30 *Years*

Previous Solved Papers

GATE 2025

Mechanical Engineering

- Fully solved with explanations
- Analysis of previous papers
- Topicwise presentation
- Thoroughly revised & updated

MADE EASY PUBLICATIONS

www.madeeasy.in

MADE EASY Publications Pvt. Ltd.

Corporate Office: 44-A/4, Kalu Sarai (Near Hauz Khas Metro Station), New Delhi-110016 | **Ph.:** 9021300500, 8860378007

Email: infomep@madeeasy.in | **Web:** www.madeeasypublications.org

GATE - 2025

Mechanical Engineering

Topicwise Previous GATE Solved Papers (1995-2024)

Editions

1st Edition	:	2008
2nd Edition	:	2009
3rd Edition	:	2010
4th Edition	:	2011
5th Edition	:	2012
6th Edition	:	2013
7th Edition	:	2014
8th Edition	:	2015
9th Edition	:	2016
10th Edition	:	2017
11th Edition	:	2018
12th Edition	:	2019
13th Edition	:	2020
14th Edition	:	2021
15th Edition	:	2022
16th Edition	:	2023
17th Edition	**:**	**2024**

Preface

B. Singh (Ex. IES)

Over the period of time the GATE examination has become more challenging due to increasing number of candidates. Though every candidate has ability to succeed but competitive environment, in-depth knowledge, quality guidance and good source of study is required to achieve high level goals.

The new edition of **GATE 2025 Solved Papers : Mechanical Engineering** has been fully revised, updated and edited. The whole book has been divided into topicwise sections.

At the beginning of each subject, analysis of previous papers are given to improve the understanding of subject.

I have true desire to serve student community by way of providing good source of study and quality guidance. I hope this book will be proved an important tool to succeed in GATE examination. Any suggestions from the readers for the improvement of this book are most welcome.

B. Singh (Ex. IES)
Chairman and Managing Director
MADE EASY Group

GATE-2025

Mechanical Engineering

CONTENTS

Engineering Mathematics

UNIT I

CONTENTS

Engineering Mathematics

Syllabus

***Linear Algebra*:** Matrix algebra, systems of linear equations, eigenvalues and eigenvectors.

***Calculus*:** Functions of single variable, limit, continuity and differentiability, mean value theorems, indeterminate forms; evaluation of definite and improper integrals; double and triple integrals; partial derivatives, total derivative, Taylor series (in one and two variables), maxima and minima, Fourier series; gradient, divergence and curl, vector identities, directional derivatives, line, surface and volume integrals, applications of Gauss, Stokes and Green's theorems.

***Differential equations*:** First order equations (linear and nonlinear); higher order linear differential equations with constant coefficients; Euler-Cauchy equation; initial and boundary value problems; Laplace transforms; solutions of heat, wave and Laplace's equations.

***Complex variables*:** Analytic functions; Cauchy-Riemann equations; Cauchy's integral theorem and integral formula; Taylor and Laurent series.

***Probability and Statistics*:** Definitions of probability, sampling theorems, conditional probability; mean, median, mode and standard deviation; random variables, binomial, Poisson and normal distributions.

***Numerical Methods*:** Numerical solutions of linear and non-linear algebraic equations; integration by trapezoidal and Simpson's rules; single and multi-step methods for differential equations.

Analysis of Previous GATE Papers

Exam Year	1 Mark Ques.	2 Marks Ques.	Total Marks
2003	3	6	**15**
2004	3	5	**13**
2005	5	10	**25**
2006	4	8	**20**
2007	4	8	**20**
2008	6	9	**24**
2009	4	8	**20**
2010	5	3	**11**
2011	5	4	**13**
2012	5	5	**15**
2013	5	5	**15**
2014 Set-1	5	4	**13**
2014 Set-2	5	4	**13**
2014 Set-3	5	4	**13**
2014 Set-4	5	4	**13**
2015 Set-1	4	3	**10**
2015 Set-2	4	4	**12**
2015 Set-3	6	5	**16**
2016 Set-1	5	4	**13**
2016 Set-2	5	4	**13**
2016 Set-3	5	4	**13**
2017 Set-1	5	4	**13**
2017 Set-2	4	4	**12**
2018 Set-1	5	4	**13**
2018 Set-2	5	4	**13**
2019 Set-1	5	4	**13**
2019 Set-2	5	4	**13**
2020 Set-1	6	4	**14**
2020 Set-2	5	4	**13**
2021 Set-1	5	3	**11**
2021 Set-2	5	4	**13**
2022 Set-1	5	4	**13**
2022 Set-2	5	4	**13**
2023	5	4	**13**
2024	5	4	**13**

Engineering Mathematics

UNIT

I

CONTENTS

Engineering Mathematics

Syllabus

Linear Algebra : Matrix algebra, systems of linear equations, eigenvalues and eigenvectors.

Calculus: Functions of single variable, limit, continuity and differentiability, mean value theorems, indeterminate forms; evaluation of definite and improper integrals; double and triple integrals; partial derivatives, total derivative, Taylor series (in one and two variables), maxima and minima, Fourier series; gradient, divergence and curl, vector identities, directional derivatives, line, surface and volume integrals, applications of Gauss, Stokes and Green's theorems.

Differential equations: First order equations (linear and nonlinear); higher order linear differential equations with constant coefficients; Euler-Cauchy equation; initial and boundary value problems; Laplace transforms; solutions of heat, wave and Laplace's equations.

Complex variables: Analytic functions; Cauchy-Riemann equations; Cauchy's integral theorem and integral formula; Taylor and Laurent series.

Probability and Statistics: Definitions of probability, sampling theorems, conditional probability; mean, median, mode and standard deviation; random variables, binomial, Poisson and normal distributions.

Numerical Methods: Numerical solutions of linear and non-linear algebraic equations; integration by trapezoidal and Simpson's rules; single and multi-step methods for differential equations.

Analysis of Previous GATE Papers

Exam Year	1 Mark Ques.	2 Marks Ques.	Total Marks
2003	3	6	**15**
2004	3	5	**13**
2005	5	10	**25**
2006	4	8	**20**
2007	4	8	**20**
2008	6	9	**24**
2009	4	8	**20**
2010	5	3	**11**
2011	5	4	**13**
2012	5	5	**15**
2013	5	5	**15**
2014 Set-1	5	4	**13**
2014 Set-2	5	4	**13**
2014 Set-3	5	4	**13**
2014 Set-4	5	4	**13**
2015 Set-1	4	3	**10**
2015 Set-2	4	4	**12**
2015 Set-3	6	5	**16**
2016 Set-1	5	4	**13**
2016 Set-2	5	4	**13**
2016 Set-3	5	4	**13**
2017 Set-1	5	4	**13**
2017 Set-2	4	4	**12**
2018 Set-1	5	4	**13**
2018 Set-2	5	4	**13**
2019 Set-1	5	4	**13**
2019 Set-2	5	4	**13**
2020 Set-1	6	4	**14**
2020 Set-2	5	4	**13**
2021 Set-1	5	3	**11**
2021 Set-2	5	4	**13**
2022 Set-1	5	4	**13**
2022 Set-2	5	4	**13**
2023	5	4	**13**
2024	5	4	**13**

1 Linear Algebra

1.1 For the matrix $\begin{bmatrix} 4 & 1 \\ 1 & 4 \end{bmatrix}$ the eigen values are
(a) 3 and –3 (b) –3 and –5
(c) 3 and 5 (d) 5 and 0 **[2003 : 1 M]**

1.2 Consider the system of simultaneous equations

$$x + 2y + z = 6$$
$$2x + y + 2z = 6$$
$$x + y + z = 5$$

This system has
(a) unique solution (b) infinite number of solutions
(c) no solution (d) exactly two solutions
[2003 : 2 M]

1.3 The sum of the eigen values of the matrix given below is

$$\begin{bmatrix} 1 & 1 & 3 \\ 1 & 5 & 1 \\ 3 & 1 & 1 \end{bmatrix}$$

(a) 5 (b) 7
(c) 9 (d) 18 **[2004 : 1 M]**

1.4 For which value of x will the matrix given below become singular?

$$\begin{bmatrix} 8 & x & 0 \\ 4 & 0 & 2 \\ 12 & 6 & 0 \end{bmatrix}$$

(a) 4 (b) 6
(c) 8 (d) 12 **[2004 : 2 M]**

1.5 A is a 3×4 real matrix and $Ax = B$ is an inconsistent system of equations. The highest possible rank of A is
(a) 1 (b) 2
(c) 3 (d) 4 **[2005 : 1 M]**

1.6 Which one of the following is an eigenvector of the matrix

$$\begin{bmatrix} 5 & 0 & 0 & 0 \\ 0 & 5 & 5 & 0 \\ 0 & 0 & 2 & 1 \\ 0 & 0 & 3 & 1 \end{bmatrix}$$

(a) $\begin{bmatrix} 1 \\ -2 \\ 0 \\ 0 \end{bmatrix}$ (b) $\begin{bmatrix} 0 \\ 0 \\ 1 \\ 0 \end{bmatrix}$

(c) $\begin{bmatrix} 1 \\ 0 \\ 0 \\ -2 \end{bmatrix}$ (d) $\begin{bmatrix} 1 \\ -1 \\ 2 \\ 1 \end{bmatrix}$ **[2005 : 2 M]**

1.7 With a 1 unit change in b, what is the change in x in the solution of the system of equations $x + y = 2$, $1.01\,x + 0.99\,y = b$?
(a) zero (b) 2 units
(c) 50 units (d) 100 units
[2005 : 2 M]

1.8 Let x denotes a real number. Find out the INCORRECT statement.
(a) $S = \{x : x > 3\}$ represents the set of all real numbers greater than 3
(b) $S = \{x : x^2 < 0\}$ represents the empty set.
(c) $S = \{x : x \in A \text{ and } x \in B\}$ represents the union of set A and set B.
(d) $S = \{x : a < x < b\}$ represents the set of all real numbers between a and b, where a and b are real numbers.
[2006 : 1 M]

1.9 Eigenvalues of a matrix $S = \begin{bmatrix} 3 & 2 \\ 2 & 3 \end{bmatrix}$ are 5 and 1. What are the eigenvalues of the matrix $S^2 = SS$?
(a) 1 and 25 (b) 6 and 4
(c) 5 and 1 (d) 2 and 10
[2006 : 2 M]

1.10 Match the items in columns I and II.

Column I
P. Singular matrix
Q. Non-square matrix
R. Real symmetric
S. Orthogonal matrix

Column II
1. Determinant is not defined
2. Determinant is always one
3. Determinant is zero
4. Eigenvalues are always real
5. Eigenvalues are not defined

(a) P–3, Q–1, R–4, S–2
(b) P–2, Q–3, R–4, S–1
(c) P–3, Q–2, R–5, S–4
(d) P–3, Q–4, R–2, S–1 [2006 : 2 M]

1.11 Multiplication of matrices E and F is G. Matrices E and G are

$$E=\begin{bmatrix}\cos\theta & -\sin\theta & 0\\ \sin\theta & \cos\theta & 0\\ 0 & 0 & 1\end{bmatrix} \text{ and } G=\begin{bmatrix}1 & 0 & 0\\ 0 & 1 & 0\\ 0 & 0 & 1\end{bmatrix}$$

What is the matrix F?

(a) $\begin{bmatrix}\cos\theta & -\sin\theta & 0\\ \sin\theta & \cos\theta & 0\\ 0 & 0 & 1\end{bmatrix}$

(b) $\begin{bmatrix}\cos\theta & \cos\theta & 0\\ -\cos\theta & \sin\theta & 0\\ 0 & 0 & 1\end{bmatrix}$

(c) $\begin{bmatrix}\cos\theta & \sin\theta & 0\\ -\sin\theta & \cos\theta & 0\\ 0 & 0 & 1\end{bmatrix}$

(d) $\begin{bmatrix}\sin\theta & -\cos\theta & 0\\ \cos\theta & \sin\theta & 0\\ 0 & 0 & 1\end{bmatrix}$ [2006 : 2 M]

1.12 If a square matrix A is real and symmetric, then the eigenvalues
(a) are always real
(b) are always real and positive
(c) are always real and non-negative
(d) occur in complex conjugate pairs
[2007 : 1 M]

1.13 The number of linearly independent eigenvectors of $\begin{bmatrix}2 & 1\\ 0 & 2\end{bmatrix}$ is
(a) 0 (b) 1
(c) 2 (d) infinite [2007 : 2 M]

1.14 The matrix $\begin{bmatrix}1 & 2 & 4\\ 3 & 0 & 6\\ 1 & 1 & p\end{bmatrix}$ has one eigenvalue equal to 3. The sum of the other two eigenvalues is
(a) p (b) $p-1$
(c) $p-2$ (d) $p-3$ [2008 : 1 M]

1.15 For what value of a, if any, will the following system of equations in x, y and z have a solution?

$$2x + 3y = 4$$
$$x + y + z = 4$$
$$x + 2y - z = a$$

(a) Any real number (b) 0
(c) 1 (d) There is no such value
[2008 : 2 M]

1.16 The eigenvectors of the matrix $\begin{bmatrix}1 & 2\\ 0 & 2\end{bmatrix}$ are written in the form $\begin{bmatrix}1\\ a\end{bmatrix}$ and $\begin{bmatrix}1\\ b\end{bmatrix}$. What is $a + b$?
(a) 0 (b) $\frac{1}{2}$
(c) 1 (d) 2 [2008 : 2 M]

1.17 For a matrix $[M] = \begin{bmatrix}\frac{3}{5} & \frac{4}{5}\\ x & \frac{3}{5}\end{bmatrix}$, the transpose of the matrix is equal to the inverse of the matrix, $[M]^T = [M]^{-1}$. The value of x is given by
(a) $-\frac{4}{5}$ (b) $-\frac{3}{5}$
(c) $\frac{3}{5}$ (d) $\frac{4}{5}$ [2009 : 1 M]

1.18 One of the eigen vectors of the matrix $A=\begin{bmatrix}2 & 2\\ 1 & 3\end{bmatrix}$ is
(a) $\begin{Bmatrix}2\\ -1\end{Bmatrix}$ (b) $\begin{Bmatrix}2\\ 1\end{Bmatrix}$
(c) $\begin{Bmatrix}4\\ 1\end{Bmatrix}$ (d) $\begin{Bmatrix}1\\ -1\end{Bmatrix}$ [2010 : 2 M]

1.19 Eigenvalues of a real symmetric matrix are always
(a) positive (b) negative
(c) real (d) complex [2011 : 1 M]

1.20 Consider the following system of equations

$$2x_1 + x_2 + x_3 = 0$$
$$x_2 - x_3 = 0$$
$$x_1 + x_2 = 0$$

This system has
(a) a unique solution
(b) no solution
(c) infinite number of solutions
(d) five solutions [2011 : 2 M]

1.21 For the matrix $A = \begin{bmatrix} 5 & 3 \\ 1 & 3 \end{bmatrix}$, ONE of the normalized eigen vectors is given as

(a) $\begin{pmatrix} \frac{1}{2} \\ \frac{\sqrt{3}}{2} \end{pmatrix}$ (b) $\begin{pmatrix} \frac{1}{\sqrt{2}} \\ \frac{-1}{\sqrt{2}} \end{pmatrix}$

(c) $\begin{pmatrix} \frac{3}{\sqrt{10}} \\ \frac{-1}{\sqrt{10}} \end{pmatrix}$ (d) $\begin{pmatrix} \frac{1}{\sqrt{5}} \\ \frac{2}{\sqrt{5}} \end{pmatrix}$ [2012 : 2 M]

1.22 $x + 2y + z = 4$
$2x + y + 2z = 5$
$x - y + z = 1$
The system of algebraic given below has
(a) A unique solution of $x = 1, y = 1$ and $z = 1$
(b) only the two solutions of ($x = 1, y = 1, z = 1$) and ($x = 2, y = 1, z = 0$)
(c) infinite number of solutions
(d) no feasible solution [2012 : 2 M]

1.23 Choose the **CORRECT** set of functions, which are linearly dependent.
(a) $\sin x$, $\sin^2 x$ and $\cos^2 x$
(b) $\cos x$, $\sin x$ and $\tan x$
(c) $\cos 2x$, $\sin^2 x$ and $\cos^2 x$
(d) $\cos 2x$, $\sin x$ and $\cos x$ [2013 : 1 M]

1.24 The eigen values of a symmetric matrix are all
(a) complex with non-zero positive imaginary part
(b) complex with non-zero negative imaginary part
(c) real
(d) pure imaginary [2013 : 1 M]

1.25 Given that the determinant of the matrix $\begin{bmatrix} 1 & 3 & 0 \\ 2 & 6 & 4 \\ -1 & 0 & 2 \end{bmatrix}$ is – 12, the determinant of the matrix $\begin{bmatrix} 2 & 6 & 0 \\ 4 & 12 & 8 \\ -2 & 0 & 4 \end{bmatrix}$ is

(a) – 96 (b) – 24
(c) 24 (d) 96 [2014 : 1 M, Set-1]

1.26 The matrix form of the linear system $\frac{dx}{dt} = 3x - 5y$ and $\frac{dy}{dt} = 4x + 8y$ is

(a) $\frac{d}{dt}\begin{Bmatrix} x \\ y \end{Bmatrix} = \begin{bmatrix} 3 & -5 \\ 4 & 8 \end{bmatrix}\begin{Bmatrix} x \\ y \end{Bmatrix}$

(b) $\frac{d}{dt}\begin{Bmatrix} x \\ y \end{Bmatrix} = \begin{bmatrix} 3 & 8 \\ 4 & -5 \end{bmatrix}\begin{Bmatrix} x \\ y \end{Bmatrix}$

(c) $\frac{d}{dt}\begin{Bmatrix} x \\ y \end{Bmatrix} = \begin{bmatrix} 4 & -5 \\ 3 & 8 \end{bmatrix}\begin{Bmatrix} x \\ y \end{Bmatrix}$

(d) $\frac{d}{dt}\begin{Bmatrix} x \\ y \end{Bmatrix} = \begin{bmatrix} 4 & 8 \\ 3 & -5 \end{bmatrix}\begin{Bmatrix} x \\ y \end{Bmatrix}$ [2014 : 1 M, Set-1]

1.27 One of the eigenvectors of matrix $\begin{bmatrix} -5 & 2 \\ -9 & 6 \end{bmatrix}$ is

(a) $\begin{Bmatrix} -1 \\ 1 \end{Bmatrix}$ (b) $\begin{Bmatrix} -2 \\ 9 \end{Bmatrix}$

(c) $\begin{Bmatrix} 2 \\ -1 \end{Bmatrix}$ (d) $\begin{Bmatrix} 1 \\ 1 \end{Bmatrix}$

[2014 : 1 M, Set-2]

1.28 Consider a 3 × 3 real symmetric matrix S such that two of its eigenvalues are $a \neq 0$, $b \neq 0$ with respective eigenvectors $\begin{bmatrix} x_1 \\ x_2 \\ x_3 \end{bmatrix}$, $\begin{bmatrix} y_1 \\ y_2 \\ y_3 \end{bmatrix}$. If $a \neq b$ then $x_1y_1 + x_2y_2 + x_3y_3$ equals
(a) a (b) b
(c) ab (d) 0 [2014 : 1 M, Set-3]

1.29 Which one of the following equations is a correct identity for arbitrary 3 × 3 real matrices P, Q and R?
(a) $P(Q + R) = PQ + RP$
(b) $(P - Q)^2 = P^2 - 2PQ + Q^2$
(c) det $(P + Q)$ = det P + det Q
(d) $(P + Q)^2 = P^2 + PQ + QP + Q^2$

[2014 : 1 M, Set-4]

1.30 If any two columns of a determinant $P = \begin{vmatrix} 4 & 7 & 8 \\ 3 & 1 & 5 \\ 9 & 6 & 2 \end{vmatrix}$ are interchanged, which one of the following statements regarding the value of the determinant is CORRECT?
(a) Absolute value remains unchanged but sign will change
(b) Both absolute value and sign will change
(c) Absolute value will change but sign will not change
(d) Both absolute value and sign will remain unchanged [2015 : 1 M, Set-1]

1.31 At least one eigenvalue of a singular matrix is
(a) positive
(b) zero
(c) negative
(d) imaginary

[2015 : 1 M, Set-3]

1.32 The lowest eigenvalue of the 2 × 2 matrix $\begin{bmatrix} 4 & 2 \\ 1 & 3 \end{bmatrix}$ is ______.

[2015 : 1 M, Set-3]

1.33 For given matrix $P = \begin{bmatrix} 4+3i & -i \\ i & 4-3i \end{bmatrix}$ where $i = \sqrt{-1}$, the inverse of matrix P is
(a) $\frac{1}{24}\begin{bmatrix} 4-3i & i \\ -i & 4+3i \end{bmatrix}$
(b) $\frac{1}{25}\begin{bmatrix} i & 4-i \\ 4+3i & -i \end{bmatrix}$
(c) $\frac{1}{24}\begin{bmatrix} 4+3i & -i \\ i & 4-3i \end{bmatrix}$
(d) $\frac{1}{25}\begin{bmatrix} 4+3i & -i \\ i & 4-3i \end{bmatrix}$

[2015 : 2 M, Set-3]

1.34 The solution to the system of equations is
$$\begin{bmatrix} 2 & 5 \\ -4 & 3 \end{bmatrix}\begin{bmatrix} x \\ y \end{bmatrix} = \begin{bmatrix} 2 \\ -30 \end{bmatrix}$$
(a) 6, 2
(b) –6, 2
(c) –6, –2
(d) 6, – 2

[2016 : 1 M, Set-1]

1.35 The condition for which the eigenvalues of the matrix $A = \begin{bmatrix} 2 & 1 \\ 1 & k \end{bmatrix}$ are positive, is
(a) $k > \frac{1}{2}$
(b) $k > -2$
(c) $k > 0$
(d) $k < -\frac{1}{2}$

[2016 : 1 M, Set-2]

1.36 A real square matrix A is called skew-symmetric if
(a) $A^T = A$
(b) $A^T = A^{-1}$
(c) $A^T = -A$
(d) $A^T = A + A^{-1}$

[2016 : 1 M, Set-3]

1.37 The number of linearly independent eigenvectors of matrix $A = \begin{bmatrix} 2 & 1 & 0 \\ 0 & 2 & 0 \\ 0 & 0 & 3 \end{bmatrix}$ is______

[2016 : 2 M, Set-3]

1.38 The product of eigenvalues of the matrix P is
$$P = \begin{bmatrix} 2 & 0 & 1 \\ 4 & -3 & 3 \\ 0 & 2 & -1 \end{bmatrix}$$
(a) – 6
(b) 2
(c) 6
(d) – 2

[2017 : 1 M, Set-1]

1.39 Consider the matrix $P = \begin{bmatrix} \frac{1}{\sqrt{2}} & 0 & \frac{1}{\sqrt{2}} \\ 0 & 1 & 0 \\ \frac{-1}{\sqrt{2}} & 0 & \frac{1}{\sqrt{2}} \end{bmatrix}$

Which one of the following statements about P is INCORRECT?
(a) Determinant of P is equal to 1.
(b) P is orthogonal.
(c) Inverse of P is equal to its transpose.
(d) All eigenvalues of P are real numbers.

[2017 : 2 M, Set-1]

1.40 The determinant of a 2 × 2 matrix is 50. If one eigenvalue of the matrix is 10, the other eigen value is ____________.

[2017 : 1 M, Set-2]

1.41 Consider the matrix $A = \begin{bmatrix} 50 & 70 \\ 70 & 80 \end{bmatrix}$ whose eigenvectors corresponding to eigenvalues λ_1 and λ_2 are $x_1 = \begin{bmatrix} 70 \\ \lambda_1 - 50 \end{bmatrix}$ and $x_2 = \begin{bmatrix} \lambda_2 - 80 \\ 70 \end{bmatrix}$, respectively. The value of $x_1^T x_2$ is __________.

[2017 : 2 M, Set-2]

1.42 The rank of the matrix $\begin{bmatrix} -4 & 1 & -1 \\ -1 & -1 & -1 \\ 7 & -3 & 1 \end{bmatrix}$ is
(a) 1
(b) 2
(c) 3
(d) 4

[2018 : 1 M, Set-1]

1.43 If $A = \begin{bmatrix} 1 & 2 & 3 \\ 0 & 4 & 5 \\ 0 & 0 & 1 \end{bmatrix}$ then $\det(A^{-1})$ is _______ (correct to two decimal places).

[2018 : 1 M, Set-2]

1.44 Consider the matrix

$$P = \begin{bmatrix} 1 & 1 & 0 \\ 0 & 1 & 1 \\ 0 & 0 & 1 \end{bmatrix}$$

The number of distinct eigen values of P is
(a) 2 (b) 0
(c) 1 (d) 3 **[2019 : 1 M, Set-1]**

1.45 The set of equations
$x + y + z = 1$
$ax - ay + 3z = 5$
$5x - 3y + az = 6$
has infinite solution if a =
(a) – 4 (b) – 3
(c) 3 (d) 4 **[2019 : 2 M, Set-1]**

1.46 In matrix equation [A]{X} = {R}

$$[A] = \begin{bmatrix} 4 & 8 & 4 \\ 8 & 16 & -4 \\ 4 & -4 & 15 \end{bmatrix}, \{X\} = \begin{Bmatrix} 2 \\ 1 \\ 4 \end{Bmatrix} \text{ and } \{R\} = \begin{Bmatrix} 32 \\ 16 \\ 64 \end{Bmatrix}$$

One of the eigen values of Matrix [A] is
(a) 4 (b) 15
(c) 8 (d) 16
[2019 : 1 M, Set-2]

1.47 Multiplication of real valued square matrices of same dimension is
(a) not always possible to compute
(b) associative
(c) always positive definite
(d) commutative **[2020 : 1 M, Set-1]**

1.48 A matrix *P* is decomposed into its symmetric part *S* and skew symmetric part *V*. If

$$S = \begin{pmatrix} -4 & 4 & 2 \\ 4 & 3 & 7/2 \\ 2 & 7/2 & 2 \end{pmatrix}, \quad V = \begin{pmatrix} 0 & -2 & 3 \\ 2 & 0 & 7/2 \\ -3 & -7/2 & 0 \end{pmatrix}$$

then matrix *P* is

(a) $\begin{pmatrix} -2 & 9/2 & -1 \\ -1 & 81/4 & 11 \\ -2 & 45/2 & 73/4 \end{pmatrix}$ (b) $\begin{pmatrix} -4 & 6 & -1 \\ 2 & 3 & 0 \\ 5 & 7 & 2 \end{pmatrix}$

(c) $\begin{pmatrix} -4 & 2 & 5 \\ 6 & 3 & 7 \\ -1 & 0 & 2 \end{pmatrix}$ (d) $\begin{pmatrix} 4 & -6 & 1 \\ -2 & -3 & 0 \\ -5 & -7 & -2 \end{pmatrix}$

[2020 : 1 M, Set-2]

1.49 Let *I* be a 100 dimensional identity matrix and *E* be the set of its distinct (no value appears more than once in *E*) real eigenvalues. The number of elements in *E* is ______.
[2020 : 1 M, Set-2]

1.50 Consider an $n \times n$ matrix *A* and a non-zero $n \times 1$ vector *p*. Their product $Ap = \alpha^2 p$, where $\alpha \in \Re$ and $\alpha \notin \{-1, 0, 1\}$. Based on the given information, the eigen value of A^2 is:
(a) $\sqrt{\alpha}$ (b) α^2
(c) α (d) α^4
[2021 : 1 M, Set-2]

1.51 Let the superscript *T* represent the transpose operation. Consider the function $f(x) = \frac{1}{2}x^T Q x - r^T x$, where x and r are $n \times 1$ vectors and *Q* is a symmetric n × n matrix. The stationary point of *f*(*x*) is
(a) $\frac{r}{r^T r}$ (b) $Q^T r$
(c) $Q^{-1} r$ (d) r **[2021 : 2 M, Set-2]**

1.52 Consider a vector *p* in 2-dimensional space. Let its direction (counter-clockwise angle with the positive *x*-axis) be θ. Let *p* be an eigen vector of a 2 × 2 matrix A with corresponding eigen value λ, λ > 0. If we denote the magnitude of a vector *v* by ||*v*||, identify the VALID statement regarding *p*′, where *p*′ = *Ap*.
(a) Direction of $p' = \theta, \|p'\| = \lambda \|p\|$
(b) Direction of $p' = \lambda\theta, \|p'\| = \|p\|$
(c) Direction of $p' = \theta, \|p'\| = \|p\|/\lambda$
(d) Direction of $p' = \lambda\theta, \|p'\| = \lambda \|p\|$
[2021 : 2 M, Set-1]

1.53 If $A = \begin{bmatrix} 10 & 2k+5 \\ 3k-3 & k+5 \end{bmatrix}$ is a symmetric matrix, the value of *k* is ______.
(a) 8 (b) 5
(c) –0.4 (d) $\frac{1+\sqrt{1561}}{12}$
[2022 : 1 M, Set-1]

1.54 The system of linear equations in real (*x*, *y*) given by

$$(x\ y)\begin{bmatrix} 2 & 5-2\alpha \\ \alpha & 1 \end{bmatrix} = (0\ 0)$$

involves a real parameter α and has infinitely

many non-trivial solutions for special value(s) of α. Which one or more among the following options is/are non-trivial solution(s) of (x, y) for such special value(s) of α?

(a) $x = 2,\ y = -2$ (b) $x = -1,\ y = 4$
(c) $x = 1,\ y = 1$ (d) $x = 4,\ y = -2$

[2022 : 2 M, Set-1]

1.55 A is a 3×5 real matrix of rank 2. For the set of homogeneous equations $Ax = 0$, where 0 is a zero vector and x is a vector of unknown variables, which of the following is/are true?

(a) The given set of equations will have a unique solution.
(b) The given set of equations will be satisfied by a zero vector of appropriate size.
(c) The given set of equations will have infinitely many solutions.
(d) The given set of equations will have many but a finite number of solutions.

[2022 : 2 M, Set-2]

1.56 If the sum and product of eigen values of a 2×2 matrix $\begin{bmatrix} 3 & p \\ p & q \end{bmatrix}$ are 4 and −1 respectively, then $|p|$ is ______ (in integer).

[2022 : 2 M, Set-2]

1.57 A linear transformation maps a point (x, y) in the plane to the point $(\hat{x}, \hat{y})$ according to the rule

$$\hat{x} = 3y,\ \hat{y} = 2x$$

Then, the disc $x^2 + y^2 \leq 1$ gets transformed to a region with an area equal to ______. (Rounded off to two decimals)

Use $\pi = 3.14$.

[2023 : 1 M]

1.58 Consider the system of linear equations

$$x + 2y + z = 5$$
$$2x + ay + 4z = 12$$
$$2x + 4y + 6z = b$$

The values of a and b such that there exists a non-trivial null space and the system admits infinite solutions are

(a) $a = 4,\ b = 12$ (b) $a = 8,\ b = 12$
(c) $a = 4,\ b = 14$ (d) $a = 8,\ b = 14$

[2024 : 1 M]

1.59 The matrix $\begin{bmatrix} 1 & a \\ 8 & 3 \end{bmatrix}$ (where $a > 0$) has a negative eigen value if a is greater than

(a) $\frac{1}{8}$ (b) $\frac{1}{4}$
(c) $\frac{3}{8}$ (d) $\frac{1}{5}$

[2024 : 2 M]

■■■■

Answers Linear Algebra

1.1	(c)	1.2	(c)	1.3	(b)	1.4	(a)	1.5	(b)	1.6	(a)	1.7	(c)	1.8	(c)
1.9	(a)	1.10	(a)	1.11	(c)	1.12	(a)	1.13	(b)	1.14	(c)	1.15	(b)	1.16	(b)
1.17	(a)	1.18	(a)	1.19	(c)	1.20	(c)	1.21	(b)	1.22	(c)	1.23	(c)	1.24	(c)
1.25	(a)	1.26	(a)	1.27	(d)	1.28	(d)	1.29	(d)	1.30	(a)	1.31	(b)	1.32	(2)
1.33	(a)	1.34	(d)	1.35	(a)	1.36	(c)	1.37	(2)	1.38	(b)	1.39	(d)	1.40	(5)
1.41	(0)	1.42	(b)	1.43	(0.25)	1.44	(c)	1.45	(d)	1.46	(d)	1.47	(b)	1.48	(c)
1.49	(1)	1.50	(d)	1.51	(c)	1.52	(a)	1.53	(a)	1.54	(a, b)	1.55	(b, c)	1.56	(2)
1.57	(18.84)	1.58	(c)	1.59	(c)										

Explanations Linear Algebra

1.1 (c)

$$A = \begin{bmatrix} 4 & 1 \\ 1 & 4 \end{bmatrix}$$

Now $|A - \lambda I| = 0$

where λ = eigen value

$$\therefore \begin{vmatrix} 4-\lambda & 1 \\ 1 & 4-\lambda \end{vmatrix} = 0$$

$$(4-\lambda)^2 - 1 = 0$$

or, $(4-\lambda)^2 - (1)^2 = 0$

or, $(4 - \lambda + 1)(4 - \lambda - 1) = 0$

or, $(5-\lambda)(3-\lambda) = 0$

$\therefore$ $\lambda = $ **3, 5**

1.2 (c)

Given equation are

$$x + 2y + z = 6$$
$$2x + y + 2z = 6$$
$$x + y + z = 5$$

Given system can be written as

$$\begin{bmatrix} 1 & 2 & 1 \\ 2 & 1 & 2 \\ 1 & 1 & 1 \end{bmatrix}\begin{bmatrix} x \\ y \\ z \end{bmatrix} = \begin{bmatrix} 6 \\ 6 \\ 5 \end{bmatrix}$$

Applying row operation

$R_2 \to R_2 - 2R_1$, $R_3 \to R_3 - R_1$

$$\begin{bmatrix} 1 & 2 & 1 \\ 0 & -3 & 0 \\ 0 & -1 & 0 \end{bmatrix}\begin{bmatrix} x \\ y \\ z \end{bmatrix} = \begin{bmatrix} 6 \\ -6 \\ -1 \end{bmatrix}$$

and applying $R_3 \to 3R_3 - R_2$

$$\begin{bmatrix} 1 & 2 & 1 \\ 0 & -3 & 0 \\ 0 & 0 & 0 \end{bmatrix}\begin{bmatrix} x \\ y \\ z \end{bmatrix} = \begin{bmatrix} 6 \\ -6 \\ 3 \end{bmatrix}$$

Since the rank of coefficient matrix is 2 and rank of argument matrix is 3, which are not equal. Hence system has **no solution**.

1.3 (b)

Sum of eigen values of given matrix
= sum of diagonal elementof given matrix
= 1 + 5 + 1 = **7**

1.4 (a)

For singularity of matrix = $\begin{vmatrix} 8 & x & 0 \\ 4 & 0 & 2 \\ 12 & 6 & 0 \end{vmatrix} = 0$

$\Rightarrow 8(0 - 12) - x(0 - 2 \times 12) = 0$

$\therefore$ $x = $ **4**

1.5 (b)

$$C = [A : B]_{3 \times 5}$$

$\therefore$ $\rho[C_{3\times5}] \le \min\{3, 5\}$

& $\because$ The system is inconsistent

$\rho(A) < \rho(C)$

$\therefore$ $\rho(A) < 3$

Hence maximum possible rank of

$A = $ **2**

1.6 (a)

Eigen vector is an independent vector. Here in option *a* last two terms are zero. Hence only this option is correct.

1.7 (c)

Given $x + y = 2$ ...(i)

$1.01\,x + 0.99\,y = b$ ...(ii)

Multiply 0.99 is equation (i), and substract from equation (ii); we get

$(1.01 - 0.99)\,x = b - 2 \times 0.99$

$0.02x = b - 1.98$

$\therefore$ $0.02\,\Delta x = \Delta b$

$\therefore$ $\Delta x = \dfrac{1}{0.02}$ = **50 unit**

1.9 (a)

If $\lambda_1, \lambda_2, \lambda_3 \ldots \lambda_4$ are the eigen values of A. Then the eigen values of

A^m are $\lambda_1^m, \lambda_2^m, \lambda_3^m \ldots$

$\Rightarrow$ S^2 are $S_1^1, S_2^2, S_3^2 \ldots$

S matrix has eigen values 1 and S.

$\Rightarrow$ S^2 matrix has eigen values 1^2 and S^2

i.e. **1 and 25**

1.11 (c)

$$E = \begin{bmatrix} \cos\theta & -\sin\theta & 0 \\ \sin\theta & \cos\theta & 0 \\ 0 & 0 & 1 \end{bmatrix}$$

and $$G = \begin{bmatrix} 1 & 0 & 0 \\ 0 & 1 & 0 \\ 0 & 0 & 1 \end{bmatrix}$$

The according to problem

$$E \times F = G$$

or $\begin{bmatrix} \cos\theta & \sin\theta & 0 \\ -\sin\theta & \cos\theta & 0 \\ 0 & 0 & 1 \end{bmatrix} \times F = \begin{bmatrix} 1 & 0 & 0 \\ 0 & 1 & 0 \\ 0 & 0 & 1 \end{bmatrix}$

Hence we see that product of $(E \times F)$ is unit matrix so F will be the inverse of E.

$$F = E^{-1} = \begin{bmatrix} \cos\theta & \sin\theta & 0 \\ -\sin\theta & \cos\theta & 0 \\ 0 & 0 & 1 \end{bmatrix}$$

1.12 (a)

The eigen values of any real and symmetric matrix **are always real.**

1.13 (b)

$$A = \begin{bmatrix} 2 & 1 \\ 0 & 2 \end{bmatrix}$$

$$[a - \lambda I] = 0$$

$$\begin{bmatrix} 2-\lambda & 1 \\ 0 & 2-\lambda \end{bmatrix} = 0$$

$\Rightarrow \quad (2-\lambda)^2 = 0$

$\Rightarrow \quad \lambda = 2$

Number of linearly independent eigen vector is one.

1.14 (c)

Sum of the eigen values of matrix is

= sum of diagonal values present in the matrix

$\therefore 1 + 0 + p = 3 + \lambda_2 + \lambda_3$

$\Rightarrow p + 1 = 3 + \lambda_2 + \lambda_3$

$\Rightarrow \lambda_2 + \lambda_3 = p + 1 - 3 =$ **$p - 2$**

1.15 (b)

Augmented matrix is

$$\left[\begin{array}{ccc|c} 2 & 3 & 0 & 4 \\ 1 & 1 & 1 & 4 \\ 1 & 2 & -1 & a \end{array}\right] \begin{matrix} R_1 \to R_1 - 2R_2 \\ R_3 \to R_3 - R_2 \end{matrix}$$

$$\sim \left[\begin{array}{ccc|c} 0 & 1 & -2 & -4 \\ 1 & 1 & 1 & 4 \\ 0 & 1 & -2 & a-4 \end{array}\right] R_1 \to R_1 - R_3$$

$$\sim \left[\begin{array}{ccc|c} 0 & 0 & 0 & -a \\ 1 & 1 & 1 & 4 \\ 0 & 1 & -2 & a-4 \end{array}\right]$$

will have solution if **$a = 0$**

as Rank A = Rank (aug. A)

1.16 (b)

$$\begin{vmatrix} (1-\lambda) & 2 \\ 0 & (2-\lambda) \end{vmatrix} = 0$$

$\Rightarrow \quad (1-\lambda)(2-\lambda) = 0$

$\therefore \quad \lambda = 1, 2$

Putting the value of $\lambda = 1$

$\Rightarrow \quad \begin{bmatrix} 0 & 2 \\ 0 & 1 \end{bmatrix}\begin{bmatrix} 1 \\ a \end{bmatrix} = 0$

$\Rightarrow \quad a = 0$

putting the value of $\lambda = 2$

$$\begin{bmatrix} -1 & 2 \\ 0 & 0 \end{bmatrix}\begin{bmatrix} 1 \\ b \end{bmatrix} = 0$$

$\Rightarrow \quad b = \frac{1}{2}$

$\therefore \quad a + b = \frac{1}{2}$

1.17 (a)

If $\quad A^T = A^{-1}$

then A is orthogonal matrix.

Therefore $A.A^{-1} = A^{-1} A = I$

and $A^T A = A A^T = I$

Since M is orthogonal matrix

$$M^T M = I$$

$$\begin{bmatrix} \frac{3}{5} & x \\ \frac{4}{5} & \frac{3}{5} \end{bmatrix}\begin{bmatrix} \frac{3}{5} & \frac{4}{5} \\ x & \frac{3}{5} \end{bmatrix} = \begin{bmatrix} 1 & 0 \\ 0 & 1 \end{bmatrix}$$

$$\Rightarrow \begin{bmatrix} \left(\frac{3}{5}\right)^2 + x^2 & \left(\frac{3}{5}\cdot\frac{4}{5}\right) + \frac{3}{5}x \\ \left(\frac{4}{5}\cdot\frac{3}{5}\right) + \frac{3}{5}\cdot x & \left(\frac{4}{5}\right)^2 + \left(\frac{3}{5}\right)^2 \end{bmatrix} = \begin{bmatrix} 1 & 0 \\ 0 & 1 \end{bmatrix}$$

$\Rightarrow$ Compare both sides a_{12}

$$a_{12} = \left(\frac{3}{5}\right)\left(\frac{4}{5}\right) + \left(\frac{3}{5}\right)x = 0$$

$\Rightarrow \quad \frac{3}{5}x = -\frac{3}{5}\cdot\frac{4}{5}$

$\Rightarrow \quad x = -\frac{4}{5}$

1.18 (a)

$$A = \begin{bmatrix} 2 & 2 \\ 1 & 3 \end{bmatrix}$$

Characteristic equation of A is

$$\begin{vmatrix} 2-\lambda & 2 \\ 1 & 3-\lambda \end{vmatrix} = 0$$

$(2-\lambda)(3-\lambda) - 2 = 0$

$\lambda^2 - 5\lambda + 4 = 0$

$\lambda = 1, 4$

The eigen value problem is $[A-\lambda I]\hat{x} = 0$

$$\begin{bmatrix} 2-\lambda & 2 \\ 1 & 3-\lambda \end{bmatrix}\begin{bmatrix} x_1 \\ x_2 \end{bmatrix} = \begin{bmatrix} 0 \\ 0 \end{bmatrix}$$

Putting $\lambda = 1$

$$\begin{bmatrix} 1 & 2 \\ 1 & 2 \end{bmatrix}\begin{bmatrix} x_1 \\ x_2 \end{bmatrix} = \begin{bmatrix} 0 \\ 0 \end{bmatrix}$$

$$x_1 + 2x_2 = 0 \quad \text{... (i)}$$

$$x_1 + 2x_2 = 0 \quad \text{... (ii)}$$

Solution is $x_2 = k$, $x_1 = -2k$

$$\hat{X}_1 = \begin{bmatrix} -2k \\ k \end{bmatrix}$$

i.e. $x_1 : x_2 = -2 : 1$

Since, choice (*A*) $\begin{bmatrix} 2 \\ -1 \end{bmatrix}$ is in same ratio of x_1 to x_2.

∴ Choice (*A*) is an eigen vector.

1.19 (c)

Eigen values of symmetric matrix are always real.

1.20 (c)

The Augmented matrix

$$[A \mid B] = \left[\begin{array}{ccc|c} 2 & 1 & 1 & 0 \\ 0 & 1 & -1 & 0 \\ 1 & 1 & 0 & 0 \end{array}\right]$$

Performing gauss elimination on $[A \mid B]$ we get

$$\left[\begin{array}{ccc|c} 2 & 1 & 1 & 0 \\ 0 & 1 & -1 & 0 \\ 1 & 1 & 0 & 0 \end{array}\right] \xrightarrow{R_3 - \frac{1}{2}R_1} \left[\begin{array}{ccc|c} 2 & 1 & 1 & 0 \\ 0 & 1 & -1 & 0 \\ 0 & \frac{1}{2} & \frac{1}{2} & 0 \end{array}\right]$$

$$\xrightarrow{R_3 - \frac{1}{2}R_2} \left[\begin{array}{ccc|c} 2 & 1 & 1 & 0 \\ 0 & 1 & -1 & 0 \\ 0 & 0 & 0 & 0 \end{array}\right]$$

Rank (*A*) = Rank (*A* | *B*) = **2 < 3**

So infinite number of solutions are obtained.

1.21 (b)

$$A = \begin{bmatrix} 5 & 3 \\ 1 & 3 \end{bmatrix}$$

Characteristic equation is

$$\begin{vmatrix} 5-\lambda & 3 \\ 1 & 3-\lambda \end{vmatrix} = 0$$

$$(5-\lambda)(3-\lambda) - 3 = 0$$

$$\lambda^2 - 8\lambda + 12 = 0$$

$$\lambda = 2, 6$$

Now, to find eigen vectors:

$$[A-\lambda I]\hat{x} = 0$$

Which is

$$\begin{bmatrix} 5-\lambda & 3 \\ 1 & 3-\lambda \end{bmatrix}\begin{bmatrix} x_1 \\ x_2 \end{bmatrix} = \begin{bmatrix} 0 \\ 0 \end{bmatrix}$$

Put $\lambda = 2$ in above equation and we get

$$\begin{bmatrix} 3 & 3 \\ 1 & 1 \end{bmatrix}\begin{bmatrix} x_1 \\ x_2 \end{bmatrix} = \begin{bmatrix} 0 \\ 0 \end{bmatrix}$$

Which gives us the equation,

$$3x_1 + 3x_2 = 0$$

and $x_1 + x_2 = 0$

Which is only one equation,

$$x_1 + x_2 = 0$$

Whose solution is

$$\begin{bmatrix} x_1 \\ x_2 \end{bmatrix} = \begin{bmatrix} -k \\ k \end{bmatrix}$$

So one eigen vector is $\hat{x}_1 = \begin{bmatrix} -k \\ k \end{bmatrix}$

Which after normalization is $= \dfrac{\hat{x}_1}{|\hat{x}_1|}$

$$= \frac{1}{\sqrt{(-k)^2 + (k^2)}}\begin{bmatrix} -k \\ k \end{bmatrix} = \begin{bmatrix} -\frac{1}{\sqrt{2}} \\ \frac{1}{\sqrt{2}} \end{bmatrix}$$

The other eigen vector is obtained by putting the other eigen value

$\lambda = 6$ in eigen value problem

$$\begin{bmatrix} 5-\lambda & 3 \\ 1 & 3-\lambda \end{bmatrix}\begin{bmatrix} x_1 \\ x_2 \end{bmatrix} = \begin{bmatrix} 0 \\ 0 \end{bmatrix}$$

Which gives,

$$\begin{bmatrix} -1 & 3 \\ 1 & -3 \end{bmatrix}\begin{bmatrix} x_1 \\ x_2 \end{bmatrix} = \begin{bmatrix} 0 \\ 0 \end{bmatrix}$$

Which gives the equation

$$-x_1 + 3x_2 = 0$$

and $x_1 - 3x_2 = 0$

Which is only one equation

$$-x_1 + 3x_2 = 0$$

Whose solution is

$$\hat{x}_2 = \begin{bmatrix} x_1 \\ x_2 \end{bmatrix} = \begin{bmatrix} 3k \\ k \end{bmatrix}$$

Which after normalization is

$$\frac{\hat{x}_2}{|\hat{x}_2|} = \frac{1}{\sqrt{((3k)^2 + k^2)}}\begin{bmatrix} 3k \\ k \end{bmatrix} = \begin{bmatrix} 3/2 \\ 1/2 \end{bmatrix}$$

Choice (b) $\begin{bmatrix} \frac{1}{\sqrt{2}} \\ -\frac{1}{\sqrt{2}} \end{bmatrix}$ is the only correct choice,

since it is a constant multiple of one the normalized vectors which is $\hat{x}_1$.

1.22 (c)

The given system is

$$x + 2y + z = 4$$
$$2x + y + 2z = 5$$
$$x - y + z = 1$$

Use Gauss elimination method as follows:

Augmented matrix is

$$[A/B] = \left[\begin{array}{ccc|c} 1 & 2 & 1 & 4 \\ 2 & 1 & 2 & 5 \\ 1 & -1 & 1 & 1 \end{array}\right] \xrightarrow[R_3 - R_1]{R_2 - 2R_1}$$

$$\left[\begin{array}{ccc|c} 1 & 2 & 1 & 4 \\ 0 & -3 & 0 & -3 \\ 0 & -3 & 0 & -3 \end{array}\right] \xrightarrow{R_3 - R_2} \left[\begin{array}{ccc|c} 1 & 2 & 1 & 4 \\ 0 & -3 & 0 & -3 \\ 0 & 0 & 0 & 0 \end{array}\right]$$

$$\text{Rank}(A) = 2$$
$$\text{Rank}(AB) = 2$$

So rank (A) = Rank (AB) = 2

System is consistent

Now system rank $r = 2$

Number of variables n = 3

$$r < n$$

So we have infinite number of solutions.

1.23 (c)

We know $\cos 2x = \cos^2 x - \sin^2 x$

$\therefore$ option (c) is correct.

1.24 (c)

(i) The Eigen values of symmetric matrix $[A^T = A]$ are purely real.

(ii) The Eigen value of skew-symmetric matrix $[A^T = -A]$ are either purely imaginary or zeros.

1.25 (a)

Let $D = -12$ for the given matrix

$$|A| = \begin{vmatrix} 2 & 6 & 0 \\ 4 & 12 & 8 \\ -2 & 0 & 4 \end{vmatrix} = (2)^3 \begin{vmatrix} 1 & 3 & 0 \\ 2 & 6 & 4 \\ -1 & 0 & 2 \end{vmatrix}$$

(Taking 2 common from each row)

$$\therefore \text{Det}(A) = (2)^3 \times D$$
$$= 8 \times -12 = \mathbf{-96}$$

1.26 (a)

$$\frac{dx}{dt} = 3x - 5y$$

$$\frac{dy}{dt} = 4x + 8y$$

$$\frac{d}{dt}\begin{Bmatrix} x \\ y \end{Bmatrix} = \begin{bmatrix} 3 & -5 \\ 4 & 8 \end{bmatrix}\begin{Bmatrix} x \\ y \end{Bmatrix} = \begin{bmatrix} 3x & -5y \\ 4x & +8y \end{bmatrix}$$

1.27 (d)

The characteristic equation $|A - \lambda I| = 0$

i.e. $\begin{vmatrix} -5-\lambda & 2 \\ -9 & 6-\lambda \end{vmatrix} = 0$

or $(\lambda - 6)(\lambda + 5) + 18 = 0$

or $\lambda^2 - 6\lambda + 5\lambda - 30 + 18 = 0$

or $\lambda^2 - \lambda - 12 = 0$

or $\lambda = \dfrac{1 \pm \sqrt{1+48}}{2} = \dfrac{1 \pm 7}{2} = 4, -3$

Corresponding to $\lambda = 4$, we have

$$[A - \lambda I]x = \begin{bmatrix} -5-\lambda & 2 \\ -9 & 6-\lambda \end{bmatrix}\begin{bmatrix} x \\ y \end{bmatrix} = 0$$

or, $\begin{bmatrix} -9 & 2 \\ -9 & 2 \end{bmatrix}\begin{bmatrix} x \\ y \end{bmatrix} = 0$

which gives only one independent equation, $-9x + 2y = 0$

$\therefore \dfrac{x}{2} = \dfrac{y}{9}$ gives eigen vector (2, 9)

Corresponding to $\lambda = -3$,

$$= \begin{bmatrix} -2 & 2 \\ -9 & 9 \end{bmatrix}\begin{bmatrix} x \\ y \end{bmatrix} = 0$$

which gives $-x + y = 0$ (only one independent equation)

$\therefore \quad \dfrac{x}{1} = \dfrac{y}{1}$ which gives (1,1)

So, the eigen vectors are $\begin{Bmatrix} 2 \\ 9 \end{Bmatrix}$ and $\begin{Bmatrix} 1 \\ 1 \end{Bmatrix}$.

1.28 (d)

3 × 3 real symmetric matrix such that two of its eigen value are $a \neq 0$ $b \neq 0$ with respective eigen vectors $\begin{bmatrix} x_1 \\ x_2 \\ x_3 \end{bmatrix}\begin{bmatrix} y_1 \\ y_2 \\ y_3 \end{bmatrix}$ if $a \neq b$ then

$x_1 y_1 + x_2 y_2 + x_3 y_3 = 0$ because they are orthogonal.

$\therefore \quad x^T y = 0 \quad (\text{since } a \neq b)$

$$[x_1 \ x_2 \ x_3]\begin{bmatrix} y_1 \\ y_2 \\ y_3 \end{bmatrix} = 0$$

$$x_1 y_1 + x_2 y_2 + x_3 y_3 = \mathbf{0}$$

1.29 (d)

$$(P+Q)^2 = P^2 + PQ + QP + Q^2$$
$$= P.P + P.Q + Q.P + Q.Q$$
$$= P^2 + PQ + QP + Q^2$$

1.30 (a)

Property of determinant : If any two row or column are interchanged, then magnitude of determinant remains same but sign changes.

1.31 (b)

For singular matrix

$$|A| = 0$$

According to properties of eigen value

Product of eigen values = $|A| = 0$

$\Rightarrow$ Atleast one of the eigen value is **zero**.

1.32 Sol.

$$A = \begin{bmatrix} 4 & 2 \\ 1 & 3 \end{bmatrix}$$

$$\therefore \quad |A - \lambda I| = 0$$

$$\begin{bmatrix} 4-\lambda & 2 \\ 1 & 3-\lambda \end{bmatrix} = (4-\lambda)(3-\lambda) - 2 = 0$$

$$(\lambda - 4)(\lambda - 3) - 2 = 0$$
$$\lambda^2 - 7\lambda + 10 = 0$$
$$\Rightarrow \quad \lambda = 5, 2$$

Minimum value = **2**

1.33 (a)

$$P = \begin{bmatrix} 4+3i & -i \\ i & 4-3i \end{bmatrix}$$

$$P^{-1} = \frac{\begin{bmatrix} 4-3i & -(-i) \\ -i & 4+3i \end{bmatrix}}{|A|} = \frac{1}{24}\begin{bmatrix} 4-3i & i \\ -i & 4+3i \end{bmatrix}$$

1.34 (d)

$$\begin{bmatrix} 2 & 5 \\ -4 & 3 \end{bmatrix}\begin{bmatrix} x \\ y \end{bmatrix} = \begin{bmatrix} 2 \\ -30 \end{bmatrix}$$

$$\begin{bmatrix} 2 & 5 & 2 \\ -4 & 3 & -30 \end{bmatrix}$$

$$R_2 + 2R_1$$

$$\begin{bmatrix} 2 & 5 & 2 \\ 0 & 13 & -26 \end{bmatrix}$$

$$13y = -26$$

or $\quad y = -2$

$$2x + 5y = 2$$
$$2x + 5(-2) = 2$$
$$2x = 2 + 10 = 12$$

or $\quad x = \mathbf{6}$

1.35 (a)

All Eigen values of $A = \begin{bmatrix} 2 & 1 \\ 1 & k \end{bmatrix}$ are positive

$$2 > 0$$

$\therefore$ 2 × 2 leading minor must be greater than zero

$$\begin{vmatrix} 2 & 1 \\ 1 & k \end{vmatrix} > 0$$
$$2k - 1 > 0$$
$$2k > 1$$
$$k > \frac{1}{2}$$

1.36 (c)

A is skew-symmetric

$\therefore \quad A^T = -A$

1.37 Sol.

Consider $A - \lambda I = \begin{bmatrix} 2-\lambda & 1 & 0 \\ 0 & 2-\lambda & 0 \\ 0 & 0 & 3-\lambda \end{bmatrix}$

Ch. equation is $|A - \lambda I| = 0$

$(2-\lambda)(2-\lambda)(3-\lambda) = 0$

$$\lambda = 2, 2, 3$$

$\lambda = 3$ there is one L.I. Eigen vector

$\lambda = 2$ Consider $(A - 2I)x = 0$

rank = 2 The equation are $x_2 = 0$

No. of variables = 3 $\quad x_3 = 0$

Let $x_1 = k$ be independent.

$\therefore$ Eigen vector is $\begin{bmatrix} k \\ 0 \\ 0 \end{bmatrix} = k\begin{bmatrix} 1 \\ 0 \\ 0 \end{bmatrix}$

Only one independent Eigen vector in the case of $\lambda = 2$

Hence finally no. of L.I. Eigen vectors = 2

1.38 (b)

Product of Eigen value = determinant value

$$= 2(3-6) + 1(8-0)$$
$$= 2(-3) + 8 = -6 + 8 = \mathbf{2}$$

1.39 (d)

Explanations: The Eigen values of an orthogonal matrix are of magnitude 1 and are real (a) complex conjugate pairs.

1.40 Sol.

The product of eigen value is always equal to the determinant value of the matrix.

$\lambda_1 = 10 \quad \lambda_2 =$ unknown $\quad |A| = 50$

$$\lambda_1 \cdot \lambda_2 = 50$$
$$10(\lambda_2) = 50$$
$$\therefore \quad \lambda_2 = 5$$

1.41 Sol.

$$A = \begin{bmatrix} 50 & 70 \\ 70 & 80 \end{bmatrix}$$

Eigen values of A are λ_1, λ_2

$$\lambda_1 + \lambda_2 = 130$$
$$\lambda_1\lambda_2 = -900$$

Given that $X_1 = \begin{bmatrix} 70 \\ \lambda_1 - 50 \end{bmatrix}$ $X_2 = \begin{bmatrix} \lambda_2 - 80 \\ 70 \end{bmatrix}$

$$X_1^T X_2 = [70 \quad \lambda_1 - 50] \begin{bmatrix} \lambda_2 - 80 \\ 70 \end{bmatrix}$$

$= 70\,\lambda_2 - 5600 + 70\,\lambda_1 - 3500$
$= 70\,(\lambda_1 + \lambda_2) - 9100 = 70\,(130) - 9100$
$= 9100 - 9100 = \mathbf{0}$

1.42 (b)

$$\begin{bmatrix} -4 & 1 & -1 \\ -1 & -1 & -1 \\ 7 & -3 & 1 \end{bmatrix}$$

$$R_1 \longleftrightarrow R_2 \quad \begin{bmatrix} -1 & -1 & -1 \\ -4 & 1 & -1 \\ 7 & -3 & 1 \end{bmatrix}$$

$$R_2 - 4R_1, R_3 + 7R_1 \quad \begin{bmatrix} -1 & -1 & -1 \\ 0 & 5 & 3 \\ 0 & -10 & -6 \end{bmatrix}$$

$$R_3 + 2R_2 \quad \begin{bmatrix} -1 & -1 & -1 \\ 0 & 5 & 3 \\ 0 & 0 & 0 \end{bmatrix}$$

No. of non zero rows = 2
rank = **2**

1.43 Sol.

$$A = a\begin{bmatrix} 1 & 2 & 3 \\ 0 & 4 & 5 \\ 0 & 0 & 1 \end{bmatrix}$$

$$|A| = 4$$

$$\left|A^{-1}\right| = \frac{1}{|A|} = \frac{1}{4}$$

1.44 (c)

$$A = \begin{bmatrix} 1 & 1 & 0 \\ 0 & 1 & 1 \\ 0 & 0 & 1 \end{bmatrix}$$

Since given matrix is U.T.M.
So, diagonal elements are the eigen values.
Hence, $\lambda = 1, 1, 1$. So matrix has only one distinct eigen value i.e.

1.45 (d)

Given system is non-homogeneous system when augmented matrix

$$c = [a/B] = \left[\begin{array}{ccc|c} 1 & 1 & 1 & 1 \\ a & -a & 3 & 5 \\ 5 & -3 & a & 6 \end{array}\right]$$

$R_2 \div a$ and $R_3 \rightarrow R_3 - 5R_1$

$$c = \left[\begin{array}{ccc|c} 1 & 1 & 1 & 1 \\ 1 & -1 & 3/a & 5/a \\ 0 & -8 & a-5 & 1 \end{array}\right]$$

$R_2 \rightarrow R_2 - R_1$,

$$c = \left[\begin{array}{ccc|c} 1 & 1 & 1 & 1 \\ 0 & -2 & \frac{3}{a}-1 & \frac{5}{a}-1 \\ 0 & -8 & a-5 & 1 \end{array}\right]$$

$R_3 \rightarrow R_3 - 4R_2$,

$$c = \left[\begin{array}{ccc|c} 1 & 1 & 1 & 1 \\ 0 & -2 & \frac{3}{a}-1 & \frac{5}{a}-1 \\ 0 & 0 & a-\frac{12}{a}-1 & 5-\frac{20}{a} \end{array}\right]$$

This system will have infinitely many solution only.

If $\quad a - \frac{12}{a} - 1 = 0$ and $5 - \frac{20}{a} = 0$

$$a = -3, 4 \text{ and } a = 4$$

for $a = 4$, the system has infinite many solution.

1.46 (d)

$$[A]\{x\} = \begin{bmatrix} 4 & 8 & 4 \\ 8 & 16 & -4 \\ 4 & -4 & 15 \end{bmatrix}\begin{bmatrix} 2 \\ 1 \\ 4 \end{bmatrix}$$

$$= \begin{bmatrix} 32 \\ 16 \\ 64 \end{bmatrix} = 16\begin{bmatrix} 2 \\ 1 \\ 4 \end{bmatrix} = 16\{x\}$$

So, one of the eigen value is 16.

1.47 (b)

Matrix multiplication is associative.

1.48 (c)

$$S = \frac{P + P^T}{2}$$

$$V = \frac{P - P^T}{2}$$

$$P = S + V = \begin{pmatrix} -4 & 2 & 5 \\ 6 & 3 & 7 \\ -1 & 0 & 2 \end{pmatrix}$$

1.49 Sol.

I_{100}

Eigen values of $I \rightarrow \underbrace{1, 1, \ldots\ldots\ldots, 1}_{100 \text{ times}}$

Set of distributed eigen value $E = \{1\}$

Number of elements in $E = 1$

1.50 (d)

Given, $AP = \alpha^2 P$

By comparison with $AX = \lambda X$

$\Rightarrow \quad \lambda = \alpha^2$

Hence, eigen value of A is α^2, so eigen value of A^2 is α^4.

1.51 (c)

For $n = 2$, matrix are

$$Q = \begin{bmatrix} a & c \\ c & b \end{bmatrix}, x = \begin{bmatrix} x_1 \\ x_2 \end{bmatrix}, R = \begin{bmatrix} r_1 \\ r_2 \end{bmatrix}$$

$$F(x) = \frac{1}{2}[x_1, x_2]\begin{bmatrix} a & c \\ c & b \end{bmatrix}\begin{bmatrix} x_1 \\ x_2 \end{bmatrix} - [r_1\ r_2]\begin{bmatrix} x_1 \\ x_2 \end{bmatrix}$$

$$= \frac{1}{2}\left[ax_1^2 + bx_2^2 + 2cx_1x_2\right] - \left[r_1x_1 + r_2x_2\right]$$

i.e. $F(x_1, x_2) = \frac{1}{2}ax_1^2 + \frac{1}{2}bx_2^2 + cx_1x_2 - r_1x_1 - r_2x_2$

Now, for critical point, $\frac{\partial f}{\partial x_1} = 0$ and $\frac{\partial f}{\partial x_2} = 0$

$\Rightarrow \quad a_1x_1 + cx_2 - r_1 = 0$

and $bx_2 + cx_1 - r_2 = 0$

In matrix form we can write it as

$$\begin{bmatrix} a & c \\ c & b \end{bmatrix}\begin{bmatrix} x_1 \\ x_2 \end{bmatrix} = \begin{bmatrix} r_1 \\ r_2 \end{bmatrix}$$

$\Rightarrow \quad Qx = R$

By multiplying both side by Q^{-1}

$$x = Q^{-1}R$$

1.52 (a)

$\because$ A is a 2×2 matrix and P is the eigen vector of matrix A with corresponding eigen value λ.

Given: $P' = AP$

$\because \quad AP = \lambda P$

Hence, $P' = \lambda P$

So, $\|P'\| = \|\lambda P\| = \lambda\|P\|$

But direction of vector P' will be same as vector P.

1.53 (a)

For the matrix to be symmetric,

$$a_{12} = a_{21}$$

$\Rightarrow \quad 2k + 5 = 3k - 3$

$\Rightarrow \quad k = 8$

1.54 (a, b)

For non-trivial solution,

$$|A| = 0$$

$$\Rightarrow \begin{vmatrix} 2 & 5-2\alpha \\ \alpha & 1 \end{vmatrix} = 0$$

$2 - \alpha(5 - 2\alpha) = 0$

$2\alpha - 5\alpha + 2\alpha^2 = 0$

$2\alpha^2 - 5\alpha + 2 = 0$

$2\alpha^2 - 4\alpha - \alpha + 2 = 0$

$2\alpha(\alpha - 2) - (\alpha - 2) = 0$

$$\alpha = 2, \frac{1}{2}$$

For $\alpha = 2$, $A = \begin{bmatrix} 2 & 1 \\ 2 & 1 \end{bmatrix}$

Thus, $[x \quad y]\begin{bmatrix} 2 & 1 \\ 2 & 1 \end{bmatrix} = [0, 0]$

$2x + 2y = 0$

$\Rightarrow \quad x + y = 0$

$x = -y$

Option (a) satisfies this condition.

For, $\alpha = \frac{1}{2}$, $A = \begin{bmatrix} 2 & 4 \\ \frac{1}{2} & 1 \end{bmatrix}$

Thus, $[x \quad y]\begin{bmatrix} 2 & 4 \\ \frac{1}{2} & 1 \end{bmatrix} = [0, 0]$

$2x + \frac{1}{2}y = 0$

and, $4x + y = 0$

$\Rightarrow \quad x = -\frac{y}{4}$

Option (b) satisfies this condition.

1.55 (b, c)

Given: $\rho(A) = 2$

Since the rank of matrix A is less than the number of variables, so there will be infinitely many solutions including a zero vector of appropriate size.

1.56 (2) (2 to 2)

From properties of eigen values,

1. Sum of eigen values = Trace of matrix

$\Rightarrow \quad 4 = 3 + q$

or $\quad q = 1$

2. Product of eigen values = Determinant

$$\Rightarrow \quad -1 = \begin{vmatrix} 3 & p \\ p & q \end{vmatrix}$$

$\Rightarrow \quad 3q - p^2 = -1$

$\Rightarrow \quad 3 - p^2 = -1$

$\Rightarrow \quad p^2 = 4$

$\Rightarrow \quad p = \pm 2$

Thus, $\quad |p| = 2$

1.57 (18.84) (18.80 to 18.90)

Given that,

$\hat{x} = 3y$ and $\hat{y} = 2x$

$\therefore y = \dfrac{\hat{x}}{3}$ and $x = \dfrac{\hat{y}}{2}$

The given region, $x^2 + y^2 \leq 1$

or, $\left(\dfrac{\hat{y}}{2}\right)^2 + \left(\dfrac{\hat{x}}{3}\right)^2 \leq 1$

$$\frac{(\hat{y})^2}{4} + \frac{(\hat{x})^2}{9} \leq 1$$

This is equation of an ellipse with semi major axis, $a = 3$ and semi-minor axis, $b = 2$, thus area of transformed region is

$A = \pi ab = 6\pi \simeq 18.84$ units

1.58 (c)

$$x + 2y + z = 5$$
$$2x + ay + 4z = 12$$
$$2x + 4y + 6z = b$$

Performing row operations on augmented matrix,

$$\left[\begin{array}{ccc|c} 1 & 2 & 1 & 5 \\ 2 & a & 4 & 12 \\ 2 & 4 & 6 & b \end{array}\right] \xrightarrow[R_3 = R_2 - 2R_1]{R_2 = R_2 - 2R_1} \left[\begin{array}{ccc|c} 1 & 2 & 1 & 5 \\ 0 & a-4 & 2 & 2 \\ 0 & 0 & 4 & b-10 \end{array}\right]$$

For infinitely many solutions,

$$\rho(A) = \rho(AB) = 2$$

$\Rightarrow \quad a - 4 = 0$

$\quad a = 4$

and, $\quad \rho(AB) = 2$

for this $\quad R_3 = 2R_2$

$\quad b - 10 = 4$

$\quad b = 14$

1.59 (c)

For a negative eigen value, product of eigen values < 0

$\Rightarrow$ Determinant < 0

$\Rightarrow \quad 3 - 8a < 0$

$\Rightarrow \quad 8a > 3$

$\Rightarrow \quad a > \dfrac{3}{8}$

■■■■

2 Calculus

2.1 $\underset{x \to 0}{Lt} \dfrac{\sin^2 x}{x}$ is equal to

(a) 0 (b) ∞
(c) 1 (d) –1 [2003 : 1 M]

2.2 The area enclosed between the parabola $y = x^2$ and the straight line $y = x$ is

(a) 1/8 (b) 1/6
(c) 1/3 (d) 1/2 [2003 : 2 M]

2.3 If $x = a(\theta + \sin\theta)$ and $y = a(1 - \cos\theta)$, then dy/dx will be equal to

(a) $\sin\left(\dfrac{\theta}{2}\right)$ (b) $\cos\left(\dfrac{\theta}{2}\right)$
(c) $\tan\left(\dfrac{\theta}{2}\right)$ (d) $\cot\left(\dfrac{\theta}{2}\right)$ [2004 : 1 M]

2.4 The angle between two unit-magnitude coplanar vectors P(0.866, 0.500, 0) and Q(0.259, 0.966, 0) will be

(a) 0° (b) 30°
(c) 45° (d) 60° [2004 : 1 M]

2.5 The volume of an object expressed in spherical coordinates is given by

$$V = \int_0^{2\pi}\int_0^{\pi/3}\int_0^1 r^2 \sin\phi \, dr\, d\phi\, d\theta$$

The value of the integral is

(a) $\dfrac{\pi}{3}$ (b) $\dfrac{\pi}{6}$
(c) $\dfrac{2\pi}{3}$ (d) $\dfrac{\pi}{4}$ [2004 : 2 M]

2.6 $\displaystyle\int_{-a}^{a} \left(\sin^6 x + \sin^7 x\right) dx$ is equal to

(a) $2\displaystyle\int_0^a \sin^6 x\, dx$ (b) $2\displaystyle\int_0^a \sin^7 x\, dx$
(c) $2\displaystyle\int_0^a \left(\sin^6 x + \sin^7 x\right) dx$ (d) Zero

[2005 : 1 M]

2.7 Changing the order of the integration in the double integral $I = \displaystyle\int_0^8 \int_{x/4}^2 f(x, y)\, dy\, dx$ leads to $I = \displaystyle\int_r^s \int_p^q f(x, y)\, dx\, dy$. What is q?

(a) $4y$ (b) $16y^2$
(c) x (d) 8 [2005 : 1 M]

2.8 By a change of variable $x(u, v) = uv$, $y(u, v) = v/u$ is double integral, the integrand $f(x, y)$ changes to $f(uv, v/u)\,\phi(u, v)$. Then, $\phi(u, v)$ is

(a) $2v/u$ (b) $2uv$
(c) v^2 (d) 1 [2005 : 2 M]

2.9 The right circular cone of largest volume that can be enclosed by a sphere of 1 m radius has a height of

(a) $\dfrac{1}{3}$m (b) $\dfrac{2}{3}$m
(c) $\dfrac{2\sqrt{2}}{3}$m (d) $\dfrac{4}{3}$m [2005 : 2 M]

2.10 Equation of the line normal to function $f(x) = (x - 8)^{2/3} + 1$ at P(0, 5) is

(a) $y = 3x - 5$ (b) $y = 3x + 5$
(c) $3y = x + 15$ (d) $3y = x - 15$

[2006 : 2 M]

2.11 If $f(x) = \dfrac{2x^2 - 7x + 3}{5x^2 - 12x - 9}$, then $\lim\limits_{x \to 3} f(x)$ will be

(a) –1/3 (b) 5/18
(c) 0 (d) 2/5 [2006 : 2 M]

2.12 The minimum value of function $y = x^2$ in the interval [1, 5] is

(a) 0 (b) 1
(c) 25 (d) undefined

[2007 : 1 M]

2.13 If, $y = x + \sqrt{x + \sqrt{x + \sqrt{x + \ldots\infty}}}$, then $y(2) =$

(a) 4 or 1 (b) 4 only
(c) 1 only (d) undefined

[2007 : 2 M]

2.14 $\lim\limits_{x \to 0} \dfrac{e^x - \left(1 + x + \dfrac{x^2}{2}\right)}{x^3} =$

(a) 0 (b) 1/6
(c) 1/3 (d) 1 [2007 : 2 M]

2.15 The value of $\lim_{x\to 8} \frac{x^{1/3}-2}{(x-8)}$

(a) $\frac{1}{16}$ (b) $\frac{1}{12}$
(c) $\frac{1}{8}$ (d) $\frac{1}{4}$ [2008 : 1 M]

2.16 Consider the shaded triangular region P shown in the figure. What is $\iint_P xy\,dx\,dy$?

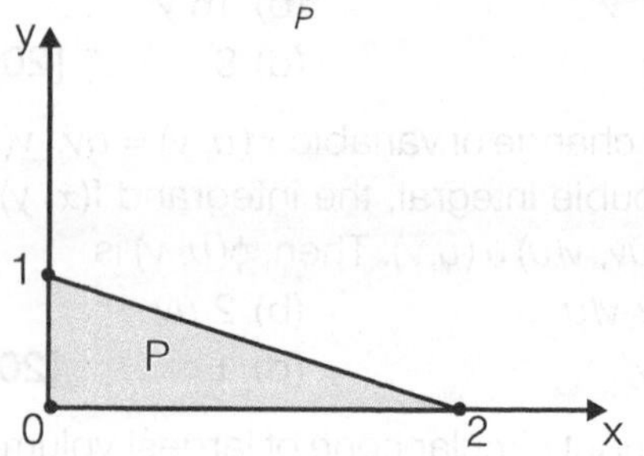

(a) $\frac{1}{6}$ (b) $\frac{2}{9}$
(c) $\frac{7}{16}$ (d) 1 [2008 : 2 M]

2.17 Which of the following integrals is unbounded ?

(a) $\int_0^{\pi/4} \tan x\,dx$ (b) $\int_0^{\infty} \frac{1}{x^2+1}dx$
(c) $\int_0^{\infty} xe^{-x}\,dx$ (d) $\int_0^{1} \frac{1}{1-x}dx$

[2008 : 2 M]

2.18 The length of the curve $y=\frac{2}{3}x^{3/2}$ between $x = 0$ and $x = 1$ is
(a) 0.27 (b) 0.67
(c) 1 (d) 1.22 [2008 : 2 M]

2.19 Let $f = y^x$. What is $\frac{\partial^2 f}{\partial x \partial y}$ at $x = 2$, $y = 1$?
(a) 0 (b) ln 2
(c) 1 (d) $\frac{1}{\ln 2}$ [2008 : 2 M]

2.20 In the Taylor series expansion of e^x about $x = 2$, the coefficient of $(x - 2)^4$ is
(a) 1/4! (b) $2^4/4!$
(c) $e^2/4!$ (d) $e^4/4!$ [2008 : 1 M]

2.21 The distance between the origin and the point nearest to it on the surface $z^2 = 1 + xy$ is
(a) 1 (b) $\frac{\sqrt{3}}{2}$
(c) $\sqrt{3}$ (d) 2 [2009 : 2 M]

2.22 The area enclosed between the curves $y^2 = 4x$ and $x^2 = 4y$ is
(a) $\frac{16}{3}$ (b) 8
(c) $\frac{32}{3}$ (d) 16 [2009 : 2 M]

2.23 The parabolic arc $y=\sqrt{x}$, $1 \le x \le 2$ is revolved around the x-axis. The volume of the solid of revolution is
(a) $\pi/4$ (b) $\pi/2$
(c) $3\pi/4$ (d) $3\pi/2$ [2010 : 1 M]

2.24 The value of the integral $\int_{-\infty}^{\infty} \frac{dx}{1+x^2}$ is
(a) $-\pi$ (b) $-\pi/2$
(c) $\pi/2$ (d) π [2010 : 1 M]

2.25 The function $y = |2 - 3x|$
(a) is continuous $\forall\, x \in R$ and differentiable $\forall\, x \in R$
(b) is continuous $\forall\, x \in R$ and differentiable $\forall\, x \in R$ except at $x = 3/2$
(c) is continuous $\forall\, x \in R$ and differentiable $\forall\, x \in R$ except at $x = 2/3$
(d) is continuous $\forall\, x \in R$ except $x = 3$ and differentiable $\forall\, x \in R$ [2010 : 1 M]

2.26 A series expansion for the function $\sin\theta$ is
(a) $1-\frac{\theta^2}{2!}+\frac{\theta^4}{4!}-\ldots$ (b) $\theta-\frac{\theta^3}{3!}+\frac{\theta^5}{5!}-\ldots$
(c) $1+\theta+\frac{\theta^2}{2!}+\frac{\theta^3}{3!}+\ldots$ (d) $\theta+\frac{\theta^3}{3!}+\frac{\theta^5}{5!}+\ldots$
[2011 : 1 M]

2.27 What is $\lim_{\theta\to 0} \frac{\sin\theta}{\theta}$ equal to?
(a) θ (b) $\sin\theta$
(c) 0 (d) 1 [2011 : 1 M]

2.28 If $f(x)$ is an even function and a is a positive real number, then $\int_{-a}^{a} f(x)dx$ equals
(a) 0 (b) a
(c) $2a$ (d) $2\int_0^a f(x)dx$
[2011 : 1 M]

2.29 Consider the function $f(x) = |x|$ in the interval $-1 < x \le 1$. At the point $x = 0$, $f(x)$ is
(a) continuous and differentiable
(b) non-continuous and differentiable
(c) continuous and non-differentiable
(d) neither continuous nor differentiable
[2012 : 1 M]

2.30 $\lim_{x\to 0}\left(\frac{1-\cos x}{x^2}\right)$ is
(a) 1/4 (b) 1/2
(c) 1 (d) 2 [2012 : 1 M]

2.31 The area enclosed between the straight line $y = x$ and the parabola $y = x^2$ in the $x - y$ plane is
(a) 1/6 (b) 1/4
(c) 1/3 (d) 1/2 [2012 : 1 M]

2.32 At $x = 0$, the function $f(x) = x^3 + 1$ has
(a) a maximum value
(b) a minimum value
(c) a singularity
(d) a point of inflection [2012 : 1 M]

2.33 The value of the definite integral $\int_1^e \sqrt{x}\ln(x)dx$ is
(a) $\frac{4}{9}\sqrt{e^3}+\frac{2}{9}$ (b) $\frac{2}{9}\sqrt{e^3}-\frac{4}{9}$
(c) $\frac{2}{9}\sqrt{e^3}+\frac{4}{9}$ (d) $\frac{4}{9}\sqrt{e^3}-\frac{2}{9}$
[2013 : 2 M]

2.34 $\text{Lim}_{x\to 0}\frac{x-\sin x}{1-\cos x}$
(a) 0 (b) 1
(c) 3 (d) not defined
[2014 : 1 M, Set-1]

2.35 $\text{Lim}_{x\to 0}\left(\frac{e^{2x}-1}{\sin(4x)}\right)$ is equal to
(a) 0 (b) 0.5
(c) 1 (d) 2 [2014 : 1 M, Set-2]

2.36 The best approximation of the minimum value attained by $e^{-x}\sin(100x)$ for $x \ge 0$ is ____.
[2014 : 1 M, Set-2]

2.37 If a function is continuous at a point,
(a) the limit of the function may not exist at the point.
(b) the function must be derivable at the point.
(c) the limit of the function at the point tends to infinity.
(d) the limit must exist at the point and the value of limit should be same as the value of the function at that point.
[2014 : 1 M, Set-3]

2.38 The value of the integral
$$\int_0^2 \frac{(x-1)^2\sin(x-1)}{(x-1)^2+\cos(x-1)}dx \text{ is}$$
(a) 3 (b) 0
(c) –1 (d) –2
[2014 : 1 M, Set-4]

2.39 The value of the integral $\int_0^2\int_0^{2x} e^{x+y}dy\,dx$
(a) $\frac{1}{2}(e-1)$ (b) $\frac{1}{2}(e^2-1)^2$
(c) $\frac{1}{2}(e^2-e)$ (d) $\frac{1}{2}\left(e-\frac{1}{e}\right)^2$
[2014 : 2 M, Set-4]

2.40 The value of $\lim_{x\to 0}\frac{1-\cos(x^2)}{2x^4}$ is
(a) 0 (b) $\frac{1}{2}$
(c) $\frac{1}{4}$ (d) undefined
[2015 : 1 M, Set-1]

2.41 At $x = 0$, the function $f(x) = |x|$ has
(a) a minimum
(b) a maximum
(c) a point of inflection
(d) neither a maximum nor minimum
[2015 : 1 M, Set-2]

2.42 The value of $\lim_{x\to 0}\left(\frac{-\sin x}{2\sin x+\cos x}\right)$ is ________.
[2015 : 1 M, Set-3]

2.43 Consider a spatial curve in three-dimensional space given in parametric form by
$x(t) = \cos t,\ y(t) = \sin t,\ z(t) = \frac{2}{\pi}t,\ 0 \le t \le \frac{\pi}{2}$
The length of the curve is ________.
[2015 : 2 M, Set-1]

2.44 Consider an ant crawling along the curve $(x-2)^2 + y^2 = 4$, where x and y are in meters. The ant starts at the point (4, 0) and moves counter-clockwise with a speed of 1.57 meters per second. The time taken by the ant to reach the point (2, 2) is (in seconds) _______.

[2015 : 2 M, Set-1]

2.45 Consider the function $f(x) = 2x^3 - 3x^2$ in the domain [–1, 2]. The global minimum of $f(x)$ is_______.

[2016 : 2 M, Set-1]

2.46 The values of x for which the function

$$f(x) = \frac{x^2 - 3x - 4}{x^2 + 3x - 4}$$

is NOT continuous are

(a) 4 and –1 (b) 4 and 1
(c) – 4 and 1 (d) – 4 and –1

[2016 : 1 M, Set-2]

2.47 $\underset{x\to 0}{Lt} \dfrac{\log_e(1+4x)}{e^{3x}-1}$ is equal to

(a) 0 (b) 1/12
(c) 4/3 (d) 1 [2016 : 1 M, Set-3]

2.48 $\lim\limits_{x\to\infty} \sqrt{x_2 + x - 1} - x$ is

(a) 0 (b) ∞
(c) 1/2 (d) – ∞

[2016 : 2 M, Set-3]

2.49 The value of $\lim_{x\to 0} \dfrac{x^3 - \sin(x)}{x}$ is

(a) 0 (b) 3
(c) 1 (d) – 1 [2017 : 2 M, Set-1]

2.50 A parametric curve defined by $x = \cos\left(\dfrac{\pi u}{2}\right), y = \sin\left(\dfrac{\pi u}{2}\right)$ in the range $0 \le u \le 1$ is rotated about the X-axis by 360 degrees. Area of the surface generated is

(a) $\dfrac{\pi}{2}$ (b) π
(c) 2π (d) 4π [2017 : 2 M, Set-1]

2.51 According to the Mean Value Theorem, for a continuous function $f(x)$ in the interval $[a, b]$, there exists a value ξ in this interval such that $\int_a^b f(x)dx =$

(a) $f(\xi)(b-a)$ (b) $f(b)(\xi - a)$
(c) $f(a)(b-\xi)$ (d) 0 [2018 : 1 M, Set-1]

2.52 A parabola $x = y^2$ with $0 \le x \le 1$ is shown in the figure. The volume of the solid of rotation obtained by rotating the shaded area by 360° around the x–axis is

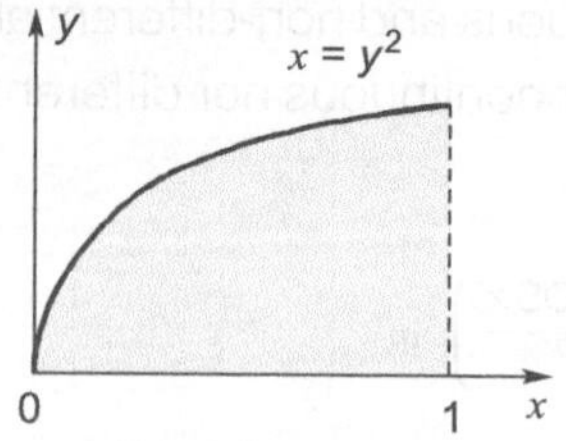

(a) π/4 (b) π
(c) π/2 (d) 2π [2019 : 1 M, Set-1]

2.53 The value of the following definite integral is _______ (round off to three decimal places)

$$\int_1^e (x \ln x)\,dx$$ [2019 : 2 M, Set-1]

2.54 The derivative of $f(x) = \cos x$ can be estimated using the approximation

$$f'(x) = \frac{f(x+h) - f(x-h)}{2h}.$$

The percentage error is calculated as $\left(\dfrac{\text{Exact value} - \text{Approx value}}{\text{Exact value}} \times 100\right)$.

The percentage error in the derivative of $f(x)$ at $x = \dfrac{\pi}{6}$ radian choosing $h = 0.1$ radian is

(a) > 1% and <5% (b) < 0.1%
(c) > 0.1% and <1% (d) > 5%

[2019 : 2 M, Set-2]

2.55 The value of $\lim\limits_{x\to 1}\left(\dfrac{1-e^{-c(1-x)}}{1-xe^{-c(1-x)}}\right)$

(a) $c + 1$ (b) $\dfrac{c+1}{c}$
(c) c (d) $\dfrac{c}{c+1}$

[2020 : 1 M, Set-1]

2.56 Let $I = \int_{x=0}^{1}\int_{y=0}^{x^2} xy^2\,dy\,dx$. then, I may also be expressed as

(a) $\int_{y=0}^{1}\int_{x=\sqrt{y}}^{1} yx^2\,dx\,dy$

(b) $\int_{y=0}^{1}\int_{x=0}^{\sqrt{y}} xy^2\,dx\,dy$

(c) $\int_{y=0}^{1}\int_{x=\sqrt{y}}^{1} xy^2 dx\, dy$

(d) $\int_{y=0}^{1}\int_{x=0}^{\sqrt{y}} yx^2 dx\, dy$ [2020 : 1 M, Set-2]

2.57 An analytic function of a complex variable $z = x + iy\left(i = \sqrt{-1}\right)$ is defined as

$$f(z) = x^2 - y^2 + i\psi(x, y)$$

where $\psi(x, y)$ is a real function. The value of the imaginary part of $f(z)$ at $z = (1 + i)$ is ______ (round off to 2 decimal places).

[2020 : 2 M, Set-1]

2.58 Let $f(x) = x^2 - 2x + 2$ be a continuous defined on $x \in [1, 3]$. The point x at which the tangent of $f(x)$ becomes parallel to the straight line joining f(1) and f(3) is

(a) 1 (b) 2

(c) 0 (d) 3 [2021 : 2 M, Set-1]

2.59 The value of $\int_{0}^{\pi/2}\int_{0}^{\cos\theta} r\sin\theta\, dr\, d\theta$ is

(a) 0 (b) $\frac{1}{6}$

(c) $\frac{4}{3}$ (d) π [2021 : 2 M, Set-2]

2.60 The value of $\lim_{x\to 0}\left(\frac{1-\cos x}{x^2}\right)$ is

(a) $\frac{1}{2}$ (b) $\frac{1}{3}$

(c) 1 (d) $\frac{1}{4}$ [2021 : 1 M, Set-1]

2.61 The limit

$$p = \lim_{x\to\pi}\left(\frac{x^2 + \alpha x + 2\pi^2}{x - \pi + 2\sin x}\right)$$

has a finite value of real α. The value of α and the corresponding limit p are

(a) $\alpha = -3\pi$, and $p = \pi$

(b) $\alpha = -2\pi$, and $p = 2\pi$

(c) $\alpha = \pi$, and $p = \pi$

(d) $\alpha = 2\pi$, and $p = 3\pi$ [2022 : 1 M, Set-1]

2.62 Solution of $\nabla^2 T = 0$ in a square domain $(0 < x < 1 \text{ and } 0 < y < 1)$ with boundary conditions: $T(x, 0) = x$; $T(0, y) = y$; $T(x, 1) = 1 + x$; $T(1, y) = 1 + y$ is

(a) $T(x, y) = x - xy + y$

(b) $T(x, y) = x + y$

(c) $T(x, y) = -x + y$

(d) $T(x, y) = x + xy + y$ [2022 : 1 M, Set-1]

2.63 Given $\int_{-\infty}^{\infty} e^{-x^2} dx = \sqrt{\pi}$. If a and b are integers, the value of $\int_{-\infty}^{\infty} e^{-a(x+b)^2} dx$ is ______.

(a) $\sqrt{a\pi}$ (b) $\sqrt{\frac{\pi}{a}}$

(c) $b\sqrt{\pi a}$ (d) $b\sqrt{\frac{\pi}{a}}$

[2022 : 1 M, Set-2]

2.64 A polynomial $\phi(s) = a_n s^n + a_{n-1}s^{n-1} + \ldots + a_1 s + a_o$ of degree $n > 3$ with constant real coefficients a_n, a_{n-1}, ... a_o has triple roots at $s = -\sigma$. Which one of the following conditions must be satisfied?

(a) $\phi(s) = 0$ at all the three values of s satisfying $s^3 + \sigma^3 = 0$

(b) $\phi(s) = 0,\ \frac{d\phi(s)}{ds} = 0, \text{ and } \frac{d^2\phi(s)}{ds^2} = 0$ at $s = -\sigma$

(c) $\phi(s) = 0,\ \frac{d^2\phi(s)}{ds^2} = 0, \text{ and } \frac{d^4\phi(s)}{ds^4} = 0$ at $s = -\sigma$

(d) $\phi(s) = 0, \text{ and } \frac{d^3\phi(s)}{ds^3} = 0$ at $s = -\sigma$

[2022 : 1 M, Set-2]

2.65 The figure shows the plot of a function over the interval [–4, 4]. Which one of the options given CORRECTLY identifies the function?

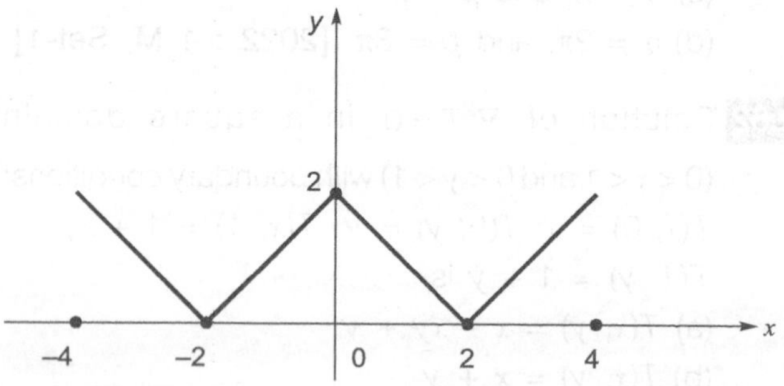

(a) $|2-x|$ (b) $|2-|x||$

(c) $|2+|x||$ (d) $2-|x|$ [2023 : 1 M]

2.66 The smallest perimeter that a rectangle with area of 4 square units can have is ________ units. (Answer in integer)

[2023: 2 M]

2.67 Let $f(.)$ be a twice differentiable function from $\mathbb{R}^2 \to \mathbb{R}$. If $p, x_o \in \mathbb{R}^2$ where $\|p\|$ is sufficiently small (here, $\|.\|$ is the Euclidean norm or distance function),

then $f(x_o + p) = f(x_o)+\nabla f(x_o)^T p+\frac{1}{2}p^T\nabla^2 f(\psi)p$

where $\psi \in \mathbb{R}^2$ is a point on the line segment joining x_o and x_o + p. If x_o is a strict local minimum of $f(x)$, then which one of the following statements is TRUE?

(a) $\nabla f(x_o)^T p=0$ and $p^T\nabla^2 f(\psi)p>0$

(b) $\nabla f(x_o)^T p=0$ and $p^T\nabla^2 f(\psi)p=0$

(c) $\nabla f(x_o)^T p>0$ and $p^T\nabla^2 f(\psi)p=0$

(d) $\nabla f(x_o)^T p=0$ and $p^T\nabla^2 f(\psi)p<0$

[2024 : 1 M]

2.68 If the value of the double integral

$$\int_{x=3}^{4}\int_{y=1}^{2}\frac{dydx}{(x+y)^2}$$

is $\log_e(a/24)$, then a is ________ (answer in integer)

[2024 : 2 M]

■■■■

Answers Calculus

2.1	(a)	2.2	(b)	2.3	(c)	2.4	(c)	2.5	(a)	2.6	(a)	2.7	(a)	2.8	(a)
2.9	(d)	2.10	(b)	2.11	(b)	2.12	(b)	2.13	(b)	2.14	(b)	2.15	(b)	2.16	(a)
2.17	(d)	2.18	(d)	2.19	(c)	2.20	(c)	2.21	(a)	2.22	(a)	2.23	(d)	2.24	(d)
2.25	(c)	2.26	(b)	2.27	(d)	2.28	(d)	2.29	(c)	2.30	(b)	2.31	(a)	2.32	(d)
2.33	(c)	2.34	(a)	2.35	(b)	2.36	(0.9844)	2.37	(d)	2.38	(b)	2.39	(b)	2.40	(c)
2.41	(a)	2.42	(0)	2.43	(1.86)	2.44	(2)	2.45	(–5)	2.46	(c)	2.47	(c)	2.48	(c)
2.49	(d)	2.50	(c)	2.51	(a)	2.52	(c)	2.53	(2.097)	2.54	(c)	2.55	(d)	2.56	(c)
2.57	(2)	2.58	(b)	2.59	(b)	2.60	(a)	2.61	(a)	2.62	(b)	2.63	(b)	2.64	(b)
2.65	(b)	2.66	(8)	2.67	(a)	2.68	(25)								

Explanations Calculus

2.1 (a)

$$\underset{x\to 0}{\text{Lt}}\frac{\sin^2 x}{x} = \underset{x\to 0}{\text{Lt}}\left(\frac{\sin x}{x}\right)^2 x = 1\times 0 = \mathbf{0}$$

2.2 (b)

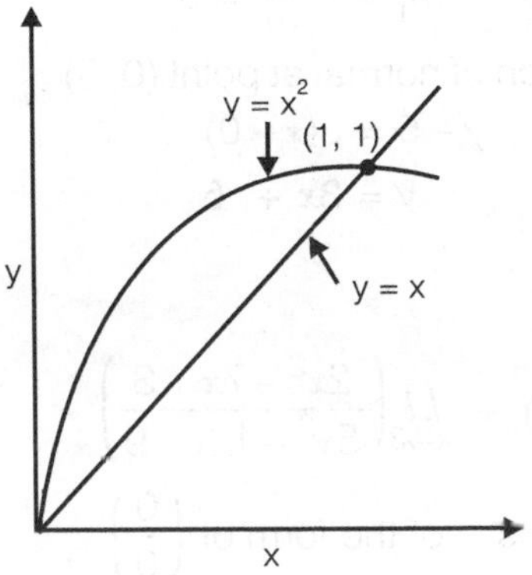

$$\text{Area enclosed} = \int_0^1 \left|\left(x^2 - x\right)\right| dx$$

$$= \left|\frac{x^3}{3} - \frac{x^2}{2}\right|_0^1 = \left|-\frac{1}{6}\right| = \frac{1}{6}$$

2.3 (c)

Given, $x = a(\theta + \sin\theta)$

$y = a(1 - \cos\theta)$

$$\therefore \quad \frac{dx}{d\theta} = a(1+\cos\theta), \quad \frac{dy}{d\theta} = a\sin\theta$$

$$\frac{dy}{dx} = \frac{dy/d\theta}{dx/d\theta} = \frac{a\sin\theta}{a(1+\cos\theta)}$$

$$= \frac{2a\sin\left(\frac{\theta}{2}\right)\cdot\cos\left(\frac{\theta}{2}\right)}{a\times 2\cos^2\left(\frac{\theta}{2}\right)} = \mathbf{\tan\theta/2}$$

2.4 (c)

$$\vec{P} = 0.866\hat{i} + 0.500\hat{j} + 0\hat{k}$$

$$\vec{Q} = 0.259\hat{i} + 0.966\hat{j} + 0\hat{k}$$

$$\therefore \quad \vec{P}\cdot\vec{Q} = |\vec{P}||\vec{Q}|\cos\theta$$

or, $(0.866\hat{i} + 0.5\hat{j} + 0\hat{k}).(0.259\hat{i} + 0.966\hat{j} + 0\hat{k})$

$$= \sqrt{(0.866)^2 + (0.5)^2} \times \sqrt{(0.259)^2 + (0.966)^2}.\cos\theta$$

$$\therefore \quad \cos\theta = \frac{0.866\times 0.259 + 0.5\times 0.966}{\sqrt{0.99}\times\sqrt{1.001}}$$

$$\cos\theta = 0.707$$

or $\theta = \mathbf{45°}$

2.5 (a)

$$V = \int_0^{2\pi}\int_0^{\pi/3}\int_0^1 r^2\sin\phi\cdot dr\cdot d\phi\cdot d\theta$$

$$= \int_0^{2\pi}\int_0^{\pi/3}\int_2^1 \sin\phi\cdot d\phi\cdot d\theta \quad = \frac{1}{2}\int_0^{2\pi}[1-\cos\phi]_0^{\pi/3}\, d\theta$$

$$= \frac{1}{3}\int_0^{2\pi}\frac{1}{2}d\theta = \frac{1}{6}\times 2\pi = \frac{\pi}{3}$$

2.6 (a)

$$I = \int_{-a}^{a}\left(\sin^6 x + \sin^7 x\right)dx = 2\int_0^a \sin^6 x\, dx + 0$$

$$= 2\int_0^a \sin^6 \boldsymbol{x}\, \boldsymbol{dx}$$

$$\left(\text{because } \int_0^a \sin^7 x\cdot dx = 0\right)$$

2.7 (a)

When $$I = \int_0^8\int_{x/4}^2 f(x\cdot y)\,dy\,dx$$

i.e.

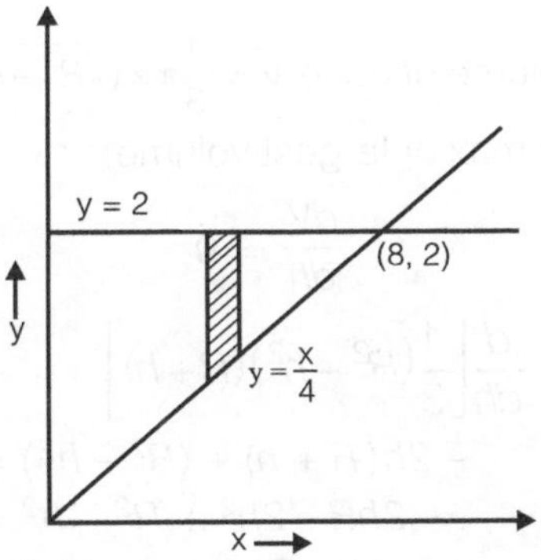

Now,

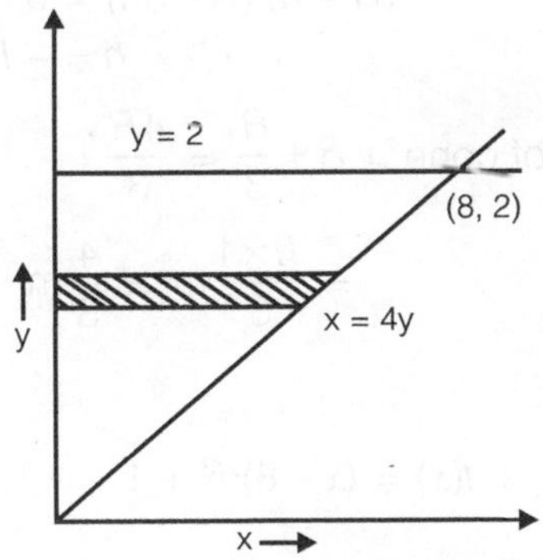

$$I = \int_0^2 \int_0^{4y} f(x,y)\,dy\,dx$$

2.8 (a)

$$\frac{\partial x}{\partial u} = v;\ \frac{\partial x}{\partial v} = u$$

and $$\frac{\partial y}{\partial u} = -\frac{v}{u^2}$$

$$\frac{\partial y}{\partial v} = \frac{1}{u}$$

and $$\phi(u,v) = \begin{vmatrix} \frac{\partial x}{\partial u} & \frac{\partial x}{\partial v} \\ \frac{\partial y}{\partial u} & \frac{\partial y}{\partial v} \end{vmatrix} = \begin{vmatrix} v & u \\ -\frac{v}{u^2} & \frac{1}{u} \end{vmatrix}$$

$$= \frac{v}{u} + \frac{v}{u} = \frac{2v}{u}$$

2.9 (d)

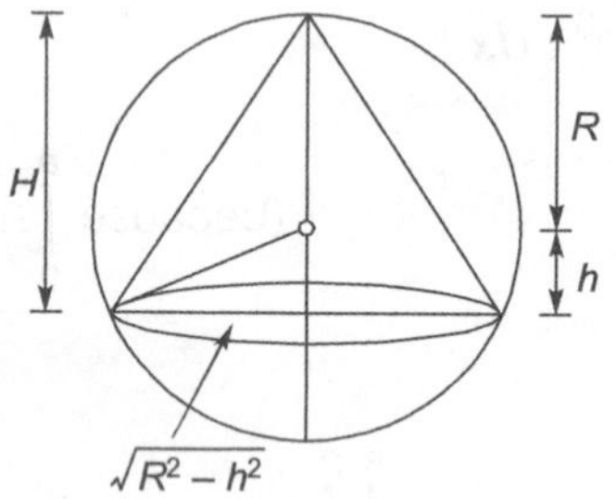

Given $R = 1$ radius of sphere. Let height of cone is $H = h + R$

Volume of cone $V = \frac{1}{3}\pi \times \left(\sqrt{R^2-h^2}\right)^2 (R+h)$

for max or largest volume,

$$\Rightarrow \frac{dV}{dh} = 0$$

$$\Rightarrow \frac{d}{dh}\left[\frac{1}{3}(R^2-h^2)(R+h)\right]$$

$$\Rightarrow -2h(R+h) + (R^2-h^2) = 0$$

$$\Rightarrow -2hR - 2h^2 + R^2 - h^2 = 0$$

$$\Rightarrow R^2 - 3h^2 - 2hR = 0$$

$$\Rightarrow (R+h)(R-3h) = 0$$

$$h = -R,\ R/3$$

Height of cone $= R + \frac{R}{3} = \frac{4R}{3}$

$$= \frac{h\times 1}{3}\text{m} = \frac{4}{3}\text{m}$$

2.10 (b)

Given $f(x) = (x-8)^{2/3} + 1$

$$f'(x) = \frac{2}{3}(x-8)^{-1/3}$$

Slope of tangent at point (0, 5)

$$m = \frac{2}{3}(0-8)^{-1/3} = -\frac{1}{3}$$

Slope of normal at point (0, 5)

$$m_1 = -\frac{1}{m} = 3$$

Equation of normal at point (0, 5)

$$y - 5 = 3(x-0)$$

$$\Rightarrow y = 3x + 5$$

2.11 (b)

$$\underset{x\to 3}{Lt} f(x) = \underset{x\to 3}{Lt}\left(\frac{2x^2-7x+3}{5x^2-12x-9}\right)$$

Here this is of the form of $\left(\frac{0}{0}\right)$

So, applying L-Hospital's rule

$$\underset{x\to 3}{Lt}\left(\frac{4x-7}{10x-12}\right) = \frac{5}{18}$$

2.12 (b)

Given $y = x^2$

At $x = 1;\ y = 1$

at $x = 5;\ y = 25$

Minimum value of function is 1.

2.13 (b)

Given $y = x + \sqrt{x + \sqrt{x + \sqrt{x + \ldots\infty}}}$

or, $(y - x) = \sqrt{x + \sqrt{x + \sqrt{x + \ldots\infty}}}$

Square both side, we get

$$(y-x)^2 = x + \sqrt{x + \sqrt{x + \ldots\infty}}$$

$$(y-x)^2 = y$$

$$y^2 - (2x+1)y + x^2 = 0$$

put $x = 2$

$\therefore$ $y^2 - 5y + 4 = 0$

$$(y-4)(y-1) = 0$$

$\therefore$ $y = 1$ or 4

But y is always greater than x.

Hence $y = 4$ only

2.14 (b)

$$\underset{x\to 0}{Lt}\ \frac{e^x - \left(1 + x + \frac{x^2}{2}\right)}{x^3}$$

This is of the form of $\left(\frac{0}{0}\right)$

Applying L' Hospital rule

$$\underset{x\to 0}{Lt}\frac{e^x-\left(1+x+\frac{x^2}{2}\right)}{x^3}\left(\frac{0}{0}\right)$$

$$\underset{x\to 0}{Lt}\frac{e^x-1}{6x}$$

$$\underset{x\to 0}{Lt}\frac{e^x}{6}=\frac{1}{6}$$

2.15 (b)

$$(x-8) = h\,(\text{say})$$

$$\Rightarrow \quad x = 8+h$$

$$\therefore \lim_{h\to 0}\frac{(8+h)^{1/3}-2}{h}$$

Above form in the $\left(\frac{0}{0}\right)$ by putting the value $h = 0$

Applying L' hospital rule

$$\lim_{h\to 0}\frac{\frac{1}{3}(8+h)^{\left(\frac{1}{3}-1\right)}}{1}=\frac{1}{3}(8)^{-\frac{2}{3}}=\frac{1}{12}$$

2.16 (a)

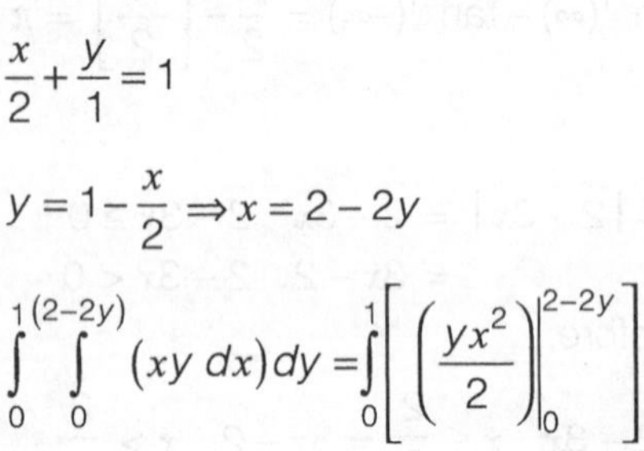

$$\frac{x}{2}+\frac{y}{1}=1$$

$$\Rightarrow \; y=1-\frac{x}{2}\Rightarrow x=2-2y$$

$$\int_0^1\int_0^{(2-2y)}(xy\,dx)dy=\int_0^1\left[\left(\frac{yx^2}{2}\right)\Bigg|_0^{2-2y}\right]dy$$

$$=\int_0^1\frac{y}{2}(2-2y)^2\,dy=\int_0^1 2y(1-y)^2\,dy=\frac{1}{6}$$

2.17 (d)

$$\int_0^{\pi/4}\tan x\,dx = \log\sqrt{2}$$

$$\int_0^{\infty}\frac{dx}{x^2+1}=\frac{\pi}{2}$$

$$\int_0^{\infty}x\,e^{-x}dx = -e^{-x}(x+1)\Big|_0^{\infty}=1$$

$$\int_0^1\frac{dx}{(x-1)}=\ln 0-\ln 1$$

(where [ln 0] is not defined)

So (d) is the answer

2.18 (d)

$$y=\frac{2}{3}x^{3/2}$$

$$\frac{dy}{dx}=x^{1/2}$$

length of the cursive is given by

$$\int_0^1\sqrt{\left(\frac{dy}{dx}\right)^2+1}\,dx=\int_0^1\sqrt{x+1}\,dx=1.22$$

2.19 (c)

$$f = y^x$$

$$\ln f = x\ln y$$

$$\Rightarrow \quad \frac{1}{f}\frac{\partial f}{\partial y}=\frac{x}{y}$$

$$\Rightarrow \quad \frac{\partial f}{\partial y}=y^x\left(\frac{x}{y}\right)=y^{x-1}x$$

$$\Rightarrow \quad \frac{\partial^2 f}{\partial x\partial y}=\frac{\partial}{\partial x}\left(y^{x-1}x\right)$$

$$= y^{x-1}+x\,y^{x-1}\ln y = 1^{(2-1)}(1+2\cdot\ln 1)=1$$

2.20 (c)

$f(x)$ in the neighbourhood of a

$f(x) = \Sigma b_n(x-a)^n$ where n from 0 to 0

$$b_n=\frac{f^n(a)}{n!}$$

$$f^4(x)=e^x;\; f^4(2)=e^2$$

$$\therefore \text{Coefficient of}(x-2)^4 = b_4=\frac{f^4(2)}{4!}=\frac{e^2}{4!}$$

2.21 (a)

Let the point be the (x, y, z)

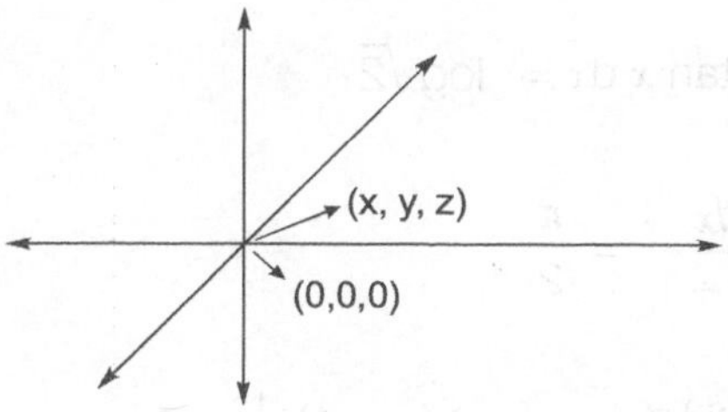

The distance between two given coordinates is given as,

$$D^2 = \sqrt{(x-0)^2+(y-0)^2-(z-0)^2}$$

$$D^2 = x^2 + y^2 + z^2$$

given $z^2 = 1 + xy$

$$D^2 = x^2 + y^2 + 1 + xy$$

Let $D^2 = U$

$$U = x^2 + y^2 + 1 + xy$$

For minimum or nearest distance

$$p = \frac{\partial u}{\partial x} = 0$$

$$\Rightarrow 2x + y = 0 \quad \text{...(i)}$$

$$\Rightarrow \quad q = \frac{\partial u}{\partial y} = 0 = 2y + x = 0 \quad \text{...(ii)}$$

By solving equation (i) and (ii)

$$x = 0$$
$$y = 0$$
$$z = 1$$

$$r = \frac{\partial^2 v}{\partial x^2} = 2; t = \frac{\partial^2 v}{\partial y^2} = 2; \; s = \frac{\partial^2 u}{\partial x \partial y} = 1$$

$$rt - s^2 = 2 \times 2 - 1^2 = 3$$

$$rt - s^2 > 0$$

$$r > 0$$

$\therefore$ distance is minimum.

Hence the distance should be

$$D = \sqrt{(0-0)^2+(0-0)^2-(1-0)^2} = 1 \text{ m}$$

Hence answer should be (a).

2.22 (a)

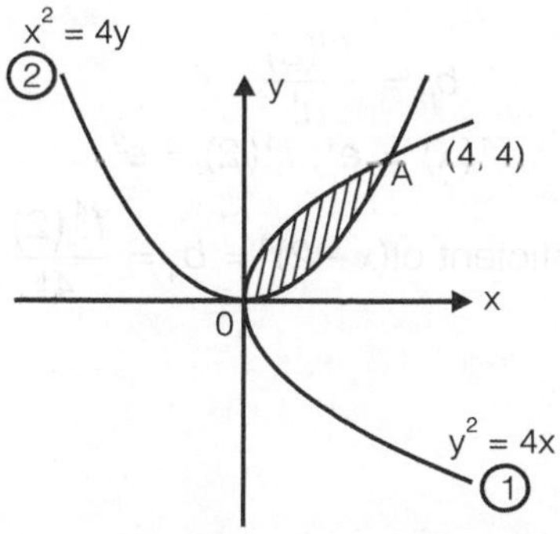

Curve 1 : $y^2 = 4x$

Curve 2 : $x^2 = 4y$

Intersection points of curve 1 and 2

$$y^2 = 4x = 4\sqrt{4y} = 8\sqrt{y}$$

$$y^4 = 8 \times 8\, y$$

$$\Rightarrow \quad y(y^3 - 64) = 0$$

Solution $\quad y = 4$ and $y = 0$

then $\quad x = 4 \quad x = 0$

Therefore intersection point are $A(4, 4)$ and $0(0, 0)$

The area enclosed between curves 1 and 2 are given by

$$\text{Area} = \int_{x_1}^{x_2} (y_1 - y_2)dx = \int_0^4 \left(\sqrt{4x} - \frac{x^2}{4}\right)dx$$

$$= 2\frac{x^{3/2}}{3/2} - \frac{x^3}{3 \times 4}\Bigg|_0^4 = \frac{4}{3}(4)^{3/2} - \frac{(4)^3}{3 \times 4} = \frac{16}{3}$$

2.23 (d)

The volume of a solid generated by revolution about the x-axis, of the area bounded by curve $y = f(x)$, the x-axis and the ordinates $x = a$, $y = b$ is

$$\text{Volume} = \int_a^b \pi y^2 dx$$

Here, $a - 1$, $b - 2$ and $y = \sqrt{x} \Rightarrow y^2 = x$

$$\therefore \text{ Volume} = \int_1^2 \pi \cdot x \cdot dx$$

$$= \pi \cdot \left[\frac{x^2}{2}\right]_1^2 = \frac{\pi}{2}\left[x^2\right]_1^2 = \frac{\pi}{2}\left[2^2 - 1^2\right] = \frac{3}{2}\pi$$

2.24 (d)

$$\int_{-\infty}^{\infty} \frac{dx}{1+x^2} = \left[\tan^{-1} x\right]_{-\infty}^{\infty}$$

$$= \tan^{-1}(\infty) - \tan^{-1}(-\infty) = \frac{\pi}{2} - \left[\frac{-\pi}{2}\right] = \pi$$

2.25 (c)

$$y = |2 - 3x| = 2 - 3x \quad 2 - 3x \geq 0$$
$$= 3x - 2 \quad 2 - 3x < 0$$

Therefore,

$$y = 2 - 3x \quad x \leq \frac{2}{3} = 3x - 2 \quad x > \frac{2}{3}$$

Since $2 - 3x$ and $3x - 2$ are polynomials, these are continuous at all points. The only concern is at

$x = \frac{2}{3}$

Left limit at $x = \frac{2}{3}$ is $2 - 3 \times \frac{2}{3} = 0.$

Right limit at $x = \frac{2}{3}$ is $3 \times \frac{2}{3} - 2 = 0.$

$$f\left(\frac{2}{3}\right) = 2 - 3 \times \frac{2}{3} = 0$$

Since, Left limit = Right limit = $f\left(\frac{2}{3}\right)$

Function is continuous at $\frac{2}{3}$.

y is therefore continuous $\forall\, x \in R$

Now since $2 - 3x$ and $3x - 2$ are polynomials, they are differentiable.

only concern is at $x = \frac{2}{3}$.

Now at $x = \frac{2}{3}$ LD = Left derivative = – 3

$$\Rightarrow \ln \int_1 \sqrt{x}\, dx - \int_1 \left(\frac{d}{dx}\right.$$

$$\Rightarrow \left[\ln x \frac{x^{3/2}}{3/2}\right]_1^e - \int_1^e \frac{1}{x}\frac{x^{3/2}}{3/2}$$

$$\Rightarrow \frac{2}{3}\left[\ln e \cdot e^{3/2} - 0\right] - \frac{4}{9}\left[x^{3/2}\right]_1^e$$

$$\Rightarrow \frac{2}{3}\sqrt{e^3} - \frac{4}{9}\sqrt{e^3} + \frac{4}{9} \Rightarrow \frac{2}{9}\sqrt{e^3} + \frac{4}{9}$$

2.34 (a)

$$\underset{x \to 0}{\text{Lt}} \frac{x - \sin x}{1 - \cos x}$$

Applying L' Hospital's rule

$$\underset{x \to 0}{\text{Lt}} \frac{\frac{d}{dx}(x - \sin x)}{\frac{d}{dx}(1 - \cos x)} = \underset{x \to 0}{\text{Lt}} \frac{(1 - \cos x)}{\sin x}$$

(It is still of $\frac{0}{0}$ form)

Again applying L' Hospital's rule

$$\underset{x \to 0}{\text{Lt}} \frac{\frac{d}{dx}(1 - \cos x)}{\frac{d}{dx}(\sin x)} = \underset{x \to 0}{\text{Lt}} \frac{\sin x}{\cos x} = \frac{0}{1} = 0$$

2.35 (b)

$\underset{x \to 0}{\text{Lt}} \frac{(e^{2x} - 1)}{\sin 4x}$, it is of $\left(\frac{0}{0}\right)$ from

$$f(0) = 0$$

Since left limit = Right limit = $f(0)$

So $|x|$ is continuous at $x = 0$

Now

LD = Left derivative (at $x = 0$) = –1

RD = Right derivative (at $x = 0$) = +1

$LD \neq RD$

So $|x|$ is not differentiable at $x = 0$

So $|x|$ is continuous and non-differentiable at $x = 0$.

2.30 (b)

$$\lim_{x \to 0} \frac{(1 - \cos x)}{x^2} = \frac{1 - \cos 0}{0^2} = \frac{0}{0}$$

So use L' hospitals rule

$$\lim_{x \to 0} \frac{(1 - \cos x)}{x^2} = \lim_{x \to 0} \frac{\sin x}{2x} = \frac{0}{0}$$

So use L'hospitals rule again

$$I = \int_0^2 \frac{(x-1)^2 \sin(x-1)}{(x-1)^2 + \cos(x-1)} dx$$

Taking $x - 1 = z \Rightarrow dx = dz$

for $x = 0$, $z \to -1$ and $x = 2$, $z \to 1$

$$\therefore \quad I = \int_{-1}^{1} \frac{z^2 \sin z}{z^2 + \cos z} dz$$

let $\quad f(z) = \frac{z^2 \sin z}{z^2 + \cos z}$

$$f(-z) = \frac{z^2 \sin z}{z^2 + \cos z}$$

$f(z) = -f(z)$ function is ODD.

$\therefore \quad I = 0$

2.39 (b)

$$I = \int_0^2 \left[\int_0^x e^{x+y} dy\right] dx$$

$$\int_0^x e^{x+y} dy = e^{x+y} dy\Big|_0^x = e^{2x} - e^x$$

$$I = \int_0^2 (e^{2x} - e^x) dx$$

$$I = \frac{1}{2} e^{2x}\Big|_0^2 - e^x\Big|_0^2 = \frac{1}{2}(e^4 - 1) - (e^2 - 1)$$

$$I = \frac{1}{2} e^4 - \frac{1}{2} - e^2 + 1$$

2.32 (d)

$$f(x) = x^3 + 1$$

Put $f'(x) = 0$

$\Rightarrow$ $3x^2 = 0$

$\Rightarrow$ $x = 0$ is the only critical point

at this critical point

$$f''(x) = 6x$$

$$f''(0) = 6 \times 0 = 0$$

Now $f'''(x) = 6$ and

so $f'''(0) = 6$ which is non zero.

Since the first non zero derivative value occurs at the third derivative which is an odd derivative, this function has a point of inflection at $x = 0$.

2.33 (c)

$$\int_1^e \sqrt{x} \ln x \, dx$$

$$\int_1^e \ldots = \int_1^e \left(\frac{d}{dx}\ln x \int x\,dx\right)dx$$

RD = Right derivative = +3

$LD \neq RD$

$\therefore$ The function y is not differentiable at

$$x = \frac{2}{3}$$

So, we can say that y is differential $\forall x \in R$,

except at $x = \frac{2}{3}$

2.26 (b)

$$\sin\theta = \theta - \frac{\theta^3}{3!} + \frac{\theta^2}{5!} \ldots$$

2.27 (d)

$$\lim_{\theta \to 0} \frac{\sin\theta}{\theta} = 1$$

2.28 (d)

If $f(x)$ is even function then

$$\int_{-a}^{a} f(x)dx = 2\int_0^a f(x)dx$$

2.29 (c)

$|x| = x \qquad x \geq 0$

$\;\;\;\; = -x \qquad x < 0$

at $x = 0$ left limit = 0

Right limit = –0 = 0

Applying L' Hospital's rule,

$$\underset{x\to 0}{\text{Lt}} \frac{2e^{2x}}{4\cos 4x} = \frac{2 \times 1}{4 \times 1} = \frac{1}{2}$$

2.36 Sol.

$$f(x) = e^{-x}\sin(100x)$$

$$f'(x) = -e^{-x}\sin(100x) + e^{-x}\cos(100x) \times 100$$

for minima $f'(x) = 0$

from here $\tan(100x) = 100$

$$x = \frac{1}{100}\tan^{-1}(100) = 0.0156$$

$$f(x) = e^{-0.0156}\sin(100 \times 0.0156) = \mathbf{0.9844}$$

2.37 (d)

$f(x)$ is continuous at any point

if $\underset{x\to a^-}{Lt} f(x) = \underset{x\to a^+}{Lt} f(x) = f(a)$

2.38 (b)

$$= \lim_{x\to 0}\left[\frac{\cos x}{2}\right] = \frac{1}{2}$$

2.31 (a)

The area enclosed is shown below as shaded:

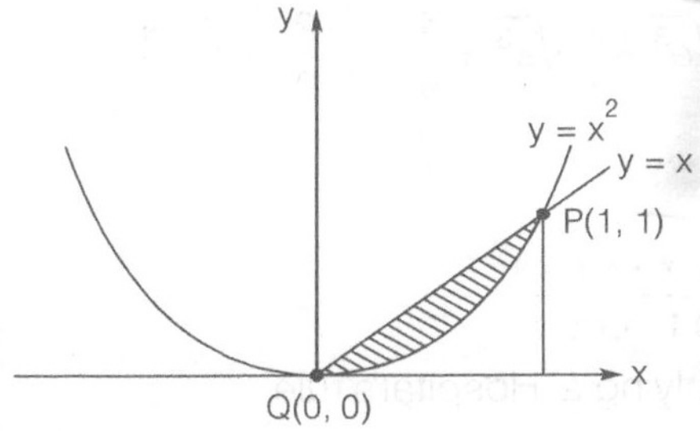

The coordinates of point P and Q is obtained by solving

$y = x$ and $y = x^2$ simultaneously,

i.e. $x = x^2$

$\Rightarrow$ $x(x - 1) = 0$

$\Rightarrow$ $x = 0, \quad x = 1$

Now, $x = 0 \Rightarrow y = 0$ which is *pt* $Q(0, 0)$

and $x = 1 \Rightarrow y = 1^2 = 1$ which is *pt* $P(1, 1)$

So required area is

$$= \int_0^1 x\,dx - \int_0^1 x^2 dx$$

$$\left[\frac{x^2}{2}\right]_0^1 - \left[\frac{x^3}{3}\right]_0^1 = \frac{1}{2} - \frac{1}{3} = \frac{1}{6}$$

$$= \frac{1}{2}e^4 - e^2 + \frac{1}{2} = \frac{1}{2}(e^4 - 2e^2 + 1) = \frac{1}{2}(e^2 - 1)^2$$

2.40 (c)

$$\lim_{x \to 0} \frac{1 - \cos(x^2)}{2x^4} \qquad \text{putting the } x \to 0$$

we get $\frac{0}{0}$ form

Applying *L'* Hospital rule

$$\Rightarrow \lim_{x \to 0} \frac{2x \sin(x^2)}{8x^3} \qquad \Rightarrow \lim_{x \to 0} \frac{\sin(x^2)}{4x^2}$$

$$\Rightarrow \frac{1}{4} \lim_{x \to 0} \frac{\sin(x^2)}{x^2} \Rightarrow \frac{1}{4} \lim_{x^2 \to 0} \frac{\sin(x^2)}{x^2} = \frac{1}{4} \times 1 = \frac{1}{4}$$

2.41 (a)

The graph of $|x|$ is

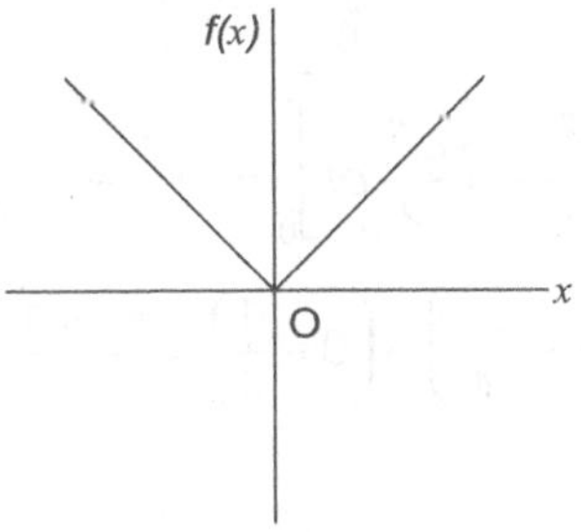

from the graph we can say that
$|x|$ has **minimum** at x = **0**.

2.42 Sol.

$$\lim_{x \to 0} \left(\frac{-\sin x}{2\sin x + \cos x} \right) = \lim_{x \to 0} \left(\frac{-\sin 0}{2\sin 0 + \cos 0} \right) = \frac{0}{1} = 0$$

(**Note:** Since the function is not evaluating to 0/0 no need to use *L*'hospital's rule)

2.43 Sol.

$$S = \int_0^{\pi/2} \sqrt{\left(\frac{dx}{dt}\right)^2 + \left(\frac{dy}{dt}\right)^2 + \left(\frac{dz}{dt}\right)^2}\, dt$$

$$S = \int_0^{\pi/2} \sqrt{(-\sin t)^2 + (\cos t)^2 + \left(\frac{2}{\pi}\right)^2}\, dt$$

$$= \int_0^{\pi/2} \sqrt{1 + \left(\frac{4}{\pi^2}\right)}\, dt = \sqrt{1 + \left(\frac{4}{\pi^2}\right)}\, [t]_0^{\pi/2}$$

$$= \sqrt{1 + \left(\frac{4}{\pi^2}\right)} \left(\frac{\pi}{2}\right) = 1.86$$

2.44 Sol.

$(x - 2)^2 + (y^2) = (2)^2$, is a circle of radius 2 m and centre at (2, 0)

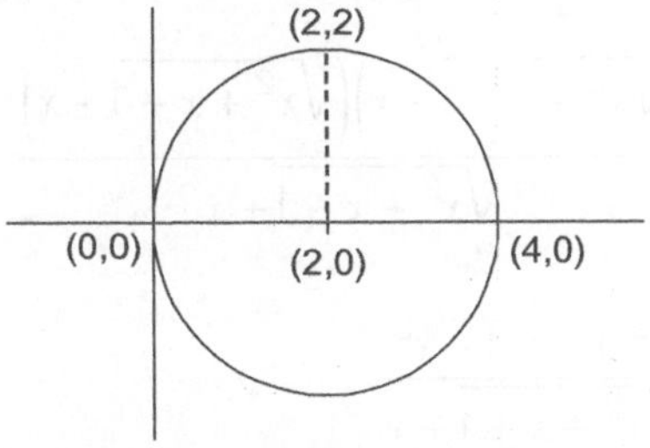

Time to reach from (4, 0) to (2, 2) is

$$\text{Time} = \frac{\text{Distance}}{\text{Speed}}$$

$$= \frac{\left(\frac{2\pi r}{4}\right)}{1.57} = \frac{\left(\frac{2\pi^2}{4}\right)}{1.57} = \frac{\pi}{1.57} = 2 \text{ s}$$

2.45 Sol.

$$f(x) = 2x^3 - 3x^2 \text{ in } [-1, 2]$$
$$f'(x) = 6x^2 - 6x$$
$$f'(x) = 0$$
$$6x^2 - 6x = 0 \qquad x = -1 \quad f(-1) = -5 \text{ G. Min.}$$
$$6x(x - 1) = 0 \qquad x = 2 \quad f(2) = 4$$
$$x = 0, 1 \qquad x = 0 \quad f(0) = 0$$
$$f''(x) = 12x - 6 \qquad x = 1 \quad f(1) = -1$$
$$f''(0) = -6 \quad \text{Max}$$
$$f''(1) = 6 \quad \text{Min}$$

G. Minima is **–5** at x = 1.

2.46 (c)

$$f(x) = \frac{x^2 - 3x - 4}{x^2 + 3x - 4} \text{ is not continuous}$$

when

$$x^2 + 3x - 4 = 0$$
$$(x + 4)(x - 1) = 0$$
$$\boldsymbol{x = -4, 1}$$

2.47 (c)

$$\underset{x \to 0}{Lt} \frac{\ln(1 + 4x)}{e^{3x} - 1} \quad 0/0 \text{ form}$$

$$\underset{x \to 0}{Lt} \frac{\frac{1}{1 + 4x} \cdot 4}{3e^{3x}} = \frac{4}{3}$$

2.48 (c)

$$\underset{x\to\infty}{Lt}\sqrt{x^2+x-1}-x$$

$$\underset{x\to\infty}{Lt}\frac{\left(\sqrt{x^2+x-1}-x\right)\left(\sqrt{x^2+x-1}+x\right)}{\sqrt{x^2+x-1}+x}$$

$$\underset{x\to\infty}{Lt}\frac{x^2+x-1-x^2}{\sqrt{x^2+x+1}+x}$$

$$\underset{x\to\infty}{Lt}\frac{x-1}{x\sqrt{1+\frac{1}{x}+\frac{1}{x^2}}+x}$$

$$\underset{x\to\infty}{Lt}\frac{1-\frac{1}{x}}{\sqrt{1+\frac{1}{x}+\frac{1}{x^2}}+1}=\frac{1}{1+1}=\frac{1}{2}$$

2.49 (d)

$$\underset{x\to 0}{Lt}\frac{x^3-\sin x}{x} = \left(\frac{0}{0}\text{form}\right)$$

$$\underset{x\to 0}{Lt}\frac{3x^2-\cos x}{1} = 0-\cos 0$$

$$= 0 - 1 = -1$$

2.50 (c)

$$x = \cos\left(\frac{\pi u}{2}\right);\quad y = \sin\left(\frac{\pi u}{2}\right)$$

$$x^2 + y^2 = 1$$

It represents a circle in x-y plane.

$0 \le u \le 1$ (given range)

$\therefore$ $0 \le x \le 1,\ 0 \le y \le 1$

so, $0 \le \theta \le \frac{\pi}{2}$

Thus, we will get as quarter circle in x-y plane.

$\therefore$ Area of hemisphere $= 2\pi(r)^2 = 2\pi \times (1)^2 = \mathbf{2\pi}$

2.51 (a)

$$\int_a^b f(x)dx = f(\xi)(b-a)$$

2.52 (c)

Since rotation is about x axis, so.

$$\text{Volume} = \int_0^1 \pi y^2 dx = \int_0^1 \pi x\,dx = \pi\left[\frac{x^2}{2}\right]_0^1 = \frac{\pi}{2}$$

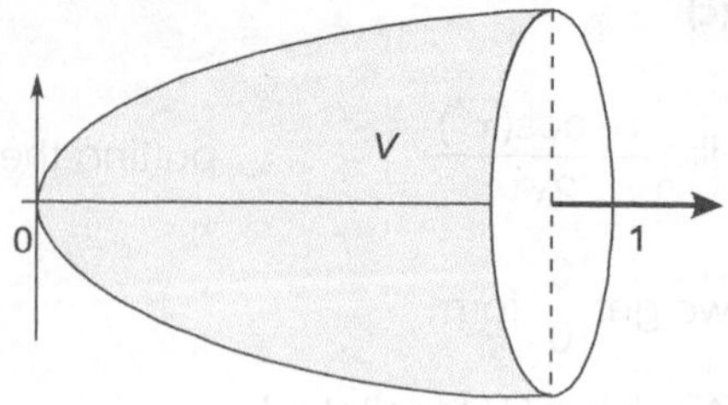

2.53 Sol.

$$I = \int_1^e x\ln x dx$$

Put $x = e^t,\ dx = e^t dt$

(when $x = 1,\ t = 0$; when $x = e,\ t = 1$)

$$I = \int_0^1 t\cdot e^t e^t dt = \int_0^1 t\cdot e^{2t}\,dt$$

$$=\left[t\frac{e^{2t}}{2}-\int\frac{e^{2t}}{2}dt\right]_0^1 = \left(t\frac{e^{2t}}{2}-\frac{e^{2t}}{4}\right)_0^1$$

$$=\left(\frac{e^2}{2}-\frac{e^2}{4}\right)-\left(0-\frac{1}{4}\right)=\frac{e^2+1}{4} = 2.097$$

2.54 (c)

Exact value

$$f'(x) = -\sin x$$

$$f'\left(\frac{\pi}{6}\right) = -\frac{1}{2}$$

Approximate value

$$f'(x) = \frac{\cos(x+0.1)-\cos(x-0.1)}{2(0.1)}$$

$$= \frac{-\sin x\cdot\sin(0.1)}{2(0.1)}$$

$$f'\left(\frac{\pi}{6}\right) = -0.49916$$

Error = Exact value − Approximate value

$= -0.5 + 0.49916 = -0.00084$

$\therefore$ %Error $= \frac{-0.00084}{0.5}\times 100 = 0.168\%$

$\therefore$ The value is > 0.1% and <1%

2.55 (d)

$$\lim_{x\to 1}\frac{1-e^{-c(1-x)}}{1-xe^{-c(1-x)}} = \frac{0}{0}\text{ form}$$

Applying 'L' Hospital rule

$$= \lim_{x\to 1}\frac{-ce^{-c(1-x)}}{-e^{-c(1-x)} - x(e)^{-c(1-x)}}$$

$$= \lim_{x\to 1}\frac{-c}{-1-cx} = \frac{-c}{-1-c} = \frac{c}{1+c}$$

2.56 (c)

$$I = \int_0^1\int_0^{x^2} xy^2\,dy\,dx$$

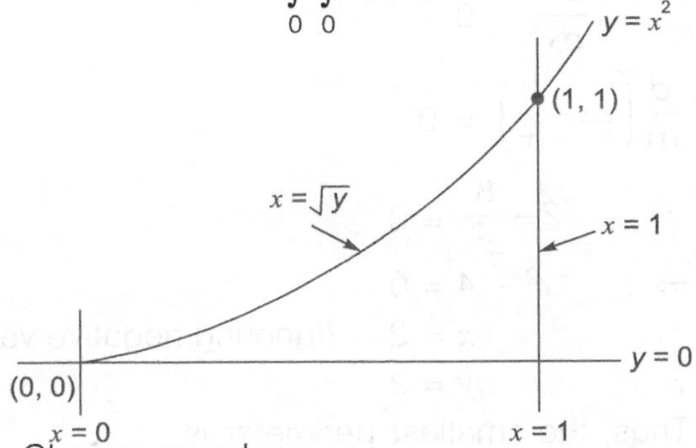

Change on rules,

$$I = \int_{y=0}^{1}\int_{x=\sqrt{y}}^{1} xy^2\,dx\,dy$$

2.57 Sol.

$f(z) = \phi + i\psi$ is analytic

$\phi = x^2 - y^2$ $\quad \dfrac{C\text{-}R\text{equations}}{\phi_x = \psi_y}$

$$\phi_x = 2x = \psi_y$$

$$\phi_y = -\psi_x$$

$$\phi_y = -2y = -\psi_x$$

$$\psi_x = 2y \Rightarrow \psi = 2xy + C_1$$

$$\psi_y = 2x \Rightarrow \psi = 2xy + C_2$$

Comparing $\psi = 2xy + C$

valid for all C put $C = 0$

$\psi(1 + i) \Rightarrow (x = 1 \quad y = 1)$

$\therefore \quad \psi = 2$

2.58 (b)

By Lagrange's mean value theorem,

$$\frac{f(3)-f(1)}{3-1} = f'(c)$$

$$\Rightarrow \quad \frac{5-1}{3-1} = 2c-2$$

$$\Rightarrow \quad 2c-2 = 2$$

$$\Rightarrow \quad c = 2 \in (1, 3)$$

2.59 (b)

$$I = \int_{\theta=0}^{\theta=\frac{\pi}{2}}\int_{r=0}^{r=\cos\theta} r\sin\theta\,dr\,d\theta$$

$$= \int_{\theta=0}^{\frac{\pi}{2}}\left[\frac{r^2}{2}\right]_0^{\cos\theta} \times \sin\theta\,d\theta = \frac{1}{2}\int_{\theta}^{\frac{\pi}{2}}\sin\theta\cdot\cos^2\theta\,d\theta$$

Let, $\quad \cos\theta = t$

$$-\sin\theta\,d\theta = dt$$

For $\quad \cos\theta = t$

At, $\quad \theta = \dfrac{\pi}{2}; t = 0$

$$\theta = 0,\ t = 1$$

$$= \int_1^0 \frac{-t^2}{2}\,dt$$

$$= \frac{-1}{2}\left[\frac{t^3}{3}\right]_1^0 = \frac{-1}{2}\times\left(\frac{-1}{3}\right) = \frac{-1}{2}\left[\frac{-1}{3}\right] = \frac{1}{6}$$

2.60 (a)

$$\lim_{x\to 0}\left(\frac{1-\cos x}{x^2}\right) \quad \left(\frac{0}{0}\text{ form}\right)$$

$$\text{Applying L-H rule} = \lim_{x\to 0}\frac{\sin x}{2x} = \frac{1}{2}\left[\lim_{x\to 0}\frac{\sin x}{x}\right] = \frac{1}{2}$$

2.61 (a)

Given, $\quad p = \lim\limits_{x\to\pi}\dfrac{x^2 + ax + 2\pi^2}{x - \pi + 2\sin x}$

For $x = \pi$, the denominator becomes zero. Thus for p to have a finite value, the numerator must also be zero for $x = \pi$.

$\Rightarrow$ At $\quad x = \pi$,

$$\pi^2 + a\pi + 2\pi^2 = 0$$

$$a\pi = -3\pi^2$$

$$a = -3\pi$$

Now, $\quad p = \lim\limits_{x\to\pi}\dfrac{x^2 - 3\pi x + 2\pi^2}{x - \pi + 2\sin x}\left(\dfrac{0}{0}\text{ form}\right)$

Using L' Hospital's rule,

$$p = \lim_{x\to\pi}\frac{2x - 3\pi}{1 + 2\cos x} = \frac{2\pi - 3\pi}{1 + 2\cos\pi} = \frac{-\pi}{1-2} = \pi$$

$\therefore p = \pi$

2.62 (b)

For boundary condition, $T(x, 1) = x + 1$

Checking options:

Option A

$$T(x, 1) = x - x\times 1 + 1 = 1 \neq x + 1$$

(not satisfy)

Option B $\quad T(x, 1) = x + 1$ (satisfy)

Option C $\quad T(x, 1) = -x + 1$ (not satisfy)

Option D $\quad T(x, 1) = x + x(1) = 2x + 1$ (not satisfy)

Hence correct option is (B).

2.63 (b)

Let $(x + b) = t \Rightarrow dx = dt$

When, $x = -\infty$; $t = -\infty$ and $x = \infty$; $t = \infty$

$$\therefore \int_{-\infty}^{\infty} e^{-a(x+b)^2} dx = \int_{-\infty}^{\infty} e^{-at^2} dt$$

Now, let, $at^2 = y^2 \Rightarrow t = \frac{y}{\sqrt{a}}$

$$\Rightarrow \quad 2at\,dt = 2y dy$$

$$\Rightarrow \quad dt = \frac{ydy}{at} = \frac{ydy}{a\frac{y}{\sqrt{a}}} = \frac{dy}{\sqrt{a}}$$

$$\therefore \quad \int_{-\infty}^{\infty} e^{-at^2} dt = \int_{-\infty}^{\infty} e^{-y^2} \cdot \frac{dy}{\sqrt{a}} = \frac{1}{\sqrt{a}} \int_{-\infty}^{\infty} e^{-y^2} dy$$

$$= \sqrt{\frac{\pi}{a}}$$

2.64 (b)

Since, $\phi(s)$ has triple roots at $s = -\sigma$, $(s + \sigma)^3$ will be one of its factors and $\phi(-\sigma) = 0$.

There will be an inflection point on the curve of $\phi(s)$ (just like in the case of x^3). Hence, the first and second order derivative at $s = -\sigma$ will be zero.

2.65 (b)

We know that the graph of $y = |x|$ is

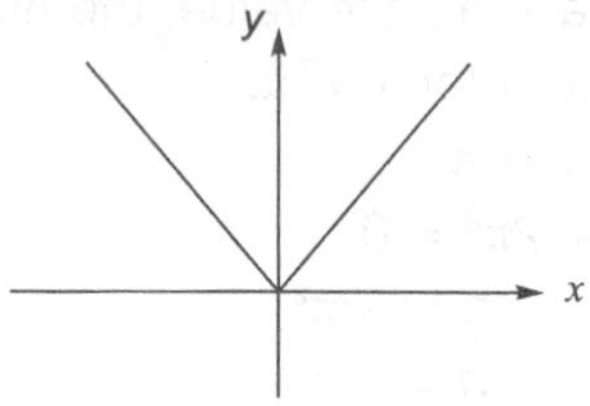

and the graph of $y = -|x|$ is

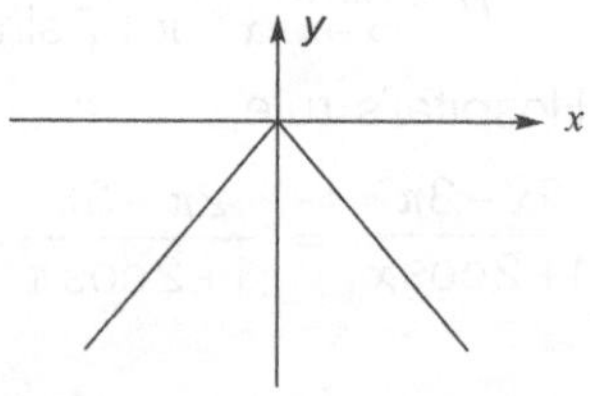

It can be observed that the given graph can be obtained by first shifting the graph of $y = -|x|$ up by 2 units and then taking the modulus of resultant function.

Shifting up by 2 units transforms the equation to,

$$y = 2 - |x|$$

and, taking modulus gives the resultant equation as

$$y = |2 - |x||$$

2.66 (8) (7.999 to 8.001)

Let x and y be the dimensions of the rectangle.

Given, $xy = 4$...(i)

Perimeter, $P = 2(x + y)$

By equation (i),

$$P = 2\left(x + \frac{4}{x}\right) = 2x + \frac{8}{x}$$

For smallest perimeter,

$$\frac{dP}{dx} = 0$$

$$\frac{d}{dx}\left(2x + \frac{8}{x}\right) = 0$$

$$2 - \frac{8}{x^2} = 0$$

$$\Rightarrow \quad x^2 - 4 = 0$$

$$x = 2 \quad \text{(ignoring negative value)}$$

$$\therefore \quad y = 2$$

Thus, the smallest perimeter is

$$P = 2(2 + 2) = 8 \text{ units}$$

2.67 (a)

It can be assumed that the function is parabolic in nature i.e. of the form $ax^2 + bx + c$. For a strictly local minimum to exist, the coefficient of x^2 i.e. 'a' should be positive.

Also, for a local minimum to exist, its first derivative should be zero while the second derivative should be positive. Similar conditions can be observed in option (a) only. Hence, option (a) is correct.

2.68 (25) (25 to 25)

To evaluate, $\int_{x=3}^{x=4}\int_{y=1}^{y=2} \frac{dx\,dy}{(x+y)^2}$

$$= \int_{x=3}^{x=4} -\left[\frac{1}{(x+y)} dx\right]_{y=1}^{y=2} = -\int_{x=3}^{x=4}\left[\frac{1}{2+x} - \frac{1}{x+1}\right]$$

$$= -[\ln(2+x) - \ln(x+1)]_{x=3}^{x=4} = -\left[\ln\left(\frac{2+x}{x+1}\right)\right]_3^4$$

$$= -\left[\ln\left(\frac{6}{5}\right) - \ln\left(\frac{5}{4}\right)\right] = -\left[\ln\left(\frac{6\times4}{5\times5}\right)\right] = -\ln\left(\frac{24}{25}\right)$$

$$= \ln\left(\frac{25}{24}\right)$$

Thus, $\log_e\left(\frac{a}{24}\right) = \ln\left(\frac{25}{24}\right)$

$$\therefore \quad a = 25$$

■■■■

3 Vector Calculus

3.1 The vector field $\vec{F} = x\hat{i} - y\hat{j}$ (where $\hat{i}$ and $\hat{j}$ are unit vector) is
(a) divergence free, but not irrotational
(b) irrotational, but not divergence free
(c) divergence free and irrotational
(d) neither divergence free nor irrotational
[2003 : 2 M]

3.2 Stokes theorem connects
(a) a line integral and a surface integral
(b) a surface integral and a volume integral
(c) a line integral and a volume integral
(d) gradient of a function and its surface integral
[2005 : 1 M]

3.3 The line integral $\int \vec{V} \cdot d\vec{r}$ of the vector $\vec{V}(\vec{r}) = 2xyz\hat{i} + x^2z\hat{j} + x^2y\hat{k}$ from the origin to the point $P(1, 1, 1)$
(a) is 1
(b) is zero
(c) is –1
(d) cannot be determined without specifying the path
[2005 : 2 M]

3.4 The area of a triangle formed by the tips of vectors $\bar{a}, \bar{b}$ and $\bar{c}$ is
(a) $\frac{1}{2}(\bar{a}-\bar{b})\cdot(\bar{a}-\bar{c})$ (b) $\frac{1}{2}|(\bar{a}-\bar{b})\times(\bar{a}-\bar{c})|$
(c) $\frac{1}{2}|\bar{a}\times\bar{b}\times\bar{c}|$ (d) $\frac{1}{2}(\bar{a}\times\bar{b})\cdot\bar{c}$
[2007 : 2 M]

3.5 The divergence of the vector field $(x-y)\hat{i} + (y-x)\hat{j} + (x+y+z)\hat{k}$ is
(a) 0 (b) 1
(c) 2 (d) 3 **[2008 : 1 M]**

3.6 The directional derivative of the scalar function $f(x, y, z) = x^2 + 2y^2 + z$ at the point $P = (1,1,2)$ in the direction of the vector $\vec{a} = 3\hat{i} - 4\hat{j}$ is
(a) – 4 (b) –2
(c) –1 (d) 1 **[2008 : 2 M]**

3.7 The divergence of the vector field $3xz\hat{i} + 2xy\hat{j} - yz^2\hat{k}$ at a point (1,1,1) is equal to
(a) 7 (b) 4
(c) 3 (d) 0 **[2009 : 1 M]**

3.8 A path *AB* in the form of one quarter of a circle of unit radius is shown in the figure. Integration of $(x + y)^2$ on path *AB* traversed in a counter-clockwise sense is

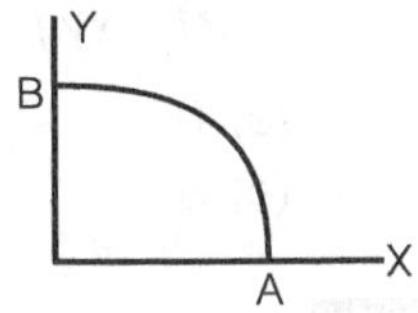

(a) $\frac{\pi}{2}-1$ (b) $\frac{\pi}{2}+1$
(c) $\frac{\pi}{2}$ (d) 1
[2009 : 2 M]

3.9 For the spherical surface $x^2 + y^2 + z^2 = 1$, the unit outward normal vector at the point $\left(\frac{1}{\sqrt{2}}, \frac{1}{\sqrt{2}}, 0\right)$ is given by
(a) $\frac{1}{\sqrt{2}}\hat{i} + \frac{1}{\sqrt{2}}\hat{j}$ (b) $\frac{1}{\sqrt{2}}\hat{i} - \frac{1}{\sqrt{2}}\hat{j}$
(c) $\hat{k}$ (d) $\frac{1}{\sqrt{3}}\hat{i} + \frac{1}{\sqrt{3}}\hat{j} + \frac{1}{\sqrt{3}}\hat{k}$
[2012 : 1 M]

3.10 The following surface integral is to be evaluated over a sphere for the given steady velocity vector field $\vec{F} = xi + yj + zk$ defined with respect to a Cartesian coordinate system having *i*, *j* and *k* as unit base vectors.

$$\iint_S \frac{1}{4}(\vec{F}\cdot n)\,dA$$

where S is the sphere, $x^2 + y^2 + z^3 = 1$ and *n* is the outward unit normal vector to the sphere. The value of the surface integral is
(a) π (b) 2π
(c) 3π/4 (d) 4π **[2013 : 2 M]**

3.11 Which one of the following describes the relationship among the three vectors, $\hat{i}+\hat{j}+\hat{k}$, $2\hat{i}+3\hat{j}+\hat{k}$ and $5\hat{i}+6\hat{j}+4\hat{k}$?
(a) The vectors are mutually perpendicular

(b) The vectors are linearly dependent
(c) The vectors are linearly independent
(d) The vectors are unit vectors

[2014 : 1 M, Set-1]

3.12 Curl of vector $\vec{F} = x^2z^2\hat{i} - 2xy^2z\hat{j} + 2y^2z^3\hat{k}$ is

(a) $(4yz^3 + 2xy^2)\hat{i} + 2x^2z\hat{j} - 2y^2z\hat{k}$
(b) $(4yz^3 + 2xy^2)\hat{i} - 2x^2z\hat{j} - 2y^2z\hat{k}$
(c) $2xz^2\hat{i} - 4xyz\hat{j} + 6y^2z^2\hat{k}$
(d) $2xz^2\hat{i} + 4xyz\hat{j} + 6y^2z^2\hat{k}$

[2014 : 1 M, Set-2]

3.13 Divergence of the vector field

$x^2z\hat{i} + xy\hat{j} - yz^2\hat{k}$ at (1, –1, 1) is

(a) 0 (b) 3
(c) 5 (d) 6 [2014 : 1 M, Set-3]

3.14 The integral $\oint_C (ydx - xdy)$ is evaluated along the circle $x^2 + y^2 = \frac{1}{4}$ traversed in counter clockwise direction. The integral is equal to

(a) 0 (b) $-\frac{\pi}{4}$
(c) $-\frac{\pi}{2}$ (d) $\frac{\pi}{4}$ [2014 : 2 M, Set-1]

3.15 Curl of vector $V(x, y, z) = 2x^2 i + 3z^2 j + y^3 k$ at $x = y = z = 1$ is

(a) $-3i$ (b) $3i$
(c) $3i - 4j$ (d) $3i - 6k$

[2015 : 1 M, Set-2]

3.16 Let ϕ be an arbitrary smooth real valued scalar function and V be an arbitrary smooth vector valued function in a three-dimensional space. Which one of the following is an identity?

(a) $Curl(\phi\vec{V}) = \nabla(\phi Div\vec{V})$
(b) $Div\vec{V} = 0$
(c) $Div\ Curl\vec{V} = 0$
(d) $Div(\phi\vec{V}) = \phi Div\vec{V}$ [2015 : 1 M, Set-3]

3.17 The surface integral

$\iint_S \frac{1}{\pi}(9xi - 3yj)\cdot ndS$ over the sphere given by $x^2 + y^2 + z^2 = 9$ is ________.

[2015 : 2 M, Set-2]

3.18 The value of $\int_C [(3x - 8y^2)dx + (4y - 6xy)dy]$, (where C is the boundary of the region boundary by $x = 0$, $y = 0$ and $x + y = 1$) is ________.

[2015 : 2 M, Set-3]

3.19 A scalar potential φ has the following gradient : $\nabla\varphi = yz\hat{i} + xz\hat{j} + xy\hat{k}$. Consider the integral $\int_C \nabla\varphi . d\vec{r}$ on the curve

$\vec{r} = x\hat{i} + y\hat{j} + z\hat{k}$.

The curve C is parameterized as follows :]

$$\begin{cases} x = t \\ y = t \quad \text{and } 1 \le t \le 3 \\ z = 3t^2 \end{cases}$$

The value of the integral is____

[2016 : 2 M, Set-2]

3.20 The value of the line integral $\oint_C \bar{F}\cdot\bar{r}'ds$, where C is a circle of radius $\frac{4}{\sqrt{\pi}}$ units is______

Here, $\bar{F}(x, y) = y\hat{i} + 2x\hat{j}$ and $\bar{r}'$ is the UNIT tangent vector on the curve C at an arc length s from a reference point on the curve $\hat{i}$ and $\hat{j}$ are the basis vectors in the x–y Cartesian reference. In evaluating the line integral, the curve has to be traversed in the counter-clockwise direction.

[2016 : 2 M, Set-3]

3.21 For the vector $\vec{V} = 2yz\hat{i} + 3xz\hat{j} + 4xy\hat{k}$, the value of $\nabla\cdot(\nabla\times\vec{V})$ is ________.

[2017 : 2 M, Set-1]

3.22 The divergence of the vector $-yi + xj$ is ________. [2017 : 1 M, Set-2]

3.23 The surface integral $\iint_S F{\cdot}ndS$ over the surface S of the sphere $x^2 + y^2 + z^2 = 9$, where $F = (x + y)i + (x + z)j + (y + z)k$ and n is the unit outward surface normal, yields ________.

[2017 : 2 M, Set-2]

3.24 The value of the integral

$$\oiint_S \vec{r}\cdot\hat{n}dS$$

over the closed surface S bounding a volume, where $\vec{r} = x\hat{i} + y\hat{i} + z\hat{k}$ is the position vector and $\hat{n}$ is the normal to the surface S, is

(a) V (b) $2V$
(c) $3V$ (d) $4V$ [2018 : 2 M, Set-1]

3.25 The divergence of the vector field $\vec{u} = e^x(\cos y\hat{i} + \sin y\hat{j})$ is

(a) 0 (b) $e^x \cos y + e^x \sin y$
(c) $2e^x \cos y$ (d) $2e^x \sin y$

[2018 : 1 M, Set-2]

3.26 For a position vector $r = x\hat{i} + y\hat{j} + z\hat{k}$ the norm of the vector can be defined as $|\vec{r}| = \sqrt{x^2 + y^2 + z^2}$. Given a function $\phi = \ln|\vec{r}|$, its gradient $\nabla\phi$ is

(a) $\vec{r}$ (b) $\frac{\vec{r}}{|\vec{r}|}$
(c) $\frac{\vec{r}}{\vec{r}\cdot\vec{r}}$ (d) $\frac{\vec{r}}{|\vec{r}|^3}$

[2018 : 2 M, Set-2]

3.27 The directional derivative of the function $f(x, y) = x^2 + y^2$ along a line directed from (0, 0) to (1, 1), evaluated at point $x = 1$, $y = 1$ is

(a) $4\sqrt{2}$ (b) 2
(c) $2\sqrt{2}$ (d) $\sqrt{2}$

[2019 : 1 M, Set-2]

3.28 Given a vector $\bar{u} = \frac{1}{3}\left(-y^3\hat{i} + x^3\hat{j} + z^3\hat{k}\right)$ and $\hat{n}$ as the unit normal vector to the surface of the hemisphere $\left(x^2 = 1;\ z \geq 0\right)$, the value of integral $\int(\nabla\times\bar{u})\cdot\hat{n}dS$ evaluated on the curved surface of the hemisphere S is

(a) π (b) $\frac{\pi}{2}$
(c) $\frac{-\pi}{2}$ (d) $\frac{\pi}{3}$ [2019 : 2 M, Set-2]

3.29 For three vectors $\vec{A} = 2\hat{j} - 3\hat{k}, \vec{B} = -2\hat{i} + \hat{k}$ and $\vec{C} = 3\hat{i} - \hat{j}$, where $\hat{i}, \hat{j}$ and $\hat{k}$ are unit vectors along the axes of a right-handed rectangular/Cartesian coordinate system, the value of $(\vec{A}\cdot(\vec{B}\times\vec{C})+6)$ is ________.

[2020 : 1 M, Set-1]

3.30 A vector field is defined as

$$\vec{f}(x,y,z) = \frac{x}{\left[x^2+y^2+z^2\right]^{3/2}}i + \frac{y}{\left[x^2+y^2+z^2\right]^{3/2}}\hat{j} + \frac{z}{\left[x^2+y^2+z^2\right]^{3/2}}\hat{k}$$

where, $\hat{i}, \hat{j}, \hat{k}$ are unit vectors along the axes of a right-handed rectangular/Cartesian coordinate system. The surface integral $\iint \vec{f}\,d\vec{S}$ (where $d\vec{S}$ is an elemental surface area vector) evaluated over the inner and outer surfaces of a spherical shell formed by two concentric spheres with origin as the center, and internal and external radii of 1 and 2, respectively, is

(a) 0 (b) 2π
(c) 8π (d) 4π

[2020 : 2 M, Set-1]

3.31 The directional derivative $f(x, y, z) = xyz$ at point (–1, 1, 3) in the direction of vector $\hat{i} - 2\hat{j} + 2\hat{k}$ is

(a) 7/3 (b) 7
(c) $-\frac{7}{3}$ (d) $3\hat{i} - 3\hat{j} - \hat{k}$

[2020 : 2 M, Set-2]

3.32 Given a function $\varphi = \frac{1}{2}\left(x^2 + y^2 + z^2\right)$ in three-dimensional Cartesian space, the value of the surface integral

$$\oiint_S \hat{n}\cdot\nabla\varphi\, dS,$$

where S is the surface of a sphere of unit radius and $\hat{n}$ is the outward unit normal vector on S, is

(a) 4π (b) 3π
(c) $\frac{4\pi}{3}$ (d) 0 [2022 : 1 M, Set-1]

3.33 Consider two vectors:

$\vec{a} = 5i + 7j + 2k$

$\vec{b} = 3i - j + 6k$

Magnitude of the component of $\vec{a}$ orthogonal to $\vec{b}$ in the plane containing the vectors $\vec{a}$ and $\vec{b}$ is ________ (round off to 2 decimal places).

[2022 : 2 M, Set-1]

3.34 Consider a cube of unit edge length and sides parallel to coordinate axes with its centroid at the point (1, 2, 3). The surface integral $\int_A \vec{F}\cdot d\vec{A}$ of a vector field, $\vec{F} = 3x\hat{i} + 5y\hat{j} + 6z\hat{k}$ over the entire surface A area of the cube is _____.

(a) 14 (b) 27
(c) 28 (d) 31 [2022 : 1 M, Set-2]

3.35 A vector field

$B(x, y, z) = x\hat{i} + y\hat{j} - 2z\hat{k}$

is defined over a conical region having height $h = 2$, base radius $r = 3$ and axis along z, as shown in the figure. The base of the cone lies in the x-y plane and is centered at the origin. If n denotes the unit outward normal to the curved surface S of the cone, the value of the integral,

$$\int_S B\cdot n dS$$

equals _____. [Answer in integer]

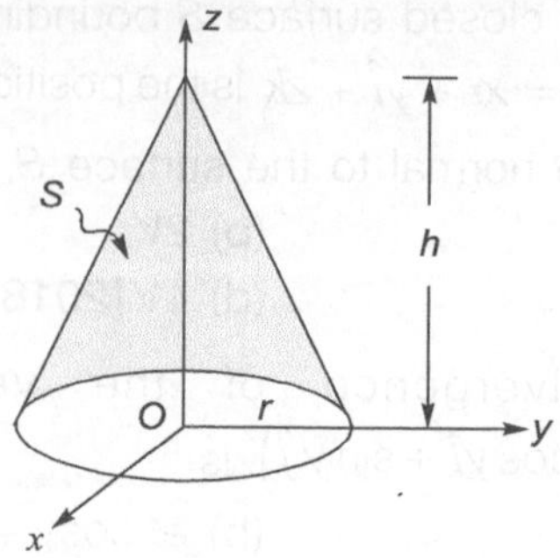

[2023 : 1 Mark]

3.36 The value of the surface integral

$$\oiint_S z dx dy$$

where S is the external surface of the sphere $x^2 + y^2 + z^2 = R^2$ is

(a) $4\pi R^3$ (b) $\frac{4\pi R^3}{3}$
(c) 0 (d) πR^3

[2024 : 1 M]

Answers Vector Calculus

3.1	(c)	3.2	(a)	3.3	(a)	3.4	(b)	3.5	(d)	3.6	(b)	3.7	(c)	3.8	(b)
3.9	(a)	3.10	(a)	3.11	(b)	3.12	(a)	3.13	(c)	3.14	(c)	3.15	(a)	3.16	(c)
3.17	(216)	3.18	(1.6666)	3.19	(726)	3.20	(16)	3.21	(0)	3.22	(0)	3.23	(226.08)	3.24	(c)
3.25	(c)	3.26	(c)	3.27	(c)	3.28	(b)	3.29	(6)	3.30	(a)	3.31	(a)	3.32	(a)
3.33	(8.32)	3.34	(a)	3.35	(0)	3.36	(b)								

Explanations Vector Calculus

3.1 (c)

$$\vec{F} = x\hat{i} - y\hat{j}$$

For divergence

Grade $\vec{F} = \nabla\cdot\vec{F} = \left(\hat{i}\frac{\partial}{\partial x} + \hat{j}\frac{\partial}{\partial y} + \hat{k}\frac{\partial}{\partial z}\right).(x\hat{i} - y\hat{j})$

$= 1 - 1 = 0$

Hence vector field is divergence free.

For irrotational

$$\text{Curl}\vec{F} = \nabla\times\vec{F} = \left(i\frac{\partial}{\partial x} + j\frac{\partial}{\partial y} + j\frac{\partial}{\partial z}\right)\times(x\hat{i} - y\hat{i})$$

$$= \begin{vmatrix} i & j & k \\ \frac{\partial}{\partial x} & \frac{\partial}{\partial y} & \frac{\partial}{\partial z} \\ x & -y & 0 \end{vmatrix} = \hat{i}(0-0) - \hat{j}(0-0) + \hat{k}(0-0) = 0$$

Hence vector field is irrotational

3.2 (a)

A line integral and a surface integral is related by stroke's theorem.

3.3 (a)

Potential function of

$$\vec{V} = x^2yz$$

∴ line integral of the vector function from origin to the point (1, 1, 1) is

$= (x^2yz)_{1,1,1} - (x^2yz)_{(0,0,0)} = 1 - 0 = 1$

3.5 (d)

div $\{(x-y)\hat{i} + (y-x)\hat{j} + (x+y+z)\hat{k}\}$

$$= \frac{\partial}{\partial x}(x-y) + \frac{\partial}{\partial y}(y-x) + \frac{\partial}{\partial z}(x+y+z) = 3$$

3.6 (b)

The directional derivative of the scalar function $f(x, y, z) = x^2 + 2y^2 + z$ in the direction of the vector $\vec{a} = 3\hat{i} - 4\hat{j}$ is

$$\left(\frac{\partial f}{\partial x}\hat{i} + \frac{\partial f}{\partial y}\hat{j} + \frac{\partial f}{\partial z}\hat{k}\right)\cdot(\hat{a})$$

$$= \left(2x\hat{i} + 4y\hat{j} + 1\hat{k}\right)\cdot\left(\frac{3\hat{i} - 4\hat{j}}{\sqrt{3^2 + 4^2}}\right)$$

$$= \left(2\hat{i} + 4\hat{j} + \hat{k}\right)\cdot\left(\frac{3\hat{i} - 4\hat{j}}{5}\right)$$

(given that $x = 1$, $y = 1$, $z = 2$)

$$= -2$$

3.7 (c)

Vector field,

$$\vec{f} = 3xz\,\hat{i} + 2xy\,\hat{j} - yz^2\,\hat{k}$$

$$\vec{f} = \left[f_1\,\hat{i} + f_2\,\hat{j} + f_3\,\hat{k}\right]$$

Divergence of vector field $\vec{f} = \nabla\cdot\vec{f}$

$$\text{Div}(f) = \left(i\frac{\partial}{\partial x} + \hat{j}\frac{\partial}{\partial y} + \hat{k}\frac{\partial}{\partial z}\right)\cdot\left(f_1\,\hat{i} + f_2\,\hat{j} + f_3\,\hat{k}\right)$$

$$\text{Div}(f) = \frac{\partial f_1}{\partial x} + \frac{\partial f_2}{\partial y} + \frac{\partial f_3}{\partial z}$$

$$\text{Div}(f) = \frac{\partial}{\partial x}[3xz] + \frac{\partial}{\partial y}[2xy] + \frac{\partial}{\partial z}\left[-2yz^2\right]$$

$$\text{Div}(f) = 3z + 2x - 2zy$$

$$\text{Div}(f)\big|_{(1,1,1)} = 3(1) + 2(1) - 2(1)(1) = \mathbf{3}$$

3.8 (b)

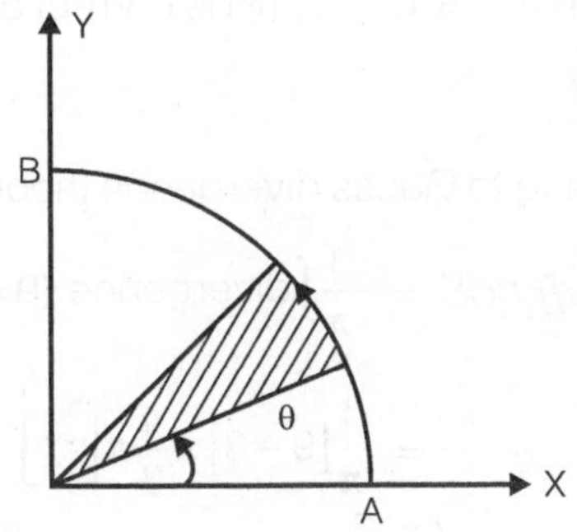

Path AB:

$$x^2 + y^2 = 1$$
$$x = \cos\theta$$
$$y = \sin\theta$$

Along path AB θ varies from 0° to 90° [0 to $\pi/2$]

$$\int_{\text{Path } AB} (x + y)^2\,(r\,d\theta)$$

$$= \int_0^{\pi/2} (\cos\theta + \sin\theta)^2\, 1\cdot d\theta$$

$$= \int_0^{\pi/2} (\sin^2\theta + \cos^2\theta + 2\sin\theta\cos\theta)d\theta$$

$$= \int_0^{\pi/2} (1 + \sin 2\theta)d\theta = \theta + \frac{(-\cos 2\theta)}{2}\Bigg|_0^{\pi/2}$$

$$= \frac{\pi}{2} - \frac{1}{2}\left[\cos 2\frac{\pi}{2} - \cos 0\right]$$

$$= \frac{\pi}{2} - \frac{1}{2}[-1 - 1] = \frac{\pi}{2} + 1$$

3.9 (a)

$$x^2 + y^2 + z^2 = 1$$

$$f(x, y, z) = x^2 + y^2 + z^2 - 1 = 0$$

$$\text{grad } f = \frac{\partial f}{\partial x}\hat{i} + \frac{\partial f}{\partial y}\hat{j} + \frac{\partial f}{\partial z}\hat{k}$$

$$= 2x\hat{i} + 2y\,j + 2z\,\hat{k}$$

at $\left(\frac{1}{\sqrt{2}}, \frac{1}{\sqrt{2}}, 0\right)$

$$\text{grad } f = \frac{2}{\sqrt{2}}\hat{i} + \frac{2}{\sqrt{2}}\hat{j} + 2\times 0\times\hat{k}$$

$$= \sqrt{2}\,i + \sqrt{2}\,j + 0\hat{k}$$

$$|\text{grad } f| = \sqrt{2+2} = \sqrt{4} = 2$$

The unit outward normal vector at point P is

$$n = \frac{1}{|\text{grad } f|}(\text{grad } f)_{\text{at } P}$$

$$= \frac{1}{2}\left(\sqrt{2}\,i + \sqrt{2}\,j\right) = \frac{1}{\sqrt{2}}\,i + \frac{1}{\sqrt{2}}\,j$$

3.10 (a)

$$\iint \frac{1}{4}(F\cdot n)dA \;;\; \iint \frac{1}{4}(F\times dA)$$

By using Gauss divergence theorem

$$\Rightarrow \iiint \frac{1}{4}(\vec{\nabla}\cdot F)dV$$

$$\Rightarrow \iiint \frac{1}{4}\left(\frac{\partial}{\partial x}\hat{i} + \frac{\partial}{\partial y}\hat{j} + \frac{\partial}{\partial y}\hat{k}\right)dV$$

$$\Rightarrow \iiint \frac{1}{4}(1+1+1)dV \;\Rightarrow\; \frac{3}{4}\times V_{\text{sphere}}$$

$$\Rightarrow \frac{3}{4}\times\frac{4}{3}\times\pi(1)^3 = \pi$$

3.11 (b)

For linear dependency, det $\begin{vmatrix} 1 & 1 & 1 \\ 2 & 3 & 1 \\ 5 & 6 & 4 \end{vmatrix}$ must be zero.

$\therefore\ \Delta = 1(12-6) - 1(8-5) + 1\,(12-15)$

$= 6 - 3 - 3 = 0$

$\therefore$ The three vectors are linearly dependent.

3.12 (a)

$$\vec{F} = x^2z^2\vec{i} - 2xy^2z\vec{j} + 2y^2z^3\vec{k}$$

$$\nabla\times\vec{F} = \begin{vmatrix} \vec{i} & \vec{j} & \vec{k} \\ \frac{\partial}{\partial x} & \frac{\partial}{\partial y} & \frac{\partial}{\partial z} \\ x^2z^2 & -2xy^2z^2 & 2y^2z^3 \end{vmatrix}$$

$$= \vec{i}\left[\frac{\partial}{\partial y}(2y^2z^3) + \frac{\partial}{\partial z}(2xy^2z)\right]$$

$$-\vec{j}\left[\frac{\partial}{\partial x}(2y^2z^3) - \frac{\partial}{\partial z}(x^2z^2)\right]$$

$$+\vec{k}\left[\frac{\partial}{\partial x}(-2xy^2z) - \frac{\partial}{\partial y}(x^2z^2)\right]$$

$$\nabla\times\vec{F} = \vec{i}[4yz^3 + 2xy^2] - \vec{j}[-2zx^2] + \vec{k}[-2y^2z - 0]$$

$$= (4yz^3 + 2xy^2)\vec{i} + (2x^2z)\vec{j} - (2y^2z)\vec{k}$$

3.13 (c)

$$\vec{F} = x^2z\hat{i} + xy\hat{j} - yz^2\hat{k}$$

$$\nabla\cdot\vec{F} = \frac{\partial}{\partial x}(x^2z) + \frac{\partial}{\partial y}(xy) - \frac{\partial}{\partial z}(yz^2)$$

$$\nabla\cdot\vec{F} = 2xz + x - 2yz$$

$$\therefore \nabla\cdot\vec{F}\Big|_{(1,-1,1)} = 2\times1\times1 + 1 - 2\times -1\times 1$$

$$= 2 + 1 + 2 = \mathbf{5}$$

3.14 (c)

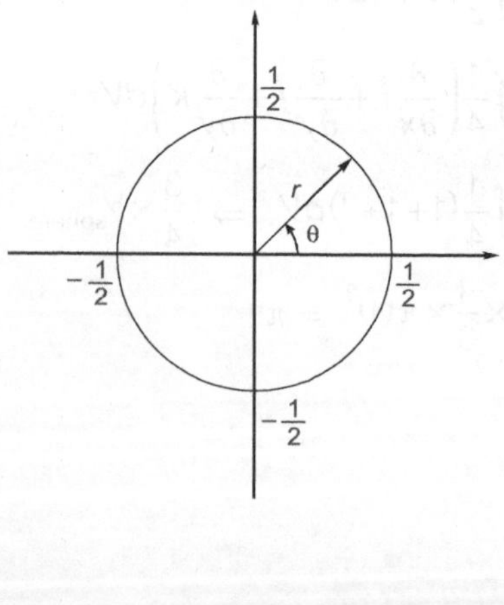

$$\int_C(\phi\,dx + \psi\,dy) = \iint_E\left(\frac{\partial\psi}{\partial x} - \frac{\partial\phi}{\partial y}\right)dxdy$$

$$\phi = y$$

$$\psi = -x$$

$$\frac{\partial\psi}{\partial x} = -1$$

$$\frac{\partial\phi}{\partial y} = 1$$

$$I = \iint_E\left(\frac{\partial\psi}{\partial x} - \frac{\partial\phi}{\partial y}\right)dxdy = \int_0^{1/2}\int_0^{2\pi}(-2)\,r\,d\theta\,dr$$

$$= -2\left(\int_0^{1/2} r\,dr\right)\left(\int_0^{2\pi} r d\theta\right) = -2\left(\frac{r^2}{2}\right)_0^{1/2} 2\pi = -\frac{\pi}{2}$$

3.15 (a)

$$\text{Curl of vector} = \begin{vmatrix} i & j & k \\ \frac{\partial}{\partial x} & \frac{\partial}{\partial y} & \frac{\partial}{\partial z} \\ 2x^2 & 3z^2 & y^3 \end{vmatrix}$$

$$= i\left[\frac{\partial}{\partial y}(y^3)\,\frac{\partial}{\partial z}(3z^2)\right] - j\left[\frac{\partial}{\partial x}(y^3)\,\frac{\partial}{\partial z}(2x^2)\right] + k\left[\frac{\partial}{\partial x}(3z^2)\,\frac{\partial}{\partial y}(2x^2)\right]$$

$= i[3y^2 - 6z] - j[0] + k[0 + 0]$

At $x = 1$, $y = 1$ and $z = 1$

Curl $= i(3\times1^2 - 6\times1) = \mathbf{-3}i$

3.16 (c)

Div Curl $\vec{V} = 0$ $\quad\therefore$ (c) is correct option.

3.17 Sol.

According to Gauss divergence theorem

$$\iint_S \frac{1}{\pi}(9xi - 3yj)\cdot n\,dS = \frac{1}{\pi}\int \text{divergence }(9xi - 3yj), dv$$

$$= \frac{1}{\pi}[9-3]\times\frac{4}{3}\pi\left[r^3\right]$$

$r = 3$ [given]

$$= \frac{1}{\pi}\times6\times\frac{4}{3}\pi\times27 = \mathbf{216}$$

3.18 Sol.

$\int_C[(3x-8y^2)dx + (4y-6xy)dy]$, C is boundary of region bounded by $x = 0$, $y = 1$, and $z + y = 1$.

Using Green's theorem

$$I = \oint_c (Pdx + Qdy) = \oint\oint_R \left(\frac{\partial Q}{\partial x} - \frac{\partial P}{\partial y}\right) dxdy$$

Here, $P = 3x - 8y^2$

$Q = 4y - 6xy$

$$\frac{\partial Q}{\partial x} = -6y$$

$$\frac{\partial P}{\partial y} = -16y$$

$$I = \iint(-6y - (-16y))dxdy = \iint 10y\,dx\,dy$$

$$I = 10\int_0^1 dx \int_0^{1-x} \frac{y^2}{2} = 5\int_0^1 dx(1-x)^2 = 5\int_0^1 (1-x)^2.dx$$

$I = 1.6666$

3.19 Sol.

$$\int_c \nabla\phi \cdot d\bar{r} = \int_c (yz\bar{i} + xz\bar{p} + xy\bar{k})$$

$$(dx\bar{i} + dx\bar{p} + dz\bar{k})$$

$$= \int_c yzdx + xzdy + xydz = \int_c d(xyz)$$

$$= (xyz)$$

Given that $x = t$, $y = t^2$, $z = 3t^2$

$$= (t \cdot t^2 \cdot 3t^2)\Big|_1^3 = 3(t^5)\Big|_1^3 = 3(3^5 - 1)$$

$$= 3^6 - 3 = 729 - 3 = \mathbf{726}$$

3.20 Sol.

$$\int_c \bar{F}\cdot\bar{r}'dx = \int_c \bar{F}\cdot d\bar{r} = \int_c F_1dx + F_2dy$$

$$= \iint_R \left(\frac{\partial F_2}{dx} - \frac{\partial F_1}{dy}\right)dxdy$$

$F_1 = y \quad F_2 = 2x$

$$= \iint_R (2-1)dxdy$$

$$\frac{\partial F_1}{\partial y} = 1 \quad \frac{\partial F_2}{\partial x} = 2$$

$$= \iint dx\,dy$$

= Area of the circle with radius $\frac{4}{\sqrt{\pi}} = \pi\left(\frac{4}{\sqrt{\pi}}\right)^2$

$$= \pi\frac{16}{\pi} = \mathbf{16}$$

3.21 Sol.

By vector identities

div. (curl $\bar{F}$) = **0**

3.22 Sol.

$$\bar{F} = -y\bar{i} + x\bar{j}$$

$$\nabla \cdot \bar{F} = \frac{\partial}{\partial x}(-y) + \frac{\partial}{\partial y}(x) = \mathbf{0 + 0 = 0}$$

3.23 Sol.

$$F = (x + y)i + (x + z)j + (y + z)k$$

$$\nabla \cdot F = \frac{\partial}{\partial x}(x+y) + \frac{\partial}{\partial y}(x+z) + \frac{\partial}{\partial z}(y+z)$$

$$= 1 + 0 + 1 = 2$$

By Gauss divergence theorem

$$\iint_s F\cdot\hat{n}dS = \iint_V \nabla\cdot\bar{F}dV = \iint_V 2dV$$

$= 2V$ where V is volume of $x^2 + y^2 + z^2$

$$= 2\left(\frac{4}{3}\pi(3)^3\right) = \mathbf{226.08}$$

3.24 (c)

By Gauss Divergence Theorem

$$\iint_s \vec{r}\cdot\hat{n}dS = \iiint_v \nabla\cdot\vec{r}dV = \iiint_v 3\,dV = \mathbf{3V}$$

3.25 (c)

$$\vec{u} = e^x\cos y\hat{i} + e^x\cdot\sin y\hat{j}$$

$$\nabla\cdot\vec{u} = \frac{\partial}{\partial x}(u_1) + \frac{\partial}{\partial y}(u_2)$$

$$= \frac{\partial}{\partial x}(e^x\cdot\cos y) + \frac{\partial}{\partial y}(e^x\cdot\sin y)$$

$$= e^x\cos y + e^x\cos y$$

$$\nabla\cdot\vec{u} = \mathbf{2e^x\cdot\cos y}$$

3.26 (c)

$$\phi = \ln r$$

$$\nabla\phi = \nabla(\ln r)$$

$$f(r) = \ln(r)$$

$$f'(r) = \frac{1}{r}$$

$$\nabla f(r) = \frac{f'(r)}{r}\cdot\vec{r} = \left(\frac{1}{r}\right)\times\left(\frac{1}{r}\right)\cdot\vec{r}$$

$$\nabla f(r) = \frac{\vec{r}}{r^2}$$

3.27 (c)

Directional derivative of $f(x, y)$ at $P(1, 1)$ along the direction of line joining (0, 0) to

$$(1, 1) = \nabla f \cdot \frac{\bar{a}}{|\bar{a}|}$$

where, $\bar{a}$ = line joining (0, 0) and (1, 1)

$$\bar{a} = (1-0)\hat{i} + (1-0)\hat{j} = \hat{i} + \hat{j}$$

and $$\nabla f = \hat{i}\frac{\partial f}{\partial x} + \hat{j}\frac{\partial f}{\partial y} + \hat{k}\frac{\partial f}{\partial z}$$

$$= 2x\hat{i} + 2y\hat{j}$$

$$(\nabla f)_{P(1,1)} = 2\hat{i} + 2\hat{j}$$

∴ Directional derivative of $f(x, y)$ at $P(1, 1)$ in the direction of $\vec{a}$

$$= \nabla f \cdot \frac{\bar{a}}{|\bar{a}|} = (2\hat{i} + 2\hat{j}) \cdot \frac{(\hat{i} + \hat{j})}{\sqrt{2}} = \frac{2+2}{\sqrt{2}} = 2\sqrt{2}$$

3.28 (b)

Given $\bar{u}$ is differentiable vector in the given open surface bounded by closed contour C i.e. C: $x^2 + y^2 = 1$, $z = 0$

By stoke's theorem

$$\iint_S \text{curl } \vec{F} \cdot \hat{n}\, ds = \oint_C \vec{F} \cdot \overline{dr}$$

$$\iint_S (\nabla \times \bar{u}) \cdot \hat{n}\, ds = \oint_C \bar{u} \cdot \overline{dr}$$

$$= \frac{1}{3}\oint -y^3 dx + x^3 dy \quad (\because z = 0)$$

C: $x = \cos\theta \Rightarrow dx = -\sin\theta\, d\theta$; $y = \sin\theta$

⇒ $dy = \cos\theta\, d\theta$; $\theta = 0$ to 2π

$$= \frac{1}{3}\int_0^{2\pi} \sin^4\theta\, d\theta + \frac{1}{3}\int_0^{2\pi} \cos^4\theta\, d\theta$$

$$= \frac{4}{3}\int_0^{\pi/2} \sin^4\theta\, d\theta + \frac{4}{3}\int_0^{\pi/2} \cos^4\theta\, d\theta$$

$$= \frac{4}{3}\left(\frac{3}{4} \times \frac{1}{2} \times \frac{\pi}{2} + \frac{3}{4} \times \frac{1}{2} \times \frac{\pi}{2}\right) = \frac{\pi}{2}$$

3.29 Sol.

Given, $\vec{A} = 2\hat{j} - 3\hat{k}, \vec{B} = -2\hat{i} + \hat{k}$ and $\vec{C} = 3\hat{i} - \hat{j}$

$$\vec{A} \cdot (\vec{B} \times \vec{C}) = \begin{vmatrix} 0 & 2 & -3 \\ -2 & 0 & 1 \\ 3 & -1 & 0 \end{vmatrix}$$

$$= 0 - 2[0 - 3] - 3[2 - 0] = 0$$

$$\vec{A} \cdot (\vec{B} \times \vec{C}) + 6 = 0 + 6 = 6$$

3.30 (a)

$$\vec{f} = \frac{\vec{r}}{r^3}$$

$$\nabla \cdot \vec{f} = \nabla \cdot \frac{\vec{r}}{r^3} = 0$$

By divergence theorem

$$\iint_S \vec{f} \cdot \vec{dS} = \iiint_R \nabla \vec{f}\, dx\, dy\, dz = 0$$

3.31 (a)

$$f = xyz$$

$$\nabla f = f_x\bar{i} + f_y\bar{j} + f_z\bar{k} = yz\bar{i} + xz\bar{j} + xy\bar{k}$$

$$(\nabla f)_{(-1,1,3)} = 3\bar{i} - 3\bar{j} - \bar{k}$$

$$\vec{a} = \bar{i} - 2\bar{j} + 2\bar{k}$$

Directional derivative of f in direction of $\vec{a}$

$$Df_{\vec{a}} = \nabla f \times \frac{\vec{a}}{|\vec{a}|} = \frac{3(1) + (-3)(-2) + (-1)(2)}{\sqrt{(1)^2 + (-2)^2 + (2)^2}} = \frac{7}{3}$$

3.32 (a)

Given, $$\varphi = \frac{1}{2}(x^2 + y^2 + z^2)$$

$$\vec{\nabla}\varphi = \frac{\partial\varphi}{\partial x}i + \frac{\partial\varphi}{\partial y}j + \frac{\partial\varphi}{\partial z}k = \frac{1}{2}(2xi + 2yj + 2zk)$$

$$\vec{\nabla}\varphi = xi + yj + zk$$

Using Gauss' divergence theorem,

$$\oiint_S \hat{n} \cdot \nabla\varphi\, ds = \iiint_V \vec{\nabla} \cdot (\vec{\nabla}\varphi)\, dV$$

$$= \iiint_V \vec{\nabla} \cdot (xi + yj + zk)\, dV$$

$$= \iiint_V (1 + 1 + 1)\, dV = 3\iiint_V dV$$

$$= 3 \times \text{volume of sphere} = 3 \times \frac{4}{3}\pi(1)^3 = 4\pi$$

3.33 (8.32) (7.90 to 8.70)

$$\vec{a} = 5\hat{i} + 7\hat{j} + 2\hat{k}; \quad \vec{b} = 3\hat{i} - \hat{j} + 6\hat{k}$$

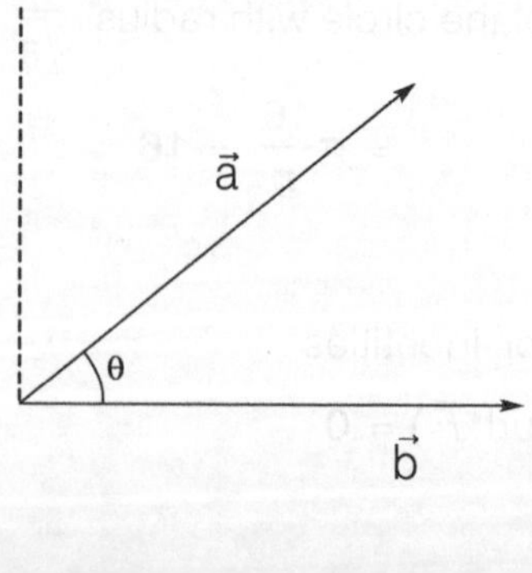

Magnitude of $\vec{a}$ orthogonal to $\vec{b}$ in plane of $\vec{a}$

and $\vec{b}$ = $\|a|\sin\theta|$

$$\cos\theta = \frac{\vec{a}\cdot\vec{b}}{|\vec{a}|\cdot|\vec{b}|}$$

$$= \frac{5\times3-7\times1+2\times6}{\sqrt{5^2+7^2+2^2}\cdot\sqrt{3^2+(-1)^2+6^2}}$$

$$= \frac{20}{\sqrt{78}\cdot\sqrt{46}}$$

$$\Rightarrow \quad \theta = 70.495°$$

$$\therefore \quad \|a|\sin\theta| = \left|\sqrt{78}\sin(70.495°)\right|$$

$$= 8.32$$

3.34 (a)

Given: $\vec{F} = 3x\hat{i} + 5y\hat{j} + 6z\hat{k}$

$$\vec{\nabla}\cdot\vec{F} = \frac{\partial}{\partial x}(3x) + \frac{\partial}{\partial y}(5y) + \frac{\partial}{\partial z}(6z)$$

$$= 3 + 5 + 6 = 14$$

$\therefore$ By Gauss' divergence theorem,

$$\int_A \vec{F}\cdot d\vec{A} = \iiint_V \vec{\nabla}\cdot\vec{F}dV = \iiint_V 14dV$$

$$= 14 \times \text{Volume of cube}$$

$$= 14 \times 1^3 = 14$$

3.35 (0) (–0.001 to 0.001)

Given, $\vec{B}(x,y,z) = x\hat{i} + y\hat{j} - 2z\hat{k}$

Since surface S is a closed surface, Gauss-divergence theorem can be applied

$$\therefore \quad \int_S \vec{B}\cdot\hat{n}ds = \iiint \vec{\nabla}\cdot\vec{B}dv$$

where,

$$\vec{\nabla}\cdot\vec{B} = \frac{\partial}{\partial x}x + \frac{\partial}{\partial y}(y) - \frac{\partial}{\partial z}(2z) = 1 + 1 - 2 = 0$$

Thus, $\int_S \vec{B}\cdot\hat{n}ds = 0$

3.36 (b)

To evaluate, $\oiint_S z\,dx\,dy$

where $\quad S \equiv x^2 + y^2 + z^2 = R^2$

$$\oiint_S z\,dxdy = \text{Volume of sphere} = \frac{4\pi}{3}R^3$$

■■■■

4 Differential Equations

4.1 The solution of the differential equation $\frac{dy}{dx} + y^2 = 0$ is

(a) $y = \frac{1}{x+c}$

(b) $y = \frac{-x^3}{3} + c$

(c) ce^x

(d) unsolvable as equation is non-linear

[2003 : 2 M]

4.2 If $x^2 \frac{dy}{dx} + 2xy = \frac{2\ln(x)}{x}$, and $y(1) = 0$, then what is $y(e)$?

(a) e (b) 1

(c) $1/e$ (d) $1/e^2$ [2005 : 2 M]

Linked Answer Questions 4.3 and 4.4

The complete solution of the ordinary differential equation $\frac{d^2y}{dx^2} + p\frac{dy}{dx} + qy = 0$ is $y = c_1e^{-x} + c_2e^{-3x}$.

4.3 Then, p and q are

(a) $p = 3, q = 3$ (b) $p = 3, q = 4$

(c) $p = 4, q = 3$ (d) $p = 4, q = 4$

[2005 : 2 M]

4.4 Which of the following is a solution of the differential equation

$$\frac{d^2y}{dx^2} + p\frac{dy}{dx} + (q+1)y = 0?$$

(a) e^{-3x} (b) $x\,e^{-x}$

(c) xe^{-2x} (d) x^2e^{-2x} [2005 : 2 M]

4.5 The solution of the differential equation $\frac{dy}{dx} + 2xy = e^{-x^2}$ with $y(0) = 1$ is

(a) $(1+x)e^{+x^2}$ (b) $(1+x)e^{-x^2}$

(c) $(1-x)e^{+x^2}$ (d) $(1-x)e^{-x^2}$ [2006 : 1 M]

4.6 For $\frac{d^2y}{dx^2} + 4\frac{dy}{dx} + 3y = 3e^{2x}$, the particular integrals is

(a) $\frac{1}{15}e^{2x}$ (b) $\frac{1}{5}e^{2x}$

(c) $3e^{2x}$ (d) $C_1e^{-x} + C_2e^{-3x}$

[2006 : 2 M]

4.7 The solution of $dy/dx = y^2$ with initial value $y(0) = 1$ bounded in the interval

(a) $-\infty \le x \le \infty$ (b) $-\infty \le x \le 1$

(c) $x < 1, x > 1$ (d) $-2 \le x \le 2$

[2007 : 2 M]

4.8 Given that $\ddot{x} + 3x = 0$, and $x(0) = 1$, $\dot{x}(0) = 0$, what is $x(1)$?

(a) –0.99 (b) –0.16

(c) 0.16 (d) 0.99 [2008 : 1 M]

4.9 It is given that $y'' + 2y' + y = 0$, $y(0) = 0$, $y(1) = 0$. What is $y(0.5)$?

(a) 0 (b) 0.37

(c) 0.62 (d) 1.13 [2008 : 2 M]

4.10 The inverse Laplace transform of $\frac{1}{(s^2+s)}$ is

(a) $1 + e^t$ (b) $1 - e^t$

(c) $1 - e^{-t}$ (d) $1 + e^{-t}$ [2009 : 1 M]

4.11 The solution of $x\frac{dy}{dx} + y = x^4$ with the condition $y(1) = \frac{6}{5}$ is

(a) $y = \frac{x^4}{5} + \frac{1}{x}$ (b) $y = \frac{4x^4}{5} + \frac{4}{5x}$

(c) $y = \frac{x^4}{5} + 1$ (d) $y = \frac{x^5}{5} + 1$

[2009 : 2 M]

4.12 The Blasius equation, $\frac{d^3f}{d\eta^3} + \frac{f}{2}\frac{d^2f}{d\eta^2} = 0$, is a

(a) second order nonlinear ordinary differential equation

(b) third order nonlinear ordinary differential equation

(c) third order linear ordinary differential equation

(d) mixed order nonlinear ordinary differential equation

[2010 : 1 M]

4.13 Consider the differential equation $\frac{dy}{dx} = (1+y^2)x$.

The general solution with constant c is

(a) $y = \tan\frac{x^2}{2} + \tan c$ (b) $y = \tan^2\left(\frac{x}{2} + c\right)$

(c) $y = \tan^2\left(\frac{x}{2}\right) + c$ (d) $y = \tan\left(\frac{x^2}{2} + c\right)$

[2011 : 2 M]

4.14 The inverse Laplace transform of the function $F(s) = \frac{1}{s(s+1)}$ is given by

(a) $f(t) = \sin t$ (b) $f(t) = e^{-t}\sin t$

(c) $f(t) = e^{-t}$ (d) $f(t) = 1 - e^{-t}$

[2012 : 2 M]

4.15 Consider the differential equation $x^2\frac{d^2y}{dx^2} + x\frac{dy}{dx} - 4y = 0$ with the boundary conditions of $y(0) = 0$ and $y(1) = 1$. The complete solution of the differential equation is

(a) x^2 (b) $\sin\left(\frac{\pi x}{2}\right)$

(c) $e^x\sin\left(\frac{\pi x}{2}\right)$ (d) $e^{-x}\sin\left(\frac{\pi x}{2}\right)$

[2012 : 2 M]

4.16 The solution to the differential equation $\frac{d^2u}{dx^2} - k\frac{du}{dx} = 0$ is where k is constant, subjected to the boundary conditions $u(0) = 0$ and $u(L) = U$, is

(a) $u = U\frac{x}{L}$ (b) $u = U\left(\frac{1-e^{kx}}{1-e^{kL}}\right)$

(c) $u = U\left(\frac{1-e^{-kx}}{1-e^{-kL}}\right)$ (d) $u = U\left(\frac{1+e^{kx}}{1+e^{kL}}\right)$

[2013 : 2 M]

4.17 The solution of the initial value problem $\frac{dy}{dx} = -2xy;\ y(0) = 2$ is

(a) $1+e^{-x^2}$ (b) $2e^{-x^2}$

(c) $1+e^{x^2}$ (d) $2e^{x^2}$

[2014 : 1 M, Set-4]

4.18 If $y = f(x)$ is the solution of $\frac{d^2y}{dx^2} = 0$, with the boundary conditions $y = 5$ at $x = 0$, and $\frac{dy}{dx} = 2$ at $x = 10$, $f(15) =$ ______.

[2014 : 2 M, Set-1]

4.19 The general solution of the differential equation $\frac{dy}{dx} = \cos(x + y)$, with c as a constant, is

(a) $y + \sin(x + y) = x + c$

(b) $\tan\left(\frac{x+y}{2}\right) = y + c$

(c) $\cos\left(\frac{x+y}{2}\right) = x + c$

(d) $\tan\left(\frac{x+y}{2}\right) = x + c$ [2014 : 2 M, Set-2]

4.20 Consider two solutions $x(t) = x_1(t)$ and $x(t) = x_2(t)$ of the differential equation $\frac{d^2x(t)}{dt^2} + x(t) = 0,\ t > 0$, such that

$x_2 = 0, \left.\frac{dx_2(t)}{dt}\right|_{t=0} = 1.$

The Wronskian $W(t) = \begin{vmatrix} x_1(t) & x_2(t) \\ \frac{dx_1(t)}{dt} & \frac{dx_2(t)}{dt} \end{vmatrix}$ at $t = \pi/2$ is

(a) 1 (b) –1

(c) 0 (d) $\pi/2$

[2014 : 2 M, Set-3]

4.21 Find the solution of $\frac{d^2y}{dx^2} = y$ which passes through the origin and the point $\left(ln\ 2, \frac{3}{4}\right)$.

(a) $y = \frac{1}{2}e^x - e^{-x}$ (b) $y = \frac{1}{2}(e^x + e^{-x})$

(c) $y = \frac{1}{2}(e^x - e^{-x})$ (d) $y = \frac{1}{2}e^x + e^{-x}$

[2015 : 2 M, Set-1]

4.22 Consider the following differential equation:

$\frac{dy}{dt} = -5y$; initial condition: $y = 2$ at $t = 0$

The value of y at $t = 3$ is

(a) $-5e^{-10}$ (b) $2e^{-10}$

(c) $2e^{-15}$ (d) $-15e^2$

[2015 : 2 M, Set-2]

4.23 If $y = f(x)$ satisfies the boundary value problem $y'' + 9y = 0$, $y(0) = 0$, $y\left(\frac{\pi}{2}\right) = \sqrt{2}$, then $y\left(\frac{\pi}{4}\right)$ is ______

[2016 : 2 M, Set-1]

4.24 The differential equation $\frac{d^2y}{dx^2} + 16y = 0$ for $y(x)$ with the two boundary conditions $\left.\frac{dy}{dx}\right|_{x=0} = 1$ and $\left.\frac{dy}{dx}\right|_{x=\frac{\pi}{2}} = -1$ has

(a) no solution (b) exactly two solutions
(c) exactly one solution (d) infinitely many solutions

[2017 : 1 M, Set-1]

4.25 Consider the differential equation $3y''(x) + 27y(x) = 0$ with initial conditions $y(0) = 0$ and $y'(0) = 2000$. The value of y at $x = 1$ is _______.

[2017 : 2 M, Set-2]

4.26 If y is the solution of the differential equation

$$y^3\frac{dy}{dx} + x^3 = 0,\ y(0) = 1$$

the value of $y(-1)$ is

(a) –2 (b) –1
(c) 0 (d) 1

[2018 : 1 M, Set-2]

4.27 Given the ordinary differential equation

$$\frac{d^2y}{dx^2} + \frac{dy}{dx} - 6y = 0$$

with $y(0) = 0$ and $\frac{dy}{dx}(0) = 1$, the value of $y(1)$ is _______ (correct to two decimal places).

[2018 : 2 M, Set-2]

4.28 For the equation, $\frac{dy}{dx} + 7x^2y = 0$, if $y(0) = \frac{3}{7}$, then the value of $y(1)$ is

(a) $\frac{3}{7}e^{-\frac{7}{3}}$ (b) $\frac{7}{3}e^{-\frac{7}{3}}$

(c) $\frac{3}{7}e^{-\frac{3}{7}}$ (d) $\frac{7}{3}e^{-\frac{3}{7}}$

[2019 : 1 M, Set-1]

4.29 The differential equation $\frac{dy}{dx} + 4y = 5$ is valid in the domain $0 \le x \le 1$ with $y(0) = 2.25$. The solution of differential equation is

(a) $y = e^{-4x} + 1.25$ (b) $y = e^{-4x} + 5$
(c) $y = e^{4x} + 5$ (d) $y = e^{4x} + 1.25$

[2019 : 1 M, Set-2]

4.30 A differential equation is given as

$$x^2\frac{d^2y}{dx^2} - 2x\frac{dy}{dx} + 2y = 4$$

The solution of differential equation in terms of arbitrary constant C_1 and C_2 is

(a) $y = \frac{C_1}{x^2} + C_2x + 2$

(b) $y = C_1x^2 + C_2x + 4$

(c) $y = C_1x^2 + C_2x + 2$

(d) $y = \frac{C_1}{x^2} + C_2x + 4$

[2019 : 2 M, Set-2]

4.31 The ordinary differential equation $\frac{dy}{dt} = -\pi y$ subject to an initial condition $y(0) = 1$ is solved numerically using the following scheme:

$$\frac{y(t_{n+1}) - y(t_n)}{h} = -\pi y(t_n)$$

where h is the time step, t_n = nh, and $n = 0, 1, 2,$ .... This numerical scheme is stable for all values of h in the interval _________.

(a) $0 < h < \frac{2}{\pi}$ (b) for all $h > 0$

(c) $0 < h < 1$ (d) $0 < h < \frac{\pi}{2}$

[2021 : 1 M, Set-1]

4.32 Consider the following differential equation

$$(1+y)\frac{dy}{dx} = y.$$

The solution of the equation that satisfies the condition $y(1) = 1$ is

(a) $ye^y = e^x$ (b) $y^2e^y = e^x$
(c) $(1 + y)e^y = 2e^x$ (d) $2ye^y = e^x + e$

[2021 : 2 M, Set-2]

4.33 If $y(x)$ satisfies the differential equation

$$(\sin x)\frac{dy}{dx} + y\cos x = 1$$

subject to the condition $y\left(\frac{\pi}{2}\right) = \frac{\pi}{2}$, then $y\left(\frac{\pi}{6}\right)$ is

(a) 0 (b) $\frac{\pi}{3}$

(c) $\frac{\pi}{6}$ (d) $\frac{\pi}{2}$ [2021 : 1 M, Set-1]

4.34 For the exact differential equation,

$$\frac{du}{dx} = \frac{-xu^2}{2 + x^2u}$$

Which one of the following is the solution?

(a) $u^2 + 2x^2$ = Constant

(b) $xu^2 + u$ = Constant

(c) $\frac{1}{2}x^2u^2 + 2u$ = Constant

(d) $\frac{1}{2}ux^2 + 2x$ = Constant [2022 : 2 M, Set-2]

4.35 Consider the second-order linear ordinary differential equation

$$x^2\frac{d^2y}{dx^2} + x\frac{dy}{dx} - y = 0,\ x \geq 1$$

with the intial conditions

$$y(x=1) = 6,\ \left.\frac{dy}{dx}\right|_{x=1} = 2$$

the value of y at $x = 2$ equals ______. [Answer in integer]

[2023 : 2 M]

4.36 If x(t) satisfies the differential equation

$$t\frac{dx}{dt} + (t - x) = 0$$

subject to the condition $x(1) = 0$, then the value of $x(2)$ is ______. (rounded off to 2 decimal places)

[2024 : 2 M]

■■■■

Answers Differential Equations

4.1	(a)	4.2	(d)	4.3	(c)	4.4	(c)	4.5	(b)	4.6	(b)	4.7	(c)	4.8	(b)
4.9	(a)	4.10	(c)	4.11	(a)	4.12	(b)	4.13	(d)	4.14	(d)	4.15	(a)	4.16	(b)
4.17	(b)	4.18	(35)	4.19	(d)	4.20	(a)	4.21	(c)	4.22	(c)	4.23	(–1)	4.24	(a)
4.25	(94.08)	4.26	(c)	4.27	(1.4678)	4.28	(a)	4.29	(a)	4.30	(c)	4.31	(a)	4.32	(a)
4.33	(b)	4.34	(c)	4.35	(9)	4.36	(–1.39)								

Explanations Differential Equations

4.1 (a)

Given differential equation

$$\frac{dy}{dx} + y^2 = 0$$

$$\Rightarrow \quad -\frac{dy}{y^2} = dx$$

On integrating, we get

$$-\int\frac{dy}{y^2} = \int dx;\quad \frac{1}{y} = x + c$$

$$\therefore \quad y = \frac{1}{x + c}$$

4.2 (d)

$$\text{Given } x^2\frac{dy}{dx} + 2xy = \frac{2\ln x}{x} \qquad ....(1)$$

$$\text{or, } \frac{dy}{dx} + \frac{2}{x}y = \frac{2\ln x}{x^3} \qquad ....(2)$$

$$\text{Integrating factor} = e^{\int\frac{2}{x}dx}$$

$$\text{I.F.} = e^{2\ln x} = x^2$$

Multiply by I.F. then integrate the equation we get

$$y \cdot x^2 = \int\frac{2\ln x}{x^3}\cdot x^2$$

$$\Rightarrow \quad x^2y = \int\frac{2\ln x}{x}$$

$$\therefore \quad x^2y = (\ln x)^2 + c$$

when $x = 1$; and $y = 0$

then $c = 0$

$$\therefore \quad x^2y = (\ln x)^2$$

Now at $x = e$

$$y = \frac{(\ln e)^2}{e^2} = \frac{1}{e^2}$$

4.3 (c)

Given equation is

$$\frac{d^2y}{dx^2} + p\frac{dy}{dx} + qy = 0$$

$$(D^2 + pD + q)y = 0$$

$$\therefore \quad D^2 + pD + q = 0$$

Its solution is

$$y = C_1 e^{-x} + C_2e^{-3x}$$

Sum of roots $-p = -1 - 3$

$= P = 4$

Product of roots $q = (-1)(-3)$

$\therefore$ $q = 3$

4.4 (c)

Given equation is

$$\frac{d^2y}{dx^2}+p\frac{dy}{dx}+(q+1)y=0$$

$\Rightarrow [D^2 + PD + (q + 1)]y = 0$

Put $P = 4$

and $q = 3$

$\therefore (D^2 + 4D + 4)y = 0$

$D^2 + 4D + 4 = 0$

$(D + 2)^2 = 0$

$\therefore D = -2, -2$

$y = \boldsymbol{x e^{-2x}}$

4.5 (b)

Given equation

$$\frac{dy}{dx}+2xy = e^{x^2}$$

Integrating factor

$$\text{I.F.} = e^{\int 2x dx} = e^{x^2}$$

Solution is $y \cdot x \cdot e^{x^2} = \int e^{-x^2} \cdot e^{x^2}\, dx$

$\therefore y \cdot x \cdot e^{x^2} = x + c$

$dt\ x = 0$

$y = 1$

$\therefore c = 1$

$y = \boldsymbol{(1 + x)e^{-x^2}}$

4.6 (b)

$$\frac{d^2y}{dx^2}+4\frac{dy}{dx}+3y = 3e^{2x}$$

$\Rightarrow (D^2 + 4D + 3)y = 3x^{e^{2x}}$

Auxiliary equation is

$D^2 + 4D + 3 = 0$

or $(D + 1)(D + 3) = 0$

$\therefore D = -1\ \&\ -3$

Particular integration

$$\text{P.I.} = \frac{1}{D^2+4D+3}\times 3e^{2x} = 3\frac{e^{2x}}{(2)^2+4(2)+3}$$

$$= \frac{3e^{2x}}{15}=\frac{e^{2x}}{5}$$

4.7 (c)

Given $\frac{dy}{dx} = y^2$

$\Rightarrow \int\frac{dy}{y^2} = \int dx$

$\Rightarrow -\frac{1}{y} = x + c$

$\therefore y = -\frac{1}{x+c}$

When $x = 0$

$y = 1$

$\therefore C = -1$

$\therefore y = -\frac{1}{x-1}$

$x - 1 \neq 0$

$\Rightarrow x \neq 1$

i.e. $\boldsymbol{x < 1}$

and $\boldsymbol{x > 1}$

4.8 (b)

$\ddot{x} + 3x = 0$

Auxiliary equation is

$m^2 + 3 = 0$

i.e. $m = \pm\sqrt{3}$

$\therefore x = A\cos\sqrt{3}t + B\sin\sqrt{3}t$

$\dot{x} = \sqrt{3}(B\cos\sqrt{3}t - A\sin\sqrt{3}t)$

At $t = 0$

$1 = A$ ...(i)

$0 = B$ ...(ii)

$\therefore x = \cos\sqrt{3}t$

$x(1) = \cos(\sqrt{3}\times 57.3) = -0.16$

4.9 (a)

$y'' + 2y' + y = 0$

$(D^2 + 2D + 1)y = 0$

$\Rightarrow D^2 + 2D + 1 = 0$

$\Rightarrow (D + 1)^2 = 0$

$\Rightarrow D = -1, -1$

$\therefore y = (C_1 + C_2x)e^{-x}$

$\Rightarrow 0 = (C_1 + C_2(0))e^{-0}$

$\Rightarrow C_1 = 0$

$0 = (C_1 + C_2)e^{-1}$

$\Rightarrow C_1 + C_2 = 0$

$\Rightarrow C_2 = 0$

$\therefore y(0.5) = \boldsymbol{0}$

4.10 (c)

$$L^{-1}\left(\frac{1}{s^2+s}\right) = ?$$

$$\frac{1}{s^2+s} = \frac{1}{s(s+1)} = \frac{1}{s} - \frac{1}{s+1}$$

$$L^{-1}\left(\frac{1}{s^2+s}\right) = L^{-1}\left(\frac{1}{s}\right) - -L^{-1}\left(\frac{1}{s+1}\right)$$

$$= 1 - e^{-t}$$

[Using standard formulae]

Standard formula:

$$L^{-1}\left(\frac{1}{s}\right) = 1$$

$$L^{-1}\left(\frac{1}{s+a}\right) = e^{-at};\ L^{-1}\left(\frac{1}{s-a}\right) = e^{at}$$

4.11 (a)

Given differential equation is

$$x\frac{dy}{dx} + y = x^4$$

$$\Rightarrow \frac{dy}{dx} + \left(\frac{1}{x}\right)y = x^3 \quad \text{... (i)}$$

Standard form

$$\Rightarrow \frac{dy}{dx} + Py = Q \quad \text{... (ii)}$$

where P and Q function of x only and solution is given by

$$y(\text{I.F.}) = \int Q\cdot(I.F.)\,dx + C$$

where, integrating factor (I.F.) = $e^{\int P dx}$

Here $P = \frac{1}{x}$ and $Q = x^3$

$$\text{I.F.} = e^{\int \frac{1}{x}dx} = e^{\ln x} = x$$

Solution $y(x) = \int x^3 \cdot x\,dx + c$

$$yx = \frac{x^5}{5} + c$$

Given condition

$$y(1) = \frac{6}{5}$$

Means at $x = 1$; $y = \frac{6}{5}$

$$\Rightarrow \frac{6}{5} \times 1 = \frac{1}{5} + c$$

$$\Rightarrow c = \frac{6}{5} - \frac{1}{5} = 1$$

Therefore $yx = \frac{x^5}{5} + 1$

$$\Rightarrow y = \frac{x^4}{5} + \frac{1}{x}$$

4.12 (b)

$\frac{d^3f}{dn^3} + \frac{f}{2}\frac{d^2f}{dn^2} = 0$ is third order $\left(\frac{d^3f}{dn^3}\right)$ and it is non linear, since the product $f \times \frac{d^2f}{dn^2}$ is not allowed in linear differential equation.

4.13 (d)

$$\frac{dy}{dx} = (1+y^2)x$$

$$\int \frac{dy}{1+y^2} = \int x dx$$

$$\tan^{-1} y = \frac{x^2}{2} + C$$

$$y = \tan\left(\frac{x^2}{2} + C\right)$$

4.14 (d)

$$F(s) = \frac{1}{s(s+1)} = \frac{A}{s} + \frac{B}{s+1} = \frac{A(s+1)+B(s)}{s(s+1)}$$

$\Rightarrow A(s+1) + B(s) = 1$

Put $s = 0$

$\Rightarrow A = 1$

and $s = -1$

$\Rightarrow B = -1$

So $F(s) = \frac{1}{s} - \frac{1}{s+1}$

Now $f(t) = L^{-1}(F(s)) = e^{0t} - e^{-t}$

$f(t) = 1 - e^{-t}$

4.15 (a)

$$x^2\frac{d^2y}{dx^2} + x\frac{dy}{dx} - 4y = 0$$

$y(0) = 0$ and $y(1) = 1$

Choice (a) satisfies the initial condition as well as equation as shown below:

If $y = x^2$

$\Rightarrow y(0) = 0^2 = 0$ and $y(1) = 1^2 = 1$

Also, $\frac{dy}{dx} = 2x$

$$\frac{d^2y}{dx^2} = 2$$

$$\Rightarrow x^2\frac{d^2y}{dx^2}+x\frac{dy}{dx}-4y = x^2\times 2+x\times 2x-4\times x^2$$

$$= 2x^2 + 2x^2 - 4x^2 = 0$$

So $y = x^2$ is the solution to this equation with given boundary conditions.

4.16 (b)

$AE = m^2 - km = 0 \Rightarrow m = 0, k.$

$$\therefore \quad u(x) = c_1 + c_2e^{kx}$$

$$u(0) = 0 \Rightarrow c_1 + c_2 = 0$$

$$u(L) = u \Rightarrow c_1 + c_2e^{KL} = U$$

$$\Rightarrow \quad c_2 = \frac{U}{e^{KL}-1} \text{ and } c_1 = \frac{-U}{e^{kL}-1}$$

$$\therefore \quad u(x) = \frac{U}{1-e^{kL}}+\frac{Ue^{kx}}{e^{kL}-1} = U\left[\frac{1-e^{kx}}{1-e^{kL}}\right]$$

Hence the correct option is (b).

4.17 (b)

$$\frac{dy}{dx} = 2xy = 0 \quad \text{...(1)}$$

$$\text{I.F.} = e^{\int 2x\,dx} = e^{x^2}$$

Multiplying I.F. to both side of equation (1)

$$e^{x^2}\left[\frac{dy}{dx}+2xy\right] = 0$$

$$\Rightarrow \quad \frac{d}{dx}\left(e^{x^2}y\right)=0$$

$$e^{x^2}y = C$$

from the given boundary condition, $C = 2$

$$\therefore e^{x^2}y = 2$$

$$y = 2e^{-x^2}$$

4.18 Sol.

$$\frac{d^2y}{dx^2} = 0$$

$$\frac{dy}{dx} = C_1$$

$$\Rightarrow \quad C_1 = 2$$

$$y = C_1x + C_2$$

at $x = 0;\ y = 5 = C_2$

$$\therefore \quad y = 2x + 5$$

at $y(15) = 2 \times 15 + 5 = \mathbf{35}$

4.19 (d)

Let $z = x + y$

$$\frac{dz}{dx} = 1+\frac{dy}{dx}$$

$$\therefore \quad \frac{dz}{dx}-1 = \cos z$$

$$\text{or} \quad \int\frac{dz}{1+\cos z} = \int dx$$

$$\text{or} \quad \frac{1}{2}\int \sec^2\left(\frac{z}{2}\right)dz = x + c$$

$$\text{or} \quad \tan\left(\frac{z}{2}\right) = x + c$$

$$\text{or } \tan\left(\frac{x+y}{2}\right) = \boldsymbol{x + c}$$

4.20 (a)

Given differential equation in symbolic form is $(D^2 + 1)\,x(t) = 0$

Its A.E. is $D^2 + 1 = 0$,

$$\therefore \quad D = \pm i$$

So, C.F. is

$$x_1(t) = C_1 \cos t,$$

$$x_2(t) = C_2 \sin t$$

$$\therefore \quad x_1(0) = C_1 = 1$$

$$\Rightarrow \quad x_1(t) = \cos t$$

$$\left[\text{it satisfies } \left.\frac{dx_1(t)}{dt}\right|_{t=0} = 0\right]$$

$$x_2(0) = 0 = C_2\sin(0) \ (\because C_2 \neq 0)$$

$$\frac{dx_2(t)}{dt} = C_2 \cos t$$

$$\Rightarrow \quad \left.\frac{dx_2(t)}{dt}\right|_{t=0} = C_2 = 1$$

$$\therefore \quad x_2(t) = \sin t$$

$$W(t) = \begin{vmatrix} x_1(t) & x_2(t) \\ \frac{dx_1(t)}{dt} & \frac{dx_2(t)}{dt} \end{vmatrix} = \begin{vmatrix} \cos t & \sin t \\ -\sin t & \cos t \end{vmatrix}$$

$$W(t) = \cos^2 t + \sin^2 t = 1$$

4.21 (c)

$$\frac{d^2y}{dx^2} = y$$

$$\Rightarrow \quad D^2y = y \quad \left(\because \frac{d}{dx} = D\right)$$

$$(D^2 - 1)y = 0$$

$$D^2 - 1 = 0$$
$$D = \pm 1$$
$$y = C_1 e^x + C_2 e^{-x}$$

Given point passes through origin

$$\Rightarrow \quad O = C_1 + C_2$$
$$C_1 = -C_2 \quad \ldots(i)$$

Also, point passes through (ln 2, 3/4)

$$\Rightarrow \quad \frac{3}{4} = C_1 e^{\ln 2} + C_2 e^{-\ln 2}$$
$$\frac{3}{4} = 2C_1 + \frac{C_2}{2}$$
$$\Rightarrow \quad C_2 + 4C_1 = 1.5 \quad \ldots(ii)$$

From (i) $C_1 = -C_2$, putting in (ii), we get

$$\Rightarrow \quad -3C_2 = 1.5$$
$$C_2 = -0.5$$
$$\therefore \quad C_1 = 0.5$$
$$\Rightarrow \quad y = 0.5(e^x - e^{-x})$$
$$y = \frac{e^x - e^{-x}}{2}$$

4.22 (c)

$$\frac{dy}{dt} = -5y$$
$$\int \frac{dy}{y} = -\int 5dt$$
$$\ln y = -5t + C$$

at $t = 0$

$$y = 2$$
$$\ln 2 = C$$

So, $\ln y = -5t + \ln 2$

$$\ln \frac{y}{2} = -5t$$
$$\frac{y}{2} = e^{-5t}$$
$$y = 2e^{-5t}$$

at $t = 3$

$$y = \mathbf{2e^{-15}}$$

4.23 Sol.

$$(D^2 + 9)y = 0$$
$$m^2 + 9 = 0$$
$$m = \pm 3i$$
$$y = C_1 \cos 3x + C_2 \sin 3x \quad \ldots(1)$$

$x = 0$ $\quad \theta = C_1$

$$y\left(\frac{\pi}{2}\right) = \sqrt{2}$$
$$\sqrt{2} = C_1 \cos\frac{3\pi}{2} + C_2 \sin\frac{3\pi}{2}$$
$$C_2 = -\sqrt{2}$$
$$\therefore \quad y = \sqrt{2}\sin 3x$$
$$y\left(\frac{\pi}{4}\right) = -\sqrt{2}\sin\frac{3\pi}{4} = \frac{-5}{4} = -1$$

4.24 (a)

$$(D^2 + 16)y = 0$$

AE is $m^2 + 16 = 0$

$$m = \pm 4i$$

Solution is $y = c_1 \cos 4x + c_2 \sin 4x$

$$y' = -4c_1 \sin 4x + 4c_2 \cos 4x$$
$$y'(0) = 1$$
$$1 = 4c_2$$
$$c_2 = 1/4$$
$$y'(\pi/2) = -1$$
$$-1 = -4c_1 \sin 2\pi + 4c_2 \cos 2\pi$$
$$-1 = 0 + 4c_2$$
$$c_2 = -1/4$$

Therefore the given differential equation has **no solution.**

4.25 Sol.

The direction is $3y''(x) + 27y(x) = 0$

The auxiliary equation is

$$3m^2 + 27 = 0$$
$$m^2 + 9 = 0$$
$$m = \pm 3i$$

Solution is $y = c_1 \cos 3x + c_2 \sin 3x$

given that $y(0) = 0$

$$\therefore \quad 0 = c_1$$
$$y' = -3c_1 \sin 3x + 3c_2 \cos 3x$$
$$y'(0) = 2000$$
$$2000 = 0 + 3c_2$$
$$c_2 = \frac{2000}{3}$$

$\therefore$ Solution is

$$y = \frac{2000}{3}\sin 3x$$

when $x = 1$ $\quad y = \frac{2000}{3}\sin 3 = 94.08$

4.26 (c)

$$y^3 \frac{dy}{dx} = -x^3$$

$$y^3dy = -x^3\ dx$$
$$\int y^3dy = -\int x^3dx$$
$$\frac{y^4}{4} = \frac{-x^4}{4}+C$$
$$\frac{x^4+y^4}{4} = C$$
$$y(0) = 1,$$
$$\frac{0+1}{4} = C$$
$$C = \frac{1}{4}$$
$$x^4 + y^4 = 1$$
$$y^4 = 1 - x^4$$
$$y = \sqrt[4]{1-x^4}$$

When, $x = -1$

$$\mathbf{y = 0}$$

4.27 Sol.

$$(D^2 + D - 6)y = 0$$
$$y(0) = 0,$$
$$y'(0) = 1$$
$$(D + 3)(D - 2)y = 0$$
$$D = 2, -3$$
$$\text{C.F.} = C_1e^{2x} + C_2e^{-3x}$$
$$y = C_1e^{2x} + C_2e^{-3x}$$
$$y(0) = 0$$

So, $0 = C_1 + C_2$ ...(i)

$$\frac{dy}{dx} = 2C_1e^{2x} - 3C_2e^{-3x}$$
$$y'(0) = 1,$$

$1 = 2C_1 - 3C_2$ ...(ii)

From equations (i) and (ii),

$$C_1 = \frac{1}{5},$$
$$C_2 = \frac{-1}{5}$$
$$y = \frac{1}{5}e^{2x} - \frac{1}{5}e^{-3x}$$

When, $x = 1$

$$y(1) = \frac{e^2 - e^{-3}}{5} = \mathbf{1.4678}$$

4.28 (a)

$$\frac{dy}{dx} + 7x^2y = 0$$
$$\int \frac{dy}{y} = \int -7x^2dx$$

$\ln y = \frac{-7x^3}{3} + c$ ...(i)

At $x = 0$, $y = \frac{3}{7}$

$\ln\left(\frac{3}{7}\right) = c$ ...(ii)

From (i) and (ii),

$$\ln\ y = \frac{-7x^3}{3} + \ln\left(\frac{3}{7}\right)$$
$$\ln\left(\frac{7y}{3}\right) = \frac{-7x^3}{3}$$
$$\Rightarrow \quad y = \frac{3}{7}e^{-\frac{7}{3}x^3}$$
$$y(1) = \frac{3}{7}e^{-\frac{7}{3}}$$

4.29 (a)

$$P(x) = 4,\ Q(x) = 5$$
$$\text{I.F} = e^{\int P(x)dx} = e^{\int 4dx} = e^{4x}$$
$$ye^{4x} = \int 5e^{4x}dx + c$$
$$ye^{4x} = \frac{5}{4}e^{4x} + c$$
$$y = \frac{5}{4} + ce^{-4x}$$

Substitute given initial condition,

$x = 0$, $y = 2.25$

We get, $c = 1$

$$y = \frac{5}{4} + e^{-4x}$$

4.30 (c)

Given D.E. is Euler-Cauchy differential equation,

Let $x = e^z$

$\Rightarrow$ $z = \ln x$

Let, $\frac{d}{dx} = D$ and $\frac{d}{dz} = \theta$

Then $x\frac{d}{dx} = \frac{d}{dz}$

and $x^2\frac{d^2}{dx^2} = \frac{d^2}{dz^2} - \frac{d}{dz}$

i.e. $xD = \theta$, $x^2D^2 = \theta(\theta - 1)$

Sub in D.E.

We get $(\theta(\theta-1) - 2\theta + 2)y = 4$

$(\theta - 1)(\theta - 2)y = 4$

It is LDE with constant coefficients

$\therefore \quad y = y_{CF} + y_{PI}$

$y_{CF} = (m-1)(m-2) = 0$

$\Rightarrow \quad m = 1, 2$

$\therefore \quad y_{CF} = C_1e^z + C_2e^{2z} = C_1x + C_2x^2$

$$y_{P.I.} = \frac{\phi(z)}{f(\theta)} = \frac{4 \times e^{0z}}{(\theta-1)(\theta-2)} = \frac{4}{2} = 2$$

Hence, $\quad y = C_1x + C_2x^2 + 2$

So accordingly option (c) is correct.

4.31 (a)

$$\frac{y(t_{n+1}) - y(t_n)}{h} = -\pi y(t_n)$$

$$y_{n+1} = -\pi h y_n + y_n = (-\pi h + 1)y_n$$

It is recursion relation between y_{n+1} and y_n so solution will be stable if

$$|-\pi h + 1| < 1$$

$$-1 < -\pi h + 1 < 1$$

$$-2 < -\pi h < 0$$

$$0 < \pi h < 2 < h < \frac{2}{\pi}$$

Theretore option (a) is correct.

4.32 (a)

$$(1+y)\frac{dy}{dx} = y$$

$$\Rightarrow \int\left(\frac{1}{y}+1\right)dy = \int dx$$

$\Rightarrow \quad \log y + y = x + c$

Using, $\quad y(1) = 1$

$\log 1 + 1 = 1 + c \quad \Rightarrow c = 0$

Hence, $\log y + y = x$

$\Rightarrow \log y + y\log e = x$

$\log_e(y \cdot e^y) = x$

$\Rightarrow \quad ye^y = e^x$

4.33 (b)

$$\frac{dy}{dx} + y\cot x = \text{cosec}\, x$$

$$\text{I.F.} = e^{\int \cot x\, dx} = e^{\log \sin x} = \sin x$$

$$\Rightarrow y(\sin x) = \int \text{cosec}\, x \sin x\, dx + c$$

$\Rightarrow \quad y\sin x = x + c$

$$\Rightarrow \frac{\pi}{2}\sin\frac{\pi}{2} = \frac{\pi}{2} + c$$

$$\Rightarrow \frac{\pi}{2} = \frac{\pi}{2} + c \qquad \Rightarrow c = 0$$

$\Rightarrow \quad y\sin x = x$

$$\Rightarrow y\sin\frac{\pi}{6} = \frac{\pi}{6}$$

$$\Rightarrow y\left(\frac{1}{2}\right) = \frac{\pi}{6}$$

$$\Rightarrow y = \frac{\pi}{3}$$

4.34 (c)

Given differential equation, $\dfrac{du}{dx} = \dfrac{-xu^2}{2 + x^2u}$

Rewriting it as, $(2 + x^2u)\,du + (xu^2)\,dx = 0$

Checking for exactness,

$$\frac{\partial M}{\partial u} = \frac{\partial}{\partial u}(xu^2) = 2xu$$

$$\frac{\partial N}{\partial x} = \frac{\partial}{\partial x}(2 + xu^2) = 2xu$$

Hence, it is an exact differential equation.

Thus, its solution is given by

$$\int_{u=C} xu^2 dx + \int 2du = C$$

$$\frac{x^2u^2}{2} + 2u = C$$

4.35 (9) (8.999 to 9.001)

It is Euler differential equation

$\therefore D(D-1)y + Dy - y = 0$

$[D(D-1) + (D-1)]y = 0$

Corresponding auxiliary equation,

$m^2 - m + m - 1 = 0$

On solving, $m^2 = 1$

$m = \pm 1$

$\therefore y = C_1e^t + C_1e^{-t}$

But, $t = \ln x$

$$\therefore y = C_1x + \frac{C_2}{x}$$

Also, $\dfrac{dy}{dx} = C_1 - \dfrac{C_2}{x^2}$

Now applying boundary conditions,

i.e. $y(x = 1) = 6; \quad C_1 + C_2 = 6 \qquad$ ...(i)

and, $\left.\dfrac{dy}{dx}\right|_{x=1} = 2; \quad C_1 - C_2 = 2 \qquad$ ...(ii)

On solving (i) and (ii), $C_1 = 4$; $C_2 = 2$

Thus, $y = 4x + \dfrac{2}{x}$

Substituting, $x = 2$, $y = 4(2) + \dfrac{2}{2} = 9$

4.36 (-1.39) (-1.40 to -1.38)

$$t\frac{dx}{dt}+t-x = 0$$

$$\frac{dx}{dt}-\frac{x}{t}+1 = 0$$

$$\frac{dx}{dt}-\frac{1}{t}x = -1$$

This is a linear DE of the form,

$$\frac{dx}{dt}+Px = Q$$

where, $P=-\frac{1}{t}$ and $Q=-1$

Integrating factor,

$$IF = e^{\int -\frac{1}{t}dx} = e^{-\ln t} = e^{\ln\left(\frac{1}{t}\right)} = \frac{1}{t}$$

Thus, the solutions is

$$x \times IF = \int Q\cdot IF dt + c$$

$$x\cdot\frac{1}{t} = \int -1\cdot\frac{1}{t}dt + c$$

$$\frac{x}{t} = -\ln t + c \Rightarrow \frac{x}{t} = \ln\left(\frac{1}{t}\right)+c$$

$$\frac{x}{t} = \ln\frac{1}{t}+c \Rightarrow x = t\cdot\ln\frac{1}{t}+ct$$

Given, at $t = 1$, $x = 0$

$\Rightarrow$ $0 = 0 + c \Rightarrow c = 0$

Thus, particular solution is,

$$x = t\cdot\ln\frac{1}{t}$$

Now, for $t = 2$,

$$x = 2\ln\frac{1}{2}\ ; \quad x = -2\ln 2$$

$$x = -2 \times 0.693 = -1.386 \simeq -1.39$$

■■■■

5 Complex Variables

5.1 Assuming $i = \sqrt{-1}$ and t is a real number, $\int_0^{\pi/3} e^{it}dt$ is

(a) $\frac{\sqrt{3}}{2} + i\frac{1}{2}$ (b) $\frac{\sqrt{3}}{2} - i\frac{1}{2}$

(c) $\frac{1}{2} + i\left(\frac{\sqrt{3}}{2}\right)$ (d) $\frac{1}{2} + i\left(1 - \frac{\sqrt{3}}{2}\right)$

[2006 : 2 M]

5.2 If $\varphi(x, y)$ and $\psi(x, y)$ are functions with continuous second derivatives, then $\varphi(x, y) + i\psi(x, y)$ can be expressed as an analytic function of $x + i\psi(i = \sqrt{-1})$, when

(a) $\frac{\partial\varphi}{\partial x} = -\frac{\partial\psi}{\partial x}, \frac{\partial\varphi}{\partial y} = \frac{\partial\psi}{\partial y}$

(b) $\frac{\partial\varphi}{\partial y} = -\frac{\partial\psi}{\partial x}; \frac{\partial\varphi}{\partial x} = \frac{\partial\psi}{\partial y}$

(c) $\frac{\partial^2\varphi}{\partial x^2} + \frac{\partial^2\varphi}{\partial y^2} = \frac{\partial^2\psi}{\partial x^2} + \frac{\partial^2\psi}{\partial y^2} = 1$

(d) $\frac{\partial\varphi}{\partial x} + \frac{\partial\varphi}{\partial y} = \frac{\partial\psi}{\partial x} + \frac{\partial\psi}{\partial y} = 0$ **[2007 : 1 M]**

5.3 The integral $\oint f(z)dz$ evaluated around the unit circle on the complex plane for $f(z) = \frac{\cos z}{z}$ is

(a) $2\pi i$ (b) $4\pi i$

(c) $-2\pi i$ (d) 0 **[2008 : 2 M]**

5.4 An analytic function of a complex variable $z = x + iy$ is expressed as $f(z) = u(x, y) + i\,v(x, y)$ where $i = \sqrt{-1}$. If $u = xy$, the expression for v should be

(a) $\frac{(x+y)^2}{2} + k$ (b) $\frac{x^2 - y^2}{2} + k$

(c) $\frac{y^2 - x^2}{2} + k$ (d) $\frac{(x-y)^2}{2} + k$

[2009 : 2 M]

5.5 The modulus of the complex number $\left(\frac{3+4i}{1-2i}\right)$ is

(a) 5 (b) $\sqrt{5}$

(c) $1/\sqrt{5}$ (d) 1/5 **[2010 : 1 M]**

5.6 The product of two complex numbers $(1 + i)$ and $2 - 5i$ is

(a) $7 - 3i$ (b) $3 - 4i$

(c) $-3 - 4i$ (d) $7 + 3i$ **[2011 : 1 M]**

5.7 The argument of the complex number $\frac{1+i}{1-i}$, where $i = \sqrt{-1}$, is

(a) $-\pi$ (b) $-\frac{\pi}{2}$

(c) $\frac{\pi}{2}$ (d) π **[2014 : 1 M, Set-1]**

5.8 An analytic function of a complex variable $z = x + i\,y$ is expressed as $f(z) = u(x, y) + i\,v(x, y)$, where $i = \sqrt{-1}$. If $u(x, y) = 2\,x\,y$, then $v(x, y)$ must be

(a) $x^2 + y^2$ + constant (b) $x^2 - y^2$ + constant

(c) $-x^2 + y^2$ + const. (d) $-x^2 - y^2$ + constant

[2014 : 2 M, Set-2]

5.9 An analytic function of a complex variable $z = x + iy$ is expressed as $f(z) = u(x, y) + iv(x, y)$, where $i = \sqrt{-1}$. If $u(x, y) = x^2 - y^2$, then expression for $v(x, y)$ in terms of x, y and a general constant c would be

(a) $xy + c$ (b) $\frac{x^2 + y^2}{2} + c$

(c) $2xy + c$ (d) $\frac{(x-y)^2}{2} + c$

[2014 : 2 M, Set-3]

5.10 If z is a complex variable, the value of $\int_5^{3i} \frac{dz}{z}$ is

(a) $-0.511 - 1.57i$ (b) $-0.511 + 1.57i$

(c) $0.511 - 1.57i$ (d) $0.511 + 1.57i$

[2014 : 2 M, Set-4]

5.11 Given two complex numbers

$z_1 = 5 + \left(5\sqrt{3}\right)i$ and $z_2 = \frac{2}{\sqrt{3}} + 2i$

the argument of $\frac{z_1}{z_2}$ in degree is

(a) 0 (b) 30

(c) 60 (d) 90 **[2015 : 1 M, Set-1]**

5.12 $f(z) = u(x, y) + iv(x, y)$ is an analytic function or complex variable $z = x + iy$ where $i = \sqrt{-1}$ $u(x, y) = 2xy$, then $v(x, y)$ may be expressed as

(a) $-x^2 + y^2$ + constant

(b) $x^2 - y^2$ + constant

(c) $x^2 + y^2$ + constant

(d) $-(x^2 + y^2)$ + constant [2016 : 1 M, Set-1]

5.13 A function f of the complex variable $z = x + iy$, is given as $f(x, y) = u(x, y) + iv(x, y)$, where $u(x, y) = 2kxy$ and $v(x, y) = x^2 - y^2$. The value of k, for which the function is analytic, is______.

[2016 : 1 M, Set-2]

5.14 The value of $\oint_\Gamma \frac{3z-5}{(z-1)(z-2)}dz$ along a closed path Γ is is equal to $(4\pi i)$, where $z = x + iy$ and $i = \sqrt{-1}$. The correct path Γ is

(a)

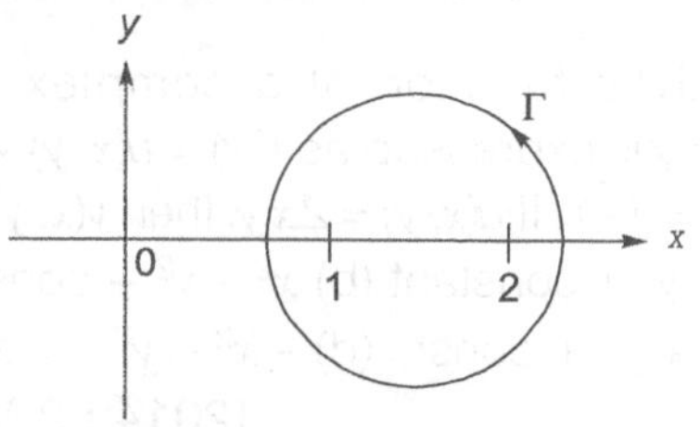

(b)

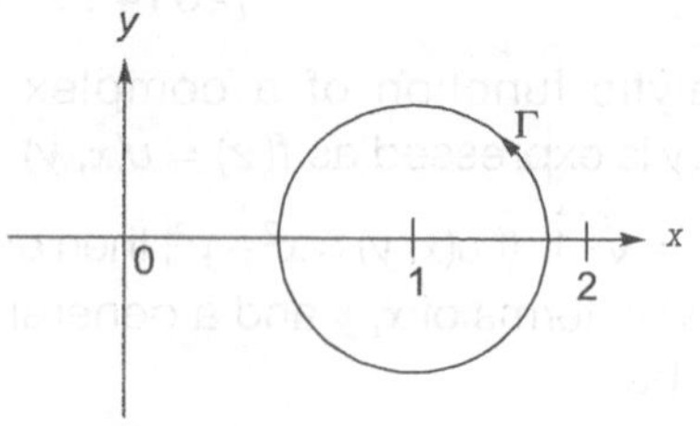

(c)

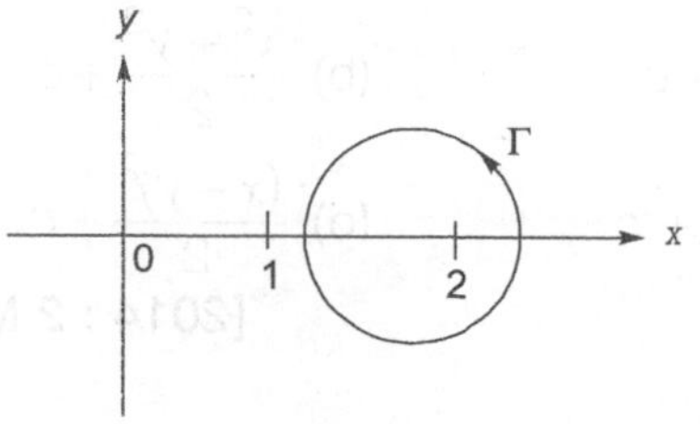

(d)

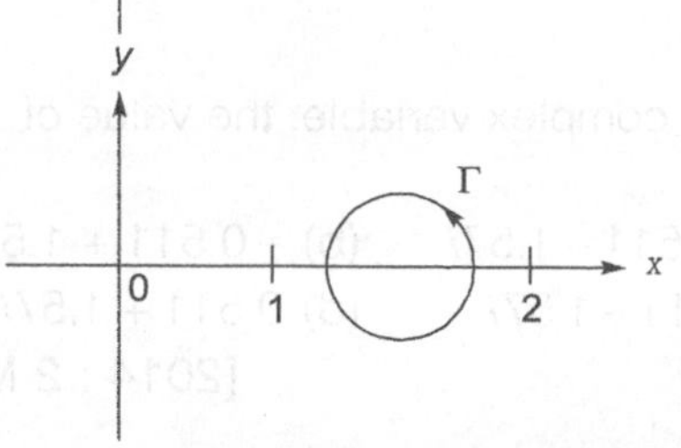

[2016 : 2 M, Set-2]

5.15 The area (in percentage) under standard normal distribution curve of random variable Z within limits from –3 to +3 is______ [2016 : 1 M, Set-3]

5.16 The value of the integral

$$\int_{-\infty}^{\infty} \frac{\sin x}{x^2 + 2x + 2} dx$$

evaluated using contour integration and the residue theorem is

(a) $\frac{-\pi \sin(1)}{e}$ (b) $\frac{-\pi \cos(1)}{e}$

(c) $\frac{\sin(1)}{e}$ (d) $\frac{\cos(1)}{e}$

[2016 : 2 M, Set-1]

5.17 Solutions of Laplace equation having continuous second-order partial derivatives are called

(a) biharmonic functions

(b) harmonic functions

(c) conjugate harmonic functions

(d) error functions [2016 : 1 M, Set-3]

5.18 If $f(z) = (x^2 + ay^2) + ibxy$ is a complex analytic function of $z = x + iy$, where $i = \sqrt{-1}$, then

(a) $a = -1, b = -1$ (b) $a = -1, b = 2$

(c) $a = 1, b = 2$ (d) $a = 2, b = 2$

[2017 : 2 M, Set-2]

5.19 $F(z)$ is a function of the complex variable $z = x + iy$ given by

$F(z) = iz + k\, Re(z) + i\, Im(z)$

For what value of k will F(z) satisfy the Cauchy-Riemann equations?

(a) 0 (b) 1

(c) –1 (d) y [2018 : 1 M, Set-1]

5.20 Let z be a complex variable. For a counterclockwise integration around a unit circle C, centered at origin,

$$\oint_C \frac{1}{5z-4} dz = A\pi i$$ the value of A is

(a) 2/5 (b) 1/2

(c) 2 (d) 4/5

[2018 : 2 M, Set-2]

5.21 A harmonic function is analytic if it satisfies the Laplace equation. If $u(x, y) = 2x^2 - 2y^2 + 4xy$ is a harmonic function, then its conjugate harmonic function $v(x, y)$ is

(a) $4y^2 - 4xy$ + constant

(b) $-4xy + 2y^2 - 2x^2$ + constant

(c) $2x^2 - 2y^2 + xy$ + constant

(d) $4xy - 2x^2 + 2y^2$ + constant

[2019 : 2 M, Set-1]

5.22 An analytic function $f(z)$ of complex variable $z = x + iy$ may be written as $f(z) = u(x, y) + iv(x, y)$. Then, $u(x, y)$ and $v(x, y)$ must satisfy,

(a) $\frac{\partial u}{\partial x} = \frac{-\partial v}{\partial y}$ and $\frac{\partial u}{\partial y} = \frac{\partial v}{\partial x}$

(b) $\frac{\partial u}{\partial x} = \frac{-\partial v}{\partial y}$ and $\frac{\partial u}{\partial y} = \frac{-\partial v}{\partial x}$

(c) $\frac{\partial u}{\partial x} = \frac{\partial v}{\partial y}$ and $\frac{\partial u}{\partial y} = \frac{-\partial v}{\partial x}$

(d) $\frac{\partial u}{\partial x} = \frac{\partial v}{\partial y}$ and $\frac{\partial u}{\partial y} = \frac{\partial v}{\partial x}$ **[2019 : 1 M, Set-2]**

5.23 Which of the following function $f(z)$, of the complex variable z, is NOT analytic at all the points of the complex plane?

(a) $f(z) = z^2$ (b) $f(z) = \log z$
(c) $f(z) = e^z$ (d) $f(z) = \sin z$

[2020 : 1 M, Set-1]

5.24 The function $f(z)$ of complex variable $z = x + iy$, where $i = \sqrt{-1}$, is given as $f(z) = (x^3 - 3xy^2) + iv(x, y)$. For this function to be analytic, $v(x, y)$ should be

(a) $(3x^2y - y^3)$ + constant
(b) $(3x^2y^2 - y^3)$ + constant
(c) $(x^2 - 3x^2y)$ + constant
(d) $(3xy^2 - y^3)$ + constant **[2020 : 2 M, Set-2]**

5.25 Value of $(1 + i)^8$, where $i = \sqrt{-1}$, is equal to

(a) $16i$ (b) 16
(c) $4i$ (d) 4 **[2021 : 1 M, Set-2]**

5.26 Let C represent the unit circle centered at origin in the complex plane, and complex variables, $z = x + iy$. The valve of the contour integral $\oint_C \frac{\cosh 3z}{2z} dz$ (where integration is taken counter clockwise) is

(a) πi (b) $2\pi i$
(c) 0 (d) 2 **[2021 : 2 M, Set-1]**

5.27 The value of the integral

$$\oint \left(\frac{6z}{2z^4 - 3z^3 + 7z^2 - 3z + 5} \right) dz$$

evaluated over a counter-clockwise circular contour in the complex plane enclosing only the pole $z = i$, where i is the imaginary unit, is

(a) $(-1 + i)\pi$ (b) $(1 + i)\pi$
(c) $2(1 - i)\pi$ (d) $(2 + i)\pi$

[2022 : 2 M, Set-1]

5.28 Given $z = x + iy$, $i = \sqrt{-1}$. C is a circle of radius 2 with the centre at the origin. If the contour C is traversed anticlockwise, then the value of the integral $\frac{1}{2\pi} \int_C \frac{1}{(z-i)(z+4i)} dz$ is_____ (round off to one decimal place).

[2022 : 2 M, Set-2]

5.29 The value of k that makes the complex-valued function
$f(z) = e^{-kx} (\cos 2y - i \sin 2y)$
analytic, where $z = x + iy$, is______. [Answer in integer]

[2023 : 1 M]

5.30 Let f(z) be an analytic function, where $z = x + iy$. If the real part of f(z) is $\cosh x \cos y$, and the imaginary part of $f(z)$ is zero for $y = 0$, then $f(z)$ is

(a) $\cosh x \exp(-iy)$ (b) $\cosh z \exp z$
(c) $\cosh z \cos y$ (d) $\cosh z$

[2024 : 1 M]

Answers Complex Variables

5.1	(a)	5.2	(b)	5.3	(a)	5.4	(c)	5.5	(b)	5.6	(a)	5.7	(c)	5.8	(c)
5.9	(c)	5.10	(b)	5.11	(a)	5.12	(a)	5.13	(–1)	5.14	(b)	5.15	(99.74)		
5.16	(a)	5.17	(b)	5.18	(b)	5.19	(b)	5.20	(a)	5.21	(d)	5.22	(a)	5.23	(b)
5.24	(a)	5.25	(b)	5.26	(a)	5.27	(a)	5.28	(0.2)	5.29	(2)	5.30	(d)		

Explanations Complex Variables

5.1 (a)

$$I = \int_0^{\pi/3} e^{it} = \left[\frac{e^{it}}{i}\right]_0^{\pi/3} = \left[\frac{\cos t + i \sin t}{i}\right]_0^{\pi/3}$$

$$= \frac{1}{i}\left[\frac{1}{2} + i\frac{\sqrt{3}}{2} - 1\right] = \left[-\frac{1}{2i} + \frac{\sqrt{3}}{2}\right] = \left[\frac{\sqrt{3}}{2} + i\frac{1}{2}\right]$$

5.2 (b)

The necessary condition for a function $f(z) = \phi(x, y) + i\,\psi(x, y)$ to be analytic

(i) $\frac{\partial \phi}{\partial x} = \frac{\partial \psi}{\partial y}$

(ii) $\frac{\partial \phi}{\partial y} = -\frac{\partial \psi}{\partial x}$

are known as Cauchy Reiman equations

Provided $\frac{\partial \phi}{\partial x}, \frac{\partial \phi}{\partial y}, \frac{\partial \psi}{\partial x}, \frac{\partial \psi}{\partial y}$ exist.

5.3 (a)

$f(z) = \frac{\cos z}{z}$ has simple pole at $z = 0$

$\therefore$ Residue of $f(z)$ at $z = 0$

$$\underset{z\to 0}{Lt}\, f(z)\cdot z = \underset{z\to 0}{Lt} \cos z = 1$$

$$\int_e f(z)dz = 2\pi i \text{ (Residue at } z = 0)$$

$$= 2\pi i \cdot 1 = \mathbf{2\pi i}$$

5.4 (c)

$$u = xy$$

$$u_x = \frac{\partial u}{\partial x} = y;\ u_y = \frac{\partial u}{\partial y} = x$$

$$f(z) = u(x, y) + i\,v(x, y)$$

$$f'(z) = u_x + i\,v_x = v_y - i\,u_y$$

Cauchy Riemann Equation

$$u_x = v_y$$

$$v_x = -u_y$$

$$f'(z) = u_x - i\,u_y$$

$$f'(z) = y - i\,x$$

Put $x = z$ and $y = 0$

$$f'(z) = 0 - i\,z$$

$$f(z) = \int(-i\,z)dz = -i\frac{z^2}{2} + c_1 + i\,c_2$$

$c_1 = 0$ (constant in real part)

$$f(z) = \frac{-i\,z^2}{2} + i\,c_2 = \frac{-i\,(x + iy)^2}{2} + i\,c_2$$

$$f(z) = xy + \frac{i(y^2 - x^2)}{2} + i\,c_2$$

$$f(z) = u(x, y) + i\,v(x, y)$$

Comparing imaginary part

$$v = \frac{y^2 - x^2}{2} + c_2 = \frac{y^2 - x^2}{2} + k$$

5.5 (b)

$$Z = \frac{3 + 4i}{1 - 2i} = \frac{(3 + 4i)(1 + 2i)}{(1 - 2i)(1 + 2i)}$$

$$= \frac{-5 + 10i}{5} = -1 + 2i$$

$$|Z| = \sqrt{(-1)^2 + (2)^2} = \sqrt{5}$$

5.6 (a)

$(1 + i)(2 - 5i) = \mathbf{7 - 3i}$

5.7 (c)

Let $z = \frac{1+i}{1-i} = \frac{(1+i)(1+i)}{(1-i)(1+i)} = \frac{1+i^2+2i}{1-i^2} = \frac{2i}{2} = i$

$z = x + iy = i$

so, $x = 0$

$y = 1$

$$\text{Arg}(z) = \tan^{-1}\left(\frac{y}{x}\right) = \tan^{-1}\left(\frac{1}{0}\right) = \tan^{-1}\infty = \frac{\pi}{2}$$

5.8 (c)

As per Cauchy-Riemann equation

$$\frac{\partial u}{\partial x} = \frac{\partial v}{\partial y}$$

and $\frac{\partial u}{\partial y} = -\frac{\partial v}{\partial x}$

$$\frac{\partial u}{\partial x} = 2y$$

and $\frac{\partial u}{\partial y} = 2x$

$$\frac{\partial v}{\partial y} = 2y$$

$\Rightarrow$ $v = y^2 + f(x)$

$$\frac{\partial v}{\partial x} = 0 + f'(x) = -2x$$

$\therefore$ $f(x) = -x^2 + \text{constant}$

$\therefore$ $\mathbf{v = y^2 - x^2 + \text{constant}}$

5.9 (c)

As per Cauchy-Riemann equations:

$$\frac{\partial u}{\partial x} = \frac{\partial v}{\partial y} \text{ and } \frac{\partial u}{\partial y} = -\frac{\partial v}{\partial x}$$

$$u = x^2 - y^2$$

$$\frac{\partial u}{\partial x} = 2x, \ \frac{\partial u}{\partial y} = -2y$$

$$\therefore \quad \frac{\partial v}{\partial x} = 2x$$

$$\Rightarrow \quad v = 2xy + f(x)$$

$$\frac{\partial v}{\partial x} = 2y + f'(x)$$

$$\Rightarrow \quad \frac{\partial u}{\partial y} = 2y + f'(x)$$

$$\Rightarrow \quad f'(x) = 0 \quad \text{i.e. } f(x) = C$$

$$\therefore \quad v = \mathbf{2xy + C}$$

5.10 (b)

If z is a complex number, the value of $\int_5^{3i} \frac{1}{z} dz$

$$= \log z \Big|_5^{3i}$$

$$\log z = \frac{1}{2}\ln(x^2 + y^2) + i\tan^{-1}\left(\frac{y}{x}\right)$$

$$= \log 3i - \log 5 = \log(0 + 3i) - \log(5 + 0i)$$

$$= \left[\frac{1}{2}\ln(0+9) + i\tan^{-1}\left(\frac{3}{0}\right)\right] - \left[\frac{1}{2}\ln(25+0) + i\tan^{-1}\left(\frac{0}{5}\right)\right]$$

$$= \frac{1}{2}\ln 9 + i\tan^{-1}\infty - \frac{1}{2}\ln 25 - 0$$

$$= \ln(9)^{1/2} + i\tan^{-1}\infty - \ln(25)^{1/2}$$

$$= \ln 3 + i\frac{\pi}{2} - \ln 5 = \ln\left(\frac{3}{5}\right) + i\frac{\pi}{2}$$

$$= -0.5108 + i\,1.57 = \mathbf{-0.511 + 1.57i}$$

5.11 (a)

$$z_1 = 5 + \left(5\sqrt{3}\right)i$$

$$z_2 = \frac{2}{\sqrt{3}} + 2i$$

$$\arg(z_1) = \theta_1 = \tan^{-1}\left(\frac{5\sqrt{3}}{5}\right)$$

$$\theta_1 = 60^\circ$$

$$\arg(z_2) = \theta_2 = \tan^{-1}\left(\frac{2}{2\sqrt{3}}\right)$$

$$\theta_2 = 60^\circ$$

$$\arg\left(\frac{z_1}{z_2}\right) = \arg(z_1) - \arg(z_2)$$

$$= 60 - 60 = \mathbf{0^\circ}$$

5.12 (a)

$$u = 2xy$$

$$u_x = 2y \qquad u_y = 2x$$

In option (a)

$$V_x = -2x \quad u_y = -V_x$$

$$V_y = 2y$$

C–R equation are satisfied only in option (a)

5.13 Sol.

Given that $f(z) = u + iv$ is analytic

$$u(x, y) = 2kxy \qquad v = x^2 - y^2$$

$$u_x = 2ky \qquad v_y = -2y$$

$$u_x = v_y$$

$$k = -1$$

$$u_y = 2kx \qquad v_x = 2x$$

$$u_y = -v_x$$

$$2kx = -2x$$

$$\mathbf{k = -1}$$

5.14 (b)

$$\oint_\Gamma \frac{3z-5}{(z-1)(z-2)} dz = 4\pi i$$

$$\oint_\Gamma \frac{3z-5}{(z-1)(z-2)} dz = 2\pi i(2)$$

Sum of residues must be equal to 2.

$$\underset{z=1}{\text{Res} f(z)} = \underset{z \to 1}{u} (z-1)\frac{3z-5}{(z-1)(z-2)} = \frac{-2}{-1} = 2$$

$$\underset{z=2}{\text{Res} f(z)} = \underset{z \to 2}{u} (z-2)\frac{3z-5}{(z-1)(z-2)} = \frac{6-5}{2-1} = 1$$

Therefore $z = 1$ must lies inside C

$z = 2$ lies outside C

then only we will get the given integral values is equal to $4\pi i$.

5.15 Sol.

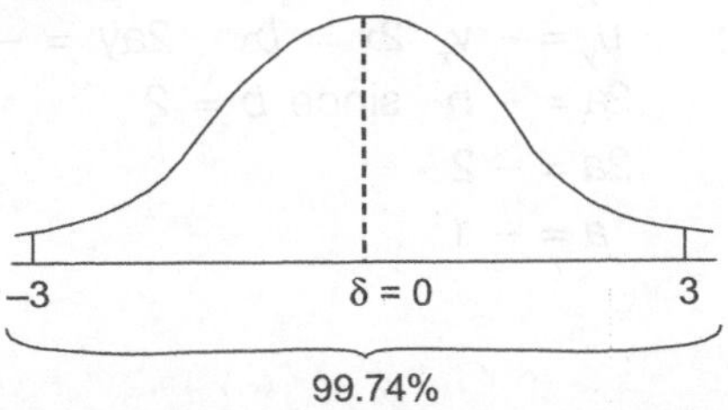

5.16 (a)

$$\int_{-\infty}^{\infty} \frac{\sin x}{x^2+2x+2}dx$$

$$I = \int_{-\infty}^{\infty} \frac{\sin z}{z^2+2z+2}dz = \int_{-\infty}^{\infty} \frac{\text{I.P of } e^{iz}}{z^2+2z+2}dz$$

Poles are $z^2 + 2z + 2 = 0$

$$z = \frac{-2 \pm \sqrt{4-8}}{2} = \frac{-2 \pm 2i}{2} = -1 \pm i$$

$z = -1 - i$

↓

Outside upper half

↓

Residue is 0

$-1 + i$

↓

inside upperhalf

Res $\phi(z)$

$z = -1 + i$

$$= \underset{z \to -1+i}{Lt} \; z-(-1+i) \times \frac{e^{iz}}{(z-(-1+i)\,z-(-1-9))}$$

$$= \frac{e^{i(-1+i)}}{(-1+i)-(-1-i)} = \frac{e^{-i-1}}{-1+i+1+i} = \frac{e^{-i-1}}{2i}$$

$$I = \text{I.P. of } 2\pi i\left(\frac{e^{-i-1}}{2i}\right) = \text{I.P of } \pi\left(e^{-i}\cdot e^{-1}\right)$$

$$= \text{I.P of } \pi e^{-1}(\cos 1^\circ - i\sin 1^\circ) = \frac{-\pi \sin 1^\circ}{e}$$

5.17 (b)

Solution of laplace equation having continuous Second order partial derivating

$\therefore \quad \nabla^2\phi = 0$

$$\frac{\partial^2\phi}{\partial x^2} + \frac{\partial^2\phi}{\partial y^2} = 0$$

$\therefore$ **ϕ is harmonic function.**

5.18 (b)

Given that the analytic function

$$f(2) = (x^2 + ay^2) + i\, bxy$$

$$u + i\,v = (x^2 + ay^2) + i\,(bxy)$$

$$u = x^2 + ay^2; \quad v = bxy$$

$u_x = 2x; \quad u_y = 2ay; \quad v_x = by; \quad v_y = bx$

$u_x = v_y; \quad u_y = -v_x \quad 2x = bx; \quad 2ay = -by$

$b = 2; \quad 2a = -b \quad \text{since } b = 2$

$2a = -2$

$a = -1$

5.19 (b)

$$F(z) = iz + k\,Re(z) + i\,Im(z)$$

$$u + iv = i(x + iy) + kx + iy$$

$$u + iv = kx - y + i(x + y)$$

$$u = kx - y, \; v = x + y$$

$$u_x = k, \; u_y = -1$$

$$v = x + y$$

$$v_x = 1$$

$$v_y = 1$$

By Cauchy-Riemann equations

$u_x = v_y; \quad k = 1$

5.20 (a)

$5z - 4 = 0$

$z = \frac{4}{5}$ lies inside circle,

$|z| = 1$

$$\int \frac{1}{(5z-4)}dz = A\pi i$$

$$\frac{1}{5}\int \frac{1}{\left(z-\frac{4}{5}\right)}dz = A\pi i$$

$$\int \frac{\left(\frac{1}{5}\right)}{\left(z-\frac{4}{5}\right)}dz = 2\pi i \cdot f\left(\frac{4}{5}\right) = 2\pi i \times \left(\frac{1}{5}\right) = \frac{2}{5}\pi i$$

$$A = \frac{2}{5} = 0.4$$

5.21 (d)

Let $f(z) = u + iv$ is an analytic function where u is harmonic then v is called its harmonic conjugate.

Here $u = 2x^2 - 2y^2 + 4xy$ i.e. real part is given, so by using Milne-Thomson method.

Step 1: $\frac{\partial u}{\partial x} = 4x + 4y = \phi_1(x, y)$

Step 2: $\phi_1(z, 0) = 4z$

Step 3: $\frac{\partial u}{\partial y} = -4y + 4x = \phi_2(x, y)$

Step 4: $\phi_2(z, 0) = 4z$

Step 5: $f(z) = \int[\phi_1(z,0) - i\phi_2(z,0)] + c$

$$= \int(4z - i4z)dz + c = 4(1-i)\frac{z^2}{2} + c$$

$$= 2(1-i)\left[x^2 - y^2 + 2ixy\right] + c$$

$$f(z) = \left(2x^2 - 2y^2 + 4xy\right) + i\left(4xy - 2x^2 + 2y^2 + c\right)$$

$$= u + iv$$

Hence harmonic conjugate,

$v = 4xy - 2x^2 + 2y^2 + \text{constant}$

5.22 (a)

C–R equation for the, $f(z) = u + iv$ to be analytic are $\frac{\partial u}{\partial x} = \frac{\partial v}{\partial y}$ and $\frac{\partial u}{\partial y} = \frac{-\partial v}{\partial x}$

5.23 (b)

$\because$ $\log z$ is not analytic at all points.

5.24 (a)

$f(z) = u + iv$

$u = x^3 - 3xy^2,\ v = v(x, y)$

For $f(z)$ to be Analytical,

$u_x = 3x^2 - 3y^2 = v_y$

$u_y = -6xy = -v_x$

$v_x = 6xy$ by integrating w.r.t x

$\Rightarrow \quad v = 3x^2y + C_1$

$v_y = 3x^2 - 3y^2$ by integrating w.r.t y

$\Rightarrow \quad v = 3x^2y - y^3 + C_2$

$v = (3x^2y - y^3) + \text{constant}(C_1 = -y^3)$

5.25 (b)

$(1 + i)^8$

$z = 1 + i;\quad r = |z| = \sqrt{2};\ \theta = \frac{\pi}{4}$

$(1 + i)^8 = \left(\sqrt{2}e^{i\frac{\pi}{4}}\right)^8$

$= 16 \times e^{i \times 2\pi} = 16(\cos 2\pi + i \sin 2\pi)$

$= 16 \times 1 = 16$

5.26 (a)

Pole of $f(c)$ is $z = 0$ simple pole.

Reduce at $z = 0$,

$$\underset{z=0}{R} f(z) = \lim_{z \to 0}(z - 0)f(z)$$

$$= \lim_{z \to 0} \frac{\cosh(3z)}{2} = \lim_{z \to 0} \frac{e^{3z} + e^{-3z}}{2 \times 2} = \frac{1}{2}$$

By Cauchy Riemann theorem, $I = 2\pi\left(\frac{1}{2}\right) = \pi i$

5.27 (a)

Let, $\quad f(z) = 2z^4 - 3z^3 + 7z^2 - 3z + 5$

Here, $\quad f(i) = 2i^4 - 3i^3 + 7i^2 - 3i + 5$

$= 2(1) - 3(-i) + 7(-1) - 3i + 5$

$= 2 + 3i - 7 - 3i + 5 = 0$

$\therefore \quad z = i$ is a singular point

Thus, $\oint \frac{6z}{2z^4 - 3z^3 + 7z^2 - 3z + 5} dz$

$= 2\pi i \left(\text{Res}|_{z=i}\right)$

where,

$$\text{Res}|_{z=i} = \lim_{z \to i}(z - i)\frac{6z}{2z^4 - 3z^3 + 7z^2 - 3z + 5}$$

$$= \lim_{z \to i} \frac{6z^2 - 6zi}{2z^4 - 3z^3 + 7z^2 - 3z + 5} \left(\frac{0}{0} \text{ form}\right)$$

$$= \lim_{z \to i} \frac{12z - 6i}{8z^3 - 9z^2 + 14z - 3} = \frac{12i - 6i}{8i^3 - 9i^2 + 14i - 3}$$

$$= \frac{6i}{8(-i) - 9(-1) + 14i - 3} = \frac{6i}{6 + 6i} = \frac{i}{i + 1}$$

$$\therefore \oint \frac{6z}{2z^4 - 3z^3 + 7z^2 - 3z + 5} dz = 2\pi i\left(\frac{i}{i + 1}\right)$$

$$= \frac{2\pi(-1)}{i + 1} = \frac{-2\pi}{i + 1} \times \frac{i - 1}{i - 1} = \frac{-2\pi(i - 1)}{i^2 - 1} = \frac{-2\pi(i - 1)}{-2}$$

$= (-1 + i)\pi$

5.28 (0.2) (0.2 to 0.2)

Let, $f(z) = \frac{1}{(z - i)(z + 4i)}$

Singular points : $z = i, -4i$

Since C is circle of radius 2, only $z = i$ will lie inside the circle.

Thus, $\oint_C f(z)dx = 2\pi i(\text{Res}|_{z=i})$

where, $\text{Res}|_{z=i} = \lim_{z \to i}(z - i)f(z)$

$$= \lim_{z \to i} \frac{1}{z + 4i} = \frac{1}{5i}$$

$$\therefore \oint_C \frac{1}{(z - i)(z + 4i)} dz = 2\pi i\left(\frac{1}{5i}\right) = \frac{2}{5}\pi$$

or, $\frac{1}{2\pi}\oint_C \frac{1}{(z - i)(z + 4i)} dz = \frac{1}{5} = 0.2$

5.29 (2) (1.999 to 2.001)

Given, $f(z) = u(x, y) + iv(x, y)$

$f(z) = e^{-kx} \cos 2y - ie^{-kx} \sin 2y$

Here, $\frac{\partial u}{\partial x} = -ke^{-kx} \cos 2y$

$\frac{\partial u}{\partial y} = -2e^{-kx} \sin 2y$

and, $\frac{\partial v}{\partial x} = ke^{-kx} \sin 2y$

$\frac{\partial v}{\partial y} = -2e^{-kx} \cos 2y$

Since $f(z)$ is analytic, it will satisfy Cauchy–Riemann equation, $\frac{\partial u}{\partial x} = \frac{\partial v}{\partial y}$

$\Rightarrow -ke^{-kx} \cos 2y = -2e^{-kx} \cos 2y$

$\Rightarrow k = 2$

5.30 (d)

Let, $f(z) = u + iv$

$u = \cosh z \cos y$

$$= \left(\frac{e^x + e^{-x}}{2}\right)\cos y$$

$$u = \left(\frac{e^x + e^{-x}}{2}\right)\cos y$$

by Milne Thomson method

$$u(x, y) = \left(\frac{e^x + e^{-x}}{2}\right)\cos y$$

Partial differentiate w.r.t. *x*, *y*

$$u_x = \left(\frac{e^x - e^{-x}}{2}\right)\cos y; \quad u_y = \left(\frac{e^x + e^{-x}}{2}\right)(-\sin y)$$

$$u_x = \left(\frac{e^x - e^{-x}}{2}\right)\cos y, \quad u_y = -\frac{(e^x + e^{-x})}{2}\sin y$$

$$u_x(z, 0) = \frac{e^z - e^{-z}}{2} = \sinh z, \; u_y(z, 0) = -0$$

$$f(z) = \int (u_x - iu_y)dz + c$$

$$= \int (\sinh z - a)dz + c$$

$$= \int \sinh z + c$$

$$= \cosh z + c$$

■■■■

6 Probability and Statistics

6.1 A box contains 5 black and 5 red balls. Two balls are randomly picked one after another from the box, without replacement. The probability for both balls being red is

(a) $\frac{1}{90}$ (b) $\frac{1}{2}$

(c) $\frac{19}{90}$ (d) $\frac{2}{9}$ **[2003 : 2 M]**

6.2 From a pack of regular playing cards, two cards are drawn at random. What is the probability that both cards will be Kings, if first card in NOT replaced?

(a) $\frac{1}{26}$ (b) $\frac{1}{52}$

(c) $\frac{1}{169}$ (d) $\frac{1}{221}$ **[2004 : 2 M]**

6.3 A lot has 10% defective items. Ten items are chosen randomly from this lot. The probability that exactly 2 of the chosen items are defective is

(a) 0.0036 (b) 0.1937

(c) 0.2234 (d) 0.3874 **[2005 : 1 M]**

6.4 A single die is thrown twice. What is the probability that the sum is neither 8 nor 9?

(a) 1/9 (b) 5/36

(c) 1/4 (d) 3/4 **[2005 : 2 M]**

6.5 A box contains 20 defective items and 80 non-defective items. If two items are selected at random without replacement, what will be the probability that both items are defective?

(a) $\frac{1}{5}$ (b) $\frac{1}{25}$

(c) $\frac{20}{99}$ (d) $\frac{19}{495}$ **[2006 : 1 M]**

6.6 Consider the continuous random variable with probability density function

$f(t) = 1 + t$ for $-1 \le t \le 0$

$= 1 - t$ for $0 \le t \le 1$

The standard deviation of the random variable is

(a) $\frac{1}{\sqrt{3}}$ (b) $\frac{1}{\sqrt{6}}$

(c) $\frac{1}{3}$ (d) $\frac{1}{6}$ **[2006 : 2 M]**

6.7 A coin is tossed 4 times. What is the probability of getting heads exactly 3 times?

(a) $\frac{1}{4}$ (b) $\frac{3}{8}$

(c) $\frac{1}{2}$ (d) $\frac{3}{4}$ **[2008 : 1 M]**

6.8 Let X and Y be two independent random variables. Which one of the relations between expectation (E), variance (Var) and covariance (Cov) given below is FALSE?

(a) $E(XY) = E(X)\,E(Y)$

(b) $\text{Cov}(X, Y) = 0$

(c) $\text{Var}(X + Y) = \text{Var}(X) + \text{Var}(Y)$

(d) $E(X^2 Y^2) = (E(X))^2 (E(Y))^2$ **[2007 : 2 M]**

6.9 If three coins are tossed simultaneously, the probability of getting at least one head is

(a) 1/8 (b) 3/8

(c) 1/2 (d) 7/8 **[2009 : 1 M]**

6.10 The standard deviation of a uniformly distributed random variable between 0 and 1 is

(a) $\frac{1}{\sqrt{12}}$ (b) $\frac{1}{\sqrt{3}}$

(c) $\frac{5}{\sqrt{12}}$ (d) $\frac{7}{\sqrt{12}}$ **[2009 : 2 M]**

6.11 A box contains 2 washers, 3 nuts and 4 bolts. Items are drawn from the box at random one at a time without replacement. The probability of drawing 2 washers first followed by 3 nuts and subsequently the 4 bolts is

(a) 2/315 (b) 1/630

(c) 1/1260 (d) 1/2520 **[2010 : 2 M]**

6.12 A box contains 4 red balls and 6 black balls. Three balls are selected randomly from the box one after another, without replacement. The probability that the selected set contains one red ball and two black balls is

(a) 1/20 (b) 1/12

(c) 3/10 (d) 1/2 **[2012 : 2 M]**

6.13 Let X be a normal random variable with mean 1 and variance 4. The probability $P\{X < 0\}$ is

(a) 0.5

(b) greater than zero and less than 0.5
(c) greater than 0.5 and less than 1.0
(d) 1.0 [2013 : 1 M]

6.14 The probability that a student knows the correct answer to a multiple choice question is 2/3. If the student does not known the answer, then the student guesses the answer. The probability of the guessed answer being correct is 1/4. Given that the student has answered the question correctly, the conditional probability that the student known the correct answer is
(a) 2/3 (b) 3/4
(c) 5/6 (d) 8/9 [2013 : 2 M]

6.15 A box contains 25 parts of which 10 are defective. Two parts are being drawn simultaneously in a random manner from the box. The probability of both the parts being good is
(a) $\frac{7}{20}$ (b) $\frac{45}{125}$
(c) $\frac{25}{29}$ (d) $\frac{5}{9}$ [2014 : 1 M, Set-2]

6.16 A group consists of equal number of men and women. Of this group 20% of the men and 50% of the women are unemployed. If a person is selected at random from this group, the probability of the selected person being employed is ______.
[2014 : 1 M, Set-3]

6.17 A nationalized bank has found that the daily balance available in its savings accounts follows a normal distribution with a mean of Rs. 500 and a standard deviation of Rs. 50. The percentage of savings account holders, who maintain an average daily balance more than Rs. 500 is ____
[2014 : 1 M, Set-4]

6.18 In the following table, x is a discrete random variable and $p(x)$ is the probability density. The standard deviation of x is

x	1	2	3
$p(x)$	0.3	0.6	0.1

(a) 0.18 (b) 0.36
(c) 0.54 (d) 0.6
[2014 : 2 M, Set-1]

6.19 Consider an unbiased cubic dice with opposite faces coloured identically and each face coloured red, blue or green such that each colour appears only two times on the dice. If the dice is thrown thrice, the probability of obtaining red colour on top face of the dice at least twice is ______.
[2014 : 2 M, Set-2]

6.20 A machine produces 0, 1 or 2 defective pieces in a day with associated probability of 1/6, 2/3 and 1/6, respectively. The mean value and the variance of the number of defective pieces produced by the machine in a day, respectively, are
(a) 1 and 1/3 (b) 1/3 and 1
(c) 1 and 4/3 (d) 1/3 and 4/3
[2014 : 2 M, Set-3]

6.21 The number of accidents occurring in a plant in a month follows Poisson distribution with mean as 5.2. The probability of occurrence of less than 2 accidents in the plant during a randomly selected month is
(a) 0.029 (b) 0.034
(c) 0.039 (d) 0.044
[2014 : 2 M, Set-4]

6.22 Three vendors were asked to supply a very high precision component. The respective probabilities of their meeting the strict design specifications are 0.8, 0.7 and 0.5. Each vendor supplies one component. The probability that out of total three components supplied by the vendors, at least one will meet the design specification is ______.
[2015 : 1 M, Set-2]

6.23 If $P(X) = \frac{1}{4}$, $P(Y) = \frac{1}{3}$, and $P(X \cap Y) = \frac{1}{12}$, the value of $P(Y/X)$ is
(a) $\frac{1}{4}$ (b) $\frac{4}{25}$
(c) $\frac{1}{3}$ (d) $\frac{29}{50}$
[2015 : 1 M, Set-3]

6.24 The chance of a student passing an exam is 20%. The chance of a student passing the exam and getting above 90% M in it is 5%, Given that a student passes the examination, the probability that the student gets above 90% M is
(a) $\frac{1}{18}$ (b) $\frac{1}{4}$
(c) $\frac{2}{9}$ (d) $\frac{5}{18}$
[2015 : 2 M, Set-2]

6.25 The probability of obtaining at least two "SIX" in throwing a fair dice 4 times is

(a) $\frac{425}{432}$ (b) $\frac{19}{144}$

(c) $\frac{13}{144}$ (d) $\frac{125}{432}$

[2015 : 2 M, Set-3]

6.26 Consider a Poisson distribution for the tossing of a biased coin. The mean for this distribution is μ. The standard deviation for this distribution is given by

(a) $\sqrt{\mu}$ (b) μ^2

(c) μ (d) $\frac{1}{\mu}$ [2016 : 1 M, Set-1]

6.27 The probability that a screw manufactured by a company is defective is 0.1. The company sells screws in packets containing 5 screws and gives a guarantee of replacement if one or more screws in the packet are found to be defective. The probability that a packet would have to be replaced is____. [2016 : 2 M, Set-2]

6.28 Three cards were drawn from a pack of 52 cards. The probability that they are a king, a queen, and a jack is

(a) $\frac{16}{5525}$ (b) $\frac{64}{2197}$

(c) $\frac{3}{13}$ (d) $\frac{8}{16575}$

[2016 : 2 M, Set-3]

6.29 A six-face fair dice is rolled a large number of times. The mean value of the outcomes is ________. [2017 : 1 M, Set-1]

6.30 Two coins are tossed simultaneously. The probability (upto two decimal points accuracy) of getting at least one head is ______________.

[2017 : 1 M, Set-2]

6.31 A six-faced fair dice is rolled five times. The probability (in%) of obtaining "ONE" at least four times is

(a) 33.3 (b) 3.33

(c) 0.33 (d) 0.0033

[2018 : 1 M, Set-1]

6.32 Four red balls, four green balls and four blue balls are put in a box. Three balls are pulled our of the box at random one after another without replacement. The probability that all the three balls are red is

(a) 1/72 (b) 1/55

(c) 1/36 (d) 1/27

[2018 : 1 M, Set-1]

6.33 Let X_1, X_2 be two independent normal random variables with means μ_1, μ_2 and standard deviations σ_1, σ_2 respectively. Consider $Y = X_1 - X_2$; $\mu_1 = \mu_2 = 1$, $\sigma_1 = 1$, $\sigma_2 = 2$, Then,

(a) Y is normal distributed with mean 0 and variance 1

(b) Y is normally distributed with mean 0 and variance 5

(c) Y has mean 0 and variance 5, but is NOT normally distributed

(d) Y has mean 0 and variance 1, but is NOT normally distributed

[2018 : 2 M, Set-1]

6.34 Let X_1 and X_2 be two independent exponentially distributed random variables with means 0.5 and 0.25 respectively. Then $Y = \min(X_1, X_2)$ is

(a) exponentially distributed with mean $\frac{1}{6}$

(b) exponentially distributed with mean 2

(c) normally distributed with mean $\frac{3}{4}$

(d) normally distributed with mean $\frac{1}{6}$

[2018 : 2 M, Set-2]

6.35 The lengths of large stock of titanium rods follow a normal distribution with a mean (μ) of 440 mm and a standard deviation (σ) of 1 mm. What is the percentage of rods whose lengths lie between 438 mm and 441 mm?

(a) 68.4% (b) 86.64%

(c) 81.85% (d) 99.75%

[2019 : 1 M, Set-1]

6.36 The variable x takes a value between 0 and 10 with uniform probability distribution. The variable y takes a value between 0 and 20 with uniform probability distribution. The probability of the sum of variables $(x + y)$ being greater than 20 is

(a) 0.25 (b) 0.33

(c) 0.50 (d) 0 [2019 : 2 M, Set-1]

6.37 If x is the mean of data 3, x, 2 and 4, then the mode is _______

[2019 : 1 M, Set-2]

6.38 The probability that a part manufactured by a company will be defective is 0.05. If 15 such parts are selected randomly and inspected, then the probability that at least two part will be defective is ______. (round off to two decimal places)

[2019 : 2 M, Set-2]

6.39 A company is hiring to fill four managerial vacancies. The candidates are five men and three women. If every candidate is equally likely to be chosen then the probability that at least one women will be selected is ______ (round off to 2 decimal places).

[2020 : 1 M, Set-1]

6.40 The sum of two normally distributed random variables X and Y is

(a) normally distributed, only if X and Y have the same standard deviation
(b) always normally distributed
(c) normally distributed, only if X and Y have the same mean
(d) normally distributed, only if X and Y are independent

[2020 : 1 M, Set-2]

6.41 Consider two exponentially distributed random variables X and Y, both having a mean of 0.50. Let $Z = X + Y$ and r be the correlation coefficient between X and Y. If the variance of Z equals 0, then the value of r is ______ (round off to 2 decimal places).

[2020 : 2 M, Set-1]

6.42 A fair coin is tossed 20 times. The probability that 'head' will appear exactly 4 times in the first ten tosses, and 'tail' will appear exactly 4 times in the next ten tosses is ______ (round off to 3 decimal places)

[2020 : 2 M, Set-2]

6.43 The mean and variance, respectively, of a binomial distribution for n independent trials with the probability of success as p, are

(a) $\sqrt{np}, \sqrt{np(1-p)}$ (b) $np, np(1-p)$
(c) np, np (d) $\sqrt{np}, np(1-2p)$

[2021 : 1 M, Set-2]

6.44 Consider a binomial random variable X. If $X_1, X_2, \ldots X_n$ are independent and identically distributed samples from the distribution of X with sum $Y = \sum_{i=1}^{n} X_i$, then the distribution of Y as $n \to \infty$ can be approximated as

(a) Exponential (b) Bernoulli
(c) Binomial (d) Normal

[2021 : 1 M, Set-1]

6.45 Let a random variable X follow Poisson distribution such that

$\text{Prob}(X = 1) = \text{Prob}(X = 2)$

The value of $\text{Prob}(X = 3)$ is (round off to 2 decimal places).

[2022 : 2 M, Set-1]

6.46 A machine produces a defective component with a probability of 0.015. The number of defective components in a packed box containing 200 components produced by the machine follows a Poisson distribution. The mean and the variance of the distribution are

(a) 3 and 3, respectively
(b) $\sqrt{3}$ and $\sqrt{3}$ respectively
(c) 0.015 and 0.015, respectively
(d) 3 and 9, respectively

[2023 : 1 M]

6.47 Let X be a continuous random variable defined on [0, 1] such that its probability density function $f(x) = 1$ for $0 \le x \le 1$ and 0 otherwise. Let $Y = \log_e(X + 1)$. Then the expected value of Y is ______. [Rounded off to 2 decimal places]

[2024 : 2 M]

Answers Probability and Statistics

6.1 (d)	6.2 (d)	6.3 (b)	6.4 (d)	6.5 (d)	6.6 (b)	6.7 (a)	6.8 (d)
6.9 (d)	6.10 (a)	6.11 (c)	6.12 (d)	6.13 (b)	6.14 (d)	6.15 (a)	6.16 (0.65)
6.17 (50%)	6.18 (d)	6.19 (0.259)	6.20 (a)	6.21 (b)	6.22 (0.97)	6.23 (c)	6.24 (b)
6.25 (b)	6.26 (a)	6.27 (0.4095)	6.28 (a)	6.29 (3.5)	6.30 (0.75)	6.31 (c)	6.32 (b)
6.33 (b)	6.34 (a)	6.35 (c)	6.36 (a)	6.37 (3)	6.38 (0.17)	6.39 (0.93)	6.40 (*)
6.41 (–1)	6.42 (0.042)	6.43 (b)	6.44 (d)	6.45 (0.18)	6.46 (a)	6.47 (0.39)	

Explanations Probability and Statistics

6.1 (d)

Probability of drawing two red balls

$$= \frac{5}{10} \times \frac{4}{9} = \frac{2}{9}$$

6.2 (d)

Probability that both cards are kings

$$= \frac{^4C_1}{^{52}C_1} \frac{^3C_1}{^{51}C_1} = \frac{4}{52} \times \frac{3}{51} = \frac{1}{221}$$

6.3 (b)

Probability of defective item

$P = 0.1$

Probability of non-defective item

$Q = 1 - p = 1 - 0.1 = 0.9$

Probability that exactly 2 of the chosen items are defective

$$= {}^{10}C_2(P)^2(Q)^8 = {}^{10}C_2(0.1)^2(0.9)^8 = \mathbf{0.1937}$$

6.4 (d)

Sample space = $(6)^2 = 36$

Total ways in which sum is either 8 or 9 is

(2, 6), (3, 5), (3, 6), (4, 4), (4, 5), (5, 3), (5, 4), (6, 2), (6, 3)

$\therefore$ Probability of coming sum 8 or 9

$$= \frac{9}{36} = \frac{1}{4}$$

So probability of not coming sum 8 or 9

$$= 1 - \frac{1}{4} = \frac{3}{4}$$

6.5 (d)

Probability of first item being defective is

$$P_1 = \frac{20}{100}$$

Probability that second item being defective is

$$P_2 = \frac{19}{99}$$

Probability that both are defective

$$P = P_1P_2 = \frac{20}{100} \times \frac{19}{99} = \frac{19}{495}$$

6.6 (b)

$$\text{Mean } (\bar{t}) = \int_{-\infty}^{\infty} t \cdot f(t) \cdot dt$$

$$= \int_{-1}^{0} t(1+t)dt + \int_{0}^{1} t(1-t)dt$$

$$= \left[\frac{t^2}{2} + \frac{t^3}{3}\right]_{-1}^{0} + \left[\frac{t^2}{2} - \frac{t^3}{3}\right]_{0}^{1}$$

$$= -\left[\frac{1}{2} - \frac{1}{3}\right] + \left[\frac{1}{2} - \frac{1}{3}\right] = -\left[\frac{1}{6} + \frac{1}{6}\right] = 0$$

$$\text{Variance} = \int_{-\infty}^{\infty} (t - \bar{t})^2 f(t) dt$$

$$= \int_{-1}^{0} t^2(1+t) \cdot dt + \int_{0}^{1} t^2 (1-t) dt$$

$$= \int_{-1}^{0} (t^2 + t^3) \cdot dt + \int_{0}^{1} t^2 (1-t) dt$$

$$= \left[\frac{t^3}{3} + \frac{t^4}{4}\right]_{-1}^{0} + \left[\frac{t^3}{3} - \frac{t^4}{4}\right]_{0}^{1}$$

$$= \frac{1}{12} + \frac{1}{12} = \frac{1}{6}$$

Standard deviation

$$= \sqrt{\text{variance}} = \frac{1}{\sqrt{6}}$$

6.7 (a)

$P(H) = 0.5$

Probability of getting head exactly 3 times is

$$^4C_3 (0.5)^3 (0.5) = \frac{1}{4}$$

6.9 (d)

Three coins are tossed simultaneously, the total number of ways = $2^3 = 8$

Number of ways to get all tails = 1

Probability to get all tails

$$= \frac{\text{Number of ways for tails}}{\text{Total number of ways}} = \frac{1}{8}$$

Therefore probability to get at least one head is

$$= 1 - \frac{1}{8} = \frac{7}{8}$$

or, Total number of ways with at least one head is

$$\left.\begin{matrix} H & T & T \\ T & H & T \\ T & T & H \\ H & H & T \\ H & T & H \\ T & H & H \\ H & H & H \end{matrix}\right\} 7 \text{ ways}$$

Therefore probability to get at least one head $= \dfrac{7}{8}$

6.10 (a)

For uniformly distributed random variable $P(x) = P$

$P(v)$ = probability density function

$$E(x^2) \text{ mean} = \frac{\int_0^1 xP(x)dx}{\int P(x)dx} = \frac{\int_0^1 xp\,dx}{\int p\,dx}$$

$$= \frac{P[x^2/2]_0^1}{P[x]_0^1} = \frac{1}{2}$$

$$\Rightarrow E(x^2) = \frac{\int x^2 P(x)\,dx}{\int P(x)\,dx} = \frac{\int_0^1 x^2 p\,dx}{\int_0^1 p\,dx} = \frac{p[x^3/3]_0^1}{p[x]_0^1}$$

$$= \frac{p(1/3-0)}{p(1-0)}$$

$$\text{variance } \sigma^2 = E(x^2) - (E(x))^2$$

$$= \frac{1}{3} - \left(\frac{1}{2}\right)^2 = \frac{1}{3} - \frac{1}{4} = \frac{1}{12}$$

Standard deviation $\sigma = \dfrac{1}{\sqrt{12}}$

Answer option (a) is correct.

Alternate

'x' defined on [0, 1]

$$\text{Variance } V(x) = \frac{(b-a)^2}{12} = \frac{(1-0)^2}{12}$$

$\therefore$ Standard deviation

$$\sigma = \frac{1}{\sqrt{12}}$$

6.11 (c)

Probability of 2 washer first, $P(A) = \dfrac{^2C_2}{^9C_2}$

Probability of 3 nuts, $P(B) = \dfrac{^3C_3}{^7C_3}$

Probability of 4 bolts, $P(C) = \dfrac{^4C_4}{^4C_4}$

Probability of given sequence of drawing

$$P = P(A) \times P(B) \times P(C)$$

$$= \frac{2\times 1}{9\times 8} \times \frac{3\times 2\times 1}{7\times 6\times 5} \times \frac{4\times 3\times 2\times 1}{4\times 3\times 2\times 1} = \frac{1}{1260}$$

Hence answer should be option (c).

6.12 (d)

The problem can be represented by the following diagram.

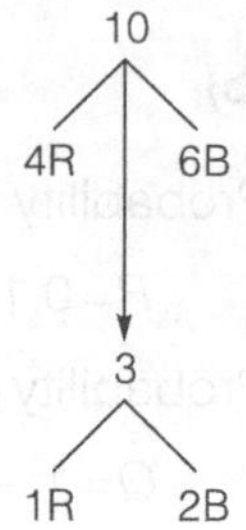

p(1R and 2B)

$$= \frac{^4C_1 \times {}^6C_2}{10C_3}$$

$$= \frac{60}{120} = \frac{1}{2}$$

6.13 (b)

By normal distribution

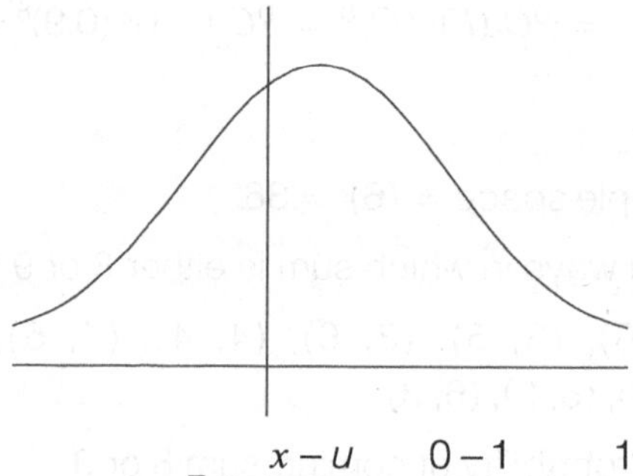

$$z = \frac{x-u}{\sigma} = \frac{0-1}{4} < -\frac{1}{4}$$

For this value of z

Probability $0 < p < 0.5$

6.15 (a)

$$\textbf{required prob} = \frac{^{15}C_2}{^{25}C_2} = \frac{14\times 15}{25\times 24} = \frac{7}{20}$$

6.16 Sol.

Total % of employed person

$$= 100 - \frac{20+50}{2}\% = 100 - 35\% = 65\%$$

$$\text{Required prob} = \frac{65}{100} = 0.65$$

6.17 Sol.

(49 to 51)

$$Z = \frac{x-\mu}{\sigma}$$

Given, $x = 500$, $\mu = 500$, $\sigma = 50$

$\therefore \quad z = \dfrac{500-500}{50} = 0$

$\therefore P(0) =$ **50%**

6.18 (d)

Mean,

$$\bar{x} = \frac{\Sigma f_i x_i}{\Sigma f_i} = \frac{1\times0.3+2\times0.8+3\times0.1}{1} = 1.8$$

Standard deviation,

$$\sigma = \left(\frac{\Sigma f_i x_i^2}{\Sigma f_i} - \bar{x}^2\right)^{1/2}$$

$$\therefore \quad \sigma = \left(\frac{0.3\times1^2+0.6\times2^2+0.1\times3^2}{1} - 1.8^2\right)^{1/2}$$

$$= (3.6-1.8^2)^{1/2} = (0.36)^{1/2} = \mathbf{0.6}$$

6.19 Sol.

x	R	B	G
P(x)	$\frac{2}{6}$	$\frac{2}{6}$	$\frac{2}{6}$

$n = 3x$: red colour

$$P = P(\text{Red}) = \frac{2}{6}$$

$$q = 1 - P = 1-\frac{2}{6} = \frac{4}{6}$$

Prob. of getting red colour on top face atleast twice is

$= P(x = 2) + P(x = 3)$

$= {}^nC_2P^2\, q^{n-2} + {}^nC_3\, P^3\, q^{n-3}$

$$= {}^nC_2\left(\frac{8}{6}\right)^2\left(\frac{1}{4}\right)^1 + \frac{3}{3}\left(\frac{2}{6}\right)^2\left(\frac{4}{6}\right)^0$$

$$= 3\cdot\frac{4}{36}\cdot\frac{4}{6}+1\cdot\frac{8}{216} = \frac{48+8}{216} = \frac{56}{216} = \mathbf{0.259}$$

6.20 (a)

x	0	1	2
p(x)	$\frac{1}{6}$	$\frac{2}{3}$	$\frac{1}{6}$

$$\text{mean} = \Sigma x\cdot p(x) = 0\left(\frac{1}{6}\right)+1\left(\frac{2}{3}\right)+2\left(\frac{1}{6}\right)$$

$$= \frac{2}{3}+\frac{2}{6} = \frac{2}{3}+\frac{1}{3} = \frac{3}{3} = 1$$

$$E(x^2) = 0\left(\frac{1}{6}\right)+1\left(\frac{2}{3}\right)+4\left(\frac{1}{6}\right)$$

$$= \frac{2}{3}+\frac{4}{6} = \frac{2}{3}+\frac{2}{3} = \frac{4}{3}$$

$$\text{Variance} = E(x^2) - (E(x)^2) = \frac{4}{3}-1 = \frac{1}{3}$$

6.21 (b)

$$P(x) = \frac{\mu^x e^{-\mu}}{x!}$$

$$P(x < 2) = P(0) + P(1)$$

$$= \frac{\mu^0 e^{-\mu}}{0!} + \frac{\mu^1 e^{-\mu}}{1!} = \frac{1}{e^{\mu}}+\frac{\mu}{e^{\mu}} = \frac{1+\mu}{e^{\mu}}$$

As μ(mean) = 5.2

$$\therefore \quad P(x<2) = \frac{6.2}{e^{5.2}} = 0.0342$$

6.22 Sol.

Probability of atleast one meet the specification

$= 1-(\bar{A}\cap\bar{B}\cap\bar{C}) = 1 - (0.2 \times 0.3 \times 0.5) =$ **0.97**

6.23 (c)

$$P\left(\frac{Y}{X}\right) = \frac{P(Y\cap X)}{P(X)} = \frac{1/12}{1/4} = \frac{1}{3}$$

6.24 (b)

Given,

p(passing the exam) = 0.2

p(passing the exam $\cap$ > 90%) = 0.05

The desired probability

$= p$(> 90% | passing the exam)

$$= \frac{p(\text{passing the exam} \cap > 90\%)}{p(\text{passing the exam})} = \frac{0.05}{0.2} = \frac{1}{4}$$

6.25 (b)

Let P be the probability that six occurs on a fair dice,

$$\therefore \quad P = \frac{1}{6}$$

$$\therefore \quad q = \frac{5}{6}$$

Let X, be the number of times 'six' occurs,

Probability of obtaining at least two 'six' in throwing a fair dice 4 times is

$= 1 - \{P(X = 0) + P(X = 1)\}$

$= 1 - \{{}^4C_0\, p^0 q^4 + {}^4C_1\, p^1 q^3\}$

$$= 1-\left\{\left(\frac{5}{6}\right)^4+\left[4\times\frac{1}{6}\times\left(\frac{5}{6}\right)^3\right]\right\}$$

$$=1-\left\{\frac{125}{144}\right\} = \frac{19}{144}$$

6.26 (a)

In poisson distribution mean = Variance

Given that mean = Variance = m

Standard deviation = $\sqrt{\text{Variance}} = \sqrt{\mu}$

6.27 Sol.

$n = 5,\ P = 0.1,\ q = 0.9$

X : no of defectives

$$P(X \geq 1) = 1 - P(X = 0)$$
$$= 1 - {}^5C_0 \times (0.1)^0 \times (0.9)^5$$
$$= 1 - (0.9)^5 = \mathbf{0.4095}$$

6.28 (a)

$$\frac{{}^4C_1 \cdot {}^4C_1 \cdot {}^4C_1}{{}^{52}C_3} = \frac{64}{\frac{52\times51\times50}{3\times2\times1}}$$

$$= \frac{64}{22{,}100} = \frac{16}{5525}$$

6.29 Sol.

Face	1	2	3	4	5	6
Prob.	1/6	1/6	1/6	1/6	1/6	1/6

mean = $E(x) = \Sigma x \cdot P(x)$

$= 1(1/6) + 2(1/6) + 3(1/6) + 4(1/6) + 5(1/6) + 6(1/6)$

$$= \frac{1}{6}(1+2+3+4+5+6) = \frac{21}{6} = 3.5$$

6.30 Sol.

Possible outcomes when 2 coins are tossed

TT, TH, HT, HH

∴ Probability of getting atleast one head

= (Total) − ^{P}(none head)

$$=1-\frac{1}{4}=\frac{3}{4} = 0.75$$

6.31 (c)

A dice is rolled 5 times

$n = 5$

$$P = (\text{Probability of getting 1}) = \frac{1}{6}$$

$$q = 1-\frac{1}{6}=\frac{5}{6}$$

Probability of getting 1 at least 4 times is

$$P(x \geq 4) = P(x = 4) + P(x = 5)$$
$$= {}^nC_4p^4q^{n-4} + {}^nC_5p^5q^{n-5}$$
$$= {}^5C_4p^4q^1 + {}^5C_5p^5q^0$$
$$= 5\times\left(\frac{1}{6}\right)^4\left(\frac{5}{6}\right)^1+1\times\left(\frac{1}{6}\right)^5\times\left(\frac{5}{6}\right)^0$$
$$= \frac{25}{(6)^5}+\frac{1}{(6)^5}=\frac{26}{(6)^5} = 0.0033$$

% probability = 0.33%

6.32 (b)

Probability that all the three balls are red is

4R (struck) 4G 4B
3R (struck)
2R

$$= \text{R·R·R.} = \frac{4}{12}\times\frac{3}{11}\times\frac{2}{10}=\frac{24}{1320}=\frac{1}{55}$$

6.33 (b)

$\mu_1 = 1,\ \mu_2 = 1,\ \sigma_1 = 1,\ \sigma_2 = 2$

x_1 and x_2 are two independent random variables

$$Y = X_1 - X_2$$
$$\mu(Y) = \mu(X_1 - X_2) = \mu(X_1) - \mu(X_2) = \mu_1 - \mu_2$$
$$= 1 - 1 = \mathbf{0}$$
$$\text{Var}(Y) = \text{Var}(X_1 - X_2)$$
$$= \text{Var}(X_1) + \text{Var}(X_2) - \text{Cov}(X_1, X_2)$$

Since X_1 and X_2 are independent variables

$$\text{Var}(Y) = \text{Var}(X_1) + \text{Var}(X_2) = \sigma_1^2+\sigma_2^2 = 1+4$$

$$\text{Var}(Y) = \mathbf{5}$$

6.34 (a)

$$\text{Mean}(x_1) = 0.5$$
$$\frac{1}{\lambda_1} = 0.5$$
$$\lambda_1 = \frac{1}{0.5} = 2$$
$$\text{Mean}(x_2) = 0.25$$

$$\frac{1}{\lambda_2} = 0.25$$

$$\lambda_2 = \frac{1}{0.25} = 4$$

$$y = \text{mean } (x_1, x_2)$$

$$\text{Mean } (y) = \frac{1}{\lambda_1 + \lambda_2} = \frac{1}{2+4} = \frac{1}{6}$$

6.35 (c)

$$\mu = 400 \text{ mm},$$

$$UL = 441 \text{ mm}$$

$$\sigma = 1 \text{ mm},$$

$$LL = 438 \text{ mm}$$

$$\text{Now, Upper limit} = \left(\frac{Z-\mu}{\sigma}\right) = \left(\frac{441-440}{1}\right) = 1$$

$$\text{Lower limit} = \left(\frac{Z-\mu}{\sigma}\right) = \left(\frac{438-440}{1}\right) = -2$$

Hence total percentage of rods in between 438 mm and 441 mm

$$= P(Z-1) - P(Z-2)$$

$$= 0.3413 + (0.5 - 0.0228)$$

$$= 0.81854 = 81.854\%$$

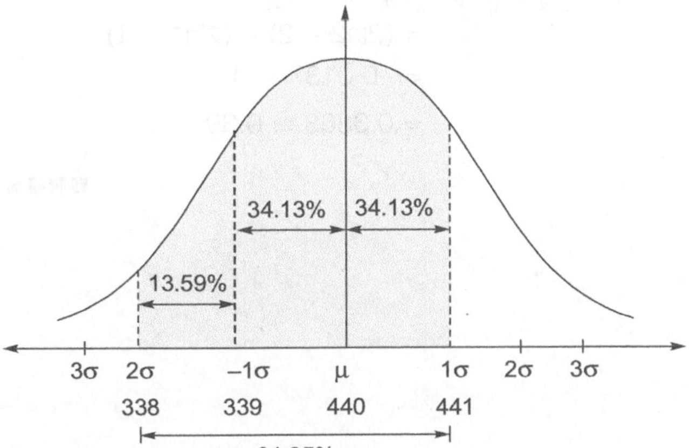

6.36 (a)

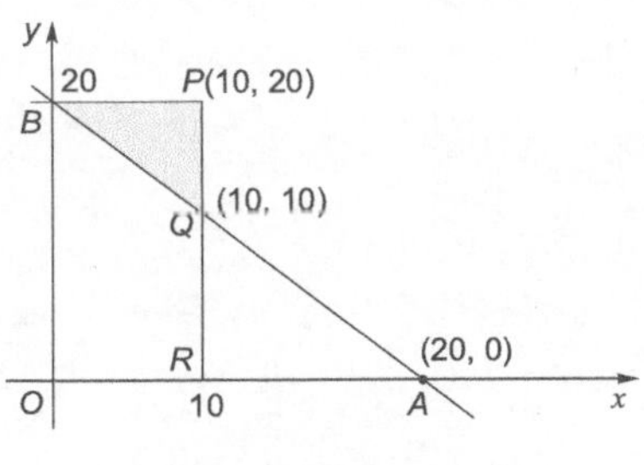

$0 \le x \le 10$ and $0 \le y \le 20$.

Here x and y together constitute a rectangle given by OBPR.

Where AB: $x + y = 20$

Req. probability,

$$P[(x+y) > 20] = \frac{\text{Shaded area}}{\text{Total area}} = \frac{50}{200}$$

$$= \frac{1}{4} = 0.25$$

6.37 Sol.

$$\text{Mean,} \quad x = \frac{3+x+2+4}{4}$$

$$4x = 9 + x$$

$$3x = 9$$

$$x = 3$$

So the given numbers are 3, 3, 2, 4

Mode = Maximum times repeated value.

= 3

6.38 Sol.

$n = 15$, $p = 0.05$, $q = 0.95$

$$x \sim B(n, p) \approx x \sim P(\lambda)$$

$$\lambda = np = 15(0.05) = 0.75$$

P (Atleast two defective)

$$P(x \ge 2) = 1 - P(x \le 1)$$

$$= 1 - (P(0) + P(1)) = 1 - e^{-0.75}(1 + (0.75))$$

$$= 1 - 1.75\,(e^{-0.75}) = 0.1733 \approx 0.17$$

6.39 Sol.

There are probability, 5 men, and 3 women

P[atleast one women selected for 4 vacancies]

$$= 1 - P[\text{none}]$$

$$= 1 - \frac{{}^5C_4\,{}^3C_0}{{}^8C_4} = 1 - \frac{5}{70}$$

$$= 1 - \frac{1}{14} = \frac{13}{14} = 0.93$$

6.41 Sol.

$$X \sim E(\lambda_1); \text{ mean} = \frac{1}{\lambda_1} = 0.5$$

$$\Rightarrow \quad \lambda_1 = 2$$

$$\text{Variance, } x = \frac{1}{\lambda_1^2} = \frac{1}{4} = 0.25$$

$$Y \sim E(\lambda_2); \text{ Mean} = \frac{1}{\lambda_2} = 0.5$$

$$\Rightarrow \quad \lambda_2 = 2$$

Variance, $y = 0.25$

Given Var (Z) = Var(x) + Var (y) + 2 COV (x, y)

$$0 = 0.25 + 0.25 + 2\text{ COV }(x, y)$$

$$\text{COV }(x, y) = -\frac{0.5}{2} = -0.25$$

Correlation,

$$\rho = \frac{COV(x,y)}{\sigma_x\sigma_y} = \frac{-0.25}{\sqrt{(0.25)}\sqrt{(0.25)}} = -1$$

6.42 Sol.

Fair coin tossed 20 times

First 10 times probability that head will appear exactly 4 times

P[4 heads in 10 tosses] · P[4 tails in 10 tosses]

$$= 10_{C_4}\left(\frac{1}{2}\right)^4\left(\frac{1}{2}\right)^6 \cdot 10_{C_4}\left(\frac{1}{2}\right)^6\left(\frac{1}{2}\right)^4$$

$$= \frac{10_{C_4}\cdot 10_{C_4}}{2^{10}\cdot 2^{10}} = \frac{210\times 210}{2^{10}\times 2^{10}} = 0.042$$

6.43 (b)

$$\text{Mean} = np$$

$$\text{Variance} = npq = np(1-p)$$

6.45 (0.18) (0.17 to 0.19)

Given that,

$$P(x = 1) = P(x = 2)$$

$$\frac{\lambda e^{-\lambda}}{1!} = \frac{\lambda^2 e^{-\lambda}}{2!}$$

$$\lambda = \frac{\lambda^2}{2}$$

$$\lambda^2 - 2\lambda = 0$$

$$\lambda(\lambda - 2) = 0$$

$\lambda = 2$ (Because λ cannot be zero)

$$\therefore \quad P(x = 3) = \frac{e^{-2}(2)^3}{3!} = \frac{8}{6}e^{-2} = 0.180$$

6.46 (a)

Given: Probability of producing a defective component, $P = 0.015$

Number of components in the box, $n = 200$

Since it is a case of Poisson's distribution,

Mean = Variance = $nP = 200 \times 0.015 = 3$

6.47 (0.39) (0.38 to 0.39)

$$f(x) = \begin{cases} 1 & 0 \le x \le 1 \\ 0 & \text{otherwise} \end{cases}$$

Expected value of Y,

$$E[Y] = \int_{-\infty}^{\infty} Y\cdot f(x)dx$$

$$= \int_0^1 1\cdot \ln(x+1)dx = \int_0^1 \ln(x+1)dx$$

$$= \int_1^2 \ln t\, dt \qquad \begin{bmatrix} \because x+1 = t \\ \therefore dx = dt \end{bmatrix}$$

$$= [t\cdot \ln t - t]_1^2$$

$$= (2\ln 2 - 2) - (1\ln 1 - 1)$$

$$= -0.6137 + 1$$

$$= 0.3863 \simeq 0.39$$

■■■■

7 Numerical Methods

7.1 The accuracy of Simpson's rule quadrature for a step size h is
(a) $O(h^2)$ (b) $O(h^3)$
(c) $O(h^4)$ (d) $O(h^5)$ **[2003 : 1 M]**

7.2 The values of a function $f(x)$ are tabulated below:

x	$f(x)$
0	1
1	2
2	1
3	10

Using Newton's forward difference formula, the cubic polynomial that can be fitted to the above data, is
(a) $2x^3 + 7x^2 - 6x + 2$ (b) $2x^3 - 7x^2 + 6x - 2$
(c) $x^3 - 7x^2 - 6x^2 + 1$ (d) $2x^3 - 7x^2 + 6x + 1$
[2004 : 2 M]

7.3 Starting from $x_0 = 1$, one step of Newton-Raphson method in solving the equation $x^3 + 3x - 7 = 0$ gives the next value (x_1) as
(a) $x_1 = 0.5$ (b) $x_1 = 1.406$
(c) $x_1 = 1.5$ (d) $x_1 = 2$ **[2005 : 2 M]**

7.4 Match the items in columns I and II.

Column I
P. Gauss-Seidel method
Q. Forward Newton-Gauss method
R. Runge-Kutta method
S. Trapezoidal Rule

Column II
1. Interpolation
2. Non-linear differential equations
3. Numerical integration
4. Linear algebraic equations

(a) P-1, Q-4, R-3, S-2 (b) P-1, Q-4, R-2, S-3
(c) P-1, Q-3, R-2, S-4 (d) P-4, Q-1, R-2, S-3
[2006 : 1 M]

7.5 A calculator has accuracy up to 8 digits after decimal place. The value of $\int_0^{2\pi} \sin x \, dx$ when evaluated using this calculator by trapezoidal method with 8 equal intervals, to 5 significant digits is
(a) 0.00000 (b) 1.0000
(b) 0.00500 (d) 0.00025 **[2007 : 2 M]**

7.6 Torque exerted on a flywheel over a cycle is listed in the table. Flywheel energy (in J per unit cycle) using Simpson's rule is

Angle (degree)	0	60	120	180	240	300	360
Torque (N m)	0	1066	–323	0	323	–355	0

(a) 542 (b) 993
(c) 1444 (d) 1986 **[2010 : 2 M]**

7.7 The integral $\int_1^3 \frac{1}{x} dx$, when evaluated by using Simpson's 1/3 rule on two equal subintervals each of length 1, equals
(a) 1.000 (b) 1.098
(c) 1.111 (d) 1.120 **[2011 : 2 M]**

7.8 Match the correct pairs

Numerical Integration Scheme	Order of Fitting Polynomial
P. Simpson's 3/8 Rule	1. First
Q. Trapezoidal Rule	2. Second
R. Simpson's 1/3 Rule	3. Third

(a) P-2, Q-1, R-3 (b) P-3, Q-2, R-1
(c) P-1, Q-2, R-3 (d) P-3, Q-1, R-2
[2013 : 1 M]

7.9 The definite integral $\int_1^3 \frac{1}{x} dx$ is evaluated using trapezoidal rule with a step size of 1. The correct answer is ________.
[2014 : 1 M, Set-3]

7.10 Using the trapezoidal rule, and dividing the interval of integration into three equal subintervals, the definite integral $\int_{-1}^{+1} |x| dx$ is ________.
[2014 : 2 M, Set-1]

7.11 The value of $\int_{2.5}^{4} \ln(x) dx$ calculated using the Trapezoidal rule with five subintervals is ___.
[2014 : 2 M, Set-2]

7.12 The real root of the equation $5x - 2\cos x - 1 = 0$ (up to two decimal accuracy) is ______.

[2014 : 2 M, Set-3]

7.13 Consider an ordinary differential equation $\frac{dx}{dt} = 4t + 4$. If $x = x_0$ at $t = 0$, the increment in x calculated using Runge-Kutta fourth order multi-step method with a step size of $\Delta t = 0.2$ is

(a) 0.22 (b) 0.44
(c) 0.66 (d) 0.88

[2014 : 2 M, Set-4]

7.14 Simpson's $\frac{1}{3}$ rule is used to integrate the function $f(x) = \frac{3}{5}x^2 + \frac{9}{5}$ between $x = 0$ and $x = 1$ using the least number of equal sub-intervals. The value of the integral is ______.

[2015 : 1 M, Set-1]

7.15 Using a unit step size, the volume of integral $\int_1^2 x\ln x\, dx$ by trapezoidal rule is ______.

[2015 : 1 M, Set-3]

7.16 The values of function $f(x)$ at 5 discrete points are given below:

x	0	0.1	0.2	0.3	0.4
$f(x)$	0	10	40	90	160

Using Trapezoidal rule step size of 0.1, the value of $\int_0^{0.4} f(x)dx$ is ______.

[2015 : 2 M, Set-2]

7.17 Newton-Raphson method is used to find the roots of the equation, $x^3 + 2x^2 + 3x - 1 = 0$. If the initial guess is $x_0 = 1$, then the value of x after 2nd iteration is ______.

[2015 : 2 M, Set-3]

7.18 Solve the equation $x = 10\cos(x)$ using the Newton-Raphson method. The initial guess is $x = \frac{\pi}{4}$. The value of the predicted root after the first iteration, up to second decimal, is______.

[2016 : 1 M, Set-1]

7.19 Gauss seidel method is used to solve the following equations (as per the given order):

$$x_1 + 2x_2 + 3x_3 = 5$$
$$2x_1 + 3x_2 + x_3 = 1$$
$$3x_1 + 2x_2 + x_3 = 3$$

Assuming initial guess as $x_1 = x_2 = x_3 = 0$, the value of x_3 after the first iteration is______

[2016 : 2 M, Set-1]

7.20 Numerical integration using trapezoidal rule gives the best result for a single variable function, which is

(a) linear (b) parabolic
(c) logarithmic (d) hyperbolic

[2016 : 1 M, Set-2]

7.21 The root of the function $f(x) = x^3 + x - 1$ obtained after first iteration on application of Newton Raphson scheme using an initial guess of $x_0 = 1$ is

(a) 0.682 (b) 0.686
(c) 0.750 (d) 1.000

[2016 : 1 M, Set-3]

7.22 The error in numerically computing the integral $\int_0^{\pi}(\sin x + \cos x)dx$ using the trapezoidal rule with three intervals of equal length between 0 and π is______.

[2016 : 2 M, Set-2]

7.23 $P(0, 3)$, $Q(0.5, 4)$ and $R(1, 5)$ are three points on the curve defined by $f(x)$. Numerical integration is carried out using both Trapezoidal rule and Simpson's rule within limits $x = 0$ and $x = 1$ for the curve. The difference between the two results will be

(a) 0 (b) 0.25
(c) 0.5 (d) 1

[2017 : 2 M, Set-1]

7.24 An explicit forward Euler method is used to numerically integrate the differential equation

$$\frac{dy}{dt} = y$$

using a time step of 0.1. With the initial condition $y(0) = 1$, the value of y(1) computed by this method is ______(correct to two decimal places).

[2018 : 2 M, Set-1]

7.25 Evaluation of $\int_2^4 x^3 dx$ using a 2-equal segment trapezoidal rule gives a value of ________.

[2019 : 1 M, Set-1]

7.26 The evaluation of the definite integral $\int_{-1}^{1.4} x|x|dx$ by using Simpson's 1/3rd (one-third) rule with step size $h = 0.6$ yields

(a) 0.592 (b) 0.581
(c) 0.914 (d) 1.248

[2020 : 2 M, Set-1]

7.27 For the integral $\int_0^{\pi/2} (8+4\cos x)dx$, the absolute percentage error in numerical evaluation with the Trapezoidal rule, using only the end points, is ________. (round off to one decimal place).

[2020 : 2 M, Set-2]

7.28 Find the positive real root of $x^3 - x - 3 = 0$ using Newton-Raphson method. If the starting guess (x_0) is 2, the numerical value of the root after two iterations (x_2) is _____ (round off to two decimal places).

[2021 : 2 M, Set-2]

7.29 Value of $\int_4^{5.2} \ln x dx$ using Simpson's one-third rule with interval size 0.3 is

(a) 1.60 (b) 1.83
(c) 1.51 (d) 1.06

[2021 : 1 M, Set-2]

7.30 Consider the definite integral

$$\int_1^2 (4x^2 + 2x + 6)dx$$

Let I_e be the exact value of the integral. If the same integral is estimated using Simpson's rule with 10 equal subintervals, the value is I_s. The percentage error is defined as $e = 100 \times \frac{(I_e - I_s)}{I_e}$. The value of e is

(a) 2.5 (b) 3.5
(c) 1.2 (d) 0 **[2022 : 1 M, Set-2]**

7.31 The initial value problem

$$\frac{dy}{dt} + 2y = 0, \quad y(0) = 1$$

is solved numerically using the forward Euler's method with a constant and positive time step of Δt.

Let y_n represent the numerical solution obtained after n steps. The condition $|y_{n+1}| \le |y_n|$ is satisfied if and only if Δt does not exceed _______. (Answer in integer)

[2023 : 2 Marks]

7.32 In order to numerically solve the ordinary differential equation $\frac{dy}{dt} = -y$ for $t > 0$, with an initial condition $y(0) = 1$, the following scheme is employed

$$\frac{y_{n+1} - y_n}{\Delta t} = -\frac{1}{2}(y_{n+1} + y_n)$$

Here, Δt is the time step and $y_n = y(n\Delta t)$ for $n = 0, 1, 2, ...$ This numerical scheme will yield a solution with non-physical oscillations for $\Delta t > h$. The value of h is

(a) 1 (b) $\frac{3}{2}$
(c) $\frac{1}{2}$ (d) 2

[2024 : 1 M]

■■■■

Answers Numerical Methods

7.1 (c)	7.2 (d)	7.3 (c)	7.4 (d)	7.5 (a)	7.6 (b)	7.7 (c)	7.8 (d)
7.9 (1.165)	7.10 (1.11)	7.11 (1.7533)	7.12 (0.54)	7.13 (d)	7.14 (2)	7.15 (0.693)	
7.16 (22)	7.17 (0.3043)	7.18 (1.5639)	7.19 (1.555)	7.20 (a)	7.21 (c)	7.22 (0.1863)	
7.23 (a)	7.24 (2.5937)	7.25 (63)	7.26 (a)	7.27 (5.2)	7.28 (1.67)	7.29 (b)	7.30 (d)
7.31 (1)	7.32 (d)						

Explanations Numerical Methods

7.2 (d)

x	$f(x)$	$\nabla f(x)$	$\nabla^2 f(x)$	$\nabla^3 f(x)$
0	1			
1	2	1	2	
2	1	–1	10	12
3	10	9		

Using Newton's forward interpolation formula we get

$$f(x) = f(0) + \frac{x}{1}\nabla f(0) + \frac{x(x-1)}{1:2}\nabla^2 f(0) + \frac{x(x-1)(x-2)}{1:2:3}\nabla^3 f(0)$$

$$= 1 + x(1) + \frac{x(x-1)}{2}(-2) + \frac{x(x-1)(x-2)}{6}(12)$$

$$= 1 + x + (x - x^2) + 2x(x^2 - 3x + 2)$$

$$= 1 + x + x - x^2 + 2x^3 - 6x^2 + 4x$$

$$= \mathbf{2x^3 - 7x^2 + 6x + 1}$$

7.3 (c)

From Newton–Raphson method

$$x_{n+1} = x_n - \frac{f(x)}{f'(x)}x_0$$

Given function is

$$f(x) = x^3 + 3x - 7$$

and $f'(x) = 3x^2 + 3$

Putting $x = 1$

$$f(1) = (1)^3 + 3 \times (1) - 7 = -3$$

$$f'(1) = 3 \times (1)^2 + 3 = 6$$

$$\therefore \quad x_1 = 1 - \left(\frac{-3}{6}\right) \times 1 = \mathbf{1.5}$$

7.5 (a)

$$h = \frac{2\pi - 0}{8} = \frac{\pi}{4}$$

$$y_0 = \sin(0) = 0$$

$$y_1 = \sin\left(\frac{\pi}{4}\right) = 0.70710$$

$$y_2 = \sin\left(\frac{\pi}{2}\right) = 1$$

$$y_3 = \sin\left(\frac{3\pi}{4}\right) = 0.7010$$

$$y_4 = \sin(\pi) = 0$$

$$y_5 = \sin\left(\frac{5\pi}{4}\right) = -0.70710$$

$$y_6 = \sin\left(\frac{6\pi}{4}\right) = -1$$

$$y_7 = \sin\left(\frac{7\pi}{4}\right) = -0.70710$$

$$y_8 = \sin\left(\frac{8\pi}{4}\right) = 0$$

Trapezoidal rule

$$\int_{x_0}^{x_0+nh} f(x)\cdot dx = \frac{h}{2}[(y_0 + y_n + 2(y_1 + y_2 + \dots + y_{n-1})]$$

$$\int_0^{2\pi} \sin x \cdot dx = \frac{h}{8}$$

$$[(0+0) + 2(0.70710 + 1 + 0.70710 + 0 - 0.70710 - 0.70710] = \mathbf{0.00000}$$

7.6 (b)

$$T = \int_0^{2\pi} T\,d\theta$$

Angle	0	60°	120°	180°	240°	300°	360°
Torque	0	1066	– 323	0	323	– 355	0

By using Simpson rule

$$\int_0^{2\pi} T d\theta = \frac{h}{3}[(y_0 + y_n) + 4(y_1 + y_3 \dots) + 2(y_2 + y_4)\dots]$$

$$= \frac{60}{3} \times \frac{\pi}{180}\ [(0+0) + 4(1066 + 0.355)] + 2(-323 + 323)$$

$$= \mathbf{992.24}$$

7.7 (c)

$$I = \int_1^3 \frac{1}{x}dx$$

x	$f(x) = \frac{1}{x}$
1	1
2	$\frac{1}{2}$
3	$\frac{1}{3}$

$$I = \frac{h}{3}(f_o + 4f_1 + f_2) = \frac{1}{3}\left(1 + 4 \times \frac{1}{2} \times \frac{1}{3}\right)$$

$$= \mathbf{1.111}$$

7.9 Sol.

x	1	2	3
y = f(x)	1	0.5	0.33
y_n	y_0	y_1	y_2

$$I = \int_1^3 \frac{1}{x} dx = \frac{h}{2}[(y_0 + y_2) + 2y_1]$$

$$= \frac{1}{2}[1 + 0.33 + 2 \times 0.5] = \frac{2.33}{2} = 1.165$$

7.10 Sol.

$$I = \int_{-1}^{1} |x| dx$$

$$n = 3,\ h = \frac{1-(-1)}{3} = \frac{2}{3}$$

x	-1	$-\frac{1}{3}$	$\frac{1}{3}$	1
$\lvert x \rvert$	1	$\frac{1}{3}$	$\frac{1}{3}$	1

By Trapezoidal rule

$$\int_{-1}^{1} |x| dx = \frac{h}{2}[(y_0 + y_3) + 2(y_1 + y_2)]$$

$$= \frac{2}{3 \times 2}\left[(1+1) + 2\left(\frac{1}{3} + \frac{1}{3}\right)\right] = \frac{1}{3}\left[2 + \frac{4}{3}\right] = 1.11$$

7.11 Sol.

x	2.5	2.8	3.1	3.4	3.7	4
$y = f(x)$	0.1963	1.0296	1.1314	1.2237	1.3083	1.3863
y_n	y_0	y_1	y_2	y_3	y_4	y_5

$$I = \int_{2.5}^{4} \ln(x) dx$$

$$= \frac{h}{2}[(y_0 + y_5) + 2(y_1 + y_2 + y_3 + y_4)]$$

$$I = \frac{0.3}{2}[(0.9163 + 1.3863) + 2(1.0296 + 1.1314 + 1.2237 + 1.3083)]$$

$$I = \frac{0.3}{2} \times 11.6886 = \mathbf{1.7533}$$

7.12 Sol.

$$f(x) = 5x - 2\cos x - 1$$

$$f'(x) = 5 + 2\sin x$$

By Newton Raphson's equation:

$$x_{n+1} = x_n - \frac{f(x_n)}{f'(x_n)}$$

Assuming $x_0 = 1$ (1 rad = 57.32°)

$$\Rightarrow x_1 = 1 - \frac{5 \times 1 - 2\cos(57.32°) - 1}{5 + 2\sin(57.32°)} = 0.5632$$

Again,

$$x_2 = 0.5632 - \frac{5 \times 0.5632 - 2\cos(32.27°) - 1}{5 + 2\sin(32.27°)}$$

$$= 0.5425$$

$$x_3 = 0.5425 - \frac{5 \times 0.5425 - 2\cos(31.09°) - 1}{5 + 2\sin(31.09°)}$$

$$= 0.5424$$

$\therefore$ Real root, $x = \mathbf{0.54}$

7.13 (d)

$$\frac{dx}{dt} = 4t + 4 = f(t_0, x_0)$$

At $t = 0$, $x = x_0$

Irrespective of values of x, $f(t_0, x_0)$ depends on t only.

$$k_1 = h f(t_0, x_0) = 0.2x \times 4 = 0.8$$

$$k_2 = hf\left(t_0 + \frac{1}{2}h, x_0 + \frac{k_1}{2}\right) = hf(0.1, x_0 + 0.4)$$

$$= 0.2(4 \times 0.1 + 4) = 0.88$$

$$k_3 = hf\left(t_0 + \frac{1}{2}h, x_0 + \frac{k_2}{2}\right)$$

$$= 0.2f(0.1, x_0 + 0.44)$$

$$= 0.2(4 \times 0.1 + 4) = 0.88$$

$$k_4 = hf(t_0 + h, x_0 + k_3)$$

$$= 0.2f(0.2, x_0 + 0.88)$$

$$= 0.2(4 \times 0.2 + 4) = 0.96$$

$$k = \frac{1}{6}(k_1 + 2k_2 + 2k_3 + k_4)$$

$$= \frac{1}{6}(0.8 + 2 \times 0.88 + 2 \times 0.88 + 0.96)$$

$$= \mathbf{0.88}$$

7.14 Sol.

$$f(x) = \frac{3}{5}x^2 + \frac{9}{5}$$

x	0	0.5	1
$f(x)$	1.8	1.95	2.4

$$\Rightarrow \int_0^1 f(x) = \frac{h}{3}[y_0 + 4y_1 + y_2]$$

$$= \frac{0.5}{3}[1.8 + 4(1.95) + 2.4] = 2$$

7.15 Sol.

x	1	2
$f(x)$	0	2(ln2)
	y_0	y_n

$$\therefore \quad I = \frac{h}{2}[y_0 + y_n]$$

$$I = \frac{1}{2}[0 + 2\ln 2] = \ln 2 = \mathbf{0.693}$$

7.16 Sol.

$$\int_0^{0.4} f(x)dx = \frac{h}{2}[y_0 + 2[y_1 + y_2 + y_3] + y_4]$$

$$= \frac{0.1}{2}[0 + 2[10 + 40 + 90] + 160] = \mathbf{22}$$

7.17 Sol.

$$f(x) = x^3 + 2x^2 + 3x - 1$$
$$f'(x) = 3x^2 + 4x + 3$$
$$x_0 = 1$$
$$x_{n+1} = x_n - \frac{f(x_n)}{f'(x_n)}$$
$$x_1 = x_0 - \frac{f(x_0)}{f'(x_0)} = 1 - \frac{[5]}{[10]} = 0.5$$
$$x_2 = 0.5 - \frac{[f(0.5)]}{[f'(0.5)]} = 0.5 - \frac{[0.125]}{[5.15]}$$
$$x_2 = \mathbf{0.3043}$$

7.18 Sol.

$$f(x) = x - 10\cos x$$
$$\therefore \quad f\left(\frac{\pi}{4}\right) = \frac{\pi}{4} - \frac{10}{\sqrt{2}} = -6.286$$
$$f'(x) = 1 + 10\sin x$$
$$f'\left(\frac{\pi}{4}\right) = 1 + \frac{10}{\sqrt{2}} = 8.07$$
$$x_1 = x_0 - \frac{f(x_0)}{f'(x_0)} = \frac{\pi}{4} - \left(\frac{-6.286}{8.07}\right)$$
$$= \frac{\pi}{4} + \frac{6.286}{8.07} = \mathbf{1.5639}$$

7.19 Sol.

The equations are

$$x_1 + 2x_2 + 3x_3 = 5 \quad ...(3)$$
$$2x_1 + 3x_2 + x_3 = 1 \quad ...(2)$$
$$3x_1 + 2x_2 + x_3 = 3 \quad ...(1)$$

By pivoting we can write

$$3x_1 + 2x_2 + x_3 = 3$$
$$2x_1 + 3x_2 + x_3 = 1$$
$$x_1 = \frac{3 - 2x_2 - x_3}{3} \quad ...(1)$$
$$x_2 = \frac{1 - 2x_1 - x_3}{3} \quad ...(2)$$
$$x_1 + 2x_2 + 3x_3 = 5$$
$$x_3 = \frac{5 - x_1 - 2x_2}{3} \quad ...(3)$$

Put $x_2 = 0$ $x_3 = 0$ in equation (1) $x_1 = 1$

Put $x_1 = 1$ $x_3 = 0$ in equation (3) $x_2 = -0.333$

Put $x_1 = 1$ $x_2 = -0.333$ in equation (3) $x_3 = 1.555$

$x_3 = \mathbf{1.555}$

7.20 (a)

Trapezoidal rule gives the best result in single variable function when the function is **linear** (degree 1)

7.21 (c)

$$f(x) = x^3 + x - 1 \qquad f(1) = 1 + 1 - 1 = 1$$
$$f'(x) = 3x^2 + 1 \qquad f'(1) = 3 + 1 = 4$$
$$x_1 = x_0 - \frac{f(x_0)}{f'(x_0)} = 1 - \frac{1}{4} = 1 - 0.25 = \mathbf{0.750}$$

7.22 Sol.

x	0	$\frac{\pi}{3}$	$\frac{2\pi}{3}$	π
$f(x)$	1 y_0	1.366 y_1	0.366 y_2	–1 y_3

By Trapezoidal

$$\int_0^{\pi}(\sin x + \cos x)dx = \frac{\pi/3}{2}(1 + (-1) + 2(1.366 + 0.366))$$
$$= \frac{\pi}{3}(1.732) = 1.8137$$

$$\int_0^{\pi}(\sin x + \cos x)dx = \int_0^{\pi/2}(\sin x + \cos x)dx + \int_{\pi/2}^{\pi}(\sin x + \cos x)dx$$
$$= (-\cos x + \sin x)\Big|_0^{\pi/2} + (-\cos x + \sin x)\Big|_{\pi/2}^{\pi}$$
$$= [(0 + 1) - (-1 + 0)] + [(1 + 0) - (0 + 1)]$$
$$= 1 + 1 + 1 - 1 = 2$$

Error = Exact value – approx value
= 2 – 1.8137 = **0.1863**

7.23 (a)

x	0	0.5	1
y	3_{y0}	4_{y1}	5_{y2}

By Trapezoidal rule

$$\int_0^1 f(x)dx = \frac{h}{2}[(y_0 + y_2) + 2y_1]$$

$$= \frac{0.5}{2}[(3+5)+2(4)] = 4$$

By Simpson rule

$$\int_0^1 f(x)dx = \frac{h}{3}[(y_0 + y_2) + 4y_1]$$

$$= \frac{0.5}{3}[(3+5)+4(4)] = 4$$

The difference between the two results will be **0**.

7.24 Sol.

$y_1 = y_0 + hf(t_0, y_0) = y_0 + hy_0 = 1 + 0.1\ (1)$

$y_1 = 1.1$

$y_2 = y_1 + hf(t_1, y_1) = y_1 + h.y_1 = 1.1 + 0.1(1.1)$

$y_2 = 1.21$

$y_3 = y_2 + hf(t_2, y_2) = y_2 + h.y_2 = 1.21 + 0.1 \times 1.21$

$y_3 = 1.331$

$y_4 = y_3 + h.f(t_3, y_3) = y_3 + h.y_3$

$= 1.331 + 0.1 \times 1.331$

$y_4 = 1.4641$

$y_5 = y_4 + h.f(t_4, y_4) = y_4 + h.y_4$

$= 1.4641 + 0.1 \times (1.4641) = 1.61051$

$y_6 = y_5 + h.f(t_5, y_5) = y_5 + h.y_5$

$= 1.61051 + 0.1 \times 1.61051$

$y_6 = 1.771561$

$y_7 = y_6 + h.f(t_6, y_6) = y_6 + h \times y_6$

$= 1.771561 + 0.1 \times 1.771561$

$= 1.9487$

$y_8 = y_7 + h.f(t_7, y_7) = y_7 + h.y_7$

$= 1.9487 + 0.1 \times (1.9487)$

$y_8 = 2.14357$

$y_9 = y_8 + h.f(t_8, y_8) = y_8 + h.y_8$

$= 2.14357 + 0.1 \times 2.14357$

$y_9 = 2.3579$

$y_{10} = y_9 + h.f(t_9, y_9) = y_9 + h.y_9$

$= 2.3579 + 0.1 \times (2.3579)$

$\mathbf{y_{10} = 2.5937}$

7.25 Sol.

x	2	3	4
y	8	27	64

Step size, $h = \frac{b-a}{n} = \frac{4-2}{2} = 1$

By trapezoidal rule,

$$\int_2^4 x^3 dx = \frac{h}{2}[(y_0 + y_2) + 2y_1]$$

$$= \frac{1}{2}[8 + 64 + 2 \times 27] = 63$$

7.26 (a)

Step size $h = 0.6$, $n = \frac{b-a}{h} = \frac{1.4+1}{0.6} = \frac{2.4}{0.6} = 4$

x	−1	−0.4	0.2	0.8	1.4
y	−1	−0.16	0.04	0.64	1.96

Simpson's 1/3rd rule:

$$\int_{-1}^{1.4} x|x|dx = \frac{h}{3}[y_0 + y_4] + 4(y_1 + y_3) + 2 \times y_2]$$

$$= \frac{0.6}{3}[(-1+1.96) + 4(-0.16+0.64) + 2(0.04)]$$

$$= 0.2[0.96 + 1.92 + 0.08] = 0.592$$

7.27 Sol.

True value

$$\int_0^{\pi/2} (8 + 4\cos x)dx = [8x + (4\sin x)]_0^{\pi/2}$$

$$= 4\pi + 4 = 16.566$$

By trapezoidal rule, (single step)

x	0	$\pi/2$
$f(x)$	12	8

$$h = \frac{\pi}{2}$$

Approx. $I = \frac{\pi}{4}[12 + 8] = 5\pi = 15.707$

Absolute error = |True value − Approximate value|

= |16.566 − 15.707| = 0.859

Absolute percentage error $= \frac{0.859}{16.566} \times 100$

$= 5.18\% \approx 5.2\%$

7.28 Sol.

Given, $f(x) = x^3 - x - 3,\ x_0 = 2$

$f'(x) = 3x^2 - 1$

Iteration 1: $x_1 = x_0 - \frac{f(x_0)}{f'(x_0)} = 2 - \frac{(8-2-3)}{3(4)-1}$

$= 1.72$

Iteration 2: $x_2 =$

$$x_1 - \frac{f(x_1)}{f'(x_1)} = 1.72 - \frac{(1.72^3 - 1.72 - 3)}{3(1.72)^2 - 1} = 1.67$$

7.29 (b)

Here $f(x) = \log x$

$a = 4, \quad b = 5.2, \; h = 0.3$

So, $n = \dfrac{b-a}{h} = \dfrac{5.2-4}{0.3} = 4$

x	4	4.3	4.6	4.9	5.2
y	log4	log4.3	log4.6	log4.9	log5.2
	y_0	y_1	y_2	y_3	y_4

As per Simpson's 1/3rd rule

$$\int_4^{5.2} \log x dx = \frac{h}{3}[y_0 + y_4 + 4(y_1 + y_3) + 2(y_2)]$$

$$= 1.8272 \simeq 1.83$$

7.30 (d)

Since the function $f(x) = 4x^2 + 2x + 6$ is a second degree polynomial and the Simpson's 1/3 rule also uses a second degree polynomial for approximation, the values of I_s and I_e will be same.

$$\therefore \quad e = \frac{I_e - I_s}{I_e} \times 100 = 0$$

7.31 (1) (0.999 to 1.001)

Given, $\dfrac{dy}{dt} + 2y = 0$

$$\frac{dy}{dt} = -2y$$

$$f(t, y) = -2y$$

$$y_{n+1} = y_n + hf(t_n, y_n)$$

$$y_{n+1} = y_n + \Delta t(-2y_n)$$

$$y_{n+1} = y_n - \Delta t 2y_n$$

$$y_{n+1} = y_n(1 - 2\Delta t)$$

$$\frac{y_{n+1}}{y_n} = 1 - 2\Delta t$$

$$\frac{|y_{n+1}|}{|y_n|} \le 1$$

$$\Rightarrow |1 - 2\Delta t| \le 1$$

$$-1 \le 1 - 2\Delta t \le 1$$

$$-2 \le -2\Delta t \le 0$$

$$0 \le 2\Delta t \le 2$$

$$0 \le \Delta t \le 1$$

Thus, Δt cannot exceed 1.

7.32 (d)

Given, $\dfrac{y_{n+1} - y_n}{\Delta t} = -\dfrac{1}{2}(y_{n+1} + y_n)$

$$y_{n+1} - y_n = -\frac{\Delta t}{2}(y_{n+1} + y_n)$$

$$y_{n+1} + \frac{\Delta t}{2} y_{n+1} = y_n - \frac{\Delta t}{2} y_n$$

$$y_{n+1} = \frac{2 - \Delta t}{2 + \Delta t} y_n$$

For stability or non physical oscillations,

$$-1 \le \frac{1 - \frac{\Delta t}{2}}{1 + \frac{\Delta t}{2}} \le 0 \text{ and } 0 \le \frac{1 - \frac{\Delta t}{2}}{1 + \frac{\Delta t}{2}} \le 1$$

$$1 + \frac{\Delta t}{2} \ge 0 \text{ and } 1 - \frac{\Delta t}{2} \le 0 \text{ and } 1 - \frac{\Delta t}{2} \ge 0$$

$$\Delta t \ge 2 \text{ and } \frac{\Delta t}{2} \le 1$$

$$\Delta t \ge 2 \text{ and } \Delta t \le 2$$

$$\therefore \quad h = 2$$

■■■■

8 Laplace Transforms & Fourier Series

Laplace Transforms

8.1 Laplace transform of the function sin ωt is

(a) $\frac{s}{s^2+\omega^2}$ (b) $\frac{\omega}{s^2+\omega^2}$

(c) $\frac{s}{s^2-\omega^2}$ (d) $\frac{\omega}{s^2-\omega^2}$

[2003 : 2 M]

8.2 A delayed unit step function is defined as $u(t-a) = \begin{cases} 0, & \text{for } t < a \\ 1, & \text{for } t \geq a \end{cases}$. Its Laplace transform is

(a) $a \cdot e^{-as}$ (b) $\frac{e^{-as}}{s}$

(c) $\frac{e^{as}}{s}$ (d) $\frac{e^{as}}{a}$ **[2004 : 2 M]**

8.3 If $F(s)$ is the Laplace transform of function $f(t)$, then Laplace transform of $\int_0^t f(\tau)d\tau$ is

(a) $\frac{1}{s}F(s)$ (b) $\frac{1}{s}F(s)-f(0)$

(c) $sF(s)-f(0)$ (d) $\int F(s)\,ds$

[2007 : 2 M]

8.4 The function $f(t)$ satisfies the differential equation $\frac{d^2f}{dt^2}+f=0$ and the auxiliary conditions, $f(0)=0$, $\frac{df}{dt}(0)=4$. The Laplace transform of $f(t)$ is given by

(a) $\frac{2}{s+1}$ (b) $\frac{4}{s+1}$

(c) $\frac{4}{s^2+1}$ (d) $\frac{2}{s^2+1}$ **[2013 : 2 M]**

8.5 Laplace transform of $\cos(\omega t)$ is $\frac{s}{s^2+\omega^2}$. The laplace transform of $e^{-2t}\cos(4t)$ is

(a) $\frac{s-2}{(s-2)^2+16}$ (b) $\frac{s+2}{(s-2)^2+16}$

(c) $\frac{s-2}{(s+2)^2+16}$ (d) $\frac{s+2}{(s+2)^2+16}$

[2014 : 1 M, Set-4]

8.6 Laplace transform of the function $f(t)$ is given by $F(s)=L\{f(t)\}=\int_0^\infty f(t)e^{-st}dt$. Laplace transform of the function shown below is given by

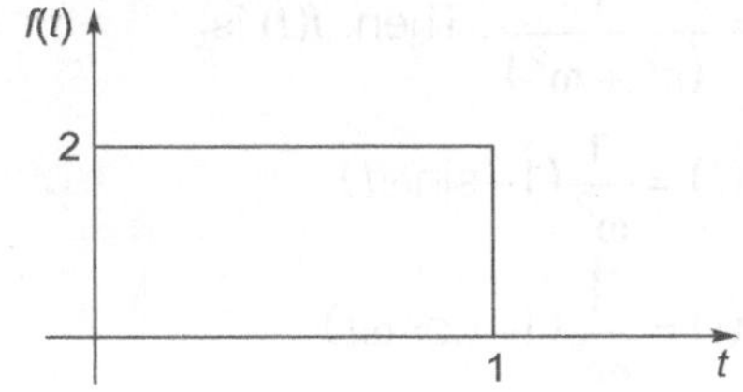

(a) $\frac{1-e^{-2s}}{s}$ (b) $\frac{1-e^{-s}}{2s}$

(c) $\frac{2-2e^{-s}}{s}$ (d) $\frac{1-2e^{-s}}{s}$

[2015 : 2 M, Set-3]

8.7 The Laplace transform of e^{i5t} where $i=\sqrt{-1}$, is

(a) $\frac{s-5i}{s^2-25}$ (b) $\frac{s+5i}{s^2+25}$

(c) $\frac{s+5i}{s^2-25}$ (d) $\frac{s-5i}{s^2+25}$

[2015 : 1 M, Set-2]

8.8 If $f(t)$ is a function defined for all $t \geq 0$, its Laplace transform $F(s)$ is defined as

(a) $\int_0^\infty e^{st}f(t)\,dt$ (b) $\int_0^\infty e^{-st}f(t)dt$

(c) $\int_0^\infty e^{ist}f(t)dt$ (d) $\int_0^\infty e^{-ist}f(t)dt$

[2016 : 1 M, Set-1]

8.9 Laplace transform of $\cos(\omega t)$ is

(a) $\frac{s}{s^2+\omega^2}$ (b) $\frac{\omega}{s^2+\omega^2}$

(c) $\frac{s}{s^2-\omega^2}$ (d) $\frac{\omega}{s^2-\omega^2}$

[2016 : 1 M, Set-2]

8.10 The Laplace transform of te^t is

(a) $\frac{s}{(s+1)^2}$ (b) $\frac{1}{(s-1)^2}$

(c) $\frac{1}{(s+1)^2}$ (d) $\frac{s}{s-1}$

[2017 : 1 M, Set-2]

8.11 $F(s)$ is the Laplace transform of the function

$f(t) = 2t^2e^{-t}$

F(1) is________(correct to two decimal places).

[2018 : 2 M, Set-1]

8.12 The Laplace transform of a function $f(t)$ is $L(f) = \frac{1}{(s^2+\omega^2)}$. Then, $f(t)$ is

(a) $f(t) = \frac{1}{\omega^2}(1-\sin\omega t)$

(b) $f(t) = \frac{1}{\omega^2}(1-\cos\omega t)$

(c) $f(t) = \frac{1}{\omega}\cos\omega t$

(d) $f(t) = \frac{1}{\omega}\sin\omega t$ [2020 : 1 M, Set-1]

8.13 The solution of $\frac{d^2y}{dt^2} - y = 1$, which additionally satisfies $y|_{t=0} = \frac{dy}{dt}\Big|_{t=0} = 0$ in the Laplace s-domain is

(a) $\frac{1}{s(s+1)}$ (b) $\frac{1}{s(s+1)(s-1)}$

(c) $\frac{1}{s(s-1)}$ (d) $\frac{1}{s-1}$

[2020 : 1 M, Set-2]

8.14 If the Laplace transform of a function $f(t)$ is given by $\frac{s+3}{(s+1)(s+2)}$, then $f(0)$ is

(a) $\frac{1}{2}$ (b) 0

(c) 1 (d) $\frac{3}{2}$ [2021 : 1 M, Set-2]

8.15 The Dirac-delta function $(\delta(t-t_0))$ for $t, t_0 \in \mathbb{R}$, has the following property

$$\int_a^b \phi(t)\delta(t-t_0)\,dt = \begin{cases} \phi(t_0) & a < t_0 < b \\ 0 & \text{otherwise} \end{cases}$$

The Laplace transform of the Dirac-delta function $\delta(t-a)$ for $a > 0$; $\mathcal{L}(\delta(t-a)) = F(s)$ is

(a) e^{sa} (b) ∞

(c) e^{-sa} (d) 0 [2021 : 1 M, Set-1]

8.16 Which one of the options given is the inverse Laplace transform of $\frac{1}{s^3-s}$?

where, $u(t)$ denotes the unit-step function.

(a) $\left(-1+\frac{1}{2}e^{-t}+\frac{1}{2}e^{t}\right)u(t)$

(b) $\left(\frac{1}{3}e^{-t}-e^{t}\right)u(t)$

(c) $\left(-1+\frac{1}{2}e^{-(t-1)}+\frac{1}{2}e^{(t-1)}\right)u(t-1)$

(d) $\left(-1-\frac{1}{2}e^{-(t-1)}-\frac{1}{2}e^{(t-1)}\right)u(t-1)$ [2023 : 2 M]

Fourier Series

8.17 The Fourier cosine series for an even function $f(x)$ is given by

$$f(x) = a_0 + \sum_{n=1}^{\infty} a_n \cos(nx)$$

The value of the coefficient a_2 for the function $f(x) = \cos^2(x)$ in $[0, \pi]$ is

(a) –0.5 (b) 0.0

(c) 0.5 (d) 1.0

[2018 : 1 M, Set-2]

8.18 The Fourier series expansion of x^3 in the interval $-1 \le x \le 1$ with periodic continuation has

(a) only sine terms

(b) only cosine terms

(c) both sine and cosine terms

(d) only sine terms and a non-zero constant

[2022 : 1 M, Set-1]

8.19 $F(t)$ is a periodic square wave function as shown. It takes only two values, 4 and 0, and stays at each of these values for 1 second before changing. What is the constant term in the Fourier series expansion of $F(t)$?

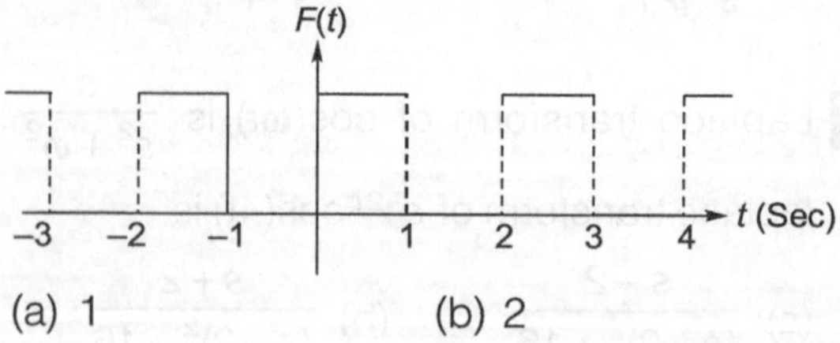

(a) 1 (b) 2

(c) 3 (d) 4 [2022 : 1 M, Set-2]

Answers Laplace Transforms & Fourier Series

8.1	(b)	8.2	(b)	8.3	(a)	8.4	(c)	8.5	(d)	8.6	(c)	8.7	(b)	8.8	(b)
8.9	(a)	8.10	(b)	8.11	(0.5)	8.12	(d)	8.13	(b)	8.14	(c)	8.15	(c)	8.16	(a)
8.17	(c)	8.18	(a)	8.19	(b)										

Explanations Laplace Transforms & Fourier Series

8.1 (b)

$$L[\sin \omega t] = \frac{\omega}{s^2 + \omega^2}$$

8.2 (b)

$$L(u(t-a)) = \int_0^\infty e^{-st} u(t-s)\,dt$$

$$= \int_0^a e^{-st} u(t-s)\,dt + \int_a^\infty e^{-st} u(t-a)\,dt$$

$$= \frac{[e^{-st}]_a^\infty}{-s} = e^{-\infty \times t} - \frac{e^{-sa}}{-s} = \frac{e^{-sa}}{s}$$

8.3 (a)

$$L\left[\int_0^t \int_0^t \dots \int_0^t f(t)\,dt^n\right] = \frac{1}{S^n} F(s)$$

8.4 (c)

Apply Laplace on both sides

$$\mathcal{L}\left\{\frac{d^2 f}{dt^2}\right\} + \mathcal{L}\{f\} = 0$$

$$\Rightarrow s^2 \mathcal{L}\{f(t)\} - sf(0) - f'(0) + \mathcal{L}\{f(t)\} = 0$$

$$\mathcal{L}\{f(t)\}\left[s^2 + 1\right] = 4$$

$$\mathcal{L}\{f(t)\} = \frac{4}{s^2 + 1}$$

8.5 (d)

Given $L\cos(\omega t) = \dfrac{s}{s^2 + \omega^2}$

$\Rightarrow L(e^{-2t}\cos 4t) = ?$

By the given formula

$$\Rightarrow L(\cos 4t) = \frac{s}{s^2 + 16}$$

$$\Rightarrow L(e^{-2t}\cos 4t) = \frac{s+2}{(s+2)^2 + 16}$$

($\because$ By using first shifting property of Laplace)

8.6 (c)

$$F(s) = \int_0^\infty f(t) e^{-st}\,dt = \int_0^1 2e^{-st} dt + \int_1^\infty 0 \cdot e^{-st} dt$$

$$= 2\left[\frac{e^{-st}}{-s}\right]_0^1 = \frac{2}{-s}[e^{-s} - 1]$$

$$= \frac{2(1 - e^{-s})}{s} = \frac{2 - 2e^{-s}}{s}$$

8.7 (b)

$$e^{i5t} = \cos 5t + i \sin 5t$$

$$L\{e^{(i5t)}\} = \frac{s}{s^2 + 25} + \frac{5i}{s^2 + 25} = \frac{s + 5i}{s^2 + 25}$$

8.8 (b)

$$L(f(t)) = \int_0^\infty e^{-st} f(t)\,dt$$

8.9 (a)

$$L(\cos \omega t) = \frac{s}{s^2 + \omega^2}$$

8.10 (b)

$$f(t) = te^t$$

$$L(t) = \frac{1}{s^2}$$

By first shifting rule

$$L(te^t) = \frac{1}{(s-1)^2}$$

8.11 Sol.

$$L(t^2) = \frac{L^2}{S^3}$$

$$L(2t^2) = \frac{4}{S^3}$$

$$F(s) = L[e^{-t} \cdot 2t^2] = \frac{4}{(S+1)}$$

$$F(1) = \frac{4}{(1+1)^3} = \frac{4}{8} = 0.5$$

8.12 (d)

$$L\{f\} = \frac{1}{s^2+\omega^2}$$

$$f(t) = L^{-1}\left\{\frac{1}{s^2+\omega^2}\right\} = \frac{1}{\omega}\sin\omega t$$

8.13 (b)

$$y'' - y = 1$$
$$y(0) = 1$$
$$y'(0) = 1$$
$$L\{y'' - y\} = L\{1\}$$

$$s^2Y(s) - sy(0) - y'(0) - y(s) = \frac{1}{s}$$

$$y(s) = \frac{1}{s(s^2-1)} = \frac{1}{s(s+1)(s-1)}$$

8.14. (c)

By using partial fraction concept.

$$f(t) = L^{-1}\left[\frac{s+3}{(s+1)(s+2)}\right] = L^{-1}\left[\frac{2}{s+1} - \frac{1}{s+2}\right]$$

$\Rightarrow$ $f(t) = 2e^{-t} - e^{-2t}$

So, $f(c) = 2e^0 - e^0 = 2 - 1 = 1$

8.15 (c)

$$\because \int_0^\infty f(t)\delta(t-a)\,dt = f(a)$$

$$\therefore L\{\delta(t-a)\} = \int_0^\infty e^{-st}\delta(t-a)\,dt = e^{-as}$$

8.16 (a)

Given,

$$F(s) = \frac{1}{s^3 - s} = \frac{1}{s(s^2-1)} = \frac{1}{s(s-1)(s+1)}$$

On partial fraction decomposition,

$$F(s) = -\frac{1}{s} + \frac{\frac{1}{2}}{s-1} + \frac{\frac{1}{2}}{s+1}$$

Thus,

$$L^{-1}(F(s)) = L^{-1}\left(-\frac{1}{s}\right) + L^{-1}\left(\frac{\frac{1}{2}}{s-1}\right) + L^{-1}\left(\frac{\frac{1}{2}}{s+1}\right)$$

$$= -L^{-1}\left(\frac{1}{s}\right) + \frac{1}{2}L^{-1}\left(\frac{1}{s-1}\right) + \frac{1}{2}L^{-1}\left(\frac{1}{s+1}\right)$$

$$= -1 + \frac{1}{2}e^t + \frac{1}{2}e^{-t}$$

8.17 (c)

$$\cos^2 x = \frac{1+\cos 2x}{2}$$

$$f(x) = \frac{1}{2} + \frac{\cos 2x}{2}$$

$$f(x) = \frac{a_0}{2} + \sum_{n=1}^{\infty} a_n \cdot \cos nx$$

$$a_0 = 1$$
$$a_1 = 0,$$
$$a_2 = \frac{1}{2}$$

8.18 (a)

Since $f(x) = x^3$ is an odd function, its Fourier series expansion consists of only sine terms.

8.19 (b)

Constant term in Fourier series is given by $\frac{a_o}{2}$.

where, $a_o = \frac{1}{L}\int_o^2 f(x)dx = \frac{1}{1}\int_o^2 f(x)dx$

[where, $L = 1$ and period (0, 2)]

$= 1 \times$ Area under the curve within [0, 2]

$= 1 \times 4 = 4$

Thus, $\frac{a_o}{2} = \frac{4}{2} = 2$

■■■■

9 Partial Differential Equations

9.1 The partial differential equation

$$\frac{\partial^2\varphi}{\partial x^2}+\frac{\partial^2\varphi}{\partial y^2}+\left(\frac{\partial\varphi}{\partial x}\right)+\left(\frac{\partial\varphi}{\partial y}\right)=0 \text{ has}$$

(a) degree 1 order 2 (b) degree 1 order 1
(c) degree 2 order 1 (d) degree 2 order 2

[2007 : 1 M]

9.2 The partial differential equation

$$\frac{\partial u}{\partial t}+u\frac{\partial u}{\partial x}=\frac{\partial^2 u}{\partial x^2} \text{ is a}$$

(a) linear equation of order 2
(b) non-linear equation of order 1
(c) linear equation of order 1
(d) non-linear equation of order 2 [2013 : 1 M]

9.3 Consider the following partial differential equation $u(x, y)$ with the constant $c > 1$:

$$\frac{\partial u}{\partial y}+c\frac{\partial u}{\partial x}=0$$

Solution of this equation is
(a) $u(x, y) = f(x + cy)$
(b) $u(x, y) = f(x - cy)$
(c) $u(x, y) = f(cx + y)$
(d) $u(x, y) = f(cx - y)$ [2017 : 1 M, Set-1]

9.4 Consider a function u which depends on position x and time t. The partial differential equation

$$\frac{\partial u}{\partial t}=\frac{\partial^2 u}{\partial x^2} \text{ is known as the}$$

(a) Wave equation (b) Heat equation
(c) Laplace's equation (d) Elasticity equation

[2018 : 1 M, Set-2]

Answers Partial Differential Equations

9.1 (a) 9.2 (d) 9.3 (b) 9.4 (b)

Explanations Partial Differential Equations

9.1 (a)

Order: The order of a differential equation is the order of the highest derivative appears in the equation.
Degree: The degree of a differential equation is the degree of the highest order differential coefficient or derivative, when the differential coefficient is free from radicals and fraction.
The general solution of a differential equation of order 'n' must involve 'n' arbitrary constant.

9.2 (d)

A differential equation in the form $\frac{dy}{dx}+Py=Q$ where, P and Q are functions of x i.e., $f(x)$ is said to be linear equation.

$$\frac{\partial u}{\partial t}+u\frac{\partial u}{\partial x}=\frac{\partial^2 u}{\partial x^2}$$

The given equation is not complying with the definition of linear equation, therefore it is a non-linear equation of order 2.

9.3 (b)

$$u = f(x-cy)$$

$$\frac{\partial u}{\partial x} = f'(x-cy)(1)$$

$$\frac{\partial u}{\partial y} = f'(x-cy)(-c)$$

$$= -c\cdot f'(x-cy) = -c\cdot\frac{\partial u}{\partial x}$$

$$\therefore \quad \frac{\partial u}{\partial y}+c\frac{\partial u}{\partial x} = 0$$

9.4 (b)

$$\frac{\partial u}{\partial t}=\frac{\partial^2 u}{\partial x^2} \text{ is known as heat equation}$$

Engineering Mechanics

UNIT

II

CONTENTS

Engineering Mechanics

Syllabus

Free-body diagrams and equilibrium; friction and its applications including rolling friction, belt-pulley, brakes, clutches, screw jack, wedge, vehicles, etc.; trusses and frames; virtual work; kinematics and dynamics of rigid bodies in plane motion; impulse and momentum (linear and angular) and energy formulations; Lagrange's equation.

Analysis of Previous GATE Papers

Exam Year	1 Mark Ques.	2 Marks Ques.	3 Marks Ques.	5 Marks Ques.	Total Marks
1995	–	1	–	–	**2**
1996	2	3	–	–	**8**
1997	–	1	–	–	**2**
1998	1	–	–	–	**1**
1999	–	1	–	–	**2**
2000	2	–	–	–	**2**
2003	5	–	–	–	**5**
2004	–	3	–	–	**6**
2005	2	4	–	–	**10**
2006	–	2	–	–	**4**
2007	1	1	–	–	**3**
2008	1	2	–	–	**5**
2009	1	1	–	–	**3**
2011	1	2	–	–	**5**
2012	2	3	–	–	**8**
2013	–	1	–	–	**2**
2014 Set-1	1	2	–	–	**5**
2014 Set-2	1	2	–	–	**5**
2014 Set-3	1	3	–	–	**7**

Exam Year	1 Mark Ques.	2 Marks Ques.	Total Marks
2014 Set-4	–	4	**8**
2015 Set-1	3	4	**11**
2015 Set-2	1	2	**5**
2015 Set-3	2	3	**8**
2016 Set-1	2	2	**6**
2016 Set-2	1	3	**7**
2016 Set-3	2	3	**8**
2017 Set-1	2	1	**4**
2017 Set-2	–	1	**2**
2018 Set-1	1	2	**5**
2018 Set-2	1	1	**3**
2019 Set-1	1	2	**5**
2019 Set-2	2	1	**4**
2020 Set-1	1	1	**3**
2020 Set-2	2	1	**4**
2021 Set-1	1	2	**5**
2021 Set-2	–	2	**4**
2022 Set-1	1	3	**7**
2022 Set-2	2	2	**6**
2023	2	2	**6**
2024	2	2	**6**

1 FBD, Equilibrium, Plane Trusses and Virtual Work

1.1 A spring scale indicates a tension T in the right hand cable of the pulley system shown in figure. Neglecting the mass of the pulleys and ignoring friction between the cable and pulley the mass m is

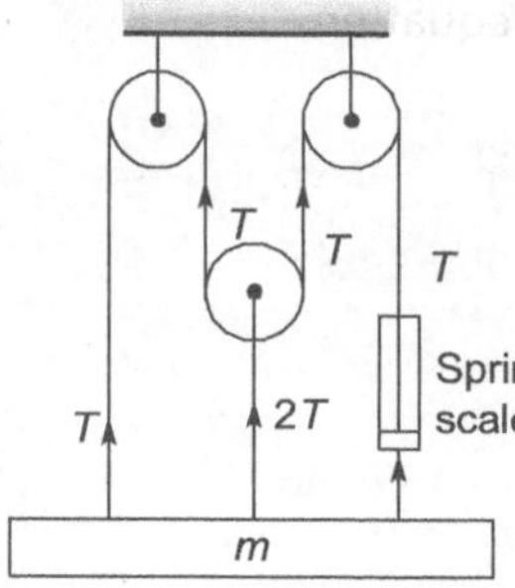

(a) $2T/g$
(b) $T(1 + e^{4\pi})/g$
(c) $4T/g$
(d) None of these

[1995 : 2 M]

1.2 A mass 35 kg is suspended from a weightless bar AB which is supported by a cable CB and a pin at A as shown in figure. The pin reactions at A on the bar AB are

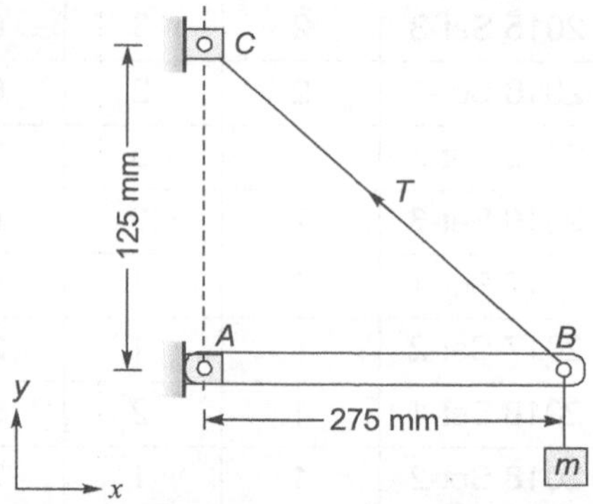

(a) $R_x = 343.4$ N, $R_y = 755.4$ N
(b) $R_x = 343.4$ N, $R_y = 0$
(c) $R_x = 755.4$ N, $R_y = 343.4$ N
(d) $R_x = 755.4$ N, $R_y = 0$

[1997 : 2 M]

1.3 A truss consists of horizontal members (AC, CD, DB and EF) and vertical members (CE and DF) having length l each. The members AE, DE and BF are inclined at 45° to the horizontal. For the uniformly distributed load p per unit length on the members EF of the truss shown in figure given below, the force in the member CD is

(a) $\frac{pl}{2}$
(b) pl
(c) 0
(d) $\frac{2pl}{3}$

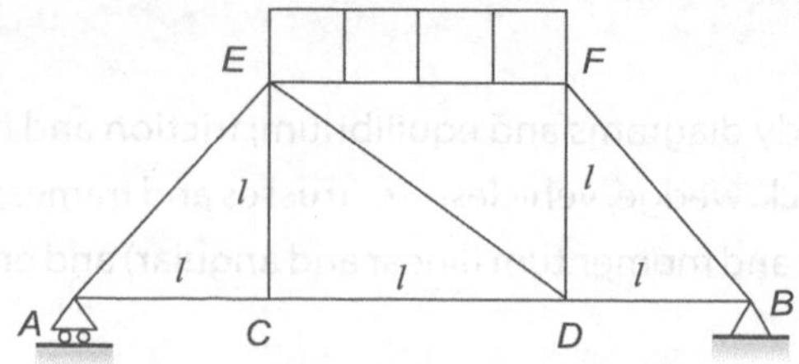

[2003 : 1 M]

1.4 The figure shows a pin-jointed plane truss loaded at the point M by hanging a mass of 100 kg. The member LN of the truss is subjected to a load of

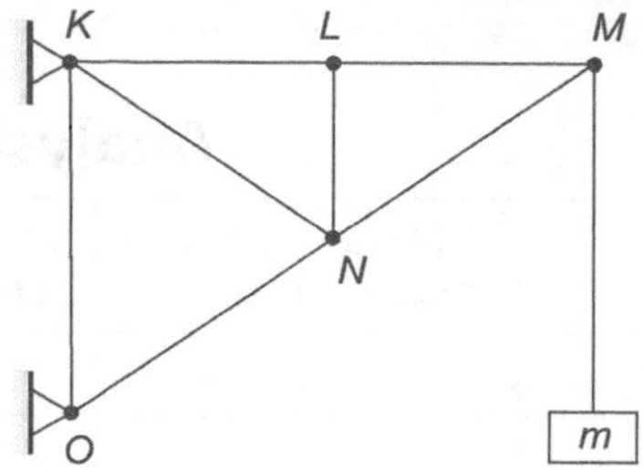

(a) 0 Newton
(b) 490 Newtons in compression
(c) 981 Newtons in compression
(d) 981 Newtons in tension

[2003 : 1 M]

1.5 If a system is in equilibrium and the position of the system depends upon many independent variables, the principle of virtual work states that the partial derivatives of its total potential energy with respect to each of the independent variable must be

(a) –1.0
(b) 0
(c) 1.0
(d) ∞

[2006 : 2 M]

1.6 If point A is in equilibrium under the action of the applied forces, the values of tensions T_{AB} and T_{AC} are respectively.

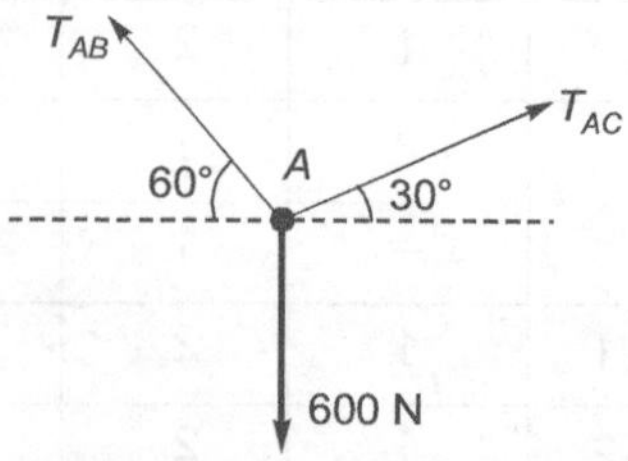

(a) 520 N and 300 N
(b) 300 N and 520 N
(c) 450 N and 150 N
(d) 150 N and 450 N

[2006 : 2 M]

1.7 Consider a truss PQR loaded at P with a force F as shown in the figure.

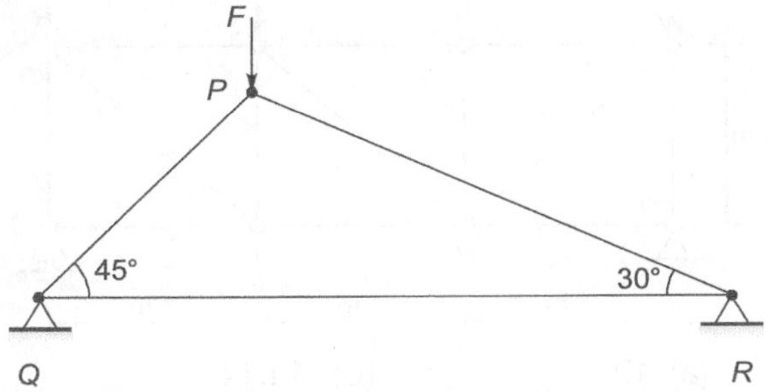

The tension in the member QR is

(a) 0.5 F (b) 0.63 F
(c) 0.73 F (d) 0.87 F **[2008 : 2 M]**

Common Data Questions (1.8 and 1.9):

Two steel truss members, AC and BC, each having cross sectional area of 100 mm^2, are subjected to a horizontal force F as shown in figure. All the joints are hinged.

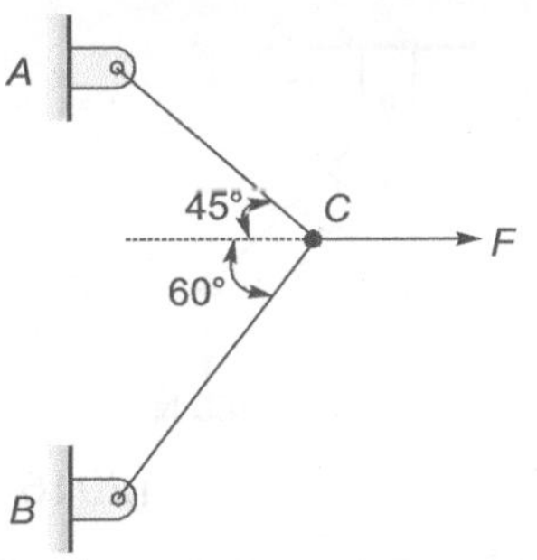

1.8 If $F = 1$ kN, the magnitude of the vertical reaction force developed at the point B in kN is

(a) 0.63 (b) 0.32
(c) 1.26 (d) 1.46 **[2012 : 2 M]**

1.9 The maximum force F in kN that can be applied at C such that the axial stress in any of the truss members DOES NOT exceed 100 MPa is

(a) 8.17 (b) 11.15
(c) 14.14 (d) 22.30 **[2012 : 2 M]**

1.10 A pin jointed uniform rigid rod of weight W and length L is supported horizontally by an external force F as shown in the figure below. The force F is suddenly removed. At the instant of force removal, the magnitude of vertical reaction developed at the support is

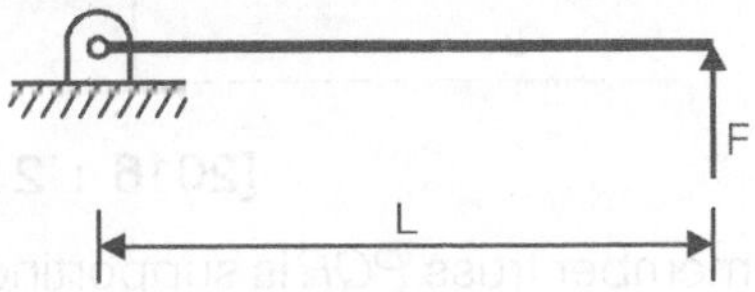

(a) zero (b) $W/4$
(c) $W/2$ (d) W **[2013 : 2 M]**

1.11 A two member truss ABC is shown in the figure. The force (in kN) transmitted in member AB is ______.

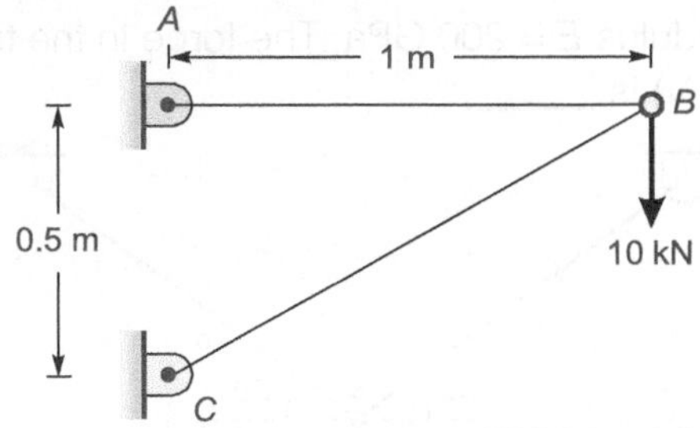

[2014 : 1 M, Set-2]

1.12 For the truss shown in the figure, the forces F_1 and F_2 are 9 kN and 3 kN, respectively. The force (in kN) in the member QS is (All dimensions are in m)

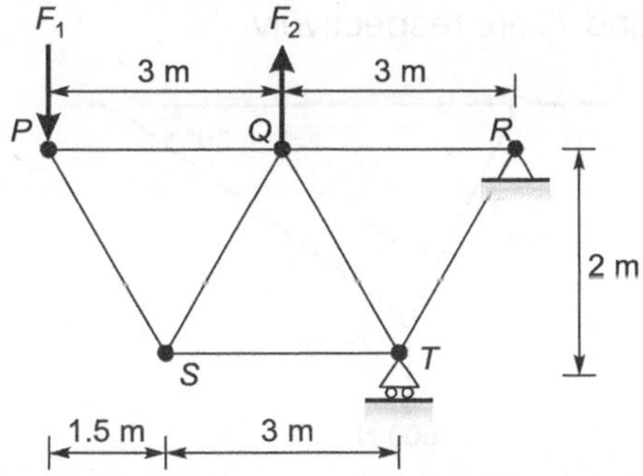

(a) 11.25 tension (b) 11.25 compression
(c) 13.5 tension (d) 13.5 compression
[2014 : 2 M, Set-4]

1.13 A ladder AB of length 5 m and weight (W) 600 N is resting against a wall. Assuming frictionless contact at the floor (B) and the wall (A), the magnitude of the force P (in newton) required to maintain equilibrium of the ladder is ______.

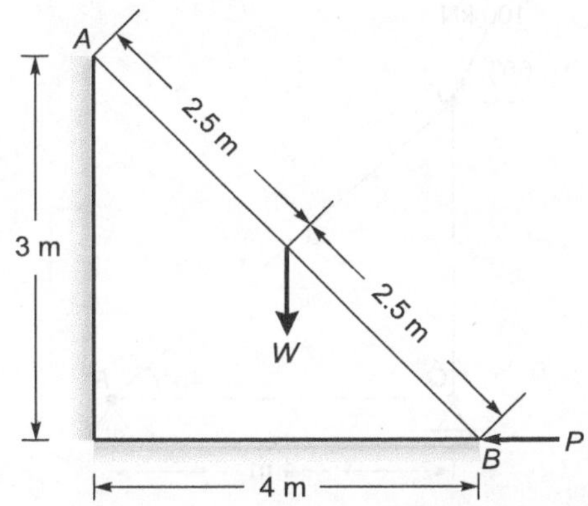

[2014 : 2 M, Set-4]

1.14 In a statically determinate plane truss, the number of joints (j) and the number of members (m) are related by

(a) $j = 2m - 3$ (b) $m = 2j + 1$
(c) $m = 2j - 3$ (d) $m = 2j - 1$
[2014 : 1 M, Set-4]

1.15 Two identical trusses support a load of 100 N as shown in the figure. The length of each truss is 1.0 m, cross-sectional area is 200 mm^2; Young's modulus E = 200 GPa. The force in the truss AB (in N) is ______

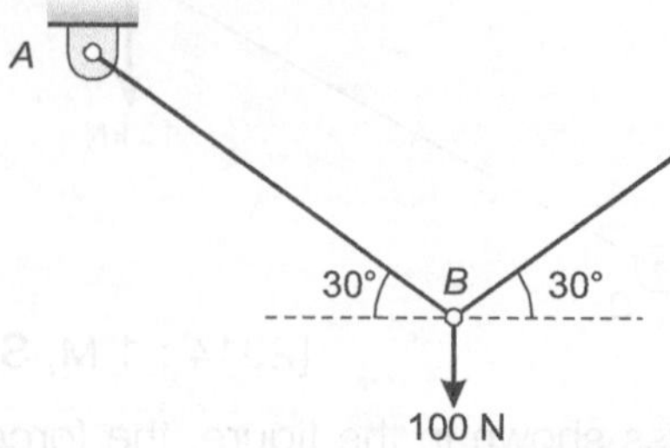

[2015 : 1 M, Set-1]

1.16 A weight of 500 N is supported by two metallic ropes as shown in the figure. The values of tensions T_1 and T_2 are respectively

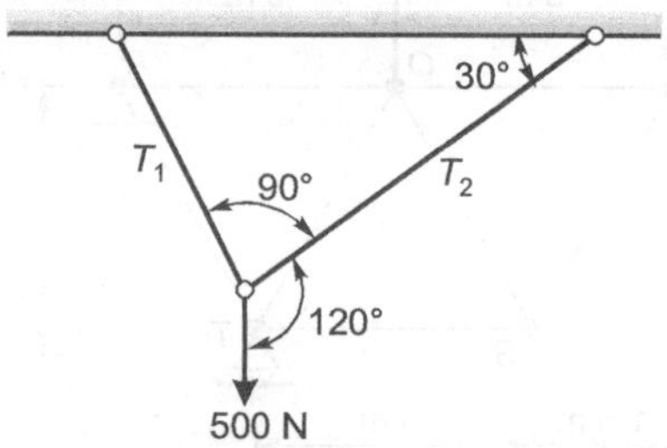

(a) 433 N and 250 N
(b) 250 N and 433 N
(c) 353.5 N and 250 N
(d) 250 N and 353.5 N [2015 : 1 M, Set-3]

1.17 For the truss shown in figure, the magnitude of the force in member PR and the support reaction at R are respectively

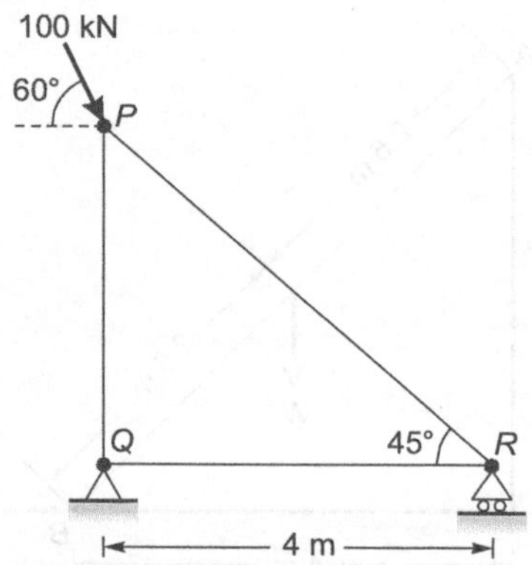

(a) 122.47 kN and 50 kN
(b) 70.71 kN and 100 kN
(c) 70.71 kN and 50 kN
(d) 81.65 kN and 100 kN [2015 : 2 M, Set-1]

1.18 For the truss shown in the figure, the magnitude of the force (in kN) in the member SR is

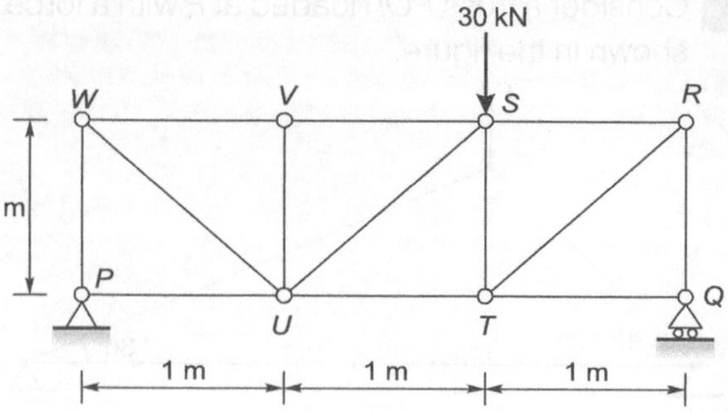

(a) 10
(b) 14.14
(c) 20
(d) 28.28

[2015 : 2 M, Set-2]

1.19 A rigid ball of weight 100 N is suspended with the help of a string. The ball is pulled by a horizontal force F such that the string makes an angle of 30° with the vertical. The magnitude of force F (in N) is________.

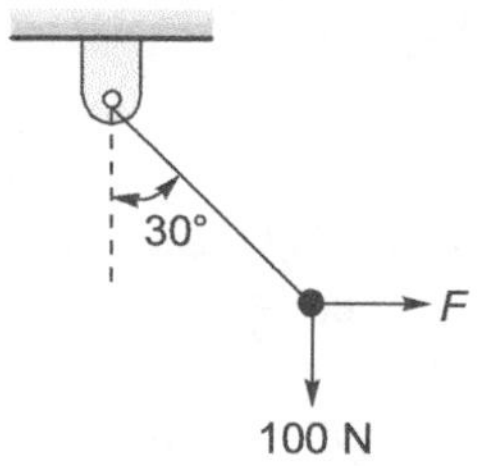

[2016 : 1 M, Set-1]

1.20 A block of mass m rests on an inclined plane and is attached by a string to the wall as shown in the figure. The coefficient of static friction between the plane and the block is 0.25. The string can withstand a maximum force of 20 N. The maximum value of the mass (m) for which the string will not break and the block will be in static equilibrium is ______ kg.

Take $\cos\theta = 0.8$ and $\sin\theta = 0.6$.

Acceleration due to gravity $g = 10$ m/s^2

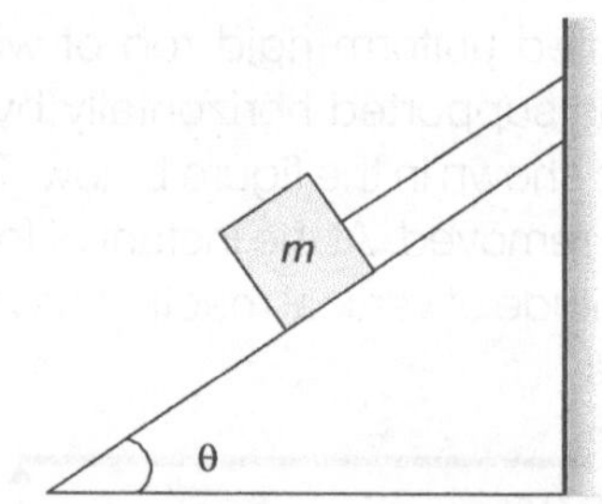

[2016 : 2 M, Set-1]

1.21 A two member truss PQR is supporting a load W. The axial forces in members PQ and QR are respectively

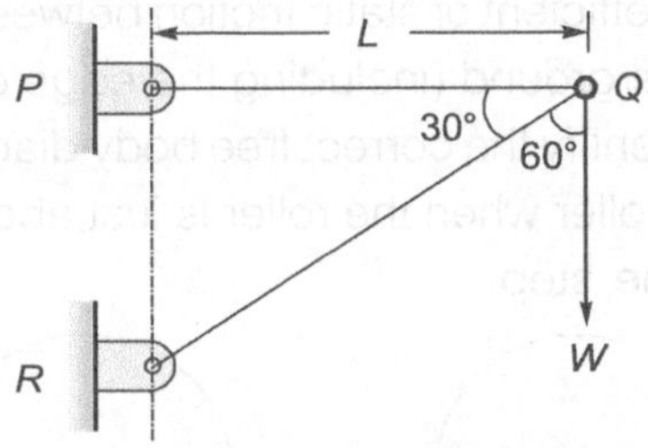

(a) $2W$ tensile and $\sqrt{3}W$ compressive
(b) $\sqrt{3}W$ tensile and $2W$ compressive
(c) $\sqrt{3}W$ compressive and $2W$ tensile
(d) 2W compressive and $\sqrt{3}W$ tensile
[2016 : 2 M, Set-1]

1.22 A force F is acting on a bent bar which is clamped at one end as shown in the figure.

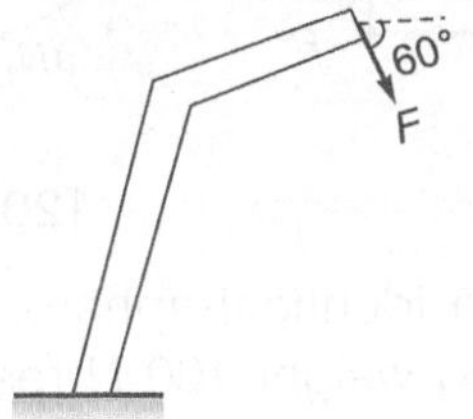

The CORRECT free body diagram is

(a)
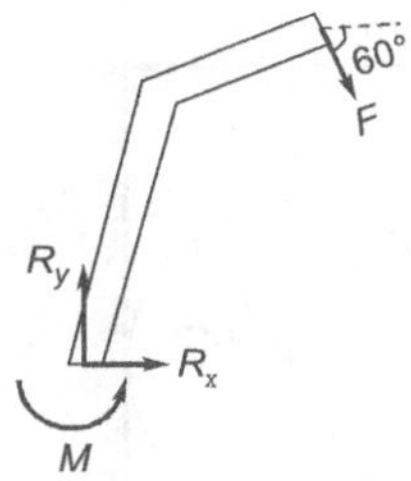

(b)
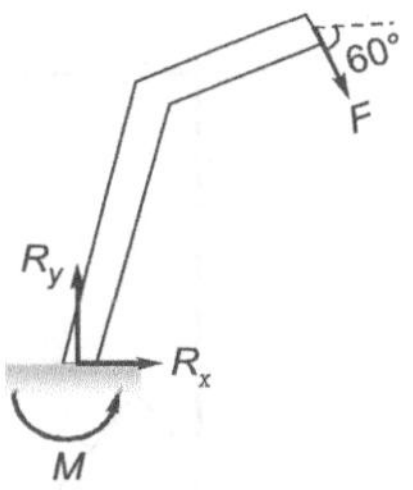

(c)
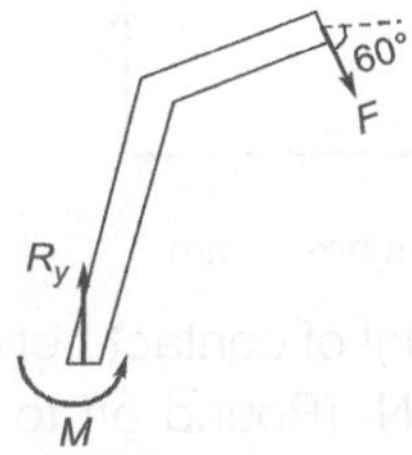

(d)
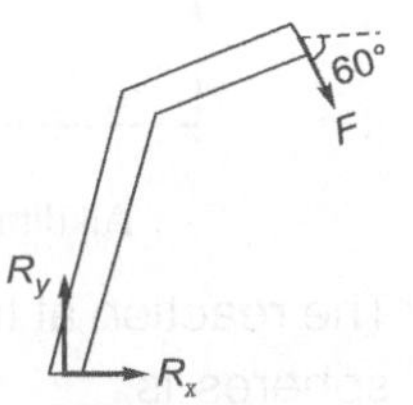

[2016 : 1 M, Set-3]

1.23 In the figure, the load P = 1 N, length L = 1 m, Young's modulus E = 70 GPa, and the cross-section of the links is a square with dimension 10 mm × 10 mm. All joints are pin joints.

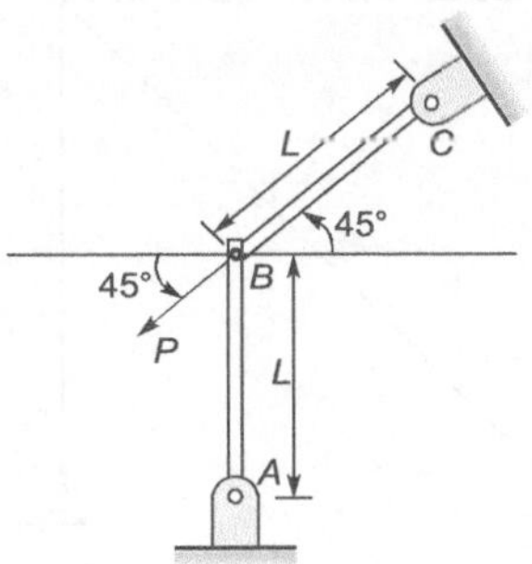

The stress (in Pa) in the AB is______
(Indicate compressive stress by a negative sign and tensile stress by a positive sign.)
[2016 : 2 M, Set-2]

1.24 A bar of uniform cross section and weighing 100 N is held horizontally using two massless and inextensible strings S_1 and S_2 as shown in the figure.

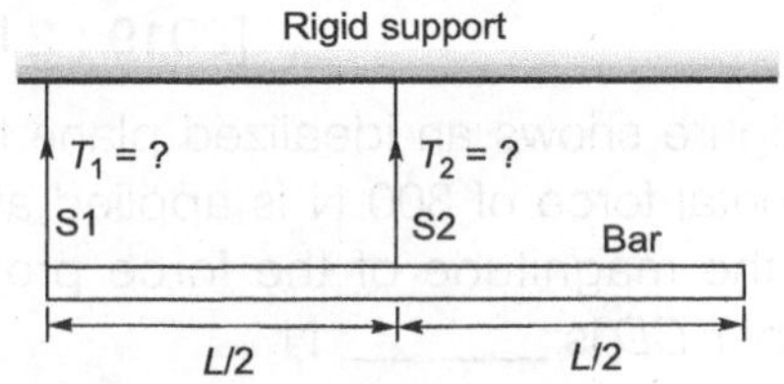

(a) T_1 = 100 N and T_2 = 0 N
(b) T_1 = 0 N and T_2 = 100 N
(c) T_1 = 75 N and T_2 = 25 N
(d) T_1 = 25 N and T_2 = 75 N [2018 : 1 M, Set-1]

1.25 A truss is composed of members AB, BC, CD, AD and BD, as shown in the figure. A vertical load of 10 kN is applied at point D. The magnitude of force (in kN) in the member BC is ______.

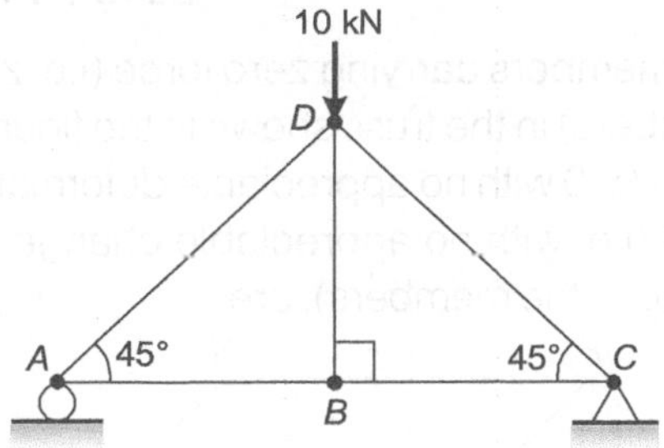

[2019 : 2 M, Set-1]

1.26 A car is having weight W is moving in the direction as shown in the figure. The centre of gravity (CG) of the car is located at height h from the ground, midway between the front and rear wheels. The distance between the front and rear wheels, is l. The acceleration of the car is a, and acceleration due to gravity is g. The reactions on the front wheels (R_f) and rear wheels (R_r) are given by

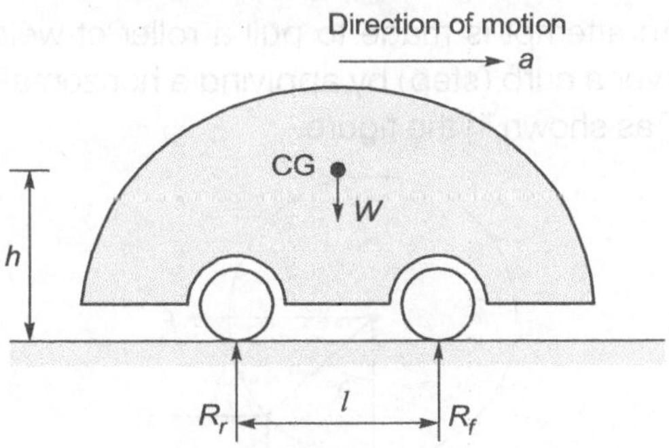

(a) $R_f = \frac{W}{2} + \frac{W}{g}\left(\frac{h}{l}\right)a; R_r = \frac{W}{2} - \frac{W}{g}\left(\frac{h}{l}\right)a$

(b) $R_f = R_r = \frac{W}{2} + \frac{W}{g}\left(\frac{h}{l}\right)a$

(c) $R_f = R_r = \frac{W}{2} - \frac{W}{g}\left(\frac{h}{l}\right)a$

(d) $R_f = \frac{W}{2} - \frac{W}{g}\left(\frac{h}{l}\right)a; R_r = \frac{W}{2} + \frac{W}{g}\left(\frac{h}{l}\right)a$

[2019 : 2 M, Set-1]

1.27 The figure shows an idealized plane truss. If a horizontal force of 300 N is applied at point A, then the magnitude of the force produced in member CD is ________ N.

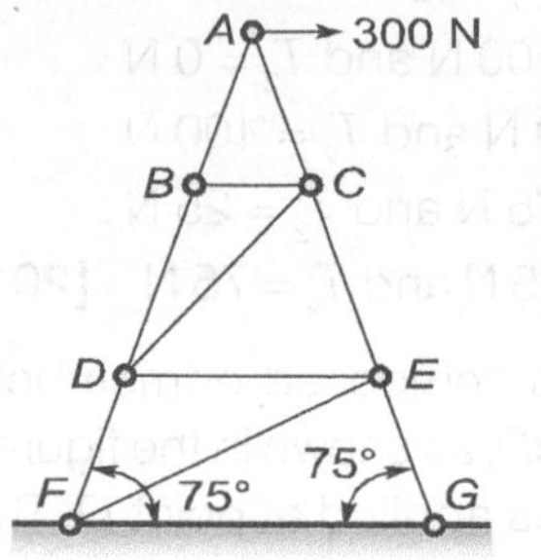

[2019 : 1 M, Set-2]

1.28 The members carrying zero force (i.e. zero-force members) in the truss shown in the figure, for any load $P > 0$ with no appreciable deformation of the truss (i.e. with no appreciable change in angles between the members), are

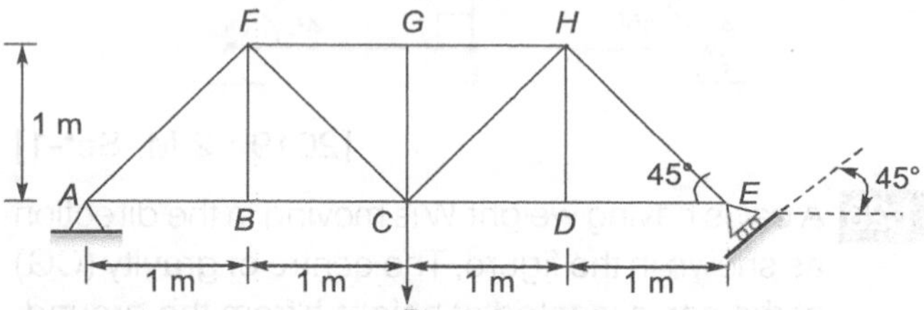

(a) BF, DH and GC only
(b) BF, DH, GC, CD and DE only
(c) BF and DH only
(d) BF, DH, GC, FG and GH only

[2020 : 1 M, Set-1]

1.29 An attempt is made to pull a roller of weight W over a curb (step) by applying a horizontal force F as shown in the figure.

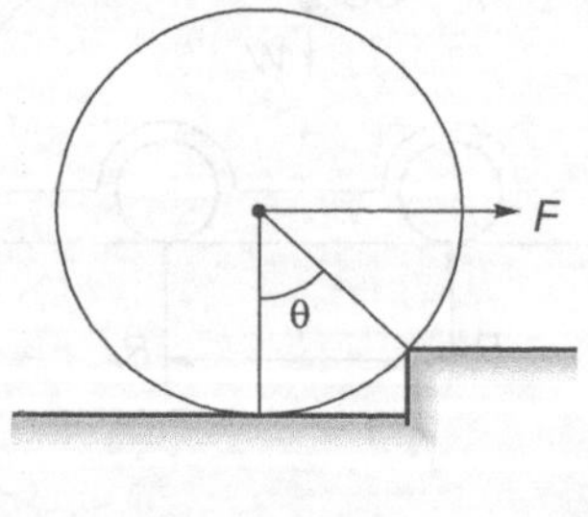

The coefficient of static friction between the roller and the ground (including the edge of the step) is μ. Identify the correct free body diagram (FBD) of the roller when the roller is just about to climb over the step.

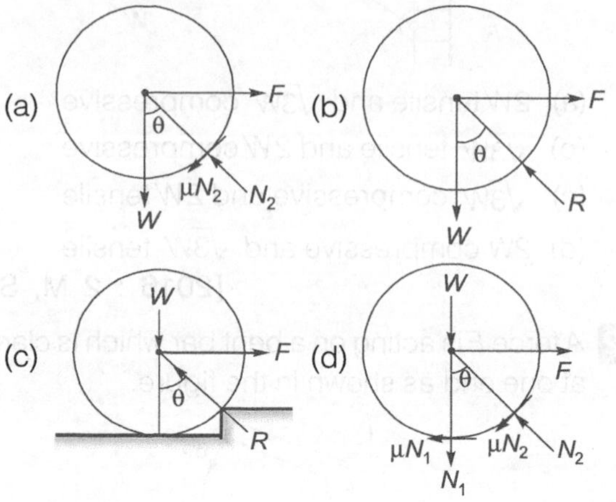

[2020 : 1 M, Set-2]

1.30 Two smooth identical spheres each of radius 125 mm and weight 100 N rest in a horizontal channel having vertical walls. The distance between vertical walls of the channel is 400 mm.

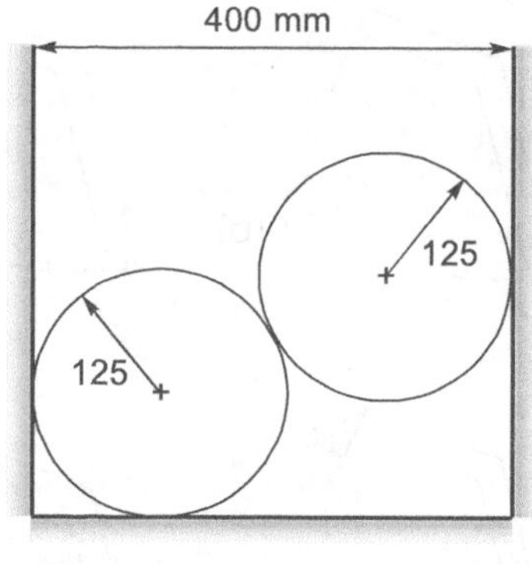

All dimensions are in mm

The reaction at the point of contact between two spheres is ________ N. [Round off to end one decimal place]

[2021 : 2 M, Set-1]

1.31 A plane truss $PQRS$ ($PQ = RS$, and $\angle PQR = 90°$) is shown in the figure.

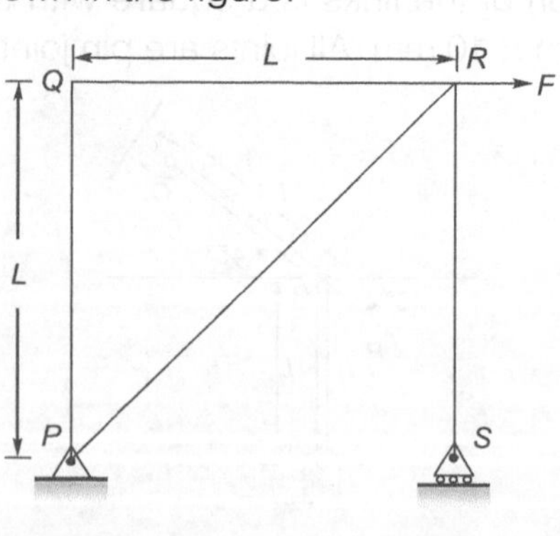

The forces in the members *PR* and *RS*, respectively, are ______.

(a) $F\sqrt{2}$ (tensile) and F(tensile)

(b) $F\sqrt{2}$ (tensile) and F (compressive)

(c) F(compressive) and $F\sqrt{2}$ (compressive)

(d) F(tensile) and $F\sqrt{2}$ (tensile)

[2021 : 1 M, Set-2]

1.32 A square plate is supported in four different ways (configurations (P) to (S) as shown in the figure). A couple moment *C* is applied on the plate. Assume all the members to be rigid and mass-less, and all joints to be frictionless. All support links of the plate are identical.

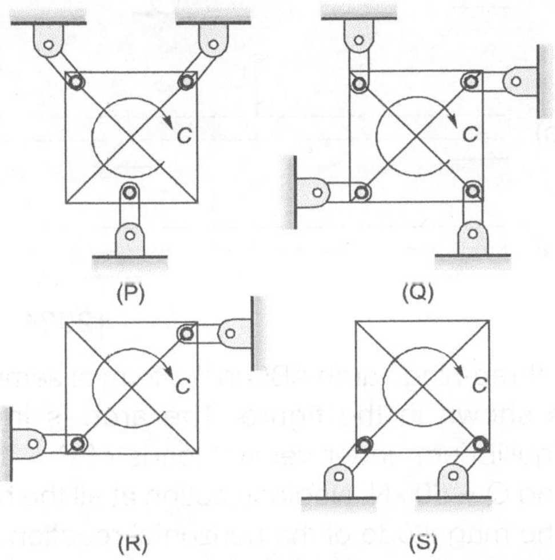

The square plate can remain in equilibrium in its initial state for which one or more of the following support configurations?

(a) Configuration (P) (b) Configuration (Q)

(c) Configuration (R) (d) Configuration (S)

[2022 : 1 M, Set-2]

1.33 A structure, along with the loads applied on it, is shown in the figure. Self-weight of all the members is negligible and all the pin joints are friction-less. *AE* is a single member that contains pin *C*. Likewise, *BE* is a single member that contains pin *D*. Members *GI* and *FH* are overlapping rigid members. The magnitude of the force carried by member *CI* is ________ kN (in integer).

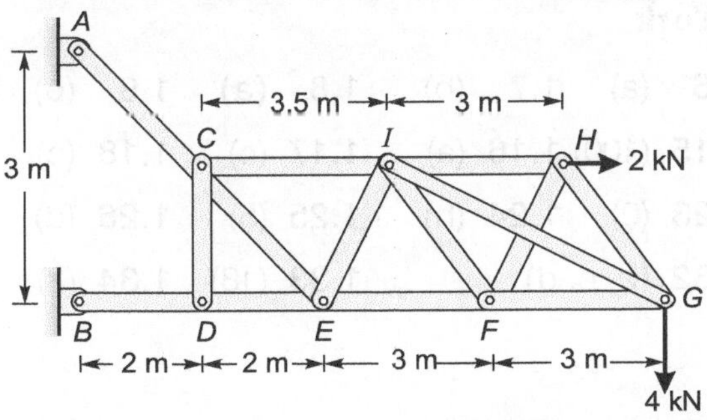

[2022 : 2 M, Set-1]

1.34 A rope with two mass-less platforms at its two ends passes over a fixed pulley as shown in the figure. Discs with narrow slots and having equal weight of 20 N each can be placed on the platforms. The number of discs placed on the left side platform is *n* and that on the right side platform is *m*.

It is found that for *n* = 5 and *m* = 0, a force *F* = 200 N (refer to part (i) of the figure) is just sufficient to initiate upward motion of the left side platform. If the force *F* is removed then the minimum value of *m* (refer to part (ii) of the figure) required to prevent downward motion of the left side platform is______ (in integer).

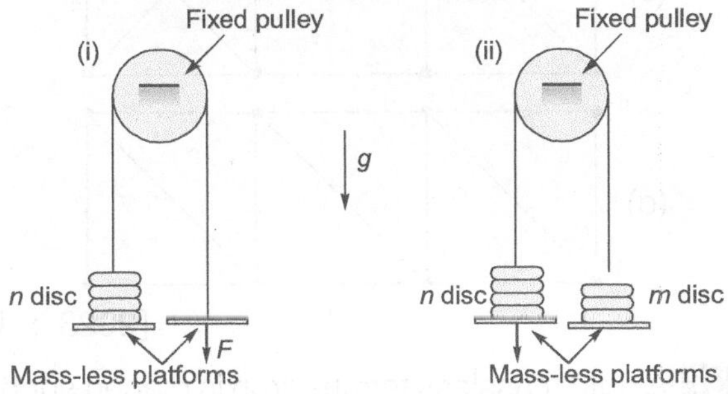

[2022 : 1 M, Set-2]

1.35 The lengths of members BC and CE in the frame shown in the figure are equal. All the members are rigid and lightweight, and the friction at the joints is negligible. Two forces of magnitude $Q > 0$ are applied as shown, each at the mid-length of the respective member on which it acts.

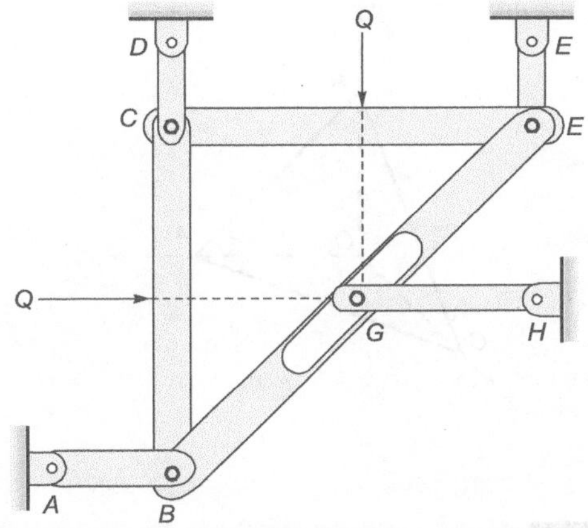

Which one or more of the following members do not carry any load (force)?

(a) AB (b) CD

(c) EF (d) GH

[2022 : 2 M, Set-2]

1.36 The options show frames consisting of rigid bars connected by pin joints. Which one of the frames is non-rigid?

(a)

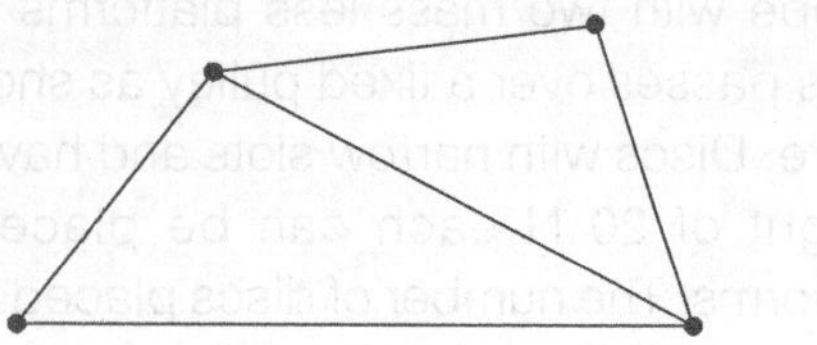

(b)

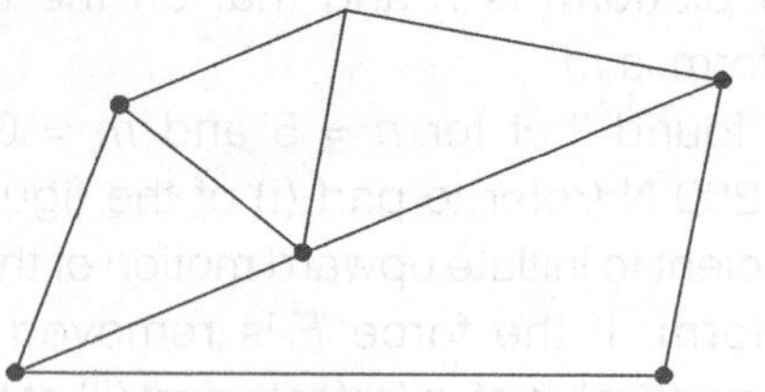

(c)

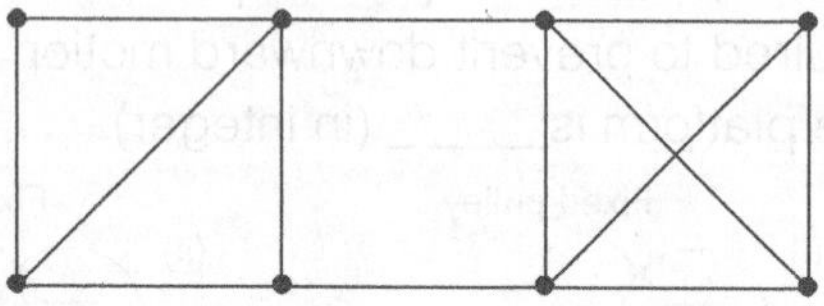

(d)

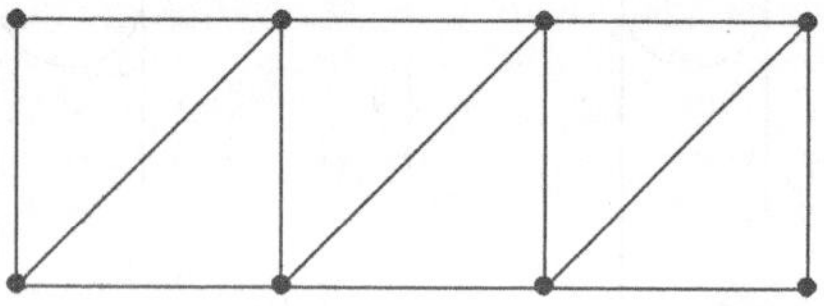

[2023 : 1 M]

1.37 A rigid massless tetrahedron is placed such that vertex *O* is at the origin and the other three vertices *A*, *B*, *C* lie on the coordinate axes as shown in the figure. The body is acted on by three point loads, of which one is acting at *A* along x-axis and another at point *B* along y-axis. For the body to be in equilibrium, the third point load acting at point *O* must be

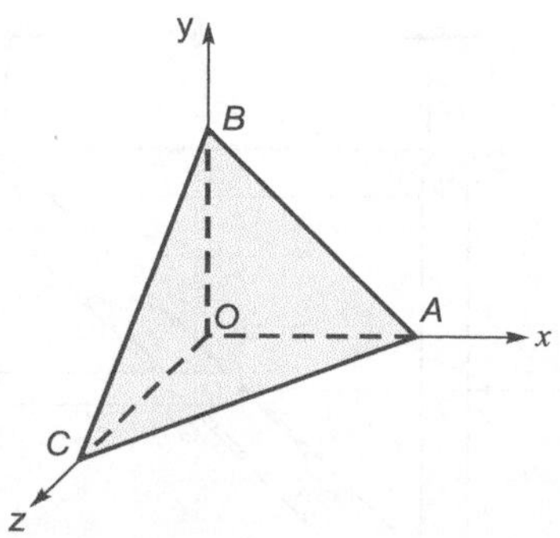

(a) In y-z plane but not along y or z axis
(b) along z-axis
(c) in z-x plane but not along z or x axis
(d) in x-y plane but not along x or y axis

[2024 : 1 M]

1.38 Which of the following beam(s) is/are statically indeterminate?

(a)

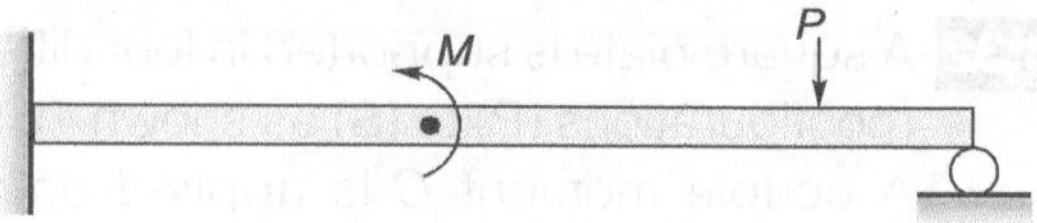

(b)

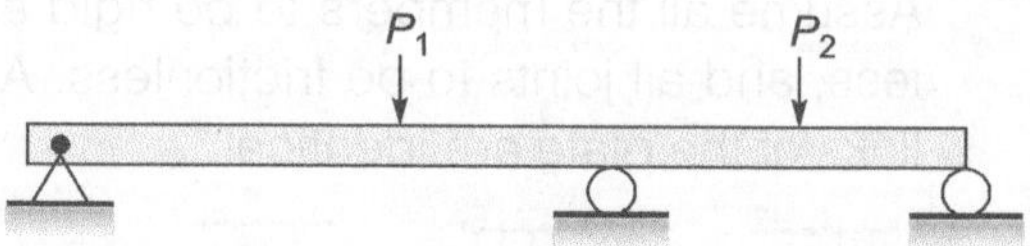

(c)

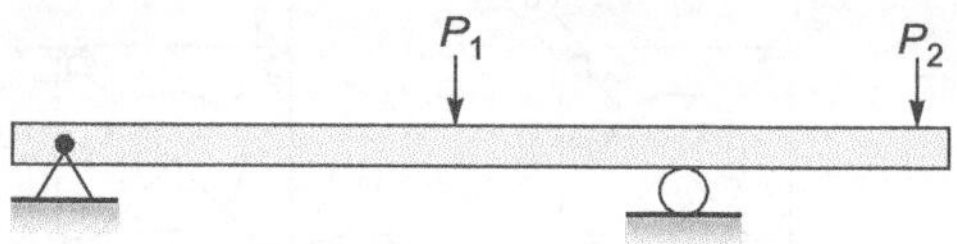

(d)

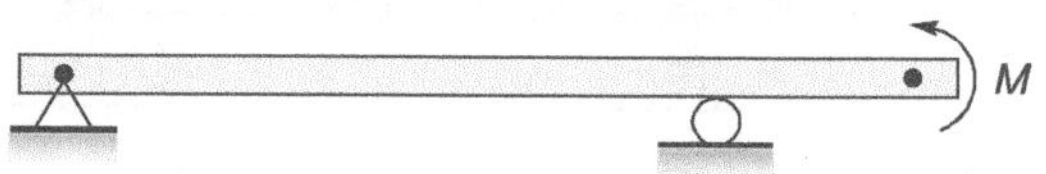

[2024 : 2 M]

1.39 A three-hinge arch ABC in the form of semi-circle is shown in the figure. The arch is in static equilibrium under vertical loads of P = 100 kN and Q = 50 kN. Neglect friction at all the hinges. The magnitude of the horizontal reaction at B is _______ kN. (rounded off to 1 decimal place)

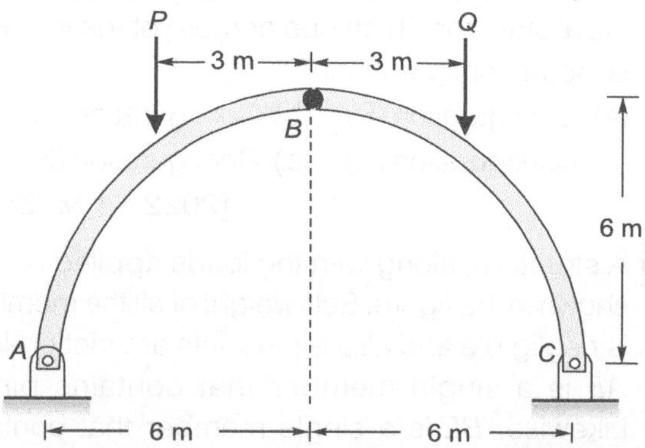

[2024 : 2 M]

■■■■

Answers FBD, Equilibrium, Plane Trusses and Virtual Work

1.1 (c)	1.2 (d)	1.3 (a)	1.4 (a)	1.5 (b)	1.6 (a)	1.7 (b)	1.8 (a)	1.9 (b)
1.10 (b)	1.11 (20)	1.12 (a)	1.13 (400)	1.14 (c)	1.15 (100)	1.16 (a)	1.17 (c)	1.18 (c)
1.19 (57.74)		1.20 (5)	1.21 (b)	1.22 (a)	1.23 (0)	1.24 (b)	1.25 (5)	1.26 (d)
1.27 (0)	1.28 (b)	1.29 (b)	1.30 (1.25)	1.31 (b)	1.32 (b, c, d)		1.33 (18)	1.34 (3)
1.35 (b, d)	1.36 (c)	1.37 (d)	1.38 (a, b)	1.39 (37.5)				

Explanations FBD, Equilibrium, Plane Trusses and Virtual Work

1.1 (c)

The free body diagram of mass m,

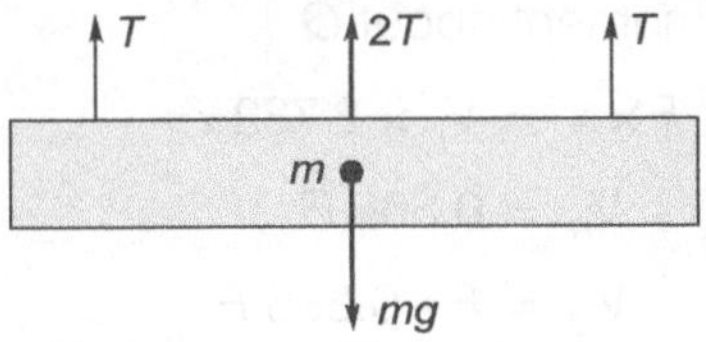

In equilibrium condition,

$$4T = mg$$

or $$m = \frac{4T}{g}$$

1.2 (d)

Since point A is hinge support, so there will be horizontal and vertical reactions at point A.

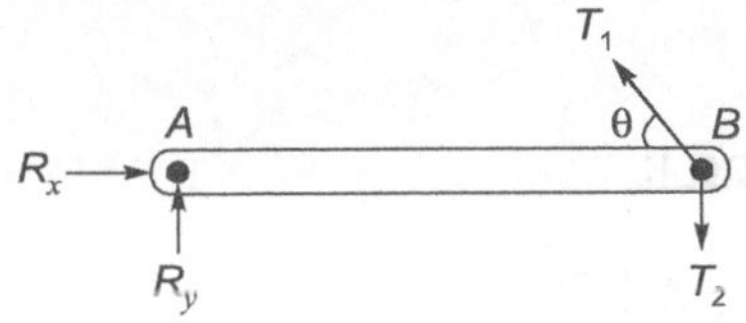

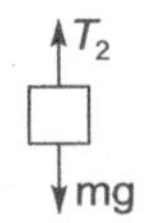

For block $T_2 = mg = 343.35 \text{ N}$ ...(i)

$$\tan\theta = \frac{125}{275} = 0.4545$$

$$\theta = \tan^{-1}(0.4545) = 24.44°$$

For bar $\Sigma M_A = 0$

$$T_1 \sin\theta \times l = T_2 \times l$$

$$\Rightarrow T_1 = \frac{T_2}{\sin\theta} = \frac{343.35}{\sin 24.44°} = 829.74 \text{ N}$$

$$\Sigma F_y = 0$$

$$+R_y - T_2 + T_1 \sin\theta = 0$$

$$\therefore \quad R_y = 0$$

$$\Sigma F_x = 0$$

$$R_x - T_1 \cos\theta = 0$$

$$R_x = \mathbf{755.39 \text{ N}}$$

1.3 (a)

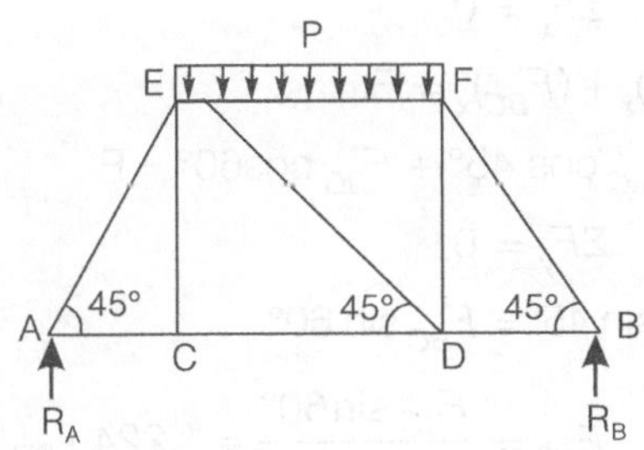

$$R_A + R_B = Pl$$

Taking moment about A

$$Pl(l + l/2) = R_B \times 3l$$

$$\therefore \quad R_B = \frac{Pl}{2}$$

$$\therefore \quad R_A = \frac{Pl}{2}$$

Joint A

Now at joint A

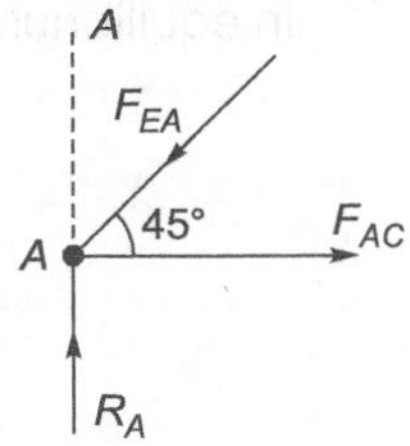

$$\Sigma F_V = 0$$

$$F_{EA} \sin 45° = \frac{Pl}{2}$$

$$\therefore \quad F_{EA} = \frac{Pl}{\sqrt{2}}$$

$$F_{AC} = F_{EA} \cos 45° = \frac{Pl}{\sqrt{2}} \times \frac{1}{\sqrt{2}}$$

$$F_{AC} = \frac{Pl}{2}$$

At joint C

$$F_{CA} = F_{CD} = \frac{Pl}{2} \qquad [\because F_{CE} = 0]$$

1.4 (a)

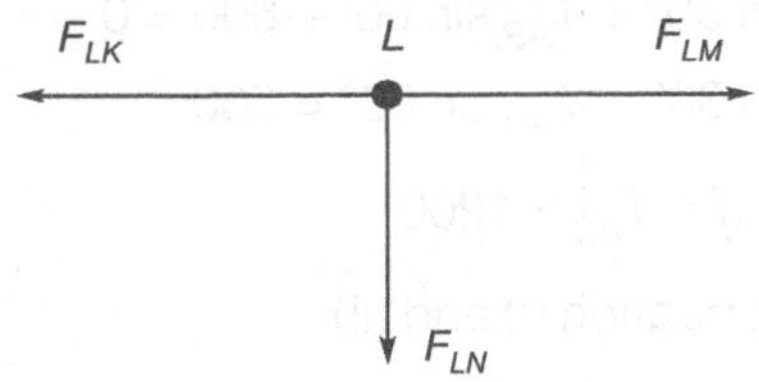

$\Sigma F_H = 0$ & $\Sigma F_V = 0$

At joint "L"

$$\therefore \quad F_{LK} - F_{LM} = 0 \quad (\Sigma F_H = 0)$$

$$F_{LN} = 0 \quad (\Sigma F_V = 0)$$

Hence no force is acting on the truss number LN.

1.5 (b)

In equilibrium, potential energy is minimum.

If any system is in equilibrium and subjected to many independent variables, partial derivatives of its total potential energy with respect to each of the independent variable must be zero.

1.6 (a)

Method I:

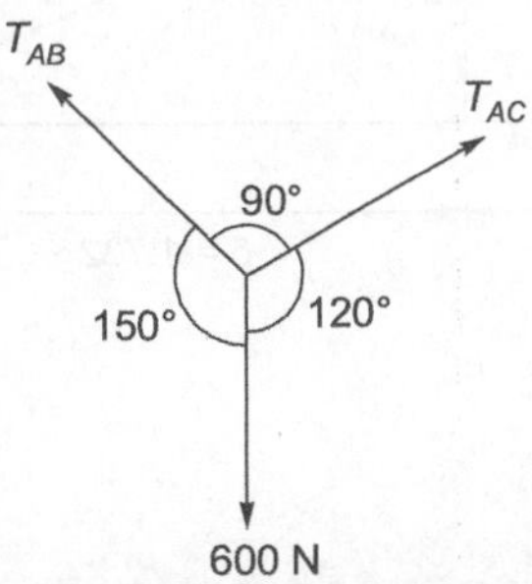

By Lami's theorem

$$\frac{T_{AB}}{\sin 120^\circ} = \frac{T_{AC}}{\sin 150^\circ} = \frac{600}{\sin 90^\circ}$$

$$\therefore \quad T_{AB} = 600 \sin 120^\circ = 519.61 \approx 520 \text{ N}$$

and $\quad T_{AC} = 600 \sin 150^\circ = \mathbf{300\ N}$

<u>Method II:</u>

In equilibrium,

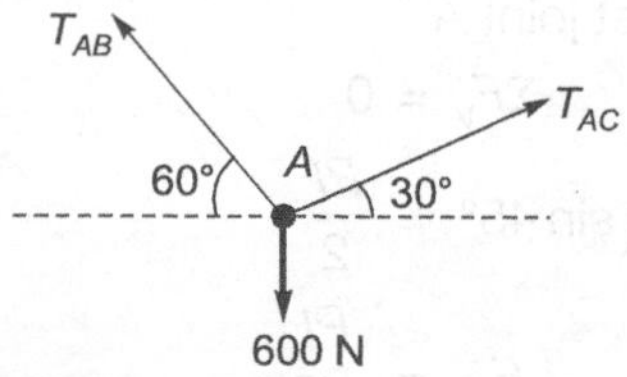

Horizontal forces,

$$\Sigma F_x = 0,$$

$$T_{AC} \cos 30^\circ - T_{AB} \cos 60^\circ = 0$$

$$T_{AC} = T_{AB} \frac{\cos 60^\circ}{\cos 30^\circ} = \frac{T_{AB}}{\sqrt{3}} \quad \text{...(i)}$$

Vertical forces,

$$\Sigma F_y = 0,$$

$$T_{AC} \sin 30^\circ + T_{AB} \sin 60^\circ - 600 = 0$$

$$T_{AC} \sin 30^\circ + T_{AB} \sin 60^\circ = 600$$

$$T_{AC} + \sqrt{3}\ T_{AB} = 1200 \quad \text{...(ii)}$$

From equation (i) and (ii)

$$\frac{T_{AB}}{\sqrt{3}} + \sqrt{3} T_{AB} = 1200$$

$$4T_{AB} = 1200\sqrt{3}$$

$$T_{AB} = 300\sqrt{3} = 519.61 \text{ N} = 520 \text{ N}$$

and $\quad T_{AC} = \dfrac{T_{AB}}{\sqrt{3}} = \dfrac{300\sqrt{3}}{\sqrt{3}} = 300 \text{ N}$

1.7 (b)

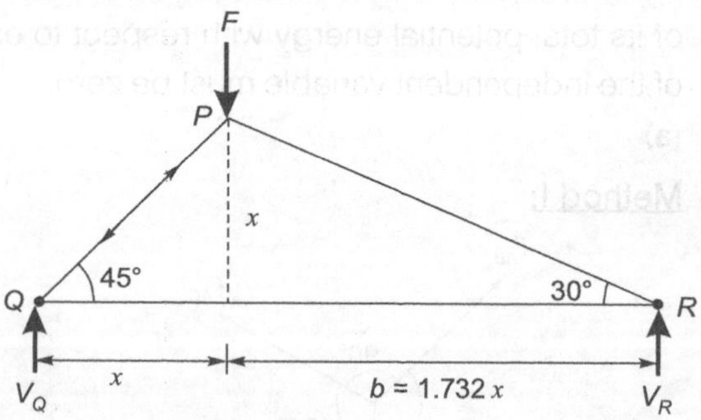

$$\tan 30^\circ = \frac{x}{b}$$

$$b = \frac{x}{\tan 30^\circ} = 1.732\,x$$

Taking moment about Q

$$F \times x = V_R \times 2.732\,x$$

$$V_R = 0.366\,F$$

$$V_Q = F - 0.366\,F$$

$$= 0.634\,F$$

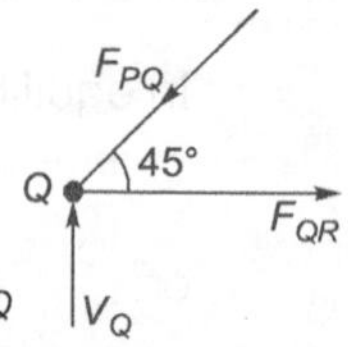

FBD of joint Q

Let force in the member PQ is F_{PQ}

$$\therefore F_{PQ} \sin 45^\circ = V_Q$$

$$\Rightarrow F_{PQ} \sin 45^\circ = 0.634 \text{ F}$$

Force in member QR

$$F_{QR} = F_{PQ} \cos 45^\circ = \mathbf{0.634F}$$

1.8 (a)

<u>Method I:</u>

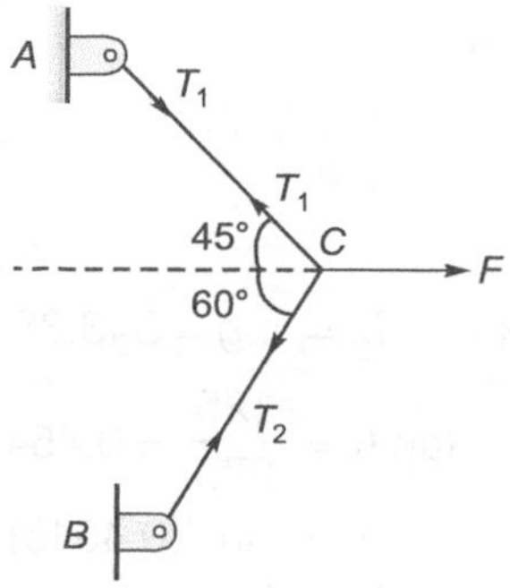

Using Lami's theorem

$$\frac{T_1}{\sin 120^\circ} = \frac{T_2}{\sin 135^\circ} = \frac{F}{\sin 105^\circ}$$

$$T_1 = 0.8965F$$

$$T_2 = 0.732F$$

Vertical reaction at B,

$$R_B = T_2 \cos 30^\circ = 0.732 \cos 30^\circ$$

$$R_B = \mathbf{0.634\ kN}$$

Method II:

$$\Sigma F_x = 0,$$

$$(F_{AC})_x + (F_{BC})_x = F \quad \text{...(i)}$$

$$\Rightarrow F_{AC} \cos 45^\circ + F_{BC} \cos 60^\circ = \text{F}$$

$$\Sigma F_y = 0,$$

$$F_{AC} \sin 45^\circ = F_{BC} \sin 60^\circ$$

$$F_{AC} = \frac{F_{BC} \sin 60^\circ}{\sin 45^\circ} = 1.224\ F_{BC}$$

$$\Rightarrow \quad 1 = 0.865\,F_{BC} + 0.5\,F_{BC}$$

$$\therefore \quad F_{BC} = \frac{1}{1.365} = 0.732 \text{ kN}$$

Vertical force at B, $(R_B)_V = F_{BC}\sin 60°$

$$= 0.732 \sin 60° = 0.634 \text{ kN}$$

1.9 (b)

Method I:

Maximum force = $0.8965F$

$$\therefore \text{ Max stress } \frac{0.8965F}{100} \le 100 \text{ MPa}$$

$$\therefore\ 100 \ge \frac{0.8965 \times F}{100}$$

$\therefore\ F \le$ **11.154 kN**

Method II:

As, $F_{AC} = 0.895$ kN

then

$$\frac{F \times F_{AC}}{A} = 100$$

$$\frac{F \times 0.895}{100} = 100$$

$$F = 11173.18 \text{ N} = 11.173 \text{ kN}$$

1.10 (b)

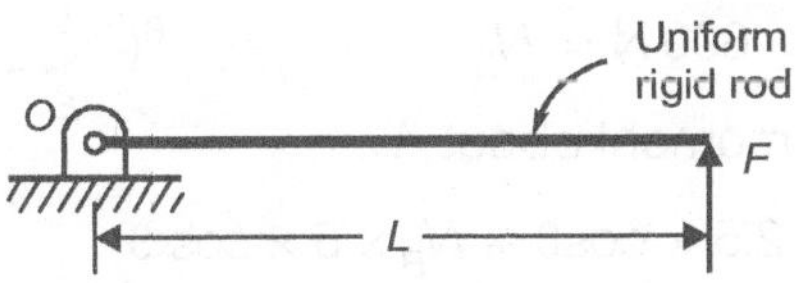

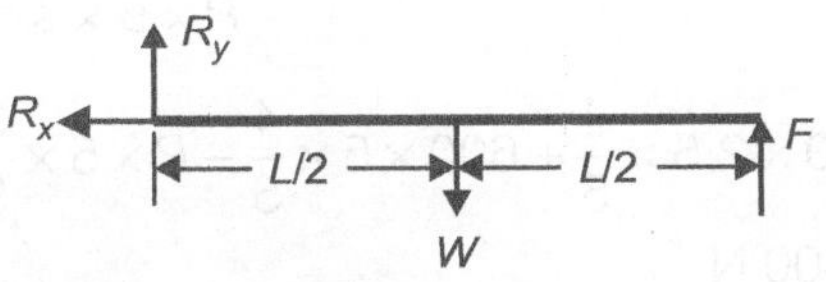

$$\Sigma R_x = 0 \Rightarrow R_x = 0$$

$$\Sigma R_y = 0 \Rightarrow R_y + F = W$$

$$R_y = (W - F) \qquad \text{...(i)}$$

If force F removes, then rod will rotate about O, then

$$T_{(\text{torque})} = I\alpha$$

$$W \times \frac{L}{2} = \left(\frac{W}{g}\right)\frac{L^3}{3} \times \alpha \quad \text{...(ii)}$$

$$W - R_y = \left(\frac{W}{g}\right)a_{cg} \qquad \text{...(iii)}$$

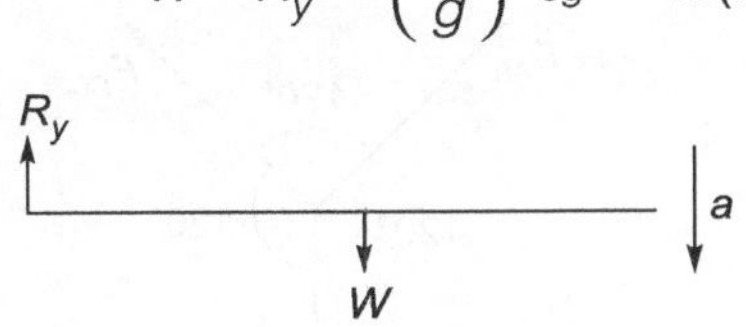

$$a_{cg} = \left(\frac{L}{2}\right)\alpha \qquad \text{...(iv)}$$

Equations (ii), (iii), (iv)

$$W \times \frac{L}{2} = \left(\frac{W}{g}\right)\frac{L^2}{3} \times \frac{a_{c.m}}{\left(\frac{L}{2}\right)}$$

$$W \times \frac{L}{2} = \left(\frac{W}{g}\right) \times \frac{L^2}{3} \times \left(\frac{2}{L}\right) \times \frac{(W - R_y)}{\left(\frac{W}{g}\right)}$$

$$\frac{3W}{4} = W - R_y$$

$$R_y = \frac{W}{4}$$

1.11 Sol.

Method I:

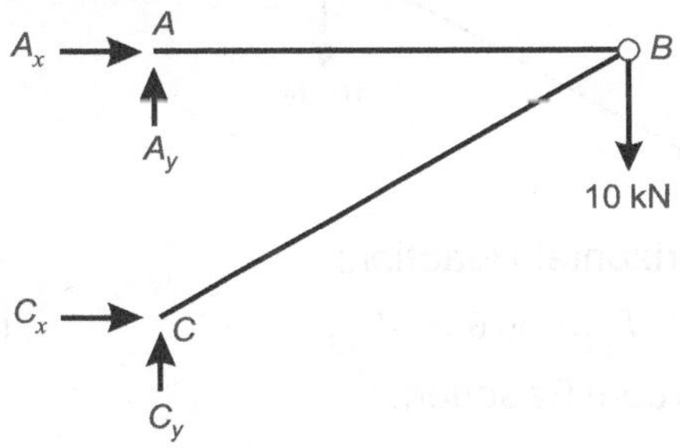

$$AB = 1 \text{ m}$$

$$AC = 0.5 \text{ m}$$

$$BC = \sqrt{1^2 + 0.5^2} = \sqrt{1.25} = 1.118 \text{ m}$$

$$A_x + C_x = 0$$

$$A_y + C_y = 10$$

(from force equilibrium)

$$\Sigma M_A = 0$$

$$C_x \times 0.5 = 10 \times 1$$

or $\quad C_x = 20$ kN

and $\quad A_x = -20$ kN

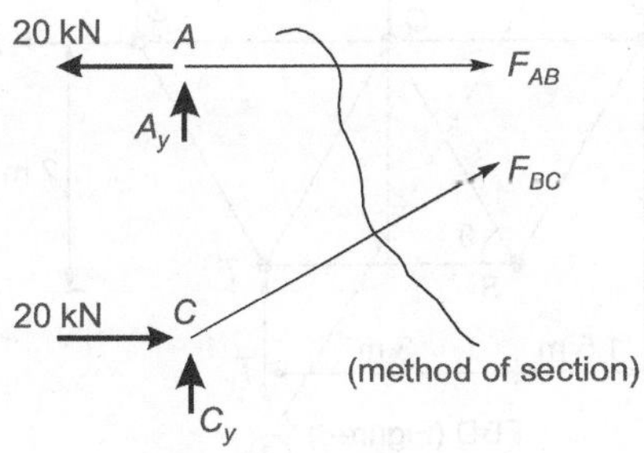

$$\Sigma M_c = 0$$

$$\Rightarrow \quad F_{AB} \times 0.5 = 20 \times 0.5$$

$$\therefore \quad F_{AB} = \mathbf{20\ kN}$$

<u>Method II:</u>

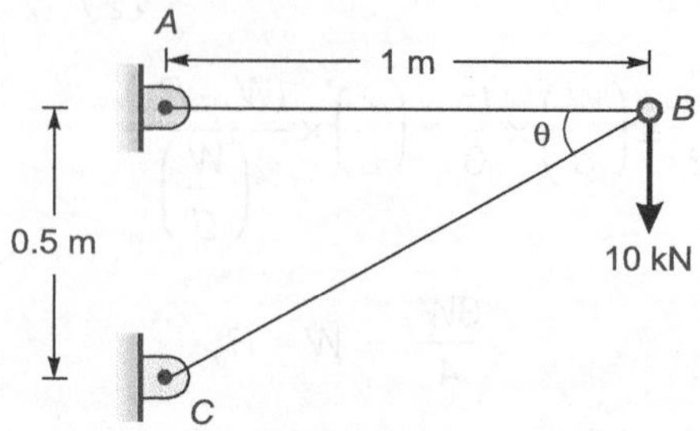

Free body diagram of point B,

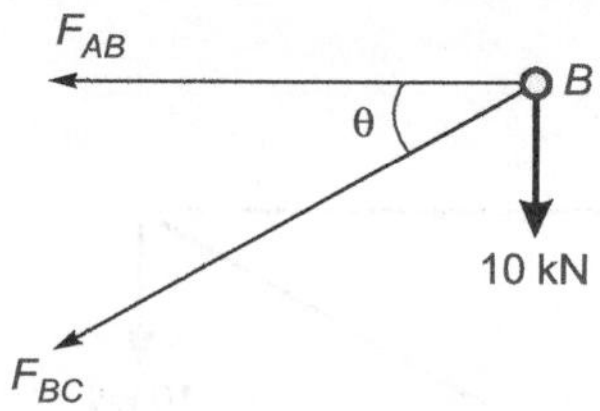

Horizontal Reaction:

$$F_{BC} \cos\theta = -F_{AB} \quad \text{...(i)}$$

Vertical Reaction:

$$F_{BC} \sin\theta = -10 \quad \text{...(ii)}$$

Putting the value of F_{BC} from equation (ii) to equation (i)

$$\frac{-10}{\sin\theta} \times \cos\theta = -F_{AB}$$

$$\therefore \quad F_{AB} = 10 \times \cot\theta = 10 \times \frac{1}{0.5} = 20\text{ kN}$$

1.12 (a)

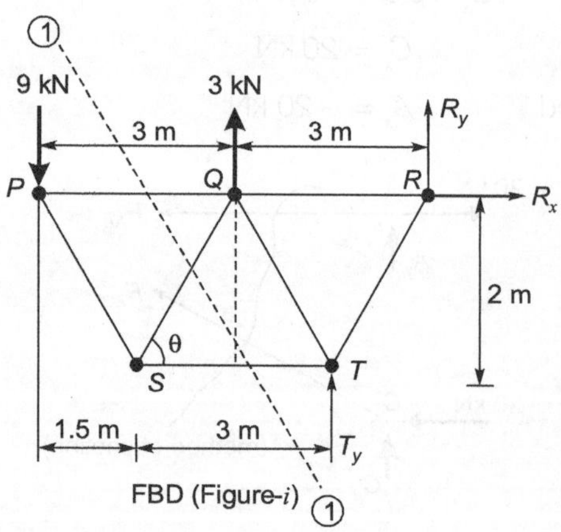

FBD (Figure-*i*)

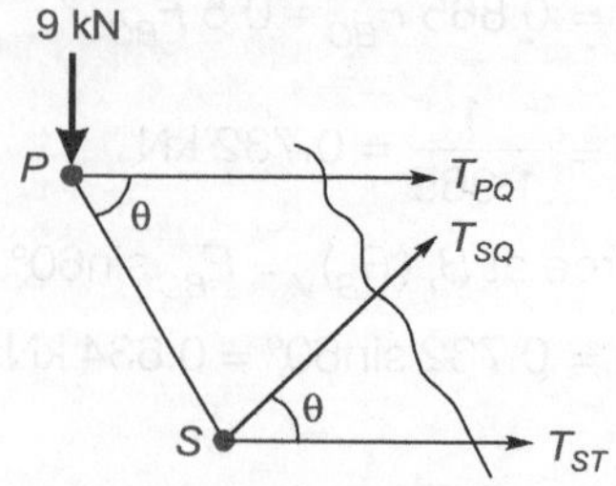

Section through *PQ*,*QS*, & ST (Figure-*ii*)

$$\tan\theta = \frac{2}{1.5}$$

$$\theta = \mathbf{53.13°}$$

Considering L.H.S. of section (1)......(1)

$$\Sigma F_v = 0$$

$$-9 + T_{SQ} \times \sin\theta = 0$$

$$T_{SQ} = \frac{9}{\sin 53.13°} = 11.25\text{ kN (Tensile)}$$

1.13 Sol.

Drawing FBD of ladder AB

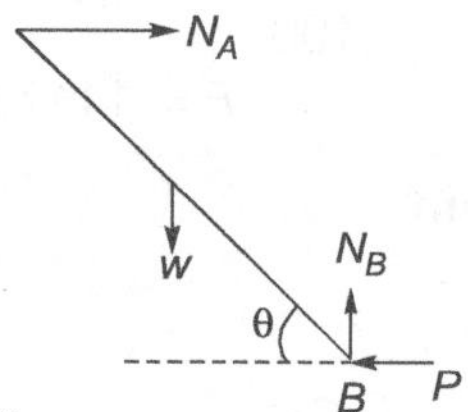

$$\tan\theta = \frac{3}{4}$$

$$(\Sigma F)_y = 0$$

$$\Rightarrow \quad 600\text{ N} = N_B$$

Taking moment about A

$$-600 \times 2.5 \times \cos\theta + N_B \times 5 \times \cos\theta - P \times 5 \times \sin\theta = 0$$

$$\Rightarrow -600 \times 2.5 \times \frac{4}{5} + 600 \times 5 \times \frac{4}{5} - P \times 5 \times \frac{3}{5} = 0$$

$$\Rightarrow P = 400\text{ N}$$

1.14 (c)

For a joint we can write 2 equilibrium equations, so for *j* joints we will have 2 *j* equations. If there are *m* members in a truss then we need to calculate *m* unknowns. Also we need to calculate 3 reactions for a plane truss. So for determinate truss $m + 3 = 2j$.

1.15 Sol.

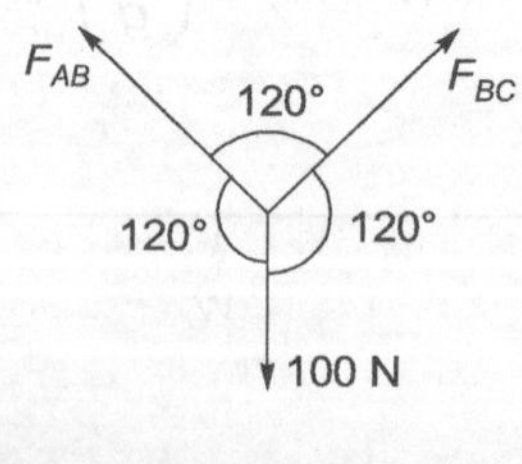

$$\frac{F_{AB}}{\sin 120°} = \frac{100}{\sin 120°} \text{ or } F_{AB} = \mathbf{100\ N}$$

1.16 (a)

Using Lami's theorem,

$$\frac{T_1}{\sin 120°} = \frac{500}{\sin 90°}$$

$$\therefore \quad T_1 = 500 \sin 120° = 433 \text{ N}$$

$$\frac{T_2}{\sin(360° - (120° + 90°))} = \frac{500}{\sin 90°}$$

$$\Rightarrow \quad T_2 = \mathbf{250\ N}$$

1.17 (c)

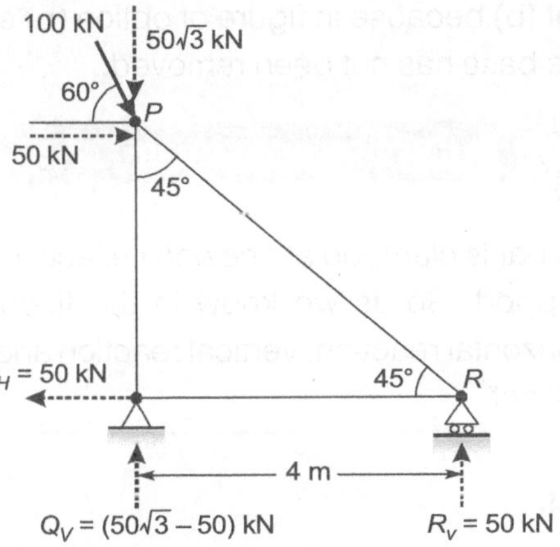

$$\Sigma M_Q = 0$$

$$R_v \times 4 - 50 \times 4 = 0$$

$$R_v = 50 \text{ kN}$$

Force in member QR = 50 kN (Tensile)

Free body diagram of joint R

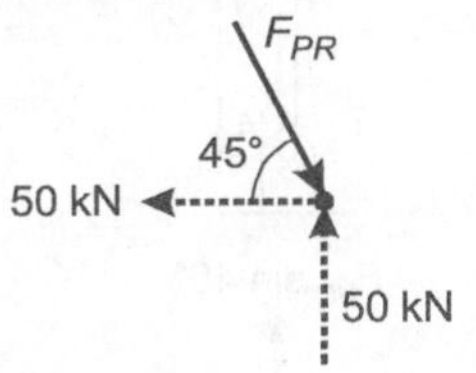

$$F_{PR} \cos 45° = 50$$

$$F_{PR} = \frac{50}{\cos 45°} = \mathbf{70.71\ kN}$$

1.18 (c)

$$R_P + R_Q = 30 \text{ kN}$$

$$\Sigma M_P = 0$$

$$30 \times 2 = R_Q \times 3$$

$$R_Q = 20 \text{ N}$$

$$R_P = 10 \text{ N}$$

Using method of section

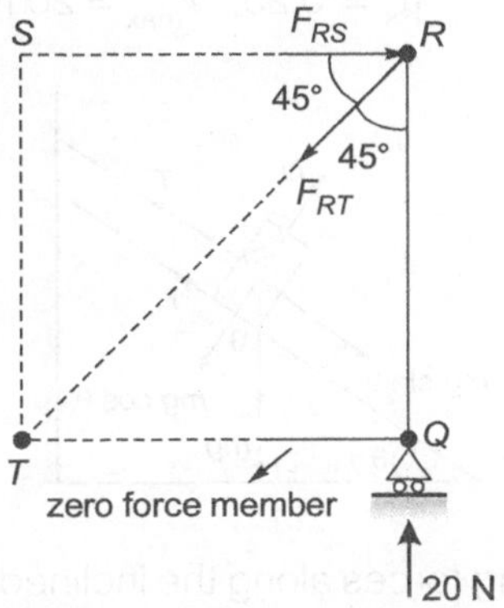

Considering moment about point T, which is zero

$$R_Q \times 1 = F_{RS} \times 1$$

$$F_{RS} = R_Q = \mathbf{20\ N\ compressive}$$

1.19 Sol.

Method I:

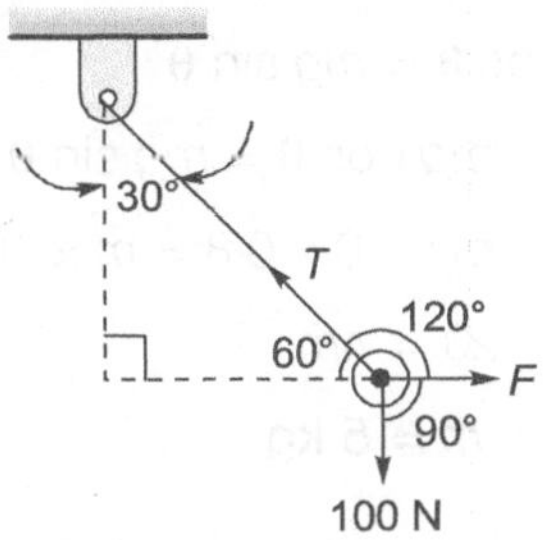

Applying Lami's theorem

$$\frac{100}{\sin 120°} = \frac{F}{\sin(60° + 90°)} = \frac{T}{\sin 90°}$$

$$F = \frac{100 \sin 150°}{\sin 120°} = \mathbf{57.74\ N}$$

Method II:

This problem can be solved without applying Lami's theorem,

As,

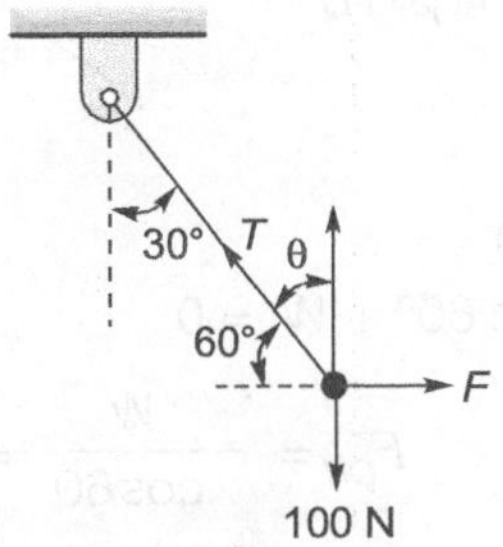

$$\therefore \quad \theta = 30°, \text{ then}$$

Vertical reactions,

$$T \cos 30° = 100 \text{ N}$$

$$T = \frac{100 \times 2}{\sqrt{3}} \text{ N} = \frac{200}{\sqrt{3}}$$

Horizontal reactions,

$$F = T \cos 60° = \frac{200}{\sqrt{3}} \times \frac{1}{2} = \frac{100}{\sqrt{3}} = 57.74 \text{ N}$$

1.20 Sol.

$$\mu_s = 0.25, \; T_{max} = 20 \text{ N}$$

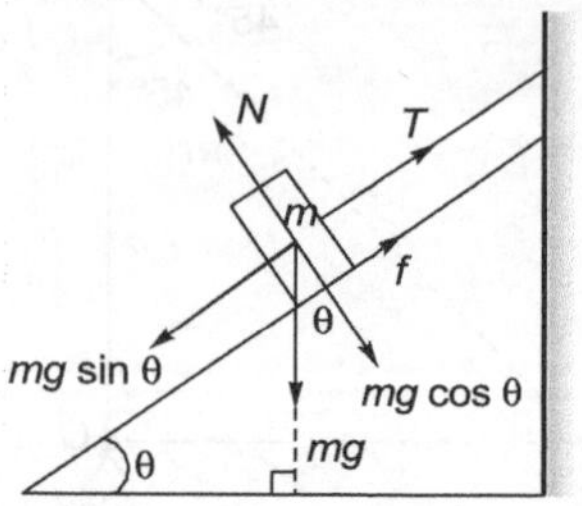

Balancing forces along the inclined plane

$$\Rightarrow \quad T + f = mg \sin\theta$$

$$f = \mu N$$

$$N = mg \cos\theta$$

(balancing forces perpendicular to the inclined plane)

$$T + \mu mg \cos\theta = mg \sin\theta$$

$$20 + 0.25 \times mg \cos\theta = mg \sin\theta$$

$$20 + 0.25 \times m \times 10 \times 0.8 = m \times 10 \times 0.6$$

$$20 = 4m$$

$$\therefore \quad m = 5 \text{ kg}$$

1.21 (b)

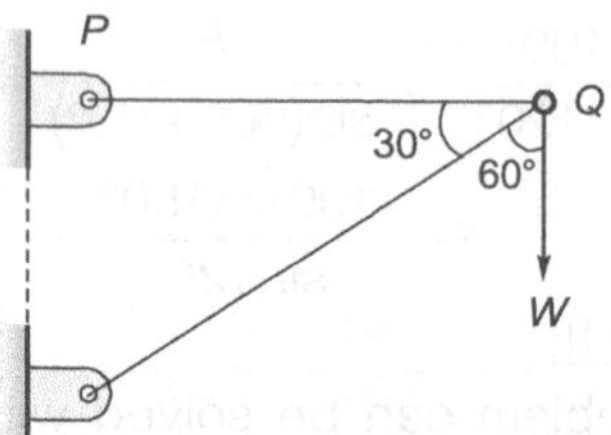

At joint Q

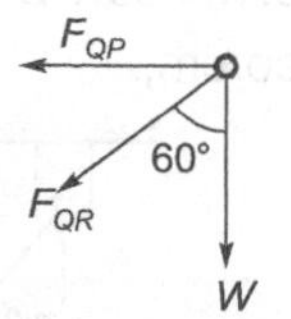

$$\Sigma F_V = 0$$

$$F_{QR} \cos 60° + W = 0$$

$$F_{QR} = -\frac{W}{\cos 60} = -2W$$

$$= 2W \text{(compressive)}$$

$$\Sigma F_H = 0$$

$$\therefore F_{QP} + F_{QR} \cos 30° = 0$$

$$F_{QP} = -F_{QR} \cos 30°$$

$$= -(-2W) \cos 30°$$

$$F_{QP} = +2W \times \frac{\sqrt{3}}{2} = +W\sqrt{3} \quad \text{(tensile)}$$

1.22 (a)

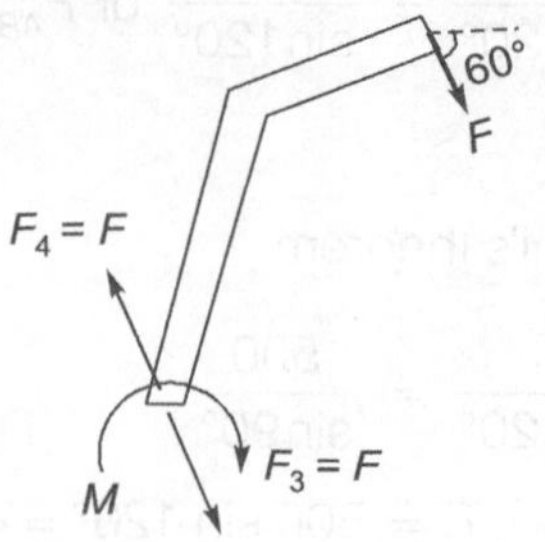

Applying equal, opposite and parallel forces at the base, F_3 and F_4.

F_4 and F will form a couple at the base whose magnitude will be M and F_3 will be force which will have a horizontal and vertical components, which are displayed by R_H and R_V in the figure of option (a). Hence correct answer will be (a) and not (b) because in figure of option (b) support at the base has not been removed.

POINTS TO REMEMBER

As bar is clamped at one end i.e. acting like fixed support. So as we know in the fixed support horizontal reaction, vertical reaction and moment will act.

1.23 Sol.

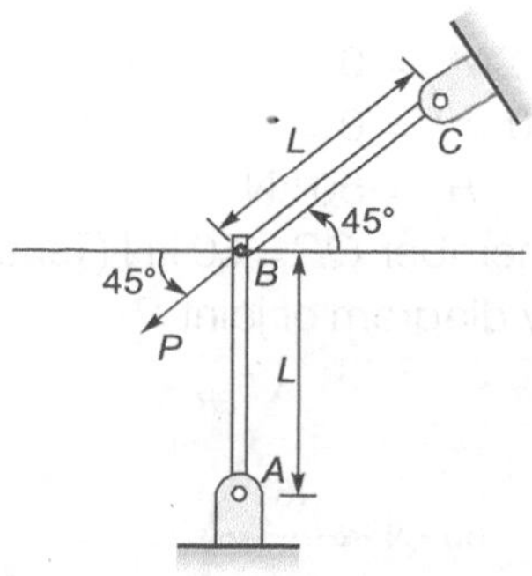

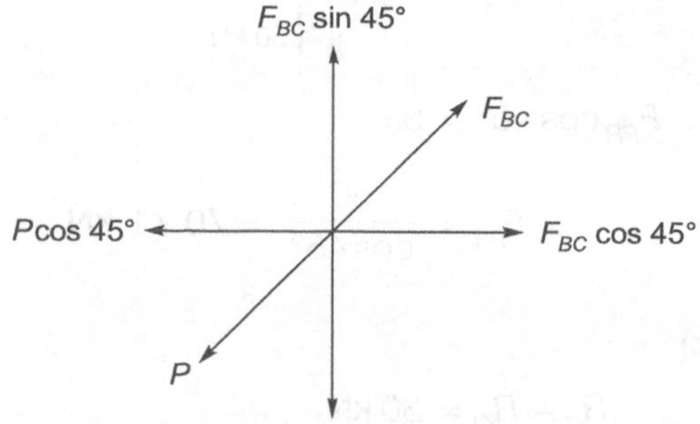

At equilibrium,

$$\Sigma F_V = 0 \text{ and } \Sigma F_H = 0$$

$$F_{BC} \sin 45° = F_{AB} + P \sin 45° \quad \text{...(i)}$$

$$F_{BC} \cos 45° = P \cos 45° \quad \text{...(ii)}$$

$$\therefore \quad F_{BC} = P$$

$$P \sin 45° = F_{AB} + P \sin 45°$$

$$\therefore \quad F_{AB} = 0$$

$$\sigma_{AB} = 0$$

1.24 (b)

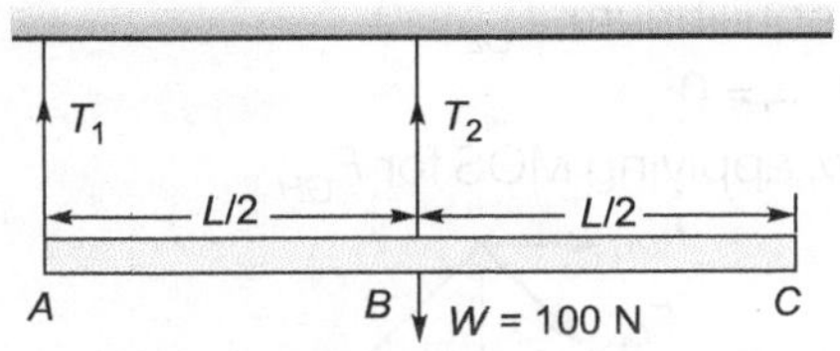

Considering equilibrium of bar,

$$\Sigma \vec{F} = 0$$

$$T_1 + T_2 = 100 \text{ N} \quad \text{...(i)}$$

and $\quad \Sigma M_A = 0$

$$T_2 \cdot \frac{L}{2} = 100 \times \frac{L}{2}$$

$$\therefore \quad T_2 = 100 \text{ N}$$

$$T_1 = 0 \text{ N}$$

1.25 (5)

By symmetry, force in member AB and CD will be same

By symmetry, $\quad R_A = R_C = 5$ kN

At joint D,

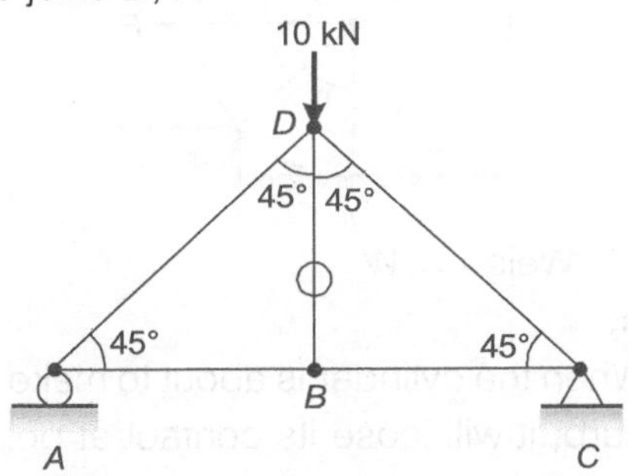

Joint D,

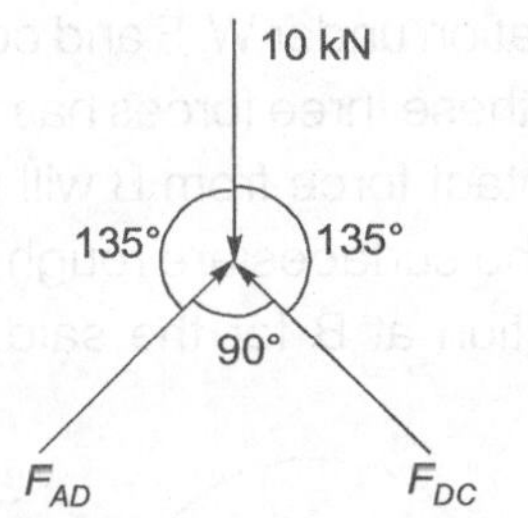

$$\frac{F_{AD}}{\sin 135°} = \frac{10}{1} = \frac{F_{DC}}{\sin 135°}$$

$$F_{DC} = 10 \sin 135° = 5\sqrt{2} \text{ kN (Compressive)}$$

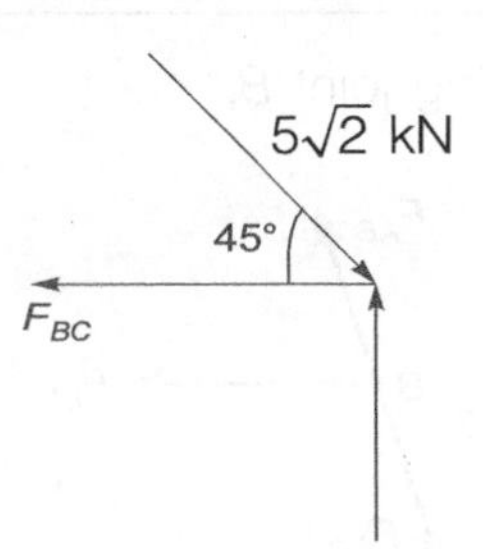

Joint C,

$$5\sqrt{2} \cos 45° = F_{BC}$$

$$F_{BC} = 5 \text{ kN (Tensile)}$$

1.26 (d)

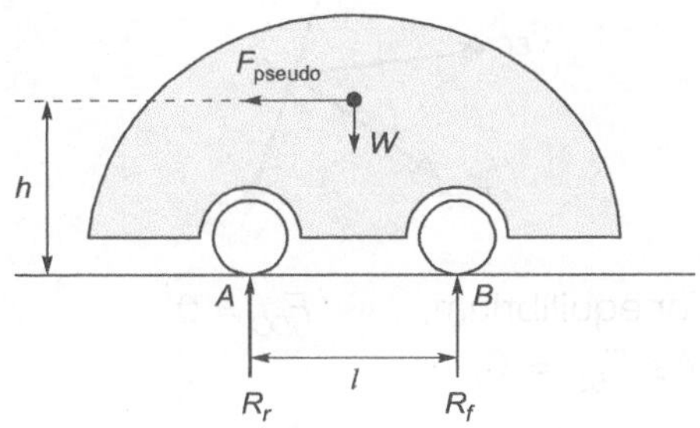

Pseudo force, $F_{\text{pseudo}} = ma$ in opposite direction of motion of car.

$$\Sigma F_V = 0$$

$$R_r + R_f = W \quad \text{...(1)}$$

Taking moment about A

$$(F_{\text{pseudo}})h + (R_f)l = \frac{Wl}{2}$$

$$m_a h + R_f l = \frac{Wl}{2}$$

$$R_f = \frac{W}{2} - \frac{W}{g}\left(\frac{h}{l}\right)a \quad \text{...(2)}$$

From (1) and (2)

$$R_r = \frac{W}{2} + \frac{W}{g}\left(\frac{h}{l}\right)a$$

1.27 (0)

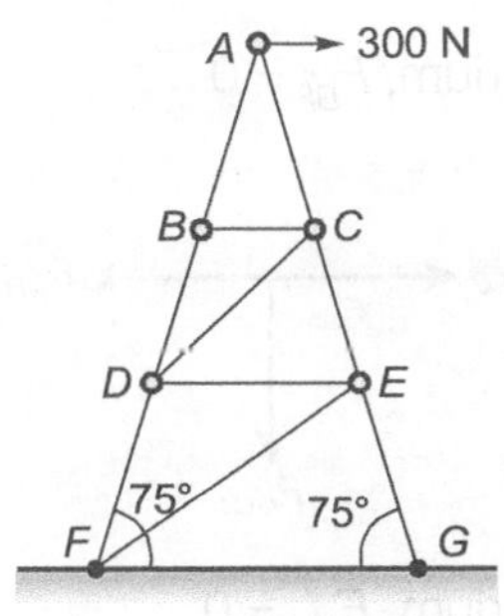

Considering joint B,

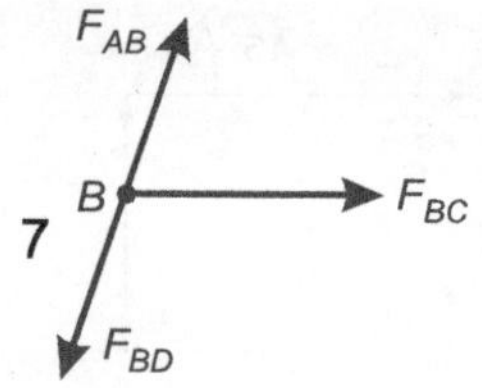

$\Rightarrow F_{BC} = 0$ (for equilibrium)

Now considering joint C,

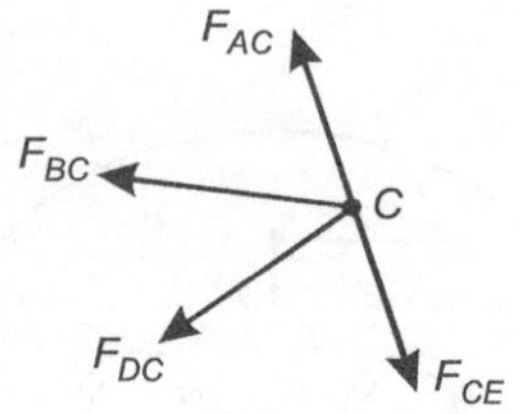

For equilibrium, $F_{DC} = 0$
(As $F_{BC} = 0$)

1.28 (b)

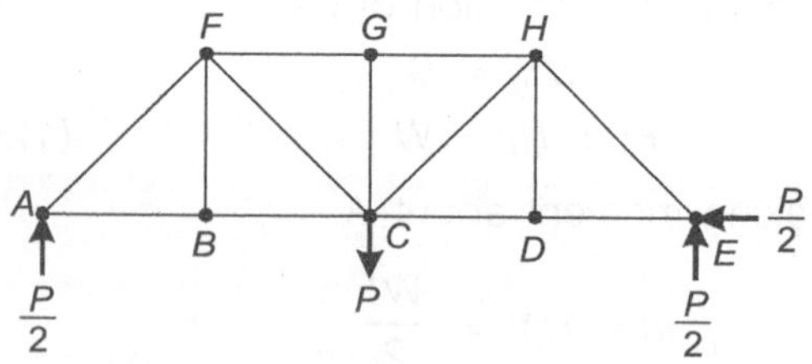

For equilibrium, vertical component of reaction at E is equal to $\frac{P}{2}$. Since the angle make by reaction at E is 45°, so both R_{HE} and R_{VE} are equal.

At joint B,

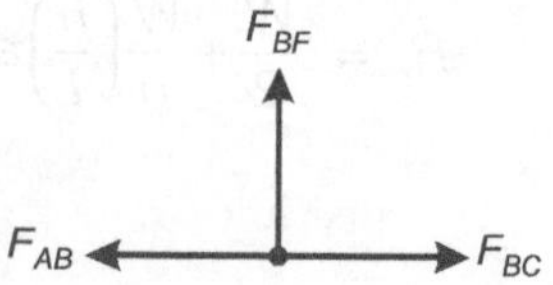

For equilibrium, $F_{BF} = 0$

At joint G

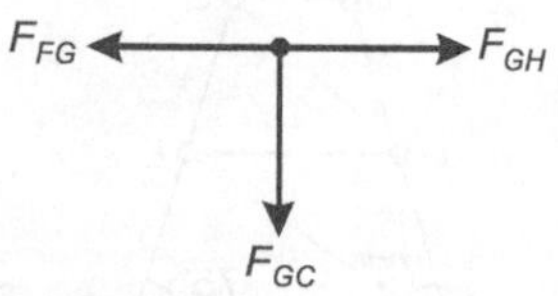

For equilibrium, $F_{GC} = 0$

Similarly, $F_{DH} = 0$

At joint E,

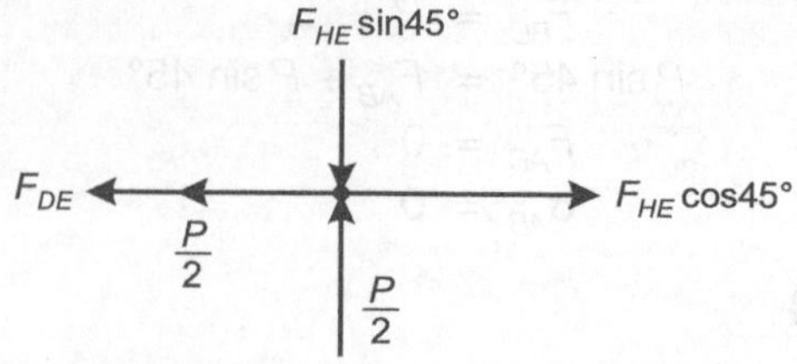

For equilibrium, $F_{DE} = 0$

$\therefore\ F_{CD} = 0$

Now, applying MOS for F_{GH}

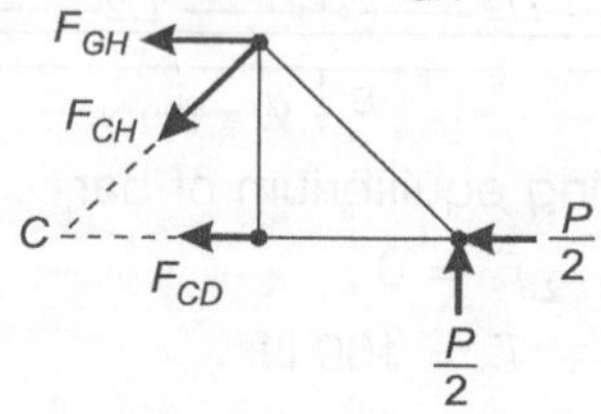

Taking moment about C,

$$F_{GH} \times 1 + \frac{P}{2} \times 2 = 0$$

$$F_{GH} = -P$$

$$\therefore \quad F_{GH} \neq 0$$

$$F_{GF} \neq 0$$

1.29 (b)

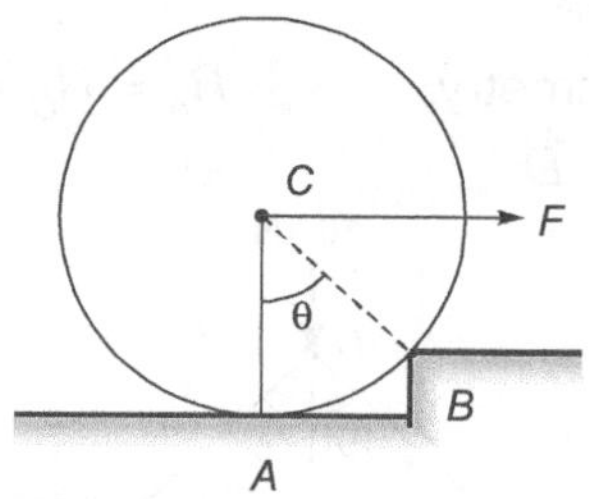

Weigh = W

Note:

(i) When the cylinder is about to make out of the curb, it will loose its contact at point A, only contact will be at it B.

(ii) At verge of moving out of curb, Roller will be in equation under W, F and contact force from B and these three forces has to be concurrent so contact force from B will pass through C.

(iii) Even the surfaces are rough but there will be no friction at B for the said condition.

FBD

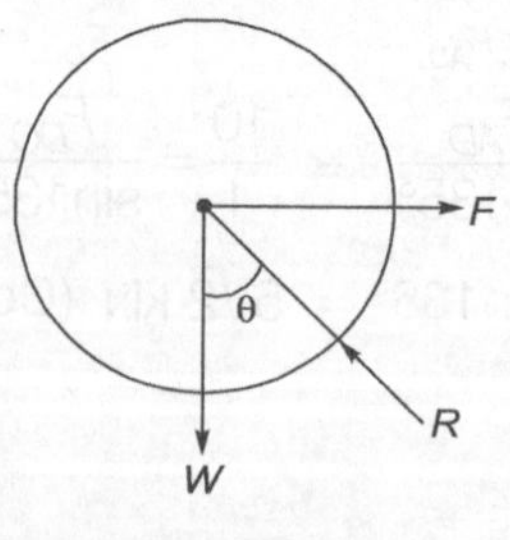

1.30 Sol.

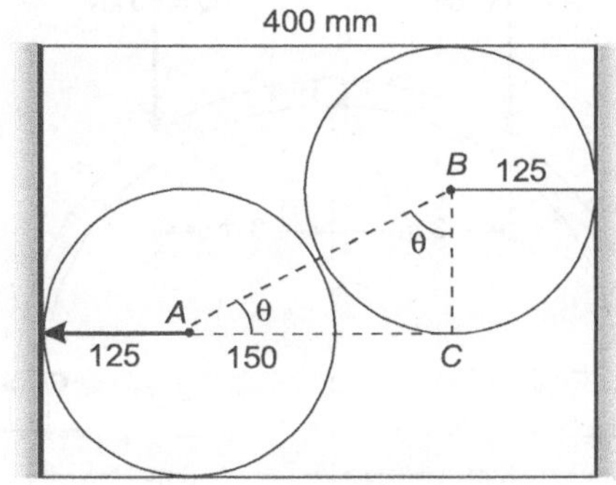

$$BC = 250^2 - 150^2$$

$$\cos\theta = \frac{200}{250}$$

$$\theta = 36.869°$$

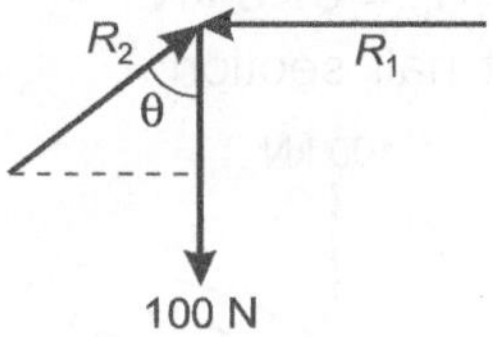

$$R_2 \cos\theta = 100$$

$$R_2 = \frac{100 \times 250}{200} = 125 \text{ N}$$

1.31 (b)

Consider joint Q,

$\Rightarrow \quad F_{QR} = 0$

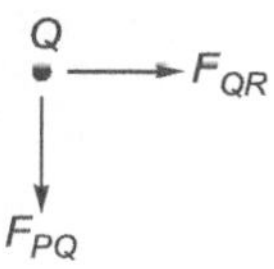

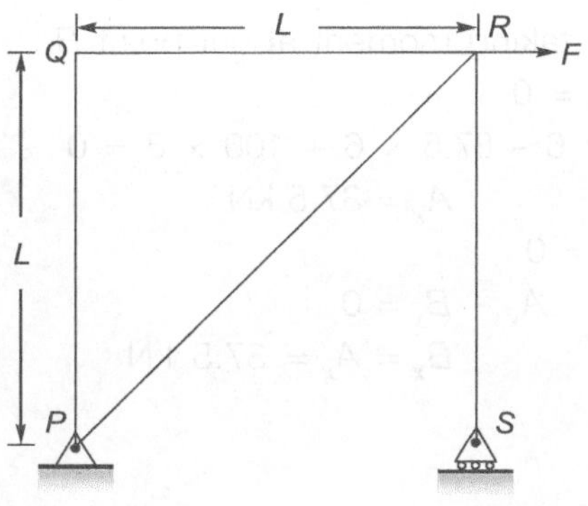

Joint R,

$\Sigma F_H = 0$

$\Rightarrow \quad F_{PR} \sin 45° = F$

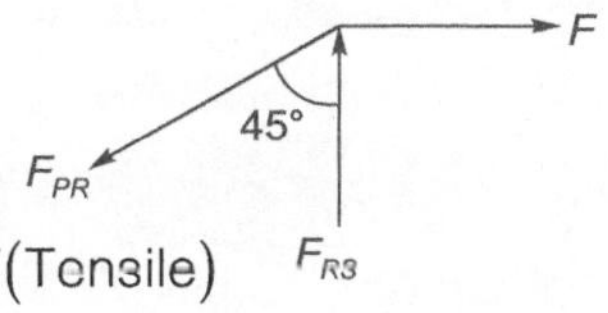

$$F_{PR} = \sqrt{2}F \text{(Tensile)}$$

$\Rightarrow \quad F_{PR} \cos 45° = F_{RS}$

$\Sigma F_V = 0$

$$F_{RS} = \sqrt{2}F \times \frac{1}{\sqrt{2}}$$

$\Rightarrow \quad F_{RS} = F \text{(Comp.)}$

1.33 (18) (18 to 18)

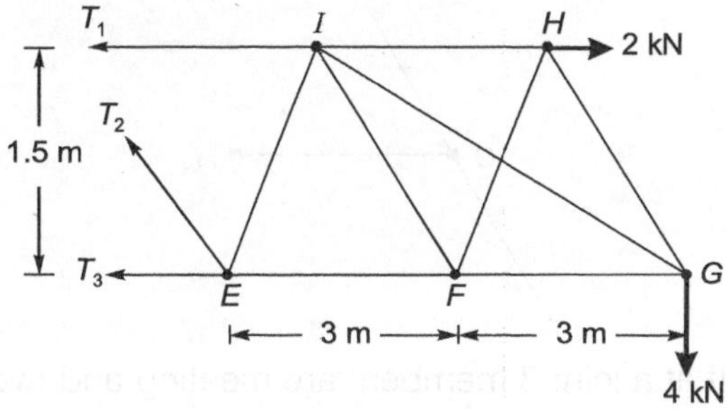

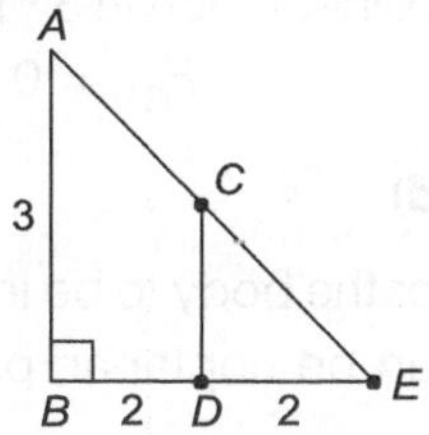

$$\frac{CD}{DE} = \frac{AB}{BE}$$

$\Rightarrow \quad CD = \frac{2 \times 3}{4} = 1.5 \text{ m}$

$\Sigma M_E = 0$

$\Rightarrow \quad 2 \times 1.5 - T_1 \times 1.5 + 4 \times 6 = 0$

$$T_1 = \frac{24 + 3}{1.5} = \frac{27}{1.5} = 18 \text{ kN}$$

Hence, magnitude of force carried by member CI is 18 kN.

1.34 (3) (3 to 3)

$$\frac{T_1}{T_2} = e^{\mu\theta}$$

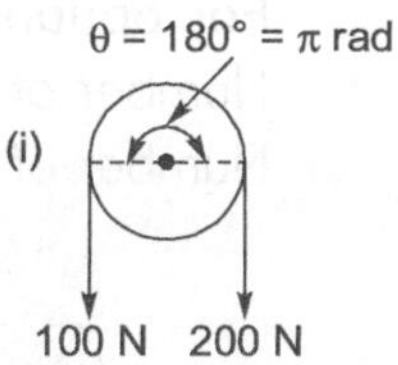

$$\frac{200}{100} = e^{\mu\theta}$$

$\Rightarrow \quad \mu = 0.22$

Now, $\quad \frac{200}{T_3} = e^{0.22\pi}$

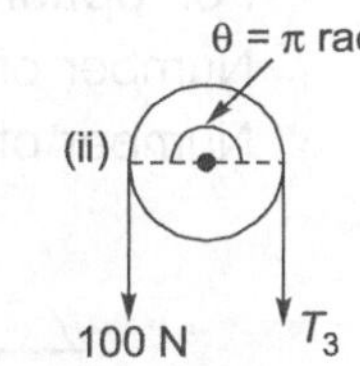

$\Rightarrow \quad T_3 = \frac{100}{2} = 50 \text{ N}$

So, for a force of 50 N, number of discs required

$= \frac{50}{20} = 2.5$

So, minimum number of discs required are 3.

1.35 (b, d)

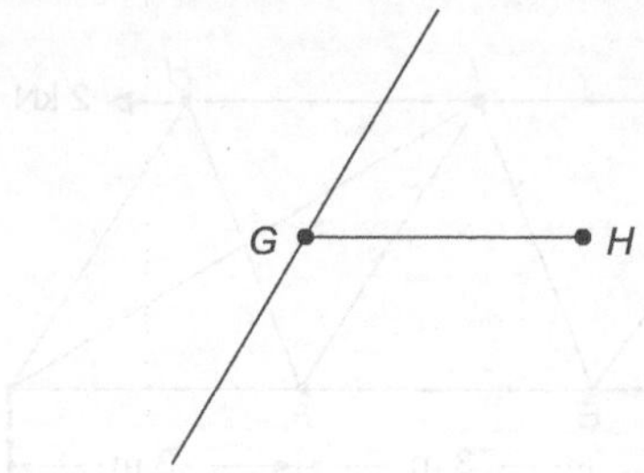

If at a joint 3 members are meeting and two are collinear then in 3rd member force will be zero.

$$F_{GH} = 0$$

1.37 (d)

For the body to be in equilibrium, the three forces can be coplanar, parallel or concurrent.

1.38 (a, b)

Statically indeterminate structures are those structures that cannot be analyzed using statics or equations of equilibrium. In such cases, the number unknowns exceed the number of equilibrium equations available.

Checking each option:

For option (a),

Number of unknown = 3

Number of equilibrium equation = 2

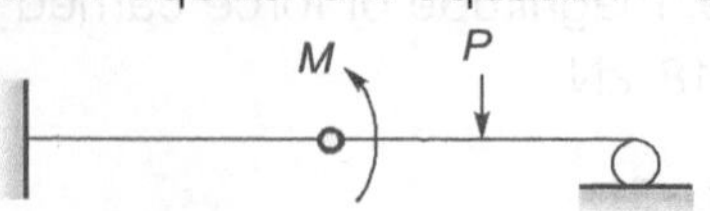

For option (b),

Number of unknown = 3

Number of equilibrium equation = 2

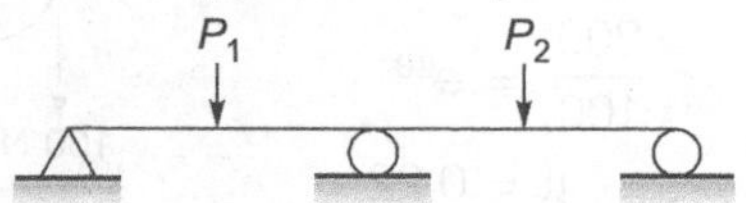

For option (c),

Number of unknown = 2

Number of equilibrium equation = 2

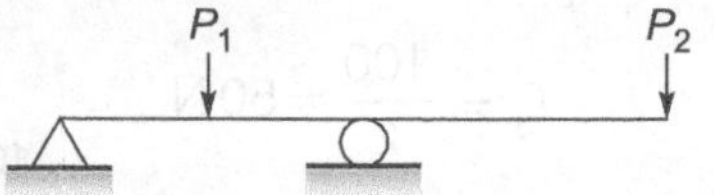

For option (d),

Number of unknown = 2

Number of equilibrium equation = 2

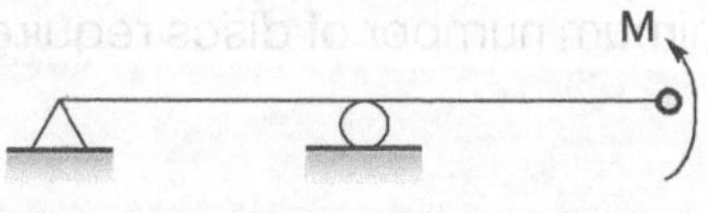

So, (a) and (b) are statically indeterminate structure.

1.39 (37.5) (37.0 to 38.0)

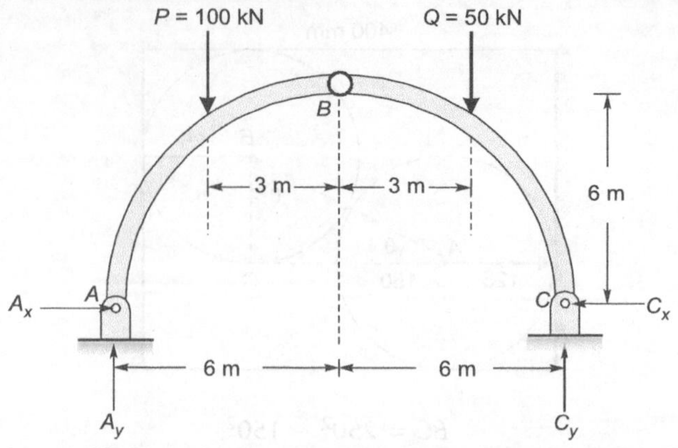

Initially, taking moment about hinged point C,

$$\Sigma M_C = 0$$

$$A_y \times 12 - 100 \times 9 - 50 \times 3 = 0$$

$$\Rightarrow \quad A_y = 87.5 \text{ kN}$$

FBD for left half section:

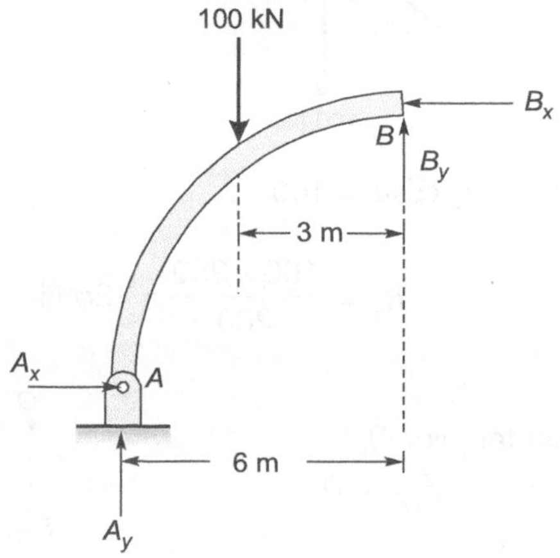

Now taking moment about point B.

$$\Sigma M_B = 0$$

$$A_x \times 6 - 87.5 \times 6 + 100 \times 3 = 0$$

$$\Rightarrow \quad A_x = 37.5 \text{ kN}$$

$$\Sigma F_x = 0$$

$$A_x - B_x = 0$$

$$\Rightarrow \quad B_x = A_x = 37.5 \text{ kN}$$

■■■■

2 Translation and Rotation

2.1 *AB* and *CD* are two uniform and identical bars of mass 10 kg each, as shown. The hinges at *A* and *B* are frictionless. The assembly is released from rest and motion occurs in the vertical plane. At the instant that the hinge *B* passes the point *B′*, the angle between the two bars will be

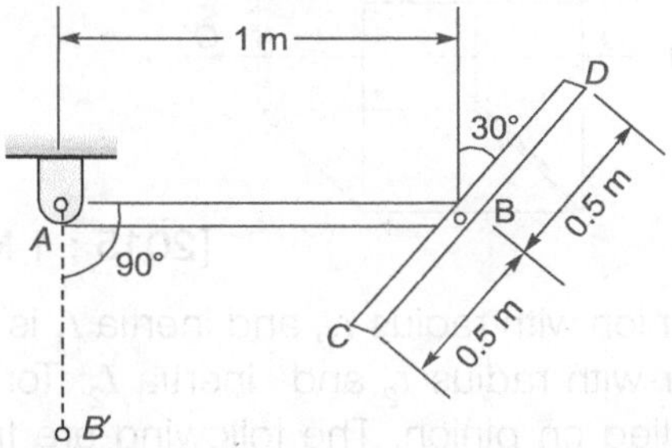

(a) 60 degrees (b) 37.4 degrees
(c) 30 degrees (d) 45 degrees

[1996 : 2 M]

2.2 A car moving with uniform acceleration covers 450 m in a 5 second interval, and covers 700 m in the next 5 second interval. The acceleration of the car is

(a) 7 m/s^2 (b) 50 m/s^2
(c) 25 m/s^2 (d) 10 m/s^2 **[1998 : 1 M]**

2.3 As shown in the figure, a person *A* is standing at the centre of a rotating platform facing person *B* who is riding a bicycle, heading East. The relevant speeds and distances are shown in the given figure person, a bicycle, heading East. At the instant under consideration, what is the apparent velocity of *B* as seen by *A*?

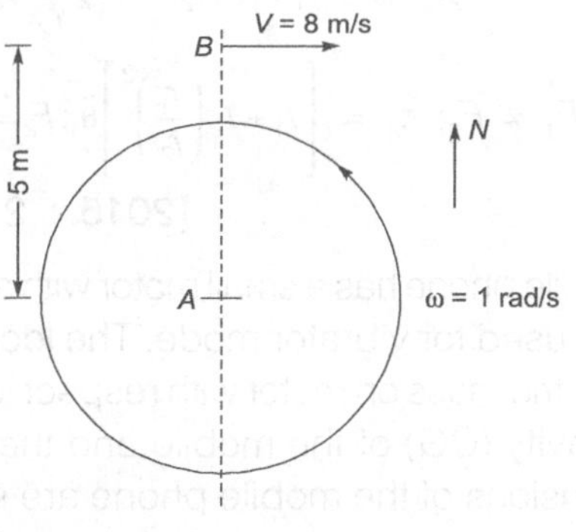

(a) 3 m/s heading east
(b) 3 m/s heading west
(c) 8 m/s heading east
(d) 13 m/s heading east **[1999 : 2 M]**

2.4 A rigid body shown in the fig. (a) has a mass of 10 kg. It rotates with a uniform angular velocity ω. A balancing mass of 20 kg is attached as shown in fig. (b). The percentage increase in mass moment of inertia as a result of this addition is

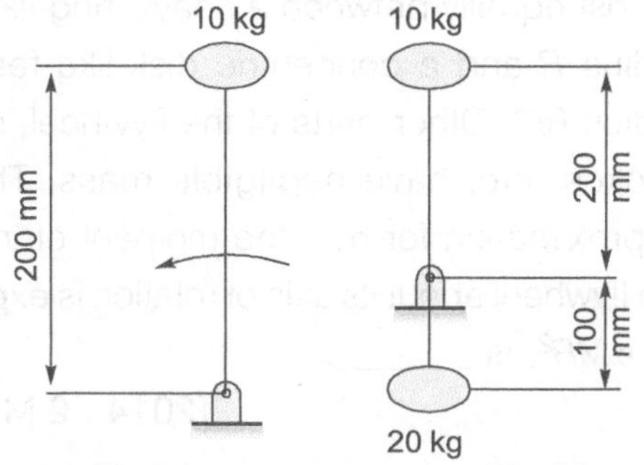

Fig. (a) Fig. (b)

(a) 25% (b) 50%
(c) 100% (d) 200% **[2004 : 2 M]**

2.5 The time variation of the position of a particle in rectilinear motion is given by $x = 2t^3 + t^2 + 2t$. If *v* is the velocity and *a* the acceleration of the particle in consistent units, the motion started with

(a) $v = 0,\ a = 0$ (b) $v = 0,\ a = 2$
(c) $v = 2,\ a = 0$ (d) $v = 2,\ a = 2$

[2005 : 1 M]

2.6 A simple pendulum of length of 5 m, with a bob of mass 1 kg, is in simple harmonic motion. As it passes through its mean position, the bob has a speed of 5 m/s. The net force on the bob at the mean position is

(a) zero (b) 2.5 N
(c) 5 N (d) 25 N **[2005 : 1 M]**

2.7 A uniform rigid rod of mass *M* and length *L* is hinged at one end as shown in the adjacent figure. A force *P* is applied at a distance of 2*L*/3 from the hinge so that the rod swings to the right. The horizontal reaction at the hinge is

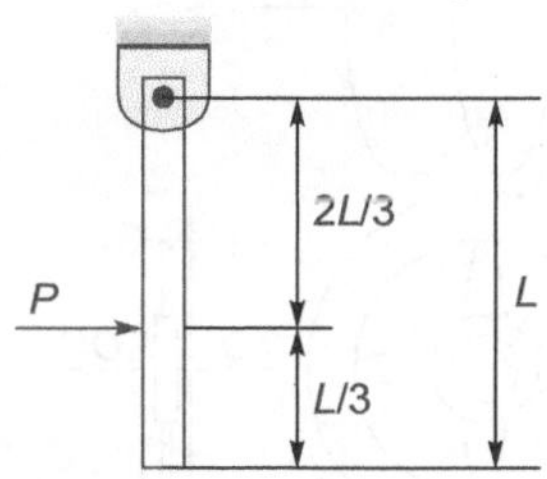

(a) −*P* (b) 0
(c) *P*/3 (d) 2*P*/3 **[2009 : 2 M]**

2.8 A circular solid disc of uniform thickness 20 mm, radius 200 mm and mass 20 kg, is used as a flywheel. If it rotates at 600 rpm, the kinetic energy of the flywheel, in Joules is
(a) 395 (b) 790
(c) 1580 (d) 3160 [2012 : 1 M]

2.9 onsider a flywheel whose mass M is distributed almost equally between a heavy, ring-like rim of radius R and a concentric disk-like feature of radius $R/2$. Other parts of the flywheel, such as spokes, etc, have negligible mass. The best approximation for α, if the moment of inertia of the flywheel about its axis of rotation is expressed as αMR^2, is ______.

[2014 : 2 M, Set-2]

2.10 A uniform slender rod (8 m length and 3 kg mass) rotates in a vertical plane about a horizontal axis 1 m from its end as shown in the figure. The magnitude of the angular acceleration (in rad/s^2) of the rod at the position shown is ______

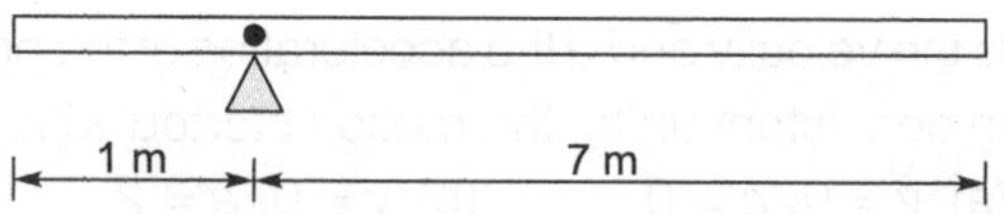

[2014 : 2 M, Set-4]

2.11 A swimmer can swim 10 km in 2 hours when swimming along the flow of a river. While swimming against the flow, she takes 5 hours for the same distance. Her speed in still water (in km/h) is ______.

[2015 : 1 M, Set-1]

2.12 For the same material and the mass, which of the following configurations of flywheel will have maximum mass moment of inertia about the axis of rotation OO' passing through the center of gravity.

(a)
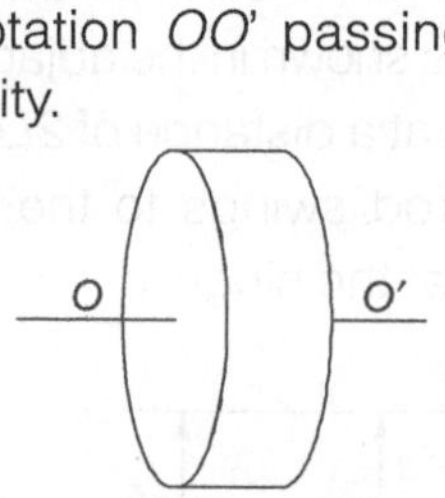

(b)
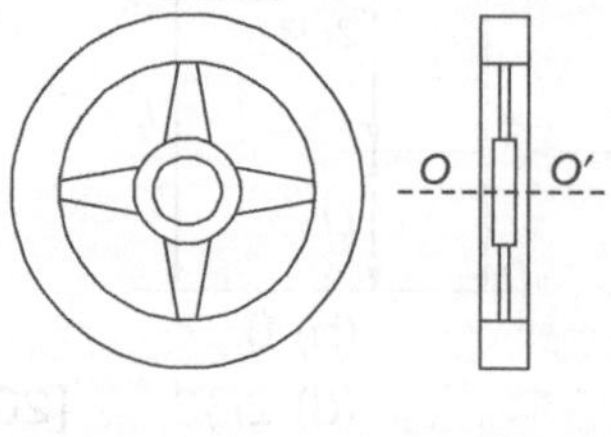

(c)
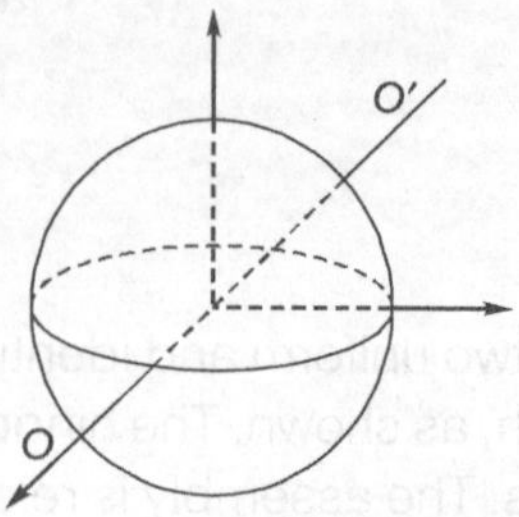

(d)
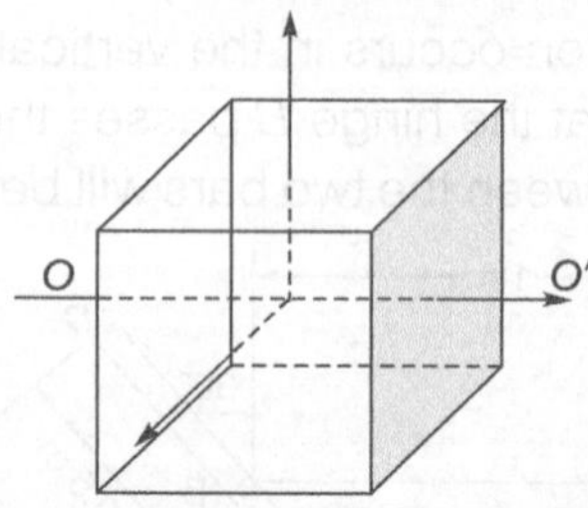

[2015 : 1 M, Set-3]

2.13 A pinion with radius r_1, and inertia I_1 is driving a gear with radius r_2 and inertia I_2. Torque τ_1 is applied on pinion. The following are free body diagrams of pinion and gear showing important forces (F_1 and F_2) of interaction. Which of the following relations hold true?

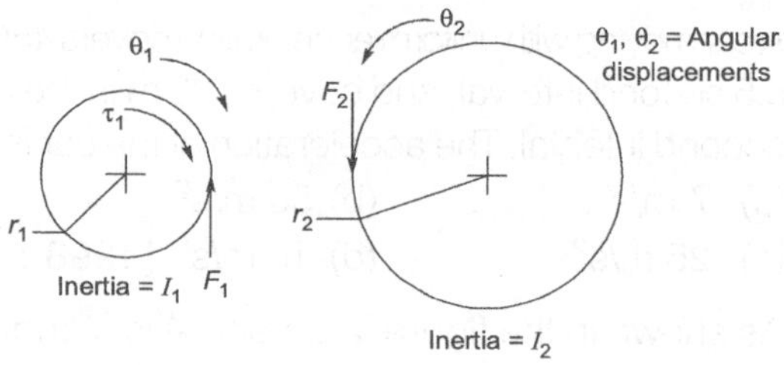

(a) $F_1 \neq F_2$; $\tau_1 = I_1\ddot{\theta}_1$; $F_2 = I_2\dfrac{r_1}{r_2^2}\ddot{\theta}_1$

(b) $F_1 = F_2$; $\tau_1 = \left[I_1 + I_2\left(\dfrac{r_1}{r_2}\right)^2\right]\ddot{\theta}_1$; $F_2 = I_2\dfrac{r_1}{r_2^2}\ddot{\theta}_1$

(c) $F_1 = F_2$; $\tau_1 = I_1\ddot{\theta}_1$; $F_2 = I_2\dfrac{1}{r_2}\ddot{\theta}_2$

(d) $F_1 \neq F_2$; $\tau_1 = \left[I_1 + I_2\left(\dfrac{r_1}{r_2}\right)^2\right]\ddot{\theta}_1$; $F_2 = I_2\dfrac{1}{r_2}\ddot{\theta}_2$

[2015 : 2 M, Set-1]

2.14 A mobile phone has a small motor with an eccentric mass used for vibrator mode. The location of the eccentric mass on motor with respect to the center of gravity (CG) of the mobile and the rest of the dimensions of the mobile phone are shown. The mobile is kept on a flat horizontal surface.

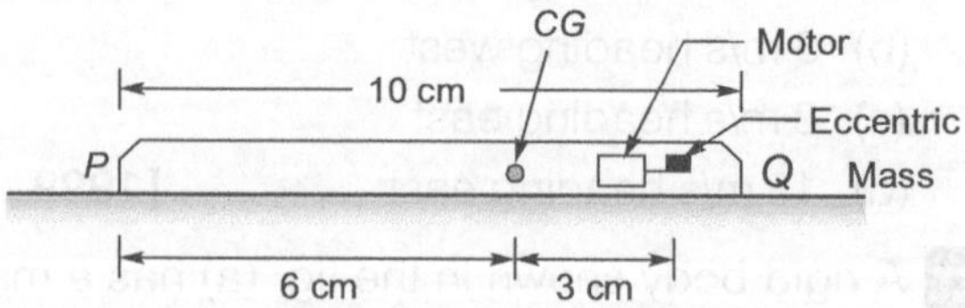

Given in addition that the eccentric mass = 2 grams, eccentricity = 2.19 mm, mass of the mobile = 90 grams, g= 9.81 m/s^2. Uniform speed of the motor in RPM for which the mobile will get just lifted off the ground at the end Q is approximately

(a) 3000 (b) 3500
(c) 4000 (d) 4500

[2015 : 2 M, Set-1]

2.15 The initial velocity of an object is 40 m/s. The acceleration a of the object is given by the following expression.

$$a = -0.1v,$$

where v is the instantaneous velocity of the object. The velocity of the object after 3 seconds will be ________.

[2015 : 2 M, Set-2]

2.16 Figure shows a wheel rotating about Q_2. Two points A and B located along the radius of wheel has speed of 80 m/s and 140 m/s respectively. The distance between the points A and B is 300 mm. The diameter of the wheel (in mm) is ________.

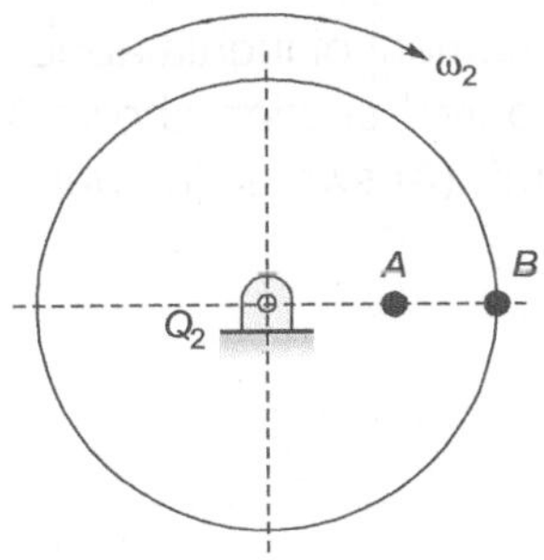

[2015 : 2 M, Set-3]

2.17 An inextensible massless string goes over a frictionless pulley. Two weights of 100 N and 200 N are attached to the two ends of the string. The weights are released from rest, and start moving due to gravity. The tension in the string (in N) is______

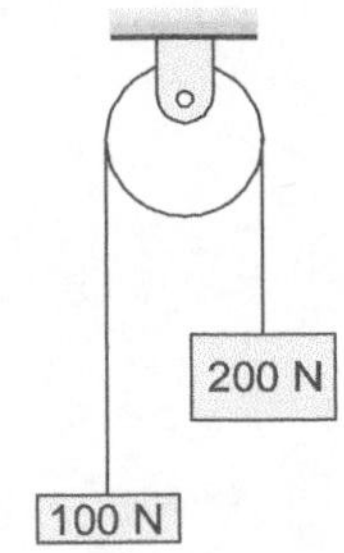

[2016 : 2 M, Set-3]

2.18 The rod PQ of length $L = \sqrt{2}$ m and uniformly distributed mass of $M = 10$ kg, is released from rest at the position shown in the figure. The ends slide along the frictionless faces OP and OQ. Assume acceleration due to gravity, g = 10 m/s^2. The mass moment of inertia of the rod about its centre of mass and an axis perpendicular to the plane of the figure is (ML^2/12). At this instant, the magnitude of angular acceleration (in radian/s^2) of the rod is __________.

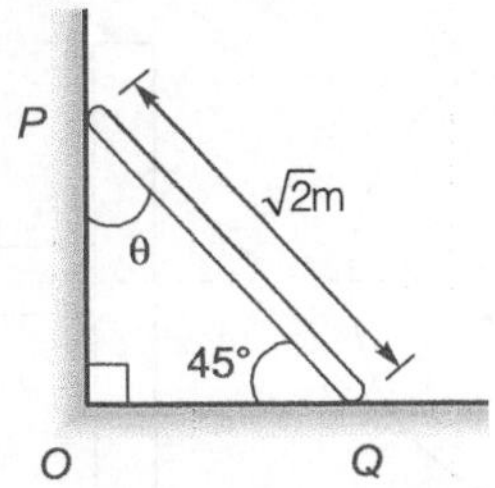

[2017 : 2 M, Set-2]

2.19 A rigid triangular body, PQR. with sides of equal length of 1 unit moves on a flat plate. At the instant shown, edge QR is parallel to the x-axis and the body moves such that velocities of points P and R are V_P and V_R in the x and y directions, respectively. The magnitude of the angular velocity of the body is

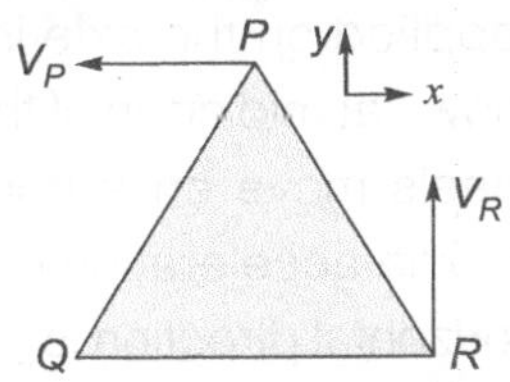

(a) $2V_P$ (b) $\frac{V_R}{\sqrt{3}}$

(c) $\frac{V_P}{\sqrt{3}}$ (d) $2\ V_R$

[2019 : 1 M, Set-2]

2.20 A circular disk of radius r is confined to roll without slipping at P and Q as shown in the figure.

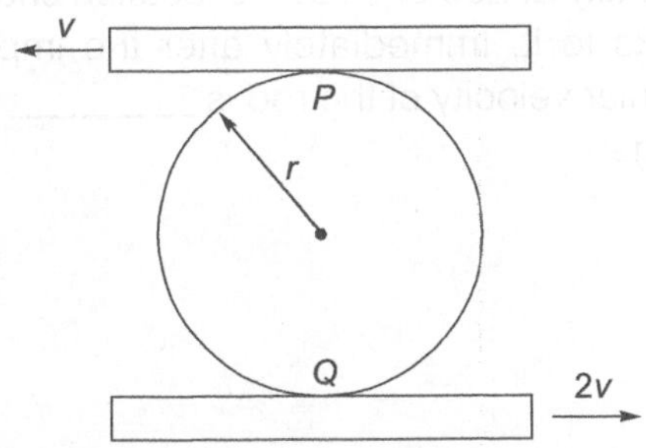

If the plates have velocities as shown, the magnitude of the angular velocity of the disk is

(a) $\frac{3v}{2r}$ (b) $\frac{v}{2r}$

(c) $\frac{v}{r}$ (d) $\frac{2v}{3r}$

[2020 : 1 M, Set-2]

2.21 The wheels and axle system lying an a rough surface is shown in the figure.

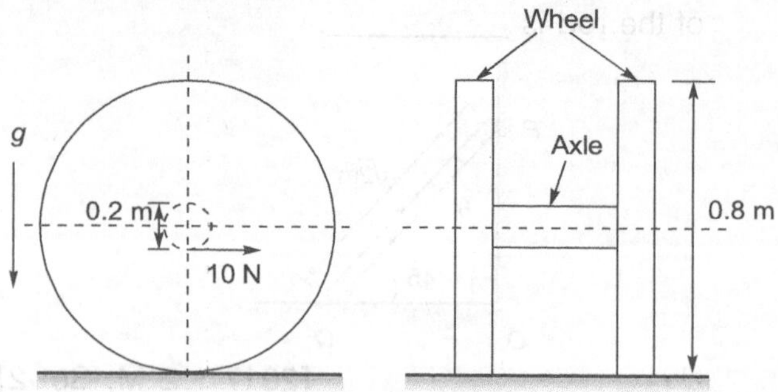

Each wheel has diameter 0,8 m and mass 1 kg. Assume that the mass of the wheel is concentrated at rim and neglect the mass of the spokes. The diameter of axle is 0.2 m and its mass is 1.5 kg. Neglect the moment of inertia of the axle and assume g = 9.8 m/s^2. An effort of 10 N is applied on the axle in the horizontal direction shown at mid span of the axle. Assume that the wheels move on a horizontal surface without slip. The acceleration of the wheel axle system in horizontal direction is_____m/s^2 (round off to one decimal place).

[2021 : 2 M, Set-2]

2.22 The plane of the figure represents a horizontal plane. A thin rigid rod at rest is pivoted without friction about a fixed vertical axis passing through *O*. Its mass moment of inertia is equal to 0.1 kg·cm^2 about *O*. A point mass of 0.001 kg hits it normally at 200 cm/s at the location shown, and sticks to it. Immediately after the impact, the angular velocity of the rod is ________ rad/s (in integer).

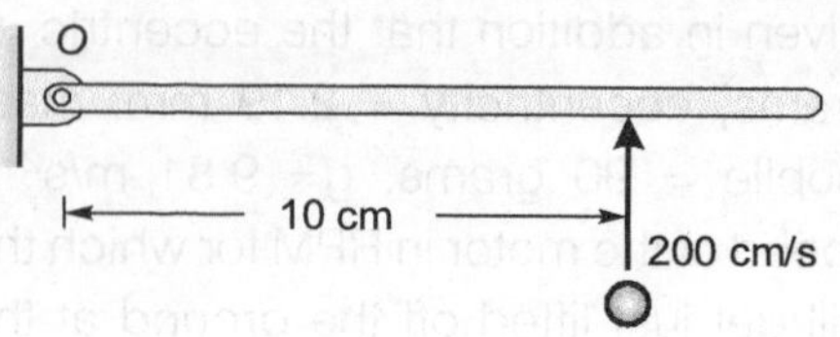

[2022 : 1 M, Set-1]

2.23 A cylindrical disc of mass $m = 1$ kg and radius $r = 0.15$ m was spinning at $\omega = 5$ rad/s when it was placed on a flat horizontal surface and released (refer to the figure). Gravity g acts vertically downwards as shown in the figure. The coefficient of friction between the disc and the surface is finite and positive. Disregarding any other dissipation except that due to friction between the disc and the surface, the horizontal velocity of the center of the disc, when it starts rolling without slipping, will be m/s (round off to 2 decimal places).

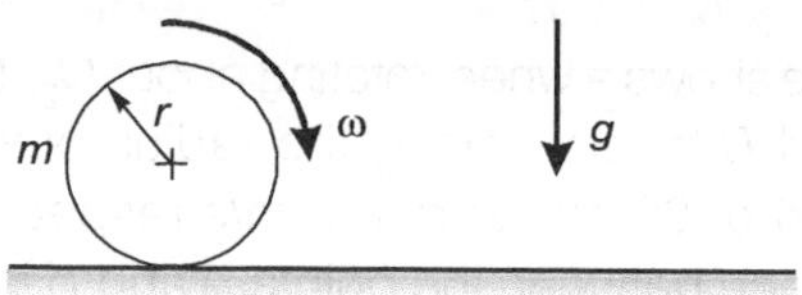

[2022 : 2 M, Set-1]

2.24 The area moment of inertia about the y-axis of a linearly tapered section shown in the figure is ________ m^4. (Answer in integer)

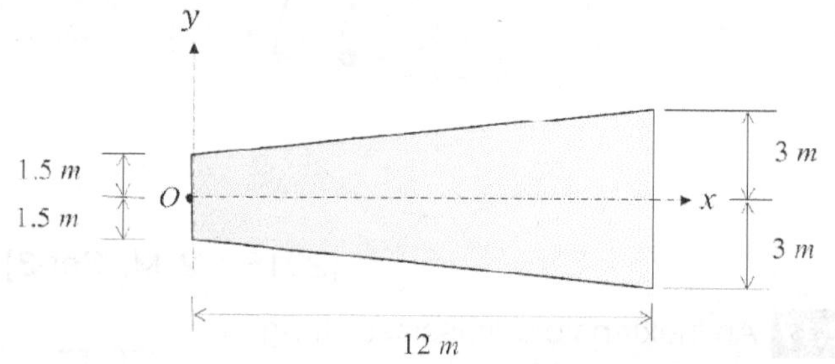

[2023 : 2 M]

■■■■

Answers Translation and Rotation

2.1 (c) 2.2 (d) 2.3 (c) 2.4 (b) 2.5 (d) 2.6 (c) 2.7 (b) 2.8 (b)

2.9 (0.5625) 2.10 (2.053) 2.11 (3.5) 2.12 (b) 2.13 (b) 2.14 (b) 2.15 (29.63)

2.16 (1400) 2.17 (133.33) 2.18 (7.5) 2.19 (d) 2.20 (a) 2.21 (1.36) 2.22 (10)

2.23 (0.25) 2.24 (3024)

Explanations Translation and Rotation

2.1 (c)

As weight and hinge reaction passes through center of mass and hinge is frictionless, the rod CD will not rotate. CD rod will make same angle 30° with vertical.

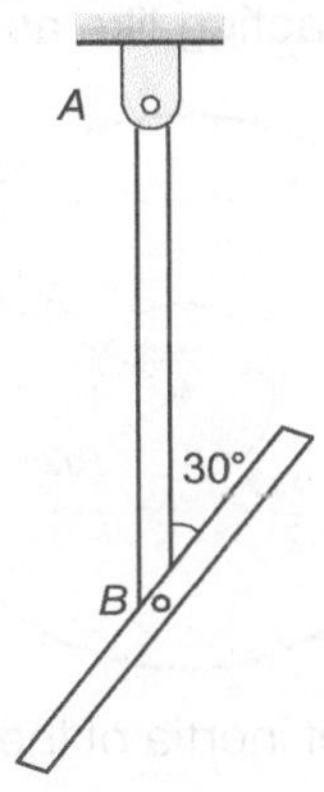

2.2 (d)

Method I:

$t = 5$ s, $s_5 = 450$ m

$t = 10$ s

$s_{10} = 450 + 700 = 1150$

$$s = ut + \frac{1}{2}at^2$$

$$450 = 5u + \frac{1}{2}a(25) \quad ...(i)$$

$$1150 = 10u + \frac{1}{2}a \times 100 \quad ...(ii)$$

Solving equations (i) and (ii), we get

$u = 65$ m/s

$a = \mathbf{10\ m/s^2}$

Method II:

Let u be the velocity at the start of first 5 second interval and v be the velocity at the start of next 5 second interval.

From I-equation of motion;

$$v = u + at$$

$$v - u = at \quad ...(i)$$

From II-equation of motion;

$$S = ut + \frac{1}{2}at^2$$

For first 5 sec. interval,

$$S_1 = ut + \frac{1}{2}at^2 \quad ...(ii)$$

For next 5 sec interval,

$$S_2 = vt + \frac{1}{2}at^2 \quad ...(iii)$$

$\therefore$ $S_2 - S_1 = (v - u)t$

Also;

$S_2 - S_1 = at \times t$ [from equation (i)]

$S_2 - S_1 = at^2$

$700 - 450 = a \times 5 \times 5$

$250 = a \times 5 \times 5$

$a = 10$ m/s^2

2.3 (c)

$$\overrightarrow{V_{B/A}} = \overrightarrow{V_B} - \overrightarrow{V_A} = 8 - 0$$

$= 8$ m/s heading east.

2.4 (b)

Moment of inertia of first $= mk^2$

$= 10 \times (0.2)^2 = 0.4$ kgm^2

Moment of inertia of second

$= m_1k_1^2 + m_2k_2^2 = 10 \times (0.2)^2 + 20 \times (0.1)^2$

$= 0.6$ kgm^2

$\therefore$ Increases in mass moment of inertia

$$= \frac{0.6 - 0.4}{0.4} \times 100 = 50\%$$

2.5 (d)

Given: $x = 2t^3 + t^2 + 2t$

Velocity: $V = \frac{dx}{dt} = 6t^2 + 2t + 2$

At $t = 0$ (start of motion)

$V = 0 + 0 + 2$

$V = 2$

Acceleration: $a = \frac{dV}{dt} = 12t + 2$

At $t = 0$

$a = 0 + 2$

$\therefore$ $a = 2$

2.6 (c)

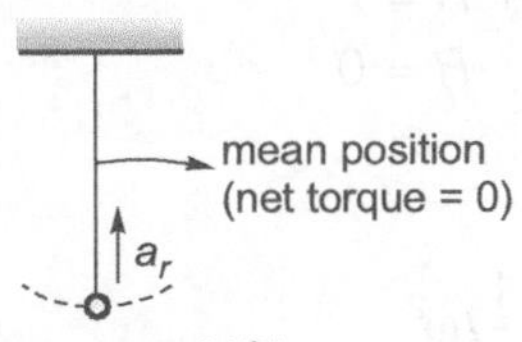

$T - mg = ma_r \rightarrow$ NSL

$$T - mg = \frac{mv^2}{r}$$

$$F_{net} = T - mg = \frac{1 \times 5^2}{5} = 5\text{ N}$$

2.7 (b)

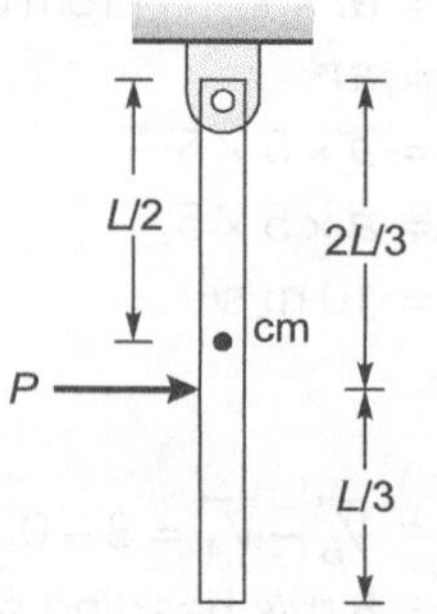

At this moment

$$\omega = 0$$

but $\alpha \neq 0$ [start of rotation]

$$\vec{a}_{CM} = \vec{a}_r + \vec{a}_t$$

(because centre of mass is in circular motion at a radius $L/2$)

$$\vec{a}_r = 0 \quad [\because \omega = 0]$$

$$\vec{a}_{CM} = \vec{a}_t = \frac{L}{2}\alpha(\hat{i})$$

Torque, $T = I\alpha$ (about hinge)

Taking moment about A,

$$P \times \frac{2L}{3} = \frac{ML^2}{3}\alpha$$

$$\alpha = \frac{2P}{ML}$$

$$\vec{a}_{CM} = \frac{L}{2} \times \frac{2P}{ML}(\hat{i})$$

$$\vec{a}_{CM} = \frac{P}{M}(\hat{i})$$

Let R be the reaction in horizontal direction at hinge:

$$\left(\Sigma f\right)_{ext} = M\vec{a}_{CM} \quad \text{[in horizontal direction]}$$

$$P + R = Ma_{CM}$$

$$P + R = M \times \frac{P}{M}$$

$$P + R = P$$

$$R = 0$$

2.8 (b)

$$KE = \frac{1}{2}I\omega^2$$

$$I = \frac{mr^2}{2} = \frac{20 \times (0.2)^2}{2} = 0.4\text{ kgm}^2$$

$$\omega = \frac{2\pi N}{60} = \frac{2 \times \pi \times 600}{60} = 62.83\text{ rad/s}$$

$$\therefore KE = \frac{1}{2} \times 0.4 \times (62.83)^2 = \mathbf{790\ J}$$

2.9 Sol.

Since half of the mass is acting like rim and half of the mass is acting like as disk

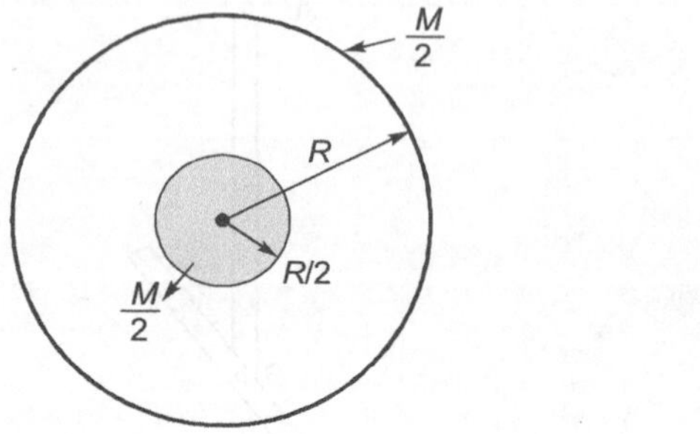

Total moment of inertia of the flywheel,

$$I = \frac{M}{2}(R)^2 + \frac{1}{2}\left(\frac{M}{2}\right)\left(\frac{R}{2}\right)^2 = \frac{MR^2}{2} + \frac{1}{2} \times \frac{M}{2} \times \frac{R^2}{4}$$

$$= \frac{MR^2}{2} + \frac{MR^2}{16} = \frac{9}{16}MR^2 = \alpha MR^2$$

$$\alpha = \frac{9}{16} = 0.5625$$

2.10 Sol.

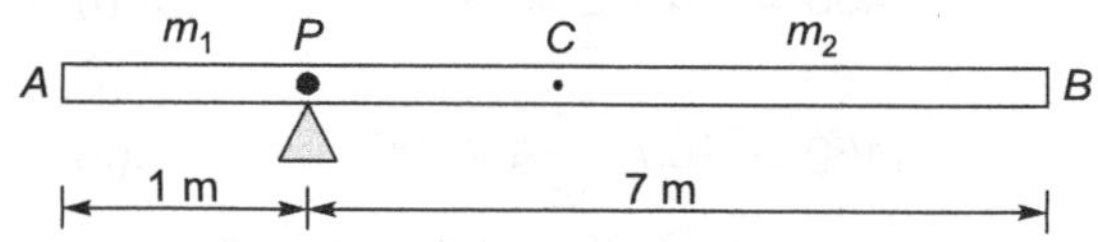

$AP = 1$ m, $AC = 4$ m, $PC = 3$ m

$$I_C = \frac{mL^2}{12} = \frac{3 \times 8 \times 8}{12} = 16\text{ kg-m}^2$$

$$I_P = I_C + m(PC)^2$$

$$= 16 + 3 \times 3^2 = 43\text{ kg-m}^2$$

$$\text{Torque} = m_2 g \times \frac{PB}{2} - m_1 g\,\frac{AP}{2}$$

$$T = m_2 \times 9.81 \times \frac{7}{2} - m_1 \times 9.81 \times \frac{1}{2}$$

$$= 34.335\,m_2 - 4.905\,m_1$$

$$m_1 = \frac{3}{8}\text{ kg}, m_2 = \frac{3 \times 7}{8} = 2.625\text{ kg}$$

$$T = 34.335 \times 2.625 - 4.905 \times 0.375$$

$$= 88.29\text{ N.m}$$

We can also calculate torque as

$$T = mg \times PC$$

$$= 3 \times 9.81 \times 3 = 88.29\text{ N.m}$$

$$T = I_P\alpha$$

$$\alpha = \frac{T}{I_P} = \frac{88.29}{43} = \mathbf{2.053\ rad/s^2}$$

2.11 Sol.

Let velocity of swimmer in still water = x
Velocity of river flow = y
(i) Swimming along the river flow
Resultant velocity of swimmer = $x + y$

$$\Rightarrow \quad 2 = \frac{10}{x+y}$$

$$\Rightarrow \quad x + y = 5 \quad \text{...(i)}$$

(ii) Swimming against the river flow
Resultant velocity of woman = $x - y$

$$\Rightarrow \quad 5 = \frac{10}{x-y}$$

$$x - y = 2 \quad \text{...(ii)}$$

Adding equations (i) and (ii)

$$2x = 7$$

$$x = \mathbf{3.5\ km/h}$$

2.12 (b)

$$I = mr^2$$

Here for the rimmed wheel, the mass m will be most far away distribution from the axis 0-0′.
∴ (b) is the correct answer.

2.13 (b)

$$\tau_1 - F_1r_1 = I_1\ddot{\theta}_1 \quad \text{...(i)}$$

$$F_1 = F_2 \quad \text{...(ii)}$$

[Action and reaction are equal by Newton's 3rd law]

$$F_2r_2 = \tau_2 = I_2\ddot{\theta}_2 \quad \text{...(iii)}$$

From equations (ii) and (iii)

$$F_2 = F_1 = \frac{I_2\ddot{\theta}_2}{r_2} \quad \text{...(iv)}$$

From equations (i) and (iv)

$$\tau_1 - \left(\frac{I_2\ddot{\theta}_2}{r_2}\right)r_1 = I_1\ddot{\theta}_1 \quad \text{...(v)}$$

We know, $r_1\omega_1 = r_2\omega_2$

$$r_1\dot{\theta}_1 = r_2\dot{\theta}_2$$

Differentiating with respect to time

$$r_1\ddot{\theta}_1 = r_2\ddot{\theta}_2$$

$$\ddot{\theta}_2 = \frac{r_1}{r_2}\ddot{\theta}_1$$

$$\tau_1 = I_1\ddot{\theta}_1 + \frac{I_2}{r_2}\left[\frac{r_1}{r_2}\ddot{\theta}_1\right]r_1 = \left[I_1 + I_2\left[\frac{r_1}{r_2}\right]^2\right]\ddot{\theta}_1$$

$$\tau_2 = I_2\ddot{\theta}_2 = F_2r_2$$

$$F_2 = \frac{I_2}{r_2}\ddot{\theta}_2 = \frac{I_2}{r_2}\left[\frac{r_1}{r_2}\ddot{\theta}_1\right] = I_2\frac{r_1}{r_2^2}\ddot{\theta}_1$$

2.14 (b)

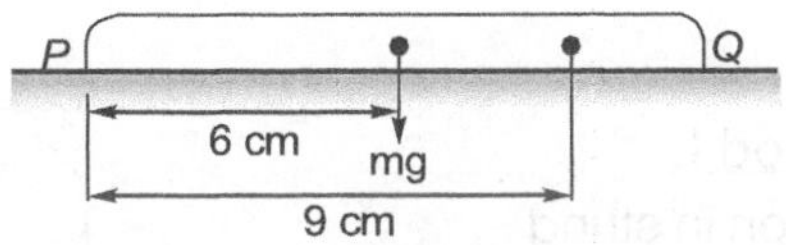

$$\Sigma M_P = 0 \quad [\text{moment about } P = 0]$$

$$mg \times 6 = m_e(2.19 \times 10^{-3})(\omega^2) \times [9]$$

$$90 \times 9.81 \times 6 = 2(2.19 \times 10^{-3})\,\omega^2 \times 9$$

or $\quad \omega = 366.58$ rad/s

also, $\quad \omega = \dfrac{2\pi N}{60}$

∴ $\quad 366.58 = \dfrac{2\pi \times N}{60}$

or $\quad N = 3500.61$ rpm ≈ 3500 rpm

2.15 Sol.

Given: $u = 40$ m/s, $a = -0.1v$, $V = ?$, $t = 3$ s

$$a = \frac{dv}{dt} = -0.1v$$

$$\int_{40}^{V}\frac{dv}{v} = \int_0^3 -0.1dt$$

$$[\ln v]_{40}^{V} = -0.1\,[t]$$

$$\ln v - \ln 40 = -0.1\,[3.0] = -0.3$$

$$\ln v = \ln 40 - 0.3$$

$$\ln v = 3.38887$$

or $\quad v = e^{3.38887} = \mathbf{29.63\ m/s}$

2.16 Sol.

Angular velocity of point A and B will be same

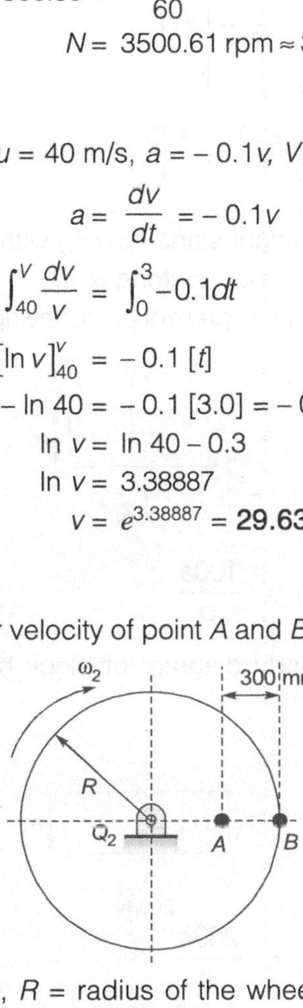

where, R = radius of the wheel

$$\omega_A = \omega_B$$

$$\frac{V_A}{R-0.3} = \frac{V_B}{R}$$

$$\frac{80}{R-0.3} = \frac{140}{R}$$

$$80R = 140R - 42$$

$$42 = 60R$$

$$R = 0.7 \text{ m}$$

$\therefore$ Diameter of the wheel,

$$D = 2R = 1.4 \text{ m} = \mathbf{1400 \text{ mm}}$$

2.17 Sol.

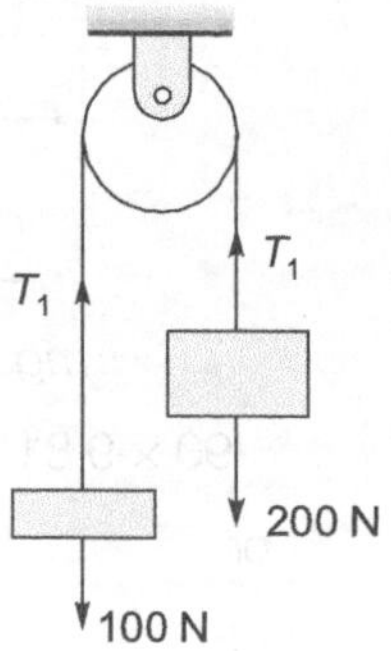

Method I:

Tension in string

$$= \left[\frac{2m_1m_2}{m_1+m_2}\right]g$$

$$= \left[\frac{2\times\frac{100}{g}\times\frac{200}{g}}{\frac{100}{g}+\frac{200}{g}}\right]g$$

$$= \left[\frac{\frac{2\times100\times200}{g^2}}{\frac{(100+200)}{g}}\right]\times g$$

$$= \frac{2\times100\times200}{300} = \frac{400}{3} = \mathbf{133.33 \text{ N.}}$$

Method II:

Assume weight starts moving with acceleration 'a', while tension in string is T_1

$\therefore$ Free body diagram of block having 10 N weight,

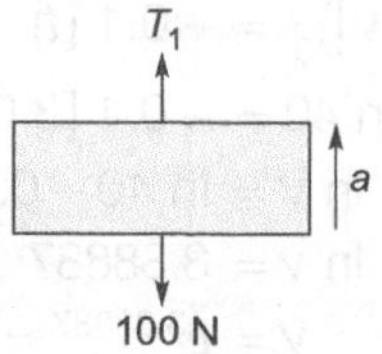

$$T_1 - 100 = \frac{100a}{g} \quad \text{...(i)}$$

$\therefore$ Free body diagram of block having 200 N weight.

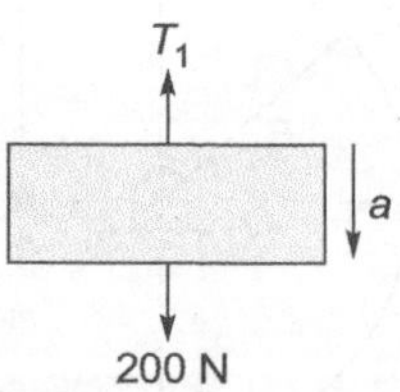

$$200 - T_1 = \frac{200a}{g} \quad \text{...(ii)}$$

Dividing equation (ii) with equation (ii)

$$\frac{T_1-100}{200-T_1} = \frac{100}{200}$$

$$2T_1 - 200 = 200 - T_1$$

$$3T_1 = 400$$

Tension in string,

$$T_1 = \frac{400}{3} = 133.33 \text{ N}$$

2.18 Sol.

$L = \sqrt{2}$ m, $M = 10$ kg, $g = 10$ m/s²

Both the surfaces are smooth

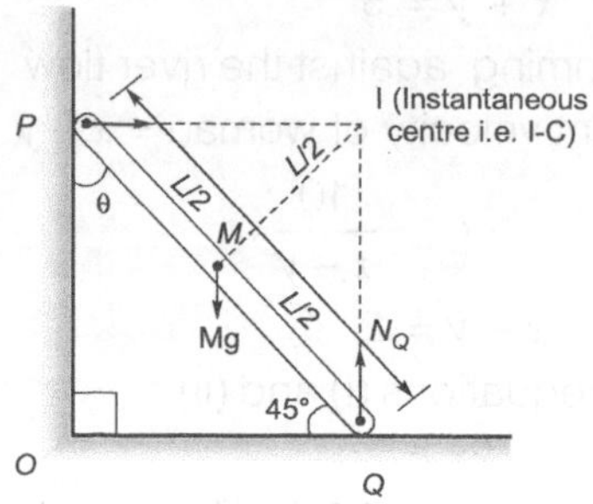

$$T_{I-C} = I_{I-C}\,\alpha$$

$$\Rightarrow Mg\times\frac{L}{2}\cos45° = \left[I_{CM} + M\times\frac{L^2}{4}\right]\alpha$$

$$\Rightarrow Mg\frac{L}{2}\cos45° = \left[\frac{ML^2}{12}+\frac{ML^2}{4}\right]\alpha$$

$$\Rightarrow \alpha = \frac{3g}{2L}\cos45° = \frac{30}{2\times\sqrt{2}}\times\frac{1}{\sqrt{2}}$$

$$= \mathbf{7.5 \text{ rad/s}^2}$$

2.19 (d)

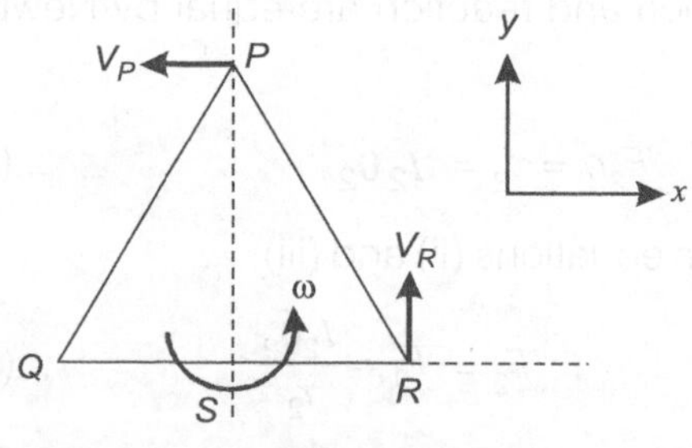

At the given instant, instantaneous centre (I_C) is at mid of Q and R, that is S.

Let's take ω to be angular velocity of body.

Then, $\quad V_P = (PS)\cdot\omega$

$$\Rightarrow \omega = \frac{V_P}{PS} = \frac{V_P}{\sqrt{3}/2} = \frac{2V_P}{\sqrt{3}} \quad \{\because PQ = PR = QR = 1\}$$

Also, $\quad V_R = SR\cdot\omega$

$$\Rightarrow \qquad \omega = \frac{V_R}{SR} = \frac{V_R}{1/2} = 2V_R$$

Observing options, $2V_R$ is correct i.e. option (d).

2.20 (a)

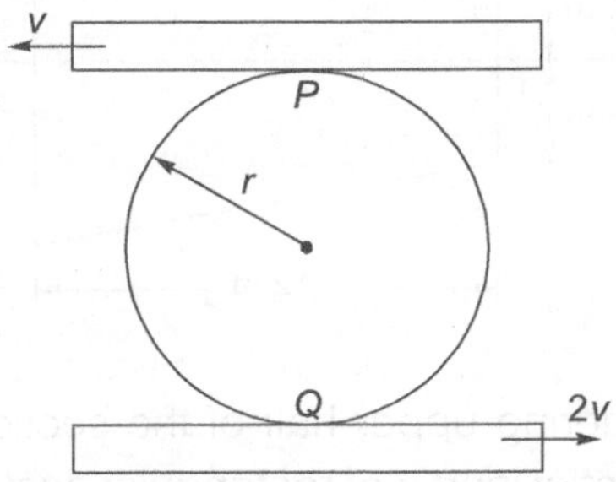

For pure rolling

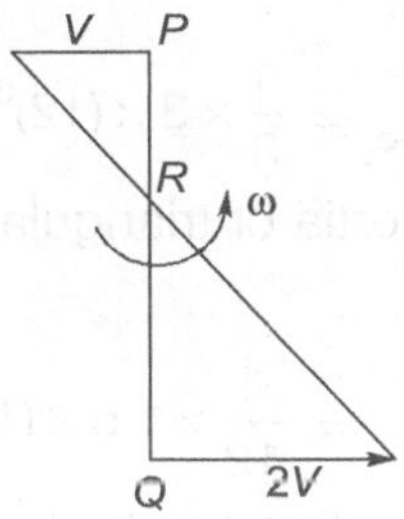

$$V_P = V = (PR)\omega \qquad \text{... (i)}$$

$$V_Q = 2V = (QR)\omega \qquad \text{... (ii)}$$

Divide by (ii) to (i),

$$2 = \frac{QR}{PR} \Rightarrow QR = 2(PR)$$

$$PR + QR = 2r$$

$$PR + 2(PR) = 2r$$

$$PR = \frac{2}{3}r$$

From equation (i),

$$V = \left(\frac{2}{3}r\right)\omega \Rightarrow \omega = \frac{3V}{2r}$$

2.21 Sol.

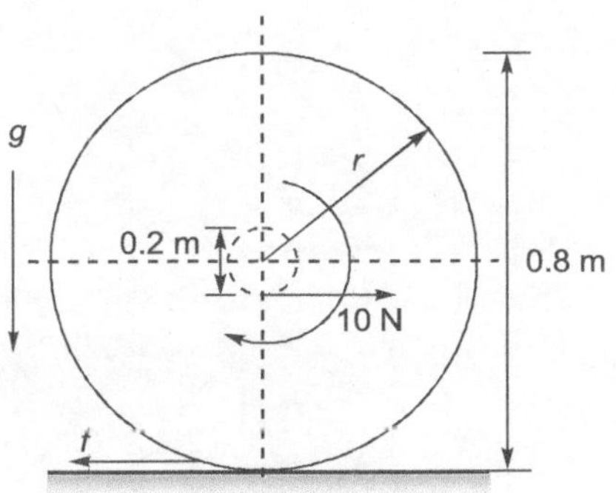

$$m = 1 \text{ kg/wheel}$$

$$m_{axle} = 1.5 \text{ kg}$$

$$10 - 2f = 3.5a \rightarrow \text{NSL}$$

$$10 - 2f = 3.5r\alpha$$

$$10 - 2f = 1.4\alpha \quad \text{...(1)} \qquad [r = 0.4 \text{ m}]$$

$$\Sigma T_{cm} = \text{T}_{cm}\alpha$$

$$2fr - 10 \times 0.1 = 2mr^2\alpha$$

$$0.8f - 1 = 2 \times 1 \times 0.4^2\alpha$$

$$0.8f - 1 = 0.32\alpha \qquad \text{...(2)}$$

from (1)

$$\frac{10 - 1.4\alpha}{2} = f$$

$$\Rightarrow \qquad f = 5 - 0.7\alpha$$

Using eq. (2)

$$0.8(5 - 0.7\alpha) - 1 = 0.32\alpha$$

$$4 - 0.56\alpha - 1 = 0.32\alpha$$

$$3 = (0.32 + 0.56)\alpha$$

$$\Rightarrow \qquad \alpha = \frac{3}{0.88}$$

$$\alpha = 3.409 \text{ rad/s}^2$$

$$a_{cm} = r\alpha = 1.36 \text{ m/s}^2$$

2.22 (10)

Taking rod and mass m as the system and using the principle of conservation of angular momentum about point O.

$$\Rightarrow \qquad L_{A,\text{ before striking}} = L_{A,\text{ after striking}}$$

$$\Rightarrow \qquad mVr = (I_{\text{combined}})\,\omega$$

$$\Rightarrow \qquad 0.001 \times 200 \times 10 = (0.1 \times 0.001 \times 10^2) \times \omega$$

$$\Rightarrow \qquad \omega = \frac{2}{0.2} = 10 \text{ rad/s}$$

2.23 (0.25) (0.24 to 0.26)

Given : Cylindrical disc, m = 1 kg, r = 0.15 m, initial angular speed, ω = 5 rad/s

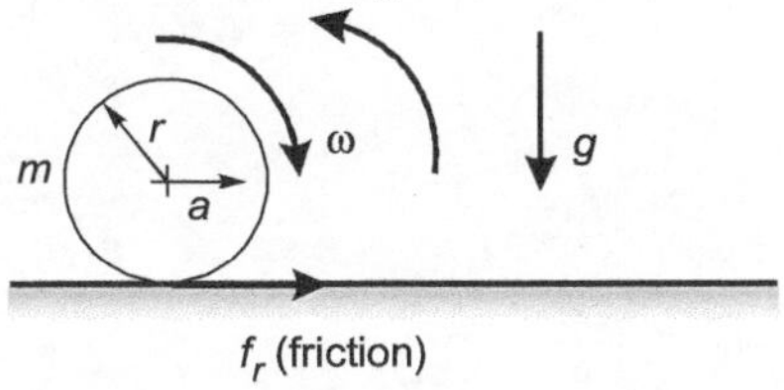

$$\Sigma F_x = ma$$

$$\Rightarrow \qquad f_r = ma \qquad \text{...(i)}$$

$$\text{and} \qquad \Sigma\text{Torque} = I\alpha$$

$$\Rightarrow \qquad f_r \times r = I\alpha \qquad \left(\because I = \frac{mr^2}{2}\right)$$

$$\Rightarrow \qquad f_r = \frac{mr\alpha}{2} \qquad \text{...(ii)}$$

From equation (i) and (ii)

$$ma = \frac{mr\alpha}{2}$$

$$a = \frac{r\alpha}{2}$$

When rolling starts

$$v = \omega' r$$

$$v = u + at \text{ and } \omega' = \omega - \alpha t$$

$$\Rightarrow \quad v = 0 + at = r(\omega - \alpha t) \quad \text{...(iii)}$$

$$at = r\omega - r\alpha t$$

$$\frac{r\alpha t}{2} = r\omega - r\alpha t$$

$$\Rightarrow \quad \alpha t = \frac{2}{3}\omega$$

From equation (iii),

$$v = r(\omega - \alpha t) = r\left(\omega - \frac{2}{3}\omega\right)$$

$$= \frac{r\omega}{3} = \frac{0.15 \times 5}{3} = 0.25 \text{ m/s}$$

2.24 (3024) (3023.999 to 3024.001)

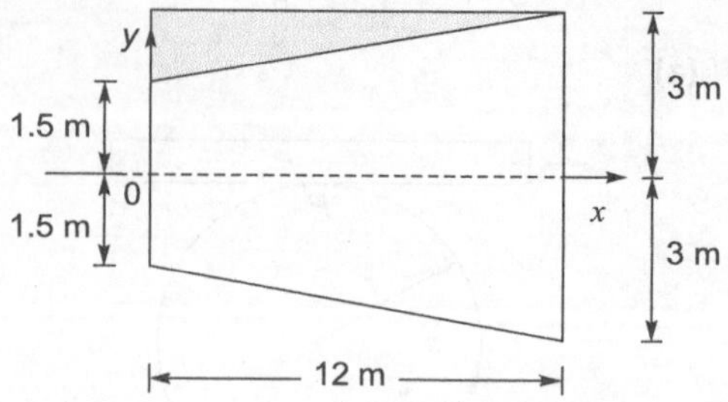

Considering upper half of the section.

Moment of inertia of rectangular secton about y-axis,

$$I_{\text{rectangular sec.}} = \frac{1}{3} \times 3 \times (12)^3 = (12)^3$$

Moment of inertia of triangular section about y-axis,

$$I_{\text{triangular sec.}} = \frac{1}{12} \times 1.5 \times (12)^3 = \frac{1}{8} \times (12)^3$$

Moment of inertia of actual section,

$$I = (12)^3 - \frac{(12)^3}{8} = \frac{7}{8} \times (12)^3$$

$\therefore$ Moment of inertia for the whole section

$$= 2 \times \frac{7}{8} \times (12)^3 = 3024 \text{ m}^3$$

3 Friction

3.1 A steel wheel of 600 mm diameter rolls on a horizontal steel rail. It carries a load of 500 N. The coefficient of rolling resistance is 0.3 mm. The force in Newton, necessary to roll the wheel along the rail is

(a) 0.5 (b) 5
(c) 1.5 (d) 150 **[2000 : 1 M]**

3.2 The ratio of tension on the tight side to that of the slack ride in a flat belt drive is

(a) proportional to the product of coefficient of friction and lap angle
(b) an exponential function of the product of coefficient of friction and lap angle
(c) proportional to lap angle
(d) proportional to the coefficient of friction

[2000 : 1 M]

3.3 The figure shows a pair of pin-jointed gripper-tongs holding an object weighing 2000 N. The coefficient of friction (μ) at the gripping surface is 0.1. *XX* is the line of action of the input force and *YY* is the line of application of gripping force. If the pin-joint is assumed to be frictionless, the magnitude of force *F* required to hold the weight is

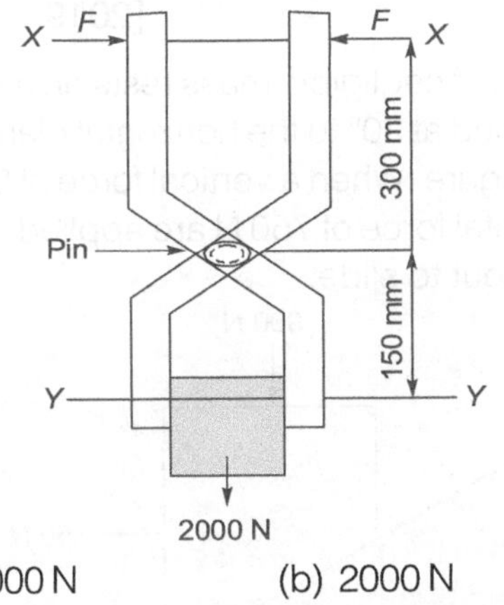

(a) 1000 N (b) 2000 N
(c) 2500 N (d) 5000 N **[2004 : 2 M]**

3.4 Two books of mass 1 kg each are kept on a table, one over the other. The coefficient of friction on every pair of contacting surfaces is 0.3. The lower book is pulled with a horizontal force *F*. The minimum value of *F* for which slip occurs between the two books is

(a) zero (b) 11.77 N
(c) 5.88 N (d) 8.83 N **[2005 : 2 M]**

3.5 A block of mass *M* is released from point *P* on a rough inclined plane with inclination angle θ, shown in the figure below. The coefficient of friction is μ. If $\mu < \tan\theta$, then the time taken by the block to reach another point *Q* on the inclined plane, where *PQ* = *s*, is

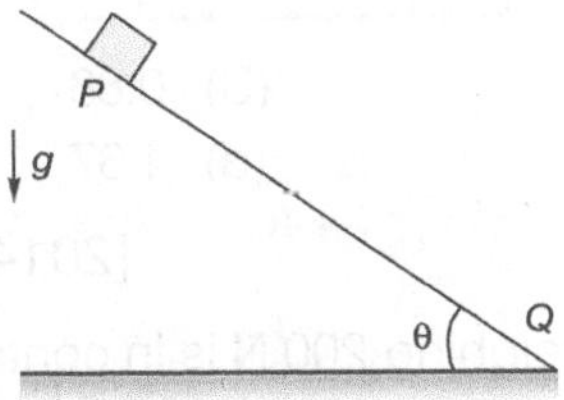

(a) $\sqrt{\dfrac{2s}{g\cos\theta(\tan\theta-\mu)}}$ (b) $\sqrt{\dfrac{2s}{g\cos\theta(\tan\theta+\mu)}}$

(c) $\sqrt{\dfrac{2s}{g\sin\theta(\tan\theta-\mu)}}$ (d) $\sqrt{\dfrac{2s}{g\sin\theta(\tan\theta+\mu)}}$

[2007 : 2 M]

3.6 A block weighing 981 N is resting on a horizontal surface. The coefficient of friction between the block and the horizontal surface is μ = 0.2. A vertical cable attached to the block provides partial support as shown. A man can pull horizontally with a force of 100 N. What will be the tension, *T* (in N) in the cable if the man is just able to move the block to the right?

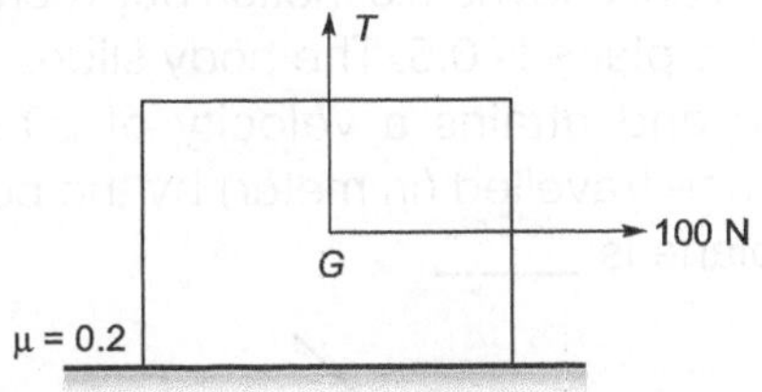

(a) 176.2 (b) 196.0
(c) 481.0 (d) 981.0 **[2009 : 1 M]**

3.7 A 1 kg block is resting on a surface with coefficient of friction μ = 0.1. A force of 0.8 N is applied to the block as shown in the figure. The friction force is

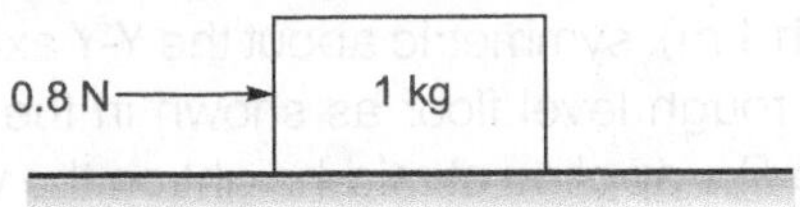

(a) 0 (b) 0.8 N
(c) 0.98 N (d) 1.2 N **[2011 : 2 M]**

3.8 A block R of mass 100 kg is placed on a block S of mass 150 kg as shown in the figure. Block R is tied to the wall by a massless and inextensible string PQ. If the coefficient of static friction for all surfaces is 0.4, the minimum force F (in kN) needed to move the block S is

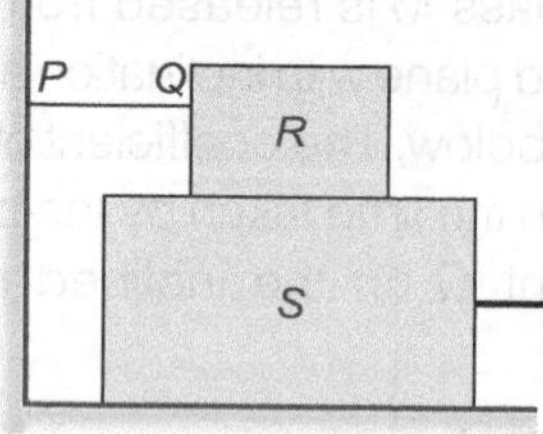

(a) 0.69 (B) 0.88
(c) 0.98 (d) 1.37

[2014 : 2 M, Set-1]

3.9 A block weighing 200 N is in contact with a level plane whose coefficients of static and kinetic friction are 0.4 and 0.2, respectively. The block is acted upon by a horizontal force (in newton) $P = 10t$, where t denotes the time in seconds. The velocity (in m/s) of the block attained after 10 seconds is ___. [2014 : 2 M, Set-1]

3.10 A truck accelerates up a 10° incline with a crate of 100 kg. Value of static coefficient of friction between the crate and the truck surface is 0.3. The maximum value of acceleration (in m/s^2) of the truck such that the crate does not slide down is ______. [2014 : 2 M, Set-2]

3.11 A body of mass (M) 10 kg is initially stationary on a 45° inclined plane as shown in figure. The coefficient of dynamic friction between the body and the plane is 0.5. The body slides down the plane and attains a velocity of 20 m/s. The distance travelled (in meter) by the body along the plane is ______.

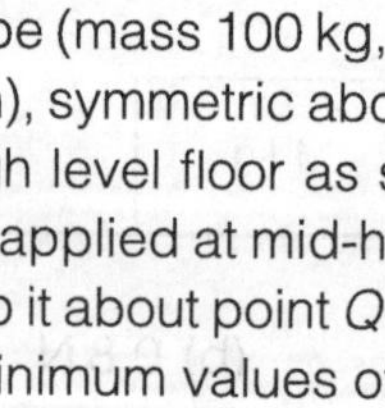

[2014 : 2 M, Set-3]

3.12 A wardrobe (mass 100 kg, height 4 m, width 2 m, depth 1 m), symmetric about the Y-Y axis, stands on a rough level floor as shown in the figure. A force P is applied at mid-height on the wardrobe so as to tip it about point Q without slipping. What are the minimum values of the force (in newton) and the static coefficient of friction μ between the floor and the wardrobe, respectively?

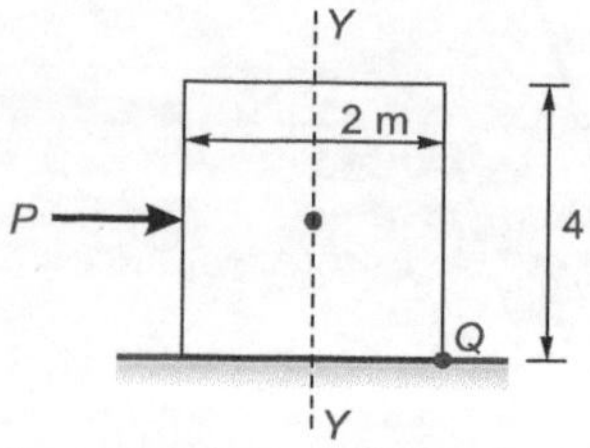

(a) 490.5 and 0.5 (b) 981 and 0.5
(c) 1000.5 and 0.15 (d) 1000.5 and 0.25

[2014 : 2 M, Set-4]

3.13 A force of 100 N is applied to the centre of a circular disc, of mass 10 kg and radius 1 m, resting on a floor as shown in the figure. If the disc rolls without slipping on the floor, the linear acceleration (in m/s^2) of the centre of the disc is ______ (correct to two decimal places).

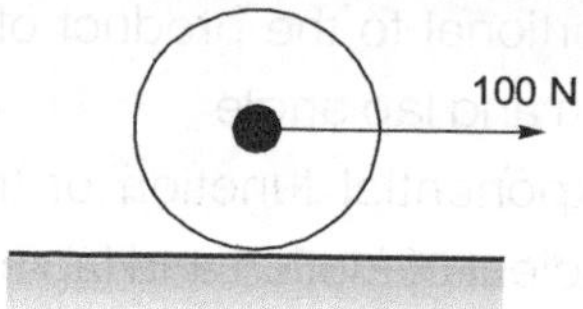

[2018 : 2 M, Set-2]

3.14 A block of mass 10 kg rests on a horizontal floor. The acceleration due to gravity is 9.81 m/s^2. The coefficient of static friction between the floor and the block is 0.2. A horizontal force of 10 N is applied on the block as shown in the figure. The magnitude of force of friction (in N) on the block is ______.

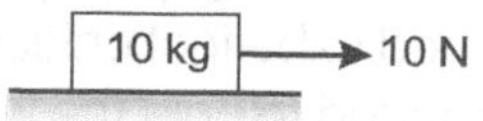

[2019 : 1 M, Set-1]

3.15 A block of negligible mass rests on a surface that is inclined at 30° to the horizontal plane as shown in the figure. When a vertical force of 900 N and a horizontal force of 750 N are applied, the block is just about to slide.

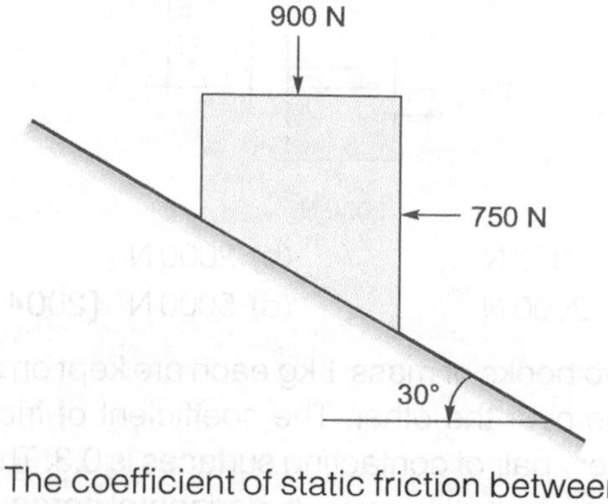

The coefficient of static friction between the block and surface is ______ (round off to two decimal places)

[2021 : 2 M, Set-2]

3.16 Two rigid massless rods PR and RQ are joined at frictionless pin-joint R and are resting on ground at P and Q, respectively, as shown in the figure. A vertical force F acts on the pin R as shown. When the included angle $\theta < 90°$, the rods remain in static equilibrium due to Coulomb friction between the rods and ground at locations P and Q. At $\theta = 90°$, impending slip occurs simultaneously at points P and Q. Then the ratio of the coefficient of friction at Q to that at $P\left(\frac{\mu_Q}{\mu_P}\right)$ is ________. (round off to two decimal places).

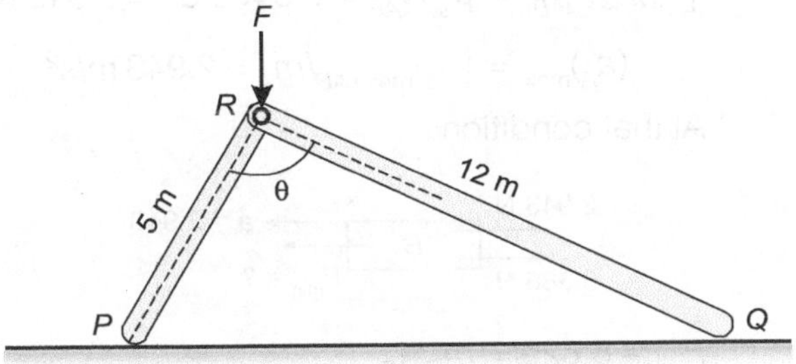

[2022 : 2 M, Set-1]

3.17 The braking system shown in the figure uses a belt to slow down a pulley rotating in the clockwise direction by the application of a force P. The belt wraps around the pulley over an angle $\alpha = 270$ degrees. The coefficient of friction between the belt and the pulley is 0.3. The influence of centrifugal forces on the belt is negligible. During braking, the ratio of the tensions T_1 to T_2 in the belt is equal to ________. (Rounded off to two decimal places)
Take $\pi = 3.14$.

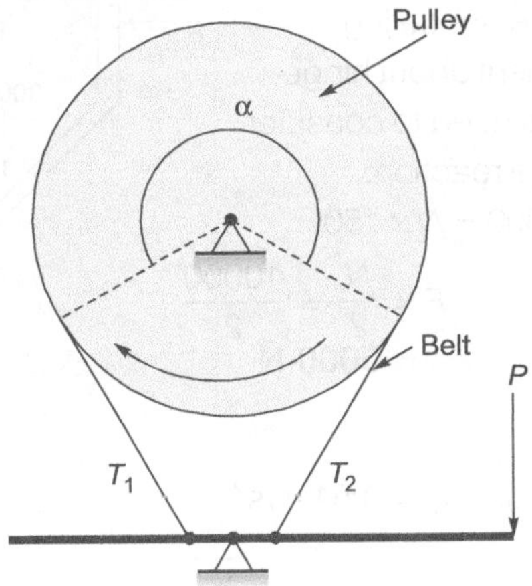

[2023 : 1 M]

Answers Friction

3.1 (a) 3.2 (b) 3.3 (d) 3.4 (b) 3.5 (a) 3.6 (c) 3.7 (b) 3.8 (d) 3.9 (4.9)
3.10 (1.1948) 3.11 (57.667) 3.12 (a) 3.13 (6.67) 3.14 (10) 3.15 (0.17) 3.16 (5.76)
3.17 (4.11)

Explanations Friction

3.1 (a)

Rolling Resistance: The minimum force required at centre of roller parallel to the contacting surface so as to start rolling.
Considering rolling resistance. There will not be point or line contact, there will be a surface contact

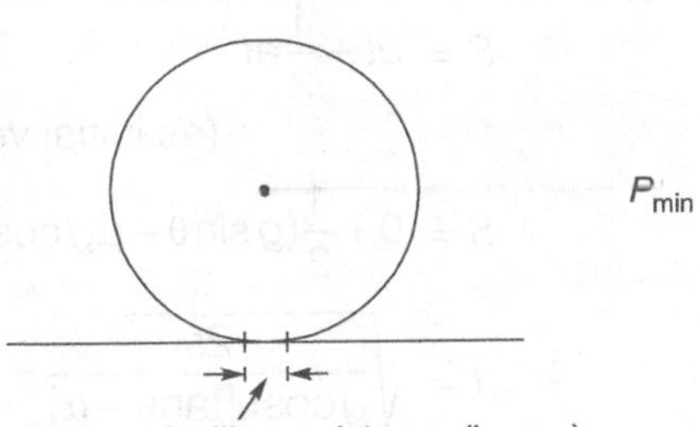

Coefficient of rolling resistance (in mm)

(this length of contact in mm is known as coefficient of rolling resistance)

$$f_R \text{ (rolling friction of wheel)} = \frac{W}{R}\mu_R$$

$\mu_R \rightarrow$ Coefficient of rolling resistance
$W \rightarrow$ Load on wheel
$R \rightarrow$ Radius of wheel

$$f_R = \frac{500 \times 0.3}{300} = 0.5 \text{ N}$$

3.2 (b)

$$\frac{T_1}{T_2} = e^{\mu\theta}$$

where θ = Lap angle and μ = Coefficient of friction.

3.3 (d)

FBD of object

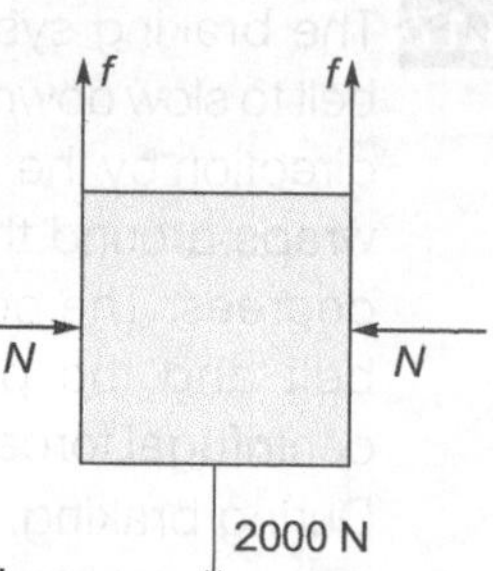

$2f = 2000$

$f = 1000$

$= \mu N$

$= 0.1 \times N$

$N = 10000$ N

Considering one tong of gripper.

Taking net moment about hinge point of gripper tong zero (because it is in equilibrium).

As we are taking moment about hinge so no need to consider hinge reaction.

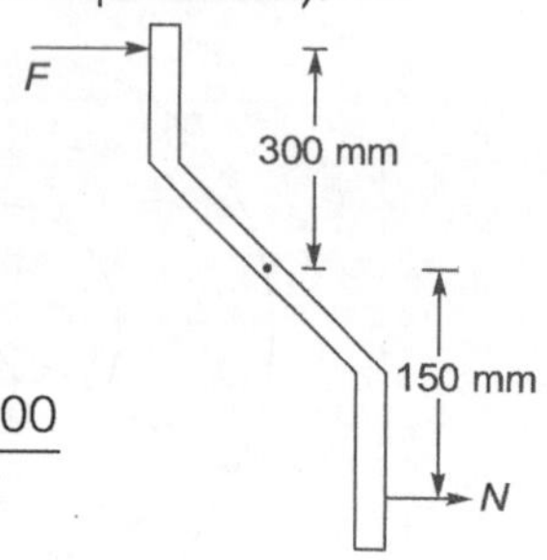

$F \times 300 = N \times 150$

$$F = \frac{N}{2} = \frac{10000}{2}$$

$= 5000$ N

3.4 (b)

Taking, $g = 9.81$ m/s^2

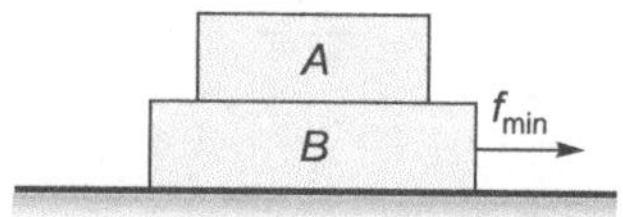

Now as we apply force on lower book B. Firstly book A and B will start moving together with same acceleration slipping from C, so F_{min} need to over come $[(f_s)_{max}]_{BC}$

As there no acceleration in vertical direction, so in vertical direction net force will be zero on A and B both.

$[(f_s)_{max}]_{BC} = \mu_s[N_{BC}]$

FBD of A

$\Rightarrow \quad N_{AB} = mg = 1 \times 9.81$

$= 9.81$ N

FBD of B

$N_{BC} = mg + N_{BA}$

$= 9.81 + 9.81 = 19.62$ N

$[(f_s)_{max}]_{BC} = 0.3 \times 19.62 = 5.886$ N

Now when we will apply force more than 5.886 N, they will start moving together with same acceleration.

Now as we keep on increasing F, acceleration of book B will go on increasing and thus increasing acceleration of book A. But acceleration of book A is increasing because of static friction and it has a maximum value. So after that acceleration of book B will be more than acceleration of A. So there will be relative slipping between B and A.

$[a_{max}]_A = ?$

$[a_{max}]_A$ will be corresponding to $[(f_s)_{max}]_{AB}$

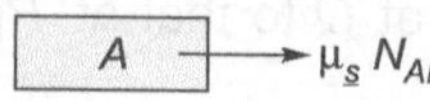

$[(f_s)_{max}]_{AB} = \mu_s N_{AB} = 0.3 \times 9.81 = 2.943$ N

$(a_A)_{max} = [(f_s)_{max}]_{AB}/m_A = 2.943$ m/s^2

At that condition

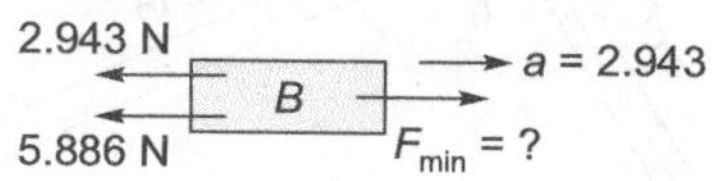

$F_{min} - 8.829 = m_B a_B$

$F_{min} = 8.829 + 1 \times 2.943 = 11.772$ N

At this point both will be moving with same acceleration but beyond this if we will increase F, acceleration of B will be more than A and there will be slipping between the books.

3.5 (a)

FBD of block :

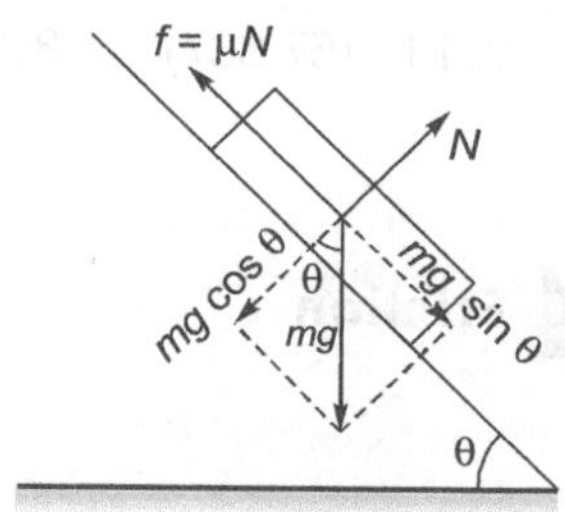

$N = mg\cos\theta$

From Newton's second law

$mg\sin\theta - \mu\, mg\cos\theta = ma$

$a = g\sin\theta - \mu\, g\cos\theta$

$$S = ut + \frac{1}{2}at^2$$

(as initial velocity, $u = 0$)

$$S = 0 + \frac{1}{2}(g\sin\theta - \mu g\cos\theta)t^2$$

$$t = \sqrt{\frac{2s}{g\cos\theta[\tan\theta - \mu]}}$$

3.6 (c)

Free body diagram

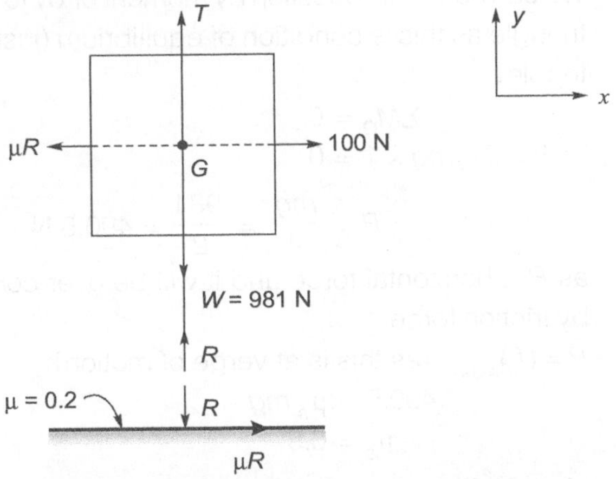

Assume normal reaction = R (Newton)

Balancing force in horizontal direction

$$(\Sigma Fx = 0)$$

$$\mu R = 100$$

$$(0.2)R = 100$$

$$R = 500\text{ N}$$

Balancing the force in vertical direction

$$(\Sigma Fy = 0)$$

$$T + R - W = 0$$

$$\Rightarrow T + 500 - 981 = 0$$

$$T = \mathbf{481\ N}$$

3.7 (b)

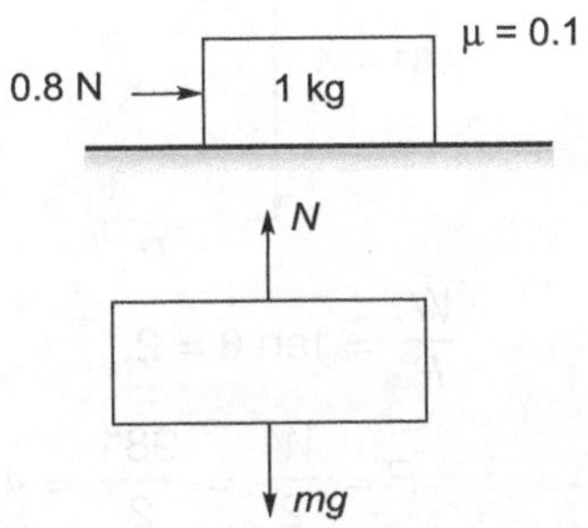

$$\text{Friction force} = \mu \times N$$

$$\Sigma F_V = 0$$

$$\therefore \quad N - mg = 0$$

$$N = mg = 1 \times 9.81$$

$$\text{Friction force} = \mu \times N = 0.1 \times 9.81$$

$$= 0.981\text{ N}$$

Since applied force is less than friction force hence friction force acting on body at equilibrium is **0.8 N.**

3.8 (d)

Free body diagram of block *R* and *S*.

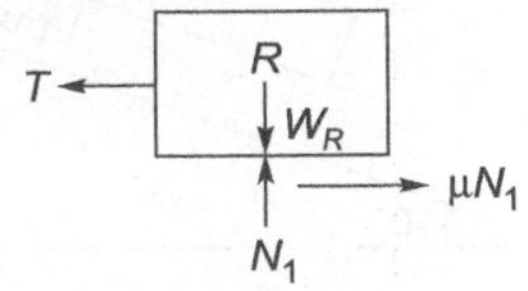

$$N_1 = W_R$$

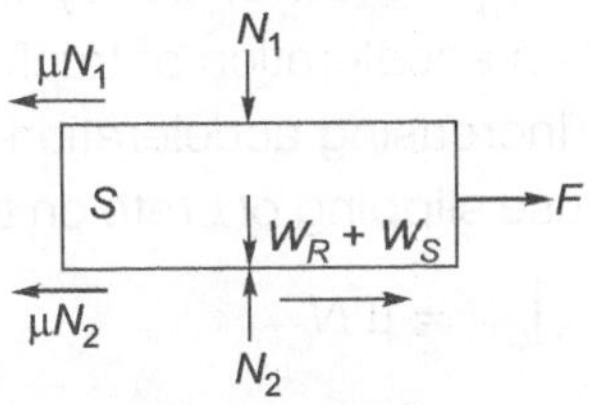

$$N_2 = W_R + W_S$$

Free body diagram of block *R* and *S*.

$$F \geq \mu N_1 + \mu N_2$$

$$F \geq \mu(N_1 + N_2)$$

$$F \geq \mu(W_R + W_R + W_S)$$

$$F \geq 0.4\,(100 + 100 + 150) \times 9.81$$

$$F \geq 1373.4\text{ N}$$

$$F \geq 1.37\text{ kN}$$

3.9 Sol.

$$N = mg = 200\text{ N}$$

$$(f_s)_{max} = \mu_s N$$

$$= 0.4 \times 200 = 80\text{ N}$$

upto 8 sec [$P = 10\,t$] $\rightarrow$ no motion

$\Rightarrow$ from 8 sec to 10 sec

$$f_k = \mu_k N = 0.2 \times 200 = 40\text{ N}$$

Average applied force

$$= \frac{P_{max} + P_{min}}{2} = \frac{100 + 80}{2} = 90\text{ N}$$

Average resultant force = Avg. applied force − kinetic friction = 90 − 40 = 50 N

$$\text{Average acceleration} = \frac{50 \times 9.8}{200}$$

$$\Rightarrow \quad V = u + at$$

$$u = 0$$

$$a = \frac{50 \times 9.8}{200}$$

$$t = 2\text{ second}$$

$$\therefore \quad V = \frac{50 \times 9.8}{200} \times 2 = \mathbf{4.9\ m/s}$$

3.10 Sol.

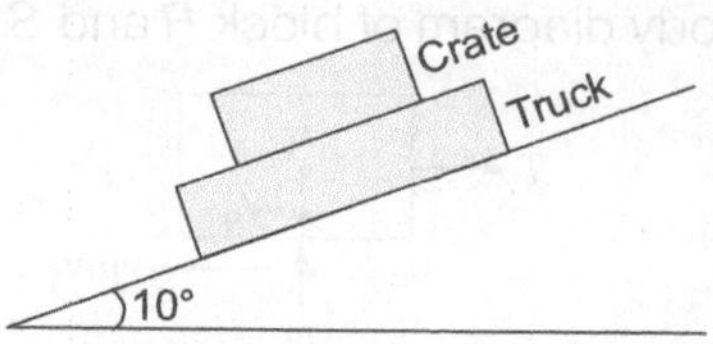

Crate will acquire max acceleration when $(f_s)_{max}$ will be acting on crate by truck and at that condition acceleration of truck and crate will be same. Increasing acceleration of truck after that will cause slipping of crate on truck.

$$[(f_s)_{max}]_{CT} = \mu N_{CT}$$

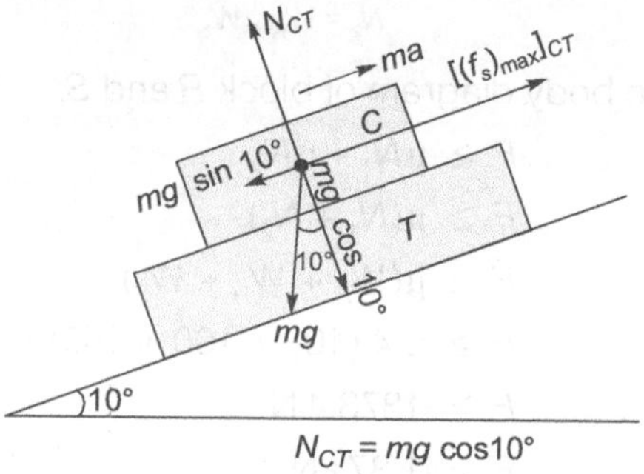

$$[(f_s)_{max}]_{CT} - mg\sin 10° = ma$$

$$\mu mg\cos 10° - mg\sin 10° = ma$$

or $\quad a = \mathbf{1.1948\ m/s^2}$

3.11 Sol.

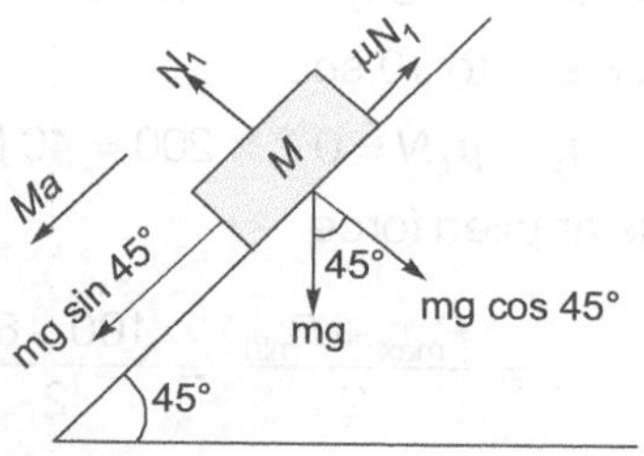

$N_1 = Mg\cos 45°$ (From Newton's first law)

$\therefore \quad Mg\sin 45° - \mu N_1 = \text{Ma}$

(From Newton's second law)

$Mg\sin 45° - \mu\ \text{Mg}\cos 45° = \text{Ma}$

$$a = g\sin 45° - \mu g\cos 45°$$
$$= 3.468\ \text{m/s}^2$$

Now, from 2nd equation of motion,

$$v^2 = u^2 + 2as$$
$$(20)^2 = 0 + 2 \times 3.468 \times s$$
$$s = 57.667\ \text{m}$$

3.12 (a)

__Method I:__

We can solve this question by moment or by force triangle as this is condition of equilibrium (just to topple)

$$\Sigma M_Q = 0$$
$$P \times 2 - mg \times 1 = 0$$
$$P = \frac{mg}{2} = \frac{981}{2} = 490.5\ \text{N}$$

as P is horizontal force and it will be over come by friction force

$P = (f_s)_{max}$ [as this is at verge of motion]

$$490.5 = \mu_s mg$$
$$\mu_s = 0.5$$

__Method II:__

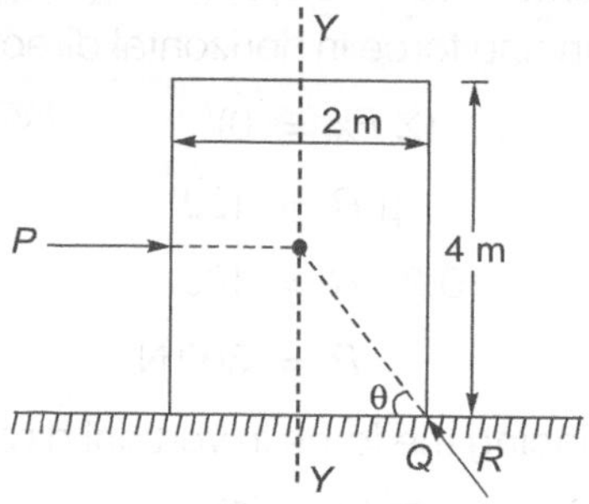

$$\tan\theta = \frac{2}{1}$$
$$\theta = 63.4349°$$

Wardrobe is about to topple about pt Q so there will be contact force from point Q on the wardrobe

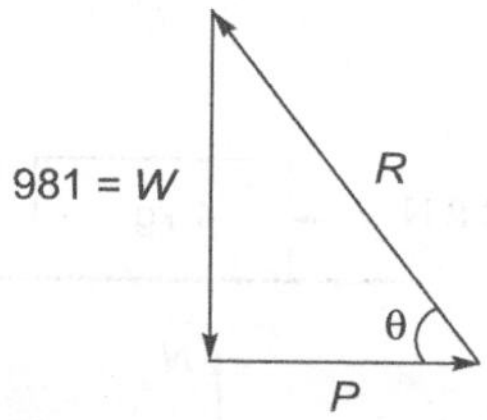

$$\frac{W}{P} = \tan\theta = 2$$

$$P = \frac{W}{2} = \frac{981}{2} = \mathbf{490.5\ N}$$

Two components of R will balance W and P and horizontal component (parallel to contacting surface is called friction and perpendicular to contacting surface is called normal reaction)

$$R\cos\theta = \text{friction force}$$
$$= P\text{(in magnitude)}$$
$$(f_s)_{max} = P = 490.5$$

and $\quad (f_s)_{max} = \mu_s mg$

$$\mu_s mg = 490.5$$
$$\mu_s = \mathbf{0.5}$$

3.13 Sol.

Given : $m = 10$ kg, $r = 1$ m

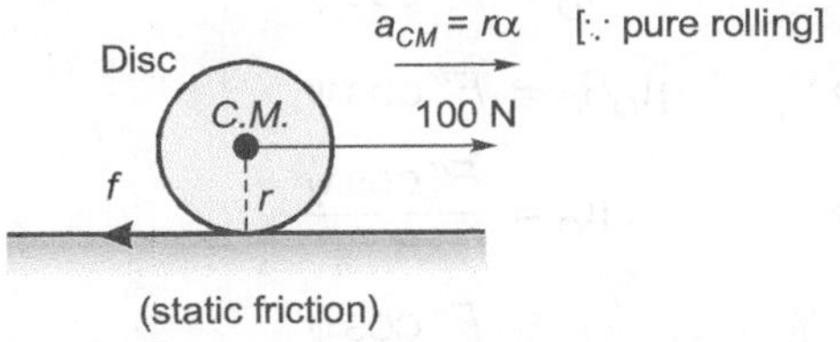

$$100 - f = ma_{CM}$$
$$= mr\alpha \text{ (Newton's II}^{nd}\text{ law)} \quad ...(i)$$
$$[\text{Torque}]_{CM} = I_{CM}\alpha$$
$$f \times r = \frac{mr^2}{2} \times \alpha$$
$$2f = mr\alpha$$
$$2f = ma_{CM} \quad [\because a_{CM} = r\alpha]$$
$$f = \frac{m}{2} \times a_{CM} \quad ...(ii)$$

By equation (i) and (ii):

$$100 - \frac{ma_{CM}}{2} = ma_{CM}$$
$$100 = \frac{3}{2} ma_{CM}$$
$$\frac{200}{3 \times 10} = a_{CM}$$
$$a_{CM} = \frac{20}{3} = \mathbf{6.67 \ m/s^2}$$

<u>Method II:</u>

Given : $F = 100$ N, $m = 10$ kg, $r = 1$ m

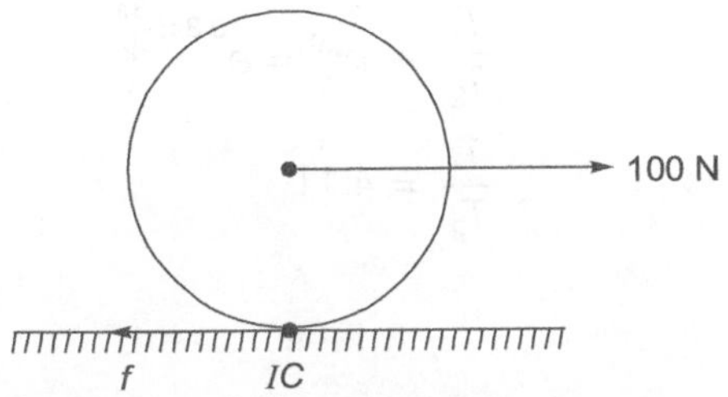

Calculating torque about I centre

$$T_{IC} = I_{IC}\alpha$$
$$T_{IC} = \frac{1}{2}mr^2 + mr^2 = \frac{3}{2}mr^2$$
$$\Rightarrow \quad 100 \times r = \frac{3}{2}mr^2\alpha$$
$$100 = \frac{3}{2} \times 10 \times 1^2 \times \alpha$$
$$\alpha = \frac{20}{3} = 6.67 \text{ rad/s}^2$$
$$a_{cm} = r\alpha = 1 \times 6.67 = 6.67 \text{ m/s}^2$$

Remember : Angular acceleration will be same i.e. $a_{cm} = \alpha_{IC}$

3.14 Sol.

Given : $m = 10$ kg, $\mu = 0.2$, $F = 10$ N

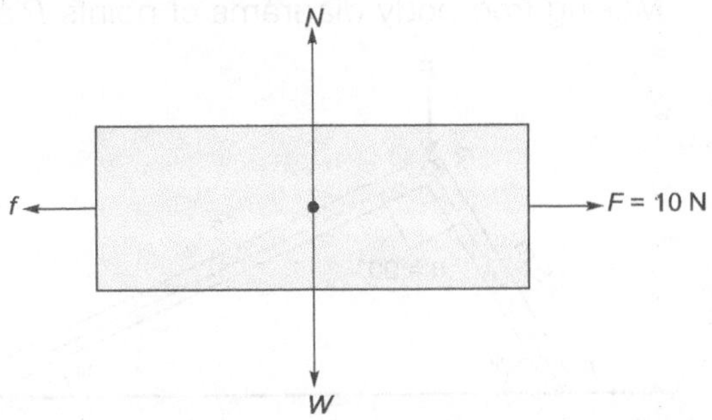

$$W = mg = 10 \times 9.81 = 98.1 \text{ N}$$

Normal force, $N = W = 98.1$ N

Now, Maximum value of static friction force

$$= \mu_S N = 0.2 \times 98.1 = 19.62 \text{ N}$$

∵ Applied force is less than static friction force, so friction force will be same as applied force.

Hence, Friction force = 10 N

3.15 Sol.

After forces are applied block is just about move (mass is negligible). Calculate coefficient of friction,

$$F_H = 750 \text{ N}$$
$$F_V = 900 \text{ N}$$
$$\theta = 30°$$

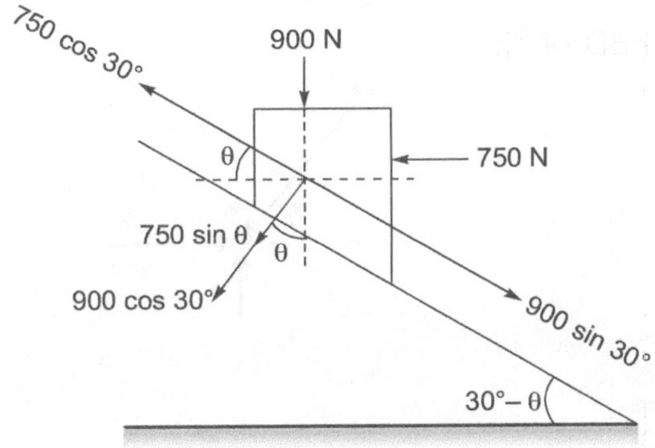

$$N = 900 \cos\theta + 750 \sin\theta$$
$$N = 900 \cos 30° + 750 \sin 30°$$
$$= 1154.4228 \text{ N}$$
$$F_{max} + 900 \sin 30° = 750 \cos 30°$$
$$\mu N = 199.519$$
$$\mu = \frac{199.519}{1154.4228}$$
$$\mu = 0.1728$$

3.16 (5.76) (5.70 to 5.80)

Considering the limiting condition at which $\theta = 90°$

Let F' be the force in member PR and F'' be the force in the member QR.

Making free-body diagrams of points P and Q.

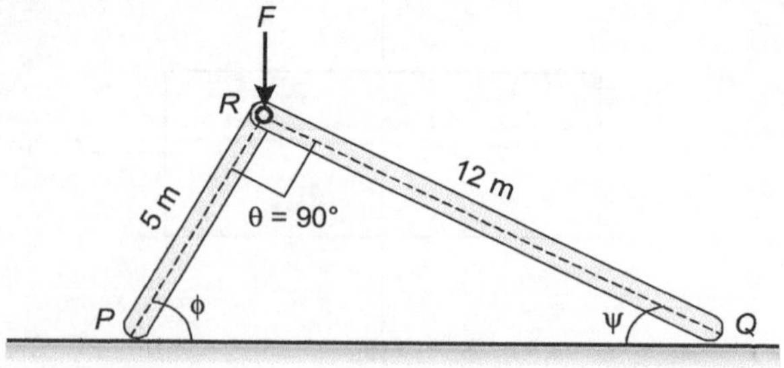

FBD of P,

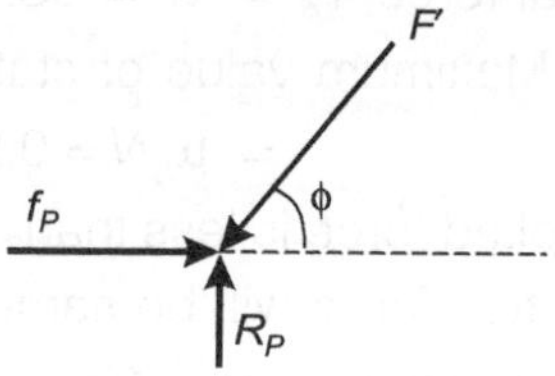

$$\Rightarrow \quad R_P = F'\sin\phi$$

$$\Rightarrow \quad f_P = F'\cos\phi$$

$$\Rightarrow \quad \mu_P R_P = F'\cos\phi$$

$$\Rightarrow \quad \mu_P = \frac{F'\cos\phi}{F'\sin\phi}$$

$$\Rightarrow \quad \mu_P = \frac{1}{\tan\phi}$$

FBD of Q,

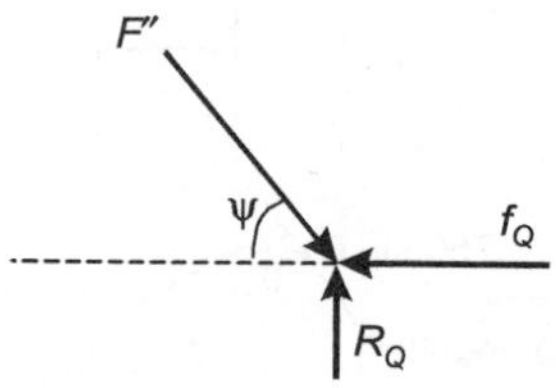

$$\Rightarrow \quad R_Q = F''\sin\psi$$

$$\Rightarrow \quad f_Q = F''\cos\psi$$

$$\Rightarrow \quad \mu_Q R_Q = F''\cos\psi$$

$$\Rightarrow \quad \mu_Q = \frac{F''\cos\psi}{R_Q}$$

$$\Rightarrow \quad \mu_Q = \frac{F''\cos\psi}{F''\sin\psi}$$

$$\Rightarrow \quad \mu_Q = \frac{1}{\tan\psi}$$

$$\therefore \quad \frac{\mu_Q}{\mu_P} = \frac{\dfrac{1}{\tan\psi}}{\dfrac{1}{\tan\phi}} = \frac{\tan\phi}{\tan\psi}$$

From the given diagram

$$\Rightarrow \quad \tan\phi = \frac{12}{5}$$

$$\Rightarrow \quad \tan\psi = \frac{5}{12}$$

$$\therefore \quad \frac{\mu_Q}{\mu_P} = \frac{\tan\phi}{\tan\psi} = \frac{12/5}{5/12} = \frac{12}{5}\times\frac{12}{5} = 5.76$$

3.17 (4.11) (4.05 to 4.15)

Given, coefficient of friction, $\mu = 0.3$

Angle of contact, $\theta = 270°$

$$\theta = \frac{3\pi}{2}$$

The ratio of tensions is given by,

$$\frac{T_1}{T_2} = e^{\mu\theta} = e^{0.3\times\frac{3\pi}{2}}$$

$$\Rightarrow \quad \frac{T_1}{T_2} = 4.11$$

■■■■

4 Impulse & Momentum, Impacts & Work-Energy

4.1 A ball A of mass m falls under gravity from a height h and strikes another ball B of mass m which is supported at rest on a spring of stiffness k. Assume perfect elastic impact. Immediately after the impact

(a) the velocity of ball A is $\frac{1}{2}\sqrt{2gh}$

(b) the velocity of ball A is zero

(c) the velocity of both balls is $\frac{1}{2}\sqrt{2gh}$

(d) None of the above **[1996 : 1 M]**

4.2 Match 4 correct pairs between **List-I** and **List-II** for questions. No credit will be given for partially correct matching.

List-I

A. Collision of particles
B. Stability
C. Satellite motion
D. Spinning top

List-II

1. Euler's equation of motion
2. Minimum kinetic energy
3. Minimum potential energy
4. Impulse momentum principle
5. Conservation of moment of momentum

Codes:

	A	B	C	D
(a)	1	3	5	1
(b)	2	4	3	5
(c)	4	3	2	5
(d)	4	3	1	5

[1996 : 2 M]

4.3 A bullet of mass m travels at a very high velocity V (as shown in the figure) and gets embedded inside the block of mass M initially at rest on a rough horizontal floor. The block with the bullet is seen to move a distance s along the floor. Assuming μ to be the coefficient of kinetic friction between the block and the floor and g the acceleration due to gravity what is the velocity V of the bullet?

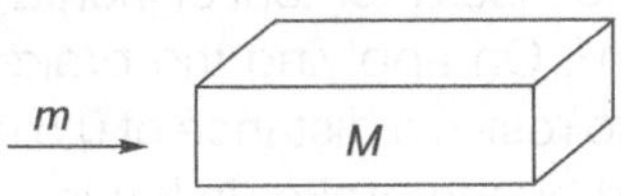

(a) $\frac{M+m}{m}\sqrt{2\mu gs}$ (b) $\frac{M-m}{m}\sqrt{2\mu gs}$

(c) $\frac{\mu(M+m)}{mm}\sqrt{2\mu gs}$ (d) $\frac{M}{m}\sqrt{2\mu gs}$

[2003 : 1 M]

4.4 An ejector mechanism consists of a helical compression spring having a spring constant of $K = 981 \times 10^3$ N/m. It is pre-compressed by 100 mm from its free state. If it is used to eject a mass of 100 kg held on it, the mass will move up through a distance of

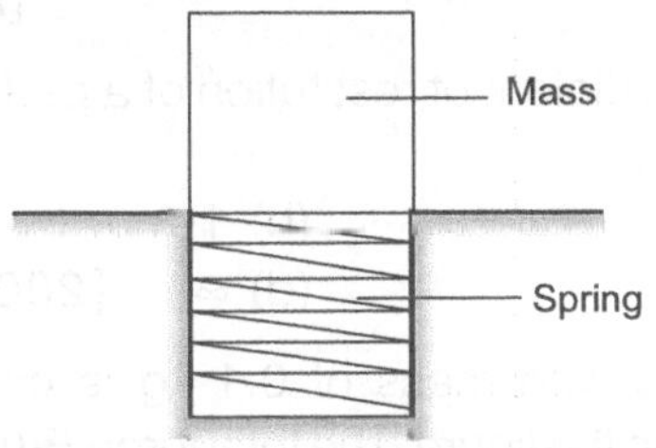

(a) 100 mm (b) 5000 mm

(c) 581 mm (d) 1000 mm

[2004 : 2 M]

4.5 An elevator (lift) consists of the elevator cage and a counter weight, of mass m each. The cage and the counterweight are connected by chain that passes over a pulley. The pulley is coupled to a motor. It is desired that the elevator should have a maximum stopping time of t seconds from a peak speed v. If the inertias of the pulley and the chain are neglected, the minimum power that the motor must have is

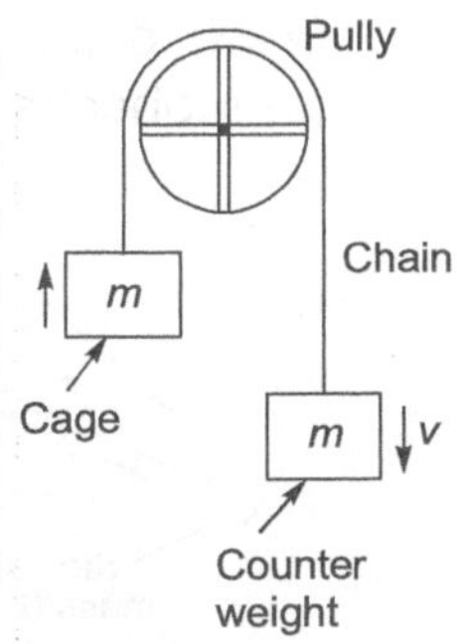

(a) $\frac{1}{2}mv^2$ (b) $\frac{mv^2}{2t}$

(c) $\frac{mv^2}{t}$ (d) $\frac{2mv^2}{t}$ **[2005 : 2 M]**

4.6 A 1 kg mass of clay, moving with a velocity of 10 m/s, strikes a stationary wheel and sticks to it. The solid wheel has a mass of 20 kg and a radius of 1 m. Assuming that the wheel is set into pure rolling motion, the angular velocity of the wheel immediately after the impact is approximately

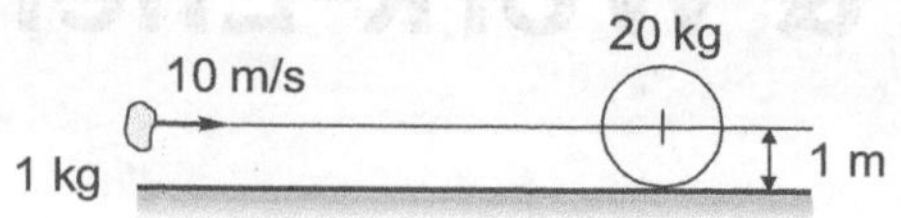

(a) zero (b) $\frac{1}{3}$ rad/s

(c) $\sqrt{\frac{10}{3}}$ rad/s (d) $\frac{10}{3}$ rad/s

[2005 : 2 M]

4.7 During inelastic collision of two particles, which one of the following is conserved?

(a) Total linear momentum only

(b) Total kinetic energy only

(c) Both linear momentum and kinetic energy

(d) Neither linear momentum nor kinetic energy

[2007 : 1 M]

4.8 The coefficient of restitution of a perfectly plastic impact is

(a) 0 (b) 1

(c) 2 (d) ∞ [20011 : 1 M]

4.9 A stone with mass of 0.1 kg is catapulted as shown in the figure. The total force F_x (in N) exerted by the rubber band as a function of distance x (in m) is given by $F_x = 300x^2$. If the stone is displaced by 0.1 m from the un-stretched position ($x = 0$) of the rubber band, the energy stored in the rubber bad is

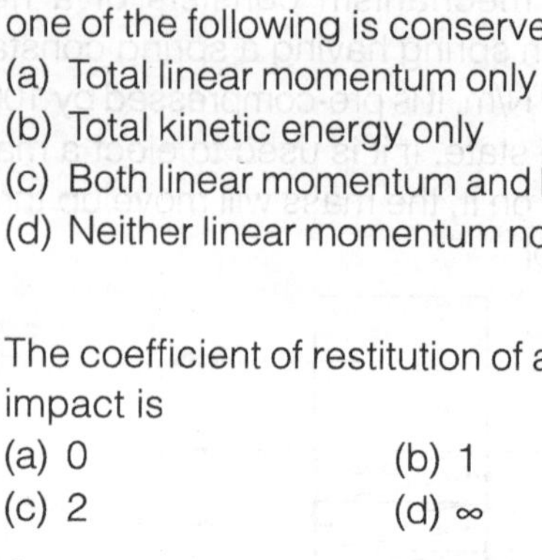

(a) 0.01 J (b) 0.1 J

(c) 1 J (d) 10 J [2011 : 2 M]

4.10 A single degree of freedom system having mass 1 kg and stiffness 10 kN/m initially at rest is subjected to an impulse force of magnitude 5 kN for 10^{-4} seconds. The amplitude in mm of the resulting free vibration is

(a) 0.5 (b) 1.0

(c) 5.0 (d) 10.0

[2013 : 2 M]

4.11 A mass m_1 of 100 kg travelling with a uniform velocity of 5 m/s along a line collides with a stationary mass m_2 of 1000 kg. After the collision, both the masses travel together with the same velocity. The coefficient of restitution is

(a) 0.6 (b) 0.1

(c) 0.01 (d) 0

[2014 : 1 M, Set-3]

4.12 An annular disc has a mass m, inner radius R and outer radius $2R$. The disc rolls on a flat surface without slipping. If the velocity of the center of mass is v, the kinetic energy of the disc is

(a) 9/16 mv^2 (b) 11/16 mv^2

(c) 13/16 mv^2 (d) 15/16 mv^2

[2014 : 2 M, Set-3]

4.13 A small ball of mass 1 kg moving with a velocity of 12 m/s undergoes a direct central impact with a stationary ball of mass 2 kg. The impact is perfectly elastic. The speed (in m/s) of 2 kg mass ball after the impact will be_____.

[2015 : 1 M, Set-2]

4.14 A ball of mass 0.1 kg, initially at rest, dropped from height of 1 m. The ball hits the ground and bounces off the ground. Upon impact with the ground, the velocity reduces by 20%. The height (in m) to which the ball will rise is ______.

[2015 : 2 M, Set-1]

4.15 A point mass M is released from rest and slides down a spherical bowl of radius R from a height H as shown in the figure below. The surface of the bowl is smooth (no friction). The velocity of the mass at the bottom of the bowl is

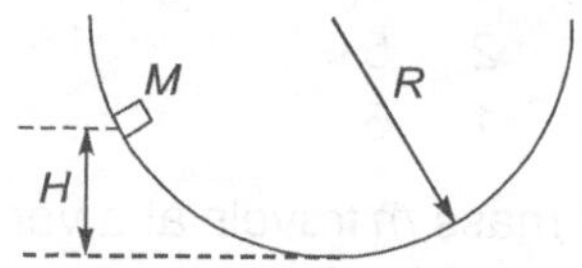

(a) $\sqrt{gH}$ (b) $\sqrt{2gR}$

(c) $\sqrt{2gH}$ (d) 0

[2016 : 1 M, Set-1]

4.16 A mass of 2000 kg is currently being lowered at a velocity of 2 m/s from the drum as shown in the figure. The mass moment of inertia of the drum is 150 kg-m^2. On applying the brake, the mass is brought to rest in a distance of 0.5 m. The energy absorbed by the brake (in kJ) is ________.

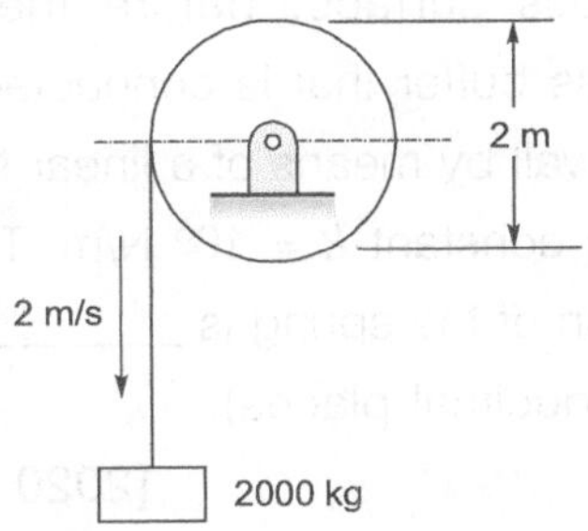

[2016 : 2 M, Set-2]

4.17 A point mass having mass M is moving with a velocity V at an angle θ to the wall as shown in the figure. The mass undergoes a perfectly elastic collision with the smooth wall and rebounds. The total change (final minus initial) in the momentum of the mass is

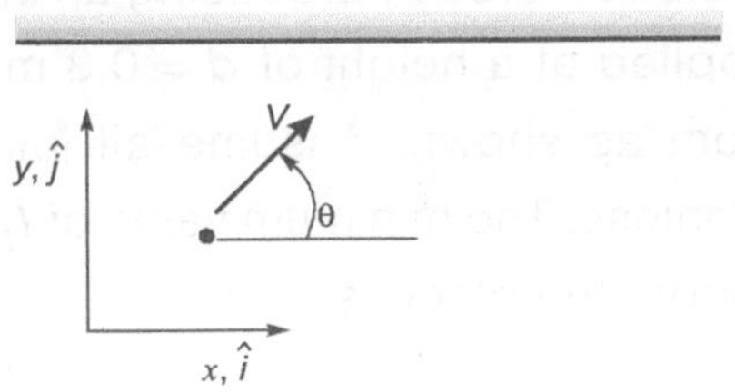

(a) $-2MV\cos\theta\,\hat{j}$ (b) $2MV\sin\theta\,\hat{j}$
(c) $2MV\cos\theta\,\hat{j}$ (d) $-2MV\sin\theta\,\hat{j}$

[2016 : 1 M, Set-2]

4.18 A circular disc of radius 100 mm and mass 1 kg, initially at rest at position A, rolls without slipping down a curved path as shown in figure. The speed v of the disc when it reaches position B is________ m/s.
Acceleration due to gravity, g = 10 m/s^2.

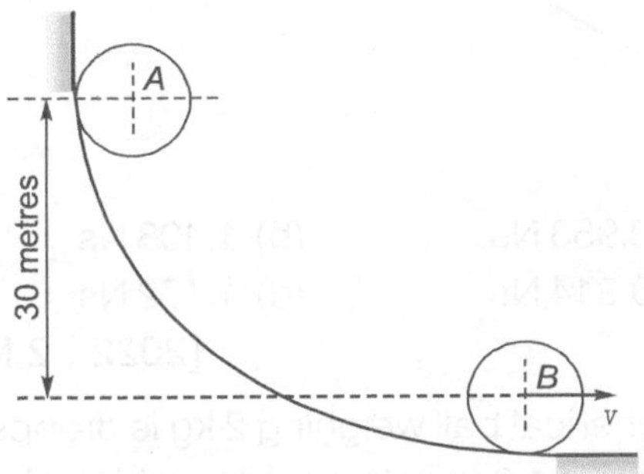

[2016 : 2 M, Set-3]

4.19 A system of particles in motion has mass center G as shown in the figure. The particle i has mass m_i and its position with respect to a fixed point O is given by the position vector r_i. The position of the particle with respect to G is given by the vector ρ_i. The time rate of change of the angular momentum of the system of particles about G is (The quantity $\ddot{\rho}_i$ indicates second derivative of ρ_i with respect to time and likewise for r_i).

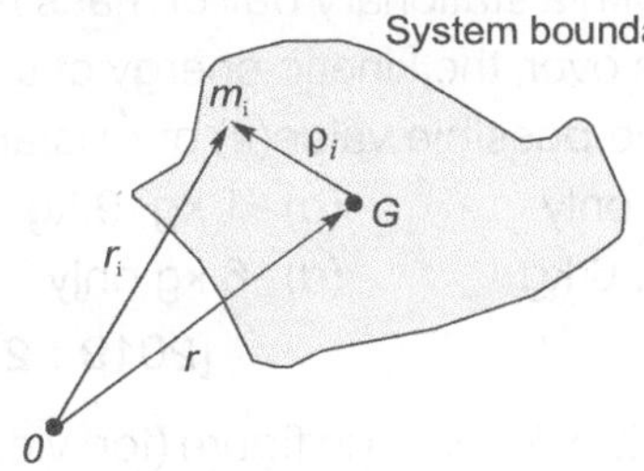

(a) $\sum_i r_i \times m_i \ddot{\rho}_i$ (b) $\sum \rho_i \times m_i \ddot{r}_i$
(c) $\sum_i r_i \times m_i \ddot{r}_i$ (d) $\sum_i \rho_i \times m_i \ddot{\rho}_i$

[2016 : 2 M, Set-2]

4.20 A particle of unit mass is moving on a plane. Its trajectory, in polar coordinates, is given by $r(t) = t^2$, $\theta(t) = t$, where t is time. The kinetic energy of the particle at time t = 2 is
(a) 4 (b) 12
(c) 16 (d) 24

[2017 : 1 M, Set-1]

4.21 Two disc A and B with identical mass (m) and radius (R) are initially at rest. They roll down from the top of identical inclined planes without slipping. Disc A has all of its mass concentrated at the rim, while Disc B has its mass uniformly distributed. At the bottom of the plane, the ratio of velocity of the center of disc A to the velocity of the center of disc B is
(a) $\sqrt{\dfrac{3}{4}}$ (b) $\sqrt{\dfrac{3}{2}}$
(c) 1 (d) $\sqrt{2}$

[2017 : 2 M, Set-1]

4.22 Block P of mass 2 kg slides down the surface and has a speed 20 m/s at the lowest point, Q where the local radius of curvature is 2 m as shown in the figure. Assuming g = 10 m/s^2, the normal force (in N) at Q is ________(correct to two decimal places).

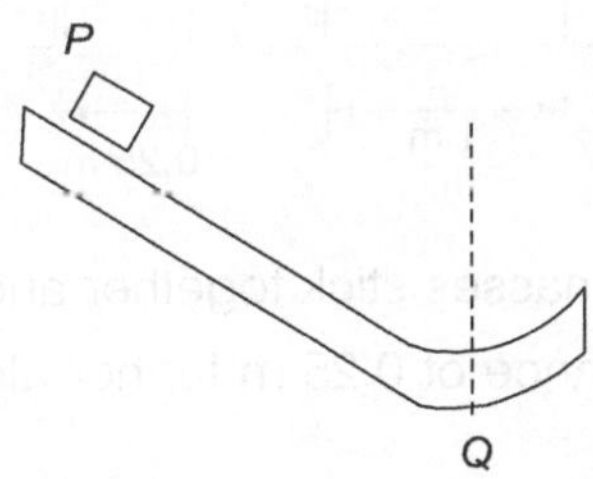

[2018 : 2 M, Set-1]

4.23 A ball of mass 3 kg moving with a velocity of 4 m/s undergoes a perfectly-elastic direct-central impact with a stationary ball of mass *m*. After the impact is over, the kinetic energy of the 3 kg ball is 6 J. The possible value(s) of *m* is/are

(a) 1 kg only (b) 1 kg, 9 kg
(c) 1 kg, 6 kg (d) 6 kg only

[2019 : 2 M, Set-2]

4.24 The 2 kg block shown in figure (top view) rests on a smooth horizontal surface and is attached to a massless elastic cord that has a stiffness 5 N/m.

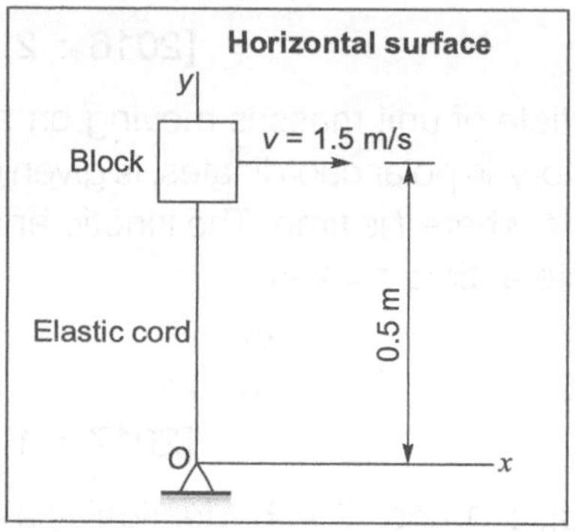

The cord hinged at *O* is initially unstretched and always remains elastic. The block is given a velocity *v* of 1.5 m/s perpendicular to the cord. The magnitude of velocity in m/s of the block at the instant the cord is stretched by 0.4 m is

(a) 1.36 (b) 1.07
(c) 0.83 (d) 1.50

[2020 : 2 M, Set-1]

4.25 A rigid block of mass m_1 = 10 kg having velocity v_0 = 2 m/s strikes a stationary block of mass m_2 = 30 kg after travelling 1 m along a frictionless horizontal surface as shown in the figure.

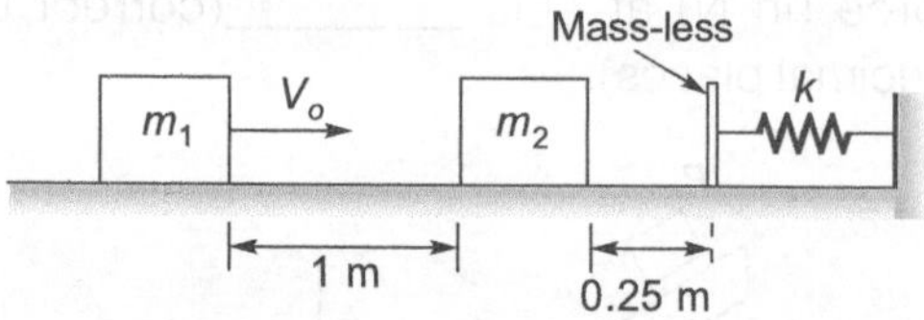

The two masses stick together and jointly move by a distance of 0.25 m further along the same frictionless surface, before they touch the mass-less buffer that is connected to the rigid vertical wall by means of a linear spring having a spring constant $k = 10^5$ N/m. The maximum deflection of the spring is ________ cm (round off to 2 decimal places).

[2020 : 2 M, Set-2]

4.26 A rigid homogeneous uniform block of mass 1 kg, height *h* = 0.4 m and width *b* = 0.3 m is pinned at one corner and placed upright in a uniform gravitational field (g = 9.81 m/s²), supported by a roller in the configuration shown in the figure. A short duration (impulsive) force *F*, producing an impulse I_F, is applied at a height of *d* = 0.3 m from the bottom as shown. Assume all joints to be frictionless. The minimum value of I_F required to topple the block is

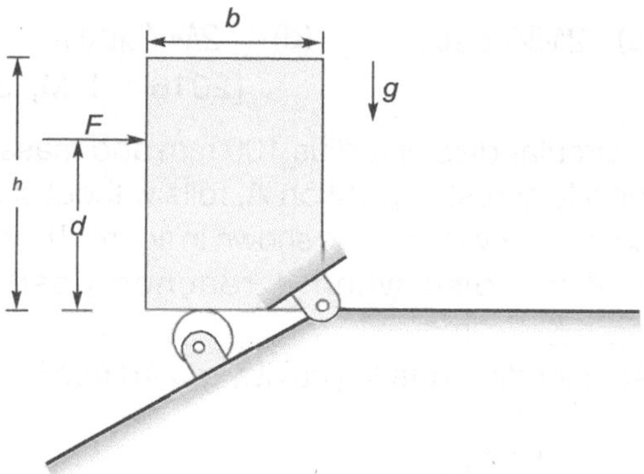

(a) 0.953 Ns (b) 1.403 Ns
(c) 0.814 Ns (d) 1.172 Ns

[2022 : 2 M, Set-2]

4.27 A spherical ball weighing 2 kg is dropped from a height of 4.9 m onto an immovable rigid block as shown in the figure. If the collision is perfectly elastic, what is the momentum vector of the ball (in kg m/s) just after impact?

Take the acceleration due to gravity to be g = 9.8 m/s². Options have been rounded off to one decimal place.

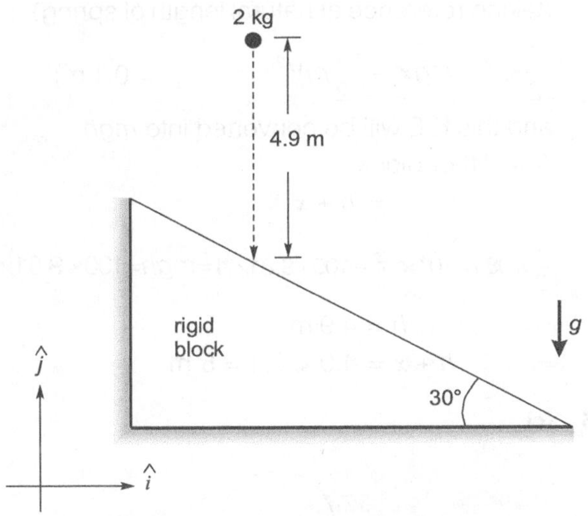

(a) $19.6\hat{i}$ (b) $19.6\hat{j}$
(c) $17.0\hat{i}+9.8\hat{j}$ (d) $9.8\hat{i}+17.0\hat{j}$

[2023 : 2 M]

4.28 A ram in the from of a rectangular body of size, l = 9 cm and b = 2 m is suspended by two parallel ropes of lengths 7 m. Assume the center-of-mass of the body is at its geometric center and g = 9.81 m/s^2. For striking the object P with horizontal velocity of 5 m/s, what is the angle θ with the vertical from which the ram should be released from rest?

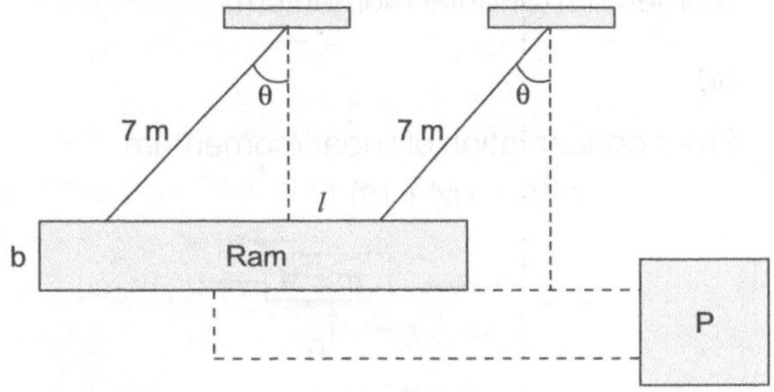

(a) 67.1° (b) 35.1°
(c) 40.2° (d) 79.5°

[2024 : 1 M]

Answers Impulse, Momentum, Impacts and Work-Energy

4.1 (b)	4.2 (d)	4.3 (a)	4.4 (b)	4.5 (c)	4.6 (b)	4.7 (a)	4.8 (a)	4.9 (b)
4.10 (c)	4.11 (d)	4.12 (c)	4.13 (8)	4.14 (0.638)	4.15 (c)	4.16 (14.11)	4.17 (d)	4.18 (20)
4.19 (b)	4.20 (c)	4.21 (a)	4.22 (420)	4.23 (b)	4.24 (a)	4.25 (1)	4.26 (a)	4.27 (c)
4.28 (b)								

Explanations Impulse, Momentum, Impacts and Work-Energy

4.1 (b)

<u>Method I:</u>

In perfectly elastic collision velocity of ball A will be zero after impact. Because in perfectly elastic impact velocities are interchanged when masses are same.

<u>Method II:</u>

$e = 1 \rightarrow$ in perfectly elastic collision.

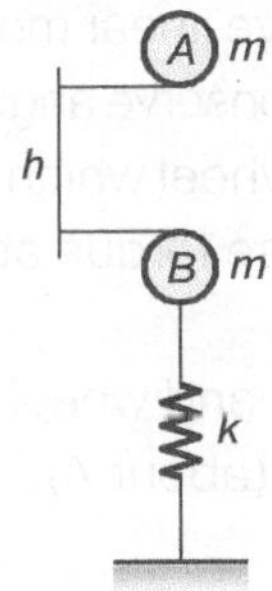

Before impact

Velocity of ball A,

$$u_A = \sqrt{2gh}$$

and $u_B = 0$

Vel. of app. $= u_A - u_B = \sqrt{2gh}$

Conserving linear momentum

$$m_A u_A + 0 = m_A V_A + m_B V_B$$

$$m_A = m_B = m$$

$$u_A = V_A + V_B \quad \text{...(i)}$$

For elastic collision

Velocity of approach = velocity of separation

$$u_A - u_B = V_B - V_A \quad \text{...(ii)}$$

Adding Eqs. (i) and (ii), we get

$$2u_A - u_B = 2V_B$$

As $u_B = 0$

$$\Rightarrow \quad 2u_A = 2V_B$$

$$u_A = V_B$$

Using value of u_A in Eq. (i), we get

$$\mathbf{V_A = 0}$$

$$V_B = u_A = \sqrt{2gh}$$

4.2 (d)

Collision of particles → Impulse momentum principle
Stability → Minimum potential energy
Satellite motion → Euler's equation of motion
(Constant K. E. not minimum K.E.)
Spinning top → Conservation of moment of momentum (angular momentum)

4.3 (a)

From conservation of linear momentum

$$mV = (M+m)\,U$$

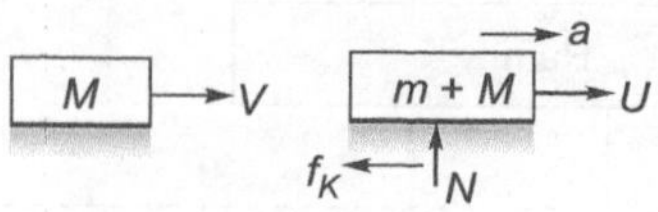

$$\therefore \quad U = \frac{mV}{M+m}$$

Kinetic friction,

$$\vec{f}_k = \mu N(-\hat{i})$$

$$(M+m)\quad \vec{a} = \vec{f}_K$$

$$\vec{f}_k = \mu(M+m)g(-\hat{i})$$

$$\vec{a} = \frac{\vec{f}_K}{(M+m)} = \frac{\mu(M+m)g(-\vec{i})}{(M+m)}$$

$$\vec{a} = \mu g(-\hat{i})$$

Now $\quad V_2^2 = U^2 - 2\mu gs$

$$\therefore \quad 0 = \left(\frac{mV}{M+m}\right)^2 - 2\mu gs$$

$$\therefore \quad V = \left(\frac{M+m}{m}\right)\sqrt{2\mu gs}$$

4.4 (b)

<u>Method I:</u>

$\frac{1}{2}kx^2$ will be converted into potential energy (taking reference as compressed condition)

$$\frac{1}{2}kx^2 = mgh$$

$$\frac{1}{2}\times 981\times 10^3\times(0.1)^2 = 100\times 9.81\,h$$

$$h = 5\text{ m} = \mathbf{5000\text{ mm}}$$

<u>Method II:</u>

When the spring will reach its natural length, K.E of the block of mass 100 kg held on it.

(taking reference at natural length of spring)

$$\frac{1}{2}kx^2 - mgx = \frac{1}{2}mV^2 \qquad [x = 0.1\text{ m}]$$

and this K.E will be converted into mgh.

Total lift of block

$$= h + x$$

$$\frac{1}{2}\times 981\times 10^3\times .1^2 - 100\times 9.81\times .1 = mgh = 100\times 9.81\times h$$

$$h = 4.9\text{ m}$$

$$\Rightarrow \quad h + x = 4.9 + 0.1 = \mathbf{5\text{ m}}$$

4.5 (c)

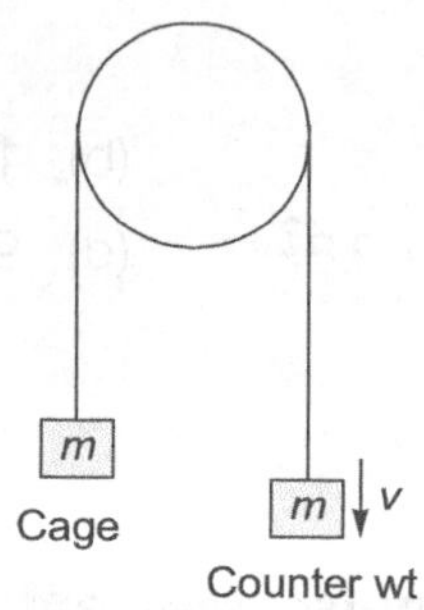

Power = Rate of energy supplied 'or' obtained.

If counter weight is moving with velocity V then cage will also be moving with velocity V.

So total K.E. $= \frac{1}{2}mv^2 + \frac{1}{2}mv^2 = mv^2$

and to absorb (or to stop) this energy time available is t second.

$$\text{Power} = \frac{\text{Total energy absorbed}}{\text{time taken for this absorption}}$$

$$= \frac{mv^2}{t}$$

4.6 (b)

In this question we need to neglect the mass of clay w.r.t. mass of the wheel.

We can't conserve linear momentum as friction is there so we will conserve angular momentum about the point of the wheel which is in contact with the surface. (because torque about A of all forces is zero)

Taking both clay and wheel as system.

$$L_i = L_f \text{(about } A)$$

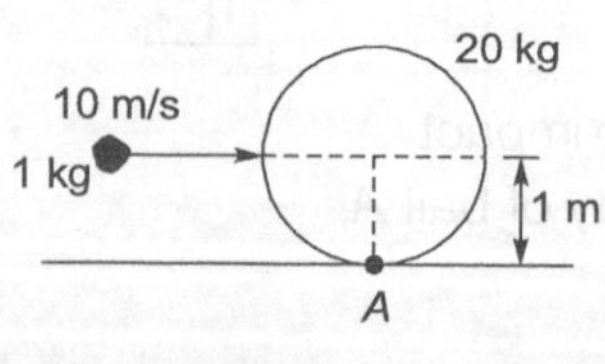

$$L_i = mvr = 1 \times 10 \times 1 = 10$$

$$L_f = mv_{cm}r + I\omega = 20 \times r\omega \times r + \frac{mr^2}{2}\omega$$

$$= 20 \times 1 \times \omega \times 1 + \frac{20}{2} \times 1^2 \times \omega$$

$$L_i = L_f$$

$$10 = 20\,\omega + 10\,\omega$$

$$\Rightarrow \quad 10 = 30\,\omega$$

or $$\omega = \frac{10}{30} = \frac{1}{3} \text{ rad/s}$$

4.7 (a)

During **inelastic** collision of two bodies only linear momentum remains conserved.

POINTS TO REMEMBER

- Conservation of momentum is applied, in both elastic and plastic provided there is no external force.
- Conservation of energy is applied, in elastic collisions only and in all other collisions there is energy loss.

4.8 (a)

Coefficient of restitution of a perfectly plastic impact or perfectly inelastic collision is zero.
Coefficient of restitution, $e = 0$

4.9 (b)

Rate of change of energy stored = $F\,dx$

$$dE = 300x^2 dx$$

$$\int dE = \int_0^{0.1} 300x^2\,dx$$

$$E = 300 \times \left.\frac{x^3}{3}\right|_0^{0.1} = \mathbf{0.1\ J}$$

4.10 (c)

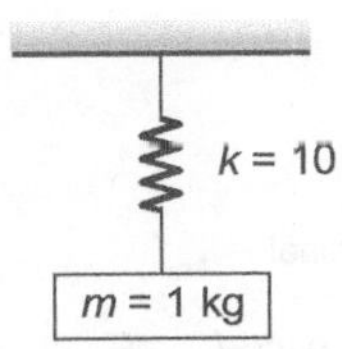

At $t = 0$ mass is at rest

$F = 5$ kN for 10^{-4} seconds

Amplitude = x

$$F = m \times dv/dt$$

$$F \cdot \int_{t_1}^{t_2} dt = m \cdot \int_{v_1}^{v_2} dV$$

$$F(t_2 - t_1) = m \times (v_2 - v_1)$$

$$5 \times 10^3 \times 10^{-4} = 1 \times (v_2 - 0)$$

$$0.5 \text{ m/s} = v_2$$

$$\frac{1}{2}mv_2^2 = \frac{1}{2}kx^2$$

$$x = \sqrt{\frac{mv_2^2}{k}} = \sqrt{\frac{1 \times 0.5 \times 0.5}{10 \times 1000}}$$

$$= \frac{0.5}{100} = 0.005 \text{ m} = \mathbf{5\ mm}$$

4.11 (d)

Coefficient of restitution,

$$e = \frac{\text{Relative speed after collision}}{\text{Relative speed before collision}}$$

After collision, both the masses move with same velocity. So, there is zero relative velocity between them,

$$\therefore \quad e = \frac{0}{5} = 0$$

4.12 (c)

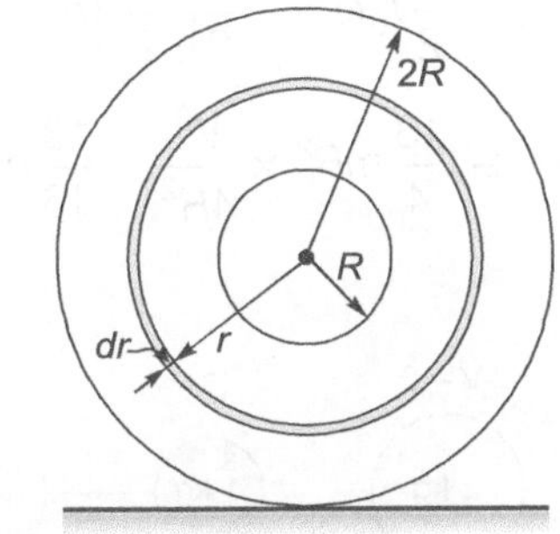

m is mass in area $3\pi R^2[4\pi R^2 - \pi R^2]$
mass in area $2\pi r dr = dm$

$$dm = \frac{2mrdr}{3R^2}$$

$$I = \int_R^{2R} dm\,r^2 = \frac{2m}{3R^2}\int_R^{2R} r^3 dr$$

$$= \frac{2m}{3R^2}\left[\frac{r^4}{4}\right]_R^{2R} = \frac{m}{6R^2}[15R^4] = \frac{5mR^2}{2}$$

K.E = (K.E)$_{translation}$ + (K.E)$_{rotation}$

$$= \frac{1}{2}mv^2 + \frac{1}{2}I\omega^2$$

where $V = 2R\omega$

$$\Rightarrow \quad \omega = \frac{V}{2R}$$

$$\therefore \quad K.E = \frac{1}{2}mv^2 + \frac{1}{2} \times \frac{5mR^2}{2} \times \frac{v^2}{4R^2}$$

$$= \frac{1}{2}mv^2 + \frac{5mv^2}{16}$$

$$K.E = \frac{8mv^2 + 5mv^2}{16} = \frac{13mv^2}{16}$$

Method II:

Also solved above,

$$I = \frac{5}{2}mR^2$$

this is the moment of inertia about the centre of mass of the disc.

∴ Again, the moment of inertia about the instantaneous axis of rotation.

$$I_0 = I + m(2R)^2$$

$$= \frac{5}{2}mR^2 + 4mR^2 = \frac{13}{2}mR^2$$

∴ Kinetic energy of the disc,

$$KE = \frac{1}{2}I_0\omega^2 = \frac{1}{2}\left(\frac{13}{2}\right)mR^2 \times \left(\frac{V}{2R}\right)^2$$

$$= \frac{13}{4}mR^2 \times \frac{V^2}{4R^2} = \frac{13}{16}mV^2$$

4.13 Sol.

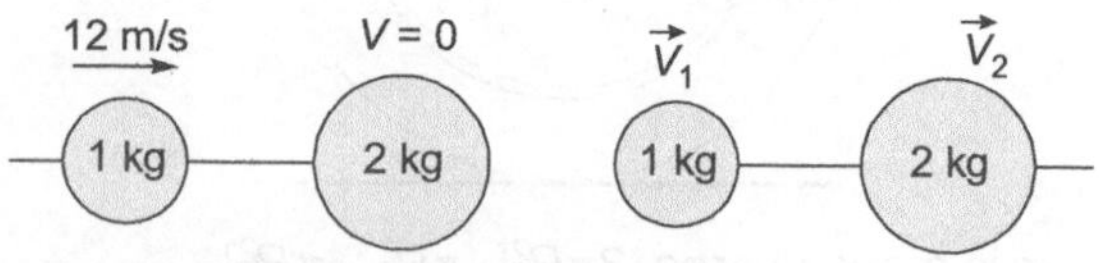

1. Conserving linear momentum

$$1 \times 12 = 1 \times V_1 + 2 \times V_2$$

$$12 = V_1 + 2V_2 \quad \text{...(i)}$$

2. Velocity of approach = Velocity of separation

$$12 - 0 = V_2 - V_1$$

$$V_2 - V_1 = 12 \quad \text{...(ii)}$$

From equations (i) and (ii), we get

$$V_2 = \mathbf{8\ m/s}$$

4.14 Sol.

Given: $m = 0.1$ kg, $h = 1$ m

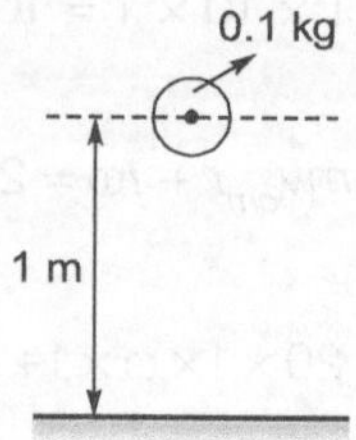

Velocity with which ball hits the ground

$$V_1 = \sqrt{2gh} = \sqrt{2 \times 9.81 \times 1}$$

$$= 4.4294 \text{ m/s}$$

and $V_2 = 0.8 \times 4.4294 = 3.54$ m/s

⇒ Now $\frac{1}{2}mV_2^2$ is kinetic energy of ball which will be decreasing as it goes upwards (because work is done against gravity) and gravitational potential energy increase

$$\Rightarrow \quad \frac{1}{2}mV_2^2 = mgh'$$

$$h' = \frac{V_2^2}{2g} = \frac{(3.54)^2}{2 \times 9.81} = \mathbf{0.638\ m}$$

4.15 (c)

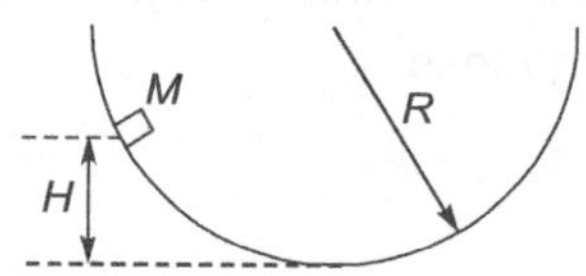

Loss of P.E. = Gain in K.E.

$$MgH = \frac{1}{2}Mv^2$$

$$v = \sqrt{2gH}$$

4.16 Sol.

Given: $m = 2000$ kg, $V = 2$ m/s, I = 150 kgm²

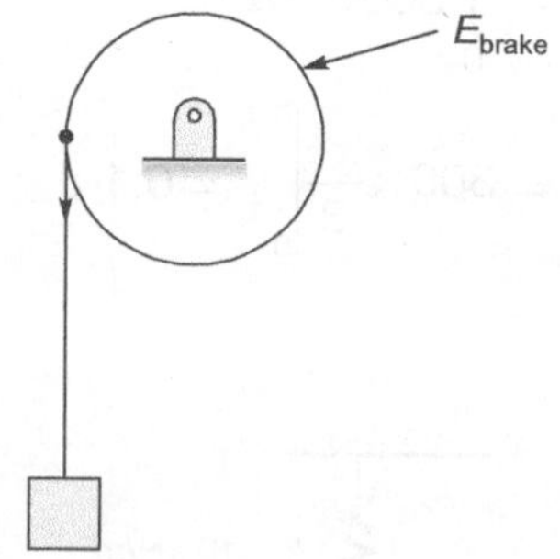

$$E_{initial} - E_{brake} = E_{final}$$

$$\frac{1}{2}I\omega_1^2 + \frac{1}{2}mv_1^2 + mgH_1 - E_{brake}$$

$$= \frac{1}{2}I\omega_2^2 + \frac{1}{2}mv_2^2 + mgH_2$$

$$E_{brake} = \frac{1}{2}I\omega_1^2 + \frac{1}{2}mv_1^2 + mg(H_1 - H_2)$$

$$h = H_1 - H_2 = 0.5 \text{ m} \quad \left(\frac{1}{2}I\omega_2^2 = 0;\ \frac{1}{2}mV_2^2 = 0\right)$$

$$E_{brake} = \frac{1}{2}I\omega_1^2 + mgh + \frac{1}{2}mv_1^2$$

$$= \frac{\frac{1}{2}\times 150\times 2^2 + 2000\times 9.81\times 0.5 + \frac{1}{2}\times 2000\times 2^2}{1000}$$

= **14.11 kJ**

4.17 (d)

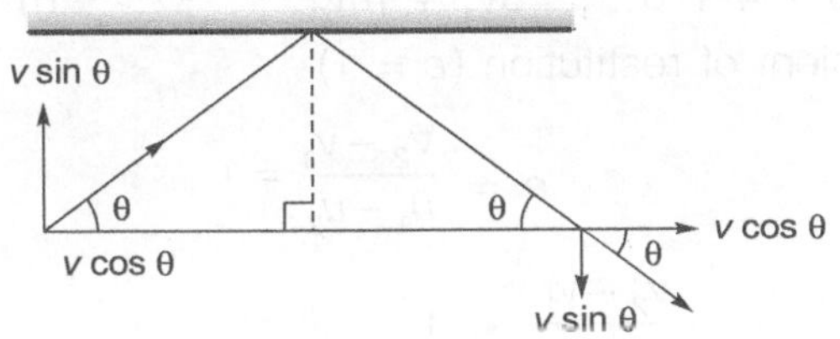

$$P_i = MV\cos\theta\,\hat{i} + MV\sin\theta\,\hat{j}$$

$$P_f = MV\cos\theta\,\hat{i} - MV\sin\theta\,\hat{j}$$

$$P_f - P_i = -2MV\sin\theta\,\hat{j}$$

4.18 Sol.

Given: $m = 1$ kg, $r = 100$ mm

Loss in P.E. = Gain in rotational K.E. + gain in translational K.E.

$$mgH = \frac{1}{2}I\omega^2 + \frac{1}{2}mv^2$$

$$v = r\omega$$

$$\omega = \frac{v}{r}$$

$$mgH = \frac{1}{2}\frac{mr^2}{2}\times\left(\frac{v}{r}\right)^2 + \frac{1}{2}mv^2$$

$$10\times 30 = \frac{v^2}{4} + \frac{v^2}{2}$$

$$10\times 30\times 4 = v^2 + 2v^2$$

$$3v^2 = 30\times 10\times 4$$

$$v^2 = 400$$

or $v =$ **20 m/s**

4.19 (b)

Angular momentum is the moment of linear momentum about given point. The moment of linear momentum is obtained by product of linear momentum and distance perpendicular from the given point.

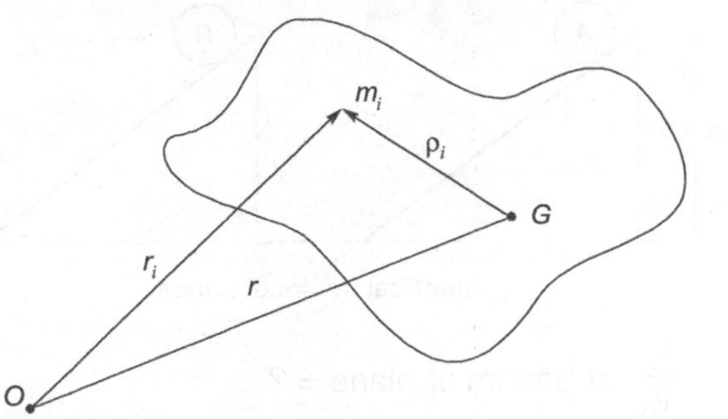

Time rate of change of angular momentum

$$\dot{L} = \frac{d\vec{L}}{dt}$$

$$L = \rho\times P$$

$$[L_i]_G = \rho_i\times P_i \text{ (For particle } m_i \text{ about G)}$$

where, $P_i = m_i V_i$, linear momentum

$$[L_i]_G = \rho_i\times m_i V_i$$

$$- \rho_i\times m_i\dot{r}_i$$

$$[\dot{L}_i]_G = \rho_i\times m_i\ddot{r}_i$$

$$[\dot{L}_{Body}] = \sum\rho_i\times m_i\ddot{r}_i$$

4.20 (c)

Given: $m = 1$ kg, $r = t^2$; $\theta = t$

$$\text{K.E.} = \frac{1}{2}mv^2 = ? \text{ at } t = 2 \text{ sec}$$

$$\vec{V} = r\omega(\hat{t}) + \frac{dr}{dt}\hat{r} = t^2\times 1(\hat{t}) + 2t\hat{r}$$

$$\left[\because \frac{d\theta}{dt} = \omega = 1\right]$$

$$\vec{V} = t^2(\hat{t}) + 2t(\hat{r})$$

at $t = 2s$

$$\vec{V} = 4(\hat{t}) + 4(\hat{r})$$

$$|\vec{V}| = \sqrt{16+16} = \sqrt{32}$$

$$\text{K.E.} = \frac{1}{2}mv^2 = \frac{1}{2}\times 1\times 32 = 16$$

4.21 (a)

$$m_A = m_B = m$$

Outer circumferential radius of both disc A and B is also same i.e. R

$$I_A = mR^2$$

(mass is concentrated at the rim)

$$I_B = \frac{mR^2}{2}$$
(mass is uniformly distributed)

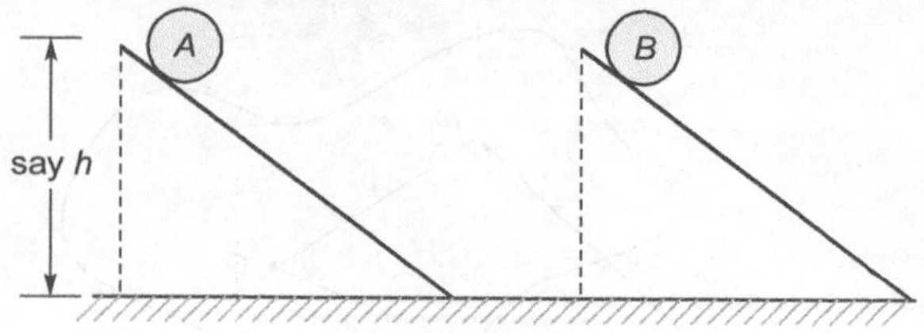

Identical inclined planes

$\frac{V_A}{V_B}$ at bottom of plane = ?

Loss in potential energy

= (Gain in rotational K.E. + Gain in translational K.E.)$_A$

(Gain in rotational K.E. + Gain in translational K.E.)$_B$

$$\therefore \quad mgh = \frac{1}{2}mV_A^2 + \frac{1}{2}I_A\omega_A^2$$

$$= \frac{1}{2}mV_B^2 + \frac{1}{2}I_B\omega_B^2$$

$$\frac{1}{2}mV_A^2 + \frac{1}{2}I_A\omega_A^2 = \frac{1}{2}mV_B^2 + \frac{1}{2}I_B\omega_B^2$$

$$\frac{1}{2}mV_A^2 + \frac{1}{2}mR^2\frac{V_A^2}{R^2} = \frac{1}{2}mV_B^2 + \frac{1}{2}\frac{mR^2}{2}\times\frac{V_B^2}{R^2}$$

$$\frac{V_A^2}{2} + \frac{V_A^2}{2} = \frac{V_B^2}{2} + \frac{V_B^2}{4}$$

$$V_A^2 = \frac{3}{4}V_B^2$$

$$\frac{V_A}{V_B} = \sqrt{\frac{3}{4}}$$

4.22 Sol.

Given: $m = 2$ kg, $g = 10$ m/s^2

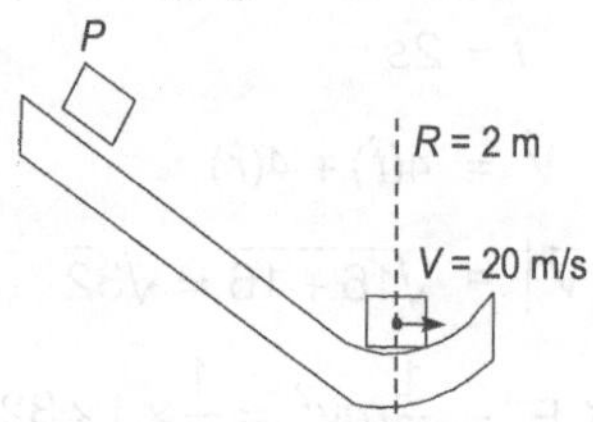

FBD of block *P* at point *Q*

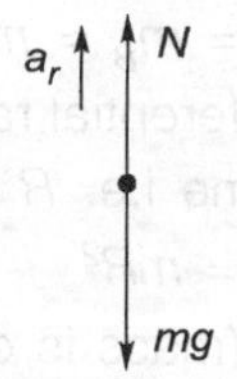

$$N - mg = ma_r \quad \text{(Newton's second law)}$$

$$N = mg + \frac{mv^2}{R}$$

$$= 2\times 10 + \frac{2\times 20\times 20}{2} = 420 \text{ N}$$

4.23 (b)

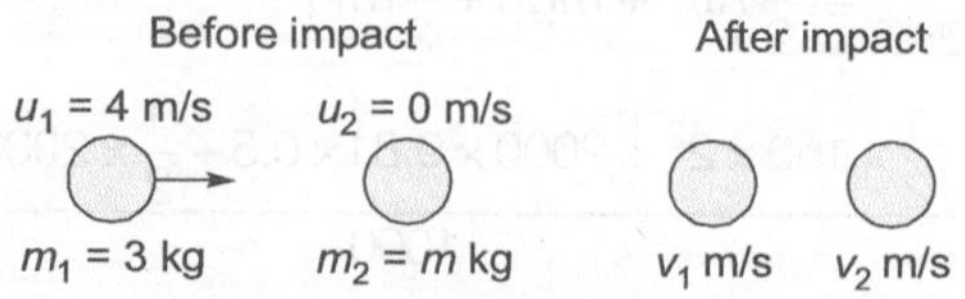

Momentum is conserved

$$m_1u_1 + m_2u_2 = m_1v_1 + m_2v_2$$

$$3\times 4 + 0 = 3v_1 + mv_2 \quad \text{...(i)}$$

Coefficient of restitution ($e = 1$)

$$e = \frac{v_2 - v_1}{u_1 - u_2} = 1$$

$$\frac{v_2 - v_1}{4} = 1$$

$$v_2 - v_1 = 4 \text{ m/s} \quad \text{...(ii)}$$

KE of m_1 after impact is 6 J

$$\frac{1}{2}m_1v_1^2 = 6 \text{ J}$$

$$v_1 = \sqrt{\frac{6\times 2}{3}} = \pm 2 \text{ m/s}$$

From equation (ii),

For $v_1 = +2$ m/s
$v_2 = 6$ m/s

For $v_1 = -2$ m/s
$v_2 = 2$ m/s

From equation (i),

For $v_2 = 6$ m/s, $v_1 = 2$ m/s
$m = 1$ kg

For $v_2 = 2$ m/s, $v_1 = -2$ m/s
$m = 9$ kg

4.24 (a)

As per given data: $m = 2$ kg, $V = 1.5$ m/s, stiffness $k = 5$ N/m

Energy conservation,

$$\frac{1}{2}mV_1^2 = \frac{1}{2}mV_0^2 + \frac{1}{2}kx^2$$

$$\Rightarrow \quad 2\times 1.5^2 = 2\times V_0^2 + 5\times 0.4^2$$

$$V_0 = 1.360 \text{ m/s}$$

4.25 Sol.

Given data: $m_1 = 10$ kg, $V_0 = 2$ m/s, $m_2 = 30$ kg

Collision Theory

From conservation of momentum,

$$m_1 \times v_0 + m_2 \times 0 = (m_1 + m_2) \times v$$

$$10 \times 2 = (10 + 30)v$$

$$20 = 40v$$

$$v = 0.5 \text{ m/s}$$

From energy conservation,

$$\text{Now, } \frac{1}{2}(m_1 + m_2)v^2 = \frac{1}{2}kx^2$$

$$\frac{1}{2} \times 40 \times (0.5)^2 = \frac{1}{2} \times 10^5 \times x^2$$

$$10 = 10^5 \times x^2$$

$$\Rightarrow \quad x^2 = \frac{1}{10^4}$$

Maximum deflection,

$$x = \frac{1}{100} \text{ m} = 1 \text{ cm}$$

4.26 (a)

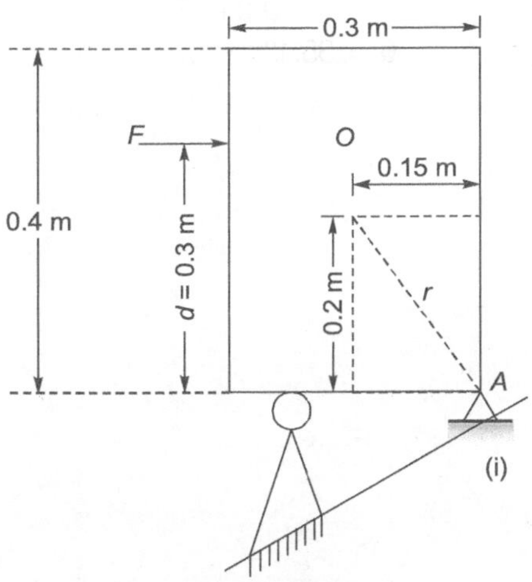

O → C.G. of block

$$r = \sqrt{0.15^2 + 0.2^2} = 0.25 \text{ m}$$

$I_F \rightarrow$ Impulse due to force F

Due to impulse provided by force F, block will topple about point A.

As we know:

Angular impulse $(I_F) = I_F \times d$ = Change in angular momentum

$$I_F \times 0.3 = L_i - L_i$$

$$L_i = 0$$

L_i = Angular momentum after application of force about point A.

$$\Rightarrow \quad I_F \times 0.3 = I_A \times W \quad \text{... (i)}$$

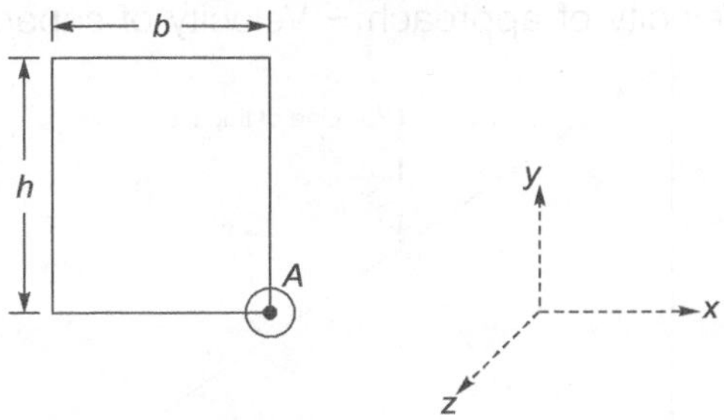

$I_A \rightarrow$ MOI about axis perpendicular to plane passing through point A.

$$I_{A,z} = I_{Ax} + I_{Ay}$$

$$I_A = I_{Az} = \frac{mh^2}{3} + \frac{mb^2}{3}$$

$$I_A = \frac{1 \times 0.4^2}{3} + \frac{1 \times 0.3^2}{3} = 0.08334 \quad \text{... (ii)}$$

Minimum impulse required just to topple the block should be able to produce sufficient kinetic energy such that at highest point (for centre of gravity) of motion velocity just becomes zero.

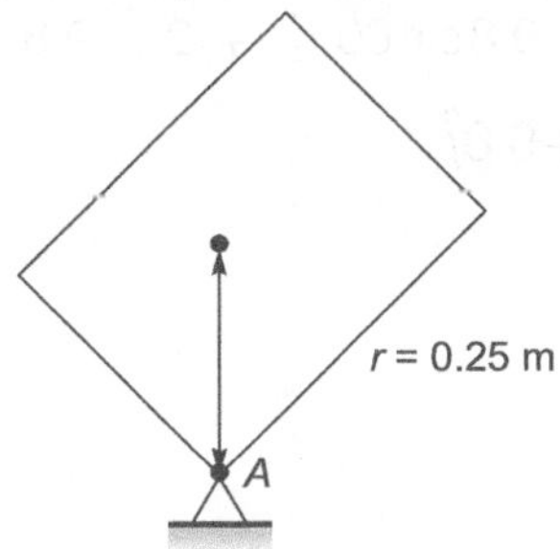

(ii)

From energy conservation:

$$(K.E.)_i + (P.E.)_i = (K.E.)_f + (P.E.)_f$$

$$\Rightarrow \quad \frac{1}{2}I_A\omega^2 + 0 = 0 + mg\left(r - \frac{h}{2}\right)$$

$$\Rightarrow \quad \frac{1}{2} \times 0.08334 \times \omega^2 = 1 \times 9.81(0.25 - 0.2)$$

$$\Rightarrow \quad \omega = 3.43 \text{ rad/s}$$

From equation (i)

$$I_F \times 0.3 = 0.08334 \times 3.43$$

$$I_F = 0.953 \text{ Ns}$$

4.27 (c)

Velocity of the ball just before making an impact with the block,

$$u = \sqrt{2gh} = \sqrt{2\times 9.8\times 4.9} = 9.8 \text{ m/s}$$

For perfectly elastic collision, along the line of impact,

Velocity of approach = Velocity of separation

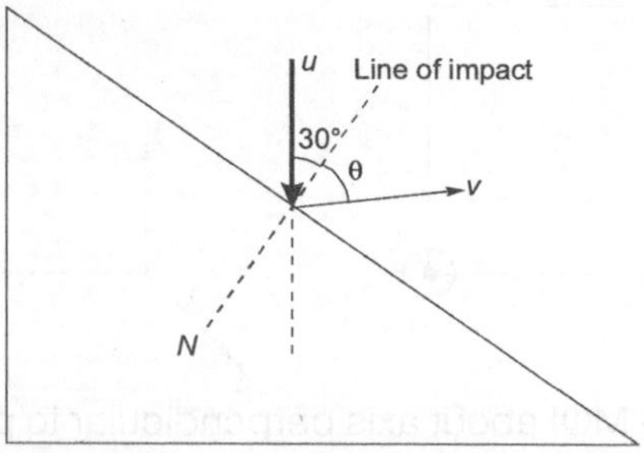

$$9.8\cos 30° = v\cos\theta \quad ...(i)$$

Now, momentum along the inclined plane will remain conserved,

$$P_i = P_f$$

$$\Rightarrow \quad 2u\sin 30° = 2v\sin\theta$$

$$\Rightarrow \quad v\sin\theta = 9.8\sin 30° = 4.9 \quad ...(ii)$$

Using equation (i) and (ii), we get

$$v = 9.8\sqrt{\sin^2 30° + \cos^2 30°} = 9.8 \text{ m/s}$$

Therefore, $\theta = 30°$

Thus, momentum vector just after the impact,

$$\vec{P}_f = 2\times 9.8\sin 60°\,\hat{i} + 2\times 9.8\cos 60°\,\hat{j}$$

$$= 17\hat{i} + 9.8\hat{j}$$

4.28 (b)

Given, $AB = 7$ m, $u = 0$, $v = 5$ m/s

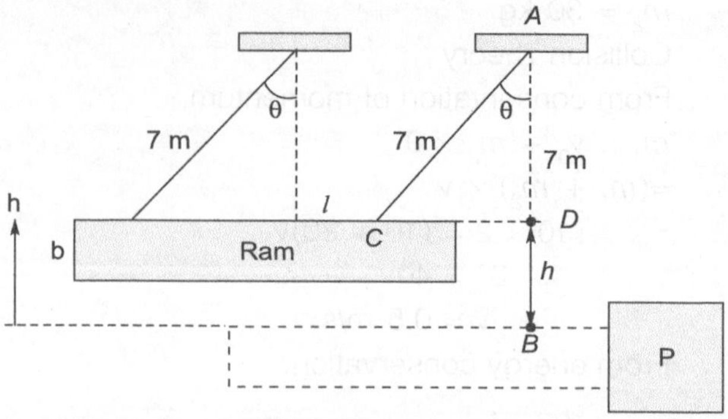

According to question, the rope will become vertical on striking the object, hence $AB = 7$ m.

From geometry, $AD = 7\cos\theta$

$$\therefore \quad h = AB - AD = 7 - 7\cos\theta$$

$$= 7(1-\cos\theta)$$

From energy conservation,

Initial potential energy = Final kinetic energy

$$\Rightarrow \quad mgh = \frac{1}{2}mv^2$$

$$9.81\times 7(1-\cos\theta) = \frac{1}{2}\times 5^2$$

$$\cos\theta = \frac{9.81\times 7 - 12.5}{9.81\times 7}$$

$$\Rightarrow \quad \theta = 35.1°$$

■■■■

5 Plane Motion

5.1 A wheel of mass m and radius r is in accelerated rolling motion without slip under a steady axle torque T. If the coefficient of kinetic friction is μ, the friction force from the ground on the wheel is

(a) μ mg (b) T/r
(c) zero (d) None of these

[1996 : 1 M]

5.2 A rod of length 1 m is sliding in a corner as shown in figure. At an instant when the rod makes an angle of 60 degrees with the horizontal plane, the velocity of point A on the rod is 1 m/s. The angular velocity of the rod at the instant is

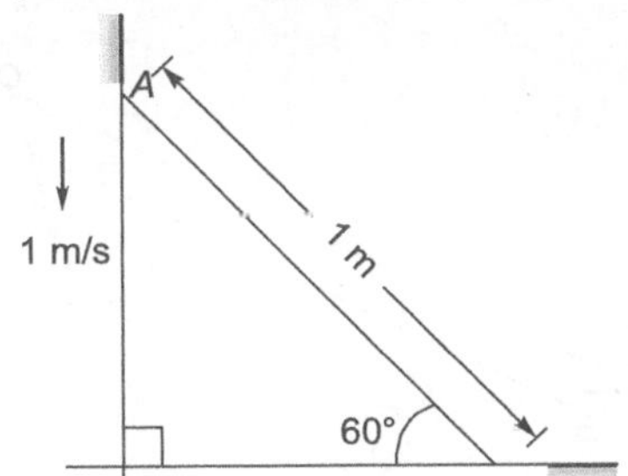

(a) 2 rad/s (b) 1.5 rad/s
(c) 0.5 rad/s (d) 0.75 rad/s

[1996 : 2 M]

Common Data Questions Q.5.3 and Q.5.4

A reel of mass m and radius of gyration k is rolling down smoothly from rest with one end of the thread wound on it held in the ceiling as depicted in the figure. Consider the thickness of the thread and its mass negligible in comparison with the radius r of the hub and the reel mass m. Symbol g represents the acceleration due to gravity.

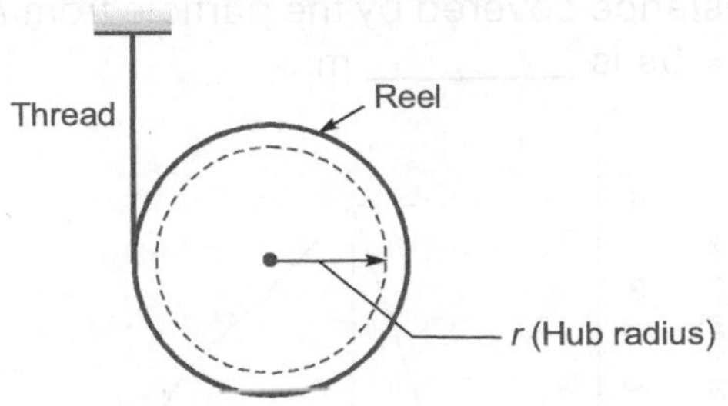

5.3 The linear acceleration of the reel is

(a) $\dfrac{gr^2}{(r^2+k^2)}$ (b) $\dfrac{gk^2}{(r^2+k^2)}$

(c) $\dfrac{grk}{(r^2+k^2)}$ (d) $\dfrac{mgr^2}{(r^2+k^2)}$

[2003 : 1 M]

5.4 The tension in the thread is

(a) $\dfrac{mgr^2}{(r^2+k^2)}$ (b) $\dfrac{mgrk}{(r^2+k^2)}$

(c) $\dfrac{mgk^2}{(r^2+k^2)}$ (d) $\dfrac{mg}{(r^2+k^2)}$

[2003 : 1 M]

5.5 A shell is fired from a cannon. At the instant the shell is just about to leave the barrel, its velocity relative to the barrel is 3 m/s, while the barrel is swinging upwards with a constant angular velocity of 2 rad/s. The magnitude of the absolute velocity of the shell is

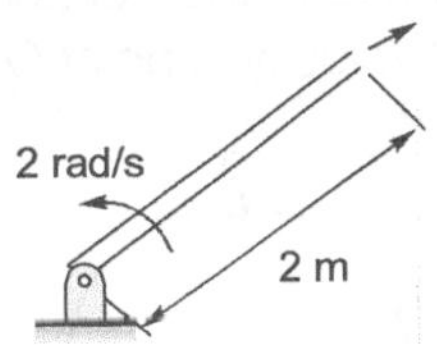

(a) 3 m/s (b) 4 m/s
(c) 5 m/s (d) 7 m/s [2005 : 2 M]

5.6 A straight rod length $L(t)$, hinged at one end freely extensible at the other end, rotates through an angle $\theta(t)$ about the hinge. At time t, $L(t)$=1 m, $\dot{L}(t)$ = 1 m/s, $\theta(t) = \pi/4$ rad and $\dot{\theta}(t)$ = 1 rad/s. The magnitude of the velocity at the other end of the rod is

(a) 1 m/s (b) $\sqrt{2}$ m/s
(c) $\sqrt{3}$ m/s (d) 2 m/s [2008 : 1 M]

5.7 A circular disk of a radius R rolls without slipping at a velocity V. The magnitude of the velocity at point P (see figure) is:

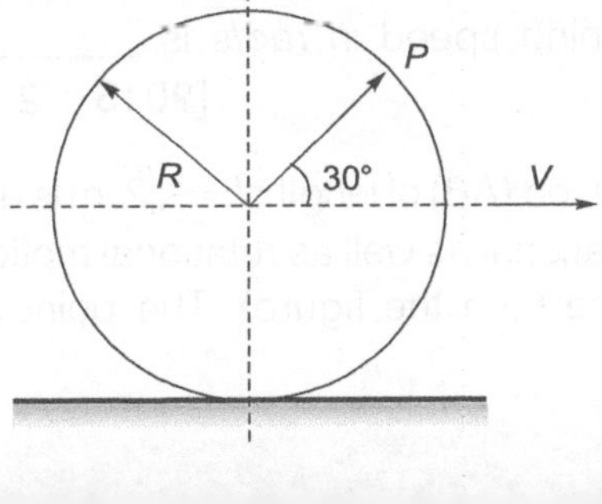

(a) $\sqrt{3}\,V$ (b) $\sqrt{3}\,V/2$

(c) $V/2$ (d) $2V/\sqrt{3}$ [2008 : 2 M]

5.8 A solid disk of radius r rolls without slipping on a horizontal floor with angular velocity ω and angular acceleration α. The magnitude of the acceleration of the point of contact on the disc is

(a) zero (b) $r\alpha$

(c) $\sqrt{(r\alpha)^2 + (r\omega^2)^2}$ (d) $r\omega^2$ [2012 : 1 M]

5.9 A circular object of radius r rolls without slipping on a horizontal level floor with the center having velocity V. The velocity at the point of contact between the object and the floor is

(a) zero

(b) V in the direction of motion

(c) V opposite to the direction of motion

(d) V vertically upward from the floor

[2014 : 1 M, Set-1]

5.10 A wheel of radius r rolls without slipping on a horizontal surface shown below. If the velocity of point P is 10 m/s in the horizontal direction, the magnitude of velocity of point Q (in m/s) is ______.

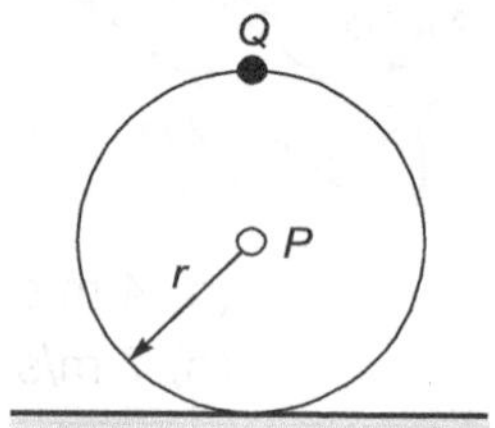

[2015 : 1 M, Set-1]

5.11 A bullet spins as the shot is fired from a gun. For this purpose, two helical slots as shown in the figure are cut in the barrel. Projections A and B on the bullet engage in each of the slots.

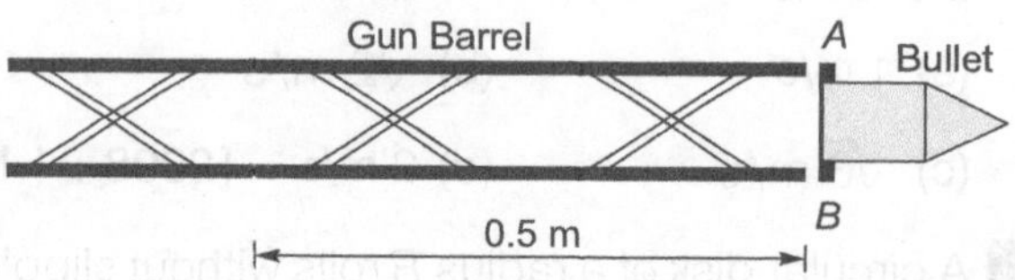

Helical slots are such that one turn of helix is completed over a distance of 0.5 m. If velocity of bullet when it exits the barrel is 20 m/s, its spinning speed in rad/s is ______.

[2015 : 2 M, Set-3]

5.12 A rigid rod (AB) of length $L = \sqrt{2}$ m is undergoing translational as well as rotational motion in the x-y plane (see the figure). The point A has the velocity $V_1 = \hat{i} + 2\hat{j}$ m/s. The end B is constrained to move only along the x direction.

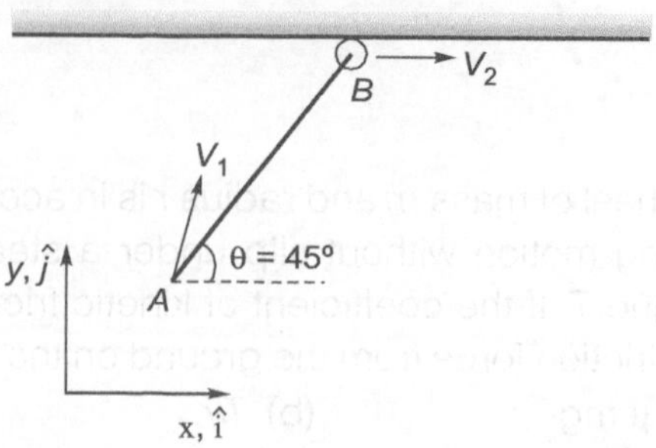

The magnitude of the velocity V_2 (in m/s) at the end B is ____ .

[2016 : 2 M, Set-3]

5.13 A rigid link PQ is undergoing plane motion as shown in the figure (V_P and V_Q are non-zero). V_{QP} is the relative velocity of point Q with respect to point P.

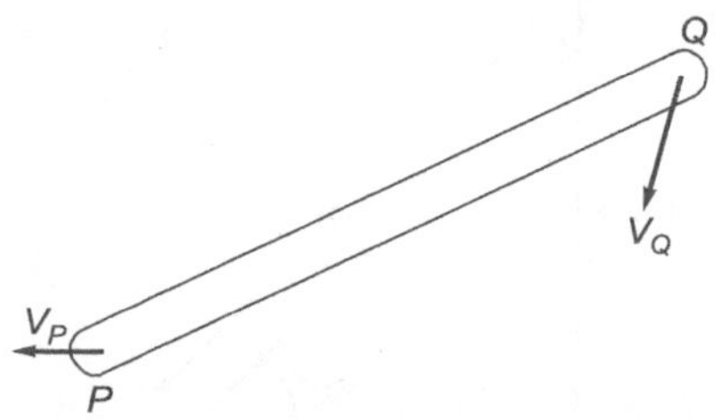

Which one of the following is TRUE?

(a) V_{QP} has components along and perpendicular to PQ

(b) V_{QP} has only one component directed from P to Q

(c) V_{QP} has only one component directed from Q to P

(d) V_{QP} has only one component perpendicular to PQ [2016 : 1 M, Set-3]

5.14 The following figure shows the velocity time plot for a particle traveling along a straight line. The distance covered by the particle from $t = 0$ to t = 5s is __________ m.

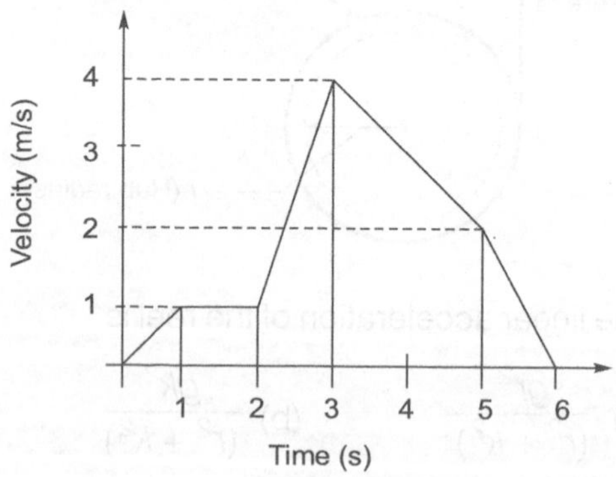

[2017 : 1 M, Set-1]

5.15 A point mass is shot vertically up from ground level with a velocity of 4 m/s at time, $t = 0$. It loses 20% of its impact velocity after each collision with the ground. Assuming that the acceleration due to gravity is 10 m/s^2 and that air resistance is negligible, the mass stops bouncing and comes to complete rest on the ground after a total time (in seconds) of

(a) 1 (b) 2
(c) 4 (d) ∞ **[2018 : 2 M, Set-1]**

5.16 A ball is dropped from rest from a height of 1 m in a frictionless tube as shown in the figure. If the tube profile is approximated by two straight lines (ignoring the curved portion), the total distance travelled (in m) by the ball is ________ (correct to two decimal places).

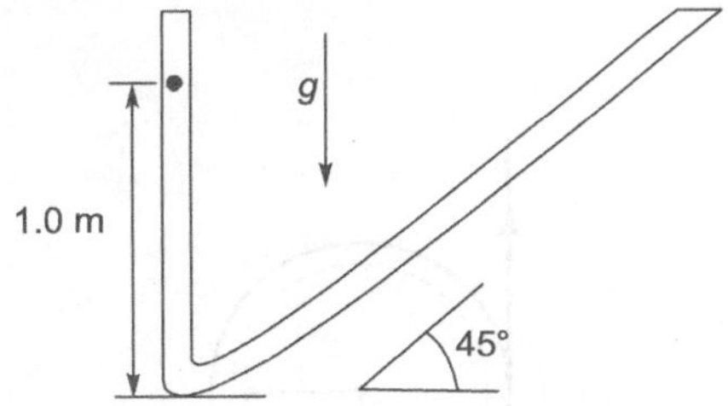

[2018 : 1 M, Set-2]

Answers Plane Motion

5.1 (d) 5.2 (a) 5.3 (a) 5.4 (c) 5.5 (c) 5.6 (b) 5.7 (a) 5.8 (d) 5.9 (a)
5.10 (20) 5.11 (251.327) 5.12 (3) 5.13 (d) 5.14 (10) 5.15 (c) 5.16 (2.414)

Explanations Plane Motion

5.1 (d)

Assuming point contact
Because coefficient of rolling resistance is not given.

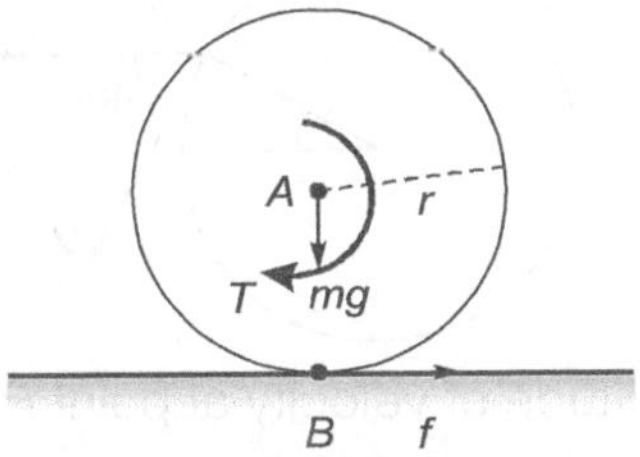

as wheel is in accelerated motion so pt A is in acceleration.

$$a_A = r\alpha \text{ (for pure rolling)} \quad \text{... (i)}$$

$T \rightarrow$ Torque given about A. (Axle)

Friction will oppose slipping of point B, so acting rightward.

Torque = $I\alpha$

Torque is about A or line perpendicular to wheel & passing through A.

$$T - fr = I\alpha = \frac{mr^2}{2}\alpha \quad \text{...(ii)}$$

[As mg passes through A]

Newton's 2nd law in horizontal direction

$$f = ma_A \quad \text{...(iii)}$$

From equations (i) and (iii)

$$f = mr\alpha \quad \text{...(iv)}$$

From equations (ii) and (iv)

$$T - mr^2\alpha = \frac{mr^2}{2}\alpha$$

$$T = \frac{3}{2}mr^2.\alpha$$

$$\alpha = \frac{2T}{3mr^2}$$

We know that

$$f = mr\,\alpha = mr \times \frac{2T}{3mr^2} = \frac{2T}{3r}$$

5.2 (a)

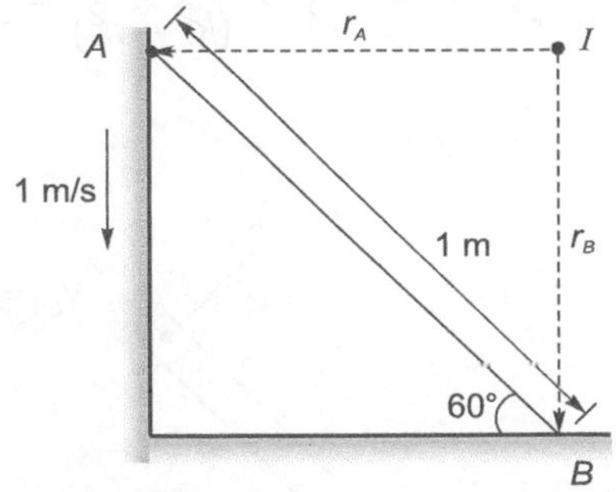

The instantaneous center of AB will be perpendicular to V_A and V_B,

$$IA = 1\cos\theta = 1\cos 60° = \frac{1}{2}\text{ m}$$

$$V_A = \omega \times IA = \omega \times r_A$$

or $$\omega = \frac{V_A}{r_A} = \frac{1}{1/2} = 2 \text{ rad/s}$$

5.3 (a)

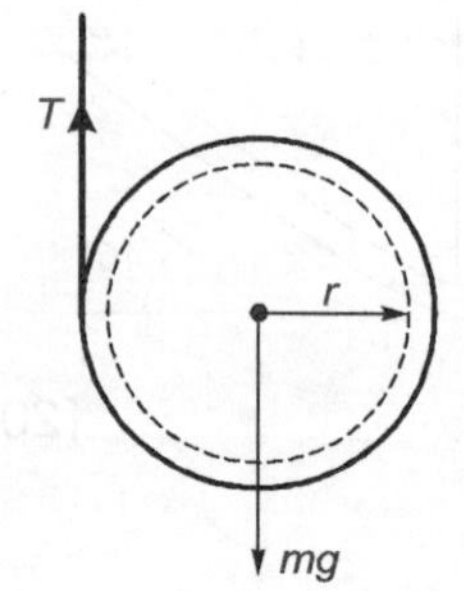

Let a = linear acceleration

For a vertical translation motion.

$$mg - T = ma \quad ...(i)$$

For rotational motion

$$T \times r = I\alpha$$

[$a = r\alpha$ (pure rolling)]

$$Tr = mk^2 \frac{a}{r}$$

$\therefore$ $$T = \frac{mk^2}{r^2}\cdot a \quad ...(ii)$$

From Eqs. (i) and (ii), we get

$$mg - \frac{mk^2}{r^2}\cdot a = ma$$

or $$a = \frac{gr^2}{k^2 + r^2}$$

5.4 (c)

$$T = \frac{mk^2}{r^2}\cdot a = \frac{mk^2}{r^2} \times \frac{gr^2}{(k^2 + r^2)} = \frac{mgk^2}{(k^2 + r^2)}$$

5.5 (c)

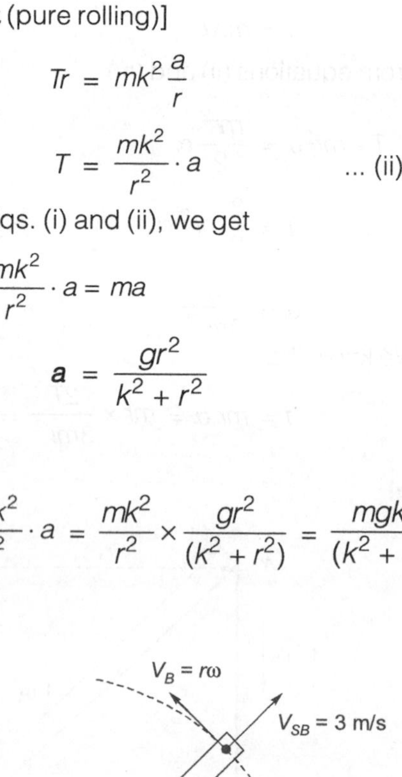

$V_B = r\omega = 2 \times 2 = 4$ m/s

V_{sB} = velocity of shell w.r.t. barrel = 3 m/s

As shell is in barrel, the net velocity of shell

$$\Rightarrow \quad \vec{V_s} = \vec{V_B} + \vec{V_{sB}}$$

$\Rightarrow$ as $\vec{V_B}$ and $\vec{V_{sB}}$ are perpendicular

$$|\vec{V_s}| = \sqrt{(3)^2 + (4)^2} = \mathbf{5\ m/s}$$

5.6 (b)

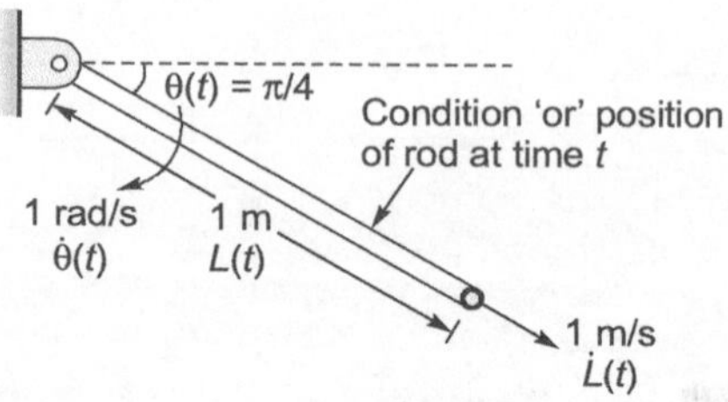

Velocity has two components

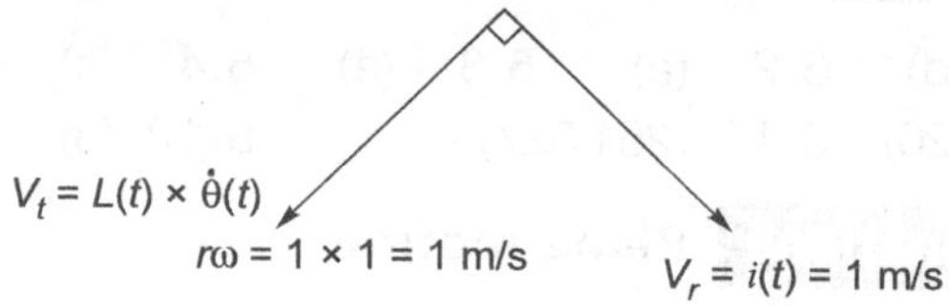

$$\vec{v} = \sqrt{(1)^2 + (1)^2} = \sqrt{2}\ \text{m/s}$$

5.7 (a)

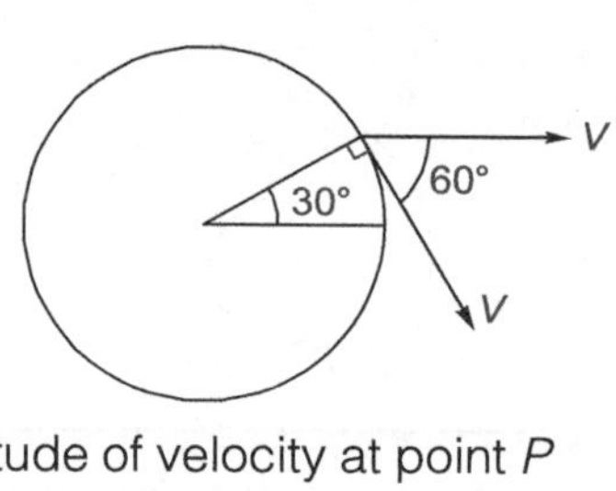

magnitude of velocity at point P

$$= \sqrt{V^2 + V^2 + 2V^2\cos 60°}$$

$$= \sqrt{2V^2 + 2V^2\cos 60°}$$

$$= V\sqrt{2 + 2\cos 60°} = V\sqrt{3}$$

5.8 (d)

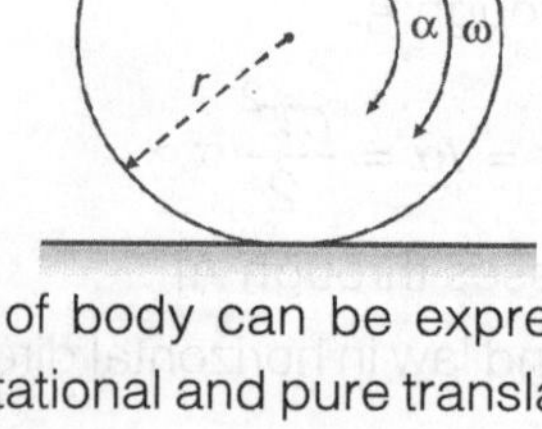

Motion of body can be expressed as sum of pure rotational and pure translation.

For no slip condition.

Translational acceleration of centre of mass (a)
$= \alpha r$
Velocity of centre of mass (v) = ωr
I Pure rotational II Pure translation

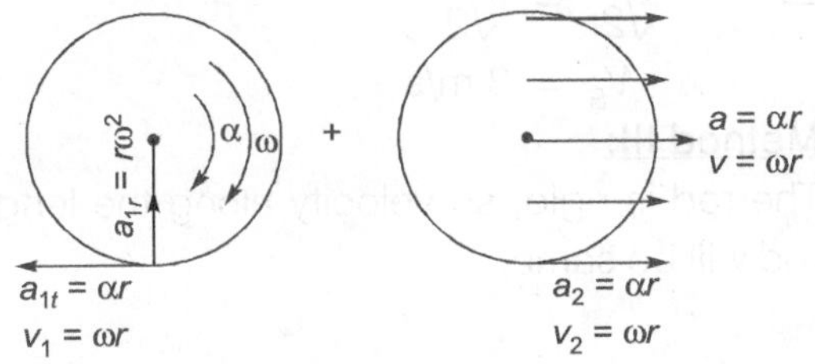

Now superimposing the two motions.

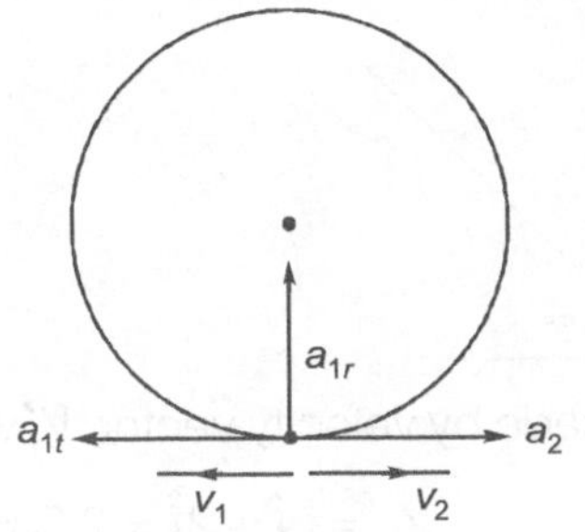

Linear acceleration of bottom point
$$= \alpha r - \alpha r + r\omega^2$$
Velocity of bottom point = $\omega r - \omega r = 0$
$\therefore$ Net acceleration of bottom point = $\boldsymbol{r\omega^2}$

5.9 (a)

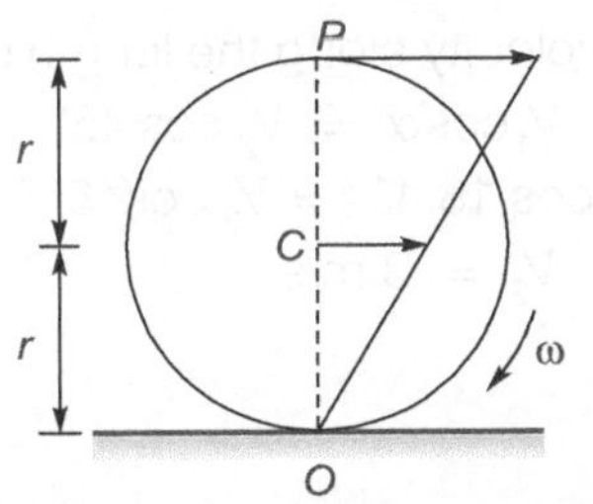

$$V_P = \omega \times \overline{OP} = 2\,\omega r$$
$$V_C = \omega \times \overline{OC} = \omega r$$
$$V_O = \omega \times 0 = 0$$

For no slipping, velocity at point of contact is zero.

5.10 Sol.

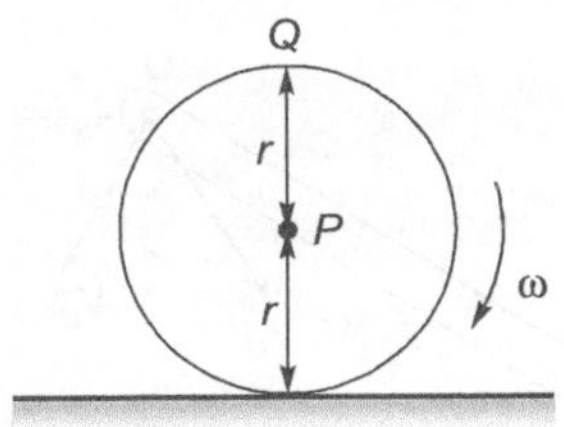

Velocity of point P,
$$V_P = r\omega = 10 \text{ m/s}$$

Velocity of point Q,
$$V_Q = 2\,r\omega = 2 \times 10 = \mathbf{20\ m/s}$$

5.11 Sol.

Method I:

One turn of helix is completed over a distance of 0.5 m
So bullet spins 2π radian angle in 0.5 m.
As speed of bullet is 20 m/s, so bullet completes a distance of 20 m in 1 sec.

In 20 m bullet spins through $\dfrac{2\pi}{0.5} \times 20$ radian

$$\text{So, angular velocity} = \frac{\dfrac{2\pi \times 20}{0.5}}{1} = 80\,\pi$$
$$= \mathbf{251.3274\ rad/s}$$

Method II:

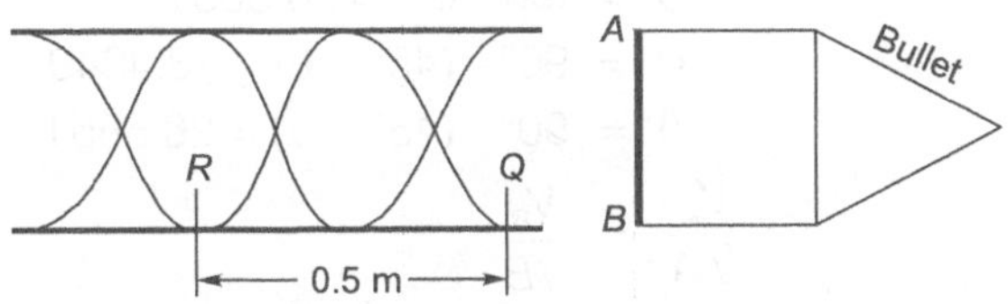

As according to the given, bullet spins due to the projections in the barrel.
Speed of the bullet,
$$V_B = 20 \text{ m/s}$$
$\because$ One revolution of bullet is completed when bullet travels from point R to Q.
$\because$ Time required by bullet to travel from point

$$R \text{ to } Q, (t) = \frac{RQ}{V} = \frac{0.5}{20} = 0.025 \text{ s}$$

$\because$ In 0.025 s, bullet will complete one revolution.
$\because$ In 1 s, revolution of bullet,

$$N = \frac{1}{0.025} \text{ rev.} = 40 \text{ rev.}$$

$\because N = 40$ revolutions per second
Spinning sped,
$$\omega = 2\pi N = 2\pi \times 40 = 80\,\pi$$
$$= 80 \times 3.14 = 251.327 \text{ rad/s}$$

5.12 Sol.

Method I:

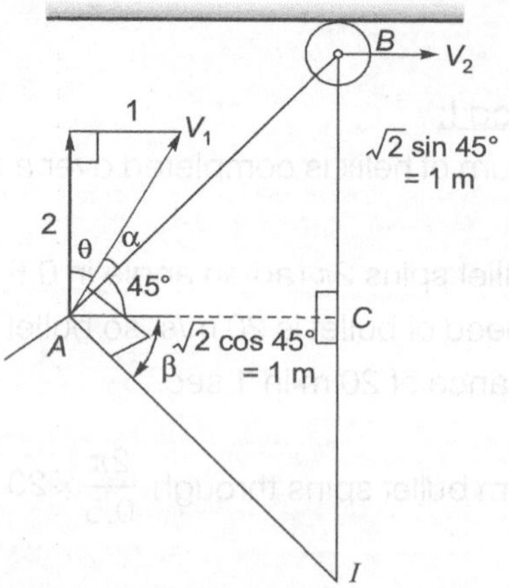

$$\sqrt{2}\sin 45° = 1 \text{ m}$$

$$\tan\theta = \frac{1}{2}$$

$$V_A = i + 2j = \sqrt{1+2^2} = \sqrt{5} \text{ m/s}$$

$$\theta = \tan^{-1} 0.5 = 26.5651°$$

$$\alpha = 90° - (45° + \theta) = 18.4349°$$

$$\beta = 90° - (45° + \alpha) = 26.5651°$$

$$\frac{V_A}{IA} = \frac{V_B}{IB}$$

$$V_B = \frac{V_A \cdot IB}{IA} = \frac{\sqrt{5}\cdot IB}{IA}$$

$$\cos\beta = \frac{1}{IA}$$

$$IA = \frac{1}{\cos 26.5651} = 1.118 \text{ m.}$$

$$IC = \sqrt{(IA)^2 - 1^2} = \sqrt{(1.118)^2 - 1}$$

$$= 0.5 \text{ m}$$

$$IB = (1 + 0.5) \text{ m}$$

$$\frac{V_A}{IA} = \frac{V_B}{IB}$$

$$V_2 = V_B = \frac{\sqrt{3}\times[1+0.5]}{1.118} = \mathbf{3 \text{ m/s}}$$

Method II:

As *AB* is rigid rod, velocity of point *A* and *B* along *AB* will be same.

Let $\vec{V}_B$ be the velocity of *B* then $\vec{V}_B = V_B\hat{i}$

$$\vec{V}_A = \hat{i} + 2\hat{j}$$

Unit vector along,

$$AB, \vec{u} = \frac{i+j}{\sqrt{2}}$$

Now $\vec{V}_A \cdot \vec{u} = \vec{V}_B \cdot \vec{u}$

$$\Rightarrow (i+2j)\cdot\left(\frac{i+j}{\sqrt{2}}\right) = (V_B\hat{i})\cdot\left(\frac{i+j}{\sqrt{2}}\right)$$

$$\Rightarrow \frac{3}{\sqrt{2}} = \frac{V_B}{\sqrt{2}}$$

$$V_B = 3 \text{ m/s}$$

Method III:

The rod is rigid, so velocity along the length of rod will be same.

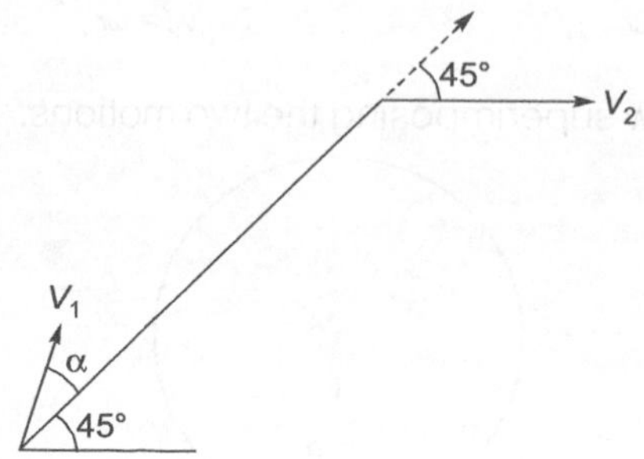

Angle made by velocity vector, V_1 with horizontal

$$V_1 = \hat{i} + 2\hat{j} = 2.236 \text{ m/s}$$

$$\tan\beta = \frac{2}{1}$$

$$\Rightarrow \beta = 63.434°$$

$$\alpha = \beta - \theta = 63.434° - 45°$$

$$= 18.434°$$

Equate velocity along the length of rod

$$V_1\cos\alpha = V_2\cos 45°$$

$$2.236 \times \cos 18.434 = V_2 \cos 45°$$

$$V_2 = 3 \text{ m/s}$$

5.13 (d)

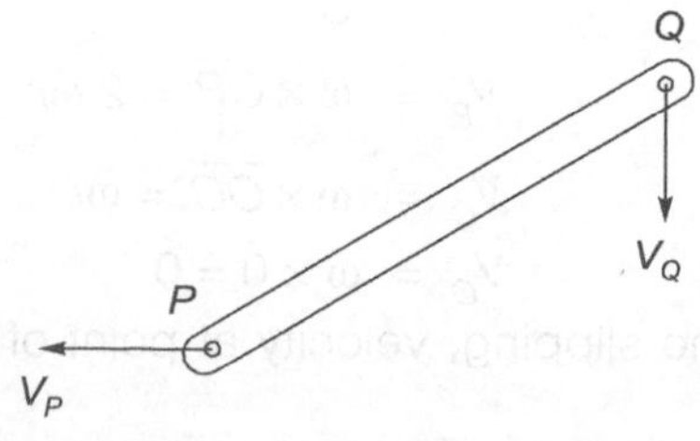

To find relative velocity direction of V_P is reversed and placed at tail of $\vec{V}_Q$

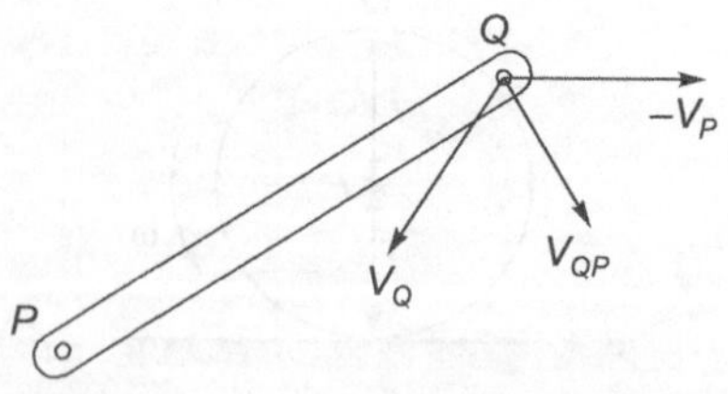

Resultant V_{Qp} is perpendicular to link *PQ* or we

can simply say as PQ is rigid component of V_P and V_Q along PQ will be same. So V_{QP} can only be perpendicular to PQ.

5.14 Sol.

V-t curve

Distance travelled is asked from t = 0 to 5s

$$V = \frac{dS}{dt}$$

$$dS = Vdt$$

$$S = \int_0^5 Vdt$$

Area of V-t graph plotted on time axis from 0 to 5s

$$S = \frac{1}{2}\times1\times1+1\times1\times\frac{1}{2}[1+4]\times1+\frac{1}{2}[4+2]\times2$$

$$= 10 \text{ m}$$

5.15 (c)

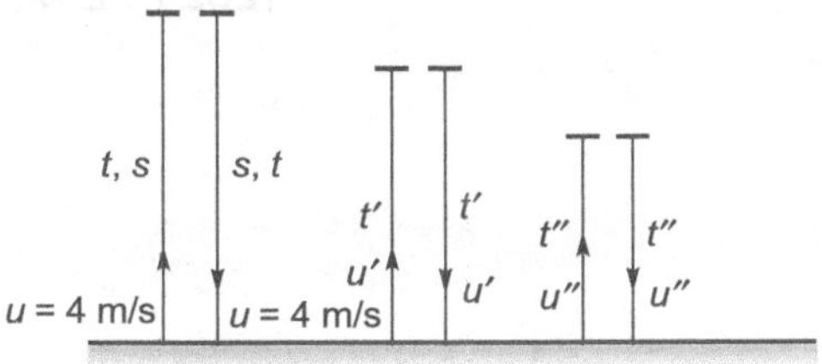

(1) → t = ?

$$v = u + at$$

$$0 = 4 - 10t$$

$$t = \frac{4}{10} = 0.4\text{s}$$

(2) → t' = ?

$$u' = 0.8 \times u = 0.8 \times 4 = 3.2 \text{ m/s}$$

$$v' = u' + at'$$

$$0 = 3.2 - 10t'$$

$$t' = \frac{3.2}{10} = 0.32\text{s}$$

(3) → t'' = ?

$$u'' = 0.8\ u' = 0.8 \times 3.2 = 2.56 \text{ m/s}$$

$$v'' = u'' + at''$$

$$0 = 2.56 - 10t''$$

$$t'' = 0.256\text{s}$$

So, t, t', t'' are forming a GP series

So, total time = $2(t + t' + t'' + \ldots\ldots 0)$

$$= 2[0.4 + 0.32 + 0.256 + \ldots\ 0]$$

$$= 2\times\frac{0.4}{1-0.8} = 2 \times 2 = \mathbf{4s}$$

5.16 Sol.

By conservation of energy (because frictionless tube) and neglecting the curved portion,

Ball will rise through the same vertical height in inclined tube as in vertical tube i.e. 1 m.

Total distance travelled by the ball = $OA + AB$ (by neglecting curve portion)

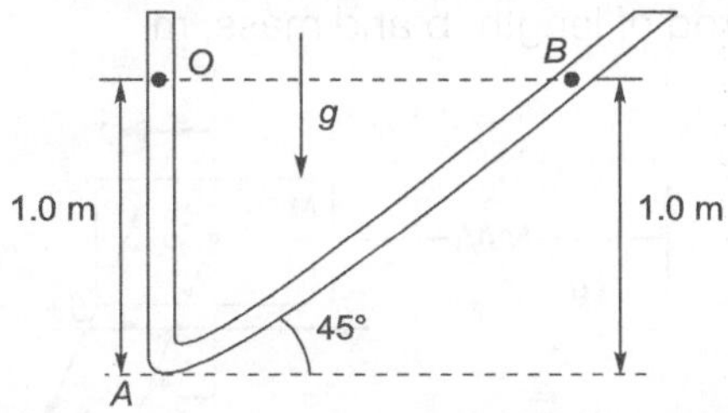

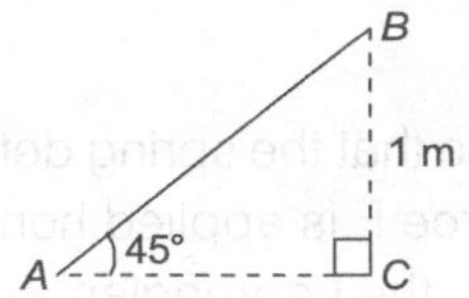

$$\frac{BC}{AB} = \sin 45°$$

$$AB = \frac{BC}{\sin 45°} = \frac{1}{\sin 45°} = 1.4142 \text{ m}$$

Total travel, $OA + AB = 1 + 1.4142$ m

$$= \mathbf{2.414\ m}$$

6 Lagrangian Equations

6.1 Consider a two degree of freedom system as shown in the figure, where PQ is a rigid uniform rod of length, b and mass, m.

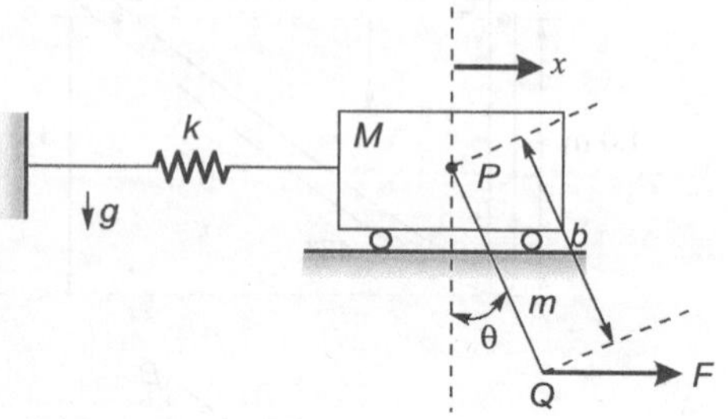

Assume that the spring deflects only horizontally and force F is applied horizontally at Q. For this system, the Lagrangian, L is

(a) $\frac{1}{2}(M+m)\dot{x}^2+\frac{1}{2}mb\dot{\theta}\dot{x}\cos\theta$
$+\frac{1}{6}mb^2\dot{\theta}^2-\frac{1}{2}kx^2+mg\frac{b}{2}\cos\theta$

(b) $\frac{1}{2}(M+m)\dot{x}^2+\frac{1}{6}mb^2\dot{\theta}^2$
$-\frac{1}{2}kx^2+mg\frac{b}{2}\cos\theta$

(c) $\frac{1}{2}M\dot{x}^2+\frac{1}{2}mb\dot{\theta}\dot{x}\cos\theta+\frac{1}{6}mb^2\dot{\theta}^2$
$-\frac{1}{2}kx^2+mg\frac{b}{2}\cos\theta+Fb\sin\theta$

(d) $\frac{1}{2}M\dot{x}^2+\frac{1}{2}mb\dot{\theta}\dot{x}\cos\theta+\frac{1}{6}mb^2\dot{\theta}^2-\frac{1}{2}kx^2$

[2021 : 2 M, Set-1]

Answers Lagrangian Equations

6.1 (a)

Explanations Lagrangian Equations

6.1 (a)

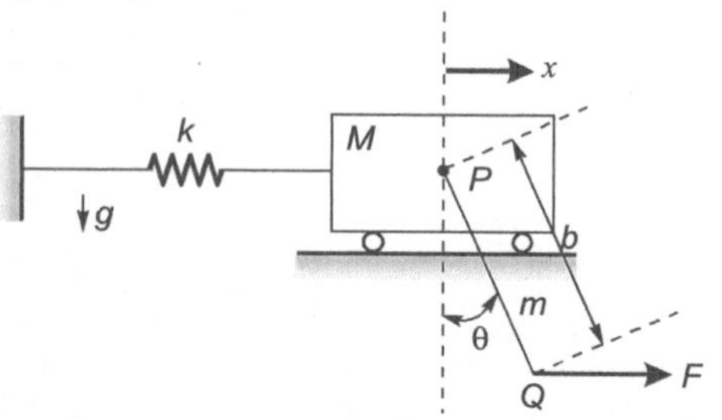

For mass, M $T=\frac{1}{2}M\dot{x}^2$

$$V=\frac{1}{2}Kx^2$$

For mass, m

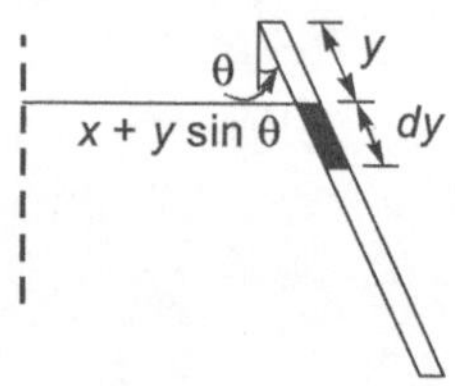

$$dm = \frac{m}{b}dy$$

Displacement $= x + y\sin\theta$

Velocity $= \dot{x}+y\cos\theta\dot{\theta}$

$$dT = \frac{1}{2}dMVel^2$$

$$= \frac{1}{2}dm\left[\dot{x}^2+y^2\dot{\theta}^2\cos^2\theta+2\dot{x}\dot{\theta}y\cos\theta\right]$$

$$T = dT = \frac{1}{2}\frac{m}{b}\int(\dot{x}^2+y^2\dot{\theta}^2\times 1+2\dot{x}\dot{\theta}y\cos\theta)dy$$

$$T = \frac{1}{2}\frac{m}{b}\left(\dot{x}^2b+\frac{b^3}{3}\dot{\theta}^2+2\dot{x}\dot{\theta}\cos\theta\frac{b^2}{2}\right)$$

$$T = \frac{1}{2}\frac{m}{b}\left(\dot{x}^2b+\frac{b^3}{3}\dot{\theta}^2+\dot{x}\dot{\theta}b^2\cos\theta\right)$$

$$T = \frac{1}{2}m\dot{x}^2+\frac{mb^2}{6}\dot{\theta}^2+\frac{m}{2}\dot{x}\dot{\theta}b\cos\theta$$

$$V = -mg\frac{b}{2}\cos\theta$$

For both masses, M and m

$$T=\frac{1}{2}M\dot{x}^2+\frac{1}{2}m\dot{x}^2+\frac{mb^2}{6}\dot{\theta}^2+\frac{m}{2}b\dot{\theta}\dot{x}^2\cos\theta$$

$$V = \frac{1}{2}kx^2-mg\frac{b}{2}\cos\theta$$

$$L = T-V$$

$$= \frac{1}{2}(M+m)\dot{x}^2+\frac{1}{2}mb\dot{\theta}\dot{x}\cos\theta$$

$$+\frac{mb^2\dot{\theta}^2}{6}-\frac{1}{2}kx^2+mg\frac{b}{2}\cos\theta$$

■■■■

Strength of Materials

UNIT

III

CONTENTS

Strength of Materials

Syllabus

Strength of Materials: Stress and strain, elastic constants, Poisson's ratio; Mohr's circle for plane stress and plane strain; thin cylinders; shear force and bending moment diagrams; bending and shear stresses; concept of shear centre; deflection of beams; torsion of circular shafts; Euler's theory of columns; energy methods; thermal stresses; strain gauges and rosettes; testing of materials with universal testing machine; testing of hardness and impact strength.

Analysis of Previous GATE Papers

Exam Year	1 Mark Ques.	2 Marks Ques.	Total Marks
1995	1	2	**5**
1996	1	1	**3**
1997	1	1	**3**
1998	–	1	**2**
2003	6	4	**14**
2004	3	4	**11**
2005	2	6	**14**
2006	2	5	**12**
2007	2	6	**14**
2008	4	8	**20**
2009	2	2	**6**
2010	1	2	**5**
2011	4	3	**10**
2012	3	3	**9**
2013	2	1	**4**
2014 Set-1	2	3	**8**
2014 Set-2	3	2	**7**
2014 Set-3	2	2	**6**
2014 Set-4	2	2	**6**

Exam Year	1 Mark Ques.	2 Marks Ques.	Total Marks
2015 Set-1	3	2	**5**
2015 Set-2	2	5	**12**
2015 Set-3	1	1	**3**
2016 Set-1	3	4	**11**
2016 Set-2	2	4	**10**
2016 Set-3	2	3	**8**
2017 Set-1	5	4	**13**
2017 Set-2	4	3	**10**
2018 Set-1	2	2	**6**
2018 Set-2	2	3	**8**
2019 Set-1	2	3	**8**
2019 Set-2	3	3	**9**
2020 Set-1	1	2	**5**
2020 Set-2	1	4	**9**
2021 Set-1	4	2	**8**
2021 Set-2	3	2	**7**
2022 Set-1	2	2	**6**
2022 Set-2	2	4	**10**
2023	4	5	**14**
2024	–	3	**6**

1 Simple Stress-Strain and Elastic Constants

1.1 free bar of length l is uniformly heated from 0°C to a temperature t°C, α is the coefficient of linear expansion and E the modulus of elasticity. The stress in the bar is

(a) αtE (b) $\alpha tE/2$
(c) Zero (d) None of these

[1995 : 1 M]

1.2 Two identical circular rods of same diameter and same length are subjected to same magnitude of axial tensile force. One of the rods is made out of mild steel having the modulus of elasticity of 206 GPa. The other rod is made out of cast iron having the modulus of elasticity of 100 GPa. Assume both the materials to be homogeneous and isotropic and the axial force causes the same amount of uniform stress in both the rods. The stresses developed are within the proportional limit of the respective materials. Which of the following observations is correct?

(a) Both rods elongate by the same amount
(b) Mild steel rod elongates more than the cast iron
(c) Cast iron rod elongates more than the mild steel rod
(d) As the stresses are equal strains are also equal in both the rods

[2003 : 1 M]

1.3 The figure below shows a steel rod of 25 mm^2 cross sectional area. It is loaded at four points. K, L, M and N. Assume E_{steel} = 200 GPa. The total change in length of the rod due to loading is

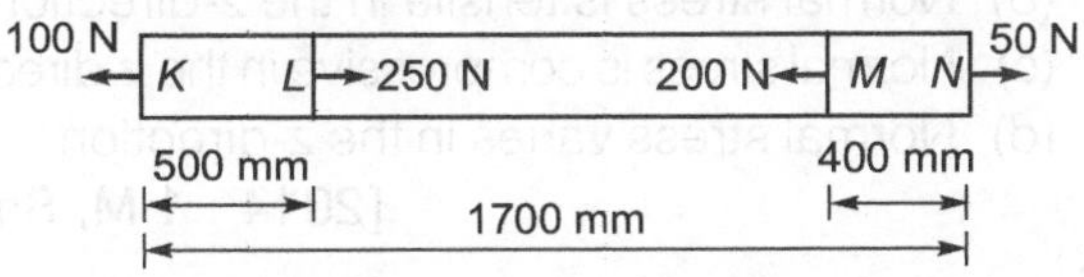

(a) 1 μm (b) −10 μm
(c) 16 μm (d) −20 μm [2004 : 2 M]

1.4 A uniform, slender cylindrical rod is made of a homogeneous and isotropic material. The rod rests on a frictionless surface. The rod is heated uniformly. If the radial and longitudinal thermal stresses are represented by σ_r and σ_z respectively, then

(a) $\sigma_r = 0, \sigma_z = 0$ (b) $\sigma_r \neq 0, \sigma_z = 0$
(c) $\sigma_r = 0, \sigma_z \neq 0$ (d) $\sigma_r \neq 0, \sigma_z \neq 0$

[2005 : 1 M]

1.5 A steel bar of 40 mm × 40 mm square cross-section is subjected to an axial compressive load of 200 kN. If the length of the bar is 2 m and E = 200 GPa, the contraction of the bar will be

(a) 1.25 mm (b) 2.70 mm
(c) 4.05 mm (d) 5.40 mm [2006 : 2 M]

1.6 A bar having a cross-sectional area of 700 mm^2 is subjected to axial loads at the positions indicated. The value of stress in the segment QR is

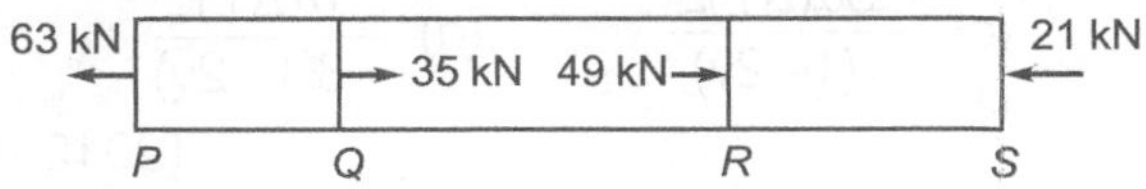

(a) 40 MPa (b) 50 MPa
(c) 70 MPa (d) 120 MPa

[2006 : 2 M]

1.7 A 200 × 100 × 50 mm steel block is subjected to a hydrostatic pressure of 15 MPa. The Young's modulus and Poisson's ratio of the material are 200 GPa and 0.3 respectively. The change in the volume of the block in mm^3 is

(a) 85 (b) 90
(c) 100 (d) 110 [2007 : 2 M]

1.8 A steel rod of length L and diameter D, fixed at both ends, is uniformly heated to a temperature rise of ΔT. The Young's modulus is E and the coefficient of linear expansion is α. The thermal stress in the rod is

(a) 0 (b) $\alpha\Delta T$
(c) $E\alpha\Delta T$ (d) $E\alpha\Delta TL$ 2007 : 1 M]

1.9 stepped steel shaft shown below is subjected to 10 N-m torque. If the modulus of rigidity is 80 GPa, the strain energy in the shaft in N-mm is

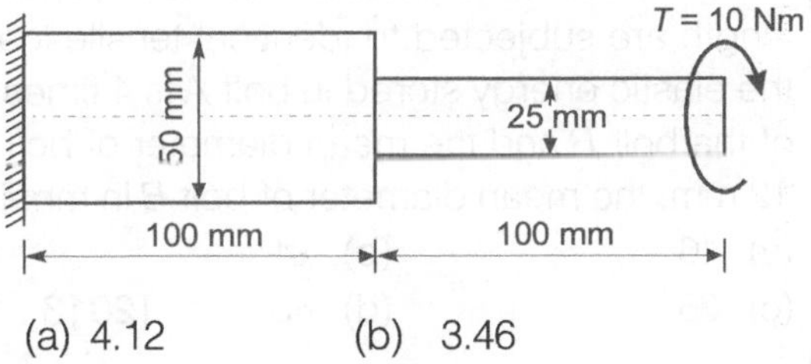

(a) 4.12 (b) 3.46
(c) 1.73 (d) 0.86 [2007 : 2 M]

1.10 A rod of length L and diameter D is subjected to a tensile load P. Which of the following is sufficient to calculate the resulting change in diameter?
(a) Young's modulus
(b) Shear modulus
(c) Poisson's ratio
(d) Both Young's modulus and shear modulus

[2008 : 1 M]

1.11 A solid steel cube constrained on all six faces is heated so that the temperature rises uniformly by ΔT. If the thermal coefficient of the material is α, Young's modulus is E and the Poisson's ratio is ν, the thermal stress developed in the cube due to heating is

(a) $-\dfrac{\alpha(\Delta T)E}{(1-2\nu)}$ (b) $-\dfrac{2\alpha(\Delta T)E}{(1-2\nu)}$

(c) $-\dfrac{3\alpha(\Delta T)E}{(1-2\nu)}$ (d) $-\dfrac{\alpha(\Delta T)E}{3(1-2\nu)}$

[2012 : 2 M]

1.12 A rod of length L having uniform cross-section area A is subjected to a tensile force P as shown in the figure below. If the Young's modulus of the material varies linearly from E_1 to E_2 along the length of the rod, the normal stress developed at the section-SS at

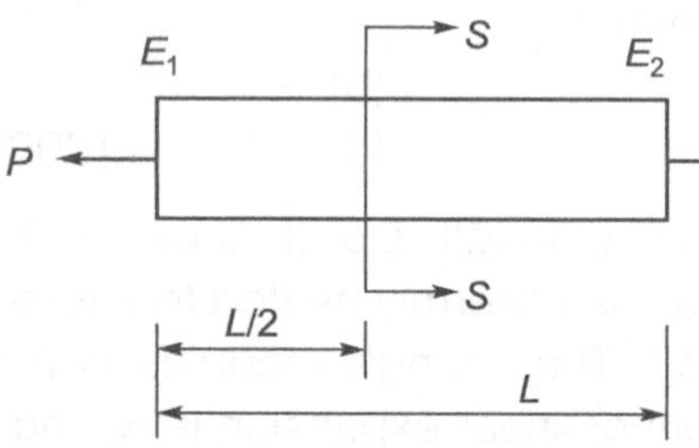

(a) $\dfrac{P}{A}$ (b) $\dfrac{P(E_1-E_2)}{A(E_1+E_2)}$

(c) $\dfrac{PE_2}{AE_1}$ (d) $\dfrac{PE_1}{AE_2}$

[2013 : 1 M]

1.13 Two threaded bolts A and B of same material and length are subjected to identical tensile load. If the elastic energy stored in bolt A is 4 times that of the bolt B and the mean diameter of bolt A is 12 mm, the mean diameter of bolt B in mm is
(a) 16 (b) 24
(c) 36 (d) 48 [2013 : 1 M]

1.14 A circular rod of length L and area of cross-section A has a modulus of elasticity E and coefficient of thermal expansion α. One end of the rod is fixed and other end is free. If the temperature of the rod is increased by ΔT, then
(a) stress developed in the rod is $E\alpha\Delta T$ and strain developed in the rod is $\alpha\Delta T$
(b) both stress and strain developed in the rod are zero
(c) stress developed in the rod is zero and strain developed in the rod is $\alpha\Delta T$
(d) stress developed in the rod is $E\alpha\Delta T$ and strain developed in the rod is zero

[2014 : 1 M, Set-1]

1.15 A steel cube, with all faces free to deform, has Young's modulus, E, Poisson's ratio, v, and coefficient of thermal expansion, α. The pressure (hydrostatic stress) developed within the cube, when it is subjected to a uniform increase in temperature, ΔT, is given by

(a) 0 (b) $\dfrac{\alpha(\Delta T)E}{1-2v}$

(c) $-\dfrac{\alpha(\Delta T)E}{1-2v}$ (d) $-\dfrac{\alpha(\Delta T)E}{3(1-2v)}$

[2014 : 1 M, Set-2]

1.16 A thin plate of uniform thickness is subject to pressure as shown in the figure below

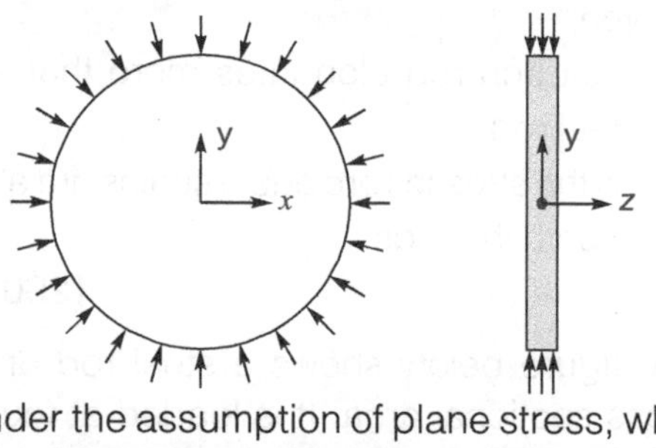

Under the assumption of plane stress, which one of the following is correct?
(a) Normal stress is zero in the z-direction
(b) Normal stress is tensile in the z-direction
(c) Normal stress is compressive in the z-direction
(d) Normal stress varies in the z-direction

[2014 : 1 M, Set-2]

1.17 A 200 mm long, stress free rod at room temperature is held between two immovable rigid walls. The temperature of the rod is uniformly raised by 250°C. If the Young's modulus and coefficient of thermal expansion are 200 GPa and 1×10^{-5}/°C, respectively, the magnitude of the longitudinal stress (in MPa) developed in the rod is ______.

[2014 : 2 M, Set-1]

1.18 A horizontal bar with a constant cross-section is subjected to loading as shown in the figure. The Young's moduli for the sections *AB* and *BC* are 3*E* and *E*, respectively.

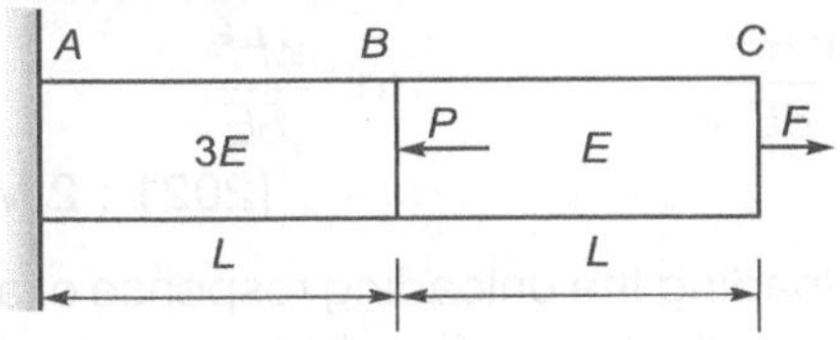

For the deflection at *C* to be zero, the ratio *P*/*F* is__________.

[2016 : 2 M, Set-1]

1.19 A circular metallic rod of length 250 mm is placed between two rigid immovable walls as shown in the figure. The rod is perfect contact with the wall on the left side and there is a gap of 0.2 mm between the rod and the wall on the right side. If the temperature of the rod is increased by 200°C, the axial stress developed in the rod is___MPa. Young's modulus of the material of the rod is 200 GPa and the coefficient of thermal expansion is 10^{-5} per°C.

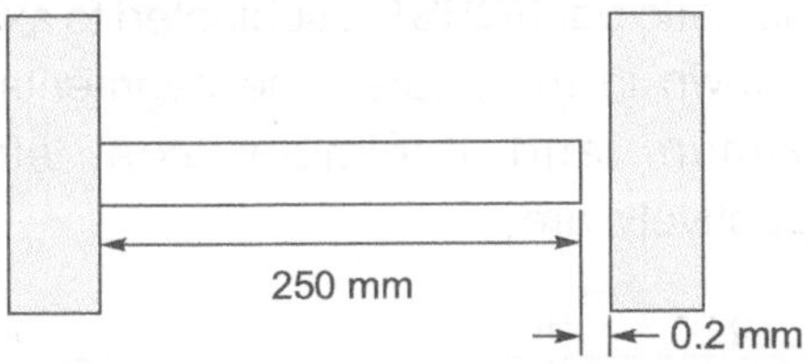

[2016 : 2 M, Set-2]

1.20 A square plate of dimension $L \times L$ is subjected to a uniform pressure load $p = 250$ MPa on its edges as shown in the figure. Assume plane stress conditions. The Young's modulus $E = 200$ GPa.

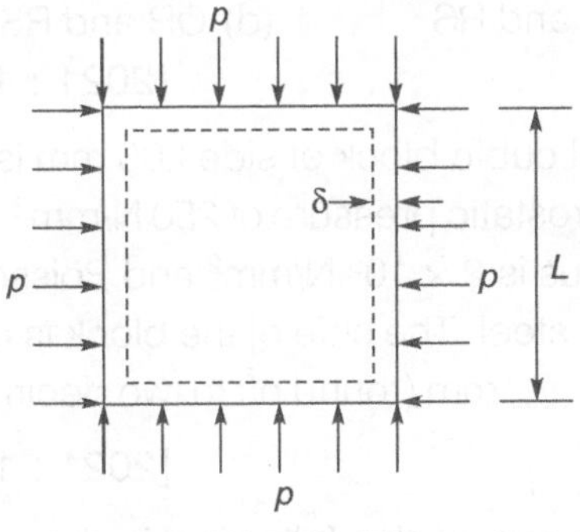

The deformed shape is a square of dimension $l - 2\delta$. If $L = 2$ m and $\delta - 0.001$ m, tho Poisson's ratio of the plate material is______

[2016 : 2 M, Set-3]

1.21 A horizontal bar, fixed at one end ($x = 0$), has a length of 1 m, and cross-sectional area of 100 mm^2. Its elastic modulus varies along its length as given by $E(x) = 100e^{-x}$ GPa, where x is the length coordinate (in m) along the axis of the bar. An axial tensile load of 10 kN is applied at the free end ($x = 1$). The axial displacement of the free end is ______ mm.

[2017 : 2 M, Set-1]

1.22 A point mass of 100 kg is dropped onto a massless elastic bar (cross-sectional area = 100 mm^2, length = 1 m, Young's modulus = 100 GPa) from a height *H* of 10 mm as shown (Figure is not to scale). If $g = 10$ m/s^2, the maximum compression of the elastic bar is ______ mm.

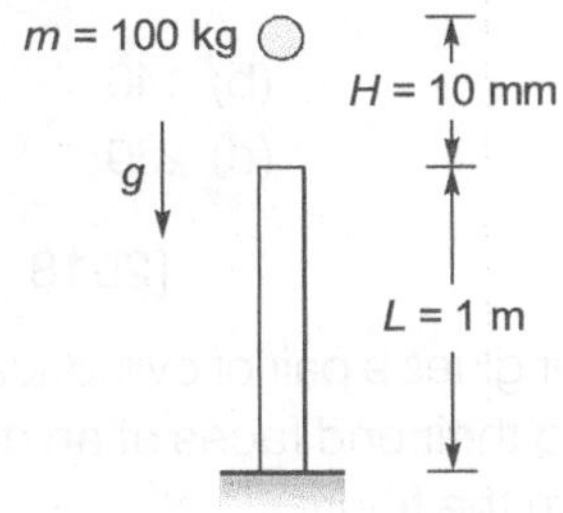

[2017 : 2 M, Set-1]

1.23 A rod of length 20 mm is stretched to make a rod of length 40 mm. Subsequently, it is compressed to make a rod of final length 10 mm. Consider the longitudinal tensile strain as positive and compressive strain as negative. The total true longitudinal strain in the rod is

(a) – 0.5 (b) – 0.69
(c) – 0.75 (d) – 1.0

[2017 : 2 M, Set-2]

1.24 A cantilever beam of length *L* and flexural modulus *EI* is subjected to a point load *P* at the free end. The elastic strain energy stored in the beam due to bending (neglecting transverse shear) is

(a) $\dfrac{P^2L^3}{6EI}$ (b) $\dfrac{P^2L^3}{3EI}$

(c) $\dfrac{PL^3}{3EI}$ (d) $\dfrac{PL^3}{6EI}$

[2017 : 1 M, Set-2]

1.25 A steel bar is held by two fixed supports as shown in the figure and is subjected to an increase of temperature $\Delta T = 100$°C. If the coefficient of thermal expansion and Young's modulus of elasticity of steel are 11×10^{-6} /°C and 200 GPa, respectively, the magnitude of thermal stress (in MPa) induced in the bar is ______.

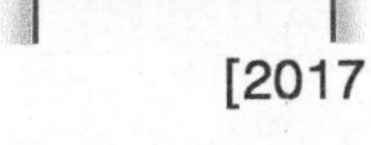

[2017 : 1 M, Set-2]

1.26 A bimetallic cylindrical bar of cross sectional area 1 m^2 is made by bonding Steel (Young's modulus = 210 GPa) and Aluminium (Young's modulus = 70 GPa) as shown in the figure. To maintain tensile axial strain of magnitude 10^{-6} in steel bar and compressive axial strain of magnitude 10^{-6} in Aluminum bar, the magnitude of the required force P (in kN) along the indicated direction is

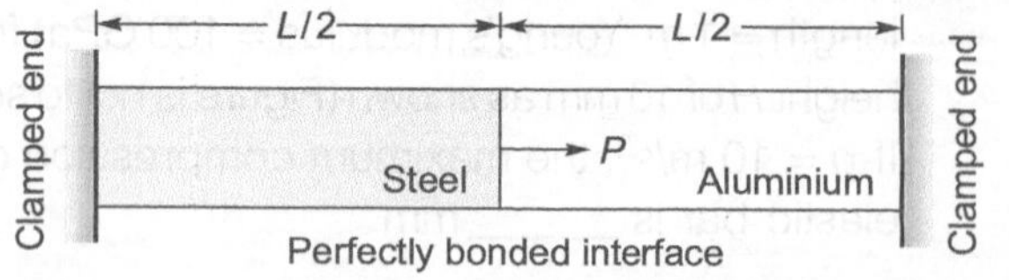

(a) 70 (b) 140
(c) 210 (d) 280

[2018 : 2 M, Set-2]

1.27 A carpenter glues a pair of cylindrical wooden logs by bonding their end faces at an angle of θ = 30° as shown in the figure.

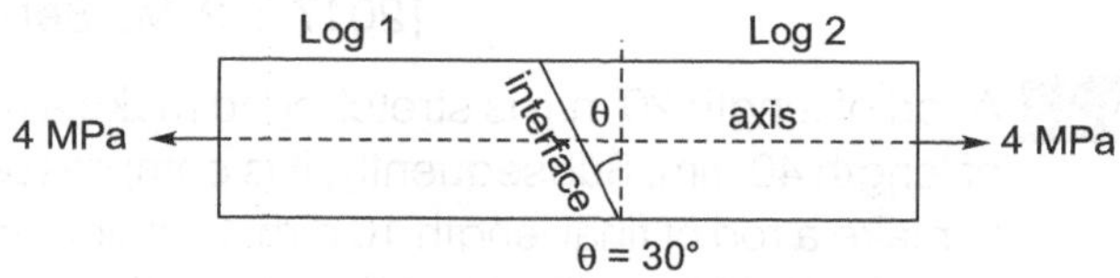

The glue used at the interface fails if

Criterion 1 : the maximum normal stress exceeds 2.5 MPa

Criterion 2 : the maximum shear stress exceeds 1.5 MPa

Assume that the interface fails before the logs fail. When a uniform tensile stress of 4 MPa is applied, the interface

(a) fails only because of criterion 1
(b) fails only because of criterion 2
(c) fails because of both criteria 1 and 2
(d) does not fail. [2018 : 2 M, Set-1]

1.28 A solid cube of side 1 m is kept at a room temperature of 32°C. The coefficient of linear thermal expansion of the cube material is 1×10^{-5}/°C and the bulk modulus is 200 GPa. If the cube is constrained all around and heated uniformly to 42°C, then the magnitude of volumetric (mean) stress induced due to heating is ________ MPa.

[2019 : 1 M, Set-1]

1.29 A right solid circular cone standing on its base on a horizontal surface is of height H and base radius R. The cone is made of a material with specific weight w and elastic modulus E. The vertical deflection at the mid-height of the cone due to self-weight is given by

(a) $\frac{wRH}{8E}$ (b) $\frac{wH^2}{8E}$

(c) $\frac{wRH}{6E}$ (d) $\frac{wH^2}{6E}$

[2021 : 2 M, Set-1]

1.30 The loading the unloading response of a metal is shown in the figure. The elastic and plastic strains corresponding to 200 MPa stress, respectively are

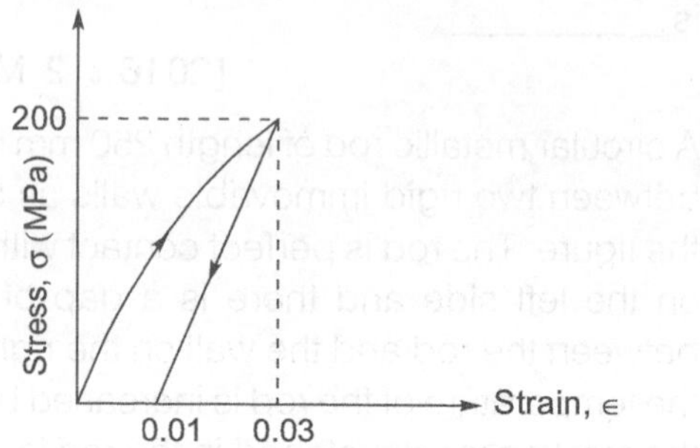

(a) 0.01 and 0.01 (b) 0.02 and 0.02
(c) 0.01 and 0.02 (d) 0.02 and 0.01

[2021 : 1 M, Set-1]

1.31 A prismatic bar PQRST is subjected to axial loads as shown in the figure. The segments having maximum and minimum axial stresses, respectively, are

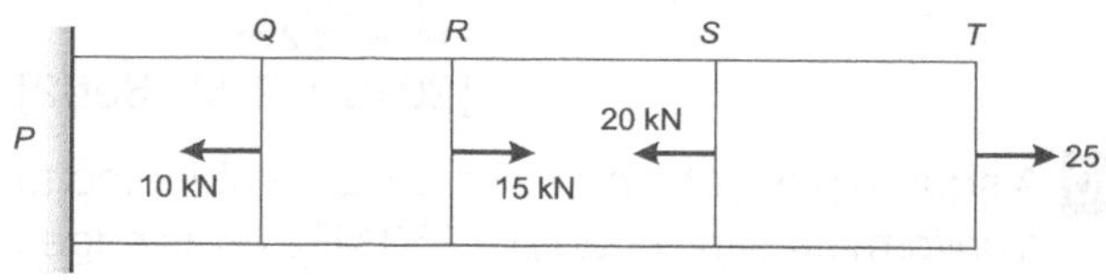

(a) ST and PQ (b) QR and PQ
(c) ST and RS (d) QR and RS

[2021 : 1 M, Set-1]

1.32 A steel cubic block of side 200 mm is subjected to hydrostatic pressure of 250 N/mm². The elastic modulus is 2×10^5 N/mm² and Poisson's ratio is 0.3 for steel. The side of the block is reduced by ________ mm (round off to two decimal places).

[2021 : 1 M, Set-2]

1.33 Which one of the following is the definition of ultimate tensile strength (UTS) obtained from a stress-strain test on a metal specimen?

(a) Stress value where the stress-strain curve transitions from elastic to plastic behavior.
(b) The maximum load attained divided by the original cross-sectional area.

(c) The maximum load attained divided by the corresponding instantaneous cross-sectional area.

(d) Stress where the specimen fractures.

[2022 : 1 M, Set-2]

1.34 A rigid beam AD of length $3a = 6$ m is hinged at frictionless pin joint A and supported by two strings as shown in the figure. String BC passes over two small frictionless pulleys of negligible radius. All the strings are made of the same material and have equal cross-sectional area. A force $F = 9$ kN is applied at C and the resulting stresses in the strings are within linear elastic limit. The self-weight of the beam is negligible with respect to the applied load. Assuming small deflections, the tension developed in the string at C is ________ kN (round off to 2 decimal places).

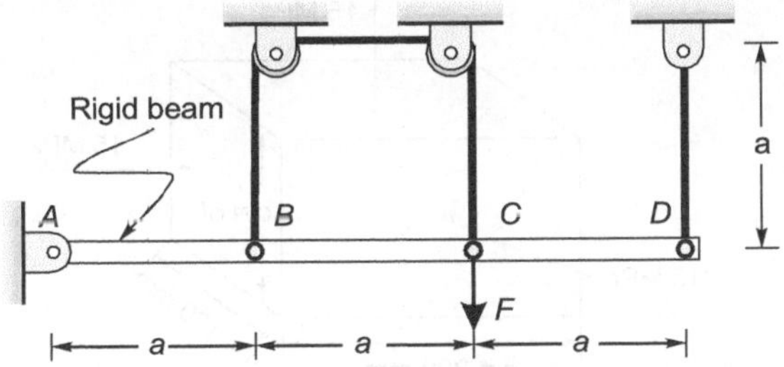

[2022 : 2 M, Set-2]

1.35 Cylindrical bars P and Q have identical lengths and radii, but are composed of different linear elastic materials. The Young's modulus and coefficient of thermal expansion of Q are twice the corresponding values of P. Assume the bars to be perfectly bonded at the interface, and their weights to be negligible.

The bars are held between rigid supports as shown in the figure and the temperature is raised by ΔT. Assume that the stress in each bar is homogeneous and uniaxial. Denote the magnitudes of stress in P and Q by σ_1 and σ_2, respectively.

Which of the statement(s) given is/are CORRECT?

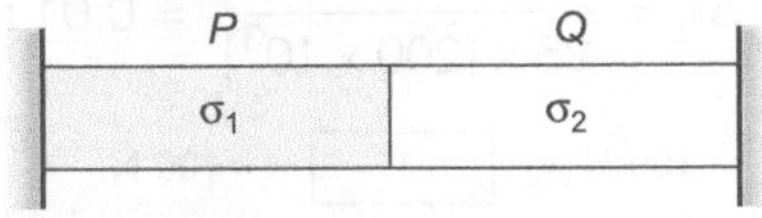

(a) The interface between P and Q moves to the left after heating.

(b) The interface between P and Q moves to the right after heating.

(c) $\sigma_1 < \sigma_2$

(d) $\sigma_1 = \sigma_2$

[2023 : 2 M], MSQ]

Answers Simple Stress-Strain and Elastic Constants

1.1 (c)	1.2 (c)	1.3 (b)	1.4 (a)	1.5 (a)	1.6 (a)	1.7 (b)	1.8 (c)
1.9 (c)	1.10 (d)	1.11 (a)	1.12 (a)	1.13 (b)	1.14 (c)	1.15 (a)	1.16 (a)
1.17 (500)	1.18 (4)	1.19 (240)	1.20 (0.2)	1.21 (1.718)	1.22 (1.517)	1.23 (b)	1.24 (a)
1.25 (220)	1.26 (d)	1.27 (c)	1.28 (60)	1.29 (b)	1.30 (d)	1.31 (c)	
1.32 (0.1)	1.33 (b)	1.34 (1.5)	1.35 (a, d)				

Explanations Simple Stress-Strain and Elastic Constants

1.1 (c)

If a bar is free, it will expand freely on heating and no thermal stress will generate. Thermal stress gets generated only when bar is constrained.

1.2 (c)

Given, $E_S = 206$ GPa and $E_I = 100$ GPa

Elongation in mild steel,

$$\Delta L_s = \frac{PL}{AE_s} \quad \text{...(i)}$$

Elongation in cast Iron,

$$\Delta L_I = \frac{PL}{AE_I} \quad \text{...(ii)}$$

Dividing Eq. (ii) by Eq. (i), we get

$$\frac{\Delta L_s}{\Delta L_I} = \frac{E_I}{E_S}$$

Since diameter and length of both shafts are equal

$$\therefore \quad \frac{\Delta L_s}{\Delta L_I} = \frac{100}{206} < 1$$

1.3 (b)

Given: Area = 25 mm²; $E_{steel} = 200$ GPa

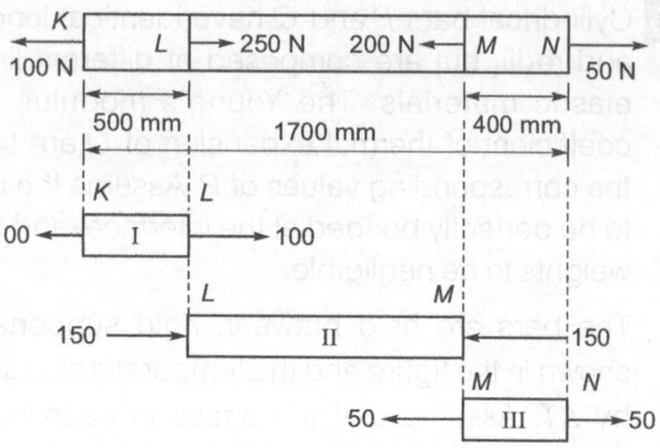

Axial deformation $= \dfrac{PL}{AE}$

$$\Delta l_{\text{total}} = \Delta l_1 + \Delta l_2 + \Delta l_3$$

$$\Delta l_1 = \frac{100 \times 500}{25 \times \{200 \times 10^3\}} = 0.01 \text{ mm}$$

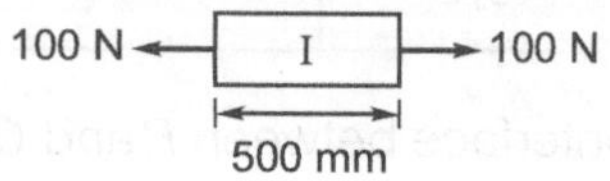

$$\Delta l_2 = \frac{-150 \times [1700 - \{500 + 400\}]}{25 \times \{200 \times 10^3\}}$$

$$= -0.024 \text{ mm}$$

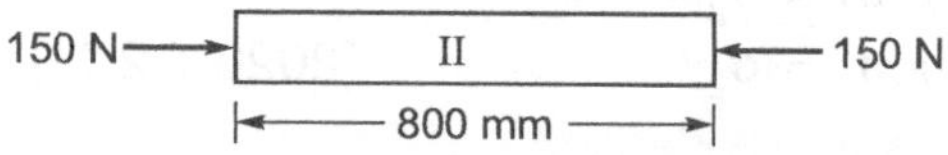

$$\Delta l_3 = \frac{50 \times 400}{25 \times \{200 \times 10^3\}} = 4 \times 10^{-3} \text{ mm}$$

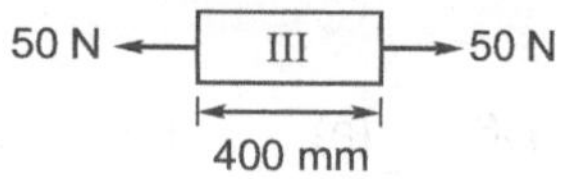

$$\Delta l_{\text{Total}} = \Delta l_1 + \Delta l_2 + \Delta l_3$$

$$= 0.01 - 0.024 + 4 \times 10^{-3}$$

$$= -0.01 \text{ mm} = 0.01 \times 10^{-3} \text{ m} = -1 \times 10^{-5} \text{ m}$$

$$= \frac{-1 \times 10^{-5} \times 10}{10} \text{ m} = -10 \times 10^{-6} \text{ m} = \mathbf{-10\ \mu m}$$

1.4 (a)

If a body is allowed to expand or contract freely with rise or fall in temperature then no stress is induced in body. i.e.

$$\sigma_r = 0 \text{ and } \sigma_z = 0$$

1.5 (a)

We know that

$$\Delta L = \frac{PL}{AE} = \frac{200 \times 10^3 \times 2 \times 10^3}{40 \times 40 \times 200 \times 10^3} = 1.25 \text{ mm}$$

1.6 (a)

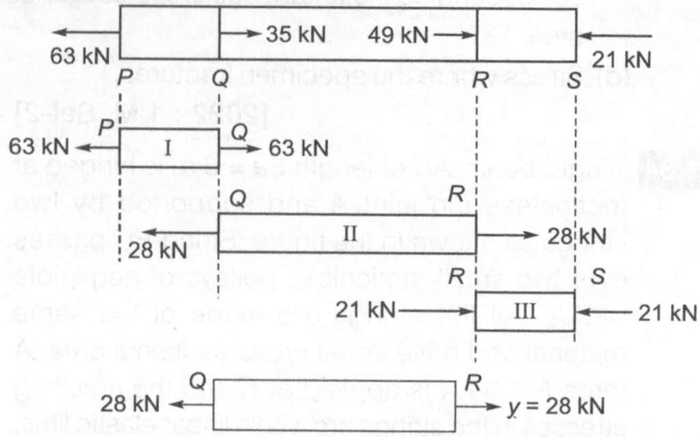

$$\sigma_{QR} = \frac{28 \times 10^3 \text{ N}}{700 \text{ mm}^2} = 40 \text{ MPa}$$

1.7 (b)

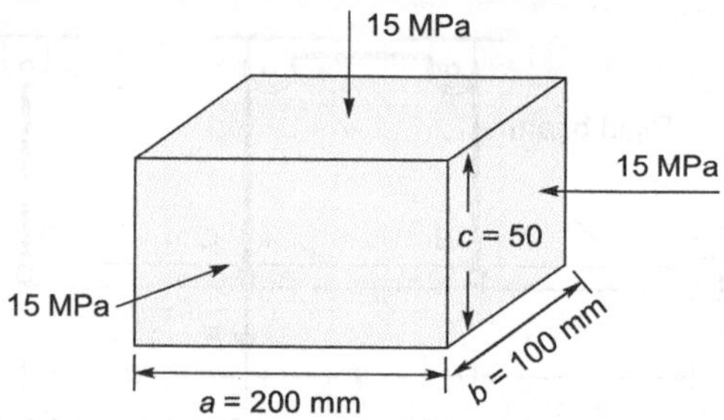

$$V = a \times b \times c$$

$$= 200 \times 100 \times 50 \text{ mm}^3$$

$$= 10^6 \text{ mm}^3$$

$$E = 200 \text{ GPa}, \mu = 0.3$$

$$\frac{dV}{V} = \frac{3\sigma}{E}\{1 - 2\mu\}$$

$$dV = \frac{3\sigma}{E}\{1 - 2\mu\} \times V$$

$$= \frac{3 \times 15}{200 \times 10^3} \times \{1 - (2 \times 0.3)\} \times 10^6$$

$$= \mathbf{90\ mm^3}$$

1.8 (c)

Thermal strain,

$$\varepsilon_t = \frac{\text{Change in length prevented}}{\text{Original length}} = \frac{L\alpha\Delta T}{L}$$

$$= \alpha\Delta T$$

Thermal stress, $\sigma_t = \varepsilon_t \times E = \alpha\,\Delta T E$

1.9 (c)

$$U_{\text{torsion}} = \frac{1}{2}T_1\theta_1 + \frac{1}{2}T_2\theta_2$$

$$= \frac{T_1}{2}\left(\frac{T_1 L}{GJ_1}\right) + \frac{T_2}{2}\left(\frac{T_2 L}{GJ_2}\right)$$

$$= \frac{T^2L}{2GJ_1} + \frac{T^2L}{2GJ_2} \quad (\text{As } T_1 = T_2 = T)$$

$$= \frac{T^2L}{2G}\left(\frac{1}{J_1} + \frac{1}{J_2}\right) = \frac{(10\times1000)^2\times100}{2\times80\times1000}$$

$$\left[\left(\frac{32}{\pi\times(50)^4} + \frac{32}{\pi\times(25)^4}\right)\right]$$

$$= \frac{10^8\times100\times32}{2\pi\times80\times1000}\left[\frac{1}{(50)^4} + \frac{1}{(25)^4}\right] = \mathbf{1.73\ N\text{-}mm}$$

1.11 (a)

Let the side of cube be a

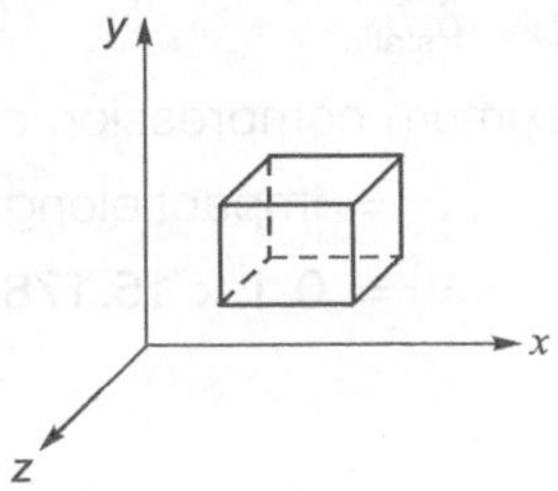

Now since the cube is uniformly constrained to expand, the stress produced in all the three directions will be same

$\therefore$ Strain in x direction $= -\alpha(\Delta T)$

$$-\frac{\sigma_x}{E} - \frac{\nu\sigma_y}{E} - \frac{\nu\sigma_z}{E}$$

$$\sigma_x = \sigma_y = \sigma_z = \sigma$$

$$\therefore \sigma = -\frac{\alpha(\Delta T)E}{(1-2\nu)}$$

1.12 (a)

Normal stress at any section is independent of modulus of elasticity.

1.13 (b)

Given:

$P_1 = P_2 = P$

(identical tensile load on bolt A & B)

$L_1 = L_2 = L$ (same length)

$d_A = 12$ mm

U_A = Strain energy in bolt A

U_B = Strain energy in bolt B

$U_A = 4U_B$ (Given) ...(i)

$\therefore$ $d_B = ?$

Strain energy is given by

$$U_A = \frac{1}{2}P\times\delta = \frac{1}{2}\frac{P^2L}{AE}$$

$\therefore$ Eq. (i)

$$\frac{1}{2}\frac{P_1^2L_1}{A_1E} = 4\times\frac{1}{2}\frac{P_2^2L_2}{A_2E}$$

$$\frac{1}{A_1} = 4\times\frac{1}{A_2}$$

$$\frac{4}{\pi(12)^2} = 4\times\frac{4}{\pi(d_B)^2}$$

$$d_B^2 = 576$$

or $\quad d_B = \mathbf{24\ mm}$

1.14 (c)

For thermal stress to be developed, there must be constraint in the system. So, although strain develops but there is no thermal stress.

1.15 (a)

As all forces are free to deform, there is no thermal stress.

1.16 (a)

Thin plate of uniform thickness pertains to plane stress condition. So, stress out of plane would be zero.

1.17 Sol.

Given,

$l = 200$ mm; $\Delta T = 250°C$; $\alpha = 1\times10^{-5}/°C$

$E = 200$ GPa $= 200\times10^3$ MPa, $\sigma = ?$

$$\sigma = \alpha\Delta TE = 10^{-5}\times250\times200\times10^3$$

$$= \mathbf{500\ MPa}$$

1.18 Sol.

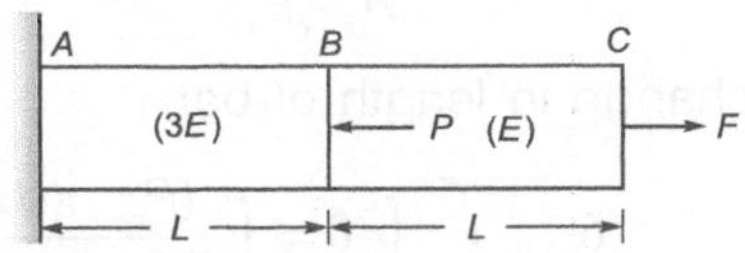

$$\Delta_C = 0$$

$$\Delta_{BC} + \Delta_{AB} = 0$$

$$\frac{FL}{AE} + \frac{(F-P)L}{A\times3E} = 0$$

$$F + \frac{F-P}{3} = 0$$

$$3F + F - P = 0$$

$$4E = P$$

or $$\frac{P}{F} = 4$$

1.19 Sol.

Free expansion of rod = $L\alpha\Delta T$

$$= 250 \times 10^{-5} \times 200 = 0.5 \text{ mm}$$

Actual expansion = 0.2 mm

∴ Expansion resisted = 0.3 mm

$$\therefore \quad \text{Strain resisted} = \frac{0.3}{250}$$

$$\text{Stress developed} = \frac{0.3}{250} \times 200 \times 10^3$$

$$= 0.24 \times 10^3 = \mathbf{240\ MPa}$$

1.20 Sol.

Strain in horizontal direction

$$\frac{-2\delta}{L} = \frac{-P}{E} + \mu\frac{P}{E}$$

$$\frac{-2\delta}{L}E = P(-1+\mu) \Rightarrow \frac{2\delta E}{LP} = 1-\mu$$

$$\therefore \quad \mu = 1 - \frac{2\delta E}{LP}$$

$$= 1 - \frac{2 \times 0.001 \times 200 \times 10^3}{2 \times 250} = 1 - 0.8 = \mathbf{0.2}$$

1.21 Sol.

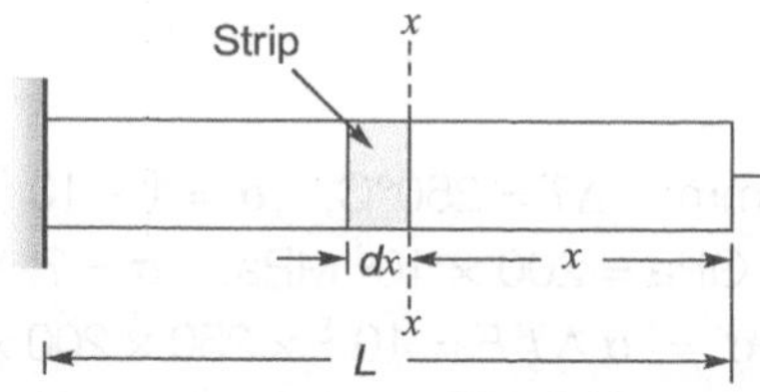

Change in length of small strip,

$$\delta = \frac{P_{x-x}(dx)}{A_{x-x}E_{xx}}$$

Total change in length of bar,

$$\delta_{total} = \int_0^L \delta = \int_0^L \frac{(P_{x-x})(dx)}{(A_{x-x})E_{x-x}}$$

$$P_{x-x} = P = \text{Constant}$$

$$A_{x-x} = A = \text{Constant}$$

$$E_{x-x} = 100\ e^{-x} \text{ GPa}$$

$$\delta_{total} = \frac{P}{A}\int_0^L \frac{dx}{100e^{-x}} = \frac{P}{10^9 \times 100\,A}\int_0^L \frac{dx}{e^{-x}}$$

$$= \frac{10000}{\frac{100}{10^6} \times 100 \times 10^9}\int_0^1 \frac{dx}{e^{-x}} = \frac{1}{1000}[1.7183]\text{ m}$$

Hence, $\delta_{total} = \mathbf{1.7183\ mm}$

1.22 Sol.

Given loading is impact loading

$$\delta_{impact} = \delta_{static} \times \text{Impact factor}$$

$$\delta_{static} = \frac{WL}{AE} = \frac{100 \times 10 \times 1000}{100 \times 100 \times 10^3}$$

$$= 0.1 \text{ mm}$$

Impact factor,

$$\text{I.F.} = 1 + \sqrt{1 + \frac{2h}{\delta_{static}}} = 1 + \sqrt{1 + \frac{2 \times 10}{0.1}} = 15.178$$

Hence maximum compression of bar

$$= \text{Impact elongation}$$

$$= \mathbf{0.1 \times 15.178 = 1.5178\ mm}$$

1.23 (b)

$$\varepsilon_E = \frac{L_f - L_0}{L_0} = \frac{10-20}{20} = -0.5$$

$$\varepsilon_T = \ln[1 + \varepsilon_E] = \ln[1 - 0.5]$$

$$= \mathbf{-0.693}$$

1.24 (a)

Let U is the S.E. in the beam due to B.M. (M)

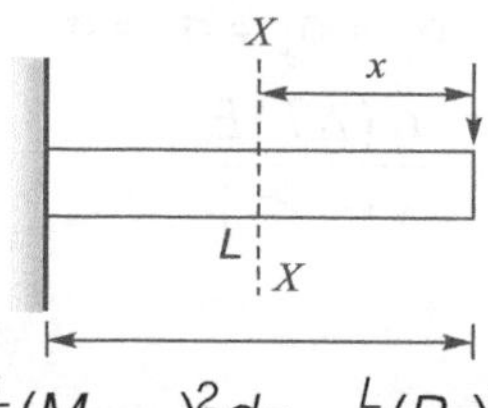

$$U = \int_0^L \frac{(M_{X-X})^2 dx}{2(EI)_{X-X}} = \int_0^L \frac{(Px)^2 dx}{2EI_{N.A}}$$

$$U = \frac{P^2}{2EI_{N.A}}\int_0^L x^2(dx) = \frac{P^2}{2EI}\left(\frac{L^3}{3}\right)$$

$$U = \frac{P^2L^3}{6EI_{N.A}}$$

1.25 Sol.

We know that for completely restricted expansion, thermal stress developed in bar is given by

$$\sigma_{th} = \alpha(\Delta T)E$$

$$= 11 \times 10^{-6} \times 100 \times 200 \times 10^3$$

$$= \mathbf{220\ MPa} \text{ (comp. in native)}$$

1.26 (d)

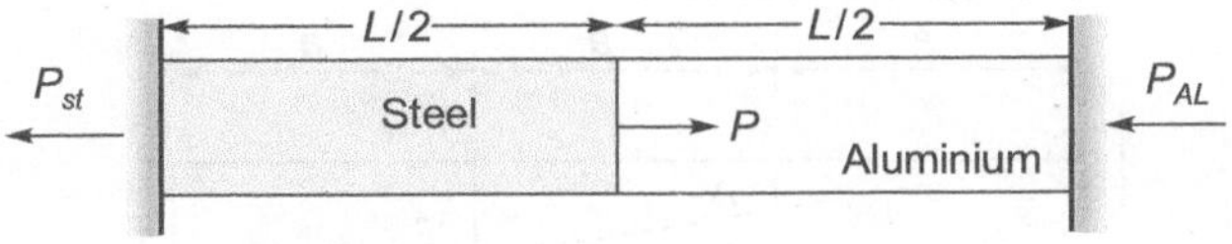

$$P_{st} + P_{Al} = P \qquad \ldots \text{(i)}$$

$$\Delta_{st} + \Delta_{Al} = 0$$

$$\frac{P_{st} \cdot \frac{L}{2}}{AE_{st}} + \frac{(P_{st} - P)\frac{L}{2}}{AE_{AL}} = 0$$

$$\frac{P_{st}}{210} + \frac{P_{st} - P}{70} = 0$$

$$\therefore \qquad P_{st} = \frac{3P}{4}$$

Now,
$$\Delta_{st} = \frac{P_{st} \cdot \frac{L}{2}}{AE_{st}}$$

$$\in_{st} = \frac{P_{st}}{AE_{st}} = \frac{3P}{4AE_{st}}$$

$$\therefore P = \frac{\in_{st} \times 4AE_{st}}{3} = \frac{10^{-6} \times 4 \times 1 \times 210 \times 10^9}{3}$$

$$= 280 \text{ kN}$$

1.27 (c)

Normal stress on inclined plane

$$\sigma' = \sigma_x \cos^2 \theta$$

$$= 4 \times \cos^2 30° = 3 \text{ MPa}$$

Shear stress on inclined plane,

$$\tau' = \frac{\sigma_x}{2} \sin 2\theta$$

$$= 2 \times \sin 60° = 1.73 \text{ MPa}$$

Since both the stress exceeds the given limits, answer is option (c).

1.28 Sol.

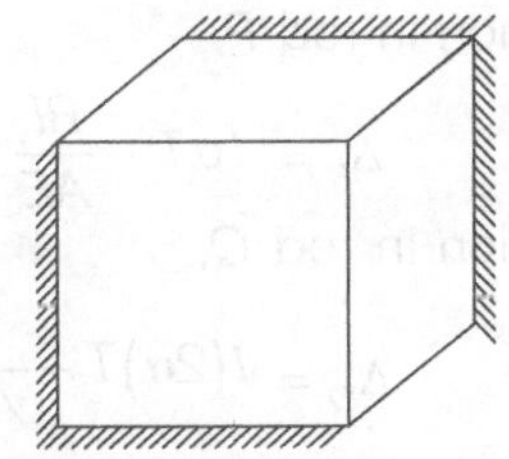

Since block is constrained in all direction and heated uniformly. So it is the case of hydrostatic state of stress.

As we know,

For hydrostatic state of stress,

Volumetric strain,

$$\varepsilon_v = \frac{\sigma_v}{K} = 3\alpha \Delta T$$

or
$$\sigma_v = 3K\alpha\Delta T$$

$$= 3 \times 200 \times 10^3 \times 10^{-5} \times (42 - 32)$$

$$= 60 \text{ MPa}$$

1.29 (b)

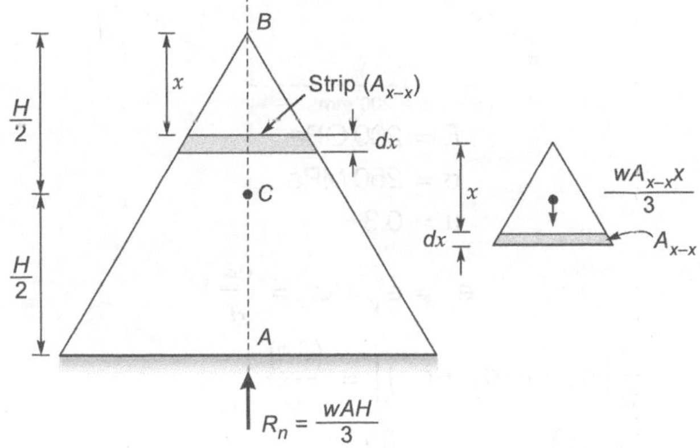

$$P_{x\text{-}x} = \frac{(-)wA_{x-x}(x)}{3}$$

Contraction of small strip = $(\delta l)_{strip}$

$$= \frac{P_{x-x}dx}{(AE)_{x-x}} = \frac{w(A_{x-x})(x)dx}{3(A_{x-x})E} = \frac{wx}{3E}dx$$

Contraction of conical bar at mid height

$$\left(\text{i.e. } x = \frac{H}{2}\right) = \int_{H/2}^{H} (\delta l)_{strip} = \frac{w}{3E}\int_{H/2}^{H} x dx$$

$$= \frac{w}{6E}\left[H^2 - \frac{H^2}{4}\right] = \frac{wH^2}{8E}$$

1.30 (d)

Elastic strain : Which can be recovered = 0.03 – 0.01 = 0.02

Plastic strain : Permanent strain = 0.01

1.31 (c)

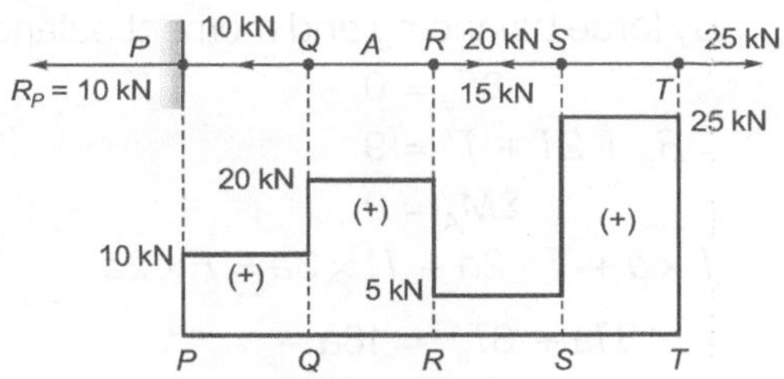

Axial Loading Diagram

$$P_{max} = P_{ST} = 25 \text{ kN}$$
$$P_{min} = P_{RS} = 5 \text{ kN}$$

Hence, maximum and minimum axial stresses are in ST and RS portions because of prismatic bar. Hence the answer is (c).

1.32 Sol.

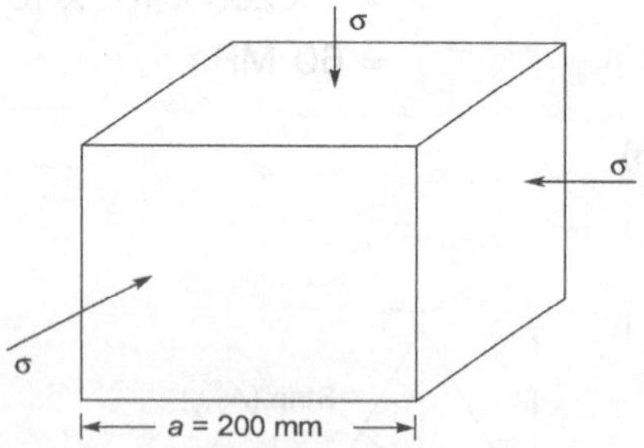

$$E = 200 \text{ GPa}$$
$$\sigma = 250 \text{ MPa}$$
$$\mu = 0.3$$
$$\epsilon_x = \epsilon_y = \epsilon_z = \frac{\delta a}{a}$$
$$\frac{1}{E}\left[\sigma_x - \mu(\sigma_y + \sigma_z)\right] = \frac{(\delta a)}{a}$$
$$\left(\frac{-\sigma}{E}\right)(1-2\mu) = \frac{\delta a}{a}$$
$$\delta a = -\frac{(250)(1-0.6)(200)}{200 \times 10^3} = (-)\ 0.10 \text{ mm}$$

Reduction in side of cube is 0.10 mm.

1.33 (b)

Ultimate tensile strength S_{ut} from tension test is given by

$$S_{ut} = \frac{\text{Maximum load}}{\text{Original cross-sectional area}}$$

1.34 (1.5) (1.48 to 1.52)

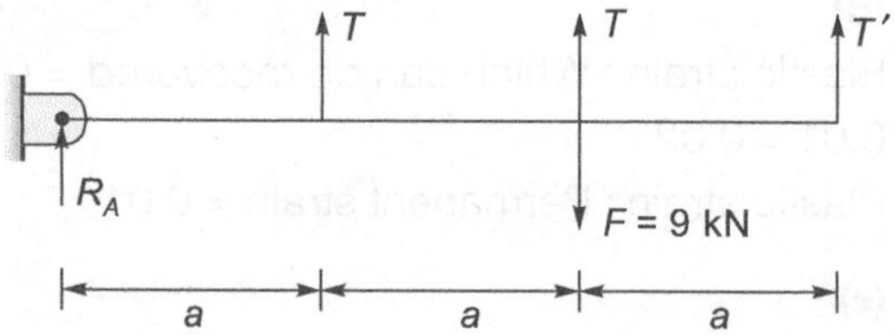

By force balancing and moment balancing:

$$\Sigma F_y = 0$$
$$R_A + 2T + T' = 9 \quad \text{...(i)}$$
$$\Sigma M_A = 0$$
$$T \times a + T \times 2a + T' \times 3a = F \times 2a$$
$$3Ta + 3T'_a = 18a$$
$$T + T' = \frac{18}{3}$$
$$T + T' = 6 \quad \text{...(ii)}$$

For small deflections

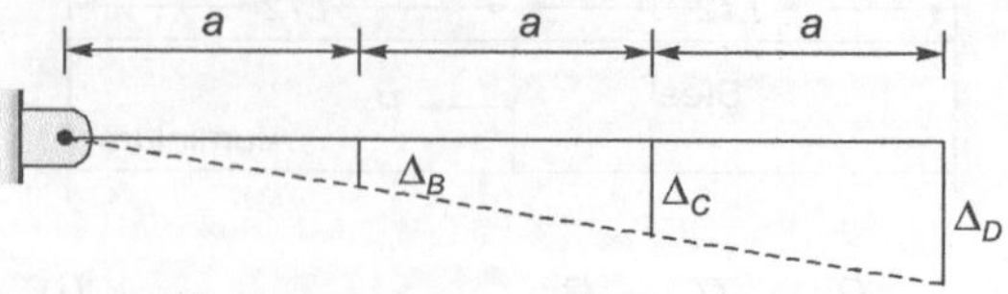

By similar triangles

$$\frac{\Delta_B}{a} = \frac{\Delta_C}{2a} = \frac{\Delta_D}{3a}$$
$$\Delta_B = \frac{\Delta_C}{2}; \quad \Delta_B = \frac{\Delta_D}{3} \quad \text{...(iii)}$$

As rope BC is stretching due to tension T, length of rope $BC\ l_{BC} = 3a$

So, $\Delta_B + \Delta_C = \dfrac{T \times 3a}{AE}$ [Due to deflection wire stretched = $\Delta_B + \Delta_C$]

$$\Delta_B + 2\Delta_B = \frac{T \times 3a}{AE}$$
$$\Delta_B = \frac{Ta}{AE}$$

By equation (iii) $\dfrac{Ta}{AE} = \dfrac{T'_a}{3AE}$

$$T = \frac{T'}{3}$$

By equation (ii)

$$\frac{T'}{3} + T' = 6$$
$$\Rightarrow \quad T' = \frac{6 \times 3}{4} = 4.5 \text{ kN}$$
$$\Rightarrow \quad T = 6 - 4.5 = 1.5 \text{ kN}$$

1.35 (a, d)

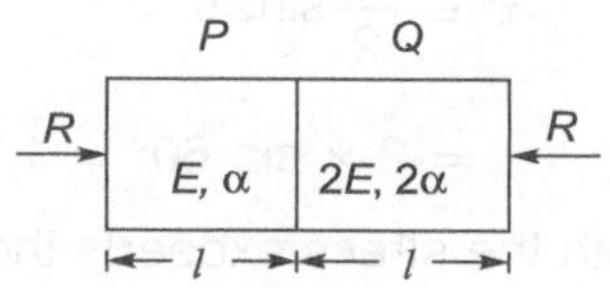

As the areas of both the bars are same, therefore, the stress developed in both the bars will remain same.

Deflection in rod P,

$$\Delta_P = l\alpha T - \frac{Rl}{AE}$$

Deflection in rod Q,

$$\Delta_Q = l(2\alpha)T - \frac{Rl}{A(2E)}$$

We can say that $\Delta_Q > \Delta_P$. Therefore, interface will move towards left side.

■■■■

2 Properties of Material

2.1 In terms of Poisson's ratio (μ) the ratio of Young's modulus (E) to Shear modulus (G) of elastic materials is

(a) $2(1+\mu)$ (b) $2(1-\mu)$

(c) $\frac{1}{2}(1+\mu)$ (d) $\frac{1}{2}(1-\mu)$ **[2004 : 1 M]**

2.2 The stress-strain curve for mild steel is shown in figure given below. Choose the correct option referring to both figure and table.

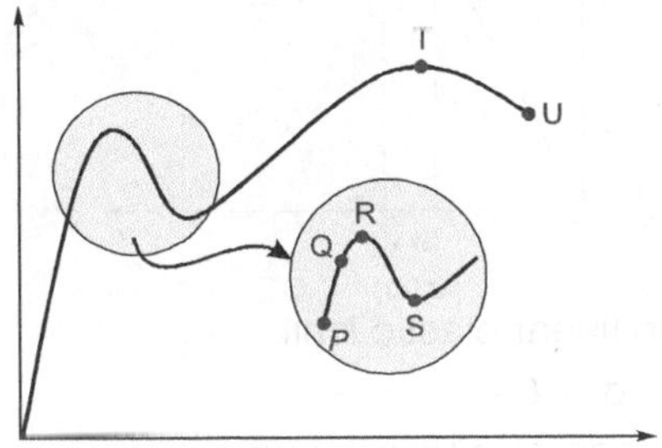

Point on the graph	Description of the point
P.	1. Upper yield point
Q.	2. Ultimate tensile strength
R.	3. Proportionality limit
S.	4. Elastic limit
T.	5. Lower yield point
U.	6. Failure

(a) P–1, Q–2, R–3, S–4, T–5, U–6

(b) P–3, Q–1, R–4, S–2, T–6, U–5

(c) P–3, Q–4, R–1, S–5, T–2, U–6

(d) P–4, Q–1, R–5, S–2, T–3, U–6

[2014 : 1 M, Set-3]

2.3 A metallic rod of 500 mm length and 50 mm diameter, when subjected to a tensile force of 100 kN at the ends, experiences an increase in its length by 0.5 mm and a reduction in its diameter by 0.015 mm. The Poisson's ratio of the rod material is _____.

[2014 : 1 M, Set-1]

2.4 If the Poisson's ratio of an elastic material is 0.4, the ratio of modulus of rigidity to Young's modulus is ______.

[2014 : 1 M, Set-4]

2.5 The number of independent elastic constants required to define the stress-strain relationship for an isotropic elastic solid is______.

[2014 : 1 M, Set-4]

2.6 Which one of the following types of stress-strain relationship best describes the behavior of brittle materials, such as ceramics and thermosetting plastics, (σ = stress and ε = strain)?

(a) 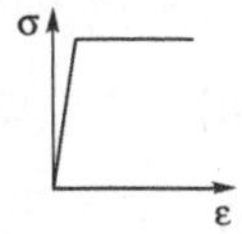

(b) 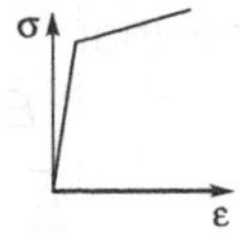

(c) 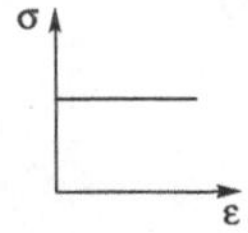

(d) 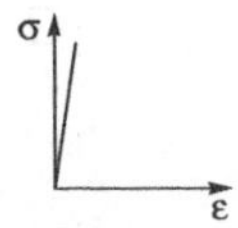

[2015 : 1 M, Set-1]

2.7 A rod is subjected to a uni-axial load within linear elastic limit. When the change in the stress is 200 MPa, the change in the strain is 0.001. If the Poisson's ratio of the rod is 0.3, the modulus of rigidity (in GPa) is ________.

[2015 : 1 M, Set-2]

2.8 A hypothetical engineering stress-strain curve shown in the figure has three straight lines PQ, QR, RS with coordinates $P(0, 0)$, $Q(0.2, 100)$, $R(0.6, 140)$ and $S(0.8, 130)$. Q is the yield point, R is the UTS point and S the fracture point.

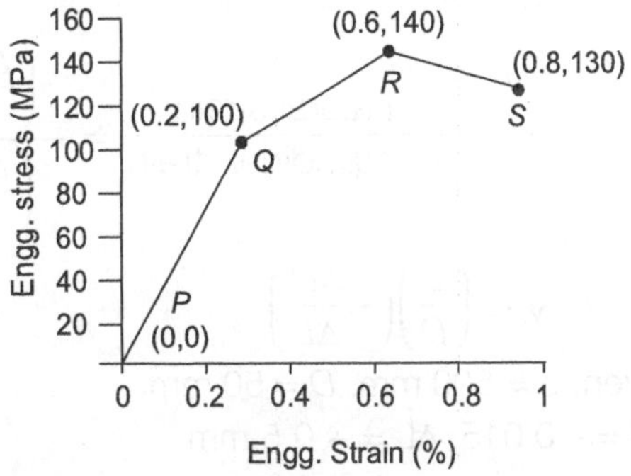

The toughness of the material (in MJ/m^3) is_______.

[2016 : 2 M, Set-1]

2.9 In the engineering stress-strain curve for mild steel, the Ultimate Tensile Strength (UTS) refers to

(a) Yield stress (b) Proportional limit

(c) Maximum stress (d) Fracture stress

[2017 : 1 M, Set-1]

2.10 The Poisson's ratio for a perfectly incompressible linear elastic material is

(a) 1 (b) 0.5

(c) 0 (d) infinity

[2017 : 1 M, Set-1]

Answers Properties of Material

2.1 (a) 2.2 (c) 2.3 (0.3) 2.4 (0.357) 2.5 (2) 2.6 (d) 2.7 (76.9230)

2.8 (0.85) 2.9 (c) 2.10 (b)

Explanations Properties of Material

2.1 (a)

We know that

$$E = 2G(1+\mu)$$

or

$$\frac{E}{G} = 2(1+\mu)$$

2.2 (c)

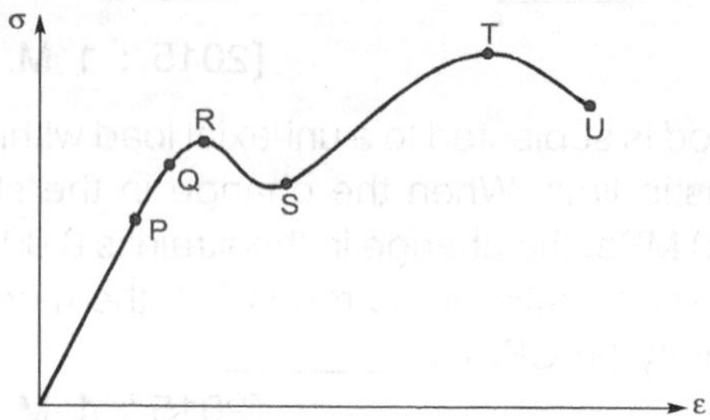

P : Proportional limit Q : Elastic limit
R : Upper yield point S : Lower yield point
T : Ultimate tensile strength
U : Failure/Rupture

2.3 Sol.

$$\nu = -\frac{\text{Lateral strain}}{\text{Longitudinal strain}} = -\frac{\left(\frac{\Delta D}{D}\right)}{\left(\frac{\Delta L}{L}\right)}$$

$$\nu = \left(\frac{L}{D}\right)\left(-\frac{\Delta D}{\Delta L}\right)$$

Given, $L = 500$ mm, $D = 50$ mm,
$\Delta D = -0.015$, $\Delta L = +0.5$ mm

$$\therefore \quad \nu = \frac{500}{50}\times\frac{0.015}{0.500} = \mathbf{0.3}$$

2.4 Sol.

$$E = 2G(1+\nu)$$

$$\frac{G}{E} = \frac{1}{2(1+\nu)} = \frac{1}{2\times 1.4} = \frac{1}{2.8} = 0.357$$

2.5 Sol.

Either E or G, 2 independent constant
Either G or K, 2 independent constant
Either E or K, 2 independent constant

2.7 Sol.

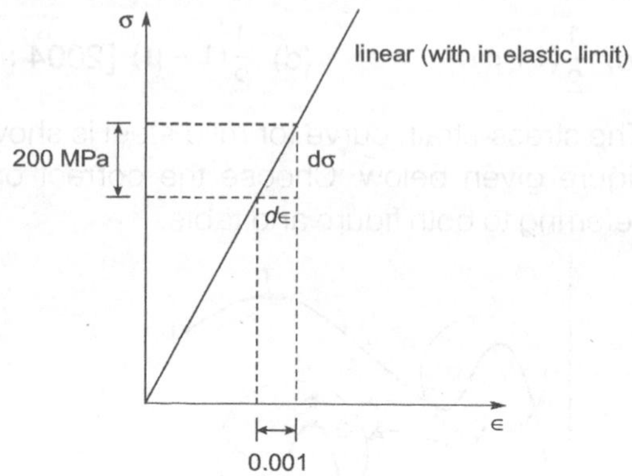

With in linear elastic limit

$$\sigma = E\epsilon$$

$E \rightarrow$ slope of σ vs ϵ curve

$$E = \frac{d\sigma}{d\epsilon} = \frac{200}{0.001} = 200 \text{ GPa}$$

$$E = 2G[1+\mu]$$

$$G = \frac{E}{2(1+\mu)} = \frac{200}{2(1+0.3)} = \frac{100}{1.3}$$

$$= \mathbf{76.9230\ GPa}$$

2.8 Sol.

Toughness is area of curve upto S on strain axis

$$\left[\frac{1}{2}\times\frac{0.2}{100}\times 100\right] + \left\{\frac{[100+140]}{2}\times\frac{0.4}{100}\right\} + \left\{\frac{140+130}{2}\right\}\times\frac{0.2}{100}$$

$$= 0.1 + 0.48 + 0.27 = \mathbf{0.85\ MJ/m^3}$$

2.9 (c)

Ultimate tensile strength represents the maximum stress that a material can withstand without fracture.

2.10 (b)

$$E = 3k(1-2\mu)$$

For perfectly incompressible material; $k = \infty$

$$\therefore \quad 1 - 2\mu = 0$$

$$\mu = 0.5$$

Points to remember: For perfectly incompressible material, Poisson's ratio = 0.5.

3 Shear Force and Bending Moment

3.1 A concentrated load P acts on a simply supported beam of span L at a distance $L/3$ from the left support. The bending moment at the point of application of the load is given by

(a) $\frac{PL}{3}$ (b) $\frac{2PL}{3}$

(c) $\frac{PL}{9}$ (d) $\frac{2PL}{9}$ **[2003 : 1 M]**

3.2 A beam is made up of two identical bars AB and BC, by hinging them together at B. The end A is built-in (cantilevered) and the end C is simply-supported. With the load P acting as shown, the bending moment at A is

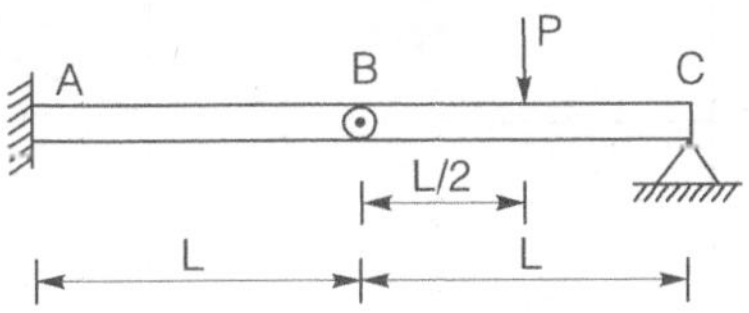

(a) Zero (b) $PL/2$

(c) 3PL/2 (d) Indeterminate

[2005 : 2 M]

3.3 In a simply-supported beam loaded as shown below, the maximum bending moment in Nm is

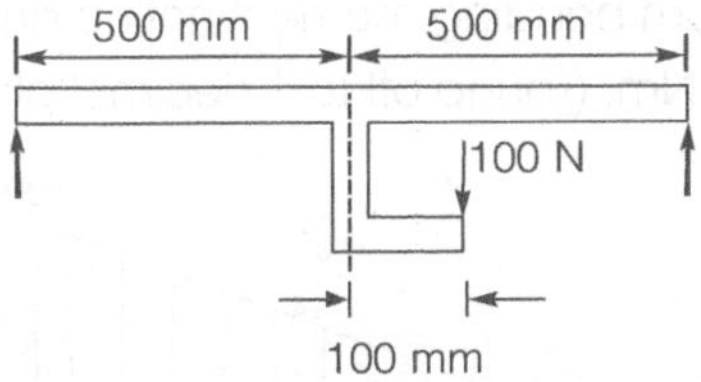

(a) 25 (b) 30

(c) 35 (d) 60 **[2007 : 1 M]**

3.4 The strain energy stored in the beam with flexural rigidity EI and loaded as shown in the figure is

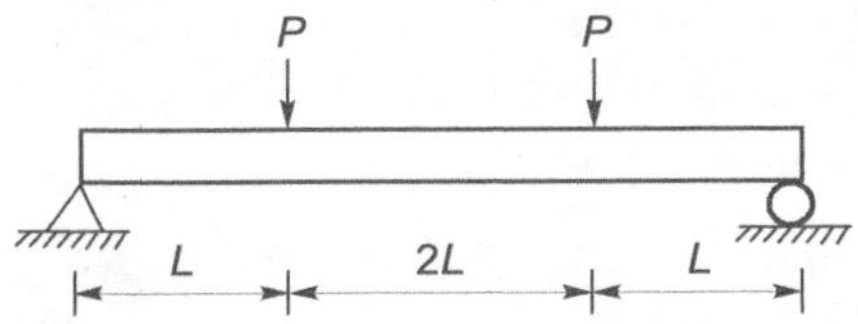

(a) $\frac{P^2L^3}{3EI}$ (b) $\frac{2P^2L^3}{3EI}$

(c) $\frac{4P^2L^3}{3EI}$ (d) $\frac{8P^2L^3}{3EI}$

[2008 : 2 M]

3.5 A simply supported beam PQ is loaded by a moment of 1 kNm at the mid-span of the beam as shown in the figure. The reaction forces R_P and R_Q at supports P and Q respectively are

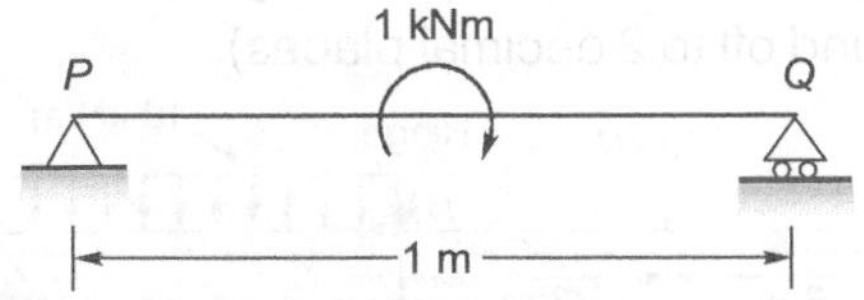

(a) 1 kN downward, 1 kN upward

(b) 0.5 kN upward, 0.5 kN downward

(c) 0.5 kN downward, 0.5 kN upward

(d) 1 kN upward, 1 kN upward **[2011 : 1 M]**

3.6 A simply supported beam of length L is subjected to a varying distributed load sin $(3\pi x/L)$ Nm^{-1}, where the distance x is measured from the left support. The magnitude of the vertical reaction force in N at the left support is

(a) zero (b) $L/3\pi$

(c) L/π (d) $2L/\pi$ **[2013 : 2 M]**

3.7 A cantilever beam OP is connected to another beam PQ with a pin joint as shown in the figure. A load of 10 kN is applied at the mid-point of PQ. The magnitude of bending moment (in kNm) at fixed end O is ______

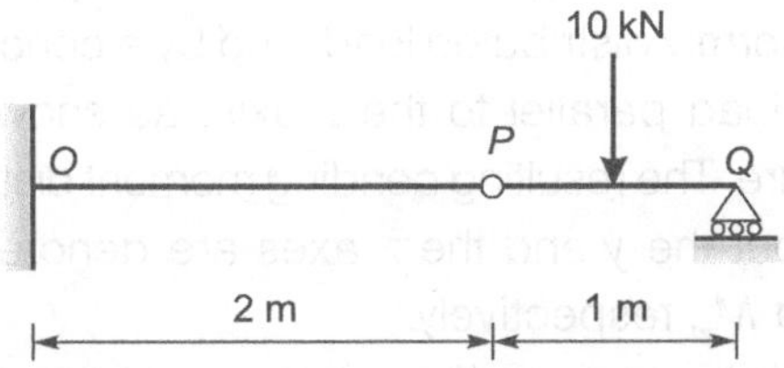

(a) 2.5 (b) 5

(c) 10 (d) 25

[2015 : 2 M, Set-2]

3.8 For the overhanging beam shown in figure, the magnitude of maximum bending moment (in kNm) is ______.

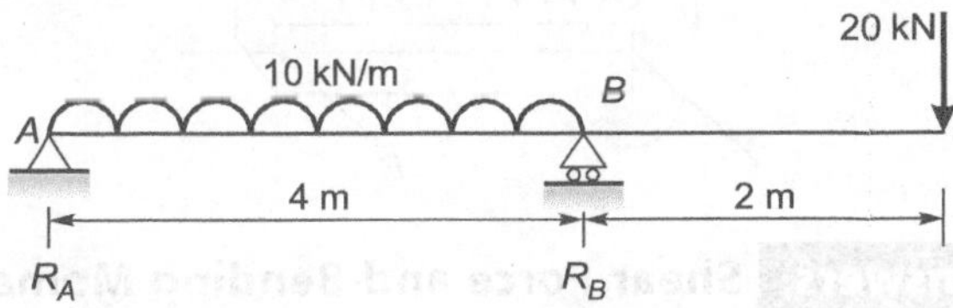

[2015 : 2 M, Set-3]

3.9 For a loaded cantilever beam of uniform cross-section, the bending moment (in Nmm) along the length is $M(x) = 5x^2 + 10x$, where x is the distance (in mm) measured from the free end of the beam. The magnitude of shear force (in N) in the cross-section at $x = 10$ mm is ______

[2017 : 1 M, Set-2]

3.10 The magnitude of reaction force at joint C of the hinge-beam shown in the figure is ______ kN (round off to 2 decimal places).

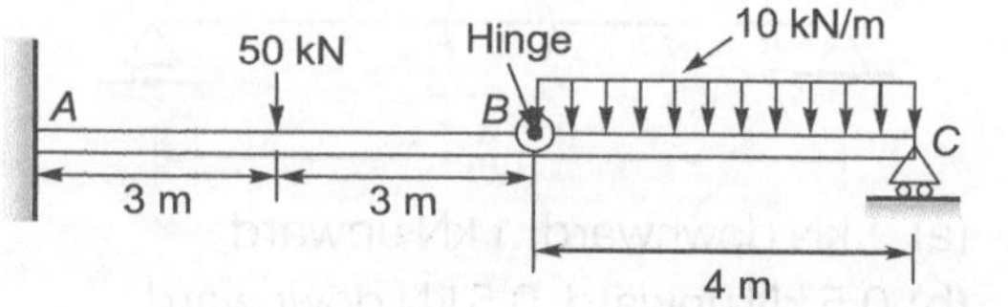

[2020 : 2 M, Set-1]

3.11 A beam of negligible mass is hinged at support P and has a roller support Q as shown in the figure.

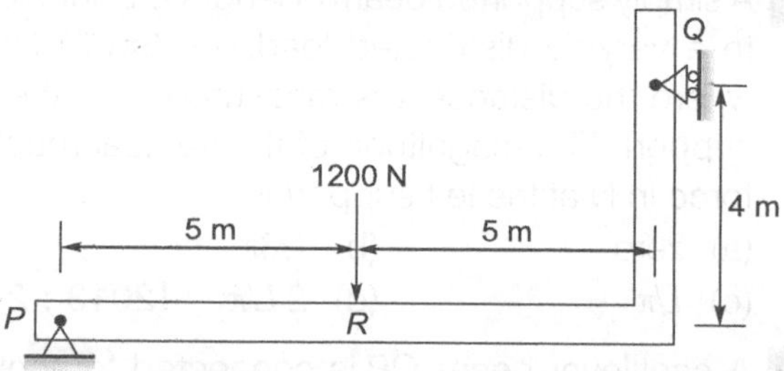

A point load of 1200 N is applied at point R. The magnitude of the reaction force at support Q is__N.

[2020 : 1 M, Set-2]

3.12 A beam of length L is loaded in the xy-plane by a uniformly distributed load, and by a concentrated tip load parallel to the z-axis, as shown in the figure. The resulting bending moment distributions about the y and the z axes are denoted by M_y and M_z, respectively.

Which one of the options given depicts qualitatively CORRECT variations of M_y and M_z along the length of the beam?

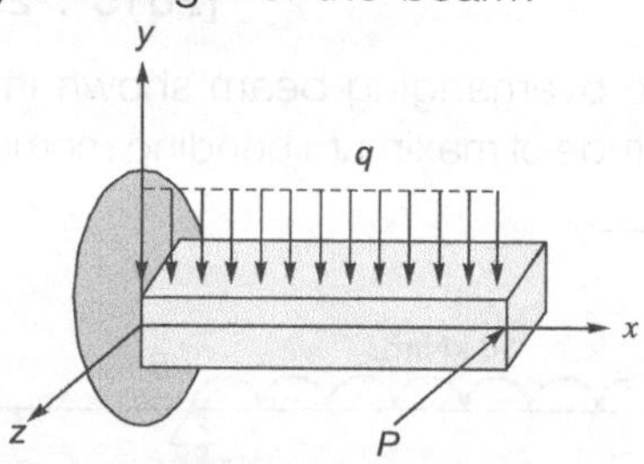

(a)

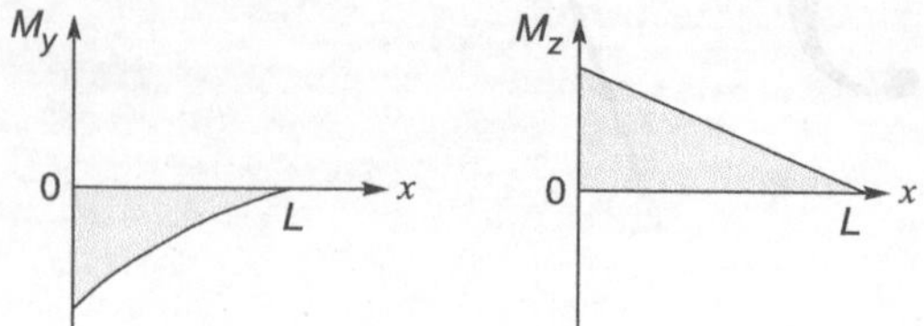

(b)

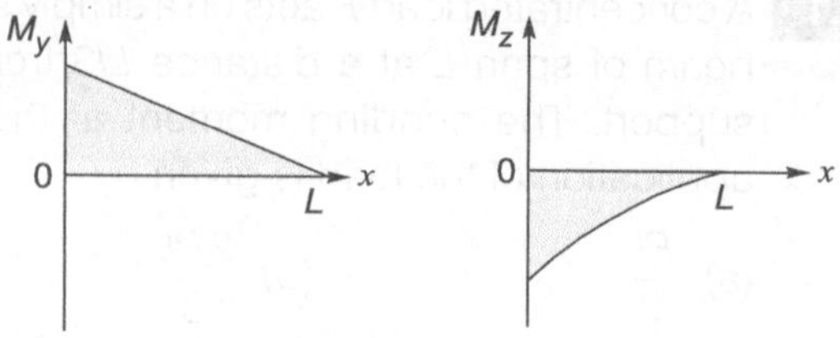

(c)

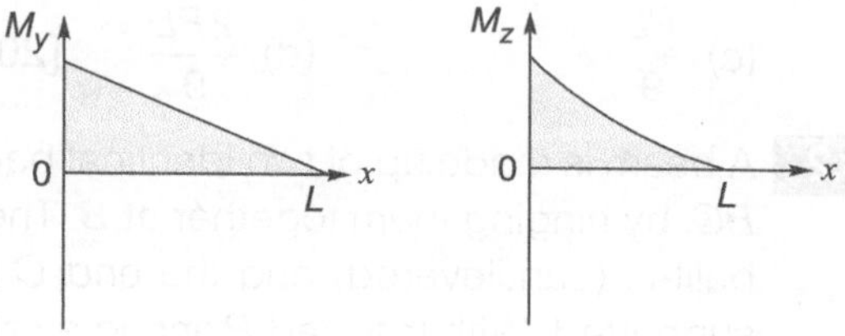

(d)

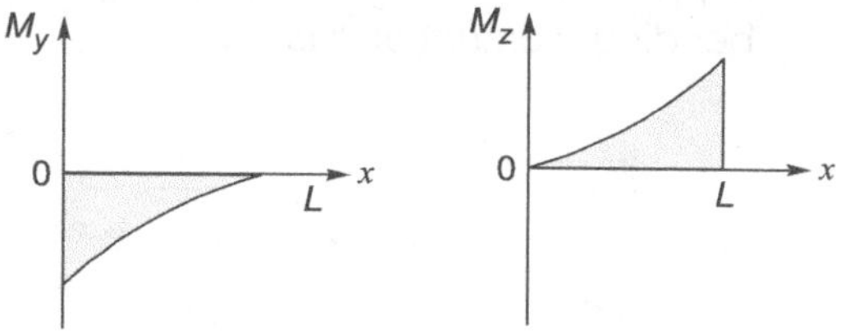

[2023 : 2 M]

3.13 A horizontal beam of length 1200 mm is pinned at the left end and is resting on a roller at the other end as shown in the figure. A linearly varying distribution load is applied on the beam. The magnitude of maximum bending moment acting on the beam is ______ Nm. (round off to 1 decimal place)

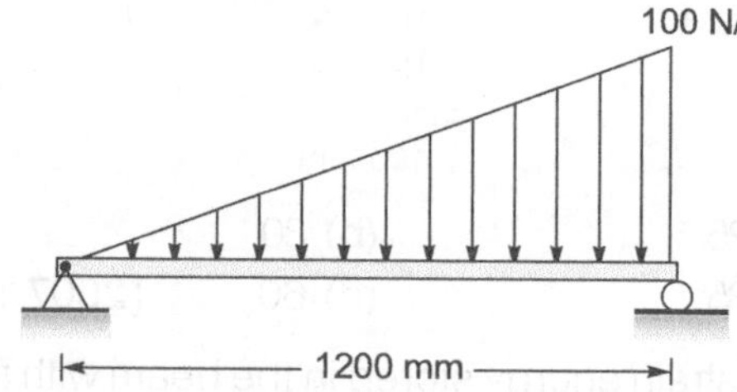

[2024 : 2 M]

■■■■

Answers Shear Force and Bending Moment

3.1 (d)	3.2 (b)	3.3 (b)	3.4 (c)	3.5 (a)	3.6 (b)	3.7 (c)	3.8 (40)
3.9 (110)	3.10 (20)	3.11 (1500)	3.12 (b)	3.13 (9.2)			

Explanations Shear Force and Bending Moment

3.1 (d)

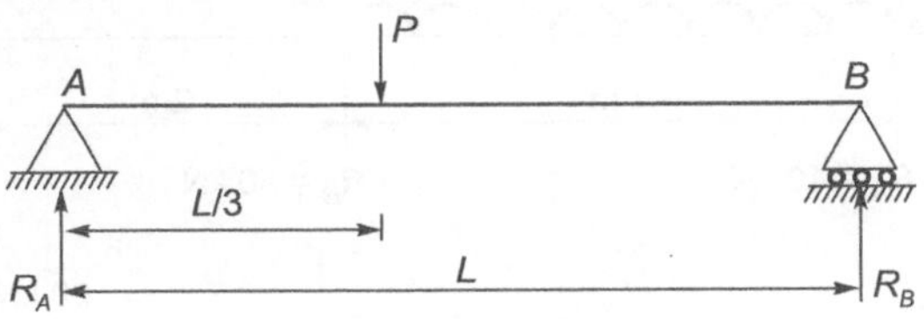

Taking moment about A, we have

$$R_B \times L = P \times \frac{L}{3}$$

$$\therefore \quad R_B = \frac{P}{3}$$

and $$R_A = P - R_B = P - \frac{P}{3} = \frac{2P}{3}$$

Bending moment at the point of application of load,

$$M = \frac{2P}{3} \times \frac{L}{3} = \frac{2PL}{9}$$

3.2 (b)

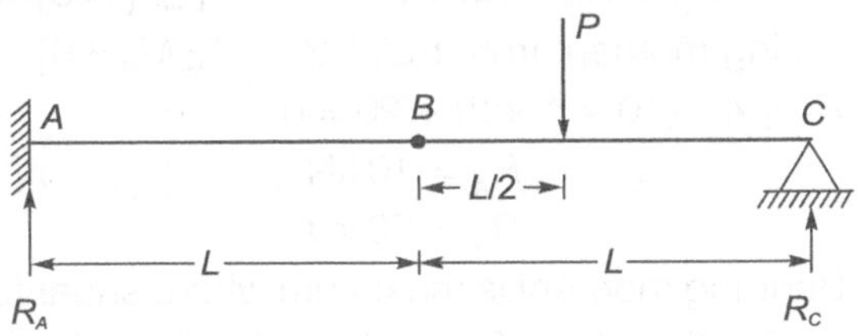

This loading diagram can be converted into

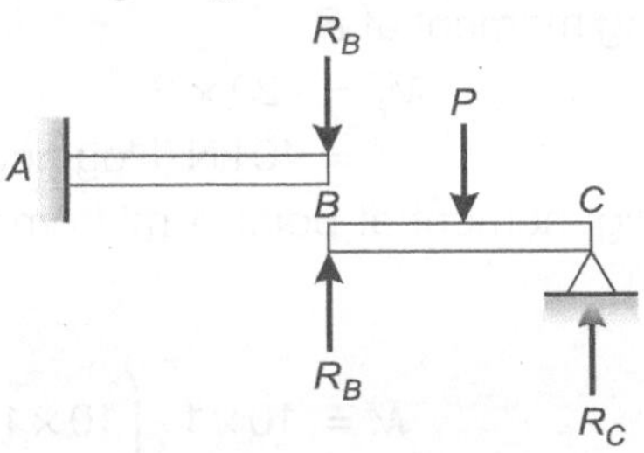

$\Sigma F_v = 0$ for member BC, $R_B + R_C = P$

$$\Sigma M_C = 0;\ R_B \times L - P \times \frac{L}{2} = 0$$

$$R_B = \frac{P}{2}$$

$$\text{Bending moment at } A = R_B \times L = \frac{P}{2} \times L = \frac{PL}{2}$$

3.3 (b)

Resultant beam diagram

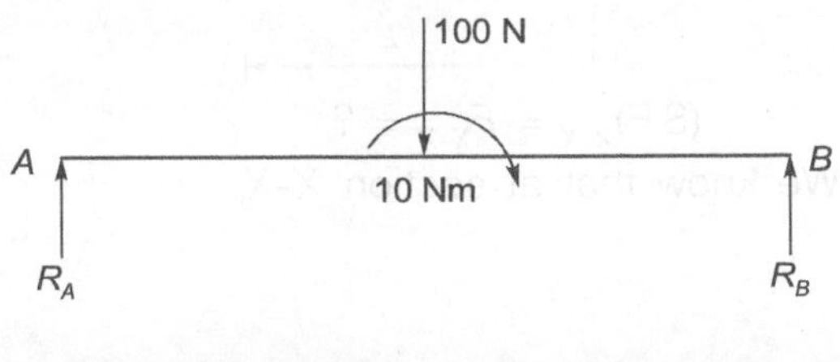

$$R_A + R_B = 100 \quad \text{...(1)}$$

Taking moment about A,

$$-100 \times 0.5 - 10 + R_B \times 1 = 0$$

$$\therefore \quad R_B = 60 \text{ N}$$

and $$R_A = 40 \text{ N}$$

Bending moment diagram

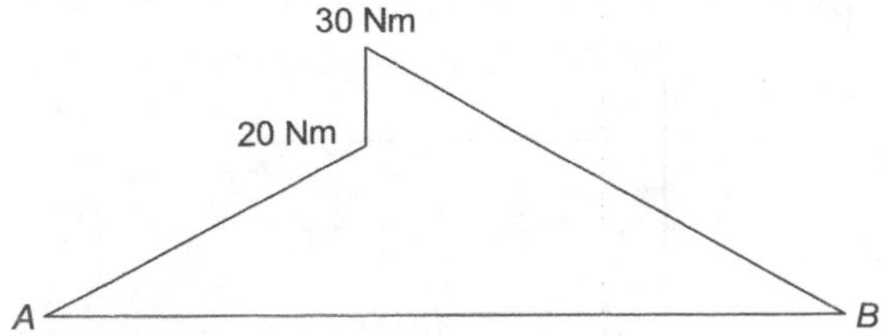

Maximum bending moment

$$= 40 \times 0.5 + 10 = \mathbf{30\ Nm}$$

3.4 (c)

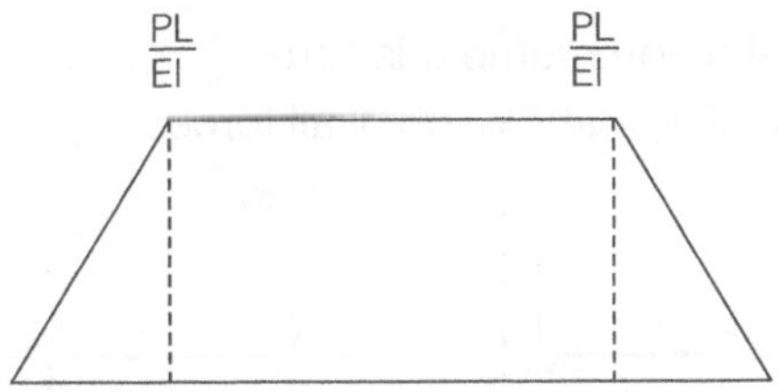

Total strain energy stored is given by

$$2\int_0^L \frac{(Px)^2\,dx}{2EI} + \frac{(PL)^2 \cdot 2L}{2EI}$$

$$= \frac{2P^2}{2EI}\int_0^L x^2 dx + \frac{P^2L^3}{EI} = \frac{P^2}{EI} \times \frac{L^3}{3} + \frac{P^2L^3}{EI}$$

$$= \frac{P^2L^3}{3EI} + \frac{P^2L^3}{EI} = \frac{4}{3}\frac{P^2L^3}{EI}$$

3.5 (a)

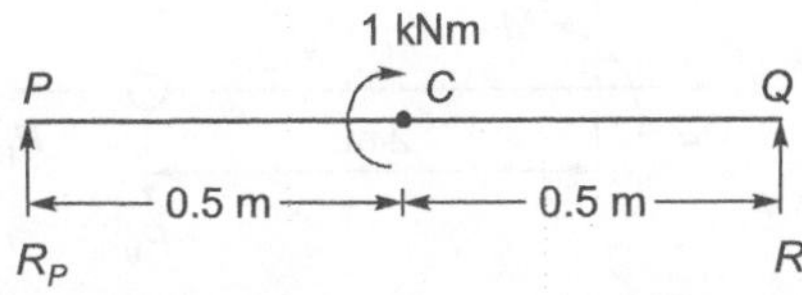

Let us assumed the direction of R_P and R_Q as upward. Now,

$$\therefore \quad R_P + R_Q = 0$$

$\therefore$ Taking moment about point Q, we get

$$R_P \times 1 + 1 = 0$$

$$R_P = -1 \text{ kN}$$

$$\therefore \quad R_Q = +1 \text{ kN}$$

Since our assume direction of R_P is wrong therefore $R_P = 1$ kN which act in the downward direction and $R_Q = 1$ kN acting in upward direction.

3.6 (b)

$$\text{Load, } W = \int_0^L \sin\left(\frac{3\pi x}{L}\right)dx = \frac{2L}{3\pi}$$

Vertical reaction,

$$R_1 = R_2 = \frac{L}{3\pi} \quad \text{(due to symmetrical loading)}$$

3.7 (c)

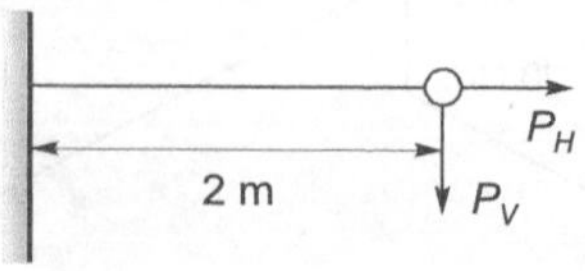

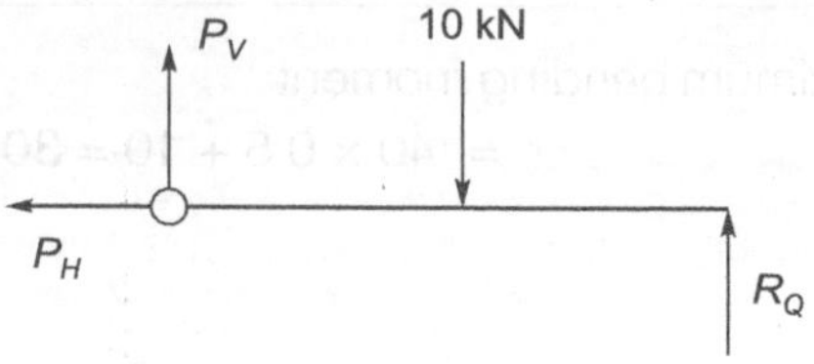

At hinge moment is zero.
Taking summation of all forces

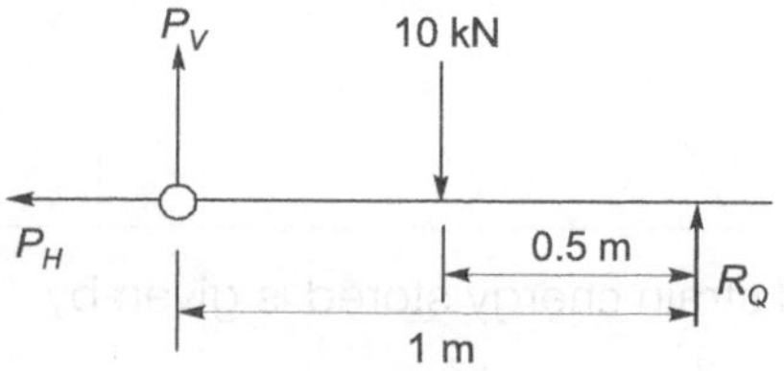

$$P_V + R_Q = 10$$

Taking moment of all forces about Q

$$M = \underset{\text{clockwise}}{(P_V \times 1)} - \underset{\text{anticlockwise}}{(10 \times 0.5)}$$

At equilibrium $M = 0$

$$\therefore \quad 0 = (P_V \times 1) - 5$$

$$P_V = \mathbf{5\ kN}$$

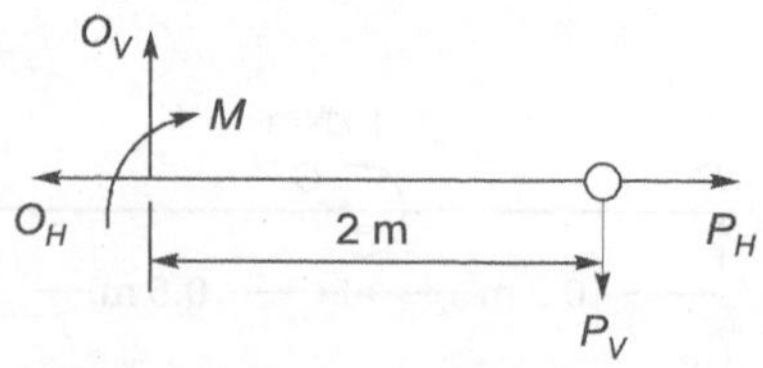

Summation of all vertical forces is zero.

$$\therefore \quad O_V + P_V = 0$$

Summation of moments of all forces at equilibrium is zero.

$$\therefore \quad \underset{\text{(clockwise)}}{+M} + \underset{\text{(clockwise)}}{P_V \times 2} = 0$$

$$\therefore \quad M = -P_V \times 2$$

where $\quad P_V = -5$ kN

$$\therefore \quad M = \mathbf{10\ kN}$$

3.8 Sol.

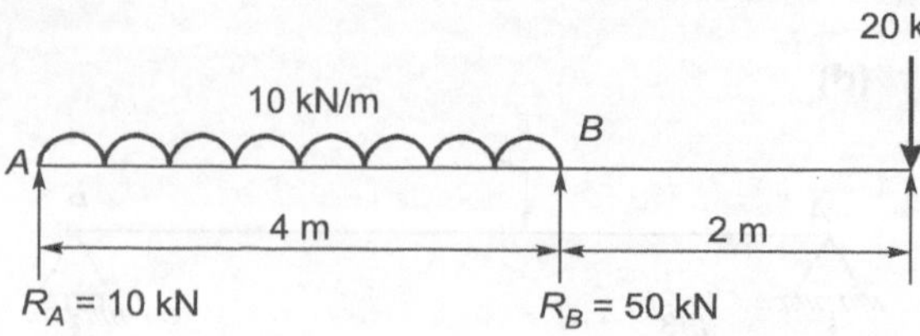

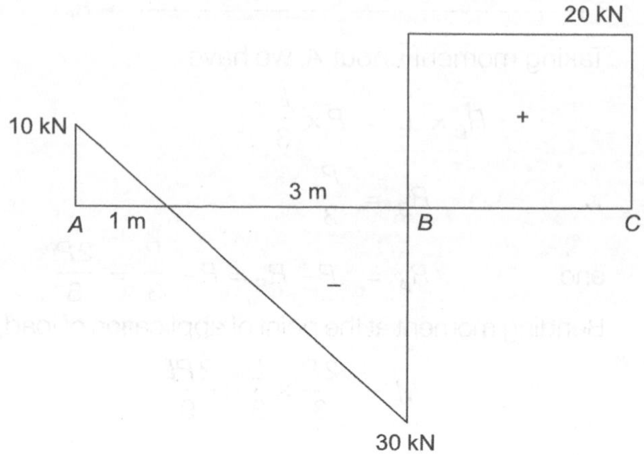

$$R_A + R_B = 60 \text{ kN} \qquad [\Sigma F_V = 0]$$

Taking moment from point B, $\quad [\Sigma M_B = 0]$

$$R_A \times 4 - (10 \times 4 \times 2) + 20 = 0$$

$$R_A = 10 \text{ kN}$$

$$R_B = 50 \text{ kN}$$

Bending moment is maximum where shear force is zero (i.e. shear force changes its sign).

Bending moment at B,

$$M_B = -20 \times 2$$

$$= 40 \text{ kN (Hogging)}$$

Bending moment at point '1 m' from the right of point A,

$$M = 10 \times 1 - \left(10 \times 1 \times \frac{1}{2}\right)$$

$$= 5 \text{ kNm (Sagging)}$$

Maximum bending moment is at point B i.e. 40 kNm.

3.9 Sol.

$$(\text{BM})_{X\text{-}X} = M_{X\text{-}X} = 5x^2 + 10x$$

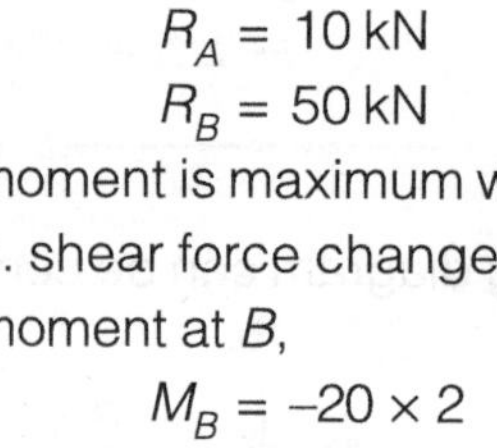

$$(\text{S.F})_{X\text{-}X} = F_{X\text{-}X} = ?$$

We know that at section X–X,

$$F_{X\text{-}X} = \frac{dM}{dx}$$

$$F_{X\text{-}X} = \frac{d}{dx}\left[5x^2 + 10x\right] = 10x + 10$$

$$(F_{X\text{-}X})_{x\,=\,10\text{ mm}} = 10(10) + 10 = \mathbf{110\ N}$$

3.10 Sol.

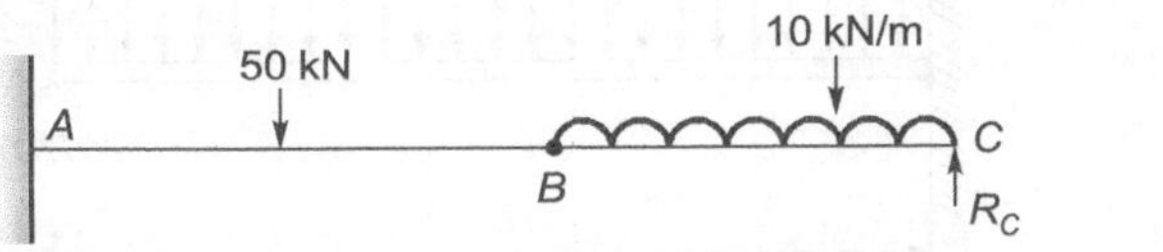

$$\Sigma M_{B_{\text{Right}}} = 0 \qquad \text{[Internal hinge at } B]$$

$$4R_C = 10 \times 4 \times 2$$

$$R_C = 20\text{ kN}$$

3.11 Sol.

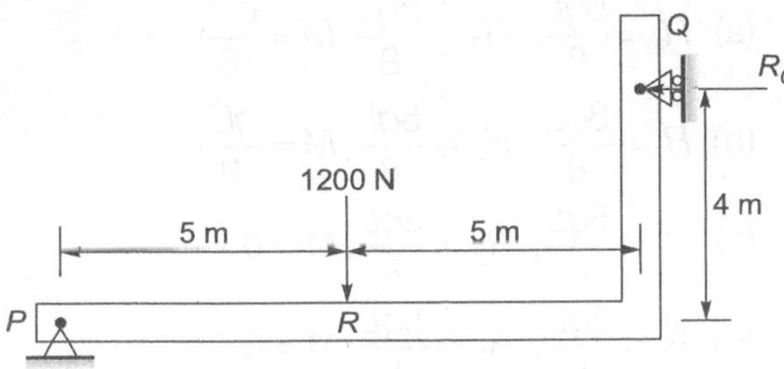

$$\Sigma \vec{M}_P = 0$$

$$1200 \times 5 - R_Q \times 4 = 0$$

$$R_Q = \frac{1200 \times 5}{4} = 1500\text{ N}$$

3.12 (b)

Due to UDL, BM is negative and varies parabolically in vertical plane i.e bending moment is acting about Z-axis (M_z).

Due to Horizontal concentrated point load, bending moment is positive and varies linearly in horizontal plane i.e. bending moment is acting about Y-axis(M_y).

3.13 (9.2) (9.0 to 9.5)

Reactions at the ends,

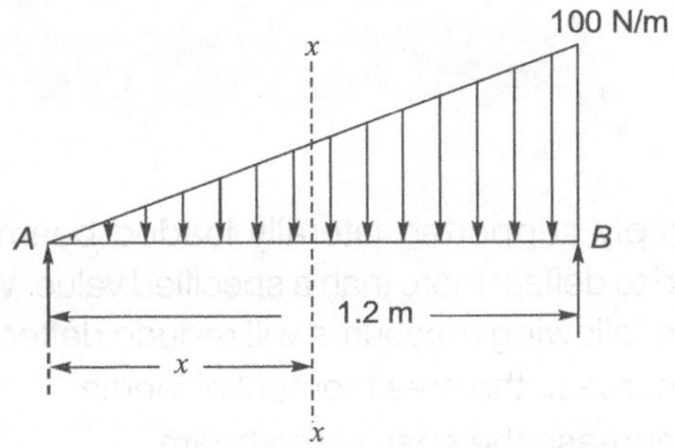

$$R_A + R_B = \frac{1}{2} \times 1.2 \times 100$$

$$\therefore \quad R_A + R_B = 60\text{ N} \quad \text{... (i)}$$

$$\Sigma M_B = 0$$

$$\therefore \quad R_A \times 1.2 = \left(\frac{1}{2} \times 1.2 \times 100\right) \times \frac{1.2}{3}$$

$$\therefore \quad R_A = 20\text{ N}$$

$$\therefore \quad R_B = 40\text{ N}$$

Now, SF at section 'x-x' at a distance of x

$$SF_x = 0$$

$$\therefore 20 - \frac{1}{2} \times x \times \frac{100x}{1.2} = 0$$

$$\therefore \quad x = 0.693\text{ m}$$

$\therefore$ At x = 0.693, BM will be maximum,

$$\therefore \quad M_{\text{max}} =$$

$$20 \times 0.693 - \frac{1}{2} \times \frac{0.693 \times 100 \times 0.693}{1.2} \times \frac{0.693}{3}$$

$$\therefore \quad M_{\text{max}} = 9.237\text{ N-m} \simeq 9.2\text{ N.m}$$

■■■■

4 Deflection of Beams

4.1 A simply supported laterally loaded beam was found to deflect more than a specified value. Which of the following measures will reduce deflection?
(a) Increase the area moment of inertia
(b) Increase the span of the beam
(c) Select a different material having lesser modulus of elasticity
(d) Magnitude of the load to be increased

[2003 : 2 M]

Common Data for Q. 4.2 and Q. 4.3

A steel beam of breadth 120 mm and height 750 mm is loaded as shown in the figure. Assume E_{steel} = 200 GPa.

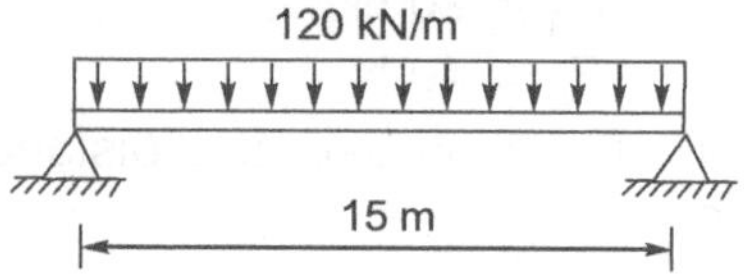

4.2 The beam is subjected to a maximum bending moment of
(a) 3375 kNm (b) 4750 kNm
(c) 6750 kNm (d) 8750 kNm

[2004 : 2 M]

4.3 The value of maximum deflection of the beam is
(a) 93.75 mm (b) 83.75 mm
(c) 73.75 mm (d) 63.75 mm

[2004 : 2 M]

4.4 Two identical cantilever beams are supported as shown, with their free ends in contact through a rigid roller. After the load P is applied, the free ends will have

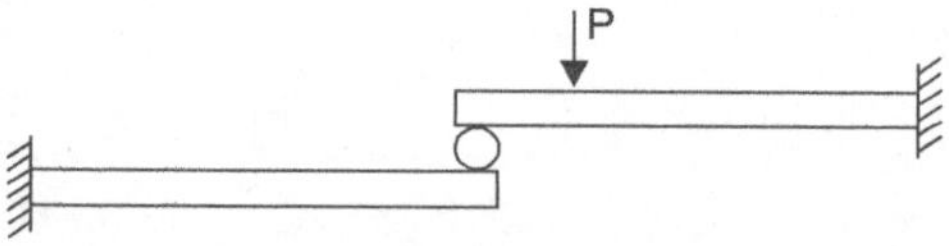

(a) equal deflections but not equal slopes
(b) equal slopes but not equal deflections
(c) equal slopes as well as equal deflections
(d) neither equal slopes nor equal deflections

[2005 : 1 M]

4.5 A uniformly loaded propped cantilever beam and its free body diagram are shown below. The reactions are

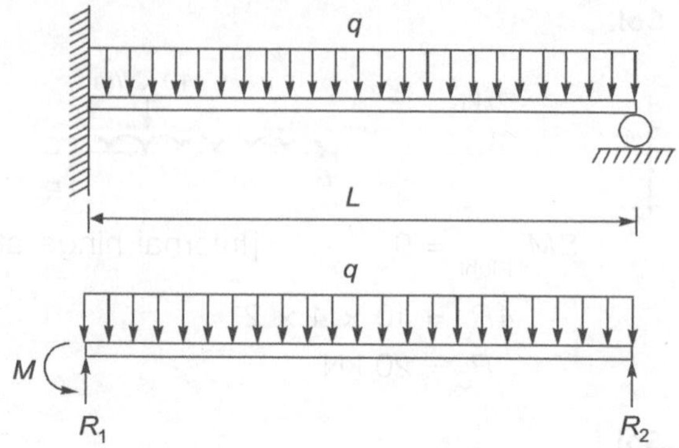

(a) $R_1 = \frac{5qL}{8}$, $R_2 = \frac{3qL}{8}$, $M = \frac{qL^2}{8}$

(b) $R_1 = \frac{3qL}{8}$, $R_2 = \frac{5qL}{8}$, $M = \frac{qL^2}{8}$

(c) $R_1 = \frac{5qL}{8}$, $R_2 = \frac{3qL}{8}$, $M = 0$

(d) $R_1 = \frac{3qL}{8}$, $R_2 = \frac{5qL}{8}$, $M = 0$

[2007 : 2 M]

4.6 A frame of two arms of equal length L is shown in the adjacent figure. The flexural rigidity of each arm of the frame is EI. The vertical deflection at the point of application of load P is

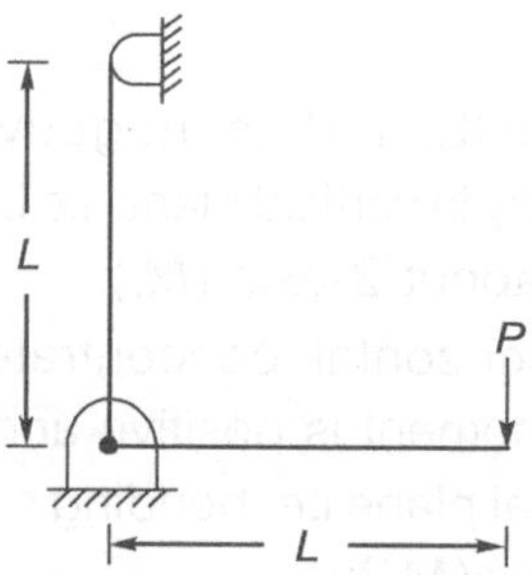

(a) $\frac{PL^3}{3EI}$ (b) $\frac{2PL^3}{3EI}$

(c) $\frac{PL^3}{EI}$ (d) $\frac{4PL^3}{3EI}$

[2009 : 2 M]

Common Data for Q. 4.7 and Q. 4.8

A triangular-shaped cantilever beam of uniform-thickness is shown in the figure. The Young's modulus of the material of the beam is E. A concentrated load P is applied at the free end of the beam.

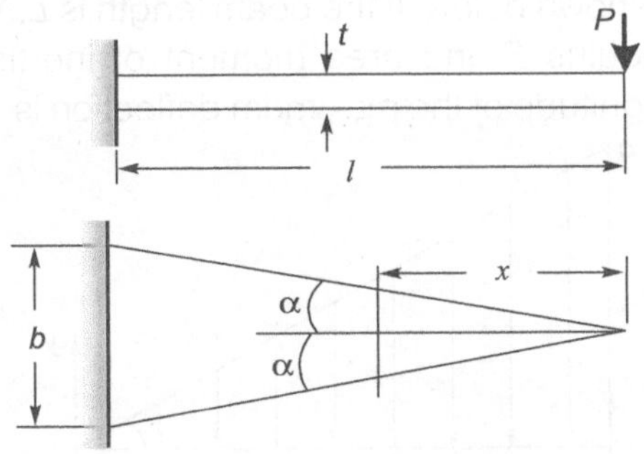

4.7 The area moment of inertia about the neutral axis of a cross-section at a distance x measured from the free end is

(a) $\frac{bxt^3}{6l}$ (b) $\frac{bxt^3}{12l}$

(c) $\frac{bxt^3}{24l}$ (d) $\frac{xt^3}{12l}$ [2011 : 2 M]

4.8 The maximum deflection of the beam is

(a) $\frac{24Pl^3}{Ebt^3}$ (b) $\frac{12Pl^3}{Ebt^3}$

(c) $\frac{8Pl^3}{Ebt^3}$ (d) $\frac{6Pl^3}{Ebt^3}$ [2011 : 2 M]

4.9 A cantilever beam of length L is subjected to a moment M at the free end. The moment of inertia of the beam cross section about the neutral axis is I and the Young's modulus is E. The magnitude of the maximum deflection is

(a) $\frac{ML^2}{2EI}$ (b) $\frac{ML^2}{EI}$

(c) $\frac{2ML^2}{EI}$ (d) $\frac{4ML^2}{EI}$ [2012 : 1 M]

4.10 The flexural rigidity (EI) of a cantilever beam is assumed to be constant over the length of the beam shown in figure. If a load P and bending moment $PL/2$ are applied at the free end of the beam then the value of the slope at the free end is

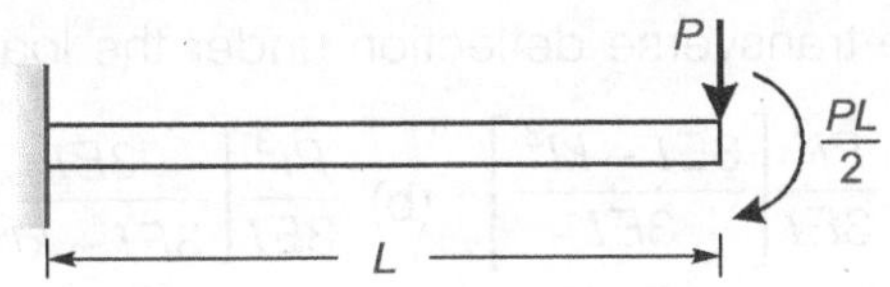

(a) $\frac{1}{2}\frac{PL^2}{EI}$ (b) $\frac{PL^2}{EI}$

(c) $\frac{3}{2}\frac{PL^2}{EI}$ (d) $\frac{5}{2}\frac{PL^2}{EI}$

[2014 : 2 M, Set-2]

4.11 A cantilever beam of length, L, with uniform cross-section and flexural rigidity, EI, is loaded uniformly by a vertical load, w per unit length. The maximum vertical deflection of the beam is given by

(a) $\frac{wL^4}{8EI}$ (b) $\frac{wL^4}{16EI}$

(c) $\frac{wL^4}{4EI}$ (d) $\frac{wL^4}{24EI}$

[2014 : 2 M, Set-2]

4.12 A force P is applied at a distance x from the end of the beam as shown in the figure. What would be the value of x so that the displacement at A is equal to zero?

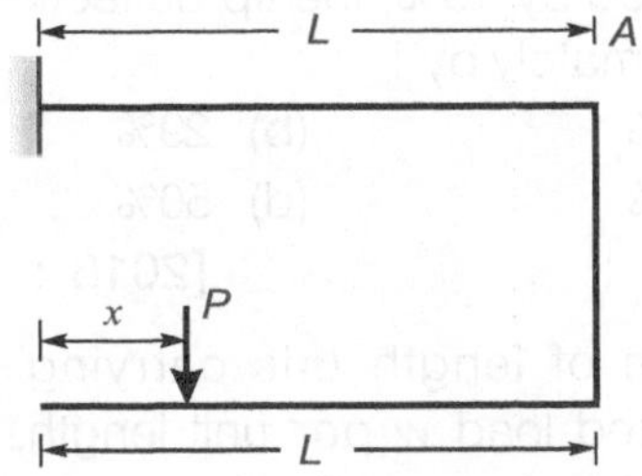

(a) 0.5 L (b) 0.25 L

(c) 0.33 L (d) 0.66 L

[2014 : 2 M, Set-3]

4.13 A frame is subjected to a load P as shown in the figure. The frame has a constant flexural rigidity EI. The effect of axial load is neglected. The deflection at point A due to the applied load P is

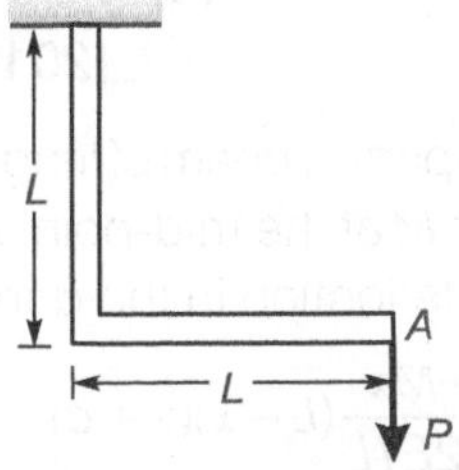

(a) $\frac{1}{3}\frac{PL^3}{EI}$ (b) $\frac{2}{3}\frac{PL^3}{EI}$

(c) $\frac{PL^4}{EI}$ (d) $\frac{4}{3}\frac{PL^3}{EI}$

[2014 : 2 M, Set-4]

4.14 A cantilever beam with flexural rigidity of 200 Nm^2 is loaded as shown in the figure. The deflection (in mm) at the tip of the beam is _____.

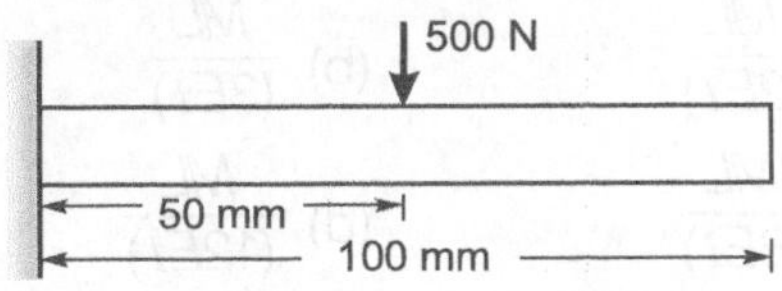

[2015 : 2 M, Set-1]

4.15 A cantilever beam with square cross-section of 6 mm side is subjected to a load of 2 kN normal to the top surface as shown in the figure. The Young's modulus of elasticity of the material of the beam is 210 GPa. The magnitude of slope (in radian) at Q (20 mm from the fixed end) is ____.

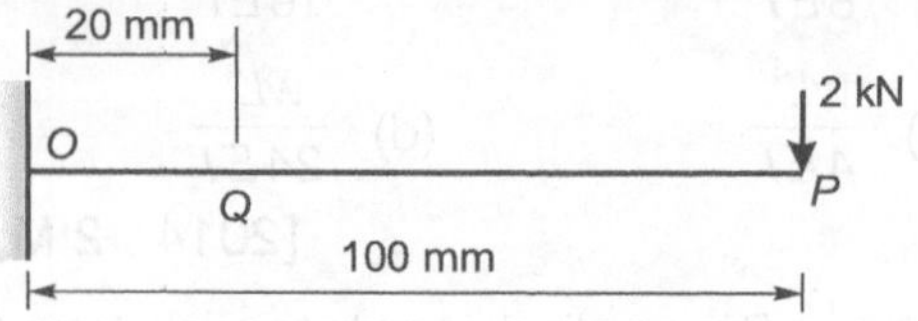

[2015 : 2 M, Set-2]

4.16 A cantilever beam having square cross-section of side a is subjected to an end load, If a is increased by 19%, the tip deflection decreases approximately by

(a) 19% (b) 29%
(c) 41% (d) 50%

[2016 : 1 M, Set-1]

4.17 A beam of length L is carrying a uniformly distributed load w per unit length. The flexural rigidity of the beam is EI. The reaction at the simple support at the right end is

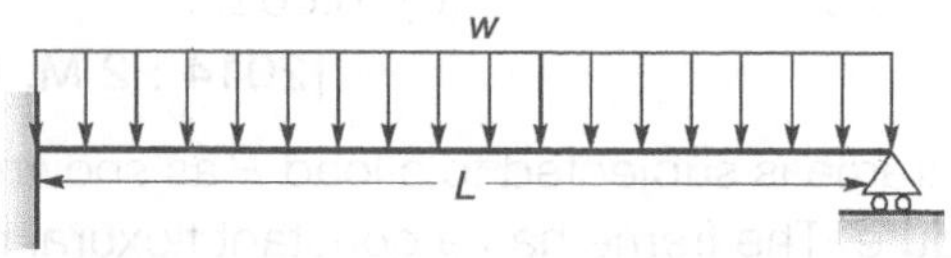

(a) $wL/2$ (b) $\frac{3wL}{8}$
(c) $wL/4$ (d) $wL/8$

[2016 : 2 M, Set-3]

4.18 A simply supported beam of length $2L$ is subjected to a moment M at the mid-point $x = 0$ as shown in figure. The deflection in the domain $0 \le x \le L$ is

$$w = \frac{-Mx}{12EIL}(L-x)(x+c)$$

where E is the Young's modulus, I is the area moment of inertia and C is a constant (to be determined).

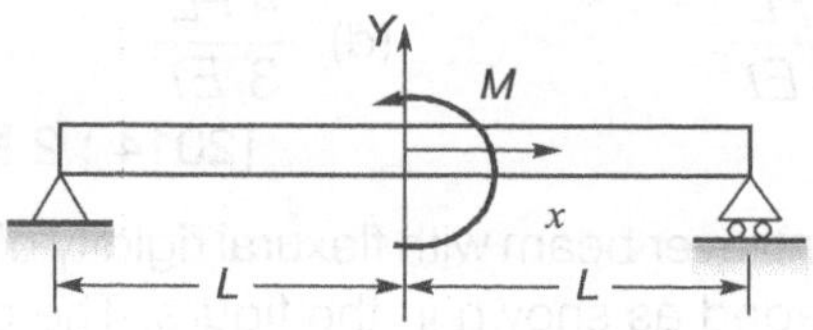

The slope at the center $x = 0$ is

(a) $\frac{ML}{(2EI)}$ (b) $\frac{ML}{(3EI)}$
(c) $\frac{ML}{(6EI)}$ (d) $\frac{ML}{(12EI)}$

[2016 : 2 M, Set-2]

4.19 A prismatic, straight, elastic, cantilever beam is subjected to a linearly distributed transverse load as shown below. If the beam length is L, Young's modulus E and area moment of inertia I, the magnitude of the maximum deflection is

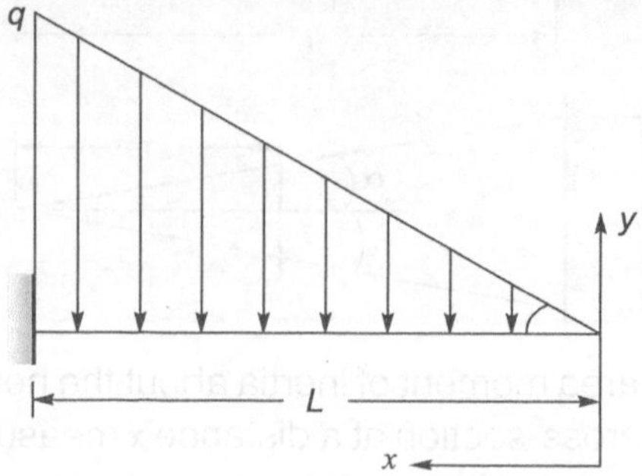

(a) $\frac{qL^4}{10EI}$ (b) $\frac{qL^4}{60EI}$
(c) $\frac{qL^4}{15EI}$ (d) $\frac{qL^4}{30EI}$

[2019 : 2 M, Set-2]

4.20 A horizontal cantilever beam of circular cross-section length = 1 m and flexural rigidity $EI = 200$ Nm^2 is subjected to an applied moment $M_A = 1.0$ Nm at the free end as shown in the figure. The magnitude of vertical deflection of the free end is ______ mm. (round off to one decimal place)

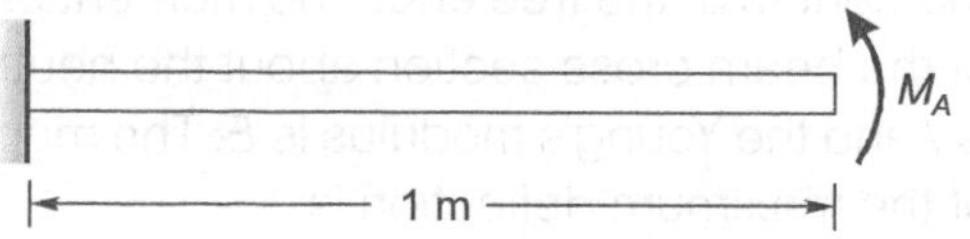

[2019 : 2 M, Set-2]

4.21 A cantilever of length l, and flexural rigidity EI, stiffened by a spring of stiffness k, is loaded by transverse force P, as shown

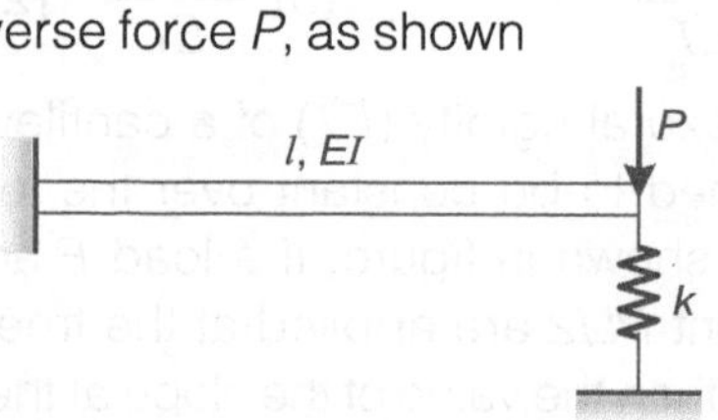

The transverse deflection under the load is

(a) $\frac{Pl^3}{3EI}\left[\frac{3EI-kl^3}{3EI}\right]$ (b) $\frac{Pl^3}{3EI}\left[\frac{3EI}{3EI+kl^3}\right]$
(c) $\frac{Pl^3}{3EI}\left[\frac{6EI-kl^3}{6EI}\right]$ (d) $\frac{Pl^3}{3EI}\left[\frac{3EI}{3EI+2kl^3}\right]$

[2020 : 2 M, Set-2]

4.22 A cantilever beam of length, L, and flexural ridigity, EI, is subjected to an end moment, M, as shown in the figure. The deflection of the beam at $x = \frac{L}{2}$ is

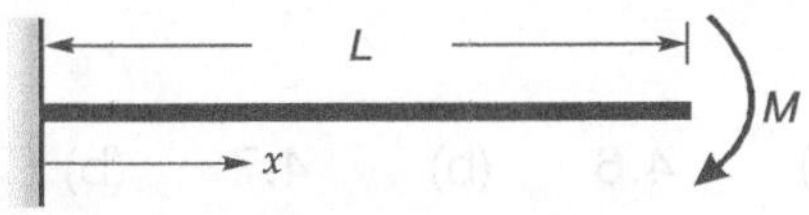

(a) $\frac{ML^2}{16EI}$ (b) $\frac{ML^2}{4EI}$

(c) $\frac{ML^2}{2EI}$ (d) $\frac{ML^2}{8EI}$

[2021 : 1 M, Set-1]

4.23 A cantilever beam with a uniform flexural rigidity ($EI = 200 \times 10^6$ N.m^2) is loaded with a concentrated force at its free end. The area of the bending moment diagram corresponding to the full length of the beam is 10000 N.m^2. The magnitude of the slope of the beam at its free end is ________ micro radian (round off to the nearest integer).

[2021 : 2 M, Set-2]

4.24 A plane frame *PQR* (fixed at *P* and free at *R*) is shown in the figure. Both members (*PQ* and *QR*) have length, *L*, and flexural rigidity, EI. Neglecting the effect of axial stress and transverse shear, the horizontal deflection at free end, *R*, is

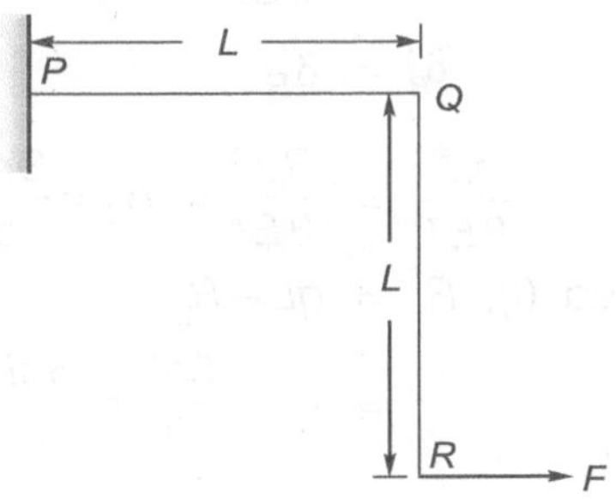

(a) $\frac{2FL^3}{3EI}$ (b) $\frac{4FL^3}{3EI}$

(c) $\frac{FL^3}{3EI}$ (d) $\frac{5FL^3}{3EI}$

[2021 : 2 M, Set-2]

4.25 An *L*-shaped elastic member *ABC* with slender arms *AB* and *BC* of uniform cross- section is clamped at end *A* and connected to a pin at end *C*. The pin remains in continuous contact with and is constrained to move in a smooth horizontal slot. The section modulus of the member is same in both the arms. The end C is subjected to a horizontal force *P* and all the deflections are in the plane of the figure. Given the length *AB* is 4α and length *BC* is α, the magnitude and direction of the normal force on the pin from the slot, respectively, are

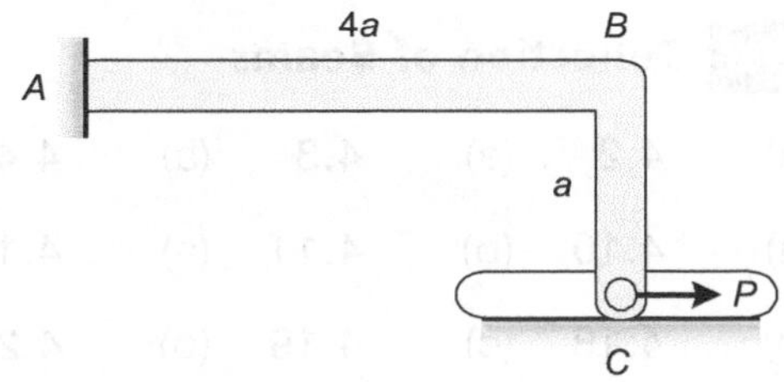

(a) $\frac{3P}{8}$, and downwards

(b) $\frac{5P}{8}$, and upwards

(c) $\frac{P}{4}$, and downwards

(d) $\frac{3P}{4}$, and upwards

[2022 : 2 M, Set-1]

4.26 The effective stiffness of a cantilever beam of length *L* and flexural rigidity *EI* subjected to a transverse tip load I is

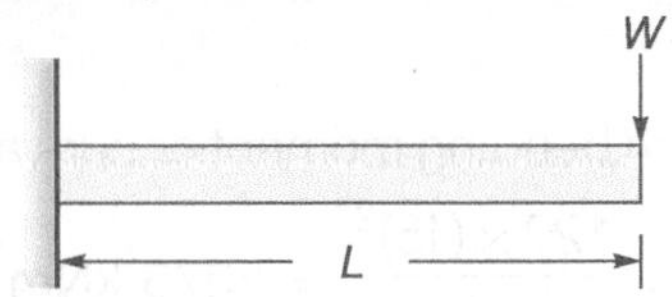

(a) $\frac{3EI}{L^3}$ (b) $\frac{2EI}{L^3}$

(c) $\frac{L^3}{2EI}$ (d) $\frac{L^3}{3EI}$ [2023 : 1 M]

4.27 A cylindrical bar has a length $L = 5$ m and cross section area $S = 10$ m^2. The bar is made of a linear elastic material with a density $\rho = 2700$ kg/m^3 and Young's modulus $E = 70$ GPa. The bar is suspended as shown in the figure and is in a state of uniaxial tension due to its self-weight. The elastic strain energy stored in the bar equals ________ J. (Rounded off to two decimal places)

Take the acceleration due to gravity as $g = 9.8$ m/s^2.

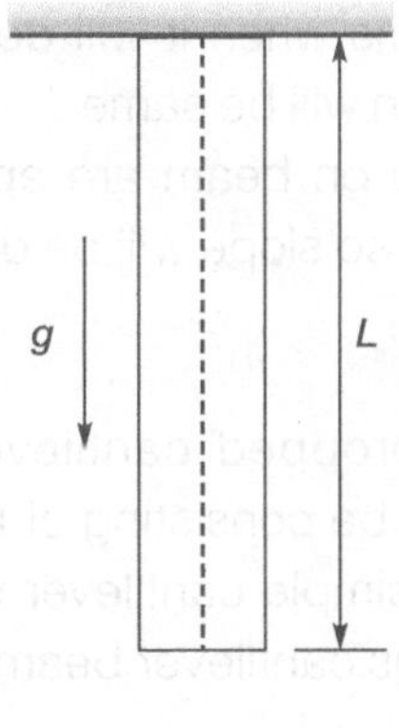

[2023 : 2 M]

■■■■

Answers Deflection of Beams

4.1	(a)	4.2	(a)	4.3	(a)	4.4	(a)	4.5	(a)	4.6	(b)	4.7	(b)	4.8	(d)
4.9	(a)	4.10	(b)	4.11	(a)	4.12	(c)	4.13	(d)	4.14	(0.26042)	4.15	(0.1587)	4.16	(d)
4.17	(b)	4.18	(c)	4.19	(d)	4.20	(2.50)	4.21	(b)	4.22	(d)	4.23	(50)	4.24	(b)
4.25	(a)	4.26	(a)	4.27	(2.09)										

Explanations Deflection of Beams

4.1 (a)

From differential equation of elastic curve

$$\frac{d^2y}{dx^2} = \frac{M}{EI}$$

$$\therefore \quad y = \frac{1}{EI}\iint M dx + C_1x + C_2$$

As area moment (I) increases, the deflection (y) will reduce.

4.2 (a)

Maximum bending moment occurs at the centre

$$= \frac{wL^2}{8} = \frac{120 \times (15)^2}{8} = \mathbf{3375\ kNm}$$

4.3 (a)

Moment of inertia,

$$I = \frac{bh^3}{12} = \frac{120 \times (750)^3}{12} = 42.1875 \times 10^8 \text{ mm}^4$$

Maximum deflection in the beam is given by

$$\delta = \frac{5}{384}\frac{wL^4}{EI}$$

$$= \frac{5}{384} \times \frac{120 \times 10^3}{10^3} \times \frac{(15 \times 10^3)^4}{200 \times 10^3 \times 42.1875 \times 10^8}$$

$$= \mathbf{93.75\ mm}$$

4.4 (a)

- Rigid roller will remain in contact with both the beams when it will deflect together. So deflection will be same.
- As leads on beam are applied at different location, so slope will be different.

4.5 (a)

The given propped cantilever beam can be assumed to be consisting of two type of loads. First part is simple cantilever with UDL and the second part is cantilever beam with point load of R_2 at end.

Balancing the forces, we get

$$R_1 + R_2 = qL \quad \text{...(i)}$$

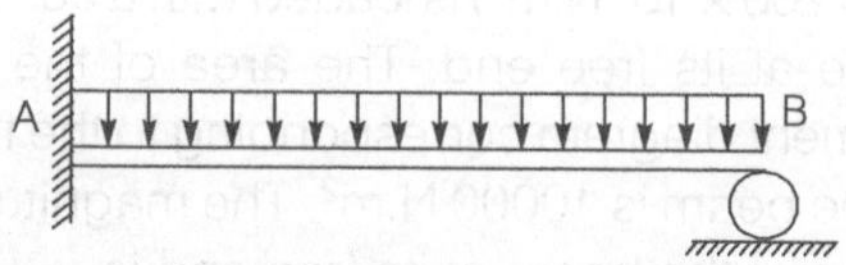

Balancing the deflection at end point as net deflection at the end is zero.

Deflection at B due to the UDL alone,

$$\delta_B = \frac{qL^4}{8EI}$$

Deflection at B due to R_2 alone,

$$\delta'_B = \frac{R_2L^3}{3EI}$$

$$\therefore \quad \delta_B = \delta'_B$$

$$\Rightarrow \quad \frac{qL^4}{8EI} = \frac{R_2L^3}{3EI} \Rightarrow R_2 = \frac{3qL}{8}$$

From eq. (i), $R_1 = qL - R_2$

$$= qL - \frac{3qL}{8} = \frac{5qL}{8}$$

Moment: $M = R_2L - qL \times \frac{L}{2}$

$$= \frac{3qL^2}{8} - \frac{qL^2}{2} = -\frac{qL^2}{8}$$

4.6 (b)

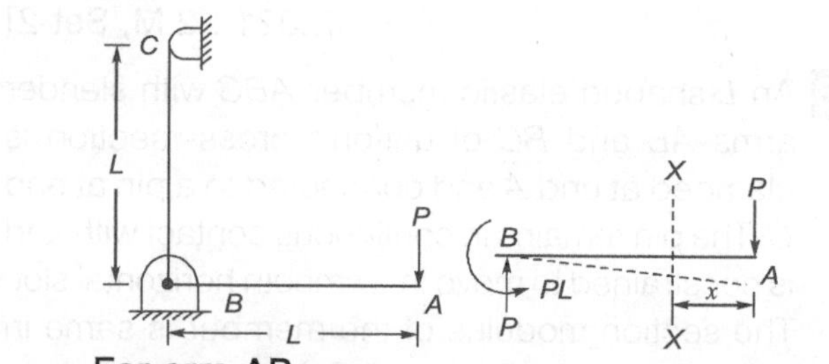

For arm AB:

At any section X-X, at a distance x from A

$$M_x = -Px$$

$$\therefore \quad U_{AB} = \int \frac{M_x^2\, dx}{2EI} = \int_0^L \frac{(-Px)^2}{2EI} dx$$

$$U_{AB} = \frac{P^2}{2EI}\left(\frac{x^3}{3}\right)_0^L = \frac{P^2L^3}{6EI}$$

For arm BC:

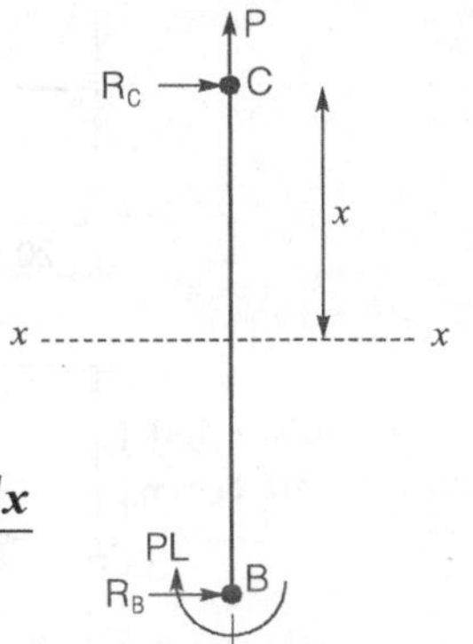

$R_B + R_C = 0$

$M_C = 0$

$R_BL - PL = 0$

$R_B = P$

$R_C = -P$

$M_x = R_Cx = Px$

$$U_{BC} = \int_0^L \frac{M_x^2 dx}{2EI} = \int_0^L \frac{P^2x^2dx}{2EI}$$

$$= \frac{P^2L^3}{6EI}$$

Total strain energy stored in the frame of two arm

$$U = U_{AB} + U_{BC} = \frac{P^2L^3}{6EI} + \frac{P^2L^3}{6EI} = \frac{P^2L^3}{3EI}$$

Therefore, vertical deflection of A is

$$\delta_A = \frac{\partial U}{\partial P} = \frac{\partial}{\partial P}\left[\frac{P^2L^3}{3EI}\right] = \frac{2PL^3}{3EI}$$

4.7 (b)

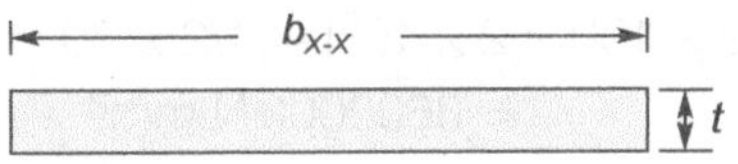

At section X-X,

$$b_{X\text{-}X} = \frac{bx}{l}$$

$$\text{MOI} = \frac{b_{X-X}t^3}{12} = \frac{bxt^3}{12l}$$

4.8 (d)

Consider an element of length 'dx' at distance x from free end.

$$U = \int_0^l \frac{(M_{X-X})dx}{2EI} = \int_0^l \frac{(Px)dx \times 12l}{2Ebxt^3}$$

$$= \frac{6P^2l}{Ebt^3}\int_0^l xdx = \frac{3P^2l^3}{Ebt^3}$$

Maximum deflection will be at the free end (i.e. point of application of load)

$$\delta = \frac{\partial U}{\partial P} = \frac{6Pl^3}{Ebt^3}$$

4.9 (a)

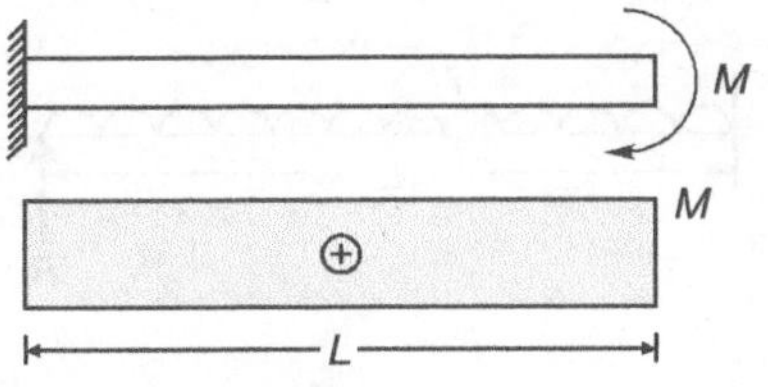

BMD

$$\text{Deflection} = \frac{1}{EI}(\text{Moment of area of BMD})$$

$$= \frac{1}{EI} \times M \times L \times \frac{L}{2} = \frac{ML^2}{2EI}$$

4.10 (b)

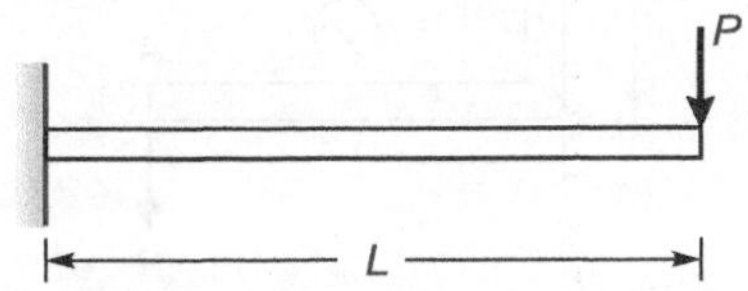

$$\theta_1 = \frac{PL^2}{2EI}$$

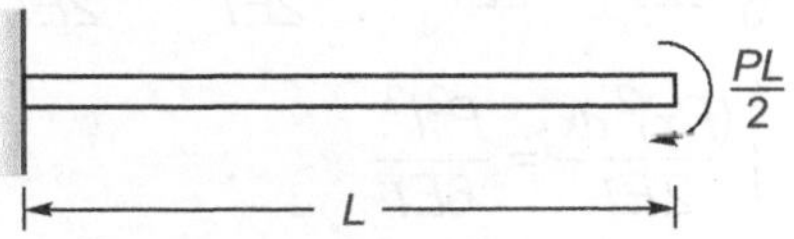

$$\theta_2 = \frac{ML}{EI} = \frac{PL^2}{2EI}$$

By superposition, $\theta = \theta_1 + \theta_2 = \dfrac{PL^2}{EI}$

4.11 (a)

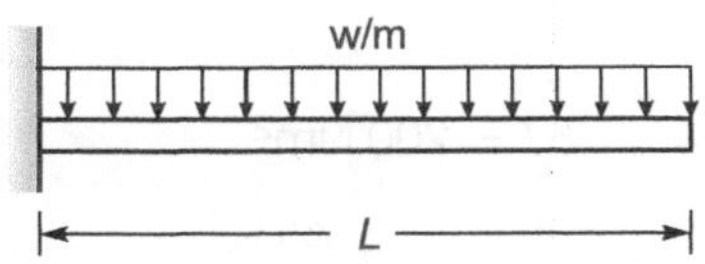

$$y = \frac{wx^2}{24EI}(x^2 + 6L^2 - 4Lx)$$

and $\qquad \delta_{max} = \dfrac{wL^4}{8EI}$

4.12 (c)

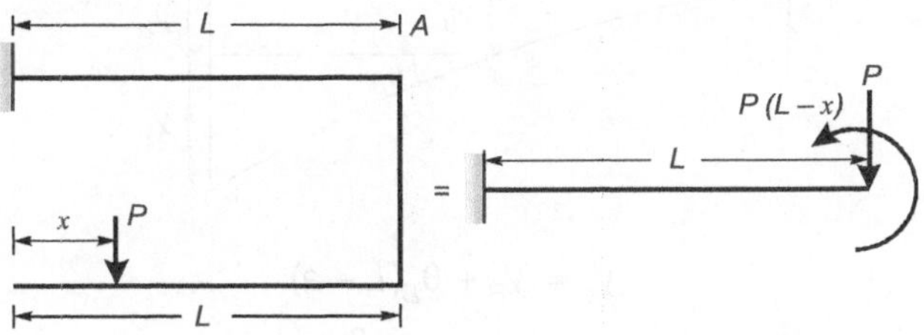

Given: $\qquad \delta_A = 0$

$$\delta_A = \frac{-PL^3}{3EI} + \frac{P(L-x)L^2}{2EI} = 0$$

$$\Rightarrow \quad \frac{P(L-x)L^2}{2EI} = \frac{PL^3}{3EI}$$

$$\Rightarrow \quad \frac{(L-x)}{2} = \frac{L}{3}$$

$$\Rightarrow \quad 3L - 3x = 2L$$

$$x = \frac{L}{3} = 0.33L$$

4.13 (d)

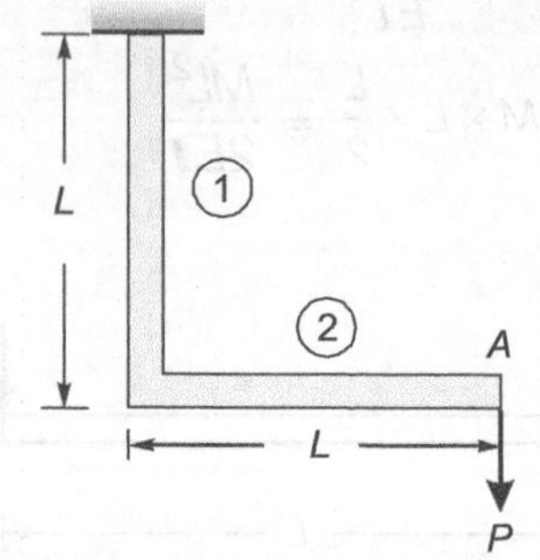

$$U_1 = \int_0^L \frac{M^2 dx}{2EI} = \frac{M^2 L}{2EI} = \frac{(PL)^2 \cdot L}{2EI} = \frac{P^2 L^3}{2EI}$$

$$U_2 = \int_0^L \frac{(Px)^2 dx}{2EI} = \frac{P^2 L^3}{6EI}$$

$$U = U_1 + U_2 = \frac{4P^2 L^3}{6EI}$$

By Castigliano's theorem,

$$\delta_A = \frac{\partial U}{\partial P} = \frac{\partial}{\partial P}\left(\frac{4P^2 L^3}{6EI}\right) = \frac{8PL^3}{6EI} = \frac{4PL^3}{3EI}$$

4.14 Sol.

$$EI = 200 \text{ Nm}^2$$

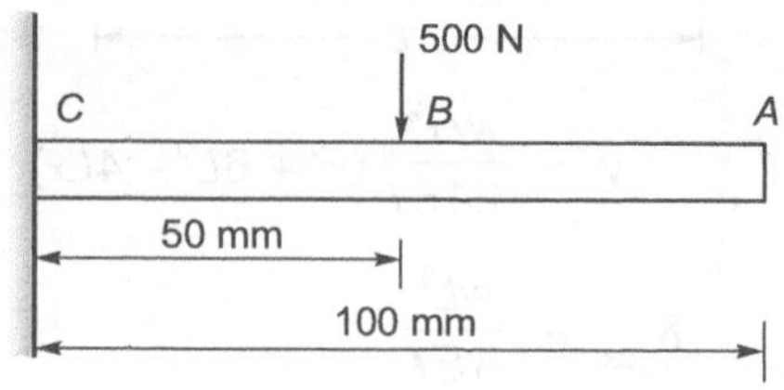

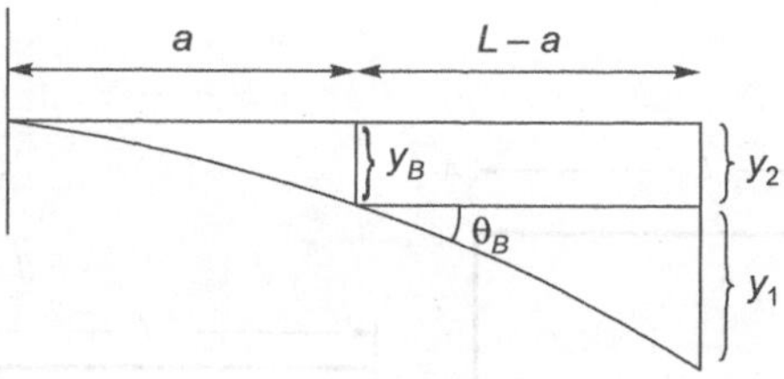

$$y_c = y_B + \theta_B(L-a)$$

$$y_B = \frac{Pa^3}{3EI} = \frac{500 \times \left\{\frac{50}{1000}\right\}^3}{3 \times 200} = 1.04167 \times 10^{-4} \text{ m}$$

$$\theta_B = \frac{Pa^2}{2EI} = \frac{500 \times \left\{\frac{50}{1000}\right\}^2}{2 \times 200} = 3.125 \times 10^{-3} \text{ rad}$$

$$y_c = y_B + \theta_B(L-a) = [1.04167 \times 10^{-4}] + \left[3.125 \times 10^{-3} \times \frac{50}{1000}\right]$$

$$= 2.6042 \times 10^{-4} \text{ m} = \mathbf{0.26042 \text{ mm}}$$

4.15 Sol.

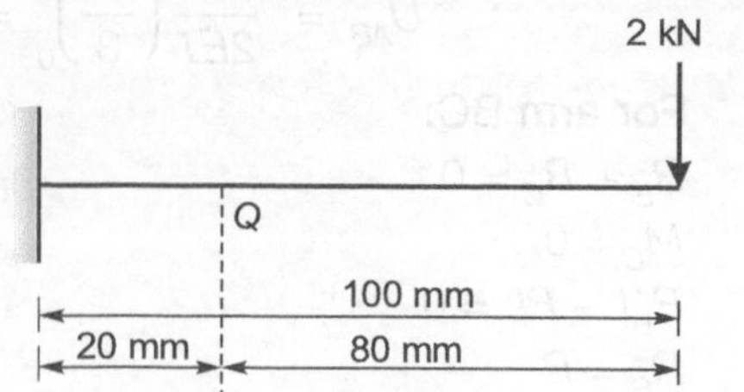

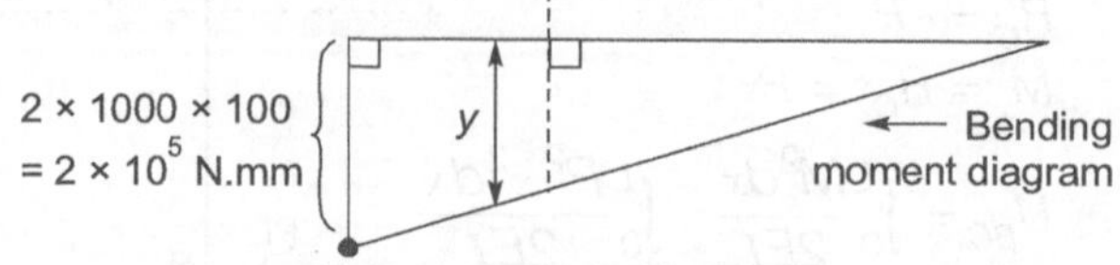

Slope $\theta_Q - \theta_0 = \frac{1}{EI}$ [Area of diagram till Q]

Area of diagram till "Q"

$$\left[\frac{1}{2} \times 100 \times 2 \times 10^5\right] - \left[\frac{1}{2} \times 80 \times y\right]$$

Calculating; y from properties of similar triangles.

$$\frac{y}{2 \times 10^5} = \frac{80}{100}$$

$$y = 160{,}000 \text{ Nmm}$$

Area of diagram till Q

$\Rightarrow$ [1/2 × 100 × 2 × 10^5] – [1/2 × 80 × 160,000]

$= 3600000 \text{ N.mm}^2$

$$\theta_Q = \frac{1}{210 \times 10^3 \times I} \times 3600000$$

$$I = \frac{bd^3}{12} = \frac{b^4}{12} = 108 \text{ mm}^4$$

$$\theta_Q = \frac{1}{210 \times 10^3 \times 108} \times 3600000 = \mathbf{0.1587 \text{ rad.}}$$

4.16 (d)

For cantilever beam,

$$\delta = \frac{wL^3}{3EI}$$

$$I = \frac{a^4}{12};\ \delta = \frac{4wL^3}{Ea^4}$$

$$\delta' = \frac{4wL^3}{E(1.19a)^4} = 0.498\delta$$

% decrease in deflection $= \frac{\delta' - \delta}{\delta} = 50.13\%$

4.17 (b)

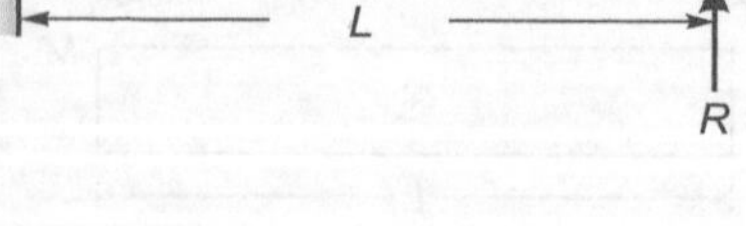

Deflection at the support is zero,

$$\delta = \frac{-wL^4}{8EI} + \frac{RL^3}{3EI} = 0$$

$$R = \frac{3wL}{8}$$

4.18 (c)

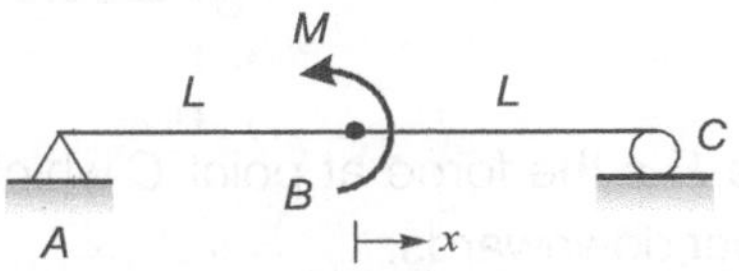

Equation of deflection

$$W = \frac{-M}{12EIL}\left[Lx^2 - x^3 + Lcx - Cx^2\right]$$

$$\frac{dW}{dx} = \frac{-M}{12EIL}\left[2Lx - 3x^2 + LC - 2Cx\right]$$

$$\frac{d^2W}{dx^2} = \frac{-M}{12EIL}\left[2L - 6x - 2C\right]$$

$$\therefore \quad EI\frac{d^2W}{dx^2} = \frac{-M}{12L}\left[2L - 6x - 2C\right]$$

$$\text{now} \quad EI\frac{d^2W}{dx^2} = \frac{-M}{12L}\left[2L - 6x - 2C\right]$$

at $\quad x = L,\ M_x = 0$

$$\therefore \quad -4L - 2C = 0$$

$$C = -2L$$

Slope

$$\frac{dW}{dx} = \frac{-M}{12EIL}\left[2Lx - 3x^2 - 2L^2 + 4Lx\right]$$

$$\therefore \quad \left|\frac{dW}{dx}\right|_{x=0} = \frac{ML}{6EI}$$

Method II:

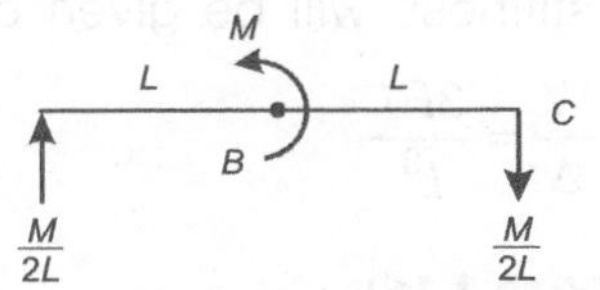

$$\text{Strain energy } (U) = 2\int_0^L \frac{\left(\frac{Mx}{2L}\right)^2 \cdot dx}{2EI}$$

$$= \frac{1}{EI}\left[\frac{M^2}{4L^2}\cdot\frac{l^3}{3}\right] = \frac{M^2 l}{12EI}$$

By Castigliano's theorem,

$$\text{Slope} = \frac{dU}{dM} = \frac{ML}{6EI}$$

4.19 (d)

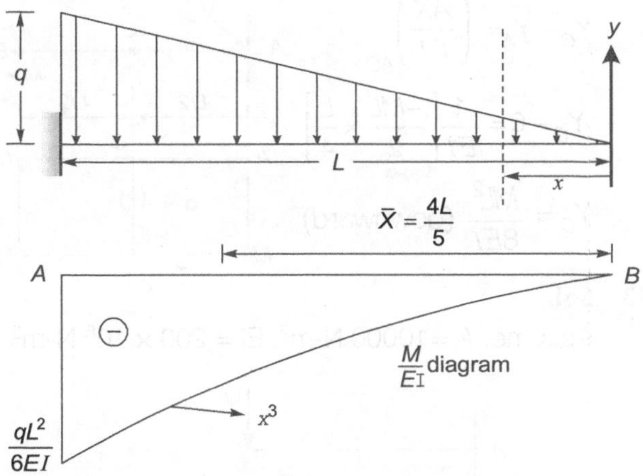

By using moment area IInd theorem

$$Y_B - Y_A = A\bar{X} \text{ of } \frac{M}{EI} \text{ diagram between } B \text{ and } A$$

$$Y_{max} - 0 = \left(\frac{1}{4}\right)\left(\frac{qL^2}{6EI}\right)(L)\left(\frac{4L}{5}\right)$$

$$y_{max} = \frac{qL^4}{30EI}\ (\downarrow)$$

4.20 Sol.

$$y_{max} = \frac{ML^2}{2EI} = \frac{1 \times 10^3 \times 10^6}{2 \times 200 \times 10^6} = 2.5\text{ mm}$$

4.21 (b)

As per given configuration diagram of beam:

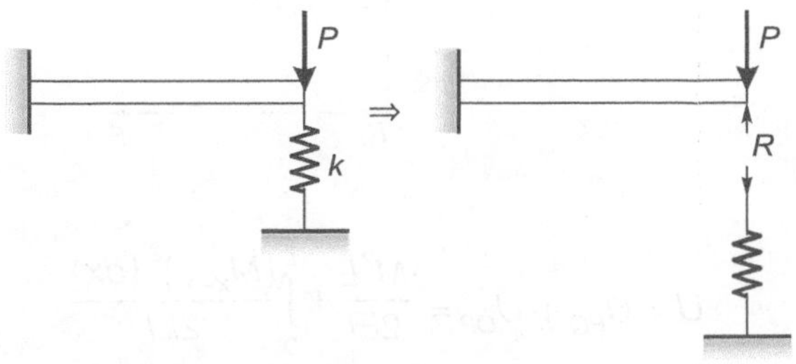

$$\Delta_{beam} = \Delta_{spring}$$

$$\frac{(P-R)L^3}{3EI} = \frac{R}{k}$$

$$P - R = \frac{3EI}{KL^3}R$$

$$\therefore \quad R = \frac{P}{1 + \frac{3EI}{KL^3}} = \frac{PL^3K}{KL^3 + 3EI}$$

Deflection of beam,

$$\Delta_{beam} = \frac{R}{K} = \frac{PL^3}{KL^3 + 3EI}$$

4.22 (d)

$$Y_C - Y_A = \left(\frac{A\bar{X}}{EI}\right)_{AC}$$

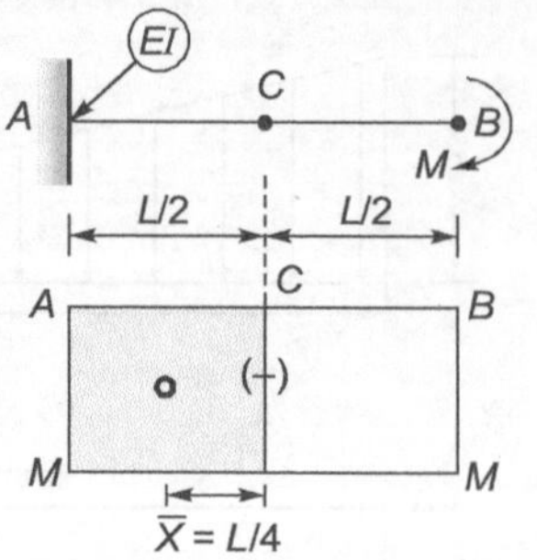

$$Y_C - 0 = \frac{1}{EI}\left[\frac{-ML}{2}\times\frac{L}{4}\right]$$

$$Y_C = \frac{ML^2}{8EI} \text{ (downward)}$$

4.23 Sol.

Assume: A = 10000 N-m^2; EI = 200 × 10^6 N-m^2

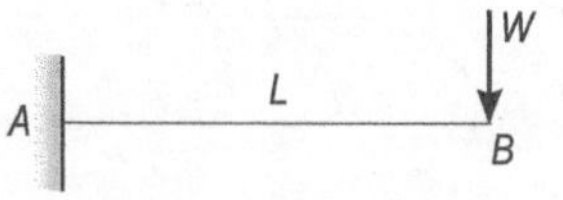

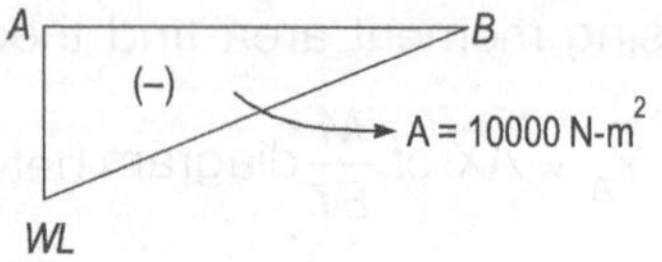

As per moment area first theorem,

$$\theta_B - \theta_A = \left(\frac{A}{EI}\right)AB$$

$$\theta_B - 0 = \frac{10000}{200\times10^6} = 0.5\times10^{-4} \text{ radians}$$

$$\theta_B = 50\,\mu \text{ radians}$$

4.24 (b)

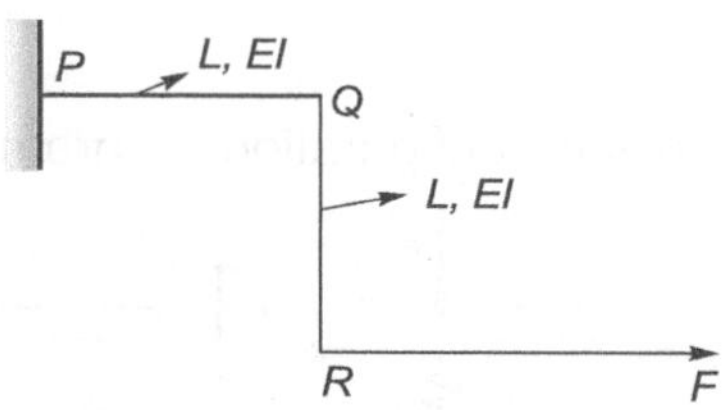

$$U = U_{PQ} + U_{QP} = \frac{M^2L}{2EI} + \int_0^L \frac{(M_{x-x})^2\,(dx)}{2EI}$$

$$U = \frac{(FL)^2 L}{2EI} + \int_0^L\left(\frac{(Fx)^2\,(dx)}{2EI}\right)$$

$$U = \frac{F^2L^3}{2EI} + \frac{F^2L^3}{6EI} = \frac{2F^2L^3}{3EI}$$

By Castigliano's theorem:

$$(\delta_H)_R = \frac{\partial U}{\partial F} = \frac{4FL^3}{3EI}$$

4.25 (a)

Free body diagram of the member is:

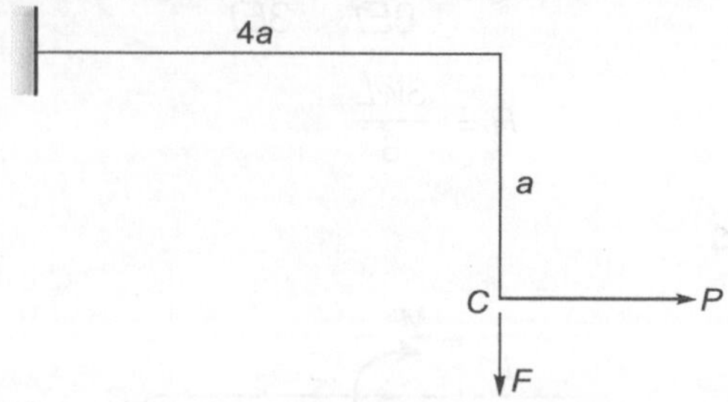

Where F is the force at point C which will move member downwards.

Let, ΔC_1 = Deflection upwards by force P

ΔC_2 = Deflection downwards by force F

$$\Delta C_1 = \frac{(Pa)(4a)^2}{2EI} = \frac{8Pa^3}{EI}$$

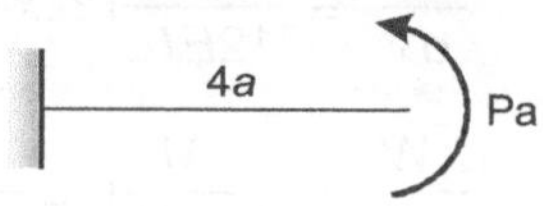

$$\Delta C_2 = \frac{F\times(4a)^3}{3EI} = \frac{64Fa^3}{3EI}$$

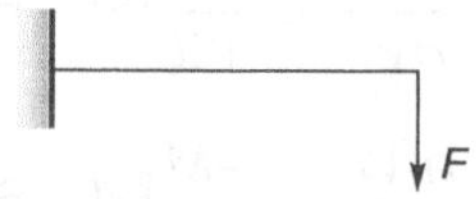

As, $\Delta C_1 = \Delta C_2$

$$\Rightarrow \quad \frac{8Pa^3}{EI} = \frac{64Fa^3}{3EI}$$

$$\Rightarrow \quad F = \frac{24P}{64} = \frac{3P}{8} \text{ (downwards)}$$

4.26 (a)

The deflection of the beam is given by,

$$\Delta = \frac{WL^3}{3EI}$$

and the stiffness will be given by,

$$K = \frac{W}{\Delta} = \frac{3EI}{L^3}$$

4.27 (2.09) (2.00 to 2.16)

Given, mass density, ρ = 2700 kg/m^3

∴ Weight density, γ = 2700 × 9.81 N/m^3

Cross-section area, A = 10 m^2

Length of bar, L = 5 m

The strain energy is given by:

$$U = \frac{W^2L}{6AE} = \frac{\gamma^2AL^3}{6E}$$

$$\Rightarrow U = \frac{(2700\times9.81)^2\times10\times5^3}{6\times70\times10^9} = 2.09 \text{ N-m}$$

■■■■

5 Bending and Shear Stress in Beams

5.1 The beams, one having square cross-section and another circular cross-section, are subjected to the same amount of bending moment. If the cross sectional area as well as the material of both the beams are the same then

(a) maximum bending stress developed in both the beams is the same

(b) the circular beam experience more bending stress than the square one

(c) the square beam experiences more bending stress than the circular one

(d) as the material is same both beams will experience same deformation

[2003 : 1 M]

5.2 The second moment of a circular area about the diameter is given by (D is the diameter)

(a) $\frac{\pi D^4}{4}$ (b) $\frac{\pi D^4}{16}$

(c) $\frac{\pi D^4}{32}$ (d) $\frac{\pi D^4}{64}$

[2003 : 1 M]

5.3 A cantilever beam has the square cross section of 10 mm × 10 mm. It carries a transverse load of 10 N. Considering only the bottom fibres of the beam, the correct representation of the longitudinal variation of the bending stress is

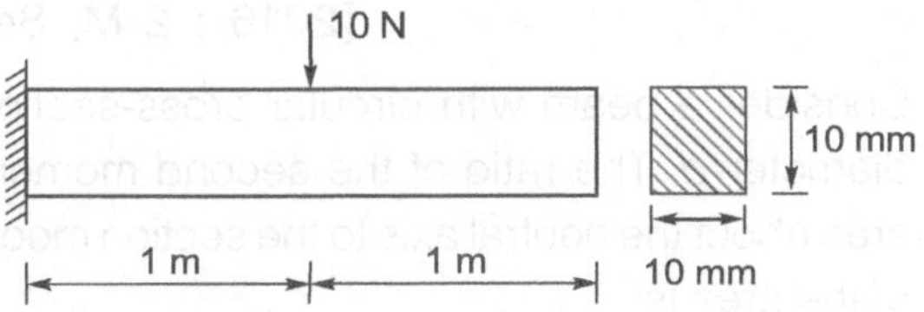

(a) 60 MPa 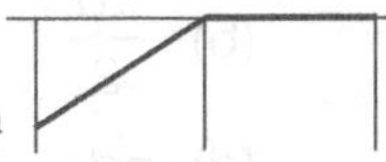

(b) 60 MPa 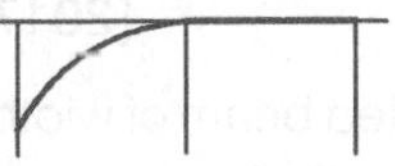

(c) 400 MPa 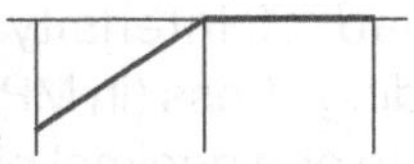

(d) 400 MPa 

[2005 : 2 M]

Linked Data Questions Q. 5.4 and Q. 5.5

A simply supported beam of span length 6 m and 75 mm diameter carries a uniformly distributed load of 1.5 kN/m.

5.4 What is the maximum value of bending moment?

(a) 9 kNm (b) 13.5 kNm

(c) 6.75 kNm (d) 125 kNm [2006 : 2 M

5.5 What is the maximum value of bending stress?

(a) 162.98 MPa (b) 325.95 MPa

(c) 625.95 MPa (d) 651.90 MPa

[2006 : 2 M]

5.6 The transverse shear stress acting in a beam of rectangular cross-section, subjected to a transverse shear load, is

(a) variable with maximum at the bottom of the beam

(b) variable with maximum at the top of the beam

(c) uniform

(d) variable with maximum on the neutral axis

[2008 : 1 M]

5.7 For the component loaded with a force F as shown in the figure, the axial stress at the corner point P is

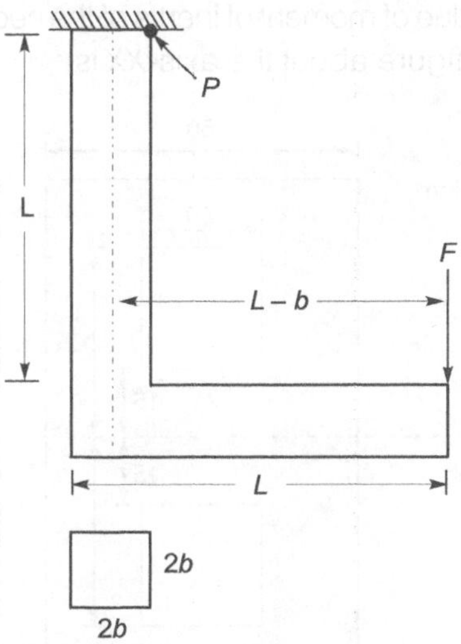

(a) $\frac{F(3L-b)}{4b^3}$ (b) $\frac{F(3L+b)}{4b^3}$

(c) $\dfrac{F(3L-4b)}{4b^3}$ (d) $\dfrac{F(3L-2b)}{4b^3}$

[2008 : 2 M]

Linked Data Questions 5.8 and 5.9.

A massless beam has a loading pattern as shown in the figure. The beam is of rectangular cross-section with a width of 30 mm and height of 100 mm.

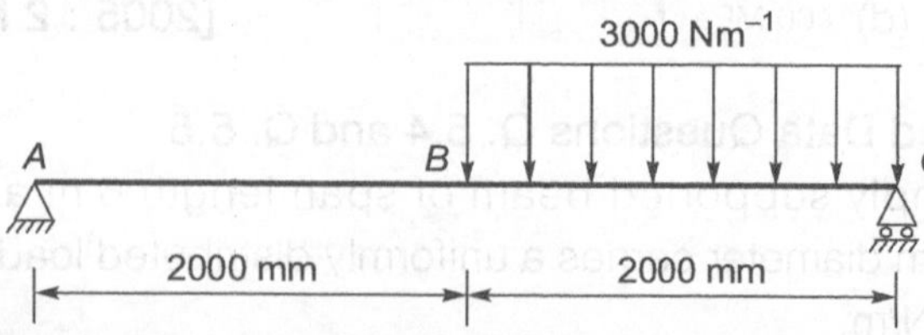

5.8 The maximum bending moment occurs at

(a) location B
(b) 2675 mm to the right of A
(c) 2500 mm to the right of A
(d) 3225 mm to the right of A [2010 : 2 M]

5.9 The maximum magnitude of bending stress (in MPa) is given by

(a) 60.0 (b) 67.5
(c) 200.0 (d) 225.0 [2010 : 2 M]

5.10 Consider a simply supported beam of length, $50h$, with a rectangular cross-section of depth, h, and width, $2h$. The beam carries a vertical point load, P, at its mid-point. The ratio of the maximum shear stress to the maximum bending stress in the beam is

(a) 0.02 (b) 0.10
(c) 0.05 (d) 0.01

[2014 : 2 M, Set-3]

5.11 The value of moment of inertia of the section shown in the figure about the axis-XX is

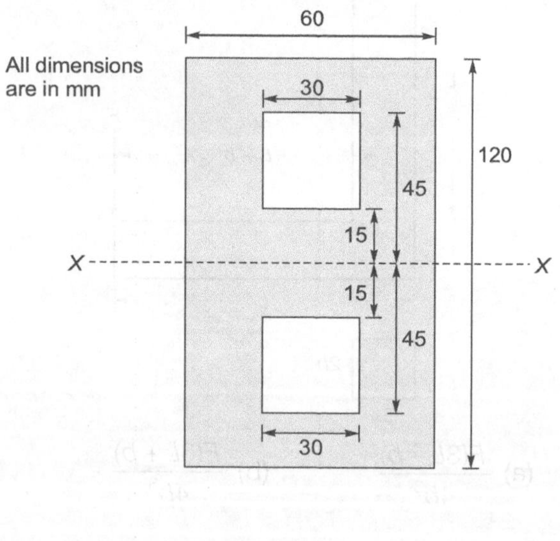

(a) 8.5050×10^6 mm^4
(b) 6.8850×10^6 mm^4
(c) 7.7625×10^6 mm^4
(d) 8.5725×10^6 mm^4 [2015 : 2 M, Set-3]

5.12 A simply supported beam of length $3L$ is subjected to the loading shown in the figure.

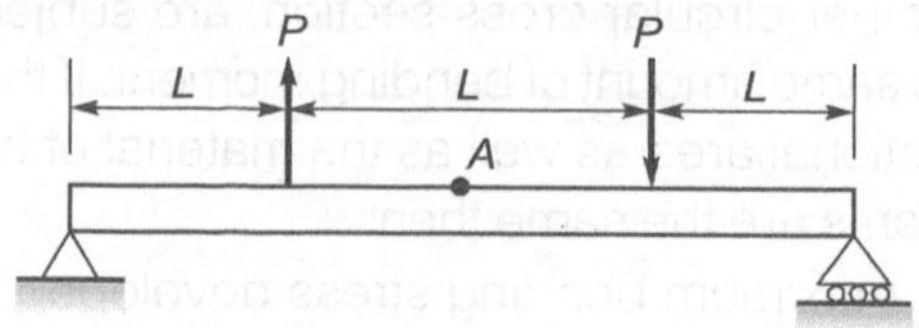

It is given that $P = 1$ N, $L = 1$ m and Young's modulus $E = 200$ GPa. The cross-section is a square with dimension 10 mm × 10 mm. The bending stress (in Pa) at the point A located at the top surface of the beam at the distance of 1.5 L from the left end is________.

[2016 : 2 M, Set-1]

5.13 The figure shows cross-section of a beam subjected to bending. The are moment of inertia (in mm^4) of this cross-section about its base is________.

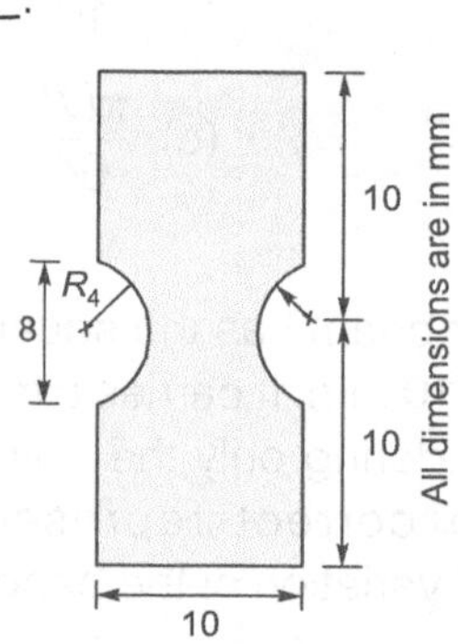

[2016 : 2 M, Set-1]

5.14 Consider a beam with circular cross-section of diameter d. The ratio of the second moment of area about the neutral axis to the section modulus of the area is

(a) $\dfrac{d}{2}$ (b) $\dfrac{\pi d}{2}$
(c) d (d) πd

[2017 : 1 M, Set-1]

5.15 A simply supported beam of width 100 mm, height 200 mm and length 4 m is carrying a uniformly distributed load of intensity 10 kN/m. The maximum bending stress (in MPa) in the beam is ______(correct to one decimal place)

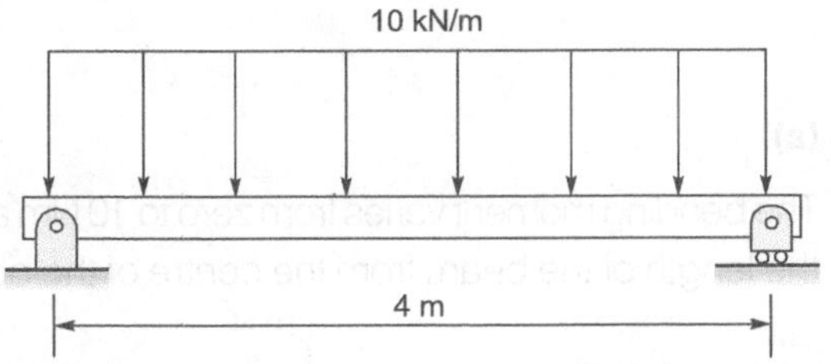

[2018 : 2 M, Set-1]

5.16 Consider an elastic straight beam of length $L = 10\pi$ m, with square cross-section of side a = 5 mm, and Young's modulus E = 200 GPa. This straight beam was bent in such a way that the two ends meet, to form a circle of mean radius R. Assuming that Euler-Bernoulli beam theory is applicable to this bending problem, the maximum tensile bending stress in the bent beam is ________ MPa.

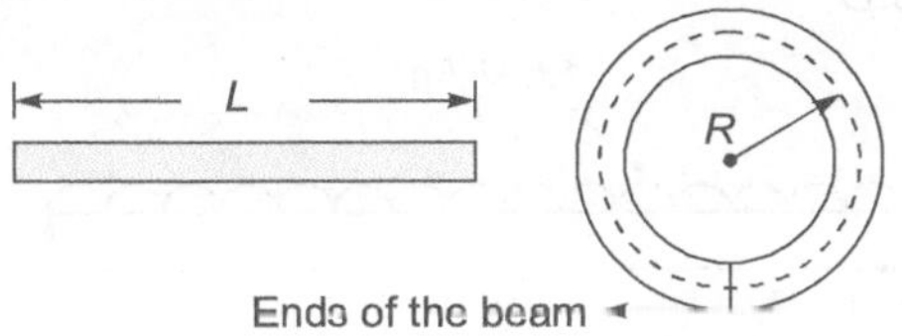

[2019 : 2 M, Set-1]

5.17 A wire of circular cross-section of diameter 1.0 mm is bent into a circular arc of radius 1.0 m by application of pure bending moments at its ends. The Young's modulus of the material of the wire is 100 GPa. The maximum tensile stress developed in the wire is _______ MPa.

[2019 : 1 M, Set-2]

5.18 An overhanging beam PQR is subjected to uniformly distributed load 20 kN/m as shown in the figure.

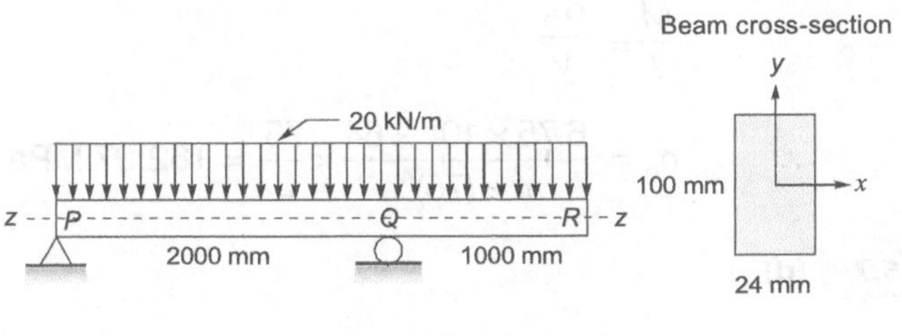

The maximum bending stress developed in the beam is _________ MPa [Round off to one decimal place]

[2021 : 2 M, Set-1]

5.19 A beam is undergoing pure bending as shown in the figure. The stress (σ)-strain (ε) curve for the material is also given. The yield stress of the material is σ_y. Which of the option(s) given represent(s) the bending stress distribution at cross-section AA after plastic yielding?

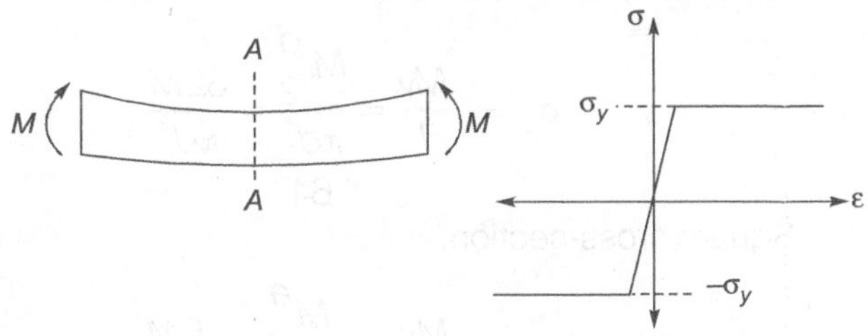

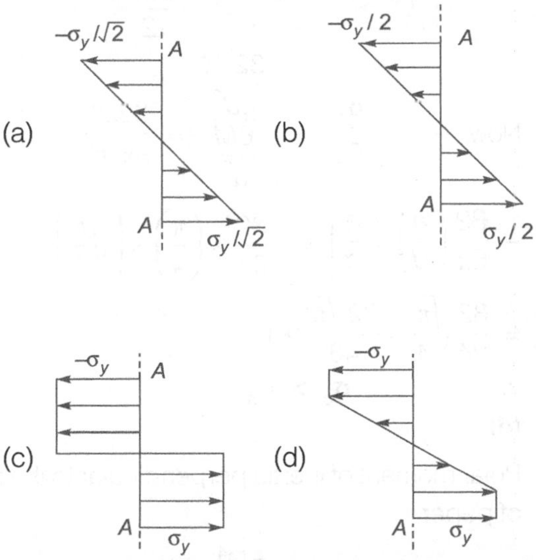

[2023 : 1 M, MSQ]

Answers Bending and Shear Stress in Beams

5.1	(b)	5.2	(d)	5.3	(a)	5.4	(c)	5.5	(a)	5.6	(d)	5.7	(d)	5.8	(c)
5.9	(b)	5.10	(d)	5.11	(b)	5.12	(0)	5.13	(21439.06)			5.14	(a)	5.15	(30)
5.16	(100)	5.17	(50)	5.18	(250)	5.19	(c, d)								

Explanations Bending and Shear Stress in Beams

5.1 (b)

Let, d = diameter of circular cross-section

a = side of square cross-section

Since cross-sectional area of square and circular cross-section are equal

$$\therefore \quad \frac{\pi}{4}d^2 = a^2 \quad ...(i)$$

For circular cross-section,

$$\sigma_c = \frac{My}{I} = \frac{M.\frac{d}{2}}{\frac{\pi d^4}{64}} = \frac{32M}{\pi d^3}$$

Square cross-section,

$$\sigma_s = \frac{My}{I} = \frac{M\frac{a}{2}}{\frac{a^4}{12}} = \frac{6M}{a^3}$$

Now,

$$\frac{\sigma_c}{\sigma_s} = \frac{\frac{32\,M}{\pi d^3}}{\frac{6M}{a^3}} = \frac{32a^3}{6\pi d^3}$$

$$= \frac{32}{6\pi}\left(\frac{a}{d}\right)^2\left(\frac{a}{d}\right) = \frac{32}{6\pi}\times\left(\frac{\pi}{4}\right)\times\left(\sqrt{\frac{\pi}{4}}\right)$$

$$= \frac{32}{24}\sqrt{\frac{\pi}{4}} = \frac{32\sqrt{\pi}}{48} > 1$$

$$\therefore \quad \boldsymbol{\sigma_c > \sigma_s}$$

5.2 (d)

Polar moment of inertia perpendicular to the plane of paper

$$I_p = \frac{\pi D^4}{32}$$

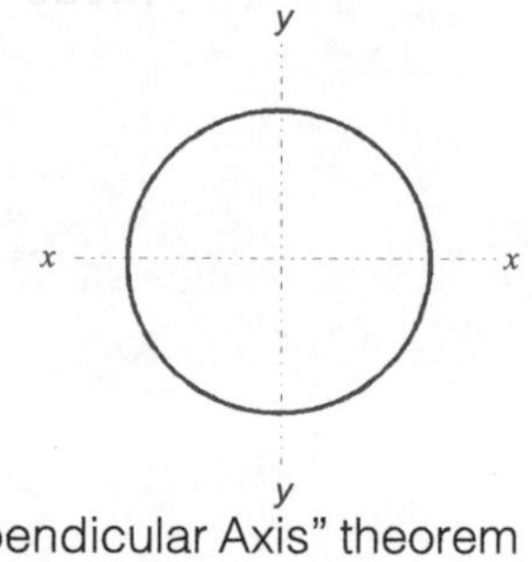

By 'Perpendicular Axis" theorem

$$I_{xx} + I_{yy} = I_p \quad [\because I_{xx} = I_{yy}]$$

$$2I_{xx} = \frac{\pi D^4}{32}$$

$$\therefore \quad I_{xx} = I_{yy} = \frac{\pi D^4}{64}$$

Here xx and yy represents diametral axis.

5.3 (a)

The bending moment varies from zero to 10 Nm along the length of the beam from the centre of the beam.

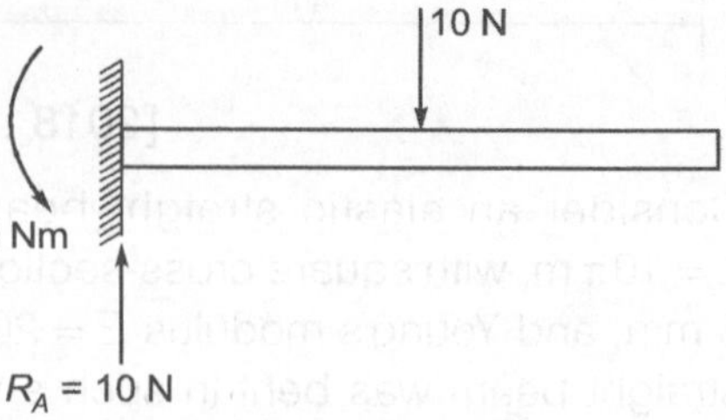

$$\therefore \sigma = \frac{M}{I}\times y_{max} = \frac{10\times10^3}{10^4/12}\times\frac{10}{2} = \mathbf{60\ MPa}$$

5.4 (c)

F.B.D

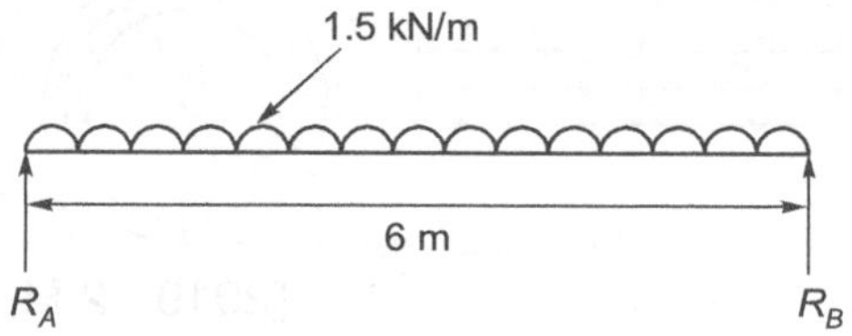

$$R_A + R_B = 9\text{ kN}$$

Taking moment about B, we get

$$\therefore \quad R_A \times 6 = 1.5\times6\times3$$

$$R_A = R_B = 4.5\text{ kN}$$

$$M_{max} = 4.5x - 1.5\frac{x^2}{2} \text{ [at } x = 3\text{m]}$$

$$= \mathbf{6.75\ kNm}$$

5.5 (a)

We know that

$$\frac{M}{I} = \frac{\sigma_b}{y}$$

$$\therefore \quad \sigma_b = \frac{6.75\times10^6\times64}{\pi\times(75)^4}\times\frac{75}{2} = \mathbf{162.97\ MPa}$$

5.7 (d)

$$\frac{M}{I} = \frac{\sigma}{y}$$

$$\frac{F(L-b)\times12}{(2b)^4} = \frac{\sigma_b}{b}$$

or $$\sigma_b = \frac{3F(L-b)}{4b^3}$$

(due to bending)

and $$\sigma_a = \frac{F}{4b^2}$$

(due to axial force)

$\therefore$ Total axial stress

$$= \frac{3F(L-b)}{4b^3} + \frac{F}{4b^2} = \frac{3F(L-b)}{4b^3} + \frac{Fb}{4b^3}$$

$$= \frac{3FL - 3Fb + Fb}{4b^3} = \frac{F(3L-2b)}{4b^3}$$

5.8 (c)

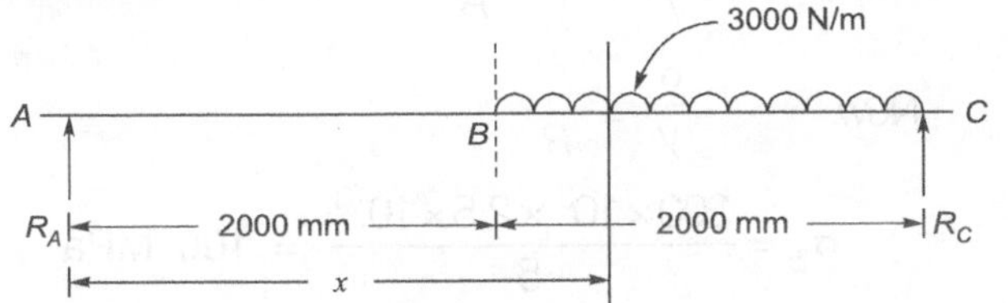

$$R_A + R_B = 6000 \text{ N}$$

Taking moment about A

$$600 \times 3 - R_C \times 4 = 0$$

$$\Rightarrow \quad R_C = 4500 \text{ N}$$

$$R_A = 1500 \text{ N}$$

Taking any section x m from A.

$$\text{Shear force} = 1500 - 3000(x-2)$$

and maximum bending moment occurs where shear force = 0.

$$\Rightarrow 1500 - 3000(x-2) = 0$$

$$x = 2.5 \text{ m}$$

or **2500 mm from A**

5.9 (b)

Maximum bending moment,

$$(\text{BM})_{\text{max.}} = 1500x - \frac{3000}{2}(x-2)^2$$

at $x = 2.5$ m

$$(\text{BM})_{\text{max}} = 1500 \times 2.5 - 1500(0.5)^2$$

$$= \mathbf{3375 \text{ Nm}}$$

Bending stress,

$$\sigma = \frac{M}{Z}$$

where $z = \frac{1}{6} bd^2$

$$\therefore \quad \sigma = \frac{6 \times 3375}{30 \times 100^2 \times 10^{-9}}$$

$$= 67.5 \times 10^6 \text{ N/m}^2 = \mathbf{67.5 \text{ MPa}}$$

5.10 (d)

$$\tau_{\text{max}} = \frac{3}{2}\frac{P_{\text{max}}}{A} = \frac{3P_{\text{max}}}{4h^2} = 0.75\frac{P_{\text{max}}}{h^2}$$

$$\tau_{\text{max}} = \frac{0.75P}{h^2 \, 2} = \frac{0.375P}{h^2} \qquad \left[\because P_{\text{max}} = \frac{P}{2}\right]$$

$$M_{\text{max}} = \frac{PL}{4} = \frac{P \times 50h}{4} = 12.5Ph$$

$$Z = \frac{bh^2}{6} = \frac{h^3}{3}$$

$$\sigma_{\text{max}} = \frac{M_{\text{max}}}{Z} = \frac{12.5Ph \times 3}{h^3} = 37.5\frac{P}{h^2}$$

$$\therefore \quad \frac{\tau_{\text{max}}}{\sigma_{\text{max}}} = \frac{0.375}{37.5} = 0.01$$

5.11 (b)

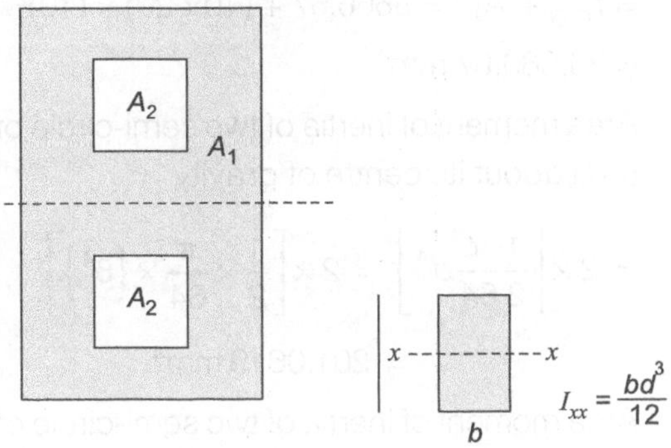

$$I_{\text{total}} = I_{A1} - 2I_{A2}$$

$$I_{A1} = 60 \times \frac{120^3}{12} = 8640000 \text{ mm}^4$$

$$I_{A2} = 30 \times \frac{30^3}{12} + 30^2 \times 30^2$$

$$I_{A2} = 877500 \text{ mm}^4$$

$$\therefore \quad I = 8640000 - 2 \times 877500$$

$$= 68850000 \text{ mm}^4$$

$$= \mathbf{6.885 \times 10^6 \text{ mm}^4}$$

5.12 Sol.

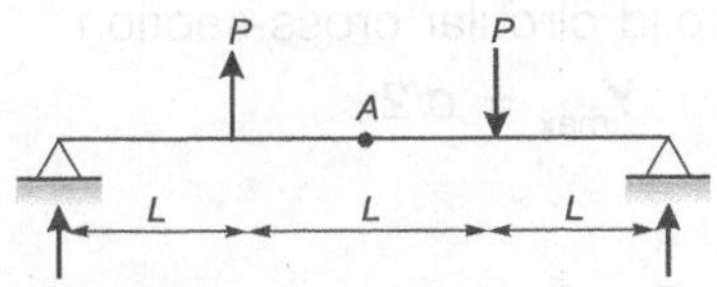

$$\Sigma F_V = 0$$

$$R_1 + R_2 = 0$$

$$\Sigma M_1 = 0$$

$$3R_2L - 2PL + PL = 0$$

$$R_2 = \frac{P}{3}; \quad R_1 = \frac{-P}{3}$$

B.M at A, $M_A = (R_1 \times 1.5L) + (P \times 0.5L)$

$$= \left(\frac{-P}{3} \times 1.5L\right) + 0.5\,PL = 0$$

$\therefore$ B.M. at $A = 0$

$\therefore$ Bending stress = **0 Pa.**

5.13 Sol.

Area moment of inertia of rectangular cross-section about its centre of gravity

$$I_{C.G} = \frac{bd^3}{12} = \frac{10 \times 20^3}{12} = 6666.67 \text{ mm}^4$$

Area moment of inertia of rectangular cross-section about its base

$= I_{C.G} + Ay^2 = 6666.67 + (10 \times 20) \times (10)^2$

$= 26{,}666.67 \text{ mm}^4$

Area moment of inertia of two semi-circle of equal radii about its centre of gravity

$$= 2 \times \left[\frac{1}{2}\frac{\pi}{64}d^4\right] = 2 \times \left[\frac{1}{2} \times \frac{\pi}{64} \times \{8^4\}\right]$$

$$= 201.0619 \text{ mm}^4$$

Area moment of inertia of two semi-circle of equal radii about base

$$= I_{C.G} + Ay^2 = 201.0619 + \left[2 \times \frac{1}{2}\pi \times 4^2\right][10]^2$$

$= \mathbf{5227.61 \text{ mm}^4}$

Area moment of inertia of configuration given in question about its base

$$= 26666.67 - 5227.61$$

$$= \mathbf{21439.06 \text{ mm}^4}$$

5.14 (a)

$$\frac{I_{N.A}}{Z_{N.A}} = \frac{I_{N.A.}}{(I_{N.A} / Y_{max})} = Y_{max}$$

For a solid circular cross-section

$$Y_{max} = d/2$$

5.15 Sol.

Maximum B.M.

$$M = \frac{wL^2}{8} = \frac{10 \times 16}{8} = 20 \text{ kNm} \quad (L = 4 \text{ m})$$

Maximum bending stress,

$$\sigma_{max} = \frac{M}{I} y_{max} = \frac{20 \times 10^3}{\left(\frac{0.1 \times 0.2^3}{12}\right)} \times 0.1$$

$$= \mathbf{30 \times 10^6 \text{ N/m}^2 = 30 \text{ MPa}}$$

5.16 Sol.

Given: $L = 10\,\pi$ m; $a = 5$ mm; $E = 200$ GPa

Perimeter of wire = Length of wire

$$2\pi R = 10\pi$$

$$R = 5 \text{ m}$$

We know that,

$$\frac{M}{I} = \frac{\sigma}{y} = \frac{E}{R}$$

Now,
$$\frac{\sigma}{y} = \frac{E}{R}$$

$$\sigma_b = \frac{200 \times 10^3 \times 2.5 \times 10^{-3}}{5} = \mathbf{100 \text{ MPa}}$$

5.17 Sol.

Given: $E = 100$ GPa, $d = 1$ mm, $R = 1$ m

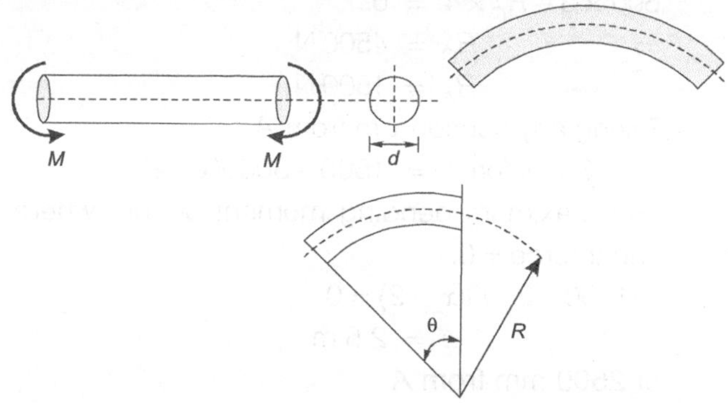

$$(\sigma_b)_{max} = \frac{EY_{max}}{R}$$

$$(\sigma_b)_{max} = \frac{E\frac{d}{2}}{R} = \frac{100 \times 10^3 \times 0.5}{1000} = 50 \text{ MPa}$$

5.18 Sol.

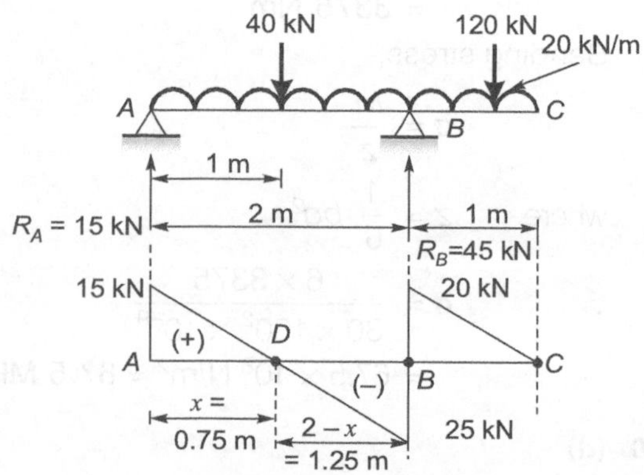

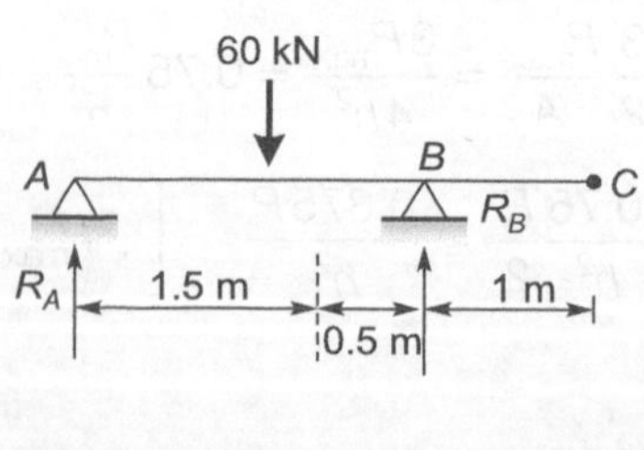

$$M_D - M_A = \frac{1}{2} \times 15 \times 0.75$$

$$\Sigma M_A = 60 \times 1.5 - R_B \times 2 = 0$$

$$M_D = 5.625 \text{ kN-m(s)}$$

$$R_B = 45 \text{ kN}(\uparrow); \; R_A = 15 \text{ kN}(\uparrow)$$

Location of D:

$$\frac{15}{x} = \frac{25}{2-x}$$

$$6 - 3x = 5x$$

$$= \frac{3}{4}\text{m}$$

$$M_C - M_B = \frac{1}{2} \times 20 \times 1$$

$$M_B = -10 \text{ kN-m}$$

$$M_B = 10 \text{ kN-m } (+1)$$

$$\text{Max. B.M.} = \text{Larger of } (M_B \text{ and } M_D)$$

$$M_B = 10 \text{ kN-m } (+1)$$

$$(\sigma_b)_{max} = \frac{M_{max}}{Z_{N.A}} = \frac{6 \times 10 \times 10^6}{24 \times 100^2} = 250 \text{ MPa}$$

5.19 (c, d)

If bending stress at every fibre is equal to yield strength of the material, then entire cross-section undergoes plastic bending. In such a case, the distribution will be as represented in option (c). If bending stress at inner fibres is less than yield strength and bending stress at extreme fibres is equal to yield strength then, the bending stress will vary for some distance and then become constant i.e. equal to yield strength as shown in option (d).

■■■■

6 Torsion in Shafts

6.1 aximum shear stress developed on the surface of a solid circular shaft under pure torsion is 240 MPa. If the shaft diameter is doubled then the maximum shear stress developed corresponding to the same torque will be

(a) 120 MPa (b) 60 MPa
(c) 30 MPa (d) 15 MPa **[2003 : 1 M]**

6.2 A torque of 10 Nm is transmitted through a stepped shaft as shown in figure. The torsional stiffnesses of individual sections of lengths *MN*, *NO* and *OP* are 20 Nm/rad, 30 Nm/rad and 60 Nm/rad respectively. The angular deflection between the ends *M* and *P* of the shaft is

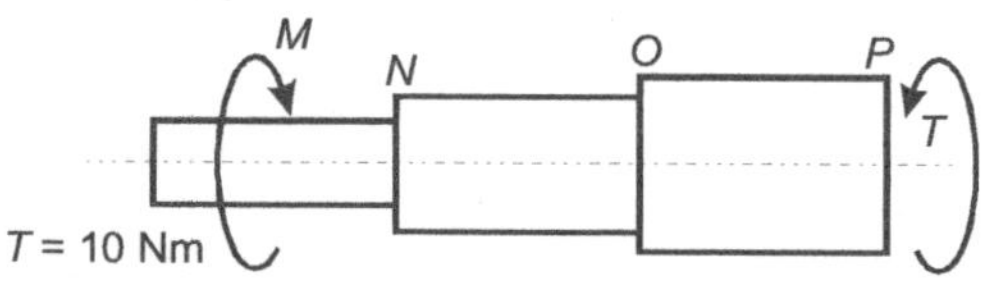

(a) 0.5 rad (b) 1.0 rad
(c) 5.0 rad (d) 10 rad **[2004 : 1 M]**

6.3 A solid circular shaft of 60 mm diameter transmits a torque of 1600 Nm. The value of maximum shear stress developed is

(a) 37.72 MPa (b) 47.72 MPa
(c) 57.72 MPa (d) 67.72 MPa

[2004 : 2 M]

6.4 Two shafts *AB* and *BC*, of equal length and diameters *d* and 2*d*, are made of the same material. They are joined at *B* through a shaft coupling, while the ends *A* and *C* are built-in (cantilevered). A twisting moment *T* is applied to the coupling. If T_A and T_C represent the twisting moments at the ends *A* and *C*, respectively, then

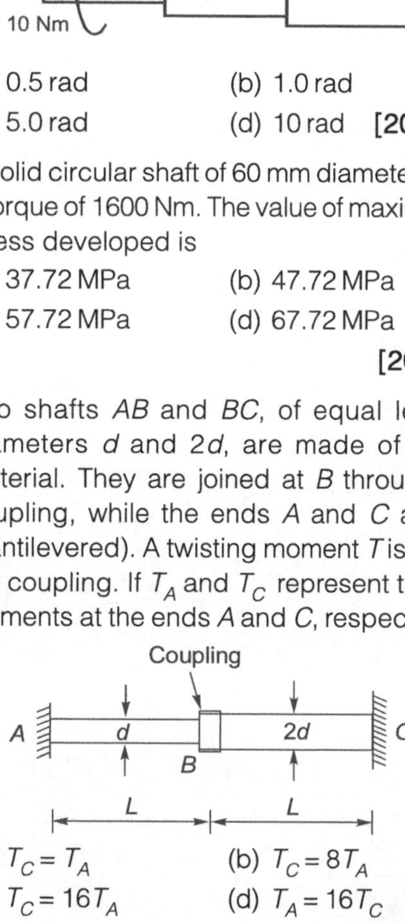

(a) $T_C = T_A$ (b) $T_C = 8T_A$
(c) $T_C = 16T_A$ (d) $T_A = 16T_C$

[2005 : 2 M]

6.5 For a circular shaft of diameter *d* subjected to torque *T*, the maximum value of the shear stress is

(a) $\frac{64T}{\pi d^3}$ (b) $\frac{32T}{\pi d^3}$
(c) $\frac{16T}{\pi d^3}$ (d) $\frac{8T}{\pi d^3}$ **[2006 : 1 M]**

6.6 A solid shaft of diameter *d* and length *L* is fixed at both the ends. A torque, T_0 is applied at a distance, *L*/4 from the left end as shown in the figure given below.

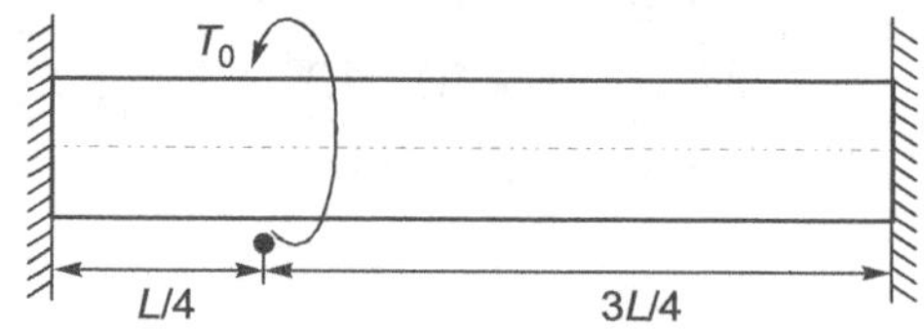

The maximum shear stress in the shaft is

(a) $\frac{16T_0}{\pi d^3}$ (b) $\frac{12T_0}{\pi d^3}$
(c) $\frac{8T_0}{\pi d^3}$ (d) $\frac{4T_0}{\pi d^3}$ **[2009 : 2 M]**

6.7 A torque *T* is applied at the free end of a stepped rod of circular cross-sections as shown in the figure. The shear modulus of the material of the rod is *G*. The expression for *d* to produce an angular twist θ at the free end is

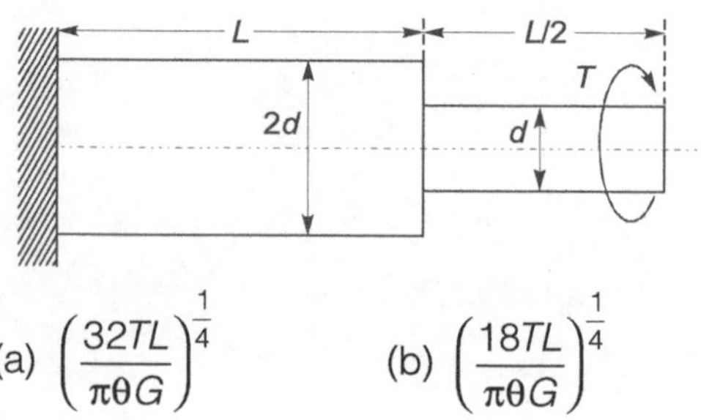

(a) $\left(\frac{32TL}{\pi\theta G}\right)^{\frac{1}{4}}$ (b) $\left(\frac{18TL}{\pi\theta G}\right)^{\frac{1}{4}}$
(c) $\left(\frac{16TL}{\pi\theta G}\right)^{\frac{1}{4}}$ (d) $\left(\frac{2TL}{\pi\theta G}\right)^{\frac{1}{4}}$ **[2011 : 2 M]**

6.8 Two solid circular shafts of radii R_1 and R_2 are subjected to same torque. The maximum shear stresses developed in the two shafts are τ_1 and τ_2. If $R_1/R_2 = 2$, then τ_2/τ_1 is _____.

[2014 : 1 M, Set-3]

6.9 onsider a stepped shaft subjected to a twisting moment applied at B as shown in the figure. Assume shear modulus, G = 77 GPa. The angle of twist at C (in degree) is ________.

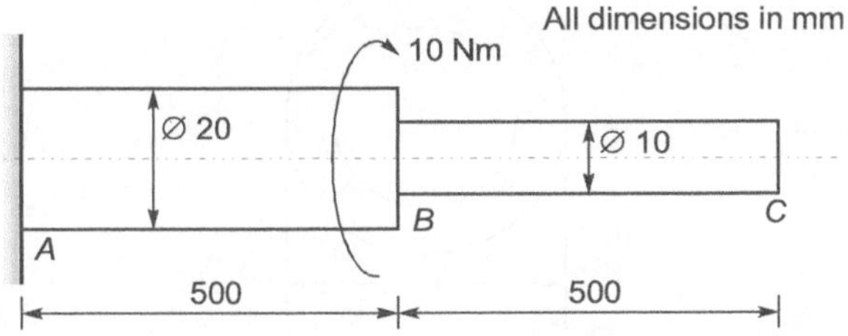

[2015 : 1 M, Set-1]

6.10 A hollow shaft ($d_o = 2d_i$ where d_o and d_i are the outer and inner diameters respectively) needs to transmit 20 kW power at 3000 RPM. If the maximum permissible shear stress is 30 MPa, d_o is

(a) 11.29 mm (b) 22.58 mm
(c) 33.87 mm (d) 45.16 mm

[2015 : 2 M, Set-2]

6.11 A hollow shaft of 1 m length is designed to transmit a power of 30 kW at 700 rpm. The maximum permissible angle of twist in the shaft is 1°. The inner diameter of the shaft is 0.7 times the outer diameter. The modulus of rigidity is 80 GPa. The outside diameter (in mm) of the shaft is ____.

[2015 : 2 M, Set-2]

6.12 The cross sections of two hollow bars made of the same material are concentric circles as shown in the figure. It is given that $r_3 > r_1$ and $r_4 > r_2$, and that the areas of the cross-sections are the same, J_1 and J_2 are the torsional rigidities of the bars on the left and right, respectively. The ratio J_2/J_1 is

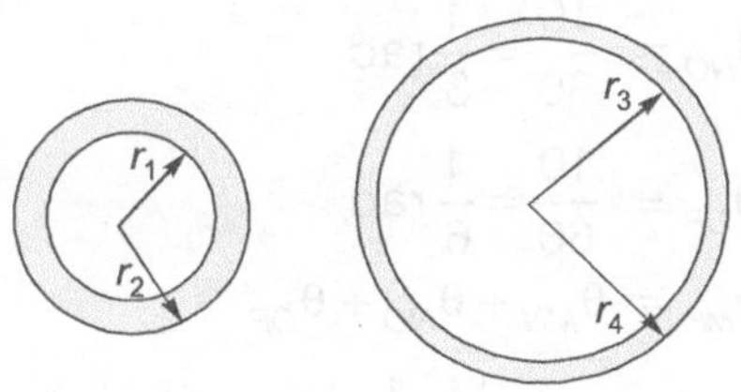

(a) > 1
(b) < 0.5
(c) =1
(d) between 0.5 and 1 [2016 : 1 M, Set-1]

6.13 Two circular shafts made of same material, one solid (S) and one hollow (H), have the same length and polar moment of inertia. Both are subjected to same torque. Here, θ_S is the twist and τ_S is the maximum shear stress in the solid shaft, whereas θ_H is the twist and τ_H is the maximum shear stress in the hollow shaft. Which one of the following is TRUE?

(a) $\theta_S = \theta_H$ and $\tau_S = \tau_H$
(b) $\theta_S > \theta_H$ and $\tau_S > \tau_H$
(c) $\theta_S < \theta_H$ and $\tau_S < \tau_H$
(d) $\theta_S = \theta_H$ and $\tau_S < \tau_H$ [2016 : 2 M, Set-3]

6.14 A rigid horizontal rod of length $2L$ is fixed to a circular cylinder of radius R as shown in the figure. Vertical forces of magnitude P are applied at the two ends as shown in the figure. The shear modulus for the cylinder is G and the Young's modulus is E.

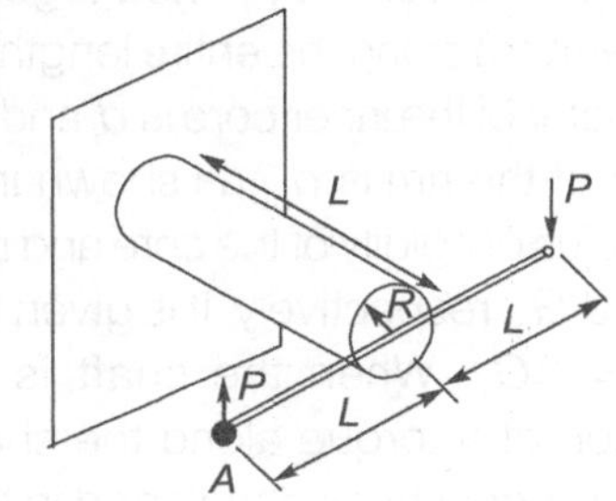

The vertical deflection at point A is

(a) $\dfrac{PL^3}{(\pi R^4 G)}$ (b) $\dfrac{PL^3}{(\pi R^4 E)}$

(c) $\dfrac{2PL^3}{(\pi R^4 E)}$ (d) $\dfrac{4PL^3}{(\pi R^4 G)}$

[2016 : 2 M, Set-2]

6.15 A motor driving a solid circular steel shaft transmits 40 kW of power at 500 rpm. If the diameter of the shaft is 40 mm, the maximum shear stress in the shaft is ________ MPa.

[2017 : 1 M, Set-1]

6.16 A hollow circular shaft of inner radius 10 mm outer radius 20 mm and length 1 m is to be used as a torsional spring. If the shear modulus of the material of the shaft is 150 GPa, the torsional stiffness of the shaft (in kN-m/rad) is ______ (correct to two decimal places).

[2018 : 1 M, Set-2]

6.17 A bar of circular cross section is clamped at ends P and Q as shown in the figure. A torsional moment T = 150 Nm is applied at a distance of 100 mm from end P. The torsional reactions (T_P, T_Q) in Nm at the ends P and Q respectively are

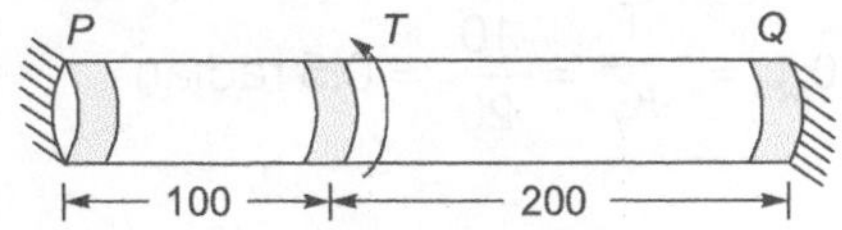

(All dimensions are in mm)

(a) (50, 100) (b) (75, 75)
(c) (100, 50) (d) (120, 30)

[2018 : 2 M, Set-2]

6.18 A cylindrical rod of diameter 10 mm and length 1.0 m is fixed at one end. The other end is twisted by an angle of 10° by applying a torque. If the maximum shear strain in the rod is $p \times 10^{-3}$, then p is equal to ________ (round off to two decimal places) [2019 : 1 M, Set-1]

6.19 A shaft of length L is made of two materials, one in the inner core and the other in the outer rim, and the two are perfectly joined together (no slip at the interface) along the entire length of the shaft. The diameter of the inner core is d_i and the external diameter of the rim is d_o, as shown in the figure. The modulus of rigidity of the core and rim materials are G_i and G_o, respectively. It is given that $d_o = 2d_i$ and $G_i = 3G_o$. When the shaft is twisted by application of a torque along the shaft axis, the maximum shear stress developed in the outer rim and the inner core turn out to be τ_o and τ_i, respectively. All the deformations are in the elastic range and stress-strain relations are linear. Then the ratio $\frac{\tau_i}{\tau_o}$ is ______ (round off to 2 decimal places).

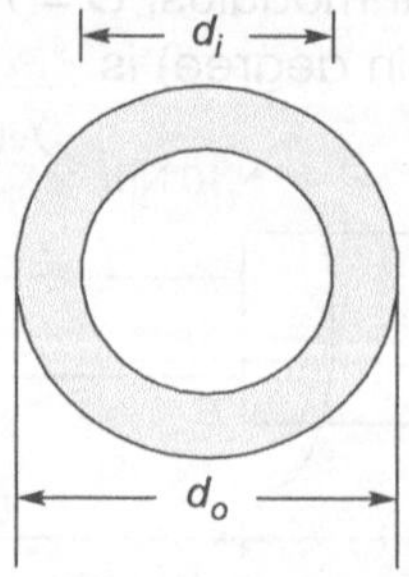

Shaft cross-section

[2022 : 2 M, Set-2]

6.20 A cylindrical transmission shaft of length 1.5 m and diameter 100 mm is made of a linear elastic material with a shear modulus of 80 GPa. While operating at 500 rpm, the angle of twist across its length is found to be 0.5 degrees. The power transmitted by the shaft at this speed is _____ kW. (Rounded off to two decimal places) Take π = 3.14.

[2023 : 2 M]

■■■■

Answers Torsion in Shafts

6.1	(c)	6.2	(b)	6.3	(a)	6.4	(c)	6.5	(c)	6.6	(b)	6.7	(b)	6.8	(8)
6.9	(0.2368°)	6.10	(b)	6.11	(44.5212)	6.12	(a)	6.13	(d)	6.14	(d)	6.15	(60.792)		
6.16	(35.343)	6.17	(c)	6.18	(0.8726)	6.19	(1.50)	6.20	(239.25)						

Explanations Torsion in Shafts

6.1 (c)

$$\tau_{max} = \frac{16T}{\pi d^3} = 240$$

When shaft diameter is doubled then

$$\tau'_{max} = \frac{16T}{\pi(2d)^3} = \frac{16T}{8\pi d^3}$$

$$= \frac{\tau_{max}}{8} = \frac{240}{8} = 30 \text{ MPa}$$

6.2 (b)

Here each shaft connected with each other in series. Torque in same in each part.

$$\theta_{MN} = \frac{T}{K_t} = \frac{10}{20} = 0.5 \text{ radian}$$

$$\theta_{NO} = \frac{10}{30} = \frac{1}{3} \text{ rad}$$

$$\theta_{OP} = \frac{10}{60} = \frac{1}{6} \text{ rad}$$

$$\theta_{MP} = \theta_{MN} + \theta_{NO} + \theta_{OP}$$

$$= 0.5 + \frac{1}{3} + \frac{1}{6} = 1.0 \text{ rad} = \mathbf{1.0 \text{ radian}}$$

6.3 (a)

$$\tau_{max} = \frac{16T}{\pi d^3} = \frac{16 \times 1600 \times 10^3}{\pi \times (60)^3} = 37.72 \text{ MPa}$$

6.4 (c)

Angular deflection in both shafts are same

$$\therefore \quad \theta_{AB} = \theta_{BC}$$

$$\frac{T_A \times L}{G \times \frac{\pi}{32} \times d^4} = \frac{T_C \times L}{G \times \frac{\pi}{32} \times (2d)^4}$$

$$T_A = \frac{T_C}{16}$$

$$\therefore \quad T_C = 16\,T_A$$

6.5 (c)

$$\frac{T}{J} = \frac{\tau}{d/2}$$

$$\Rightarrow \quad \frac{T}{\frac{\pi}{32} d^4} = \frac{\tau}{d/2}$$

$$\therefore \quad \tau = \frac{16T}{\pi d^3}$$

6.6 (b)

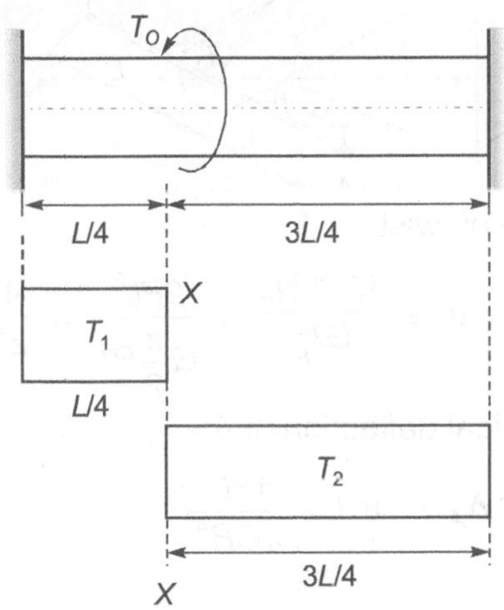

$$T_1 + T_2 = T_o \qquad \text{...(i)}$$

Angle of twist for both shafts at section X-X will be same.

Therefore,

$$\frac{T_1 L}{4GJ} = \frac{T_2 \times 3L}{4GJ}$$

$$\therefore \quad T_1 = 3T_2 \qquad \text{...(ii)}$$

Solving equations (i) and (ii)

$$T_2 = \frac{T_o}{4}, \; T_1 = \frac{3T_o}{4}$$

$$\text{Shear stress in 1} = \frac{16T_1}{\pi d^3} = \frac{16}{\pi d^3} \times \frac{3T_o}{4}$$

$$= \frac{12T_o}{\pi d^3}$$

$$\text{Shear stress in 2} = \frac{16T_2}{\pi d^3} = \frac{16}{\pi d^3} \times \frac{T_o}{4} = \frac{4T_o}{\pi d^3}$$

Hence maximum shear stress in shaft is $\frac{12T_o}{\pi d^3}$

6.7 (b)

By using torsion equation we have

$$\frac{T}{J} = \frac{\tau}{r} = \frac{G\theta}{L}$$

$$\therefore \quad \Sigma\theta = \Sigma \frac{TL}{GJ}$$

$$\therefore \quad \theta = \frac{T\left(\frac{L}{2}\right)}{G\left(\frac{\pi}{32}\right)d^4}\left[\frac{1}{2} + \frac{1}{16}\right]$$

$$\theta = \frac{TL}{G\left(\frac{\pi}{32}d^4\right)} \times \left(\frac{9}{16}\right) = \frac{18TL}{G\pi d^4}$$

$$\text{or} \quad d = \left(\frac{18TL}{\pi\theta G}\right)^{1/4}$$

6.8 Sol.

$$\tau_{max} = \frac{2T}{\pi R^3}$$

$$\frac{\tau_2}{\tau_1} = \left(\frac{R_1}{R_2}\right)^3 = (2)^3 = 8 \quad (\because T_1 = T_2 = T)$$

6.9 Sol.

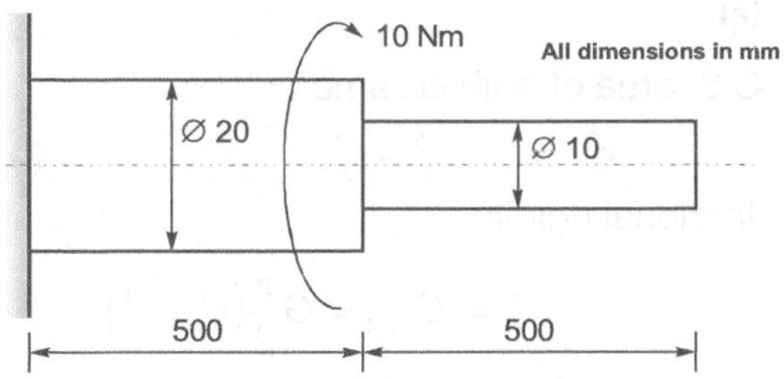

$$\theta_C = \theta_B = \frac{TL}{GJ} = \frac{10 \times 10^3 \times 500 \times 32}{\pi \times (20)^4 \times 77 \times 10^3}$$

$$= 4.1338946 \times 10^{-3} \text{ radian}$$

$$= 4.134 \times 10^{-3} \times \frac{180}{11} = 0.2368°$$

6.10 (b)

Given data; $d_o = 2d_i$; $k = \frac{d_i}{d_o} = 0.5$

$P = 20$ kW

$N = 3000$ RPM

$$\frac{2\pi NT}{60} = P$$

$$\text{or} \quad T = \frac{60 \times 20 \times 10^3}{2\pi \times 3000} = 63.66197 \text{ Nm}$$

$$\tau = \frac{16T}{\pi d_o^3\left[1-\left(\frac{d_i}{d_o}\right)^4\right]}$$

$$\Rightarrow \quad \tau_{max} = \frac{16T}{\pi d_o^3(1-k^4)}$$

$$30 = \frac{16 \times 63.66197 \times 10^3}{\pi \times d_o^3(1-0.5^4)}$$

or d_o = **22.59 mm**

6.11 Sol.

Given data: P = 30 kW, N = 700 rpm

$$\frac{2\pi NT}{60} = 30 \times 10^3 = P$$

or T = 409.25 Nm

$$\theta = \frac{TL}{GJ} \quad (\theta \text{ in radian})$$

$$\therefore \frac{1\times\pi}{180} = \frac{409.25\times10^3\times10^3\times32}{\pi\times(d_o)^4(1-0.7^4)\times80\times10^3}$$

or d_o = **44.5212 mm**

6.12 (a)

C.S. area of both is same

$$r_2^2 - r_1^2 = r_4^2 - r_3^2 \quad \text{...(i)}$$

Torsional rigidity,

$$J_1 = GI_{p1} = G\frac{\pi}{2}(r_2^4 - r_1^4)$$

$$J_2 = GI_{p2} = G\frac{\pi}{2}(r_4^4 - r_3^4)$$

$$\therefore \frac{J_2}{J_1} = \frac{r_4^4 - r_3^4}{r_2^2 - r_1^2} = \frac{(r_4^2 - r_3^2)(r_4^2 + r_3^2)}{(r_2^2 - r_1^2)(r_2^2 + r_1^2)}$$

$$\therefore \frac{J_2}{J_1} = \frac{r_4^2 + r_3^2}{r_2^2 + r_1^2} \quad \text{[From equation (i)]}$$

$$\therefore r_4 > r_2,\ r_3 > r_1$$

$$\therefore \frac{J_2}{J_1} > 1$$

6.13 (d)

$$L_s = L_H,\ T_s = T_H = T,$$

$$I_{p_s} = I_{p_H}$$

$$\therefore \frac{\pi}{32}d^4 = \frac{\pi}{32}\left[d_o^4 - d_i^4\right]$$

$$\therefore d_o > d$$

$$\therefore r_o > r$$

Angle of twist,

$$\theta = \frac{TL}{GI_p}$$

$$\therefore \theta_s = \theta_H$$

$$\tau_{max_s} = \frac{I}{I_p}\cdot r$$

$$\tau_{max_H} = \frac{I}{I_p}\cdot r_o$$

$$\therefore \tau_{max_H} > \tau_{max_s}$$

6.14 (d)

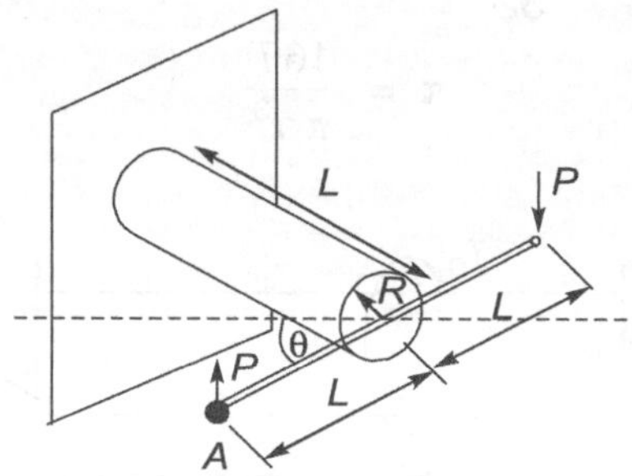

Angle of twist,

$$\theta = \frac{(2PL)\cdot L}{GI_p} = \frac{2PL^2}{G\frac{\pi}{2}R^4} = \frac{4PL^2}{\pi GR^4}$$

∴ Vertical deflection of A,

$$\Delta_A = \theta\cdot L = \frac{4PL^3}{\pi GR^4}$$

6.15 Sol.

$$T = \frac{P\times60\times10^6}{2\pi N} = \frac{40\times60\times10^6}{2\pi(500)}$$

$$= 763943.7268 \text{ N-mm}$$

$$\tau_{max} = \frac{16T}{\pi d^3} = \frac{16\times763943.7268}{\pi(40)^3}$$

= **60.792 MPa**

6.16 Sol.

$$\text{Torsional stiffness} = \frac{GI_P}{L}$$

$$= \frac{150\times10^9\times\frac{\pi}{32}[0.04^4 - 0.02^4]}{1}$$

= 35343 Nm/rad

= **35.343 kNm/rad**

6.17 (c)

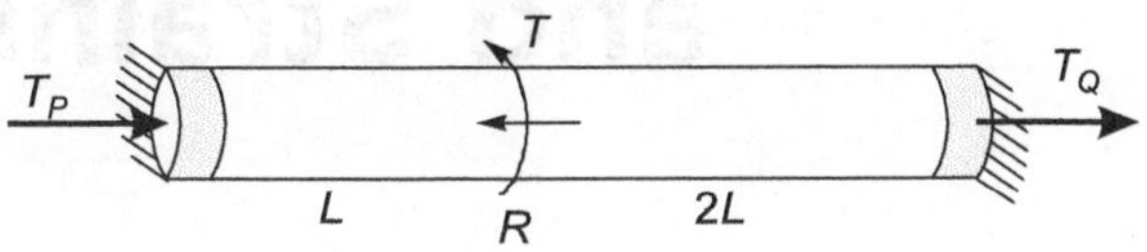

$$T_P + T_Q = T \quad ...(i)$$

$$Q_{PR} + Q_{RQ} = 0$$

$$\frac{T_P \cdot L}{GI_P} + \frac{(T_P - T)2L}{GI_P} = 0$$

$$T_P + 2T_P = 2T$$

$$T_P = \frac{2T}{3} = 100 \text{ Nm}$$

From equation (i), we get,

$$T_Q = \frac{T}{3} = 50 \text{ Nm}$$

6.18 Sol.

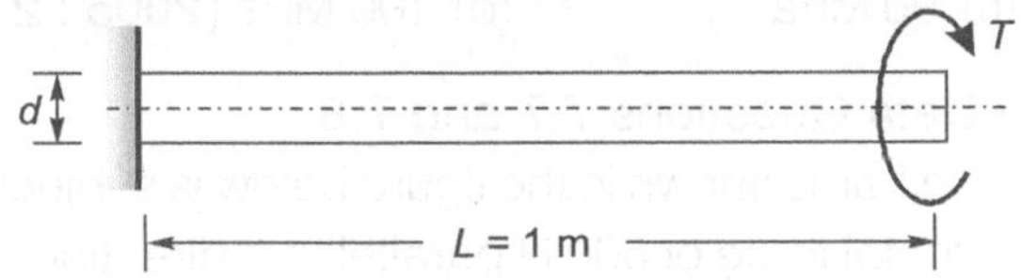

d = 10 mm; θ = 10°

L = 1 m, $\phi = P \times 10^{-3}$

From torsion eq.

$$\frac{T}{J} = \frac{\tau_{max}}{R} = \frac{G\theta}{L}$$

$$\frac{\tau_{max}}{G} = \frac{R\theta}{L}$$

$$\Rightarrow \quad \phi_{max} = \frac{R\theta}{L}$$

$$p \times 10^{-3} = \frac{5 \times 10^{\circ}}{1000} \times \frac{\pi}{180} = 0.8726 \times 10^{-3}$$

$$p = 0.8726$$

6.19 (1.50) (1.48 to 1.52)

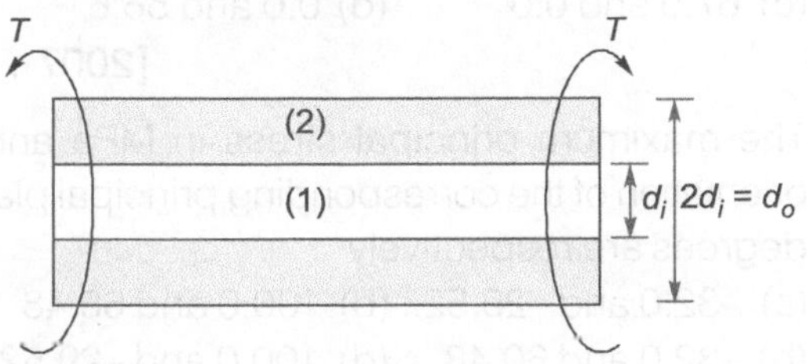

Given: $\frac{G_1}{G_2} = 3$, Let $d_i = d_1$

$$\Rightarrow \quad d_o = 2d_1 \text{ and } \frac{(\tau_{max})_1}{(\tau_{max})_2} = ?$$

As shaft are connected parallely,

Angle of twist, $\theta_1 = \theta_2$ (Compatibility equation)

$$\frac{32T_1L_1}{G_1 \times \pi d_1^4} = \frac{32T_2L_2}{G_2 \times \pi d_2^4(1-k^4)}$$

$$\frac{T_1}{T_2} = \frac{G_1}{G_2}\left(\frac{d_1}{d_2}\right)^4\left(\frac{1}{1-k^4}\right)$$

$$= 3 \times \left(\frac{1}{2}\right)^4 \times \frac{1}{1-\left(\frac{1}{2}\right)^4}$$

$$= \frac{3}{16} \times \frac{1}{1-\frac{1}{16}} = \frac{3}{16} \times \frac{16}{15}$$

$$\Rightarrow \quad \frac{T_1}{T_2} = \frac{3}{15}$$

Now, $$\frac{\tau_1}{\tau_2} = \frac{16T_1}{\pi d_1^3} \times \frac{\pi d_2^3}{16T_2}(1-k^4)$$

$$= \frac{3}{15} \times \left(\frac{2d}{d}\right)^3 \times \left(1-\left(\frac{1}{2}\right)^4\right)$$

$$= \frac{3}{15} \times 8 \times \frac{15}{16} = \frac{3}{2} = 1.5$$

6.20 (239.25) (237 to 240)

Using the equation, $\theta = \frac{TL}{GJ}$

$$\therefore T = \frac{GJ\theta}{L} = \frac{80 \times 10^3 \times \frac{\pi}{32} \times (100)^4 \times 0.5 \times \frac{\pi}{180}}{1500}$$

$$\Rightarrow \quad T = 4569261.297 \text{ N-mm}$$

$$\Rightarrow \quad T = 4569.261 \text{ N-m}$$

$$\therefore \quad P = \frac{2\pi NT}{60} = \frac{2\pi \times 500 \times 4569.261}{60}$$

$$\Rightarrow \quad P = 239245.96 \text{ W} = 239.25 \text{ kW}$$

■■■■

7 Principal Stresses and Strains

7.1 If the two principal strains at a point are 1000×10^{-6} and -600×10^{-6}, then the maximum shear strain is

(a) 800×10^{-6} (b) 500×10^{-6}
(c) 1600×10^{-6} (d) 200×10^{-6}

[1996 : 1 M]

7.2 A shaft subjected to torsion experiences a pure shear stress τ on the surface. The maximum principal stress on the surface which is at 45° to the axis will have a value

(a) $\tau \cos 45°$ (b) $2\tau \cos 45°$
(d) $\tau \cos^2 45°$ (d) $2\tau \sin 45° \cos 45°$

[2003 : 2 M]

Common Data Questions Q.7.3 and Q.7.4.

The state of stress at a point P in a two dimensional loading is such that the Mohr's circle is a point located at 175 MPa on the positive normal stress axis.

7.3 The maximum and minimum principal stresses respectively from the Mohr's circle are

(a) +175 MPa, –175 MPa
(b) +175 MPa, +175 MPa
(c) 0, –175 MPa
(d) 0, 0

[2003 : 2 M]

7.4 The directions of maximum and minimum principal stresses at the point P from the Mohr's circle are

(a) 0, 90° (b) 90°, 0
(c) 45°, 135° (d) All directions

[2003 : 2 M]

7.5 The figure shows the state of stress at a certain point in a stressed body. The magnitudes of normal stresses in the x and y directions are 100 MPa and 20 MPa respectively. The radius of Mohr's stress circle representing this state of stress is

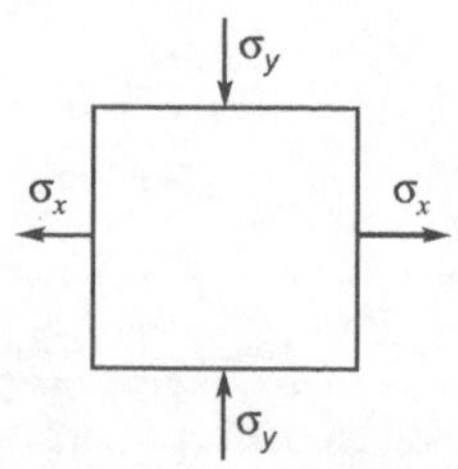

(a) 120 MPa (b) 80 MPa
(c) 60 MPa (d) 40 MPa

[2004 : 1 M]

7.6 The Mohr's circle of plane stress for a point in a body is shown. The design is to be done on the basis of the maximum shear stress theory for yielding. Then, yielding will just begin if the designer chooses a ductile material whose yield strength is

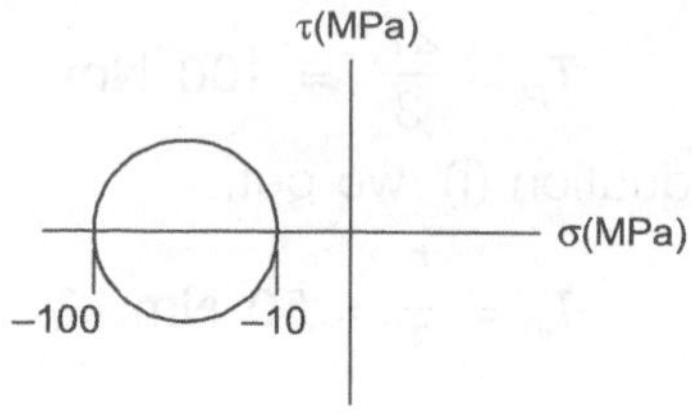

(a) 45 MPa (b) 50 MPa
(c) 90 MPa (d) 100 MPa **[2005 : 2 M]**

Linked Data Questions 7.7 and 7.8

A machine frame shown in the figure below is subjected to a horizontal force of 600 N parallel to z-direction.

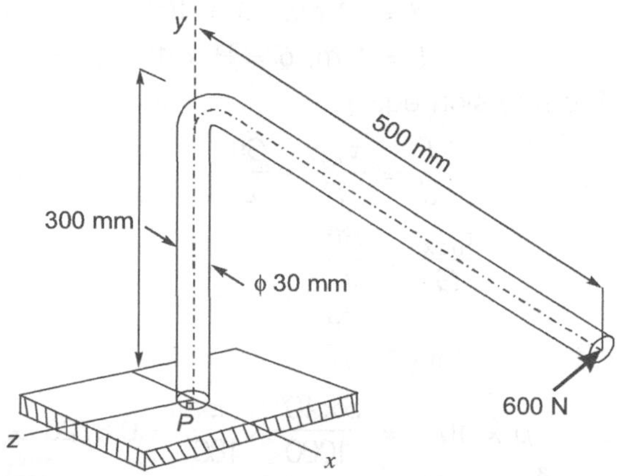

7.7 The normal and shear stresses in MPa at point P are respectively

(a) 67.9 and 56.6 (b) 56.6 and 67.9
(c) 67.9 and 0.0 (d) 0.0 and 56.6

[2007 : 2 M]

7.8 The maximum principal stress in MPa and the orientation of the corresponding principal plane in degrees are respectively

(a) –32.0 and –29.52 (b) 100.0 and 60.48
(b) –32.0 and 60.48 (d) 100.0 and –29.52

[2007 : 2 M]

7.9 A two dimensional fluid element rotates like a rigid body. At a point within the element, the pressure is 1 unit . Radius of the Mohr's circle, characterizing the state at that point, is

(a) 0.5 unit (b) 0 unit
(c) 1 unit (d) 2 units [2008 : 2 M]

7.10 A solid circular shaft of diameter 100 mm is subjected to an axial stress of 50 MPa. It is further subjected to a torque of 10 kNm. The maximum principal stress experienced on the shaft is closest to
(a) 41 MPa (b) 82 MPa
(c) 164 MPa (d) 204 MPa [2008 : 2 M]

7.11 If the principal stresses in a plane stress problem are $\sigma_1 = 100$ MPa, $\sigma_2 = 40$ MPa, the magnitude of the maximum shear stress (in MPa) will be
(a) 60 (b) 50
(c) 30 (d) 20 [2009 : 1 M]

7.12 A solid circular shaft of diameter d is subjected to a combined bending moment M and torque, T. The material property to be used for designing the shaft using the relation $\frac{16}{\pi d^3}\sqrt{M^2+T^2}$ is
(a) ultimate tensile strength (S_u)
(b) tensile yield strength (S_y)
(c) torsional yield strength (S_{sy})
(d) endurance strength (S_e) [2009 : 1 M]

7.13 The state of plane-stress at a point is given by $\sigma_x = 200$ MPa, $\sigma_y = 100$ MPa and $\tau_{xy} = 100$ MPa. The maximum shear stress (in MPa) is
(a) 111.8 (b) 150.1
(c) 180.3 (d) 223.6 [2010 : 1 M]

7.14 The state of stress at a point under plane stress condition is
$\sigma_{xx} = 40$ MPa, $\sigma_{yy} = 100$ MPa and $\tau_{xy} = 40$ MPa
The radius of Mohr's circle representing the given state of stress in MPa is
(a) 40 (b) 50
(c) 60 (d) 100 [2012 : 2 M]

7.15 The state of stress at a point is given by $\sigma_x = -6$ MPa, $\sigma_y = 4$ MPa, and $\tau_{xy} = -8$ MPa. The maximum tensile stress (in MPa) at the point is ________. [2014 : 2 M, Set-1]

7.16 In a plane stress condition, the components of stress at a point are $\sigma_x = 20$ MPa, $\sigma_y = 80$ MPa and $\tau_{xy} = 40$ MPa. The maximum shear stress (in MPa) at the point is
(a) 20 (b) 25
(c) 50 (d) 100
[2015 : 2 M, Set-2]

7.17 A shaft with a circular cross-section is subjected to pure twisting moment. The ratio of the maximum shear stress to the largest principal stress is
(a) 2.0 (b) 1.0
(c) 0.5 (d) 0
[2016 : 1 M, Set-2]

7.18 The state of stress at a point on an element is shown in figure (a). The same state of stress is shown in another coordinate system in figure (b).

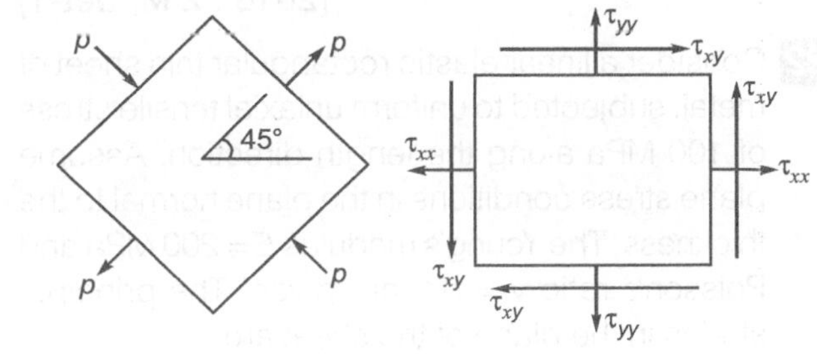

(a) (b)

The components $(\tau_{xx}, \tau_{yy}, \tau_{xy})$ are given by
(a) $\left(\frac{p}{\sqrt{2}}, -\frac{p}{\sqrt{2}}, 0\right)$ (b) $(0, 0, p)$
(c) $\left(p, -p, \frac{p}{\sqrt{2}}\right)$ (d) $\left(0, 0, \frac{p}{\sqrt{2}}\right)$
[2016 : 1 M, Set-3]

7.19 A rectangular region in a solid is in a state of plane strain. The (x, y) coordinates of the corners of the undeformed rectangle are given by $P(0, 0)$, $Q(4, 0)$, $R(4, 3)$, $S(0, 3)$. The rectangle is subjected to uniform strain $\varepsilon_{xx} = 0.001$, $\varepsilon_{yy} = 0.002$, $\gamma_{xy} = 0.003$. The deformed length of the elongated diagonal, upto three decimal places, is ______ units.
[2017 : 2 M, Set-1]

7.20 The state of stress at a point is $\sigma_x = \sigma_y = \sigma_z = \tau_{xz} = \tau_{zx} = \tau_{yz} = \tau_{zy} = 0$ and $\tau_{xy} = \tau_{yx} = 50$ MPa. The maximum normal stress (in MPa) at that point is ______.
[2017 : 1 M, Set-2]

7.21 The state of stress at a point, for a body in place stress, is shown in the figure below. If the minimum principal stress is 10 kPa, then the normal stress σ_y (in kPa) is

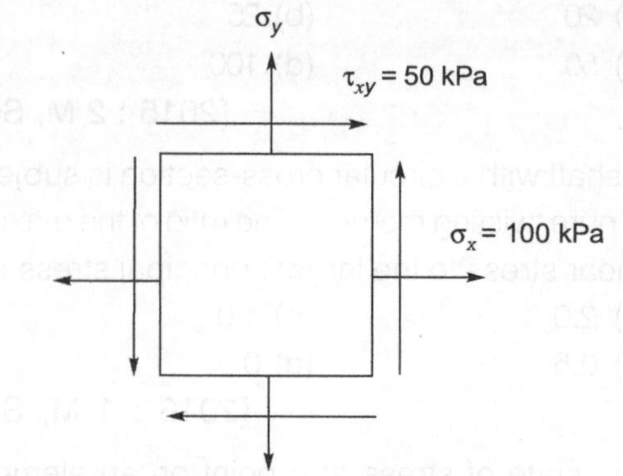

(a) 9.45 (b) 18.88
(c) 37.78 (d) 75.50

[2018 : 2 M, Set-1]

7.22 Consider a linear elastic rectangular thin sheet of metal, subjected to uniform uniaxial tensile stress of 100 MPa along the length direction. Assume plane stress conditions in the plane normal to the thickness. The Young's modulus $E = 200$ MPa and Poisson's ratio $\nu = 0.3$ are given. The principal strains in the plane of the sheet are

(a) (0.5, –0.5) (b) (0.5, –0.15)
(c) (0.35, –0.15) (d) (0.5, 0.0)

[2019 : 1 M, Set-2]

7.23 The state of stress at a point in a component is represented by a Mohr's circle of radius 100 MPa centered at 200 MPa on the normal stress axis. On a plane passing through the same point, the normal stress is 260 MPa. The magnitude of the shear stress on the same plane at the same point is _______ MPa.

[2019 : 1 M, Set-2]

7.24 The stress state at a point in a material under plane stress condition is equi-biaxial tension with a magnitude of 10 MPa. If one unit on the $\sigma - \tau$ plane is 1 MPa, the Mohr's circle representation of the state-of-stress is given by

(a) a circle with a radius of 10 units on the $\sigma - \tau$ plane
(b) a circle with a radius equal to principal stress and its center at the origin of the $\sigma - \tau$ plane
(c) a point on the τ axis at a distance of 10 units from the origin
(d) a point on the σ axis at a distance of 10 units from the origin

[2020 : 1 M, Set-1]

7.25 A structural member under loading has a uniform state of plane stress which in usual notations is given by $\sigma_x = 3P$, $\sigma_y = -2P$ and $\tau_{xy} = \sqrt{2}P$, where $P > 0$. The yield strength of the material is 350 MPa. If the member is designed using the maximum distortion energy theory, then the value of P at which yielding starts (according to the maximum distortion energy theory) is

(a) 70 MPa (b) 90 MPa
(c) 120 MPa (d) 75 MPa

[2022 : 1 M, Set-2]

7.26 A linear elastic structure under plane stress condition is subjected to two sets of loading, I and II. The resulting states of stress at a point corresponding to these two loadings are as shown in the figure below. If these two sets of loading are applied simultaneously, then the net normal component of stress σ_{xx} is _____.

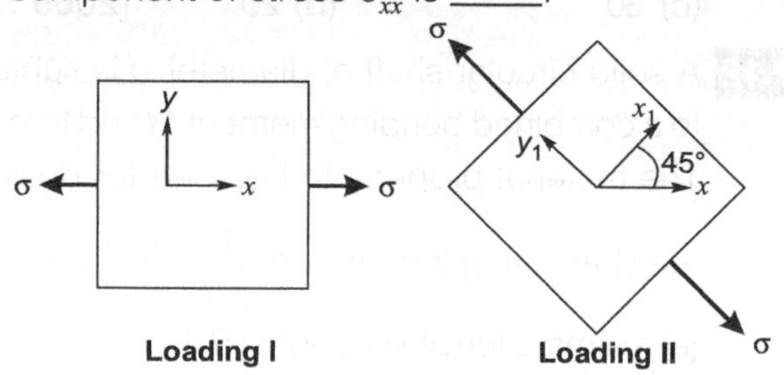

(a) $\dfrac{3\sigma}{2}$ (b) $\sigma\left(1+\dfrac{1}{\sqrt{2}}\right)$
(c) $\dfrac{\sigma}{2}$ (d) $\sigma\left(1-\dfrac{1}{\sqrt{2}}\right)$

[2022 : 2 M, Set-2]

7.27 Which of the plot(s) shown is/are valid Mohr's circle representations of a plane stress state in a material? (The center of each circle is indicated by O)

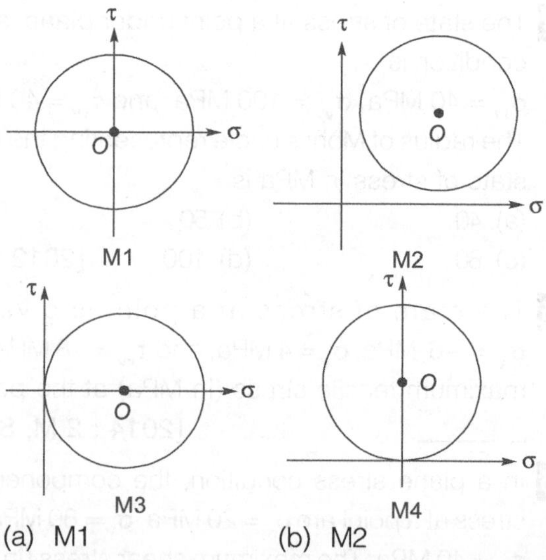

(a) M1 (b) M2
(c) M3 (d) M4

[2023 : 1 M, MSQ]

■■■■

Answers Principal Stresses and Strains

7.1	(c)	7.2	(d)	7.3	(b)	7.4	(d)	7.5	(c)	7.6	(d)	7.7	(a)	7.8	(b)
7.9	(b)	7.10	(b)	7.11	(b)	7.12	(c)	7.13	(a)	7.14	(b)	7.15	(8.43)	7.16	(c)
7.17	(b)	7.18	(b)	7.19	(5.015)	7.20	(50)	7.21	(c)	7.22	(b)	7.23	(80)	7.24	(d)
7.25	(a)	7.26	(a)	7.27	(a, c)										

Explanations Principal Stresses and Strains

7.1 (c)

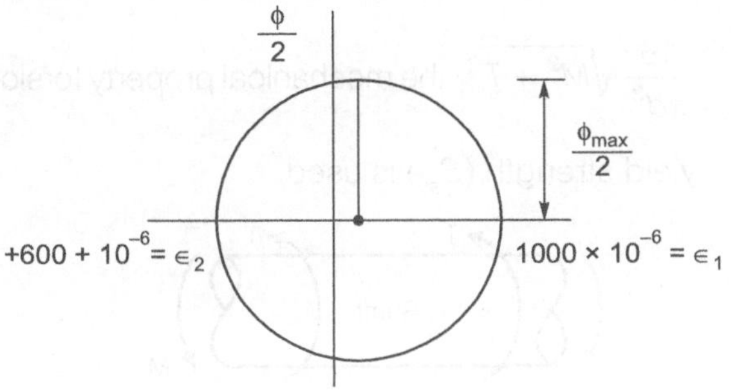

$$\frac{\phi_{max}}{2} = \text{radius} = \frac{\epsilon_1 - \epsilon_2}{2}$$

$$\phi_{max} = \epsilon_1 - \epsilon_2 = [1000 - (-600)] \times 10^{-6}$$

$$= \mathbf{1600 \times 10^{-6}}$$

7.2 (d)

$$\sigma_{1/2} = \frac{\sigma_x + \sigma_y}{2} \pm \sqrt{\left(\frac{\sigma_x - \sigma_y}{2}\right)^2 + \tau_{xy}^2}$$

For pure shear,

σ_x and $\sigma_y = 0$ and $\tau_{xy} = \tau$

$$\Rightarrow \quad \sigma_{1/2} = 0 \pm \sqrt{\tau^2} = \pm\tau$$

∴ Maximum principal stress = τ

It can be written as **2τ sin 45° cos45°**

$$= 2 \times \tau \times \frac{1}{\sqrt{2}} \times \frac{1}{\sqrt{2}} = \tau$$

7.3 (b)

Mohr's circle is a point,

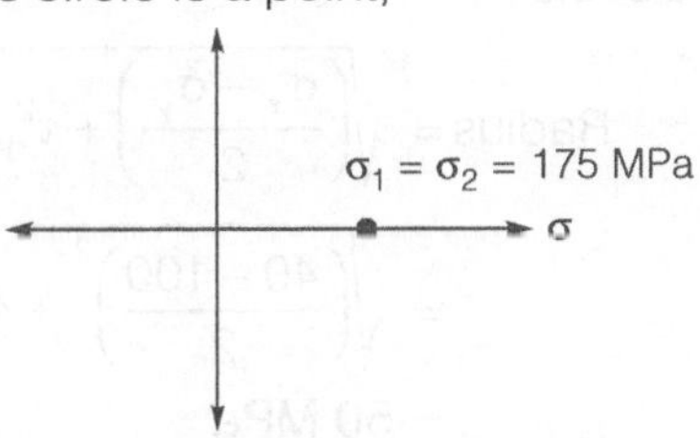

i.e., $\sigma_1 = \sigma_2 = $ **175 MPa**

and $\tau_{max} = 0$

7.4 (d)

Mohr's circle is a point, hence maximum and minimum stress at point *P* is same in all directions.

7.5 (c)

$$\text{Radius} = \sqrt{\left(\frac{\sigma_x - \sigma_y}{2}\right)^2 + \tau_{xy}}$$

Given, $\sigma_x = 100$ MPa, $\sigma_y = -20$ MPa

$\tau_{xy} = 0$

$$\therefore \text{Radius} = \sqrt{\left(\frac{100+20}{2}\right)^2 + 0} = \mathbf{60\ MPa}$$

7.6 (d)

As per maximum shear stress theory,

$$(\tau_{max})_{absolute} \le \left(\frac{S_{yt}}{2}\right)_{TT}$$

and when σ_1 and σ_2 are like in nature

$$\sigma_1 \le S_{yt}$$

$$S_{yt} = \mathbf{100\ MPa}$$

7.7 (a)

Normal stress:

$$\sigma = \frac{32M}{\pi d^3} = \frac{32 \times 600 \times 300}{\pi \times (30)^3} = 67.9 \text{ MPa}$$

Shear stress:

$$\tau = \frac{16T}{\pi d^3} = \frac{16 \times 600 \times 500}{\pi \times (30)^3} = \mathbf{56.6\ MPa}$$

7.8 (b)

The stress element

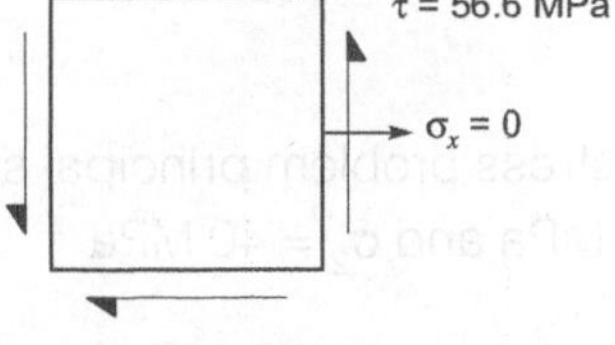

Centre of Mohr circle

$$=\frac{\sigma_x+\sigma_y}{2}=33.95\text{ MPa}$$

Radius of Mohr's circle

$$=\left[\left(\frac{\sigma_x-\sigma_y}{2}\right)^2+\tau_{xy}^2\right]^{\frac{1}{2}}=\left[\left(\frac{0-67.9}{2}\right)^2+(56.6)^2\right]^{\frac{1}{2}}$$

= 66 MPa

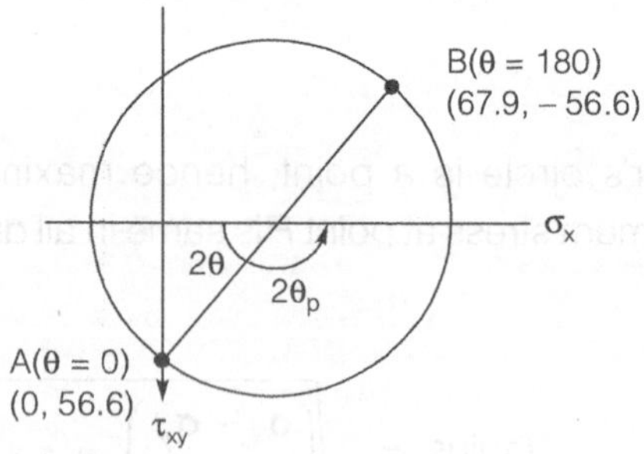

$$\text{Principal stress} = 66+33.95$$
$$= 99.95 \approx \mathbf{100\ MPa}$$

$$\sin 2\theta = \frac{56.6}{66}$$

$$\Rightarrow \quad 2\theta = 59.04$$
$$\therefore \quad 2\theta_p = 120.95$$
$$\theta_p = \mathbf{60.47°}$$

7.9 (b)

Since the fluid element will be subjected to hydrostatic loading, Mohr circle will reduce into a point on σ-axis.

∴ Radius of Mohr's circle = 0 unit

7.10 (b)

$$\frac{T}{I_P}=\frac{\tau}{r}$$

$$\frac{10\times10^6}{\frac{\pi}{32}(100)^4}=\frac{\tau}{\frac{100}{2}}$$

or $\tau = 50.95\text{ N/mm}^2$

Maximum principal stress

$$=\frac{\sigma}{2}+\sqrt{\left(\frac{\sigma}{2}\right)^2+\tau^2}=\frac{50}{2}+\sqrt{\left(\frac{50}{2}\right)^2+(50.95)^2}$$

$$=25+\sqrt{25^2+50.95^2}=\mathbf{82\ MPa}$$

7.11 (b)

In plane stress problem principal stresses, $\sigma_1 = 100$ MPa and $\sigma_2 = 40$ MPa

Magnitude of the maximum shear stress in plane stress case is given by

$$\tau_{max}=\text{Maximum of}\left[\left|\frac{\sigma_1}{2}\right|,\left|\frac{\sigma_2}{2}\right|,\left|\frac{\sigma_1-\sigma_2}{2}\right|\right]$$

$$=\text{Maximum of}\left[\left|\frac{100}{2}\right|,\left|\frac{40}{2}\right|,\left|\frac{100-40}{2}\right|\right]$$

= 50 MPa

7.12 (c)

A solid circular shaft of diameter d is subjected to a combined bending moment *M* and torque *T*. For design of shaft using the relation $\frac{16}{\pi d^3}\sqrt{M^2+T^2}$ the mechanical property torsional yield strength (S_{sy}) is used.

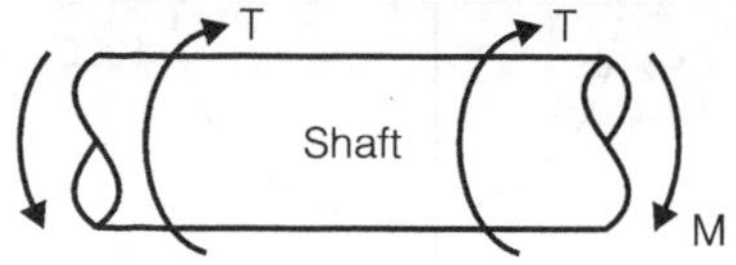

$$T_{eq}=\sqrt{M^2+T^2}$$

$$\frac{\tau}{r}=\frac{T_{eq}}{I_P}$$

$$\tau=\frac{16}{\pi d^3}\sqrt{M^2+T^2}$$

$$\le\frac{S_{sy}}{N}\ \text{(for safe design)}$$

7.13 (a)

$$\text{In plane, } \tau_{max}=\frac{1}{2}\sqrt{(\sigma_x-\sigma_y)^2+4\tau_{xy}^2}$$

$$=\frac{1}{2}\sqrt{100^2+4\times100^2}$$

$$=\mathbf{111.80\ MPa}$$

7.14 (b)

Mohr's circle

$$\text{Radius}=\sqrt{\left(\frac{\sigma_x-\sigma_y}{2}\right)+\tau_{xy}^2}$$

$$=\sqrt{\left(\frac{40-100}{2}\right)^2+40^2}$$

$$=\mathbf{50\ MPa}$$

7.15 Sol.

$$\sigma_1 = \frac{\sigma_x + \sigma_y}{2} + \sqrt{\left(\frac{\sigma_x - \sigma_y}{2}\right)^2 + \tau_{xy}^2}$$

$$= \frac{-6+4}{2} + \sqrt{\left(\frac{-6-4}{2}\right)^2 + (-8)^2}$$

$$= -1 + \sqrt{25+64}$$

$$= -1 + \sqrt{89} = \mathbf{8.43\ MPa}$$

7.16 (c)

$$\sigma_{1,2} = \frac{1}{2}\left[(\sigma_x + \sigma_y) \pm \sqrt{(\sigma_x - \sigma_y)^2 + 4\tau_{xy}^2}\right]$$

$$= \frac{1}{2}\left[100 \pm \sqrt{(60)^2 + 4 \times 40^2}\right]$$

$$\sigma_1 = 100\ \text{MPa}$$
$$\sigma_2 = 0$$

$$\tau_{max} = \frac{\sigma_1}{2} = \frac{100}{2} = \mathbf{50\ MPa}$$

7.17 (b)

For the case of pure shear

$$\sigma_1 = \tau_{max}$$
$$\sigma_2 = -\tau_{max}$$

Required ratio $= \tau_{max}/\sigma_1 = \sigma_1/\sigma_1 = 1$

7.18 (b)

The given plane is principal plane ($\sigma_1 = p$, $\sigma_2 = -p$). At 45° from principal plane, plane of maximum shear occurs. On the plane of maximum shear.

both, $\tau_{xy} = \tau_{yy} = \dfrac{\sigma_1 + \sigma_2}{2} = 0$

$$\tau_{xy} = \frac{\sigma_1 + \sigma_2}{2} = p$$

7.19 Sol.

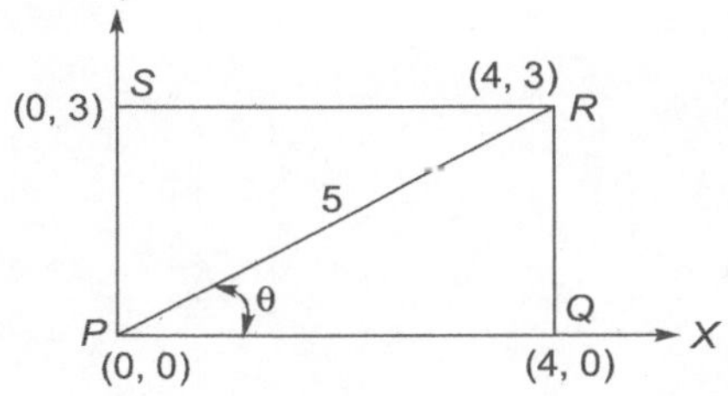

$\varepsilon_x = 0.001$
$\varepsilon_y = 0.002$
$\gamma_{xy} = 0.003$

$$\theta = \tan^{-1}\left(\frac{3}{4}\right) = 36.87°$$

$$\varepsilon_\theta = \frac{(\varepsilon_x + \varepsilon_y)}{2} + \frac{(\varepsilon_x - \varepsilon_y)}{2}\cos 2\theta + \frac{\gamma_{xy}}{2}(\sin 2\theta)$$

$$\varepsilon_\theta = 2.8 \times 10^{-3}$$

$$\frac{\Delta L_{\text{diagonal(PR)}}}{L_{\text{diagonal(PR)}}} = 2.8 \times 10^{-3}$$

$$\Delta L_{\text{diagonal(PR)}} = 0.014 \text{ units}$$

So, $L'_{\text{diagonal(PR)}} = L_{\text{diagonal(PR)}} + \Delta L_{\text{diagonal(PR)}}$
$= 5 + 0.014$
$= \mathbf{5.014\ units}$

7.20 Sol.

Given state of stress condition indicates pure shear state of stress.

For pure shear state of stress,

Max. tensile stress = Maximum comp. stress
= Maximum shear stress
$= \tau_{xy} = 50$ MPa

Hence, maximum normal stress = **50 MPa**

7.21 (c)

Given $\sigma_x = 100$ kPa, $\tau_{xy} = 50$ kPa, $\sigma_2 = 10$ kPa

Minimum principal stress,

$$\sigma_2 = \frac{\sigma_x + \sigma_y}{2} - \sqrt{\left(\frac{\sigma_x - \sigma_y}{2}\right)^2 + \tau_{xy}^2}$$

$$10 = \frac{100 + \sigma_y}{2} - \sqrt{\left(\frac{100 - \sigma_y}{2}\right)^2 + 50^2}$$

$$\therefore \quad \sqrt{\left(50 - \frac{\sigma_y}{2}\right)^2 + 50^2} = 50 + \frac{\sigma_y}{2} - 10 = 40 + \frac{\sigma_y}{2}$$

By squaring

$$2500 + \frac{\sigma_y^2}{4} - 50\sigma_y + 2500 = 1600 + \frac{\sigma_y^2}{4} + 40\sigma_y$$

$$\therefore \quad 90\,\sigma_y = 3400$$

$$\boldsymbol{\sigma_y = 37.78\ \text{kPa}}$$

7.22 (b)

Given, $E = 200$ MPa, $\nu = 0.3$

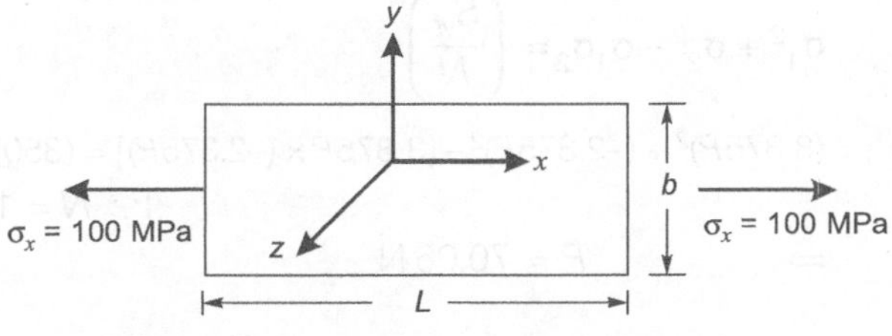

Assume plane stress condition, $\sigma_z = 0$

Given, $\sigma_y = 0$ and $\sigma_x \neq 0$[Uniaxial state of stress condition]

As no shear stress acting on x and y face,

So, $\sigma_x = \sigma_1 = 100$ MPa

$\sigma_y = \sigma_2 = 0$

$$\varepsilon_1 = \frac{\sigma_1}{E} - \nu\frac{\sigma_2}{E} = \frac{100}{200} - 0 = 0.5$$

$$\varepsilon_2 = \frac{\sigma_2}{E} - \nu\frac{\sigma_1}{E} = 0 - 0.3\left(\frac{100}{200}\right)$$

$$= \mathbf{-0.15}$$

Hence, option (b) is correct.

7.23 Sol.

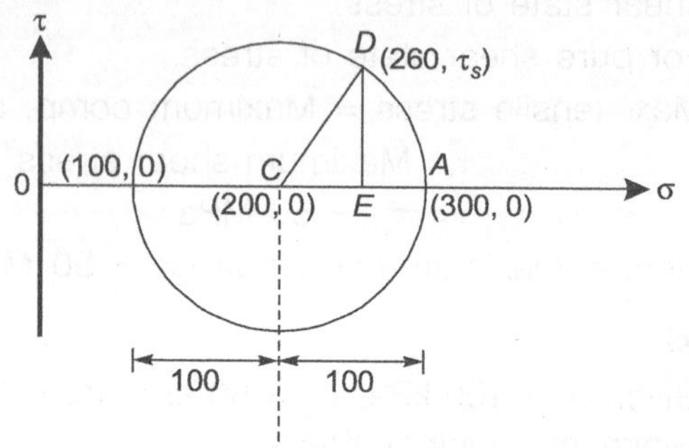

In ΔCDE, $(CD)^2 = (CE)^2 + (DE)^2$

$(100)^2 = (60)^2 + (\tau_s)^2$

or $\tau_s = \mathbf{80\ MPa}$

7.24 (d)

Given $\sigma_1 = 100 = \sigma_2 = 10$ MPa

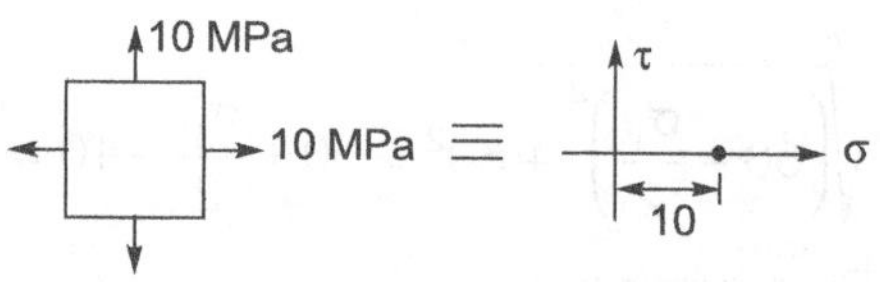

7.25 (a)

Principal stresses,

$$\sigma_{1,2} = \frac{1}{2}\left[(3P - 2P) \pm \sqrt{(3P + 2P)^2 + 4 \times \left(\sqrt{2}P\right)^2}\right]$$

$$\sigma_{1,2} = 3.375\,P, -2.375\,P$$

So by MDET

$$\sigma_1^2 + \sigma_2^2 - \sigma_1\sigma_2 = \left(\frac{S_{yt}}{N}\right)^2$$

$$(3.375P)^2 + (-2.375P)^2 - [3.375P \times (-2.375P)] = (350)^2$$

$[\because N = 1]$

$\Rightarrow$ $P = 70.06$ N

7.26 (a)

By Mohr's circle method,

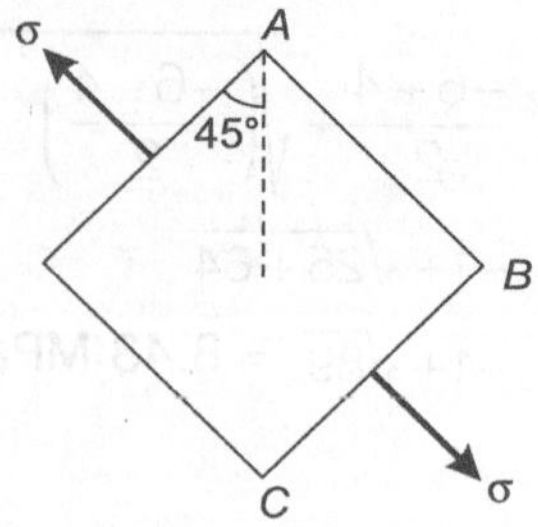

$AB = (0, 0)$

$BC = (+\sigma, 0)$

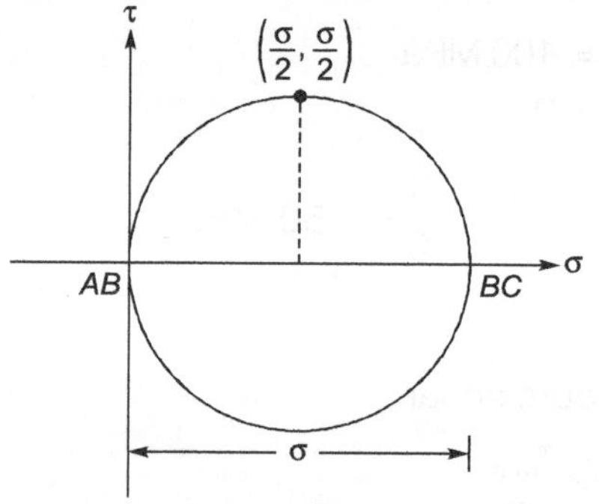

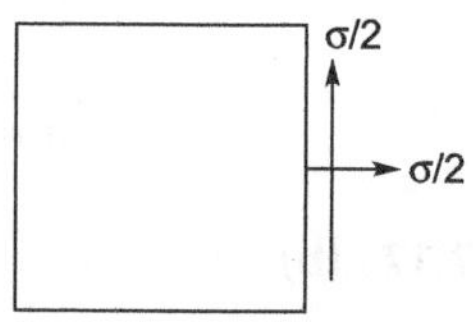

We need to find σ_{xx}, so this should be algebraic sum of σ and $\dfrac{\sigma}{2}$

$$\sigma_{xx} = \sigma + \frac{\sigma}{2} = \frac{3\sigma}{2}$$

■■■■

8 Theories of Failure

8.1 According to Von-Mises' distortion energy theory, the distortion energy under three dimensional stress state is represented by

(a) $\frac{1}{2E}\left[\sigma_1^2+\sigma_2^2+\sigma_3^2-2\mu(\sigma_1\sigma_2+\sigma_3\sigma_2+\sigma_1\sigma_3)\right]$

(b) $\frac{1-2\mu}{6E}\left[\sigma_1^2+\sigma_2^2+\sigma_3^2+2(\sigma_1\sigma_2+\sigma_3\sigma_2+\sigma_1\sigma_3)\right]$

(c) $\frac{1+\mu}{3E}\left[\sigma_1^2+\sigma_2^2+\sigma_3^2-(\sigma_1\sigma_2+\sigma_3\sigma_2+\sigma_1\sigma_3)\right]$

(d) $\frac{1}{3E}\left[\sigma_1^2+\sigma_2^2+\sigma_3^2-\mu(\sigma_1\sigma_2+\sigma_3\sigma_2+\sigma_1\sigma_3)\right]$

[2006 : 2 M]

8.2 Match the following criteria of material failure, under biaxial stresses σ_1 and σ_2 and yield stress σ_y, with their corresponding graphic representations:

Criterion	Graph
P. Maximum-normal-stress criterion	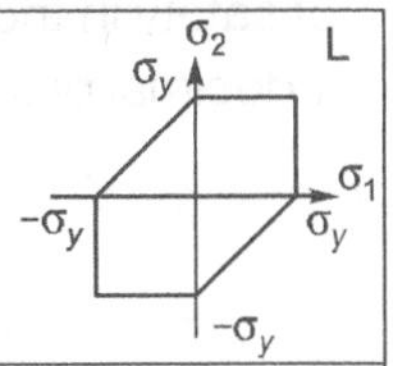
Q. Maximum-distortion-energy criterion	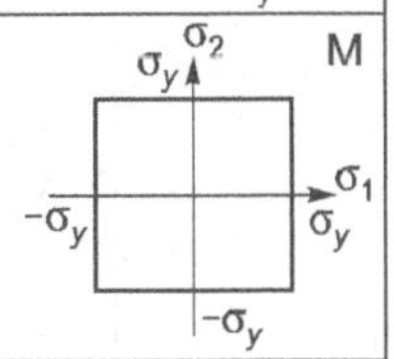
R. Maximum-shear-stress criterion	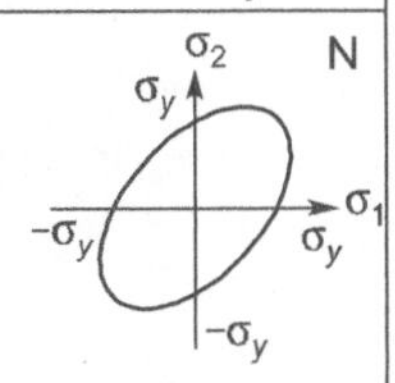

(a) P–M, Q–L, R–N (b) P–N, Q–M, R–L
(c) P–M, Q–N, R–L (d) Q–N, Q–L, R–M

[2011 : 1 M]

8.3 The homogenous state of stress for a metal part undergoing plastic deformation is

$$T=\begin{pmatrix}10 & 5 & 0\\ 5 & 20 & 0\\ 0 & 0 & -10\end{pmatrix}$$

Where the stress component values are in MPa. Using Von Mises yield criterion, the value of estimated shear yield stress, in MPa is

(a) 9.50 (b) 16.07
(c) 28.52 (d) 49.41 **[2012 : 2 M]**

8.4 The principal stresses at a point inside a solid object are σ_1 = 100 MPa, σ_2 = 100 MPa and σ_3 = 0 MPa. The yield strength of the material is 200 MPa. The factor of safety calculated using Tresca (maximum shear stress) theory is n_T and the factor of safety calculated using Von Mises (maximum distortional energy) theory is n_V. Which one of the following relations is TRUE?

(a) $n_T=\left(\frac{\sqrt{3}}{2}\right)n_V$ (b) $n_T=\left(\sqrt{3}\right)n_V$

(c) $n_T=n_V$ (d) $n_V=\left(\sqrt{3}\right)n_T$

[2016 : 2 M, Set-1]

8.5 In a metal forming operation when the material has just started yielding, the principal stresses are σ_1 = +180 MPa, σ_2 = – 100 MPa, σ_3 = 0. Following von Mises criterion, the yield stress is ______ MPa.

[2017 : 1 M, Set-1]

8.6 The principal stresses at a point in a critical section of a machine component are σ_1 = 60 MPa, σ_2 = 5 MPa and σ_3 = – 40 MPa. For the material of the component, the tensile yield strength is σ_y = 200 MPa. According to the maximum shear stress theory, the factor of safety is

(a) 1.67 (b) 2.00
(c) 3.60 (d) 4.00

[2017 : 2 M, Set-2]

8.7 At a critical point in a component, the state of stress is given as σ_{xx} = 100 MPa, σ_{yy} = 220 MPa, $\sigma_{xy}=\sigma_{yx}$ = 80 MPa and all other stress components are zero. The yield strength of the material is 468 MPa. The factor of safety on the basis of maximum shear stress theory is ______ (round off to one decimal place).

[2019 : 2 M, Set-1]

8.8 The Von-Mises stress at a point in a body subjected to forces is proportional to the square root of the

(a) distortional strain energy per unit volume
(b) dilatational strain energy per unit volume
(c) plastic strain energy per unit volume
(d) total strain energy per unit volume

[2021 : 1 M, Set-2]

8.9 Assuming the material considered in each

statement is homogeneous, isotropic, linear elastic, and the deformations are in the elastic range, which one or more of the following statement(s) is/are TRUE?

(a) A body subjected to hydrostatic pressure has no shear stress.
(b) If a long solid steel rod is subjected to tensile load, then its volume increases.
(c) Maximum shear stress theory is suitable for failure analysis of brittle materials.
(d) If a portion of a beam has zero shear force, then the corresponding portion of the elastic curve of the beam is always straight.

[2022 : 1 M, Set-1]

8.10 A shaft AC rotating at a constant speed carries a thin pulley of radius $r = 0.4$ m at the end C which drives a belt. A motor is coupled at the end A of the shaft such that it applies a torque M_z about the shaft axis without causing any bending moment. The shaft is mounted on narrow frictionless bearings at A and B where $AB = BC = L = 0.5$ m. The taut and slack side tensions of the belt are $T_1 = 300$ N and $T_2 = 100$ N, respectively. The allowable shear stress for the shaft material is 80 MPa. The self-weights of the pulley and the shaft are negligible. Use the value of π available in the on-screen virtual calculator. Neglecting shock and fatigue loading and assuming maximum shear stress theory, the minimum required shaft diameter is ______ mm (round off to 2 decimal places).

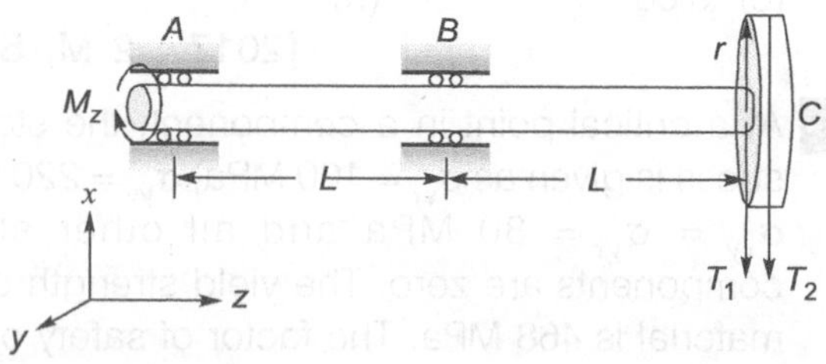

[2022 : 2 M, Set-2]

8.11 The principal stresses at a point P in a solid are 70 MPa, –70 MPa and 0. The yield stress of the material is 100 MPa. Which prediction(s) about material failure at P is/are CORRECT?

(a) Maximum normal stress theory predicts that the material fails
(b) Maximum shear stress theory predicts that the material fails
(c) Maximum normal stress theory predicts that the material does not fail
(d) Maximum shear stress theory predicts that the material does not fail

[2023 : 1 M, MSQ]

8.12 A solid massless cylindrical member of 50 mm diameter is rigidly attached at one end, and is subjected to an axial force P = 100 kN and a torque T = 600 Nm at the other end as shown. Assume that the axis of the cylinder is normal to the support. Considering distortion energy theory with allowable yield stress as 300 MPa, the factor of safety in the design is ______ (rounded off to 1 decimal place)

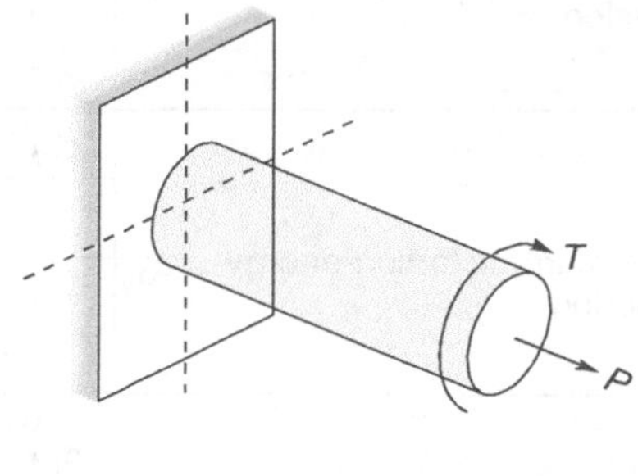

[2024 : 2 M]

Answers Theories of Failure

8.1 (c) 8.2 (c) 8.3 (b) 8.4 (c) 8.5 (245.7641) 8.6 (b) 8.7 (1.8)
8.8 (a) 8.9 (a, b) 8.10 (23.94) 8.11 (b, c) 8.12 (4.5)

Explanations Theories of Failure

8.3 (b)

Method I :

$$T = \begin{bmatrix} 10 & 5 & 0 \\ 5 & 20 & 0 \\ 0 & 0 & -10 \end{bmatrix} = \begin{bmatrix} \sigma_x & \tau_{xy} & \tau_{xz} \\ \tau_{yx} & \sigma_y & \tau_{yz} \\ \tau_{zx} & \tau_{zy} & \sigma_z \end{bmatrix}$$

$\sigma_x = 10,\ \sigma_y = 20,\ \tau_{xy} = \tau_{yx} = 5,\ \tau_{xz} = 0, \sigma_z = -10$

$$\sigma_{1,2} = \frac{1}{2}\left[(\sigma_x + \sigma_y) \pm \sqrt{(\sigma_x - \sigma_y)^2 + 4\tau_{xy}^2}\right]$$

$$\sigma_{1,2} = \frac{1}{2}\left[30 \pm \sqrt{(-10)^2 + 4 \times 5^2}\right]$$

$\sigma_1 = 22.07,\ \sigma_2 = 7.929$

Rewritting principal stresses, $\sigma_1 = 22.07$, $\sigma_2 = -10$, $\sigma_3 = 7.929$

From Von-Mises yield criterion,

$$\left[(\sigma_1 - \sigma_2)^2 + (\sigma_2 - \sigma_3)^2 + (\sigma_3 - \sigma_1)^2\right] \le 2\left(\frac{S_{yt}}{N}\right)^2$$

Take $N = 1$

$$\left[(32.07)^2 + (17.929)^2 + (14.141)^2\right] = 2(s_{yt})^2$$

$s_{yt} = 27.838$ MPa

$$s_{ys} = \frac{s_{yt}}{\sqrt{3}} = \frac{27.838}{\sqrt{3}} = 16.07 \text{ MPa}$$

Method II:

$$\begin{bmatrix} (\sigma_x - \sigma_y)^2 + (\sigma_y - \sigma_z)^2 + (\sigma_z - \sigma_x)^2 \\ +6(\tau_{xy}^2 + \tau_{yz}^2 + \tau_{xz}^2) \end{bmatrix} = 2\left(\frac{S_{yt}}{N}\right)^2$$

$$\left[(-10)^2 + (30)^2 + (-20)^2 + 6(5)^2\right] = 2(s_{yt})^2$$

$s_{yt} = 27.838$ MPa

$$s_{ys} = \frac{s_{yt}}{\sqrt{3}} = \frac{27.838}{\sqrt{3}} = 16.07 \text{ MPa}$$

8.4 (c)

$\sigma_1 = 100$ MPa, $\sigma_2 = 100$ MPa,
$\sigma_3 = 0$; $\sigma_y = 200$ MPa

According to Tresca theory

$$\tau_{max} \le \frac{\sigma_y}{2FOS} = \frac{\sigma_y}{2n_T}$$

now $\tau_{max\,abs} = \dfrac{\sigma_1 - \sigma_3}{2} = 50$ MPa

$\therefore\quad 50 = \dfrac{200}{2n_T}$

$\therefore\quad n_T = 2$

According to Von-Mises theory

$$\sigma_1^2 + \sigma_2^2 - \sigma_1\sigma_2 \le \left(\frac{\sigma_y}{n_V}\right)^2$$

$$100^2 + 100^2 - 100 \times 100 \le \left(\frac{200}{n_V}\right)^2$$

$\therefore\quad 100 = \dfrac{200}{n_V}$

or $\quad n_v = 2$

$\therefore\quad n_T = n_V$

8.5 Sol.

Safe condition for design as per von-Mises criterion,

$$\sigma_1^2 + \sigma_2^2 - \sigma_1\sigma_2 \le \left(\frac{S_{yt}}{N}\right)^2$$

When yielding occurs, $N = 1$

$$\sigma_1^2 + \sigma_2^2 - \sigma_1\sigma_2 = (S_{yt})^2$$

$$(180)^2 + (-100)^2 - (180)(-100) = (S_{yt})^2$$

$S_{yt} =$ **245.7641 MPa**

8.6 (b)

As per maximum shear stress theory, safe condition for design,

$$\text{Absolute } \tau_{max} = \frac{S_{YS}}{N} \text{ or } \frac{S_{Yt}}{2N}$$

Absolute τ_{max} = larger of

$$\left[\left|\frac{\sigma_1 - \sigma_2}{2}\right|, \left|\frac{\sigma_2 - \sigma_3}{2}\right|, \left|\frac{\sigma_3 - \sigma_1}{2}\right|\right]$$

$$= \left|\frac{\sigma_3 - \sigma_1}{2}\right|$$

[$\because$ σ_1 and σ_3 are unlike in nature]

$$= \left|\frac{-40 - 60}{2}\right| = 50 \text{ MPa}$$

Hence, $\quad 50 = \dfrac{200}{2(N)}$

$N = 2$

8.7 Sol.

As per given data

$\sigma_{xx} = 100$ MPa,

$\sigma_{yy} = 220$ MPa

$$\sigma_{xy} = \sigma_{yx} = 80 \text{ MPa}$$

Principal stresses,

$$\sigma_{1,2} = \left(\frac{\sigma_x + \sigma_y}{2}\right) \pm \sqrt{\left(\frac{\sigma_x - \sigma_y}{2}\right)^2 + \sigma_{xy}^2}$$

$$= \left(\frac{100+220}{2}\right) \pm \sqrt{\left(\frac{100-220}{2}\right)^2 + 80^2}$$

$$= 160 \pm 100$$

$$\sigma_1 = 160 + 100 = 260 \text{ MPa}$$

$$\sigma_2 = 160 - 100 = 60 \text{ MPa}$$

Both principal stresses are like in nature.

According to maximum shear stress theory.

Maximum shear stress

$$= \left[\max\left(\frac{\sigma_1 - \sigma_2}{2}\right), \left(\frac{\sigma_1}{2}\right), \left(\frac{\sigma_2}{2}\right)\right]$$

$$\tau_{max} = \frac{\sigma_1}{2}$$

$$\frac{\sigma_{yt}}{FOS \times 2} = \frac{\sigma_1}{2}$$

$$\text{FOS} = \frac{\sigma_{yt}}{\sigma_1} = \frac{468}{260} = 1.8$$

8.8 (a)

Condition for failure as per M.D.E.T.

Distortion energy per unit volume under tri-axial state of stress > Distortion energy per unit volume under uni-axial state of stress.

Hence,

$$\left(\frac{1+\mu}{6E}\right)\left[(\sigma_1-\sigma_2)^2+(\sigma_2-\sigma_3)^2+(\sigma_1-\sigma_3)^2\right] > \left(\frac{1+\mu}{3E}\right)(S_{yt})^2$$

$$(\sigma_1-\sigma_2)^2+(\sigma_2-\sigma_3)^2+(\sigma_1-\sigma_3)^2 > 2(S_{yt})^2$$

$$\sqrt{\sigma_1^2+\sigma_2^2+\sigma_3^2+\sigma_1\sigma_2-\sigma_2\sigma_3-\sigma_1\sigma_3} > (S_{yt})$$

(Von Mises effective stress)

S_{yt} = Von Mises effective stress is defined as the uni-axial yield stress that would create same distortion energy created by the tri-axial state of stress.

8.9 (a, b)

Under hydrostatic stress state condition, Mohr's circle will become a point. Every plane will become a principal plane and on all the principal planes, shear stress is zero. So, the statement is correct.

Under axial tensile load, volume of rod increases because longitudinal and lateral strains are tensile and compressive in nature respectively and longitudinal strain is more than lateral strain. So this statement is correct.

Maximum shear stress theory is more suitable for ductile materials, not for brittle materials. For brittle materials, the maximum principle stress theory is suitable. So, this statement is incorrect.

When shear force and bending moment both are zero, then elastic curve will be a straight line. When shear force is zero and bending moment has a constant non-zero value, then elastic curve will be a circular arc. Hence, the last statement is not correct.

8.10 (23.94) (23.60 to 24.20)

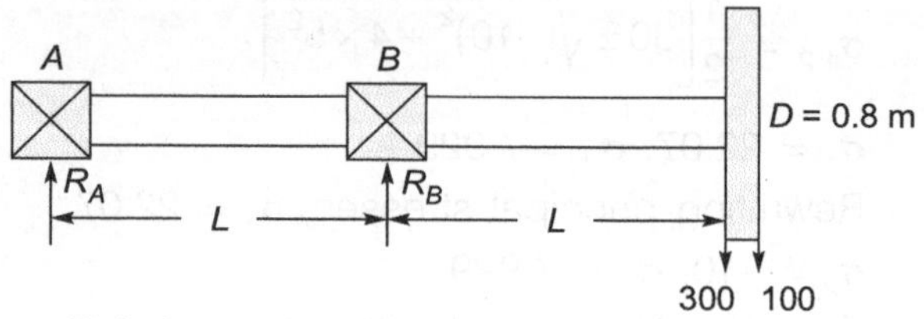

Twisting moment,

$$T = (T_1 - T_2)r = (300 - 100) \times 0.4$$

$$= 80 \text{ Nm} = 80 \times 10^3 \text{ N.mm}$$

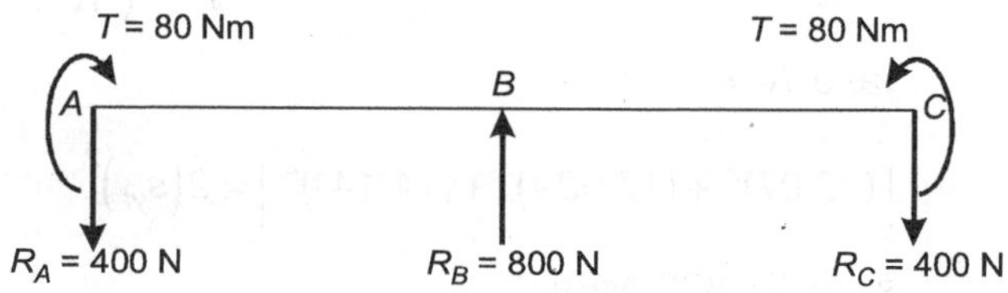

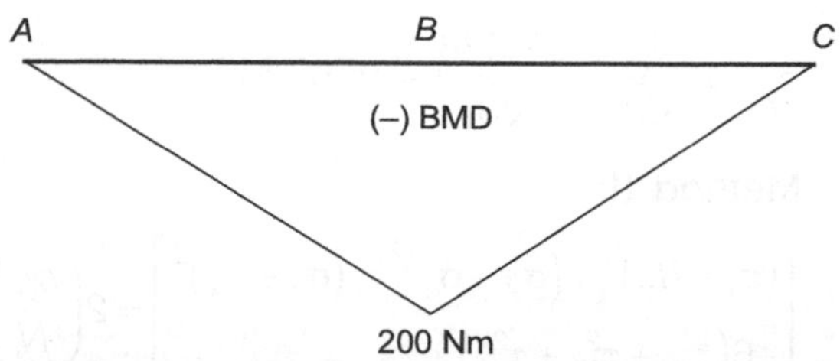

***Note** : As given in the question the bending moment due to motor is to be neglected but the bending moment due to reactions is to be considered.*

$$\Sigma M_A = 0$$

$$\Rightarrow \quad R_B \times (L) = 400 \times 2L$$

$$\Rightarrow \quad R_B = 800 \text{ N} (\uparrow)$$

$$\Sigma V = 0$$

$$\Rightarrow \quad R_A = 400 (\uparrow)$$

As per M.S.S.T.,

$$T_e \le \left[T_R = \frac{\pi}{16} d^3 \tau_{per}\right]$$

$$\sqrt{(k_b M)^2 + (k_t T)^2} \le \left(\frac{\pi}{16}\right)(d^3)(80 \times 10^6)$$

$$\sqrt{(1\times 200)^2+(1\times 80)^2} \le \frac{\pi}{16}\times d^3(80\times 10^6)$$

$$215.407 \ge \left(\frac{\pi}{16}\right)(d^3)(80\times 10^6)$$

$$d \ge \sqrt[3]{\frac{16\times 215.407}{\pi\times 180\times 10^6}}$$

$$\ge 0.023935\text{ m} \ge 23.935\text{ mm}$$

So, minimum required shaft diameter will be 23.94 mm.

8.11 (b, c)

Given, $\sigma_1 = 70$ MPa. $\sigma_2 = -70$ MPa

If, $S_{yt} = 100$ MPa

Maximum shear stress,

$$\tau_{max} = 100\text{ MPa}$$

As per maximum normal stress theory,

$$\sigma_1 < S_{yt}$$

Since this condition is satisfied for the given material, the maximum normal stress theory predicts that the material does not fails.

As per maximum shear stress theory,

$$\tau_{max} < S_{ys}\left(\frac{S_{yt}}{2}\right)$$

Since thus condition is not satisfied, the maximum shear stress theory predicts that the material fail.

8.12 (4.5) (4.3 to 4.7)

Given : $D = 50$ mm; $S_{yt} = 300$ MPa

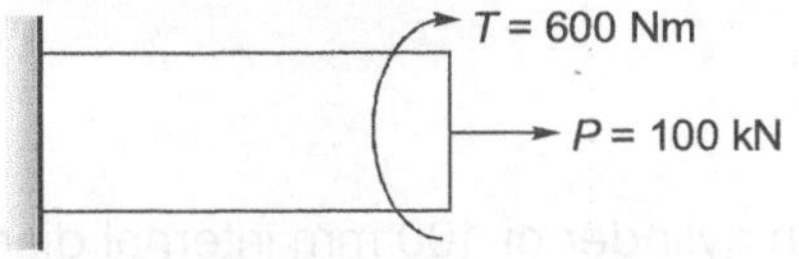

Axial stress due to P,

$$\sigma_x = \sigma_a = \frac{P}{A} = \frac{100\times 10^3}{\left(\frac{\pi\times 50^2}{4}\right)} = 50.93\text{ MPa},\ \sigma_y = 0$$

Shear stress due to T,

$$\tau_{xy} = (\tau_t)_{max} = \frac{16\times T}{\pi D^3} = \frac{16\times 600\times 10^3}{\pi\times 50^3} = 24.446\text{ MPa}$$

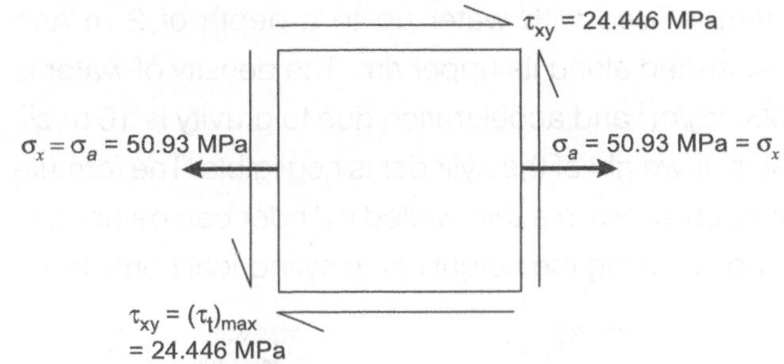

From distortion energy theory,

$$(\sigma_t)_{per} = \sqrt{\sigma_x^2+3\tau_{xy}^2}$$

$$\frac{300}{N} = \sqrt{50.93^2+3\times 24.446^2}$$

$$\left[\because (\sigma_t)_{per} = \frac{S_{yt}}{N}\right]$$

$$N = \frac{300}{66.232}$$

$$N = 4.529 \simeq 4.5$$

9 Thick and Thin Cylinders and Spheres

9.1 A thin cylinder of 100 mm internal diameter and 5 mm thickness is subjected to an internal pressure of 10 MPa and a torque of 2000 Nm. Calculate the magnitudes of the principal stresses (in MPa)

(a) 109.8, 45.2 (b) 109.8, 40.2
(c) 109.8, 31 (d) 109.8, 50 **[1996 : 2 M]**

Linked answer questions Q.9.2 and Q.9.3

A cylindrical container of radius $R = 1$ m, wall thickness 1 mm is filled with water up to a depth of 2 m and suspended along its upper rim. The density of water is 1000 kg/m^3 and acceleration due to gravity is 10 m/s^2. The self-weight of the cylinder is negligible. The formula for hoop stress in a thin-walled cylinder can be used at all points along the height of the cylindrical container.

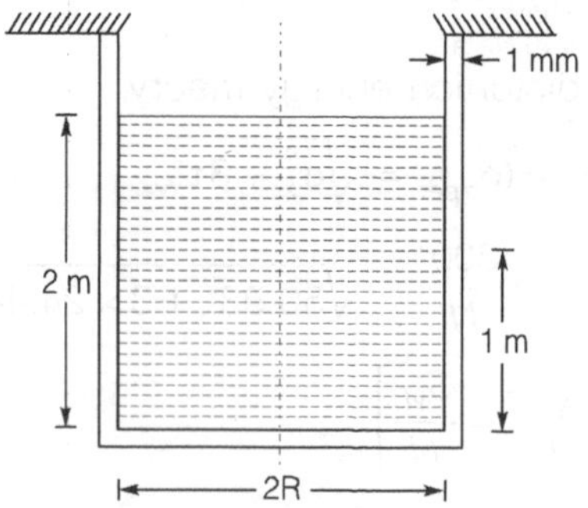

9.2 The axial and circumferential stress (σ_a, σ_c) experienced by the cylinder wall at mid-depth (1 m as shown) are

(a) (10,10) MPa (b) (5, 10) MPa
(c) (10, 5) MPa (d) (5, 5) MPa **[2008 : 2 M]**

9.3 If the Young's modulus and Poisson's ratio of the container material are 100 GPa and 0.3, respectively, the axial strain in the cylinder wall at mid-depth is

(a) 2×10^{-5} (b) 6×10^{-5}
(c) 7×10^{-5} (d) 1.2×10^{-4} **[2008 : 2 M]**

9.4 A thin cylinder of inner radius 500 mm and thickness 10 mm subjected to an internal pressure of 5 MPa. The average circumferential (hoop) stress in MPa is

(a) 100 (b) 250
(c) 500 (d) 1000 **[2011 : 1 M]**

9.5 A thin walled spherical shell is subjected to an internal pressure. If the radius of the shell is increased by 1% and the thickness is reduced by 1%, with the internal pressure remaining the same, the percentage change in the circumferential (hoop) stress is

(a) 0 (b) 1
(c) 1.08 (d) 2.02 **[2012 : 1 M]**

9.6 A long thin walled cylindrical shell, closed at both the ends, is subjected to an internal pressure. The ratio of the hoop stress (circumferential stress) to longitudinal stress developed in the shell is

(a) 0.5 (b) 1.0
(c) 2.0 (d) 4.0 **[2013 : 1 M]**

9.7 A thin gas cylinder with an internal radius of 100 mm is subject to an internal pressure of 10 MPa. The maximum permissible working stress is restricted to 100 MPa. The minimum cylinder wall thickness (in mm) for safe design must be ______

[2014 : 2 M, Set-4]

9.8 A gas is stored in a cylindrical tank of inner radius 7 m and wall thickness 50 mm. The gauge pressure of the gas is 2 MPa. The maximum shear stress (in MPa) in the wall is

(a) 35 (b) 70
(c) 140 (d) 280

[2015 : 1 M, Set-2]

9.9 A cylindrical tank with closed ends is filled with compressed air at a pressure of 500 kPa. The inner radius of the tank is 2 m, and it has wall thickness of 10 mm. The magnitude of maximum in-plane shear stress (in MPa) is _______.

[2015 : 1 M, Set-3]

9.10 A thin cylindrical pressure vessel with closed-ends is subjected to internal pressure. The ratio of circumferential (hoop) stress to the longitudinal stress is

(a) 0.25 (b) 0.50
(c) 1.0 (d) 2.0 **[2016: 1 M, Set-2]**

9.11 A thin-walled cylindrical can with rigid end caps has a mean radius R = 100 mm and a wall thickness of t = 5 mm. The can is pressurized and an additional tensile stress of 50 MPa is imposed along the axial direction as shown in the figure. Assume that the state of stress in the wall is uniform along its length. If the magnitudes of axial and circumferential components of stress in the can are equal, the pressure (in MPa) inside the can is ________ (correct to two decimal places).

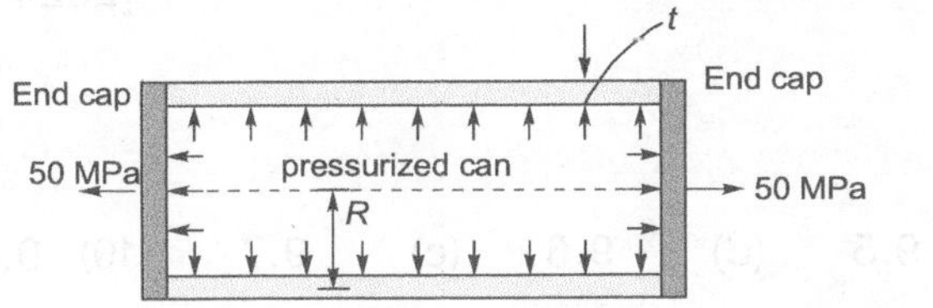

[2018 : 2 M, Set-2]

9.12 Consider two concentric circular cylinders of different materials M and N in contact with each other at $r = b$, as shown below. The interface at $r = b$ is frictionless. The composite cylinder system is subjected to internal pressure P. Let $\left(u_r^M, u_\theta^M\right)$ and $\left(\sigma_{rr}^M, \sigma_{\theta\theta}^M\right)$ denote the radial and tangential displacement and stress components, respectively, in material M. Similarly, $\left(u_r^N, u_\theta^N\right)$ and $\left(\sigma_{rr}^N, \sigma_{\theta\theta}^N\right)$ denote the radial and tangential displacement and stress components, respectively, in material N. The boundary conditions that need to be satisfied at the frictionless interface between the two cylinders is:

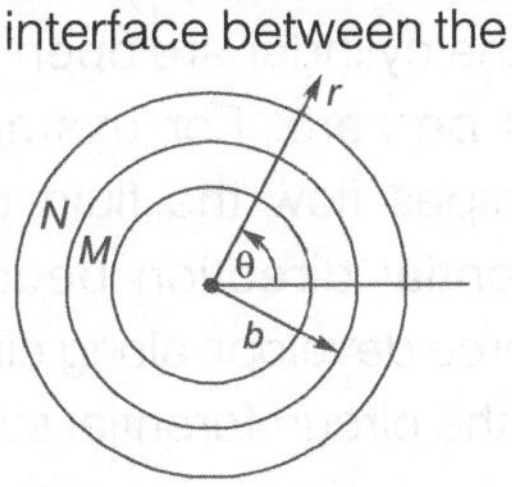

(a) $u_\theta^M = u_\theta^N$ and $\sigma_{\theta\theta}^M = \sigma_{\theta\theta}^N$ only
(b) $u_r^M = u_r^N$ and $\sigma_{rr}^M = \sigma_{rr}^N$ only
(c) $\sigma_{rr}^M = \sigma_{rr}^N$ and $\sigma_{\theta\theta}^M = \sigma_{\theta\theta}^N$ only
(d) $u_r^M = u_r^N$ and $\sigma_{rr}^M = \sigma_{rr}^N$ and $u_\theta^M = u_\theta^N$ and $\sigma_{\theta\theta}^M = \sigma_{\theta\theta}^N$

[2019 : 2 M, Set-2]

9.13 A thin-walled cylinder of radius r and thickness t is open at both ends, and fits snugly between two rigid walls under ambient conditions, as shown in the figure

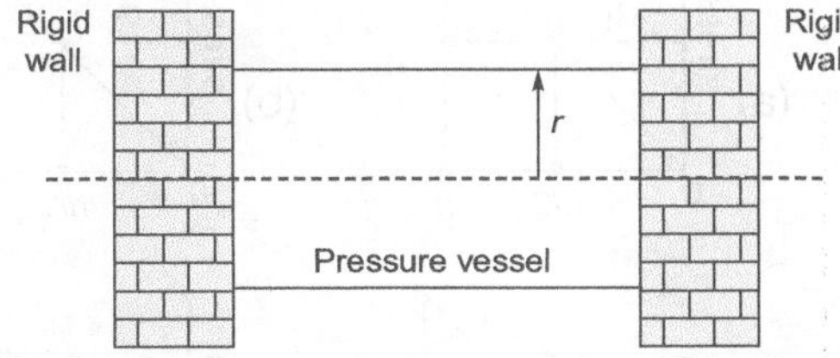

The material of the cylinder has Young's modulus E, Poisson's ratio ν, and coefficient of thermal expansion α. What is the minimum rise in temperature ΔT of the cylinder (assume uniform cylinder temperature with no buckling of the cylinder) required to prevent gas leakage if the cylinder has to store the gas at an internal pressure of p above the atmosphere?

(a) $\Delta T = \left(\nu - \frac{1}{4}\right)\frac{pr}{\alpha t E}$ (b) $\Delta T = \left(\nu + \frac{1}{2}\right)\frac{pr}{\alpha t E}$
(c) $\Delta T = \frac{\nu pr}{\alpha t E}$ (d) $\Delta T = \frac{3\nu pr}{2\alpha t E}$

[2020 : 2 M, Set-2]

9.14 A thin-walled cylindrical pressure vessel has mean wall thickness of t and nominal radius of r. The Poisson's ratio of the wall material is $\frac{1}{3}$. When it was subjected to some internal pressure, its nominal perimeter in the cylindrical portion increased by 0.1% and the corresponding wall thickness became $\bar{t}$. The corresponding change in the wall thickness of the cylindrical portion, i.e. $100 \times \frac{(\bar{t} - t)}{t}$, is ________ % (round off to 3 decimal places).

[2022 : 2 M, Set-1]

9.15 The figure shows a thin-walled open-top cylindrical vessel of radius r and wall thickness t. The vessel is held along the brim and contains a constant-density liquid to height h from the base. Neglect atmospheric pressure, the weight of the vessel and bending stresses in the vessel walls.

Which one of the plots depicts qualitatively correct dependence of the magnitude of axial wall stress (σ_1) and circumferential wall stress (σ_2) on y?

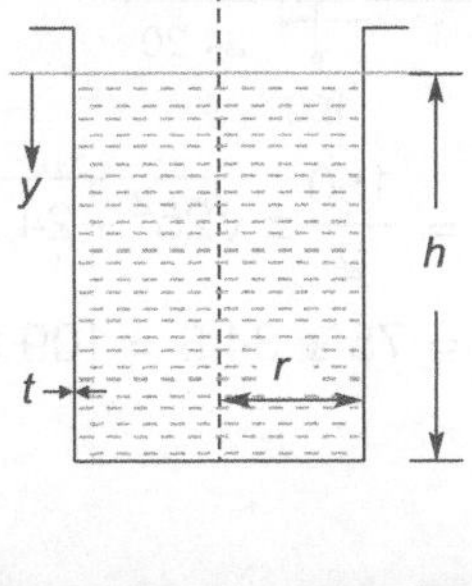

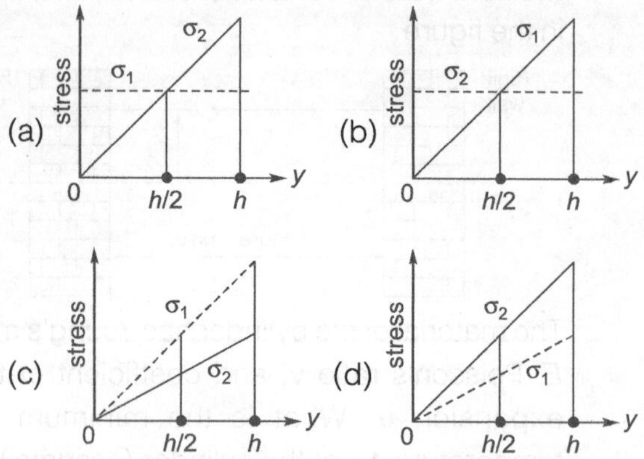

[2023 : 2 M]

9.16 The figure shows a thin cylindrical pressure vessel constructed by welding plates together along a line that makes an angle $\alpha = 60°$ with the horizontal. The closed vessel has a wall thickness of 10 mm and diameter of 2 m. When subjected to an internal pressure of 200 kPa, the magnitude of the normal stress acting on the weld is ________ MPa. (rounded off to 1 decimal place)

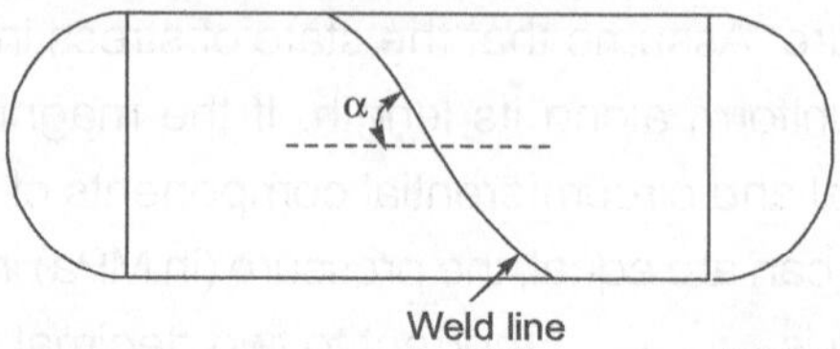

[2024 : 2 M]

Answers Thick and Thin Cylinders and Spheres

9.1	(b)	9.2	(a)	9.3	(c)	9.4	(b)	9.5	(d)	9.6	(c)	9.7	(10)	9.8	(c)
9.9	(25)	9.10	(d)	9.11	(5.128)	9.12	(b)	9.13	(c)	9.14	(–0.06)	9.15	(a)		
9.16	(12.5)														

Explanations Thick and Thin Cylinders and Spheres

9.1 (b)

Given: $D_i = 100$ mm, $D_o = 110$ mm, $D = 105$ mm

Thin cylinder

$$J = \pi D t\left(\frac{D^2}{4}\right) = \pi \times 105 \times 5 \times (52.5)^2$$

$$= 4545982 \text{ mm}^4 = 4.545 \times 10^6 \text{ mm}^4$$

$$\tau_{max} = \frac{T \times r_{max}}{J}$$

where $r_{max} = \dfrac{D_o}{2} = 55$ mm

$$\therefore \quad \tau_{max} = \frac{2000 \times 10^3 \times 55}{4.545 \times 10^6} = 24.20 \text{ MPa}$$

$$\sigma_H = \sigma_c = \frac{pd_i}{2t} = 100 \text{ MPa}$$

$$\sigma_L = 50 \text{ MPa}$$

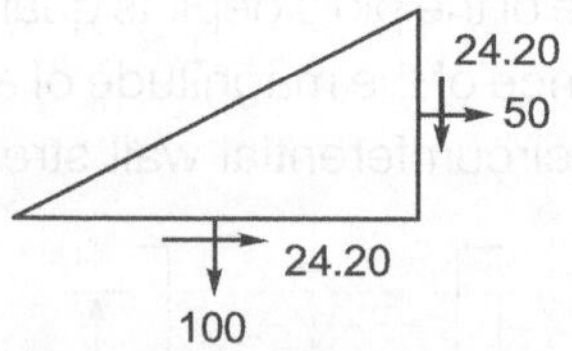

$$\sigma_1, \sigma_2 = \frac{150}{2} \pm \sqrt{25^2 + 24.20^2}$$

$$= 75 \pm 34.8 = \mathbf{109.8,\ 40.2\ MPa}$$

9.2 (a)

Note: Why $\dfrac{pd}{2t}$ and $\dfrac{pd}{4t}$ not applicable for the given case?

- In case of closed end cylinder when pressure acts along all direction the resisting force developed in longitudinal as well as circumferential direction that's why the both stresses come into picture.
- This is only applicable for thin cylinder with internal gas pressure and both ends closed.
- If ends of the cylinder are open then longitudinal stress will be zero. For example in pipe flow case. In pipes flow the fluid pressure acts in circumferential direction because of this the resisting force develops along circumference and that's why the circumferential stresses come into picture.

In our problem

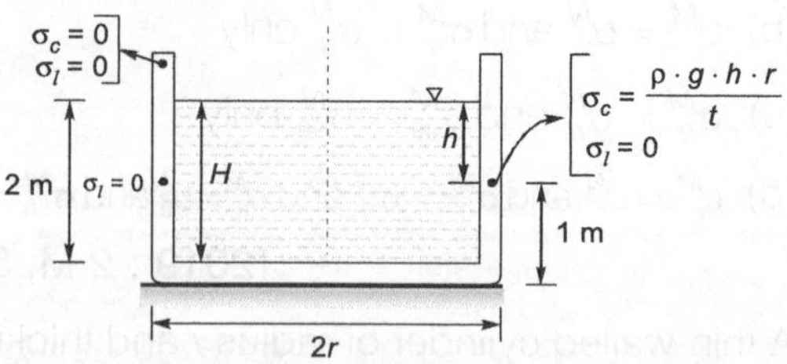

- When this container which is hold at the bottom side then longitudinal stress will not exist at any point along longitudinal direction. Because the weight of water which is hold by bottom support and the net resisting force along longitudinal direction is zero. But circumferential stress comes because of fluid pressure which is dependent from height of water column.

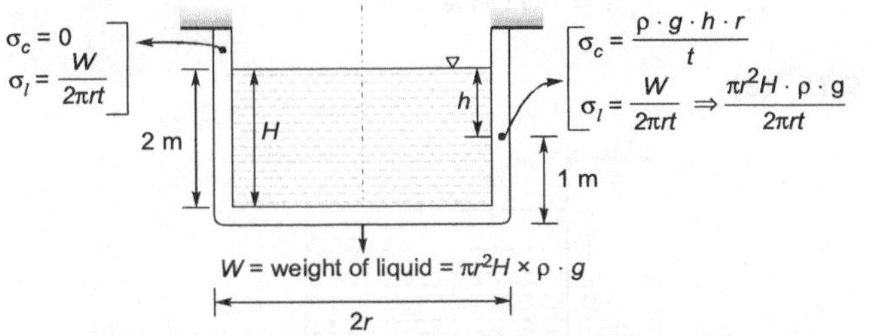

- When this container which is hold at the top side then longitudinal stress comes into picture because the weight of water develops the resisting force along the longitudinal direction which is equal in magnitude at every point along longitudinal direction and circumferential stresses varying point to point if the height of water column varying.
- So this is the special case in thin cylinder.

Special Case: When thin cylinder containing liquid with open end.

As per given data

$$\sigma_l = \frac{\rho g H r}{2t} = \frac{1 \times 2 \times 1000 \times 10}{2 \times 0.001} = 10 \text{ MPa}$$

$$\sigma_c = \frac{\rho g h r}{t} = \frac{1000 \times 10 \times 1 \times 1}{0.001} = 10 \text{ MPa}$$

9.3 (c)

$\because$ Axial strain is given by, $\frac{\sigma_l}{E} - \mu \frac{\sigma_c}{E}$

Axial strain at mid-depth

$$= \frac{10 - 0.3 \times 10}{100 \times 10^3} = 7 \times 10^{-5}$$

9.4 (b)

Circumferential or hoop stress, $\sigma_h = \frac{pd}{2t}$

where p = 5 MPa, d = 1000 mm, t = 10 mm

$$\therefore \quad \sigma_H = \frac{5 \times 1000}{2 \times 10} = 250 \text{ MPa}$$

9.5 (d)

Hoop stress $= \frac{pd}{4t}$

where, p = Internal pressure; d = Diameter of shell; t = Thickness of shell

Now $\sigma' = \frac{pd'}{4t'}$; $d' = 1.01d$ and $t' = 0.99t$

$$\therefore \quad \sigma' = \frac{p \times 1.01d}{4 \times 0.99t} = 1.0202\sigma$$

$$\% \text{ change} = \frac{\sigma' - \sigma}{\sigma} \times 100$$

$$= \frac{(1.0202 - 1)\sigma}{\sigma} \times 100 = \mathbf{2.02\%}$$

9.6 (c)

Hoop stress: $\sigma_h = \frac{pd}{2t}$

Longitudinal stress: $\sigma_L = \frac{pd}{4t}$

$$\therefore \quad \frac{\sigma_h}{\sigma_L} = 2$$

9.7 Sol.

$$\sigma_{working} = \frac{pd}{2t} \quad (d = 2r = 200 \text{ mm})$$

$$t = \frac{pd}{2\sigma_w} = \frac{10 \times 200}{2 \times 100} = \mathbf{10 \text{ mm}}$$

9.8 (c)

Maximum shear stress in the wall

$$= \frac{\sigma_1}{2} = \frac{pd}{4t} = \frac{2 \times 14 \times 1000}{4 \times 50} = \mathbf{140 \text{ MPa}}$$

9.9 Sol.

Maximum in plane shear stress (in MPa)

$$\tau = \frac{\sigma_1 - \sigma_2}{2} = \frac{\frac{pd}{2t} - \frac{pd}{4t}}{2} = \frac{500 \times 10^{-3} \times 4000}{8 \times 10}$$

$= \mathbf{25 \text{ MPa}}$

9.10 (d)

For thin cylinder: Circumferential stress,

$$\sigma_h = \frac{pd}{2t}$$

Longitudinal stress,

$$\sigma_L = \frac{pd}{4t} \quad \therefore \quad \frac{\sigma_h}{\sigma_L} = 2$$

9.11 Sol.

NOTE: This answer is obtained by considering inner radius.

R_m = 100 mm, t = 5 mm,

$R_i = R_m - \frac{t}{2}$ = 97.5 mm, D_i = 195 mm

σ_h = Hoop stress

$$= \frac{pD_i}{2t} = \frac{(p)(195)}{2 \times 5} = 19.5p \text{ MPa}$$

σ_L = Longitudinal or axial stress

$$= \frac{\sigma_h}{2} = 9.75p \text{ MPa}$$

σ_a = Axial stress due to axial load, σ_a = 50 MPa

Net axial stress = $\sigma_L + \sigma_a = 9.75p + 50$

As per given data:

$$\sigma_h = (\sigma_{axial})_{net}$$
$$19.5p = 9.75p + 50$$
$$9.75p = 50$$
$$p = 5.128 \text{ MPa}$$

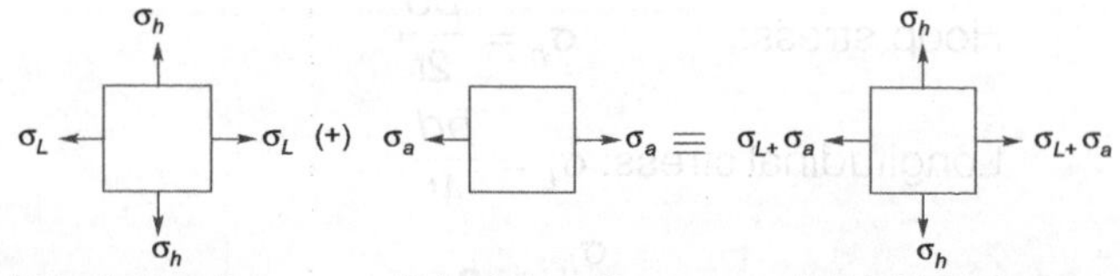

State of stress due to pressure — State of stress due to axial load — Actual state of stress

9.12 (b)

Due to fluid pressure both the cylinder will remain in contact, so radial displacement and radial stress will be same at interface. But due to frictionless surface both the cylinder can slip tangentially, therefore tangential displacement and tangential stress will be different.

9.13 (c)

Since cylinder is open at both ends.

Longitudinal stress, $\sigma_L = 0$

For no leakage, $\varepsilon_{L_{P_r}} = \varepsilon_{L_{temp}}$

$$\frac{\nu\sigma_h}{E} = \alpha\Delta T; \quad \frac{\nu Pr}{tE} = \alpha\Delta T; \quad \Delta T = \frac{\nu pr}{\alpha tE}$$

9.14 (-0.06) (-0.062 to -0.058)

Given: ϵ_H = Hoop strain, $\frac{\delta D}{D} = \frac{1}{\epsilon}(\sigma_H - \mu\sigma_L)$

As $\sigma_H = \frac{Pd}{2t}$, $\sigma_L = \frac{\sigma_H}{2}$

Putting these values of stress in the hoop strain equation, we get

$$\epsilon_H = \frac{PD}{4tE}[2-\mu] = \frac{0.1}{100} \quad \text{... (i)}$$

$\epsilon_3 = \epsilon_r$ = Radial strain

$$= \frac{\delta t}{t} = \frac{1}{E}[\sigma_3 - \mu(\sigma_1 + \sigma_2)]$$

$$\epsilon_r = \frac{\delta t}{t} = \frac{1}{E}[0 - \mu(\sigma_1 + \sigma_2)]$$

$$\epsilon_r = \frac{\delta t}{t} = \frac{1}{E}\left[-\mu\frac{3\sigma_1}{2}\right] = -3\mu\left[\frac{PD}{4tE}\right] \quad \text{...(ii)}$$

From equation (i),

$$\frac{PD}{4tE} = \frac{0.001}{2-\mu} = \frac{0.001}{2-\frac{1}{3}}$$

Putting the value of $\frac{PD}{4tE}$ in equation (ii)

$$\Rightarrow \quad \frac{\delta t}{t} = -3 \times \frac{1}{3} \times \frac{0.001}{2-\frac{1}{3}} = -0.0006$$

$$\therefore \quad \frac{\delta t}{t} = -0.06\%$$

9.15 (a)

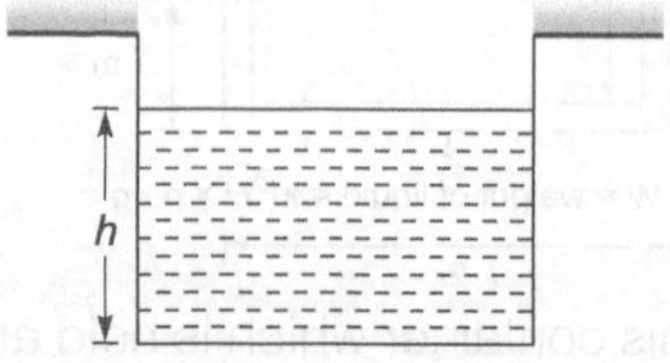

As we move from the free surface towards bottom, the pressure variation can be written as:

$$P = \rho g h$$

Now, $\sigma_2 = \frac{PD}{2t} = \rho g h\left(\frac{D}{2t}\right)$

$\therefore \quad \sigma_2 \propto h$

Longitudinal stress will be developed because of the total amount of fluid and since this amount is constant, the longitudinal stress will be constant.

9.16 (12.5) (12.3 to 12.7)

Given: P = 0.2 MPa, D = 2 m, t = 10 mm

$$\sigma_x = \sigma_L = \frac{PD}{4t} = \frac{0.2 \times 2000}{4 \times 10} = 10 \text{ MPa}$$

$$\sigma_y = \sigma_h = \frac{PD}{2t} = \frac{0.2 \times 2000}{2 \times 10} = 20 \text{ MPa}$$

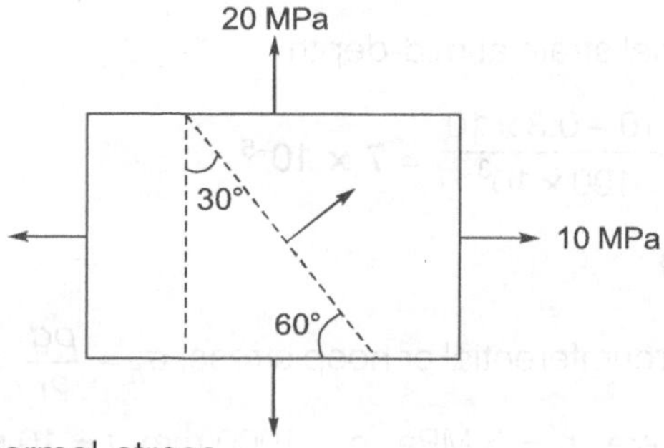

Normal stress,

$$(\sigma_n)_\theta = \left(\frac{\sigma_x + \sigma_y}{2}\right) + \left(\frac{\sigma_x - \sigma_y}{2}\right)\cos 2\theta + \tau_{xy}\sin 2\theta$$

$$(\sigma_n)_{\theta=30^\circ} = \left(\frac{10+20}{2}\right) + \left(\frac{10-20}{2}\right)\cos(2 \times 30^\circ)$$

= 12.5 MPa

■■■■

10 Theory of Column

10.1 A pin-ended column of length L, modulus of elasticity E and second moment of the cross-sectional area is I loaded concentrically by a compressive load P. The critical buckling load (P_{cr}) is given by

(a) $P_{cr} = \dfrac{EI}{\pi^2 L^2}$ (b) $P_{cr} = \dfrac{\pi^2 EI}{3L^2}$

(c) $P_{cr} = \dfrac{\pi EI}{L^2}$ (d) $P_{cr} = \dfrac{\pi^2 EI}{L^2}$

[2006 : 1 M]

10.2 The rod PQ of length L and with flexural rigidity EI is hinged at both ends. For what minimum force F is it expected to buckle?

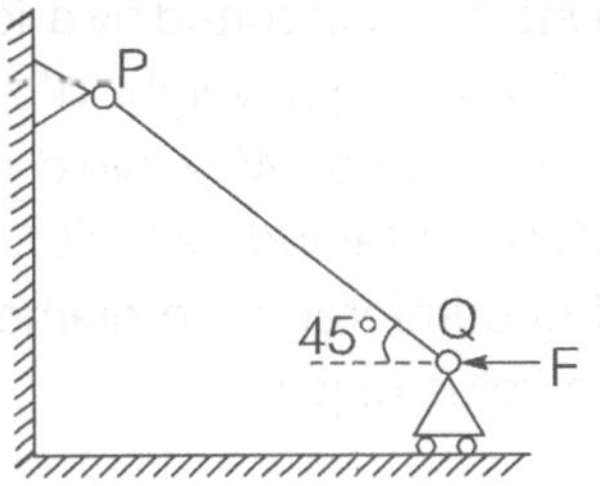

(a) $\dfrac{\pi^2 EI}{L^2}$ (b) $\dfrac{\sqrt{2}\pi^2 EI}{L^2}$

(c) $\dfrac{\pi^2 EI}{\sqrt{2}L^2}$ (d) $\dfrac{\pi^2 EI}{2L^2}$ **[2008 : 2 M]**

10.3 A column has a rectangular cross-section of 10 mm × 20 mm and a length of 1 m. The slenderness ratio of the column is close to

(a) 200 (b) 346

(c) 477 (d) 1000

[2011 : 1 M]

10.4 For a long slender column of uniform cross section, the ratio of critical buckling load for the case with both ends clamped to the case with both ends hinged is

(a) 1 (b) 2

(c) 4 (d) 8 **[2012 : 1 M]**

10.5 Consider a steel (Young's modulus E = 200 GPa) column hinged on both sides. Its heights is 1.0 m and cross-section is 10 mm × 20 mm. The lowest Euler critical buckling load (in N) is ______.

[2015 : 1 M, Set-1]

10.6 An initially stress-free massless elastic beam of length L and circular cross-section with diameter d($d << L$) is held fixed between two walls as shown. The beam material has Young's modulus E and coefficient of thermal expansion α.

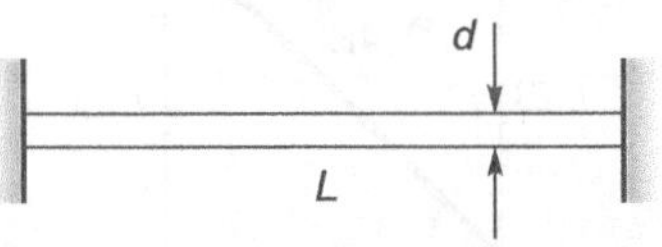

If the beam is slowly and uniformly heated, the temperature rise required to cause the beam to buckle is proportional to

(a) d (b) d^2

(c) d^3 (d) d^4

[2017 : 2 M, Set-1]

10.7 A steel column of rectangular section (15 mm × 10 mm) and length 1.5 m is simply supported at both ends. Assuming modulus of elasticity, E = 200 GPa for steel, the critical axial load (in kN) is________(correct to two decimal places)

[2018 : 1 M, Set-1]

10.8 The minimum axial compressive load, P required to initiate buckling for a pinned-pinned slender column with bending stiffness EI and length L is

(a) $P = \dfrac{\pi^2 EI}{4L^2}$ (b) $P = \dfrac{\pi^2 EI}{L^2}$

(c) $P = \dfrac{3\pi^2 EI}{4L^2}$ (d) $P = \dfrac{4\pi^2 EI}{L^2}$

[2018 : 1 M, Set-2]

10.9 Consider a prismatic straight beam of length L = π m, pinned at the two ends as shown in the figure. The beam has a square cross-section of side p = 6 mm. The Young's modulus E = 200 GPa, and the coefficient of thermal expansion $\alpha = 3 \times 10^{-6}$ K^{-1}. The minimum temperature rise required to cause Euler buckling of the beam is _____K.

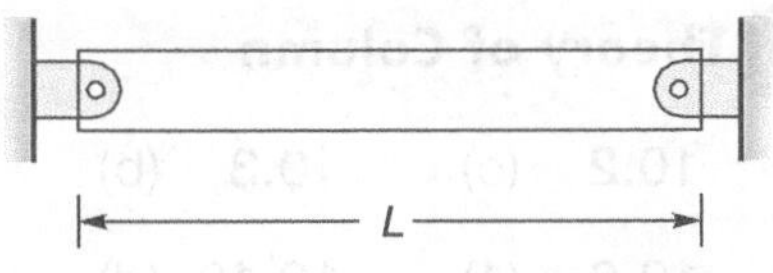

[2019 : 2 M, Set-1]

10.10 The truss shown in the figure has four members of length l and flexural rigidity EI, and one member of length $l\sqrt{2}$ and flexural rigidity 4EI. The truss is loaded by a pair of forces of magnitude P, as shown in the figure.

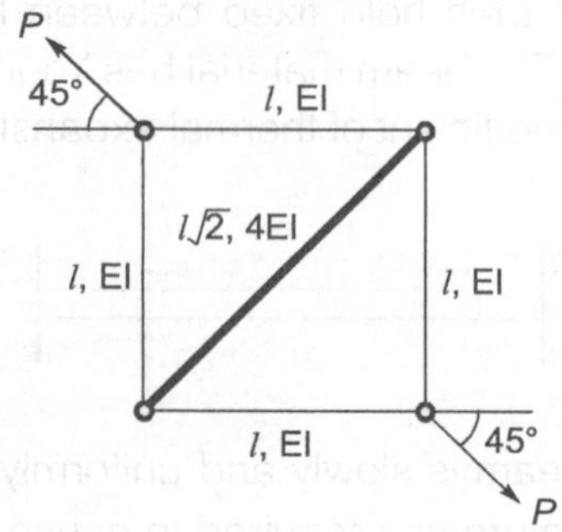

The smallest value of P, at which any of the truss members will buckle is

(a) $\frac{\pi^2 EI}{l^2}$ (b) $\frac{\sqrt{2}\pi^2 EI}{l^2}$

(c) $\frac{\pi^2 EI}{2l^2}$ (d) $\frac{2\pi^2 EI}{l^2}$

[2020 : 2 M, Set-1]

10.11 Uniaxial compression test data for a solid metal bar of length 1 m is shown in the figure.

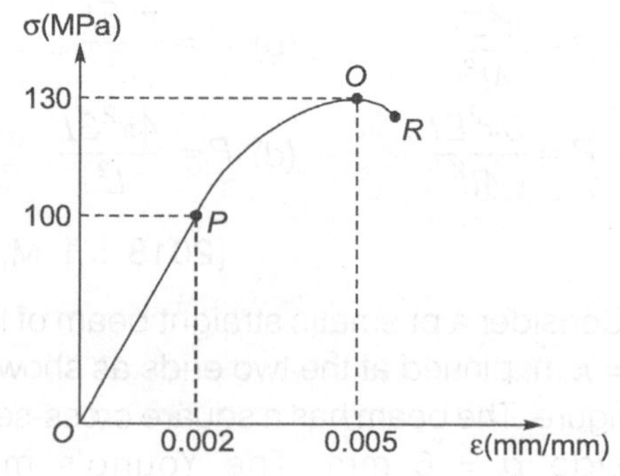

The bar material has a linear elastic response from O to P followed by a non-linear response. The point P represents the yield point of the material. The rod is pinned at both the ends. The minimum diameter of the bar so that it does not buckle under axial loading before reaching the yield point is ______ mm (round off to one decimal place).

[2020 : 2 M, Set-2]

10.12 A column with one end fixed and one end free has a critical buckling load of 100 N. For the same column, if the free end is replaced with a pinned end then the critical buckling load will be ________ N (round off to the nearest integer).

[2021 : 1 M, Set-2]

10.13 A uniform light slender beam AB of section modulus EI is pinned by a frictionless joint A to the ground and supported by a light inextensible cable CB to hang a weight W as shown. If the maximum value of W to avoid buckling of the beam AB is obtained as $\beta\pi^2 EI$, where π is the ratio of circumference to diameter of a circle, then the value of β is

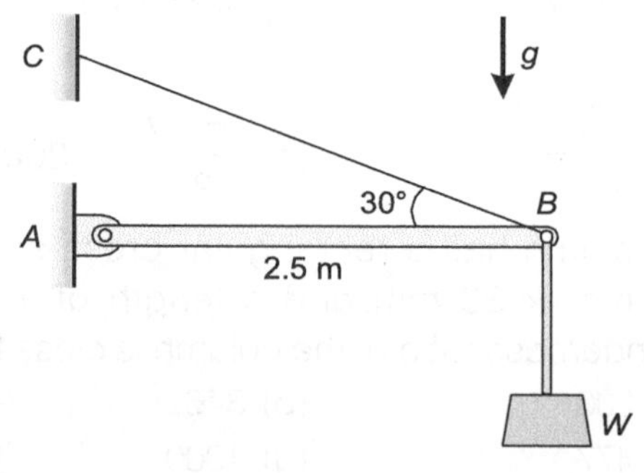

(a) 0.0924 m^{-2} (b) 0.0713 m^{-2}

(c) 0.1261 m^{-2} (d) 0.1417 m^{-2}

[2022 : 1 M, Set-1]

Answers Theory of Column

10.1 (d) 10.2 (c) 10.3 (b) 10.4 (c) 10.5 (3289.88) 10.6 (b) 10.7 (1.097)

10.8 (b) 10.9 (1) 10.10 (d) 10.11 (56.94) 10.12 (800) 10.13 (a)

Explanations Theory of Column

10.1 (d)

$$P_E = \frac{n\pi^2 EI}{L^2}$$

For pin-ended column, $n = 1$

Here
n = End fixity coefficient
E = Modulus of elasticity
I = Second moment of area
L = Actual length of column

10.2 (c)

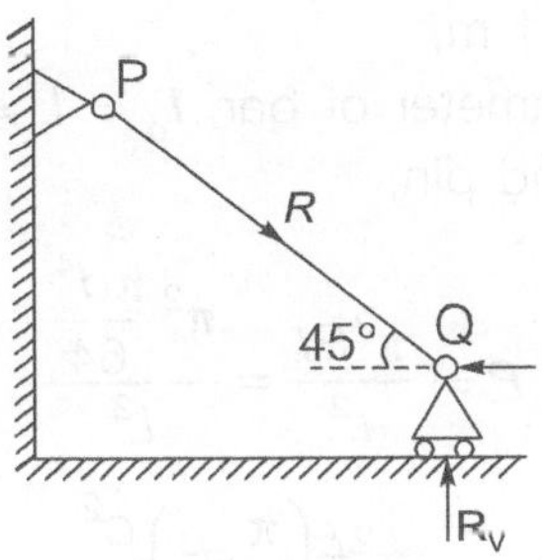

Let the axial compressive load acting on the rod is R.

$$R_H = R\cos 45°$$
$$R_V = R\sin 45°$$

Hence, $$R = \frac{F}{\cos 45°} = \sqrt{2}F$$

Condition for buckling, $R > P_e$

$$\sqrt{2}F > \frac{\pi^2 EI_{min}}{L_e^2} \quad \text{where } L_e = \alpha L$$

$$F > \frac{\pi^2 EI_{min}}{\sqrt{2}\,\alpha^2 L^2}$$

$$\alpha = 1 \quad (\because \text{both ends hinged})$$

Hence $$F_{min} = \frac{\pi^2 EI_{min}}{\sqrt{2}\,L^2}$$

10.3 (b)

Assuming both ends of column to be hinged.

Given:

$L_{eq} = 1$ m

$$\text{Slenderness ratio} = \frac{L_{eq}}{k_{min}}$$

$$k_{min} = \sqrt{\frac{I_{min}}{A}} = \sqrt{\frac{\frac{1}{12}\times(10)^3\times 20\times 10^{-12}}{10\times 20\times 10^{-6}}}$$

$$= 2.88\times 10^{-3}\text{ m}$$

$$\text{Slenderness ratio} = \frac{1}{2.88\times 10^{-3}} = 347.22 \approx \mathbf{346}$$

10.4 (c)

Euler's buckling load,

$$P_e = \frac{n\pi^2 EI_{min}}{L^2}$$

$n = 4$ for both end clamped (fixed)

$n = 1$ for both end hinged

$\therefore$ Ratio = **4**

10.5 Sol.

$L_e = L$ for both end hinged

$$F = \frac{\pi^2 EI}{L^2} = \frac{\pi^2\times 200\times 10^3\times 20\times 10^3}{12\times(1000)^2}$$

$$= \mathbf{3289.868\ N}$$

10.6 (b)

Condition for buckling,

(Reaction offered by the fixed support) > Buckling load

i.e. $$R > P_e$$

$$\sigma_{Th} A > \frac{\pi^2 EI_{min}}{L_e^2}$$

$$\alpha\,(\Delta T)\,EA > \frac{\pi^2 EI_{min}}{(L_e)^2}$$

For circular cross-section

$$\alpha(\Delta T)E\frac{\pi}{4}d^2 > \frac{\pi^2 E}{(L_e)^2}\frac{\pi}{64}d^4$$

$$\Delta T > \frac{\pi^2 d^2}{16\alpha(L_e)^2}$$

Hence, $$\Delta T \propto d^2$$

10.7 Sol.

$$\text{Buckling load} = \frac{\pi^2 EI_{min}}{L^2}$$

$$= \frac{\pi^2\times 200\times 10^3\times\frac{15\times 10^3}{12}}{1500^2}$$

$$= 1096.62\text{ N} = \mathbf{1.097\ kN}$$

10.8 (b)

For both ends hinged column buckling load,

$$P = \frac{\pi^2 EI}{L^2}$$

10.9 Sol.

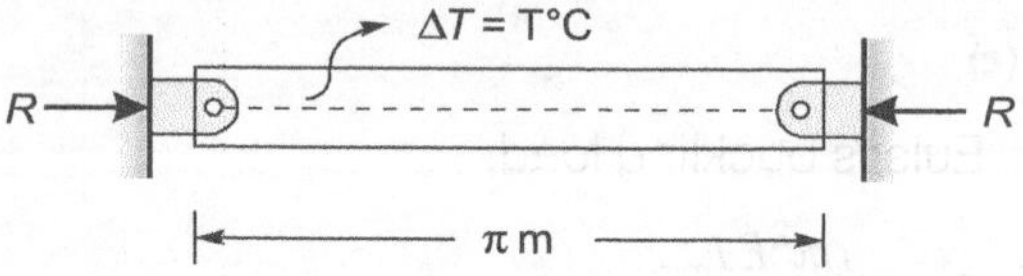

$$L = \pi \text{ m}$$

Side of square, $p = 6$ mm

$$E = 200 \text{ GPa}$$

$$\alpha = 3 \times 10^{-6}/\text{K}$$

For completely restricted expansion,

$$R = \sigma_{Th}$$

$$A = \alpha TEA \quad \text{...(i)}$$

$$\text{Buckling load, } P_e = \left(\frac{n\pi^2 EI_{min}}{L^2}\right)$$

$$P_e = (1)\frac{\pi^2 Ep^4}{12L^2} \quad \text{...(ii)}$$

Condition for buckling,

$$R > P_e$$

$$\alpha \text{TE}(A) > \frac{\pi^2 Ep^4}{12L^2}$$

$$\alpha \text{T}(p^2) > \frac{\pi^2 p^4}{12L^2}$$

$$T > \frac{\pi^2 p^2}{(\alpha)(12)\pi^2(1000)^2}$$

$$T > \frac{6^2}{(12)\times 3\times 10^{-6}\times 10^6}$$

$$T > 1\text{K}$$

or Change in temperature is 1K.

10.10 (d)

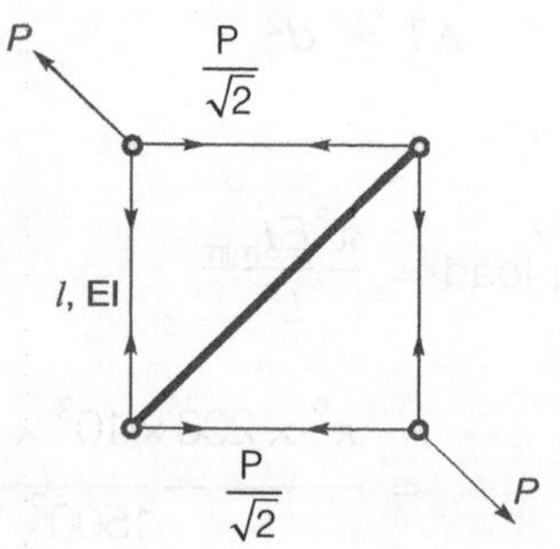

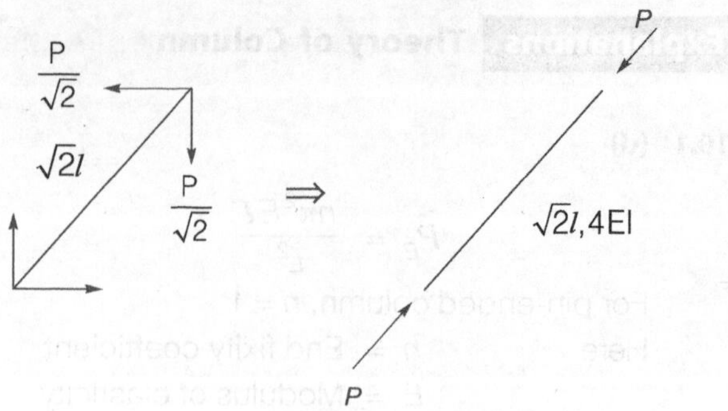

$$P = \frac{\pi^2(4EI)}{(\sqrt{2}l)^2} = \frac{2\pi^2 EI}{l^2}$$

10.11 Sol.

Given $L = 1$ m,

Let d = diameter of bar, $l_c = L = 1$ mm

For both end pin,

$$P = \frac{\pi^2 EI}{L^2} = \frac{\pi^2 \frac{\pi d^4}{64}}{L^2}$$

$$= \frac{\pi^2 E\left(\frac{\pi}{4}d^2\right)\frac{d^2}{16}}{L^2} = \frac{\frac{\pi^2 EAd^2}{16}}{L^2}$$

$$\frac{P}{AE} = \frac{\pi^2 d^2}{16L^2}$$

$$\varepsilon_y = \frac{\pi^2 d^2}{16L^2} = 0.002 \quad \text{[at point } P\text{]}$$

$$d = \sqrt{\frac{16L^2 \varepsilon_y}{\pi^2}}$$

$$= \sqrt{\frac{16\times 1000^2 \times 0.002}{\pi^2}}$$

$$= 56.94 \text{ mm}$$

10.12 Sol.

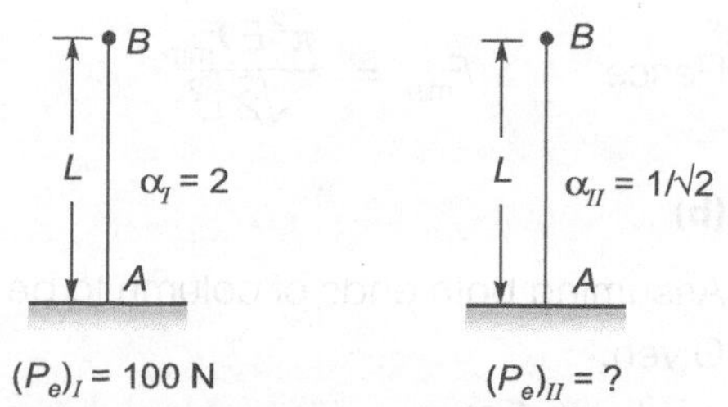

For a given material in 0.1 and length,

$$\text{Pe} \propto \frac{1}{\alpha^2}$$

where, α = length fixity coefficient

Pe = Buckling or critical load.

$$\frac{(Pe)_{II}}{(Pe)_I} = \left(\frac{\alpha_I}{\alpha_{II}}\right)^2 = \left(\frac{2L}{1/\sqrt{2}}\right)^2 = 8$$

$$(P_e)_{II} = 8(P_e)_I = 800\text{ N}$$

10.13 (a)

Drawing free body diagram of point B,

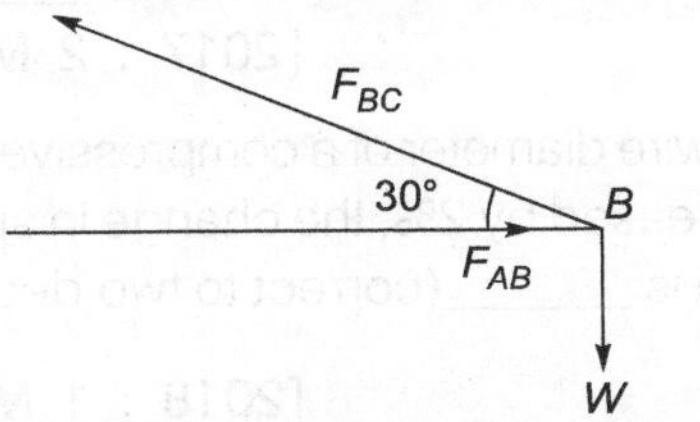

Writing two equations for equilibrium at point B.

$\Rightarrow \quad F_{BC}(\sin 30°) = W$

$$F_{BC} = 2W$$

$$F_{AB} = F_{BC}\cos 30°$$

$$F_{AB} = 2W\cos 30°$$

$$F_{AB} = 2W \times \frac{\sqrt{3}}{2}$$

$$F_{AB} = 1.732W$$

Using the standard formula for buckling load,

$$P_C = \frac{\pi^2 EI}{L_C^2} = \frac{\pi^2 EI}{(2.5)^2}$$

$$F_{AB} = 1.732W = \frac{\pi^2 EI}{(2.5)^2}$$

$$W = \frac{\pi^2 EI}{(2.5)^2 \times 1.732} = \beta\pi^2 EI$$

$$\beta = \frac{1}{1.732 \times 6.25} = 0.092$$

■■■■

11 Springs

11.1 The deflection of a spring with 20 active turns under a load of 1000 N is 10 mm. The spring is made into two pieces each of 10 active coils and placed in parallel under the same load. The deflection of this system is

(a) 20 mm (b) 10 mm
(c) 5 mm (d) 2.5 mm

[1995 : 2 Marks]

11.2 A weighing machine consists of a 2 kg pan resting on a spring. In this condition, with the pan resting on the spring, the length of the spring is 200 mm. When a mass of 20 kg is placed on the pan, the length of the spring becomes 100 mm. For the spring, the undeformed length L and the spring constant k (stiffness) are

(a) L = 220 mm, k = 1862 N/m
(b) L = 210 mm, k = 1960 N/m
(c) L = 200 mm, k = 1960 N/m
(d) L = 200 mm, k = 2156 N/m **[2005 : 2 Marks]**

11.3 A compression spring is made of music wire of 2 mm diameter having a shear strength and shear modulus of 800 MPa and 80 GPa respectively. The mean coil diameter is 20 mm, free length is 40 mm and the number of active coils is 10. If the mean coil diameter is reduced to 10 mm, the stiffness of the spring is approximately

(a) decreased by 8 times
(b) decreases by 2 times
(c) increased by 2 times
(d) increased by 8 times **[2008 : 2 Marks]**

11.4 The spring constant of a helical compression spring DOES NOT depend on

(a) coil diameter
(b) material strength
(c) number of active turns
(d) wire diameter **[2016 : 1 Mark, Set-1]**

11.5 A helical compression spring made of a wire of circular cross-section is subjected to a compressive load. The maximum shear stress induced in the cross-section of the wire is 24 MPa. For the same compressive load, if both the wire diameter and the mean coil diameter are doubled, the maximum shear stress (in MPa) induced in the cross-section of the wire is _________.

[2017 : 2 Marks, Set-2]

11.6 If the wire diameter of a compressive helical spring is increased by 2%, the change in spring stiffness (in %) is _______(correct to two decimal places.)

[2018 : 1 Marks, Set-1]

11.7 A helical spring has spring constant k. If the wire diameter, spring diameter and the number of coils are all doubled then the spring constant of the new spring becomes

(a) $8k$ (b) $16k$
(c) k (d) $k/2$

[2020 : 2 Marks, Set-2]

11.8 Shear stress distribution on the cross-section of the coil wire in a helical compression spring is shown in the figure. This shear stress distribution represents

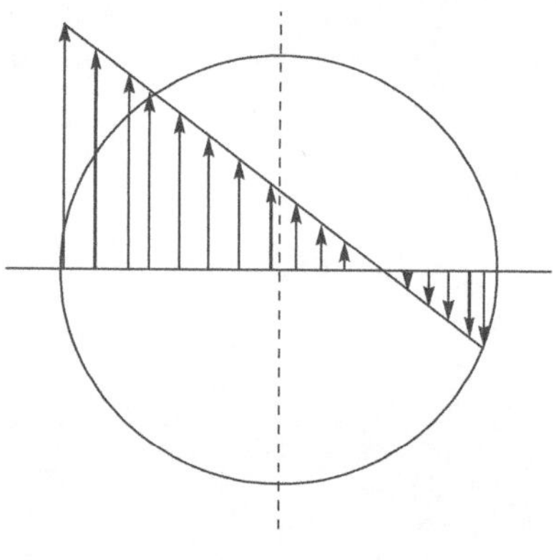

(a) Combined direct shear and torsional shear stress in the coil wire cross-section.
(b) Combined direct shear and torsional shear stress along with the effect of stress concentration at inside edge of the coil wire cross-section.
(c) Torsional shear stress in the coil wire cross-section.
(d) Direct shear stress in the coil wire cross-section.

[2021 : 1 Mark, Set-1]

Answers Springs

11.1 (d) 11.2 (b) 11.3 (d) 11.4 (b) 11.5 (6) 11.6 (8.243) 11.7 (c) 11.8 (a)

Explanations Springs

11.1 (d)

$\Delta x = 10$ mm, $N = 20$ turns, $F = 1000$ N

$$k = \frac{F}{\Delta x} = \frac{1000}{10} = 100 \text{ N/mm}$$

On cutting spring in two equal parts, new k' will equal to $2k$

Under parallel arrangement, k of the system will become $2k + 2k = 4k$ *i.e.*, 400 N/mm

Under same force F, we get

$$\Delta x' = \frac{1000}{400} = 2.5 \text{ mm}$$

11.2 (b)

$$F = k\Delta x$$

$$\Rightarrow \quad 2g = k(L_0 - 0.2) \quad \text{...(i)}$$

When mass of 20 kg is placed on the pan

$$22g = k(L_0 - 0.1) \quad \text{...(ii)}$$

Solving equations (i) and (ii), we get

$$\mathbf{L_0 = 210 \text{ mm}}$$

$$\mathbf{k = 1960 \text{ N/m}}$$

11.3 (d)

$$k = \frac{Gd^4}{8D^3N}$$

$$\frac{k_2}{k_1} = \left(\frac{D_1}{D_2}\right)^3 = \left(\frac{20}{10}\right)^3 = 8$$

$\therefore$ Stiffness of spring is increased by 8 times.

11.4 (b)

Stiffness of helical spring,

$$k = \frac{Gd^4}{64R^3n}$$

It depends on wire diameter (d), shear modulus (G), Coil radius(R) and number of active turns (n). Does not depend on strength.

11.5 Sol.

Maximum shear stress induced in spring wire of a helical compression spring is given by,

$$\tau_{max} = \frac{8WD}{\pi d^3}k_W \quad \text{or} \quad \frac{8WC}{\pi d^2}k_W$$

$$\tau_{max} \propto \frac{1}{d^2}$$

[$\because W_1 = W_2 = W;\ C_1 = C_2$ (i.e., $C = D/d$) $(k_W)_1 = (k_W)_2 = k_W$]

$$\frac{\tau_2}{\tau_1} = \left(\frac{d_1}{d_2}\right)^2$$

$$\frac{\tau_2}{24} = \left(\frac{d}{2d}\right)^2$$

$$\Rightarrow \quad \mathbf{\tau_2 = 6 \text{ MPa}}$$

11.6 Sol.

Stiffness of helical spring,

$$k = \frac{Gd^4}{64R^3n}$$

d = Spring wire diameter; R = Mean coil radius

n = Number of turns

$$\therefore \quad k \propto d^4$$

$$\frac{k'}{k} = \left(\frac{d'}{d}\right)^4$$

$$k' = \left(\frac{1.02d}{d}\right)^4 k = 1.08243\ k$$

$$\% \text{ increase in stiffness} = \frac{k' - k}{k} \times 100\%$$

$$= \left(\frac{1.08243k - k}{k}\right) \times 100\% = \mathbf{8.243\%}$$

11.7 (c)

As per given data: Spring constant,

$$k_{(spring)} = \frac{Gd^4}{8D^3n}$$

Spring constant,

$$k_{new} = \frac{G(2d)^4}{8(2D)^3(2n)} = \frac{Gd^4}{8D^3n}$$

Hence, $k_{new} = k$

11.8 (a)

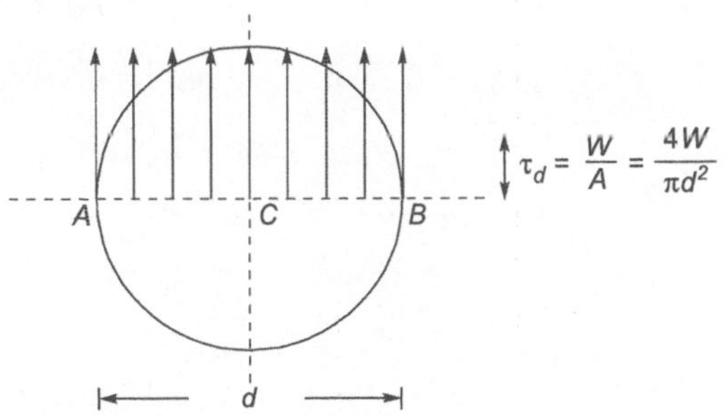

Fig. : Direct shear stress distribution across the cross-section of wire

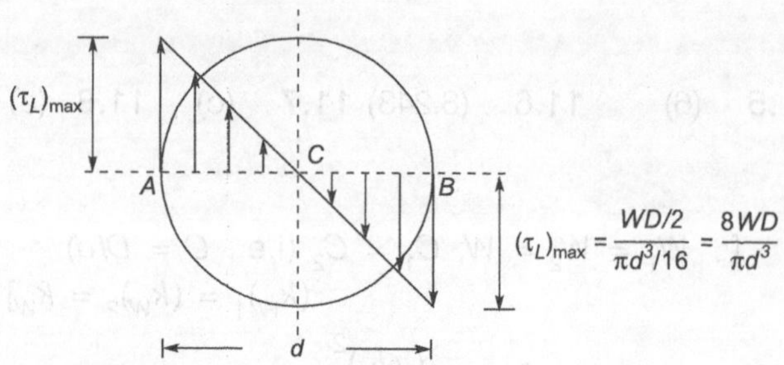

Fig. : Torsional shear stress distribution across the cross-section of wire

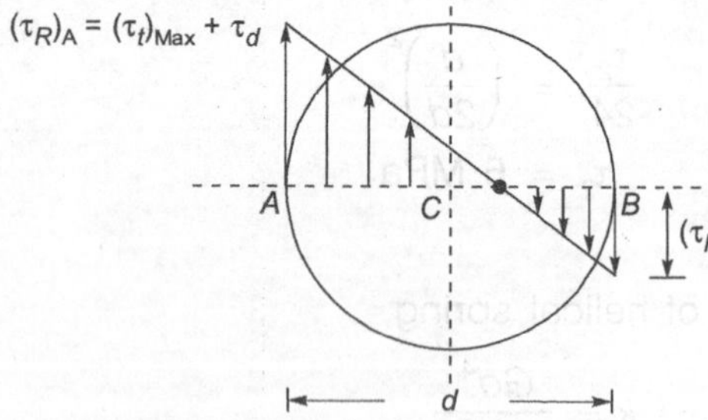

Fig. : Resultant shear stress variation across the cross-section of wire (without considering stress concentration effect)

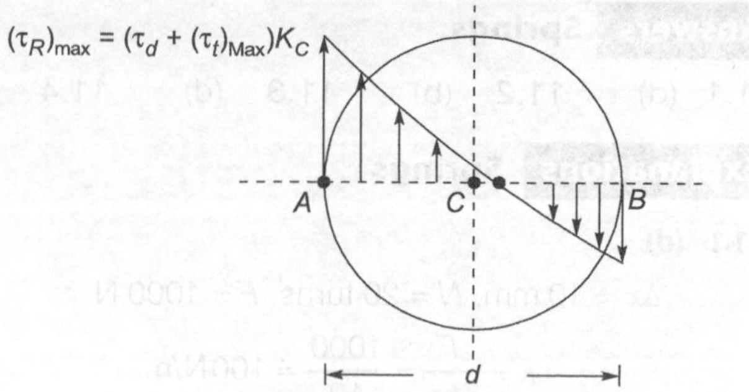

Fig. : Resultant shear stress variation considering stress concentration effect

UNIT IV

Theory of Machines

CONTENTS

Theory of Machines

Syllabus

Displacement, velocity and acceleration analysis of plane mechanisms; dynamic analysis of linkages; cams; gears and gear trains; flywheels and governors; balancing of reciprocating and rotating masses; gyroscope.

Vibrations : Free and forced vibration of single degree of freedom systems, effect of damping; vibration isolation; resonance; critical speeds of shafts.

Analysis of Previous GATE Papers

Exam Year	1 Mark Ques.	2 Marks Ques.	Total Marks
1995	–	3	**6**
1996	–	2	**14**
1997	1	–	**6**
1998	1	1	**3**
1999	1	2	**10**
2000	1	1	**3**
2001	2	1	**9**
2002	–	–	**–**
2003	3	6	**15**
2004	2	8	**18**
2005	2	6	**14**
2006	3	9	**21**
2007	1	6	**13**
2008	1	3	**7**
2009	2	4	**10**
2010	5	3	**11**
2011	1	3	**7**
2012	2	1	**4**
2013	3	2	**7**
2014 Set-1	2	3	**8**
2014 Set-2	2	3	**8**
2014 Set-3	2	4	**10**
2014 Set-4	2	3	**8**
2015 Set-1	1	2	**5**

Exam Year	1 Mark Ques.	2 Marks Ques.	Total Marks
2015 Set-2	2	2	**6**
2015 Set-3	3	3	**9**
2016 Set-1	2	3	**8**
2016 Set-2	1	2	**5**
2016 Set-3	3	3	**9**
2017 Set-1	1	3	**7**
2017 Set-2	2	4	**10**
2018 Set-1	3	3	**9**
2018 Set-2	2	3	**8**
2019 Set-1	3	3	**9**
2019 Set-2	1	5	**11**
2020 Set-1	4	2	**8**
2020 Set-2	2	3	**8**
2021 Set-1	3	2	**7**
2021 Set-2	2	4	**10**
2022 Set-1	2	3	**8**
2022 Set-2	2	3	**8**
2023	2	1	**4**
2024	2	3	**8**

1 Displacement, Velocity and Acceleration

1.1 A rod of length 1 m is sliding in a corner as shown in figure. At an instant when the rod makes an angle of 60 degrees with the horizontal plane., the velocity of point *A* on the rod is 1 m/s. The angular velocity of the rod at this instant is

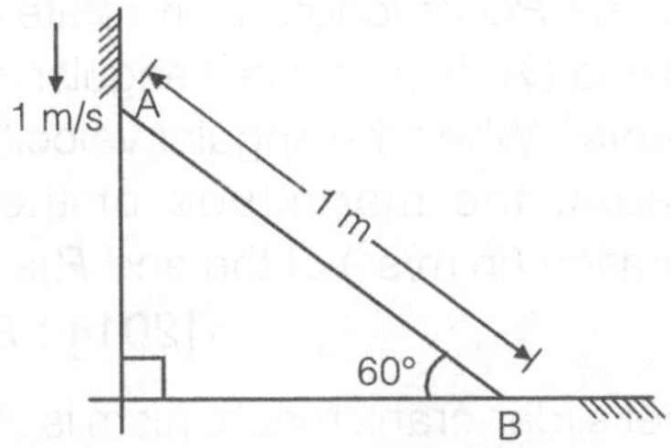

(a) 2 rad/s (b) 1.5 rad/s
(c) 0.5 rad/s (d) 0.75 rad/s **[1996 : 2 M]**

Common Data Questions Q.1.2 and Q.1.3

The circular disc shown in its plan view in the figure rotates in a plane parallel to the horizontal plane about the point *O* at a uniform angular velocity ω. Two other points *A* and *B* are located on the line *OZ* at distances r_A and r_B from *O* respectively.

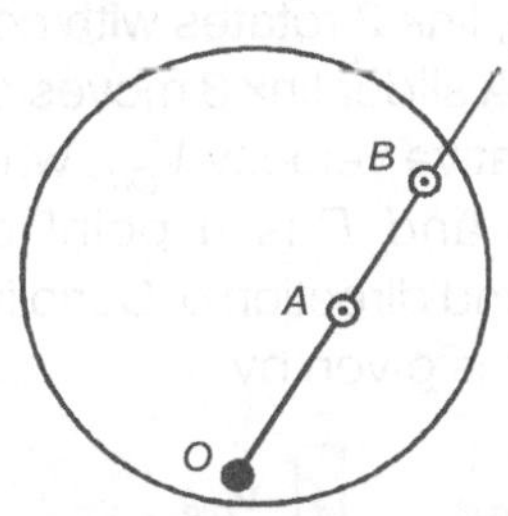

1.2 The velocity of point *B* with respect to point *A* is a vector of magnitude

(a) 0
(b) $\omega\,(r_B - r_A)$ and direction opposite to the direction of motion of point *B*
(c) $\omega\,(r_B - r_A)$ and direction same as the direction of motion of point *B*.
(d) $\omega\,(r_B - r_A)$ and direction being from *O* to *Z*.

[2003 : 2 M]

1.3 The acceleration of point *B* with respect to point *A* is a vector of magnitude

(a) 0
(b) $\omega\,(r_B - r_A)$ and direction same as the direction of motion of point *B*.
(c) $\omega\,(r_B - r_A)$ and direction opposite to be direction of motion of point *B*.
(d) $\omega^2\,(r_B - r_A)$ and direction being from *Z* to *O*.

[2003 : 2 M]

1.4 In the figure shown, the relative velocity of link 1 with respect of link 2 is 12 m/s. Link 2 rotates at a constant speed of 120 rpm. The magnitude of Coriolis component of acceleration of link 1 is

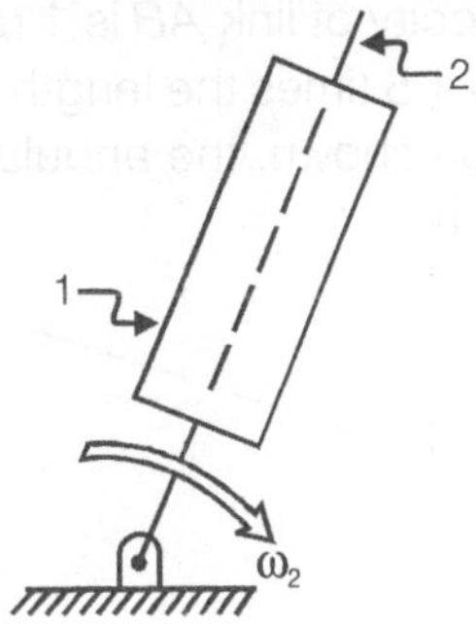

(a) 302 m/s² (b) 604 m/s²
(c) 906 m/s² (d) 1208 m/s² **[2004 : 2 M]**

1.5 The input link O_2P of a four bar linkage is rotated at 2 rad/s in counter clockwise direction as shown below. The angular velocity of the coupler *PQ* in rad/s, at an instant when $\angle O_4O_2P = 180°$, is

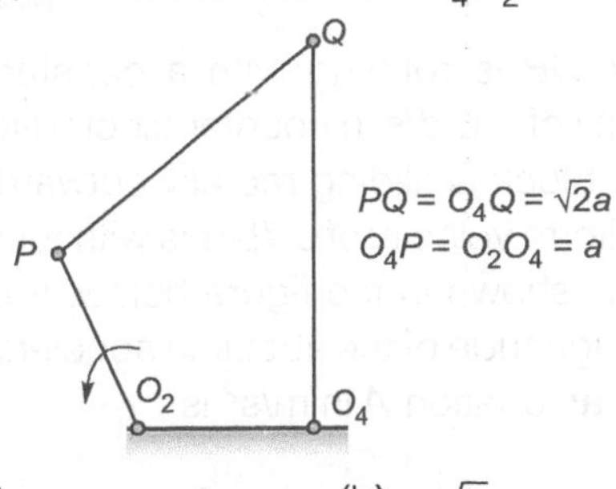

(a) 4 (b) $2\sqrt{2}$
(c) 1 (d) $1/\sqrt{2}$ **[2007 : 2 M]**

1.6 For the configuration shown, the angular velocity of link *AB* is 10 rad/s counter clockwise. The magnitude of the relative sliding velocity (in ms^{-1}) of slider *B* with respect to rigid link *CD* is

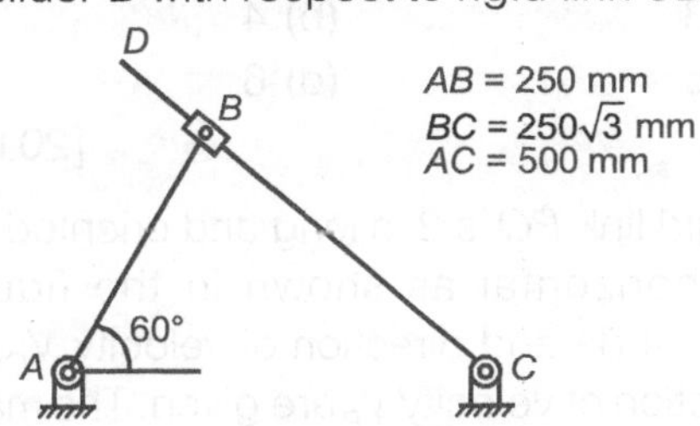

(a) 0 (b) 0.86
(c) 1.25 (d) 2.50

[2010 : 2 M]

1.7 There are two points P and Q on a planar rigid body. The relative velocity between the two points

(a) should always be along PQ
(b) can be oriented along any direction
(c) should always be perpendicular to PQ
(d) should be along QP when the body undergoes pure translation

[2010 : 1 M]

1.8 For the four-bar linkage shown in the figure, the angular velocity of link AB is 1 rad/s. The length of link CD is 1.5 times the length of link AB. In the configuration shown, the angular velocity of link CD in rad/s is

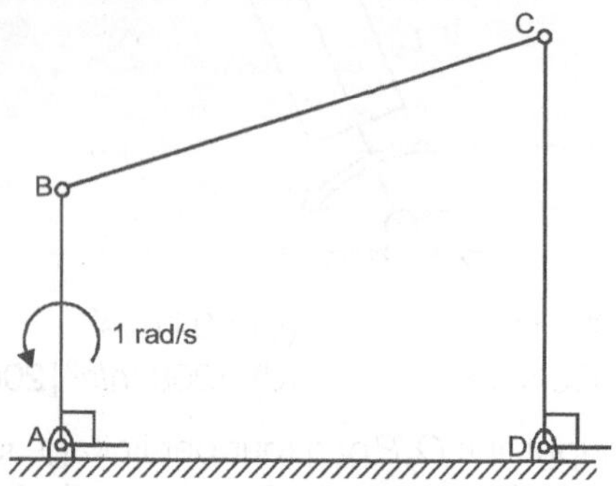

(a) 3
(b) 3/2
(c) 1
(d) 2/3

[2011 : 2 M]

1.9 A link OB is rotating with a constant angular velocity of 2 rad/s in counter clockwise direction and a block is sliding radially outward on it with an uniform velocity of 0.75 m/s with respect to the rod, as shown in the figure below. If OA = 1 m, the magnitude of the absolute acceleration of the block at location A in m/s^2 is

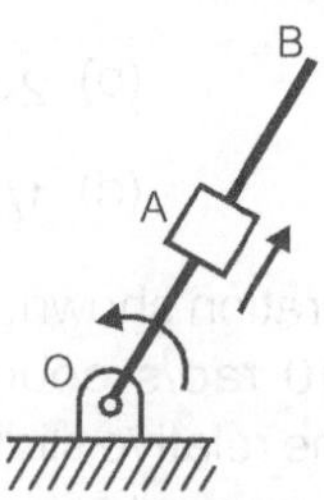

(a) 3
(b) 4
(c) 5
(d) 6

[2013 : 1 M]

1.10 A rigid link PQ is 2 m long and oriented at 20° to the horizontal as shown in the figure. The magnitude and direction of velocity V_Q, and the direction of velocity V_P are given. The magnitude of V_P (in m/s) at this instant is

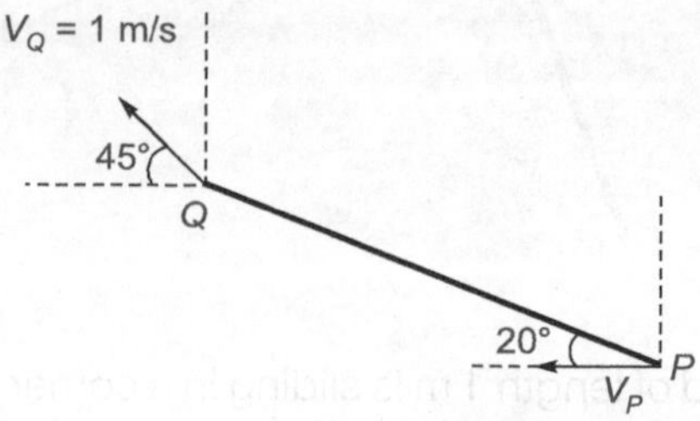

(a) 2.14
(b) 1.89
(c) 1.21
(d) 0.96

[2014 : 1 M, Set-1]

1.11 A rigid link PQ of length 2 m rotates about the pinned end Q with a constant angular acceleration of 12 rad/s^2. When the angular velocity of the link is 4 rad/s, the magnitude of the resultant acceleration (in m/s^2) of the end P is _______.

[2014 : 2 M, Set-2]

1.12 An offset slider-crank mechanism is shown in the figure at an instant. Conventionally, the Quick Return Ratio (QRR) is considered to be greater than one. The value of QRR is _______.

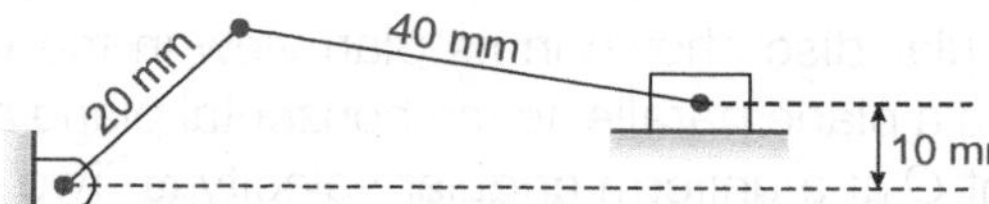

[2014 : 2 M, Set-1]

1.13 In the figure, link 2 rotates with constant angular velocity ω_2. A slider link 3 moves outwards with a constant relative velocity $V_{Q/P}$, where Q is a point on slider 3 and P is a point on link 2. The magnitude and direction of Coriolis component of acceleration is given by

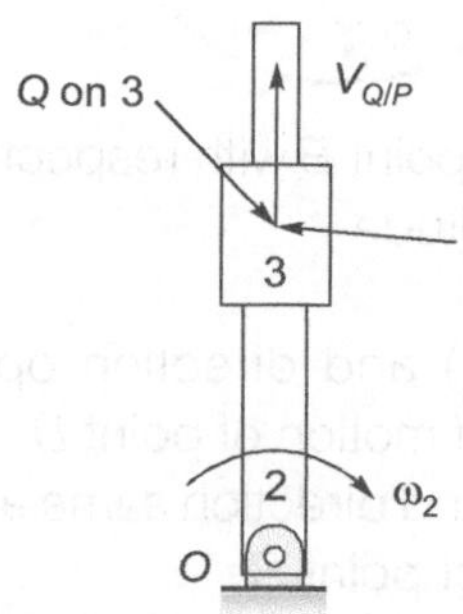

(a) $2\omega_2 V_{Q/P}$; direction of $V_{Q/P}$ rotated by 90° in the direction ω_2
(b) $\omega_2 V_{Q/P}$; direction of $V_{Q/P}$ rotated by 90° in the direction ω_2
(c) $2\omega_2 V_{Q/P}$; direction of $V_{Q/P}$ rotated by 90° opposite to the direction of ω_2
(d) $\omega_2 V_{Q/P}$; direction of $V_{Q/P}$ rotated by 90° opposite to the direction ω_2

[2015 : 1 M, Set-3]

1.14 The rod AB, of length 1 m, shown in the figure is connected to two sliders at each end through pins. The sliders can slide along QP and QR. If the velocity V_A of the slider at A is 2 m/s, the velocity of the midpoint of the rod at this instant is ________ m/s.

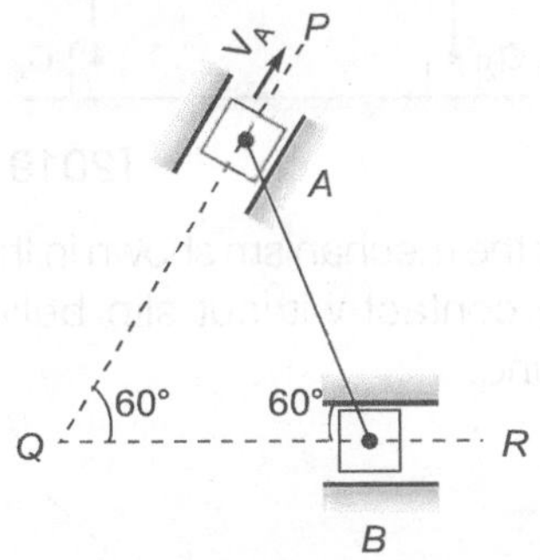

[2016 : 2 M, Set-2]

1.15 The number of degrees of freedom in a planar mechanism having n links and j simple hinge joints is

(a) $3(n - 3) - 2j$ (b) $3(n - 1) - 2j$
(c) $3n - 2j$ (d) $2j - 3n + 4$

[2016 : 1 M, Set-3]

1.16 A rigid link PQ is undergoing plane motion as shown in the figure (V_P and V_Q are non-zero). V_{QP} is the relative velocity of point Q with respect to point P.

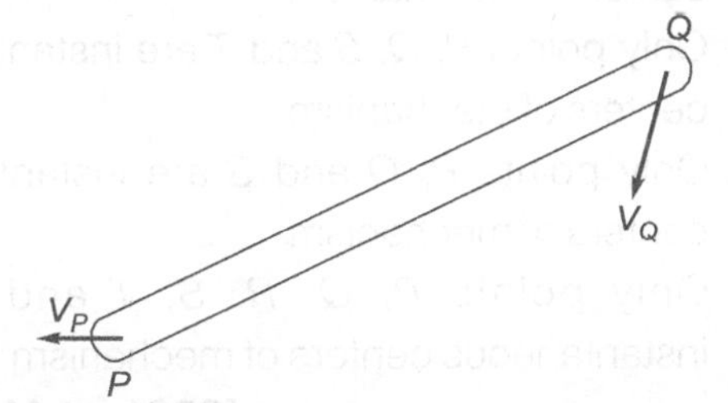

Which one of the following is TRUE?

(a) V_{QP} has components along and perpendicular to PQ
(b) V_{QP} has only one component directed from P to Q
(c) V_{QP} has only one component directed from Q to P
(d) V_{QP} has only one component perpendicular to PQ

[2016 : 1 M, Set-3]

1.17 Block 2 slides outward on link 1 at a uniform velocity of 6 m/s as shown in the figure. Link 1 is rotating at a constant angular velocity of 20 radian/s counterclockwise. The magnitude of the total acceleration (in m/s²) of point P of the block with respect to fixed point O is ________.

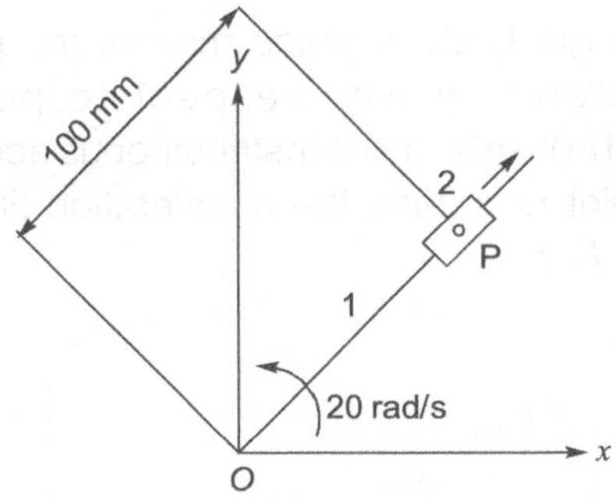

[2017 : 2 M, Set-2]

1.18 For an inline slider-crank mechanism, the length of the crank and connecting rod are 3 m and 4 m, respectively. At the instant when the connecting rod is perpendicular to the crank, if the velocity of the slider is 1 m/s, the magnitude of angular velocity (upto 3 decimal points accuracy) of the crank is ________ radian/s.

[2017 : 2 M, Set-1]

1.19 In a slider-crank mechanism, the lengths of the crank and the connecting rod are 100 mm and 160 mm, respectively. The crank is rotating with an angular velocity of 10 radian/s counter clockwise. The magnitude of linear velocity (in m/s) of the piston at the instant corresponding to the configuration shown in the figure is ______.

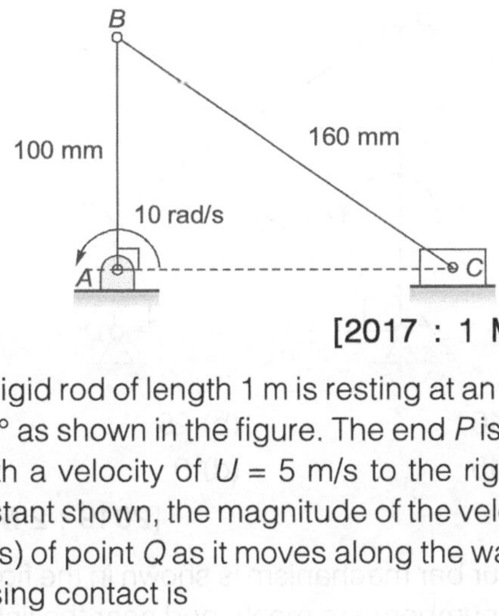

[2017 : 1 M, Set-2]

1.20 A rigid rod of length 1 m is resting at an angle θ = 45° as shown in the figure. The end P is dragged with a velocity of U = 5 m/s to the right. At the instant shown, the magnitude of the velocity V (in m/s) of point Q as it moves along the wall without losing contact is

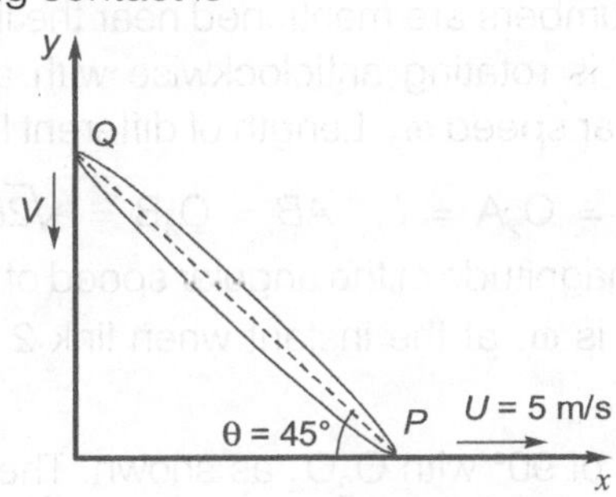

(a) 5 (b) 6
(c) 8 (d) 10 [2018 : 2 M, Set-2]

1.21 In a rigid body in plane motion, the point R is accelerating with respect to point P at $10\angle 180°$ m/s^2. If the instantaneous acceleration of point Q is zero, the acceleration (in m/s^2) of point R is

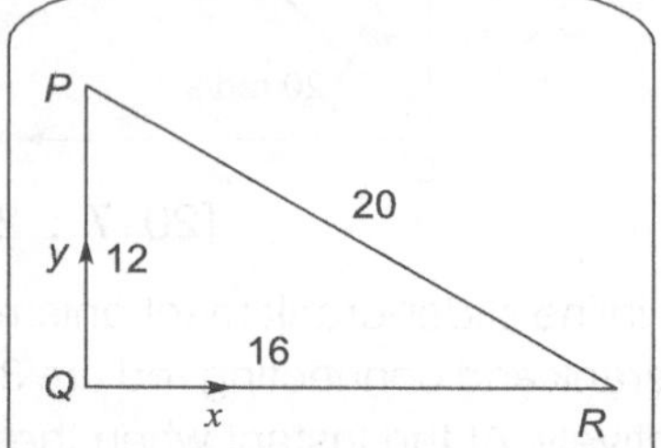

(a) $8\angle 233°$ (b) $10\angle 225°$
(c) $10\angle 217°$ (d) $8\angle 217°$

[2018 : 2 M, Set-2]

1.22 In a four bar planar mechanism shown in the figure, AB = 5 cm, AD = 4 cm and DC = 2 cm. In the configuration shown, both AB and DC are perpendicular to AD. The bar AB rotates with an angular velocity of 10 rad/s. The magnitude of angular velocity (in rad/s) of bar DC at this instant is

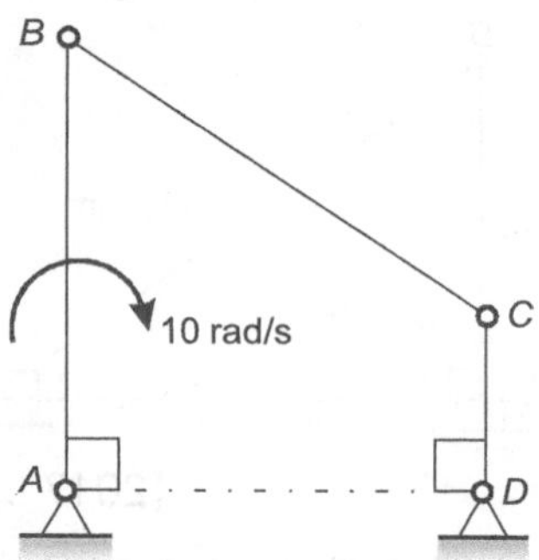

(a) 10 (b) 25
(c) 15 (d) 0

[2019 : 2 M, Set-1]

1.23 A four bar mechanism is shown in the figure. The link numbers are mentioned near the links. Input link 2 is rotating anticlockwise with a constant angular speed ω_2. Length of different links are:

$$O_2O_4 = O_2A = L,\ AB = O_4B = \sqrt{2}L$$

The magnitude of the angular speed of the output link 4 is ω_4 at the instant when link 2 makes an angle of 90° with O_2O_4 as shown. The ratio $\frac{\omega_4}{\omega_2}$ is ______ (round off to two decimal places).

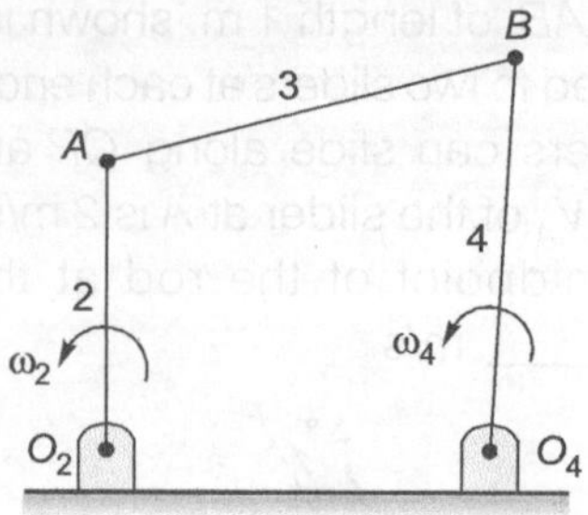

[2019 : 2 M, Set-2]

1.24 Consider the mechanism shown in the figure. There is rolling contact without slip between the disc and ground.

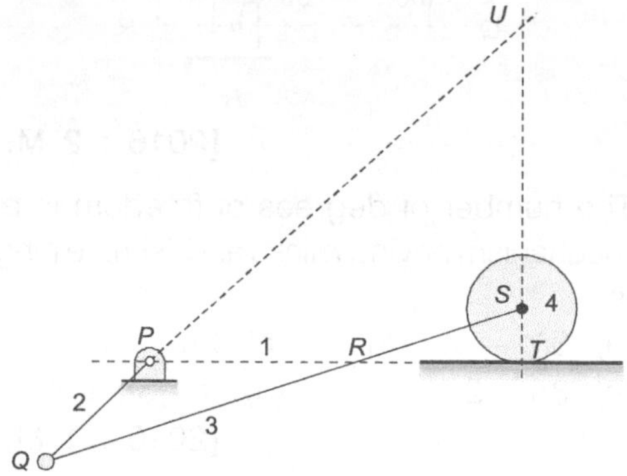

Select the correct statement about instantaneous centers in the mechanism.

(a) Only points P, Q, R, S and U are instantaneous centers of mechanism
(b) Only points P, Q, S and T are instantaneous centers of mechanism
(c) Only points P, Q and S are instantaneous centers of mechanism
(d) Only points P, Q, R, S, T and U are instantaneous centers of mechanism

[2021 : 1 M, Set-2]

1.25 A planar four-bar linkage mechanism with 3 revolute kinematic pairs and 1 prismatic kinematic pair is shown in the figure, where $AB \perp CE$ and $FD \perp CE$. The T-shaped link $CDEF$ is constructed such that the slider B can cross the point D, and CE is sufficiently long. For the given lengths as shown, the mechanism is

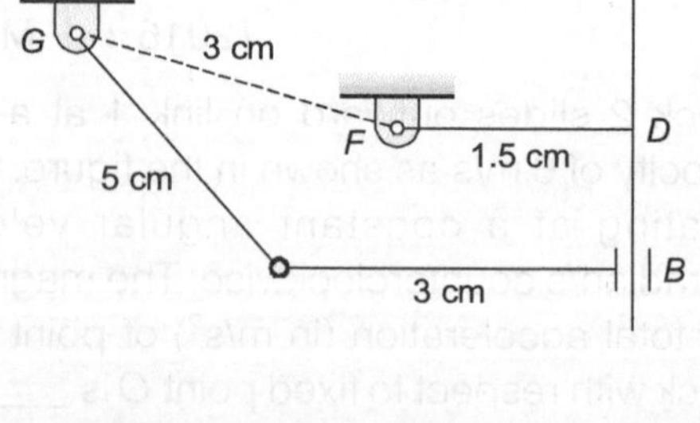

(a) a Grashof chain with links *AG, AB,* and *CDEF* completely rotatable about the ground link *FG*
(b) a non-Grashof chain with all oscillating links
(c) a Grashof chain with *AB* completely rotatable about the ground link *FG*, and oscillatory links *AG* and *CDEF*
(d) on the border of Grashof and non-Grashof chains with uncertain configuration(s)

[2022 : 2 M, Set-1]

1.26 In the configuration of the planar four-bar mechanism at a certain instant as shown in the figure, the angular velocity of the 2 cm long link is $\omega_2 = 5$ rad/s. Given the dimensions as shown, the magnitude of the angular velocity ω_4 of the 4 cm long link is given by _____ rad/s (round off to 2 decimal places).

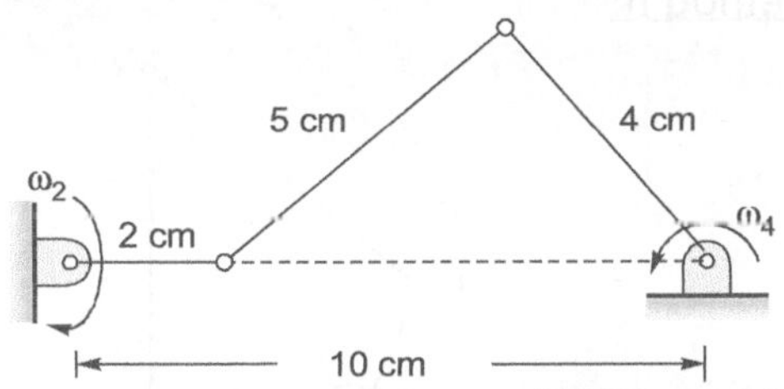

[2022 : 2 M, Set-2]

1.27 The figure shows a wheel rolling without slipping on a horizontal plane with angular velocity ω_1. A rigid bar *PQ* is pinned to the wheel at *P* while the end *Q* slides on the floor. What is the angular velocity ω_2 of the bar *PQ*?

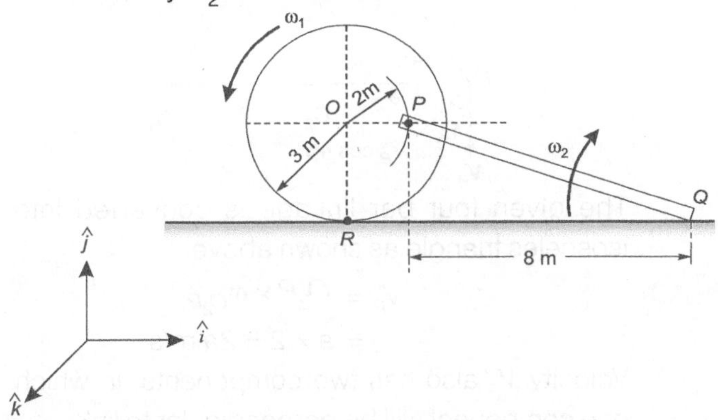

(a) $\omega_2 = 2\omega_1$ (b) $\omega_2 = \omega_1$
(c) $\omega_2 = 0.5\omega_1$ (d) $\omega_2 = 0.25\omega_1$

[2023 : 2 M]

1.28 At the instant when *OP* is vertical and AP is horizontal, the link *OD* is rotating counter clockwise at a constant rate $\omega = 7$ rad/s. Pin *P* on link *OD* slides in the slot *BC* of link *ABC* which is hinged at *A*, and causes a clockwise rotation of the link *ABC*. The magnitude of angular velocity of link *ABC* for this instant is _______ rad/s. (rounded off to 2 decimal places)

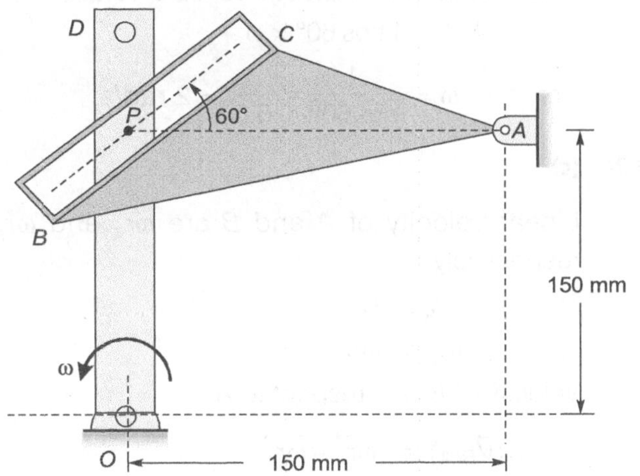

[2024 : 2 M]

Answers Displacement, Velocity and Acceleration

1.1 (a) 1.2 (c) 1.3 (d) 1.4 (a) 1.5 (c) 1.6 (d) 1.7 (c) 1.8 (d) 1.9 (c)
1.10 (d) 1.11 (40) 1.12 (1.25) 1.13 (a) 1.14 (1) 1.15 (b) 1.16 (d) 1.17 (243.310)
1.18 (0.266) 1.19 (1) 1.20 (a) 1.21 (d) 1.22 (b) 1.23 (0.79) 1.24 (d) 1.25 (a)
1.26 (1.25) 1.27 (d) 1.28 (12.12)

Explanations Displacement, Velocity and Acceleration

1.1 (a)

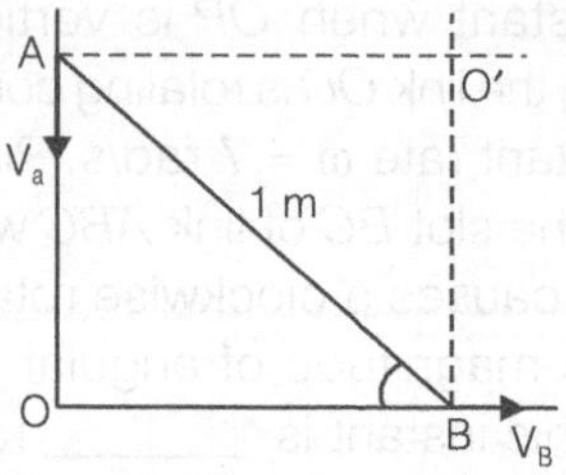

$$V_a = O'A \times \omega$$

Here, O' is instantaneous centre of rotation.

$$1 = 1\cos 60° \times \omega$$

or $$\omega = \frac{1}{\cos 60°} = \frac{1}{0.5} = \mathbf{2\ rad/s}$$

1.2 (c)

Linear velocity of A and B are ωr_A and ωr_B respectively.

$\because$ $r_B > r_A$

$\therefore$ $\omega r_B > \omega r_A$

Velocity of B with respect to A

$$(\vec{V}_{B/A}) = \overrightarrow{\omega r_B} - \overrightarrow{\omega r_A}$$

$\omega(r_B - r_A)$ in the direction of motion of point B.

1.3 (d)

Acceleration of point B with respect to point A.

$$\overrightarrow{a_{B/A}} = \vec{a}_B - \vec{a}_A$$

$$= \overrightarrow{\omega^2 r_B} - \overrightarrow{\omega^2 r_A}$$

$$= \omega^2 (r_B - r_A) \text{ in direction from } Z \text{ to } O.$$

1.4 (a)

Velocity of link 1 with respect to 2

$$V_{1/2} = 12 \text{ m/s}$$

$$\omega = \frac{2\pi N}{60} = \frac{2\pi \times 120}{60} = 12.566 \text{ rad/s}$$

$\therefore$ Corioli's component of acceleration

$$= 2V_{1/2}\omega = 2 \times 12 \times 12.566 \approx \mathbf{302\ m/s^2}$$

1.5 (c)

Method I:

When

$\angle O_4 O_2 P = 180°$

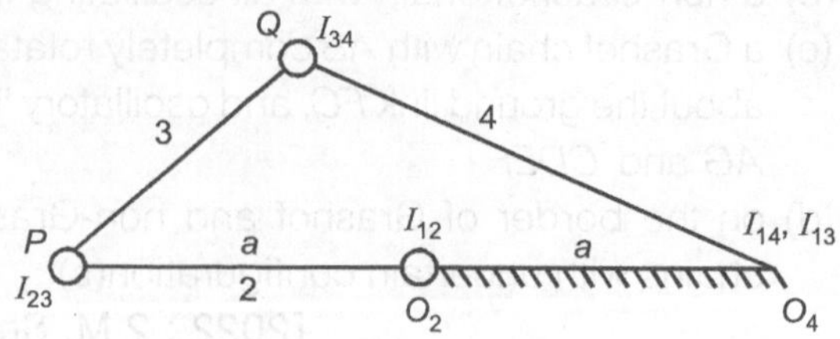

$$I_{13} \begin{cases} I_{12}\, I_{23} \\ I_{14}\, I_{34} \end{cases}$$

Applying angular velocity theorem at I_{23}.

$$\omega_2(I_{23}I_{12}) = \omega_3(I_{23}I_{13})$$

$$2(a) = \omega_3(2a)$$

$$\omega_3 = 2/2 = \mathbf{1\ rad/s}$$

Method II:

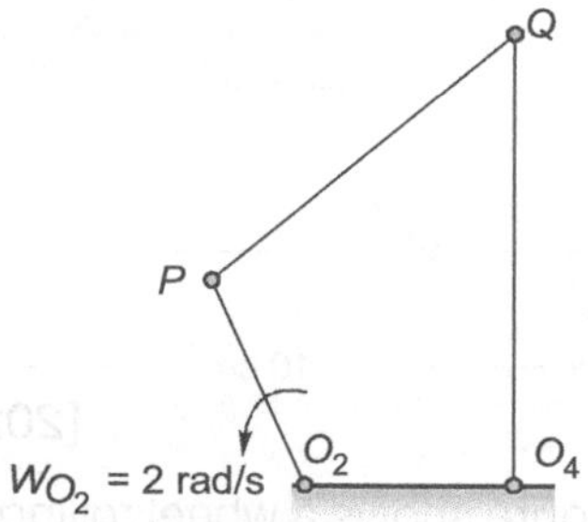

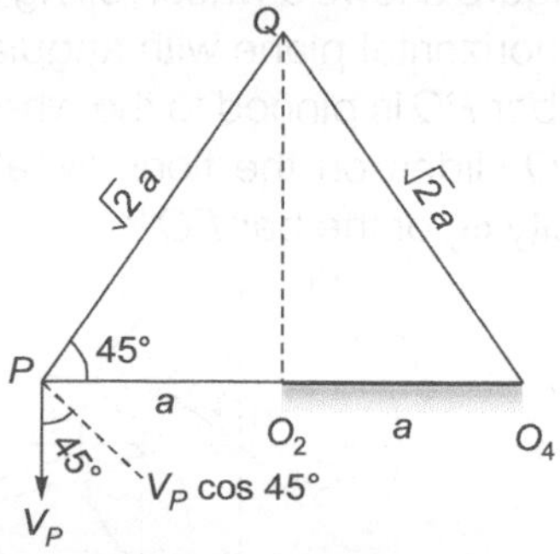

The given four bar linkage is converted into isosceles triangle as shown above.

$$V_P = O_2P \times \omega_{O_2P}$$

$$= a \times 2 = 2a \text{ m/s}$$

Velocity V_P also has two components, in which one component will be perpendicular to link,

$$PQ = V_P \cos 45°$$

$$= 2a \times \frac{1}{\sqrt{2}} = \sqrt{2}a \text{ m/s}$$

$\because$ Angular velocity of link PQ,

$$\omega_{PQ} = \frac{V_P \cos 45°}{PQ} = \frac{\sqrt{2}\,a}{\sqrt{2}\,a}$$

$$= 1 \text{ rad/s}$$

1.6 (d)

As AB is perpendicular to slider for given condition, so sliding velocity of slider equals to velocity of link AB

$$= \omega \times AB = 10 \times 0.25 = \mathbf{2.5\ m/s}$$

1.7 (c)

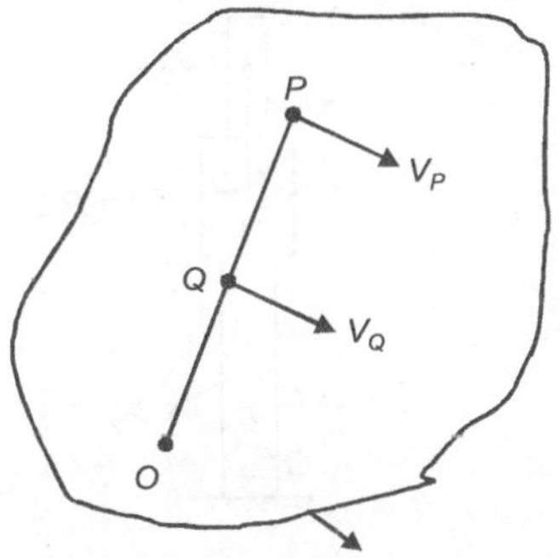

Planar rigid body.

V_{PQ} = Relative velocity between P and Q.

$V_{PQ} = V_P - V_Q$ Always perpendicular to PQ.

1.8 (d)

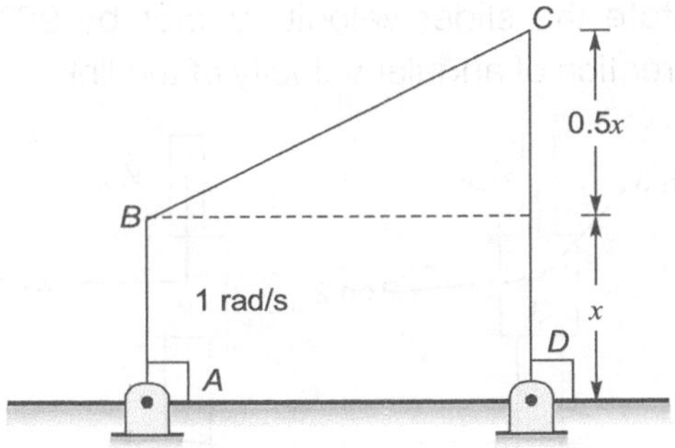

Let length of link AB is x velocity diagram;

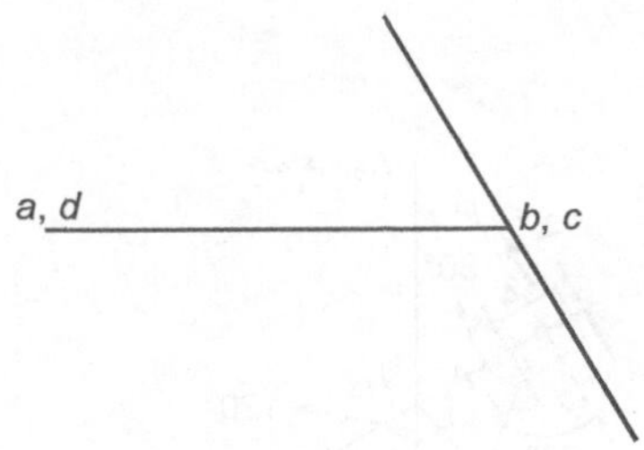

$$V_b = V_c$$

(As per from velocity diagram)

$$AB \times \omega_{AB} = CD \times \omega_{CD}$$

$$1 = \left(\frac{CD}{AB}\right) \times \omega_{CD}$$

$$1 = 1.5 \times \omega_{CD}$$

$$\omega_{CD} = \frac{2}{3}\ \mathbf{rad/s}$$

1.9 (c)

Acceleration of the block at $A = \vec{a}_{cr} + \vec{a}_r$

$$= \overrightarrow{2\omega V} + \overrightarrow{\omega^2 r}$$

$$\Rightarrow \quad r = OA$$

$$\Rightarrow \quad 2\omega V = 2 \times 2 \times 0.75 = 3\ \text{m/s}^2$$

$$\Rightarrow \quad \omega^2 r = 2^2 \times 1 = 4\ \text{m/s}^2$$

As Corioli's component and radial component are perpendicular to each other.

So, $\quad |\vec{a}_{abs}| = \sqrt{3^2 + 4^2} = \mathbf{5\ m/s^2}$

1.10 (d)

<u>Method I:</u>

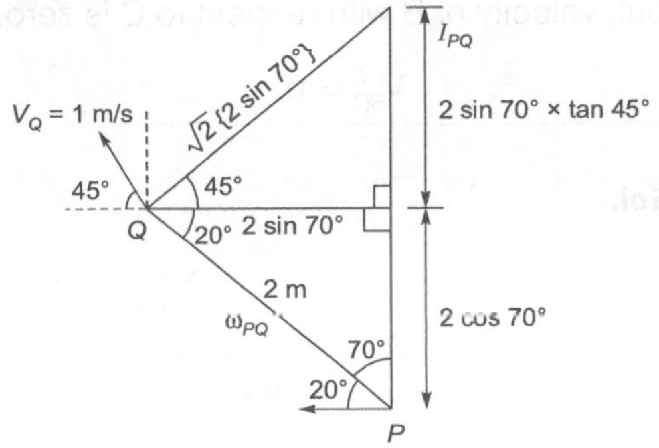

$$\therefore \quad \omega_{PQ} = \frac{V_P}{PI_{PQ}} = \frac{V_Q}{QI_{PQ}}$$

$P.I_{PQ}$ = Distance between P and I_{PQ}

$= 2 \cos 70° + 2 \sin 70° \times \tan 45°$

$= 2.56$ m

$Q.I_{PQ}$ = Distance between Q and I_{PQ}

$= \sqrt{2} \times 2 \times \sin 70° = 2.66$ m

$$\therefore \quad \frac{V_P}{2.56} = \frac{1}{2.66}$$

or $\quad V_P = \dfrac{2.56}{2.66} = \mathbf{0.96\ m/s}$

<u>Method II:</u>

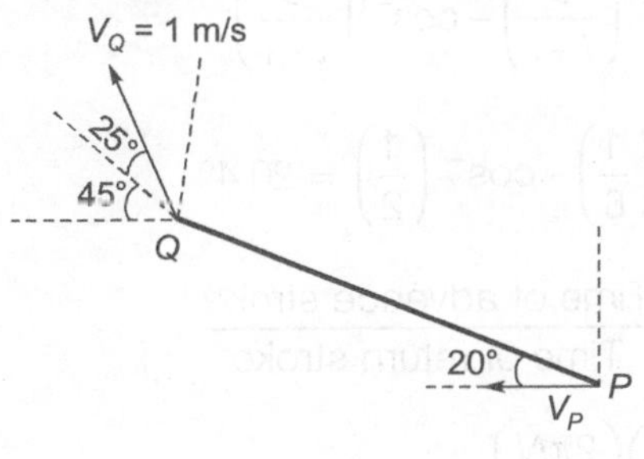

As, according to the data, link PQ is rigid link, so the velocity at very along the link will be same. Therefore:

$$V_p \cos 20° = V_Q \cos 25°$$

$$V_P = V_Q \frac{\cos 25°}{\cos 20°} = 1 \times 0.9644$$

$$V_p = 0.96 \text{ m/s}$$

POINTS TO REMEMBER

In this problem, as we have ssen that

$$V_B = V_C$$

but, velocity of B with respect to C is zero. i.e.

$$V_{BC} = 0$$

1.11 Sol.

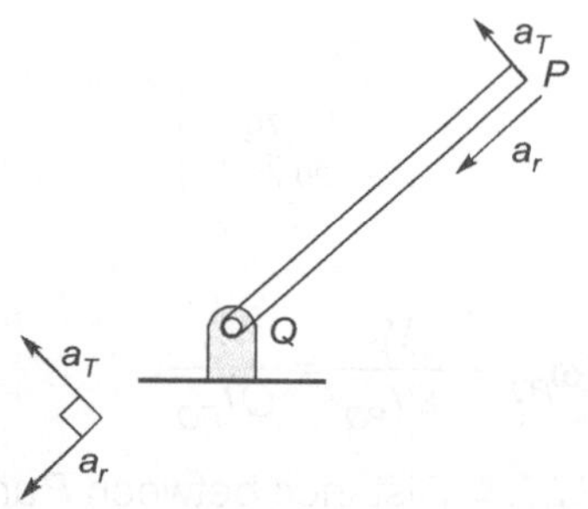

$$a_T = r\alpha = 2 \times 12 = 24 \text{ m/s}^2$$

$$a_R = \omega^2 r = 16 \times 2 = 32 \text{ m/s}^2$$

$$a = \sqrt{a_T^2 + a_R^2} = \sqrt{24^2 + 32^2}$$

$$= \sqrt{1600} = \mathbf{40 \text{ m/s}^2}$$

1.12 Sol.

Length of connecting rod: l_r = 40 mm

Crank radius: r = 20 mm

Eccentricity: e = 10 mm

$$\phi = \cos^{-1}\left(\frac{e}{l+r}\right) - \cos^{-1}\left(\frac{e}{l-r}\right)$$

$$= \cos^{-1}\left(\frac{1}{6}\right) - \cos^{-1}\left(\frac{1}{2}\right) = 20.4°$$

$$\text{QRR} = \frac{\text{Time of advance stroke}}{\text{Time of return stroke}}$$

$$= \left(\frac{\theta_C}{2\pi N}\right)\left(\frac{2\pi N}{\theta_R}\right)$$

$$\text{QRR} = \frac{\theta_C}{\theta_R} = \frac{180° + \phi}{180° - \phi}$$

$$= \frac{180° + 20.4°}{180° - 20.4°} = 1.25$$

1.13 (a)

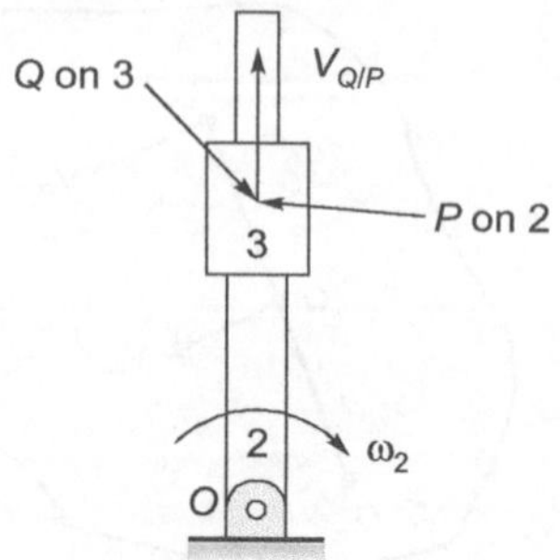

Coriolis acceleration

$$= 2 \times \text{velocity of slide} \times \text{angular velocity of link 2.}$$

$$a_c = 2 \times V_{a/p} \times \omega_2$$

To determine the direction of coriolis component, rotate the slider velocity vector by 90° in the direction of angular velocity of the link.

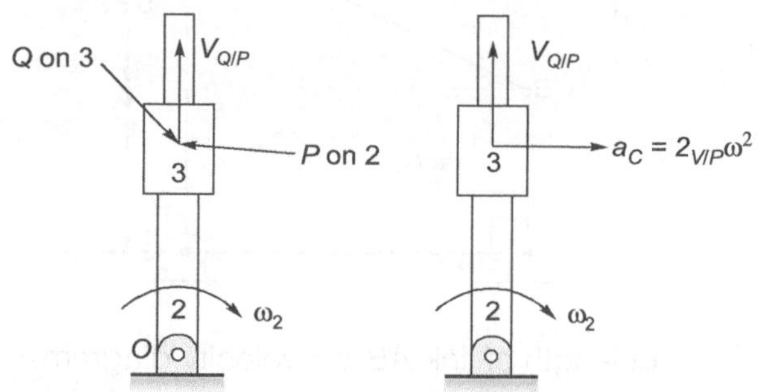

1.14 Sol.

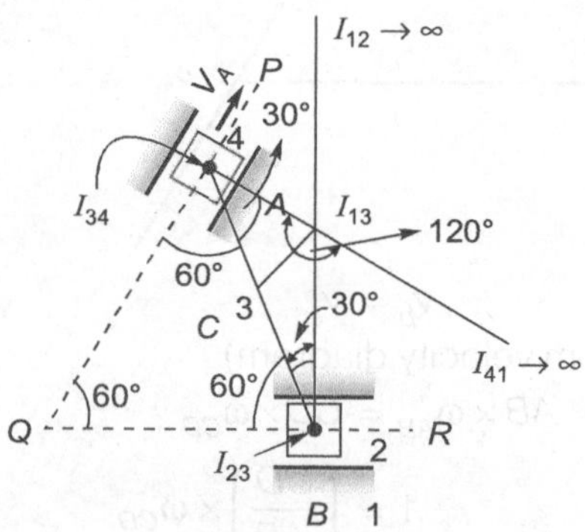

$$\omega_3 = \omega_{AB} = \frac{V_A}{I_{13A}} = \frac{V_B}{I_{13B}} = \frac{V_C}{I_{13C}}$$

I_{13A} = Distance between I_{13} and point A

I_{13B} = Distance between I_{13} and point B

I_{13C} = Distance between I_{13} and point C

$$\frac{V_A}{I_{13A}} = \frac{V_C}{I_{13C}}; \qquad \frac{2}{I_{13A}} = \frac{V_C}{I_{13C}}$$

Applying sine law in triangle ABI_{13}

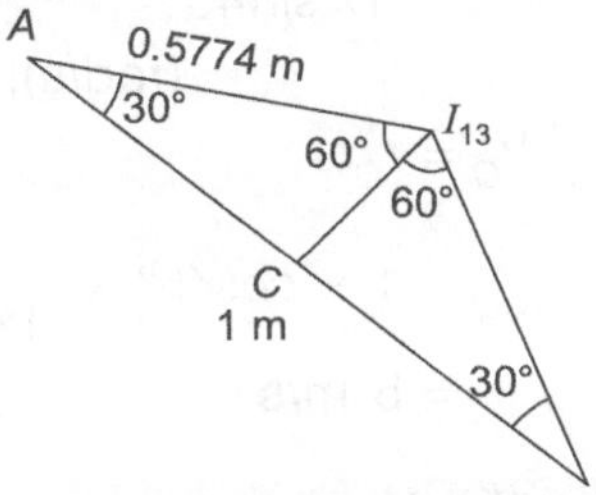

$$\frac{1}{\sin 120°} = \frac{I_{13A}}{\sin 30°}$$

$$I_{13A} = \frac{\sin 30°}{\sin 120°}$$

Applying sine law in ΔACI_{13} = **0.5774**

$$\frac{I_{13C}}{\sin 30°} = \frac{I_{13A}}{\sin 90°}$$

$$I_{13C} = \frac{I_{13A}\sin 30°}{\sin 90°}$$

$$= 0.5774 \times 0.5 = 0.2887 \text{ m}$$

$$V_C = \frac{V_A \cdot I_{13C}}{I_{13a}} = \frac{2 \times 0.2887}{0.5774} = 1 \text{ m/s}$$

1.15 (b)

The relation between the no. of links 'n', lower pair 'j' and higher pair 'h' gives degre of freedom. For planar mechanism.

$$F = 3(n-1) - 2j - h$$

Here, no higher pair, so, $h = 0$

$$F = 3(n-1) - 2j$$

1.16 (d)

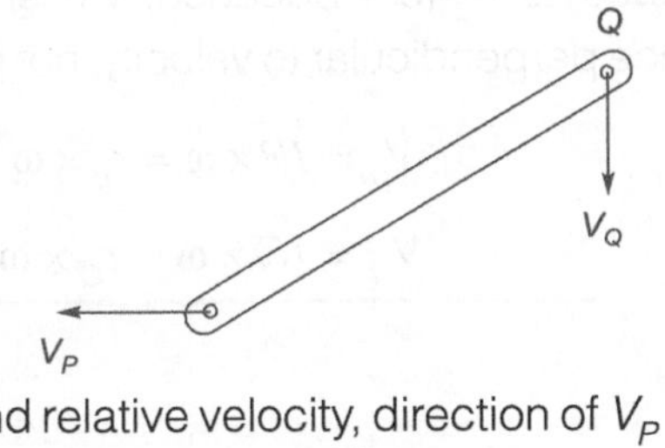

To find relative velocity, direction of V_P is reversed and placed at fail of $\vec{V}_Q$

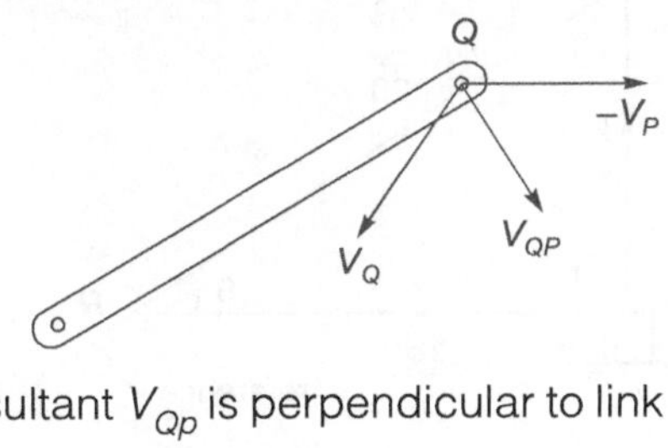

Resultant V_{Qp} is perpendicular to link PQ.

1.17 Sol.

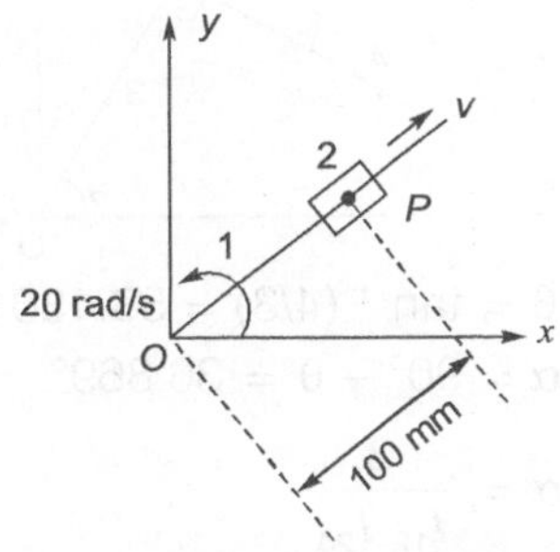

V = 6 m/s, ω = 20 rad/s

Here the centripetal and Corioli's acceleration will be there

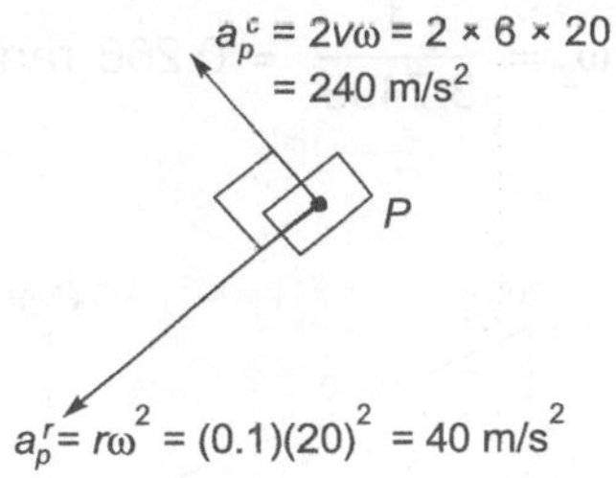

Total acceleration:

$$a_p = \sqrt{(40)^2 + (240)^2} = \mathbf{243.310 \text{ m/s}^2}$$

1.18 Sol.

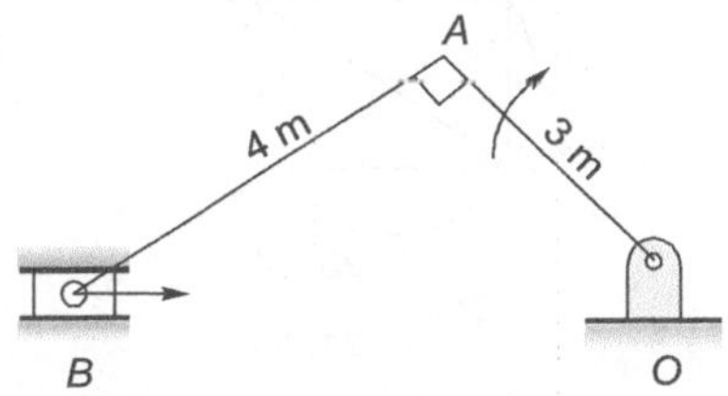

Here V_B = 1 m/s, ω_{crank} = ?

Let us draw the I-centres

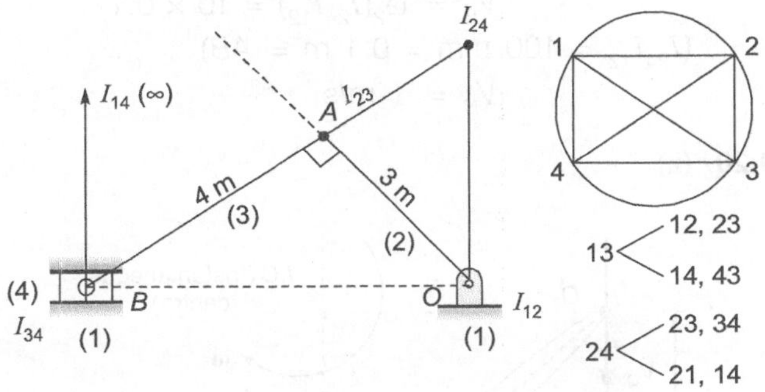

Applying angular velocity theorem at I_{24}

$$\therefore \omega_2(I_{24}I_{12}) = V_4 = V_B = 1$$

$$\omega_2(I_{24}I_{12}) = 1$$

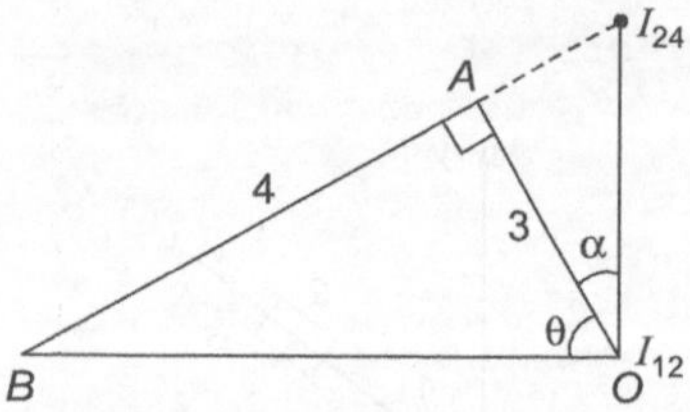

$$\theta = \tan^{-1}(4/3) = 53.13010°$$

$$\alpha = 90° - \theta = 36.869°$$

$$\cos\alpha = \frac{3}{I_{12}I_{24}}$$

$$I_{12}I_{24} = \frac{3}{\cos 36.869°} = 3.7499 \text{ meter}$$

Now, $\omega_2(3.7499) = 1$

$$\omega_2 = \frac{1}{3.7499} = 0.266 \text{ rad/s}$$

1.19 Sol.

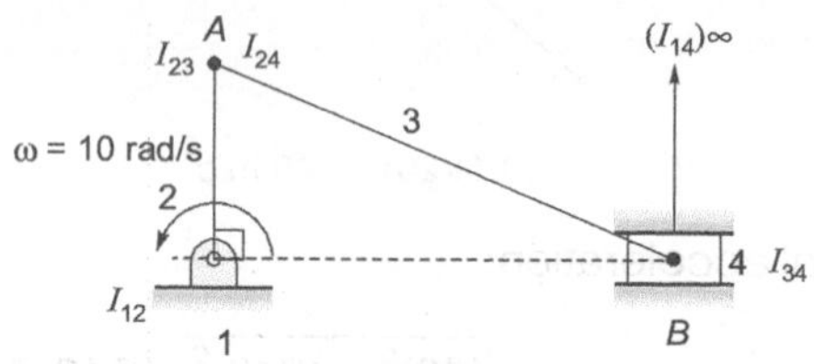

After plotting I-centres

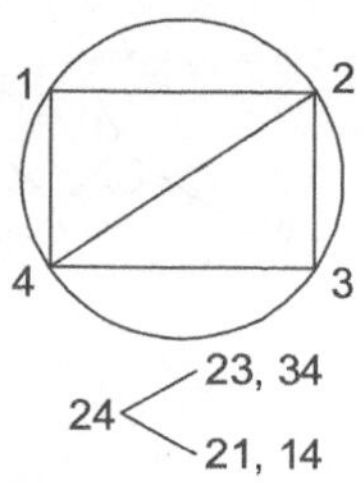

Here, I_{23} and I_{24} will come at same point

Applying angular Velocity Theorem at I_{24}

$$\therefore \quad \omega_2(I_{24}I_{12}) = V_4 = V_B$$

$$V_B = \omega_2(I_{24}I_{12}) = 10 \times 0.1$$

$(I_{24}I_{12} = 100 \text{ mm} = 0.1 \text{ m} = AB)$

$$V_B = 1 \text{ m/s}$$

1.20 (a)

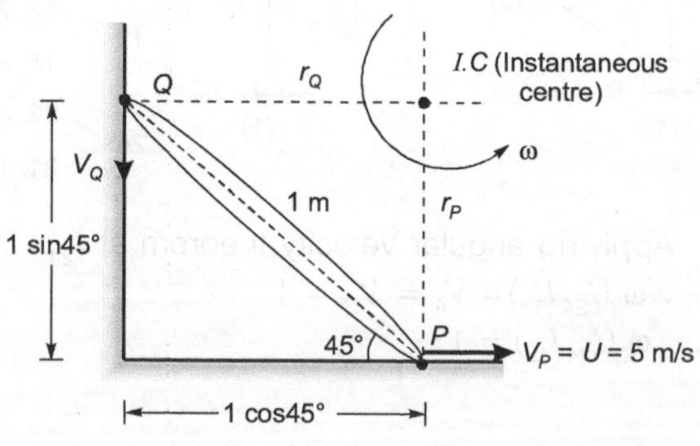

$$V_P = r_P\omega$$

$$\Rightarrow \quad 5 = 1 \times \sin 45° \times \omega$$

$$\Rightarrow \quad \omega = \frac{5}{1 \times \sin 45°}$$

(rad/s); anticlockwise

$$V_Q = r_Q\omega$$

$$\Rightarrow \quad V_Q = 1 \times \cos 45° \times \frac{5}{1 \times \sin 45°}$$

$$= 5 \text{ m/s}$$

AVOID MISTAKE

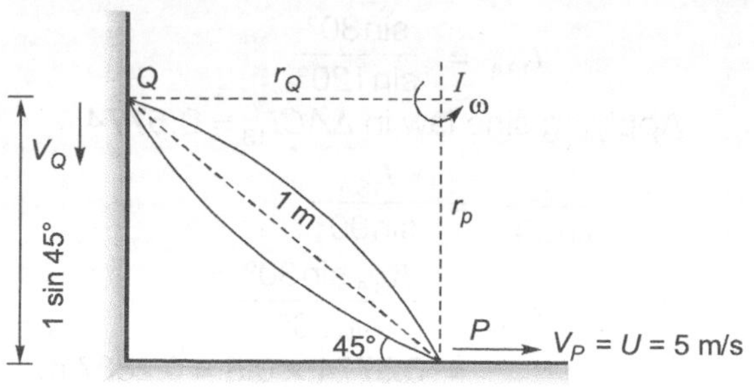

1 cos 45°

$$V_P = IQ \times \omega$$

$$5 = 1 \cos 45° \times \omega$$

$$\omega = \frac{5}{\cos 45°}$$

$$V_Q = IP \times \omega$$

$$= 1 \sin 45° \times \frac{5}{\cos 45°}$$

$$= \frac{5}{\tan 45°} = 5 \text{ m/s}$$

Note: Here student gets right answer by chance, (because ∠ = 45°) because, we should take distance perpendicular to velocity, not parallel, so

$$V_p = IP \times \omega = r_p \times \omega$$

and $$V_Q = IQ \times \omega = r_Q \times \omega$$

1.21 (d)

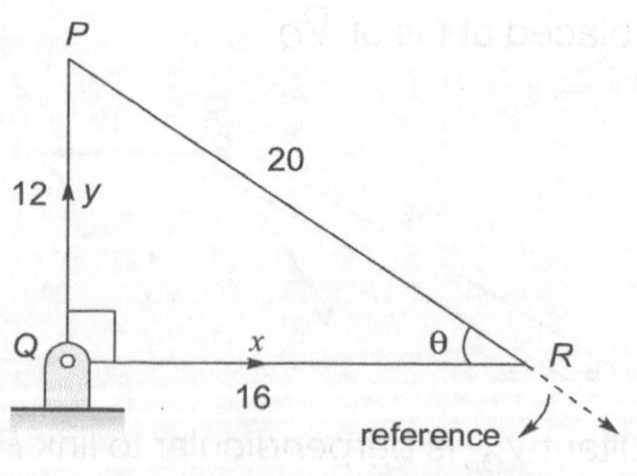

As acceleration of point Q is zero, therefore we can assume this rigid body PQR is hinged at Q at this instant.

$$\vec{a}_{RP} = \vec{a}_R - \vec{a}_P$$

is given as 10 m/s^2 at an angle of 180°, that means only radial acceleration is there at that instant and reference is PR

$$a_{RP} = (RP)\omega^2 = 10$$

$$\Rightarrow \quad 20\omega^2 = 10$$

$$\Rightarrow \quad \omega = \frac{1}{\sqrt{2}}$$

Since point R has only radial acceleration, so total acceleration is only radial, therefore tangential acceleration is zero.

Therefore $\alpha_{body} = 0$

$$a_R = QR(\omega^2) = 16 \times \frac{1}{2} = 8 \text{ m/s}^2$$

and will be in the horizontal backward direction, but our reference is along PR, so the angle of it from reference is $(180 + \theta)$

from ΔPQR

$$\tan\theta = \frac{12}{16}$$

$$\Rightarrow \quad \theta = 36.8698°$$

So, $180 + 36.8698 = 216.8698 \simeq 217°$

So answer is **8∠217°**

1.22 (b)

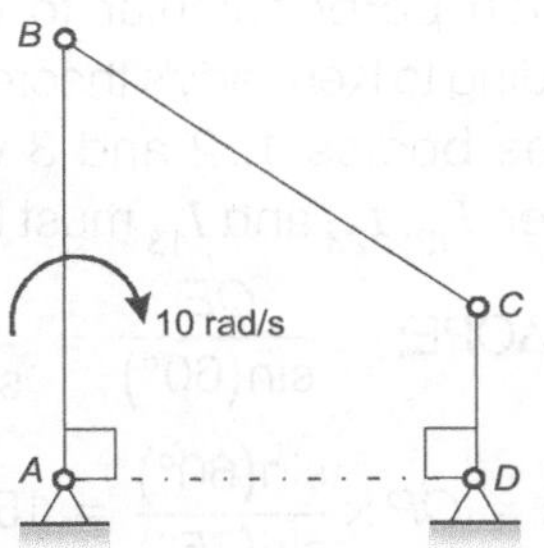

Velocity diagram:

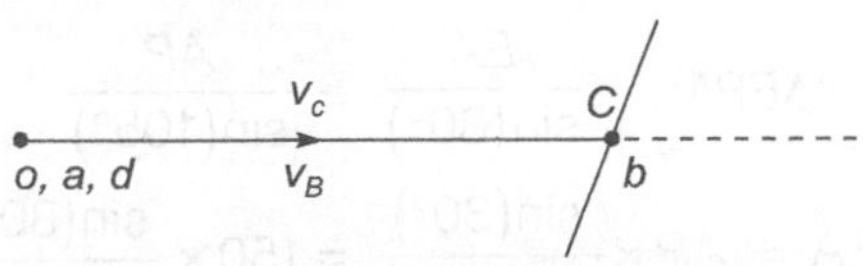

$$V_C = V_B = (AB) \times \omega_{AB}$$

$$= 0.05 \times 10) = 0.5 \text{ m/s}$$

$$\omega_{CD} = \frac{V_C}{CD} = \frac{0.5}{0.02} = 25 \text{ rad/s}$$

1.23 Sol.

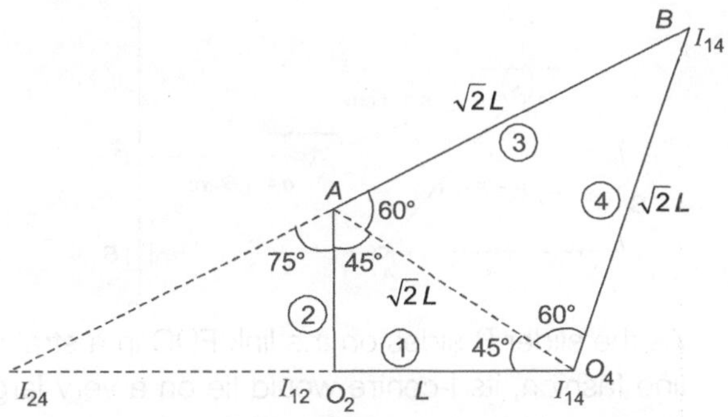

According to Arnold's Kennedy theorem,

$$\Rightarrow \quad \tan 75° = \frac{(I_{24}I_{12})}{(L)}$$

$$(I_{24}I_{12}) = (L \tan 75°) = 3.732\, L$$

$$(I_{24}I_{14}) = 4.732\, L$$

$$\omega_2 \times (I_{24}I_{12}) = \omega_4 (I_{24}I_{14})$$

$$\frac{\omega_4}{\omega_2} = \frac{I_{24}I_{12}}{I_{24}I_{14}}$$

$$\frac{\omega_4}{\omega_2} = \left(\frac{3.732}{4.732}\right) = 0.7886$$

1.24 (d)

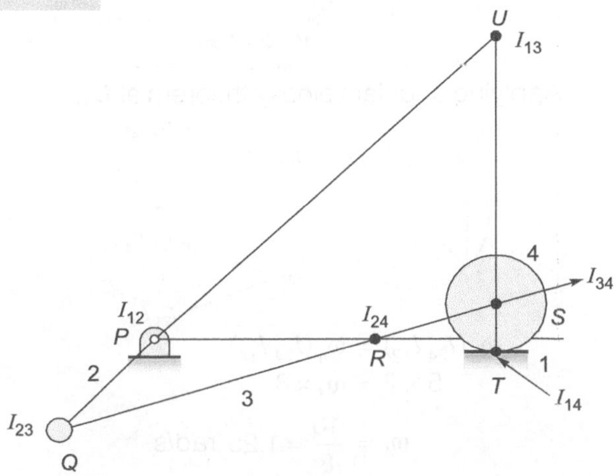

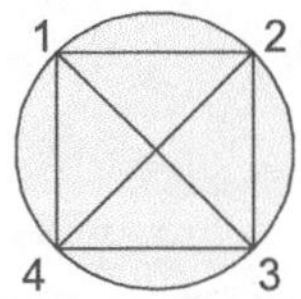

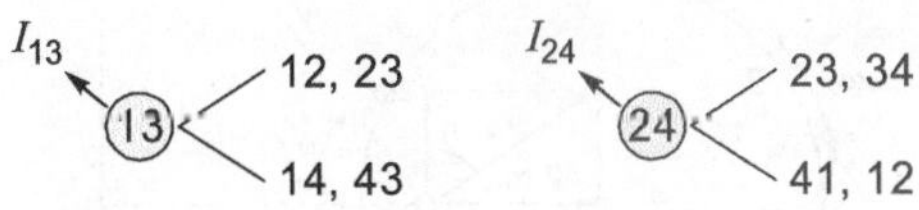

Points P, Q, R, S, T and U are instantaneous centers of mechanism.

1.25 (a)

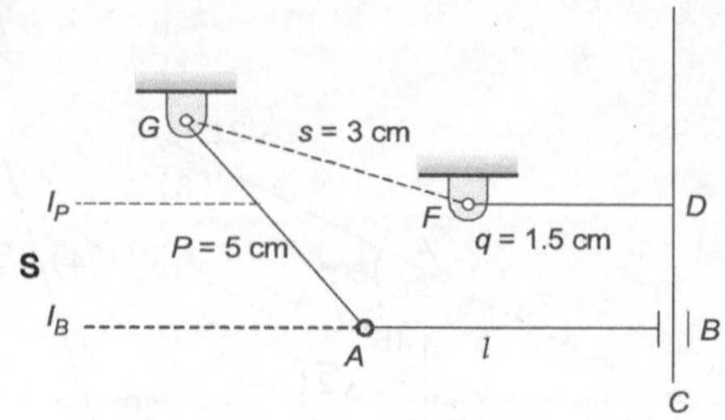

As the slider B sides on the link FDC in a straight line fashion, its I-centre would lie on a very large distance. Similarly for the link FDC, its I-centre would also be on a large distance. We may consider that

$$Q - q = 3 - 1.5 = 1.5 \text{ cm}$$

$$P - s = 5 - 3 = 2$$

$\therefore$ Since Grashof's law is, $s + l < P + q$

$\therefore \quad l - q < P - P$

$\therefore \quad 1.5 < 2$ (Grashof's law is satisfied)

As the shortest link (s = 3 cm) is fixed, so all other links would rotate and acts as cranks.

1.26 (1.25) (1.24 to 1.26)

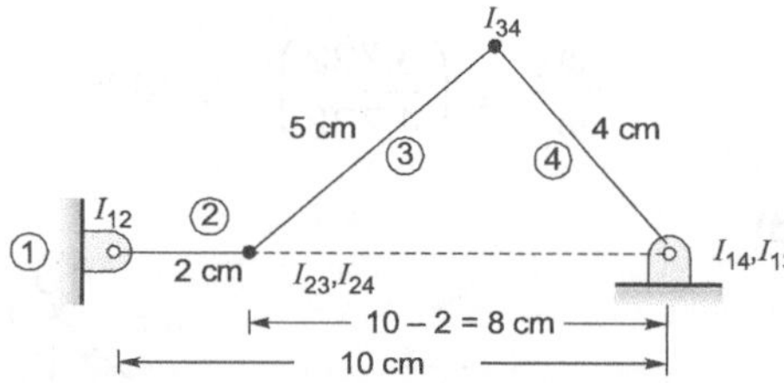

Applying angular velocity theorem at I_{24}:

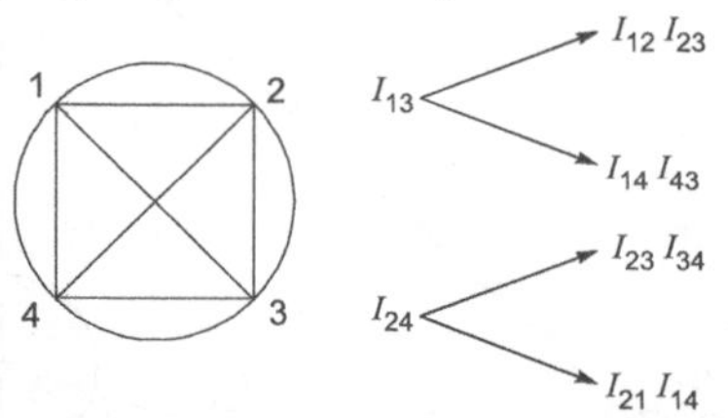

$$\omega_2 (I_{24} I_{12}) = \omega_4 (I_{24} I_{14})$$

$$5 \times 2 = \omega_4 \times 8$$

$$\omega_4 = \frac{10}{8} = 1.25 \text{ rad/s}$$

1.27 (d)

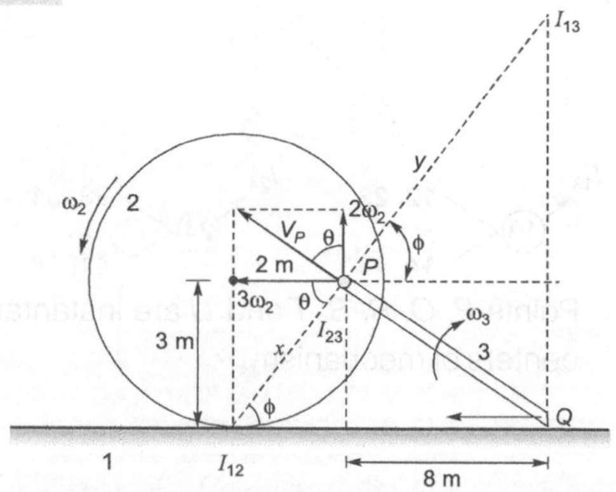

Let us assume fixed link as 1, disc as link 2 and rod as link 3.

From geometry, $\tan\theta = \dfrac{3}{2}$ and $\tan\phi = \dfrac{3}{2}$

$\Rightarrow \quad \theta = \phi$

$\therefore \quad x\cos\phi = 2 \Rightarrow y\cos\phi = 8$

Applying Kennedy theorem,

$$\omega_2 (I_{12} I_{23}) = \omega_3 (I_{13} I_{23})$$

Since '1' is used for disc and '2' is used for rod in the question,

$$\therefore \quad \omega_1\left(\frac{2}{\cos\phi}\right) = \omega_2\left(\frac{8}{\cos\phi}\right)$$

$$\Rightarrow \quad \omega_1 = 4\omega_2 \Rightarrow \omega_2 = 0.25\omega_1$$

1.28 (12.124) (11.30 to 12.70)

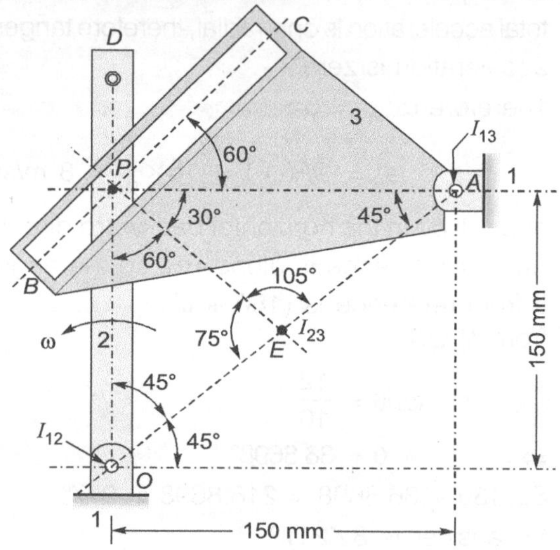

$\omega_2 = \omega = 7$ rad/s

Pin P slides in direction BC, so I_{23} must lie in direction perpendicular to direction BC. Also, according to Kennedy's theorem for relative motion of three bodies 1, 2 and 3 with respect to one another, I_{12}, I_{23} and I_{13} must lie on a straight line.

From ΔOPE; $\quad \dfrac{OE}{\sin(60°)} = \dfrac{OP}{\sin(75°)}$

$$\Rightarrow OE = OP \times \frac{\sin(60°)}{\sin(75°)} = 150 \times \frac{\sin(60°)}{\sin(75°)}$$

$$= 134.486 \text{ mm}$$

From ΔEPA; $\quad \dfrac{EA}{\sin(30°)} = \dfrac{AP}{\sin(105°)}$

$$\Rightarrow EA = AP \times \frac{\sin(30°)}{\sin(105°)} = 150 \times \frac{\sin(30°)}{\sin(105°)}$$

$$= 77.646 \text{ mm}$$

Now, $\quad V_{I_{23}} = \omega_2 (I_{12} I_{23}) = \omega_3 (I_{13} I_{23})$

$$\Rightarrow \quad \omega_3 = \omega_2 \times \frac{I_{12} I_{23}}{I_{13} I_{23}} = \omega_2 \times \frac{OE}{EA}$$

$$= 7 \times \frac{134.486}{77.646} = 12.124 \text{ rad/s}$$

■■■■

2 Planar Mechanisms

2.1 Figure shows a quick return mechanism. The cranks OA rotates clockwise uniformly. OA = 2 cm, OO' = 4 cm. The ratio of time for forward motion to that for return motion is

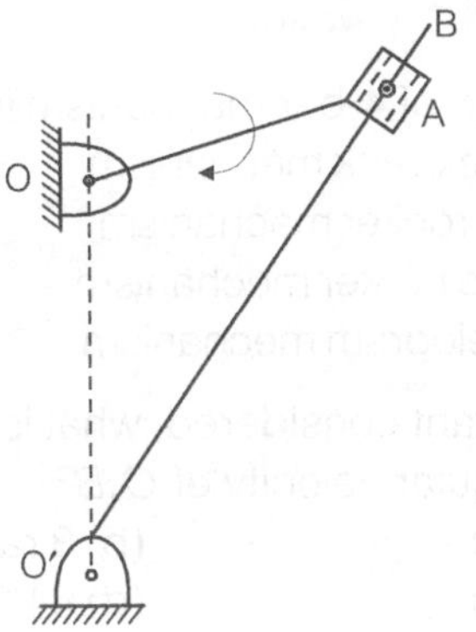

(a) 0.5 (b) 2.0

(c) $\sqrt{2}$ (d) 1 **[1995 : 2 M]**

2.2 For the planar mechanism shown in figure select the most appropriate choice for the motion of link 2 when link 4 is moved upwards.

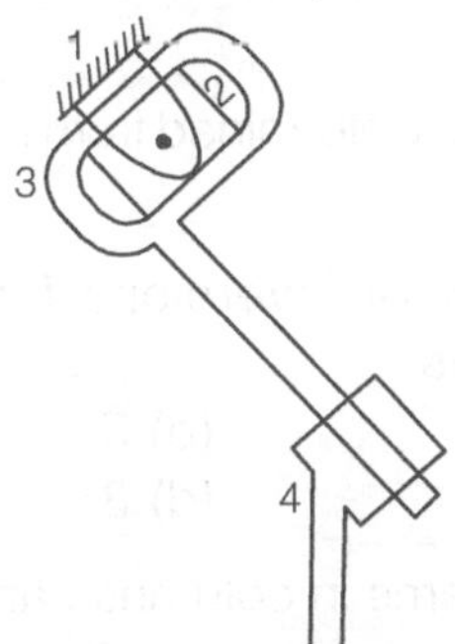

(a) Link 2 rotates clockwise
(b) Link 2 rotates counter clockwise
(c) Link 2 does not move
(d) Link 2 motion cannot be determined

[1999 : 1 M]

2.3 For the audio Cassette mechanism shown in figure given below where is the instantaneous centre of rotation (P) of the two spools?

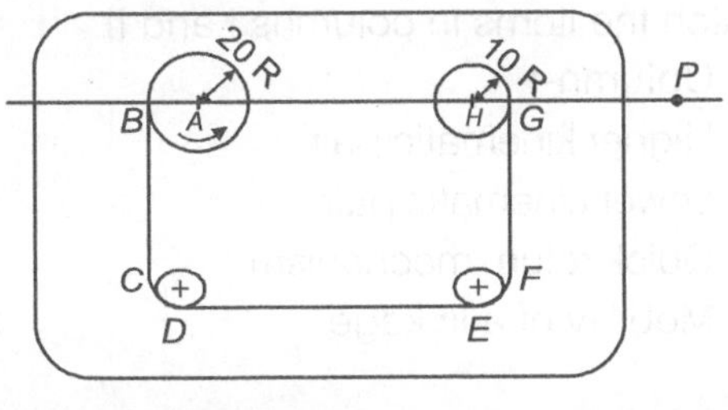

(a) Point P lies to the left of both the spools but at infinity along the line joining A and H
(b) Point P lies in between the two spools on the line joint A and H, such that $PH = 2AP$
(c) Point P lies to the right of both the spools on the line joining A and H, such that $AH = HP$
(d) Point P lies at the intersection of the line joining B and C and the line joining G and F

[1999 : 2 M]

2.4 The mechanism used in a shaping machine is
(a) a closed 4-bar chain having 4 revolute pairs
(b) a closed 6-bar chain having 6 revolute pairs
(c) a closed 4-bar chain having 2 revolute and 2 sliding pairs
(d) an inversion of the single slider-crank chain

[2003 : 1 M]

2.5 The lengths of the links of a 4-bar linkage with revolute pairs only are p, q, r and s units. Given that $p < q < r < s$. Which of these links should be the fixed one, for obtaining a "double crank" mechanism?

(a) links of length p (b) links of length q

(c) link of length r (d) link of length s

[2003 : 1 M]

2.6 The figure below shows a planar mechanism with single degree of freedom. The instant center 24 for the given configuration is located at a position

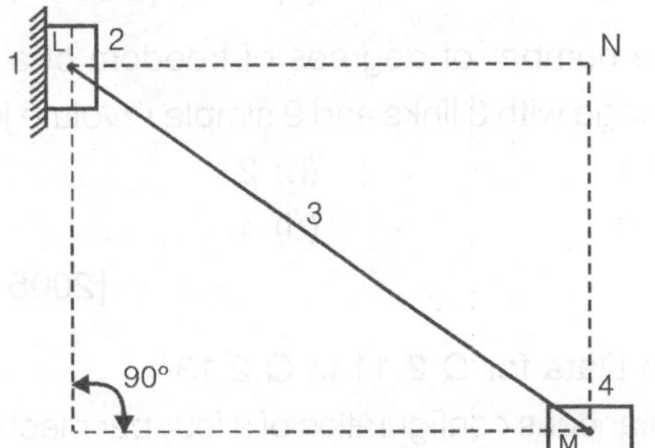

(a) L (b) M

(c) N (d) ∞

[2004 : 2 M]

2.7 Match the following:

Type of Mechanism

P. Scott - Russel mechanism
Q. Geneva mechanism
R. Off-set slider-crank mechanism
S. Scotch Yoke mechanism

Motion achieved

1. Intermittent motion
2. Quick return motion
3. Simple harmonic motion
4. Straight line motion

(a) P-2 Q-3 R-1 S-4 (b) P-3 Q-2 R-4 S-1
(c) P-4 Q-1 R-2 S-3 (d) P-4 Q-3 R-1 S-2

[2004 : 2 M]

2.8 Match the following with respect to spatial mechanisms.

Type of Joint	Degrees of constraint
P. Revolute	1. Three
Q. Cylindrical	2. Five
R. Spherical	3. Four
	4. Two
	5. Zero

(a) P-1 Q-3 R-1 (b) P-5 Q-4 R-3
(c) P-2 Q-3 R-1 (d) P-4 Q-5 R-3

[2004 : 2 M]

2.9 For a mechanism shown below, the mechanical advantage for the given configuration is

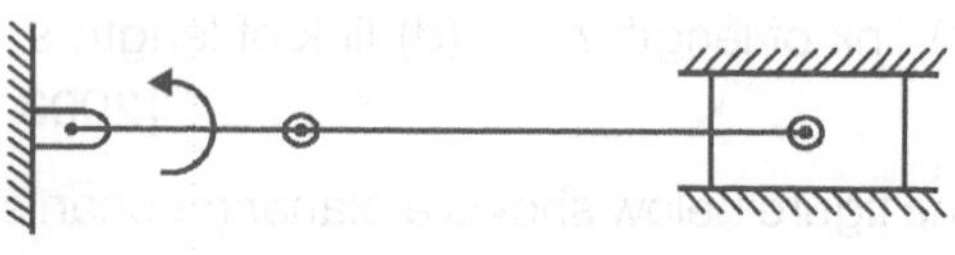

(a) 0 (b) 0.5
(c) 1.0 (d) ∞ [2004 : 1 M]

2.10 The number of degrees of freedom of a planar linkage with 8 links and 9 simple revolute joints is

(a) 1 (b) 2
(c) 3 (d) 4

[2005 : 1 M]

Common Data for Q 2.11 to Q.2.13

An instantaneous configuration of a four-bar mechanism, whose plane is horizontal, is shown in the figure below. At this instant, the angular velocity and angular acceleration of link O_2A are $\omega = 8$ rad/s and $\alpha = 0$, respectively, and the driving torque (τ) is zero. The link O_2A is balanced so that its centre of mass falls at O_2.

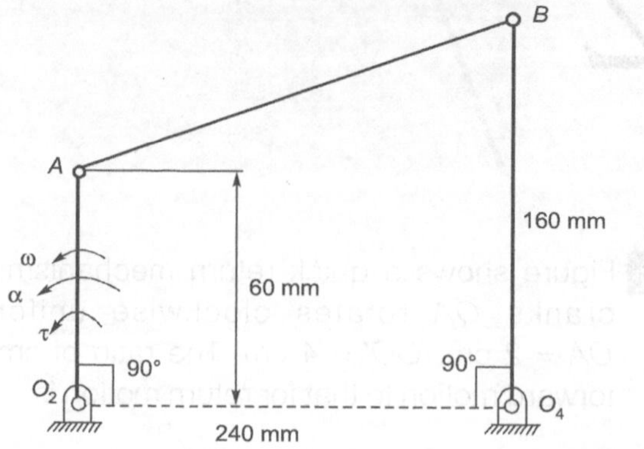

2.11 Which kind of 4-bar mechanism is O_2ABO_4?

(a) Double crank mechanism
(b) Crank-rocker mechanism
(c) Double rocker mechanism
(d) Parallelogram mechanism [2005 : 2 M]

2.12 At the instant considered, what is the magnitude of the angular velocity of O_4B?

(a) 1 rad/s (b) 3 rad/s
(c) 8 rad/s (d) 64/3 rad/s

[2005 : 2 M]

2.13 At the same instant, if the component of the force at joint A along AB is 30 N, then the magnitude of the joint reaction at O_2

(a) is zero
(b) is 30 N
(c) is 78 N
(d) cannot be determined from the given data

[2005 : 2 M]

2.14 The number of inversions for a slider crank mechanism is

(a) 6 (b) 5
(c) 4 (d) 3 [2006 : 1 M]

2.15 Match the items in columns I and II.

Column I	Column II
P. Addendum	1. Cam
Q. Instantaneous center of velocity	2. Beam
R. Section modulus	3. Linkage
S. Prime circle	4. Gear

(a) P-4, Q-2, R-3, S-1
(b) P-4, Q-3, R-2, S-1
(c) P-3, Q-2, R-1, S-4
(d) P-3, Q-4, R-1, S-2 [2006 : 2 M]

2.16 Match the items in columns I and II

Column-I

P. Higher kinematic pair
Q. Lower kinematic pair
R. Quick return mechanism
S. Mobility of a linkage

Column-II
1. Grubler's equation
2. Line contact
3. Euler's equation
4. Planer
5. Shaper
6. Surface contact

(a) P–2, Q–6, R–4, S–3
(b) P–6, Q–2, R–4, S–1
(c) P–6, Q–2, R–5, S–3
(d) P–2, Q–6, R–5, S–1 [2006 : 2 M]

2.17 In a four-bar linkage, S denotes the shortest link length, L is the longest link length, P and Q are the lengths of other two links. At least one of the three moving links will rotate by 360° if

(a) $S + L \leq P + Q$ (b) $S + L > P + Q$
(c) $S + P \leq L + Q$ (d) $S + P > L + Q$
[2006 : 2 M]

2.18 For a four bar linkage in toggle position, the value of mechanical advantage is

(a) 0.0 (b) 0.5
(c) 1.0 (d) ∞ [2006 : 1 M]

Linked Answer Questions 2.19 to 2.20:

A quick return mechanism is shown below. The crank OS is driven at 2 rev/s in counter clockwise direction.

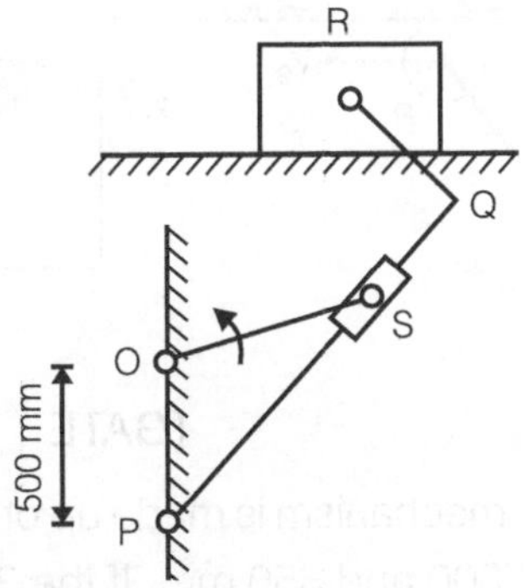

2.19 If the quick return ratio is 1 : 2, then the length of the crank in mm is

(a) 250 (b) $250\sqrt{3}$
(c) 500 (d) $500\sqrt{3}$
[2007 : 2 M]

2.20 The angular speed of PQ in rev/s when the block R attains maximum speed during forward stroke (stroke with slower speed) is (in rad/s)

(a) 1/3 (b) 2/3
(c) 2 (d) 3 [2007 : 2 M]

2.21 A planar mechanism has 8 links and 10 rotary joints. The number of degrees of freedom of the mechanism, using Grubler's criterion, is

(a) 0 (b) 1
(c) 2 (d) 3 [2008 : 1 M]

2.22 A simple quick return mechanism is shown in the figure. The forward to return ratio of the quick return mechanism is 2 : 1. If the radius of the crank O_1P is 125 mm, then the distance d (in mm) between the crank centre to lever pivot centre point should be

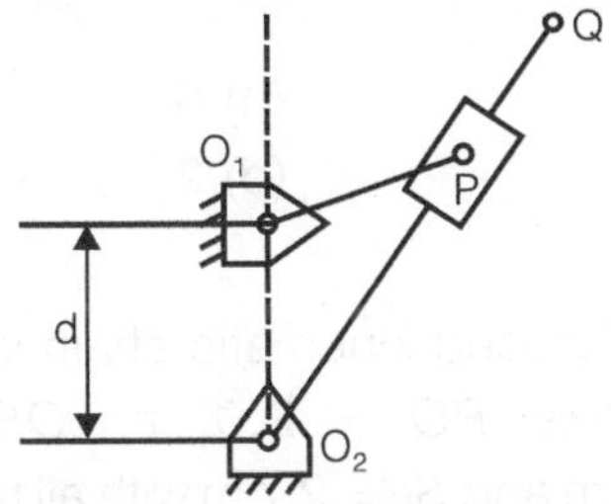

(a) 144.3 (b) 216.5
(c) 240.0 (d) 250.0 [2009 : 1 M]

2.23 Match the approaches given below to perform stated kinematics/dynamics analysis of machine.

Analysis	Approach
P. Continuous relative rotation	1. D' Alembert's principle
Q. Velocity and acceleration	2. Grubler's criterion
R. Mobility	3. Grashof's law
S. Dynamic – static analysis	4. Kennedy's theorem

(a) P–1, Q–2, R–3, S–4
(b) P–3, Q–4, R–2, S–1
(c) P–2, Q–3, R–4, S–1
(d) P–4, Q–2, R–1, S–3 [2009 : 2 M]

2.24 Mobility of a statically indeterminate structure is

(a) ≤ –1 (b) 0
(c) 1 (d) ≥ 2 [2010 : 1 M]

2.25 Which of the following statements is INCORRECT?

(a) Grashof's rule states that for a planar crank-rocker four bar mechanism, the sum of the shortest and longest link lengths cannot be less than the sum of the remaining two link lengths.
(b) Inversions of a mechanism are created by fixing different links one at a time.
(c) Geneva mechanism is an intermittent motion device.
(d) Grubler's criterion assumes mobility of a planar mechanism to be one.

[2010 : 1 M]

2.26 A double parallelogram mechanism is shown in the figure. Note that PQ is a single link. The mobility of the mechanism is

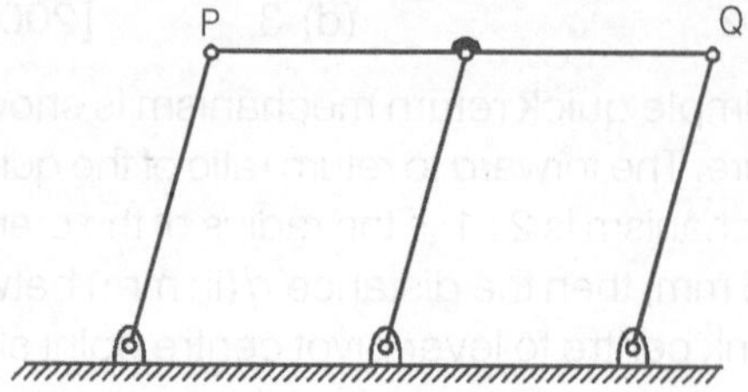

(a) −1 (b) 0
(c) 1 (d) 2

[2011 : 1 M]

2.27 A planar closed kinematic chain is formed with rigid links PQ = 2.0 m, QR = 3.0 m, RS = 2.5 m and SP = 2.7 m with all revolute joints. The link to be fixed to obtain rocker (rocker-rocker) mechanism is

(a) PQ (b) QR
(c) RS (d) SP

[2013 : 1 M]

2.28 A 4-bar mechanism with all revolute pairs has link lengths l_f = 20 mm, l_{in} = 40 mm, l_{co} = 50 mm and l_{out} = 60 mm. The suffixes '*f*', 'in', 'co' and 'out' denote the fixed link, the input link, the coupler and output link respectively. Which one of the following statements is true about the input and output links?

(a) Both links can execute full circular motion
(b) Both links cannot execute full circular motion
(c) Only the output link cannot execute full circular motion
(d) Only the input link cannot execute full circular motion

[2014 : 1 M, Set-2]

2.29 The number of degrees of freedom of the linkage shown in the figure is

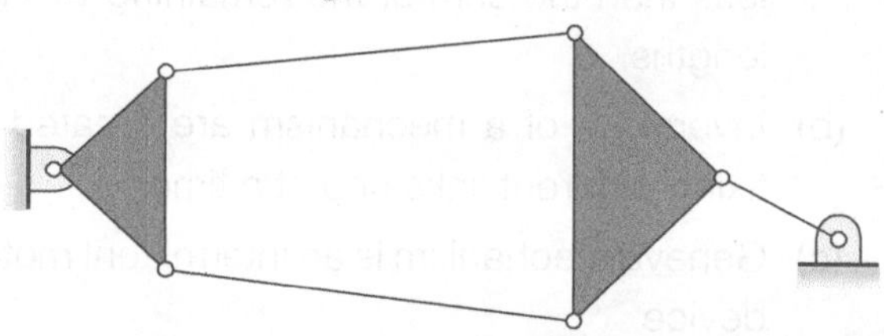

(a) − 3 (b) 0
(c) 1 (d) 2

[2015 : 2 M, Set-3]

2.30 The number of degree of freedom of the planetary gear train shown in the figure is

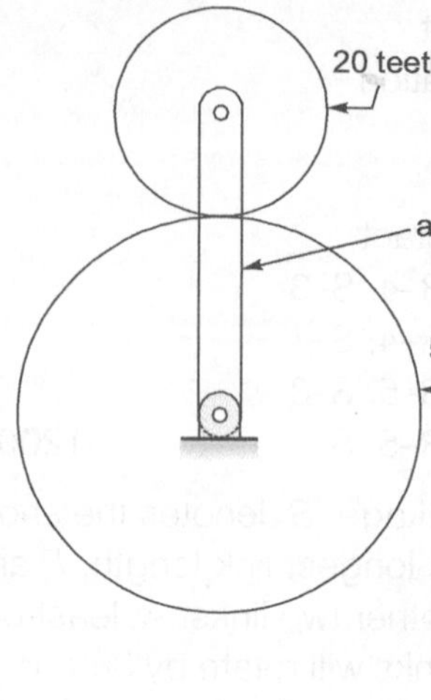

(a) 0 (b) 1
(c) 2 (d) 3 [2015 : 1 M, Set-2]

2.31 A rigid (AB) of length $L = \sqrt{2}$ is undergoing translational as well as rotational motion in the x–y plane (see the figure). The point A has the velocity $V_1 = i + 2j$ m/s. The end B is constrained to move only along the x-direction. The magnitude of velocity V_2 (in m/s) at the end B is_____.

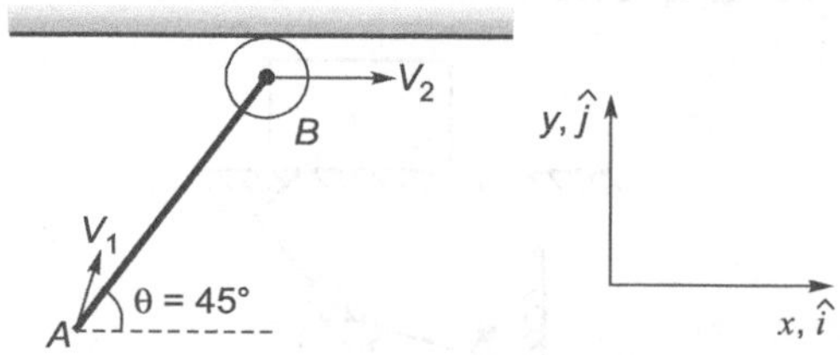

[GATE : 2016 : 2 M]

2.32 A four bar mechanism is made up of links of length 100, 200, 300 and 350 mm. If the 350 mm link is fixed, the number of links that can rotate fully is________.

[2018 : 1 M, Set-1]

2.33 For an Oldham coupling used between two shafts, which among the following statements are correct?

I. Torsional load is transferred along shaft axis.
II. A velocity ratio of 1 : 2 between shaft is obtained without using gears.
III. Bending load is transferred transverse to shaft axis.
IV. Rotation is transferred along shaft axis.

(a) I and III (b) I and IV
(c) II and III (d) II and IV

[2018 : 1 M, Set-1]

2.34 The crank of a slider-crank mechanism rotates counterclockwise (CCW) with a constant angular velocity ω, as shown. Assume the length of the crank to be *r*.

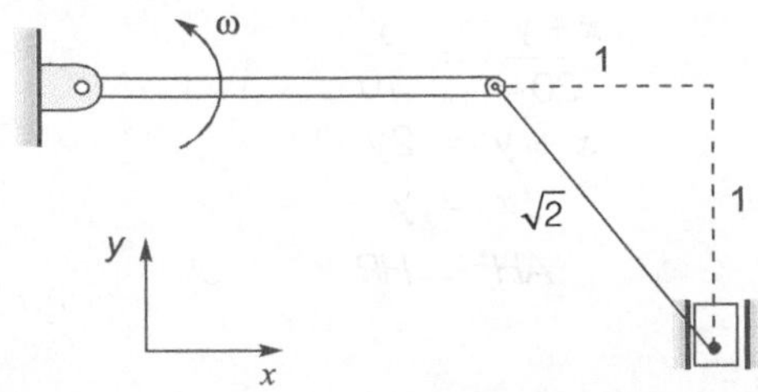

Using exact analysis, the acceleration of the slider in the *y*-direction, at the instant shown, where the crank is parallel to x-axis, is given by

(a) $\omega^2 r$ (b) $2\omega^2 r$
(c) $-2\omega^2 r$ (d) $-\omega^2 r$

[2019 : 2 M, Set-2]

2.35 A four bar mechanism is shown below.

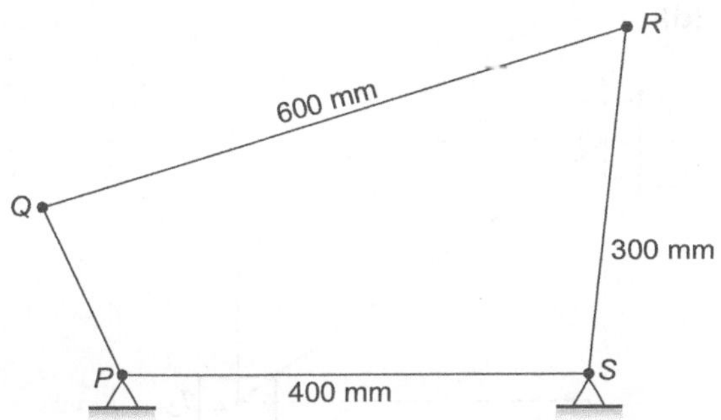

For the mechanism to be a crank-rocker mechanism, the length of the link *PQ* can be

(a) 350 mm (b) 200 mm
(c) 300 mm (d) 80 mm

[2020 : 1 M, Set-1]

2.36 The number of qualitatively distinct kinematic inversions possible for a Grashof chain with four revolute pairs is

(a) 4 (b) 2
(c) 1 (d) 3

[2020 : 1 M, Set-2]

2.37 The Whitworth quick return mechanism is shown in the figure with link lengths as follows: OP = 300 mm, OA = 150 mm, AR = 160 mm, RS = 450 mm.

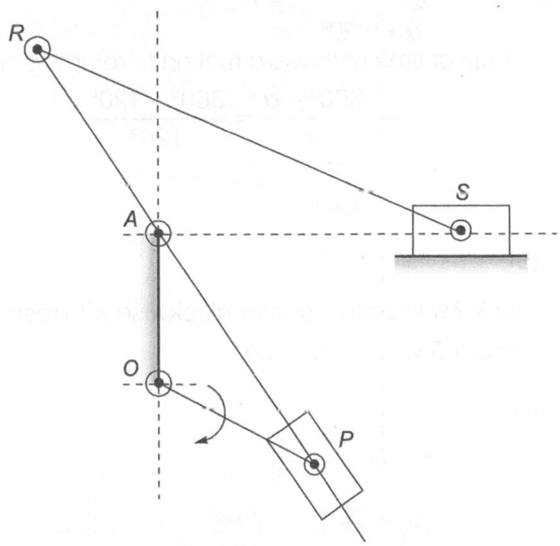

The quick return ratio for the mechanism is ________. [Round off to one decimal place]

[2021 : 2 M, Set-1]

Answers Planar Mechanisms

2.1	(b)	2.2	(b)	2.3	(c)	2.4	(d)	2.5	(a)	2.6	(d)	2.7	(c)	2.8	(c)
2.9	(d)	2.10	(c)	2.11	(b)	2.12	(b)	2.13	(b)	2.14	(c)	2.15	(b)	2.16	(d)
2.17	(a)	2.18	(d)	2.19	(a)	2.20	(b)	2.21	(b)	2.22	(d)	2.23	(b)	2.24	(a)
2.25	(a)	2.26	(c)	2.27	(c)	2.28	(a)	2.29	(c)	2.30	(c)	2.31	(3)	2.32	(1)
2.33	(b)	2.34	(a)	2.35	(d)	2.36	(d)	2.37	(2)						

Explanations Planar Mechanisms

2.1 (b)

Given, $OA = 2$ cm, $OO' = 4$ cm

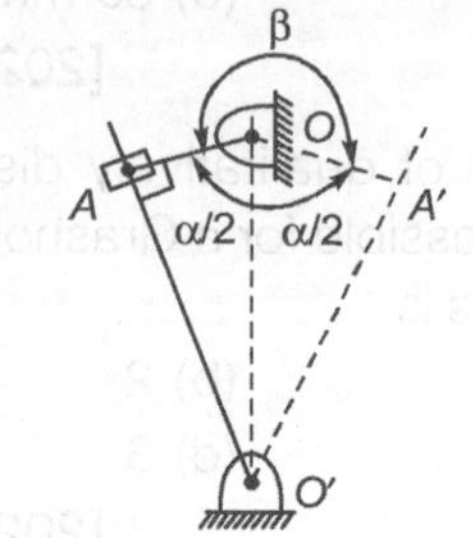

$$\cos\frac{\alpha}{2} = \frac{OA'}{OO'}$$

$$\frac{\alpha}{2} = \cos^{-1}\left(\frac{2}{4}\right)$$

$$\alpha = 120°$$

Ratio of time of forward motion to return motion

$$= \frac{360° - \alpha}{\alpha} = \frac{360° - 120°}{120°}$$

$$= \frac{240°}{120°} = 2$$

2.2 (b)

Link 2 will rotate, counter-clockwise with respect to link 3 when is moved.

2.3 (c)

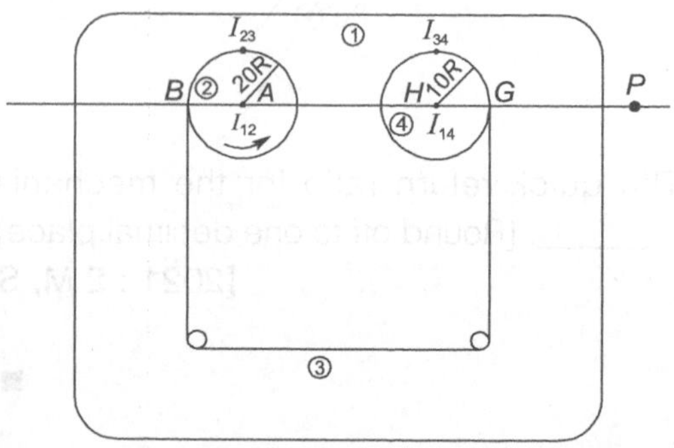

$$I_{24} = ?$$

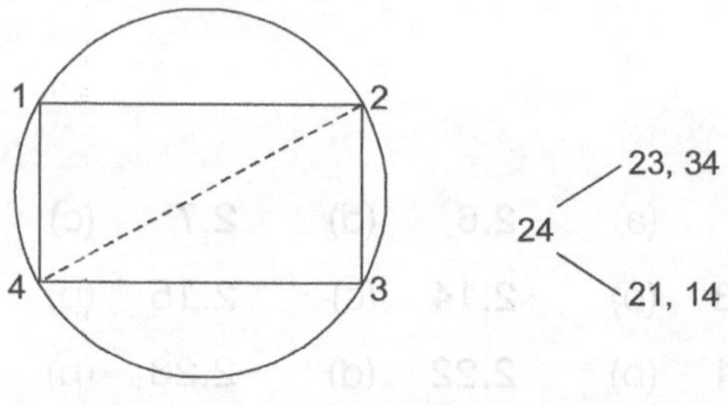

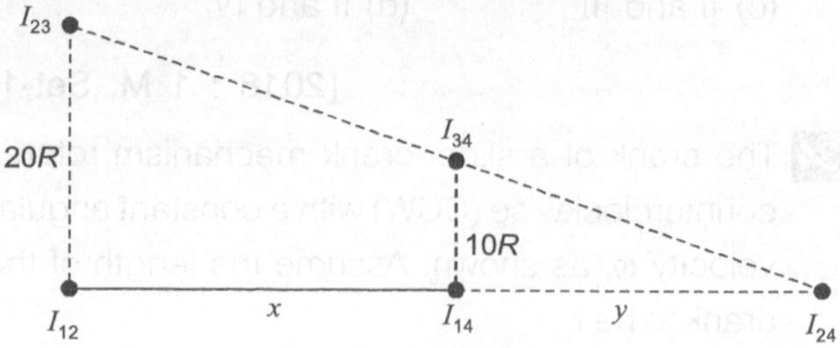

$$\frac{x+y}{20} = \frac{y}{10}$$

$$x + y = 2y$$

$$x = y$$

$$\Rightarrow \quad AH = HP$$

2.4 (d)

Crank and slotted lever mechanism is mostly used in shaping machines, which is the inversion of single slider-crank chain.

2.5 (a)

Double crank mechanism is obtained when the shortest link is fixed. Here, the shortest link is p. Therefore link of length p should be fixed.

2.6 (d)

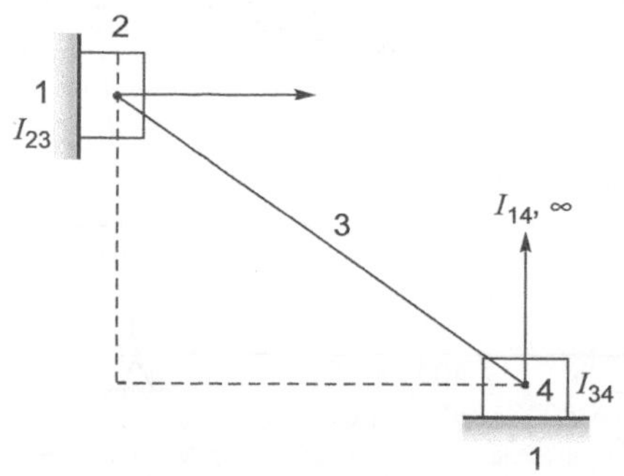

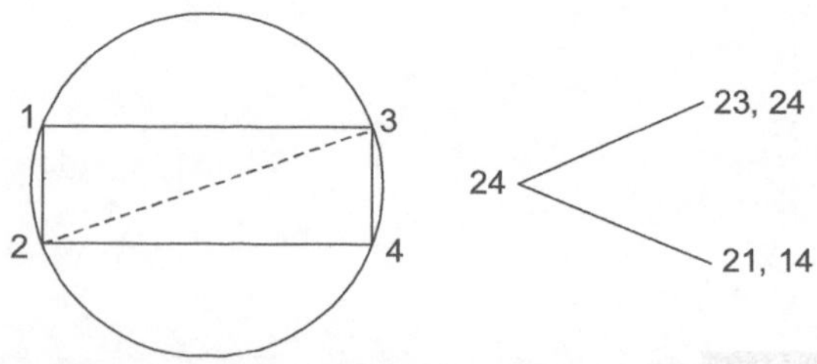

The only condition when I_{23}, I_{34}, and I_{21}, I_{14} will meet, is at ∞.

So, I_{24} will lies at infinity.

2.7 (c)

(P) Scott - Russel mechanism - Straight line motion

(Q) Geneva mechanism - Intermittent motion

(R) Off - set slider crank mechanism - Quick return motion

(S) Scotch Yoke mechanism - Simple harmonic motion

2.8 (c)

(P) Revolute — Five
(Q) Cylinder — Four
(R) Spherical — Three

2.9 (d)

$$\text{Mechanical advantage} = \frac{T_{output}}{T_{input}} = \frac{\omega_{input}}{\omega_{output}}$$

Since output link is a slider

Hence, $\omega_{output} = 0$

$\therefore$ Mechanical advantage $= \infty$

2.10 (c)

No. of links: $l = 8$

No. of revolute joints,

$j = 9$

No. of higher pair,

$h = 0$

$\therefore$ Number of degree of freedom,

$$n = 3(l-1) - 2j - h$$
$$= 3(8-1) - 2 \times 9 - 0$$
$$= 3$$

2.11 (b)

Length of link,

$$AB = \sqrt{(240)^2 + (160-60)^2}$$
$$= l_{AB} = 260 \text{ mm}$$
$$= l_{AO_2} + l_{O_2O_4} < l_{AB} + l_{BO_4}$$

and $O_2 O_4$ is fixed link. Hence it will acts as crank-rocker mechanism.

2.12 (b)

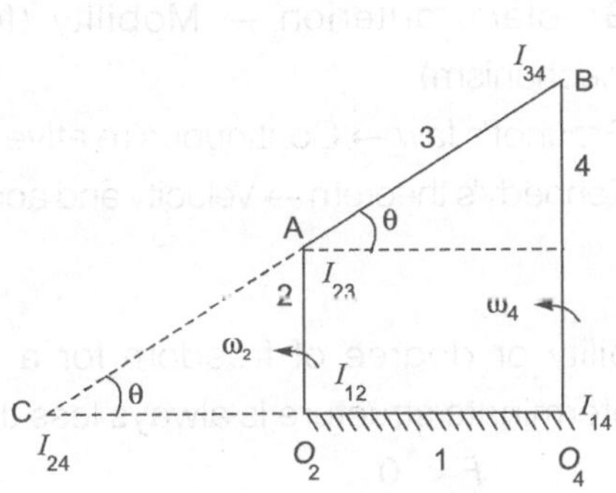

$$\theta = \frac{160-60}{240} = \frac{60}{CO_2}$$

$\therefore$ $CO_2 = 144$ mm

Now, $$\frac{\omega_4}{\omega_2} = \frac{I_{12}I_{24}}{I_{14}I_{24}} = \frac{144}{(240+144)}$$

$$\frac{\omega_4}{8} = \frac{144}{384}$$

$\therefore$ $$\omega_4 = \frac{144}{384} \times 8 = \mathbf{3\ rad/s}$$

2.13 (b)

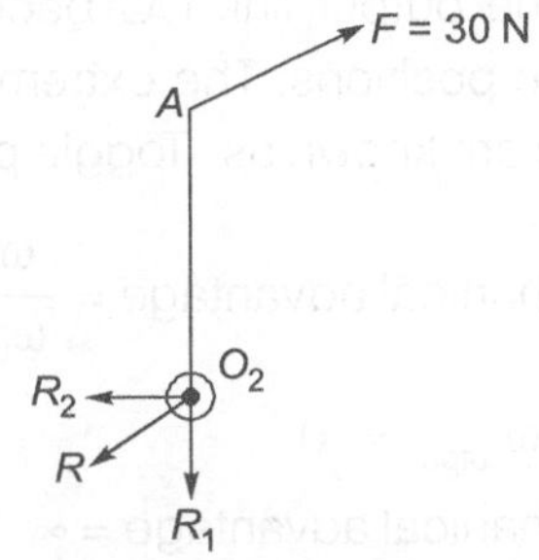

Since, according to the given data, only one force, $F = 30$ N is acting at point A along line AB, So

$$F = R = 30 \text{ N}$$

R, reaction at O_2,

$$R = \sqrt{R_1^2 + R_2^2} = 30 \text{ N}$$

2.14 (c)

No. of links of a slider crank mechanism = 4

So there are four inversion of slider crank mechanism.

2.15 (b)

(P) Addendum — Gear
(Q) Instantaneous centre of velocity — Linkage
(R) Section modulus — Beam
(S) Prime circle — Cam.

2.16 (d)

(P) Higher kinematic pair — Line contact
(Q) Lower kinematic pair — Surface contact
(R) Quick return mechanism — Shaper
(S) Mobility of a linkage — Grubler's equation

2.17 (a)

According to Grashof's law for a four bar mechanism. The sum of shortest and longest link lengths should not be greater than the sum of the remaining two link length.

i.e., $S + L \leq P + Q$

2.18 (d)

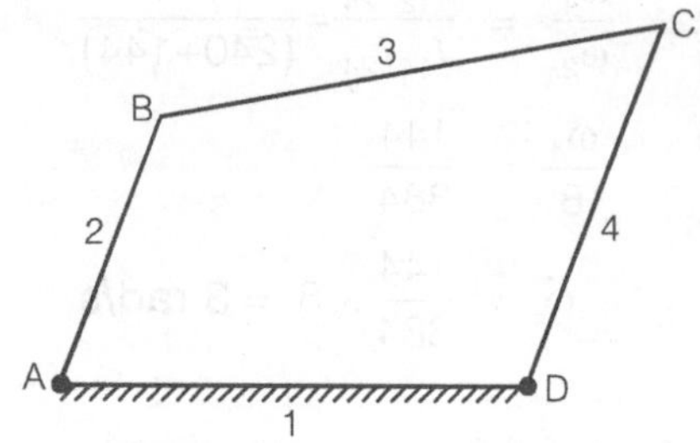

ω_4 of the output link DC becomes zero at the extreme positions. The extreme positions of the linkage are known as "Toggle position".

$$\therefore \text{ Mechanical advantage} = \frac{\omega_{input}}{\omega_{output}} \times \eta_{mechanism}$$

$$\because \quad \omega_{output} = 0$$

$\therefore$ Mechanical advantage = ∞

2.19 (a)

$$\text{Given : } \frac{2\theta}{2\pi - 2\theta} = \frac{1}{2}$$

$$\therefore \quad \theta = \frac{\pi}{3}$$

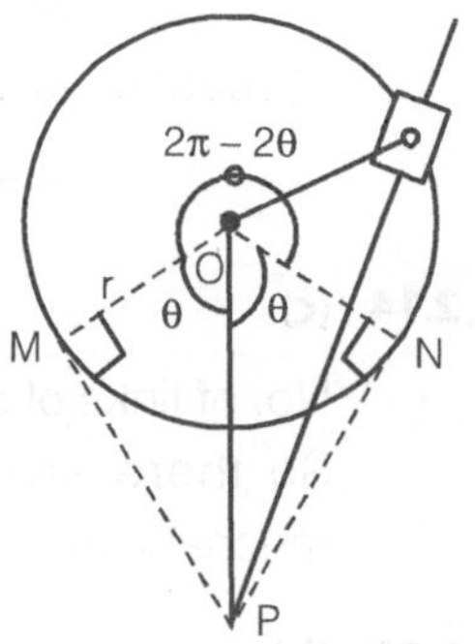

In triangle *OMP*

$$\cos\theta = \frac{r}{OP},$$

$$\cos\left(\frac{\pi}{3}\right) = \frac{r}{500}$$

or $\quad r =$ **250 mm**

POINTS TO REMEMBER

While solving these type of questions, students should keep in mind that quick-return ratio is (> 1) or (< 1) and then according to the given data, take the ratio of forward and return stroke.

2.20 (b)

Maximum speed during forward stroke occur when *QR* and *QS* are perpendicular.

$$V = 250 \times 2 = 750 \times \omega_{PQ}$$

$$\omega_{PQ} = \frac{2}{3} \text{ rad/s}$$

2.21 (b)

Degree of freedom is given by

$$= 3(l - 1) - 2j - h$$

Here, $\quad h =$ No. of higher pair = 0

$$= 3(8 - 1) - (2 \times 10) = 1$$

2.22 (d)

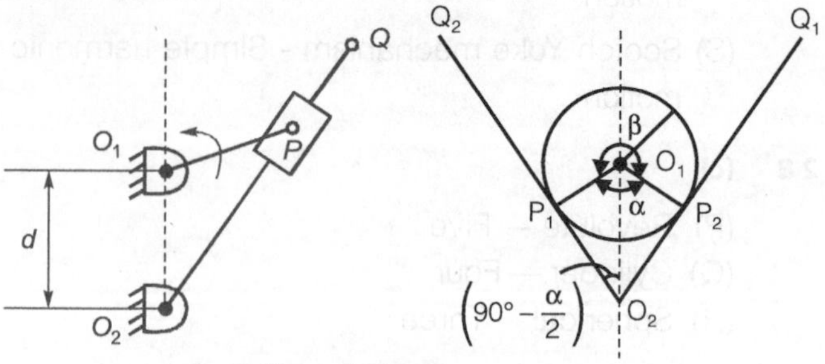

$$O_1P = 125 \text{ mm}$$

Quick return mechanism

$$\frac{\text{Time of working (forward) stroke}}{\text{Time of return stroke}} = \frac{\beta}{\alpha}$$

$$= \frac{360 - \alpha}{\alpha}$$

$$\frac{2}{1} = \frac{360° - \alpha}{\alpha}$$

$$\Rightarrow \quad 2\alpha = 360° - \alpha$$

$$\Rightarrow \quad 3\alpha = 360° \quad \Rightarrow \alpha = 120°$$

$$\Rightarrow \quad \frac{\alpha}{2} = 60°$$

The extreme position of the crank (O_1 P) are shown in figure.

From right angle triangle $O_2\, O_1\, P_1$, we find that

$$\sin(90° - \alpha/2) = \frac{O_1\,P_1}{O_1\,O_2}$$

$$\sin(90° - 60°) = \frac{125}{O_1\,O_2} = \frac{125}{d}$$

$$\sin 30° = \frac{125}{d}$$

or $\quad d = \dfrac{125}{\sin 30°} =$ **250 mm**

2.23 (b)

1. D'Alembert's principle → Dynamic-static analysis
2. Grubler's criterion → Mobility (for plane mechanism)
3. Grashof's law → Continuous relative rotation
4. Kennedy's theorem → Velocity and acceleration

2.24 (a)

Mobility or degree of freedom for a statically indeterminate structure is always less than zero.

i.e., $\quad F < 0$

$\therefore$ only option (a) is negative value

$$\therefore \quad F \leq -1$$

2.25 (a)

Grashof's rule states that for a planar crank rocker four bar mechanism, the sum of shortest and longest link length is less than the sum of remaining two link length i.e.

$$s + l \le p + q$$

2.26 (c)

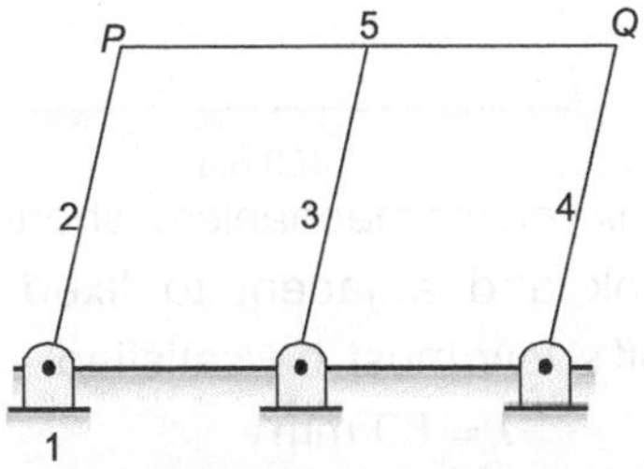

In this case there is one link is redundant i.e. function of mechanism is not affected even if any one of thinks, (2), (3) or (4) are removed.

So, effective number of link.

$$l = 4, j = 4, h = 0$$

$$\text{DOF} = 3(l-1) - 2j - h$$

$$= 3(4-1) - 2 \times 4 - 0 = 1$$

2.27 (c)

PQ = 2.0 m, QR = 3.0 m, RS = 2.5 m, SP = 2.7 m

(i) If shortest line PQ = 2.0 m is fixed, then we will give double crank.

(ii) If link adjacent to shortest is fixed, then we will get crank-rocker mechanism.

(iii) If we fix coupler 'RS', we will get double rocker mechanism.

POINTS TO REMEMBER

This problem is based on Grashoff's law, for continuous relative motion.

2.28 (a)

$$S + L < P + Q$$

In this case, shortest link is fixed and hence, the resulting mechanism is double crank mechanism.

2.29 (c)

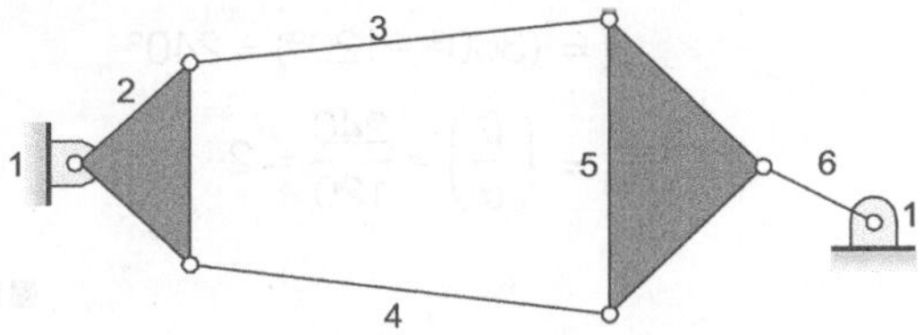

$$F = 3(6-1) - 2 \times 7 - 0$$

$$= 15 - 14 = 1$$

2.30 (c)

Degree of freedom: $F = 3(l-1) - 2j - h$

$$= 3(4-1) - 2 \times 3 - 1 = 2$$

2.31 Sol.

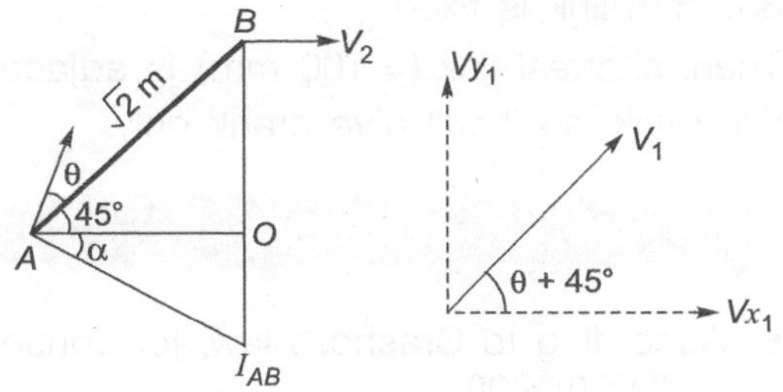

Given: $V_1 = \hat{i} + 2\hat{j}$ m/s $= V_{x_1}\hat{i} + V_{y_1}\hat{j}$

So, V_{x_1} = 1 m/s and V_{y_1} = 2 m/s

$$\tan(\theta + 45°) = \frac{V_{y_1}}{V_{x_1}} = \frac{2}{1}$$

$$\theta + 45° = 63.435° = 18.435°$$

$$\theta = 18.435°$$

According to geometry, $\theta + 45° + \alpha = 90°$

$$\alpha = 90° - 45° - \theta = 90° - 45 - 18.435°$$

$$= 26.565°$$

In ΔABO, $AB = \sqrt{2}\,m$, then $AO = BO$ = 1 m

In ΔOAI_{AB}, $\cos\alpha = \frac{AO}{AI_{AB}} \Rightarrow AI_{AB}$

$$= \frac{AO}{\cos 26.565} = \frac{1}{\cos 26.565}$$

$$\tan\alpha = \frac{OI_{AB}}{AO}$$

$$OI_{AB} = (\tan\alpha)(AO) = 1 \times \tan 26.565$$

$$= 0.4999 \simeq 0.5 \text{ m}$$

So, $BI_{AB} = BO + OI_{AB} = 1 + 0.5 = 1.5$ m

So, for velocity of rod, V_2;

$$\omega_{AB} = \frac{V_1}{AI_{AB}} = \frac{V_2}{BI_{AB}}$$

$$\frac{\sqrt{1+4}}{1.118} = \frac{V_2}{1.5}$$

$$V_2 = \frac{(\sqrt{5}) \times 1.5}{1.118} = 3 \text{ m/s}$$

2.32 Sol.

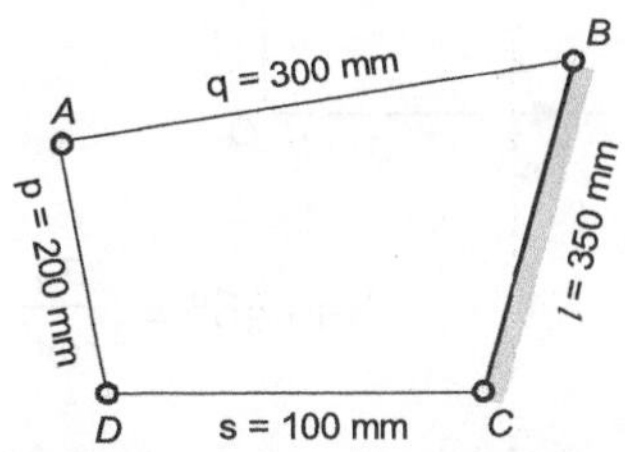

$s = 100$, $p = 200$, $l = 350$, $q = 300$

$$(s + l) = 350 + 100 = 450 < (p + q)$$
$$450 < 200 + 300$$
$$450 < 500$$

Grashof law is satisfied.

350 mm link is fixed.

Then, shortest link (= 100 mm) is adjacent to fixed link, so it will give crank only.

POINTS TO REMEMBER

- According to Grashof's law, for continuous relative motion,
$$s + l \leq p + q$$
So, in this problem, $s + l < p + q$ is satisfied
- If shortest link, s is fixed, mechanism will be double crank.
- If link adjacent of shortest link is fixed, mechanism will be crank-rocker type.

2.34 (a)

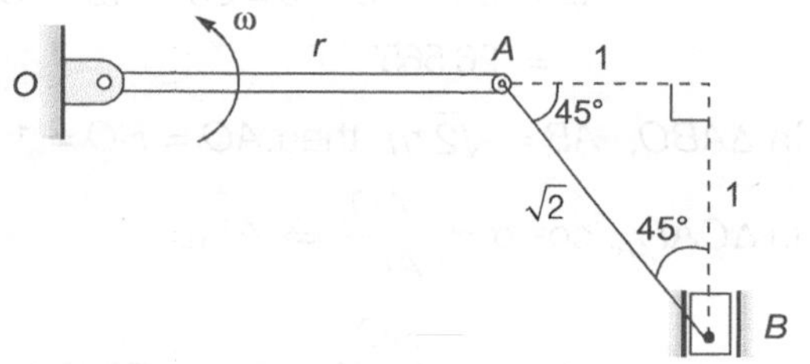

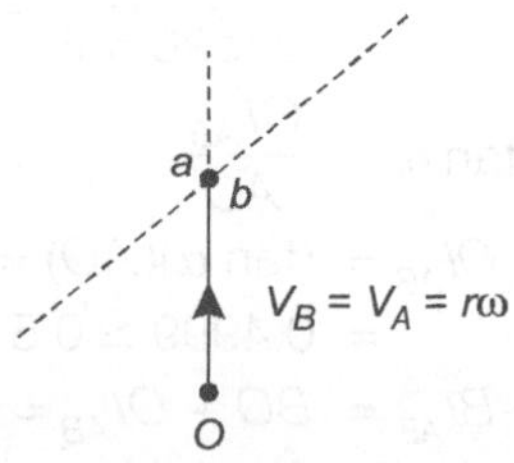

$$V_{BA} = 0$$
$$\Rightarrow \quad \omega_{BA} = 0$$

Acceleration of $BA = 0$

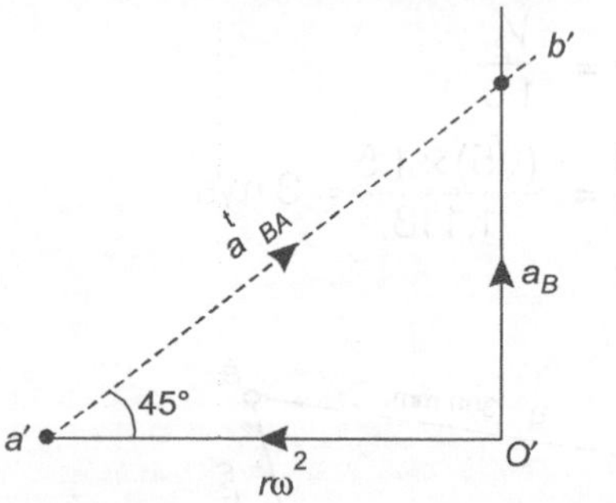

$(a^r_{BA} = 0, V_{BA} = 0)$

$$\tan 45° = \frac{a_B}{r\omega^2}$$

$$\Rightarrow \quad a_B = r\omega^2$$

2.35 (d)

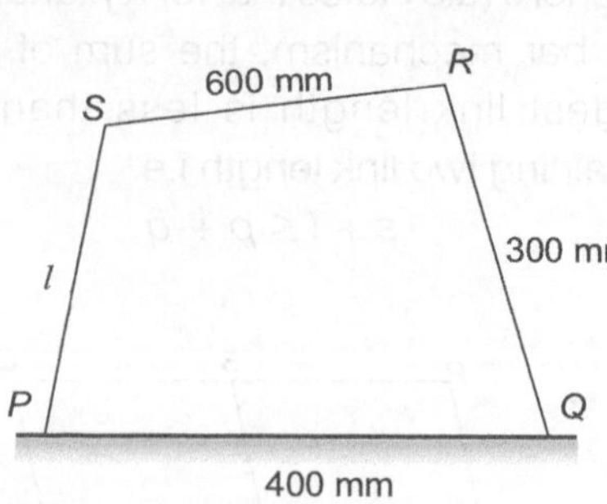

For crank-rocker mechanism, shortest link must be crank and adjacent to fixed as well as Grashoff's law must be satisfied.

If $\quad l = 80$ mm

then shortest will be = 80 mm

as well as(80 + 600) < (400 + 300)

$$680 < 700$$

Therefore law is satisfied.

$$\Rightarrow \quad l = 80 \text{ mm}$$

2.36 (d)

The number of qualitatively distinct kinematic inversion of Grashof chain with four revolute pair are:

1. Double crank mechanism
2. Crank-rocker mechanism
3. Double rocker mechanism

2.37 Sol.

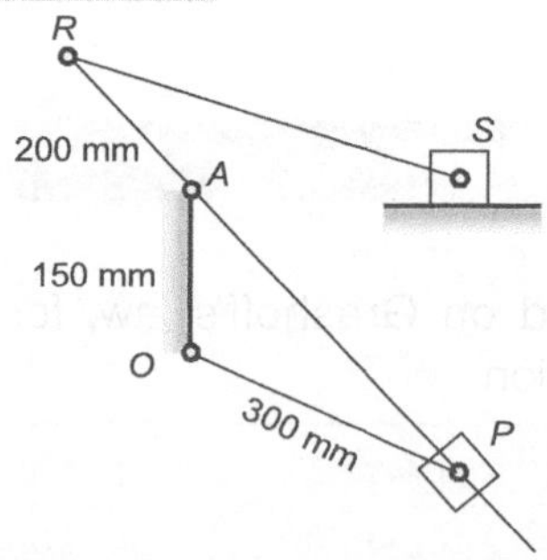

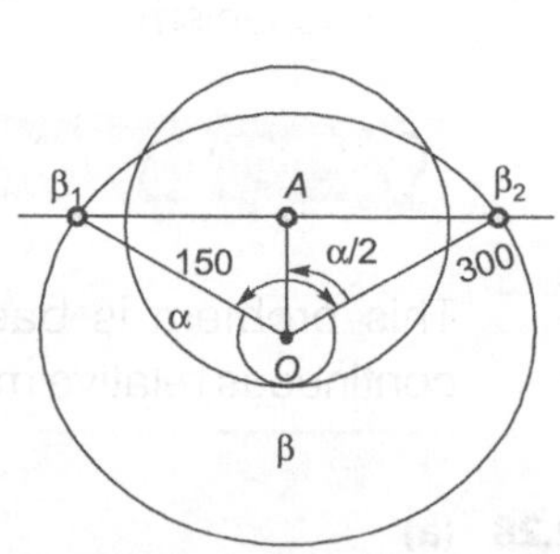

$$\cos\frac{\alpha}{2} = \frac{150}{300} = \frac{1}{2}$$

$$\Rightarrow \quad \frac{\alpha}{2} = 60°$$

$$\alpha = 120°$$

$$\beta = (360° - 120°) = 240°$$

$$\text{QRR} = \left(\frac{\beta}{\alpha}\right) = \frac{240}{120} = 2$$

■■■■

3 Cams

3.1 In a cam-follower mechanism, the follower needs to rise through 20 mm during 60° of cam rotation, the first 30° with a constant acceleration and then with a deceleration of the same magnitude. The initial and final speeds of the follower are zero. The cam rotates at uniform speed of 300 rpm. The maximum speed of the follower is

(a) 0.60 m/s (b) 1.20 m/s
(c) 1.68 m/s (d) 2.40 m/s **[2005 : 2 M]**

3.2 In a cam design, the rise motion is given by a Simple Harmonic Motion (SHM) $s = \frac{h}{2}\left(1-\cos\frac{\pi\theta}{\beta}\right)$ where h is total rise, θ is camshaft angle, β is the total angle of the rise interval. The jerk is given by

(a) $\frac{h}{2}\left(1-\cos\frac{\pi\theta}{\beta}\right)$ (b) $\frac{\pi}{\beta}\frac{h}{2}\sin\left(\frac{\pi\theta}{\beta}\right)$

(c) $\frac{\pi^2}{\beta^2}\frac{h}{2}\cos\left(\frac{\pi\theta}{\beta}\right)$ (d) $-\frac{\pi^3}{\beta^3}\frac{h}{2}\sin\left(\frac{\pi\theta}{\beta}\right)$

[2008 : 2 M]

3.3 In the mechanism given below, if the angular velocity of the eccentric circular disc is 1 rad/s, the angular velocity (rad/s) of the follower link for the instant shown in the figure is

[Note : All dimensions are in mm.]

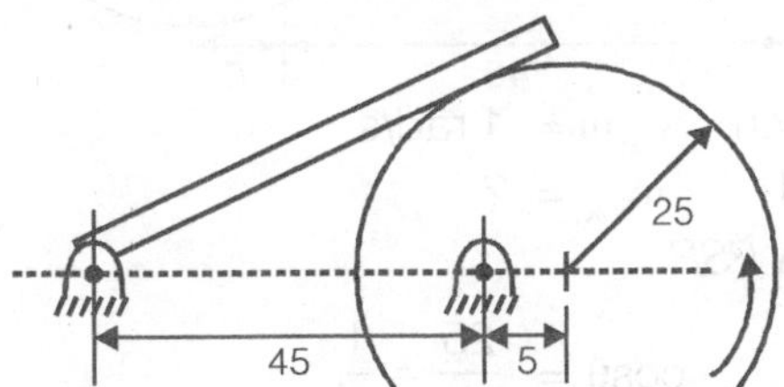

(a) 0.05 (b) 0.1
(c) 5.0 (d) 10.0 **[2012 : 1 M]**

3.4 Consider a rotating disk cam and a translating roller follower with zero offset. Which one of the following pitch curves, parameterized by t, lying in the interval 0 to 2π, is associated with the maximum translation of the follower during one full rotation of the cam rotating about the center at $(x, y) = (0, 0)$?

(a) $x(t) = \cos t$, $y(t) = \sin t$

(b) $x(t) = \cos t$, $y(t) = 2\sin t$

(c) $x(t) = \frac{1}{2} + \cos t$, $y(t) = 2\sin t$

(d) $x(t) = \frac{1}{2} + \cos t$, $y(t) = \sin t$

[2014 : 2 M, Set-3]

3.5 In a cam-follower, the follower rises by h as the cam rotates by δ (radians) at constant angular velocity ω (radians/s). The follower is uniformly accelerating during the first half of the rise period and it is uniformly decelerating in the later half of the rise period. Assuming that the magnitudes of the acceleration and deceleration are same, the maximum velocity of the follower is

(a) $\frac{4h\omega}{\delta}$ (b) $h\omega$

(c) $\frac{2h\omega}{\delta}$ (d) $2h\omega$

[2018 : 2 M, Set-2]

3.6 A flat-faced follower is driven using a circular eccentric cam rotating at a constant angular velocity ω. At time $t = 0$, vertical position of follower is $y(0) = 0$, and the system is in the configuration shown below.

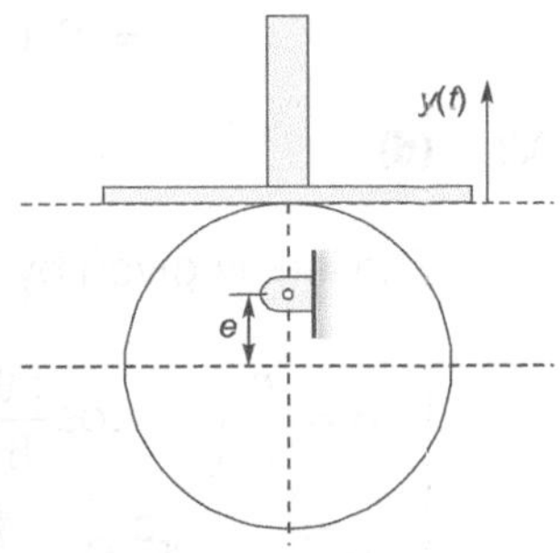

Then vertical position of the follower face, $y(t)$ is given by

(a) $e(1 - \cos \omega t)$ (b) $e(1 + \cos 2\omega t)$
(c) $e \sin \omega t$ (d) $e \sin 2\omega t$

[2019 : 1 M, Set-1]

3.7 A cam with a translating flat-face follower is desired to have the follower motion

$$y(\theta) = 4[2\pi\theta - \theta^2], \qquad 0 \le \theta \le 2\pi$$

Contact stress considerations dictate that the radius of curvature of the cam profile should not be less than 40 mm anywhere. The minimum permissible base circle radius is _____ mm (round off to one decimal place).

[2020 : 2 M, Set-1]

■■■■

Answers Cams

3.1 (b) 3.2 (d) 3.3 (b) 3.4 (c) 3.5 (c) 3.6 (a) 3.7 (48)

Explanations Cams

3.1 (b)

Angular velocity,

$$\omega = \frac{2\pi N}{60} = \frac{2\pi 300}{60} = 10\pi$$

Time taken to move 30°

$$= \frac{\frac{\pi}{180}\times 30}{\omega} = \frac{\frac{\pi}{6}}{10\pi} = \frac{1}{60}\text{ s}$$

During this time, follower moves by distance of 10 mm with initial velocity

$$u = 0$$

Now; $S = ut + \frac{1}{2}at^2$

$$0.01 = 0 + \frac{1}{2}\times a\times\left(\frac{1}{60}\right)^2$$

$\therefore$ $a = 0.01 \times 2 \times (60)^2 = 72 \text{ m/s}^2$

$$V_{max} = u + at$$

$$= 0 + 72 \times \frac{1}{60} = \mathbf{1.20 \text{ m/s}}$$

3.2 (d)

The jerk is given by $\frac{d^3s}{dt^3}$

if $S = \frac{h}{2}\left(1-\cos\frac{\pi\theta}{\beta}\right)$

$$\therefore \frac{d^3s}{dt^3} = -\frac{\pi^3}{\beta^3}\frac{h}{2}\sin\left(\frac{\pi\theta}{\beta}\right)$$

where $\theta = \omega t$

3.3 (b)

Method I:

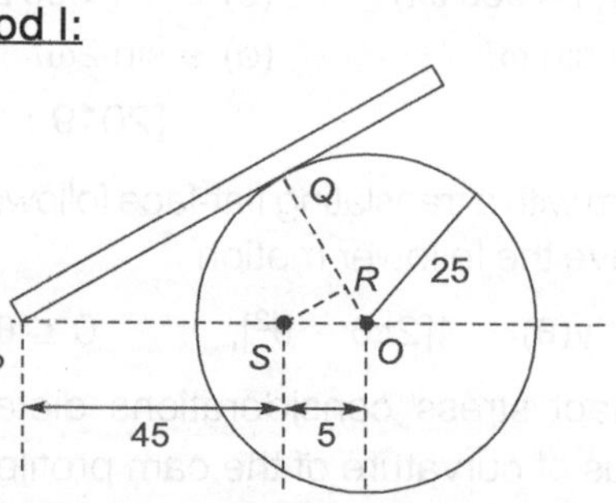

$\Delta PQO \sim \Delta SRO$

$$\therefore \frac{PQ}{SR} = \frac{PO}{SO} \quad ...(1)$$

Applying Pythagoras theorem in ΔPQO.

$$(PO)^2 = (PQ)^2 + (OQ)^2$$

$$\therefore (PQ) = \sqrt{(PO)^2 - (OQ)^2}$$

$$= \sqrt{(50)^2 - (25)^2} = 43.3 \text{ mm}$$

Putting value of PQ in Eq. (1), we get

$$\frac{43.3}{SR} = \frac{45+5}{5}$$

$$SR = \frac{43.3\times 5}{50} = 4.33 \text{ mm}$$

Velocity of Q is equal to velocity of R.

$$V_Q = V_R = SR\times\omega$$

$$= 4.33 \times 1 = 4.33 \text{ mm/s}$$

$$\omega_{PQ} = \frac{V_{PQ}}{PQ} = \frac{4.33}{43.3} = \mathbf{0.1 \text{ rad/s}}$$

Method II:

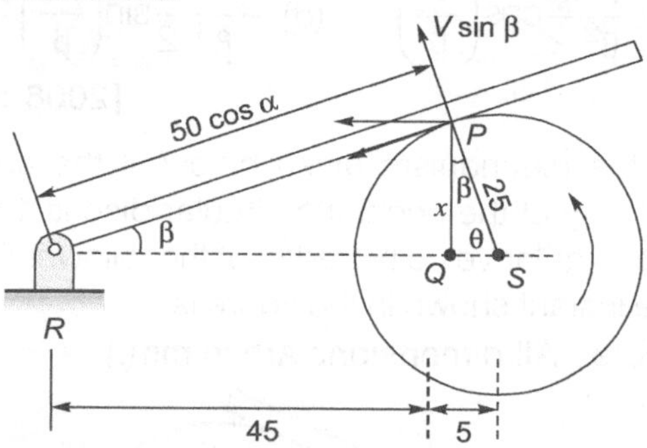

Given, $\omega = 1$ rad/s

and $\omega_{link} = ?$

In ΔPSR,

$$\cos\theta = \frac{25}{50} = \frac{1}{2}$$

$$\theta = 60°$$

So, $\beta = 30°$

Applying sine rule, ΔPQS,

$$\frac{5}{\sin\beta} = \frac{x}{\sin\theta}$$

$$x = \frac{5\sin\theta}{\sin\beta} = \frac{5\sin 60°}{\sin\beta}$$

Velocity at point P,

$$V = PQ\times\omega = x\times\omega = x\times 1 = x$$

$$V = x = \frac{5\sin 60°}{\sin\beta}$$

$$V\sin\beta = 5\sin 60°$$

Angular velocity of follower link;

$$\omega_{link} = \frac{V\sin\beta}{50\cos\beta} = \frac{5\sin 60°}{50\cos 30°} = \frac{1}{10}$$

$$= 0.1 \text{ rad/s}$$

3.5 (c)

Here, outstroke angle

$$\theta_0 = \delta$$

and stroke length $= h$

Angular velocity $= \omega$

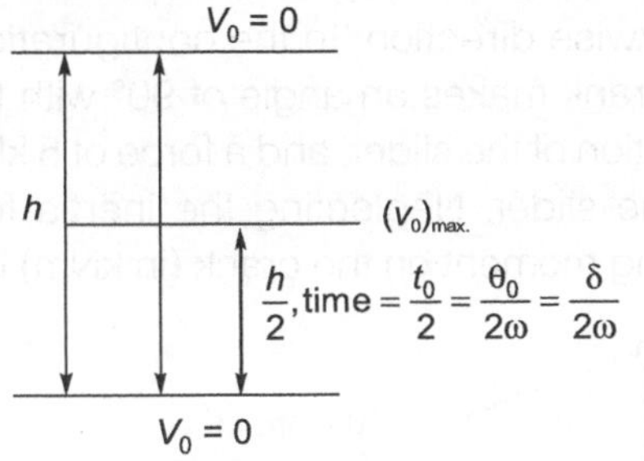

$$V = u + at$$

$$(V_0)_{max} = 0 + a \times \frac{t_0}{2} = a \times \frac{\delta}{2\omega} \quad \ldots \text{(i)}$$

$$\frac{h}{2} = 0 + \frac{1}{2}a\left(\frac{t_0}{2}\right)^2$$

$$h = \frac{at_0^2}{4} = \frac{a\delta^2}{4\omega^2}$$

$$a = \frac{4\omega^2 . h}{\delta^2} \quad \ldots \text{(ii)}$$

By equation (i) and (ii), we get

$$(V_0)_{max} = \frac{a\delta}{2\omega} = \frac{4\omega^2 . h}{\delta^2} \times \frac{\delta}{2\omega} = \frac{2\omega . h}{\delta}$$

POINTS TO REMEMBER

- For the case of constant acceleration and retardation, and also for the case of cycloidal motion

$$V_{max} = \frac{2h\omega}{\delta}$$

where, $h = 0.02$ m

$$\omega = \frac{2\pi N}{60} = \frac{2\pi \times 300}{60} = 10\pi$$

$$\delta = 60° = \frac{\pi}{N}$$

Maximum speed,

$$V_{max} = \frac{2h\omega}{\delta} = \frac{2 \times 0.02 \times 10\pi}{\pi/3}$$

$$= 1.2 \text{ m/s}$$

3.6 (a)

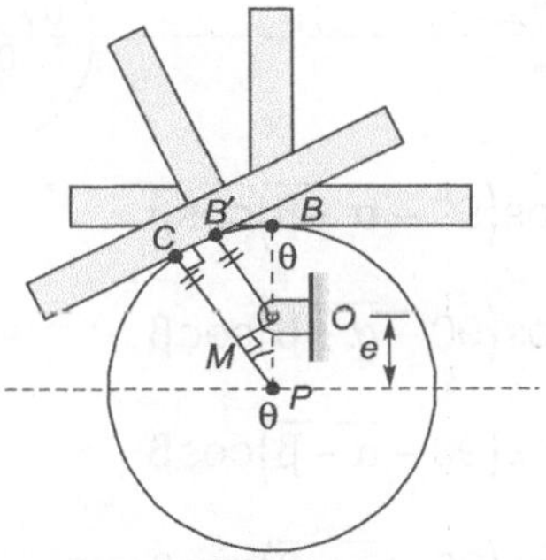

Displacement of follower

$y = OB' - OB$ (OB' will be parallel to PC)

$= CM - OB$

$= (PC - PM) - OB(OB' = CM)$

$= R - e\cos\theta - (R - e)$

$= R - e\cos\theta - R + e$

$= e(1 - \cos\theta)$(In time(t) $\theta = \omega t$)

$= e(1 - \cos\omega t)$

3.7 Sol.

$(R_{curvature})_{min} = 40$ mm

For flat face follower,

Displacement equation:

$$y = 4(2\pi\theta - \theta^2)$$

$$\frac{dy}{d\theta} = V = 4(2\pi - 2\theta)$$

$$= 8(\pi - \theta)$$

(For y to be max $\frac{dy}{d\theta} = 0 \Rightarrow \theta = \pi$)

$$a = \frac{dv}{d\theta} = -8$$

$(R_{curvature})_{Min} = R_{Base} + (y + a)_{min}$

(y_{min} is 0 at $\theta = 0, 2\pi$)

$$40 = R_{Base} + [0 - 8]_{Min}$$

$$40 = R_{Base} + [-8]$$

$$R_{Base} = 40 - (-8) = 40 + 8 = 48 \text{ mm}$$

4 Dynamic Analysis of Slider-crank

4.1 The cross head velocity in the slider crank mechanism, for the position shown in the figure

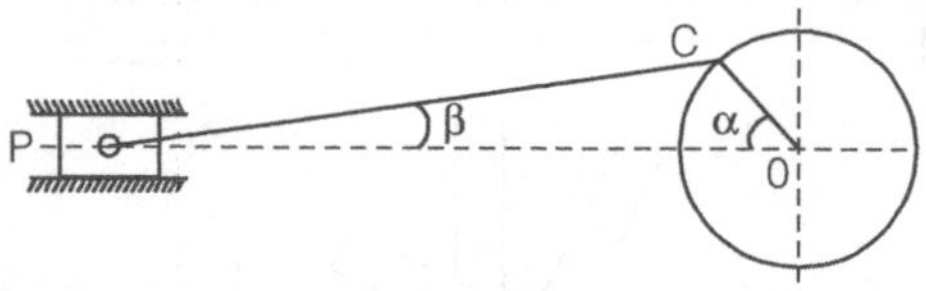

(a) $V_c \cos\left(90 - \overline{\alpha + \beta}\right)\cos\beta$

(b) $V_c \cos\left(90 - \overline{\alpha + \beta}\right)\sec\beta$

(c) $V_c \cos\left(90 - \overline{\alpha - \beta}\right)\cos\beta$

(d) $V_c \cos\left(90 - \overline{\alpha - \beta}\right)\sec\beta$

[1997 : 1 M]

4.2 Consider the triangular formed by the connecting rod and the crank of an IC engine as the two sides of the triangle. If the maximum area of this triangle occurs when the crank is 75°, the ratio of connecting rod length to crank radius to

(a) 5 (b) 4

(c) 3.73 (d) 3 [1998 : 1 M]

4.3 A slider-crank mechanism with crank radius 60 mm and connecting rod length 240 mm is shown in the figure. The crank is rotating with a uniform angular speed of 10 rad/s, counter clockwise. For the given configuration, the speed (in m/s) of the slider is ______.

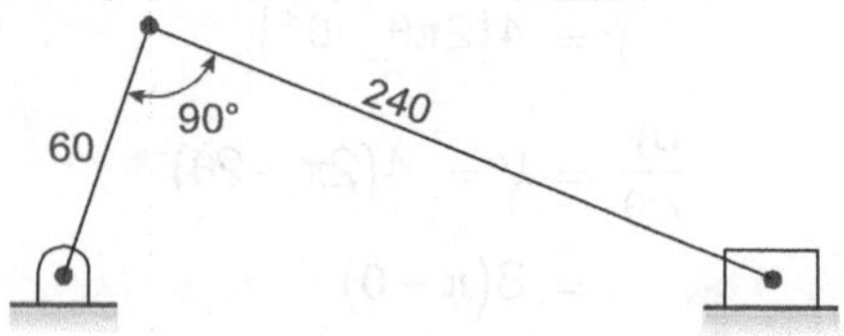

[2014 : 2 M, Set-3]

4.4 In a certain slider-crank mechanism, lengths of crank and connecting rod are equal. If the crank rotates with a uniform angular speed of 14 rad/s and the crank length is 300 mm, the maximum acceleration of the slider (in m/s^2) is ______.

[2015 : 2 M, Set-2]

4.5 A slider crank mechanism with crank radius 200 mm and connecting rod length 800 mm is shown. The crank is rotating at 600 rpm in the counter clockwise direction. In the configuration shown, the crank makes an angle of 90° with the sliding direction of the slider, and a force of 5 kN is acting on the slider. Neglecting the inertia forces, the turning moment on the crank (in kNm) is______.

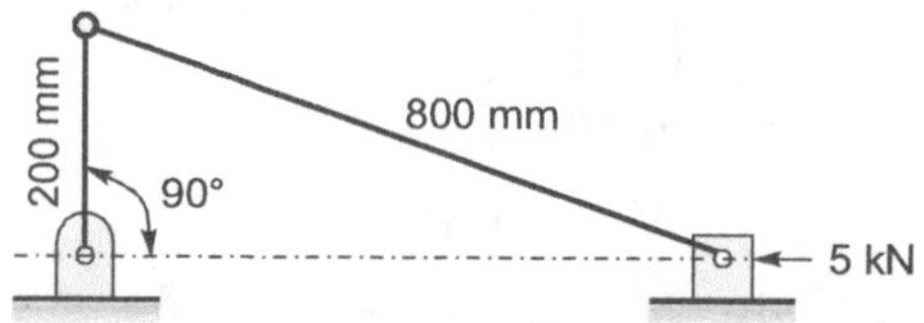

[2016 : 2 M, Set-1]

4.6 A slider crank mechanism is shown in the figure. At some instant, the crank angle is 45° and a force of 40 N is acting towards the left on the slider. The length of the crank is 30 mm and the connecting rod is 70 mm. Ignoring the effect of gravity, friction and inertial forces, the magnitude of the crankshaft torque (in Nm) needed to keep the mechanism in equilibrium is__________(correct to two decimal places).

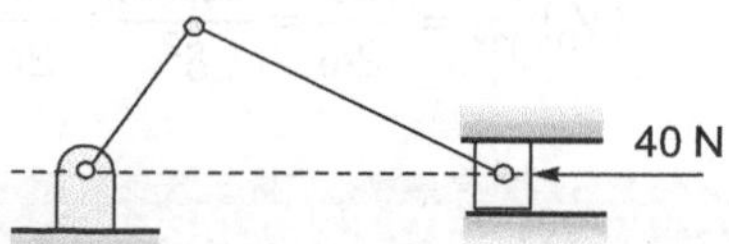

[2018 : 2 M, Set-1]

■■■■

Answers **Dynamic Analysis of Slider-crank**

4.1 (b) 4.2 (c) 4.3 (0.6) 4.4 (117.6) 4.5 (1) 4.6 (1.12)

Explanations Dynamic Analysis of Slider-crank

4.1 (b)

Method I:

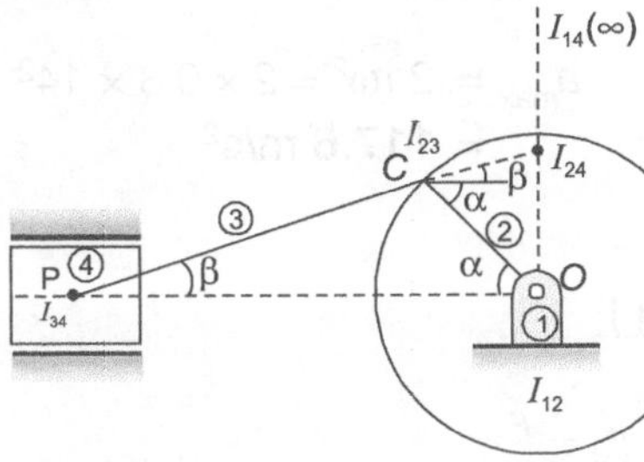

Cross head is 4th link

$$V_4 = ?$$
$$V_4 = \omega_2(I_{24}I_{12})$$
$$V_4 = \omega_2(I_{24}I_{12})$$
$$\Rightarrow \quad V_4 = \frac{V_c}{r}(I_{24}I_{12})$$

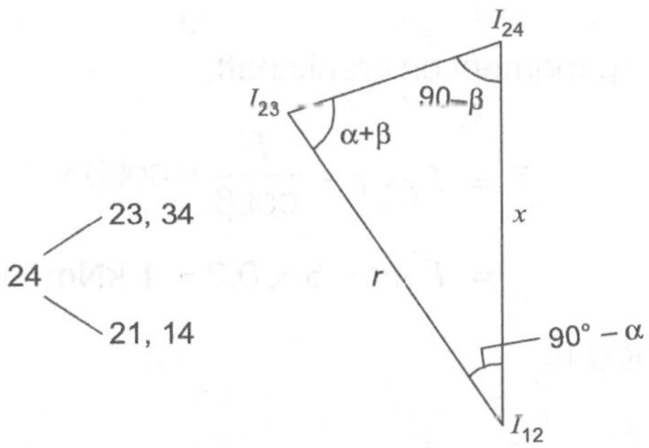

$$\Rightarrow \quad \frac{r}{\sin(90° - \beta)} = \frac{x}{\sin(\alpha + \beta)} \quad [\text{Let } I_{24}I_{12} = x]$$

$$x = \frac{r\sin(\alpha+\beta)}{\cos\beta}$$

$$\Rightarrow \quad V_4 = \frac{V_c}{r} \times \frac{r\cos[90 - (\alpha+\beta)]}{\cos\beta}$$

$$\Rightarrow \quad \mathbf{V_4 = V_c \cos[90° - (\alpha + \beta)]\sec\beta}$$

Method II:

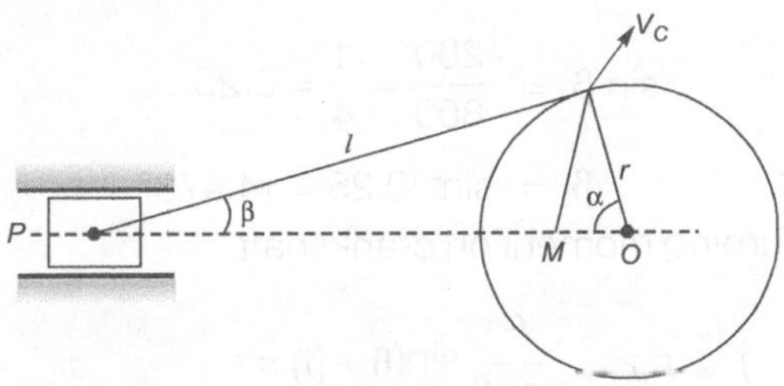

Considering ΔCPM and ΔCOM,

$$CM = l\sin\beta = r\sin\alpha$$

$$\sin\beta = \frac{r}{l}\sin\alpha$$

$$\cos\beta = \sqrt{1 - \frac{r^2\sin^2\alpha}{l^2}}$$

Velocity of cross-head,

$$V = r\omega\left[\sin\alpha + \frac{r\sin\alpha\cos\alpha}{l^2 - r^2\sin^2\alpha}\right]$$

$$= V_C\left[\sin\alpha + \frac{r\sin\alpha\cos\alpha}{l^2\cos^2\beta}\right]$$

$$= V_C\left[\sin\alpha\frac{\cos\beta}{\cos\beta} + \frac{r\sin\alpha\cos\alpha}{l\cos\beta}\right]$$

$$= V_C\left[\frac{\sin\alpha\cos\beta}{\cos\beta} + \frac{\sin\beta\cos\alpha}{\cos\beta}\right]$$

$$= V_C\sec\beta\,[\sin\alpha\cos\beta + \cos\alpha\sin\beta]$$

$$= V_C\sec\beta\sin(\alpha + \beta)$$

$$= V_C\sec\beta\cos(90 - \alpha + \beta)$$

$$= V_C\cos(90 - \overline{\alpha + \beta})\sec\beta$$

4.2 (c)

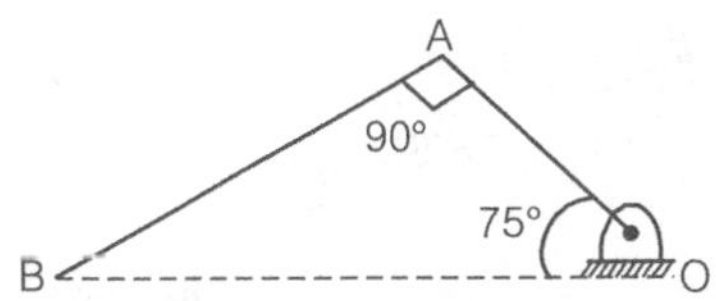

$$\text{Area} = \frac{1}{2}(AB)(AO)\sin A°$$

Area will be maximum, when

$$\angle A = 90°$$

when ΔOAB is a right-angled triangle.

$$\frac{l}{r} = \tan 75° = \frac{AB}{OA} = 3.73$$

$$\frac{l}{r} = 3.73$$

4.3 Sol.

Method I:

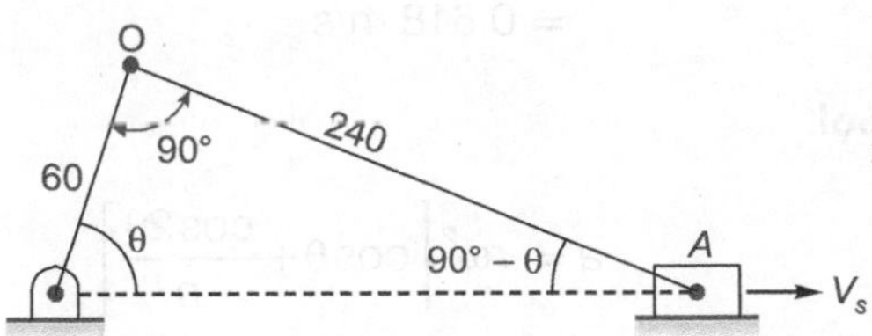

L = 240 mm

$r = 60$ mm

$n = \frac{L}{r} = \frac{240}{60} = 4$

$\omega = 10$ rad/s

From the given configuration,

$\cos\theta = \frac{\sin\theta}{n}$ or $\tan\theta = n$

or $\theta = \tan^{-1} 4 = 75.96°$

$\therefore \quad V_s = \omega r\left[\sin\theta + \frac{\sin 2\theta}{2n}\right]$

$= 10\times 0.06\left[\sin 75.96° + \frac{\sin 151.92°}{2\times 4}\right]$

$= 0.6[0.97 + 0.0588]$

$= 0.6 \times 1.0288 = 0.617$ m/s

or $V_0 = \omega r$

Since slider is ⊥ar to the crank

so, $V_0 = \omega r = 10 \times 0.06$

$= \mathbf{0.6\ m/s}$

Method II:

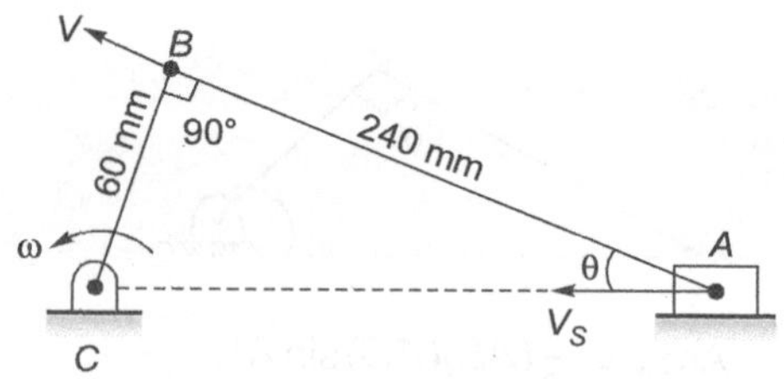

V_S = Velocity of slider

$AC^2 = 60^2 + 240^2$

$AC = 247.386$ mm

Velocity, $V = (BC)\times\omega$

$= 0.06 \times 10 = 0.6$ m/s

Since, crank and connecting rod is perpendicular.

$V_S\cos\theta = V$

$$V_S = \frac{V}{\cos\theta} = \frac{0.6}{\left(\frac{240}{297.386}\right)}$$

$= 0.618$ m/s

4.4 Sol.

$$a = r\omega^2\left[\cos\theta + \frac{\cos 2\theta}{n}\right]$$

at $\theta = 0$

$a \Rightarrow a_{max}$

$$a_{max} = r\omega^2\left[1 + \frac{1}{n}\right]$$

$n = \frac{l}{r} = 1$

$a_{max} = 2r\omega^2 = 2 \times 0.3 \times 14^2$

$= \mathbf{117.6\ m/s^2}$

4.5 Sol.

Method I:

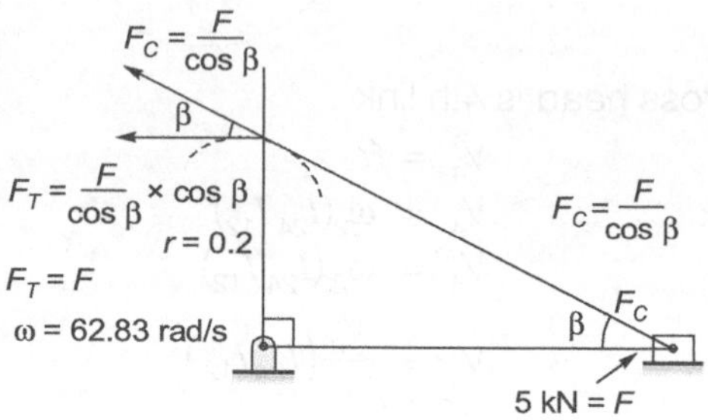

Turning moment on crankshaft,

$$T = F_T \times r = \frac{F}{\cos\beta}\times\cos\beta\times r$$

$= F \times r = 5 \times 0.2 = \mathbf{1\ kNm}$

Method II:

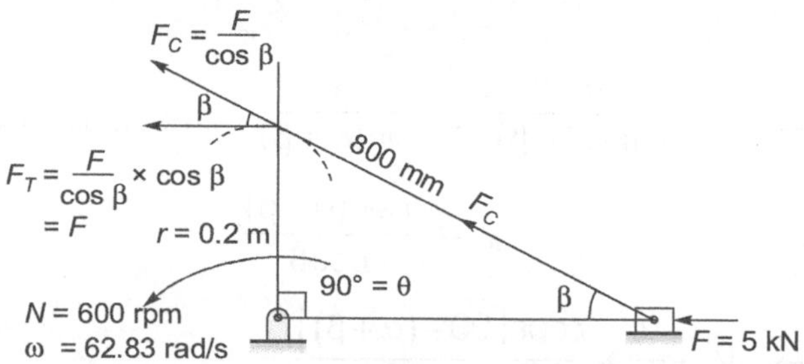

$$\sin\beta = \frac{\text{Perpendicular}}{\text{Hypotenuse}}$$

$$\sin\beta = \frac{200}{800} = \frac{1}{4} = 0.25$$

$\beta = \sin^{-1}0.25 = 14.478°$

Turning moment on crankshaft,

$$T = F_t r = \frac{F}{\cos\beta}\sin(\theta+\beta)\times r$$

$$= \frac{5}{\cos 14.478}\sin[90° + 14.478°]\times 0.2$$

$= \mathbf{1\ kNm}$

4.6 Sol.

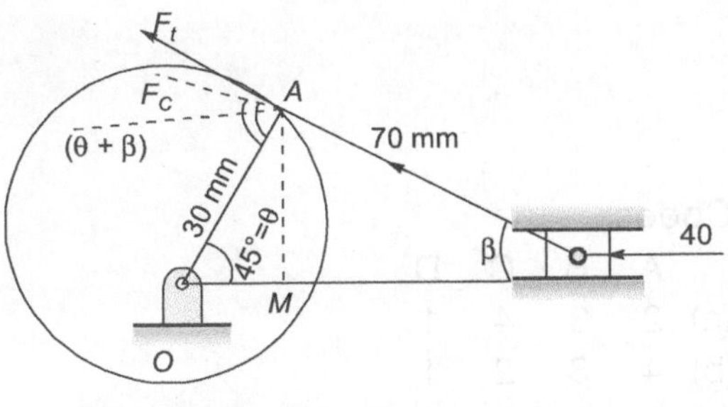

$$30 \sin 45° = 70 \sin \beta$$

$$\sin \beta = \frac{30 \sin 45°}{70} = 0.303045$$

$$\beta = 17.6406°$$

$$F_C \cos \beta = 40$$

$$F_C = \frac{40}{\cos\beta} = 41.9737 \text{ N}$$

$$F_t = F_C \sin(\theta + \beta)$$

$$= 41.9737 \times \sin(17.6406° + 45°)$$

$$F_t = 37.2785 \text{ N}$$

$$T = F_t \times r = 37.2785 \times 0.030$$

$$= 1.1183 \text{ Nm} = 1.118 \text{ Nm}$$

$$= \mathbf{1.12 \text{ Nm}}$$

■■■■

5 Gear and Gear Train

5.1 **List-I (Gear types)**
A. Worm gears
B. Cross helical gears
C. Bevel gears
D. Spur gears

List-II (Application)
1. Parallel shafts
2. Nonparallel, intersecting shafts
3. Nonparallel, nonintersecting shafts
4. Large speed ratios

Codes:

	A	B	C	D
(a)	2	3	1	4
(b)	4	1	3	2
(c)	4	3	2	1
(d)	3	1	2	4

[1995 : 2 M]

5.2 The arm *OA* of a epicyclic gear train shown in figure revolves counter clockwise about *O* with an angular velocity of 4 rad/s. Both gear are of same size. The angular velocity of gears *C*, if the sun gear *B* is fixed is

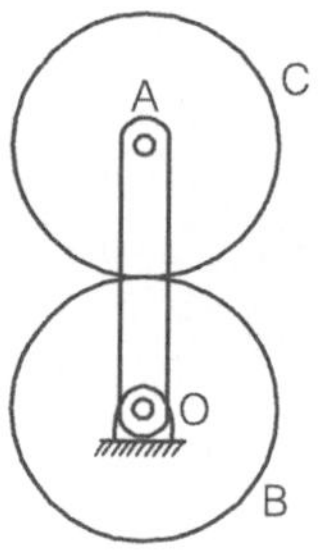

(a) 4 rad/s (b) 8 rad/s
(c) 10 rad/s (d) 12 rad/s

[1995 : 2 M]

5.3 Interference in a pair of gears is avoided, if the addendum circles of both the gears intersect common tangent to the base circles with in the points of tangency.
(a) True
(b) False [1995 : 1 M]

5.4 Match the Following
List-I (Gear Types)
A. Worm gears
B. Cross helical gears
C. Bevel gears
D. Spur gears

List-II (Applications
1. Parallel shafts
2. Nonparallel, intersecting shafts
3. Nonparallel, non-intersecting shafts
4. Large speed ratios

Codes:

	A	B	C	D
(a)	2	3	4	1
(b)	4	3	2	1
(c)	3	4	2	1
(d)	1	3	4	2

[1995 : 1 M]

5.5 In spur gears, the circle on which the involute is generated is called the
(a) pitch circle
(b) clearance angle
(c) base circle
(d) addendum circle [1996 : 1 M]

5.6 To make a worm drive reversible, it is necessary to increase
(a) centre distance
(b) worm diameter factor
(c) number of starts
(d) reduction ratio [1997 : 1 M]

5.7 A 1.5 kW motor is running at 1440 rev/min. It is to be connected to a stirrer running at 36 rev/min. The gearing arrangement suitable for this application is
(a) differential gear (b) helical gear
(c) spur gear (d) worm gear

[2000 : 1 M]

5.8 A 1.5 kW motor is running at 1440 rev/min. It is to be connected to a stirrer running at 36 rev/min. The gearing arrangement suitable for this application is
(a) differential gear (b) helical gear
(c) spur gear (d) worm gear [2000 : 1 M]

5.9 arge speed reductions (greater than 20) in one stage of a gear train are possible through
(a) Spur gearing (b) Worm gearing
(c) Bevel gearing (d) Helical gearing

[2002 : 1 M]

5.10 The minimum number of teeth on the pinion to operate without interference in standard full height involute teeth gear mechanism with 20° pressure angle is
(a) 14 (b) 12
(c) 18 (d) 32 [2002 : 1 M]

Common Data Questions Q.5.11 and Q.5.12

A compacting machine shown in the figure below is used to create a desired thrust force by using a rack and pinion arrangement. The input gear is mounted on the motor shaft. The gears have involute teeth of 2 mm module.

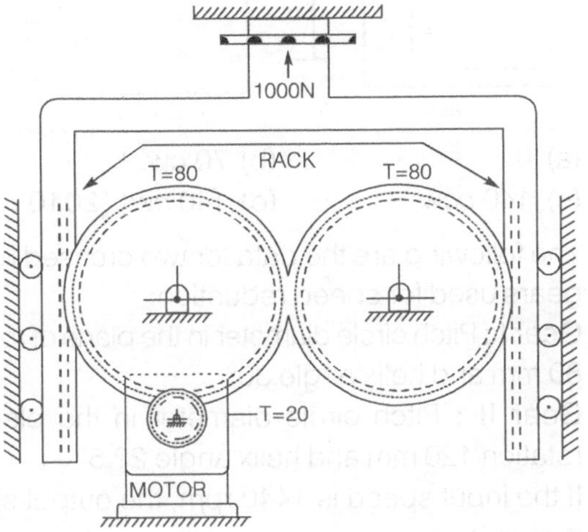

5.11 If the drive efficiency is 80%, the torque required on the input shaft to create 1000 N output thrust is

(a) 20 Nm (b) 25 Nm
(c) 32 Nm (d) 50 Nm [2004 : 2 M]

5.12 If the pressure angle of the rack is 20°, the force acting along the line of action between the rack and the gear teeth is

(a) 250 N (b) 342 N
(c) 532 N (d) 600 N [2004 : 2 M]

5.13 Match the following

Type of gears
P. Bevel gears
Q. Worm gears
R. Herringbone gears
S. Hypoid gears

Arrangement of shafts
1. Nonparallel offset shafts
2. Nonparallel intersecting shafts
3. Nonparallel, non-intersecting shafts
4. Parallel shafts

(a) P - 4, Q - 2, R - 1, S - 3
(b) P - 2, Q - 3, R - 4, S - 1
(c) P - 3, Q - 2, R - 1, S - 4
(d) P - 1, Q - 3, R - 4, S - 2 [2004 : 2 M]

Common Data for Questions 5.14 and 5.15:

A planetary gear train has four gears and one carrier. Angular velocities of the gears are ω_1, ω_2, ω_3, and ω_4, respectively. The carrier rotates with angular velocity ω_5.

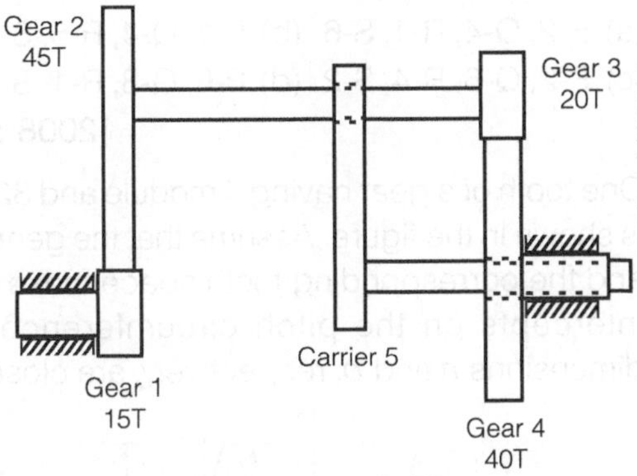

5.14 What is the relation between the angular velocities of Gear 1 and Gear 4?

(a) $\frac{\omega_1 - \omega_5}{\omega_4 - \omega_5} = 6$ (b) $\frac{\omega_4 - \omega_5}{\omega_1 - \omega_5} = 6$

(c) $\frac{\omega_1 - \omega_2}{\omega_4 - \omega_5} = -\left(\frac{2}{3}\right)$ (d) $\frac{\omega_2 - \omega_5}{\omega_4 - \omega_5} = \frac{8}{9}$

[2006 : 2 M]

5.15 For ω_1 = 60 rpm clockwise (*cw*) when viewed from the left, what is the angular velocity of the carrier and its direction so that Gear 4 rotates in counterclockwise (*ccw*) direction at twice the angular velocity of Gear 1 when viewed from the left?

(a) 130 rpm, *cw* (b) 223 rpm, *ccw*
(c) 256 rpm, *cw* (d) 156 rpm, *ccw*

[2006 : 2 M]

5.16 Twenty degree full depth involute profiled 19-teeth pinion and 37-teeth gear are in mesh. If the module is 5 mm, the center distance between the gear pair will be

(a) 140 mm (b) 150 mm
(c) 280 mm (d) 300 mm [2006 : 2 M]

5.17 Match the type of gears with their most appropriate description.

Type of gear
P. Helical
Q. Spiral Bevel
R. Hypoid
S. Rack and pinion

Description
1. Axes non parallel and non intersecting
2. Axes parallel and teeth are inclined to the axis
3. Axes parallel and teeth are parallel to the axis
4. Axes are perpendicular and intersecting, and teeth are inclined to the axis
5. Axes are perpendicular and used for large speed reduction
6. Axes parallel and one of the gears has infinite radius

(a) P-2, Q-4, R-1, S-6 (b) P-1, Q-4, R-5, S-6
(c) P-2, Q-6, R-4, S-2 (d) P-6, Q-3, R-1, S-5

[2008 : 2 M]

5.18 One tooth of a gear having 4 module and 32 teeth is shown in the figure. Assume that the gear tooth and the corresponding tooth space make equal intercepts on the pitch circumference. The dimensions *a* and *b*, respectively, are closest to.

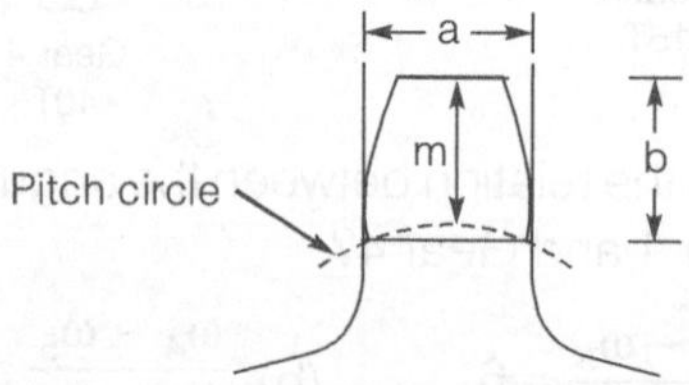

(a) 6.08 mm, 4 mm (b) 6.48 mm, 4.2 mm
(c) 6.28 mm, 4.3 mm (d) 6.28 mm, 4.1 mm

[2008 : 2 M]

5.19 An epicyclic gear train is shown schematically in the adjacent figure. The sun gear 2 on the input shaft is a 20 teeth external gear. The planet gear 3 is a 40 teeth external gear. The ring gear 5 is a 100 teeth internal gear. The ring gear 5 is fixed and the gear 2 is rotating at 60 rpm *ccw* (*ccw* = counterclockwise and *cw* = clockwise)

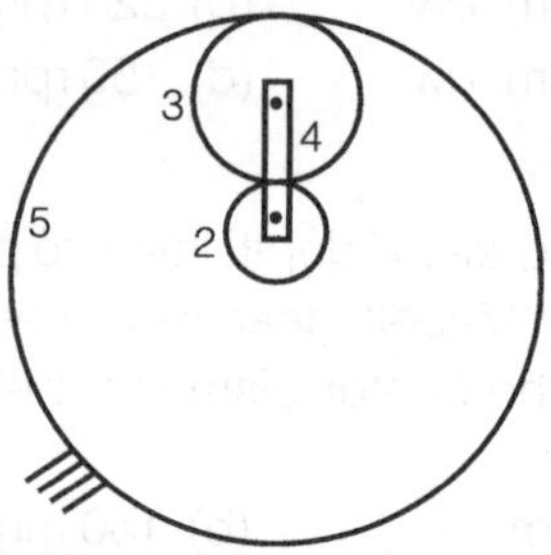

The arm 4 attached to the output shaft will rotate at

(a) 10 rpm *ccw* (b) 10 rpm *cw*
(c) 12 rpm *cw* (d) 12 rpm *ccw*

[2009 : 2 M]

5.20 Tooth interference in an external involute spur gear pair can be reduced by

(a) decreasing center distance between gear pair
(b) decreasing module
(c) decreasing pressure angle
(d) increasing number of gear teeth

[2010: 1 M]

5.21 For the epicyclic gear arrangement shown in the figure, ω_2 = 100 rad/s clockwise (*cw*) and ω_{arm} = 80 rad/s counter clockwise (*ccw*). The angular velocity ω_5 (in rad/s) is

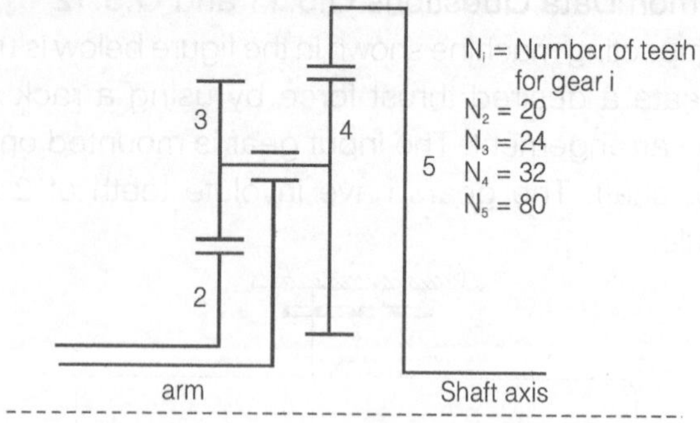

(a) 0 (b) 70 *cw*
(c) 140 *ccw* (d) 140 *cw* [2010 : 2 M]

5.22 The following are the data for two crossed helical gears used for speed reduction:

Gear I : Pitch circle diameter in the plane of rotation 80 mm and helix angle 30°

Gear II : Pitch circle diameter in the plane of rotation 120 mm and helix angle 22.5°

If the input speed is 1440 rpm, the output speed in rpm is

(a) 1200 (b) 900
(c) 875 (d) 720 [2012 : 1 M]

5.23 A compound gear train with gears *P*, *Q*, *R* and *S* has number of teeth 20, 40, 15 and 20 respectively. Gears *Q* and *R* are mounted on the same shaft as shown in the figure below. The diameter of the gear *Q* is twice that of the gear *R*. IF the module of the gear *R* is 2 mm, the centre distance in mm between gears *P* and *S* is

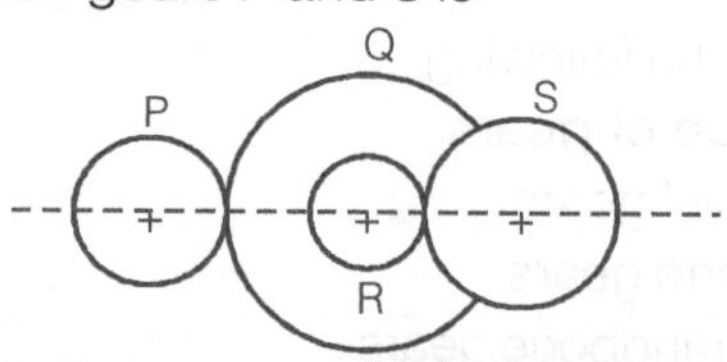

(a) 40 (b) 80
(c) 120 (d) 160 [2013 : 2 M]

5.24 For the given statements:

I. Mating spur gear teeth is an example of higher pair
II. A revolute joint is an example of lower pair

Indicate the correct answer.

(a) Both I and II are false
(b) I is true and II is false
(c) I is false and II is true
(d) Both I and II are true [2014 : 1 M, Set-1]

5.25 Which one of the following is used to convert a rotational motion into a translational motion?

(a) Bevel gears
(b) Double helical gears
(c) Worm gears

(d) Rack and pinion gears [2014 : 1 M, Set-4]

5.26 A pair of spur gears with module 5 mm and a center distance of 450 mm is used for a speed reduction of 5 :1. The number of teeth on pinion is ________.

[2014 : 2 M, Set-1]

5.27 Gear 2 rotates at 1200 rpm is counter clockwise direction and engages with Gear 3 and Gear 4 mounted on the same shaft. Gear 5 engages with Gear 4. The numbers of teeth on Gears 2, 3, 4 and 5 is 20, 40, 15 and 30, respectively. The angular speed of Gear 5 is

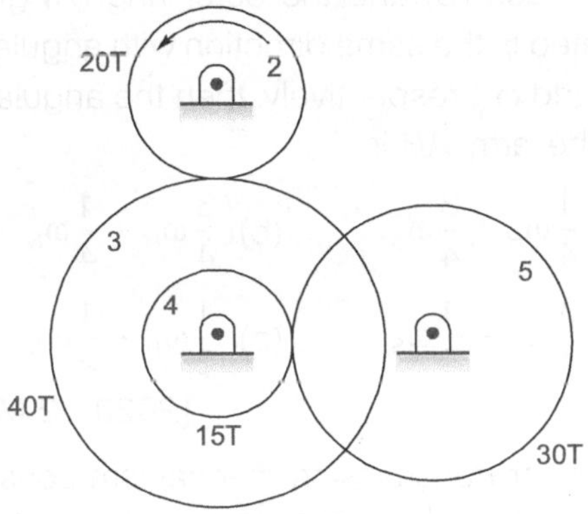

(a) 300 rpm counter clockwise
(b) 300 rpm clockwise
(c) 4800 rpm counter clockwise
(d) 4800 rpm clockwise [2014 : 1 M, Set-3]

5.28 It is desired to avoid interference in a pair of spur gears having a 20° pressure angle. With increase in pinion to gear speed ratio, the minimum number of teeth on the pinion
(a) increases
(b) decreases
(c) first increases and then decreases
(d) remains unchanged [2014 : 2 M, Set-4]

5.29 A gear train is made up of five spur gears as shown in the figure. Gear 2 is driver and gear 6 is driven member. N_2, N_3, N_4, N_5 and N_6 represent the number of teeth on gears 2, 3, 5 and 6 respectively. The gear(s) which act(s) as idler(s) is/are

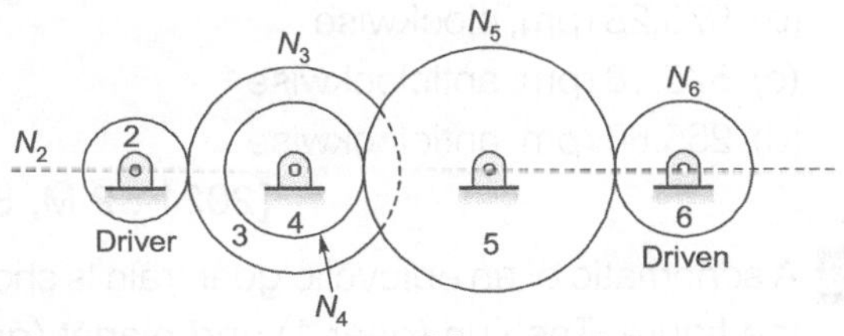

(a) only 3 (b) only 4
(c) only 5 (d) Both 3 and 5
[2015 : 1 M, Set-3]

5.30 In the gear train shown gear 3 is carried on arm 5, Gear 3 meshes with gear 2 and gear 4. The number of teeth on gear 2, 3 and 4 are 60, 20, and 100, respectively. If gear 2 is fixed and gear 4 rotates with an angular velocity of 100 rpm in the counterclockwise direction, the angular speed of arm 5 (in rpm) is

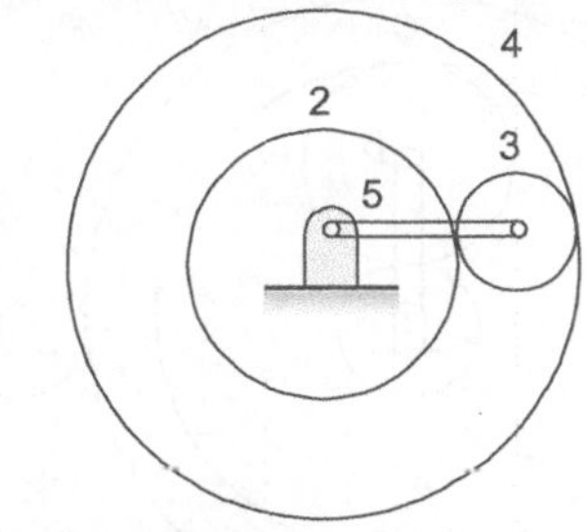

(a) 166.7 counterclockwise
(b) 166.7 clockwise
(c) 62.5 counterclockwise
(d) 62.5 clockwise [2016 : 2 M, Set-1]

5.31 In an epicyclic gear train, shown in the figure, the outer ring gear is fixed, while the sun gear rotates counterclockwise at 100 rpm. Let the number of teeth on the sun planet and outer gears to be 50, 25 and 100, respectively. The ratio of magnitude of angular velocity of the planet gear to the angular velocity of the carrier arm is ______.

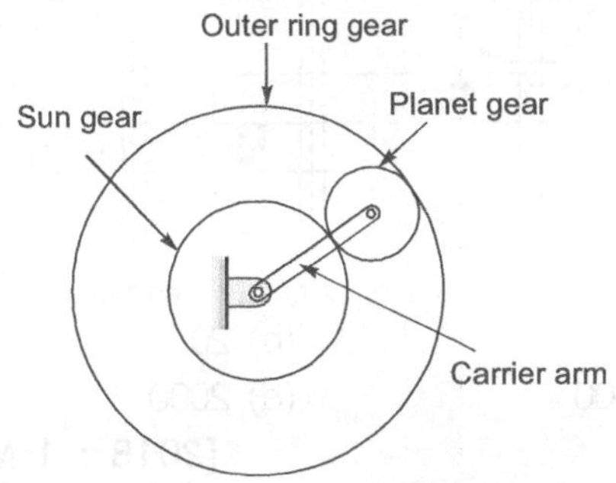

[2017 : 2 M, Set-1]

5.32 A gear train shown in the figure consists of gears *P*, *Q*, *R* and *S*. Gear *Q* and gear *R* are mounted on the same shaft. All the gears are mounted on parallel shafts and the number of teeth of *P*, *Q*, *R* and *S* are 24, 45, 30 and 80, respectively. Gear *P* is rotating at 400 rpm. The speed (in rpm) of the gear *S* is ________.

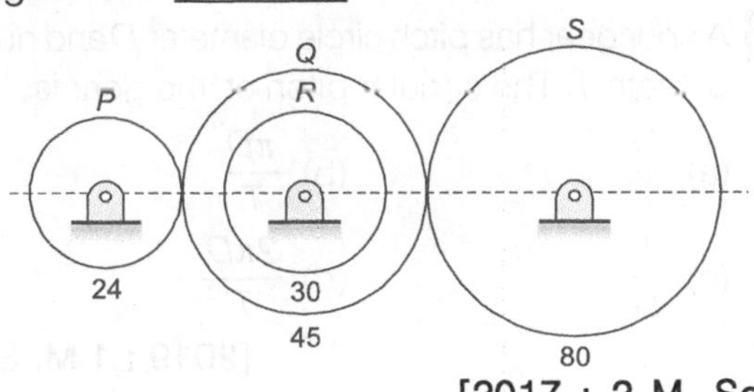

[2017 : 2 M, Set-2]

5.33 An epicyclic gear train is shown in the figure below. The number of teeth on the gears A, B and D are 20, 30 and 20 respectively. Gear C has 80 teeth on the inner surface and 100 teeth on the outer surface. If the carrier arm AB is fixed and the sun gear A rotates at 300 rpm in the clockwise direction, then the rpm of D in the clockwise direction is

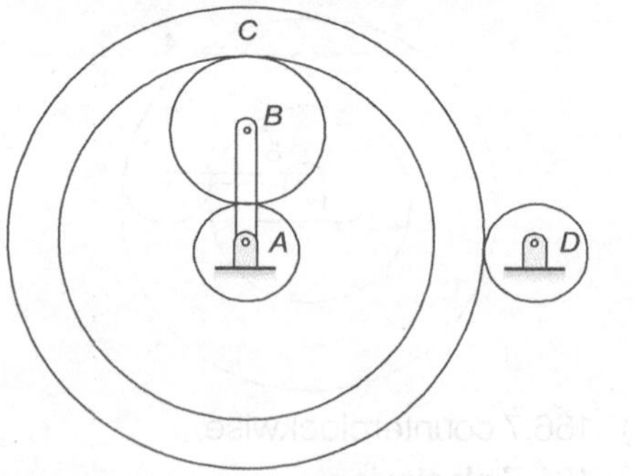

(a) 240 (b) –240
(c) 375 (d) –375

[2018 : 2 M, Set-1]

5.34 A frictionless gear train is shown in the figure. The left most 12-teeth gear is given a torque of 100 Nm. The output torque from the 60-teeth gear on the right in Nm is

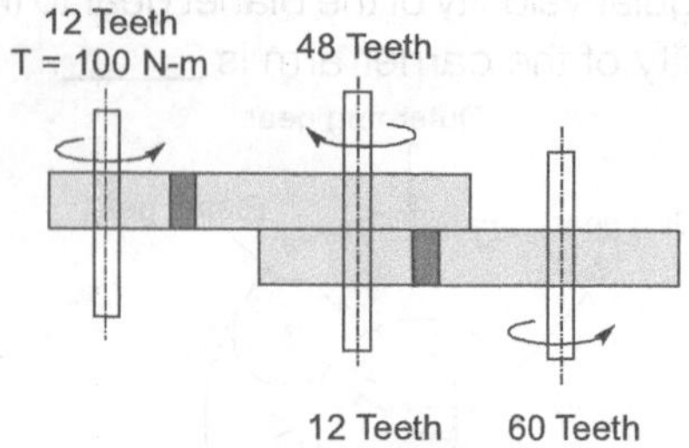

(a) 5 (b) 20
(c) 500 (d) 2000

[2018 : 1 M, Set-2]

5.35 A spur gear with 20° full depth teeth is transmitting 20 kW at 200 rad/s. The pitch circle diameter of the gear is 100 mm. The magnitude of the force applied on the gear in the radial direction is

(a) 0.73 kN (b) 2.78 kN
(c) 1.39 kN (d) 0.36 kN

[2019 : 1 M, Set-1]

5.36 A spur gear has pitch circle diameter D and number of teeth T. The circular pitch of the gear is

(a) $\frac{D}{T}$ (b) $\frac{\pi D}{T}$
(c) $\frac{T}{D}$ (d) $\frac{2\pi D}{T}$

[2019 : 1 M, Set-2]

5.37 The sun (S) and the planet (P) of an epicyclic gear train shown in the figure have identical number of teeth

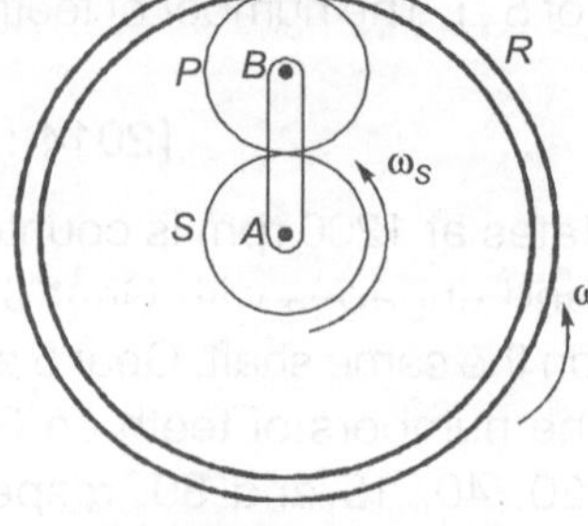

If the sun (S) and the outer ring (R) gears are rotated in the same direction with angular speed ω_S and ω_R, respectively, then the angular speed of the arm AB is

(a) $\frac{1}{4}\omega_R + \frac{3}{4}\omega_S$ (b) $\frac{3}{4}\omega_R + \frac{1}{4}\omega_S$
(c) $\frac{3}{4}\omega_R - \frac{1}{4}\omega_S$ (d) $\frac{1}{2}\omega_R - \frac{1}{2}\omega_S$

[2020 : 2 M, Set-2]

5.38 A power transmission mechanism consists of a belt drive and a gear train as shown in the figure.

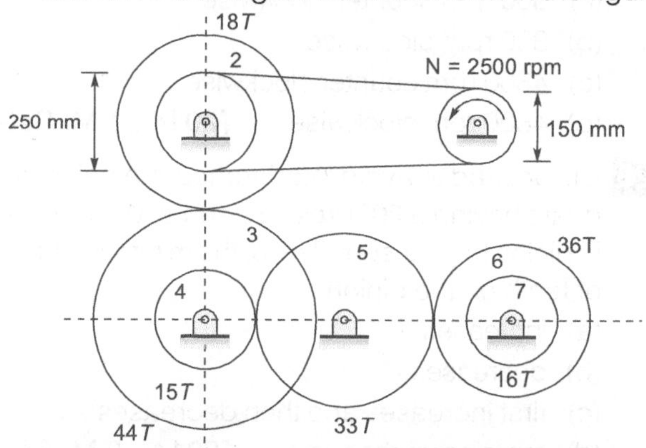

Diameters of pulleys of belt drive and number of teeth (T) on the gears 2 to 7 are indicated in the figure. The speed and direction of rotation of gear 7, respectively, are

(a) 255.68 rpm; clockwise
(b) 575.28 rpm; clockwise
(c) 575.28 rpm; anticlockwise
(d) 255.68 rpm; anticlockwise

[2021 : 2 M, Set-2]

5.39 A schematic of an epicyclic gear train is shown in the figure. The sun (gear 1) and planet (gear 2) are external, and the ring gear (gear 3) is internal. Gear 1, gear 3 and arm OP are pivoted to the ground at O. Gear 2 is carried on the arm OP via the pivot joint at *P*, and is in mesh with the other two gears. Gear 2 has 20 teeth and gear 3 has 80 teeth. If gear 1 is kept fixed at 0 rpm and gear 3 rotates at 900 rpm counter clockwise (ccw), the magnitude of angular velocity of arm *OP* is ________ rpm (in integer).

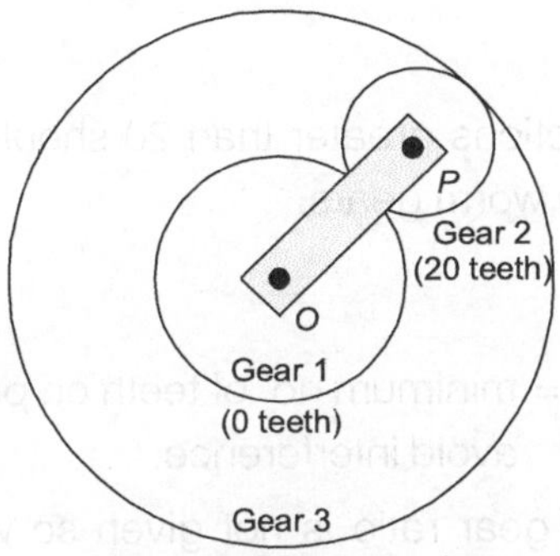

[2022 : 2 M, Set-1]

5.40 Two meshing spur gears 1 and 2 with diametral pitch of 8 teeth per mm and an angular velocity ratio $\frac{|\omega_2|}{|\omega_1|} = \frac{1}{4}$, have their centers 30 mm apart.

The number of teeth on the driver (gear 1) is _______. (Answer in integer)

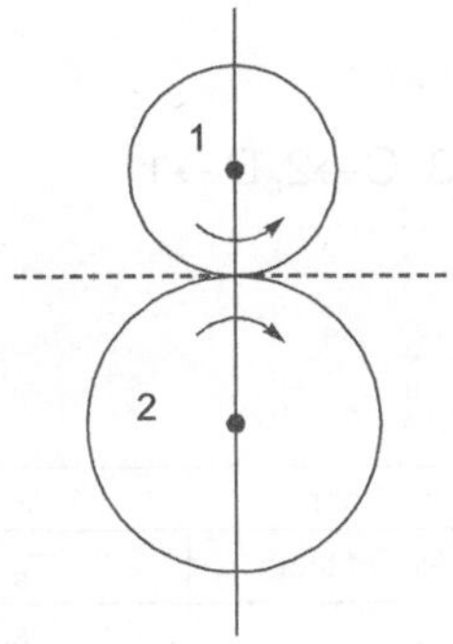

[2023 : 1 M]

5.41 The Levai type-A train illustrated in the figure has gears with module $m = 8$ mm/tooth. Gears 2 and 3 have 19 and 24 teeth respectively. Gear 2 is fixed and internal gear 4 rotates at 20 rev/min counter-clockwise. The magnitude of angular velocity of the arm is _____ rev/min. (rounded off to 2 decimal places)

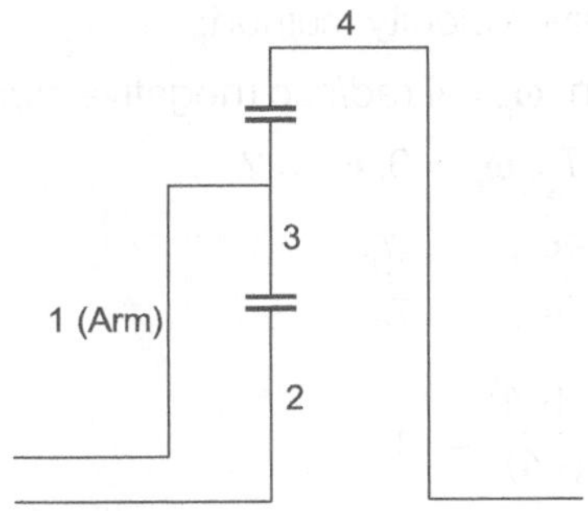

[2024 : 2 M]

Answers Gear and Gear Train

5.1 (c)	5.2 (b)	5.3 (a)	5.4 (b)	5.5 (c)	5.6 (c)	5.7 (d)	5.8 (d)
5.9 (b)	5.10 (c)	5.11 (b)	5.12 (c)	5.13 (b)	5.14 (a)	5.15 (d)	5.16 (a)
5.17 (a)	5.18 (d)	5.19 (a)	5.20 (d)	5.21 (c)	5.22 (b)	5.23 (b)	5.24 (d)
5.25 (d)	5.26 (30)	5.27 (a)	5.28 (b)	5.29 (c)	5.30 (c)	5.31 (3)	
5.32 (120)	5.33 (c)	5.34 (d)	5.35 (a)	5.36 (b)	5.37 (b)	5.38 (a)	
5.39 (600)	5.40 (96)	5.41 (15.58)					

Explanations Gear and Gear Train

5.1 (c)

A→4, B→3, C→2, D→1

5.2 (b)

Method I:

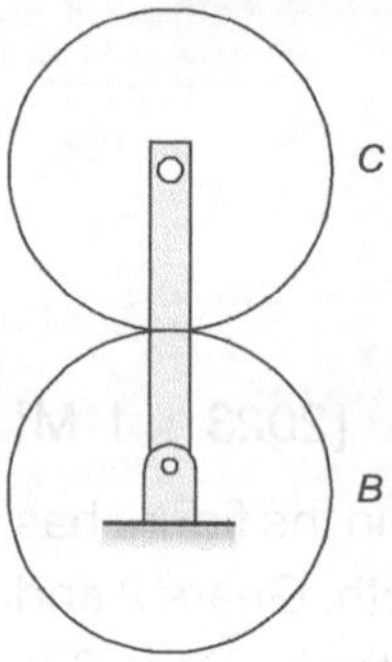

Gear	B	C
No. of teeth	T_B	$T_B = T_C$
Speed N in rpm	+ N positive sign means clockwise	$\frac{-N \cdot T_B}{T_C} = -N$
Speed after adding speed of arm, let speed of arm be $+y$ clockwise	$N + y$	$-N + y$

Given: $y = -4$ [counter clockwise direction]

$N + y = 0$ [sun gear B is fined]

$\therefore \quad N = -y$

$= -1\{-4\} = +4$ rad/s

Speed of gear $C = -4 - 4 =$ **−8 rad/s**

Method II:

Relative velocity method:

Given: $\omega_a = 4$ rad/sec (negative sign CCW)

$T_B = T_C, \omega_B = 0, \omega_C = ?$

$$\frac{\omega_C - \omega_a}{\omega_B - \omega_a} = \frac{-T_B}{T_C}$$

$$\frac{\omega_C - (-4)}{0 - (-4)} = -1$$

$$\omega_C + 4 = -4$$

$$\omega_C = -8 \text{ rad/s}$$

$$= 8 \text{ rad/s}$$

5.4 (b)

A – 4, B – 3, C – 2, D – 1

5.7 (d)

Power of motor = 1.5 kW

$N = 1440$ rev/min

Speed of stirrer = 36 rev/min

$$\text{Speed reduction ratio} = \frac{1440}{36} = \mathbf{40}$$

For such high reduction, **worm gear** is most suitable.

5.8 (d)

$$\text{Velocity reduction} = \frac{36}{1440} = \frac{1}{40}$$

For this kind of reduction, worm gear arrangement is most suitable.

5.9 (b)

Reductions greater than 20 should be carried out by worm gears.

5.10 (c)

Z_{min} = minimum no. of teeth on pinion to avoid interference.

Since gear ratio is not given so we assumed that arrangement is rack & pinion arrangement.

$$Z_{min} = \frac{2 \times A_w}{\sin^2 \phi} = \frac{2 \times 1}{\sin^2 20} = 17.09$$

$$\simeq \mathbf{18 \text{ teeth}}$$

5.11 (b)

F_l = Force acting along the line of action.

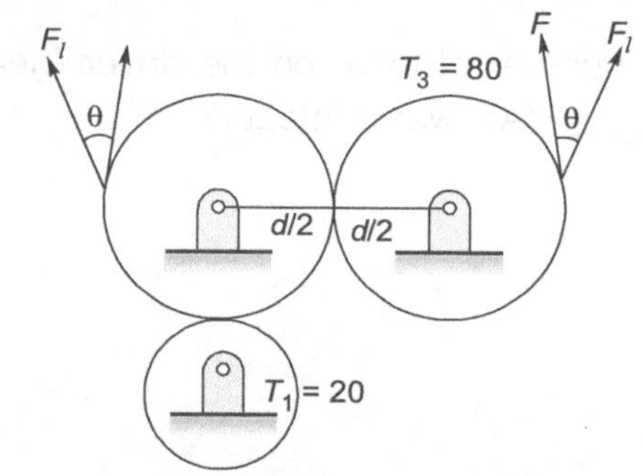

$$2F = 1000$$

$$F = 500 \text{ N}$$

$$m = 2 \text{ mm}$$

$$R = \frac{mT}{2} = \frac{2 \times 80}{2} = 40 \text{ mm} = 0.040 \text{ m}$$

$$\eta = 0.80 = \frac{T_2 \times \omega_2}{T_1 \times \omega_1} = \frac{(500 \times 0.040 \times 2)}{T_1} \times \frac{20}{80}$$

$$\left[\frac{\omega_2}{\omega_1} = \frac{T_1}{T_2}\right]$$

$$T_1 = 25 \text{ N-m}$$

5.12 (c)

$$F_l \cos\theta = F$$

$$F_l = \frac{F}{\cos 20°}$$

(pressure angle, $\theta = 20$)

$$F_l = 532 \text{ N}$$

POINTS TO REMEMBER

In this problem, we have taken output power as $2T_2\omega_2$ because output power is transmitting the both gears.

5.14 (a)

Arm	Gears	1	2	3	4
0	No. of teeth	15	45	20	40
0	Velocities	N_1	$\frac{15 N_1}{45}$	$\frac{-15N_1}{45}$	$\frac{+20 \times 15N_1}{40 \times 45}$
+y	Velocities after adding velocity of arm "y"	$N_1 + y$	$\frac{-15N_1}{45} + y$	$\frac{-15N_1}{45} + y$	$\frac{+20 \times 15N_1}{45 \times 40} + y$

Therefore

$$\omega_1 = N_1 + y$$

$$\omega_2 = \frac{-15N_1}{45} + y = \frac{-N_1}{3} + y$$

$$\omega_3 = \frac{-15N_1}{45} + y = \frac{-N_1}{3} + y$$

$$\omega_4 = \frac{20 \times 15 \times N_1}{45 \times 40} + y = \frac{N_1}{6} + y$$

$$\therefore \quad \frac{\omega_1 - \omega_5}{\omega_4 - \omega_5} = \frac{N_1 + y - y}{\frac{N_1}{6} + y - y} = \frac{N_1}{N_{1/6}} = 6$$

$$\frac{\omega_1 - \omega_5}{\omega_4 - \omega_5} = 6$$

5.15 (d)

Given:

$$N_1 + y = +60 \quad \text{...(i)} \quad [\omega_1 = 60]$$

Also

$$\frac{N_1}{6} + y = -2(N_1 + y) \qquad [\omega_4 = -2\omega_1]$$

$$\therefore \quad \frac{N_1}{6} + y = -2N_1 - 2y$$

$$\frac{N_1}{6} + 2N_1 = -3y$$

$$\frac{13N_1}{6} = -3y$$

$$N_1 = (60 - y) \qquad \text{from equation (i)}$$

$$\frac{13}{6}[60 - y] = -3y$$

$$780 - 13y = -18y$$

$$780 = -18y + 13y$$

$$780 = -5y$$

$$y = -\frac{780}{5} = \mathbf{-156 \text{ rpm}}$$

Negative sign represents counter clockwise direction.

5.16 (a)

Given $T_p = 19$, $T_a = 37$, $m = 5$

$$m = D/T$$

$$D_P = 95 \text{ mm}$$

$$D_G = 185 \text{ mm}$$

$$\text{Centre distance} = \frac{95 + 185}{2} = \mathbf{140 \text{ mm}}$$

5.17 (a)

Helical gear – Axes parallel and teeth are inclined to the axis.

Spiral Bevel – Axes are perpendicular and interesting and teeth are inclined to the axis.

Hypoid – Axis nonparallel and non-interesting.

Rack and pinion – Axes parallel and one of the gears has infinite radius.

5.18 (d)

Given: $m = 4$ mm and $T = 32$

So $D = m \times T = 4 \times 32$

$= 128$ mm

$$P_c = \frac{\pi D}{T} = \frac{\pi \times 128}{32}$$

$$P_c = 12.566 \text{ mm}$$

but $P_c = a + a = 2a$

(because tooth thickness = tooth space)

$$2a = 12.566$$

$$\mathbf{a = 6.28 \text{ mm}}$$

and $b = 4.1$ mm

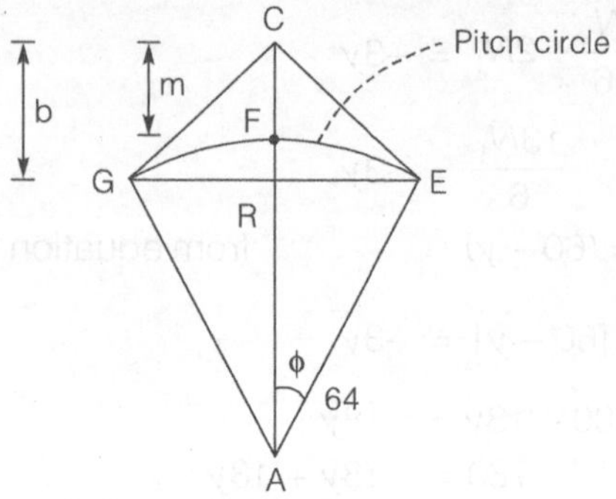

$\because \quad \phi = \frac{360°}{32 \times 4} = 2.8°$

$\therefore \quad AR = 64 \times \cos 2.8° = 63.9°$

$\Rightarrow \quad b = m + FR [\because FR = FA - AR]$

$= m + (64 - 63.9)$

$\Rightarrow \quad b = 4.1 \text{ mm}$

Addendum = module

Dedendum = 1.1 × module

5.19 (a)

Method I:

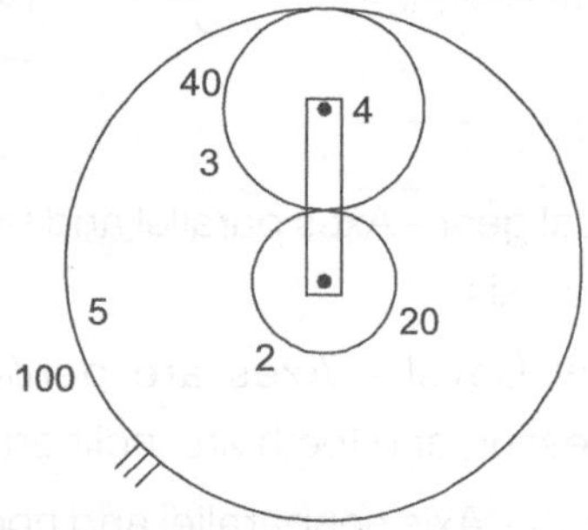

Arm– 4	Gear– 2	Gear– 3	Gear– 5
0	$+x$	$-\frac{T_2}{T_3}x$	$-\frac{T_2}{T_3}x \times \frac{T_3}{T_5}$
$+y$	$+y$	$+y$	$+y$
y	$x + y$	$y - \frac{T_2}{T_3}x$	$y - \frac{T_2}{T_5}x$

Gear 5 is fixed

$$y - \frac{T_2}{T_5}x = 0$$

$$y = \frac{20}{100}x$$

$$y = 0.2x$$

$$x + y = -60$$

[Gear 2 rotates counterclockwise]

$$x + 2x = -60$$

$$1.2x = -60$$

$$x = -\frac{60}{1.2} = -50 \text{ rpm}$$

$$y = -10 \text{ rpm}$$

Method II:

Relative velocity method:

Given: $T_2 = 20$, $T_3 = 40$, $T_5 = 100$, $N_5 = 0$

$N_2 = 60$ rpm (CCW) = –60 rpm

$$\frac{N_2 - N_4}{N_5 - N_4} = \frac{-T_5}{T_3} \times \frac{T_3}{T_2} = \frac{-T_5}{T_2}$$

$$\frac{-60 - N_4}{0 - N_4} = \frac{-100}{20} = -5$$

$$-60 - N_4 = +5 N_4$$

$$6 N_4 = -60$$

$$N_4 = -10 \text{ rpm} = 10 \text{ rpm (ccw)}$$

5.20 (d)

Methods to prevent interference in gears are:

1. Stubbing the teeth, decreases the path of contact and contact ratio, thereby decreases interference.
2. Undercutting of gear, eliminates interference however gear becomes weaker.
3. Increasing pressure angle, decreases interference.
4. Increasing number of teeth decreases interference.
5. Addendum of gear can be modified.
6. Circular pitch can be increased.
7. Decreasing the addendum of gear and by increasing the same for pinion.

5.21 (c)

Method I:

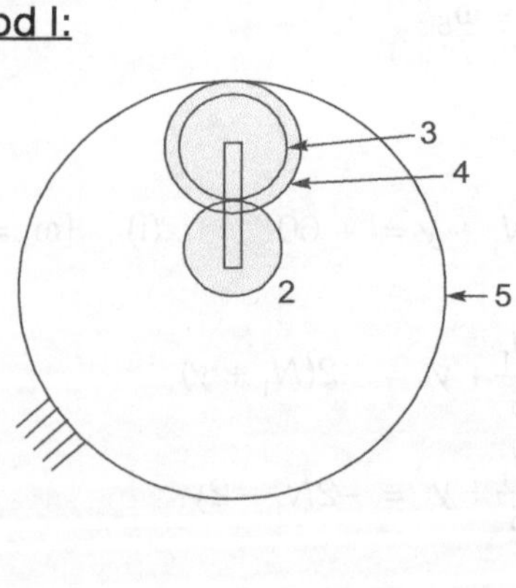

Gears	2	Compound gear 3	4	5
Teeth	20	24	32	80
Speed	$+N$	$\frac{-20N}{24}$	$\frac{-20N}{24}$	$\frac{-20N}{24}\times\frac{32}{80}$
	$+N$	$\frac{-5N}{6}$	$\frac{-5N}{6}$	$\frac{-5N}{6}\times\frac{2}{5}$
Adding speed of arm in clockwise direction	$N+y$	$\frac{-5N}{6}+y$	$\frac{-5N}{6}+y$	$\frac{-N}{3}+y$

Given, $N + y = +100$ (c.w)

$$y = -80$$

$$\therefore \quad N - 80 = 100$$

$$N = 180$$

$$\therefore \quad \omega_s = \frac{-N}{3} + y = \frac{-180}{3} - 80$$

$$= -60 - 80 = \mathbf{-140\ rad/s}$$

Method II:

Relative velocity method:

Given: ω_2 100 rad/s (cw),

ω_a = 80 rad/s (ccw) = –80 rad/s

$$\frac{\omega_5 - \omega_a}{\omega_2 - \omega_a} = -\frac{T_2}{T_3}\times\frac{T_4}{T_5}$$

$$\frac{\omega_5 - (-80)}{100 - (-80)} = \frac{-20}{24}\times\frac{32}{80}$$

$$\frac{\omega_5 + 80}{180} = -\frac{1}{3}$$

$$\omega_5 = -140 \text{ rad/s} = 140 \text{ rad/s (ccw)}$$

5.22 (b)

For helical gear

$$\text{Velocity ratio} = \frac{N_2}{N_1} = \frac{D_1\cos\phi}{D_2\cos\phi} = \frac{80\cos 30^\circ}{120\cos 22.5^\circ}$$

$$\therefore \quad N_2 = 1440 \times 0.625$$

$$= \mathbf{900\ rpm}$$

5.23 (b)

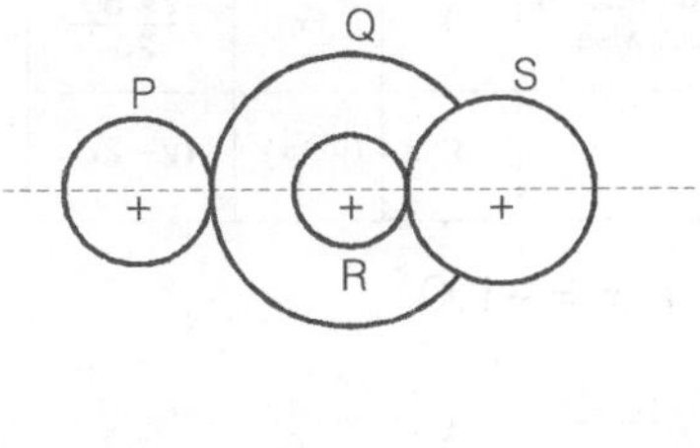

Centre distance between P and S.

$(\text{Radius})_P + (\text{Radius})_Q + (\text{Radius})_R + (\text{Radius})_S$

$$\Rightarrow \quad D_Q = 2 \times D_R$$

$$M_R = 2 \text{ mm}$$

$$\therefore \quad M_R = \frac{D_R}{T_R}$$

$$\therefore \quad D_R = T \times M_R$$

$$= 15 \times 2 = 30 \text{ mm}$$

$$D_Q = 2 \times 30 = 60 \text{ mm}$$

and $\quad T_Q = 40$

$$\therefore \quad M_Q = \frac{D_Q}{T_Q} = \frac{60}{40} = 1.5$$

Since Q meshes with P module of gear Q is equal to module of gear P.

$$D_P = M_P \times T_P = 1.5 \times 20 = 30.0 \text{ mm}$$

Since S meshes with R module of S and R will be same

$\therefore$ Module of S = 2 mm

$$D_s = M_s \times T_s = 2 \times 20 = 40 \text{ mm}$$

$\therefore$ Centre distance

$$= 15 + 30 + 15 + 20 = \mathbf{80\ mm}$$

5.24 (d)

Example of Lower Pair :

Revolute pair, Prismatic joint, Screw pair, cylindrical joint, spherical joint.

Example of higher pair :

Wheel rolling an surface, cam and follower contact, meshing teeth of two gears.

5.25 (d)

Bevel gears: Rotational motion transfer between axes at right angle.

Worm gears: For large reduction ratio in a single stage.

Double helical gears: Rotational motion transfer between parallel axes.

Rack and Pinion gears: Rotational to linear motion conversion.

5.26 Sol.

$$\frac{N_P}{N_G} = 5 = \frac{T_G}{T_P}$$

$$T_G = 5\,T_P$$

$$C = \frac{m}{2}(T_P + T_G)$$

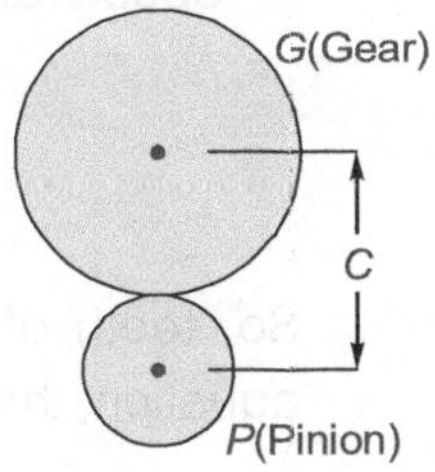

$$T_P + T_G = \frac{2C}{m} = \frac{2 \times 450}{5} = 180$$

or $\quad 6T_P = 180$

or $\quad T_P = \mathbf{30}$

5.27 (a)

<u>Method I:</u>

Gears	2	3	4	5
Teeth on gears	20	40	15	30
Speed of gears in RPM	–1200	$\frac{+20 \times 1200}{40}$	$\frac{+20 \times 1200}{40}$	$\frac{-20 \times 1200}{40} \times \frac{15}{30}$

∴ Angular speed of gear 5

$$= \frac{-20 \times 1200 \times 15}{40 \times 30} = -300 \text{ RPM}$$

$$= \mathbf{300\ rpm} \qquad \{\text{counter clockwise}\}$$

Method II:

Given: $N_2 = -1200$ rpm = 1200 rpm (ccw)

$T_2 = 20$, $T_3 = 40$, $T_4 = 15$, $T_5 = 30$, $N_5 = ?$

Speed ratio;

$$\frac{N_5}{N_2} = \frac{T_2 \times T_4}{T_5 \times T_3}$$

$$N_5 = N_2\left(\frac{T_2 \times T_4}{T_5 \times T_3}\right)$$

$$= -1200\left(\frac{20 \times 15}{40 \times 30}\right) = -300 \text{ rpm}$$

$$= 300 \text{ rpm (ccw)}$$

5.29 (c)

Let y revolution per minute is given to driver gear 2. Gear 3 and gear 4 are compound gears.

Gears	2	3	4	5	6
No. of teeth	T_2	T_3	T_4	T_5	T_6
Speed	y	$-y\frac{T_2}{T_3}$	$-y\frac{T_2}{T_3}$	$+y\frac{T_2}{T_3} \times \frac{T_4}{T_5}$	$-y \times \frac{T_2}{T_3} \cdot \frac{T_4}{T_5} \cdot \frac{T_5}{T_6}$

$$\text{Speed ratio} = \frac{\text{Speed of driver}}{\text{Speed of driven}}$$

$$= \frac{y}{-y \times \frac{T_2}{T_3} \times \frac{T_4}{T_5} \times \frac{T_5}{T_6}} = \frac{-T_3 \times T_6}{T_2 \times T_4}$$

So, teeth of gear 5 is not in the speed ratio equation, then gear 5 will act as an idler year.

POINTS TO REMEMBER

Idler gears give the following advantages:

- The number of intermediate gears controls directin of rotation of output gear.
- Idler gear serve the purpose or bridging the gap between the input and output shaft, when this centre distance is large.
- Simple way to judge idler is that gear which is connecting left and right both will be idler.

5.30 (c)

Gears	2	3	4
Number of teeth	60	20	100
Speed in rpm	$+N$ (clockwise)	$\frac{-N \times 60}{20}$ anticlockwise	$\frac{-N \times 60 \times 20}{20 \times 100}$ (anticlockwise)
Let speed of arm by y rpm ($+y$ clockwise)	$N + y$	$\frac{-N \times 60}{20} + y$	$\frac{-N \times 60}{100} + y$

Given : Gear 2 is fixed

$$\therefore \quad N + y = 0 \qquad [\therefore N = -y]$$

Gear 4 rotates 100 rpm is counter clockwise.

$$-\frac{N \times 60}{100} + y = -100$$

$$+ y \times 0.6 + y = -100$$

$$y = \frac{-100}{1.6} = \mathbf{-62.5\ rpm}$$

Hence angular speed of arm is 62.5 rpm (counterclockwise).

5.31 Sol.

We take convention clockwise (+ve)

$N_D = 0$, $N_S = -100$, $T_S = 50$, $T_P = 25$, $T_D = 100$

$$\frac{N_P}{N_{ARM}} = ?$$

Motions	Arm	S (50)	P (25)	D (100)
1. Arm fixed let sun rotates by $+x$ r.p.m. clockwise	0	$+x$	$\frac{-x \times 50}{25}$	$-x \times \left(\frac{50}{25}\right) \times \left(\frac{25}{100}\right)$
2. Amr free	y	$(y + x)$	$(y - 2x)$	$\left(y - \frac{x}{2}\right)$

$$y + x = -100 \qquad \ldots \text{(i)}$$

$$y - \frac{x}{2} = 0 \qquad \dots \text{(ii)}$$

By Equation (i) – Equation (ii),

$$\frac{3x}{2} = -100 = x = \frac{-200}{3}$$

$$y = -100 - x = -100 + \frac{200}{3} = \frac{-100}{3}$$

$$N_P = y - 2x = \frac{-100}{3} + \frac{400}{3} = \frac{300}{3} = 100$$

$$N_{ARM} = y = \frac{-100}{3}$$

$$\frac{N_P}{N_{ARM}} = \frac{100}{\frac{-100}{3}} = -3$$

$$\left|\frac{N_P}{N_{ARM}}\right| = 3$$

5.32 Sol.

$T_P = 24$

$T_Q = 45$ $\quad N_P = 400$ rpm

$T_R = 30$ $\quad N_S = ?$

$T_S = 80$

Here gear R is not meshing at all.

$$\frac{N_P}{N_Q} = \frac{T_Q}{T_P} = \frac{45}{24} \qquad \dots(1)$$

$$\frac{N_Q}{N_S} = \frac{T_S}{T_Q} = \frac{80}{45} \qquad \dots(2)$$

By (1) × (2)

$$\frac{N_P}{N_S} = \frac{45}{24} \times \frac{80}{45}$$

$$N_S = N_P \times \frac{24}{80} = \frac{400 \times 24}{80}$$

$$\mathbf{N_S = 120 \text{ rpm}}$$

5.33 (c)

$T_A = 20$, $T_B = 30$, $T_D = 20$, $T_C = 80$ (Inner), $T_C = 100$ (Outer)

Arm is fixed, no epicyclic nature. Taking clockwise direction as positive.

$$N_A = +300$$

$$(A, B) \quad N_B = -\frac{300 \times 20}{30} = -200$$

$$(B, C) \quad N_C = -200 \times \frac{30}{80} = -75$$

$$(C, D) \quad N_D = +75 \times \frac{100}{20} = +375$$

5.34 (d)

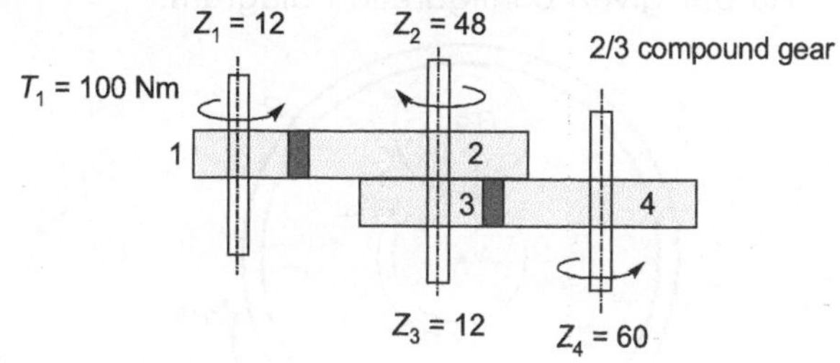

$$T_1 = 100 \text{ Nm}$$

Let speed of 1 is N_1

$$(1, 2): \quad N_2 = N_1 \times \frac{Z_1}{Z_2} = N_1 \times \frac{12}{48} = \frac{N_1}{4}$$

$$N_3 = N_2 = \frac{N_1}{4}$$

$$(3, 4): \quad N_4 = N_3 \times \frac{Z_3}{Z_4} = \frac{N_1}{4} \times \frac{12}{60} = \frac{N_1}{20}$$

By Power conservation

(Assume η (efficiency) = 1)

$$T_1 \times N_1 = T_4 \times N_4$$

$$100 \times N_1 = T_4 \times \frac{N_1}{20}$$

$$\mathbf{T_4 = 2000 \text{ Nm}}$$

5.35 (a)

Given, $\phi = 20°$, $P = 20$ kW, $\omega = 200$ rad/s

$D = 100$ mm

We know that,

Power, $\quad P = T \times \omega$

$$20 \times 10^3 = T \times 200$$

$$T = 100 \text{ N-m}$$

We know that,

$$T = F_t \times r$$

$$F_t = \frac{2T}{D} = \frac{2 \times 100}{100 \times 10^{-3}} = 2 \text{ kN}$$

Radial force,

$$F_r = F_t(\tan\phi)$$

$$= 2 \times \tan 20° = 0.72794 \text{ kN}$$

$$= 0.728 \text{ kN} = 0.73 \text{ kN}$$

5.36 (b)

Pitch circle diameter = D

Number of teeth = T

$$\text{Circular pitch, } P = \frac{\pi D}{T}$$

5.37 (b)

As per given configuration diagram:

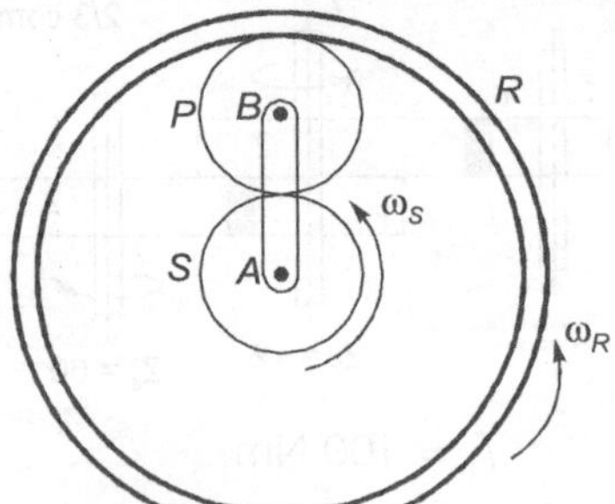

$$r_s + 2r_P = r_R$$

$$\Rightarrow T_S + 2T_P = T_R \quad (T_P = T_S)$$

$$3T_P = T_R$$

$$\Rightarrow 3T_P = T_R$$

Motion	Arm	$S(T_P)$	$P(T_P)$	$R(3T_P)$
1. Let arm fixed SUN → +x rpm clockwise	0	$+x$	$-x\dfrac{T_P}{T_P}$	$-x\dfrac{T_P}{3T_P}$
2. Arm free	y	$(y+x)$	$(y-x)$	$\left(y-\dfrac{x}{3}\right)$

$$y + x = \omega_S$$

$$y - \frac{x}{3} = \omega_R$$

Substract by,

$$\frac{4x}{3} = (\omega_S - \omega_R)$$

$$x = \frac{3}{4}(\omega_S - \omega_R)$$

$$y = \omega_S - x = \omega_S - \frac{3}{4}(\omega_S - \omega_R)$$

$$= \omega_S - \frac{3\omega_S}{4} + \frac{3\omega_R}{4} = \frac{\omega_S}{4} + \frac{3\omega_R}{4}$$

5.38 (a)

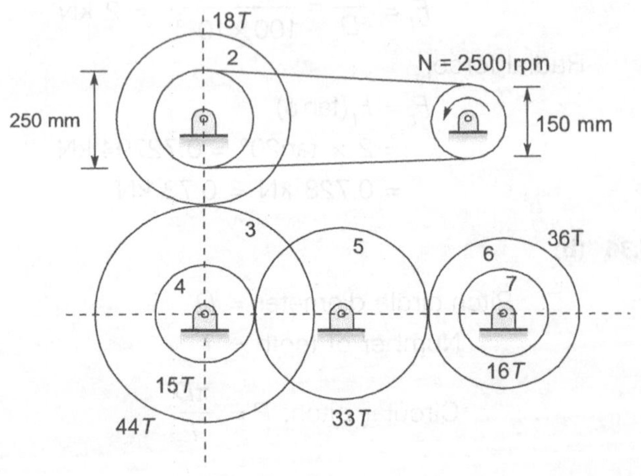

$T_3 = 44, \quad T_6 = 36, \quad T_2 = 18, \quad T_4 = 15$

$$\frac{N_1}{N_0} = \frac{d_0}{d_1}$$

$$\Rightarrow N_1 = \frac{150}{250} \times 2500 = 15{,}00 = N_2$$

$$N_3 = \frac{N_2 \times T_2}{T_3} = \frac{1500 \times 18}{44}$$

$$= 613.63636 = N_4$$

$$N_6 = \frac{N_4 \times T_4}{T_6} = \frac{613.63636 \times 15}{36}$$

$$= 255.6818 = N_7 \text{(Clockwise)}$$

Gear 5 is idler.

5.39 (600) (600 to 600)

Here the sun gear is fixed and the number of teeth for following gears have been given

$$T_2 = 20$$

$$T_3 = 80$$

$$T_1 = ?$$

∵ Here, modulus of all the gears are same, so the radius relation is given by

$$\Rightarrow r_1 + 2r_2 = r_3$$

$$\Rightarrow \frac{mT_1}{2} + \frac{2mT_2}{2} = \frac{mT_3}{2}$$

$$\Rightarrow T_1 + 2T_2 = T_3$$

$$\Rightarrow T_1 = T_3 - 2T_2 = 80 - 2 \times 20$$

$$\Rightarrow T_1 = 40 \text{ teeth}$$

Motions	Arm	1(40)	2(20)	3(80)
1 let arm be fixed (w.r.t. arm) Sun speed is $+x$ clockwise	0	$+x$	$-x\times\dfrac{40}{20}$	$-x\times\dfrac{40}{20}\times\dfrac{20}{80}$
2 Arm effect consideration	y	$y+x$	$y-2x$	$y-\dfrac{x}{2}$

From the data given, where sun's speed is zero.

$$\Rightarrow N_1 = 0$$

$$\Rightarrow y + x = 0 \quad \text{...(i)}$$

Speed of 3 is given as

$N_3 = 900$ [Consider counter-clockwise as positive]

$$\Rightarrow y - \frac{x}{2} = 900 \quad \text{...(ii)}$$

Solving these two equations, we get

$$x = -600 \text{ and } y = 600 \text{ rpm}$$

So speed of arm is y which is equal to 600 rpm (counter-clockwise).

5.40 (96) (95.999 to 96.001)

Given:

Diametral pitch, $P_d = \dfrac{8 \text{ teeth}}{\text{mm}} = \dfrac{T}{D}$

$\Rightarrow$ Module, $m = \dfrac{D(\text{mm})}{T} = \dfrac{1}{P_d}$

$\therefore \quad m = \dfrac{1}{8}\text{mm}$

$$\frac{\omega_2}{\omega_1} = \frac{T_1}{T_2} = \frac{1}{4} \quad \text{(Given)}$$

$$\Rightarrow \quad T_2 = 4T_1 \quad \text{(i)}$$

If r and R be the radius of gears 1 and 2 respectively then,

$$\because \quad R + r = 30$$

$$\Rightarrow \frac{mT_2}{2} + \frac{mT_1}{2} = 30 \quad \left\{\therefore m = \frac{D}{T} = \frac{2R}{T}\right\}$$

$$\Rightarrow \frac{m}{2}(T_2 + T_1) = 30$$

$$\Rightarrow \frac{1}{8 \times 2}(4T_1 + T_1) = 30 \quad \text{[Using (i)]}$$

$$\Rightarrow \quad 5T_1 = 30 \times 16$$

$$\therefore \quad T_1 = \frac{30 \times 16}{5} = 96$$

5.41 (15.58) (15.00 to 16.00)

Gear train:

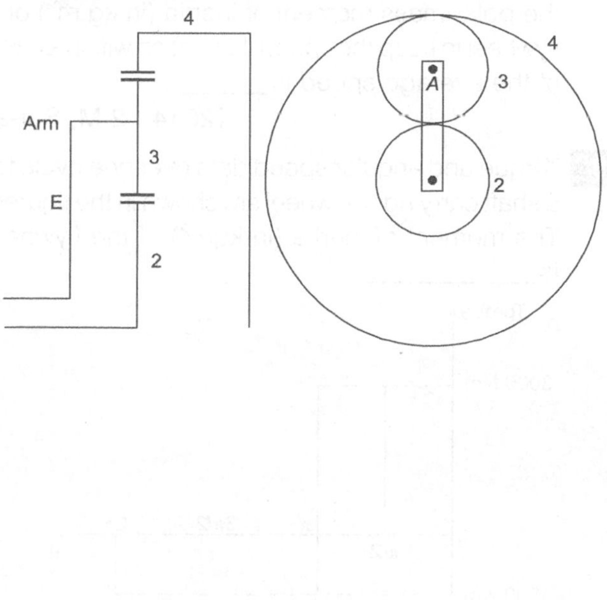

Given: $T_2 = 19$, $T_3 = 24$, $N_4 = 0$, $N_4 = 20$ rpm (CCW)

As $R_2 + 2R_3 = R_4$

$\Rightarrow \quad T_2 + 2T_3 = T_4$

$\Rightarrow \quad T_4 = 19 + 2 \times 24 = 67$

	A	Gear 2	Gear 3	Gear 4
Arm is fixed	0	$+x$	$-x\times\frac{T_2}{T_3}$ $=-x\times\frac{19}{24}$	$-x\times\frac{T_2}{T_3}\times\frac{T_3}{T_4}$ $=-x\times\frac{19}{27}\times\frac{27}{67}$ $=-x\times\frac{19}{67}$
Arm is free	$+y$	$+y+x$	$+y-x\times\frac{19}{24}$	$+y-x\times\frac{19}{67}$

$N_2 = y + x \Rightarrow x + y = 0$ ... (i)

and $\quad N_4 = 20$ rpm

$$\Rightarrow \quad y - x \times \frac{19}{67} = 20 \quad \text{... (ii)}$$

On solving equation (i) and (ii);

$x = -15.58$ rpm $= 15.58$ rpm (CW)

$y = 15.58$ (CCW)

■■■■

6 Flywheel and Governor

6.1 A flywheel of moment of inertia 9.8 kg m^2 fluctuates by 30 rpm for a fluctuation in energy of 1936 joules. The mean speed of the flywheel is (in rpm)

(a) 600 (b) 900
(c) 968 (d) 29470 **[1998 : 2 M]**

6.2 Which of the following statements is correct?

(a) Flywheel reduces speed fluctuations during a cycle for a constant load, but flywheel does not control the mean speed of the engine if the load changes.
(b) Flywheel does not reduce speed fluctuations during a cycle for a constant load, but flywheel does controls the mean speed of the engine if the load changes
(c) Governor controls a speed fluctuations during a cycle for a constant load, but governor does not control the mean speed of the engine if the load changes.
(d) Governor controls speed fluctuations during a cycle for a constant load, and governor also controls the mean speed of the engine if the load changes.

[2001 : 1 M]

6.3 For a certain engine having an average speed of 1200 rpm, a flywheel approximated as a solid disc, is required for keeping the fluctuation of speed within 2% about the average speed. The fluctuation of kinetic energy per cycle is found to be 2 kJ. What is the least possible mass of the flywheel if its diameter is not to exceed 1 m?

(a) 40 kg (b) 51 kg
(c) 62 kg (d) 73 kg **[2003 : 2 M]**

6.4 If C_f is the coefficient of speed fluctuation of a flywheel then the ratio of $\omega_{max}/\omega_{min}$ will be

(a) $\frac{1-2C_f}{1+2C_f}$ (b) $\frac{2-C_f}{2+C_f}$

(c) $\frac{1+2C_f}{1-2C_f}$ (d) $\frac{2+C_f}{2-C_f}$ **[2006 : 2 M]**

6.5 The speed of an engine varies from 210 rad/s to 190 rad/s. During cycle the change in kinetic energy is found to be 400 Nm. The inertia of the flywheel in kgm^2 is

(a) 0.10 (b) 0.20
(c) 0.30 (d) 0.40

[2007 : 2 M]

6.6 A flywheel connected to a punching machine has to supply energy of 400 Nm while running at a mean angular speed of 20 rad/s. If the total fluctuation of speed is not to exceed ±2% the mass moment of inertia of the flywheel in kg-m^2 is

(a) 25 (b) 50
(c) 100 (d) 125

[2013 : 2 M]

6.7 Maximum fluctuation of kinetic energy in an engine has been calculated to be 2600 J. Assuming that the engine runs at an average speed of 200 rpm, the polar mass moment of inertia (in kg.m^2) of a flywheel to keep the speed fluctuation within ±0.5% of the average speed is _______.

[2014 : 2 M, Set-2]

6.8 Torque and angular speed data over one cycle for a shaft carrying a flywheel are shown in the figures. The moment of inertia (in kgm^2) of the flywheel is_____________.

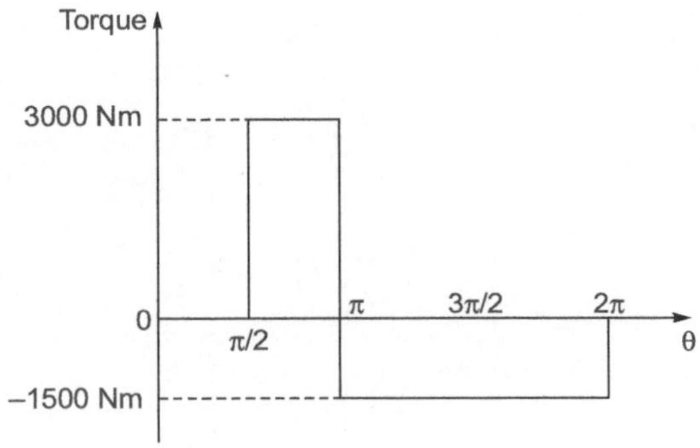

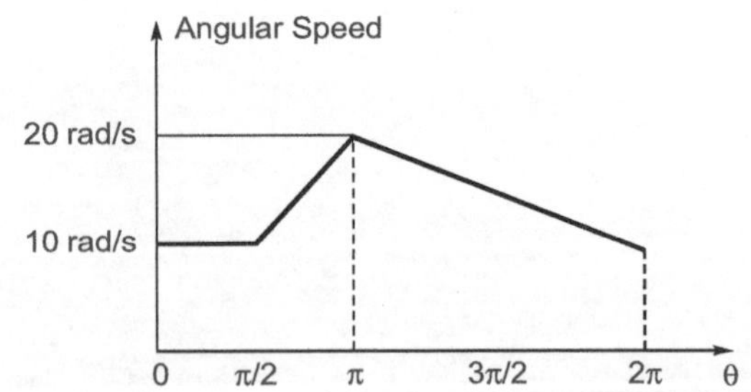

[2014 : 2 M, Set-4]

6.9 Consider a flywheel whose mass M is distributed almost equally between a heavy, ring-like rim of radius R and a concentric dislike feature of radius $R/2$. Other parts of the flywheel, such as spokes, etc, have negligible mass. The best approximation for α, if the mass moment of inertia of the flywheel about its axis of rotation is expressed as αMR^2 is ________.

[GATE : 2014]

6.10 The torque (in Nm) exerted on the crank shaft of a two stroke engine can be described as $T = 10000 + 1000 \sin 2\theta - 1200 \cos 2\theta$, where θ is the crank angle as measured from inner dead center position. Assuming the resisting torque to be constant, the power (in kW) developed by the engine at 100 rpm is ________.

[2015 : 2 M, Set-3]

6.11 Consider a slider crank mechanism with nonzero masses and inertia. A constant torque τ is applied on the crank as shown in the figure. Which of the following plots best resembles variation of crank angle, θ versus time

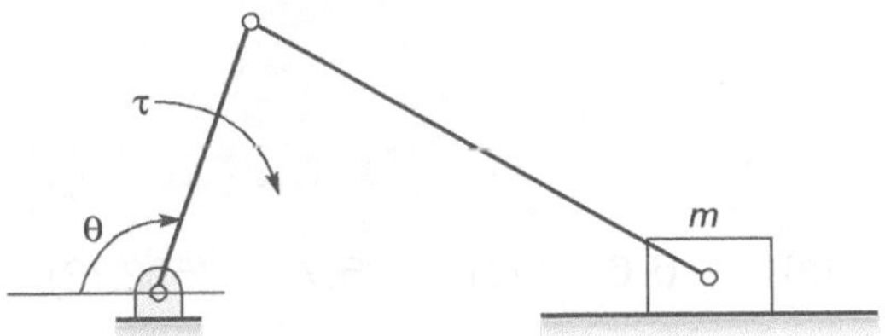

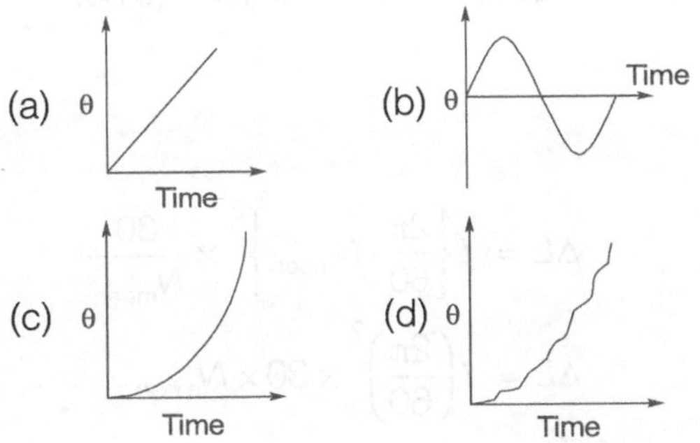

[2015 : 1 M, Set-1]

6.12 A flywheel is attached to an engine to keep its rotational speed between 100 rad/s and 110 rad/s. If the energy fluctuation in the flywheel between these two speeds is 1.05 kJ then the moment of inertia of the flywheel is ________ kg.m^2 (round off to 2 decimal places).

[2020 : 1 M, Set-1]

6.13 The turning moment diagram of a flywheel fitted to a fictitious engine is shown in the figure.

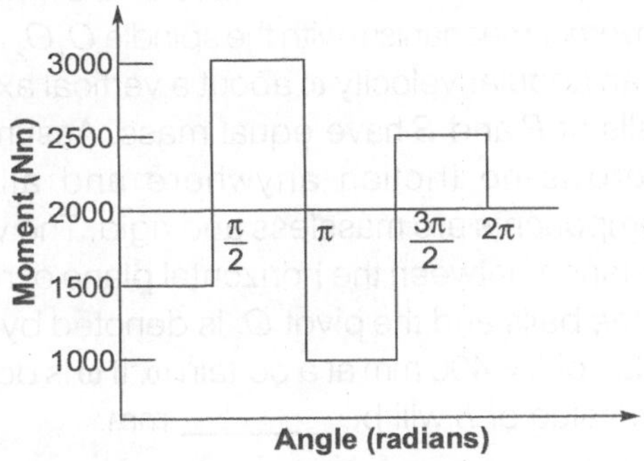

The mean turning moment is 2000 Nm. The average engine speed is 1000 rpm. For fluctuation in the speed to be within ±2% of the average speed, the mass moment of inertia of the flywheel is ________ kgm^2.

[2020 : 2 M, Set-2]

6.14 The controlling force curves P, Q and R for a spring controlled governor are shown in the figure, where r_1 and r_2 are any two radii of rotation,

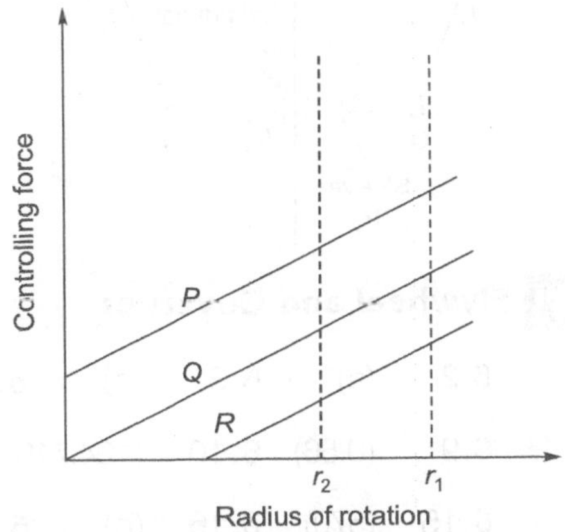

The characteristics shown by the curves are

(a) P - Unstable: Q - Isochronous: R - Stable
(b) P- Stable: Q - Isochronous; R - Unstable
(c) P - Stable: Q - Unstable: R - Isochronous
(d) P - Unstable: Q - Stable: R - Isochronous

[2021 : 1 M, Set-2]

6.15 The torque provided by an engine is given by $T(\theta) = 12000 + 2500 \sin(2\theta)$ N.m, where θ is the angle turned by the crank from inner dead center. The mean speed of the engine is 200 rpm and it drives a machine that provides a constant resisting torque. If variation of the speed from the mean speed is not to exceed ± 0.5%, the minimum mass moment of inertia of the flywheel should be______kg.m^2 (round off to the nearest integer).

[2021 : 2 M, Set-2]

6.16 The figure shows a schematic of a simple Watt governor mechanism with the spindle O_1O_2 rotating at an angular velocity ω about a vertical axis. The balls at P and S have equal mass. Assume that there is no friction anywhere and all other components are massless and rigid. The vertical distance between the horizontal plane of rotation of the balls and the pivot O_1 is denoted by h. The value of $h = 400$ mm at a certain ω. If ω is doubled, the value of h will be ________ mm.

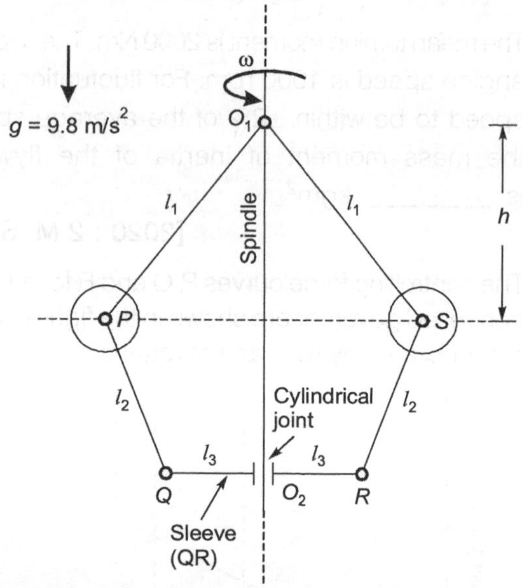

(a) 50
(b) 100
(c) 150
(d) 200

[2022 : 1 M, Set-1]

6.17 The change in kinetic energy ΔE of an engine is 300 J, and minimum and maximum shaft speeds are ω_{min} = 220 rad/s and ω_{max} = 280 rad/s, respectively. Assume that the torque-time function is purely-harmonic. To achieve a coefficient of fluctuation of 0.05, the moment of inertia (in kg.m²) of the flywheel to be mounted on the engine shaft is

(a) 0.053
(b) 0.113
(c) 0.096
(d) 0.071

[2024 : 1 M]

Answers Flywheel and Governor

6.1	(a)	6.2	(a)	6.3	(b)	6.4	(d)	6.5	(a)	6.6	(a)	6.7	(592.73)
6.8	(31.26)	6.9	(0.56)	6.10	(104.719)			6.11	(d)	6.12	(1)	6.13	(3.58)
6.14	(a)	6.15	(570)	6.16	(b)	6.17	(c)						

Explanations Flywheel and Governor

6.1 (a)

$$I = 9.8 \text{ kgm}^2$$
$$N_1 - N_2 = 30 \text{ rpm}$$
$$\Delta E = 1936 \text{ J}$$
$$N_{mean} = ?$$
$$C_s = \frac{N_1 - N_2}{\frac{N_1 + N_2}{2}} = \frac{N_1 - N_2}{N_{mean}} = \frac{30}{N_{mean}}$$
$$\Delta E = I\omega_{mean}^2 C_s$$
$$\omega_{mean} = \frac{2\pi N_{mean}}{60}$$
$$\Delta E = I\left\{\frac{2\pi N_{mean}}{60}\right\}^2 C_s$$
$$\Delta E = I\left\{\frac{2\pi}{60}\cdot N_{mean}\right\}^2 \times \frac{30}{N_{mean}}$$
$$\Delta E = I\left(\frac{2\pi}{60}\right)^2 \times 30 \times N_{mean}$$
$$N_{mean} = \frac{\Delta E}{I\left(\frac{2\pi}{60}\right)^2 \times 30} = \frac{1936}{9.8 \times \left\{\frac{2\pi}{60}\right\}^2 \times 30}$$
$$= \mathbf{600.48 \text{ rpm}}$$

6.2 (a)

Flywheel raduces speed fluctuations during a cycle for a constant load because flywheel stores energy during the period when supply of energy is more than the requirement, and release it during the period when the requirement of energy is more than the supply i.e. Flywheel controls inside cycle fluctuation only due to variation in turnig moment. Governor central the speed variations caused by the varying load i.e. outside cycle fluctuations.

6.3 (b)

Average speed,

$$\omega = \frac{2\pi \times 1200}{60} = 125.66 \text{ rad/s}$$

we know that

$$\Delta E = I\omega^2 C_S$$

and

$$\Delta E = \frac{MR^2}{2} \times \omega^2 \times C_S$$

$$\therefore M = \frac{2\Delta E}{R^2 \times \omega^2 \times C_S} = \frac{2 \times 2000}{\left(\frac{1}{2}\right)^2 \times (125.66)^2 \times 0.02}$$

$$= 50.66 \approx \mathbf{51\ kg}$$

POINTS TO REMEMBER

It is important to note that here value of $C_s = 0.02$, not $C_s = 0.04$, because it is given that fluctuation of speed within 2% about the average speed.

6.4 (d)

$$C_f = \frac{\omega_{max} - \omega_{min}}{\frac{\omega_{max} + \omega_{min}}{2}} = \frac{2(\omega_{max} - \omega_{min})}{\omega_{max} + \omega_{min}}$$

$$\frac{C_f}{2} = \frac{\omega_{max} - \omega_{min}}{\omega_{max} + \omega_{min}}$$

$$C_f\omega_{max} + C_f\omega_{min} = 2\omega_{max} - 2\omega_{min}$$

$$C_f\omega_{min} + 2\omega_{min} = 2\omega_{max} - C_f\omega_{max}$$

$$(C_f + 2)\omega_{min} = \omega_{max}(2 - C_s)$$

$$\frac{C_f + 2}{2 - C_f} = \frac{\omega_{max}}{\omega_{min}}$$

or

$$\frac{\omega_{max}}{\omega_{min}} = \frac{2 + C_f}{2 - C_f}$$

6.5 (a)

$$\omega_{min} = 190 \text{ rad/s}, \ \omega_{max} = 210 \text{ rad/s}$$

$$\Delta E = 400 \text{ Nm}$$

$$C_s = \frac{\omega_{max} - \omega_{min}}{\frac{(\omega_{max} + \omega_{min})}{2}} = \frac{210 - 190}{\frac{(210 + 190)}{2}} = \frac{20}{200} = 0.1$$

$$\omega_{mean} = \frac{\omega_{min} + \omega_{max}}{2} = \frac{190 + 210}{2}$$

$$= 200 \text{ rad/s}$$

$$\Delta E = I\omega_{mean}^2 C_S$$

$$400 = I \times (200)^2 \times 0.1$$

$$I = \frac{400}{200^2 \times 0.1} = \mathbf{0.1\ kgm^2}$$

6.6 (a)

$$\Delta E = 400 \text{ Nm}, \ \omega_{mean} = 20 \text{ rad/s}$$

$$C_s = \pm 2\% = 4\% = \frac{4}{100}$$

$$\Delta E = I\omega_{mean}^2 C_S$$

$$400 = I \times \{20\}^2 \times \{0.04\}$$

or

$$I = \mathbf{25\ kgm^2}$$

6.7 Sol.

$$\Delta E = 2600 \text{ J}, \ N_{average} = 200 \text{ rpm}, \ I = ?$$

$$C_s = \pm 0.5\% = 1\% = \frac{1}{100} = 0.01$$

$$\Delta E = I\omega^2 C_s$$

$$2600 = I \times \left\{\frac{2\pi}{60} \times 200\right\}^2 \times 0.01$$

$$2600 = I \times 438.6491 \times 0.01$$

$$I = \frac{2600}{4.3865} = \mathbf{592.73\ kgm^2}$$

6.8 Sol.

$$I = ?$$

Work done in cycle,

$$E = 3000 \times \left[\pi - \frac{\pi}{2}\right] - 1500 \times [2\pi - \pi]$$

$$= 3000 \times \frac{\pi}{2} - 1500\pi = 0$$

Resisting torque,

$$T_{mean} = \frac{\text{Work done in cycle } (E)}{2\pi}$$

$$T_{mean} \times 2\pi = E$$

$$T_{mean} = \frac{E}{2\pi} = 0$$

$$\omega_{min} = 10 \text{ rad/s at } 0 = \pi/2$$

$$\omega_{max} = 20 \text{ rad/s at } 0 = \pi$$

$$C_s = \frac{\omega_{max} - \omega_{min}}{\frac{\omega_{max} + \omega_{min}}{2}}$$

$$= \frac{2\{\omega_{max} - \omega_{min}\}}{\omega_{max} + \omega_{min}} = \frac{2 \times \{20 - 10\}}{20 + 10}$$

$$= 0.67$$

$$\Delta E = 3000 \times \frac{\pi}{2} = 1500\pi = I\omega^2 C_s$$

$$1500\pi = I \times \omega_{mean}^2 \times 0.67$$

$$1500\pi = I \times \left\{\frac{20+10}{2}\right\}^2 \times 0.67 = \frac{1500\pi}{15^2 \times 0.67}$$

$$= \mathbf{31.26\ kg\ m^2}$$

POINTS TO REMEMBER

In this problem, value of θ is taken as 2π because one cycle is cmpleting in 2π angle. But it may differ in other problems depending on the cycle i.e. value of θ may be π, 4π, 6π....

6.9 Sol.

Given: Radius of rim = R

Radius of disk = $R/2$

Average radius of flywheel

$$= \frac{\left(R + \frac{R}{2}\right)}{2} = \frac{\left(\frac{3R}{2}\right)}{2} = \frac{3R}{4}$$

Mass moment of inertia,

$$I_0 = M\left(\frac{3R}{4}\right)^2$$

$$I_0 = \frac{9}{16}MR^2 \quad \text{...(i)}$$

Given, $I_0 = \alpha MR^2$ ...(ii)

On comparing equation (i) and (ii), we get

$$\alpha = \frac{9}{16} = 0.5625 \approx 0.56$$

6.10 Sol.

$T = 10000 + 1000 \sin 2\theta - 1200 \cos 2\theta$

$N = 100$ rpm

$$\omega = \frac{2\pi \times 100}{60} = 10.472 \text{ rad/s}$$

$$T_{mean} = \frac{1}{2\pi}\int_0^{2\pi} [10{,}000 + 1000 \sin 2\theta - 1200 \cos 2\theta] d\theta$$

$$= 10{,}000[2\pi - 0] + \frac{1000}{2} \times [\cos 2\theta]_0^{2\pi} - \frac{1200}{2}[\sin 2\theta]_0^{2\pi}$$

$$= \frac{1}{2\pi}[10{,}000 \cdot 2\pi + 500 \cdot [\cos 4\pi - \cos 0]$$

$$-600[\sin 4\pi - \sin 0]$$

$$= (10{,}000 \times 2\pi) + 500\,[1 - 1] - 600\,[0 - 0]$$

$$= \frac{10000 \times 2\pi}{2\pi} = 10000 \text{ Nm}$$

$$\text{Power} = 10000 \times \frac{2\pi \times 100}{60} = 104719.75 \text{ W}$$

$$= \mathbf{104.719\ kW}$$

6.11 (d)

$$\tau(\text{torque}) = I\ddot{\theta}$$

$$\tau = \text{constant}$$

$$I = \text{constant}$$

$$\tau = I\frac{d^2\theta}{dt^2}$$

$$\frac{d^2\theta}{dt^2} = \frac{\tau}{I}$$

$$\frac{d\theta}{dt} = \frac{\tau}{I}t + C_1$$

$$\theta = \frac{\tau}{I}\frac{t^2}{2} + C_1 t + C_2$$

$$\theta = f(t) = \text{parabolic function}$$

But practically graph will be fluctuating about ideal parabolic curve because, inertial torque, gives fluctuation.

6.12 Sol.

Given, $\Delta E_{max} = 1.05$ kJ, $\omega_{max} = 110$ rad/s; $\omega_{min} = 100$ rad/s

$$\Delta E_{max} = \frac{1}{2}I\left(\omega_{max}^2 - \omega_{min}^2\right)$$

$$\therefore 1.05 \times 10^3 \times 2 = I\left(110^2 - 100^2\right)$$

$$I = 1 \text{ kg.m}^2$$

6.13 Sol.

Given (T-θ) diagram:

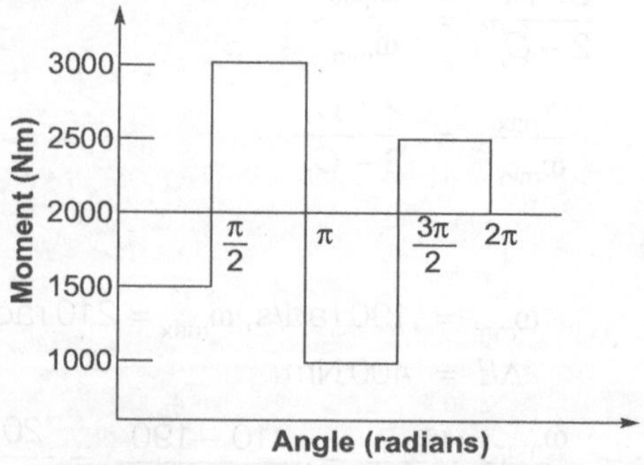

$N = 1000$ rpm

$$\omega = \frac{2\pi \times 1000}{60} = 104.7195 \text{ rad/s}$$

$$\Delta E_{max} = \left(\pi - \frac{\pi}{2}\right) \cdot (3000 - 2000)$$

$$= \left(\frac{\pi}{2} \times 1000\right) \text{J}$$

$$C_s = \pm 2\% = 4\% = 0.04$$

$$\Delta E_{max} = I \cdot \omega^2 \cdot C_s$$

$$\frac{\pi}{2} \times 1000 = I \times (104.7195)^2 \times 0.04$$

$$I = \frac{1570.795}{(104.7195)^2 \times 0.04} = 3.58 \text{ kg-m}^2$$

6.15 Sol.

T_{mean} = 12000 Nm

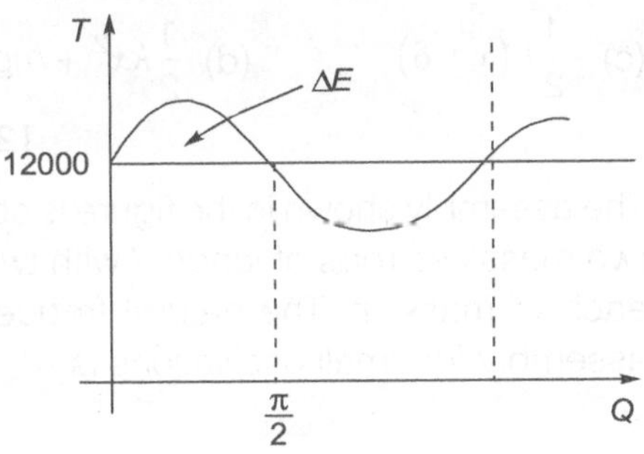

$$\omega = \frac{\pi \times 200}{30} = 20.9439 \text{ rad/s}$$

$$\Delta E = \int_0^{\frac{\pi}{2}} (T - T_{mean}) d\theta = 2500 \int_0^{\frac{\pi}{2}} \sin 2\theta \, d\theta$$

$$= 2500 \times 1 = 2500 \text{ J}$$

$$\Delta E = I\omega^2 C_s$$

$$2500 = I \times 20.9439^2 \times 0.01$$

$$\Rightarrow \quad I = 569.934 \text{ kgm}^2 \simeq 570 \text{ kg.m}^2$$

6.16 (b)

For a Watt governor:

$$\Rightarrow \quad \omega^2 = \frac{g}{h}$$

$$\Rightarrow \quad h \propto \frac{1}{\omega^2}$$

$$\Rightarrow \quad \frac{h_1}{h_2} = \frac{\omega_2^2}{\omega_1^2}$$

$$\Rightarrow \quad \frac{400}{h_2} = \frac{(2\omega_1)^2}{\omega_1^2} = 4$$

$$\Rightarrow \quad h_2 = \frac{400}{4} = 100 \text{ mm}$$

6.17 (c)

Given: ΔE = 300 J, ω_{min} = 220 rad/s, ω_{max} = 280rad/s, C_s = 0.05, I = ?

Energy fluctuation, $\Delta E = I\omega^2 C_s$

$$\Rightarrow \quad \Delta E = I \times \left(\frac{\omega_{max} + \omega_{min}}{2}\right)^2 \times C_s$$

$$\Rightarrow \quad 300 = I \times \left(\frac{280 + 220}{2}\right)^2 \times 0.05$$

$$\Rightarrow \quad I = 0.096 \text{ kgm}^2$$

■■■■

7 Vibration

7.1 A mass of 1 kg is suspended by means of 3 springs as shown in the figure. The spring constant k_1, k_2 and k_3 are respectively 1 kN/m, 3 kN/m and 2 kN/m. The natural frequency of the system is approximately

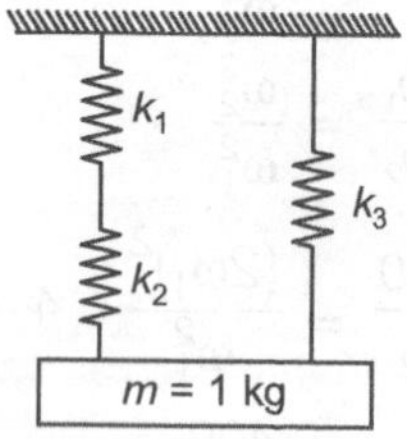

(a) 46.90 rad/s (b) 52.44 rad/s
(c) 60.55 rad/s (d) 77.46 rad/s

[1996 : 2 M]

7.2 Consider the system of two wagons shown in the figure. The natural frequencies of this system are

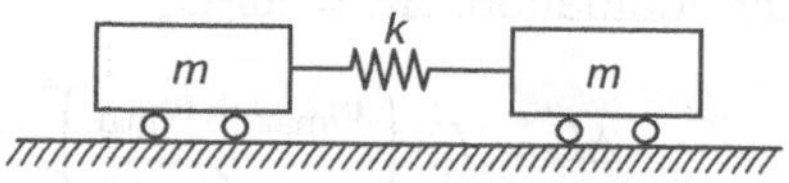

(a) $0, \sqrt{\frac{2k}{m}}$ (b) $\sqrt{\frac{k}{m}}, \sqrt{\frac{2k}{m}}$

(c) $\sqrt{\frac{k}{m}}, \sqrt{\frac{k}{2m}}$ (d) $0, \sqrt{\frac{k}{2m}}$ [1999 : 2 M]

7.3 As shown in the figure, a mass of 100 kg is held between two springs. The natural frequency of vibration of the system in cycle/s, is

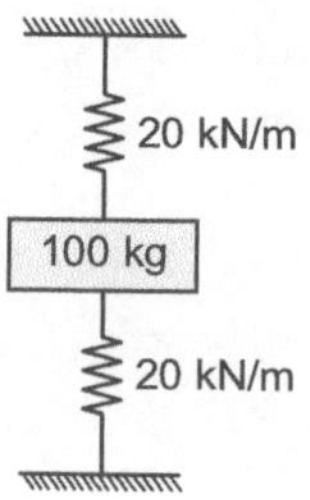

(a) $\frac{1}{2\pi}$ (b) $\frac{5}{\pi}$

(c) $\frac{10}{\pi}$ (d) $\frac{20}{\pi}$ [2000 : 2 M]

7.4 In the figure shown, the spring deflects by δ to position A (the equilibrium position) when a mass m is kept on it. During free vibration, the mass is at position B at some instant. The change in potential energy of the spring mass system from position A to position B is

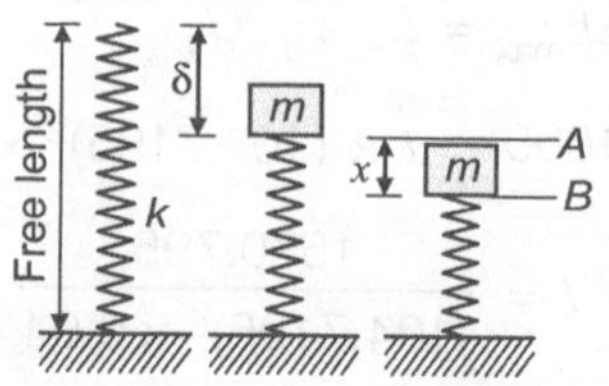

(a) $\frac{1}{2}kx^2$ (b) $\frac{1}{2}kx^2 - mgx$

(c) $\frac{1}{2}k(x+\delta)$ (d) $\frac{1}{2}kx^2 + mgx$

[2001 : 1 M]

7.5 The assembly shown in the figure is composed of two massless rods of length l with two particles, each of mass m. The natural frequency of this assembly for small oscillations is

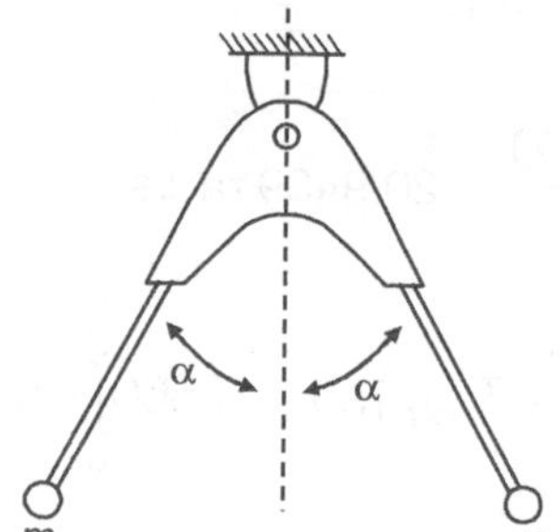

(a) $\sqrt{g/l}$ (b) $\sqrt{2g/l(\cos\alpha)}$

(c) $\sqrt{g/l(\cos\alpha)}$ (d) $\sqrt{(g\cos\alpha)/l}$

[2001 : 2 M]

7.6 Consider the arrangement shown in the figure below where J is the combined polar mass moment of inertia of the disc and the shafts. k_1, k_2, k_3 are the torsional stiffness of the respective shafts. The natural frequency of torsional oscillation of the disc is given by

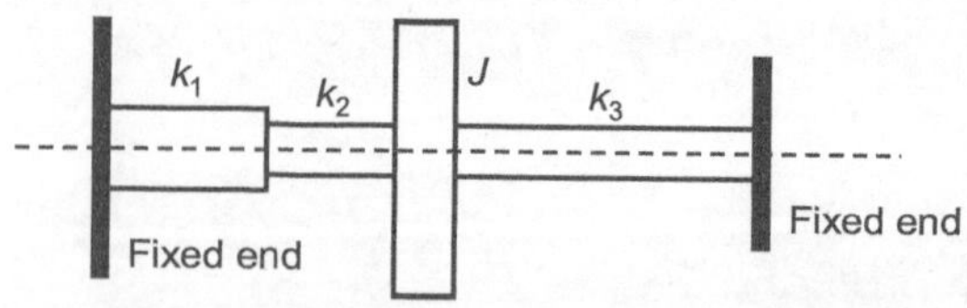

(a) $\sqrt{\frac{k_1 + k_2 + k_3}{J}}$ (b) $\sqrt{\frac{k_1k_2 + k_2k_3 + k_3k_1}{J(k_1 + k_2)}}$

(c) $\sqrt{\dfrac{k_1+k_2+k_3}{J(k_1k_2+k_2k_3+k_3k_1)}}$

(d) $\sqrt{\dfrac{k_1k_2+k_2k_3+k_3k_1}{J(k_2+k_3)}}$ [2003 : 1 M]

7.7 A flexible rotor-shaft system comprises a 10 kg rotor disc placed in the middle of a massless shaft of diameter 30 mm and length 500 mm between bearings (shaft is being taken massless as the equivalent mass of the shaft is included in the rotor mass) mounted at the ends. The bearings are assumed to simulate simply supported boundary conditions. The shaft is made of steel for which the value of E is 2.1×10^{11} Pa. What is the critical speed of rotation of the shaft?

(a) 60 Hz (b) 90 Hz
(d) 135 Hz (d) 180 Hz [2003 : 2 M]

Common Data Questions Q.7.8 and Q.7.9

A uniform rigid slender bar of mass 10 kg, hinged at the left end is suspended with the help of spring and damper arrangement as shown in the figure where $K = 2$ kN/m, $C = 500$ Ns/m and the stiffness of the torsional spring K_θ is 1 kN /m/rad. Ignore the hinge dimensions.

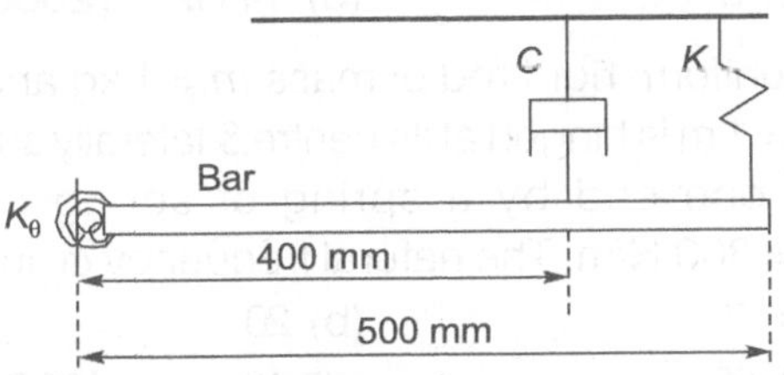

7.8 The undamped natural frequency of oscillations of the bar about the hinge point is

(a) 42.43 rad/s (b) 30 rad/s
(c) 17.32 rad/s (d) 14.14 rad/s
[2003 : 2 M]

7.9 The damping coefficient in the vibration equation is given by

(a) 500 Nms/rad (b) 500 N/(m/s)
(c) 80 Nms/rad (d) 80 N/(m/s) [2003 : 2 M]

7.10 A vibrating machine is isolated from the floor using springs. If the ratio of excitation frequency of vibration of machine to the natural frequency of the isolation system is equal to 0.5, the transmissibility ratio of isolation is

(a) $\frac{1}{2}$ (b) $\frac{3}{4}$

(c) $\frac{4}{3}$ (d) 2 [2004 : 1 M]

7.11 A uniform stiff rod of length 300 mm and having a weight of 300 N is pivoted at one end and connected to a spring at the other end. For keeping the rod vertical in a stable position the minimum value of spring constant k needed is

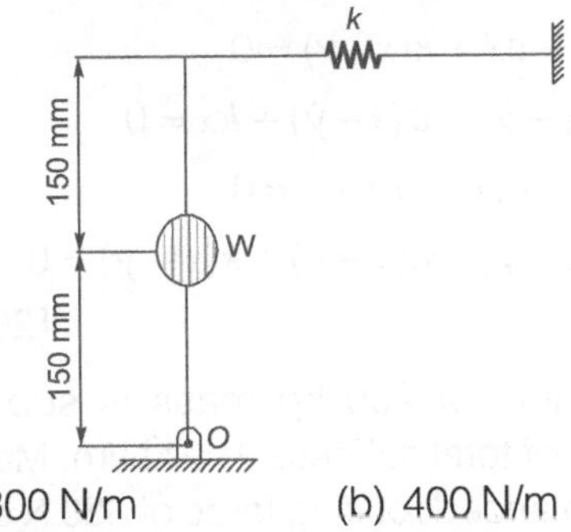

(a) 300 N/m (b) 400 N/m
(c) 500 N/m (d) 1000 N/m
[2004 : 2 M]

7.12 A mass m, of 20 kg is attached to the free end of a steel cantilever beam of length 1000 mm having a cross-section of 25× 25 mm. Assume the mass of the cantilever to be negligible and E_{steel} = 200 GPa. If the lateral vibration of this system is critically damped using a viscous damper, the damping constant of the damper is

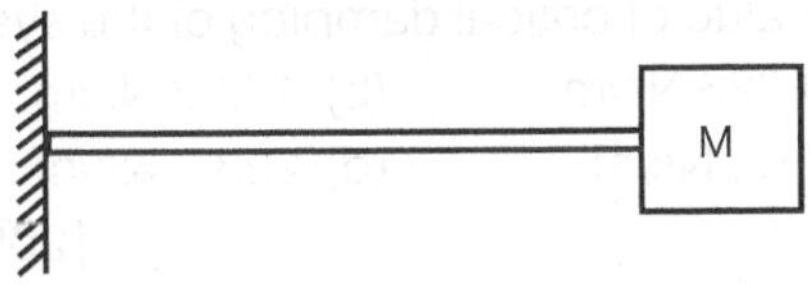

(a) 1250 Ns/m (b) 625 Ns/m
(c) 312.50 Ns/m (d) 156.25 Ns/m
[2004 : 2 M]

7.13 There are four samples P, Q, R and S with natural frequencies 64, 96, 128 and 256 Hz, respectively. They are mounted on test setups for conducting vibration experiments. If a loud pure note of frequency 144 Hz is produced by some instrument, which of the samples will show the most perceptible induced vibration?

(a) P (b) Q
(c) R (d) S [2005 : 1 M]

7.14 In a spring-mass system, the mass is 0.1 kg and the stiffness of the spring is 1 kN/m. By introducing a damper, the frequency of oscillation is found to be 90% of the original value. What is the damping coefficient of the damper?

(a) 1.2 Ns/m (b) 3.4 Ns/m
(c) 8.7 Ns/m (d) 12.0 Ns/m [2005 : 2 M]

7.15 The differential equation governing the vibrating system is

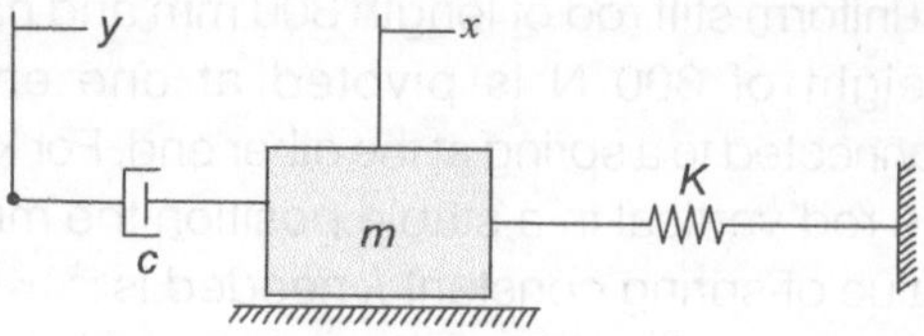

(a) $m\ddot{x} + c\ddot{x} + k(x-y) = 0$
(b) $m(\ddot{x}-\ddot{y}) + c(\dot{x}-\dot{y}) + kx = 0$
(c) $m\ddot{x} + c(\dot{x}-\dot{y}) + kx = 0$
(d) $m(\ddot{x}-\ddot{y}) + c(\dot{x}-\dot{y}) + k(x-y) = 0$

[2006 : 1 M]

7.16 A machine of 250 kg mass is supported on springs of total stiffness 100 kN/m. Machine has an unbalanced rotating force of 350 N at speed of 3600 rpm. Assuming a damping factor of 0.15, the value of transmissibility ratio is
(a) 0.0531 (b) 0.9922
(c) 0.0162 (d) 0.0028 [2006 : 2 M]

Linked Answer Questions 7.17 and 7.18

A vibratory system consists of a mass 12.5 kg, a spring of stiffness 1000 N/m, and a dashpot with damping coefficient of 15 Ns/m.

7.17 The value of critical damping of the system is
(a) 0.223 Ns/m (b) 17.88 Ns/m
(c) 71.4 Ns/m (d) 223.6 Ns/m

[2006 : 2 M]

7.18 The value of logarithmic decrement is
(a) 1.35 (b) 1.32
(c) 0.68 (d) 0.66 [2006 : 2 M]

7.19 For an under damped harmonic oscillator, resonance
(a) occurs when excitation frequency is greater than undamped natural frequency
(b) occurs when excitation frequency is less than undamped natural frequency
(c) occurs when excitation frequency is equal to undamped natural frequency
(d) never occurs [2007 : 1 M]

7.20 The natural frequency of the system shown below is

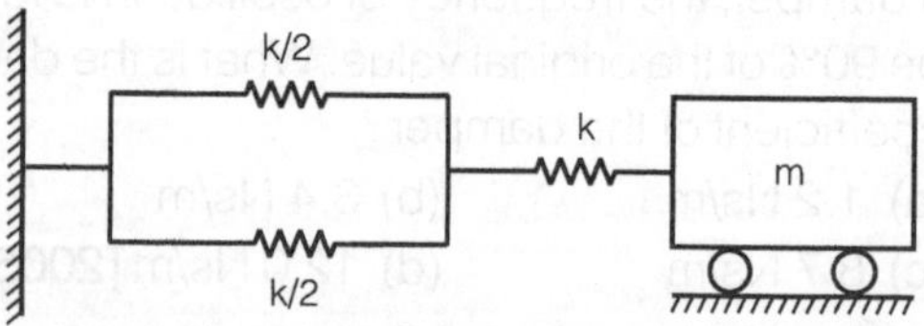

(a) $\sqrt{\frac{k}{2m}}$ (b) $\sqrt{\frac{k}{m}}$
(c) $\sqrt{\frac{2k}{m}}$ (d) $\sqrt{\frac{3k}{m}}$

[2007 : 2 M]

7.21 The equation of motion of a harmonic oscillator is given by

$$\frac{d^2x}{dt^2} + 2\zeta\omega_n\frac{dx}{dt} + \omega_n^2 x = 0$$

and the initial conditions at $t = 0$ are $x(0) = X$, $\frac{dx}{dt}(0) = 0$. The amplitude of $x(t)$ after n complete cycles is
(a) $Xe^{-2n\pi\left(\frac{\zeta}{\sqrt{1-\zeta^2}}\right)}$ (b) $Xe^{2n\pi\left(\frac{\zeta}{\sqrt{1-\zeta^2}}\right)}$
(c) $Xe^{-2n\pi\left(\frac{\sqrt{1-\zeta^2}}{\zeta}\right)}$ (d) X [2007 : 2 M]

7.22 The natural frequency of the spring mass system shown in the figure is closest to

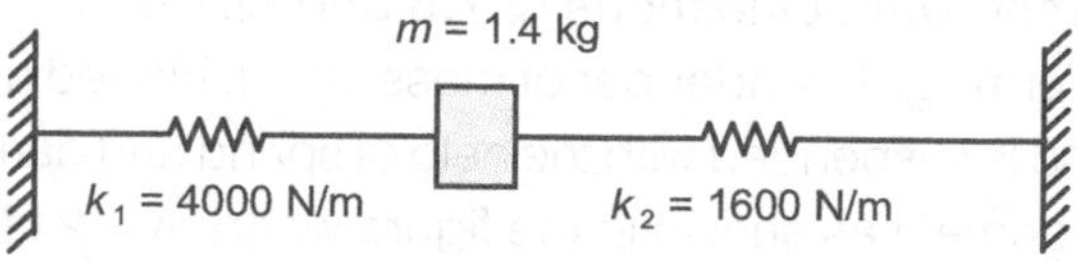

(a) 8 Hz (b) 10 Hz
(c) 12 Hz (d) 14 Hz [2008 : 2 M]

7.23 A uniform rigid rod of mass $m = 1$ kg and length $L = 1$ m is hinged at its centre & laterally supported at one end by a spring of spring constant $k = 300$ N/m. The natural frequency ω_n in rad/s is
(a) 10 (b) 20
(c) 30 (d) 40 [2008 : 2 M]

7.24 The rotor shaft of a large electric motor supported between short bearings at both the ends shows a deflection of 1.8 mm in the middle of the rotor. Assuming the rotor to be perfectly balanced and supported at knife edges at both the ends, the likely critical speed (in rpm) of the shaft is
(a) 350 (b) 705
(c) 2810 (d) 4430 [2009 : 1 M]

7.25 An automotive engine weighing 240 kg is supported on four springs with linear characteristics. Each of the front two springs have a stiffness of 16 MN/m while the stiffness of each rear spring is 32 MN/m. The engine speed (in rpm), at which resonance is likely to occur, is
(a) 6040 (b) 3020
(c) 1424 (d) 955 [2009 : 2 M]

7.26 A vehicle suspension system consists of a spring and a damper. The stiffness of the spring is

3.6 kN/m and the damping constant of the damper is 400 Ns/m. If the mass is 50 kg, then the damping factor (d) and damped natural frequency (f_n), respectively, are
(a) 0.471 and 1.19 Hz (b) 0.471 and 7.48 Hz
(c) 0.666 and 1.35 Hz (d) 0.666 and 8.50 Hz
[2009 : 2 M]

7.27 The natural frequency of a spring-mass system on earth is ω_n. The natural frequency of this system on the moon ($g_{moon} = g_{earth}/6$) is
(a) ω_n (b) $0.408\ \omega_n$
(c) $0.204\ \omega_n$ (d) $0.167\ \omega_n$ [2010 : 1 M]

7.28 A mass m attached to a spring is subjected to a harmonic force as shown in figure. The amplitude of the forced motion is observed to be 50 mm. The value of m (in kg) is

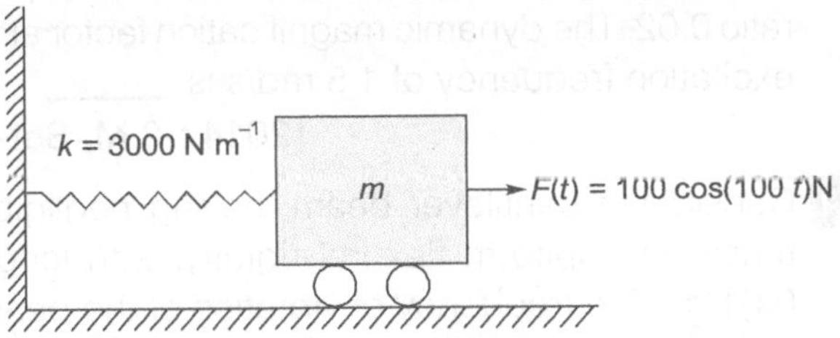

(a) 0.1 (b) 1.0
(c) 0.3 (d) 0.5 [2010 : 2 M]

7.29 A mass of 1 kg is attached to two identical springs each with stiffness k = 20 kN/m as shown in the figure. Under frictionless condition, the natural frequency of the system in Hz is close to

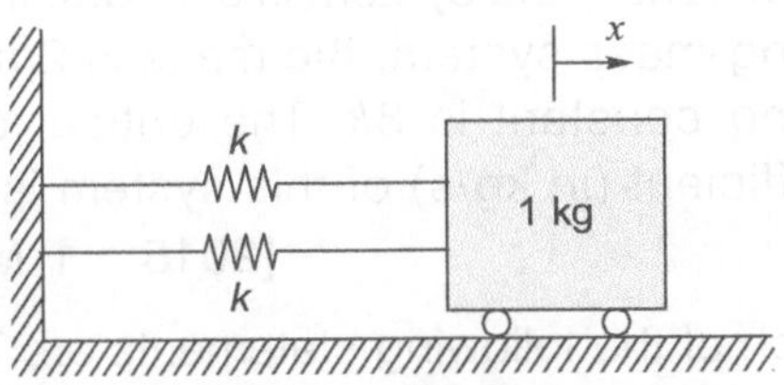

(a) 32 (b) 23
(c) 16 (d) 11 [2011 : 2 M]

7.30 A disc of mass m is attached to a spring of stiffness k as shown in the fig. The disc rolls without slipping on a horizontal surface. The natural frequency of vibration of the system is

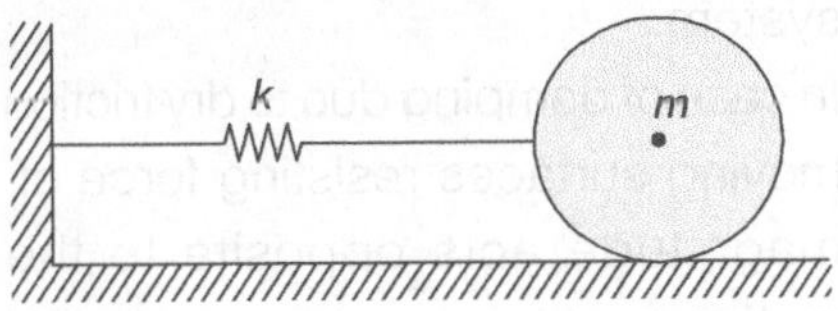

(a) $\frac{1}{2\pi}\sqrt{\frac{k}{m}}$ (b) $\frac{1}{2\pi}\sqrt{\frac{2k}{m}}$

(c) $\frac{1}{2\pi}\sqrt{\frac{2k}{3m}}$ (d) $\frac{1}{2\pi}\sqrt{\frac{3k}{2m}}$ [2011 : 2 M]

7.31 A concentrated mass m is attached at the centre of a rod of length $2L$ as shown in the figure. The rod is kept in a horizontal equilibrium position by a spring of stiffness k. For very small amplitude of vibration, neglecting the weights of the rod and spring, the undamped natural frequency of the system is

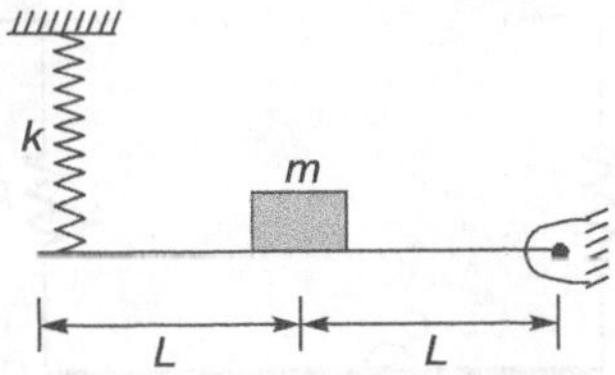

(a) $\sqrt{\frac{k}{m}}$ (b) $\sqrt{\frac{2k}{m}}$

(c) $\sqrt{\frac{k}{2m}}$ (d) $\sqrt{\frac{4k}{m}}$ [2012 : 2 M]

7.32 Critical damping is the
(a) largest amount of damping for which no oscillation occurs in free vibration
(b) smallest amount of damping for which no oscillation occurs in free vibration
(c) largest amount of damping for which the motion is simple harmonic in free vibration
(d) smallest amount of damping for which the motion is simple harmonic in free vibration
[2014 : 1 M, Set-1]

7.33 In vibration isolation, which one of the following statements is **NOT** correct regarding Transmissibility (T)?
(a) T is nearly unity at small excitation frequencies
(b) T can be always reduced by using higher damping at any excitation frequency
(c) T is unity at the frequency ratio of $\sqrt{2}$
(d) T is infinity at resonance for undamped systems
[2014 : 1 M, Set-2]

7.34 Consider a single degree of freedom system with viscous damping excited by a harmonic force. At resonance, the phase angle (in degree) of the displacement with respect to the exciting force is
(a) 0 (b) 45
(c) 90 (d) 135
[2014 : 1 M, Set-3]

7.35 A point mass is executing simple harmonic motion with an amplitude of 10 mm and frequency of 4 Hz. The maximum acceleration (m/s^2) of the mass is ______

[2014 : 1 M, Set-4]

7.36 A rigid uniform rod AB of length L and mass m is hinged at C such that $AC = L/3$, $CB = 2L/3$. Ends A and B are supported by springs of spring constant k. The natural frequency of the system is given by

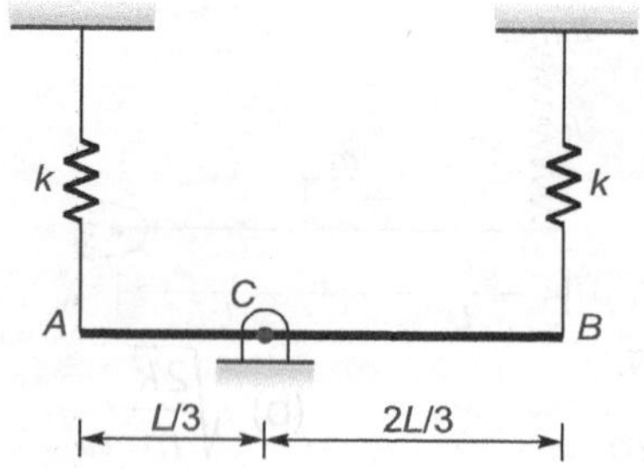

(a) $\sqrt{\dfrac{k}{2m}}$ (b) $\sqrt{\dfrac{k}{m}}$

(c) $\sqrt{\dfrac{2k}{m}}$ (d) $\sqrt{\dfrac{5k}{m}}$

[2014 : 2 M, Set-1]

7.37 What is the natural frequency of the spring mass system shown below? The contact between the block and the inclined plane is frictionless. The mass of the block is denoted by m and the spring constants are denoted by k_1 and k_2 as shown below

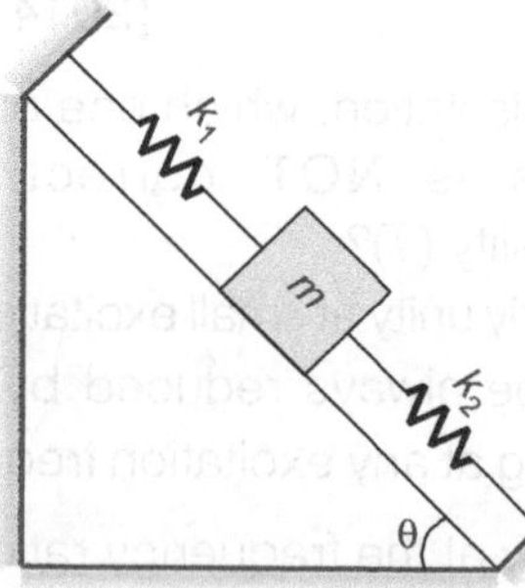

(a) $\sqrt{\dfrac{k_1+k_2}{2m}}$ (b) $\sqrt{\dfrac{k_1+k_2}{4m}}$

(c) $\sqrt{\dfrac{k_1-k_2}{m}}$ (d) $\sqrt{\dfrac{k_1+k_2}{m}}$

[2014 : 2 M, Set-2]

7.38 The damping ratio of a single degree of freedom spring mass damper system with mass of 1 kg, stiffness 100 N/m and viscous damping coefficient of 5 Ns/m is _____.

[2014 : 2 M, Set-3]

7.39 A mass-spring-dashpot system with mass m = 10 kg, spring constant k = 6250 N/m is excited by a harmonic excitation of 10 cos(25t) N. At the steady state, the vibration amplitude of the mass is 40 mm. The damping coefficient (c, in Ns/m) of the dashpot is ______.

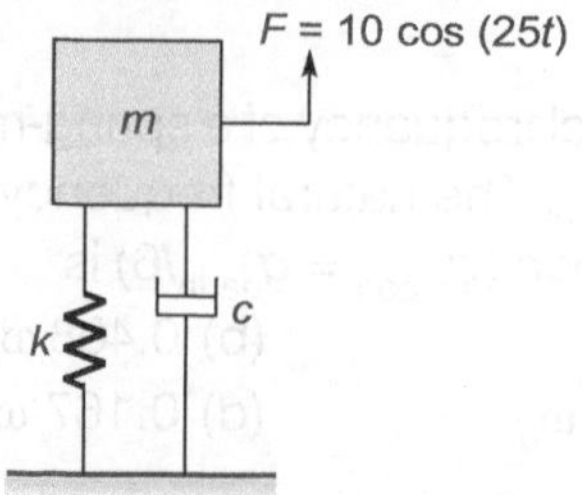

[2014 : 2 M, Set-3]

7.40 A single degree of freedom system has a mass of 2 kg, stiffness 8 N/m and viscous damping ratio 0.02. The dynamic magnification factor at an excitation frequency of 1.5 rad/s is ______

[2014 : 2 M, Set-4]

7.41 Consider a cantilever beam, having negligible mass and uniform flexural rigidity, with length 0.01 m. The frequency of vibration of the beam, with a 0.5 kg mass attached at the free tip, is 100 Hz. The flexural rigidity (in Nm^2) of the beam is ______.

[2014 : 2 M, Set-1]

7.42 In a spring-mass system, the mass is m and the spring constant is k. The critical damping coefficient of the system is 0.1 kg/s. In another spring-mass system, the mass is $2m$ and the spring constant is $8k$. The critical damping coefficient (in kg/s) of this system is _____.

[2015 : 1 M, Set-2]

7.43 Which of the following statements are TRUE for damped vibrations?

P. For a system having critical damping, the value of damping ratio is unity and system does not undergo a vibratory motion.

Q. Logarithmic decrement method is used to determine the amount of damping in a physical system.

R. In case of damping due to dry friction between moving surfaces resisting force of constant magnitude acts opposite to the relative motion.

S. For the case of viscous damping, drag force is directly proportional to the square of relative velocity.

(a) P and Q only (b) P and S only
(c) P, Q and R only (d) Q and S only

[2015 : 1 M, Set-3]

7.44 Considering massless rigid rod small oscillations, the natural frequency (in rad/s) of vibration of the system shown in the figure is

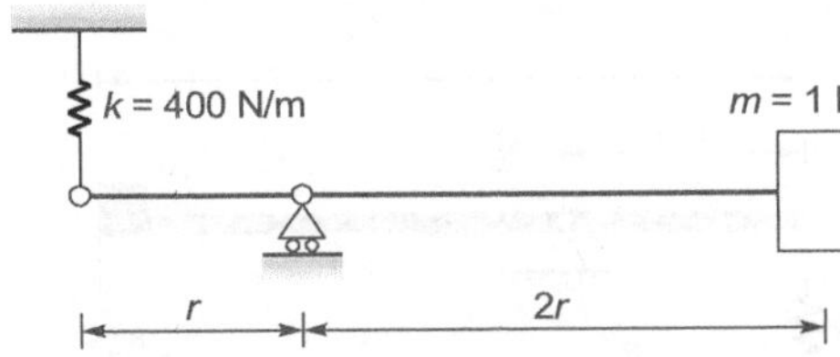

(a) $\sqrt{\frac{400}{1}}$ (b) $\sqrt{\frac{400}{2}}$

(c) $\sqrt{\frac{400}{3}}$ (d) $\sqrt{\frac{400}{4}}$

[2015 : 2 M, Set-1]

7.45 A precision instrument package ($m = 1$ kg) needs to be mounted on a surface vibrating at 60 Hz. It is desired that only 5% of the base surface vibration amplitude be transmitted to the instrument. Assuming that the isolation is designed with its natural frequency significantly lesser than 60 Hz, so that the effect of damping may be ignored. The stiffness (in N/m) of the required mounting pad is ______.

[2015 : 2 M, Set-1]

7.46 A single-degree-freedom spring-mass system is subjected to a sinusoidal force of 10 N amplitude and frequency ω along the axis of the spring. The stiffness of the spring is 150 N/m, damping factor is 0.2 and the undamped natural frequency is 10ω. At steady state, the amplitude of vibration (in m) is approximately

(a) 0.05 (b) 0.07
(c) 0.70 (d) 0.90

[2015 : 2 M, Set-2]

7.47 Figure shows a single degree of freedom system. The system consists of a massless rigid bar OP hinged at O and a mass m at end P. The natural frequency of vibration of the system is

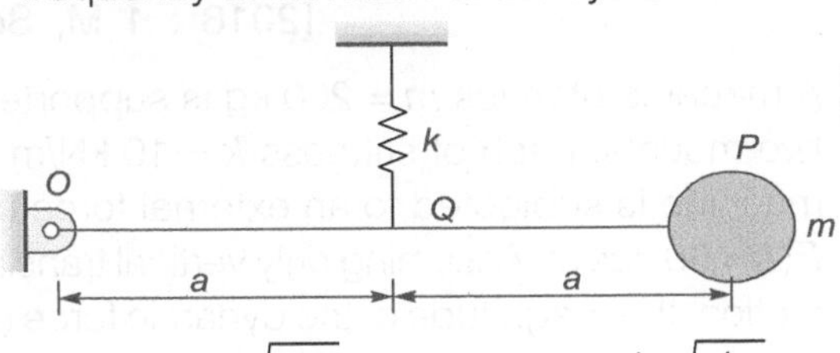

(a) $f_n = \frac{1}{2\pi}\sqrt{\frac{k}{4m}}$ (b) $f_n = \frac{1}{2\pi}\sqrt{\frac{k}{2m}}$

(c) $f_n = \frac{1}{2\pi}\sqrt{\frac{k}{m}}$ (d) $f_n = \frac{1}{2\pi}\sqrt{\frac{2k}{m}}$

[2015 : 2 M, Set-3]

7.48 A single degree of freedom spring-mass system is subjected to a harmonic force of constant amplitude. For an excitation frequency of $\sqrt{\frac{3k}{m}}$, the ratio of the amplitude of steady state response to the static deflection of the spring is ______

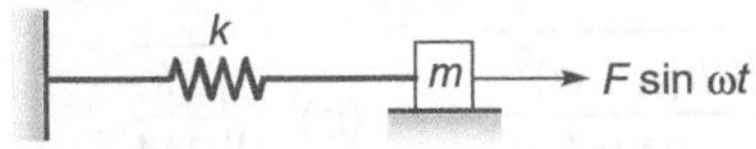

[2016 : 2 M, Set-3]

7.49 A single degree of freedom mass-spring-viscous damper system with mass m, spring constant k and viscous damping coefficient q is critically damped. The correct relation among m, k, and q is

(a) $q = \sqrt{2km}$ (b) $q = 2\sqrt{km}$

(c) $q = \sqrt{\frac{2k}{m}}$ (d) $q = 2\sqrt{\frac{k}{m}}$

[2016 : 1 M, Set-2]

7.50 The system shown in the figure consists of block A of mass 5 kg connected to a spring through a massless rope passing over pulley B of radius r and mass 20 kg. The spring constant k is 1500 N/m. If there is no slipping of the rope over the pulley, the natural frequency of the system is______ rad/s.

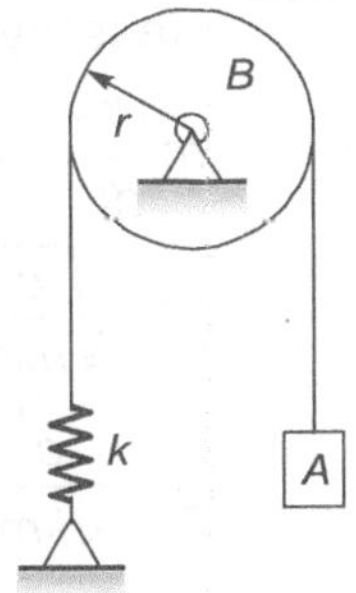

[2016 : 2 M, Set-2]

7.51 A single degree of freedom spring mass system with viscous damping has a spring constant of 10 kN/m. The system is excited by a sinusoidal force of amplitude 100 N. If the damping factor (ratio) is 0.25, the amplitude of steady state oscillation at resonance is______mm.

[2016 : 1 M, Set-1]

7.52 A solid disc with radius a is connected to a spring at a point d above the center of the disc. The other end of the spring is fixed to the vertical wall. The disc is free to roll without slipping on the ground. The mass of the disc is M and the spring constant is k. The polar moment of inertia for the disc about its centre is $J = \frac{Ma^2}{2}$

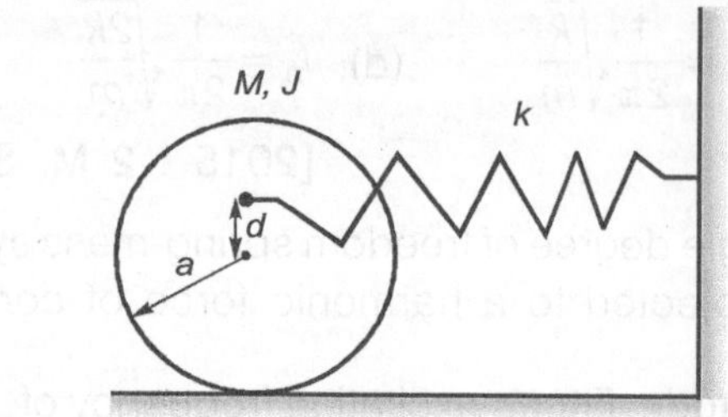

The natural frequency of this system in rad/s is given by

(a) $\sqrt{\dfrac{2k(a+d)^2}{3Ma^2}}$ (b) $\sqrt{\dfrac{2k}{3M}}$

(c) $\sqrt{\dfrac{2k(a+d)^2}{Ma^2}}$ (d) $\sqrt{\dfrac{k(a+d)^2}{Ma^2}}$

[2016 : 2 M, Set-1]

7.53 The static deflection of a spring under gravity, when a mass of 1 kg is suspended from it, is 1 mm. Assume the acceleration due to gravity $g = 10$ m/s^2. The natural frequency of this spring-mass system (in rad/s) is________

[2016 : 1 M, Set-3]

7.54 The damping ratio for a viscously damped spring mass system, governed by the relationship $m\dfrac{d^2x}{dt^2} + c\dfrac{dx}{dt} + kx = F(t)$, is given by

(a) $\sqrt{\dfrac{c}{mk}}$ (b) $\dfrac{c}{2\sqrt{km}}$

(c) $\dfrac{c}{\sqrt{km}}$ (d) $\sqrt{\dfrac{c}{\sqrt{2mk}}}$

[2017 : 1 M, Set-1]

7.55 A mass m is attached to two identical springs having constant k as shown in the figure. The natural frequency ω of this single degree of freedom system is

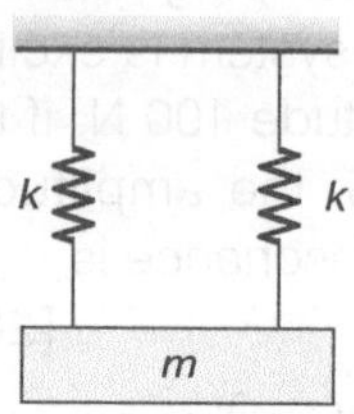

(a) $\sqrt{\dfrac{2k}{m}}$ (b) $\sqrt{\dfrac{k}{m}}$

(c) $\sqrt{\dfrac{k}{2m}}$ (d) $\sqrt{\dfrac{4k}{m}}$

[2017 : 1 M, Set-2]

7.56 A thin uniform rigid bar of length L and M is hinged at point O, located at a distance of $L/3$ from one of its ends. The bar is further supported using springs, each of stiffness k, located at the two ends. A particle of mass $m = M/4$ is fixed at one end of the bar, as shown in the figure. For small rotations of the bar about O, the natural frequency of the system is

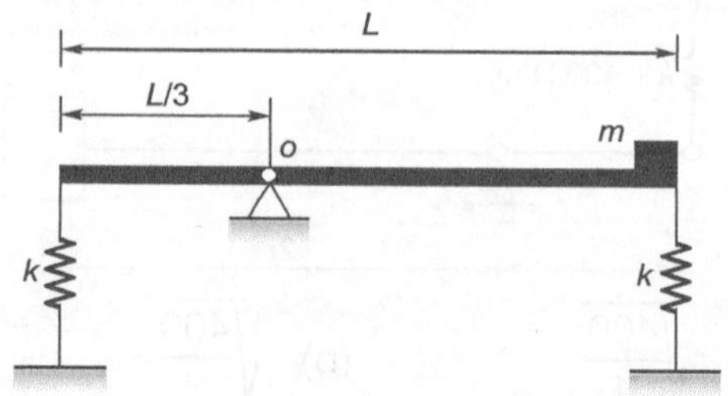

(a) $\sqrt{\dfrac{5k}{M}}$ (b) $\sqrt{\dfrac{5k}{2M}}$

(c) $\sqrt{\dfrac{3k}{2M}}$ (d) $\sqrt{\dfrac{3k}{M}}$

[2017 : 2 M, Set-1]

7.57 The radius of gyration of a compound pendulum about the point of suspension is 100 mm. The distance between the point of suspension and the centre of mass is 250 mm. Considering the acceleration due to gravity is 9.81 m/s^2, the natural frequency (in radian/s) of the compound pendulum is ________.

[2017 : 2 M, Set-2]

7.58 The equation of motion for a spring-mass system excited by a harmonic force is

$M\ddot{x} + kx = F\cos(\omega t)$

where M is the mass, k is the spring stiffness, F is the force amplitude and ω is the angular frequency of excitation. Resonance occurs when ω is equal to

(a) $\sqrt{\dfrac{M}{k}}$ (b) $\dfrac{1}{2\pi}\sqrt{\dfrac{k}{M}}$

(c) $2\pi\sqrt{\dfrac{k}{M}}$ (d) $\sqrt{\dfrac{k}{M}}$

[2018 : 1 M, Set-1]

7.59 A machine of mass $m = 200$ kg is supported on two mounts, each of stiffness $k = 10$ kN/m. The machine is subjected to an external force (in N) $F(t) = 50\cos 5t$. Assuming only vertical translatory motion, the magnitude of the dynamic force (in N) transmitted from each mount to the ground is ________(correct to two decimal places).

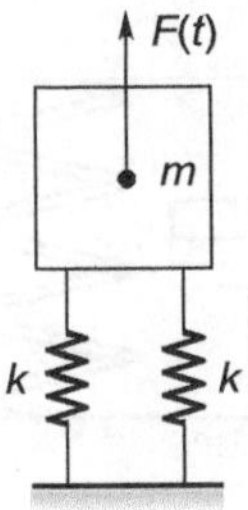

[2018 : 2 M, Set-1]

7.60 In a single degree of freedom underdamped spring-mass-damper system as shown in the figure, an additional damper, is added in parallel such that the system still remains underdamped. Which one of the following statements is ALWAYS true?

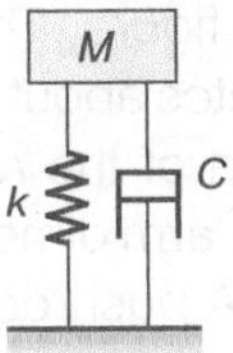

(a) Transmissibility will increase
(b) Transmissibility will decrease
(c) Time period of free oscillations will increase
(d) Time period of free oscillations will decrease

[2018 : 1 M, Set-2]

7.61 The natural frequencies corresponding to the spring-mass systems I and II are ω_I and ω_{II}, respectively. The ratio $\dfrac{\omega_I}{\omega_{II}}$ is

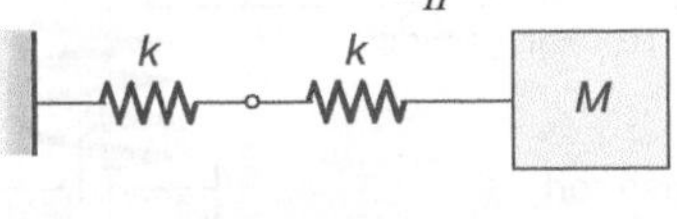

System I

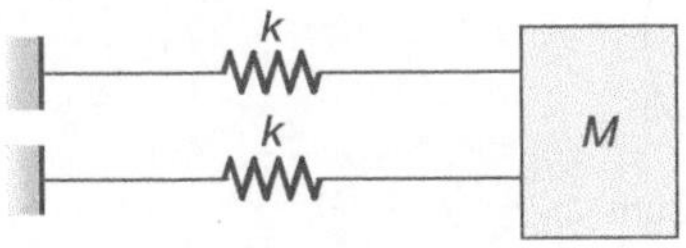

System II

(a) 2 (b) $\dfrac{1}{2}$

(c) $\dfrac{1}{4}$ (d) 4 [2019 : 1 M, Set-1]

7.62 A uniform thin disk of mass 1 kg and radius 0.1 m is kept on a surface as shown in the figure. A spring of stiffness k_1 = 400 N/m is connected to the disk center A and another spring of stiffness k_2 = 100 N/m is connected at point B just above point A on the circumference of the disk. Initially, both the springs are unstretched. Assume pure rolling of the disk. For small disturbance from the equilibrium, the natural frequency of vibration of the system is ______ rad/s (round off to one decimal place)

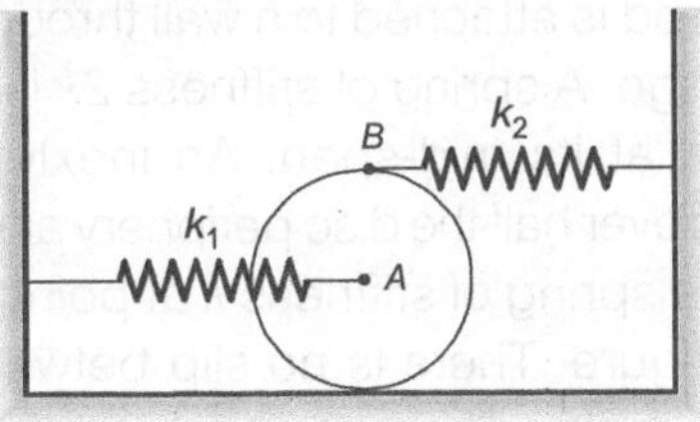

[2019 : 2 M, Set-1]

7.63 A slender uniform rigid bar of mass 'm' is hinged at 'O' and supported by two springs with stiffness $3k$ and k and a damper with damping coefficient c as shown in figure. For the system to be critically damped, the ratio of $\dfrac{c}{\sqrt{km}}$ should be ____

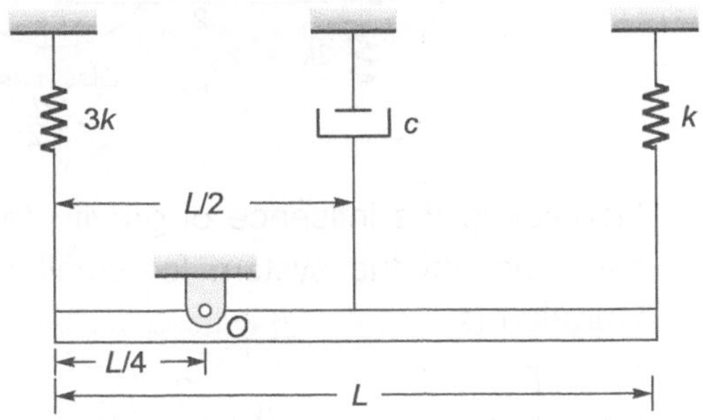

(a) $4\sqrt{7}$ (b) 2
(c) 4 (d) $2\sqrt{7}$

[2019 : 2 M, Set-2]

7.64 A single-degree-of-freedom oscillator is subjected to harmonic excitation $F(t) = F_0\cos(\omega t)$ as shown in the figure.

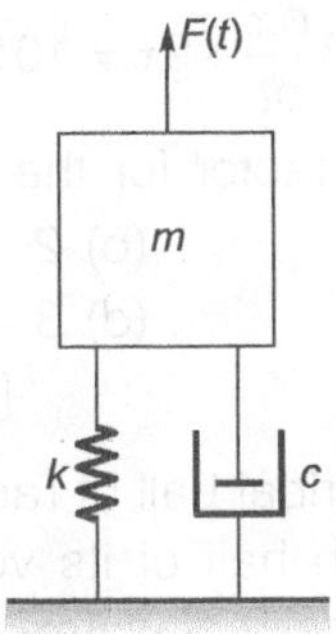

The non-zero value of ω, for which the amplitude of the force transmitted to the ground will be F_0, is

(a) $\sqrt{\dfrac{k}{2m}}$ (b) $\sqrt{\dfrac{2k}{m}}$

(c) $\sqrt{\dfrac{k}{m}}$ (d) $2\sqrt{\dfrac{k}{m}}$

[2020 : 1 M, Set-1]

7.65 A rigid mass-less rod of length L is connected to a disc (pulley) of mass m and radius $r = L/4$ through a friction-less revolute joint. The other end of that rod is attached to a wall through a friction-less hinge. A spring of stiffness $2k$ is attached to the rod at its mid-span. An inextensible rope passes over half the disc periphery and is securely tied to a spring of stiffness k at point C as shown in the figure. There is no slip between the rope and the pulley. The system is in static equilibrium in the configuration shown in the figure and the rope is always taut.

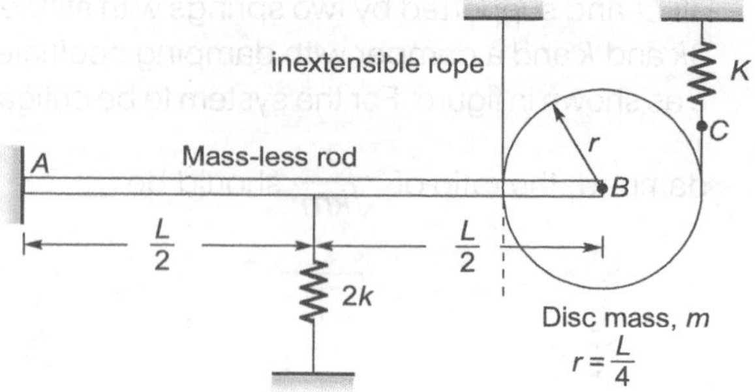

Neglecting the influence of gravity, the natural frequency of the system for small amplitude vibration is

(a) $\sqrt{\frac{k}{m}}$
(b) $\frac{3}{\sqrt{2}}\sqrt{\frac{k}{m}}$
(c) $\sqrt{3}\sqrt{\frac{k}{m}}$
(d) $\sqrt{\frac{3}{2}}\sqrt{\frac{k}{m}}$

[2020 : 2 M, Set-1]

7.66 The equation of motion of a spring-mass-damper system is given by

$$\frac{d^2x}{dt^2} + 3\frac{dx}{dt} + 9x = 10\sin(5t)$$

The damping factor for the system is

(a) 0.25
(b) 2
(c) 0.5
(d) 3

[2020 : 1 M, Set-2]

7.67 A hollow spherical ball of radius 20 cm floats in still water, with half of its volume submerged. Taking the density of water as 1000 kg/m^3, and the acceleration due to gravity as 10 m/s^2, the natural frequency of small osc illations of the ball, normal to the water surface is ________ radians/s (roundoff to 2 decimal places).

[2020 : 2 M, Set-2]

7.68 Consider a single degree of freedom system comprising a mass M, supported on a spring and a dashpot as shown in the figure.

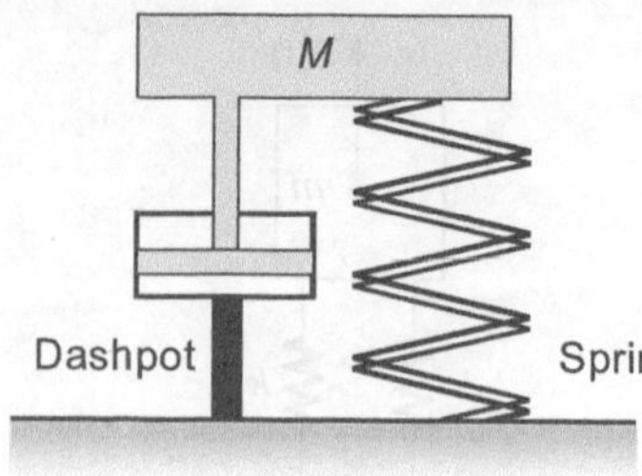

If the amplitude of the free vibration response reduces from 8 mm to 1.5 mm in 3 cycles, the damping ratio of the system is ________ [Round off to three decimal places]

[2021 : 1 M, Set-1]

7.69 A tappet valve mechanism in an IC engine comprises a rocker arm ABC that is hinged at B as shown in the figure. The rocker is assumed rigid and it oscillates about the hinge B. The mass moment of inertia of the rocker about B is 10^{-4} kgm^2. The rocker arm dimensions are a = 3.5 cm and b = 2.5 cm. A pushrod pushes the rocker at location A, when moved vertically by a cam that rotates at N rpm. The pushrod pushes the rocker at location A, when moved vertically by a cam that rotates at N rpm. The pushrod is assumed massless and has a stiffness of 15 N/mm. At the other end C, the rocker pushes a valve against a spring of stiffness 10 N/mm. The valve is assumed massless and rigid.

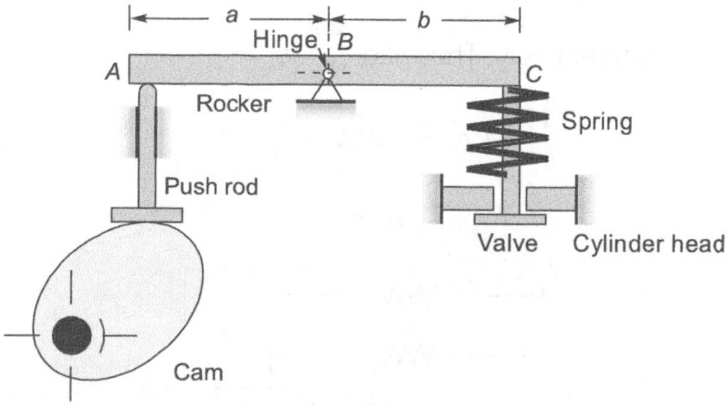

Resonance in the rocker system occurs when the cam shaft runs at a speed of ________ rpm. [Round off to the nearest integer]

(a) 2369
(b) 790
(c) 496
(d) 4739

[2021 : 2 M, Set-1]

7.70 Consider the system shown in the figure. A rope goes over a pulley. A mass, m, is hanging from the rope. A spring of stiffness, k, is attached at one end of the rope. Assume rope is inextensible, massless and there is no slip between pulley and rope.

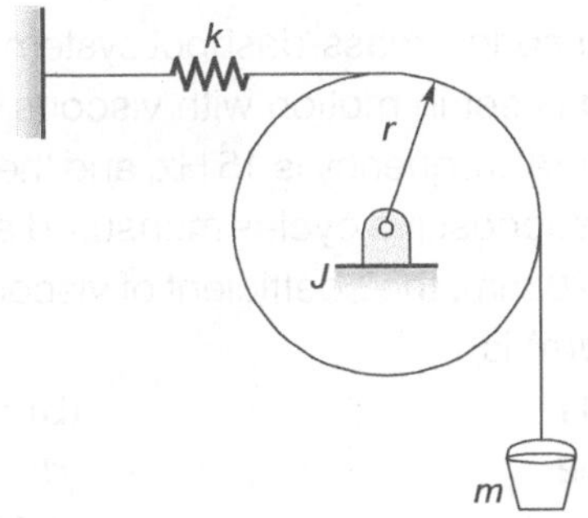

The pulley radius is r and its mass moment of inertia is J. Assume that the mass is vibrating harmonically about its static equilibrium position. The natural frequency of the system is

(a) $\sqrt{\dfrac{kr^2}{J+mr^2}}$ (b) $\sqrt{\dfrac{kr^2}{J}}$

(c) $\sqrt{\dfrac{kr^2}{J-mr^2}}$ (d) $\sqrt{\dfrac{k}{m}}$

[2021 : 2 M, Set-2]

7.71 A machine of mass 100 kg is subjected to an external harmonic force with a frequency of 40 rad/s. The designer decides to mount the machine on an isolator to reduce the force transmitted to the foundation. The isolator can be considered as a combination of stiffness (K) and damper (damping factor, ξ) in parallel. The designer has the following four isolators:

1. K = 640 kN/m, ξ = 0.70
2. K = 640 kN/m, ξ = 0.07
3. K = 22.5 kN/m, ξ = 0.70
4. K = 22.5 kN/m, ξ = 0.07

Arrange the isolators in the ascending order of the force transmitted to the foundation.

(a) 1 – 3 – 4 – 2 (b) 1 – 3 – 2 – 4

(c) 3 – 1 – 2 – 4 (d) 4 – 3 – 1 – 2

[2021 : 2 M, Set-2]

7.72 A rigid uniform annular disc is pivoted on a knife edge A in a uniform gravitational field as shown, such that it can execute small amplitude simple harmonic motion in the plane of the figure without slip at the pivot point. The inner radius r and outer radius R are such that $r^2 = \dfrac{R^2}{2}$, and the acceleration due to gravity is g. If the time period of small amplitude simple harmonic motion is given by $T = \beta\pi\sqrt{\dfrac{R}{g}}$, where π is the ratio of circumference to diameter of a circle, then β = ________. (round off to 2 decimal places).

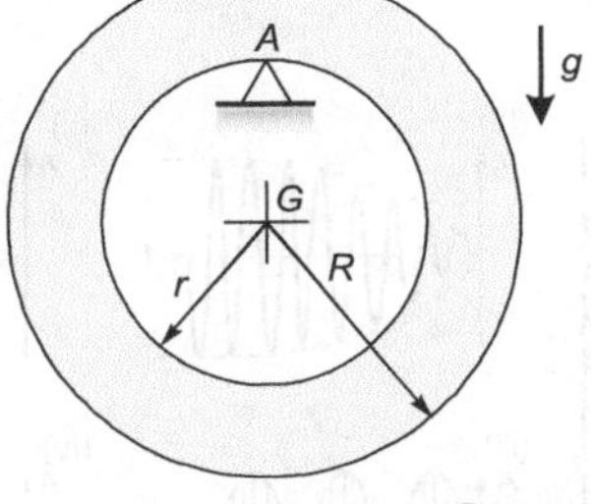

[2022 : 1 M, Set-1]

7.73 Consider a forced single degree-of-freedom system governed by $\ddot{x}(t) + 2\zeta\omega_n\dot{x}(t) + \omega_n^2 x(t) = \omega_n^2\cos(\omega t)$, where ζ and ω_n are the damping ratio and undamped natural frequency of the system, respectively, wile ω is forcing frequency. The amplitude of the forced steady state response of this system is given by $\left[(1-r^2)^2 + (2\zeta r)^2\right]^{-\frac{1}{2}}$, where $r = \dfrac{\omega}{\omega_n}$. The peak amplitude of this response occurs at a frequency $\omega = \omega_p$. If ω_d denotes the damped natural frequency of this system, which one of the following options is true?

(a) $\omega_p < \omega_d < \omega_n$ (b) $\omega_p = \omega_d < \omega_n$

(c) $\omega_d < \omega_n = \omega_p$ (d) $\omega_d < \omega_n < \omega_p$

[2022 : 2 M, Set-1]

7.74 For a dynamical system governed by the equation,

$$\ddot{x}(t) + (2\xi\omega_n)\dot{x}(t) + (\omega_n^2)x(t) = 0$$

the damping ratio ξ is equal to $\dfrac{1}{2\pi}\log_e(2)$. The displacement x of this system is measured during a hammer test. A displacement peak in the positive displacement direction is measured to be 4 mm. Neglecting higher powers (>1) of the damping ratio, the displacement at the next peak in the positive direction will be_____ mm (in integer).

[2022 : 1 M, Set-2]

7.75 A spring mass damper system (mass m, stiffness k, and damping coefficient c) excited by a force $F(t) = B\sin\omega t$, where B, ω, and t are the amplitude, frequency and time, respectively, is shown in the figure. Four different responses of the system (marked as (i) to (iv)) are shown just to the right of the system figure. In the figures of the responses, A is the amplitude of response shown in red color and the dashed lines indicate its envelope. The responses represent only the qualitative trend and those are not drawn to any specific scale.

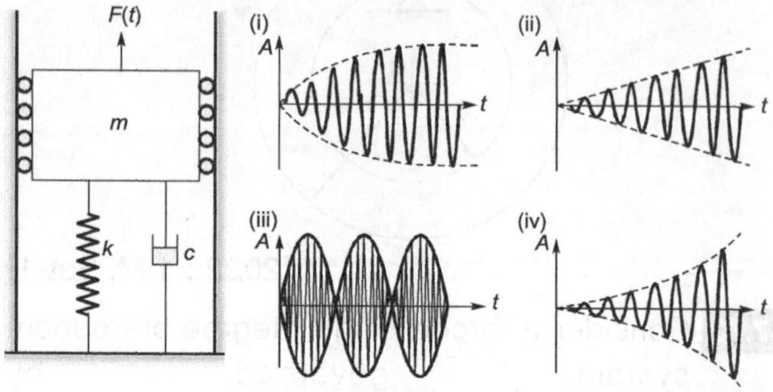

Four different parameter and forcing conditions are mentioned below.

P. $c > 0$ and $\omega = \sqrt{\frac{k}{m}}$

Q. $c < 0$ and $\omega \neq 0$

R. $c = 0$ and $\omega = \sqrt{\frac{k}{m}}$

S. $c = 0$ and $\omega \cong \sqrt{\frac{k}{m}}$

(a) P-(i), Q-(iii), R-(iv), S-(ii)
(b) P-(ii), Q-(iii), R-(iv), S-(i)
(c) P-(i), Q-(iv), R-(ii), S-(iii)
(d) P-(iii), Q-(iv), R-(ii), S-(i) [2022 : 2 M, Set-2]

7.76 The figure shows a block of mass m = 20 kg attached to a pair of identical linear springs, each having a spring constant k = 1000 N/m. The block oscillates on a frictionless horizontal surface. Assuming free vibrations, the time taken by the block to complete ten oscillations is ________ seconds. (Rounded off to two decimal places) Take π = 3.14.

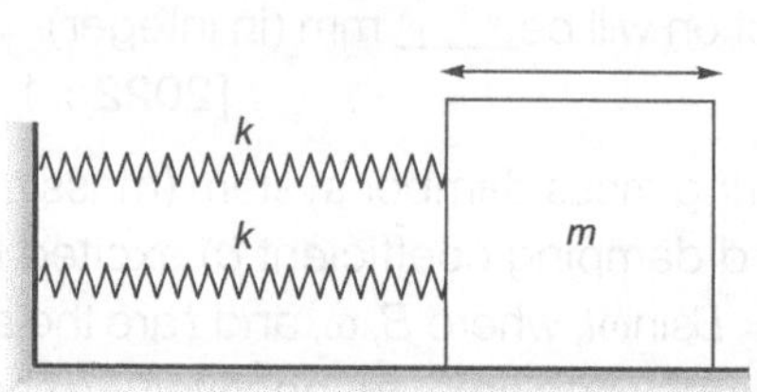

[2023 : 1 M]

7.77 A linear spring-mass-dashpot system with a mass of 2 kg is set in motion with viscous damping. If the natural frequency is 15 Hz, and the amplitudes of two successive cycles measured are 7.75 mm and 7.20 mm, the coefficient of viscous damping (in N.s/m) is

(a) 4.41 (b) 6.11
(c) 2.52 (d) 7.51

[2024 : 1 M]

7.78 A vibratory system consists of mass m, a vertical spring of stiffness $2k$ and a horizontal spring of stiffness k. The end A of the horizontal spring is given a horizontal motion $x_A = a \sin \omega t$. The other end of the spring is connected to an inextensible rope that passes over two massless pulleys as shown. Assume m = 10 kg, k = 1.5 kN/m, and neglect friction. The magnitude of critical driving frequency for which the oscillations of mass m tend to become excessively large is ______ rad/s. (answer in integer)

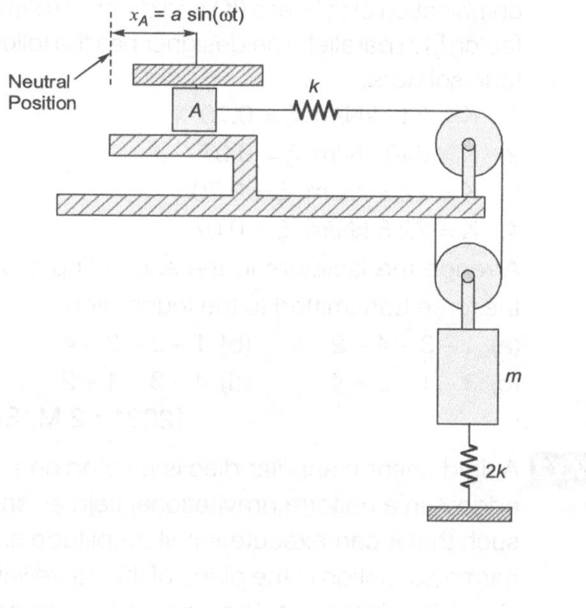

[2024 : 2 M]

■■■■

Answers Vibration

7.1	(b)	7.2	(a)	7.3	(c)	7.4	(a)	7.5	(d)	7.6	(b)	7.7	(b)	7.8	(a)
7.9	(c)	7.10	(c)	7.11	(c)	7.12	(a)	7.13	(c)	7.14	(c)	7.15	(c)	7.16	(c)
7.17	(d)	7.18	(*)	7.19	(c)	7.20	(a)	7.21	(a)	7.22	(b)	7.23	(c)	7.24	(b)
7.25	(a)	7.26	(a)	7.27	(a)	7.28	(a)	7.29	(a)	7.30	(c)	7.31	(d)	7.32	(b)
7.33	(b)	7.34	(c)	7.35	(6.31)	7.36	(d)	7.37	(d)	7.38	(0.25)	7.39	(10)		
7.40	(2.28)	7.41	(0.06573)			7.42	(0.4)	7.43	(c)	7.44	(d)	7.45	(6767.72)		
7.46	(b)	7.47	(a)	7.48	(0.5)	7.49	(b)	7.50	(10)	7.51	(20)	7.52	(a)		
7.53	(100)	7.54	(b)	7.55	(a)	7.56	(b)	7.57	(15.66)	7.58	(d)	7.59	(33.33)		
7.60	(c)	7.61	(b)	7.62	(23.094)	7.63	(a)	7.64	(b)	7.65	(c)	7.66	(c)		
7.67	(8.66)	7.68	(0.088)	7.69	(d)	7.70	(a)	7.71	(d)	7.72	(2.66)	7.73	(a)		
7.74	(2)	7.75	(c)	7.76	(6.28)	7.77	(a)	7.78	(30)						

Explanations Vibration

7.1 (b)

k_1 and k_2 are in series, therefore k_{eq}, will be,

$$\frac{1}{k_{eq_1}} = \frac{1}{k_1}+\frac{1}{k_2} = \frac{1}{1}+\frac{1}{3} = \frac{4}{3}$$

Now, k_{eq_1} and k_3 are in parallel arrangement

$$k_{eq} = k_{eq_1} + k_3 = \frac{3}{4}+2 = \frac{11}{4} \text{ kN/m}$$

Natural frequency

$$= \sqrt{\frac{k_{eq}}{m}} = \sqrt{\frac{11\times 10^3}{4\times 1}} = \mathbf{52.44 \text{ rad/s}}$$

7.2 (a)

The system is a unconstrained system with 2 degrees of freedom. In one case it will behave like rigid body and natural frequency in that case will be **equal to zero.** (About its centre of mass).

$$x_1 = x_2$$

For second case, there will be node which will lie in between the spring since both masses are same.

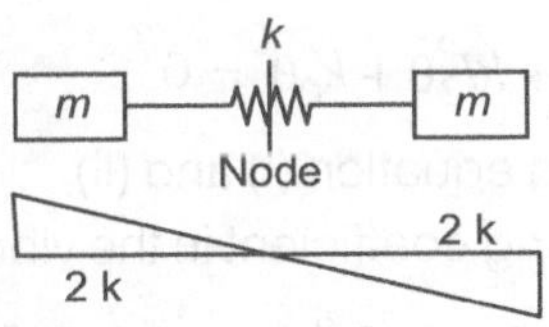

We can take system from node and it as shown below.

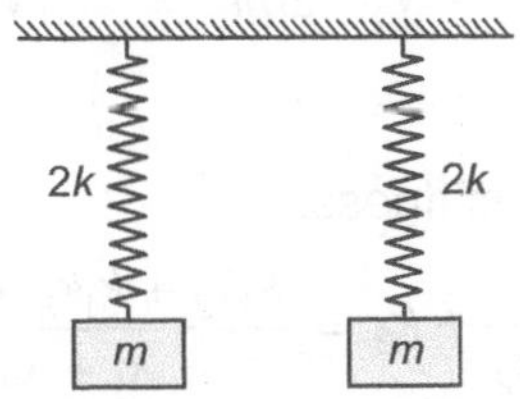

In this case, value of spring coefficient become $2k$ because spring of coefficient k is divided into two identical parts.

$$x_1 = x_2 = x$$

Sprin force on block = $2k \cdot x$

$$m\ddot{x} + 2kx = 0$$

Natural frequency in this case is $\sqrt{\frac{2k}{m}}$.

7.3 (c)

Springs are in parallel arrangement,

$$k_{eq} = k_1 + k_2 = 20 + 20 = 40 \text{ kN/m}$$

$$\text{Natural frequency} = \frac{1}{2\pi}\sqrt{\frac{40\times 10^3}{100}} = \frac{10}{\pi}$$

7.4 (a)

$$\Delta(PE) = (PE)_B-(PE)_A$$

$$= \frac{1}{2}k(x+\delta)^2 + 0 - \left[\frac{1}{2}k\delta^2 + mgx\right]$$

Taking reference datum at position B.

At equilibrium position *i.e.*, at A.

$$mg = k\delta$$

$$\Delta(PE) = \frac{1}{2}kx^2 + \frac{1}{2}k\delta^2 + xk\delta - \frac{1}{2}k\delta^2 - mgx$$

as $\quad mg = k\delta$

$$\Delta(PE) = \frac{1}{2}kx^2 + \frac{1}{2}k\delta^2 + mgx - \frac{1}{2}k\delta^2 - mgx$$
$$= \frac{1}{2}kx^2$$

7.5 (d)

$$I = ml^2 + ml^2 = 2ml^2$$

Considering moment equilibrium condition about axis of rotation.

$$2ml^2\ddot{\theta} + mgl\sin(\alpha+\theta) - mgl\sin(\alpha-\theta) = 0$$
$$\Rightarrow 2ml^2\ddot{\theta} + mlg\,[\sin\alpha\cos\theta + \cos\alpha\sin\theta - \sin\alpha\cos\theta + \cos\alpha\cos\theta]$$
$$\Rightarrow 2ml^2\ddot{\theta} + 2mgl\cos\alpha\sin\theta = 0$$

For very small, θ, $\sin\theta \approx \theta$

Hence, we get

$$2ml^2\ddot{\theta} + 2mgl\cos\alpha = 0$$
$$\omega = \sqrt{\frac{2mgl\cos\alpha}{2ml^2}} = \sqrt{\frac{g\cos\alpha}{l}}$$

7.6 (b)

Equivalent stiffness,

$$k_{eq} = \frac{k_1k_2}{k_1+k_2} + k_3 = \frac{k_1k_2 + k_1k_3 + k_2k_3}{k_1+k_2}$$

Now natural frequency

$$= \sqrt{\frac{k_{eq}}{J}} = \sqrt{\frac{k_1k_2 + k_1k_3 + k_2k_3}{J(k_1+k_2)}}$$

7.7 (b)

For shaft loaded at middle with mass M

Deflection: $\Delta = \dfrac{Mgl^3}{48EI}$

Moment of inertia,

$$I = \frac{\pi}{64}d^4 = \frac{\pi}{64}\times\left(30\times10^{-3}\right)^4$$
$$= 3.976\times10^{-8}\ \text{m}^4$$

So; deflection of shaft,

$$\Delta = \frac{Mgl^3}{48EI} = \frac{10\times9.81\times(500\times10^{-3})^3}{48\times2.1\times10^{11}\times3.976\times10^{-8}}$$
$$= 3.06\times10^{-5}\ \text{m}$$
$$\omega_n = \sqrt{\frac{g}{\Delta}} = \sqrt{\frac{9.81}{3.06\times10^{-5}}} = 566.24\ \text{rad/s}$$

Natural frequency,

$$f_n = \frac{\omega_n}{2\pi} = \frac{566.24}{2\pi} = \mathbf{90.11\ Hz}$$

7.8 (a)

For small deflection, after equilibrium

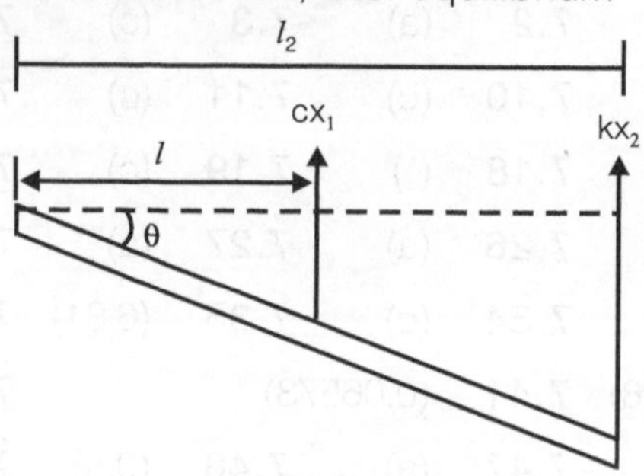

Now, $\theta = \dfrac{x_1}{0.4} = \dfrac{x_2}{0.5}$

$\therefore\quad x_1 = 0.4\,\theta$

and $\quad x_2 = 0.5\,\theta$

Moment of inertia

$$= \frac{ml_2^2}{3} = \frac{10\times(0.5)^2}{3} = 0.833\ \text{kgm}^2$$

Now, from Newton's law of motion

$$c\dot{x}_1l_1 + kx_2l_2 + k_\theta.\theta + I\ddot{\theta} = 0$$
$$\Rightarrow Cl_1^2\dot{\theta} + kl_2^2\,\theta + k_\theta.\theta + I\ddot{\theta} = 0$$
$$\Rightarrow 500\times(0.4)^2\dot{\theta} + (2000\times(0.5)^2 + 1000)\theta + 0.833\ddot{\theta} = 0$$
$$\Rightarrow 0.833\ddot{\theta} + 80\dot{\theta} + 1500\theta \qquad \text{....(i)}$$

Equation can be written as :

$$I\ddot{\theta} + C_T\dot{\theta} + K_T\theta = 0$$

$\therefore\quad I = 0.833$

$C_T = 80$

$K_T = 1500$

Now natural frequency,

$$\omega_n = \sqrt{\frac{K_T}{I}} = \sqrt{\frac{1500}{0.833}} = \mathbf{42.43\ rad/s}$$

7.9 (c)

Equivalent can be written as:

$$I\ddot{\theta} + C_T\dot{\theta} + k_T\theta = 0 \qquad \text{...(i)}$$
$$Cl_1^2\dot{\theta} + I\ddot{\theta} + kl_2^2\theta + k_0\theta = 0 \qquad \text{...(ii)}$$

Comparing equation (i) and (ii)

The damping coefficient in the vibration equation,

$$C_T = Cl_1^2 = 500\,(0.4)^2 = 0.16\times500$$
$$= 80\ \text{Nms/rad}$$

7.10 (c)

Given $r = 0.5$

Due to isolation

$$\xi_S = 0$$

Transmissibility ratio,

$$TR = \frac{\sqrt{1+(2\xi r)^2}}{\sqrt{(1-r^2)^2+(2\xi r)^2}}$$

$$= \frac{1}{\sqrt{\left(1-(0.5)^2\right)^2}} = \frac{1}{1-0.25} = \frac{4}{3}$$

7.11 (c)

Let the weight be displaced from equilibrium position, then

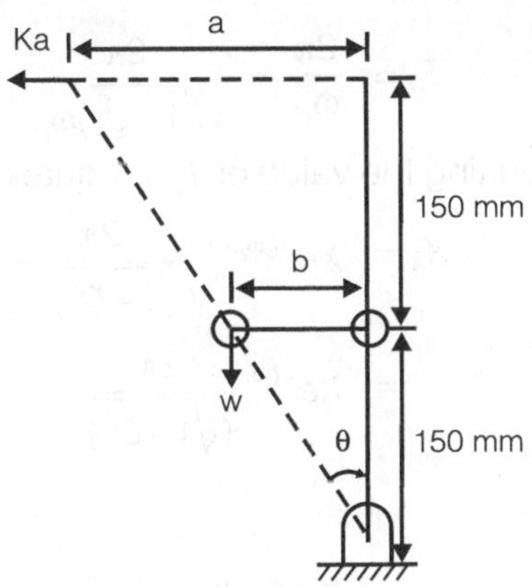

For small displacement

$$\theta = \frac{a}{300} = \frac{b}{150}$$

Taking moment about hinged point 'O'

$$k.a.(0.3) = W.b$$

$$\Rightarrow \quad k.(0.3)^2 = 300\,(0.150)$$

$$\therefore \quad k = \frac{300 \times 0.150}{(0.30)^2} = \mathbf{500\ N/m}$$

7.12 (a)

Moment of inertia of the shaft (I)

$$= \frac{1}{12} \times \left(25 \times 10^{-3}\right)^4$$

$$= 3.25 \times 10^{-8}\ m^4$$

Now, deflection at free end of cantilever beam

$$\delta = \frac{WL^3}{3EI}$$

$$= \frac{20 \times 9.81 \times (1)^3}{3 \times 200 \times 10^9 \times 3.25 \times 10^{-8}}$$

$$= 0.01\ m$$

Now, we known that

$$\omega_n = \sqrt{\frac{g}{\delta}} = \sqrt{\frac{9.81}{0.01}} = 31.32\ rad/s$$

Critical damping constant,

$$C_C = 2\,m\omega_n = 2 \times 20 \times 31.32$$

$$\approx \mathbf{1250\ Ns/m}$$

POINTS TO REMEMBER

For critically damped system, damping factor is one. As we know,

$$\text{Damping factor, } \xi = \frac{C}{2\sqrt{km}}$$

Fro critical damnped system,

$$\xi = 1,\ C = C_c$$

i.e.

$$1 = \frac{C_c}{2\sqrt{km}}$$

$$C_c = 2\sqrt{Km}$$

This problem is also based on deflection of beam.

7.13 (c)

Execution frequency;

$$f = 144\ Hz$$

Natural frequency,

$$f_n = 64, 96, 128 \text{ and } 256$$

$$\therefore \quad \frac{\text{Execution frequency}}{\text{Natural frequency}}$$

$$= \frac{f}{f_n} = \underset{(P)}{\frac{144}{64}}, \underset{(Q)}{\frac{144}{96}}, \underset{(R)}{\frac{144}{128}}, \underset{(S)}{\frac{144}{256}}$$

$$= \mathbf{2.25,\ 1.5,\ 1.25,\ 0.5625}$$

Since 1.125 is closed to 1, so the sample 'R' will show the most perceptible induced vibration.

7.14 (c)

Given $\omega_d = 0.9\omega_n$

we know that

$$\omega_d = \omega_n\sqrt{1-\xi^2}$$

$$\Rightarrow \quad 0.9\omega_n = \omega_n\sqrt{1-\xi^2}$$

$$\therefore \quad \xi = 0.436$$

Now

$$\xi = \frac{c}{2\sqrt{km}}$$

$$\therefore \quad c = 2 \times 0.436 \times \sqrt{1000 \times 0.1}$$

$$= \mathbf{8.71\ Ns/m}$$

7.15 (c)

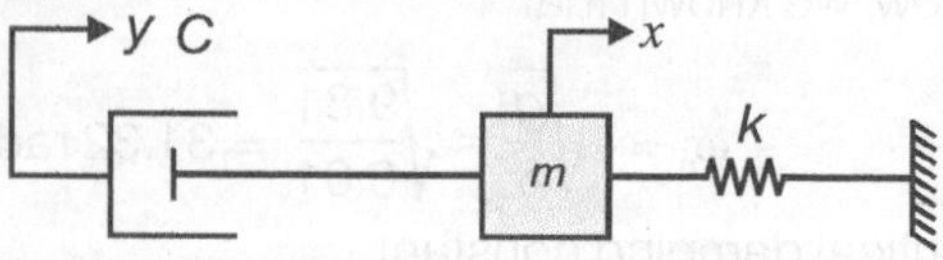

Differential equation governing the above vibration system is given by

$$\Rightarrow \frac{md^2x}{dt^2} + C\left(\frac{dx}{dt} - \frac{dy}{dt}\right) + kx = 0$$

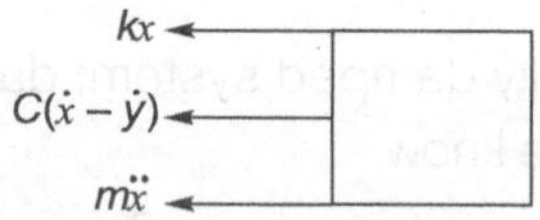

$$\Rightarrow m\ddot{x} + c(\dot{x} - \dot{y}) + kx = 0$$

7.16 (c)

$$\omega = \frac{2\pi \times 3600}{60} = 377 \text{ rad/s}$$

Natural frequency,

$$\omega_n = \sqrt{\frac{k}{m}} = \sqrt{\frac{100 \times 1000}{250}} = 20 \text{ rad/s}$$

Now, $r = \frac{\omega}{\omega_n} = \frac{377}{20} = 18.85$

Transmissibility ratio,

$$\text{TR} = \frac{\sqrt{1 + (2\xi r)^2}}{\sqrt{(1-r^2)^2 + (2\xi r)^2}}$$

$$= \frac{\sqrt{1 + (2 \times 0.15 \times 18.85)^2}}{\sqrt{1-(18.85)^2) + (2 \times 0.15 \times 18.85)^2}}$$

$$= 0.0162$$

7.17 (d)

Given : $m = 12.5$ kg, $k = 1000$ N/m, $c = 15$ Ns/m

Critical damping $= 2\sqrt{km} = 2\sqrt{1000 \times 12.5}$

$$= 223.6 \text{ Ns/m}$$

7.18 (*)

$$\xi = \frac{c}{2\sqrt{km}} = \frac{15}{223.6} = 0.0671$$

Logarithmic decrement

$$= \frac{2\pi\xi}{\sqrt{1-\xi^2}} = \frac{2\pi \times 0.0671}{\sqrt{(1-(0.0671)^2}}$$

$$= 0.422$$

7.19 (c)

Resonance is

$$\omega = \omega_n$$

7.20 (a)

Equivalent stiffness

$$\frac{1}{k_{eq}} = \frac{1}{\frac{k}{2} + \frac{k}{2}} + \frac{1}{k}$$

$$\therefore \quad k_{eq} = k/2$$

Now natural frequency

$$= \sqrt{\frac{k_{eq}}{m}} = \sqrt{\frac{k}{2m}}$$

7.21 (a)

$$X_1 = Xe^{-\zeta\omega_n T_d}$$

After n complete cycles,

$$X_n = Xe^{-\zeta\omega_n T_d \cdot n} \quad \text{...(i)}$$

We know that,

$$T_d = \frac{2\pi}{\omega_d} = \frac{2\pi}{(\sqrt{1-\zeta^2})\omega_n}$$

Now putting the value of T_d in equation (i)

$$\therefore \quad X_n = Xe^{-\zeta\omega_n \cdot n \frac{2\pi}{(\sqrt{1-\zeta^2})\omega_n}}$$

$$= Xe^{-\zeta \cdot n \frac{2\pi}{(\sqrt{1-\zeta^2})}}$$

7.22 (b)

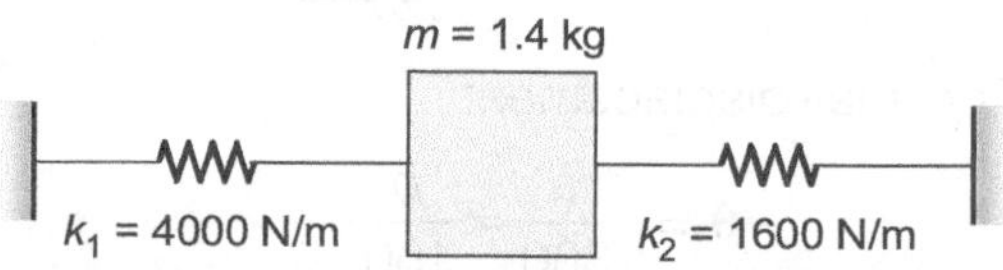

Springs are in parallel connection

$\therefore k_{eq} = k_1 + k_2 = 4000 + 1600 = 5600$ N/m

$$\omega = \sqrt{\frac{5600}{1.4}} = \sqrt{4000}$$

$$f = \frac{\omega}{2\pi} = \frac{\sqrt{4000}}{2\pi} = 10 \text{ Hz}$$

7.23 (c)

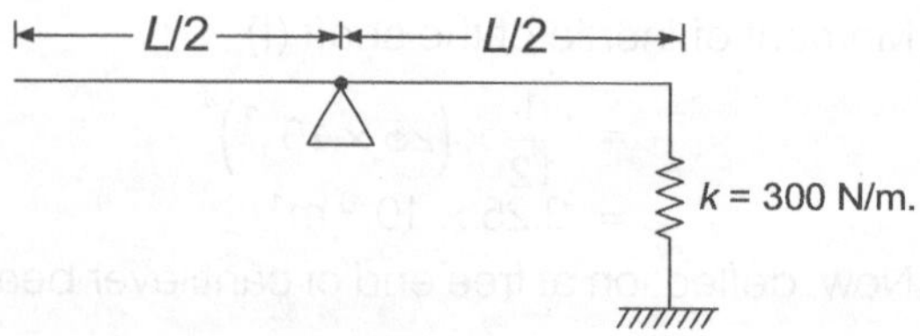

$$\frac{1}{2}I\dot{\theta}^2 + \frac{1}{2}k\,x^2 = \text{constant}$$

$$\frac{1}{2}I\dot{\theta}^2 + \frac{1}{2}k\left(\frac{L}{2}\theta\right)^2 = \text{constant}$$

Differentiating the whole expression w.r.t. t

$$\frac{1}{2}I\,2\dot{\theta}\ddot{\theta}+\frac{k}{2}\times\frac{L^2}{4}\times 2\theta\dot{\theta}=0$$

$$\Rightarrow I\ddot{\theta}+\frac{KL^2}{4}\theta=0$$

Putting the value of $I=\frac{mL^2}{12}$

$$\Rightarrow \frac{mL^2}{12}\ddot{\theta}+\frac{kL^2}{4}\theta=0 \Rightarrow \ddot{\theta}+\frac{3k}{m}\theta=0$$

$\therefore$ Natural frequency: $\omega_x=\sqrt{\frac{3k}{m}}$

$$=\sqrt{\frac{3\times300}{1}}=\textbf{30 rad/s}$$

7.24 (b)

Critical or whirling speed,

$$\omega_c=\omega_n=\sqrt{\frac{K}{m}}=\sqrt{\frac{g}{\delta}}\text{ Hz}$$

If N_c is the critical or whirling speed in rpm then

$$\frac{2\pi N_c}{60}=\sqrt{\frac{g}{\delta}}$$

$$\Rightarrow \frac{2\pi N_c}{60}=\sqrt{\frac{9.81\,m/s^2}{1.8\times10^{-3}\,m}}$$

$$\Rightarrow N_c=705.32\text{ rpm}\approx\textbf{705 rpm}$$

7.25 (a)

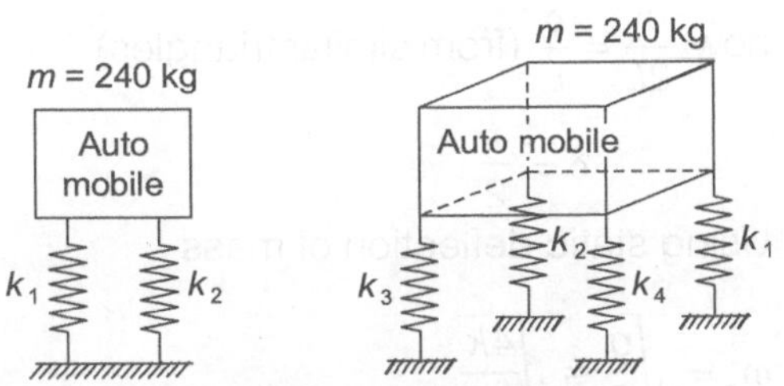

$k_1=k_2=16$ MN/m $=16\times10^6$ N/m

$k_3=k_4=32$ MN/m $=32\times10^6$ N/m

All the springs are in parallel combination. Therefore,

Equivalent stiffness,

$$k_{eq.}=k_1+k_2+k_3+k_4$$
$$=16+16+32+32$$
$$=96\text{ MN/m}$$

At resonance

$$\omega=\omega_n=\sqrt{\frac{k}{m}}$$

$$\frac{2\pi N}{60}=\sqrt{\frac{k}{m}}=\sqrt{\frac{96\times10^6\text{ N/m}}{240\text{ kg}}}$$

$$N=6042.5\text{ rpm}\approx\textbf{6040 rpm}$$

7.26 (a)

Stiffness of spring: $k=3.6$ kN/m

Damping constant of the damper

$$C=400\text{ Ns/m}$$

Mass, $m=50$ kg

$$m\ddot{x}+C\dot{x}+kx=0$$

[Suspension system consisting spring and damper]

Damping factor,

$$(d)\text{ or }\xi=\frac{C}{C_C}=\frac{C}{2\sqrt{km}}$$

$$=\frac{400\text{ N.s/m}}{2\sqrt{3.6\times10^3\times50\text{ N/m}\times\text{kg}}}$$

$$=\textbf{0.471}$$

Natural frequency,

$$\omega_n=\sqrt{\frac{k}{m}}=\sqrt{\frac{3.6\times10^3}{50}}$$

$$=8.485\text{ rad/s}$$

Damped natural frequency,

$$\omega_d=\omega_n\sqrt{1-\xi^2}=8.485\sqrt{1-(0.471)^2}$$

$$2\pi f_d=\omega_d=7.48\text{ rad/s}$$

$$f_d=\textbf{1.19 Hz}$$

7.27 (a)

As we know that, ω_n doesn't depend on gravity, it depends on the mass

$$\omega_n\propto\frac{1}{\sqrt{m}}$$

And again,

$$\omega_n=\sqrt{\frac{g}{\delta}}$$

Since it is spring mass-system,

$$\delta=\frac{mg}{k}$$

$$\omega_n=\sqrt{\frac{g}{\frac{mg}{k}}}=\sqrt{\frac{k}{m}}$$

So, ω_n will be constant every where.

7.28 (a)

Method I:

Given, $F(t) = 100\cos(100t)$ N

$F_0 = 100$ N

$\omega = 100$ rad/s

$k = 3000$ N/m

Amplitude: $X = 50 \times 10^{-3}$ m

Now by $X = \dfrac{F_0}{(k - m\omega^2)}$

as damping is zero.

$$k - m\omega^2 = \frac{100}{50 \times 10^{-3}} = 2000$$

$$m\omega^2 = 3000 - 2000$$

$$m = \frac{1000}{(100)^2} = \mathbf{0.1\ kg}$$

Method II:

$$\text{Amplitude, } A = \frac{\frac{F_0}{k}}{\left[1 - \left(\frac{\omega}{\omega_n}\right)^2\right]^2 + \left(\frac{2\xi\omega}{\omega_n}\right)^2}$$

For no damping,

$\xi = 0$

$$A = \frac{\frac{F_0}{k}}{\left[1 - \left(\frac{\omega}{\omega_n}\right)^2\right]^2} = \frac{\frac{F_0}{k}}{1 - \left(\frac{\omega}{\omega_n}\right)^2}$$

$$0.05 = \frac{\frac{100}{300}}{1 - \left(\frac{100}{\omega_n}\right)^2}$$

$\therefore$ $\omega_n = 100\sqrt{3}$

But, we know that,

$$\omega_n = \sqrt{\frac{k}{m}}$$

$\because$ $m = \dfrac{K}{\omega_n^2}$

$$= \frac{3000}{3 \times 100 \times 100} = 0.1 \text{ kg}$$

7.29 (a)

Since springs are in parallel hence

$k_{\text{equivalent}} = k + k = 2k$

$\therefore$ Frequency $= \dfrac{1}{2\pi}\sqrt{\dfrac{2k}{m}}$

$$= \frac{1}{2\pi}\sqrt{\frac{2 \times 20 \times 10^3}{1}} = \mathbf{31.83\ Hz}$$

7.30 (c)

$$I\ddot{\theta} + Kxr = 0$$

$$\left(\frac{mr^2}{2} + mr^2\right)\ddot{\theta} + K(r \cdot \theta)r = 0$$

$$\frac{3}{2}mr^2\ddot{\theta} + Kr^2\theta = 0$$

$$\ddot{\theta} + \frac{2K}{3m}\theta = 0$$

$$\text{Natural frequency} = \frac{1}{2\pi}\sqrt{\frac{2k}{3m}}$$

7.31 (d)

Method I:

Taking moments about O

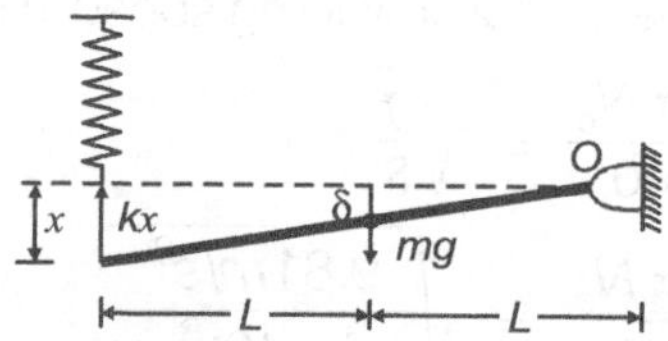

$kx \times 2L = mg \times L$

$$x = \frac{mg}{2k}$$

now $\dfrac{x}{2L} = \dfrac{\delta}{L}$ (from similar triangles)

$$\delta = \frac{x}{2}$$

Using static deflection of mass

$$\omega_n = \sqrt{\frac{g}{\delta}} = \sqrt{\frac{4k}{m}}$$

Method II:

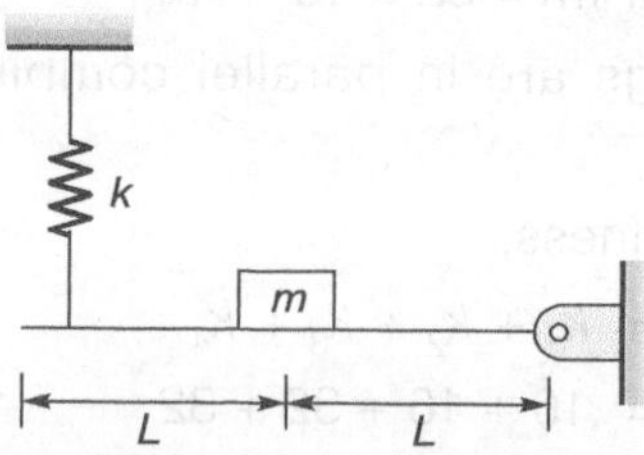

By equation of motion,

$$mL^2\ddot{\theta} + k(2L)^2\theta = 0$$

$$mL^2\ddot{\theta} + 4L^2k\theta = 0$$

$$\ddot{\theta} + \frac{4kL^2}{mL^2}\theta = 0$$

We know that,

$$\frac{4kL^2}{mL^2} = \frac{4k}{m}$$

$$\omega_n = \sqrt{\frac{4k}{m}} \text{ rad/s}$$

7.32 (b)

The system with critical damping returns to equilibrium in fastest time without any oscillation. In critically damped free vibrations, the damping force is just sufficient to dissipate the energy within one cycle of motion. The system never executes a cycle, it approaches equilibrium with exponentially decaying displacement.

7.33 (b)

$$\text{Transmissibility, } T = \frac{\sqrt{1+\left(\frac{2\xi\omega}{\omega_n}\right)^2}}{\sqrt{\left(1-\frac{\omega^2}{\omega_n^2}\right)+\left(\frac{2\xi\omega_n}{\omega_n}\right)^2}}$$

If $\quad \frac{\omega}{\omega_n} < \sqrt{2}$ "T" reduce as damping increases.

If $\quad \frac{\omega}{\omega_n} = \sqrt{2}$ then $T = 1$

If $\quad \frac{\omega}{\omega_n} < \sqrt{2}$

"T" increases with increased in damping but remains less than 1.

7.34 (c)

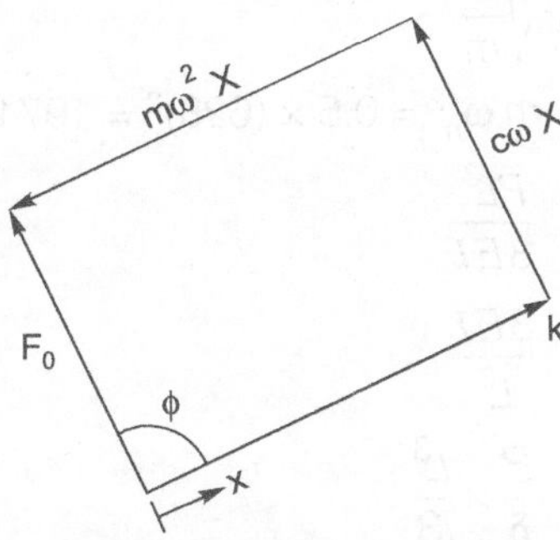

Phase diagram for $\frac{\omega}{\omega_n} = 1$

X : Displacement
F_0 : Excitation force
$c\omega X$: Damping force
kX : Spring force
$m\omega^2 X$: Inertia force

7.35 Sol.

$$f_n = \frac{\omega_n}{2\times\pi}$$

$$\Rightarrow \omega_n = 2\pi f = 2\times 3.14\times 4 = 25.12 \text{ rad/s}$$

$$a_{max} = \omega_n^2 x = (25.12)^2\times 10\times 10^{-3} = \mathbf{6.31\ m/s^2}$$

7.36 (d)

Method I:

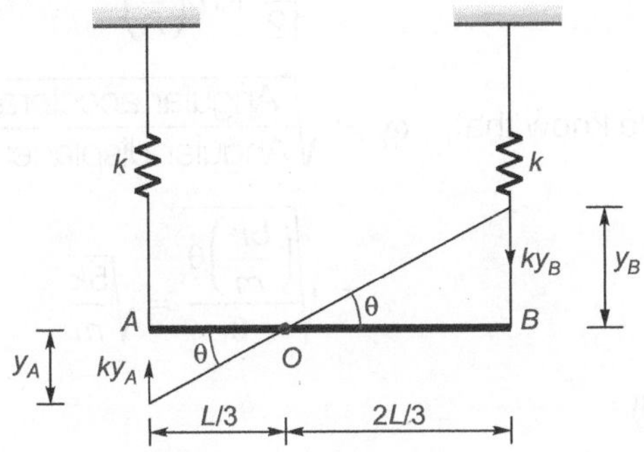

$$\theta = \frac{y_A}{\left(\frac{L}{3}\right)} = \frac{y_B}{\left(\frac{2L}{3}\right)}$$

$$I_0 = I_C + m\left(\frac{2L}{3}-\frac{L}{2}\right)^2$$

$$I_C = \frac{mL^2}{12}$$

$$\therefore \quad I_0 = \frac{mL^2}{12}+\frac{mL^2}{36} = \frac{mL^2}{9}$$

Taking, $\Sigma M_0 = 0$

$$\Rightarrow \left(ky_A\times\frac{L}{3}\right)+\left(ky_B\times\frac{2L}{3}\right)+I_0\ddot{\theta} = 0$$

$$\Rightarrow k\theta\left(\frac{L}{3}\right)^2 + k\theta\left(\frac{2L}{3}\right)^2 + \frac{mL^2}{9}\ddot{\theta} = 0$$

$$\Rightarrow \left(\frac{mL^2}{9}\right)\ddot{\theta}+\left(\frac{5kL^2}{9}\right)\theta = 0$$

$$\Rightarrow \ddot{\theta}+\left(\frac{5k}{m}\right)\theta = 0$$

$$\therefore \omega_n^2 = \frac{5k}{m}; \quad \omega_n = \sqrt{\frac{5k}{m}}$$

Method II:

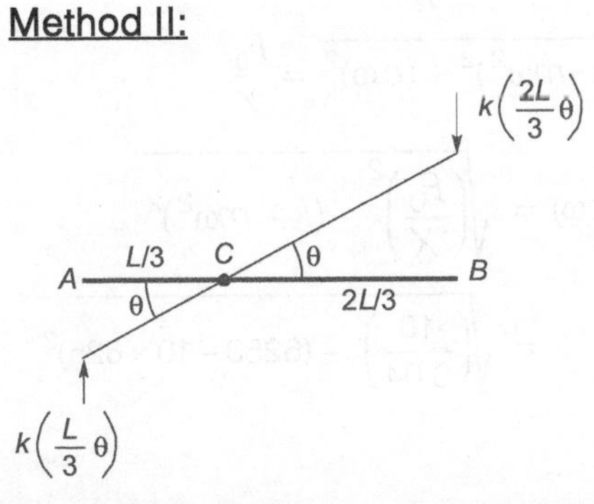

Restoring torque:

$$T = \left(k\frac{L}{3}\theta\right)\frac{L}{3} + \frac{2L}{3}\left(k\frac{2L}{3}\theta\right)$$

$$I_c\alpha = \left(\frac{5kL^2}{9}\right)\theta$$

$$\alpha = \frac{\left(\frac{5kL^2}{9}\right)0}{\frac{mL}{12} + m\left(\frac{L}{6}\right)^2} = \left(\frac{5k}{m}\right)\theta$$

We know that, $\omega_n = \sqrt{\dfrac{\text{Angular acceleration}}{\text{Angular displacement}}}$

$$= \sqrt{\frac{\left(\frac{5k}{m}\right)\theta}{\theta}} = \sqrt{\frac{5k}{m}}$$

7.37 (d)

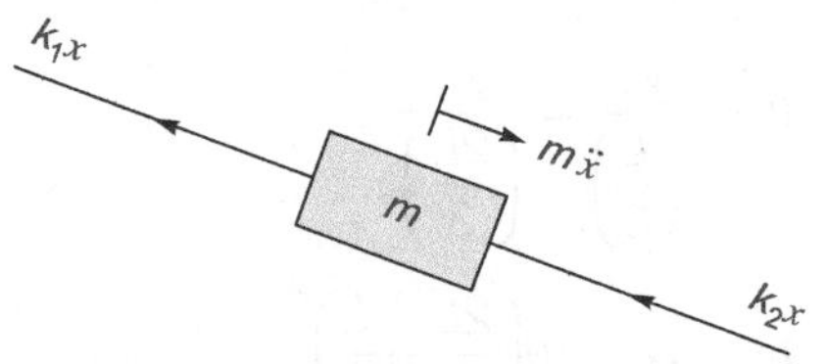

$$m\ddot{x} + (k_1 + k_2)x = 0$$

$$\ddot{x} + \left(\frac{k_1 + k_2}{m}\right)x = 0$$

$$\therefore \quad \omega_n = \sqrt{\frac{k_1 + k_2}{m}}$$

7.38 Sol.

Damping ratio,

$$\zeta = \frac{C}{2\sqrt{km}} = \frac{5}{2\sqrt{1 \times 100}} = \frac{5}{20} = 0.25$$

7.39 Sol.

Given, $m = 10$ kg, $F_0 = 10$ N,
$\omega = 25$ rad/s, $k = 6250$ N/m

$$X = \frac{F_0}{\sqrt{(k - m\omega^2)^2 + (c\omega)^2}}$$

or $\sqrt{(k - m\omega^2)^2 + (c\omega)^2} = \dfrac{F_0}{X}$

$$(c\omega) = \sqrt{\left(\frac{F_0}{X}\right)^2 - (k - m\omega^2)^2}$$

$$= \sqrt{\left(\frac{10}{0.04}\right)^2 - (6250 - 10 \times 625)^2}$$

$$c\omega = \frac{10}{0.04} = 250$$

$$\therefore \quad c = \frac{250}{\omega} = \frac{250}{25} = 10 \text{ Ns/m}$$

7.40 Sol.

$$MF = \frac{1}{\sqrt{\left[1 - \left(\frac{\omega}{\omega_n}\right)^2\right]^2 + \left[2\zeta\frac{\omega}{\omega_n}\right]^2}}$$

Given, $\omega = 1.5$ rad/s

$$\omega_n = \sqrt{\frac{k}{m}} = \sqrt{\frac{8}{2}} = 2 \text{ rad/s}$$

$$\zeta = 0.02, \frac{\omega}{\omega_n} = \frac{1.5}{2} = 0.75$$

$$\therefore \quad MF = \frac{1}{\sqrt{[1 - (0.75)^2]^2 + (2 \times 0.02 \times 0.75)^2}}$$

$$\therefore \quad MF = \frac{1}{\sqrt{(0.4375)^2 + (0.03)^2}} = \frac{1}{\sqrt{0.1923}}$$

$\therefore$ $MF =$ **2.28**

7.41 Sol.

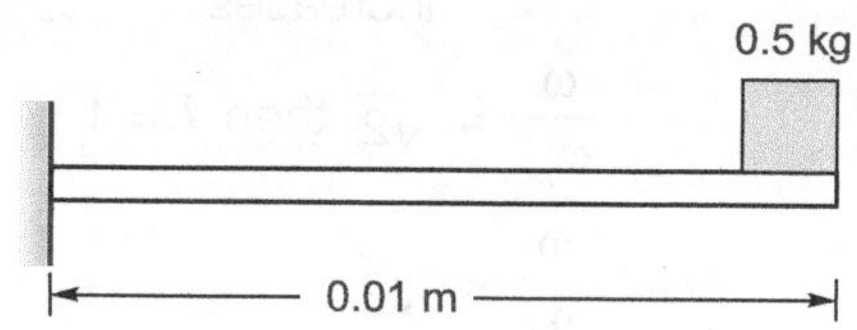

$P = mg = 0.5 \times 9.81 = 4.905$ N
$L = 0.01$ m, $f_n = 100$ Hz
$\omega_n = 2\pi f_n = 628$ rad/s

$$\omega_n = \sqrt{\frac{k}{m}}$$

or $k = m\omega_n^2 = 0.5 \times (628)^2 = 197192$ N/m

$$\delta = \frac{PL^3}{3EI}$$

$$\Rightarrow \quad \frac{P}{\delta} = \frac{3EI}{L^3}$$

or $EI = \dfrac{P}{\delta} \times \dfrac{L^3}{3}$

$$= 197192 \times \frac{(0.01)^3}{3} = \mathbf{0.06573 \text{ Nm}^2}$$

7.42 Sol.

$$\xi = \frac{C}{2\sqrt{mk}}; \; C = 2\xi\sqrt{mk}$$

$$\frac{C_2}{C_1} = \frac{2\xi_2\sqrt{m_2k_2}}{2\xi_1\sqrt{m_1k_1}}$$

$$\frac{C_2}{0.1} = \frac{2\times1\sqrt{2m\times8k}}{2\times1\sqrt{m\times k}}$$

$$C_2 = \mathbf{0.4\ kg/s}$$

POINTS TO REMEMBER

If $\xi > 1 \Rightarrow$ Overdamped system $\Rightarrow$ No vibration
If $\xi = 1 \Rightarrow$ Critically damped system $\Rightarrow$ No vibration.
If $\xi < 1 \Rightarrow$ Under damped system $\Rightarrow$ Vibration occur.

7.43 (c)

P, Q, R are right option.
S: for the viscous damping, drag force is directly proportional to the relative velocity.

$$f_D = C \cdot V$$

7.44 (d)

Method I:

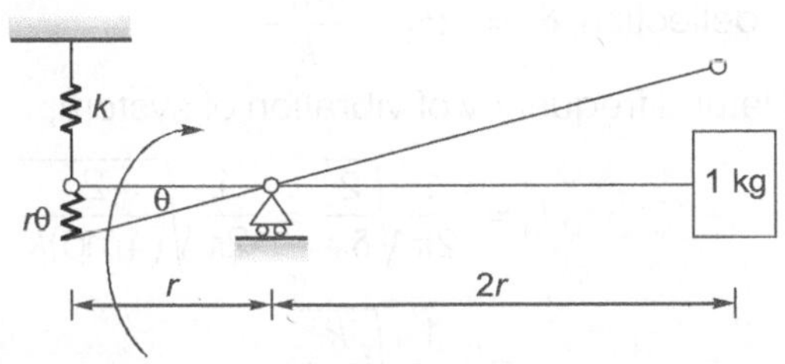

$Q \rightarrow$ Change in angular positions of rod in anticlockwise direction.
Restoring torque = $(kr\theta)r$ will act in clockwise direction
So,

$$\tau = -kr^2\theta$$
$$I\alpha = -kr^2\theta$$
$$\alpha = -\frac{kr^2}{I}\theta$$
$$\omega_n = \sqrt{\frac{kr^2}{I}}$$
$$I = m(2r)^2 = 1\,(4r^2)$$
$$\omega_n = \sqrt{\frac{400\times r^2}{1\times 4r^2}} = \sqrt{\frac{400}{4}} = \mathbf{10\ rad/s}$$

Method II:

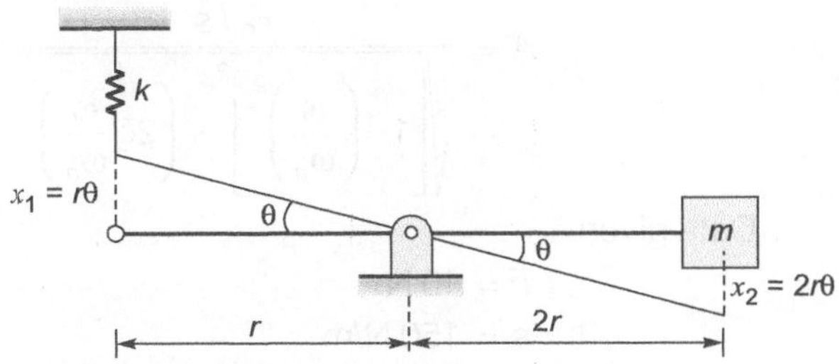

Given: $k = 400$, $m = 1$ kg
When the system is disturbed from its equilibrium position,

$$m\frac{d^2x_2}{dt^2}2r + kx_1r = 0$$
$$2mr\frac{d^2x_2}{dt^2} + k(r\theta)r = 0$$
$$2mr\frac{d^2(2r\theta)}{dt^2} + kr^2\theta = 0$$
$$4mr^2\frac{d^2\theta}{dt^2} + kr^2\theta = 0$$
$$\frac{d^2\theta}{dt^2} + \frac{kr^2}{4mr^2}\theta = 0$$

So, $$\omega_n = \sqrt{\frac{kr^2}{4mr^2}} = \sqrt{\frac{k}{4\,m}} = \sqrt{\frac{400}{4}}$$

7.45 Sol.

Given, $\omega = 2\pi f = 2\pi \times 60 = 120\pi$, $m = 1$ kg

$$\frac{F_{TR}}{F_0} = \frac{\sqrt{1+(2\xi r)^2}}{\sqrt{(1-r^2)^2+(2\xi r)^2}}$$

No damping,

$$\therefore \quad \xi = 0$$
$$\frac{F_{TR}}{F_0} = 0.05$$
$$0.05 = \frac{1}{r^2-1}$$
$$r^2 - 1 = 20$$
$$r = \sqrt{21}$$
$$\frac{\omega}{\omega_n} = \sqrt{21};\quad \frac{120\pi}{\omega_n} = \sqrt{21}$$
$$\omega_n = 82.266 \text{ rad/s}$$
$$\sqrt{\frac{k}{m}} = 82.266$$
$$\sqrt{\frac{k}{1}} = 6767.728$$
$$k = \mathbf{6767.72\ N/m}$$

7.46 (b)

$$A = \frac{F_0/s}{\sqrt{\left[1-\left(\frac{\omega}{\omega_n}\right)^2\right]^2 + \left(2\xi\frac{\omega}{\omega_n}\right)^2}}$$

Data given

$$F = 10\text{ N}$$
$$s = 150\text{ N/m}$$
$$\frac{\omega}{\omega_n} = \frac{\omega}{10\omega} = 0.1$$
$$\varepsilon = 0.2$$

$$\Rightarrow A = \frac{10/150}{\sqrt{\left[1-(0.1)^2\right]^2 + [2\times 0.2\times 0.1]^2}} = \mathbf{0.07\ m}$$

7.47 (a)

Method I:

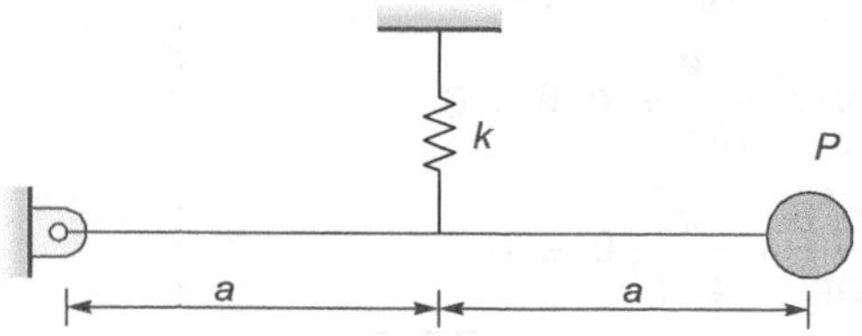

$\therefore$ moving mass m by angle θ

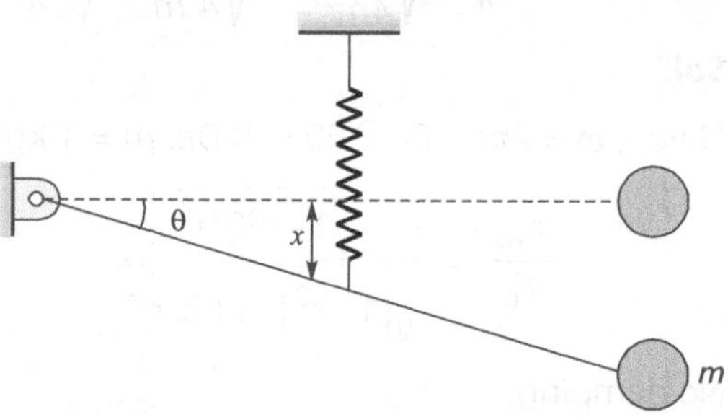

$\therefore$ restoring toque = $kx \times a = k \times (a\theta) \times a$

Restoring torque = inertia torque

$$k(a^2\theta) = -I\alpha$$
$$k(a^2\theta) = -m(2a)^2\ddot{\theta}$$
$$k\theta = -m\times 4a^2\ddot{\theta}$$
$$4\ddot{\theta} + k\theta = 0$$
$$\ddot{\theta} + \frac{k}{4m}\theta = 0$$

$\therefore$ $$\omega_n = \sqrt{\frac{k}{4m}};\quad \frac{2\pi}{T} = \sqrt{\frac{k}{4m}}$$

$$T = 2\pi\sqrt{\frac{4m}{k}}$$

$\therefore$ $$f_n = \frac{1}{T} = \frac{1}{2\pi}\sqrt{\frac{k}{4m}}$$

Method II:

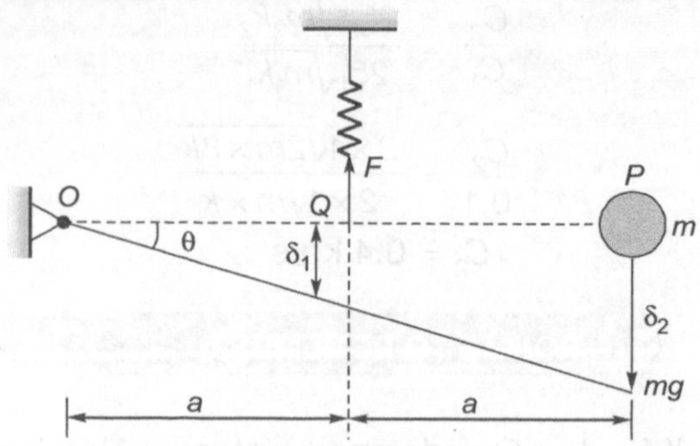

$\therefore$ rod is rigid, so it will not deform, for the triangle similarity.

$$\frac{\delta_1}{a} = \frac{\delta_2}{2a}$$
$$\frac{\delta_1}{\delta_2} = \frac{1}{2}$$
$$\delta_2 = 2\delta_1$$

Taking moment about point O,

$F\times a - mg\times 2a = 0$

$$F = 2\text{ mg}$$

Now,

$$\text{Deflection, } \delta_1 = \frac{F}{k} = \frac{2mg}{k}$$

and

$$\text{deflection, } \delta_2 = 2\delta_1 = \frac{4mg}{k}$$

Natural frequency of vibration of system,

$$f_n = \frac{1}{2\pi}\sqrt{\frac{2}{\delta_2}} = \frac{1}{2\pi}\sqrt{\frac{2}{(4\text{mg})k}}$$
$$= \frac{1}{2\pi}\sqrt{\frac{k}{4\text{ m}}}$$

7.48 Sol.

$$\frac{A}{x_0} = \text{Magnification factor}$$

$$\frac{A}{F_o/k} = \frac{1}{\sqrt{\left(1-\left(\frac{\omega}{w_n}\right)^2\right)^2 + \left(2\zeta\frac{\omega}{\omega_n}\right)^2}}$$

$\because$ $$\zeta = 0$$

$$\frac{A}{F_o/k} = \frac{1}{\left(\frac{\omega}{\omega_n}\right)^2 - 1}$$

$$\omega = \sqrt{\frac{3k}{m}},\quad \omega_n = \sqrt{\frac{k}{m}}$$

$$\frac{\omega}{\omega_n} = \sqrt{\frac{3k}{m}} \times \sqrt{\frac{m}{k}} = \sqrt{3}$$

$$\frac{A}{F_o/k} = \frac{1}{(\sqrt{3})^2 - 1} = \frac{1}{2}$$

7.49 (b)

Given :

1. Single degree of freedom
2. Mass = m
3. Spring constant = k
4. Damping coefficient = q
5. Critically damped

Damping factor = 1

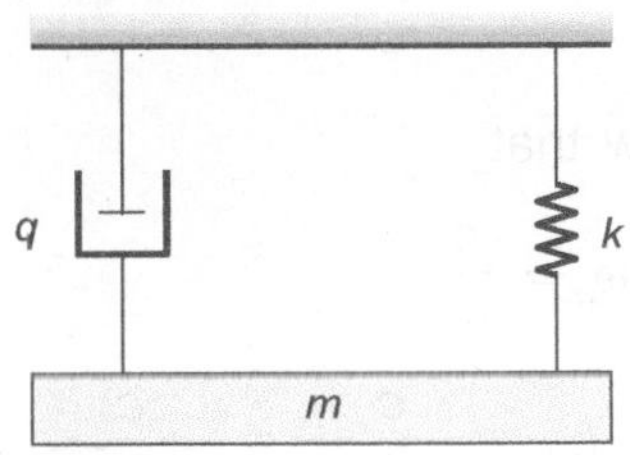

Correct relation:

Damping factor=

$$\frac{\text{Damping coefficient}}{2 \times \text{Mass} \times \text{Natural frequency of vibration}}$$

$$1 = \frac{q}{2 \times m \times \sqrt{\frac{k}{m}}}$$

$$q = 2\sqrt{mk}$$

7.50 Sol.

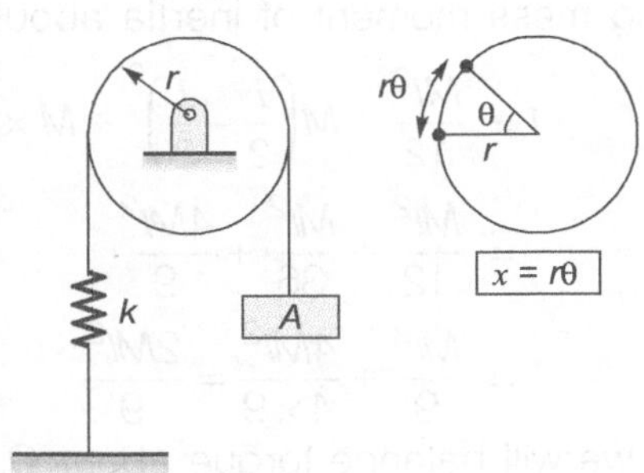

Total energy of system

$$= \frac{1}{2}mv^2 + \frac{1}{2}kx^2 + \frac{1}{2}I\omega^2$$

$$= \frac{1}{2}m\left(r \times \frac{d\theta}{dt}\right)^2 + \frac{1}{2}k \times (r\theta)^2 + \frac{1}{2} \times \frac{mr^2}{2} \times \left[\frac{d\theta}{dt}\right]^2$$

$$= \frac{1}{2}mr^2 \cdot (\dot{\theta})^2 + \frac{1}{2}k \times r^2\theta^2 + \frac{mr^2}{4}(\dot{\theta})^2$$

Since total energy remains constant with respect to time.

$$\frac{dE}{dt} = 0$$

$$\frac{dE}{dt} = \frac{1}{2}mr^2 2 \cdot \dot{\theta}\ddot{\theta} + \frac{kr^2}{2} \cdot 2\theta \cdot \dot{\theta} + \frac{Mr^2}{4} 2\dot{\theta}\ddot{\theta}$$

$$0 = mr^2\dot{\theta}\ddot{\theta} + kr^2\theta \cdot \dot{\theta} + \frac{Mr^2}{2}\dot{\theta}\ddot{\theta}$$

Divide both sides by $\dot{\theta}$

$$0 = mr^2\ddot{\theta} + kr^2\theta + \frac{Mr^2}{2}\ddot{\theta}$$

Divide both sides by r^2

$$0 = m\ddot{\theta} + k\theta + \frac{M\ddot{\theta}}{2}$$

$$0 = \left(m + \frac{M}{2}\right)\ddot{\theta} + k\theta$$

$$0 = \ddot{\theta} + \left[\frac{k}{m + \frac{M}{2}}\right]\theta$$

$$\omega_n^2 = \frac{k}{m + \frac{M}{2}}$$

$$\omega_n = \sqrt{\frac{2k}{M + 2m}} = \sqrt{\frac{2 \times 1500}{20 + (2 \times 5)}}$$

$$= \sqrt{\frac{3000}{20 + 10}} = \sqrt{100} = \mathbf{10\ rad/s}$$

7.51 Sol.

Forcing function = $F_0 \sin \omega t$

Given, F_0 = 100 N

Damping factor,

$$\xi = 0.25$$

Spring constant,

$$k = 10\text{ kN/m} = 10 \times 1000 = 10000\text{ N/m}$$

$$A = \frac{F_0 / k}{\sqrt{\left[\left\{1 - \left(\frac{\omega}{\omega_n}\right)^2\right\}^2 + \left(2\xi\frac{\omega}{\omega_n}\right)^2\right]}}$$

At resonance

$$\omega = \omega_n$$

$$\therefore \quad A = \frac{\frac{100}{10000}}{2\xi} = \frac{0.01}{2 \times 0.25} = 0.02\text{ m}$$

$$= \mathbf{20\ mm}$$

7.52 (a)

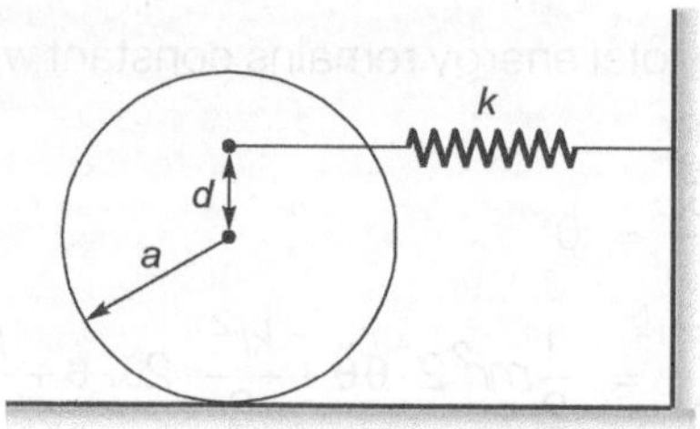

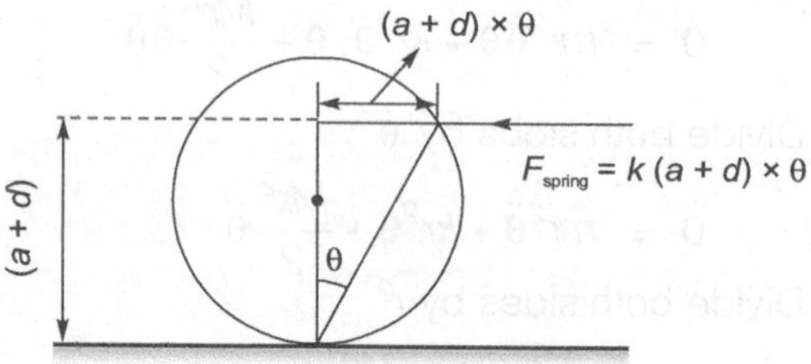

Total energy of system

$$\Rightarrow \quad \frac{1}{2}Mv^2 + \frac{1}{2}kx^2 + \frac{1}{2}I\omega^2$$

$$\Rightarrow \quad \frac{1}{2}M\times\left(a\times\frac{d\theta}{dt}\right)^2 + \frac{1}{2}k\times \{(a+d)\theta\}^2 + \frac{1}{2}\times\frac{Ma^2}{2}\times\left(\frac{d\theta}{dt}\right)^2$$

Let $\frac{d\theta}{dt} = \dot{\theta}$

$$E = \frac{1}{2}Ma^2\dot{\theta}^2 + \frac{k}{2}(a+d)^2\cdot\theta^2 + \frac{Ma^2}{4}\dot{\theta}^2$$

Since energy of system remains concerned with respect to time.

$$\frac{dE}{dt} = 0$$

$$\frac{dE}{dt} = \frac{Ma^2}{2}2\cdot\dot{\theta}\cdot\ddot{\theta} + \frac{k}{2}(a+d)^2 \cdot 2\theta\cdot\dot{\theta} + \frac{Ma^2}{4}\cdot 2\dot{\theta}\ddot{\theta}$$

$$0 = Ma^2\dot{\theta}\ddot{\theta} + \frac{Ma^2}{2}\cdot\dot{\theta}\,\ddot{\theta} + k(a+d)^2\theta\cdot\dot{\theta}$$

Divide both sides by $\dot{\theta}$

$$0 = Ma^2\ddot{\theta} + \frac{Ma^2}{2}\cdot\ddot{\theta} + k(a+d)^2\cdot\theta$$

$$0 = \left(Ma^2 + \frac{Ma^2}{2}\right)\ddot{\theta} + k(a+d)^2\theta$$

$$0 = \frac{3}{2}Ma^2\cdot\ddot{\theta} + k(a+d)^2\theta$$

Divide both sides by $\frac{3}{2}ma^2$

$$0 = \ddot{\theta} + \frac{k(a+d)^2}{\frac{3Ma^2}{2}}\theta$$

$$\ddot{\theta} + \omega_n^2\theta = 0$$

Comparing above two equation

$$\omega_n^2 = \frac{2k(a+d)^2}{3Ma^2};\ \omega_n = \sqrt{\frac{2k(a+d)^2}{3Ma^2}}$$

7.53 Sol.

$g = 10\ \text{m/s}^2$, $m = 1$ kg

Deflection: $\Delta = 1$ mm

$$\omega_n = \sqrt{\frac{g}{\Delta}} = \sqrt{\frac{10}{1\times10^{-3}}} = \textbf{100 rad/s}$$

7.54 (b)

We know that:

$$2\xi\omega_n = \frac{c}{m}$$

$$\xi = \frac{c}{m\times 2\omega_n} = \frac{c}{2m\sqrt{\frac{k}{m}}} = \frac{c}{2\sqrt{km}}$$

7.55 (a)

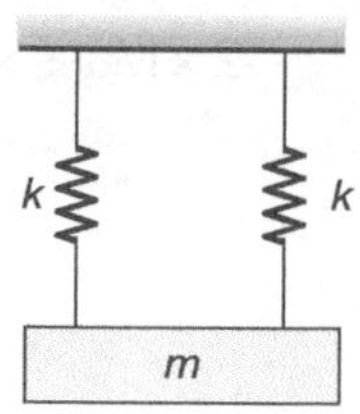

Equivalent stiffness,

$$k_{eq} = k + k = 2k$$

Natural frequency,

$$\omega_n = \sqrt{\frac{k_{eq}}{m}} = \sqrt{\frac{2k}{m}}$$

7.56 (b)

Taking mass moment of inertia about point O,

$$I = \frac{Ml^2}{12} + M\left(\frac{l}{2} - \frac{l}{3}\right)^2 + M\times\left(\frac{2l}{3}\right)^2$$

$$= \frac{Ml^2}{12} + \frac{Ml^2}{36} + \frac{4Ml^2}{9}$$

$$= \frac{Ml^2}{9} + \frac{4Ml^2}{4\times9} = \frac{2Ml^2}{9}$$

Now we will balance torque about O,

$$I\alpha = k\times\frac{2l}{3}\times\left(\frac{2l}{3}\theta\right) + K\times\frac{l}{3}\times\left(\frac{l}{3}\theta\right)$$

$$I\alpha = \frac{5kl^2}{9}\theta$$

$$\frac{2Ml^2}{9}\frac{d^2\theta}{dt^2} = \frac{5kl^2}{9}\theta$$

$$\Rightarrow \quad \frac{d^2\theta}{dt^2} = \frac{5kl^2}{2ml^2}$$

$$\omega_n^2\theta = \frac{5k}{2M}$$

$$\therefore \quad \omega_n = \sqrt{\frac{5k}{2M}}$$

7.57 Sol.

If we displace this by an angle θ clockwise:

Compound pendulum

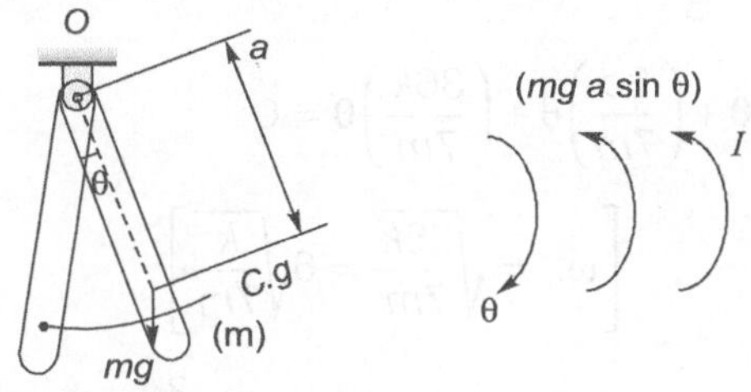

$$I\ddot{\theta} + mg.a\sin\theta = 0$$

Taking θ very small:

$\sin\theta \approx \theta$ (k = 100 mm = 0.1 m)

$$I\ddot{\theta} + mg.a.\theta = 0$$

$$\ddot{\theta} + \left(\frac{mg}{I}a\right)\theta = 0$$

Here $I = mk^2 = m(0.1)^2$

(I is the MI about the point of hinge)

$I = m(0.01)$ kg-m²

a = 250 mm = 0.250 m

g = 9.81 m/s²

$$\ddot{\theta} + \left(\frac{m \times 9.81 \times 0.250}{(0.01)m}\right)\theta = 0$$

$$\ddot{\theta} + (245.25)\theta = 0$$

$$\omega_n = \sqrt{245.25} = \mathbf{15.660\ rad/s}$$

7.59 Sol.

m = 200 kg, k = 10 kN/m = 10000 N/m

$k_{eq} = k + k$ = 10000 + 10000 = 20000 N/m

$F_t = 50\cos 5t$

F_0 = 50 N

ω = 5 rad/s

$$\omega_n = \sqrt{\frac{k_{eq}}{m}} = \sqrt{\frac{20000}{200}} = \sqrt{100}$$

= 10 rad/s

$$\frac{\omega}{\omega_n} = \frac{5}{10} = \frac{1}{2}$$

No damping ⇒ $c = 0 \Rightarrow \xi = 0$

$$\text{Transmissibility, } \epsilon = \frac{\sqrt{1+\left(\frac{2\xi\omega}{\omega_n}\right)^2}}{\sqrt{\left\{1-\left(\frac{\omega}{\omega_n}\right)^2\right\}^2 + \left\{\frac{2\xi\omega}{\omega_n}\right\}^2}}$$

(ξ = 0)

$$= \frac{1}{\sqrt{\left\{1-\left(\frac{\omega}{\omega_n}\right)^2\right\}^2}} = \frac{1}{\sqrt{\left\{1-\left(\frac{1}{2}\right)^2\right\}^2}}$$

$$= \frac{1}{1-\frac{1}{4}} = \frac{1}{\frac{3}{4}} = \left(\frac{4}{3}\right)$$

$$\epsilon = \frac{F_t}{F_0}$$

$$\frac{4}{3} = \frac{F_t}{50}$$

$$F_t = \frac{200}{3} = 66.666 \text{ N}$$

Force from each mount

$$= \frac{F_t}{2} = \frac{66.666}{2} = 33.333 \text{ N}$$

7.60 (c)

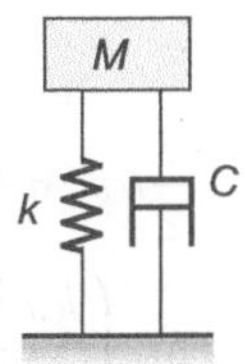

Additional damper is added into parallel.

Then damping will increase.

But it is still under damped.

But ξ will increase.

Here no unbalance force is there.

But $\omega_d = \sqrt{1-\xi^2}\cdot\omega_n$ $\left\{\omega_n = \sqrt{\frac{k}{M}}\right\}$

But as ξ increases

then ξ^2 will increase

Then ω_d will decrease

Then $T_d = \frac{2\pi}{\omega_d}$ **will increase**

7.61 (b)

$$\frac{1}{k_{eg}} = \frac{1}{k} + \frac{1}{k} = \frac{2}{k}$$

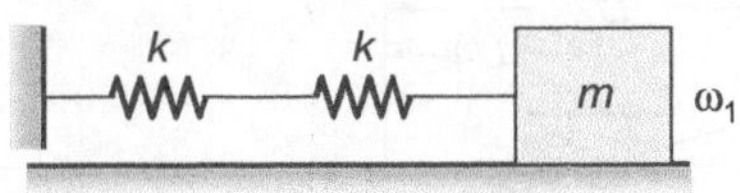

$$k_{eq} = \frac{k}{2}$$

$$\Rightarrow \quad \omega_1 = \sqrt{\frac{(k/2)}{m}} = \sqrt{\frac{k}{2m}}$$

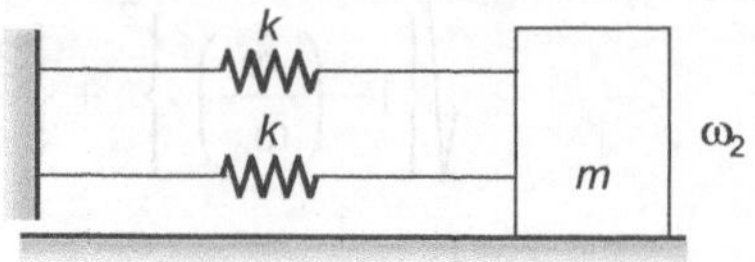

$$\omega_2 = \sqrt{\frac{2k}{m}}$$

$$\frac{\omega_1}{\omega_2} = \sqrt{\frac{k/2m}{2k/m}} = \sqrt{\frac{1}{4}} = \frac{1}{2}$$

$$\frac{\omega_1}{\omega_2} = \frac{1}{2}$$

7.62 Sol.

As per given data

Disc mass,

$$m = 1 \text{ kg}$$

Radius, $r = 0.1$m

torque equation about point 'o'

$$I\ddot{\theta} + (k_1 \times r \times \theta)r + (k_2 \times 2r \times \theta)2r = 0$$

$$I \text{ about 'o'} = \frac{mr^2}{2} + mr^2$$

$$I = \frac{3}{2}mr^2$$

$$\frac{3}{2}mr^2\ddot{\theta} + \left(k_1r^2 + k_2\left(4r^2\right)\right)\theta = 0$$

$$\ddot{\theta} + \left(\frac{k_1r^2 + 4k_2r^2}{\frac{3}{2}mr^2}\right)\theta = 0$$

$$\omega_n = \sqrt{\frac{k_1r^2 + 4k_2r^2}{\frac{3}{2}mr^2}}$$

$$= \sqrt{\frac{400(0.2)^2 + 4 \times 100 \times 0.2^2}{1.5 \times 1 \times 0.2^2}}$$

$$\omega_n = 23.094 \text{ rad/s}$$

7.63 (a)

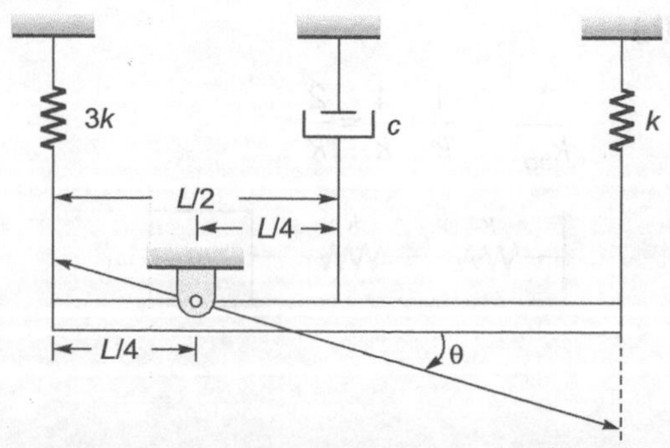

$$I = \frac{ml^2}{12} + m\left(\frac{l}{4}\right)^2 = \frac{ml^2}{12} + \frac{ml^2}{16} = \frac{7ml^2}{48}$$

Total torque about hinge 'O

$$I\ddot{\theta} + c\left(\frac{l}{4}\right)^2\dot{\theta} + \left[k\frac{9l^2}{16} + 3k\frac{l^2}{16}\right]\theta = 0$$

$$\frac{7ml^2}{48}\ddot{\theta} + \frac{cl^2}{16}\dot{\theta} + \frac{12kl^2}{16}\theta = 0$$

$$\frac{7ml^2}{3}\ddot{\theta} + \left(cl^2\right)\dot{\theta} + \left(12kl^2\right)\theta = 0$$

$$\ddot{\theta} + \left(\frac{3c}{7m}\right)\dot{\theta} + \left(\frac{36k}{7m}\right)\theta = 0$$

$$\left[\omega_n = \sqrt{\frac{36k}{7m}} = 6\sqrt{\frac{k}{7m}}\right]$$

$$2\xi\omega_n = \frac{3c}{7m}$$

For critical damping $\xi = 1$

$$\Rightarrow \quad 2\omega_n = \frac{3c}{7m}$$

$$\Rightarrow \quad 2 \times 6\sqrt{\frac{k}{7m}} = \frac{3c}{7m}$$

$$\Rightarrow \quad 12 = \frac{3c}{\sqrt{km}\sqrt{7}}$$

$$\Rightarrow \quad \frac{c}{\sqrt{km}} = 4\sqrt{7}$$

7.64 (b)

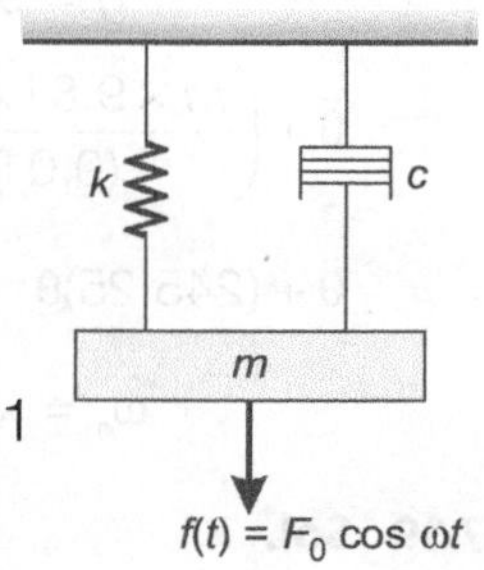

Given,

$$F_T = F_0$$

Then transmissibility

$$\epsilon = \frac{F_T}{F_0} = \frac{F_0}{F_0} = 1$$

$$\frac{\sqrt{1 + \left(\frac{2\xi\omega}{\omega_n}\right)^2}}{\sqrt{\left(1 - \left(\frac{\omega}{\omega_n}\right)^2\right)^2 + \left(\frac{2\xi\omega}{\omega_n}\right)^2}} = 1$$

$$1 + \left(\frac{2\xi\omega}{\omega_n}\right)^2 = \left(1 - \left(\frac{\omega}{\omega_n}\right)^2\right)^2 + \left(\frac{2\xi\omega}{\omega_n}\right)^2$$

$$1 - \left(\frac{\omega}{\omega_n}\right)^2 = \pm 1$$

Taking (–ve) sign

$$\Rightarrow 1-\left(\frac{\omega}{\omega_n}\right)^2 = -1$$

$$\Rightarrow \left(\frac{\omega}{\omega_n}\right)^2 = \sqrt{2}$$

$$\Rightarrow \frac{\omega}{\omega_n} = \sqrt{2} = \sqrt{2}\omega_n = \sqrt{\frac{2k}{m}}$$

7.65 (c)

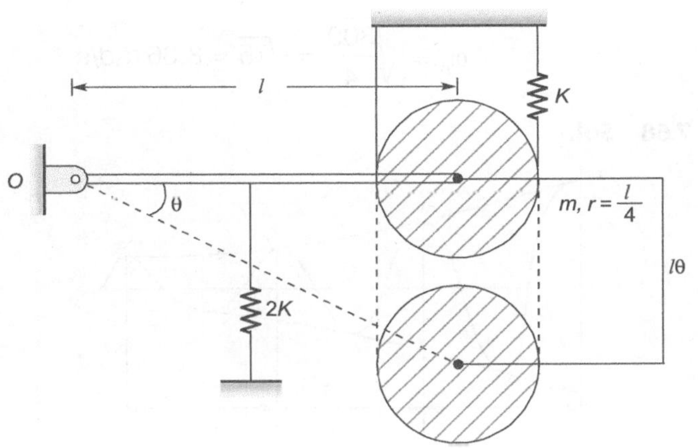

MI about point O:

$$I = \frac{mr^2}{2} + ml^2$$

$$= \frac{m}{2}\left(\frac{l}{4}\right)^2 + ml^2 = \frac{33ml^2}{32}$$

MI about disc centre:

$$I_{disc} = \frac{mr^2}{2}$$

When rod rotates by β,

$$\beta \cdot r = l \cdot \theta$$

$$\beta = \frac{l \cdot \theta}{r}$$

(If disc is also rotating about its own centre due to static friction).

Energy Method:

E = (Rotational KE)$_{\text{system about } O}$
+ (Rotational KE)$_{\text{disc about its own centre}}$
+ (PE of spring of 2K)
+ (PE of spring of K)

$$E = \frac{1}{2}I(\dot{\theta})^2 + \frac{1}{2}I_{disc}(\dot{\beta})^2 + \frac{1}{2}(2K)\times\left(\frac{l}{2}\cdot\theta\right)^2 + \frac{1}{2}K(2l\cdot\theta)^2$$

$$= \frac{1}{2}\cdot\frac{33ml^2}{32}\dot{\theta}^2 + \frac{1}{2}\cdot\frac{mr^2}{2}\left(\frac{l^2}{r^2}\right)\dot{\theta}^2 + \frac{1}{2}\cdot 2K\cdot\frac{l^2}{4}\cdot\theta^2 + \frac{1}{2}K\cdot 4l^2\theta^2$$

$$E = \frac{1}{2}\left(\frac{33ml^2}{32} + \frac{ml^2}{2}\right)\dot{\theta}^2 + \frac{1}{2}\left(\frac{Kl^2}{2} + 4Kl^2\right)\theta^2$$

$$= \frac{1}{2}\left(\frac{49ml^2}{32}\right)\dot{\theta}^2 + \frac{1}{2}\left(\frac{9Kl^2}{2}\right)\theta^2$$

$$\frac{dE}{dt} = 0$$

$$\frac{1}{2}\left[\frac{49ml^2}{32}\cdot 2\dot{\theta}\ddot{\theta} + \frac{9Kl^2}{2}2\cdot\theta\cdot\dot{\theta}\right] = 0$$

$$\frac{49ml^2}{16}\ddot{\theta} + 9Kl^2\theta = 0$$

$$\frac{49m}{16}\ddot{\theta} + 9K\cdot\theta = 0$$

$$\ddot{\theta} + \frac{9K\times 16}{49m}\cdot\theta = 0$$

$$\omega_n = \frac{12}{7}\sqrt{\frac{k}{m}}$$

This is exact solution.

But this solution is very close to $\sqrt{3}\sqrt{\frac{k}{m}}$

Because, $\frac{12}{7}\sqrt{\frac{k}{m}} = (1.714)\sqrt{\frac{k}{m}}$

$$\sqrt{3}\sqrt{\frac{k}{m}} = (1.732)\sqrt{\frac{k}{m}}$$

Therefore, $\omega_n = \frac{12}{7}\sqrt{\frac{k}{m}} = \sqrt{3}\sqrt{\frac{k}{m}}$

Note: If we take an approximation in moment of inertia about hinge axis.

That, $I = \frac{mr^2}{2} + ml^2 \quad (r = l/4)$

$$= \left(\frac{ml^2}{32} + ml^2\right)$$

If we neglect $\frac{ml^2}{32}$ because $\frac{ml^2}{32}$ <<<< ml^2

If we take, $I = ml^2$

Then we get, $\omega_n = \sqrt{3}\sqrt{\frac{k}{m}}$

But if we take exact inertia,

$$I = \left(\frac{ml^2}{32} + ml^2\right) = \frac{33ml^2}{32}$$

Then exact answer is $\omega_n = \frac{12}{7}\sqrt{\frac{k}{m}}$.

7.66 (c)

$$\frac{d^2x}{dt^2}+3\left(\frac{dx}{dt}\right)+9x = 10 \sin 5t$$

Comparing with standard equation:

$$\ddot{x}+(2\xi\omega_n)\dot{x}+(\omega_n^2)x = \left(\frac{F_o}{m}\right)\sin\omega t$$

$$2\xi\omega_n = 3 \qquad \begin{bmatrix}\omega_n^2 = 9\\ \omega_n = 3\end{bmatrix}$$

$$2\xi \times 3 = 3$$

$$\xi = \frac{1}{2} = 0.5$$

7.67 Sol.

Given data: R = 20 cm = 0.20 m

ρ = 1000 kg/m^3

g = 10 m/s^2

$$\left[\forall = \text{Volume displaced} = \frac{V_s}{2}\right]$$

$$mg = F_B = \rho\forall\cdot g$$

$$m = \rho\cdot\frac{V_s}{2}$$

$$V_s\rho_s = \rho\cdot\frac{V_s}{2}$$

$$\rho_s = \frac{\rho}{2} = \frac{1000}{2} = 500 \text{ kg/m}^3$$

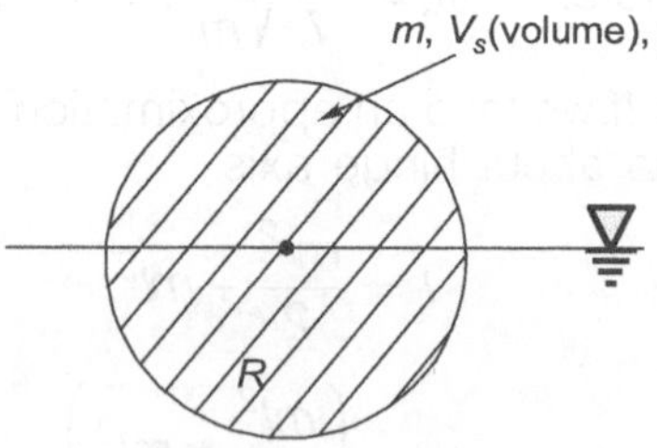

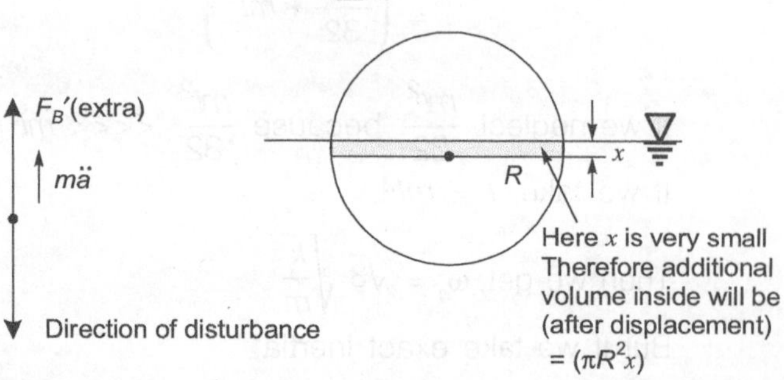

$$m\ddot{x}+f'_{B\,\text{extra}} = 0$$

$$m\ddot{x}+V_{\text{extra}}\cdot\rho\cdot g = 0$$

$$\rho\cdot\frac{V_s}{2}\ddot{x}+(\pi R^2\cdot x)\rho\cdot g = 0$$

$$\frac{4}{3}\frac{\pi R^3}{2}\hat{x}+\pi R^2 xg = 0$$

$$\ddot{x}+\frac{3g}{2R}x = 0$$

$$\omega_n = \sqrt{\frac{3g}{2R}} = \sqrt{\frac{3\times 10}{2\times 0.20}} = \sqrt{\frac{30}{0.4}}$$

$$\omega_n = \sqrt{\frac{300}{4}} = \sqrt{75} = 8.66 \text{ rad/s}$$

7.68 Sol.

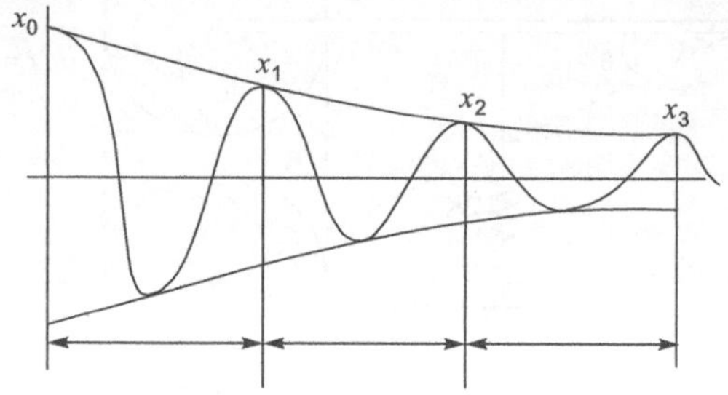

$$\frac{x_0}{x_3} = \frac{8}{1.5} = \frac{80}{15}$$

$$\frac{x_0}{x_1}\cdot\frac{x_1}{x_2}\cdot\frac{x_2}{x_3} = \frac{80}{15}$$

$$e^{\delta}\cdot e^{\delta}\cdot e^{\delta} = \frac{80}{15}$$

$$e^{3\delta} = \ln\left(\frac{80}{15}\right) = 1.67397$$

$$\delta = 0.55799$$

$$\frac{2\pi\xi}{\sqrt{1-\xi^2}} = 0.55799$$

$$\xi = 0.088$$

7.69 (d)

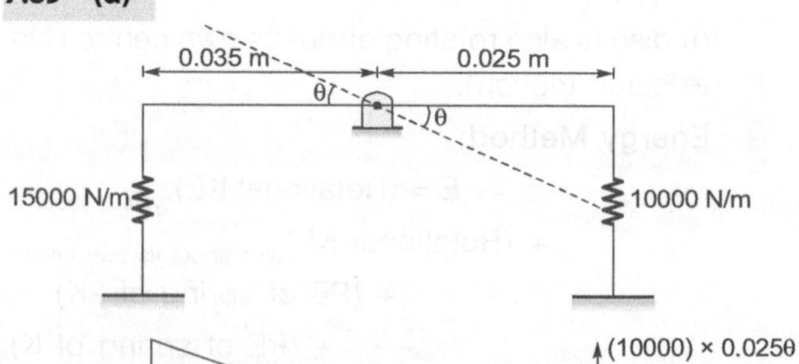

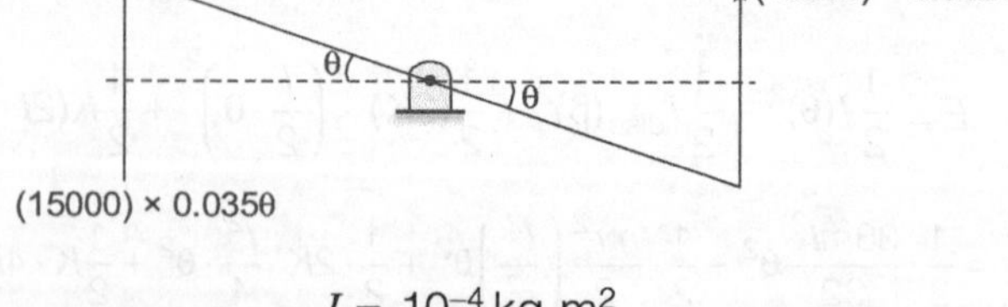

$$I = 10^{-4}\text{ kg-m}^2$$

By D'Alembert Principle

$$I\ddot{\theta} + [10000 \times (0.025)^2 + 15000 \times (0.035)^2]\theta = 0$$

θ

$(10000)(0.025)^2\theta$

$$(10^{-4})\ddot{\theta} + (24.625)\theta = 0$$

$(15000)(0.035)^2\theta$

$$\ddot{\theta} + \left(\frac{24.625}{10^{-4}}\right)\theta = 0 \qquad I\ddot{\theta}$$

$$\Rightarrow \quad \omega_n^2 = (246250)$$

$$\omega_n = 496.2358 \text{ rad/s}$$

$$\Rightarrow N_C = \frac{496.2358 \times 60}{2\pi} = 4738.7031 \text{ r.p.m.}$$

7.70 (a)

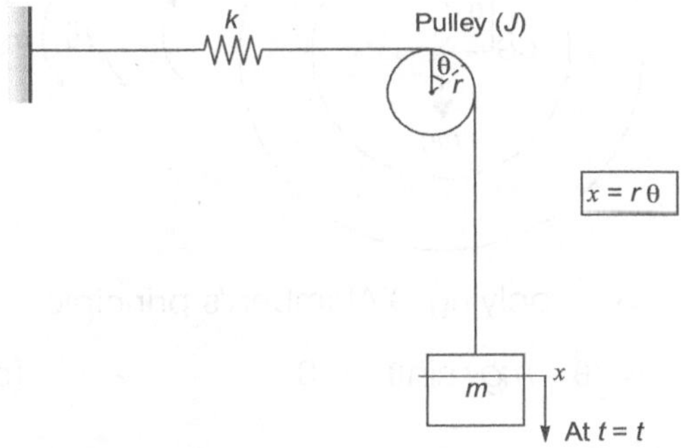

$$E = \frac{1}{2}m\dot{x}^2 + \frac{1}{2}kx^2 + \frac{1}{2}I\omega^2$$

$$= \frac{1}{2}mr^2\dot{\theta}^2 + \frac{1}{2}kr^2\theta^2 + \frac{1}{2}J\dot{\theta}^2$$

$$= \frac{1}{2}\left[\left(J + mr^2\right)\dot{\theta}^2 + kr^2\theta^2\right] = 0$$

$$\frac{dE}{dt} = 0$$

$$\left(J + mr^2\right) \times 2\dot{\theta}\ddot{\theta} + kr^2 2\theta\dot{\theta} = 0$$

$$\left(\ddot{\theta} + \frac{kr^2}{J + mr^2}\right)\theta = 0$$

$$\omega_n = \sqrt{\frac{kr^2}{J + mr^2}}$$

7.71 (d)

Transmitted force problem:

$$\omega = 40 \text{ rad/s (force frequency)}$$

$$m = 100 \text{ kg}$$

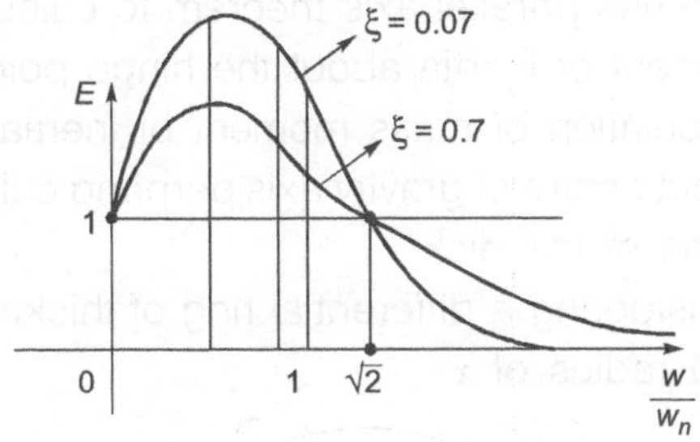

$$\in = \frac{\sqrt{1 + \left(\frac{2\xi\omega}{\omega_n}\right)^2}}{\sqrt{\left[\left(1 - \left(\frac{\omega}{\omega_n}\right)^2\right)^2 + \left(\frac{2\xi\omega}{\omega_n}\right)^2\right]}}$$

$$\omega_n = \sqrt{\frac{640 \times 10^3}{100}} = 80 \Rightarrow \frac{\omega}{\omega_n} = 0.5$$

$$\omega_n = \sqrt{\frac{22.5 \times 10^3}{100}} = 47.434164$$

$$\Rightarrow \quad \frac{\omega}{\omega_n} = \frac{40}{15} = 2.666$$

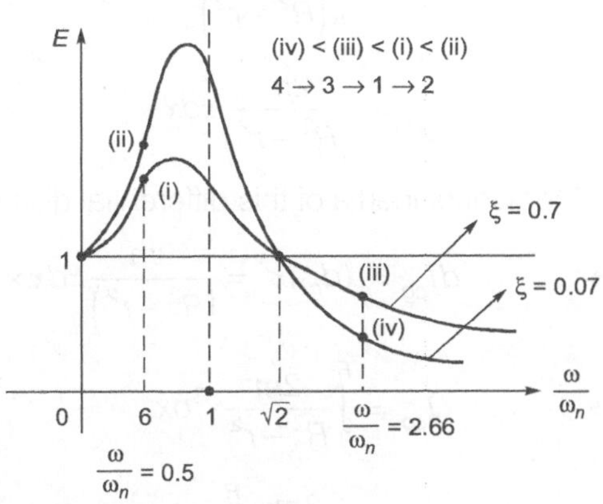

7.72 (2.66) (2.66 to 2.70)

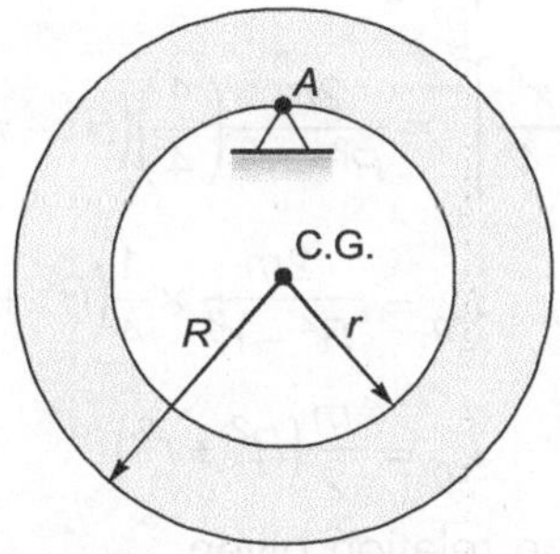

Here the disk will oscillate about the hinge which is on the smaller radius of the body.

To calculate the time period, we need to first calculate the mass moment of inertia about the hinge point A. For that, we need to calculate moment of inertia about centre of gravity and then use parallel axis theorem to calculate the moment of inertia about the hinge point.

Calculation of mass moment of inertia of disk about centre of gravity axis perpendicular to the plane of the disk.

Considering a differential ring of thickness 'dx' at a radius of x.

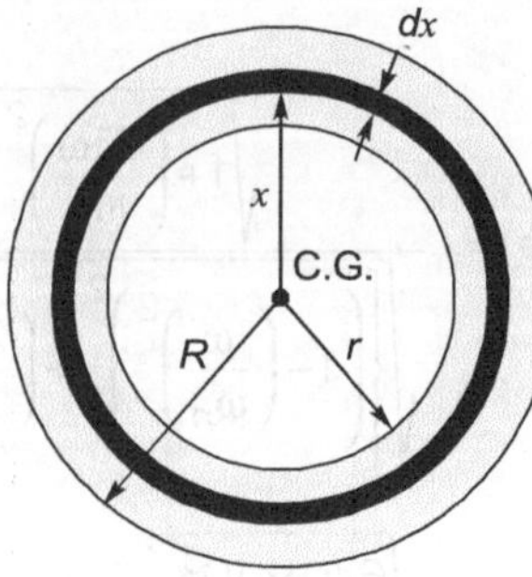

$\Rightarrow$ Suppose 'm' is the mass of the disk.

$\Rightarrow$ Area of the disk will be $\pi(R^2 - r^2)$

$\Rightarrow$ Mass per unit area will be $\dfrac{m}{\pi(R^2-r^2)}$

$\therefore$ So, mass of the differential element will be

$$dM = \frac{m}{\pi(R^2-r^2)} \times 2\pi x dx$$

$$= \frac{2m}{(R^2-r^2)} x dx$$

$\therefore$ Moment of inertia of this differential disk will be

$$\Rightarrow \quad dI_{cg} = (d_m)x^2 = \frac{2m}{(R^2-r^2)} x dx \times x^2$$

$$\Rightarrow \quad I_{cg} = \int_r^R \frac{2m}{R^2-r^2} x^3 dx$$

$$\Rightarrow \quad I_{cg} = \frac{2m}{R^2-r^2}\int_r^R x^3 dx$$

$$\Rightarrow \quad I_{cg} =$$

$$\frac{2m}{R^2-r^2}\left[\frac{x^4}{4}\right]_r^R = \frac{2m}{R^2-r^2}\left(\frac{1}{4}\right)(R^4-r^4)$$

$$\Rightarrow \quad I_{cg} = \frac{2m}{R^2-r^2} \times \frac{1}{4}(R^2+r^2)(R^2-r^2)$$

$$\Rightarrow \quad I_{cg} = \frac{m}{2}(R^2+r^2)$$

Putting the relation given

$$r = \frac{R}{\sqrt{2}}$$

We get, $\quad I_{cg} = \dfrac{3}{4}mR^2$

$\Rightarrow$ Using parallel axis theorem to get mass moment of inertia about the hinge point A, perpendicular to the plane.

$$\Rightarrow \quad I_A = I_{cg} + mr^2$$

$$\Rightarrow \quad I_A = \frac{3}{4}mR^2 + mr^2$$

$$= \frac{3}{4}mR^2 + \frac{mR^2}{2} = \frac{5}{4}mR^2$$

$\therefore$ Applying torque method to calculate the time period. Let the disk be disturbed by a small amount (θ) in anti-clockwise direction. So, there would be a torque due to mg in clockwise direction.

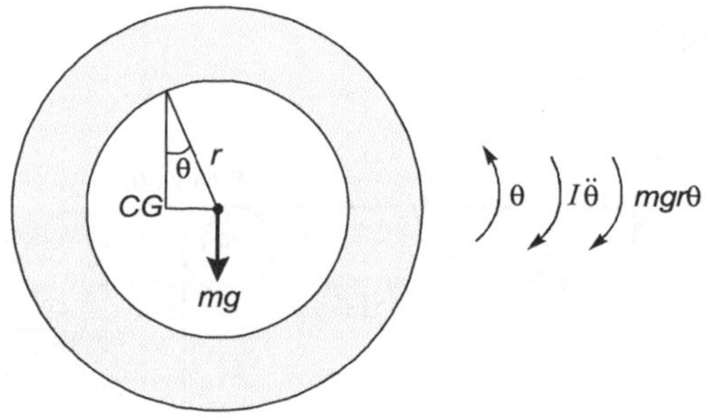

$\Rightarrow$ Applying D'Alambert's principle

$$\Rightarrow I\ddot{\theta} + mgr\sin\theta = 0 \qquad [\sin\theta \approx \theta]$$

$$\Rightarrow \frac{5}{4}mR^2\ddot{\theta} + mg\frac{R}{\sqrt{2}}\theta = 0$$

$$\Rightarrow \ddot{\theta} + \frac{4g}{5\sqrt{2}R}\theta = 0 \qquad \text{Equation of motion}$$

$\Rightarrow$ Standard form of oscillation is given as:

$$\ddot{\theta} + \omega_n^2\theta = 0$$

Comparing the equations, we get

$$\omega_n^2 = \frac{4g}{5\sqrt{2}R}$$

$$\omega_n = \frac{2}{\sqrt{5\sqrt{2}}}\sqrt{\frac{g}{R}}$$

Time period is given as

$$T = \frac{2\pi}{\omega_n} = \frac{2\pi}{\dfrac{2}{\sqrt{5\sqrt{2}}}\sqrt{\dfrac{g}{R}}} = \sqrt{5\sqrt{2}}\,\pi\sqrt{\frac{R}{g}}$$

⇒ Comparing the equation with the time period given in the question, we get

$$\beta = \sqrt{5\sqrt{2}}$$
$$= 2.659 \simeq 2.66$$

7.73 (a)

$$A = \frac{(f_0/s)}{\sqrt{\left\{1-\left(\frac{\omega}{\omega_n}\right)^2\right\}^2 + \left\{\frac{2\zeta\omega}{\omega_n}\right\}^2}} \quad \left(\frac{\omega}{\omega_n}\right) = r$$

$$= \frac{(f_0/s)}{\sqrt{(1-r^2)^2 + (2\zeta r)^2}}$$

Let: $y = (1-r^2)^2 + (2\zeta r)^2$

For A_{max} : ⇒ $y \to$ min

For y_{min} :

$$\frac{dy}{dr} = 0$$

$$2(1-r^2)(-2r) + (4\zeta^2)\cdot 2r = 0$$

$$-2 + 2r^2 + 4\zeta^2 = 0$$
$$-2r^2 = 4\zeta^2 - 2$$
$$-2r^2 = 2 - 4\zeta^2 = 2(1-2\zeta^2)$$
$$r^2 = (1-2\zeta^2)$$
$$r = \sqrt{1-2\zeta^2}$$
$$\frac{\omega_p}{\omega_n} = \sqrt{1-2\zeta^2}$$
$$\omega_p = \left(\sqrt{1-2\zeta^2}\right)\cdot\omega_n$$

Now,

$$\omega_p = \left(\sqrt{1-2\zeta^2}\right)\cdot\omega_n$$

And,

$$\omega_d = \left(\sqrt{1-2\zeta^2}\right)\cdot\omega_n$$

Therefore,

In general,

$$\omega_d < \omega_n$$
$$\omega_p < \omega_n$$

and $\omega_p < \omega_d$

⇒ $\omega_p < \omega_d < \omega_n$

Therefore, option (a) is correct answer.

7.74 (2) (2 to 2)

Equation of damped system,

$$\ddot{x} + (2\xi\omega_n)\dot{x} + (\omega_n^2)x = 0$$

where, ξ = Damping factor

Given: $\xi = \frac{1}{2\pi}\log_e 2 = 0.1103178$

Initial peak displacement = 4 mm

As $\xi < 1$

⇒ This is an under damped system.

Logarithmic decrement,

$$\delta = \frac{2\pi\xi}{\sqrt{1-\xi^2}} = 0.6974$$

Decrement ratio:

$$\frac{x_o}{x_1} = \frac{x_1}{x_2} = \frac{x_2}{x_3} .... = e^{\delta}$$

$$\frac{x_o}{x_1} = e^{0.6974}$$

$$x_1 = \frac{x_o}{e^{0.6974}} = \frac{4}{2.0085} = 1.991 \text{ mm}$$
$$\simeq 2 \text{ mm}$$

7.75 (c)

$$x = (A_1 \sin\omega_n t + A_i \cos\omega_n t) + \frac{(F_o/k)}{\left(1-\left(\frac{\omega}{\omega_n}\right)^2\right)}\cos\omega t$$

At $\omega = \sqrt{\frac{k}{m}} = \omega_n$

$$\frac{\omega}{\omega_n} = 1$$

$$x = (A_1 \sin\omega_n t + A_2 \cos\omega_n t) + \frac{(F_o/k)}{2m\omega_n}\sin\omega_n t$$

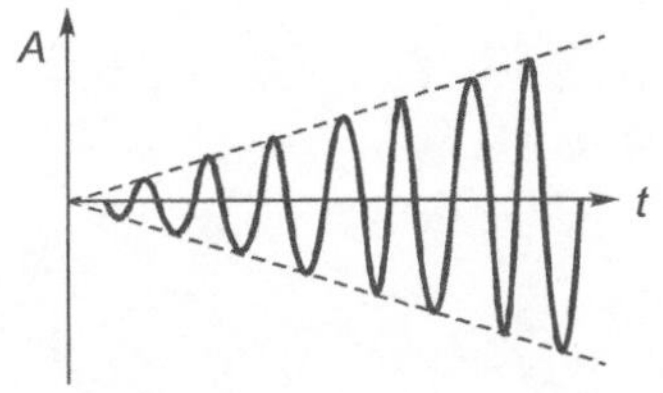

R - (ii)

S - (iii)

This implies option (c) is correct.

7.76 (6.28) (6.27 to 6.29)

Since the springs are arranged parallely, the equivalent stiffness is given by

$$k_{eq} = k + k = 2k = 2 \times 1000$$

∴ k_{eq} = 2000 N/m

Natural frequency,

$$\omega_n = \sqrt{\frac{k_{eq}}{m}} = \sqrt{\frac{2000}{20}} = 10 \text{ rad/s}$$

Time period of oscillation is,

$$T_n = \frac{2\pi}{\omega_n}$$

Thus, time taken to complete 10 oscillations is

$$10T_n = 10\times\frac{2\pi}{\omega_n} = 10\times\frac{2\pi}{10}$$

$$= 2\pi = 2 \times 3.14 = 6.28 \text{ sec}$$

7.77 (a)

Given: $m = 2$ kg, $f_n = 15$ Hz, $x_n = 7.75$ mm, $x_{n+1} = 7.20$ mm

$$\omega_n = 2\pi(f_n) = 2\pi \times 15 = 30\pi \text{ rad/s}$$

Decrement ratio, $\frac{x_n}{x_{n+1}} = \frac{7.75}{7.20} = 1.07638$

Logarithmic decrement,

$$\delta = \log_e\left(\frac{x_n}{x_{n+1}}\right) = \log_e(1.07638)$$

$$\delta = 0.0736$$

$$\frac{2\pi\xi}{\sqrt{1-\xi^2}} = 0.0736$$

$$\Rightarrow \quad \xi = 0.0117$$

But, $2\xi\omega_n = \frac{C}{m}$

$$\Rightarrow 2 \times 0.0117 \times 30\pi = \frac{C}{2}$$

$$\Rightarrow \quad C = 4.41 \text{ N/m/s}$$

7.78 (30) (29 to 31)

Given : $m = 10$ kg; $k = 1.5$ kN/m $= 1500$ N/m.

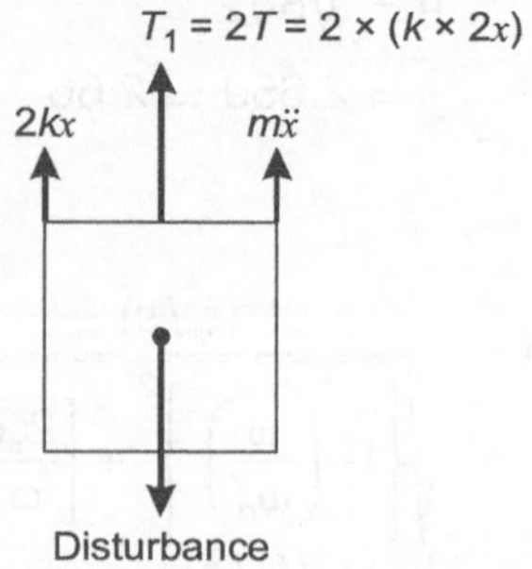

At time

Using D'Alembert's principle;

$$m\ddot{x} + 4kx + 2kx = 0$$

$$m\ddot{x} + 6kx = 0$$

$$\Rightarrow \quad \ddot{x} + \left(\frac{6k}{m}\right)x = 0;$$

Compairing with general equation of motion,

So, $\omega_n = \sqrt{\frac{6k}{m}} = \sqrt{\frac{6\times1500}{10}}$

$$= 30 \text{ rad/s}$$

■■■■

8 Balancing and Gyroscope

8.1 A rotating disc of 1 m diameter has two eccentric masses of 0.5 kg each at radii of 50 mm and 60 mm at angular positions of 0° and 150°, respectively. A balancing mass of 0.1 kg is to be used to balance the rotor. What is the radial position of the balancing mass?

(a) 50 mm (b) 120 mm
(c) 150 mm (d) 280 mm **[2005 : 2 M]**

8.2 A cantilever type gate hinged at Q is shown in the figure. P and R are the centers of gravity of the cantilever part and the counterweight respectively. The mass of the cantilever part is 75 kg. The mass of the counterweight, for static balance, is

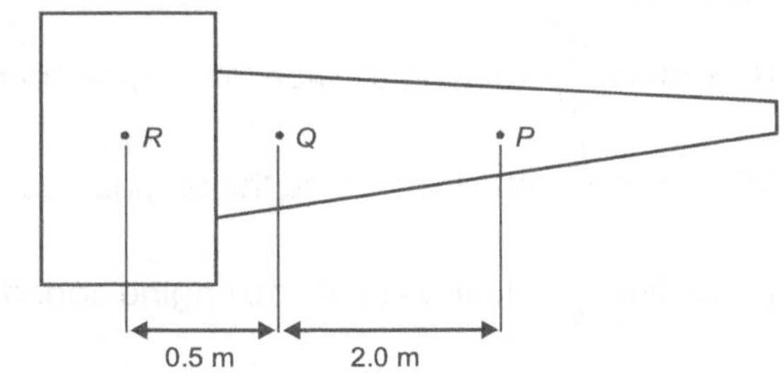

(a) 75 kg (b) 150 kg
(c) 225 kg (d) 300 kg **[2008 : 1 M]**

8.3 A slider crank mechanism has slider mass of 10 kg, stroke of 0.2 m and rotates with a uniform angular velocity of 10 rad/s. The primary inertia forces of the slider are partially balanced by a revolving mass of 6 kg at the crank, placed at a distance equal to crank radius. Neglect the mass of connecting rod and crank. When the crank angle (with respect to slider axis) is 30°, the unbalanced force (in newton) normal to the slider axis is_____.

[2014 : 2 M, Set-1]

8.4 Two masses m are attached to opposite sides of a rigid rotating shaft in the vertical plane. Another pair of equal masses m_1 is attached to the opposite sides of the shaft in the vertical plane as shown in figure. Consider $m = 1$ kg, $e = 50$ mm, $e_1 = 20$ mm, $b = 0.3$ m, $a = 2$ m and $a_1 = 2.5$ m. For the system to be dynamically balanced, m_1 should be _______ kg.

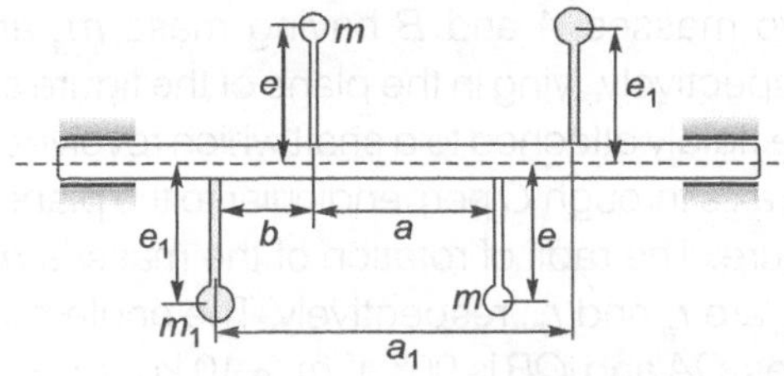

[2016 : 2 M, Set-3]

8.5 Three masses are connected to a rotating shaft supported on bearing A and B as shown in the figure. The system is in a space where the gravitational effect is absent. Neglect the mass of shaft and rods connecting the masses. For $m_1 = 10$ kg, $m_2 = 5$ kg and $m_3 = 2.5$ kg and for a shaft angular speed of 1000 radian/s, the magnitude of the bearing reaction (in N) at location B is ________.

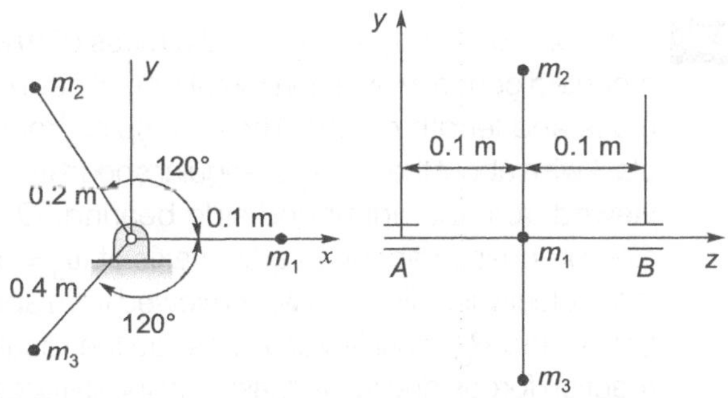

[2017 : 2 M, Set-2]

8.6 A car is moving on a curved horizontal road of radius 100 m with a speed of 20 m/s, The rotating masses of the engine have an angular speed of 100 rad/s in clockwise direction when viewed from the front of the car. The combined moment of inertia of the rotating masses is 10 kg-m^2. The magnitude of the gyroscopic moment in (N-m) is_____.

[2016 : 1 M, Set-1]

8.7 he rotor of a turbojet engine of an aircraft has a mass 180 kg and polar moment of inertia 10 kg m^2 about the rotor axis. The rotor rotates at a constant speed of 1100 rad/s in the clockwise direction when viewed from the front of the aircraft. The aircraft while flying at a speed of 800 km per hour takes a turn with a radius of 1.5 km to the left. The gyroscopic moment exerted by the rotor on the aircraft structure and the direction of motion of the nose when the aircraft turns, are

(a) 162.9 N.m, the nose goes up
(b) 1629.6 N.m, the nose goes down
(c) 1629.6 N.m, and the nose goes up
(d) 162.9 N.m, and the nose goes down

[2019 : 2 M, Set-1]

8.8 Two masses *A* and *B* having mass m_a and m_b respectively, lying in the plane of the figure shown, are rigidly attached to a shaft which revolves about an axis through *O* perpendicular to the plane of the figure. The radii of rotation of the masses m_a and m_b are r_a and r_b, respectively. The angle between lines *OA* and *OB* is 90°. If $m_a = 10$ kg, $m_b = 20$ kg, $r_a = 200$ mm and $r_b = 400$ mm, then the balance mass to be placed at a radius of 200 mm is ______kg (round off to two decimal places).

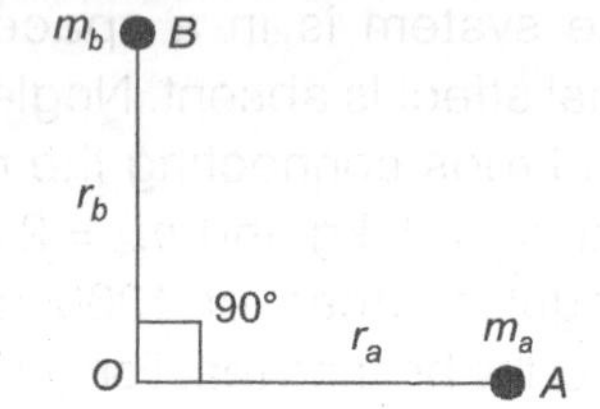

[2019 : 2 M, Set-2]

8.9 A uniform disc with radius r and a mass of *m* kg is mounted centrally on a horizontal axle of negligible mass and length of 1.5*r*. The disc spins counter-clockwise about the axle with angular speed ω, when viewed from the right-hand side bearing, *Q*. The axle precesses about a vertical axis at $\omega_p = \omega/10$ in the clockwise direction when viewed from above. Let R_P and R_Q (positive upwards) be the resultant reaction forces due to the mass and the gyroscopic effect, at bearings *P* and *Q*, respectively. Assuming $\omega^2 r = 300$ m/s^2 and $g = 10$ m/s^2, the ratio of the larger to the smaller bearing reaction force (considering appropriate signs) is ____.

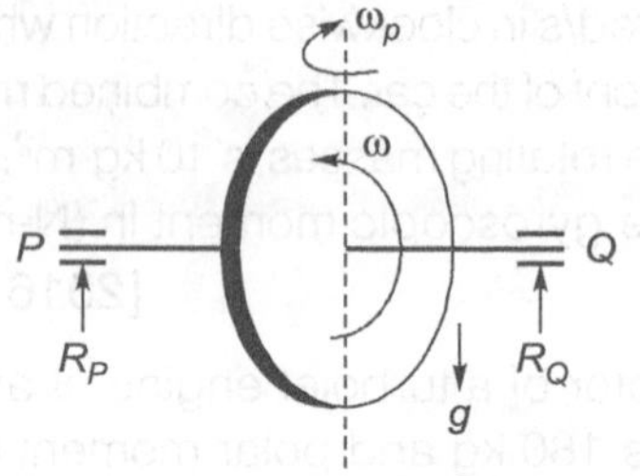

[2019 : 2 M, Set-2]

8.10 A balanced rigid disc mounted on a rigid rotor has four identical point masses, each of 10 grams, attached to four points on the 100 mm radius circle shown in the figure

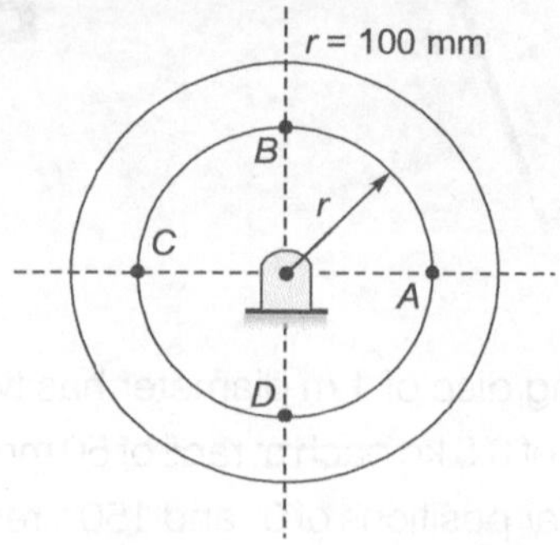

The rotor is driven by a motor at uniform angular speed of 10 rad/s. If one of the masses gets detached then the magnitude of the resultant unbalance force on the rotor is _____ N. (round off to 2 decimal places).

[2020 : 1 M, Set-1]

8.11 Consider a reciprocating engine with crank radius R and connecting rod of length L. The secondary unbalance force for this case is equivalent to primary unbalance force due to a virtual crank of ________.

(a) radius $\frac{R^2}{4L}$ rotating at twice the engine speed
(b) radius $\frac{L}{2}$ rotating at twice the engine speed
(c) radius $\frac{R}{4}$ rotating at half the engine speed
(d) radius $\frac{L^2}{4R}$ rotating at half the engine speed

[2021 : 1 M, Set-1]

8.12 The figure shows an arrangement of a heavy propeller shaft in a ship. The combined polar mass moment of inertia of the propeller and the shaft is 100 kg.m^2. The propeller rotates at $\omega = 12$ rad/s. The waves acting on the ship hull induces a rolling motion as shown in the figure with an angular velocity of 5 rad/s. The gyroscopic moment generated on the shaft due to the motion described is ______ N.m. [Round off to the nearest integer]

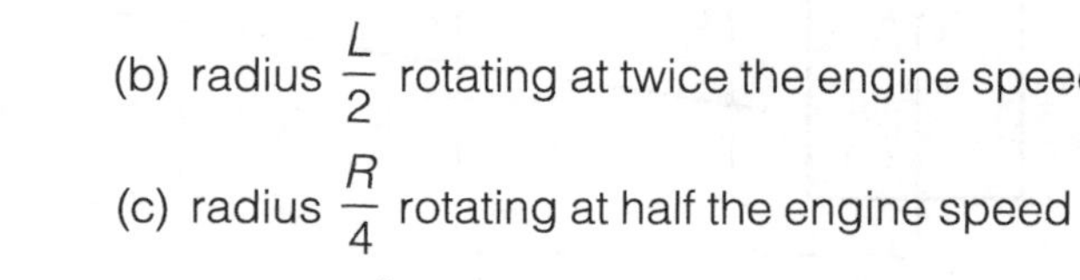

[2021 : 1 M, Set-1]

8.13 A massive uniform rigid circular disc is mounted on a frictionless bearing at the end *E* of a massive uniform rigid shaft AE which is suspended horizontally in a uniform gravitational field by two identical light inextensible strings AB and CD as shown, where G is the center of mass of the shaft-disc assembly and g is the acceleration due to gravity. The disc is then given a rapid spin ω about its axis in the positive x-axis direction as shown, while the shaft remains at rest. The direction of rotation is defined by using the right-hand thumb rule. If the string AB is suddenly cut, assuming negligible energy dissipation, the shaft AE will

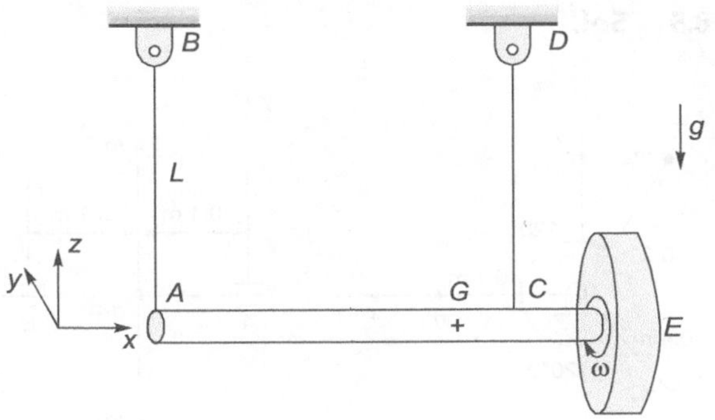

(a) rotate slowly (compared to ω) about the negative z-axis direction
(b) rotate slowly (compared to ω) about the positive z-axis direction
(c) rotate slowly (compared to ω) about the positive y-axis direction
(d) rotate slowly (compared to ω) about the negative y-axis direction

[2022 : 1 M, Set-2]

8.14 A rigid body in the X-Y plane consists of two point masses (1 kg each) attached to the ends of two massless rods, each of 1 cm length, as shown in the figure. It rotates at 30 RPM counter-clockwise about the Z-axis passing through point O. A point mass of $\sqrt{2}$ kg, attached to one end of a third massless rod, is used for balancing the body by attaching the free end of the rod to point O. The length of the third rod is ____ cm.

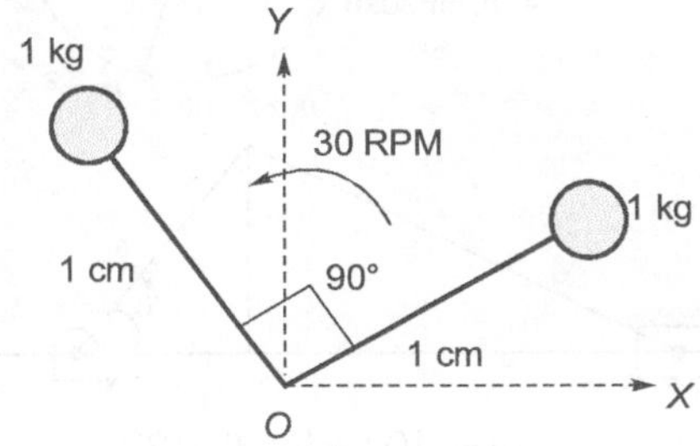

(a) 1 (b) $\sqrt{2}$
(c) $\frac{1}{\sqrt{2}}$ (d) $\frac{1}{2\sqrt{2}}$

[2022 : 2 M, Set-2]

Answers Balancing and Gyroscope

8.1	(c)	8.2	(d)	8.3	(30)	8.4	(2)	8.5	(0)	8.6	(200)	8.7	(b)
8.8	(41.23)	8.9	(–3)	8.10	(0.1)	8.11	(a)	8.12	(0)	8.13	(a)	8.14	(a)

Explanations Balancing and Gyroscope

8.1 (c)

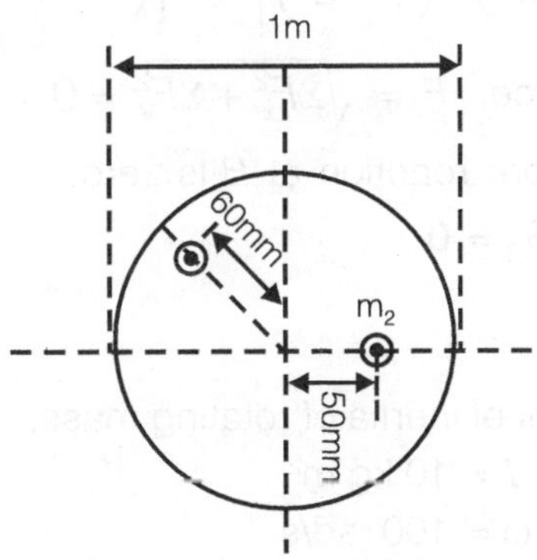

Let x and y is position of balancing mass along x and y directions. Resolving forces along x-axis

$$0.5(-60 \times 10^{-3}\cos 30° + 50 \times \cos 0 \times 10^{-3})\omega^2 = 0.1 \times x \times 10^{-3}$$

$$\therefore \quad x = -9.8 \text{ mm}$$

$$0.5 \times (60 \times 10^{-3} \times \sin 30)\omega^2 + 0.5\,(50 \times 10^{-3} \times \sin 0)\omega^2 = 0.1\omega^2 y$$

$$\therefore \quad y = 150 \text{ mm}$$

Position of balancing mass,

$$r = \sqrt{x^2 + y^2} = \sqrt{(-9.8)^2 + (150)^2}$$

$$= \mathbf{150.3 \text{ mm}}$$

8.2 (d)

Let counterweight at R is m kg

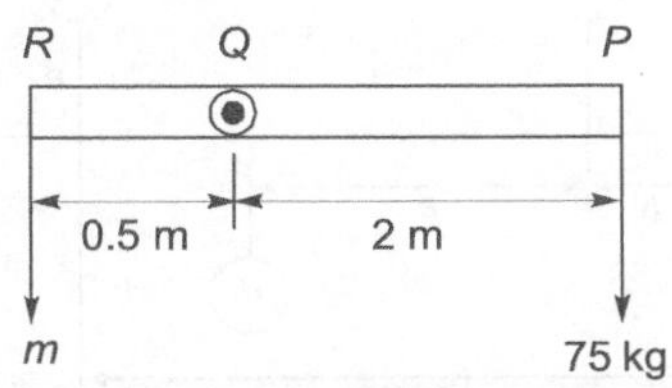

For static condition

$$\Sigma M_Q = 0$$

$$m \times 0.5 = 75 \times 2$$

$$m = \mathbf{300\ kg}$$

8.3 Sol.

Given: $m_p = 10$ kg, $m = 6$ kg, $r = 0.1$ m, $\omega = 10$ rad/s, $\theta = 30°$

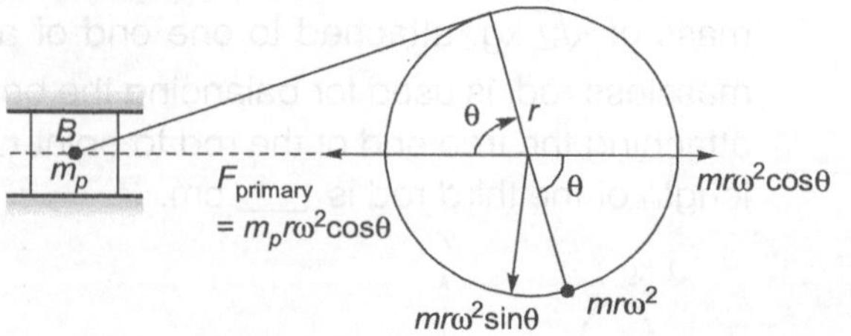

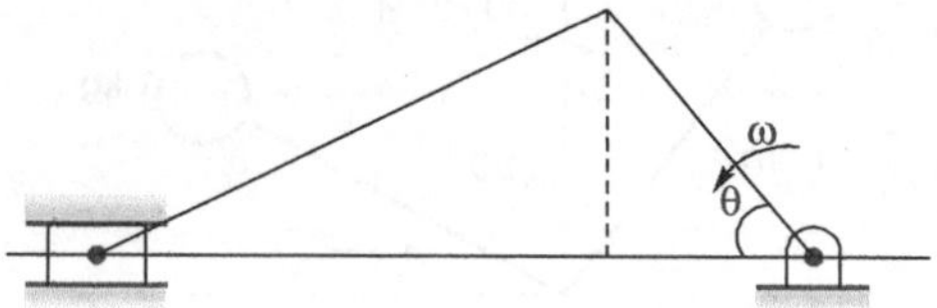

$$\omega = 10 \text{ rad/s}, \theta = 30°$$

$$r = \frac{s}{2} = 0.1 \text{ m}$$

Unbalanced force $= m\omega^2 r \sin 30°$

$$= 6 \times 100 \times 0.1 \times \frac{1}{2} = \mathbf{30\ N}$$

8.4 Sol.

Method I:

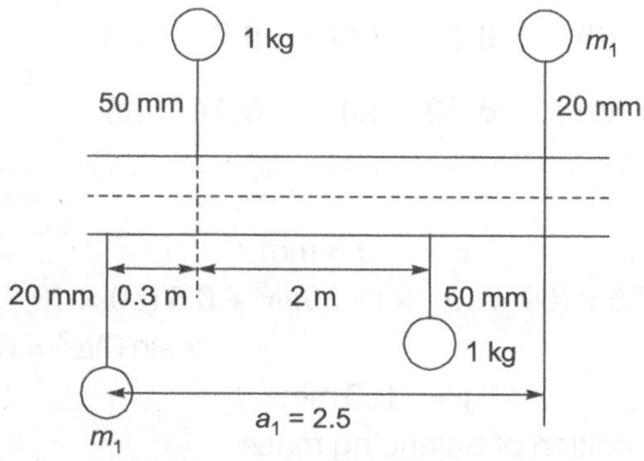

Balance moment of all forces

$$\omega^2(m_1 \times 0.02 \times 2.5) + (1 \times 0.05 \times 0.3)\omega^2 = (m_1 \times 0.02 \times 0)\,\omega^2 + (1 \times 0.05 \times 2.3)\omega^2$$

$$m_1 = \mathbf{2\ kg}$$

Method II:

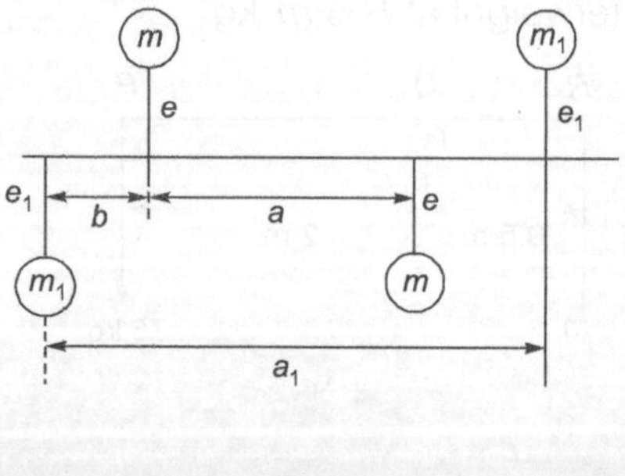

Let reference PQ, passes through m_1 mass on left side.

Taking moment about PQ,

$m_1e_1 \times 0 \times \cos 180° + meb \cos 0° + (a + b) \cos 180° + m_1e_1a_1 \cos 0° = 0$

$$0 + meb - me(a + b) + m_1e_1a_1 = 0$$

$$m_1e_1a_1 = mea$$

$$m_1 = \frac{mea}{e_1a_1} = \frac{1 \times 50 \times 2}{20 \times 2.5} = 2 \text{ kg}$$

8.5 Sol.

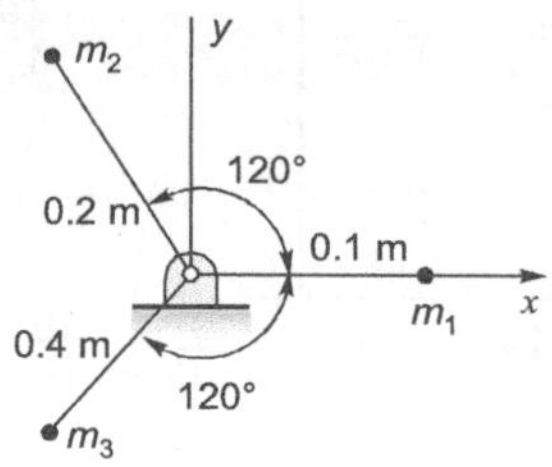

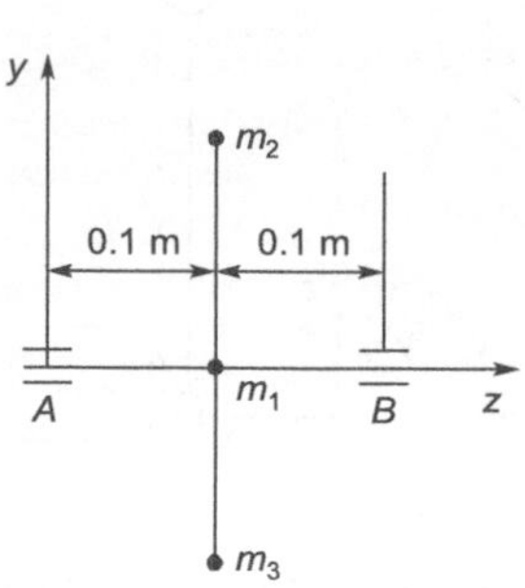

It means all three masses are in same plane let us calculates the net force

$$\Sigma F_x = \Sigma mr\omega^2\cos\theta$$

$$= (10 \times 0.1 \times \omega^2) - (5 \times 0.2 \times \omega^2 \cos 60°) - (2.5 \times 0.4 \times \omega^2 \cos 60°)$$

$$= \left[1 - \left(5 \times 0.2 \times \frac{1}{2}\right) - \left(2.5 \times 0.4 \times \frac{1}{2}\right)\right]\omega^2 = 0$$

$$\Sigma F_y = \Sigma mr\omega^2\sin\theta = (5 \times 0.2 \times \omega^2 \sin 60°) - (2.5 \times 0.4 \times \omega^2 \sin 60°)$$

$$= \left[\left(1 \times \frac{\sqrt{3}}{2}\right) - \left(1 \times \frac{\sqrt{3}}{2}\right)\right]\omega^2 = \left[\left(\frac{\sqrt{3}}{2} - \frac{\sqrt{3}}{2}\right)\right]\omega^2 = 0$$

Net force, $F = \sqrt{\Sigma F_x^2 + \Sigma F_y^2} = 0$

Therefore reaction at B is zero.

$$R_B = 0$$

8.6 Sol.

Moment of inertia of rotating mass,

$$I = 10 \text{ kg m}^2$$

$$\omega = 100 \text{ rad/s}$$

$$\omega_p = \frac{v}{r} = \frac{20}{100} \text{ rad/s}$$

Gyroscopic moment

$$= I \times \omega \times \omega_p = 10 \times 100 \times \frac{20}{100} = \mathbf{200\ Nm}$$

8.7 (b)

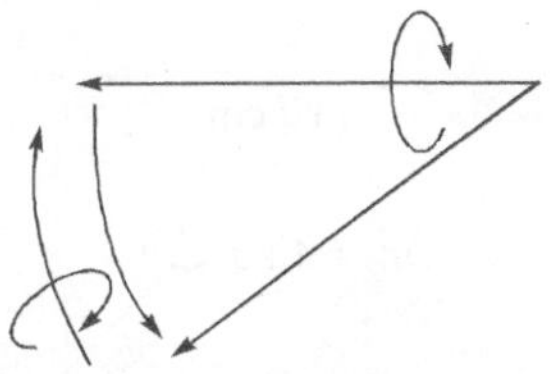

$m = 180$ kg

$I = 10$ kg-m^2

$\omega = 1100$ rad/s

$R = 1.5$ km $= 1500$ m

$v = 800$ km/hr $= 800 \times \dfrac{5}{18} = 222.222$ m/s

$\omega_p = \dfrac{v}{R} = 0.14814$ rad/s

Gyroscopic couple

$= I.\omega.\omega._p = 10 \times 1100 \times 0.14814$

$= 1629.629$ Nm

Active gyroscopic couple direction clockwise in vertical plane looking from the left side of the aircraft.

8.8 Sol.

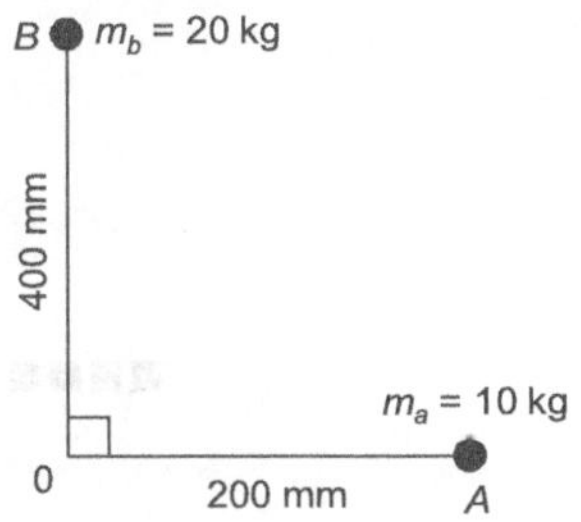

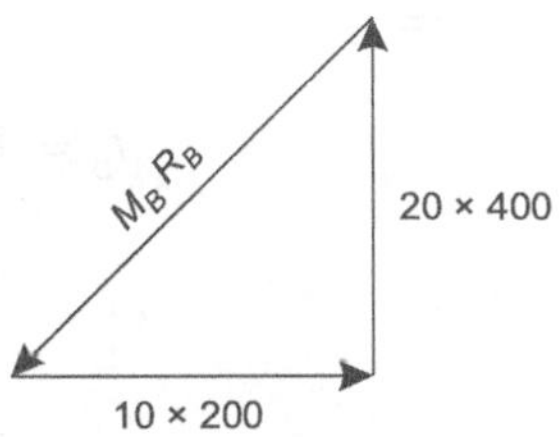

Force polygon

Balance mass radius, $R_B = 200$ mm

$M_B = ?$

$$M_B R_B = \sqrt{(10 \times 200)^2 + (20 \times 400)^2}$$

$M_B \times 200 = 8246.211$

$M_B = 41.23$ kg

8.9 Sol.

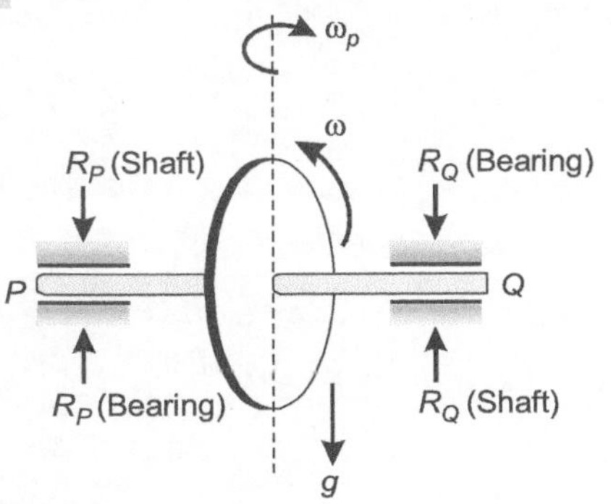

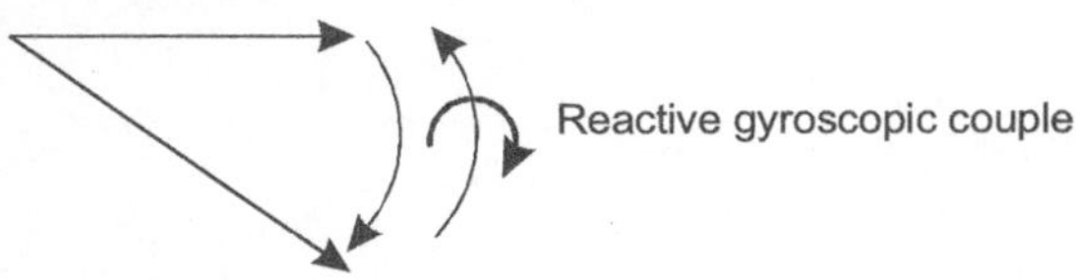

Active gyroscopic couple $= I\omega\omega_p$

$$= \frac{mr^2}{2} \times \omega \times \frac{\omega}{10} = \left(\frac{m\omega^2 r^2}{20}\right)$$

Reaction force at bearing P due to gyroscopic couple is (R_{P_m})

$$R_{P_m} = \frac{m\omega^2 r^2}{20 \times 1.5r}$$

$$= \frac{m\omega^2 r^2}{20 \times 1.5r} = \frac{m\omega^2 r}{30} = 10m \text{ N (upward)}$$

Reaction force at bearing Q due to gyroscopic couple is R_{Q_m}

$$R_{Q_m} = \frac{m\omega^2 r^2}{20 \times 1.5r} = \frac{m\omega r}{30}$$

$= 10m$ N (downward)

Reaction force at bearing P due to gravity

$$= \frac{mg}{2} \text{(downward)} = 5m \text{ N}$$

Reaction force at bearing Q due to gravity

$$= \frac{mg}{2} = 5m \text{ N (upward)}$$

Reaction at bearing P,

$R_P = 10m\,\text{N} - 5m\,\text{N} = 5m\,\text{N}$ [Upward]

Reaction at bearing Q,

$R_Q = -10m\,\text{N} - 5m\,\text{N} = -15m\,\text{N}$ [Upward]

$$\frac{R_Q}{R_P} = \frac{-15}{5} = -3$$

8.10 Sol.

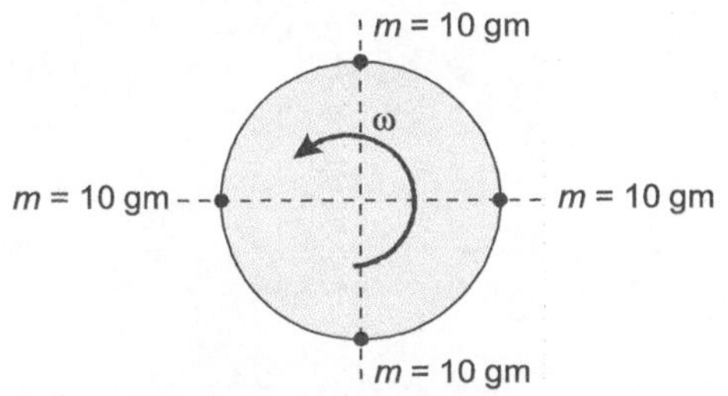

$\omega = 10$ rad/s, $r = 100$ mm $= 0.1$ m

If one mass is detached then

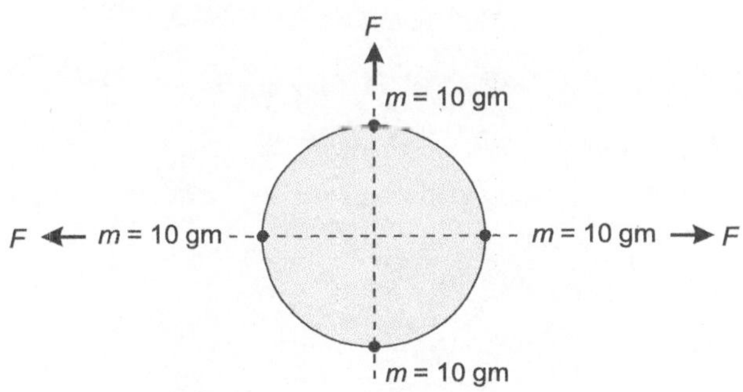

Now, unbalance force,

$$F = mr\omega^2$$

$$= \frac{10}{1000} \times 0.1 \times (10)^2 = 0.1 \text{ N}$$

8.13 (a)

As centre of mass is in between strings AB and CD so due to gyroscopic effect it rotates about the negative z-axis direction as it is rotating clockwise when viewed from the top.

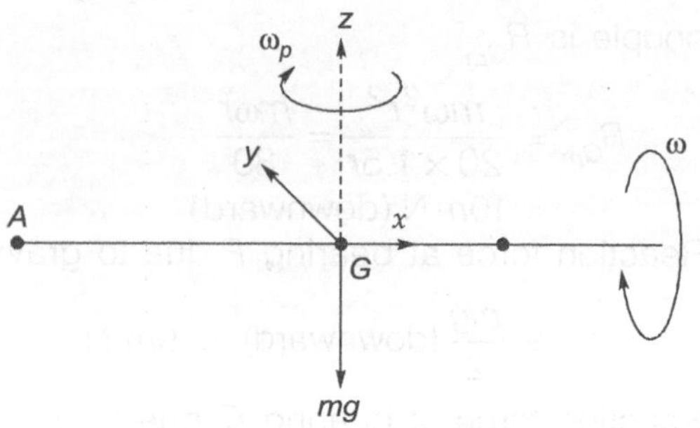

8.14 (a)

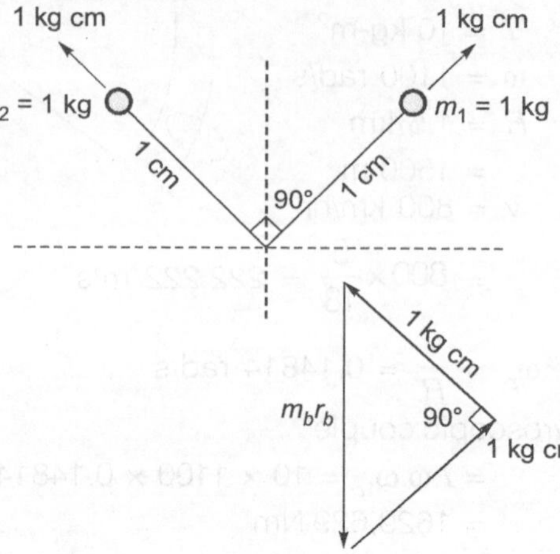

Resultant balancing force = $m_b r_b$

$$\Rightarrow \quad m_b r_b = \sqrt{1^2 + 1^2}$$

$$\sqrt{2} \times r_b = 1\sqrt{2}$$

$$\Rightarrow \quad r_b = \frac{1\sqrt{2}}{\sqrt{2}} \text{ cm}$$

$$r_b = 1 \text{ cm}$$

■■■■

Machine Design

UNIT

V

CONTENTS

Machine Design

Syllabus

Design for static and dynamic loading; failure theories; fatigue strength and the S-N diagram; principles of the design of machine elements such as bolted, riveted and welded joints; shafts, gears, rolling and sliding contact bearings, brakes and clutches, springs.

Analysis of Previous GATE Papers

Exam Year	1 Mark Ques.	2 Marks Ques.	Total Marks
1995	2	–	**2**
1996	5	–	**5**
1997	2	2	**6**
1998	2	–	**2**
1999	–	–	**–**
2000	2	–	**2**
2001	–	–	**–**
2002	3	–	**3**
2003	1	4	**9**
2004	3	2	**7**
2005	2	2	**6**
2006	–	4	**8**
2007	1	8	**17**
2008	–	7	**14**
2009	–	3	**6**
2010	–	3	**6**
2011	–	1	**2**
2012	–	2	**4**
2013	3	5	**13**
2014 Set-1	1	2	**5**
2014 Set-2	–	8	**8**
2014 Set-3	2	2	**6**
2014 Set-4	2	3	**8**
2015 Set-1	1	2	**5**

Exam Year	1 Mark Ques.	2 Marks Ques.	Total Marks
2015 Set-2	1	–	**1**
2015 Set-3	1	2	**5**
2016 Set-1	–	1	**2**
2016 Set-2	2	–	**2**
2016 Set-3	–	1	**2**
2017 Set-1	1	1	**3**
2017 Set-2	1	3	**7**
2018 Set-1	1	3	**7**
2018 Set-2	2	1	**4**
2019 Set-1	1	1	**3**
2019 Set-2	1	1	**3**
2020 Set-1	1	3	**7**
2020 Set-2	1	1	**3**
2021 Set-1	–	3	**6**
2021 Set-2	–	1	**2**
2022 Set-1	2	1	**4**
2022 Set-2	–	–	**–**
2023	1	–	**1**
2024	2	1	**4**

1 Static, Dynamic Loading and Failure Theories

1.1 A solid shaft can resist a bending moment of 3.0 kNm and a twisting moment of 4.0 kNm together, then the maximum torque that can be applied is

(a) 7.0 kNm (b) 3.5 kNm
(c) 4.5 kNm (d) 5.0 kNm **[1996 : 1 M]**

1.2 A small element at the critical section of a component in biaxial state of stress with the two principal stresses being 360 MPa and 140 MPa. The maximum working stress according to distortion energy theory is

(a) 220 MPa (b) 110 MPa
(c) 314 MPa (d) 330 MPa **[1997 : 2 M]**

1.3 Which theory of failure will you use for aluminium components under steady loading?

(a) Principal stress theory
(b) Principal strain theory
(c) Strain energy theory
(d) Maximum shear stress theory **[1998 : 1 M]**

1.4 A thin spherical pressure vessel of 200 mm diameter and 1 mm thickness is subjected to an internal pressure varying from 4 to 8 MPa. Assume that the yield, ultimate, and endurance strength of material are 600, 800 and 400 MPa respectively. The factor of safety as per Goodman's relation is

(a) 2.0 (b) 1.6
(c) 1.4 (d) 1.2 **[2007 : 2 M]**

1.5 The piston rod of diameter 20 mm and length 700 mm in a hydraulic cylinder is subjected to a compressive force of 10 kN due to the internal pressure. The end conditions for the rod may be assumed as guided at the piston end and hinged at the other end. The Young's modulus is 200 GPa. The factor of safety for the piston rod is

(a) 0.68 (b) 2.75
(c) 6.32 (d) 11.0 **[2007 : 2 M]**

1.6 A forged steel link with uniform diameter of 30 mm at the centre is subjected to an axial force that varies from 40 kN in compression to 160 kN in tension. The tensile (S_u), yield (S_y) and corrected endurance (S_e) strengths of the steel material are 600 MPa, 420 MPa and 240 MPa respectively. The factor of safety against fatigue endurance as per Soderberg's criterion is

(a) 1.26 (b) 1.37
(c) 1.45 (d) 2.00 **[2009 : 2 M]**

1.7 A bar is subjected to fluctuating tensile load from 20 kN to 100 kN. The material has yield strength of 240 MPa and endurance limit in reversed bending is 160 MPa. According to the Soderberg principle, the area of cross-section in mm^2 of the bar for a factor of safety of 2 is

(a) 400 (b) 600
(c) 750 (d) 1000 **[2013 : 2 M]**

1.8 A rotating steel shaft is supported at the ends. It is subjected to a point load at the center. The maximum bonding stress developed is 100 MPa. If the yield, ultimate and corrected endurance strength of the shaft material are 300 MPa, 500 MPa and 200 MPa, respectively, then the factor of safety for the shaft is ______.

[2014 : 1 M, Set-3]

1.9 Which one of the following is not correct?

(a) Intermediate principal stress is ignored when applying the maximum principal stress theory.
(b) The maximum shear stress theory gives the most accurate results amongst all the failure theories.
(c) As per the maximum strain energy theory, failure occurs when the strain energy per unit volume exceeds a critical value.
(d) As per the maximum distortion energy theory, failure occurs when the distortion energy per unit volume exceeds a critical value.

[2014 : 1 M, Set-3]

1.10 Consider the two states of stress as shown in configurations I and II in the figure below. From the standpoint of distortion energy (von-Mises) criterion, which one of the following statements is true?

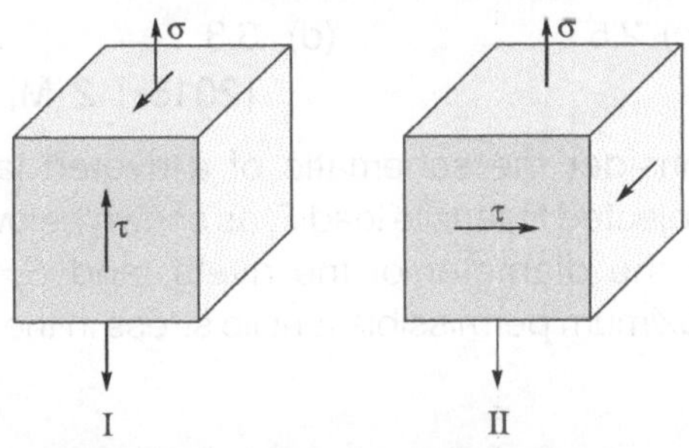

(a) I yields after II
(b) II yields after I
(c) Both yield simultaneously
(d) Nothing can be said about their relative yielding

[2014 : 2 M, Set-2]

1.11 A shaft is subjected to pure torsional moment. The maximum shear stress developed in the shaft is 100 MPa. The yield and ultimate strengths of the shaft material in tension are 300 MPa and 450 MPa, respectively. The factor of safety using maximum distortion energy (Von-Mises) theory is ______.

[2014 : 2 M, Set-4]

1.12 The uniaxial yield stress of a material is 300 MPa. According to Von Mises criterion, the shear yield stress (in MPa) of the material is ______.

[2015 : 1 M, Set-2]

1.13 In a structure subjected to fatigue loading, the minimum and maximum stresses developed in a cycle are 200 MPa and 400 MPa respectively. The value of stress amplitude (in MPa) is ______.

[2014 : 1 M, Set-2]

1.14 For the given fluctuating fatigue load, the values of stress amplitude and stress ratio are respectively

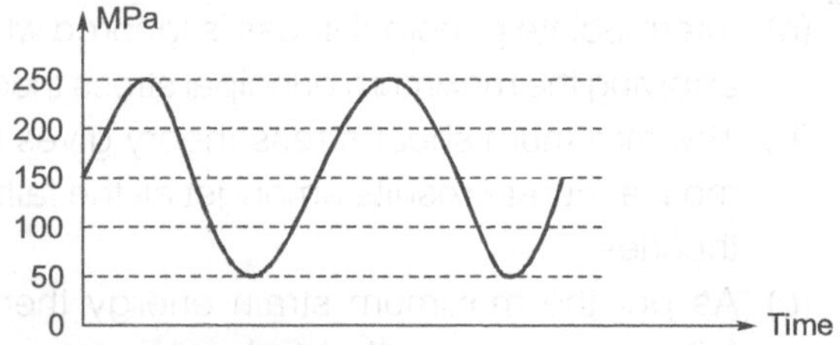

(a) 100 MPa and 5
(b) 250 MPa and 5
(c) 100 MPa and 0.20
(d) 250 MPa and 0.20

[2015 : 1 M, Set-3]

1.15 A machine element is subjected to the following bi-axial state of stress; $\sigma_x = 80$ MPa; $\sigma_y = 20$ MPa; $\tau_{xy} = 40$ MPa. If the shear strength of the materials is 100 MPa, the factor of safety as per Tresca's maximum shear stress theory is

(a) 1.0
(b) 2.0
(c) 2.5
(d) 3.3

[2015 : 2 M, Set-1]

1.16 Consider the schematic of a riveted lap joint subjected to tensile load F, as shown below. Let d be the diameter of the rivets, and S_f be the maximum permissible tensile stress in the plates. What should be the minimum value for the thickness of the plates to guard against tensile failure of the plates? Assume the plates to be identical.

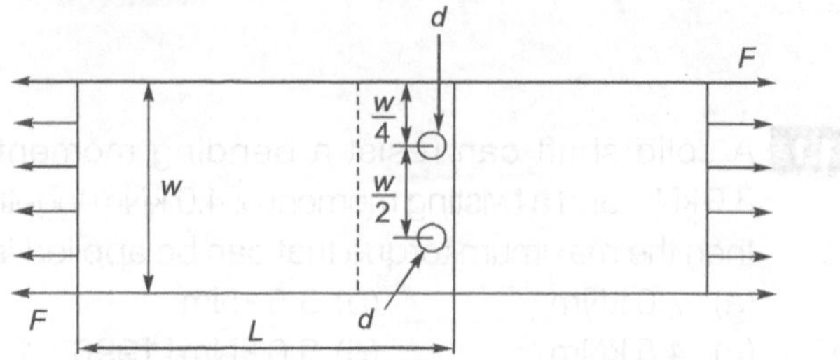

(a) $\frac{F}{S_f(w-2d)}$
(b) $\frac{F}{S_f w}$
(c) $\frac{F}{S_f(w-d)}$
(d) $\frac{2F}{S_f w}$

[2017 : 1 M, Set-1]

1.17 A machine component made of a ductile material is subjected to a variable loading with $\sigma_{min} = -50$ MPa and $\sigma_{max} = 50$ MPa. If the corrected endurance limit and the yield strength for the material are $\sigma_e' = 100$ MPa and $\sigma_y = 300$ MPa, respectively, the factor of safety is ______.

[2017 : 1 M, Set-2]

1.18 If σ_1 and σ_3 are the algebraically largest and smallest principal stresses respectively, the value of the maximum shear stress is

(a) $\frac{\sigma_1+\sigma_3}{2}$
(b) $\frac{\sigma_1-\sigma_3}{2}$
(c) $\sqrt{\frac{\sigma_1+\sigma_3}{2}}$
(d) $\sqrt{\frac{\sigma_1-\sigma_3}{2}}$

[2018 : 1 M, Set-1]

1.19 Bars of square and circular cross-section with 0.5 m length are made of a material with shear strength of 20 MPa. The square bar cross-section dimension is 4 cm x 4 cm and the cylindrical bar cross-section diameter is 4 cm. The specimens are loaded as shown in the figure.

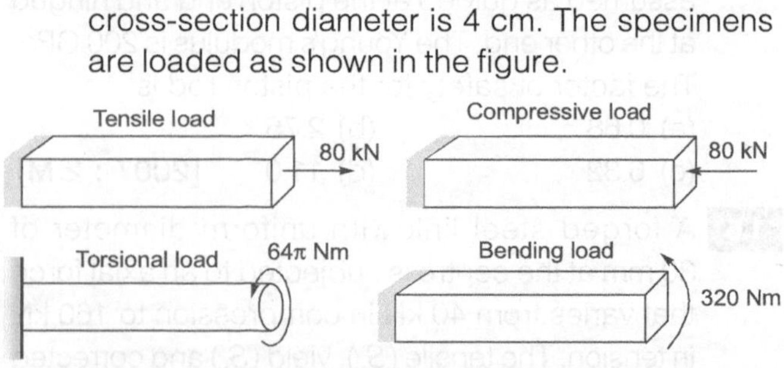

Which specimen(s) will fail due to the applied load as per maximum shear stress theory?

(a) Torsional load specimen
(b) Bending load specimen
(c) None of the specimens
(d) Tensile and compressive load specimens

[2020 : 2 M, Set-1]

1.20 A machine part in the form of cantilever beam is subjected to fluctuating load as shown in the figure. The load varies from 800 N to 1600 N. The modified endurance, yield and ultimate strengths of the material are 200 MPa, 500 MPa and 600 MPa, respectively

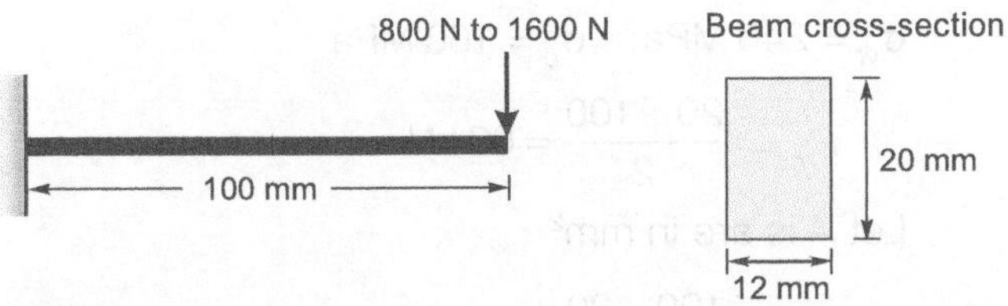

The factor of safety of the beam using modified Goodman criterion is ________ [round off to one decimal place]

[2021 : 2 M, Set-1]

1.21 Which one of the following failure theories is the most conservative design approach against fatigue failure?
(a) Modified Goodman line
(b) Yield line
(c) Gerber line
(d) Soderberg line

[2024 : 1 M]

■■■■

Answers Static, Dynamic Loading and Failure Theories

1.1	(d)	1.2	(c)	1.3	(d)	1.4	(b)	1.5	(c)	1.6	(a)	1.7	(d)	1.8	(2)
1.9	(b)	1.10	(c)	1.11	(1.732)	1.12	(173.1)	1.13	(100)	1.14	(c)	1.15	(b)	1.16	(a)
1.17	(2)	1.18	(b)	1.19	(d)	1.20	(2.5)	1.21	(d)						

Explanations Static, Dynamic Loading and Failure Theories

1.1 (d)

Maximum torque $= \sqrt{M^2 + T^2}$ = Equivalent torque

$$= \sqrt{3^2 + 4^2} = 5 \text{ kNm}$$

1.2 (c)

According to distortion energy theory

$$\tau = \sqrt{\sigma_1^2 + \sigma_2^2 - \sigma_1\sigma_2} = 314 \text{ MPa}$$

1.3 (d)

Since, aluminium is a ductile material therefore we will use maximum shear stress theory.

1.4 (b)

Stress induced

$$\sigma_1 = \sigma_2 = \frac{pr}{2t}$$

$$\sigma_{1max} = \frac{8 \times 100}{2 \times 1} = 400 \text{ MPa}$$

$$\sigma_{1min} = \frac{4 \times 100}{2 \times 1} = 200 \text{ MPa}$$

$$\sigma_{2max} = 400 \text{ MPa}$$

$$\sigma_{2min} = 200 \text{ MPa}$$

$$\sigma_{1m} = 300 \text{ MPa}$$

$$\sigma_{1a} = 100 \text{ MPa}$$

$$\sigma_{2m} = 300 \text{ MPa}$$

$$\sigma_{2a} = 100 \text{ MPa}$$

Equivalent Stresses

$$\sigma_{me} = \sqrt{\sigma_{1m}^2 + \sigma_{2m}^2 - \sigma_{1m}\sigma_{2m}}$$

$$= \sqrt{300^2 + 300^2 - 300 \times 300}$$

$$= 300 \text{ MPa}$$

Similarly,

$$\sigma_{ae} = 100 \text{ MPa}$$

Goodman equation,

$$\frac{\sigma_{ae}}{S_e} + \frac{\sigma_{me}}{S_{ut}} = \frac{1}{n}$$

$$\Rightarrow \frac{100}{400} + \frac{300}{800} = \frac{1}{n}$$

$$n = 1.6$$

1.5 (c)

Assuming guided end to be fixed and other end given as hinged. The crippling load according to Euler's equation,

$$P_{cr} = \frac{2\pi^2 EI}{L^2}$$

$$I = \frac{\pi}{64}(20)^4 = 7853.9 \text{ mm}^4$$

$$E = 200 \times 10^9 \text{ N/m}^2 = 200 \times 10^3 \text{ N/mm}^2$$

$$= \frac{2\pi^2 \times 200 \times 10^3 \times 7853.9}{(700)^2} = 63.27 \text{ kN}$$

Factor of safety

$$= \frac{\text{Crippling load}}{\text{Working load}} = \frac{63.27}{10} = 6.32$$

1.6 (a)

Diameter: $d = 30$ mm

$F_{max} = +160$ kN (Tension)

$F_{min} = -40$ kN (Compression)

Tensile strength: $S_u = 600$ MPa

Yield strength: $S_y = 430$ MPa

Corrected endurance,

$S_e = 240$ MPa

Maximum stress,

$$\sigma_{max} = \frac{F_{max}}{A} = \frac{160 \times 10^3 \text{ N}}{\frac{\pi}{4}(30)^2 \text{ mm}^2}$$

$= 226.47$ MPa (Tensile)

Minimum stress,

$$\sigma_{min} = \frac{F_{min}}{A} = \frac{-40 \times 10^3 \text{ N}}{\frac{\pi}{4}(30)^2 \text{ mm}^2}$$

$= -56.62$ MPa (Compression)

Stress amplitude,

$$\sigma_a = \frac{1}{2}(\sigma_{max} - \sigma_{min}) = \frac{1}{2}[226.47 - (-56.62)]$$

$= 141.54$ MPa

Mean stress, $\sigma_m = \frac{1}{2}(\sigma_{max} + \sigma_{min})$

$$= \frac{1}{2}[226.47 + (-56.62)]$$

$= 84.925$ MPa

Assume factor of safety is n then

$$S_a = n\sigma_a = 141.54\, n$$

$$S_m = n\sigma_m = 84.925\, n$$

The equation of Soderberg line is as follows

$$\frac{S_a}{S_e} + \frac{S_m}{S_{yt}} = 1$$

$$\Rightarrow \frac{141.54\, n}{240} + \frac{84.925\, n}{420} = 1$$

$$\Rightarrow n = \mathbf{1.26}$$

1.7 (d)

$P_1 = 20$ kN; $P_2 = 100$ kN

$\sigma_{yt} = 240$ MPa; $\sigma_e = 160$ MPa

$$P_m = \frac{20 + 100}{2} = 60 \text{ kN}$$

Let A is are in mm²

$$P_v = \frac{100 - 20}{2} = 40 \text{ kN}$$

$$\sigma_v = \frac{40}{A} \text{ kN/mm}^2$$

By Soderberg principle

$$\frac{1}{\text{FOS}} = \frac{\sigma_m}{\sigma_{yt}} + \frac{\sigma_v}{\sigma_e}$$

$$\frac{1}{2} = \frac{60 \times 10^3}{A\sigma_{yt}} + \frac{40 \times 10^3}{A\sigma_v}$$

$$A = \left[\frac{60}{240} + \frac{40}{160}\right] 2 \times 10^3$$

$$A = \left[\frac{2}{4}\right] \times 2 \times 10^3 = \mathbf{1000 \text{ mm}^2}$$

1.8 Sol.

Same fiber under point load at the center of a rotating steel shaft supported at ends is subjected to reversal of stresses. Here only maximum stress is given, its sign (nature) is not given

$\therefore \quad \sigma_{min} = -100$ MPa; $\sigma_{max} = 100$ MPa

$$\sigma_m = \frac{\sigma_{max} + \sigma_{min}}{2} = 0$$

$$\sigma_a = \frac{\sigma_{max} - \sigma_{min}}{2} = \frac{100 + 100}{2} = 100 \text{ MPa}$$

As the steel is ductile material, Soderberg criteria will be applied.

$$\therefore \quad \frac{\sigma_m}{\sigma_y} + \frac{\sigma_a}{\sigma_e} = \frac{1}{n}$$

$$\text{or} \quad 0 + \frac{100}{200} = \frac{1}{n} \Rightarrow \mathbf{n = 2}$$

1.9 (b)

The maximum shear stress theory predicts that shear yield value τ_y is 0.5 times the tensile yield value. This is about 15% less than the value predicted by the distortion energy (or the octahedral shear) theory. The maximum shear stress theory gives values for design on the safe side. Also, because of its simplicity, this theory is widely used in machine design dealing with ductile materials.

1.10 (c)

Simplified Von Mises equation for general load scenario is given by

$$\sigma_v = \sqrt{\frac{1}{2}[(\sigma_{11}-\sigma_{22})^2+(\sigma_{22}-\sigma_{33})^2 + (\sigma_{33}-\sigma_{11})^2+6(\sigma_{12}^2+\sigma_{23}^2+\sigma_{31}^2)]}$$

Case-I

$\sigma_{11} = 0;\quad \sigma_{22} = \sigma;\quad \sigma_{33} = 0;\quad \sigma_{12} = 0$

$\sigma_{23} = \tau;\quad \sigma_{31} = 0$

$$\therefore\ \sigma_v = \sqrt{\frac{1}{2}(2\sigma^2+6\tau^2)} = \sqrt{\sigma^2+3\tau^2}$$

Case-II

$\sigma_{11} = 0;\quad \sigma_{22} = \sigma;\quad \sigma_{33} = 0;\quad \sigma_{12} = 0$

$\sigma_{23} = 0;\quad \sigma_{31} = \tau$

$$\therefore\ \sigma_v = \sqrt{\frac{1}{2}(2\sigma^2+6\tau^2)} = \sqrt{\sigma^2+3\tau^2}$$

In both the cases, σ_v is same, so both yield simultaneously.

1.11 Sol.

$\sigma_1 = 100$ MPa, $\sigma_2 = -100$ MPa

As per Von-Mises yield theory:

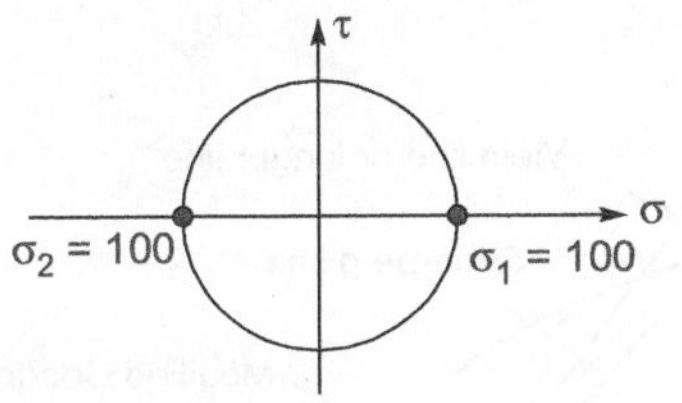

$$\frac{1}{2}[(\sigma_1-\sigma_2)^2+(\sigma_2-\sigma_3)^2+(\sigma_1-\sigma_3)^2] \le \left(\frac{\sigma_y}{N}\right)^2$$

In this case, $\sigma_3 = 0$

$$[\sigma_1^2+\sigma_2^2-\sigma_1\sigma_2] \le \left(\frac{\sigma_y}{N}\right)^2$$

$$\Rightarrow\ [(100)^2+(100)^2+(100)^2] \le \left(\frac{300}{N}\right)^2$$

$$\Rightarrow\ N^2 = \frac{9}{3} \Rightarrow N \le \sqrt{3}$$

$$\therefore\ N = 1.732$$

1.12 Sol.

According to Von-Mises criterion,

$$\tau_y = \frac{\sigma_y}{\sqrt{3}} = \frac{300}{\sqrt{3}} = 173.1 \text{ MPa}$$

1.13 Sol.

$$\sigma_{min} = 200 \text{ MPa},\ \sigma_{max} = 400 \text{ MPa}$$

$$\sigma_a = \frac{\sigma_{max}-\sigma_{min}}{2} = \frac{400-200}{2} = 100 \text{ MPa}$$

1.14 (c)

Stress amplitude

$$\sigma_a = \frac{\sigma_{max}-\sigma_{min}}{2} = \frac{250-50}{2} = 100 \text{ MPa}$$

$$\text{Stress ratio} = \frac{\sigma_{min}}{\sigma_{max}} = \frac{50}{250} = 0.2$$

1.15 (b)

$$\sigma_{1,2} = \frac{1}{2}\left[(\sigma_x+\sigma_y) \pm \sqrt{(\sigma_x-\sigma_y)^2+4\tau_{xy}^2}\right]$$

$$= \frac{1}{2}\left[(80+20) \pm \sqrt{(60)^2+4\times 40^2}\right]$$

$$\sigma_1 = 100 \text{ MPa},\ \sigma_2 = 0 \text{ MPa}$$

$$(\tau_{max}) = \frac{\sigma_1}{2} = 50 \text{ MPa}$$

$$\text{FOS (Factor of safety)} = \frac{\tau_{ys}}{\tau_{max}} = \frac{100}{50} = 2$$

1.16 (a)

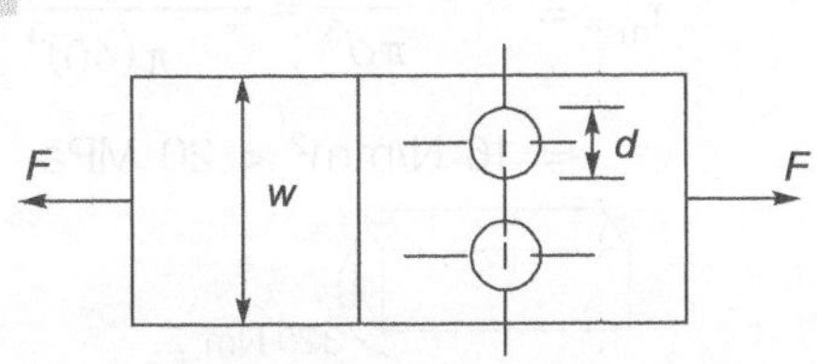

Safe condition to avoid tearing failure of the joint along the row of rivets, as it has minimum area to resist external load F.

$$(\sigma_{max})_{tensile} \le (\sigma_t)_{permissible}$$

$$\frac{F}{(w-2d)t} = S_t$$

$$t \geq \frac{F}{S_f(w-2d)}$$

1.17 Sol.

Given variable loading is a completely reversed fatigue or variable loading because

$$\sigma_{max} = -\sigma_{min}$$

Hence, $\sigma_{mean} = \sigma_m = 0$; $\sigma_{variable} = \sigma_v = \sigma_{max}$

For completely reversed fatigue loading Soderberg, Goodman, Gerber and strength criterion will give same results.

As per strength criterion,

$$\sigma_{max} \leq \sigma_{per} \quad \text{or} \quad \frac{\text{failure stress}}{\text{F.O.S.}}$$

$$\sigma_{max} = \frac{\text{Endurance limit}}{N}$$

$$N = \frac{\text{Endurance limit}}{\text{Maximum stress}} = \frac{100}{50} = 2$$

1.18 (b)

$$\text{Maximum shear stress} = \frac{\sigma_1 - \sigma_3}{2}$$

1.19 (d)

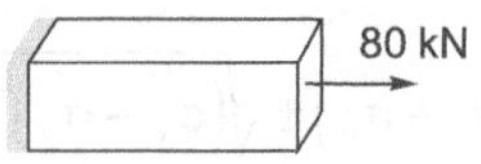

$$\sigma = \frac{80\times10^3}{40^2} = 50 \text{ N/mm}^2$$

$$\tau_{max} = \frac{\sigma}{2} = 25 \text{ N/mm}^2 > 20 \text{ MPa}$$

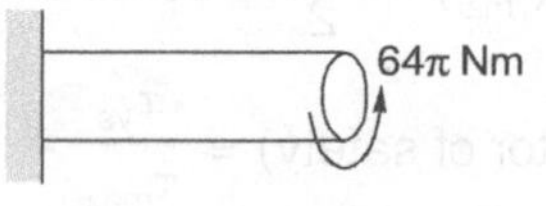

$$\tau_{max} = \frac{16T}{\pi d^3} = \frac{16\times64\pi\times10^3}{\pi(40)^3}$$

$$= 16 \text{ N/mm}^2 < 20 \text{ MPa}$$

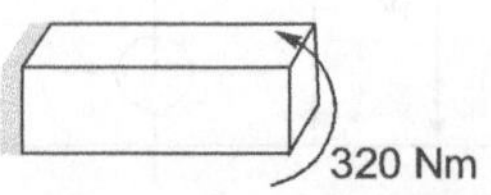

$$\sigma = \frac{\mu}{a^3/6} = \frac{320\times10^3}{\frac{40^3}{6}} = 30 \text{ N/mm}^2$$

$$\tau_{max} = \frac{\sigma}{2} = 15 \text{ N/mm}^2 < 20 \text{ MPa}$$

1.20 Sol.

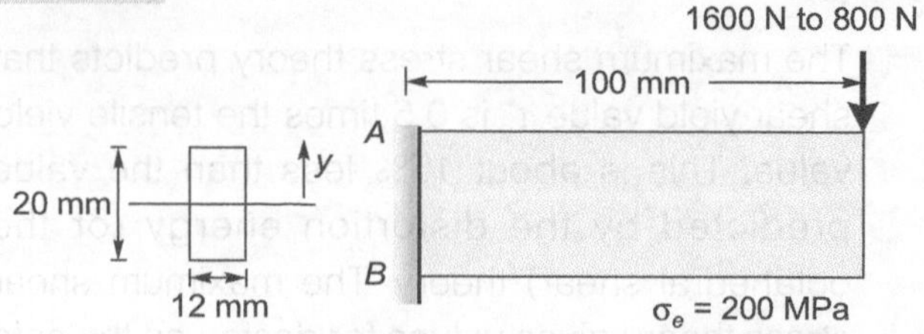

$\sigma_e = 200$ MPa
$S_{yt} = 600$ MPa
$S_{yt} = 500$ MPa

A : Critical Point

$$\sigma_{max,A} = \sigma_{b,max} \text{ at } A \text{ due to 1600 N}$$

$$= \frac{6M}{bd^2} = \frac{6\times1600\times100}{12\times(20)^2}$$

$$\sigma_{max} = 200 \text{ MPa}$$

$$\sigma_{min,A} = \sigma_{b,max} \text{ at } A \text{ due to 800 N}$$

$$= \frac{6\times800\times100}{12\times(20)^2} = 100 \text{ MPa}$$

$$\text{Modified Goodman} = \frac{\sigma_m}{S_{yt}} + \frac{\sigma_a}{\sigma_e} \leq \frac{1}{N}$$

$$\sigma_m = \left|\frac{\sigma_{max}+\sigma_{min}}{2}\right| = 150 \text{ MPa}$$

$$\sigma_a = \left|\frac{\sigma_{max}-\sigma_{min}}{2}\right| = 50 \text{ MPa}$$

$$\frac{150}{600} + \frac{50}{200} \leq \frac{1}{N}$$

$$N \leq 2$$

$$N \approx 2$$

$$\text{Langer, } \frac{\sigma_m}{S_{yt}} + \frac{\sigma_a}{S_{yt}} \leq \frac{1}{N}$$

$$\frac{150}{500} + \frac{50}{500} \leq \frac{1}{N} \qquad (N \leq 2.5)$$

$$N = 2.5$$

Modified Goodman = Safe result of [Goodman or Langer]

1.21 (d)

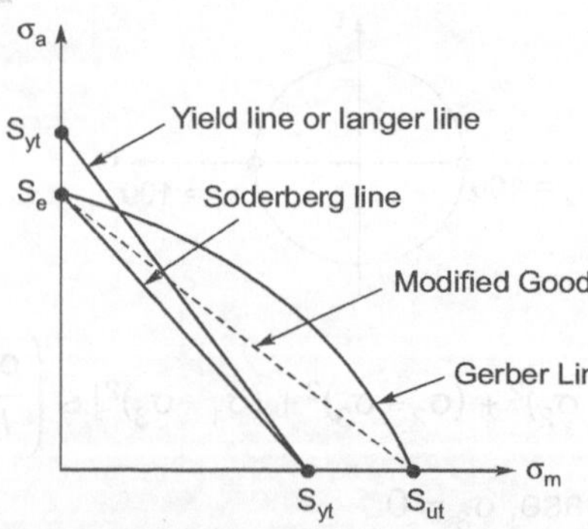

Soderberg line is the most conservative fatigue failure criterion.

■■■■

2 Fatigue Strength and S-N Diagram

2.1 static load is mounted at the centre of a shaft rotating at uniform angular velocity. This shaft will be designed for
(a) the maximum compressive stress (static)
(b) the maximum tensile stress (static)
(c) the maximum bending moment (static)
(d) fatigue loading [2002 : 1 M]

2.2 terms of theoretical stress concentration factor (k_t) and fatigue stress concentration factor (k_f), the notch sensitivity q is expressed as

(a) $\dfrac{(k_f - 1)}{(k_t - 1)}$ (b) $\dfrac{(k_f - 1)}{(k_t + 1)}$

(c) $\dfrac{(k_t - 1)}{(k_f - 1)}$ (d) $\dfrac{(k_f + 1)}{(k_t + 1)}$ [2004 : 1 M]

2.3 he S-N curve for steel becomes asymptotic nearly at
(a) 10^3 cycles (b) 10^4 cycles
(c) 10^6 cycles (d) 10^9 cycles
[2004 : 1 M]

2.4 cylindrical shaft is subjected to an alternating stress of 100 MPa. Fatigue strength to sustain 1000 cycles is 490 MPa. If the corrected endurance strength is 70 MPa, estimated shaft life will be
(a) 1071 cycles (b) 15000 cycles
(c) 281914 cycles (d) 928643 cycles
[2006 : 2 M]

2.5 hich one of the following is the most conservative fatigue failure criterion?
(a) Soderberg (b) Modified Goodman
(c) ASME Elliptic (d) Gerber
[2015 : 1 M, Set-1]

2.6 A machine element has an ultimate strength (σ_u) of 600 N/mm^2, and endurance limit (σ_{en}) of 250 N/mm^2. The fatigue curve for the element on a log-log plot is shown below. If the element is to be designed for a finite life of 10000 cycles, the maximum amplitude of a completely reversed operating stress is _______ N/mm^2.

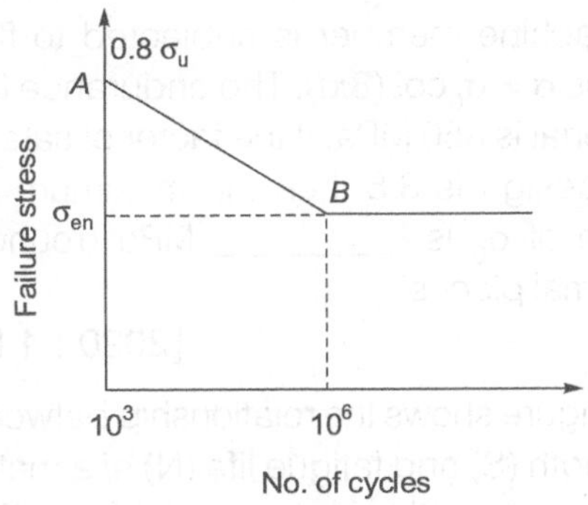

[2017 : 2 M, Set-1]

2.7 Fatigue life of a material for a fully reversed loading condition is estimated from
$\sigma_a = 1100\, N^{-0.15}$
where σ_a is the stress amplitude in MPa and N is the failure life in cycles. The maximum allowable stress amplitude (in MPa) for a life of 1×10^5 cycles under the same loading condition is _______ (correction to two decimal places).
[2018 : 1 M, Set-2]

2.8 A bar is subjected to a combination of a steady load of 60 kN and a load fluctuating between – 10 kN to 90 kN. The corrected endurance limit of the bar is 150 MPa. the yield strength of the material is 480 MPa and the ultimate strength of the material is 600 MPa. The bar cross-section is square with side a. If the factor of safety is 2, the value of a (in mm), according to the modified Goodman's criterion, is ______ (correct to two decimal places).
[2018 : 2 M, Set-2]

2.9 During a high cycle fatigue test, a metallic specimen is subjected to cyclic loading with a mean stress of +140 MPa, and a minimum stress of –70 MPa. The R-ratio (minimum stress to maximum stress) for this cyclic loading is ______ (round off to one decimal place)
[2019 : 1 M, Set-1]

2.10 Endurance limit of a beam subjected to pure bending decreases with
(a) decrease in the surface roughness and increase in the size of the beam
(b) decrease in the surface roughness and decrease in the size of the beam

(c) increase in the surface roughness and increase in the size of the beam
(d) increase in the surface roughness and decrease in the size of the beam

[2019 : 1 M, Set-2]

2.11 A machine member is subjected to fluctuating stress $\sigma = \sigma_0 \cos(8\pi t)$. The endurance limit of the material is 350 MPa. If the factor of safety used in the design is 3.5 then the maximum allowable value of σ_0 is ________ MPa (round off to 2 decimal places).

[2020 : 1 M, Set-2]

2.12 The figure shows the relationship between fatigue strength (S) and fatigue life (N) of a material. The fatigue strength of the material for a life of 1000 cycles is 450 MPa, while its fatigue strength for a life of 10^6 cycles is 150 MPa.

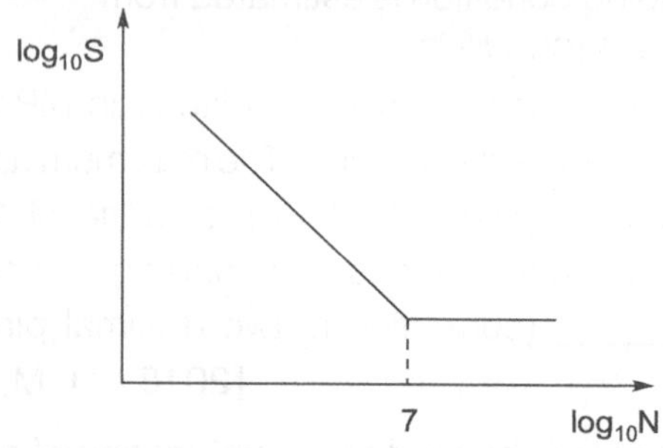

The life of a cylindrical shaft made of this material subjected to an alternating stress of 200 MPa will then be ____ cycles (round off to the nearest integer).

[2021 : 2 M, Set-2]

2.13 Which of the following methods can improve the fatigue strength of a circular mild steel (MS) shaft?
(a) Enhancing surface finish
(b) Shot peening of the shaft
(c) Increasing relative humidity
(d) Reducing relative humidity

[2022 : 1 M, Set-1]

2.14 The S-N curve from a fatigue test for steel is shown. Which one of the options gives the endurance limit?

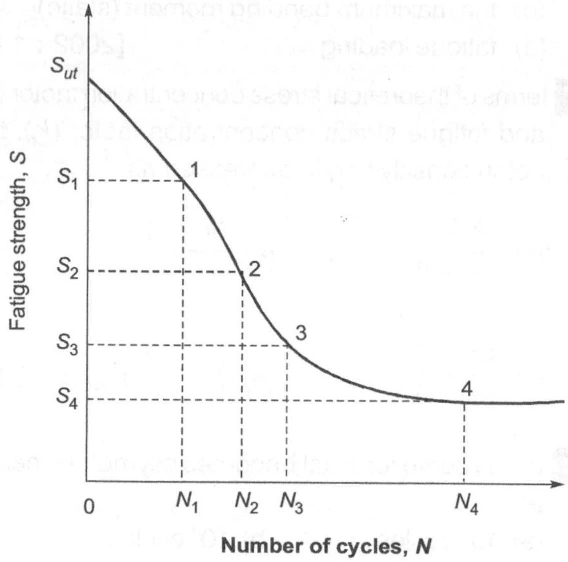

(a) S_{ut}
(b) S_2
(c) S_3
(d) S_4

[2023 : 1 M]

Answers Fatigue Strength and S-N Diagram

2.1 (d) 2.2 (a) 2.3 (c) 2.4 (c) 2.5 (a) 2.6 (386.195) 2.7 (195.61)
2.8 (31.62) 2.9 (–0.2) 2.10 (c) 2.11 (100) 2.12 (163841) 2.13 (a, b, d) 2.14 (d)

Explanations Fatigue Strength and S-N Diagram

2.1 (d)

The elements of shaft will be under fatigue load due to its rotation.

2.2 (a)

We know that

$$K_f = 1 + q(K_t - 1), \quad q = \frac{K_f - 1}{K_t - 1}$$

2.3 (c)

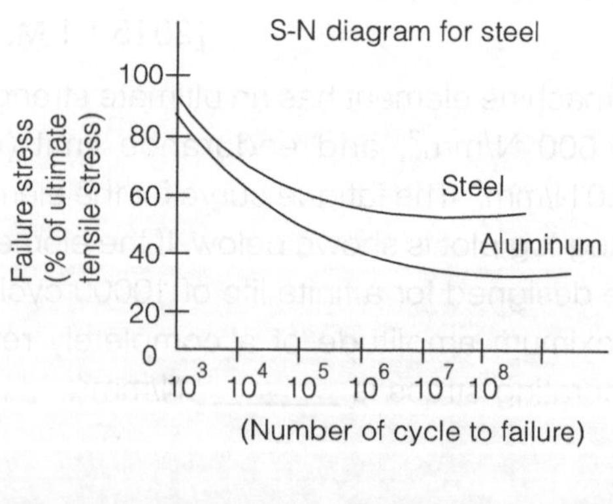

2.4 (c)

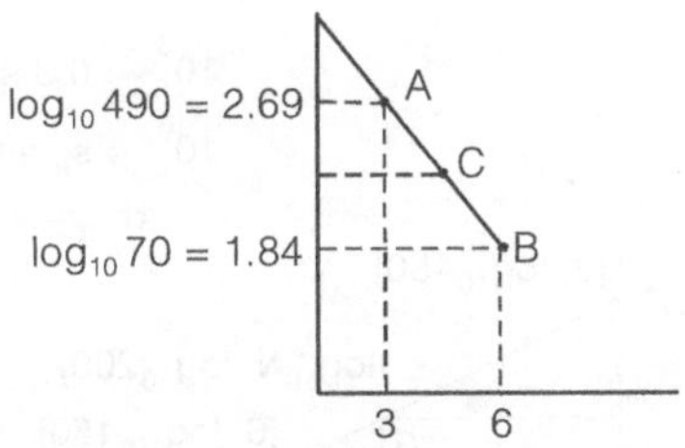

Equation of straight line connecting $(3, \log_{10} 490)$ and $(6, \log_{10} 70)$

$$\frac{y - 1.8451}{x - 6} = \frac{1.84 - 2.6902}{6 - 3}$$

$$y - 1.8451 = -0.2817\,(x - 6)$$

At $\quad y = \log_{10} 100 = 2$

$$\Rightarrow 2 - 1.8451 = -0.2817\,(x - 6)$$

$$x = 5.4501$$

$$\log_{10} N = 5.4501, \quad N = \mathbf{281914}$$

2.6 Sol.

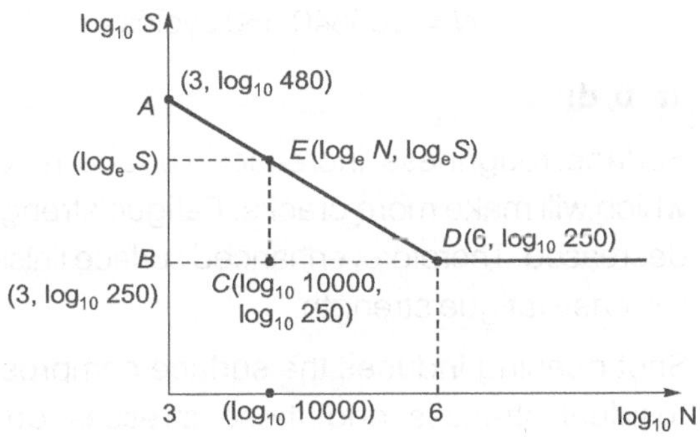

From two similar triangles,
(i.e. ΔABD and ΔECD)

$$\frac{AB}{EC} = \frac{BD}{CD}$$

$$\frac{(\log_{10} 480 - \log_{10} 250)}{(\log_{10} S - \log_{10} 250)} = \frac{6 - 3}{6 - \log_{10} 10000}$$

$$\frac{2.68124 - 2.39794}{\log_{10} S - 2.39794} = 1.5$$

$$S = \mathbf{386.195\ N/mm^2}$$

2.7 Sol.

$$\frac{\sigma_{max} - \sigma_{min}}{2} = 1100\, N^{-0.15}$$

$$\frac{\sigma_{max} - (-\sigma_{max})}{2} = 1100\, N^{-0.15}$$

$$\frac{2\sigma_{max}}{2} = 1100\, N^{-0.15}$$

$$\sigma_{max} = 1100\, N^{-0.15} = 1100 \times (10^5)^{-0.15}$$

$$= 1100 \times (10)^{-0.75} = \frac{1100}{5.62}$$

$$\boldsymbol{\sigma_{max} = 195.61\ \text{MPa}}$$

2.8 Sol.

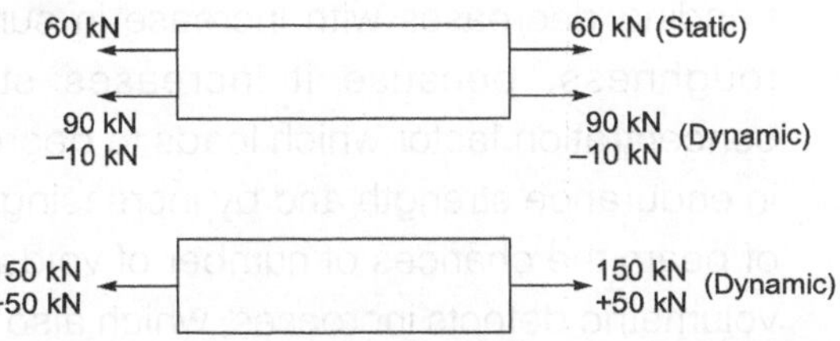

$$P_m = \frac{P_{max} + P_{min}}{2}$$

$$P_a = \frac{P_{max} - P_{min}}{2}$$

$$P_m = 100 \text{ kN}$$

$$P_a = 50 \text{ kN}$$

$$\sigma_m = \frac{100 \times 10^3}{a^2} \text{MPa}$$

$$\sigma_a = \frac{50 \times 10^3}{a^2} \text{MPa}$$

Solution by Goodman's equation,

$$\frac{\sigma_m}{S_{ut}} + \frac{\sigma_a}{\sigma_e} = \frac{1}{N}; \; 1000\left[\frac{100}{a^2 \times 600} + \frac{50}{150a^2}\right] = \frac{1}{2}$$

$$a^2 = 1000$$

$$a = 31.62 \text{ mm}$$

Solution by yield (Langer's) line equation,

$$\frac{\sigma_m}{S_{yt}} + \frac{\sigma_a}{S_{yt}} = \frac{1}{N}; 1000\left[\frac{100}{480a^2} + \frac{50}{480a^2}\right] = \frac{1}{2}$$

$$a^2 = 625$$

$$a = 25 \text{ mm}$$

Hence final answer by modified **Goodman's criterion is 31.62 mm**

2.9 Sol.

$$\sigma_{min} = 70 \text{ MPa} \quad \text{(Compressive)},$$

$$\sigma_{mean} = 140 \text{ MPa}$$

$$\frac{\sigma_{min}}{\sigma_{max}} = ?$$

We know that,

$$\sigma_{mean} = \frac{\sigma_{max} + \sigma_{min}}{2}$$

$$140 = \frac{\sigma_{max} - 70}{2}$$

$$280 + 70 = \sigma_{max}$$

$$\sigma_{max} = 350 \text{ MPa (Tensile)}$$

$$\frac{\sigma_{min}}{\sigma_{max}} = -\frac{70}{350} = -0.2$$

2.10 (c)

Endurance limit of beams subjected to pure bending decreases with increase in surface roughness, because it increases stress concentration factor which leads to decrease in endurance strength and by increasing size of beam the chances of number of voids and volumetric defects increases, which also lead to decrease in endurance limit.

2.11 Sol.

Fluctuating stress,

$$\sigma = \sigma_o \cos(8\pi t)$$

Hence, $\sigma_{max} = \sigma_o$ (at $\cos(8\pi t) = 1$)

$\sigma_{min} = -\sigma_o$ (at $\cos(8\pi t) = -1$)

$$\sigma_{mean} = \frac{\sigma_{max} + \sigma_{min}}{2} = 0$$

$$\sigma_v = \frac{\sigma_{max} - \sigma_{min}}{2} = \frac{2\sigma_o}{2} = \sigma_o$$

$\sigma_e = 350$ MPa (given)

FOS = 3.5

From strength criteria, $\frac{\sigma_v}{\sigma_e} \le \frac{1}{FOS}$

$$\frac{\sigma_o}{350} \le \frac{1}{3.5}$$

$$\sigma_o \le 100 \text{ MPa}$$

2.12 Sol.

$10^3 \rightarrow 0.9\, s_{ut} = 450$ Mpa

$10^6 \rightarrow s_e = 150$ MPa

$\sigma_{act} = \pm 200$ MPa

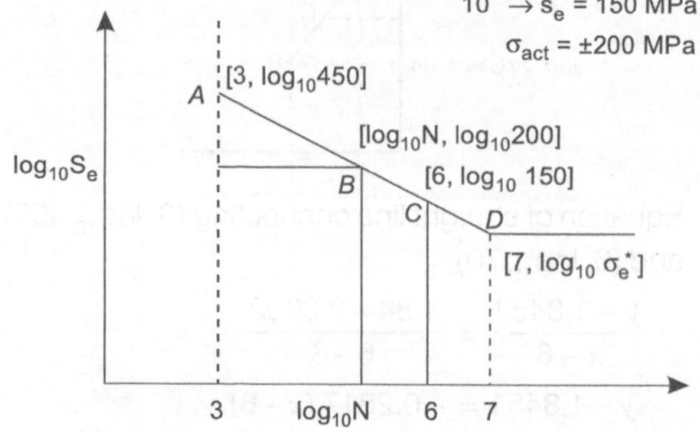

Equation of line *AB*:

$$y - y_1 = \frac{(y_2 - y_1)}{(x_2 - x_1)}[x - x_1]$$

$$\log_{10}200 - \log_{10}450 = \frac{\log_{10}150 - \log_{10}450}{(6-3)}[\log_{10} N - 3]$$

$$N = 163840.580 \text{ cycles}$$

2.13 (a, b, d)

Surface roughness increases surface notches which will make more cracks. Fatigue strength is decreased. Therefore, enhanced surface finish will increase fatigue strength.

Shot peening induces the surface compressive residual stresses and these stresses do not allow crack propagation to take place. Shot peening increases fatigue strength.

Humidity can cause corrosion, so if humidity is there, it may lead to more crack formation. Therefore, fatigue strength increases when relative humidity decreases.

2.14 (d)

Endurance limit is the fatigue strength corresponding to infinite number of cycles. Thus, S_4 gives the endurance limit.

■■■■

3 Bolted Riveted and Welded Joint

3.1 If the ratio of the diameter of rivet hole to the pitch of rivets is 0.25, then the tearing efficiency of the joint is

(a) 0.50 (b) 0.75
(c) 0.25 (d) 0.87 **[1996 : 1 M]**

3.2 The bolts in a rigid flanged coupling connecting two shafts transmitting power are subjected to

(a) shear force and bending moment
(b) axial force
(c) torsion
(d) torsion and bending moment **[1996 : 1 M]**

3.3 Bolts in the flanged end of pressure vessel are usually pre-tensioned. Indicate which of the following statements is true

(a) Pre-tensioning helps to seal the pressure vessel
(b) Pre-tensioning increase the fatigue life of the bolts
(c) Pre-tensioning reduces the maximum tensile stress in the bolts
(d) Pre-tensioning helps to reduce the effect of pressure pulsations in the pressure vessels

[1998 : 1 M]

3.4 In a bolted joint two smooth members are connected with an axial tightening force of 2200 N. If the bolt used has metric threads of 4 mm pitch, the torque required for achieving the tightening force is

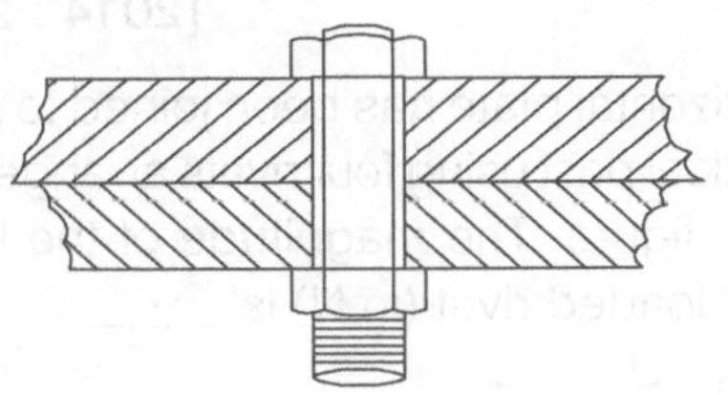

(a) 0.7 Nm (b) 1.0 Nm
(c) 1.4 Nm (d) 2.8 Nm **[2004 : 2 M]**

3.5 A 60 mm long and 6 mm thick fillet weld carries a steady load of 15 kN along the weld. The shear strength of the weld material is equal to 200 MPa. The factor of safety is

(a) 2.4 (b) 3.4
(c) 4.8 (d) 6.8 **[2006 : 2 M]**

3.6 A bolted joint is shown below. The maximum shear stress, in MPa, in the bolts at *A* and *B*, respectively are

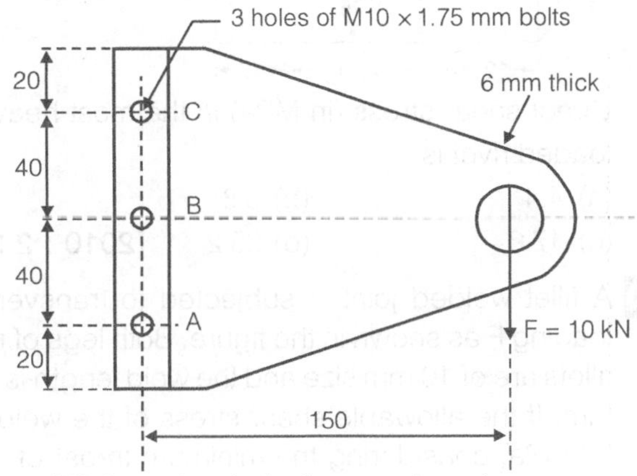

(All dimensions in the figure are in mm)

(a) 242.6, 42.5 (b) 42.5, 242.6
(c) 42.5, 42.5 (d) 60.15, 343.64

[2007 : 2 M]

Linked Answer Questions 3.7 and 3.8

A steel bar of 10 × 50 mm is cantilevered with two *M*12 bolts (*P* and *Q*) to support a static load of 4 kN as shown in the figure.

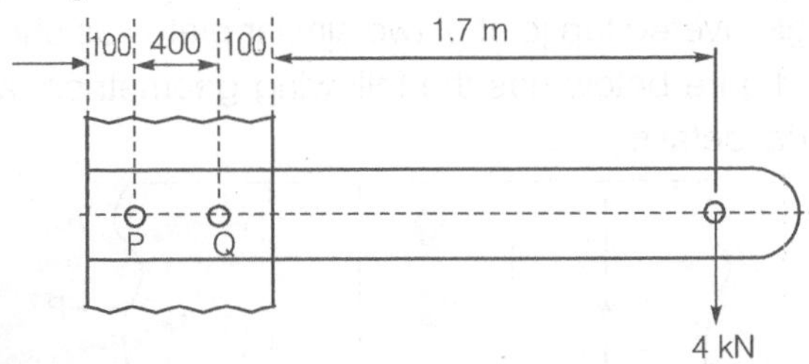

3.7 The primary and secondary shear loads on bolt *P*, respectively, are

(a) 2 kN, 20 kN (b) 20 kN, 2 kN
(c) 20 kN, 0 kN (d) 0 kN, 20 kN

[2008 : 2 M]

3.8 The resultant shear stress on bolt *P* is closest to:

(a) 132 MPa (b) 159 MPa
(c) 178 MPa (d) 195 MPa

[2008 : 2 M]

3.9 A bracket (shown in figure) is rigidly mounted on wall using four rivets. Each rivet is 6 mm in diameter and has an effective length of 12 mm.

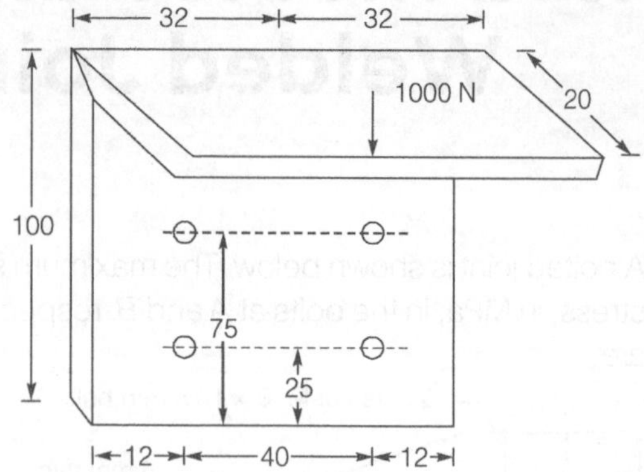

Direct shear stress (in MPa) in the most heavily loaded rivet is

(a) 4.4 (b) 8.8

(c) 17.6 (d) 35.2 [2010 : 2 M]

3.10 A fillet welded joint is subjected to transverse loading F as shown in the figure. Both legs of the fillets are of 10 mm size and the weld length is 30 mm. If the allowable shear stress of the weld is 94 MPa, considering the minimum throat of the weld, the maximum allowable transverse load in kN is

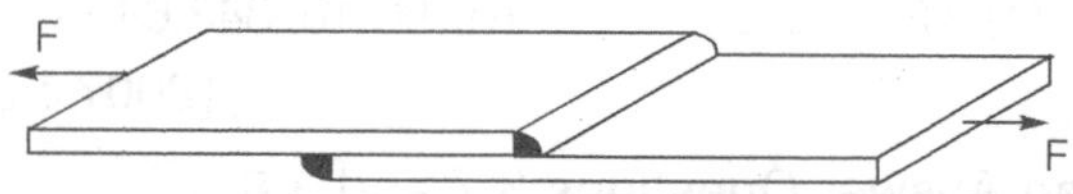

(a) 14.44 (b) 17.92

(c) 19.93 (d) 22.16 [2012 : 2 M]

Common Data Questions 3.11 and 3.12

A single riveted lap joint of two similar plates as shown in the figure below has the following geometrical and material details.

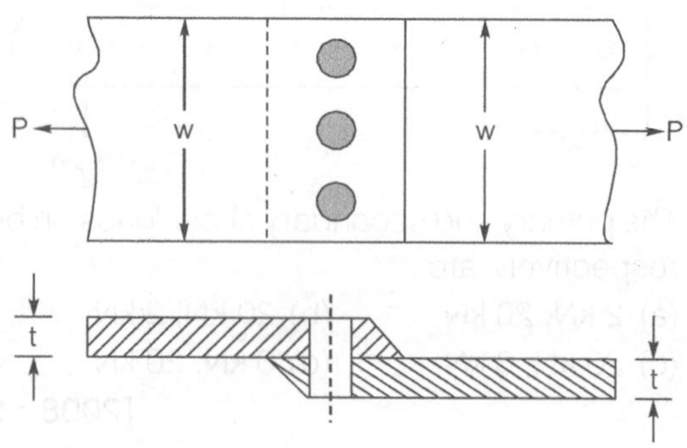

width of the plate w = 200 mm, thickness of the plate t = 5 mm, number of rivets n = 3, diameter of the rivet d_r = 10 mm, diameter of the rivet hole d_h = 11 mm, allowable tensile stress of the plate σ_p = 200 MPa, allowable shear stress of the rivet σ_s = 100 MPa and allowable bearing stress of the rivet σ_c = 150 MPa.

3.11 If the rivets are to be designed to avoid crushing failure, the maximum permissible load P in kN is

(a) 7.50 (b) 15.00

(c) 22.50 (d) 30.00 [2013 : 2 M]

3.12 If the plates are to be designed to avoid tearing failure, the maximum permissible load P in kN is

(a) 83 (b) 125

(c) 167 (d) 501 [2013 : 2 M]

3.13 For the three bolt system shown in the figure, the bolt material has shear yield strength of 200 MPa. For a factor of safety of 2, the minimum metric specification required for the bolt is

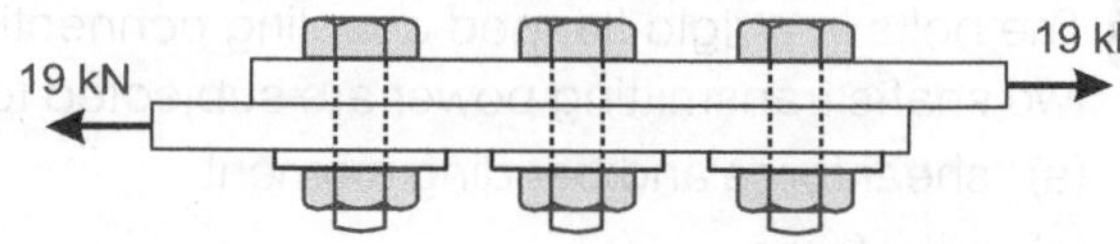

(a) $M8$ (b) $M10$

(c) $M12$ (d) $M16$

[2014 : 2 M, Set-2]

3.14 A bolt of major diameter 12 mm is required to clamp two steel plates. Cross sectional area of the threaded portion of the bolt is 84.3 mm^2. The length of the threaded portion in grip is 30 mm, while the length of the unthreaded portion in grip is 8 mm. Young's modulus of material is 200 GPa. The effective stiffness (in MN/m) of the bolt in the clamped zone is ______

[2014 : 2 M, Set-4]

3.15 A butt weld joint is developed on steel plates having yield and ultimate tensile strength of 500 MPa and 700 MPa, respectively. The thickness of the plates is 8 mm and width is 20 mm. Improper selection of welding parameters caused an undercut of 3 mm depth along the weld. The maximum transverse tensile load (in kN) carrying capacity of the developed weld joint is ______.

[2014 : 2 M, Set-4]

3.16 A horizontal plate has been joined to a vertical to a vertical post using four rivets arranged as shown in the figure. The magnitude of the load on the worst loaded rivet (in N) is ______.

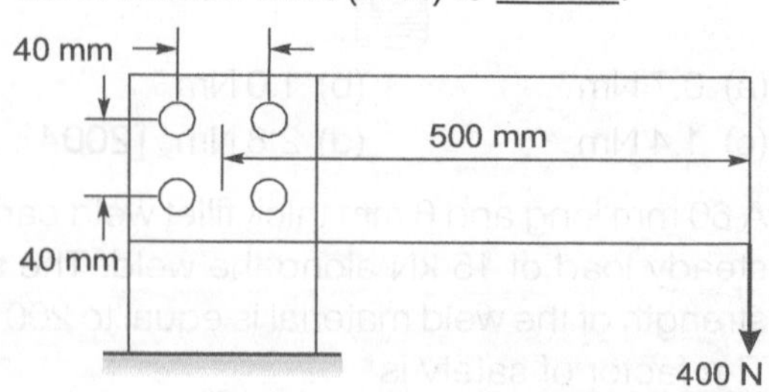

[2015 : 2 M, Set-1]

3.17 A cantilever bracket is bolted to a column using three $M12 \times 1.75$ bolts, P, Q and R. The value of maximum shear stress developed in the bolt P (in MPa) is ______.

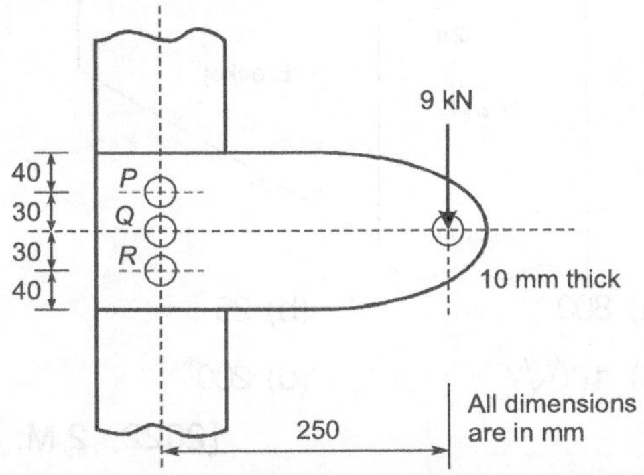

[2015 : 2 M, Set-3]

3.18 A bolted joint has four bolts arranged as shown in figure. The cross sectional area of each bolt is 25 mm^2. A torque T = 200 Nm is acting on the joint. Neglecting friction due to clamping force, maximum shear stress in a bolt is ______ MPa.

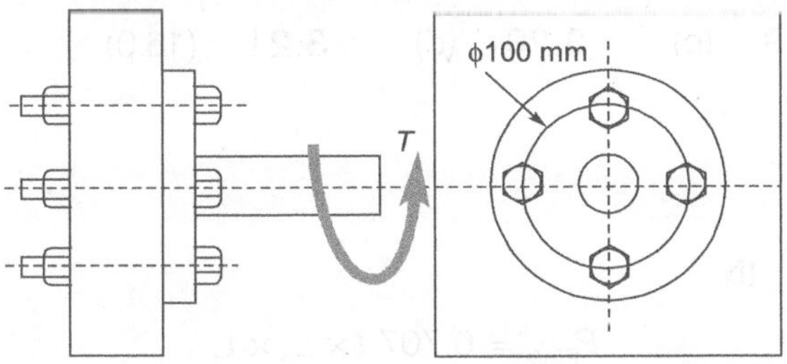

[2016 : 2 M, Set-3]

3.19 A steel plate, connected to a fixed channel using three identical bolts A, B and C, carries a load of 6 kN as shown in the figure. Considering the effect of direct load and moment, the magnitude of resultant shear force (in kN) on bolt C is

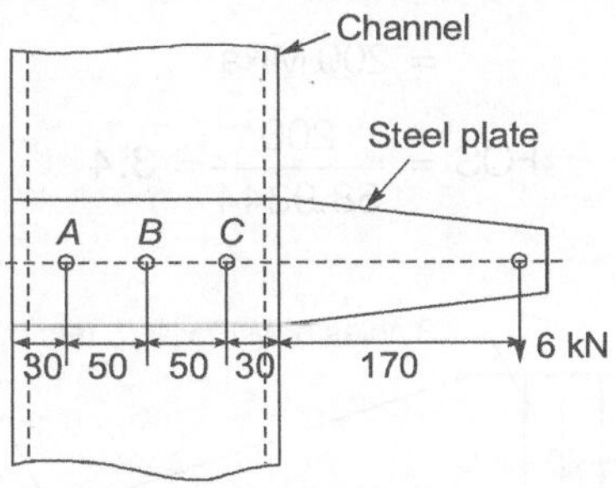

All dimensions are in mm

(a) 13 (b) 15
(c) 17 (d) 30 **[2017 : 2 M, Set-2]**

3.20 Pre-tensioning of a bolted joint is used to
(a) strain harden the bolt head
(b) decrease stiffness of the bolted joint
(c) increase stiffness of the bolted joint
(d) prevent yielding of the thread root

[2018 : 1 M, Set-2]

3.21 A rectangular steel bar of length 500 mm, width 100 mm, and thickness 15 mm is cantilevered to a 200 mm steel channel using 4 bolts, as shown.

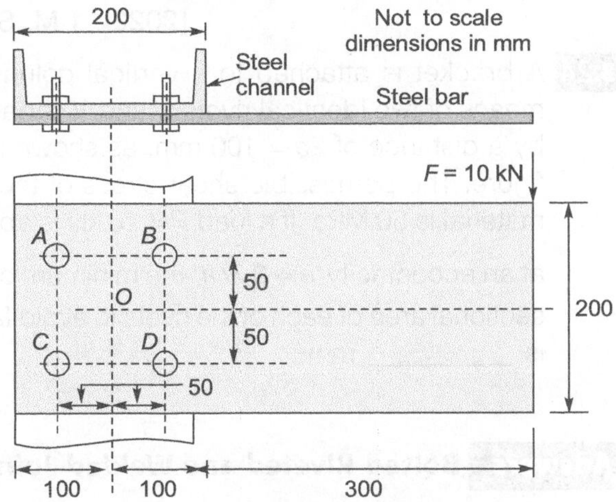

For an external load of 10 kN applied at the tip of the steel bar, the resultant shear load on the bolt at B, is ________ kN (round off to one decimal place).

[2020 : 2 M, Set-1]

3.22 A cantilever beam of rectangular cross-section is welded to a support by means of two fillet welds as shown in figure. A vertical load of 2 kN acts at free end of the beam.

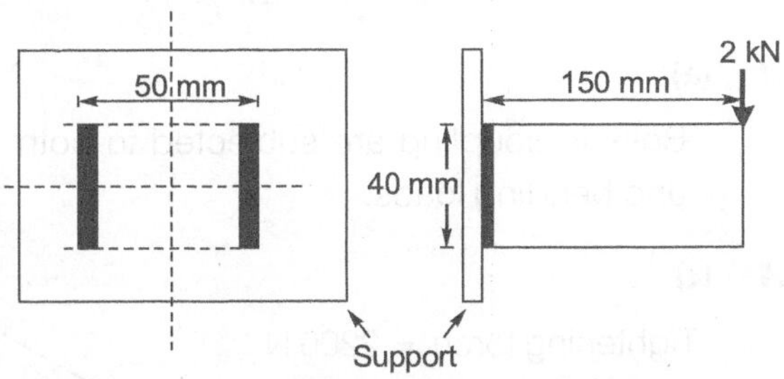

Considering that the allowable shear stress in weld is 60 N/mm^2, the minimum size (leg) of the weld required is ________ mm [Round off to one decimal place]

[2021 : 2 M, Set-1]

3.23 A square threaded screw is used to lift a load W by applying a force F. Efficiency of square threaded screw is expressed as
(a) The ratio of work done by W per revolution to work done by F per revolution
(b) $\frac{W}{F}$

(c) $\frac{F}{W}$

(d) The ratio of work done by F per revolution to work done by W per revolution

[2022 : 1 M, Set-1]

3.24 A bracket is attached to a vertical column by means of two identical rivets U and V separated by a distance of $2a$ = 100 mm, as shown in the figure. The permissible shear stress of the rivet material is 50 MPa. If a load P = 10 kN is applied at an eccentricity $e = 3\sqrt{7}a$, the minimum cross-sectional area of each of the rivets to avoid failure is ________ mm^2.

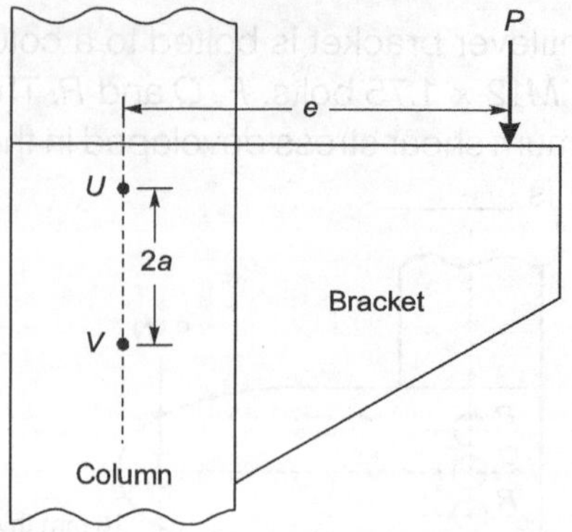

(a) 800
(b) 25
(c) $100\sqrt{7}$
(d) 200

[2022 : 2 M, Set-1]

Answers Bolted Riveted and Welded Joint

3.1 (b)	3.2 (a)	3.3 (b)	3.4 (c)	3.5 (b)	3.6 (a)	3.7 (a)	3.8 (b)
3.9 (b)	3.10 (c)	3.11 (c)	3.12 (c)	3.13 (b)	3.14 (468.77)	3.15 (70)	
3.16 (1839.83)	3.17 (332.6321)	3.18 (40)	3.19 (c)	3.20 (c)	3.21 (16.0)		
3.22 (6.65)	3.23 (a)	3.24 (a)					

Explanations Bolted Riveted and Welded Joint

3.1 (b)

$$\eta = \frac{p-d}{p} = 1 - \frac{d}{p} = 1 - 0.25 = \mathbf{0.75}$$

3.2 (a)

Bolts in coupling are subjected to both shear and bending loads.

3.4 (c)

Tightening force = 2200 N

Pitch = 0.004 m

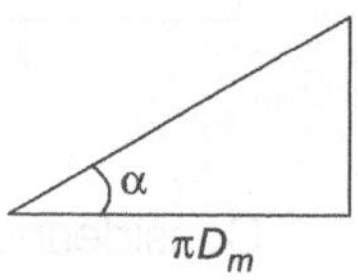

For single start thread

Lead = Pitch

$$\tan\alpha = \frac{L}{\pi D_m} = \frac{\text{pitch}}{\pi D_m}$$

$$T = \frac{WD_m}{2} \times \tan(\phi + \alpha)$$

For smooth thread, $\phi = 0$ [given in question]

$$T = \frac{WD_m}{2} \times \frac{\text{Pitch}}{\pi D_m} = W \times \frac{\text{Pitch}}{2\pi}$$

$$= 2200 \times \frac{0.004}{2\pi} = 1.4 \text{ Nm}$$

3.5 (b)

$$P_{PFW} = 0.707\, t \times L_e \times \tau_s$$

{here PFW → parallel filter weld}

$$\frac{15 \times 10^3}{0.707 \times 6 \times 60} = \tau_s$$

or τ_s = 58.9344 MPa

Given shear strength of material

= 200 MPa

$$\text{FOS} = \frac{200}{58.9344} = \mathbf{3.4}$$

3.6 (a)

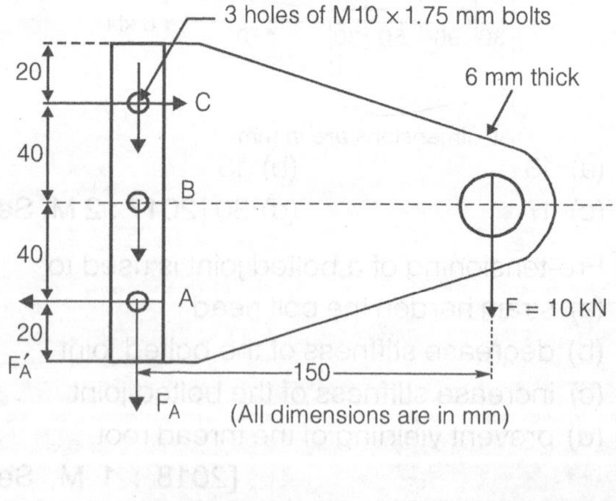

$$F_A = FB = FC = \frac{10}{3}\text{ kN}$$

$$F'_A = \frac{Per}{r_A^2 + r_B^2 + r_C^2}$$

$$= \frac{10 \times 10^3 \times 150 \times 40}{40^2 + 0^2 + 40^2}$$

$$F'_A = 18.75\text{ kN} = F'_C$$

Shear load at A and C

$$F_R = \sqrt{F_A^2 + F_A'^2} \Rightarrow \sqrt{3.33_A^2 + 18.75^2}$$

$$F_R = 19.04\text{ kN}$$

$$\tau_{max} = \frac{19.04 \times 10^3}{\frac{\pi}{4} \times 10^2}$$

$$(\tau_{max})_{A,C} = 242.5\text{ MPa}$$

$$(\tau_{max})_B = \mathbf{42.44\ MPa}$$

3.7 (a)

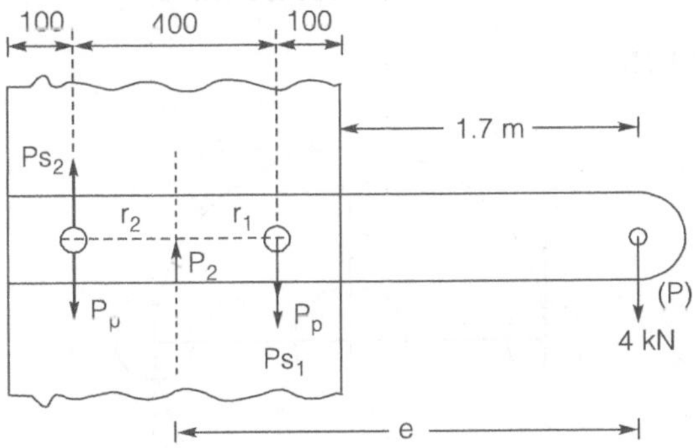

⇒ Primary shear load due to P_1

$$P_p = \frac{P_1}{n} = \frac{4}{2} = \mathbf{2\ kN}$$

Now for secondary shear load effect of P and P_2

$$TM = P \times e = 8 \times 10^6\text{ N-mm.}$$

$$Ps_1 r_1 + Ps_2 r_2 = P \times e$$

$$\frac{Ps_1}{r_1}\left[r_1^2 + r_1^2\right] = P \times e$$

$$Ps_1 = Ps_2$$

$$r_1 = r_2$$

$$\frac{Ps_1}{200}[2 \times (200)^2] = 8 \times 10^6$$

$$Ps_1 = Ps_2 = 20{,}000\text{ N} = \mathbf{20\ kN.}$$

3.8 (b)

Resultant shear load on bolt P

$$R_p = 20\text{ kN} - 2\text{ kN}$$

$$= 18\text{ kN (here d = 12 mm)}$$

$$\tau_{max} = \frac{18 \times 10^3}{\frac{\pi}{4}(12)^2} = \mathbf{159\ MPa}$$

3.9 (b)

There are four rivets, direct force on each rivet $= \frac{1000}{4} = 250$ N.

Direct shear stress,

$$\tau = \frac{250}{\text{Area of rivet}} = \frac{250}{\frac{\pi}{4} \times (6 \times 10^{-3})^2}$$

$$= \mathbf{8.8\ MPa}$$

Here, direct shear stress will be same for each bolts.

3.10 (c)

The shear strength of the joint for single parallel fillet weld is

P = Throat area × Allowable stress

$= t \times l \times \tau$

t = leg size × sin 45° = 0.707 × leg size

For double parallel fillet weld

$P = 2 \times 0.707 \times \text{leg size} \times l \times \tau$

$= 0.707 \times \text{leg size} \times (2 \times l) \times \tau$ [$2l = 30$]

$= 0.707 \times 0.01 \times 0.03 \times 94 \times 10^6 = 19{,}937.4$ N

$P = 19.93$ kN

3.11 (c)

$$P_{crushing} = 3(d \times t \times \sigma_c) = 3(10 \times 5 \times 150)$$

$$= \mathbf{22.50\ kN}$$

3.12 (c)

Tearing load

$$= (W - 3d)t \times \sigma_p = (200 - 3 \times 11) \times 5 \times 200$$

$$= \mathbf{167\ kN}$$

3.13 (b)

$$\text{Design stress} = \frac{\tau_y}{\text{FOS}} = \frac{200}{2} = 100\text{ MPa}$$

Load = 19 kN

$$\text{Total shear area} = \frac{19\times10^3}{100\times10^6} = 1.9\times10^{-4}\ \text{m}^2$$

$$\text{Shear area for one bolt} = \frac{A}{3} = 6.33\times10^{-5}\ \text{m}^2$$

$$\therefore \quad d = \left(\frac{4\times6.33\times10^{-5}}{\pi}\right)^{1/2} = 8.982\times10^{-3}\ \text{m} = 8.982\ \text{mm}$$

So, $M10$

3.14 Sol.

$$k_t = \frac{A_t E}{L_t} \quad \text{(Stiffness in threaded portion)}$$

$$k_t = \frac{84.3\times200\times10^3}{30\times10^{-3}} = 562\times10^6\ \text{N/m} = 562\ \text{MN/m}$$

$$k_d = \frac{A_d E}{L_d} \quad \text{(Stiffness in unthreaded region)}$$

A_d (major diameter c/s area)

$$= \frac{\pi}{4}d^2 = 0.785\times144 = 113.04\ \text{mm}^2$$

L_d (length of unthreaded portion) = 8 mm

$$\therefore \quad k_d = \frac{113.04\times200\times10^3}{8\times10^{-3}} = 2826\times10^6\ \text{N/m} = 2826\ \text{MN/m}$$

$$\frac{1}{k} = \frac{1}{k_t}+\frac{1}{k_d} \Rightarrow k = \frac{k_d k_t}{k_d+k_t} = \frac{2826\times562}{2826+562}$$

$$= \mathbf{468.77\ MN/m}$$

3.15 Sol.

$$P_t = S_{ut}(t - t_{undercut})L = 700\times(8-3)\times20 = 700\times100\ \text{N} = \mathbf{70\ kN}$$

3.16 Sol.

Primary force (P_1) = 100 N

$$P_1'' = P_2'' = P_3'' = P_4'' = \frac{P_e\times r}{4r^2}$$

$$= \frac{400\times500\times\left(40\sqrt{2}/2\right)}{4\times\left(\frac{40\sqrt{2}}{2}\right)^2}$$

$$P_1'' = \frac{400\times500\times20\sqrt{2}}{4\times\left(20\sqrt{2}\right)^2} = \frac{400\times500}{4\times20\sqrt{2}}$$

$$P_1'' = 1767.766\ \text{N}$$

(1 and 4) are worst loaded

$$\therefore P_{net}^2 = P_1 + P_1''^2 + 2P_1P_1''\cos45°$$

$$P_{net} = \mathbf{1839.83\ N}$$

3.17 Sol.

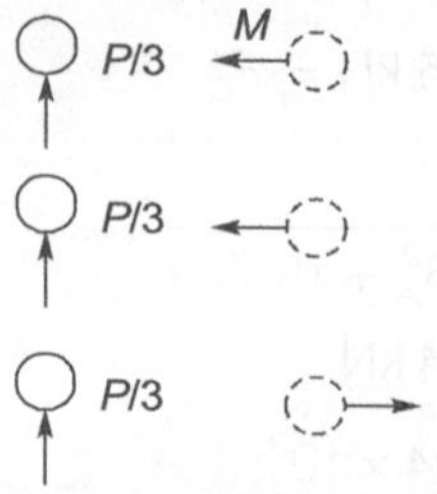

$$M = \frac{Pe\times r_1}{2\times r_1^2}$$

$$M = \frac{9000\times250\times30}{2\times30^2}$$

$$M = 37500$$

Total shear force on P,

$$F_s = \sqrt{(37500)^2+(3000)^2} = 37619.808\ \text{N}$$

$$\therefore \quad \tau_{max} = \frac{F_s}{A} = \frac{37619.808}{\pi\times0.25\times12^2} = \mathbf{332.6321\ MPa}$$

3.18 Sol.

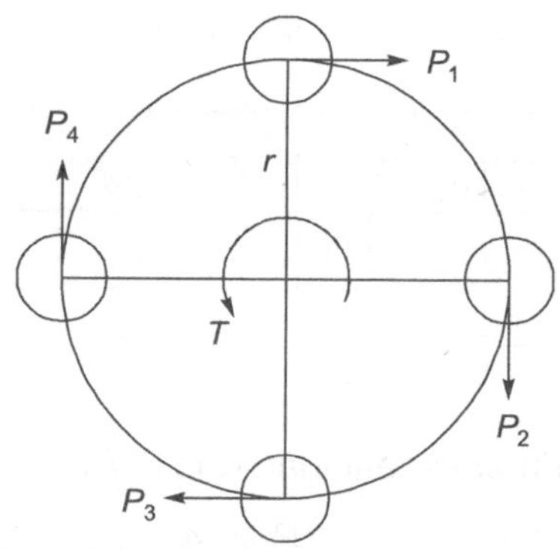

$$\therefore \quad P_1r + P_2r + P_3r + P_4r - T = 0$$

$P_1 = P_2 = P_3 = P_4$ [By symmetry]

$$4Pr = T$$

$$P = \frac{T}{4r} = \frac{200}{4\times\frac{50}{1000}} = 1000\ \text{N}$$

$$\text{Shear stress} = \frac{T}{\text{Area}} = \frac{1000}{25} = 40\ \text{N/mm}^2 = \mathbf{40\ MPa}$$

3.19 (c)

$$P_P = \frac{P}{n} = \frac{P}{3} = 2\ \text{kN}$$

$$r_A = r_C = 50;\ (P_S)_A = (P_S)_C$$

$$r_B = 0 \Rightarrow (P_S)_B = 0$$

$$\frac{(P_S)_C}{r_C}\left[r_C^2 + r_B^2 + r_A^2\right] = P\times e$$

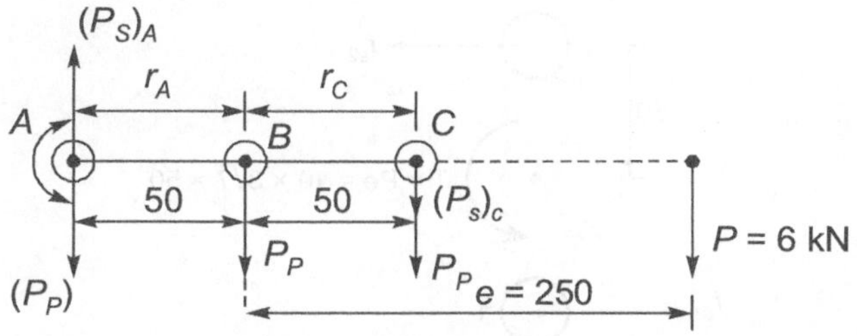

$$(P_S)_C \times 2 \times r_C = P \times e$$

$$(P_S)_C = \frac{6 \times 250}{2 \times 50} = 15\text{ kN}$$

$$\theta_A = 180°;\ \theta_C = 0°$$

Resultant shear force $(R_C) = P_P + (P_S)_C$

$$[\because \theta_C = 0°]$$

$$\mathbf{R_C = 2 + 15 = 17\ kN}$$

$$R_A = (P_S)A - P_P = 13\text{ kN}$$

$$R_B = P_P = 2\text{ kN}$$

3.20 (c)

Pretension increases stiffness of system.

3.21 Sol.

Primary tensile force,

$$F_p = \frac{10}{4} = 2.5\text{ kN}$$

Ecentricity, $e = 400$ mm

$$r = \sqrt{50^2 + 50^2} = 50\sqrt{2}\text{ mm}$$

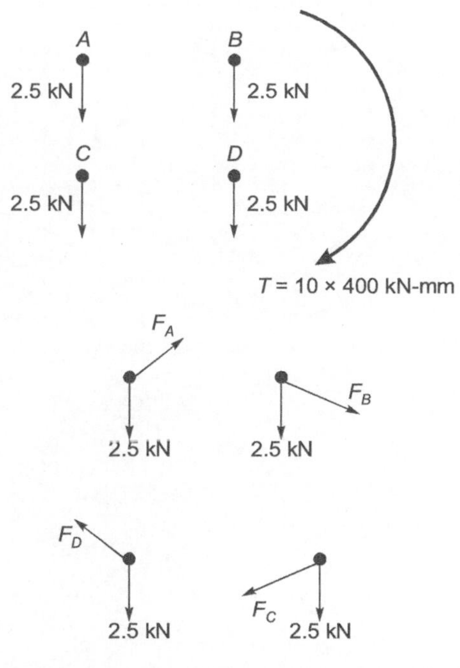

Secondary tensile force,

$$F_A = F_B = F_C = F_D$$

$$= \frac{P \times e}{4 \times r^2} = \frac{10 \times 400}{4 \times (50\sqrt{2})^2} = 14.14\text{ kN}$$

Resultant force at B,

$$B = \sqrt{F_p^2 + F_B^2 + 2f_B \cdot f_p \times \cos 45°}$$

$$= \sqrt{14.14^2 + 2.5^2 + 2(14.14)(2.5)\cos 45}$$

$$= 16.005\text{ kN}$$

3.22 Sol.

$$\tau_{max} = \frac{2 \times 10^3}{0.707t(40) \times 2} = \frac{35.36}{t}\text{ MPa}$$

$$\sigma_{max} = \frac{M_{max}}{I_{NA}} \cdot y_{max} = \frac{2000 \times 150 \times 20}{\frac{0.707t(40)^3 \times 2}{12}}$$

$$\sigma_{max} = \frac{795.615}{t}\text{ MPa}$$

$$\text{MSST, } \sqrt{\sigma_{max}^2 + 4\tau^2} \le 2\left(\frac{S_{ys}}{N}\right)$$

$$\sqrt{\left(\frac{795.615}{t}\right)^2 + 4\left(\frac{35.36}{t}\right)^2} \le 2 \times 60$$

$$\frac{798.752}{t} \le 2(60)$$

$$t = 6.65\text{ mm}$$

3.23 (a)

$$\text{Velocity ratio} = \frac{\text{Velocity of } P}{\text{Velocity of } W} = \frac{x}{y}\text{(say)}$$

Mechanical advantage,

$$MA = \frac{F_{output}}{F_{input}} = \frac{W}{F}$$

$$\eta = \frac{\text{Work done by machine on load}}{\text{Work done by effort}}$$

$$= \frac{W \cdot y}{F \cdot x}$$

$$\eta = \frac{\text{Work done by } W \text{ per revolution}}{\text{Work done by } F \text{ per revolution}}$$

3.24 (a)

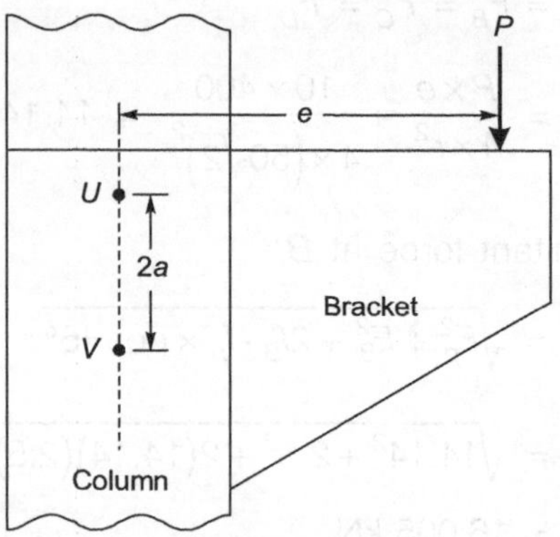

$$2a = 100 \text{ mm}$$

$$\Rightarrow \quad a = 50 \text{ mm}$$

Primary forces

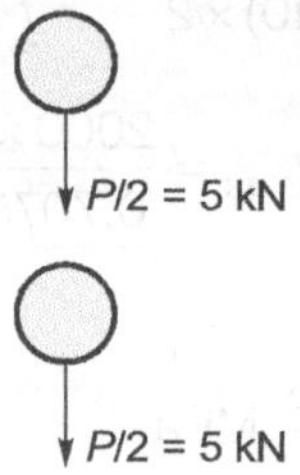

Secondary forces

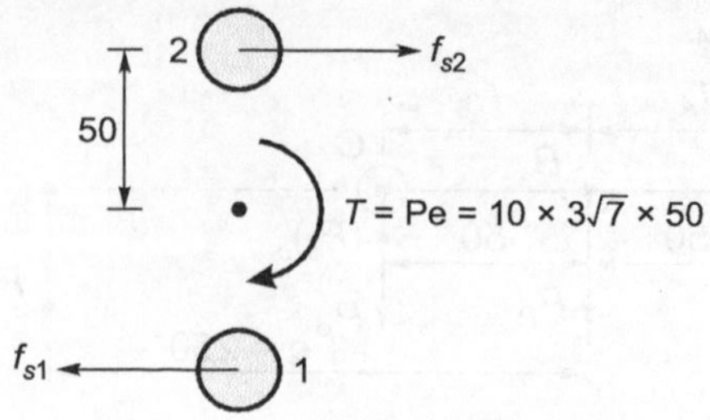

$$f_s \propto r$$

and $f_{s1} = f_{s2} = \dfrac{Pe}{2r} = \dfrac{10\times 3\sqrt{7}\times 50}{2\times 50} = 39.68 \text{ kN}$

Critical rivet both 1 and 2.

∴ Maximum shear force = Resultant of force

$$R = \sqrt{(39.68)^2 + 5^2} = 40 \text{ kN}$$

Safe condition:

$$\frac{R}{A} \le \tau_{per}$$

$$\frac{40\times 10^3}{A} \le 50$$

$$A = 800 \text{ mm}^2$$

■■■■

4 Bearings, Shafts and Keys

4.1 In thick film hydrodynamic journal bearings, the coefficient of friction
(a) increase with increase in load
(b) is independent on load
(c) decreases with increase in load
(d) may increase or decrease with increase in load
[1996 : 1 M]

4.2 To restore stable operating condition in a hydrodynamic journal bearing, when it encounters higher magnitude of loads
(a) oil viscosity is to be decreased
(b) oil viscosity is to be increased
(c) oil viscosity index is to be increased
(d) oil viscosity index is to be decreased
[1997 : 1 M]

4.3 The dynamic load capacity of 6306 bearing is 22 kN. The maximum radial load it can sustain to operate at 600 rev/min, for 2000 hours is
(a) 4.16 kN (b) 3.60 kN
(c) 6.25 kN (d) 5.29 kN **[1997 : 2 M]**

4.4 Square key of side *d*/4 each and length *l* is used to transmit torque *T* from the shaft of diameter *d* to the hub of a pulley. Assuming the length of the key to be equal to the thickness of the pulley, the average shear stress developed in the key is given by
(a) $\frac{4T}{ld}$ (b) $\frac{16T}{ld^2}$
(c) $\frac{8T}{ld^2}$ (d) $\frac{16T}{\pi d^3}$ **[2003 : 2 M]**

4.5 The basic load rating of a ball bearing is
(a) the maximum static radial load that can be applied without causing any plastic deformation of bearing components
(b) the radial load at which 90% of the group of apparently identical bearings run for on million revolutions before the first evidence of failure
(c) the maximum radial load that can be applied during operation without any plastic deformation of bearing components
(d) a combination of radial and axial loads that can be applied without any plastic deformation
[2005 : 1 M]

4.6 Which one of the following is criterion in the design of hydrodynamic journal bearings?
(a) Sommerfeld number
(b) Rating life
(c) Specific dynamic capacity
(d) Rotation factor **[2005 : 1 M]**

4.7 A ball bearing operating at a load *F* has 8000 hours of life. The life of the bearing, in hours, when the load is doubled to 2*F* is
(a) 8000 (b) 6000
(c) 4000 (d) 1000 **[2007 : 1 M]**

4.8 A natural feed journal bearing of diameter 50 mm and length 50 mm operating at 20 revolution/sec. carries a load of 2.0 kN. The lubricant used has a viscosity of 20 mPa.s. The radial clearance is 50 μm. The Sommerfeld number for the bearings is
(a) 0.062 (b) 0.125
(c) 0.250 (d) 0.785 **[2007 : 2 M]**

4.9 A journal bearing has a shaft diameter of 40 mm and a length of 40 mm. The shaft is rotating at 20 rad/s and the viscosity of the lubricant is 20 mPa.s. The clearance is 0.020 mm. The loss of torque due to the viscosity of the lubricant is approximately
(a) 0.040 Nm (b) 0.252 Nm
(c) 0.400 Nm (d) 0.652 Nm **[2008 : 2 M]**

4.10 A lightly loaded full journal bearing has journal diameter of 50 mm, bush bore of 50.05 mm and bush length of 20 mm. If rotational speed of journal is 1200 rpm and average viscosity of liquid lubricant is 0.03 Pa s, the power loss (in W) will be
(a) 37 (b) 74
(c) 118 (d) 237 **[2010 : 2 M]**

4.11 Two identical ball bearings *P* and *Q* are operating at loads 30 kN and 45 kN respectively. The ratio of the life of bearing *P* to the life of bearing Q is
(a) 81/16 (b) 27/8
(c) 9/4 (d) 3/2 **[2011 : 2 M]**

4.12 A solid circular shaft needs to be designed to transmit a torque of 50 N.m. If the allowable shear stress of the material is 140 MPa, assuming a factor of safety of 2, minimum allowable design diameter in mm is

(a) 8 (b) 16
(c) 24 (d) 34 [2012 : 2 M]

4.13 Ball bearings are rated by a manufacturer for a life of 10^6 revolutions. The catalogue rating of a particular bearing is 16 kN. If the design load is 2 kN, the life of the bearing will be $P \times 10^6$ revolutions, where P is equal to ______

[2014 : 1 M, Set-4]

4.14 A hydrodynamic journal bearing is subjected to 2000 N load at a rotational speed of 2000 rpm. Both bearing bore diameter and length are 40 mm. If radial clearance is 20 μm and bearing is lubricated with an oil having viscosity 0.03 Pa.s, the Sommerfeld number of the bearing is ____ .

[2014 : 2 M, Set-1]

4.15 For ball bearings, the fatigue life L measured in number of revolutions and the radial load F are related by $FL^{1/3} = K$, where K is a constant. It withstands a radial load of 2 kN for a life of 540 million revolutions. The load (in kN) for a life of one million revolutions is ______.

[2015 : 2 M, Set-3]

4.16 Which of the bearings given below SHOULD NOT be subjected to a thrust load?
(a) Deep groove ball bearing
(b) Angular contact ball bearing
(c) Cylindrical (straight) roller bearing
(d) Single row tapered roller bearing

[2016 : 1 M, Set-3]

4.17 A self-aligning ball bearing has a basic dynamic load rating (C_{10}, for 10^6 revolutions) of 35 kN. If the equivalent radial load on the bearing is 45 kN, the expected life (in 10^6 revolutions) is
(a) below 0.5 (b) 0.5 to 0.8
(c) 0.8 to 1.0 (d) above 1.0

[2018 : 2 M, Set-1]

4.18 A helical gear with 20° pressure angle and 30° helix angle mounted at the mid-span of a shaft that is supported between two bearings at the ends. The nature of the stresses induced in the shaft is
(a) normal stress due to bending in one plane and axial loading; shear stress due to torsion
(b) normal stress due to bending in two planes; shear stress due to torsion
(c) normal stress due to bending only
(d) normal stress due to bending in two planes and axial loading; shear stress due to torsion

[2020 : 1 M, Set-1]

4.19 For a ball bearing the fatigue life in millions of revolutions is given by $L = \left(\frac{C}{P}\right)^n$, where P is the constant applied load and C is the basic dynamic load rating. Which one of the following statements is TRUE?
(a) n = 3, assuming that the inner racing is fixed and outer racing is revolving.
(b) n = 1/3, assuming that the inner racing is fixed and outer racing is revolving.
(c) n = 3, assuming that the outer racing is fixed and inner racing is revolving.
(d) n = 1/3, assuming that the outer racing is fixed and inner racing is revolving.

[2024 : 1 M]

■■■■

Answers Bearing, Shafts and Keys

4.1	(c)	4.2	(b)	4.3	(d)	4.4	(c)	4.5	(b)	4.6	(a)	4.7	(d)	4.8	(b)
4.9	(a)	4.10	(a)	4.11	(b)	4.12	(b)	4.13	(512)	4.14	(0.8)	4.15	(16.2865)		
4.16	(c)	4.17	(a)	4.18	(c, d)	4.19	(c)								

Explanations Bearing, Shafts and Keys

4.1 (c)

$$f \propto \frac{ZN}{P}$$

where, Z is viscosity, N is speed and P is load.

4.2 (b)

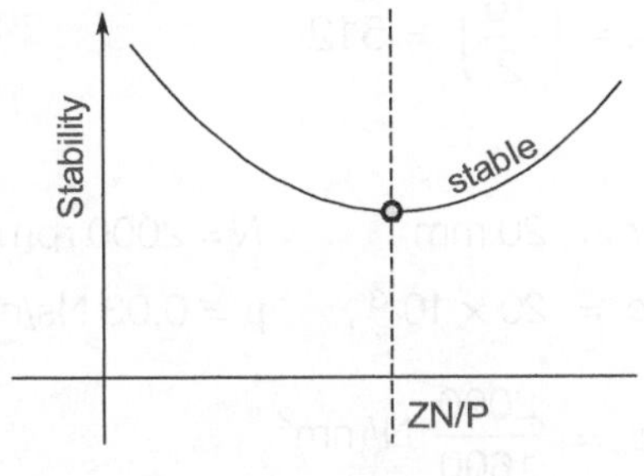

For stability $\frac{ZN}{P}$ increases and for that Z should be increased and P should be decreased.

4.3 (d)

$$L = 600 \times 60 \times 2000 = 72 \times 10^6 \text{ rev}$$

$$L = \left(\frac{C}{P}\right)^3 \times 10^6 \text{rev}$$

$$72 \times 10^6 = \left(\frac{22}{P}\right)^3 \times 10^6$$

$$P = \mathbf{5.29\ kN}$$

4.4 (c)

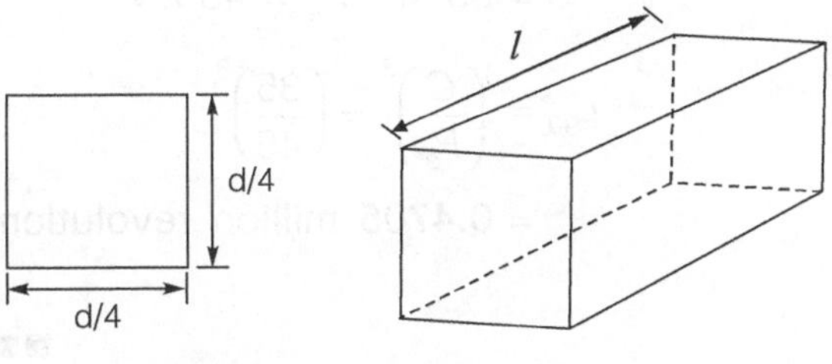

Diameter of shaft $= d$

Torque transmitted $= T$

Force acting on circumference of shaft

$$F = \frac{T}{d/2} = \frac{2T}{d}$$

$$\text{Resisting area} = \frac{l \times d}{4}$$

Average shear stress,

$$\tau = \frac{F}{A} = \frac{2T}{d} \times \frac{4}{ld} = \frac{8T}{ld^2}$$

4.6 (a)

Sommerfeld number,

$$S = \frac{ZN}{P}\left(\frac{D}{C}\right)^2$$

It is also called bearing characteristic number

where Z = Absolute viscosity of lubricant

N = Speed (rps)

P = Bearing pressure on projected bearing area (N/mm²)

D = Diameter of the journal

C = Diametral clearance

4.7 (d)

$$\text{Life of bearing}, L = \left(\frac{C}{W}\right)^K \times 10^6$$

where K = 3 for ball bearing

$$\text{at load } F, \quad L_1 = \left(\frac{C}{F}\right)^K \Rightarrow C = FL^{\frac{1}{K}}$$

$$\text{at load } 2F \quad L_2 = \left(\frac{C}{2F}\right)^3 = \left(\frac{FL^{1/3}}{2F}\right)^3 = \left(\frac{L^{1/3}}{2}\right)^3$$

$$= \mathbf{1000\ hours.}$$

4.8 (b)

Sommerfeld number,

$$S = \left(\frac{r}{c}\right)^2 \times \frac{\mu N}{p}$$

Therefore,

$$p = \frac{W}{d \times l} = \frac{2000}{50 \times 50} = 0.8\ \text{N/mm}^2$$

$$S = \left(\frac{25}{50 \times 10^{-3}}\right)^2 \times \frac{20 \times 20 \times 10^{-3}}{0.8 \times 10^6} = \mathbf{0.125}$$

4.9 (a)

$$\tau = \frac{\mu v}{c} = \frac{20 \times 10^{-3} \times 20 \times 20 \times 10^{-3}}{0.02 \times 10^{-3}}$$

$$= 400\ \text{N/m}^2$$

Force $= \tau A = 400 \times 3.14 \times 0.04 \times 0.04\ N$

$(\because F = \tau \cdot \pi dl)$

Torque $= 400 \times 3.14 \times 0.04 \times 0.04 \times 0.02$

$(\because \text{Torque} = F \times r)$

$= \mathbf{0.040\ Nm}$

4.10 (a)

Tangential velocity of shaft,

$$u = \frac{\pi DN}{60} = \frac{\pi \times 50 \times 10^{-3} \times 1200}{60}$$
$$= 3.14 \text{ m/s}$$

Clearance,

$$y = \frac{50.05 - 50}{2} = 0.025 \text{ mm}$$

By $\tau = \mu \frac{du}{dy}$ shear stress on shaft

$$\tau = 0.03 \times \frac{3.14}{0.025 \times 10^{-3}} = 3768 \text{ N/m}^2$$

Shear force on shaft,

$$F = \tau \times \text{Area} = 3768 \times \pi D \times L$$
$$= 3768 \times \pi \times 50 \times 10^{-3} \times 20 \times 10^{-3}$$
$$= 11.83 \text{ N}$$

$$\text{Torque: } T = F \times \frac{D}{2} = 11.83 \times \frac{50 \times 10^{-3}}{2}$$
$$= 0.2953 \text{ Nm}$$

$$\text{Power loss} = \frac{2\pi NT}{60} = \frac{2\pi \times 1200 \times 0.2953}{60}$$
$$= \mathbf{37.1 \text{ W}}$$

4.11 (b)

$$L = \left(\frac{C}{P}\right)^3$$

where L = Life of bearing
P = Load

$$\frac{(\text{Life})_P}{(\text{Life})_Q} = \frac{(\text{Load})_Q^3}{(\text{Load})_P^3} = \left[\frac{45}{30}\right]^3 = 3.375 \text{ or } \frac{27}{8}$$

4.12 (b)

$$T = 50 \text{ Nm}$$
$$\tau_{\text{allowable/permissible}} = 140 \text{ MPa}$$

$$\therefore \quad \tau_{per} = \frac{16T}{\pi d^3}$$

$$d \geq \left(\frac{16T}{\pi \tau_{per}}\right)^{1/3} \geq \left(\frac{16 \times 50}{\pi \times 140 \times 10^6}\right)^{1/3}$$
$$d \geq 0.0122 \text{ m}$$
$$d \geq 12.2 \text{ mm}$$

Hence from options, d = 16 mm

4.13 Sol.

$$L_{10} = \left(\frac{C}{W}\right)^3$$

where

L_{10} = Basic life rating in millions of revolutions

C = Basic dynamic load rating

P = Equivalent dynamic radial load

$$\therefore \ L_{10} = \left(\frac{16}{2}\right)^3 = 512$$

4.14 Sol.

r = 20 mm, N = 2000 rpm

$c = 20 \times 10^{-3}$, $\mu = 0.03 \text{ Ns/m}^2$

$$p = \frac{2000}{1600} \text{ N/mm}^2 \qquad \left(p = \frac{W}{A}\right)$$
$$= 1.25 \text{ MPa}$$

$$S = \frac{ZN}{60p}\left(\frac{r}{c}\right)^2$$

$$= \frac{0.03 \times 2000}{60 \times 1.25 \times 10^6}\left(\frac{20}{20 \times 10^{-3}}\right)^2 = \mathbf{0.8}$$

4.15 Sol.

$$FL^{1/3} = k$$
$$2 \text{ kN} \times (540)^{1/3} = F \times (1)^{1/3}$$
$$F = \mathbf{16.2865 \text{ kN}}$$

4.17 Sol.

$$C = 35 \text{ kN}, \ P_e = 45 \text{ kN}$$

$$L_{90} = \left(\frac{C}{P_e}\right)^3 = \left(\frac{35}{45}\right)^3$$
$$= \mathbf{0.4705 \text{ million revolution}}$$

■■■■

5 Gears

Common Data Questions 5.1 to 5.2

The overall gear ratio in a 2 stage speed reduction gear box (with all spur gears) is 12. The input and output shafts of the gear box are collinear. The countershaft which is parallel to the input and output shafts has a gear (Z_2 teeth) and pinion (Z_3 = 15 teeth) to mesh with pinion (Z_1= 16 teeth) on the input shaft and gear (Z_4 teeth) on the output shaft respectively. It was decided to use a gear ratio of 4 with 3 module in the first stage and 4 module in the second stage.

5.1 Z_2 and Z_4 are

(a) 64 and 45 (b) 45 and 64
(c) 48 and 60 (d) 60 and 48

[2003 : 2 M]

5.2 The centre distance in the second stage is

(a) 90 mm (b) 120 mm
(c) 160 mm (d) 240 mm **[2003 : 2 M]**

5.3 Two mating spur gears have 40 and 120 teeth respectively. The pinion rotates at 1200 rpm and transmits a torque of 20 N.m. The torque transmitted by gear is

(a) 6.6 Nm (b) 20 Nm
(c) 40 Nm (d) 60 Nm **[2004 : 1 M]**

Linked Answer Questions 5.4 to 5.6

A gear set has a pinion with 20 teeth and a gear with 40 teeth. The pinion runs at 30 rev/s and transmits a power of 20 kW. The teeth are on the 20° full-depth system and have a module of 5 mm. The length of the line of action is 19 mm.

5.4 The center distance for the above gear set in mm is

(a) 140 (b) 150
(c) 160 (d) 170 **[2007 : 2 M]**

5.5 The contact ratio of the contacting tooth is

(a) 1.21 (b) 1.25
(c) 1.29 (d) 1.33 **[2007 : 2 M]**

5.6 The resultant force on the contacting gear tooth in *N* is

(a) 77.23 (b) 212.20
(c) 2258.1 (d) 289.43 **[2007 : 2 M]**

5.7 A spur gear has a module of 3 mm, the number of teeth 16, a face width of 36 mm and a pressure angle of 20°. It is transmitting a power of 3 kW at 20 rev/s. Taking a velocity factor of 1.5 and a form factor of 0.3, the stress in the gear tooth is about

(a) 32 MPa (b) 46 MPa
(c) 58 MPa (d) 70 MPa **[2008 : 2 M]**

Linked Answer Questions 5.8 and 5.9

A 20° full depth involute spur pinion of 4 mm module and 21 teeth is to transmit 15 kW at 960 rpm. Its face width is 25 mm.

5.8 The tangential force transmitted (in N) is

(a) 3552 (b) 2611
(c) 1776 (d) 1305 **[2009 : 2 M]**

5.9 Given that the tooth geometry factor is 0.32 and the combined effect of dynamic load and allied factors intensifying the stress is 1.5; the minimum allowable stress (in MPa) for the gear material is

(a) 242.0 (b) 166.5
(c) 121.0 (d) 74.0 **[2009 : 2 M]**

5.10 A spur pinion of pitch diameter 50 mm rotates at 200 rad/s and transmits 3 kW power. The pressure angle of the tooth of the pinion is 20°. Assuming that only one pair of the teeth is in contact, the total force (in Newton) exerted by a tooth of the pinion on the tooth on a mating gear is ______.

[2014 : 2 M, Set-2]

5.11 A steel spur pinion has a module (m) of 1.25 mm, 20 teeth and 20° pressure angle. The pinion rotates at 1200 rpm and transmits power to a 60 teeth gear. The face width (F) is 50 mm, Lewis form factor Y = 0.322 and a dynamic factor K_v = 1.26. The bending stress (σ) induced in a tooth can be calculated by using the Lewis formula given below.

If the maximum bending stress experienced by the pinion is 400 MPa. the power transmitted is ________ kW (round off to one decimal place),

Lewis formula: $\sigma = \frac{K_v W^t}{FmY}$, where W^t is the tangential load acting on the pinion.

[2020 : 2 M, Set-2]

Answers Gears

5.1	(a)	5.2	(b)	5.3	(d)	5.4	(b)	5.5	(c)	5.6	(c)	5.7	(b)	5.8	(a)
5.9	(b)	5.10	(638.5)	5.11	(10)										

Explanations Gears

5.1 (a)

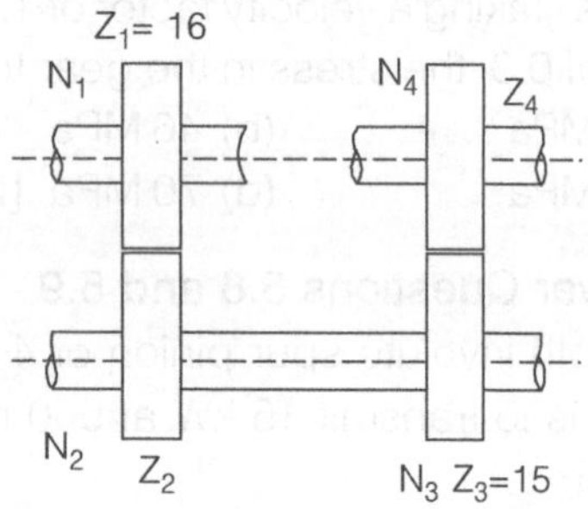

$$\frac{N_1}{N_4} = 12$$

$$\frac{N_1}{N_2} = 4 = \frac{D_2}{D_1}$$

$$m_1 = 3,\ m_2 = 4$$

Since, $\dfrac{D_1}{Z_1} = \dfrac{D_2}{Z_2}$

$$\Rightarrow \quad \frac{Z_1}{Z_2} = \frac{D_1}{D_2} = \frac{N_2}{N_1} = \frac{1}{4}$$

$$Z_2 = Z_1 \times 4 = \mathbf{64}$$

and $\dfrac{N_1}{N_4} = \dfrac{D_2 \times D_4}{D_1 \times D_3}$

$$\Rightarrow \quad 12 = \frac{D_4}{D_3} \times 4$$

$$\Rightarrow \quad \frac{D_4}{D_3} = 3$$

Since, $\dfrac{Z_3}{Z_4} = \dfrac{D_3}{D_4}$

$$Z_4 = Z_3 \frac{D_4}{D_3} = 15 \times 3 = 45$$

5.2 (b)

Centre distance in second stage

$$= \frac{D_3 + D_4}{2}$$

$$m = \frac{D_3}{Z_3} = \frac{D_4}{Z_4}$$

$$\Rightarrow \quad 4 = \frac{D_4}{45}$$

$$\Rightarrow \quad D_4 = 180 \text{ mm}$$

$$D_3 = \frac{180}{3} = 60 \text{ mm}$$

$$\text{Centre distance} = \frac{180 + 60}{2} = 120 \text{ mm}$$

5.3 (d)

Power transmitted,

$$P = \frac{2\pi N_G T_G}{60} = \frac{2\pi N_P T_P}{60}$$

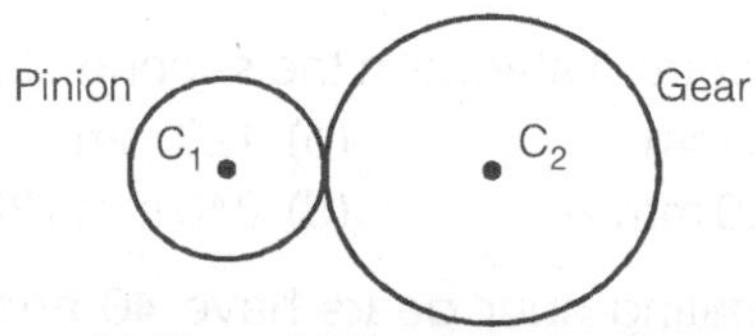

$$\Rightarrow \quad N_G T_G = N_P T_P \quad \text{(where } T \text{ is torque)}$$

and, $\dfrac{N_P}{N_G} = \dfrac{Z_G}{Z_P}$

(where Z is number of teeths)

$$\Rightarrow \quad \frac{1200}{N_G} = \frac{120}{40}$$

$$N_G = 400 \text{ rpm}$$

Hence,

$$400 \times T_G = 1200 \times 20$$

$$T_G = \mathbf{60\ N.m}$$

5.4 (b)

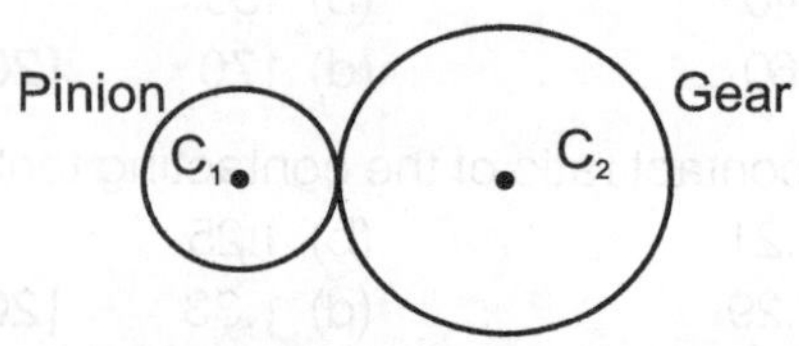

$Z_p = 20$ $\qquad Z_G = 40$

$N_p = 30$ rev/s $\qquad N_G = ?$

Pressure angle $= 20°$

Module: $m = 5$ mm

$$m = \frac{d_p}{Z_p} = \frac{d_G}{Z_G}$$

$d_p = 5 \times 20 = 100$ mm

$d_G = 5 \times 40 = 200$ mm

Centre distance $= \dfrac{d_p + d_G}{2} = 150$ mm

5.5 (c)

Length of line of action = 19 mm

Length of arc of contact

$= \dfrac{19 \text{ mm}}{\cos 20°} = 20.21 \text{mm}$

Circular pitch $= \dfrac{\pi d}{z} = \pi \times 5 = 5\pi$ mm

Contact ratio $= \dfrac{\text{Length of arc}}{\text{Circular pitch}} = \dfrac{20.21}{5\pi} = 1.29$

5.6 (c)

For resultant force on contacting gear tooth:

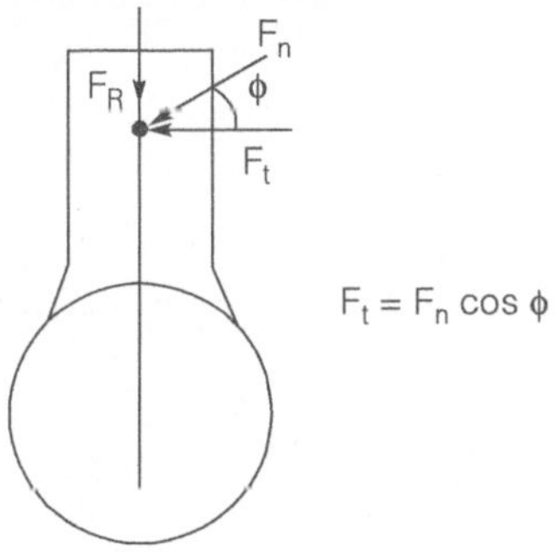

Torque: $T = \dfrac{60 \times P}{2 \times \pi \times N}$ $(N = 1800 \text{ rpm})$

$= \dfrac{60 \times 20{,}000}{2 \times \pi \times 1800} = 106.10$ Nm

$m = \dfrac{d_p}{z_p}$

$d_p = 5 \times 20 = 0.1$ m

$F_t \times R = 106.10$

$F_t = \dfrac{106.10}{0.05} = 2122$ N

$F_n = \dfrac{F_t}{\cos 20°} = \dfrac{2122}{\cos 20°}$

$= \mathbf{2258.1 \ N}$

5.7 (b)

Given:

$P = 3$ kW = 3000 W, $N = 20$ rps

Form factor, $y = 0.3$

Face width, $b = 36$ mm

Velocity factor, $C_v = 1.5$

Module, $m = 3$ mm

Teeth, $Z = 16$

$F_t = \dfrac{2T}{D_1}$

$D = mZ = 3 \times 16 = 48$ mm

$P = 2\pi NT$ where N is in rps

$T = \dfrac{P}{2\pi N} = \dfrac{3000}{2\pi \times 20} = 23.8732$ Nm

$= 23.8732 \times 10^3$ Nmm

We know

$(F_d)_{\text{Lewis}} \le F_s$

$F_t \times C_v \le (\sigma_b \cdot y\, b\, m)$

$\dfrac{2T}{D} \times C_v = (\sigma_b)\, y\, b\, m$

$\dfrac{2 \times 23.8732 \times 10^3 \times 1.5}{48} = (\sigma_b) 0.3 \times 36 \times 3$

$\sigma_b = 46 \text{ N/mm}^2 = \mathbf{46 \ MPa}$

5.8 (a)

Module:	$m = 4$ mm
Number of teeth:	$T = 21$
Speed:	$N = 960$ rpm
Pressure angle:	$\phi = 20°$
Face width:	$b = 25$ mm

Full depth involute spur pinion

Transmission of power,

$P = 15$ kW at 960 rpm

Power = Torque × Angular speed

$P = T\omega = F_t \times r \times \omega$

= Tangential force × Radius × ω

Radius: $r = \dfrac{D}{2} = \dfrac{mT}{2} = \dfrac{4 \times 21}{2} = 42$ mm

$P = F_t \times r \times \omega$

$\Rightarrow 15 \times 1000 = F_t \times 42 \times 10^{-3} \times \dfrac{2\pi \times 960}{60}$

$\Rightarrow \quad F_t = \mathbf{3552.56 \ N}$

5.9 (b)

Tooth geometry factor, $Y = 0.32$

Combined effect of dynamic load and allied factors intensifying the stress is $f_s = 1.5$

$F_t = \dfrac{\sigma_b\, b\, Y_m}{f_s}$

$\Rightarrow \ 3552\, N = \dfrac{\sigma_b \,(25 \text{ mm})\,(0.32)\,(4 \text{ mm})}{1.5}$

$\Rightarrow \quad \sigma_b = \mathbf{166.5 \ MPa}$

5.10 Sol.

Given, $d = 50$ mm; $\omega = 200$ rad/s

$P = 3000$ W; $\phi = 20°$

$$T = \frac{P}{\omega} = \frac{3000}{200} = 15 \text{ N.m}$$

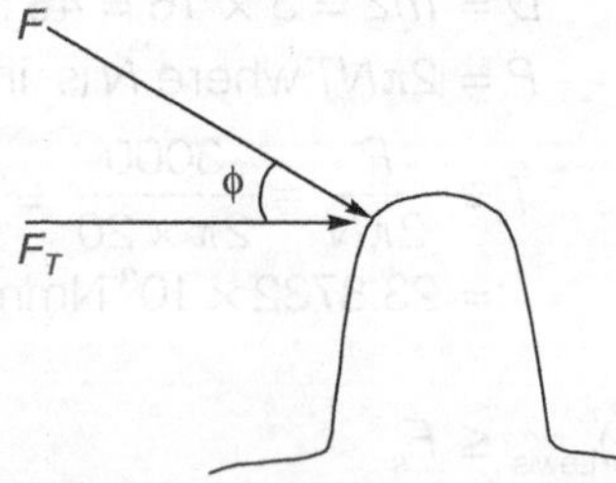

$$F_T \times r = T$$

$$\therefore \quad F_T = \frac{T}{r} = \frac{15}{0.025} = 600 \text{ N}$$

$$\therefore \quad F_T = F\cos\phi$$

$$F = \frac{600}{\cos 20°} = \mathbf{638.5\ N}$$

5.11 Sol.

As per given data: $m = 1.25$ mm

$Z_p = 20$ teeth; $\phi = 20°$

$N_p = 1200$ rpm; $Z_G = 60$ teeth

$$F_t \times c_v s = bmy[\sigma_b]_{max}$$

$$c_v = 1.26$$

$$F_t \times 1.26 \times 1 = 50 \times 1.25 \times 0.322 \times 400$$

$$F_t = 6388.88 \text{ N}$$

$$\text{Power Transmitted} = F_t \times v = \frac{F_t \times \pi D_p N}{60}$$

$$= \frac{6388.88 \times \pi \times 1.25 \times 20 \times 1200}{60} = 10 \text{ kW}$$

■■■■

6 Brakes, Clutches and Ropes

6.1 A wire rope is designated as 6 × 19 standard hoisting. The numbers 6 × 19 represent

(a) diameter in millimeter × length in meter
(b) diameter in centimeter × length in meter
(c) number of strands × number of wires in each strand
(d) number of wires in each strand × number of strands

[2003 : 1 M]

6.2 In a band brake the ratio of tight side band tension to the tension on the slack side is 3. If the angle of overlap of band on the drum is 180° the coefficient of friction required between drum and the band is

(a) 0.20 (b) 0.25
(c) 0.30 (d) 0.35 **[2003 : 2 M]**

Linked Answer Questions 6.3 and 6.4

A band brake consists of a lever attached to one end of the band. The other end of the band is fixed to the ground. The wheel has a radius of 200 mm and the wrap angle of the band is 270°. The braking force applied to the lever is limited to 100 N, and the coefficient of friction between the band and the wheel is 0.5. No other information is given.

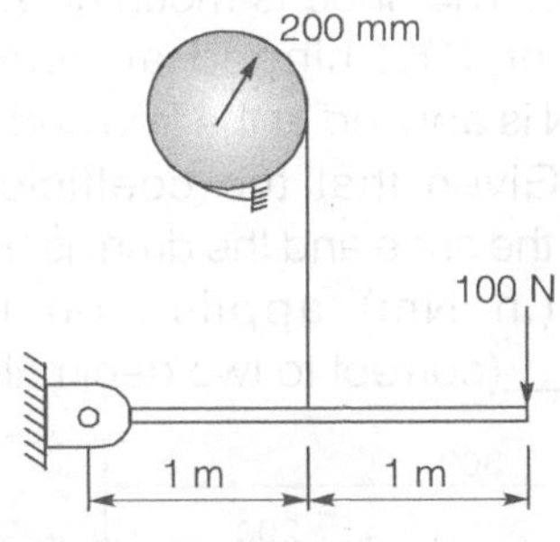

6.3 The maximum tension that can be generated in the band during braking is

(a) 1200 N (b) 2110 N
(c) 3224 N (d) 4420 N **[2005 : 2 M]**

6.4 The maximum wheel torque that can be completely braked is

(a) 200 N.m (b) 382 N.m
(c) 604 N. m (d) 844 N.m **[2005 : 2 M]**

6.5 A disk clutch is required to transmit 5 kW at 2000 rpm. The disk has a friction lining with coefficient of friction equal to 0.25. Bore radius of friction lining is equal to 25 mm. Assume uniform contact pressure of 1 MPa. The value of outside radius of the friction lining is

(a) 39.4 mm (b) 49.5 mm
(c) 97.9 mm (d) 142.9 mm

[2006 : 2 M]

6.6 A block-brake shown below has a face width of 300 mm and a mean coefficient of friction of 0.25. For an activating force of 400 N, the braking torque in Nm is

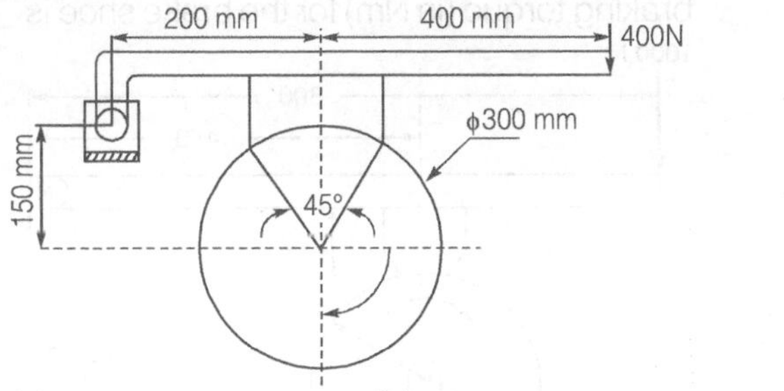

(a) 30 (b) 40
(c) 45 (d) 60 **[2007 : 2 M]**

6.7 A clutch has outer and inner diameters 100 mm and 40 mm respectively. Assuming a uniform pressure of 2 MPa and coefficient of friction of liner material 0.4, the torque carrying capacity of the clutch is

(a) 148 Nm (b) 196 Nm
(c) 372 Nm (d) 490 Nm **[2008 : 2 M]**

6.8 A band brake having bandwidth of 80 mm, drum diameter of 250 mm, coefficient of friction of 0.25 and angle of wrap of 270 degrees is required to exert a friction torque of 1000 N m. The maximum tension (in kN) developed in the band is

(a) 1.88 (b) 3.56
(c) 6.12 (d) 11.56 **[2010 : 2 M]**

6.9 A force of 400 N is applied to the brake drum of 0.5 m diameter in a band brake system as shown in figure, where the wrapping angle is 180°. If the coefficient of friction between the drum and the band is 0.25, the braking torque applied, in Nm is

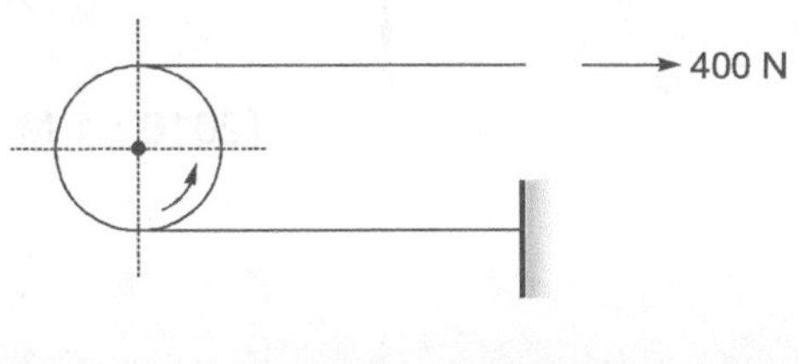

(a) 100.6 (b) 54.4
(c) 22.1 (d) 15.7 [2012 : 2 M]

6.10 A disc clutch with a single friction surface has coefficient of friction equal to 0.3. The maximum pressure which can be imposed on the friction material is 1.5 MPa. The outer diameter of the clutch plate is 200 mm and its internal diameter is 100 mm. Assuming uniform wear theory for the clutch plate, the maximum torque (in Nm) that can be transmitted is ______.

[2014 : 2 M, Set-2]

6.11 A drum brake is shown in the figure. The drum is rotating in anticlockwise direction. The coefficient of friction between drum and shoe is 0.2. The dimensions shown in the figure are in mm. The braking torque (in Nm) for the brake shoe is _____.

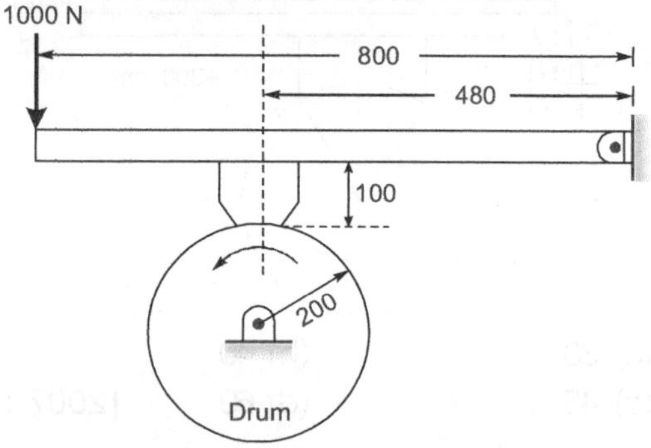

[2014 : 2 M, Set-3]

6.12 A four-wheel vehicle of mass 1000 kg moves uniformly in a straight line with the wheels revolving at 10 rad/s. The wheels are identical, each with a radius of 0.2 m. Then a constant braking torque is applied to all the wheels and the vehicle experience a uniform deceleration. For the vehicle to stop in 10 s, the braking torque (in Nm) on each wheel is ________. [2014 : 2 M, Set-3]

6.13 The forces F_1 and F_2 in a brake band and the direction of rotation of the drum are as shown in the figure. The coefficient of friction is 0.25. The angle of wrap is $3\pi/2$ radians. It is given that $R = 1$ m and $F_2 = 1$ N. The torque (in Nm) exerted on the drum is ____.

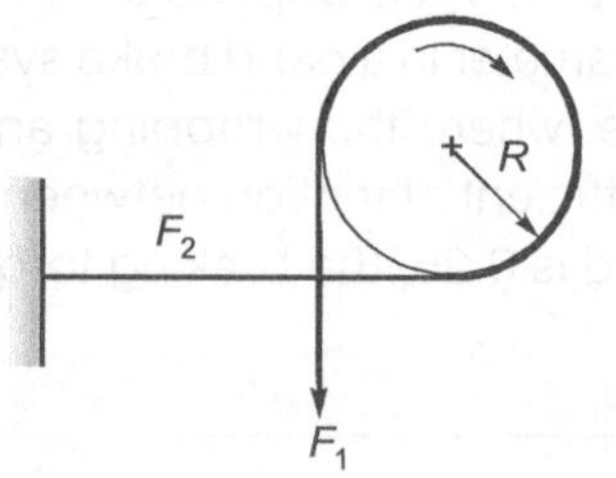

[2016 : 1 M, Set-2]

6.14 For the brake shown in the figure, which one of the following is TRUE?

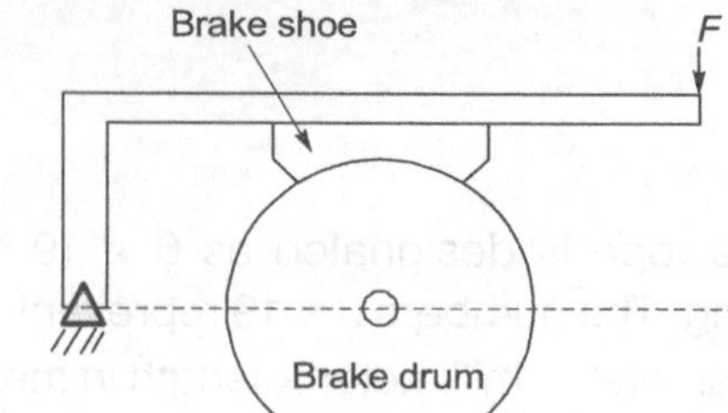

(a) Self energizing for clockwise rotation of the drum
(b) Self energizing for anti-clockwise rotation of the drum
(c) Self energizing for rotation in either direction of the drum
(d) Not of the self energizing type

[2016 : 1 M, Set-2]

6.15 A single-plate clutch has a friction disc with inner and outer radii of 20 mm and 40 mm, respectively. The friction lining in the disc is made in such a way that the coefficient of friction μ varies radially as $\mu = 0.01r$, where r is in mm. The clutch needs to transmit a friction torque of 18.85 kN mm. As per uniform pressure theory, the pressure (in MPa) on the disc is ________.

[2017 : 2 M, Set-2]

6.16 The schematic of an external drum rotating clockwise engaging with a short shoe is shown in the figure. The shoe is mounted at point Y on a rigid lever XYZ hinged at point X. A force $F = 100$ N is applied at the free end of the lever as shown. Given that the coefficient of friction between the shoe and the drum is 0.3, the braking torque (in Nm) applied on the drum is ________(correct to two decimal places).

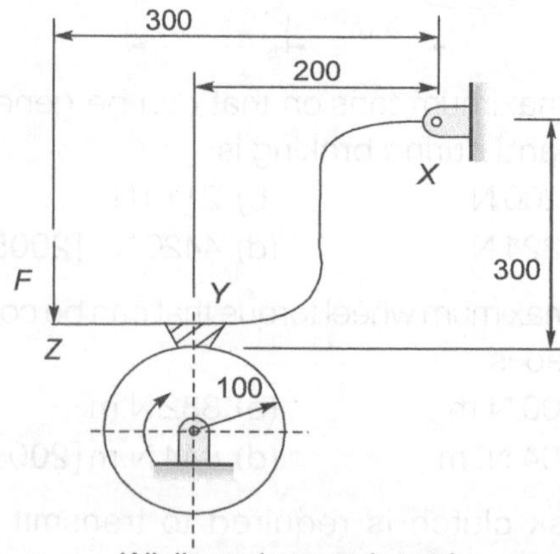

(All dimensions are in mm)

[2018 : 2 M, Set-1]

6.17 A single block brake with a short shoe and torque capacity of 250 Nm is shown. The cylindrical brake drum rotates anticlockwise at 100 rpm and the coefficient of friction is 0.25. The value of *a*, in mm (round off to one decimal place), such that the maximum actuating force *P* is 2000 N, is ____.

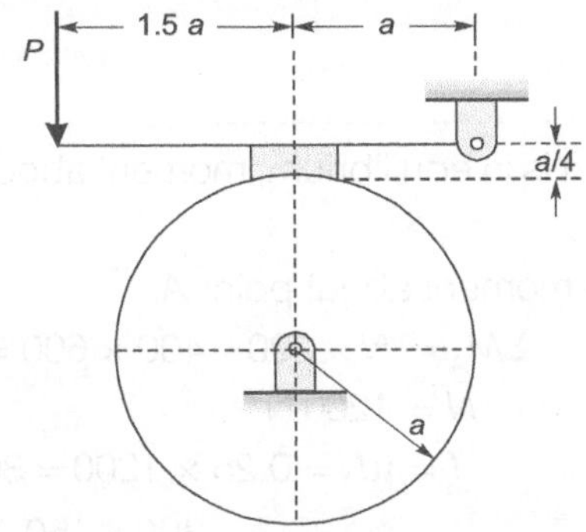

[2019 : 2 M, Set-1]

6.18 A short shoe external drum brake is shown in the figure. The diameter of the brake drum is 500 mm. The dimensions a = 1000 mm, b = 500 mm and c = 200 mm. The coefficient of friction between the drum and the shoe is 0.35. The force applied on the lever F = 100 N as shown in the figure. The drum is rotating anti-clockwise. The braking torque on the drum is______Nm (round off to two decimal places).

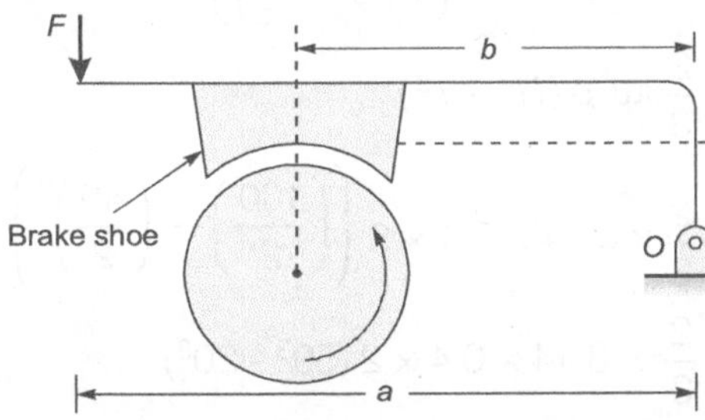

[2019 : 2 M, Set-2]

6.19 In a disc-type axial clutch, the frictional contact takes place within an annular region with outer and inner diameters 250 mm and 50 mm, respectively. An axial force F_1 is needed to transmit a torque by a new clutch. However, to transmit the same torque, one needs an axial force F_2 when the clutch wears out. If contact pressure remains uniform during operation of a new clutch while the wear is assumed to be uniform for an old clutch and the coefficient of friction does not change, then the ratio F_1/F_2 is________ (round off to 2 decimal places).

[2020 : 2 M, Set-1]

6.20 A short shoe drum (radius 260 mm) brake is shown in the figure. A force of 1 kN is applied to the lever. The coefficient of friction is 0.4.

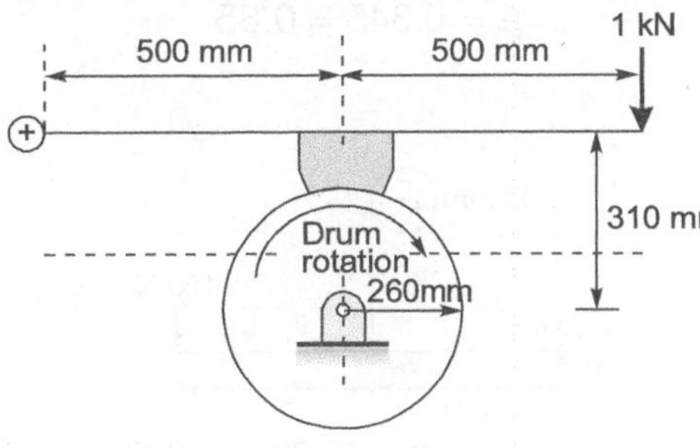

The magnitude of the torque applied by the brake is _________ N.m. [Round off to one decimal place]

[2021 : 2 M, Set-1]

6.21 A band brake shown in the figure has a coefficient of friction of 0.3. The band can take a maximum force of 1.5 kN. The maximum braking force (F) that can be safely applied is _______ N (rounded off to the nearest integer)

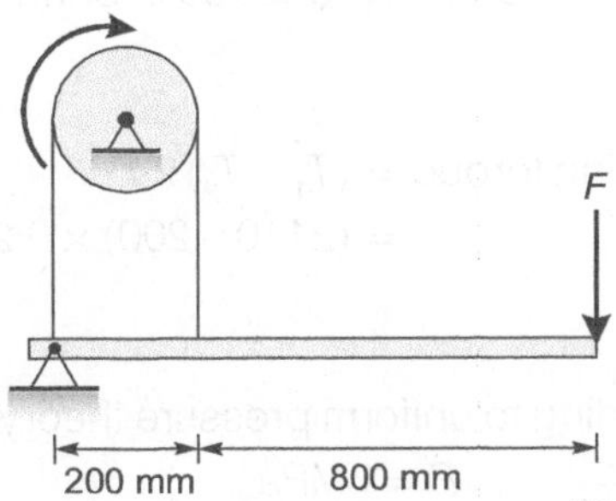

[2024 : 2 M]

Answers Brakes, Clutches and Ropes

6.1 (c) 6.2 (d) 6.3 (b) 6.4 (b) 6.5 (a) 6.6 (c) 6.7 (b) 6.8 (d)
6.9 (b) 6.10 (530.14) 6.11 (64) 6.12 (10) 6.13 (2.2482) 6.14 (a) 6.15 (0.5)
6.16 (8.18) 6.17 (212.5) 6.18 (20.35) 6.19 (0.870) 6.20 (200) 6.21 (117)

Explanations Brakes, Clutches and Ropes

6.2 (d)

Given data:

$$\frac{T_1}{T_2} = 3,\ \theta = 180° = \frac{\pi \times 180}{180} = \pi \text{ rad}$$

$$\frac{T_1}{T_2} = e^{\mu\theta}$$

$$\Rightarrow \quad 3 = e^{\mu\pi}$$

$$\Rightarrow \quad \ln 3 = \mu\pi$$

$$\Rightarrow \quad \mu = 0.345 = \mathbf{0.35}$$

6.3 (b)

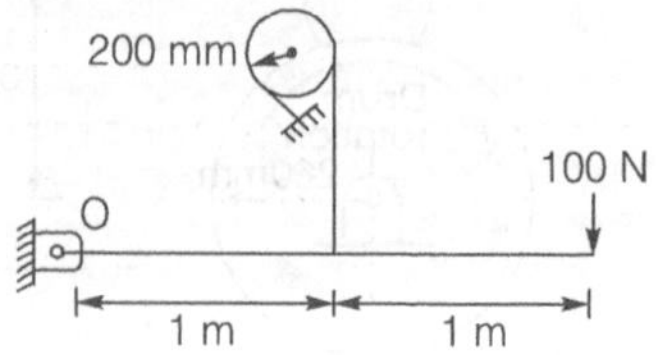

$$\theta = 270°,\ \mu = 0.5$$

Taking moment about O,

$$SM_o = 0$$

$$\Rightarrow \quad T \times 1 - 100 \times 2 = 0$$

$$T = 200 \text{ N}$$

$$\frac{T_1}{T_2} = e^{\mu\theta} = e^{0.5 \times \frac{3\pi}{2}}$$

$$T_1 = \mathbf{2110\ N}$$

Maximum tension produced in band is 2110 N

6.4 (b)

$$\text{Braking torque} = (T_1 - T_2) \times r$$

$$= (2110 - 200) \times 0.2 = \mathbf{382\ Nm}$$

6.5 (a)

According to uniform pressure theory,

$$P = 1 \text{ MPa}$$

Power transmitted,

$$P = T\omega = \frac{T \times 2\pi N}{60}$$

$$\Rightarrow \quad T = \frac{60 \times 5 \times 10^3}{2\pi \times 2000} = 23.87 \text{ Nm}$$

$$T = \frac{2}{3} \times \mu P \pi (r_1^3 - r_2^3)$$

$$\Rightarrow \quad 23.81 \times 10^3 = \frac{2}{3} \times 0.25 \times 1 \times (r_1^3 - 25^3) \times \pi$$

$$r_1 = \mathbf{39.4\ mm}$$

6.6 (c)

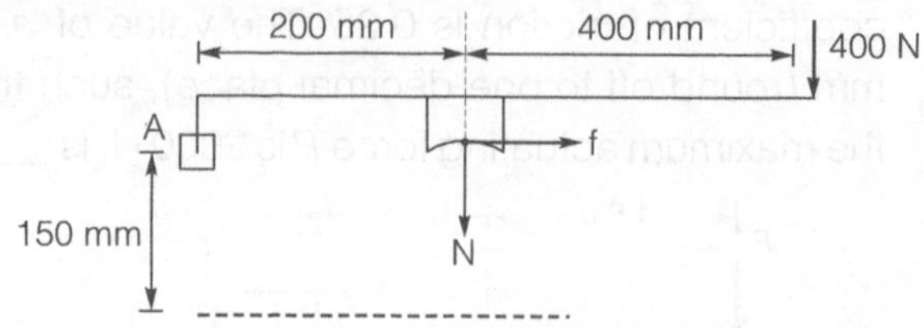

Since, it is in equilibrium, moment about A should be zero.

Taking moment about point A,

$$\Sigma M_A = N \times 200 - 400 \times 600 = 0$$

$$N = 1200 \text{ N}$$

$$f = \mu N = 0.25 \times 1200 = 300 \text{ N}$$

$$\text{Braking torque} = f \times 150 = \frac{300 \times 150}{1000} = \mathbf{45\ Nm}$$

6.7 (b)

Using uniform pressure theory torque carrying capacity of clutch is given by

$$T = \frac{2}{3}\mu W\left(\frac{R^3 - r^3}{R^2 - r^2}\right)$$

$$= \frac{2}{3}\pi\mu(R^3 - r^3)\left(\frac{W}{\pi(R^2 - r^2)}\right)$$

$$= \frac{2}{3}\pi\mu\, p\,(R^3 - r^3)$$

$$= \frac{2}{3} \times 3.14 \times 0.4 \times 2\left(\left(\frac{100}{2}\right)^3 - \left(\frac{40}{2}\right)^3\right)$$

$$= \frac{2}{3} \times 3.14 \times 0.4 \times 2\,(50^3 - 20^3)$$

$$= 195936 \text{ Nmm} = \mathbf{195.936\ Nm}$$

6.8 (d)

$$M_t = (T_1 - T_2) \cdot R$$

Given, friction torque,

$$M_t = 1000 \text{ Nm}$$

$$1000 = (T_1 - T_2) \times 0.125$$

$$\Rightarrow \quad T_1 - T_2 = 8000 \text{ N} \quad \text{... (i)}$$

By,

$$\frac{T_1}{T_2} = e^{\mu\theta} = e^{\left(0.25 \times \frac{270 \times \pi}{180}\right)} = 3.248$$

$$T_1 = 3.248\, T_2 \quad \text{... (ii)}$$

Putting T_1 from equation (ii) into equation (i)

$$T_2 = \frac{8000}{2.248} = 3558.7 \text{ N}$$

and

$$T_1 = 3.248 \times 3558.7 = \mathbf{11.56\ kN}$$

6.9 (b)

Braking torque = $(T_1 - T_2)R$

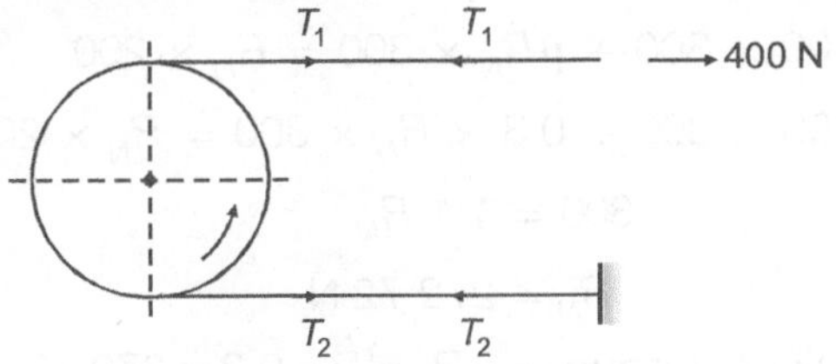

$$T_1 = 400\text{ N}$$

$$\frac{T_1}{T_2} = e^{\mu\theta}$$

$$\mu = 0.25$$

$$\theta = \pi$$

$$\therefore \quad T_2 = 182.375\text{ N}$$

$$\therefore \quad BT = (400 - 182.375) \times \frac{0.5}{2} = \mathbf{54.4\ Nm}$$

6.10 Sol.

$$r_1 = 0.1\text{ m}; \qquad r_2 = 0.05\text{ m}$$

$$P_{max} \times r_2 = C$$

$$\Rightarrow \quad C = 1.5 \times 10^6 \times 0.05 = 75000$$

$$W = 2\pi C(r_1 - r_2)$$

$$= 2\pi \times 75000 \times (0.1 - 0.05)$$

$$= 23561.94\text{ N}$$

$$T = n\mu WR = 1 \times 0.3 \times 23561.94 \times 0.075$$

$$= \mathbf{530.14\ Nm}$$

$$\left(R = \frac{r_1 + r_2}{2} = \frac{0.1 + 0.05}{2} = 0.075\text{ m}\right)$$

6.11 Sol.

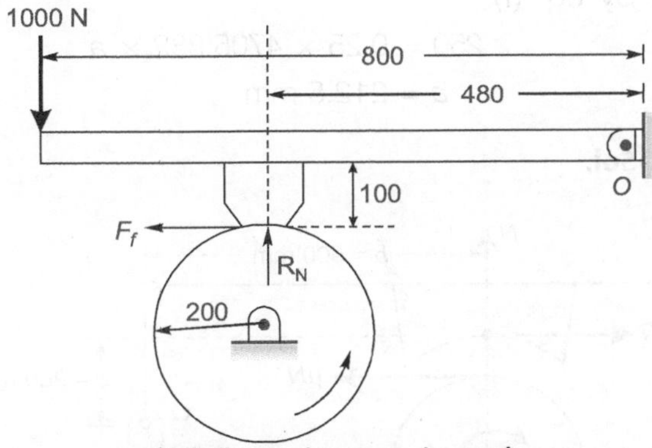

(All dimensions are in mm)

Taking $\Sigma M_o = 0$

$$\Rightarrow \quad 1000 \times 800 = F_f \times 100 + R_N \times 480$$

$$\Rightarrow \quad 8000 = F_f + 4.8\,R_N$$

$$F_f = \mu R_N = 0.2\,R_N$$

$$\therefore\ 0.2\,R_N + 4.8\,R_N = 8000$$

or $\quad R_N = \dfrac{8000}{5} = 1600\text{ N}$

$$\therefore \quad F_f = \mu R_N = 0.2 \times 1600 = 320\text{ N}$$

Braking torque,

$$T = F_f \times r = 320 \times 0.2 = \mathbf{64\ Nm}$$

6.12 Sol.

Initial velocity, $u = \omega r = 10 \times 0.2 = 2$ m/s

For vehicle to stop,

$$v = u - at \Rightarrow 0 = u - at$$

$$\therefore \quad a = \frac{u}{t} = \frac{2}{10} = 0.2\text{ m/s}^2$$

Braking force = Total mass × deceleration

$$F = 1000 \times 0.2 = 200\text{ N}$$

$$F_{\text{each wheel}} = \frac{200}{4} = 50\text{ N}$$

Braking torque on each wheel

$$= 50 \times 0.2 = \mathbf{10\ Nm}$$

6.13 Sol.

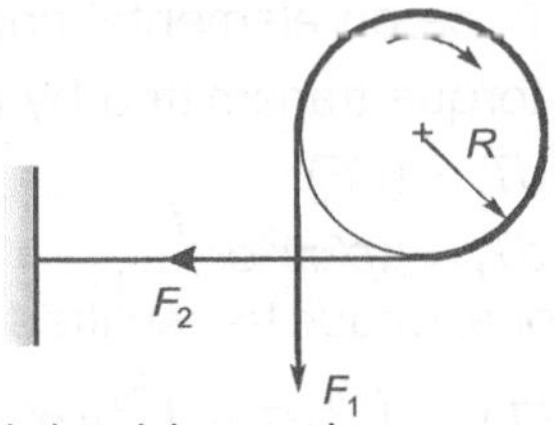

i.e. F_1 is tight side tension.

$$F_1 > F_2$$

$$\frac{F_1}{F_2} = e^{\mu\theta} = e^{0.25 \times \frac{3\pi}{2}}$$

$$F_1 = F_2 \cdot e^{0.25 \times \frac{3\pi}{2}}$$

$$= 1 \times e^{0.25 \times \frac{3 \times \pi}{2}} = 3.2482\text{ N}$$

$$\text{Torque} = (F_1 - F_2) \times r$$

$$= (3.2482 - 1) \times 1 = \mathbf{2.2482\ Nm}$$

6.14 (a)

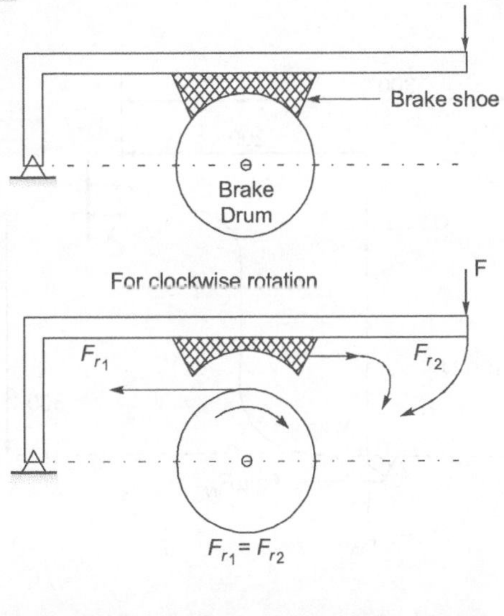

For clockwise direction rotation of drum direction of moment of forces F_{r2} and F will be same. Hence self energizing for clockwise rotation of the drum.

6.15 Sol.

Given: $R_i = 20$ mm, $R_0 = 40$ mm, $\mu = 0.01\ r$, $T_f = 18.85$ kN/mm

Let p = pressure on the disc

$$d_A = 2\pi r(dr)$$

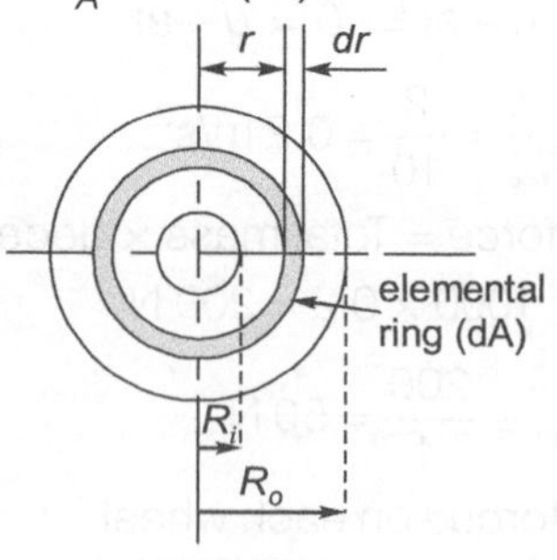

Clutch plate

Normal load on elemental ring $(dW) = p2\pi r dr$

Frictional force on elemental ring $(dF_f) = \mu dW$

Frictional torque transmitted by elemental ring,

$$dT_f = (dF_f)r$$

Hence, $dT_f = \mu p 2\pi r^2 dr$

Total frictional torque transmitted by clutch plate

$$(T_f) = \int_{R_i}^{R_0} dT_f = \int_{R_i}^{R_0} \mu p 2\pi r^2 dr$$

As per U.P.T.

$$T_f = 2\pi p\int_{R_i}^{R_0} \mu r^2 dr = 2\pi p\int_{R_i}^{R_0} (0.01r)r^2 dr$$

$$(T_f)_{U.P.T.} = (2\pi p)(0.01)\left[\frac{R_0^4 - R_i^4}{4}\right]$$

$$18.85 \times 10^3 = 2\pi p(0.01)\left[\frac{40^4 - 20^4}{4}\right]$$

$$p = \mathbf{0.5\ MPa}$$

6.16 Sol.

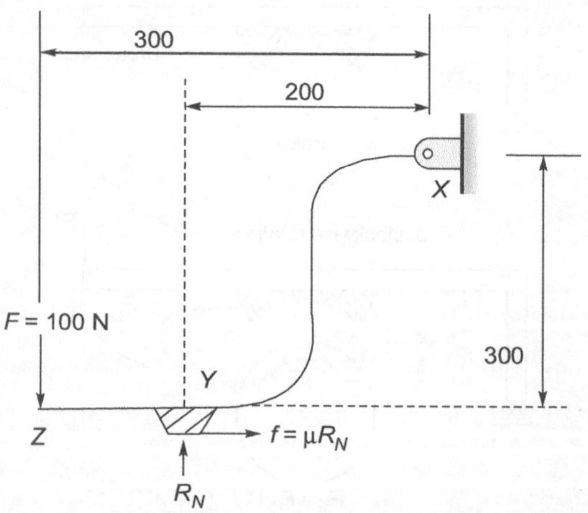

$$\Sigma M_x = 0$$

$$F \times 300 + f \times 300 = R_N \times 200$$

$$100 \times 300 + \mu R_N \times 300 = R_N \times 200$$

$$100 \times 300 + 0.3 \times R_N \times 300 = R_N \times 200$$

$$300 = 1.1\ R_N$$

$$R_N = 272.72 \text{ N}$$

$$\text{Braking torque} = \mu R_N \times R = 0.3 \times 272.72 \times 0.100$$

$$T = 8.18 \text{ Nm}$$

6.17 Sol.

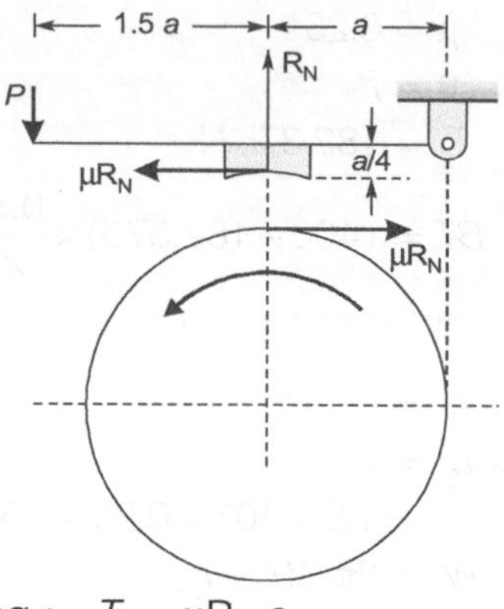

Drum eq.: $T_f = \mu R_N . a$

$$250 = 0.25 \times R_N \times a \qquad ...(i)$$

Lever eq.

$$P \times 2.5a - \mu R_N \times \frac{a}{4} - R_N \times a = 0$$

$$R_N\left(1 + \frac{0.25}{4}\right) = 2000 \times 2.5$$

$$R_N = 4705.882 \text{ N} \qquad ...(ii)$$

By eq. (i)

$$250 = 0.25 \times 4705.882 \times a$$

$$a = 212.5 \text{ mm}$$

6.18 Sol.

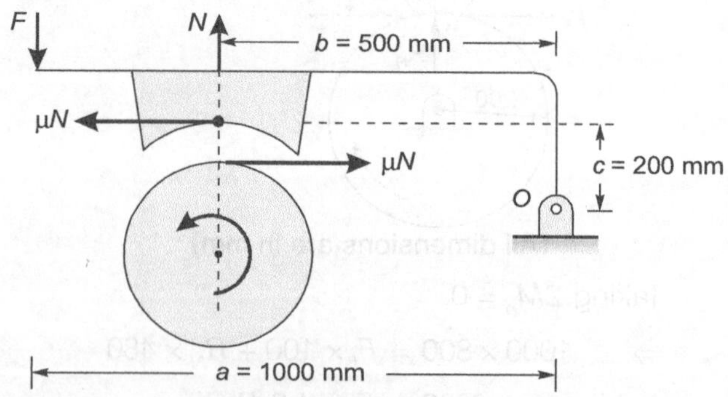

Free body diagram

By considering the moment about point 'O'

$$F \times a = N \times b - \mu N \times c$$

$$100 \times 1000 = N \times 500 - 0.35N \times 200$$
$$N = 232.558 \text{ Newton}$$

$$\text{Braking torque} = \mu N \times \frac{D}{2}$$
$$= 0.35 \times 232.558 \times \frac{500}{2} \times 10^{-3}$$
$$= 20.35 \text{ Nm}$$

6.19 Sol.

$$T = \mu W \times R_m$$
$$T_{new} = \mu \times W_{new} \times R_{new}$$
$$T_{new} = \mu \times W_{new} \times \frac{2}{3}\frac{(R_0^3 - R_i^3)}{R_0^2 - R_i^2}$$
$$T_{old} = \mu \times W \times R_{old}$$
$$T_{new} = T_{old}$$
$$W_{new} \times \frac{2}{3}\frac{(R_0^3 - R_i^3)}{R_0^2 - R_i^2} = W_{old} \times \frac{(R_0 + R_i)}{2}$$
$$\frac{W_{new}}{W_{old}} = 0.871$$

6.20 Sol.

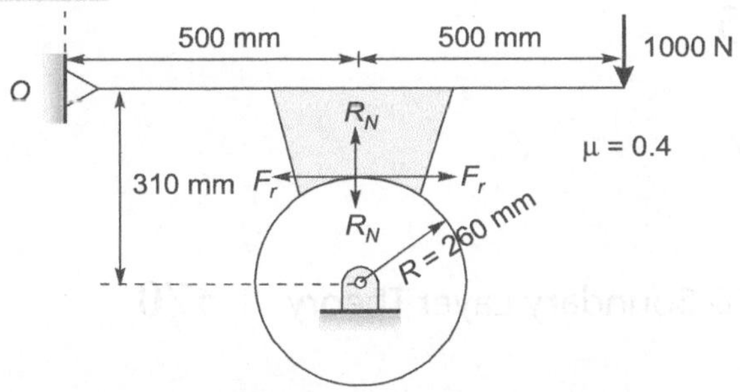

Taking moment about 'O'

$$R_N(500) + F_Y[310 - 260] - 1000 \times 1000 = 0$$
$$R_N(500) + 0.4(R_N)(50) - 1000 \times 1000 = 0$$
$$R_N = 1923.076 \text{ N}$$
$$F_r = \mu R_N = 769.23 \text{ N}$$
$$T_f = F_r \times R = 200 \text{ N-m}$$

6.21 (117) (115 to 119)

Given: $\mu = 0.3$, $T_1 = 1.5$ kN

Taking moment about hinge point O.

$$\Sigma M_o = 0$$
$$\Rightarrow F \times 1000 - T_2 \times 200 = 0$$
$$\Rightarrow \quad F = \frac{T_2 \times 200}{1000} \quad \text{... (i)}$$

Also, $$\frac{T_1}{T_2} = e^{\mu\theta} \Rightarrow \frac{1.5}{T_2} = e^{0.3\times\pi}$$

$$\Rightarrow \quad T_2 = \frac{1.5 \times 1000}{e^{0.3\times\pi}} = 584.49 \text{ N}$$

From equation (i), we have

$$F = \frac{200 \times 584.49}{1000} = 116.89 \text{ N} \simeq 117 \text{ N}$$

■■■■

Fluid Mechanics and Hydraulic Machines

UNIT

VI

CONTENTS

Fluid Mechanics and Hydraulic Machines

Syllabus

Fluid properties; fluid statics, manometry, buoyancy, forces on submerged bodies, stability of floating bodies; control-volume analysis of mass, momentum and energy; fluid acceleration; differential equations of continuity and momentum; Bernoulli's equation; dimensional analysis; viscous flow of incompressible fluids, boundary layer, elementary turbulent flow, flow through pipes, head losses in pipes, bends and fittings.

Analysis of Previous GATE Papers

Exam Year	1 Mark Ques.	2 Marks Ques.	3 Marks Ques.	5 Marks Ques.	Total Marks
1995	2	2	–	–	**6**
1996	3	1	–	1	**10**
1997	2	1	–	1	**9**
1998	–	1	–	1	**7**
1999	2	1	–	–	**4**
2000	1	1	–	–	**3**
2001	2	2	–	1	**11**
2002	3	–	–	–	**3**
2003	1	6	–	–	**13**
2004	2	9	–	–	**20**
2005	1	3	–	–	**7**
2006	3	7	–	–	**17**
2007	3	7	–	–	**17**
2008	1	5	–	–	**11**
2009	–	7	–	–	**14**
2010	4	3	–	–	**10**
2011	1	3	–	–	**7**
2012	3	2	–	–	**7**
2013	2	2	–	–	**6**
2014 Set-1	1	5	–	–	**11**

Exam Year	1 Mark Ques.	2 Marks Ques.	Total Marks
2014 Set-2	2	3	**8**
2014 Set-3	3	3	**9**
2014 Set-4	2	2	**6**
2015 Set-1	1	4	**9**
2015 Set-2	2	2	**6**
2015 Set-3	2	2	**6**
2016 Set-1	3	3	**9**
2016 Set-2	2	3	**8**
2016 Set-3	3	3	**9**
2017 Set-1	4	3	**10**
2017 Set-2	2	3	**8**
2018 Set-1	3	4	**11**
2018 Set-2	1	3	**7**
2019 Set-1	2	2	**6**
2019 Set-2	2	3	**8**
2020 Set-1	2	4	**10**
2020 Set-2	2	2	**6**
2021 Set-1	2	2	**6**
2021 Set-2	3	3	**9**
2022 Set-1	1	2	**5**
2022 Set-2	1	3	**7**
2023	1	3	**7**
2024	1	2	**5**

1 Fluid Properties

1.1 A fluid is said to be Newtonian fluid when the shear stress is
(a) directly proportional to the velocity gradient
(b) inversely proportional to the velocity gradient
(c) independent of the velocity gradient
(d) none of these **[1995 : 1 M]**

1.2 A fluid is one which can be defined as a substance that
(a) has that same shear stress at all points
(b) can deform indefinitely under the action of the smallest shear force
(c) has the small shear stress in all directions
(d) is practically incompressible **[1996 : 1 M]**

1.3 The dimension of surface tension is
(a) ML^{-1} (b) L^2T^{-1}
(c) $ML^{-1}T^{-1}$ (d) MT^{-2} **[1996 : 1 M]**

1.4 The dimension of surface tension is
(a) N/m^2 (b) J/m
(c) J/m^2 (d) W/m **[1997 : 1 M]**

1.5 If P is the gauge pressure within a spherical droplet, then gauge pressure within a bubble of the same fluid and of same size will be
(a) $P/4$ (b) $P/2$
(c) P (d) $2P$ **[1999 : 1 M]**

1.6 Kinematic viscosity of air at 20°C is given to be $1.6 \times 10^{-5}\ m^2/s$. Its kinematic viscosity at 70°C will be varying approximately
(a) $2.2 \times 10^{-5}\ m^2/s$ (b) $1.6 \times 10^{-5}\ m^2/s$
(c) $1.2 \times 10^{-5}\ m^2/s$ (d) $3.2 \times 10^{-5}\ m^2/s$
[1999 : 1 M]

1.7 The SI unit of kinematic viscosity (ν) is
(a) m^2/s (b) kg/ms
(c) m/s^2 (d) m^3/s^2 **[2001 : 1 M]**

1.8 For a Newtonian fluid
(a) shear stress is proportional to shear strain
(b) rate of shear stress is proportional to shear strain
(c) shear stress is proportional to rate of shear strain
(d) rate of shear stress is proportional to rate of shear strain
[2006 : 1 M]

1.9 Assuming constant temperature condition and air to be an ideal gas, the variation in atmospheric pressure with height calculated from fluid statics is
(a) linear (b) exponential
(c) quadratic (d) cubic
[2016 : 1 M; Set-2]

1.10 Two immiscible, incompressible, viscous fluids having same densities but different viscosities are contained between two infinite horizontal parallel plates, 2 m apart as shown below. The bottom plate is fixed and the upper plate moves to the right with a constant velocity of 3 m/s. With the assumptions of Newtonian fluid, steady, and fully developed laminar flow with zero pressure gradient in all directions, the momentum equations simplify to

$$\frac{d^2u}{dy^2} = 0$$

If the dynamic viscosity of the lower fluid, μ_2, is twice that of the upper fluid, μ_1, then the velocity at the interface (round off to two decimal places) is _______ m/s.

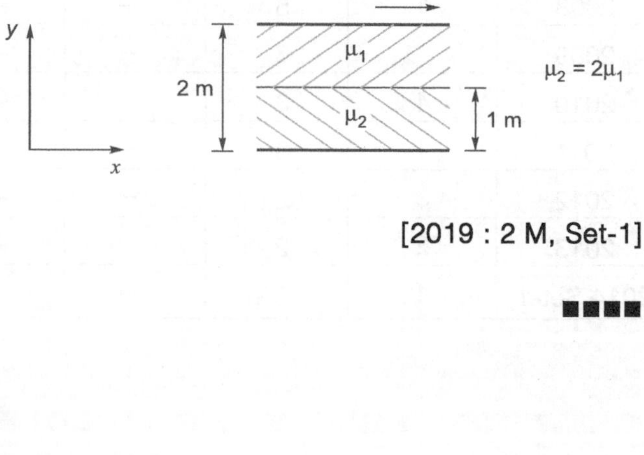

[2019 : 2 M, Set-1]

■■■■

Answers Fluid Properties

1.1 (a) 1.2 (b) 1.3 (d) 1.4 (c) 1.5 (d) 1.6 (a) 1.7 (a) 1.8 (c)
1.9 (b) 1.10 (1)

Explanations Fluid Properties

1.1 (a)

Newtonian fluid: A real fluid which obeys the Newton's law of viscosity is known as Newtonian fluid.

We know that Newton's law of viscosity,

Shear stress: $\tau \propto \frac{du}{dy}$, velocity gradient

$$\tau = \mu \frac{du}{dy}$$

where μ = Dynamic viscosity of fluid

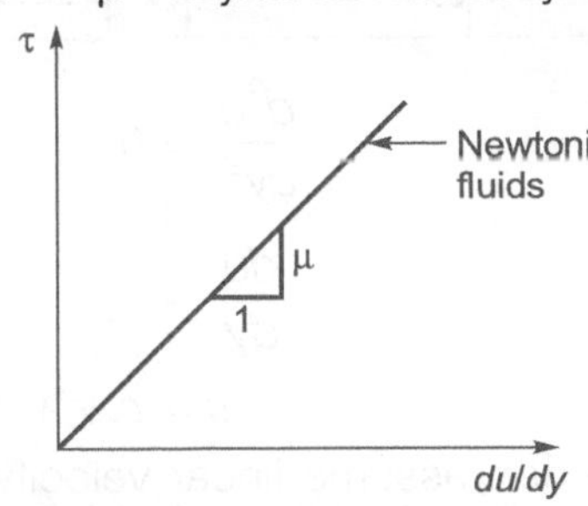

Examples of Newtonian fluids:
Water, alcohol, gasoline, air, kerosene

POINTS TO REMEMBER

For Newtonian fluid, coefficient of viscosity remain constant.

1.2 (b)

Fluid has almost negligible resistance to shear force hence deforms indefinitely under action of shear force.

1.3 (d)

Surface tension:

$$\sigma = \frac{\text{Force} : F}{\text{Length} : l} = \frac{\text{N}}{\text{m}} = \frac{\text{kgm/s}^2}{\text{m}} = \text{kg/s}^2 = \text{MT}^{-2}$$

1.4 (c)

Surface tension: It is defined as force per unit length.

Surface tension is also defined as the surface energy per unit area.

SI unit of surface tension: 1. N/m, 2. J/m^2
Dimension of surface tension: MT^{-2}

1.5 (d)

P = Gauge pressure within a spherical droplet.

$$P = \frac{4\sigma}{d} \text{ for droplet}$$

$$P_{\text{bubble}} = \frac{8\sigma}{d} \text{ for bubble}$$

$$P_{\text{bubble}} = 2P_{\text{droplet}}$$

Gauge pressure within a bubble of the same fluid and of same size will be two times the gauge pressure within a spherical droplet.

1.6 (a)

Dynamic viscosity of gases increase with increase in temperature,

i.e., $\mu \propto \sqrt{T}$

and density of gases decrease with increase in temperature at constant pressure.

i.e., $\rho \propto \frac{1}{T}$

Kinematic viscosity,

$$\nu = \frac{\mu}{\rho}; \ \nu \propto \frac{\sqrt{T}}{1/T} \propto T^{3/2}; \ \frac{\nu}{T^{3/2}} = C$$

$$\frac{\nu_1}{T_1^{3/2}} = \frac{\nu_2}{T_2^{3/2}}$$

where T_1 = 20°C = (20 + 273) K = 293 K
$\nu_1 = 1.6 \times 10^{-5}$ m^2/s
T_2 = 70°C = (70 + 273) K = 343 K
ν_2 = ?

$$\therefore \frac{1.6 \times 10^{-5}}{(293)^{3/2}} = \frac{\nu_2}{(343)^{3/2}}$$

or $\nu_2 = \mathbf{2.02 \times 10^{-5}}$ **m^2/s**

POINTS TO REMEMBER

According to kinetic theory of gases, dynamic viscosity of gases,

$$\mu = \frac{2}{3\pi}\sqrt{\frac{\pi mkT}{\pi d^2}} \quad ; m \propto \sqrt{T}$$

where, πd^2 = Collision cross-section of molecules

m = Mass of molecule

k = Boltzmann constant

1.7 (a)

Kinematic viscosity:

$$\nu = \frac{\text{Dynamic viscosity} : \mu}{\text{Density} : \rho} = \frac{\text{Ns/m}^2}{\text{kg/m}^3}$$

$$= \frac{\text{kg m}}{\text{s}^2} \times \frac{\text{s}}{\text{m}^2} \times \frac{1}{\text{kg/m}^3}$$

$\because$ 1 N = 1 kgm/s^2

$$= \text{m}^2/\text{s}$$

1.8 (c)

For a Newtonian fluid,

$$\tau \propto \frac{du}{dy}; \quad \tau = \mu\frac{du}{dy}$$

where τ = Shear stress

$\frac{du}{dy}$ = Rate of shear strain

1.9 (b)

Air pressure falls with altitude exponentially

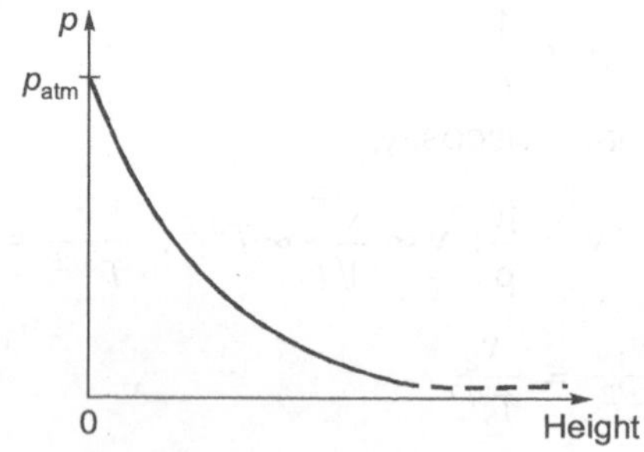

Pressure as a function of the height above the sea level.

POINTS TO REMEMBER

At isothermal condition, the variation in pressure with height is given as:

$$P = P_o e^{\frac{-zg}{RT}}$$

where, z = Height

P_o = atmospheric pressure at earth surface

1.10 Sol.

Velocity profile is laminar in both fluids

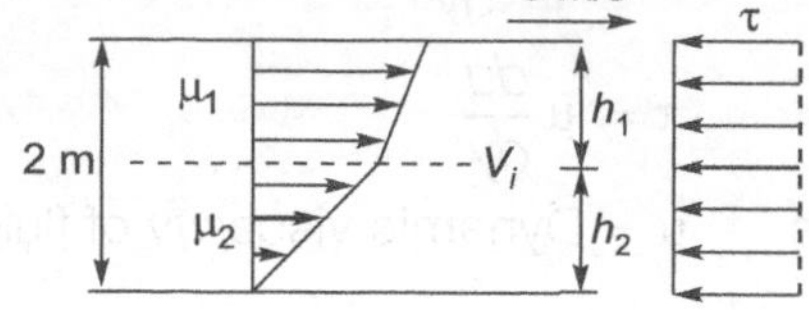

$$\frac{d^2u}{dy^2} = 0$$

$$\frac{du}{dy} = C_1$$

$$u = C_1 y + C_2$$

i.e. we can assume linear velocity profile.

If velocity profile is linear shear stress will be constant in gap everywhere i.e. in fluid (1) and fluid (2)

Also at the interface shear stress will be constant.

$$\tau_1 = \tau_2$$

$$\mu_2 \frac{V_i}{h_2} = \mu_1 \frac{(V - V_i)}{h_1}$$

where V_i is velocity at the interface.

$$2\mu_1 \frac{V_i}{1} = \frac{\mu_1(3 - V_i)}{1}$$

$$2V_i = 3 - V_i$$

$$3V_i = 3$$

$$V_i = 1 \text{ m/s}$$

■■■■

2 Fluid Statics

2.1 A mercury manometer is used to measure the static pressure at a point in a water pipe as shown in figure. The level difference of mercury in the two limbs is 10 mm. The gauge pressure at that point is

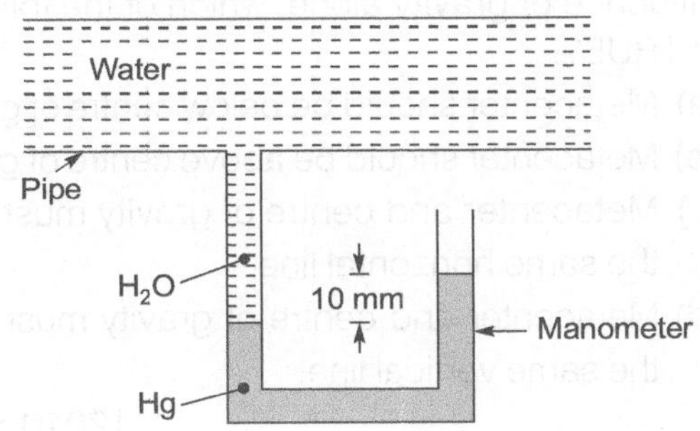

(a) 1236 Pa (b) 1333 Pa
(c) Zero (d) 98 Pa

[1996 : 1 M]

2.2 The force F needed to support the liquid of density ρ and the vessel on top (fig.) is

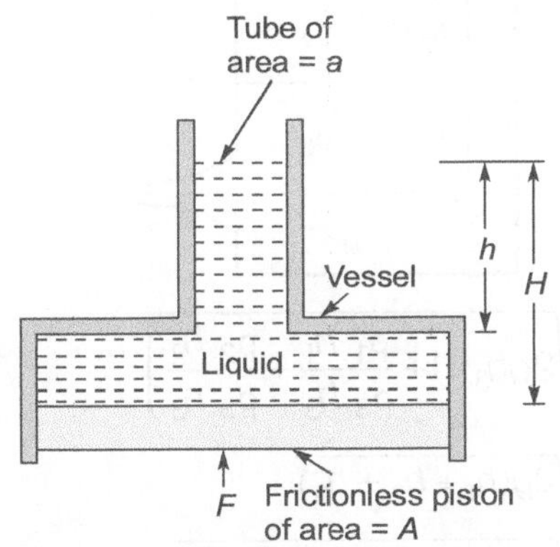

(a) $\rho g[ha-(H-h)A]$ (b) ρgHa
(c) ρgHA (d) $\rho g(H-h)A$

[1995 : 2 M]

2.3 Refer to figure, the absolute pressure of gas A in the bulb is

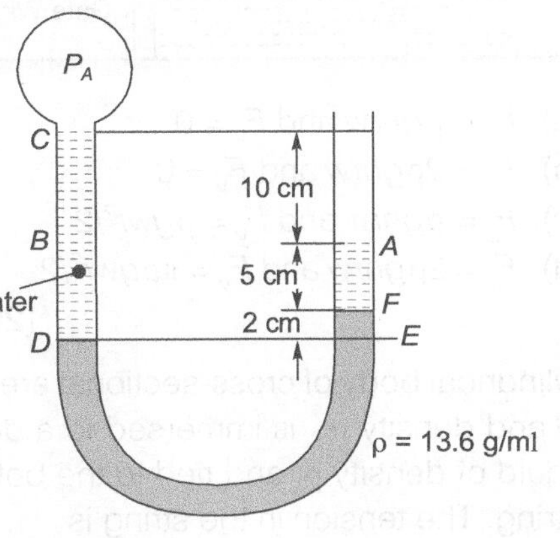

(a) 771.2 mm Hg (b) 752.65 mm Hg
(c) 767.35 mm Hg (d) 748.8 mm Hg

[1997 : 2 M]

2.4 In given figure, if the pressure of gas in bulb A is 50 cm Hg vacuum and p_{atm} = 76 cm Hg, then height of column H is equal to

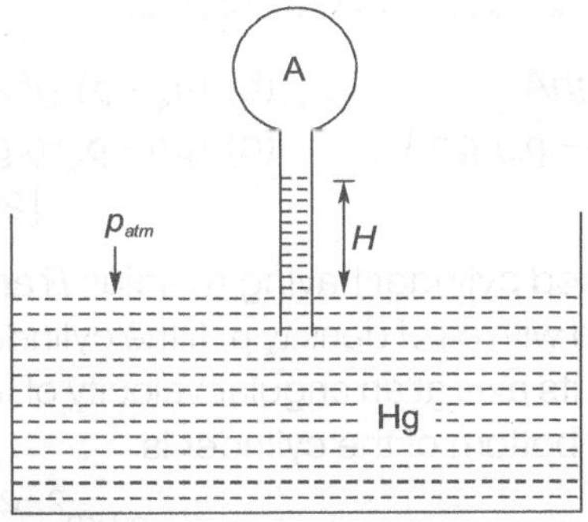

(a) 26 cm (b) 50 cm
(c) 76 cm (d) 126 cm

[2000 : 2 M]

2.5 A static fluid can have

(a) Non-zero normal and shear stress
(b) Negative normal stress and zero shear stress
(c) Positive normal stress and zero shear stress
(d) Zero normal stress and non-zero shear stress

[2001 : 1 M]

2.6 The horizontal and vertical hydrostatic forces F_x and F_y on the semicircular gate, having a width w into the plane of figure, are

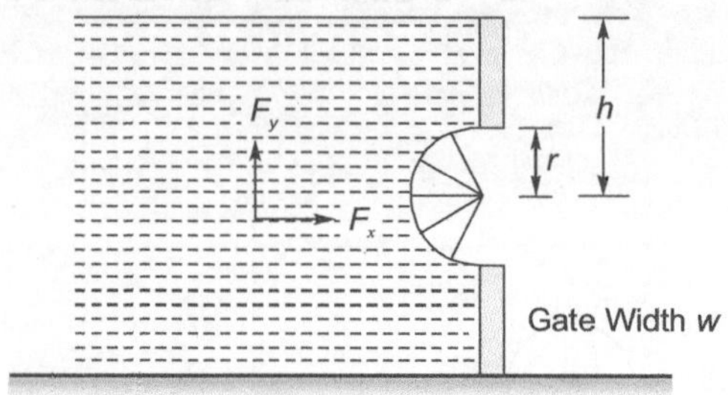

(a) $F_x = \rho ghrw$ and $F_y = 0$
(b) $F_x = 2\rho ghrw$ and $F_y = 0$
(c) $F_x = \rho ghrw$ and $F_y = \rho gwr^2/2$
(d) $F_x = 2\rho ghrw$ and $F_y = \pi\rho gwr^2/2$

[2001 : 2 M]

2.7 cylindrical body of cross-sectional area A, height H and density ρ_s, is immersed to a depth h in a liquid of density ρ, and tied to the bottom with a string. The tension in the string is

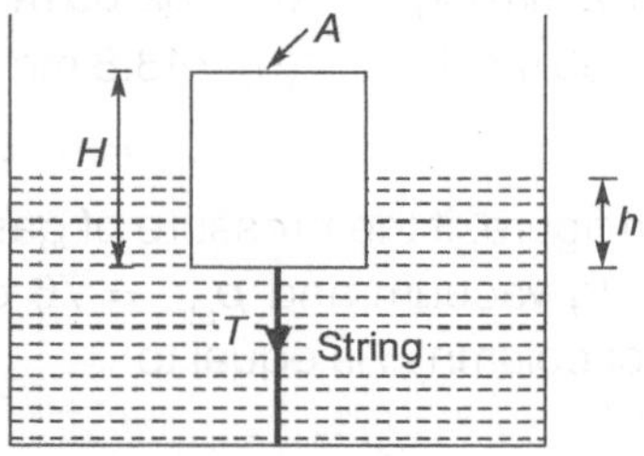

(a) ρghA (b) $(\rho_s - \rho)\, ghA$
(c) $(\rho - \rho_s)\, ghA$ (d) $(\rho h - \rho_s H)\, gA$

[2003 : 1 M]

2.8 A closed cylinder having a radius R and height H is filled with oil of density ρ. If the cylinder is rotated about its axis at an angular velocity of ω, the thrust at the bottom of the cylinder is

(a) $\pi R^2 \rho gH$ (b) $\pi R^2 \dfrac{\rho\omega^2 R^2}{4}$

(c) $\pi R^2(\rho\omega^2 R^2 + \rho gH)$ (d) $\pi R^2\left(\dfrac{\rho\omega^2 R^2}{4} + \rho gH\right)$

[2004 : 2 M]

2.9 he pressure gauges G_1 and G_2 installed on the system show pressure of P_{G1} = 5.00 bar and P_{G2} = 1.00 bar. The value of unknown pressure P is [Atmospheric pressure = 1.01 bar]

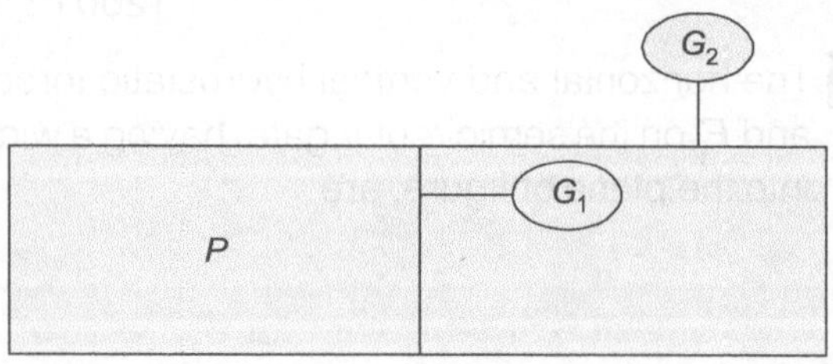

(a) 1.01 bar (b) 2.01 bar
(c) 5.00 bar (d) 7.01 bar [2004 : 2 M]

2.10 Oil in a hydraulic cylinder is compressed from an initial volume 2 m^3 to 1.96 m^3. If the pressure of oil in the cylinder changes from 40 MPa to 80 MPa during compression, the bulk modulus of elasticity of oil is

(a) 1000 MPa (b) 2000 MPa
(c) 4000 MPa (d) 8000 MPa [2007 : 2 M]

2.11 For the stability of a floating body, under the influence of gravity alone, which of the following is TRUE?

(a) Metacenter should be below centre of gravity.
(b) Metacenter should be above centre of gravity.
(c) Metacenter and centre of gravity must lie on the same horizontal line.
(d) Metacenter and centre of gravity must lie on the same vertical line.

[2010 : 1 M]

2.12 A large tank with a nozzle attached contains three immiscible, inviscid fluids as shown. Assuming that the changes in h_1, h_2 and h_3 are negligible, the instantaneous discharge velocity is

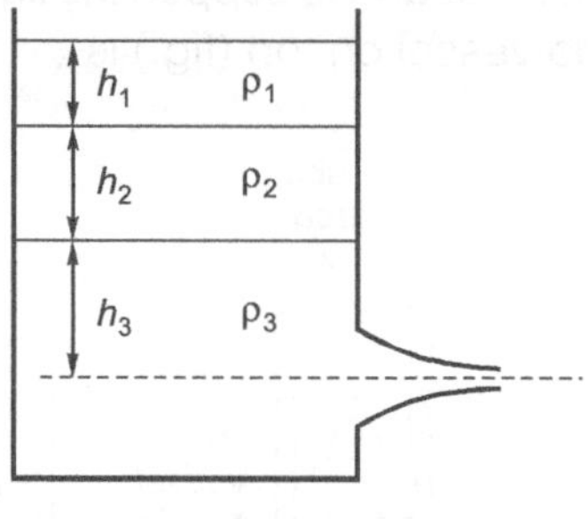

(a) $\sqrt{2gh_3\left(1 + \dfrac{\rho_1}{\rho_3}\dfrac{h_1}{h_3} + \dfrac{\rho_2}{\rho_3}\dfrac{h_2}{h_3}\right)}$

(b) $\sqrt{2g(h_1 + h_2 + h_3)}$

(c) $\sqrt{2g\left(\dfrac{\rho_1 h_1 + \rho_2 h_2 + \rho_3 h_3}{\rho_1 + \rho_2 + \rho_3}\right)}$

(d) $\sqrt{2g\left(\dfrac{\rho_3 h_2 h_3 + \rho_2 h_3 h_1 + \rho_3 h_1 h_2}{\rho_1 h_1 + \rho_2 h_2 + \rho_3 h_3}\right)}$ [2012 : 2 M]

2.13 A hinged gate of length 5 m, inclined at 30° with the horizontal and with water mass on its left, is shown in figure below. Density of water is 1000 kg/m^3. The minimum mass of the gate in kg per unit width (perpendicular to the plane of paper), required to keep it closed

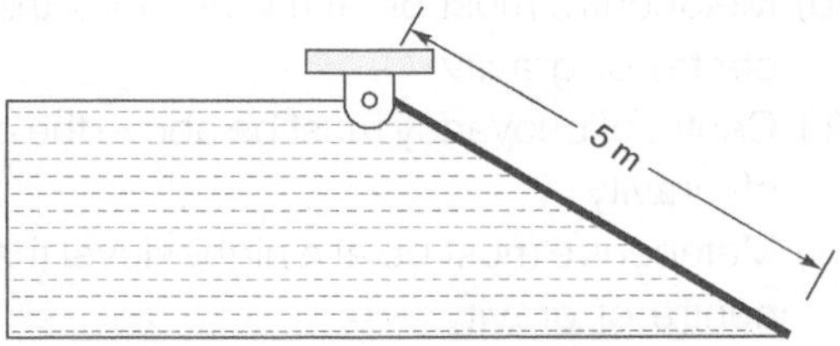

(a) 5000 (b) 6600
(c) 7546 (d) 9623 [2013 : 2 M]

2.14 For a completely submerged body with centre of gravity *G* and centre of buoyancy *B*, the condition of stability will be
(a) *G* is located below *B*
(b) *G* is located above *B*
(c) *G* and *B* are coincident
(d) independent of the locations of *G* and *B*
[2014 : 1 M, Set-1]

2.15 The difference in pressure (in N/m^2) across an air bubble of diameter 0.001 m immersed in water (surface tension = 0.072 N/m) is ______.
[2014 : 1 M, Set-2]

2.16 An aluminium alloy (density 2600 kg/m^3) casting is to be produced. A cylindrical hole of 100 mm diameter and 100 mm length is made in the casting using sand core (density 1600 kg/m^3). The net buoyancy force (in Newton) acting on the core is ____.
[2014 : 2 M, Set-1]

2.17 A spherical balloon with a diameter of 10 m, shown in the figure below is used for advertisements. The balloon is filled with helium (R_{He} = 2.08 kJ/kgK) at ambient conditions of 15°C and 100 kPa. Assuming no disturbances due to wind, the maximum allowable weight (in Newton) of balloon material and rope required to avoid the fall of the balloon (R_{air} = 0.289 kJ/kgK) is ______.

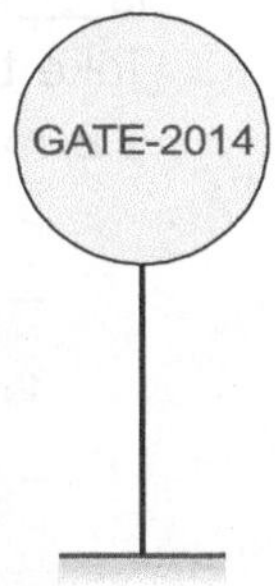

[2014 : 2 M, Set-2]

2.18 For a floating body, buoyant force acts at the
(a) centroid of the floating body
(b) center of gravity of the body
(c) centroid of the fluid vertically below the body
(d) centroid of the displaced fluid
[2016 : 1 M, Set-1]

2.19 An inverted U-tube manometer is used to measure the pressure difference between two pipes *A* and *B*, as shown in the figure. Pipe *A* is carrying oil (specific gravity = 0.8) and pipe *B* is carrying water. The densities of air and water are 1.16 kg/m^3 and 1000 kg/m^3, respectively. The pressure difference between pipes *A* and *B* is ______kPa. Acceleration due to gravity: g = 10 m/s^2

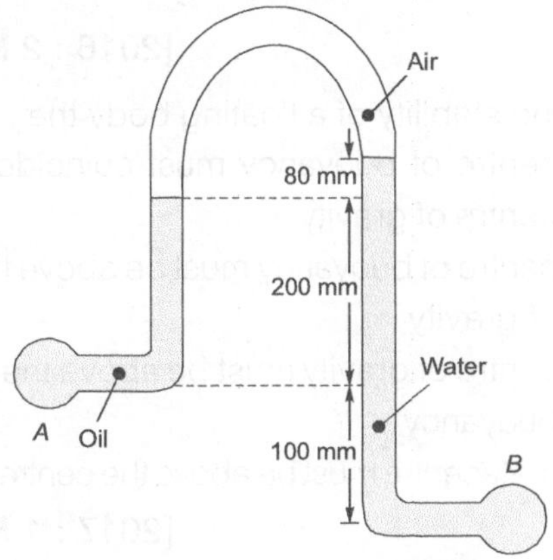

[2016 : 2 M, Set-1]

2.20 The large vessel shown in the figure contains oil and water. A body is submerged at the interface of oil and water such that 45 percent of its volume is in oil while the rest is in water. The density of the body is ________ kg/m^3.

The specific gravity of oil is 0.7 and density of water is 1000 kg/m^3.

Acceleration due to gravity: g = 10 m/s^2.

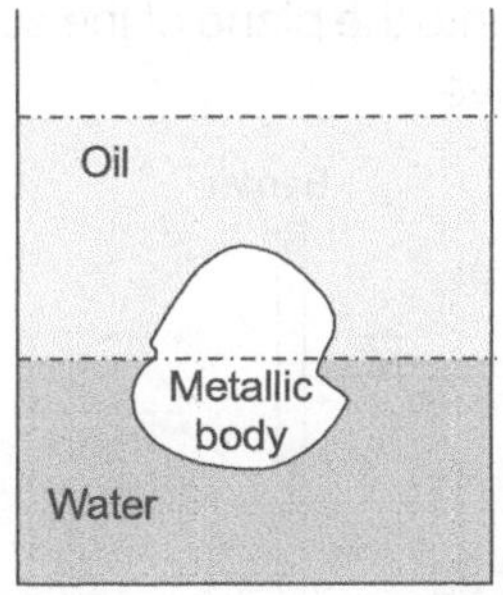

[2016 : 2 M, Set-2]

2.21 Consider a frictionless, massless and leak-proof plug blocking a rectangular hole of dimensions $2R \times L$ at the bottom of an open tank as shown in the figure. The head of the plug has the shape of a semi-cylinder of radius *R*. The tank is filled with a liquid of density ρ up to the tip of the plug. The gravitational acceleration is *g*. Neglect the effect of the atmospheric pressure.

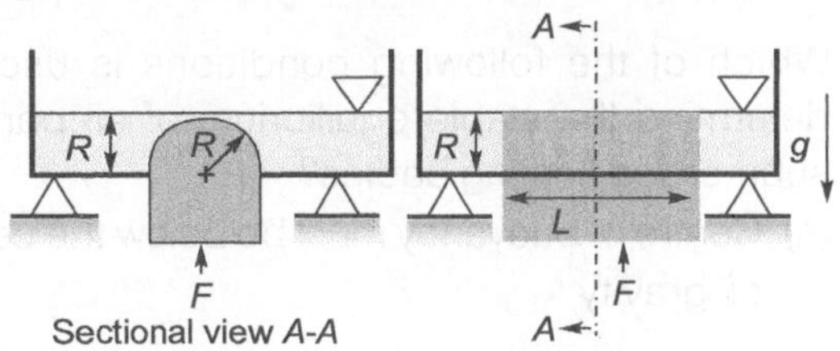

The force F required to hold the plug in its position is

(a) $2\rho R^2 gL\left(1-\frac{\pi}{4}\right)$ (b) $2\rho R^2 gL\left(1+\frac{\pi}{4}\right)$

(c) $\pi R^2 \rho gL$ (d) $\frac{\pi}{2}\rho R^2 gL$

[2016 : 2 M, Set-2]

2.22 For the stability of a floating body the

(a) centre of buoyancy must coincide with the centre of gravity

(b) centre of buoyancy must be above the centre of gravity

(c) centre of gravity must be above the centre of buoyancy

(d) metacentre must be above the centre of gravity

[2017 : 1 M, Set-2]

2.23 A cube of side 100 mm is placed at the bottom of an empty container on one of its faces. The density of the material of the cube is 800 kg/m^3. Liquid of density 1000 kg/m^3 is now poured into the container. The minimum height to which the liquid needs to be poured into the container for the cube to just lift up is ________ mm.

[2019 : 2 M, Set-1]

2.24 The barrier shown between two water tanks of unit width (1 m) into the plane of the screen is modeled as a cantilever.

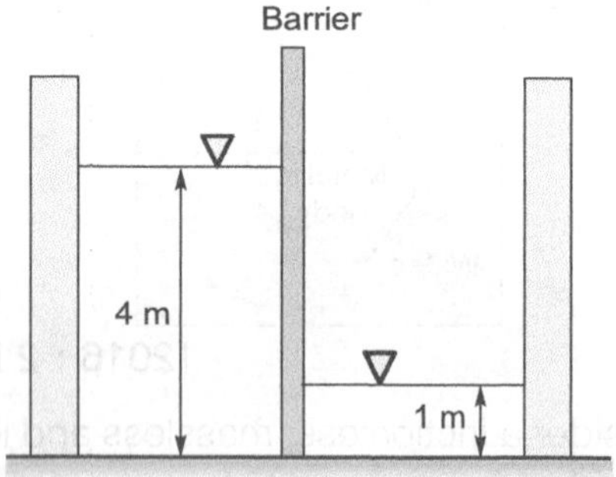

Taking the density of water as 1000 kg/m^3, and the acceleration due to gravity as 10 m/s^2, the maximum absolute bending moment developed in the cantilever is _____ kNm (round off to the nearest integer).

[2020 : 2 M, Set-1]

2.25 Which of the following conditions is used lo determine the stable equilibrium of all partially submerged floating bodies?

(a) Centre of buoyancy must be below the centre of gravity

(b) Metacentre must be at a lower level than the centre of gravity

(c) Centre of buoyancy must be above the centre of gravity

(d) Metacentre must be at a higher level than the centre of gravity

[2020 : 1 M, Set-2]

2.26 In the space above the mercury column in a barometer tube, the gauge pressure of the vapour is

(a) positive, but more than one atmosphere

(b) negative

(c) zero

(d) positive, but less than one atmosphere

[2020 : 1 M, Set-2]

2.27 A pressure measurement device fitted on the surface of a submarine, located at a depth H below the surface of an ocean, reads an absolute pressure of 4.2 MPa. The density of sea water is 1050 kg/m^3, the atmospheric pressure is 101 kPa, and the acceleration due to gravity is 9.8 m/s^2. The depth H is _______ m. [Round off to the nearest integer]

[2021 : 1 M, Set-1]

2.28 The figure shows two fluids held by a hinged gate. The atmospheric pressure is P_a = 100 kPa. The moment per unit width about the base of the hinge is ___ kNm/m. (Rounded off to one decimal place) Take the acceleration due to gravity to be g = 9.8 m/s^2.

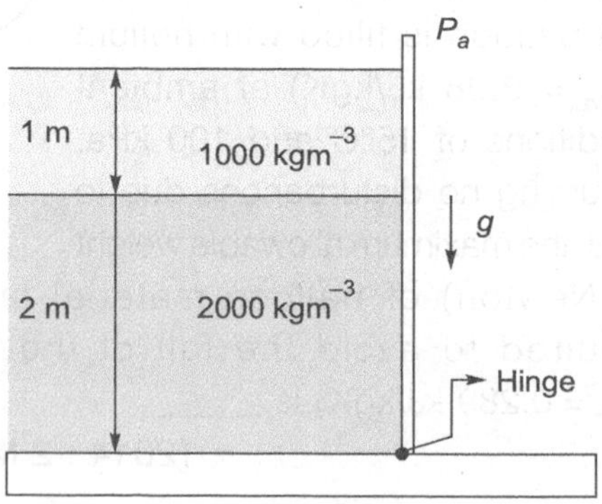

[2023 : 2 M]

■■■■

Answers Fluid Statics

2.1 (a)	2.2 (c)	2.3 (a)	2.4 (b)	2.5 (c)	2.6 (d)	2.7 (d)	2.8 (d)
2.9 (d)	2.10 (b)	2.11 (b)	2.12 (a)	2.13 (d)	2.14 (a)	2.15 (288)	
2.16 (7.7)	2.17 (5303.688)	2.18 (d)	2.19 (–2.199)	2.20 (865)	2.21 (a)	2.22 (d)	
2.23 (80)	2.24 (105)	2.25 (d)	2.26 (b)	2.27 (398)	2.28 (57.225)		

Explanations Fluid Statics

2.1 (a)

Method I:

Considering density of water be 1000 kg/m^3 and that of mercury be 13600 kg/m^3.

Gauge pressure $= (\rho_m - \rho_w)gh$

$= (13600 - 1000) \times 9.81 \times 10 \times 10^{-3}$

$\approx$ **1236 Pa**

Method II:

Given: Level difference of mercury in the two limbs, $x = 10$ mm

$$\text{Now,} \quad h_w = x\left(\frac{\rho_{Hg}}{\rho_w} - 1\right) = 10\left(\frac{13600}{1000} - 1\right)$$

$= 10 \times 12.6 = 126$ mm of water

$= 0.126$ m of water

Gauge pressure, $P = \rho_w g h_w = 1000 \times 9.81 \times 0.126$

$= 1236.06 \approx 1236$ Pa.

2.2 (c)

Height of water column above plate is H.

Force = Pressure × Area

$= \rho g H \times A = \boldsymbol{\rho g H A}$

2.3 (a)

Method I:

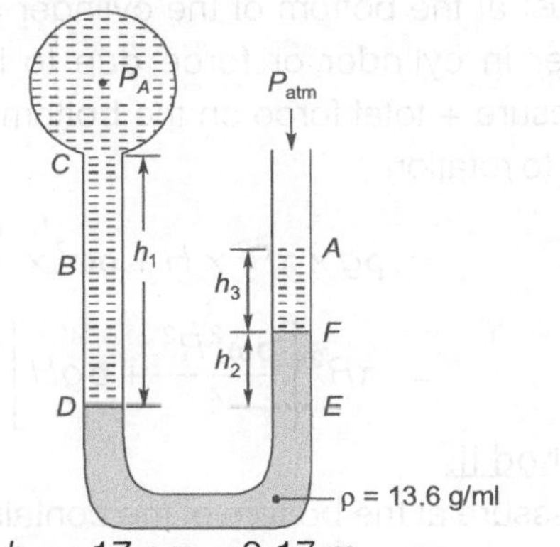

$h_1 = 17$ cm $= 0.17$ m

$h_2 = 2$ cm $= 0.02$ m

$h_3 = 5$ cm $= 0.05$ m

$\rho_{Hg} = \rho = 13.6$ g/ml $= 13600$ kg/m^3

For atmosphere $h_{atm} = 760$ mm of mercury

$P_{atm} = \rho_{Hg} \times g \times h_{atm}$

$= 13600 \times 9.81 \times 0.760 = 101396.16$ Pa

$\rho_{water} = 1000$ kg/m^3

Absolute pressure at point D

= Absolute pressure at point E

$P_A + \rho_{water} g h_1 = \rho_{Hg} g h_2 + \rho_{water} g h_3 + P_{atm}$

$P_A + 1000 \times 9.81 \times 0.17$

$= 13600 \times 9.81 \times 0.02 + 1000 \times 9.81 \times 0.05 + 101396.16$

$P_A = 102887.28$ Pa (absolute)

also, $P_A = \rho_{Hg} g h_A$

$\therefore 102887.28 = 13600 \times 9.81 \times h_A$

or $h_A = 0.77117$ m of Hg

$=$ **771.17 mm of Hg (absolute)**

Method II:

Head in terms of Hg column is given by:

$$h_{Hg} = \frac{P_A}{\rho_{Hg} \cdot g} + \frac{\Sigma h_i S_i}{S_{Hg}}$$

$$= 760 + \frac{5 \times 10 \times 1}{13.6} + \frac{2 \times 10 \times 13.6}{13.6} - \frac{17 \times 10 \times 1}{13.6}$$

$$= 760 + \frac{20 \times 13.6 - 120}{13.6}$$

$= 771.176$ mm of Hg $= 771.2$ mm of Hg

2.4 (b)

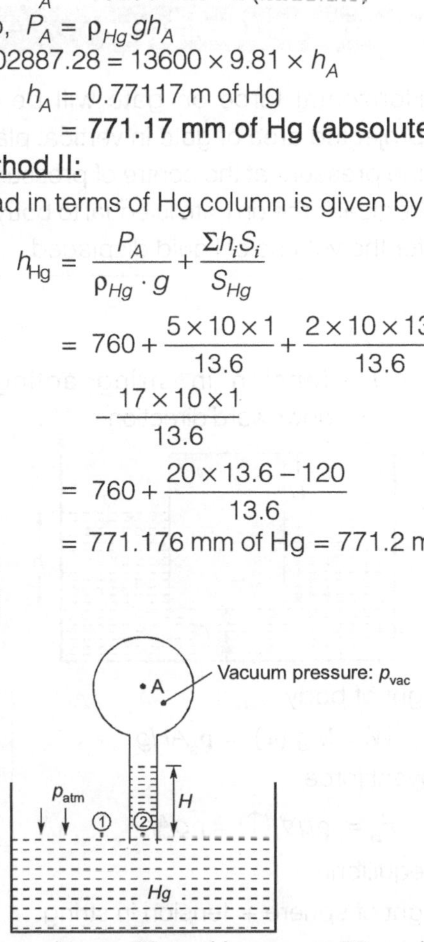

$h_{vac} = 50$ cm Hg, $p_{atm} = 76$ cm Hg

Absolute pressure at point 1

= Absolute pressure of point 2

$$(p_{abs})_1 = (p_{abs})_2$$
$$(p_{atm} + p_G)_1 = (p_{atm} - p_{vac})_2 + \rho gH$$
$$p_{atm} + 0 = p_{atm} - p_{vac} + \rho gH$$
or $\rho gH = p_{vac}$
$$\rho gH = \rho g h_{vac}$$
or $H = h_{vac} = 50 \text{ cm Hg}$

2.5 (c)

A static fluid can have positive normal stress and zero shear stress.

2.6 (d)

F_H = Pressure × Projected area of gate
$= \rho gh \times (2r \times w)$
$= 2\rho ghrw$

$F_v = \rho g \times \pi r^2 w/2$

F_v = Weight of liquid above the semi-circular gate
$= Mg$

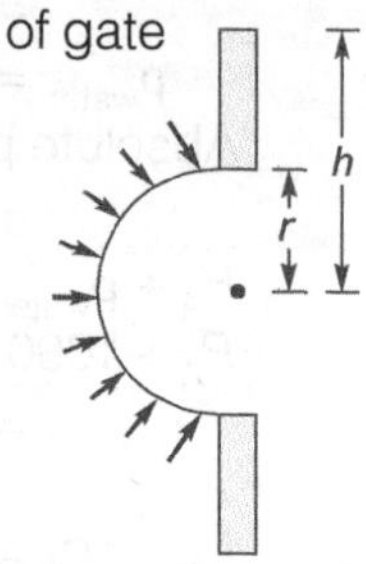

POINTS TO REMEMBER

- Horizontal force on gate will be equal to projected area of gate in vertical plane times the pressure at the centre of pressure.
- Vertical force on gate is equal to bouyant force for the volume of liquid displaced.

2.7 (d)

Let T = Tension in string acting in the downward direction

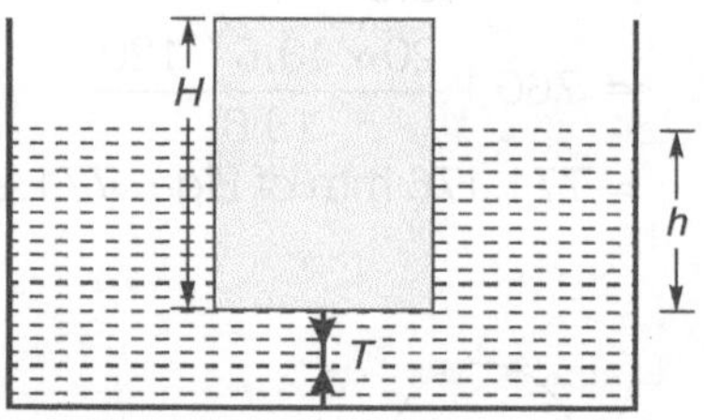

Weight of body,
$$W = Mg\ (\downarrow) = \rho_s AHg$$
Buoyant force,
$$F_B = \rho g \forall\ (\uparrow) = \rho gAh$$

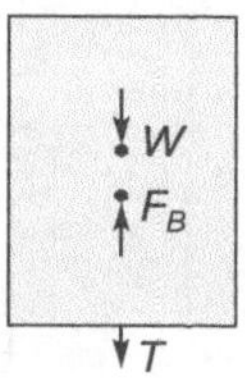

For equilibrium,

Weight of sphere + Tension in string
= Buoyant force
$$W + T = F_B$$
or $T = F_B - W = \rho gAh - \rho_s AHg$
$$= (\rho h - \rho_s H)gA$$

2.8 (d)

Method I:

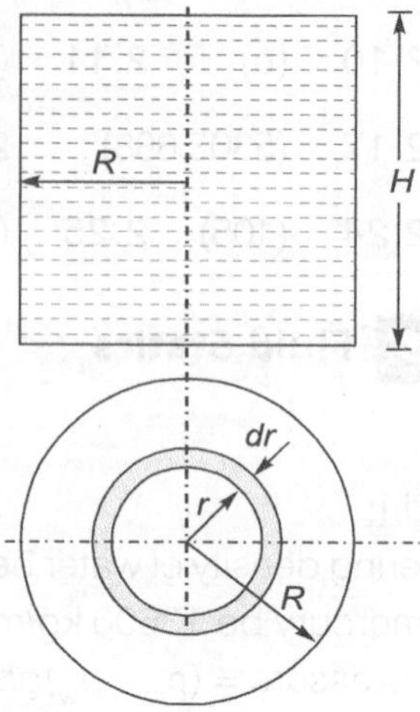

We know that
$$\frac{\partial p}{\partial r} = \frac{\rho V^2}{r} = \rho\omega^2 r$$
$$\Rightarrow \quad \int_0^p \partial p = \int_0^r \rho\omega^2 r dr$$
$$\therefore \quad p = \frac{\rho\omega^2 r^2}{2}$$

Area of circular ring = $2\pi r dr$

Force on elementary ring
= Intensity of pressure × Area of ring
$$= \frac{\rho}{2}\omega^2 r^2 2\pi r dr$$

∴ Total force on the bottom of the cylinder due to rotation
$$\int_0^R \frac{\rho\omega^2 r^2}{2}.2\pi r dr = \frac{\rho}{2}\omega^2.2\pi\int_0^R r^3 dr$$
$$= \pi\omega^2\rho\left[\frac{r^4}{4}\right]_0^R = \rho\omega^2.\frac{\pi R^4}{4}$$

Thrust at the bottom of the cylinder = weight of water in cylinder or force due to hydrostatic pressure + total force on the bottom of cylinder due to rotation
$$= \rho g \times \pi R^2 \times H + \rho\omega^2 \times \frac{\pi R^4}{4}$$
$$= \pi R^2\left[\frac{\rho\omega^2 R^2}{4} + \rho gH\right]$$

Method II:

Presssure at the bottom of the container is due to two pressure heads:

(i) Pressure in radial direction due to rotation (variable)

(ii) Weight of the liquid (constant)

Pressure head in radial direction due to rotation will be:

$$Z = \frac{\omega^2 r^2}{2g}$$

$\therefore$ Total pressure at radius r,

$$P = \left(\frac{\rho r^2 \omega^2}{2}\right) + H\rho g$$

Thrust force,

$$F = \int_0^R (P) \cdot 2\pi r \cdot dr = \int_0^R \left[\frac{\rho r^2 \omega^2}{2} + \rho g H\right] 2\pi r \cdot dr$$

$$= \frac{\rho\pi R^4 \omega^2}{4} + \rho g H \cdot \pi R^2 = \rho\pi R^2\left(\frac{R^2\omega^2}{4} + gH\right)$$

2.9 (d)

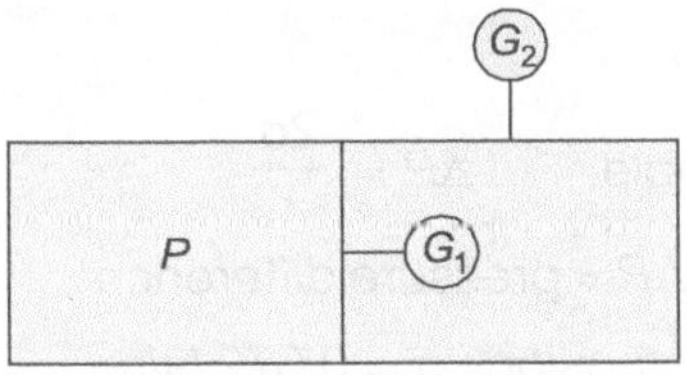

$$P_{G1} = 5 \text{ bar}, P_{G2} = 1 \text{ bar}, P_{atm} = 2.01 \text{ bar}$$

For pressure gauge G_2,

Absolute pressure

= Atmospheric pr. + Gauge pr.

$$P_{abs} = P_{atm} + P_{G2} = 1.01 + 1 = 2.01 \text{ bar}$$

= Local atmospheric pr. for pr. gauge G_1

$= (P_{atm})_1$

For pressure gauge G_1,

Absolute pressure = Local atm. pr. + Gauge pr.

$$P = (P_{atm})_1 + P_{G1}$$

$= 2.01 + 5 =$ **7.01 bar**

2.10 (b)

Given data: $V_1 = 2 \text{ m}^3$, $V_2 = 1.96 \text{ m}^3$

$\therefore \quad dV = V_2 - V_1$

$= 1.96 - 2 = -0.04 \text{ m}^3$

$p_1 = 40$ MPa, $p_2 = 80$ MPa

$\therefore \quad dp = p_2 - p_1 = 80 - 40 = 40$ MPa

Bulk modulus of elasticity,

$$K = -\frac{dp}{dV/V_1} = -V_1\frac{dp}{dV} = \frac{-2\times 40}{-0.04}$$

= **2000 MPa**

2.11 (b)

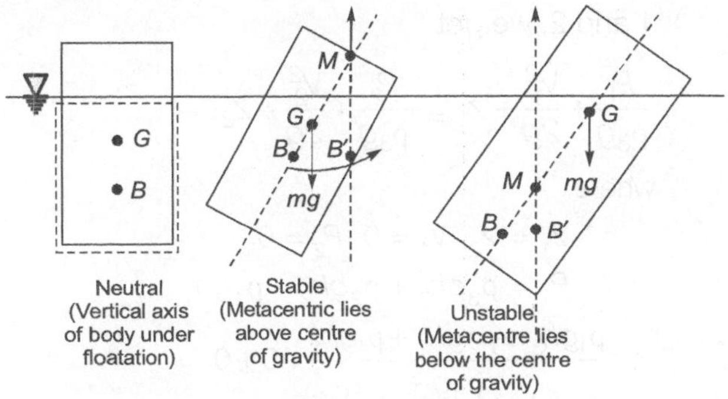

POINTS TO REMEMBER

- The distance between the centre of gravity and metacentre i.e. GM is known as metacentric height.
- If GM = 0, neutral equilibrium
  If GM > 0, stable equilibrium
  If GM < 0, unstable equilibrium

2.12 (a)

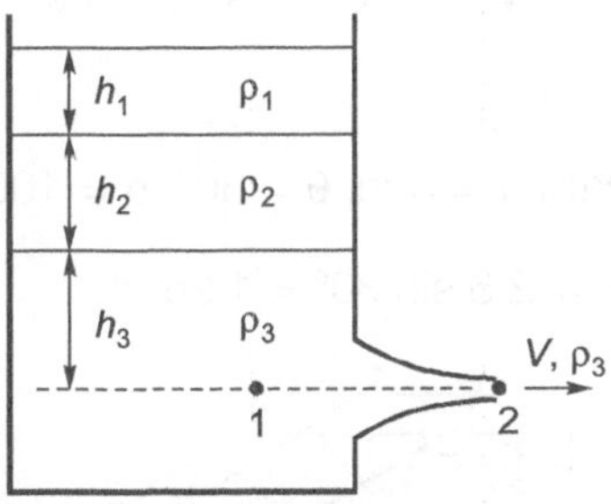

Total pressure at state 1,

$$P_1 = \rho_1 g h_1 + \rho_2 g h_2 + \rho_3 g h_3$$

Total pressure at state 2,

$$P_2 = \frac{\rho_3 V^2}{2}$$

Thus, $P_2 = P_1$

$$\frac{\rho_3 V^2}{2} = \rho_1 g h_1 + \rho_2 g h_2 + \rho_3 g h_3$$

$$V^2 = 2g\left(\frac{\rho_1}{\rho_3}h_1 + \frac{\rho_2}{\rho_3}h_2 + \frac{\rho_1}{\rho_3}h_1\right)$$

$$V = \sqrt{2g\left(h_3 + \frac{\rho_2}{\rho_3}h_2 + \frac{\rho_1}{\rho_3}h_1\right)}$$

$$= \sqrt{2gh_3\left(1 + \frac{\rho_2}{\rho_3}\frac{h_2}{h_3} + \frac{\rho_1}{\rho_3}\frac{h_1}{h_3}\right)}$$

Alternatively

Applying the Bernoulli's equation between states 1 and 2, we get

$$\frac{P_1}{\rho_3 g}+\frac{V_1^2}{2g}+z_1=\frac{P_2}{\rho_3 g}+\frac{V_2^2}{2g}+z_2$$

where,

$$z_1 = z_2,\ V_1 = 0,\ P_2 = 0$$

$$P_1 = \rho_3 g h_3 + \rho_2 g h_2 + \rho_1 g h_1$$

$$\therefore \quad \frac{\rho_1 g h_3+\rho_2 g h_2+\rho_1 g h_1}{\rho_3 g}+0+0$$

$$= 0+\frac{V_2^2}{2g}+0$$

$$h_3+\frac{\rho_2}{\rho_3}h_2+\frac{\rho_1}{\rho_3}h_1 = \frac{V_2^2}{2g}$$

or $$V_2^2 = 2g\left(h_3+\frac{\rho_2}{\rho_3}h_2+\frac{\rho_1}{\rho_3}h_1\right)$$

or $$V_2 = \sqrt{2g\left(h_3+\frac{\rho_2}{\rho_3}h_2+\frac{\rho_1}{\rho_3}h_1\right)}$$

$$V_2 = \sqrt{2gh_3\left(1+\frac{\rho_2}{\rho_3}\frac{h_2}{h_3}+\frac{\rho_1}{\rho_3}\frac{h_1}{h_3}\right)}$$

2.13 (d)

Given data: l = 5 m, θ = 30°, ρ = 1000 kg/m³

$$\bar{h} = 2.5 \sin 30° = 1.25 \text{ m}$$

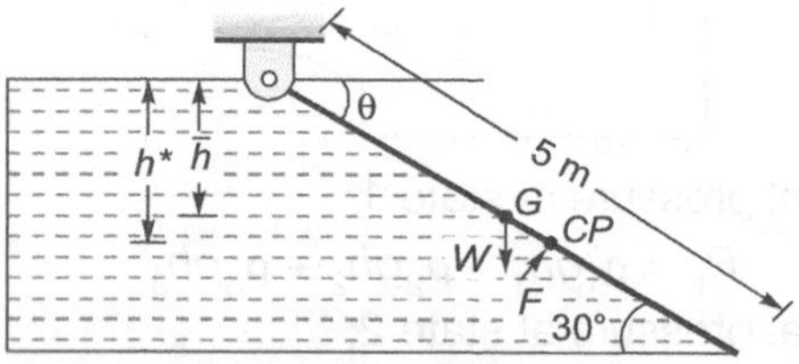

$$A = bl = 1 \times 5 = 5 \text{ m}^2$$

$$I_G = \frac{bl^3}{12} = \frac{1\times(5)^3}{12} = 10.41 \text{ m}^4$$

$$F = \rho g A\bar{h} = 1000 \times 9.81 \times 5 \times 1.25$$

$$= 61312.5 \text{ N}$$

$$h^* = \bar{h}+\frac{I_G}{A\bar{h}}\sin^2\theta$$

$$= 1.25+\frac{10.41}{5\times 1.25}\sin^2 30° = 1.666 \text{ m}$$

Taking moment about hinge,

Moment due F = Moment due to W

$$F\times\frac{h^*}{\sin 30°} = W \times 2.5 \times \cos 30°$$

$$61312.5\times\frac{1.666}{0.5} = \text{Mg} \times 2.165$$

or $$M = 9618.93 \text{ kg} \simeq \mathbf{9623\ kg}$$

2.14 (a)

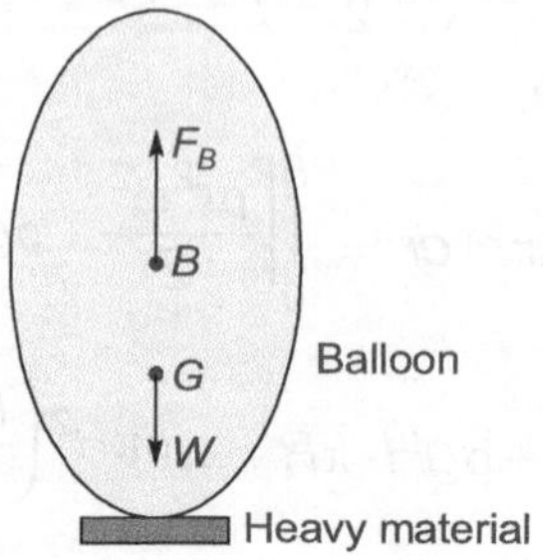

(A submerged body in stable equilibrium)

2.15 Sol.

For bubble, $\Delta P = \frac{2\sigma}{r}$

where, ΔP = pressure difference

Surface tension : σ = 0.072 N/m

Radius of bubble: $r = \frac{0.001}{2}$ m = 0.0005 m

$$\therefore \quad \Delta P = \frac{2\times 0.072}{0.0005} = \mathbf{288\ N/m^2}$$

2.16 Sol.

$$\text{Volume of core} = \frac{\pi}{4}D^2H = 0.785 \times 0.1^2 \times 0.1$$

$$= 7.85 \times 10^{-4} \text{ m}^3$$

$$\text{Net buoyancy force} = (\rho_{Al} - \rho_{core})\, V_{core}\, g$$

$$= 1000 \times 9.81 \times 7.85 \times 10^{-4}$$

$$= \mathbf{7.7\ N}$$

2.17 Sol.

Step-I : Mass of balloon

Volume: $$V = \frac{4}{3}\pi r^3 = \frac{4\times 3.14\times 5^3}{3}$$

$$= 523.33 \text{ m}^3$$

$$\rho_{He} = \frac{P}{RT} = \frac{100\times 10^3}{2080\times 288}$$

$$= 0.16693 \text{ kg/m}^3$$

$$\text{Total mass} = 0.16693 \times 523.33$$

$$= 87.359 \text{ kg}$$

Step-II : Mass of air

$$\rho = \frac{P}{RT} = \frac{100 \times 10^3}{289 \times 288}$$

$$= 1.2 \text{ kg/m}^3$$

Mass of displaced volume of air

$$= 1.2 \times 523.33 = 628 \text{ kg}$$

Step-III : Net buoyant mass

$$= 628 - 87.359 = 540.64 \text{ kg}$$

$\therefore$ Maximum allowable weight $= 540.64 \times 9.81$

$$= \mathbf{5303.688 \text{ N}}$$

2.18 (d)

Either a submerged body or a floating body, buoyant force always acts at the centroid of the displaced fluid.

2.19 Sol.

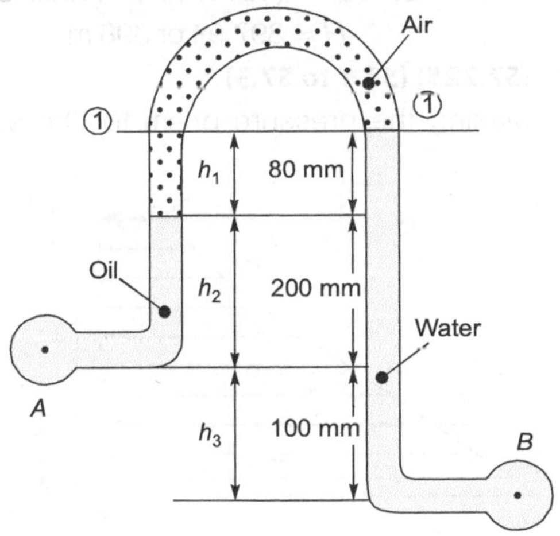

Given data:

Specific gravity of oil,

$$S_{oil} = 0.8$$

$\therefore \quad \rho_{oil} = 0.8 \times 1000 = 800 \text{ kg/m}^3$

Density of air : $\rho_{air} = 1.16 \text{ kg/m}^3$

Density of water, $\rho_w = 1000 \text{ kg/m}^3$

Acceleration due to gravity, $g = 10 \text{ m/s}^2$

$h_1 = 80 \text{ mm} = 0.08 \text{ m}$; $h_2 = 200 \text{ mm} = 0.2 \text{ m}$

$h_3 = 100 \text{ mm} = 0.1 \text{ m}$

Pressure at section 1 on left limb = Pressure at section 1 on right limb

$$p_A - \rho_{oil}\, gh_2 - \rho_{air}\, gh_1 = p_B - \rho_w\, g(h_1 + h_2 + h_3)$$

$$p_A - 800 \times 10 \times 0.2 - 1.16 \times 10 \times 0.08 = p_B - 1000 \times 10\,(0.08 + 0.2 + 0.1)$$

$$p_A - 1600 - 0.928 = p_B - 3800$$

or $\quad p_A - p_B = -2199.07 \text{ Pa} = \mathbf{-2.199 \text{ kPa}}$

2.20 Sol.

Let V = Volume of metallic body

45% of volume is in oil

i.e., $\quad V_{oil} = 45\% \text{ of } V = 0.45V$

and $\quad V_{water} = 0.55V$

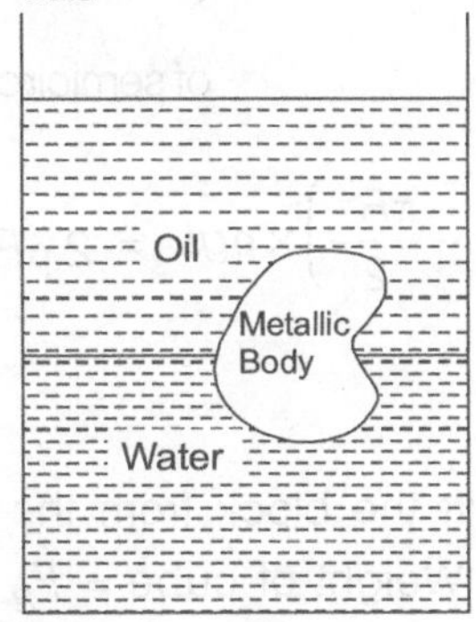

For equilibrium condition,

Net buoyant force $= (F_B)_w + (F_B)_{oil}$

$$= \text{Weight of body}$$

$$M_b g = \rho_w V_w g + \rho_{oil} V_{oil} g$$

$$M_b = \rho_w V_w + \rho_{oil} V_{oil}$$

$$\rho_b V = \rho_w \times 0.55V + \rho_{oil} \times 0.45V$$

or $\quad \rho_b = \rho_w \times 0.55 + \rho_{oil} \times 0.45$

$$= 1000 \times 0.55 + 700 \times 0.45$$

$$= 550 + 315 = \mathbf{865 \text{ kg/m}^3}$$

2.21 (a)

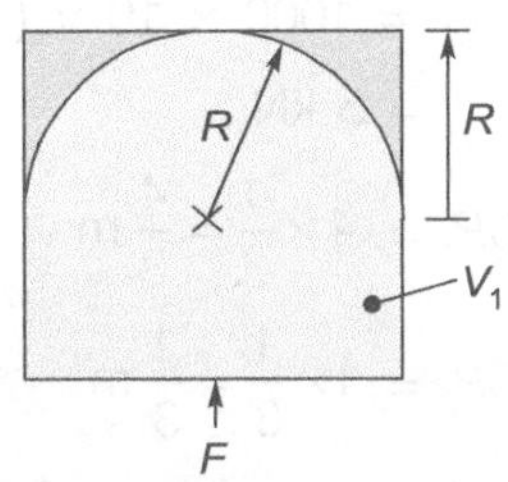

Volume, $V_{total} = 2R \times L \times R$

$$= 2R^2L \quad \text{...(i)}$$

Volume: $\quad V_1 = \frac{1}{2} \times \pi R^2 \times L \quad \text{...(ii)}$

F = Weight of fluid

$$= [V_{total} - V_1]\rho g$$

$$= \left[2R^2L - \frac{\pi R^2 L}{2}\right]\rho g = \left[1 - \frac{\pi}{4}\right] 2R^2 L g \rho$$

$$= 2\rho R^2 g L\left[1 - \frac{\pi}{4}\right]$$

POINTS TO REMEMBER

This problem can also be solved by considering.

Force, F = (Area of rectangle – Area of semicircle) × length × ρg

$$= \left(R \times 2R - \frac{\pi R^2}{2}\right) \times \rho g L = 2\rho R^2 g L\left(1 - \frac{\pi}{4}\right)$$

2.23 Sol.

To just lift the block from the bottom,

Weight of cube = F_B

$$\Rightarrow \quad \rho g \times V = \rho_l g \times \forall$$

$$\Rightarrow \quad 800 \times a^3 = 1000 \times a^2 \times h$$

$$\Rightarrow \quad h = 0.8 \times a$$

$$= 0.8 \times 100 = \mathbf{80\ mm}$$

2.24 Sol.

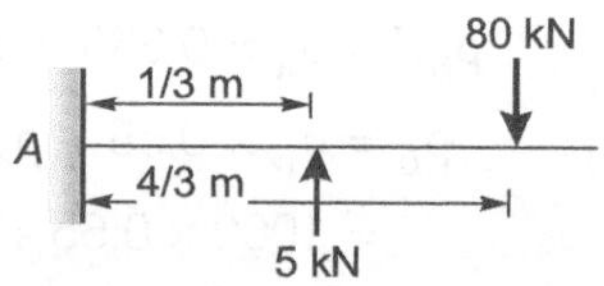

$$F_1 = \rho g A \cdot \bar{x}_1 = 1000 \times 10 \times (1 \times 4) \times 2$$

$$= 80\ \text{kN}$$

$$F_2 = \rho g A \cdot \bar{x}_2 = 1000 \times 10 \times (1 \times 1) \times 0.5$$

$$= 5\ \text{kN}$$

$$CP_1 = 4 \times \frac{1}{3} = \frac{4}{3}\ \text{m}$$

$$CP_2 = 1 \times \frac{1}{3} = \frac{1}{3}\ \text{m}$$

So, $$M_A = 80 \times \frac{4}{3} - 5 \times \frac{1}{3} = 105\ \text{kN-m}$$

2.25 (d)

Metacentre must be higher level than the centre of gravity.

2.26 (b)

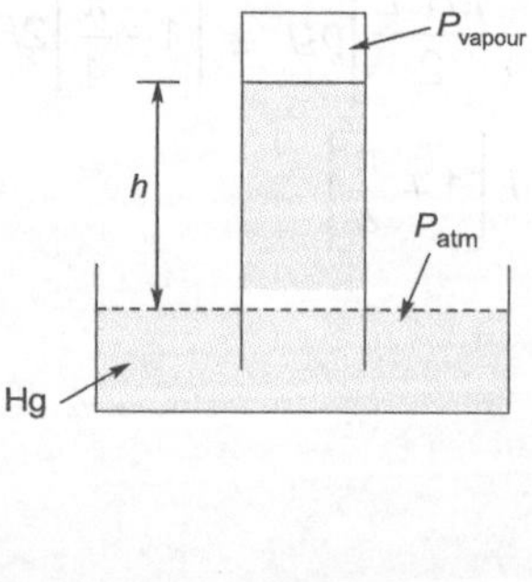

$$P_{vap} + \rho_{Hg} g h = P_{atm}$$

$$(P_{vap}) = P_{atm} - \rho_{Hg} g h$$

For gauge pressure $P_{atm} = 0$

$$(P_{vap})_{gauge} = -\rho_{Hg} g h$$

So, –ve gauge pressure.

2.27 Sol.

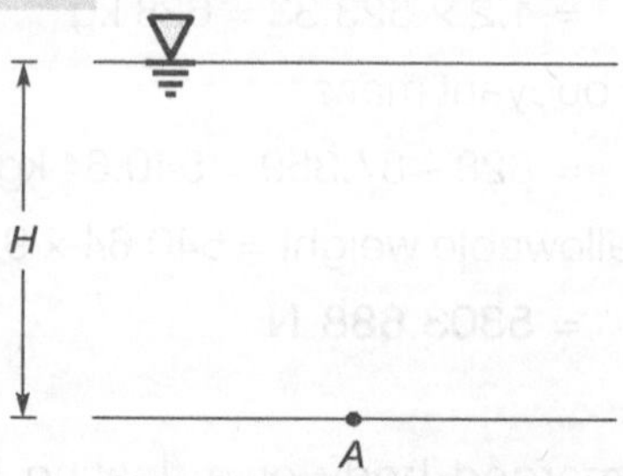

$P_A = 4.2$ MPa [Absolute pressure]

$P_{atm} = 101$ kPa; $\rho = 1050$ kg/m^3; $g = 9.8$ m/s^2

$$P_A = P_{atm} + \rho g H$$

$$4.2 \times 10^6 = (101 \times 10^3) + [1050 \times 9.81 \times H]$$

$$H = 397.94 \text{ or } 398\ \text{m}$$

2.28 (57.225) (57.1 to 57.3)

Making the pressure prism for the system.

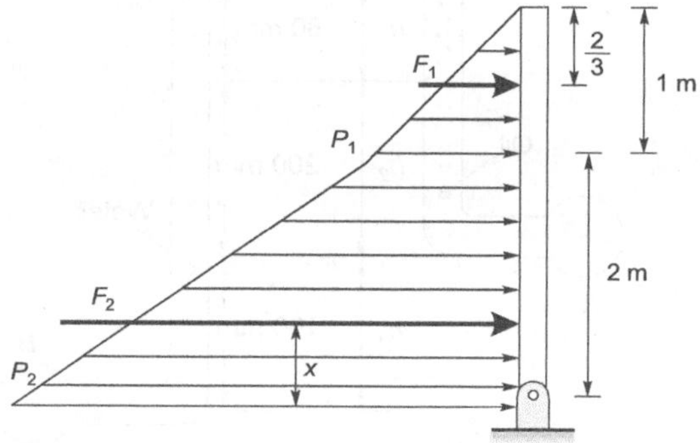

$$P_1 = 10^3 \times 9.81 \times 1 = 9.81\ \text{kPa}$$

$$P_2 = P_1 + [2000 \times 9.81 \times 2] = 49.05\ \text{kPa}$$

The resultant forces F_1 and F_2 due to pressure P_1 and P_2 respectively are given as:

F_1 = Volume of triangular pressure prism

$$= \frac{1}{2} \times P_1 \times 1 \times 1 = 4.9\text{kN}$$

F_2 = Volume of trapezoidal pressure prism

$$= \frac{1}{2} \times [P_1 + P_2] \times 2 \times 1 = 58.86\ \text{kN}$$

Now, the moment due to forces is given as

$$M = F_1 \times \left[2 + \frac{1}{3}\right] + F_2 \times \left[\frac{2P_1 + P_2}{P_1 + P_2}\right] \times \frac{2}{3}$$

Putting the values, we get

$$M = 57.225\ \text{kNm}$$

■■■■

3 Fluid Kinematics

3.1 The velocity components in the x and y directions are given by $u = \lambda xy^3 - x^2y$, $v = xy^2 - \frac{3}{4}y^4$. The value of λ for a possible flow field involving an incompressible fluid is

(a) $-\frac{3}{4}$ (b) $-\frac{4}{3}$

(c) $\frac{4}{3}$ (d) 3

[1995 : 2 M]

3.2 The 2-D flow with, velocity $\vec{V} = (x + 2y + 2)\hat{i} + (4 - y)\hat{j}$ is

(a) Compressible and irrotational
(b) Compressible and not irrotational
(c) Incompressible and irrotational
(d) Incompressible and not irrotational

[2001 : 2 M]

3.3 A fluid flow is represented by the velocity field $\vec{V} = ax\hat{i} + ay\hat{j}$, where a is a constant. The equation of stream line passing through a point (1, 2) is

(a) $x - 2y = 0$ (b) $2x + y = 0$
(c) $2x - y = 0$ (d) $x + 2y = 0$

[2004 : 1 M]

3.4 The velocity components in the x and y directions of a two dimensional potential flow are u and v, respectively. Then $\frac{\partial u}{\partial x}$, is equal to

(a) $\frac{\partial v}{\partial x}$ (b) $-\frac{\partial v}{\partial x}$

(c) $\frac{\partial v}{\partial y}$ (d) $-\frac{\partial v}{\partial y}$

[2005 : 1 M]

3.5 A leaf is caught in a whirlpool. At a given instant, the leaf is at a distance of 120 m from the centre of the whirlpool. The whirlpool can be described by the following velocity distribution:

$$V_r = -\left(\frac{60 \times 10^3}{2\pi r}\right) \text{m/s} \quad \text{and} \quad V_\theta = \frac{300 \times 10^3}{2\pi r} \text{m/s},$$

where r (in meters) is the distance from the centre of the whirlpool. What will be the distance of the leaf from the centre when it has moved through half a revolution?

(a) 48 m (b) 64 m
(c) 120 m (d) 142 m

[2005 : 2 M]

3.6 A two-dimensional flow field has velocities along the x and y directions given by $u = x^2t$ and $v = -2xyt$ respectively, where t is time. The equation of streamline is

(a) x^2y = constant
(b) xy^2 = constant
(c) xy = constant
(d) not possible to determine

[2006 : 2 M]

3.7 a two-dimensional velocity field with velocities u and v along the x and y directions respectively, the convective acceleration along the x-direction is given by

(a) $u\frac{\partial v}{\partial x} + v\frac{\partial u}{\partial y}$ (b) $u\frac{\partial u}{\partial x} + v\frac{\partial v}{\partial y}$

(c) $u\frac{\partial u}{\partial x} + v\frac{\partial u}{\partial y}$ (d) $v\frac{\partial u}{\partial x} + u\frac{\partial u}{\partial y}$

[2006 : 1 M]

3.8 a steady flow through a nozzle, the flow velocity on the nozzle axis is given by $v = u_0(1 + 3x/L)i$, where x is the distance along the axis of the nozzle from its inlet plane and L is the length of the nozzle. The time required for a fluid particle on the axis to travel from the inlet to the exit plane of the nozzle is

(a) $\frac{L}{u_0}$ (b) $\frac{L}{3u_0}\ln 4$

(c) $\frac{L}{4u_0}$ (d) $\frac{L}{2.5u_0}$

[2007 : 1 M]

3.9 Which combination of the following statements about steady incompressible forced vortex flow is correct?

P : Shear stress is zero at all points in the flow.
Q : Vorticity is zero at all points in the flow.
R : Velocity is directly proportional to the radius from the centre of the vortex.
S : Total mechanical energy per unit mass is constant in the entire flow field.

Select the correct answer using the codes given below:

(a) P and Q (b) R and S
(c) P and R (d) P and S

[2007 : 2 M]

3.10 For the continuity equation given by $\vec{\nabla}\cdot\vec{V}=0$ to be valid, where $\vec{V}$ is the velocity vector, which one of the following is a necessary condition?

(a) Steady flow
(b) Irrotational flow
(c) Inviscid flow
(d) Incompressible flow

[2008 : 1 M]

Linked Answer Questions 3.11 and 3.12: The gap between a moving circular plate and a stationary surface is being continuously reduced, as the circular plate comes down at a uniform speed V towards the stationary bottom surface, as shown in the figure. In the process, the fluid contained between the two plates flows out radially. The fluid is assumed to be incompressible and inviscid.

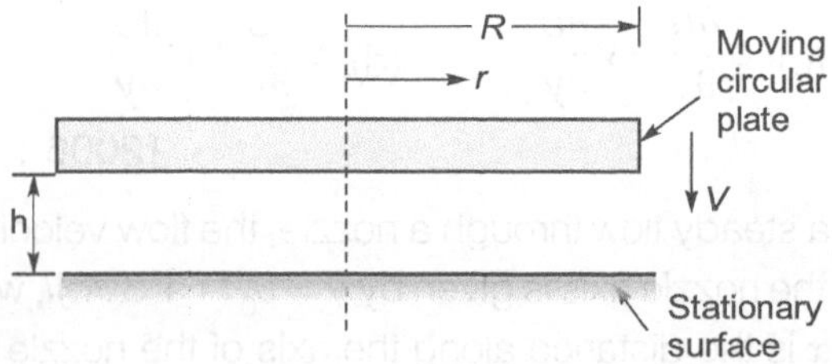

3.11 The radial velocity V_r at any radius r, when the gap width is h, is

(a) $V_r=\frac{Vr}{2h}$ (b) $V_r=\frac{Vr}{h}$

(c) $V_r=\frac{2Vh}{r}$ (d) $V_r=\frac{Vh}{r}$ [2008 : 2 M]

3.12 The radial component of the fluid acceleration at $r = R$ is

(a) $\frac{3V^2R}{4h^2}$ (b) $\frac{V^2R}{4h^2}$

(c) $\frac{V^2R}{2h^2}$ (d) $\frac{V^2h}{2R^2}$ [2008 : 2 M]

3.13 You are asked to evaluate assorted fluid flows for their suitability in a given laboratory application. The following three flow choices, expressed in terms of the two-dimensional velocity fields in the *xy*-plane, are made available.

P. $u = 2y,\ v = -3x$
Q. $u = 3xy,\ v = 0$
R. $u = -2x,\ v = 2y$

Which flow(s) should be recommended when the application requires the flow to be incompressible and irrotational?

(a) P and R (b) Q
(c) Q and R (d) R [2009 : 2 M]

3.14 Velocity vector of a flow field is given as $\vec{V}=2xy\hat{i}-x^2z\hat{j}$. The vorticity vector at (1, 1, 1) is

(a) $4\hat{i}-\hat{j}$ (b) $4\hat{i}-\hat{k}$

(c) $\hat{i}-4\hat{j}$ (d) $\hat{i}-4\hat{k}$ [2010 : 2 M]

3.15 A streamline and an equipotential line in a flow field

(a) are parallel to each other
(b) are perpendicular to each other
(c) intersect at an acute angle
(d) are identical [2011 : 1 M]

3.16 For an incompressible flow field, V, which one of the following conditions must be satisfied?

(a) $\nabla\cdot\vec{V}=0$ (b) $\nabla\times\vec{V}=0$

(c) $(\vec{V}\cdot\nabla)\vec{V}=0$ (d) $\frac{\partial\vec{V}}{\partial t}+(\vec{V}\cdot\nabla)\vec{V}=0$

[2014 : 1 M, Set-2]

3.17 A flow field which has only convective acceleration is

(a) a steady uniform flow
(b) an unsteady uniform flow
(c) a steady non-uniform flow
(d) an unsteady non-uniform flow

[2014 : 1 M, Set-4]

3.18 Consider the following statements regarding streamline(s):

(i) It is a continuous line such that the tangent at any point on it shows the velocity vector at that point

(ii) There is no flow across streamlines

(iii) $\frac{dx}{u} = \frac{dy}{v} = \frac{dz}{w}$ is the differential equation of a streamline, where u, v and w are velocities in directions x, y and z, respectively

(iv) In an unsteady flow, the path of a particle is a streamline

Which one of the following combinations of the statements is true?

(a) (i), (ii), (iv) (b) (ii), (iii), (iv)
(c) (i), (iii), (iv) (d) (i), (ii), (iii)

[2014 : 2 M, Set-4]

3.19 Consider a velocity field $\vec{V} = K(y\hat{i} + x\hat{k})$, where K is a constant. The vorticity, Ω_Z, is

(a) $-K$ (b) K
(c) $-K/2$ (d) $K/2$

[2014 : 2 M, Set-4]

3.20 If the fluid velocity for a potential flow is given by $V(x, y) = u(x, y)\,i + v(x, y)\,j$ with usual notations, then the slope of the potential line at (x, y) is

(a) $\frac{v}{u}$ (b) $-\frac{u}{v}$
(c) $-\frac{v^2}{u^2}$ (d) $\frac{u}{v}$ [2015 : 1 M, Set-2]

3.21 Match the following pairs:

	Equation	Physical Interpretation
P	$\nabla \times \vec{V} = 0$	I. Incompressible continuity equation
Q	$\nabla \cdot \vec{V} = 0$	II. Steady flow
R	$\frac{\Delta \vec{V}}{\Delta t} = 0$	III. Irrotational flow
S	$\frac{\partial \vec{V}}{\partial t} = 0$	IV. Zero acceleration of fluid particle

(a) P-IV, Q-I, R-II, S-III
(b) P-IV, Q-III, R-I, S-II
(c) P-III, Q-I, R-IV, S-II
(d) P-III, Q-I, R-II, S-IV [2015 : 2 M, Set-1]

3.22 The velocity field of an incompressible flow is given by
$V = (a_1x + a_2y + a_3z)i + (b_1x + b_2y + b_3z)j + (c_1x + c_2y + c_3z)k$, where $a_1 = 2$ and $c_3 = -4$. The value of b_2 is _______.

[2015 : 2 M, Set-1]

3.23 The volumetric flow rate (per unit depth) between two streamlines having stream function ψ_1 and ψ_2 is

(a) $|\psi_1 + \psi_2|$ (b) $\psi_1\psi_2$
(c) ψ_1/ψ_2 (d) $|\psi_1 - \psi_2|$

[2016 : 1 M, Set-2]

3.24 For a certain two-dimensional incompressible flow, velocity field is given by $2xy\hat{i} - y^2\hat{j}$. The streamlines for this flow are given by the family of curves

(a) x^2y^2 = constant (b) xy^2 = constant
(c) $2xy - y^2$ = constant (d) xy = constant

[2016 : 1 M, Set-3]

3.25 For a two-dimensional flow, the velocity field is $\vec{u} = \frac{x}{x^2+y^2}\hat{i} + \frac{y}{x^2+y^2}\hat{j}$, where $\hat{i}$ and $\hat{j}$ are the basis vectors in the x-y Cartesian coordinate system. Identify the CORRECT statements from below.

1. The flow is incompressible
2. The flow is unsteady
3. y-component of acceleration,
$$a_y = \frac{-y}{(x^2+y^2)^2}$$
4. x-component of acceleration,
$$a_x = \frac{-(x+y)}{(x^2+y^2)^2}$$

(a) 2 and 3 (b) 1 and 3
(c) 1 and 2 (d) 3 and 4

[2016 : 2 M, Set-3]

3.26 Consider the two-dimensional velocity field given by $\vec{V} = (5 + a_1x + +b_1y)\hat{i} + (4 + a_2x + b_2y)\hat{j}$, where a_1, b_1, a_2 and b_2 are constants. Which one of the following conditions needs to be satisfied for the flow to be incompressible?

(a) $a_1 + b_1 = 0$ (b) $a_1 + b_2 = 0$
(c) $a_2 + b_2 = 0$ (d) $a_2 + b_1 = 0$

[2017 : 1 M, Set-1]

3.27 For a steady flow, the velocity field is $\vec{V} = (-x^2 + 3y)\hat{i} + (2xy)\hat{j}$. The magnitude of the acceleration of a particle at (1, – 1) is

(a) 2 (b) 1
(c) $2\sqrt{5}$ (d) 0

[2017 : 2 M, Set-1]

3.28 For a two-dimensional incompressible flow field given by $\vec{u} = A(x\hat{i} - y\hat{j})$, where $A > 0$, which one of the following statements is FALSE?

A. It satisfies continuity equation
B. It is unidirectional when $x \to 0$ and $y \to \infty$.
C. Its streamlines are given by $x = y$.
D. It is irrotational

(a) A (b) B
(c) C (d) D [2018 : 1 M, Set-1]

3.29 A flat plate of width $L = 1$ m is pushed down with a velocity $U = 0.01$ m/s towards a wall resulting in the drainage of the fluid between the plate and the wall as shown in the figure. Assume two-dimensional incompressible flow and that the plate remains parallel to the wall. The average velocity, u_{avg} of the fluid (in m/s) draining out at the instant shown in the figure is ______(correct to three decimal places).

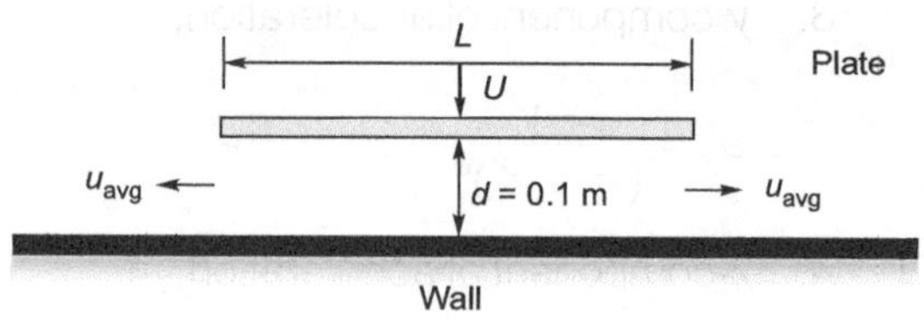

[2018 : 1 M, Set-1]

3.30 In a Lagrangian system, the position of a fluid particle in a flow is described as $x = x_0e^{-kt}$ and $y = y_0e^{kt}$ where t is the time while x_0, y_0, and k are constants. The flow is

(a) unsteady and one-dimensional
(b) steady and two-dimensional
(c) steady and one-dimensional
(d) unsteady and two-dimensional

[2018 : 2 M, Set-1]

3.31 Water flows through a pipe with a velocity given by $\vec{V} = \left(\frac{4}{t} + x + y\right)\hat{j}$ m/s, where $\hat{j}$ is the unit vector in the y direction, t (> 0) is in second, and x and y are in meters. The magnitude of total acceleration at the point $(x, y) = (1, 1)$ at $t = 2$ s is ______ m/s^2.

[2019 : 1 M, Set-1]

3.32 A two-dimensional incompressible frictionless flow field is given by $\vec{u} = x\hat{i} - y\hat{j}$. If ρ is the density of the fluid, the expression for pressure gradient vector at any point in the flow field is given as

(a) $-\rho(x\hat{i} + y\hat{j})$ (b) $-\rho(x^2\hat{i} + y^2\hat{j})$
(c) $\rho(x\hat{i} - y\hat{j})$ (d) $\rho(x\hat{i} + y\hat{j})$

[2019 : 1 M, Set-2]

3.33 The velocity field of an incompressible flow in a Cartesian system is represented by

$$\vec{V} = 2(x^2 - y^2)i + v\hat{j} + 3\hat{k}$$

Which one of the following expressions for $\vec{V}$ is valid?

(a) $-4xy - 4xz$ (b) $4xy - 6xz$
(c) $4xy + 4xz$ (d) $-4xz + 6xz$

[2020 : 1 M, Set-1]

3.34 A two dimensional flow has velocities in x and y directions given by $u = 2xyt$ and $v = -y^2t$, where t denotes time. The equation for streamline passing through $x = 1$, $y = 1$ is

(a) $x^2y = 1$ (b) $\frac{x}{y^2} = 1$
(c) $x^2y^2 = 1$ (d) $xy^2 = 1$

[2021 : 1 M, Set-2]

3.35 For a two-dimensional, incompressible flow having velocity components u and v in the x and y directions, respectively, the expression

$$\frac{\partial(u^2)}{\partial x} + \frac{\partial(uv)}{\partial y}$$

can be simplified to

(a) $2u\frac{\partial u}{\partial x} + u\frac{\partial v}{\partial y}$ (b) $u\frac{\partial u}{\partial x} + u\frac{\partial v}{\partial y}$
(c) $u\frac{\partial u}{\partial x} + v\frac{\partial u}{\partial y}$ (d) $2u\frac{\partial u}{\partial x} + v\frac{\partial u}{\partial y}$

[2021 : 1 M, Set-2]

3.36 A tiny temperature probe is fully immersed in a flowing fluid and is moving with zero relative velocity with respect to the fluid. The velocity field in the fluid is $\vec{V} = (2x)i + (y + 3t)\hat{j}$, and the temperature field in the fluid is $T = 2x^2 + xy + 4t$, where x and y are the spatial coordinates, and t is the time. The time rate of change of temperature recorded by the probe at (x = 1, y = 1, t = 1) is ___.

(a) 4 (b) 0
(c) 18 (d) 14 **[2022 : 1 M, Set-1]**

3.37 A steady two-dimensional flow field is specified by the stream function $\psi = kx^3y$, where x and y are in meter and the constant $k = 1\ m^{-2}s^{-1}$.

The magnitude of acceleration at a point (x, y) = (1 m, 1 m) is _______ m/s^2 (round off to 2 decimal places).

[2022 : 2 M, Set-1]

3.38 The velocity field in a fluid is given to be $\vec{V} = (4xy)\hat{i} + 2(x^2 - y^2)\hat{j}$. Which of the following statement(s) is/are correct?

(a) The velocity field is one-dimensional.
(b) The flow is incompressible.
(c) The flow is irrotational.
(d) The acceleration experienced by a fluid particle is zero at (x = 0, y = 0).

[2022 : 1 M, Set-2]

3.39 The velocity field of a two-dimentional, incompressible flow is given by

$\vec{V} = 2\sinh x\ \hat{i} + v(x,y)\hat{j}$ where $\hat{i}$ and $\hat{j}$ denote the unit vector in x and y direction, respectively. If $v(x,0) = \cosh x$, then $v(0, -1)$ is

(a) 1 (b) 2
(c) 4 (d) 3

[2024 : 1 M]

Answers Fluid Kinematics

3.1 (d)	3.2 (d)	3.3 (c)	3.4 (d)	3.5 (b)	3.6 (a)	3.7 (c)	3.8 (b)
3.9 (b)	3.10 (d)	3.11 (a)	3.12 (a)	3.13 (d)	3.14 (d)	3.15 (b)	3.16 (a)
3.17 (c)	3.18 (d)	3.19 (a)	3.20 (b)	3.21 (c)	3.22 (2)	3.23 (d)	3.24 (b)
3.25 (b)	3.26 (b)	3.27 (c)	3.28 (c)	3.29 (0.05)	3.30 (b)	3.31 (3)	3.32 (a)
3.33 (a)	3.34 (d)	3.35 (c)	3.36 (c)	3.37 (4.242)	3.38 (b,c,d)	3.39 (d)	

Explanations Fluid Kinematics

3.1 (d)

$$u = \lambda xy^3 - x^2y$$

$$v = xy^2 - \frac{3}{4}y^4$$

Continuity equation for 2-D incompressible flow,

$$\frac{\partial u}{\partial x} + \frac{\partial v}{\partial y} = 0$$

where $\frac{\partial u}{\partial x} = \lambda y^3 - 2xy$

and $\frac{\partial v}{\partial y} = 2xy - 3y^3$

$$\Rightarrow \lambda y^3 - 2xy + 2xy - 3y^3 = 0$$

$$\lambda y^3 = 3y^3$$

or $\lambda = 3$

3.2 (d)

$$\vec{V} = (x + 2y + 2)i + (4 - y)\hat{j}$$

$$\frac{\partial u}{\partial x} = 1; \quad \frac{\partial v}{\partial y} = -1$$

$$\frac{\partial u}{\partial x} + \frac{\partial v}{\partial y} = 0$$

Hence, flow is **incompressible.**

$$\omega_z = \frac{1}{2}\left(\frac{\partial v}{\partial x} - \frac{\partial u}{\partial y}\right) = \frac{1}{2}(0 - 2) = -1$$

($\because$ Rotational)

Hence, flow is **non-irrotational.**

3.3 (c)

Given: $\vec{V} = ax\hat{i} + ay\hat{j}$

i.e., $u = ax$ and $v = ay$

Equation of stream line is given by

$$\frac{dx}{u} = \frac{dy}{v}$$

$$\therefore \quad \frac{dx}{ax} = \frac{dy}{ay}$$

$$\text{or} \quad \frac{dx}{x} = \frac{dy}{y}$$

Integrating both sides, we get

$$\ln x = \ln y + \ln c$$

(where c is constant of integration)

$$\therefore \quad x = cy$$

Since the stream line passed through point (1, 2), therefore

$$1 = 2c$$

$$c = 1/2$$

Hence equation of stream line is

$$x = \frac{1}{2}y$$

or $\mathbf{2x - y = 0}$

3.4 (d)

For two dimensional flow, continuity equation should be satisfied.

$$\text{i.e.,} \quad \frac{\partial u}{\partial x} + \frac{\partial v}{\partial y} = 0$$

$$\therefore \quad \frac{\partial u}{dx} = -\frac{\partial v}{\partial y}$$

3.5 (b)

Radial distance = 120 m

$$V_r = -\frac{60\times 10^3}{2\pi r} \text{ m/s}$$

$$\text{and} \quad V_\theta = \frac{300\times 10^3}{2\pi r} \text{ m/s}$$

Let θ = Angle turned by a leaf

For $\theta = \frac{1}{2}$ revolution

$$\text{i.e.,} \quad \theta = \frac{2\pi}{2}$$

$$\theta = \pi \text{ radian}$$

$$\frac{V_r}{V_\theta} = -\frac{60\times 10^3}{2\pi r}\times\frac{2\pi r}{300\times 10^3} = -\frac{1}{5}$$

or $V_\theta = -V_r$

We know that

$$V_r = \frac{dr}{dt} \quad \text{...(i)}$$

$$\text{and} \quad V_\theta = r\omega = r\frac{d\theta}{dt}$$

$$-5V_r = r\frac{d\theta}{dt}$$

$$\text{or} \quad V_r = -\frac{r}{5}\frac{d\theta}{dt} \quad \text{...(ii)}$$

Equating Eq. (i) and Eq. (ii), we get

$$\frac{dr}{dt} = -\frac{r}{5}\frac{d\theta}{dt} \quad \text{or} \quad \frac{dr}{r} = -\frac{d\theta}{5}$$

Integrating both sides, we get

$$\int_{120}^{r}\frac{dr}{r} = -\frac{1}{5}\int_0^{\pi} d\theta$$

$$[\log_e r]_{120}^{r} = -\frac{1}{5}[\theta]_0^{\pi}$$

$$\log_e r - \log_e 120 = -\frac{\pi}{5}$$

$$\log_e \frac{r}{120} = -\frac{\pi}{5}$$

$$\text{or} \quad \frac{r}{120} = e^{-\frac{\pi}{5}}$$

$$\text{or} \quad r = 120\times e^{-\frac{\pi}{5}} = \frac{120}{e^{3.14/5}} = \mathbf{64\ m}$$

3.6 (a)

Equation of streamline,

$$\frac{dx}{u} = \frac{dy}{v}$$

$$\frac{dx}{xy} = \frac{dy}{-2xy}$$

$$\frac{dx}{x} = \frac{dy}{-2y}$$

$$2\frac{dx}{x} + \frac{dy}{y} = 0$$

$$2\log_e x + \log_e y = \log_e c$$

$$\log_e yx^2 = \log_e c$$

$$\mathbf{x^2 y = c}$$

3.7 (c)

Convective acceleration is defined as the rate of change of velocity due to change of position of fluid particle in a fluid flow.

Convective acceleration along the x-direction is given by

$$a_x = \frac{\partial u}{\partial t} = u\frac{\partial u}{\partial x} + v\frac{\partial u}{\partial y} + w\frac{\partial u}{\partial z}$$

for 3-*D* velocity field

$$= u\frac{du}{dx} + v\frac{du}{dy}$$

for 2-*D* (*x* and *y* direction) velocity field

3.8 (b)

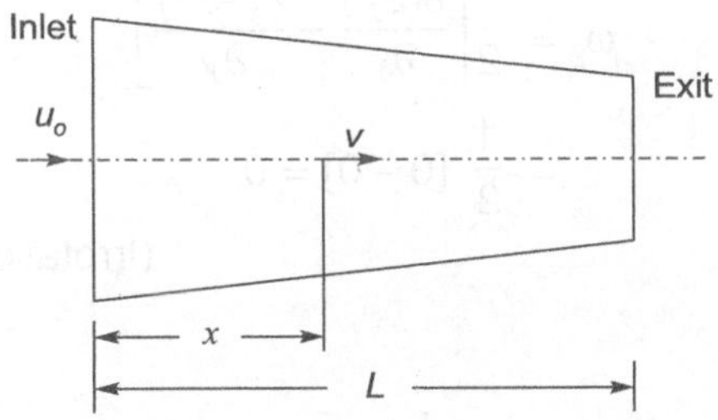

$$v = u_o\left(1 + \frac{3x}{L}\right)$$

We know that

Velocity: $v = \frac{dx}{dt}$

$$\therefore \quad \frac{dx}{dt} = u_0\left(1 + \frac{3x}{L}\right)$$

or $$u_0 . dt = \frac{dx}{\left(1 + \frac{3x}{L}\right)}$$

Integrating both sides, we get

$$u_0\int_0^t dt = \int_0^L \frac{dx}{\left(1 + \frac{3x}{L}\right)}$$

$$u_0 t = \frac{L}{3}\left[\ln\left(1 + \frac{3x}{L}\right)\right]_0^L = \frac{L}{3}\ln 4$$

$$\therefore \quad t = \frac{L}{3u_0}\ln 4$$

3.9 (b)

For steady, incompressible forced, vortex flow;

P. Shear stress is zero at all points in the flow. — Incorrect

Shear stress is always exit due to external torque acting on mass of fluid.

Q. Vorticity is zero at all points in the flow. — Incorrect

Velocity exists at all points in forced vortex flow field.

R. Velocity is directly proportional to the radius from the centre of the vortex. — Correct.

For forced vortex flow,

$$\frac{V}{r} = \text{Constant}$$

or $V \propto r$

S. Total mechanical energy per unit mass is constant in the entire flow field. — Correct

$$\frac{\rho\omega^2}{2}(r_2^2 - r_1^2) - \rho g(z_2 - z_1) = 0$$

or $$\frac{\omega^2}{2}(r_2^2 - r_1^2) - g(z_2 - z_1) = 0$$

ΔK.E. – ΔP.E. = 0

where,

$$\Delta\text{K.E.} = \frac{\omega^2}{2}(r_2^2 - r_1^2) - (z_2 - z_1)$$

change in K.E. per unit mass

ΔP.E.= $g(z_2 - z_1)$

change in P.E. per unit mass

$\therefore$ ΔK.E. = ΔP.E.

Hence, total mechanical energy per unit mass in constant.

3.10 (d)

$$\vec{\nabla} \cdot \vec{V} = 0$$

$$\frac{\partial u}{\partial x} + \frac{\partial v}{\partial y} + \frac{\partial w}{\partial z} = 0$$

$$\rho\left(\frac{\partial u}{\partial x} + \frac{\partial v}{\partial y} + \frac{\partial w}{\partial z}\right) = 0$$

So, the above equation represents **incompressible flow.**

3.11 (a)

Let us consider for small time *dt*, then for any cylindrical control volume between 2 plates of radius *r* volume reduced by the movement of upper plate in time *dt* is equal to the volume of fluid moving out

$$\therefore \quad V \times \pi r^2 = V_r \times 2\pi r h$$

or $$V_r = \frac{Vr}{2h}$$

3.12 (a)

Radial acceleration,

$$a_r = V_r \times \frac{\partial V_r}{\partial r} + \frac{\partial V_r}{\partial t}$$

$$a_r = \frac{Vr}{2h} \times \frac{\partial}{\partial r}\left(\frac{Vr}{2h}\right) + \frac{\partial}{\partial t}\left(\frac{Vr}{2h}\right)$$

For convective part r is variable and for local acceleration h is variable

$$\therefore \quad a_r = \frac{Vr}{2h} \times \frac{V}{2h} + \frac{Vr}{2}\left(\frac{-1}{h^2} \times \frac{\partial h}{\partial t}\right)$$

We know that, $\left(\frac{-\partial h}{\partial t} = V\right)$

So, $$a_r = \frac{V^2 r}{4h^2} + \frac{Vr}{2}\left(\frac{-1}{h^2}\frac{\partial h}{\partial t}\right)$$

$$a_r = \frac{V^2 r}{4h^2} + \frac{Vr}{2}\left(\frac{V}{h^2}\right) = \frac{3V^2 r}{4h^2}$$

At $r = R$

$$a_R = \frac{3V^2 R}{4h^2}$$

3.13 (d)

Incompressible flow satisfy continuity equation

$$\frac{\partial u}{\partial x} + \frac{\partial v}{\partial y} = 0$$

P. $u = 2y,\ v = -3x$

$$\frac{\partial(2y)}{\partial x} + \frac{\partial(-3x)}{\partial y}$$

Incompressible flow $\Rightarrow 0 + 0 = 0$

Q. $u = 3xy,\ v = 0$

$$\frac{\partial(3xy)}{\partial x} + \frac{\partial(0)}{\partial y}$$

Compressible flow

$\Rightarrow \ 3y + 0 \neq 0$

R. $u = -2x,\ v = 2y$

$$\frac{\partial(-2x)}{\partial x} + \frac{\partial(2y)}{\partial y}$$

Incompressible flow

$-2 + 2 = 0$

For irrotational flow condition is

$$\omega_z = \frac{1}{2}\left(\frac{\partial v}{\partial x} - \frac{\partial u}{\partial y}\right) = 0$$

For P : $u = 2y,\ v = -3x$

$$\omega_z = \frac{1}{2}\left[\frac{\partial(-3x)}{\partial x} - \frac{\partial(2y)}{\partial y}\right]$$

$$= \frac{1}{2}\,[-3 - 2] \neq 0$$

(Rotational flow)

For Q : $u = 3xy,\ v = 0$

$$\omega_z = \frac{1}{2}\left[\frac{\partial(0)}{\partial x} - \frac{\partial(3xy)}{\partial y}\right]$$

$$= \frac{1}{2}\,[0 - 3x] \neq 0$$

(Rotational flow)

For R : $u = -2x,\ v = 2y$

$$\omega_z = \frac{1}{2}\left[\frac{\partial(2y)}{\partial x} - \frac{\partial(-2x)}{\partial y}\right]$$

$$= \frac{1}{2}\,[0 - 0] = 0$$

(Irrotational flow)

3.14 (d)

$$\text{Vorticity vector} = \begin{vmatrix} i & j & k \\ \frac{\partial}{\partial x} & \frac{\partial}{\partial y} & \frac{\partial}{\partial z} \\ 2xy & -x^2 z & 0 \end{vmatrix}$$

$$= \hat{i}\left[0 - \frac{\partial}{\partial z}(-x^2 z)\right] - \hat{j}\left[0 - \frac{\partial}{\partial z}(2xy)\right]$$

$$+\hat{k}\left[\frac{\partial}{\partial x}(-x^2 z) - \frac{\partial}{\partial y}(2xy)\right]$$

$$= \hat{i}(x^2) + \hat{k}(-2xz - 2x) - 0$$

$$= i(x^2) + \hat{k}(-2xz - 2x) - 0$$

$\therefore$ At (1, 1, 1)

Vorticity vector = $\hat{i} - 4\hat{k}$

3.15 (b)

Stream line and equipotential line in a flow field are perpendicular to each other.

3.16 (a)

For an incompressible flow field, divergence must be zero *i.e.*, $\nabla . \vec{V} = 0$.

3.17 (c)

Total acceleration, $\frac{\Delta \vec{V}}{\Delta t} = \frac{\partial V}{\partial t} + V \cdot \nabla V$

$V \cdot \nabla V \rightarrow$ Convective acceleration

$\frac{\partial V}{\partial t} \rightarrow$ Local acceleration

In steady non-uniform flow, conditions change from point to point in the stream but do not change with time.

POINTS TO REMEMBER

Type of flow	Convective acceleration	Temporal or local acceleration
Steady and uniform	0	0
Steady and Non-uniform	Exists	0
Unsteady and uniform	0	Exists
Unsteady and non-uniform	Exists	Exists

3.19 (a)

Velocity vector,

$$\vec{V} = K\left(y\hat{i} + x\hat{k}\right) = Ky\hat{i} + Kx\hat{k}$$

where, $u = Ky$, $v = 0$ and $w = Kx$

Vorticity about z-axis,

$$\Omega_Z = \frac{\partial v}{\partial x} - \frac{\partial u}{\partial y}$$

$$\frac{\partial v}{\partial x} = 0, \quad \frac{\partial u}{\partial y} = K$$

$$\therefore \quad \Omega_Z = 0 - K = -\mathbf{K}$$

3.20 (b)

$$d\phi = \frac{\partial \phi}{\partial x}dx + \frac{\partial \phi}{\partial y}dy$$

$$\therefore \quad \frac{\partial \phi}{\partial x} = -u$$

$$\frac{\partial \phi}{\partial y} = -v$$

$$\therefore \quad d\phi = -udx - vdy$$

If ϕ = constant, $d\phi = 0$

$$\therefore \quad \left(\frac{dy}{dx}\right)_\phi = -\frac{u}{v}$$

3.21 (c)

- $\nabla \times \vec{V} = 0$ Irrotational flow
- $\nabla.\vec{V} = 0$ Incompressible continuity equation
- $\frac{\Delta \vec{V}}{\Delta t} = 0$ Zero acceleration of fluid particle
- $\frac{\partial \vec{V}}{dt} = 0$ Steady flow

3.22 Sol.

From continuity equation

$$\frac{\partial u}{\partial x} + \frac{\partial v}{\partial y} + \frac{\partial w}{\partial z} = 0$$

$$a_1 + b_2 + c_3 = 0$$

$$2 + b_2 - 4 = 0$$

or $\quad b_2 = 4 - 2 = \mathbf{2}$

3.24 (b)

$$\vec{V} = 2xy\hat{i} - y^2\hat{j}$$

where $u = 2xy$ and $v = -y^2$

Equation of stream line,

$$\frac{dx}{u} = \frac{dy}{v}$$

$$\frac{dx}{2xy} = \frac{dy}{-y^2}$$

or $\quad \frac{dx}{2x} = -\frac{dy}{y}$

On integrating, we get

$$\frac{1}{2}\log_e x = -\log_e y + \log_e c$$

$$\log_e x^{1/2} + \log_e y = \log_e c$$

$$\log_e x^{1/2} y = \log_e c$$

$$x^{1/2}y = \mathbf{c}$$

or $\quad xy^2 = c_1 = \text{constant}$

3.25 (b)

For 2D-flow velocity field is:

$$\vec{V} = \frac{x}{x^2+y^2}\hat{i} + \frac{y}{x^2+y^2}\hat{j}$$

so $\quad u = \frac{x}{x^2+y^2}$ and $v = \frac{y}{x^2+y^2}$

$$a_x = u\frac{\partial u}{\partial x} + v\frac{\partial u}{\partial y}$$

$$= \frac{x}{x^2+y^2}\frac{\partial}{\partial x}\left[\frac{x}{x^2+y^2}\right] + \frac{y}{x^2+y^2}\frac{\partial}{\partial y}\left[\frac{x}{x^2+y^2}\right]$$

$$= \frac{x}{x^2+y^2}\left[\frac{x(-2x)}{(x^2+y^2)^2} + \frac{1}{x^2+y^2}\right] + \frac{y}{x^2+y^2}\left[\frac{-x(2y)}{(x^2+y^2)^2}\right]$$

$$= \frac{-2x^3}{(x^2+y^2)^3} + \frac{x}{(x^2+y^2)^2} - \frac{2xy^2}{(x^2+y^2)^3}$$

$$= \frac{-2x^3 + x(x^2+y^2) - 2xy^2}{(x^2+y^2)^3}$$

$$= \frac{-2x^3 + x^3 + xy^2 - 2xy^2}{(x^2+y^2)^3} = \frac{-x^3 - xy^2}{(x^2+y^2)^3}$$

$$= \frac{-x(x^2+y^2)}{(x^2+y^2)^3}$$

$$a_x = \frac{x}{(x^2+y^2)^2}$$

$$a_y = u\frac{\partial v}{\partial x} + v\frac{\partial v}{\partial y}$$

$$= \frac{x}{x^2+y^2}\frac{\partial}{\partial x}\left[\frac{y}{x^2+y^2}\right] + \frac{y}{x^2+y^2}\frac{\partial}{\partial y}\left[\frac{y}{x^2+y^2}\right]$$

$$= \frac{x}{x^2+y^2}\frac{(-2xy)}{(x^2+y^2)^2} + \frac{y}{x^2+y^2} \times \left[\frac{-2y^2}{(x^2+y^2)^2} + \frac{1}{x^2+y^2}\right]$$

$$= \frac{-2x^2y}{(x^2+y^2)^3} - \frac{2y^3}{(x^2+y^2)^3} + \frac{y}{(x^2+y^2)^2}$$

$$= \frac{-2x^2y - 2y^3 + y(x^2+y^2)}{(x^2+y^2)^3}$$

$$= \frac{-x^2y - y^3}{(x^2+y^2)^3} = \frac{-y(x^2+y^2)}{(x^2+y^2)^3}$$

$$a_y = \frac{-y}{(x^2+y^2)^2}$$

3.26 (b)

Given that the flow is incompressible.

$$\therefore \quad \text{div}\,\bar{V} = 0$$

$$\frac{\partial}{\partial x}(5 + a_1x + b_1y) + \frac{\partial}{\partial y}(8 + a_2x + b_2y) = 0$$

$$a_1 + b_2 = 0$$

3.27 (c)

Given velocity field,

$$\vec{V} = (-x^2 + 3y)\hat{i} + (2xy)\hat{j}$$

where $u = -x^2 + 3y$ and $v = 2xy$

The acceleration components along x and y-axis.

$$a_x = u\frac{\partial u}{\partial x} + v\frac{\partial u}{\partial y}$$

and $\quad a_y = u\dfrac{\partial v}{\partial x} + v\dfrac{\partial v}{\partial y}$

$$= (-x^2 + 3y) \times (-2x) + 2xy \times 3$$
$$= 2x^3 - 6xy + 6xy = 2x^3$$

and $\quad a_y = (-x^2 + 3y) \times 2y + 2xy \times 2x$

$$= -2yx^2 + 6y^2 + 4x^2y = 2yx^2 + 6y^2$$

At point (1, −1),

$$a_x = 2$$

and $\quad a_y = 2 \times (-1) \times 1 + 6 \times (-1)^2$

$$= -2 + 6 = 4$$

$$\vec{a} = a_x\hat{i} + a_y\hat{j} = 2i + 4\hat{j}$$

Resultant acceleration,

$$a = \sqrt{4+16} = \sqrt{20}$$
$$= \sqrt{4 \times 5} = 2\sqrt{5}$$

3.28 (c)

C is the false statement

2D incompressible flow continuity equation.

$$\frac{\partial u}{\partial x} + \frac{\partial v}{\partial y} = 0$$

$$\frac{\partial(Ax)}{\partial x} + \frac{\partial(-Ay)}{\partial y} = 0$$

$A - A = 0$ it satisfies continuity equation.

⇒ As $\quad \vec{u} = Ax\,\hat{i} - Ay\hat{j}$

where, u = Ax and $v = -Ay$

⇒ Stream line equation for 2D

$$\frac{dx}{u} = \frac{dy}{v}$$

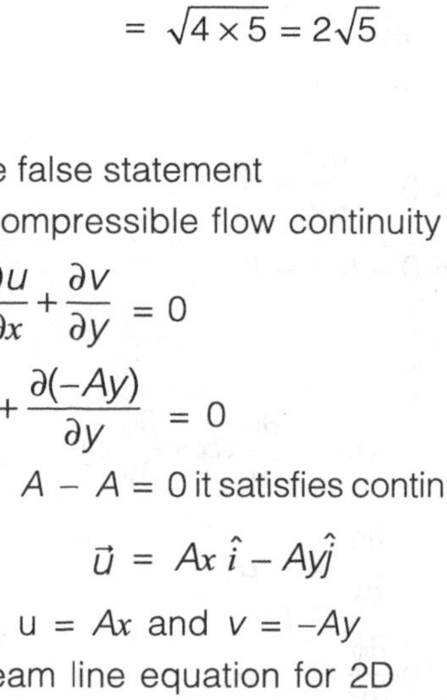

$$\frac{dx}{Ax} = \frac{dy}{-Ay}$$

$$\ln x = -\ln y + \ln c$$

$$\ln xy = \ln c$$

$xy = c \rightarrow$ streamline equation

$x \to 0$, Flow will be along y-axis

$y \to 0$, Flow will be along x-axis

3.29 Sol.

Assuming length of plate as B

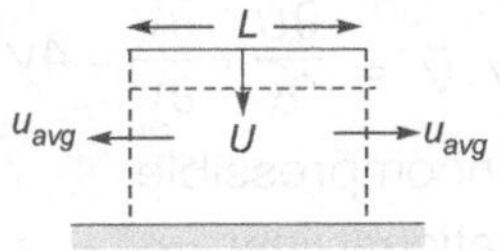

As per continuity:

Rate of mass displaced through plate

= Rate of mass displaced between plate and wall

$$\rho(U \times B \times L) = \rho(u_{avg}) \times (2L + 2B)d$$

$$UBL = u_{avg}\ 2[L + B]d$$

Assuming B>>>>L

$$UBL = 2u_{avg}\ B \times d$$

$$u_{avg} = \frac{U \times L}{2 \times d} = \frac{0.01 \times 1}{2 \times 0.1} = 0.05 \text{ m/s}$$

3.30 (b)

x direction scalar of velocity field,

$$u = \frac{dx}{dt}$$

$$u = -kx_0 e^{-kt}$$

y direction scalar of velocity field

$$v = \frac{dy}{dt} = ky_0 e^{kt}$$

$$\vec{V} = u\hat{i} + v\hat{j}$$

$$\vec{v} = -k(x_0 e^{-kt})\hat{i} + k(y_0 e^{kt})\hat{j} = -kx\hat{i} + ky\hat{j}$$

So flow is steady $\vec{V} \neq f(t)$

u & v are non zero scalar $t \geq 0$ so it is 2D flow. 2D possible flow field.

$$\frac{\partial u}{\partial x} + \frac{\partial v}{\partial y} = 0$$

$$\frac{\partial}{\partial x}(-kx_0 e^{-kt}) + \frac{\partial}{\partial y}(ky_0 e^{kt}) = 0$$

0 + 0 = 0 **continuity satisfied.**

POINTS TO REMEMBER

There is two types of approach in the fluid kinematics:

(i) Lagrangian (ii) Eularian

- Lagrangian concept is for single fluid particle.
- Eulerian concept is for particular section or point.

3.31 Sol.

$$\vec{V} = \left(\frac{4}{t} + x + y\right)\hat{j}$$

$$a = \sqrt{a_x^2 + a_y^2 + a_z^2} = a_y$$

$$a_y = v\frac{\partial v}{\partial y} + \frac{\partial v}{\partial t} = a$$

at $x = 1;\ y = 1;\ t = 2$

$$a = 3 \text{ m/s}^2$$

3.32 (a)

Euler's equation for vector form (incompressible and frictionless)

$$\vec{F}_p + \vec{F}_g = \vec{F}_i$$

$$\vec{F}_p + \vec{F}_g = m\vec{a}$$

$$\vec{g} = g_x\hat{i} + g_y\hat{j} + g_z\hat{k}$$

$$\vec{g} = 0\hat{i} + 0\hat{j} + (-g)\hat{k}$$

$$-\nabla p + \rho\vec{g} = \rho\frac{D\vec{V}}{Dt}$$

$$-\nabla p + \rho\vec{g} = \rho\left[u\frac{\partial\vec{V}}{\partial x}i + v\frac{\partial\vec{V}}{\partial y}\hat{j} + w\frac{\partial\vec{V}}{\partial z}\hat{k} + \frac{\partial\vec{V}}{\partial t}\right]$$

$$-\nabla p + \rho\vec{g} = \rho\left[x\hat{i} + (-y)(-\hat{j}) + 0 + 0\right]$$

$$-\nabla p + \rho\left[0\hat{i} + 0\hat{j}\right] = \rho\left[x\hat{i} + y\hat{j}\right]$$

$$\nabla p = -\rho\left[x\hat{i} + y\hat{j}\right]$$

3.33 (a)

$$\vec{V} = (2x^2 - 2y^2)\hat{i} + v\hat{j} + 3\hat{k}$$

For Incompressible flow, continuity equation:

$$\frac{\partial u}{\partial x} + \frac{\partial v}{\partial y} + \frac{\partial w}{\partial z} = 0$$

$$4x + \frac{\partial v}{\partial y} = 0$$

$$\frac{\partial v}{\partial y} = -4x$$

$$v = -4xy + f(x, z)$$

3.34 (d)

$$u = 2xyt$$

$$v = -y^2 t$$

$$\frac{dx}{u} = \frac{dy}{v} = \frac{dz}{w}$$

$$\frac{dx}{2xyt} = \frac{dy}{-y^2t}$$

$$-ydx = 2xdy$$

$$\ln xy^2 = C$$

$$xy^2 = 1$$

3.35 (c)

$$\frac{\partial(u^2)}{\partial x} + \frac{\partial(uv)}{\partial y}$$

By differentiating:

$$\Rightarrow \quad 2u\left[\frac{\partial u}{\partial x}\right] + u\frac{\partial v}{\partial y} + v\frac{\partial u}{\partial y}$$

$$\Rightarrow \quad u\frac{\partial u}{\partial x} + \left[u\frac{\partial u}{\partial x} + u\frac{\partial v}{\partial y}\right] + v\frac{\partial u}{\partial y}$$

According to contínity eq. : $\frac{\partial u}{\partial x} + \frac{\partial v}{\partial y} = 0$

So, $\quad u\frac{\partial u}{\partial x} + v\frac{\partial u}{\partial y}$

3.36 (c)

$$\vec{V} = (2x)i + (y + 3t)\hat{j}, \quad T = 2x^2 + xy + 4t$$

$$\frac{DT}{Dt} = \frac{\partial T}{\partial x}\cdot\frac{dx}{dt} + \frac{\partial T}{\partial y}\cdot\frac{dy}{dt} + \frac{\partial T}{\partial t}\frac{dt}{dt}$$

$$= u\frac{\partial T}{\partial x} + v\frac{\partial T}{\partial y} + \frac{\partial T}{\partial t}$$

$$= 2x(4x + y) + (y + 3t)x + 4$$

Put $x = 1,\ y = 1;\ t = 1$

$$\left(\frac{DT}{Dt}\right)_{\begin{pmatrix}x=1\\ y=1\\ t=1\end{pmatrix}} = 2(4 + 1) + (1 + 3)1 + 4$$

$$= 10 + 4 + 4 = 18$$

3.37 (4.242) (4.20 to 4.28)

$$\psi = x^3y$$

$$U = \frac{\partial\psi}{\partial y} \qquad V = \frac{-\partial\psi}{\partial x}$$

$$U = x^3 \qquad V = -3x^2y$$

$$a_x = U\frac{\partial U}{\partial x} + V\frac{\partial U}{\partial y} \qquad a_y = U\frac{\partial V}{\partial x} + V\frac{\partial V}{\partial y}$$

$$a_x = x^3(3x^2)$$

$$= x^3(6xy) + (-3x^2y)(-3x^2)$$

$$a_x = 3 \qquad a_y = (-6) + (9) = 3$$

$$|a| = \sqrt{3^2 + 3^2} = 3\sqrt{2} = 4.242$$

3.38 (b, c, d)

Given: $u = 4xy,\ v = 2(x^2 - y^2)$

(i) For incompressible flow,

$$\vec{\nabla}\cdot\vec{V} = 0$$

Here, $\quad \nabla\cdot\vec{V} = \frac{\partial u}{\partial x} + \frac{\partial v}{\partial y} = 4y - 4y = 0$

So, flow is incompressible.

(ii) For irrotational flow,

$$\omega_z = 0$$

Here, $\quad \omega_z = \frac{1}{2}\left(\frac{\partial v}{\partial x} - \frac{\partial u}{\partial y}\right) = \frac{1}{2}(4x - 4x) = 0$

So, flow is irrotational.

(iii) Acceleration at $x = 0,\ y = 0$

$$a_x = \frac{Du}{Dt} = \left[u\frac{\partial u}{\partial x} + v\frac{\partial u}{\partial y}\right]_{(x=0,y=0)}$$

$$= [0\times0 + 0\times0] = 0$$

$$a_y = \frac{Dv}{Dt} = \left[u\frac{\partial v}{\partial x} + v\frac{\partial v}{\partial y}\right]_{(x=0,y=0)}$$

$$= [0\times0 + 0\times0] = 0$$

So, acceleration of fluid at $x = 0,\ y = 0$ is zero.

It is a 2-D flow.

3.39 (d)

The velocity field of a two-dimensional, incompressible flow is given by,

$$\vec{V} = 2\sinh x\,\hat{i} + v(x, y)\hat{j}$$

and $\quad v(x, 0) = \cosh x,$

For an incompressible flow, $\nabla\cdot\vec{V} = 0$

$$\Rightarrow \quad \frac{\partial u}{\partial \_} + \frac{\partial v}{\partial y} = 0$$

$$2\cosh\_ + \frac{\partial v}{\partial y} = 0$$

$$\Rightarrow \quad \frac{\partial v}{\partial y} = -2\cosh x$$

$$\Rightarrow \quad \int \partial v = \int -2\cosh\_\cdot\partial y$$

$$v = -2y\cdot\cosh x + f(x)$$

For, $\quad v(x, 0) = \cosh x$

$$\Rightarrow -2(0)\cdot\cosh x + f(x) = \cosh x$$

$$\Rightarrow \quad f(x) = \cosh x$$

$$\Rightarrow \quad v = -2\cdot y\cdot\cosh x + \cosh x$$

$$v = (1 - 2y)\cosh x$$

$$v(0, -1) = [1 - 2\times(-1)]\times\cosh(0)$$

$$= 3$$

■■■■

4 Fluid Dynamics

4.1 Water flows though a vertical contraction from a pipe of diameter d to another of diameter $d/2$ (see fig.)

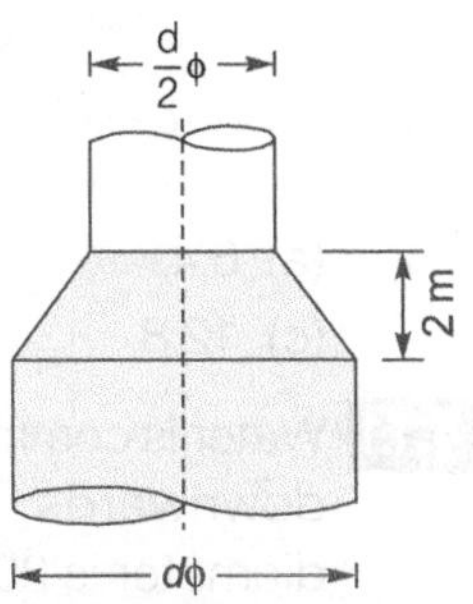

The flow velocity at the inlet to the contraction is 2 m/s and pressure 200 kN/m². If the height of the contraction measures 2 m, then pressure at the exit of the contraction will be very nearly

(a) 168 kN/m² (b) 192 kN/m²
(c) 150 kN/m² (d) 174 kN/m²

[1999 : 2 M]

4.2 Navier Stoke's equation represents the conservation of

(a) Energy (b) Mass
(c) Pressure (d) Momentum

[2000 : 1 M]

4.3 A water container is kept on a weighing balance. Water from a tap is falling vertically into the container with a volume flow rate of Q; the velocity of the water when it hits the water surface is U. At a particular instant of time the total mass of the container and water is m. The force registered by the weighing balance at this instant of time is

(a) $mg + \rho QU$ (b) $mg + 2\rho QU$
(c) $mg + \rho QU^2/2$ (d) $\rho QU^2/2$ [2003 : 2 M]

4.4 The following data about the flow of liquid was observed in a continuous chemical process plant:

Flow rate (litres /sec.)	7.5 to 7.7	7.7 to 7.9	7.9 to 8.1	8.1 to 8.3	8.3 to 8.5	8.5 to 8.7
Frequency	1	5	35	17	12	10

Mean flow rate of the liquid is

(a) 8.00 liters/s (b) 8.06 liters/s
(c) 8.16 liters/s (d) 8.26 liters/s

[2004 : 2 M]

4.5 Air flows through a venturi and into atmosphere. Air density is ρ; atmospheric pressure is p_a; throat diameter is D_t; exit diameter is D and exit velocity is U. The throat is connected to a cylinder containing a frictionless piston attached to a spring. The spring constant is k. The bottom surface of the piston is exposed to atmosphere. Due to the flow, the piston moves by distance x. Assuming incompressible frictionless flow, x is

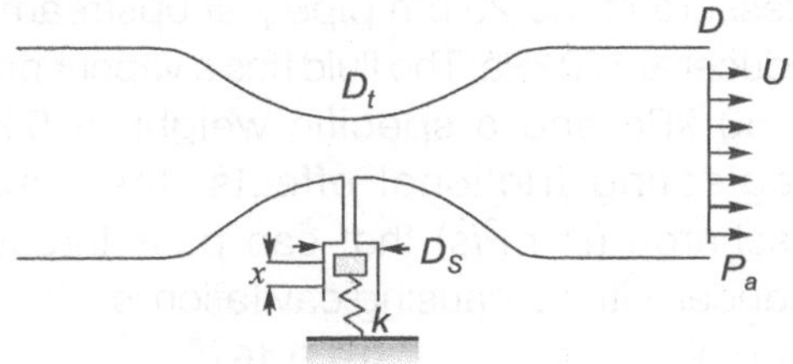

(a) $(\rho U^2/2k)\pi D_s^2$

(b) $(\rho U^2/8k)\left(\dfrac{D^2}{D_t^2}-1\right)\pi D_s^2$

(c) $(\rho U^2/2k)\left(\dfrac{D^2}{D_t^2}-1\right)\pi D_s^2$

(d) $(\rho U^2/8k)\left(\dfrac{D^4}{D_t^4}-1\right)\pi D_s^2$ [2003 : 2 M]

4.6 A venturimeter of 20 mm throat diameter is used to measure the velocity of water in a horizontal pipe of 40 mm diameter. If the pressure difference between the pipe and throat sections is found to be 30 kPa then, neglecting frictional losses, the flow velocity is

(a) 0.2 m/s (b) 1.0 m/s
(c) 1.4 m/s (d) 2.0 m/s [2005 : 2 M]

4.7 A U-tube manometer with a small quantity of mercury is used to measure the static pressure difference between two locations A and B in a conical section through which an incompressible fluid flows. At a particular flow rate, the mercury column appears as shown in the figure. The density of mercury is 13600 kg/m³ and g = 9.81 m/s². Which of the following is correct?

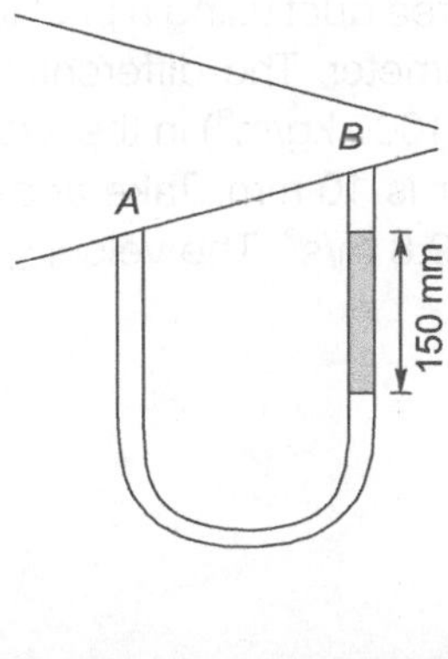

(a) Flow direction is A to B and $p_A - p_B = 20$ kPa
(b) Flow direction is B to A and $p_A - p_B = 1.4$ kPa
(c) Flow direction is A to B and $p_B - p_A = 20$ kPa
(d) Flow direction is B to A and $p_B - p_A = 1.4$ kPa
[2005 : 2 M]

4.8 Consider steady, incompressible and irrotational flow through a reducer in a horizontal pipe where the diameter is reduced from 20 cm to 10 cm. The pressure in the 20 cm pipe just upstream of the reducer is 150 kPa. The fluid has a vapour pressure of 50 kPa and a specific weight of 5 kN/m^3. Neglecting frictional effects, the maximum discharge (in m^3/s) that can pass through the reducer without causing cavitation is
(a) 0.05 (b) 0.16
(c) 0.27 (d) 0.38 [2009 : 2 M]

4.9 Consider steady flow of water in a situation where two pipe lines (pipe 1 and pipe 2) combine into a single pipe line (pipe 3) as shown in figure. The cross-sectional area of all three pipelines are constant. The following data is given

Pipe number	Area (m^2)	Velocity (m/s)
1	1	1
2	2	2
3	2.5	?

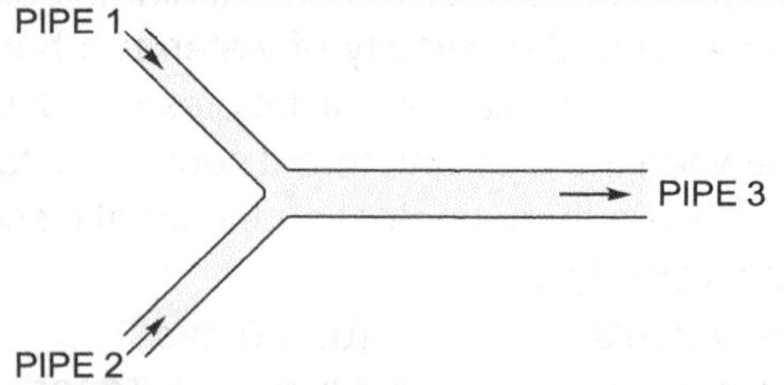

Assuming the water properties and the velocities to be uniform across the cross-section of the inlets and the outlet, the exit velocity (in m/s) in pipe 3 is
(a) 1 (b) 1.5
(c) 2 (d) 2.5 [2009 : 2 M]

4.10 Figure shows the schematic for the measurement of velocity of air (density = 1.2 kg/m^3) through a constant-area duct using a pitot tube and a water tube manometer. The differential head of water (density = 1000 kg/m^3) in the two columns of the manometer is 10 mm. Take acceleration due to gravity as 9.8 m/s^2. The velocity of air in m/s is

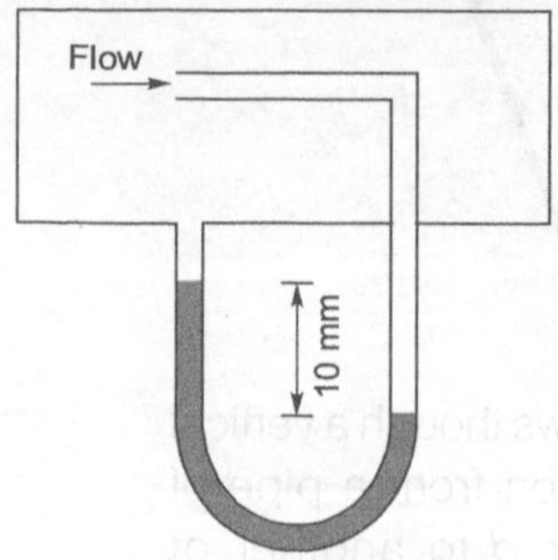

(a) 6.4 (b) 9.0
(c) 12.8 (d) 25.6 [2011 : 2 M]

4.11 Water is coming out from a tap and a falls vertically downwards. At the tap opening, the stream diameter is 20 mm with uniform velocity of 2 m/s. Acceleration due to gravity is 9.81 m/s^2. Assuming steady, inviscid flow, constant atmospheric pressure everywhere and neglecting curvature and surface tension effects, the diameter in mm of the stream 0.5 m below the tap is approximately
(a) 10 (b) 15
(c) 20 (d) 25 [2013 : 2 M]

4.12 Within a boundary layer for a steady incompressible flow, the Bernoulli equation
(a) holds because the flow is steady
(b) holds because the flow is incompressible
(c) holds because the flow is transitional
(d) does not hold because the flow is frictional
[2015 : 1 M, Set-2]

4.13 A Prandtl tube (Pitot-static tube with C = 1) is used to measure the velocity of water. The differential manometer reading is 10 mm of liquid column with a relative density of 10. Assuming g = 9.8 m/s^2, the velocity of water (in m/s) is _______.
[2015 : 2 M, Set-3]

4.14 Water (ρ = 1000 kg/m^3) flows through a venturimeter with inlet diameter 80 mm and throat diameter 40 mm. The inlet and throat gauge pressures are measured to be 400 kPa and 130 kPa respectively. Assuming the venturimeter to be horizontal and neglecting friction, the inlet velocity (in m/s) is _______.
[2015 : 2 M, Set-1]

4.15 The water jet exiting from a stationary tank through a circular opening of diameter 300 mm impinges on a rigid wall as shown in the figure. Neglect all minor losses and assume the water level in the tank to remain constant. The net horizontal force experienced by the wall is _________ kN.

Density of water is 1000 kg/m^3.
Acceleration due to gravity g = 10 m/s^2.

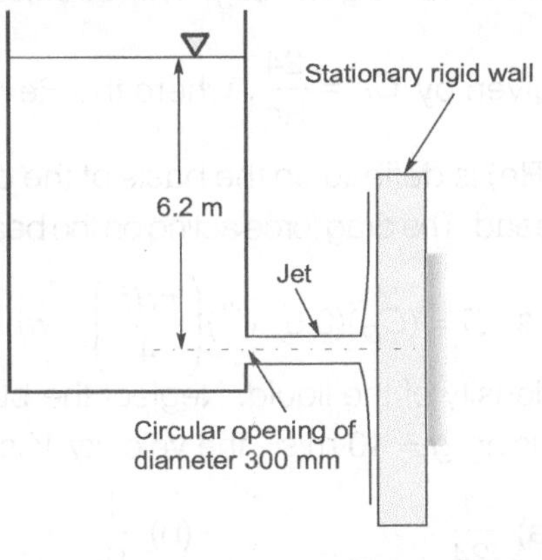

[2016 : 2 M, Set-3]

4.16 The arrangement shown in the figure measures the velocity V of a gas of density 1 kg/m^3 flowing through a pipe. The acceleration due to gravity is 9.81 m/s^2. If the manometric fluid is water (density 1000 kg/m^3) and the velocity V is 20 m/s, the differential head h (in mm) between the two arms of the manometer is ______.

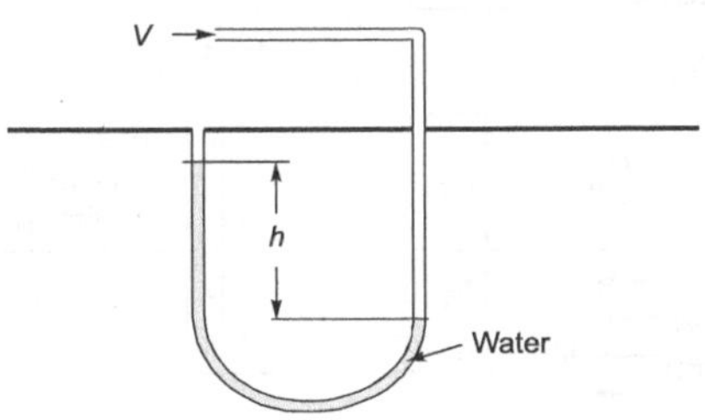

[2017 : 2 M, Set-2]

4.17 A tank open at the top with a water level of 1 m, as shown in the figure, has a hole at a height of 0.5 m. A free jet leaves horizontally from the smooth hole. The distance X (in m) where the jet strikes the floor is

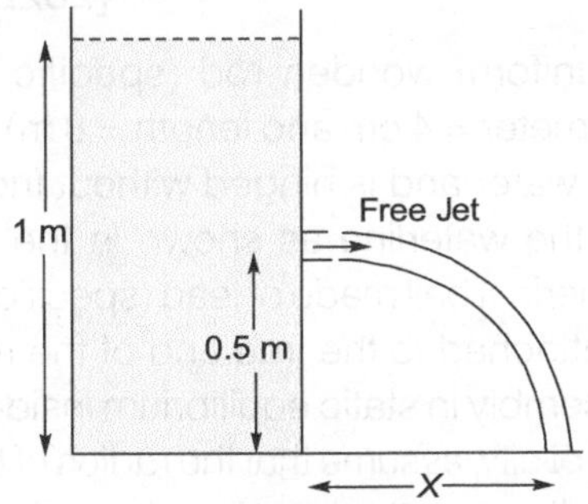

(a) 0.5 (b) 1.0
(c) 2.0 (d) 4.0

[2018 : 2 M, Set-1]

4.18 A frictionless circular piston of area 10^{-2} m^2 and mass 100 kg sinks into a cylindrical container of the same area filled with water of density 1000 kg/m^3 as shown in the figure.

The container has a hole of area 10^{-3} m^2 at the bottom that is open to the atmosphere. Assuming there is no leakage from the edges of the piston and considering water to be incompressible, the magnitude of the piston velocity (in m/s) at the instant shown is ______ (correct to three decimal places).

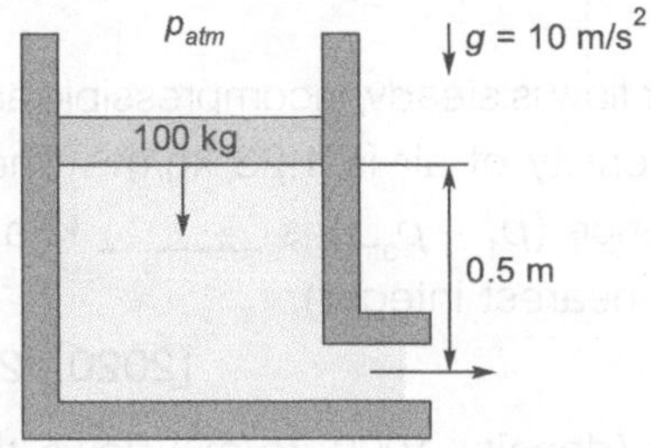

[2018 : 2 M, Set-2]

4.19 Air flows at the rate of 1.5 m^3/s through a horizontal pipe with a gradually reducing cross-section as shown in the figure. The two cross-sections of the pipe have diameters of 400 mm and 200 mm. Take the air density as 1.2 kg/m^3 and assume inviscid incompressible flow. The change in pressure ($p_2 - p_1$) (in kPa) between sections 1 and 2 is

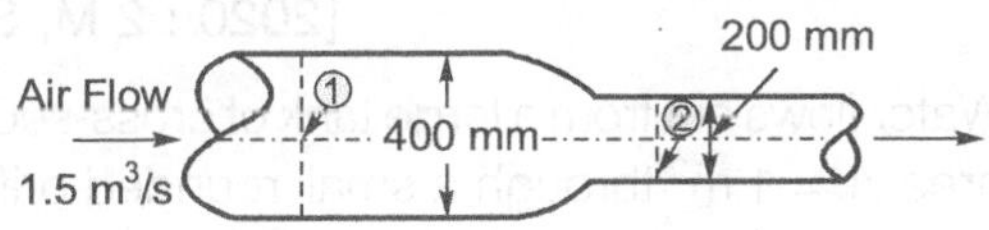

(a) −1.28 (b) 2.56
(c) −2.13 (d) 1.28

[2018 : 2 M, Set-2]

4.20 Air discharges steadily through a horizontal nozzle and impinges on a stationary vertical plate as shown in figure.

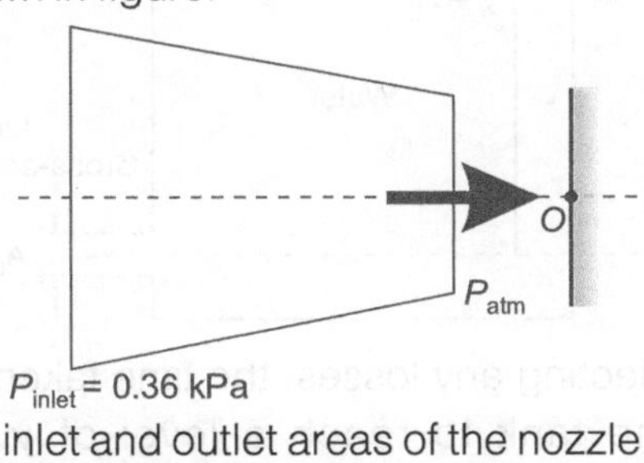

The inlet and outlet areas of the nozzle are 0.1 m^2 and 0.02 m^2, respectively. Take air density as constant and equal to 1.2 kg/m^3. If the inlet gauge pressure of air is 0.36 kPa, the gauge pressure at point O on the plate is ______ kPa (round off to two decimal places).

[2020 : 2 M, Set-1]

4.21 Consider a flow through a nozzle, as shown in the figure below:

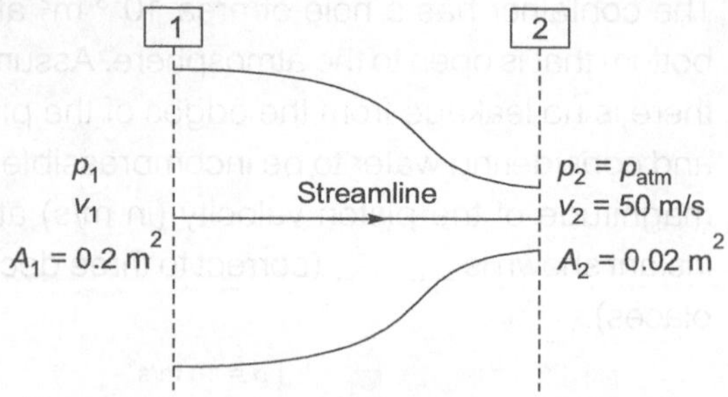

The air flow is steady, incompressible and inviscid. The density of air is 1.23 kg/m^3. The pressure difference $(p_1 - p_{atm})$ is ______ kPa (round off to the nearest integer).

[2020 : 2 M, Set-2]

4.22 Water (density 1000 kg/m^3) flows through an inclined pipe of uniform diameter. The velocity, pressure and elevation at section A are V_A = 3.2 m/s, p_A = 186 kPa and z_A = 24.5 m respectively, and those at section B are V_B = 3.2 m/s, p_B = 260 kPa and z_B = 9.1 m, respectively. If acceleration due to gravity is 10 m/s^2 then the head lost due to friction is ______ m (round off to one decimal place).

[2020 : 2 M, Set-2]

4.23 Water flows out from a large tank of cross-sectional area A_t = 1 m^2 through a small rounded orifice of cross-sectional area A_0 = 1 cm^2, located at y = 0. Initially the water level (H), measured from y = 0, is 1 m. The acceleration due to gravity is 9.8 m/s^2.

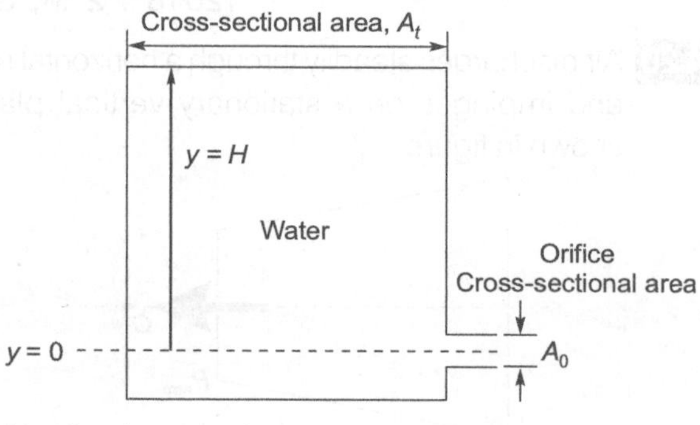

Neglecting any losses, the time taken by water in the tank to reach a level of y = H/4 is ______ seconds (round off to one decimal place).

[2021 : 2 M, Set-2]

4.24 A solid spherical bead of lead (uniform density = 11000 kg/m^3) of diameter d = 0.1 mm sinks with a constant velocity V in a large stagnant pool of a liquid (dynamic viscosity = 1.1 × 10^{-3} kg·m^{-1}s^{-1}). The coefficient of drag is given by $C_D = \frac{24}{\text{Re}}$, where the Reynolds number (Re) is defined on the basis of the diameter of the bead. The drag force acting on the bead is expressed as $D = (C_D)(0.5\rho V^2)\left(\frac{\pi d^2}{4}\right)$, where ρ is the density of the liquid. Neglect the buoyancy force. Using g = 10 m/s^2, the velocity V is ___ m/s.

(a) $\frac{1}{24}$ (b) $\frac{1}{6}$

(c) $\frac{1}{18}$ (d) $\frac{1}{12}$

[2022 : 2 M, Set-1]

4.25 A tube of uniform diameter D is immersed in a steady flowing inviscid liquid stream of velocity V, as shown in the figure. Gravitational acceleration is represented by g. The volume flow rate through the tube is ______.

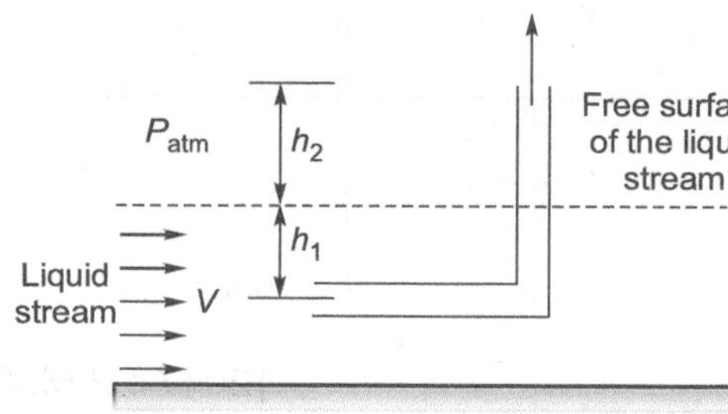

(a) $\frac{\pi}{4}D^2V$ (b) $\frac{\pi}{4}D^2\sqrt{2gh_2}$

(c) $\frac{\pi}{4}D^2\sqrt{2g(h_1+h_2)}$ (d) $\frac{\pi}{4}D^2\sqrt{V^2-2gh_2}$

[2022 : 2 M, Set-2]

4.26 A uniform wooden rod (specific gravity = 0.6, diameter = 4 cm and length = 8 m) is immersed in the water and is hinged without friction at point A on the waterline as shown in the figure. A solid spherical ball made of lead (specific gravity = 11.4) is attached to the free end of the rod to keep the assembly in static equilibrium inside the water. For simplicity, assume that the radius of the ball is much smaller than the length of the rod.

Assume density of water = 10^3 kg/m^3 and π = 3.14.

Radius of the ball is ______ cm (round off to 2 decimal places).

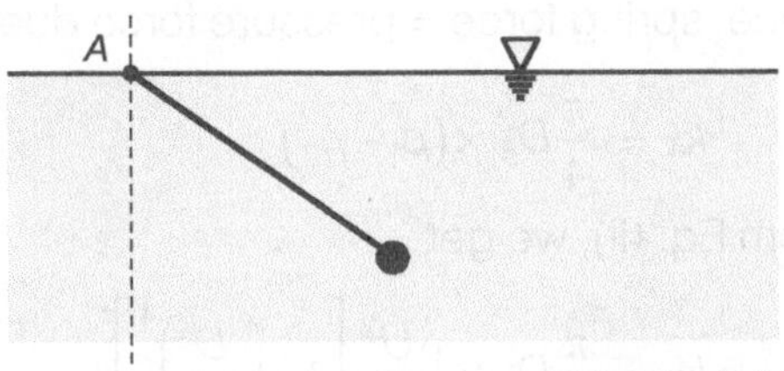

[2022 : 2 M, Set-2]

4.27 The velocity field of a certain two-dimensional flow is given by

$V(x,y) = k(x\hat{i} - y\hat{j})$

where $k = 2\ s^{-1}$. The coordinates x and y are in meters. Assume gravitational effects to be negligible. If the density of the fluid is 1000 kg/m³ and the pressure at the origin is 100 kPa, the pressure at the location (2 m, 2 m) is ______ kPa.

(Answer in integer) [2023 : 2 M]

4.28 Consider a unidirectional fluid flow with the velocity field given by

$V(x,y,z,t) = u(x,t)\hat{i}$

where $u(0, t) = 1$. If the spatially homogeneous density field varies with time t as

$\rho(t) = 1 + 0.2e^{-t}$

the value of $u(2,1)$ is ______. (Rounded off to two decimal places)

Assume all quantities to be dimensionless.

[2023 : 2 M]

Answers Fluid Dynamics

4.1 (c)	4.2 (d)	4.3 (a)	4.4 (c)	4.5 (d)	4.6 (d)	4.7 (a)	4.8 (b)
4.9 (c)	4.10 (c)	4.11 (b)	4.12 (d)	4.13 (1.328)	4.14 (6)	4.15 (8.76)	
4.16 (20.40)	4.17 (b)	4.18 (1.456)	4.19 (a)	4.20 (0.375)	4.21 (1.52)	4.22 (8)	
4.23 (2258.8)	4.24 (c)	4.25 (d)	4.26 (3.58)	4.27 (84)	4.28 (1.137)		

Explanations Fluid Dynamics

4.1 (c)

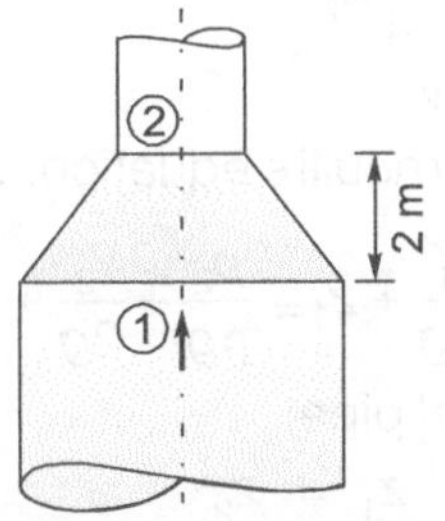

From equation of continuity, we get

$$A_1V_1 = A_2V_2$$

$$\frac{\pi}{4}d^2 \times 2 = \frac{\pi}{4}\times\left(\frac{d}{2}\right)^2 \times V_2$$

or $V_2 = 8$ m/s

Using Bernoulli's equations, we get

$$\frac{p_1}{\rho g} + \frac{V_1^2}{2g} + z_1 = \frac{p_2}{\rho g} + \frac{V_2^2}{2g} + z_2$$

$$\frac{200\times10^3}{10^3\times9.81} + \frac{(2)^2}{2\times9.81} + 0 = \frac{p_2}{10^3\times9.81} + \frac{(8)^2}{2\times9.81} + 2$$

$$20.38 + 0.20 = \frac{p_2}{9810} + 3.26 + 2$$

or $$\frac{p_2}{9810} = 15.32$$

or $$p_2 = 15.32\times9810 = 150289.2\ \text{N/m}^2$$
$$= \mathbf{150.289\ kN/m^2 \approx 150\ kN/m^2}$$

4.2 (d)

Bernoulli's equation: Law of conservation of energy

Continuity equation: Law of conservation of mass

Navier-Stoke's equation: Law of conservation of momentum

4.3 (a)

Mass of water striking per second = ρQ

Force on weighing balance due to water striking

= Initial momentum – Final momentum

= $\rho QU - 0 = \rho QU$

Weight of container and water = mg

Now total force on weighing balance

= $\boldsymbol{mg + \rho QU}$

4.4 (c)

Flow rate	Mean value of flow rate (x)	Frequency (f)	fx
7.5–7.7	7.6	1	7.6
7.7–7.9	7.8	5	39
7.9–8.1	8.0	35	280
8.1–8.3	8.2	17	139.4
8.3–8.5	8.4	12	100.8
8.5–8.7	8.6	10	86

$$\Sigma f = 80$$

$$\Sigma fx = 652.8$$

$$\therefore \text{ Mean flow rate} = \frac{\Sigma fx}{\Sigma x} = \frac{652.8}{80} = \mathbf{8.16}$$

4.5 (d)

Applying continuity equation, we get

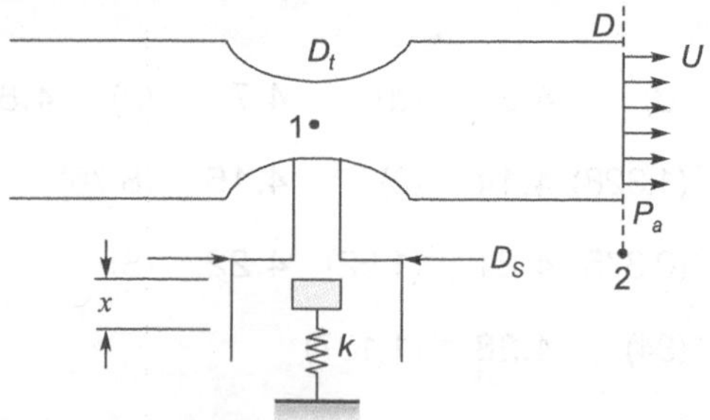

$$A_1V_1 = A_2V_2$$

$$V_1 = \left(\frac{A_2}{A_1}\right)V_2 = \left(\frac{\frac{\pi}{4}D^2}{\frac{\pi}{4}D_t^2}\right)U$$

$$V_1 = \left(\frac{D}{D_t}\right)^2 U \quad \text{...(i)}$$

Applying Bernoulli's equation

$$\frac{p_1}{\rho g} + \frac{V_1^2}{2g} = \frac{p_2}{\rho g} + \frac{V_2^2}{2g}$$

[$z_1 = z_2$ because venturi is horizontal]

$$\frac{p_1 - p_2}{\rho g} = \frac{V_2^2 - V_1^2}{2g}$$

$$p_1 - p_2 = \frac{\rho}{2}(V_2^2 - V_1^2)$$

From Eq. (i), we get

$$p_1 - p_2 = \frac{\rho}{2}\left[U^2 - \left(\frac{D}{D_t}\right)^4 U^2\right]$$

$$= \frac{\rho}{2}U^2\left[1 - \left(\frac{D}{D_t}\right)^4\right] \quad \text{...(ii)}$$

Since, spring force = pressure force due to air

$$-kx = \frac{\pi}{4}D_s^2 \times (p_1 - p_2)$$

From Eq. (ii), we get

$$-kx = \frac{\pi}{4}D_s^2 \times \frac{\rho U^2}{2}\left[1 - \left(\frac{D}{D_t}\right)^4\right]$$

$$\therefore \quad \boldsymbol{x} = \frac{\rho U^2}{8k}\left[\left(\frac{D}{D_t}\right)^4 - 1\right]\pi D_s^2$$

4.6 (d)

Given data:

$d_2 = 20$ mm $= 0.02$ m, $d_1 = 40$ mm $= 0.04$ m

$\Delta p = p_1 - p_2 = 30$ kPa $= 30 \times 10^3$ Pa

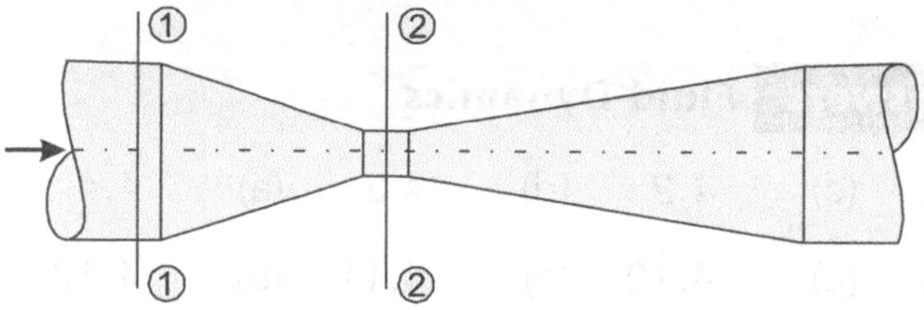

Venturimeter

For continuity equation, we get

$$A_1V_1 = A_2V_2$$

$$V_2 = \frac{A_1}{A_2}V_1 = \frac{\frac{\pi}{4}d_1^2}{\frac{\pi}{4}d_2^2} \times V_1 = \left(\frac{d_1}{d_2}\right)^2 V_1 = \left(\frac{0.04}{0.02}\right)^2 V_1$$

$$V_2 = 4V_1 \quad \text{...(i)}$$

Applying Bernoulli's equation, we get

$$\frac{p_1}{\rho g} + \frac{V_1^2}{2g} + z_1 = \frac{p_2}{\rho g} + \frac{V_2^2}{2g} + z_2$$

For horizontal pipe,

$$z_1 = z_2$$

$$\frac{p_1 - p_2}{\rho g} = \frac{V_2^2 - V_1^2}{2g}$$

$$\frac{\Delta p}{\rho} = \frac{1}{2}(V_2^2 - V_1^2)$$

$$\frac{30 \times 10^3}{1000} = \frac{1}{2}\left[(4V_1)^2 - V_1^2\right]$$

$$\frac{15V_1^2}{2} = 30$$

$$V_1^2 = 4$$

or $\quad V_1 = \mathbf{2\ m/s}$

4.7 (a)

Flow direction is A to B because pressure at section A is more than at section B.

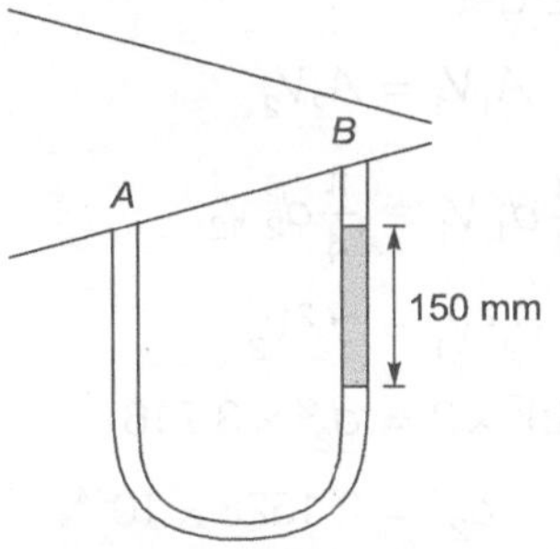

$$p_A - p_B = (\rho gh)_{Hg} = \rho_{Hg}\, gh_{Hg}$$

where,

$$\rho_{Hg} = 13600 \text{ kg/m}^3,\ g = 9.81 \text{ m/s}^2$$

$$h_{Hg} = 150 \text{ mm} = 0.15 \text{ m}$$

$$\therefore p_A - p_B = 13600 \times 9.81 \times 0.15$$

$$= 20012.4 \text{ Pa} = \mathbf{20.01 \text{ kPa}}$$

4.8 (b)

Given data:

$$d_1 = 20 \text{ cm} = 0.2 \text{ m}$$

$$d_2 = 10 \text{ cm} = 0.1 \text{ m}$$

$$p_1 = 150 \text{ kPa} = 150000 \text{ Pa}$$

Vapour pressure,

$$p_v = 50 \text{ kPa}$$

Specific weight,

$$w = 5 \text{ kN/m}^3 = 5000 \text{ N/m}^3 = \rho g$$

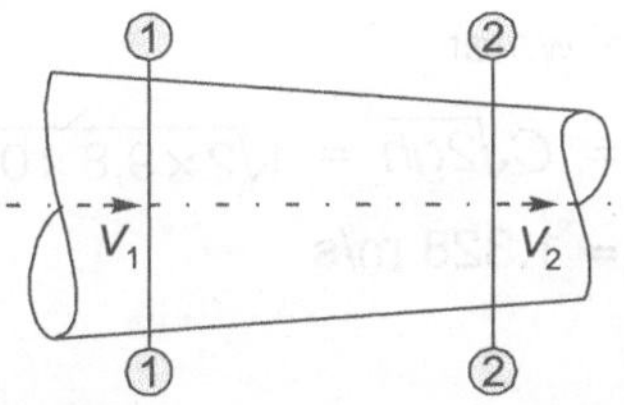

Applying continuity equation at sections (1)-(1) and (2)-(2), we get

$$A_1V_1 = A_2V_2 \quad \text{for incompressible flow}$$

$$\frac{\pi}{4}d_1^2 V_1 = \frac{\pi}{4}d_2^2 V_2$$

$$d_1^2 V_1 = d_2^2 V_2$$

or $$V_1 = \left(\frac{d_2}{d_1}\right)^2 V_2 = \left(\frac{10}{20}\right)^2 V_2 = 0.25\ V_2$$

The maximum pressure in downstream of reducer should be greater or equal to vapour pressure (p_v) to avoid cavitation. Therefore for maximum discharge,

$$p_2 = p_v = 50 \text{ kPa} = 50000 \text{ Pa}$$

Applying Bernoulli's equation at sections (1)-(1) and (2)-(2), we get

$$\frac{p_1}{\rho g} + \frac{V_1^2}{2g} + z_1 = \frac{p_2}{\rho g} + \frac{V_2^2}{2g} + z_2$$

$$\frac{p_1}{\rho g} + \frac{V_1^2}{2g} = \frac{p_2}{\rho g} + \frac{V_2^2}{2g} \quad \because z_1 = z_2$$

$$\frac{150000}{5000} + \frac{(0.25)^2 V_2^2}{2\times 9.81} = \frac{50000}{5000} + \frac{V_2^2}{2\times 9.81}$$

$$30 + 0.00318\ V_2^2 = 10 + 0.0509\ V_2^2$$

$$V_2^2 = 418.51$$

or $$V_2 = 20.45 \text{ m/s}$$

$$\therefore \quad Q = A_2V_2 = \frac{\pi}{4}d_2^2 V_2$$

$$= \frac{3.14}{4}\times(0.1)^2 \times 20.45 = \mathbf{0.160 \text{ m}^3\text{/s}}$$

4.9 (c)

$$Q_1 + Q_2 = Q_3$$

$$A_1V_1 + A_2V_2 = A_3V_3$$

$$1 + 4 = 2.5 \times V_3 \text{ or } \mathbf{V_3 = 2 \text{ m/s}}$$

4.10 (c)

Given data:

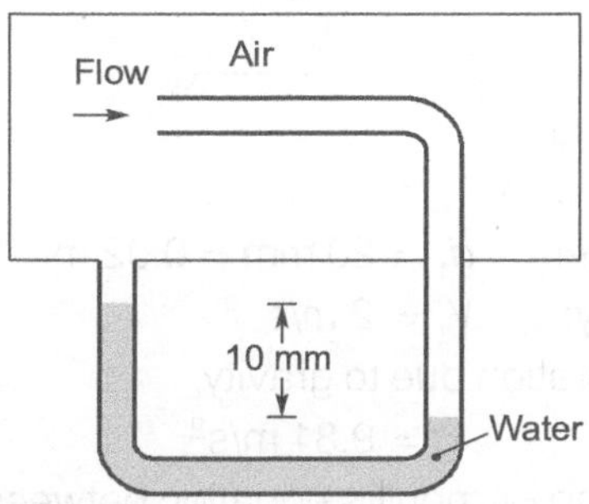

Density of air,

$$\rho_a = 1.2 \text{ kg/m}^3$$

Density of water,

$$\rho_w = 1000 \text{ kg/m}^3$$

Differential head in manometer,

$$h = 10 \text{ mm of water} = 0.01 \text{ m of water}$$

This reading is the dynamic pressure head. Hence dynamic pressure,

$$p_{dyn} = (\rho gh)_{water} = \rho_w gh$$

$$= 1000 \times 9.81 \times 0.01 = 98.1 \text{ N/m}^2$$

also $p_{dyn} = \left(\frac{1}{2}\rho V^2\right)_{air} = \frac{1}{2}\rho_a V^2$

$$\therefore \quad 98.1 = \frac{1}{2}\times 1.2\times V^2$$

or $V^2 = 163.5$

$V = \mathbf{12.78\ m/s}$

Alternatively

$$V = \sqrt{2gh}$$

$$h = x\left(\frac{\rho_m}{\rho}-1\right)$$

where

x = Reading of differential manometer $= 10\times 10^{-3}$ m

ρ_m = Density of manometric fluid $= 1000$ kg/m^3

ρ = Density of following fluid = 1.2 kg/m^3

$$\therefore \quad V = \sqrt{2\times 9.81\times 10\times 10^{-3}\left(\frac{1000}{1.2}-1\right)}$$

$V = 12.779$ m/s $\approx$ **12.8 m/s**

4.11 (b)

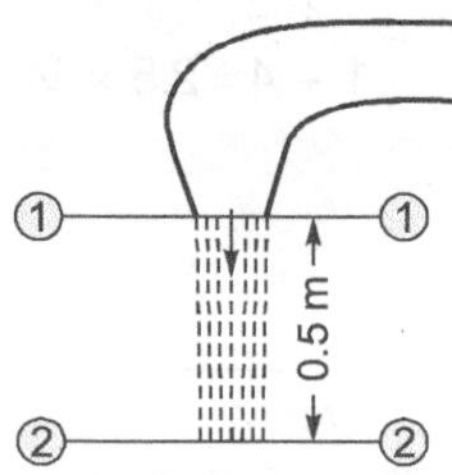

Given data:

At section 1-1,

Diameter: d_1 = 20 mm = 0.02 m

Velocity: V_1 = 2 m/s

Acceleration due to gravity,

$g = 9.81$ m/s^2

Applying Bernoulli's equation between sections 1 and 2, we get

$$\frac{p_1}{\rho g}+\frac{V_1^2}{2g}+z_1 = \frac{p_2}{\rho g}+\frac{V_2^2}{2g}+z_2$$

$$0+\frac{V_1^2}{2g}+0.5 = 0+\frac{V_2^2}{2g}+0$$

$$\frac{V_1^2}{2g}+0.5 = \frac{V_2^2}{2g}$$

or $V_2^2 = V_1^2 + 0.5\times 2g$

or $V_2 = \sqrt{V_1^2+0.5\times 2\times g}$

$= \sqrt{(2)^2+0.5\times 2\times 9.81}$

$= \sqrt{13.81} = 3.716$ m/s

Applying continuity equation between sections 1 and 2, we get

$$A_1V_1 = A_2V_2$$

$$\frac{\pi}{4}d_1^2 V_1 = \frac{\pi}{4}d_2^2 V_2$$

or $d_1^2V_1 = d_2^2V_2$

$(0.02)^2\times 2 = d_2^2\times 3.716$

or $d_2^2 = 2.1528\times 10^{-4}$

or $d_2 = 0.01467$ m

$= 14.67$ mm $\approx$ **15 mm**

4.12 (d)

Within boundary layer, Bernoulli's equation is not valid because due to friction, flow is irrotational.

4.13 Sol.

Given data:

$C = 1$

x = 10 mm of liquid = 0.01 m of liquid

$S = 10$

$\rho_{mano} = 10\times 1000 = 10000$ kg/m^3

$g = 9.8$ m/s^2

$$h = x\left[\frac{\rho_{mano}}{\rho_{water}}-1\right] = 0.01\left[\frac{10000}{1000}-1\right]$$

= 0.09 m of water

Velocity of water,

$$V = C\sqrt{2gh} = 1\sqrt{2\times 9.8\times 0.09}$$

= **1.328 m/s**

4.14 Sol.

Given data:

$\rho = 1000$ kg/m^3

d_1 = 80 mm = 0.08 m

d_2 = 40mm = 0.04 m

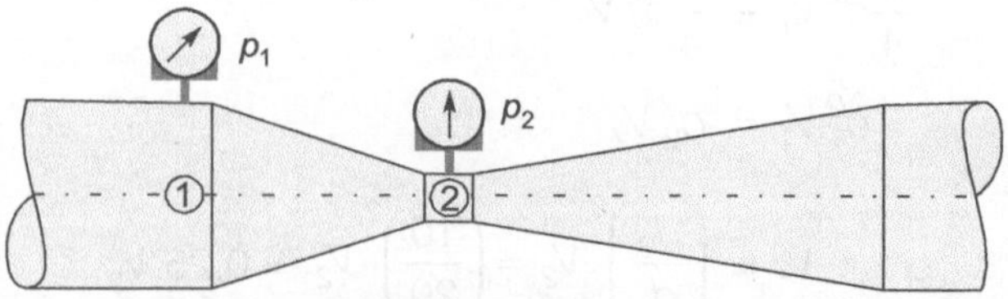

Venturimeter

$p_1 = 400$ kPa $= 400\times 10^3$ Pa

$p_2 = 130$ kPa $= 130\times 10^3$ Pa

Applying Bernoulli's equation between sections 1 and 2,

$$\frac{p_1}{\rho g}+\frac{V_1^2}{2g}+z_1 = \frac{p_2}{\rho g}+\frac{V_2^2}{2g}+z_2$$

where $z_1 = z_1$ for horizontal pipe

$$\therefore \quad \frac{p_1}{\rho g}+\frac{V_1^2}{2g} = \frac{p_2}{\rho g}+\frac{V_2^2}{2g}$$

or $$\frac{p_1-p_2}{\rho g} = \frac{V_2^2-V_1^2}{2g}$$

or $$\frac{p_1-p_2}{\rho} = \frac{V_2^2-V_2^2}{2} \qquad ...(i)$$

Applying continuity equation between sections 1 and 2,

$$A_1V_1 = A_2V_2$$

$$\frac{\pi}{4}d_1^2 V_1 = \frac{\pi}{2}d_2^2 V_2$$

or $$d_1^2V_1 = d_2^2V_2$$

or $$V_2 = \left(\frac{d_1}{d_2}\right)^2 V_1 = \left(\frac{0.08}{0.04}\right)^2 V_1$$

$$V_2 = 4V_1$$

Substituting $V_2 = 4V_1$ in Eq. (i), we get

$$\frac{400\times10^3-130\times10^3}{1000} = \frac{(4V_1)^2-V_1^2}{2}$$

$$540 = 16V_1^2 - V_1^2$$

$$540 = 15V_1^2$$

or $$V_1^2 = 36$$

or $$V_1 = \mathbf{6\ m/s}$$

<u>**Method II:**</u>

Given: P_1 = 400 kPa, = 400 × 10^3 Pa

P_2 = 130 kPa = 130 × 10^3 Pa

$Z_1 = Z_2$ (Venturimeter is horizontal)

Discharge of venturimeter,

$$Q = \frac{a_1a_2\sqrt{2gh}}{\sqrt{a_1^2-a_2^2}}$$

$$h = \left(\frac{p_1}{\rho g}+z_1\right)-\left(\frac{p_2}{\rho g}+z_2\right)$$

$$= \left(\frac{400\times10^3}{9810}-\frac{130\times10^3}{9810}\right) = 27.52\text{ m}$$

$$\text{Inlet velocity, } V_1 = \frac{Q}{a_1} = \frac{a_2\sqrt{2gh}}{\sqrt{a_1^2-a_2^2}}$$

$$= \frac{\frac{\pi}{4}(0.04)^3\sqrt{2\times9.81\times27.52}}{\sqrt{\left[\frac{\pi}{4}(0.08)\right]^2\left[\frac{\pi}{4}(0.04)^2\right]^2}} = 6\text{ m/s}$$

POINTS TO REMEMBER

- Here, we have taken, discharge of venturimeter, Q as the ideal discharge, that's why we applied:
$$Q = \frac{a_1a_2\sqrt{2gh}}{\sqrt{a_1^2-a_2^2}}$$
- If we take actual discharge, then
$$Q_{actual} = C_DQ$$
where, C_D = coefficient of discharge = 0.94 to 0.98
- Then in the problem,
$$V_{actual} = 0.98\times6 = 5.88\text{ m/s}$$

4.15 Sol.

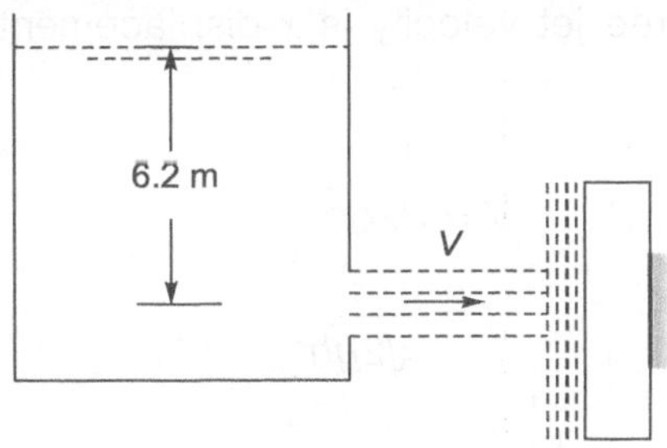

Diameter of jet,

$$d = 300\text{ mm} = 0.30\text{ m}$$

∴ Cross-section area,

$$a = \frac{\pi}{4}d^2 = \frac{3.14}{4}\times(0.30)^2$$

$$= 0.07065\text{ m}^2$$

Jet velocity: $V = \sqrt{2gh} = \sqrt{2\times10\times6.2}$

$$= 11.135\text{ m/s}$$

Net horizontal force,

$$F = m[V-0] = mV = \rho aV\times V$$

$$= \rho aV^2 = 1000\times0.07065\times(11.135)^2$$

$$= 8759.76\text{ N} = \mathbf{8.76\ kN}$$

4.16 Sol.

Method : (i)

Given data: ρ_{gas} = 1 kg/m^3; g = 9.81 m/s^2, ρ_m = 1000 kg/m^3, V = 20 m/s

Dynamic pressure of gas = $(\rho gh)_{water}$

$$\frac{1}{2}\rho_{gas}V^2 = \rho_m\times9.81\times h$$

$$\frac{1}{2}\times 1\times(20)^2 = 1000 \times 9.81 \times h$$

or $\quad h = 0.02038$ m of water

$\quad = 20.38$ mm of water

Method : (ii)

$\rho_m = 1000\,\text{kg/m}^3,\ \rho_{gas} = 1\ \text{kg/m}^3$

$V = 20$ m/s

$$V = \sqrt{2gh\left(\frac{\rho_m}{\rho_{gas}}-1\right)}$$

$$20 = \sqrt{2\times 9.81\times h\left(\frac{1000}{1}-1\right)}$$

or $\quad h = $ **20.40 mm of water**

4.17 (b)

Let free jet velocity is x displacement

$$t = \frac{x}{v}$$

$$V = \sqrt{2gh}$$

$$t = \frac{x}{\sqrt{2gh}} \qquad \ldots\text{(i)}$$

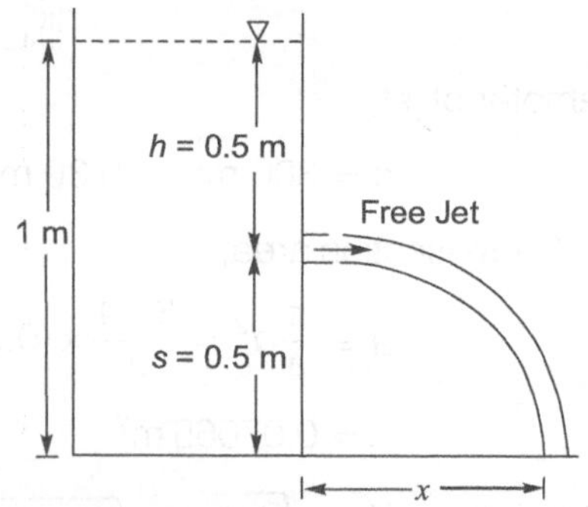

By motion equation,

$$s = ut + \frac{1}{2}gt^2$$

For free fall: $u = 0,\ s = \frac{1}{2}gt^2$

$t = \sqrt{\frac{2s}{g}}$ ...(ii) is the time of free fall of object

By equations (i) and (ii)

$$\frac{x}{\sqrt{2gh}} = \sqrt{\frac{2s}{g}}$$

$$x = \sqrt{4hs} = \sqrt{4\times 0.5\times 0.5} = 1\,\text{m}$$

4.18 Sol.

$$A_1V_1 = A_2V_2$$

$$V_2 = \left(\frac{A_1}{A_2}\right)V_1$$

Applying Bernoulli's equation between points 1 and 2, we get

$$\frac{p_1}{\rho g}+\frac{V_1^2}{2g}+Z_1 = \frac{p_2}{\rho g}+\frac{V_2^2}{2g}+Z_2$$

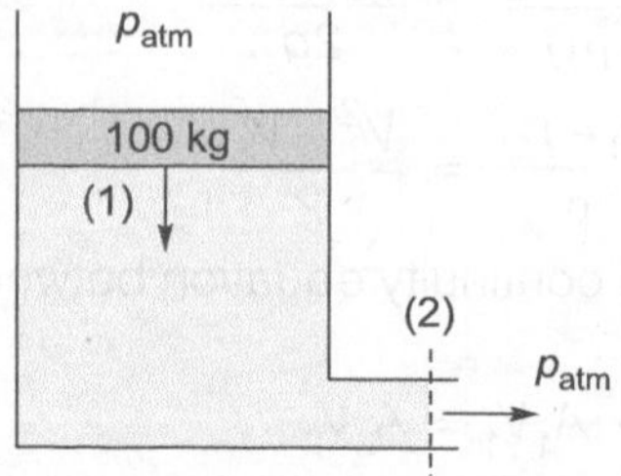

$$\left[\frac{p_{atm}+\dfrac{100\times 10}{10^{-2}}}{\rho\times g}\right]+\frac{V_1^2}{2g}+0.5$$

$$= \frac{p_{atm}}{\rho g}+\frac{A_1^2}{A_2^2}\times\frac{V_1^2}{2g}$$

$$\frac{100\times 10}{10^{-2}\times 1000\times 10}+0.5 = \left(\frac{A_1^2}{A_2^2}-1\right)\frac{V_1^2}{2g}$$

$$10 + 0.5 = (10^2-1)\frac{V_1^2}{2g}$$

$$V_1^2 = \frac{10.5\times 2\times 10}{99} = 2.12$$

$$V_1 = \mathbf{1.456\ m/s}$$

4.19 (a)

$$\frac{p_1}{\rho g}+\frac{V_1^2}{2g}+Z_1 = \frac{p_2}{\rho g}+\frac{V_2^2}{2g}+Z_2$$

{where, Z = constant}

$$A_1 = \frac{\pi}{4}\times 0.4^2 = 0.1256\ \text{m}^2$$

$$A_2 = \frac{\pi}{4}\times 0.2^2 = 0.0314\ \text{m}^2$$

$$\frac{p_2-p_1}{\rho g} = \frac{V_1^2-V_2^2}{2g}$$

$$\frac{p_2-p_1}{1.2} = \frac{1}{2}\left[\frac{Q^2}{A_1^2}-\frac{Q^2}{A_2^2}\right]$$

$$p_2 - p_1 = 0.6\times(1.5)^2\left[\frac{1}{A_1^2}-\frac{1}{A_2^2}\right]$$

$$p_2 - p_1 = 1.35\left[\frac{1}{0.01577}-\frac{1}{0.000985}\right]$$

$$= 1.35[63.41 - 1015.22]$$

$$= -1284.94\ \text{Pa} = \mathbf{-1.28\ kPa}$$

4.20 Sol.

On applying continuity equation,

$$\dot{m} = \rho_1 \cdot A_1 \cdot v_1 = \rho_2 \cdot A_2 \cdot v_2$$

$$\Rightarrow \quad 0.1 V_1 = 0.02 V_2$$

$$\Rightarrow \quad V_2 = \frac{10}{2} V_1 = 5V_1$$

Now on applying Bernoulli's between sections 1 and 2

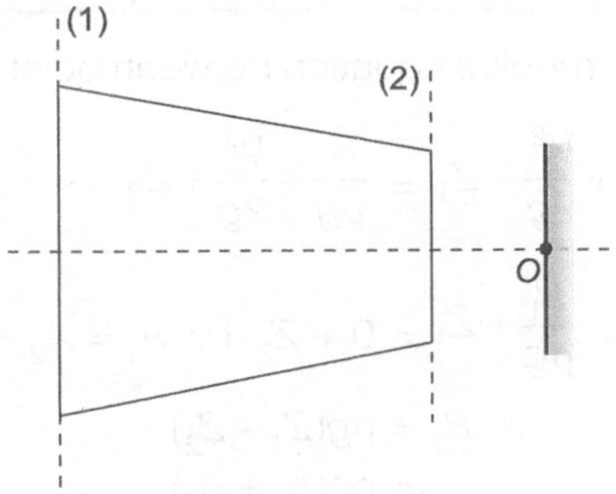

$$\frac{P_1}{\rho g} + \frac{V_1^2}{2g} + z_1 = \frac{P_2}{\rho g}^{\,0} + \frac{V_2^2}{2g} + z_2$$

$$\Rightarrow \quad \frac{0.36 \times 10^3}{1.21 \times 9.81} + \frac{V_1^2}{2g} = \frac{25 V_1^2}{2g}$$

$$\Rightarrow \quad V_1 = 4.98 \text{ m/s}$$

$$\Rightarrow \quad V_2 = 24.89 \text{ m/s}$$

On applying Bernoulli's between section 2 and point O

$$\frac{P_2}{\rho g}^{\,0} + \frac{V_2^2}{2g} + z_2 = \frac{P_O}{\rho g} + \frac{V_O^2}{2g}^{\,0} + z_O$$

$$P_O = \frac{\rho g \cdot V_2^2}{2g} = \frac{1.21 \times (24.89)^2}{2} = 375 \text{ Pa}$$

$$= 0.375 \text{ kPa (gauge)}$$

4.21 Sol.

As per given data, $\rho_{air} = 1.23$ kg/m^3

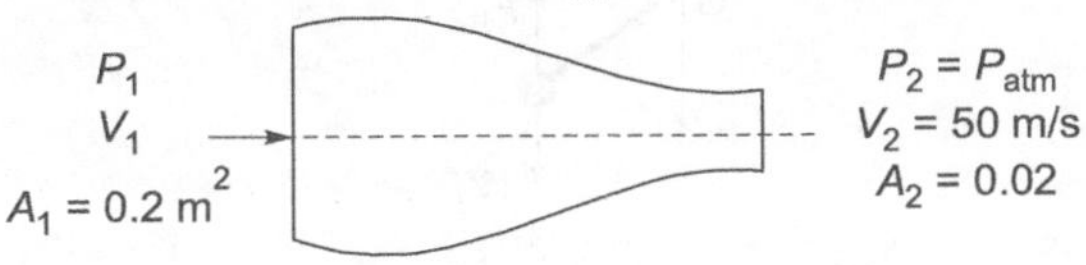

$$A_1 V_1 = A_2 V_2$$

$$0.2 \times V_1 = 0.02 \times 50$$

$$V_1 = \frac{1}{10} \times 50 = 5 \text{ m/s}$$

Applying BE

$$\frac{P_1}{w} + \frac{V_1^2}{2g} + z_1 = \frac{P_2}{w} + \frac{V_2^2}{2g} + z_2$$

$(\because z_1 = z_2)$

$$\frac{P_1 - P_2}{\rho_{air} g} = \frac{V_2^2 - V_1^2}{2g}$$

Pressure difference,

$$P_1 - P_2 = \left(\frac{50^2 - 5^2}{2}\right) \times 1.23$$

$$= 1522.125 \text{ Pa}$$

$$= 1.52 \text{ kPa}$$

4.22 Sol.

Given: $v_A = 3.2$ m/s, $p_A = 186$ kPa
$z_A = 24.5$ m, $v_B = 32$ m/s,
$p_B = 260$ kPa, $z_B = 9.1$ m

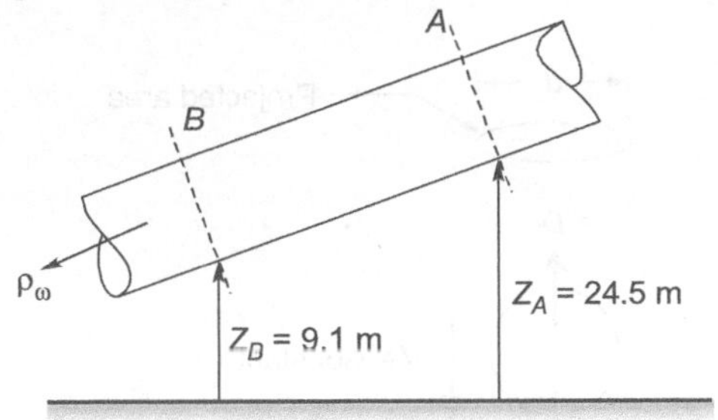

$$\text{Energy at 'A' head} = \frac{P_A}{\rho g} + \frac{V_A^2}{2g} + Z_A$$

$$= \frac{186 \times 10^3}{1000 \times 10} + \frac{3.2^2}{2 \times 10} + 24.5$$

$$= 18.6 + 0.512 + 24.5$$

$$= 43.612$$

$$\text{Energy at 'b' head} = \frac{P_b}{\rho g} + \frac{V_b^2}{2g} + Z_b$$

$$= \frac{286 \times 10^3}{1000 \times 10} + \frac{3.2^2}{2 \times 10} + 9.1$$

$$= 26 + 0.512 + 9.1 = 35.612$$

$E_A > E_B$, so flow from 'A' to 'B'.

Heat loss $= E_A - E_B = 43.612 - 35.612$
$= 8$ m of water head

4.23 Sol.

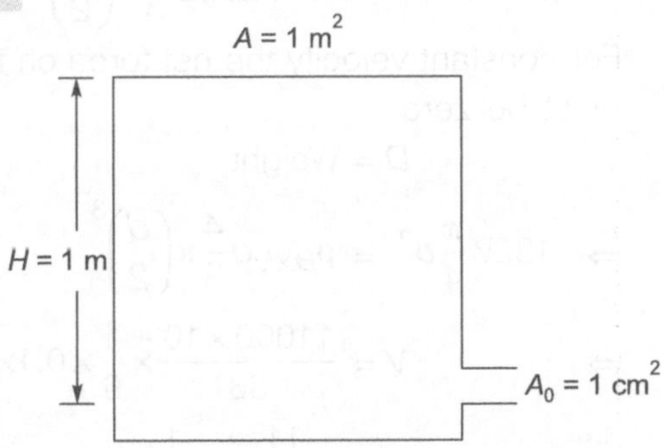

$$\dot{m}_{in} - \dot{m}_{out} = \frac{dm_w}{dt} \qquad (\dot{m}_{in} = 0)$$

$$-\rho A_o\sqrt{2gH} = \frac{d}{dt}[A\times H\times\rho]$$

$$dt = -\frac{A}{A_o}\times\frac{1}{\sqrt{2g}}\times\frac{dH}{\sqrt{H}}$$

$$t = -\frac{A}{A_o}\times\frac{1}{\sqrt{2g}}\int_1^{0.25}\frac{dH}{\sqrt{H}}$$

$$t = \frac{A}{A_o}\times\frac{1}{\sqrt{2g}}(2\sqrt{H})_{0.25}^{1}$$

$$t = \frac{1}{10^{-4}}(2-1)\times\frac{1}{\sqrt{2\times 9.81}}$$

$$t = 2258.8 \text{ s}$$

4.24 (c)

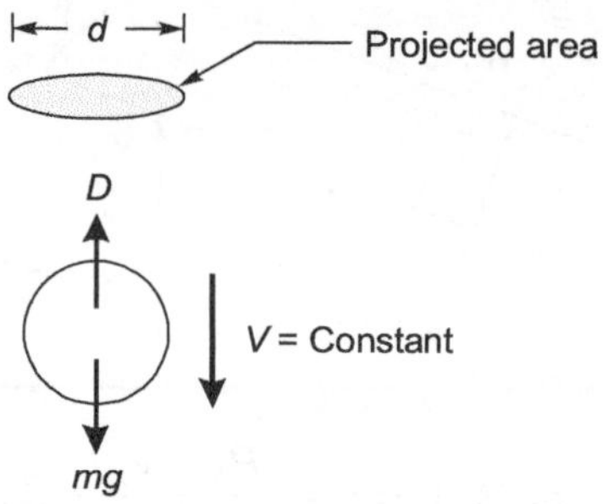

$$\text{Re} = \frac{\rho V d}{\mu} = \frac{\rho V\times 0.1\times 10^{-3}}{1.1\times 10^{-3}}$$

$$\text{Re} = \frac{\rho V}{11}$$

$\Rightarrow$ Drag force on sphere $= C_D\left(0.5\rho V^2\right)\left(\frac{\pi d^2}{4}\right)$

$$= \frac{24}{\left(\frac{\rho V}{11}\right)}\left(0.5\rho V^2\right)\left(\frac{\pi d^2}{4}\right)$$

$$D = 132V\left(\frac{\pi}{4}d^2\right) \quad \text{...(i)}$$

Weight of sphere $= \rho_{Body}g\frac{4}{3}\pi\left(\frac{d}{2}\right)^3$ ...(ii)

For constant velocity the net force on the body must be zero.

$$D = \text{Weight}$$

$$\Rightarrow \quad 132V\frac{\pi}{4}d^2 = \rho_{Body}g\frac{4}{3}\pi\left(\frac{d}{2}\right)^3$$

$$\Rightarrow \quad V = \frac{11000\times 10}{33}\times\frac{1}{6}\times 0.1\times 10^{-3}$$

$$\therefore \quad V = \frac{11}{33\times 6} = \frac{1}{18}$$

4.25 (d)

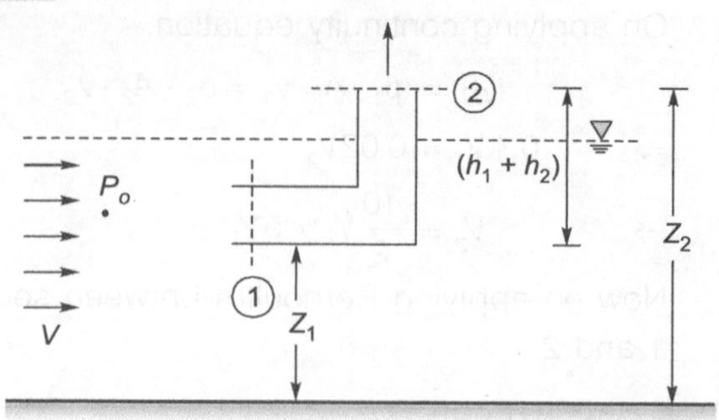

By Bernoulli's equation between point 1 and 2

$$\frac{P_1}{\rho g}+\frac{V_1^2}{2g}+Z_1 = \frac{P_2}{\rho g}+\frac{V_2^2}{2g}+Z_2$$

$$\frac{P_1}{\rho g}+Z_1 = 0 + Z_2 \quad [\because A_1 = A_2 \Rightarrow V_1 = V_2]$$

$$P_1 = \rho g(Z_1 - Z_2)$$

$$= \rho g(h_1 + h_2)$$

$$\because \quad P_o = \rho g h_1$$

$$\frac{P_o}{\rho g}+\frac{V^2}{2g}+Z_o = \frac{P_1}{\rho g}+\frac{V_1^2}{2g}+Z_1$$

As $Z_o = Z_1$

$$\frac{\rho g h_1}{\rho g}+\frac{V^2}{2g} = \frac{\rho g(h_1+h_2)}{\rho g}+\frac{V_1^2}{2g}$$

$$\frac{V_1^2}{2g} = \frac{V^2}{2g} - h_2$$

$$V_1 = \sqrt{V^2 - 2gh_2}$$

$$\therefore \quad Q = V_1 A_1 = \frac{\pi}{4}D^2\sqrt{V^2 - 2gh_2}$$

4.26 (3.58) (3.48 to 3.70)

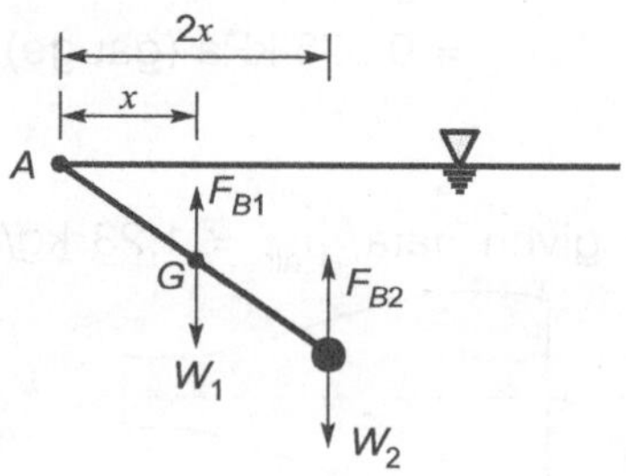

$$W_1 = \rho_1 g\pi r_1^2 l; \quad W_2 = \rho_2 g\frac{4}{3}\pi r_2^3$$

where, r_1 = Radius of rod

r_2 = Radius of sphere

ΣM_A = 0 for equilibrium

$$W_1 x + W_2(2x) = F_{B1}\times x + F_{B2}\times 2x$$

$$\rho_1 g\pi r_1^2 l + 2\rho_2 g\frac{4}{3}\pi r_2^3 = \rho_w g\left[\pi r_1^2 l + \frac{4}{3}\pi r_2^3\times 2\right]$$

$$S_1 l r_1^2 + 2S_2 \frac{4}{3} r_2^3 = r_1^2 l + \frac{4}{3} r_2^3 \times 2$$

[Here, $S_1 = \frac{\rho_1}{\rho_w}$ and $S_2 = \frac{\rho_2}{\rho_w}$]

$$(2S_2 - 2)\frac{4}{3} r_2^3 = (-S_1 + 1) r_1^2 l$$

$$(2 \times 11.4 - 2)\frac{4}{3} r_2^3 = (1 - 0.6) \times 8 \times (0.02)^2$$

$$r_2 = 0.0358 \text{ m} = 3.58 \text{ cm}$$

4.27 (84) (83.999 to 84.001)

Given $V(x,y) = k(x\hat{i} - y\hat{j})$

$$V_x = kx = u$$

$$V_y = -ky = v$$

Using Navier-Stokes equation in x-direction,

$$-\frac{\partial P}{\partial x} + \mu \frac{\partial^2 u}{\partial x^2} = \rho\left(u\frac{\partial u}{\partial x} + v\frac{\partial u}{\partial y}\right)$$

$$-\frac{\partial P}{\partial x} = \rho(kx \times k + (-ky)0)$$

$$-\frac{\partial P}{\partial x} = k^2 \rho x$$

$$P(x) = \frac{-k^2 \rho x^2}{2} + C_1 \qquad \text{...(i)}$$

Using Navier-Stokes equation in y-direction,

$$-\frac{\partial P}{\partial y} + \mu \frac{\partial^2 u}{\partial y^2} = \rho\left(u\frac{\partial v}{\partial x} + v\frac{\partial v}{\partial y}\right)$$

$$-\frac{\partial P}{\partial y} = \rho(kx \times 0 + (-ky)(-k))$$

$$-\frac{\partial P}{\partial y} = k^2 \rho y$$

$$P(y) = \frac{-k^2 \rho y^2}{2} + C_2 \qquad \text{...(ii)}$$

From equation (i) and (ii), we get

$$P(x, y) = -\frac{k^2 \rho x^2}{2} - \frac{k^2 \rho y^2}{2} + C$$

Now, it is given that

$$P(0, 0) = 100 \text{ kPa}$$

$$100 = 0 + c$$

$$c = 100$$

then,

$$P(x, y) = \frac{-\rho k^2}{2}(x^2 + y^2) + 100$$

$$P(2, 2) = \frac{-4}{2}(2^2 + 2^2) + 100$$

$$P(2, 2) = -16 + 100 = 84 \text{ kPa}$$

4.28 (1.137) (1.13 to 1.15)

Given: $u(0, t) = 1$

$\rho(t) = 1 + 0.2e^{-t}$

Using continuity equation

$$\frac{\partial \rho}{\partial t} + \frac{\partial}{\partial x}(\rho u) + \cancelto{o}{\frac{\partial}{\partial y}(\rho v)} + \cancelto{o}{\frac{\partial}{\partial y}(\rho \omega)} = 0$$

$$\frac{\partial \rho}{\partial t} + \frac{\partial}{\partial x}(\rho u) = 0 \qquad \text{...(i)}$$

$$\frac{\partial \rho}{\partial t} = \frac{\partial(1 + 0.2e^{-t})}{\partial t} = -0.2e^{-t}$$

$$\frac{\partial}{\partial x}(\rho u) = \frac{\partial}{\partial x}((1 + 0.2e^{-t})u(x,t))$$

$$= (1 + 0.2c^{-t}) \cdot \frac{\partial u}{\partial x}$$

Now, put in equation (i)

$$-0.2e^{-t} + (1 + 0.2e^{-t}) \cdot \frac{\partial u}{\partial x} = 0$$

$$\frac{\partial u}{\partial x} = \frac{0.2e^{-t}}{1 + 0.2e^{-t}}$$

$$u = \left(\frac{0.2e^{-t}}{1 + 0.2e^{-t}}\right) \cdot x + c \qquad \text{...(ii)}$$

Using boundary condition,

$$u(0, t) = 1$$

$$c = 1$$

From equation (ii),

$$u = \frac{0.2e^{-t}}{1 + 0.2e^{-t}} x + 1$$

$$u(2, 1) = \frac{0.2e^{-1}}{1 + 0.2e^{-1}} \times 2 + 1$$

$$u(2, 1) = 1.137 \text{ m/s}$$

■■■■

5 Viscous, Turbulent Flow & Boundary Layer Theory

5.1 In fully developed laminar flow in the circular pipe, the head loss due to friction is directly proportional to .... (mean velocity/square of the mean velocity)

[1995 : 1 M]

5.2 For laminar flow through a long pipe, the pressure drop per unit length increases
(a) in linear proportion to the cross-sectional area
(b) in proportion to the diameter of the pipe
(c) in inverse proportion to the cross-sectional area
(d) in inverse proportional to the square of cross-sectional area

[1996 : 2 M]

5.3 The discharge velocity at the pipe exit in figure is

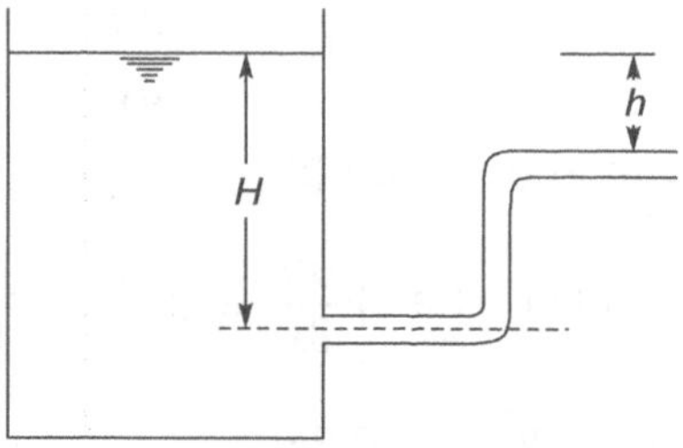

(a) $\sqrt{2gH}$ (b) $\sqrt{2gh}$
(c) $\sqrt{g(H+h)}$ (d) 0 [1998 : 2 M]

5.4 Flow separations flow past a solid object is caused by
(a) A reduction of pressure to vapour pressure
(b) A negative pressure gradient
(c) A positive pressure gradient
(d) The boundary layer thickness reducing to zero

[2002 : 1 M]

5.5 If x is the distance measured from the leading edge of a flat plate, then laminar boundary layer thickness varies as
(a) $1/x$ (b) $x^{4/5}$
(c) x^2 (d) $x^{1/2}$ [2002 : 1 M]

Common Data Questions 5.6 to 5.7: A syringe with a frictionless plunger contains water and has at its end a 100 mm long needle of 1 mm diameter. The internal diameter of the syringe is 10 mm. Water density is 1000 kg/m^3. The plunger is pushed in at 10 mm/s and the water comes out as a jet.

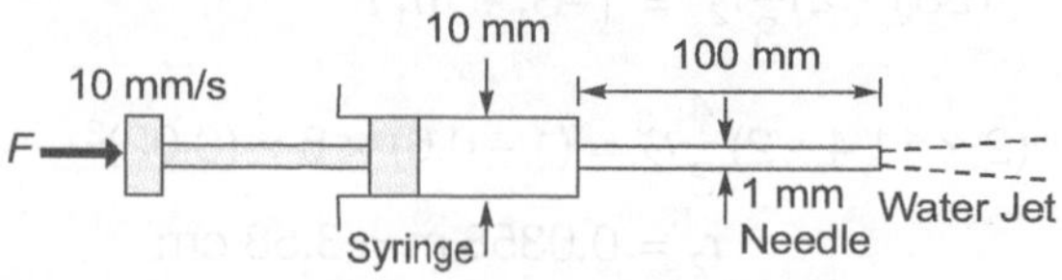

5.6 Assuming ideal flow, the force F in newtons required on the plunger to push out the water is
(a) 0 (b) 0.04
(c) 0.13 (d) 1.15 [2003 : 2 M]

5.7 Neglect losses in the cylinder and assume fully developed laminar viscous flow throughout the needle; the Darcy friction factor is $64/Re$. where Re is the Reynolds number. Given that the viscosity of water is 1.0×10^{-3} kg/ms, the force F in newtons required on the plunger is
(a) 0.13 (b) 0.16
(c) 0.3 (d) 4.4 [2003 : 2 M]

5.8 A cubic block of side L and mass M is dragged over an oil film across table by a string connects to a hanging block of mass m as shown in figure. The Newtonian oil film of thickness h has dynamic viscosity μ and the flow condition is laminar. The acceleration due to gravity is g. The steady state velocity V of block is

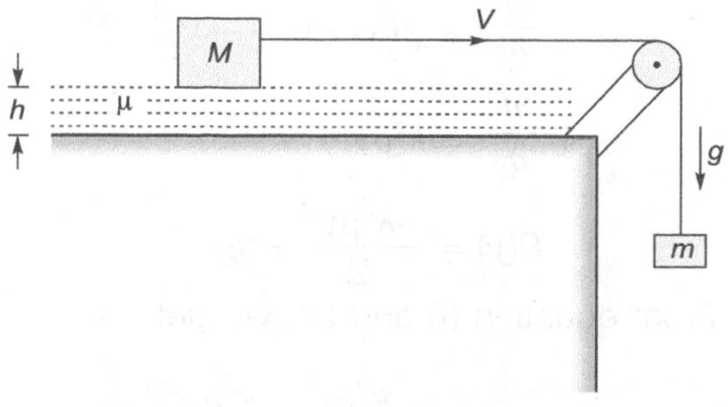

(a) $\frac{Mgh}{\mu L^2}$ (b) $\frac{Mgh}{\mu}$
(c) $\frac{mgh}{\mu L^2}$ (d) $\frac{mgh}{\mu}$ [2004 : 2 M]

5.9 An incompressible fluid (kinematic viscosity, 7.4×10^{-7} m^2/s, specific gravity, 0.88) is held between two parallel plates. If the top plate is moved with a velocity of 0.5 m/s while the bottom one is held stationary, the fluid attains a linear velocity profile in the gap of 0.5 mm between these plates; the shear stress in Pascals on the surface of top plate is

(a) 0.651×10^{-3} (b) 0.651
(c) 6.51 (d) 0.651×10^{3}
[2004 : 1 M]

5.10 For air flow over a flat plate, velocity (U) and boundary layer thickness (δ) can be expressed respectively, as

$$\frac{u}{U_\infty} = \frac{3}{2}\frac{y}{\delta} - \frac{1}{2}\left(\frac{y}{\delta}\right)^3; \quad \delta = \frac{4.64x}{\sqrt{Re_x}}$$

If the free stream velocity is 2 m/s, and air has kinematic viscosity of 1.5×10^{-5} m²/s and density of 1.23 kg/m³, the wall shear stress at x = 1 m, is
(a) 2.36×10^{2} N/m² (b) 43.6×10^{-3} N/m²
(c) 4.36×10^{-3} N/m² (d) 2.18×10^{-3} N/m²
[2004 : 2 M]

5.11 The velocity profile in fully developed laminar flow in a pipe of diameter D is given by $u = u_0(1 - 4r^2/D^2)$, where r is the radial distance from the center. If the viscosity of the fluid is μ, the pressure drop across a length L of the pipe is

(a) $\frac{\mu u_0 L}{D^2}$ (b) $\frac{4\mu u_0 L}{D^2}$

(c) $\frac{8\mu u_0 L}{D^2}$ (d) $\frac{16\mu u_0 L}{D^2}$ [2006 : 2 M]

Linked Answer Questions 5.12 and 5.13: A smooth flat plate with a sharp leading edge is placed along a gas stream flowing at U = 10 m/s. The thickness of the boundary layer at section r–s is 10 mm, the breadth of the plate is 1 m (into the paper) and the density of the gas, ρ = 1.0 kg/m³. Assume that the boundary layer is thin, two-dimensional, and follows a linear velocity distribution, $u = U(y/\delta)$, at the section r–s, where y is the height from plate.

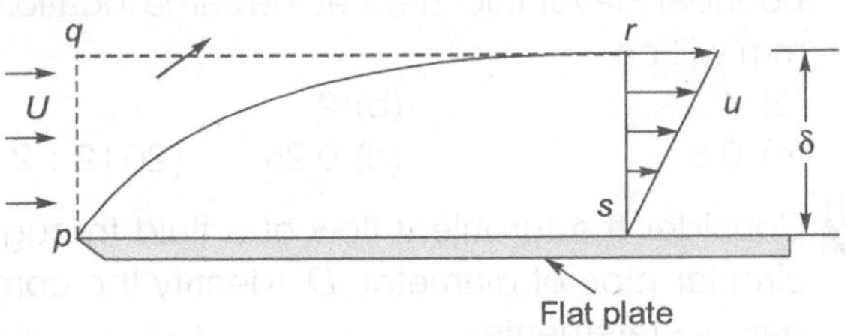

5.12 The mass flow rate (in kg/s) across the section q–r is
(a) zero (b) 0.05
(c) 0.10 (d) 0.15 [2006 : 2 M]

5.13 The integrated drag force (in N) on the plate, between p-s, is
(a) 0.67 (b) 0.33
(c) 0.17 (d) zero [2006 : 2 M]

5.14 Consider an incompressible laminar boundary layer flow over a flat plate of length L, aligned with the direction of an incoming uniform free stream. If F is the ratio of the drag force on the front half of the plate to the drag force on the rear half, then
(a) $F < 1/2$ (b) $F = 1/2$
(c) $F = 1$ (d) $F > 1$ [2007 : 1 M]

5.15 Consider steady laminar incompressible axi-symmetric fully developed viscous flow through a straight circular pipe of constant cross-sectional area at a Reynolds number of 5. The ratio of inertia force to viscous force on a fluid particle is
(a) 5 (b) 1/5
(c) 0 (d) ∞ [2007 : 1 M]

Linked Answer Questions 5.16 and 5.17: Consider a steady incompressible flow through a channel as shown below:

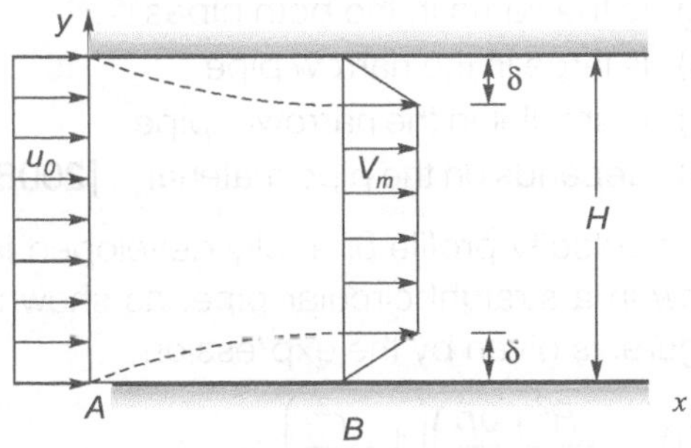

The velocity profile is uniform with a value of u_0 at the inlet section A. The velocity profile at section B downstream is

$$u = \begin{cases} V_m \frac{y}{\delta}, & 0 \le y \le \delta \\ V_m, & \delta \le y \le H - \delta \\ V_m \frac{H-y}{\delta}, & H - \delta \le y \le H \end{cases}$$

5.16 The ratio V_m/u_0 is

(a) $\frac{1}{1-2(\delta/H)}$ (b) 1

(c) $\frac{1}{1-(\delta/H)}$ (d) $\frac{1}{1+(\delta/H)}$ [2007 : 2 M]

5.17 The ratio $\frac{p_A - p_B}{\frac{1}{2}\rho u_0^2}$ (where p_A and p_B are the pressures at section A and B, respectively, and ρ is the density of the fluid) is

(a) $\dfrac{1}{[1+(\delta/H)]^2}$ (b) $\dfrac{1}{[1-(\delta/H)]^2}$

(c) $\dfrac{1}{(1-(2\delta/H))^2}-1$ (d) $\dfrac{1}{1+(\delta/H)}$

[2007 : 2 M]

5.18 Oil is being pumped through a straight pipe, the pipe length, diameter and volumetric flow rate are all doubled in a new arrangement. The pipe friction factor, however, remains constant. The ratio of pipe frictional losses in the new arrangement to that in the original configuration would be

(a) $\frac{1}{4}$ (b) $\frac{1}{2}$

(c) 2 (d) 4 [2008 : 2 M]

5.19 Two pipes of uniform section but different diameters carry water at the same flow rate. Water properties are the same in the two pipes. The Reynolds number, based on the pipe diameter

(a) is the same in the both pipes

(b) is large in the narrow pipe

(c) is smaller in the narrower pipe

(d) depends on the pipe material [2008 : 2 M]

5.20 The velocity profile of a fully developed laminar flow in a straight circular pipe, as shown in the figure, is given by the expression

$$u(r) = -\frac{R^2}{4\mu}\left(\frac{\partial p}{\partial x}\right)\left(1-\frac{r^2}{R^2}\right)$$

where $\frac{\partial p}{\partial x}$ is a constant

The average velocity of fluid in the pipe is

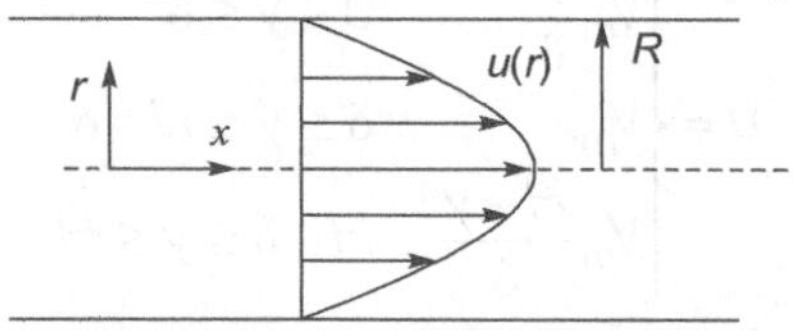

(a) $-\frac{R^2}{8\mu}\left(\frac{\partial p}{\partial x}\right)$ (b) $-\frac{R^2}{4\mu}\left(\frac{\partial p}{\partial x}\right)$

(c) $-\frac{R^2}{2\mu}\left(\frac{\partial p}{\partial x}\right)$ (d) $-\frac{R^2}{\mu}\left(\frac{\partial p}{\partial x}\right)$

[2009 : 2 M]

5.21 The pressure drop for laminar flow of a liquid in a smooth pipe at normal temperature and pressure is

(a) directly proportional to density

(b) inversely proportional to density

(c) independent of density

(d) proportional to $(\text{density})^{0.75}$ [2009 : 2 M]

5.22 A phenomenon is modeled using n dimensional variables with k primary dimensions. The number of non-dimensional variables is

(a) k (b) n

(c) $n-k$ (d) $n+k$ [2010 : 1 M]

5.23 The maximum velocity of a one-dimensional incompressible fully developed viscous flow, between two fixed parallel plates, is 6 ms^{-1}. The mean velocity (in ms^{-1}) of the flow is

(a) 2 (b) 3

(c) 4 (d) 5 [2010 : 1 M]

5.24 Match the following

P : Compressible flow	U : Reynolds number
Q: Free surface flow	V: Nusselt number
R: Boundary layer flow	W: Weber number
S: Pipe flow	X: Froude number
T: Heat convection	Y: Mach number
	Z: Skin friction coefficient

(a) P–U; Q–X; R–V; S–Z; T–W

(b) P–W; Q–X; R–Z; S–U; T–V

(c) P–Y; Q–W; R–Z; S–U; T–X

(d) P–Y; Q–X; R–Z; S–U; T–V [2010 : 2 M]

5.25 Water is flowing through a horizontal pipe of constant diameter and the flow is laminar. If the diameter of the pipe is increased by 50% keeping the volume flow rate constant, then the pressure drop in the pipe due to friction will decrease by

(a) 33% (b) 50%

(c) 70% (d) 80% [2011 : 2 M]

5.26 An incompressible fluid flows over a flat plate with zero pressure gradient. The boundary layer thickness is 1 mm at a location where the Reynolds number is 1000. If the velocity of the fluid alone is increased by a factor of 4, then the boundary layer thickness at the same location, in mm will be

(a) 4 (b) 2

(c) 0.5 (d) 0.25 [2012 : 2 M]

5.27 Consider the turbulent flow of a fluid through a circular pipe of diameter, D. Identify the correct pair of statements.

I. The fluid is well-mixed

II. The fluid is unmixed

III. $Re_D < 2300$

IV. $Re_D > 2300$

(a) I and III (b) II and IV

(c) II and III (d) I and IV

[2014 : 1 M, Set-3]

5.28 Water flows through a pipe having an inner radius of 10 mm at the rate of 36 kg/hr at 25°C. The viscosity of water at 25°C is 0.001 kg/ms. The Reynolds number of the flow is ______.

[2014 : 2 M, Set-1]

5.29 For a fully developed flow of water in a pipe having diameter 10 cm, velocity 0.1 m/s and kinematic viscosity 10^{-5} m²/s, the value of Darcy friction factor is ______.

[2014 : 2 M, Set-1]

5.30 In a simple concentric shaft-bearing arrangement, the lubricant flows in the 2 mm gap between the shaft and the bearing. The flow may be assumed to be a plane Couette flow with zero pressure gradient. The diameter of the shaft is 100 mm and its tangential speed is 10 m/s. The dynamic viscosity of the lubricant is 0.1 kg/ms. The frictional resisting force (in Newton) per 100 mm length of the bearing is ______.

[2014 : 2 M, Set-1]

5.31 Consider laminar flow of water over a flat plate of length 1 m. If the boundary layer thickness at a distance of 0.25 m from the leading edge of the plate is 8 mm, the boundary layer thickness (in mm), at a distance of 0.75 m, is ______

[2014 : 2 M, Set-2]

5.32 Couette flow is characterized by

(a) steady, incompressible, laminar flow through a straight circular pipe

(b) fully developed turbulent flow through a straight circular pipe

(c) steady, incompressible, laminar flow between two fixed parallel plates

(d) steady, incompressible, laminar flow between one fixed plate and the other moving with a constant velocity

[2015 : 1 M, Set-3]

5.33 Air (ρ = 1.2 kg/m³ and kinematic viscosity, ν = 2 × 10^{-5} m²/s) with a velocity of 2 m/s flows over the top surface of a flat plate of length 2.5 m. If the average value of friction coefficient is $C_f = \dfrac{1.328}{\sqrt{Re_x}}$, the total drag force (in N) per unit width of the plate is ______.

[2015 : 2 M, Set-1]

5.34 For a fully developed laminar flow of water (dynamic viscosity 0.001 Pa-s) through a pipe radius 5 cm, the axial pressure gradient is – 10 Pa/m. The magnitude of axial velocity (in m/s) at a radial location of 0.2 cm is ______.

[2015 : 2 M, Set-2]

5.35 The instantaneous stream wise velocity of a turbulent flow is given as follows:

$$u(x, y, z, t) = \bar{u}(x, y, z) + u'(x, y, z, t)$$

The time average of the fluctuating velocity $u'(x, y, z, t)$ is

(a) $\dfrac{u'}{2}$ (b) $-\dfrac{\bar{u}}{2}$

(c) zero (d) $\dfrac{\bar{u}}{2}$ [2016 : 1 M, Set-1]

5.36 Oil (kinematic viscosity, ν_{oil} = 1.0 × 10^{-5} m²/s) flows through a pipe of 0.5 m diameter with a velocity of 10 m/s. Water (kinematic viscosity, ν_w = 0.89 × 10^{-6} m²/s) is flowing through a model pipe of diameter 20 mm. For satisfying the dynamic similarity, the velocity of water (in m/s) is ______.

[2016 : 2 M, Set-1]

5.37 A steady laminar boundary layer is formed over a flat plate as shown in the figure. The free stream velocity of the fluid is U_0. The velocity profile at the inlet *a-b* is uniform, while that at a downstream location *c-d* is given by $u = U_0\left[2\left(\dfrac{y}{\delta}\right) - \left(\dfrac{y}{\delta}\right)^2\right]$.

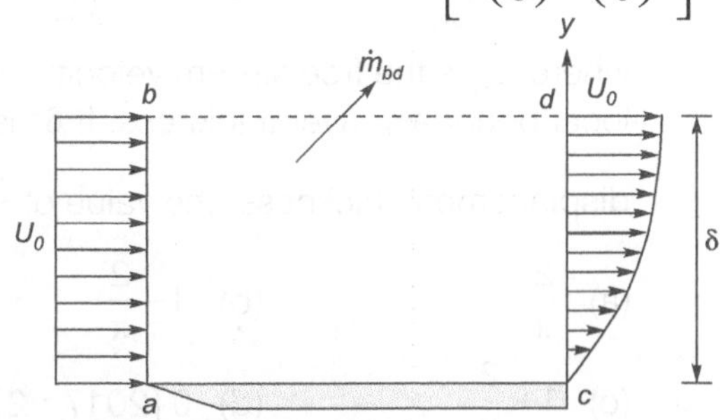

The ratio of the mass flow rate, $\dot{m}_{bd}$, leaving through the horizontal section *b-d* to that entering through the vertical section *a-b* is ______

[2016 : 2 M, Set-1]

5.38 Consider fluid flow between two infinite horizontal plates which are parallel (the gap between them being 50 mm). The top plate is sliding parallel to the stationary bottom plate at a speed of 3 m/s. The flow between the plates is solely due to the motion of the top plate. The force per unit area (magnitude) required to maintain the bottom plate stationary is ______ N/m².

Viscosity of the fluid $\mu = 0.44$ kg/ms and density $\rho = 888$ kg/m^3.

[2016 : 2 M, Set-2]

5.39 Consider a fully developed steady laminar flow of an incompressible fluid with viscosity μ through a circular pipe of radius R. Given that the velocity at a radial location of $R/2$ from the centerline of the pipe is U_1, the shear stress at the wall is $K\mu U_1/R$, where K is ________.

[2016 : 2 M, Set-3]

5.40 For steady flow of a viscous incompressible fluid through pipe of constant of diameter, the average velocity in the fully developed region is constant. Which one of the following statements about the average velocity in the developing region is TRUE?

(a) It increases until the flow is fully developed.
(b) It is constant and is equal to the average velocity in the fully developed region.
(c) It decreases until the flow is fully developed.
(d) It is constant but is always lower than the average velocity in the fully developed region.

[2017 : 1 M, Set-1]

5.41 The velocity profile inside the boundary layer for flow over a plate is given as $\frac{u}{U_\infty} = \sin\left(\frac{\pi}{2}\frac{y}{\delta}\right)$, where U_∞ is the free stream velocity and δ is the local boundary layer thickness. If δ^* is the local displacement thickness, the value of $\frac{\delta^*}{\delta}$ is

(a) $\frac{2}{\pi}$ (b) $1-\frac{2}{\pi}$
(c) $1+\frac{2}{\pi}$ (d) 0 [2017 : 2 M, Set-1]

5.42 Consider a laminar flow at zero incidence over a flat plate. The shear stress at the wall is denoted by τ_w. The axial position x_1 and x_2 on the plate are measured from the leading edge in the direction of flow. If $x_2 > x_1$, then

(a) $\tau_w|_{x_1} = \tau_w|_{x_2} = 0$ (b) $\tau_w|_{x_1} = \tau_w|_{x_2} \neq 0$
(c) $\tau_w|_{x_1} > \tau_w|_{x_2}$ (d) $\tau_w|_{x_1} < \tau_w|_{x_2}$

[2017 : 1 M, Set-2]

5.43 For the laminar flow of water over a sphere, the drag coefficient C_F is defined as $C_F = F/(\rho U^2 D^2)$, where F is the drag force, ρ is the fluid density, U is the fluid velocity and D is the diameter of the sphere. The density of water is 1000 kg/m^3. When the diameter of the sphere is 100 mm and the fluid velocity is 2 m/s, the drag coefficient is 0.5. If water now flows over another sphere of diameter 200 mm under dynamically similar conditions, the drag force (in N) on this sphere is ________.

[2017 : 2 M, Set-2]

5.44 A solid block of 2.0 kg mass slides steadily at a velocity V along a vertical wall as shown in the figure below. A thin oil film of thickness $h = 0.15$ mm provides lubrication between the block and the wall. The surface area of the face of the block in contact with the oil film is 0.04 m^2. The velocity distribution within the oil film gap is linear as shown in the figure. Take dynamic viscosity of oil as 7×10^{-3} Pa-s and acceleration due to gravity as 10 m/s^2. Neglect weight of the oil. The terminal velocity V (in m/s) of the block is ________ (correct to one decimal place).

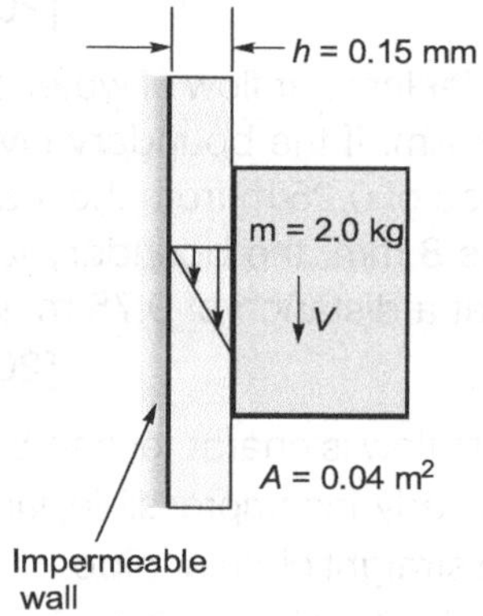

[2018 : 2 M, Set-1]

5.45 The viscous laminar flow of air over a flat plate results in the formation of a boundary layer. The boundary layer thickness at the end of the plate of length L is δ_L. When the plate length is increased to twice of its original length, the percentage change in laminar boundary layer thickness at the end of the plate (with respect to δ_L) is ______ (correct to two decimal places).

[2018 : 1 M, Set-2]

5.46 Water enters a circular pipe of length $L = 5.0$ m and diameter $D = 0.20$ m with Reynolds number $Re_D = 500$. The velocity profile at the inlet of the pipe is uniform while it is parabolic at the exit. The Reynolds number at the exit of the pipe is ______.

[2019 : 1 M, Set-2]

5.47 The aerodynamic drag on a sports car depends on its shape. The car has a drag coefficient of 0.1 with the windows and the roof closed. With the windows and the roof open, the drag coefficient becomes 0.8. The car travels at 44 km/h with the windows and roof closed for the same amount of power needed to overcome the aerodynamic drag, the speed of the car with the windows and roof open (round off to two decimal places), is_______ km/h. (density of air and the frontal area may be assumed to be constant).

[2019 : 2 M, Set-2]

5.48 Froude number is the ratio of
(a) inertia forces to viscous forces
(b) buoyancy forces to viscous forces
(c) buoyancy forces to inertia forces
(d) inertia forces to gravity forces

[2020 : 1 M, Set-1]

5.49 Consider fully developed, steady state incompressible laminar flow of a viscous fluid between two large parallel horizontal plates. The bottom plate is fixed and the top plate moves with a constant velocity of U = 4 m/s. Separation between the plates is 5 mm. There is no pressure gradient in the direction of flow. The density of fluid is 800 kg/m^3, and the kinematic viscosity is 1.25 $\times 10^{-4}$ m^2/s. The average shear stress in the fluid is ________ Pa. [Round off to the nearest integer]

[2021 : 1 M, Set-1]

5.50 A high velocity water jet of cross section area = 0.01 m^2 and velocity = 35 m/s enters a pipe filled with stagnant water. The diameter of the pipe is 0.32 m. This high velocity water jet entrains additional water from the pipe and the total water leaves the pipe with a velocity 6 m/s as shown in the figure.

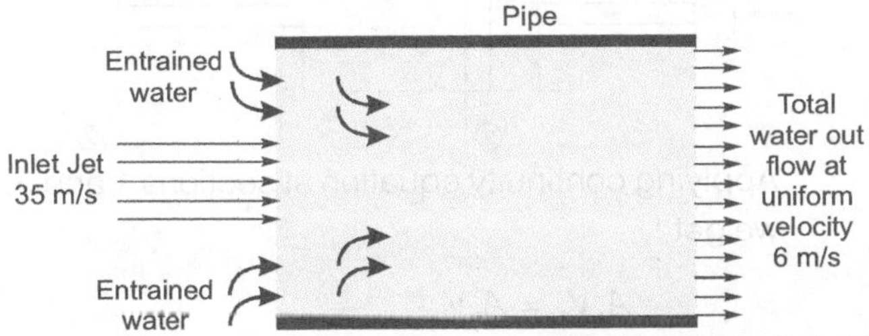

[2021 : 2 M, Set-2]

5.51 Which of the following is responsible for eddy viscosity (or turbulent viscosity) in a turbulent boundary layer on a flat plate?
(a) Nikuradse stresses
(b) Boussinesq stresses
(c) Reynolds stresses
(d) Prandtl stresses

[2021 : 1 M, Set-2]

5.52 The steady velocity field in an inviscid fluid of density 1.5 is given to be

$$\vec{V} = (y^2 - x^2)\hat{i} + (2xy)\hat{j}$$

Neglecting body forces, the pressure gradient at (x = 1, y = 1) is ________.

(a) $10\hat{j}$ (b) $20\hat{i}$
(c) $-6\hat{i} - 6\hat{j}$ (d) $-4\hat{i} - 4\hat{j}$

[2022 : 2 M, Set-2]

5.53 Air (density = 1.2 kg/m^3, kinematic viscosity = 1.5 × 10^{-5} m^2/s) flows over a flat plate with a free-stream velocity of 2 m/s. The wall shear stress at a location 15 mm from the leading edge is τ_w. What is the wall shear stress at a location 30 mm from the leading edge?

(a) $\frac{\tau_w}{2}$ (b) $\sqrt{2}\tau_w$
(c) $2\tau_w$ (d) $\frac{\tau_w}{\sqrt{2}}$

[2023 : 1 Mark]

5.54 A liquid fills a horizontal capillary tube whose one end is dipped in a large pool of the liquid. Experiments show that the distance L travelled by the liquid meniscus inside the capillary in time t is given by

$$L = k\gamma^{\alpha}R^{b}\mu^{c}\sqrt{t},$$

where γ is the surface tension, R is the inner radius of the capillary, and μ is the dynamic viscosity of the liquid. If k is a dimensionless constant, then the exponent a is ______ (rounded off to 1 decimal place)

[2024 : 2 M]

■■■■

Answers Viscous, Turbulent Flow and Boundary Layer Theory

5.1 (Mean velocity) 5.2 (c) 5.3 (b) 5.4 (c) 5.5 (d) 5.6 (b) 5.7 (c)
5.8 (c) 5.9 (b) 5.10 (c) 5.11 (d) 5.12 (b) 5.13 (c) 5.14 (d) 5.15 (a)
5.16 (c) 5.17 (b) 5.18 (a) 5.19 (b) 5.20 (a) 5.21 (c) 5.22 (c) 5.23 (c)
5.24 (d) 5.25 (d) 5.26 (c) 5.27 (d) 5.28 (636.94) 5.29 (0.064) 5.30 (15.7)
5.31 (13.8564) 5.32. (d) 5.33 (0.0159) 5.34 (6.24) 5.35 (c) 5.36 (22.25) 5.37 (0.333)
5.38 (26.4) 5.39 (2.66) 5.40 (b) 5.41 (b) 5.42 (c) 5.43 (20) 5.44 (10.714)
5.45 (41.42) 5.46 (500) 5.47 (22) 5.48 (d) 5.49 (80) 5.50 (132.55) 5.51 (c) 5.52 (c)
5.53 (d) 5.54 (0.5)

Explanations Viscous, Turbulent Flow and Boundary Layer Theory

5.1 Sol.

Hagen-Poiseuille equation

Head loss: $$h_f = \frac{32\mu\bar{u}L}{\rho g D^2}$$

$$h_f \propto \bar{u}$$

where h_f = Head loss due to friction

$\bar{u}$ = Mean velocity

5.2 (c)

Loss in pressure in due to head loss due to friction,

$$p_1 - p_2 = \frac{32\mu\bar{u}L}{D^2}$$

5.3 (b)

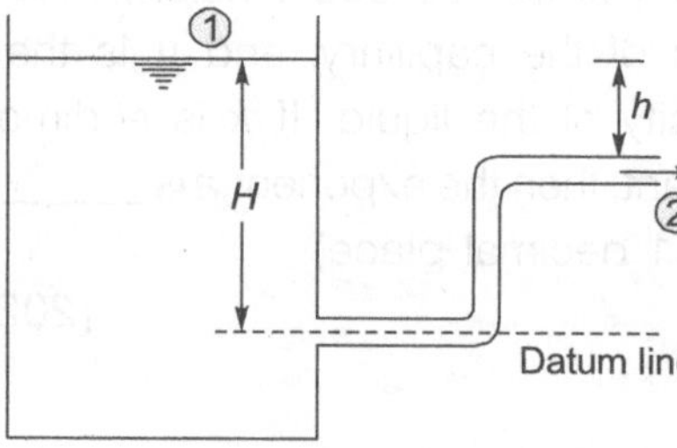

Applying Bernoulli's between points (1) and (2), we get

$$\frac{p_1}{\rho g} + \frac{V_1^2}{2g} + z_1 = \frac{p_2}{\rho g} + \frac{V_2^2}{2g} + z_2$$

where,

$p_1 = p_2 = 0$, $V_1 = 0$, $z_1 = H$, $z_2 = H - h$

$$\therefore\ 0 + 0 + H = 0 + \frac{V_2^2}{2g} + H - h$$

or $$V_2^2 = 2gh$$

or $$V_2 = \sqrt{2gh}$$

5.4 (c)

Positive pressure gradient.

5.5 (d)

Boundary layer thickness,

$$\delta = \frac{5x}{\sqrt{Re_x}} \text{ for laminar boundary layer}$$

where $Re_x = \frac{Vx}{\nu}$

$$\therefore\ \delta = \frac{5x}{\sqrt{Vx/\nu}} = \frac{5\sqrt{x}}{\sqrt{V/\nu}} \propto x^{1/2}$$

5.6 (b)

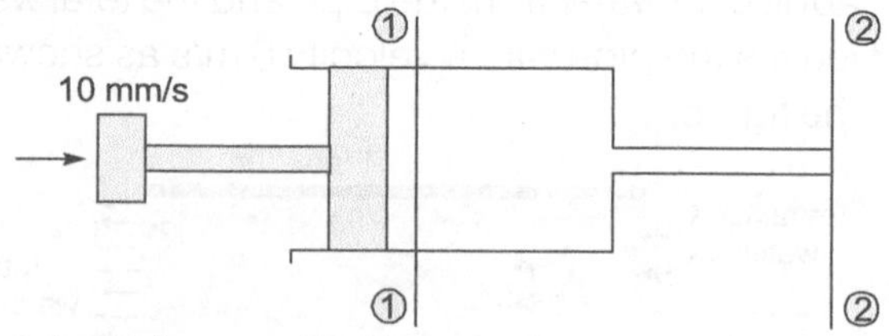

Applying continuity equation at sections 1 and 2, we get

$$A_1V_1 = A_2V_2$$

$$V_2 = \left(\frac{A_1}{A_2}\right)V_1 = \frac{\frac{\pi}{4}\times(0.01)^2}{\frac{\pi}{4}\times(0.001)^2}\times 0.01$$

$$= 1 \text{ m/s}$$

Applying Bernoulli's equation at sections 1 and 2,

we get

$$\frac{p_1}{\rho g}+\frac{V_1^2}{2g}+z_1=\frac{p_2}{\rho g}+\frac{V_2^2}{2g}+z_2$$

Since $z_1 = z_2$

and taking gauge pressure

$$p_2 = 0 \quad \text{(atmospheric)}$$

$$\therefore \quad \frac{p_1}{\rho g} = \frac{V_2^2 - V_1^2}{2g}$$

$$p_1 = \frac{\rho}{2}\left(V_2^2-V_1^2\right)$$

$$= \frac{1000}{2}\left[(1)^2-(0.01)^2\right]$$

$$= 499.95 \text{ N/m}^2$$

Force required on plunger,

$$= p_1 \times A_1 = 499.95 \times \frac{\pi}{4} \times (0.01)^2$$

$$= \mathbf{0.039\ N \approx 0.04\ N}$$

5.7 (c)

$$Re = \frac{\rho V_2 d_2}{\mu} = \frac{1000 \times 1 \times 0.001}{1 \times 10^{-3}}$$

$$= 1000 \quad \text{(flow is laminar)}$$

Now Darcy's friction factor,

$$f = \frac{64}{\text{Re}} = \frac{64}{1000} = 0.064$$

Head loss in needle,

$$h_f = \frac{fLV_2^2}{2gd_2} = \frac{0.064 \times 0.1 \times (1)^2}{2 \times 9.81 \times 0.001}$$

$$= 0.3262 \text{ m of water}$$

Applying Bernoulli's equation at points 1 and 2, we get

$$\frac{p_1}{\rho g}+\frac{V_1^2}{2g}+z_1 = \frac{p_2}{\rho g}+\frac{V_2^2}{2g}+z_2+h_f$$

Since $z_1 = z_2$

and $p_2 = 0$ (atmospheric)

$$\frac{p_1}{\rho g} = \left(\frac{V_2^2 - V_1^2}{2g}\right) + h_f$$

$$p_1 = \frac{\rho}{2}\left(V_2^2 - V_1^2\right) + \rho g h_f$$

$$= \frac{1000}{2}\left[(1)^2 - (0.01)^2\right] + 1000 \times 9.81 \times 0.3262$$

$$= 3699.97 \text{ N/m}^2$$

Force required on plunger,

$$= p_1 \times A_1 = 3699.97 \times \frac{\pi}{4} \times (0.01)^2$$

$$= \mathbf{0.29\ N \approx 0.3\ N}$$

5.8 (c)

Shear stress: $\tau = \mu\left(\frac{du}{dy}\right)$

where $du = V - 0 = V$

$dy = h$

$$\therefore \quad \tau = \mu\frac{V}{h} \quad \text{...(i)}$$

also $\tau = \frac{F}{A}$

where $F = mg$

$A = L^2$

$$\therefore \quad \tau = \frac{mg}{L^2} \quad \text{...(ii)}$$

Equating Eqs. (*i*) and (*ii*), we get

$$\mu\frac{V}{h} = \frac{mg}{L^2} \text{ or } V = \frac{mg\,h}{\mu L^2}$$

5.9 (b)

Given data:

$\nu = 7.4 \times 10^{-7}$ m²/s

$S = 0.88$

$\therefore$ $\rho = 1000 \times 0.88 = 880$ kg/m³

$U = 0.5$ m/s

$h = 0.5$ mm $= 0.0005$ m

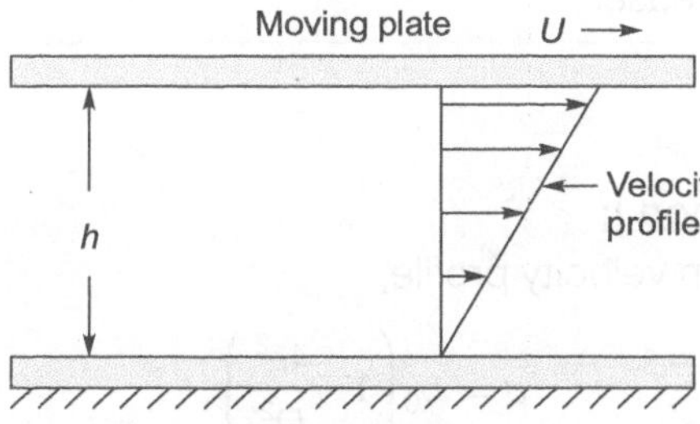

Dynamic viscosity,

$$\mu = \rho\nu = 880 \times 7.4 \times 10^{-7}$$

$$= 6.512 \times 10^{-3} \text{ Ns/m}^2$$

Shear stress: $\tau = \mu\frac{du}{dy}$

where $du = U - 0 = 0.5$ m/s

$dy = h = 0.0005$ m

$$\therefore \quad \tau = \frac{6.512 \times 10^{-3} \times 0.5}{0.0005} = \mathbf{0.651\ Pa}$$

5.10 (c)

Reynold number,

$$\text{Re}_x = \frac{u_\infty x}{\nu} = \frac{2\times 1}{1.5\times 10^{-5}} = 1.33\times 10^5$$

$$\delta = \frac{4.64x}{\sqrt{\text{Re}_x}} = \frac{4.64\times 1}{\sqrt{1.33\times 10^5}} = 0.0127$$

Now, $\frac{du}{dy} = u_\infty\left[\frac{3}{2}\cdot\frac{1}{\delta} - \frac{3}{2}\left(\frac{y^2}{\delta^3}\right)\right]$

$$\left.\frac{du}{dy}\right|_{y=0} = \frac{3u_\infty}{2\delta}$$

Now, shear stress,

$$\tau_0 = \mu\left(\frac{du}{dy}\right)_{y=0} = \mu\times\frac{3u_\infty}{2\delta}$$

$$= \frac{3u_\infty\times\nu\times\rho}{2\delta} \qquad \because \mu = \nu\rho$$

$$= \frac{3\times 2\times 1.5\times 10^{-5}\times 1.23}{2\times 0.0127}$$

$$= \mathbf{4.36\times 10^{-3}\ N/m^2}$$

POINTS TO REMEMBER

It is important to note that, shear stress for laminar

$$\tau = \mu\frac{3}{2\delta}u_\infty \quad \text{i.e.} \quad \tau\propto\frac{1}{\sqrt{x}}$$

So, as the distance from the leading edge (x) increases, shear stress on the plate surface decreases.

5.11 (d)

Method I:

Given velocity profile,

$$u = u_0\left(1-\frac{4r^2}{D^2}\right)$$

At $r = 0$, $u = u_{max}$

$$\therefore \quad u_{max} = u_0$$

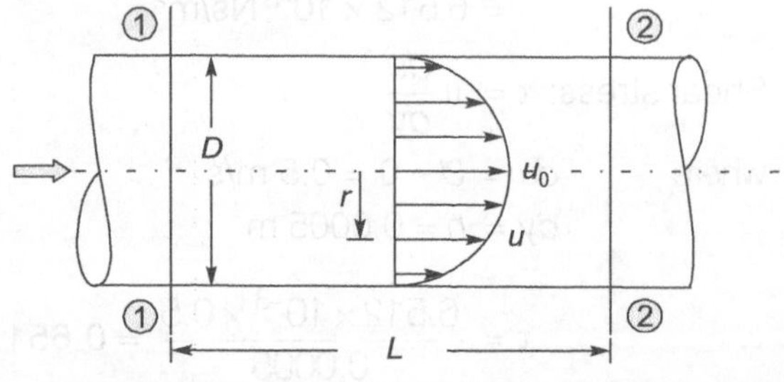

We know that the pressure drop across a length L of the pipe,

$$p_1 - p_2 = \frac{32\mu\bar{u}L}{D^2}$$

where $\bar{u}$ = Average velocity

$$\bar{u} = \frac{u_{max}}{2} = \frac{u_0}{2}$$

$$\therefore \quad p_1 - p_2 = \frac{32\mu u_0 L}{2\times D^2} = \frac{16\mu u_0 L}{D^2}$$

Method II:

For the laminar flow through circular pipe,

Shear stress, $\tau = -\frac{r}{2}\left(\frac{\partial p}{\partial x}\right)$

$$\mu\frac{\partial u}{\partial r} = -\frac{r}{2}\frac{\Delta p}{L}$$

$$-\frac{r}{2}\frac{\Delta p}{L} = \mu\frac{\partial}{\partial r}\left(u_0\left(1-\frac{4r^2}{D^2}\right)\right)$$

$$\because \quad \frac{r}{2}\frac{\Delta p}{L} = m\left(\frac{8u_0 r}{D^2}\right)$$

$$\frac{r}{2}\frac{\Delta p}{L} = \frac{8\mu u_0 r}{D^2}$$

Pressure drop,

$$\Delta p = \frac{16\mu u_0 L}{D^2}$$

5.12 (b)

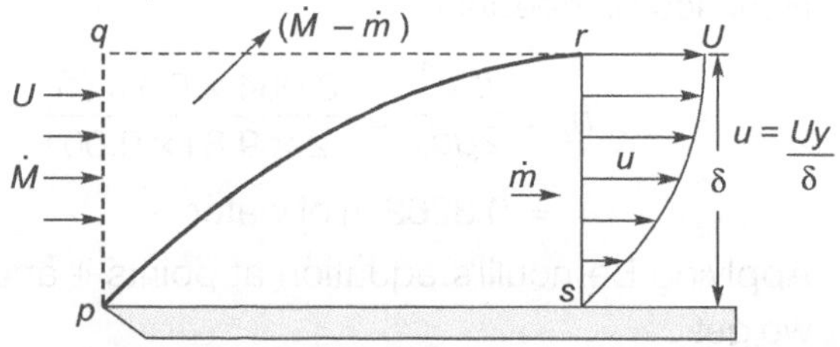

Velocity of free stream,

$$U = 10\ \text{m/s}$$

Boundary layer thickness at section r-s,

$$\delta = 10\ \text{mm} = 0.01\ \text{m}$$

Width of the plate: $b = 1$ m

Density of the gas: $\rho = 1$ kg/m³

Linear velocity distribution,

$$u = U\frac{y}{\delta}$$

Mass flow rate entering through section pq

$$\dot{M} = \rho AU = \rho b\delta U$$

$$= 1\times 1\times 0.01\times 10 = 0.1\ \text{kg/s}$$

Mass flow rate entering through section *s-r*.

$$\dot{m} = \int_0^\delta \rho b dy u = \rho b \int_0^\delta u dy = \rho b \int_0^\delta \frac{Uy}{\delta} dy \qquad \left(\because\ u = \frac{Uy}{\delta}\right)$$

$$= \frac{\rho b U}{\delta}\int_0^\delta y dy = \frac{\rho b U}{\delta}\left[\frac{y^2}{2}\right]_0^\delta$$

$$= \frac{\rho b U}{\delta} \times \frac{\delta^2}{2} = \frac{\rho b U \delta}{2}$$

$$= \frac{1 \times 1 \times 10 \times 0.01}{2} = \mathbf{0.05\ kg/s}$$

Mass flow rate through section *q-r*,

$$\dot{M} - \dot{m} = 0.1 - 0.05 = \mathbf{0.05\ kg/s}$$

5.13 (c)

Method I:

Drag force on plate = rate of change of momentum of the flowing fluid.

$$\text{Momentum thickness: } \theta = \int_0^\delta \frac{u}{U}\left(1 - \frac{u}{U}\right) dy$$

$$\theta = \int_0^\delta \frac{u}{U}\left(1 - \frac{u}{U}\right) dy = \int_0^\delta \left(\frac{y}{\delta} - \frac{y^2}{\delta^2}\right) dy$$

$$= \left.\frac{y^2}{2\delta} - \frac{y^3}{3\delta^2}\right|_0^\delta = \frac{\delta}{2} - \frac{\delta}{3} = \frac{\delta}{6}$$

$\therefore$ Force on plate per unit width, $F = \rho U^2 \times (\theta \times 1)$

Here, $U = 10$ m/s

$$\therefore \quad F = 1 \times (10)^2 \times \left(\frac{10 \times 10^{-3}}{6} \times 1\right)$$

$$= \mathbf{0.1666\ N}$$

Method II:

Integrated drag force on the plate between, (*p-s*),

F_D = Rate of change in momentum of volume (*pqrs*)

Momentum rate entering through (*p-q*) per sec

$$= \dot{m}_{pq} \times \text{velocity} = 0.1 \times U$$

$$= 0.1 \times 10 = 1\ \text{N}$$

Momentum leaving through (*q-r*) per sec

$$= \dot{m}_{qr} \times \text{velocity} = 0.05 \times 10$$

$$= 0.5\ \text{N}$$

Momentum leaving through (*r-s*) per sec

$$= \int_0^\delta \rho \times dy \times 1 \times u \times u$$

$$= \int_0^\delta \rho u^2 dy = \int \rho \frac{U^2 y^2}{\delta^2} dy \qquad \left[\text{where, } u = U\left(\frac{y}{\delta}\right)\right]$$

$$= \frac{\rho U^2 \delta}{3} = \frac{1 \times 10^2 \times 0.1}{3} = \frac{1}{3}$$

$$= 0.333\ \text{N}$$

So drag force,

$$F_D = 1 - (0.5 + 0.333)$$

$$= 0.1667 = 0.17\ \text{N}$$

5.14 (d)

Drag force,

$$F_D = C_f \times \frac{1}{2}\rho A V^2 = \frac{1.328}{\sqrt{Re_L}} \times \frac{1}{2}\rho A V^2$$

$$= \frac{1.328}{\sqrt{\dfrac{\rho L V}{\mu}}} \times \frac{1}{2}\rho \times A \times L \times V^2$$

i.e. $F_D \propto \sqrt{L}$

Now drag force on front half,

$$F_{D/2} \propto \sqrt{\frac{L}{2}}$$

$$\therefore \quad F_{D/2} = \frac{F_D}{\sqrt{2}}$$

Drag force on rear half,

$$F'_{D/2} = F_D - F_{D/2} = \left(1 - \frac{1}{\sqrt{2}}\right) F_D$$

$$\text{Now } F = \frac{F_{D/2}}{F'_{D/2}} = \frac{\dfrac{F_D}{\sqrt{2}}}{\left(1 - \dfrac{1}{\sqrt{2}}\right) F_D} = \frac{1}{\sqrt{2} - 1} > 1$$

$\therefore \quad F > 1$

5.15 (a)

Reynolds number,

$$Re = \frac{\text{Inertia force}}{\text{Viscous force}} = 5$$

5.16 (c)

Applying continuity equation at sections A and B, mass flow rate at inlet section A = Mass rate at section B.

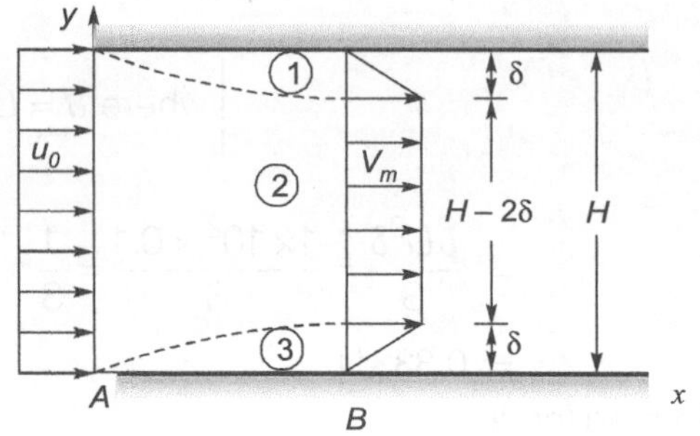

$$m_A = m_B$$

Hence, per unit width of the channel, we get

$$\rho u_0 H \times 1 = \rho V_m(H-2\delta)\times 1 + 2\rho\int_0^\delta u\,dy \times 1$$

$$u_0 H = V_m(H-2\delta) + 2\int_0^\delta \frac{V_m y}{\delta}dy$$

$$\left(\because\ u = \frac{V_m y}{\delta}\right)$$

$$u_0 H = V_m(H-2\delta) + 2\frac{V_m}{\delta}\left[\frac{y^2}{2}\right]_0^\delta$$

$$u_0 H = V_m(H-2\delta) + 2\frac{V_m}{\delta}\times\frac{\delta^2}{2}$$

$$u_0 H = V_m(H-2\delta) + V_m\delta$$

$$u_0 H = V_m H - 2V_m\delta + V_m\delta$$

$$u_0 H = V_m H - V_m\delta$$

$$u_0 H = V_m(H-\delta)$$

or $$\frac{V_m}{u_0} = \frac{H}{H-\delta} = \frac{1}{1-(\delta/H)}$$

POINTS TO REMEMBER

Here, due to symmetry of velocity profile, we can say that,

$$\dot m_1 = \dot m_3$$

$$= \int_0^\delta \rho u(dy\times 1)$$

$$= \int_0^\delta \rho V_m\frac{y}{\delta}dy = \frac{\rho V_m\delta}{2}$$

5.17 (b)

Applying Bernoulli's equation at sections A and B, we get

$$\frac{p_A}{\rho g} + \frac{u_o^2}{2g} + z_A = \frac{p_B}{\rho g} + \frac{V_m^2}{2g} + z_B$$

$$\frac{p_A}{\rho g} + \frac{u_o^2}{2g} = \frac{p_B}{\rho g} + \frac{V_m^2}{2g} \qquad \because\ z_A = z_B$$

or $$\frac{p_A}{\rho g} - \frac{p_B}{\rho g} = \frac{V_m^2}{2g} - \frac{u_o^2}{2g}$$

or $$p_A - p_B = \frac{\rho V_m^2}{2} - \frac{\rho u_o^2}{2}$$

Now, the ratio

$$\frac{p_A - p_B}{\frac{1}{2}\rho u_o^2} = \frac{\frac{\rho V_m^2}{2} - \frac{\rho u_o^2}{2}}{\frac{1}{2}\rho u_o^2}$$

$$= \frac{V_m^2 - u_o^2}{u_o^2} = \left(\frac{V_m}{u_o}\right)^2 - 1$$

Substituting the value of $\frac{V_m}{u_o} = \frac{1}{1-(\delta/H)}$

From Q.5.26, we get

$$\frac{p_A - p_B}{\frac{1}{2}\rho u_o^2} = \frac{1}{[1-(\delta/H)]^2} - 1$$

5.18 (a)

For original pipe	For new pipe
$l_1 = l$	$l_2 = 2l$
$d_1 = d$	$d_2 = 2d$
$Q_1 = Q$	$Q_2 = 2Q$

Head loss due to friction,

$$h_f = \frac{flV^2}{2gd}$$

where, f = Friction factor, l = Length of pipe, V = Average velocity, d = Diameter of pipe

Discharge: $Q = AV = \frac{\pi}{4}d^2V$

or $V = \frac{4Q}{\pi d^2}$

$$h_f = \frac{fl}{2gd} \times \left(\frac{4Q}{\pi d^2}\right)^2 = \frac{8flQ^2}{\pi^2 g d^5}$$

For original pipe,

$$h_{f_1} = \frac{8fl_1Q_1^2}{\pi^2 g d_1^5} = \frac{8flQ^2}{\pi^2 g d^5}$$

For new pipe,

$$h_{f_2} = \frac{8fl_1Q_2^2}{\pi^2 g d_2^5} = \frac{8f \times 2l \times 4Q^2}{\pi^2 g \times 2^5 \times d^5} = \frac{2flQ^2}{\pi^2 g d^5}$$

$$\frac{h_{f_1}}{h_{f_2}} = \frac{2flQ^2}{\pi^2 g d^5} \times \frac{\pi^2 g d^5}{2flQ^2} = \frac{1}{4}$$

5.19 (b)

Reynolds number,

$$Re = \frac{Vd}{\nu}$$

Discharge: $Q = AV = \frac{\pi}{4}d^2V$

or $V = \frac{4Q}{\pi d^2}$

$\therefore$ Reynolds number,

$$Re = \frac{4Q}{\pi d^2} \times \frac{d}{\nu}$$

$$Re = \frac{4Q}{\pi \nu d} \quad \ldots(i)$$

Equation (i) indicates, $Re \propto \frac{1}{d}$

Hence, the Reynolds number is large in the narrow pipe.

5.20 (a)

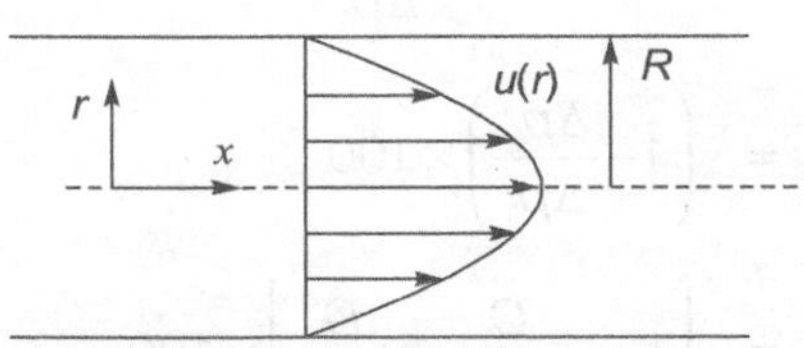

The velocity profile of a fully developed laminar flow in a straight circular pipe, as shown in figure, is given by the expression

$$u(r) = -\frac{R^2}{4\mu}\left(\frac{dp}{dx}\right)\left[1 - \frac{r^2}{R^2}\right]$$

$$= U_{max}\left[1 - \frac{r^2}{R^2}\right]$$

where $U_{max} = -\frac{R^2}{4\mu}\left(\frac{dp}{dx}\right)$

where $\frac{dp}{dx}$ is a constant

The average velocity of fluid in the pipe,

$$V_{avg} = \frac{U_{max}}{2} = -\frac{R^2}{8\mu}\left(\frac{dp}{dx}\right)$$

Alternatively

Given velocity profile expression,

$$u(r) = \frac{-R^2}{4\mu}\left(\frac{dp}{dx}\right)\left[1 - \frac{r^2}{R^2}\right]$$

Consider a small circular ring element of radius r and thickness dr as shown in figure.

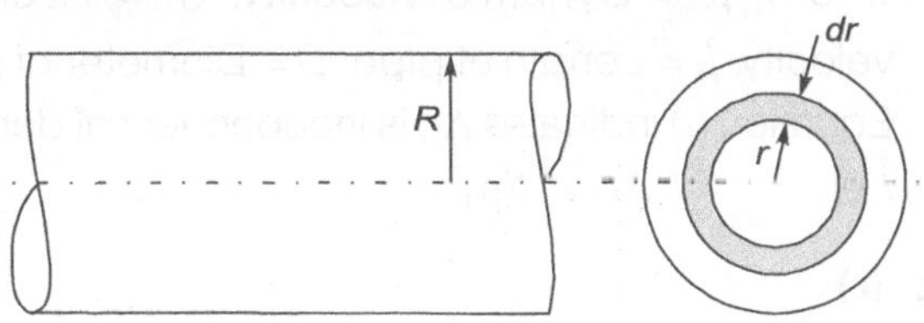

The fluid flowing per second through small elementary ring,

dQ = Velocity at radius r × Area of a small ring element)

$$= u \times 2\pi r dr$$

$$= \frac{-1}{4\mu}R^2\left(\frac{dp}{dx}\right)\left[1 - \frac{r^2}{R^2}\right] \times 2\pi r dr$$

$$= \frac{-\pi}{2\mu}R^2\left(\frac{dp}{dx}\right)\left[r - \frac{r^3}{R^2}\right]dr$$

Net discharge flow through pipe,

$$Q = \int_0^R \frac{-\pi}{2\mu}R^2\left(\frac{dp}{dx}\right)\left[r - \frac{r^3}{R^2}\right]dr$$

$$= \frac{-\pi}{2\mu}R^2\left(\frac{dp}{dx}\right)\int_0^R\left[r - \frac{r^3}{R^2}\right]dr$$

$$= \frac{-\pi}{2\mu}R^2\left(\frac{dp}{dx}\right)\left[\frac{r^2}{2} - \frac{r^4}{4R^2}\right]_0^R$$

$$= \frac{-\pi}{2\mu}R^2\left(\frac{dp}{dx}\right)\left[\frac{R^2}{2} - \frac{R^4}{4R^2}\right]$$

$$= \frac{-\pi}{2\mu}R^2\left(\frac{dp}{dx}\right)\left(\frac{R^2}{2}-\frac{R^2}{4}\right)$$

$$= \frac{-\pi}{2\mu}R^2\left(\frac{dp}{dx}\right)\times\frac{R^2}{4} = \frac{-\pi}{8\mu}R^4\left(\frac{dp}{dx}\right)$$

∴ Average velocity,

$$\bar{u} = \frac{Q}{\pi R^2} = \frac{\frac{-\pi}{8\mu}R^4\left(\frac{dp}{dx}\right)}{\pi R^2}$$

$$= \frac{-1}{8\mu}\left(\frac{dp}{dx}\right)R^2$$

5.21 (c)

Pressure drop,

$$\Delta p = \frac{32\mu\bar{u}L}{D^2} \quad ...(i)$$

where, μ = Dynamic viscosity, $\bar{u}$ = Average velocity, L = Length of pipe, D = Diameter of pipe

Equation (*i*) indicates Δp is independent of density.

i.e., $\Delta p \neq f(\rho)$

5.22 (c)

According to Buckingham's π-theorem, if there are "*n*" number of total variables (independent and dependent) in any functional relationship and "k" number of fundamental quantities, there the given system can be grouped into "*n–k*" non-dimensional π-terms.

5.23 (c)

$$U_{max} = 6 \text{ m/s}$$

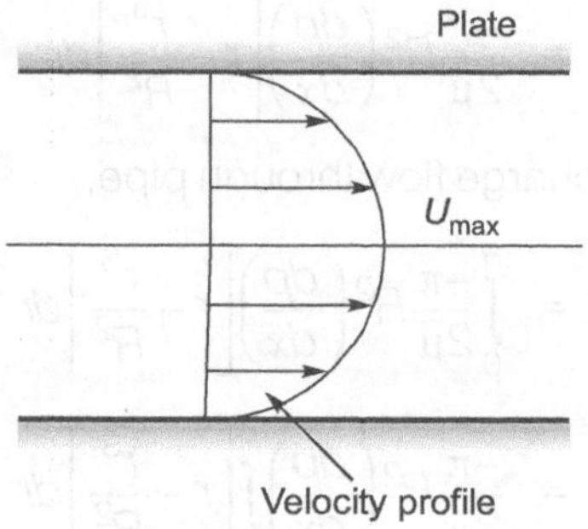

Since,

$$\frac{U_{max}}{U_{mean}} = \frac{3}{2}$$

$$\frac{6}{U_{mean}} = \frac{3}{2}$$

$$U_{mean} = 4 \text{ m/s}$$

POINTS TO REMEMBER

- In case of flow between two fixed paralle plates.
$$\frac{U_{max}}{U_{avg}} = \frac{3}{2}$$
- In case of flow in the circular pipe,
$$\frac{U_{max}}{U_{avg}} = 2$$

5.25 (d)

Pressure drop in the pipe,

$$\Delta p = \frac{32\mu\bar{u}L}{D^2}$$

Discharge: $Q = AV = \frac{\pi}{4}d^2V$

or $V = \frac{4Q}{\pi d^2}$

∴ $\Delta p = \frac{32\mu}{D^2}\times\frac{4Q}{\pi D^2}L$

$$\Delta p = \frac{128\mu QL}{\pi D^4}$$

or $\Delta p = \frac{C}{D^4}$

where $C = \frac{128\mu QL}{\pi}$

$$\Delta p_1 = \frac{C}{D_1^4} = \frac{C}{D^4} \text{ where } D_1 = D$$

and $\Delta p_2 = \frac{C}{D_2^4} = \frac{C}{(1.5D)^4}$ where $D_2 = 1.5D$

% change in pressure drop,

$$= \frac{\Delta p_1 - \Delta p_2}{\Delta p_1}\times 100$$

$$= \left(1-\frac{\Delta p_2}{\Delta p_1}\right)\times 100$$

$$= \left[1-\frac{C}{(1.5D)^4}\times\frac{D^4}{C}\right]\times 100$$

$$= \left(1-\frac{1}{(1.5)^4}\right)\times 100 = \left(1-\frac{1}{5.06}\right)\times 100$$

$$= (1-0.1976)\times 100 = 80.24\% \approx \mathbf{80\%}$$

5.26 (c)

Boundary layer thickness,

$$\delta \propto \frac{1}{Re_x^{1/2}} \text{ for laminar flow}$$

and $$Re_x = \frac{Vx}{\nu}$$

i.e., $$Re_x \propto V$$

$$\therefore \quad \delta \propto \frac{1}{V^{1/2}}$$

$$\delta_1 V_1^{1/2} = \delta_2 V_2^{1/2}$$

$$1 \times V^{1/2} = \delta_2 (4V)^{1/2}$$

$$V^{1/2} = \delta_2 \times 2V^{1/2}$$

or $$\delta_2 = \frac{1}{2} = 0.5 \text{ mm}$$

5.27 (d)

For turbulent flow, $Re > 2300$

Turbulent flow is considered as well mixed flow.

5.28 Sol.

Radius: $r = 10$ mm

$\therefore$ Diameter: $d = 2r = 2 \times 10 = 20$ mm $= 0.02$ m

$$\dot{m} = 36 \text{ kg/hr} = \frac{36}{3600} = 0.01 \text{ kg/s}$$

$$\mu = 0.001 \text{ kg/ms}$$

Reynolds number: $Re = \dfrac{\rho V d}{\mu}$

From continuity equation, $\dot{m} = \rho A V$

or $$V = \frac{\dot{m}}{\rho A} = \frac{\dot{m}}{\rho \frac{\pi d^2}{4}} = \frac{4\dot{m}}{\rho \pi d^2}$$

$$\therefore \quad Re = \frac{\rho d}{\mu} \times \frac{4\dot{m}}{\rho \pi d^2} = \frac{4\dot{m}}{\pi \mu d}$$

$$= \frac{4 \times 0.01}{3.14 \times 0.001 \times 0.02} = \mathbf{636.94}$$

5.29 Sol.

Given : $d = 10$ cm, $V = 0.1$ m/s, $\nu = 10^{-5}$ m²/s

Reynold's number,

$$Re = \frac{Vd}{\nu} = \frac{0.1 \times 0.1}{10^{-5}} = 1000$$

Friction factor, $f = \dfrac{64}{Re} = \dfrac{64}{1000} = \mathbf{0.064}$

5.30 Sol.

Given data:

$d = 100$ mm $= 0.1$ m

$L = 100$ mm $= 0.1$ m

$u = 10$ m/s, $\mu = 0.1$ kg/ms

$dy = 2$ mm $= 2 \times 10^{-3}$ m

$$\tau = \mu \frac{du}{dy} = 0.1 \times \frac{10}{2 \times 10^{-3}}$$

$$= 500 \text{ N/m}^2$$

Frictional area $= \pi d L = 3.14 \times 0.1 \times 0.1$

$= 0.0314$ m²

Frictional resistance force $= \tau \times \pi d L$

$= 500 \times 0.0314 = \mathbf{15.7\ N}$

5.31 Sol.

$$\delta \propto \sqrt{x} \text{ for laminar flow.}$$

$$\frac{\delta}{\sqrt{x}} = \text{constant}$$

$$\frac{\delta_2}{\delta_1} = \sqrt{\frac{x_2}{x_1}} = \sqrt{\frac{0.75}{0.25}} = \sqrt{3}$$

$$\therefore \quad \delta_2 = \delta_1\sqrt{3} = 1.732 \times 8 = \mathbf{13.8564\ mm}$$

5.33 Sol.

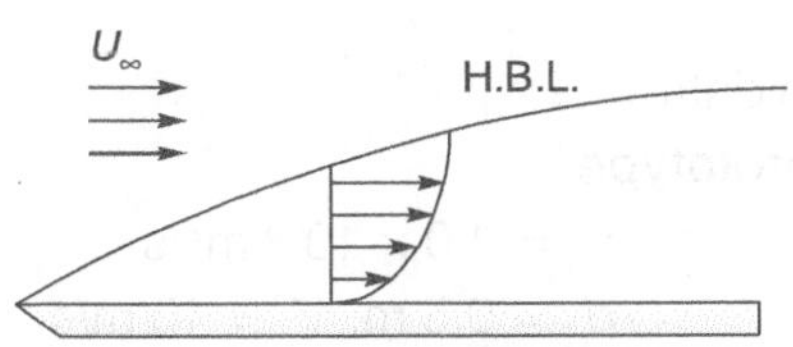

Drag force = Average shear stress × area

where $\bar{\tau}$ = Average shear stress

$$= \bar{C}_f \times \frac{\rho U_\infty^2}{2} = \frac{1.328}{\sqrt{Re_L}} \times \frac{\rho U_\infty^2}{2}$$

Re_L = Local Reynold's number at trailing edge,

$$= \frac{U_\infty L \rho}{\mu} = \frac{2 \times 2.5 \times 1.2}{1.2 \times 2 \times 10^{-5}}$$

$= 2.5 \times 10^5$ (laminar flow)

Drag force: $F_D = \bar{\tau} \times A$

$$= \frac{1.328}{\sqrt{2.5 \times 10^5}} \times \frac{1.2 \times 2^2}{2} \times (2.5 \times 1)$$

$= \mathbf{0.0159\ N}$

5.34 Sol.

$\mu = 0.001$ Pa-s

$R = 5$ cm $= 0.05$ m

$$\frac{\partial p}{\partial x} = -10 \text{ Pa/m}$$

$$r = 0.2 \text{ cm} = 0.002 \text{ m}$$

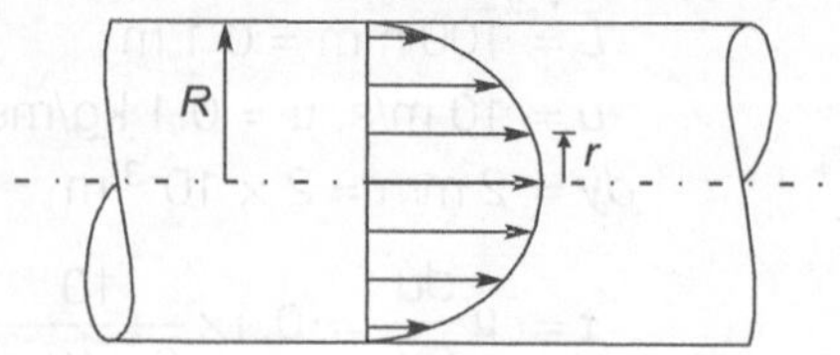

For incompressible laminar flow through pipe,

Local velocity,

$$u = -\frac{1}{4\mu}\frac{\partial p}{\partial x}[R^2 - r^2]$$

$$= -\frac{1}{4 \times 0.001} \times (-10)[(0.05)^2 - (0.002)^2]$$

$$= \mathbf{6.24 \text{ m/s}}$$

5.35 (c)

Instantaneous velocity : $u = \bar{u} + u'$

The time-average of the fluctuating velocity

$$\bar{u}' = \frac{1}{T}\int_0^T u'dt = \frac{1}{T}\int_0^T (u - \bar{u})dt$$

$$= \frac{1}{T}\int_0^T udT - \frac{1}{T}\bar{u}\int_0^T dt = \bar{u} - \frac{\bar{u}}{T}T$$

$$= \bar{u} - \bar{u} = \mathbf{0}$$

5.36 Sol.

Given data:

For prototype

$$\nu_p = 1.0 \times 10^{-5} \text{ m}^2\text{/s}$$

$$d_p = 0.5 \text{ m}, \; V_p = 10 \text{ m/s}$$

For model

$$\nu_m = 0.89 \times 10^{-6} \text{ m}^2\text{/s}$$

$$d_m = 20 \text{ mm} = 0.02 \text{ m}$$

$$V_m = ?$$

We know that

Reynolds number,

$$Re = \frac{Vd}{\nu}$$

$$(Re)_m = (Re)_p$$

$$\frac{V_m d_m}{\nu_m} = \frac{V_p d_p}{\nu_p}$$

$$\frac{V_m \times 0.02}{0.89 \times 10^{-6}} = \frac{10 \times 0.5}{1.0 \times 10^{-5}}$$

or $\quad V_m = \mathbf{22.25 \text{ m/s}}$

The velocity of water is 22.25 m/s

5.37 Sol.

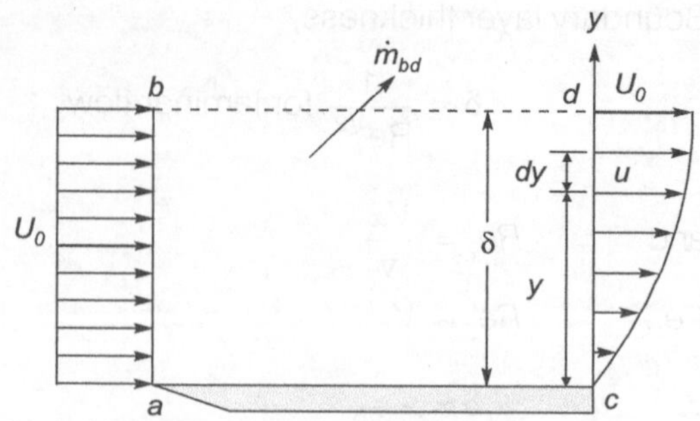

U_0 = Velocity of free stream

δ = Boundary layer thickness at section c-d

b = Width of the plate

Velocity profile,

$$u = U_0\left[2\left(\frac{y}{\delta}\right) - \left(\frac{y}{\delta}\right)^2\right]$$

Mass flow rate through section a-b,

$$\dot{m}_{ab} = \rho A U_0 = \rho b \delta U_0$$

Mass flow rate through section c-d,

$$\dot{m}_{cd} = \int_0^\delta \rho b dy u$$

$$= \int_0^\delta \rho b U_0\left[2\frac{y}{\delta} - \left(\frac{y}{\delta}\right)^2\right]dy$$

$$\dot{m}_{cd} = \rho b U_0\left[\frac{2y^2}{2\delta} - \frac{y^3}{3\delta^2}\right]_0^\delta$$

$$= \rho b U_0\left[\frac{\delta^2}{\delta} - \frac{\delta^3}{3\delta^2}\right]$$

$$= \rho b U_0\left[\delta - \frac{\delta}{3}\right] = \rho b U_0 \times \frac{2}{3}\delta$$

$$= \frac{2}{3}\rho b \delta U_0$$

Mass flow rate through section b-d,

$$\dot{m}_{bd} = \dot{m}_{ab} - \dot{m}_{cd}$$

$$= \rho b \delta U_0 - \frac{2}{3}\rho b \delta U_0$$

$$= \frac{\rho b \delta U_0}{3}$$

$$\frac{\dot{m}_{bd}}{\dot{m}_{ab}} = \frac{\rho b \delta U_0}{3 \times \rho b \delta U_0} = \frac{1}{3} = \mathbf{0.333}$$

5.38 Sol.

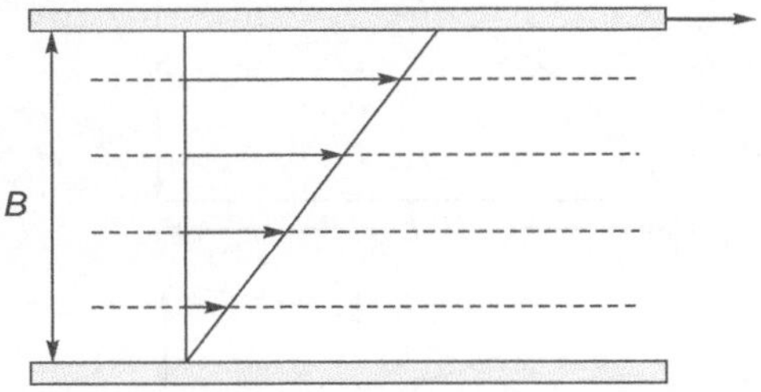

$$B = 50 \text{ mm} = 0.050 \text{ mm}$$
$$U = 3 \text{ m/s}, \mu = 0.44 \text{ kg/ms}$$
$$\rho = 888 \text{ kg/m}^3$$

Newton's law of viscosity,

$$\tau = \mu\frac{du}{dy} = \frac{\mu U}{B}$$
$$= \frac{0.44 \times 3}{0.050} = \mathbf{26.4 \text{ N/m}^2}$$

5.39 Sol.

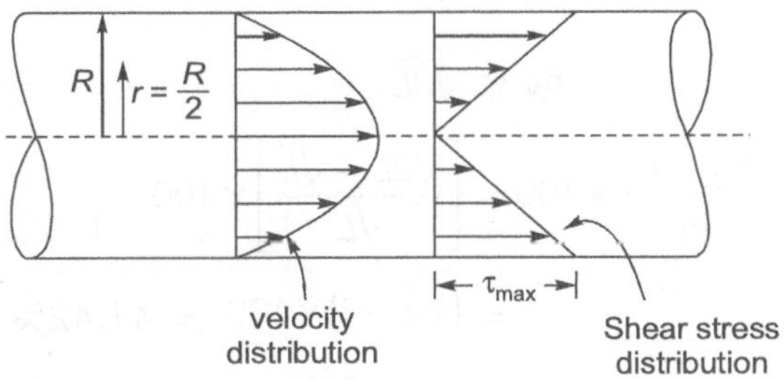

Shear stress at the wall,

$$\tau_0 = \tau_{max} = \frac{K\mu U_1}{R}$$

We know that the velocity at any point for laminar flow through pipe,

$$u = -\frac{1}{4\mu}\left(\frac{\partial p}{\partial x}\right)\left(R^2 - r^2\right)$$

At $\quad r = \frac{R}{2}, u = U_1$

$$\therefore \quad U_1 = -\frac{1}{4\mu}\left(\frac{\partial p}{\partial x}\right)\left(R^2 - \frac{R^2}{4}\right)$$
$$= -\frac{1}{4\mu}\left(\frac{\partial p}{\partial x}\right)\frac{3}{4}R^2$$
$$= -\frac{3}{16\mu}\left(\frac{\partial p}{\partial x}\right)R^2$$
$$\therefore \quad \tau_0 = -\frac{K\mu}{R} \times \frac{3}{16\mu}\left(\frac{\partial p}{\partial x}\right)R^2$$
$$= -\frac{3}{16}K\left(\frac{\partial p}{\partial x}\right)R$$

We know that maximum shear stress at wall,

$$\tau_0 = -\frac{R}{2}\frac{\partial p}{\partial x}$$
$$\therefore \quad -\frac{3}{16}K\left(\frac{\partial p}{\partial x}\right)R = -\frac{R}{2}\frac{\partial p}{\partial x}$$

or $\quad K = \frac{8}{3} = \mathbf{2.66}$

5.41 (b)

Given $\quad \frac{u}{U_\infty} = \sin\left(\frac{\pi}{2}\frac{y}{\delta}\right)$

where $\quad \delta$ = Boundary layer thickness

Local displacement thickness,

$$\delta^* = \int_0^\delta\left(1 - \frac{u}{U_\infty}\right)dy = \int_0^\delta\left[1 - \sin\left(\frac{\pi y}{2\delta}\right)\right]dy$$
$$= \left[y + \frac{2\delta}{\pi} \times \cos\left(\frac{\pi y}{2\delta}\right)\right]_0^\delta = \left[\delta + 0 - 0 - \frac{2\delta}{\pi}\right]$$
$$\delta^* = \delta\left(1 - \frac{2}{\pi}\right)$$

or, $\quad \frac{\delta^*}{\delta} = 1 - \frac{2}{\pi}$

5.42 (c)

Laminar flow over plate

As per Blasius equation

$$C_{fx} = \frac{0.664}{\sqrt{R_{ex}}}$$

and $\quad C_{fx} = \frac{\tau_w}{\frac{1}{2}\rho u_\infty^2}$

$$\frac{0.664}{\sqrt{\frac{u_\infty x}{\nu}}} = \frac{\tau_w}{\frac{1}{2}\rho u_\infty^2}$$
$$\tau_w \propto \frac{1}{\sqrt{x}}$$

as $\quad x_2 > x_1$

$$\boldsymbol{\tau_{w2} < \tau_{w1}}$$

5.43 Sol.

Drag coefficient,

$$C_F = \frac{F}{\rho U^2 D^2}$$

or Drag force,

$$F = C_F \rho U^2 D^2$$

Given data for sphere-1

$$\rho_1 = 1000 \text{ kg/m}^3$$
$$D_1 = 100 \text{ mm} = 0.1 \text{ m}$$
$$U_1 = 2 \text{ m/s}$$
$$C_F = 0.5$$

For sphere-2

$$\rho_2 = 1000 \text{ kg/m}^3$$
$$D_2 = 200 \text{ mm} = 0.2 \text{ m}$$

For dynamically similar conditions,

$$(\text{Re})_1 = (\text{Re})_2$$

$$\frac{U_1 D_1}{\nu_1} = \frac{U_2 D_2}{\nu_2}$$

or $U_1 D_1 = U_2 D_2 \qquad \because \nu_1 = \nu_2$

$$2 \times 0.1 = U_2 \times 0.2$$

or $U_2 = 1$ m/s

$$F_2 = C_F \cdot \rho U_2^2 D_2^2$$
$$= 0.5 \times 1000 \times (1)^2 \times (0.2)^2$$
$$= \mathbf{20\ N}$$

5.44 Sol.

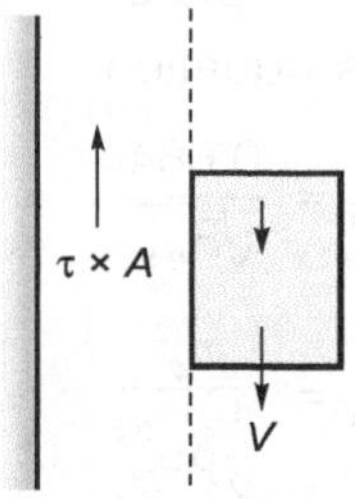

Terminal velocity is a constant velocity i.e. the net acceleration is zero.

So, $\Sigma F_{net} = ma$

$$mg - \tau A = 0$$
$$\tau A = mg$$
$$\mu \frac{V}{h} A = mg$$
$$7 \times 10^{-3} \times \frac{V}{0.15 \times 10^{-3}} \times 0.04 = 2 \times 10$$
$$\mathbf{V = 10.714\ m/s}$$

5.45 Sol.

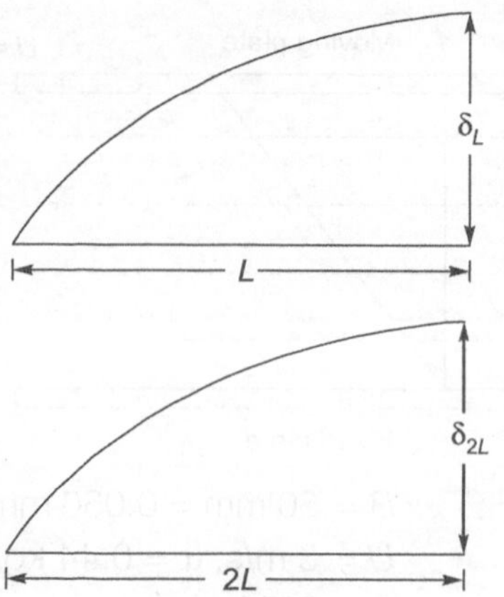

For laminar flow over flat plate,
As per Blasius equation,

$$\delta = \frac{5x}{\sqrt{\dfrac{\rho V_\infty x}{\mu}}}$$

$$\delta \propto \sqrt{x}$$
$$\delta_L \propto \sqrt{L}$$
$$\delta_{2L} \propto \sqrt{2L}$$

$$\frac{\delta_{2L} - \delta_L}{\delta_L} \times 100 = \left[\frac{\sqrt{2L} - \sqrt{L}}{\sqrt{L}}\right] \times 100$$
$$= \left(\sqrt{2} - 1\right) \times 100 = \mathbf{41.42\%}$$

5.46 Sol.

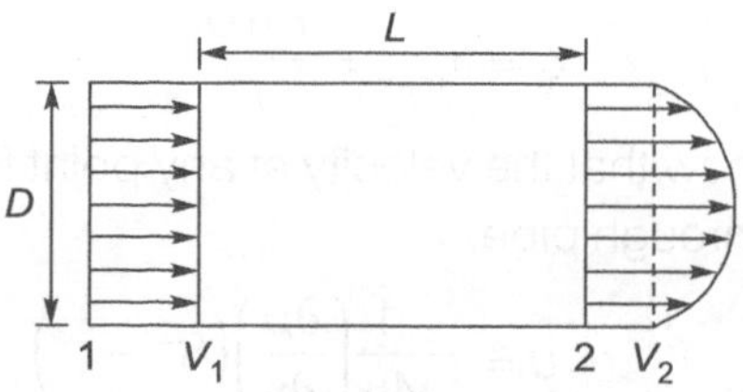

$$\text{Re} = \frac{\rho V D}{\mu} \text{ for pipe flow}$$

V = Mean velocity

Exit mean velocity V_2

$$\dot{m}_{in} = \dot{m}_{out}$$
$$\rho A_1 V_1 = \rho A_2 V_2$$

{∵ Water is incompressible fluid, so density will remain same}

Given, $A_1 = A_2$

∴ $V_1 = V_2$

So, Reynolds number is constant i.e. 500 at exit.

5.47 Sol.

Case-1:

$$C_{D1} = 0.1$$
$$U_{\infty 1} = 44 \text{ km/h}$$

Case-2:

$$C_{D2} = 0.8$$
$$U_{\infty 2} = ?$$

As we know, Power, $P = C_D \times \frac{1}{2}\rho U_\infty^3 \times A$

For same power, density and frontal area,

$$(\text{Power})_1 = (\text{Power})_2$$

$$C_{D1}\frac{1}{2}\rho_1 A_1 U_{\infty 1}^3 = C_{D2}\frac{1}{2}\rho_2 A_2 U_{\infty 2}^3 \quad (\rho_1 = \rho_2)$$

$$U_\infty^3 \propto \frac{1}{C_D}$$

$$\Rightarrow \left(\frac{U_{\infty 2}}{U_{\infty 1}}\right)^3 = \frac{C_{D1}}{C_{D2}}$$

$$\left(\frac{U_{\infty 2}}{44}\right)^3 = \frac{0.1}{0.8}$$

$$U_{\infty 2} = \frac{44}{2} = 22 \text{ km/h}$$

5.49 Sol.

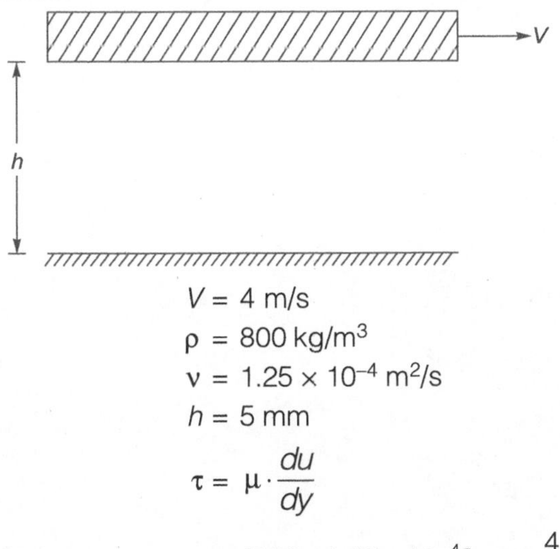

$$V = 4 \text{ m/s}$$
$$\rho = 800 \text{ kg/m}^3$$
$$\nu = 1.25 \times 10^{-4} \text{ m}^2\text{/s}$$
$$h = 5 \text{ mm}$$
$$\tau = \mu \cdot \frac{du}{dy}$$
$$\tau = [800 \times 1.25 \times 10^{-4}] \times \frac{4}{5 \times 10^{-3}}$$
$$\tau = 80 \text{ N/m}^2 = 80 \text{ Pa}$$

5.50 Sol.

Applying continuity equation :

$$35 \times 0.01 + \dot{q} = 6 \times \frac{\pi}{4} \times 0.32^2$$
$$\dot{q} = 0.13255 \text{ m}^3\text{/s}$$
$$\dot{q} = 132.55 \ l\text{/s}$$

5.52 (c)

Given: $V = (y^2 - x^2)\hat{i} + (2xy)\hat{j}, \rho = 1.5 \text{ kg/m}^3$

By Navier stokes equation,

In x direction:

$$\rho\frac{\Delta u}{\Delta t} = -\frac{\partial P}{\partial x} + \mu(\nabla^2 u) + \rho g_x$$

For inviscid fluid and negligible body forces

$$\rho\left[u\frac{\partial u}{\partial x} + v\frac{\partial u}{\partial y}\right] = -\frac{\partial P}{\partial x}$$

$$\Rightarrow \frac{\partial P}{\partial x} = -1.5\,[(y^2 - x^2)(-2x) + (2xy)(2y)]$$

$$\Rightarrow \left.\frac{\partial P}{\partial x}\right|_{(1,1)} = -1.5\,[(1-1)(-2 \times (1)) + 4 \times (1) \times (1)^2]$$

$$\left.\frac{\partial P}{\partial x}\right|_{(1,1)} = -6$$

In y direction:

$$\rho\frac{\Delta v}{\Delta t} = -\frac{\partial P}{\partial y} + \mu(\nabla^2 v) + \rho g_y$$

For inviscid fluid and negligible body forces

$$\rho\left[u\frac{\partial v}{\partial x} + v\frac{\partial v}{\partial y}\right] = -\frac{\partial P}{\partial y}$$

$$\frac{\partial P}{\partial y} = -1.5\,[(y^2 - x^2)(2y) + (2xy)(2x)]$$

$$\left.\frac{\partial P}{\partial y}\right|_{(1,1)} = -1.5\,[(1-1)(2 \times 1) + 4 \times (1)^2 \times (1)]$$

$$= -6$$

So, Pressure gradient $= -6\hat{i} - 6\hat{j}$

5.53 (d)

Given for air, $\rho = 1.2 \text{ kg/m}^3$, $\nu = 1.5 \times 10^{-5}$ m²/s, $U_\infty = 2$ m/s

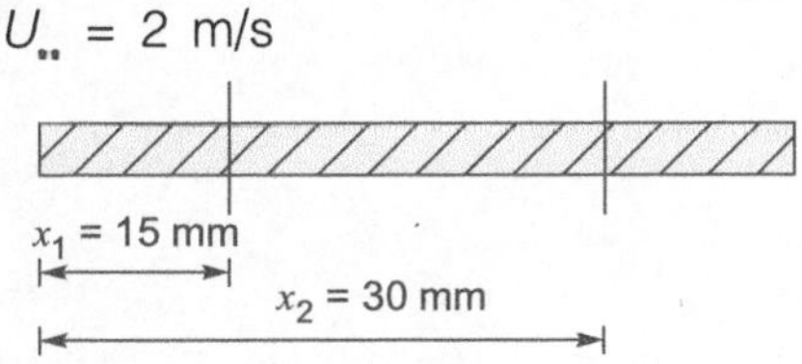

At $x_1 = 15$ mm $\tau_1 = \tau_w$

At x_2 = 30 mm,

$$\text{Reynolds number} = \frac{U_\infty x}{\nu} = \frac{2\times 30\times 10^{-3}}{1.5\times 10^{-5}} = 4000$$

As Reynolds number is lesser than the critical value, the flow is laminar.

For laminar flow over a flat plate,

$$\tau_w \propto \frac{1}{\sqrt{x}}$$

$$\therefore \quad \frac{\tau_1}{\tau_2} = \sqrt{\frac{x_2}{x_1}}$$

$$\frac{\tau_w}{\tau_2} = \sqrt{\frac{30}{15}}$$

$$\tau_2 = \frac{\tau_w}{\sqrt{2}}$$

5.54 (0.5) (0.5 to 0.5)

Dimension of surface tension = $[MT^{-2}]$

Dimension of radius = $[L]$

Dimension of dynamic viscosity = $[ML^{-1}T^{-1}]$

Dimension of time = $[T]$

Now, $L = k\gamma^a R^b \mu^c \sqrt{t}$

$$\Rightarrow [L] = [M^0L^0T^0][MT^{-2}]^a [L]^b [ML^{-1}T^{-1}]^c [T]^{1/2}$$

$$\Rightarrow \quad a + c = 0 \quad \text{...(i)}$$

$$\Rightarrow \quad b - c = 1 \quad \text{...(ii)}$$

$$-2a - c + \frac{1}{2} = 0 \quad \text{...(iii)}$$

On solving equation (i), (ii) and (iii), we get

$a = 0.5$, $b = 0.5$ and $c = -0.5$

■■■■

6 Flow Through Pipes

6.1 For a fluid flow through a divergent pipe of length L having inlet and outlet radii of R_1 and R_2 respectively and a constant flow rate of Q, assuming the velocity to be axial and uniform at any cross-section, the acceleration at the exit is

(a) $\frac{2Q(R_1 - R_2)}{\pi L R_2^3}$ (b) $\frac{2Q^2(R_1 - R_2)}{\pi L R_2^3}$

(c) $\frac{2Q^2(R_1 - R_2)}{\pi^2 L R_2^5}$ (d) $\frac{2Q^2(R_2 - R_1)}{\pi^2 L R_2^5}$

[2004 : 2 M]

6.2 A siphon draws water from a reservoir and discharge it out at atmospheric pressure. Assuming ideal fluid and the reservoir is large, the velocity at point P in the siphon tube is

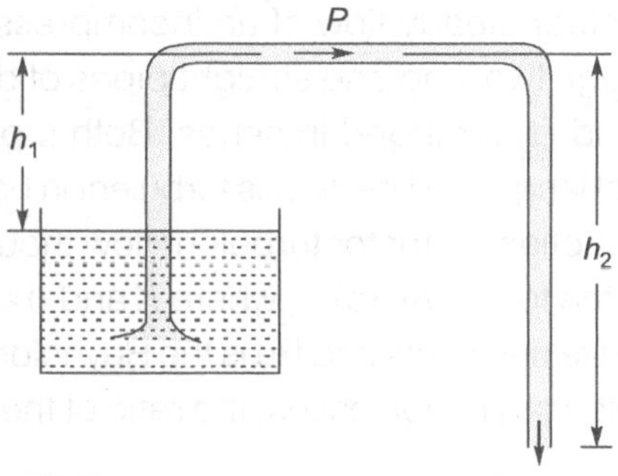

(a) $\sqrt{2gh_1}$ (b) $\sqrt{2gh_2}$

(c) $\sqrt{2g(h_2 - h_1)}$ (d) $\sqrt{2g(h_2 + h_1)}$

[2006 : 2 M]

6.3 Water at 25°C is flowing through a 1.0 km long G.I. pipe of 200 mm diameter at the rate of 0.07 m^3/s. If value of Darcy friction factor for this pipe is 0.02 and density of water is 1000 kg/m^3, the pumping power (in kW) required to maintain the flow is

(a) 1.8 (b) 17.4

(c) 20.5 (d) 41.0 **[2009 : 2 M]**

6.4 A smooth pipe of diameter 200 mm carries water. The pressure in the pipe at section $S1$ (elevation: 10 m) is 50 kPa. At section $S2$ (elevation : 12 m) the pressure is 20 kPa and velocity is 2 ms^{-1}. Density of water is 1000 kgm^{-3} and acceleration due to gravity is 9.8 ms^{-2}. Which of the following is TRUE

(a) flow is from $S1$ to $S2$ and head loss is 0.53 m

(b) flow is from $S2$ to $S1$ and head loss is 0.53 m

(c) flow is from $S1$ to $S2$ and head loss is 1.06 m

(d) flow is from $S2$ to $S1$ and head loss is 1.06 m

[2010 : 2 M]

6.5 Oil flows through a 200 mm diameter horizontal cast iron pipe (friction factor, f = 0.0225) of length 500 m. The volumetric flow rate is 0.2 m^3/s. The head loss (in m) due to friction is (assume g = 9.81 m/s^2)

(a) 116.18 (b) 0.116

(c) 18.22 (d) 232.36 **[2012 : 1 M]**

6.6 For steady, fully developed flow inside straight pipe of diameter D, neglecting gravity effects, the pressure drop Δp over a length L and the wall shear stress τ_w are related by

(a) $\tau_w = \frac{\Delta p D}{4L}$ (b) $\tau_w = \frac{\Delta p D^2}{4L^2}$

(c) $\tau_w = \frac{\Delta p D}{2L}$ (d) $\tau_w = \frac{4\Delta p L}{D}$

[2013 : 1 M]

6.7 Water flows through a 10 mm diameter and 250 m long smooth pipe at an average velocity of 0.1 m/s. The density and the viscosity of water are 997 kg/m^3 and 855 × 10^{-6} Ns/m^2, respectively. Assuming fully-developed flow, the pressure drop (in Pa) in the pipe is ______.

[2014 : 2 M, Set-2]

6.8 A fluid of dynamic viscosity 2 × 10^{-5} kg/ms and density 1 kg/m^3 flows with an average velocity of 1 m/s through a long duct of rectangular (25 mm × 15 mm) cross-section. Assuming laminar flow, the pressure drop (in Pa) in the fully developed region per meter length of the duct is ______.

[2014 : 2 M, Set-3]

6.9 A siphon is used to drain water from a large tank as shown in figure below. Assume that the level of water is maintained constant. Ignore frictional effect due to viscosity and losses at entry and exit. At the exit of the siphon, the velocity of water is

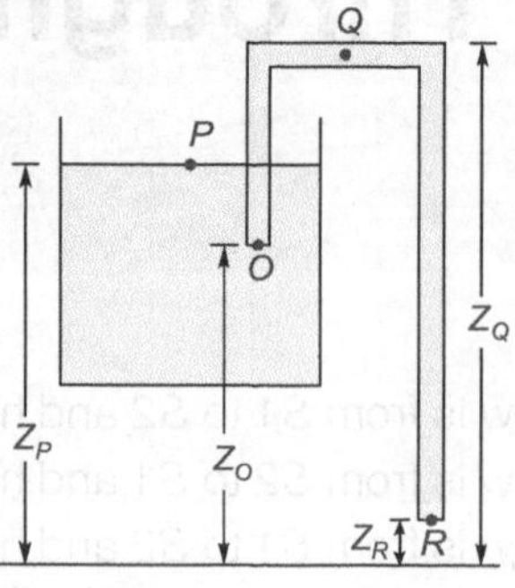

(a) $\sqrt{2g(Z_Q - Z_R)}$ (b) $\sqrt{2g(Z_P - Z_R)}$

(c) $\sqrt{2g(Z_O - Z_R)}$ (d) $\sqrt{2gZ_Q}$

[2014 : 2 M, Set-3]

6.10 Consider fully developed flow in a circular pipe with negligible entrance length effects. Assuming the mass flow rate, density and friction factor to be constant, if the length of the pipe is doubled and the diameter is halved, the head loss due to friction will increase by a factor of

(a) 4 (b) 16
(c) 32 (d) 64 [2015 : 1 M, Set-1]

6.11 Three parallel pipes connected at the two ends have flow-rates Q_1, Q_2 and Q_3 respectively, and the corresponding frictional head losses are h_{L1}, h_{L2} and h_{L3} respectively. The correct expressions for total flow rate (Q) and frictional head loss across the two ends (h_L) are

(a) $Q = Q_1 + Q_2 + Q_3;\ h_L = h_{L1} + h_{L2} + h_{L3}$
(b) $Q = Q_1 + Q_2 + Q_3;\ h_L = h_{L1} = h_{L2} = h_{L3}$
(c) $Q = Q_1 = Q_2 = Q_3;\ h_L = h_{L1} + h_{L2} + h_{L3}$
(d) $Q = Q_1 = Q_2 = Q_3;\ h_L = h_{L1} = h_{L2} = h_{L3}$

[2015 : 1 M, Set-3]

6.12 The head loss for a laminar incompressible flow through a horizontal circular pipe is h_1. Pipe length and fluid remaining the same, if the average flow velocity doubles and the pipe diameter reduces to half its previous value, the head loss is h_2. The ratio h_2/h_1 is

(a) 1 (b) 4
(c) 8 (d) 16 [2015 : 2 M, Set-2]

6.13 A channel of width 450 mm branches into two sub-channels having width 300 mm and 200 mm as shown in figure. If the volumetric flow rate (taking unit depth) of an incompressible flow through the main channel is 0.9 m³/s and the velocity in the sub-channel of width 200 mm is 3 m/s, the velocity in the sub-channel of width 300 mm is _____ m/s. Assume both inlet and outlet to be at the same elevation.

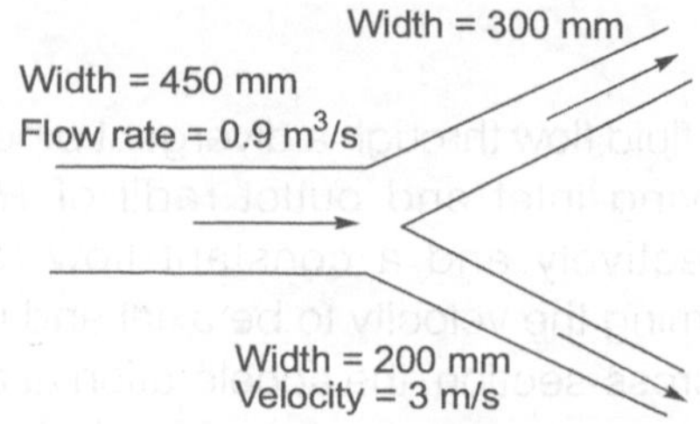

[2016 : 1 M, Set-3]

6.14 Water (density = 1000 kg/m³) at ambient temperature flows through a horizontal pipe of uniform cross-section at the rate of 1 kg/s. If the pressure drop across the pipe is 100 kPa, the minimum power required to pump the water across the pipe, in Watt, is _______.

[2017 : 1 M, Set-1]

6.15 Consider steady flow of an incompressible fluid through two long and straight pipes of diameters d_1 and d_2 arranged in series. Both pipes are of equal length and the flow is turbulent in both pipes. The friction factor for turbulent flow though pipes is of the for, $f = K(\mathrm{Re})^{-n}$, where K and n are known positive constants and Re is the Reynolds number. Neglecting minor losses, the ratio of the frictional pressure drop in pipe 1 to that in pipe 2. $\left(\frac{\Delta P_1}{\Delta P_2}\right)$, is given by

(a) $\left(\frac{d_2}{d_1}\right)^{(5-n)}$ (b) $\left(\frac{d_2}{d_1}\right)^{5}$

(c) $\left(\frac{d_2}{d_1}\right)^{(3-n)}$ (d) $\left(\frac{d_2}{d_1}\right)^{(5+n)}$

[2017 : 2 M, Set-1]

6.16 Water flows through two different pipes A and B of the same circular cross-section but at different flow rates. The length of pipe A is 1.0 m and that of pipe B is 2.0 m. The flow in both the pipes is laminar and fully developed. If the frictional head loss across the length of the pipes is same, the ratio of volume flow rates Q_B/Q_A is _____ (round off to two decimal places).

[2019 : 2 M, Set-2]

6.17 Consider steady, viscous, fully developed flow of a fluid through a circular pipe of internal diameter D. We know that the velocity profile forms a paraboloid about the pipe centre line, given by: $V = -C\left(r^2 - \frac{D^2}{4}\right)$ m/s, where C is a constant. The rate of kinetic energy (in J/s) at the control surface A-B, as shown in the figure, is proportional to D^n. The value of n is______.

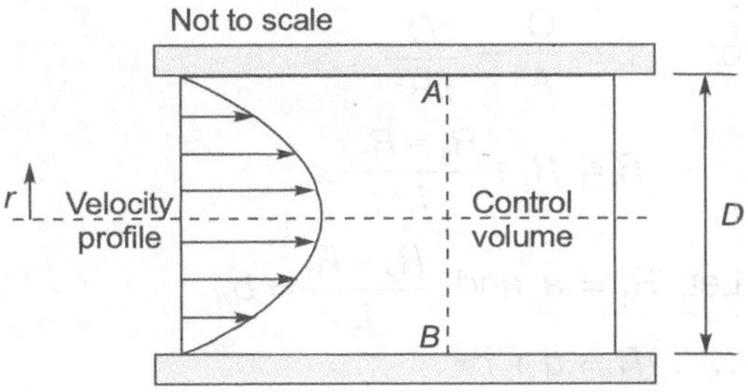

[2020 : 2 M, Set-1]

6.18 In the pipe network shown in the figure, all pipes have the same cross-section and can be assumed to have the same friction factor. The pipes, connecting points W, N, and S with point J have an equal length L. The pipe connecting points J and E has a length $10L$. The pressure at the ends N, E, and S are equal. The flow rate in the pipe connecting W and J is Q. Assume that the fluid flow is steady, incompressible, and the pressure losses at the pipe entrance and junction are negligible. Consider the following statements:

I. The flow rate in pipe connecting J and E is $Q/21$.

II. The pressure difference between J and N is equal to the pressure difference between J and E.

Which one of the following option is CORRECT?

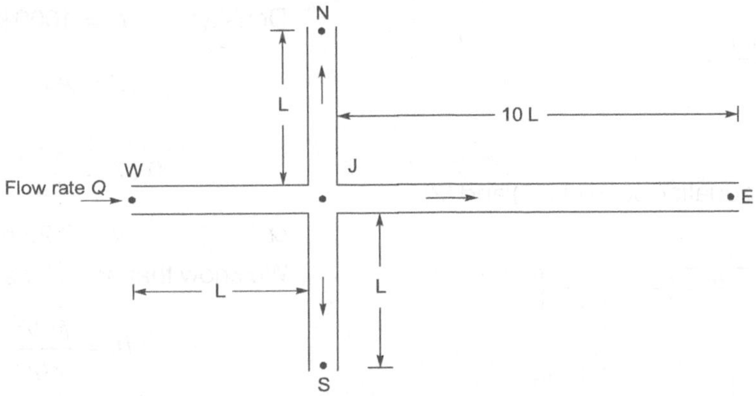

(a) I is False and II is True
(b) I is True and II is False
(c) Both I and II are False
(d) Both I and II are True

[2024 : 2 M]

Answers Flow Through Pipes

6.1	(c)	**6.2**	(c)	**6.3**	(b)	**6.4**	(c)	**6.5**	(a)	**6.6**	(a)	**6.7**	(6840)	**6.8**	(1.83)
6.9	(b)	**6.10**	(d)	**6.11**	(b)	**6.12**	(c)	**6.13**	(1)	**6.14**	(100)	**6.15**	(a)	**6.16**	(0.5)
6.17	(8)	**6.18**	(a)												

Explanations Flow Through Pipes

6.1 (c)

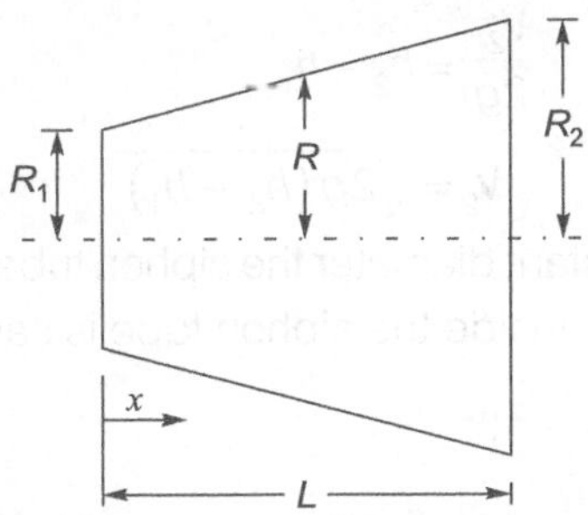

$$a_x = u\frac{\partial u}{\partial x} + v\frac{\partial u}{\partial y} + w\frac{\partial u}{\partial z} + \frac{\partial u}{\partial t}$$

$$\therefore \quad a_x = u\frac{\partial u}{\partial x}$$

$$Q = Au$$

or $u = \frac{Q}{A} = \frac{Q}{\pi R^2}$

$$R = R_1 + \frac{R_2 - R_1}{L} \cdot x$$

Let, $R_1 = a$ and $\frac{R_2 - R_1}{L} = b$

$\therefore \quad R = a + bx$

$$u = \frac{Q}{\pi(a+bx)^2}$$

$$\frac{\partial u}{\partial x} = \frac{-2bQ}{\pi(a+bx)^3}$$

$$a_x = u\frac{\partial u}{\partial x}$$

$$a_{x=L} = u\frac{\partial u}{\partial x}\bigg|_{x=L}$$

and also replacing a and b

$$a_{x=L} = \frac{2Q^2(R_1 - R_2)}{\pi^2 L R_2^5} =$$

6.2 (c)

Applying Bernoulli's equation at points (1) and (2), we get

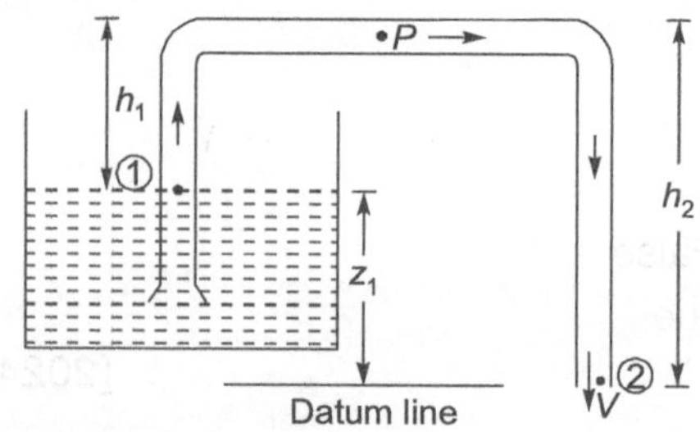

$$\frac{p_1}{\rho g} + \frac{V_1^2}{2g} + z_1 = \frac{p_2}{\rho g} + \frac{V_2^2}{2g} + z_1$$

where $\frac{p_1}{\rho g} = \frac{p_2}{\rho g} = \frac{p_{atm}}{\rho g}$, $z_1 = h_2 - h_1$

$z_2 = 0$, $V_1 = 0$

$$\therefore \frac{p_{atm}}{\rho g} + 0 + h_2 - h_1 = \frac{p_{atm}}{\rho g} + \frac{V_2^2}{2g}$$

or $\frac{V_2^2}{2g} = h_2 - h_1$

or $V_2 = \sqrt{2g(h_2 - h_1)}$

As constant diameter the siphon tube, the velocity of water inside the siphon tube is same.

POINTS TO REMEMBER

- A pipe which raises above its hydraulic grade line has negative pressure and is known as Siphon pipe.
- In Siphon pipe, the summit pressure (p_s) should not be less than vapour pressure of liquid to avoid air lock formation at summit.

6.3 (b)

Given data:

Length: $L = 1$ km $= 1000$ m

Diameter: $d = 200$ mm $= 0.2$ m

Discharge: $Q = 0.07$ m³/s

Friction factor,

$f = 0.02$

Density: $\rho = 1000$ kg/m³

$$Q = AV = \frac{\pi}{4}d^2V$$

$$0.07 = \frac{3.14}{4} \times (0.2)^2 \times V$$

or $V = 2.23$ m/s

We know that head loss due to friction,

$$h_f = \frac{fLV^2}{2gd} = \frac{0.02 \times 1000 \times (2.23)^2}{2 \times 9.81 \times 0.2}$$

$= 25.34$ m

Pumping power required to maintain the flow,

$$P = \rho Q g h_f$$

$= 1000 \times 0.07 \times 9.81 \times 25.34$

$= 17400.9$ W $=$ **17.4 kW**

6.4 (c)

Given data:

At section $S1$

$d_1 = 200$ mm $= 0.2$ m

$$\therefore \quad A_1 = \frac{\pi}{4}d^2 = \frac{3.14}{4} \times (0.2)^2$$

$= 0.0314$ m²

$z_1 = 10$ m, $p_1 = 50$ kPa

$V_1 = V_2$

$\rho = 1000$ kg/m³

and $g = 9.81$ m/s²

At section $S2$

$$d_2 = d_1$$

$$\therefore \quad A_2 = A_1$$

$$z_2 = 12 \text{ m}$$

$$p_2 = 20 \text{ kPa} = 20 \times 10^3 \text{ Pa}$$

$$V_2 = 2 \text{ m/s}$$

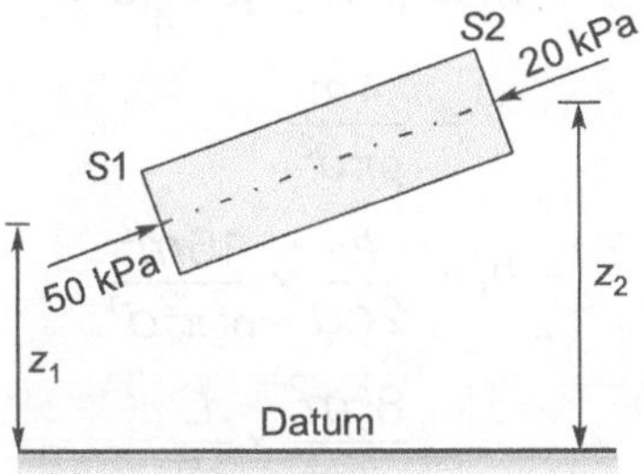

Net head at section $S1$:

$$H_1 = \frac{p_1}{\rho g} + \frac{V_1^2}{2g} + z_1$$

$$= \frac{50 \times 10^3}{1000 \times 9.81} + \frac{(2)^2}{2 \times 9.81} + 10$$

$$= 5.09 + 0.20 + 10 = 15.29 \text{ m}$$

Net head at section $S2$:

$$H_2 = \frac{p_2}{\rho g} + \frac{V_2^2}{2g} + z_1$$

$$= \frac{20 \times 10^3}{1000 \times 9.81} + \frac{(2)^2}{2 \times 9.81} + 12$$

$$= 2.03 + 0.20 + 12 = 14.23 \text{ m}$$

Since net head at section $S1$ is greater than at section $S2$, so the flow direction is from section $S1$ to section $S2$.

Head loss: $h_f = H_1 - H_2$

$$= 15.29 - 14.23 = \mathbf{1.06 \text{ m}}$$

6.5 (a)

Given data:

$$d = 200 \text{ mm} = 0.2 \text{ m}$$

$$\therefore \quad A = \frac{\pi}{4}d^2 = \frac{3.14}{4} \times (0.2)^2$$

$$= 0.0314 \text{ m}^2$$

$$f = 0.0225, L = 500 \text{ m}, Q = 0.2 \text{ m}^3/\text{s}$$

also $Q = AV$

$\therefore \quad 0.2 = 0.0314 \times V$

or $V = 6.369$ m/s

We know that the Darcy's formula, head loss due to friction,

$$h_f = \frac{fLV^2}{2gd} = \frac{0.0225 \times 500 \times (6.369)^2}{2 \times 9.81 \times 0.2}$$

$$= \mathbf{116.29 \text{ m of oil}}$$

6.6 (a)

Shear stress on wall,

$$\tau_w = -\frac{\partial p}{\partial x}\frac{R}{2}$$

where $-\dfrac{\partial p}{\partial x} = \dfrac{\Delta p}{L}$

Δp = Pressure drop

and $R = \dfrac{D}{2}$

$$\therefore \quad \tau_w = \frac{\Delta p}{L} \times \frac{D}{2 \times 2} = \frac{\Delta p D}{4L}$$

POINTS TO REMEMBER

- The negative sign on $\dfrac{\partial p}{\partial x}$ indicates decrease in pressure in the direction of flow. The pressure must decrease because pressure force is the only means available to compensate for resistance of flow.
- The maximum value of shear stress occur at the wall of the pipe.

6.7 Sol.

$$D = 10 \text{ mm} = 0.01 \text{ m}, L = 250 \text{ m}$$

$$\bar{u} = 0.1 \text{ m/s}, \rho = 997 \text{ kg/m}^3$$

$$\mu = 855 \times 10^{-6} \text{ Ns/m}^2$$

$$\text{Re} = \frac{\rho \bar{u} D}{\mu} = \frac{997 \times 0.1 \times 0.01}{855 \times 10^{-6}}$$

$$= 1166.082 < 2000$$

So flow is laminar

Pressure drop in the pipe,

$$\Delta p = \frac{32 \mu \bar{u} L}{D^2}$$

$$= \frac{32 \times 855 \times 10^{-6} \times 0.1 \times 250}{(0.01)^2}$$

$$= \mathbf{6840 \text{ Pa}}$$

6.8 Sol.

Given data: $\mu = 2 \times 10^{-5}$ kg/ms, $\rho = 1$ kg/m^3

$$\bar{u} = 1 \text{ m/s}$$

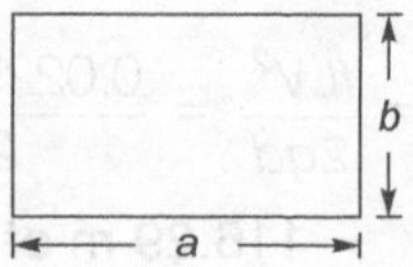

$$a = 25 \text{ mm} = 0.025 \text{ m}$$
$$b = 15 \text{ mm} = 0.015 \text{ m}$$

Characteristic diameter,

$$D_c = \frac{4A}{P}$$

where $A = a \times b$, cross-section area

$$= 0.025 \times 0.015$$
$$= 3.75 \times 10^{-4} \text{ m}^2$$

and $P = 2(a+b)$, wetted perimeter

$$= 2(0.025+0.015) = 0.08 \text{ m}$$

$$\therefore \quad D_c = \frac{4 \times 3.74 \times 10^{-4}}{0.08} = 0.0187 \text{ m}$$

Pressure drop,

$$\Delta p = \frac{32 \mu \bar{u} L}{D_c^2}$$

$$\frac{\Delta p}{L} = \frac{32 \mu \bar{u}}{D_c^2} = \frac{32 \times 2 \times 10^{-5} \times 1}{(0.0187)^2}$$

$$= \mathbf{1.83 \text{ Pa/m}}$$

POINTS TO REMEMBER

Here, in this problem characteristic diameter is the main concern to remember. As in the circular pipe, characteristic diameter is same as the diameter of pipe but in case of rectangular, triangular shape of pipes characteristic diameter will be different.

6.9 (b)

Applying Bernoulli's equation between P and R,

$$\frac{p_{atm}}{\rho g} + \frac{V_P^2}{2g} + Z_P = \frac{V_R^2}{2g} + Z_R + \frac{p_{atm}}{\rho g}$$

or

$$\frac{V_R^2}{2g} = (Z_P - Z_R)$$

($\because$ $V_P = 0$ for still water)

$$\therefore \quad V_R = \sqrt{2g(Z_P - Z_R)}$$

6.10 (d)

<u>**Method I:**</u>

Head loss due to friction,

$$h_f = \frac{fLV^2}{2gd}$$

Mass flow rate,

$$m = \rho AV = \rho \times \frac{\pi}{4} d^2 V$$

or

$$V = \frac{4m}{\rho \pi d^2}$$

$$\therefore \quad h_f = \frac{fL}{2gd} \times \frac{16m^2}{\rho^2 \pi^2 d^4}$$

$$= \frac{8fm^2}{\rho^2 \pi^2} \times \frac{L}{d^5}$$

$$= C \times \frac{L}{d^5} \quad \text{where } C = \frac{8fm^2}{\rho^2 \pi^2}$$

$$h_{f_1} = C \times \frac{L}{d^5}$$

and

$$h_{f_2} = C \times \frac{2L}{(d/2)^5} = \frac{C \times 2L}{\frac{d^5}{2^5}}$$

$$= \frac{C \times 2 \times 2^5 L}{d^5} = \frac{C \times 64 L}{d^5}$$

$$\mathbf{h_{f_2} = 64 h_{f_1}}$$

The head loss due to friction will increase by a factor of 64.

<u>**Method II:**</u>

Head loss due to friction,

$$m = \rho Q$$

(where m, ρ and Q are constant)

$$h_f = \frac{fLV^2}{2gD} = \frac{fLQ^2}{12D^5}$$

$$\frac{h_{f_2}}{h_{f_1}} = \frac{f(2L)Q^2}{12(D/2)^5} \times \frac{12D^2}{fLQ^2}$$

$$= 2 \times 2^5 = 64$$

$$h_{f_2} = 64 h_{f_1}$$

So, head loss due to friction increases by a factor of 64.

6.11 (b)

Total discharge,

$$Q = Q_1 + Q_2 + Q_2$$

Head loss: $h_L = h_{L_1} = h_{L_2} = h_{L_3}$

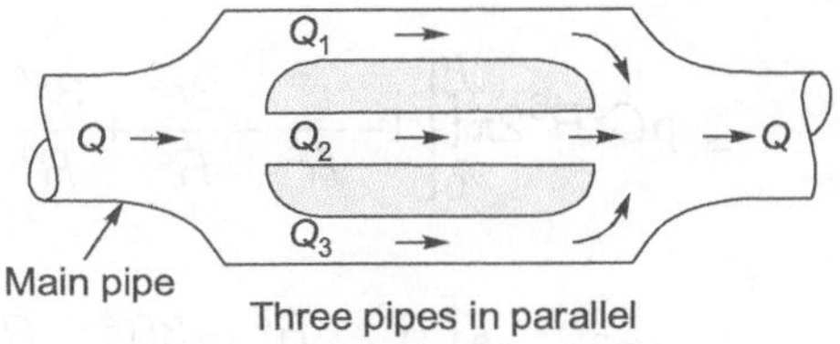

Three pipes in parallel

For the pipe connected in series,
Total discharge,

$$Q = Q_1 = Q_2 = Q_3$$

Head loss; $h_L = h_{L_1} + h_{L_2} + h_{L_3}$

6.12 (c)

$$\text{Head loss:} \quad h_f = \frac{32\mu\bar{u}L}{\rho g D^2}$$

$$h_1 = \frac{32\mu\bar{u}L}{\rho g D^2}$$

and
$$h_2 = \frac{32\mu \times 2\bar{u}L}{\rho g (D/2)^2}$$

$$= \frac{32\mu 2\bar{u}L \times 4}{\rho g D^2} = \frac{8 \times 32\mu\bar{u}L}{\rho g D^2}$$

$$\frac{h_2}{h_1} = \frac{8 \times 32\mu\bar{u}L}{\rho g D^2} \times \frac{\rho g D^2}{32\mu\bar{u}L} = 8$$

6.13 Sol.

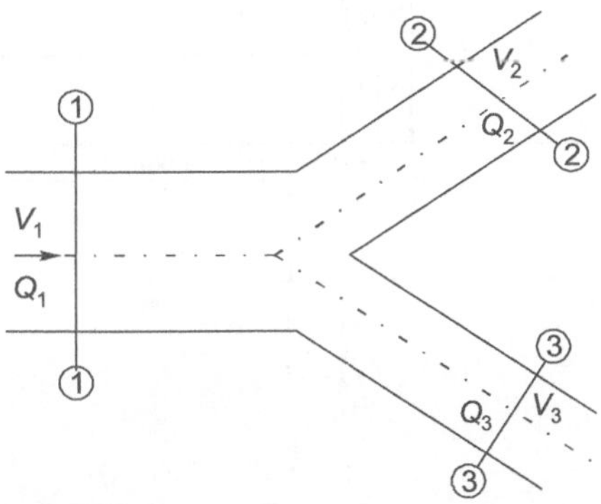

Given data: b_1 = 450 mm = 0.45 m
b_2 = 300 mm = 0.30 m; b_3 = 200 mm = 0.20 m
Depth: h = 1 m

$$Q_1 = 0.9 \text{ m}^3/\text{s}$$
$$V_3 = 3 \text{ m/s}$$
$$V_2 = ?$$

Apply equation of continuity, we get

$$Q_1 = Q_2 + Q_3$$
$$Q_1 = A_2V_2 + A_3V_3$$
$$Q_1 = b_2hV_2 + b_3hV_3$$
$$0.9 = 0.30 \times 1 \times V_2 + 0.20 \times 1 \times 3$$
$$0.9 = 0.30\, V_2 + 0.60$$

or $\quad 0.30\, V_2 = 0.9 - 0.6 = 0.30$

or $\quad V_2 = \frac{0.30}{0.30} = 1 \text{ m/s}$

6.14 Sol.

Given: ρ = 1000 kg/m³; m = 1 kg/s
Δp = 100 kPa = 100 × 10³ Pa

$$P = \rho Q g h_f = Q\Delta p$$
$$= \frac{m}{\rho}\Delta p = \frac{1}{1000} \times 100 \times 10^3$$
$$= 100 \text{ W}$$

6.15 (a)

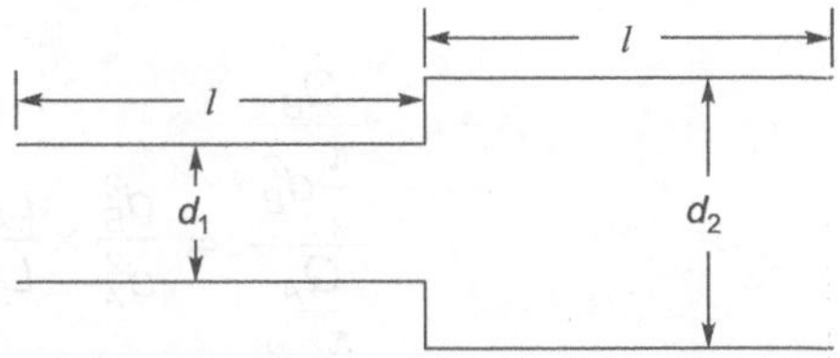

For series connection, $Q_1 = Q_2 = Q$
As given $\quad l_1 = l_2 = l$

For horizontal pipe, $h_f = \frac{\Delta P}{\rho g}$

$$h_{f_1} = \frac{8f_1LQ^2}{\pi^2 \times g \times d_1^5}; \quad h_{f_2} = \frac{8f_2LQ^2}{\pi^2 \times g \times d_2^5}$$

$$\frac{\Delta P_1}{\Delta P_2} = \frac{f_1}{d_1^5} \times \frac{d_2^5}{f_2} \quad \text{...(i)}$$

$$f = \frac{K}{\text{Re}^n} = \frac{K}{\left(\frac{Vd}{\nu}\right)^n}$$

(where Q and ν are constant)

$$Q = \frac{\pi}{4}d^2 \times V$$

$$V \propto \frac{1}{d^2}$$

$$f \propto \frac{1}{\left(\frac{1}{d^2}d\right)^n}$$

By equation (i) $f \propto D^n$

$$\frac{\Delta P_1}{\Delta P_2} = \frac{d_1^n}{d_1^5} \times \frac{d_2^5}{d_2^n} = \frac{d_2^{5-n}}{d_1^{5-n}} = \left(\frac{d_2}{d_1}\right)^{5-n}$$

6.16 Sol.

Pipe-A $\quad L_A = 1$ m
Pipe-B $\quad L_B = 2$ m
$d_A = d_B$; $h_{fA} = h_{fB}$

$$\frac{32\mu V_A L_A}{w d_A^2} = \frac{32\mu V_B L_B}{w d_B^2}$$

$$\frac{V_B}{V_A} = \frac{d_B^2}{d_A^2} \times \frac{L_A}{L_B}$$

$$Q = AV$$

$$V = \frac{Q}{\frac{\pi}{4}d^2}$$

$$\frac{\frac{Q_B}{\frac{\pi}{4}d_B^2}}{\frac{Q_A}{\frac{\pi}{4}d_A^2}} = \frac{d_B^2}{d_A^2} \times \frac{L_A}{L_B}$$

$$\frac{Q_B}{Q_A} = \frac{d_B^4}{d_A^4} \times \frac{L_A}{L_B} = \frac{1}{2}$$

{$\because$ Given $d_A = d_B$}

6.17 Sol.

Given, flow is steady, fully developed and viscous.

Velocity profile, $V = -C\left[r^2 - \frac{D^2}{4}\right] = -C\left[r^2 - R^2\right]$

$$V = C\left[R^2 - r^2\right] = CR^2\left[1 - \frac{r^2}{R^2}\right]$$

$$\dot{KE} = \frac{1}{2}\int_0^R \rho V^3 dA$$

$$= \rho\int_0^R C^3R^6\left[1 - \frac{r^2}{R^2}\right]^3 \cdot 2\pi r \cdot dr$$

$$= \rho C^3 R^6 2\pi \int_0^R \left[1 - \frac{r^2}{R^2}\right]^3 \cdot r \cdot dr$$

$$= \rho C^3 R^6 2\pi \int_0^R \left[1 - \frac{r^6}{R^6} - \frac{3r^2}{R^2}\left[1 - \frac{r^2}{R^2}\right]\right] \cdot r \cdot dr$$

$$= \rho C^3 R^6 2\pi \int_0^R \left[1 - \frac{r^6}{R^6} - \frac{3r^2}{R^2} + \frac{3r^4}{R^4}\right] \cdot r \cdot dr$$

$$= \rho C^3 2\pi R^6\left[\frac{R^2}{2} - \frac{R^2}{8} - \frac{3R^2}{4} + \frac{R^2}{2}\right]$$

$$= \rho C^3 2\pi R^6\left[R^2 - \frac{R^2}{8} - \frac{3R^2}{4}\right]$$

$$= \rho C^3 2\pi R^6\left[\frac{R^2}{8}\right]$$

$$= \frac{\rho C^3 2\pi}{8} R^8 = \frac{\rho C^3 2\pi}{8 \times (2)^8} \times D^8$$

$KE \propto D^8$

$n = 8$

6.18 (a)

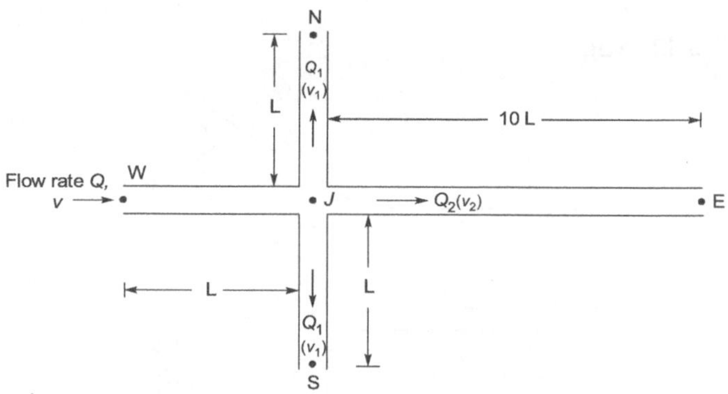

$$P_N = P_E = P_S = P$$

$$Q = Q_1 + Q_2 + Q_1 = 2Q_1 + Q_2$$

$$P_J - P_N = P_J - P_E = P_J - P_S$$

$$\Rightarrow \quad \frac{P_J - P_N}{\rho g} = \frac{P_J - P_E}{\rho g} = \frac{P_J - P_S}{\rho g}$$

$$\Rightarrow \quad (h_f)_{JN} = (h_f)_{JE} = (h_f)_{JS}$$

$$\Rightarrow \quad \frac{fLQ_1^2}{(12.1)d^5} = \frac{f(10L)Q_2^2}{(12.1)d^5} = \frac{fLQ_1^2}{(12.1)d^5}$$

$$\Rightarrow \quad Q_1^2 = 10Q_2^2$$

$$\Rightarrow \quad Q_1 = \sqrt{10}Q_2 \quad ...(iii)$$

Using equation (ii) and (iii)

$$Q = 2\left(\sqrt{10}Q_2\right) + Q_2$$

$$Q = 2\left(\sqrt{10} + 1\right)Q_2$$

$$\Rightarrow \quad Q_2 = \frac{Q}{\left(2\sqrt{10} + 1\right)} = 0.1365\,Q$$

and
$$Q_1 = \sqrt{10}Q_2 = \sqrt{10} \times \frac{Q}{\left(2\sqrt{10} + 1\right)}$$

$$= 0.4317Q$$

Hence, the flow rate in pipe connecting J and E is

$$Q_2 = 0.1365\,Q > \frac{Q}{21}$$

So, statement I is false

and$(\text{Pressure})_N = (\text{Pressure})_E$

$$\Rightarrow \quad P_N = P_E$$

$$\Rightarrow \quad P_J - P_N = P_J - P_E$$

Hence, the pressure difference between J and N is equal to the pressure difference between J and E.

So, statement II is true.

■■■■

7 Turbines and Pumps

7.1 Kaplan turbine is
(a) a high head mixed flow turbine
(b) a low head axial flow turbine
(c) an outward flow reaction turbine
(d) an impulse inward flow turbine **[1997 : 1 M]**

7.2 If there are m physical quantities and n fundamental dimensions in a particular process, the number of non-dimensional parameters is
(a) $m + n$ (b) $m \times n$
(c) $m - n$ (d) m/n **[2002 : 1 M]**

7.3 A centrifugal pump running at 500 rpm and at its maximum efficiency is delivering a head of 30 m at a flow rate of 60 liters per minute. If the rpm is changed to 1000, then the head H in meters and flow rate Q in liters per minute at maximum efficiency are estimated to be
(a) $H = 60$, $Q = 120$ (b) $H = 120$, $Q = 120$
(c) $H = 60$, $Q = 480$ (d) $H = 120$, $Q = 30$
[2003 : 2 M]

7.4 Match **List-I** with **List-II** and select the correct answer using the codes given below the lists:

List-I	List-II
P. Curtis	1. Reaction steam turbine
Q. Rateau	2. Gas turbine
R. Kaplan	3. Velocity compounding
S. Francis	4. Pressure compounding
	5. Impulse water turbine
	6. Axial turbine
	7. Mixed flow turbine
	8. Centrifugal pump

Codes:

	P	Q	R	S
(a)	2	1	1	6
(b)	3	4	6	7
(c)	1	3	1	5
(d)	3	4	7	6

[2003 : 2 M]

7.5 A centrifugal pump is required to pump water to an open water tank situated 4 km away from the location of the pump through a pipe of diameter 0.2 m having Darcy's friction factor of 0.01. The average speed of water in the pipe is 2 m/s. If it is to maintain a constant head of 5 m in the tank, neglecting other minor losses, the absolute discharge pressure at the pump exit is
(a) 0.449 bar (b) 5.503 bar
(c) 44.911 bar (d) 55.203 bar
[2004 : 2 M]

7.6 At a hydro electric power plant site, available head and flow rate are 24.5 m and 10.1 m^3/s respectively. If the turbine to be installed is required to run at 4.0 revolution per second (rps) with an overall efficiency of 90%, the suitable type of turbine for this site is
(a) Francis (b) Kaplan
(c) Pelton (d) Propeller **[2004 : 2 M]**

7.7 Match List-I with List-II and select the correct answer using the codes given below the lists:

List-I
P. Reciprocating pump
Q. Axial flow pump
R. Microhydel plant
S. Backward curved vanes

List-II
1. Plant with power output below 100 kW
2. Plant with power output between 100 kW to 1 MW
3. Positive displacement
4. Draft tube
5. High flow rate, low pressure ratio
6. Centrifugal pump impeller

Codes:

	P	Q	R	S
(a)	3	5	6	2
(b)	3	5	2	6
(c)	3	5	1	6
(d)	4	5	1	6

[2004 : 2 M]

7.8 In a Pelton wheel, the bucket peripheral speed is 10 m/s, the water jet velocity is 25 m/s and volumetric flow rate of the jet is 0.1 m^3/s. If the jet deflection angle is 120° and the flow is ideal, the power developed is
(a) 7.5 kW (b) 15.0 kW
(c) 22.5 kW (d) 37.5 kW **[2006 : 1 M]**

7.9 A large hydraulic turbine is to generate 300 kW at 1000 rpm under a head of 40 m. For initial testing, a 1 : 4 scale model of the turbine operates under

a head of 10 m. The power generated by the model (in kW) will be
(a) 2.34 (b) 4.68
(c) 9.38 (d) 18.75 [2006 : 2 M]

7.10 A horizontal – shaft centrifugal pump lifts water at 65°C. The suction nozzle is one meter below pump centerline. The pressure at this point equals 200 kPa gauge and velocity is 3 m/s. Steam tables show saturation pressure at 65°C is 25 kPa, and specific volume of the saturated liquid is 0.001020 m^3/kg. The pump Net Positive Suction Head (NPSH) in meters is

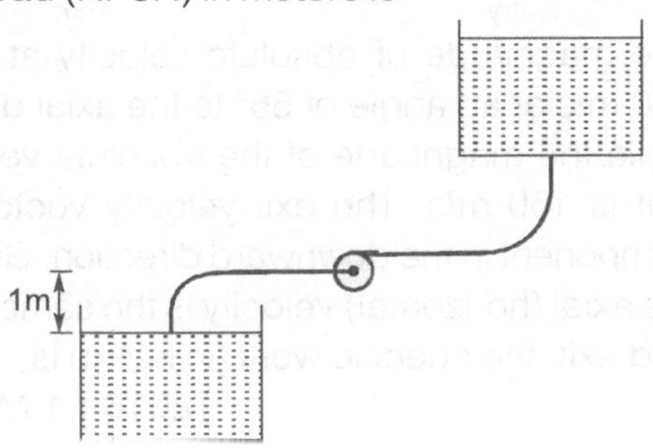

(a) 24 (b) 26
(c) 28 (d) 30 [2006 : 2 M]

7.11 The inlet angle of runner blades of a Francis turbine is 90°. The blades are so shaped that the tangential component of velocity at blade outlet is zero. The flow velocity remains constant throughout the blade passage and is equal to half of the blade velocity at runner inlet. The blade efficiency of the runner is
(a) 25% (b) 50%
(c) 80% (d) 89% [2007 : 2 M]

7.12 A model of a hydraulic turbine is tested at a head of 1/4th of that under which the full scale turbine works. The diameter of the model is half of that of the full scale turbine. If N is the RPM of the full scale turbine, then the RPM of the model will be
(a) N/4 (b) N/2
(c) N (d) 2N [2007 : 2 M]

7.13 Match **List-I** with **List-II** and select the correct answer using the codes given below the lists:

List-I
P. Centrifugal compressor
Q. Centrifugal pump
R. Pelton wheel
S. Kaplan turbine

List-II
1. Axial flow
2. Surging
3. Priming
4. Pure impulse

Codes:

	P	Q	R	S
(a)	2	3	4	1
(b)	2	3	1	4
(c)	3	4	1	2
(d)	1	2	3	4

[2007 : 2 M]

7.14 Water, having a density of 1000 kg/m^3, issues from a nozzle with a velocity of 10 m/s and the jet strikes a bucket mounted on a Pelton wheel. The wheel rotates at 10 rad/s. The mean diameter of the wheel is 1 m. The jet is split into two equal streams by the bucket, such that each stream is deflected by 120° as shown in the figure. Friction in the bucket may be neglected. Magnitude of the torque exerted by the water on the wheel, per unit mass flow rate of the incoming jet, is

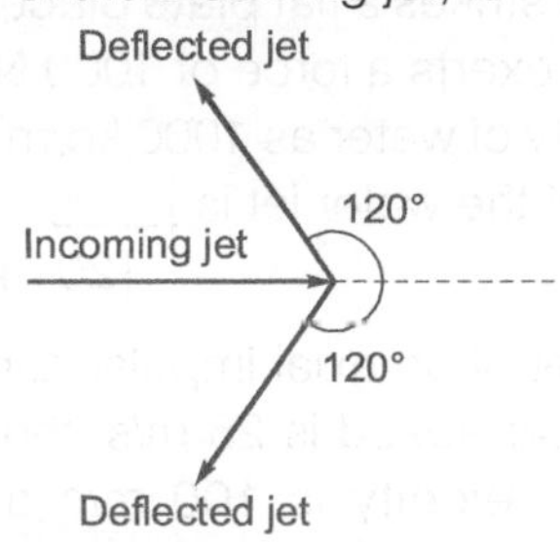

(a) 0 (Nm)/(kg/s) (b) 1.25 (Nm)/(kg/s)
(c) 2.5 (Nm)/(kg/s) (d) 3.75 (Nm)/(kg/s)
[2008 : 2 M]

7.15 A hydraulic turbine develops 1000 kW power for a head of 40 m. If the head is reduced to 20 m, the power developed (in kW) is
(a) 177 (b) 354
(c) 500 (d) 707 [2010 : 1 M]

7.16 A pump handling a liquid raises its pressure from 1 bar to 30 bar. Take the density of the liquid as 990 kg/m^3. The isentropic specific work done by the pump in kJ/kg is
(a) 0.10 (b) 0.30
(c) 2.50 (d) 2.93 [2011 : 2 M]

7.17 The velocity triangles at the inlet and exit of the rotor of a turbomachine are shown. V denotes the absolute velocity of the fluid, W denotes the relative velocity of the fluid and U denotes the blade velocity. Subscripts 1 and 2 refer to inlet and outlet respectively If $V_2 = W_1$ and $V_1 = W_2$, then the degree of reaction is
(a) 0
(b) 1
(c) 0.5
(d) 0.25

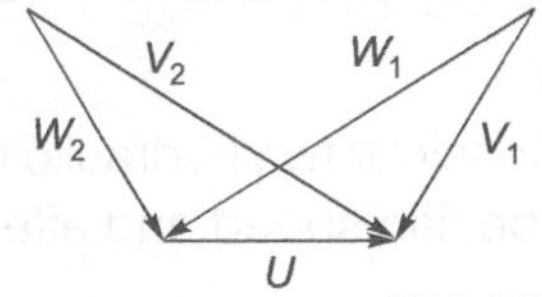

[2012 : 1 M]

7.18 In order to have maximum power form a Pelton turbine, the bucket speed must be
(a) equal to the jet speed
(b) equal to half of the jet speed
(c) equal to twice the jet speed
(d) independent of the jet speed [2013 : 1 M]

7.19 Kaplan water turbine is commonly used when the flow through its runner is
(a) axial and the head available is more than 100 m
(b) axial and the head available is less than 10 m
(c) radial and the head available is more than 100 m
(d) mixed and the head available is about 50 m
[2014 : 1 M, Set-4]

7.20 An ideal water jet with volume flow rate of 0.05 m^3/s strikes a flat plate placed normal to its path and exerts a force of 1000 N. Considering the density of water as 1000 kg/m^3, the diameter (in mm) of the water jet is _____.
[2014 : 2 M, Set-1]

7.21 At the inlet of an axial impulse turbine rotor, the blade linear speed is 25 m/s, the magnitude of absolute velocity is 100 m/s and the angle between them is 25°. The relative velocity and the axial component of velocity remain the same between the inlet and outlet of the blades. The blade inlet and outlet velocity triangles are shown in figure. Assuming no losses, the specific work (in J/kg) is ______.

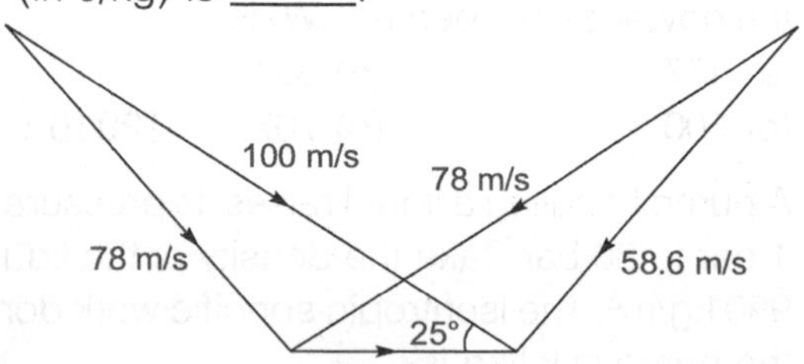

[2014 : 2 M, Set-3]

7.22 Which of the following statements are TRUE, when the cavitation parameter σ = 0?
i. the local pressure is reduced to vapor pressure
ii. cavitation starts
iii. boiling of liquid starts
iv cavitation stops
(a) i, ii and iv (b) only ii and iii
(c) only i and iii (d) i, ii and iii
[2015 : 2 M, Set-3]

7.23 Consider two hydraulic turbines having identical specific speed and effective head at the inlet. If the speed ratio (N_1/N_2) of the two turbines is 2, then the respective power ratio (P_1/P_2) is_________.
[2016 : 1 M, Set-1]

7.24 The blade and fluid velocities for an axial turbine are as shown in the figure.

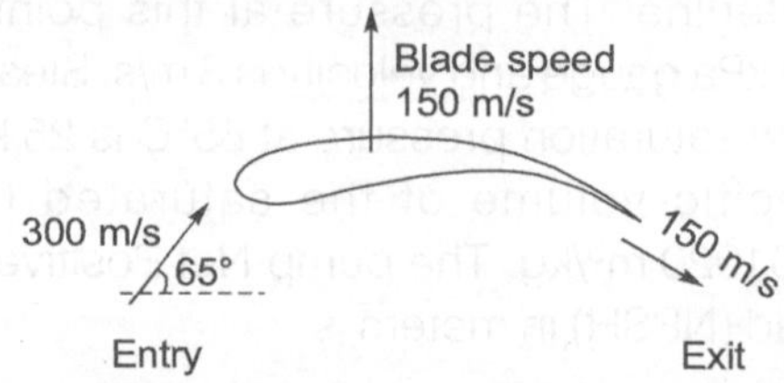

The magnitude of absolute velocity at entry is 300 m/s at an angle of 65° to the axial direction, while the magnitude of the absolute velocity at exit is 150 m/s. The exit velocity vector has a component in the downward direction. Given that the axial (horizontal) velocity is the same at entry and exit, the specific work (in kJ/kg) is_______.
[2016 : 1 M, Set-3]

7.25 Which one of the following is NOT a rotating machine?
(a) Centrifugal pump (b) Gear pump
(c) Jet pump (d) Vane pump
[2017 : 1 M, Set-1]

7.26 Which one of the following statements is TRUE?
(a) Both Pelton and Francis turbines are impulse turbines.
(b) Francis turbine is a reaction turbine but Kaplan turbine is an impulse turbine.
(c) Francis turbine is an axial-flow reaction turbine.
(d) Kaplan turbine is an axial-flow reaction turbine.
[2017 : 1 M, Set-2]

7.27 A 60 mm diameter water jet strikes a plate containing a hole of 40 mm diameter as shown in the figure. Part of the jet passes through the hole horizontally, and the remaining is deflected vertically. The density of water is 1000 kg/m^3. If velocities are as indicated in the figure, the magnitude of horizontal force (in N) required to hold the plate is ________.

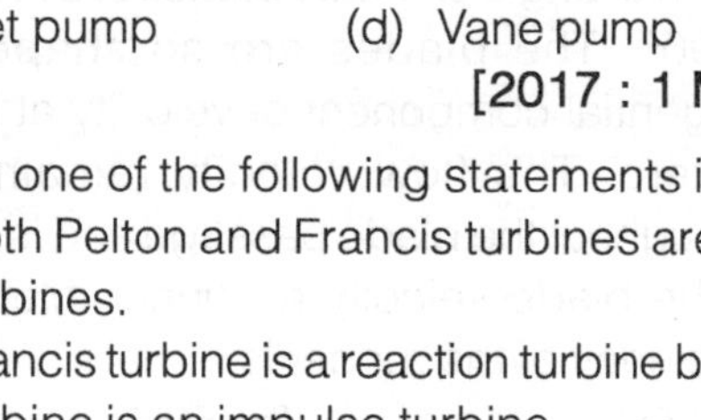

[2017 : 2 M, Set-2]

7.28 For a Pelton wheel with a given water jet velocity, the maximum output power from the Pelton wheel is obtained when the ratio of the bucket speed to the water jet speed is ____(correct to two decimal places). [2018 : 1 M, Set-1]

7.29 A sprinkler shown in the figure rotates about its hinge point in a horizontal plane due to water flow discharged through its two exit nozzles.

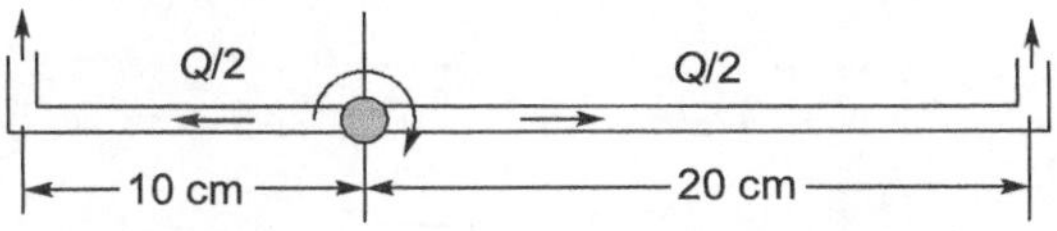

The total flow rate Q through the sprinkler is 1 litre/sec and the cross-sectional area of each exit nozzle is 1 cm^2. Assuming equal flow rate through both arms and a frictionless hinge, the steady state angular speed of rotation (rad/s) of the sprinkler is _____(correct to two decimal places).

[2018 : 2 M, Set-1]

7.30 A test is conducted on a one-fifth scale model of a Francis turbine under a head of 2 m and volumetric flow rate of 1 m^3/s at 450 rpm. Take the water density and the acceleration due to gravity as 10^3 kg/m^3 and 10 m/s^2, respectively. Assume no losses both in model and prototype turbines. The power (in MW) of a full sized turbine while working under a head of 30 m is _______ (correct to two decimal places).

[2018 : 2 M, Set-2]

7.31 As per common design practice, the three types of hydraulic turbine, in descending order of flow rate, are

(a) Pelton, Kaplan, Francis
(b) Francis, Kaplan, Pelton
(c) Kaplan, Francis, Pelton
(d) Pelton, Francis, Kaplan [2019 : 1 M, Set-1]

7.32 An idealized centrifugal pump (blade outer radius of 50 mm) consumes 2 kW power while running at 3000 rpm. The entry of the liquid into the pump is axial and exit from the pump is radial with respect to impeller. If the losses are neglected, then the mass flow rate of the liquid through the pump is _______ kg/s (round off to two decimal places).

[2019 : 2 M, Set-2]

7.33 For a Kaplan (axial flow) turbine, the outlet blade velocity diagram at a section is shown in figure.

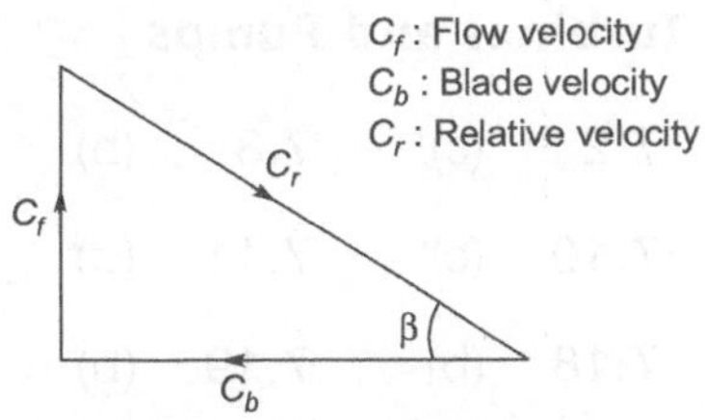

The diameter at this section is 3 m. The hub and tip diameters of the blade are 2 m and 4 m, respectively. The water volume flow rate is 100 m^3/s. The rotational speed of the turbine is 300 rpm. The blade outlet angle β is ___ degrees (round off to one decimal place). [2020 : 2 M, Set-1]

7.34 A cylindrical jet of water (density = 1000 kg/m^3) impinges at the center of a flat, circular plate and spreads radially outwards, as shown in the figure. The plate is resting on a linear spring with a spring constant k = 1 kN/m. The incoming jet diameter is D = 1 cm.

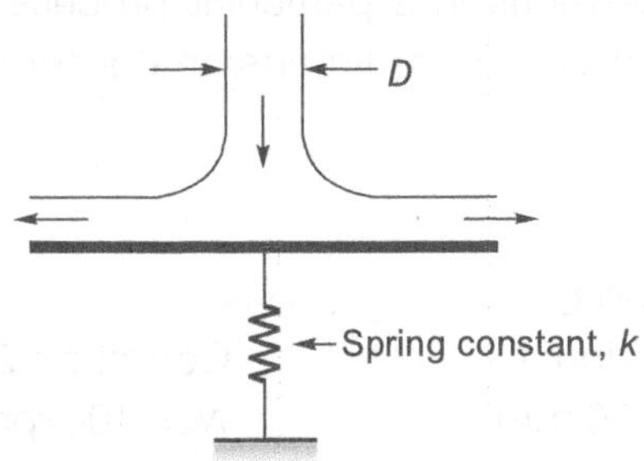

If the spring shows a steady deflection of 1 cm upon impingement of jet, then the velocity of the incoming jet is ________ m/s. [Round off to one decimal place] [2021 : 2 M, Set-1]

7.35 A pressure measurement device fitted on the surface of a submarine, located at a depth H below the surface of an ocean, reads an absolute pressure of 4.2 MPa. The density of sea water is 1050 kg/m^3, the atmospheric pressure is 101 kPa, and the acceleration due to gravity is 9.8 m/s^2. The depth H is _______ m. [Round off to the nearest integer] [2021 : 1 M, Set-1]

7.36 A vertical shaft Francis turbine rotates at 300 rpm. The available head at the inlet to the turbine is 200 m. The tip speed of the rotor is 40 m/s. Water leaves the runner of the turbine without whirl. Velocity at the exit of the draft tube is 3.5 m/s. The head losses in different components of the turbine are: (i) stator and guide vanes: 5.0 m. (ii) rotor: 10 m, and (iii) draft tube: 2 m. Flow rate through the turbine is 20 m^3/s. Take g = 9.8 m/s^2. The hydraulic efficiency of the turbine is___% (round off to one decimal place). [2021 : 2 M, Set-2]

■■■■

Answers Turbines and Pumps

7.1 (b) 7.2 (c) 7.3 (b) 7.4 (b) 7.5 (b) 7.6 (a) 7.7 (b) 7.8 (c)

7.9 (a) 7.10 (c) 7.11 (d) 7.12 (c) 7.13 (a) 7.14 (d) 7.15 (b) 7.16 (d)

7.17 (c) 7.18 (b) 7.19 (b) 7.20 (56.43) 7.21 (3280.5) 7.22 (d) 7.23 (1/4)

7.24 (52.80) 7.25 (c) 7.26 (d) 7.27 (628.32) 7.28 (0.50) 7.29 (10) 7.30 (29.05)

7.31 (c) 7.32 (8.11) 7.33 (12.7) 7.34 (11.3) 7.35 (398) 7.36 (91.2)

Explanations Turbines and Pumps

7.1 (b)

Kaplan turbines are axial flow turbines.

7.2 (c)

According to Buckingham-π theorem, if there are m physical quantities and n fundamental dimensions in a particular process, then the number of non-dimensional π-terms will be $m - n$.

7.3 (b)

Method I:

Condition-1	Condition-2
$N_1 = 500$ rpm	$N_2 = 100$ rpm
$H_1 = 30$ m	$H_2 = ?$
$Q_1 = 60$ litre/min	$Q_2 = ?$

Unit speed: $N_u = \dfrac{N}{\sqrt{H}}$

$$(N_u)_1 = (N_u)_2$$

$$\frac{N_1}{\sqrt{H_1}} = \frac{N_2}{\sqrt{H_2}}$$

$$\frac{500}{\sqrt{30}} = \frac{1000}{\sqrt{H_2}}$$

or $\sqrt{H_2} = \dfrac{1000}{500} \times \sqrt{30} = 2 \times \sqrt{30}$

or $H_2 = 4 \times 30 = 120$ m

Unit speed: $Q_u = \dfrac{Q}{\sqrt{H}}$

$$(Q_u)_1 = (Q_u)_2$$

$$\frac{60}{\sqrt{30}} = \frac{Q_2}{\sqrt{120}}$$

or $Q_2 = 60 \times \sqrt{\dfrac{120}{30}} \Rightarrow 60 \times \sqrt{4}$

or $Q_2 = 120$ litre per minute

Method II:

Head coefficient of pump;

$$\left(\frac{H}{N^2D^2}\right)_1 = \left(\frac{H}{N^2D^2}\right)_2$$

$$\left(\frac{30}{500^2 \times D^2}\right)_1 = \left(\frac{H}{1000^2 \times D^2}\right)_2$$

(diameter $D_1 = D_2$)

$$\frac{30}{500^2} = \frac{H_2}{1000^2}$$

$$H_2 = 120 \text{ m}$$

Flow coefficient of pump;

$$\left(\frac{Q}{ND^3}\right)_1 = \left(\frac{Q}{ND^3}\right)_2$$

$$\frac{60}{500} = \frac{Q_2}{1000}$$

$$Q_2 = 120 \text{ litres/minute}$$

7.4 (b)

Curtis = Velocity compounding

Rateau = Pressure compounding

Kaplan = Axial flow turbine

Francis = Mixed flow turbine

7.5 (b)

Given, $l = 4$ km $= 4000$ m

$d = 0.2$ m

$f = 0.01$

$V = 2$ m/s

$h = 5$ m

$$h_f = \frac{flV^2}{2gd} = \frac{0.01 \times 4000 \times (2)^2}{2 \times 9.81 \times 0.2}$$

$= 40.77$ m

Head produced by the pump,

$$H = h + h_f$$
$$= 5 + 40.77 = 45.77 \text{ m of water}$$

Absolute discharge pressure at the pump exit

$$p_{abs} = \rho g H + p_{atm}$$
$$= 1000 \times 9.81 \times 45.77 + 101325$$
$$= 5.503 \times 105 \text{ Pa}$$
$$= \mathbf{5.503 \text{ bar}}$$

7.6 (a)

Given data:

$$H = 24.5 \text{ m}$$
$$Q = 10.1 \text{ m}^3/\text{s}$$

Speed: $N = 4 \text{ rps} = 4 \times 60 \text{ rpm} = 240 \text{ rpm}$

Overall efficiency,

$$\eta_o = 90\% = 0.90$$

also $$\eta_o = \frac{P}{\rho Q g H}$$

$$0.90 = \frac{P}{1000 \times 10.1 \times 9.81 \times 24.5}$$

or $P = 2184736.05 \text{ W} = 2184.73 \text{ kW}$

Specific speed,

$$N_s = \frac{N\sqrt{P}}{H^{5/4}}$$

where P is in kW

N is in rpm

H is in m

$$\therefore \quad N_s = \frac{240 \times \sqrt{2184.76}}{(24.5)^{5/4}} = \mathbf{205.80 \text{ (SI unit)}}$$

As the specific speed lies between 51–255, the turbine is **Francis**.

7.7 (b)

(P) Reciprocating pump—Positive displacement

(Q) Axial flow pump—High flow rate, low pressure ratio

(R) Microhydel plant — Power output between 100 kW to 1 MW

(S) Backward curved vanes—centrifugal pump impeller

7.8 (c)

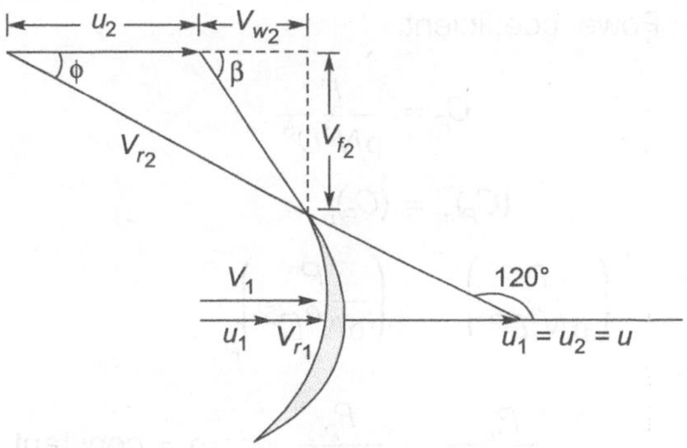

Given data: $u = 10$ m/s; $V_1 = 25$ m/s

$$Q = 0.1 \text{ m}^3/\text{s}$$

Jet deflection angle = 120°

$$\therefore \quad \phi = 180° - 120° = 60°$$

Power developed by Pelton wheel,

$$P = \rho Q (V_1 - u)(1 + \cos\phi)\, u$$
$$= 1000 \times 0.1\,(25 - 10)(1 + \cos 60°) \times 10$$
$$= 22500 \text{ W} = \mathbf{22.5 \text{ kW}}$$

Since, flow is ideal $V_{r_1} = V_{r_2}$

7.9 (a)

Given data:

For Prototype

Power generated, $P_p = 300$ kW

Speed: $N_p = 1000$ rpm

Head: $H_p = 40$ m

For Model

Power generated, $P_m = ?$

Head: $H_m = 10$ m

Scale ratio: $\frac{D_p}{D_m} = 4$

We know that the head coefficient,

$$C_H = \frac{gH}{N^2 D^2}$$

$$(C_H)_m = (C_H)_p$$

$$\left(\frac{gH}{N^2 D^2}\right)_m = \left(\frac{gH}{N^2 D^2}\right)_p$$

$$\frac{H_m}{N_m^2 D_m^2} = \frac{H_p}{N_p^2 D_p^2}$$

$$\frac{H_m}{H_p}\left(\frac{D_p}{D_m}\right)^2 = \frac{(N_m)^2}{N_p^2}$$

$$\frac{10}{40}(4)^2 = \frac{N_m^2}{(100)^2}$$

or $$N_m^2 = 4000000$$

or $N_m = 2000$ rpm

Power coefficient,

$$C_P = \frac{P}{\rho N^3 D^5}$$

$$(C_P)_m = (C_P)_p$$

$$\left(\frac{P}{\rho N^3 D^5}\right)_m = \left(\frac{P}{\rho N^3 D^5}\right)_p$$

$$\frac{P_m}{N_m^3 D_m^5} = \frac{P_p}{N_p^3 D_p^5} \quad \because \rho = \text{constant}$$

$$P_m = P_p\left(\frac{N_m}{N_p}\right)^3\left(\frac{D_m}{D_p}\right)^5$$

$$= 300\left(\frac{2000}{1000}\right)^3\left(\frac{1}{4}\right)^5 = \mathbf{2.34\ kW}$$

<u>**Method II:**</u>

According to model laws of turbine:

$$\left(\frac{P}{\frac{H^3}{2D^2}}\right)_m = \left(\frac{P}{H^{3/2}D^2}\right)_p$$

$$\because \quad P_m = P_P\left(\frac{D_m}{D_p}\right)^2\left(\frac{H_m}{H_p}\right)^{3/2}$$

$$= 300 \times \left(\frac{1}{4}\right)^2 \times \left(\frac{10}{40}\right)^{3/2} = 2.34\ \text{kW}$$

7.10 (c)

Given data:

Temperature of water,

$$T = 65°C$$

Section head,

$$h_s = 1\ \text{m}$$

Gauge pressure at section nozzle inlet,

$$p_{1,g} = 200\ \text{kPa}$$

∴ Absolute pressure,

$$p_{1,abs} = p_{1,g} + p_{atm} = 200 + 101.325$$

$$= 301.325\ \text{kPa} = 301.325 \times 10^3\ \text{Pa}$$

Velocity at section nozzle inlet,

$$V_1 = 3\ \text{m/s}$$

Saturated pressure at 65°C,

$$p_{v,abs} = 25\ \text{kPa, vapour pressure}$$

$$= 25 \times 10^3\ \text{Pa}$$

Specific volume,

$$v = 0.001020\ \text{m}^3/\text{kg}$$

$$\therefore \quad \text{Density: } \rho = \frac{1}{0.001020} = 980.39\ \text{kg/m}^3$$

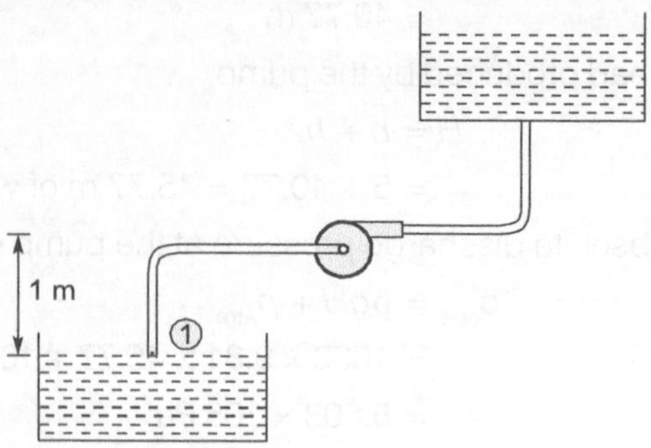

Net positive section head,

$$\text{NPSH} = \frac{p_{s,abs}}{\rho g} + \frac{V_s^2}{2g} - \frac{p_{v,abs}}{\rho g}$$

Applying Bernoulli's equation at section nozzle inlet and section of the pump

$$\frac{p_{1,abs}}{\rho g} + \frac{V_1^2}{2g} + z_1 = \frac{p_{s,abs}}{\rho g} + \frac{V_s^2}{2g} + h_s$$

$$\frac{p_{1,abs}}{\rho g} + \frac{V^2}{2g} + 0 = \frac{p_{s,abs}}{\rho g} + \frac{V_s^2}{2g} + h_s$$

$$\text{or} \quad \frac{p_{s,abs}}{\rho g} + \frac{V_s^2}{2g} = \frac{p_{1,abs}}{\rho g} + \frac{V_1^2}{2g} - h_s$$

$$\therefore \quad \text{NPSH} = \frac{p_{s,abs}}{\rho g} + \frac{V_1^2}{2g} - h_s - \frac{p_{v,abs}}{\rho g}$$

$$= \frac{301.325 \times 10^3}{980.39 \times 9.81} + \frac{(3)^2}{2 \times 9.81} - 1 - \frac{25 \times 10^3}{980.39 \times 9.81}$$

$$= 31.33 + 0.45 - 1 - 2.56$$

$$= 28.22\ \text{m} \approx \mathbf{28\ m}$$

7.11 (d)

Given data: $\beta_i = 90°$; $V_{wo} = 0$

$$V_{f_i} = V_{f_o} = \frac{u_i}{2}$$

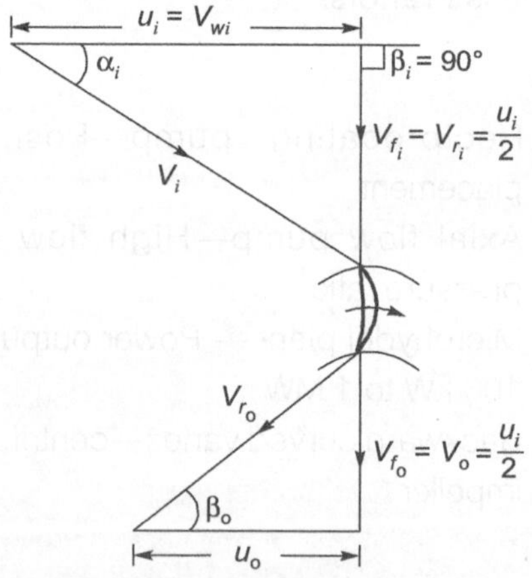

Blade efficiency of the runner,

$$\eta_H = \frac{V_{wi}u_i}{gH} = \frac{u_i^2}{gH} \quad \because V_{wi} = u_i$$

By energy balance equation,

$$\rho QgH = \rho QV_{wi}u_i + \frac{\rho QV_o^2}{2}$$

$$gH = u_i^2 + \frac{V_{fo}^2}{2} = u_i^2 + \frac{u_i^2}{8}$$

$$gH = \frac{9}{8}u_i^2$$

$$\therefore \quad \eta_H = \frac{u_i^2}{\frac{9}{8}u_i^2} = \frac{8}{9} = 0.8888$$

$$= 88.88\% \approx \mathbf{89\%}$$

7.12 (c)

Given data:

Model

$$H_m = \frac{H_p}{4};\quad D_m = \frac{D_p}{2}$$

$$N_m = ?$$

Prototype (full scale turbine)

$$N_p = N$$

We know that head coefficient,

$$C_H = \frac{H}{(ND)^2}$$

$$(C_H)_{\text{model}} = (C_H)_{\text{prototype}}$$

$$\frac{H_m}{N_m^2 D_m^2} = \frac{H_p}{N_p^2 D_p^2}$$

$$\frac{H_p/4}{N_m^2 \times \frac{D_p^2}{4}} = \frac{H_p}{N^2 \times D_p^2}$$

or $\quad N_m^2 = N^2$

or $\quad N_m = \mathbf{N}$

7.13 (a)

(P) Centrifugal compressor — Surging
(Q) Centrifugal pump — Priming
(R) Pelton wheel — Pure impulse
(Q) Kaplan turbine — Axial flow

7.14 (d)

Given data:

$$\rho = 1000 \text{ kg/m}^3$$

$$V_i = 10 \text{ m/s}$$

$$\omega = 10 \text{ rad/s}$$

$$D = 1 \text{ m}$$

$$u = \omega R = \frac{\omega D}{2} = 10 \times \frac{1}{2} = 5 \text{ m/s}$$

Angle of deflection = 120°

$$\therefore \quad \beta_o = 180° - 120° = 60°$$

Force exerted by the water of the wheel per unit mass flow rate,

$$F = (V_i - u)(1 + \cos\beta_o) \quad \text{N/(kg/s)}$$

$$= (10 - 5)(1 + \cos 60°)$$

$$= \mathbf{7.5\ N/kg/s}$$

Torque exerted by the water on the wheel per unit mass flow rate,

$$T = F\frac{D}{2} = 7.5 \times \frac{1}{2}$$

$$= \mathbf{3.75\ Nm/(kg/s)}$$

7.15 (b)

Condition-1 $P_1 = 1000$ kW; $H_1 = 40$ m

Condition-2 $P_2 = ?$; $H_2 = 20$ m

We know that the unit power,

$$P_u = \frac{P}{H^{3/2}}$$

$$(P_u)_1 = (P_u)_2$$

$$\frac{P_1}{H_1^{3/2}} = \frac{P_2}{H_2^{3/2}}$$

$$\frac{1000}{(40)^{3/2}} = \frac{P_2}{(20)^{3/2}}$$

or $\quad P_2 = 1000 \times \dfrac{(20)^{3/2}}{(40)^{3/2}}$

$$= \frac{1000}{(2)^{3/2}} = \frac{1000}{2^{1.5}} = 353.55 \text{ kW}$$

$$= \mathbf{354\ kW}$$

7.16 (d)

Given data:

Pressure of liquid at pump inlet,

$$p_1 = 1 \text{ bar} = 100 \text{ kPa}$$

Pressure of liquid at pump,

$$p_2 = 30 \text{ bar} = 3000 \text{ kPa}$$

Density of liquid,

$$\rho = 990 \text{ kg/m}^3$$

Specific pump input work,

$$w_p = v(p_2 - p_1)$$

$$= \frac{(p_2 - p_1)}{\rho} \quad \because v = 1/\rho$$

$$= \frac{3000 - 100}{990} = 2.929 \text{ kJ/kg}$$

$$= \mathbf{2.93\ kJ/kg}$$

7.17 (c)

Degree reaction,

$$R = \frac{\text{Static pressure change in the runner}}{\text{Total energy transfer in the runner}}$$

where static pressure change,

$$= \frac{W_2^2 - W_1^2}{2} + \frac{U_1^2 - U_2^2}{2}$$

$$= \frac{W_2^2 - W_1^2}{2} \qquad \because U_1 = U_2$$

Total energy transfer,

$$= \frac{V_1^2 - V_2^2}{2} + \frac{W_2^2 - W_1^2}{2} + \frac{U_1^2 - U_2^2}{2}$$

$$= \frac{W_2^2 - W_1^2}{2} + \frac{W_2^2 - W_1^2}{2} = W_2^2 - W_1^2$$

$$\therefore \qquad R = \frac{\frac{W_2^2 - W_1^2}{2}}{W_2^2 - W_1^2} = \frac{1}{2} = 0.5$$

7.18 (b)

Condition for maximum power form a Pelton turbine

Bucket speed = Half of the jet speed

$$u = \frac{V_i}{2}$$

7.19 (b)

Kaplan water turbine is commonly used when the flow through its runner is axial and the head available is less than 10 m.

Kaplan/Propeller turbine:

- Axial flow turbine
- Low head: $3 < H < 80$ m

Francis turbine:

- Mixed flow turbine
- Medium head: $80 < H < 600$ m

Pelton turbine:

- Tangential flow turbine
- High head: $600 < H \leq 2000$ m

7.20 Sol.

$$Q = 0.05 \text{ m}^3/\text{s}$$

$$F = 1000 \text{ N}$$

$$\rho = 1000 \text{ kg/m}^3$$

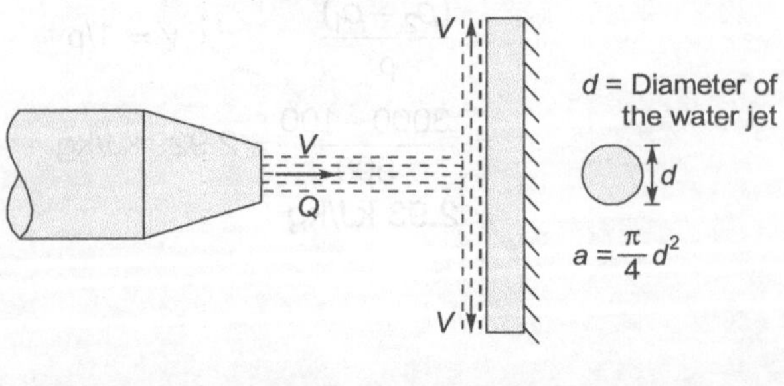

$$F = \rho a V^2$$

where a = cross-section area of the jet

$$Q = aV$$

$$\therefore \qquad F = \rho QV$$

$$1000 = 1000 \times 0.05 \times V$$

or $$V = 20 \text{ m/s}$$

$$Q = aV = \frac{\pi}{4} d^2 V$$

$$0.05 = \frac{3.14}{4} \times d^2 \times 20$$

or $$d^2 = 0.00318$$

or $$d = 0.05643 \text{ m}$$

$$= \mathbf{56.43 \text{ mm}}$$

7.21 Sol.

Given data: $u = 25$ m/s; $V_i = 100$ m/s; $\alpha_i = 25°$

$$V_{r_i} = V_{r_o} = 78 \text{ m/s}$$

$$V_o = 58.6 \text{ m/s}$$

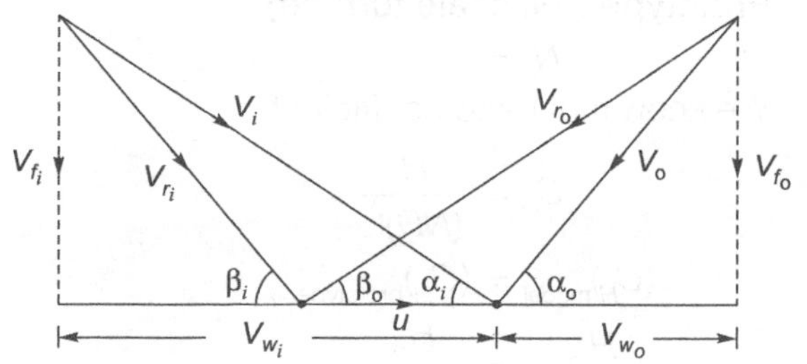

$$\cos \alpha_i = \frac{V_{wi}}{V_i}$$

$$\cos 25° = \frac{V_{wi}}{100}$$

or $$V_{w_i} = 100 \times \cos 25° = 90.36 \text{ m/s}$$

$$\sin \alpha_i = \frac{V_{f_i}}{V_i}$$

$$\sin 25° = \frac{V_{f_i}}{100}$$

or $$V_{f_i} = 100 \times \sin 25° = 42.26 \text{ m/s}$$

$$V_{wo} = \sqrt{V_o^2 - V_{fo}^2}$$

$$= \sqrt{(58.6)^2 - (42.26)^2} = 40.59 \text{ m/s}$$

Specific work,

$$w = (V_{w_i} + V_{wo})u$$

$$= (90.63 + 40.59) \times 25$$

$$= \mathbf{3280.5 \text{ J/kg}}$$

7.22 (d)

Cavitation parameter: $\sigma = 0$

It means:

(i) The local pressure is reduce to vapour pressure

(ii) Cavitation starts

(iii) Boiling of liquid starts

7.23 Sol.

Given data:

$$N_{s1} = N_{s2}$$
$$H_1 = H_2$$
$$\frac{N_1}{N_2} = 2$$

We know that specific speed of the turbine,

$$N_s = \frac{N\sqrt{P}}{H^{5/4}}$$

$$(N_s)_1 = (N_s)_2$$

$$\frac{N_1\sqrt{P_1}}{H_1^{5/4}} = \frac{N_2\sqrt{P_2}}{H_2^{5/4}}$$

$$N_1\sqrt{P_1} = N_2\sqrt{P_2} \qquad \because H_1 = H_2$$

or $$\sqrt{\frac{P_1}{P_2}} = \frac{N_2}{N_1}$$

or $$\frac{P_1}{P_2} = \left(\frac{N_2}{N_1}\right)^2 = \left(\frac{1}{2}\right)^2 = \frac{1}{4}$$

7.24 Sol.

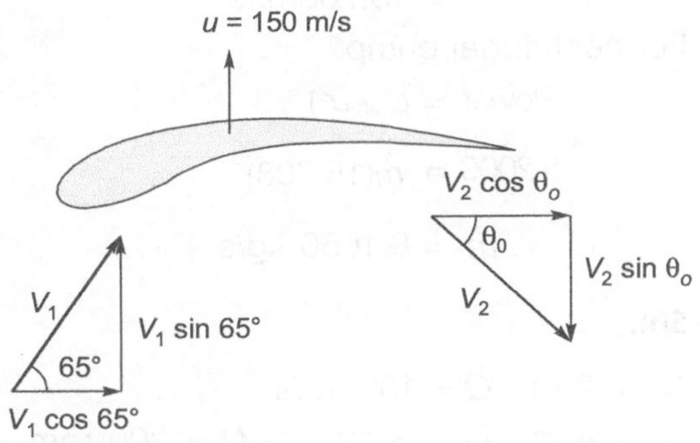

Given data:

Axial (horizontal) velocity is the same at entry and exit.

$$V_1 = 300 \text{ m/s}$$
$$\theta_i = 65°$$
$$V_2 = 150 \text{ m/s}, \; u = 150 \text{ m/s}$$

Given condition,

$$V_1 \cos 65° = V_2 \cos \theta_o$$
$$300 \cos 65° = 150 \times \cos \theta_o$$

or $$\cos \theta_o = 0.8452$$

or $$\theta_o = 32.30°$$

Specific work,

$$w = [V_1 \sin 65° - (-V_2 \sin 32.30°)]u$$
$$= [300 \sin 65° + 150 \sin 32.30°] \times 150$$
$$= 52806.77 \text{ J/kg} = \mathbf{52.80 \text{ kJ/kg}}$$

7.25 (c)

- Jet pumps, also known as ejector pump are devices capable of handling and transporting al forms of motive fluid including gas, steam or liquids.
- Jet pumps are less efficient than typical centrifugal pumps.

7.27 Sol.

$$\rho = 1000 \text{ kg/m}^3$$

Force, F = rate of change of momentum

$$F = (\dot{m}V)_{in} - (\dot{m}V)_{out}$$

$$F = \rho\left[\frac{\pi}{4}(D_1^2 - D_2^2)\right] \times V^2$$

$$= 1000\frac{\pi}{4} \times \left[0.06^2 - 0.04^2\right] \times 20^2$$

$$= \mathbf{628.32 \text{ N}}$$

7.28 Sol.

In Pelton wheel turbine for maximum output power,

$$\frac{u}{V_1} = \frac{1}{2} = 0.50$$

7.29 Sol.

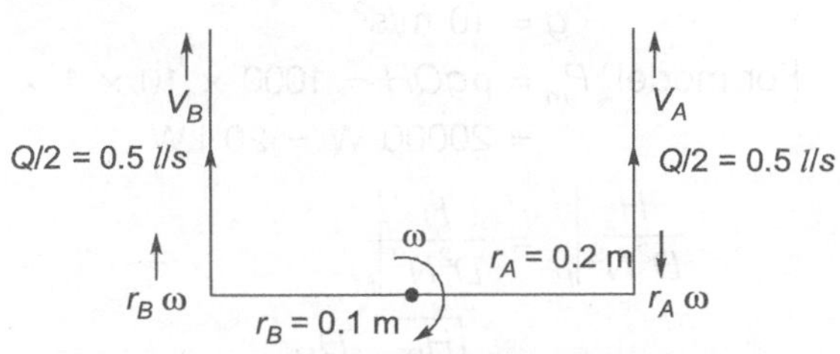

Relative velocities of water with sprinkler

$$V_A = \frac{Q/2}{A} = \frac{1 \times 10^{-3}}{2 \times 10^{-4}} = 5 \text{ m/s}$$

$$V_B = 5 \text{ m/s}$$

Absolute velocity from B side

$$V'_{Abs} - (+r_B\omega) = V_B$$

$$V'_{Abs} = V_B + r_B\omega = 5 + 0.1\,\omega$$

Absolute velocity from A side

$$V_{Abs} - (-r_A\omega) = V_A$$

$$V_{Abs} = V_A - r_A\omega = 5 - 0.2\,\omega$$

The external torque to the sprinkler is zero.

So, $\Sigma T = 0$

$$\dot{m}_A V_{Abs} r_A - \dot{m}_B V'_{Abs} r_B = 0$$

$$\rho\left(\frac{Q}{2}\right)\{5-0.2\,\omega\}0.2-\rho\frac{Q}{2}\{5+0.1\,\omega\}0.1=0$$

$$1 - 0.04\,\omega - 0.5 - 0.01\,\omega = 0$$

$$0.05\,\omega = 0.5$$

$$\omega = \mathbf{10\ rad/s}$$

POINTS TO REMEMBER

Sprinkler is a type of reaction machine, in which water comes out at a high velocity from the rotor in a tangential direction. The essential feature of the rotor is that water enters at high pressure and this pressure energy is transfomed into kinetic energy by a nozzle which is a part of the rotor itself.

7.30 Sol.

Method I:

$$\frac{D_m}{D_p} = \frac{1}{5}$$

Model	Prototype
$H = 2$ m	$H = 30$ m
$Q = 1$ m³/s;	$P = ?$

$$N = 450 \text{ rpm}$$

$$\rho = 1000 \text{ kg/m}^3$$

$$g = 10 \text{ m/s}^2$$

For model $P_m = \rho gQH = 1000 \times 10 \times 1 \times 2$

$$= 20000 \text{ W} = 20 \text{ kW}$$

$$\left.\frac{H}{D^2N^2}\right|_P = \left.\frac{H}{D^2N^2}\right|_M$$

$$N_P = \sqrt{\frac{H_P}{H_M}} \times \frac{D_M}{D_P} \times N_M$$

$$= \sqrt{\frac{30}{2}} \times \frac{1}{5} \times 450 = 348.56 \text{ rpm}$$

$$\left.\frac{P}{D^5N^3}\right|_P = \left.\frac{P}{D^5N^3}\right|_M$$

$$P_P = P_M \times \frac{D_P^5}{D_M^5} \times \frac{N_P^3}{N_M^3}$$

$$P_P = 20 \times 5^5 \times \frac{348.56^3}{450^3}$$

$$= 29045.24 \text{ kW} = \mathbf{29.05\ MW}$$

Method II:

For model, $P_m = \rho gQH = 1000 \times 10 \times 1 \times 2$

$$= 20000 \text{ W} = 20 \text{ kW}$$

From the model laws of turbine;

$$\left(\frac{P}{H^{3/2}D^2}\right)_m = \left(\frac{P}{H^{3/2}D^2}\right)_p$$

$$P_p = p_m \times \left(\frac{H_p}{H_m}\right)^{3/2}\left(\frac{D_p}{D_m}\right)^2$$

$$= 20000 \times \left(\frac{30}{2}\right)^{3/2} (5)^2$$

$$= 29047375.1 \text{ W}$$

$$= 29.04 \text{ MW}$$

7.31 (c)

Pelton turbine is high head and low mass flow rate turbine. Kaplan turbine is low head and high mass flow rate turbine. Francis turbine is medium head, medium mass flow rate turbine.

7.32 Sol.

$$\text{Power} = 2 \text{ kW} = 2000 \text{ W}$$

$$N = 3000 \text{ rpm}$$

$$D = 0.05 \text{ m} \times 2 = 0.1 \text{ m}$$

$$u = \frac{\pi DN}{60} = \frac{\pi \times 0.1 \times 3000}{60}$$

$$= 15.708 \text{ m/s}$$

For centrifugal pump,

$$\text{Power} = \rho Q(u^2)$$

$$2000 = \dot{m}(15.708)^2$$

$$\dot{m} = 8.1056 \text{ kg/s}$$

7.33 Sol.

$D_b = 2$ m $\quad Q = 100$ m³/s

$D_o = 4$ m $\quad D = 3$ m; $\quad N = 300$ rpm,

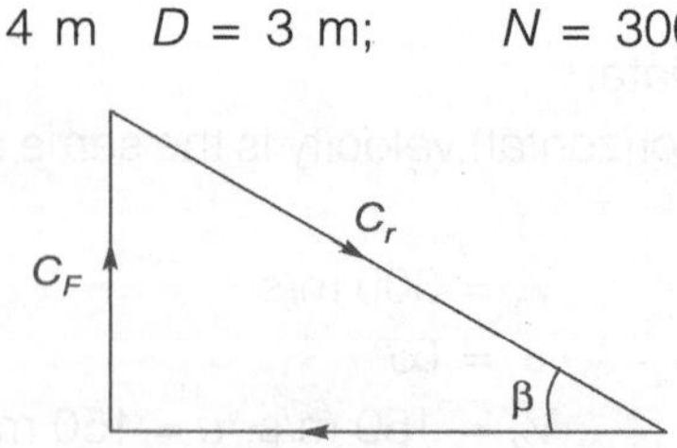

$$\tan\beta = \frac{C_F}{C_b} = \frac{V_{F2}}{u_2}$$

$$u_2 = C_b = \frac{\pi DN}{60} = \frac{\pi \times 3 \times 300}{60}$$

$$u_2 = C_b = 47.12 \text{ m/s}$$

$$Q = \frac{\pi}{4}(4^2 - 2^2) \times V_{F2}$$

$$V_{F2} = C_F = 10.61 \text{ m/s}$$

$$\tan\beta = \frac{10.61}{47.12}$$

$$\Rightarrow \quad \beta = 12.69° \simeq 12.7°$$

7.34 Sol.

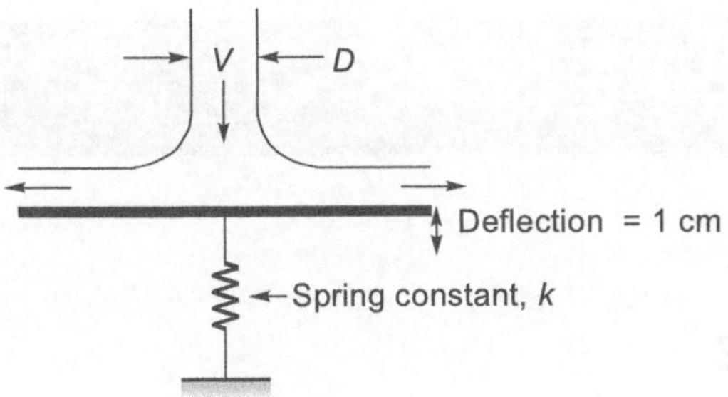

$$\delta = 1 \text{ cm}$$

$$D = 1 \text{ cm}$$

$$\rho = 1000 \text{ kg/m}^3$$

$$K = 1 \text{ kN-m}$$

Force due to jet = Spring force

$$\rho A V^2 = kx$$

$$1000 \times \frac{\pi}{4} \times (10^{-2})^2 V^2 = 1000 \times 10^{-2}$$

$$V = 11.28 \text{ m/s}$$

$$V = 11.3 \text{ m/s}$$

7.35 Sol.

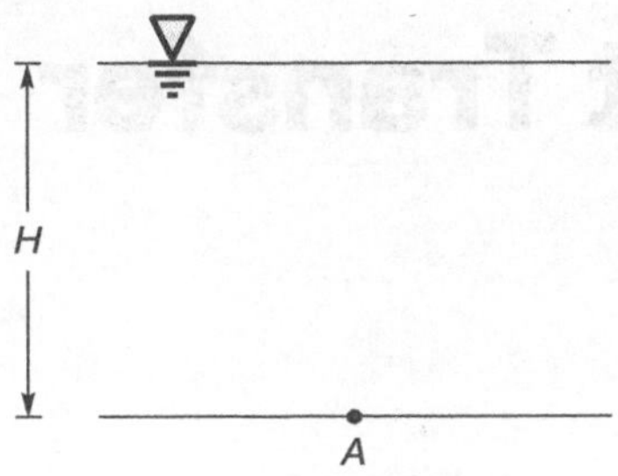

$$P_A = 4.2 \text{ MPa} \quad \text{[Absolute pressure]}$$

$$P_{atm} = 101 \text{ kPa}$$

$$\rho = 1050 \text{ kg/m}^3$$

$$g = 9.8 \text{ m/s}^2$$

$$P_A = P_{atm} + \rho g H$$

$$4.2 \times 10^6 = (101 \times 10^3) + [1050 \times 9.81 \times H]$$

$$H = 397.94 \text{ or } 398 \text{ m}$$

7.36 Sol.

$$H_{inlet} = H_{turbine} + H_{L,\text{stator + Guide}} + H_{rotor} + H_{draftuen} + \frac{V_3^2}{2g}$$

$$200 = H_{turbine} + 5 + 10 + 2 + \frac{3.5^2}{2 \times 9.8}$$

$$H_{turbine} = 182.375 \text{ m}$$

$$\eta_H = \frac{H_{turbine}}{H_{inlet}} = \frac{182.375}{200}$$

$$= 0.9118 = 91.187\%$$

■■■■

Heat Transfer

UNIT

VII

CONTENTS

Heat Transfer

Syllabus

Modes of heat transfer; one dimensional heat conduction, resistance concept and electrical analogy, heat transfer through fins; unsteady heat conduction, lumped parameter system, Heisler's charts; thermal boundary layer, dimensionless parameters in free and forced convective heat transfer, heat transfer correlations for flow over flat plates and through pipes, effect of turbulence; heat exchanger performance, LMTD and NTU methods; radiative heat transfer, Stefan- Boltzmann law, Wien's displacement law, black and grey surfaces, view factors, radiation network analysis.

Analysis of Previous GATE Papers

Exam Year	1 Mark Ques.	2 Marks Ques.	3 Marks Ques.	5 Marks Ques.	Total Marks
1995	–	–	–	2	10
1996	1	2	–	–	5
1997	–	1	–	1	7
1998	1	1	–	4	23
1999	–	1	–	1	7
2000	–	1	–	1	7
2001	2	2	–	1	11
2002	2	2	–	1	11
2003	1	4	–	–	9
2004	1	3	–	–	7
2005	3	6	–	–	15
2006	1	2	–	–	5
2007	–	7	–	–	14
2008	1	4	–	–	9
2009	1	4	–	–	9
2010	–	2	–	–	4
2011	2	3	–	–	8
2012	3	4	–	–	11
2013	3	4	–	–	11
2014 Set-1	2	2	–	–	6

Exam Year	1 Mark Ques.	2 Marks Ques.	Total Marks
2014 Set-2	2	3	8
2014 Set-3	1	3	7
2014 Set-4	2	3	8
2015 Set-1	2	2	6
2015 Set-2	2	3	8
2015 Set-3	2	2	6
2016 Set-1	1	3	7
2016 Set-2	2	2	6
2016 Set-3	3	2	7
2017 Set-1	–	2	4
2017 Set-2	1	1	3
2018 Set-1	–	1	2
2018 Set-2	1	2	5
2019 Set-1	2	2	6
2019 Set-2	3	2	7
2020 Set-1	2	1	4
2020 Set-2	1	2	5
2021 Set-1	2	2	6
2021 Set-2	1	2	5
2022 Set-1	2	3	8
2022 Set-2	2	3	8
2023	3	3	9
2024	3	3	9

1 Conduction

1.1 For a current carrying wire of 20 mm diameter exposed to air (h = 20 W/m^{2}K), maximum heat dissipation occurs when thickness of insulation (0.5 W/mK) is
(a) 30 mm (b) 25 mm
(c) 20 mm (d) 15 mm **[1996 : 2 M]**

1.2 For a given heat flow and for the same thickness, the temperature drop across the material will be maximum for
(a) Copper (b) Steel
(c) Glass wool (d) Refractory brick
[1996 : 1 M]

1.3 The temperature variation under steady heat conduction across a composite slab of two materials with thermal conductivities k_1 and k_2 is shown in figure then, which one of the following statements holds?

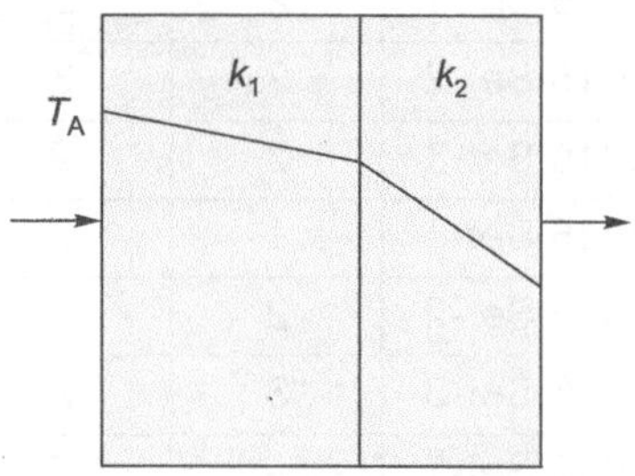

(a) $k_1 > k_2$ (b) $k_1 = k_2$
(c) $k_1 = 0$ (d) $k_1 < k_2$ **[1998 : 2 M]**

1.4 It is proposed to coat a 1 mm diameter wire with enamel paint (k = 0.1 W/mK) to increase heat transfer with air. If the air side heat transfer coefficient is 100 W/m^{2}K , the optimum thickness of enamel paint should be
(a) 0.25 mm (b) 0.5 mm
(c) 1 mm (d) 2 mm **[1999 : 2 M]**

1.5 In descending order of magnitude, the thermal conductivity (a) Pure iron (b) Liquid water (c) Saturated water vapour (d) Pure aluminum can be arranged as
(a) abcd (b) bcad
(c) dabc (d) dcba **[2001 : 1 M]**

1.6 One dimensional unsteady state heat transfer equation for a sphere with heat generation at the rate of q can be written

(a) $\frac{1}{r}\frac{\partial}{\partial r}\left(r\frac{\partial T}{\partial r}\right)+\frac{q}{k}=\frac{1}{\alpha}\frac{\partial T}{\partial t}$

(b) $\frac{1}{r^2}\frac{\partial}{\partial r}\left(r^2\frac{\partial T}{\partial r}\right)+\frac{q}{k}=\frac{1}{\alpha}\frac{\partial T}{\partial t}$

(c) $\frac{\partial^2 T}{\partial r^2}+\frac{q}{k}=\frac{1}{\alpha}\frac{\partial T}{\partial t}$

(d) $\frac{\partial^2}{\partial r^2}(rT)+\frac{q}{k}=\frac{1}{\alpha}\frac{\partial T}{\partial t}$ **[2004 : 1 M]**

1.7 A stainless steel tube (k_s = 19 W/mK) of 2 cm *ID* and 5 cm *OD* is insulated with 3 cm thick asbestos (k_a = 0.2 W/mK). If the temperature difference between the inner most and outermost surfaces is 600°C, the heat transfer rate per unit length is
(a) 0.94 W/m (b) 9.44 W/m
(c) 944.72 W/m (d) 9447.21 W/m
[2004 : 2 M]

1.8 A well machined steel plate of thickness L is kept such that the wall temperature are T_h and T_c as seen in the figure below. A smooth copper plate of the same thickness L is now attached to the steel plate without any gap as indicated in the figure below. The temperature at the interface is T_i. The temperatures of the outer walls are still the same at T_h and T_c. The heat transfer rates are q_1 and q_2 per unit area in the two cases respectively in the direction shown. Which of the following statements is correct?

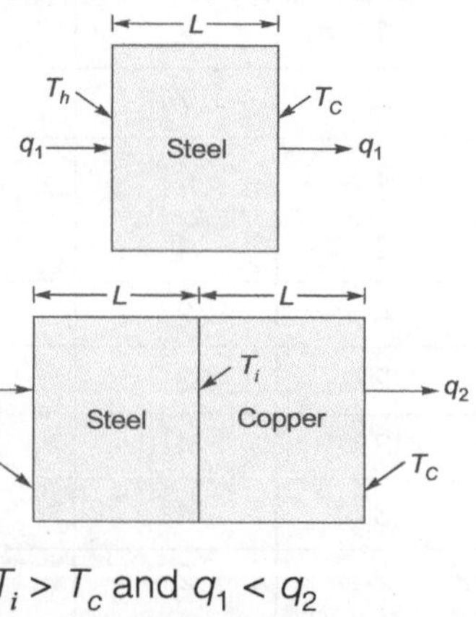

(a) $T_h > T_i > T_c$ and $q_1 < q_2$
(b) $T_h < T_i < T_c$ and $q_1 = q_2$
(c) $T_h = (T_i + T_c)/2$ and $q_1 > q_2$
(d) $T_i < (T_h + T_c)/2$ and $q_1 > q_2$ **[2005 : 1 M]**

1.9 a case of one dimensional heat conduction in a medium with constant properties, T is the temperature at position x, at time t. Then $\frac{\partial T}{\partial t}$ is proportional to

(a) $\frac{T}{x}$ (b) $\frac{\partial T}{\partial x}$

(c) $\frac{\partial^2 T}{\partial x \partial t}$ (d) $\frac{\partial^2 T}{\partial x^2}$ [2005 : 1 M]

1.10 Heat flows through a composite slab, as shown below. The depth of the slab is 1 m. The k values are in W/mK. The overall thermal resistance in K/W is

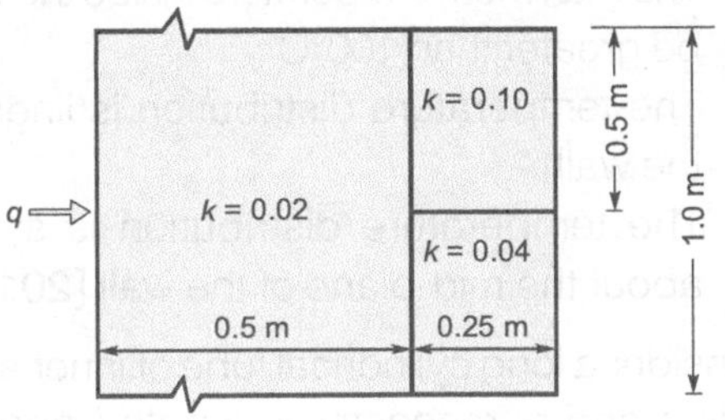

(a) 17.2 (b) 21.9
(c) 28.6 (d) 39.2 [2005 : 2 M]

1.11 In a composite slab, the temperature at the interface (T_{inter}) between two materials is equal to the average of the temperature at the two ends. Assuming steady one dimensional heat conduction, which of the following statements is true about the respective thermal conductivities?

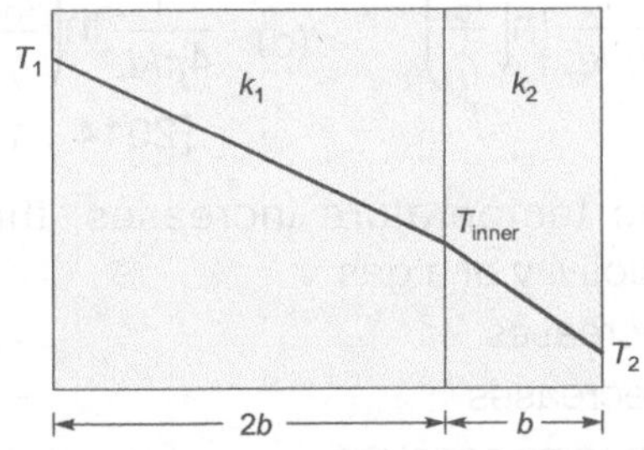

(a) $2k_1 = k_2$ (b) $k_1 = k_2$
(c) $2k_1 = 3k_2$ (d) $k_1 = 2k_2$
[2006 : 1 M]

1.12 With an increase in the thickness of insulation around a circular pipe, heat loss to surroundings due to
(a) convection increases, while that due to conduction decreases
(b) convection decreases, while that due to conduction increases
(c) convection and conduction decreases
(d) convection and conduction increases
[2006 : 2 M]

1.13 A long glass cylinder of inner diameter = 0.03 m and outer diameter = 0.05 m carries hot fluid inside. If the thermal conductivity of glass = 1.05 W/mK, the thermal resistance (K/W) per unit length of the cylinder is
(a) 0.031 (b) 0.077
(c) 0.17 (d) 0.34 [2007 : 2 M]

1.14 Heat is being transferred conductively from a cylindrical nuclear reactor fuel rod of 50 mm diameter to water at 75°C, under steady state condition, the rate of heat generation within the fuel element is 10^6 W/m^3 and the convective heat transfer coefficient is 1 kW/m^{2}K, the outer surface temperature of the fuel element would be
(a) 700 K (b) 625 K
(c) 360 K (d) 400 K [2007 : 2 M]

Linked Answer Questions Q.1.15 and Q.1.16:
Consider steady one-dimensional heat flow in a plate of 20 mm thickness with a uniform heat generation of 80 MW/m^3. The left and right faces are kept at constant temperatures of 160°C and 120°C respectively. The plate has a constant thermal conductivity of 200 W/mK.

1.15 The location of maximum temperature within the plate from its left face is
(a) 15 mm (b) 10 mm
(c) 5 mm (d) 0 mm [2007 : 2 M]

1.16 The maximum temperature within the plate in °C is
(a) 160 (b) 165
(c) 200 (d) 250 [2007 : 2 M]

1.17 Steady two-dimensional heat conduction takes place in the body shown in the figure below. The normal temperature gradients over surface P and Q can be considered to be uniform. The temperature gradient $\frac{\partial T}{\partial x}$ at surface Q is equal to 10 K/m. Surfaces P and Q are maintained at constant temperatures as shown in the figure, while the remaining part of the boundary is insulated. The body has a constant thermal conductivity of 0.1 W/mK. The values of $\frac{\partial T}{\partial y}$ and $\frac{\partial T}{\partial x}$ at surface P are

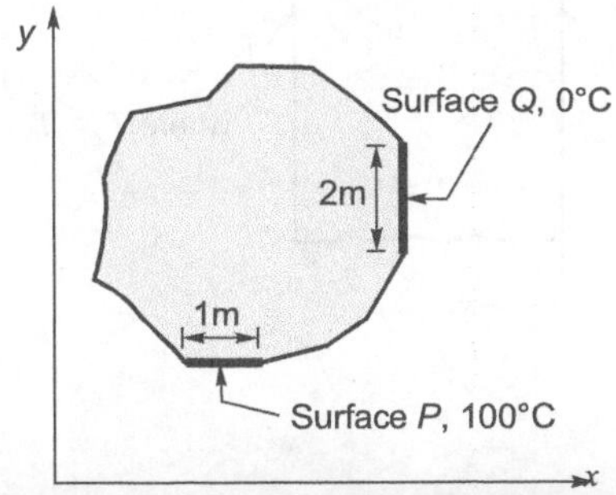

(a) $\frac{\partial T}{\partial x} = 20$ K/m, $\frac{\partial T}{\partial y} = 0$ K/m

(b) $\frac{\partial T}{\partial x} = 0$ K/m, $\frac{\partial T}{\partial y} = 10$ K/m

(c) $\frac{\partial T}{\partial x} = 10$ K/m, $\frac{\partial T}{\partial y} = 10$ K/m

(d) $\frac{\partial T}{\partial x} = 0$ K/m, $\frac{\partial T}{\partial y} = 20$ K/m **[2008 : 2 M]**

1.18 A coolant fluid at 30°C flows over a heated flat plate maintained at a constant temperature of 100°C. The boundary layer temperature distribution at a given location on the plate may be approximated as $T = 30 + 70\exp(-y)$ where y (in m) is the distance normal to the plate and T is in °C. If thermal conductivity of the fluid is 1.0 W/mK, the local convective heat transfer coefficient (in W/m²K) at that location will be

(a) 0.2 (b) 1

(c) 5 (d) 10 **[2009 : 1 M]**

1.19 A pipe of 25 mm outer diameter carries steam. The heat transfer coefficient between the cylinder and surroundings is 5 W/m²K. It is proposed to reduce the heat loss from the pipe by adding insulation having a thermal conductivity of 0.05 W/mK. Which one of the following statements is TRUE?

(a) The outer radius of the pipe is equal to the critical radius.

(b) The outer radius of the pipe is less than the critical radius.

(c) Adding the insulation will reduce the heat loss.

(d) Adding the insulation will increase the heat **[2011 : 1 M]**

1.20 Consider one-dimensional steady state heat conduction, without heat generation, in a plane wall; with boundary conditions as shown in the figure below. The conductivity of the wall is given by $k = k_0 + bT$; where k_0 and b are positive constant, and T is temperature.

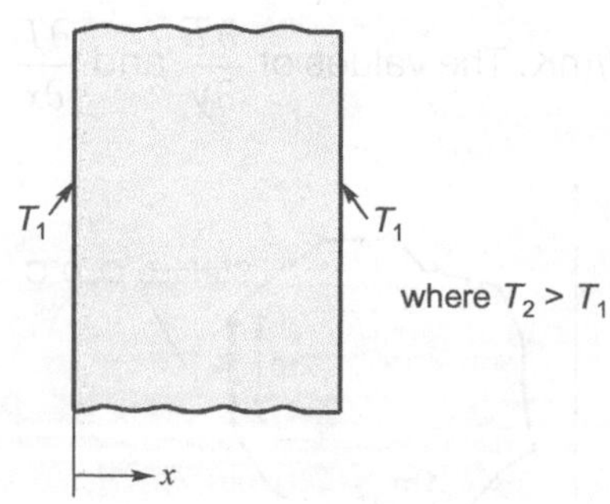

As x increases, the temperature gradient (dT/dx) will

(a) remains constant (b) be zero

(c) increase (d) decrease **[2013 : 1 M]**

1.21 Consider one-dimensional steady state heat conducting along x-axis ($0 \le x \le L$), through a plane wall with the boundary surfaces ($x = 0$ and $x = L$) maintained at temperature of 0°C and 100°C. Heat is generated uniformly through out the wall. Choose the CORRECT statement

(a) The direction of heat transfer will be from the surface at 100°C to the surface of 0°C.

(b) The maximum temperature inside the wall must be greater than 100°C

(c) The temperature distribution is linear within the wall

(d) The temperature distribution is symmetric about the mid-plane of the wall **[2013 : 1 M]**

1.22 Consider a long cylindrical tube of inner and outer radii, r_i and r_o, respectively, length L and thermal conductivity, k. Its inner and outer surfaces are maintained at T_i and T_o, respectively ($T_i > T_o$). Assuming one-dimensional steady state heat conduction in the radial direction, the thermal resistance in the wall of the tube is

(a) $\frac{1}{2\pi kL}\ln\left(\frac{r_i}{r_o}\right)$ (b) $\frac{L}{2\pi r_i k}$

(c) $\frac{1}{2\pi kL}\ln\left(\frac{r_o}{r_i}\right)$ (d) $\frac{1}{4\pi kL}\ln\left(\frac{r_o}{r_i}\right)$

[2014 : 1 M, Set-3]

1.23 As the temperature increases, the thermal conductivity of a gas

(a) increases

(b) decreases

(c) remains constant

(d) increases up to a certain temperature and then decreases **[2014 : 1 M, Set-4]**

1.24 Match **Group *A*** with **Group *B***:

Group A

A : Biot number

B : Grashof number

C : Prandtl number

D : Reynolds number

Group B

1 : Ratio of buoyancy to viscous force

2 : Ratio of inertia force to viscous force

3 : Ratio of momentum to thermal diffusivities

4 : Ratio of internal thermal resistance to boundary layer thermal resistance

Codes:

	A	B	C	D
(a)	4	1	3	2
(b)	4	3	1	2
(c)	3	2	1	4
(d)	2	1	3	4

[2014 : 1 M, Set-4]

1.25 Consider one dimensional steady state heat conduction across a wall (as shown in figure below) of thickness 30 mm and thermal conductivity 15 W/mK. At $x = 0$, a constant heat flux, $q'' = 1 \times 10^5$ W/m^2 is applied. On the other side of the wall, heat is removed from the wall by convection with a fluid at 25°C and heat transfer coefficient of 250 W/m^{2}K. The temperature (in °C), at $x = 0$ is _____.

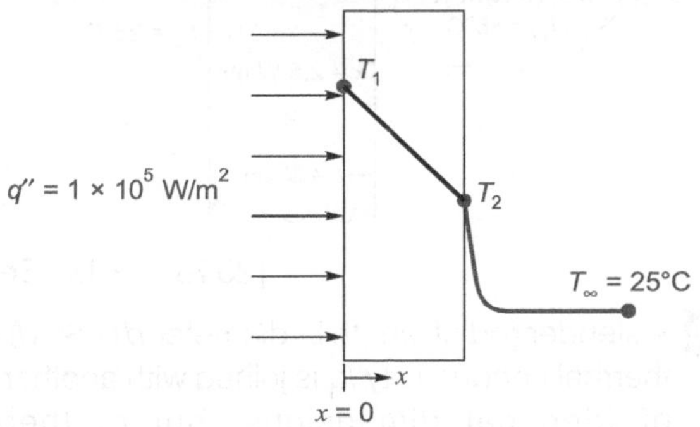

[2014 : 2 M, Set-1]

1.26 A material P of thickness 1 mm is sandwiched between two steel slabs, as shown in the figure below. A heat flux 10 kW/m^2 is supplied to one of the steel slabs as shown. The boundary temperatures of the slabs are indicated in the figure. Assume thermal conductivity of this steel is 10 W/mK. Considering one-dimensional steady state heat conduction for the configuration, the thermal conductivity (k, in W/mK) of material P is ____.

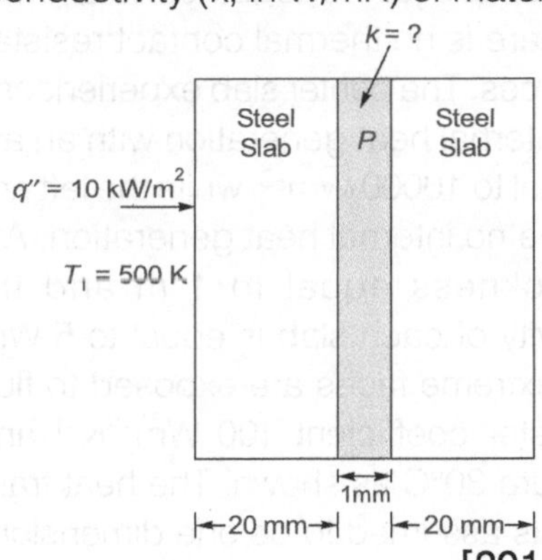

[2014 : 2 M, Set-2]

1.27 Heat transfer through a composite wall is shown in figure. Both the sections of the wall have equal thickness (l). The conductivity of one section is k and that of the other is $2k$. The left face of the wall is at 600 K and the right face is at 300 K.

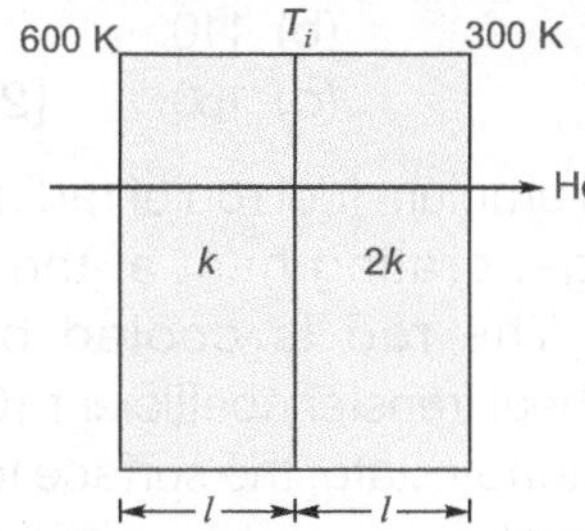

The interface temperature T_i (in K) of the composite wall is _____. [2014 : 2 M, Set-3]

1.28 A plane wall has a thermal conductivity of 1.15 W/mK. If the inner surface is at 1100°C and the outer surface is at 350°C, then the design thickness (in meter) of the wall to maintain a steady heat flux of 2500 W/m^2 should be ____. [2014 : 2 M, Set-4]

1.29 If a foam Insulation is added to a 4 cm outer diameter pipe as shown in the figure, the critical radius of insulation (in cm) is ________.

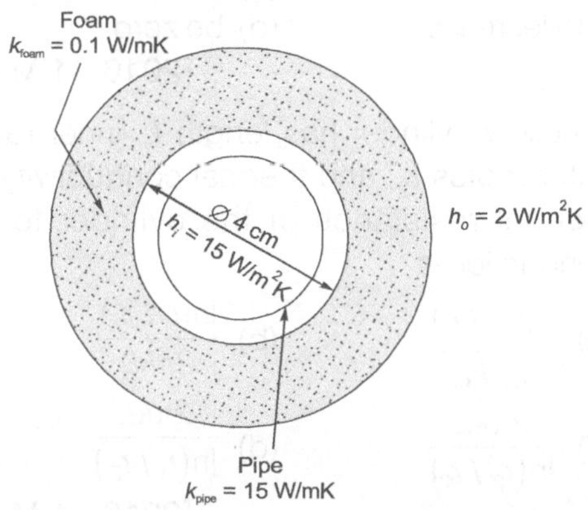

[2015 : 1 M, Set-2]

1.30 A 10 mm diameter electrical conductor is covered by an insulation of 2 mm thickness. The conductivity of the insulation is 0.08 W/mK and the convection coefficient at the insulation surface is 10 W/m^{2}K. Addition of further insulation of the same material will

(a) increase heat loss continuously

(b) decrease heat loss continuously

(c) increase heat loss to a maximum and then decrease heat loss

(d) decrease heat loss to a minimum and then increase heat loss [2015 : 2 M, Set-1]

1.31 A brick wall ($k = 0.9$ W/mK) of thickness 0.18 m separates the warm air in a room from the cold ambient air. On a particular winter day, the outside air temperature is – 5°C and the room needs to be maintained at 27°C. The heat transfer coefficient associated with outside air is 20 W/m^{2}K. Neglecting the convective resistance of the air inside the room, the heat loss, in (W/m^2), is

(a) 88 (b) 110
(c) 128 (d) 160 [2015 : 2 M]

1.32 A cylindrical uranium fuel rod of radius 5 mm in a nuclear is generating heat at the rate of 4×10^7 W/m^3. The rod is cooled by a liquid (convective heat transfer coefficient 1000 W/m^{2}K) at 25°C. At steady state, the surface temperature (in K) of the rod is

(a) 308 (b) 398
(c) 418 (d) 448 **[2015 : 2 M, Set-2]**

1.33 A plastic sleeve of outer radius r_o = 1 mm covers a wire (radius r = 0.5 mm) carrying electric current. Thermal conductivity of the plastic is 0.15 W/mK. The heat transfer coefficient on the outer surface of the sleeve exposed to air is 25 W/m^{2}K. Due to the addition of the plastic cover, the heat transfer from the wire to the ambient will

(a) increase (b) remain the same
(c) decrease (d) be zero

[2016 : 1 M, Set-1]

1.34 A hollow cylinder has length L, inner radius r_1, outer radius r_2, and thermal conductivity k. The thermal resistance of the cylinder for radial conduction is

(a) $\dfrac{\ln(r_2/r_1)}{2\pi kL}$ (b) $\dfrac{\ln(r_1/r_2)}{2\pi kL}$

(c) $\dfrac{2\pi kL}{\ln(r_2/r_1)}$ (d) $\dfrac{2\pi kL}{\ln(r_1/r_2)}$

[2016 : 1 M, Set-2]

1.35 Steady one-dimensional heat conduction takes place across the faces 1 and 3 of a composite slab consisting of slabs A and B in perfect contact as shown in the figure, where k_A, k_B denote the respective thermal conductivities. Using the data as given in the figure, the interface temperature T_2 (in °C) is ________

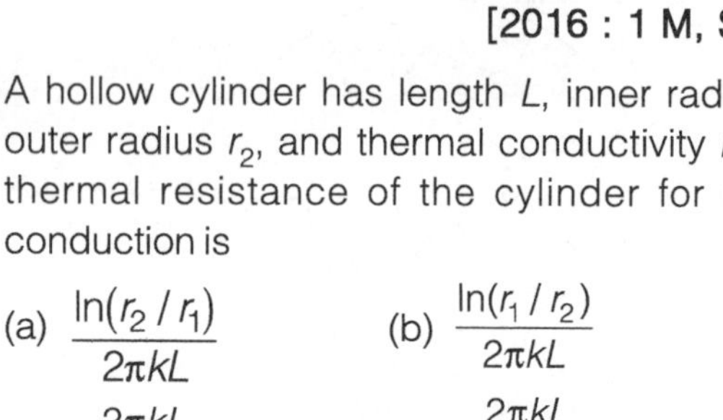

[2016 : 1 M, Set-3]

1.36 Heat is generated uniformly in a long solid cylindrical rod (diameter = 10 mm) at the rate of 4×10^7 W/m^3. The thermal conductivity of the rod material is 25 W/mK. Under steady state conditions, the temperature difference between the centre and the surface of the rod is ________ °C.

[2017 : 2 M, Set-1]

1.37 A plane slab of thickness L and thermal conductivity k is heated with a fluid on one side (P), and the other side (Q) is maintained at a constant temperature, T_Q of 25°C, as shown in the figure. The fluid is at 45°C and the surface heat transfer coefficient h, is 10 W/m^{2}K. The steady state temperature. T_P (in °C) of the side which is exposed to the fluid is ________ (correct to two decimal places).

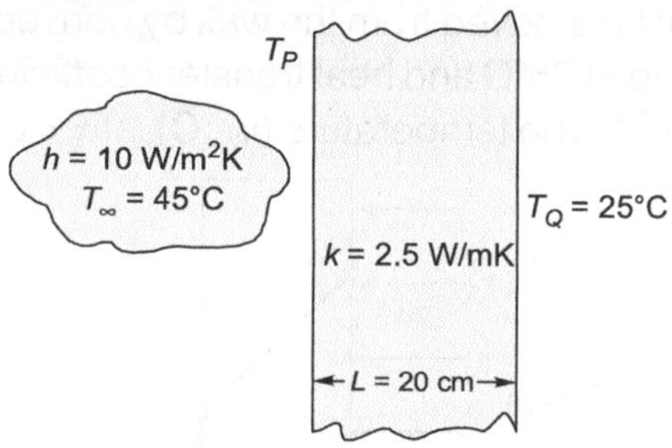

[2018 : 2 M, Set-1]

1.38 A slender rod of length L, diameter d ($L>>d$) and thermal conductivity k_1 is joined with another rod of identical dimensions, but of thermal conductivity k_2, to from a composite cylindrical rod of length 2L. The heat transfer in radial direction and contact resistance are negligible. The effective thermal conductivity of the composite rod is

(a) $\dfrac{k_1k_2}{k_1+k_2}$ (b) $\dfrac{2k_1k_2}{k_1+k_2}$

(c) $k_1 + k_2$ (d) $\sqrt{k_1k_2}$

[2019 : 1 M, Set-1]

1.39 Three slabs are joined together as shown in the figure. There is no thermal contact resistance at the interfaces. The center slab experiences a non-uniform internal heat generation with an average value equal to 10000 Wm^{-3}, while the left and right slabs have no internal heat generation. All slabs have thickness equal to 1 m and thermal conductivity of each slab is equal to 5 Wm^{-1}K^{-1}. The two extreme faces are exposed to fluid with heat transfer coefficient 100 Wm^{-2}K^{-1} and bulk temperature 30°C as shown. The heat transfer in the slabs is assumed to be one dimensional and steady, and all properties are constant. If the left extreme face temperature T_1 is measured to be

100°C, the right extreme face temperature T_2 is ________ °C.

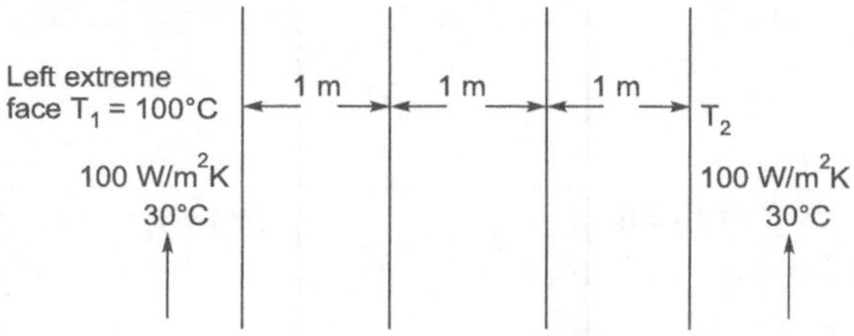

[2019 : 2 M, Set-1]

1.40 One-dimensional steady state heat conduction takes place through a solid whose cross-sectional area varies linearly in the direction of heat transfer. Assume there is no heat generation in the solid and the thermal conductivity of the material is constant and independent of temperature. The temperature distribution in the solid is

(a) Quadratic (b) Logarithmic
(c) Linear (b) Exponential

[2019 : 1 M, Set-2]

1.41 In a furnace, the inner and outer sides of the brick wall (k_1 = 2.5 W/mK) are maintained at 1100°C and 700°C respectively as shown in figure.

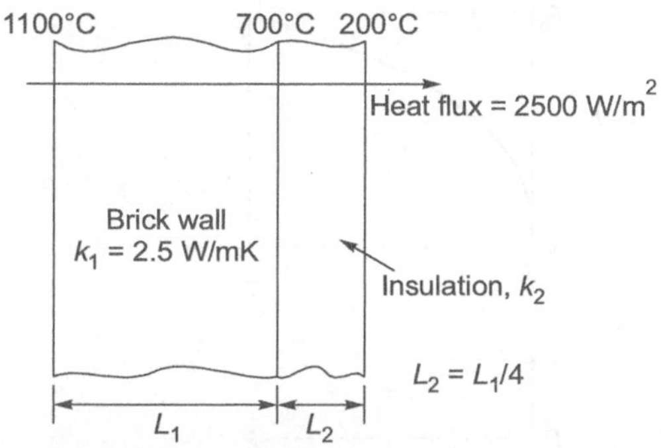

The brick wall is covered by an insulating material of thermal conductivity k_2. The thickness of the insulation is 1/4th of the thickness of the brick wall. The outer surface of the insulation is at 200°C. The heat flux through the composite walls is 2500 W/m^2. The value of k_2 is _______ W/mK (round off to 2 decimal places). [2020 : 1 M, Set-2]

1.42 Ambient air flows over a heated slab having flat, top surface at y = 0. The local temperature (in Kelvin) profile within the thermal boundary layer is given by $T(y) = 300 + 200\exp(-5y)$, where y is the distance measured from the slab surface in meters. If the thermal conductivity of air is 1.0 W/m.K and that of the slab is 100 W/m.K, then the magnitude of temperature gradient $\left|\dfrac{dT}{dy}\right|$ within the slab at $y = 0$ is _______K/m (round off to the nearest integer).

[2021 : 2 M, Set-2]

1.43 An uninsulated cylindrical wire of radius 1.0 mm produces electric heating at the rate of 5.0 W/m. The temperature of the surface of the wire is 75°C when placed in air at 25°C. When the wire is coated with PVC of thickness 1.0 mm, the temperature of the surface of the wire reduces to 55°C. Assume that the heat generation rate from the wire and the convective heat transfer coefficient are same for both uninsulated wire and the coated wire. The thermal conductivity of PVC is _________ W/mK (Round off to two decimal places)

[2021 : 2 M, Set-1]

1.44 Consider a rod of uniform thermal conductivity whose one end (x = 0) is insulated and the other end (x = L) is exposed to flow of air at temperature T_∞ with convective heat transfer coefficient h. The cylindrical surface of the rod is insulated so that the heat transfer is strictly along the axis of the rod. The rate of internal heat generation per unit volume inside the rod is given as

$$\dot{q} = \cos\frac{2\pi x}{L}$$

The steady state temperature at the mid-location of the rod is given as T_A. What will be the temperature at the same location, if the convective heat transfer coefficient increases to $2h$?

(a) $T_A + \dfrac{\dot{q}L}{2h}$ (b) $2T_A$

(c) T_A (d) $T_A\left(1-\dfrac{\dot{q}L}{4\pi h}\right)+\dfrac{\dot{q}L}{4\pi h}T_\infty$

[2022 : 2 M, Set-1]

1.45 Consider steady state, one-dimensional heat conduction in an infinite slab of thickness $2L$ (L = 1 m) as shown in the figure. The conductivity (k) of the material varies with temperature as $k = CT$, where T is the temperature in K, and C is a constant equal to 2 W·m^{-1}·K^{-2}. There is a uniform heat generation of 1280 kW/m^3 in the slab. If both faces of the slab are maintained at 600 K, then the temperature at x = 0 is _____ K (in integer).

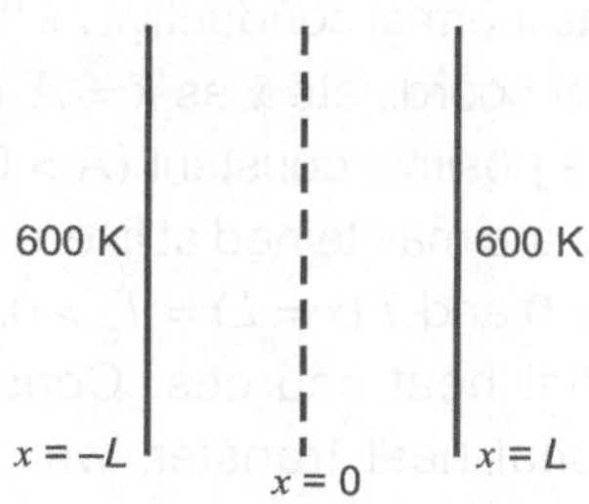

[2022 : 2 M, Set-2]

1.46 Consider a laterally insulated rod of length L and constant thermal conductivity. Assuming one-dimensional heat conduction in the rod, which of the following steady-state temperature profile(s) can occur without internal heat generation?

(a)
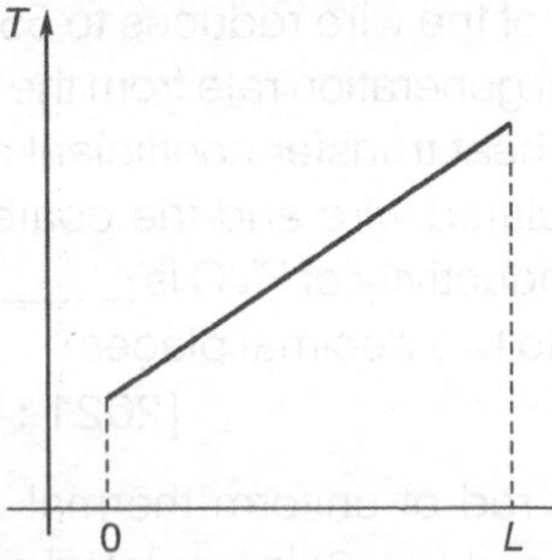

(b)
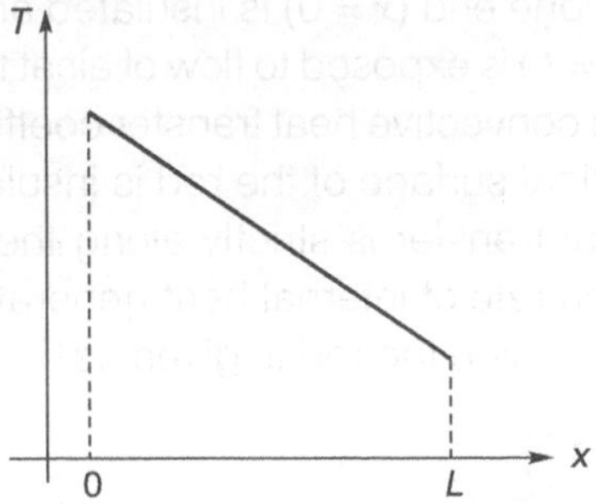

(c)
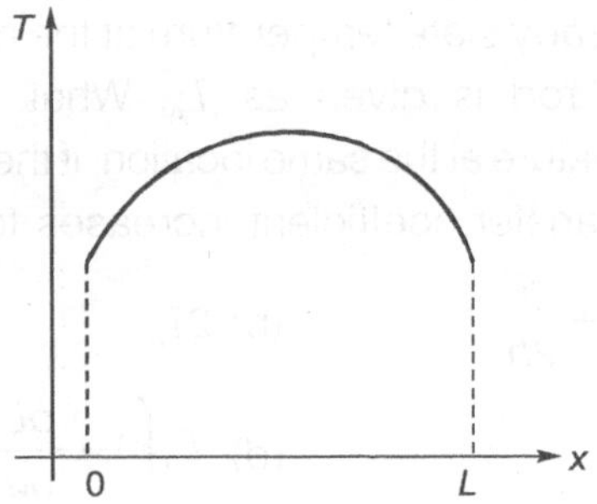

(d)
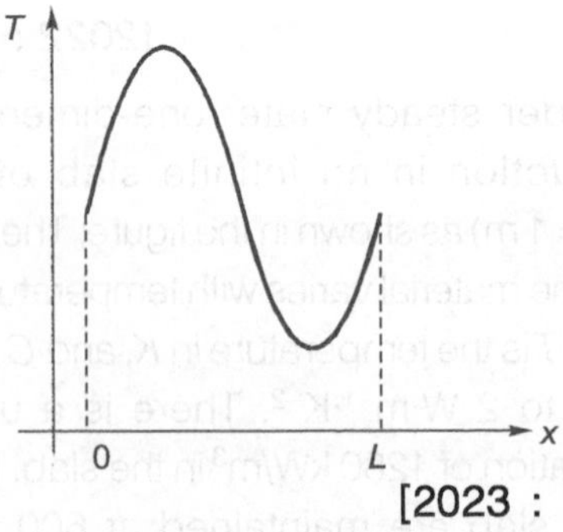

[2023 : 1 M, MSQ]

1.47 A plane, solid slab of thickness L, shown in the figure, has thermal conductivity k that varies with the spatial coordinate x as $k = A + Bx$, where A and B are positive constant ($A > 0$, $B > 0$). The slab walls are maintained at fixed temperature of $T(x = 0) = 0$ and $T(x = L) = T_o > 0$. The slab has no internal heat sources. Considering one-dimensional heat transfer, which one of the following plots qualitatively depicts the steady-state temperature distribution within the slab?

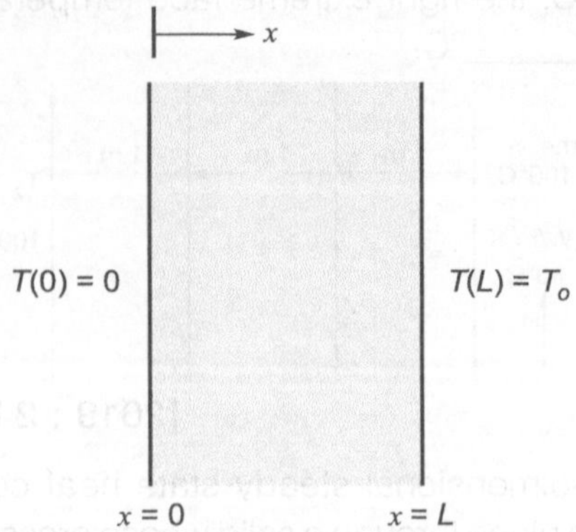

(a)
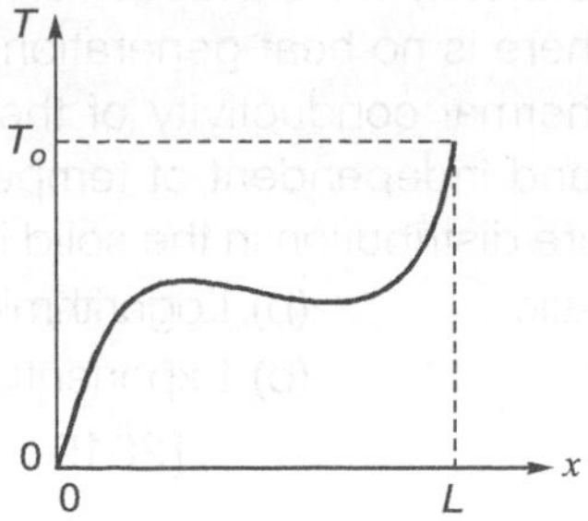

(b)
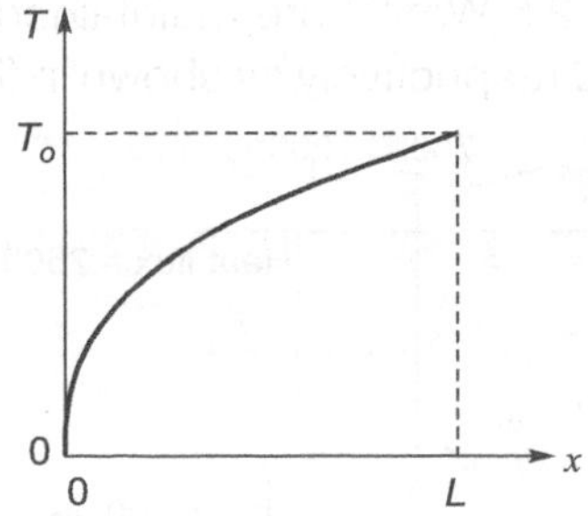

(c)
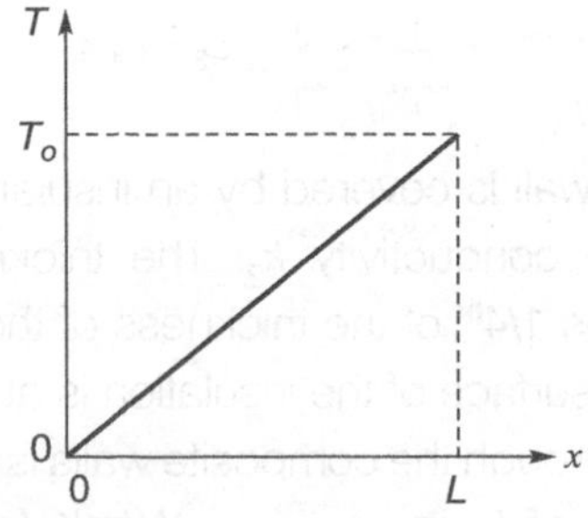

(d)
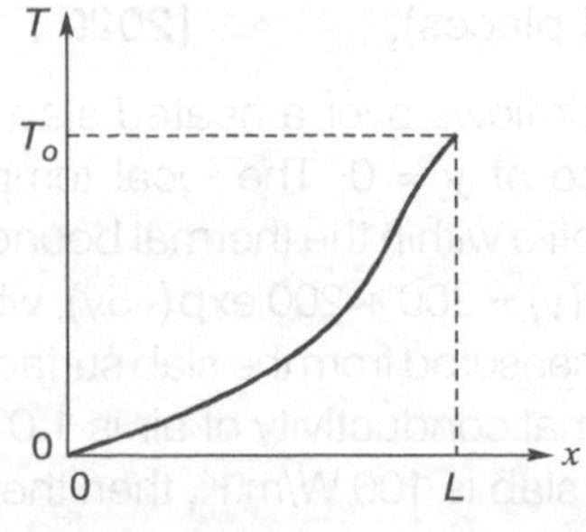

[2024 : 1 M]

1.48 Consider a slab of 20 mm thickness. There is a uniform heat generation of $\dot{q} = 100$ MW/m^3 inside the slab. The left right faces of the slab are maintained at 150°C and 110°C, respectively. The plate has a constant thermal conductivity of 200 W/mK. Considering a 1-D steady state heat conduction, the location of the maximum temperature from the left face will be at ______ mm. (answer in integer)

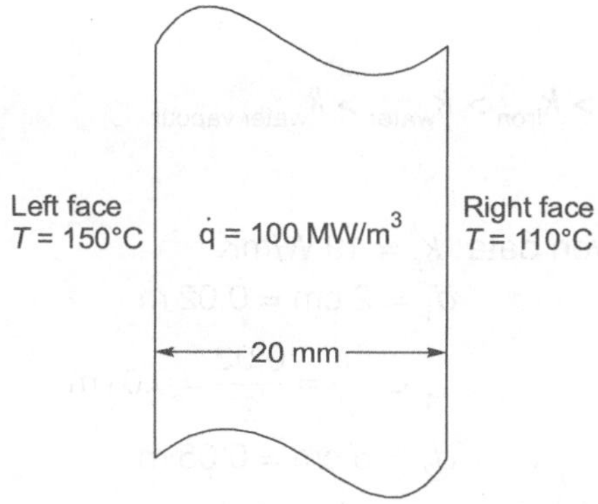

[2024 : 2 M]

Answers Conduction

1.1 (d)	1.2 (c)	1.3 (a)	1.4 (b)	1.5 (c)	1.6 (b)	1.7 (c)	1.8 (d)
1.9 (d)	1.10 (c)	1.11 (d)	1.12 (a)	1.13 (b)	1.14 (c)	1.15 (c)	1.16 (b)
1.17 (d)	1.18 (b)	1.19 (c)	1.20 (d)	1.21 (b)	1.22 (c)	1.23 (a)	1.24 (a)
1.25 (625)	1.26 (0.1)	1.27 (400)	1.28 (0.345)	1.29 (5)	1.30 (c)	1.31 (c)	1.32 (b)
1.33 (a)	1.34 (a)	1.35 (67.5)	1.36 (10)	1.37 (33.89)	1.38 (b)	1.39 (60)	1.40 (b)
1.41 (0.5)	1.42 (10)	1.43 (0.11)	1.44 (c)	1.45 (1000)	1.46 (a, b)	1.47 (b)	1.48 (6)

Explanations Conduction

1.1 (d)

Given data:

$$d = 20 \text{ mm}$$

$$\therefore \quad r = \frac{d}{2} = \frac{20}{2} = 10 \text{ mm}$$

$$h_o = 20 \text{ W/m}^2\text{K}$$

$$k = 0.5 \text{ W/mK}$$

For maximum heat dissipation,

Critical radius,

$$r_c = \frac{k}{h} \quad \text{for cylinder or wire}$$

$$= \frac{0.5}{20} = 0.025 \text{ m} = 25 \text{ mm}$$

Thickness of insulation,

$$= r_c - r = 25 - 10 = \mathbf{15\ mm}$$

1.2 (c)

Fourier's law,

$$Q = -kA\frac{dT}{dx}$$

At constant Q, A and dx,

$$dT \propto \frac{1}{k}$$

Temperature drop (dT) is inversely proportional to the thermal conductivity.

Material	Thermal conductivity : k W/mK
Copper	385
Steel	45
Glass wool	0.0372
Refractory brick	1.50

Hence, lower the thermal conductivity, higher temperature drop. So, **glass wool** has minimum thermal conductivity and maximum temperature drop.

1.3 (a)

Since, drop of temperature in case of 2nd slab is more rapid (slope in 2nd case is more than that of 1st). Therefore, conductivity of I-slab is more than II-slab.

1.4 (b)

$$d = 1 \text{ mm}$$

$$\therefore \quad r = \frac{d}{2} = \frac{1}{2} = 0.5 \text{ mm}$$

$$k = 0.1 \text{ W/mK}$$

$$h_o = 100 \text{ W/m}^2\text{K}$$

Critical thickness,

$$r_c = \frac{k}{h_o} = \frac{0.1}{100} = 0.001 \text{ m} = 1 \text{ mm}$$

Optimum thickness of enamel paint,

$$t_{opt} = r_c - r = 1 - 0.5 = \mathbf{0.5\ mm}$$

1.5 (c)

$k_{Al} > k_{iron} > k_{water} > k_{water\ vapour}$

1.7 (c)

Given data: $k_s = 19$ W/mK

$$d_1 = 2 \text{ cm} = 0.02 \text{ m}$$

$$\therefore \quad r_1 = \frac{d_1}{2} = \frac{0.02}{2} = 0.01 \text{ m}$$

$$d_2 = 5 \text{ cm} = 0.05 \text{ m}$$

$$\therefore \quad r_2 = \frac{d_2}{2} = \frac{0.05}{2} = 0.025 \text{ m}$$

$$t = 3 \text{ cm} = 0.03 \text{ m}$$

$$\therefore \quad r_3 = r_2 + t = 0.025 + 0.03 = 0.055 \text{ m}$$

$$k_a = 0.2 \text{ W/mK}$$

$$T_i - T_o = 600°\text{C}$$

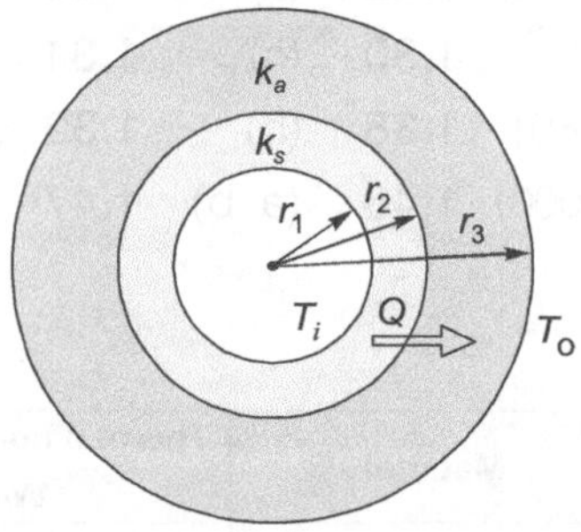

Heat transfer,

$$Q = \frac{T_i - T_o}{\dfrac{1}{2\pi k_s l}\log_e \dfrac{r_2}{r_1} + \dfrac{1}{2\pi k_a l}\log_e \dfrac{r_3}{r_2}}$$

$$\frac{Q}{l} = \frac{T_i - T_o}{\dfrac{1}{2\pi k_s}\log_e \dfrac{r_2}{r_1} + \dfrac{1}{2\pi k_a}\log_e \dfrac{r_3}{r_2}}$$

$$q = \frac{600}{\dfrac{1}{2\times 3.14\times 19}\log_e \dfrac{0.025}{0.01} + \dfrac{1}{2\times 3.14\times 0.2}\log_e \dfrac{0.055}{0.025}}$$

$$= \frac{600}{0.007679 + 0.627752} = \mathbf{944.24\ W/m}$$

1.8 (d)

Case-I:

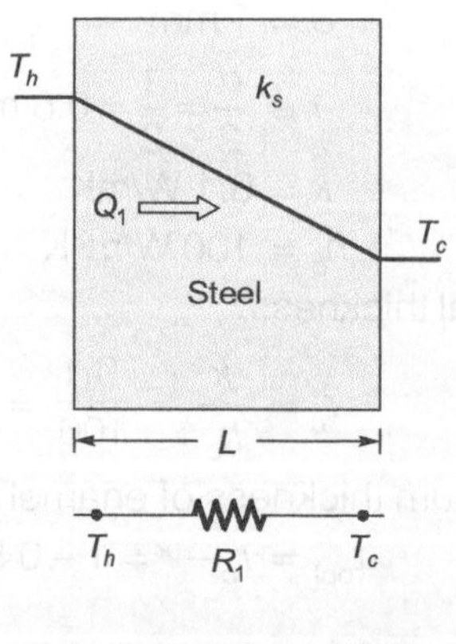

Case-II:

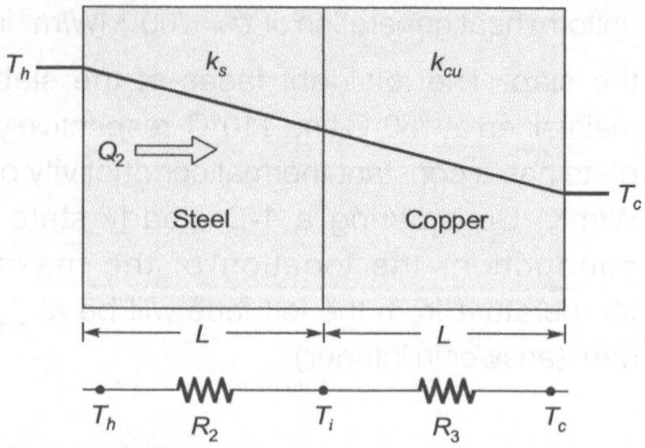

For case-I:

$$Q_1 = \frac{T_h - T_c}{R_1}$$

where $R_1 = \dfrac{L}{k_s A}$

For Case-II:

$$Q_2 = \frac{T_h - T_c}{R_2 + R_3}$$

where $R_2 = \dfrac{L}{k_s A}$ and $R_3 = \dfrac{L}{k_{cu} A}$

As $R_2 + R_3 > R_1$ at constant temperature difference in both cases.

Then $q_1\left(\text{or } \dfrac{Q_1}{A}\right) > q_2\left(\text{or } \dfrac{Q_2}{A}\right)$

For Case-II:

$$R_2 > R_3 \qquad \because k_{cu} > k_s$$

$$Q_2 = \frac{T_h - T_i}{R_2}$$

or $$R_2 = \frac{T_h - T_i}{Q_2}$$

also $$Q_2 = \frac{T_i - T_c}{R_3}$$

or $$R_3 = \frac{T_i - T_c}{Q_2}$$

$$\therefore \quad \frac{T_h - T_i}{Q_2} > \frac{T_i - T_c}{Q_2}$$

or $T_h - T_i > T_i - T_c$

or $T_h + T_c > 2T_i$

or $$T_i < \frac{T_h + T_c}{2}$$

1.9 (d)

The equation of heat transfer in Cartesian coordinates without heat generation is

$$\frac{\partial^2 T}{\partial x^2} + \frac{\partial^2 T}{\partial y^2} + \frac{\partial^2 T}{\partial z^2} = \frac{\rho c_p}{k}\frac{\partial T}{\partial t}$$

For one dimensional heat transfer,

$$\frac{\partial^2 T}{\partial x^2} = \frac{\rho c_p}{k}\frac{\partial T}{\partial t}$$

i.e.,
$$\frac{\partial T}{\partial t} \propto \frac{\partial^2 T}{\partial x^2}$$

1.10 (c)

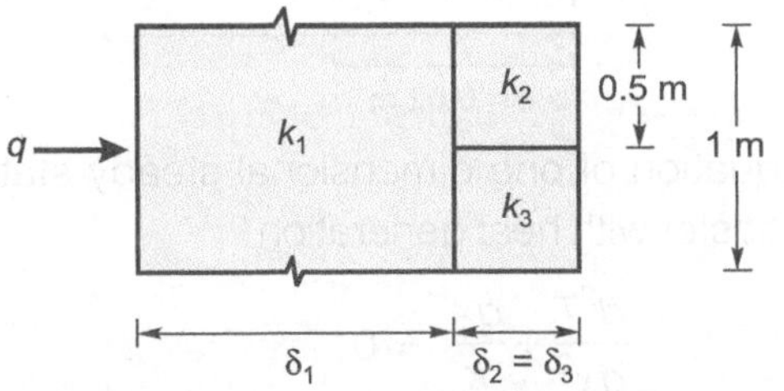

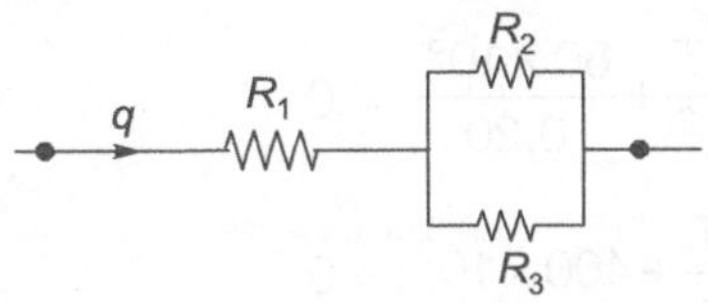

$k_1 = 0.02$ W/mK; $k_2 = 0.10$ W/mK; $k_3 = 0.04$ W/mK

$\delta_1 = 0.5$ m; $\delta_2 = \delta_3 = 0.25$ m; Depth: $w = 1$ m

Total thermal resistance,

$$R_t = R_1 + \frac{R_2 R_3}{R_2 + R_3}$$

where $$R_1 = \frac{\delta_1}{A_1 k_1} = \frac{0.5}{1\times1\times0.02} = 25 \text{ K/W}$$

$$R_2 = \frac{\delta_2}{A_2 k_2} = \frac{0.25}{0.5\times1\times0.10} = 5 \text{ K/W}$$

and $$R_3 = \frac{\delta_3}{A_3 k_3} = \frac{0.25}{0.5\times0.04} = 12.5 \text{ K/W}$$

$\therefore$ $$R_t = 25 + \frac{5\times12.5}{5+12.5} = 25 + 3.57$$
$$= 28.57 \text{ K/W} \simeq \mathbf{28.6 \text{ K/W}}$$

1.11 (d)

For steady state heat transfer through composite slab,

Heat transfer through slab-1

= Heat transfer through slab-2

$$\frac{k_1 A(T_1 - T_{inter})}{2b} = \frac{k_2 A(T_{inter} - T_2)}{b}$$

or $k_1(T_1 - T_{inter}) = 2k_2(T_{inter} - T_2)$

Given condition,

$$T_{inter} = \frac{T_1 + T_2}{2} = \frac{T_1}{2} + \frac{T_2}{2}$$

Therefore, $$k_1\left(T_1 - \frac{T_1}{2} - \frac{T_2}{2}\right) = 2k_2\left(\frac{T_1}{2} + \frac{T_2}{2} - T_2\right)$$

$$k_1\left(\frac{T_1 - T_2}{2}\right) = 2k_2\left(\frac{T_1 - T_2}{2}\right)$$

or $$\mathbf{k_1 = 2k_2}$$

1.12 (a)

An insulation is added to the circular pipe, the outer surface temperature will decrease because of higher thermal resistance but at the same time the surface area for convective heat dissipation will increase causing more heat loss. These two opposing effect lead to an optimum insulation thickness.

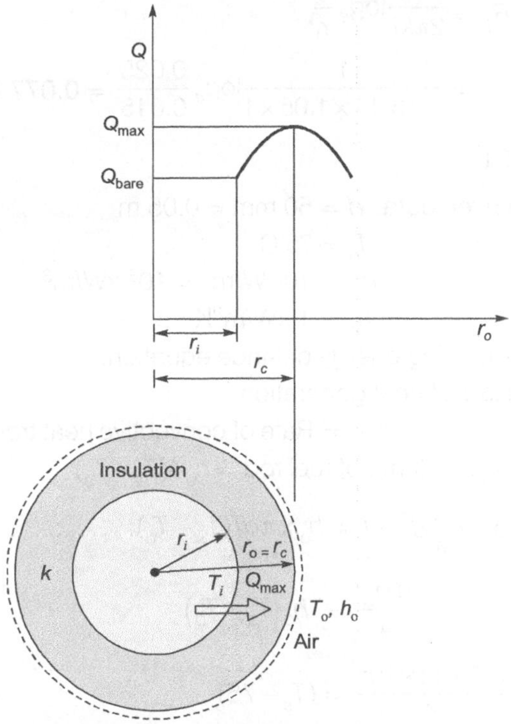

As shown in figure, the rate of heat transfer increases then decreases with increasing the thickness of insulation. The radius of insulation at which maximum rate of heat loss to surroundings is called **critical radius of insulation**. It is denoted by r_c. The difference between the critical radius of insulation (r_c) and the inner radius of a hollow cylinder is called **critical thickness of insulation**.

1.13 (b)

Given data:

$$d_1 = 0.03 \text{ m}$$

$\therefore$ $$r_1 = \frac{d_1}{2} = \frac{0.03}{2} = 0.015 \text{ m}$$

$$d_2 = 0.05 \text{ m}$$

$\therefore$ $$r_2 = \frac{d_2}{2} = \frac{0.05}{2} = 0.025 \text{ m}$$

$$k = 1.05 \text{ W/mK}$$

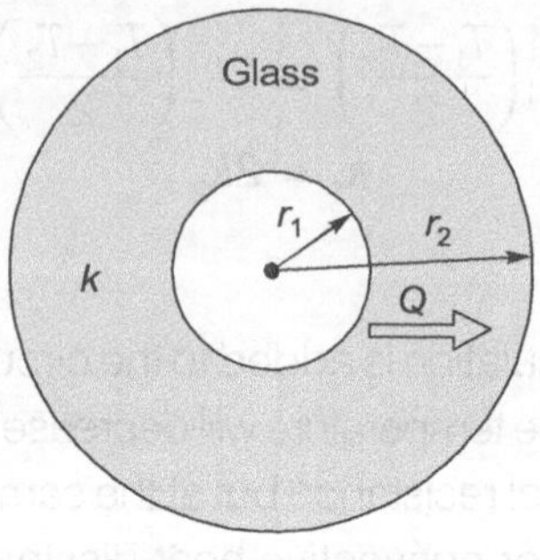

Long glass cylinder

Thermal resistance,

$$R_t = \frac{1}{2\pi k l}\log_e \frac{r_2}{r_1}$$

$$= \frac{1}{2\times 3.14\times 1.05\times 1}\log_e \frac{0.025}{0.015} = \mathbf{0.077\ K/W}$$

1.14 (c)

Given data: d = 50 mm = 0.05 m

T_o = 75°C

$q_G = 10^6\ W/m^3 = 10^3\ kW/m^3$

$h_o = 1\ kW/m^2K$

Applying energy balance equation,

Rate of heat generation

= Rate of convective heat transfer

$q_G \times$ Volume of fuel rod = $h_o A(T_s - T_o)$

$$q_G \times \frac{\pi}{4} d^2 \times l = h_o \times \pi d l (T_s - T_o)$$

$$\frac{d\,q_G}{4} = h_o(T_s - T_o)$$

$$\frac{0.05\times 1\times 10^3}{4} = (T_s - 75)$$

or $T_s - 75 = 12.5$

or $T_s = 12.5 + 75 = 87.5°C$

$= (273 + 87.5)K$

$= 360.5\ K \simeq \mathbf{360\ K}$

1.15 (c)

Given data : δ = 20 mm = 0.02 m

$q_G = 80\ MW/m^3 = 80\times 10^6\ W/m^3$

$= 80\times 10^3\ kW/m^3$

$T_{x=0} = 160°C$

$T_{x=\delta} = 120°C$

$k = 200\ W/mK = 0.20\ kW/mK$

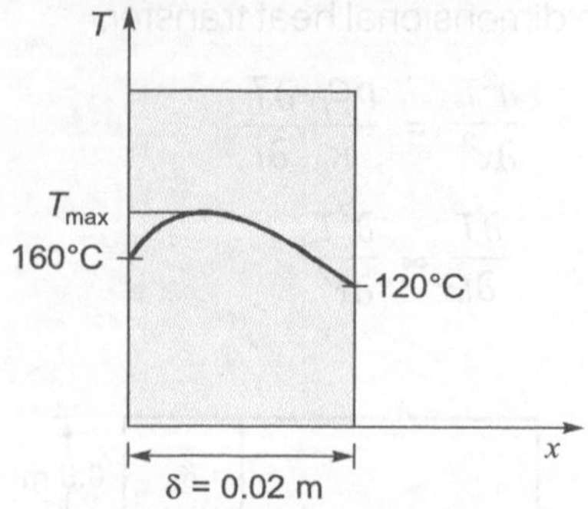

Equation of one dimensional steady state heat, transfer with heat generation

$$\frac{d^2T}{dx^2} + \frac{q_G}{k} = 0$$

$$\frac{d^2T}{dx^2} + \frac{80\times 10^3}{0.20} = 0$$

$$\frac{d^2T}{dx^2} + 400\times 10^3 = 0$$

On integrating, we get

$$\frac{dT}{dx} + 400\times 10^3 x = C_1$$

Again integrating, we get

$$T + 400\times 10^3 \frac{x^2}{2} = C_1 x + C_2$$

$$T + 200\times 10^3 x^2 = C_1 x + C_2 \quad \text{...(i)}$$

where C_1 and C_2 are constant of integration that can be determined from the boundary conditions.

At $x = 0$, $T = 160°C$

$\therefore$ $160 = C_2$

or $C_2 = 160$

At $x = \delta = 0.02$ m, $T = 120°C$

$\therefore\ 120 + 200\times 10^3 \times (0.02)^2$

$= C_1 \times 0.02 + 160$

$200 = 0.02 C_1 + 160$

$0.02 C_1 = 40$

$C_1 = 2000$

Substituting the values of C_1 and C_2 in Eq. (i), we get

$T + 200\times 10^3 x^2 = 2000x + 160$

$T = 2000x + 160 - 200\times 103x^2$ ...(ii)

The condition for the maximum temperature,

$$\frac{dT}{dx} = 0$$

$200 - 400\times 10^3 x = 0$

or $x = 0.005\ m = \mathbf{5\ mm}$

Substituting the value of x = 0.005 m in Eq. (ii), we get

$$T_{max} = 2000 \times 0.005 + 160 - 200 \times 10^3 \times (0.005)^2$$

$$= 10 + 160 - 5 = \textbf{165°C}$$

1.17 (d)

Heat flows always normal to the surface so

$$\left.\frac{\partial T}{\partial x}\right|_P = 0 \text{ and } \left.\frac{\partial T}{\partial y}\right|_Q = 0$$

For constant surface temperatures

$$q_{cond,\,Q} = q_{cond,\,P}$$

i.e., $$kA_P \left.\frac{\partial T}{\partial y}\right|_P = kA_Q \left.\frac{\partial T}{\partial x}\right|_Q$$

$$\left.\frac{\partial T}{\partial y}\right|_P = \frac{A_Q}{A_P}\left.\frac{\partial T}{\partial x}\right|_Q$$

$$= \frac{2}{1} \times 10 = \textbf{20 K/m}$$

1.18 (b)

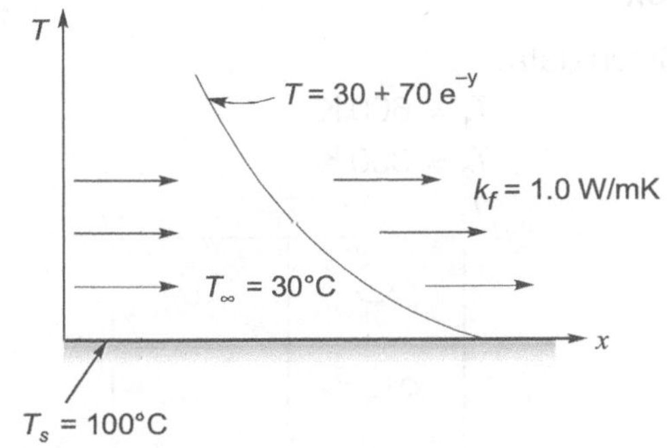

At $y = 0$; $q_{cond} = q_{conv}$

$$-k_f \left.\frac{\partial T}{\partial y}\right|_{y=0} = h\Delta T = h(T_s - T_\infty)$$

$$-(1)\frac{\partial}{\partial y}\left[30 + 70\,e^{-y}\right]\Big|_{y=0} = h(100 - 30)$$

$$-\left[0 + 70(-1)\,e^{-y}\right]\Big|_{y=0} = h(70)$$

$$70\,e^{-0} = h(70)$$

or $$h = \textbf{1 W/m}^2\textbf{K}$$

Alternative Solution

Given data: T_∞ = 30°C; T_s = 100°C

$$k_f = 1 \text{ W/mK}$$

$$T = 30 + 70 \exp(-y)$$

Differentiating w.r.t y, we get

$$\frac{dT}{dy} = -70e^{-y}$$

At $y = 0$

$$\left(\frac{dT}{dy}\right)_{y=0} = -70$$

We know that local convective heat transfer coefficient:

$$h_x = \frac{-k_f\left(\frac{dT}{dy}\right)_{y=0}}{T_s - T_\infty} = \frac{-1\times(-70)}{100-30} = \textbf{1 W/m}^2\textbf{K}$$

1.19 (c)

Given data:

Outer diameter of pipe,

$$d_o = 25 \text{ mm}$$

∴ Radius: $$r_o = \frac{d_o}{2} = \frac{25}{2} = 12.5 \text{ mm}$$

$$h = 5 \text{ W/m}^2\text{K};\; k = 0.05 \text{ W/mK}$$

Critical radius of insulation,

$$r_c = \frac{k}{h} = \frac{0.05}{5} = 0.01\,\text{m} = \textbf{10 mm}$$

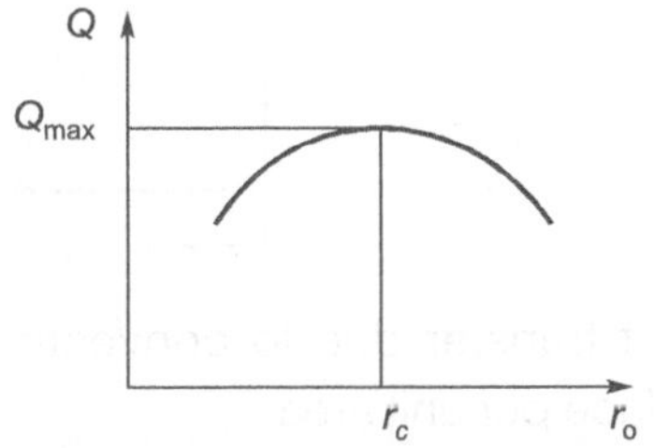

As $r_o > r_c$, then adding the insulation will reduce the heat loss.

1.20 (d)

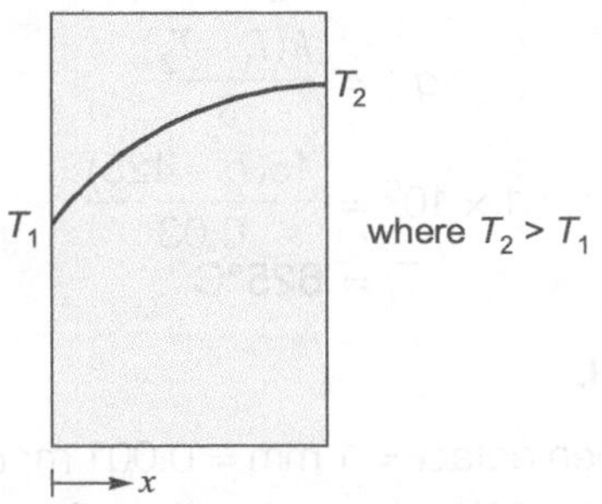

$$Q = -kA\frac{dT}{dx} \qquad \text{...(i)}$$

where $k = k_0 + bT$

As x increases, T increases, then k also increases.

i.e., k is +ve

Equation (i), indicates $\frac{dT}{dx}$ **decreases.**

1.21 (b)

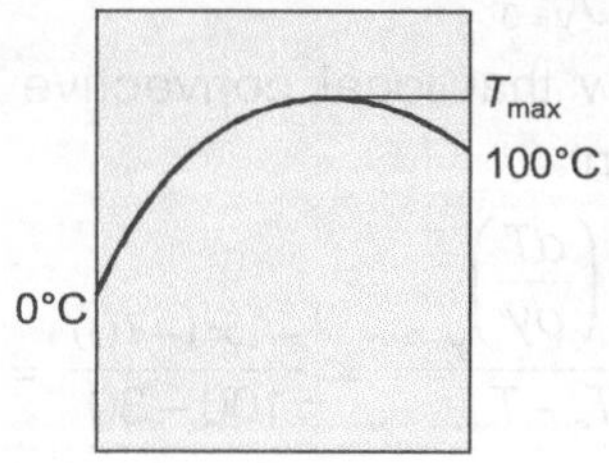

1.25 Sol.

Given data: δ = 30 mm = 0.03 m;
k = 15 W/mK
q'' = 1 × 10^5 W/m²;
T_∞ = 25°C
h = 250 W/m²K

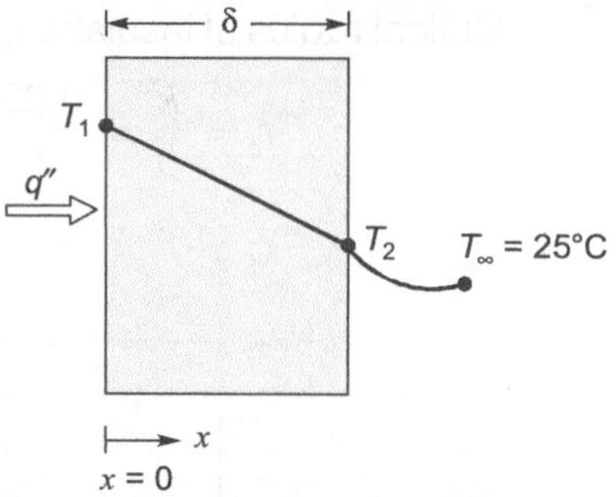

Heat transfer due to convection from outside surface per unit area:

$$q'' = \frac{Q}{A} = h(T_2 - T_\infty)$$

$$1 \times 10^5 = 250(T_2 - 25)$$

or $$T_2 = 425°C$$

Heat transfer through wall per unit area,

$$q'' = \frac{k(T_1 - T_2)}{\delta}$$

$$1 \times 10^5 = \frac{15(T_1 - 425)}{0.03}$$

or $$T_1 = \mathbf{625°C}$$

1.26 Sol.

Given data: x = 1 mm = 0.001 m; q'' = 10 kW/m² = 10000 W/m²; k_s = 10 W/mK; δ = 20 mm = 0.02 m
T_1 = 500 K; T_2 = 360 K

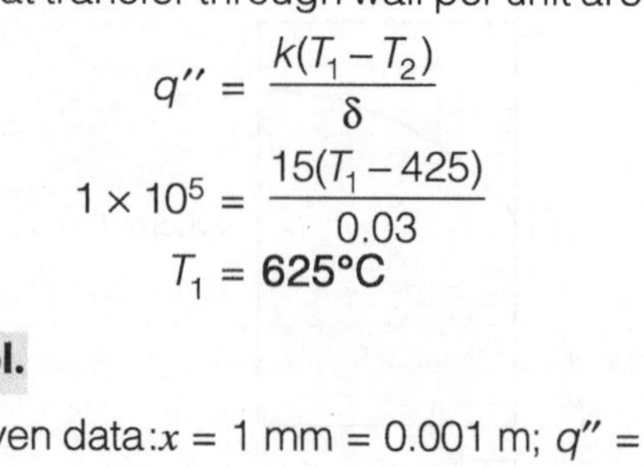

Heat flow:

$$Q = \frac{T_1 - T_2}{\frac{\delta}{k_sA} + \frac{x}{kA} + \frac{\delta}{k_sA}} = \frac{T_1 - T_2}{\frac{2\delta}{k_sA} + \frac{x}{kA}}$$

$$\frac{Q}{A} = \frac{T_1 - T_2}{\frac{2\delta}{k_s} + \frac{x}{k}}$$

$$q'' = \frac{T_1 - T_2}{\frac{2\delta}{k_s} + \frac{x}{k}}$$

or $$q''\left(\frac{2\delta}{k_s} + \frac{x}{k}\right) = T_1 - T_2$$

$$10000\left(\frac{2 \times 0.02}{10} + \frac{0.001}{k}\right) = 500 - 360$$

$$40 + \frac{10}{k} = 140$$

or $$\frac{10}{k} = 140 - 40 = 100$$

or $$k = \frac{10}{100} = \mathbf{0.1\ W/mK}$$

1.27 Sol.

Given data:

$$T_1 = 600\ K$$
$$T_2 = 300\ K$$

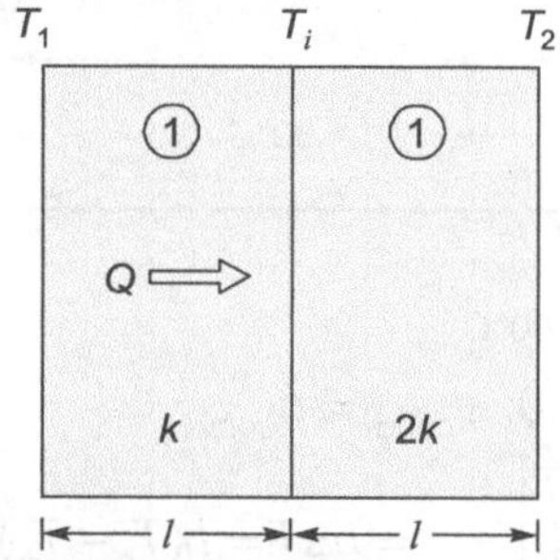

For steady state heat transfer,
Heat flow through layer-1
= Heat flow through layer-2

$$Q = Q$$

$$\frac{T_1 - T_i}{l/kA} = \frac{T_i - T_2}{l/2kA}$$

$$\frac{kA(T_1 - T_i)}{l} = \frac{2kA(T_i - T_2)}{l}$$

or $$T_1 - T_i = 2T_i - 2T_2$$

or $$T_1 + 2T_2 = 3T_i$$

or $$T_i = \frac{T_1 + 2T_2}{3} = \frac{600 + 2 \times 300}{3}$$

$$= \mathbf{400\ K}$$

1.28 Sol.

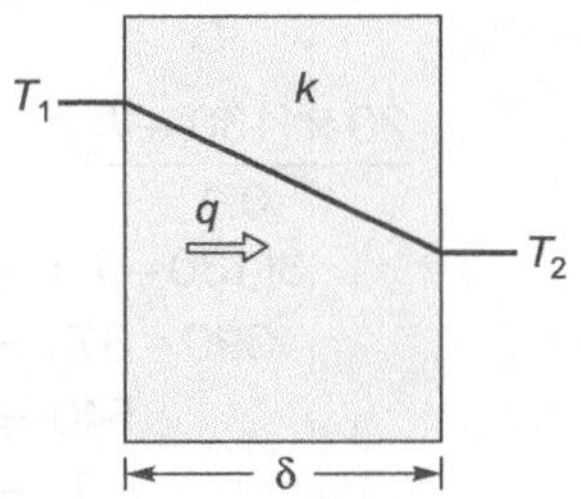

$k = 1.15$ W/mK;
$T_1 = 1100°C$;
$T_2 = 350°C$
$q = 2500$ W/m^2

$$Q = \frac{T_1 - T_2}{R_t}$$

$$\frac{Q}{A} = \frac{k(T_1 - T_2)}{\delta}$$

$$q = \frac{k(T_1 - T_2)}{\delta}$$

$$2500 = \frac{1.15(1100 - 350)}{\delta}$$

or $\delta = \mathbf{0.345\ m}$

1.29 Sol.

$$k_{foam} = 0.1 \text{ W/mK}$$
$$h_o = 2 \text{ W/m}^2\text{K}$$

Critical radius,

$$r_c = \frac{k_{foam}}{h_o} \quad \text{for a pipe}$$

$$= \frac{0.1}{2} = 0.05 \text{ m} = \mathbf{5\ cm}$$

1.30 (c)

$$d_i = 10 \text{ mm}$$

$$\therefore \quad r_i = \frac{d}{2} = \frac{10}{2} = 5 \text{ mm}$$

$$t = 2 \text{ mm}$$

$$\therefore \quad r_o = r_i + t$$
$$= 5 + 2 = 7 \text{ mm}$$
$$k = 0.08 \text{ W/mK}$$
$$h_o = 10 \text{ W/m}^2\text{K}$$

Critical radius,

$$r_c = \frac{k}{h_o} = \frac{0.08}{10}$$
$$= 0.008 \text{ m} = \mathbf{8\ mm}$$

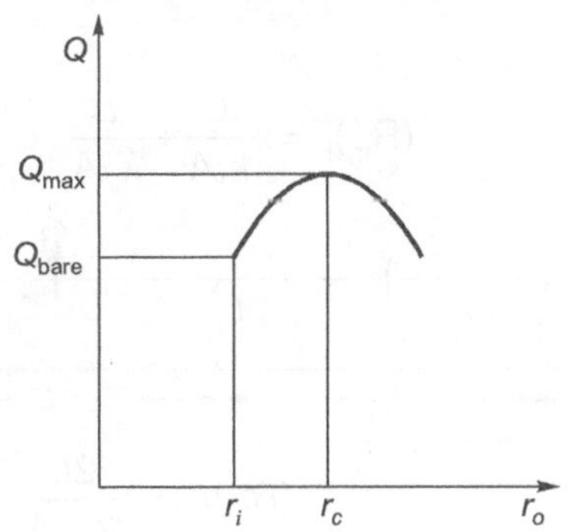

As $r_o < r_c$, the heat loss increase to a maximum and then decrease heat loss by addition of insulation.

1.31 (c)

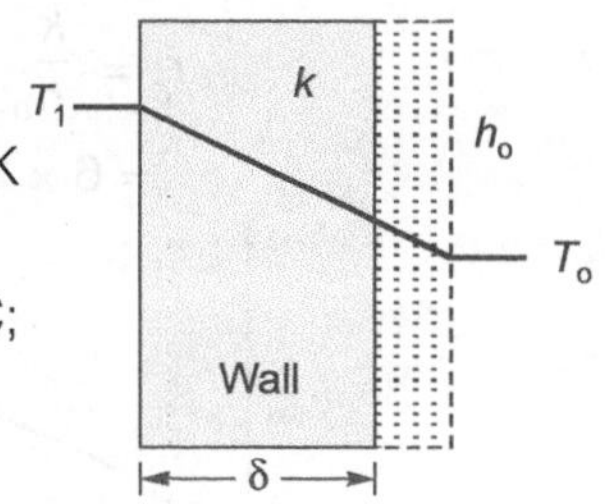

Given data:

$$k = 0.9 \text{ W/mK}$$
$$\delta = 0.18 \text{ m}$$

$T_o = -5°C$; $T_1 = 27°C$;
$h_o = 20$ W/m^{2}K

Heat loss:

$$Q = \frac{T_1 - T_o}{\frac{\delta}{kA} + \frac{1}{h_o A}}$$

$$\frac{Q}{A} = \frac{T_1 - T_0}{\frac{\delta}{k} + \frac{1}{h_0}}$$

$$q = \frac{27 - (-5)}{\frac{0.18}{0.9} + \frac{1}{20}} \quad \because q = \frac{Q}{A}$$

$$= \frac{32}{0.2 + 0.05} = \frac{32}{0.25} = \mathbf{128\ W/m^2}$$

1.32 (b)

Given data: $r = 5 \text{ mm} = 0.005 \text{ m}$

$$\therefore \quad d = 2r = 2 \times 0.005 = 0.010 \text{ m}$$
$$q_G = 4 \times 10^7 \text{ W/m}^3$$
$$h_o = 1000 \text{ W/m}^2\text{K}$$
$$T_o = 25°C$$

For steady state heat transfer,
Rate of heat generation in the rod
= rate of convection heat transfer from rod surface to fluid

$q_G \times$ volume of rod $= h_o A(T_s - T_o)$

$$q_G \times \frac{\pi}{4} d^2 l = h_o \times \pi d l (T_s - T_o)$$

$$\frac{d q_G}{4} = h_o(T_s - T_o)$$

$$\frac{0.010 \times 4 \times 10^7}{4} = 1000(T_s - 25)$$

or $T_s - 25 = 100$

or $T_s = 100 + 25 = 125°C$

$= (125 + 273)$ K $= \mathbf{398\,K}$

1.33 (a)

Given data: $r_o = 1$ mm

Inner radius: $r_i = 0.5$ mm

$k = 0.15$ W/mK

$h_o = 25$ W/m^{2}K

Critical radius,

$$r_c = \frac{k}{h_o} = \frac{0.15}{25}$$

$$= 6 \times 10^{-3} \text{ m} = 6 \text{ mm}$$

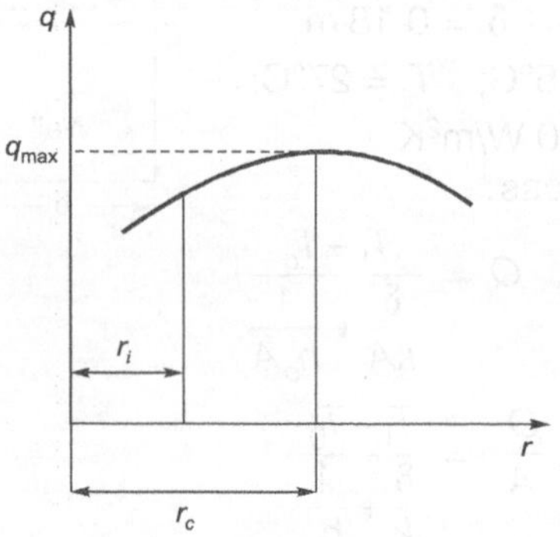

As $r_o < r_c$, due to addition of plastic cover heat transfer will increase till $r_o = r_c$ beyond that heat transfer will decrease.

1.34 (a)

$$Q = \frac{\Delta T}{\text{Thermal resistance}}$$

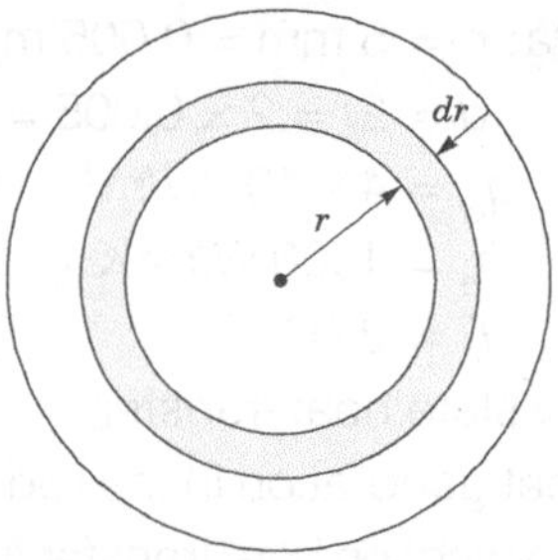

$$\text{Thermal resistance} = \int_{r_1}^{r_2} \frac{L}{kA} = \int_{r_1}^{r_2} \frac{dr}{k \times 2\pi r l}$$

$$= \frac{1}{2\pi l k}\int_{r_1}^{r_2} \frac{dr}{r} = \frac{1}{2\pi l k}[\ln r]_{r_1}^{r_2} = \frac{1}{2\pi l k}\ln\frac{r_2}{r_1}$$

1.35 Sol.

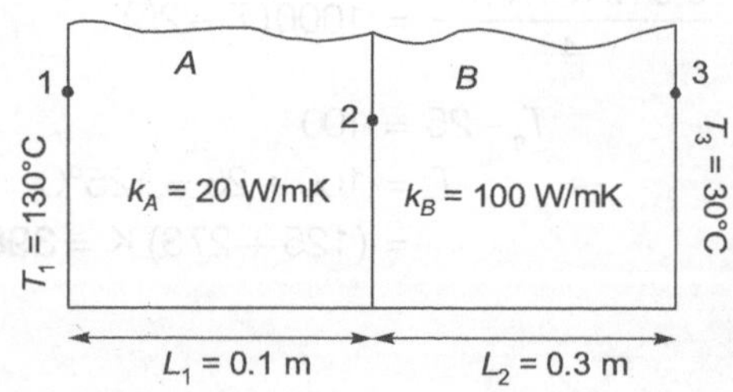

At steady state

$$Q_A = Q_B$$

$$\frac{k_A[T_1 - T_2]A}{L_1} = \frac{k_B[T_2 - T_3]A}{L_2}$$

$$\frac{20 \times [130 - T_2]}{0.1} = \frac{100 \times [T_2 - 30]}{0.3}$$

$$3[130 - T_2] = 5T_2 - 150$$

$$390 - 3T_2 = 5T_2 - 150$$

$$540 = 8T_2$$

$$T_2 = \mathbf{67.5°C}$$

1.36 Sol.

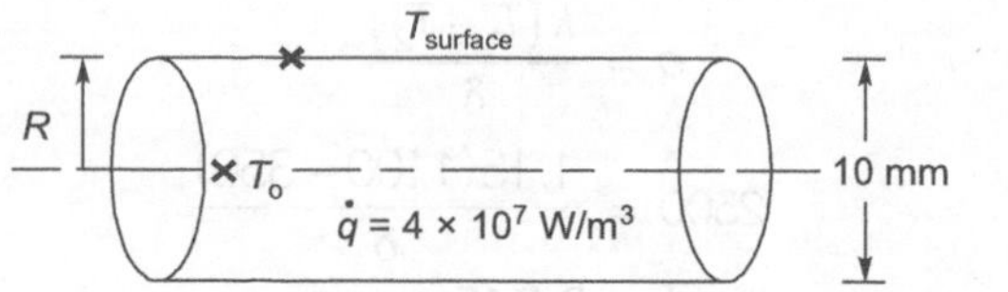

T_o = Temp. of cylinder at the axis

T_S = Surface temp. of Rod (cylinder)

$$T_o - T_S = \frac{\dot{q}}{4k}R^2$$

$\dot{q}$ = uniform heat generation rate per unit volume (W/m^3)

$$T_o - T_S = \frac{4 \times 10^7}{4 \times 25} \times \left(\frac{5}{1000}\right)^2 = 10°C$$

1.37 Sol.

Assuming steady state conditions

Thermal circuit

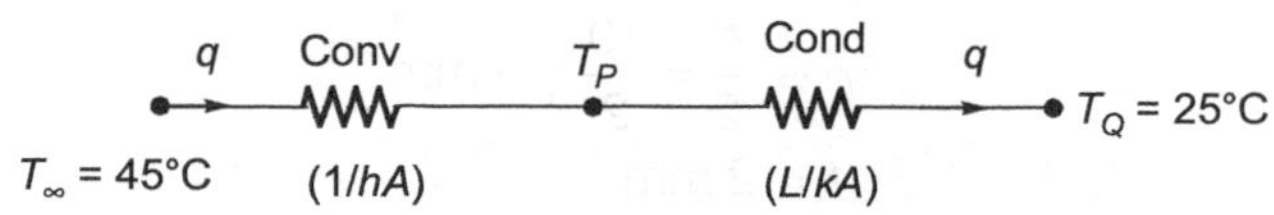

q = Rate of heat transfer

$$= \frac{45 - T_P}{1/hA} = \frac{T_P - 25}{L/kA}$$

$$\Rightarrow \frac{45 - T_P}{\left(\frac{1}{10}\right)} = \frac{T_P - 25}{\frac{0.2}{2.5}}$$

$$T_P = \mathbf{33.89°C}$$

1.38 (b)

$$(R_{th})_I = \frac{L}{k_1 A} + \frac{L}{k_2 A}$$

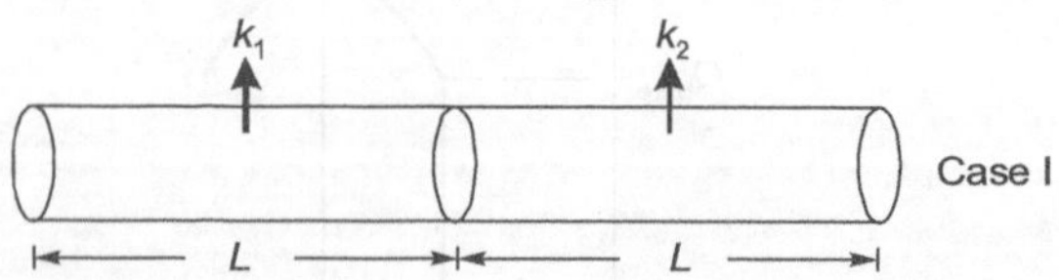

$$(R_{th})_{II} = \frac{2L}{k_{eq}A}$$

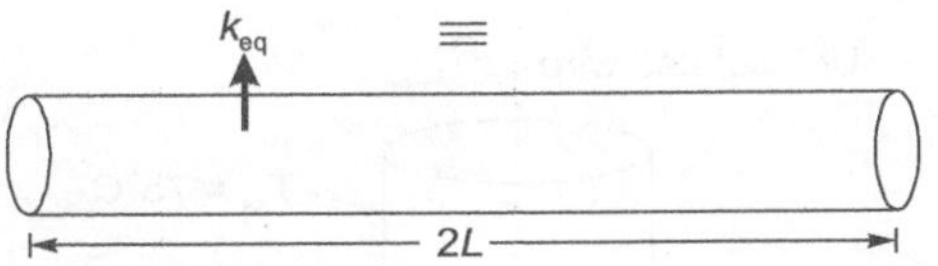

Equivalent resistance will be same in both cases,

$$(R_{th})_{II} = (R_{th})_{I}$$

$$\Rightarrow \quad \frac{2L}{k_{eq}A} = \frac{L}{k_1 A} + \frac{L}{k_2 A}$$

$$\Rightarrow \quad k_{eq} = \frac{2k_1 k_2}{k_1 + k_2}$$

1.39 Sol.

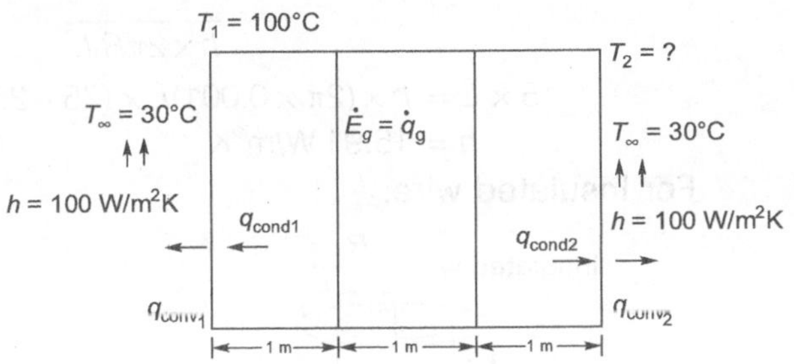

$$\dot{E}_g = 10000 \times (L \times A)$$

$$= 10000 \times 1 \times 1 = 10000\,\text{W}$$

$$hA(T_1 - T_\infty) = q_{conv_1} = q_{cond_1}$$

$$100 \times (100 - 30) = q_{cond_1}$$

$$q_{cond_1} = 7000 \text{ W}$$

$$q_{conv_2} = hA(T_2 - T_\infty)$$

$$= 100 \times 1 \times (T_2 - 30)$$

$$q_{conv_2} = 100(T_2 - 30) = q_{cond_2}$$

At steady state:

$$\dot{E}_g = \dot{E}_{out}$$

$$\dot{E}_g = q_{cond_1} + q_{cond_2}$$

$$10000 = 7000 + 100(T_2 - 30)$$

$$T_2 = \left(\frac{3000}{100}\right) + 30 = 60°\text{C}$$

1.40 (b)

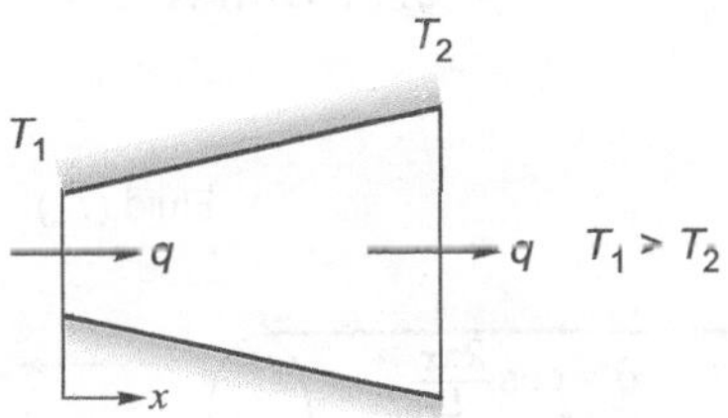

$$A = f(x)$$

Let, $\quad A = a + bx$

$$\dot{E}_{in} + \dot{E}_{out} + \not{\dot{E}}_g = \not{\dot{E}}_{st}$$

($\because$ steady state and no heat generation)

$$\dot{E}_{in} = \dot{E}_{out}$$

$q \neq f(x)$ i.e. is constant

$$q = -k A_{cs} \frac{dT}{dx} = \text{constant}$$

$$A_{cs} \frac{dT}{dx} = \text{constant}$$

($\because$ k will be constant for a material)

$$\int dT = \int \frac{C \cdot dx}{(a + bx)}$$

$$T = \frac{C}{b} \ln(a + bx) + C_1$$

Hence, temperature distribution in the solid is logarithmic.

1.41 Sol.

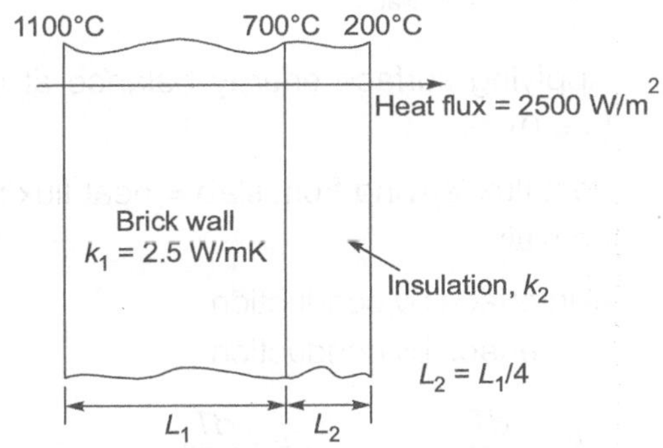

Given, $L_2 = \frac{L_1}{4}$

Assuming steady state, one-dimensional conduction heat transfer through composite slab,

Thermal circuit:

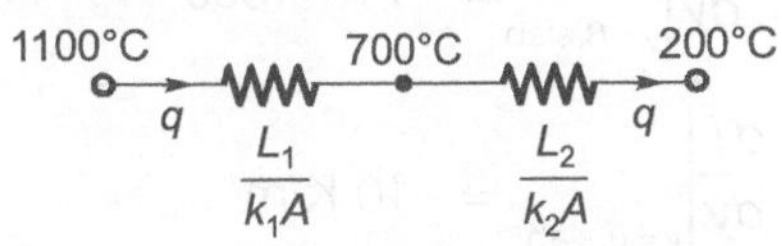

$$\Rightarrow \quad q = \frac{1100 - 700}{\frac{L_1}{k_1 A}} = \frac{700 - 200}{\frac{L_2}{k_2 A}}$$

$$\Rightarrow \quad \frac{400}{\frac{L_2}{2.5}} = \frac{500}{\frac{L_2}{k_2}}$$

$$\Rightarrow \quad k_2 = 0.5 \text{ W-mK}$$

1.42 Sol.

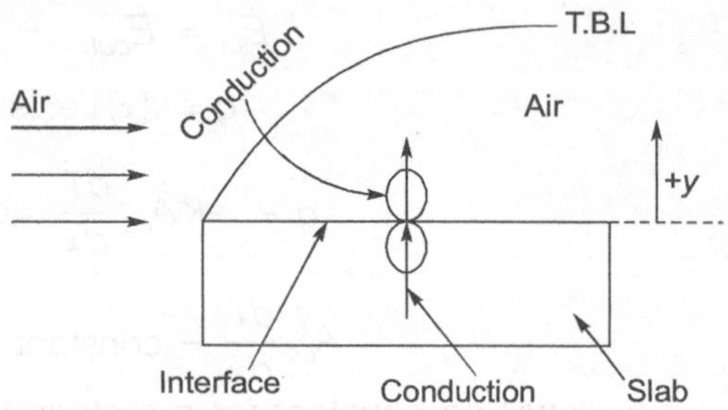

Given that

Temp. variation of air,

$$T = 300 + 200\,e^{-5y}$$

$$K_{air} = 1 \text{ W/m-K}$$

$$K_{slab} = 100 \text{ W/m-K}$$

$$\left.\frac{dT}{dy}\right|_{y=0,\ slab} = ?$$

Applying surface energy balance at interface ($y = 0$)

heat flux leaving from slab = heat flux received from air

at interface by conduction

at interface by conduction

$$-k_{slab}\left.\frac{dT}{dy}\right|_{y=0,\,slab} = -k_{air}\left.\frac{dT}{dy}\right|_{y=0,\ air}$$

$$\left.\frac{dT}{dy}\right|_{y=0,\ air} = 0 + 200 \times -5 \times 1 = -1000 \text{ K/m}$$

$$-100\left.\frac{dT}{dy}\right|_{y=0,\,slab} = -1 \times -1000$$

$$\left.\frac{dT}{dy}\right|_{y=0,\,slab} = -10 \text{ K/m}$$

$$\text{Magnitude of } \left.\frac{dT}{dy}\right|_{y=0,\,slab} = 10 \text{ K/m}$$

1.43 Sol.

For uninsulated wire:

Given: $T_\infty = 25°C$, $R_1 = 1$ mm, $\dot{q}_{gen} = 5$ W/m, $T_{s_1} = 7.5°C$

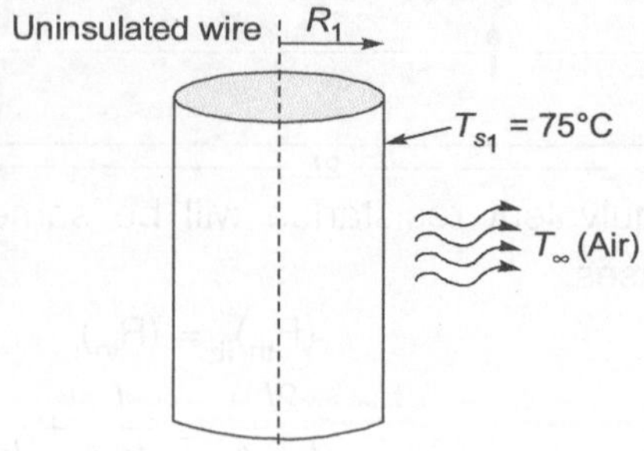

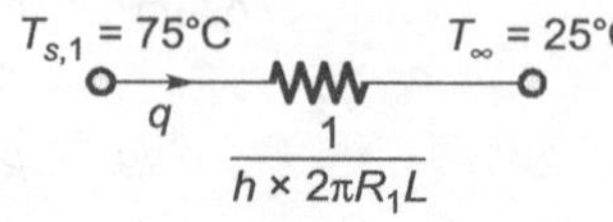

$$q = \dot{q}_{gen} \times L = \frac{T_{s1} - T_\infty}{\dfrac{1}{h \times 2\pi R_1 L}}$$

$$5 \times L = h \times (2\pi \times 0.001)L \times (75 - 25)$$

$$h = 15.91 \text{ W/m}^2\text{K}$$

For Insulated wire:

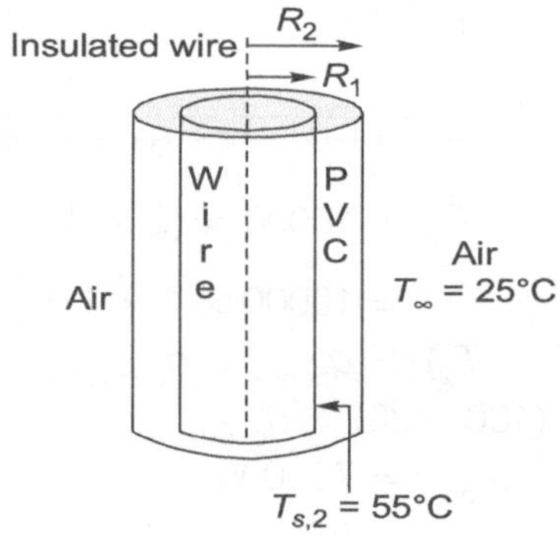

$$R_2 = 2 \text{ mm}$$

Heat transfer rate after insulation kept on wire,

$$q = q_{gen,\ wire} \times L = 5 \times L$$

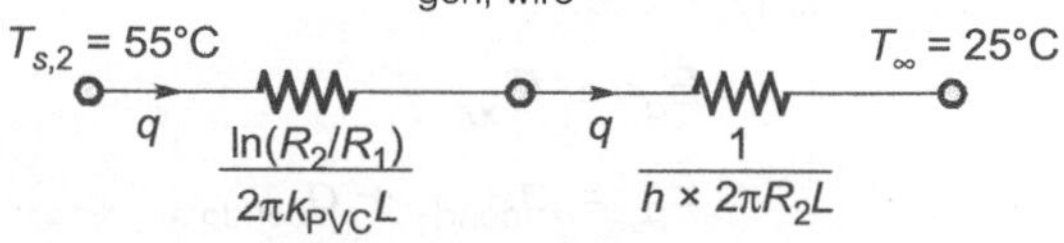

$$q = 5 \times L = \frac{55 - 25}{\dfrac{\ln(2/1)}{2\pi k_{PVC} L} + \dfrac{1}{15.91 \times 2 \times \pi \times 0.002 \times L}}$$

$$k_{PVC} = 0.1103 \text{ W/m-K}$$

$$= 0.11 \text{ W/m-K}$$

1.44 (c)

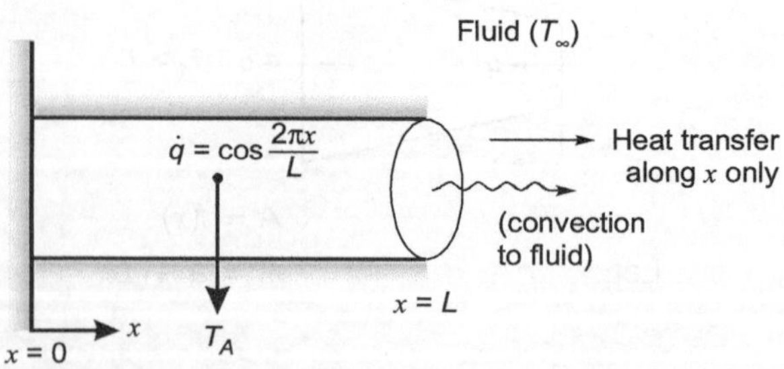

For steady state, 1-d heat flow with non-uniform heat generation.

$$\Rightarrow \quad \frac{d^2T}{dx^2} + \frac{\dot{q}}{k} = 0$$

$$\Rightarrow \quad \frac{d^2T}{dx^2} = -\frac{\dot{q}}{k}$$

Integrating the equation,

$$\Rightarrow \quad \frac{dT}{dx} = \int -\frac{\dot{q}}{k}dx + c_1$$

$$\Rightarrow \quad \frac{dT}{dx} = -\frac{1}{k}\int \cos\frac{2\pi x}{L}dx + c_1$$

$$\Rightarrow \quad \frac{dT}{dx} = -\frac{1}{k}\left(\frac{L}{2\pi}\right)\sin\frac{2\pi x}{L} + c_1 \quad \text{...(i)}$$

At $x = 0$, $q_{cond} = 0$,

$$\frac{dT}{dx} = 0$$

$$\therefore \quad c_1 = 0$$

Integrating (i) equation,

$$\Rightarrow \quad T = -\frac{L}{2\pi k} \times \frac{-L}{2\pi}\cos\frac{2\pi x}{L} + c_2$$

$$\Rightarrow \quad T = \frac{L^2}{4\pi^2 k}\cos\left(\frac{2\pi x}{L}\right) + c_2$$

At $x = L$,

$\Rightarrow$ Heat conducted = Heat convected

$$\Rightarrow -kA\left(\frac{dT}{dx}\right)_{at\,x=L} = hA(T_{x=L} - T_\infty)$$

$$\Rightarrow -k\left[-\frac{1}{k} \times \frac{L}{2\pi} \times \sin\frac{2\pi L}{L}\right] = hA(T_{x=L} - T_\infty)$$

$$\Rightarrow -k\left[-\frac{1}{k} \times \frac{L}{2\pi} \times 0\right] = hA(T_{x=L} - T_\infty)$$

$$\Rightarrow \quad T_{x=L} = T_\infty$$

$$\Rightarrow \quad \frac{L^2}{4\pi^2 k}\cos\left(\frac{2\pi \times L}{L}\right) + c_2 = T_\infty$$

$$\Rightarrow \quad \frac{L^2}{4\pi^2 k} + c_2 = T_\infty$$

Therefore, c_2 is not a function of h.

$\Rightarrow$ T = Function of x only, not h

$\therefore$ T_A remains same even if 'h' is doubled.

1.45 (1000)

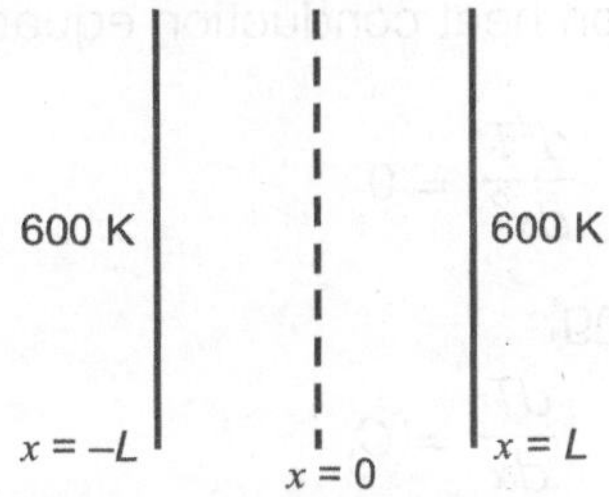

Given: $K = CT$, where $C = 2$

Steady state, one-dimensional heat conduction

$$\frac{\partial}{\partial x}\left(k\frac{\partial T}{\partial x}\right) = -\dot{q}$$

On integrating,

$$k\frac{dT}{dx} = -\dot{q}x + C_1$$

$$2T\frac{dT}{dx} = -\dot{q}x + C_1$$

$$TdT = \left[\frac{-qx + C_1}{2}\right]dx$$

On integrating,

$$\frac{T^2}{2} = \frac{1}{2}\left[-\frac{\dot{q}x^2}{2} + C_1 x\right] + C_2 \quad \text{... (i)}$$

Boundary conditions:

At $x = +L = +1$ m $\Rightarrow T = 600$ K

$$600^2 = \frac{-1280 \times 10^3 \times 1^2}{2} + C_1 \times 1 + C_2$$

$$\Rightarrow \quad C_1 + C_2 = 10^6 \quad \text{... (ii)}$$

At $x = -L = -1 \Rightarrow T = 600$ K

$$600^2 = \frac{-1280 \times 10^3 \times (-1)^2}{2} + C_1 \times (-1) + C_2$$

$$\Rightarrow \quad -C_1 + C_2 = 10^6 \quad \text{... (iii)}$$

By equation (ii) and (iii) we get,

$$C_1 = 0$$

and $C_2 = 10^6$

So by equation (i) we get

$$T^2 = -\frac{\dot{q}x^2}{2} + C_2$$

So, at $x = 0$

$$[T(0)]^2 = C_2$$

$$[T(0)] = \sqrt{C_2} = \sqrt{10^6}$$

$$\Rightarrow \quad T = 10^3 = 1000 \text{ K}$$

$$T_{x=0} = 1000 \text{ K}$$

1.46 (a, b)

For 1-D steady state without internal heat generation heat conduction equation is

$$\frac{d^2T}{dx^2} = 0$$

Integrating,

$$\frac{dT}{dx} = C_1$$

and, $T(x) = C_1x + C_2$

Thus the temperature varies linearly along its length. It may decrease with x as shown in option (b) or increase with x as shown in option (a) depending upon the value of constant C_1.

1.47 (b)

Given: $k = A + Bx$

where A and B are positive real number

So with increase with x, k increases

From energy equation;

$$\frac{\partial}{\partial x}\left(k\frac{\partial T}{\partial x}\right) + \cancel{q_g}^{\,0} = 0$$

$$\frac{\partial}{\partial x}\left((A+Bx)\cdot\frac{\partial T}{\partial x}\right) = 0$$

On integration, $(A+Bx)\cdot\dfrac{\partial T}{\partial x} = C_1$

$$\frac{\partial T}{\partial x} = \frac{C_1}{A+Bx}$$

On further integration, we will get temperature (T) as logarithmic function of x.

Hence, variation of T with x is given as;

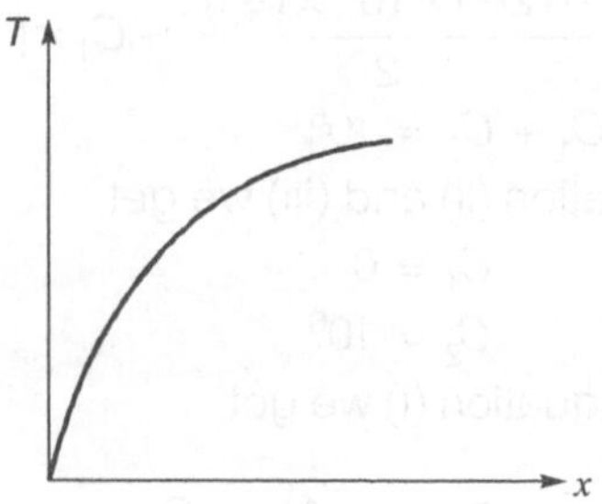

1.48 (6)

For uniform internal heat generation and constant thermal conductivity in a rectangular slab,

$$\frac{\partial^2 T}{\partial x^2} + \frac{q_g}{k} = 0$$

$$\Rightarrow \quad T = -\frac{q_g x^2}{2k} + C_1 x + C_2$$

Temperature variation is parabolic inside the slab

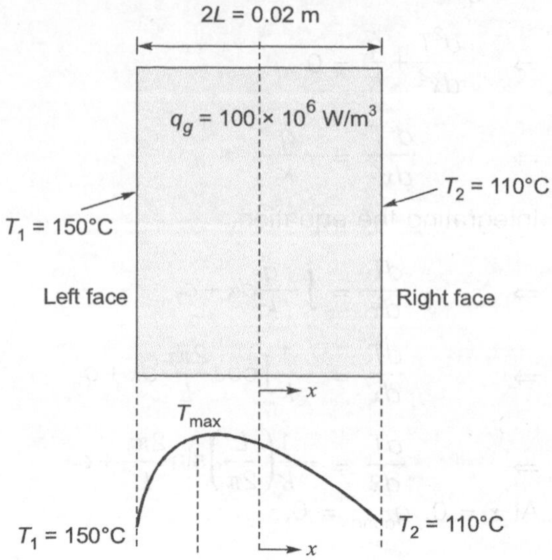

At $x = -L,\ T = T_1 = 150°C$

$$\Rightarrow \quad 150 = -\frac{q_g x^2}{2k} - C_1 L + C_2 \quad \text{... (i)}$$

and at $x = L,\ T = T_2 = 110°C$

$$110 = -\frac{q_g L^2}{2k} + C_1 L + C_2 \quad \text{... (ii)}$$

Using equation (i) and (ii)

$$C_1 = \frac{T_2 - T_1}{2L} = \frac{110-150}{2\times(0.01)} = -2000$$

At T_{max}; $\dfrac{dT}{dx} = 0$

$$\Rightarrow \quad \frac{dT}{dx} = -\frac{q_g x}{k} + C_1$$

$$\Rightarrow \quad 0 = -\frac{(100\times10^6)x}{200} - 2000$$

$$\Rightarrow \quad x = -4 \times 10^{-3}\text{ m} = -4\text{ mm}$$

Distance of location of the maximum temperature from the left face

$$= L - |x| = 10 - 4 = 6\text{ mm}$$

■■■■

2 Free and Forced Convection

2.1 Match List-I with List-II and select the correct answer using the code given below the lists:

List-I

A. Grashof nu mber B. Schmidt number

C. Weber number D. Fourier number

List-II

1. Mass diffusion 2. Transient heat conduction

3. Free convection 4. Forced convection

5. Surface tension 6. Radiation

Codes:

	A	B	C	D
(a)	4	2	3	5
(b)	3	1	5	2
(c)	5	1	2	4
(d)	1	2	5	4

[1996 : 2 M]

2.2 Heat transfer coefficients for free convection in gases, forced convection in gases and vapours, and for boiling water lie, respectively, in the range of

(a) 5–15, 20–200 and 3000–50000 W/m^{2}K

(b) 20–50, 200–500 and 50000–100000 W/m^{2}K

(c) 50–100, 500–1000 and 100000–100000 W/m^{2}K

(d) 20–100, 200–1000 and a constant 1000000 Wm^{2}K [1998 : 1 M]

2.3 The properties of mercury at 300 K are: Density = 13529 kg/m^3, c_p = 0.1393 kJ/kgK, dynamic viscosity = 0.1523 × 10^{-2} Ns/m^2 and thermal conductivity = 8.540 W/mK. The Prandtl number of the mercury at 300 K is

(a) 0.0248 (b) 2.48

(c) 24.8 (d) 248 [2002 : 2 M]

Common Data Questions Q.2.4–Q.2.5:

Heat is being transferred by convection from water at 48°C to a glass plate whose surface temp. is exposed to the water is at 40°C. The thermal conductivity of water is 0.6 W/mK and the thermal conductivity of glass is 1.2 W/mK. The spatial gradient of temperature in the water at the water-glass interface is dT/dy = 1 × 10^4 K/m.

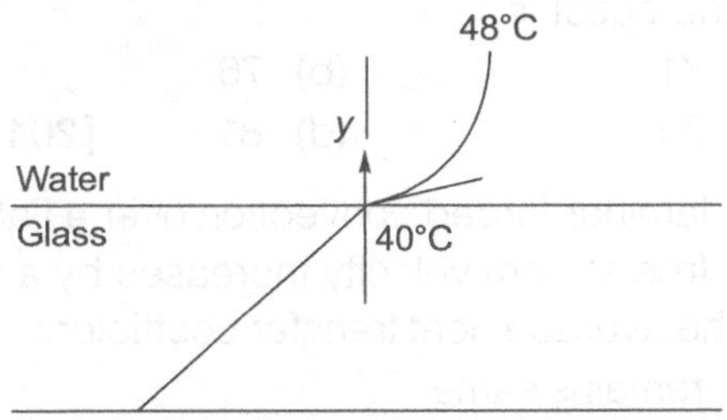

2.4 The value of the temperature gradient in the glass at the water-glass interface in K/m is

(a) – 2 × 10^4 (b) 0.0

(c) 0.5 × 10^4 (d) 2 × 10^4 [2003 : 2 M]

2.5 The heat transfer coefficient h in W/m^{2}K is

(a) 0.0 (b) 4.8

(c) 6 (d) 750 [2003 : 2 M]

2.6 For the three-dimensional object shown in the figure below, five faces are insulated. The sixth face (PQRS), which is not insulated, interacts thermally with the ambient, with a convective heat transfer coefficient of 10 W/m^{2}K. The ambient temperature is 30°C. Heat is uniformly generated inside the object at the rate of 100 W/m^3. Assuming the face PQRS to be at uniform temperature, its steady state temperature is

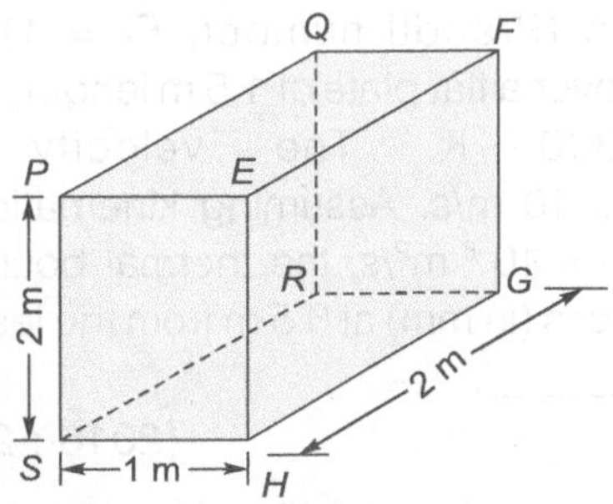

(a) 10°C (b) 20°C

(c) 30°C (d) 40°C [2008 : 2 M]

Common Data Questions Q.2.7 and Q.2.8:

Water (specific heat, c_p = 4.18 kJ/kgK) enters a pipe at a rate 0.01 kg/s and a temperature of 20°C. The pipe, of diameter 50 mm and length 3 m, is subjected to a wall heat flux q''_w in W/m^2.

2.7 If q''_w = 2500x, where x is in m and in the direction of flow (x = 0 at the inlet), the bulk means temperature of the water leaving the pipe in °C is

(a) 42 (b) 62
(c) 74 (d) 104 [2013 : 2 M]

2.8 If q''_w = 5000 and the convection heat transfer coefficient at the pipe outlet is 1000 W/m²K, the temperature in °C at the inner surface of the pipe at the outlet is
(a) 71 (b) 76
(c) 79 (d) 81 [2013 : 2 M]

2.9 For laminar forced convection over a flat plate, if the free stream velocity increases by a factor of 2, the average heat transfer coefficient
(a) remains same
(b) decreases by a factor of $\sqrt{2}$
(c) rises by a factor of $\sqrt{2}$
(d) rises by a factor of 4 [2014 : 1 M, Set-2]

2.10 In the laminar flow of air (Pr = 0.7) over a heated plate, if δ and δ_T denote, respectively, the hydrodynamic and thermal boundary layer thicknesses, then
(a) $\delta = \delta_T$ (b) $\delta > \delta_T$
(c) $\delta < \delta_T$ (d) $\delta = 0$ but $\delta_T \neq 0$
[2015 : 1 M, Set-2]

2.11 The ratio of momentum diffusivity (ν) to thermal diffusivity (α), is called
(a) Prandtl number (b) Nusselt number
(c) Biot number (d) Lewis number
[2015 : 1 M, Set-3]

2.12 A fluid (Prandtl number, Pr = 1) at 500 K flows over a flat plate of 1.5 m length, maintained at 300 K. The velocity of the fluid is 10 m/s. Assuming kinematic viscosity. $\nu = 30 \times 10^{-6}$ m²/s, the thermal boundary layer thickness (in mm) at 0.5 m from the leading edge is________.
[2016 : 2 M, Set-1]

2.13 Grashof number signifies the ratio of
(a) inertia force to viscous force
(b) buoyancy force to viscous force
(c) buoyancy force to inertia force
(d) inertia force to surface tension force
[2016 : 1 M, Set-3]

2.14 For a hydrodynamically and thermally fully developed laminar flow through a circular pipe of constant cross section, the Nusselt number at constant wall heat flux (Nu_q) and that at constant wall temperature (Nu_T) are related as
(a) $Nu_q = Nu_T$ (b) $Nu_q < Nu_T$
(c) $Nu_q > Nu_T$ (d) $Nu_q = (Nu_T)^2$
[2019 : 1 M, Set-1]

2.15 The wall of a constant diameter pipe of length 1 m is heated uniformly with flux q'' by wrapping a heater coil around it. The flow at the inlet to the pipe is hydrodynamically fully developed. The fluid is incompressible and the flow is assumed to be laminar and steady all through the pipe. The bulk temperature of the fluid is equal to 0°C at the inlet and 50°C at the exit. The wall temperatures are measured at three locations, P, Q and R, as shown in the figure. The flow thermally develops after some distance from the inlet. The following measurements are made:

Point	P	Q	R
Wall Temperature(°C)	50	80	90

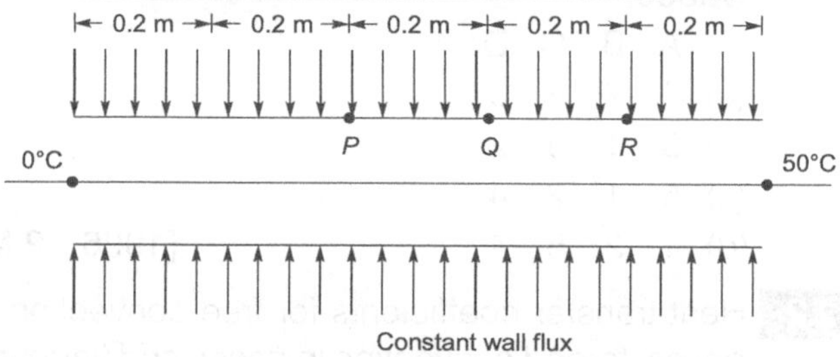

Among the locations P, Q and R the flow is thermally developed at
(a) P, Q and R (b) Q and R only
(c) R only (d) P and Q only
[2019 : 2 M, Set-1]

2.16 A thin vertical flat plate of height L, and infinite width perpendicular to the plane of the figure, is losing heat to the surroundings by natural convection. The temperatures of the plate and the surroundings and the properties of the surrounding fluid are constant. The relationship between the average Nusselt and Rayleigh numbers is given as $Nu = K\,Ra^{1/4}$, where K is a constant. The length scales for Nusselt and Rayleigh numbers are the height of the plate. The height of the plate is increased to 16 L keeping all other factors constant.

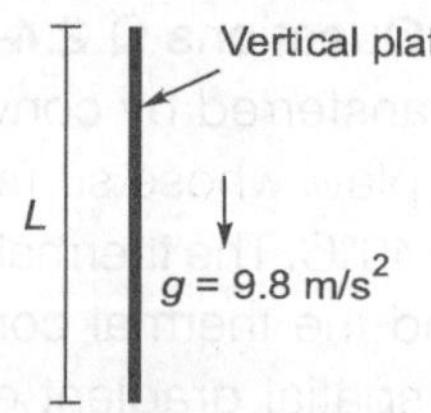

If the average heat transfer coefficient for the first plate is h_1 and that for the second plate is h_2, the value of the ratio h_1/h_2 is ______
[2019 : 1 M, Set-2]

2.17 Match the following non-dimensional numbers with the corresponding definitions:

	Non-dimensional number		Definition
P.	Reynolds number	1.	$\frac{\text{Buoyancy force}}{\text{Viscous force}}$
Q.	Grashof number	2.	$\frac{\text{Momentum diffusivity}}{\text{Thermal diffusivity}}$
R.	Nusselt number	3.	$\frac{\text{Inertia force}}{\text{Viscous force}}$
S.	Prandtl number	4.	$\frac{\text{Convective heat transfer}}{\text{Conduction heat transfer}}$

(a) P-3, Q-1, R-2, S-4
(b) P-3, Q-1, R-4, S-2
(c) P-4, Q-3, R-1, S-2
(d) P-1, Q-3, R-2, S-4 **[2020 : 1 M, Set-1]**

2.18 Water flows through a tube of 3 cm internal diameter and length 20 m, The outside surface of the tube is heated electrically so that it is subjected to uniform heat flux circumferentially and axially. The mean inlet and exit temperatures of the water are 10°C and 70°C, respectively. The mass flow rate of the water is 720 kg/h. Disregard the thermal resistance of the tube wall. The internal heat transfer coefficient is 1697 W/m^{2}K. Take specific heat C_p of water as 4.179 kJ/kgK. The inner surface temperature at the exit section of the tube is ________ °C (round off to one decimal place).

[2020 : 2 M, Set-2]

2.19 In the following two-dimensional momentum equation for natural convection over a surface immersed in a quiescent fluid at temperature T_∞ (g is the gravitational acceleration, β is the volumetric thermal expansion coefficient, ν is the kinematic viscosity, u and v are the velocities in x and y directions, respectively, and T is the temperature)

$$u\frac{\partial u}{\partial x}+v\frac{\partial u}{dy}=g\beta(T-T_\infty)+\nu\frac{\partial^2 u}{\partial y^2},$$

the term $g\beta(T - T_\infty)$ represents
(a) Ratio of inertial force to viscous force.
(b) Ratio of buoyancy force to viscous force.
(c) Viscous force per unit mass.
(d) Buoyancy force per unit mass.

[2022 : 1 M, Set-1]

2.20 Which of the following non-dimensional terms is an estimate of Nusselt number?
(a) Ratio of internal thermal resistance of a solid to the boundary layer thermal resistance.
(b) Ratio of the rate at which internal energy is advected to the rate of conduction heat transfer.
(c) Non-dimensional temperature gradient.
(d) Non-dimensional velocity gradient multiplied by Prandtl number.

[2022 : 1 M, Set-2]

2.21 Consider a hydrodynamically and thermally fully-developed, steady fluid flow of 1 kg/s in a uniformly heated pipe with diameter of 0.1 m and length of 40 m. A constant heat flux of magnitude 15000 W/m^2 is imposed on the outer surface of the pipe. The bulk-mean temperature of the fluid at the entrance to the pipe is 200°C. The Reynolds number (Re) of the flow is 85000, and the Prandtl number (Pr) of the fluid is 5. The thermal conductivity and the specific heat of the fluid are 0.08 W·m^{-1}·K^{-1} and 2600 J·kg^{-1}·K^{-1}, respectively. The correlation Nu = 0.023 Re$^{0.8}$Pr$^{0.4}$ is applicable, where the Nusselt Number (Nu) is defined on the basis of the pipe diameter. The pipe surface temperature at the exit is ________ °C (round off to the nearest integer).

[2022 : 2 M, Set-2]

Answers Free and Forced Convection

2.1 (b) 2.2 (a) 2.3 (a) 2.4 (c) 2.5 (d) 2.6 (d) 2.7 (b) 2.8 (d)
2.9 (c) 2.10 (c) 2.11 (a) 2.12 (5.97) 2.13 (b) 2.14 (c) 2.15 (b) 2.16 (2)
2.17 (b) 2.18 (85.7) 2.19 (d) 2.20 (c) 2.21 (321)

Explanations Free and Forced Convection

2.3 (a)

ρ = 13529 kg/m^3; c_p = 0.1393 kJ/kgK
= 139.3 J/kgK

$\mu = 0.1523 \times 10^{-2}$ Ns/m^2; k_f = 8.540 W/mK

Prandtl number,

$$\text{Pr} = \frac{\mu c_p}{k_f} = \frac{0.1523 \times 10^{-2} \times 139.3}{8.540}$$

$$= \mathbf{0.0248}$$

2.4 (c)

$T_w = 48°C$; $T_g = 40°C$

k_w = 0.6 W/mK; k_g = 1.2 W/mK

$$\left(\frac{dT}{dy}\right)_w = 1 \times 10^4 \text{ K/m}$$

At water-glass interface,

$$-k_w A\left(\frac{dT}{dy}\right)_w = -k_g A\left(\frac{dT}{dy}\right)_g$$

or $$k_w\left(\frac{dT}{dy}\right)_w = k_g\left(\frac{dT}{dy}\right)_g$$

$$0.6 \times 1 \times 10^4 = 1.2 \times \left(\frac{dT}{dy}\right)_g$$

$$\left(\frac{dT}{dy}\right)_g = \mathbf{0.5 \times 10^4 \text{ K/m}}$$

2.5 (d)

For water,

$$(Q_{\text{conv}})_{\text{water}} = (Q_{\text{cond}})_{\text{water}}$$

$$h_w A(T_w - T_g) = k_w A\left(\frac{dT}{dy}\right)_w$$

or $$A(T_w - T_g) = k_w\left(\frac{dT}{dy}\right)_w$$

$$h_w \times (48 - 40) = 0.6 \times 1 \times 10^4$$

or $$h_w = \mathbf{750 \text{ W/m}^2\text{K}}$$

2.6 (d)

Given data: h_o = W/m^{2}K; $T_o = 30°C$

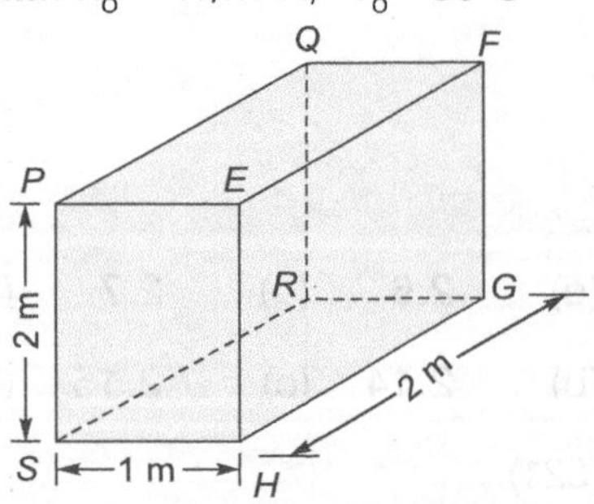

$$q_G = 100 \text{ W/m}^3$$

Volume: $V = 2 \times 1 \times 2 = 4$ m^3

Heat generated,

$$Q = q_G \times V = 100 \times 4 = 400 \text{ W}$$

Convection heat transfer from face *PQRS*,

$$Q = h_o A(T_s - T_o)$$

$$400 = 10 \times 2 \times 2(T_s - 30)$$

or $$T_s - 30 = 10$$

or $$T_s = 10 + 30 = \mathbf{40°C}$$

2.7 (b)

$$c_p = 4.18 \text{ kJ/kgK} = 4180 \text{ J/kgK}$$

$$m = 0.01 \text{ kg/s}; \; T_i = 20°C$$

$$d = 50 \text{ mm} = 0.05 \text{ m}; \; l = 3 \text{ m}$$

$$q_w'' = 2500x \text{ W/m}^2$$

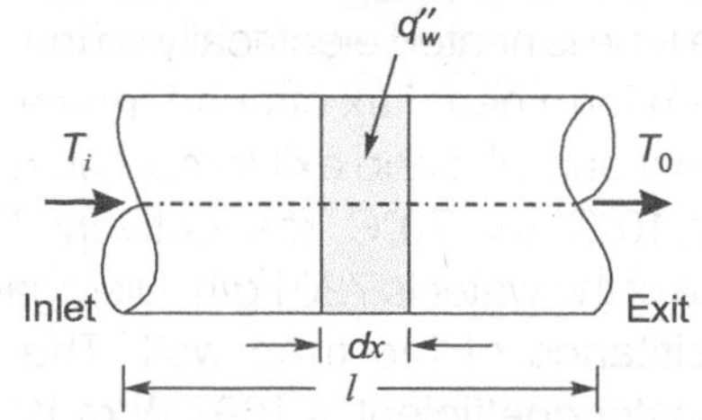

Heat transfer through small length dx,

$$\delta Q = q_w'' \times \pi d \times dx$$

$$= \pi d q_w'' \, dx$$

$$= 3.14 \times 0.05 \times 2500x\,dx$$

$$= 392.5x\,dx$$

Net heat transfer through whole length l,

$$Q = \int_0^l 392.5x\,dx = 392.5\int_0^3 x\,dx$$

$$= 392.5\left[\frac{x^2}{2}\right]_0^3$$

$$= 392.5 \times \frac{3 \times 3}{2} = 1766.25 \text{ W}$$

also $Q = mc_p(T_o - T_i)$, heat gained by the water

$\therefore$ $$1766.25 = 0.01 \times 4180(T_o - 20)$$

or $$T_o - 20 = 42.25$$

or $$T_o = 42.25 + 20$$

$$= 62.25°C \simeq \mathbf{62°C}$$

Alternatively

At pipe inlet,

$$x = 0$$

$$q_{wi}'' = 2500 \times 0 = 0$$

Art pipe exit,

$$x = l$$

$$q''_{wo} = 2500 \times l \times 3 = 7500 \text{ W/m}^2$$

Average heat transfer,

$$q''_{avg} = \frac{q''_{wi} + q''_{wo}}{2}$$

$$= \frac{0 + 7500}{2} = 3750 \text{ W/m}^2$$

Net heat transfer,

$$Q = q''_{avg} \times \text{surface area}$$

$$= 3750 \times \pi dl$$

$$= 3750 \times 3.14 \times 0.05 \times 3$$

$$= 1766.25 \text{ W}$$

also $Q = mc_p(T_o - T_i)$, heat gained by the water

$$\therefore \quad 1766.25 = 0.01 \times 4180(T_o - 20)$$

or $T_o - 20 = 42.25$

or $T_o = 42.25 + 20$

$$= 62.25°\text{C} \simeq \mathbf{62°C}$$

2.8 (d)

$$q''_w = 5000 \text{ W/m}^2;\ h = 1000 \text{ W/m}^2\text{K}$$

Heat transfer through pipe wall,

$$Q = q'_w \times \text{surface area} = 5000 \times \pi dl$$

$$= 5000 \times 3.14 \times 0.05 \times 3$$

$$= 2355 \text{ W}$$

also $Q = mc_p(T_o - T_i)$, heat gained by the water

$$2355 = 0.01 \times 4180(T_o - 20)$$

$$56.33 = T_o - 20$$

$$T_o = 76.33°\text{C}$$

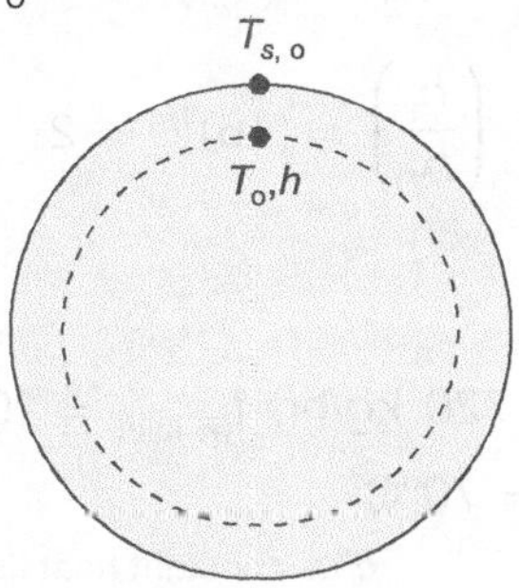

At pipe outlet

Heat flux q''_w is same at any two sections,

$$q''_w = h(T_{s,o} - T_o)$$

$$5000 = 1000(T_{s,o} - 76.33)$$

or $5 = T_{s,o} - 76.33$

or $T_{s,o} = 76.33 + 5$

$$= 81.33°\text{C} \simeq \mathbf{81°C}$$

2.9 (c)

For the laminar flow over flat plate,

$$\text{Re} < 3 \times 10^5$$

Nusselt number,

$$N_u = \frac{hl}{k} = 0.664\, \text{Re}^{1/2}\, \text{Pr}^{1/3}$$

where $\text{Re} = \dfrac{Ul}{\nu}$

$$\therefore \quad \frac{hl}{k} = 0.664 \times \left(\frac{Ul}{\nu}\right)^{1/2} \times \text{Pr}^{1/3}$$

$$h \propto U^{1/2}$$

$$\frac{h_2}{h_1} = \left(\frac{U_2}{U_1}\right)^{1/2} = \left(\frac{2U_2}{U_1}\right)^{1/2} = \sqrt{2}$$

or $h_2 = \sqrt{2}\, h_1$

2.10 (c)

Prandtl number,

$$\text{Pr} = 0.7$$

also $\text{Pr} = \left(\dfrac{\delta}{\delta_T}\right)^3$

$$\therefore \quad (0.7)^{1/3} = \frac{\delta}{\delta_T}$$

or $\delta = 0.888\, \delta_T$

$$\therefore \quad \boldsymbol{\delta < \delta_T}$$

2.12 Sol.

Given, $Pr = 1$; $T_\infty = 500$ K; Length = 1.5 m

$$T_{plate} = 300 \text{ K};\ U_\infty = 10 \text{ m/s}$$

$$\nu = 30 \times 10^{-6} \text{ m}^2/\text{s}$$

Thermal boundary layer at 0.5 m from leading edge = δ_t

$$\text{Re}_{x=0.5} = \frac{Ux}{\nu} = \frac{10 \times 0.5}{30 \times 10^{-6}} = 166{,}666.67$$

Since $\text{Re}_{x=0.5} < 5 \times 10^5$. Hence flow is laminar

$$\therefore \quad \delta = \frac{5x}{\sqrt{\text{Re}_x}} = \frac{5 \times 0.5}{\sqrt{166666.67}}$$

$$= 6.1237 \times 10^{-3} \text{ m}$$

$$\frac{\delta_t}{\delta} = \frac{1}{1.026}(Pr)^{-1/3}$$

$$\delta_t = 5.97 \text{ mm}$$

2.14 (c)

For fully developed laminar flow through pipes, Nusselt number value corresponding to uniform heat flux is 4.36.

Nusselt number value corresponding to uniform wall temperature is 3.66.

Hence, $Nu_q > Nu_T$

2.15 (b)

Internal force convection

Boundary condition: Constant Heat Flux Case

T_m(Bulk mean temperature of fluid)

At inlet, $T_{m,i} = 0°C$

At outlet, $T_{m,\theta} = 50°C$

Variation of Bulk Mean Temperature (T_m)

$$T_m = T_{m,i} + \frac{q''P}{\dot{m}c_p}x \quad \text{... linear with 'x'}$$

at $x = 1$ m; $T_{m,\theta} = 50°C$

$$50°C = 0°C + \frac{q''P}{\dot{m}c_p} \times 1$$

$$\Rightarrow \frac{q''P}{\dot{m}c_p} = 50$$

$$\Rightarrow T_m = 0 + 50 \times x$$

$$\Rightarrow T_m = 50x$$

at $x = 0.2$ m, $T_m = 10°C$

$x = 0.4$ m, $T_m = 20°C$ (P)

$x = 0.6$ m, $T_m = 30°C$ (Q)

$x = 0.8$ m, $T_m = 40°C$ (R)

$x = 1.0$ m, $T_m = 50°C$

When flow is thermally developed 'h' becomes constant.

From Newton's law of cooling,

$$q'' = h(T_s - T_m)$$

[$T_S \rightarrow$ Surface temperature of pipe]

$$\Rightarrow T_S - T_m = \text{constant}$$

Difference between pipe surface and fluid temperature is constant when flow is thermally developed. Also, when flow is thermally developed, pipe surface temperature 'T_S' varies linearly.

Now, given that, T_S at location P, Q, R,

$$T_{S,P} = 50°C$$

$$T_{S,Q} = 80°C$$

$$T_{S,R} = 90°C$$

Temperature difference between pipe surface and fluid,

at P, $T_S - T_m = 50 - 20 = 30°C$

at Q, $T_S - T_m = 80 - 30 = 50°C$

at R, $T_S - T_m = 90 - 40 = 50°C$

Now, it is clear that, from Q to R, $T_S - T_m$ is constant and T_S varies linearly. So, flow is thermally developed at "Q and R".

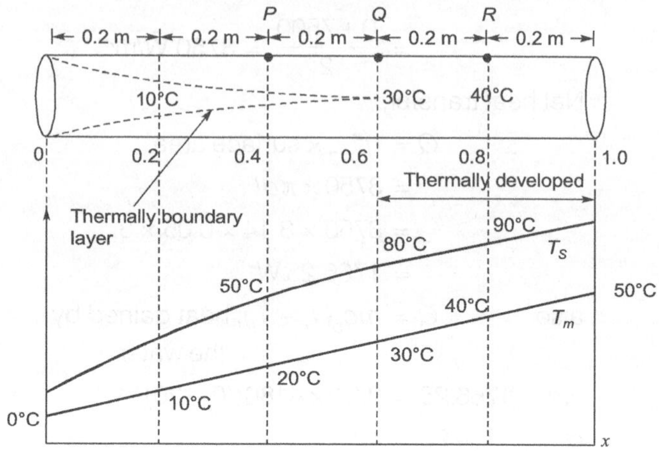

2.16 Sol.

$$Nu = K(\text{Ra})^{1/4}$$

$$Nu = \frac{hD}{K}$$

In this case of vertical plate $D = L$

$$Nu = \left(\frac{hL}{K}\right)$$

$$\frac{hL}{K} = K(\text{Ra})^{1/4} = K\left[\frac{g\beta L^3 \Delta T}{\nu^2} \times \text{Pr}\right]^{1/4}$$

So, $$h \propto \frac{1}{(L)^{1/4}}$$

$$\frac{h_1}{h_2} = \left(\frac{L_2}{L_1}\right)^{1/4}$$

$$\left(\frac{h_1}{h_2}\right) = (16)^{1/4} = 2$$

2.18 Sol.

$\dot{m}_w = 720$ kg/hr, $t_{w,\text{ inlet}} = 10°C$

$t_{w,\text{ exit}} = 70°C$

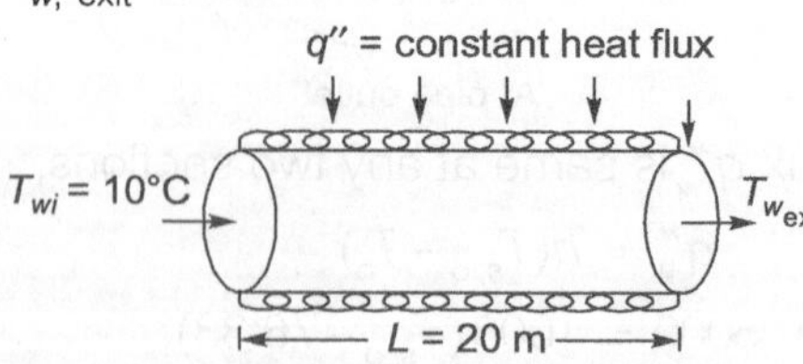

$h = 1697$ W/m²K

From energy balance equation,

Heat flux × Area of HT = $\dot{m}_w \times c_{pw}(T_{w\,\text{exit}} - T_{w\,\text{inlet}})$

$$q'' \times \pi DL = \dot{m}_w \times c_{pw}(70-10)$$

$$q'' = \frac{720}{3600} \times \frac{4.179\times10^3\times60}{\pi\times\left(\frac{3}{100}\right)\times20}\ \text{W/m}^2$$

$$q'' = 26604.34\ \text{W/m}^2$$

Applying Newton's law of cooling at exit

$$q'' = h\times(T_{\text{tube at exit}} - T_{\text{water at exit}})\ \text{W/m}^2$$

$$26604.34 = 1697\times(T_{\text{tube at exit}} - 70)\ \text{W/m}^2$$

$$T_{\text{tube at exit}} = 85.67°\text{C}$$

2.19 (d)

$$\beta = \frac{1}{V}\frac{\partial V}{\partial T}$$

β is the volumetric thermal expansion which, signify the buoyancy forces

the term $g\beta(T - T_\infty)$ represents Buoyant force per unit mass.

2.20 (c)

Nusselt number,

$$\text{Nu} = \frac{hD}{k_{\text{fluid}}}$$

Also, $$\text{Nu} = \frac{Q_{\text{conv.}}}{Q_{\text{cond.}}}$$

$$= \frac{\text{Conduction thermal resistance offered by fluid if it were stationary}}{\text{Surface convection thermal resistance}}$$

2.21 (321) (317 to 324)

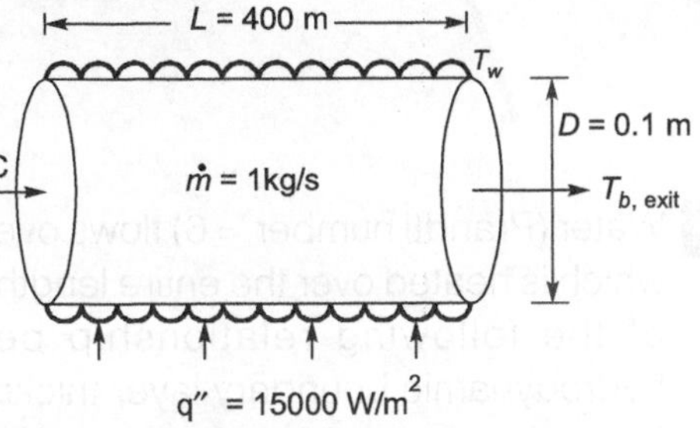

Given: Re = 85000, Pr = 5

To get $T_{b\,\text{exit}}$:

$$q''\times\pi DL = \dot{m}_f\times C_P\left(T_{b\,exit} - T_{b,inlet}\right)$$

$$15000\times\pi\times0.1\times40 = 1\times2600\times(T_{b\,\text{exit}} - 200)$$

$$\Rightarrow\quad T_{b\,\text{exit}} = 272.498°\text{C}$$

Nusselt number, $\text{Nu} = \frac{hD}{k} = 0.023\text{Re}^{0.8}\text{Pr}^{0.4}$

So, $$\frac{h\times0.1}{0.08} = 0.023\times(85000)^{0.8}\times5^{0.4}$$

$$h = 307.56\ \text{W/m}^2\text{K}$$

Now for T_w (Wall temperature):

By Newton's law of cooling,

$$q'' = h\left(T_{w,exit} - T_{b,exit}\right)$$

$$15000 = (T_{w,\text{exit}} - 272.498)$$

$$T_{w,\text{exit}} = \frac{15000}{307.56} + 272.498$$

$$= 48.7 + 6769 + 272.498$$

$$T_{w,\text{exit}} = 321.26°\text{C} \approx 321°\text{C}$$

■■■■

3 Heat Transfer in Flow Over Plates and Pipes

3.1 Water (Prandtl number = 6) flows over a flat plate which is heated over the entire length. Which one of the following relationship between the hydrodynamic boundary layer thickness (δ) and thermal boundary layer thickness (δ_t) is true

(a) $\delta_t = \delta$ (b) $\delta_t < \delta$

(c) $\delta_t > \delta$ (d) Can not be predicted

[2001 : 2 M]

3.2 Consider a laminar boundary layer over a heated flat plate. The free stream velocity is U_∞. At some distance x from the leading edge the velocity boundary layer thickness is δ_V and the thermal boundary layer thickness is δ_T. If the Prandtl number is greater than 1, then

(a) $\delta_V > \delta_T$ (b) $\delta_T > \delta_V$

(c) $\delta_V \approx \delta_T \sim (U_\infty x)^{-1/2}$ (d) $\delta_V \approx \delta_T \sim x^{-1/2}$

[2003 : 2 M]

Linked Answer Questions Q.3.3 and Q.3.4: An un-insulated air conditioning duct of rectangular cross section 1 m × 0.5 m, carrying air at 20°C with a velocity of 10 m/s, is exposed to an ambient of 30°C. Neglect the effect of duct construction material. For air in the range of 20–30°C, data is as follows : thermal conductivity = 0.025 W/mK; viscosity = 18 μPa.s; Prandtl number = 0.73; density = 1.2 kg/m³. For laminar flow Nusselt number is 3.4 for constant wall temperature conditions and , for turbulent flow, Nu = 0.023 $Re^{0.8}$ $Pr^{0.33}$.

3.3 The Reynolds number for the flow is

(a) 444 (b) 890

(c) 4.44×10^5 (d) 5.33×10^5 **[2005 : 2 M]**

3.4 The heat transfer per meter length of the duct, in watts, is

(a) 3.8 (b) 5.3

(c) 89 (d) 769 **[2005 : 2 M]**

3.5 The temperature distribution within the thermal boundary layer over a heated isothermal flat plate is given by $\dfrac{T - T_W}{T_\infty - T_W} = \dfrac{3}{2}\left(\dfrac{y}{\delta_t}\right) - \dfrac{1}{2}\left(\dfrac{y}{\delta_t}\right)^3$, where T_W and T_∞ are the temperatures of plate and free stream respectively, and y is the normal distance measured from the plate. The local Nusselt number based on the thermal boundary layer thickness δ_t is given by

(a) 1.33 (b) 1.50

(c) 2.0 (d) 4.64 **[2007 : 2 M]**

3.6 For flow of fluid over a heated plate, the following fluid properties are known: viscosity = 0.001 Pa.s; specific heat at constant pressure = 1 kJ/kgK; thermal conductivity = 1 W/mK.
The hydrodynamic boundary layer thickness at a specified location on the plate is 1 mm. The thermal boundary layer thickness at the same location is

(a) 0.001 mm (b) 0.01 mm

(c) 1 mm (d) 1000 mm **[2008 : 1 M]**

3.7 The ratios of the laminar hydrodynamic boundary layer thickness to thermal boundary layer thickness of flows of two fluids P and Q on a flat plate are 1/2 and 2 respectively. The Reynolds number based on the plate length for both the flows is 10^4. The Prandtl and Nusselt numbers for P are 1/8 and 35 respectively. The Prandtl and Nusselt numbers for Q are respectively

(a) 8 and 140 (b) 8 and 70

(c) 4 and 70 (d) 4 and 35 **[2011 : 2 M]**

3.8 Consider a two-dimensional laminar flow over a long cylinder as shown in the figure below.

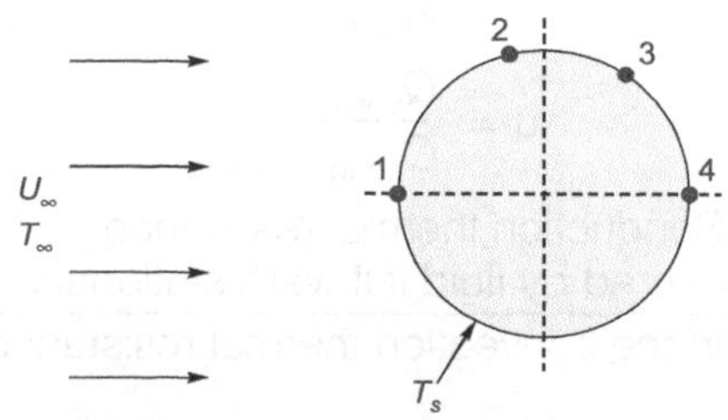

The free stream velocity is U_∞ and the free stream temperature T_∞ is lower than the cylinder surface temperature T_s. The local heat transfer coefficient is minimum at point

(a) 1 (b) 2

(c) 3 (d) 4 **[2014 : 1 M, Set-1]**

3.9 The non-dimensional fluid temperature profile near the surface of a convectively cooled flat plate is given by $\dfrac{T_W - T}{T_W - T_\infty} = a + b\dfrac{y}{L} + c\left(\dfrac{y}{L}\right)^2$, where y is measured perpendicular to the plate, L is the plate length, and a, b and c are arbitrary constants. T_W and T_∞ are wall and ambient temperatures, respectively. If the thermal conductivity of the fluid is k and the wall heat flux is q'', the Nusselt number

$$Nu = \frac{q''}{T_w - T_\infty}\frac{L}{k}$$ is equal to

(a) a (b) b

(c) $2c$ (d) $(b + 2c)$

[2014 : 2 M, Set-1]

3.10 Water flows through a tube of diameter 25 mm at an average velocity of 1.0 m/s. The properties of water are ρ = 1000 kg/m^3, $\mu = 7.25 \times 10^{-4}$ Ns/m^2, k = 0.625 W/mK, Pr = 4.85. Using Nu = 0.023 Re$^{0.8}$ Pr$^{0.4}$, the convective heat transfer coefficient (in W/m^{2}K) is ______ [2014 : 2 M, Set-4]

3.11 For flow of viscous fluid over a flat plate, if the fluid temperature is the same as the plate temperature, the thermal boundary layer is

(a) thinner than the velocity boundary layer

(b) thicker than the velocity boundary layer

(c) of the same thickness as the velocity boundary layer

(d) not formed at all [2015 : 1 M, Set-1]

3.12 The Blasius equation related to boundary layer theory is a

(a) third-order linear partial differential equation

(b) third-order nonlinear partial differential equation

(c) second-order nonlinear ordinary differential equation

(d) third-order nonlinear ordinary differential equation

[2015 : 1 M, Set-1]

3.13 For flow through a pipe of radius R, the velocity and temperature distribution are as follows:

$u(r, x) = C_1$, and $T(r, x) = C_2\left[1-\left(\frac{r}{R}\right)^3\right]$, where C_1 and C_2 are constants.

The bulk temperature is given by $T_m = \frac{2}{U_m R^2}\int_0^R u(r,x)T(r,x)r dr$ with U_m being the mean velocity of flow. The value of T_m is

(a) $\frac{0.5C_2}{U_m}$ (b) $0.5C_2$

(c) $0.6C_2$ (d) $\frac{0.6C_2}{U_m}$

[2015 : 2 M, Set-1]

3.14 Consider incompressible laminar flow of a constant property Newtonian fluid in an isothermal circular tube. The flow is steady with fully-developed temperature and velocity profiles. The Nusselt number for this flow depends on

(a) neither the Reynolds number nor the Prandtl number

(b) both the Reynolds and Prandtl numbers

(c) the Reynolds number but not the Prandtl number

(d) the Prandtl number but not the Reynolds number

[2023 : 1 M]

3.15 A very large metal plate of thickness d and thermal conductivity k is cooled by a stream of air at temperature T_∞ = 300 K with a heat transfer coefficient h, as shown in the figure. The centerline temperature of the plate is T_P. In which of the following case(s) can the lumped parameter model be used to study the heat transfer in the metal plate?

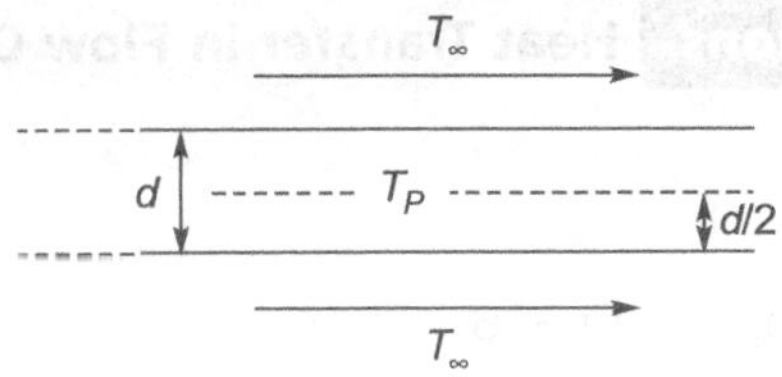

(a) h = 10 Wm^{-2}K^{-1}, k = 100 Wm^{-1}K^{-1}, d = 1 mm, T_P = 350 K

(b) h = 100 Wm^{-2}K^{-1}, k = 100 Wm^{-1}K^{-1}, d = 1 m, T_P = 325 K

(c) h = 100 Wm^{-2}K^{-1}, k = 1000 Wm^{-1}K^{-1}, d = 1 mm, T_P = 325 K

(d) h = 1000 Wm^{-2}K^{-1}, k = 1 Wm^{-1}K^{-1}, d = 1 m, T_P = 350 K

[2023 : 2 M, MSQ]

3.16 Consider a hydrodynamically fully developed laminar flow through a circular pipe with the flow along the axis (i.e. z direction). In the following statements, T is the temperature of the fluid, T_w is the wall temperature and T_m is the bulk mean temperature of the fluid. Which one of the following statements is TRUE?

(a) Nusselt number varies linearly along the z-direction for a thermally fully developed flow.

(b) For a thermally fully developed flow, $\frac{\partial T}{\partial z} = 0$, always.

(c) For constant wall temperature of the duct, $\frac{dT_m}{dz}$ = constant.

(d) For constant wall temperature ($T_w > T_m$) of the duct, $\frac{dT_m}{dz}$ increase exponentially with distance along z-direction.

[2024 : 1 M]

3.17 Consider incompressible laminar flow over a flat plate with freestream velocity of u_∞. The Nusselt number corresponding to this flow velocity is Nu_1. If the freestream velocity is doubled, the Nusselt number changes to Nu_2. Choose the correct option for $\frac{Nu_2}{Nu_1}$.

(a) 1 (b) $\sqrt{2}$
(c) 2 (d) 1.26

[2024 : 1 M]

■■■■

Answers Heat Transfer in Flow Over Plates and Pipes

3.1 (b) 3.2 (a) 3.3 (c) 3.4 (d) 3.5 (b) 3.6 (c) 3.7 (a) 3.8 (b)
3.9 (b) 3.10 (4613.66) 3.11 (d) 3.12 (d) 3.13 (c) 3.14 (a)
3.15 (a, c) 3.16 (*) 3.17 (b)

Explanations Heat Transfer in Flow Over Plates and Pipes

3.1 (b)

Given, $Pr = 6$

$$\Rightarrow \quad \frac{\delta}{\delta_t} = Pr^{1/3} = (6)^{1/3} = 1.82$$

Hence $\boldsymbol{\delta > \delta_t}$

3.2 (a)

We know that

$$\frac{\delta_V}{\delta_T} = (Pr)^{1/3}$$

here $Pr > 1$

$\therefore$ $\boldsymbol{\delta_v > \delta_T}$

3.3 (c)

Width: $a = 1$ m
Depth: $b = 0.5$ m

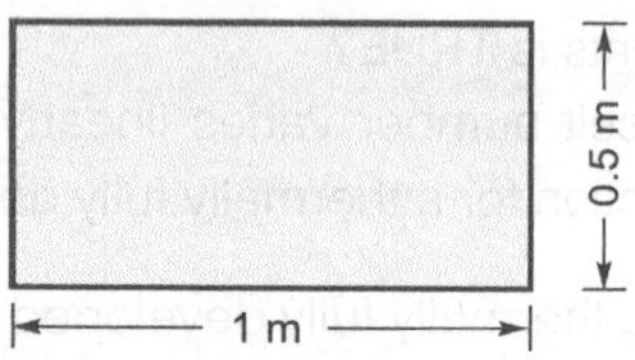

Velocity: $V = 10$ m/s
Air temperature,

$$T = 20°C$$

Ambient temperature,

$$T_0 = 30°C$$

Properties of air,

$$k = 0.025 \text{ W/mK}$$

$$\mu = 1.8\ \mu\text{Pa.s} = 18 \times 10^{-6} \text{ Pa.s}$$

$$Pr = 0.73$$

$$\rho = 1.2 \text{ kg/m}^3$$

Equivalent diameter of duct,

$$D = \frac{4 \times \text{Cross-sectional area}}{\text{Wetted perimeter}}$$

$$= \frac{4A}{P} = \frac{4 \times ab}{2(a+b)} = \frac{4 \times 1 \times 0.5}{2(1+0.5)} = 0.667 \text{ m}$$

Reynolds number,

$$Re = \frac{\rho VL}{\mu} = \frac{1.2 \times 10 \times 0.667}{18 \times 10^{-6}}$$

$$= \mathbf{4.44 \times 10^5}$$

3.4 (d)

As Re > 4000, the flow is turbulent
Relation for turbulent flow,

$$Nu = 0.023\, Re^{0.8}\, Pr^{0.33}$$

$$\frac{hL}{k} = 0.023 \times (4.44 \times 10^5)^{0.8} \times (0.73)^{0.33}$$

$$\frac{h \times 0.667}{0.025} = 683.173$$

or $h = 25.60$ W/m²K

Surface area of duct,

$$A = 2 \times al + 2 \times bl$$

where l = length of duct

$$\therefore \quad A = 2 \times 1 \times l + 2 \times 0.5 \times l$$

$$= (2 + 1)l = 3l$$

Rate of heat transfer,

$$Q = hA(T_0 - T)$$

$$Q = 25.60 \times 3l \times (30-20)$$

or $$\frac{Q}{l} = 768 \text{ W} \simeq \mathbf{769\ W}$$

3.5 (b)

$$\text{Nu} = \frac{hL}{k_f} = \left.\frac{\partial T^*}{\partial y^*}\right|_{y^*=0}$$

where $$T^* = \frac{T - T_w}{T_\infty - T_w}$$

and $$y^* = \frac{y}{\delta}$$

where T_w is surface temperature and T_∞ is free-stream temperature.

$$\Rightarrow \text{Nu} = \left.\frac{3}{2} - \frac{3}{2}\left(\frac{y}{\delta_t}\right)\right|_{y^*=0} = \left.\frac{3}{2} - \frac{3}{2}y^*\right|_{y^*=0}$$

$$= 1.5$$

3.6 (c)

Given data:

$$\mu = 0.001 \text{ Pa.s}$$
$$c_p = 1 \text{ kJ/kgK} = 1000 \text{ J/kgK}$$
$$k_f = 1 \text{ W/mK}$$

Hydrodynamic boundary layer,

$$\delta = 1 \text{ mm}$$

We know that

Prandtl number,

$$\text{Pr} = \frac{\mu c_p}{k_f} = \frac{0.001 \times 1000}{1} = 1$$

Thermal boundary layer,

$$\delta_t = \delta\, \text{Pr}^{-1/3} = 1 \times (1)^{-1/3} = \mathbf{1\ mm}$$

3.7 (a)

Given data:

For fluid P

$$\frac{\delta}{\delta_t} = \frac{1}{2}$$

$$(\text{Re})_P = 1000$$

$$(\text{Pr})_P = \frac{1}{8}$$

$$(\text{Nu})_P = 35$$

$$(\text{Nu})_P = 0.664(\text{Re})_P^{0.5}(\text{Pr})_P^{0.333}$$

For fluid Q

$$\frac{\delta}{\delta_t} = 2$$

$$(\text{Re})_Q = 10000$$
$$(\text{Pr})_Q = ?$$
$$(\text{Nu})_Q = ?$$

$$\frac{\delta}{\delta_t} = (\text{Pr})_Q^{1/3}$$

$$2 = (\text{Pr})_Q^{1/3}$$

or $$(\text{Pr})_Q = (2)^3 = \mathbf{8}$$

For laminar boundary layer on a flat plate, Nusselt number,

$$\text{Nu} = 0.664(\text{Re})^{0.5}(\text{Pr})^{0.333}$$

Ratio $$\frac{(\text{Nu})_P}{(\text{Nu})_Q} = \frac{0.664(\text{Re})_P^{0.5}(\text{Pr})_P^{0.333}}{0.664(\text{Re})_Q^{0.5}(\text{Pr})_Q^{0.333}}$$

$$\frac{(\text{Nu})_P}{(\text{Nu})_Q} = \frac{(\text{Re})_P^{0.5}(\text{Pr})_P^{0.333}}{(\text{Re})_Q^{0.5}(\text{Pr})_Q^{0.333}}$$

$$\frac{35}{(\text{Nu})_Q} = \frac{(10000)^{0.5} \times \left(\frac{1}{8}\right)^{0.333}}{(10000)^{0.5} \times (8)^{0.333}}$$

$$\frac{35}{(\text{Nu})_Q} = \frac{1}{(8)^{0.333} \times (8)^{0.333}}$$

$$= \frac{1}{(8)^{0.666}} = \frac{1}{4}$$

$$(N_u)_Q = 35 \times 4 = \mathbf{140}$$

3.8 (b)

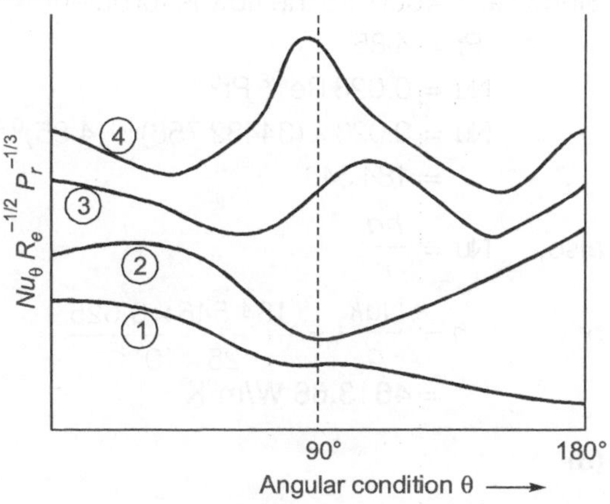

1. Re = 500
2. Re = 10^4
3. Re = 2×10^5
4. Re = 2×10^6

Experimental results for the variation of the local Nusselt number with θ are shown in above figure for the cylinder in cross flow of air. The results are strongly influenced by the nature of boundary layer development on the surface. Starting at the stagnation point, Nu_θ decreases with increasing θ as a result of laminar boundary layer development. Nu_θ as a result of laminar boundary layer development. Nu_θ reaches minimum at 80° where separation occurs. So, at **point 2**, the local heat transfer coefficient is minimum.

3.9 (b)

$$q'' = -k\left.\frac{\partial T}{\partial y}\right|_{y=0} = h(T_w - T_\infty)$$

$$T_w - T = (T_w - T_\infty)\left[a + \frac{by}{L} + c\left(\frac{y}{L}\right)^2\right]$$

$$\frac{\partial T}{\partial y} = (T_w - T_\infty)\left[\frac{b}{L} + \frac{2c}{L^2}y\right]$$

$$-\left.\frac{\partial T}{\partial y}\right|_{y=0} = (T_w - T_\infty)\frac{b}{L}$$

or $$q'' = k(T_w - T_\infty)\frac{b}{L}$$

$\therefore$ $$\text{Nu} = \frac{q''}{T_w - T_\infty}\frac{L}{k} = b$$

3.10 Sol.

Reynolds number,

$$\text{Re} = \frac{\rho V d}{\mu} = \frac{1000 \times 1 \times 25 \times 10^{-3}}{7.25 \times 10^{-4}}$$
$$= 34482.75$$

Since Re > 4000, so the flow is turbulent.

$$\text{Pr} = 4.85$$
$$\text{Nu} = 0.023\,\text{Re}^{0.8}\,\text{Pr}^{0.4}$$
$$\text{Nu} = 0.023 \times (34482.758)^{0.8}(4.85)^{0.4}$$
$$= 184.546$$

also, $$\text{Nu} = \frac{hd}{k}$$

or $$h = \frac{\text{Nu}k}{d} = \frac{184.546 \times 0.625}{25 \times 10^{-3}}$$
$$= \mathbf{4613.66\ W/m^2K}$$

3.12 (d)

$$2\frac{d^3f}{d\eta^3} + f\frac{d^2f}{d\eta^2} = 0$$

third-order non-linear differential equation.

3.13 (c)

$$u(r,x) = C_1$$

$$T(r,x) = C_2\left[1 - \left(\frac{r}{R}\right)^3\right]$$

$$T_m = \frac{2}{U_m R^2}\int_0^R u(r,x)T(r,x)\,r\,dr$$

$$= \frac{2}{U_m R^2}\int_0^R C_1 \times C_2\left[1 - \left(\frac{r}{R}\right)^3\right] r\,dr$$

$$= \frac{2C_1C_2}{U_m R^2}\int_0^R\left[1 - \left(\frac{r}{R}\right)^3\right] r\,dr = \frac{2C_1C_2}{U_m R^2}\int_0^R\left(r - \frac{r^4}{R^3}\right)dr$$

$$= \frac{2C_1C_2}{U_m R^2}\left[\frac{r^2}{2} - \frac{r^5}{5R^3}\right]_0^R = \frac{2C_1C_2}{U_m R^2}\left[\frac{R^2}{2} - \frac{R^5}{5R^3}\right]$$

$$= \frac{2C_1C_2}{U_m R^2}\left[\frac{R^2}{2} - \frac{R^2}{5}\right] = \frac{2C_1C_2}{U_m R^2}\left[\frac{5R^2 - 2R^2}{10}\right]$$

$$= \frac{2C_1C_2}{U_m R^2} \times \frac{3}{10}R^2 = \frac{0.6\,C_1C_2}{U_m} \quad \text{...(i)}$$

Since $u(r, x) = C_1$, constant

$$U_m = C_1$$

Substituting $U_m = C_1$ in above Eq. (i), we get

$$T_m = \frac{0.6C_1C_2}{C_1} = \mathbf{0.6C_2}$$

3.15 (a, c)

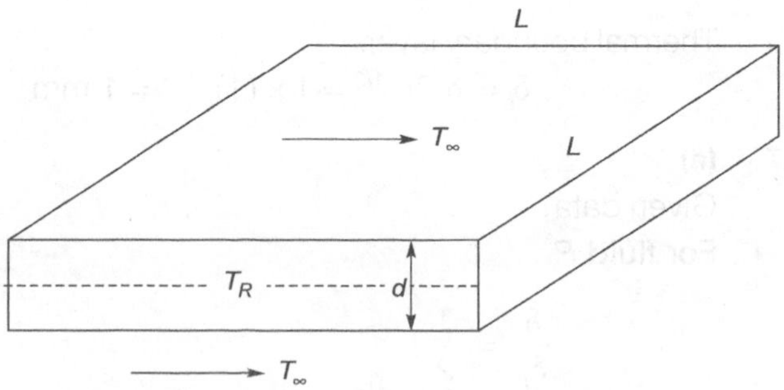

Characteristic length,

$$L_C = \frac{V}{A_s} = \frac{L \times L \times d}{(L \times L) + (L \times L)} = \frac{d}{2}$$

As edge area is negligible

Lumped analysis is used when Biot number is less than 0.1.

i.e. $$\text{Bi} = \frac{hL_C}{k} < 0.1$$

Checking each option,

(A) $\text{Bi} = \dfrac{10 \times 0.5}{100 \times 1000} = 5 \times 10^{-5} < 0.1$

(B) $\text{Bi} = \dfrac{100 \times 0.5}{100} = 0.5 > 0.1$

(C) $\text{Bi} = \dfrac{100 \times 0.5}{1000 \times 1000} = 5 \times 10^{-5} < 0.1$

(D) $\text{Bi} = \dfrac{1000 \times 0.5}{1} = 500 > 0.1$

3.17 (b)

For laminar flow over flat plate,
$\text{Nu} \propto \text{Re}^{0.5}\, \text{Pr}^{1/3}$, where Re is Reynolds number and Pr is the Prandtl number.

$$\because \quad \text{Re} = \frac{u_\infty x}{\nu}$$

$$\therefore \quad \text{Nu} \propto u_\infty^{0.5}$$

$$\text{or,} \quad \frac{Nu_2}{Nu_1} = \sqrt{\frac{u_{\infty,2}}{u_{\infty,1}}} = \sqrt{2}$$

■■■■

4 Radiation

4.1 For the circular tube of equal length and diameter shown in figure below, the view factor F_{13} is 0.17. The view factor F_{12} in this case will be

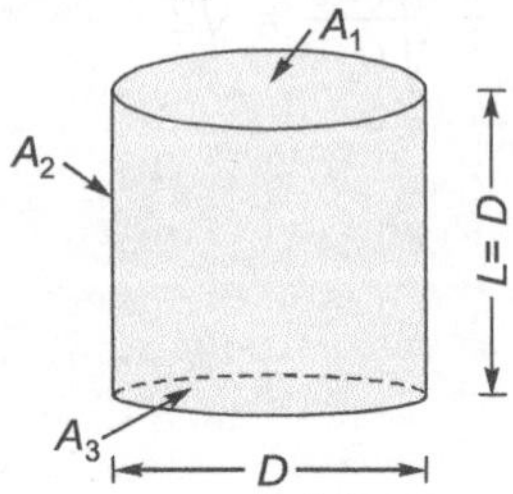

(a) 0.17 (b) 0.21
(c) 0.79 (d) 0.83 **[2001 : 2 M]**

4.2 What is the value of the view factor for two inclined flat plates having common edge of equal length and with an angle of 20 degrees?

(a) 0.83 (b) 1.17
(c) 0.66 (d) 1.34 **[2002 : 2 M]**

4.3 A plate having 10 cm^2 area each side is hanging in the middle of a room of 100 m^2 total surface area. The plate temperature and emissivity are respectively 800 K and 0.6. The temperature and emissivity values for the surfaces of the room are 300 K and 0.3 respectively. Boltzmann's constant $\sigma = 5.67 \times 10^{-8}$ W/m^{2}K^4. The total heat loss from the two surfaces of the plate is

(a) 13.66 W (b) 27.32 W
(c) 27.87 W (d) 13.66 MW **[2003 : 1 M]**

4.4 The following figure was generated from experimental data relating spectral black body emissive power to wave length at the three temperatures T_1, T_2 and T_3($T_1 > T_2 > T_3$).

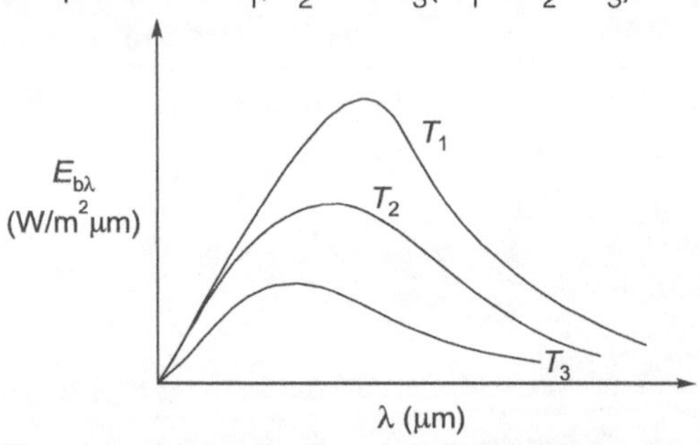

The conclusion in that the measurements are

(a) correct because the maxima in $E_{b\lambda}$ show the correct trend
(b) correct because Planck's law is satisfied
(c) wrong because the Stefan Boltzmann law in not satisfied
(d) wrong because Wien's displacement law is not satisfied **[2005 : 1 M]**

4.5 A solid cylinder (surface 2) is located at the centre of a hollow sphere (surface 1). The diameter of the sphere is 1 m, while the cylinder has a diameter and length of 0.5 m each. The radiation configuration factor F_{11} is

(a) 0.375 (b) 0.625
(c) 0.75 (d) 1 **[2005 : 2 M]**

4.6 A 100 W electric bulb was switched on in a 2.5 m × 3 m × 3 m size thermally insulated room having a temperature of 20°C. The room temperature at the end of 24 hours will be

(a) 321°C (b) 341°C
(c) 450°C (d) 470°C **[2006 : 2 M]**

4.7 A hollow enclosure is formed between two infinitely long concentric cylinders of radii 1 m and 2 m, respectively. Radiative heat exchange takes place between the inner surface of the larger cylinder (surface-2) and the outer surface of the smaller cylinder (surface-1). The radiating surface are diffuse and the medium in the enclosure is non-participating. The fraction of the thermal radiation leaving the larger surface and striking itself is

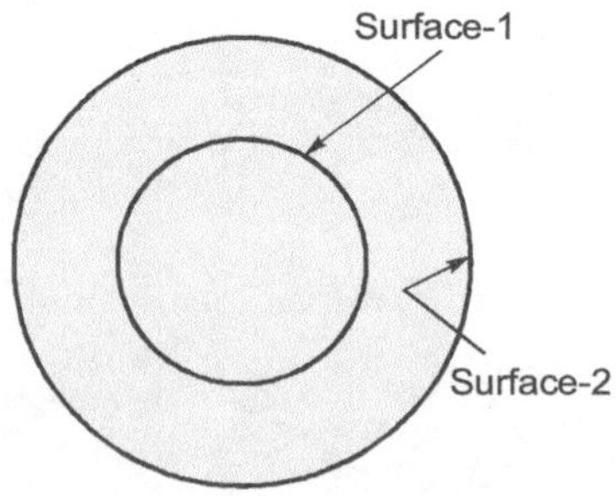

(a) 0.25 (b) 0.5
(c) 0.75 (d) 1 **[2008 : 2 M]**

Common Data for Qs.4.8 and Qs.4.9: Radiative heat transfer is intended between the inner surfaces of two very large isothermal parallel metal plates. While the upper plate (designated as plate 1) is a black surface and is the warmer one being maintained at 727°C, the lower plate (plate 2) is a diffuse and gray surface with an emissivity of 0.7 and is kept at 227°C. Assume that the surfaces are sufficiently large to form a two-surface enclosure and steady-state conditions to exist. Stefan-Boltzmann constant is given as 5.67×10^{-8} W/m^{2}K^4.

4.8 The irradiation (in kW/m^2) for the plate (plate 1) is

(a) 2.5 (b) 3.6
(c) 17.0 (d) 19.5 **[2009 : 2 M]**

4.9 If plate 1 is also a diffuse and gray surface with an emissivity value of 0.8, the net radiation heat exchange (in kW/m^2) between plate 1 and plate 2 is

(a) 17.0 (b) 19.5
(c) 23.0 (d) 31.7 **[2009 : 2 M]**

4.10 For an opaque surface, the absorptivity (α), transmissivity (τ) and reflectivity (ρ) are related by the equation

(a) $\alpha + \rho = \tau$ (b) $\rho + \alpha + \tau = 0$
(c) $\alpha + \rho = 1$ (d) $\alpha + \rho = 0$ **[2012 : 1 M]**

4.11 Consider two infinitely long thin concentric tubes of circular cross section as shown in figure. If D_1 and D_2 are the diameters of the inner and outer tubes respectively, then the view factor F_{22} is given by

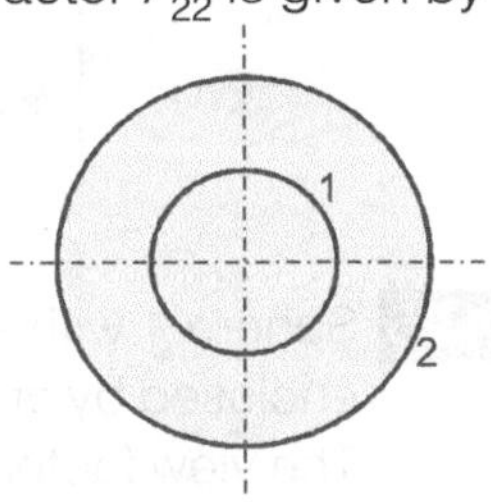

(a) $\left(\frac{D_2}{D_1}\right) - 1$

(b) Zero

(c) $\left(\frac{D_1}{D_2}\right)$

(d) $1 - \left(\frac{D_1}{D_2}\right)$ **[2012 : 2 M]**

4.12 Two large diffuse gray parallel plates, separated by a small distance, have surface temperature of 400 K and 300 K. If the emissivities of the surfaces are 0.8 and the Stefan-Boltzmann constant is 5.67×10^{-8} W/m^{3}K^4, the net radiation heat exchange rate in kW/m^2 between the two plates is

(a) 0.66 (b) 0.79
(c) 0.99 (d) 3.96 **[2013 : 2 M]**

4.13 A hemispherical furnace of 1 m radius has the inner surface (emissivity, $\varepsilon = 1$) of its roof maintained at 800 K, while its floor ($\varepsilon = 0.5$) is kept at 600 K. Stefan–Boltzmann constant is 5.668×10^{-8} W/m^{2}K^4. The net radiative heat transfer (in kW) from the roof to the floor is _____. **[2014 : 2 M, Set-2]**

4.14 A solid sphere of radius r_1 = 20 mm is placed concentrically inside a hollow sphere of radius r_2 = 30 mm as shown in the figure.

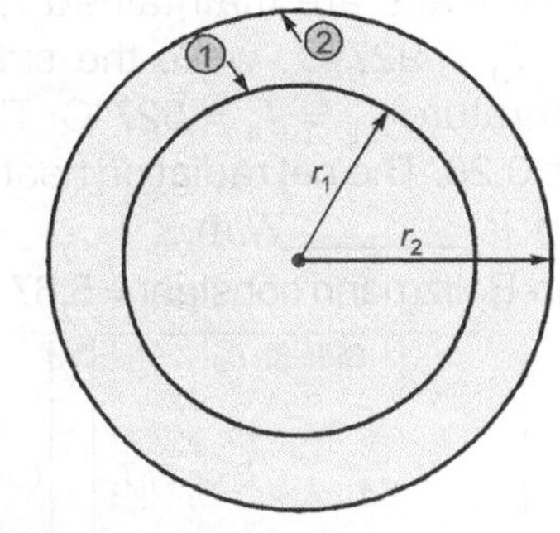

The view factor F_{21} for radiation heat transfer is

(a) $\frac{2}{3}$ (b) $\frac{4}{9}$

(c) $\frac{8}{27}$ (d) $\frac{9}{4}$ **[2014 : 2 M, Set-3]**

4.15 Two infinite parallel plates are placed at a certain distance apart. An infinite radiation shield is inserted between the plates without touching any of them to reduce heat exchange between the plates. Assume that the emissivities of plates and radiation shield are equal. The ratio of the net heat exchange between the plates with and without the shield is

(a) 1/2 (b) 1/3
(c) 1/4 (d) 1/8

[2014 : 2 M, Set-4]

4.16 The total emissive power of a surface is 500 W/m^2 at temperature T_1 and 1200 W/m^2 at a temperature T_2, where the temperature are in Kelvin. Assuming the emissivity of the surface to be constant, the ratio of the temperature T_1/T_2 is

(a) 0.0308 (b) 0.416
(c) 0.803 (d) 0.874

[2015 : 2 M, Set-2]

4.17 A solid sphere 1 of radius r is placed inside a hollow, closed hemispherical surface 2 of radius $4r$. The shape factor $F_{2\text{-}1}$ is

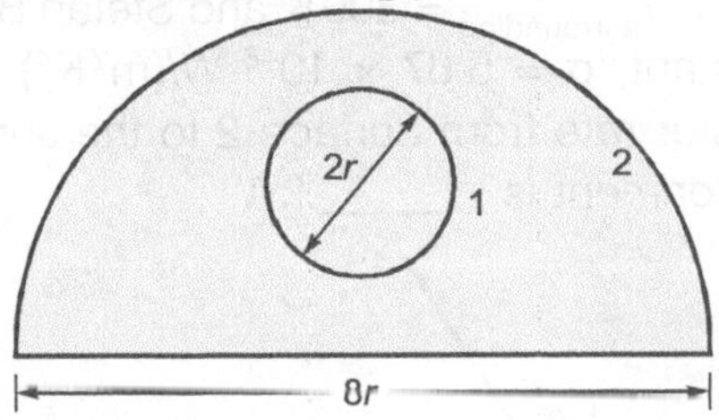

(a) $\frac{1}{12}$ (b) $\frac{1}{2}$

(c) 2 (d) 12 **[2015 : 2 M, Set-3]**

4.18 An infinitely long function 0.5 m × 0.4 m cross-section is shown in the figure below. Consider all surfaces of the furnace to be black. The top and bottom walls are maintained at temperature $T_1 = T_3 = 927°C$ while the side walls are at temperature $T_2 = T_4 = 527°C$. The view factor, $F_{1\text{-}2}$ is 0.26. The net radiation heat loss or gain on side 1 is _______W/m.

Stefan-Boltzmann constant = 5.67×10^{-8} W/m^{2}K^4.

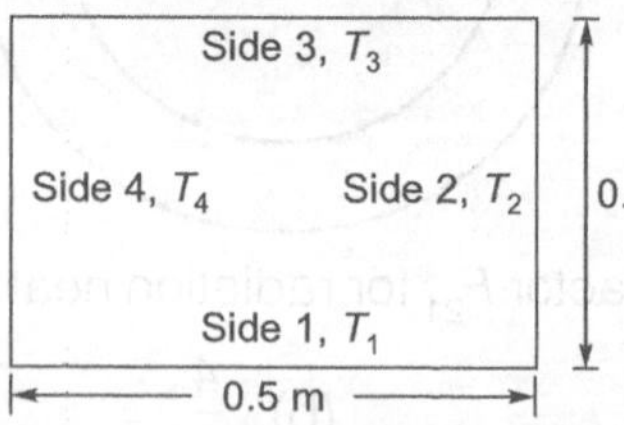

[2016 : 2 M, Set-1]

4.19 Consider the radiation heat exchange inside an annulus between two very long concentric cylinders. The radius of the outer cylinder is R_0 and that of the inner cylinder is R_i. The radiation view factor of the outer cylinder onto itself is

(a) $1-\sqrt{\dfrac{R_i}{R_0}}$ (b) $\sqrt{1-\dfrac{R_i}{R_0}}$

(c) $1-\left(\dfrac{R_i}{R_0}\right)^{1/3}$ (d) $1-\dfrac{R_i}{R_0}$

[2016 : 1 M, Set-2]

4.20 Two large parallel plates having a gap of 10 mm in between them are maintained at temperatures $T_1 = 1000$ K and $T_2 = 400$ K. Given emissivity values, $\varepsilon_1 = 0.5$, $\varepsilon_2 = 0.25$ and Stefan-Boltzmann constant $\sigma = 5.67 \times 10^{-8}$ W/m^{2}K^4, the heat transfer between the plates (in kW/m^2) is_______

[2016 : 2 M, Set-3]

4.21 Two black surfaces, *AB* and *BC*, of length 5 m and 6 m, respectively, are oriented as shown. Both surfaces extend infinitely into the third dimension. Given that view factor $F_{12} = 0.5$, $T_1 = 800$ K, $T_2 = 600$ K, $T_{surrounding} = 300$ K and Stefan Boltzmann constant, $\sigma = 5.67 \times 10^{-8}$ W/(m^{2}K^4), the heat transfer rate from Surface 2 to the surrounding environment is _______ kW.

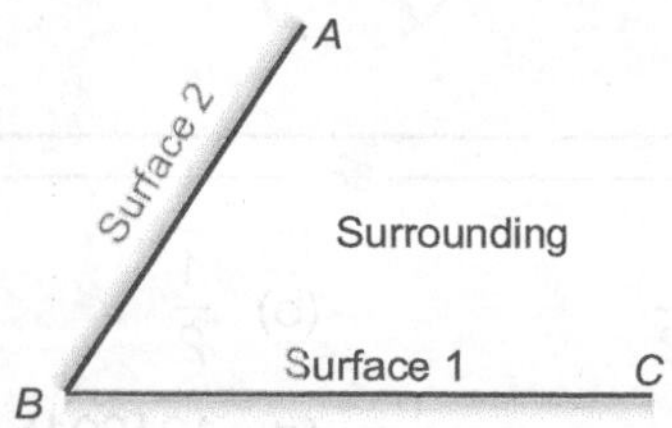

[2017 : 2 M, Set-1]

4.22 The emissive power of a blackbody is *P*. If its absolute temperature is doubled, the emissive power becomes

(a) 2*P* (b) 4*P*

(c) 8*P* (d) 16*P* [2017 : 1 M, Set-2]

4.23 The peak wavelength of radiation emitted by a black body at a temperature of 2000 K is 1.45 μm. If the peak wavelength of emitted radiation changes to 2.90 μm, then the temperature (in K) of the black body is

(a) 500 (b) 1000

(c) 4000 (d) 8000

[2018 : 1 M, Set-2]

4.24 A 0.2 m thick infinite black plate having a thermal conductivity of 3.96 W/m-K is exposed to two infinite black surfaces at 300 K and 400 K as shown in the figure. At steady state the surface temperature of the plate facing the cold side is 350 K. The value of Stefan-Boltzmann constant σ is

$$5.67 \times 10^{-8} \text{ W/m}^2\text{ K}^4$$

Assuming 1-D heat conduction, the magnitude of heat flux through the plate (in W/m^2) is _______ (correct to two decimal places).

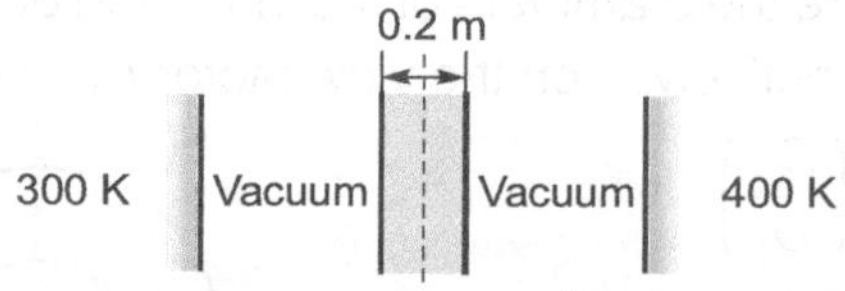

[2018 : 2 M, Set-2]

4.25 Sphere 1 with a diameter of 0.1 m is completely enclosed by another sphere 2 of diameter 0.4 m. The view factor F_{12} is

(a) 0.25 (b) 0.5

(c) 0.0625 (d) 1.0 [2019 : 1 M, Set-2]

4.26 Three sets of parallel plates LM, NR and PQ are given in figure 1, 2 and 3. The view factor F_{IJ} is defined as the fraction of radiation leaving plate *I* that is intercepted by plate *J*. Assume that the values of F_{LM} and F_{NR} are 0.8 and 0.4, respectively. The value of F_{PQ} (round off to one decimal place) is _______ .

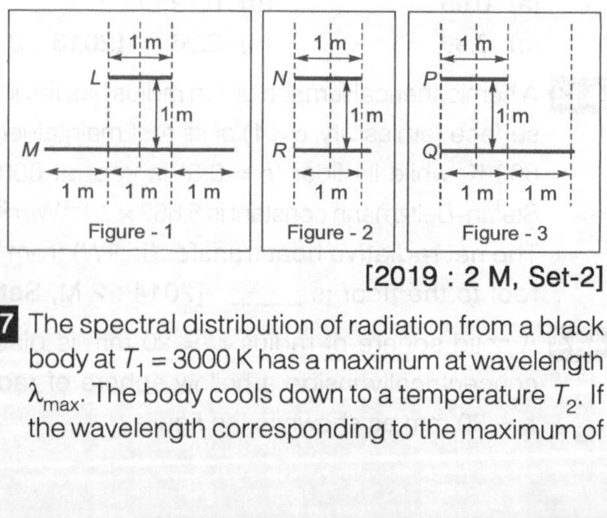

[2019 : 2 M, Set-2]

4.27 The spectral distribution of radiation from a black body at $T_1 = 3000$ K has a maximum at wavelength λ_{max}. The body cools down to a temperature T_2. If the wavelength corresponding to the maximum of the spectral distribution at T_2 is 1.2 times of the original wavelength λ_{max}, then the temperature T_2 is ________ K (round off to the nearest integer).

[2020 : 2 M, Set-2]

4.28 A solid sphere of radius 10 mm is placed at the centroid of a hollow cubical enclosure of side length 30 mm. The outer surface of the sphere is denoted by 1 and the inner surface of the cube is denoted by 2. The view factor F_{22} for radiation heat transfer is ____________ [round off to two decimal places]

[2021 : 2 M, Set-1]

4.29 A flat plate made of cast iron is exposed to a solar flux of 600 W/m^2 at an ambient temperature of 25°C. Assume that the entire solar flux is absorbed by the plate.

Cast iron has a low temperature absorptivity of 0.21. Use Stefan-Boltzmann constant = 5.669 × 10^{-8} W/m^2-K^4. Neglect all other modes of heat transfer except radiation.

Under the aforementioned conditions, the radiation equilibrium temperature of the plate is ______°C (round off to the nearest integer).

[2022 : 1 M, Set-1]

4.30 Consider a solid slab (thermal conductivity, k = 10 W·m^{-1}K^{-1}) with thickness 0.2 m and of infinite extent in the other two directions as shown in the figure. Surface 2, at 300 K, is exposed to a fluid flow at a free stream temperature (T_∞) of 293 K, with a convective heat transfer coefficient (h) of 100 W·m^{-2}K^{-1}. Surface 2 is opaque, diffuse and gray with an emissivity (ε) of 0.5 and exchanges heat by radiation with very large surroundings at 0 K. Radiative heat transfer inside the solid slab is neglected. The Stefan-Boltzmann constant is 5.67 × 10^{-8} W·m^{-2}K^{-4}. The temperature T_1 of Surface 1 of the slab, under steady-state conditions, is ________ K (round off to the nearest integer).

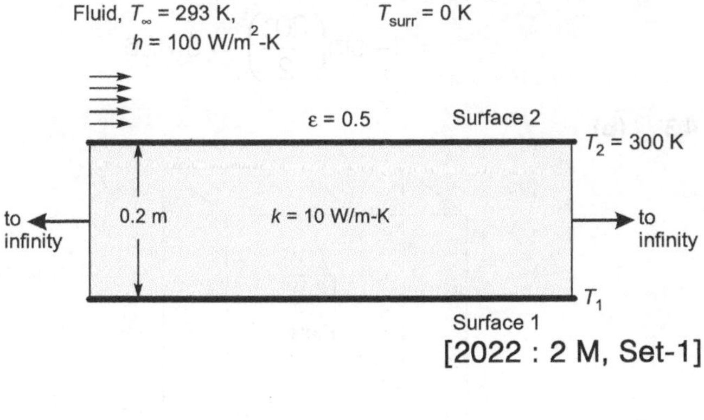

[2022 : 2 M, Set-1]

4.31 Wien's law is stated as follows: $\lambda_m T = C$, where C is 2898 μm.K and λ_m is the wavelength at which the emissive power of a black body is maximum for a given temperature T. The spectral hemispherical emissivity (ε_λ) of a surface is shown in the figure below (1Å = 10^{-10}m). The temperature at which the total hemispherical emissivity will be highest is ____________ K (round off to the nearest integer).

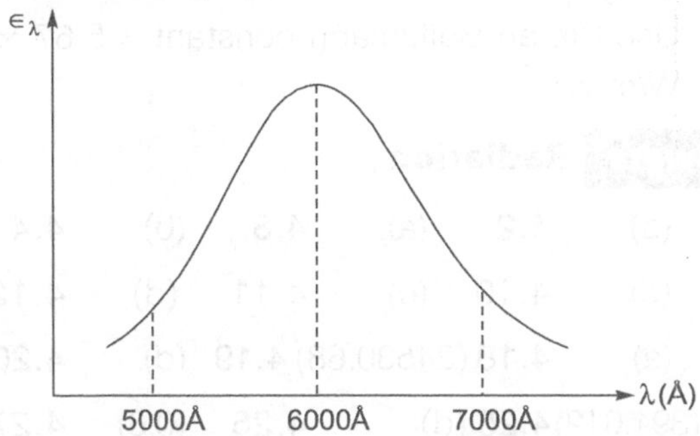

[2022 : 1 M, Set-2]

4.32 A cylindrical rod of length h and diameter d is placed inside a cubic enclosure of side length L, S denotes the inner surface of the cube. The view-factor F_{S-S} is

(a) 0
(b) 1
(c) $\dfrac{\left(\pi dh + \dfrac{\pi d^2}{2}\right)}{6L^2}$
(d) $1 - \dfrac{\left(\pi dh + \dfrac{\pi d^2}{2}\right)}{6L^2}$

[2023 : 2 M]

4.33 An explosion at time t = 0 releases energy E at the origin in a space filled with a gas of density ρ. Subsequently, a hemispherical blast wave propagates radially outwards as shown in the figure.

Let R denote the radius of the front of the hemispherical blast wave. The radius R follows the relationship $R = kt^a E^b \rho^c$, where k is a dimensionless constant. The value of exponent a is__________. (Rounded off to one decimal place)

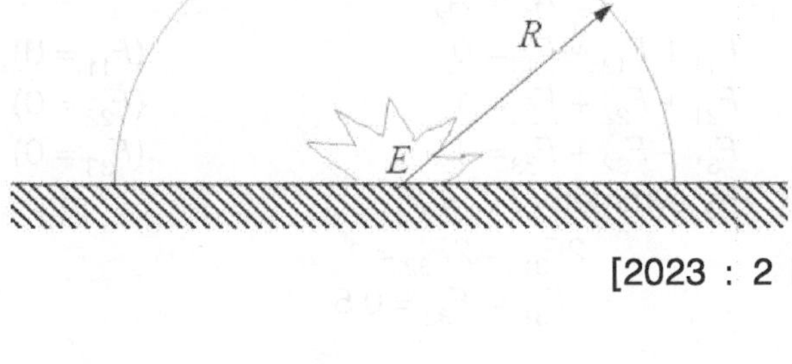

[2023 : 2 M]

4.34 Consider a hemispherical furnace of diameter D = 6 m with a flat base. The dome of the furnace has an emissivity of 0.7 and the flat base is a blackbody. The base and the dome are maintained at uniform temperature of 300 K and 1200 K, respectively. Under steady state conditions, the rate of radiation heat transfer from the dome to the base is ______ kW. (rounded off to the nearest integer).

Use Stefan-Boltzmann constant = 5.67×10^{-8} W/m^{2}K^4

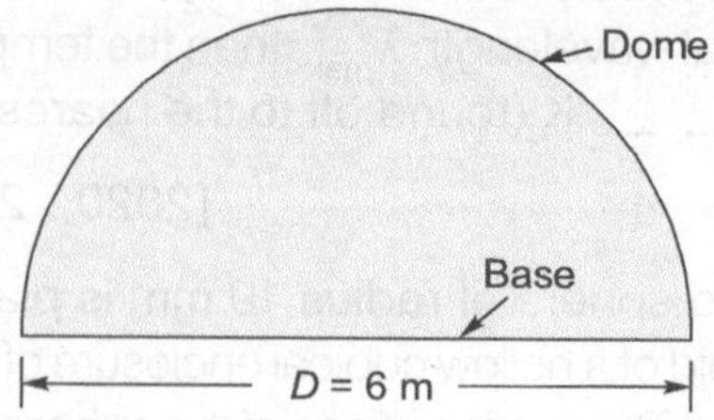

[2024 : 2 M]

■■■■

Answers Radiation

4.1	(d)	4.2	(a)	4.3	(b)	4.4	(d)	4.5	(b)	4.6	(d)	4.7	(b)	4.8	(d)
4.9	(d)	4.10	(c)	4.11	(d)	4.12	(a)	4.13	(24.929)	4.14	(b)	4.15	(a)	4.16	(c)
4.17	(a)	4.18	(24530.68)	4.19	(d)	4.20	(11.05)	4.21	(13.778)	4.22	(d)	4.23	(b)		
4.24	(391.612)	4.25	(d)	4.26	(0.6)	4.27	(2500)	4.28	(0.77)	4.29	(218)	4.30	(319)		
4.31	(4830)	4.32	(d)	4.33	(0.4)	4.34	(2727)								

Explanations Radiation

4.1 (d)

$F_{11} + F_{12} + F_{13} = 1$

For flat surfaces 1 and 3

$F_{11} = F_{33} = 0$

and given $F_{13} = 0.17$

So, we get

$0 + F_{12} + 0.17 = 1$

$F_{12} = \mathbf{0.83}$

4.2 (a)

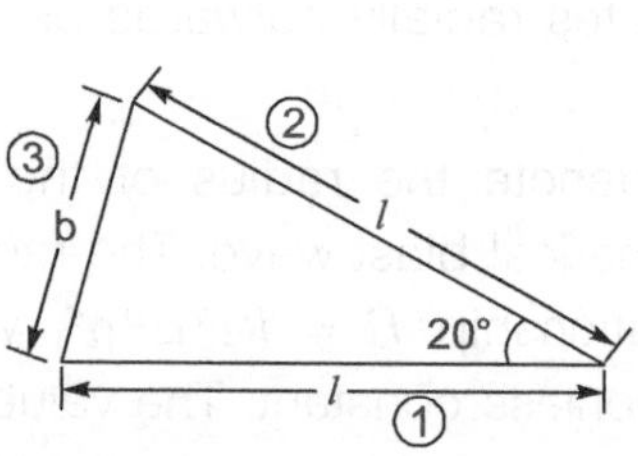

Let l be the length of both plates and assumed that surrounding between two large plate behaves as surface 3.

$A_1 = A_2$

$F_{11} + F_{12} + F_{13} = 0 \quad (F_{11} = 0)$

$F_{21} + F_{22} + F_{23} = 1 \quad (F_{22} = 0)$

$F_{31} + F_{32} + F_{33} = 1 \quad (F_{33} = 0)$

By symmetry $F_{31} = F_{32}$

$2F_{31} = 2F_{32} = 1$

$F_{31} = F_{32} = 0.5$

$A_1 F_{13} = A_3 F_{31}$

$(l \times w) \times F_{13} = (2l \sin 10° \times w) \times 0.5$

$F_{13} = \sin 10°$

By Cosine law

$b^2 = l^2 + l^2 - 2l^2 \cos 20°$

$\Rightarrow \quad b^2 = 2l^2(1 - \cos 20°)$

$= 2l^2 \times 2\sin^2 10° = 4l^2 \sin^2 10°$

$b = 2l \sin 10°$

$F_{12} = 1 - F_{13} = 1 - \sin 10°$

By symmetry

$F_{12} = F_{21} = 1 - \sin 10° = \mathbf{0.826}$

Alternatively:

$F_{12} = F_{21} = 1 - \sin\frac{\theta}{2}$

$= 1 - \sin\left(\frac{20°}{2}\right) = \mathbf{0.826}$

4.3 (b)

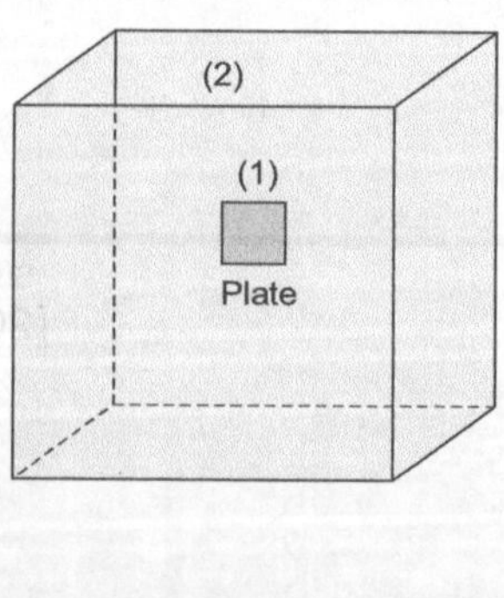

Heat loss from one surface of the plate

$$Q = \frac{\sigma\left(T_1^4 - T_2^4\right)}{\left(\frac{1-\varepsilon_1}{A_1\varepsilon_1}\right) + \frac{1}{A_1} + \frac{1-\varepsilon_2}{A_2\varepsilon_2}} = \frac{\sigma\left(T_1^4 - T_2^4\right)}{\left(\frac{1}{A_1\varepsilon_1}\right) + \left(\frac{1-\varepsilon_2}{A_2\varepsilon_2}\right)}$$

Given: $A_1 = 10 \times 10^{-4}\ m^2 = 10^{-3}\ m^2$

$T_1 = 800$ K, $\varepsilon_1 = 0.6$, $T_2 = 300$ K, $\varepsilon_2 = 0.3$

$A_2 = 100\ m^2$

$$\therefore \quad Q = \frac{5.67 \times 10^{-8}\left[(800)^4 - (300)^4\right]}{\frac{1}{10^{-3}} \times \frac{1}{0.6} + \frac{1}{100}\left(\frac{1-0.3}{0.3}\right)}$$

$$= 13.66 \text{ W}$$

$\therefore$ Heat loss from both surface = $2Q$

$$= 2 \times 13.66 = \mathbf{27.32\ W}$$

4.4 (d)

The correct diagram according to Wien's displacement law

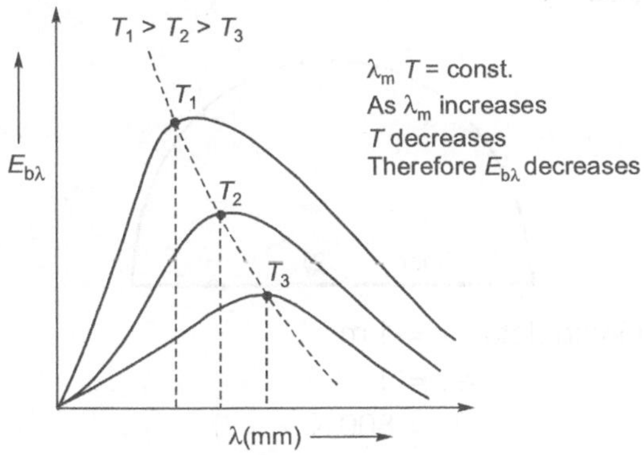

4.5 (b)

Diameter of sphere = 1 m

Diameter of cylinder = 0.5 m

Length of cylinder = 0.5 m

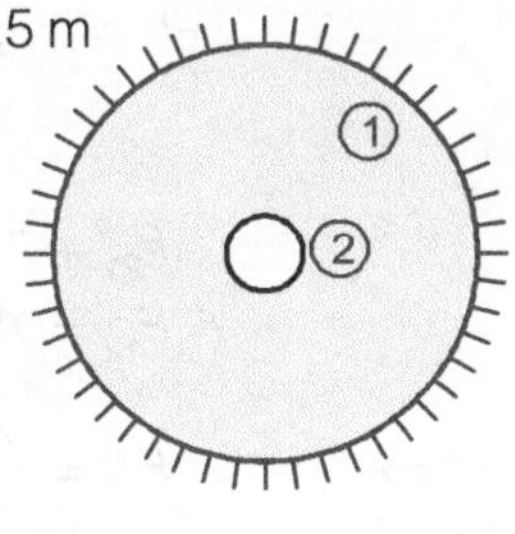

We know that

$$F_{21} + F_{22} = 1 \quad \ldots(i)$$

$$\because \quad F_{22} = 0$$

$$\therefore \quad F_{21} = 1$$

We know that

$$A_1F_{12} = A_2F_{21}$$

$$F_{12} = \frac{A_2}{A_1}$$

Now, $F_{11} + F_{12} = 1$

$$\therefore \quad F_{11} = 1 - F_{12} = 1 - \frac{A_2}{A_1}$$

$$= 1 - \frac{\left[(\pi \times 0.5 \times 0.5) + \left(2 \times \frac{\pi}{4} \times 0.5 \times 0.5\right)\right]}{4\pi \times 0.5 \times 0.5} = \mathbf{0.625}$$

4.6 (d)

During 24 hours heat liberated by bulb is

$$Q = 100 \times 24 \times 3600 = 8.64 \times 10^6 \text{ J}$$

Density of air, $\rho = 1.2$ kg/m^3

Volume of room, $V = 2.5 \times 3 \times 3 = 22.5$ m^3

Mass of air, $m = 1.2 \times 22.5 = 27$ kg

Let T_2 is final temperature of gas,

c_v of air = 0.717 kJ/kgK

Because volume of room is constant,

$$Q = mc_v(T_2 - T_1)$$

$$8.64 \times 10^6 = 27 \times 0.717 \times 1000(T_2 - 20)$$

or $\quad T_2 = \mathbf{470°C}$

4.7 (b)

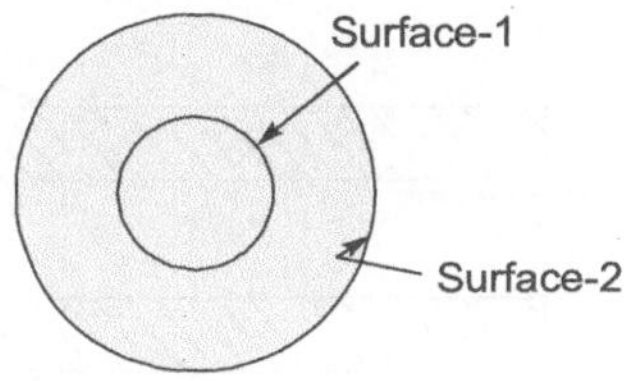

$$R_1 = 1 \text{ m}$$

$$R_2 = 2 \text{ m}$$

For surface-1,

$$F_{11} + F_{12} = 1$$

$$F_{12} = 1 \qquad \because F_{11} = 0$$

For surface -2,

$$F_{21} + F_{22} = 1 \qquad \ldots(i)$$

By reciprocity theorem,

$$A_1F_{12} = A_2F_{21}$$

$$2\pi R_1 l \times 1 = 2\pi R_2 l \times F_{21}$$

$$R_1 = R_2F_{21}$$

or $\quad F_{21} = \frac{R_1}{R_2} = \frac{1}{2} = 0.5$

Substituting $F_{21} = 0.5$ in Eq. (i), we get

$$0.5 + F_{22} = 1$$

or $\quad F_{22} = 1 - 0.5 = \mathbf{0.5}$

4.8 (d)

$$T_1 = 727°C = 1000 \text{ K}$$

$$T_2 = 227°C = 500 \text{ K}$$

$$\epsilon = 0.7$$

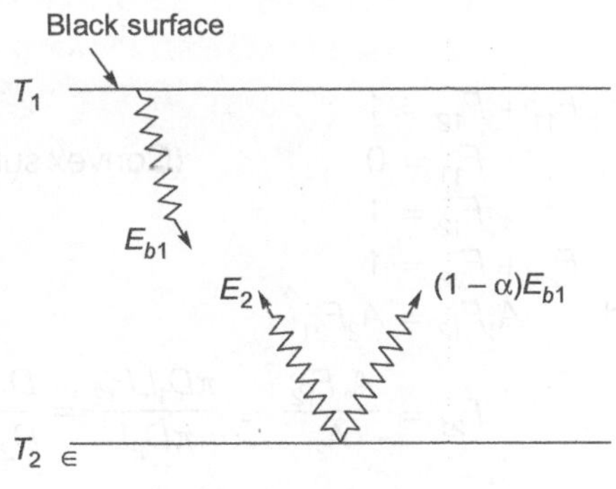

Energy incident upon plate 1(irradiation)
= Radiosity of plate 2, since $F_{12} = F_{21} = 1$
Now energy leaving from surface 2 is

$$= E_2 + (1 - \epsilon_2)E_{b1}$$

E_2 = Energy emitting due to own temperature

$$= \epsilon\sigma T_2^4 = 0.7 \times 5.67 \times 10^{-8} \times 500^4$$
$$= 2480.62 \text{ W/m}^2$$

and $(1 - \epsilon_2)E_{b1}$
= Fraction of energy reflected from surface 2.

$$= (1 - 0.7) \times 5.67 \times 10^{-8} \times 1000^4 = 17010 \text{ W/m}^2$$

∴ Total energy incident upon plate 1
= **19.49 kW/m²**

4.9 (d)

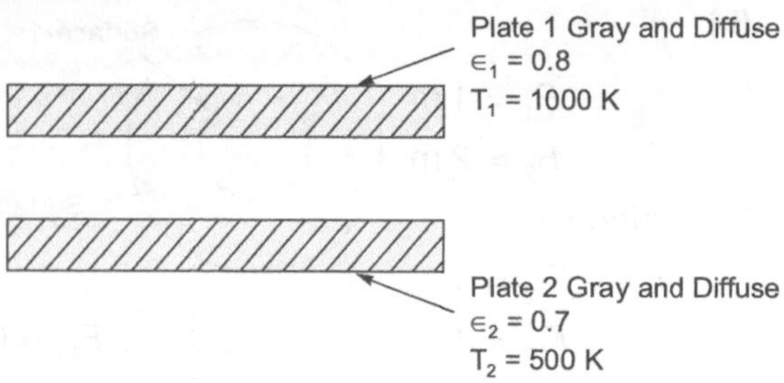

The net radiation heat exchange between plate 1 and plate 2 [i.e. infinite parallel gray planes]

$$Q_{12} = \frac{A_1 \sigma (T_1^4 - T_2^4)}{\frac{1}{\epsilon_1} + \frac{1}{\epsilon_2} - 1}$$

$$\frac{Q_{12}}{A_1} = \frac{\sigma(T_1^4 - T_2^4)}{\frac{1}{\epsilon_1} + \frac{1}{\epsilon_2} - 1} = \frac{5.67 \times 10^{-8}[(1000)^4 - (500)^4]}{\frac{1}{0.8} + \frac{1}{0.7} - 1}$$

$= 31678.34 \text{ W/m}^2$ = **31.68 kW/m²**

4.10 (c)

For any surface

$$\alpha + \tau + \rho = 1$$

But for opaque surface, transmissivity $(\tau) = 0$

$$\therefore \quad \alpha + \rho = 1$$

4.11 (d)

$$F_{11} + F_{12} = 1$$
$$F_{11} = 0 \qquad \text{(Convex surface)}$$
$$\therefore \quad F_{12} = 1$$
$$F_{21} + F_{22} = 1$$

and $\quad A_1F_{12} = A_2F_{21}$

or $\quad F_{21} = \frac{A_1F_{12}}{A_2} = \frac{\pi D_1 L F_{12}}{\pi D_2 L} = \frac{D_1}{D_2}$

$$F_{22} = 1 - F_{21} = 1 - \frac{D_1}{D_2}$$

4.12 (a)

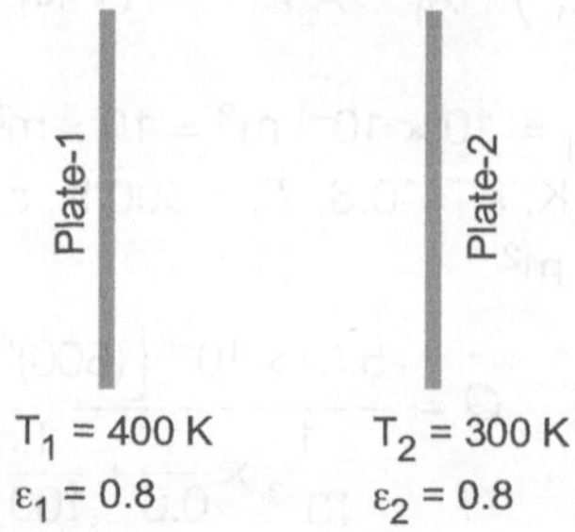

$$\sigma = 5.67 \times 10^{-8} \text{ W/m}^2\text{K}^4$$

$$\frac{Q}{A} = \frac{\sigma(T_1^4 - T_2^4)}{\frac{1}{\varepsilon_1} + \frac{1}{\varepsilon_2} - 1} = \frac{5.67 \times 10^{-8}(400^4 - 300^4)}{\frac{1}{0.8} + \frac{1}{0.8} - 1}$$

$= 661.5 \text{ W/m}^2$ = **0.66 kW/m²**

4.13 Sol.

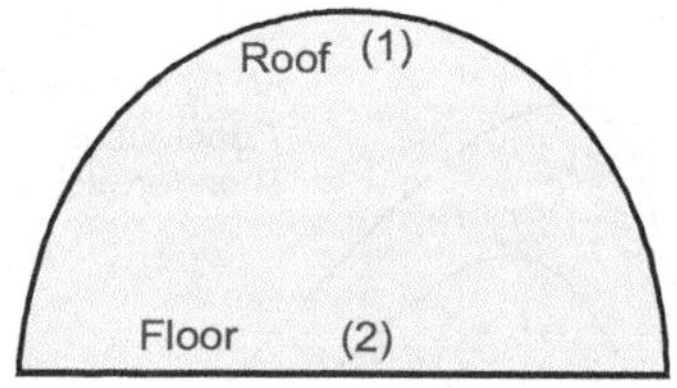

Given data: $r = 1$ m

$$\epsilon_1 = 1$$
$$T_1 = 800 \text{ K}$$
$$\epsilon_2 = 0.5$$
$$T_2 = 600 \text{ K}$$
$$\sigma = 5.668 \times 10^{-8} \text{ W/m}^2\text{K}^4$$

$$(F_g)_{1-2} = \frac{1}{\frac{1-\epsilon_1}{\epsilon_1} + \frac{1}{F_{12}} + \frac{A_1}{A_2}\left(\frac{1-\epsilon_1}{\epsilon_2}\right)}$$

$$F_{21} = 1$$
$$A_1F_{12} = A_2F_{21}$$

$$\Rightarrow \quad F_{12} = \frac{A_2}{A_1} = \frac{\pi r^2}{2\pi r^2} = \frac{1}{2}$$

$$\therefore \quad (F_g)_{1-2} = \frac{1}{\left(\frac{1-1}{1}\right) + 2 + 2\left(\frac{1-0.5}{0.5}\right)}$$

$$= \frac{1}{4} = 0.25$$

$$Q_{1-2} = \sigma A_1 (F_g)_{1-2} [T_1^4 - T_2^4]$$

$$Q_{12} = 5.668 \times 2 \times \pi \times (1)^2 \times 0.25\,(800^4 - 600^4) \times 10^{-8}$$

$$Q_{12} = 24929.1666\ \text{W} = \mathbf{24.929\ kW}$$

4.14 (b)

$$F_{11} + F_{12} = 1$$

$$F_{11} = 0$$

$$\therefore \quad F_{12} = 1$$

Applying reciprocity theorem,

$$A_1 F_{12} = A_2 F_{21}$$

or

$$F_{21} = \frac{A_1}{A_2} F_{12} = \frac{4\pi r_1^2}{4\pi r_2^2} F_{12}$$

$$= \left(\frac{r_1}{r_2}\right)^2 = \left(\frac{2}{3}\right)^2 = \frac{4}{9}$$

4.15 (a)

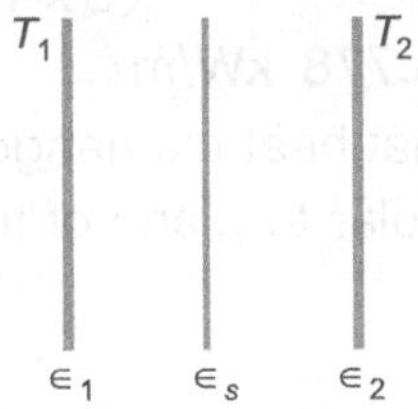

$$\in_1 = \in_s = \in_2 = \in$$

$$\text{R(with } N \text{ shields)} = (N+1)\left(\frac{2}{\in} - 1\right)$$

$$\text{R(without shield)} = \frac{2}{\in} - 1$$

$$\left(\frac{Q}{A}\right)_{\text{with } N \text{ shield}} = \frac{1}{N+1}\left(\frac{Q}{A}\right)_{\text{without shield}}$$

$$\therefore \quad \frac{Q_{\text{with 1 shield}}}{Q_{\text{without shield}}} = \frac{1}{N+1} = \frac{1}{2} \qquad (\because N = 1)$$

4.16 (c)

$$\text{E(emmisive power)} \propto T^4 \qquad (\because \in = \text{Constant})$$

$$\frac{E_1}{E_2} = \left(\frac{T_1}{T_2}\right)^4$$

or

$$\frac{T_1}{T_2} = \left(\frac{500}{1200}\right)^{1/4} = 0.803$$

4.17 (a)

Applying reciprocity theorem,

$$A_1 F_{1-2} = A_2 F_{2-1} \qquad (\because F_{1-2} = 1)$$

$$4\pi R_1^2 = (2\pi R_2^2 + \pi R_2^2) F_{2-1}$$

$$R_1 = r$$

$$R_2 = 4r$$

$$4\pi \times r^2 = 3\pi \times (4r)^2 F_{2-1}$$

$$F_{2-1} = \frac{4\pi r^2}{48\pi r^2} = \frac{1}{12}$$

4.18 Sol.

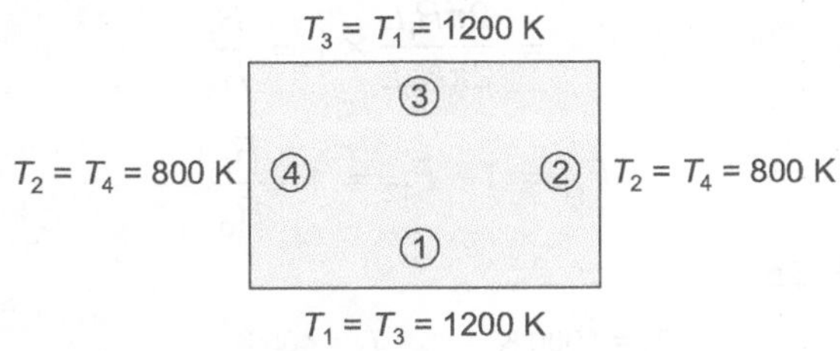

Given: $F_{12} = 0.26$

$F_{12} = F_{14} = 0.26$ [symmetry]

Given $\sigma = 5.67 \times 10^{-8}\ \text{W/m}^2\text{K}^4$

sides (1) and (3) will not exchange any heat being at same temperature

sides (2) and (4) will not exchange any heat being at same temperature

Hence side (1) and side (2) and side (1) and side (4) will exchange heat

$$Q_{1-2} = A_1 F_{12} \sigma \left[T_1^4 - T_2^4\right]$$

$$= 0.5 \times 0.26 \times 5.67 \times 10^{-8} \times [1200^4 - 800^4]$$

$$= 12265.34\ \text{W/m}$$

$$Q_{1-4} = 0.5 \times 0.26 \times 5.67 \times 10^{-8} \times [1200^4 - 800^4]$$

$$= 12265.34\ \text{W/m}$$

$$Q_{net} = Q_{1-2} + Q_{1-4} = 12265.34 + 12265.34$$

$$= \mathbf{24530.68\ W/m}$$

4.19 (d)

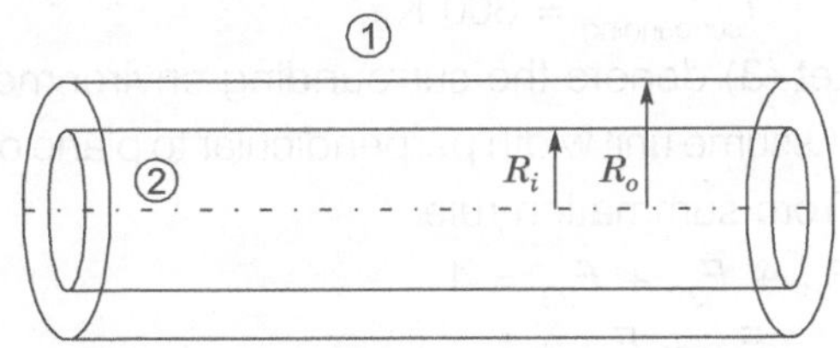

To find : Radiation view factor of the outer cylinder onto itself.

Hence F_{11} is required.

$\Rightarrow \quad A_1 F_{12} = A_2 F_{21}$ [Reciprocity]

$F_{11} + F_{12} = 1$ [Summation]

$F_{21} + F_{22} = 1$

$\therefore \quad F_{11} = 1 - F_{12}$

$F_{22} = 0$ [No radiation leaving surface 2 reaches back to 2]

$\therefore \quad F_{21} = 1$

$$F_{12} = \frac{A_2}{A_1} \times F_{21} = \frac{A_2}{A_1} \times 1$$

$$= \frac{2\pi R_i L}{2\pi R_o L} \times 1 = \frac{R_i}{R_o}$$

$$F_{11} = 1 - F_{12} = 1 - \frac{R_i}{R_o}$$

4.20 Sol.

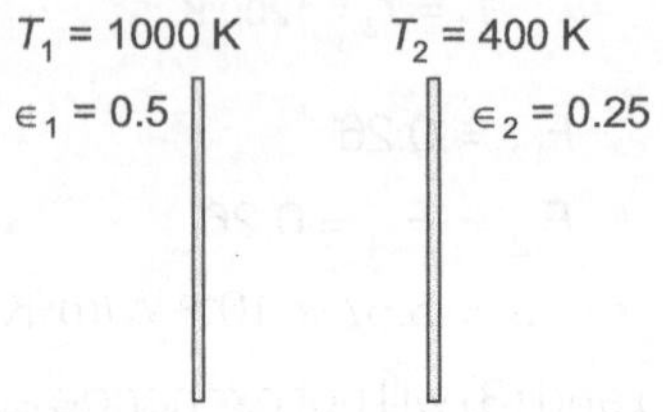

$$Q = \frac{\sigma\left[T_1^4 - T_2^4\right]}{\frac{1}{\epsilon_1} + \frac{1}{\epsilon_2} - 1} = \frac{5.67 \times 10^{-8} \times (1000^4 - 400^4)}{\frac{1}{0.5} + \frac{1}{0.25} - 1}$$

$= 11049.7 \text{ W/m}^2 =$ **11.05 kW/m²**

4.21 Sol.

$F_{12} = 0.5$

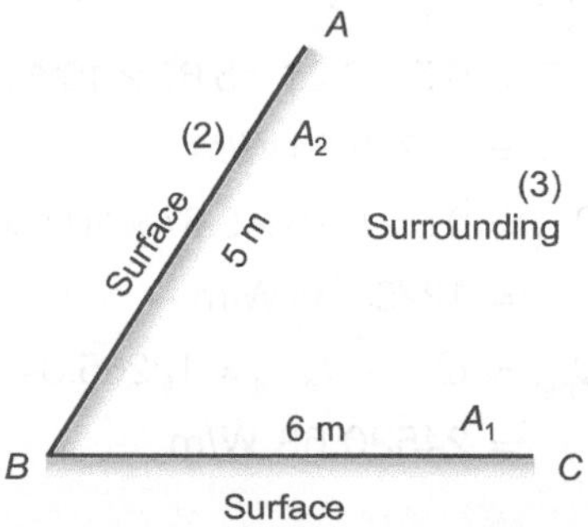

$T_1 = 800 \text{ K}, \; T_2 = 600 \text{ K}$

$T_{surrounding} = 300 \text{ K}$

Let (3) denote the surrounding environment (Assume unit width perpendicular to plane of Fig.)

From summation rule:

$F_{21} + F_{22} + F_{23} = 1$

$F_{21} + F_{23} = 1$

$\because \quad F_{22} = 0$

or $\quad F_{23} = 1 - F_{21}$

$A_1F_{12} = A_2F_{21} = 6 \times 1 \times 0.5$

$= (5 \times 1) \times F_{21}$

$$F_{21} = \frac{6 \times 0.5}{5} = 0.6$$

$\therefore \quad F_{23} = 1 - F_{21} = 0.4$

Since surrounding environment being large, No surface has a surface resistance

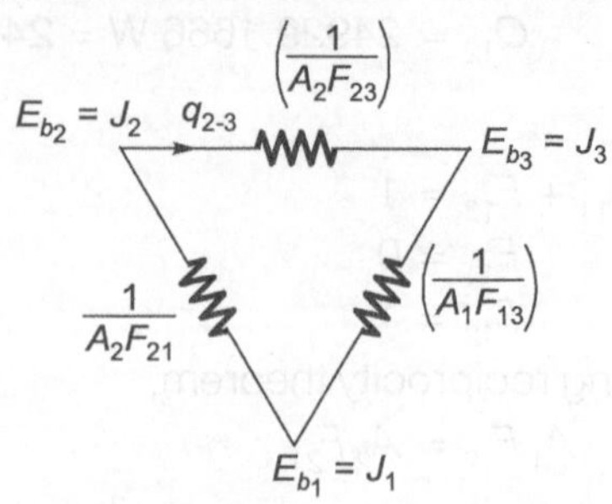

$$\text{Heat transfer rate from (2) and (3)} = \frac{E_{b_2} - E_{b_3}}{\frac{1}{A_2F_{23}}}$$

$$= \frac{\sigma\left(T_2^4 - T_3^4\right)}{\frac{1}{A_2F_{23}}} = \frac{5.67 \times 10^{-8}(600^4 - 300^4)}{\left(\frac{1}{5 \times 1 \times 0.4}\right)} \text{ W/m}$$

= 13.778 kW/m

Assume that heat exchange is per unit width perpendicular to plane of figure.

4.22 (d)

$E_b \propto T^4$ (Stefan Boltzmann law of radiation)

$$\frac{E_{b_2}}{E_{b_1}} = \left(\frac{2T}{T}\right)^4 \Rightarrow E_{b_2} = 16P$$

4.23 (b)

From Wein's displacement law

For black body,

$\lambda_M T = \text{constant}$

$\lambda_{M1} T_1 = \lambda_{M2} T_2$

$1.45 \times 2000 = 2.90 \times T_2$

$$\therefore \quad T_2 = \left(\frac{1.45}{2.90} \times 2000\right) = \mathbf{1000\ K}$$

4.24 Sol.

Under steady state condition, all rate of heat transfer i.e. from surface at 400 K to black plate (via radiation), inside black plate (via conduction) and from black plate to surface at 300 K (via radiation) are equal.

So, heat flux through wall = Radiation flux from wall to surface at 300 K

$$= \frac{\sigma(350^4 - 300^4)}{\frac{1}{1} + \frac{1}{1} - 1} = \mathbf{391.612\ W/m^2}$$

4.25 (d)

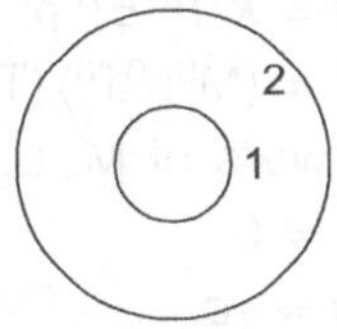

$$F_{11} = 0$$

We know that, $F_{11} + F_{12} = 1$

$$F_{12} = 1$$

4.26 Sol.

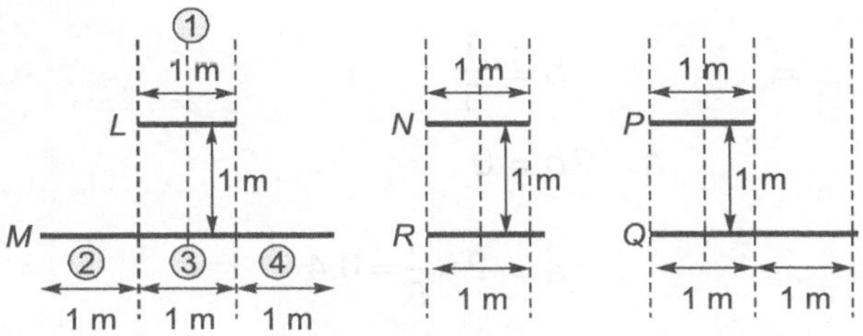

As $\quad F_{NR} = 0.4$

$$F_{LM} = F_{12} + F_{13} + F_{14}$$

$$F_{13} = F_{NR} = 0.4$$

$$F_{12} = F_{14}$$

$$F_{LM} = 0.8 = 0.4 + 2F_{12}$$

$$0.2 = F_{12}$$

$$F_{PQ} = F_{12} + F_{13} = 0.2 + 0.4 = 0.6$$

4.27 Sol.

Given, $T_1 = 8000$ K

From Wien's Displacement law,

$$\lambda_m T = \text{constant} \Rightarrow \lambda_{m1} T_1 = \lambda_{m2} T_2$$

$$\lambda_{m1} \times 3000 = 1.2\lambda_{m1} \times T_2$$

$$\text{Temperature, } T_2 = \left(\frac{3000}{1.2}\right) \text{K} = 2500 \text{ K}$$

4.28 Sol.

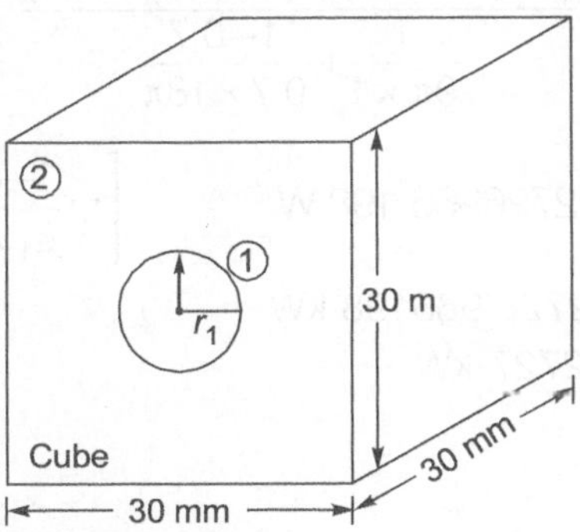

$$r_1 = 10 \text{ mm}$$

$$A_1 = 4\pi r_1^2$$

$$A_2 = 6 \times (30^2)$$

$$F_{12} = 1$$

$$A_1 F_{12} = A_2 F_{21}$$

$$\Rightarrow \quad F_{21} = \frac{A_1}{A_2} = \frac{4\pi \times (10)^2}{6 \times (30)^2} = 0.2327$$

$$F_{22} = 1 - F_{21} = 0.7672 \simeq 0.77$$

4.29 (218) (210 to 225)

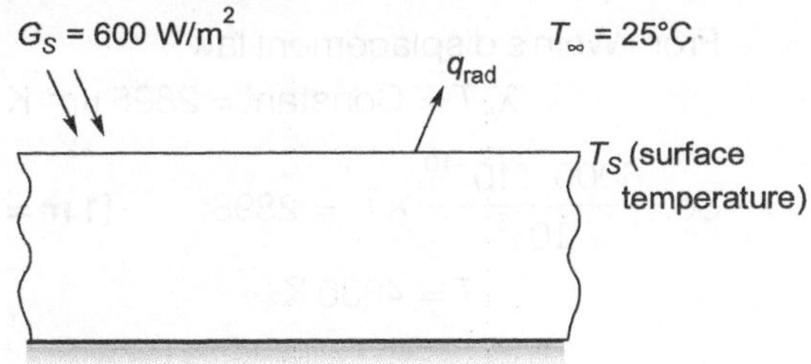

For steady state of slab,

Heat absorbed by plate = Net radiation heat transfer between plate and ambient

As per Kirchoff's law,

$$\alpha_{Sky} = \varepsilon_{Sky}$$

$$\varepsilon = 0.21$$

For equilibrium temperature :

Net heat transfer at surface = 0

$$q''_{rad} - G_s = 0$$

$$\varepsilon\sigma\left(T^4 - T_\infty^4\right) = G_s$$

$$0.21 \times 5.67 \times 10^{-8}\left[T^4 - 298^4\right] = 600$$

$$T = 491.33 \text{ K}$$

$$T = 218°\text{C}$$

4.30 (319) (315 to 320)

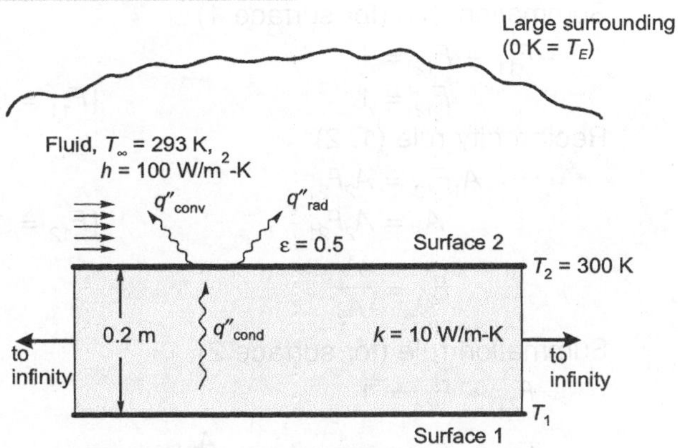

Applying energy balance on surface (2)

$$q''_{cond} = q''_{conv} + q''_{rad}$$

$$\frac{K(T_1 - T_2)}{L} = h(T_2 - T_\infty) + \sigma\varepsilon(T_2^4 - T_E^4)$$

$$\frac{10 \times (T_1 - 300)}{0.2} = 100 \times (300 - 293) + 0.5 \times 5.67 \times 10^{-8} \times (300^4 - 0^4)$$

$$T_1 = 318.59 \text{ K}$$
$$T_1 = 319 \text{ K}$$

4.31 (4830) (4825 to 4835)

As we know that $\epsilon_\lambda = \dfrac{E_\lambda}{E_{b\lambda}}$

From Wein's displacement law

$$\lambda_m T = \text{Constant} = 2898\ \mu\text{m-K}$$

So, $\dfrac{6000 \times 10^{-10}}{10^{-6}} \times T = 2898$ **[1 m = 10⁻⁶ m]**

$$T = 4830 \text{ K}$$

4.32 (d)

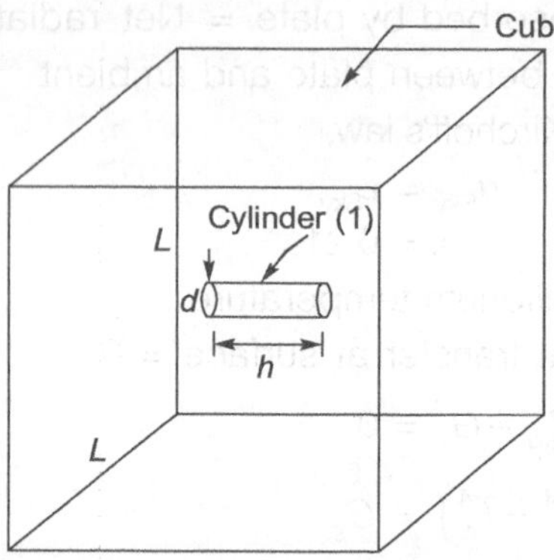

$$A_1 = \pi dh + \frac{\pi}{4}d^2 + \frac{\pi}{4}d^2$$

$$A_1 = \left(\pi dh + \frac{\pi d^2}{2}\right);\ A_2 = 6L^2$$

Summation rule (for surface 1)

$$F_{11} + F_{12} = 1$$
$$F_{12} = 1 \qquad \{F_{11} = 0\}$$

Reciprocity rule (1, 2):

$$A_1F_{12} = A_2F_{21}$$
$$A_1 = A_2F_{21} \qquad \{F_{12} = 1\}$$
$$F_{21} = \frac{A_1}{A_2}$$

Summation rule (for surface 2)

$$F_{21} + F_{22} = 1$$

$$F_{22} \text{ or } F_{ss} = 1 - F_{21} = 1 - \frac{A_1}{A_2}$$

$$F_{22} = 1 - \frac{\left(\pi dh + \dfrac{\pi d^2}{2}\right)}{6L^2}$$

4.33 (0.4) (0.39 to 0.41)

$$R = k\, t^a E^b \rho^c$$

$$M^0L^1T^0 = (M^0L^0T^0)(T)^a(ML^2T^{-2})^b(ML^{-3})^c$$

Comparing powers of M, L, T

$$b + c = 0$$
$$\Rightarrow \quad b = -c$$
$$2b - 3c = 1$$
$$-2c - 3c = 1$$
$$c = -\frac{1}{5}$$
$$\Rightarrow \quad b = \frac{1}{5}$$
$$a - 2b = 0$$
$$a = 2 \times \frac{1}{5} = 0.4$$

4.34 (2727) (2720 to 2740)

Given: $T_1 = 300$ K, $T_2 = 1200$ K, $D = 6$ m, $F_{11} = 0$, $F_{12} = 1$, $\epsilon_1 = 1$, $\epsilon_2 = 0.7$

$$A_1 = \pi R^3 = \pi(3)^2 = 9\pi \text{ m}^2$$
$$A_2 = 2\pi R^2 = 2\pi(3)^2 = 18\pi \text{ m}^2$$

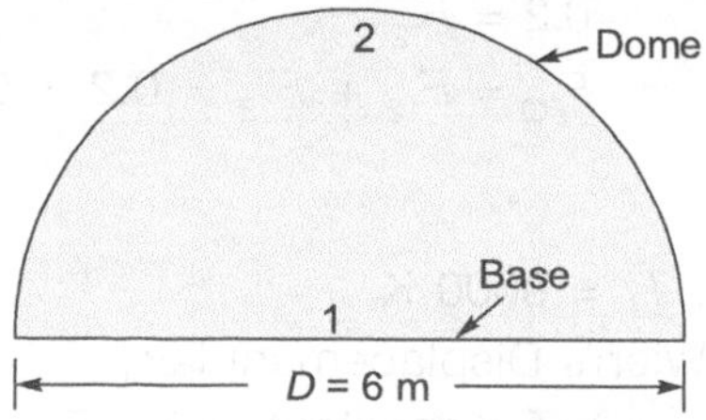

$$\text{Heat transfer, } q = \frac{\sigma\left(T_2^4 - T_1^4\right)}{\dfrac{1-\epsilon_1}{\epsilon_1 A_1} + \dfrac{1}{A_1F_{12}} + \dfrac{1-\epsilon_2}{\epsilon_2 A_2}}$$

$$= \frac{\left(5.67 \times 10^{-8}\right) \times \left(1200^4 - 300^4\right)}{\dfrac{1}{9\pi \times 1} + \dfrac{1-0.7}{0.7 \times 18\pi}}$$

$$= 2726966.198 \text{ W} \qquad \left[\because \frac{1-\epsilon_1}{\epsilon_1 A_1} = 0\right]$$

$$= 2726.966198 \text{ kW}$$
$$\simeq 2727 \text{ kW}$$

■■■■

5 Heat Exchanger

5.1 In certain heat exchanger, both the fluids have identical mass flow rate specific heat product. The hot fluid enters at 76°C and leaves at 47°C, and the cold fluid entering at 26°C leaves at 55°C, the effectiveness of the HE is

(a) 0.16 (b) 0.58
(c) 0.72 (d) 1.0 **[1997 : 2 M]**

5.2 Air enters a counter flow HE at 70°C and leaves at 40°C. Water enters at 30°C and leaves at 50°C, the LMTD in degree °C is

(a) 5.65 (b) 14.43
(c) 19.52 (d) 20.17 **[2000 : 2 M]**

5.3 For the same inlet and exit temperatures of the hot and cold fluids, the log mean temperature difference (LMTD) is

(a) Greater for parallel flow heat exchanger than for counter flow heat exchanger
(b) Greater for counter flow heat exchanger than for parallel flow heat exchanger
(c) Same for both parallel and counter flow heat exchangers
(d) Depending on the properties of fluid

[2002 : 1 M]

5.4 In a counter flow heat exchanger, for the hot fluid the heat capacity = 2 kJ/kgK, mass flow rate = 5 kg/s, inlet temperature = 150°C, outlet temperature = 100°C. For the cold fluid, heat capacity = 4 kJ/kgK, mass flow rate = 10 kg/s, inlet temperature = 20°C. Neglecting heat transfer to the surroundings, the outlet temperature of the cold fluid in °C is

(a) 7.5 (b) 32.5
(c) 45.5 (d) 70.0 **[2003 : 2 M]**

5.5 In a condenser, water enters at 30°C and flows at the rate 1500 kg/hr. The condensing steam is at a temperature of 120°C and cooling water leaves the condenser at 80°C. Specific heat of water is 4.187 kJ/kgK. If the overall heat transfer coefficient is 2000 W/m^{2}K, the heat transfer area is

(a) 0.707 m^2 (b) 7.07 m^2
(c) 70.7 m^2 (d) 141.4 m^2 **[2004 : 2 M]**

5.6 Hot oil is cooled from 80 to 50°C in an oil cooler which uses air as the coolant. The air temperature rises from 30 to 40°C. The designer uses a LMTD value of 26°C. The type of heat exchanger is

(a) parallel flow (b) double pipe
(c) counter flow (d) cross flow **[2005 : 2 M]**

5.7 In a counterflow heat exchanger, hot fluid enters at 60°C and cold fluid leaves at 30°C. Mass flow rate of the hot fluid is 1 kg/s and that of the cold fluid is 2 kg/s. Specific heat of the hot fluid is 10 kJ/kgK and that of the cold fluid is 5 kJ/kgK. The Log Mean Temperature Difference (LMTD) for the heat exchanger in °C is

(a) 15 (b) 30
(c) 35 (d) 45 **[2007 : 2 M]**

5.8 The logarithmic mean temperature difference (LMTD) of a counter flow heat exchanger is 20°C. The cold fluid enters at 20°C and the hot fluid enters at 100°C. Mass flow rate of the cold fluid is twice that of the hot fluid. Specific heat at constant pressure of the hot fluid is twice that of the cold fluid. The exit temperature of the cold fluid

(a) is 40°C
(b) is 60°C
(c) is 80°C
(d) cannot be determined **[2008 : 2 M]**

5.9 a parallel flow heat exchanger operating under steady state, the heat capacity rates (product of specific heat at constant pressure and mass flow rate) of the hot and cold fluid are equal. The hot fluid, flowing at 1 kg/s with c_p = 4 kJ/kgK, enters the heat exchanger at 102°C while the cold fluid has an inlet temperature of 15°C. The overall heat transfer coefficient for the heat exchanger is estimated to be 1 kW/m^{2}K and the corresponding heat transfer surface area is 5 m^2. Neglect heat transfer between the heat exchanger and the ambient. The heat exchanger is characterized by the following relation:

$2\varepsilon = 1 - \text{expt}(-2\text{NTU})$. The exit temperature (in °C) for the cold fluid is

(a) 45 (b) 55
(c) 65 (d) 75 **[2009 : 2 M]**

5.10 An industrial gas (c_p = 1 kJ/kgK) enters a parallel flow heat exchanger at 250°C with a flow rate of

2 kg/s to heat a water stream. The water stream (c_p = 4 kJ/kgK) enters the heat exchanger at 50°C with a flow rate of 1 kg/s. The heat exchanger has an effectiveness of 0.75. The gas stream exit temperature will be

(a) 75°C (b) 100°C
(c) 125°C (d) 150°C [2010 : 2 M]

5.11 Cold water flowing at 0.1 kg/s is heated from 20°C to 70°C in a counter flow type heat exchanger by a hot water stream flowing at 0.1 kg/s and entering at 90°C. The specific heat of water is 4200 J/(kgK) and density is 1000 kg/m^3. If the overall heat transfer coefficient U for the heat exchanger is 2000 W/(m^{2}K), the required heat exchange area (in m^2) is

(a) 0.052 (b) 0.525
(c) 0.151 (d) 0.202 [2011 : 2 M]

5.12 Water (c_p = 4.18 kJ/kgK) at 80°C enters a counter flow heat exchanger with a mass flow rate of 0.5 kg/s. Air (c_p = 1 kJ/kgK) enter at 30°C with a mass flow rate 2.09 kg/s. If the effectiveness of the heat exchanger is 0.8, the LMTD (in °C) is

(a) 40 (b) 20
(c) 10 (d) 5 [2012 : 2 M]

5.13 In a condenser of a power plant, the steam condenses at a temperature of 60°C. The cooling water enters at 30°C and leaves at 45°C. The Logarithmic Mean Temperature Difference (LMTD) of the condenser is

(a) 16.2°C (b) 21.6°C
(c) 30°C (d) 37.5°C [2011 : 1 M]

5.14 In a heat exchanger, it is observed that $\Delta T_1 = \Delta T_2$, where ΔT_1 is the temperature difference between the two single phase fluid streams at one end and ΔT_2 is the temperature difference at the other end. This heat exchanger is

(a) a condenser
(b) an evaporator
(c) a counter flow heat exchanger
(d) a parallel flow heat exchanger

[2014 : 1 M, Set-2]

5.15 In a concentric counter flow heat exchanger, water flows through the inner tube at 25°C and leaves at 42°C. The engine oil enters at 100°C and flows in the annular flow passage. The exit temperature of the engine oil is 50°C. Mass flow rate of water and the engine oil are 1.5 kg/s and 1 kg/s, respectively. The specific heat of water and oil are 4178 J/kgK and 2130 J/kgK, respectively. The effectiveness of this heat exchanger is ______.

[2014 : 2 M, Set-2]

5.16 A double pipe counter flow heat exchanger transfers heat between two water streams. Tube side water at 19 litre/s is heated from 10°C to 38°C. Shell side water at 25 litre/s is entering at 46°C. Assume constant properties of water, density is 1000 kg/m^3 and specific heat is 4186 J/kgK. The LMTD (in °C) is ______.

[2014 : 2 M, Set-3]

5.17 Saturated vapor is condensed to saturated liquid in a condenser. The heat capacity ratio is $C_r = \dfrac{C_{min}}{C_{max}}$. The effectiveness($\varepsilon$) of the condenser is

(a) $\dfrac{1-\exp[-NTU(1+C_r)]}{1+C_r}$

(b) $\dfrac{1-\exp[-NTU(1-C_r)]}{1+C_r[-NTU(1-C_r)]}$

(c) $\dfrac{NTU}{1+NTU}$

(d) $1-\exp(-NTU)$ [2015 : 1 M, Set-3]

5.18 A balanced counter flow heat exchanger has a surface area of 20 m^2 and overall heat transfer coefficient of 20 W/m^{2}K. Air (c_p = 1000 J/kgK) entering at 0.4 kg/s and 280 K is to be preheated by the air leaving the system at 0.4 kg/s and 300 K. The temperature (in K) of the preheated air is

(a) 290 (b) 300
(c) 320 (d) 350 [2015 : 2 M, Set-2]

5.19 Consider a parallel-flow heat exchanger with area A_p and a counter-flow heat exchanger with area A_c. In both the heat exchangers, the hot stream flowing at 1 kg/s cools from 80°C to 50°C. For the cold stream in both the heat exchangers, the flow rate and the inlet temperature are 2 kg/s and 10°C, respectively. The hot and cold streams in both the heat exchangers are of the same fluid. Also, both the heat exchangers have the same overall heat transfer coefficient. The ratio A_c/A_p is ______.

[2016 : 2 M, Set-2]

5.20 For a heat exchanger, ΔT_{max} is the maximum temperature difference and ΔT_{min} is the minimum temperature difference between the two fluids. $LMTD$ is the log mean temperature difference. C_{min} and C_{max} are the minimum and the maximum heat capacity rates. The maximum possible heat transfer (Q_{max}) between the two fluids is

(a) C_{min} LMTD (b) $C_{min} \Delta T_{max}$
(c) $C_{max} \Delta T_{max}$ (d) $C_{max} \Delta T_{min}$

[2016 : 1 M, Set-3]

5.21 Saturated steam at 100°C condenses on the outside of a tube. Cold enters the tube at 20°C and exits at 50°C. The value of the Log Mean Temperature Difference (LMTD) is ________ °C.

[2017 : 1 M, Set-1]

5.22 In a counter-flow heat exchanger, water is heated at the rate of 1.5 kg/s from 40°C to 80°C by an oil entering at 120°C and leaving at 60°C. The specific heats of water and oil are 4.2 kJ/kgK and 2 kJ/kgK, respectively. The overall heat transfer coefficient is 400 W/m^{2}K. The required heat transfer surface area (in m^2) is
(a) 0.104 (b) 0.022
(c) 10.4 (d) 21.84

[2017 : 2 M, Set-2]

5.23 Steam in the condenser of a thermal power plant is to be condensed at a temperature of 30°C with cooling water which enters the tubes of the condenser at 14°C and exits at 22°C. The total surface area of the tubes is 50 m^2. and the overall heat transfer coefficient is 2000 W/m^2. The heat transfer (in MW) to the condenser is ________ (correct to two decimal places).

[2018 : 2 M, Set-2]

5.24 Hot and cold fluids enter a parallel flow double tube heat exchanger at 100°C and 15°C respectively. The heat capacity rates of hot and cold fluids are C_h = 2000 W/K and C_c = 1200 W/K respectively. If the outlet temperature of the cold fluid is 45 °C, the log mean temperature difference (LMTD) of the heat exchanger is ________ K (round off to two decimal places).

[2019 : 2 M, Set-2]

5.25 In a concentric tube counter-flow heat exchanger, hot oil enters at 102°C and leaves at 65°C. Cold water enters at 25°C and leaves at 42°C. The log mean temperature difference (LMTD) is ________ °C (round off to one decimal place).

[2020 : 1 M, Set-1]

5.26 A shell and tube heat exchanger is used as a steam condenser. Coolant water enters the tube at 300 K at a rate of 100 kg/s. The overall heat transfer coefficient is 1500 W/m^2.K, and total heat transfer area is 400 m^2. Steam condenses at a saturation temperature of 350 K. Assume that the specific heat of coolant water is 4000 J/kg.K. The temperature of the coolant water coming out of the condenser is ________ K (round off to the nearest integer).

[2021 : 2 M, Set-2]

5.27 During open-heart surgery, a patient's blood is cooled down to 25°C from 37°C using a concentric tube counter-flow heat exchanger. Water enters the heat exchanger at 4°C and leaves at 18°C. Blood flow rate during the surgery is 5 L/minute.

Use the following fluid properties:

Fluid	Density (kg/m^3)	Specific heat (J/kg-K)
Blood	1050	3740
Water	1000	4200

Effectiveness of the heat exchanger is (round off to 2 decimal places).

[2022 : 2 M, Set-1]

5.28 Saturated vapor at 200 °C condenses to saturated liquid at the rate of 150 kg/s on the shell side of a heat exchanger (enthalpy of condensation h_{fg} = 2400 kJ/kg). A fluid with c_p = 4 kJ·kg^{-1}·K^{-1} enters at 100°C on the tube side. If the effectiveness of the heat exchanger is 0.9, then the mass flow rate of the fluid in the tube side is _____ kg/s (in integer).

[2022 : 2 M, Set-2]

5.29 Consider a counter-flow heat exchanger with the inlet temperatures of two fluids (1 and 2) being $T_{1,in}$ = 300 K and $T_{2,in}$ = 350 K. The heat capacity rates of the two fluids are C_1 = 1000 W/K and C_2 = 400 W/K, and the effectiveness of the heat exchanger is 0.5. The actual heat transfer rate is _____ kW. (Answer in integer) [2023 : 1 M]

5.30 A condenser is used as a heat exchanger in a large steam power plant in which steam is condensed to liquid water. The condenser is a shell and tube heat exchanger which consists of 1 shell and 20000 tubes. Water flows through each of the tubes at a rate of 1 kg/s with an inlet temperature of 30°C. The steam in the condenser shell condenses at the rate of 430 kg/s at a temperature of 50°C. If the heat of vaporization is 2.326 MJ/kg and specific heat of water is 4 kJ/(kg.K), the effectiveness of the heat exchanger is ________ (rounded off to 3 decimal places)

[2024 : 2 M]

■■■■

Answers Heat Exchanger

5.1	(b)	5.2	(b)	5.3	(b)	5.4	(b)	5.5	(a)	5.6	(d)	5.7	(b)	5.8	(c)
5.9	(b)	5.10	(b)	5.11	(b)	5.12	(c)	5.13	(b)	5.14	(c)	5.15	(0.6676)		
5.16	(11.02)	5.17	(d)	5.18	(a)	5.19	(0.927)	5.20	(b)	5.21	(63.82)	5.22	(d)		
5.23	(1.154)	5.24	(57.71)	5.25	(49.3)	5.26	(339)	5.27	(0.42)	5.28	(1000)	5.29	(10)		
5.30	(0.625)														

Explanations Heat Exchanger

5.1 (b)

$$\varepsilon = \frac{C_h\left(T_{h_1} - T_{h_2}\right)}{C_{min}\left(T_{h_1} - T_{c_1}\right)}$$

Here, $C_h = C_c$

$$\varepsilon = \frac{T_{h_1} - T_{h_2}}{T_{h_1} - T_{c_1}} = \frac{76-47}{76-26} = \mathbf{0.58}$$

5.2 (b)

$$\text{LMTD} = \frac{\theta_1 - \theta_2}{\ln\left(\frac{\theta_1}{\theta_2}\right)}$$

For counter flow HE

$$\theta_1 = 70 - 50 = 20°\text{C}$$

$$\theta_2 = 40 - 30 = 10°\text{C}$$

$$\text{LMTD} = \frac{20-10}{\ln\left(\frac{20}{10}\right)} = \mathbf{14.43°C}$$

5.4 (b)

$c_{ph} = 2000$ J/kgK; $m_h = 5$ kg/s;

$c_{pc} = 4000$ J/kgK; $T_{hi} = 150°$C; $m_c = 10$ kg/s

$T_{ci} = 20°$C; $T_{co} = ?$

Heat lost by hot fluid = Heat gained by cold fluid

$$m_h c_{ph} (T_{hi} - T_{ho}) = m_c c_{pc} (T_{co} - T_{ci})$$

$$5 \times 2000 \times (150 - 100) = 10 \times 4000\,(T_{co} - T_{ci})$$

$$\therefore \quad T_{co} = \mathbf{32.5°C}$$

5.5 (a)

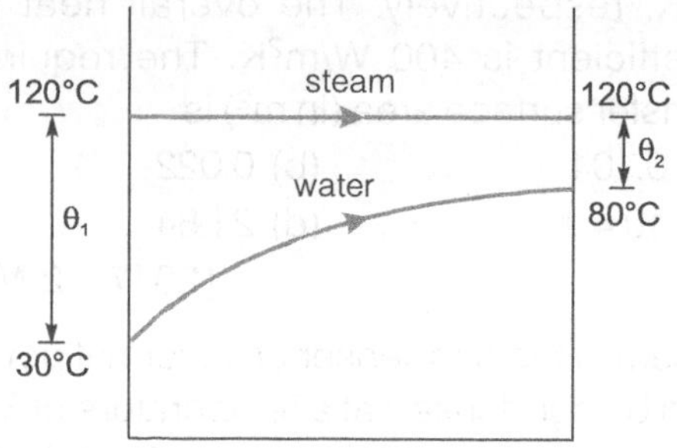

$$\therefore \quad \theta_1 = 120° - 30° = 90°\text{C}$$

$$\theta_2 = 120° - 80° = 40°\text{C}$$

Mass flow rate,

$$m = \frac{1500}{3600} = 0.4167 \text{ kg/s}$$

Log mean temperature difference,

$$\theta_m = \frac{\theta_1 - \theta_2}{\ln\left(\frac{\theta_1}{\theta_2}\right)} = \frac{90-40}{\ln\left(\frac{90}{40}\right)} = 61.65°\text{C}$$

From energy balance,

$$Q = mc_{pw}\Delta T = UA\theta_m$$

$$0.4167 \times 4.187 \times 1000 \times 50 = 2000 \times A \times 61.65$$

or $$A = \mathbf{0.707\ m^2}$$

5.6 (d)

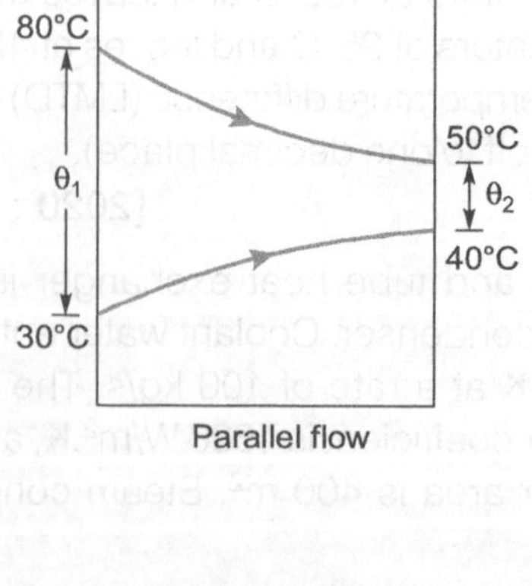

Parallel flow

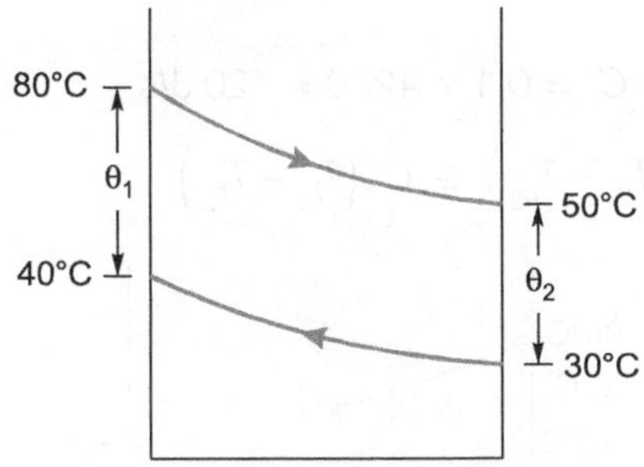

Counter flow heat exchanger

For parallel flow

$$\theta_{mp} = \frac{\theta_1 - \theta_2}{\ln\left(\frac{\theta_1}{\theta_2}\right)} = \frac{(80-30)-(50-40)}{\ln\left(\frac{80-30}{50-40}\right)}$$

$$\theta_{mp} = 24.85°C$$

which is less than consider LMTD *i.e.*, 26°C

For counter flow

$$\theta_{mc} = \frac{\theta_1 - \theta_2}{\ln\left(\frac{\theta_1}{\theta_2}\right)} = \frac{(80-40)-(50-30)}{\ln\left(\frac{80\ \ 40}{50-30}\right)}$$

$$= \frac{40-20}{\ln\left(\frac{40}{20}\right)}$$

$$\theta_{mc} = \mathbf{28.85°C}$$

which is greater than considered LMTD *i.e.*, 26°C.

Hence **cross flow** is better for given problem.

5.7 (b)

Given data:

$T_{h1} = 60°C$; $T_{c2} = 30°C$; $m_h = 1$ kg/s
$m_c = 2$ kg/s; $c_h = 10$ kJ/kgK; $c_c = 5$ kJ/kgK

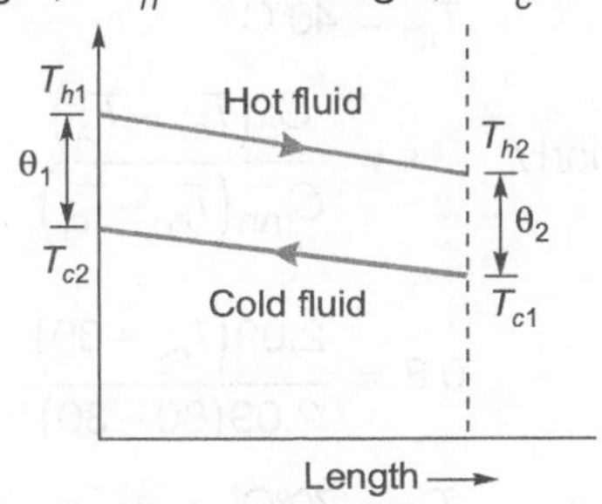

Heat lost by hot fluid = Heat gained by cold fluid

$$m_h c_h (T_{h1} - T_{h2}) = m_c c_c (T_{c2} - T_{c1})$$

$$1 \times 10(60 - T_{h2}) = 2 \times 5\,(30 - T_{c1})$$

or $\quad 60 - T_{h2} = 30 - T_{c1}$

$T_{h2} - T_{c1} = 60 - 30 = 30°C$

If $\quad \theta_1 = \theta_2$

Then $\quad$ LMTD $= \theta_1 = \theta_2 = \mathbf{30°C}$

5.8 (c)

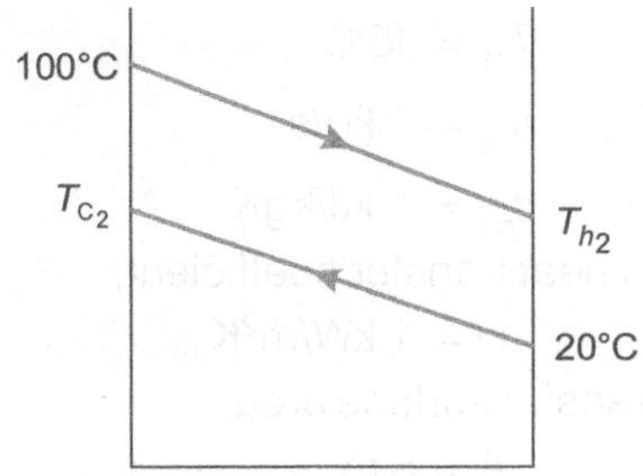

$$m_c = 2m_h$$

$$c_h = 2c_c$$

By energy balance

$$m_h c_h \left(T_{h_1} - T_{h_2}\right) = m_c c_c \left(T_{c_2} - T_{c_1}\right)$$

$$T_{h_1} - T_{h_2} = T_{c_2} - T_{c_1}$$

$$T_{h_1} - T_{c_2} = T_{h_2} - T_{c_1}$$

i.e. $\quad \theta_1 = \theta_2$

$$\text{LMTD} = \frac{\left(T_{h_1} - T_{c_2}\right) - \left(T_{h_2} - T_{c_1}\right)}{\ln\left(\frac{T_{h_1} - T_{c_2}}{T_{h_2} - T_{c_1}}\right)}$$

$$\text{LMTD} = \frac{\theta_1 - \theta_2}{\ln\left(\frac{\theta_1}{\theta_2}\right)} = \theta_m$$

Let, $\quad \frac{\theta_1}{\theta_2} = x$

$$\theta_m = \frac{\theta_2(x-1)}{\ln x}$$

$$\theta_m = \lim_{x \to 1} \frac{\theta_2(x-1)}{\ln x} \quad \text{...(i)}$$

Since $\frac{0}{0}$,

Now applying L'Hospital rule

$$\theta_m = \theta_2 = \theta_1$$

Given $\quad \theta_m = 20°C$

$$T_{h_1} - T_{c_2} = 20°C$$

$$100 - T_{c_2} = 20°C$$

or $\quad T_{c_2} = \mathbf{80°C}$

5.9 (b)

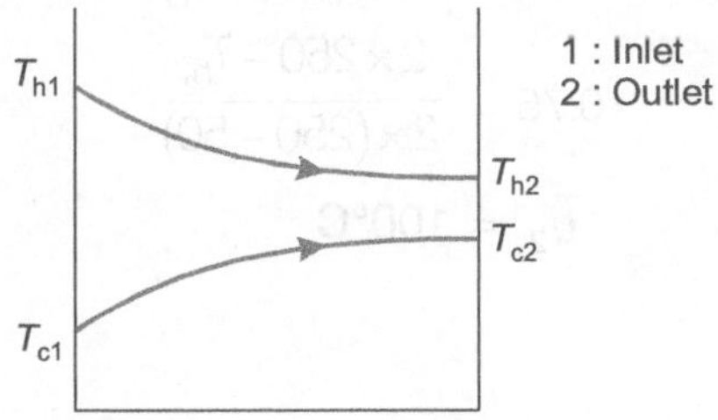

Parallel flow heat exchanger

$$T_{h1} = 102°C$$
$$T_{c1} = 15°C$$
$$m_h = 1 \text{ kg/s}$$
$$c_{ph} = 4 \text{ kJ/kgK}$$

Overall heat transfer coefficient,

$$U = 1 \text{ kW/m}^2\text{K}$$

Heat transfer surface area,

$$A = 5 \text{ m}^2$$

Given:

$$m_c c_{pc} = m_h c_{ph} = 1 \times 4 = 4 \text{ kJ/sK}$$

Neglect heat transfer between the heat exchanger and the ambient.

The heat exchanger is characterized by the following relation

$$2\epsilon = 1 - \exp(-2 \text{ NTU})$$

or

$$\epsilon = \frac{1-\exp(-2\text{ NTU})}{2}$$

$$\text{NTU} = \frac{UA}{C_{min}} = \frac{1\times 5}{4} = 1.25$$

$$\epsilon = \frac{1-\exp(-2\text{ NTU})}{2}$$

$$= \frac{1-\exp(-2\times 1.25)}{2} = 0.46$$

The maximum possible heat transfer rate,

$$Q_{max} = C_{min}(T_{h1} - T_{c1})$$
$$= 4 \times (102 - 15) = 348 \text{ kW}$$

Actual rate of heat transfer,

$$Q = \epsilon Q_{max}$$
$$= 0.46 \times 348 = 160 \text{ kW}$$

also $\quad Q = C_c(T_{c2} - T_{c1})$

$\therefore \quad 160 = 4 \times (T_{c2} - 15)$

or $\quad T_{c2} = \mathbf{55°C}$

5.10 (b)

$$C_h = m_c c_{ph} = 2 \times 1 \times 10^3 = 2 \text{ kJ/Ks}$$
$$C_c = m_c c_{pc} = 1 \times 4 \times 10^3 = 4 \text{ kJ/Ks}$$
$$C_{min} = C_h = 2 \text{ kJ/Ks}$$

$$\varepsilon = \frac{C_h(T_{h_1} - T_{h_2})}{C_{min}(T_{h_1} - T_{c_1})}$$

$$0.75 = \frac{2\times 250 - T_{h_2}}{2\times(250-50)}$$

or $\quad T_{h_2} = \mathbf{100°C}$

5.11 (b)

$$C_h = C_c = 0.1 \times 4200 = 420 \text{ J/s}$$

$$C_c(T_{c_1} - T_{c_2}) = C_h(T_{h_1} - T_{h_2})$$

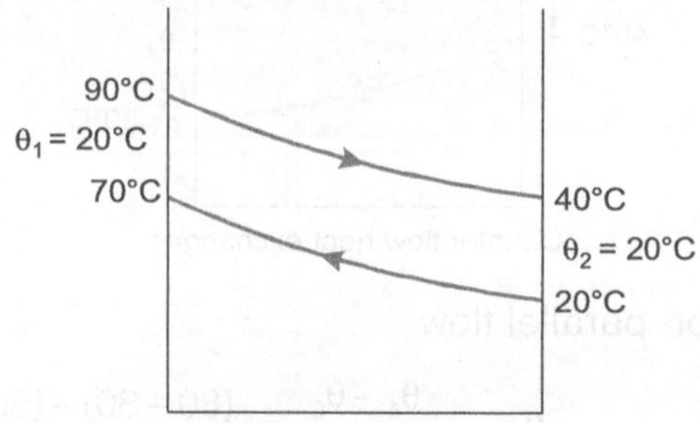

Counter flow heat exchanger

$$(70 - 20) = 90 - T_{h_2}$$

$$T_{h_2} = 40°C$$

$$\text{LMTD} = \theta_1 = \theta_2$$

$$C_c(T_{c_2} - T_{c_1}) = UA\text{ (LMTD)}$$

$$420 \times 50 = A \times 2000 \times 20$$

or $\quad A = \mathbf{0.525 \text{ m}^2}$

5.12 (c)

$$C_c = 2.09 \times 1 = 2.09 \text{ kJ/Ks}$$
$$C_h = 0.5 \times 4.18 = 2.09 \text{ kJ/Ks}$$

$$\varepsilon = \frac{C_h(T_{h_1} - T_{h_2})}{C_{min}(T_{h_1} - T_{c_1})}$$

$$0.8 = \frac{C_h(80 - T_{h_2})}{C_{min}(80-30)}$$

$$T_{h2} = 40°C$$

Similarly, $\quad \varepsilon = \dfrac{C_c(T_{c_2} - T_{c_1})}{C_{min}(T_{h_1} - T_{c_1})}$

$$0.8 = \frac{2.09(T_{c_2} - 30)}{2.09(80-30)}$$

or $\quad T_{c2} = 70°C$

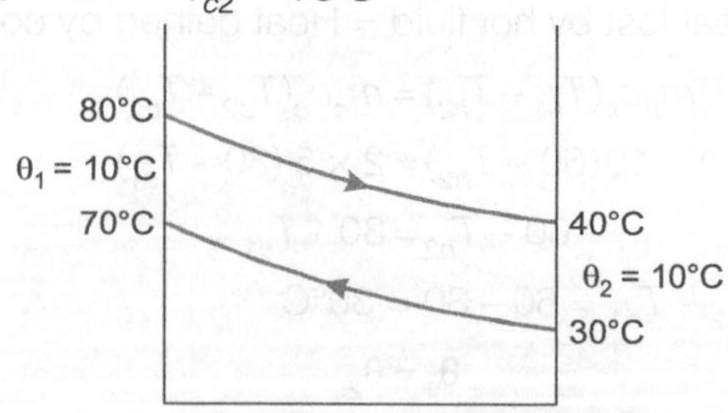

$$\theta_1 = \theta_2 = \text{LMTD} = \mathbf{10°C}$$

5.13 (b)

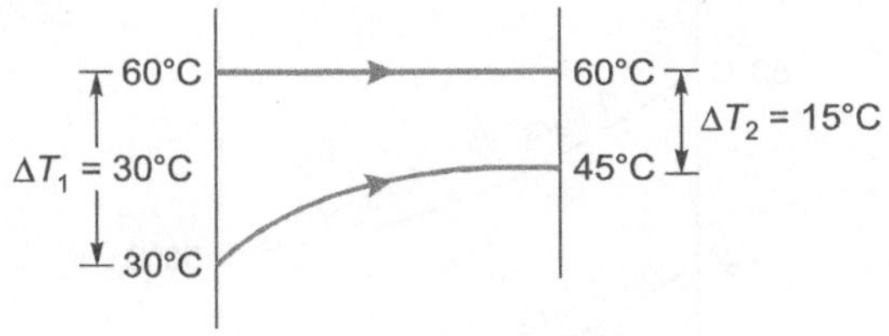

$$\therefore \quad \text{LMTD} = \frac{\Delta T_1 - \Delta T_2}{\ln\left(\dfrac{\Delta T_1}{\Delta T_2}\right)} = \frac{30-15}{\ln(2)} = \mathbf{21.64°C}$$

5.15 Sol.

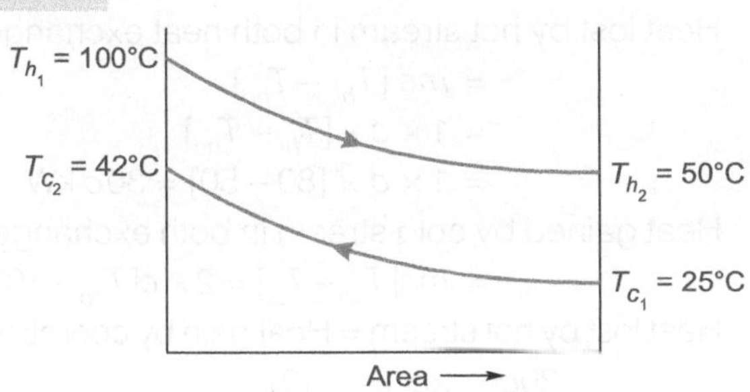

Given data:

c_c = 4178 J/kgK; c_h = 2130 J/kgK

m_c = 1.5 kg/s; m_h = 1 kg/s

$$C_{min} = c_h m_h = 2130 \times 1 = 2130 \text{ W/K}$$

$$C_{max} = c_c m_c = 4178 \times 1.5 = 6267 \text{ W/K}$$

$$\Delta T_1 = 100 - 42 = 58°C$$

$$\Delta T_2 = 50 - 25 = 25°C$$

$$\Delta T_{lm} = \frac{\Delta T_1 - \Delta T_2}{\ln\left(\dfrac{\Delta T_1}{\Delta T_2}\right)} = \frac{58-25}{\ln\left(\dfrac{58}{25}\right)} = 39.21°C$$

$$Q = UA\,\Delta T_{lm}$$

also, $$Q = m_c c_c (T_{c2} - T_{c1})$$

$$\therefore \quad UA\,\Delta T_{lm} = m_c c_c (T_{c2} - T_{c1})$$

$$UA = \frac{4178 \times 1.5(42-25)}{39.21}$$

$$= 2717.138 \text{ W/K}$$

$$\text{NTU} = \frac{\text{UA}}{\text{C}_{\text{min}}} = \frac{2717}{2130} = 1.2756$$

$$C = \frac{C_{\text{min}}}{C_{\text{max}}} = \frac{2130}{6267} = 0.34$$

$$\epsilon = \frac{1-\exp[-NTU\,(1-C)]}{1-C\exp[-NTU\,(1-C)]}$$

$$= \frac{1-\exp[-1.2756 \times 0.66]}{1-0.34\exp[1-2756 \times 0.66]}$$

$$= \frac{1-0.43}{1-0.34 \times 0.43} = \mathbf{0.6676}$$

Alternatively:

$$\epsilon = \frac{T_{h_1} - T_{h_2}}{T_{h_1} - T_{c_1}} = \frac{100-50}{100-25}$$

$$= \frac{50}{75} = \frac{2}{3} = \mathbf{0.6667}$$

5.16 Sol.

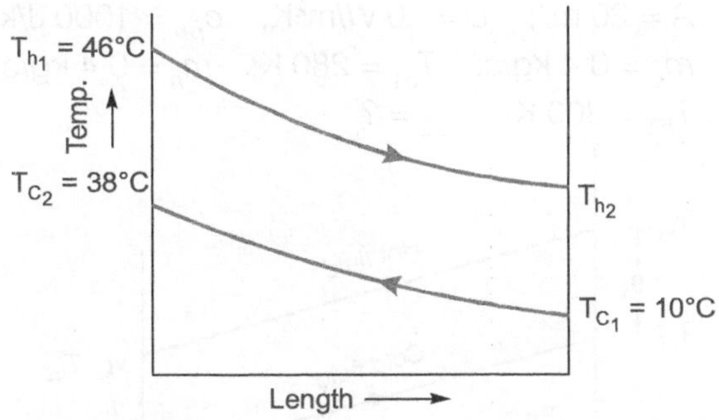

Given data:

T_{h1} = 46°C; T_{h2} = ?; T_{c1} = 10°C; T_{c2} = 38°C

$$\rho_{water} = 1000 \text{ kg/m}^3$$

$$m_c = 19 \text{ L/s}, \; m_h = 25 \text{ L/s}$$

From energy balance,

$$m_c c_c (T_{c2} - T_{c1}) = m_h c_h (T_{h1} - T_{h2})$$

$$19 \times (38 - 10) = 25(46 - T_{h_2})$$

or $$T_{h_2} = 46 - \frac{19 \times 28}{25} = 24.72°C$$

$$\Delta T_1 = T_{h_1} - T_{c_2} = 46 - 38 = 8°C$$

$$\Delta T_2 = T_{h_2} - T_{c_1} = 24.72 - 10 = 14.72°C$$

$$\text{LMTD} = \frac{\Delta T_1 - \Delta T_2}{\ln\left(\dfrac{\Delta T_1}{\Delta T_2}\right)}$$

$$= \frac{8-14.72}{\ln\left(\dfrac{8}{14.72}\right)} = \frac{6.72}{0.60976} = \mathbf{11.02°C}$$

5.17 (d)

For parallel

$$\varepsilon = \frac{1-\exp[-\text{NTU}(1+C_r)]}{1+C_r}$$

For condenser,

$$C_{\text{max}} = \infty$$

$$C_r = 0$$

$$\boldsymbol{\varepsilon} = \frac{1-\exp[-\text{NTU}]}{1}$$

∴ (d) is correct option.

Considering counter-flow heat exchanger

$$\frac{1-\exp[-NTU(1+C_r)]}{1-C_r\exp[-NTU(1+C_r)]}$$

$\therefore \quad C_r = 0$

1 – exp (– NTU)

$\therefore$ Again (d) is correct option.

5.18 (a)

Given data:

$A = 20\text{ m}^3$; $U = 20\text{ W/m}^2\text{K}$; $c_{ph} = 1000\text{ J/kgK}$

$m_c = 0.4\text{ kg/s}$; $T_{c1} = 280\text{ K}$; $m_h = 0.4\text{ kg/s}$

$T_{h1} = 300\text{ K}$; $T_{c2} = ?$

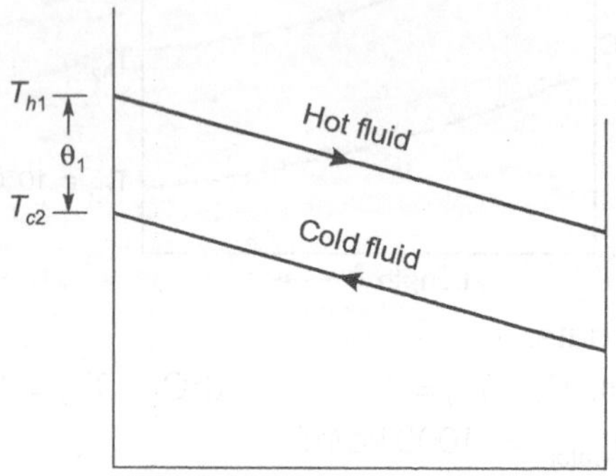

By energy balance,

Heat lost by hot fluid = Heat gained by cold fluid

$$m_h c_{ph}(T_{h1}-T_{h2}) = m_c c_{pc}(T_{c2}-T_{c1})$$

or $\quad T_{h1}-T_{h2} = T_{c2}-T_{c1} \qquad (\because m_h c_{ph} = m_c c_{pc})$

$$T_{h1}-T_{c2} = T_{h2}-T_{c1}$$

$$\theta_1 = \theta_2$$

If $\theta_1 = \theta_2$, then LMTD $= \theta_1 = \theta_2$

$$Q = A\times U\times \text{LMTD}$$

$$Q = A\times U\times\theta_1 = 20\times 20\times(T_{h1}-T_{c2})$$

$$= 400\,(300-T_{c2})$$

also $\quad Q = m_c c_{pc}(T_{c2}-T_{c1})$

$\therefore \quad 400(300-T_{c2}) = 0.4\times 1000\,(T_{c2}-280)$

$$300-T_{c2} = T_{c2}-280$$

or $\quad 2T_{c2} = 580$

or $\quad T_{c2} = 290°\text{C}$

5.19 Sol.

Parallel flow

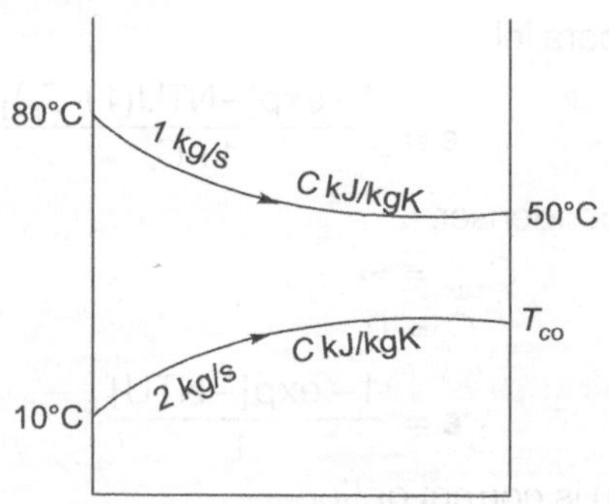

A_P = Area of parallel flow heat exchanger

Counter flow

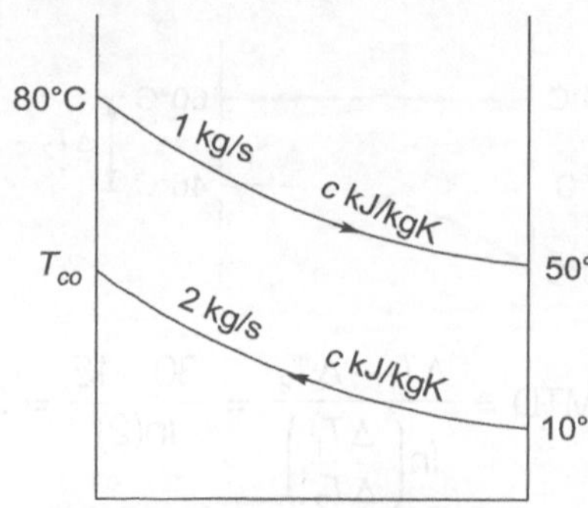

A_C = Area of counter flow heat exchanger

Heat lost by hot stream in both heat exchanger

$$= mc\,[T_{hi}-T_{ho}]$$

$$= 1\times c\times[T_{hi}-T_{ho}]$$

$$= 1\times c\times[80-50] = 30c\text{ kW}$$

Heat gained by cold stream in both exchange

$$= mc[T_{co}-T_{ci}] = 2\times c[T_{co}-10]$$

Heat lost by hot stream = Heat gain by cool stream

$$30c = 2c(T_{co}-10)$$

$$T_{co} = \frac{30}{2}+10 = 15+10 = 25°\text{C}$$

For counter flow,

$$LMTD = \frac{\Delta T_i-\Delta T_e}{\ln(\Delta T_i/\Delta T_e)}$$

$$= \frac{70-25}{\ln[70/25]} = 43.705°\text{C}$$

$$Q_H = UA_P(43.705)$$

For counter flow,

$$LMTD = \frac{\Delta T_i-\Delta T_e}{\ln(\Delta T_i/\Delta T_e)}$$

$$= \frac{55-40}{\ln(55/40)} = 47.103°\text{C}$$

$$Q_H = UA_C(47.103)$$

In order to compare area, we assume both heat exchanger have same heat transfer rate.

$$\therefore UA_P(43.705) = UA_C(47.103)$$

$$\frac{43.705}{47.103} = \frac{A_C}{A_P} = \mathbf{0.927}$$

5.20 (b)

Maximum heat transfer possible

$$= C_{\min}\times(\Delta T)_{\max}$$

5.21 Sol.

Temperature profiles are

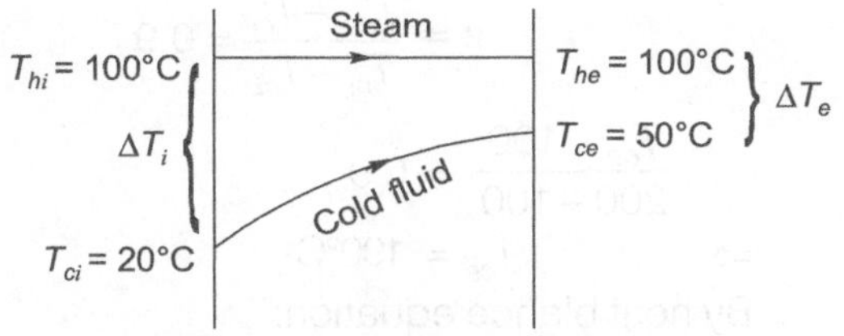

$$\Delta T_i = 100 - 20 = 80°C$$
$$\Delta T_e = 100 - 50 = 50°C$$

(LMTD) (parallel or counter HE)

$$= \frac{\Delta T_i - \Delta T_e}{\ln \frac{\Delta T_i}{\Delta T_e}} = \frac{80 - 50}{\ln \frac{80}{50}}$$

$$= \mathbf{63.82°C}$$

5.22 (d)

H.E. = Counterflow type

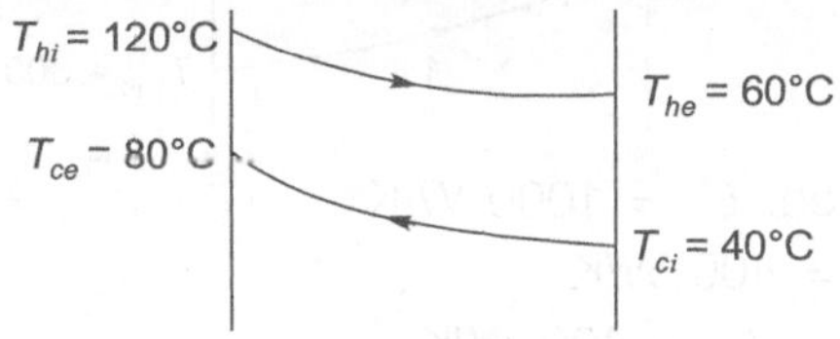

$$\Delta T_i = 120 - 80 = 40°C$$
$$\Delta T_e = 60 - 40 = 20°C$$

$$LMTD = \frac{\Delta T_i - \Delta T_e}{\ln \frac{\Delta T_i}{\Delta T_e}} = \frac{40 - 20}{\ln \frac{40}{20}} = 28.86°C$$

$$Q = \dot{m}_C c_{p_C}(T_{ce} - T_{ci})$$
$$= 1.5 \times 4.2 \times 10^3 \,(80 - 40) \text{ W}$$
$$Q = 252000 \text{ W}$$

also $Q = UA\Delta T_m$

$$A = \frac{Q}{U\Delta T_m} = \frac{1.5 \times 4.2 \times 10^3 \times 40}{400 \times 28.86}$$
$$= \mathbf{21.83\ m^2}$$

5.23 Sol.

Since steam is condensing, the temperature of hot fluid remains constant.

∴ Temperature profiles of hot and cold fluids are

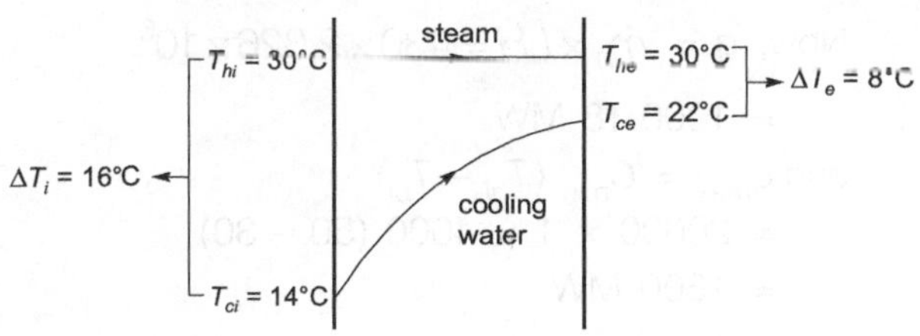

∴ (LMTD) of *HE*

$$= \frac{\Delta T_i - \Delta T_e}{\ln \frac{\Delta T_i}{\Delta T_e}} = \frac{16 - 8}{\ln\left(\frac{16}{8}\right)} = 11.54°C$$

Total heat transfer rate between steam and cooling water.

$$Q = UA(\text{LMTD}) = 2000 \times 50 \times 11.54$$
$$= 11.54 \times 10^5 \text{ W}$$
$$= \mathbf{1.154\ MW}$$

5.24 Sol.

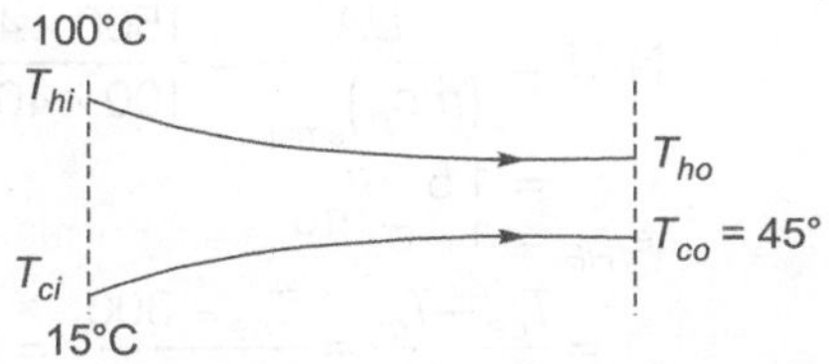

$$C_h = 2000 \text{ W/K}$$
$$C_c = 1200 \text{ W/K}$$

By energy equation,

$$C_h(\Delta T_h) = C_c(\Delta T_c)$$
$$2000\,(100 - T_{ho}) = 1200\,(30)$$
$$T_{ho} = 82°C$$

$$\text{LMTD} = \frac{(100-15)-(82-45)}{\ln\left(\frac{100-15}{82-45}\right)} = 57.71°C$$

Since LMTD is temperature difference so the answer will be 57.71 K

5.25 Sol.

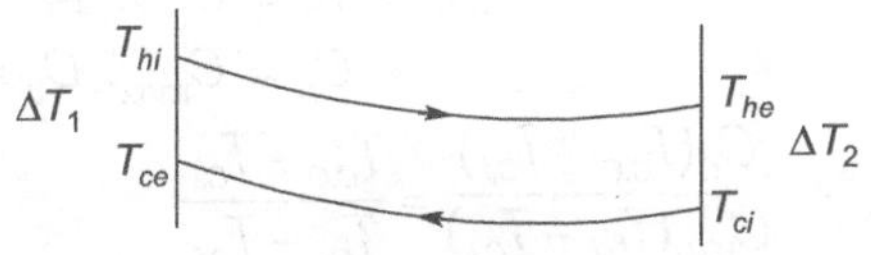

$$T_{hi} = 102°C, \quad T_{he} = 65°C$$
$$T_{c,i} = 25°C, \quad T_{ce} = 42°C$$

$$\text{LMTD} = \frac{\Delta T_1 - \Delta T_2}{\ln\left(\frac{\Delta T_1}{\Delta T_2}\right)};$$

For counter flow heat exchanger:

$$\Delta T_1 = 102°C - 42°C = 60°C;$$
$$\Delta T_2 = 65°C - 25°C = 40°C;$$

$$= \frac{60 - 40}{\ln\left(\frac{60}{40}\right)} = \frac{20}{0.4054}$$

$$= 49.3°C$$

5.26 Sol.

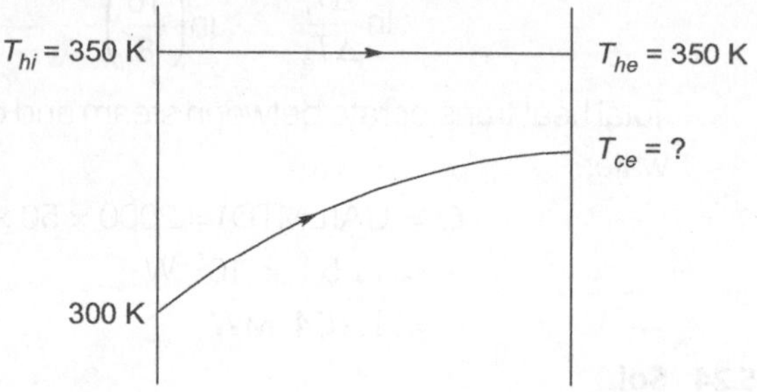

$$NTU = \frac{UA}{(\dot{m}c_p)_{small}} = \frac{1500 \times 400}{100 \times 4000} = 1.5$$

$$\in_{HE} = 1 - e^{-NTU}$$

$$= \frac{T_{ce} - T_{ci}}{T_{hi} - T_{ci}} = \frac{T_{ce} - 300}{350 - 300} = 0.7768$$

$$T_{ce} = 338.84 \text{ K} \simeq 339 \text{ K}$$

5.27 (0.42) (0.40 to 0.44)

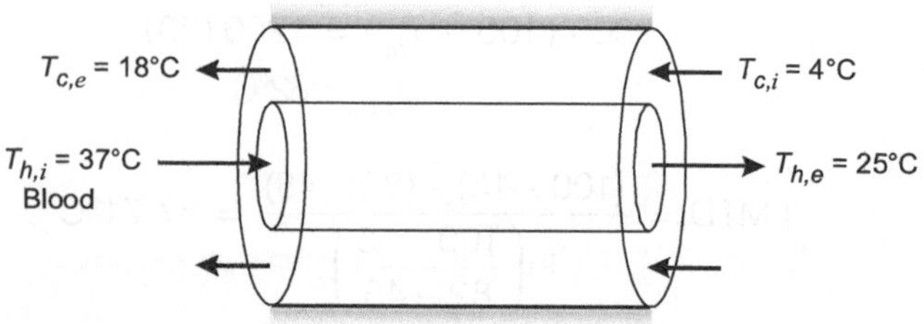

Applying energy balance

$$C_h(T_{h,i} - T_{h,e}) = C_c(T_{c,e} - T_{c,i})$$

$$C_h(37 - 25) = C_c(18 - 4)$$

$$C_h \times 12 = C_c \times 14$$

$$C_h = C_{max};\ C_c = C_{min}$$

$$\varepsilon = \frac{C_c(T_{c,e} - T_{c,i})}{C_{min}(T_{h,i} - T_{c,i})} = \frac{T_{c,e} - T_{c,i}}{T_{h,i} - T_{c,i}}$$

$$\varepsilon = \frac{18-4}{37-4} = \frac{14}{33} = 0.4242 = 0.42$$

5.28 (1000) (1000 to 1000)

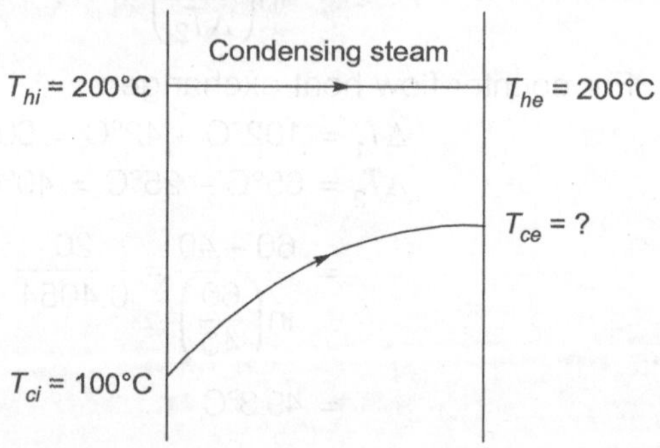

For exit temperature of cold fluid:

$$(\dot{m}C_P)_{\text{hot fluid}} \to \infty$$

$$\therefore \quad C = 0$$

$$\varepsilon = \frac{T_{ce} - T_{ci}}{T_{hi} - T_{ci}} = 0.9$$

$$\frac{T_{ce} - 100}{200 - 100} = 0.9$$

$$\Rightarrow \quad T_{ce} = 190°C$$

By heat blance equation:

$$\dot{m}_{\text{condensing}} \times h_{fg} = \dot{m}_c C_{Pc} \times (T_{ce} - T_{ci})$$

$$150 \times 2400 = \dot{m}_c \times 4.0 \times (190 - 100)$$

Mass flow rate of cold fluid, $\dot{m}_c$ = 1000 kg/s

5.29 (10) (9.999 to 10.001)

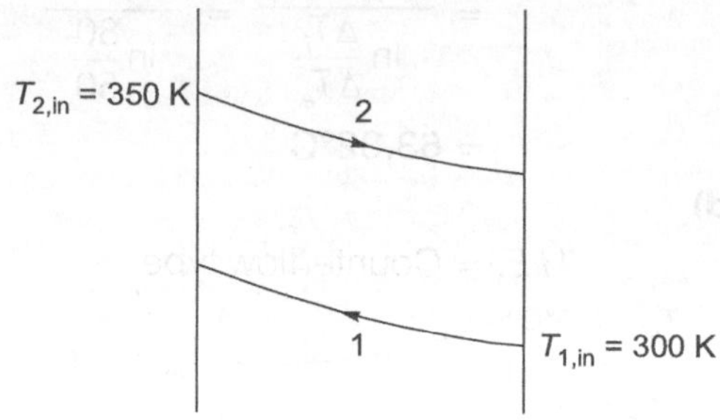

Given: C_1 = 1000 W/K

C_2 = 400 W/K

$\therefore$ $C_{min} = C_2$ = 400 W/K

Effectiveness, $\in = \frac{\dot{q}_{actual}}{\dot{q}_{max}}$

$$\in = \frac{\dot{q}_{actual}}{C_{min}(T_{2,in} - T_{1,in})} = 0.5$$

$$q_{actual} = 0.5 \times 400\ (350 - 300)$$

$$= 10000 \text{ Watt} = 10 \text{ kW}$$

5.30 (0.625) (0.590 to 0.633)

Given: $\dot{m}_h$ = 430 kg/s, LH = 2.326 MJ/kg

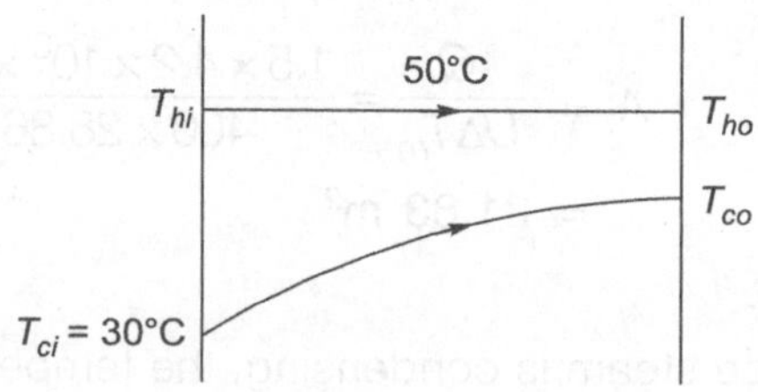

Effectiveness, $\in = \frac{q}{q_{max}}$

Now, $q = \dot{m}_h \times LH = 430 \times 2.326 \times 10^6$

$= 1000.18$ MW

and $q_{max} = C_{min}\ (T_{hi} - T_{ci})$

$= 20000 \times 1 \times 4000\ (50 - 30)$

$= 1600$ MW

$$\therefore \in = \frac{1000.18}{1600} = 0.625$$

■■■■

6 Fins and Unsteady Heat Transfer (Conduction and Convection)

6.1 Lumped heat transfer analysis of a solid object suddenly exposed to a fluid medium at a different temp is valid when
(a) Biot number < 0.1
(b) Biot number > 0.1
(c) Fourier number < 0.1
(d) Fourier number > 0.1 **[2001 : 1 M]**

6.2 The value of Biot number is very small (less than 0.01), when
(a) The convective resistance of fluid is negligible
(b) The conductive resistance of fluid is negligible
(c) The conductive resistance of solid is negligible
(d) None of the above **[2002 : 1 M]**

6.3 A spherical thermocouple junction of diameter 0.706 mm is to be used for the measurement of temperature of a gas stream. The convective heat transfer coefficient on bead surface is 400 W/m^{2}K. Thermophysical properties of thermocouple material are k = 20 W/mK, c = 400 J/kgK and ρ = 8500 kg/m^3. If the thermocouple initially at 30°C is placed in a hot stream of 300°C, the time taken by the bead to reach 298°C, is
(a) 2.35 s (b) 4.9 s
(c) 14.7 s (d) 29.4 s **[2004 : 2 M]**

6.4 A small copper ball of 5 mm diameter at 500 K is dropped into an oil bath whose temperature is 300 K. The thermal conductivity of copper is 400 W/mK, its density 9000 kg/m^3 and its specific heat 385 J/kgK. If the heat transfer coefficient is 250 W/m^{2}K and lumped analysis is assumed to be valid, the rate of fall of the temperature of the ball at the beginning of cooling will be, in K/s,
(a) 8.7 (b) 13.9
(c) 17.3 (d) 27.7 **[2005 : 2 M]**

6.5 The average heat transfer coefficient on a thin hot vertical plate suspended in still air can be determined from observations of the change in plate temperature with time as it cools. Assume the plate temperature to be uniform at any instant of time and radiation heat exchange with the surroundings negligible. The ambient temperature is 25°C, the plate has a total surface area of 0.1 m^2 and a mass of 4 kg. The specific heat of the plate material is 2.5 kJ/kgK. The convective heat transfer coefficient in W/m^{2}K, at the instant when the plate temperature is 225°C and the change in plate temperature with time dT/dt = – 0.02 K/s, is
(a) 200 (b) 20
(c) 15 (d) 10 **[2007 : 2 M]**

6.6 Consider steady-state heat conduction across the thickness in a plane composite wall (as shown in the figure) exposed to convection conditions on both sides.
Given: h_i = 20 W/m^{2}K; h_o = 50 W/m^{2}K; $T_{\infty,i}$ = 20°C; $T_{\infty,o}$ = –2°C; k_1 = 20 W/mK; k_2 = 50 W/mK; L_1 = 0.30 m and L_2 = 0.15 m
Assuming negligible contact resistance between the wall surfaces, the interface temperature, T (in °C), of the two walls will be

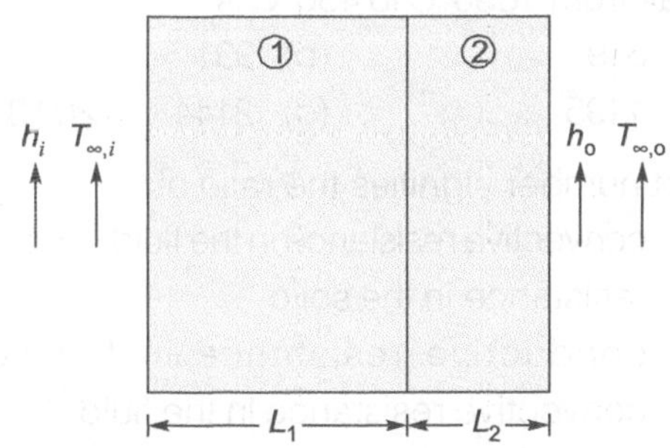

(a) – 0.50 (b) 2.75
(c) 3.75 (d) 4.50 **[2009 : 2 M]**

6.7 A fin has 5 mm diameter and 100 mm length. The thermal conductivity of fin material is 400 Wm^{-1}K^{-1}. One end of the fin is maintained at 130°C and its remaining surface is exposed to ambient air at 30°C. If the convective heat transfer coefficient is 40 Wm^{-2}K^{-1}, the heat loss (in W) from the fin is
(a) 0.08 (b) 5.0
(c) 7.0 (d) 7.8 **[2010 : 2 M]**

6.8 A spherical steel ball of 12 mm diameter is initially at 1000 K. It is slowly cooled in a surrounding of 300 K. The heat transfer coefficient between the steel ball and the surroundings is 5 W/m^{2}K. The thermal conductivity of steel is 20 W/mK. The temperature difference between the centre and the surface of the steel ball is

(a) large because conduction resistance is far higher than the convective resistance.
(b) large because conduction resistance is far less than the convective resistance.
(c) small because conduction resistance is far higher than the convective resistance.
(d) small because conduction resistance is far less than the convective resistance. [2011 : 2 M]

6.9 Which one of the following configurations has the highest fin effectiveness?
(a) Thin, closely spaced fins
(b) Thin, widely spaced fins
(c) Thick, widely spaced fins
(d) Thick, closely spaced fins [2012 : 1 M]

6.10 A steel ball of diameter 60 mm is initially in thermal equilibrium at 1030°C in a furnace. It is suddenly removed from the furnace and cooled in ambient air at 30°C, with convective heat transfer coefficient $h = 20$ W/m^{2}K. The thermo-physical properties of steel are : density ρ = 7800 kg/m^2, conductivity k = 40 W/mK and specific heat c = 600 J/kgK. The time required in seconds to cool the steel ball in air from 1030°C to 430°C is
(a) 519 (b) 931
(c) 1195 (d) 2144 [2013 : 2 M]

6.11 Biot number signifies the ratio of
(a) convective resistance in the fluid to conductive resistance in the solid
(b) conductive resistance in the solid to convective resistance in the fluid
(c) inertia force to viscous force in the fluid
(d) buoyancy force to viscous force in the fluid [2014 : 1 M, Set-1]

6.12 A steel ball of 10 mm diameter at 1000 K is required to be cooled to 350 K by immersing it in a water environment at 300 K. The convective heat transfer coefficient is 1000 W/m^{2}K. Thermal conductivity of steel is 40 W/mK. The time constant for the cooling process t is 16 s. The time required (in s) to reach the final temperature is_______. [2016 : 2 M, Set-1]

6.13 Two cylindrical shafts A and B at the same initial temperature are simultaneously placed in a furnace. The surfaces of the shafts remain at the furnace gas temperature at all times after they are introduced into the furnace. The temperature variation in the axial direction of the shafts can be assumed to be negligible. The data related to shafts A and B is given in the following Table.

Quantity	Shaft A	Shaft B
Diameter (m)	0.4	0.1
Thermal Conductivity (W/mK)	40	20
Volumetric heat capacity (J/m^{3}K)	2×10^6	2×10^7

The temperature at the centerline of the shaft A reaches 400°C after two hours. The time required (in hours) for the centerline of the shaft B to attain the temperature of 400°C is _______. [2016 : 2 M, Set-2]

6.14 A cylindrical steel rod, 0.01 m in diameter and 0.2 m in length is first heated to 750°C and then immersed in a water bath at 100°C. The heat transfer coefficient is 250 W/m^{2}K. The density, specific heat and thermal conductivity of steel are ρ = 7801 kg/m^3, c = 473 J/kgK, and k = 43 W/mK, respectively. The time required for the rod to reach 300°C is _______ seconds. [2016 : 2 M, Set-3]

6.15 The heat loss from a fin is 6 W. The effectiveness and efficiency of the fin are 3 and 0.75 respectively. The heat loss (in W) from the fin, keeping the entire fin surface at base temperature is _______. [2017 : 1 M, Set-2]

6.16 A metal ball of diameter 60 mm is initially at 220°C. The ball is suddenly cooled by an air jet of 20°C. The heat transfer coefficient is 200 W/m^{2}K. The specific heat, thermal conductivity and density of the metal ball are 400 J/kgK, 400 W/mK and 9000 kg/m^3, respectively. The ball temperature (in °C) after 90 seconds will be approximately
(a) 141 (b) 163
(c) 189 (d) 210 [2017 : 2 M, Set-2]

6.17 A small metal bead (radius 0.5 mm), initially at 100°C, when placed in a stream of fluid at 20°C, attains a temperature of 28°C in 4.35 seconds. The density and specific heat of the metal are 8500 kg/m^3 and 400 J/kgK, respectively. If the bead is considered as lumped system, the convective heat transfer coefficient (in W/m^{2}K) between the metal bead and the fluid stream is
(a) 149.9 (b) 449.7
(c) 283.3 (d) 299.8 [2020 : 2 M, Set-1]

6.18 An infinitely long pin fin, attached to an isothermal hot surface, transfers heat at a steady rate of $\dot{Q}_1$ to the ambient air. If the thermal conductivity of the fin material is doubled, while keeping everything else constant, the rate of steady-rate heat transfer from the fin becomes $\dot{Q}_2$. The ratio $\frac{\dot{Q}_2}{\dot{Q}_1}$ is

(a) 2 (b) $\frac{1}{2}$

(c) $\frac{1}{\sqrt{2}}$ (d) $\sqrt{2}$

[2021 : 1 M, Set-1]

6.19 A hot steel spherical ball is suddenly dipped into a low temperature oil bath.
Which of the following dimensionless parameters are required to determine instantaneous center temperature of the ball using a Heisler chart?

(a) Nusselt number and Grashoff number
(b) Reynolds number and Prandtl number
(c) Biot number and Fourier number
(d) Biot number and Froude number

[2021 : 1 M, Set-1]

Answers Fins and Unsteady Heat Transfer (Conduction and Convection)

6.1	(a)	6.2	(c)	6.3	(b)	6.4	(c)	6.5	(d)	6.6	(c)	6.7	(b)	6.8	(d)
6.9	(a)	6.10	(d)	6.11	(b)	6.12	(42.23)	6.13	(2.5)	6.14	(43.5)	6.15	(8)	6.16	(a)
6.17	(d)	6.18	(d)	6.19	(c)										

Explanations Fins and Unsteady Heat Transfer (Conduction and Convection)

6.2 (c)

Biot number is defined as

$$\text{Bi} = \frac{R_{cond}}{R_{conv}}$$

The value of Biot number is very small (less than 0.01), when the conductive resistance of solid is negligible.

6.3 (b)

Given data:
$h = 400$ W/m²K; $k = 20$ W/mK; $c = 400$ J/kgK
$\rho = 8500$ kg/m³; $T(t) = 298$°C

Now,
$$L = \frac{V}{A} = \frac{1/6\pi D^3}{\pi D^2} = \frac{D}{6} = \frac{0.706\times10^{-3}}{6}$$
$$= 1.76 \times 10^{-4} \text{ m}$$

Biot number,
$$\text{Bi} = \frac{hL}{k} = \frac{400\times1.76\times10^{-4}}{20}$$
$$= 0.0023$$

Since Bi < 0.1

Lumped system analysis can be used

$$\frac{T-T_\infty}{T_i-T_\infty} = e^{-\left(\frac{hAt}{\rho cV}\right)}$$

$$\frac{298-300}{30-300} = e^{-\left(\frac{400\times6\times t}{8500\times0.706\times10^{-3}\times400}\right)}$$
$$7.407\times10^{-3} = e^{-t}$$

Taking $\log_e$ both sides, we get

$\log_e 7.407 \times 1 = -t$

or $t = \mathbf{4.90\ seconds}$

6.4 (c)

Given data:
$k = 400$ W/mK; $T_i = 500$ K; $\rho = 9000$ kg/m³
$T_\infty = 300$ K; $c_p = 385$ J/kgK; $h = 250$ W/m²K

$$L_c = \frac{V}{A} = \frac{D}{6} = \frac{0.005}{6}$$
$$= 8.33\times10^{-4} \text{ m}$$

$$T = T_\infty + (T_i - T_\infty)\,e^{\frac{-hAt}{\rho cV}}$$

$$\frac{dT}{dt} = \frac{-hA}{\rho cV}(T_i - T_\infty)e^{\frac{-hAt}{\rho cV}}$$

$$\left.\frac{dT}{dt}\right|_{t=0} = \frac{hA}{\rho cV}(T_i - T_\infty) \quad \text{[magnitude]}$$

$$= \frac{250 \times 200}{9000 \times 385 \times 8.33 \times 10^{-4}}$$

$$\left.\frac{dT}{dt}\right|_{t=0} = \mathbf{17.31\ K/s}$$

6.5 (d)

Given data:

$$A = 0.1\ m^2; \quad m = 4\ kg$$

$$c_p = 2.5\ kJ/kgK$$

Net rate of heat transfer

= Rate of change of internal energy

$$hA(T - T_f) = -mc_p \frac{dT}{dt}$$

$$h \times 0.1\,(225 - 25)$$

$$= -4 \times 2.5 \times 1000 \times (-0.02)$$

$$h = \frac{4 \times 2.5 \times 1000 \times 0.02}{0.1 \times 200}$$

$$= \mathbf{10\ W/m^2K}$$

6.6 (c)

Given data:

$h_i = 20\ W/m^2K$; $h_o = 50\ W/m^2K$; $T_{\infty, o} = 20°C$

$k_1 = 20\ W/mK$; $k_2 = 50\ W/mK$;

$L_1 = 0.30\ m$; $L_2 = 0.15\ m$

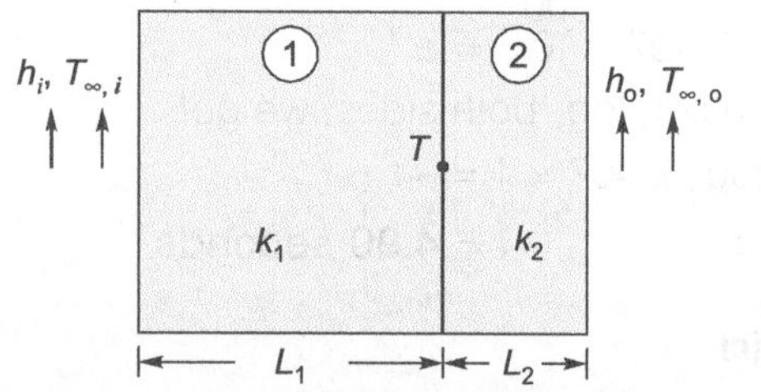

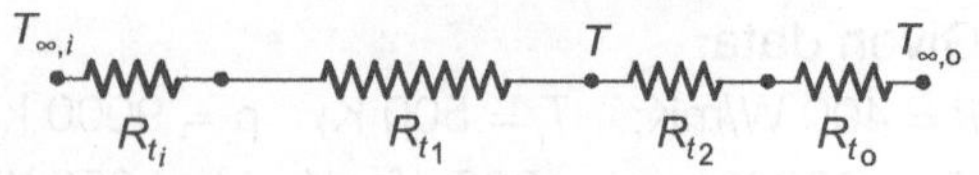

Heat flow:

$$Q = \frac{T_{\infty,i} - T_{\infty,O}}{R_{t_i} + R_{t_1} + R_{t_2} + R_{t_o}}$$

$$Q = \frac{T_{\infty,i} - T_{\infty,O}}{\frac{1}{h_i A} + \frac{L_1}{k_1 A} + \frac{L_2}{k_2 A} + \frac{1}{h_o A}}$$

$$\frac{Q}{A} = \frac{T_{\infty,i} - T_{\infty,O}}{\frac{1}{h_i} + \frac{L_1}{k_1} + \frac{L_2}{k_2} + \frac{1}{h_o}}$$

$$= \frac{20 - (-2)}{\frac{1}{20} + \frac{0.30}{20} + \frac{0.15}{50} + \frac{1}{50}}$$

$$= \frac{20 + 2}{\frac{5 + 1.50 + 0.3 + 2}{100}} = \frac{22 \times 100}{8.8} = 250\ W$$

also $$\frac{Q}{A} = \frac{T_{\infty,i} - T}{\frac{1}{h_i} + \frac{L_1}{k_1}}$$

$$\therefore \quad 250 = \frac{20 - T}{\frac{1}{20} + \frac{0.3}{20}} = \frac{20 - T}{\frac{0.3}{20}}$$

$$250 \times \frac{0.3}{20} = 20 - T$$

$$16.25 = 20 - T$$

or $$T = 20 - 16.25 = \mathbf{3.75°C}$$

6.7 (b)

Given data:

Diameter of fin, $d = 5\ mm = 0.005\ m$

$\therefore$ Cross-sectional area,

$$A = \frac{\pi}{4} d^2 = \frac{3.14}{4} \times (0.005)^2$$

$$= 1.96 \times 10^{-5}\ m^2$$

Length of fin,

$$L = 100\ mm = 0.1\ m$$

Thermal conductivity,

$$k = 400\ W/mK$$

$$T_0 = 130°C$$

$$T_\infty = 30°C$$

$$h = 40\ W/m^2K$$

Perimeter: $P = \pi d = 3.14 \times 0.005 = 0.0157\ m$

$$m = \sqrt{\frac{hP}{kA}} = \sqrt{\frac{40 \times 0.0157}{400 \times 1.95 \times 10^{-5}}}$$

$$= 8.97\ m^{-1}$$

$$mL = 8.97 \times 0.1 = 0.897$$

For long fin, $mL > 2.64$

So, the fin is insulated tip,

$$Q_{fin} = \sqrt{hPkA}\,(T_0 - T_\infty) \tanh mL$$

$$= \sqrt{40 \times 0.0157 \times 400 \times 1.95 \times 10^{-5}}\ (130 - 30) \tanh (8.97 \times 0.1)$$

$$= 0.0699 \times 100 \times \tanh 0.897$$

$$= 0.0699 \times 100 \times 0.7148 = 4.99\ W \simeq \mathbf{5\ W}$$

6.8 (d)

Biot number,

$$Bi = \frac{\text{Conduction resistance: } \frac{l}{kA}}{\text{Convection resistance: } \frac{1}{hA}} = \frac{hl}{k}$$

where $$l = \frac{V}{A} = \frac{\pi}{6}\frac{d^3}{\pi d^2} = \frac{d}{6}$$

$$\therefore \quad Bi = \frac{h}{k}\frac{d}{6} = \frac{5\times 0.012}{20\times 6} = \mathbf{0.0005}$$

It means that the conduction resistance is far less than convective resistance

6.9 (a)

Fin effectiveness,

$$\varepsilon = \frac{\text{Heat transfer with fin: } Q_{Fin}}{\text{Heat transfer without fin: } Q} = \frac{Q_{Fin}}{Q}$$

where $Q_{Fin} = \sqrt{kPhP}\,\theta_o$ for long fin

$$Q = hA\theta_o$$

$$\therefore \quad \varepsilon = \frac{\sqrt{kAhP}\,\theta_o}{hA\theta_o} = \frac{\sqrt{kAhP}}{hA} = \sqrt{\frac{kP}{hA}}$$

$$\varepsilon \propto \sqrt{P}; \quad \propto \frac{1}{\sqrt{A}}$$

For highest fin effectiveness, perimeter of fin should be large and cross-sectional area should be low. It means, for large value of perimeter, fins should be closely spaced (to increase the total number of fins) and for lower value of cross-sectional area, it should be thin.

6.10 (d)

Diameter of ball,

$$d = 60 \text{ mm} = 0.06 \text{ m}$$

Initial temperature,

$$T_i = 1030°C$$

Ambient temperature,

$$T_o = 30°C$$

Convective heat transfer coefficient,

$$h = 20 \text{ W/m}^2\text{K}$$

Density: $\rho = 7800 \text{ kg/m}^3$

Conductivity: $k = 40$ W/mK

Specific heat: $c = 600$ J/kgK

Final temperature of steel ball,

$$T = 430°C$$

Time: $t = ?$

The temperature distribution relation is given by

$$\frac{T - T_o}{T_i - T_o} = e^{-\frac{ht}{\rho c l}} \quad \text{...(i)}$$

where $l = \frac{V}{A}$ characteristic length of balls

$$= \frac{\pi}{6}\frac{d^3}{\pi d^2}$$

$$\because \quad V = \frac{\pi}{6}d^3, A = \pi d^2$$

$$\therefore \quad l = \frac{d}{6} = \frac{0.06}{6} = 0.01 \text{ m}$$

Substituting the values of T, T_o, T_i, h, ρ, c and l in Eq. (i), we get

$$\frac{430-30}{1030-30} = e^{\frac{-20\times t}{7800\times 600\times 0.01}}$$

$$0.4 = e^{-4.273\times 10^{-4}\,t}$$

Taking $\log_e$ both sides, we get

$$\log_e 0.4 = -4.273\times 10^{-4}\times t\times 1$$

or $$t = \frac{0.9162}{4.273\times 10^{-4}} = \mathbf{2144.16\ s}$$

6.11 (b)

Biot number,

$$Bi = \frac{hl}{k} = \frac{\frac{l}{kA}}{\frac{1}{hA}}$$

$$Bi = \frac{\text{Conductive resistance of solid}}{\text{Convective resistance of fluid}}$$

6.12 Sol.

Diameter of ball = 10 mm

$$T_{initial} = 1000 \text{ K}$$
$$T_{final} = 350 \text{ K}$$
$$T_\infty = 300 \text{ K}$$
$$h = 1000 \text{ W/m}^2\text{K}$$
$$k = 40 \text{ W/mK}$$

Time constant = 16 s

Time required = t s

$$\frac{T - T_\infty}{T_i - T_\infty} = e^{-\frac{hAt}{\rho vc}}$$

But $$\frac{\rho vc}{hA} = \text{Time constant}$$
$$= 16 \text{ second}$$

$$\frac{350-300}{1000-300} = e^{-\frac{1}{16}\times t}$$

$$0.07143 = e^{-t/16}$$

Taking $\log_e$ both sides, we get

$$\log_e 0.07143 = \frac{-t}{16}\log_e e \qquad (\because \log_e e = 1)$$

$$-2.639 = \frac{-t}{16}$$

or $$t = \mathbf{42.23\ s}$$

6.13 Sol.

$$T_{(r,t)} = T_0 + (T_\infty - T_0)\,\text{erf}(z)$$

$$\frac{T_{(r,t)} - T_0}{T_\infty - T_0} = \text{erf}(z)$$

$$\frac{T_{(r,t)} - T_0}{T_\infty - T_0} = \text{erf(z)}$$

For shaft A $\quad \dfrac{T_{0.2,t} - T_i}{400 - T_i}$

For shaft B $\quad \dfrac{T_{0.05,t} - T_i}{400 - T_i}$

Hence error function of both shafts will be same.

$$\therefore \quad z_A = z_B$$

$$\frac{r_A}{2\sqrt{\alpha_A t_A}} = \frac{r_B}{2\sqrt{\alpha_B t_B}}$$

$$\frac{r_A}{\sqrt{\alpha_A t_A}} = \frac{r_B}{\sqrt{\alpha_B t_B}}$$

$$\frac{r_A^2}{\alpha_A t_A} = \frac{r_B^2}{\alpha_B t_B}$$

$$t_B = \left(\frac{r_B}{r_A}\right)^2 \times \frac{\alpha_A}{\alpha_B} \times t_A$$

$$\alpha_A = \frac{k}{\rho c_p} = \frac{40}{2\times10^6} = \frac{20}{10^6}\ \text{m}^2/\text{s}$$

$$\alpha_B = \frac{k}{\rho c_p} = \frac{20}{2\times10^7} = \frac{1}{10^6}\ \text{m}^2/\text{s}$$

$$t_B = \left(\frac{r_B}{r_A}\right)^2 \times \frac{\alpha_A}{\alpha_B} \times t_A$$

$$= \left(\frac{0.1}{0.4}\right)^2 \times \frac{\dfrac{20}{10^6}}{\dfrac{1}{10^6}} \times 2$$

$$= \frac{1}{16} \times 20 \times 2 = \textbf{2.5 hrs.}$$

6.14 Sol.

Given data:

$D = 0.01$ m; $\quad L = 0.2$ m; $\quad T_i = 750°\text{C}$

$T_\infty = 100°\text{C}$; $\quad h = 250\ \text{W/m}^2\text{K}$; $\quad \rho = 7801\ \text{kg/m}^3$

$c = 473$ J/kgK

$$\frac{T - T_\infty}{T_i - T_\infty} = e^{-\frac{hAt}{\rho Vc}}$$

where $\quad \dfrac{V}{A} = \dfrac{\pi D^2 L}{4 \times \pi D L} = \dfrac{D}{4} = \dfrac{0.01}{4}$

$$= 2.5 \times 10^{-3}\ \text{m}$$

$$\therefore \quad \frac{300 - 100}{750 - 100} = e^{\frac{-250 \times t}{7801 \times 2.5 \times 10^{-3} \times 473}}$$

$$\frac{200}{650} = e^{-0.0271t}$$

$$0.30769 = e^{-0.0271t}$$

Taking $\log_e$ both sides, we get

$$\log_e 0.30769 = -0.0271t \log_e e$$

$$(\because \log_e e = 1)$$

$$-1.1786 = -0.027t$$

or $\quad t = \dfrac{1.1786}{0.027} = \textbf{43.5 s}$

6.15 Sol.

q_{actual} from fin = 6 watt

$\eta_{\text{fin}} = \dfrac{q_{\text{actual}}}{q_{\text{max. possible}}}$ i.e. when entire fin at base temperature

$\therefore$ q from fin if entire fin at base temperature

$$= \frac{q_{\text{act.}}}{0.75} = \frac{6}{0.75} = \textbf{8 W}$$

6.16 (a)

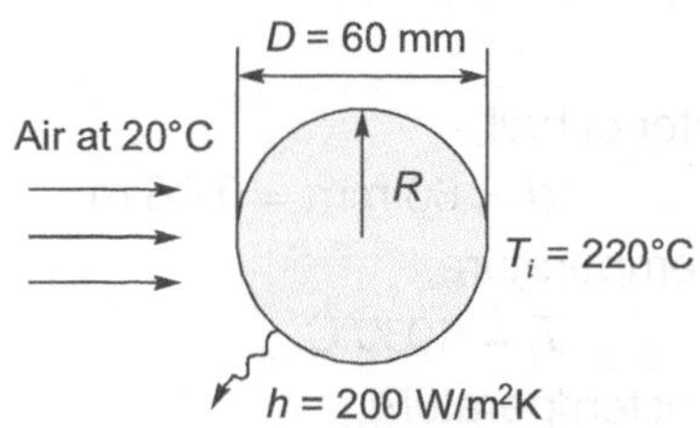

$c_p = 400$ J/kgK
$k = 400$ W/mK
$\rho = 9000\ \text{kg/m}^3$
Time $(\tau) = 90$ sec

Since K being high and size of ball being small, lumped heat analysis is valid.

$$\Rightarrow \quad e^{\left(\frac{hA}{\rho V c_p}\right)\tau} = \frac{T_i - T_\infty}{T - T_\infty}$$

$$\Rightarrow \quad \left(\frac{hA}{\rho V c_p}\right)\tau = \ln\left(\frac{T_i - T_\infty}{T - T_\infty}\right)$$

Put $\quad \dfrac{A}{V} = \dfrac{3}{R} = \dfrac{3}{\left(\dfrac{30}{1000}\right)}$

$$\left(\frac{hA}{\rho v c_p}\right)\tau = \left(200 \times \frac{3}{0.03} \times \frac{1}{9000} \times \frac{1}{400}\right) \times 90$$

$$= 0.5$$

$$\frac{220-20}{T-20} = e^{0.5} = 1.6487$$

$$T = \mathbf{141.3°C}$$

6.17 (d)

r = 0.5 mm; c_p = 400 J/kgK;
ρ = 8500 kg/m^3; t = 4.35 sec
T_0 = 100°C, T_∞ = 20°C
T = 28°C, τ = 4.35 seconds

$$\because \quad \frac{V}{A} = \frac{\frac{4}{3}\pi R^3}{4\pi R^2} = \frac{R}{3}$$

$$\because \quad \frac{\theta}{\theta_0} = \frac{T-T_\infty}{T_i - T_\infty} = e^{\frac{-hA\tau}{\rho c_p V}}$$

$$\Rightarrow \quad \frac{28-20}{100-20} = e^{-\left(\frac{h \cdot 3000}{0.5 \times 8500 \times 400}\right) \times 4.35}$$

$$\Rightarrow \quad h = 299.95 \text{ W/m}^2\text{K}$$

6.18 (d)

Fin problem:

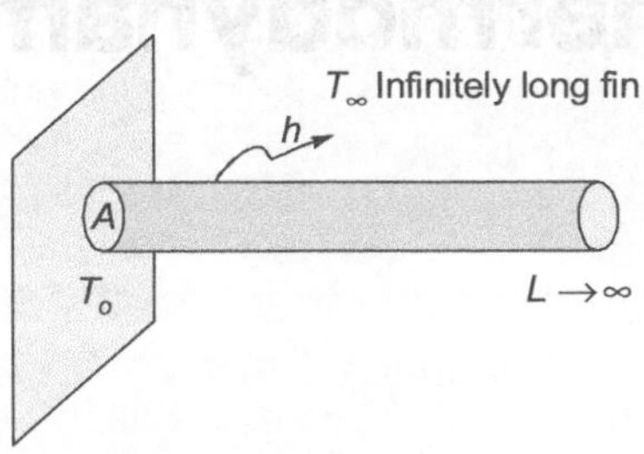

$$q = \sqrt{hPkA}\,(T_o - T_\infty) \text{ wait}$$

If k gets doubled q increases by $\sqrt{2}$ times.

6.19 (c)

Unsteady state

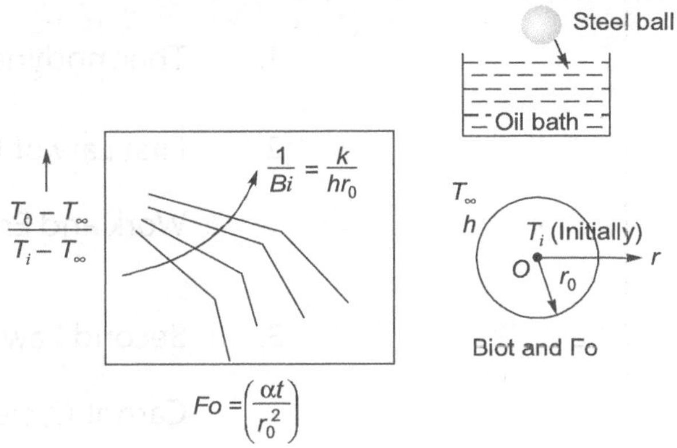

Thermodynamics

UNIT VIII

CONTENTS

Thermodynamics

Syllabus

Thermodynamic systems and processes; properties of pure substances, behaviour of ideal and real gases; zeroth and first laws of thermodynamics, calculation of work and heat in various processes; second law of thermodynamics; thermodynamic property charts and tables, availability and irreversibility; thermodynamic relations.

Analysis of Previous GATE Papers

Exam Year	1 Mark Ques.	2 Marks Ques.	3 Marks Ques.	5 Marks Ques.	Total Marks
1995	3	3	–	2	19
1996	4	3	–	1	15
1997	2	1	–	3	19
1998	–	1	–	2	12
1999	3	–	–	1	8
2000	2	1	–	2	14
2001	–	3	–	2	16
2002	–	1	–	2	12
2003	1	3	–	–	7
2004	2	2	–	–	7
2005	1	4	–	–	9
2006	–	4	–	–	8
2007	2	4	–	–	10
2008	1	8	–	–	17
2009	3	3	–	–	9
2010	1	2	–	–	5
2011	2	4	–	–	10
2012	1	2	–	–	5
2013	1	1	–	–	3
2014 Set-1	1	2	–	–	5
2014 Set-2	1	–	–	–	1

Exam Year	1 Mark Ques.	2 Marks Ques.	Total Marks
2014 Set-3	–	2	4
2014 Set-4	–	2	4
2015 Set-1	3	2	7
2015 Set-2	1	2	5
2015 Set-3	1	3	7
2016 Set-1	1	2	5
2016 Set-2	2	2	6
2016 Set-3	1	1	3
2017 Set-1	2	1	4
2017 Set-2	1	3	7
2018 Set-1	2	2	6
2018 Set-2	2	1	4
2019 Set-1	2	2	6
2019 Set-2	2	1	4
2020 Set-1	2	1	4
2020 Set-2	2	2	6
2021 Set-1	2	2	6
2021 Set-2	1	1	3
2022 Set-1	2	1	4
2022 Set-2	2	2	6
2023	1	3	7
2024	1	2	5

1 Thermodynamic System and Processes

1.1 An isolated thermodynamic system executes a process. Choose the correct statement (s) from the following

(a) No heat is transferred
(b) No work is done
(c) No mass flows across the boundary of the system
(d) No chemical reaction takes place within the system **[1999 : 2 M]**

1.2 Heat and work are

(a) intensive properties
(b) extensive properties
(c) point functions
(d) path functions **[2011 : 1 M]**

1.3 A certain amount of an ideal gas is initially at a pressure p_1 and temperature T_1. First, it undergoes a constant pressure process 1-2 such that $T_2 = 3T_1/4$. Then, it undergoes a constant volume process 2-3 such that $T_3 = T_1/2$. The ratio of the final volume to the initial volume of the ideal gas is

(a) 0.25 (b) 0.75
(c) 1.0 (d) 1.5

[2014 : 2 M, Set-3]

1.4 Which of the following statements are **TRUE** with respect to heat and work?

(i) They are boundary phenomena
(ii) They are exact differentials
(iii) They are path functions

(a) both (i) and (ii) (b) both (i) and (iii)
(c) both (ii) and (iii) (d) only (iii)

[2016 : 1 M, Set-1]

1.5 The volume and temperature of air (assumed to be an ideal gas) in a closed vessel is 2.87 m^3 and 300 K, respectively. The gauge pressure indicated by a manometer fitted to the wall of the vessel is 0.5 bar. If the gas constant of air is $R = 287$ J/kgK and the atmospheric pressure is 1 bar, the mass of air (in kg) in the vessel is

(a) 1.67 (b) 3.33
(c) 5.00 (d) 6.66

[2017 : 2 M, Set-2]

1.6 During a non-flow thermodynamic process (1 – 2) executed by a perfect gas, the heat interaction is equal to the work interaction ($Q_{1-2} = W_{1-2}$) when the process is

(a) Isentropic (b) Polytropic
(c) Adiabatic (d) Isothermal

[2019 : 1 M, Set-1]

1.7 Which one of the following is an intensive property of a thermodynamic system?

(a) Mass (b) Density
(c) Energy (d) Volume

[2022 : 1 M, Set-2]

■■■■

Answers Thermodynamic System and Processes

1.1 (a, b & c) 1.2 (d) 1.3 (b) 1.4 (b) 1.5 (c) 1.6 (d) 1.7 (b)

Explanations Thermodynamic System and Processes

1.1 (a, b, c)

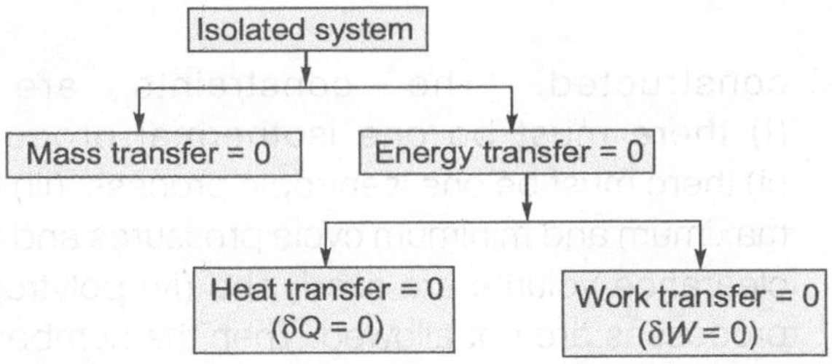

1.2 (d)

Heat and work are not properties and are path function.

1.3 (b)

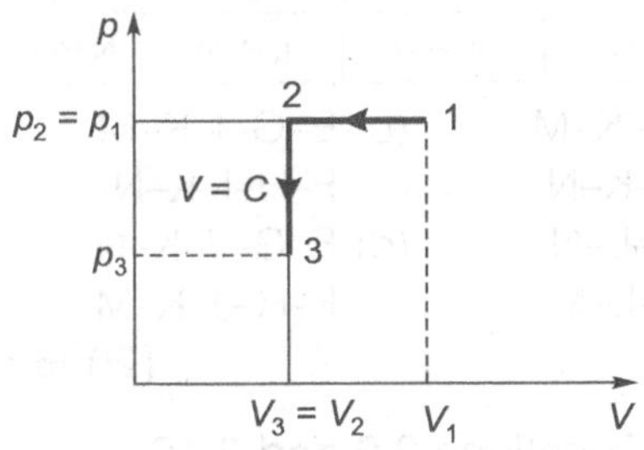

Process 1 – 2: At $p = C$

$$T_2 = \frac{3T_1}{4} \quad \text{(given condition)}$$

or $$\frac{T_2}{T_1} = \frac{3}{4}$$

According to Charle's law

$$\frac{V_2}{V_1} = \frac{T_2}{T_1} = \frac{3}{4}$$

Process 2 – 3: At $V = C$

$$T_3 = \frac{T_1}{2} \text{ given condition}$$

$$\frac{\text{Final volume}}{\text{Initial volume}} = \frac{V_3}{V_1} = \frac{V_2}{V_1} = \frac{T_2}{T_1} \quad (\because V_3 = V_2)$$

$$= \frac{3}{4} = 0.75$$

1.4 (b)

Heat:

1. When heat is added to the system, it is taken as positive.
2. Heat transfer is path function.
3. Heat is not a property.
4. Heat is inexact differential
5. Heat is boundary phenomena.
6. Heat is energy in transit.

Works:

1. Work transfer is path function.
2. It is not a property.
3. It is inexact differential.
4. It is boundary phenomena.
5. When work is added to the system, it is taken negative.
6. Work is energy in transit.

1.5 (c)

$$P_g = 0.5 \text{ bar} = 50 \text{ kPa};$$
$$R = 287 \text{ J/kgK} = 0.287 \text{ kJ/kgK}$$

Ideas gas equation:

$$PV = mRT$$

Here P is absolute pressure:

$$P = P_{abs} = P_{\text{gauge}} + P_{\text{atm}}$$
$$= 50 + 100 = 150 \text{ kPa}$$

$$m = \frac{PV}{RT} = \frac{150 \times 2.87}{0.287 \times 300} = \mathbf{5 \text{ kg}}$$

1.7 (b)

As Density, $\rho = \dfrac{\text{Mass}}{\text{Volume}}$

■■■■

2 First Law of Thermodynamics, Heat, Work & Energy Interactions

2.1 For reversible adiabatic compression in a steady flow process, the work transfer per unit mass is

(a) $\int pdv$ (b) $\int vdp$

(c) $\int Tds$ (d) $\int sdT$ **[1996 : 1 M]**

2.2 The specific heats of an ideal gas depend on its

(a) Temperature
(b) Pressure
(c) Volume
(d) Molecular weight and structure **[1996 : 1 M]**

2.3 A steam turbine receives steam steadily at 10 bar with a enthalpy of 3000 kJ/kg and discharges at 1 bar with an enthalpy of 2700 kJ/kg. The work output is 250 kJ/kg. The changes in kinetic and potential energies are negligible The heat transfer from the turbine casing to the surroundings is equal to

(a) 0 kJ (b) 50 kJ
(c) 150 kJ (d) 250 kJ **[2000 : 1 M]**

2.4 A small steam whistle (perfectly insulated and doing no shaft work) causes a drop of 0.8 kJ/kg in enthalpy of steam from entry to exit. If the kinetic energy of the steam at entry is negligible, the velocity of the steam at exit is

(a) 4 m/s (b) 40 m/s
(c) 80 m/s (d) 120 m/s **[2001 : 2 M]**

2.5 A 2 kW, 40 litres water heater is switched on for 20 minutes. The heat capacity c_p for water is 4.2 kJ/kgK. Assuming all the electrical energy has gone into heating the water, increase of the water temperature in degree centigrade is

(a) 2.7 (b) 4.0
(c) 14.3 (d) 25.25 **[2003 : 1 M]**

2.6 A gas contained in a cylinder is compressed, the work required for compression being 5000 kJ. During the process, heat interaction of 2000 kJ causes the surroundings to be heated. The changes in internal energy of the gas during the process is

(a) –7000 kJ (b) –3000 kJ
(c) +3000 kJ (d) +7000 kJ **[2004 : 1 M]**

2.7 A reversible thermodynamic cycle containing only three processes and producing work is to be constructed. The constraints are : (i) there must be one isothermal process, (ii) there must be one isentropic process, (iii) the maximum and minimum cycle pressures and the clearance volume are fixed, and (iv) polytropic processes are not allowed. Then the number of possible cycles are

(a) 1 (b) 2
(c) 3 (d) 4 **[2005 : 2 M]**

2.8 Match items from groups I, II, III, IV and V.

Group I	Group II	Group III	Group IV	Group V
	When added to the system, is	Differential	Function	Phenomenon
E Heat	G Positive	I Exact	K Path	M Transient
F Work	H Negative	J Inexact	L Point	N Boundary

(a) F–G–J–K–M
E–G–I–K–N
(b) E–G–I–K–M
F–H–I–K–N
(b) F–H–J–L–N
E–H–I–L–M
(d) E–G–J–K–N
F–H–J–K–M
[2006 : 2 M]

Linked Answer Questions 2.9 and 2.10

A football was inflated to a gauge pressure of 1 bar when the ambient temperature was 15°C. When the game started next day, the air temperature at the stadium was 5°C. Assume that the volume of the football remains constant at 2500 cm^3.

2.9 The amount of heat lost by the air in the football and the gauge pressure of air in the football at the stadium respectively equal

(a) 30.6 J, 1.94 bar (b) 21.8 J, 0.93 bar
(c) 61.1 J, 1.94 bar (d) 43.7 J, 0.93 bar
[2006 : 2 M]

2.10 Gauge pressure of air to which the ball must have been originally inflated so that it would be equal 1 bar gauge at the stadium is

(a) 2.23 bar (b) 1.94 bar
(c) 1.07 bar (d) 1.00 bar **[2006 : 2 M]**

2.11 Which of the following relationships is valid only for reversible processes undergone by a closed system of simple compressible substance (neglect changes in kinetic and potential energy)?

(a) $\delta Q = dU + \delta W$ (b) $TdS = dU + pdV$
(c) $TdS = dU + \delta W$ (d) $\delta Q = dU + pdV$
[2007 : 1 M]

2.12 A balloon containing an ideal gas is initially kept in an evacuated and insulated room. The balloon ruptures and the gas fills up the entire room. Which one of the following statements is TRUE at the end of above process?

(a) The internal energy of the gas decreases from its initial value, but the enthalpy remains constant.
(b) The internal energy of the gas increases from its initial value, but the enthalpy remains constant.
(c) Both internal energy and enthalpy of the gas remain constant
(d) Both internal energy and enthalpy of the gas increase **[2008 : 2 M]**

2.13 A rigid, insulated tank is initially evacuated. The tank is connected with a supply line through which air (assumed to be ideal gas with constant specific heats) passes at 1 MPa, 350°C.

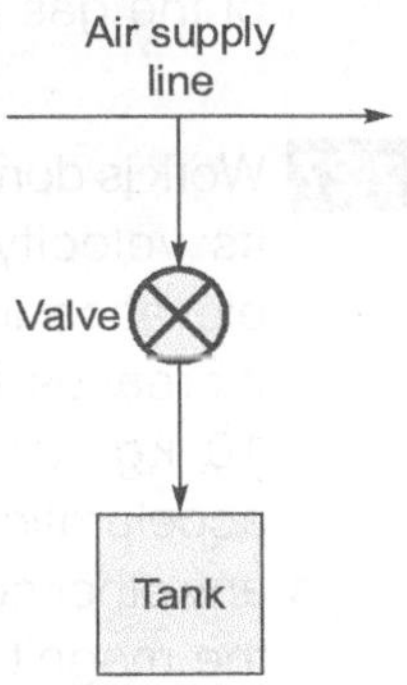

A valve connected with the supply line is opened and the tank is charged with air until the final pressure inside the tank reaches 1 MPa. The final temperature inside the tank

(a) is greater than 350°C
(b) is less than 350°C
(c) is equal to 350°C
(d) may be greater than, less than, or equal to 350°C, depending on the volume of the tank. **[2008 : 2 M]**

2.14 In a steady state steady flow process taking place in a device with a single inlet and a single outlet, the work done per unit mass flow rate is given by

$w = -\int_{inlet}^{outlet} v dp$, where v is the specific volume and p is the pressure, The expression for w given above

(a) is valid only if the process is both reversible and adiabatic
(b) is valid only if the process is both reversible and isothermal
(c) is valid for any reversible process.
(d) is incorrect; it must be $w = \int_{inlet}^{outlet} p dv$

[2008 : 2 M]

2.15 A frictionless piston-cylinder device contains a gas initially at 0.8 MPa and 0.015 m^3. It expands quasistatically at constant temperature to a final volume of 0.030 m^3. The work output (in kJ) during this process will be

(a) 8.32 (b) 12.00
(c) 554.67 (d) 8320.00 **[2009 : 1 M]**

2.16 A compressor undergoes a reversible, steady flow process. The gas at inlet and outlet of the compressor is designated as state 1 and state 2 respectively. Potential and kinetic energy changes are to be ignored. The following notations are used: v = specific volume and P = pressure of the gas. The specific work required to be supplied to the compressor for this gas compression process is

(a) $\int_1^2 P dv$ (b) $\int_1^2 v dP$
(c) $v_1 (P_2 - P_1)$ (d) $-P_2(v_1 - v_2)$

[2009 : 1 M]

Common Data for Question 2.17 and 2.18.

The inlet and the outlet conditions of steam for an adiabatic steam turbine are as indicated in the notations are as usually followed.

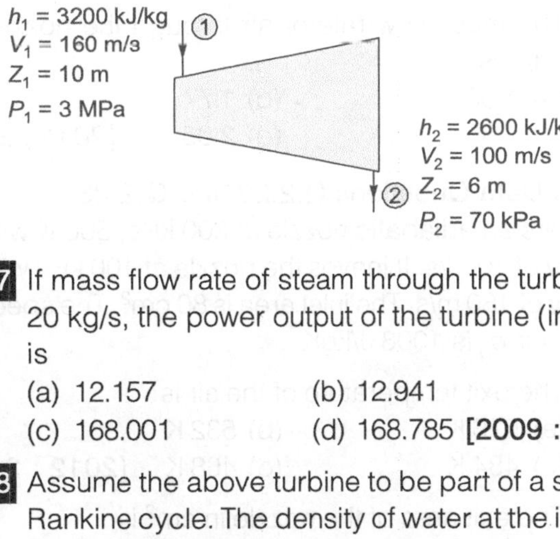

2.17 If mass flow rate of steam through the turbine is 20 kg/s, the power output of the turbine (in MW) is

(a) 12.157 (b) 12.941
(c) 168.001 (d) 168.785 **[2009 : 2 M]**

2.18 Assume the above turbine to be part of a simple Rankine cycle. The density of water at the inlet to the pump is 1000 kg/m^3. Ignoring kinetic and potential energy effects, the specific work (in kJ/kg) supplied to the pump is

(a) 0.293 (b) 0.351
(c) 2.930 (d) 3.510 **[2009 : 2 M]**

2.19 The contents of a well-insulated tank are heated by a resistor of 23 Ω in which 10 A current is flowing. Consider the tank along with its contents as a thermodynamic system. The work done by the system and the heat transfer to the system are positive. The rates of heat (Q), work (W) and

change in internal energy (ΔU) during the process in kW are
(a) $Q = 0,\ W = -2.3,\ \Delta U = +2.3$
(b) $Q = +2.3,\ W = 0,\ \Delta U = +2.3$
(c) $Q = -2.3,\ W = 0,\ \Delta U = -2.3$
(d) $Q = 0,\ W = +2.3,\ \Delta U = -2.3$ [2011 : 1 M]

Linked Data Questions 2.20 and 2.21

The temperature and pressure of air in a large reservoir are 400 K and 3 bar respectively. A converging diverging nozzle of exit area 0.005 m^2 is fitted to the wall of the reservoir as shown in the figure. The static pressure of air at the exit section for isentropic flow through the nozzle is 50 kPa. The characteristic gas constant and the ratio of specific heats of air are 0.287 kJ/kgK and 1.4 respectively.

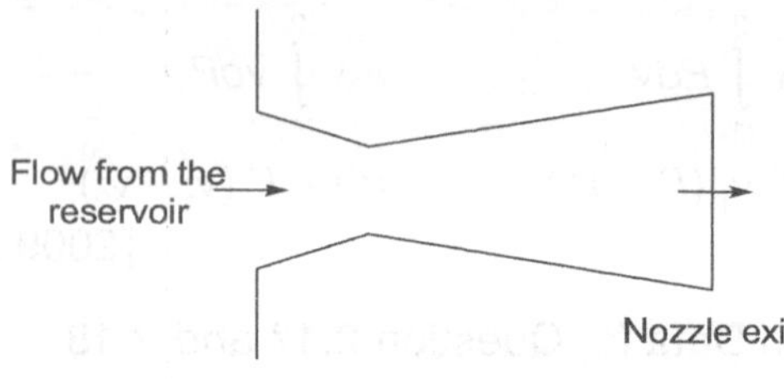

2.20 The density of air in kg/m^3 at the nozzle exit is
(a) 0.560 (b) 0.600
(c) 0.727 (d) 0.800 [2011 : 2 M]

2.21 The mass flow rate of air through the nozzle in kg/s is
(a) 1.30 (b) 1.77
(c) 1.85 (d) 2.06 [2011 : 2 M]

Linked Data Questions Q.2.22 and Q.2.23

Air enters an adiabatic nozzle at 300 kPa, 500 K with a velocity of 10 m/s. It leaves the nozzle at 100 kPa with a velocity of 180 m/s. The inlet area is 80 cm^2. The specific heat of air c_p is 1008 J/kgK

2.22 The exit temperature of the air is
(a) 516 K (b) 532 K
(c) 484 K (d) 468 K [2012 : 2 M]

2.23 The exit area of the nozzle in cm^2 is
(a) 90.1 (b) 56.3
(c) 4.4 (d) 12.9 [2012 : 2 M]

2.24 For an ideal gas with constant values of specific heats, for calculation of the specific enthalpy,
(a) it is sufficient to know only the temperature
(b) both temperature and pressure are required to be known
(c) both temperature and volume are required to be known
(d) both temperature and mass are required to be known [2015 : 1 M, Set-1]

2.25 Temperature of nitrogen in a vessel of volume 2 m^3 is 288 K. A *U*-tube manometer connected by the vessel shows a reading of 70 cm of mercury (level higher in the end open to atmosphere). The universal gas constant is 8314 J/kmol-K, atmospheric pressure is 1.01325 bar, acceleration due to gravity is 9.81 m/s^2 and density of mercury is 13600 kg/m^3. The mass of nitrogen (in kg) in the vessel is ______. [2015 : 2 M, Set-1]

2.26 A well insulated rigid container of volume 1 m^3 contains 1.0 kg of an ideal gas [c_p = 1000 J/(kgK) and c_v = 800 J/(kgK)] at a pressure of 10^5 Pa. A stirrer is rotated at constant rpm in the container for 1000 rotations and the applied torque is 100 Nm. The final temperature of the gas (in K) is ______. [2015 : 2 M, Set-1]

2.27 Work is done on an adiabatic system due to which its velocity changes from 10 m/s to 20 m/s, elevation increases by 20 m and temperature increases by 1 K. The mass of the system is 10 kg. c_v = 100 J/(kgK) and gravitational acceleration is 10 m/s^2. If there is no change in any other component of the energy of the system, the magnitude of total work done (in kJ) on the system is ______. [2015 : 2 M, Set-2]

2.28 Steam enters a turbine at 30 bar, 300°C (u = 2750 kJ/kg, h = 2993 kJ/kg) and exits the turbine as saturated liquid at 15 kPa (u = 225 kJ/kg, h = 226 kJ/kg). Heat loss to the surrounding is 50 kJ/kg of steam flowing through the turbine. Neglecting changes in kinetic energy and potential energy, the work output of the turbine (in kJ/kg of steam) is ______. [2015 : 2 M, Set-3]

2.29 An ideal gas undergoes a reversible process in which the pressure varies linearly with volume. The conditions at the start (subscript 1) and at the end (subscript 2) of the process with usual notation are : p_1 = 100 kPa, V_1 = 0.2 m^3 and p_2 = 200 kPa, V_2 = 0.1 m^3 and the gas constant, R = 0.275 kJ/kgK. The magnitude of the work required for the process (in kJ) is______. [2016 : 2 M, Set-1]

2.30 The internal energy of an ideal gas is a function of
(a) temperature and pressure
(b) volume and pressure
(c) entropy and pressure
(d) temperature only [2016 : 1 M, Set-2]

2.31 A piston-cylinder device initially contains 0.4 m^3 of air (to be treated as an ideal gas) at

100 kPa and 80°C. The air is now isothermally compressed to 0.1 m^3. The work done during this process is ______ kJ.
(Take the sign convention such that work done on the system is negative) [2016 : 2 M, Set-2]

2.32 Steam at an initial enthalpy of 100 kJ/kg and inlet velocity of 100 m/s, enters an insulated horizontal nozzle. It leaves the nozzle at 200 m/s. The exit enthalpy (in kJ/kg) is ___. [2016 : 2 M, Set-3]

2.33 The molar specific heat at constant volume of an ideal gas is equal to 2.5 times the universal gas constant (8.314 J/mol-K). When the temperature increases by 100 K, the change in molar specific enthalpy is ______ J/mol. [2017 : 1 M, Set-1]

2.34 A mass m of a perfect gas at pressure p_1 and volume V_1 undergoes an isothermal process. The final pressure is p_2 and volume is V_2. The work done on the system is considered positive. If R is the gas constant and T is the temperature, then the work done in the process is

(a) $p_1V_1 \ln\frac{V_2}{V_1}$ (b) $-p_1V_1 \ln\frac{p_1}{p_2}$

(c) $RT \ln\frac{V_2}{V_1}$ (d) $-mRT \ln\frac{p_2}{p_1}$

[2017 : 1 M, Set-2]

2.35 A calorically perfect gas (specific heat at constant pressure 1000 J/kgK) enters and leaves a gas turbine with the same velocity. The temperature of the gas at turbine entry and exit are 1100 K and 400 K, respectively. The power produced is 4.6 MW and heat escapes at the rate of 300 kJ/s through the turbine casing. The mass flow rate of the gas (in kg/s) through the turbine is

(a) 6.14 (b) 7.00
(c) 7.50 (d) 8.00 [2017 : 2 M, Set-2]

2.36 Steam flows through a nozzle at mass flow rate of $\dot{m}$ = 0.1 kg/s with a heat loss of 5 kW. The enthalpies at inlet and exit are 2500 kJ/kg and 2350 kJ/kg, respectively. Assuming negligible velocity at inlet ($C_1 \approx 0$), the velocity (C_2) of steam (in m/s) at the nozzle exit is ____(correct to two decimal places)

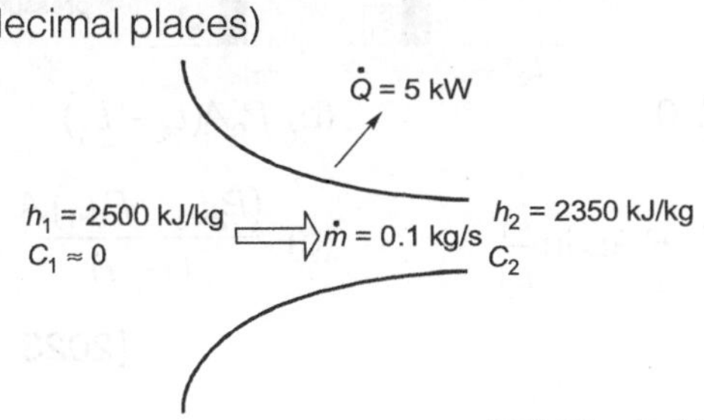

[2018 : 2 M, Set-1]

2.37 An engine operates on the reversible cycles as shown in the figure. The work output from the engine (in kJ/cycle) is ______ (correct to two decimal places).

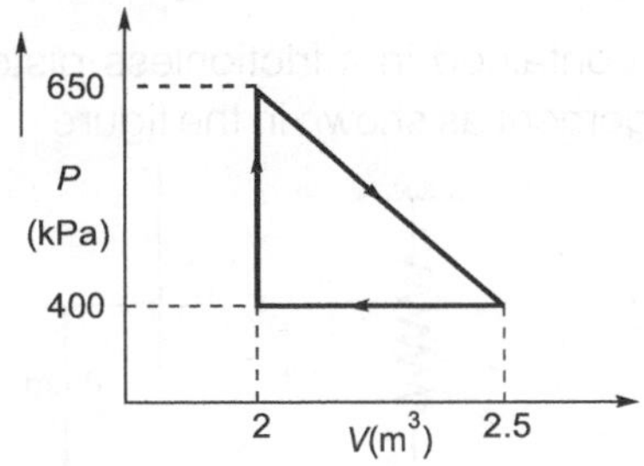

[2018 : 1 M, Set-2]

2.38 Air is held inside a non-insulated cylinder using a piston (mass M = 25 kg and area A = 100 cm^2) and stoppers (of negligible area), as shown in the figure. The initial pressure P_1 and temperature T_1 of air inside the cylinder are 200 kPa and 400° C, respectively. The ambient pressure P_∞ and temperature T_∞ are 100 kPa and 27° C respectively. The temperature of the air inside the cylinder (°C) at which the piston will begin to move is ______ (correct to two decimal places).

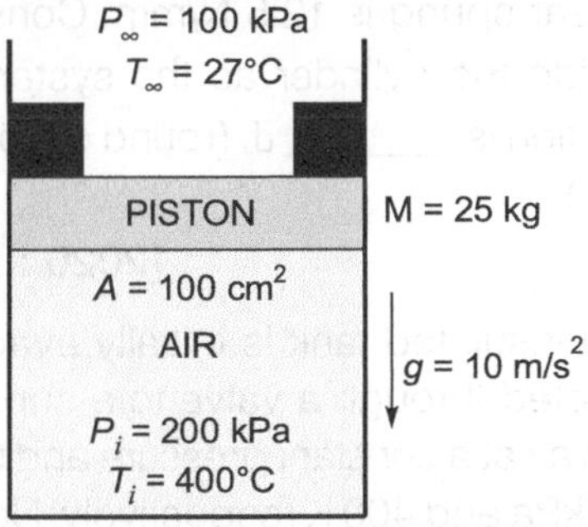

[2018 : 2 M, Set-2]

2.39 A gas is heated in a duct as it flows over a resistance heater. Consider a 101 kW electric heating system. The gas enters the heating section of the duct at 100 kPa and 27°C with a volume flow rate of 15 m^3/s. If heat is lost from the gas in the duct to the surroundings at a rate of 51 kW, the exit temperature of the gas is
(Assume constant pressure, ideal gas, negligible change in kinetic and potential energies and constant specific heat; C_p = 1 kJ/kgK; R = 0.5 kJ/kgK)

(a) 32°C (b) 37°C
(c) 76°C (d) 53°C

[2019 : 2 M, Set-1]

2.40 One kg of air, initially at a temperature of 127°C, expands reversibly at a constant pressure until the volume is doubled. If the gas constant of air is 287 J/kg.K, the magnitude of work transfer is ________ kJ (round off to 2 decimal places).

[2020 : 2 M, Set-1]

2.41 Air is contained in a frictionless piston-cylinder arrangement as shown in the figure.

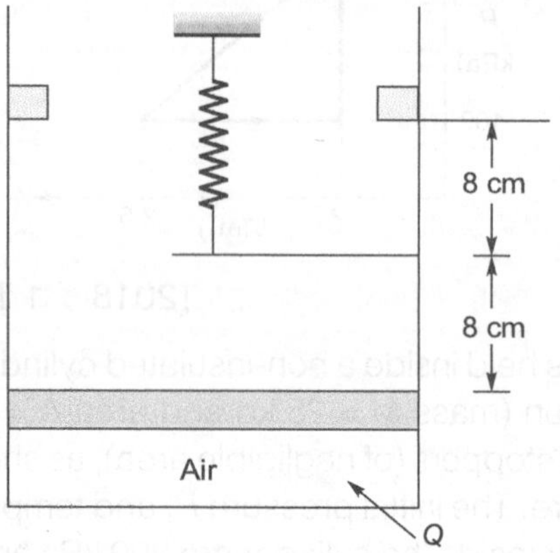

The atmospheric pressure is 100 kPa and the initial pressure of air in the cylinder is 105 kPa. The area of piston is 300 cm^2. Heat is now added and the piston moves slowly from its initial position until it reaches the stops. The spring constant of the linear spring is 12.5 N/mm. Considering the air inside the cylinder as the system, the work interaction is ________ J. (round off to the nearest integer).

[2020 : 2 M, Set-2]

2.42 A rigid insulated tank is initially evacuated. It is connected through a valve to a supply line that carries air at a constant pressure and temperature of 250 kPa and 400 K respectively. Now the valve is opened and air is allowed to flow into the tank until the pressure inside the tank reaches to 250 kPa at which point the valve is closed. Assume that the air behaves as a perfect gas with constant properties (c_p = 1.005 kJ/kgK, c_v = 0.718 kJ/kgK, R = 0.287 kJ/kgK). Final temperature of the air inside the tank is ________ K. [Round off to one decimal place]

[2021 : 1 M, Set-1

2.43 A polytropic process is carried out from an initial pressure of 110 kPa and volume of 5 m^3 to a final volume of 2.5 m^3. The polytropic index is given by n = 1.2. The absolute value of the work done during the process is ________ kJ (round off to 2 decimal places).

[2022 : 1 M, Set-1]

2.44 Consider 1 kg of an ideal gas at 1 bar, 300 K contained in a rigid and perfectly insulated container. The specific heat of the gas at constant volume, C_V is equal to 750 $Jkg^{-1}K^{-1}$. A stirrer performs 225 kJ of work on gas. Assume that the container does not participate in the thermodynamic interaction. The final pressure of the gas will be ____ bar (in integer).

[2022 : 1 M, Set-2]

2.45 A rigid tank of volume of 8 m^3 is being filled up with air from a pipeline connected through a valve. Initially the valve is closed and the tank is assumed to be completely evacuated. The air pressure and temperature inside the pipeline are maintained at 600 kPa and 306 K, respectively. The filling of the tank begins by opening the valve and the process ends when the tank pressure is equal to the pipeline pressure. During the filling process, heat loss to the surrounding is 1000 kJ. The specific heats of air at constant pressure and at constant volume are 1.005 kJ/kg.K and 0.718 kJ/kg.K, respectively. Neglect changes in kinetic energy and potential energy.

The final temperature of the tank after the completion of the filling process is ____ K (round off to the nearest integer).

[2022 : 2 M, Set-2]

2.46 Consider a fully adiabatic piston-cylinder arrangement as shown in the figure. The piston is massless and cross-sectional area of the cylinder is A. The fluid inside the cylinder is air (considered as a perfect gas), with γ being the ratio of the specific heat at constant pressure to the specific heat at constant volume for air. The piston is initially located at a position L_1. The initial pressure of the air inside the cylinder is $P_1 >> P_0$, where P_0 is the atmospheric pressure. The stop S_1 is instantaneously removed and the piston moves to the position L_2, where the equilibrium pressure of air inside the cylinder is $P_2 >> P_0$.

What is the work done by the piston on the atmosphere during this process?

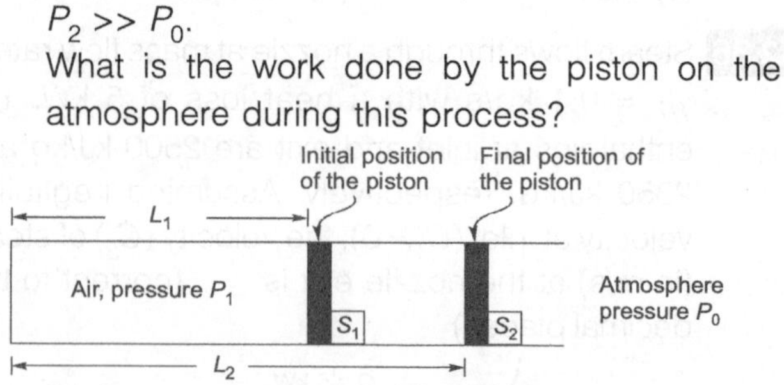

(a) 0

(b) $P_0A(L_2 - L_1)$

(c) $P_1AL_1 \ln\frac{L_1}{L_2}$

(d) $\frac{(P_2L_2 - P_1L_1)A}{(1-\gamma)}$

[2023 : 2 M]

2.47 A piston-cylinder arrangement shown in the figure has a stop located 2 m above the base. The cylinder initially contains air at 140 kPa and 350°C and the piston is resting in equilibrium at a position which is 1 m above the stops. The system is now cooled to the ambient temperature of 25°C. Consider air to be an ideal gas with a value of gas constant R = 0.287 kJ/kgK.

The absolute value of specific work done during the process is _______ kJ/kg. (rounded off to 1 decimal place)

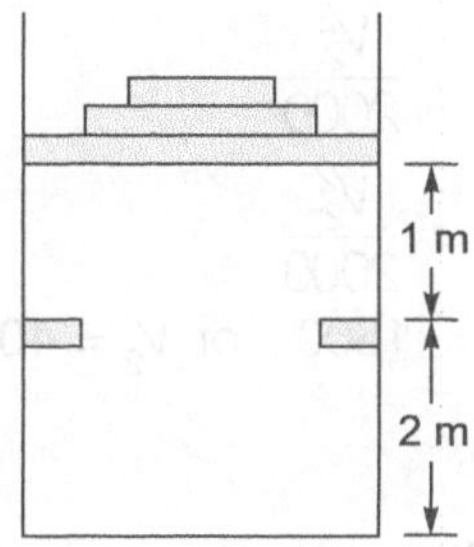

[2024 : 2 M]

Answers First Law, Heat, Work and Energy

2.1	(b)	2.2	(d)	2.3	(b)	2.4	(b)	2.5	(c)	2.6	(c)	2.7	(d)	2.8	(d)
2.9	(d)	2.10	(c)	2.11	(d)	2.12	(c)	2.13	(a)	2.14	(c)	2.15	(a)	2.16	(b)
2.17	(a)	2.18	(c)	2.19	(a)	2.20	(c)	2.21	(d)	2.22	(c)	2.23	(d)	2.24	(a)
2.25	(4.56)	2.26	(1285.39)	2.27	(4.5)	2.28	(2717)	2.29	(15)	2.30	(d)	2.31	(–55.45)		
2.32	(85)	2.33	(2909.9)	2.34	(b)	2.35	(b)	2.36	(447.213)	2.37	(62.5)	2.38	(117.625)		
2.39	(a)	2.40	(114.8)	2.41	(544)	2.42	(559.9)	2.43	(408.92)	2.44	(2)	2.45	(395.388)		
2.46	(b)	2.47	(59.6)												

Explanations First Law, Heat, Work and Energy

2.2 (d)

Spefic heat of an ideal gas,

$$c_p = \frac{\gamma R}{\gamma - 1} \text{ and, } c_v = \frac{R}{\gamma - 1}$$

where, R = Gas constant and depends upon molecular weight as

$$R = \frac{\bar{R}}{M} \quad (M = \text{Molecular weight})$$

and, γ- depends upon the molecular structure i.e. monoatomic, diatomic and polyatomic.

2.3 (b)

Given data:

p_1 = 10 bar; h_1 = 3000 kJ/kg; p_2 = 1 bar

h_2 = 2700 kJ/kg; w = 250 kJ/kg

q = ?

The steady flow energy equation for unit mass flow rate,

$$h_1 + \frac{V_1^2}{2} + gz_1 + q = h_2 + \frac{V_2^2}{2} + gz_2 + w$$

Change in KE and PE are negligible

$$h_1 + q = h_2 + w$$

$$3000 + q = 2700 + 250$$

or $\quad q = -$ **50 kJ/kg**

The heat transfer from the turbine casing to the surroundings is equal to 50 kJ/kg.

2.4 (b)

Given data:

$$\Delta h = -0.8 \text{ kJ/kg}$$
$$h_2 - h_1 = -0.8 \text{ kJ/kg}$$
$$h_1 - h_2 = 0.8 \text{ kJ/kg}$$

Applying steady flow energy equation,

$$h_1 + \frac{V_1^2}{2} + gz_1 + q = h_2 + \frac{V_2^2}{2} + gz_2 + w$$

Assumptions:

(i) $q = 0$ adiabatic flow

(ii) $w = 0$

(iii) Change K.E. and P.E. are neglect

(iv) Neglect inlet velocity.

$$\therefore \quad h_1 = h_2 + \frac{V_2^2}{2}$$

$$h_1 = h_2 + \frac{V_2^2}{2000}$$

$$h_1 - h_2 = \frac{V_2^2}{2000}$$

or $$8 = \frac{V_2^2}{2000}$$

$$V_2^2 = 1600 \quad \text{or } V_2 = \mathbf{40\ m/s}$$

2.5 (c)

Given data:

$$P = 2 \text{ kW}$$

$$V = 40 \text{ litres} = \frac{40}{1000} = 0.04 \text{ m}^3$$

$$t = 20 \text{ minutes} = 20 \times 60 \text{ s}$$
$$= 1200 \text{ s}$$
$$c_p = 4.2 \text{ kJ/kg K}$$

Heat supplied:

$$Q = \text{Power} \times \text{Time}$$
$$= 2 \times 1200 = 2400 \text{ kJ}$$

also $$Q = mc_p\Delta T = \rho V c_p \Delta T \quad \because m = \rho V$$

$\therefore$ $$2400 = 1000 \times 0.04 \times 4.2 \times \Delta T$$

or $$\Delta T = 14.28°\text{C} \approx \mathbf{14.3°C}$$

2.6 (c)

Given data: Work done on gas in cylinder during compression,

$$W = -5000 \text{ kJ}$$

Heat, $$Q = -2000 \text{ kJ}$$

According to first flow for process,

$$Q = dU + W$$
$$-2000 = dU - 5000$$

or $$dU = -2000 + 5000 = \mathbf{+3000\ kJ}$$

2.7 (d)

Four clockwise cycles can be formed, out of which 2 cycles are formed having constant volume process and 2 cycles are formed having constant pressure process.

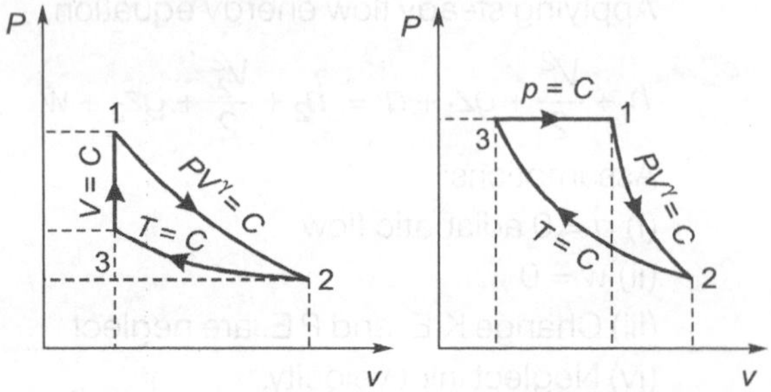

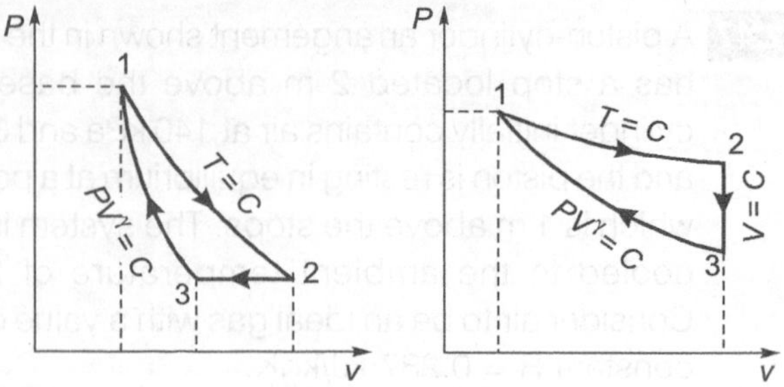

2.8 (d)

Heat:

1. When heat is added to the system, it is positive.
2. Heat transfer is path function.
3. Heat is not a property.
4. Heat is inexact differential
5. Heat is boundary phenomena.
6. Heat is energy in transit.

Works:

1. Work transfer is path function.
2. It is not a property.
3. It is inexact differential.
4. It is boundary phenomena.
5. When work is added to the system, it is negative.
6. Work is energy in transit.

2.9 (d)

Given data:

Gauge pressure, p_{G1} = 1 bar = 100 kPa

Ambient temperature,

$$T_1 = 15°\text{C} = (15 + 273) \text{ K} = 288 \text{ K}$$

Absolute pressure,

$$p_1 = p_{G1} + p_{atm} = 100 + 101.325$$
$$= 201.325 \text{ kPa}$$

Volume: $V = 2500 \text{ cm}^3 = 2500 \times 10^{-6} \text{ m}^3$

Temperature, $T_2 = 5°\text{C} = (5 + 273) \text{ K} = 278 \text{ K}$

From equation of state, $p_1V_1 = mRT_1$

$$201.325 \times 2500 \times 10^{-6} = m \times 0.287 \times 288$$

or $$m = 6.089 \times 10^{-3} \text{ kg}$$

Heat loss: $Q = mc_v(T_1 - T_2)$

$$= 6.089 \times 10^{-3} \times 0.718 \times (288 - 278)$$
$$= 0.0437 \text{ kJ} = \mathbf{43.7\ J}$$

Method I:

At $V = C$

$$\frac{p_1}{T_1} = \frac{p_2}{T_2}$$

$$\frac{201.325}{288} = \frac{p_2}{278}$$

or $p_2 = 194.33$ kPa (absolute)

also $p_2 = p_{G2} + p_{atm}$

$\therefore$ $194.33 = p_{G2} + 101.325$

or $p_{G2} = 93$ kPa = **0.93 bar**

Method II:

$$p_2 V = mRT_2$$

$$p_2 \times 2500 \times 10^{-6} = 6.089 \times 10^{-3} \times 0.287 \times 278$$

or $p_2 = 194.32$ kPa (absolute)

also $p_2 = p_{G2} + p_{atm}$

$\therefore$ $194.32 = p_{G2} + 101.325$

or $p_{G2} = 92.99$ kPa $\approx$ 93 kPa = **0.93 bar**

2.10 (c)

If $p_{G1} = 1$ bar = 100 kPa

Absolute pressure,

$$p_1 = p_{G1} + p_{atm} = 100 + 101.325 = 201.325 \text{ kPa}$$

$$T_1 = 278 \text{ K}$$

$$T_2 = 288 \text{ K}$$

At $V = C$

$$\frac{p_2}{T_2} = \frac{p_1}{T_1}$$

$$\frac{p_2}{288} = \frac{201.325}{278}$$

or $p_2 = 208.56$ kPa

also $p_2 = p_{G2} + p_{atm}$

$\therefore$ $208.56 = p_{G2} + 101.325$

or $p_{G2} = 107.235$ kPa = **1.07 bar**

2.11 (d)

$$\delta Q = dU + pdV$$

This equation holds good for a closed system when only pdV work is present. This is true only for a reversible (quasistatic) process.

2.12 (c)

The balloon ruptures in an evacuated and insulated room, the gas fills up the entire room. This case related to free expansion.

i.e., $\delta Q = 0$, $\delta W = 0$

According to first law of thermodynamics,

$$\delta Q = dU + \delta W$$

$$0 = dU + 0$$

$$\mathbf{dU = 0}$$

$$U_2 - U_1 = 0$$

$$\mathbf{U_2 = U_1}$$

$$mc_v T_2 = mc_v T_1$$

or $mc_p T_2 = mc_p T_1$

$$H_2 = H_1$$

2.13 (a)

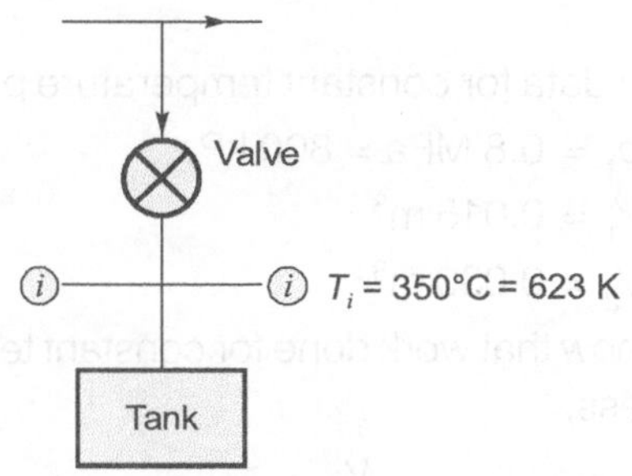

Applying unsteady flow energy equation,

$$Q = m_2 u_2 - m_1 u_1 + m_e\left(h_e + \frac{V_e^2}{2} + gz_e\right) - m_i\left(h_i + \frac{V_i^2}{2} + gz_i\right) + W \quad \text{... (i)}$$

For given problem,

$m_1 = 0$ evacuated tank

$m_e = 0$ no exit

V_i is neglected

$z_i = 0$; $Q = 0$; $W = 0$

Above Eq. (i) becomes

$$0 = m_2 u_2 - m_i h_i$$

or $m_i h_i = m_2 u_2$

$h_i = u_2$ $\because m_i = m_2$

$$c_p T_i = c_v T_2$$

$$\frac{c_p}{c_v} T_i = T_2 \quad \text{where } \gamma = \frac{c_p}{c_v}$$

$$\gamma T_i = T_2$$

or $T_2 = \gamma T_i = 1.4 \times 623$ K $= 872.3$ K $= 599.2$°C

The final temperature, T_2 = **599.2°C** inside the tank is greater than 350°C.

POINTS TO REMEMBER

This is an important formula i.e. $T_2 = \gamma T_i$ (in case of insulated and evacuated tank which is charging through a pipe line) because problems have been asked many times based on this only.

2.14 (c)

The expression, $w = -\int_{\text{inlet}}^{\text{outlet}} vdp$ is valid for reversible process, for open system.

2.15 (a)

Given data for constant temperature process:

$p_1 = 0.8 \text{ MPa} = 800 \text{ kPa}$

$V_1 = 0.015 \text{ m}^3$

$V_2 = 0.030 \text{ m}^3$

We know that work done for constant temperature process,

$$W = p_1V_1 \log_e \frac{V_2}{V_1}$$

$$= 800 \times 0.015 \log_e \frac{0.030}{0.015}$$

$$= 8.317 \text{ kJ} \approx \mathbf{8.32 \text{ kJ}}$$

2.16 (b)

Reversible steady flow process.

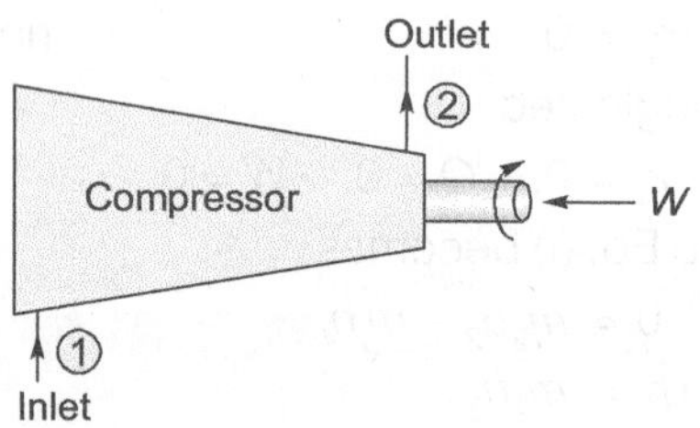

State 1: Inlet

State 2: Outlet

Potential and kinetic energy change are to be ignored.

v = Specific volume of the gas

P = Pressure of the gas

Specific work required to be supplied to the compressor for this gas compression process is

$$w = \int_1^2 v\,dp \quad \text{(Open system flow in compressor)}$$

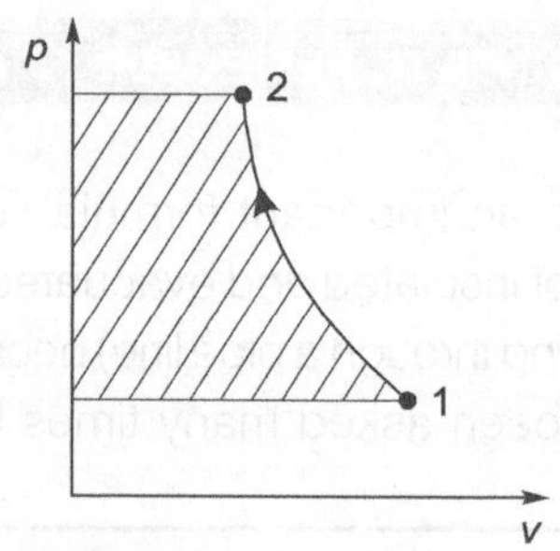

2.17 (a)

Applying the steady flow energy equation,

$$\dot{m}\left(h_1 + \frac{V_1^2}{2} + gz_1\right) + Q = \dot{m}\left(h_2 + \frac{V_2^2}{2} + gz_2\right) + W$$

where $Q = 0$ for adiabatic flow,

$$\dot{m}\left(h_1 + \frac{V_1^2}{2} + gz_1\right) = \dot{m}\left(h_2 + \frac{V_2^2}{2} + gz_2\right) + W$$

On units balancing, we get

$$\dot{m}\left(h_1 + \frac{V_1^2}{2000} + \frac{gz_1}{1000}\right) = \dot{m}\left(h_2 + \frac{V_2^2}{2000} + \frac{gz_2}{1000}\right) + W$$

$$20\left(3200 + \frac{(160)^2}{2000} + \frac{9.81 \times 10}{1000}\right) = 20\left(2600 + \frac{(100)^2}{2000} + \frac{9.81 \times 6}{1000}\right) + W$$

$$64257.96 = 52101.17 + W$$

or $W = 12156.79 \text{ kW}$

$= 12.1568 \text{ MW}$

$\approx \mathbf{12.157 \text{ MW}}$

2.18 (c)

Specific work supplied to the pump,

$$w_p = v(p_1 - p_2) \text{ kJ/kg}$$

where v is in m³/kg

p_1 and p_2 are in kPa

$$\therefore \quad w_p = \frac{p_1 - p_2}{\rho} = \frac{3000 - 70}{1000}$$

$$= \mathbf{2.93 \text{ kJ/kg}}$$

2.19 (a)

Heat transfer,

$Q = 0$, tank is well insulated

Increase in internal energy,

ΔU = rate of heat dissipated by a resistor

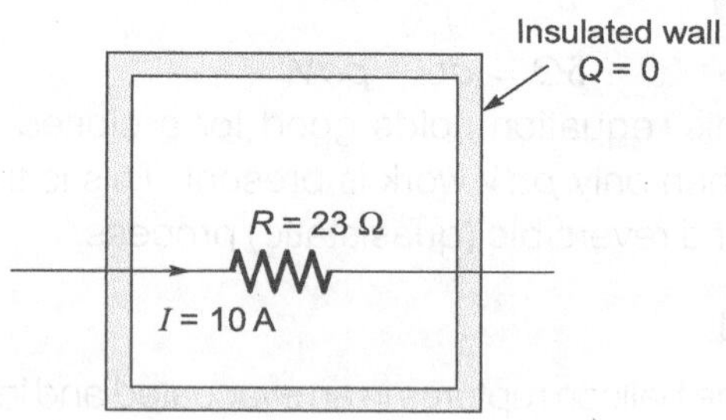

$$\Delta U = I^2R = (10)^2 \times 23$$

$$= 2300 \text{ W} = 2.3 \text{ kW}$$

According to first law of thermodynamic for any process,

$$\delta Q = dU + \delta W$$

$$0 = 2.3 + \delta W$$

or $\delta W = \mathbf{-2.3 \text{ kW}}$

Negative sign indicates that work is done on the system.

2.20 (c)

Density of air in (kg/m³) at nozzle exit is

$$\frac{p_1}{\rho_1^\gamma} = \frac{p_2}{\rho_2^\gamma}$$

where ρ = Density

$$\rho_2 = \left(\frac{p_2}{p_1}\right)^{\frac{1}{\gamma}} \rho_1$$

Also,

$$\because \quad p_1V_1 = mRT_1$$

$$\frac{p_1}{RT_1} = \frac{m}{V_1}; \quad \rho_1 = \frac{p_1}{RT_1}$$

$$\therefore \quad \rho_2 = \left(\frac{50}{300}\right)^{1/1.4} \times \frac{3\times10^5}{0.287\times1000\times400}$$

$$= \mathbf{0.727\ kg/m^3}$$

2.21 (d)

Mass flow rate of air through nozzle

= mass entering nozzle

= mass exit from nozzle

$$\dot{m} = \rho_i A_i V_i = \rho_e A_e V_e$$

where i = inlet; e = exit; ρ = density

A = Area; V = velocity

$$p = \rho RT$$

$$\dot{m} = \frac{A_eV_ep_e}{RT_e}$$

Given: $R = 0.287$ kJ/kgK

$p_e = 50$ kPa

For isentropic process

$$\left(\frac{T_e}{T_i}\right) = \left(\frac{p_e}{p_i}\right)^{(\gamma-1)/\gamma}$$

$$\frac{T_e}{400} = \left(\frac{50}{300}\right)^{0.4/1.4}$$

or $\quad T_e = 239.73$ K

Mass flow rate,

$$\dot{m} = \frac{50\times0.005\times V_e}{0.287\times239.73} \quad \text{... (1)}$$

Using steady flow energy equation

$$h_i + \frac{V_i^2}{2} + q = h_e + \frac{V_e^2}{2} + w$$

where $V_i = 0$

$q = 0$ for adiabatic process

$w = 0$

$$\therefore \quad h_i = h_e + \frac{V_e^2}{2}$$

On units balancing, we get

$$h_i = h_e + \frac{V_e^2}{2000}$$

or

$$h_i - h_e = \frac{V_e^2}{2000}$$

$$c_p(T_i - T_e) = \frac{V_e^2}{2000}$$

$$1\times(400-239.73) = \frac{V_e^2}{2000}$$

$$V_e = 567.57 \text{ m/s}$$

From Eq. (1), we get

$$\dot{m} = \frac{50\times0.005\times567.57}{0.287\times239.73} = \mathbf{2.06\ kg/s}$$

2.22 (c)

Applying the steady flow energy equation

$$h_1 + \frac{V_1^2}{2} + gz_1 + q = h_2 + \frac{V_2^2}{2} + gz_2 + w$$

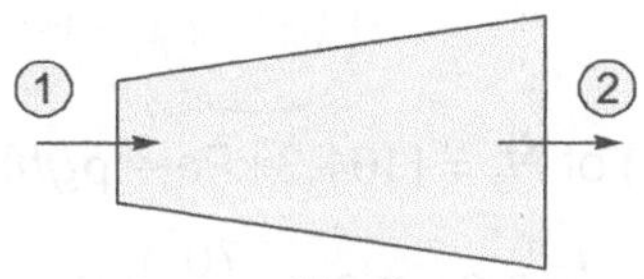

Nozzle

Assumptions:

$z_1 = z_2$

$q = 0$, adiabatic flow

$w = 0$, always for nozzle

$$h_1 + \frac{V_1^2}{2} = h_2 + \frac{V_2^2}{2}$$

$$c_pT_1 + \frac{V_1^2}{2} = c_pT_2 + \frac{V_2^2}{2}$$

$$1008\times500 + \frac{(10)^2}{2} = 1008\times T_2 + \frac{(180)^2}{2}$$

$$504050 = 1008\,T_2 + 16200$$

or $\quad T_2 = 483.97$ K ≈ **484 K**

2.23 (d)

Continuity equation for compressible flow

$$\rho_1A_1V_1 = \rho_2A_2V_2$$

and,

$$\frac{A_1V_1}{v_1} = \frac{A_2V_2}{v_2}$$

$$\frac{A_1V_1}{\frac{RT_1}{p_1}} = \frac{A_2V_2}{\frac{RT_2}{p_2}} \quad (\because pv = RT)$$

$$\frac{A_1V_1p_1}{T_1} = \frac{A_2V_2p_2}{T_2}$$

$$\frac{80\times10\times300}{500} = \frac{A_2\times180\times100}{484}$$

or $$A_2 = \mathbf{12.90\ cm^2}$$

2.24 (a)

In case of ideal gas and any process, specific enthalpy is given as

$$dh = c_p dT$$

Also, enthalpy is a function of temperature i.e.

$$h = f(T).$$

So, for the calculation of the specific enthalpy, it is sufficient to know only the temperature.

2.25 Sol.

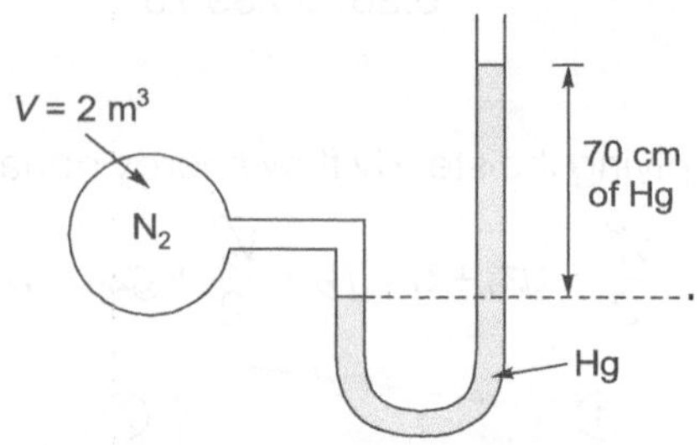

$(P_{absolute})$ of N_2 = [101.3 kPa + ρgh] kPa

$$= 101.3+\left(13.6\times9.81\times\frac{70}{100}\right)\text{kPa}$$

= 194.6 kPa

As N_2 being ideal gas,

$$pV = mRT$$

$$m = \left(\frac{pV}{RT}\right)\text{kg}$$

where $$R = \frac{8.314}{28} = 0.296 \text{ kJ/kgK}$$

$$\therefore \quad m = \frac{194.6\times2}{0.296\times288} = \mathbf{4.56\ kg}$$

2.26 Sol.

Given: c_p = 1000 J/kgK, c_v = 800 J/kgK

Rotation of stirrers = 1000

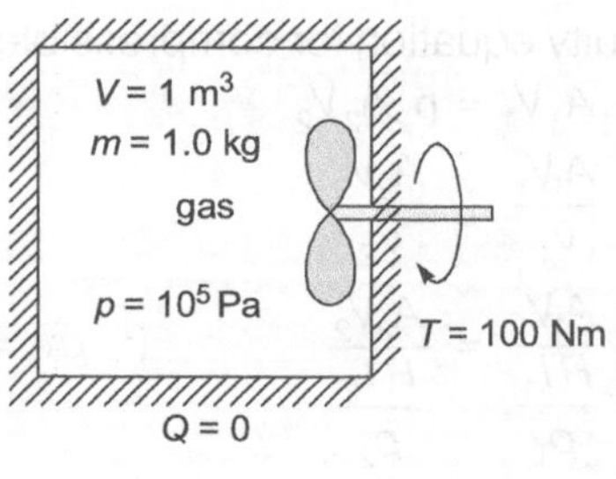

Work done on the gas,

= Torque × angular displacement

= 100 × 2π × 1000 = – 628.318 kJ

(Negative sign shows work is done on the gas)

$$p_1V_1 = mRT_1$$

$$R = c_p - c_v = 0.2 \text{ kJ/kgK}$$

$$T_1 = \frac{p_1V_1}{mR} = \frac{10^2\times1}{0.2\times1} = 500 \text{ K}$$

According to 1st law of thermodynamic

$$\delta Q - \delta W = \Delta U$$

$$0-(-628.318)\text{ kJ} = mc_v(T_2-T_1) \qquad (\delta Q = 0)$$

or $$T_2 = 500+\frac{628.318}{1\times0.8} = \mathbf{1285.39\ K}$$

2.27 Sol.

Ist law for process:

$$\delta Q = dE + \delta W$$

$$0 = \Delta U + \Delta K.E + \Delta P.E + \delta W$$

$$0 = mc_v dT + \frac{1}{2}m(V_2^2 - V_1^2) + mg(z_2 - z_1) + W_{1-2}$$

$$0 = 10\times0.1\times1 + \frac{1}{2}\times10((20)^2-(10)^2)\times10^{-3} + 10\times10\times20\times10^{-3} + W_{1-2}$$

$$\therefore \quad W_{1-2} = -4.5 \text{ kJ}$$

= **4.5 kJ**, work done on the system

2.28 Sol.

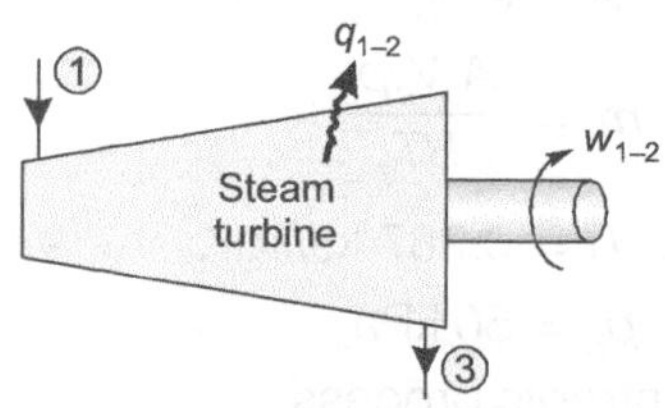

Given data:

At inlet, u_1 = 2750 kJ/kg; h_1 = 2993 kJ/kg

At exit, u_2 = 225 kJ/kg; h_2 = 226 kJ/kg

$$q_{1-2} = -50 \text{ kJ/kg}$$

Specific work done,

$$w_{1-2} = (h_1 - h_2) + q_{1-2}$$

= (2993 – 226) – 50

= **2717 kJ/kg**

2.29 Sol.

Given: p_1 = 100 kPa, V_1 = 0.2 m³

p_2 = 200 kPa, V_2 = 0.1 m³, R = 0.275 kJ/kgK

$p \propto V$

$\therefore$ p-V diagram will be a straight line

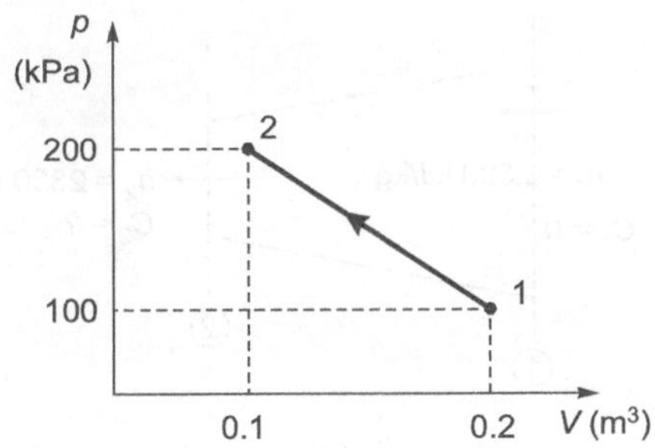

Work done is equal to area under p-V diagram,

$$\therefore \text{ Work} = \frac{1}{2}(100 + 200) \times (0.2 - 0.1)$$

$$= \mathbf{15\ kJ}$$

The work is done on the gas, since the process is compression.

Since only magnitude is asked hence final answer is without $-ve$ sign.

POINTS TO REMEMBER

Whenever it is given in the problem that pressure varies linearly with volume,

Always use,

$$\text{Work done, } W = \frac{1}{2}(P_1 + P_2)(V_2 - V_1)$$

2.31 Sol.

$V_1 = 0.4\ m^3$; $p_1 = 100$ kPa

$T_1 = 80°C = (80 + 273)\ K = 353\ K$

$V_2 = 0.1\ m^3$

For isothermal process

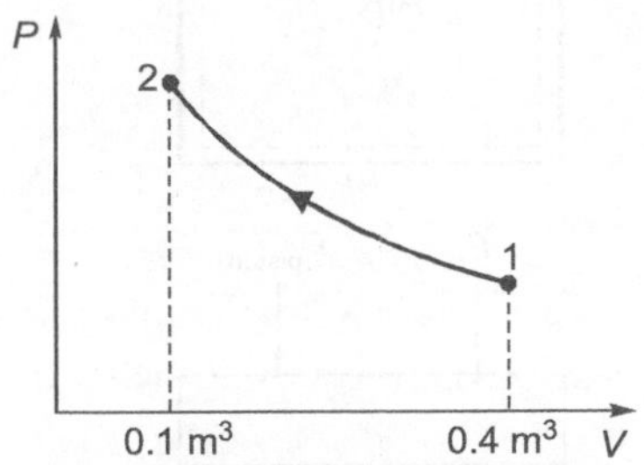

P-V diagram of isothermal process

$$W = p_1 V_1 \log_e \frac{V_2}{V_1}$$

$$= 100 \times 0.4 \log_e \frac{0.1}{0.4}$$

$$= \mathbf{-55.45\ kJ}$$

2.32 Sol.

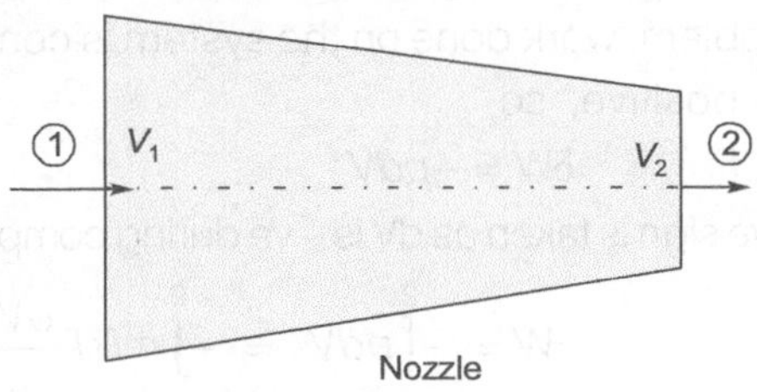

Nozzle

Applying steady flow energy equation between section 1 and 2.

$$h_1 + \frac{V_1^2}{2} + gz_1 + q = h_2 + \frac{V_2^2}{2} + gz_2 + w$$

where $q = 0$ insulated nozzle

$z_1 = z_2$ horizontal nozzle

$w = 0$ always for nozzle

$$h_1 + \frac{V_1^2}{2} = h_2 + \frac{V_2^2}{2}$$

$$h_1 + \frac{V_1^2}{2000} = h_2 + \frac{V_2^2}{2000}$$

where h_1 and h_2 are in kJ/kg and V_1 and V_2 are in m/s

$$\therefore 100 + \frac{(100)^2}{2000} = h_2 + \frac{(200)^2}{2000}$$

$$100 + 5 = h_2 + 20$$

or $h_2 = \mathbf{85\ kJ/kg}$

2.33 Sol.

Given that: $\bar{C}_v = 2.5\bar{R}$

$\bar{C}_v \rightarrow$ Molar specific heat at constant volume

$$\bar{C}_p - \bar{C}_v = \bar{R}$$

$$\Rightarrow \quad \bar{C}_p = \bar{R} + \bar{C}_v = \bar{R} + 2.5\bar{R}$$

$$\bar{C}_p = 3.5\bar{R}$$

Molar specific enthalpy change of ideal gas ($\Delta\bar{H}$)

$$\Delta H = n\bar{C}_p \Delta T$$

$$\overline{\Delta H} = \frac{\Delta H}{n} = \overline{\Delta H} = \bar{C}_p \Delta T$$

$$= 3.5 \times 8.314 \times 100$$

$$\overline{\Delta H} = 2909.9 \text{ J/mole or kJ/kmole}$$

2.34 (b)

Since, according to the statement given to problem, work done on the system is considered as positive, so

$$\delta W = -pdV$$

(–ve sign is taken as dV is – ve during compression)

$$W = -\int_1^2 pdV = -\int_1^2 mRT\frac{dV}{V}$$

$$= -mRT\int_1^2 \frac{dV}{V} = -mRT\ln\frac{V_2}{V_1}$$

$$= -mRT\ln\frac{p_1}{p_2} = -p_1V_1\ln\frac{p_1}{p_2}$$

2.35 (b)

$$c_p = 1 \text{ kJ/kg-K}$$

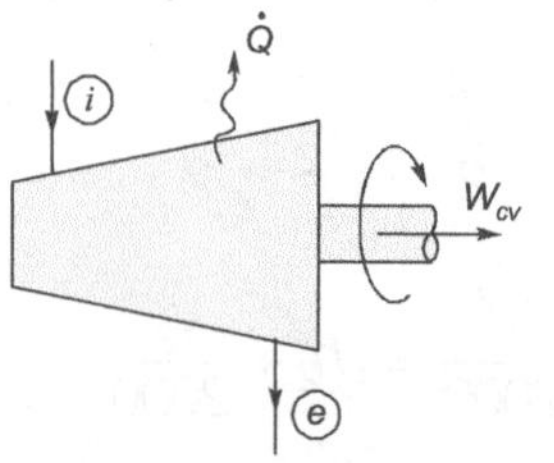

$$V_i = V_e \text{ (Inlet and exit velocity same)}$$

$$\dot{W}_{C.V} = 4.6 \text{ MW} = 4.6 \times 10^3 \text{ kW}$$

$$\dot{Q} = 300 \text{ kW}$$

Applying SFEE,

$$\dot{m}(h_i + \frac{1}{2}V_i^2 + gz_i) + \dot{Q} = \dot{m}(h_e + \frac{1}{2}V_e^2 + gz_e) + \dot{W}_{C.V.}$$

$$\dot{m}\left[c_p(T_i - T_e)\right] + \dot{Q} = \dot{W}_{C.V.} \quad \text{(Let } z_i = z_e)$$

$$\dot{m}[1\times(1100-400)] + (-300) = 4.6 \times 10^3$$

$$\Rightarrow \quad \dot{m} = \mathbf{7\ kg/s}$$

2.36 Sol.

$$\dot{m} = 0.1 \text{ kg/s},$$

$$\dot{Q} = 5 \text{ kW (heat loss)}$$

Applying SFEE

$$\dot{m}\left(h_1 + \frac{1}{2}C_1^2 + gz_1\right) + \dot{Q} = \dot{m}\left(h_2 + \frac{1}{2}C_2^2 + gz_2\right) + \dot{W}_{cv}$$

$$C_1 = 0 \text{ and } \dot{W}_{cv} = 0$$

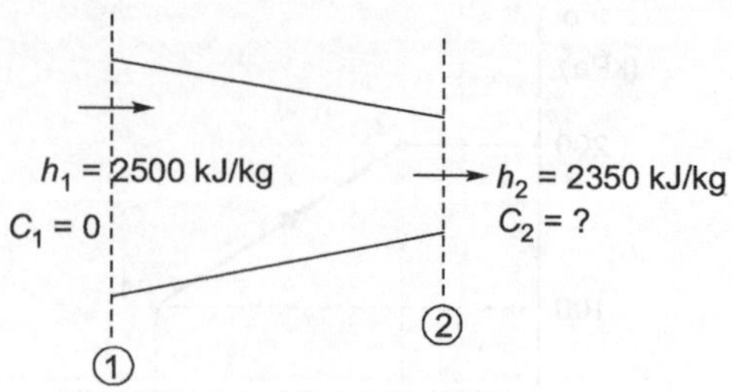

$$z_1 = z_2 \text{ (assume)}$$

$$\dot{m}h_1 + \dot{Q} = \dot{m}h_2 + \dot{m}\frac{1}{2}C_2^2$$

$$\Rightarrow \quad \dot{m}\times\frac{1}{2}C_2^2 = \dot{m}(h_1 - h_2) + \dot{Q}$$

$$\Rightarrow \quad 0.1\times\frac{1}{2}C_2^2\times 10^{-3} = 0.1(2500 - 2350) - 5$$

$$C_2 = \mathbf{447.213\ m/s}$$

2.37 Sol.

Since it is a case of reversible cycle, so work done will be equal to the area of P-v cycle.

Work output=Area enclosed on P-v diagram

$$= \frac{1}{2}(650-400)(2.5-2)$$

$$= \mathbf{62.5\ (kJ/cycle)}$$

2.38 Sol.

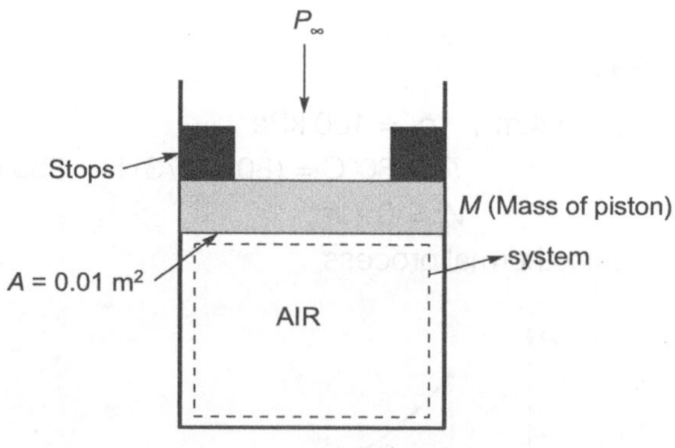

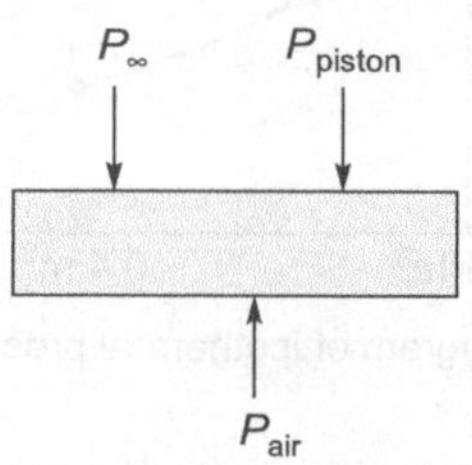

$$P_\infty = 100 \text{ kPa}$$

$$P_{piston} = \frac{Mg}{A} = \frac{25\times10\times10^{-3}}{0.01} = 25 \text{ kPa}$$

Total external pressure $= P_\infty + P_{piston} = 125$ kPa.

Initial pressure of air, P_i = 200 kPa.

Piston will be about to come down if system pressure is equal to total pressure i.e. 125 kPa.

⇒ Final pressure of air at which it is about to move is P_f = 125 kPa

Applying mass conservation at two states i and f

$$m_i = m_f$$

$$\frac{P_i V_i}{RT_i} = \frac{P_f V_f}{RT_f} \qquad [\because V_i = V_f]$$

$$\frac{P_i}{T_i} = \frac{P_f}{T_f}$$

$$\frac{200}{673} = \frac{125}{T_f}$$

$$T_f = 420.625 \text{ K}$$

$$T_f = \mathbf{147.625°C}$$

2.39 (a)

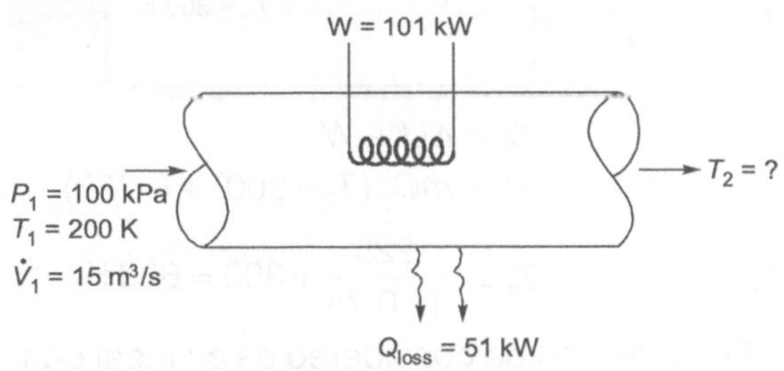

Since it is given an ideal gas

and c_p = 1 kJ/kgK

R = 0.5 kJ/kgK

at inlet $P_1\dot{V}_1 = \dot{m}RT_1$

$$\dot{m} = \frac{P_1\dot{V}_1}{RT_1} = \frac{100\times 15}{0.5\times 300} = 10 \text{ kg/s}$$

Using SFEE

$$\dot{m}h_1 + Q = \dot{m}h_2 + W$$

$$h_1 + \frac{Q}{\dot{m}} = h_2 + \frac{W}{\dot{m}}$$

$$c_pT_1 - \left(\frac{51}{10}\right) = c_pT_2 - \frac{101}{10}$$

$$(T_2 - T_1)c_p = \frac{101-51}{10} = 5$$

$$T_2 - T_1 = \frac{5}{c_p} = 5\text{K}$$

$$T_2 = 300 + 5 = 305\text{K or } 32°\text{C}$$

2.40 Sol.

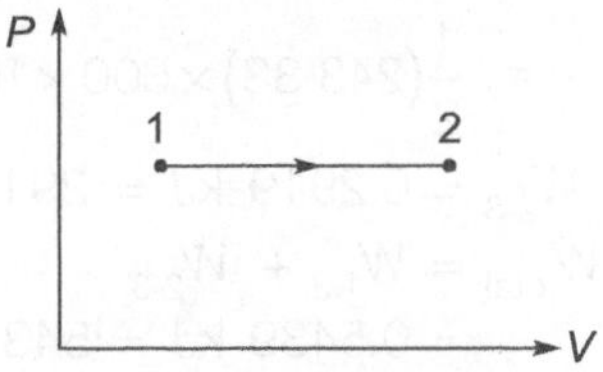

m = 1 kg; T_1 = 127°C = 400 K

$P = C$; $V_2 = 2V_1$; R = 0.287 kJ/kgK

$$W = P_1(V_2 - V_1) = P_1(2V_1 - V_1)$$
$$= P_1\cdot V_1 = mRT_1 = 1 \times 0.287 \times 400$$
$$= 114.8 \text{ kJ}$$

2.41 Sol.

As per given information

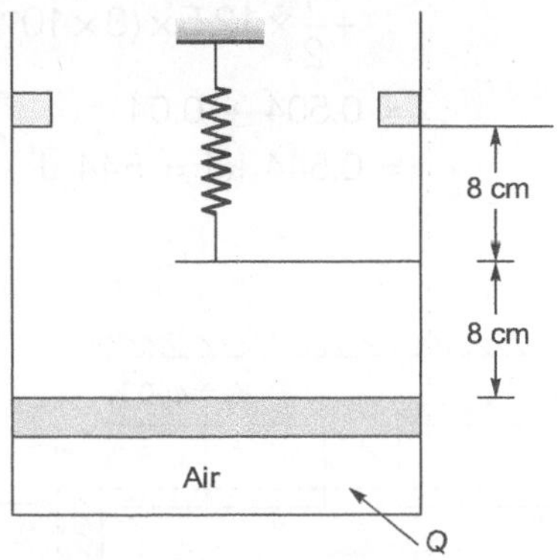

P_0 = 100 kPa, P_1 = 105 kPa,

K = 12.5 N/mm = 12.5 kN/m

$$A = 300 \text{ cm}^2 = 300 \times 10^{-4} \text{ m}^2$$

$$x = 8 \text{ cm} = 8 \times 10^{-2} \text{ m}$$

1-2 constant pressure

$$W_{1\text{-}2} = P_1 \times A \times x$$
$$= 105 \times 300 \times 10^{-4} \times 8 \times 10^{-2}$$
$$= 0.252 \text{ kJ} = 252 \text{ J}$$

$$P_2 = 105 \text{ kPa}$$

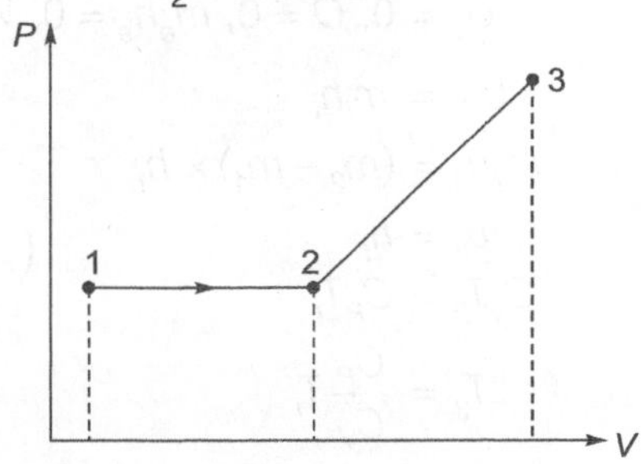

At 3, $P_3 \times A = P_2A + Kx$

$$P_3 = P_2 + \frac{Kx}{A} = 105 + \frac{12.5\times 8\times 10^{-2}}{300\times 10^{-4}}$$
$$= 138.33 \text{ kPa}$$

$$W_{2\text{-}3} = \frac{1}{2}(P_2 + P_3)\times(V_3 - V_2)$$

$$= \frac{1}{2}(105+138.33)A \times x$$

$$= \frac{1}{2}(243.33)\times 300\times 10^{-4}\times 8\times 10^{-2}$$

$$W_{2\text{-}3} = 0.2919 \text{ kJ} = 291.9 \text{ J}$$

$$\therefore \quad W_{total} = W_{1\text{-}2} + W_{2\text{-}3}$$

$$= 0.5439 \text{ kJ} = 543.91 \approx 544 \text{ J}$$

Alternate Solution:

Total work = Work done because of 105 kPa pressure + Work done against spring which is equal to energy stored in spring

$$\text{Work done} = P_1 \times A \times 2x + \frac{1}{2}k\cdot x^2$$

$$= 105\times 300\times 10^{-4}\times 2\times 8\times 10^{-2} + \frac{1}{2}\times 12.5\times (8\times 10^{-2})^2$$

$$= 0.504 + 0.04$$

$$= 0.544 \text{ kJ} = 544 \text{ J}$$

2.42 Sol

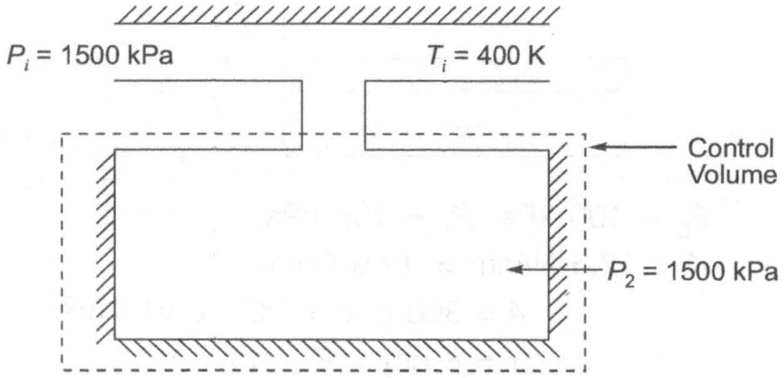

Let, 1 = Initial condition of tank

2 = Final condition of tank

Applying unsteady state energy equation

$$U_2 - U_1 = m_i h_i + Q - m_e h_e - W_{CV}$$

$$U_1 = 0,\ Q = 0,\ m_e h_e = 0,\ W_{CV} = 0$$

$$U_2 = m_i h_i$$

$$m_2 u_2 = (m_2 - m_1)\times h_i$$

$$u_2 = h_i \qquad (\because m_1 = 0)$$

$$C_V T_2 = C_P T_i$$

$$T_2 = \frac{C_P}{C_V}T_i$$

$$T_2 = \left(\frac{1.005}{0.718}\right)\times 400$$

$$T_2 = 559.9 \text{ K}$$

2.43 (408.92) (404.00 to 414.00)

$$P_1 V_1^{1.2} = P_2 V_2^{1.2}$$

$$110(5)^{1.2} = P_2(2.5)^{1.2}$$

$$P_2 = 252.7136 \text{ kPa}$$

$$W = \frac{P_1V_1 - P_2V_2}{n-1}$$

$$= \frac{110\times 5 - 252.7136\times 2.5}{1.2-1}$$

$$W = -408.92 \text{ kJ}$$

$\therefore$ Magnitude is 408.92 kJ.

2.44 (2) (2 to 2)

Given: C_V = 750 J/kgK = 0.75 kJ/kgK, m = 1 kg

By first law of thermodynamics

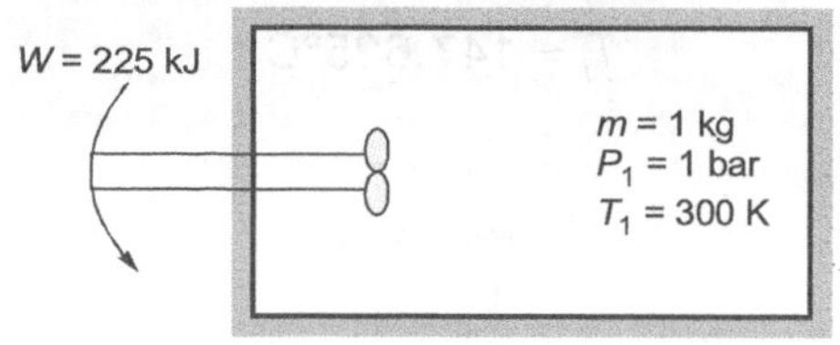

$$Q = \Delta U + W$$

$$0 = mC_V(T_2 - 300) + (-225)$$

$$T_2 = \frac{225}{1\times 0.75} + 300 = 600 \text{ K}$$

Since air can be considered as an ideal gas,

$$PV = mRT$$

$$P \propto T \qquad [\because V = \text{Constant}]$$

$$\frac{P_2}{P_1} = \frac{T_2}{T_1}$$

$$P_2 = \frac{600}{300}\times 1 \text{ bar} = 2 \text{ bar}$$

2.45 (395.388) (385 to 405)

Here, i = Inlet, e = Exit,

Initial condition of CV ⇒ 1

Final condition of CV ⇒ 2

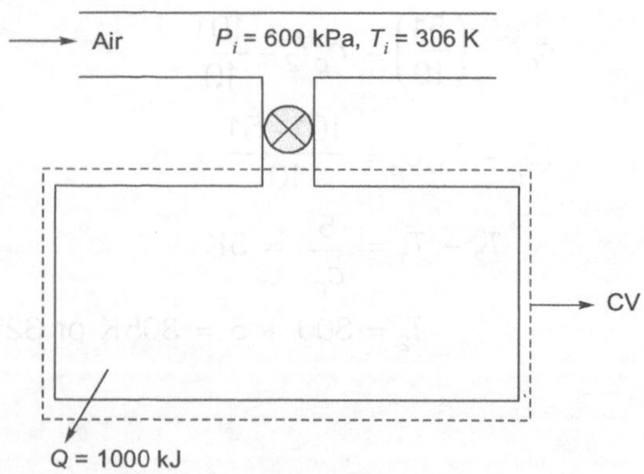

From conservation of mass:

$$m_2 - \cancel{m_1}^{\,0} = m_i - \cancel{m_e}^{\,0}$$

$$m_2 = m_e$$

Conservation of energy:

$$U_2 - \cancel{U_1}^{\,0} = m_i h_i + Q - \cancel{m_e h_e}^{\,0} - \cancel{W_{CV}}^{\,0}$$

$$m_2 u_2 = m_2 \times C_P T_i + (-1000)$$

$$[\because u_2 = C_V T_2 \text{ and } m_i = m_2]$$

$$m_2 C_V T_2 = m_2 C_V T_i - 1000$$

$$\frac{P_2 V}{R T_2} \times C_V \times T_2 = \frac{P_2 V}{R T_2} \times C_V \times T_i - 1000$$

[By Ideal gas equation]

$$\frac{600 \times 8}{0.287} \times 0.718 = \frac{600 \times 8}{0.287 \times T_2} \times 1.005 \times 306 - 1000$$

$$\Rightarrow \quad T_2 = 395.388 \text{ K}$$

2.46 (b)

From the given data, $V_1 = L_1 \times A$

$$V_2 = L_2 \times A$$

$$W_{atm} = P_0(V_2 - V_1)$$

$$= P_0(L_2 A - L_1 A)$$

$$= P_0 A(L_2 - L_1)$$

2.47 ((59.6) (58.0 to 61.0)

Let A_p be the area of piston.

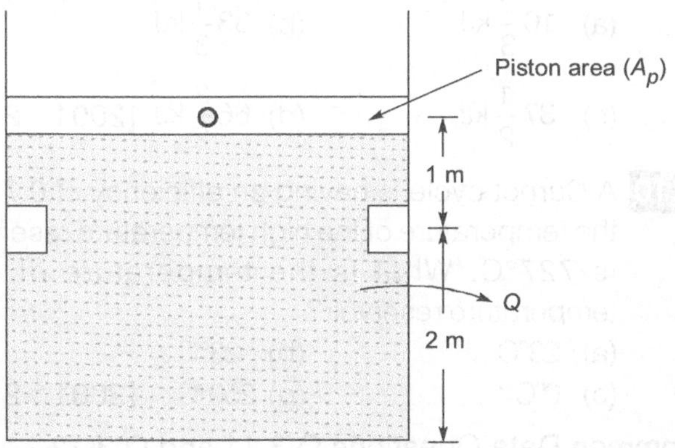

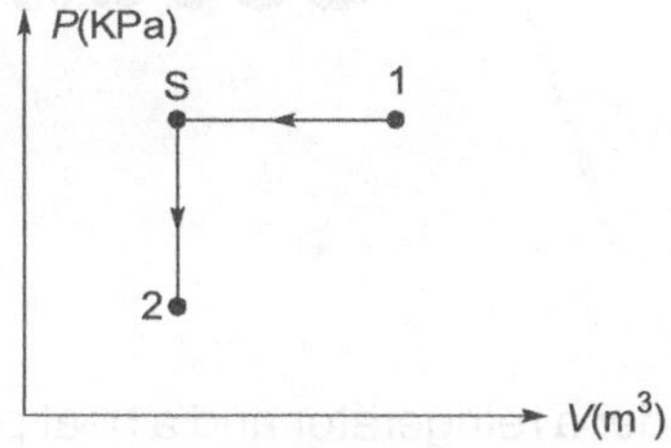

Given: $P_1 = 140$ KPa

$$T_1 = 350°C = 350 + 273 = 623 \text{ K}$$

$$T_2 = 25°C = 25 + 273 = 298 \text{ K}$$

$$W_{1-s} = P_1 V_1 - P_s V_s$$

$$= mR(T_1 - T_s)$$

For constant pressure process:

$$V \propto T$$

$$\Rightarrow \quad \frac{V_1}{V_s} = \frac{T_1}{T_s}$$

$$\Rightarrow \quad \frac{A_p \times 3}{A_p \times 2} = \frac{623}{T_s} \Rightarrow T_s = 415.33 \text{ K}$$

$$W_{1-s} = 1 \times 0.287 \times (623 - 415.33)$$

$$= 59.60 \text{ kJ/kg}$$

$$W_{s-2} = 0$$

[$\because$ Constant volume process]

$$\therefore \quad W_{1-2} = W_{1-s} + W_{s-2}$$

$$= 59.60 + 0 = 59.60 \text{ kJ/kg}$$

■■■■

3 Second Law of Thermodynamics, Carnot Cycle and Entropy

3.1 Consider a refrigerator and a heat pump working on the reversed Carnot cycle between the same temperature limits. Which of the following is correct?
(a) COP of refrigerator = COP of heat pump
(b) COP of refrigerator = COP of heat pump + 1
(c) COP of refrigerator = COP of heat pump – 1
(d) COP of refrigerator = Inverse of the COP of heat pump **[1995 : 1 M]**

3.2 In the case of a refrigeration system undergoing an irreversible cycle, $\oint \frac{\delta Q}{T}$ is.... (< 0 / = 0 / > 0)
[1995 : 1 M]

3.3 One kilomole of an ideal gas is throttled from an initial pressure of 0.5 MPa to 0.1 MPa. The initial temperature is 300 K. The entropy change of the universe is
(a) 13.38 kJ/K (b) 4014.3 kJ/K
(c) 0.0446 kJ/K (d) 0.0445 kJ/K
[1995 : 2 M]

3.4 A solar energy based heat engine which receives 80 kJ of heat at 100°C and rejects 70 kJ of heat to the ambient at 30°C is to be designed. The thermal efficiency of the heat engine is
(a) 70% (b) 18.8%
(c) 12.5% (d) Indetermine
[1996 : 2 M]

3.5 For two cycles coupled in series, the topping cycle has an efficiency of 30% and the bottoming cycle has an efficiency of 20%. The overall combined cycle efficiency is
(a) 50% (b) 44%
(c) 38% (d) 55% **[1996 : 2 M]**

3.6 For an ideal gas the expression $\left[T\left(\frac{\delta s}{\delta T}\right)_p - T\left(\frac{\delta s}{\delta T}\right)_v\right]$ is always equal to
(a) Zero (b) $\frac{c_p}{c_v}$
(c) R (d) RT **[1997 : 1 M]**

3.7 system undergoes a state change from 1 to 2. According to the second law of thermodynamics, for the process to be feasible, the entropy change, $S_2 - S_1$ of the system
(a) is positive or zero
(b) is negative or zero
(c) is zero
(d) can be positive, negative or zero
[1997 : 1 M]

3.8 hen an ideal gas with constant specific heat is throttled adiabatically, with negligible changes in kinetic and potential energies.
(a) $\Delta h = 0, \Delta T = 0$ (b) $\Delta h > 0, \Delta T = 0$
(c) $\Delta h > 0, \Delta S > 0$ (d) $\Delta h = 0, \Delta S > 0$
[2000 : 2 M]

3.9 cycle heat engine does 50 kJ of work per cycle. If the efficiency of the heat engine is 75%. The heat rejected per cycle is
(a) $16\frac{2}{3}$ kJ (b) $33\frac{1}{3}$ kJ
(c) $37\frac{1}{2}$ kJ (d) $66\frac{2}{2}$ kJ **[2001 : 2 M]**

3.10 A Carnot cycle is having an efficiency of 0.75. If the temperature of the high temperature reservoir is 727°C. What is the temperature of low temperature reservoir?
(a) 23°C (b) –23°C
(c) 0°C (d) 250°C **[2001 : 2 M]**

Common Data Questions Q.3.11 and Q.3.12

Nitrogen gas (molecular weight 28) is enclosed in a cylinder by a piston, at the initial condition of 2 bar, 298 K and 1 m^3. In a particular process, the gas slowly expands under isothermal condition, until the volume becomes 2 m^3. Heat exchange occurs with the atmosphere at 298 K during this process.

3.11 The work interaction for the nitrogen gas is
(a) 200 kJ (b) 138.6 kJ
(c) 2 kJ (d) –200 kJ **[2003 : 2 M]**

3.12 The entropy changes for the universe during the process in kJ/K is
(a) 0.4652 (b) 0.0067
(c) 0 (d) – 0.6711 **[2003 : 2 M]**

3.13 Considering the relationship $TdS = dU + pdV$ between the entropy (S), internal energy (U), pressure (p), temperature (T) and volume (V), which of the following statements is correct?
(a) It is applicable only for a reversible process
(b) For an irreversible process, $TdS > dU + pdV$
(c) It is valid only for an ideal gas
(d) It is equivalent to 1st law, for a reversible process **[2003 : 2 M]**

3.14 A solar collector receiving solar radiation at the rate of 0.6 kW/m^2 transforms it to the internal energy of a fluid at an overall efficiency of 50%. The fluid heated to 350 K is used to run a heat engine which rejects heat at 313 K. If the heat engine is to deliver 2.5 kW power, the minimum area of the solar collector required would be
(a) 83.33 m^2 (b) 16.66 m^2
(c) 39.68 m^2 (d) 79.36 m^2 **[2004 : 2 M]**

3.15 The following four figures have been drawn to represent a fictitious thermodynamic cycle, on the *P-v* and *T-s* planes.

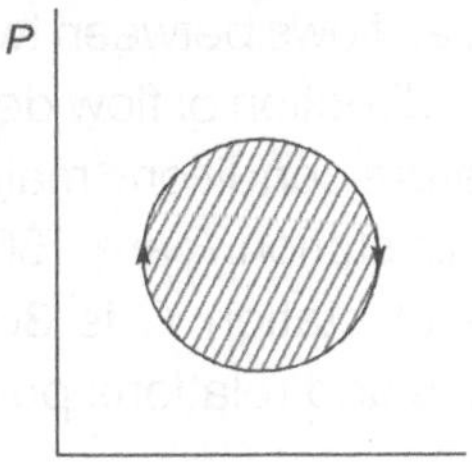
Fig. 1

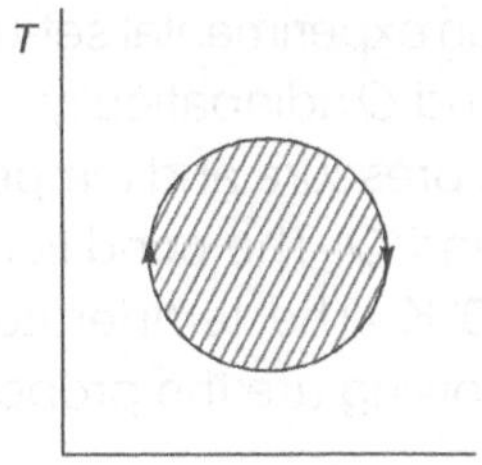
Fig. 2

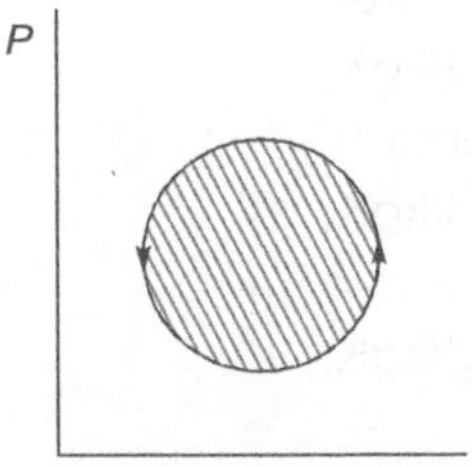
Fig. 3

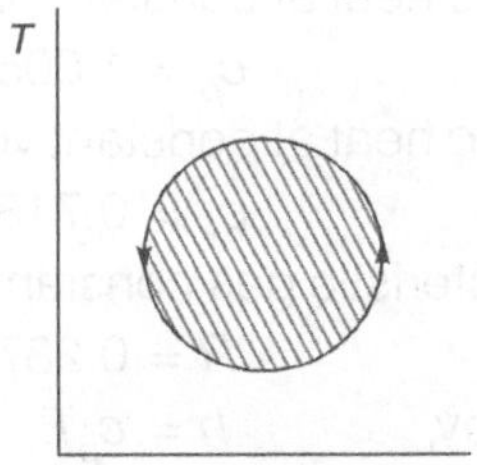
Fig. 4

According to the first law of thermodynamics, equal areas are enclosed by
(a) Figures 1 and 2 (b) Figures 1 and 3
(c) Figures 1 and 4 (d) Figures 2 and 3
[2005 : 1 M]

3.16 A heat transformer is device that transfers a part of the heat, supplied to it at an intermediate temperature, to a high temperature reservoir while rejecting the remaining part to a low temperature heat sink. In such a heat transformer, 100 kJ of heat is supplied at 350 K. The maximum amount of heat in kJ that can be transferred to 400 K, when the rest is rejected to a heat sink at 300 K is
(a) 12.50 (b) 14.29
(c) 33.33 (d) 57.14 **[2007 : 2 M]**

Common Data for Questions 3.17 and 3.18

A thermodynamic cycle with an ideal gas as working fluid is shown below.

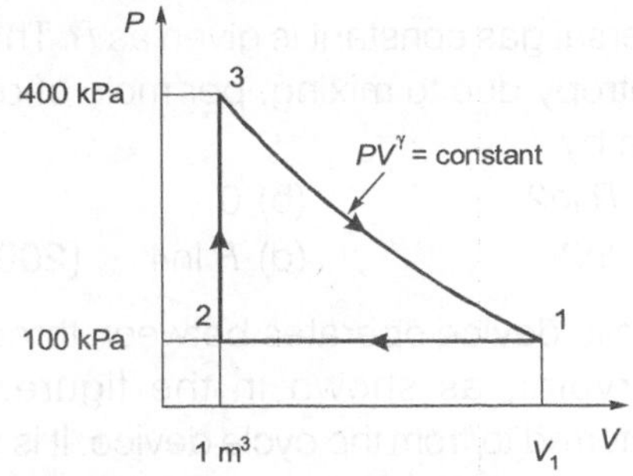

3.17 The above cycle is represented on *T-s* plane by

(a)
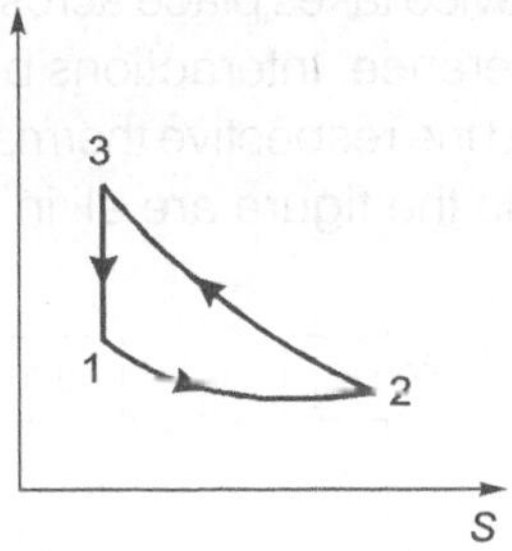

(b)
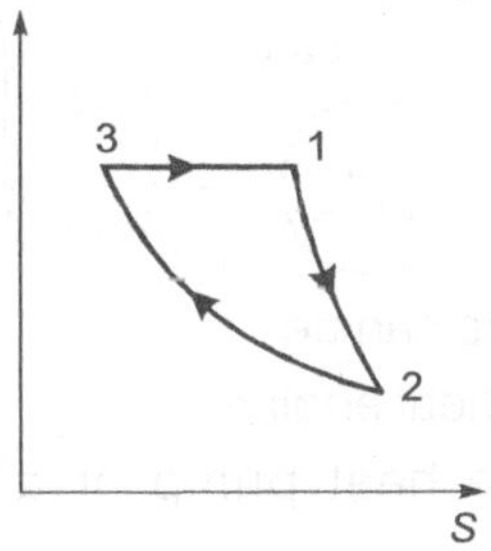

(c)
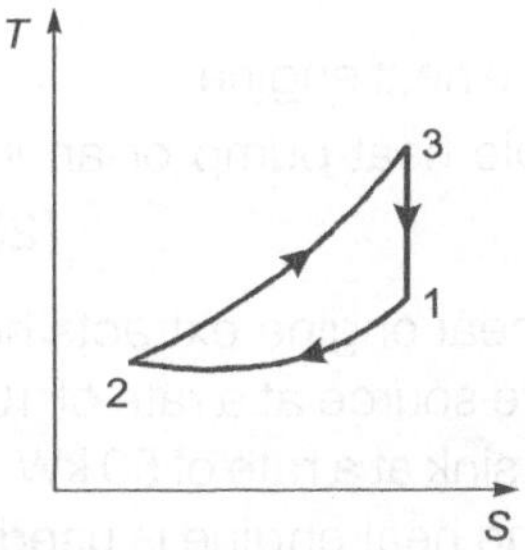

(d)
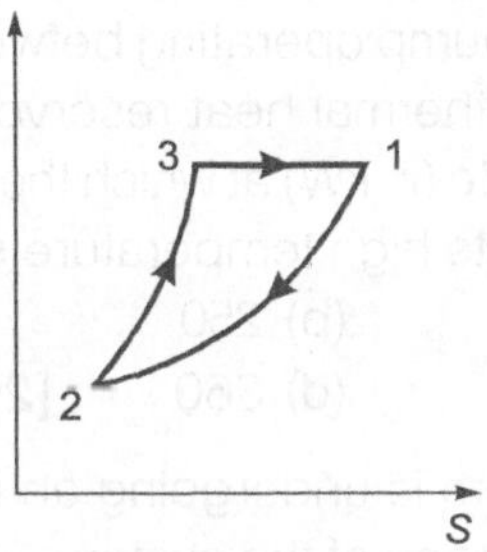

[2007 : 2 M]

3.18 If the specific heats of the working fluid are constant and the value of specific heat ratio γ is 1.4, the thermal efficiency (%) of the cycle is
(a) 21 (b) 40.9
(c) 42.6 (d) 59.7 **[2007 : 2 M]**

3.19 2 moles of oxygen are mixed adiabatically with another 2 moles of oxygen in a mixing chamber, so that the final total pressure and temperature of the mixtures become same as those of the individual constituents at their initial states. The universal gas constant is given as R. The change in entropy due to mixing, per mole of oxygen, is given by

(a) $-R \ln 2$ (b) 0
(c) $R \ln 2$ (d) $R \ln 4$ [2008 : 1 M]

3.20 A cyclic device operates between three thermal reservoirs, as shown in the figure. Heat is transferred to/from the cycle device. It is assumed that heat transfer between each thermal reservoir and the cyclic device takes place across negligible temperature difference. Interactions between the cyclic device and the respective thermal reservoirs that are shown in the figure are all in the form of heat transfer.

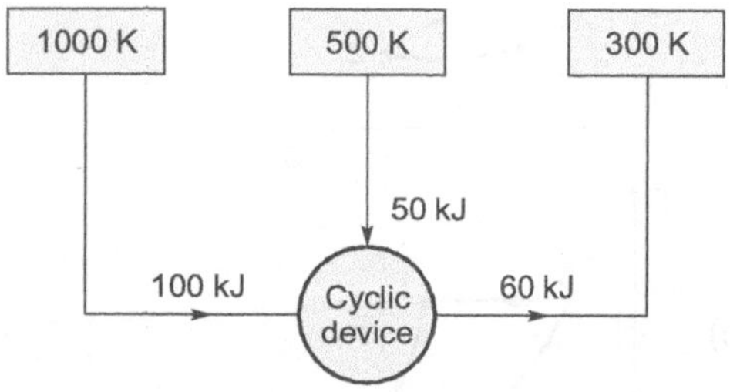

The cyclic device can be
(a) a reversible heat engine
(b) a reversible heat pump or a reversible refrigerator
(c) an irreversible heat engine
(d) an irreversible heat pump or an irreversible refrigerator. [2008 : 2 M]

3.21 An irreversible heat engine extracts heat from a high temperature source at a rate of 100 kW and rejects heat to a sink at a rate of 50 kW. The entire work output of the heat engine is used to drive a reversible heat pump operating between a set of independent isothermal heat reservoirs at 17°C and 75°C. The rate (in kW) at which the heat pump delivers heat to its high temperature sink is

(a) 50 (b) 250
(c) 300 (d) 360 [2009 : 2 M]

3.22 If a closed system is undergoing an irreversible process, the entropy of the system
(a) must increase
(b) always remains constant
(c) must decrease
(d) can increase, decrease or remain constant [2009 : 1 M]

3.23 One kilogram of water at room temperature is brought into contact with a high temperature thermal reservoir. The entropy change of the universe is
(a) equal to entropy change of the reservoir
(b) equal to entropy change of water
(c) equal to zero
(d) always positive [2010 : 1 M]

3.24 Consider the following two processes;
I. A heat source at 1200 K loses 2500 kJ of heat to a sink at 800 K
II. A heat source at 800 K loses 2000 kJ of heat to a sink at 500 K

Which of the following statements is true?
(a) Process I is more irreversible than Process II
(b) Process II is more irreversible than Process I
(c) Irreversibility associated in both the processes are equal
(d) Both the processes are reversible [2010 : 2 M]

Common Data Question 3.25 and 3.26

In an experimental set-up, air flows between two stations P and Q adiabatically. The direction of flow depends on the pressure and temperature conditions maintained at P and Q. The conditions at station P are 150 kPa and 350 K. The temperature at station Q is 300 K. The following are the properties and relations pertaining to air:

Specific heat at constant pressure,
$$c_p = 1.005 \text{ kJ/kgK};$$
Specific heat at constant volume,
$$c_v = 0.718 \text{ kJ/kgK};$$
Characteristic gas constant,
$$R = 0.287 \text{ kJ/kgK}.$$
Enthalpy, $h = c_p T$.
Internal energy, $u = c_v T$.

3.25 If the air has to flow from station P to station Q, the maximum possible value of pressure in kPa at station Q is close to

(a) 50 (b) 87
(c) 128 (d) 150 [2011 : 2 M]

3.26 If the pressure at station Q is 50 kPa, the change in entropy $(s_Q - s_P)$ in kJ/kgK is

(a) -0.155 (b) 0
(c) 0.160 (d) 0.355 [2011 : 2 M]

3.27 An ideal gas of mass m and temperature T_1 undergoes a reversible isothermal process from an initial pressure P_1 to final pressure P_2. The heat loss during the process is Q. The entropy change ΔS of the gas is

(a) $mR\ln\left(\frac{P_2}{P_1}\right)$ (b) $mR\ln\left(\frac{P_1}{P_2}\right)$

(c) $mR\ln\left(\frac{P_2}{P_1}\right)-\frac{Q}{T_1}$ (d) zero [2012 : 1 M]

3.28 Which one of the following pairs of equations describes an irreversible heat engine?

(a) $\oint \delta Q > 0$ and $\oint \frac{\delta Q}{T} < 0$

(b) $\oint \delta Q < 0$ and $\oint \frac{\delta Q}{T} < 0$

(c) $\oint \delta Q > 0$ and $\oint \frac{\delta Q}{T} > 0$

(d) $\oint \delta Q < 0$ and $\oint \frac{\delta Q}{T} > 0$ [2014 : 1 M, Set-3]

3.29 A source at a temperature of 500 K provides 1000 kJ of heat. The temperature of environment is 27°C. The maximum useful work (in kJ) that can be obtained from the heat source is _______.

[2014 : 1 M, Set-3]

3.30 A reversible heat engine receives 2 kJ of heat from a reservoir at 1000 K and a certain amount of heat from a reservoir at 800 K. It rejects 1 kJ of heat to a reservoir at 400 K. The net work output (in kJ) of the cycle is

(a) 0.8 (b) 1.0

(c) 1.4 (c) 2.0

[2014 : 2 M, Set-1]

3.31 An amount of 100 kW of heat is transferred through a wall in steady state. One side of the wall is maintained at 127°C and the other side at 27°C. The entropy generated (in W/K) due to the heat transfer through the wall is _____.

[2014 : 2 M, Set-3]

3.32 A closed system contains 10 kg of saturated liquid ammonia at 10°C. Heat addition required to convert the entire liquid into saturated vapour at a constant pressure is 16.2 MJ. If the entropy of the saturated liquid is 0.88 kJ/kgK, the entropy (in kJ/kgK) of saturated vapour is _____

[2014 : 2 M, Set-4]

3.33 Two identical metal blocks L and M (specific heat = 0.4 kJ/kg K), each having a mass of 5 kg, are initially at 313 K. A reversible refrigerator extracts heat from block L and rejects heat to block M until the temperature of block L reaches 293 K. The final temperature (in K) of block M is _____

[2014 : 2 M, Set-4]

3.34 A Carnot engine (CE-1) works between two temperature reservoirs A and B, where $T_A = 900$ K and $T_B = 500$ K. A second Carnot engine (CE-2) works between temperature reservoirs B and C, where $T_C = 300$ K. In each cycle of CE-1 and CE-2, all the heat rejected by CE-1 to reservoir B is used by CE-2. For one cycle of operation, if the net Q absorbed by CE-1 from reservoir A is 150 MJ, the net heat rejected to reservoir C by CE-2 (in MJ) is _____.

[2015 : 1 M, Set-1]

3.35 One kg of air (R = 287 J/kgK) undergoes an irreversible process between equilibrium state 1 (20°C, 0.9 m^3) and equilibrium state 2 (20°C, 0.6 m^3). The change in entropy $s_2 - s_1$ (in J/kgK) is _______.

[2015 : 2 M, Set-2]

3.36 A reversible cycle receives 40 kJ of heat from one heat source at a temperature of 127°C and 37 kJ from another heat source at 97°C. The heat rejected (in kJ) to the heat sink at 47°C is____

[2016 : 2 M, Set-2]

3.37 A heat pump absorbs 10 kW of heat from outside environment at 250K while absorbing 15 kW of work. It delivers the heat to a room that must be kept warm at 300 K. The Coefficient of Performance (COP) of the heat pump is _______.

[2017 : 1 M, Set-1]

3.38 One kg of an ideal gas (gas constant, R = 400 J/kgK; specific heat at constant volume, c_v = 1000 J/kgK) at 1 bar, and 300 K is contained in a sealed rigid cylinder. During an adiabatic process, 100 kJ of work is done on the system by a stirrer. The increase in entropy of the system is _______ J/K. [2017 : 2 M, Set-1]

3.39 One kg of an ideal gas (gas constant R = 287 J/kgK) undergoes an irreversible process from state-1 (1 bar, 300 K) to state-2 (2 bar, 300 K). The change in specific entropy ($s_2 - s_1$) of the gas (in J/kgK) in the process is _______.

[2017 : 2 M, Set-2]

3.40 An ideal gas undergoes a process from state 1 (T_1 = 300 K, p_1 = 100 kPa) to state 2 (T_2 = 600 K, p_2 = 500 kPa). The specific heats of the ideal gas are: c_p = 1 kJ/kg-K and c_v = 0.7 kJ/kg-K. The change in specific entropy of the ideal gas from state 1 to state 2 (in kJ/kg-K) is_____(correct to two decimal places) [2018 : 1 M, Set-1]

3.41 For an ideal gas with constant properties undergoing a quasi-static process, which one of the following represents the change of entropy (Δs) from state 1 to 2?

(a) $\Delta s = c_p \ln\left(\frac{T_2}{T_1}\right) - R\ln\left(\frac{P_2}{P_1}\right)$

(b) $\Delta s = c_v \ln\left(\frac{T_2}{T_1}\right) - c_p \ln\left(\frac{V_2}{V_1}\right)$

(c) $\Delta s = c_p \ln\left(\frac{T_2}{T_1}\right) - c_v \ln\left(\frac{P_2}{P_1}\right)$

(d) $\Delta s = c_v \ln\left(\frac{T_2}{T_1}\right) + R\ln\left(\frac{V_1}{V_2}\right)$

[2018 : 1 M, Set-2]

3.42 Air of mass 1 kg, initially at 300 K and 10 bar, is allowed to expand isothermally till it reaches a pressure of 1 bar. Assuming air as an ideal gas with gas constant of 0.287 kJ/kgK, the change in entropy of air (in kJ/kgK, round off to two decimal places) is ________. [2019 : 1 M, Set-1]

3.43 The figure shows a heat engine (HE) working between two reservoirs. The amount of heat (Q_2) rejected by the heat engine is drawn by a heat pump (HP). The heat pump receives the entire work output (W) of the heat engine. If temperature, $T_1 > T_3 > T_2$, then the relation between the efficiency (η) of the heat engine and the coefficient of performance (COP) of the heat pump is

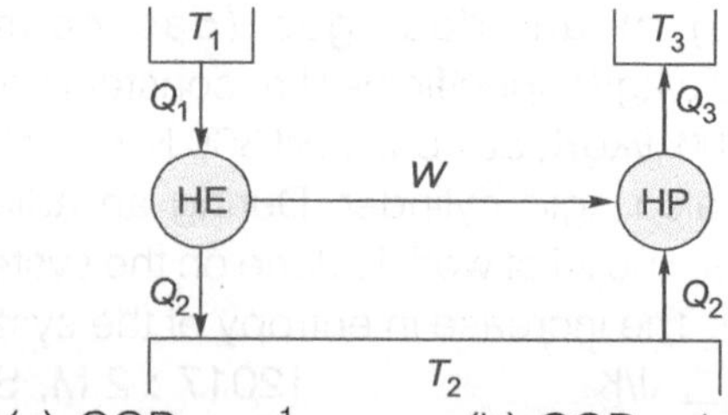

(a) COP = η^{-1} (b) COP = $1 + \eta$
(c) COP = $\eta^{-1} - 1$ (d) COP = η

[2019 : 2 M, Set-2]

3.44 For an ideal gas, a constant pressure line and a constant volume line intersect at a point, in the temperature (T) versus specific entropy (s) diagram. C_P is the specific heat at constant pressure and C_V is the specific heat at constant volume. The ratio of the slopes of the constant pressure and constant volume lines at the point of intersection is

(a) $\frac{C_P - C_V}{C_P}$ (b) $\frac{C_P - C_V}{C_V}$

(c) $\frac{C_P}{C_V}$ (d) $\frac{C_V}{C_P}$ [2020 : 1 M, Set-1]

3.45 One kg of air in a closed system undergoes an irreversible process from an initial state of p_1 = 1 bar (absolute) and T_1 = 27°C, to a final state of p_2 = 3 bar (absolute) and T_2 = 127°C. If the gas constant of air is 287 J/kgK and the ratio of the specific heats γ = 1.4, then the change in the specific entropy (in J/kgK) of the air in the process is

(a) 172.0
(b) –26.3
(c) indeterminate, as the process is irreversible
(d) 28.4 [2020 : 2 M, Set-2]

3.46 An adiabatic vortex tube, shown in the figure given below is supplied with 5 kg/s of air (inlet 1) at 500 kPa and 300 K. Two separate streams of air are leaving the device from outlets 2 and 3. Hot air leaves the device at a rate of 3 kg/s from outlet 2 at 100 kPa and 340 K, while 2 kg/s of cold air stream is leaving the device from outlet 3 at 100 kPa and 240 K.

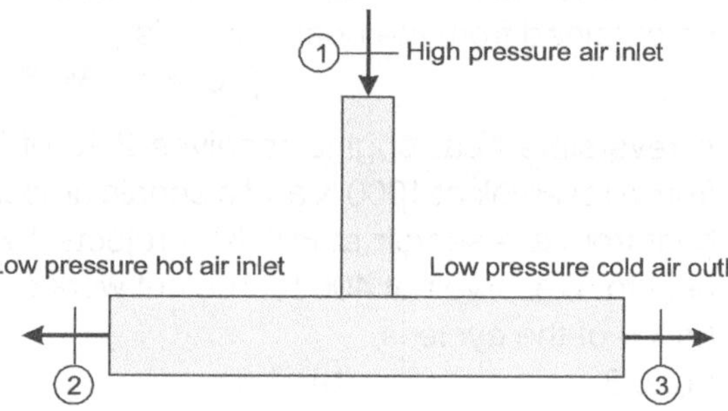

Consider constant specific heat of air is 1005 J/kg.K and gas constant is 287 J/kg.K. There is no work transfer across the boundary of this device. The rate of entropy generation is ________ kW/K (round off to one decimal place).

[2021 : 2 M, Set-2]

3.47 The Clausius inequality holds good for

(a) any process
(b) any cycle
(c) only reversible process
(d) only reversible cycle [2022 : 1 M, Set-1]

3.48 Consider a one-dimensional steady heat conduction process through a solid slab of thickness 0.1 m. The higher temperature side A has a surface temperature of 80°C, and the heat transfer rate per unit area to low temperature side B is 4.5 kW/m^2. The thermal conductivity of the slab is 15 W/m.K. The rate of entropy generation per unit area during the heat transfer process is ________ W/m^2.K (round off to 2 decimal places).

[2022 : 2 M, Set-1]

3.49 At steady state, 500 kg/s of steam enters a turbine with specific enthalpy equal to 3500 kJ/kg and specific entropy equal to 6.5 $kJ \cdot kg^{-1} \cdot K^{-1}$. It expands reversibly in the turbine to the condenser pressure. Heat loss occurs reversibly in the turbine at a temperature of 500 K. If the exit specific enthalpy and specific entropy are 2500 kJ/kg and 6.3 $kJ \cdot kg^{-1} \cdot K^{-1}$, respectively, the work output from the turbine is ________ MW (in integer). [2022 : 2 M, Set-2]

3.50 A heat engine extracts heat (Q_H) from a thermal reservoir at a temperature of 1000 K and rejects heat (Q_L) to a thermal reservoir at a temperature of 100 K, while producing work (W). Which one of the combinations of Q_H, Q_L and W given is allowed?

(a) $Q_H = 2000$ J, $Q_L = 500$ J, $W = 1000$ J
(b) $Q_H = 2000$ J, $Q_L = 750$ J, $W = 1250$ J
(c) $Q_H = 6000$ J, $Q_L = 500$ J, $W = 5500$ J
(d) $Q_H = 6000$ J, $Q_L = 600$ J, $W = 5500$ J

[2023 : 1 M]

3.51 A furnace can supply heat steadily at 1200 K at a rate of 24000 kJ/min. The maximum amount of power (in kW) that can be produced by using the heat supplied by this furnace in an environment at 300 K is

(a) 0 (b) 150
(c) 300 (d) 18000

[2024 : 1 M]

3.52 A heat pump (HP) is driven by the work output of a heat engine (HE) as shown in the figure. The heat engine extracts 150 kJ of heat from the source at 1000 K. The heat pump absorbs heat from the ambient at 280 K and delivers heat to the room which is maintained at 300 K. Considering the combined system to be ideal, the total amount of heat delivered to the room together by the heat engine and heat pump is_____ kJ. [Answer in integer]

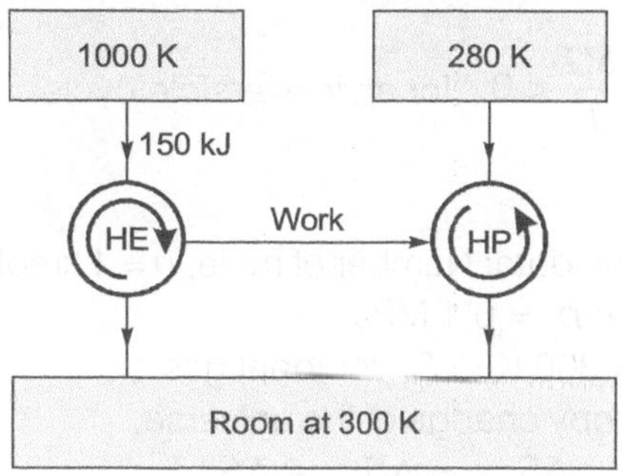

[2024 : 2 M]

Answers Second Law, Carnot Cycle and Entropy

3.1 (c)	3.2 (< 0)	3.3 (a)	3.4 (c)	3.5 (b)	3.6 (c)	3.7 (d)	3.8 (a, d)
3.9 (a)	3.10 (b)	3.11 (b)	3.12 (c)	3.13 (d)	3.14 (d)	3.15 (a)	3.16 (d)
3.17 (c)	3.18 (a)	3.19 (b)	3.20 (a)	3.21 (c)	3.22 (d)	3.23 (d)	3.24 (b)
3.25 (b)	3.26 (c)	3.27 (b)	3.28 (a)	3.29 (400)	3.30 (c)	3.31 (83.33)	
3.32 (6.604)	3.33 (334.36)	3.34 (50)	3.35 (−116.368)	3.36 (64)	3.37 (1.67)	3.38 (287.682)	
3.39 (−198.93)		3.40 (0.21)	3.41 (a)	3.42 (0.66)	3.43 (a)	3.44 (d)	3.45 (b)
3.46 (2.2)	3.47 (b)	3.48 (1.184)	3.49 (450)	3.50 (b)	3.51 (c)	3.52 (1620)	

Explanations Second Law, Carnot Cycle and Entropy

3.1 (c)

Refrigerator:

Coefficient of performance of refrigerator,

$$(COP)_R = \frac{\text{Cooling effect: } Q_2}{\text{Work input: } W} = \frac{Q_2}{W}$$

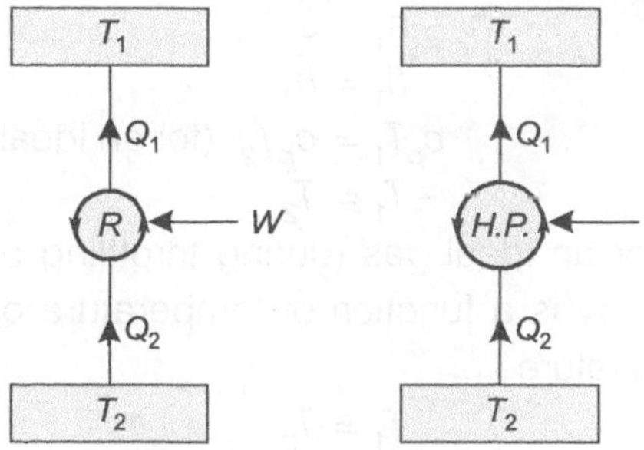

(a) Refrigerator (b) Heat Pump

Heat pump:

Coefficient of performance of heat pump,

$$(COP)_{HP} = \frac{\text{Heating effect: } Q_1}{\text{Work input: } W} = \frac{Q_1}{W}$$

$$(COP)_{HP} = \frac{Q_1 - Q_2 + Q_2}{W}$$

$$(COP)_{HP} = \frac{W + Q_2}{W} \quad \because W = Q_1 - Q_2$$

$$(COP)_{HP} = 1 + \frac{Q_2}{W} = 1 + (COP)_R$$

or $(COP)_R = \mathbf{(COP)_{HP} - 1}$

3.2 Sol.

$$\oint \frac{\delta Q}{T} < 0 \quad \text{for an irreversible cycle.}$$

3.3 (a)

Given data: Number of mole, $n = 1$ kmol; $p_1 = 0.5$ MPa; $p_2 = 0.1$ MPa;

$T_1 = 300\text{ K} = T_2$ for ideal gas

Entropy change of the universe,

$$\Delta S_{univ} = \Delta S_{sys} + \Delta S_{surr}$$

$$= \Delta S_{sys} \quad \because \Delta S_{surr} = 0$$

$$= mc_p \log_e \frac{T_2}{T_1} - mR \log_e \frac{p_2}{p_1}$$

$$= -mR \log_e \frac{p_2}{p_1} \quad \because T_1 = T_2$$

(Throttling process)

$$= -m\frac{\bar{R}}{M} \log_e \frac{p_2}{p_1} = -n\bar{R} \log_e \frac{p_2}{p_1}$$

$$= -1 \times 8.314 \log_e \frac{0.1}{0.5}$$

$$= \mathbf{13.38\ kJ/K}$$

POINTS TO REMEMBER

During throttling process,

Enthalpy change, $\Delta h = 0$

$$\because \quad h_2 - h_1 = 0$$

$$h_1 = h_2$$

$$c_p T_1 = c_p T_2 \quad \text{(for an ideal gas)}$$

$$T_1 = T_2$$

So, for an ideal gas (during throttling process) enthalpy is a function of temperature only and temperature.

$$T_1 = T_2$$

3.4 (c)

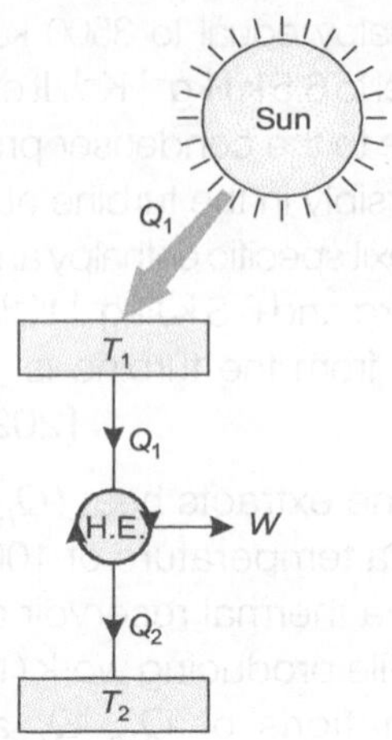

Given data:

$$Q_1 = 80\text{ kJ}$$

$$T_1 = 100°\text{C} = 373\text{ K}$$

$$Q_2 = 70\text{ kJ}$$

$$T_2 = 30°\text{C} = (30 + 273)\text{ K} = 303\text{ K}$$

Thermal efficiency,

$$\eta = \frac{W}{Q_1} = \frac{Q_1 - Q_2}{Q_1} = \frac{80 - 70}{80}$$

$$= 0.125 = \mathbf{12.5\%}$$

3.5 (b)

$$\eta_{overall} = \eta_1 + \eta_2 - \eta_1\eta_2$$

$$= 0.3 + 0.2 - 0.3 \times 0.2$$

$$= 0.5 - 0.06 = 0.44 = \mathbf{44\%}$$

POINTS TO REMEMBER

Formula for overall thermal efficiency is calculated as,

$$\text{Overall efficiency} = \frac{\text{Work output}}{\text{Heat input}}$$

$$= \frac{\eta_1 Q + \eta_2(1 - \eta_2)Q}{Q}$$

$$= \eta_1 + \eta_2(1 - \eta_2)$$

$$= \eta_1 + \eta_2 - \eta_1\eta_2$$

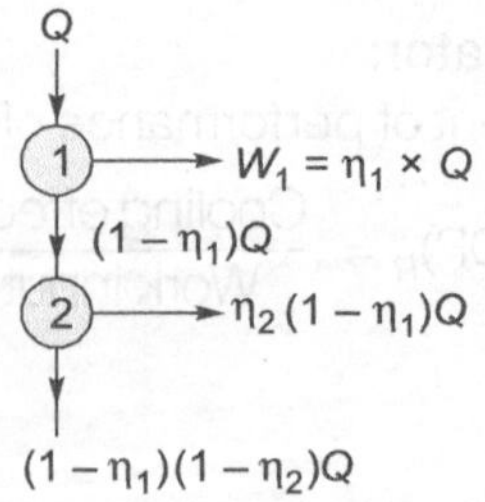

3.6 (c)

$$T\left(\frac{\partial s}{\partial T}\right)_p - T\left(\frac{\partial s}{\partial T}\right)_v = T \times \frac{c_p}{T} - T\frac{c_v}{T}$$

$$= c_p - c_v = R$$

where $\left(\frac{\partial s}{\partial T}\right)_p = \frac{c_p}{T}$ and $\left(\frac{\partial s}{\partial T}\right)_v = \frac{c_v}{T}$

3.7 (d)

For the process to be feasible, the entropy change can be positive, negative or zero.
When entropy change,
$(\Delta S) > 0$ i.e., Heat is added to the sytem.
$(\Delta S) < 0$ i.e., Heat is rejected from the system.
$(\Delta S) = 0$ i.e., Reversible adiabatic process.

3.8 (a)

Option (b) and (c) are wrong because in throttling process

$$\Delta h = 0$$

Option (d) is right for throttling process for both the real and ideal gas.
But option (a) is right only for ideal gas.

3.9 (a)

$$\eta = \frac{W}{Q}$$

$$0.75 = \frac{50}{Q}$$

$$Q = 66.67$$

$$\therefore \quad Q_R = 66.67 - 50 = 16.67 \text{ kJ}$$

$$= 16\frac{2}{3} \text{ kJ}$$

3.10 (b)

Given data:

$$\eta_{Carnot} = 0.75$$

$$T_1 = 727°\text{C} = 1000 \text{ K}$$

$$T_2 = ?$$

$$\therefore \quad \eta_{Carnot} = 1 - \frac{T_2}{T_1}$$

$$0.75 = 1 - \frac{T_2}{1000}$$

or $$\frac{T_2}{1000} = 0.25$$

or $$T_2 = 0.25 \times 1000 = 250 \text{ K}$$

$$= (250 - 273)°\text{C} = \mathbf{-23°C}$$

3.11 (b)

For isothermal process

$$\text{Work} = p_1V_1 \ln\frac{V_2}{V_1} = (2 \times 10^5) \times (1) \times \ln\frac{2}{1}$$

$$= 138629.4 \text{ J} = 138.63 \text{ kJ}$$

3.12 (c)

Given process is isothermal, so change in entropy of universe

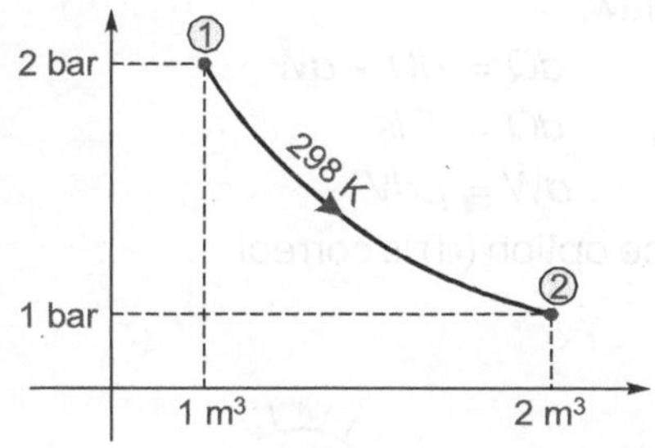

$$(dS)_{universe} = (dS)_{system} + (dS)_{surrounding}$$

Work done in isothermal process

$$= P_1V_1 \ln(V_2/V_1)$$

$$= 200 \times 1 \ln(2/1)$$

$$= 138.62 \text{ kJ}$$

$$\delta Q = dU + \delta W$$

$$\delta Q = 138.62 \text{ kJ} \quad (\because dU = 0)$$

Heat taken by surroundings to the system

$$= 138.62 \text{ kJ}$$

Change in entropy of surrounding

$$(dS)_{surr} = \frac{-138.62}{298} = -0.465 \text{ kJ/K}$$

$$\text{Mass of nitrogen gas} = \frac{PV}{RT} = \frac{200 \times 1}{\frac{8.314}{28} \times 298}$$

or $$m = 2.26 \text{ kg}$$

Change in entropy of system

$$= mc_v \ln\left(\frac{T_2}{T_1}\right) + mR\ln\left(\frac{V_2}{V_1}\right)$$

$$= mR\ln\left(\frac{V_2}{V_1}\right) \quad \because T_1 = T_2$$

$$= 2.26 \times \frac{8.314}{28} \ln\left(\frac{2}{1}\right)$$

$$= +0.465 \text{ kJ/K}$$

$$(dS)_{universe} = 0.465 - 0.465$$

$$= 0$$

3.13 (d)

$TdS = dU + pdV$

This equation holds good for any process reversible or irreversible, undergone by a closed system, since it is a relation among properties which are independent of the both.

$(dS)_{process}$ can be positive, negative or zero, hence option (b) is wrong.

For ideal gas,

First law,

$\Rightarrow \quad dQ = dU + dW$

Also, $\quad dQ = Tds$

$dW = pdV$

Hence option (d) is correct.

3.14 (d)

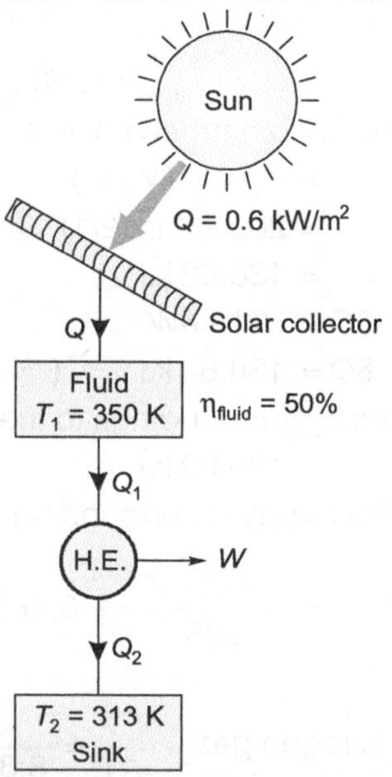

Given data:

$Q = 0.6 \text{ kW/m}^2$

Overall efficiency of the fluid system,

$\eta_{fluid} = 50\% = 0.50$

$T_1 = 350 \text{ K}$

$T_2 = 313 \text{ K}$

$W = 2.5 \text{ kW}$

Overall efficiency of the fluid system,

$$\eta_{fluid} = \frac{Q_1}{Q}$$

$$0.50 = \frac{Q_1}{0.6}$$

or $\quad Q_1 = 0.50 \times 0.6 = 0.3 \text{ kW/m}^2$

Let $\quad A$ = Minimum area of the solar collector

$\therefore \quad Q_1 = 0.3A \text{ kW}$

Minimum area of the solar collector is when, the heat engine is reversible

$\therefore$ Thermal efficiency,

$$\eta = 1 - \frac{T_2}{T_1} = 1 - \frac{313}{350} = 1 - 0.895$$

$$= 0.105$$

also $\quad \eta = \dfrac{W}{Q_1}$

$$\therefore \quad 0.105 = \frac{2.5}{0.3A}$$

$$A = \mathbf{79.36 \text{ m}^2}$$

3.15 (a)

We know that

For closed cycle, change in internal energy is zero.

First law of thermodynamic for cyclic process

$\Sigma W_{cycle} = \Sigma Q_{cycle}$

So area enclosed by PV diagram and Ts diagram must be equal, as well as the direction of the cycle.

3.16 (d)

<u>Method I:</u>

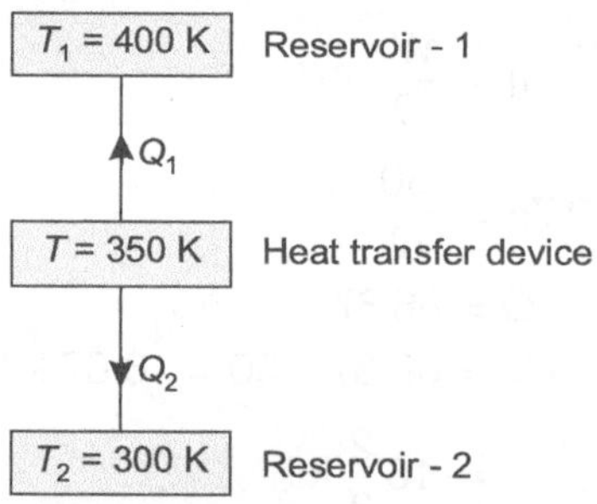

$Q_1 + Q_2 = 100 \text{ kJ}$

$T = 350 \text{ K}; \; T_1 = 400 \text{ K}$

$T_2 = 300 \text{ K}$

Entropy change of the heat temperature device :

$$\Delta S = -\frac{(Q_1 + Q_2)}{350} = -\frac{100}{350} = -\frac{2}{7} \text{ kJ/K}$$

Entropy change of reservoir-1,

$$\Delta S_1 = \frac{Q_2}{T_2} = \frac{100 - Q_1}{300} \text{ kJ/K}$$

Entropy change of reservoir-2,

$$\Delta S_2 = \frac{Q_1}{T_1} = \frac{Q_1}{400} \text{ kJ/K}$$

Total entropy,

$$\Delta S_{total} = 0$$

$$\Delta S + \Delta S_1 + \Delta S_2 = 0$$

$$-\frac{2}{7}+\frac{100-Q_1}{300}+\frac{Q_1}{400}=0$$

$$-\frac{2}{7}+\frac{1}{3}-\frac{Q_1}{300}+\frac{Q_1}{400}=0$$

$$-\frac{400Q_1+300Q_1}{120000}=\frac{2}{7}-\frac{1}{3}=\frac{6-7}{21}=-\frac{1}{21}$$

$$-\frac{100Q_1}{120000}=-\frac{1}{21}$$

or $\quad Q_1 = 57.14$ kJ

Method II:

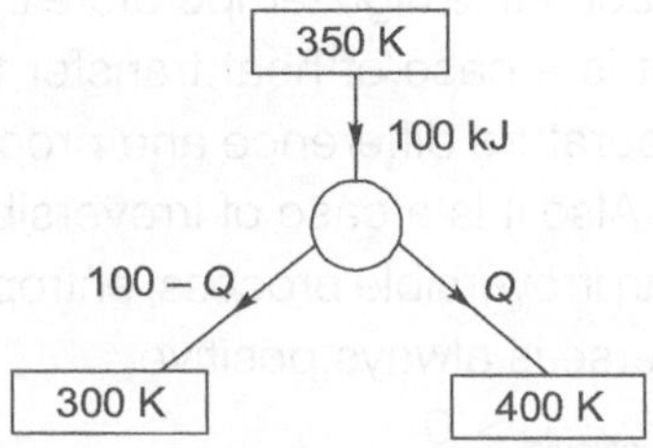

Where, Q is the maximum amount of heat rejected.

∴maximum heat transfer, transofrmer must bc reversible.

∴ Clausius inequality,

$$\oint\frac{\delta Q}{T}=0$$

$$\frac{100}{350}-\frac{100-Q}{300}-\frac{Q}{400}=0$$

$$\frac{10}{35}=\frac{100-Q}{300}+\frac{Q}{400}$$

After solving, $\quad Q = 57.142$ kJ

3.17 (c)

We can observe in the P-V diagram that temperature is not constant during any stage hence options (b) and (d) are rejected as temperature is constant during the stage 3-1 in both the options which is not possible option (a) is rejected because clockwise process in P-V diagram cannot have anticlockwise T-S diagram. Hence the correct option is (c)

3.18 (a)

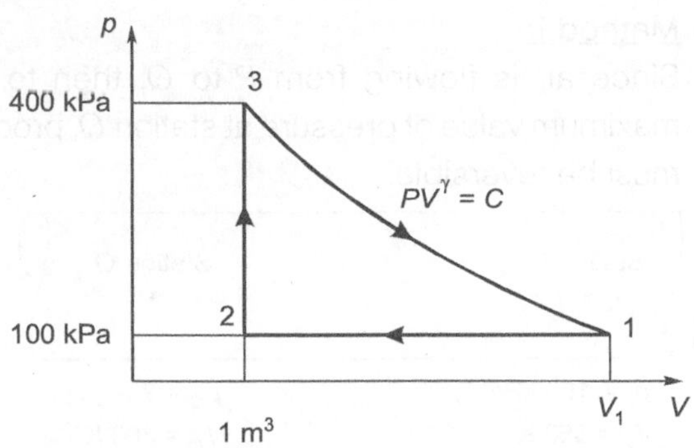

Process 2—3

$$Q_{2\text{-}3} = (U_3 - U_2) + W_{2\text{-}3}$$

$$Q_{2\text{-}3} = (U_3 - U_2) + 0$$

$$= mc_v(T_3 - T_2)$$

$$= mc_v\left(\frac{p_3V_3}{mR}-\frac{p_2V_2}{mR}\right)$$

$$= \frac{c_v}{R}(p_3V_3 - p_2V_2)$$

$$= \frac{c_v}{c_p - c_v}(400\times1 - 100\times1)$$

$$Q_{2\text{-}3} = \frac{1}{\gamma-1}(400-100)$$

$$= \frac{300}{1.4-1}=\frac{300}{0.4}=750\text{ kJ}$$

$$Q_{1\text{-}2} = (U_2 - U_1) + W_{1\text{-}2}$$

$$= mc_v(T_2 - T_1) + p(V_2 - V_1)$$

$$= mc_v\left[\frac{p_2V_2}{mR}-\frac{p_1V_1}{mR}\right]+p(V_2-V_1)$$

$$= \frac{c_v}{R}[p(V_2-V_1)]+p(V_2-V_1)$$

$$= p(V_2-V_1)\left\{\frac{c_v}{R}+1\right\}$$

$$= p(1-V_1)\left\{\frac{c_v}{c_p-c_v}+1\right\} \quad [\because V_2 = 1\text{ m}^3]$$

$$= 100(1-V_1)\left\{\frac{1}{\gamma-1}+1\right\}$$

Calculating V_1

$$p_3V_3^\gamma = p_1V_1^\gamma$$

$$\therefore \quad V_1 = \left[\frac{p_3\cdot V_3^\gamma}{p_1}\right]^{\frac{1}{\gamma}}$$

$$= \left[\frac{400}{100}\times1\right]^{\frac{1}{1.4}} = 2.692\text{ m}^3$$

$$Q_{1\text{-}2} = 100[1-2.692]\times\left\{\frac{1}{0.4}+1\right\}$$

$$= -592.2\text{ kJ}$$

(minus sign represent heat is rejected)

There is no heat transfer in process 3-1 as it is reversible adiabatic process.

Efficiency of cycle,

$$\eta = 1-\frac{Q_{reject}}{Q_{added}} = 1-\frac{592.2}{750}$$

$$= 0.2104 = \mathbf{21.04\%}$$

3.19 (b)

$$dS = mc_v \ln\frac{T_2}{T_1} - mR\ln\frac{P_1}{P_2}$$

Since $P_1 = P_2$ and $T_2 = T_1$

$\therefore \quad dS = 0$

3.20 (a)

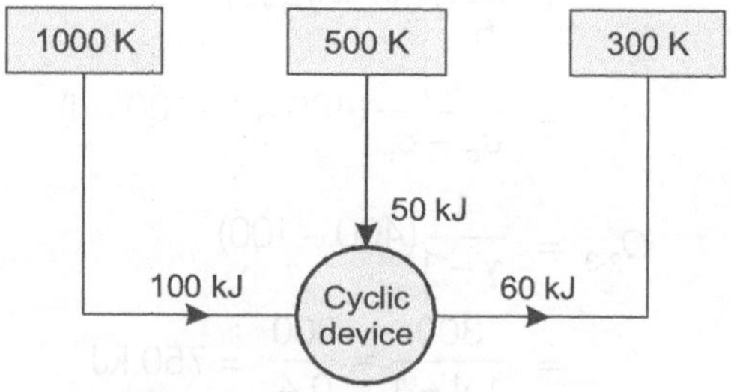

According to Clausius's inequality,

$$\oint \frac{\delta Q}{T} \leq 0$$

$$\frac{100}{1000} + \frac{50}{500} - \frac{60}{300} \leq 0$$

$$0.1 + 0.1 - 0.2 = 0$$

$$0 = 0$$

Hence, the cyclic device can be a reversible heat engine.

3.21 (c)

Given data: $Q_1 = 100$ kW; $Q_2 = 50$ kW;
$T_3 = 290$ K; $T_4 = 348$ K

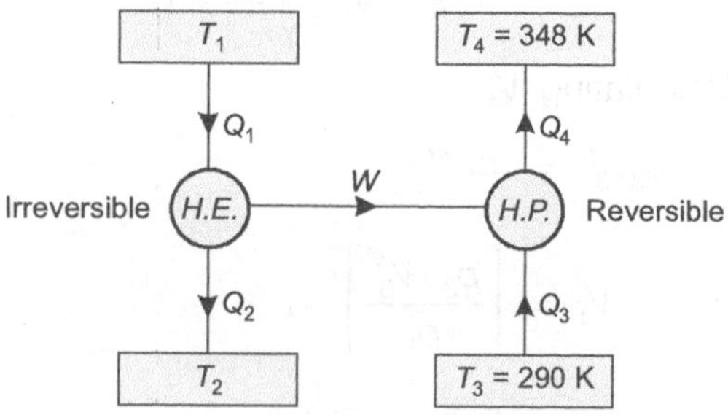

Heat engine work output,

$$W = Q_1 - Q_2 = 100 - 50 = 50 \text{ kW}$$

= Pump input work

$$(COP)_{HP} = \frac{Q_4}{W}$$

also $$(COP)_{HP} = \frac{T_4}{T_4 - T_3}$$

$$\therefore \quad \frac{Q_4}{W} = \frac{T_4}{T_4 - T_3}$$

$$\frac{Q_4}{50} = \frac{348}{348-290} = \frac{348}{58}$$

or $Q_4 = $ **300 kW**

3.22 (d)

If a closed system is undergoing an irreversible process, the entropy of the system can increase, decrease or remain constant.

$$ds = \frac{dQ}{T} + \delta s_{gen}$$

[δs_{gen} is always positive but dQ can be positive, negative or zero.]

3.23 (d)

Since, water at room temperature is brought into contact with a high temperature thermal reservoir. So, it is a case of heat transfer through a large temperature difference and process will be very fast. Also it is a case of irreversible process.

For an irreversible process, entropy change of the universe is always positive.

$(\Delta S)_{universe} > 0$

3.24 (b)

Process-I

Entropy change of universe process-1,

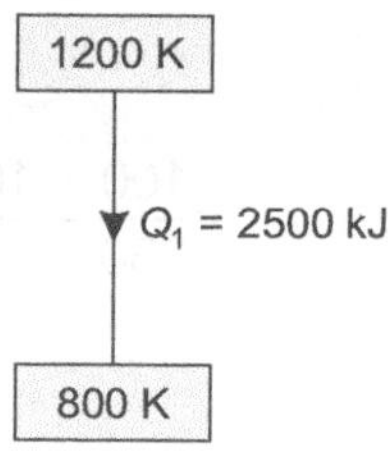

$$\Delta S_1 = -\frac{Q_1}{T_1} + \frac{Q_1}{T_2}$$

$$= -\frac{2500}{1200} + \frac{2500}{800}$$

$$= -2.08 + 3.12$$

$$= 1.04 \text{ kJ/K}$$

Process-II

Entropy change of process 2,

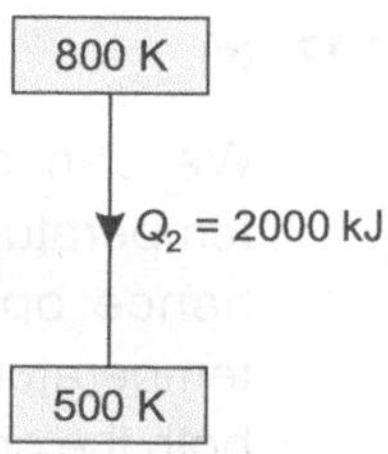

$$\Delta S_2 = -\frac{Q_2}{T_2} + \frac{Q_2}{T_2}$$

$$= -\frac{2000}{800} + \frac{2000}{500}$$

$$= -2.5 + 4$$

$$= 1.5 \text{ kJ/K}$$

Thus, $\boldsymbol{\Delta S_2 > \Delta S_1}$, process-II is more irreversible than process-I.

3.25 (b)

<u>**Method I:**</u>

Since air is flowing from P to Q, then to get maximum value of pressure at station Q, process must be reversible.

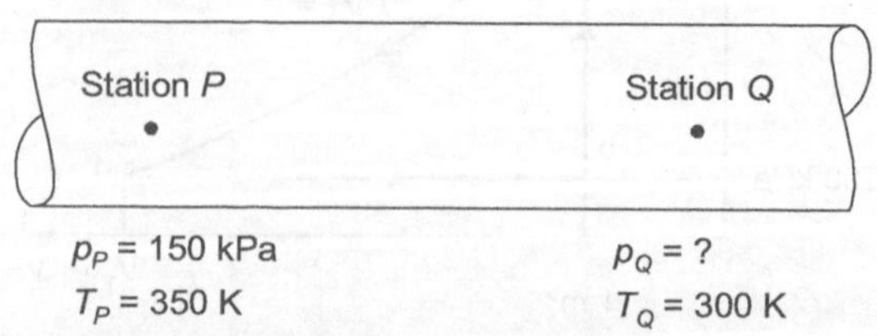

$$c_p = 1.005 \text{ kJ/kgK}$$
$$c_v = 0.718 \text{ kJ/kgK}$$
$$R = 0.287 \text{ kJ/kgK}$$

$$\gamma = \frac{c_p}{c_v} = \frac{1.005}{0.718} = 1.4$$

For reversible adiabatic process *P–Q*,

$$\frac{T_Q}{T_P} = \left(\frac{p_Q}{p_P}\right)^{\frac{\gamma-1}{\gamma}}$$

$$\frac{300}{350} = \left(\frac{p_Q}{150}\right)^{\frac{1.4-1}{1.4}} = \left(\frac{p_Q}{150}\right)^{0.285}$$

or $$\frac{p_Q}{150} = \left(\frac{300}{350}\right)^{3.5} = 0.583$$

or $$p_Q = \mathbf{87.45\ kPa}$$

<u>**Method II:**</u>

Since air is flowing *P* to *Q*, then to get maximum value of pressure at station *Q*,

Change in entropy,

$$S_Q - S_P \geq 0$$

$$c_p \ln\left(\frac{T_Q}{T_P}\right) - R\ln\left(\frac{P_Q}{P_P}\right) \geq 0$$

$$1.005\ln\left(\frac{300}{350}\right) - 0.287\ln\left(\frac{P_Q}{150}\right) \geq 0$$

Now, $$\ln\left(\frac{P_Q}{150}\right) \leq -0.537$$

then, $$P_Q \geq 87.665 \text{ kPa}$$
$$P_Q = 87 \text{ kPa}$$

3.26 (c)

If $p_Q = 50$ kPa

Change in entropy,

$$s_Q - s_P = c_p \ln\frac{T_Q}{T_P} - R\ln\frac{p_Q}{p_P}$$
$$= 1.005\ln\frac{300}{350} - 0.287\ln\frac{50}{150}$$
$$= 1.005(-0.1541) - 0.287(-1.098)$$
$$= -0.1548 + 0.3151 = \mathbf{0.160\ kJ/kgK}$$

3.27 (b)

We know that

$$dS = \frac{\delta Q}{T}$$

Integrating between states 1 and 2 gives

$$\int_1^2 \Delta S = \int_1^2 \frac{\delta Q}{T}$$

$$S_2 - S_1 = \frac{1}{T}\int_1^2 \delta Q \qquad \because T = C$$

$$= \frac{1}{T} \times Q_{1-2} = \frac{mRT\ln\frac{V_2}{V_1}}{T}$$

$$= mR\ln\frac{V_2}{V_1} = mR\ln\frac{p_1}{p_2}$$

where $\frac{p_1}{p_2} = \frac{V_2}{V_1}$ for isothermal process

Alternatively

We know that the entropy change in terms of temperature and pressure ratios,

$$S_2 - S_1 = mc_p \ln\frac{T_2}{T_1} - mR\ln\frac{p_2}{p_1}$$

where $T_1 = T_2$ for isothermal process

$$\therefore \quad S_2 - S_1 = -mR\ln\frac{p_2}{p_1} = mR\ln\frac{p_1}{p_2}$$

3.28 (a)

1. Clausius inequality,

$$\oint \frac{\delta Q}{T} \leq 0$$

(i) $\oint \frac{\delta Q}{T} < 0$ for irreversible heat engine.

(ii) $\oint \frac{\delta Q}{T} = 0$ for reversible heat engine.

(iii) $\oint \frac{\delta Q}{T} > 0$ for impossible heat engine

2. $\oint \delta Q \geq 0$ both for reversible and irreversible heat engine.

(i). $\oint \delta Q > 0$ for irreversible or reversible heat engine.

3.29 Sol.

<u>**Method I:**</u>

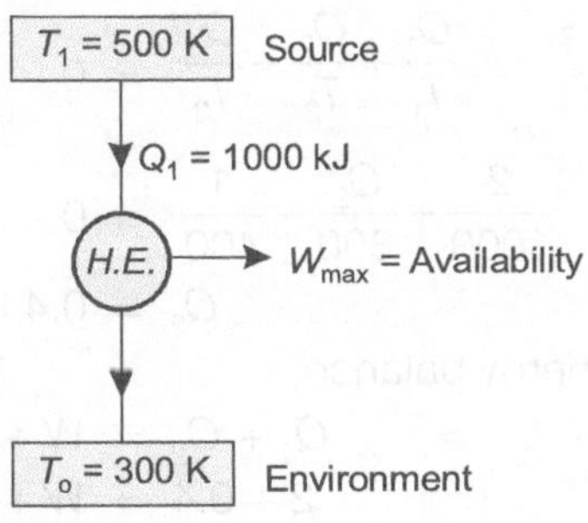

Given data:

$$T_1 = 500 \text{ K}$$
$$Q_1 = 1000 \text{ kJ}$$
$$T_o = 27°\text{C} = (27 + 273) \text{ K} = 300 \text{ K}$$

$$\eta = 1 - \frac{T_o}{T_1} = 1 - \frac{300}{500} = 1 - 0.6 = 0.4$$

also $$\eta = \frac{W_{max}}{Q_1}$$

$$\therefore \quad 0.4 = \frac{W_{max}}{1000}$$

or $$W_{max} = 0.4 \times 1000 = \mathbf{400\ kJ}$$

<u>Method II:</u>

As we know, maximum useful work is possible, when the device will operate in a reversible cycle.

For reversible cycle,

$$\frac{Q_1}{T_1} = \frac{Q_0}{T_0}$$

$$\therefore \quad Q_0 = T_0 \cdot \frac{Q_1}{T_1} = \frac{300 \times 1000}{500} = 600 \text{ kJ}$$

Maximum useful work,

$$(W_{useful})_{max} = Q_1 - Q_0 = 1000 - 600 = 400 \text{ kJ}$$

3.30 (c)

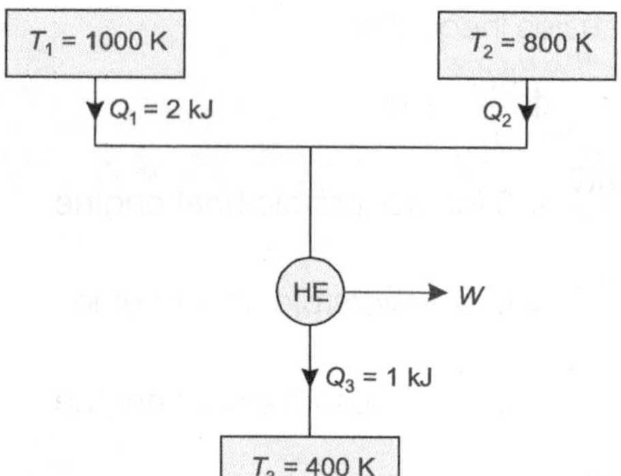

For reversible heat engine,

$$\oint \frac{\delta Q}{T} = 0$$

i.e. $$(\Delta S)_{\text{Reversible cycle}} = 0$$

or $$\frac{Q_1}{T_1} + \frac{Q_2}{T_2} - \frac{Q_3}{T_3} = 0$$

or, $$\frac{2}{1000} + \frac{Q_2}{800} - \frac{1}{400} = 0$$

or $$Q_2 = 0.4 \text{ kJ}$$

From energy balance,

$$Q_1 + Q_2 = W + Q_3$$

or $$2 + 0.4 = W + 1$$

or $$W = \mathbf{1.4\ kJ}$$

3.31 Sol.

As it is given that process is steady so heat transfer through the wall will be:

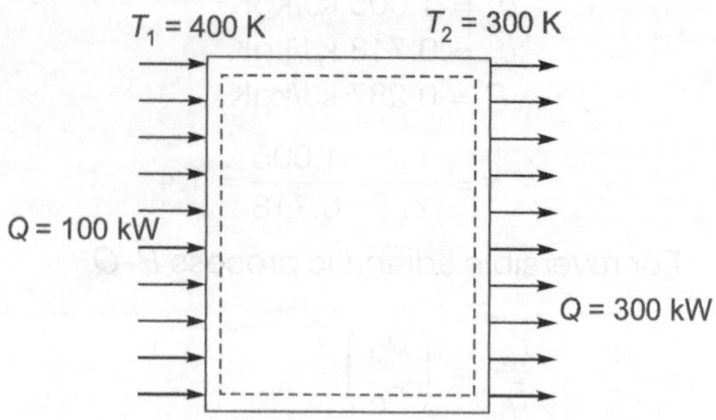

$$S_2 - S_1 = \int_1^2 \frac{dQ}{T} + S_{gen} = \frac{Q}{T_1} - \frac{Q}{T_2} + S_{gen}$$

$$\therefore \quad 0 = \frac{100}{400} - \frac{100}{300} + S_{gen}$$

or $$S_{gen} = \frac{1}{3} - \frac{1}{4} = \frac{1}{12} \text{ kW/K}$$

$$\therefore \quad S_{gen} = \frac{1000}{12} \text{ W/K} = 83.33 \text{ W/K}$$

3.32 Sol.

Given, $m = 10$ kg, $T = 10°C$

For closed system, $\delta Q = TdS$

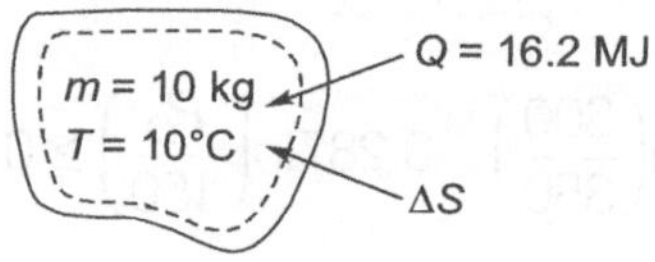

Closed system

$$\therefore \quad \delta Q = mT(s_g - s_f)$$

$$\therefore \quad s_g = \frac{\delta Q}{mT} + s_f = \frac{16.2 \times 10^3}{10 \times (273 + 10)} + 0.88$$

$$= 5.724 + 0.88$$

$$= \mathbf{6.604\ kJ/kgK}$$

3.33 Sol.

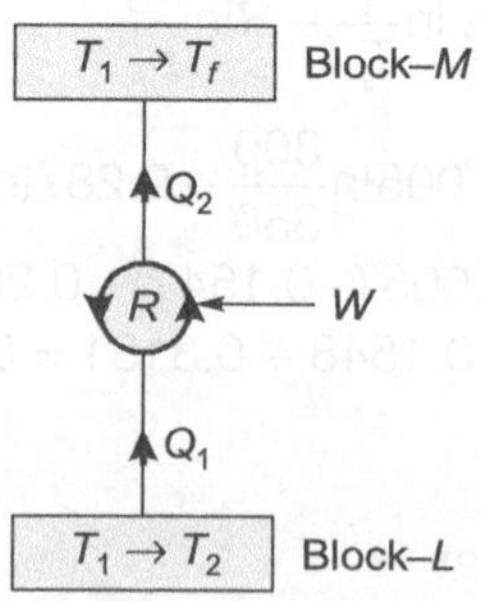

Reversible refrigerator

Given data:

Specified heat,

$c = 0.4$ kJ/kgK each body

Mass: $m = 5$ kg each body

Initial temperature,

$T_1 = 313$ K each body

Final temperature of block L,

$$T_2 = 293\text{ K}$$

Let T_f = Final temperature of block M

Entropy change of block L,

$$\Delta S_L = mc\log_e \frac{T_2}{T_1}$$

Entropy change of block M,

$$\Delta S_M = mc\log_e \frac{T_f}{T_1}$$

Entropy change of universe,

$$\Delta S_{\text{univ}} = \Delta S_L + \Delta S_M$$

$$= mc\log_e \frac{T_2}{T_1} + mc\log_e \frac{T_f}{T_1}$$

For a reversible refrigeration,

$$\Delta S_{\text{univ}} = 0$$

$$mc\log_e \frac{T_2}{T_1} + mc\log_e \frac{T_f}{T_1} = 0$$

$$mc\log_e \frac{T_2T_f}{T_1^2} = 0$$

or $$\log_e \frac{T_2T_f}{T_1^2} = 0$$

or $$\frac{T_2T_f}{T_1^2} = e^0 = 1$$

$$T_f = \frac{T_1^2}{T_2} = \frac{(313)^2}{293} = \mathbf{334.36\ K}$$

3.34 Sol.

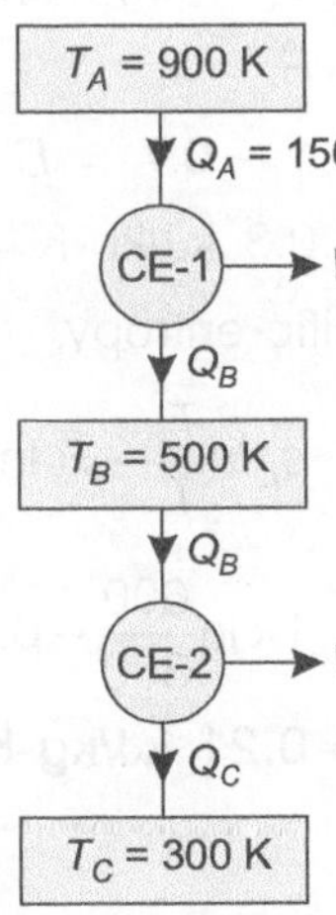

Carnot engines (CE-1, CE-2) are in series.

For Carnot engine (CE-1)

$$\eta_1 = 1 - \frac{T_B}{T_A}$$

also $$\eta_1 = 1 - \frac{Q_B}{Q_A}$$

$\therefore$ $$1 - \frac{T_B}{T_A} = 1 - \frac{Q_B}{Q_A}$$

or $$\frac{T_B}{T_A} = \frac{Q_B}{Q_A}$$

$$\frac{500}{900} = \frac{Q_B}{150}$$

or $$Q_B = \frac{5 \times 150}{9}\text{ MJ}$$

For Carnot engine (CE-2)

$$\eta_2 = 1 - \frac{T_C}{T_B}$$

also $$\eta_2 = 1 - \frac{Q_C}{Q_B}$$

$\therefore$ $$1 - \frac{T_C}{T_B} = 1 - \frac{Q_C}{Q_B}$$

or $$\frac{T_C}{T_B} = \frac{Q_C}{Q_B}$$

$$\frac{300}{500} = \frac{Q_C}{\frac{5 \times 150}{9}}$$

or $$Q_c = \mathbf{50\ MJ}$$

3.35 Sol.

Change in entropy,

$$s_2 - s_1 = c_v \ln\frac{T_2}{T_1} + R\ln\frac{V_2}{V_1} = R\ln\frac{V_2}{V_1}$$

Since internal energy,

$$du = 0,$$

So, $$T_1 = T_2$$

$\therefore$ $$s_2 - s_1 = 287\ln\frac{0.6}{0.9} = \mathbf{-\ 116.368\ J/kgK}$$

3.36 Sol.

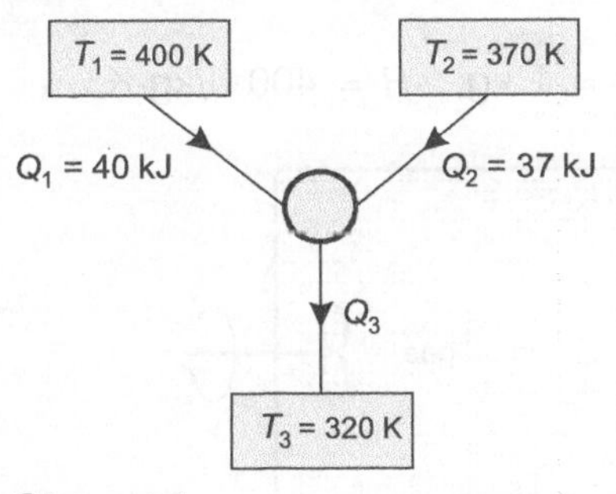

Given data:

$$T_1 = 127°\text{C} = (127 + 273)\text{ K}$$

$= 400\text{ K}$

$T_2 = 97°C = (97 + 273)K$

$= 370\text{ K}$

$T_3 = 47°C = (47 + 273)K$

$= 320\text{ K}$

For reversible cycle

Clausius inequality, $\oint \frac{\delta Q}{T} = 0$

$$\therefore \frac{40}{400} + \frac{37}{370} - \frac{Q_3}{320} = 0$$

$$0.1 + 0.1 = \frac{Q_3}{320}$$

or $Q_3 = \mathbf{64\ kJ}$

3.37 Sol.

Given that: $\dot{Q}_L = 10$ kW

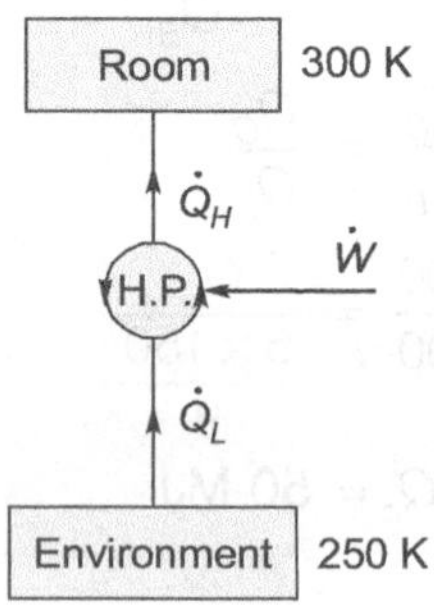

$\dot{W} = 15$ kW

Applying first law on heat pump

$$\dot{Q}_L + \dot{W} = \dot{Q}_H$$

$$10\text{ kW} + 15\text{ kW} = \dot{Q}_H$$

$$\dot{Q}_H = 25\text{ kW}$$

$$(COP)_{H.P.} = \frac{\dot{Q}_H}{\dot{W}} = \frac{25\text{ kW}}{15\text{ kW}} = \mathbf{1.67}$$

3.38 Sol.

$m = 1$ kg; $R = 400$ J/kg-K

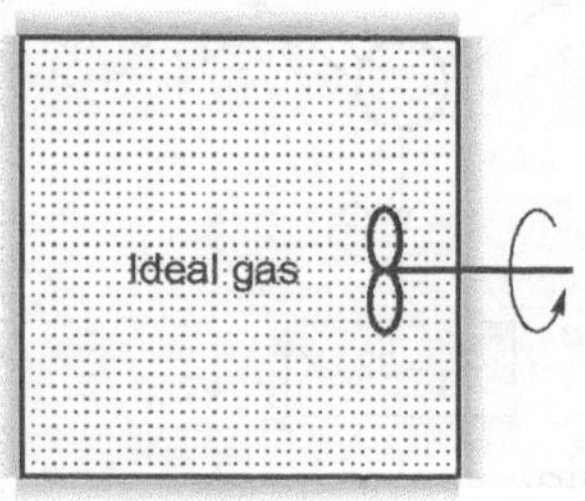

$c_v = 1000$ J/kg-K

$T_i = 300$ K; $P_i = 100$ kPa

$W = 100$ kJ

Applying first law:

$$\delta Q = \Delta U + \delta W$$

$$\Rightarrow \quad 0 = mc_v(T_f - T_i) + (-100) \times 10^3 \text{ J}$$

$$\Rightarrow \quad 0 = 1 \times 1000 \times (T_f - 300) - 100 \times 10^3$$

$$\Rightarrow \quad T_f = 400\text{ K}$$

Entropy change for ideal gas

$$\Delta S = m\left[c_v \ln\frac{T_f}{T_i} + R\ln\frac{V_f}{V_i}\right]$$

(As volume is constant ($\because$ $V_i = V_f$))

$$= 1 \times \left[1000\ln\frac{400}{300} + 0\right] = \mathbf{287.682\ J/K}$$

3.39 Sol.

Change in specific entropy of ideal gas

$$s_f - s_i = c_p \ln\frac{T_f}{T_i} - R\ln\frac{P_f}{P_i}$$

$$= 1005\ln\frac{300}{300} - 287\ln\frac{2}{1}$$

$$= 0 - 287\ln 2 = \mathbf{-198.93\ J/kg\text{-}K}$$

3.40 Sol.

Ideal gas

State–1: $T_1 = 300$ K, $p_1 = 100$ kPa

State–2: $T_2 = 600$ K, $p_2 = 500$ kPa,

$c_p = 1$ kJ/kg-K, $c_v = 0.7$ kJ/kg-K,

$$c_p - c_v = R,$$

$$\Rightarrow \quad c_p - c_v = 1 - 0.7 = R$$

$$R = 0.3\text{ kJ/kg-K}$$

Change in specific entropy,

$$s_2 - s_1 = c_p \ln\frac{T_2}{T_1} - R\ln\frac{p_2}{p_1}$$

$$= 1 \times \ln\frac{600}{300} - 0.3\ln\frac{500}{100}$$

$$= \mathbf{0.21\ kJ/kg\text{-}K}$$

3.42 Sol.

$m = 1$ kg; $P_1 = 10$ bar $= 1000$ kPa; $T_1 = 300$K

$T_2 = 300$K, $P_2 = 1$ bar

$$S_2 - S_1 = mc_p \ln\frac{T_2}{T_1} - mR\ln\frac{P_2}{P_1}$$

$$= -mR\ln\frac{P_2}{P_1} = -1\times 0.287\ln\frac{1}{10}$$
$$= 0.66 \text{ kJ/kgK}$$

3.43 (a)

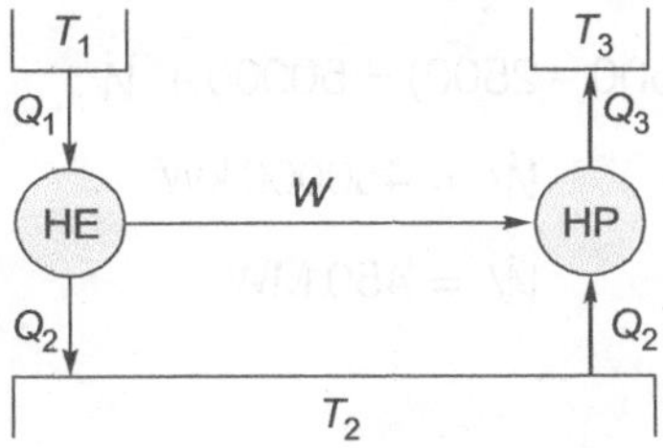

$$\eta = \frac{W}{Q_1} = \frac{W}{Q_2+W}$$
$$\text{COP} = \frac{Q_3}{W} = \frac{Q_2+W}{W}$$
$$\text{COP} = \frac{1}{\eta} = \eta^{-1}$$

3.44 (d)

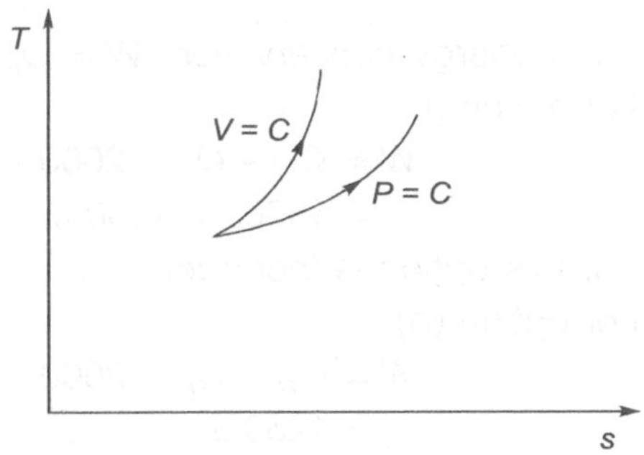

If $v = C$

$$Tds = du + pdv$$
$$Tds = du$$
$$du = C_V dT \quad [\because V = C]$$
$$Tds = C_V dT$$
$$\left(\frac{dT}{ds}\right)_V = \frac{T}{C_V}$$

If $P = C$

$$Tds = dh - vdP$$
$$dh = C_P dT$$

$[\because P = C]$

$$Tds = C_P dT$$
$$\left(\frac{dT}{ds}\right)_P = \frac{T}{C_P}$$

$$\text{Ratio, } \frac{(dT/ds)_P}{(dT/ds)_V} = \frac{(T/C_P)_P}{(T/C_V)_V} = \frac{C_V}{C_P}$$

3.45 (b)

P_1 = 1 bar, P_2 = 3 bar, T_1 = 300 K, T_2 = 400 K,

$$c_p = \frac{\gamma R}{\gamma - 1} = \frac{1.4\times 287}{0.4}$$
$$= 1005 \text{ J/kgK}$$
$$\Delta S = c_p \ln\frac{T_2}{T_1} - R\ln\frac{P_2}{P_1}$$
$$= 1005\ln\frac{400}{300} - 287\ln\frac{3}{1}$$
$$= -26.32 \text{ J/kgK}$$

3.46 Sol.

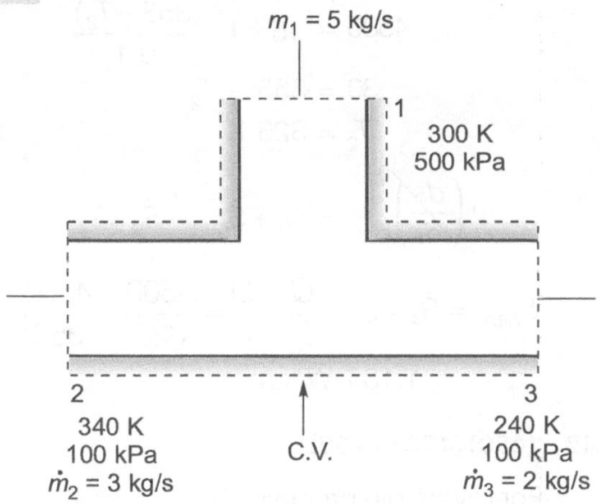

$$\left(\frac{dS}{dt}\right)_{C.V.} = \dot{S}_i + \dot{S}_{gen} - \dot{S}_e$$
$$\because \left(\frac{dS}{dt}\right)_{C.V.} = 0 \text{ [Steady state]}$$
$$\dot{S}_{gen} = \dot{S}_e - \dot{S}_i = \underset{3}{\dot{m}_2} s_2 + \underset{2}{\dot{m}_3} s_3 - \underset{5}{\dot{m}_1} s_1$$

$$= 3(s_2 - s_1) + 2(s_3 - s_1) = 3\times 0.587 + 2(0.237)$$
$$= 2.235 \text{ kW/K} \simeq 2.2 \text{ kW/K}$$

3.47 (b)

Clasius inequality is given as:

$$\oint\frac{dQ}{T} \le 0$$

It is given as

$$\oint\frac{dQ}{T} = 0; \text{ Valid for reversible cycle}$$
$$\oint\frac{dQ}{T} < 0; \text{ Valid for irreversible cycle}$$
$$\oint\frac{dQ}{T} > 0; \text{ Not possible}$$

So, the inequality is valid for any cycle.

3.48 (1.184) (1.12 to 1.24)

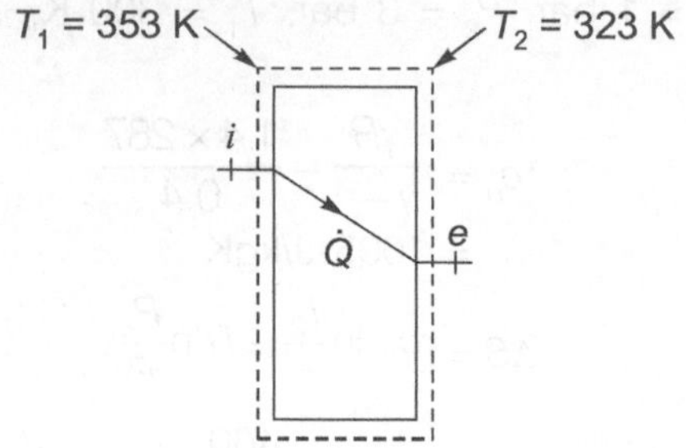

$$\dot{Q} = KA\frac{dT}{dx}$$

$$\dot{Q} = KA\frac{dT}{dx}$$

$$4500 = 15\times 1\times\frac{(353-T_2)}{0.1}$$

$$30 = 353 - T_2$$

$$T_2 = 323\text{ K}$$

$$\left(\cancel{\frac{ds}{dt}}\right)^{0}_{\text{C.V.}} = \dot{S}_i + \dot{S}_{\text{gen}} - \dot{S}_e$$

$$\dot{S}_{\text{gen}} = \dot{S}_e - \dot{S}_i = \frac{\dot{Q}}{T_e} - \frac{\dot{Q}}{T_i} = \frac{4500}{323} - \frac{4500}{353}$$

$$= 1.184\text{ W/k.m}^2$$

3.49 (450) (450 to 450)

For reversible process

$$\dot{S}_{\text{gen}} = 0$$

$$\underbrace{\left(\cancel{\frac{dS}{dt}}\right)_{CV}}_{0\text{(Steady state)}} = \dot{S}_i + \overbrace{\cancel{\dot{S}_{\text{gen}}}}^{0\text{(Reversible)}} - \dot{S}_e$$

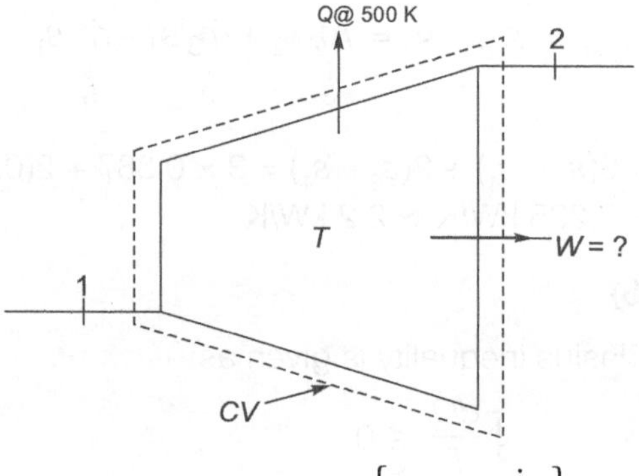

$$0 = \dot{m}s_1 - \left\{\dot{m}s_2 + \frac{\dot{Q}}{500}\right\}$$

$$0 = 500\times\{6.5-6.3\} - \frac{\dot{Q}}{500}$$

$$\dot{Q} = 50000\text{ kW}$$

$$\dot{Q} = 50\text{ MW}$$

Conservation of energy:

$$\dot{E}_{in} = \dot{E}_{exit}$$

$$\dot{m}h_1 = \dot{m}h_2 + \dot{Q} + \dot{W}$$

$$\dot{m}(h_1 - h_2) = \dot{Q} + \dot{W}$$

$$500\,(3500-2500) = 50000 + \dot{W}$$

$$\dot{W} = 450000\text{ kW}$$

$$\dot{W} = 450\text{ MW}$$

3.50 (b)

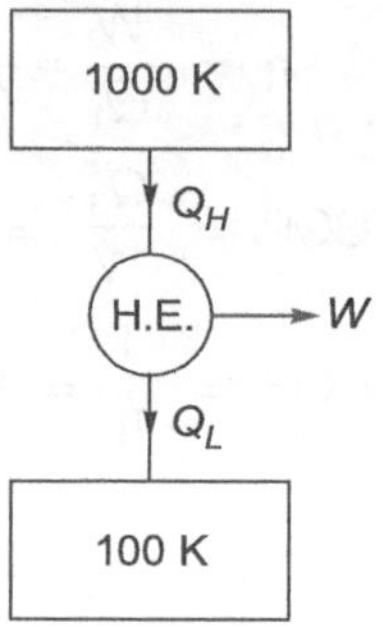

From energy conservation, $W = Q_H - Q_L$

For option (a)

$$W = Q_H - Q_L = 2000 - 500$$
$$= 1500\text{J} \neq 1000\text{ J}$$

So, this option is incorrect.

For option (b)

$$W = Q_H - Q_L = 2000 - 750$$
$$= 1250\text{ J}$$

Checking the possibility of cyclic device,

$$\oint\frac{\delta Q}{T} = \frac{2000}{1000} - \frac{750}{100} = 2 - 7.5$$
$$= -5.5 < 0 \quad \text{(Possible)}$$

So, this option is correct.

For option (c)

$$W = Q_H - Q_L = 6000 - 500 = 5500\text{ J}$$

Checking the possibility of cyclic device,

$$\oint\frac{\delta Q}{T} = \frac{6000}{1000} - \frac{500}{100} = 6 - 5 = 1 > 0$$

(Not possible)

So, this option is incorrect.

For option (d)

$W = Q_H - Q_L = 6000 - 600 = 5400 \neq 5500$

So, this option is incorrect.

3.51 (c)

$Q_{in} = 24000 \text{ kJ/min} = 400 \text{ kW}$

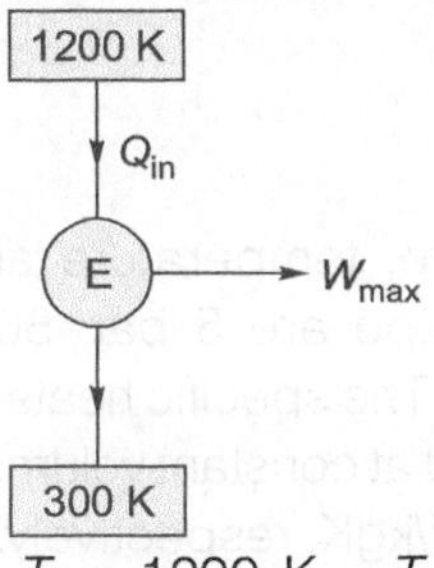

$T_1 = 1200 \text{ K}, \quad T_2 = 300 \text{ K}$

Maximum power can be obtained when the heat supplied

in used in Carnot engine,,

$$\therefore \quad W_{max} = Q_{in} \times \eta_{carnot}$$

$$= 400\left(1 - \frac{300}{1200}\right)$$

$$W_{max} = 300 \text{ kW}$$

3.52 (1620)

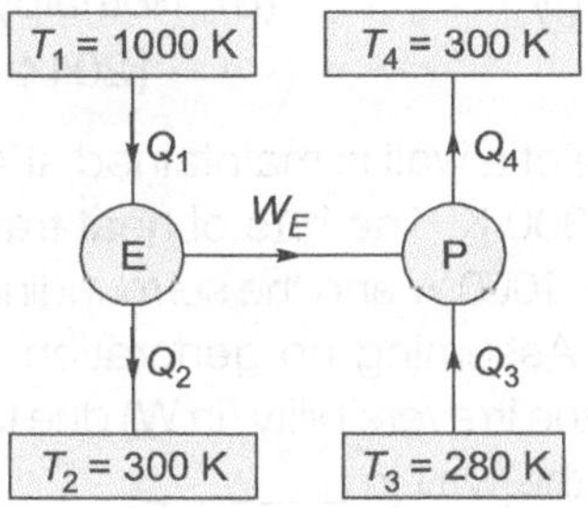

Given: $T_1 = 1000 \text{ K}, \quad T_2 = 300 \text{ K} = T_4$

$Q_1 = 150 \text{ kJ}$

$T_3 = 280 \text{ K}$

For heat engine,

$$\eta = 1 - \frac{T_2}{T_1} = 1 - \frac{300}{1000} = 1 - \frac{Q_2}{Q_1}$$

or $Q_2 = 0.3Q_1 = 0.3 \times 150 = 45 \text{ kJ}$

Now, $Q_1 - Q_2 = 150 - 45$

$\Rightarrow W_E = 105 \text{ kJ}$

Now, $\text{COP}_{HP} = \dfrac{Q_4}{W_E} = \dfrac{T_4}{T_4 - T_3}$

$$\therefore \quad \frac{Q_4}{105} = \frac{300}{300 - 280}$$

$$\therefore \quad Q_4 = 1575 \text{ kJ}$$

$$\therefore \quad Q_2 + Q_4 = 1575 + 45 = 1620 \text{ kJ}$$

Hence, total amount of heat delivered to the room together by the heat engine and heat pump is 1620 kJ.

■■■■

4 Availability and Irreversibility

4.1 A heat reservoir at 900 K is brought into contact with the ambient at 300 K for a short time. During this period 9000 kJ of heat is lost by the heat reservoir. The total loss in availability due to this process is

(a) 18000 kJ (b) 9000 kJ
(c) 6000 kJ (d) None of these

[1995 : 2 M]

4.2 Availability of a system at any given state is

(a) a property of the system
(b) the maximum work obtainable as the system goes to dead state
(c) the total energy of the system
(d) the maximum useful work obtainable as the system goes to dead state

[2000 : 1 M]

4.3 A steel billet of 2000 kg mass is to be cooled from 1250 K to 450 K. The heat released during this process is to be used as a source of energy. The ambient temperature is 303 K and specific heat of steel is 0.5 kJ/kgK. The available energy of this billet is

(a) 490.44 MJ (b) 30.95 MJ
(c) 10.35 MJ (d) 0.10 MJ

[2004 : 2 M]

4.4 A gas expands in a frictionless piston-cylinder arrangement. The expansion process is very slow, and is resisted by an ambient pressure of 100 kPa. During the expansion process, the pressure of the system (gas) remains constant at 300 kPa. The change in volume of the gas is 0.01 m^3. The maximum amount of work that could be utilized from the above process is.

(a) 0 kJ (b) 1 kJ
(c) 2 kJ (d) 3 kJ

[2008 : 2 M]

4.5 The pressure, temperature and velocity of air flowing in pipe are 5 bar, 500 K and 50 m/s, respectively. The specific heats of air at a constant pressure and at constant volume are 1.005 kJ/kgK and 0.718 kJ/kgK, respectively. Neglect potential energy. If the pressure and temperature of the surroundings are 1 bar and 300 K, respectively, the available energy in kJ/kg of the air stream is

(a) 170 (b) 187
(c) 191 (d) 213

[2013 : 2 M]

4.6 The maximum theoretical work obtainable, when a system interacts to equilibrium with a reference environment, is called

(a) Entropy (b) Enthalpy
(c) Exergy (d) Rothalpy

[2014 : 1 M, Set-1]

4.7 One side of a wall is maintained at 400 K and the other at 300 K. The rate of heat transfer through the wall is 1000 W and the surrounding temperature is 25°C. Assuming no generation of heat within the wall, the irreversibility (in W) due to heat transfer through the wall is _______.

[2015 : 2 M, Set-3]

4.8 Water flowing at the rate of 1 kg/s through a system is heated using an electric heater such that the specific enthalpy of the water increases by 2.50 kJ/kg and the specific entropy increases by 0.007 kJ/kgK. The power input to the electric heater is 2.50 kW. There is no other work or heat interaction between the system and the surroundings. Assuming an ambient temperature of 300 K, the irreversibility rate of the system is _________ kW. (round off to two decimal places).

[2019 : 2 M, Set-2]

Answers Availability and Irreversibility

4.1 (c) 4.2 (d) 4.3 (a) 4.4 (c) 4.5 (b) 4.6 (c) 4.7 (248.23)
4.8 (2.10)

Explanations Availability and Irreversibility

4.1 (c)

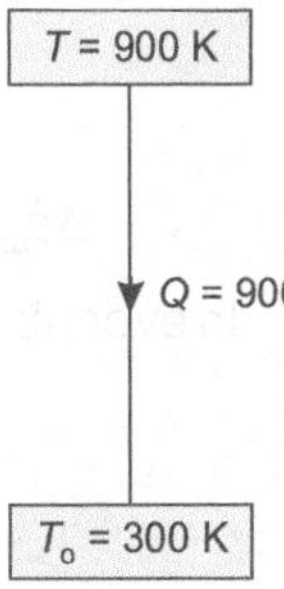

Given data:

$$T = 900 \text{ K}$$
$$T_0 = 300 \text{ K}$$
$$Q = 9000 \text{ kJ}$$

Change in universe entropy,

$$\Delta S_{uni} = -\frac{Q}{T} + \frac{Q}{T_0}$$
$$= -\frac{9000}{900} + \frac{9000}{300}$$
$$= -10 + 30 = 20 \text{ kJ/K}$$

Total loss in availability (= Irreversibility),

$$I = T_0 \Delta S_{uni}$$
$$= 300 \times 20 = \mathbf{6000 \text{ kJ}}$$

4.2 (d)

Availability of a system at any given state is defined as the maximum useful work obtainable as the system goes to dead state.

4.3 (a)

Given: Ambient temperture, $T_0 = 303$ K

$$Q = mc_p\Delta T$$
$$= 2000 \times (0.5 \text{ kJ/kgK}) \times (1250 - 450)$$
$$= 800000 \text{ kJ}$$

$$\Delta S = mc_p \ln\frac{T_1}{T_2} = 2000 \times 0.5 \times \ln\frac{1250}{450}$$
$$= 1021.165 \text{ kJ/K}$$

Available energy for the billet

$$\text{A.E.} = Q - T_0 \Delta S = 800000 - 303 \times 1021.165$$
$$= 490587.005 \text{ kJ} = \mathbf{490.58 \text{ MJ}}$$

4.4 (c)

Given data:

Pressure of gas in a cylinder, $p = 300$ kPa

Ambient pressure, $p_{atm} = 100$ kPa

Change in volume,

$$dV = 0.01 \text{ m}^3$$

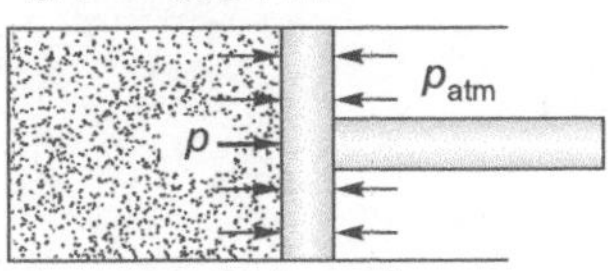

Maximum amount of work,

$$W_{max} = (p - p_{atm})\, dV$$
$$= (300 - 100) \times 0.01 = \mathbf{2 \text{ kJ}}$$

4.5 (b)

$$P_0 = 100 \text{ kPa}, \; T_0 = 300 \text{ K}$$

Available energy

$$= (h_1 - h_0) - T_0(s_1 - s_0) + \left(\frac{V_1^2}{2} - \frac{V_2^2}{2}\right)$$

where $s_1 - s_0 = c_p \ln\frac{T_1}{T_0} - R\ln\frac{p_1}{p_0}$ $\quad (V_2 = 0)$

$$= 1.005\ln\left(\frac{500}{300}\right) - 0.287\ln\left(\frac{5}{1}\right)$$
$$= 0.05147 \text{ kJ/kgK}$$

$\therefore$ Available energy

$$= 1.005(500 - 300) - 300(0.05147) + \frac{(50)^2}{2} \times 10^{-3}$$
$$= 186.8 \text{ kJ/kg} \approx \mathbf{187 \text{ kJ/kg}}$$

4.6 (c)

The maximum theoretical work obtainable, when a system interacts to equilibrium with a reference envirionment, is called as exergy.

4.7 Sol.

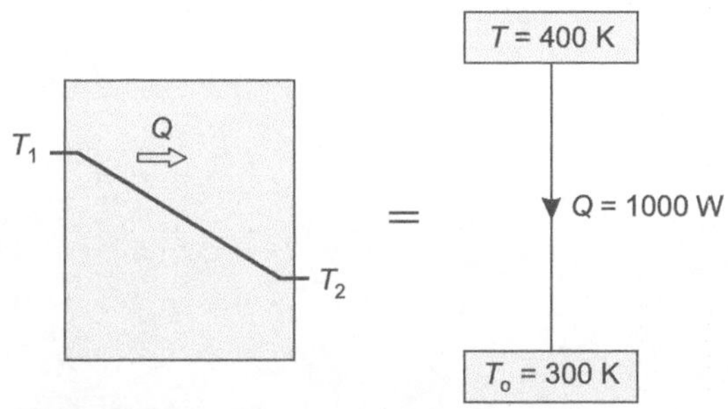

Given data:

$T_1 = 400$ K; $\quad T_2 = 300$ K; $\quad Q = 1000$ W

$$T_0 = 25°\text{C} = (25 + 273) \text{ K} = 298 \text{ K}$$

$$S_2 - S_1 = \int_1^2 \frac{dQ}{T} + S_{gen}$$
$$0 = \frac{Q}{T_1} - \frac{Q}{T_2} + S_{gen}$$
$$0 = \frac{1000}{400} - \frac{1000}{300} + S_{gen}$$
$$S_{gen} = 0.833 \text{ W/K}$$

Irreversibility,

$$I = T_0 S_{gen}$$
$$= 298 \times (0.833) = \mathbf{248.23 \text{ W}}$$

4.8 Sol.

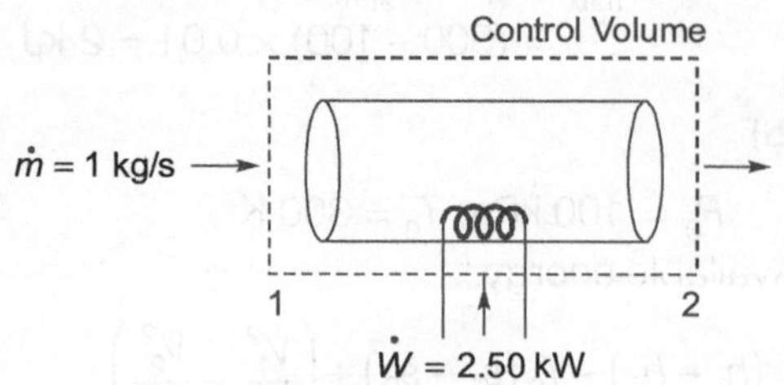

$$h_2 - h_1 = 2.50 \text{ kJ/kg}$$

$$s_2 - s_1 = 0.007 \text{ kJ/kgK}$$

$$\dot{I} = T_0 \dot{S}_{gen} = 300 \times \dot{S}_{gen}$$

$$\dot{S}_{gen} = \dot{S}_{exit} - \dot{S}_{inlet} = (\dot{m}s_2) - (\dot{m}s_1 + 0)$$

(Entropy associated with work is zero)

$$= \dot{m}(s_2 - s_1) = 1 \times 0.007 = 0.007$$

$$\therefore \quad \dot{I} = 300 \times 0.007 = 2.10 \text{ kW}$$

Alternate Solution:

$$\Delta\dot{s}_{univ} = \Delta\dot{s}_{sys} + \Delta\dot{s}_{surr} = \Delta\dot{s}_{sys}$$

$$= \dot{m}(s_2 - s_1) = 1(0.007)$$

$$\Delta\dot{s}_{univ} = 0.007 \text{ kW/K}$$

Irreversibility rate,

$$\dot{I} = T_0 \Delta\dot{s}_{univ} = 300 \times 0.007 = 2.1 \text{ kW}$$

■■■■

5 Properties of Pure Substances

5.1 Constant pressure lines in the superheated region of the Mollier diagram will have
(a) A positive slope
(b) A negative slope
(c) Zero slope
(d) Both positive and negative slope
[1995 : 1 M]

5.2 When wet steam flows through a throttle valve and remains wet at exit
(a) Its temperature and quality increase
(b) Its temperature decreases but quality increases
(c) Its temperature increases but quality decreases
(d) Its temperature and quality decrease
[1996 : 1 M]

Linked Answer Questions 5.3 and 5.5

The following table of properties was printed out for saturated liquid and saturated vapour of ammonia. The title for only the first two columns are available. All that we know that the other columns (columns 3 to 8) contain data on specific properties, namely, internal energy (kJ/kg), enthalpy (kJ/kg) and entropy (kJ/kgK).

t°C	p (kPa)						
–20	190.2	88.76	0.3657	89.05	5.6155	1299.5	1418.0
0	429.6	179.69	0.7114	180.36	5.3309	1318.0	1442.2
20	587.5	272.89	1.0408	274.30	5.0860	1332.2	1460.2
40	1554.9	368.74	1.3574	371.43	4.8662	1341.0	1470.2

5.3 The specific enthalpy data are in columns
(a) 3 and 7 (b) 3 and 8
(c) 5 and 7 (d) 5 and 8 **[2005 : 2 M]**

5.4 When saturated liquid at 40°C is throttled to –20°C, the quality at exit will be
(a) 0.189 (b) 0.212
(c) 0.231 (d) 0.788 **[2005 : 2 M]**

5.5 Given below is an extract from steam tables.

Temperature in °C	P_{sat} (bar)	Specific volume (m³/kg)		Enthalpy (kJ/kg)	
		Saturated liquid	Saturated vapour	Saturated liquid	Saturated vapour
45	0.09593	0.001010	15.26	188.45	2394.8
342.24	150	0.001658	0.010337	1610.5	2610.5

Specific enthalpy of water in kJ/kg at 150 bar and 45°C is
(a) 203.60 (b) 200.53
(c) 196.38 (d) 188.45 **[2006 : 2 M]**

5.6 Which combination of the following statements is correct?
P : A gas cools upon expansion only when its Joule-Thomson coefficient is positive in the temperature range of expansion.
Q : For a system undergoing a process, its entropy remains constant only when the process is reversible.
R : The work done by a closed system in an adiabatic is a point function.
S : A liquid expands upon freezing when the slope of its fusion curve on pressure-Temperature diagram is negative.
(a) R and S (b) P and Q
(c) Q, R and S (d) P, Q and R
[2007 : 2 M]

5.7 Water has a critical specific volume of 0.003155 m³/kg. A closed and rigid steel tank of volume 0.025 m³ contains a mixture of water and steam at 0.1 MPa. The mass of the mixture is 10 kg. The tank is now slowly heated. The liquid level inside the tank
(a) will rise
(b) will fall
(c) will remain constant
(d) may rise or fall depending on the amount of heat transferred **[2007 : 1 M]**

Common Data for Q. 5.8, 5.9 and 5.10

In the figure shown, the system is a pure substance kept in a piston-cylinder arrangement. The system is initially a two-phase mixture containing 1 kg of liquid and 0.03 kg of vapour at a pressure of 100 kPa. Initially, the piston rests on a set of stops, a shown in the figure. A pressure of 200 kPa is required to exactly balance the weight of the piston and the outside atmospheric pressure. Heat transfer takes place into the system until its volume increases by 50%. Heat transfer to the system occurs in such a manner that the piston, when allowed to move, does so in a very slow (quasi-static/quasi-equilibrium) process. The thermal reservoir from which heat is transferred to the system has a temperature of 400°C. Average temperature of the system boundary can be taken as 175°C. The heat transfer to the system is 1 kJ, during which its entropy increases by 10 J/K. Specific volumes of liquid (v_f) and vapour (v_g) phases, as well as values of saturation temperatures, are given in the table below.

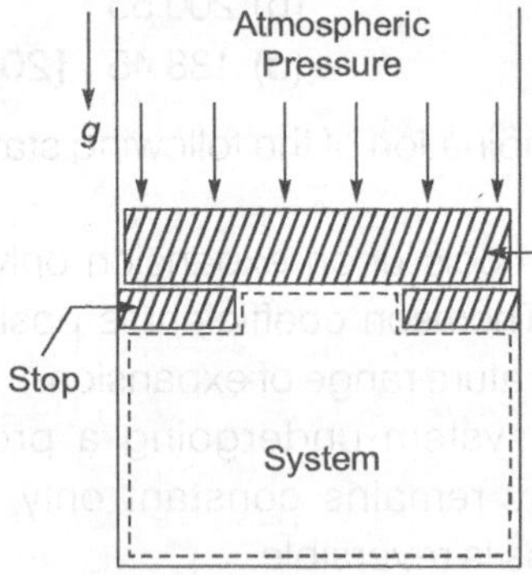

Pressure (kPa)	Saturation temp. T_{sat} (°C)	$v_f(m^3/kg)$	$v_g(m^3/kg)$
100	100	0.001	0.1
200	200	0.0015	0.002

5.8 At the end of the process, which one of the following situations will be true?
(a) superheated vapour will be left in the system
(b) no vapour will be left in the system
(c) a liquid + vapour mixture will be left in the system
(d) the mixture will exist at a dry saturated vapour state [2008 : 2 M]

5.9 The work done by the system during the process is
(a) 0.1 kJ (b) 0.2 kJ
(c) 0.3 kJ (d) 0.4 kJ [2008 : 2 M]

5.10 The net entropy generation (considering the system and the thermal reservoir together) during the process is closest to
(a) 7.5 J/K (b) 7.7 J/K
(c) 8.5 J/K (d) 10 J/K [2008 : 2 M]

5.11 A pure substance at 8 MPa and 400°C is having a specific internal energy of 2864 kJ/kg and a specific volume of 0.03432 m³/kg. Its specific enthalpy (in kJ/kg) is___. [2014 : 1 M, Set-2]

5.12 1.5 kg of water is in saturated liquid state at 2 bar (v_f = 0.001061 m³/kg, u_f = 504.0 kJ/kg, h_f = 505 kJ/kg). Heat is added in a constant pressure process till the temperature of water reaches 400°C (v = 1.5493 m³/kg, u = 2967.0 kJ/kg, h = 3277.0 kJ/kg). The heat added (in kJ) in the process is ______. [2014 : 2 M, Set-1]

5.13 The Van der Waals equation of state is $\left(p+\frac{a}{v^2}\right)(v-b)=RT$, where p is pressure, v is specific volume, T is temperature and R is characteristic gas constant. The SI unit of a is
(a) J/kgK (b) m³/kg
(c) m⁵/kg-s² (d) Pa/kg
[2015 : 1 M, Set-2]

5.14 A rigid container of volume 0.5 m³ contains 1.0 kg of water at 120°C (v_f = 0.00106 m³/kg, v_g = 0.8908 m³/kg). The state of water is
(a) compressed liquid
(b) saturated liquid
(c) a mixture of saturated liquid and saturated vapor
(d) superheated vapor [2015 : 1 M, Set-3]

5.15 A mixture of ideal gases has the following composition by mass:

N_2	O_2	CO_2
60%	30%	10%

If the universal gas constant is 8314 J/kmol-K, the characteristic gas constant of the mixture (in J/kgK) is ____. [2015 : 2 M, Set-3]

5.16 For water at 25°C, dp_s/dT_s = 0.189 kPa/K (p_s is the saturation pressure in kPa and T_s is the saturation temperature in K) and the specific volume of dry saturated vapour is 43.38 m³/kg. Assume that the specific volume of liquid is negligible in comparison with that of vapour. Using the Clausius-Clapeyron equation, an estimate of the enthalpy of evaporation of water at 25°C (in kJ/kg) is_________. [2016 : 2 M, Set-1]

5.17 The INCORRECT statement about the characteristics of critical point of a pure substance is that
(a) there is no constant temperature vaporization process
(b) it has point of inflection with zero slope
(c) the ice directly converts from solid phase to vapor phase
(d) saturated liquid and saturated vapor states are identical [2016 : 1 M, Set-3]

5.18 Which one of the following statements is correct for a superheated vapour?
(a) Its pressure is less than the saturation pressure at a given temperature.

(b) Its temperature is less than the saturation temperature at a given pressure.
(c) Its volume is less than the volume of the saturated vapour at a given temperature.
(d) Its enthalpy is less than enthalpy of the saturated vapour at a given pressure.
[2018 : 1 M, Set-1]

5.19 A tank of volume 0.05 m^3 contains a mixture of saturated water and saturated steam at 200°C. The mass of the liquid present is 8 kg. The entropy (in kJ/kgK) of the mixture is_____________(correct of two decimal places)
Property data for saturated steam and water are:
At 200°C, p_{sat} = 1.5538 MPa
v_f = 0.001157 m^3/kg, v_g = 0.12736 m^3/kg
s_{fg} = 4.1014 kJ/kgK, s_f = 2.3309 kJ/kgK
[2018 : 2 M, Set-1]

5.20 If one mole of H_2 gas is occupies a rigid container with a capacity of 1000 litres and the temperature is raised from 27°C to 37°C, the change in pressure of the contained gas (round off to two decimal places), assuming ideal gas behaviour, is______ Pa. (R = 8.314 J/mol-K) [2019 : 2 M, Set-1]

5.21 A closed vessel contains pure water, in thermal equilibrium with its vapour at 25°C (Stage #1), as shown.

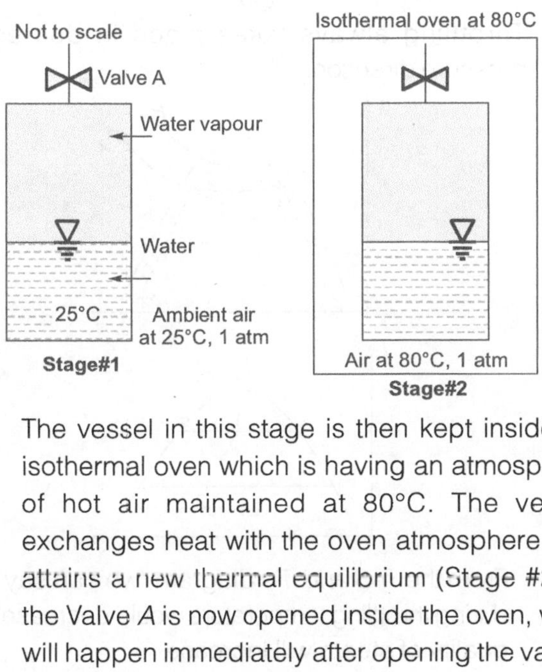

The vessel in this stage is then kept inside an isothermal oven which is having an atmosphere of hot air maintained at 80°C. The vessel exchanges heat with the oven atmosphere and attains a new thermal equilibrium (Stage #2). If the Valve A is now opened inside the oven, what will happen immediately after opening the valve?
(a) Hot air will go inside the vessel through Valve A
(b) Water vapor inside the vessel will come out of the Valve A
(c) All the vapor inside the vessel will immediately condense
(d) Nothing will happen - the vessel will continue to remain in equilibrium [2020 : 1 M, Set-2]

5.22 A rigid tank of volume 50 m^3 contains a pure substance as a saturated liquid vapour mixture at 400 kPa. Of the total mass of the mixture, 20% mass is liquid and 80% mass is vapour. Properties at 400 kPa are: Saturation temperature, T_{sat} = 143.61°C; Specific volume of saturated liquid. v_f = 0.001084 m^3/kg; Specific volume of saturated vapour. v_g = 0.46242 m^3/kg. The total mass of liquid vapour mixture in the tank is ________ kg (round off to the nearest integer).
[2021 : 1 M, Set-2]

5.23 Superheated steam at 1500 kPa, has a specific volume of 2.75 m^3/kmol and compressibility factor (Z) of 0.95. The temperature of steam is ______ °C [Round off to the nearest integer]
(a) 249
(b) 471
(c) 522
(d) 198 [2021 : 1 M, Set-1]

5.24 In the vicinity of the triple point, the equation of liquid-vapour boundary in the P-T phase diagram for ammonia is ln P = 24.38 – 3063/T, where P is pressure (in Pa) and T is temperature (in K). Similarly, the solid-vapour boundary is given by ln P = 27.92 – 3754/T. The temperature at the triple point is_____ K. [Round off to one decimal place]
[2021 : 2 M, Set-1]

5.25 Which one of the following statements is FALSE?
(a) For an ideal gas, the enthalpy is independent of pressure.
(b) For a real gas going through an adiabatic reversible process, the process equation is given by PV^g = constant, where P is the pressure, V is the volume and g is the ratio of the specific heats of the gas at constant pressure and constant volume.
(c) For an ideal gas undergoing a reversible polytropic process $PV^{1.5}$ = constant, the equation connecting the pressure, volume and temperature of the gas at any point along the process is $\frac{P}{R} = \frac{mT}{V}$, where R is the gas constant and m is the mass of the gas.
(d) Any real gas behaves as an ideal gas at sufficiently low pressure or sufficiently high temperature. [2023 : 2 M]

5.26 Consider a mixture of two ideal gases, X and Y, with molar masses $\overline{M}_X = 10$ kg/kmol and $\overline{M}_Y = 20$ kg/kmol, respectively, in a container. The total pressure in the container is 100 kPa, the total volume of the container is 10 m^3 and the temperature of the contents of the container is 300 K. If the mass of gas-X in the container is 2 kg, then the mass of gas-Y in the container is ______ kg. (Rounded off to one decimal place) Assume that the universal gas constant is 8314 J kmol^{-1}K^{-1}.

[2023 : 2 M]

Answers Properties of Pure Substances

5.1 (a)	5.2 (b)	5.3 (d)	5.4 (b)	5.5 (d)	5.6 (a)	5.7 (a)	5.8 (a)
5.9 (d)	5.10 (c)	5.11 (3138.56)	5.12 (4158.97)	5.13 (c)	5.14 (c)		
5.15 (274.86)	5.16 (2443.248)	5.17 (c)	5.18 (a)	5.19 (2.4884)	5.20 (83.14)	5.21 (a)	
5.22 (135)	5.23 (a)	5.24 (195.2)	5.25 (b)	5.26 (4)			

Explanations Properties of Pure Substances

5.1 (a)

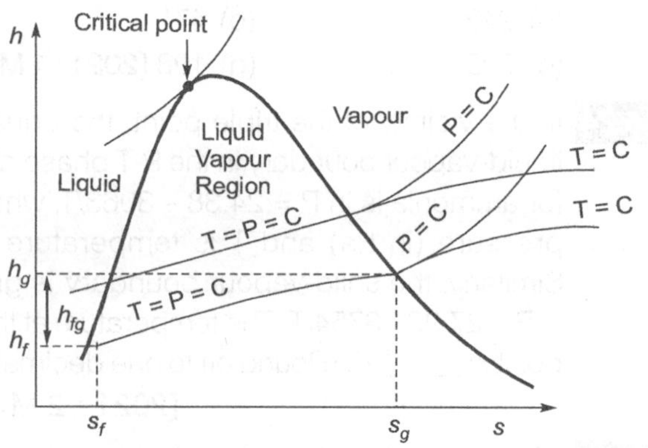

Figure 7.9 *h-s* diagram

As we know,

$$Tds = dh - vdp$$

$$Tds = dh$$

($dp = 0$, constant pressure lines)

$$\frac{dh}{dS} = T$$

$$\because \quad T > 0$$

$$\frac{dh}{dS} > 0$$

Mollier diagram slope > 0

POINTS TO REMEMBER

In superheated region, constant pressure lines diverge, because with increasing temperature, the slope increases.

5.2 (b)

Throttling process

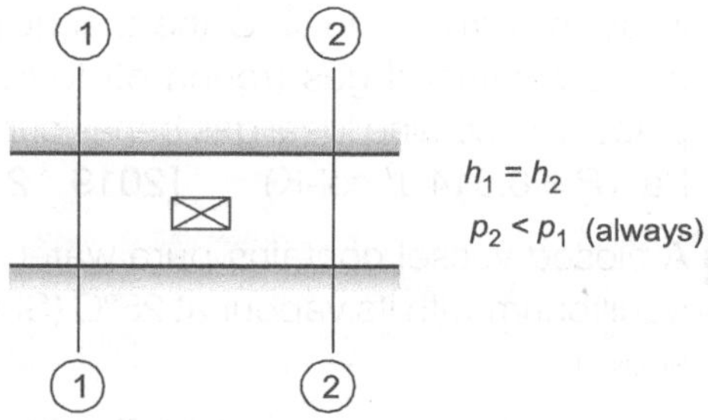

Throttling always takes place in decresing pressure direction.

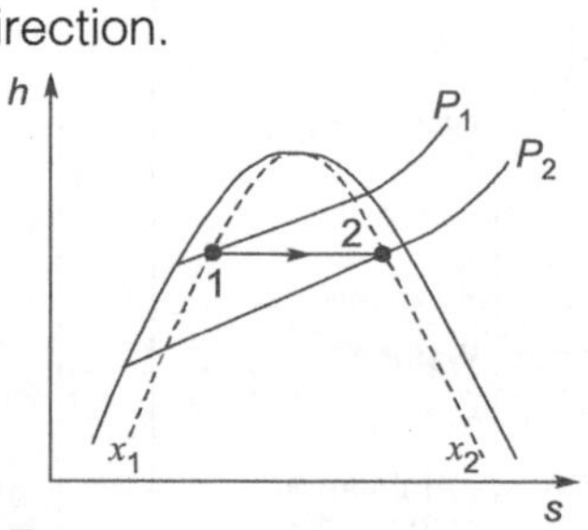

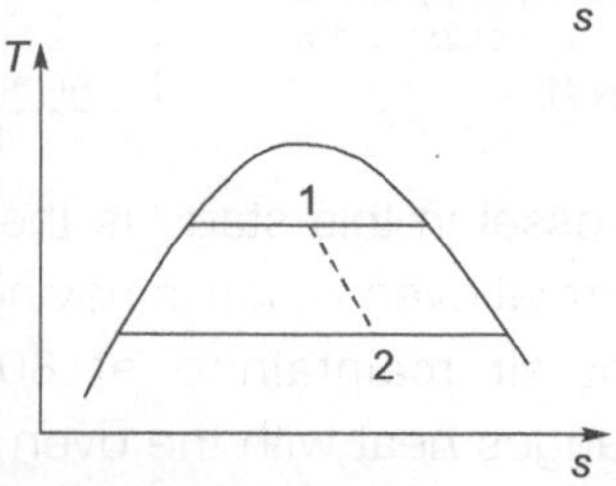

From the *h-S* and *T-s* diagram, we can say that during throttling temperature would decrease and quality of steam will increase.

5.3 (d)

Specific enthalpy,

$$h = u + pv$$

i.e., $h > u$ and the value of entropy is less than u.

5.4 (b)

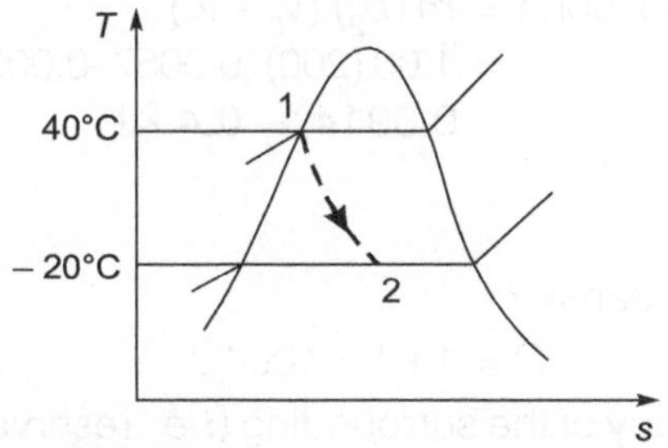

For throttling process,

Enthalpy of state 1 = Enthalpy at state 2

$$h_1 = h_2$$

where $h_1 = h_f = 371.43$ kJ/kg at 40°C

$$h_2 = h_f + x_2\,(h_g - h_f) \text{ at } 20°C$$

$$h_2 = 89.05 + x_2\,(1418 - 89.05)$$

$$h_2 = 89.05 + 1328.95\,x_2$$

also $h_2 = h_1$

$\therefore$ $89.05 + 1328.95\,x_2 = 371.43$

or $x_2 = \mathbf{0.212}$

5.5 (d)

Given data:

$$p_3 = 150 \text{ bar} = p_2$$

$$T_3 = 45°C$$

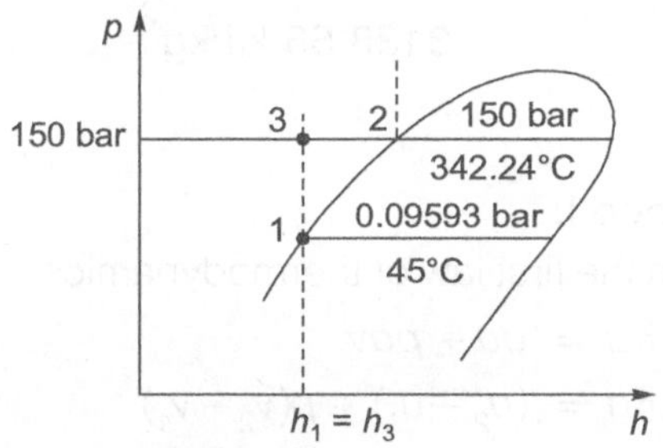

At saturated point 2,

$$p_2 = 150 \text{ bar}$$

$$T_2 = T_{sat} = 342.24°C$$

At saturated point 1,

$$p_1 = 0.09594 \text{ bar}$$

$$T_1 = T_3 = 45°C$$

As $T_3 < T_2$, the given condition of water is sub-cooled liquid.

and $T_3 = T_1 = 45°C$

$\therefore$ The specific enthalpy at point 3 ≈ the specific enthalpy at saturated liquid point 1

$$h_3 = h_1 = 188.45 \text{ kJ/kg}$$

5.6 (a)

P:

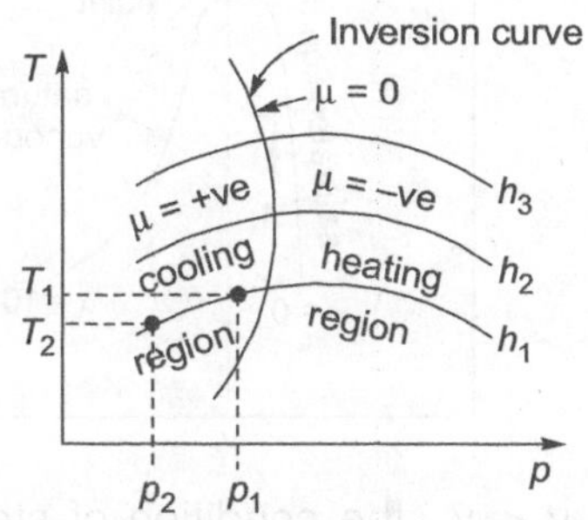

From the figure, we can say that Joule-Thomson coefficeint is positive in the temperature range of expansion. So statement is correct.

Q: Entropy is constant when process is reversible adiabatic, not only reversible, so statement is wrong.

R: From first law of thermodynamics,

$$\delta Q = dU + \delta W$$

For adibatic process,

$$\delta Q = 0$$

$$dU = -\delta W$$

Since dU is a point function, so we can say work done by a closed system in an adiabatic is a point function. So statement is correct.

S:

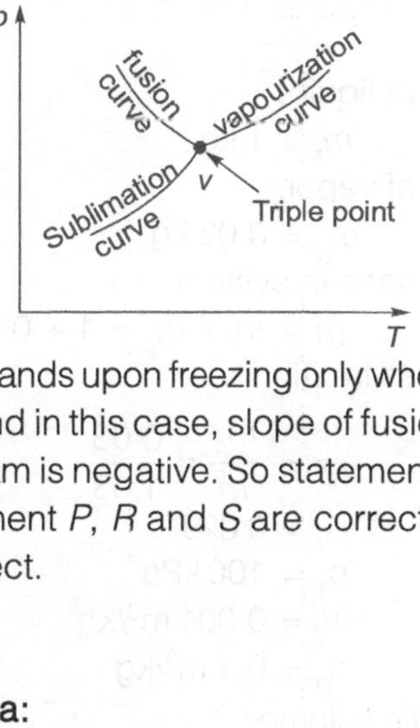

Liquid expands upon freezing only when the liquid is water and in this case, slope of fusion curve on P-T diagram is negative. So statement is correct. So, statement P, R and S are correct. So option (a) is correct.

5.7 (a)

Given data:

Critical specific volume of water,

$$v_c = 0.003155 \text{ m}^3/\text{kg}$$

Volume of steel tank,

$$V = 0.025 \text{ m}^3$$

Pressure of mixture of water and steam,

$$p = 0.1 \text{ MPa} = 1 \text{ bar}$$

Mass of mixture of water and steam,

$$m = 10 \text{ kg}$$

Specific volume of mixture,

$$v = \frac{V}{m} = \frac{0.025}{10} = 0.0025 \text{ m}^3/\text{kg}$$

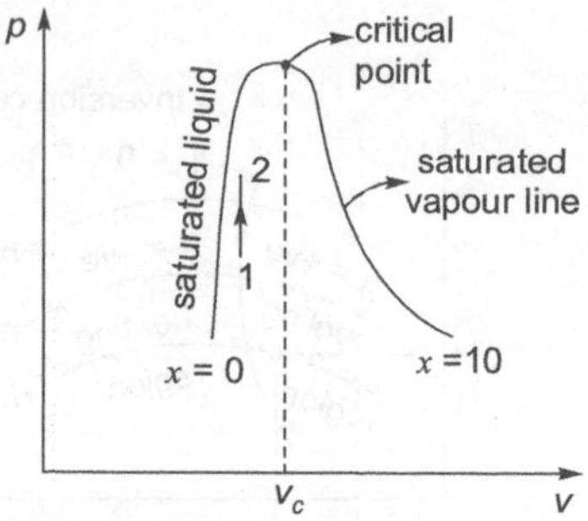

As $v < v_c$, the condition of steam lies near to saturated liquid line and the liquid level inside the tank **will rise** with heating.

5.8 (a)

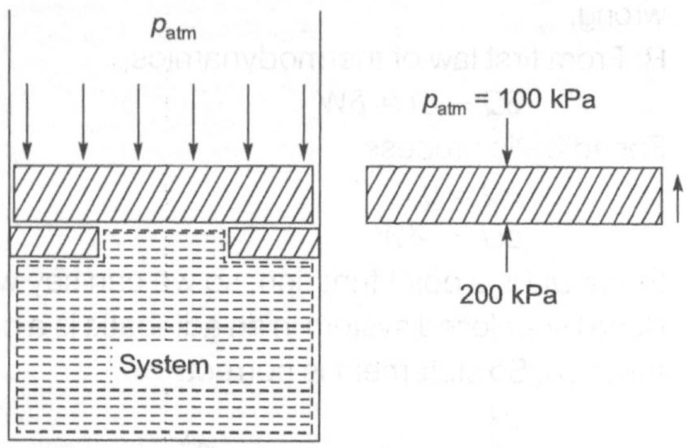

Mass of liquid,

$$m_f = 1 \text{ kg}$$

Mass of vapor,

$$m_g = 0.03 \text{ kg}$$

Total mass in system,

$$m = m_f + m_g = 1 + 0.03 = 1.03 \text{ kg}$$

Dryness fraction,

$$x = \frac{m_g}{m} = \frac{0.03}{1.03} = 0.029$$

At $p_1 = 100$ kPa

$$v_f = 0.001 \text{ m}^3/\text{kg}$$

$$m_g = 0.1 \text{ m}^3/\text{kg}$$

$\therefore$ Specific volume,

$$v_1 = v_f + x(v_g - v_f) = 0.001 + 0.029(0.1 - 0.001) = 0.0038 \text{ m}^3/\text{kg}$$

As pressure increases from 100 kPa to 200 kPa at constant temperature (quasi-static), volume increases by 50%.

i.e., $v_2 = 1.5\, v_1 = 1.5 \times 0.0038 = 0.0057 \text{ m}^3/\text{kg}$

As $v_2 = v_{g,\,200°C}$, the steam is superheated vapor in the system

5.9 (d)

$$\text{Work done} = m(p_2)(v_2 - v_1) = 1.03(200)(0.0057 - 0.0038) = \mathbf{0.3914 \approx 0.4 \text{ kJ}}$$

5.10 (c)

Heat transfer:

$$Q = 1 \text{ kJ} = 1000 \text{ J}$$

Entropy of the surrounding (*i.e.*, reservoir),

$$(\Delta S)_{surr} = -\frac{1000}{673} = -1.485 \text{ J/K}$$

Entropy of the system increases,

$$(\Delta S)_{sys} = 10 \text{ J/K}$$

Net entropy generation,

$$(\Delta S)_{uni} = (\Delta S)_{sys} + (\Delta S)_{surr} = 10 - 1.485 = \mathbf{8.51 \text{ J/K}}$$

5.11 Sol.

$$p = 8 \text{ MPa} = 8 \times 10^3 \text{ kPa}$$

$$T = 400°\text{C}$$

$$u = 2864 \text{ kJ/kg}$$

$$v = 0.03432 \text{ m}^3/\text{kg}$$

By definition of specific enthalpy,

$$h = u + pv = 2864 + 8 \times 10^3 \times 0.03432 = \mathbf{3138.56 \text{ kJ/kg}}$$

5.12 Sol.

Method I:

From the first law of thermodynamics

$$\delta q = du + pdv$$

$$\delta q = (u_2 - u_1) + p(v_2 - v_1)$$

$$\delta q = (2967 - 504) + 2 \times 10^2 (1.5493 - 0.001061)$$

$$\delta q = 2463 + 309.6478 = 2772.6478 \text{ kJ/kg}$$

For 1.5 kg of water,

$$Q = 2772.6478 \times 1.5 = \mathbf{4158.97 \text{ kJ}}$$

Method II:

As we know, heat transfer in a constant pressure system is equal to change in enthalpy.

$$(\delta Q) = m(h_2 - h_1) = 1.5(3277 - 505) = \mathbf{4158 \text{ kJ}}$$

5.13 (c)

The Vander waals equation,

$$\left(p + \frac{a}{v^2}\right)(v - b) = RT$$

As according to the equation, the unit of pressure and (a/v^2) should be same,

$$\therefore \quad \text{Unit of } p = \text{Unit of } \frac{a}{v^2}$$

$$\frac{\text{kg}}{\text{m-sec}^2} = \frac{(\text{kg})^2 \times a}{m^6}$$

$$a = \text{m}^5/\text{kg-sec}^2$$

5.14 (c)

Method I:

Volume of container, $V = 0.5$ m^3

Mass of water, $m = 1$ kg

Specific volume of water,

$$v = \frac{V}{m} = \frac{0.5}{1} = 0.5 \text{ m}^3/\text{kg}$$

$$v_f = 0.00106 \text{ m}^3/\text{kg}$$

$$v_g = 0.8908 \text{ m}^3/\text{kg}$$

As, x is the dryness fraction,

$$v = v_f + x(v_g - v_f)$$

$$0.5 = 0.00106 + x(0.8908 - 0.00106)$$

$$x = 0.5608$$

As, value of x is greater than zero and less than 1 i.e. $0 < x < 1$,

So state of water is a mixture of saturated liquid and saturated vapour.

Method II:

Specific volume of water, $v = \frac{V}{m} = \frac{0.5}{1} = 0.5 \text{ m}^3/\text{kg}$

$\therefore$ It is lying between v_f and v_g, so state of water is a mixture of saturated liquid and saturated vapour.

5.15 Sol.

Equation state for N_2,

$$p_{N_2}V = m_{N_2}R_{N_2}T$$

or

$$\frac{p_{N_2}V}{T} = m_{N_2}R_{N_2}$$

where

$$R_{N_2} = \frac{\bar{R}}{M_{N_2}}$$

$$\therefore \quad \frac{p_{N_2}V}{T} = \frac{m_{N_2}\bar{R}}{M_{N_2}} \quad \text{... (i)}$$

Similarly for O_2 and CO_2

$$\frac{p_{O_2}V}{T} = \frac{m_{O_2}\bar{R}}{M_{O_2}} \quad \text{... (ii)}$$

and

$$\frac{p_{CO_2}V}{T} = \frac{m_{CO_2}\bar{R}}{M_{CO_2}} \quad \text{... (iii)}$$

Adding Eqs. (i), (ii) and (iii), we get

$$\frac{V}{T}(p_{N_2} + p_{O_2} + p_{CO_2}) = \frac{m_{N_2}\bar{R}}{M_{N_2}} + \frac{m_{O_2}\bar{R}}{M_{O_2}} + \frac{m_{CO_2}\bar{R}}{M_{CO_2}}$$

$$\frac{V}{T}p = \left(\frac{m_{N_2}}{M_{N_2}} + \frac{m_{O_2}}{M_{O_2}} + \frac{m_{CO_2}}{M_{CO_2}}\right)\bar{R} \quad \text{... (iv)}$$

where, pressure of mixture

$$p = p_{N_2} + p_{O_2} + p_{CO_2}$$

Equation of state for mixture of ideal gas

$$pV = mRT$$

or

$$\frac{pV}{T} = mR \quad \text{...(v)}$$

Equating Eqs. (iv) and (v), we get

$$mR = \left(\frac{m_{N_2}}{M_{N_2}} + \frac{m_{O_2}}{M_{O_2}} + \frac{m_{CO_2}}{M_{CO_2}}\right)\bar{R}$$

$$mR = \left(\frac{0.60\,m}{M_{N_2}} + \frac{0.30\,m}{M_{O_2}} + \frac{0.1\,m}{M_{CO_2}}\right)\bar{R}$$

or

$$R = \left(\frac{0.60}{M_{N_2}} + \frac{0.30}{M_{O_2}} + \frac{0.10}{M_{CO_2}}\right)\bar{R}$$

$$= \left(\frac{0.60}{28} + \frac{0.30}{32} + \frac{0.10}{44}\right) \times 8314$$

$$= (0.02142 + 0.00937 + 0.00227) \times 8314$$

$$= 0.03306 \times 8134$$

$$= \mathbf{274.86\ J/kgK}$$

5.16 Sol.

At 25°C, $\frac{dp_s}{dT_s} = 0.189$ kPa/K

Specific volume of dry saturated vapour,

$$v_s = 43.38 \text{ m}^3/\text{kg}$$

From Clausius-Clapeyron equation, neglecting specific volume of liquid ($v_2 >> v_f$)

$$\left(\frac{dp}{dT}\right)_S = \frac{h_{fg}}{T(v_g - v_\delta)} \approx \frac{h_{fg}}{Tv_g}$$

$$\frac{dp_s}{dT_s} = \frac{h_{fg}}{T_s \times v_s}$$

h_{fg} = Enthalpy of vapourisation

$$0.189 = \frac{h_{fg}}{(25 + 273) \times 43.38}$$

$$\therefore \quad h_{fg} = \mathbf{2443.248\ kJ/kg}$$

5.17 (c)

The characteristics of critical point of a pure substance.

(i) There is no constant temperature vapourization because heat of vapourization is zero.

(ii) Critical point is point of inflection i.e. isothermal curve changes its curvature.

(iii) Liquid directly converts into vapour and when ice directly converts into vapour, it is known as sublimation.

(iv) Saturated liquid and saturated vapour states are identical.

5.18 (a)

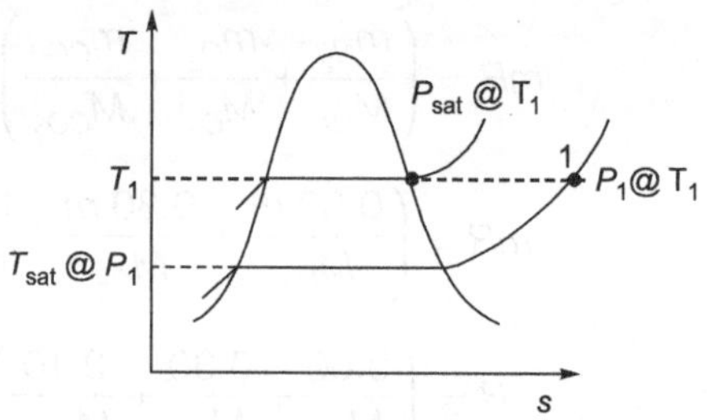

P_{sat} @ $T_1 \rightarrow$ saturation pressure at T_1 temperature

$P_1 \rightarrow$ pressure of superheated vapour at state 1.

So, $P_1 < P_{sat@T_1}$ i.e. pressure less than the saturation pressure at a given temperature.

5.19 Sol.

Total volume of tank (V) = 0.05 m^3

Mass of liquid (m_2) = 8 kg

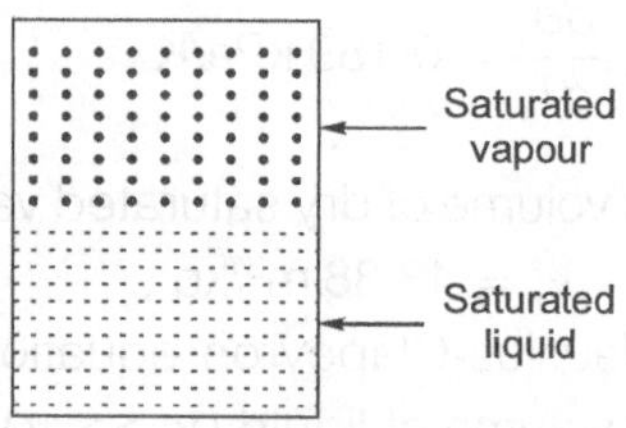

$$x = \frac{m_V}{m_V + m_L} \text{ (dryness fraction)}$$

$$V = V_L + V_V$$

$$\begin{bmatrix} V_L - \text{Volume of liquid in tank} \\ V_V - \text{Volume of vapour in tank} \end{bmatrix}$$

$$V_V = m_V v_V$$

$$v_V = v_g @ 200°C = 0.12736 \text{ m}^3/\text{kg}$$

$$V_L = m_L v_L$$

$$v_L = v_f @ 200°C = 0.001157 \text{ m}^3/\text{kg}$$

$$V_L = 8 \times 0.001157 = 9.256 \times 10^{-3} \text{ m}^3$$

$$0.05 = 9.256 \times 10^{-3} + V_V$$

$$V_V = 0.0474 \text{ m}^3$$

So, $0.04074 = m_V \times 0.12736$

$$m_V = 0.3198 \text{ kg}$$

$$x = \frac{0.3198}{0.3198 + 8} \Rightarrow x = 0.0384$$

Specific entropy of mixture(s),

$$s = s_f + x s_{fg}$$

$$= 2.3309 + 0.0384 \times 4.1014$$

$$= \mathbf{2.4884 \text{ kJ/kg-K}}$$

5.20 Sol.

V = Volume of rigid container

T_1 = Initial temperature of the gas

T_2 = Final temperature of the gas

$$P_1 V = n\bar{R}T_1$$

$$P_2 V = n\bar{R}T_2$$

$$(P_2 - P_1)V = n\bar{R}(T_2 - T_1)$$

$$\Delta P \times 1 = 1 \times 8.314(37 - 27)$$

$$\Delta P = 83.14 \text{ Pa}$$

5.21 (a)

Initially when water and water vapour mixture is at 25°C then the maximum vapour pressure that can be at 25°C in the saturation pressure of vapour at 25°C.

The saturation press at 25°C is very less than 1 atm (101.3 kPa). It is around 3.17 kPa.

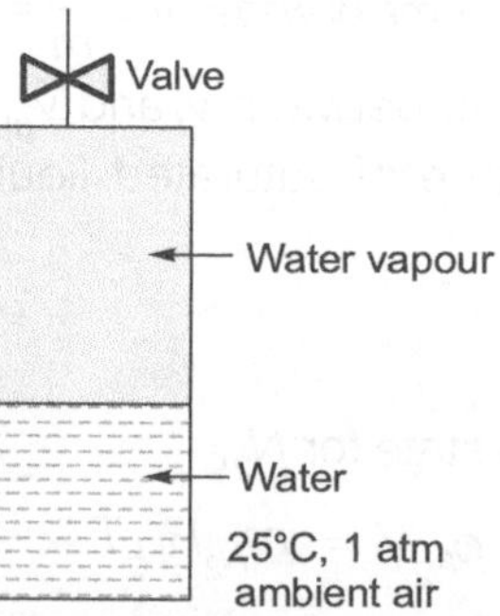

Now when this vessel will be kept in oven at 80°C then at 80°C the saturation pressure of water is still less than 1 atm. It is around 47.2 kPa.

The vapour pressure will reach 1 atm when temperature is 100°C. Hence at 80°C also the pressure will be the than 1 atm 80 when valve is opened air will enter the valve.

5.22 Sol.

$$V_{total} = V_l + V_v$$

$$= v_f \times m_l + v_g \times m_v$$

$$50 = 0.001084 \times 0.2\,m + 0.46242 \times 0.8\,m$$

$$m = 135.08 \text{ kg} = 135 \text{ kg}$$

5.23 (a)

Given: $P = 1500$ kPa,

$V = 2.75$ m3/k-mol, $Z = 0.95$

$$P\bar{V} = Z \times n\bar{R}T$$

$$P\frac{\bar{V}}{n} = Z\bar{R}T$$

$$P\bar{V} = Z\bar{R}T$$

1500 KPa × 2.75 m^3/K-mol = 0.95 × 8.314 kJ/K-molK × T

$$T = 522.26 \text{ K}$$

$$T = 522.26 - 273 = 249.26°\text{C}$$

5.24 Sol.

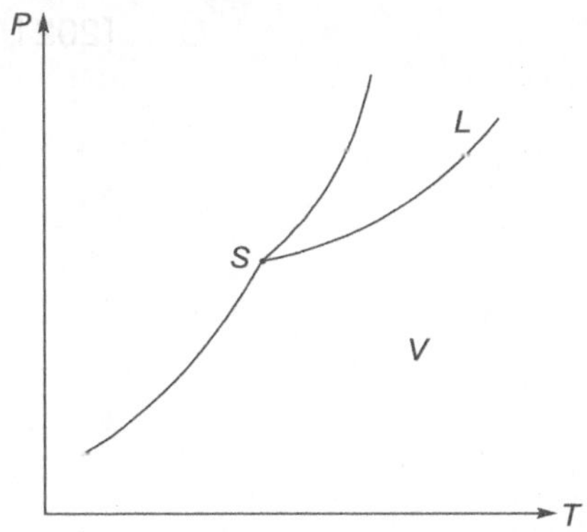

Liquid vapour,

$$\ln P = 24.38 - \frac{3063}{T}$$

Solid vapour,

$$\ln P = 27.92 - \frac{3754}{T}$$

At triple point temperature of solid, liquid and vapour is same.

∴ equating

$$24.38 - \frac{3063}{T} = 27.92 - \frac{3754}{T}$$

Multiplying by T on both sides

$$24.38T - 3063 = 27.92T - 3754$$

$$3.54T = 691$$

$$T = 195.197 \text{ K}$$

5.25 (b)

The process $PV^{\gamma} = C$ is applicable for an ideal gas going through an adiabatic reversible process.

5.26 (4) (3.9 to 4.1)

We are given with two gases:

$$P_t = 100 \text{ kPa}$$

$$V = 10 \text{ m}^3$$

$$T = 300 \text{ K}$$

Using the equation, $PV = n\bar{R}T$

$\Rightarrow$ $100 \times 10 = n \times 8.314 \times 300$

$\Rightarrow$ $n = 0.4$

Now, $n = n_x + n_y$

$$n = \frac{m_x}{M_x} + \frac{m_y}{M_y}$$

$\Rightarrow$ $0.4 = \frac{2}{10} + \frac{m_y}{20}$

$\Rightarrow$ $m_y = 4$ kg

■■■■

6 Thermodynamics Relations

6.1 For a simple compressible system, v, s, p and T are specific volume, specific entropy, pressure and temperature, respectively. As per Maxwell's relations, $\left(\frac{\partial v}{\partial s}\right)_p$ is equal to

(a) $\left(\frac{\partial T}{\partial p}\right)_s$ (b) $\left(\frac{\partial p}{\partial v}\right)_T$

(c) $\left(\frac{\partial s}{\partial T}\right)_p$ (d) $-\left(\frac{\partial T}{\partial v}\right)_p$

[2019 : 1 M, Set-2]

6.2 For an ideal gas, the value of the Joule-Thomson coefficient is

(a) positive (b) zero

(c) negative (d) indeterminate

[2020 : 1 M, Set-1]

6.3 The fundamental thermodynamic relation for a rubber band is given by $dU = TdS + \tau dL$, where T is the absolute temperature, S is the entropy, τ is the tension in the rubber band, and L is the length of the rubber band. Which one of the following relations is CORRECT?

(a) $T = \left(\frac{\partial U}{\partial S}\right)_\tau$ (b) $T = \left(\frac{\partial U}{\partial S}\right)_L$

(c) $\left(\frac{\partial T}{\partial S}\right)_L = \left(\frac{\partial \tau}{\partial L}\right)_S$ (d) $\left(\frac{\partial T}{\partial L}\right)_S = \left(\frac{\partial \tau}{\partial S}\right)_L$

[2021 : 2 M, Set-1]

Answers Thermodynamics Relations

6.1 (a) 6.2 (b) 6.3 (d)

Explanations Thermodynamics Relations

6.1 (a)

As per Maxwell's relation,

$$\left(\frac{\partial v}{\partial s}\right)_p = \left(\frac{\partial T}{\partial p}\right)_s$$

6.2 (b)

Value of Joule-Thomson coefficient for ideal gas,

$$\mu = \left(\frac{\partial T}{\partial P}\right)_h = 0$$

6.3 (d)

$$dU = Tds + \tau dL$$

T = Absolute temperature,
S = Entropy, τ = Tension, L = Length
Comparing with $dz = Mdx + Ndy$

$$M = \left(\frac{\partial z}{\partial x}\right)_y$$

$$N = \left(\frac{\partial z}{\partial y}\right)_x$$

Since z is exact differential

$$\left(\frac{\partial M}{\partial y}\right)_x = \left(\frac{\partial N}{\partial x}\right)_y$$

Since U is exact differential

$$T = \left(\frac{\partial U}{\partial S}\right)_L \quad \text{Since } M = \left(\frac{\partial z}{\partial x}\right)_y$$

$$T = \left(\frac{\partial U}{\partial L}\right)_S \quad \text{Since } N = \left(\frac{\partial z}{\partial y}\right)_x$$

$$\left(\frac{\partial T}{\partial L}\right)_S = \left(\frac{\partial \tau}{\partial S}\right)_L$$

Power Plant Engineering

UNIT

IX

CONTENTS

Power Plant Engineering

Syllabus

Power Engineering: Air and gas compressors; vapour and gas power cycles, concepts of regeneration and reheat.

Analysis of Previous GATE Papers

Exam Year	1 Mark Ques.	2 Marks Ques.	Total Marks
1995	2	–	**2**
1996	–	1	**2**
1997	1	3	**7**
1998	2	1	**4**
1999	–	1	**2**
2000	1	1	**3**
2001	1	1	**3**
2002	–	1	**2**
2003	2	1	**4**
2004	1	2	**5**
2005	1	3	**7**
2006	–	3	**6**
2007	–	1	**2**
2008	–	1	**2**
2009	–	–	**–**
2010	–	2	**4**
2011	–	2	**4**
2012	–	–	**–**
2013	–	3	**6**
2014 Set-1	1	1	**3**
2014 Set-2	1	2	**5**

Exam Year	1 Mark Ques.	2 Marks Ques.	Total Marks
2014 Set-3	1	–	**1**
2014 Set-4	–	1	**2**
2015 Set-1	–	1	**2**
2015 Set-2	1	1	**3**
2015 Set-3	–	–	**–**
2016 Set-1	1	1	**3**
2016 Set-2	1	–	**1**
2016 Set-3	–	1	**2**
2017 Set-1	–	1	**2**
2017 Set-2	–	1	**2**
2018 Set-1	–	–	**–**
2018 Set-2	1	–	**1**
2019 Set-1	–	2	**4**
2019 Set-2	1	–	**1**
2020 Set-1	1	3	**7**
2020 Set-2	1	1	**3**
2021 Set-1	–	1	**2**
2021 Set-2	2	1	**4**
2022 Set-1	1	2	**5**
2022 Set-2	1	–	**1**
2023	1	–	**1**
2024	1	2	**5**

1 Compressible Flow

1.1 In adiabatic flow with friction, the stagnation temperature along a streamline .... (increase/ remain constant) **[1995 : 1 M]**

1.2 For a compressible fluid, sonic velocity is
(a) a property of the fluid
(b) always given by $(\gamma RT)^{1/2}$ where γ, R and T are respectively the ratio of specific heats, gas constant and temperature in K
(c) always given by $\left(\frac{\partial p}{\partial \rho}\right)_s^{1/2}$, where p, ρ and s are respectively pressure, density and entropy
(d) always greater then the velocity of fluid at any location

[2000 : 2 M]

1.3 Consider adiabatic flow of air through a duct. At a given point in the duct, velocity of air is 300 m/s, temperature is 330 K and pressure is 180 kPa. Assume that the air behaves as a perfect gas with constant c_p = 1.005 kJ/kgK. The stagnation temperature at this point is ______ K (round off to two decimal places.)

[2021 : 1 M, Set-2]

1.4 An object is moving with a Mach number of 0.6 in an ideal gas environment, which is at a temperature of 350 K. The gas constant is 320 J/kg.K and ratio of specific heats is 1.3. The speed of object is_________ m/s (round off to that nearest integer).

[2021 : 1 M, Set-2]

1.5 The figure shows a purely convergent nozzle with a steady, inviscid compressible flow of an ideal gas with constant thermophysical properties operating under choked condition. The exit plane shown in the figure is located within the nozzle. If the inlet pressure (P_0) is increased while keeping the back pressure (P_{back}) unchanged, which of the following statements is/are true?

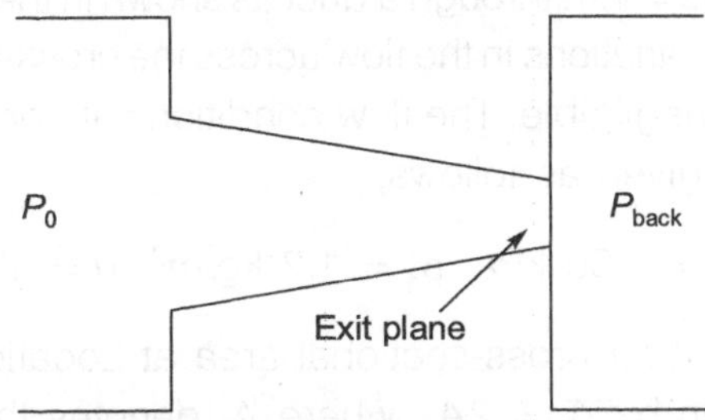

(a) Mass flow rate through the nozzle will remain unchanged.
(b) Mach number at the exit plane of the nozzle will remain unchanged at unity.
(c) Mass flow rate through the nozzle will increase.
(d) Mach number at the exit plane of the nozzle will become more than unity.

[2022 : 1 M, Set-1]

1.6 Consider steady, one-dimensional compressible flow of a gas in a pipe of diameter 1 m. At one location in the pipe, the density and velocity are 1 kg/m^3 and 100 m/s, respectively. At a downstream location in the pipe, the velocity is 170 m/s. If the pressure drop between these two locations is 10 kPa, the force exerted by the gas on the pipe between these two locations is ________ N.

(a) $350\,\pi^2$ (b) $750\,\pi$
(c) $1000\,\pi$ (d) 3000

[2022 : 2 M, Set-1]

1.7 Consider a steady flow through a horizontal divergent channel, as shown in the figure, with supersonic flow at the inlet. The direction of flow is from left to right.

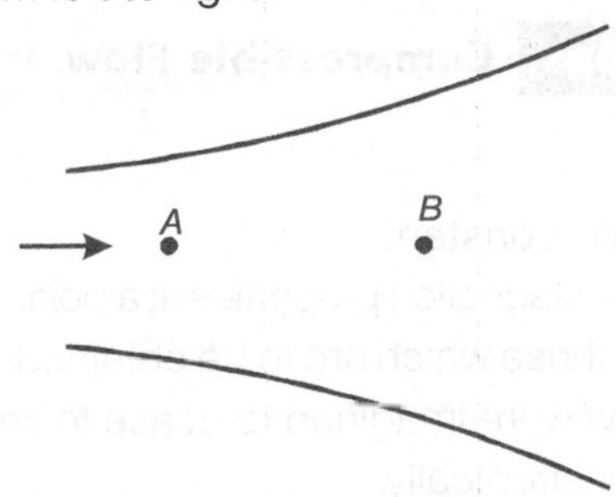

Pressure at location B is observed to be higher than that at an upstream location A. Which among the following options can be the reason?
(a) Since volume flow rate is constant, velocity at B is lower than velocity at A

(b) Normal shock
(c) Viscous effect
(d) Boundary layer separation **[2022 : 1 M, Set-2]**

1.8 Consider an isentropic flow of air (ratio of specific heats = 1.4) through a duct as shown in the figure. The variations in the flow across the cross-section are negligible. The flow conditions at Location 1 are given as follows:

P_1 = 100 kPa, ρ_1 = 1.2 kg/m^3, u_1= 400 m/s

The duct cross-sectional area at Location 2 is given by $A_2 = 2A_1$, where A_1 denotes the duct cross-sectional area at Location 1. Which one of the given statements about the velocity u_2 and pressure P_2 at Location 2 is TRUE?

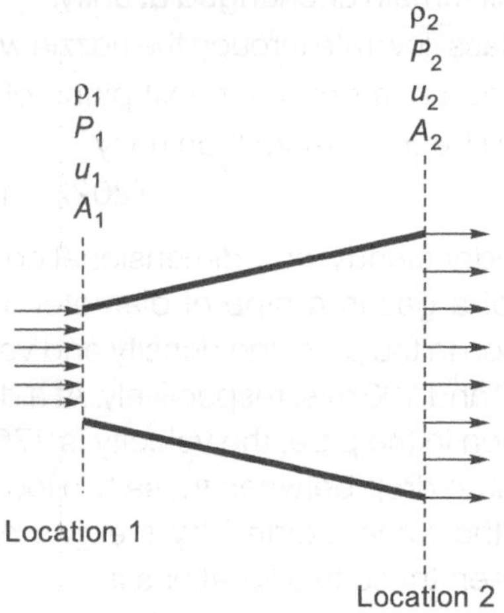

(a) $u_2 < u_1$, $P_2 < P_1$
(b) $u_2 < u_1$, $P_2 > P_1$
(c) $u_2 > u_1$, $P_2 < P_1$
(d) $u_2 > u_1$, $P_2 > P_1$ **[2023 : 1 M]**

1.9 Steady, compressible flow of air takes place through an adiabatic converging-diverging nozzle, as shown in the figure. For a particular value of pressure difference across the nozzle, a stationary normal shock wave forms in the diverging section of the nozzle. If *E* and *F* denote the flow conditions just upstream and downstream of the normal shock, respectively, which of the following statement(s) is/are TRUE?

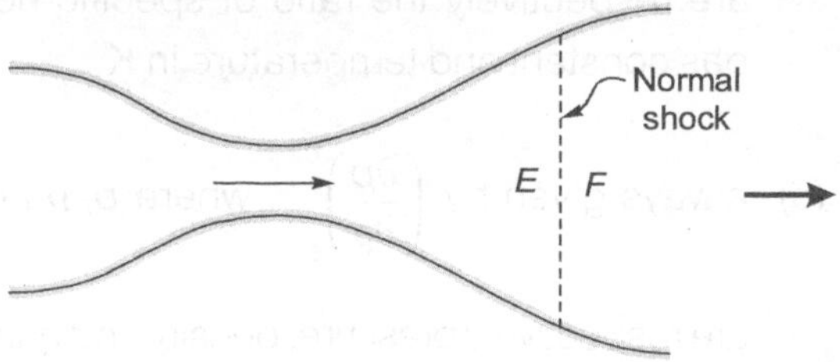

(a) Mach number at *E* is lower than the Mach number at *F*
(b) Density at *E* is lower than the density at *F*
(c) Specific entropy at *E* is lower than the specific entropy at *F*
(d) Static pressure at *E* is lower than the static pressure at *F*

[2024 : 2 M]

■■■■

Answers Compressible Flow

1.1	(Remain constant)	1.2	(c)	1.3	(374.70)	1.4	(229)	1.5	(b, c)	1.6	(b)		
1.7	(b)	1.8	(c)	1.8	(c)	1.9	(b, c, d)						

Explanations Compressible Flow

1.1 Sol.

Remain constant

- The stagnation properties at a point are defined as those which are to be obtained. If the local flow were imagined to cease to zero velocity isenstropically.
- The stagnation values are denoted by a subscript zero.
- If the flow is adiabatic, the stagnation temperature(T) along a streamline is constant, even in the presence of friction.

1.2 (c)

Sonic velocity or velocity of sound :

- The velocity of sound depends upon the temperatur and nature of steam.
- In case of compressible fluid, sonic velocity,

$$C = \sqrt{\frac{dp}{d\rho}}$$

- In case of Ideal gas, sonic velocity,

$$C = \sqrt{\gamma RT}$$

1.3 Sol.

$$M_1 = \frac{V_1}{\sqrt{\gamma R T_1}} = \frac{300}{\sqrt{1.4 \times 287 \times 330}}$$

$$= 0.823$$

$$\frac{T_{o1}}{T_1} = 1 + \frac{\gamma - 1}{2} M_1^2 = 1 + \frac{1.4 - 1}{2}(0.823)^2$$

$$\frac{T_{o1}}{T_1} = 1.154$$

$$T_{01} = 374.7037 \text{ K}$$

1.4 Sol.

$$M = \frac{V}{C} = \frac{V}{\sqrt{\gamma R T}}$$

$$0.6 = \frac{V}{\sqrt{1.3 \times 350 \times 320}}$$

$$V = 228.94 \text{ m/s} \simeq 229 \text{ m/s}$$

1.5 (b, c)

When, P_0 is increased and P_b is unchanged, due to choked condition, mass flow rate through exit plane will first increase and then will become maximum.

Mach number corresponding to maximum mass flow will be unity.

1.6 (b)

Given : Diameter of duct, $D = 1$ m; $V_1 = 100$ m/s; $V_2 = 170$ m/s;

Change in pressure, $P = 10$ kPa

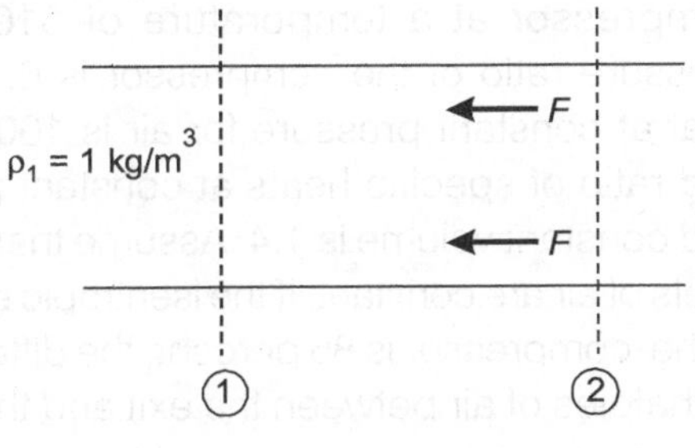

Using the momentum equation,

$$(P_1 - P_2)A - F = \dot{m}(V_2 - V_1)$$

$$\Rightarrow 10 \times 10^3 \times \frac{\pi \times 1^2}{4} - F = \rho_1 A_1 V_1 (V_2 - V_1)$$

$$\Rightarrow 2500\,\pi - F = 1 \times \frac{\pi}{4} \times 1^2 \times 100 \times (70)$$

$$\Rightarrow \quad F = 750\,\pi \text{ Newtons}$$

1.7 (b)

Since this is a horizontal divergent channel with supersonic flow at inlet so if normal flow occurs, it acts as a nozzle.

So, V_B should be greater than V_A *and* P_B should be lesser than P_A.

If we found out $P_B > P_A$ so there should be a normal shock available.

1.8 (c)

Given: $P_1 = 100$ kPa, $\rho_1 = 1.2$ kg/m^3, $u_1 = 400$ m/s.

The Mach number is given by

$$M = \frac{V}{\sqrt{\gamma R T_1}} \quad ...(1)$$

From ideal gas equation,

$$P_1 = \rho R T_1$$

$$\Rightarrow T_1 = \frac{P_1}{\rho R} = \frac{100}{1.2 \times 0.287} = 290.36 \text{ K}$$

$$\therefore \quad M = \frac{400}{\sqrt{1.4 \times 0.287 \times 10^3 \times 290.36}} = 1.17$$

Hence, the flow is supersonic.

Therefore, it will work as a nozzle, and the velocity will increase and pressure will decrease along the flow.

1.9 (b, c, d)

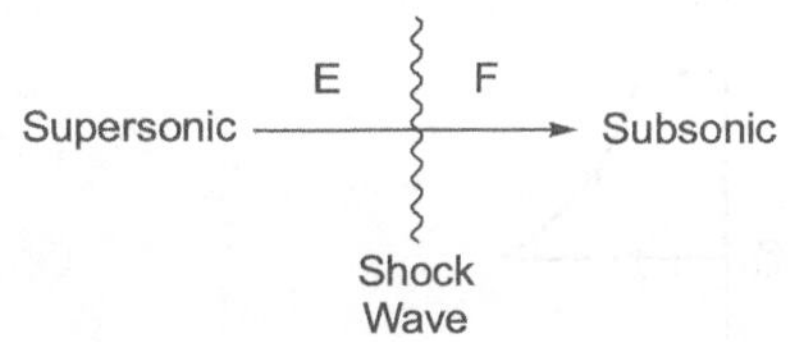

After normal shock:

- Stagnation temperature remains the same.
- Density after shock will increase, i.e., $\rho_E < \rho_F$.
- Mach number will decrease, i.e., $M_E > M_F$.
- Entropy across a shockwave always increases as it is a highly irreversible process.

■■■■

2 Air Compressors (Reciprocating & Rotary)

2.1 Select statement from List-II matching the processes in List-I. Enter your answer as *A*, *B* is the correct choice for (1) is (*A*) and (2) and (*B*).

List-I
1. Intercooling
2. Isothermal compression

List-II
- A. No heat transfer during compression
- B. Reduce low pressure compressor work
- C. Heat rejection during compression
- D. Reduce high pressure compressor work

[1999 : 2 M]

2.2 A single acting two stage compressor with complete inter cooling delivers air at 16 bar. Assuming an intake state of 1 bar at 15°C, the pressure ratio per stage is

(a) 16 (b) 8
(c) 4 (d) 2

[2001 : 2 M]

2.3 A *p-V* diagram has been obtained from a test on a reciprocating compressor. Which of the following represents that diagram?

(a)
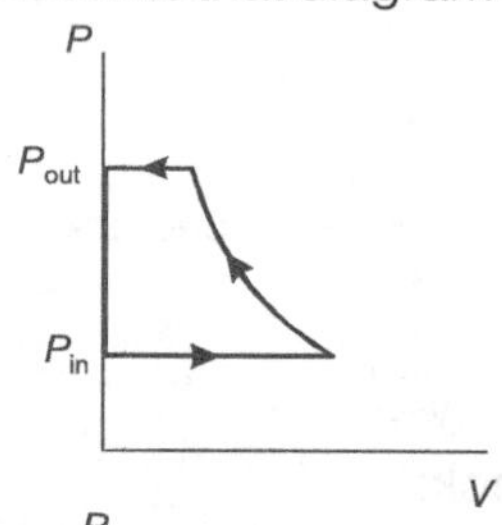

(b)
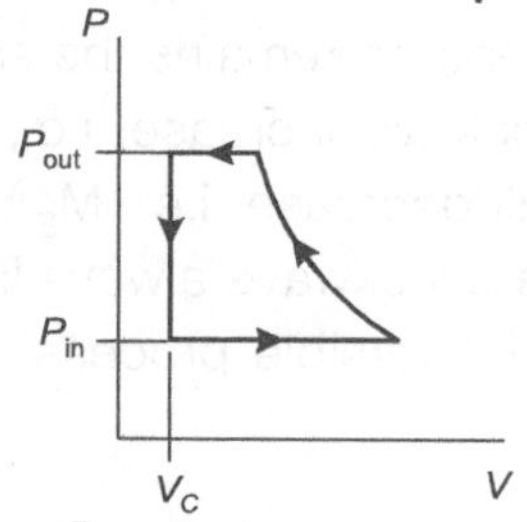

(c)
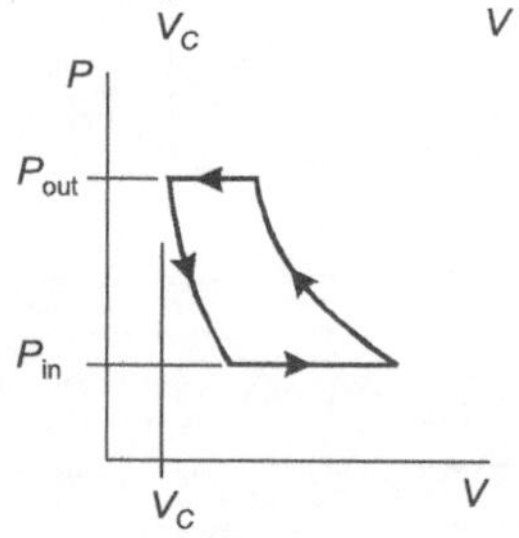

(d)
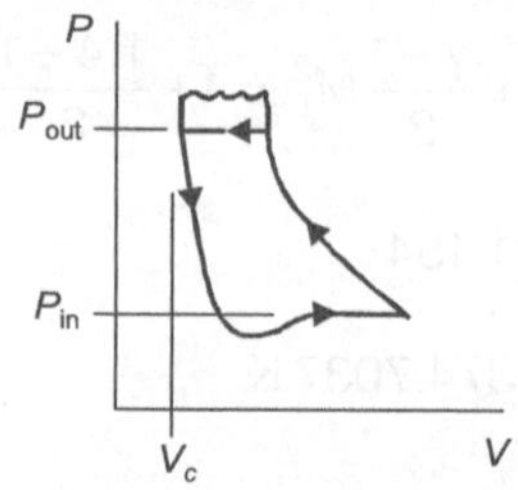

[2005 : 1 M]

2.4 In a 3-stage air compressor, the inlet pressure is p_1, discharge pressure is p_4 and the intermediate pressures are p_2 and p_3 ($p_2 < p_3$). The total pressure ratio of the compressor is 10 and the pressure ratios of the stages are equal. If p_1 = 100 kPa, the value of the pressure p_3 (in kPa) is______

[2016 : 2 M, Set-3]

2.5 The values of enthalpies at the stator inlet and rotor outlet of a hydraulic turbomachine stage are h_1 and h_3 respectively. The enthalpy at the stator outlet (or, rotor inlet) is h_2. The condition $(h_2 - h_1) = (h_3 - h_2)$ indicates that the degree of reaction of this stage is

(a) 75% (b) 100%
(c) 50% (d) zero

[2020 : 1 M, Set-2]

2.6 Air (ideal gas) enters a perfectly insulated compressor at a temperature of 310 K. The pressure ratio of the compressor is 6. Specific heat at constant pressure for air is 1005 J/kg.K and ratio of specific heats at constant pressure and constant volume is 1.4. Assume that specific heats of air are constant. If the isentropic efficiency of the compressor is 85 percent, the difference in enthalpies of air between the exit and the inlet of the compressor is _______ kJ/kg (round off to nearest integer).

[2020 : 2 M, Set-1]

■■■■

Answers Air Compressors (Reciprocating & Rotary)

2.1 (1–C, 2–D) 2.2 (c) 2.3 (d) 2.4 (464.16) 2.5 (c) 2.6 (245)

Explanations Air Compressors (Reciprocating & Rotary)

2.2 (c)

Given:
Intake pressure, $p_1 = 1$ bar
Delivery pressure, $P_{N+n} = 16$ bar
As, we know

$$P_{N+1} = K^N P_1$$

where, N = Number of stage

$$K = \frac{P_2}{P_1} = \text{pressure ratio across a stage}$$

$$\therefore \quad 16 = K^2(1)$$

Pressure ratio across a stage,

$$K = \sqrt{\frac{16}{1}} = 4$$

2.3 (d)

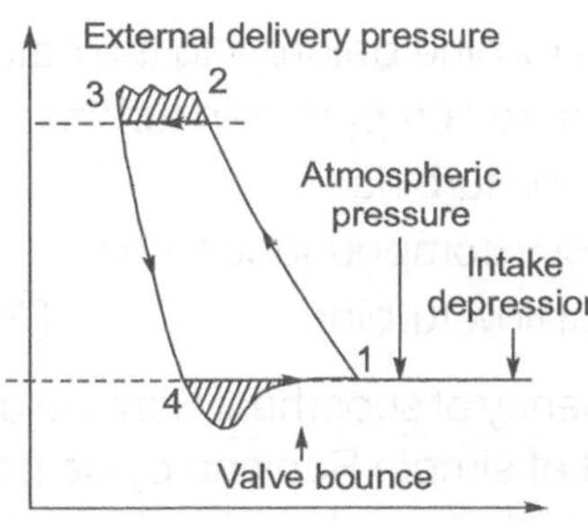

At point 4, when the clearance air has reduced to atmospheric pressure. The inlet valve in practice will not open due to mainly two reason.

A. There must be a pressure difference across the inlet valve in order to move it.

B. Inlet valve inertia.

Thus, the pressure drops away until the valve is forced off its seat, thus valve bounce will occur shown by way line.

2.4 Sol.

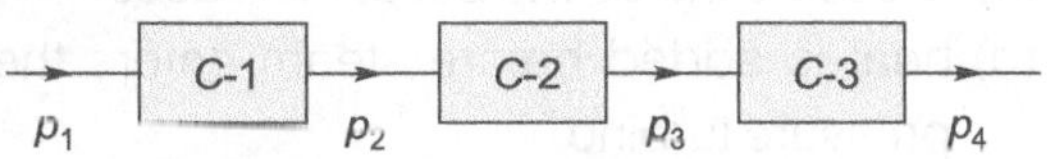

$p_1 = 100$ kPa

Total pressure ratio,

$$\frac{p_4}{p_1} = 10$$

Pressure ratio of each stages (r_p),

$$\frac{p_2}{p_1} = \frac{p_3}{p_2} = \frac{p_4}{p_3}$$

To calculate $p_3 = ?$

$$\frac{p_4}{p_1} = \frac{p_4}{p_3} \times \frac{p_3}{p_2} \times \frac{p_2}{p_1} = 10$$

$$r_p \times r_p \times r_p = 10$$

$$r_p^3 = 10$$

$$r_p = (10)^{1/3}$$

$$p_2 = r_p p_1$$

$$p_3 = r_p p_2 = r_p^2 p_1 = (10)^{2/3} \times 100$$

$$= \mathbf{464.16\ kPa}$$

2.5 (c)

As enthalpy across stator and rotor is equal it is 50% reaction stage.

2.6 Sol.

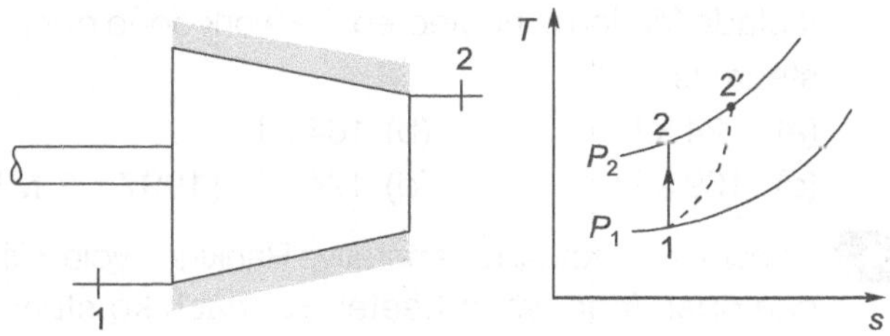

$$T_1 = 310\,\text{K}$$

$$r_p = \frac{P_2}{P_1} = 6 \qquad \eta_s = 0.85$$

$$c_p = 1.005\ \text{kJ/kgK}$$

$$\gamma = 1.4$$

$$\therefore \quad \frac{T_2}{T_1} = \left(\frac{P_2}{P_1}\right)^{\frac{\gamma-1}{\gamma}}$$

$$\Rightarrow \quad \frac{T_2}{310} = (6)^{(1.4-1)/1.4}$$

$$\Rightarrow \quad T_2 = 517.225\ \text{K}$$

Now,

$$\eta_{isen} = \frac{W_{isen}}{W_{actual}} = \frac{h_2 - h_1}{h_2' - h_1}$$

$$0.85 = \frac{c_p(T_2 - T_1)}{h_2' - h_1}$$

$$\Rightarrow \quad h_2' - h_1 = \frac{1.005(517.22 - 310)}{0.85}$$

$$= 245\ \text{kJ/kg}$$

■■■■

3 Steam Power Cycles & Steam Turbines

3.1 Consider a Rankine cycle with superheat. If the maximum pressure in the cycle is increased without changing the maximum temperature and the minimum pressure, the dryness fraction of steam after the isentropic expansion will increase [TRUE/FALSE] **[1995 : 1 M]**

3.2 A steam power plant has the boiler efficiency of 92%, turbine efficiency (mechanical) of 94%, generator efficiency of 95% and cycle efficiency of 44%. If 6% of the generated power is used to run the auxiliaries, the overall plant efficiency is

(a) 34% (b) 39%
(c) 45% (d) 30% **[1996 : 2 M]**

3.3 The following data pertain to a single stage impulse steam turbine:
Nozzle angle = 20°; Blade velocity = 200 m/s
Relative steam velocity at entry = 350 m/s
Blade inlet angle = 30°; Blade exit angle = 25°
If blade friction is neglected the work done per kg steam is

(a) 124 kJ (b) 164 kJ
(c) 169 kJ (d) 174 kJ **[1997 : 2 M]**

3.4 Consider an actual regenerative Rankine cycle with one open feed water heater. For each kg steam entering the turbine, if m kg steam with a specific enthalpy of h_1 is bladed from the turbine, and the specific enthalpy of liquid water entering the heater is h_2, then h_3 the specific enthalpy of saturated liquid leaving the heater is equal to

(a) $mh_1 - (h_2 - h_1)$ (b) $h_1 - m(h_2 - h_1)$
(c) $h_2 - m(h_2 - h_1)$ (d) $mh_2 - (h_2 - h_1)$
[1997 : 2 M]

3.5 If V_N and α are the nozzle exit velocity and nozzle angle in an impulse turbine, the optimum blade velocity is given by

(a) $V_N \cos 2\alpha$ (b) $V_N \sin 2\alpha$
(c) $V_N \cos\left(\frac{\alpha}{2}\right)$ (d) $\frac{V_N \cos\alpha}{2}$ **[1998 : 1 M]**

3.6 A Curtis stage, Rateau stage and a 50% reaction stage in a steam turbine are examples of

(a) different types of impulse stages
(b) different types of reaction stages
(c) a simple impulse stage, a velocity compounded impulse stage and reaction stage
(d) a velocity compounded impulse stage, a pressure compounded impulse stage and a reaction stage **[1998 : 1 M]**

3.7 The isentropic heat drop in the nozzle of an impulse steam turbine with a nozzle efficiency 0.9, blade velocity ratio 0.5 and mean blade velocity 150 m/s in kJ/kg is

(a) 50 (b) 40
(c) 60 (d) 75 **[1998 : 2 M]**

3.8 hich of the following is a pressure compounded turbine?

(a) Parsons (b) Curtis
(c) Rateau (d) All of three
[2000 : 1 M]

3.9 e Rateau turbine belongs to the category of

(a) pressure compounded turbine
(b) reaction turbine
(c) velocity compounded turbine
(d) radial flow turbine **[2001 : 1 M]**

3.10 The efficiency of superheat Rankine cycle is higher than that of simple Rankine cycle because

(a) the enthalpy of main steam is higher for superheat cycle
(b) the mean temperature of heat addition is higher for superheat cycle
(c) the temperature of steam in the condenser is high
(d) the quality of steam in the condenser is low
[2002 : 2 M]

3.11 In Rankine cycle, regeneration results in higher efficiency because

(a) pressure inside the boiler increases
(b) heat is added before steam enters the low pressure turbine
(c) average temperature of heat addition in the boiler increases
(d) total work delivered by the turbine increases
[2003 : 1 M]

3.12 Considering the variation of static pressure and absolute velocity in an impulse steam turbine, across one row of moving blades

(a) both pressure and velocity decrease
(b) pressure decreases but velocity increases
(c) pressure remains constant, while velocity increases
(d) pressure remains constant, while velocity decreases [2003 : 1 M]

Common Data Questions Q. 3.13 and Q.3.14

Consider a steam power plant using a reheat cycle as shown. Steam leaves the boiler and enters the turbine at 4 MPa, 350°C (h_3 = 3095 kJ/kg). After expansion in the turbine to 400 kPa (h_4 = 2609 kJ/kg), the steam is reheated to 350°C (h_5 = 3170 kJ/kg), and then expanded in a low pressure turbine to 10 kPa (h_6 = 2165 kJ/kg) the specific volume of liquid handled by the pump can be assumed to be 0.0010025 m^3/kg.

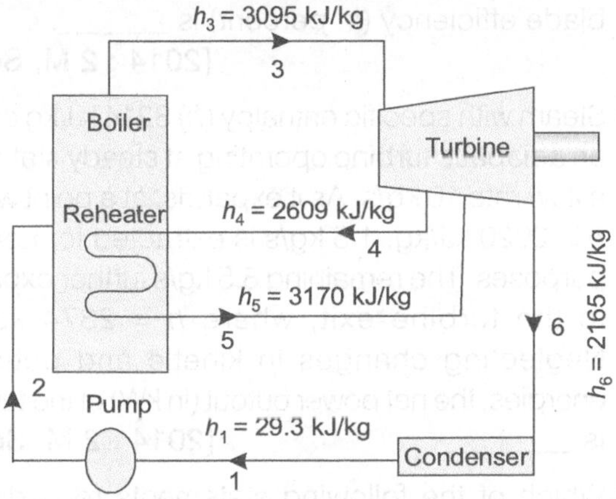

3.13 The thermal efficiency of the plant neglecting pump work is
(a) 15.8% (b) 41.1%
(c) 48.5% (d) 58.6% [2004 : 2 M]

3.14 The enthalpy at the pump discharge (h_2) is
(a) 0.33 kJ/kg (b) 3.33 kJ/kg
(c) 4.0 kJ/kg (d) 33.3 kJ/kg [2004 : 2 M]

3.15 In the velocity diagram shown below, u = blade velocity, C = absolute fluid velocity and W = relative velocity of fluid and the subscripts 1 and 2 refer to inlet and outlet. This diagram is for

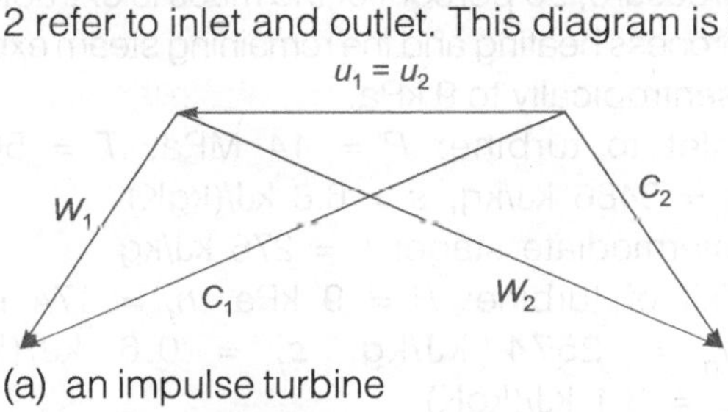

(a) an impulse turbine
(b) a reaction turbine
(c) a centrifugal compressor
(d) an axial flow compressor [2005 : 2 M]

3.16 Determine the correctness or otherwise **Assertion (A)** and the **Reason (R)**.
Assertion (A) : In a power plant working on a Rankine cycle, the regenerative feed water heating improves the efficiency of the steam turbine.
Reason (R) : The regenerative feed water heating raises the average temperature of heat addition in the Rankine cycle.
(a) Both (A) and (R) are true and (R) is the correct reason for (A)
(b) Both (A) and (R) are true but (R) is NOT the correct reason for (A)
(c) Both (A) and (R) are false
(d) (A) is false but (R) is true [2006 : 2 M]

3.17 Which combination of the following statements is correct?
The incorporation of reheater in a steam power plant:
P : always increases the thermal efficiency of the plant.
Q : Always increases the dryness fraction of steam at condenser inlet.
R : Always increases the mean temperature of heat addition.
S : Always increases the specific work output.
(a) P and S (b) Q and S
(c) P, R and S (d) P, Q, R and S [2007 : 2 M]

3.18 A thermal power plant operates on a regenerative cycle with a single open feedwater heater, as shown in the figure. For the state points shown, the specific enthalpies are: h_1 = 2800 kJ/kg and h_2 = 200 kJ/kg. The bleed to the feedwater heater is 20% of the boiler steam generation rate. The specific enthalpy at state 3 is

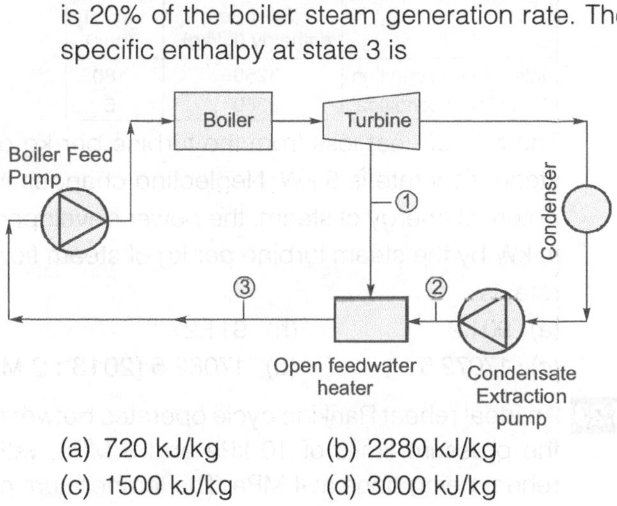

(a) 720 kJ/kg (b) 2280 kJ/kg
(c) 1500 kJ/kg (d) 3000 kJ/kg

[2008 : 2 M]

Common Data for Questions 3.19 and 3.20.

In a steam power plant operating on the Rankine cycle, steam enters the turbine at 4 MPa, 350°C and exists at a pressure of 15 kPa. Then it enters the condenser and exists as saturated water. Next, a pump feeds back the water to the boiler. The adiabatic efficiency of the turbine is 90%. The thermodynamic states of water and steam are given in the table.

State	h (kJkg^{-1})		s (kJkg^{-1}K^{-1})		v(m^{3}kg^{-1})	
Steam : 4 MPa, 350°C	3092.5		6.5821		0.06645	
Water : 15 kPa	h_f	h_g	s_f	s_g	v_f	v_g
	225.94	2599.1	0.7549	8.0085	0.001014	10.02

h is specific enthalpy, s is specific entropy and v the specific volume; subscript f and g denote saturated liquid state and saturated vapour state.

3.19 The net work output (kJ kg^{-1}) of the cycle is
(a) 498 (b) 775
(c) 860 (d) 957 [2010 : 2 M]

3.20 Heat supplied (kJ kg^{-1}) to the cycle is
(a) 2372 (b) 2576
(c) 2863 (d) 3092 [2010 : 2 M]

3.21 The values of enthalpy of steam at the inlet and outlet of a steam turbine in a Rankine cycle are 2800 kJ/kg and 1800 kJ/kg respectively. Neglecting pump work, the specific steam consumption in kg/kW hour is
(a) 3.60 (b) 0.36
(c) 0.06 (d) 0.01 [2011 : 2 M]

3.22 Specific enthalpy and velocity of steam at inlet and exit of a steam turbine, running under steady state, are as given below:

	Specific enthalpy (kJ/kg)	Velocity (m/s)
Inlet steam condition	3250	180
Exit steam condition	2360	5

The rate of heat loss from the turbine per kg of steam flow rate is 5 kW. Neglecting changes in potential energy of steam, the power developed in kW by the steam turbine per kg of steam flow rate, is
(a) 901.2 (b) 911.2
(c) 17072.5 (d) 17082.5 [2013 : 2 M]

3.23 An ideal reheat Rankine cycle operates between the pressure limits of 10 kPa and 8 MPa, with reheat being done at 4 MPa. The temperature of steam at the inlets of both turbines is 500°C and the enthalpy of steam is 3185 kJ/kg at the exit of the high pressure turbine and 2247 kJ/kg at the exit of low pressure turbine. The enthalpy of water at the exit from the pump is 191 kJ/kg. Use the following table for relevant data.

Super heated steam temperature (°C)	Pressure (MPa)	v (m^3/kg)	h (kJ/kg)	s (kJ/kgK)
500	4	0.08644	3446	7.0922
500	8	0.04177	3399	6.7266

Disregarding the pump work, the cycle efficiency (in percentage) is ______. [2014 : 2 M, Set-1]

3.24 Steam at a velocity of 10 m/s enters the impulse turbine stage with symmetrical blading having blade angle 30°. The enthalpy drop in the stage is 100 kJ. The nozzle angle is 20°. The maximum blade efficiency (in percent) is ______.
[2014 : 2 M, Set-2]

3.25 Steam with specific enthalpy (h) 3214 kJ/kg enters an adiabatic turbine operating at steady state with a flow rate 10 kg/s. As it expands, at a point where h is 2920 kJ/kg, 1.5 kg/s is extracted for heating purposes. The remaining 8.5 kg/s further expands to the turbine exit, where h = 2374 kJ/kg. Neglecting changes in kinetic and potential energies, the net power output (in kW) of the turbine is ______. [2014 : 2 M, Set-4]

3.26 Which of the following statements regarding a Rankine cycle with reheating are TRUE?
(i) increase in a average temperature of heat addition
(ii) reduction in thermal efficiency
(iii) drier steam at the turbine exit
(a) only (i) and (ii) are correct
(b) only (ii) and (iii) are correct
(c) only (i) and (iii) are correct
(d) (i), (ii) and (iii) are correct [2015 : 1 M, Set-2]

3.27 Steam enters a well insulated turbine and expands isentropically throughout. At an intermediate pressure, 20 percent of the mass is extracted for process heating and the remaining steam expands isentropically to 9 kPa.
Inlet to turbine: P = 14 MPa, T = 560°C, h = 3486 kJ/kg, s = 6.6 kJ/(kgK)
Intermediate stage: h = 276 kJ/kg
Exit of turbine: P = 9 kPa, h_f = 174 kJ/kg, h_g = 2574 kJ/kg, s_f = 0.6 kJ/(kgK), s_g = 8.1 kJ/(kgK)
If the flow rate of steam entering the turbine is 100 kg/s, then the work output (in MW) is ______.
[2015 : 2 M, Set-1]

3.28 In a Rankine cycle, the enthalpies at turbine entry and outlet are 3159 kJ/kg and 2187 kJ/kg, respectively. If the specific pump work is 2 kJ/kg, the specific steam consumption (in kg/kWh) of the cycle based on net output is ________.

[2015 : 2 M, Set-2]

3.29 The **INCORRECT** statement about regeneration in vapor power cycle is that

(a) it increases the irreversibility by adding the liquid with higher energy content to the steam generator
(b) heat is exchanged between the expanding fluid in the turbine and the compressed fluid before heat addition
(c) the principle is similar to the principle of Stirling gas cycle
(d) it is practically implemented by providing feed water heaters

[2016 : 1 M, Set-1]

3.30 In a steam power plant operating on a ideal Rankine cycle, superheated steam enters the turbine at 3 MPa and 350°C. The condenser pressure is 75 kPa. The thermal efficiency of the cycle is ______%.

Given data :

For saturated liquid, at p = 75 kPa, h_f = 384.39 kJ/kg, v_f = 0.001037 m^3/kg, s_f = 1.213 kJ/kgK.

At 75 kPa, h_{fg} = 2278.6 kJ/kg,

s_{fg} = 6.2434 kJ/kgK

At P = 3 MPa and T = 350°C (superheated steam), h = 3115.3 kJ/kg, s = 6.7428 kJ/kgK.

[2016 : 2 M, Set-1]

3.31 In the Rankine cycle for a steam power plant the turbine entry and exit enthalpies are 2803 kJ/kg and 1800 kJ/kg, respectively. The enthalpies of water at pump entry and exit are 121 kJ/kg and 124 kJ/kg respectively. The specific steam consumption (in kg/kWh) of the cycle is ______.

[2017 : 2 M, Set-2]

3.32 Select the correct statement for 50% reaction stage in a steam turbine.

(a) The rotor blade is symmetric
(b) The stator blade is symmetric
(c) The absolute inlet flow angle is equal to absolute exit flow angle
(d) The absolute exit flow angle is equal to inlet angle of rotor blade.

[2018 : 1 M, Set-2]

3.33 A steam power cycle with regeneration as shown below on the T-s diagram employs a single open feedwater heater for efficiency improvement. The fluids mix with each other in an open feedwater heater. The turbine is isentropic and the input (bleed) to the feedwater heater from the turbine is at state 2 as shown in the figure. Process 3-4 occurs in the condenser. The pump work is negligible. The input to the boiler is at state 5. The following information is available from the steam tables:

State	1	2	3	4	5	6
Enthalpy (kJ/kg)	3350	2800	2300	175	700	1000

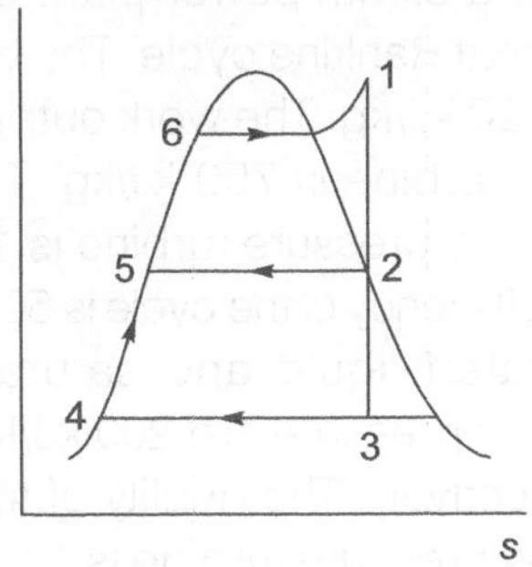

The mass flow rate of steam bled from the turbine as a percentage of the total mass flow rate at the inlet to the turbine at state 1 is ________.

[2019 : 2 M, Set-1]

3.34 Which one of the following modifications, of the simple ideal Rankine cycle increases the thermal efficiency and reduces the moisture content of the steam at the turbine outlet?

(a) Decreasing the boiler pressure.
(b) Decreasing the condenser pressure.
(c) Increasing the boiler pressure.
(d) Increasing the turbine inlet temperature.

[2019 : 1 M, Set-2]

3.35 For an ideal Rankine cycle operating between pressures of 30 bar and 0.04 bar, the work output from the turbine is 903 kJ/kg and the work input to the feed pump is 3 kJ/kg. The specific steam consumption is ____________ kg/kW.h (round off to 2 decimal places).

[2020 : 2 M, Set-1]

3.36 In a steam power plant, superheated steam at 10 MPa and 500°C, is expanded isentropically in a turbine until it becomes a saturated vapour. It is then reheated at constant pressure to 500°C. The steam is next expanded isentropically in another turbine until it reaches the condenser pressure of 20 kPa. Relevant properties of steam are given in the following two tables. The work done by both the turbines together is___ kJ/kg (roundoff to the nearest integer).

Superheated Steam Table:

Pressure, p (MPa)	Temp, p (°C)	Enthalpy, h (kJ/kg)	Entropy, s (kJ/kg.K)
10	500	3373.6	6.5965
1	500	3478.4	7.7621

Saturated Steam Table:

Pressure, p	Sat. Temp, T_{sat}(°C)	Enthalpy, h (kJ/kg) h_f	h_g	Entropy, s (kJ/kgK) s_f	s_g
1 MPa	179.91	762.9	2778.1	2.1386	6.5965
20 kPa	60.06	251.38	2609.7	0.8319	7.9085

[2020 : 2 M, Set-2]

3.37 Consider a steam power plant operating on an ideal reheat Rankine cycle. The work input to the pump is 20 kJ/kg. The work output from the high pressure turbine is 750 kJ/kg. The work output from the low pressure turbine is 1500 kJ/kg. The thermal efficiency of the cycle is 50%. The enthalpy of saturated liquid and saturated vapour at condenser pressure are 200 kJ/kg and 2600 kJ/kg, respectively. The quality of steam at the exit of the low pressure turbine is ______ % [Round off to the nearest integer] [2021 : 2 M, Set-1]

3.38 Consider the open feed water heater (FWH) shown in the figure given below:

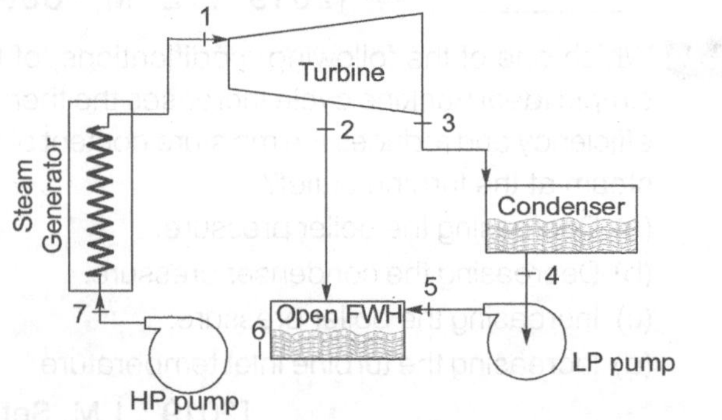

Specific enthalpy of steam at location 2 is 2624 kJ/kg, specific enthalpy of water at location 5 is 226.7 kJ/kg and specific enthalpy of saturated water at location 6 is 708.6 kJ/kg. If the mass flow rate of water entering the open feed water heater (at location 5) is 100 kg/s then the mass flow rate of steam at location 2 will be____________kg/s (round off to one decimal place).

[2021 : 2 M, Set-2]

3.39 In a steam power plant based on Rankine cycle, steam is initially expanded in a high-pressure turbine. The steam is then reheated in a reheater and finally expanded in a low-pressure turbine. The expansion work in the high-pressure turbine is 400 kJ/kg and in the low-pressure turbine is 850 kJ/kg, whereas the pump work is 15 kJ/kg. If the cycle efficiency is 32%, the heat rejected in the condenser is ______ kJ/kg (round off to 2 decimal places).

[2022 : 2 M, Set-1]

3.40 Which one of the following statements regarding a Rankine cycle is FALSE?

(a) Cycle efficiency increases as boiler pressure decreases.

(b) Superheating the steam in the boiler increases the cycle efficiency.

(c) The pressure at the turbine outlet depends on the condenser temperature.

(d) Cycle efficiency increases as condenser pressure decreases.

[2024 : 1 M]

■■■■

Answers Steam Power Cycles & Steam Turbines

3.1	(False)	3.2	(a)	3.3	(a)	3.4	(c)	3.5	(d)	3.6	(d)	3.7	(a)	3.8	(c)
3.9	(a)	3.10	(b)	3.11	(c)	3.12	(d)	3.13	(b)	3.14	(d)	3.15	(b)	3.16	(a)
3.17	(b)	3.18	(a)	3.19	(c)	3.20	(c)	3.21	(a)	3.22	(a)	3.23	(40.73)		
3.24	(88.3)	3.25	(7581)	3.26	(c)	3.27	(125.56)	3.28	(3.711)			3.29	(a)		
3.30	(26.01)	3.31	(3.6)	3.32	(d)	3.33	(20)	3.34	(d)	3.35	(4)	3.36	(1513)		
3.37	(93)	3.38	(25.2)	3.39	(2624.37)			3.40	(a)						

Explanations — Steam Power Cycles & Steam Turbines

3.1 Sol.

False

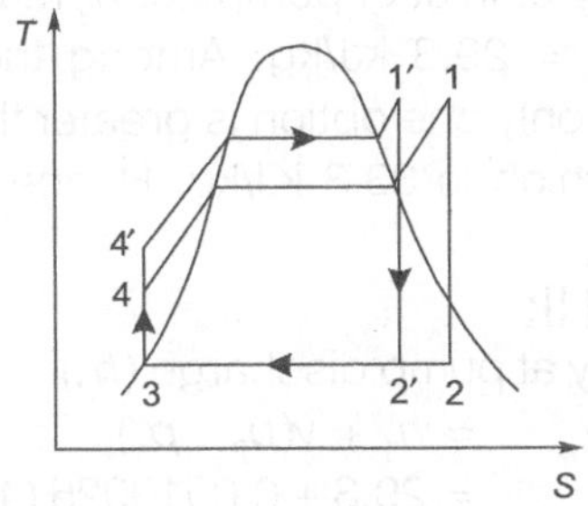

3.2 (a)

$$\eta_0 = \eta_b \times \eta_{T(R)} \times \eta_{mech(T)} \times \eta_{gen} \times \eta_{Aux}$$
$$= 0.92 \times 0.44 \times 0.94 \times 0.96 \times 0.94$$
$$= 0.34 = \mathbf{34\%}$$

3.3 (a)

Given data: $\alpha_1 = 20°$; $u = 200$ m/s; $V_{r1} = 350$ m/s

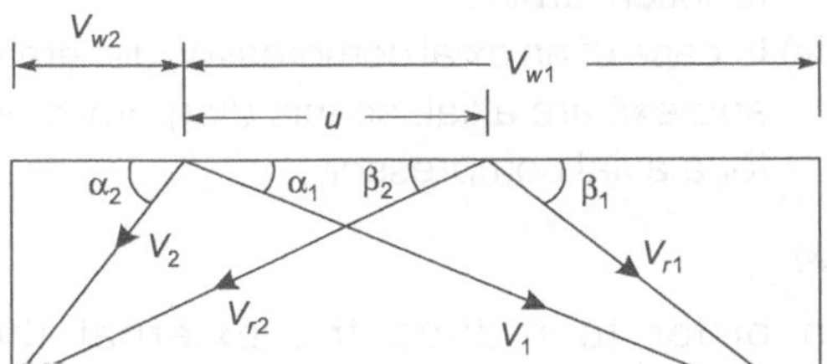

$$\beta_1 = 30°; \quad \beta_2 = 25°$$
$$K = 1; \quad V_{r1} = V_{r2}$$
$$V_{w1} = u + V_{r1}\cos\beta_1$$
$$= 200 + 350\cos 30° = 503.10 \text{ m/s}$$
$$V_{w2} = V_{r2}\cos\beta_2 - u$$
$$= 350\cos 25° - 200 = 117.20 \text{ m/s}$$

Work done/kg,

$$w = (V_{w1} + V_{w2})\,u$$
$$= (503.10 + 117.20) \times 200$$
$$= 124060 \text{ J/kg} = \mathbf{124.06\ kJ/kg}$$

POINTS TO REMEMBER

V_{w_1} and V_{w_2} made opposite direction intendedly to get more work done.

3.4 (c)

$$mh_1 + (1 - m)h_2 = h_3 \quad \text{(by heat balance)}$$
$$h_3 = h_2 - m(h_2 - h_1)$$

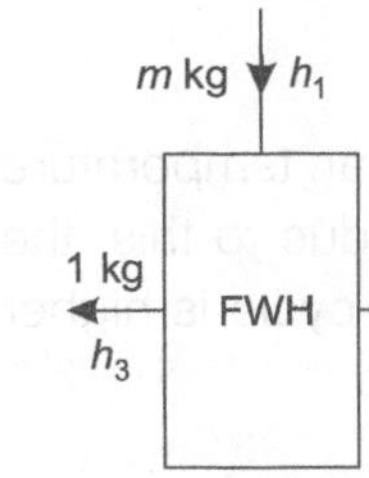

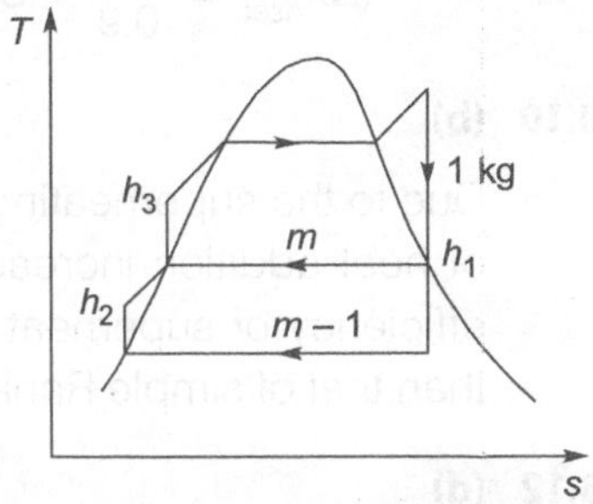

POINTS TO REMEMBER

- Regeneration in Rankine cycle result in increases in mean temperature of heat addition which inturn results in increase of thermal efficiency of cycle.
- Higher the number of bleeding point with feed water heaters, higher is the increasing in thermal efficiency.

3.5 (d)

Optimum velocity ratio

$$= \frac{u}{V_N} = \frac{\cos\alpha}{2}$$

$\therefore$ Optimum blade velocity,

$$u = \frac{V_N \cos\alpha}{2}$$

3.7 (a)

Given data:

$$\eta_N = 0.9,$$
$$\frac{u}{V_2} = 0.5$$
$$u = 150 \text{ m/s}$$
$$V_2 = \frac{u}{0.5} = \frac{150}{0.5} = 300 \text{ m/s}$$
$$(\Delta h)_{act} = \frac{V_2^2}{2000} = \frac{(300)^2}{2000}$$
$$= 45 \text{ kJ/kg}$$

Nozzle efficiency,

$$\eta_N = \frac{(\Delta h)_{act}}{(\Delta h)_{isen}}$$
$$0.9 = \frac{45}{(\Delta h)_{isen}}$$

or $(\Delta h)_{isen} = \frac{45}{0.9} = 50$ **kJ/kg**

3.10 (b)

Due to the superheating, the mean temperature of heat addition increases and due to this, the efficiency of superheat Rankine cycle is higher than that of simple Rankine cycle.

3.12 (d)

In impulse turbine, the steam pressure remains constant while it flows through the moving blades, where only the kinetic energy converts into mechanical energy.

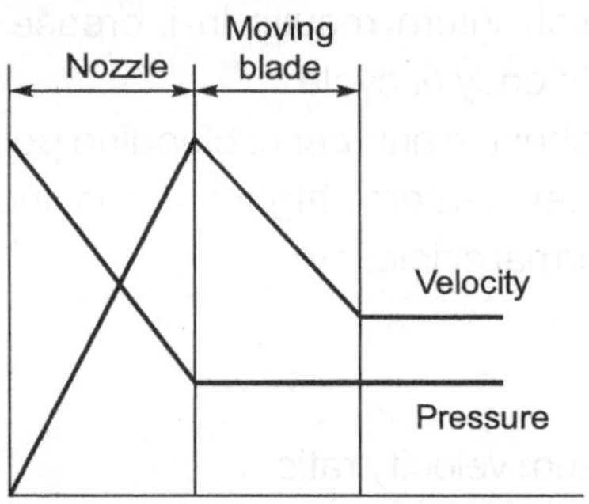

3.13 (b)

Given data:

$h_3 = 3095$ kJ/kg; $h_4 = 2604$ kJ/kg

$h_5 = 3170$ kJ/kg; $h_6 = 2165$ kJ/kg

$h_1 = 29.3$ kJ/kg

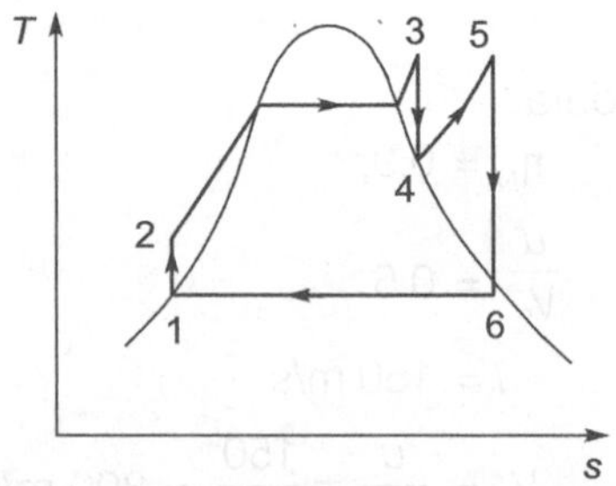

Net work output per kg,

$$w_{net} = (h_3 - h_4) + (h_5 - h_6)$$
$$= (3095 - 2609) + (3170 - 2165)$$
$$= 1491 \text{ kJ/kg}$$

Net heat supplied,

$$q_s = (h_3 - h_2) + (h_5 - h_4)$$
$$= h_3 - h_1 + h_5 - h_4 \qquad \because h_2 \approx h_1$$
$$= 3095 - 29.3 + 3170 - 2609$$
$$= 3626.7 \text{ kJ/kg}$$

Thermal efficiency,

$$\eta = \frac{w_{net}}{q_s} = \frac{1491}{3626.7} = 0.4111 = \mathbf{41.11\%}$$

3.14 (d)

Method I:

Enthalpy at exit of pump must be greater than enthalpy at inlet of pump i.e. h_2 must be greater then $h_1 = 29.3$ kJ/kg. Among the given four options only one option is greater than $h_1 = 29.3$ kJ/kg which is 33.3 kJ/kg. Hence option (d) is correct.

Method II:

Enthalpy at pump discharge (h_2)

$$= h_1 + v(p_2 - p_1)$$
$$= 29.3 + 0.0010025\,(400 - 10)$$
$$= 33.2999 \text{ kJ/kg} = 33.3 \text{ kJ/kg}$$

3.15 (b)

(i) In case of an impulse turbine, $W_1 = W_2$, so this diagram is not for impulse turbine.

(ii) In case of a reaction turbine, velocity of flow, $u_1 = u_2$ = constant, so this diagram is true fo reaction turbine.

(iv) In case of an axial compressor, generaly entry and exit are axial, so this diagram is not true for a axial compressor.

3.16 (a)

In order to reduce the external thermal irreversibility, increase in the mean temperature of heat addition is required. Mean temperature of heat addition can also be increased by reducing the amount of heat added at low temperature in liquid phase *i.e.,* why regenerative feed water heaters are required.

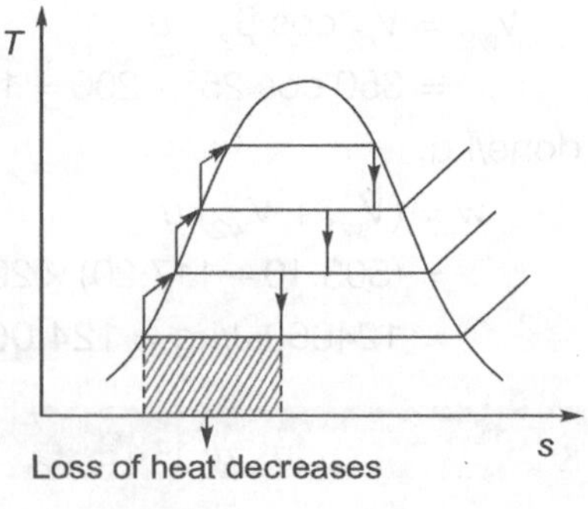

3.17 (b)

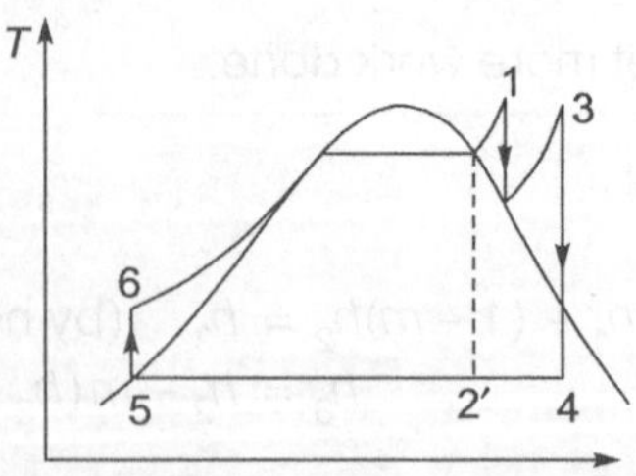

Fig: Rankine cycle with reheat

Above figure shows the Rankine cycle with reheat and due to the reheat:

1. Work done due to turbine increases $(W_T)\uparrow$
2. Pump work remains the same (W_P)
3. Net work output due to Rankine cycle increases $(W_{net})\uparrow$
4. Dryness fraction increases $(x)\uparrow$

POINTS TO REMEMBER

By reheating, the efficiency of steam power plant increases only when

$$\frac{p_{(reheater)}}{p_{(at\,inlet\,of\,HP\,turbine)}} = 0.20 \text{ to } 0.25$$

Thus, the optimum reheat pressure for power plant is 0.20 to 0.25 of the initial steam pressure.

3.18 (a)

Given data: $h_1 = 2800$ kJ/kg; $h_2 = 200$ kJ/kg

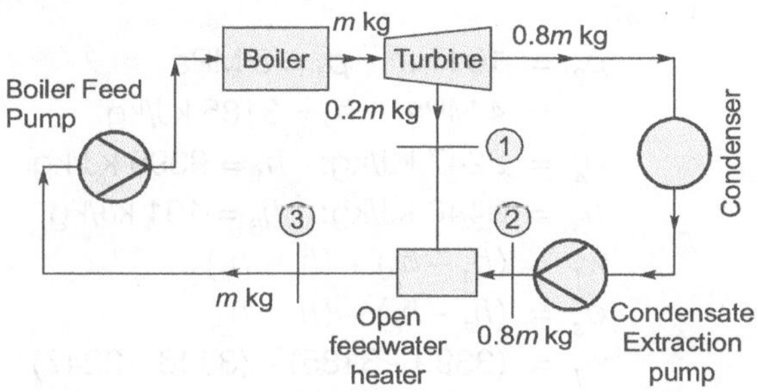

Using energy balance equation for open feedwater heater

$$mh_3 = 0.2\,mh_1 + 0.8\,mh_2$$

or $$h_3 = 0.2\,h_1 + 0.8\,h_2$$
$$= 0.2 \times 2800 + 0.8 \times 200$$
$$= 560 + 160 = \mathbf{720\ kJ/kg}$$

3.19 (c)

Given data:

$$p_1 = p_4 = 4 \text{ MPa} = 4000 \text{ kPa}$$
$$T_1 = 350°\text{C}$$
$$p_2 = p_3 = 15 \text{ kPa}$$
$$\eta_T = 90\% = 0.90$$

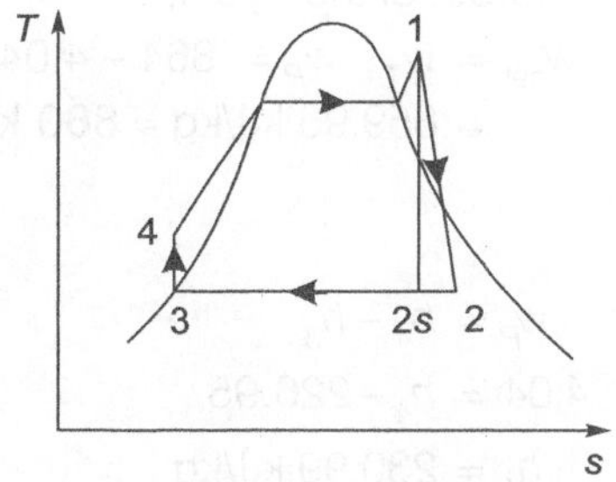

From given saturated steam table at

$$p_1 = 4 \text{ MPa}, T_1 = 350°\text{C}$$

we get, $$h_1 = 3092.5 \text{ kJ/kg}$$
$$s_1 = 6.5821 \text{ kJ/kg/K}$$

From given saturated steam table (pressure based)

$$h_f = h_3 = 226.95 \text{ kJ/kg}$$
$$h_g = 2599.1 \text{ kJ/kg}$$
$$s_f = 0.7549 \text{ kJ/kgK}$$
$$s_g = 8.0085 \text{ kJ/kgK}$$
$$v_f = v_3 = 0.001014 \text{ m}^3\text{kg}$$
$$s_1 = s_{2s} = s_f + x_{2s}(s_g - s_f)$$
$$6.5821 = 0.7549 + x_{2s}(8.0085 - 0.7549)$$
$$6.5821 = 0.7549 + 7.2536\,x_{2s}$$

or $$7.2536\,x_{2s} = 5.8272$$

or $$x_{2s} = 0.8033$$
$$h_{2s} = h_f + x_{2s}(h_g - h_f)$$
$$= 226.95 + 0.8033(2599.1 - 226.95)$$
$$= 226.95 + 1905.54 = 2132.49 \text{ kJ/kg}$$
$$\eta_T = \frac{(\Delta h)_{act}}{(\Delta h)_{Isen}} = \frac{h_1 - h_2}{h_1 - h_{2s}}$$
$$0.90 = \frac{3092.5 - h_2}{3092.5 - 2132.49}$$
$$0.90 = \frac{3092.5 - h_2}{960.01}$$

or $$960.01 \times 0.90 = 3092.5 - h_2$$

or $$864 = 3092.5 - h_2$$

or $$h_2 = 2228.5 \text{ kJ/kg}$$

Turbine work,

$$W_T = h_1 - h_2 = 3092.5 - 2228.5$$
$$= 864 \text{ kJ/kg}$$

Pump work,

$$W_P = v_3(p_1 - p_2) \text{ kJ/kg}$$

where v_3 in m³/kg and p_1 and p_2 in kPa

$\therefore$ $$W_P = 0.001014(4000 - 15)$$
$$= 4.04 \text{ kJ/kg}$$

Net work output of the cycle,

$$w_{net} = w_T - w_P = 864 - 4.04$$
$$= 859.96 \text{ kJ/kg} \approx \mathbf{860 \text{ kJ/kg}}$$

3.20 (c)

also, $w_P = h_4 - h_3$

$\therefore$ $4.04 = h_4 - 226.95$

or $h_4 = 230.99$ kJ/kg

Heat supplied to the cycle,

$$q_{4-1} = h_1 - h_4 = 3092.5 - 230.99$$
$$= 2861.51 \text{ kJ/kg} \approx \mathbf{2863 \text{ kJ/kg}}$$

3.21 (a)

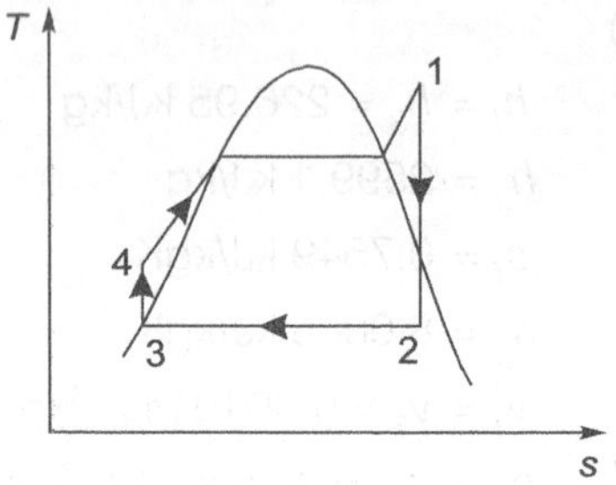

Given data:

Specific enthalpy at turbine inlet,

$h_1 = 2800$ kJ/kg

Specific enthalpy at turbine outlet: $h_2 = 1800$ kJ/kg

Specific work output by the turbine,

$$w_T = h_1 - h_2 = 2800 - 1800$$
$$= 1000 \text{ kJ/kg}$$

Net work output,

$$w_{net} = w_T - w_P$$

where pump work,

$$w_P = 0$$

$\therefore$ $w_{net} = w_T = 1000$ kJ/kg

Specific steam consumption,

$$ssc = \frac{3600}{1000} \text{ kg/kWh} = \mathbf{3.6 \text{ kg/kWh}}$$

POINTS TO REMEMBER

- Specific fuel consumption is the amount of fuel required for producing unit kWh power.
- Air rate is the air flow required per kWh output and the reciprocal of air rate is termed as specific power.
- Air rate is actually the orientation of the size of the plant i.e. the lows the air rate, the smaller the plant.

3.22 (a)

Applying steady flow energy equation

$$h_1 + \frac{V_1^2}{2} + q = h_2 + \frac{V_2^2}{2} + w$$

$$3250 + \frac{180^2}{2000} - 5 = 2360 + \frac{5^2}{2000} + w$$

or $w = \mathbf{901.2 \text{ kW/kg}}$

3.23 Sol.

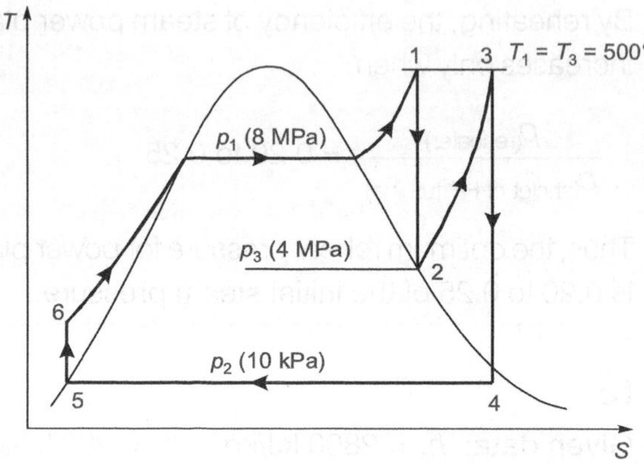

Given,

$$p_2 = 10 \text{ kPa}; \quad p_1 = 8 \text{ MPa}$$
$$p_3 = 4 \text{ MPa}; \quad h_2 = 3185 \text{ kJ/kg}$$
$$h_4 = 2247 \text{ kJ/kg}; \quad h_1 = 3399 \text{ kJ/kg}$$
$$h_3 = 3446 \text{ kJ/kg}; \quad h_6 = 191 \text{ kJ/kg}$$
$$w_T = (h_1 - h_2) + (h_3 - h_4)$$
$$q_s = (h_1 - h_6) + (h_3 - h_2)$$

$\therefore$ $w_T = (3399 - 3185) + (3446 - 2247)$

$$= 1413 \text{ kJ/kg}$$
$$q_s = (3399 - 191) + (3446 - 3185)$$
$$= 3469 \text{ kJ/kg}$$
$$\eta = \frac{w_T}{q_s} = \frac{1413}{3469} = 0.4073$$
$$= \mathbf{40.73\%}$$

3.24 Sol.

$$\eta_{b,\max} = \cos^2\alpha = (\cos 20°)^2 = 0.883$$
$$= \mathbf{88.3\%}$$

POINTS TO REMEMBER

- Max blade efficiency of reaction turbine is,

$$(\eta_b)_{\max} = \frac{2\cos^2\alpha}{1 + \cos^2\alpha}$$

- It is important to remember that efficiency of reaction turbine is greater than the impulse turbine.

- Optimum speed ratio, for impulse turbine,

$$\frac{u}{V} = \frac{\cos\alpha}{2}$$

For reaction turbine,

$$\frac{u}{V} = \cos\alpha$$

3.25 Sol.

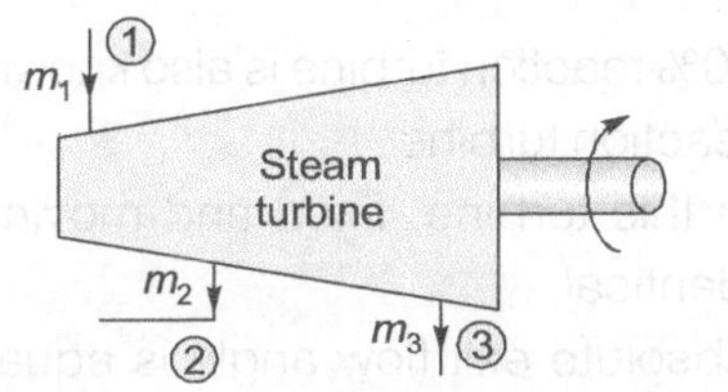

$m_1 = 10$ kg/s; $m_2 = 1.5$ kg/s; $m_3 = 8.5$ kg/s

$h_1 = 3214$ kJ/kg; $h_2 = 2920$ kJ/kg

$h_3 = 2374$ kJ/kg

$$W_T = m_1(h_1 - h_2) + m_3(h_2 - h_3)$$
$$= 10(3214 - 2920) + 8.5(2920 - 2374)$$
$$= \mathbf{7581\ kW}$$

3.26 (c)

Rankine cycle with reheating:

Advantages:

(i) Increase in a average temperature of heat addition.

(ii) Increase in thermal efficiency.

(iii) Drier steam at the turbine exit *i.e.*, decrease in moisture content of the steam at the low-pressure turbine exit, reducing the turbine blade erosion.

3.27 Sol.

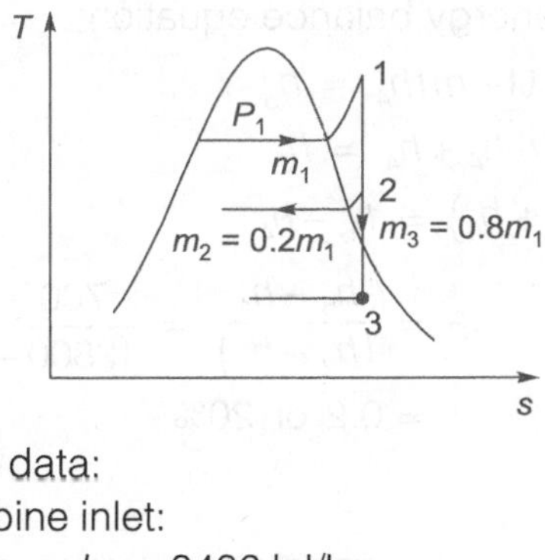

Given data:

At turbine inlet:

$$h_1 = 3486 \text{ kJ/kg}$$
$$s_1 = 6.6 \text{ kJ/kgK}$$
$$m_1 = 100 \text{ kg/s}$$

At intermediate stage:

$$h_2 = 2776 \text{ kJ/kg}$$
$$m_2 = 0.2\ m_1 = 0.2 \times 100 = 20 \text{ kg/s}$$

At turbine exit:

$$P_3 = 9 \text{ kPa}$$
$$m_3 = 0.8\ m_1 = 0.8 \times 100 = 80 \text{ kg/s}$$

At $P_3 = 9$ kPa

$$h_f = 174 \text{ kJ/kg}$$
$$h_g = 2574 \text{ kJ/kg}$$
$$s_f = 0.6 \text{ kJ/kgK}$$
$$s_g = 8.1 \text{ kJ/kgK}$$
$$s_3 = s_1 = s_f + x_3(s_g - s_f)$$
$$6.6 = 0.6 + x_3(8.1 - 0.6)$$
$$6 = 7.5\,x_3$$

or $x_3 = 0.8$

$$h_3 = h_f + x_3(h_g - h_f)$$
$$= 174 + 0.8(2574 - 174)$$
$$= 2094 \text{ kJ/kg}$$

Work output,

$$W = m_1(h_1 - h_2) + m_3(h_2 - h_3)$$
$$= 100(3486 - 2776) + 80(2776 - 2094)$$
$$= 71000 + 54560$$
$$= 125560 \text{ kW} = \mathbf{125.56\ MW}$$

3.28 Sol.

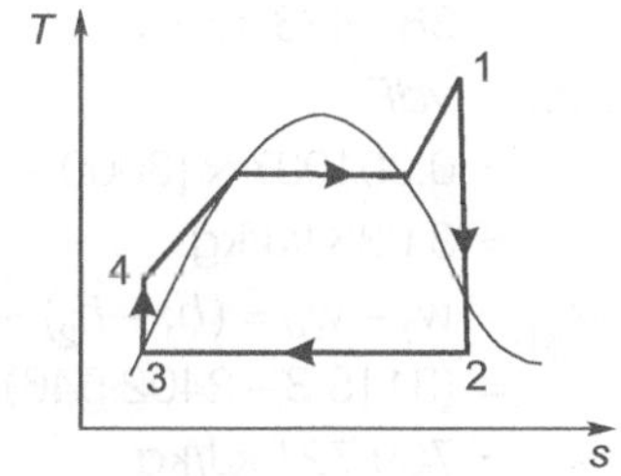

$$w_{turbine} = h_1 - h_2$$
$$= (3159 - 2187) \text{ kJ/kg} = 972 \text{ kJ/kg}$$
$$w_{net} = w_{turbine} - w_{pump}$$
$$= 972 - 2 = 970 \text{ kJ/kg}$$

Specific steam consumption

$$= \frac{3600}{970} \text{ kg/kWh}$$
$$= \mathbf{3.711\ kg/kWh}$$

Specific steam consumption is amount of steam required to produce 1 kWh (or) 3600 kJ of work.

3.29 (a)

Regeneration in vapour power cycle shall result in increase mean temperature of heat addition and it is achieved by transfering heat from exapanding steam to liquid. The temperature difference between liquid water and steam in the vapur power cycle is approximately same, so irreversibility involved is not much more.

3.30 Sol.

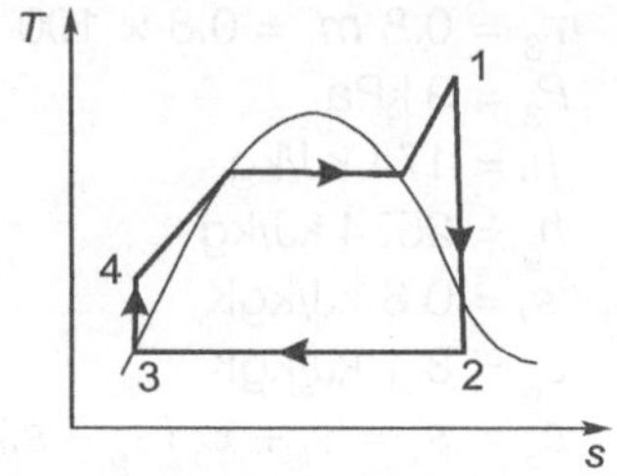

Calculating dryness fraction at point 2

$$s_1 = s_2$$

$$6.7428 = s_f + x_2 s_{fg} = 1.213 + x_2 \times 6.2434$$

or $$x_2 = \frac{6.7428 - 1.213}{6.2434} = 0.8857$$

Calculating enthalpy at point 2,

$$h_2 = h_f + x_2 h_{fg}$$
$$= 384.39 + 0.8857 \times 2278.6$$
$$= 2402.546 \text{ kJ/kg}$$

Given: $h_1 = 3115.3$ kJ/kg

$$h_4 = h_3 + vdP$$
$$= 384.39 + 0.001037 \times [3000 - 75]$$
$$= 387.423 \text{ kJ/kg}$$

$$\text{Pump work} = vdP$$
$$= 0.001037 \times [3000 - 75]$$
$$= 3.033 \text{ kJ/kg}$$

$$w_{net} = w_T - w_P = (h_1 - h_2) - 3.033$$
$$= (3115.3 - 2402.546) - 3.033$$
$$= 709.721 \text{ kJ/kg}$$

$$\text{Heat input} = h_1 - h_4$$
$$= 3115.3 - 387.423$$
$$= 2727.877 \text{ kJ/kg}$$

$$\text{Efficiency of cycle} = \frac{\text{Work delivered}}{\text{Heat input}}$$
$$= \frac{709.721}{2727.877} = 0.2601 = \mathbf{26.01\%}$$

3.31 Sol.

Given data $h_1 = 2803$ kJ/kg

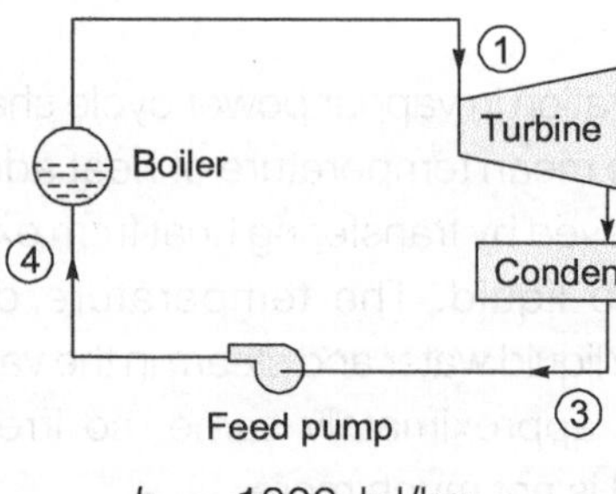

$$h_2 = 1800 \text{ kJ/kg}$$
$$h_3 = 121 \text{ kJ/kg}$$
$$h_4 = 124 \text{ kJ/kg}$$
$$w_T = h_1 - h_2$$
$$= 2803 - 1800 = 1003 \text{ kJ/kg}$$
$$w_P = h_4 - h_3 = 124 - 121 = 3 \text{ kJ/kg}$$
$$w_{net} = w_T - w_P = 1003 - 3 = 1000 \text{ kJ/kg}$$
$$ssc = \frac{3600}{w_{net}} = \frac{3600}{1000} = \mathbf{3.6 \text{ kg/kWh}}$$

3.32 (d)

(i) 50% reaction turbine is also known as Parson's reaction turbine.

(ii) In this turbine, fixed and moving blade are identical.

(iii) Absolute exit flow angle is equal to the inlet angle of rotor blade.

POINTS TO REMEMBER

In practical working, 50% reaction turbine i.e. Parson's turbine is used because Axial thrust get utilized, blade manufacturing becomes very simple and efficiency obtained is near maximum.

3.33 Sol.

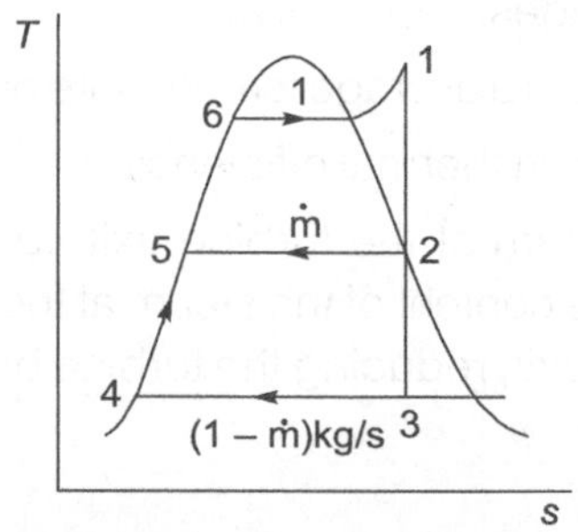

From energy balance equation

$$\dot{m}h_2 + (1 - \dot{m})h_4 = h_5$$
$$\dot{m}h_2 + \dot{m}h_4 + h_4 = h_5$$
$$\dot{m}(h_2 + h_4) = h_5 - h_4$$
$$\dot{m} = \frac{h_5 - h_4}{(h_2 - h_4)} = \frac{700 - 175}{(2800 - 175)} \times 100$$
$$= 0.2 \text{ or } 20\%$$

3.35 Sol.

$$W_T = 903 \text{ kJ/kg}; \; W_P = 3 \text{ kJ/kg}$$

Specific steam consumption = ? (kg/kW-hr)

$$\text{SSC} = \frac{3600}{W_T - W_P} = \frac{3600}{903 - 3} = 4 \text{ kg/kW-hr}$$

3.36 Sol.

Given data:
at 10 MPa, 500°C, h_1 = 3373.6 kJ/kg,
at 1 MPa, 500°C, h_3 = 3478.4 kJ/kg,
s_3 = 7.7621 kJ/kgK
at 1 MPa, $h_2 = h_g$ = 2778.1 kJ/kg

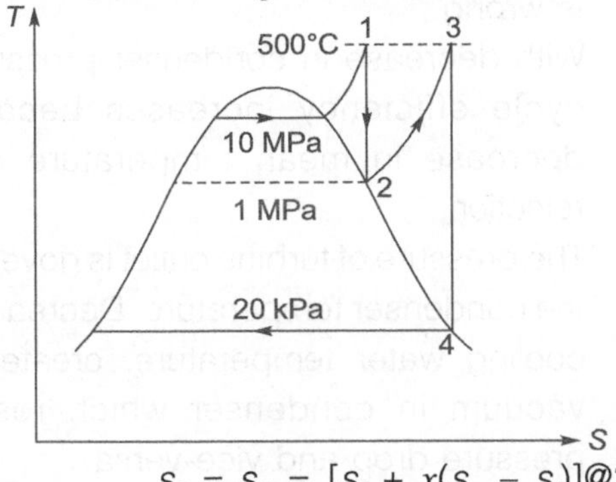

$$s_3 = s_4 = [s_f + x(s_g - s_f)]@20 \text{ kPa}$$
$$s_3 = 7.7621$$
$$= 0.8319 + x + (7.9085 - 0.8319)$$
$$x_4 - 0.9793$$
at 20 kPa, $h_4 = h_f + x_4 \times (h_g - h_f)$
$$= 2560.91 \text{ kJ/kg}$$
workdone by turbine, $W_T = (h_1 - h_2) + (h_3 - h_4)$
$$= 1512.95 \text{ kJ/kg}$$

3.37 Sol.

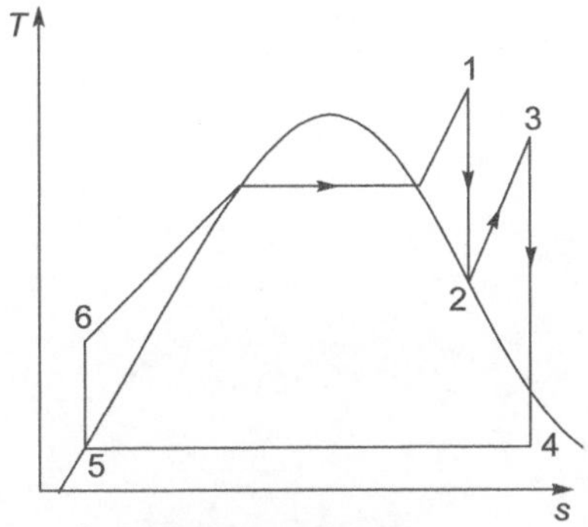

$$h_f = 200 \text{ kJ/kg}$$
$$h_g = 2600 \text{ kJ/kg}$$
$$w_p = 20 \text{ kJ/kg} = h_6 - h_5$$
$$h_1 - h_2 = 750 \text{ kJ/kg}$$
$$h_3 - h_4 = 1500 \text{ kJ/kg}$$
$$\eta = 0.5 = \frac{W_{NET}}{Q_s} = \frac{W_T - W_P}{Q_s}$$
$$0.5 = \frac{750 + 1500 - 20}{Q_s}$$
$$Q_s = 4460 \text{ kJ/kg}$$
$$\eta = 1 - \frac{Q_R}{Q_S}$$
$$\Rightarrow \frac{Q_R}{Q_S} = 0.5$$
$$Q_R = 2230 \text{ kJ/kg}$$
$$Q_R = h_4 - h_5$$
$$2230 = h_4 - 200$$
$$\Rightarrow h_4 = 2430 \text{ kJ/kg}$$
$$h_4 = h_f + x(h_g - h_f)$$
$$2430 = 200 + x(2600 - 200)$$
$$x = 0.9291$$
$$x = 93\%$$

3.38 Sol.

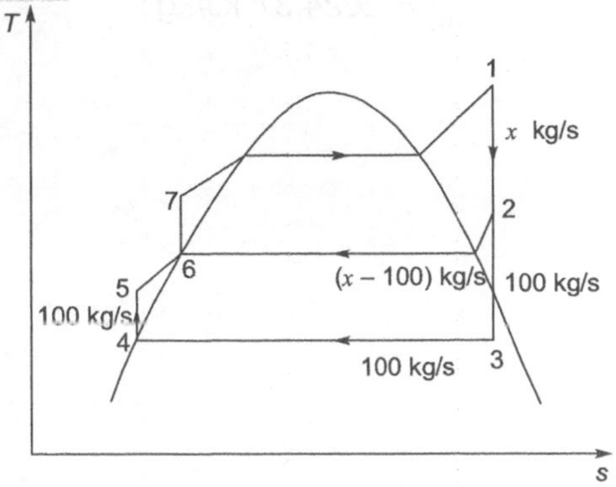

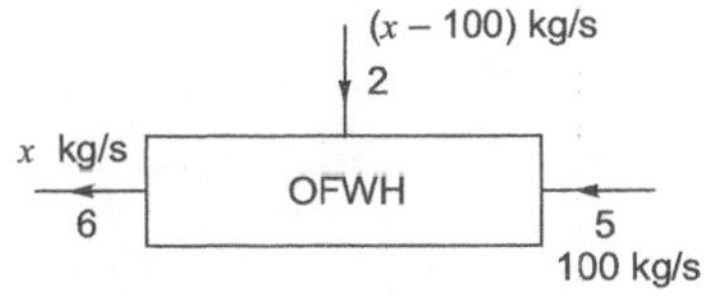

$$100h_5 + (x - 100)h_2 = xh_6$$
$$100 \times 226.7 + (x - 100)2624 = 708.6x$$
$$22670 + 2624x - 262400 = 708.6x$$
$$2624x - 708.6x = 239730$$
$$1915.4x = 239730$$
$$x = 125.159 \simeq 125.2 \text{ kg/s}$$
Mass flow rate at state 2$(x - 100)$ = 25.2 kg/s

3.39 (2624.37) (2620.00 to 2630.00)

Given data : W_{T_1} = 400 kJ/kg;
W_{T_2} = 850 kJ/kg; W_P = 15 kJ/kg; η = 0.32

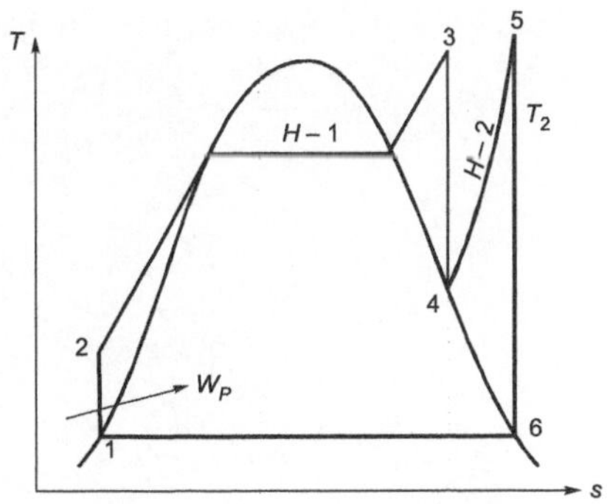

∴ Efficiency of the cycle will be given as

$$\eta = \frac{w_{net}}{Q_s} = \frac{w_{T_1} + w_{T_2} - w_p}{Q_s}$$

$$\Rightarrow \quad 0.32 = \frac{400 + 850 - 15}{Q_s}$$

$$\Rightarrow \quad Q_s = 3859.37 \text{ kJ/kg}$$

∴ Efficiency can also be written as

$$\eta = 1 - \frac{Q_{\text{Rej}}}{Q_s}$$

$$\Rightarrow \quad 0.32 = 1 - \frac{Q_{\text{Rej}}}{3859.37}$$

$$\Rightarrow \quad Q_{\text{Rej}} = 2624.37 \text{ kJ/kg}$$

3.40 (a)

- Superheating in Rankine cycle increases the cycle efficiency because of increase in mean temperature of heat addition.
- With increase in pressure of boiler, the cycle efficiency increases. So, the given statement is wrong.
- With decrease in condenser pressure, the cycle efficiency increases because of decrease in mean temperature of heat rejection.
- The pressure of turbine outlet is governed by the condenser temperature. Decreasing the cooling water temperature, creates more vacuum in condenser which results in pressure drop and vice-versa.

■■■■

4 Gas Turbines

4.1 In a gas turbine, hot combustion products with the specific heats c_p= 0.98 kJ/kgK, and c_v = 0.7538 kJ/kgK enters the turbine at 20 bar, 1500 K and exits at 1 bar. The isentropic efficiency of the turbine is 0.94. The work developed by the turbine per kg of gas flow is

(a) 689.64 kJ/kg (b) 794.66 kJ/kg
(c) 1009.72 kJ/kg (d) 1312.00 kJ/kg

[2003 : 2 M]

4.2 The compression ratio of a gas power plant cycle corresponding to maximum work output for the given temperature limits of T_{min} and T_{max} will be

(a) $\left(\frac{T_{max}}{T_{min}}\right)^{\frac{\gamma}{2(\gamma-1)}}$ (b) $\left(\frac{T_{min}}{T_{max}}\right)^{\frac{\gamma}{2(\gamma-1)}}$

(c) $\left(\frac{T_{max}}{T_{min}}\right)^{\frac{\gamma-1}{\gamma}}$ (d) $\left(\frac{T_{min}}{T_{max}}\right)^{\frac{\gamma-1}{\gamma}}$ **[2004 : 1 M]**

Common Data for Questions 4.3 and 4.4

In two air standard cycles - one operating in the Otto and the other on the Brayton cycle - air is isentropically compressed from 300 to 450 K. Heat is added to raise the temperature to 600 K in the Otto cycle and to 550 K in the Brayton cycle.

4.3 In η_o and η_B are the efficiencies of the Otto and Brayton cycles, then

(a) η_o = 0.25, η_B = 0.18
(b) η_o = η_B = 0.33
(c) η_o = 0.5, η_B = 0.45
(d) it is not possible to calculate the efficiencies unless the temperature after the expansion is given

[2005 : 2 M]

4.4 If W_o and W_B are work outputs per unit mass, then

(a) $W_o > W_B$ (b) $W_o < W_B$
(c) $W_o = W_B$
(d) it is not possible to calculate the work outputs unless the temperature after the expansion is given

[2005 : 2 M]

4.5 An ideal Brayton cycle, operating between the pressure limits of 1 bar and 6 bar, has minimum and maximum temperatures of 300 K and 1500 K. The ratio of specific heats of the working fluid is 1.4. The approximate final temperature in Kelvin at the end of the compression and expansion processes are respectively

(a) 500 and 900 (b) 900 and 500
(c) 500 and 5400 (d) 900 and 900

[2011 : 2 M]

Common Data Questions 4.6 and 4.7

In a simple Brayton cycle, the pressure ratio is 8 and temperature at the entrance of compressor and turbine are 300 K and 1400 K, respectively. Both compressor and gas turbine have isentropic efficiencies equal to 80%. For the gas, assume a constant value of c_p(specific heat at constant pressure) equal to 1 kJ/kgK and ratio of specific heats as 1.4. Neglect changes in kinetic and potential energies.

4.6 The power required by the compressor in kW/kg of gas flow are is

(a) 194.7 (b) 243.4
(c) 304.3 (d) 378.5 **[2013 : 2 M]**

4.7 The thermal efficiency of the cycle in percentage (%) is

(a) 24.8 (b) 38.6
(c) 44.8 (d) 53.1 **[2013 : 2 M]**

4.8 The thermal efficiency of an air-standard Brayton cycle in terms of pressure ratio r_p and $\gamma(= c_p/c_v)$ is given by

(a) $1-\frac{1}{r_p^{\gamma-1}}$ (b) $1-\frac{1}{r_p^{\gamma}}$

(c) $1-\frac{1}{r_p^{1/\gamma}}$ (d) $1-\frac{1}{r_p^{(\gamma-1)/\gamma}}$

[2014 : 1 M, Set-2]

4.9 For a gas turbine power plant, identify the correct pair of statements.

P. Smaller in size compared to steam power plant for same power output
Q. Starts quickly compared to steam power plant
R. Works on the principle of Rankine cycle
S. Good compatibility with solid fuel

(a) P and Q (b) R and S
(c) Q and R (d) P and S

[2014 : 1 M, Set-3]

4.10 In an ideal Brayton cycle, atmospheric air (ratio of specific heats, $c_p/c_v = 1.4$, specific heat at constant pressure = 1.005 kJ/kgK) at 1 bar and 300 K is compressed to 8 bar. The maximum temperature in the cycle is limited to 1280 K. If the heat is supplied at the rate of 80 MW, the mass flow rate (in kg/s) of air required in the cycle is ______.
[2014 : 2 M, Set-2]

4.11 Consider a simple gas turbine (Brayton) cycle and a gas turbine cycle with perfect regeneration. In both the cycles, the pressure ratio is 6 and the ratio of the specific heats of the working medium is 1.4. The ratio of minimum to maximum temperatures is 0.3 (with temperatures expressed in K) in the regenerative cycle. The ratio of the thermal efficiency of the simple cycle to that of the regenerative cycle is__. [2016 : 1 M, Set-2]

4.12 The pressure ratio across a gas turbine (for air, specific heat at constant pressure, c_p = 1040 J/kgK and ratio of specific heats γ = 1.4) is 10. If the inlet temperature to the turbine is 1200 K and the isentropic efficiency is 0.9, the gas temperature at turbine exit is ______ K. [2017 : 2 M, Set-1]

4.13 A gas turbine with air as the working fluid has an isentropic efficiency of 0.70 when operating at a pressure ratio of 3. Now, the pressure ratio of the turbine is increased to 5, while maintaining the same inlet conditions. Assume air as a perfect gas with specific heat ratio γ = 1.4. If the specific work output remains the same for both the cases, the isentropic efficiency of the turbine at the pressure ratio of 5 is ______ (round off to two decimal places) [2019 : 2 M, Set-1]

4.14 The compressor of a gas turbine plant, operating on an ideal intercooled Brayton cycle, accomplishes an overall compression ratio of 6 in a two-stage compression process, Intercooling is used to cool the air coming out from the first stage to the inlet temperature of the first stage, before its entry to the second stage. Air enters the compressor at 300 K and 100 kPa. If the properties of gas are constant, the intercooling pressure for minimum compressor work is_______ kPa (round off to 2 decimal places).
[2020 : 1 M, Set-1]

4.15 Consider an air-standard Brayton cycle with adiabatic compressor and turbine, and a regenerator, as shown in the figure. Air enters the compressor at 100 kPa and 300 K and exits the compressor at 600 kPa and 550 K. The air exits the combustion chamber at 1250 K and exits the adiabatic turbine at 100 kPa and 800 K. The exhaust air exits the regenerator (state 6) at 600 K. There is no pressure drop across the regenerator and the combustion chamber. Also, there is no heat loss from the regenerator to the surroundings. The ratio of specific heats at constant pressure and volume is c_p/c_v = 1.4. The thermal efficiency of the cycle is __________ %. (answer in integer)

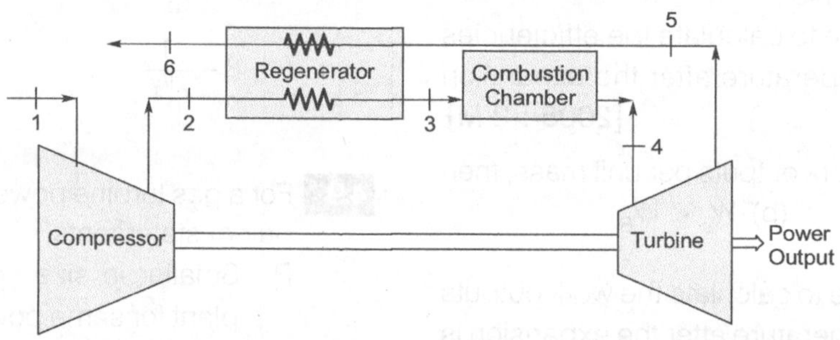

[2024 : 2 M]

Answers Gas Turbines

4.1 (a) 4.2 (a) 4.3 (b) 4.4 (a) 4.5 (a) 4.6 (c) 4.7 (a) 4.8 (d)
4.9 (a) 4.10 (108.07) 4.11 (0.8141) 4.12 (679.4) 4.13 (0.514) 4.14 (244.94) 4.15 (40)

Explanations Gas Turbines

4.1 (a)

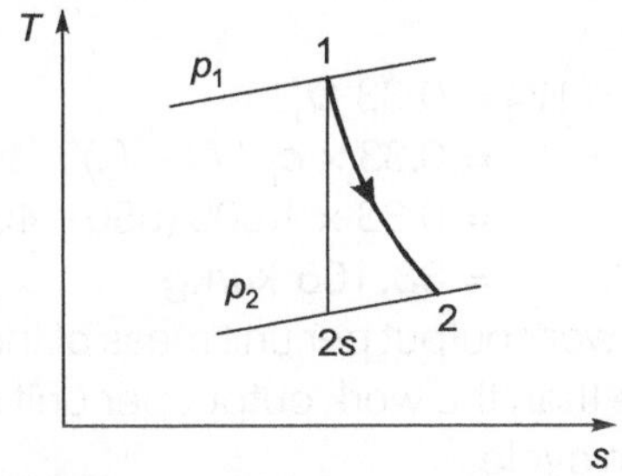

Given data:

$c_p = 0.98$ kJ/kgK; $c_v = 0.7538$ kJ/kgK

$p_1 = 20$ bar; $T_1 = 1500$ K; $p_2 = 1$ bar

$$\eta_{isen} = 0.94$$

$$\gamma = \frac{c_p}{c_v} = \frac{0.98}{0.7538} = 1.3$$

For process 1–2s,

$$\frac{T_1}{T_{2s}} = \left(\frac{p_1}{p_2}\right)^{\frac{\gamma-1}{\gamma}}$$

$$\frac{1500}{T_{2s}} = \left(\frac{20}{1}\right)^{\frac{1.3-1}{1.3}} = (20)^{0.2307} = 1.996$$

or $$T_{2s} = 751.50 \text{ K}$$

$$\eta_{isen} = \frac{T_1 - T_2}{T_1 - T_{2s}}$$

$$0.94 = \frac{1500 - T_2}{1500 - 751.50} = \frac{1500 - T_2}{748.5}$$

or $$1500 - T_2 = 0.94 \times 748.5 = 703.59$$

or $$T_2 = 796.41 \text{ K}$$

Work developed by the turbine per kg of gas,

$$w = c_p(T_1 - T_2)$$
$$= 0.98(1500 - 796.41)$$
$$= \mathbf{689.51 \text{ kJ/kg}}$$

4.2 (a)

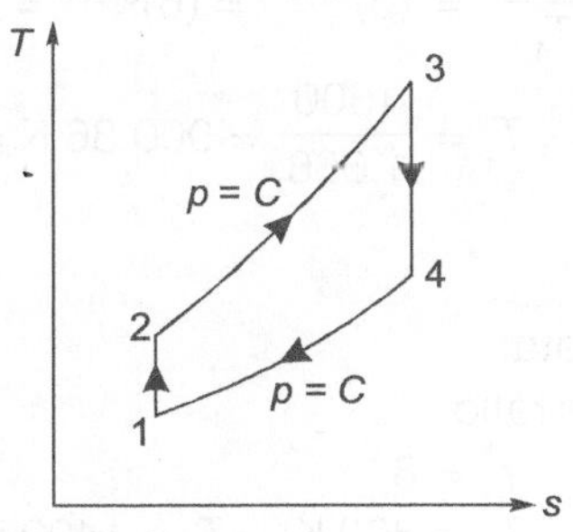

Work output during the cycle

$$= mc_p(T_3 - T_2) - mc_p(T_4 - T_1)$$
$$= mc_p(T_3 - T_4) - mc_p(T_2 - T_1)$$
$$= mc_p\left[T_3\left(1 - \frac{T_4}{T_3}\right) - T_1\left(\frac{T_2}{T_1} - 1\right)\right]$$

Since

$$\frac{T_3}{T_4} = \frac{T_2}{T_1} = (r_p)^{\frac{\gamma-1}{\gamma}} \quad \left(\text{Let } \frac{\gamma-1}{\gamma} = x,\ mc_p = K\right)$$

$\therefore$ Work/cycle

$$= K\left[T_3\left(1 - \frac{1}{r_p^x}\right) - T_1\left(r_p^x - 1\right)\right]$$

For maximum power, differentiating with respect to r_p

$$\Rightarrow \frac{dW}{dr_p} = K\left[T_3 \times \frac{x}{r_p^{(x+1)}} - T_1 \times r_p^{(x-1)}\right] = 0$$

(For max.)

$$\Rightarrow \frac{xT_3}{r_p^{(x+1)}} = T_1 \times (r_p)^{x-1}$$

$$\Rightarrow r_p^{2x} = \frac{T_3}{T_1}$$

$$\therefore r_p = \left(\frac{T_3}{T_1}\right)^{\frac{1}{2x}}$$

or $$r_p = \left(\frac{T_{max}}{T_{min}}\right)^{\frac{\gamma}{2(\gamma-1)}}$$

4.3 (b)

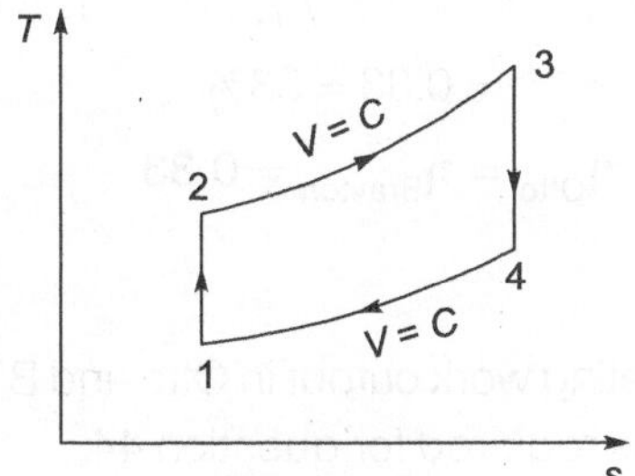

(i) Otto cycle

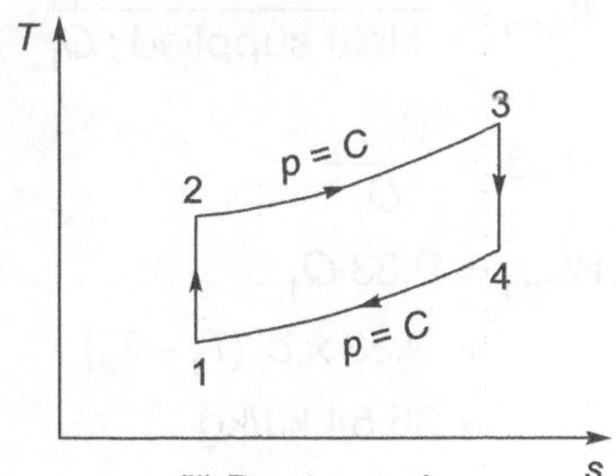

(ii) Brayton cycle

For Otto cycle

$$T_1 = 300 \text{ K}; \quad T_2 = 450 \text{ K}$$
$$T_3 = 600 \text{ K}$$

For process 1–2,

$$\frac{T_2}{T_1} = \left(\frac{V_1}{V_2}\right)^{\gamma-1}$$

or

$$\frac{V_1}{V_2} = \left(\frac{T_2}{T_1}\right)^{\frac{1}{\gamma-L}} = \left(\frac{450}{300}\right)^{\frac{1}{1.4-1}}$$
$$= (1.5)^{2.5} = 2.75$$

i.e., Compression ratio,

$$r = \frac{V_1}{V_2} = 2.75$$

$\therefore$

$$\eta_{Otto} = 1 - \frac{1}{r^{\gamma-1}} = 1 - \frac{1}{(2.75)^{1.4-1}}$$
$$= 1 - 0.667 = 0.33 = 33\%$$

For Brayton cycle

$$T_1 = 300 \text{ K}; \quad T_2 = 450 \text{ K}$$
$$T_3 = 550 \text{ K}$$

For process 1–2,
Pressure ratio,

$$r_p = \frac{p_2}{p_1} = \left(\frac{T_2}{T_1}\right)^{\frac{\gamma}{\gamma-1}}$$

$$\eta_{Brayton} = 1 - \frac{1}{(r_p)^{\frac{\gamma-1}{\gamma}}} = 1 - \frac{1}{\left[\left(\frac{T_2}{T_1}\right)^{\frac{\gamma}{\gamma-1}}\right]^{\frac{\gamma-1}{\gamma}}}$$

$$= 1 - \frac{1}{T_2/T_1} = 1 - \frac{T_1}{T_2} = 1 - \frac{300}{450}$$
$$= 0.33 = 33\%$$

Thus, $\eta_{Otto} = \eta_{Brayton} = 0.33$

4.4 (a)

Calculating work output in Otto and Brayton cycle which is required for question 44

also

$$\eta_{Otto} = \frac{\text{Work output} : W_{Otto}}{\text{Heat supplied} : Q_1}$$

$$0.33 = \frac{W_{Otto}}{Q_1}$$

or

$$W_{Otto} = 0.33\, Q_1$$
$$= 0.33 \times c_v (T_3 - T_2)$$
$$= 35.54 \text{ kJ/kg}$$

$$\eta_{Brayton} = \frac{\text{Work output} : W_B}{\text{Heat supplied}}$$

$$0.33 = \frac{W_B}{Q_1}$$

or

$$W_B = 0.33\, Q_1$$
$$= 0.33 \times c_p (T_3 - T_2)$$
$$= 0.33 \times 1.005\,(550 - 450)$$
$$= \mathbf{33.165 \text{ kJ/kg}}$$

Hence work output per unit mass of the Otto cycle is more than the work output per unit mass of the Brayton cycle.

4.5 (a)

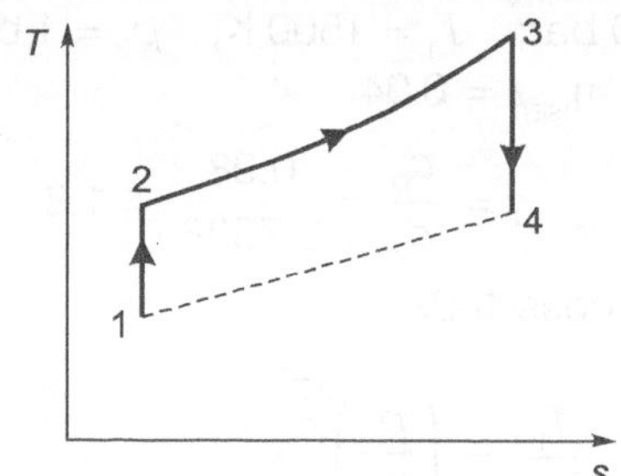

Brayton Cycle

Given data:

$$p_1 = 1 \text{ bar}; \quad p_2 = 6 \text{ bar}$$
$$T_1 = 300 \text{ K}; \quad T_3 = 1500 \text{ K}$$

For process 1–2

$$\frac{T_2}{T_1} = \left(\frac{p_2}{p_1}\right)^{\frac{\gamma-1}{\gamma}}$$

$$\frac{T_2}{300} = \left(\frac{6}{1}\right)^{\frac{1.4-1}{1.4}} = (6)^{0.285} = 1.666$$

or

$$T_2 = 300 \times 1.666$$
$$= 499.8 \text{ K} \approx \mathbf{500 \text{ K}}$$

For process 3–4

$$\frac{T_3}{T_4} = \left(\frac{p_3}{p_4}\right)^{\frac{\gamma-1}{\gamma}} = \left(\frac{p_2}{p_1}\right)^{\frac{\gamma-1}{\gamma}}$$

$$\because p_3 = p_2,\ p_4 = p_1$$

$$\frac{1500}{T_4} = (6)^{\frac{1.4-1}{1.4}} = (6)^{0.285} = 1.666$$

or

$$T_4 = \frac{1500}{1.666} = 900.36 \text{ K} \approx \mathbf{900 \text{ K}}$$

4.6 (c)

Given data:
Pressure ratio,

$$r_p = 8$$
$$T_1 = 300 \text{ K}; \quad T_3 = 1400 \text{ K}$$

$$\eta_c = \eta_T = 0.8$$
$$c_p = 1 \text{ kJ/kgK}$$
$$\gamma = 1.4$$

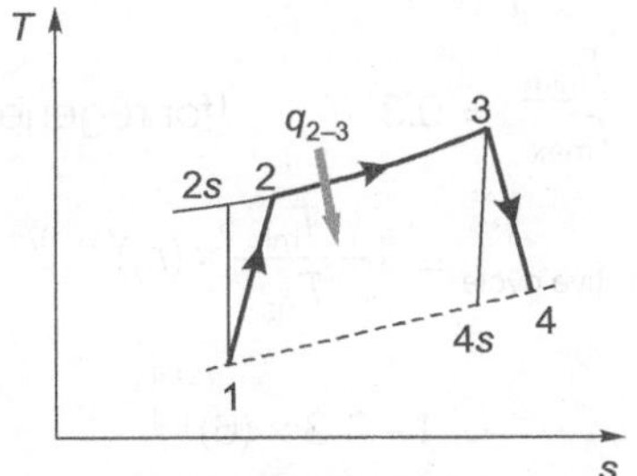

Actual Brayton Cycle

For adiabatic isentropic process 1–2s,

$$\frac{T_{2s}}{T_1} = \left(\frac{p_2}{p_1}\right)^{\frac{\gamma-1}{\gamma}}$$

$$\frac{T_{2s}}{300} = (r_p)^{\frac{\gamma-1}{\gamma}} = (8)^{\frac{1.4-1}{1.4}} = (8)^{0.2857}$$

$$\frac{T_{2s}}{300} = 1.811$$

or $$T_{2s} = 300 \times 1.811 = 543.3 \text{ K}$$

$$\eta_c = \frac{\text{Isentropic rise in temperature}}{\text{Actual rise in temperature}}$$

$$= \frac{T_{2s} - T_1}{T_2 - T_1}$$

∴ $$0.8 = \frac{543.3 - 300}{T_2 - 300}$$

or $$T_2 - 300 = \frac{243.3}{0.8} = 304.12$$

or $$T_2 = 604.12 \text{ K}$$

Power required by compressor,

$$w_C = c_p(T_2 - T_1)$$
$$= 1(604.12 - 300)$$
$$= \mathbf{304.12 \text{ kJ/kg}}$$

4.7 (a)

For adiabatic process 3–4s

$$\frac{T_3}{T_{4s}} = \left(\frac{p_2}{p_1}\right)^{\frac{\gamma-1}{\gamma}} = (8)^{\frac{1.4-1}{1.4}} = (8)^{0.2857} = 1.811$$

or $$T_{4s} = \frac{T_3}{1.811} = \frac{1400}{1.811} = 773.05 \text{ K}$$

$$\eta_T = \frac{\text{Actual fall in temperature}}{\text{Isentropic fall in temperature}}$$

$$\eta_T = \frac{T_3 - T_4}{T_3 - T_{4s}}$$

$$0.8 = \frac{1400 - T_4}{1400 - 773.05} = \frac{1400 - T_4}{626.95}$$

or $$1400 - T_4 = 0.8 \times 626.95$$
$$1400 - T_4 = 501.56$$

or $$T_4 = 898.44 \text{ K}$$

Power developed by the turbine,

$$w_T = c_p(T_3 - T_4)$$
$$= 1(1400 - 898.44) = 501.56 \text{ kJ/kg}$$

Net power output,

$$w_{net} = w_T - w_C = 501.56 - 304.12$$
$$= 197.44 \text{ kJ/kg}$$

Heat supplied,

$$q_{2-3} = c_p(T_3 - T_2) = 1(1400 - 604.12)$$
$$= 795.88 \text{ kJ/kg}$$

Thermal efficiency of the cycle,

$$\eta_{th} = \frac{\text{Net power output} : w_{net}}{\text{Heat supplied} : q_{2-3}}$$

$$= \frac{197.44}{795.88} = 0.2480 = \mathbf{24.80\%}$$

4.8 (d)

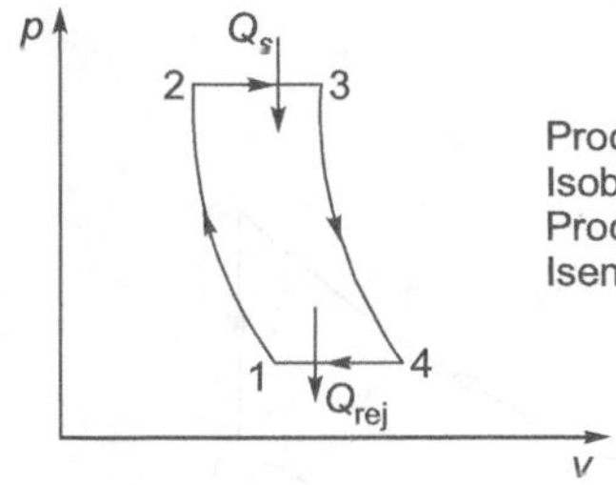

Process 4–1 & 2–3 : Isobaric process
Process 1–2 & 3–4 : Isentropic process

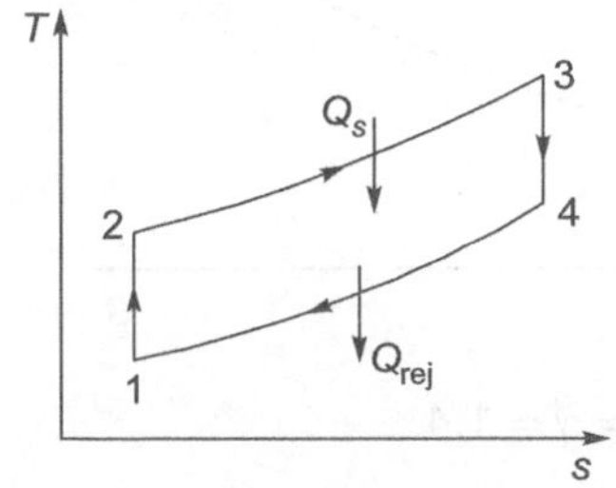

$$\frac{T_2}{T_1} = \frac{T_3}{T_4} \Rightarrow \frac{T_4}{T_1} = \frac{T_3}{T_2}$$

Thermal efficiency

$$= \frac{\text{Work done}}{\text{Heat supplied}} = 1 - \frac{\text{Heat rejected}}{\text{Heat supplied}}$$

$$= 1 - \frac{c_p(T_4 - T_1)}{c_p(T_3 - T_2)} = 1 - \frac{T_1\left(\frac{T_4}{T_1} - 1\right)}{T_2\left(\frac{T_3}{T_2} - 1\right)}$$

$$\eta_{th} = 1 - \frac{T_1}{T_2} = 1 - \frac{1}{(r_p)^{\frac{\gamma-1}{\gamma}}}$$

$$\left[\frac{T_1}{T_2} = \left(\frac{p_1}{p_2}\right)^{\frac{\gamma-1}{\gamma}}\right]$$

POINTS TO REMEMBER

As the pressure ratio of the Brayton cycle increases, the ideal air standard efficiency of the brayton cycle increases.

4.9 (a)

(i) A gas turbine plant can be started up as well as shutdown quickly, like a diesel engine.
(ii) The weight of the gas turbine plant per kW output is low, which is a favourable feature in all vehicle (land, air and sea).
(iii) A gas turbine plant works on the principle of Brayton cycle.
(iv) Any hydrocarbon fuel from high octane gasoline to heavy diesel oil and pulverized coal can be used effectively.

4.10 Sol.

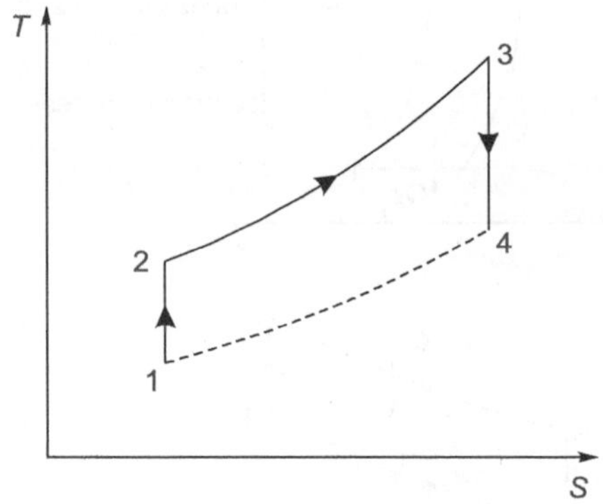

Given, $\frac{c_p}{c_v} = \gamma = 1.4$

$c_p = 1.005$ kJ/kgK; $p_1 = 1$ bar

$T_1 = 300$ K; $p_2 = 8$ bar

$T_3 = T_{max} = 1280$ K

$Q_s = 80$ MW $= 80 \times 10^3$ kW

For process 1–2,

$$\frac{T_2}{T_1} = \left(\frac{p_2}{p_1}\right)^{\frac{\gamma-1}{\gamma}} = 8^{\frac{0.4}{1.4}} = 1.8114$$

or $T_2 = 300 \times 1.8114 = 543.43$ K

$Q_s = mc_p(T_3 - T_2)$

$80 \times 10^3 = m \times 1.005\,(1280 - 543.43)$

or $m =$ **108.07 kg/s**

4.11 Sol.

Given: $r_p = 6$

$\gamma = 1.4$

$\frac{T_{min}}{T_{max}} = 0.3$ [for regenerative cycle]

$$\eta_{\text{regenerative cycle}} = 1 - \frac{T_{min}}{T_{max}} \times (r_p)^{(\gamma-1)/\gamma}$$

$$= 1 - 0.3 \times (6)^{\frac{0.4}{1.4}}$$

$$= 0.49945$$

$$\eta_{\text{simple cycle}} = 1 - \frac{1}{(r_p)^{(\gamma-1)/\gamma}} = 1 - \frac{1}{(6)^{\frac{1.4-1}{1.4}}}$$

$$= 0.4066$$

$$\text{Required ratio} = \frac{\eta_{\text{simple cycle}}}{\eta_{\text{regenerative cycle}}} = \frac{0.40666}{0.49945}$$

$$= \mathbf{0.8141}$$

4.12 Sol.

Given data: $c_p = 1040$ J/kgK; $\gamma = 1.4$; $r_p = 10$; $T_3 = 1200$ K; $\eta_T = 0.9$

$$\frac{T_3}{T_{4s}} = (r_p)^{\frac{\gamma-1}{\gamma}}$$

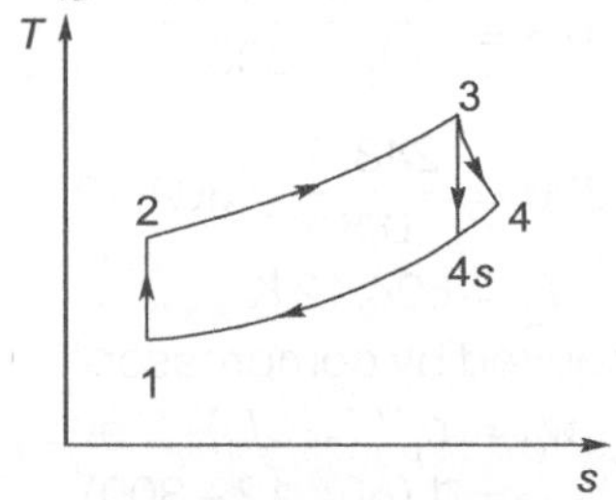

$$\frac{1200}{T_{4s}} = (10)^{(1.4-1)/1.4}$$

or $T_{4s} = 621.53$ K

$$\eta_T = \frac{T_3 - T_4}{T_3 - T_{4s}}$$

$$0.9 = \frac{1200 - T_4}{1200 - 621.53}$$

or $T_4 =$ **679.4 K**

4.13 Sol.

$$\frac{P_3}{P_{4s}} = 3$$

$$\frac{P_3'}{P_{4s}} = 5$$

$$W_{net_1} = c_pT_3\left\{1-\frac{1}{3^{\gamma-1/\gamma}}\right\}0.7$$

$$W_{net_1} = c_pT_3\left\{1-\frac{1}{5^{\gamma-1/\gamma}}\right\}\eta$$

$$W_{net_1} = W_{net2}$$

$$\Rightarrow \quad \eta = \frac{(0.7)\left\{1-\frac{1}{3^{\gamma-1/\gamma}}\right\}}{\left\{1-\frac{1}{5^{\gamma-1/\gamma}}\right\}} = 0.514$$

Consider it is asked only for turbine portion as for full cycle, data is not sufficient.

4.14 Sol.

$$\frac{P_3}{P_1} = 6 \qquad \Rightarrow P_3 = 600 \text{ kPa}$$

$$\text{Intermediate pressure} = \sqrt{P_3P_1}$$
$$= \sqrt{600\times 100}$$
$$= 244.94 \text{ kPa}$$

4.15 (40)

Given: T_1 = 300 K, T_2 = 550 K, T_4 = 1250 K, T_5 = 800 K, T_6 = 600 K,

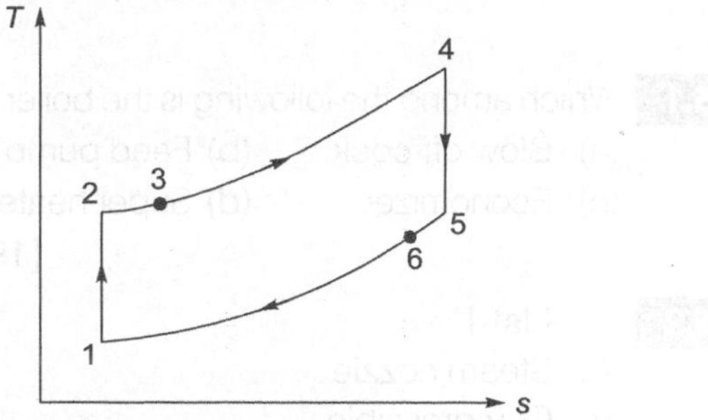

From energy balance across regenerator

$$T_3 - T_2 = T_5 - T_6$$
$$T_3 - 550 = 800 - 600$$
$$T_3 = 750 \text{ K}$$

Now, $$\eta = 1-\frac{Q_R}{Q_s}$$

$$= 1-\frac{(T_6-T_1)}{(T_4-T_3)} = 1-\frac{(600-300)}{(1250-750)}$$

$$\eta = 1-\frac{300}{500} = 0.4 \text{ or } 40\%$$

■■■■

5 Miscellaneous

5.1 Which among the following is the boiler mounting?
(a) Blow off cock (b) Feed pump
(c) Economizer (d) Super heater
[1997 : 1 M]

5.2 List-I
A. Steam nozzle
B. Compressible
C. Surface tension
D. Heat conduction
List-II
1. Mach number
2. Reaction turbine
3. Biot number
4. Nusselt number
5. Super saturation
6. Weber number **[1997 : 2 M]**

5.3 Determine the correctness or otherwise of the following Assertion (A) and the Reason (R).
Assertion (A) : Condenser is an essential equipment in a steam power plant.
Reason (R) : For the same mass flow rate and the same pressure rise, a water pump requires substantially less power than a steam compressor.
(a) Both (A) and (R) are true and (R) is the correct reason for (A)
(b) Both (A) and (R) are true but (R) is NOT the correct reason for (A)
(c) Both (A) and (R) are false
(d) (A) is false but (R) is true **[2006 : 2 M]**

5.4 Group-I shows different heat addition processes in power cycles. Likewise, Group-II shows different heat removal processes. Group-III lists power cycles. Match items from Groups I, II and III.

Group I	Group II	Group III
P. Pressure constant	S. Pressure constant	1. Rankine cycle
Q. Volume constant	T. Volume constant	2. Otto cycle
R. Temperature	U. Temperature	3. Carnot cycle
		4. Diesel cycle
		5. Brayton cycle

(a) P–S–5, R–U–3, P–S–1, Q–T–2
(b) P–S–1, R–U–3, P–S–4, P–T–2
(c) R–T–3, P–S–1, P–T–4, Q–S–5
(d) P–T–4, R–S–3, P–S–1, P–S–5
[2006 : 2 M]

5.5 In a power plant, water (density = 1000 kg/m^3) is pumped from 80 kPa to 3 MPa. The pump has an isentropic efficiency of 0.85. Assuming that the temperature of the water remains the same, the specific work (in kJ/kg) supplied to the pump is
(a) 0.34 (b) 2.48
(c) 2.92 (d) 3.43
[2014 : 1 M, Set-1]

Answers Miscellaneous

5.1 (a) 5.2 (A-5, B-1, C-6, D-3) 5.3 (b) 5.4 (a) 5.5 (d)

Explanations Miscellaneous

5.1 (a)

Boiler mounting : Provided on the body of boiler for the safety and proper control of boiler.
Ex: Safety valve, water level indicator, fusible plug, blow-off cock, pressure gauge, steam stop valve etc.

POINTS TO REMEMBER

Machine component installed either inside or outside the boiler to increase the performance and efficiency of plant are called accessories.
Ex: Air preheater, economiser, super heater, feed pump, steam trap.

5.2 Sol.

Steam nozzle → Super saturation
Compressible → Mach number
Surface tension → Weber number
Heat conduction → Biot number

5.3 (b)

Condenser is an essential equipment in a steam power plant because it is not economical to feed steam directly to the boiler.
For same mass flow rate and the same pressure rise, a water pump require very less power because the specific volume of liquid is very less as compare to specific volume of vapour.

5.4 (a)

For Rankine cycle:
Constant pressure heat addition.
Constant pressure heat rejection.
For Otto cycle:
Constant volume heat addition.
Constant volume heat rejection.
For Carnot cycle:
Constant temperature heat addition.
Constant temperature heat rejection.
For Diesel cycle:
Constant pressure heat addition.
Constant volume heat rejection.
For Brayton cycle:
Constant pressure heat addition.
Constant pressure heat rejection.

5.5 (d)

Ideal work input: $w = \nu_f \Delta p$

$$= \frac{1}{\rho_{water}}(p_2 - p_1) = \frac{(3000-80)}{1000} = \frac{2920}{1000}$$

$= 2.92$ kJ/kg

$$\text{Actual input work} = \frac{w}{\eta} = \frac{2.92}{0.85} = \mathbf{3.43\ kJ/kg}$$

■■■■

IC Engines

UNIT

X

CONTENTS

IC Engines

Syllabus

***I.C. Engines:** Air-standard Otto, Diesel and dual cycles.*

Analysis of Previous GATE Papers

Exam Year	1 Mark Ques.	2 Marks Ques.	Total Marks
1995	–	1	2
1996	–	–	–
1997	–	–	–
1998	1	2	5
1999	–	1	2
2000	–	–	–
2001	–	1	2
2002	–	1	2
2003	2	2	6
2004	–	2	4
2005	–	–	–
2006	–	–	–
2007	–	1	2
2008	1	–	1
2009	–	1	2
2010	1	–	1
2011	–	1	2
2012	–	–	–
2013	–	–	–
2014 Set-1	–	1	2
2014 Set-2	–	–	–

Exam Year	1 Mark Ques.	2 Marks Ques.	Total Marks
2014 Set-3	–	1	2
2014 Set-4	–	1	2
2015 Set-1	1	–	1
2015 Set-2	1	1	3
2015 Set-3	1	–	1
2016 Set-1	1	–	1
2016 Set-2	–	–	–
2016 Set-3	–	–	–
2017 Set-1	–	1	2
2017 Set-2	–	–	–
2018 Set-1	–	1	2
2018 Set-2	–	1	2
2019 Set-1	–	–	–
2019 Set-2	–	1	2
2020 Set-1	–	1	2
2020 Set-2	1	1	3
2021 Set-1	1	–	1
2021 Set-2	–	–	–
2022 Set-1	–	1	2
2022 Set-2	–	–	–
2023	–	–	–
2024	–	–	–

1 Air Standard Cycles

1.1 With increasing temperature of intake air, IC engine efficiency
(a) decreases
(b) increases
(c) remains same
(d) depends on other factors **[1998 : 1 M]**

1.2 An IC engine has a bore and stroke of 2 units each. The area to calculate heat loss can be taken as
(a) 4π (b) 5π
(c) 6π (d) 8π **[1998 : 2 M]**

1.3 In a spark ignition engine working on the ideal Otto cycle, the compression ratio is 5.5. The work output per cycle (*i.e.*, area of the *P-v* diagram) is equal to $23.625 \times 10^5 \times V_c$ J, where V_c is the clearance volume in m^3. The indicated mean effective pressure is
(a) 4.295 bar (b) 5.250 bar
(c) 86.870 bar (d) 106.300 bar
[2001 : 2 M]

1.4 An ideal air standard Otto cycle has a compression ratio of 8.5. If the ratio of the specific heats of air (γ) is 1.4, what is the thermal efficiency (in percentage) of the Otto cycle?
(a) 57.5 (b) 45.7
(c) 52.5 (d) 95 **[2002 : 2 M]**

1.5 For an engine operating on air standard Otto cycle, the clearance volume is 10% of the swept volume. The specific heat ratio of air is 1.4. The air standard cycle efficiency is
(a) 38.3% (b) 39.8%
(c) 60.2% (d) 61.7% **[2003 : 2 M]**

1.6 An engine working on air standard Otto cycle has a cylinder diameter of 10 cm and stroke length of 15 cm. The ratio of specific heats for air is 1.4. If the clearance volume is 196.3 cc and the heat supplied per kg of air per cycle is 1800 kJ/kg, the work output per cycle per kg of air is
(a) 879.1 kJ (b) 890.2 kJ
(c) 895.3 kJ (d) 973.5 kJ **[2004 : 2 M]**

1.7 The stroke and bore of a four stroke spark ignition engine are 250 mm and 200 mm respectively. The clearance volume is 0.001 m^3. If the specific heat ratio $\gamma = 1.4$, the air-standard cycle efficiency of the engine is
(a) 46. 40% (b) 56.10%
(c) 58.20% (d) 62.80% **[2007 : 2 M]**

1.8 Which one of the following is NOT a necessary assumption for the air-standard Otto cycle?
(a) All processes are internally reversible.
(b) Intake and exhaust processes are constant volume heat rejection processes.
(c) The combustion process is a constant volume heat addition process.
(d) The working fluid is an ideal gas with constant specific heats. **[2008 : 1 M]**

1.9 In an air-standard Otto cycle, the compression ratio is 10. The condition at the beginning of the compression process is 100 kPa and 27°C. Heat added at constant volume is 1500 kJ/kg, while 700 kJ/kg of heat is rejected during the other constant volume process in the cycle. Specific gas constant for air = 0.287 kJ/kgK. The mean effective pressure (in kPa) of the cycle is
(a) 103 (b) 310
(c) 515 (d) 1032 **[2009 : 2 M]**

1.10 In an air-standard Otto cycle, air is supplied at 0.1 MPa and 308 K. The ratio of the specific heats (γ) and the specific gas constant (R) of air are 1.4 and 288.8 J/kgK, respectively. If the compression ratio is 8 and the maximum temperature in the cycle is 2660 K, the heat (in kJ/kg) supplied to the engine is ______. **[2014 : 2 M, Set-1]**

1.11 A diesel engine has a compression ratio of 17 and cut-off takes place at 10% of the stroke. Assuming ratio of specific heats (γ) as 1.4, the air-standard efficiency (in percent) is ______.
[2014 : 2 M, Set-3]

1.12 In a compression ignition engine, the inlet air pressure is 1 bar and the pressure at the end of isentropic compression is 32.42 bar. The expansion ratio is 8. Assuming ratio of specific heats (γ) as 1.4, the air standard efficiency (in percent) is _____. **[2014 : 2 M, Set-4]**

1.13 An air-standard Diesel cycle consists of the following processes:

1-2 : Air is compressed isentropically.

2-3 : Heat is added at constant pressure.

3-4 : Air expands isentropically to the original volume.

4-1 : Heat is rejected at constant volume

If γ and T denote the specific heat ratio and temperature, respectively, the efficiency of the cycle is

(a) $1-\frac{T_4-T_1}{T_3-T_2}$ (b) $1-\frac{T_4-T_1}{\gamma(T_3-T_2)}$

(c) $1-\frac{\gamma(T_4-T_1)}{T_3-T_2}$ (d) $1-\frac{T_4-T_1}{(\gamma-1)(T_3-T_2)}$

[2015 : 1 M, Set-3]

1.14 For the same values of peak pressure, peak temperature and heat rejection, the correct order of efficiencies for Otto, Duel and Diesel cycles is

(a) $\eta_{Otto} > \eta_{Dual} > \eta_{Diesel}$

(b) $\eta_{Diesel} > \eta_{Dual} > \eta_{Otto}$

(c) $\eta_{Dual} > \eta_{Diesel} > \eta_{Otto}$

(d) $\eta_{Diesel} > \eta_{Otto} > \eta_{Dual}$

[2015 : 2 M, Set-2]

1.15 A vehicle powered by a spark ignition engine follows air standard Otto cycle (γ = 1.4). The engine generates 70 kW while consuming 10.3 kg/hr of fuel. The calorific value of fuel is 44000 kJ/kg. The compression ratio is _______ (correct to two decimal places). [2018 : 2 M, Set-2]

1.16 An engine working on air standard Otto cycle is supplied with air at 0.1 MPa and 35°C. The compression ratio is 8. The heat supplied is 500 kJ/kg. Property data for air: c_p = 1.005 kJ/kgK, c_v = 0.718 kJ/kgK, R = 0.287 kJ/kgK. The maximum temperature (in K) of the cycle is_______(correct to one decimal place).

[2018 : 2 M, Set-1]

1.17 An air standard Otto cycle has thermal efficiency of 0.5 and the mean effective pressure of the cycle is 1000 kPa. For air assume specific heat ratio γ = 1.4 and specific gas constant R = 0.287 kJ/kg.K. If the pressure and temperature at the beginning of the compression stroke are 100 kPa and 300 K respectively, then the specific net work output of the cycle is______kJ/kg (round off to two decimal places).

[2019 : 2 M, Set-2]

1.18 For an air-standard Diesel cycle,

(a) heat addition is at constant volume and heat rejection is at constant volume

(b) heat addition is at constant pressure and heat rejection is at constant volume

(c) heat addition is at constant pressure and heat rejection is at constant pressure

(d) heat addition is at constant volume and heat rejection is at constant pressure

[2020 : 1 M, Set-2]

1.19 Keeping all other parameters identical, the Compression Ratio (CR) of an air standard diesel cycle is increased from 15 to 21. Take ratio of specific heats = 1.3 and cut-off ratio of the cycle r_c = 2. The difference between the new and the old efficiency values, in percentage, $(\eta_{new}|_{CR=21}) - (\eta_{old}|_{CR=15})$ = _______ %. (round off to one decimal place).

[2020 : 2 M, Set-2]

1.20 In which of the following pairs of cycle, both cycles have at least one isothermal process?

(a) Brayton cycle and Rankine cycle

(b) Bell-Coleman cycle and Vapour compression refrigeration cycle

(c) Carnot cycle and Stirling cycle

(d) Diesel cycle and Otto cycle

[2021 : 1 M, Set-1]

1.21 An engine running on an air standard Otto cycle has a displacement volume 250 cm^3 and a clearance volume 35.7 cm^3. The pressure and temperature at the beginning of the compression process are 100 kPa and 300 K, respectively. Heat transfer during constant-volume heat addition process is 800 kJ/kg. The specific heat at constant volume is 0.718 kJ/kg.K and the ratio of specific heats at constant pressure and constant volume is 1.4. Assume the specific heats to remain constant during the cycle. The maximum pressure in the cycle is _______ kPa (round off to the nearest integer).

[2022 : 2 M, Set-1]

■■■■

Answers Air Standard Cycles

1.1	(a)	1.2	(b)	1.3	(b)	1.4	(a)	1.5	(d)	1.6	(d)	1.7	(c)	1.8	(b)
1.9	(d)	1.10	(1409.6)	1.11	(59.6)	1.12	(59.6)	1.13	(b)	1.14	(b)	1.15	(7.61)		
1.16	(1403.97)	1.17	(708.60)	1.18	(b)	1.19	(4.8)	1.20	(c)	1.21	(4809.136)				

Explanations Air Standard Cycles

1.1 (a)

Increase in temperature of intake air will cause less volumetric efficiency as the density of air intake goes down. This will cause an overall decrease in efficiency.

1.3 (b)

Compression ratio,

$$r = \frac{V_1}{V_c} = 5.5$$

$$\therefore \quad V_1 = 5.5\,V_c$$

$$mep = \frac{W}{V_s} = \frac{W}{V_1 - V_c}$$

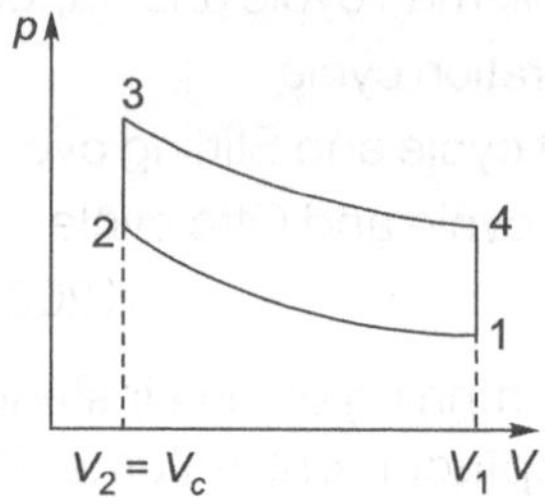

$$= \frac{23.625 \times 10^5 \times V_c}{5.5V_c - V_c}$$

$$= \frac{23.625 \times 10^5 V_c}{4.5V_c} = 5.25 \times 10^5 \text{ Pa}$$

$$= \mathbf{5.250 \text{ bar}}$$

1.4 (a)

$$\eta = 1 - \frac{1}{r^{\gamma-1}} = 1 - \frac{1}{(8.5)^{1.4-1}}$$

$$= 0.5751 = \mathbf{57.51\%}$$

1.5 (d)

$$V_c = 10\% V_s = 0.10 V_s$$

or $$\frac{V_s}{V_c} = \frac{1}{0.10} = 10$$

$$\gamma = 1.4$$

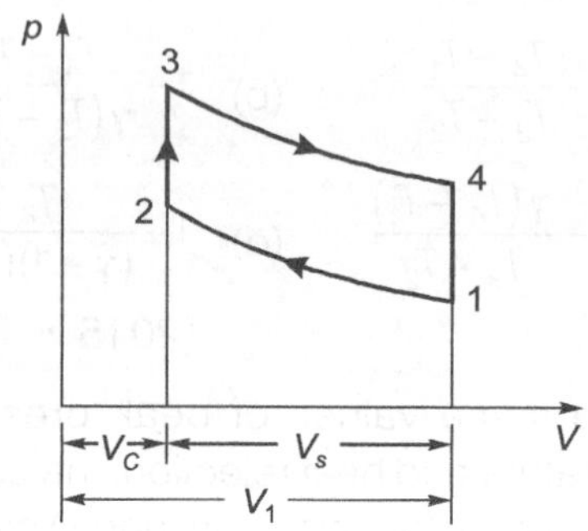

Compression ratio,

$$r = \frac{V_1}{V_c} = \frac{V_c + V_s}{V_c}$$

$$= 1 + \frac{V_s}{V_c} = 1 + 10 = 11$$

$$\eta_{Otto} = 1 - \frac{1}{r^{\gamma-1}} = 1 - \frac{1}{11^{1.4-1}}$$

$$= 1 - \frac{1}{11^{0.4}} = 1 - \frac{1}{2.609}$$

$$= 1 - 0.3832 = 0.6168 = \mathbf{61.67\%}$$

1.6 (d)

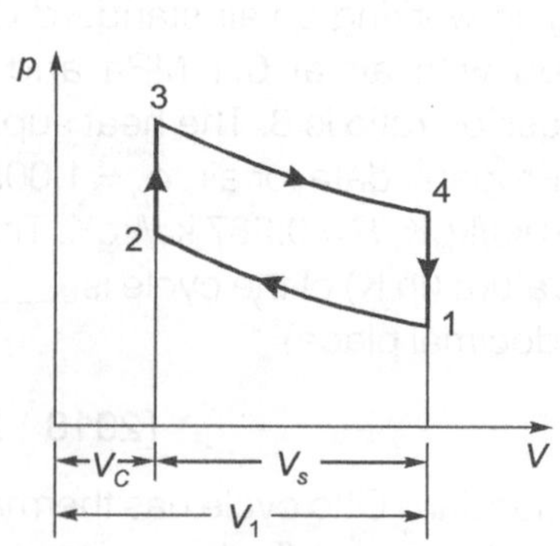

Given data:

$d = 10$ cm; $l = 15$ cm; $\gamma = 1.4$

$V_c = 196.3$ cc $= 196.3$ cm^3

$q_s = 1800$ kJ/kg

Swept volume,

$$V_s = \frac{\pi}{4} d^2 l = \frac{3.14}{4}(10)^2 \times 15$$

$$= 1177.5 \text{ cm}^3$$

Compression ratio,

$$r = \frac{V_1}{V_c} = \frac{V_c + V_s}{V_c} = \frac{196.3 - 1177.5}{196.3} = 6.99$$

$$\eta_{Otto} = 1 - \frac{1}{r^{\gamma-1}} = 1 - \frac{1}{(6.99)^{1.4-1}}$$

$$= 1 - 0.4594 = 0.5406$$

also $$\eta_{Otto} = \frac{w}{q_s}$$

$\therefore$ $$0.5406 = \frac{w}{1800}$$

or $$w = \textbf{973.08 kJ/kg}$$

1.7 (c)

Stroke length,

$$l = 250 \text{ mm} = 0.25 \text{ m}$$

Bore dia: $d = 200 \text{ mm} = 0.2 \text{ m}$

Clearance volume,

$$V_c = 0.001 \text{ m}^3$$
$$\gamma = 1.4$$

Displacement volume,

$$V_s = \frac{\pi}{4} d^2 \times l = \frac{3.14}{4} \times (2)^2 \times 0.25$$
$$= 7.85 \times 10^{-3} \text{ m}^3$$

Total volume in the cylinder,

$$V_1 = V_c + V_s = 0.001 + 7.85 \times 10^{-3}$$
$$= 8.85 \times 10^{-3} \text{ m}^3$$

Compression ratio,

$$r = \frac{V_1}{V_c} = \frac{8.85 \times 10^{-3}}{0.001} = 8.85$$

$\therefore$ Air-standard cycle efficiency,

$$\eta = 1 - \frac{1}{r^{\gamma-1}} = 1 - \frac{1}{(8.85)^{1.4-1}}$$
$$= 1 - \frac{1}{8.85^{0.4}}$$
$$= 0.5819 \approx \textbf{58.2\%}$$

1.8 (b)

Assumption of air-standard Otto cycle

1. All process are both internally as well as externally reversible.
2. Air behaves as ideal gas.
3. Specific heat remains constant.
4. Intake process is constant volume heat addition process and exhaust process is constant volume heat rejection process.

1.9 (d)

Given data for Otto cycle:

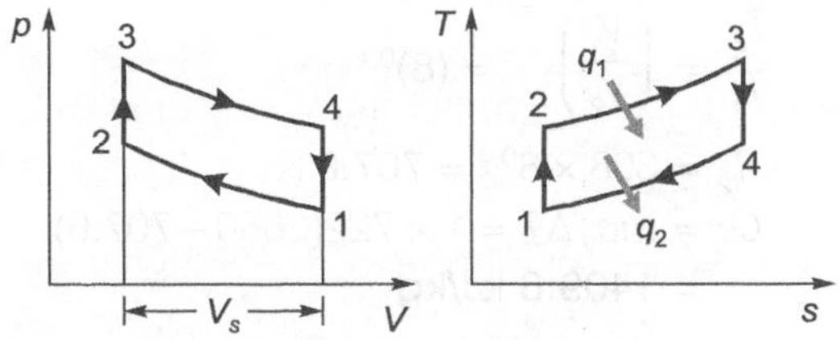

Compression ratio,

$$r = \frac{v_1}{v_2} = 10$$
$$p_1 = 100 \text{ kPa}$$
$$T_1 = 27°\text{C} = (27 + 273) \text{ K}$$
$$= 300 \text{ K}$$
$$q_1 = 1500 \text{ kJ/kg}$$
$$q_2 = 700 \text{ kJ/kg}$$
$$R = 0.287 \text{ kJ/kgK}$$
$$p_1 v_1 = RT_1$$
$$100 \times v_1 = 0.287 \times 300$$

or $$v_1 = 0.861 \text{ m}^3/\text{kg}$$

also $$v_1 = 10\, v_2$$

$\therefore$ $$0.861 = 10\, v_2$$

or $$v_2 = 0.0861 \text{ m}^3/\text{kg}$$

Specific work done per cycle,

$$w = q_1 - q_2$$
$$= 1500 - 700 = 800 \text{ kJ/kg}$$

We know that mean effective pressure,

$$p_m = \frac{w}{v_s} = \frac{800}{v_1 - v_2}$$
$$= \frac{800}{0.861 - 0.0861} = \textbf{1032.39 kPa}$$

1.10 Sol.

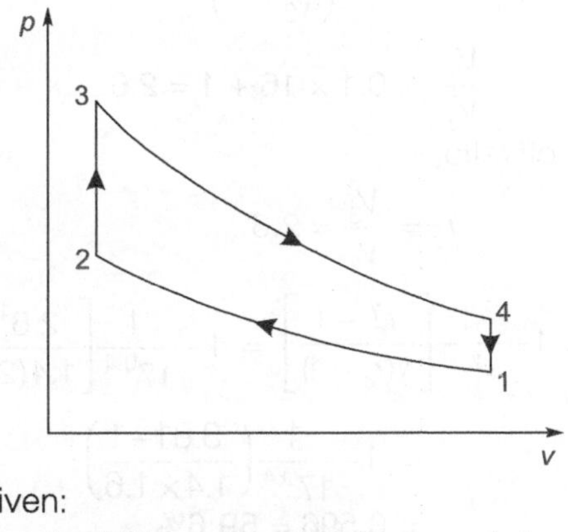

Given:

$p_1 = 0.1$ MPa; $T_1 = 308$ K; $\gamma = 1.4$

$R = 288.8$ J/kgK; $r = 8$

$T_s = 2660$ K

$$Q_s = mc_v (T_3 - T_2)$$

$$c_v = \frac{R}{\gamma - 1} = \frac{288.8}{0.4} = 722 \text{ J/kgK}$$

$$\frac{T_2}{T_1} = \left(\frac{V_1}{V_2}\right)^{\gamma-1} = (8)^{0.4}$$

$$T_2 = 308 \times 8^{0.4} = 707.6 \text{ K}$$

$$Q_s = mc_v\Delta T = 1 \times 722(2660 - 707.6)$$

$$= \mathbf{1409.6 \text{ kJ/kg}}$$

1.11 Sol.

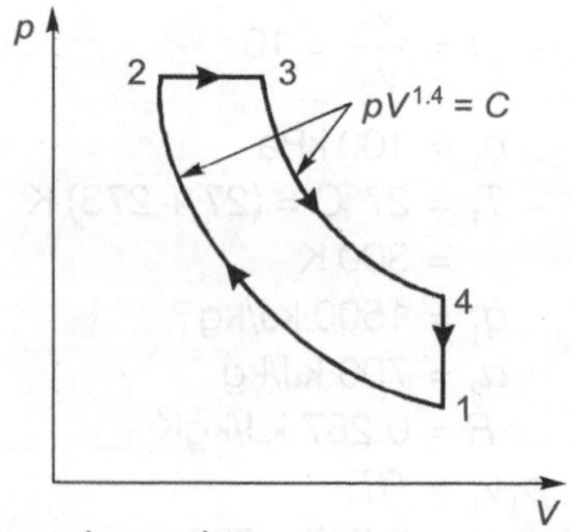

Compression ratio,

$$r = \frac{V_1}{V_2}$$

Ratio of specific heats,

$$\frac{c_p}{c_v} = \gamma = 1.4$$

Cut-off = 0.1 × stroke volume

$$V_3 - V_2 = 0.1 \times V_s$$

$$V_3 - V_2 = 0.1 \times (V_1 - V_2)$$

$$\left(\frac{V_3}{V_2} - 1\right)V_2 = 0.1\times\left(\frac{V_1}{V_2} - 1\right)V_2$$

$$\frac{V_3}{V_2} - 1 = 0.1\left(\frac{V_1}{V_2} - 1\right)$$

or $$\frac{V_3}{V_2} = 0.1 \times 16 + 1 = 2.6$$

Cut-off ratio,

$$r_c = \frac{V_3}{V_2} = 2.6$$

$$\eta_{Diesel} = 1 - \frac{1}{r^{\gamma-1}}\left[\frac{r_c^{\gamma} - 1}{\gamma(r_c - 1)}\right] = 1 - \frac{1}{17^{0.4}}\left[\frac{2.6^{1.4} - 1}{1.4(2.6 - 1)}\right]$$

$$= 1 - \frac{1}{17^{0.4}}\left(\frac{3.81 - 1}{1.4 \times 1.6}\right)$$

$$= 0.596 = \mathbf{59.6\%}$$

1.12 Sol.

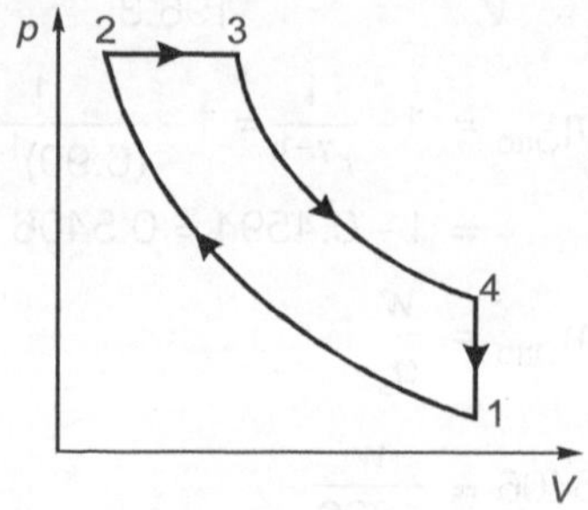

Given, $p_1 = 1$ bar, $p_2 = 32.42$ bar

$$\gamma = \frac{c_p}{c_v} = 1.4$$

$$\frac{V_4}{V_3} = \frac{V_1}{V_3} = 8$$

For process 1–2,

$$p_1 V_1^{\gamma} = p_2 V_2^{\gamma}$$

$$\therefore \left(\frac{V_1}{V_2}\right)^{\gamma} = \frac{p_2}{p_1} = 32.42$$

or $$\frac{V_1}{V_2} = (32.42)^{1/1.4} = 11.999 \approx 12$$

∴ Compression ratio, $r = 12$

Cut-off ratio,

$$r_c = \frac{V_3}{V_2} = \frac{V_1}{8} \times \frac{12}{V_1} = 1.5$$

$$\eta_{Diesel} = 1 - \frac{1}{r^{\gamma-1}}\left[\frac{r_c^{\gamma} - 1}{\gamma(r_c - 1)}\right] = 1 - \frac{1}{12^{0.4}}\left[\frac{1.5^{1.4} - 1}{1.4 \times 0.5}\right]$$

$$= 0.596 = \mathbf{59.6\%}$$

POINTS TO REMEMBER

- Here, in this problem, cut-off ratio,

$$r_c = \frac{V_3}{V_2} = \frac{V_3}{V_2} \times \frac{V_1}{V_1} = \frac{V_1}{V_2} \times \frac{V_3}{V_1}$$

So, cut-off ratio, $r_c = r \times \frac{1}{r_e} \quad \left(r_e = \frac{V_4}{V_3} = \frac{V_1}{V_3}\right)$

$$\therefore \quad r_c = \frac{r}{r_e}$$

- So, it is important to note that compression ratio is equal to the multiplicationof cut-off ratio and expansion ratio. And the value of cut-off ratio, expansion ratio and compression ratio are always greater than 1.

1.13 (b)

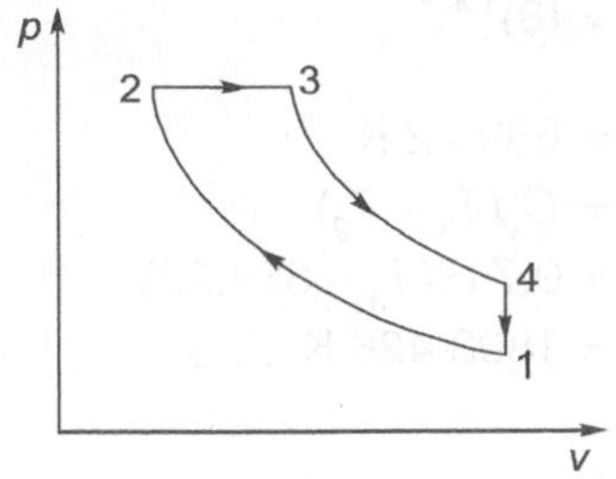

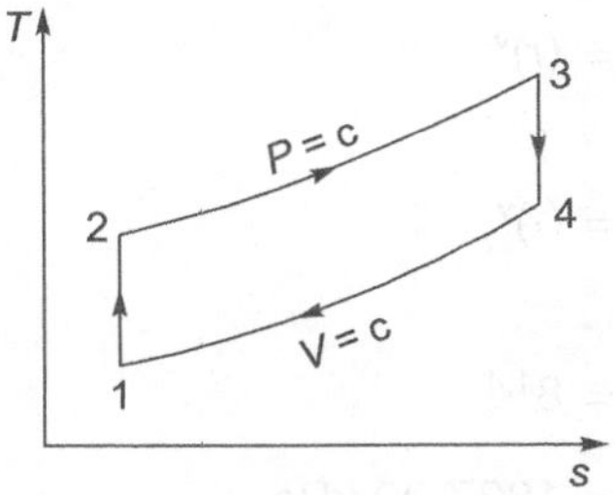

Efficiency of diesel cycle,

$$\eta = 1 - \frac{\text{Heat rejected}}{\text{Heat supplied}}$$

Heat is added in the constant pressure process,

$$Q_{2\text{-}3} = h_3 - h_2$$
$$= c_p(T_3 - T_2)$$

Heat is reject in the constant volume process,

$$Q_{4\text{-}1} = h_4 - h_1 = c_v(T_4 - T_1)$$

$$\text{efficiency, } \eta = \frac{Q_{4\text{-}1}}{Q_{2\text{-}3}} = 1 - \frac{c_v(T_4 - T_1)}{c_p(T_3 - T_2)}$$

$$= 1 - \frac{1}{\gamma}\frac{(T_4 - T_1)}{(T_3 - T_2)}$$

1.14 (b)

For same compression ratio and heat addition

$\eta_{otto} > \eta_{dual} > \eta_{diesel}$

For same compression ratio and heat rejection,

$\eta_{otto} > \eta_{dual} > \eta_{diesel}$

For same peak temperatur and heat rejection

$\eta_{diesel} > \eta_{dual} > \eta_{otto}$

For same peak pressure and heat rejection

$\eta_{diesel} > \eta_{dual} > \eta_{otto}$

1.15 Sol.

$$\gamma = 1.4$$
$$B.P. = 70 \text{ kW}$$
$$\dot{m}_f = 10.3 \text{ kg/hr}$$
$$CV = 44000 \text{ kJ/kg}$$
$$\eta = \frac{B.P.}{\dot{m}_f \times CV} = \frac{70}{\frac{10.3}{3600} \times 44000}$$

$$\eta = 0.556$$

$$\eta_{otto} = 1 - \frac{1}{(r)^{\gamma-1}} = 0.556$$

$$\Rightarrow \quad 1 - \frac{1}{(r)^{1.4-1}} = 0.556$$

$$\frac{1}{(r)^{0.4}} = 0.4439$$

$$r = \mathbf{7.61}$$

1.16 Sol.

$$P_1 = 0.1 \text{ MPa}, \; T_1 = 35°\text{C} = 308 \text{ K}$$
$$\frac{V_1}{V_2} = r = 8$$
$$Q_s = 500 \text{ kJ/kg}$$
$$c_p = 1.005 \text{ kJ/kgK}$$
$$c_v = 0.718 \text{ kJ/kgK}$$
$$R = 0.287 \text{ kJ/kgK}$$
$$\gamma = \frac{c_p}{c_v} = 1.399 \simeq 1.40$$
$$T_3 = T_{max} = ?$$

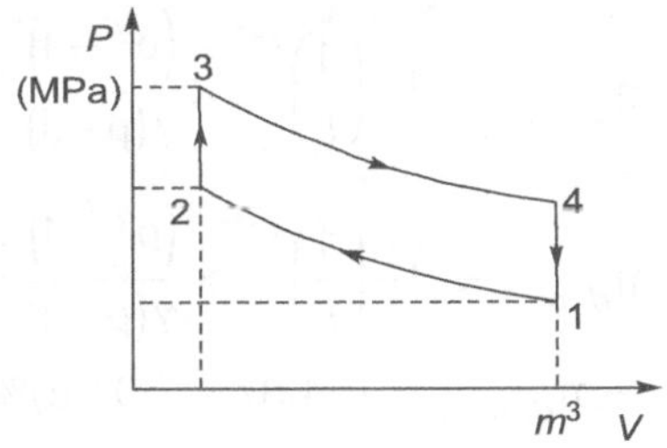

For process 1–2

$$P_1V_1^{\gamma} = P_2V_2^{\gamma}$$

$$P_2 = P_1\left(\frac{V_1}{V_2}\right)^{\gamma} = 0.1 \times (8)^{1.4}$$

$$P_2 = 1.8379 \text{ MPa}$$

and $\quad \frac{P_1V_1}{T_1} = \frac{P_2V_2}{T_2} \Rightarrow T_2 = \frac{P_2V_2}{P_1V_1}T_1$

$$T_2 = \frac{1.8379 \times 1}{0.1 \times 8} \times 308$$

$$T_2 = 707.6 \text{ K}$$

For process 2 → 3

$$Q_s = c_v(T_3 \quad T_2) = 500 \text{ kJ/ kg}$$

$$0.718(T_3 - 707.6) = 500$$

$$T_3 = 1403.97 \text{ K}$$

1.17 Sol.

As per given data,

$$\eta_{otto} = 0.5$$

$$v_1 = \frac{RT_1}{P_1} = \frac{0.287 \times 300}{100}$$
$$= 0.861 \text{ m}^3/\text{kg}$$
$$\eta_{otto} = 1 - \frac{1}{(r)^{\gamma-1}}$$
$$0.5 = 1 - \frac{1}{(r)^{\gamma-1}}$$
$$(r)^{0.4} = 2$$
$$r = 5.65$$
$$\frac{v_1}{v_2} = 5.65$$
$$v_2 = \frac{0.861}{5.65} = 0.1524 \text{ m}^3/\text{kg}$$

Specific work output $= P_{mep} \times v_S$
$$= 1000 \text{ kPa} \times (v_1 - v_2)$$
$$= 1000 \times (0.861 - 0.1524)$$
$$= 708.6 \text{ kJ/kg}$$

1.19 Sol.

As given compression ratio (*CR*) diesel 15 to 21

$r = 1.3$, cut off ratio $r_c = 2$

$$\eta_{d,\, r=21} = 1 - \left(\frac{1}{r}\right)^{\gamma-1} \times \frac{(\rho^\gamma - 1)}{\gamma(\rho - 1)} = 54.87\%$$
$$\eta_{d,\, r=15} = 1 - \left(\frac{1}{r}\right)^{\gamma-1} \times \frac{(\rho^\gamma - 1)}{\gamma(\rho - 1)} = 50.08\%$$
$$\eta_{d,\, r=21} - \eta_{d,\, r=15} = (54.87 - 50.08)\% = 4.8\%$$

1.21 (4809.136) (4780 to 4825)

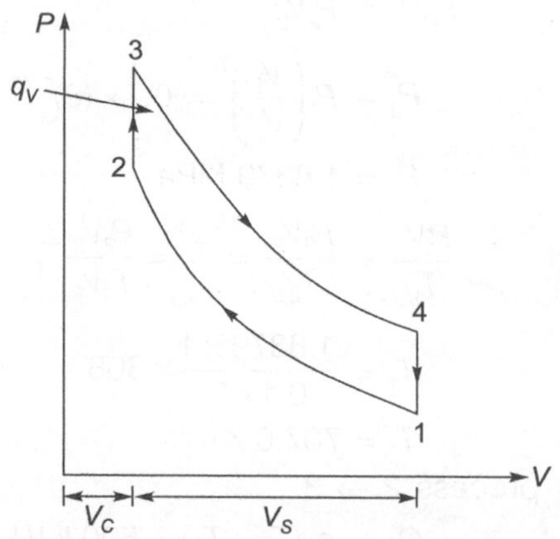

$$\frac{T_2}{T_1} = \left(\frac{P_2}{P_1}\right)^{\frac{\gamma-1}{\gamma}} = \left(\frac{V_1}{V_2}\right)^{\gamma-1} = (r)^{\gamma-1}$$
$$r = \frac{V_1}{V_2} = \frac{V_C + V_S}{V_C} = 1 + \frac{V_S}{V_C} = 1 + \frac{250}{35.7}$$
$$r = 8$$
$$\frac{T_2}{300} = (8)^{1.4-1}$$
$$T_2 = 689.22 \text{ K}$$
$$q_V = C_V(T_3 - T_2)$$
$$800 = 0.718(T_3 - 689.22)$$
$$T_3 = 1803.426 \text{ K}$$

Also, $$\left(\frac{P_2}{P_1}\right)^{\frac{\gamma-1}{\gamma}} = (r)^{\gamma-1}$$
$$\frac{P_2}{P_1} = (r)^\gamma$$
$$\frac{P_2}{100} = 8^{1.4}$$
$$P_2 = 1837.92 \text{ kPa}$$
$$P_3 = P_{max}$$
$$PV = mRT$$
$$P \propto T$$
$$\frac{P_2}{P_3} = \frac{T_2}{T_3}$$
$$\frac{1837.92 \text{ kPa}}{P_3} = \frac{689.22 \text{ K}}{1803.426 \text{ K}}$$
$$P_3 = 4809.136 \text{ kPa} \approx 4809 \text{ kPa}$$

■■■■

2 Analysis and Injection of Fuel & Fuel Emission

2.1 In order to burn 1 kilogram of CH_4 completely, the minimum number of kilograms of oxygen needed is (take atomic weights of H, C and O as 1, 12 and 16 respectively)

(a) 3 (b) 4

(c) 5 (d) 6 [1995 : 2 M]

2.2 An air breathing air craft is flying at an altitude where the air density is half the value at ground level. With reference to the ground level, the air-fuel ratio at this altitude will be

(a) $\sqrt[3]{2}$ (b) $\sqrt{\frac{1}{2}}$

(c) 2 (d) 4 [1998 : 2 M]

2.3 The silencer of an internal combustion engine

(a) reduces noise

(b) decrease break specific fuel consumption (BSFC)

(c) increase BSFC

(d) has no effect on its efficiency [1999 : 2 M]

2.4 For a spark ignition engine, the equivalence ratio (ϕ) of mixture entering the combustion chamber has values

(a) $\phi < 1$ for idling and $\phi > 1$ for peak power conditions

(b) $\phi > 1$ for both idling and peak power conditions

(c) $\phi > 1$ for idling and $\phi < 1$ for peak power conditions

(d) $\phi < 1$ for both idling and peak power conditions [2003 : 1 M]

2.5 A diesel engine is usually more efficient than a spark ignition engine because

(a) diesel being a heavier hydrocarbon, releases more heat per kg than gasoline

(b) the air standard efficiency of diesel cycle is higher than the Otto cycle, at a fixed compression ratio

(c) the compression ratio of a diesel engine is higher than that of an SI engine

(d) self ignition temperature of diesel is higher than that of gasoline [2003 : 1 M]

2.6 Propane (C_3H_8) is burned in an oxygen atmosphere with 10% deficit oxygen with respect to the stoichiometric requirement. Assuming no hydrocarbons in the products, the volume percentage of CO in the products is______.

[2016 : 1 M, Set-1]

2.7 Air contains 79% N_2 and 21% O_2 on a molar basis. Methane (CH_4) is burned with 50% excess air than required stoichiometrically. Assuming complete combustion of methane, the molar percentage of N_2 in the products is __________.

[2017 : 2 M, Set-1]

■■■■

Answers Analysis and Injection of Fuel & Fuel Emission

2.1	(b)	2.2	(b)	2.3	(a)	2.4	(b)	2.5	(c)	2.6	(14.28)	2.7	(73.83)

Explanations Analysis and Injection of Fuel & Fuel Emission

2.1 (b)

$$\underset{16\text{ kg}}{CH_4} + \underset{64\text{kg}}{2O_2} \longrightarrow CO_2 + 2H_2O$$

$$1\text{ kg}\ \frac{64}{16}\text{ kg}$$

So, to burn 1 kg of CH_2, minimum number of kilograms of oxygen needed is equal to

$$= \frac{64}{16}\text{ kg} = 4\text{ kg}$$

POINTS TO REMEMBER

- Surface area of the cylinder $= \pi DL + 2 \times \frac{\pi}{4}D^2$

  But in case of IC engines for calculating heat loss, then the area of bottom and top of the cylinder is not considered.
- Only surface area is considered for water cooled IC engines. For air cooled IC engines, piston head area is also considered if mention.

2.2 (b)

$$\frac{(AFR)_{altitude}}{(AFR)_{groundlevel}} = \sqrt{\frac{\rho_{altitude}}{\rho_{ground\ level}}} = \sqrt{\frac{1}{2}}$$

$$(AFR)_{altitude} = \sqrt{\frac{1}{2}}(AFR)_{groundlevel}$$

So with reference to the ground level, the air-fuel ratio at this altitude will be $\sqrt{\frac{1}{2}}$.

2.3 (a)

A muffler (silencer in British language) is a device for reducing the noise which is emitted by the exhaust of an internal combustion engine.

2.4 (b)

Equivalence ratio is defined as the actual fuel air ratio to stoichiometric fuel air ratio

$$\phi = \frac{\left(\frac{F}{A}\right)_{Actual}}{\left(\frac{F}{A}\right)_{Stoich}}$$

For both idling and peak power condition. Fuel-air ratio required is **more** than the stoichiometric fuel-air ratio.

2.5 (c)

For same compression ratio and the same heat supplied, Otto cycle is most efficient and diesel cycle is least efficient.

In practice, however, the compression ratio of the Diesel engine ranges between 14 and 25 whereas that of the Otto cycle engine ranges between 6 and 12. Because of diesel engine has higher efficiency than the petrol engine.

2.6 Sol.

$C_3H_8 + aO_2 \rightarrow bCO_2 + cH_2O$

Balancing carbon atoms, we get

$$3 = b \text{ or } b = 3$$

Balancing hydrogen atoms, we get

$$8 = 2c \text{ or } c = 4$$

Balancing oxygen atoms, we get

$$2a = 2b + c = 2 \times 3 + 4 = 10$$
$$a = 5$$

For stoichiometric burning (i.e., chemically correct), number of moles of O_2 required = 5.

As it is burnt with 10% deficient O_2, it will generate CO.

That is, the new equation is

$C_3H_8 + 0.9 \times 5O_2 \rightarrow \alpha CO_2 + \beta CO + \gamma H_2O$

Balancing carbon atoms, we get

$$3 = \alpha + \beta \quad \text{...(i)}$$

Balancing hydrogen atoms, we get

$$8 = 2\gamma \text{ or } \gamma = 4$$

Balancing oxygen atoms, we get

$$0.9 \times 10 = 2\alpha + \beta + \gamma$$
$$9 = 2\alpha + \beta + 4$$
$$5 = 2\alpha + \beta \quad \text{...(ii)}$$

Eq. (ii) – Eq. (i), we get

$$2 = \alpha \text{ or } \alpha = 2$$

Substituting $\alpha = 2$ in Eq. (ii), we get

$$5 = 2 \times 2 + \beta \text{ or } \beta = 1$$

$\therefore C_3H_8 + 4.5\, O_2 \rightarrow 2CO_2 + CO + 4H_2O$

The volume percentage of CO in the product

$$= \left(\frac{\beta}{\alpha + \beta + \gamma}\right) \times 100$$
$$= \left(\frac{1}{2+1+4}\right) \times 100 = \mathbf{14.28\%}$$

2.7 Sol.

Stoichiometric reaction

$$CH_4 + 2 \cdot \left[O_2 + \frac{79}{21}N_2\right] \rightarrow 2H_2O + CO_2 + 2 \times \frac{79}{21} \times N_2$$

50% excess air:

$$CH_4 + 3 \cdot \left[O_2 + \frac{79}{21}N_2\right] \rightarrow 2H_2O + CO_2 + 3 \times \frac{79}{21} \times N_2 + O_2$$

$$\%N_2 = \frac{\frac{3 \times 79}{21}}{2+1+1+3 \times \frac{79}{21}} \times 100$$

$$= \mathbf{73.83\%}$$

POINTS TO REMEMBER

If student doesn't know stoichiometric reaction for CH_4, then they can easily find by:

$$CH_4 + a(O_2 + \frac{79}{21}N_2) \longrightarrow bCO_2 + CH_2O + dN_2$$

C-balance: 1 = b

H_2-balance : $4 = 2C \Rightarrow C = 2$

O_2-balance : $2a = 2b + c = 2 \times 1 + 2 = 4 \Rightarrow a = 2$

N_2-balance: $a \times \frac{79}{21} = d \Rightarrow d = 2 \times \frac{72}{21}$

So stoichimetric reaction will be:

$$CH_4 + 2(O_2 + \frac{79}{21}N_2) \longrightarrow CO_2 + 2H_2O + 2 \times \frac{79}{21}N_2$$

■■■■

3 Testing and Performance

3.1 An automobile engine operates at a fuel air ratio of 0.05, volumetric efficiency of 90% and indicated thermal efficiency of 30%. Given that the calorific value of the fuel is 45 MJ/kg and the density of air at intake is 1 kg/m^3, the indicated mean effective pressure for the engine is

(a) 6.075 bar (b) 6.75 bar
(c) 67.5 bar (d) 243 bar **[2003 : 2 M]**

3.2 During a Morse test on a 4 cylinder engine, the following measurements of brake power were taken at constant speed.

All cylinders firing	3037 kW
Number 1 cylinder not firing	2102 kW
Number 2 cylinder not firing	2102 kW
Number 3 cylinder not firing	2100 kW
Number 4 cylinder not firing	2098 kW

The mechanical efficiency of the engine is

(a) 91.53% (b) 85.07%
(c) 81.07% (d) 61.22% **[2004 : 2 M]**

3.3 A turbocharged four-stroke direct injection diesel engine has a displacement volume of 0.0259 m^3 (25.9 liters). The engine has an output of 950 kW at 2200 rpm. The mean effective pressure (in MPa) is closest to

(a) 2 (b) 1
(c) 0.2 (d) 0.1 **[2010 : 1 M]**

3.4 The crank radius of a single-cylinder IC engine is 60 mm and the diameter of the cylinder is 80 mm. The swept volume of the cylinder is cm^3 is

(a) 48 (b) 96
(c) 302 (d) 603 **[2011 : 2 M]**

3.5 Air enters a diesel engine with a density of 1.0 kg/m^3. The compression ratio is 21. At steady state, the air intake is 30 × 10^{-3} kg/s and the net work output is 15 kW. The mean effective pressure (kPa) is _______. **[2015 : 1 M, Set-1]**

3.6 A rope-brake dynamometer attached to the crank shaft of an I.C. engine measures a brake power of 10 kW when the speed of rotation of the shaft is 400 rad/s. The shaft torque (in Nm) sensed by the dynamometer is _______ **[2015 : 1 M, Set-2]**

3.7 The indicated power developed by an engine with compression ratio of 8, is calculated using an air-standard Otto cycle {constant properties). The rate of heat addition is 10 kW. The ratio of specific heats at constant pressure and constant volume is 1.4. The mechanical efficiency of the engine is 80 percent. The brake power output of the engine is _______ kW (round off to one decimal place). **[2020 : 2 M, Set-1]**

Answers Testing and Performance

3.1 (a) **3.2** (c) **3.3** (a) **3.4** (d) **3.5** (525) **3.6** (25) **3.7** (4.5)

Explanations Testing and Performance

3.1 (a)

Volumetric efficiency

$$= \frac{\text{Actual volume}}{\text{Swept volume}} = \frac{V_a}{V_s} = 0.9$$

or $V_a = 0.9V_s$

Mass of air,

$$m_a = \rho_{air} V_a = 0.9V_s$$
$$m_f = 0.05 \times 0.9V_s = 0.045V_s$$

$$\eta_{thermal} = \frac{p_m \times V_S}{m_f \times C.V.}$$

$$0.3 = \frac{p_m \times V_s}{0.045\, V_s \times 45 \times 10^6}$$

or $p_m = 0.6075 \times 10^6$ Pa = **6.075 bar**

3.2 (c)

<u>Method I:</u>

$$(B.P.)_{1,2,3,4} = 3037 \text{ kW}$$
$$(I.P.)_{1,2,3,4} = (B.P.)_{1,2,3,4} + (F.P.)_{1,2,3,4} \quad \text{... (i)}$$

Number 1 cylinder not firing,

$$(B.P.)_{2,3,4} = 2102 \text{ kW}$$
$$(I.P.)_{2,3,4} = (B.P.)_{2,3,4} + (F.P.)_{1,2,3,4} \quad \text{... (ii)}$$

Eq. (ii) – Eq. (i), we get

$$(I.P.)_{1,2,3,4} - (I.P.)_{2,3,4} = (B.P.)_{1,2,3,4} - (B.P.)_{2,3,4}$$
$$(I.P.)_1 = 3037 - 2102 = 935 \text{ kW}$$

Similarly, number 2 cylinder not firing,

$$(B.P.)_{1,3,4} = 2102 \text{ kW}$$

$$\therefore \quad (I.P.)_2 = (B.P.)_{1,2,3,4} - (B.P.)_{1,3,4} = 3037 - 2102 = 935 \text{ kW}$$

Number 3 cylinder not firing,

$$(B.P.)_{1,2,4} = 2100 \text{ kW}$$

$$\therefore \quad (I.P.)_3 = (B.P.)_{1,2,3,4} - (B.P.)_{1,2,4} = 3037 - 2100 = 937 \text{ kW}$$

Number 4 cylinder not firing,

$$(B.P.)_{1,2,3} = 2098 \text{ kW}$$

$$\therefore \quad (I.P.)_4 = (B.P.)_{1,2,3,4} - (B.P.)_{1,2,3} = 3037 - 2098 = 939 \text{ kW}$$

Total I.P., $(I.P.)_{1,2,3,4} = (I.P.)_1 + (I.P.)_2 + (I.P.)_3 + (I.P.)_1$

$$= 935 + 935 + 937 + 939 = 3746 \text{ kW}$$

Mechanical efficiency,

$$\eta_m = \frac{(B.P.)_{1,2,3,4}}{(I.P.)_{1,2,3,4}} = \frac{3037}{3746} = 0.8107 = \mathbf{81.07\%}$$

Method II:

Given:

Brake power with 4-cylinder, 4B = 3037 kW

Brake power with 3-cylinder,

$$3B = \frac{2102 + 2102 + 2100 + 2098}{4} = 2100.5 \text{ kW}$$

Indicated power, I.P.

$$= 4(4B - 3B) = 4(3037 - 2100.5) = 4 \times 936.5 = 3746 \text{ kW}$$

Mechanicakl efficiency,

$$= \frac{B.P}{I.P.} = \frac{3037}{3746} = 0.8107 = 81.07\%$$

3.3 (a)

Given data: $n = \frac{N}{2}$ for four-stroke engine

Stroke volume, $V_s = 0.0259 \text{ m}^3$

Power, $P = 950$ kW; Speed, $N = 2200$ rpm

We know that power output,

$$P = \frac{p_m A l n x}{60} \text{ kW} = \frac{p_m V_s n x}{60}$$

where P is in kW; p_m is in kPa; V_s is in m^3

$$n = \frac{N}{2} \text{ rpm}$$

$$x = 1, \text{ number of cylinder}$$

$$\therefore \quad 950 = \frac{p_m \times 0.0259}{60} \times \frac{N}{2} \times 1$$

$$950 = \frac{p_m \times 0.0259 \times 2200}{120}$$

or $\quad p_m = 2000 \text{ kPa} = \mathbf{2\ MPa}$

3.4 (d)

Crank radius, $r = 60$ mm; Diameter of cylinder, $d = 80$ mm; Swept volume = ?

$$\therefore \quad l = 2r = 2 \times 60 \text{ mm} = 120 \text{ mm}$$

Swept volume,

$$V_s = \frac{\pi}{4} d^2 l = \frac{\pi}{4} \times (80)^2 \times 120 \times 10^{-3} = \mathbf{602.88\ cm^3}$$

3.5 Sol.

Given: $\rho = 1 \text{ kg/m}^3$; $r = \frac{V_1}{V_c} = 21$

$m = 30 \times 10^{-3}$ kg/s; $W = 15$ kW

$$V_1 = \frac{m}{\rho} = \frac{30 \times 10^{-3}}{1} \text{ m}^3/\text{s} = 30 \times 10^{-3} \text{ m}^3/\text{s}$$

$$V_c = \frac{V_1}{r} = \frac{30 \times 10^{-3}}{21} = 1.428 \times 10^{-3} \text{ m}^3/\text{s}$$

$$V_s = V_1 - V_c = 30 \times 10^{-3} - 1.428 \times 10^{-3} = 28.572 \times 10^{-3} \text{ m}^3/\text{s}$$

Mean effective pressure,

$$p_m = \frac{W}{V_s} = \frac{15}{28.572 \times 10^{-3}} = 524.99 \text{ kPa} \approx \mathbf{525\ kPa}$$

3.6 Sol.

Brake power = Brake torque × angular velocity

$$P = T\omega$$

or $$T = \frac{P}{\omega} = \frac{P}{\left(\frac{2\pi N}{60}\right)} = \frac{10{,}000}{\frac{2\pi \times N}{60}}$$

$$= \frac{10000}{400} = \mathbf{25\ Nm}$$

3.7 Sol.

$r = 8$, $Q_1 = 10$ kW, $r = 1.4$, $\eta_{mech} = 0.8$

$$\eta = 1 - \frac{1}{r_c^{\gamma-1}} = 1 - \frac{1}{8^{0.4}} = 0.5647$$

$$\eta = \frac{W}{Q_1}$$

$$W = \eta \times Q_1$$

$$W = \frac{10 \times 5647}{1000} = 5.647 \text{ kW}$$

$$BP = \eta_{mech} \times W = 0.8 \times 5.647 = 4.5175 \text{ kW}$$

4 Miscellaneous

4.1 At the time of starting, idling and low speed operation, the carburettor supplies a mixture which can be termed as
(a) Lean
(b) slightly leaner than stoichiometric
(c) stoichiometric
(d) rich [2004 : 1 M]

Answers Miscellaneous of I.C. Engines

4.1 (d)

Explanations Miscellaneous of I.C. Engines

4.1 (d)

At the time of starting, idling and low speed operation, mixture should be rich for smooth engine operation.

The richening of mixture increases the probability of contact between fuel and air particles and thus improves combustion.

■■■■

Refrigeration and Air-conditioning

UNIT

XI

CONTENTS

Refrigeration and Air-conditioning

Syllabus

Refrigeration and air-conditioning : Vapour and gas refrigeration and heat pump cycles; properties of moist air, psychrometric chart, basic psychrometric processes.

Analysis of Previous GATE Papers

Exam Year	1 Mark Ques.	2 Marks Ques.	3 Marks Ques.	5 Marks Ques.	Total Marks
1995	–	–	–	2	10
1996	–	–	–	1	5
1997	–	2	–	1	9
1998	–	–	–	–	–
1999	–	1	–	1	7
2000	2	–	–	1	7
2001	1	1	–	1	8
2002	1	–	–	–	1
2003	2	2	–	–	6
2004	3	3	–	–	9
2005	2	3	–	–	8
2006	1	2	–	–	5
2007	–	2	–	–	4
2008	–	2	–	–	4
2009	1	–	–	–	1
2010	–	1	–	–	2
2011	1	–	–	–	1
2012	–	3	–	–	6
2013	1	–	–	–	1
2014 Set-1	1	–	–	–	1
2014 Set-2	–	2	–	–	2

Exam Year	1 Mark Ques.	2 Marks Ques.	Total Marks
2014 Set-3	1	–	1
2014 Set-4	2	–	2
2015 Set-1	1	–	1
2015 Set-2	1	–	1
2015 Set-3	1	2	5
2016 Set-1	–	–	–
2016 Set-2	–	2	4
2016 Set-3	–	2	4
2017 Set-1	–	1	2
2017 Set-2	1	–	1
2018 Set-1	–	–	–
2018 Set-2	–	2	4
2019 Set-1	1	–	1
2019 Set-2	–	–	–
2020 Set-1	–	–	–
2020 Set-2	1	1	3
2021 Set-1	1	1	3
2021 Set-2	1	1	3
2022 Set-1	–	–	–
2022 Set-2	–	1	2
2023	–	–	–
2024	–	–	–

1 Vapour Refrigeration

1.1 In a vapour compression refrigeration system, liquids to suction heat exchanger is used to
(a) Keep the COP constant
(b) Prevent the liquid refrigerant from entering the compressor
(c) Sub-cool the liquid refrigerant leaving the condenser
(d) Sub-cool the vapour refrigerant from the evaporator [2000 : 1 M]

1.2 Global warming is caused by
(a) Ozone (b) Carbon dioxide
(c) Nitrogen (d) Carbon monoxide
[2000 : 1 M]

Common Data for Q. 1.3 & Q.1.4

A refrigerator based on ideal vapour compression cycle operates between the temperature limits of –20°C and 40°C. The refrigerant enters the condenser as saturated vapour and leaves as saturated liquid. The enthalpy and entropy values for saturated liquid and vapour at these temperatures are given in the table below.

T (°C)	h_f (kJ/kg)	h_g (kJ/kg)	s_f (kJ/kgK)	s_g (kJ/kgK)
–20	20	180	0.07	0.7366
40	80	200	0.3	0.67

1.3 If refrigerant circulation rate is 0.025 kg/s, the refrigeration effect is equal to
(a) 2.1 kW (b) 2.5 kW
(c) 3.0 kW (d) 4.0 kW [2003 : 2 M]

1.4 The COP of the refrigerator is
(a) 2.0 (b) 2.33
(c) 5.0 (d) 6.0 [2003 : 2 M]

1.5 In the window air conditioner, the expansion device used is
(a) capillary tube
(b) thermostatic expansion valve
(c) automatic expansion valve
(d) float valve [2004 : 1 M]

1.6 Environment friendly refrigerant R134*a* is used in the new generation domestic refrigerators. Its chemical formula is
(a) $CH\,Cl\,F_2$ (b) $C_2\,Cl_3\,F_3$
(c) $C_2\,Cl_2\,F_4$ (d) $C_2\,H_2\,F_4$ [2004 : 1 M]

1.7 A *R*-12 refrigerant reciprocating compressor operates between the condensing temperature of 30°C and evaporator temperature of –20°C. The clearance volume ratio of the compressor is 0.03. Specific heat ratio of the vapour is 1.15 and the specific volume at the suction is 0.1089 m³/kg. Other properties at various states are given in the figure. To realize 2 tons of refrigeration, the actual volume displacement rate considering the effect of clearance is

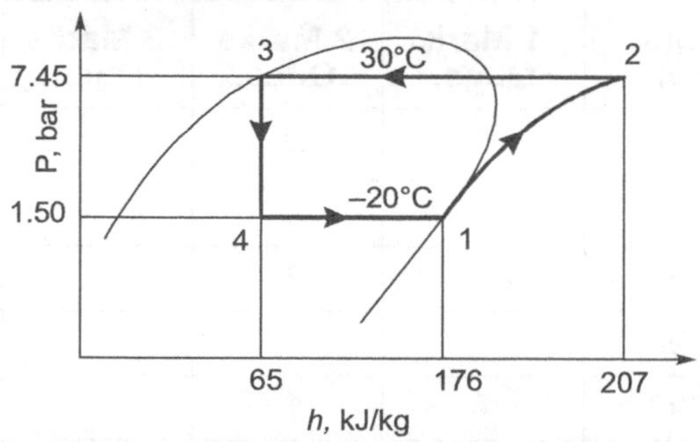

(a) 6.35×10^{-3} m³/s (b) 63.5×10^{-3} m³/s
(c) 635×10^{-3} m³/s (d) 4.88×10^{-3} m³/s
[2004 : 2 M]

1.8 The vapour compression refrigeration cycle is represented as shown in the figure below, with state 1 being the exit of the evaporator. The coordinate system used in this figure is

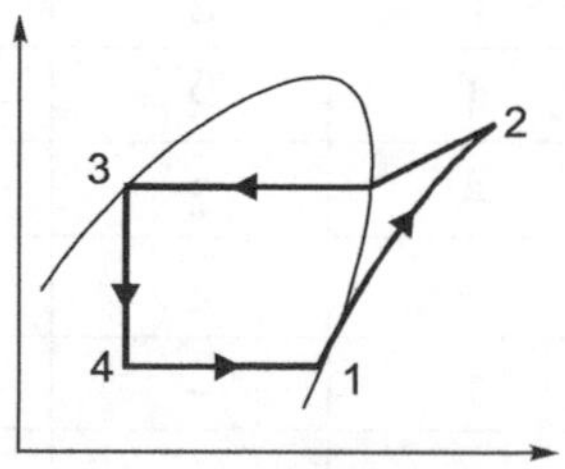

(a) *p–h* (b) *T–s*
(c) *p–s* (d) *T–h* [2005 : 2 M]

1.9 In an ideal vapour compression refrigeration cycle, the specific enthalpy of refrigerant (in kJ/kg) at the following states is given as:
Inlet of condenser : 283
Exit of condenser : 116
Exit of evaporator : 232
The COP of this cycle is
(a) 2.27 (b) 2.75
(c) 3.27 (d) 3.75 [2009 : 1 M]

Common Data Questions 1.10 and 1.11:

A refrigerator operates between 120 kPa and 800 kPa in an ideal vapor compression cycle with R–134*a* as the refrigerant. The refrigerant enters the compressor as saturated vapor and leaves the condenser as saturated liquid. The mass flow rate of the refrigerant is 0.2 kg/s. Properties for R–134*a* are as follows:

Saturated R-134*a*

P(kPa)	T(°C)	h_f(kJ/kg)	h_g(kJ/kg)	s_f(kJ/kgK)	s_g(kJ/kgK)
120	–22.32	22.5	237	0.093	0.95
800	31.31	95.5	267.3	0.354	0.918

Superheated R-134*a*

P(kPa)	T(°C)	h(kJ/kg)	s(kJ/kgK)
800	40	276.45	0.95

1.10 The rate at which heat is extracted, in kJ/s from the refrigerated space is
(a) 28.3 (b) 42.9
(c) 34.4 (d) 14.6 [2012 : 2 M]

1.11 The power required for compressor in kW is
(a) 5.94 (b) 1.83
(c) 7.9 (d) 39.5 [2012 : 2 M]

1.12 Which one of the following is a CFC refrigerant?
(a) R744 (b) R290
(c) R502 (d) R718
[2014 : 1 M, Set-1]

1.13 Refrigerant vapor enters into the compressor of a standard vapor compression cycle at – 10°C (h = 402 kJ/kg) and leaves the compreser at 50°C (h = 432 kJ/kg). It leaves the condenser at 30°C (h = 237 kJ/kg). The COP of the cycle is ________.
[2015 : 2 M, Set-3]

1.14 A refrigerator uses R-134a as its refrigerant and operates on a ideal vapour-compression refrigeration cycle between 0.14 MPa and 0.8 MPa. If the mass flow rate of the refrigerant is 0.05 kg/s, the rate of heat rejection to the environment is ____kW.
Given data :
At P = 0.14 MPa, h = 236.04 kJ/kg, s = 0.9322 kJ/kgK
At P = 0.8 MPa, h = 272.05 kJ/kg (superheated vapour)
At P = 0.8 MPa, h = 93.42 kJ/kg(saturated liquid)
[2016 : 2 M, Set-2]

1.15 In the vapour compression cycle shown in the figure, the evaporating and condensing temperatures are 260 K and 310 K, respectively. The compressor takes in liquid-vapour mixture (state 1) and isentropically compresses it to a dry saturated vapour condition (state 2). The specific heat of the liquid refrigerant is 4.8 kJ/kgK and may be treated as constant. The enthalpy of evaporation for the refrigerant at 310 K is 1054 kJ/kg.

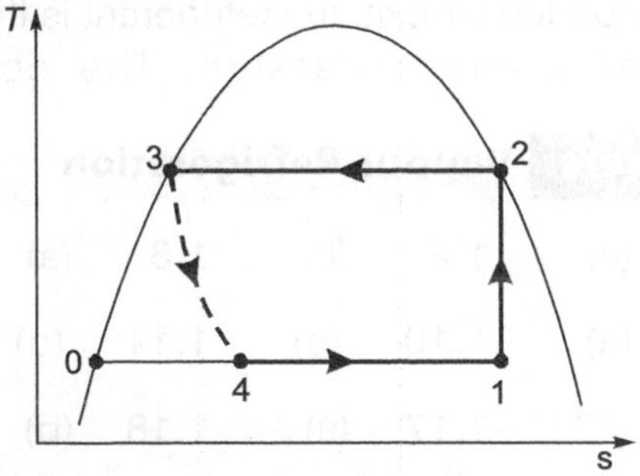

The difference between the enthalpies at state points 1 and 0 (in kJ/kg) is ________
[2016 : 2 M, Set-3]

1.16 A standard vapour compression refrigeration cycle operating with a condensing temperature of 35° C and an evaporating temperature of –10° C develops 15 kW of cooling. The P-h diagram shows the enthalpies at various states. If the isentropic efficiency of the compressor is 0.75, the magnitude of compressor power (in kW) is ______ (correct to two decimal places).

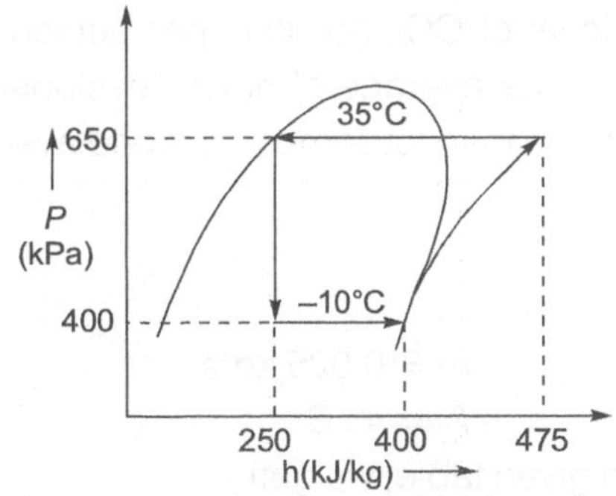

[2018 : 2 M, Set-2]

1.17 Consider an ideal vapour compression refrigeration cycle. If the throttling process is replaced by an isentropic expansion process, keeping all the other processes unchanged, which one of the following statements is true for the modified cycle?
(a) Coefficient of performance is higher than that of the original cycle
(b) Coefficient of performance is the same as that of the original cycle
(c) Refrigerating effect is lower than that of the original cycle
(d) Coefficient of performance is lower than that of the original cycle
[2019 : 1 M, Set-1]

1.18 In a vapour compression refrigeration cycle, the refrigerant enters the compressor in saturated vapour state at evaporator pressure, with specific enthalpy equal to 250 kJ/kg. The exit of the compressor is superheated at condenser pressure with specific enthalpy equal to 300 kJ/kg. At the condenser exit, the refrigerant is throttled to the evaporator pressure. The coefficient of performance (COP) of the cycle is 3. If the specific enthalpy of the saturated liquid at evaporator pressure is 50 kJ/kg, then the dryness fraction of the refrigerant at entry to evaporator is ______.

(a) 0.2 (b) 0.25

(c) 0.3 (d) 0.35

[2022 : 2 M, Set-2]

Answers Vapour Refrigeration

1.1	(c)	1.2	(b)	1.3	(a)	1.4	(b)	1.5	(a)	1.6	(d)	1.7	(*)	1.8	(d)
1.9	(a)	1.10	(a)	1.11	(c)	1.12	(c)	1.13	(5.5)	1.14	(8.93)	1.15	(1103.51)		
1.16	(10)	1.17	(a)	1.18	(b)										

Explanations Vapour Refrigeration

1.2 (b)

Global warming is caused by human activities such as burning fossil fuels and forming. When we burn fossil fuels like coal, and gas to create electricity or power our cars, we release CO_2 polution into the atmosphere.

POINTS TO REMEMBER

Our level of CO_2 pollution per person is nearly double the average of other developed nations and more than four times the world average.

1.3 (a)

$$m = 0.025 \text{ kg/s}$$

At $T_C = 40°C$

From given table, we get

$$h_2 = h_g = 200 \text{ kJ/kg}$$
$$s_2 = s_g = 0.67 \text{ kJ/kgK}$$
$$h_3 = h_f = 80 \text{ kJ/kg} = h_4$$

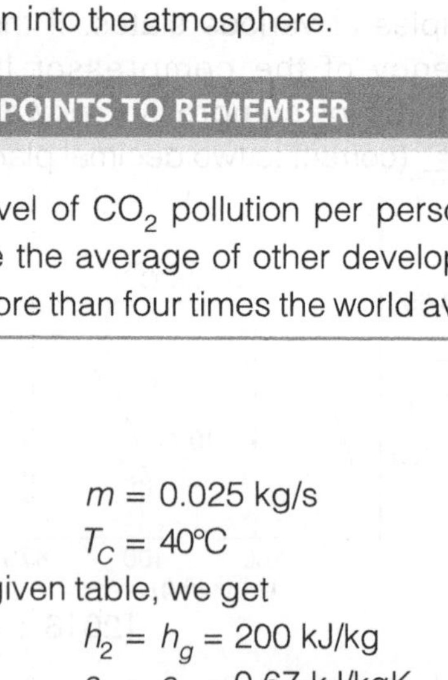

At $T_e = -20°C$

From given table, we get

$$s_f = 0.07 \text{ kJ/kgK},$$
$$s_g = 0.7366 \text{ kJ/kgK}$$
$$h_f = 20 \text{ kJ/kg}, \; h_g = 180 \text{ kJ/kg}$$
$$s_1 = s_2 = s_f + x_1(s_g - s_f)$$
$$0.67 = 0.07 + x_1(0.7366 - 0.07)$$

or $x_1 = 0.9$

Specific enthalpy at point 1,

$$h_1 = h_f + x_1(h_g - h_f)$$
$$= 20 + 0.9(180 - 20) = 164 \text{ kJ/kg}$$

Refrigeration effect, $RE = m(h_1 - h_4)$

$$= 0.025\,(164 - 80) = \mathbf{2.1\ kW}$$

1.4 (b)

$$\text{COP} = \frac{\text{Refrigeration effect}}{\text{Work input}}$$
$$= \frac{RE}{W} = \frac{2.1}{m(h_2 - h_1)}$$
$$= \frac{2.1}{0.025(200 - 164)} = 2.33$$

1.5 (a)

The capillary tube is used as an expansion device in small capacity hermetic sealed refrigeration units such as in domestic refrigerators, water coolers, room air-conditioners and freezers.

1.6 (d)

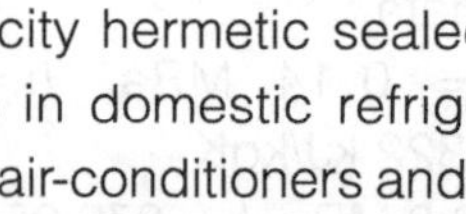

Number of carbon atoms

= 1st digit + 1 = 1 + 1 = 2

Number of hydrogen atoms

= 2nd digit – 1 = 3 – 1 = 2

Number of fluorine atoms

= 3rd digit = 4

So, its chemical formula is $C_2H_2F_4$

POINTS TO REMEMBER

We can also write R-134a like as CF_3CH_2F i.e. $C_2H_2F_4$. R-134a refrigerant has zero amount of chlorine and due to this ozone layer is not depleted and that's why R-134a is known as environment friendly refrigerant.

1.7 (No option matches)

Given data:

$C = 0.03$

$\gamma = 1.15$

$v_1 = 0.1089 \text{ m}^3/\text{kg}$

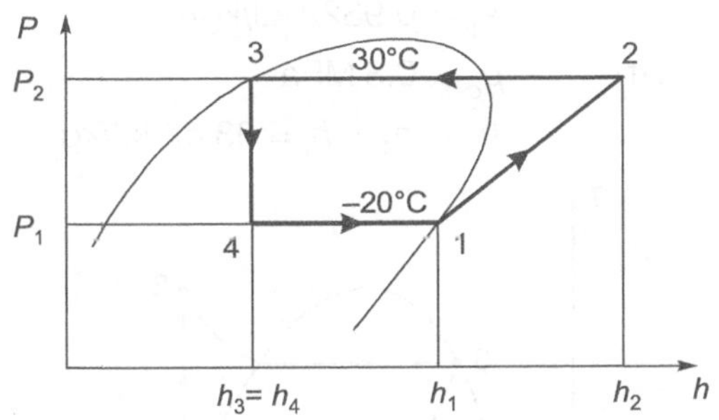

Capacity: $Q = 2 \text{ tons} = 2 \times 3.5 \text{ kW} = 7 \text{ kW}$

$P_1 = 1.50 \text{ bar}; \quad P_2 = 7.45 \text{ bar}$

$h_1 = 176 \text{ kJ/kg}; \quad h_2 = 207 \text{ kJ/kg}$

$h_3 = h_4 = 65 \text{ kJ/kg}$

Refrigerating effect,

$$RE = h_1 - h_4 = 176 - 65 = 111 \text{ kJ/kg}$$

Mass flow rate,

$$m = \frac{\text{Refrigerating capacity}(Q)}{\text{Refrigerating effect }(RE)}$$

$$= \frac{7}{111} = 0.063 \text{ kg/s}$$

We know that volumetric efficiency of the compressor,

$$\eta_v = 1 + C - C\, r_p^{1/\gamma} = 1 + C - C\left(\frac{P_2}{P_1}\right)^{1/\gamma}$$

$$= 1 + 0.03 - 0.03\left(\frac{7.45}{1.50}\right)^{1/1.15}$$

$$= 0.9091$$

also $\quad \eta_v = \dfrac{\text{Actual suction volume}}{\text{Displacement volume}}$

$$\eta_v = \frac{\dot{m}v_1}{V_s}$$

$$0.9091 = \frac{0.063 \times 0.1089}{\dot{V}_s} = \frac{0.00686}{\dot{V}_s}$$

or $\quad \dot{V}_s = 0.00754 \text{ m}^3/\text{s}$

$$= \mathbf{7.54 \times 10^{-3} \text{ m}^3/\text{s}}$$

1.8 (d)

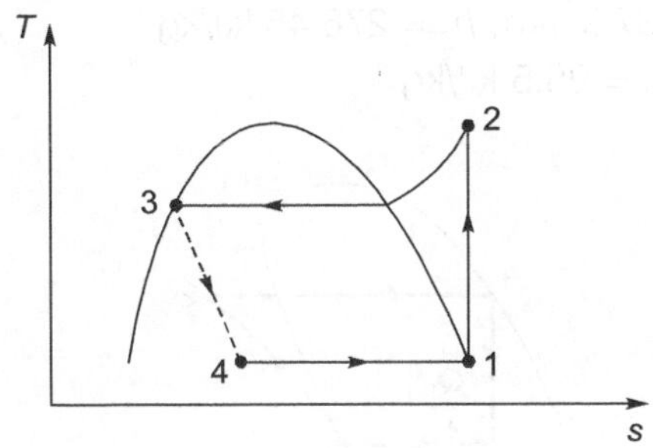

Process:

1–2: Isentropic (Reversible adiabatic) compression

2–3: Constant pressure heat rejection.

3–4: Throttling (Isenthalpic) expansion

4–1: Constant pressure heat absorption

As in VCRS process 3-4 is an adiabatic throttling process in which enthalpy remains constant ($h_3 = h_4$). So option (b) and (c) is ruled out. As process 2-3 is reversible constant pressure process first desuperheated and then condensed, ending with saturated liquid. Heat Q_1 is transferred out, so option (a) is ruled out so, we will be left with option (d) only.

1.9 (a)

Given data:

$h_2 = 283 \text{ kJ/kg}$

$h_3 = 116 \text{ kJ/kg}$

$= h_4$

$h_1 = 232 \text{ kJ/kg}$

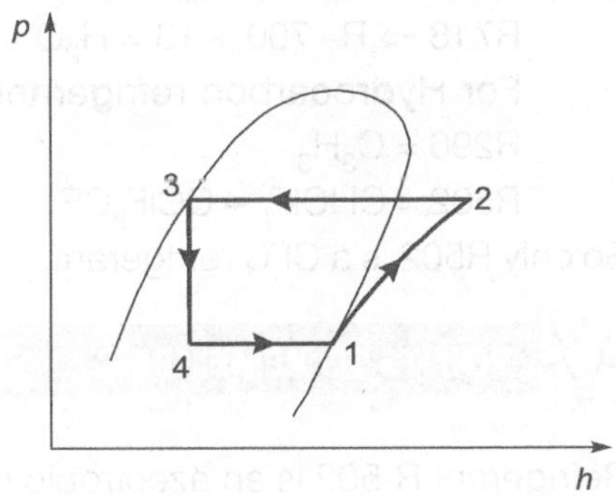

Coefficient of performance,

$$COP = \frac{\text{Refrigerating effect}(RE)}{\text{Work input } (w)}$$

where $RE = h_1 - h_4 = 232 - 116 = 116$ kJ/kg

and $w = h_2 - h_1 = 283 - 232 = 51$ kJ/kg

$\therefore$ $COP = \frac{116}{51} = \mathbf{2.27}$

1.10 (a)

From given R–134*a* tables, we get

$h_1 = 237$ kJ/kg; $h_2 = 276.45$ kJ/kg

$h_3 = h_4 = 95.5$ kJ/kg

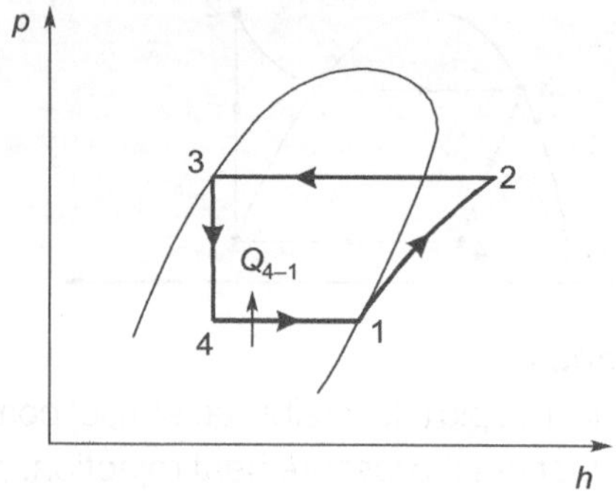

Mass flow rate: $\dot{m} = 0.2$ kg/s

Rate of heat extracted:

$$Q_{4\text{-}1} = \dot{m}(h_1 - h_4) = 0.2(237 - 95.5)$$
$$= \mathbf{28.3\ kW}$$

1.11 (c)

Power required for the compressor,

$$P = \dot{m}(h_2 - h_1) = 0.2(276.45 - 237)$$
$$= 7.89 \text{ kW} \approx \mathbf{7.9\ kW}$$

1.12 (c)

There are 3 cases for the designation of the refrigerant:

Case (i): Saturated Hydrocarbon refrigerant.

Case (ii): Unsaturated Hydrocarbon refrigerant.

Case (iii): Inorganic refrigerant.

$\therefore$ **For Inorganic refrigerant:**

R–700 + Molecular weight

i.e. R744 ⇒ R–700 + 44 = CO_2

R718 ⇒ R– 700 + 18 = H_2O

For Hydrocarbon refrigerant :

R290 = C_3H_8

R502 = $CHClF_2 + CClF_2CF_3$

So only R502 is a CFC refrigerant.

POINTS TO REMEMBER

Refrigerant R-502 is an azeotropic mixture of R-22 and R-115

1.13 Sol.

Given data:

$h_1 = 402$ kJ/kg, $h_2 = 432$ kJ/kg

$h_3 = h_4 = 237$ kJ/kg

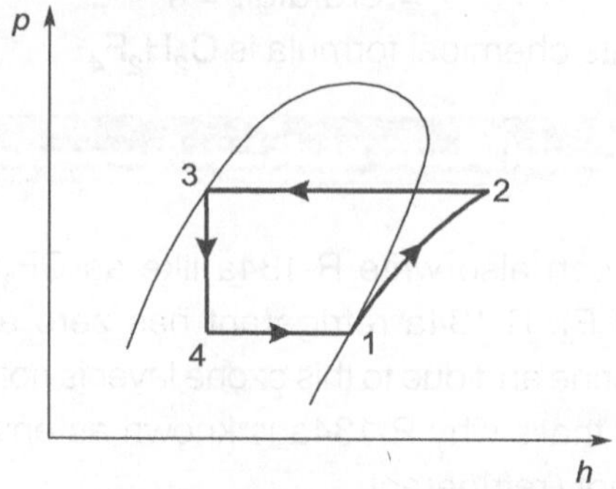

$$COP = \frac{h_1 - h_4}{h_2 - h_1} = \frac{402 - 237}{432 - 402} = \frac{165}{30}$$
$$= \mathbf{5.5}$$

1.14 Sol.

Given data: $m = 0.05$ kg/s

At $p_e = 0.14$ MPa; $h_1 = 236.04$ kJ/kg

$s_1 = 0.9322$ kJ/kgK

At $p_c = 0.8$ MPa

$h_3 = h_4 = h_f = 93.42$ kJ/kg

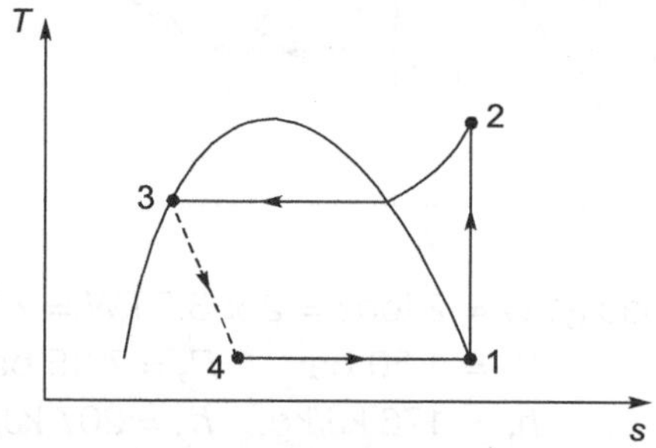

$h_2 = 272.05$ kJ/kg

Rate of heat rejection to the environment,

$$Q_{2\text{-}3} = \dot{m}[h_2 - h_3] = 0.05 \times [272.05 - 93.42]$$
$$= \mathbf{8.93\ kW}$$

1.15 Sol.

Given data: $T_e = 260$ K, $T_c = 310$ K

$c_{pl} = 4.8$ kJ/kgK, $h_{fg} = 1054$ kJ/kg at $T_c = 310$ K

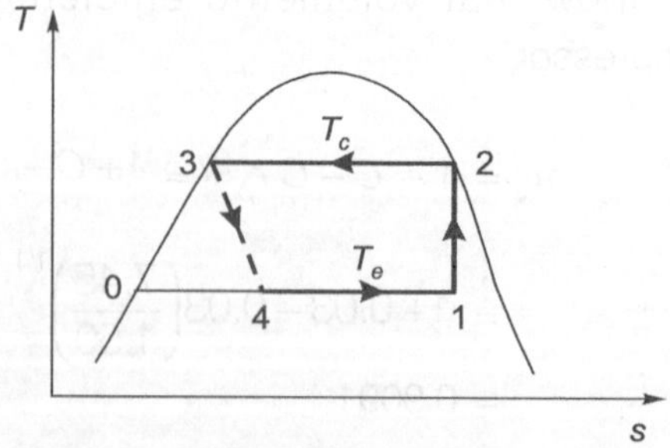

$$\frac{h_1 - h_0}{260} = s_1 - s_0 \qquad (\because s_1 = s_2)$$

$$\frac{h_1 - h_0}{260} = s_2 - s_0 \qquad ...(i)$$

$$s_2 - s_3 = \frac{1054}{310} \qquad ...(ii)$$

$$s_3 - s_0 = c_{liquid} \ln\frac{310}{260} \qquad ...(iii)$$

$$c_{liquid} = 4.8 \text{ kJ/kgK}$$

Adding equation (ii) and (iii)

$$(s_2 - s_3) + (s_3 - s_0) = s_2 - s_0$$

$$= 4.8\ln\frac{310}{260} + \frac{1054}{310} = 4.2442 \text{ kJ/kgK}$$

$\therefore$ From equation (i)

$$\frac{h_1 - h_0}{260} = 4.2442$$

$$h_1 - h_0 = \mathbf{1103.51 \text{ kJ/kg}}$$

1.16 Sol.

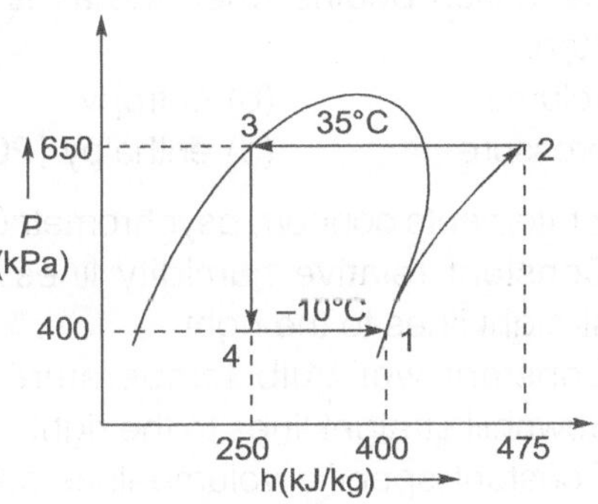

$$RC = 15 \text{ kW}$$

$$RC = \dot{m} \times (h_1 - h_4)$$

$$15 = \dot{m} \times (400 - 250)$$

$$\dot{m} = 0.1 \text{ kg/s}$$

$$w_{isentropic} = (h_2 - h_1) = (475 - 400)$$

$$= 75 \text{ kJ/kg}$$

$$\eta_C = \frac{w_{isentropic}}{w_{actual}}$$

$$w_{actual} = \frac{75}{0.75} = 100 \text{ kJ/kg}$$

$$P_{in} = \dot{m} \times w_{actual} = 0.1 \times 100$$

$$P_{in} = 10 \text{ kW}$$

1.17 (a)

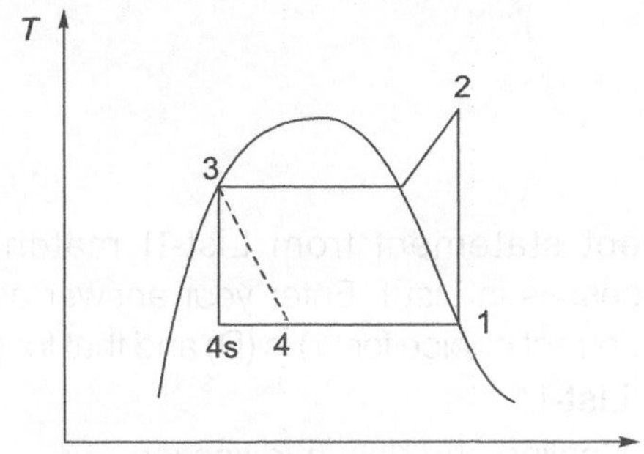

- R.E. increases.
- Work input decreases.
- COP increases.

1.18 (b)

Given: $h_1 = 250$ kJ/kg, $h_2 = 300$ kJ/kg, $h_{f\,evap} = 50$ kJ/kg

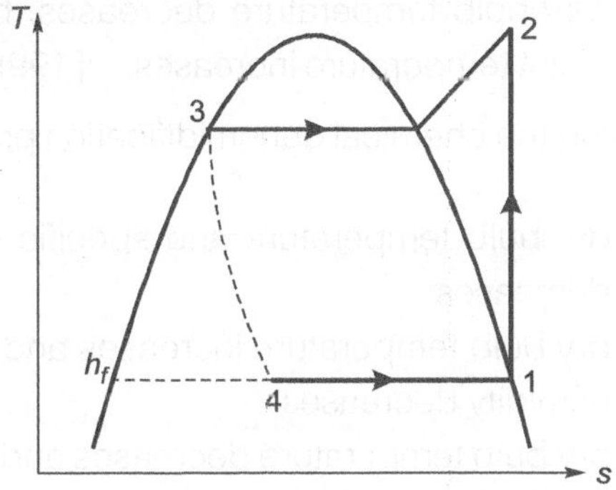

$$\text{COP} = \frac{h_1 - h_4}{h_2 - h_1} = \frac{250 - h_4}{300 - 250}$$

$$\Rightarrow \qquad h_4 = 100 \text{ kJ/kg}$$

For dryness fraction at point 4,

$$h_4 = h_{f\,evap} + x(h_1 - h_2)$$

$$100 = 50 + x(250 - 50)$$

$$x = \frac{50}{200} \Rightarrow x = 0.25$$

■■■■

2 Psychrometric Process

2.1 Select statement from List-II matching the processes in List-I. Enter your answer as D, C if the correct choice for (1) is (D) and that for (2) is (C)

List-I
1. Cooling and dehumidification
2. Chemical dehumidification

List-II
A. Dry-bulb temperature increase, but dew-point temperature decreases.
B. Dew-point temperature increase and dry bulb temperature remains unchanged.
C. Dry-bulb and wet-bulb temperature decrease
D. Dry-bulb temperature decreases, but dew-point temperature increases. **[1999 : 2 M]**

2.2 During the chemical dehumidification process of air

(a) dry bulb temperature and specific humidity decreases
(b) dry bulb temperature increases and specific humidity decreases
(c) dry bulb temperature decreases and specific humidity increases
(d) dry bulb temperature and specific humidity increases **[2004 : 1 M]**

2.3 Water at 42°C is sprayed into a stream of air at atmospheric pressure, dry bulb temperature of 40°C and a wet bulb temperature of 20°C. The air leaving the spray humidifier is not saturated. Which of the following statements is true?

(a) Air gets cooled and humidified
(b) Air gets heated and humidified
(c) Air gets heated and dehumidified
(d) Air gets cooled and dehumidified

[2005 : 1 M]

2.4 Various psychrometric processes are shown in the figure below.

Process in Figure	Name of the process
P. 0–1	i. Chemical dehumidification
Q. 0–2	ii. Sensible heating
R. 0–3	iii. Cooling and dehumidification
S. 0–4	iv. Humidification with steam injection
T. 0–5	v. Humidification with water injection

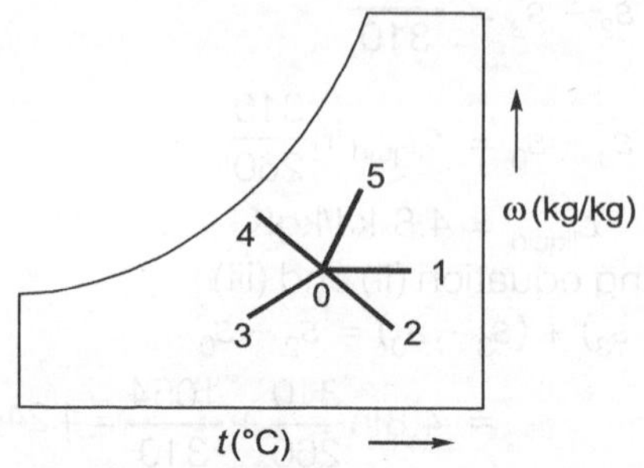

The matching pair are

(a) P-i, Q-ii, R-iii, S-iv, T-v
(b) P-ii, Q-i, R-iii, S-v, T-iv
(c) P-ii, Q-i, R-iii, S-iv, T-v
(d) P-iii, Q-iv, R-v, S-i, T-ii **[2005 : 2 M]**

2.5 Dew point temperature is the temperature at which condensation begins when the air is cooled at constant

(a) volume
(b) entropy
(c) pressure
(d) enthalpy **[2006 : 1 M]**

2.6 The statements concern psychrometric chart.

1. Constant relative humidity lines are uphill straight lines to the right
2. Constant wet bulb temperature lines are downhill straight lines to the right
3. Constant specific volume lines are downhill straight lines to the right
4. Constant enthalpy lines are coincident with constant wet bulb temperature lines

Which of the statements are correct?

(a) 2 and 3
(b) 1 and 2
(c) 1 and 3
(d) 2 and 4 **[2006 : 2 M]**

2.7 Atmospheric air at a flow rate of 3 kg/s (on dry basis) enters a cooling and dehumidifying coil with an enthalpy of 85 kJ/kg of dry air and a humidity ratio of 19 grams/kg of dry air. The air leaves the coil with an enthalpy of 43 kJ/kg of dry air and a humidity ratio of 8 grams/kg of dry air. If the condensate water leaves the coil with an enthalpy of 67 kJ/kg, the required cooling capacity of the coil in kW is

(a) 75.0
(b) 123.8
(c) 128.2
(d) 159.0 **[2007 : 2 M]**

2.8 Air (at atmospheric pressure) at a dry bulb temperature of 40°C and wet bulb temperature of 20°C is humidified in an air washer operating with continuous water recirculation. The wet bulb depression(i.e. the difference between the dry and wet bulb temperatures) at the exit is 25% of that at the inlet. The dry bulb temperature at the exit of the air washer is closest to

(a) 10°C (b) 20°C
(c) 25°C (d) 30°C [2008 : 2 M]

2.9 A sample of moist air at a total pressure of 85 kPa has a dry bulb temperature of 30°C (saturation vapour pressure of water = 4.24 kPa). If the air sample has a relative humidity of 65%, the absolute humidity (in gram) to water vapour per kg of dry air is _____. [2014 : 1 M, Set-3]

2.10 Moist air at 35°C and 100% relative humidity is entering a psychrometric device and leaving at 25°C and 100% relative humidity. The name of the device is

(a) Humidifier (b) Dehumidifier
(c) Sensible heater (d) Sensible cooler
[2014 : 1 M, Set-4]

2.11 Air in a room is at 35°C and 60% relative humidity (RH). The pressure in the room is 0.1 MPa. The saturation pressure of water at 35°C is 5.63 kPa. The humidity ratio of the air (in gram/kg of dry air) is _____. [2015 : 2 M, Set-3]

2.12 The partial pressure of water vapour in a moist air sample of relative humidity 70% is 1.6 kPa, the total pressure being 101.325 kPa. Moist air may be treated as an ideal gas mixture of water vapour and dry air. The relation between saturation temperature (T_s in K) and saturation pressure (P_s in kPa) for water is given by $\ln\left(\frac{p_s}{p_0}\right) = 14.317 - \frac{5304}{T_s}$

where p_0 = 101.325 kPa. The dry bulb temperature of the moist air sample (in °C) is____. [2016 : 2 M, Set-2]

2.13 In a mixture of dry air and water vapor at a total pressure of 750 mm of Hg, the partial pressure of water vapor is 20 mm of Hg. The humidity ratio of the air in grams of water vapor per kg of dry air (g_w/kg_{da}) is _________. [2016 : 2 M, Set-3]

2.14 If a mass of moist air contained in a closed metallic vessel is heated, then its

(a) relative humidity decreases
(b) relative humidity increases
(c) specific humidity increases
(d) specific humidity decreases
[2017 : 1 M, Set-2]

2.15 Ambient air is at a pressure of 100 kPa, dry bulb temperature of 30° C and 60% relative humidity, The saturation pressure of water at 30° C is 4.24 kPa. The specific humidity of air (in g/kg of dry air) is ____ (correct to two decimal places).

[2018 : 2 M, Set-2]

Answers Psychrometric Process

2.1 (1-C, 2-A)	2.2 (b)	2.3 (b)	2.4 (b)	2.5 (c)	2.6 (a)	2.7 (b)	2.8 (c)
2.9 (20.84)	2.10 (b)	2.11 (21.74)	2.12 (19.89)	2.13 (17.04)	2.14 (a)	2.15 (16.23)	

Explanations Psychrometric Process

2.1 Sol.

1–C, 2–A

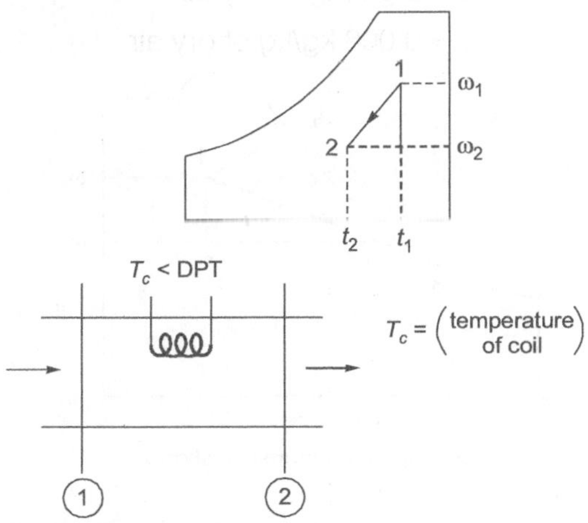

This process is possible when the coil temperature is less than dew point temperature of air. This process is generally followed in summer air conditioning.

Adiabatic chemical dehumidifications: Certain chemicals like silica gel and alumina absorbs moisture from air and hence its 'ω' value decreases, but at the same time latent heat of condensation is added to air and hence its dry bulb temperature increases.

DBT ↑ = ω↓

2.2 (b)

Chemical dehumidification process 1–2

DBT↑, ω↓

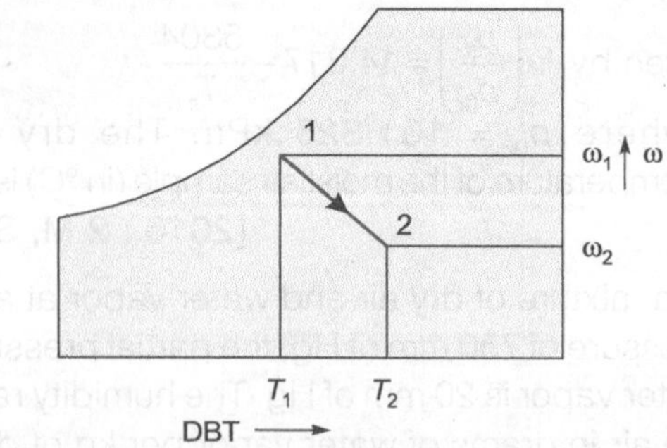

In chemical dehumidification, dry-bulb temperature increases and specific humidity decreases.

2.3 (b)

Given:

Spray water temperature = 42°C

Dry bulb temperature, DBT = 40°C

Wet bulb temperature, WBT = 20°C

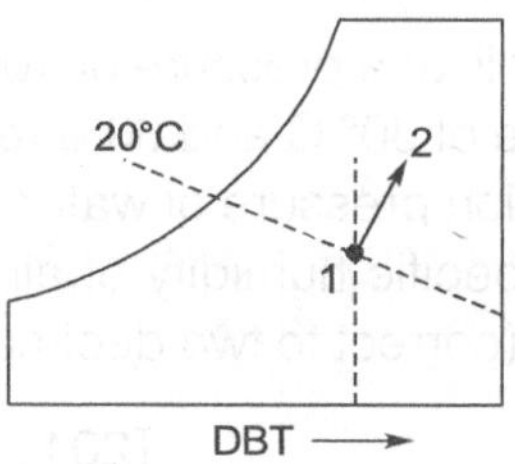

Here we see that, water spraying temperature is greater than that of air steam temperature. So air gets heated and humidified.

2.4 (b)

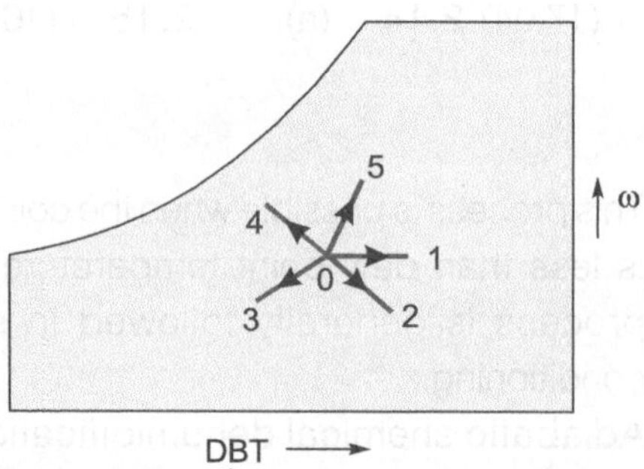

Process 0–1: DBT↑, ω = C; Sensible heating

Process 0–2: DBT↑, ω↓ : Chemical dehumidification

Process 0–3: DBT↓, ω↓: Cooling and dehumidification

Process 0–4: DBT↓, ω↑: Humidification with water injection

Process 0–5: DBT↑, ω↑: Humidification with steam injection

2.5 (c)

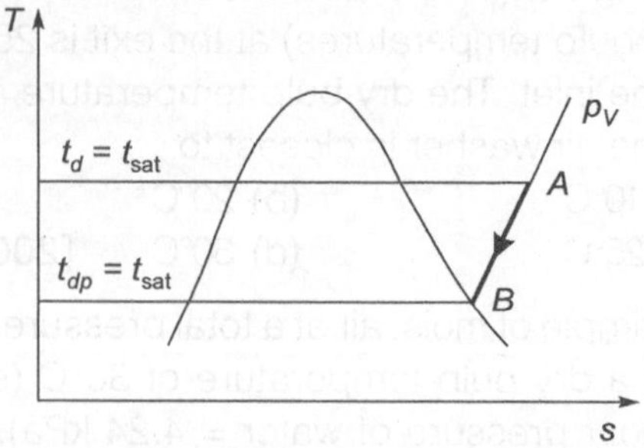

If a sample of unsaturated air, containing superheated water vapour, is cooled at constant pressure. The partial pressure (p_V) of each constituent remains constant until the water vapour reaches the saturated state as shown by point B. At this point B, the first drop of dew will be formed and hence the temperature at point B is called **dew point temperature**.

2.6 (a)

1. Constant wet bulb temperature line and specific volume lines both are straight lines but different slopes.
2. Constant relative humidity lines are curved and uphill to right.
3. Constant enthalpy and constant wet bulb temperature lines have little difference in slope.

2.7 (b)

Given data:

Mass flow rate of dry air,

$$\dot{m}_a = 3 \text{ kg/s}$$

$$h_1 = 85 \text{ kJ/kg of dry air}$$

$$\omega_1 = 19 \text{ gm/kg of dry air} = 0.019 \text{ kg/kg of dry air}$$

$$h_2 = 43 \text{ kJ/kg of dry air}$$

$$\omega_2 = 8 \text{ gm/kg of dry air} = 0.008 \text{ kg/kg of dry air}$$

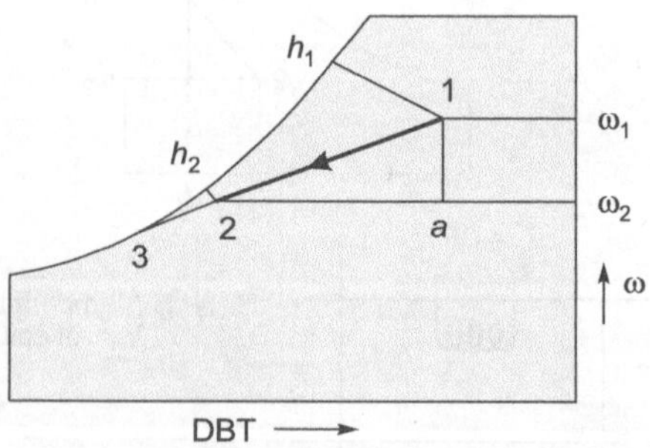

Cooling and dehumidification

Enthalpy of condensate water leaves the cooling coil: h_{fg} = 67 kJ/kg

$$\omega_1 = \frac{\dot{m}_{v1}}{\dot{m}_a}$$

or $\dot{m}_{v1} = \omega_1 \dot{m}_a = 0.019 \times 3$ kg/s

= 0.057 kg/s, mass of water vapour at inlet

Similarly,

$\dot{m}_{v2} = \omega_2 \dot{m}_a = 0.008 \times 3$

= 0.024 kg/s, mass of water vapour at outlet

Mass flow rate of condensate,

$\dot{m}_w$ = mass of water vapour at inlet – mass of water vapour at outlet

$\dot{m}_w = \dot{m}_{v1} + \dot{m}_{v2} = 0.057 - 0.024$

= 0.033 kg/s

Cooling capacity of the coil,

$$Q = \dot{m}_a (h_1 - h_2) - \dot{m}_w h_{fg}$$

where $\dot{m}(h_1 - h_2)$ = sensible heat removed

$\dot{m}_w h_{fg}$ = Latent heat removed

$\therefore \quad Q = 3(85 - 43) - 0.033 \times 67$

= **123.79 kW**

Alternatively

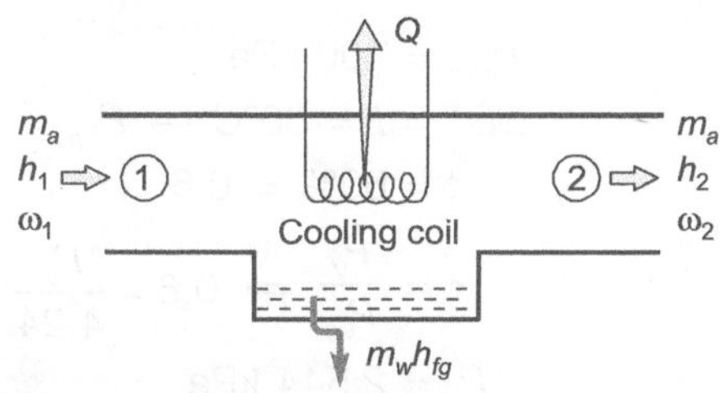

Mass flow rate of condensate,

$\dot{m}_w = \dot{m}_a(\omega_1 - \omega_2) = 3 \times (0.019 - 0.008)$

= 0.033 kg/s

Applying energy balance equation, we get

$\dot{m}_a h_1 = \dot{m}_a h_2 + \dot{m}_w h_{fg} + Q$

$3 \times 85 = 3 \times 43 + 0.033 \times 67 + Q$

Q = **123.8 kW**

2.8 (c)

Given data:

$(DBT)_i$ = 40°C

$(WBT)_i$ = 20°C

Web bulb depression,

$(WBD)_i = (DBT)_i - (WBT)_i$

= 40 – 20 = 20°C

Wet bulb depression at outlet,

$(WBD)_o = (DBT)_o - (WBT)_o$

By given condition, $(WBD)_o = 25\%(WBD)_i$

$\therefore (DBT)_o - (WBT)_o = 0.25 \times 20$

$(DBT)_o = 0.25 \times 20 + (WBT)_o = 5 + (WBT)_i$

where $(WBT)_o = (WBT)_i$ for humidified in an air washer

$\therefore \quad (DBT)_o = 5 + 20 =$ **25°C**

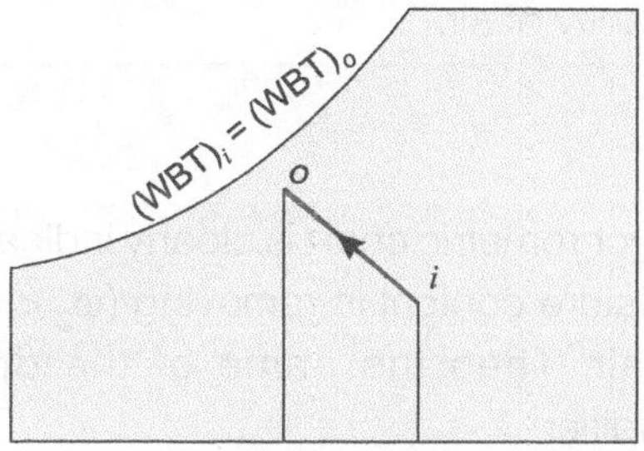

Cooling and humidification

POINTS TO REMEMBER

Air washer: An air washer is a device for conditioning air. In an air washer, air comes in direct contact with a spray of water and there will be an exchange of heat and mass (water vapour) between air and water. The outlet condition of air depends upon the temperature of water sprayed in the air washer. Hence, by controlling the water temperature exernally, it is possible to control the outlet conditions of air, which then can be used for air-conditioning purposes.

2.9 Sol.

Total pressure: p = 85 kPa

Dry bulb temperature,

T_{db} = 30°C

p_s = 4.24 kPa

ϕ = 65% = 0.65

Relative humidity,

$$\phi = \frac{p_v}{p_s} ; 0.65 = \frac{p_v}{4.24}$$

or $\quad p_v = 0.65 \times 4.24 = 2.756$ kPa

Absolute humidity,

$$\omega = \frac{0.622\, p_v}{p - p_v} = \frac{0.622 \times 2.756}{85 - 2.756}$$

= 0.02084 kg of w.v/kg of dry air

= **20.84 gram of w.v./kg of dry air**

POINTS TO REMEMBER

- For saturated air, relative humidity is 1 or 100%.
- Absolute humidity, ω gives actual mass of vapour.
- Relative humidity, ϕ gives moisture absorbing capacity of air.

2.10 (b)

The psychrometric chart is clearly indicating that the moisture content is removing ($\omega_2 < \omega_1$) from moist air. Then the name of the device is dehumidifier.

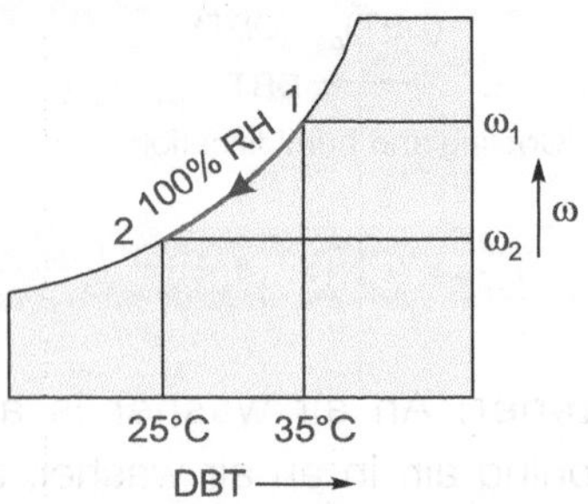

2.11 Sol.

Given data: $T_{db} = 35°C, \phi = 60\% = 0.60$

$p = 0.1$ MPa = 100 kPa, $p_s = 5.63$ kPa at 35°C

Relative humidity, $\phi = \frac{p_v}{p_s}$; $0.60 = \frac{p_v}{5.63}$

$p_v = 0.60 \times 5.63 = 3.378$ kPa

Humidity ratio, $\omega = \frac{0.622\, p_v}{p - p_v} = \frac{0.622 \times 3.378}{100 - 3.378}$

= 0.02174 kg of *w.v.*/kg of dry air

= **21.74 gram of *w.v.*/kg of dry air**

2.12 Sol.

Given data:

Relative humidity, $\phi = 70\% = 0.70$

Partial pressure, $p_v = 1.6$ kPa

Total pressure, $p_0 = 101.325$ kPa

We know that,

$$\phi = \frac{p_v}{p_s}$$

(where P_s = saturation pressure)

$$0.70 = \frac{1.6}{p_s} \text{ or } p_s = \frac{1.6}{0.70} = 2.2857 \text{ kPa}$$

Temperature corresponding to saturation pressure is dry bulb temperature,

$$\ln\left\{\frac{p_s}{p_o}\right\} = 14.317 - \frac{5304}{T_s}$$

$$\ln\left\{\frac{2.2857}{101.325}\right\} = 14.317 - \frac{5304}{T_s}$$

$$-3.79166 = 14.317 - \frac{5304}{T_s}$$

$$\frac{5304}{T_s} = 14.317 + 3.79166$$

$$\frac{5304}{T_s} = 18.10866$$

or $T_s = 292.89 \text{ K} = (292.89 - 273)°C$

$= \mathbf{19.89°C}$

2.13 Sol.

Given data:

Total pressure, p = 750 mm of Hg

Partial pressure, p_v = 20 mm of Hg

We know that humidity ratio,

$$\omega = \frac{0.622\, p_v}{p - p_v} \text{ kg } w.v./\text{kg } d.a.$$

$$= \frac{0.622 \times 20}{750 - 20}$$

= 0.01704 kg *w.v.*/kg of *d.a.*

= **17.04 g of *w.v.*/kg of *d.a.***

2.15 Sol.

$P_{atm} = 100$ kPa

$DBT = t = 30°C \rightarrow P_{vs} = 4.24$ kPa

$\phi = 60\% = 0.6$

$$\phi = \frac{P_v}{P_{vs}} \Rightarrow 0.6 = \frac{P_v}{4.24}$$

$P_V = 2.544$ kPa

$$\omega = 0.622 \times \frac{P_v}{P_{atm} - P_v}$$

$$\omega = 0.622 \times \frac{2.544}{100 - 2.544}$$

= 0.01623 kg/kg of dry air.

ω = **16.23 gram/kg of dry air**

■■■■

3 Heat Pumps and Cycles

3.1 Clearance volume of a reciprocating compressor is 100 ml, and the volume of the cylinder at bottom dead centre is 1.0 litre. The clearance ratio of the compressor is

(a) 1/11 (b) 1/10
(c) 1/9 (d) 1/1.1 **[1997 : 2 M]**

3.2 **List-I**

A. Liquid to suction heat exchanger
B. Constant volume heat addition
C. Normal shock
D. Ammonia–water

List-II

1. Vapour absorption refrigeration
2. Vapour compression refrigeration
3. Diesel cycle
4. Otto cycle
5. Converging nozzle
6. Converging diverging nozzle **[1997 : 2 M]**

3.3 An industrial heat pump operates between the temperatures of 27°C and –13°C. The rates of heat addition and heat rejection are 750 W and 1000 W, respectively. The COP for the heat pump is

(a) 7.5 (b) 6.5
(c) 4.0 (d) 3.0 **[2003 : 1 M]**

3.4 A heat engine having an efficiency of 70% is used to drive a refrigerator having a coefficient of performance of 5. The energy absorbed from low temperature reservoir by the refrigerator for each kJ of energy absorbed from high temperature source by the engine is

(a) 0.14 kJ (b) 0.71 kJ
(c) 3.5 kJ (d) 7.1 kJ **[2004 : 2 M]**

3.5 A vapour absorption refrigeration system is a heat pump with three thermal reservoirs as shown in the figure. A refrigeration effect of 100 W is required at 250 K when the heat source available is at 400 K. Heat rejection occurs at 300 K. The minimum value of heat required (in W) is

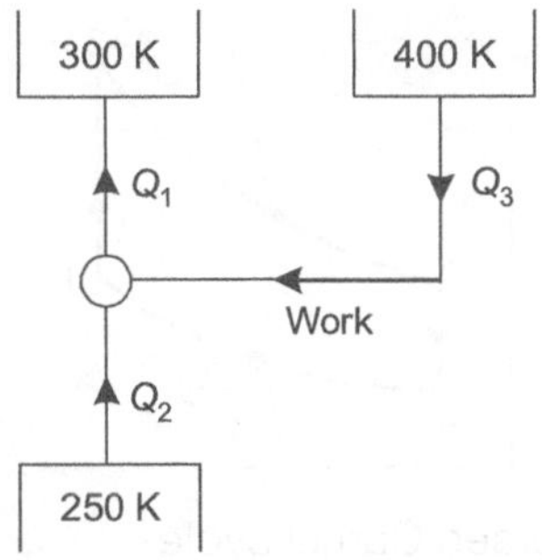

(a) 167 (b) 100
(c) 80 (d) 20 **[2005 : 2 M]**

3.6 A thin layer of water in a field is formed after a farmer has watered it. The ambient air conditions are: temp. 20°C and relative humidity 5%. An extract of steam tables is given below.

Temp.(°C)	–15	–10	–5	0.01	5	10	15	20
Saturation Pressure(kPa)	0.10	0.26	0.40	0.61	0.87	1.23	1.71	2.34

Neglecting the heat transfer between the water and the ground, the water temperature in the field after phase equilibrium is reached equals

(a) 10.3°C (b) – 10.3°C
(c) –14.5°C (d) 14.5°C **[2006 : 2 M]**

3.7 A reversed Carnot cycle refrigerator maintains a temperature of – 5°C. The ambient air temperature is 35°C. The heat gained by the refrigerator at a continuous rate is 2.5 kJ/s. The power (in watt) required to pump this heat out continuously is ____

[2014 : 1 M, Set-4]

3.8 A heat pump with refrigerant R22 is used for space heating between temperature limits of – 20°C and 25°C. The heat required is 200 MJ/h. Assume specific heat of vapour at the time of discharge as 0.98 kJ/kgK. Other relevant properties are given below. The enthalpy (in kJ/kg) of the refrigerant at isentropic compressor discharge is ______.

Saturation temperature	Pressure	Specific enthalpy		Specific entropy	
T_{sat}(°C)	P(MN/m^2)	h_f(kJ/kg)	h_g(kJ/kg)	s_f(kJ/kgK)	s_g(kJ/kgK)
– 20	0.2448	177.21	397.53	0.9139	1.7841
25	1.048	230.27	413.02	1.1047	1.7183

[2014 : 2 M, Set-2]

3.9 The COP of a Carnot heat pump operating between 6°C and 37°C is ______.

[2015 : 1 M, Set-2]

3.10 The thermodynamic cycle shown in figure (*T-s* diagram) indicates

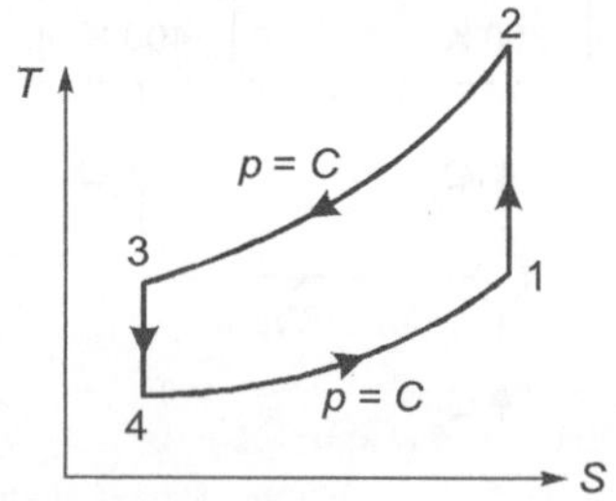

(a) reversed Carnot cycle
(b) reversed Brayton cycle
(c) vapor compression cycle
(d) vapor absorption cycle

[2015 : 1 M, Set-3]

3.11 The heat removal rate from a refrigerated space and the power input to the compressor are 7.2 kW and 1.8 kW, respectively. The coefficient of performance (COP) of the refrigerator is______.

[2016 : 1 M, Set-2]

3.12 If a reversed Carnot cycle operates between the temperature limits of 27°C and –3°C, then the ratio of the COP of a refrigerator to that of a heat pump (COP of refrigerator/COP of heat pump) based on the cycle is ________ (round off to 2 decimal places).

[2020 : 1 M, Set-2]

3.13 Consider an ideal vapour compression refrigeration cycle working on R-134a refrigerant. The COP of the cycle is 10 and the refrigeration capacity is 150 kJ/kg. The heat rejected by the refrigerant in the condenser is________ kJ/kg (round off to the nearest integer).

[2021 : 1 M, Set-2]

Answers Heat Pumps and Cycles

3.1 (c) 3.2 (A-2,B-4,C-6,D-1) 3.3 (c) 3.4 (c) 3.5 (c) 3.6 (c)
3.7 (373.13) 3.8 (433.3) 3.9 (10) 3.10 (b) 3.11 (4) 3.12 (0.9) 3.13 (165)

Explanations Heat Pumps and Cycles

3.1 (c)

Given data:
Clearance volume: $V_c = 100$ ml

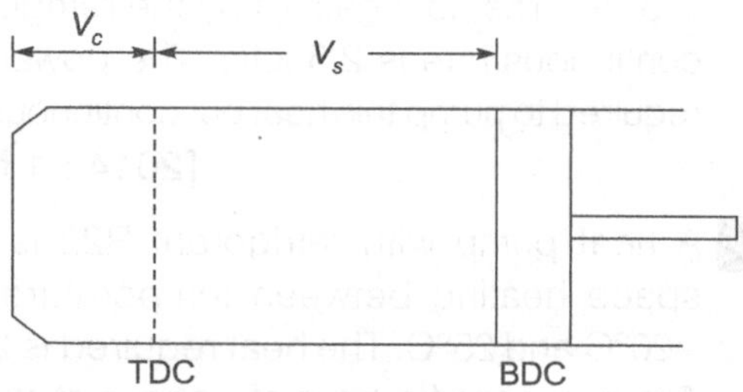

Volume of the cylinder at BDC,

$$V_c + V_s = 1 \text{ litre}$$
$$V_c + V_s = 1000 \text{ ml}$$
$$100 + V_s = 1000$$

or $$V_s = 1000 - 100 = 900 \text{ ml}$$

Clearance ratio,

$$C = \frac{V_c}{V_s} = \frac{100}{900} = \frac{1}{9}$$

3.2 Sol.

Liquid to suction heat exchanger – vapour compression refrigeration
Constant volume heat addition – Otto cycle
Normal shock – converging – diverging nozzle
Ammonia–water – vapour absorption refigeration.

POINTS TO REMEMBER

In vapour absorption refrigeration system, LiBr–H_2O can also be used, in which H_2O acts as a refrigerant and LiBr acts as a absorbent.

3.3 (c)

Given data:

$$T_1 = 27°C = (27 + 273) \text{ K} = 300 \text{ K}$$
$$T_2 = -13°C = (-13 + 273) \text{ K} = 260 \text{ K}$$
$$Q_2 = 750 \text{ W}$$
$$Q_1 = 1000 \text{ W}$$

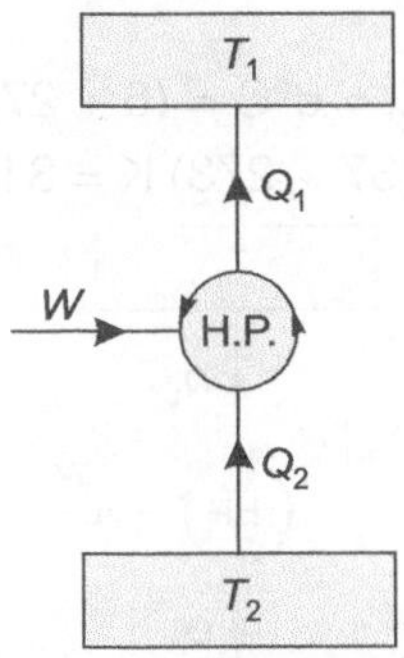

Heat pump

$$(COP)_{HP} = \frac{Q_1}{Q_1 - Q_2} \text{ for actual heat pump}$$

$$= \frac{1000}{1000 - 750} = \frac{1000}{250} = 4$$

3.4 (c)

Given data: $\eta = 70\% = 0.70$

$(COP)_R = 5$

$Q_3 - ?$

$Q_1 = 1$ kJ

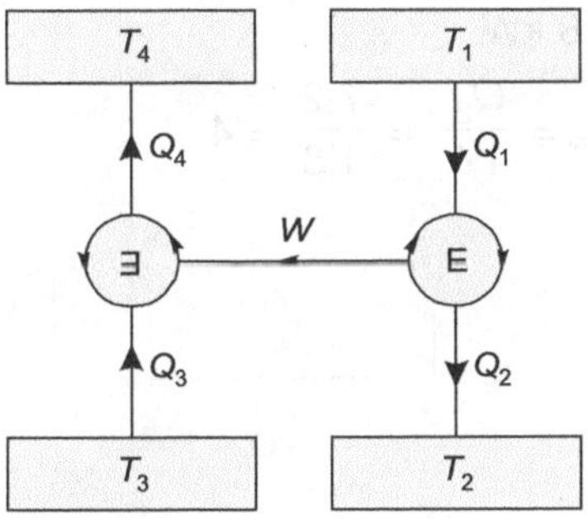

For heat engine,

$$\eta = 1 - \frac{Q_2}{Q_1}$$

$$0.70 = 1 - \frac{Q_2}{1}$$

or $Q_2 = 0.3$ kJ

$W = Q_1 - Q_2 = 1 - 0.3 = 0.7$ kJ

For refrigerator,

$$(COP)_R = \frac{Q_3}{W}$$

$$5 = \frac{Q_3}{0.7}$$

or $Q_3 = 0.7 \times 5 = \mathbf{3.5\ kJ}$

3.5 (c)

Method I:

Given data: $T_1 = 300$ K; $T_2 = 250$ K; $T_3 = 400$ K

Refrigerating effect,

$Q_2 = 100$ W

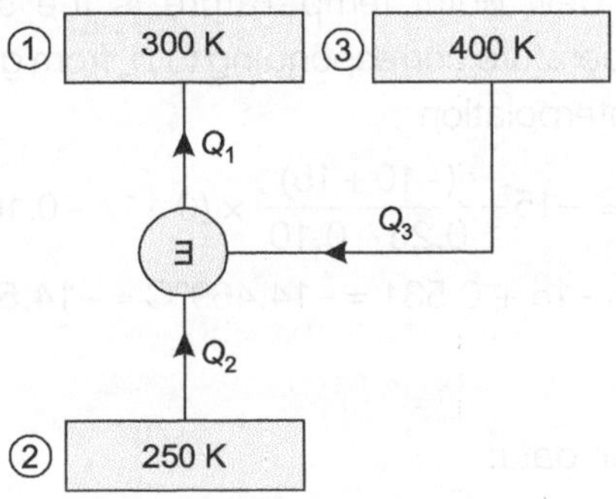

$$COP = \left(\frac{T_2}{T_1 - T_2}\right)\left(\frac{T_3 - T_1}{T_3}\right)$$

$$= \left(\frac{250}{300 - 250}\right)\left(\frac{400 - 300}{400}\right)$$

$$= 1.25$$

also $COP = \dfrac{Q_2}{Q_3}$

$\therefore \quad 1.25 = \dfrac{100}{Q_3}$

or $Q_3 = \dfrac{100}{1.25} = \mathbf{80\ W}$

Method II:

Using Clausius inequality;

$$\oint \frac{\delta Q}{T} \le 0$$

For the minimum value of heat required, the system must be reversible, i.e.

$$\oint \frac{\delta Q}{T} = 0$$

$$\therefore \quad \frac{Q_2}{250} + \frac{Q_3}{400} = \frac{Q_1}{300} \quad \text{...(i)}$$

Also, $Q_3 + Q_2 = Q_1$ ...(ii)

From equation (i) and (ii)

After solving;

$Q_3 = \mathbf{80\ W}$ and $Q_1 = 180$ W

3.6 (c)

Given data: $\phi = 5\% = 0.05$, $T_{db} = 20°C$

From table,

At 20°C, $p_s = 2.34$ kPa

$$\phi = \frac{p_v}{p_s}$$

$$0.05 = \frac{p_v}{2.34}$$

or $p_v = 0.05 \times 2.34 = 0.117$ kPa

The dew point temperature is the saturation temperature corresponding to p_v from given table by interpolation

$$T_{dp} = -15 + \frac{(-10+15)}{0.26-0.10} \times (0.117-0.10)$$

$$= -15 + 0.531 = -14.469°C \approx \mathbf{-14.5°C}$$

3.7 Sol.

Given data:

$T_2 = -5°C = (-5 + 273)$ K = 268 K

$T_1 = 35°C = (35 + 273)$ K = 308 K

$Q_2 = 2.5$ kJ/s

= 2.5 kW = 2500 W

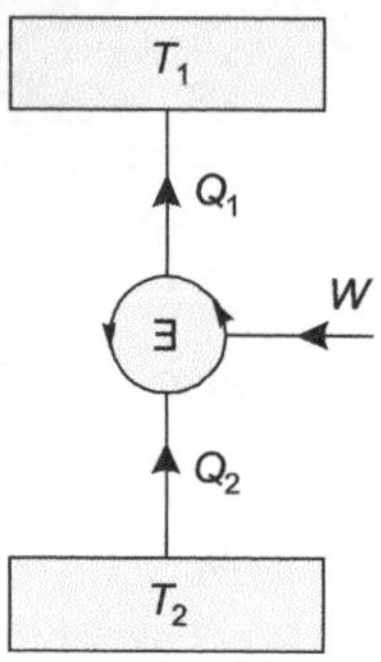

Refrigerator

$$(COP)_R = \frac{T_2}{T_1 - T_2} = \frac{268}{308-268} = 6.7$$

also $(COP)_R = \dfrac{Q_2}{W}$

$\therefore$ $6.7 = \dfrac{2500}{W}$

or $W = \dfrac{2500}{6.7} = \mathbf{373.13\ W}$

3.8 Sol.

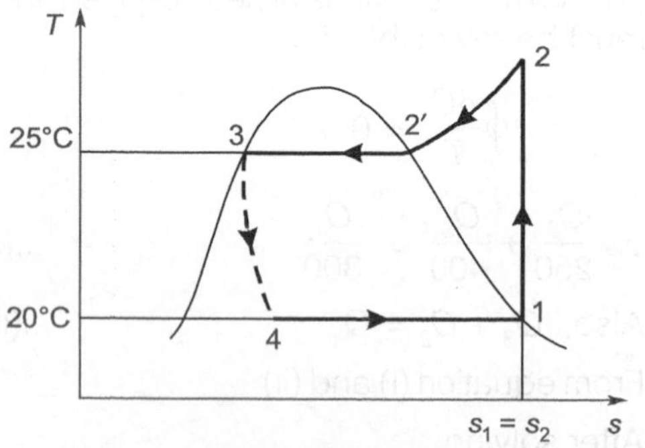

$$s_2 - s_{2'} = c_p \log_e \frac{T_2}{T_{2'}}$$

$s_1 = s_2$ (for isentropic process)

$s_2 = 1.7841$ kJ/kgK; $s_{2'} = 1.7183$ kJ/kgK

$c_p = 0.98$ kJ/kgK

$$\therefore \log_e \frac{T_2}{T_{2'}} = \frac{s_2 - s_{2'}}{c_p} = \frac{1.7841-1.7183}{0.98} = 0.0671$$

or $T_2 = 1.0694 \times T_{2'} = 1.0694 \times 298$

$= 318.695$ K

$h_2 - h_{2'} = c_p(T_2 - T_{2'})$

$h_2 = 413.02 + 0.98\,(318.695 - 298)$

$h_2 = \mathbf{433.3\ kJ/kg}$

3.9 Sol.

Given data: $T_2 = 6°C = (6 + 273)$ K = 279 K

$T_1 = 37°C = (37 + 273)$ K = 310 K

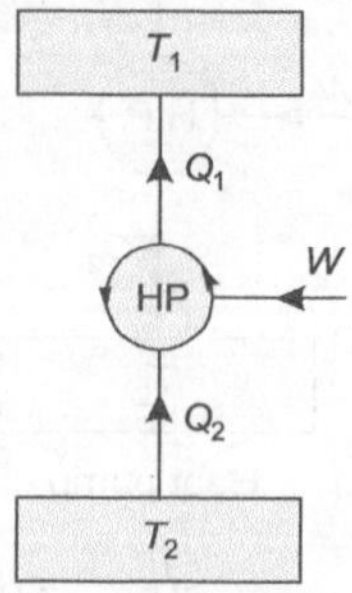

Heat pump

$$(COP)_{HP} = \frac{T_1}{T_1 - T_2} = \frac{310}{310-279} = \mathbf{10}$$

3.10 (b)

Reversed Brayton cycle is shown in the figure in question.

3.11 Sol.

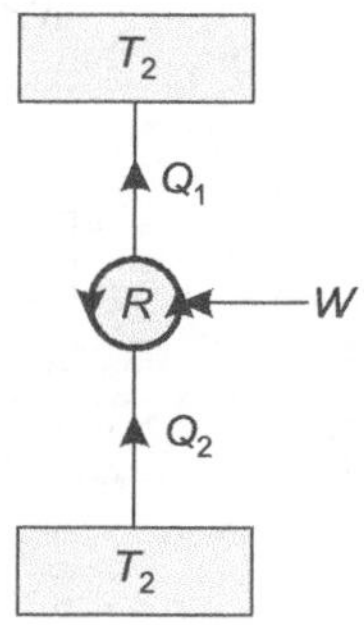

Given data: $Q_2 = 7.2$ kW;

$W = 1.8$ kW

$$(COP)_R = \frac{Q_2}{W} = \frac{7.2}{1.8} = \mathbf{4}$$

3.12 Sol.

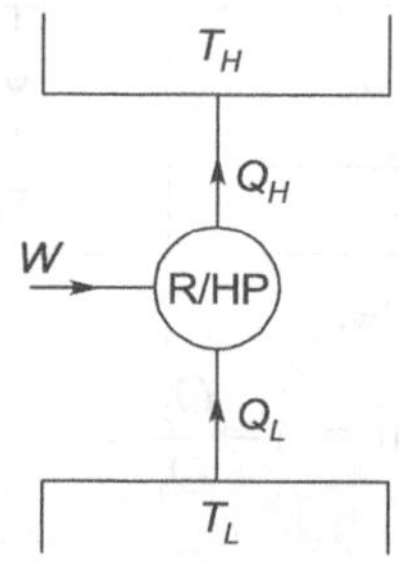

Given: $T_H = 27°C = 300$ K, $T_L = -3°C = 270$ K

$$\frac{COP_{ref}}{COP_{HP}} = \frac{\dfrac{T_L}{T_H - T_L}}{\dfrac{T_H}{T_H - T_L}} = \frac{T_L}{T_H} = \frac{270}{300} = 0.9$$

3.13 Sol.

RE = 150 kJ/kg

COP = 10

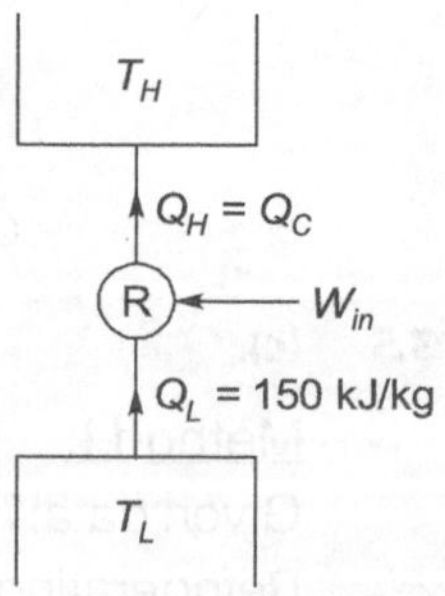

$$COP = \frac{Q_L}{W_{in}}$$

$\Rightarrow$ $W_{in} = 15$ kJ/kg

$Q_H = Q_c = Q_L + W_{in}$

$= 150 + 15$

$= 165$ kJ/kg

4 Properties of Moist Air

4.1 For air at a given temperature, as the relative humidity is increased isothermally,
(a) the wet bulb temperature and specific enthalpy increase
(b) the wet bulb temperature and specific enthalpy decrease
(c) the wet bulb temperature increases and specific enthalpy decreases
(d) the wet bulb temperature decreases and specific enthalpy increases **[2001 : 2 M]**

4.2 For air with a relative humidity of 80%
(a) the dry bulb temperature is less than the wet bulb temperature
(b) the dew point temperature is less than wet bulb temperature
(c) the dew point and web bulb temperature are equal
(d) the dry bulb and dew point temperatures are equal **[2003 : 1 M]**

4.3 Dew point temperature of air at one atmospheric pressure (1.013 bar) is 18°C. The air dry bulb temperature is 30°C. The saturation pressure of water at 18°C and 30°C are 0.02062 bar and 0.04241 bar respectively. The specific heat of air and water vapour respectively are 1.005 and 1.88 kJ/kgK and the latent heat of vaporization water of 0°C is 2500 kJ/kg. The specific humidity (kg/kg of dry air) and enthalpy (kJ/kg of dry air) of this moist air respectively, are
(a) 0.01051, 52.64 (b) 0.01291, 63.15
(c) 0.01481, 78.60 (d) 0.01532, 81.40
[2004 : 2 M]

4.4 For a typical sample of ambient air (at 30°C, 75% relative humidity and standard atmospheric pressure), the amount of moisture in kg per kg of dry air will be approximately.
(a) 0.002 (b) 0.027
(c) 0.25 (d) 0.75 **[2005 : 1 M]**

4.5 A building has to be maintained at 21°C (dry bulb) and 14.5°C (wet bulb). The dew point temperature under these conditions is 10.17°C. The outside temperature is –23°C (dry bulb) and the internal and external surface heat transfer coefficients are 8 W/m^2K and 23 W/m^2K respectively. If the building wall has a thermal conductivity of 1.2 W/mK, the minimum thickness (in m) of the wall required to prevent condensation is
(a) 0.471 (b) 0.407
(c) 0.321 (d) 0.125 **[2007 : 2 M]**

4.6 Moist air at a pressure of 100 kPa is compressed to 500 kPa and then cooled to 35°C in an aftercooler. The air at the entry to the aftercooler is unsaturated and becomes just saturated at the exit of the aftercooler. The saturation pressure of water at 35°C is 5.628 kPa. The partial pressure of water vapour (in kPa) in the moist air entering the compressor is closest to
(a) 0.57 (b) 1.13
(c) 2.26 (d) 4.52 **[2008 : 2 M]**

4.7 A moist air sample has dry bulb temperature of 30°C and specific humidity of 11.5 g water vapour per kg dry air. Assume molecular weight of air as 28.93. If the saturation vapour pressure of water at 30°C is 4.24 kPa and the total pressure is 90 kPa, then the relative humidity (in %) of air sample is
(a) 50.5 (b) 38.5
(c) 56.5 (d) 68.5 **[2010 : 2 M]**

4.8 If a mass of moist air in an airtight vessel is heated to a higher temperature, then
(a) specific humidity of the air increases
(b) specific humidity of the air decreases
(c) relative humidity of the air increases
(d) relative humidity of the air decreases
[2011 : 1 M]

4.9 A room contains 35 kg of dry air and 0.5 kg of water vapor. The total pressure and temperature of air in the room are 100 kPa and 25°C respectively. Given that the saturation pressure for water at 25°C, is 3.17 kPa, the relative humidity of the air in the room is
(a) 67% (b) 55%
(c) 83% (d) 71% **[2012 : 2 M]**

4.10 The pressure, dry bulb temperature and relative humidity of air in a room are 1 bar, 30°C and 70%, respectively. If the saturated steam pressure at 30°C is 4.25 kPa, the specific humidity of the room air in kg of water vapour/kg dry air is
(a) 0.0083 (b) 0.0101
(c) 0.0191 (d) 0.0232 **[2013 : 1 M]**

4.11 A stream of moist air (mass flow rate = 10.1 kg/s) with humidity ratio of 0.01 kg/kg dry air mixes with a second stream of superheated water vapour flowing at 0.1 kg/s. Assuming proper and uniform mixing with no condensation, the humidity ratio of the final stream in (kg/kg dry air) is ______.

[2015 : 1 M, Set-1]

4.12 Moist air is treated as an ideal gas mixture of water vapor and dry air (molecular weight of air = 28.84 and molecular weight of water = 18). At a location, the total pressure is 100 kPa, the temperature is 30°C and the relative humidity is 55%. Given that the saturation pressure of water at 30°C is 4246 Pa, the mass of water vapor per kg of dry air is ______ grams.

[2017 : 2 M, Set-1]

4.13 Moist air at 105 kPa, 30°C and 80% relative humidity flows over a cooling coil in an insulated air-conditioning duct. Saturated air exits the duct at 100 kPa and 15°C. The saturation pressure of water at 30°C and 15°C are 4.24 kPa and 1.7 kPa respectively. Molecular weight of water is 18 g/mol and that of air is 28.94 g/mol. The mass of water condensing out from the duct is ______ g/kg of dry air (round off to 2 decimal places).

[2020 : 2 M, Set-2]

4.14 An air-conditioning system provides a continuous flow of air to a room using an intake duct and an exit duct, as shown in the figure. To maintain the quality of the indoor air, the intake duct supplies a mixture of fresh air with a cold air stream. The two stream are mixed in an insulated mixing chamber located upstream of the intake duct. Cold air enters the mixing chamber at 5°C, 105 kPa with a volume flow rate of 1.25 m³/s during steady state operation. Fresh air enters the mixing chamber at 34°C and 105 kPa. The mass flow rate of the fresh air is 1.6 times of the cold air stream. Air leaves the room through the exit duct at 24°C.

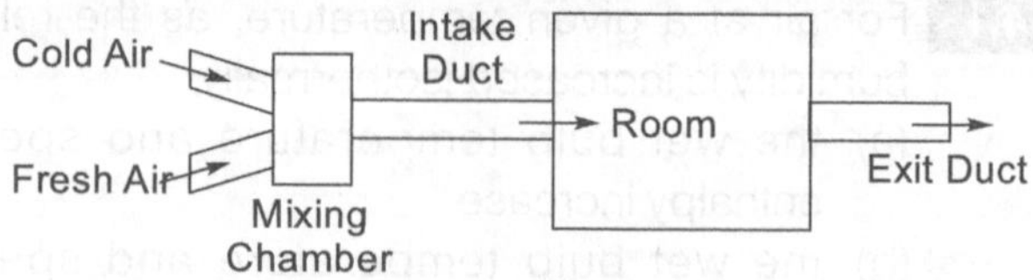

Assuming the air behaves as an ideal gas with c_p = 1.005 kJ/kgK and R = 0.287 kJ/kgK, the rate of heat gain by the air from the room is ______ kW. (Round off to two decimal places)

[2021 : 2 M, Set-1]

4.15 The relative humidity of ambient air at 300 K is 50% with a partial pressure of water vapour equal to p_v. The saturation pressure of water at 300 K is p_{sat}. The correct relation for the air-water mixture is

(a) $p_v = 0.622\, p_{sat}$ (b) $p_v = 2\, p_{sat}$
(c) $p_v = 0.5\, p_{sat}$ (d) $p_v = p_{sat}$

[2021 : 1 M, Set-1]

4.16 Ambient pressure, temperature, and relative humidity at a location are 101 kPa, 300 K, and 60%, respectively. The saturation pressure of water at 300 K is 3.6 kPa. The specific humidity of ambient air is______ g/kg of dry air.

(a) 13.6 (b) 35.1
(c) 21.4 (d) 21.9

[2021 : 2 M, Set-2]

Answers Properties of Moist Air

4.1 (a)	4.2 (b)	4.3 (b)	4.4 (b)	4.5 (b)	4.6 (b)	4.7 (b)	4.8 (d)
4.9 (d)	4.10 (c)	4.11 (0.02)	4.12 (14.92)	4.13 (10)	4.14 (4.96)	4.15 (c)	4.16 (a)

Explanations Properties of Moist Air

4.1 (a)

Relative humidity is increased isothermally,

$\phi\uparrow, T_{wb}\uparrow$

As the relative humidity is increased isothermally from state 1 to state 2 for the air, wet bulb temperature and specific enthalpy both increases.

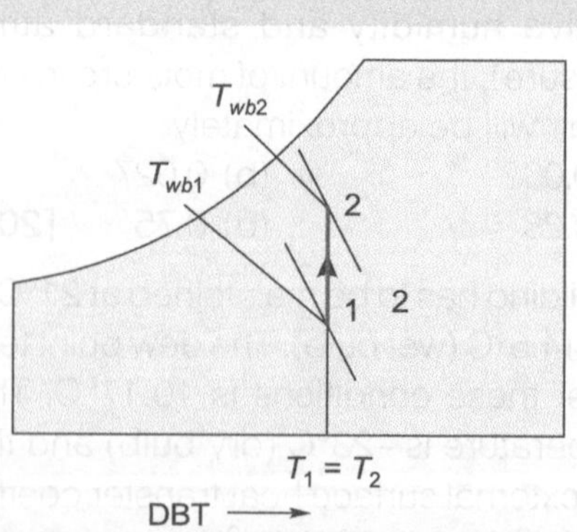

Relative humidity is increased isothermally

4.2 (b)

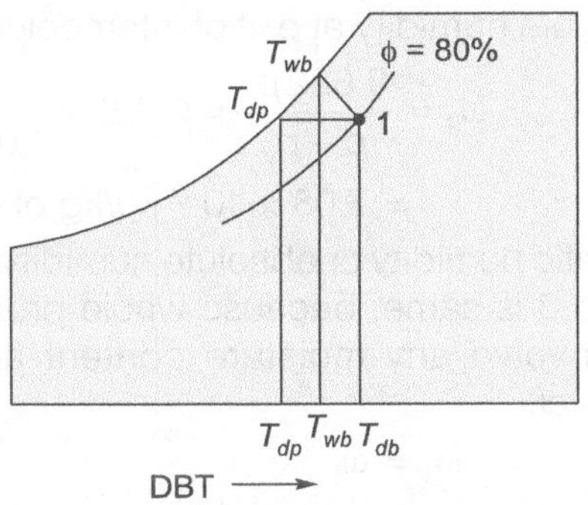

For unsaturated air *i.e.*, $\phi < 100\%$

Dry bulb temperature > Wet bulb temperature > Dew point temperature

i.e., $T_{db} > T_{wb} > T_{dp}$

POINTS TO REMEMBER

- When a volume of air at a given temperature holds the maximum amount of water vapour, the air is said to be saturated.
- Relative humidity is the water-vapour content of air relative to its content at saturation.

4.3 (b)

Given, $p = 1.013$ bar

$p_v = 0.02062$ bar (at dew point)

We know that,

Specific humidity,

$$\omega = \frac{0.622 p_v}{p - p_v} = \frac{0.622 \times 0.02062}{1.013 - 0.02062}$$

$$= \mathbf{0.012924\ kg/kg\ of\ dry\ air}$$

Enthalpy: $h = 1.005\, t_d + \omega(h_{fg,dp} + 1.88\, t_{dp})$

$= 1.005 \times 30 + 0.01291\,(2500 + 1.88 \times 30)$

$= 1.005 \times 30 + 33.00$

$= \mathbf{63.15\ kJ/kg\ of\ dry\ air}$

POINTS TO REMEMBER

- It is very important to keep in mind that the temperature, used in the formula $h = c_{pa}t + \omega(2500 + 1.88t)$, is the Dry bulb temperature.
- Enthalpy of moist air is the summation of enthalpy of dry air and enthalpy of vapour.

4.4 (b)

Since, in the problem, value of saturated vapour is not given. So taken the value of saturated vapour pressure at atmospheric condition is about 5 kPa. i.e. $p_{vs} = 5$ kPa

Relative humidity,

$$\phi = 75\% = 0.75 = \frac{p_v}{p_{vs}}$$

Partial pressure of water in mixture,

$$p_v = 5 \times 0.75 = 3.75 \text{ kPa}$$

Amount of mixture in kg per kg of dry air i.e. specific humidity,

$$\omega = \frac{0.622\, p_v}{p - p_v} = \frac{0.622 \times 3.75}{101.325 - 3.75}$$

$$= \mathbf{0.0239\ kg/kg\ of\ dry\ air.}$$

4.5 (b)

Given: $T_i = 21°C$; $T_{dp} = 10.17°C$; $T_o = -23°C$

$h_i = 8$ W/m²K, $h_o = 23$ W/m²K; $k = 1.2$ W/mK

The heat transfer rate per unit area in the slab can be calculated by considering convection at inner side.

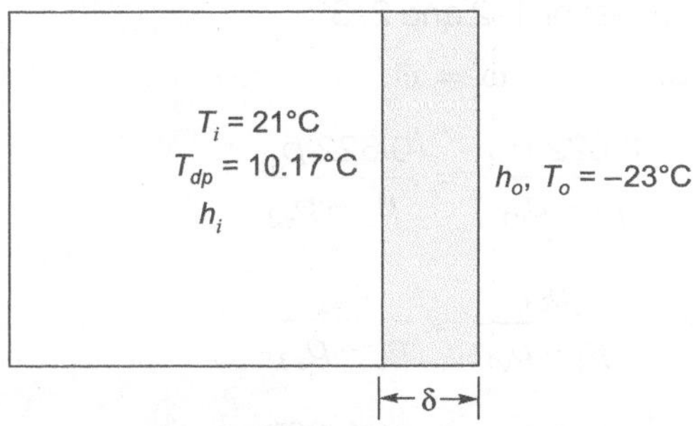

$$q = h_i(T_i - T_{dp})$$

$$= 8(21 - 10.17) = 86.64 \text{ W/m}^2$$

Further this heat is conducted through wall,

$$\therefore \quad q = \frac{T_{dp} - T_o}{\frac{x}{k} + \frac{1}{h_o}}$$

$$86.64 = \frac{10.17 - (-23)}{\frac{x}{1.2} + \frac{1}{23}} = \frac{33.17}{\frac{x}{1.2} + \frac{1}{23}}$$

$$\text{or} \quad \frac{x}{1.2} + \frac{1}{23} = \frac{33.17}{86.64} = 0.3828$$

$$\text{or} \quad \frac{x}{1.2} = 0.3393$$

$$x = 1.2 \times 0.3393 = \mathbf{0.4072\ m}$$

4.6 (b)

Method I:

Process 1–2: Isentropic compression

Process 2–3: Cooling at constant pressure

$$p_1 = 100 \text{ kPa}$$
$$p_2 = p_3 = 500 \text{ kPa}$$
$$T_3 = 35°C$$

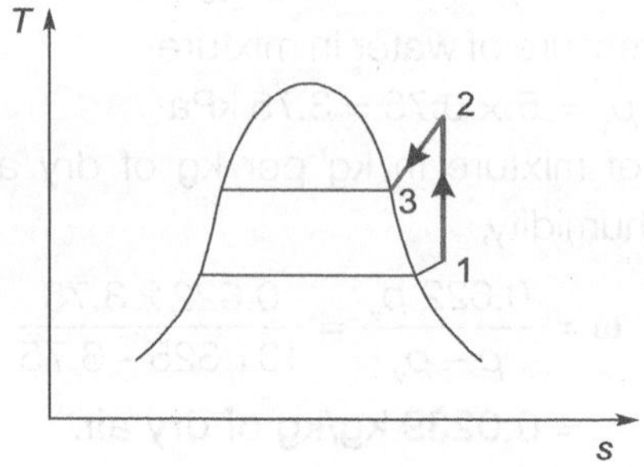

Air at points 1 and 2 is unsaturated and air at point 3 is saturated.

Saturated partial vapour pressure at point 3,

$$p_{s3} = 5.628 \text{ kPa}$$

Specific humidity at points 1, 2 and 3 is same, because no addition or removal water during processor 1–2 and 2–3

i.e., $\omega_1 = \omega_2$

$$\frac{0.622\, p_{v1}}{p_1 - p_{v1}} = \frac{0.622\, p_{s3}}{p_3 - p_{s3}}$$

or $$\frac{p_{v1}}{p_1 - p_{v1}} = \frac{p_{s3}}{p_3 - p_{s3}}$$

$$\frac{p_{v1}}{100 - p_{v1}} = \frac{5.628}{500 - 5.628} = 0.01138$$

or $$p_{v1} = 0.01138\,(100 - p_{v1})$$
$$= 1.138 - 0.01138\, p_{v1}$$

or $$1.01138\, p_{v1} = 1.138$$

or $$p_{v1} = \frac{1.38}{1.01138} = \mathbf{1.125 \text{ kPa}}$$

Method II:

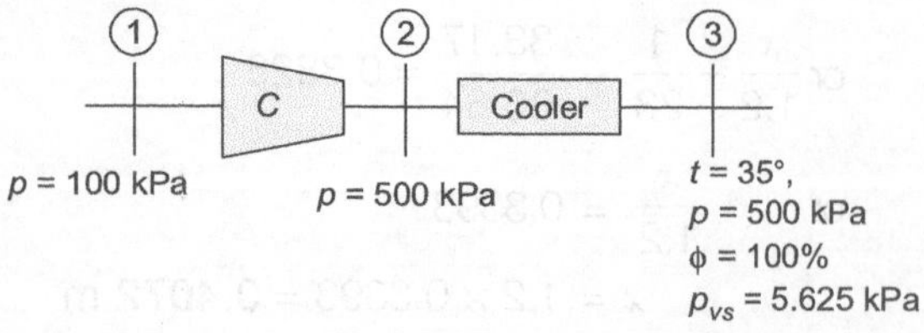

Let mixture of air water as ideal gas. At exit of intercooler, the mixture will be saturated.

Also, $\phi = 1$

$$p_v = p_{vs} = 5.628 \text{ kPa}$$

Absolute humidity at exit of intercooler,

$$\omega_3 = \frac{0.622\, p_v}{p - p_v} = 0.622 \times \frac{5.628}{500 - 5.628}$$
$$= 7.08 \times 10^{-3} \text{ kg/kg of dry air}$$

Specific humidity or absolute humidity at point 1, 2 and 3 is same, because whole process does not involve any moisture content addition or removal,

$$\therefore \quad \omega_3 = \omega_1$$

$$7.08 \times 10^{-3} = 0.622 \times \frac{p_v}{100 - p_v}$$

$$p_v = \mathbf{1.13 \text{ kPa}}$$

4.7 (b)

Relative humidity,

$$\phi = \frac{p_v}{p_{vs}}$$

Specific humidity,

$$\omega = 0.622\left(\frac{p_v}{p - p_v}\right)$$

$$11.5 \times 10^{-3} = 0.622\left(\frac{p_v}{90 - p_v}\right)$$

$$p_v = 1.63378 \text{ KPa}$$

Relative humidity,

$$\phi = \frac{p_v}{p_{vs}} = \frac{1.63378}{4.24} = 0.385$$
$$= \mathbf{38.5\%}$$

4.8 (d)

The moist air in an airtight vessel is heated to a higher temperature. This process is treated as **sensible heating**. In sensible heating process, the relative humidity of the air decreases with increase in dry bulb temperature at constant specific humidity.

This process is shown on psychrometric chart in Fig.

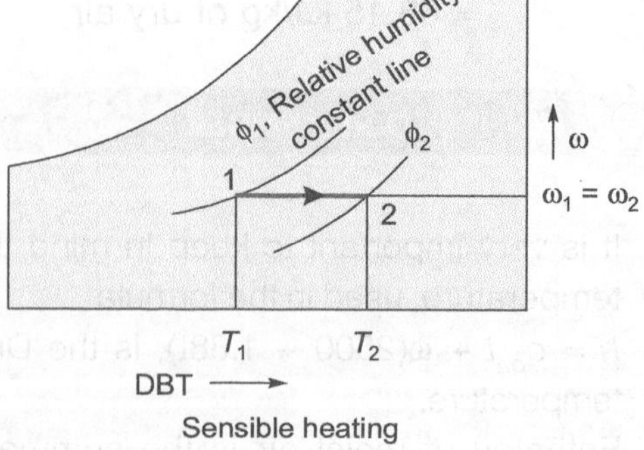

Sensible heating

4.9 (d)

$m_a = 35$ kg; $m_v = 0.5$ kg

$$\omega = \frac{m_v}{m_a} = \frac{0.5}{35} = 0.01428$$

also $\omega = 0.622\dfrac{p_v}{p_a}$

$p_t = 100$ kPa

$$\therefore \quad \omega = 0.622\frac{p_v}{p_t - p_v}$$

Now,

$$0.01428 = 0.622\frac{p_v}{100 - p_v}$$

$p_v = 2.238$ kPa

Relative humidity,

$$\phi = \frac{p_v}{p_{vs}} = \frac{2.238}{3.17} \times 100$$

$= 70.6\% \approx$ **71%**

4.10 (c)

$p_{atm} = 1$ bar $= 100$ kPa

DBT = 30°C

Relative humidity,

$\phi = 70\% = 0.7$

$p_{va} = 4.25$ kPa

Specific humidity: $\omega = ?$

$$\phi = \frac{p_v}{p_{vs}}$$

$$0.7 = \frac{p_v}{4.25}$$

$p_v = 2.975$ kPa

Specific humidity,

$$\omega = 0.622 \times \frac{p_v}{p - p_v} = 0.622 \times \frac{2.975}{100 - 2.975}$$

= **0.0191 kg *w.v.*/kg *d.a.***

4.11 Sol.

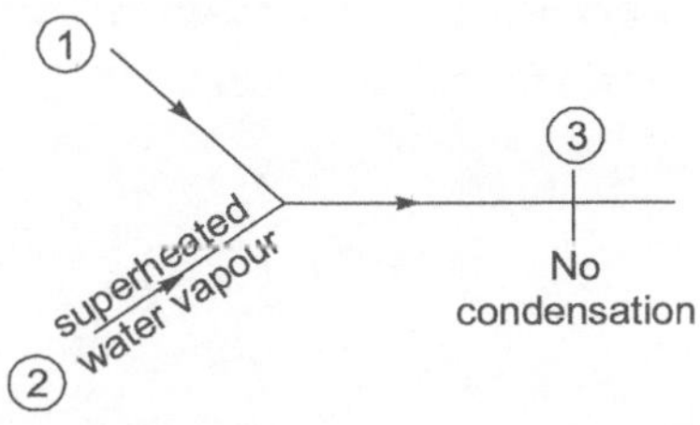

Mass flow rate of moist air = 10.1 kg/sec

Absolute humidity,

$$\omega = \frac{m_v}{m_a} = 0.01 \text{ kg/kg of dry air.}$$

Mass of water vapour,

$m_v = 0.01 \times m_a$

$10.1 = m_a + 0.01 \times m_a$

Mass of moist air = mass of dry air + mass of water vapour

$1.01\, m_a = 10.1$

$$\text{Mass of dry air: } m_a = \frac{10.1}{1.01} = 10 \text{ kg/s}$$

Mass of water vapour,

$m_{v_1} = 10.1 - 10 = 0.1$ kg/s

$m_{v_2} = 0.1$ kg/s

$(m_v)_{total} = m_{v_1} + m_{v_2} = 0.2$ kg/s

$$\text{Humidity ratio: } \omega_{final} = \frac{m_v}{m_a} = \frac{0.2}{10}$$

= **0.02 kg/kg of dry air**

4.12 Sol.

Relative humidity,

$$\phi = \frac{p_v}{p_{vs}}$$

$$0.55 = \frac{p_v}{4246}$$

or $p_v = 2335.3$ Pa $= 2.335$ kPa

$$\therefore \quad \omega = \frac{(\text{Mol. wt})_{H_2O}}{(\text{Mol. wt})_{air}} \times \left(\frac{p_v}{p - p_v}\right)$$

$$= \frac{18}{28.84} \times \left(\frac{2.335}{100 - 2.335}\right)$$

$= 0.01492$ kg/kg of d.a

also $\omega = \dfrac{m_v}{m_a}$

$$\therefore \quad 0.01492 = \frac{m_v}{1}$$

or **$m_v = 0.01492$ kg = 14.92 gm**

4.13 Sol.

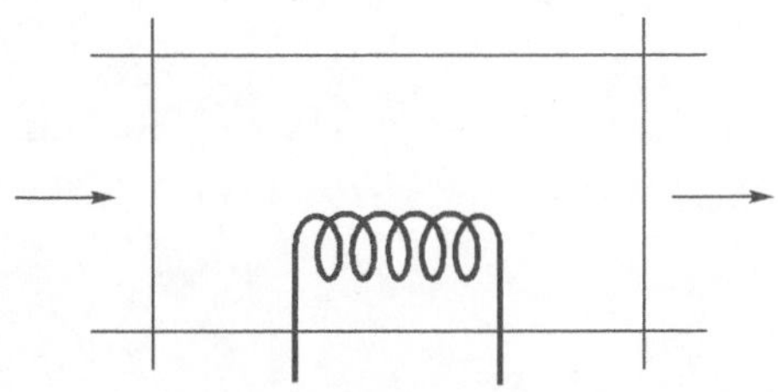

$P_{t1} = 105$ kPa, $DBT_1 = 30°C$, $\phi_1 = 0.8$

$P_{t2} = 100$ kPa, $DBT_2 = 15°C$, $\phi_2 = 1$

$P_{vs1} = 4.24$ kPa

$P_{vs2} = 1.7$ kPa

$M_{water} = 18$ g/mol.

$M_{air} = 28.94$ g/mol.

$$\omega_1 = \frac{18}{28.94} \times \frac{P_{v1}}{P_{t1} - P_{v1}}$$

$$\phi_1 = \frac{P_{v1}}{P_{vs1}}$$

$$0.8 = \frac{P_{v1}}{4.24}$$

$$\therefore \quad P_{v1} = 3.392$$

$$\therefore \quad \omega_1 = \frac{18}{28.94} \times \frac{3.392}{105 - 3.392}$$

$$= 0.02076 \text{ kgV/kgd.a}$$

$$= 20.76 \text{ gv/kgd.a}$$

$$\omega_2 = \frac{18}{28.94} \times \frac{P_{v2}}{P_{t2} - P_{v2}}$$

$$\phi_2 = \frac{P_{v2}}{P_{vs2}}$$

$$1 = \frac{P_{v2}}{1.7}$$

$$\therefore \quad P_{v2} = 1.7$$

$$\therefore \quad \omega_2 = \frac{18}{28.94} \times \frac{1.7}{100 - 1.7}$$

$$= 0.01075 \text{ Kgv/kgda}$$

Mass of water condensing $= \omega_1 - \omega_2$

$$= 20.76 - 10.75$$

$$= 10.01 \text{ g/kgd.a}$$

4.14 Sol.

1 : Cold air

2 : Hot air

$$P_1V_1 = \dot{m}_1RT_1$$

$$105 \times 1.25 = \dot{m}_1 \times 0.287 \times 278$$

$$\dot{m}_1 = 1.645 \text{ kg/sec}$$

$$\dot{m}_2 = 1.6 \times 1.645 = 2.632 \text{ kg/sec}$$

$$\dot{m}_3 = 4.277 \text{ kg/sec}$$

After mixing : 0

$$\dot{m}_1t_1 + \dot{m}_2t_2 = \dot{m}_3t_3$$

$$1.645 \times 5 + 2.632 \times 34 = 4.277t_3$$

$$t_3 = 22.85 \text{ °C}$$

Heat gain : $= h_4 - h_3$

$$= \dot{m}_3c_p(t_4 - t_3)$$

$$= 4.277 \times 1.005(24 - 22.85)$$

$$= 4.963 \text{ kW}$$

4.15 (c)

Relative humidity,

$$\phi = \frac{p_v}{p_{sat}}$$

$$0.5 = \frac{p_v}{p_{sat}}$$

$$p_v = 0.5\ p_{sat}$$

4.16 (a)

$$\phi = 0.6$$

$$\frac{P_v}{P_{vs}} = 0.6$$

$$P_{vs} = 3.6 \text{ kPa}$$

$$P_v = 0.6 \times 3.6 \text{ kPa}$$

$$P_v = 2.16 \text{ kPa}$$

$$\omega = 0.622\left(\frac{P_v}{P - P_v}\right)$$

$$= 0.622\left(\frac{2.16}{101 - 2.16}\right)$$

$= 0.01358$ kg of water vapour/kg of dry air

$= 13.58$ g/kg of dry air

■■■■

Manufacturing Engineering

UNIT

XII

CONTENTS

Manufacturing Engineering

Syllabus

Engineering Materials : Structure and properties of engineering materials, phase diagrams, heat treatment, stress-strain diagrams for engineering materials.

Casting, Forming and Joining Processes : Different types of castings, design of patterns, moulds and cores; solidification and cooling; riser and gating design. Plastic deformation and yield criteria; fundamentals of hot and cold working processes; load estimation for bulk (forging, rolling, extrusion, drawing) and sheet (shearing, deep drawing, bending) metal forming processes; principles of powder metallurgy. Principles of welding, brazing, soldering and adhesive bonding.

Machining and Machine Tool Operations : Mechanics of machining; basic machine tools; single and multi-point cutting tools, tool geometry and materials, tool life and wear; economics of machining; principles of non-traditional machining processes; principles of work holding, jigs and fixtures; abrasive machining processes; NC/CNC machines and CNC programming.

Metrology and Inspection : Limits, fits and tolerances; linear and angular measurements; comparators; interferometry; form and finish measurement; alignment and testing methods; tolerance analysis in manufacturing and assembly; concepts of coordinate-measuring machine (CMM).

Computer Integrated Manufacturing : Basic concepts of CAD/CAM and their integration tools; additive manufacturing.

UNIT XII

Manufacturing Engineering

Analysis of Previous GATE Papers

Exam Year	1 Mark Ques.	2 Marks Ques.	3 Marks Ques.	4 Marks Ques.	5 Marks Ques.	6 Marks Ques.	Total Marks
1995	7	3	–	–	5	–	**38**
1996	5	4	–	–	4	–	**33**
1997	3	4	–	–	5	–	**36**
1998	–	6	–	–	2	–	**22**
1999	4	2	–	–	5	–	**33**
2000	7	3	–	1	4	–	**37**
2001	5	4	–	–	1	–	**18**
2002	12	2	–	–	2	–	**26**
2003	8	12	–	–	–	–	**32**
2004	6	12	–	–	–	–	**30**

Exam Year	1 Mark Ques.	2 Marks Ques.	5 Marks Ques.	Total Marks
2005	7	7	–	**21**
2006	4	12	–	**28**
2007	6	14	–	**34**
2008	2	12	–	**26**
2009	4	7	–	**18**
2010	4	5	–	**14**
2011	6	4	–	**14**
2012	6	5	–	**16**
2013	4	5	–	**14**

Exam Year	1 Mark Ques.	2 Marks Ques.	4 Marks Ques.	5 Marks Ques.	6 Marks Ques.	Total Marks
2014 Set-1	6	2	–	–	–	**10**
2014 Set-2	6	3	–	–	–	**12**
2014 Set-3	6	4	–	–	–	**14**
2014 Set-4	6	4	–	–	–	**14**
2015 Set-1	5	6	–	–	–	**17**
2015 Set-2	3	6	–	–	–	**15**
2015 Set-3	5	6	–	–	–	**17**
2016 Set-1	6	5	–	–	–	**16**
2016 Set-2	4	5	–	–	–	**14**
2016 Set-3	5	7	7	–	–	**19**

Exam Year	1 Mark Ques.	2 Marks Ques.	Total Marks
2017 Set-1	3	6	**15**
2017 Set-2	3	4	**11**
2018 Set-1	7	6	**19**
2018 Set-2	6	7	**20**
2019 Set-1	5	7	**19**
2019 Set-2	6	6	**18**
2020 Set-1	4	6	**16**

Exam Year	1 Mark Ques.	2 Marks Ques.	Total Marks
2020 Set-2	5	6	**17**
2021 Set-1	4	6	**16**
2021 Set-2	5	6	**17**
2022 Set-1	5	5	**15**
2022 Set-2	6	5	**16**
2023	3	8	**19**
2024	6	5	**16**

1 Engineering Materials

1.1 The iron carbon diagram and the *TTT* curves are determined under
(a) Equilibrium and non-equilibrium conditions respectively
(b) Non-equilibrium and equilibrium conditions respectively
(c) Equilibrium conditions for both
(d) Non-equilibrium conditions for both
[ME 1996 : 1 M]

1.2 During heat treatment of steel, the hardness of various structures in increasing order is
(a) martensite, fine pearlite, coarse pearlite, spheroidite
(b) fine pearlite, martensite, spheroidite, coarse pearlite
(c) martensite, coarse pearlite, fine pearlite, spheroidite
(d) spheroidite, coarse pearlite, fine pearlite, martensite [ME 2003 : 1 M]

1.3 ardness of steel greatly improves with
(a) annealing (b) cyaniding
(c) normalising (d) tempering
[ME 2003 : 2 M]

1.4 The percentage of carbon in grey cast iron is in the range of
(a) 0.25 to 0.75 percent
(b) 1.25 to 1.75 percent
(c) 3 to 4 percent
(d) 8 to 10 percent [ME 2004 : 1 M]

1.5 From the lists given below, choose the most appropriate set of heat treatment process and the corresponding process characteristics

Process
P. Tempering
Q. Austempering
R. Martempering

Characteristics
1. Austenite is converted into bainite
2. Austenite is converted into martensite
3. Cementite is converted into globular structure
4. Both hardness and brittleness are reduced
5. Carbon is absorbed into the metal

(a) P–3, Q–1, R–5 (b) P–4, Q–3, R–2
(c) P–4, Q–1, R–2 (d) P–1, Q–5, R–4
[ME 2004 : 2 M]

1.6 When the temperature of a solid metal increases,
(a) strength of the metal decreases but ductility increases
(b) both strength and ductility of the metal decreases
(c) both strength and ductility of the metal increases
(d) strength of the metal increases but ductility decreases [ME 2005 : 1 M]

1.7 The main purpose of spheroidising treatment is to improve
(a) hardenability of low carbon steels
(b) machinability of low carbon steels
(c) hardenability of high carbon steels
(d) machinability of high carbon steels
[ME 2006 : 1 M]

1.8 atch the items in columns I and II.

	Column-I		Column-II
P.	Charpy	1.	Fluidity
Q.	Knoop	2.	Microhardness
R.	Spiral test	3.	Formability
S.	Cupping test	4.	Toughness
		5.	Permeability

(a) P–4, Q–5, R–3, S–2
(b) P–3, Q–5, R–1, S–4
(c) P–2, Q–4, R–3, S–5
(d) P–4, Q–2, R–1, S–3 [ME 2006 : 2 M]

1.9 If a particular Fe-C alloy contains less than 0.83% carbon, it is called
(a) high speed steel
(b) hypoeutectoid steel
(c) hypereutectoid steel
(d) cast iron [ME 2007 : 1 M]

1.10 The effective number of lattice points in the unit cell of simple cubic, body centered cubic, and face centered cubic space lattices, respectively, are
(a) 1, 2, 2 (b) 1, 2, 4
(c) 2, 3, 4 (d) 2, 4, 4
[ME 2009 : 1 M]

1.11 The material property which depends only on the basic crystal structure is
(a) fatigue strength (b) work hardening
(c) fracture strength (d) elastic constant
[ME 2010 : 1 M]

1.12 The crystal structure of austenite is
(a) body centered cubic
(b) face centered cubic
(c) hexagonal closed packed
(d) body centered tetragonal [ME 2011 : 1 M]

1.13 During normalizing process of steel, the specimen is heated
(a) between the upper and lower critical temperature and cooled in still air
(b) above the upper critical temperature and cooled in furnace
(c) above the upper critical temperature and cooled in still air
(d) between the upper and lower critical temperature and cooled in furnace
[ME 2012 : 1 M]

1.14 For a ductile material, toughness is a measure of
(a) resistance to scratching
(b) ability to absorb energy up to fracture
(c) ability to absorb energy till elastic limit
(d) resistance to indentation [ME 2013 : 1 M]

1.15 The process of reheating the martensitic steel to reduce its brittleness without any significant loss in its hardness is
(a) normalising (b) annealing
(c) quenching (d) tempering
[ME 2014 : 1 M, Set-1]

1.16 Match the heat treatment processes (Group A) and their associated effects on properties (Group B) of medium carbon steel
Group A
P : Tempering
Q : Quenching
R : Annealing
S : Normalizing
Group B
1 : Strengthening and grain refinement
2 : Inducing toughness
3 : Hardening
4 : Softening
Codes:

	P	Q	R	S
(a)	3	4	2	1
(b)	2	3	4	1
(c)	3	2	4	1
(d)	2	3	1	4

[ME 2014 : 1 M, Set-4]

1.17 The atomic packing factor for a material with body centered cubic structure is ______.
[2015 : 1 M, Set-2]

1.18 In a binary system of *A* and *B* a liquid of 20% *A* (80% *B*) is coexisting with a solid of 70% *A*(30%*B*). For an overall composition having 40% *A*, the fraction of solid is
(a) 0.40 (b) 0.50
(c) 0.60 (d) 0.75
[2016 : 2 M, Set-2]

1.19 The "Jominy test" is used to find
(a) Young's modulus (b) hardenability
(c) yield strength (d) thermal conductivity
[2016 : 1 M, Set-1]

1.20 In the phase diagram shown in the figure, four samples of the same composition are heated to temperatures Med by *a*, *b*, *c* and *d*.

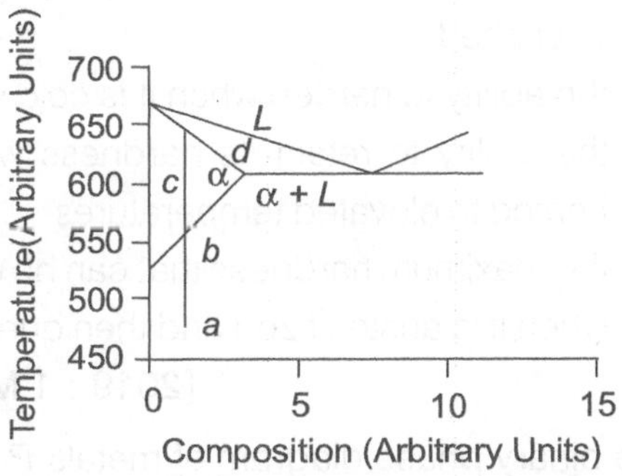

At which temperature will a sample get solutionized the fastest?
(a) *a* (b) *b*
(c) *c* (d) *d*
[2016 : 1 M, Set-2]

1.21 Given the atomic weight of Fe is 56 and that of *C* is 12, the weight percentage of carbon in cementite (Fe_3C) is ______. [2017 : 1 M, Set-2]

1.22 The crystal structure of aluminium is
(a) body-centred cubic
(b) face-centred cubic
(c) close-packed hexagonal
(d) body-centred tetragonal [2017 : 1 M, Set-2]

1.23 The number of atoms per unit cell and the number of slip systems, respectively, for a face-centered cubic (FCC) crystal are

(a) 3, 3 (b) 3, 12
(c) 4, 12 (d) 4, 48
[2018 : 1 M, Set-1]

1.24 Denoting L as liquid and M as solid in a phase-diagram with the subscripts representing different phases, a *eutectoid* reaction is described by
(a) $M_1 \rightarrow M_2 + M_3$ (b) $L_1 \rightarrow M_1 + M_2$
(c) $L_1 \rightarrow M_1 + M_2$ (d) $M_1 + M_2 \rightarrow M_3$
[2018 : 1 M, Set-2]

1.25 In UTM experiment, a sample of length 100 mm, was loaded in tension until failure. The failure load was 40 kN. The displacement, measured using the cross-head motion, at failure, was 15 mm. The compliance of the UTM is constant and is given by 5×10^{-8} m/N. The strain at failure in the sample is ______%. [2019 : 2 M, Set-1]

1.26 Hardenability of steel is a measure of
(a) the depth to which required hardening is obtained when it is austenitized and then quenched.
(b) the ability to harden when it is cold worked
(c) the ability to retain its hardness when it is heated to elevated temperatures
(d) the maximum hardness that can be obtained when it is austenitized and then quenched
[2019 : 1 M, Set-2]

1.27 The binary phase diagram of metals P and Q is shown in the figure. An alloy X containing 60% P and 40% Q (by weight) is cooled from liquid to solid state. The fractions of solid and liquid (in weight percent) at 1250° C, respectively, will be

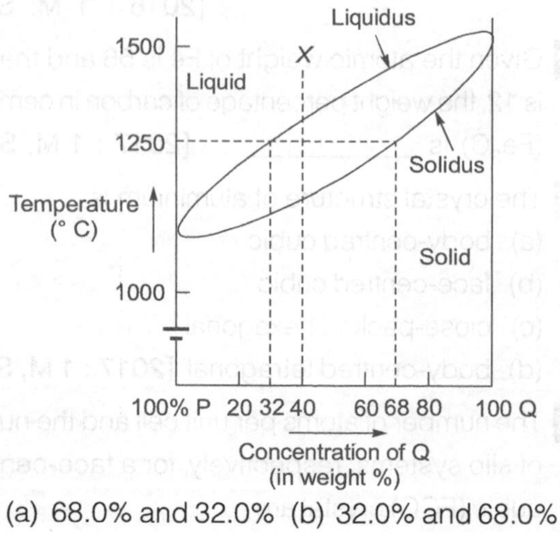

(a) 68.0% and 32.0% (b) 32.0% and 68.0%
(c) 22.2% and 77.8% (d) 77.8% and 22.2%
[2019 : 2 M, Set-2]

1.28 The crystal structure of γ iron (austenite phase) is
(a) BCC (b) BCT
(c) HCP (d) FCC
[2020 : 1 M, Set-1]

1.29 Match the following.

	Heat treatment process		Effect
P.	Tempering	1.	Strengthening
Q.	Quenching	2.	Toughening
R.	Annealing	3.	Hardening
S.	Normalizing	4.	Softening

(a) P-4, Q-3, R-2, S-1
(b) P-1, Q-1, R-3, S-2
(c) P-3, Q-3, R-1, S-3
(d) P-2, Q-3, R-4, S-1 [2020 : 1 M, Set-1]

1.30 Which one of the following statements about a phase diagram is INCORRECT?
(a) Relative amount of different phases can be found under given equilibrium conditions
(b) It gives information on transformation rates
(c) It indicates the temperature at which different phases start to melt
(d) Solid solubility limits are depicted by it
[2020 : 1 M, Set-2]

1.31 The Cast Iron which possesses all the carbon in the combined form as cementite is known as
(a) Malleable Cast Iron
(b) White Cast Iron
(c) Grey Cast Iron
(d) Spheroidal Cast Iron [2021 : 1 M, Set-2]

1.32 Which of the following heat treatment processes is/are used for surface hardening of steels?
(a) Carburizing (b) Cyaniding
(c) Annealing (d) Carbonitriding
[2022 : 1 M, Set-1]

1.33 Which of the following additive manufacturing technique(s) can use a wire as a feedstock material?
(a) Stereolithography
(b) Fused deposition modeling
(c) Selective laser sintering
(d) Directed energy deposition processes
[2022 : 1 M, Set-1]

1.34 In Fe-Fe_3C phase diagram, the eutectoid composition is 0.8 weight % of carbon at 725°C. The maximum solubility of carbon in α-ferrite phase is 0.025 weight % of carbon. A steel sample, having no other alloying element except 0.5 weight % of carbon, is slowly cooled from 1000°C to room temperature. The fraction of pro-eutectoid α-ferrite in the above steel sample at room temperature is

(a) 0.387 (b) 0.864
(c) 0.475 (d) 0.775

[2022 : 2 M, Set-1]

1.35 Fluidity of a molten alloy during sand casting depends on its solidification range. The phase diagram of a hypothetical binary alloy of components A and B is shown in the figure with its eutectic composition and temperature. All the lines in this phase diagram, including the solidus and liquidous lines, are straight lines. If this binary alloy with 15 weight % of B is poured into a mould at a pouring temperature of 800 °C, then the solidification range is

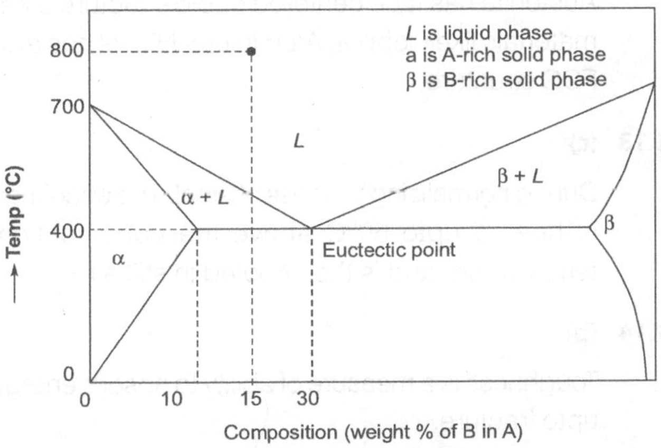

(a) 400°C (b) 250°C
(c) 800°C (d) 150°C

[2022 : 1 M, Set-2]

1.36 The atomic radius of a hypothetical face-centered cubic (FCC) metal is $\left(\frac{\sqrt{2}}{10}\right)$nm. The atomic weight of the metal is 24.092 g/mol. Taking Avogadro's number to be 6.023×10^{23} atoms/mol, the density of the metal is_____kg/m^3. (Answer in integer)

[2023 : 2 M]

1.37 A steel sample with 1.5 wt.% carbon (no other alloying elements present) is slowly cooled from 1100 °C to just below the eutectoid temperature (723 °C). A part of the iron-cementite phase diagram is shown in the figure. The ratio of the pro-eutectoid cementite content to the total cementite content in the microstructure that develops just below the eutectoid temperature is ______. (Rounded off to two decimal places)

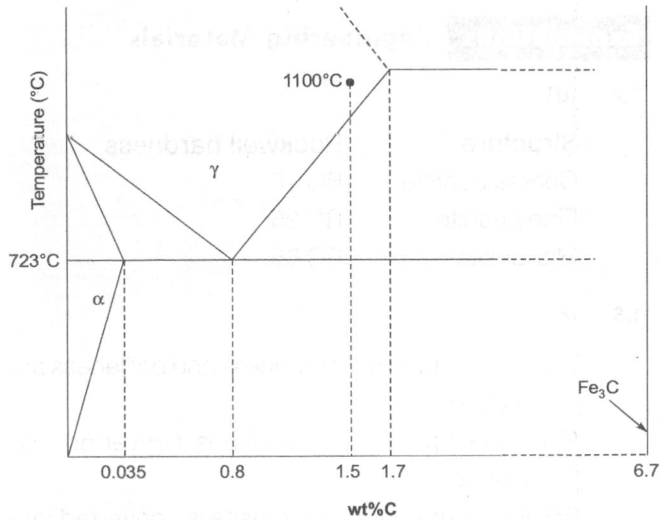

[2023 : 2 M]

1.38 The phases present in pearlite are

(a) cementite and austenite
(b) ferrite and cementite
(c) austenite and ferrite
(d) martensite and ferrite

[2024 : 1 M]

■■■■

Answers Engineering Materials

1.1	(a)	1.2	(d)	1.3	(b)	1.4	(c)	1.5	(c)	1.6	(a)	1.7	(d)	1.8	(d)
1.9	(b)	1.10	(b)	1.11	(d)	1.12	(b)	1.13	(c)	1.14	(b)	1.15	(d)	1.16	(b)
1.17	(0.68)	1.18	(a)	1.19	(b)	1.20	(c)	1.21	(6.67)	1.22	(b)	1.23	(c)	1.24	(a)
1.25	(13)	1.26	(a)	1.27	(c)	1.28	(d)	1.29	(d)	1.30	(b)	1.31	(b)		
1.32	(a,b,d)	1.33	(b, d)	1.34	(a)	1.35	(d)	1.36	(2500)	1.37	(0.54)	1.38	(b)		

Explanations Engineering Materials

1.2 (d)

Structure	Rockwell hardness
Coarse pearlite	RC 15
Fine pearlite	RC 25
Martensite	RC 65

1.5 (c)

(P) Tempering—Both hardness and brittleness are reduced

(Q) Austempering—Austenite is converted into bainite

(R) Martempering—Austensite is converted into martensite

1.8 (d)

(P) Charpy test—Toughness

(Q) Knoop—Microhardness

(R) Spiral test—Fluidity

(S) Cupping test—Formability

1.10 (b)

Space Lattice Lattice points in	Effective Number of unit cell
Simple Cubic	1/8 × 8 = 1
Body Centered Cubic (B.C.C.)	1/8 × 8 + 1 = 2
Face Centered Cubic (F.C.C.)	1/8 × 8 + 1/2 × 6 = 4

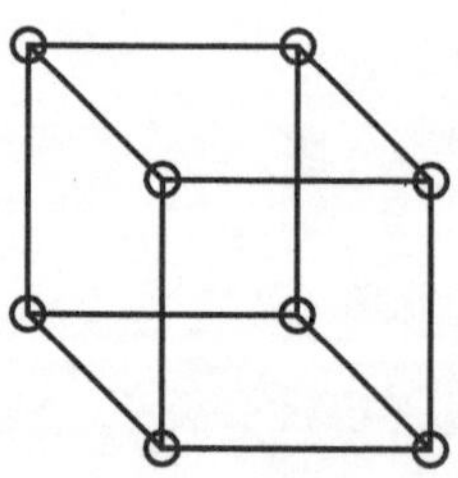

Simple Cubic

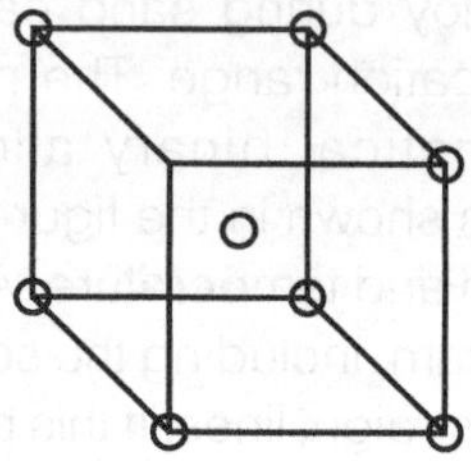

B. C. C.

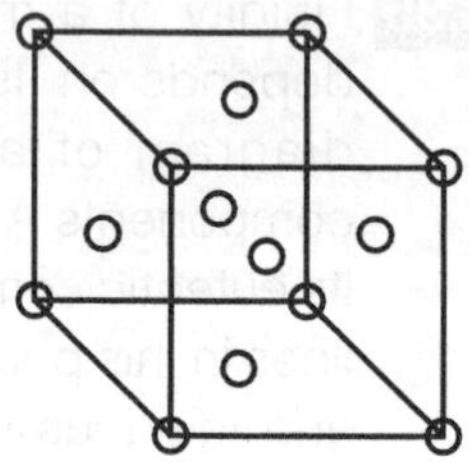

F. C. C

1.12 (b)

Austenite has face centered cubic structure other materials like Copper, Aluminium, Nickel has also FCC structure.

1.13 (c)

During normalizing process of steel. The specimen in heated upto 50°C above the upper critical temperature and is then cooled in still air.

1.14 (b)

Toughness is a measure of ability to absorb energy upto fracture.

1.17 Sol.

Atomic packing factor

$$= \frac{\text{Volume occupied by } N_{\text{average}}}{\text{Volume of unit cell}}$$

$$= \frac{N_{\text{average}} \times \frac{4}{3}\pi r^3}{a^3}$$

For BCC structure,

$$a = \frac{4r}{\sqrt{3}} \text{ and } N_{\text{average}} = 2 \Rightarrow \text{A.P.F.} = \mathbf{0.68}$$

1.19 (b)

Jominy test is used to find hardenability of metal alloy (steel) through a heat treatment process.

1.21 Sol.

Weight percentage of Carbon is cementric

$$\frac{12}{56\times3+12\times1}\times100\% = \mathbf{6.67\%}$$

1.22 (b)

Crystal Structure of materials

- **FCC:** Ni, Cu, Ag, Pt, Au, Pb, Al (soft)
- **BCC:** V, Mo, Ta, W (hard material)
- **HCP:** Mg, Zn
- Cobalt HCP < 420°C, FCC > 420°C
- Chromium HCP < 20°C , BCC >20°C
- **Glass:** Amorphous
- **BCC:** Ferrite or α-iron δ-ferrite or δ-iron
- **FCC:** Austenite or γ-iron

1.23 (c)

Number of slip systems = Number of close packed plane(CPP) × close packed direction (CPD)

For FCC, CPP are (1 1 1) × 4 Numbers

CPD are : $[1\ \bar{1}\ 0]$ × 3 Numbers

hence, number of slip systems are 12.

1.24 (a)

M_1	→	M_2	+	M_3
γ - austenite		α - ferrite		cementite (Fe_3C)

1.25 Sol.

Total compliance,

$$C_T = \frac{1}{K_T} = \frac{\delta}{P} = \frac{15\times10^{-3}}{P}$$

Machine compliance,

$$\frac{\delta}{P} = 5\times10^{-8}\ \text{N/m}$$

Metal compliance (C_{metal}) = $\frac{\delta}{P}$

According to Torsion test

$$C_T = C_m + C_{metal}$$

$$\frac{15\times10^{-3}}{40\times10^3} = 5\times10^{-8} + \frac{\delta}{P}$$

or $$\frac{15\times10^{-3}}{40\times10^3} = 5\times10^{-8} + \frac{\delta}{40\times10^3}$$

or δ = 0.013 m = 13 mm

∴ The strain at failure, $\frac{\delta}{L}\times100\%$

$$= \frac{13\ \text{mm}}{100\ \text{mm}}\times100\% = 13\%$$

1.26 (a)

The depth and hardness achieved by quenching is called hardenability. Hardenability should not be confused with hardness. Hardenability can be defined as the depth to which a certain hardness level can be obtained by the quenching process.

1.25 (c)

According to lever rule:

Amount of solid phase

$$= \frac{(40-32)}{(68-32)}\times100 = 22.2\%$$

Amount of liquid phase

$$= \frac{(68-40)}{(68-32)}\times100 = 77.8\%$$

1.32 (a, b, d)

We know that annealing reduces the hardness, but processes like carburising, carbonitriding and nitriding increases the surface hardness.

1.33 (b, d)

In stereolithography, liquid form of the polymer material is there and ultraviolet radiation falls upon the material. Due to this, one layer of material would be deposited. So, layer by layer deposition takes place. Here, wire feed is not used.

In fused deposition modelling (FDM), spool of wire of ABS material is passed through heated rollers.

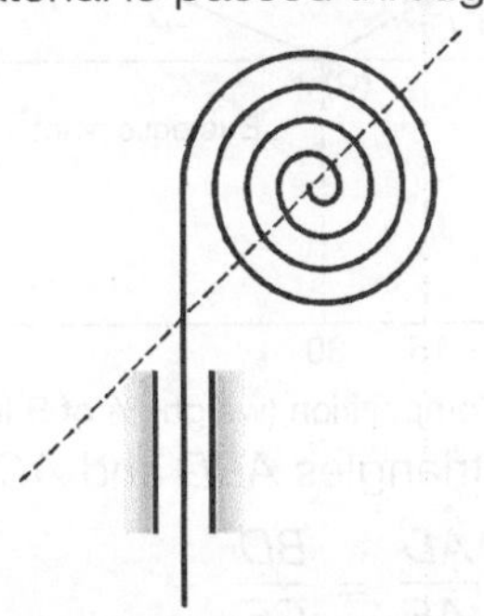

Now, this semi-solid material gets deposited. In selective laser sintering, a bed of powder polymer, resin or metal is targeted partially (sintering) or fully (melting) by a high-power directional heating source such as laser that result to a solidified layer of fused powder.

In directed energy deposition process, feed stock material is in the form of wire or powder form is used.

1.34 (a)

Below eutectoid temperature, there will not be any change in the mass fraction of Pro-eutectoid phase till room temperature.

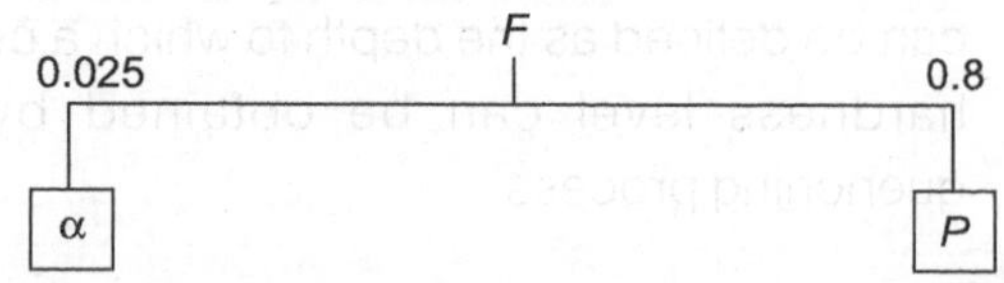

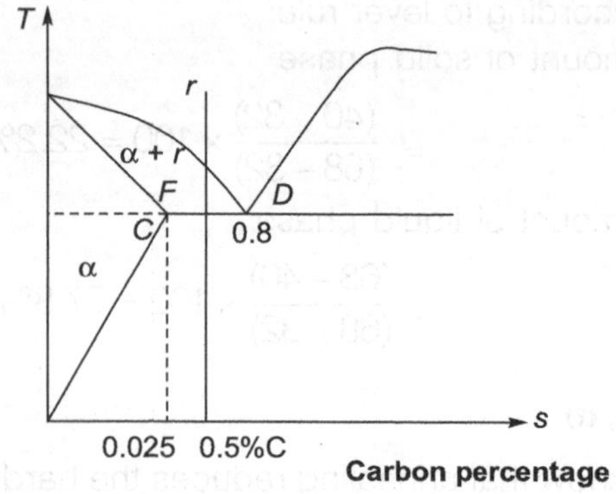

Mass fraction of pro-eutectoid a is given by $\frac{FD}{CD}$

$$\therefore \quad \frac{FD}{CD} = \frac{0.8-0.5}{0.8-0.025} = 0.387$$

1.35 (d)

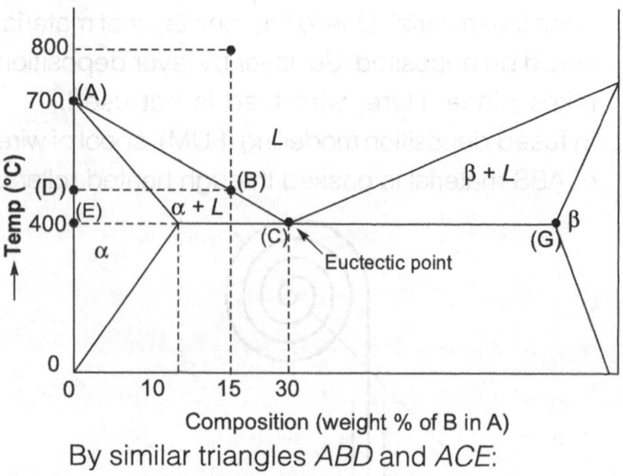

By similar triangles *ABD* and *ACE*:

$$\frac{AD}{AE} = \frac{BD}{CE}$$

$$\Rightarrow \quad AD = \frac{15}{30} \times 300 = 150$$

$$\Rightarrow \quad DE = 300 - 150 = 150°C$$

1.36 (2500) (2499.999 to 2500.001)

Given:

Radius of atom, $r = \left(\frac{\sqrt{2}}{10}\right)$nm

Atomic weight of metal, M = 24.092 g/mol

Avogadro's number,

N_A = 6.023 × 10^{23} atoms/mol

Volume of unit cell,

$$V = a^3 = \left(\frac{4r}{\sqrt{2}}\right)^3 = \left(\frac{4}{\sqrt{2}} \times \frac{\sqrt{2}}{10}\right)^3 = 0.064 \text{ nm}^3$$

Density is given by, $\rho = \frac{nM}{VN_A}$

where, n : Number of effective atoms in unit cell

M : Atomic weight; V : Volume of unit cell

N_A : Avagadro's Number

$$\therefore \rho = \frac{4 \times 24.092 \times (10^9)^3}{6.023 \times 10^{23} \times 1000 \times 0.064} = 2500 \text{ kg/m}^3$$

1.37 (0.54) (0.53 to 0.55)

Mass fraction of proeutectoid cementite,

$$m_1 = \frac{1.5-0.8}{6.7-0.8} = 0.1186$$

Mass fraction of total cementite,

$$m_2 = \frac{1.5-0.035}{6.7-0.035} = 0.2198$$

$$\frac{m_1}{m_2} = \frac{0.1186}{0.2198} = 0.54$$

■■■■

2 Metal Casting

2.1 Light impurities in the molten metal are prevented from reaching the mould cavity by providing
(a) Strainer (b) Bottom well
(c) Skim bob (d) All of the above
[1996 : 1 M]

2.2 Match the following
A. Sand casting
B. Plaster mould casting
C. Shell mould casting
D. Investment casting
1. Symmetrical and circular shapes only
2. Parts have hardened skins and soft interior
3. Minimum post casting process
4. Parts have soft skin and hard interior
5. Parts have a tendency to warp
6. Suitable only for nonferrous metals
[1998 : 2 M]

2.3 Which of the following materials requires the largest shrinkage allowance, while making a pattern for casting
(a) Aluminium
(b) Brass
(c) Cast iron
(d) Plain carbon steel [1999 : 1 M]

2.4 Disposable patterns are made of
(a) wood (b) rubber
(c) metal (d) polystyrene
[2000 : 1 M]

2.5 Shrinkage allowance on pattern is provided to compensate for shrinkage when
(a) the temperature of liquid metal drops from pouring to freezing temp
(b) the metal changes from liquid to solid state at freezing temperature
(c) the temperature of solid phase drops from freezing to room temperature
(d) the temp of metal drops from pouring to room temperature [2001 : 1 M]

2.6 In centrifugal casting, the lighter impurities are
(a) uniformly distributed
(b) forced towards the outer surface
(c) trapped near the mean radius of the casting
(d) collected at the centre of the casting
[2002 : 1 M]

2.7 The primary purpose of sprue in casting mould is to
(a) feed the casting at rate consistent with the rate of solidification
(b) Act as a reservoir for molten metal
(c) feed molten metal from the pouring basin to the gate
(d) help feed the casting until all solidification takes place [2002 : 1 M]

2.8 Hardness of green sand mould increases with
(a) increase in moisture content beyond 6 percent
(b) increase in permeability
(c) decrease in permeability
(d) increases in both moisture content and permeability [2003 : 1 M]

2.9 With a solidification factor of 0.97×10^6 s/m^2, the solidification time (in seconds) for a spherical casting of 200 mm diameter is
(a) 539 (b) 1078
(c) 4311 (d) 3233 [2003 : 2 M]

2.10 Misrun is a casting defect which occurs due to
(a) very high pouring temperature of the metal
(b) insufficient fluidity of the molten metal
(b) absorption of gases by the liquid metal
(d) improper alignment of the mould flasks
[2004 : 1 M]

2.11 Gray cast iron blocks $200 \times 100 \times 10$ mm are to be cast in sand moulds. Shrinkage allowance for pattern making is 1%. The ratio of the volume of pattern to that of the casting will be
(a) 0.97 (b) 0.99
(c) 1.01 (d) 1.03 [2004 : 2 M]

2.12 Match the items of **List-I** (Equipment) with the items of **List-II** (Process) the correct answer using the given codes.
Equipment
P. Hot chamber machine
Q. Muller
R. Dielectric baker
S. Sand blaster

Process

1. Cleaning
2. Core making
3. Die casting
4. Annealing
5. Sand mixing

(a) P–2, Q–1, R–4, S–5
(b) P–4, Q–2, R–3, S–5
(c) P–4, Q–5, R–1, S–2
(d) P–3, Q–5, R–2, S–1 [2005 : 2 M]

2.13 A mould has downsprue whose length is 20 cm and the cross sectional area at the base of the downsprue is 1 cm^2. The downsprue feeds a horizontal runner leading into the mould cavity of volume 1000 cm^3. The time required to fill the mould cavity will be

(a) 4.05 s (b) 5.05 s
(c) 6.05 s (d) 7.25 s [2005 : 2 M]

2.14 An expendable pattern is used in

(a) slush casting
(b) squeeze casting
(c) centrifugal casting
(d) investment casting [2006 : 1 M]

2.15 In a sand casting operation, the total liquid head is maintained constant such that it is equal to the mould height. The time taken to fill the mould with a top gate is t_A. If the same mould is filled with a bottom gate, then the time taken is t_B. Ignore the time required to fill the runner and frictional effects. Assume atmospheric pressure at the top molten metal surfaces. The relation between t_A and t_B is

(a) $t_B = \sqrt{2}t_A$ (b) $t_B = 2t_A$
(c) $t_B = \dfrac{t_A}{\sqrt{2}}$ (d) $t_B = 2\sqrt{2}t_A$

[2006 : 2 M]

2.16 Which of the following engineering materials is the most suitable candidate for hot chamber die casting?

(a) low carbon steel (b) titanium
(c) copper (d) tin [2007 : 1 M]

2.17 Volume of a cube of side l and volume of a sphere of radius r are equal. Both the cube and the sphere are solid and of same material. They are being cast. The ratio of the solidification time of the cube to the same of sphere is

(a) $\left(\frac{4\pi}{6}\right)^3\left(\frac{r}{l}\right)^6$ (b) $\left(\frac{4\pi}{6}\right)\left(\frac{r}{l}\right)^2$
(c) $\left(\frac{4\pi}{6}\right)^2\left(\frac{r}{l}\right)^3$ (d) $\left(\frac{4\pi}{6}\right)^2\left(\frac{r}{l}\right)^4$

[2007 : 1 M]

2.18 A 200 mm long down sprue has an area of cross-section of 650 mm^2 where the pouring basin meets the down sprue (*i.e.* at the beginning of the down sprue). A constant head of molten metal is maintained by the pouring basin. The molten metal flow rate is 6.5×10^5 mm^3/s. Considering the end of down sprue to be open to atmosphere and acceleration due to gravity of 10^4 mm/s^2, the area of the down sprue in mm^2 at its end (avoiding aspiration effect) should be

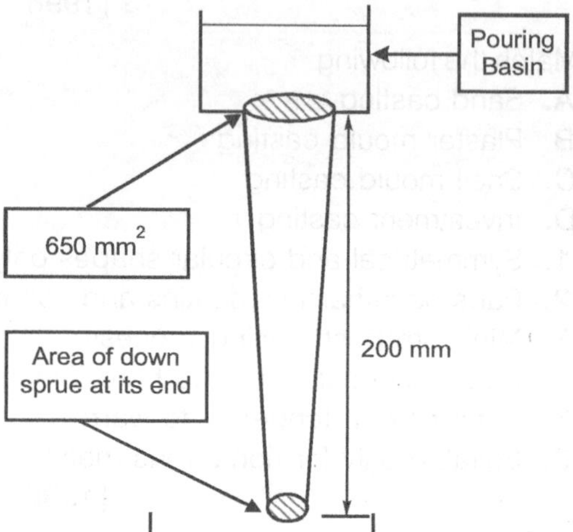

(a) 650.0 (b) 350.0
(c) 290.7 (d) 190.0 [2007 : 2 M]

2.19 While cooling, a cubical casting of side 40mm undergoes 3%, 4% and 5% volume shrinkage during the liquid state, phase transition and solid state, respectively. The volume of metal compensated from the riser is

(a) 2% (b) 7%
(c) 8% (d) 9% [2008 : 2 M]

2.20 Two streams of liquid metal which are not hot enough to fuse properly result into a casting defect known as

(a) cold shut (b) swell
(c) sand wash (d) scab [2009 : 1 M]

2.21 Match the items in **Column I** and **Column II**.

Column I	Column II
P. Metallic Chills	1. Support for the core
Q. Metallic Chaplets	2. Reservoir of the molten metal
R. Riser	3. Control cooling of critical sections
S. Exothermic Padding	4. Progressive solidification

(a) P–1, Q–3, R–2, S–4
(b) P–1, Q–4, R–2, S–3
(c) P–3, Q–4, R–2, S–1
(d) P–4, Q–1, R–2, S–3 [2009 : 2 M]

2.22 In a gating system, the ratio 1 : 2 : 4 represents
(a) sprue base area : runner area : ingate area
(b) pouring basin area : ingate area : runner area
(c) sprue base area : ingate area : casting area
(d) runner area : ingate area : casting area
[2010 : 1 M]

2.23 Green sand mould indicates that
(a) polymeric mould has been cured
(b) mould has been totally dried
(c) mould is green in colour
(d) mould contains moisture [2011 : 1 M]

2.24 A cubic casting of 50 mm side undergoes volumetric solidification shrinkage and volumetric solid contraction of 4% and 6% respectively. No riser is used. Assume uniform cooling in all directions. The side of the cube after solidification and contraction is
(a) 48.32 mm (b) 49.90 mm
(c) 49.94 mm (d) 49.96 mm
[2011 : 2 M]

2.25 A cube shaped casting solidifies in 5 min. The solidification time in min for a cube of the same material, which is 8 times heavier than the original casting, will be
(a) 10 (b) 20
(c) 24 (d) 40 [2013 : 1 M]

2.26 Match the casting defects **(Group A)** with the probable causes **(Group B)**:

Group-A
P. Hot tears Q. Shrinkage
R. Blow holes S. Cold shunt

Group-B
1. Improper fusion of two streams of liquid metal
2. Low permeability of the sand mould
3. Volumetric contraction both in liquid and solid stage
4. Differential cooling rate

(a) P–1, Q–3, R–2, S–4
(b) P–4, Q–3, R–2, S–1
(c) P–3, Q–4, R–2, S–1
(d) P–1, Q–4, R–4, S–3 [2014 : 1 M, Set-3]]

2.27 The hot tearing in a metal casting is due to
(a) high fluidity
(b) high melt temperature
(c) wide range of solidification temperature
(d) low coefficient of thermal expansion
[2014 : 1 M, Set-3]

2.28 A cylindrical blind riser with diameter d and height h, is placed on the top of the mold cavity of a closed type sand mould as shown in the figure. If the riser is of constant volume, then the rate of solidification in the riser is the least when the ratio $h : d$ is

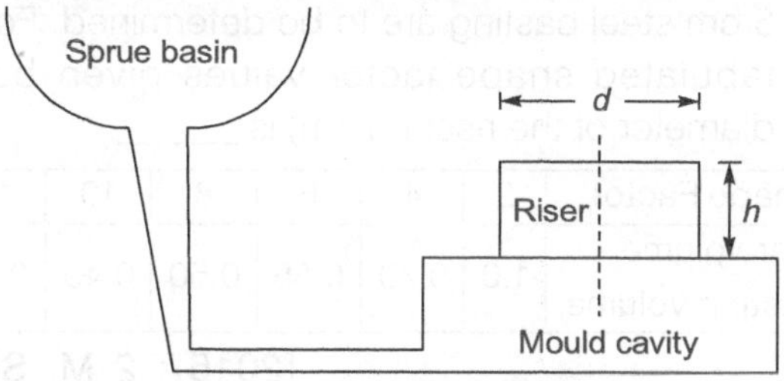

(a) 1 : 2 (b) 2 : 1
(c) 1 : 4 (d) 4 : 1
[2014 : 2 M, Set-3]

2.29 A cylindrical riser of 6 cm diameter and 6 cm height has to be designed for a sand casting mould for producing a steel rectangular plate casting of 7 cm × 10 cm × 2 cm dimensions having the total solidification time of 1.36 minute. The total solidification time (in minute) of the riser is _____
[2014 : 2 M, Set-4]

2.30 In full mould (cavity-less) casting process, the pattern is made of
(a) expandable polystyrene
(b) wax
(c) epoxy
(d) plaster of Paris [2015 : 1 M, Set-3]

2.31 The solidification time of a casting is proportional to $\left(\frac{V}{A}\right)^2$, where V is the volume of the casting and A is the total casting surface area losing heat. Two cubes of same material and size are cast using sand casting process. The top face of one of the cubes is completely insulated. The ratio of the solidification time for the cube with top face insulated to that of the other cube is
(a) $\frac{25}{36}$ (b) $\frac{36}{25}$
(c) 1 (d) $\frac{6}{5}$
[2015 : 2 M, Set-1]

2.32 A cube and a sphere made of cast iron (each of volume 1000 cm^3) were cast under identical conditions. The time taken for solidifying the cube was 4 s. The solidification time (in s) for the for the sphere is _____. [2015 : 2 M, Set-2]

2.33 Ratio of solidification time of a cylindrical casting (height = radius) of that of a cubic casting of side two times the height of cylindrical casting is ______.
[2015 : 2 M, Set-3]

2.34 The dimensions of a cylindrical side riser (height = diameter) for a 25 cm × 15 cm × 5 cm steel casting are to be determined. For the tabulated shape factor values given below, diameter of the riser (in cm) is ______.

Shape Factor	2	4	6	8	10	12
Riser volume /Casting volume	1.0	0.70	0.55	0.50	0.40	0.35

[2015 : 2 M, Set-3]

2.35 Equal amounts of a liquid metal at the same temperature are poured into three moulds made of steel, copper and aluminum. The shape of the cavity is a cylinder with 15 mm diameter. The size of the moulds are such that the outside temperature of the moulds do not increase appreciably beyond the atmospheric temperature during solidification. The sequence of solidification in the mould from the fastest to slowest is (Thermal conductivities of steel, copper and aluminum are 60.5, 401 and 237 W/mK, respectively. Specific heats of steel, copper and aluminum are 434, 385 and 903 J/kgK, respectively. Densities of steel, copper and aluminum are 7854, 8933 and 2700 kg/m^3, respectively.)
(a) Copper – Steel – Aluminium
(b) Aluminum – Steel – Copper
(c) Copper – Aluminium – Steel
(d) Steel – Copper – Aluminum
[2016 : 1 M, Set-3]

2.36 Gray cast iron blocks of size 100 mm × 50 mm × 10 mm with a central spherical cavity of diameter 4 mm are sand cast. The shrinkage allowance for the pattern is 3%. The ratio of the volume of the pattern to volume of the casting is______.
[2016 : 2 M, Set-2]

2.37 The part of a gating system which regulates the rate of pouring of molten metal is
(a) pouring basin (b) runner
(c) choke (d) ingate
[2016 : 1 M, Set-1]

2.38 A cylindrical job with diameter of 200 mm and height of 100 mm is to be cast using modulus method of riser design. Assume that the bottom surface of cylindrical riser does not contribute as cooling surface. If the diameter of the riser is equal to its height, then the height of the riser (in mm) is
(a) 150 (b) 200
(c) 100 (d) 125
[2016 : 2 M, Set-1]

2.39 A sprue in a sand mould has a top diameter of 20 mm and height of 200 mm. The velocity of the molten metal at the entry of the sprue is 0.5 m/s. Assume acceleration due to gravity as 9.8 m/s^2 and neglect all losses. If the mould is well ventilated, the velocity (upto 3 decimal points accuracy) of the molten metal at the bottom of the sprue is ______ m/s. [2017 : 2 M, Set-1]

2.40 During solidification of a pure molten metal, the grains in the casting near the mould wall are
(a) coarse and randomly oriented
(b) fine and randomly oriented
(c) fine and ordered
(d) coarse and ordered [2018 : 1 M, Set-2]

2.41 For sand-casting a steel rectangular plate with dimensions 80 mm × 120 mm × 20 mm, a cylindrical riser has to be designed. The height of the riser is equal to its diameter. The total solidification time for the casting is 2 minutes. In Chvorinov's law for the estimation of the total solidification time, exponent is to be taken as 2. For a solidification time of 3 minutes in the riser, the diameter (in mm) of the riser is ______ (correct to two decimal places). [2018 : 2 M, Set-2]

2.42 In a casting process, a vertical channel through which molten metal flows downward , from pouring basin to runner for reaching the mould cavity is called
(a) sprue (b) blister
(c) pinhole (d) riser
[2019 : 1 M, Set-1]

2.43 Match the following sand mold casting defects with their respective causes.

	Defect		Cause
P.	Blow hole	1.	Poor collapsibility
Q.	Misrun	2.	Mold erosion
R.	Hot tearing	3.	Poor permeability
S.	Wash	4.	Insufficient fluidity

(a) P-3, Q-4, R-1, S-2
(b) P-4, Q-3, R-1, S-2
(c) P-3, Q-4, R-2, S-1
(d) P-2, Q-4, R-1, S-3 [2019 : 2 M, Set-1]

2.44 The fluidity of molten metal of cast alloys (without any addition of fluxes) increases with increase in
(a) freezing range
(b) surface tension
(c) degree of superheat
(d) viscosity [2019 : 1 M, Set-2]

2.45 The figure shows a pouring arrangement for casting of a metal block. Frictional losses are negligible. The acceleration due to gravity is 9.81 m/s^2. The time (in s, round off to two decimal places) to fill up the mold cavity (of size 40 cm x 30 cm x 15 cm) is ______

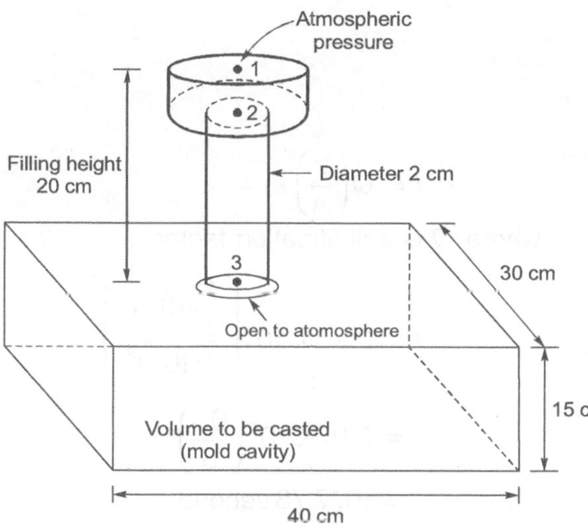

[2019 : 2 M, Set-2]

2.46 A mould cavity of 1200 cm^3 volume has to be filled through a sprue of 10 cm length feeding a horizontal runner. Cross-sectional area at the base of the sprue is 2 cm^2. Consider acceleration due to gravity as 9.81 m/s^2. Neglecting frictional losses due to molten metal flow, the time taken to fill the mould cavity is ______ seconds (round off to 2 decimal places). [2020 : 2 M, Set-2]

2.47 A true centrifugal casting operation needs to be performed horizontally to make copper tube sections with outer diameter of 250 mm and inner diameter of 230 mm. The value of acceleration due to gravity, g = 10 m/s^2. If a G-factor (ratio of centrifugal force to weight) of 60 is used for casting the tube, the rotational speed required is ______ rpm. [Round off to the nearest integer]
[2021 : 2 M, Set-1]

2.48 A cast product of a particular material has dimensions 75 mm × 125 mm × 20 mm. The total solidification time for the cast product is found to be 2.0 minutes as calculated using Chvorinov's rule having the index, n = 2. If under the identical casting conditions, the cast product shape is changed to a cylinder having diameter = 50 mm and height = 50 mm, the total solidification time will be ______ minutes (round off to two decimal places). [2021 : 2 M, Set-2]

2.49 Consider sand casting of a cube of edge length a. A cylindrical riser is placed at the top of the casting. Assume solidification time, $\tau_s \propto V/A$, where V is the volume and A is the total surface area dissipating heat. If the top of the riser is insulated, which of the following radius/radii of riser is/are acceptable?
(a) $\frac{a}{3}$
(b) $\frac{a}{2}$
(c) $\frac{a}{4}$
(d) $\frac{a}{6}$ [2022 : 1 M, Set-2]

2.50 In a metal casting process to manufacture parts, both patterns and moulds provide shape by dictating where the material should or should not go. Which of the option(s) given correctly describe(s) the mould and the pattern?
(a) Mould walls indicate boundaries within which the molten part material is allowed, while pattern walls indicate boundaries of regions where mould material is not allowed.
(b) Moulds can be used to make patterns.
(c) Pattern walls indicate boundaries within which the molten part material is allowed, while mould walls indicate boundaries of regions where mould material is not allowed.
(d) Patterns can be used to make moulds.
[2023 : 1 M, MSQ]

2.51 The allowance provided to a pattern for easy withdrawal from a sand mold is
(a) finishing allowance
(b) shake allowance
(c) shrinkage allowance
(d) distortion allowance
[2024 : 1 M]

2.52 Aluminium is casted in a cube-shaped mold having dimensions as 20 mm × 20 mm × 20 mm. Another mold of the same mold material is used to cast a sphere of aluminium having a diameter of 20 mm. The pouring temperature for both cases is the same. The ratio of the solidification times of the cube-shaped mold to the spherical mold is ______. (answer in integer)
[2024 : 2 M]

■■■■

Answers Metal Casting

2.1	(c)	2.2	(A–5, B–6, C–2, D–3)	2.3	(d)	2.4	(d)	2.5	(c)	2.6	(d)	2.7	(c)
2.8	(c)	2.9	(b)	2.10	(b)	2.11	(d)	2.12	(d)	2.13	(b)	2.14	(d)	2.15	(b)
2.16	(d)	2.17	(d)	2.18	(c)	2.19	(b)	2.20	(a)	2.21	(d)	2.22	(a)	2.23	(d)
2.24	(a)	2.25	(b)	2.26	(b)	2.27	(c)	2.28	(a)	2.29	(3)	2.30	(a)	2.31	(b)
2.32	(6.157355)	2.33	(0.5625)	2.34	(10.60)	2.35	(c)	2.36	(1.09)	2.37	(c)	2.38	(a)
2.39	(2.043)	2.40	(b)	2.41	(51.84)	2.42	(a)	2.43	(a)	2.44	(c)	2.45	(28.92)
2.46	(4.28)	2.47	(662)	2.48	(2.83)	2.49	(a, b)	2.50	(a,b,d)	2.51	(b)	2.52	(1)

Explanations Metal Casting

2.1 (c)

Skim bob prevents the slag and dirt which floats on the top of molten metal and allows only clean metal underneath to pass.

2.2 (A–5, B–6, C–2, D–3)

A–5, B–6, C–2, D–3

2.3 (d)

Among the shown material, plain carbon steel, has greatest shrinkage allowance is provided on the pattern hence size of the pattern is more.

2.4 (d)

Disposable patterns are made of polystyrene which are used in expandable mould casting.

2.5 (c)

Shrinkage allowance is provided to compensate for shrinkage when the temperature of solid phase drops from freezing to room temperature.

2.6 (d)

Lighter impurities with lower densities stay at the centre, while high density material moves away from centre due to centrifugal action.

2.7 (c)

Sprue is a tapered connecting passage between pouring basin and runner for molten metal.

2.9 (b)

Given, d = 200 mm, r = 100 mm = 0.1 m

According to Caine's relation:

Solidification time,

$$t = Q\left(\frac{V}{A}\right)^2$$

where Q is solidification factor

$$t = 0.97 \times 10^6 \left(\frac{\frac{4}{3}\pi(0.1)^3}{4\pi(0.1)^2}\right)^2$$

$$t = 0.97 \times 10^6 \left(\frac{0.1}{3}\right)^2$$

$$= 1077.78 \text{ second}$$

$$\approx \mathbf{1078 \text{ second}}$$

2.10 (b)

If molten metal is too cold, entire mould cavity may not be filled during pouring before the metal starts solidifying and the result is 'misrun'

Causes of Misrun

(i) Too cold molten metal.

(ii) Too small gates.

(iii) Too thin casting section

(iv) Too many restrictions in gating system

(v) Metal lacking fluidity

2.11 (d)

Since cast iron contracts during cooling from solid at freezing point to solid at room temperature

∴ Shrinkage allowance = +1%

Hence, shrinkage allowance given along length

$$= 200 \times 0.01 = 2 \text{ mm}$$

Shrinkage allowance given along breadth

$$= 100 \times 0.01 = 1 \text{ mm}$$

Shrinkage allowance given along height

$$= 10 \times 0.01 = 0.1 \text{ mm}$$

$\therefore$ Volume of pattern,

$$V_P = (200 + 2)(100 + 1)(10 + 0.1)$$
$$= 2.06 \times 10^5 \text{ mm}^3$$

Volume of casting,

$$V_C = 200 \times 100 \times 10 = 2 \times 10^5 \text{ mm}^3$$

$$\therefore \quad \frac{V_P}{V_C} = \frac{2.06\times10^5}{2\times10^5} = 1.03$$

2.12 (d)

(P) Hot chamber machine—Die casting
(Q) Muller—Sand mixing
(R) Dielectric baker—Core making
(S) Sand blaster—Cleaning

2.13 (b)

Length of sprue,

$$h = 20 \text{ cm} = 0.2 \text{ m}$$

Area of down sprue,

$$A = 1 \text{ cm}^2 = 10^{-4} \text{ m}^2$$

Volume of mould cavity

$$= 1000 \text{ cm}^3 = 10^{-3} \text{ m}^3$$

Velocity at base of sprue,

$$V = \sqrt{2gh} = \sqrt{2\times9.81\times0.2} = 1.98 \text{ m/s}$$

Let t is time of filling the mould cavity,

$$t_f = \frac{V_m}{A_g V_g} = \frac{10^{-3}}{10^{-4}\times\sqrt{2\times9.81\times0.2}} = 5.05 \text{ s}$$

2.15 (b)

Time taken to fill the mould with top gate

$$t_A = \frac{AH}{A_g\sqrt{2gh_m}} \quad \text{...(i)}$$

where A = Area of mould
H = Height of mould
A_g = Area of Gate
h_m = Gate height

Given that, $h_m = H$, so Eq. (i) become:

$$t_A = \frac{A\sqrt{h_m}}{A_g\sqrt{2g}} \quad \text{...(ii)}$$

Time taken to fill the mould with bottom gate,

$$t_B = \frac{2A}{A_g\sqrt{2g}}\left(\sqrt{h_m} - \sqrt{h_m - H}\right)$$

$$t_B = \frac{2A}{A_g\sqrt{2g}}\sqrt{h_m} \quad \text{...(iii)}$$

From Eqs. (ii) and (iii), we get

$$\frac{t_B}{t_A} = 2$$

or $\quad t_B = 2t_A$

2.16 (d)

The hot chamber process is used for most of the low melting point alloys, such as zinc, lead and tin.

2.17 (d)

Surface area of cube = $6l^2$
Surface area of sphere = $4\pi r^2$
According to Chvorinov's relation

$$\text{Solidification time} \propto \left(\frac{\text{volume}}{\text{surface area}}\right)^2$$

as volume of cube and sphere are equal

$$\frac{t_c}{t_s} = \left(\frac{A_s}{A_c}\right)^2 = \left(\frac{4\pi r^2}{6l^2}\right)^2$$

$$\frac{t_c}{t_s} = \left(\frac{4\pi}{6}\right)^2\left(\frac{r}{l}\right)^4$$

2.18 (c)

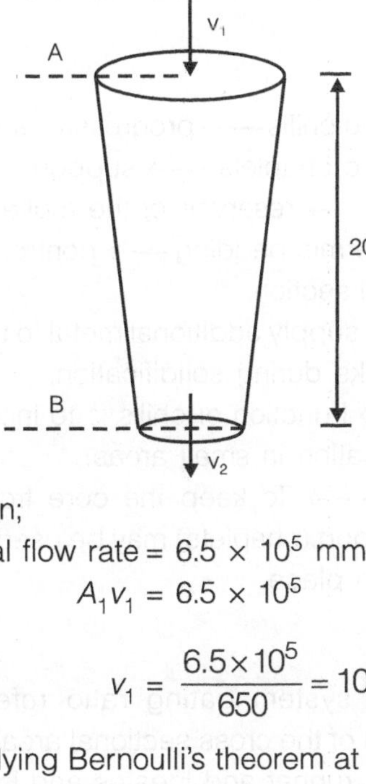

Given;
Metal flow rate = 6.5×10^5 mm^3/s

$$\therefore \quad A_1 V_1 = 6.5 \times 10^5$$

$$\therefore \quad V_1 = \frac{6.5\times10^5}{650} = 1000 \text{ mm/s}$$

Applying Bernoulli's theorem at points A and B, we get

$$h + \frac{V_1^2}{2g} = \frac{V_2^2}{2g} \quad [P_A - P_B = P_{atm}]$$

$$200 + \frac{(1000)^2}{2\times10^4} = \frac{V_2^2}{2\times10^4}$$

$\therefore \quad v_2 = 2236.068$ mm/s

Now, area of down sprue at its end,

$$A_2 = \frac{Q}{v_2} = \frac{6.5\times10^5}{2236.068} = \mathbf{290.7\ mm^2}$$

2.19 (b)

The volume of metal compensated by riser does not constitute cooling during solid state. Riser can compensate for volume shrinkage only in liquid stage and transition stage.

$\therefore$ Volume compensated by riser

$= 3 + 4 = \mathbf{7\%}$

2.20 (a)

Cold Shut:

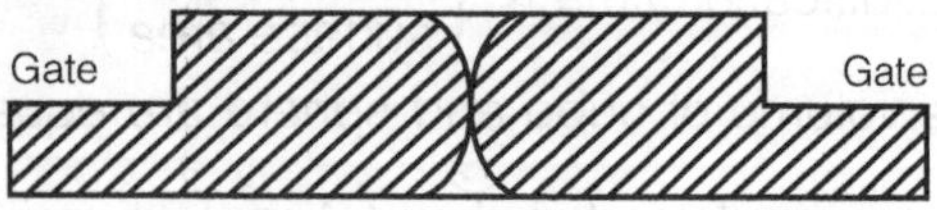

Two stream of liquid metal which are not hot enough to fuse properly result into a casting defect known as cold shut.

or

For a casting with gates at its two sides, the misrun may show up at the centre of the casting. When this happens, the defect is called a cold shut.

2.21 (d)

P. Metallic chills ⟶ progressive solidification
Q. Metallic chaplets ⟶ support for the core
R. Riser ⟶ reservoir of the molten metal
S. Exothermic padding ⟶ control cooling of critical section

Riser ⟶ supply additional metal to the casting as it shrinks during solidification.

Chills ⟶ Function of chills is to increase rate of solidification in small areas.

Chaplets ⟶ To keep the core from shifting metal support (chaplets) may be used to anchor the core in place.

2.22 (a)

In gating system gating ratio refers to the proportion of the cross sectional areas between the sprue, runner and ingates and is generally denoted as sprue area : runner area : ingates area.

2.23 (d)

Green sand mould contains 70–85% sand, 10–20% clay, 3–6% water and 1–6% additives.

2.24 (a)

Side of cube after solidification

$$= \sqrt[3]{0.96\times(50)^3} = 49.32 \text{ mm}$$

Side of cube after contraction

$$= \sqrt[3]{0.94\times(49.32)^3} = \mathbf{48.32\ mm}$$

2.25 (b)

Solidification time,

$$t = k\left(\frac{V}{A}\right)^2$$

For cube, $\left(\frac{V}{A}\right) = \frac{a}{6}$

$$\frac{t_1}{t_2} = \frac{a_1^2}{a_2^2}$$

$$t_2 = 5\left(\frac{a_2}{a_1}\right)^2 \quad \text{...(i)}$$

$$m_1 = V_1 \times \rho_1; m_2 = V_2 \times \rho_2$$

$$V_2 = 8V_1$$

$$a_2^3 = 8a_1^3$$

$\Rightarrow \quad a_2 = 2a_1$

or $\quad \frac{a_2}{a_1} = 2$

Substituting $\frac{a_2}{a_1} = 2$ in Eq. (1), we get

$$t_2 = 5(2)^2 = 5 \times 4 = \mathbf{20\ min}$$

2.27 (c)

Hot tear is a defect that occurs during solidification of a casting. Though steel is poured at high temperature, this temperature is low relative to high freezing temperature range of steel.

2.28 (a)

For least solidification time, surface area should be minimum

$$A = \pi dh + \frac{\pi d^2}{4} \quad \text{(for top rises)}$$

$$V = \frac{\pi}{4}d^2 h$$

$$\Rightarrow \quad h = \frac{4V}{\pi d^2}$$

$$A = \frac{4V}{d} + \frac{\pi d^2}{4}$$

For area to be minimum, $\frac{\partial A}{\partial d} = 0$

$$\Rightarrow -\frac{4V}{d^2}+\frac{\pi d}{2}=0$$

$$\Rightarrow \quad d^3 = \frac{8V}{\pi}=\frac{8}{\pi}\times\frac{\pi}{4}d^2\times h$$

$$\Rightarrow \quad \frac{h}{d} = \frac{1}{2}=1:2$$

2.29 (3)

$$V_{casting} = 7\times 10\times 2 = 140\text{ cm}^3$$

$$A_{casting} = 2(7\times 10+7\times 2+2\times 10)$$

$$= 2\times 104 = 208\text{ cm}^2$$

$$\left(\frac{V}{A}\right)_{casting} = \frac{140}{208}=0.673\text{ cm}$$

$$\left(\frac{V}{A}\right)_{\text{Side Riser}} = \frac{\frac{\pi}{4}d^3}{\frac{\pi}{2}d^2+\pi d^2}=\frac{d}{6}=1\text{ cm}$$

$$\frac{t_R}{t_C} = \frac{(V/A)^2_{\text{Side Riser}}}{(V/A)^2_{\text{Casting}}}=\left(\frac{1}{0.673}\right)^2$$

$\Rightarrow$ t_R = 3 minute (for bottom riser)

2.31 (b)

$$T_{solidification} = k\left(\frac{V}{A}\right)^2$$

$$\Rightarrow \quad \frac{T_1}{T_2} = \frac{k\left(\frac{V_1}{A_1}\right)^2}{k\left(\frac{V_2}{A_2}\right)^2}$$

$$V_1 = V_2$$

$$\Rightarrow \quad \frac{T_1}{T_2} = \frac{A_2^2}{A_1^2}=\frac{(6a^2)^2}{(5a^2)^2}=\frac{36}{25}$$

2.32 (6.157355)

$$\text{Time of solidification }(T) = k\left[\frac{\text{volume}}{\text{surface area}}\right]^2$$

$$\left[\frac{T_{sphere}}{T_{cube}}\right] = \left[\frac{A_{cube}}{A_{sphere}}\right]^2=\left[\frac{6a^2}{4\pi r^2}\right]^2 \quad -(1)$$

$\because [V_{sphere} = V_{cube}]$

$$V_{cube} = a^3 = 1000\text{ cm}^3$$

$$\Rightarrow \quad a = 10\text{ cm}$$

Similarly,

$$V_{sphere} = \frac{4}{3}\pi r^3 = 1000$$

$$\Rightarrow \quad r = 6.2035\text{ cm}$$

Putting in Eq. (1), we get

$$\frac{T_{sphere}}{T_{cube}} = \left[\frac{6\times 10^2}{4\pi\times(6.2035)^2}\right]^2$$

$$T_{sphere} = 6.157355\text{ second}$$

2.33 (0.5625)

$$T_{cy} = k\left[\frac{V}{A}\right]^2 = k\left[\frac{\pi r^2\times h}{2\pi r^2+2\pi rh}\right]^2$$

Given, $h = r$

$$T_{cy} = k\left[\frac{\pi r^3}{4\pi r^2}\right]^2 \Rightarrow T = k\left[\frac{r^2}{16}\right]$$

$$T_{cu} = k\left[\frac{V}{A}\right]^2 = k\left[\frac{a^3}{6a^2}\right]^2$$

But $a = 2h = 2r$

$$T_{cu} = k\left[\frac{4r^2}{36}\right]$$

$$T_{cy}/T_{cu} = \left[k\left[\frac{r^2}{16}\right]\right]\div\left[k\frac{r^2}{9}\right]=\frac{kr^2}{16}\times\frac{9}{kr^2}$$

$$= \frac{9}{16}=0.5625$$

2.34 (10.60)

$$Z = \frac{L+w}{t}$$

Z = shape factor, L = length, w = width, t = thickness

$$Z = \frac{25+15}{5}=8$$

$$\frac{\text{Riser volume}}{\text{Casting volume}} = 0.5$$

Given casting volume = 25 × 15 × 5

$$\frac{\frac{\pi}{4}d^2\times h}{25\times 15\times 5} = 0.5 \qquad \text{Given } (d = h)$$

$$\frac{\pi}{4}d^3 = 0.5\times 25\times 15\times 5$$

$$\Rightarrow \quad d = 10.60\text{ cm}$$

2.36 (1.09)

Ratio of volume of pattern to casting

$$\frac{[(1.03\times 100)\times(1.03\times 50)\times(1.03\times 10)]-\frac{4}{3}\pi(2.06)^3}{(100\times 50\times 10)-\frac{4}{3}\pi(2)^3}$$

$$= 1.09$$

2.37 (c)

Choke area regulates the rate of pouring of molten metal, maintains the required rate of liquid metal flow.

2.38 (a)

According to modulus method

$$M_R = 1.2\, M_C = \left(\frac{V}{A}\right)_R = \left[\frac{\frac{\pi}{4}d^2h}{\frac{\pi}{4}d^2 + \pi dh}\right]$$

If $h = d$

$$M_R = \frac{d}{5}$$

Now, $M_R = 1.2\, M_C$

$$\frac{d}{5} = 1.2\times\left[\frac{\frac{\pi}{4}\times 200^2\times 100}{2\times\frac{\pi}{4}\times 200^2 + \pi\times 200\times 100}\right]$$

$$d = 6\times 25 = 150 \text{ mm}$$

∵ For riser, $d = h =$ **150 mm**

2.39 (2.043)

Given : h_s = 200 mm, u_2 = 0.5 m/s, g = 9.8 m/s²; d_2 = 20 mm

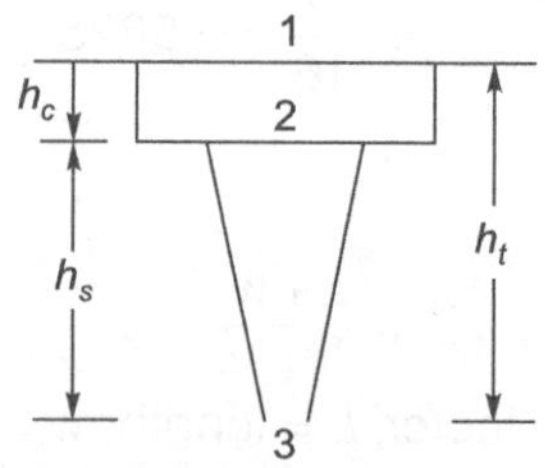

u_3 = ?, $\quad u_2 = \sqrt{2gh_c}$

$$0.5 = \sqrt{2\times 9.81\times h_c}$$

⇒ $\quad h_c = 0.01274$ m

$$h_t = h_s + h_c = 0.200 + 0.01274 = 0.2127 \text{ m}$$

$$u_3 = \sqrt{2gh_t} = \sqrt{2\times 9.81\times 0.2127}$$

$$\mathbf{u_3 = 2.043 \text{ m/s}}$$

2.40 (b)

During solidification of pure metal at the surface of the mould due to fast rate of solidification randomly oriented fine grains are produced.

2.41 (51.84)

Given: Casting size = 80 × 120 × 20 mm³

$(t_s)_{casting}$ = 2 min; $(t_s)_{riser}$ = 3 min

$$(t_s)_{casting} = k\left(\frac{V}{A}\right)^2$$

$$2 = k\left(\frac{80\times 120\times 20}{2(80\times 120 + 120\times 20 + 20\times 80)}\right)^2$$

$$k = 0.040138 \text{ min/mm}^2$$

Riser : $(t_s)_{riser} = k\left(\frac{V}{A}\right)^2$

$$3 = 0.040138\left(\frac{V}{A}\right)^2$$

$$\left(\frac{V}{A}\right) = 8.64$$

$$\frac{V}{A} = \frac{\frac{\pi}{4}d^2h}{2\frac{\pi}{4}d^2 + \pi dh}$$

For riser if $h = d$ then $\frac{V}{A} = \frac{d}{6}$

$$\frac{d}{6} = 8.64$$

$$\mathbf{h = d = 51.84 \text{ mm}}$$

2.43 (a)

Blow holes → Poor permeability
Misrun → Insufficient fluiditity
Hot tears → Poor collapsibility
Wash → Mold erosion

2.44 (c)

Fluidity:

- The ability of a metal of flow and fill a mold is known as fluidity.

Pouring temperature:

- The most important controlling factor of fluidity is the pouring temperature or the amount of superheat.
- Higher the pouring temperature, higher the fluidity.

2.45 (28.92)

Volume of mould cavity = 40 × 30 × 15 cm³
= 18000 cm³

Apply Bernoulli's equation at point 1 and 3

$$\frac{P_{atm}}{\rho g} + \frac{V_1^2}{2g} + z_1 = \frac{P_{atm}}{\rho g} + \frac{V_3^2}{2g} + 0$$

Assume $V_1 = 0$

$$V_3 = \sqrt{2gz_1} = \sqrt{2\times 981\times 20} = 198.09 \text{ cm/s}$$

Mould filling time $(t_f) = \frac{\text{Volume}}{V_2\times A_2}$

$$= \frac{18000 \text{ cm}^3}{198.09 \text{ cm/s}\times\frac{\pi}{4}(2)^2\text{cm}^2} = \mathbf{28.92 \text{ second}}$$

2.46 (4.28)

Given : Volume of mould cavity (V) = 1200 cm^3
Height of sprue (h_s) = 10 cm
Area of sprue at the bottom (A_s) = 2 cm^2
g = 9.81 m/s^2
Sprue is feed a horizontal runner: Filling time required (t_f) = ?

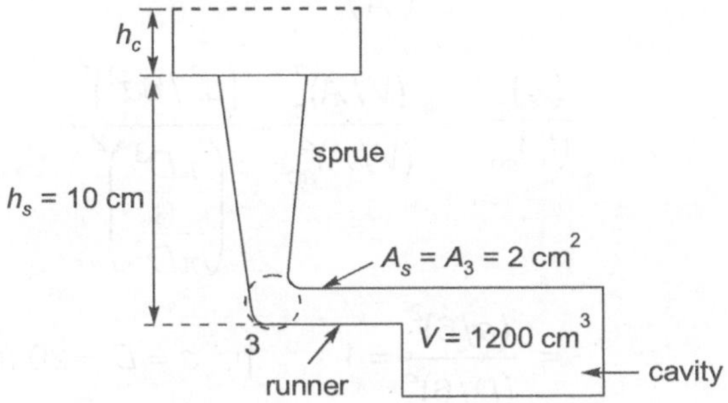

By assuming top gate, $A_g = A_s = A_3 = 2$ cm^2

$$\text{Filling time, } t_f = \frac{V}{A_g v_g} = \frac{1200}{2\sqrt{2\times 981}} = 4.28 \text{ s}$$

2.47 (662)

Given: D_0 = 250 mm; D_i = 230 mm;
g = 10 m/s^2; G-factor = 60

$$F_c = mr\omega^2$$
$$F_g = mg$$

$$\text{G-factor} = \frac{F_c}{F_g} = 60$$

$$\text{G-factor} = \frac{F_c}{F_g} = \frac{mr\omega^2}{mg} = \frac{D_0}{2g}\left(\frac{2\pi N}{60}\right)^2$$

$$N = \sqrt{\frac{2\times \text{G-Factor}\times g}{D_0}}\times\left(\frac{60}{2\pi}\right)$$

$$N = \sqrt{\frac{2\times 60\times 10}{0.25}}\times\left(\frac{60}{2\pi}\right)$$

N = 661.93 = **662 rpm**

2.48 (2.83)

Casting dimensions = 75 mm × 125 mm × 20mm
Total solidification time $(t_s)_{cu}$ = 2.0 min.
Cylindrical casting = H = D = 50 min.
Total solidification Time $(t_s)_{cy}$ = ?
Chvorinov's rule,

$$t_s \propto \left(\frac{V}{A}\right)^n \qquad n=2$$

$$\frac{(t_s)_{cu}}{(t_s)_{cy}} = \frac{\left(\frac{V}{A}\right)^2_{cu}}{\left(\frac{V}{A}\right)^2_{cy}}$$

$$\left(\frac{V}{A}\right)_{cy} = \frac{\frac{\pi}{4}d^2h}{2\frac{\pi}{4}d^2h+\pi dh}$$

$$h = d$$

$$\left(\frac{V}{A}\right)_{cy} = \frac{d}{6}$$

$$\frac{(t_s)_{cu}}{(t_s)_{cy}} = \frac{\left[\frac{75\times 125\times 20}{2(75\times 125+125\times 20+20\times 75)}\right]^2}{\left(\frac{50}{6}\right)^2}$$

$$\frac{2}{(t_s)_{cy}} = \frac{49.131}{69.44} = 0.7075$$

$(t_s)_{cy}$ = **2.83 min.**

2.49 (a, b)

Given: $t_s \propto \frac{V}{A}$

For a cylindrical top riser,
Optimum condition,

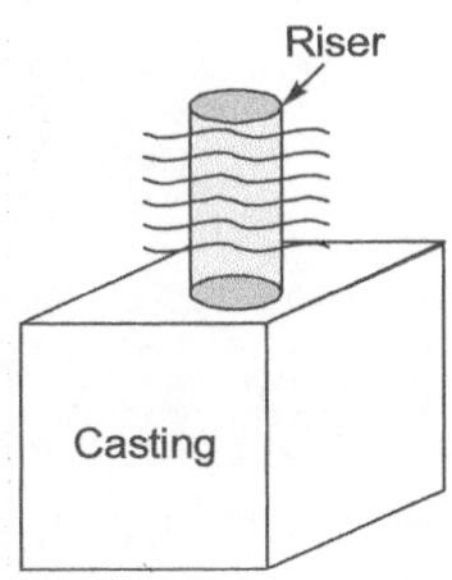

$$h = \frac{d}{2} = r$$

$$(t_s)_r \geq (t_s)_c$$

$$\left(\frac{V}{A}\right)_r \geq \left(\frac{V}{A}\right)_c$$

$$\left(\frac{\pi r^2 h}{2\pi rh}\right) \geq \frac{a}{6}$$

$$\frac{r}{2} \geq \frac{a}{6}$$

$$\Rightarrow \qquad r \geq \frac{a}{3}$$

Thus, (a) and (b) both are correct.

2.50 (a, b, d)

In sand casting process, patterns are used to make the mould cavity while in some other moulding processes like investment and full moulding, permanent moulds can be used for preparing the pattern.

2.51 (b)

Shake allowance is also known as rapping allowance. Before the withdrawal from the sand mould, the pattern is rapped/shaken all around the vertical faces to enlarge the mould cavity slightly, which facilitates its easy removal.

2.52 (1)

Given:
Cube shaped mould = 20 mm × 20 mm × 20 mm
Sphere shaped mould, D = 20 mm
By Chvorinov's rule, solidification time (t_s) is proportional to square of ratio of volume to surface area i.e..

$$t_s \propto \left(\frac{V}{A}\right)^2$$

$$\frac{(t_s)_c}{(t_s)_{sp}} = \frac{(V/A)_c^2}{(V/A)_{sp}^2} = \frac{(a^3/6a^2)^2}{\left(\dfrac{\dfrac{\pi D^3}{6}}{\pi D^2}\right)^2}$$

$$= \frac{(a/6)^2}{(D/6)^2} = 1 \qquad [\because a = D = 20 \text{ mm}]$$

■■■■

3 Metal Forming

3.1 A test specimen is stressed slightly beyond the yield point and then unloaded. Its yield strength

(a) Decreases (b) Increases
(c) remains same (d) become equal to UTS

[1995 : 1 M]

3.2 A wire of 0.1 mm dia is drawn from a rod of 15 mm diameter dies giving reductions of 20%, 40% and 80% are available. For minimum error in the final size, the number of stages and reduction at each stage respectively would be

(a) 3 stages and 80% reduction for all three stages
(b) 4 stages and 80% reduction for first three stages followed by a finishing stage of 20% reduction
(c) 5 stage and reduction of 80%, 80%, 40%, 40%, 20% in sequence
(d) None of these [1996 : 2 M]

3.3 Match the following:

List-I

A. Rivets for aircraft body
B. Carburetor body
C. Crank shafts
D. Nails

List-II

1. Forging
2. Cold heading
3. Aluminum base alloy
4. Pressure die casting
5. Investment casting [1996 : 2 M]

3.4 A strip with cross-sectional area 150 mm × 4.5 mm is being rolled with 20% reduction of area using 450 mm diameter rollers. The angle subtended by the deformation zone at the roll centre is (in radians)

(a) 0.01 (b) 0.02
(c) 0.03 (d) 0.06 [1998 : 2 M]

3.5 Hot rolling of mild steel is carried out

(a) at recrystallization temperature
(b) between 100°C to 150°C
(c) below recrystallization temperature
(d) above recrystallization temperature

[2002 : 1 M]

3.6 Ductility of material with work hardening

(a) Increases (b) Decreases
(c) Remains same (d) Unpredictable

[2002 : 1 M]

3.7 A brass billet is to be extruded from its initial diameter of 100 mm to a final diameter of 50 mm. The working temperature of 700°C and the extrusion constant is 250 MPa. The force required for extrusion is

(a) 5.44 MN (b) 2.72 MN
(c) 1.36 MN (d) 0.36 MN [2003 : 2 M]

3.8 Cold working of steel is defined as working

(a) at its recrystallisation temperature
(b) above its recrystallisation temperature
(c) below its recrystallisation temperature
(d) at two thirds of the melting temperature of the metal [ME 2003 : 1 M]

3.9 a rolling process, sheet of 25 mm thickness is rolled to 20 mm thickness. Roll is of diameter 600 mm and it rotates at 100 rpm. The roll strip contact length will be

(a) 5 mm (b) 39 mm
(c) 78 mm (d) 120 mm [2004 : 2 M]

3.10 Match the following

Product

P. Moulded luggage
Q. Packaging containers for liquid
R. Long structural shapes
S. Collapsible tubes

Process

1. Injection moulding
2. Hot rolling
3. Impact extrusion
4. Transfer moulding
5. Blow moulding
6. Coining

(a) P–1, Q–4, R–6, S–3
(b) P–4, Q–5, R–2, S–3
(c) P–1, Q–5, R–3, S–2
(d) P–5, Q–1, R–2, S–2 [2004 : 2 M]

3.11 A 2 mm thick metal sheet is to be bent at an angle of one radian with a bend radius of 100 mm. If the stretch factor is 0.5, the bend allowance is

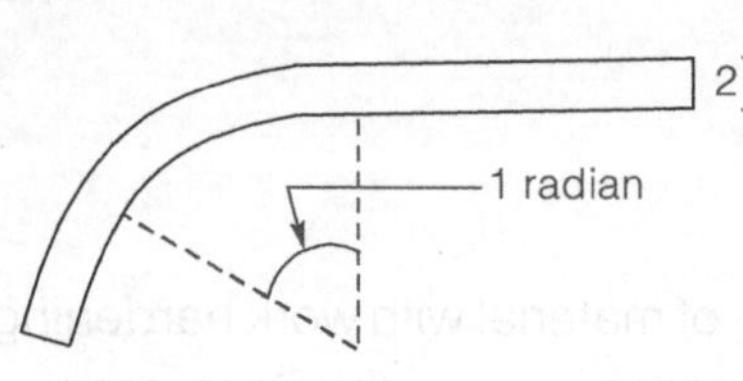

(a) 99 mm (b) 100 mm
(c) 101 mm (d) 102 mm [2005 : 2 M]

3.12 A 4 mm thick sheet is rolled with 300 mm diameter rolls to reduce thickness without any change in its width. The friction coefficient at the work-roll interface is 0.1. The minimum possible thickness of the sheet that can be produced in a single pass is
(a) 1.0 mm (b) 1.5 mm
(c) 2.5 mm (d) 3.7 mm [2006 : 2 M]

3.13 In a wire drawing operation, diameter of a steel wire is reduced from 10 mm to 8 mm. The mean flow stress of the material is 400 MPa. The ideal force required for drawing (ignoring friction and redundant work) is
(a) 4.48 kN (b) 8.97 kN
(c) 20.11 kN (d) 31.41 kN [2006 : 2 M]

3.14 Match the items in columns I and II

Column-I
P. Wrinkling
Q. Orange peel
R. Stretcher strains
S. Earing

Column-II
1. Yield point elongation
2. Anisotropy
3. Large grain
4. Insufficient blank holding force
5. Fine grain size
6. Excessive blank holding force

(a) P–6, Q–3, R–1, S–2
(b) P–4, Q–5, R–6, S–1
(c) P–2, Q–5 R–3, S–1
(d) P–4, Q–3, R–1, S–2 [2006 : 2 M]

3.15 The ultimate tensile strength of a material is 400 MPa and the elongation up to maximum load is 35%. If the material obeys power law of hardening, then the true stress-true strain relation (stress in MPa) in the plastic deformation range is
(a) $\sigma = 540\varepsilon^{0.30}$ (b) $\sigma = 775\ \varepsilon^{0.30}$
(c) $\sigma = 540\ \varepsilon^{0.35}$ (d) $\sigma = 775\ \varepsilon^{0.35}$
[ME 2006 : 2 M]

3.16 In open-die forging, a disc of diameter 200 mm and height 60 mm is compressed without any barreling effect. The final diameter of the disc is 400 mm. The true strain is
(a) 1.986 (b) 1.686
(c) 1.386 (d) 0.602
[2007 : 2 M]

3.17 The thickness of a metallic sheet is reduced from an initial value of 16 mm to a final value of 10 mm in one single pass rolling with a pair of cylindrical rollers each of diameter of 400 mm. The bite angle in degree will be
(a) 5.936 (b) 7.936
(c) 8.936 (d) 9.936 [2007 : 2 M]

3.18 Match the correct combination for following metal working processes.

Processes
P. Blanking
Q. Stretch Forming
R. Coining
S. Deep Drawing

Associated state of stress
1. Tension
2. Compression
3. Shear
4. Tension and Compression
5. Tension and shear

(a) P–2, Q–1, R–3, S–4
(b) P–3, Q–4, R–1, S–5
(c) P–5, Q–3, R–3, S–1
(d) P–3, Q–1, R–2, S–4 [2007 : 2 M]

3.19 The force requirement in a blanking operation of low carbon steel sheet is 5.0 kN. The thickness of the sheet is *t* and diameter of the blanked part is *d*. For the same work material, if the diameter of the blanked part is increased to 1.5*d* and thickness is reduced to 0.4*t*, the new blanking force in kN is
(a) 3.0 (b) 4.5
(c) 6.7 (d) 8.0 [2007 : 2 M]

3.20 In a single pass rolling operation, a 20 mm thick plate with plate width of 100 mm, is reduced to 18 mm. The roller radius is 250 mm and rotational speed is 10 rpm. The average flow stress for the plate material is 300 MPa. The power required for the rolling operation in kW is closest to
(a) 15.2 (b) 18.2
(c) 30.4 (d) 45.6 [2008 : 2 M]

3.21 The maximum possible draft in cold rolling of sheet increases with the
(a) increase in coefficient of friction
(b) decrease in coefficient of friction
(c) decrease in roll radius
(d) increase in roll velocity [2011 : 1 M]

3.22 The operation in which oil is permeated into the pores of a powder metallurgy product is known as
(a) mixing (b) sintering
(c) impregnation (d) infiltration
[2011 : 1 M]

3.23 Match the following metal forming processes with their associated stresses in the workpiece

Metal forming process
A. Coining
B. Wire Drawing
C. Blanking
D. Deep Drawing

Type of stress
1. Tensile
2. Shear
3. Tensile and compressive
4. Compressive

Codes:

	A	B	C	D
(a)	4	1	2	3
(b)	4	1	3	2
(c)	1	2	4	3
(d)	1	3	2	4

[2012 : 1 M]

3.24 A solid cylinder of diameter 100 mm and height 50 mm is forged between two frictionless flat dies to a height of 25 mm. The percentage change in diameter is
(a) 0 (b) 2.07
(c) 20.7 (d) 41.4 [2012 : 1 M]

3.25 In a single pass rolling process using 410 mm diameter steel rollers, a strip of width 140 mm and thickness 8 mm undergoes 10% reduction of thickness. The angle of bite in radians is
(a) 0.006 (b) 0.031
(c) 0.062 (d) 0.600 [2012 : 2 M]

3.26 In a rolling process, the state of stress of the material undergoing deformation is
(a) pure compression
(b) pure shear
(c) compression and shear
(d) tension and shear [2013 : 1 M]

3.27 The relationship between true strain (ε_T) and engineering strain (ε_E) in a uniaxial tension test is given as
(a) $\varepsilon_E = \ln(1 + \varepsilon_T)$ (b) $\varepsilon_E = \ln(1 - \varepsilon_T)$
(c) $\varepsilon_T = \ln(1 + \varepsilon_E)$ (d) $\varepsilon_T = \ln(1 - \varepsilon_E)$
[2014 : 1 M, Set-2]

3.28 With respect to metal working, match Group *A* with Group *B*:

Group A
P. Defect in extrusion
Q. Defect in rolling
R. Product of skew rolling
S. Product of rolling through cluster mill

Group B
I. alligatoring
II. scab
III. fish tail
IV. seamless tube
V. thin sheet with tight tolerance
VI. semi-finished balls of ball bearing

(a) *P*–II, *Q*–III, *R*–VI, *S*–V
(b) *P*–III, *Q*–I, *R*–VI, *S*–V
(c) *P*–III, *Q*–I, *R*–IV, *S*–VI
(d) *P*–I, *Q*–II, *R*–V, *S*–VI [2014 : 1 M, Set-2]

3.29 A metal rod of initial length L_0 is subjected to a drawing process. The length of the rod at any instant is given by the expression, $L(t) = L_0(1 + t^2)$, where t is the time in minutes. The true strain rate (in min^{-1}) at the end of one minute is ______.
[2014 : 2 M, Set-1]

3.30 Match the following products with preferred manufacturing processes:

	Product	Process
P	Rails	1. Blow molding
Q	Engine crankshaft	2. Extrusion
R	Aluminium channels	3. Forging
S	PET water bottles	4. Rolling

(a) *P*–4, *Q*–3, *R*–1, *S*–2
(b) *P*–4, *Q*–3, *R*–2, *S*–1
(c) *P*–2, *Q*–4, *R*–3, *S*–1
(d) *P*–3, *Q*–4, *R*–2, *S*–1 [2015 : 1 M, Set-1]

3.31 The strain hardening exponent *n* of stainless steel SS 304 with distinct yield and UTS values undergoing plastic deformation is
(a) $n < 0$ (b) $n = 0$
(c) $0 < n < 1$ (d) $n = 1$
[2015 : 1 M, Set-3]

3.32 In a slab rolling operation, the maximum thickness reduction (Δh_{max}) is given by $\Delta h_{max} = \mu^2 R$, where R is the radius of the roll and μ is the coefficient of friction between the roll and the sheet. If $\mu = 0.1$, the maximum angle subtended by the deformation zone at the centre of the roll (bite angle in degree) is _____. **[2015 : 2 M, Set-1]**

3.33 In a two-stage wire drawing operation, the fractional reduction (ratio of change in cross-sectional area to initial cross-sectional area) in the first stage is 0.4. The fractional reduction in the second stage is 0.3. The overall fractional reduction is

(a) 0.24 (b) 0.58
(c) 0.60 (d) 1.00

[2015 : 2 M, Set-2]

3.34 The flow stress (in MPa) of a material is given by

$$\sigma = 500\varepsilon^{0.1},$$

where ε is true strain. The Young's modulus of elasticity of the material is 200 GPa. A block of thickness 100 mm made of this material is compressed to 95 mm thickness and then the load is removed. The final dimension of the block (in mm) is____. **[2015 : 2 M, Set-2]**

3.35 In a rolling operation using rolls of diameter 500 mm if a 25 mm thick plate cannot be reduced to less than 20 mm in one pass, the coefficient of friction between the roll and the plate is _____. **[2015 : 2 M, Set-3]**

3.36 Engineering strain of a mild steel sample is recorded as 0.100%. The true strain is

(a) 0.010% (b) 0.055%
(c) 0.099% (d) 0.101%

[2016 : 1 M, Set-3]

3.37 A 10 mm deep cylindrical cup with diameter of 15 mm is drawn from a circular blank. Neglecting the variation in the sheet thickness, the diameter (upto 2 decimal points accuracy) of the blank is _______ mm. **[2017 : 2 M, Set-1]**

3.38 It is desired to make a product having T-shaped cross-section from a rectangular aluminium block. Which one of the following processes is expected to provide the highest strength of the product?

(a) Welding (b) Casting
(c) Metal forming (d) Machining

[2017 : 1 M, Set-2]

3.39 A strip of 120 mm width and 8 mm thickness is rolled between two 300 mm diameter rolls to get a strip of 120 mm width and 7.2 mm thickness. The speed of the strip at the exit is 30 m/min. There is no front or back tension. Assuming uniform roll pressure of 200 MPa in the roll bite and 100% mechanical efficiency, the minimum total power (in kW) required to drive the two rolls is ________. **[2017 : 2 M, Set-2]**

3.40 In a linearly hardening plastic material, the true stress beyond initial yielding

(a) increases linearly with the true strain
(b) decreases linearly with the true strain
(c) first increases linearly and then decreases linearly with the true strain
(d) remain constant **[2018 : 1 M, Set-1]**

3.41 A bar is compressed to half of its original length. The magnitude of true strain produced in the deformed bar is ________(correct to two decimal places). **[2018 : 2 M, Set-1]**

3.42 The maximum reduction in cross-sectional area per pass (R) of a cold wire drawing process is

$$R = 1 - e^{-(n+1)}$$

where n represents the strain hardening coefficient. For the case of a perfectly plastic material, R is

(a) 0.865 (b) 0.826
(c) 0.777 (d) 0.632

[2018 : 2 M, Set-1]

3.43 The true stress (σ), true strain (ε) diagram of a strain hardening material is shown in figure. First, there is loading up to point A, i.e. up to stress of 500 MPa and strain of 0.5. then from point A, there is unloading up to point B, i.e. to stress of 100 MPa, Given that the Young's modulus E = 200 GPa, the natural strain at point B (ε_B)________(correct to two decimal places).

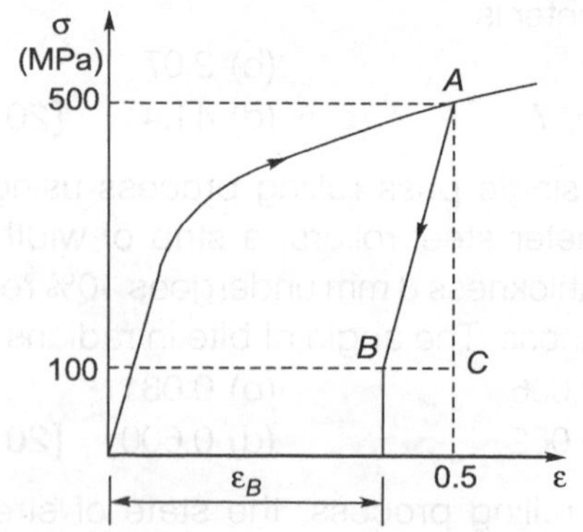

[2018 : 2 M, Set-1]

3.44 Match the following products with the suitable manufacturing process

	Product		Manufacturing Process
P	Toothpaste tube	1	Centrifugal casting
Q	Metallic pipes	2	Blow moulding
R	Plastic bottles	3	Rolling
S	Threaded bolts	4	Impact extrusion

(a) P-4, Q-3, R-1, S-2 (b) P-2, Q-1, R-3, S-4
(c) P-4, Q-1, R-2, S-3 (d) P-1, Q-3, R-4, S-2

[2018 : 1 M, Set-2]

3.45 A steel wire is drawn from an initial diameter (d_i) of 10 mm to a final diameter (d_f) of 7.5 mm. The half cone angle (α) of the die is 5° and the coefficient of friction (μ) between the die and the wire is 0.1. The average of the initial and final yield stress $[(\sigma_Y)_{avg}]$ is 350 MPa. The equation for drawing stress σ_f, (in MPa) is given as.

$$\sigma_f = (\sigma_Y)_{avg}\left\{1+\frac{1}{\mu\cot\alpha}\right\}\left[1-\left(\frac{d_f}{d_i}\right)^{2\mu\cot\alpha}\right]$$

The drawing stress (in MPa) required to carry out this operation is ______ (correct to two decimal places).

[2018 : 2 M, Set-2]

3.46 The true stress (in MPa) versus true strain relationship for a metal is given by
$\sigma = 1020\,\varepsilon^{0.4}$
The cross-sectional area at the start of a test (when the stress and strain values are equal to zero) is 100 mm^2. The cross-sectional area at the time of necking (in mm^2) is ________ (correct to two decimal places).

[2018 : 2 M, Set-2]

3.47 Consider the stress-strain curve for an ideal elastic-plastic strain hardening metal as shown in the figure. The metal was loaded in uniaxial tension starting from O. Upon loading, the stress-strain curve passes through initial yield point at P, and then strain hardens to point Q, where the loading was stopped. From point Q, the specimen was unloaded to point R, where the stress is zero. If the same specimen is reloaded in tension from point R, the value of stress at which the material yields again is ______MPa.

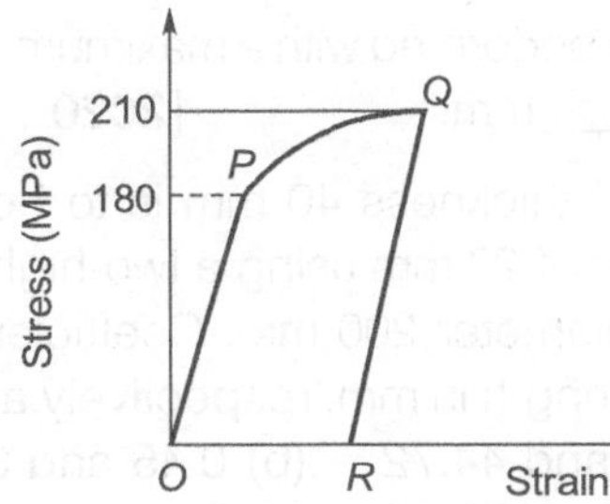

[2019 : 1 M, Set-1]

3.48 A plane-strain compression (forging) of a block is shown in the figure. The strain in the z-direction is zero. The yield strength (S_y) in uniaxial tension/compression of the material of the block is 300 MPa and it follows the Tresca (maximum shear stress) criterion. Assume that the entire block has started yielding. At a point where σ_x = 40 MPa (compressive) and τ_{xy} = 0, the stress component σ_y is

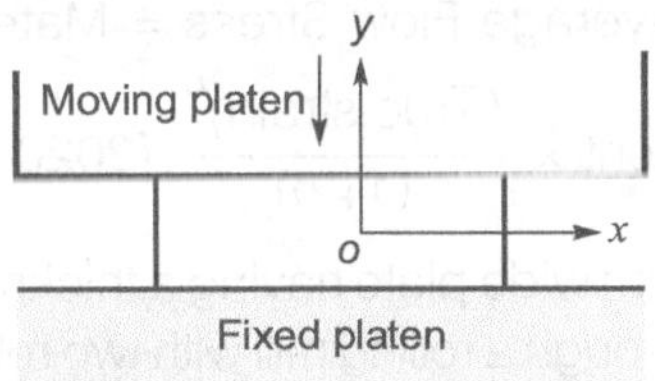

(a) 340 MPa (compressive)
(b) 260 MPa (compressive)
(c) 340 MPa (tensile)
(d) 260 MPa (tensile) [2019 : 2 M, Set-1]

3.49 The cold forming process in which a hardened tool is pressed against a workpiece (when there is relative motion between the tool and the workpiece) to produce a roughened surface with a regular pattern is
(a) Roll forming (b) Chamfering
(c) Knurling (d) Strip rolling

[2019 : 1 M, Set-2]

3.50 The thickness of a sheet is reduced by rolling (without any change in width) using 600 mm diameter rolls. Neglect elastic deflection of the rolls and assume that the coefficient of friction at the roll-workpiece interface is 0.05. The sheet enters the rotating rolls unaided. If the initial sheet thickness is 2 mm, the minimum possible final thickness that can be produced by this process in a single pass is______mm (round off to two decimal places). [2019 : 2 M, Set-2]

3.51 A bolt head has to be made at the end of a rod of diameter d = 12 mm by localized forging (upsetting) operation. The length of the unsupported portion of the rod is 40 mm. To avoid buckling of the rod, a closed forging operation has to be performed with a maximum die diameter of_______ mm. [2020 : 1 M, Set-2]

3.52 A strip of thickness 40 mm is to be rolled to a thickness of 20 mm using a two-high mill having rolls of diameter 200 mm. Coefficient of friction and arc length in mm, respectively are
(a) 0.45 and 44.72 (b) 0.45 and 38.84
(c) 0.39 and 38.84 (d) 0.39 and 44.72
[2020 : 2 M, Set-1]

3.53 The thickness of a steel plate with material strength coefficient of 210 MPa, has to be reduced from 20 mm to 15 mm in a single pass in a two-high rolling mill with a roll radius of 450 mm and rolling velocity of 28 m/min. If the plate has a width of 200 mm and its strain hardening exponent, n is 0.25, the rolling force required for the operation is__________kN (round off to 2 decimal places).
Note: Average Flow Stress = Material Strength Coefficient × $\frac{(\text{True strain})^n}{(1+n)}$ [2020 : 2 M, Set-1]

3.54 A 200 mm wide plate having a thickness of 20 mm is fed through a rolling mill with two rolls. The radius of each roll is 300 mm. The plate thickness is to be reduced to 18 mm in one pass using a roll speed of 50 rpm. The strength coefficient (K) of the work material flow curve is 300 MPa and the strain hardening exponent, n is 0.2. The coefficient of friction between the rolls and the plate is 0.1. If the friction is sufficient to permit the rolling operation then the roll force will be___kN. [Round off to the nearest integer] [2021 : 2 M, Set-1]

3.55 The size distribution of the powder particles used in Powder Metallurgy process can be determined by
(a) Laser absorption (b) Laser penetration
(c) Laser scattering (d) Laser reflection
[2021 : 1 M, Set-2]

3.56 The thickness, width and length of a metal slab are 50 mm, 250 mm and 3600 mm, respectively. A rolling operation on this slab reduces the thickness by 10% and increases the width by 3%. The length of the rolled slab is _____ mm (round off to one decimal place). [2021 : 2 M, Set-2]

3.57 A 4 mm thick aluminum sheet of width w = 100 mm is rolled in a two-roll mill of roll diameter 200 mm each. The workpiece is lubricated with a mineral oil, which gives a coefficient of friction, μ = 0.1. The flow stress (σ) of the material in MPa is $\sigma = 207 + 414\,\varepsilon$, where ε is the true strain. Assuming rolling to be a plane strain deformation process, the roll separation force (F) for maximum permissible draft (thickness reduction) is _____ kN (round off to the nearest integer).

Use:

$F = 1.15\bar{\sigma}\left(1+\frac{\mu L}{2\bar{h}}\right)wL$, where $\bar{\sigma}$ is average flow stress, L is roll-workpiece contact length, and $\bar{h}$ is the average sheet thickness.
[2022 : 2 M, Set-1]

3.58 A cylindrical billet of 100 mm diameter and 100 mm length is extruded by a direct extrusion process to produce a bar of L-section. The cross sectional dimensions of this L-section bar are shown in the figure. The total extrusion pressure (p) in MPa for the above process is related to extrusion ratio (r) as

$$p = K_s\sigma_m\left[0.8 + 1.5\ln(r) + \frac{2l}{d_o}\right]$$

where σ_m is the mean flow strength of the billet material in MPa, l is the portion of the billet length remaining to be extruded in mm, d_o is the initial diameter of the billet in mm, and K_s is the die shape factor.
If the mean flow strength of the billet material is 50 MPa and the die shape factor is 1.05, then the maximum force required at the start of extrusion is___ kN (round off to one decimal place).

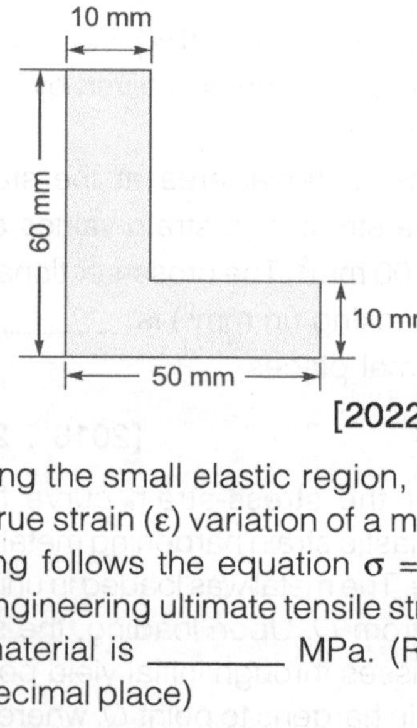

[2022 : 2 M, Set-2]

3.59 Ignoring the small elastic region, the true stress (σ) – true strain (ε) variation of a material beyond yielding follows the equation $\sigma = 400\varepsilon^{0.3}$ MPa. The engineering ultimate tensile strength value of this material is _______ MPa. (Rounded off to one decimal place) [2023 : 2 M]

3.60 The "Earing" phenomenon in metal forming is associated with
(a) deep drawing (b) extrusion
(c) rolling (d) forging [2024 : 1 M]

■■■■

Answers Metal Forming

3.1	(b)	**3.2**	(b)	**3.3**	(A–3, B–4, C–1, D–2)	**3.4**	(d)	**3.5**	(d)	**3.6**	(b)	**3.7**	(b)
3.8	(c)	**3.9**	(b)	**3.10**	(b)	**3.11**	(c)	**3.12**	(c)	**3.13**	(b)	**3.14**	(d)	**3.15**	(b)
3.16	(c)	**3.17**	(d)	**3.18**	(d)	**3.19**	(a)	**3.20**	(a)	**3.21**	(a)	**3.22**	(c)	**3.23**	(a)
3.24	(d)	**3.25**	(c)	**3.26**	(c)	**3.27**	(c)	**3.28**	(b)	**3.29**	(1)	**3.30**	(b)	**3.31**	(c)
3.32	(5.73)	**3.33**	(b)	**3.34**	(95.167)	**3.35**	(0.1414)	**3.36**	(c)	**3.37**	(28.723)	**3.38**	(c)
3.39	(9.12)	**3.40**	(a)	**3.41**	(0.693)	**3.42**	(d)	**3.43**	(0.498)	**3.44**	(c)	**3.45**	(316.25)
3.46	(67.032)	**3.47**	(210)	**3.48**	(a)	**3.49**	(c)	**3.50**	(1.25)	**3.51**	(18)	**3.52**	(a)
3.53	(1167.26)			**3.54**	(780.309)			**3.55**	(c)	**3.56**	(3883.5)	**3.57**	(351)
3.58	(2429.3)	**3.59**	(206.55)	**3.60**	(a)								

Explanations Metal Forming

3.1 (b)

Due to strain hardening phenomenon, yield strength will increase after loading a material slightly beyond yield point.

3.2 (b)

$$\% \text{ reduction} = \frac{\text{Diameter reduced in draw}}{\text{Diameter before draw}}$$

$$\text{for option (a)} = \frac{d_0 - d_1}{d_0} = 0.8$$

$$d_1 = 0.2\, d_0$$

Given, $d_0 = 15$ mm, we get $d_1 = 3$ mm

Similarly, after 2nd pass, we get $d_2 = 0.6$ mm and after 3rd pass, we get $d_3 = 0.12$ mm

For option (b), 1st 3 stages are same as in option (a), in 4th pass we have

$$\frac{d_3 - d_4}{d_3} = 0.2$$

$$\Rightarrow \quad d_4 = 0.8 d_3$$

$$\Rightarrow \quad d_4 = 0.8 \times 0.12 = \mathbf{0.096\ mm}$$

For option (c), first two stages are same as in option (a), in 3rd pass we have

$$\frac{d_2 - d_3}{d_3} = 0.4$$

$$\Rightarrow \quad d_3 = 0.6\, d_2 = 0.6 \times 0.6 = \mathbf{0.36\ mm}$$

For 4th pass we have

$$\frac{d_3 - d_4}{d_3} = 0.4$$

$$\Rightarrow \quad d_4 = 0.6\, d_3 = 0.6 \times 0.36 = \mathbf{0.216\ mm}$$

For 5th stage, we get

$$\frac{d_5 - d_4}{d_4} = 0.2$$

$$\Rightarrow \quad d_5 = 0.8\, d_4 = 0.8 \times 0.216 = \mathbf{0.1728\ mm}$$

Output in option (b) is closest to the desired diameter of 0.1 mm

3.4 (d)

$$\text{Bite angle: } \theta = \tan^{-1}\left(\frac{\sqrt{R\Delta h}}{R - \frac{\Delta h}{2}}\right)$$

From question, we have 20% reduction in cross-sectional area

$$150 \times 4.5 \times 0.8 = 150 \times h$$

$$h = 3.6 \text{ mm}$$

$$\Delta h = 4.5 - 3.6 = 0.9 \text{ mm}$$

$$\text{Given, } R = \frac{450}{2} = 225 \text{ mm}$$

$$\theta = \tan^{-1}\left(\frac{\sqrt{225 \times 0.9}}{225 - \frac{0.9}{2}}\right) = 3.62° = \mathbf{0.06\ rad}$$

3.5 (d)

Any hot working process is done at a tempeature greater than recrystalline temperature of material.

3.6 (b)

On strain hardening or work hardening, ductility of material decreases.

3.7 (b)

Force required for extrusion

$$= k\, A_i \ln\left(\frac{A_i}{A_f}\right)$$

where k = extrusion constant = 250×10^6 N/m²

A_i = initial area

$= \frac{\pi}{4} \times (0.1)^2 = 7.854 \times 10^{-3}$ m²

A_f = final area = $\frac{\pi}{4} \times (0.05)^2$

$= 1.96 \times 10^{-3}$ m²

Extrusion force,

$$P = 250 \times 10^6 \times \frac{\pi}{4} \times (0.1)^2 \ln\left(\frac{7.854 \times 10^{-3}}{1.96 \times 10^{-3}}\right)$$

$= 2.72 \times 10^6$ N = **2.72 MN**

3.8 (c)

Cold working—when material is deformed below its recrystallization temperature. If working temperature is higher than the recrystallization temperature it is called hot working.

3.9 (b)

Given; t_i = 25 mm t_f = 20 mm

R = 300 mm N = 100 rpm

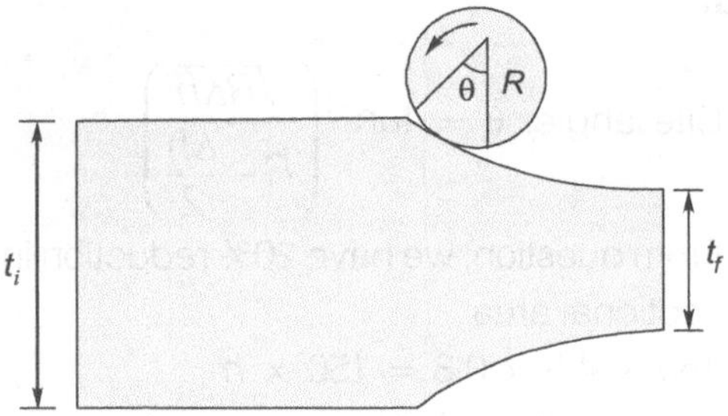

Angle subtended by the deformation zone at the roll centre,

$$\cos\theta = 1 - \frac{\Delta h}{D} = 1 - \frac{5}{600}$$

θ = 0.129 radian

Roll strip contact length = $\theta.R$ = 0.129 × 300

= **38.7 mm**

3.11 (c)

We know that

Bend allowance = α ($R + kt$)

Where R = inside radius of the bend

= 100 mm

α = bend angle = 1 radian

t = thickness of metal sheet

k = Location of neutral axis from the bottom

= 0.33 where $R < 2t$

= 0.50 where $R > 2t$

∴ Bend allowance = 1 × (100 + 0.5 × 2)

= **101 mm**

3.12 (c)

Given,

Initial thickness of sheet,

t_i = 4 mm = 4×10^{-3} m

Diameter of roll,

D = 300 mm = 0.3 m

Frictional coefficient at work–roll interface,

μ = 0.1

Let t_f = Final thickness of sheet after rolling

For single pass without slipping minimum possible thickness,

$$t_i - t_f = \mu^2 R$$

$$\mu = \sqrt{\frac{t_i - t_f}{R}}$$

$$0.1 = \sqrt{\frac{4 \times 10^{-3} - t_f}{0.15}}$$

or t_f = **2.5 mm**

3.13 (b)

Given, D_i = 10 mm, D_f = 8 mm

σ_0 = 400 MPa

Ignoring friction and redundant work means

$\mu = 0, \beta = 0$

Hence $\sigma_d = 2\sigma_0 \ln\left(\frac{r_i}{r_f}\right) = 2 \times 400 \ln\left(\frac{5}{4}\right)$

= 178.51 MPa

Ideal force = $178.51 \times 10^6 \times \pi r_f^2$

= $178.51 \times 10^6 \times 3.14 \times (0.004)^2$

= 8968 N = **8.97 kN**

3.15 (b)

True strain and engineering strain are related as:

$\epsilon = \ln(1 + \epsilon_0)$

$= \ln(1 + 0.35) = 0.3$

But at UTS $n = \epsilon$

Hence $n = \epsilon = 0.3$

Also $\sigma_t = \sigma_0(1 + \epsilon_0) = 400\ (1 + 0.35)$

= 540 MPa

Now, $\sigma_t = K\epsilon^n$ (Power law equation)

$540 = K(0.3)^{0.3}$

or K = 774.97

Hence $\sigma_f = 775\epsilon^{0.3}$

Where, σ_f = true stress

ϵ = true strain

3.16 (c)

Since volume of disc remains constant during the process, Hence

$$\frac{\pi D_1^2}{4} \times l_1 = \frac{\pi D_2^2}{4} l_2$$

or $D_1^2 l_1 = D_2^2 l_2$

$$200^2 \times 60 = 400^2 \times l_2$$

or $l_2 = 15$ mm

Change in length $= l_1 - l_2 = 60 - 15 = 45$ mm

Strain: $\varepsilon = \frac{l_1 - l_2}{l_2} = \frac{45}{15} = 3$

$\therefore$ True strain $= \ln(1 + \varepsilon) = \ln(1+3)$

$= \ln 4 = $ **1.386**

3.17 (d)

Initial sheet thickness, $t_1 = 16$ mm

Final sheet thicknesss, $t_2 = 10$ mm

Roll radius, $R = \frac{400}{2} = 200$ mm

$$\tan\alpha = \sqrt{\frac{t_1 - t_2}{R}} = \sqrt{\frac{16-10}{200}}$$

$\tan\alpha = 0.173$

$\therefore$ Bite angle, $\alpha = \tan^{-1}(0.173) =$ **9.936°**

3.19 (a)

Let τ be the shear stress

Blanking force: $F_b = \tau d t$

We know that

$$F_b \propto d t$$

$$\therefore \quad \frac{F_{b2}}{F_{b1}} = \frac{1.5d \times 0.4t}{dt}$$

$\therefore$ $F_{b2} = 0.6 F_{b1} = 0.6 \times 5 =$ **3 kN**

3.20 (a)

$h_1 = 20$ mm; $h_2 = 18$ mm; $R = 250$ mm

Width = 100 mm, $N = 10$ rpm

$\sigma_0 = 300$ MPa

$\therefore$ Load $= \sigma_0 \times L \times W$

when $L = \sqrt{R\Delta h}$

$$\text{Load} = 300 \times 10^6 \times \sqrt{0.25 \times (0.02 - 0.018)} \times 0.1$$

$= 670$ kN

Power $= 2T \times \omega$

$$= 2 \times 670 \times 0.5 \times \sqrt{0.25 \times (0.02 - 0.018)} \times 2\pi \times \frac{10}{60}$$

$= 15.68$ kW $\approx$ **15.2 kW**

3.24 (d)

Since volume remains constant

$\therefore$ $V_1 = V_2$

$$\frac{\pi}{4} d_1^2 h_1 = \frac{\pi}{4} d_2^2 h_2$$

$$(100)^2(50) = d_2^2(25)$$

or $d_2 = 141.4$ m

$$\therefore \quad \%\text{ change} = \frac{d_2 - d_1}{d_1} \times 100$$

$$= \left(\frac{141.4 - 100}{100}\right) \times 100$$

= **41.4%**

3.25 (c)

$$\Delta H = D(1 - \cos\alpha)$$

$$\cos\alpha = 1 - \frac{\Delta H}{D} = 1 - \frac{0.1 \times 8}{410} = 0.99804$$

or $\alpha = \cos^{-1}(0.99804) = 3.58°$

= **0.0625 radians**

3.27 (c)

True strain: $d\epsilon_T = \frac{dL}{L}$

$$\epsilon_T = \int_{L_0}^{L} \frac{dL}{L} = \ln\frac{L}{L_0}$$

As $dV = 0$

$$AL = A_0 L_0$$

$$\frac{L}{L_0} = \frac{A_0}{A}$$

$$\epsilon_E = \frac{L - L_0}{L_0}$$

$$\Rightarrow \quad \frac{L}{L_0} = 1 + \epsilon_E$$

$$\therefore \quad \epsilon_T = \ln(1 + \epsilon_E)$$

3.29 (1)

$$L(t) = L_0(1 + t^2)$$

$$\Rightarrow \quad dL = L_0 \times 2t\,dt$$

As $d\epsilon_T = \frac{dL}{L}$

$$d\epsilon_T = \frac{L_0 2t\,dt}{L_0(1 + t^2)}$$

$$\frac{d\in_T}{dt} = \frac{2t}{1+t^2}$$

at $t = 1$ min

$$\frac{d\in_T}{dt} = \frac{2\times1}{1+1^2} = 1\ \text{min}^{-1}$$

3.32 (5.73)

$$\Delta h_{max} = \mu^2 R$$
$$\Delta h = D(1-\cos\alpha)$$
$$\Rightarrow \mu^2 R = D(1-\cos\alpha)$$
$$\Rightarrow (0.1)^2 = 2(1-\cos\alpha)$$
or $\alpha = \mathbf{5.73°}$

3.33 (b)

Fractional reduction in first stage

$$= \frac{\text{Change in area in first stage}}{\text{Original area }(A)\text{ before reduction}}$$

$$0.4 = \frac{\text{Change in area in first stage}}{A}$$

Change in area in first stage = $0.4A$

Area after first redirection = $A - 0.4A = 0.6A$

Fraction reduction in second stage

$$= \frac{\text{Change in area in second stage}}{\text{Original area before second stage}}$$

$$0.3 = \frac{\text{Change in area in second stage}}{0.6A}$$

Change in after 2nd stage = $0.18A$

Remaining area after 2nd reduction

$= 0.6A - 0.18A = 0.42A$

Overall reduction in area

$= A - 0.42A = 0.58A$

Overall fractional reduction

$$= \frac{\text{Overall reduction in area}}{\text{Original area}} = \frac{0.58A}{A} = \mathbf{0.58}$$

3.34 (95.167)

Given: $\sigma = 500 \in^{0.1}$, $E = 200$ GPa, $h_o = 100$ mm, $h_f = 95$ mm

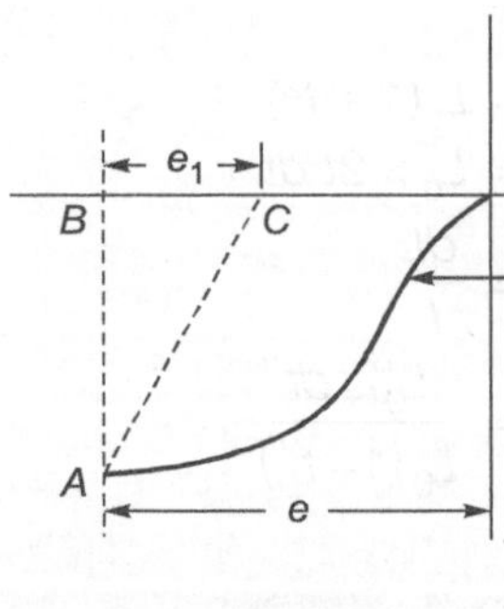

$$\in = \ln\left(\frac{h_i}{h_f}\right) = \ln\left(\frac{100}{95}\right) = 0.051$$

$$\sigma_f = 500\,(0.051)^{0.1} = 371.3\ \text{MPa}$$
$$\in = \ln(1+e)$$
$$0.051 = \ln(1+e)$$
$$e = 0.05232$$

$\in$ = True strain, e = Total engg. strain,

e_1 = Elastic strain

$$\sigma_f = \sigma(1+e)$$
$$371.3 = \sigma(1+0.05232)$$
$$\sigma = 352.83\ \text{MPa}$$

$$e_1 = \frac{\sigma}{E} = \frac{352.8394}{200\times10^3} = 0.00176$$

$$e_1 = \frac{\Delta h}{h} \Rightarrow \Delta h = e_1 \times h = 0.00176\times95$$

$$\Delta h = 95 + 0.167 = 95.167\ \text{mm}$$

3.35 (0.1414)

$D = 500$ mm; $R = 250$ mm

$(\Delta h)_{max} = \mu^2 R$; $5 = \mu^2 \times 250$

$5/250 = \mu^2$

or $\mu = \mathbf{0.1414}$

3.36 (c)

Given data:

Engineering strain,

$$\in = 0.100\% = 0.001$$

True strain : $\bar{\in} = \ln(1+\in) = \ln(1+0.001)$

$= 0.00099 = \mathbf{0.099\%}$

3.37 (28.723)

Given : $d = 15$ mm, $h = 10$ mm

$$D = \sqrt{d^2 + 4dh} = \sqrt{15^2 + 4\times15\times10}$$
$$= 28.723\ \text{mm}$$

3.38 (c)

In metal forming there will be increased mechanical properties due to change in grain size and work hardening.

3.39 (9.12)

Method-I:

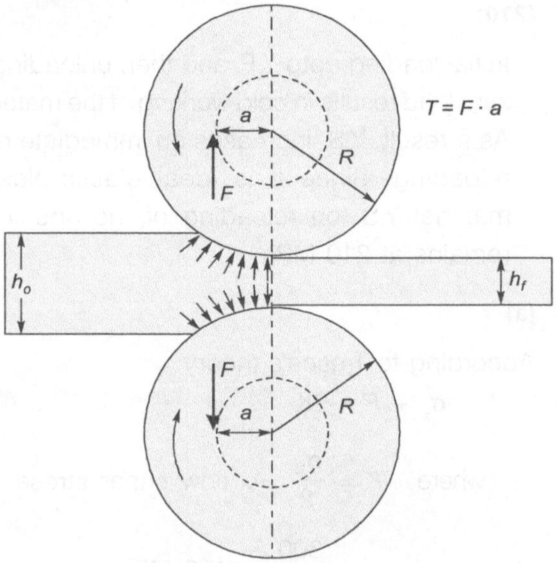

Projected length,

$$L_p = R\sin\alpha \approx \sqrt{R\Delta h}$$
$$= \sqrt{150 \times (8 - 7.2)} = 10.954 \text{ mm}$$

Projected area,

$$A_p = L_p \times b = 10.954 \times 120$$
$$= 1314.48 \text{ mm}^2$$

Roll Separating Force,

$$F = \sigma_o \times L_p \times b \text{ N}$$
$$= 200 \times 1314.48 \text{ N} = 262.9 \text{ kN}$$

[σ_o in N/mm² i.e. MPa]

Arm length(a in mm) = 0.5 L_p for hot rolling

$$= 0.5 \times 10.954 \text{ mm} = 5.477 \text{ mm}$$
$$= 0.45\ L_p \text{ for cold rolling}$$

Torque per roller,

$$T = F \times \frac{a}{1000} \text{ Nm}$$
$$= 262.9 \text{ kN} \times 5.477 \text{ mm}$$
$$= 1439.9 \text{ Nm}$$

Total power for two roller,

$$P = 2T\omega = 2T \times \frac{2\pi N}{60}$$

For the calculation of N we need velocity of neutral plane

Continuity equation

$$h_o\, b_o\, v_o = h_f\, b_f\, v_f$$
$$8 \times 120 \times v_o = 7.2 \times 120 \times 30$$

or $v_o = 27$ m/min

Assuming velocity of neutral plane average velocity,

$$V = \frac{V_0 + V_f}{2} = \frac{27 + 30}{2}$$
$$= 28.5 \text{ m/min}$$

Now, $V = \pi DN$

or $28.5 = \pi \times 0.300 \times N$

or $N = 30.24$ rpm

∴ Total power for two roller,

$$P = 2T \times \frac{2\pi N}{60}$$
$$= 2 \times 1439.9 \times \frac{2 \times \pi \times 30.24}{60} \text{ W}$$
$$= \mathbf{9.1195\ kW \approx 9.12\ kW}$$

Method-II :

Velocity at the neutral plane = ?

$$H_n V_n = H_2 V_2 \quad \text{(Applying continuity)}$$
$$H_n = H_2 + 0.25\ \Delta H$$

(Standard relation)

$$= 7.2 + 0.25 \times 0.8 = 7.4 \text{ mm}$$
$$V_n = \frac{H_2 V_2}{H_n} = 29.19 \text{ m/min}$$
$$N = \frac{29.19 \times 1000}{\pi \times 300} = 30.98 \text{ rpm}$$

Total power = $2T\omega$

$$= 2 \times 1439.9 \times 2\pi\frac{30.98}{60}$$
$$= \mathbf{9.34\ kW}$$

3.40 (a)

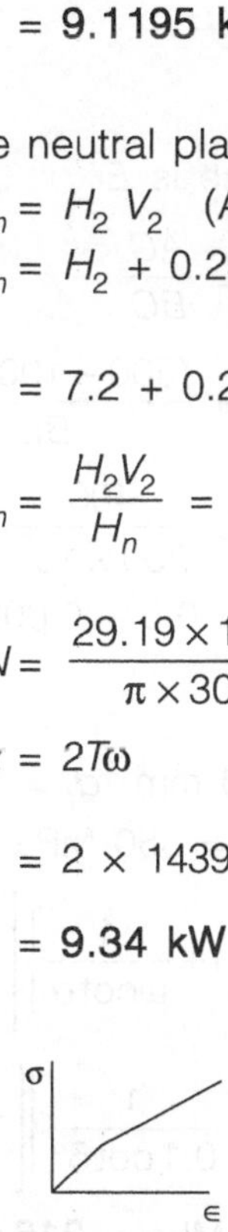

3.41 (0.693)

L_0 → Initial length

$L = \frac{L_0}{2}$ → Final length

$$\text{True strain, } \epsilon_T = \ln\frac{L}{L_0} = \ln\left[\frac{\left(\frac{L_0}{2}\right)}{L_0}\right]$$
$$= \ln\frac{1}{2} = -0.693$$

As examiner mentioned "magnitude" only magnitude will be given **0.693**.

3.42 (d)

Given:

Maximum reduction in cross sectional area per pass,

$$R = 1 - e^{-(n+1)}$$

For perfectly plastic material, strain hardening coefficient n will be zero.

So, $R = 1 - e^{-(0+1)}$

$= 1 - e^{-1} = 0.632$

Maximum reduction per pass,

$$R = \mathbf{0.632}$$

3.43 (0.498)

We know that

Slope of AB line is E

$$E = \frac{AC}{BC}$$

$$200 \times 10^3 \text{ MPa} = \frac{(500-100)\text{MPa}}{BC}$$

$$BC = \frac{400}{200 \times 10^3} = 0.002$$

$\therefore$ $\epsilon_E = 0.5 - 0.002 = \mathbf{0.498}$

3.45 (316.25)

Given: $d_i = 10$ mm, $d_f = 7.5$ mm, $\alpha = 5°$

$\mu = 0.1$, $(\sigma_y)_{avg} = 350$ MPa

$$\sigma_f = (\sigma_y)_{avg} \times \left\{1 + \frac{1}{\mu \cot\alpha}\right\}\left[1 - \left(\frac{d_f}{d_i}\right)^{2\mu\cot\alpha}\right]$$

$$= 350 \times \left[1 + \frac{1}{0.1\cot 5°}\right]\left[1 - \left(\frac{7.5}{10}\right)^{2\times 0.1\times\cot 5°}\right]$$

$= 316.2472$ MPa $\approx$ **316.25 MPa**

3.46 (67.032)

Given, $\sigma = 1020\,\varepsilon^{0.4}$

at UTS, $\varepsilon_T = n = 0.4$

at UTS, neck formation starts,

$$\varepsilon_T = 0.4 = \ln\left(\frac{A_0}{A_f}\right)$$

$$0.4 = \ln\left(\frac{100}{A_f}\right)$$

$$\frac{100}{A_f} = e^{0.4}$$

$$A_f = \frac{100}{e^{0.4}}$$

Cross-section area at the time of necking,

$$A_f = \mathbf{67.032\ mm^2}$$

3.47 (210)

Initial loading upto Y.P. and then unloading to zero load results in cold working of the material. As a result, Y.S. increases on immediate next reloading. Since it is ideal elastic-plastic, material Y.S. on reloading of the specimen remains at 210 MPa.

3.48 (a)

According to Tresca's theory

$$\sigma_x + P = 2K$$

where $K = \frac{\sigma_0}{2}$ = flow shear stress

$= \frac{300}{2} = 150$ MPa

and $\sigma_x = -40$ MPa

Now, $\sigma_x + P = 2K$

$-40 + P = 2 \times 150$

or $P = 340$ MPa

P is pressure. If $P = 340$ MPa, stress (σ_y) will be -340 MPa. i.e. 340 MPa (compressive)

3.50 (1.25)

$D = 600$ mm

$\mu = 0.05$

$h_0 = 2$ mm

$h_{f\,min} = ?$

$h_0 - h_{f\,min} = \mu^2 R$

$2 - h_{f\,min} = 0.05^2 \times 300$

$h_{f\,min} = 1.25$ mm

3.51 (18)

If $l > 3d$ then

Die dia $= 1.5d$

$= 1.5(12)$

$= 18$ mm

3.52 (a)

$h_0 = 40$ mm, $h_f = 20$ mm,

$\Delta h = 40 - 20 = 20$ mm, $D = 200$ mm,

$R = 100$ mm

Projected length,

$$L = \sqrt{R\Delta h} = \sqrt{100 \times 20}$$

$$= \sqrt{2000} = 44.7213 \text{ mm}$$

$$\Delta h = \mu^2 R$$

$$\therefore \quad 20 = \mu^2 \times 100$$

$$\therefore \quad 0.2 = \mu^2$$

$$\text{or,} \quad \mu = 0.4472$$

3.53 (1167.26)

$$\text{True strain, } (\in_T) = \ln\left(\frac{h_f}{h_o}\right) = \ln\left(\frac{15}{20}\right) = -0.28768$$

$$\bar{\sigma}_o = k\left(\frac{\in_T^n}{1+n}\right) = 210 \times \frac{(0.28768)^{0.25}}{1+0.25} = 123.04 \text{ MPa}$$

$$F = \bar{\sigma}_o \times \sqrt{R\Delta h} \times b$$

$$= 123.04 \times \sqrt{450 \times (20-15)} \times 200$$

$$= 1167259.9 \text{ N} = 1167.26 \text{ kN}$$

3.54(780.309)

$b = 200$ mm, $h_I = 20$ mm, $h_F = 18$ mm, $r = 300$ mm

$$\therefore \quad \sigma_{mf} = \frac{k\varepsilon^n}{n+1}$$

$$\because \varepsilon = \ln\frac{h_F}{h_I} = \ln\left(\frac{18}{20}\right) = -0.105 = 0.105$$

$$\because \quad \sigma_{mf} = \frac{300(0.105)^{0.2}}{0.2+1} = 159.28 \text{ MPa}$$

$$\because L_P = \sqrt{R\Delta h} = \sqrt{300 \times 2} = 24.49 \text{ mm}$$

$$A_P = b \times L_P = 200 \times 24.49$$

$$= 4898.97 \text{ mm}^2$$

$$F = \sigma_{mf} \times A_P$$

$$= 159.28 \times 4898.97$$

$$= 780309.45 \text{ N}$$

$$= \mathbf{780.309 \text{ kN}}$$

3.58 (3883.5)

$h_1 = 50$ mm; $h_2 = 0.9h_1$; $b_1 = 250$ mm; $b_2 = 1.03\,b_1$; $L_1 = 3600$ mm; $L_2 = ?$

Volume remains constant in theory of plasticity

$$h_1 b_1 L_1 = h_2 b_2 L_2 = 0.9h_1 \times 1.03 b_1 \times L_2$$

$$L_2 = \frac{3600}{0.9 \times 1.03} = 3883.5 \text{ mm}$$

3.57 (351) (340 to 360)

Given : $w = 100$ mm, $D = 200$ mm, $h_0 = 4$ mm, $\mu = 0.1$

$$\sigma_f = \sigma_t = 207 + 414\in$$

Roll separating force, $F = ?$

$$F = 1.15\bar{\sigma}\left(1 + \frac{\mu L}{2\bar{h}}\right)wL$$

$$\mu^2 R = (\Delta h)_{max}$$

$$h_0 - h_f = (0.1)^2 \times 100$$

$$h_0 - h_f = 1$$

$$h_f = 3 \text{ mm}$$

$$(\Delta h)_{max} = 4 - 3 = 1 \text{ mm}$$

Contact length, $L = \sqrt{R\Delta h}$

Average sheet thickness,

$$\bar{h} = \frac{h_0 + h_f}{2}$$

$$L = \sqrt{R \times \Delta h} = \sqrt{100 \times 1} = 10 \text{ mm}$$

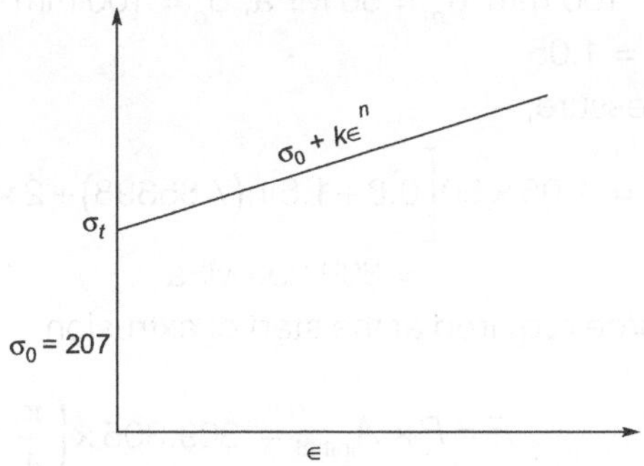

$$\bar{h} = \frac{4+3}{2} = 3.5 \text{ mm}$$

$$\sigma_f = \sigma_t = 207 + 414\in$$

$$\sigma_{avg} = \bar{\sigma} = \frac{1}{\in}\int_0^{\in} \sigma_t d\in = \frac{1}{\in}\int_0^{\in}\left(\sigma_0 + k\in^n\right)d\in$$

$$= \frac{1}{\in}\left[\sigma_0 \in + \frac{k\in^{n+1}}{n+1}\right]$$

$$\sigma_{avg} = \sigma_0 + \frac{k\in}{2} \quad [\text{if } n = 1]$$

$$\sigma_{avg} = 207 + \frac{414\in}{2}$$

$$\text{True strain, } \in = \ln\left(\frac{h_0}{h_f}\right) = \ln\left(\frac{4}{3}\right) = 0.2876$$

$$\sigma_{avg} = \bar{\sigma} = 207 + \frac{414 \times 0.2876}{2} = 266.53 \text{ MPa}$$

$$F = 1.15\bar{\sigma}\left(1 + \frac{\mu L}{2\bar{h}}\right)wL$$

$$F = 1.15 \times 266.53\left(1 + \frac{0.1 \times 10}{2 \times 3.5}\right)100 \times 10$$

$$= 351 \text{ kN}$$

3.58 (2429.3) (2426.0 to 2432.0)

Diameter of cylindrical billet $d_i = 100$ mm

Length of cylindrical billet l_i = 100 mm

So, Area $A_i = \frac{\pi}{4} \times 100^2 = 7853.98 \text{ mm}^2$

After extrusion area of final L section

$$A_f = 50 \times 10 + 50 \times 10 = 1000 \text{ mm}^2$$

$$\text{Extrusion ratio, } r = \frac{A_i}{A_f} = \frac{7853.98}{1000} = 7.85398$$

As given in question

Total extrusion pressure,

$$P = k_s \sigma_m \left[0.8 + 1.5 \ln(r) + \frac{2l}{d_o}\right]$$

At start of extrusion, $r = 7.85398$

l = 100 mm; σ_m = 50 MPa; d_o = 100 mm

k_s = 1.05

Pressure,

$$P = 1.05 \times 50 \left[0.8 + 1.5 \ln(7.85398) + 2 \times \frac{100}{100}\right] = 309.296 \text{ MPa}$$

Force required at the start of extrusion,

$$F = P \times A_{\text{initial}} = 309.305 \times \left(\frac{\pi}{4} \times 100^2\right) = 2429278.724 \text{ N} = 2429.278 \text{ kN}$$

3.59 (206.55) (206.4 to 206.6)

Given: $\sigma_T = 400 \varepsilon_T^{0.3}$

At ultimate tensile strength

$$n = \varepsilon_T = 0.3$$

So, true stress at ultimate tensile strength

$$\sigma_T = 400 \times 0.3^{0.3} = 278.74 \text{ MPa}$$

True strain, $\varepsilon_T = \ln(1 + \varepsilon)$

$\Rightarrow$ $1 + \varepsilon = e^{0.3} = 1.35$

Engineering ultimate tensile strength

$$\sigma = \frac{\sigma_T}{1+\varepsilon} = \frac{278.74}{1.35} = 206.5 \text{ MPa}$$

3.60 (a)

Anisotropy plays an important role in the performance of deep drawing processes. The anisotropy is of two types. In normal anisotropy, the properties differ in the thickness direction. In planar anisotropy, the properties vary with the orientation in the plane of the sheet. Whereas deep drawability of sheets increases with normal anisotropy, anisotropy leads to the formation of ears in cup drawing. Ears cause the wavy edge of a drawn cup

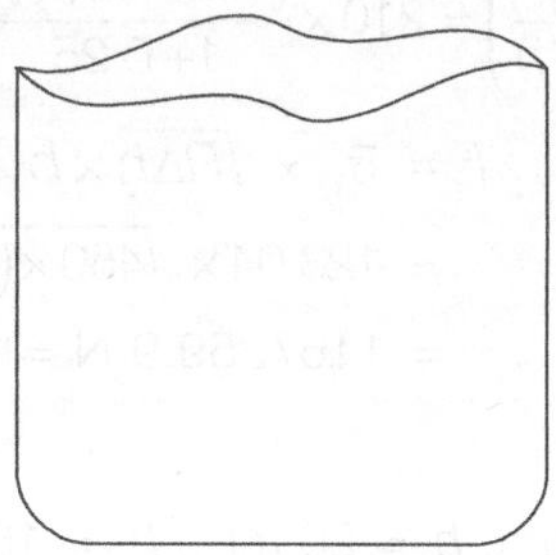

An earing defect showing the formation of ears

■■■■

4 Sheet Metal Operations

4.1 A 50 mm diameter disc is to be punched out from a carbon steel sheet 1.0 mm thick. The diameter of the die should be (Take clearance as 3% of sheet thickness)

(a) 49.925 mm (b) 50.00 mm
(c) 50.075 mm (d) None of these

[1996 : 2 M]

4.2 Identify the stress strain in flange portion of a partially drawn cylindrical cup when deep drawing without a blank holder

(a) tensile in all three directions
(b) no stress in the flange at all, because there is no blank holder
(c) tensile stress in one direction and compressive stress in other direction
(d) tensile in two directions and compressive in third direction **[1999 : 2 M]**

4.3 The cutting force in punching and blanking operations mainly depends on

(a) The modulus of elasticity of the material
(b) The shear strength of the materials
(c) The bulk modulus of the material
(d) The yield strength of the material

[2001 : 1 M]

4.4 In blanking operation the clearance is provided on

(a) The die
(b) The punch
(c) Both die and punch equally
(d) Neither the punch nor the die **[2002 : 1 M]**

4.5 A shell of 100 mm diameter and 100 mm height with the corner radius of 0.4 mm is to be produced by cup drawing. The required blank diameter is

(a) 118 mm (b) 161 mm
(c) 224 mm (d) 312 mm **[2003 : 2 M]**

4.6 A metal disc of 20 mm diameter is to be punched from a sheet of 2 mm thickness. The punch and the dia clearance is 3%. The required punch diameter is

(a) 19.88 mm (b) 19.94 mm
(c) 20.06 mm (d) 20.12 mm

[2003 : 2 M]

4.7 10 mm diameter holes are to be punched in a steel sheet of 3 mm thickness. Shear strength of the material is 400 N/mm^2 and penetration is 40%. Shear provided on the punch is 2 mm. The blanking force during the operation will be

(a) 22.6 kN (b) 37.7 kN
(c) 61.6 kN (d) 94.3 kN **[2004 : 2 M]**

4.8 In the deep drawing of cups, blanks show a tendency to wrinkle up around the periphery (flange). The most likely cause and remedy of the phenomenon are, respectively.

(a) Buckling due to circumferential compression; Increase blank holder pressure.
(b) High blank holder pressure and high friction; Reduce blank holder pressure and apply lubricant
(c) High temperature causing increase in circumferential length; Apply coolant to blank
(d) Buckling due to circumferential compression; decrease blank holder pressure.

[2008 : 2 M]

Linked Answer Questions 4.9 and 4.10

In shear cutting operation, a sheet of 5 mm thickness is cut along a length of 200 mm. The cutting blade is 400 mm long (see figure) and zero-shear (S = 0) is provided on the edge. The ultimate shear strength of the sheet is 100 MPa and penetration to thickness ratio is 0.2. Neglect friction.

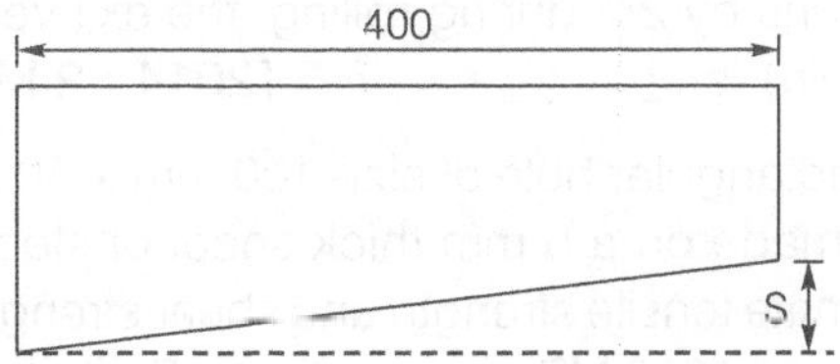

4.9 ssuming force vs displacement curve to be rectangular, the work done (in J) is

(a) 100 (b) 200
(c) 250 (d) 300 **[2010 : 2 M]**

4.10 A shear of 20 mm (S = 20 mm) is now provided on the blade. Assuming force vs displacement curve to be trapezoidal, the maximum force (in kN) exerted is
(a) 5 (b) 10
(c) 20 (d) 40 **[2010 : 2 M]**

4.11 The shear strength of a sheet metal is 300 MPa. The blanking force required to produce a blank of 100 mm diameter from a 1.5 mm thick sheet is close to
(a) 45 kN (b) 70 kN
(c) 141 kN (d) 3500 kN **[2011 : 2 M]**

4.12 Calculate the punch size in mm, for a circular blanking operation for which details

Size of blank	25 mm
Thickness of the sheet	2 mm
Radial clearance between punch and die	0.06 mm
Die allowance	0.05 mm

(a) 24.83 (b) 24.89
(c) 25.01 (d) 25.17 **[2012 : 2 M]**

4.13 In a rolling process, the maximum possible draft, defined as the difference between the initial and the final thickness of the metal sheet, mainly depends on which pair of the following parameters?
P : Strain
Q : Strength of the work material
R : Roll diameter
S : Roll velocity
T : Coefficient of friction between roll and work
(a) Q, S (b) R, T
(c) S, T (d) P, R
[2014 : 1 M, Set-4]

4.14 A mild steel plate has to be rolled in one pass such that the final plate thickness is 2/3rd of the initial thickness, with the entrance speed of 10 m/min and roll diameter of 500 mm. If the plate widens by 2% during rolling, the exit velocity (in m/min) is ______. **[2014 : 2 M, Set-2]**

4.15 A rectangular hole of size 100 mm × 50 mm is to be made on a 5 mm thick sheet of steel having ultimate tensile strength and shear strength of 500 MPa and 300 MPa, respectively. The hole is made by punching process. Neglecting the effect of clearance, the punching force (in kN) is
(a) 300 (b) 450
(c) 600 (d) 750
[2014 : 2 M, Set-2]

4.16 In a sheet metal of 2 mm thickness a hole of 10 mm diameter needs to be punched. The yield strength in tension of the sheet material is 100 MPa and its ultimate shear strength is 80 MPa. The force required to punch the hole (in kN) is _____
[2016 : 2 M, Set-3]

4.17 The value of true strain produced in compressing a cylinder to half its original length is
(a) 0.69 (b) –0.69
(c) 0.5 (d) –0.5
[2016 : 1 M, Set-2]

4.18 A 300 mm thick slab is being cold rolled using roll of 600 mm diameter. If the coefficient of friction is 0.08, the maximum possible reduction (in mm) is__________. **[2016 : 2 M, Set-1]**

4.19 The percentage scrap in a sheet metal blanking operation of a continuous strip of sheet metal as shown in the figure _______(correct to two decimal places)

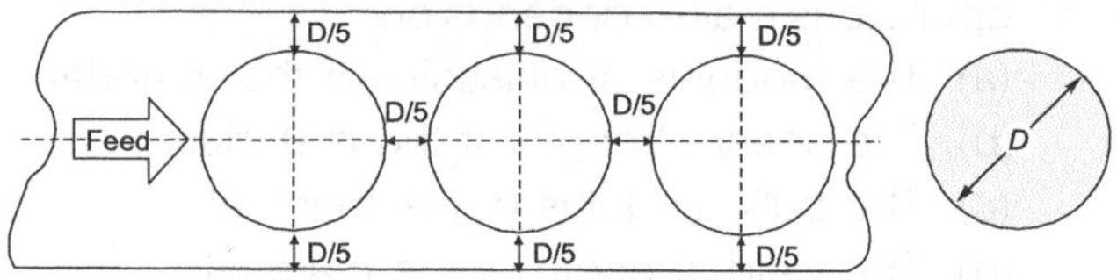

[2018 : 2 M, Set-1]

4.20 A sheet metal with a stock hardness of 250 HRC has to be sheared using a punch and a die having a clearance of 1 mm between them. If the stock hardness of the sheet metal increases to 400 HRC, the clearance between the punch and the die should be________ mm.
[2020 : 1 M, Set-1]

4.21 A blanking operation is performed on C20 steel sheet to obtain a circular disc having a diameter of 20 mm and a thickness of 2 mm. An allowance of 0.04 is provided. The punch size used for the operation is______mm (rounded off to 2 decimal places)
[2024 : 2 M]

Answers Sheet Metal Operations

4.1	(b)	4.2	(b)	4.3	(b)	4.4	(b)	4.5	(c)	4.6	(a)	4.7	(a)	4.8	(a)
4.9	(a)	4.10	(b)	4.11	(c)	4.12	(a)	4.13	(b)	4.14	(14.7)	4.15	(b)		
4.16	(5.024)	4.17	(b)	4.18	(1.92)	4.19	(53.25)	4.20	(1.26)	4.21	(19.84)				

Explanations Sheet Metal Operations

4.1 (b)

Since, it is a blanking operation, die size = blank size = 50 mm

4.2 (b)

Since there is no blank holder. Hence no stress will be developed.

4.3 (b)

During punching and blanking, shearing action of sheet is taking place, therefore cutting force depends on shear strength.

4.4 (b)

In blanking, size is controlled by die hence clearance is provided on punch.

4.5 (c)

Given, Shell diameter,

$d = 100$ mm

Shell height,

$h = 100$ mm

Corner radius,

$r = 0.4$ mm

Here $d > 20r$

∴ Blank diameter,

$$D = \sqrt{d^2 + 4hd}$$

$$= \sqrt{(100)^2 + 4 \times 100 \times 100} = \mathbf{224\ mm}$$

4.6 (a)

Given, Punched diameter,

$d = 20$ mm

Sheet thickness,

$t = 2$ mm

Clearance: $c = 3\%$ of t

Punch diameter,

$$D = d - 2c$$

$$= 20 - 2 \times \frac{3}{100} \times 2 = \mathbf{19.88\ mm}$$

4.7 (a)

Shear strength: $\tau = 400$ N/mm²

Sheet thickness: $t = 3$ mm

Diameter of hole: $d = 10$ mm

Shear on punch: $t_1 = 2$ mm

Penetration of punch in fraction,

$p = 0.4$

Blanking force when shear is applied on the punch

$$= \pi dt\left(\frac{tp}{t_1}\right)\tau = \pi \times 10 \times 3 \times \left(\frac{3 \times 0.4}{2}\right) \times 400$$

$= \mathbf{22.6\ kN}$

4.9 (a)

Maximum force: $F = L(t) . \tau$

L = Perimeter of cut, t = Thickness of cut

P = Penetration

$F = 200 \times (5) \times 100 = 100$ kN

Work done $= F \times pt$

$= 100 \times 10^3 \times 5 \times 10^{-3} \times 0.2$

$= 100$ Nm $= \mathbf{100\ J}$

4.10 (b)

For 400 mm length shear is 20 mm, therefore for 200 mm length it becomes 10 mm. Only 200 mm length is effective.

$$F = \frac{F_{max} \times pt}{s} = \frac{100}{10 \times 10^{-3}} = \mathbf{10\ kN}$$

4.11 (c)

Blanking force = Shear strength × Shear area

$= 300 \times 10^6 \times \pi \times 100 \times 1.5 \times 10^{-6} = \mathbf{141.37\ kN}$

4.12 (a)

Diameter of punch

= Diameter of blank – 2 × radial clearance – die allowance

$= 25 - 2 \times 0.06 - 0.05 = \mathbf{24.83\ mm}$

4.13 (b)

$$(t_i - t_f)_{max} = \mu^2 R$$

where μ is coefficient of friction

R is roll radius.

4.14 (14.7)

Since there is no change in metal volume at a given point per unit time throughout the process, therefore

$b_i h_i v_i = b_f h_f v_f$

Given, $\frac{h_f}{h_i} = \frac{2}{3}$; $\frac{b_f}{b_i} = 1.02$

$$\therefore v_f = \frac{b_i}{b_f} \times \frac{h_i}{h_f} \times v_i = \frac{1}{1.02} \times \frac{3}{2} \times 10$$

$= \mathbf{14.7\ m/min}$

4.15 (b)

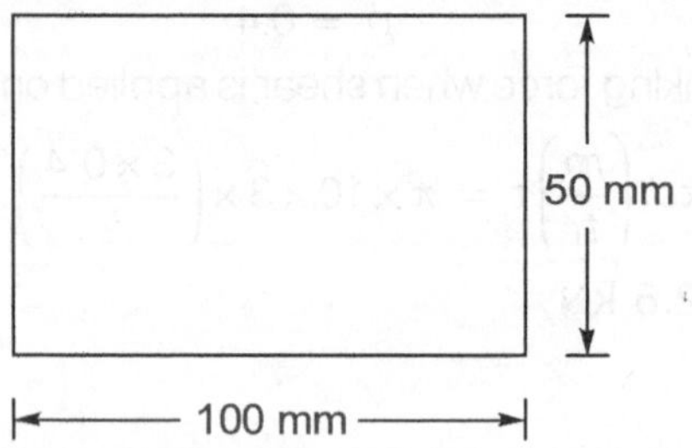

Shear area = $P \times t$

$= 2\,(100 + 50) \times 5 = 1500\ \text{mm}^2$

Punching force:

$$F = \tau \times A_s = 300 \times 1500$$

$$= 450 \times 10^3\ \text{N} = \mathbf{450\ kN}$$

4.16 (5.024)

Punching force,

$$F_{max} = \pi d t \tau_{pers.}$$

$$= \pi \times 10 \times 2 \times 80 = \mathbf{5.024\ kN}$$

4.17 (b)

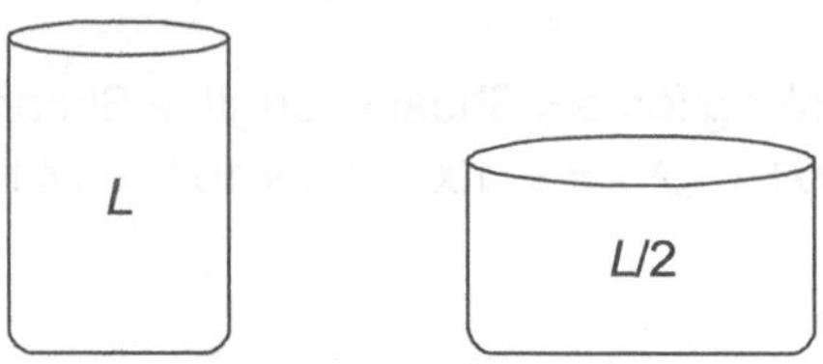

$$\in_T = \ln\left(\frac{L_f}{L_i}\right) = \ln\left(\frac{(L/2)}{L}\right) = \ln 1/2 = \mathbf{-0.69}$$

4.18 (1.92)

$$\Delta H_{max} = \mu^2 R = (0.08)^2 \times 300 = \mathbf{1.92\ mm}$$

4.19 (53.25)

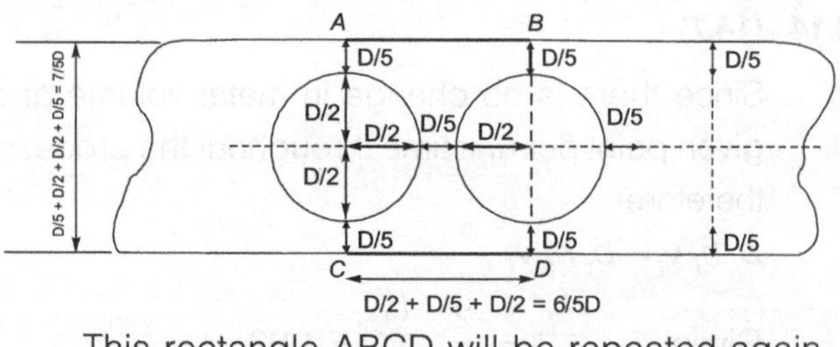

This rectangle ABCD will be repeated again and again.

$$A_t = \text{Total area} = \frac{7}{5}D \times \frac{6}{5}D = \frac{42}{25}D^2$$

$$A_u = \text{Area of blanking disc}$$

$$= \frac{\pi D^2}{4}$$

$$\%\ \text{of scrap} = \frac{A_t - A_u}{A_t} \times 100\%$$

$$= \left[1 - \frac{\left(\frac{\pi}{4}\right)}{\left(\frac{42}{25}\right)}\right] \times 100\% = 53.25\%$$

4.20 (1.26)

Tensile strength $\propto$ Hardness number

Shear strength = 0.5 tensile strength (Tresca theory)

= 0.577 tensile strength (Von-Mises theory)

Shear strength $\propto$ Hardness number and clearance,

$$C = 0.0032 t \sqrt{\tau}$$

$$C \propto \sqrt{\text{Hardness number}}$$

$$\frac{C_2}{C_1} = \sqrt{\frac{(\text{Hardness number})_2}{(\text{Hardness number})_1}}$$

or $$C_2 = 1 \times \sqrt{\frac{400}{250}} = 1.265\ \text{mm}$$

4.21 (19.84) (19.84 to 19.92)

Diameter of blank = 20 mm

Sheet thickness, t = 2 mm

Clearance allowance, $A_c = 0.04$

$\therefore$ Clearance, $C = A_c t = 0.04 \times 2 = 0.08$

Blank size = Die size = 20 mm

Punch size = Die size − 2 × C

= 20 − 2 × 0.08 = 19.84 mm

■■■■

5 Joining/Welding

5.1 Preheating before welding is done to
(a) Make the steel softer
(b) Burn away oil, grease etc. from the plate surfaces
(c) Prevent cold cracks
(d) Prevent plate distortion [1996 : 1 M]

5.2 Match the following

List-I
A. Welding of aluminum alloy
B. Ship building
C. Joining of HSS drill bit to shank
D. Deep penetration precision welds

List-II
1. Submerged arc welding
2. Electron beam welding
3. TIG welding
4. Friction welding
5. Gas welding [ME 1996 : 2 M]

5.3 For butt welding 40 mm thick steel plates, when the expected quantity of such jobs is 5000 per month over a period of 10 years, choose the best suitable welding process out of the following available alternatives.
(a) SAW
(b) Oxyacetylene gas welding
(c) EBM
(d) MIG [ME 1999 : 1 M]

5.4 Resistance spot welding is performed on two plates of 1.5 mm thickness with 6 mm diameter electrode, using 15000 A current for a time duration of 0.25 sec. Assuming the interface resistance to be 0.0001 Ω. The heat generated to form the weld is
(a) 5625 W-sec (b) 8437 W-sec
(c) 22500 W-sec (d) 33750 W-sec
[ME 2001 : 2 M]

5.5 Two plates of the same metal having equal thickness are to be butt welded with electric arc. When the plate thickness changes, welding is achieved by
(a) Adjusting the current
(b) Adjusting the duration of the current
(c) Changing the electrode size
(d) Changing the electrode coating
[ME 2002 : 1 M]

5.6 The temperature of a carburizing flame in gas welding that of a neutral or an oxidizing flame
(a) Lower than (b) Higher than
(c) Equal to (d) Unrelated to
[ME 2002 : 1 M]

5.7 Which of the following arc welding processes does **not** use consumable electrode
(a) GMAW (b) GTAW
(c) SAW (d) None [ME 2002 : 1 M]

5.8 Oxyacetylene gas welding, temperature at the inner cone of the flame is around
(a) 3500°C (b) 3200°C
(c) 2900°C (d) 2550°C
[ME 2003 : 1 M]

5.9 atch the following

Work material
P. Aluminium
Q. Die steel
R. Copper Wire
S. Titanium sheet

Type of joining
1. Submerged Arc Welding
2. Soldering
3. Thermit Welding
4. Atomic Hydrogen Welding
5. Gas Tungsten Arc Welding
6. Laser Beam Welding
(a) P–2, Q–5, R–1, S–3
(b) P–6, Q–3, R–4, S–4
(c) P–4, Q–1, R–6, S–2
(d) P–5, Q–4, R–2, S–6 [ME 2003 : 2 M]

5.10 Two 1 mm thick steel sheets are to be spot welded at a current of 5000 A. Assuming effective resistance to be 200 micro-ohms and current flow time of 0.2 second, heat generated during the process will be
(a) 0.2 Joule (b) 1 Joule
(c) 5 Joule (d) 1000 Joule
[ME 2004 : 1 M]

5.11 The strength of a brazed joint
(a) decreases with increase in gap between the two joining surfaces
(b) increases with increase in gap between the two joining surfaces
(c) decreases up to certain gap between the two joining surfaces beyond which it increases
(d) increases up to certain gap between the two joining surfaces beyond which it decreases
[ME 2005 : 1 M]

5.12 Spot welding of two 1 mm thick sheets of steel (density = 8000 kg/m^3) is carried out successfully by passing a certain amount of current for 0.1 second through the electrodes. The resultant weld nugget formed is 5 mm in diameter and 1.5 mm thick. If the latent heat of fusion of steel is 1400 kJ/kg and the effective resistance in the welding operation is 200 μΩ, the current passing through the electrodes is approximately
(a) 1480 A (b) 3300 A
(c) 4060 A (d) 9400 A
[ME 2005 : 2 M]

5.13 In an arc welding process, the voltage and current are 25 V and 300 A respectively. The arc heat transfer efficiency is 0.85 and welding speed is 8 mm/s. The net heat input (in J/mm) is
(a) 64 (b) 797
(c) 1103 (d) 79700
[ME 2006 : 2 M]

5.14 Which one of the following is a solid state joining process?
(a) gas tungsten arc welding
(b) resistance spot welding
(c) friction welding
(d) submerged arc welding [ME 2007 : 1 M]

5.15 A direct current welding machine with a linear power source characteristic provides open circuit voltage of 80 V and short circuit current of 800 A. During welding with the machine, the measured arc current is 500 A corresponding to an arc length of 5.0 mm and the measured arc current is 460 A corresponding to an arc length of 7.0 mm. The linear voltage (E) – arc length (L) characteristic of the welding arc can be given as (where E is in Volt and L is in mm)
(a) $E = 20 + 2L$ (b) $E = 20 + 8L$
(c) $E = 80 + 2L$ (d) $E = 80 + 8L$
[ME 2007 : 2 M]

5.16 Two metallic sheets, each of 2.0 mm thickness, are welded in a lap joint configuration by resistance spot welding at a welding current of 10 kA and welding time of 10 millisecond. A spherical fusion zone extending up to the full thickness of each sheet is formed and properties of the metallic sheets are given as:
Ambient temperature = 293 K
Melting temperature = 1793 K
Density = 7000 kg/m^3
Latent heat of fusion = 300 kJ/kg
Specific heat = 800 J/kgK

Assume:
(i) contact resistance along sheet-interface is 500 micro-ohm and along electrode-sheet interface is zero;
(ii) no conductive heat loss through the bulk sheet materials; and
(iii) the complete weld fusion zone is at the melting temperature.

The melting efficiency (in %) of the process is
(a) 50.37 (b) 60.37
(c) 70.37 (d) 80.37
[ME 2007 : 2 M]

5.17 In arc welding of a butt joint, the welding speed is to be selected such that highest cooling rate is achieved. Melting efficiency and heat transfer efficiency are 0.5 and 0.7, respectively. The area of the weld cross section is 5 mm^2 and the unit energy required to melt the metal is 10 J/mm^3. If the welding power is 2 kW, the welding speed in mm/s is closest to
(a) 4 (b) 14
(c) 24 (d) 34 [ME 2008 : 2 M]

5.18 Two pipes of inner diameter 100 mm and outer diameter 110 mm each are joined by flash butt welding using 30 V power supply. At the interface, 1 mm of material melts from each pipe which has a resistance of 42.4 Ω. If the unit melt energy is 64.4 MJm^{-3}, then time required for welding (in s) is
(a) 1 (b) 5
(c) 10 (d) 20 [ME 2010 : 2 M]

5.19 Which one among the following welding processes uses non-consumable electrode?
(a) Gas metal arc welding
(b) Submerged arc welding
(c) Gas tungsten arc welding
(d) Flux coated arc welding [ME 2011 : 1 M]

5.20 In a DC arc welding operation, the voltage-arc length characteristic was obtained as $V_{arc} = 20 + 5l$ where the arc length l was varied between 5 mm and 7 mm. Here V_{arc} denotes the arc voltage in Volts. The arc current was varied from 400 A to 500 A. Assuming linear power source characteristic, the open circuit voltage and the short circuit current for the welding operation are

(a) 45 V, 450 A (b) 75 V, 750 A
(c) 95 V, 950 A (d) 150 V, 1500 A

[ME 2012 : 2 M]

5.21 Filler material is .... A ... in resistance welding and the heat generated in the process is directly proportional to .... B....

(a) used/not used
(b) square of the current/cube of the current

[ME 1992 : 1 M]

5.22 Match the **CORRECT** pairs

Processes	Characteristics/ Applicaitons
P. Friction Welding	1. Non-consumbale electrode
Q. Gas Metal Arc Welding	2. Joining of thick plates
R. Electroslag Welding	3. Consumable electrode wire
S. Tungsten Inert Gas Welding	4. Joining of cylindrical dissimilar materials

(a) P–4, Q–3, R–2, S–1
(b) P–4, Q–2, R–3, S–1
(c) P–2, Q–3, R–4, S–1
(d) P–2, Q–4, R–1, S–3 [ME 2013 : 1 M]

5.23 The major difficulty during welding of aluminium is due to its

(a) high tendency of oxidation
(b) high thermal conductivity
(c) low melting point
(d) low density [ME 2014 : 1 M, Set-1]

5.24 In solid-state welding, the contamination layers between the surfaces to be welded are removed by

(a) alcohol (b) plastic deformation
(c) water jet (d) sand blasting

[ME 2014 : 1 M, Set-1]

5.25 Within the Heat Affected Zone (HAZ) in a fusion welding process, the work material undergoes

(a) microstructural changes but does not melt
(b) neither melting nor microstructural changes
(c) both melting and microstructural changes after solidification
(d) melting and retains the original microstructure after solidification [ME 2014 : 1 M, Set-4]

5.26 For spot welding of two steel sheets (base metal) each of 3 mm thickness, welding current of 10000 A is applied for 0.2 s. The heat dissipated to the base metal is 1000 J. Assuming that the heat required for melting 1 mm^3 volume of steel is 20 J and interfacial contact resistance between sheets is 0.0002 Ω, the volume (in mm^3) of weld nugget is _____. [ME 2014 : 2 M, Set-3]

5.27 In a linear arc welding process, the heat input per unit length is inversely proportional to

(a) welding current
(b) welding voltage
(c) welding speed
(d) duty cycle of the power source

[2015 : 1 M, Set-1]

5.28 Which two of the following joining processes are autogenous?

i. Diffusion welding
ii. Electroslag welding
iii. Tungsten inert gas welding
iv. Friction welding

(a) i and iv (b) ii and iii
(c) ii and iv (d) i and iii

[2015 : 1 M, Set-3]

5.29 A DC welding power source has a linear voltage-current (V-I) characteristic with open circuit voltage of 80 V and a short circuit current of 300 A. For maximum arc power, the current (in Ampere) should be set as _____. [2015 : 2 M, Set-1]

5.30 During a TIG welding process, the arc current and arc voltage were 50 A and 60 V, respectively, when the welding speed was 150 mm/min. In another process, the TIG welding is carried out at a welding speed of 120 mm/min at the same arc voltage and heat input to the material so that weld quality remains the same. The welding current (in A) for this process is

(a) 40.00 (b) 44.72
(c) 55.90 (d) 62.25

[2015 : 2 M, Set-2]

5.31 Spot welding of two steel sheets each 2 mm thick is carried out successfully by passing 4 kA of current for 0.2 seconds through the electrodes. The resulting weld nugget formed between the sheets is 5 mm in diameter. Assuming cylindrical shape for the nugget, the thickness of the nugget is________mm.

Latent heat of fusion for steel	1400 kJ/kg
Effective resistance of the weld joint	200 μΩ
Density of steel	8000 kg/m³

[2016 : 2 M, Set-3]

5.32 The welding process which uses a blanket of fusible granular flux is
(a) tungsten inert gas welding
(b) submerged arc welding
(c) electroslag welding
(d) thermit welding [2016 : 1 M, Set-2]

5.33 The voltage-length characteristic of a direct arc in an arc welding process is $V = (100 + 40l)$, where l is the length of the arc in mm and V is arc voltage in volts. During a welding operation, the arc length varies between 1 and 2 mm and the welding current is in the range 200-250 A. Assuming a linear power source, the short circuit current is_____A. [2016 : 2 M, Set-2]

5.34 Under optimal conditions of the process the temperatures experienced by a copper work piece in fusion welding, brazing and soldering are such that
(a) $T_{welding} > T_{soldering} > T_{brazing}$
(b) $T_{soldering} > T_{welding} > T_{brazing}$
(c) $T_{brazing} > T_{welding} > T_{soldering}$
(d) $T_{welding} > T_{brazing} > T_{soldering}$ [2016 : 1 M, Set-1]

5.35 Heat is removed from a molten metal of mass 2 kg at a constant rate of 10 kW till it is completely solidified. The cooling curve is shown in the fig.

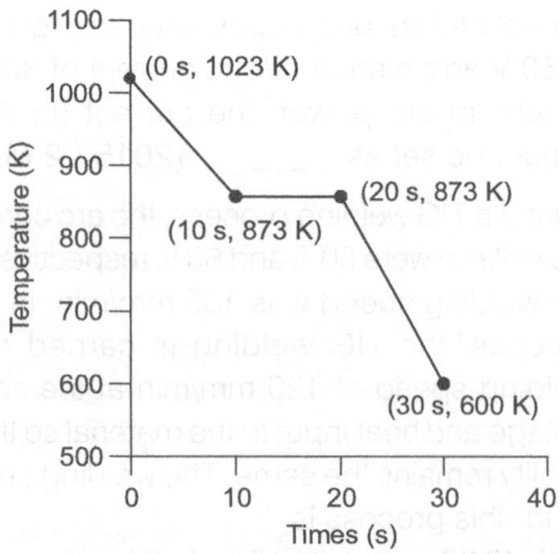

Assuming uniform temperature throughout the volume of the metal during solidifications, the latent heat of fusion of the metal (in kJ/kg) is____________ [2016 : 2 M, Set-1]

5.36 In an arc welding process, welding speed is doubled. Assuming all other process parameters to be constant, the cross sectional area of the weld bead will
(a) increase by 25% (b) increase by 50%
(c) reduce by 25% (d) reduce by 50%
[2017 : 1 M, Set-1]

5.37 The type of weld represented by the shaded region in the figure is

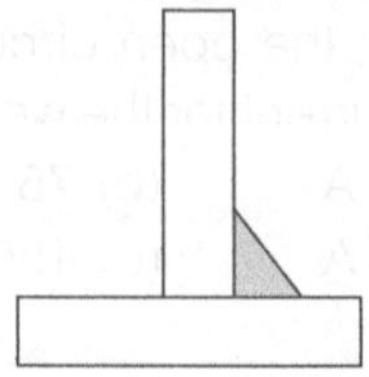

(a) groove (b) spot
(c) fillet (d) plug
[2018 : 1 M, Set-1]

5.38 A welding operation is being performed with voltage = 30 V and current = 100 A. The cross-sectional area of the weld bead is 20 mm^2. The work-piece and filler are of titanium for which of the specific energy of melting is 14 J/mm^3. Assuming a thermal efficiency of the welding process 70%, the welding speed (in mm/s) is ________ (correct to two decimal places).
[2018 : 2 M, Set-2]

5.39 Which one of the following welding methods provides the highest heat flux (W/mm^2)?
(a) Oxy-acetylene gas welding
(b) Tungsten inert gas welding
(c) Plasma arc welding
(d) Laser beam welding [2019 : 1 M, Set-1]

5.40 A gas tungsten arc welding operation is performed using a current of 250 A and an arc voltage of 20 V at a welding speed of 5 mm/s. Assuming that the arc efficiency is 70%, the net heat input per unit length of the weld will be____kJ/mm (round off to one place). [2019 : 2 M, Set-2]

5.41 Two plates, each of 6 mm thickness, are to be butt-welded. Consider the following processes and select the correct sequence in increasing order of size of the heat affected zone
1. Arc welding 2. MIG welding
3. Laser beam welding
4. Submerged arc welding
(a) 3-4-2-1 (b) 3-2-4-1
(c) 4-3-2-1 (d) 1-4-2-3
[2020 : 1 M, Set-2]

5.42 The resistance spot welding of two 1.55 mm thick metal sheet is performed using welding current of 10000 A for 0.25s. The contact resistance at the interface of the metal sheets is 0.0001 Ω. The volume of weld nugget formed after welding is 70 mm^3. Considering the heat required to melt unit volume of metal is 12 J/mm^3, the thermal efficiency of the welding process is ______%. [Round off to one decimal place] [2021 : 2 M, Set-1]

5.43 A spot welding operation performed on two pieces of steel yielded a nugget with a diameter of 5 mm and a thickness of 1 mm. The welding time was 0.1 s. The melting energy for the steel is 20 J/mm^3. Assuming the heat conversion efficiency as 10%, the power required for performing the spot welding operation is _______ kW (round off to two decimal places). [2021 : 2 M, Set-2]

5.44 Two mild steel plates of similar thickness, in butt-joint configuration, are welded by gas tungsten arc welding process using the following welding parameters.

Welding voltage	20 V
Welding current	150 A
Welding speed	5 mm/s

A filler wire of the same mild steel material having 3 mm diameter is used in this welding process. The filler wire feed rate is selected such that the final weld bead is composed of 60% volume of filler and 40% volume of plate material. The heat required to melt the mild steel material is 10 J/mm^3. The heat transfer factor is 0.7 and melting factor is 0.6. The feed rate of the filler wire is _______ mm/s (round off to one decimal place). [2022 : 2 M, Set-1]

5.45 In a direct current arc welding process, the power source has an open circuit voltage of 100 V and short circuit current of 1000 A. Assume a linear relationship between voltage and current. The arc voltage (V) varies with the arc length (l) as $V = 10 + 5l$, where V is in volts and l is in mm. The maximum available arc power during the process is ________ kVA (in integer). [2022 : 2 M, Set-2]

5.46 Two surfaces P and Q are to be joined together. In which of the given joining operation(s), there is no melting of the two surfaces P and Q for creating the joint?
(a) Arc welding
(b) Brazing
(c) Adhesive bonding
(d) Spot welding
[2023 : 1 M, MSQ]

5.47 The most suitable electrode materials used for joining low alloy steels using Gas Metal Arc Welding (GMAW) process is
(a) Copper (b) Cadmium
(c) Low alloy steel (d) tungsten
[2024 : 1 M]

5.48 In an arc welding process, the voltage and current are 30 V and 200 A, respectively. The cross-sectional area of the joint is 20 mm^2 and the welding speed is 5 mm/s. The heat required to melt the material is 20 J/s. The percentage of heat lost to the surrounding during the welding process is ______. (rounded off to 2 decimal places) [2024 : 2 M]

Answers Joining/Welding

5.1 (c) 5.2 (A–3, B–1, C–4, D–2) 5.3 (a) 5.4 (a) 5.5 (a) 5.6 (a) 5.7 (b)
5.8 (b) 5.9 (d) 5.10 (d) 5.11 (d) 5.12 (c) 5.13 (b) 5.14 (c) 5.15 (a)
5.16 (c) 5.17 (b) 5.18 (c) 5.19 (c) 5.20 (c) 5.21 (A) not used (B) square of current
5.22 (a) 5.23 (a) 5.24 (b) 5.25 (a) 5.26 (150) 5.27 (c) 5.28 (a)
5.29 (150) 5.30 (a) 5.31 (2.91) 5.32 (b) 5.33 (425) 5.34 (d) 5.35 (50) 5.36 (d)
5.37 (c) 5.38 (7.5) 5.39 (d) 5.40 (0.7) 5.41 (b) 5.42 (33.6) 5.43 (39.25)
5.44 (10.71) 5.45 (25) 5.46 (b, c) 5.47 (c) 5.48 (*)

Explanations Joining/Welding

5.3 (a)

As the no. of jobs to be done per month is very high, it requires a welding process which is capable of joining 40 mm sheets in one pass. So the answer will be SAW.

5.4 (a)

Heat generated in resistance welding = I^2RT.
= $(15000)^2 \times 0.0001 \times 0.25$ = **5625 W-sec**

5.5 (a)

On increasing current, heat generation could be increased resulting into penetration to a greater depth.

5.6 (a)

Due to lesser amount of oxygen, temperature of carburizing flame is lower than that of neutral or oxidizing flame.

5.7 (b)

In gas tungsten arc welding (GTAW), non-consumable tungsten electrode is used.

5.8 (b)

Oxyacetylene welding can be used for engineering practice. The temperature produced by inner cone of the neutral flame is around 3150°C.

5.10 (d)

Heat generated during the process = $I^2 R\Delta t$
= $(5000)^2 \times (200 \times 10^{-6}) \times 0.2$ = **1000 J**

5.12 (c)

$$\text{Mass} = \frac{\pi}{4} \times 25 \times 1.5 \times 8000 \times 10^{-9}$$
$$= 2.356 \times 10^{-4} \text{ kg}$$
$$\text{Heat} = 1400 \times 1000 \times 2.356 \times 10^{-4}$$
$$= 329.8 \text{ J}$$
$$329.8 = (A)^2 \times 200 \times 10^{-6} \times 0.1$$
$$A = \mathbf{4060 \text{ amp.}}$$

5.13 (b)

Given,
Voltage: V = 25 volt
Current: I = 300 amp
Efficiency: η = 0.85
Weld speed: v = 8 mm/s
Power generated due to arc = VI
Power utilized as heat = ηVI

$$= 0.85 \times 25 \times 300 = 6375 \text{ J/s}$$
$$= \frac{6375}{v} \text{J/mm} = \frac{6375}{8}$$
$$= \mathbf{796.8 \text{ J/mm} \approx 797 \text{ J/mm}}$$

5.15 (a)

The power source characteristic can be written analytically as

$$E = 80 - \frac{80}{800} I \qquad \text{(a)}$$

The arc characteristic is given as

$$I = aL + b$$

where a and b are constant

Given,
when I = 500 A
then L = 5.00 mm
∴ 500 = 5a + b (i)
when I = 460 A
then L = 7.00 mm
∴ 460 = 7a + b (ii)

Solving Eqs. (i) and (ii), we get
a = – 20
and b = 600

Then arc characteristic equation

$$I = -20L + 600 \qquad \text{(b)}$$

From equations (a) and (b)

$$E = 80 - \frac{80}{800}(-20L + 600)$$
$$= 80 - 0.1(-20L + 600)$$
$$= 80 + 2L - 60$$
$$= \mathbf{20 + 2L}$$

5.16 (c)

Given,
Current: $I = 10^4$ A
Time: $t = 10^{-2}$ s
Resistance: $R = 500 \times 10^{-6}\ \Omega$
Ambient temperature,
T_∞ = 293 K
Melting temperature,
T_m = 1793 K
Latent heat of fusion,
L = 300 kJ/kg

Density: $\rho = 7000 \text{ kg/m}^3$
Specific heat, $c_p = 800$ J/kgK
Radius of sphere,
$r = 2\times 10^{-3}$ m
Total heat supplied,

$$H_S = I^2Rt$$
$$= (10^4)^2 \times 500 \times 10^{-6} \times 10 \times 10^{-3}$$
$$= 500 \text{ J}$$

$$\text{Mass} = \frac{4}{3}\pi r^3 \times \rho = \frac{4}{3}\pi\left(2\times10^{-3}\right)^3 \times 7000$$
$$= 2.346 \times 10^{-4} \text{ kg}$$

Total heat for melting, $H_i = mL + m\, c_p\,(T_m - T_0)$

$$= 2.346 \times 10^{-4}\,[300 \times 1000 + 800(1793 - 293)]$$
$$H_i = 351.9 \text{ J}$$
$$\text{Efficiency} = \frac{H_i}{H_S}\times 100 = \frac{351.9}{500}\times 100$$
$$= \mathbf{70.38\%}$$

5.17 (b)

Let the welding speed is V mm/s
Volume of weld/sec = $5V$ mm^3/s

$$(5V) \times 10 = 2000 \times 0.5 \times 0.7$$
$$\Rightarrow \quad V = \frac{700}{50} = \mathbf{14\ mm/s}$$

5.18 (c)

Unit melting energy = 64.4 MJ/m^3
Voltage = 30 V
Resistance = 42.4 ohm

$$\text{Melting energy} = \frac{\pi}{4}(d_0^2 - d_i^2)t \times 64.4 \times 10^6 \text{ J}$$
$$= 212.32 \text{ J}$$

$d_0 = 110$ mm
$d_i = 100$ mm
$t = 2 \times 1 = 2$ mm

Now, I^2Rt = Melting energy

$$\frac{V^2}{R}t = \text{Melting energy}$$
$$\text{Time required} = \frac{212.32 \times 42.4}{30^2} \text{ second} = \mathbf{10\ s}$$

5.19 (c)

TIG *i.e.*, Gas Tungsten Arc welding or Tungsten Inert Gas welding uses non-consumable electrode and purpose of it is to create an arc only. It is used to weld Aluminium.

5.20 (c)

$$V_{arc} = 20 + 5l$$
$$l = 5 \text{ mm},\ V_{arc} = 45\ V$$
$$l = 7 \text{ mm},\ V_{arc} = 55\ V$$
$$V = V_0 - \frac{V_0}{I_s}I$$

V_0 = Open circuit voltage
I_s = Short circuit current

$$45 = V_0 - \frac{V_0}{I_s}\times 500$$
$$55 = V_0 - \frac{V_0}{I_s}\times 400$$
$$\Rightarrow \quad \mathbf{V_0 = 95\ volts}$$
$$\mathbf{I_s = 950\ amp.}$$

5.21 Sol. (A) not used (B) square of current

For resistance welding, $H \propto I^2$ and no filler material is used in it.

5.22 (a)

Friction welding: Joining of cylindrical dissimilar materials.
Gas metal arc welding: Consumable electrode wire.
Tungsten inert gas welding: Non-consumable electrode.
Electroslag welding: Joining of thick plates.

5.26 (150)

Total heat generation

$$= I^2Rt = (10000)^2 \times 0.0002 \times 0.2$$
$$= 4000 \text{ J}$$

Heat required to melt the nugget

$$= 4000 - 1000 = 3000 \text{ J}$$
$$\text{Volume of nugget} = \frac{3000}{20} = 150 \text{ mm}^3$$

5.28 (a)

(i) and (iv)
All solid state welding is autogenous welding.

5.29 (150)

$$V = V_{OC} - \left(\frac{V_{OC}}{I_{SC}}\right)I = 80 - \left(\frac{80}{300}\right)I$$
$$\Rightarrow \quad P = VI = (80 - 0.2666\,I)I$$
$$= 80\,I - 0.2666\,I^2$$

For power to be maximum

$$\left(\frac{dP}{dI}\right) = 0$$

$$\Rightarrow 80 - 2 \times 0.2666I = 0$$

$$\Rightarrow \quad I = 150 \text{ A}$$

5.30 (a)

$$\frac{50 \times 60}{150} = \frac{I \times 60}{120}$$

$$I = \frac{50 \times 12}{155} = 40 \text{ A}$$

5.31(2.91)

Given: $t = 2$ mm, $I = 4$ kA, Time = 0.2 s

$$D_N = 5 \text{ mm}$$

$$I^2Rt = \text{Volume} \times \rho \times \text{H.R/kg}$$

$$4000^2 \times 200 \times 10^{-6} \times 0.2 = \text{Vol.} \times 8000 \times 1400 \times 10^3$$

$$h = \mathbf{2.91 \text{ mm}}$$

5.33. (425)

$$V = 100 + 40L$$

When $\quad l = 1, \quad I = 250$

$$V = (100 + 40 \times 1) = 140$$

$$= V_0 - \frac{V_0}{I_0} \times 250$$

$$l = 2, \quad I = 200$$

$$V = (100 + 40 \times 2)$$

$$180 = V_0 - \frac{V_0}{I_0} \times 200$$

$$V_0 = 140 + \frac{4}{5} \times 250 = 340$$

$$I = \frac{V_0 \times 5}{4} = \frac{340 \times 5}{4} = \mathbf{425 \text{ A}}$$

5.35 (50)

$$Q = m \times \text{L.H}$$

Q = Total heat removed, m = Mass,

L.H. = Latent heat of fusion

$$Q = \dot{Q} \times t$$

$\dot{Q}$ = Heat removal rate, t = Time

$\therefore$ 10 kJ/s × (20 – 10) s = 2 × L.H.

or L.H. = **50 kJ/kg**

Alternatively

$$Q = m \times \text{L.H.}$$

Q = Total heat removed

m = Mass,

L.H. = Latent heat of fusion

$$Q = \dot{Q} \times t$$

$\dot{Q}$ = Heat removal rate, t = Time

$\therefore$ 10 kJ/s × (20 – 10) s = 2 × L.H.

or L.H. = **50 kJ/kg**

5.36 (d)

$$H_s = \frac{VI}{vA_b}$$

If v is doubled ($2v$) to make H_s constant A_b will be reduced by 50%.

$$H_s = \frac{VI}{2v\frac{A_b}{2}} = \frac{VI}{vA_b}$$

5.37 (c)

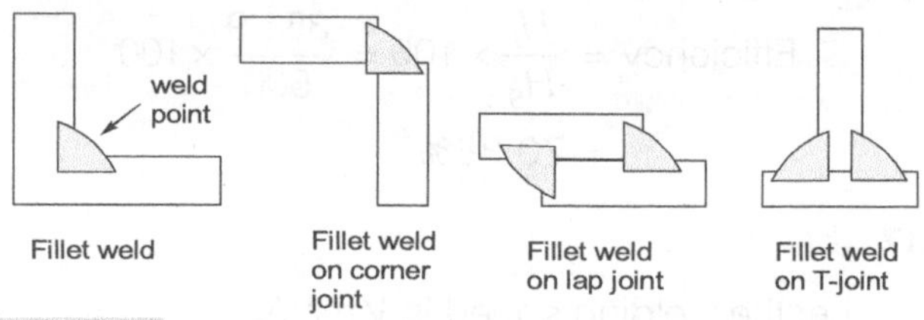

5.38 (7.5)

Given: V = 30V, I = 100A, A = 200 mm²;

η_{th} = 0.7

Heat required (H_R) = 14 J/mm³

$$\text{Heat supplied } (H_S) = \frac{VI}{Av} = \frac{3000}{20v} = \frac{150}{v}$$

$$\eta_{th} = \frac{H_R}{H_S}$$

$$\Rightarrow \quad 0.7 \times \frac{150}{v} = 14$$

$$v = 7.5 \text{ mm/s}$$

5.40 (0.7)

Given data, Welding current, I = 250 A

Voltage, V = 20 V

Welding speed, v = 5 mm per second

Arc heat transfer efficiency, η = 70%

Heat required per unit length,

$$H_s = \frac{VI}{v} \times \eta = \frac{250 \times 20}{5 \times 1000} \times 0.7$$

$$= \frac{700}{1000} = 0.7 \text{ kJ/mm}$$

5.41 (b)

Processes with low rate of heat input (slow heating) tends to produce high total heat content within the metal, slow cooling rates, and large heat-affected zones. high heat input process, has low total heats, fast cooling rates and small heat affected zones.

5.42 (33.6)

Thickness = 1.55 mm; I = 10000 A, t = 0.25 s;
R = 0.0001 Ω, V_n = 70 mm³; H_m = 12 J/mm³; η_m = ?

$$\eta_m = \frac{H_m}{H_s}$$

Melting efficiency = Thermal efficiency

$$\eta_m = \frac{H_m}{I^2Rt} = \frac{12\times 70}{(10000)^2\times 0.0001\times 0.25} = 0.336$$

$$\eta_m = 33.6\%$$

5.43 (39.25)

Given: d = 5 mm, h = 1 mm, t = 00.1s,
H_m = 2 J/mm³
Heat conversion efficiency or melting efficiency,

$$\eta_m = \frac{\eta_m}{H_s}$$

$$H_m = \frac{\pi}{4}d^2h\cdot H_{m/vol} = \frac{\pi}{4}\times(5)^2\times 1\times 20 = 392.5\text{ J}$$

$$H_s = I^2Rt = (VI)t$$

$$\eta_m = \frac{\eta_m}{H_s}$$

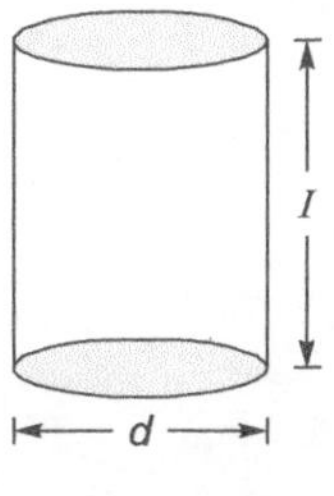

$$0.1 = \frac{392.5}{(VI)0.1}$$

$$\Rightarrow \quad P = VI = 39.25\text{ kW}$$

5.44 (10.71) (10.5 to 10.9)

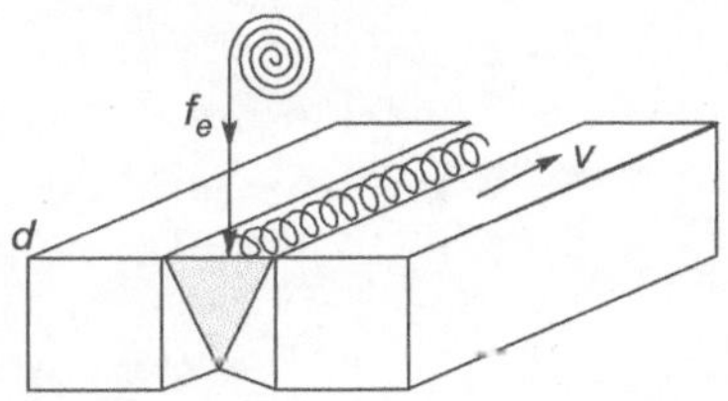

For welding,

$$\eta_h = \frac{H_m}{H_s} = \frac{H_m}{\dfrac{VI}{Av}\eta_m}$$

$$0.6 = \frac{10}{\dfrac{20\times 150}{5\times A}\times 0.7}$$

Area of weld bead,

$$A = 25.2\text{ mm}^2$$

Volume of metal from electrode/time = Volume of metal in the joint/time

$$A_e\times f_e = A\times 0.6\times v$$

$$\Rightarrow \frac{\pi}{4}\times(3)^2\times f_e = 25.2\times 0.6\times 5$$

$$f_e = 10.71\text{ mm/sec}$$

5.45 (25)

Given: V_o = 100 V, I_s = 1000 A
Max. power: $V_a = 10 + 5l_a$ ... (i)

$$V_t = V_o - \left(\frac{I_t}{I_s}\right)V_o = 100 - \left(\frac{I_t}{1000}\right)\times 100 \quad \text{... (ii)}$$

For stable arc:

$$V_t = V_a$$

$$\Rightarrow \quad 10 + 5l_a = 100 - \frac{I_t}{10}$$

$$I_t = (900 - 50l_a)$$

$$\text{Power} = V_t \times I_t$$

$$P = (10 + 5l_a)(900 - 50l_a)$$

$$\frac{\partial P}{\partial l_a} = 0$$

$$\Rightarrow \quad (l_a) = 8\text{ mm}$$

$$P_{max} = 25\text{ kVA} = 25\text{ kW}$$

$$P = 25\text{ kW}$$

Alternative solution:

For linear power source

$$V_t = \frac{V_o}{2} = \frac{100}{2} = 50\text{ V}$$

$$I_t = \frac{I_s}{2} = \frac{1000}{2} = 500\text{ A}$$

$$P = V_t\times I_t = 50\times 500 = 25\text{ kW}$$

5.46 (b, c)

- In brazing, melting of only filler metal takes place and not of parent material.
- Adhesive bonding is a solid state welding process.

5.47 (c)

In GMAW, consumable electrodes are used which has to be same composition as base metal. Thus, low alloy steels electrodes are the most suitable electrode material.

5.48 (*)

Given : $V = 30$ V; $I = 200$ Amp, $A_c = 20$ mm^2, $v = 5$ mm/sec

Heat required to melt (H_m) = 20 J/s

Heat supplied (H_s) = VI = 30 × 200 = 6000 W = 6000 J/s

$$\text{Heat lost to surrounding} = \left(\frac{H_s - H_m}{H_s}\right) \times 100$$

$$= \left(\frac{6000 - 20}{6000}\right) \times 100 = 99.67\%$$

Note: Here the question should mention the heat required to melt the material as 20 J/mm^3. Therefore we are getting such unrealistic value.

■■■■

6 Machining and Machine Tool Operation

6.1 Cutting power consumption in turning can be significantly reduced by
(a) Increasing rake angle of the tool
(b) Increasing cutting angle of the tool
(c) Widening the nose radius of the tool
(d) Increasing the clearance angle
[ME 1995 : 1 M]

6.2 Among the conventional machining processes, maximum specific energy is consumed in
(a) Turning (b) Drilling
(c) Planning (d) Grinding
[ME 1995 : 1 M]

6.3 Plain milling of mild steel plates produces
(a) irregular shaped discontinuous chips
(b) regular shaped discontinuous chips
(c) continuous chips without BUE
(d) jointed chips **[ME 1995 : 1 M]**

6.4 Diamond wheels should not be used for grinding steel components (T/F) **[ME 1995 : 1 M]**

6.5 Machine tool, structures are made ... for high process capability (tough/rigid/strong)
[ME 1995 : 2 M]

6.6 Match the following

Manufacturing process
A. Finish turning
B. Forming
C. Thread cutting
D. Down milling

Conditions
1. Backlash eliminator
2. Zero rake
3. Nose radius
4. Low speed **[ME 1995 : 2 M]**

6.7 Determine the transformation ratio (TR) of the gear train connecting the spindle and the lead screw of 4TPI required to cut 1.0 mm pitch screw threads in a centre lathe. **[ME 1995 : 2 M]**

6.8 he rake angle in drill
(a) Increase from centre to periphery
(b) Decreases from centre to periphery
(c) Remains constant
(d) Is irrelevant to the drilling operation
[ME 1996 : 1 M]

6.9 elix angle of fast helix drill is normally
(a) 35 deg (b) 60 deg
(c) 90 deg (d) 5 deg
[ME 1997 : 1 M]

6.10 In a typical metal cutting operation, using a cutting tool a positive rake $\alpha = 10°$, it is as observed that the shear angle was 20°. The friction angle is
(a) 45° (b) 30°
(c) 60° (d) 40° **[ME 1997 : 2 M]**

6.11 A cutting tool has a nose radius of 1.8 mm the feed rate for a theoretical surface roughness of R_t = 5 microns is
(a) 0.268 mm/rev (b) 0.187 mm/rev
(c) 0.036 mm/rev (d) 0.0187 mm/rev
[ME 1997 : 2 M]

6.12 In orthogonal machining operation, the chip thickness and the uncut chip thickness are equal to 0.45 mm. If the tool rake and is 0 deg. The shear plane angle is
(a) 45° (b) 30°
(c) 18° (d) 60° **[ME 1998 : 2 M]**

6.13 What is the approximate % change in the life, t, of the tool with zero rake angle used in orthogonal cutting when its clearance angle, α, is changed from 10 to 7 deg?
(Hints : Flank wear rate is proportional to cot α)
(a) 30% increase (b) 30% decrease
(c) 70% increase (d) 70% decrease
[ME 1999 : 2 M]

6.14 A built up edge is formed while machining
(a) ductile material at high speed
(b) ductile materials at low speed
(c) brittle materials at high speed
(d) brittle materials at low speed
[ME 2000 : 1 M]

6.15 Abrasive material used in grinding wheel selected for grinding ferrous alloys is
(a) SiC (b) Diamond
(c) Al_2O_3 (d) Boron carbide
[ME 2000 : 1 M]

6.16 A conventional lathe and a CNC lathe are under consideration for machining a given part. The relevant data are shown below:

	Preparation Cost (₹)	Production time per part (min)	Machine and operator cost (₹)
Conventional Lathe	30	30	75
CNC Lathe	150	15	120

The machine preferred for producing 100 pieces is
(a) conventional lathe
(b) CNC lathe
(c) any of the above
(d) None of these **[ME 2000 : 2 M]**

6.17 During orthogonal cutting of MS with a 10 deg rake angle tool, the chip thickness ratio was obtained as 0.4. The shear angle (in degrees) evaluated from this data is
(a) 6.53 (b) 20.22
(c) 22.94 (d) 50.00
[ME 2001 : 2 M]

6.18 For turning NiCr alloy steel at cutting speeds of 64 m/min and 100 m/min, the respective tool lives are 15 min and 12 min. The tool life for a cutting speed of 144 m/min is
(a) 8 min (b) 9 min
(c) 10 min (d) 11.5 min
[ME 2001 : 2 M]

6.19 A lead screw with half nuts in a lathe, free to rotate in both directions has
(a) V-threads
(b) Whitworth threads
(c) Buttress threads
(d) Acme threads **[ME 2002 : 1 M]**

6.20 The hardness of a grinding wheel is determined by the
(a) hardness of abrasive grains
(b) ability of the bond to retain abrasives
(c) hardness of the bond
(d) ability of the grinding wheel to penetrate the workpiece **[ME 2002 : 1 M]**

6.21 The time taken to drill a hole through a 25 mm thick plate with the drill rotating at 300 rpm and moving at a feed rate of 0.25 mm/rev is
(a) 10 sec (b) 20 sec
(c) 60 sec (d) 100 sec
[ME 2002 : 2 M]

6.22 Trepanning is performed for
(a) finishing a drilled hole
(b) producing a large hole without drilling
(c) truing a hole for alignment
(d) enlarging a drilled hole **[ME 2003 : 1 M]**

6.23 Strong screw threads are produced by
(a) thread milling
(b) thread chasing
(c) thread cutting with single point tool
(d) thread rolling **[ME 2003 : 1 M]**

6.24 A batch of 10 cutting tools could produce 500 components while working at 50 rpm with a tool feed of 0.25 mm/rev and depth of cut of 1 mm. A similar batch of 10 tools of the same specification could produce 122 components while working at 80 rpm with a feed of 0.25 mm/rev and 1 mm depth of cut. How many components can be produced with one cutting tool at 60 rpm?
(a) 29 (b) 31
(c) 37 (d) 42 **[ME 2003 : 2 M]**

Common Data for Q. 6.25 and 6.26

A cylinder is turned on a lathe with orthogonal machining principle. Spindle rotates at 200 rpm. The axial feed rate is 0.25 mm per revolution. Depth of cut is 0.4 mm. The rake angle is 10°. In the analysis it is found that the shear angle is 27.75°.

6.25 The thickness of the produced chip is
(a) 0.511 mm (b) 0.528 mm
(c) 0.818 mm (d) 0.846 mm
[ME 2003 : 2 M]

6.26 In the above problem, the coefficient of friction at the chip tool interface obtained using Earnest and Merchant theory is
(a) 0.18 (b) 0.36
(c) 0.71 (d) 0.98
[ME 2003 : 2 M]

6.27 Through holes of 10 mm diameter are to be drilled in a steel plate of 20 mm thickness. Drill spindle speed is 300 rpm, feed 0.2 mm/rev and drill point angle is 120°. Assuming drill overtravel of 2 mm, the time for producing a hole will be
(a) 4 seconds (b) 25 seconds
(c) 100 seconds (d) 110 seconds
[ME 2004 : 2 M]

6.28 In an orthogonal cutting test on mild steel, the following data were obtained

Cutting speed	:	40 m/min
Depth of cut	:	0.3 mm
Tool rake angle	:	+5°
Chip thickness	:	1.5 mm
Cutting force	:	900 N
Thrust force	:	450 N

Using Merchant's analysis, the friction angle during the machining will be

(a) 26.6° (b) 31.5°
(c) 45° (d) 63.4° [ME 2004 : 2 M]

6.29 In a machining operation, doubling the cutting speed reduces the tool life to 1/8th of the original value. The exponent n in Taylor's tool life equation $VT^n = C$, is

(a) $\frac{1}{8}$ (b) $\frac{1}{4}$
(c) $\frac{1}{3}$ (d) $\frac{1}{2}$ [ME 2004 : 2 M]

6.30 The figure below shows a graph which qualitatively relates cutting speed and cost per piece produced.

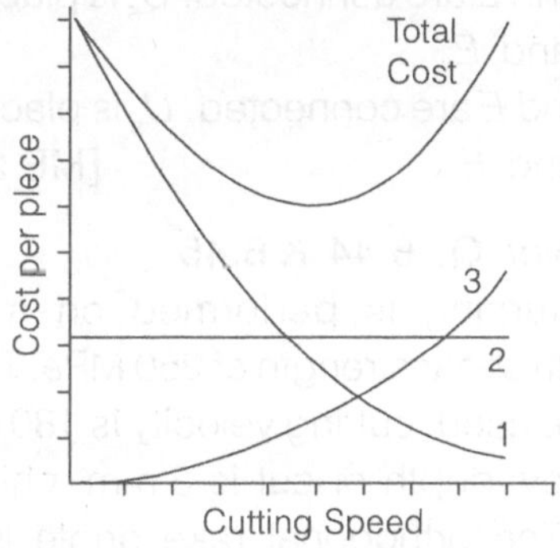

The three curves 1, 2 and 3 respectively represent

(a) machining cost, non-productive cost, tool changing cost
(b) non-productive cost, machining cost, tool changing cost
(c) tool changing cost, machining cost, non-productive cost
(d) tool changing cost, non-productive cost, machining cost [ME 2005 : 1 M]

6.31 A 600 mm x 30 mm flat surface of a plate is to be finish machined on a shaper. The plate has been fixed with the 600 mm side along the tool travel direction. If the tool over-travel at each end of the plate is 20 mm, average cutting speed is 8 m/min, feed rate is 0.3 mm/stroke and the ratio of return time to cutting time of the tool is 1 : 2, the time required for machining will be

(a) 8 minutes (b) 12 minutes
(c) 16 minutes (d) 20 minutes
[ME 2005 : 2 M]

6.32 Two tools P and Q have signatures 5°–5°–6°–6°–8°–30°–0 and 5°–5°–7°–7°–8°–15°–0 (both ASA) respectively. They are used to turn components under the same machining conditions. If h_P and h_Q denote the peak-to-valley heights of surfaces produced by the tools P and Q, the ratio h_P/h_Q will be

(a) $\frac{\tan 8° + \cot 15°}{\tan 8° + \cot 30°}$ (b) $\frac{\tan 15° + \cot 8°}{\tan 30° + \cot 8°}$
(c) $\frac{\tan 15° + \cot 7°}{\tan 30° + \cot 7°}$ (d) $\frac{\tan 7° + \cot 15°}{\tan 7° + \cot 30°}$
[ME 2005 : 2 M]

Common Data for Q. 6.33, 6.34 and 6.35

In an orthogonal machining operation:

Uncut thickness = 0.5 mm
Cutting speed = 20 m/min
Rake angle = 15°
Width of cut = 5 mm
Chip thickness = 0.7 mm
Thrust force = 200 N
Cuttting force = 1200 N

Assume Merchant's theory.

6.33 The values of shear angle and shear strain, respectively, are

(a) 30.3° and 1.98 (b) 30.3° and 4.23
(c) 40.2° and 2.97 (d) 40.2° and 1.65
[ME 2006 : 2 M]

6.34 The coefficient of friction at the tool-chip interface is

(a) 0.23 (b) 0.46
(c) 0.85 (d) 0.95 [ME 2006 : 2 M]

6.35 The percentage of total energy dissipated due to friction at the tool-chip interface is

(a) 30% (b) 42%
(c) 58% (d) 70%
[ME 2006 : 2 M]

6.36 In orthogonal turning of a low carbon steel bar of diameter 150 mm with uncoated carbide tool, the cutting velocity is 90 m/min. The feed is 0.24 mm/rev and the depth of cut is 2 mm. The chip thickness obtained is 0.48 mm. If the orthogonal rake angle is zero and the principal cutting edge angle is 90°, the shear angle in degree is

(a) 20.56 (b) 26.56
(c) 30.56 (d) 36.56
[ME 2007 : 1 M]

6.37 In orthogonal turning of medium carbon steel, the specific machining energy is 2.0 J/mm³. The cutting velocity, feed and depth of cut are 120 m/min, 0.2 mm/rev and 2 mm respectively. The main cutting force in N is

(a) 40 (b) 80
(c) 400 (d) 800
[ME 2007 : 2 M]

6.38 In orthogonal turning of low carbon steel pipe with principal cutting edge angle of 90°, the main cutting force is 1000 N and the feed force is 800 N. The shear angle is 25° and orthogonal rake angle is zero. Employing Merchant's theory, the ratio of friction force to normal force acting on the cutting tool is
(a) 1.56 (b) 1.25
(c) 0.80 (d) 0.64
[ME 2007 : 2 M]

Linked Answer Q. 6.39 and 6.40

A low cabon steel bar of 147 mm diameter with a length of 630 mm is being turned with uncoated carbide insert. The observed tool lives are 24 min and 12 min for cutting velocities of 90 m/min and 120 m/min respectively. The feed and depth of cut are 0.2 mm/rev and 2 mm respectively. Use the unmachined diameter to calculate the cutting velocity.

6.39 When tool life is 20 min, the cutting velocity in m/min is
(a) 87 (b) 97
(c) 107 (d) 114 [ME 2007 : 2 M]

6.40 Neglect over-travel or approach of the tool. When tool life is 20 min, the machining time in min for a single pass is
(a) 5 (b) 10
(c) 15 (d) 20 [ME 2007 : 2 M]

6.41 Internal gear cutting operation can be performed by
(a) milling
(b) shaping with rack cutter
(c) shaping with pinion cutter
(d) hobbing [ME 2008 : 1 M]

6.42 In a single point turning tool, the side rake angle and orthogonal rake angle are equal. ϕ is the principal cutting edge angle and its range is $0° \le \phi \le 90°$. The chip flows in the orthogonal plane. The value of ϕ is closest to
(a) 0° (b) 45°
(c) 60° (d) 90° [ME 2008 : 2 M]

6.43 The figure shows an incomplete schematic of a conventional lathe to be used for cutting threads with different pitches. The speed gear box U_v is shown and the feed gear box U_s is to be placed. *P*, *Q*, *R* and *S* denote locations and have no other significance. Changes in U_v should NOT affect the pitch of the thread being cut and changes in U_s should NOT affect the cutting speed.

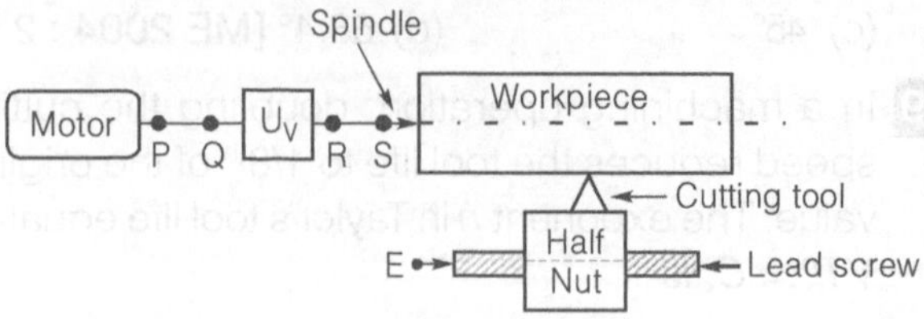

The correct connections and the correct placement of U_s are given by
(a) *Q* and *E* are connected. U_s is placed between *P* and *Q*.
(b) *S* and *E* are connected, U_s is placed between *R* and *S*.
(c) *Q* and *E* are connected. U_s is placed between *Q* and *E*.
(d) *S* and *E* are connected. U_s is placed between *S* and *E*. [ME 2008 : 2 M]

Linked Answer Q. 6.44 & 6.45

Orthogonal turning is performed on a cylindrical workpiece with shear strength of 250 MPa. The following conditions are used: cutting velocity is 180 m/min, feed is 0.20 mm/rev, depth of cut is 3 mm, chip thickness ratio = 0.5. The orthogonal rake angle is 7°. Apply Merchant's theory for analysis.

6.44 The shear plane angle (in degrees) and the shear force respectively are
(a) 52; 320 N (b) 52; 400 N
(c) 28; 400 N (d) 28; 320 N
[ME 2008 : 2 M]

6.45 The cutting and thrust forces, respectively, are
(a) 568 N; 387 N (b) 565 N; 381 N
(c) 440 N; 342 N (d) 480 N; 356 N
[ME 2008 : 2 M]

6.46 Friction at the tool-chip interface can be reduced by
(a) decreasing the rake angle
(b) increasing the depth of cut
(c) decreasing the cutting speed
(d) increasing the cutting speed
[ME 2009 : 1 M]

6.47 Minimum shear strain in orthogonal turning with a cutting tool of zero rake angle is
(a) 0.0 (b) 0.5
(c) 1.0 (d) 2.0 [ME 2009 : 2 M]

Linked Answer Q. 6.48 and 6.49

In a machining experiment, tool life was found to vary with the cutting speed in the following manner:

Cutting speed (m/min)	Tool life (minutes)
60	81
90	36

6.48 The exponent (n) and constant (k) of the Taylor's tool life equation are
(a) $n = 0.5$ and $k = 540$
(b) $n = 1$ and $k = 4860$
(c) $n = -1$ and $k = 0.74$
(d) $n = -0.5$ and $k = 1.155$ [ME 2009 : 2 M]

6.49 What is the percentage increase in tool life when the cutting speed is halved?
(a) 50% (b) 200%
(c) 300 % (d) 400%
[ME 2009 : 2 M]

6.50 For tool *A*, Taylor's tool life exponent (n) is 0.45 and constant (K) is 90. Similarly for tool *B*, $n = 0.3$ and $K = 60$. The cutting speed (in m/min) above which tool *A* will have a higher tool life than tool *B* is
(a) 26.7 (b) 42.5
(c) 80.7 (d) 142.9
[ME 2010 : 2 M]

6.51 A single-point cutting tool with 12° rake angle is used to machine a steel work-piece. The depth of cut, i.e. uncut thickness is 0.81 mm. The chip thickness under orthogonal machining condition is 1.8 mm. The shear angle is approximately
(a) 22° (b) 26°
(c) 56° (d) 76° [ME 2011 : 2 M]

6.52 In a single pass drilling operation, a through hole of 15 mm diameter is to be drilled in a steel plate of 50 mm thickness. Drill spindle speed is 500 rpm, feed is 0.2 mm/rev and drill point angle is 118°. Assuming 2 mm clearance at approach and exit, the total drill time (in seconds) is
(a) 35.1 (b) 32.4
(c) 31.2 (d) 30.1 [ME 2012 : 2 M]

6.53 Details pertaining to an orthogonal metal cutting process are given below:

Chip thickness ratio	0.4
Undeformed thickness	0.6 mm
Rake angle	+10°
Cutting speed	2.5 m/s
Mean thickness of primary shear zone	25 microns

The shear strain rate in s^{-1} during the process is
(a) 0.1781×10^5 (b) 0.7754×10^5
(c) 1.0104×10^5 (d) 4.397×10^5
[ME 2012 : 2 M]

6.54 Tool life of a 10 hours is obtained when cutting with a single point tool at 63 m/min. If Taylor's constant $C = 257.35$, tool life on doubling the velocity will be
(a) 5 hours (b) 25.7 min
(c) 38.3 min (d) unchanged
[ME 1996 : 2 M]

6.55 Tool life testing on a lathe under dry cutting conditions gauge *n* and *C* of Taylor tool life equation as 0.12 and 130 m/min. respectively. When a coolant was used, *C* increased by 10%. The increased tool life with the use of coolant at a cutting speed of 90 m/min is
(a) 47.4 min (b) 37.4 min
(c) 27.4 min (d) 17.4 min
[ME 2001 : 2 M]

Common Data Q. 6.56 and 6.57

In orthogonal turning of a bar of 100 mm diameter with a feed of 0.25 mm/rev, depth of cut of 4 mm and cutting velocity of 90 m/min, it is observed that the main (tangential) cutting force is perpendicular to the friction force acting at the chip-tool interface. The main (tangential) cutting force is 1500 N.

6.56 The orthogonal rake angle of the cutting toll in degree is
(a) zero (b) 3.58
(c) 5 (d) 7.16
[ME 2013 : 2 M]

6.57 The normal force acting the chip-tool interface in N is
(a) 1000 (b) 1500
(c) 2000 (d) 2500
[ME 2013 : 2 M]

6.58 A steel bar 200 mm in diameter is turned at a feed of 0.25 mm/rev with a depth of cut of 4 mm. The rotational speed of the workpiece is 160 rpm. The material removal rate in mm^3/s is
(a) 160 (b) 167.6
(c) 1600 (d) 1675.5 [2013 : 1 M]

6.59 The main cutting force acting on a tool during the turning (orthogonal cutting) operation of a metal is 400 N. The turning was performed using 2 mm depth of cut and 0.1 mm/rev feed rate. The specific cutting pressure (in N/mm^2) is
(a) 1000 (b) 2000

(c) 3000 (d) 4000

[ME 2014 : 1 M, Set-1]

6.60 Match the Machine Tools (Group A) with the probable Operations (Group B):

Group A	Group B
P. Centre Lathe	1. Slotting
Q. Milling	2. Counter-boring
R. Grinding	3. Knurling
S. Drilling	4. Dressing

(a) P–1, Q–2, R–4, S–3
(b) P–2, Q–1, R–4, S–3
(c) P–3, Q–1, R–4, S–2
(d) P–3, Q–4, R–2, S–1 [ME 2014 : 1 M, Set-2]

6.61 A straight turning operation is carried out using a single point cutting tool on an AISI 1020 steel rod. The feed is 0.2 mm/rev and the depth of cut is 0.5 mm. The tool has a side cutting edge angle of 60°. The uncut chip thickness (in mm) is ______.

[ME 2014 : 1 M, Set-3]

6.62 Cutting tool is much harder than the workpiece. Yet the tool wears out during the tool-work interaction, because

(a) extra hardness is imparted to the workpiece due to coolant used
(b) oxide layers on the workpiece surface impart extra hardness to it
(c) extra hardness is imparted to the workpiece due to severe rate of strain
(d) vibration of induced in the machine tool

[ME 2014 : 1 M, Set-3]

6.63 Two separate slab milling operations, 1 and 2, are performed with identical milling cutters. The depth of cut in operation 2 is twice that in operation 1. The other cutting parameters are identical. The ratio of maximum uncut chip thicknesses in operations 1 and 2 is ______

[ME 2014 : 1 M, Set-4]

6.64 Better surface finish is obtained with a large rake angle because

(a) the area of shear plane decreases resulting in the decrease in shear force and cutting force
(b) the tool becomes thinner and the cutting force is reduced
(c) less heat is accumulated in the cutting zone
(d) the friction between the chip and the tool is less [ME 2014 : 1 M, Set-4]

6.65 During pure orthogonal turning operation of a hollow cylindrical pipe, it is found that the thickness of the chip produced is 0.5 mm. The feed given to the zero degree rake angle tool is 0.2 mm/rev. The shear strain produced during the operation is ______.

[ME 2014 : 2 M, Set-1]

6.66 If the Taylor's tool life exponent n is 0.2, and the tool changing time is 1.5 min, then the tool life (in min) for maximum production rate is ______.

[ME 2014 : 2 M, Set-1]

6.67 A hole of 20 mm diameter is to be drilled in a steel block of 40 mm thickness. The drilling is performed at rotational speed of 400 rpm and feed rate of 0.1 mm/rev. The required approach and over run of the drill together is equal to the radius of drill. The drilling time (in minute) is

(a) 1.00 (b) 1.25
(c) 1.50 (d) 1.75

[ME 2014 : 2 M, Set-2]

6.68 Which pair of following statements is correct for orthogonal cutting using a single-point cutting tool?

P. Reduction in friction angle increases cutting force
Q. Reduction in friction angle decreases cutting force
R. Reduction in friction angle increases chip thickness
S. Reduction in friction angle decreases chip thickness

(a) P and R (b) P and S
(c) Q and R (d) Q and S

[ME 2014 : 2 M, Set-3]

6.69 A cast iron block of 200 mm length is being shaped in a shaping machine with a depth of cut of 4 mm, feed of 0.25 mm/stroke and the tool principal cutting edge angle of 30°. Number of cutting strokes per minute is 60. Using specific energy for cutting as 1.49 J/mm^3, the average power consumption (in watt) is ______

[ME 2014 : 2 M, Set-4]

6.70 Under certain cutting conditions, doubling the cutting speed reduces the tool life to $\left(\frac{1}{16}\right)^{th}$ of the original. Taylor's tool life index(n) for this tool-workpiece combination will be ______.

[2015 : 1 M, Set-1]

6.71 In a machining operation, if the generatrix and directrix both are straight lines, the surface obtained is

(a) cylindrical

(b) helical
(c) plane
(d) surface of revolution [2015 : 1 M, Set-3]

6.72 An orthogonal turning operation is carried out under the following conditions; rake angle = 5°, spindle rotational speed = 400 rpm; axial feed = 0.4 m/min and radial depth of cut = 5 mm. The chip thickness t_c, is found to be 3 mm. The shear angle (in degree) in this turning process is ____. [2015 : 2 M, Set-1]

6.73 A single point cutting tool with 0° rake angle is used in an orthogonal machining process. At a cutting speed of 180 m/min, the thrust force is 490 N. If the coefficient of friction between the tool and the chip is 0.7, then the power consumption (in kW) for the machining operation is ______. [2015 : 2 M, Set-2]

6.74 Orthogonal turning of a mild steel tube with a tool of rake angle 10° carried out at a feed of 0.14 mm/rev. If the thickness of the chip produced is 0.28 mm, the values of shear angle and shear strain will be respectively
(a) 28°20′ and 2.19 (b) 22°20′ and 3.53
(c) 24°30′ and 3.53 (d) 37°20′ and 5.19
[2015 : 2 M, Set-3]

6.75 A shaft of length 90 mm has a tapered portion of length 55 mm. The diameter of the taper is 80 mm at one end and 65 mm at the other. If the taper is made by tailstock set over method, the taper angle and the set over respectively are
(a) 15°32′ and 12.16 mm
(b) 15°32′ and 15.66 mm
(c) 11°22′ and 10.26 mm
(d) 10°32′ and 14.46 mm [2015 : 2 M, Set-3]

6.76 For an orthogonal cutting operation, tool material is HSS, rake angle is 22°, chip thickness is 0.8 mm, speed is 48 m/min and feed is 0.4 mm/rev. The shear plane angle (in degrees) is
(a) 19.24 (b) 29.70
(c) 56.00 (d) 68.75
[2016 : 2 M, Set-3]

6.77 A firm uses a turning center, a milling center and a grinding machine to produce two parts. The table below provides the machining time required for each part and the maximum machining time available on each machine. The profit per unit on parts I and II are ₹ 40 and ₹ 100, respectively. The maximum profit per week of the firm is ₹ ____.

Type of machine	Machining time required for the machine part (minutes)		Maximum machining time available per week (minutes)
	I	II	
Turning Center	12	6	6000
Milling Center	4	10	4000
Grinding Machine	2	3	1800

[2016 : 2 M, Set-3]

6.78 The following data is applicable for a turning operation. The length of job is 900 mm, diameter of job is 200 mm, feed rate is 0.25 mm/rev and optimum cutting speed is 300 m/min. The machining time (in min) is____.
[2016 : 1 M, Set-2]

6.79 For a certain job, the cost of metal cutting is ₹ 18 C/V and the cost of tooling is ₹ 270C/(TV), where C is a constant, V is the cutting speed in m/min and T is the tool life in minutes. The Taylor's tool life equation is $VT^{0.25} = 150$. The cutting speed (in m/min) for the minimum total cost is______.
[2016 : 2 M, Set-2]

6.80 In an orthogonal cutting process the tool used has rake angle of zero degree. The measured cutting force and thrust force are 500 N and 250 N, respectively. The coefficient of friction between the tool and the chip is___.
[2016 : 1 M, Set-1]

6.81 The tool life equation for HSS tool is $VT^{0.14}f^{0.7}d^{0.4}$ = Constant. The tool life (T) of 30 min is obtained using the following cutting conditions :
V = 45 m/min, f = 0.35 mm, d = 2.0 mm
If speed (V), feed (f) and depth of cut (d) are increased individually by 25%, the tool life (in min) is
(a) 0.15 (b) 1.06
(c) 22.50 (d) 30.0
[2016 : 2 M, Set-1]

6.82 Internal gears are manufactured by
(a) hobbing
(b) shaping with pinion cutter
(c) shaping with rack cutter
(d) milling [2016 : 1 M, Set-3]

6.83 In a single point turning operation with cemented carbide tool and steel work piece, it is found that the Taylor's exponent is 0.25. If the cutting speed is reduced by 50% then the tool life changes by ______ times. [2016 : 2 M, Set-3]

6.84 Two cutting tools with tool life equations given below are being compared:

Tool 1 : $VT^{0.1} = 150$

Tool 2 : $VT^{0.3} = 300$

where V is cutting speed in m/minute and T is tool life in minutes. The breakeven cutting speed beyond which Tool 2 will have a higher tool life is_________ m/minute. [2017 : 2 M, Set-1]

6.85 A block of length 200 mm is machined by a slab milling cutter 34 mm in diameter. The depth of cut and table feed are set at 2 mm and 18 mm/minute, respectively. Considering the approach and the over travel of the cutter to be same, the minimum estimated machining time per pass is___ minutes.

[2017 : 2 M, Set-1]

6.86 In an orthogonal machining with a tool of 9° orthogonal rake angle, the uncut chip thickness is 0.2 mm. The chip thickness fluctuates between 0.25 mm and 0.4 mm. The ratio of the maximum shear angle to the minimum shear angle during machining is ____.

[2017 : 2 M, Set-2]

6.87 During the turning of a 20 mm diameter steel bar at a spindle speed of 400 rpm, a tool life of 20 minute is obtained. When the same bar is turned at 200 rpm, the tool life becomes 60 minute. Assume that Taylor's tool life equation is valid. When the bar is turned at 300 rpm, the tool life (in minute) is approximately

(a) 25 (b) 32
(c) 40 (d) 50

[2017 : 2 M, Set-2]

6.88 Using the Taylor's tool life equation with exponent n = 0.5, if the cutting speed is reduced by 50%, the ratio of new tool life to original tool life is

(a) 4 (b) 2
(c) 1 (d) 0.5

[2018 : 1 M, Set-1]

6.89 A grinding ratio of 200 implies that the

(a) grinding wheel wears 200 times the volume of the material removed.
(b) grinding wheel wears 0.005 times the volume of the material removed
(c) aspect ratio of abrasive particles used in the grinding wheel is 200
(d) ratio of volume of abrasive particle to that of grinding wheel is 200 [2018 : 1 M, Set-1]

6.90 An orthogonal cutting operations is being carried out in which uncut thickness is 0.010 mm, cutting speed is 130 m/min, rake angle is 15° and width of cut is 6 mm. It is observed that the chip thickness is 0.015 mm, the cutting force is 60 N and the thrust force is 25 N. The ratio of friction energy to total energy is _________(correct to two decimal places) [2018 : 2 M, Set-1]

6.91 The preferred option for holding an odd-shaped workpiece in a centre lathe is

(a) live and dead centres
(b) three jaw chuck
(c) lathe dog
(d) four jaw chuck [2018 : 1 M, Set-2]

6.92 Feed rate in slab milling operation is equal to

(a) rotation per minute (rpm)
(b) product of rpm and number of teeth in the cutter
(c) product of rpm, feed per tooth and number of teeth in the cutter
(d) product of rpm, feed per tooth and number of teeth in contact [2018 : 1 M, Set-2]

6.93 Taylor's tool life equation is used to estimate the life of a batch of identical HSS twist drills by drilling through holes at constant feed in 20 mm thick mild steel plates. In test 1, a drill lasted 300 holes at 150 rpm while in test 2, another drill lasted 200 holes at 300 rpm. The maximum number of holes that can be made by another drill from the above batch at 200 rpm is _______ (correct to two decimal places).

[2018 : 2 M, Set-2]

6.94 Following data correspond to an orthogonal Turning of a 100 mm diameter rod on a lathe. Rake angle: +15°; Uncut chip thickness 0.5 mm, nominal chip thickness after the cut 1.25 mm. The shear angle (in degrees) for this process is _____(correct to two decimal places). [2018 : 2 M, Set-2]

6.95 The length, width and thickness of a steel sample are 400 mm, 40 mm and 20 mm, respectively. Its thickness needs to be uniformly reduced by 2 mm in a single pass by using horizontal slab milling. The milling cutter (diameter: 100 mm, width: 50 mm) has 20 teeth and rotates at 1200 rpm. The feed per tooth is 0.05 mm. The feed direction is along the length of the sample. If the over travel distance is the same as the approach distance, the approach distance and time taken to complete the required machining task are

(a) 14 mm, 21.4 s (b) 21 mm, 39.4 s
(c) 14 mm, 18.4 s (d) 21 mm, 28.9 s

[2019 : 1 M, Set-1]

6.96 In orthogonal turning of a cylindrical tube of wall thickness 5 mm, the axial and tangential cutting forces were measured as 1259 N and 1601 N, respectively. The measured chip thickness after machining was found to be 0.3 mm. The rake angle was 10° and axial feed was 100 mm/min. The rotational speed of the spindle was 1000 rpm. Assuming the material to be perfectly plastic and Merchant's first solution, the shear strength of the material is closest to

(a) 875 MPa (b) 920 MPa
(c) 722 MPa (d) 200 MPa

[2019 : 2 M, Set-1]

6.97 In ASA system, the side cutting and end cutting edge angles of a sharp turning tool are 45° and 10°, respectively. The feed during cylindrical turning is 0.1 mm/rev. The centre line average surface roughness (in μm, round off to one decimal place) of the generated surface is _____.

[2019 : 2 M, Set-1]

6.98 Taylor's tool life equation is given by $VT^n = C$, where V is in m/min and T is in min. In a turning operation, two tools X and Y are used. For tool X, $n = 0.3$ and C = 60 and for tool Y, $n = 0.6$ and C = 90. Both the tools will have the same tool life for the cutting speed (in m/min, round off to one decimal place) of ___. [2019 : 2 M, Set-1]

6.99 In an orthogonal machining with a single point cutting tool of rake angle 10°, the uncut chip thickness and the chip thickness are 0.125 mm and 0.22 mm respectively. Using Merchant's first solution for the condition of minimum cutting force, the coefficient of friction at the chip-tool interface is _______ (round off to two decimal places).

[2019 : 2 M, Set-2]

6.100 A through hole is drilled in an aluminum alloy plate of 15 mm thickness with a drill bit of diameter 10 mm, at a feed of 0.25 mm/rev and a spindle speed of 1200 rpm. If the specific energy required for cutting this material is 0.7 N-m/mm^3, the power required is _____W (round off to two decimal places). [2019 : 2 M, Set-2]

6.101 The base of a brass bracket needs rough grinding. For this purpose, the most suitable grinding wheel grade specification is

(a) A50G8V (b) C30Q12V
(c) A30D12V (d) C90J4B

[2020 : 1 M, Set-1]

6.102 A slot of 25 mm x 25 mm is to be milled in a workpiece of 300 mm length using a side and face milling cutter of diameter 100 mm, width 25 mm and having 20 teeth. For a depth of cut 5 mm, feed per tooth 0.1 mm, cutting speed 35 m/min and approach and over travel distance of 5 mm each, the time required for milling the slot is_______ minutes (round off to one decimal place). [2020 : 2 M, Set-1]

6.103 In a turning process using orthogonal tool geometry, a chip length of 100 mm is obtained for an uncut chip length of 250 mm.

The cutting conditions are cutting speed = 30 m/min, rake angle = 20°.

The shear plane angle is ________ degrees (round off to one decimal place).

[2020 : 2 M, Set-1]

6.104 A cylindrical bar with 200 mm diameter is being turned with a tool having geometry 0° - 9° - 7° - 8° - 15° - 30° - 0.05 inch (Coordinate system, ASA) resulting in a cutting force F_{c1}. If the tool geometry is changed to 0° - 9° - 7° - 8° - 15° - 0° - 0.05 inch (Coordinate system. ASA) and all other parameters remain unchanged, the cutting force changes to F_{c2}. Specific cutting energy (in J/mm^3) is $U_c = U_0 (t_1)^{-0.4}$, where U_0 is the specific energy coefficient, and t_1 is the uncut thickness in mm. The value of percentage change in cutting force F_{c2}. i.e., $\left(\frac{F_{c2} - F_{c1}}{F_{c1}}\right) \times 100$, is ______ (round off to one decimal place). [2020 : 2 M, Set-2]

6.105 There are two identical shaping machines S_1 and S_2. In machine S_2, the width of the workpiece is increased by 10% and the feed is decreased by 10%, with respect to that of S_1. If all other conditions remain the same then the ratio of total time per pass in S_1 and S_2 will be ______ (roundoff to one decimal place).

[2020 : 2 M, Set-2]

6.106 Bars of 250 mm length and 25 mm diameter are to be turned on a lathe with a feed of 0.2 mm/rev. Each regrinding of the tool costs Rs. 20. The time required for each tool change is 1 min. Tool life equation is given as $VT^{0.2} = 24$

(where cutting speed V is in m/min and tool life T is in min). The optimum tool cost per piece for maximum production rate is Rs. _______ (round off to 2 decimal places). [2020 : 2 M, Set-2]

6.107 In a grinding operation of a metal, specific energy consumption is 15 J/mm^3. If a grinding wheel with a diameter of 200 mm is rotating at 3000 rpm to obtain a material removal rate of 6000 mm^3/min, then the tangential force on the wheel is _________ N. [Round off to two decimal places] [2021 : 2 M, Set-1]

6.108 An orthogonal cutting operation is performed using a single point cutting tool with a rake angle of 12° on a lathe. During turning, the cutting force and the friction force are 1000 N and 600 N, respectively. If the chip thickness and the uncut thickness during turning are 1.5 mm and 0.75 mm, respectively, then the shear force is ________ N. [Round off to two decimal places]

[2021 : 2 M, Set-1]

6.109 In a machining operation, if a cutting tool traces the workpiece such that the directrix is perpendicular to the plane of the generatrix as shown in figure, the surface generated is ____

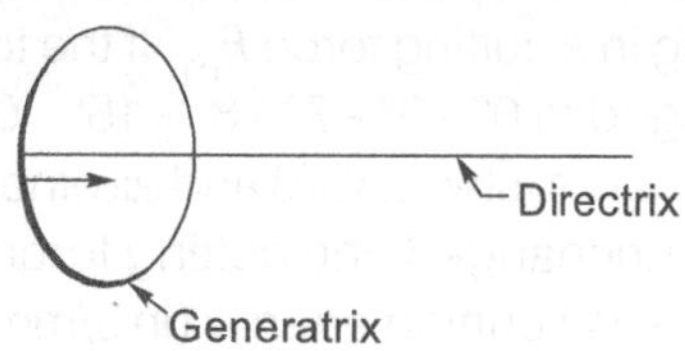

(a) spherical
(b) cylindrical
(c) surface of revolution
(d) plane [2021 : 1 M, Set-1]

6.110 The correct sequence of machining operations to be performed to finish a large diameter through hole is

(a) boring, drilling, reaming
(b) boring, reaming, drilling
(c) drilling, reaming, boring
(d) drilling, boring, reaming [2021 : 1 M, Set-1]

6.111 n a pure orthogonal turning by a zero rake angle single point carbide cutting tool, the shear force has been computed to be 400 N. If the cutting velocity, V_c = 100 m/min, depth of cut, t = 2.0 mm, feed, s_0 = 0.1 mm/revolution and chip velocity. V_f = 20 m/min, then the shear strength τ_s of the material will be _________ MPa (round off to two decimal places). [2021 : 2 M, Set-2]

6.112 A surface grinding operation has been performed on a Cast Iron plate having dimensions 300 mm (length) × 10 mm (width) × 50 mm (height). The grinding was performed using an alumina wheel having a wheel diameter of 150 mm and wheel width of 12 mm. The grinding velocity used is 40 m/s, table speed is 5 m/min, depth of cut per pass is 50 μm and the number of grinding passes is 20. The average tangential and average normal forces for each pass are found to be 40 N and 60 N respectively. The value of the specific grinding energy under the aforesaid grinding conditions is________ J/mm^3 (round off to one decimal place). [2021 : 2 M, Set-2]

6.113 Under orthogonal cutting condition, a turning operation is carried out on a metallic workpiece at a cutting speed of 4 m/s. The orthogonal rake angle of the cutting tool is 5°. The uncut chip thickness and width of cut are 0.2 mm and 3 mm, respectively. In this turning operation, the resulting friction angle and shear angle are 45° and 25°, respectively. If the dynamic yield shear strength of the workpiece material under this cutting condition is 1000 MPa, then the cutting force is _________ N (round off to one decimal place).

[2022 : 2 M, Set-1]

6.114 A 1 mm thick cylindrical tube, 100 mm in diameter, is orthogonally turned such that the entire wall thickness of the tube is cut in a single pass. The axial feed of the tool is 1 m/minute and the specific cutting energy (u) of the tube material is 6 J/mm^3. Neglect contribution of feed force towards power. The power required to carry out this operation is _______ kW (round off to one decimal place). [2022 : 2 M, Set-1]

6.115 A straight-teeth horizontal slab milling cutter is shown in the figure. It has 4 teeth and diameter (D) of 200 mm. The rotational speed of the cutter is 100 rpm and the linear feed given to the workpiece is 1000 mm/minute. The width of the workpiece (w) is 100 mm, and the entire width is milled in a single pass of the cutter. The cutting force/tooth is given by $F = Kt_c w$, where specific cutting force K = 10 N/mm^2, w is the width of cut, and t_c is the uncut chip thickness.

The depth of cut (d) is $D/2$, and hence the assumption of $\frac{d}{D} << 1$ is invalid. The maximum cutting force required is _______ kN (round off to one decimal place).

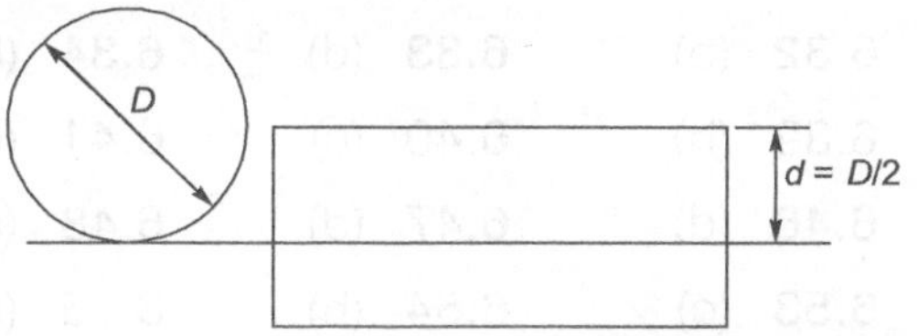

[2022 : 2 M, Set-2]

6.116 In an orthogonal machining operation, the cutting and thrust forces are equal in magnitude. The uncut chip thickness is 0.5 mm and the shear angle is 15°. The orthogonal rake angle of the tool is 0° and the width of cut is 2 mm. The workpiece material is perfectly plastic and its yield shear strength is 500 MPa. The cutting force is _______ N (round off to the nearest integer).

[2022 : 2 M, Set-2]

6.117 In an ideal orthogonal cutting experiment (see figure), the cutting speed V is 1 m/s, the rake angle of the tool $\alpha = 5°$, and the shear angle, ϕ, is known to be 45°.

Applying the ideal orthogonal cutting model, consider two shear planes PQ and RS close to each other. As they approach the thin shear zone (shown as a thick line in the figure), plane RS gets sheared with respect to PQ (point R1 shears to R2, and S1 shears to S2).

Assuming that the perpendicular distance between PQ and RS is $\delta = 25$ μm, what is the value of shear strain rate (in s^{-1}) that the material undergoes at the shear zone?

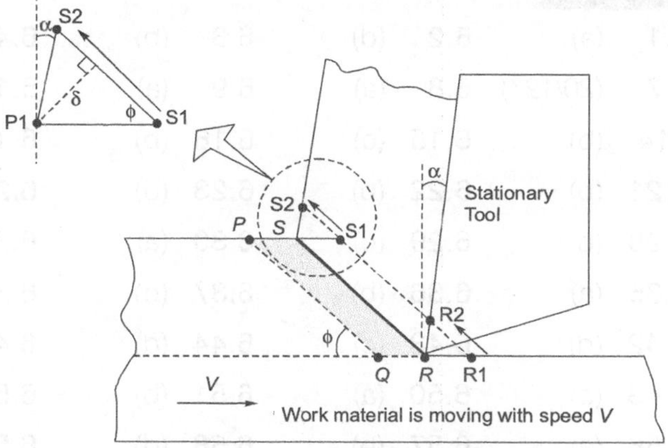

(a) 1.84×10^4
(b) 5.20×10^4
(c) 0.71×10^4
(d) 1.30×10^4

[2023 : 2 M]

6.118 The grinding wheel used to provide the best surface finish is

(a) A60L5V
(b) A80L5V
(c) A36L5V
(d) A54L5V

[2024 : 1 M]

6.119 A flat surface of a C60 steel having dimensions of 100 mm (length) × 200 mm (width) is produced by a HSS slab mill cutter. The 8-toothed cutter has 100 mm diameter and 200 mm width. The feed per tooth is 0.1 mm, cutting velocity is 20 m/min and depth of cut is 2 mm. The machining time required to remove the entire stock is_____ minutes. [Rounded off to 2 decimal places]

[2024 : 2 M]

6.120 A cutting tool provides a tool life of 60 minutes while machining with the cutting speed of 60 m/min. When the same tool is used for machining the same material, it provides a tool life of 10 minutes for a cutting speed of 100 m/min. If the cutting speed is changed to 80 m/min for the same tool and work material combination, the tool life computed using Taylor's tool life model is ____ minutes. (rounded off to 2 decimal places)

[2024 : 2 M]

■■■■

Answers Machining and Machine Tool Operation

6.1 (a)	6.2 (d)	6.3 (b)	6.4 (True)	6.5 (rigid)	6.6 (A–3, B–4, C–2, D–1)	
6.7 (20/127)	6.8 (a)	6.9 (a)	6.10 (c)	6.11 (a)	6.12 (a)	6.13 (b)
614 (b)	6.15 (c)	6.16 (b)	6.17 (c)	6.18 (c)	6.19 (d)	6.20 (b)
6.21 (b)	6.22 (b)	6.23 (d)	6.24 (a)	6.25 (a)	6.26 (d)	6.27 (b)
6.28 (b)	6.29 (c)	6.30 (a)	6.31 (b)	6.32 (b)	6.33 (d)	6.34 (b)
6.35 (a)	6.36 (b)	6.37 (d)	6.38 (c)	6.39 (b)	6.40 (c)	6.41 (c)
6.42 (d)	6.43 (c)	6.44 (d)	6.45 (b)	6.46 (d)	6.47 (d)	6.48 (a)
6.49 (c)	6.50 (a)	6.51 (b)	6.52 (a)	6.53 (c)	6.54 (b)	6.55 (a)
6.56 (a)	6.57 (b)	6.58 (d)	6.59 (b)	6.60 (c)	6.61 (0.1)	6.62 (c)
6.63 (0.707)	6.64 (a)	6.65 (2.9)	6.66 (6)	6.67 (b)	6.68 (d)	6.69 (298)
6.70 (0.25)	6.71 (c)	6.72 (18.88)	6.73 (2.1)	6.74 (a)	6.75 (a)	6.76 (b)
6.77 (40000)	6.78 (7.54)	6.79 (57.91)	6.80 (0.5)	6.81 (b)	6.82 (b)	6.83 (16)
6.84 (106.066)	6.85 (12)	6.86 (1.494)	6.87 (b)	6.88 (a)	6.89 (b)	6.90 (0.4408)
6.91 (d)	6.92 (c)	6.93 (253.53)	6.94 (23.32)	6.95 (a)	6.96 (c)	6.97 (3.75)
6.98 (40)	6.99 (0.74)	6.100 (274.89)	6.101 (b)	6.102 (8.1)	6.103 (23.5)	6.104 (–5.6)
6.105 (0.8)	6.106 (26.985)	6.107 (47.746)	6.108 (685.77)	6.109 (b)	6.110 (d)	
6.111 (392.23)	6.112 (38.40)	6.113 (2573.40)	6.114 (31.40)	6.115 (2.5)	6.116 (2732.05)	6.117 (b)
6.118 (b)	6.119 (2.24)	6.120 (21.87)				

Explanations Machining and Machine Tool Operation

6.1 (a)

With increase of rake angle, the forces induced during machining will be reduced hence the power consumption reduces, but the tool becomes weaker.

6.5 (rigid)

Machine tool, structures are made **rigid** for high process capability.

6.7 (20/127)

Transformation ratio = train value

= pitch of job threads/pitch of lead screw threads

= 1.0 mm/(1/4 inch) = (1.0 ×4)/(127/5)

= 1 ×4 ×5/127 = 20/127. (The term 127/5 is used for converting inches into mm).

6.10 (c)

$\alpha = 10°, \phi = 20°$

Here typical cutting operation means that Merchants theory is valid

$$2\phi + \beta - \alpha = 90°$$

$$\beta = 90° + \alpha - 20° = \mathbf{60°}$$

6.11 (a)

$r = 1.8$ mm, $f = ?$

$$R_t = 5 \text{ microns} = 0.005 = \frac{f^2}{8r}$$

$$f = \sqrt{0.005 \times 8 \times 1.8} = \mathbf{0.268\ mm/rev}$$

6.12 (a)

$$t_1 = t_2 = 0.45$$

$$\Rightarrow \quad r = \frac{t_1}{t_2} = 1$$

$$\alpha = 0$$

$$\phi = \tan^{-1}\left(\frac{1\cos 0°}{1 - 1\times\sin 0°}\right)$$

$$= \tan^{-1}(1) = \mathbf{45°}$$

6.13 (b)

$$\text{Tool life, } T \propto \frac{1}{\text{Flank wear}}$$

$$T \propto \frac{1}{\cot\alpha}$$

$$T \propto \tan\alpha$$

$$\frac{T_2}{T_1} = \frac{\tan\alpha_2}{\tan\alpha_1} = \frac{\tan 7°}{\tan 10°} = 0.7$$

$$T_2 = 0.7T_1$$

Tool life decreases by **30%**.

6.16 (b)

Let Q = No. of parts produced.
Total cost (TC) on conventional lathe

$$= 30 + \frac{30}{60} \times 75 \times Q = 30 + 37.5Q$$

TC on CNC lathe

$$= 150 + \frac{15}{60} \times 120 \times Q$$

$$= 150 + 30Q$$

If Q = 100 parts
TC on conventional lathe

$$= 30 + 37.5 \times 100 = 3780$$

TC on CNC lathe

$$= 150 + 30 \times 100 = 3150$$

Due to low cost CNC lathe is preferable.

6.17 (c)

$\alpha = 10°$, $r = 0.4$

$$\phi = \tan^{-1}\left(\frac{0.4\cos 10°}{1 - 0.4\sin 10°}\right) = \mathbf{22.94°}$$

6.18 (c)

$$V_1 = 64 \text{ min}, V_2 = 100 \text{ m/min}$$

$$T_1 = 15 \text{ min}, T_2 = 12 \text{ min}$$

$$V_1 T_1^n = V_2 T_2^n$$

$$\Rightarrow \quad V_1/V_2 = [T_2/T_1]^n$$

$\Rightarrow$ Taking $\log_e$ on both side, we get

$$n = \frac{\ln\frac{V_2}{V_1}}{\ln\frac{T_1}{T_2}} = \frac{\ln\frac{100}{64}}{\ln\frac{15}{12}} = \frac{0.4463}{0.2231} = 2$$

$$V_1 T_1^n = V_3 T_3^n$$

$$T_3 = T_1\left(\frac{V_1}{V_3}\right)^{1/n} = 15\left(\frac{64}{144}\right)^{0.5} = \mathbf{10 \text{ min}}$$

Note: Strictly speaking the value of $n < 1$

6.21 (b)

Length of drill hole, $L = 25$ mm, $N = 300$ rpm, $f = 0.25$ mm/rev

$$\text{drill time, } T = \frac{L}{fN} = \frac{25}{0.25 \times 300} = 0.333 \text{ min}$$

$$= \mathbf{20 \text{ sec}}$$

6.24 (a)

Velocity of first cutting tool

$$V_1 = 50 \times 0.25 = 12.5 \text{ mm/min}$$

Velocity of second cutting tool

$$V_2 = 0.25 \times 80 = 20 \text{ mm/min}$$

since tool life = no. of component produce × tool constant

$\therefore$ Tool life for first tool, $T_1 = 500$ K

Tool life for second tool, $T_2 = 122$ K

Now, $V_1 T_1^n = V_2 T_2^n$

$\Rightarrow (12.5) \times (500 \text{ K})^n = 20 \times (122 \text{ K})^n$

$$\therefore \quad \left(\frac{500}{122}\right)^n = 1.6$$

$$\therefore \quad n = 0.333$$

Hence number of components produced by one cutting tool at 60 rpm

$$\Rightarrow 12.5 \times \left(\frac{500k}{10}\right)^{0.333} = 60 \times 0.25 \times (xk)^{0.333}$$

$$\left(\frac{x}{50}\right)^{0.333} = \frac{12.5}{60 \times 0.25}$$

$$\left(\frac{x}{50}\right)^{0.333} = \frac{12.5}{60 \times 0.25}$$

$$\therefore \quad \mathbf{x = 29}$$

6.25 (a)

Given, $N = 200$ rpm
Feed (t_1) = uncut chip thickness = 0.25 mm/rev
Depth = 0.4 mm
Rake angle, $\alpha = 10°$
Shear angle, $\phi = 27.75°$
We Know

$$\tan\phi = \frac{\cos\alpha}{\frac{t_2}{t_1} - \sin\alpha}$$

$$\Rightarrow \quad t_2 = t_1\left[\frac{\cos\alpha}{\tan\phi} + \sin\alpha\right]$$

$$= 0.25\left[\frac{\cos 10°}{\tan 27.75°} + \sin 10°\right]$$

$$= \mathbf{0.511 \text{ mm}}$$

6.26 (d)

From Earnest and Merchant theory

$$2\phi + \beta - \alpha = 90^\circ$$

$$\beta = 90^\circ + 10 - 2 \times 27.75^\circ$$

$$\therefore \quad \beta = 44.5^\circ$$

$$\mu = \tan(44.5^\circ) = \mathbf{0.98}$$

6.27 (b)

Diameter of hole,

$$d = 10 \text{ mm}$$

thickness of steel plate,

$$t = 20 \text{ mm}$$

$$\alpha = \frac{120}{2} = 60^\circ$$

Break through distance,

$$A = \frac{d}{2\tan\alpha} = \frac{10}{2\tan 60^\circ}$$

$$= 2.8867 \text{ mm}$$

Total length of tool travel,

$$l = t + A + 2$$

$$= 20 + 2.8867 + 2 = 24.88 \text{ mm}$$

$$\text{Time for drilling} = \frac{l}{f \times N}$$

$$= \frac{24.88}{0.2 \times 300} = 0.4147 \text{ min}$$

$$= 0.4147 \times 60 \text{ s}$$

$$= 24.882 \text{ s} \approx \mathbf{25\ s}$$

6.28 (b)

$$\mu = \frac{F}{N} = \frac{F_c \sin\alpha + F_T \cos\alpha}{F_c \cos\alpha - F_T \sin\alpha}$$

$$= \frac{900\sin 5^\circ + 450\cos 5^\circ}{900\sin 5^\circ - 450\cos 5^\circ}$$

$$= \frac{526.72}{857.35} = 0.614$$

$$\therefore \quad \beta = \tan^{-1}(\mu) = \mathbf{31.5^\circ}$$

6.29 (c)

Taylor's tool life equation,

$$VT^n = C \quad \text{...(i)}$$

When cutting speed is doubled and tool life reduces to 1/8th, then

$$2v \times \left(\frac{1}{8}T\right)^n = C \quad \text{...(ii)}$$

from equation (i) and (ii), we get

$$2v\left(\frac{T}{8}\right)^n = VT^n$$

$$\left(\frac{1}{8}\right)^n = \frac{1}{2}$$

$$3n = 1$$

$$\therefore \quad n = \mathbf{1/3}$$

6.31 (b)

Length travelled in forward stroke

$$= 600 + 20 + 20$$

$$= 640 \text{ mm}$$

$$\text{Total strokes} = \frac{30}{0.3} = 100$$

$$\text{Total travel} = 640 \times 100$$

$$= 64000 \text{ mm} = 64 \text{ m}$$

$$\text{Time of strokes} = \frac{64}{8} = 8 \text{ minutes}$$

$$\text{Return time} = \frac{\text{cutting time}}{2} = \frac{8}{2} = 4 \text{ min}$$

Total time = 8 + 4 min = **12 min**

6.32 (b)

We know that

$$h = \frac{f}{\tan(\text{SCEA}) + \cot(\text{ECEA})}$$

where,

f - Feed rate

SCEA - Side cutting edge angle

ECEA - End cutting edge angle

Given,

$$h_p = \frac{f}{\tan 30^\circ + \cot 8^\circ}$$

and

$$h_Q = \frac{f}{\tan 15^\circ + \cot 8^\circ}$$

$$\therefore \quad \frac{h_p}{h_Q} = \frac{\tan 15^\circ + \cot 8^\circ}{\tan 30^\circ + \cot 8^\circ}$$

6.33 (d)

Given;

Uncut thickness: $t_1 = 0.5$ mm

Cutting speed: $v = 20$ m/min

Rake angle: $\alpha = 15^\circ$

Width of cut: $w = 5$ mm

Chip thickness: $t_2 = 0.7$ mm

Thrust force: $F_T = 200$ N

Cutting force: $F_C = 1200$ N

$$\text{Cutting ratio:} \quad r = \frac{t_1}{t_2} = \frac{0.5}{0.7} = 0.714$$

We know that,

$$\tan\phi = \frac{r\cos\alpha}{1-r\sin\alpha} = \frac{0.714\cos 15°}{1-0.714\sin 15°}$$
$$= 0.847$$
$$\therefore \quad \phi = \tan^{-1}(0.847) = 40.2°$$

Shear strain $= \cot\phi + \tan(\phi-\alpha)$
$$= \cot(40.2°) + \tan(40.2° - 15°)$$
$$= 1.181 + 0.4716 = \mathbf{1.65}$$

6.34 (b)

We know that

$$\mu = \frac{F}{N} = \frac{F_C\sin\alpha + F_T\cos\alpha}{F_C\cos\alpha - F_T\sin\alpha}$$
$$= \frac{1200\sin 15° + 200\cos 15°}{1200\cos 15° - 200\sin 15°}$$
$$\mu = \frac{F}{N} = \frac{F_C\sin\alpha + F_T\cos\alpha}{F_C\cos\alpha - F_T\sin\alpha}$$
$$= \frac{310.6+193.2}{1159.1-51.76} = \mathbf{0.46}$$

6.35 (a)

Frictional force

$$F_R = F_C\sin\alpha + F_T\cos\alpha$$
$$= 1200\sin(15°) + 200\cos(15°)$$
$$= 503.8$$

Velocity of chip,

$$V_C = \frac{V\sin\phi}{\cos(\phi-\alpha)} = \frac{V\sin(40.2°)}{\cos(40.2-15)} = 0.713V$$

Total power consumption

$$F_C V = 1200V$$

Power consumption due to friction

$$= F_R V_C = 503.8 \times 0.713V$$
$$= 359.21V$$

Percentage of energy dissipated due to friction

$$= \frac{359.21V}{1200\,V} \times 100 = \mathbf{30\%}$$

6.36 (b)

Given, Feed rate: $f = 0.24$ mm/rev

Chip thickness: $t_2 = 0.48$ mm

Principal cutting edge angle,

$$\Psi = 90°$$

Rake angle: $\alpha = 0$

We know that

Uncut chip thickness,

$$t_1 = f\sin\Psi = 0.24\sin 90°$$
$$t_1 = 0.24\text{ mm}$$

Cutting ratio: $r = \dfrac{t_1}{t_2} = \dfrac{0.24}{0.48} = 0.5$

$$\tan\phi = \frac{r\cos\alpha}{1-r\sin\alpha} = \frac{0.5\cos 0°}{1-0.5\sin 0°} = 0.5$$

Shear angle: $\phi = \tan^{-1}(0.5) = \mathbf{26.56°}$

6.37 (d)

The energy consumption per unit volume of material removal, commonly known as specific energy.

Cutting force $= 2 \times 1000 \times 0.2 \times 2 = \mathbf{800\ N}$

6.38 (c)

Given, Cutting force: $F_C = 1000$ N

Feed force: $F_T = 800$ N

Rake angle: $\alpha = 0°$

Shear angle: $\phi = 25°$

We know that, $\mu = \dfrac{F}{N} = \dfrac{F_C\sin\alpha + F_T\cos\alpha}{F_C\cos\alpha - F_T\sin\alpha}$

$$= \frac{1000\sin 0° + 800\cos 0°}{1000\cos 0° - 800\sin 0°} = \frac{800}{1000} = \mathbf{0.8}$$

6.39 (b)

We know that

$$v_1 T_1^n = v_2 T_2^n$$
$$\Rightarrow 90 \times (24)^n = 120 \times (12)^n$$
$$\left(\frac{24}{12}\right)^n = \frac{120}{90}$$
$$\Rightarrow \quad 2^n = 1.33$$
$$\therefore \quad n = 0.415$$

Now $90 \times (24)^n = v \times (20)^n$

$$\therefore \quad v = 90 \times \left(\frac{24}{20}\right)^{0.415} = \mathbf{97\ m/min}$$

6.40 (c)

Diameter = 147 mm = 0.147 m

Length = 630 mm = 0.63 m

Feed = 0.2 mm/rev = 0.2×10^{-3} m/rev

Velocity = 97 m/min

Required time $= \dfrac{L}{f \times N}$

and also $V = \pi DN$ m/min

where N is in RPM

$\therefore$ Required time $= \dfrac{\pi DL}{v \times f}$

$$= \frac{\pi \times 0.147 \times 0.630}{97 \times (0.2 \times 10^{-3})} = 14.99 \approx \mathbf{15\ min.}$$

6.44 (d)

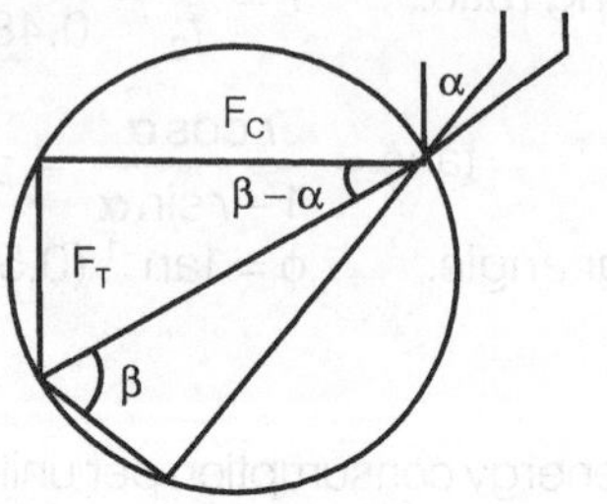

$$2\phi + \beta - \alpha = 90°$$

$$\Rightarrow \quad \tan\phi = \frac{\cos\alpha}{\frac{t_2}{t_1} - \sin\alpha}$$

$$\Rightarrow \quad \tan\phi = \frac{\cos 7°}{2 - \sin 7°}$$

$$\boldsymbol{\phi = 28°}$$

From Merchant first analysis we know that,

$$2\phi + \beta - \alpha = 90°$$

$$\Rightarrow (2 \times 28°) + \beta - 7° = 90°$$

$$\Rightarrow \quad \beta = 41°$$

$$F_s = \frac{\tau\, w\, t_1}{\sin\phi} = \frac{250 \times 3 \times 0.2}{\sin 28°} = \mathbf{319.5\ N}$$

6.45 (b)

$$\frac{F_c}{F_s} = \frac{\cos(\beta - \alpha)}{\cos(\phi + \beta - \alpha)}$$

$$\Rightarrow \quad F_c = 320 \times \frac{\cos(41° - 7°)}{\cos(28° + 41° - 7°)}$$

$$F_c = \mathbf{565.08\ N}$$

$$\tan(\beta - \alpha) = \frac{F_T}{F_c}$$

$$\Rightarrow \quad F_T = F_c \tan(\beta - \alpha) = 565 \tan(41° - 7°)$$

$$= 565 \tan 34° = \mathbf{381.55\ N}$$

6.47 (d)

During orthogonal machining.

Shear strain: $\gamma = \cot\phi + \tan(\phi - \alpha)$

Cutting tool having rake angle: $\alpha = 0°$

then $\gamma = \cot\phi + \tan\phi$

For minimum value of shear strain

$$\frac{d\gamma}{d\phi} = \frac{d}{d\phi}(\cot\phi + \tan\phi) = 0$$

$$\frac{d\gamma}{d\phi} = -\text{cosec}^2\phi + \sec^2\phi = 0$$

$$\Rightarrow \quad -\frac{1}{\sin^2\phi} + \frac{1}{\cos^2\phi} = 0$$

$$\Rightarrow \quad \frac{\cos^2\phi - \sin^2\phi}{\cos^2\phi \sin^2\phi} = 0$$

$$\Rightarrow \quad \cos 2\phi = 0 = \cos\frac{\pi}{2} = \cos\frac{3\pi}{2}$$

$$\Rightarrow \quad \phi = \frac{\pi}{4}$$

$$\frac{d^2\gamma}{d\phi^2} = -2\text{cosec}\ \phi.(-\text{cosec}\ \phi\ \cot\ \phi) + 2\sec\phi.(\sec\phi\tan\phi)$$

$$\frac{d^2\gamma}{d\phi^2} = 2\text{cosec}^2\phi\cot\phi + 2\sec^2\phi\tan\phi$$

$$\left.\frac{d^2\gamma}{d\phi^2}\right|_{at\ \phi = \frac{\pi}{4}} > 0$$

Therefore γ is minimum at $\phi = \frac{\pi}{4}$

Therefore minimum shear strain γ is given by

$$\gamma = \cot\frac{\pi}{4} + \tan\frac{\pi}{4}$$

$$= \mathbf{2°}$$

6.48 (a)

Cutting Speed (m/min)	Tool Life (minutes)
60	81
90	36

Taylor's tool life equations,

$$VT^n = \text{constant} = K$$

$$(60)(81)^n = K \quad \text{... (i)}$$

$$(90)(36)^n = K \quad \text{... (ii)}$$

Divide Eq. (i) by Eq. (ii)

$$\frac{(60)(81)^n}{(90)(36)^n} = 1$$

$$\Rightarrow \quad \left(\frac{81}{36}\right)^n = \left(\frac{90}{60}\right)$$

$$\Rightarrow \quad (2.25)^n = 1.5$$

$$\Rightarrow \quad n \ln(2.25) = \ln 1.5$$

$$\Rightarrow \quad \boldsymbol{n = 0.5}$$

Keep the value of n in Eq. (i)

$$(60)(81)^{0.5} = K$$

$$\Rightarrow \quad K = \mathbf{540}$$

6.49 (c)

Taylor's tool life equation

$$VT^{0.5} = K = 540$$

if cutting speed is halved.

$$V_1 = V, \quad T_1 = T, V_2 = V/2, T_2 = ?$$

$$V_1 T_1^{0.5} = V_2 T_2^{0.5}$$

$$T_2 = T_1\left(\frac{V_1}{V_2}\right)^2 = T_1\left(\frac{V}{V/2}\right)^2$$

$$T_2 = 4T_1 = 4T$$

Percentage increase in tool life

$$= \frac{T_2 - T_1}{T_1} \times 100 = \frac{4T - T}{T} \times 100 = \mathbf{300\%}$$

6.50 (a)

By Taylor's tool life equation $VT^n = K$.

Let V is the speed above which tool A will have a higher tool life than tool B, but at V, $T_A = T_B$.

then
$$VT_A^{0.45} = 90$$
$$VT_B^{0.3} = 60$$

Taking $\log_e$ both sides

$$\log_e V + 0.45 \log_e T_A = \log_e 90$$
$$\log_e V + 0.3 \log_e T_B = \log_e 60$$

$$\log T_A - \log_e T_B = \frac{\log_e 90 - \log_e V}{0.45} - \frac{\log_e 60 - \log_e V}{0.3}$$

as $T_A = T_B$

$$0 = \frac{0.3\log_e 90 - 0.3\log_e V - 0.45\log_e 60 + 0.45\log_e V}{0.45 \times 0.3}$$

$$0.4925 = 0.15 \log V$$

$$\mathbf{V = 26.67\ m/min}$$

6.51 (b)

Uncut chip thickness = 0.81 mm

Chip thickness = 1.8 mm

$$\tan\theta = \frac{r\cos\alpha}{1 - r\sin d\,\alpha} = \frac{\cos\alpha}{\frac{1}{r} - \sin\alpha}$$

$$= \frac{\cos 12^\circ}{\frac{t_2}{t_1} - \sin 12^\circ}$$

$$\tan\theta = \frac{\cos 12^\circ}{\left(\frac{1.8}{0.81}\right) - \sin 12^\circ}$$

$$\theta = \mathbf{25.9^\circ}.$$

6.52 (a)

$$\text{Drilling time} = \frac{\text{Distance covered}}{\text{Speed}}$$

$$\text{Distance covered} = \frac{50 + 2\times 2 + 4.5}{0.2 \times \frac{500}{60}}$$

$$= \mathbf{35.1\ s}$$

6.53 (c)

$$r = 0.4; \quad t_1 = 0.6\text{ mm}; \quad \alpha = 10^\circ$$
$$V_c = 2.5\text{ m/s}; \quad t_m = 25\ \mu\text{m}$$

$$\tan\phi = \frac{r\cos\alpha}{1 - r\sin\alpha} = 0.4233$$

$$\Rightarrow \phi = 22.9^\circ$$

Shear strain rate

$$= \frac{V_c\cos\alpha}{\cos(\phi - \alpha) \times t_m}$$

$$= \frac{2.5\cos 10}{\cos 12.9 \times 25 \times 10^{-6}}$$

$$= \mathbf{1.0104 \times 10^5/s}$$

6.54 (b)

$$V_1 = 63\text{ m/min}, \ T_1 = 10\text{ hrs}$$
$$C = 257.35$$
$$V_1 T_1^n = C$$
$$T_1^n = \frac{C}{V_1}$$

$$\Rightarrow n = \frac{\ln\left(\frac{C}{V_1}\right)}{\ln T_1} = \frac{\ln\frac{257.35}{63}}{\ln 600} = 0.22$$

$$V_2 = 2 \times V_1 = 126\text{ m/min}$$

$$T_2 = \left(\frac{257.35}{126}\right)^{1/0.22}$$

$$= \left(\frac{257.35}{126}\right)^{4.54} = \mathbf{25.6\ min}$$

6.55 (a)

$$n = 0.12, \ C = 130$$
$$C_1 = 1.1 \times 130 = 143$$
$$V = V_1 = 90\text{ m/min}$$
$$V_1 T_1^n = C_1$$

$$\Rightarrow T_1 = \left(\frac{143}{90}\right)^{1/0.12} = \mathbf{47.4\ min}$$

6.56 (a)

Diameter of bar, $d_i = 100$ mm
Feed, $f = 0.25$ mm/rev
Depth of cut, $d = 4$ mm
$V = 90$ m/min
Cutting force = 1500 N
Since cutting force, is perpendicular to the friction force,
By Merchant circle

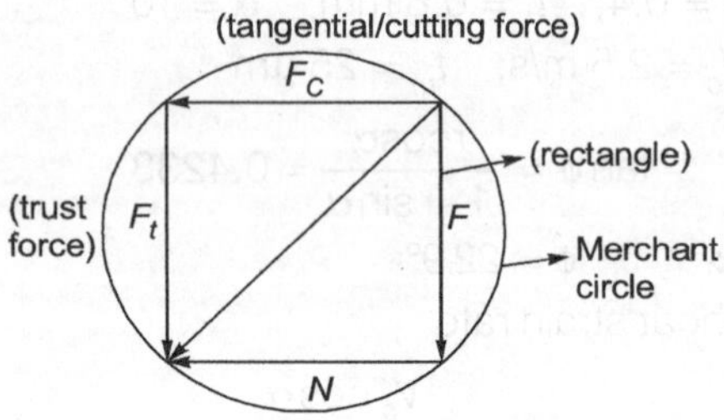

This is the case of, when $\alpha = 0$

6.57 (b)

Normal force, $N = F_C =$ **1500 N**

6.58 (d)

$$\text{MRR} = vfd$$
$$= \frac{\pi DN}{60} \times fd = \frac{\pi \times 200 \times 160 \times 0.25 \times 4}{60}$$
$$= \mathbf{1675.5\ mm^3/s}$$

6.59 (b)

Specific cutting area per revolution = wt
$= 0.2$ mm²/rev
Specific cutting pressure,
$$= \frac{F}{wt} = \frac{400}{0.2} = \mathbf{2000\ N/mm^2}$$

6.61 (0.1)

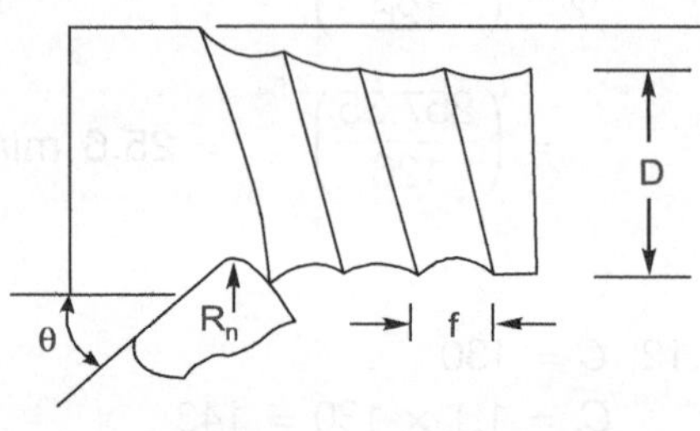

$\theta + \Psi = 90°$
$\theta = 90° - 60° = 30°$
f' (uncut chip thickness)
$= f\sin\theta = 0.2 \times \sin 30° =$ **0.1 mm**

6.63 (0.707)

$$t_{max} = \frac{2f}{NZ}\sqrt{\frac{d}{D}}$$

Formula for maximum uncut thickness

$$\therefore \quad \frac{t_1}{t_2} = \sqrt{\frac{d_1}{d_2}} = \sqrt{\frac{1}{2}} = \frac{1}{\sqrt{2}} = 0.707$$

$(\because \ D_1 = D_2)$

6.65 (2.9)

$$\gamma = \cot\phi + \tan(\phi - \alpha)$$

Given, $\alpha = 0$
$t = 0.2$ mm
$t_c = 0.5$ mm

$$r = \frac{t}{t_c} = \frac{0.2}{0.5} = 0.4$$

$$\tan\phi = \frac{r\cos\alpha}{1 - r\sin\alpha} = r$$

$(\because \alpha = 0)$

$$\therefore \gamma\,(\text{shear strain}) = \cot\phi + \tan(\phi - \alpha)$$
$$= \frac{1 + \tan^2\phi}{\tan\phi} = \frac{1 + 0.4^2}{0.4}$$
$$= \frac{1.16}{0.4} = \mathbf{2.9}$$

6.66 (6)

Given, $n = 0.2$
$T_{ct} = 1.5$ minute
Tool life for maximum production rate,

$$= \left(\frac{1}{n} - 1\right) T_{ct}$$

$$\therefore \quad \text{Tool life} = \left(\frac{1}{0.2} - 1\right) \times 1.5$$
$$= 4 \times 1.5 = \mathbf{6\ minutes}$$

6.67 (b)

Drilling time: $t = \dfrac{L}{fN}$

L = thickness + Approach + Overrun
$= 40 + 10 = 50$ mm
$f = 0.1$ mm/rev
$N = 400$ rpm

$$\therefore \quad t = \frac{50}{0.1 \times 400} = \frac{500}{400} = \mathbf{1.25\ minutes}$$

6.69 (298)

The volume rate of material removal is:
$Q = fdLN$

where f (feed) = 0.25 mm/stroke

d (depth of cut) = 4 mm

L (length of block) = 200 mm

N(number of cutting stroke per minute) = 60

$$Q = \frac{0.25 \times 4 \times 200 \times 60}{60}\left(\frac{mm^3}{s}\right)$$

$= 200\ mm^3/s$

$U_c = 1.49\ J/mm^3$

$P_{average} = U_c \times Q = 1.49 \times 200\ J/s =$ **298 W**

6.70 (0.25)

$$V_1T_1^n = V_2T_2^n$$

$$\Rightarrow \quad V_1(T_1)^n = 2V_1\left[\frac{1}{16}T_1\right]^n$$

$n =$ **0.25**

6.72 (18.88)

Rake angle (α) = 5°

N = 400 rpm

Axial feed = 0.4 m/min

$$= \frac{0.4 \times 1000}{400} = 1\ \text{mm/rev.}$$

Radial depth of cut = 5 mm

Chip thickness: t_c = 3 mm

r = chip thickness ratio

$$= \frac{t}{t_c} = \left(\frac{1}{3}\right)$$

$$\Rightarrow \quad \tan\phi = \frac{r\cos\alpha}{1 - r\sin\alpha} = \frac{\frac{1}{3}\cos(5°)}{1 - \frac{1}{3}\sin(5°)}$$

$\tan\phi = 0.342$

or $\phi = \tan^{-1}(0.342) =$ **18.88°**

6.73 (2.1)

$$\frac{F}{N} = \frac{F_T}{F_C} = 0.7$$

$$\Rightarrow \quad \frac{490}{F_C} = 0.7$$

F_C = 700 N

$$\text{Power} = F_C \times V = \frac{700 \times 180}{60} = \textbf{2.1 kW}$$

6.74 (a)

Feed = unit thickness of orthogonal turning

$$r = \frac{t_1}{t_2} = \frac{0.14}{0.28} = 0.5$$

$$\tan\phi = \frac{r\cos\alpha}{1 - r\sin\alpha}$$

$$\tan\phi = \frac{0.5\cos 10°}{1 - (0.5\sin 10°)}$$

$\phi = 28.33° =$ **28° 20′**

$\gamma = \cot\phi + \tan(\phi - \alpha)$

$\gamma = 2.1858 \approx$ **2.19**

6.75 (a)

Shaft length, L = 90 mm

Taper length, l = 55 mm

Big end diameter, D = 80 mm

Small end diameter, d = 65 mm

half tapper angle

$$\tan\alpha = \left(\frac{R - r}{l}\right) = \frac{(80 - 65)}{2 \times 55}$$

$\alpha = 7.765°$

taper angle,

$2\alpha = 15°31′$

$$\text{set over, } S = \frac{L(D - d)}{2l}$$

$$= \frac{90(80 - 65)}{2 \times 55} = 12.27\ \text{mm}$$

6.76 (b)

$\alpha = 22° \quad t_2 = 0.8 \quad v = 48 \quad t_1 = 0.4$

$r = t_1/t_2 = 0.5$

$$\phi = \tan^{-1}\left(\frac{r\cos\alpha}{1 - r\sin\alpha}\right) = \tan^{-1}\left(\frac{0.5\cos 22°}{1 - 0.5\sin 22°}\right)$$

= **29.7°**

6.77 (40000)

The maximum profit per week of the firm is **₹ 40,000.**

6.78 (7.54)

$$t = \frac{L}{fN} \text{ and } V = \pi DN$$

$$\therefore \quad t = \frac{L\pi D}{fV} = \frac{900 \times \pi \times 200}{0.25 \times 300 \times 1000}$$

$$= \frac{9\pi \times 2}{30 \times 0.25}\ \text{min} = \frac{6\pi}{2.5}\ \text{min}$$

= **7.54 min**

6.79 (57.91)

Total cost = metal cutting cost + tooling cost

$$= \frac{18C}{V} + \frac{270C}{VT}$$

Here, $$T = \left(\frac{150}{V}\right)^{\frac{1}{0.25}} = \left(\frac{15}{V}\right)^4$$

$$C_t = \frac{18C}{V} + \frac{270C}{V.\left(\frac{150}{V}\right)^4}$$

$$\frac{dC_t}{dV} = 0$$

$$-\frac{18C}{V^2} + \frac{270C}{(150)^4} \times 3v^2 = 0$$

or $$V^4 = \frac{150^4 \times 18}{270 \times 3}$$

$$V = \textbf{57.91 m/min}$$

6.80 (0.5)

$\because$ $$\tan\beta = \frac{F_T / F_C + \tan\alpha}{1 - (\tan\alpha) F_T / F_C}$$

$\because$ $$\alpha = 0$$

$$\tan\beta = \frac{F_T}{F_C} = \frac{250}{500} = \textbf{0.5}$$

6.81 (b)

$$45000 \times 30^{0.14} \times 0.35^{0.7} \times 2^{0.4} = K$$

$$K = 45{,}842.58$$

$$\Rightarrow T^{0.14} = \frac{K}{(1.25\,V)(1.25\,f)^{0.7}(1.25 \times d)^{0.4}}$$

$$\Rightarrow T^{0.14} = \frac{K}{(1.25 \times 45000)(1.25 \times 0.35)^{0.7}(1.25 \times 2)^{0.4}}$$

$$\Rightarrow T = \textbf{1.06 min}$$

6.83 (16)

$$V_2 = 0.5\,V_1$$

$$V_1 T_1^n = V_2 T_2^n \quad \text{[Taylor's tool life equation]}$$

$$\left(\frac{T_2}{T_1}\right)^n = \left(\frac{V_1}{V_2}\right); \quad \frac{T_2}{T_1} = (2)^4 = \textbf{16}$$

6.84 (106.066)

Let us assume the speed is x m/min.

$$xT_1^{0.1} = 150$$

$$\Rightarrow T_1 = \left(\frac{150}{x}\right)^{\frac{1}{0.1}} \quad \text{and} \quad xT_2^{0.3} = 300$$

$$T_2 = \left(\frac{300}{x}\right)^{\frac{1}{0.3}}$$

Now $T_2 > T_1$

or $$\left(\frac{300}{x}\right)^{\frac{1}{0.3}} = \left(\frac{150}{x}\right)^{\frac{1}{0.1}}$$

For solving inequality

$$\left(\frac{300}{x}\right)^{\frac{1}{0.3}} = \left(\frac{150}{x}\right)^{\frac{1}{0.1}}$$

or $$\left(\frac{300}{x}\right)^{0.1} = \left(\frac{150}{x}\right)^{0.3}$$

or $$\frac{x^{0.3}}{x^{0.1}} = \frac{150^{0.3}}{300^{0.1}}$$

or $$x = \textbf{106.066 m/min}$$

6.85 (12)

Feed rate F = 18 mm/min

Block length L = 200 mm

Approach length =over travel length = L_i

$$= \sqrt{d(D-d)} = \sqrt{2 \times (34-2)}$$

$$= 8 \text{ mm}$$

Machining time per pass,

$$t = \frac{L + 2L_i}{F} = \frac{200 + 2 \times 8}{18}$$

$$= \textbf{12 min.}$$

6.86 (1.494)

Given $\alpha = 9°$

t = 0.2 mm

$t_{c\,max}$ = 0.4 mm

$t_{c\,min}$ = 0.25 mm

$$\tan\phi_{max} = \frac{r_{max}\cos\alpha}{1 - r_{max}\sin\alpha} = \frac{\frac{0.2}{0.25}\cos 9°}{1 - \frac{0.2}{0.25}\sin 9°}$$

or $$\phi_{max} = 42.09°$$

$$\tan\phi_{min} = \frac{r_{min}\cos\alpha}{1 - r_{min}\sin\alpha} = \frac{\frac{0.2}{0.4}\cos 9°}{1 - \frac{0.2}{0.4}\sin 9°}$$

or $$\phi_{min} = 28.18°$$

$$\text{Ratio} = \frac{\phi_{max}}{\phi_{min}} = \frac{42.09}{28.18} = \textbf{1.494}$$

6.87 (b)

$V_1 = \pi D \times 400$ m/min; T_1 = 20 min

$V_2 = \pi D \times 200$ m/min; T_2 = 60 min

Now using Taylor's equation

$$V_1 T_1^n = V_2 T_2^n$$

or $\pi D \times 400 \times (20)^n = \pi D \times 200 \times 60^n$

or $n = 0.6309$

and $V_3 = \pi D \times 300$ m/min; $T_3 = ?$

$$V_1T_1^n = V_2T_2^n = V_3T_3^n$$

$$\pi D \times 400 \times (20)^{0.6309} = \pi D \times 300 \times T_3^{0.6309}$$

$$\therefore \quad T_3 = 31.55 \text{ min} \approx \mathbf{32\ min}$$

6.88 (a)

$$V_2 = \frac{V_1}{2}$$

$$V_1T_1^{0.5} = V_2T_2^{0.5}$$

$$V_1T_1^{0.5} = \frac{V_1}{2} \cdot T_2^{0.5}$$

$$\left(\frac{T_2}{T_1}\right)^{0.5} = 2$$

$$\frac{T_2}{T_1} = 2^{\frac{1}{0.5}} = 4$$

6.89 (b)

Grinding ratio is defined as

$$= \frac{\text{Volume of work material removed } (V_m)}{\text{Volume of wheel wear } (V_w)}$$

$$200 = \frac{\text{Volume of work material removed } (V_m)}{\text{Volume of wheel wear } (V_w)}$$

Grinding wheel wears 0.005 times the volume of material removed.

6.90 (0.4408)

Given: $t = 0.010$ mm, $v = 130$ m/min

$\alpha = 15°$, $b = 6$ mm, $t_c = 0.015$ mm

$F_c = 60$ N, $F_t = 25$ N

$$F = F_c \sin\alpha + F_t \cos\alpha$$

$$= 60 \sin 15° + 25 \cos 15°$$

$$= 39.6773 \text{ N}$$

Ratio of frictional energy to total energy

$$= \frac{F}{F_c} \cdot \frac{V_c}{V} = \frac{F}{F_c}\left(\frac{t}{t_c}\right) \left[\because \frac{t}{t_c} = \frac{V_c}{V} = r\right]$$

$$= \frac{39.6773}{60} \times \frac{0.010}{0.015} = 0.4408$$

6.91 (d)

- 4 jaw independently adjusted chuck are used to hold and support a wide variety of workpiece including odd shaped workpiece.
- Practically odd shaped job can be mounted on a face plate. **But this is not given as option.**
- Lathe dog may be used on round, square, rectangular and odd-shaped workpieces but most appropriate is option (d).

6.92 (c)

Feed = fZN

(f → Feed per tooth; Z → No. of teeth in the cutter; N → rpm)

Feed rate in slab milling operation = f_m

$$f_m = f_t Z N$$

where, f_t = feed per tooth

Z = number of teeth in cutter

N = rpm

6.93 (253.53)

Time required for drilling a hole

$$t = \frac{L}{fN}$$

Here L and f are constant and N is variable.

$$V_1 = \pi D \times 150 \text{ m/min}$$

$$T_1 = 300 \times \frac{L}{f \times 150} \text{ min}$$

$$V_2 = \pi D \times 300 \text{ min}$$

$$T_2 = 200 \times \frac{L}{f \times 300} \text{ min}$$

$$V_3 = \pi D \times 200 \text{ min}$$

$$T_3 = x \times \frac{L}{f \times 200} \text{ min}$$

[No. of hole is 'x']

$$V_1T_1^n = V_2T_2^n$$

$$\pi D \times 150 \times \left(300 \times \frac{L}{f \times 150}\right)^n = \pi D \times 300 \times \left(200 \times \frac{L}{f \times 300}\right)^n$$

or, $$2^n = 2\left(\frac{2}{3}\right)^n$$

or, $$3^n = 2$$

$$n = 0.63093$$

$$V_1T_1^n = V_3T_3^n$$

$$\pi D \times 150 \times \left(300 \times \frac{L}{f \times 150}\right)^{0.63093} = \pi D \times 200 \times \left(x \times \frac{L}{f \times 200}\right)^{0.63093}$$

$$3 \times 2^{0.63093} = 4\times\left(\frac{x}{200}\right)^{0.63093}$$

$$\Rightarrow \quad x = 253.53$$

6.94 (23.32)

$$\alpha = 15^\circ$$
$$t = 0.5 \text{ mm}$$
$$t_c = 1.25 \text{ mm}$$
$$r = \frac{t}{t_c} = \frac{0.5}{1.25} = 0.4$$
$$\tan\phi = \frac{r\cos\alpha}{1 - r\sin\alpha} = \frac{0.4\cos 15^\circ}{1 - 0.4\sin 15^\circ}$$
$$\phi = 23.3155^\circ \approx \mathbf{23.32^\circ}$$

6.95 (a)

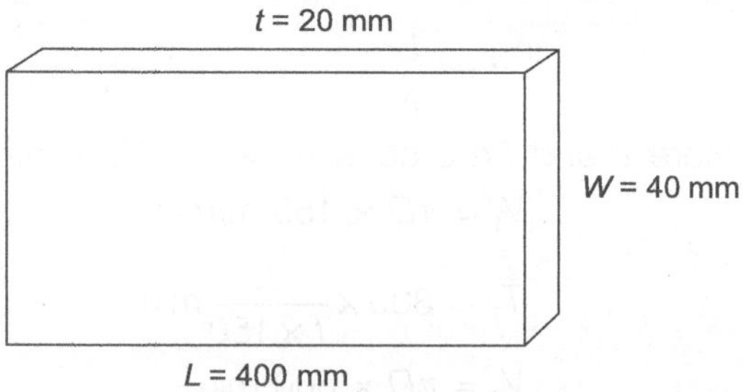

Cutter diameter,

$$D = 100 \text{ mm}$$

Cutter width = 50 mm that means single pass is sufficient.

Number of teeth,

$z = 20$, Depth of cut, $d = 2$ mm

Rotational speed,

$$N = 1200 \text{ rpm},$$
$$f = 0.05 \text{ mm per tooth}$$

Approach distance,

$$A = \sqrt{\left(\frac{D}{2}\right)^2 - \left(\frac{D}{2} - d\right)^2}$$
$$= \sqrt{d(D-d)}$$
$$= \sqrt{2(100-2)} \text{ mm} = 14 \text{ mm}$$

Time for one pass in slab or slot milling

$$= \frac{L+A}{fzN} \text{ for rough milling}$$
$$= \frac{L+2A}{fzN} \text{ for finish milling}$$

As length of approach and over travel is same

$$= \frac{L+A+A}{fzN} = \frac{400+14+14}{0.05\times 20\times 1200} \text{ min} = 21.4 \text{ s}$$

6.96 (c)

Orthogonal turning, $\lambda = 90^\circ$,
Wall thickness = depth of cut, (d) = 5 mm (assumed)

$$\text{Axial force, } F_x = 1259 \text{ N}, \; F_t = \frac{F_x}{\sin 90}$$
$$= F_x = 1259 \text{ N}$$

Tangential force, $F_z = 1601$ N, $F_c = F_t = 1601$ N

$$t_c = 0.3 \text{ mm}, \; \alpha = 10^\circ$$
$$fN = 100 \text{ mm/min}$$
$$N = 1000 \text{ rpm or } f \times 1000 = 100$$

or
$$f = 0.1 \text{ mm/rev.}$$
$$t = f\sin\lambda = 0.1 \sin 90^\circ = 0.1 \text{ mm}$$
$$b = \frac{d}{\sin\lambda} = \frac{5}{\sin 90} = 5 \text{ mm}$$
$$r = \frac{t}{t_c} = \frac{0.1}{0.3} = 0.33$$
$$\tan\phi = \frac{r\cos\alpha}{1 - r\sin\alpha} = \frac{0.33\cos 10^\circ}{1 - 0.33\sin 10^\circ}$$

or
$$\phi = 19.02^\circ$$
$$\tau_s = \frac{F_s}{A_s} = \frac{F_s \sin\phi}{bt}$$
$$= \frac{(F_c\cos\phi - F_t\sin\phi)\times\sin\phi}{bt}$$
$$= \frac{(1601\cos 19.02 - 1259\sin 19.02)\times\sin 19.02}{5\times 0.1}$$
$$= 719.12 \text{ MPa} \simeq 722 \text{ MPa}$$

Option (c) is correct.

6.97 (3.75)

Side cutting edge angle (SCEA) = 45°
End cutting edge angle (ECEA) = 10°
Feed (f) = 0.1 mm/rev
We have to consider nose radius is zero i.e. Ideal surface peak to valley surface roughness,

$$h = \frac{f}{\tan \text{SCEA} + \cot \text{ECEA}}$$

The centre line average (CLA)

$$= \frac{h}{4} = \frac{f}{4(\tan SCEA + \cot ECEA)}$$
$$= \frac{0.1}{4(\tan 45^\circ + \cot 10^\circ)} \text{ mm} = 0.00375 \text{ mm}$$
$$= 3.75 \; \mu\text{m}$$

6.98 (40)

For tool x, $VT^{0.3} = 60$
For tool y, $VT^{0.6} = 90$

Let x is the cutting speed for same tool life in m/min.

Then $xT_x^{0.3} = 60$ or $T_x = \left(\frac{60}{x}\right)^{1/0.3}$

$xT_y^{0.6} = 90$ or $T_y = \left(\frac{90}{x}\right)^{1/0.6}$

Now $T_x = T_y$

$$\left(\frac{60}{x}\right)^{1/0.3} = \left(\frac{90}{x}\right)^{1/0.6}$$

or $$\left(\frac{60}{x}\right)^{0.6} = \left(\frac{90}{x}\right)^{0.3}$$

$$\frac{x^{0.3}}{x^{0.6}} = \frac{90^{0.3}}{60^{0.6}}$$

or $x^{-0.3} = 0.33066$

or $x = (0.33066)^{1/-0.3} = 40$ m/min

6.99 (0.74)

$$\alpha = 10^\circ$$
$$t = 0.125 \text{ mm}$$
$$t_c = 0.22 \text{ mm}$$
$$r = \frac{t}{t_c} = \frac{0.125}{0.22} = 0.5682$$
$$\tan\phi = \frac{r\cos\alpha}{1 - r\sin\alpha} = \frac{0.5682\cos 10^\circ}{1 - 0.5682\sin 10^\circ}$$

or $\phi = 31.83^\circ$

Using Merchant's thoery,

$$\phi = 45^\circ + \frac{\alpha}{2} - \frac{\beta}{2}$$

or, $$31.83^\circ = 45^\circ + \frac{10^\circ}{2} - \frac{\beta}{2}$$

or, $\beta = 36.34^\circ$

$$\mu = \tan\beta = \tan 36.34^\circ$$
$$= 0.7356 \approx 0.74$$

6.100 (274.89)

Power required =Specific energy consumption × MRR

Specific energy consumption

$$= 0.7 \text{ Nm/mm}^3 = 0.7 \text{ J/mm}^3$$

$$\text{MRR} = \frac{\pi d^2}{4} \times fN$$
$$= \frac{\pi \times 10^2}{4} \times 0.25 \times 1200 \text{ mm}^3/\text{min}$$
$$= 23561.94 \text{ mm}^3/\text{min}$$
$$= 392.7 \text{ mm}^3/\text{s}$$

∴ Power, $P = 0.7 \text{ J/mm}^3 \times 392.7 \text{ mm}^3/\text{s}$
$= 274.89 \text{ J/s} = 274.89$ W

6.101 (b)

For brass 'Silicon carbide' is the abrasive. As brass is soft material, we need hard wheel. Therefore 'Q' is the best choice.

$$\underset{\text{soft}}{A} \rightarrow \underset{\text{hard}}{Z}$$

So, 'J' is in the softer side.
So best option is (b).

6.102 (8.1)

$L = 300$ mm, $D = 100$ mm, $Z = 20$ teeth, $d = 5$ mm, $f = 0.1$ mm, $V = 35$ m/min
$A = O = 5$ mm

$$V = \pi DN$$
$$35 = \pi \times 0.100 \times N$$
$$N = \pi \times 111.408 \text{ rpm}$$

$$t = \frac{L + \frac{D}{2} + A + O}{fzN} = \frac{300 + \frac{100}{2} + 5 + 5}{0.1 \times 20 \times 111.408}$$

= 1.6157 min per pass

For 25 mm cuts min 5 mm depth of cut 5 pass needed.

Total machining time = 5 × 1.6157
= 8.078 min ≃ 8.1 min

6.103 (23.5)

Given, $l_c = 100$ mm, $l = 250$ mm, $\alpha = 20^\circ$

$$r = \frac{l_c}{l} = \frac{100}{250} = 0.4$$

$$\tan\phi = \frac{r\cos\alpha}{1 - r\sin\alpha} = \frac{0.4\cos 20^\circ}{1 - 0.4 \times \sin 20^\circ} = 0.4354$$

$$\phi = \tan^{-1}(0.4354) = 23.5^\circ$$

6.104 (–5.6)

Given data, D = 200 mm,
From tool geometry,

$$C_{s1} = 30^\circ$$

∴ $\lambda_1 = 90 - 30 = 60^\circ$

From tool geometry,

$$C_{s2} = 0^\circ$$

∴ $\lambda_2 = 90 - 0 = 90^\circ$

We know that specific energy consumption

$$U_c = \frac{F_c}{1000fd} = U_0(t_1)^{-0.4} \text{ (given)}$$

∴ $F_c = U_0(f\sin\lambda)^{-0.4} \times 1000fd$

$\therefore \quad F_c \propto (\sin \lambda)^{-0.4}$

Percentage change in cutting force,

$$= \frac{F_{c2} - F_{c1}}{F_{c1}} \times 100 = \left[\left(\frac{\sin\lambda_2}{\sin\lambda_1}\right)^{-0.4} - 1\right] \times 100$$

$$= \left[\left(\frac{\sin 90^\circ}{\sin 60^\circ}\right)^{-0.4} - 1\right] \times 100 = -5.59\%$$

6.105 (0.8)

Given feed far machine 1 is f_1 and feed far machine 2 f_2

$$\text{Cutting time, } t_1 = \frac{B_1}{f_1 N_1}$$

$$\text{Cutting time, } t_2 = \frac{B_2}{f_2 N_2} = \frac{1.1 B_1}{0.9 f_1 \times N_2}$$

Ratio of cutting time,

$$\frac{t_1}{t_2} = \frac{B_1 / f_1 N}{\frac{1.1 B_1}{0.9 f_1 N}} = \frac{0.9}{1.1} = 0.8$$

6.106 (26.985)

Given L = 250 mm, D = 25 mm,

Feed, f = 0.2 mm/rev; Tool change time = 1 min

$$\text{Optimum tool life } (T_o) = \frac{T_c(1-n)}{n} = \frac{1(1-0.2)}{0.2}$$

$$= 4 \text{ min}$$

$$V_o T_o^n = C \text{ or } V_o = \frac{C}{T_o^n} = \frac{24}{4^{0.2}}$$

$$= 18.19 \text{ m/min}$$

$$V = \pi D N$$

$$\Rightarrow \quad 18.19 = \pi \times 0.025 \times N$$

$$\Rightarrow \quad N = 231.6 \text{ rpm}$$

Machining time per piece

$$(T_m) = \frac{L}{fN} = \frac{250}{0.2 \times 231.6} = 5.397 \text{ min}$$

Number of tool needed per piece work

$$= \frac{5.397}{4} \text{ piece}$$

$\therefore$ The optimum tool cost per piece

$$= \frac{5.397}{4} \times 20 = 26.985$$

6.107 (47.746)

$$E_{sp} = 15 \text{ J/mm}^3$$

$$\text{MRR} = 6000 \text{ mm}^3\text{/min}$$

$$N = 3000 \text{ rpm}$$

$$D = 200 \text{ mm}$$

$$E_{sp} = \frac{F_c V}{L d V}$$

$$15 = \frac{F_c V}{MRR} = \frac{F}{600}$$

$$V = \frac{\pi D N}{1m} = \frac{\pi(200)(3000)}{1000} = 600\,\pi \text{ m/min}$$

$$15 = \frac{F_c \times 600 \times 3\pi y}{6000}$$

$$F_c = \frac{150}{\pi} = 47.746 \text{ N} \simeq 47.75 \text{ N}$$

6.108 (685.77)

F_c = 1000 N, F = 600 N, t = 0.75 mm, t_c = 1.5 mm, $\alpha = 12^\circ$

This is orthogonal turning operation

$$r = \frac{t}{t_c} = \frac{0.75}{1.5} = 0.5$$

$$\tan\phi = \frac{0.5\cos 12^\circ}{1 - 0.5\sin 12^\circ}$$

$$\Rightarrow \quad \phi = 28.63^\circ$$

Method - 1 : Using force relations

$$F = F_c \sin\alpha + F_t \cos\alpha$$

$$600 = 1000 \sin 12^\circ + F_t \cos 12^\circ$$

$$\text{or} \quad F_t = 400.85 \text{ N}$$

$$\therefore \quad F_s = F_c \cos\phi - F_t \sin\phi$$

$$= 1000 \cos 28.63^\circ - 400.85 \sin 28.63^\circ$$

$$= 685.66 \text{ N}$$

Method - 2 : Using Merchant circle

$$F_c = R\cos(\beta - \alpha)$$

$$F = R \sin\beta$$

$$\text{or} \quad \frac{F}{F_c} = \frac{R\sin\beta}{R\cos(\beta - \alpha)}$$

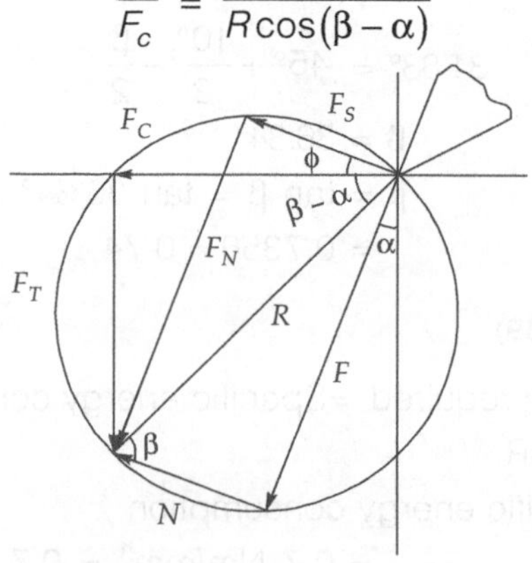

$$\frac{600}{1000} = \frac{\sin\beta}{\cos(\beta - 12)}$$

$$0.6 \cos\beta \cos 12^\circ + 0.6 \sin\beta \sin 12^\circ = \sin\beta$$

$$0.586888 \cos\beta = (1 - 0.6 \sin 12^\circ) \sin\beta = 0.875253 \sin\beta$$

$$\tan\beta = \frac{0.586888}{0.875253}$$

$$\Rightarrow \quad \beta = 33.84^\circ$$

$$\therefore \quad F = R\sin\beta$$

$$600 = R\sin 33.84^\circ$$

$$\Rightarrow \quad R = 1077.44 \text{ N}$$

$$\therefore \quad F_s = R\cos(\phi + \beta - \alpha)$$

$$= 1077.44\cos(28.63^\circ + 33.84^\circ - 12^\circ)$$

$$= 685.77 \text{ N}$$

6.109 (b)

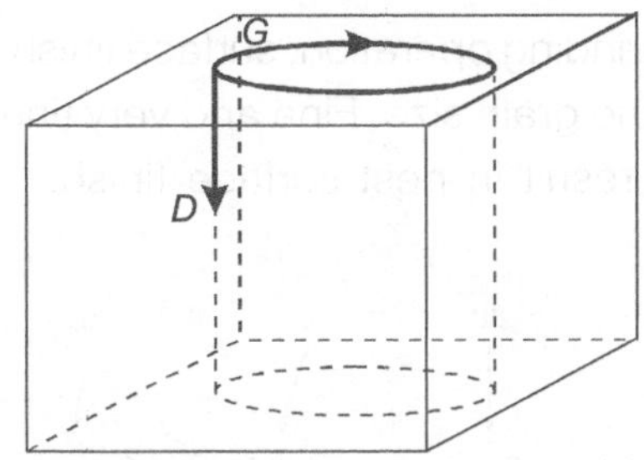

6.111 (392.23)

$$\alpha = 0^\circ$$

$$F_c = 400 \text{ N}$$

Cutting velocity (V) = 100 m/min (V_c)

$$d = 2.0 \text{ mm}(t)$$

$$f = 0.1 \text{ mm/rev}(S_o)$$

Chip velocity (V_c) = 20 m/min (V_f)

6.112 (38.40)

$$\text{Power} = F_c \cdot V$$

$$= 40\text{N} \times 40 \text{ m/s} = 1600 \text{ W}$$

$$\text{MRR} = 10 \times 0.050 \times \frac{5000}{60} \text{ mm}^3/\text{s}$$

$$= 41.667 \text{ mm}^3/\text{s}$$

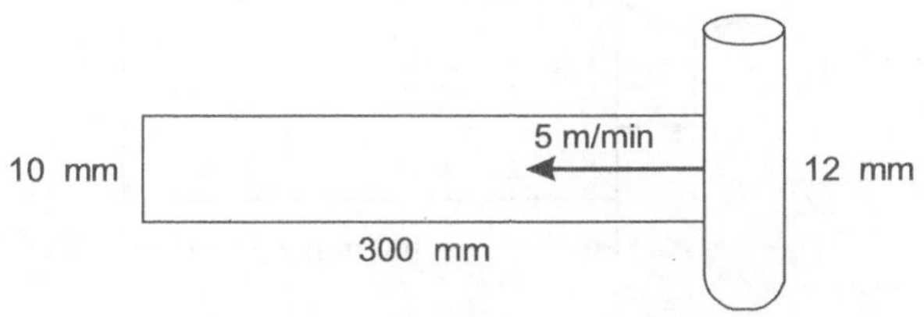

$$e = \frac{\text{Power}}{\text{MRR}} = \frac{1600 \text{J/s}}{41.667 \text{mm}^3/\text{s}}$$

$$= 38.40 \text{ J/mm}^3$$

6.113 (2573.40) (2570.0 to 2576.0)

Shear strength, τ_s = 1000 MPa, Uncut chip thickness, t_1 = 0.2 mm, Width of cut, ω = 3 mm, Velocity of cutting, V = 4 m/sec, Friction angle, β = 45°, Shear angle , ϕ = 25°, Rake angle, α = 5° From the Merchant's analysis

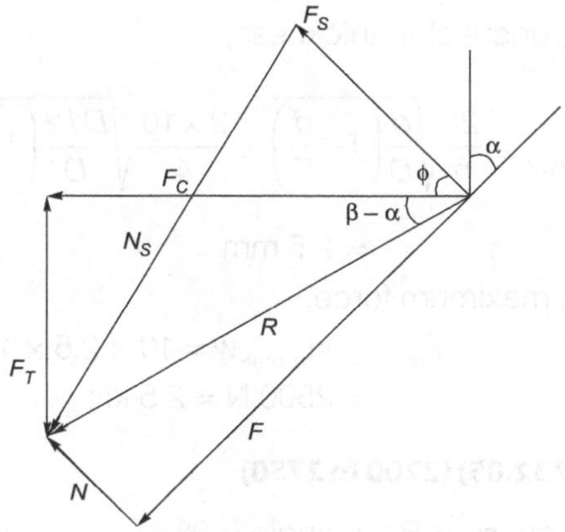

$$\Rightarrow \quad \frac{F_C}{F_S} = \frac{R(\cos\beta - \alpha)}{R\cos(\phi + \beta - \alpha)} = \frac{(\cos\beta - \alpha)}{\cos(\phi + \beta - \alpha)}$$

$$F_C = \frac{F_S(\cos\beta - \alpha)}{\cos(\phi + \beta - \alpha)} \left[F_S = \tau_s \times A = \tau_S \times \frac{\omega t_1}{\sin\phi}\right]$$

$$F_C = \tau_S \times \frac{\omega t_1}{\sin\phi} \times \frac{(\cos\beta - \alpha)}{\cos(\phi + \beta - \alpha)} = 2568.62 \text{ N}$$

6.114 (31.40) (31.2 to 31.6)

Solving the question through the dimensions of the given quantities

Power = Specific energy (J/mm³) × Area of cut (mm²) × Axial feed (mm/s)

$$= 6 \text{ J/mm}^3 \times (\pi D t)\text{mm}^2 \times \frac{1000}{60} \text{mm/sec}$$

Power = 31.40 kW

6.115 (2.5) (2.4 to 2.6)

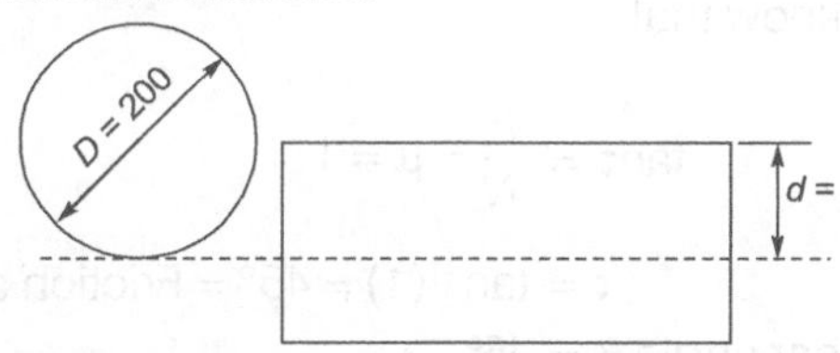

Given: Number of teeth, n = 4; Diameter of cutter, D = 200 mm; Rotational speed, N = 100 rpm; Linear feed to workpiece = 1000 mm/min; Width of workpiece, w = 100 mm, Cutting force/tooth, $F = kt_c w$, Specific cutting force, k = 10 N/mm²

$$\text{Depth of cut, } d = \frac{D}{2}$$

$$\text{Feed } (f) = \frac{1000}{100} = 10 \text{ mm/rev}$$

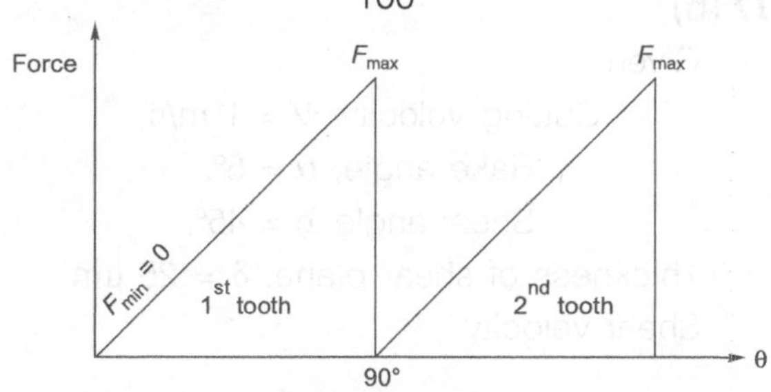

Torque diagram

So, uncut chip thickness,

$$t_{c,max} = \frac{2f}{n}\sqrt{\frac{d}{D}\left(1-\frac{d}{D}\right)} = \frac{2\times10}{4}\sqrt{\frac{D/2}{D}\left(1-\frac{D/2}{D}\right)}$$

$$= 2.5 \text{ mm}$$

So, maximum force,

$$F_{max} = kt_{c,max}w = 10\times2.5\times100$$
$$= 2500 \text{ N} = 2.5 \text{ kN}$$

6.116 (2732.05) (2700 to 2750)

Given: α_o = Rake angle = 0°

$$F_C = F_T$$

$$\Rightarrow \quad F = N; \text{ then, } \mu = \frac{F}{N}$$

$$\Rightarrow \quad \mu = 1$$

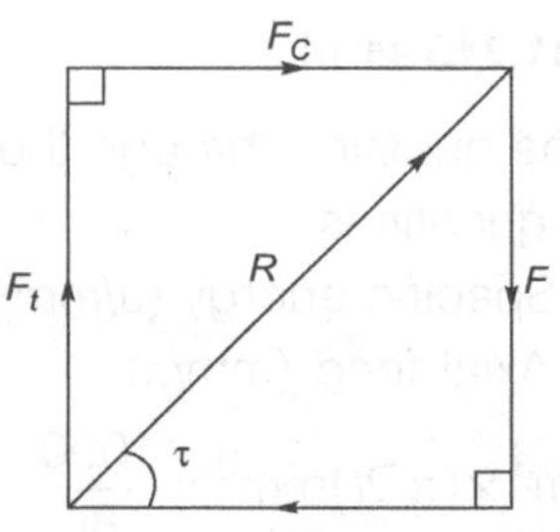

We know that

$$\tan\tau = \frac{F}{N} = \mu = 1$$

$$\tau = \tan^{-1}(1) = 45° = \text{Friction angle}$$

Shear angle θ = 15°

$$F_C = \tau_s\frac{bt_1}{\sin\theta}\times\frac{\cos(\tau-\alpha)}{\cos(\theta+\tau-\alpha_o)}$$

$$F_C = 500\times\frac{2\times0.5}{\sin15°}\times\frac{\cos(45-0)}{\cos(15+45-0)}$$

$$F_C = 2732.05 \text{ N} \approx 2732 \text{ N}$$

6.117 (b)

Given:

Cutting velocity, V = 1 m/s,

Rake angle, α = 5°,

Shear angle, ϕ = 45°,

Thickness of shear plane, δ = 25 μm

Shear velocity,

$$V_s = \frac{V\cos\alpha}{\cos(\phi-\alpha)} = \frac{1\times\cos5°}{\cos(45°-5°)} = 1.3 \text{ m/s}$$

$$\text{Shear strain rate} = \frac{V_s}{\delta} = \frac{1.3}{25\times10^{-6}}$$

$$= 5.2\times10^4 \text{s}^{-1}$$

6.118 (b)

In grinding operation, surface finish is decided by the grain size. Fine and very fine grain size will result in best surface finish.

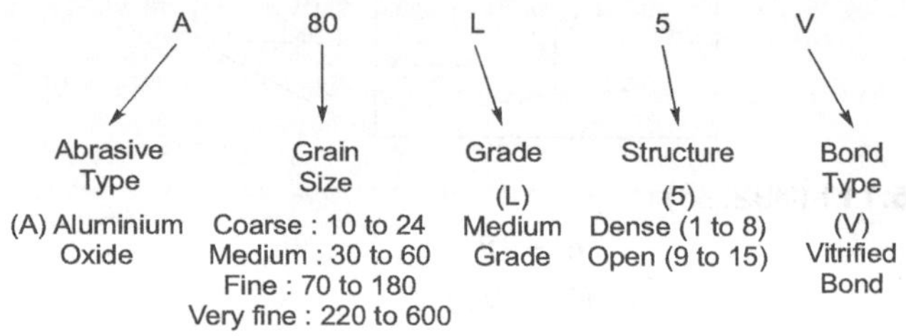

6.118 (2.24) (2.20 to 2.65)

Given: $L\times B$ = 100 mm × 200 mm, Z = 8, D = 100 mm, W = 200 mm, f_t = 0.1 mm/tooth, V = 20 m/min, d = 2 mm

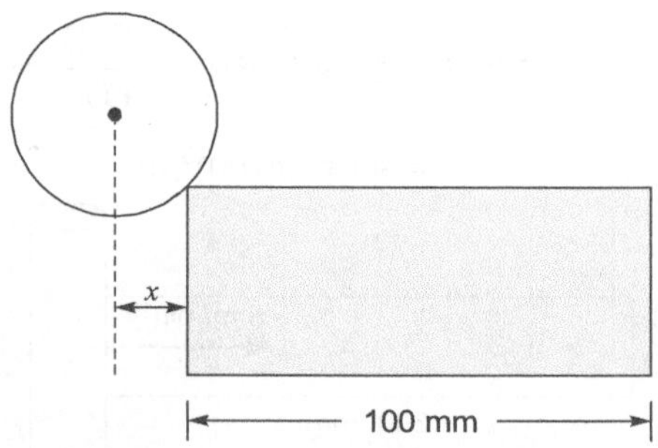

Approach length, x =

$$\sqrt{d(D-d)} = \sqrt{2\times(100-2)} = 14 \text{ mm}$$

∴ The effective length, $L_e = L + x$ = 100 + 14 = 114 mm

$$N = \frac{V\times1000}{\pi\times D} = \frac{20\times1000}{\pi\times100} = 63.662 \text{ rpm}$$

$F = f_t\times NZ$ = 0.1 × 63.662 × 8 = 50.93 mm/min

∴ The machining time required to remove the entire stock;

$$t_{m/c} = \frac{L_e}{F} = \frac{114}{50.93} = 2.238 \text{ minutes}$$

$$\simeq 2.24 \text{ minutes}$$

6.118 (21.87) (21.00 to 22.50)

Given : $V_1 = 60$ m/min; $V_2 = 100$ m/min; $V_3 = 80$ m/min

$T_1 = 60$ min; $T_2 = 10$ min; $T_3 = ?$

By using Taylor's equation,

$$V_1T_1^n = V_2T_2^n$$

$$\Rightarrow \quad 60 \times (60)^n = 100 \times (10)^n$$

$$\Rightarrow \quad \left(\frac{60}{10}\right)^n = \frac{100}{60}$$

$$\Rightarrow \quad n = \frac{\ln\left(\frac{100}{60}\right)}{\ln\left(\frac{60}{10}\right)} = 0.285$$

Also,

$$V_1T_1^n = V_3T_3^n$$

$$\Rightarrow \quad 60 \times (60)^{0.285} = 80 \times (T_3)^{0.285}$$

$$\Rightarrow \quad T_3 = 21.87 \text{ min.}$$

■■■■

7 Metrology and Inspection

7.1 Slip gauges are calibrated by outside micrometer (T/F) **[1995 : 1 M]**

7.2 Checking the diameter of a hole using GO and NOGO gauges in an example of inspection by .... Variable/attributes) **[1995 : 1 M]**

7.3 The fit on a hole-shaft pair system is specified GO/NO GO are **attributes** of a dimensions.as H7-h6, the type of fit is

(a) Clearance fit
(b) Running fit (Sliding fit)
(c) Push fit (transition fit)
(d) Force fit (Interference fit) **[1996 : 1 M]**

7.4 3 blocks of B_1, B_2 and B_3 are to be inserted in a channel of width S. maintaining a min gap of width $T = 0.125$ mm, as shown in figure. For $P = 18.75 \pm 0.08$, $Q = 25.00 \pm 0.12$, $R = 28.125 \pm 0.1$ and $S = 72.35 + X$. The tolerance X is

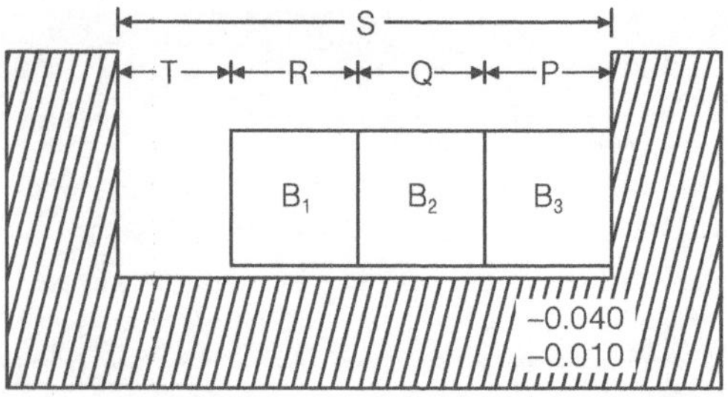

[ME 1997 : 2 M]

7.5 Allowance in limits and fits refers to

(a) Maximum clearance between the shaft and hole
(b) Minimum clearance between the shaft and hole
(c) Difference between the maximum and minimum size of hole
(d) Difference between the maximum and minimum size of shaft **[ME 2001 : 1 M]**

7.6 The dimensional limits on a shaft of 25h 7 are

(a) 25.000, 25.021 mm
(b) 25.000, 24.979 mm
(c) 25.000, 25.007 mm
(d) 25.000, 24.993 mm **[ME 2003 : 1 M]**

7.7 A threaded nut of M 16, ISO metric type, having 2 mm pitch with a pitch diameter of 14.701 mm is to be checked for its pitch diameter using two or three number of balls or rollers of the following sizes

(a) Rollers of 2 mm φ
(b) Rollers of 1.155 mm φ
(c) Balls of 2 mm φ
(d) Balls of 1.155 mm φ **[ME 2003 : 2 M]**

7.8 Two slip gauges of 10 mm width measuring 1.000 mm and 1.002 mm are kept side by side in contact with each other lengthwise. An optical flat is kept resting on the slip gauges as shown in the figure. Mono-chromatic light of wavelength 0.0058928 mm is used in the inspection. The total number of straight fringes that can be observed on both slip gauges is

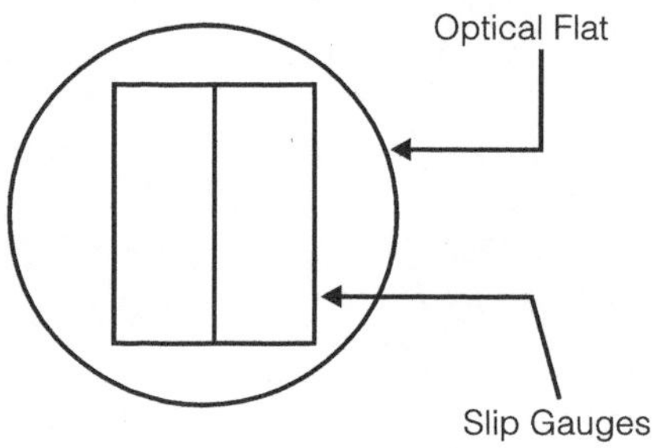

(a) 2 (b) 6
(c) 8 (d) 13 **[ME 2003 : 2 M]**

7.9 A part shown in the figure is machined to the sizes given below

$P = 35.00 \pm 0.08$ mm
$Q = 12.00 \pm 0.02$ mm
$R = 13.00^{+0.04}_{-0.02}$ mm

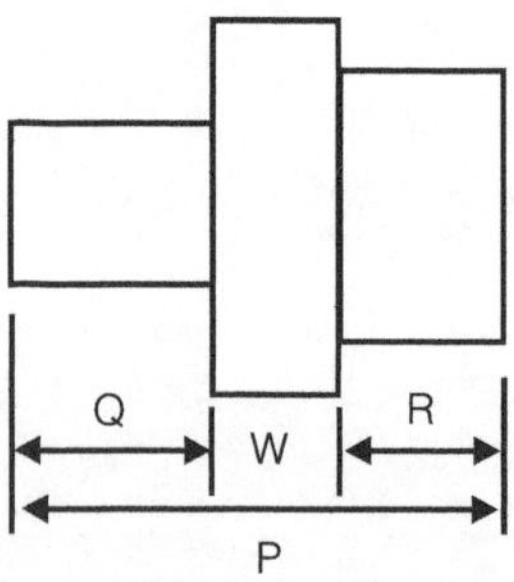

With 100% confidence, the resultant dimension W will have the specification

(a) 9.99 ± 0.03 mm
(b) 9.99 ± 0.13 mm
(c) 10.00 ± 0.03 mm
(d) 10.00 ± 0.13 mm **[ME 2003 : 2 M]**

7.10 In an interchangeable assembly, shafts of size $25.000^{-0.040}_{-0.0100}$ mm mate with holes of size $25.000^{+0.020}_{-0.000}$ mm. The maximum possible clearance in the assembly will be

(a) 10 microns (b) 20 microns
(c) 30 microns (d) 60 microns

[ME 2004 : 1 M]

7.11 GO and NO-GO plug gauges are to be designed for a hole $20.000^{+0.050}_{+0.010}$ mm. Gauge tolerances can be taken as 10% of the hole tolerance. Following ISO system of gauge design, sizes of GO and NO-GO gauge will be respectively.

(a) 20.010 mm and 20.050 mm
(b) 20.014 mm and 20.046 mm
(c) 20.006 mm and 20.054 mm
(d) 20.014 mm and 20.054 mm

[ME 2004 : 2 M]

7.12 Match the following

Feature to be inspected

P. Pitch and Angle errors of screw thread
Q. Flatness error of a surface
R. Alignment error of a machine sideway
S. Profile of a cam

Instrument

1. Auto Collimator
2. Optical interferometer
3. Dividing Head and Dial Gauge
4. Spirit Level
5. Sine bar
6. Tool maker's Microscope

(a) P–6, Q–2, R–4, S–6
(b) P–5, Q–2, R–1, S–6
(c) P–6, Q–4, R–1, S–3
(d) P–1, Q–4, R–4, S–2 [ME 2004 : 2 M]

7.13 In order to have interference fit, it is essential that the lower limit of the shaft should be

(a) greater than the upper limit of the hole
(b) lesser than the upper limit of the hole
(c) greater than the lower limit of the hole
(d) lesser than the lower limit of the hole

[ME 2005 : 1 M]

7.14 A ring gauge is used to measure

(a) outside diameter but not roundness
(b) roundness but not outside diameter
(c) both outside diameter and roundness
(d) only external threads

[ME 2006 : 1 M]

7.15 A hole is specified as $40^{0.050}_{0.000}$ mm. The mating shaft has a clearance fit with minimum clearance of 0.01 mm. The tolerance on the shaft is 0.04 mm. The maximum clearance in mm between the hole and the shaft is

(a) 0.04 (b) 0.05
(c) 0.10 (d) 0.11 **[ME 2007 : 2 M]**

7.16 A displacement sensor (a dial indicator) measures the lateral displacement of a mandrel mounted on the taper hole inside a drill spindle. The mandrel axis is an extension of the drill spindle taper hole axis and the protruding portion of the mandrel surface is perfectly cylindrical. Measurements are taken with the sensor placed at two positions *P* and *Q* as shown in the figure. The readings are recorded as R_x = maximum deflection minus minimum deflection, corresponding to sensor position at *X*, over one rotation.

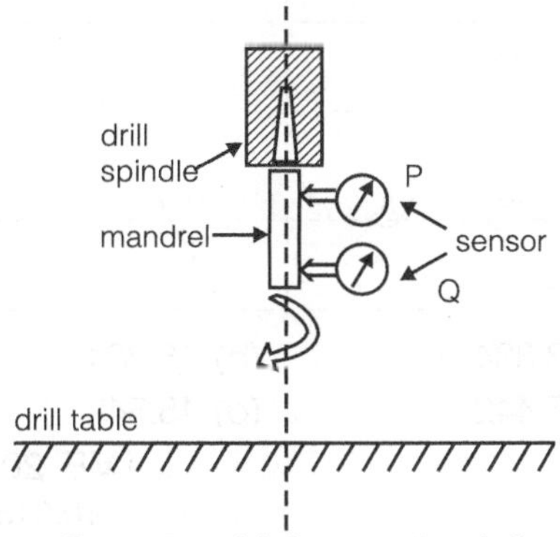

If $R_P = R_Q > 0$, which one the following would be consistent with the observation?

(a) The drill spindle rotational axis is coincident with the drill spindle taper hole axis.
(b) The drill spindle rotational axis intersects the drill spindle taper hole axis at point *P*.
(c) The drill spindle rotational axis is parallel to the drill spindle taper hole axis.
(d) The drill spindle rotational axis intersects the drill spindle taper hole axis at point *Q*.

[2008 : 2 M]

7.17 What are the upper and lower limits of the shaft represented by 60 f8?

Use the following data :

Diameter 60 lies in the diameter step of 50-80 mm. Fundamental tolerance unit, *i*, in μm = $0.45\ D^{1/3} + 0.001D$, where *D* is the representative size in mm; Tolerance value for IT8 = 25*i*. Fundamental deviation for *f* shaft= $-5.5\ D^{0.41}$

(a) lower lim. = 59.924 mm, Upper lim. = 59.970 mm

(b) lower lim. = 59.954 mm, Upper lim. = 60.000 mm
(c) lower lim. = 59.970 mm, Upper lim. = 60.016 mm
(d) lower limit = 60.000 mm, Upper limit = 60.046 mm
[2009 : 2 M]

7.18 A shaft has a dimension, $\phi 35^{-0.009}_{-0.025}$. The respective values of fundamental deviation and tolerance are
(a) –0.025, ± 0.008 (b) –0.025, 0.016
(c) –0.009, ± 0.008 (d) –0.009, 0.016
[ME 2010 : 1 M]

7.19 A taper hole is inspected using a CMM, with a probe of 2 mm diameter. At a height, $Z = 10$ mm from the bottom, 5 points are touched and a diameter of circle (not compensated for probe size) is obtained as 20 mm. Similarly, a 40 mm diameter is obtained at a height $Z = 40$ mm. The smaller diameter (in mm) of hole at $Z = 0$ is

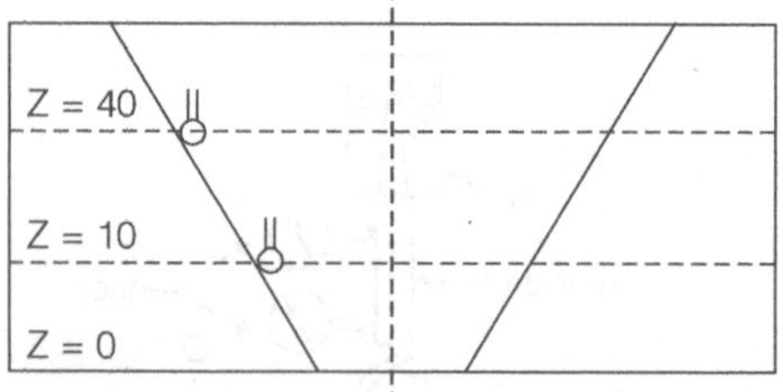

(a) 13.334 (b) 15.334
(c) 15.442 (d) 15.542
[ME 2010 : 2 M]

7.20 A hole is of dimension $\phi 9^{+0.015}_{+0}$ mm. The corresponding shaft is of dimension $\phi 9^{+0.010}_{+0.001}$ mm. The resulting assembly has
(a) loose running fit (b) close running fit
(c) transition fit (d) interference fit
[ME 2011 : 1 M]

7.21 In an interchangeable assembly, shafts of size $25.000^{+0.040}_{-0.010}$ mm mate with holes of size $25.000^{+0.030}_{+0.020}$ mm. The maximum interference (in microns) in the assembly is
(a) 40 (b) 30
(c) 20 (d) 10 **[ME 2012 : 1 M]**

7.22 Cylindrical pins of $25^{+0.020}_{+0.010}$ mm diameter are electroplated in a shop. Thickness of the plating is $30^{\pm 2.0}$ micron. Neglecting gage tolerance, the sized of the GO gage in mm to inspect the plated component is
(a) 25.042 (b) 25.052
(c) 25.074 (d) 25.084
[ME 2013 : 2 M]

7.23 A metric thread of pitch 2 mm and thread angle 60° is inspected for its pitch diameter using 3-wire method. The diameter of the best size wire in mm is
(a) 0.866 (b) 1.000
(c) 1.154 (d) 2.000 **[2013 : 1 M]**

7.24 Which one of the following instruments is widely used to check and calibrate geometric features of machine tools during their assembly?
(a) Ultrasonic probe
(b) Coordinate Measuring Machine (CMM)
(c) Laser interferometer
(d) Vernier calipers **[2014 : 1 M, Set-1]**

7.25 The flatness of a machine bed can be measured using
(a) Vernier calipers
(b) Auto collimator
(c) Height gauge
(d) Tool maker's microscope **[2014 : 1 M, Set-2]**

7.26 For the given assembly: 25 H7/g8, match Group A with Group B

Group - A		Group - B	
P.	H	I.	Shaft Type
Q.	IT8	II.	Hole Type
R.	IT7	III.	Hole Tolerance Grade
S.	g	IV.	Shaft Tolerance Grade

(a) P-I, Q-III, R-IV, S-II
(b) P-I, Q-IV, R-III, S-II
(c) P-II, Q-III, R-IV, S-I
(d) P-II, Q-IV, R-III, S-I **[ME 2014 : 2 M, Set-1]**

7.27 GO and GO-NO plug gauges is to be designed for measuring a hole of nominal diameter 25 mm with a hole tolerance of ± 0.015 mm. Considering 10% of work tolerance to be the gauge tolerance and no wear condition, the dimension (in mm) of the GO plug gauge as per the unilateral tolerance system is
(a) $24.985^{+0.003}_{-0.003}$ (b) $25.015^{+0.000}_{-0.006}$
(c) $24.985^{+0.03}_{-0.03}$ (d) $24.985^{+0.003}_{-0.000}$
[ME 2014 : 2 M, Set-4]

7.28 The diameter of a recessed ring was measured by using two spherical balls of diameter $d_1 = 60$ mm and $d_2 = 40$ mm as shown in the figure.

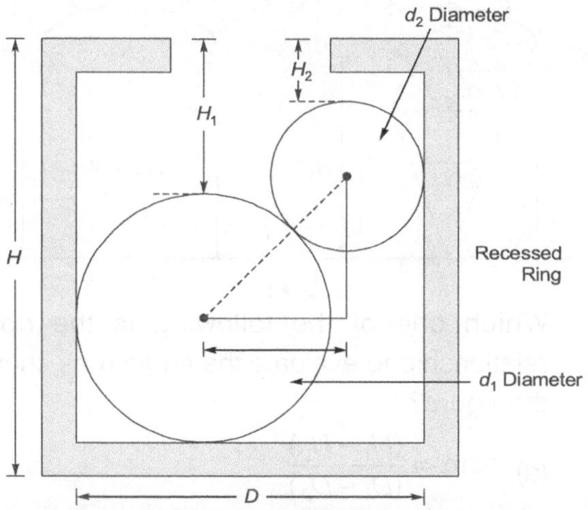

The distance $H_1 = 35.55$ mm and $H_2 = 20.55$ mm. The diameter (D, in mm) of the ring gauge is ___.

[2014 : 2 M, Set-3]

7.29 Holes of diameter $25.0^{+0.040}_{+0.020}$ mm are assembled interchangeably with the pins of diameter $25.0^{+0.005}_{-0.008}$ mm. The minimum clearance in the assembly will be

(a) 0.048 mm (b) 0.015 mm
(c) 0.005 mm (d) 0.008 mm

[2015 : 1 M, Set-1]

7.30 Which one of the following statements is TURE?

(a) The 'GO' gauge controls the upper limit of a hole
(b) The 'NO GO' gauge controls the lower limit of a shaft
(c) The 'Go' gauge controls the lower limit of a hole
(d) The 'NO GO' gauge control the lower limit of a hole

[2015 : 1 M, Set-2]

7.31 In the assembly shown below, the part dimensions are:

$L_1 = 22.0^{\pm 0.01}$ mm,

$L_2 = L_3 = 10.0^{\pm 0.005}$ mm

Assuming the normal distribution of part dimensions, the dimension L_4 in mm for assembly condition would be:

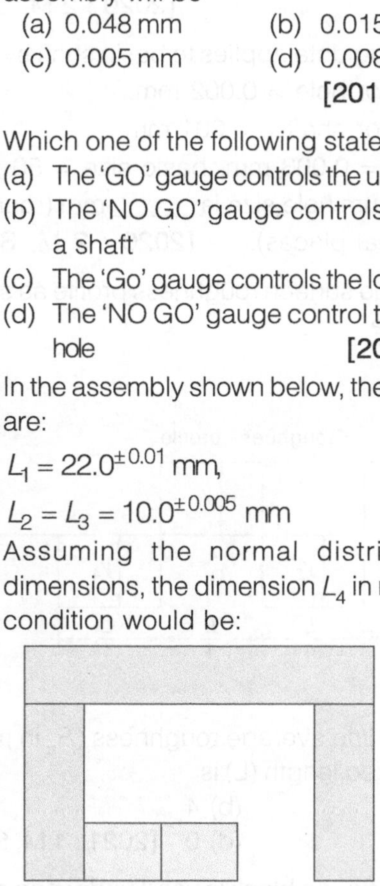

(a) $2.0^{\pm 0.008}$ (b) $2.0^{\pm 0.012}$
(c) $2.0^{\pm 0.016}$ (d) $2.0^{\pm 0.020}$

[2015 : 2 M, Set-1]

7.32 Two optically flat plates of glass are kept at a small angle θ as shown in the figure. Monochromatic light is incident vertically.

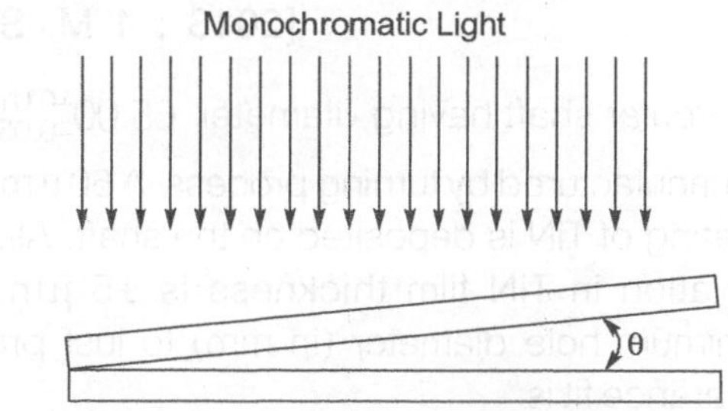

If the wavelength of light used to get a fringe spacing of 1 mm is 450 nm, the wavelength of light (in nm) to get a fringe spacing of 1.5 mm is ________.

[2016 : 2 M, Set-3]

7.33 Match the following :

P. Feeler gauge Q. Fillet gauge
R. Snap gauge S. Cylindrical plug gauge

(i) Radius of an object
(ii) Diameter within limits by comparison
(iii) Clearance or gap between components
(iv) Inside diameter of straight hole

(a) P–(iii), Q–(i), R–(ii), S–(iv)
(b) P–(iii), Q–(ii), R–(i), S–(iv)
(c) P–(iv), Q–(ii), R–(i), S–(iii)
(d) P–(iv), Q–(i), R–(ii), S–(iii)

[2016 : 1 M, Set-1]

7.34 Metric thread of 0.8 mm pitch is to be cut on a lathe. Pitch of the lead screw is 1.5 mm. If the spindle rotates at 1500 rpm, the speed of rotation of the lead screw (rpm) will be ________.

[2017 : 1 M, Set-1]

7.35 Assume that the surface roughness profile is triangular as shown schematically in the figure. If the peak to valley height is 20 μm, the central line average surface roughness R_o (in μm) is

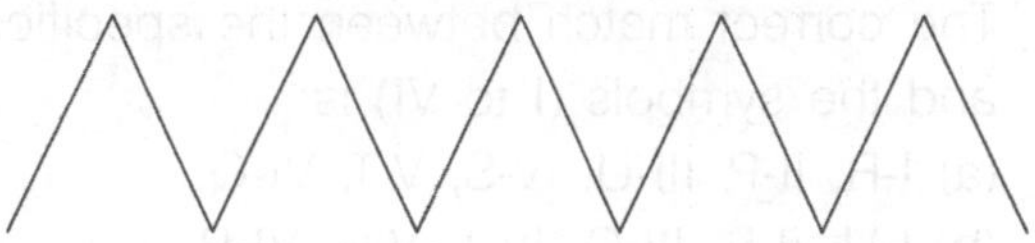

(a) 5 (b) 6.67
(c) 10 (d) 20

[2017 : 2 M, Set-1]

7.36 A cylindrical pin of $25^{+0.020}_{+0.010}$ mm diameter is electroplated, plating thickness is $2.0^{\pm 0.005}$ mm. Neglecting the gauge tolerance, the diameter (in mm, up to 3 decimal points accuracy) of the GO ring gauge to inspect the plated pin is ______.

[2017 : 2 M, Set-2]

7.37 The height (in mm) for a 125 mm sine bar to measure a taper of 27°32′ on a flat work piece is_____(correct to three decimal places).

[2018 : 1 M, Set-1]

7.38 A circular shaft having diameter $65.00^{+0.01}_{-0.05}$ mm is manufactured by turning process. A 50 μm thick coating of TiN is deposited on the shaft. Allowed variation in TiN film thickness is ±5 μm. The minimum hole diameter (in mm) to just provide clearance fit is

(a) 64.95 (b) 65.01
(c) 65.12 (d) 65.10 [2019 : 2 M, Set-1]

7.39 The most common limit gage used for inspecting the hole diameter is

(a) Master gage (b) Plug gage
(c) Ring gage (d) Snap gage

[2019 : 1 M, Set-2]

7.40 The figure below shows a symbolic representation of the surface texture in a perpendicular lay orientation with indicative values (I through VI) Ming the various specifications whose definitions are listed below.

P: Maximum Waviness Height (mm);
Q: Maximum Roughness Height (mm);
R: Minimum Roughness Height (mm);
S: Maximum Waviness Width (mm);
T: Maximum Roughness Width (mm);
U: Roughness Width (mm);

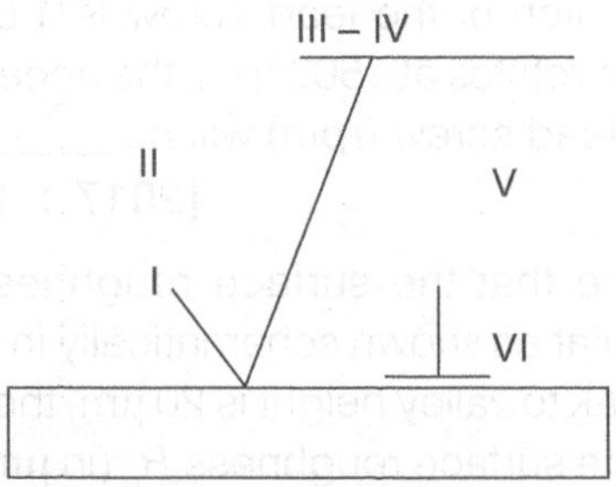

The correct match between the specifications and the symbols (I to VI) is:

(a) I-R, II-P, III-U, IV-S, V-T, VI-Q
(b) I-U, II-S, III-Q, IV-T, V-R, VI-P
(c) I-R, II-Q, III-P, IV-S, V-U, VI-T
(d) I-Q, II-U, III-R, IV-T, V-S, VI-P

[2020 : 1 M, Set-2]

7.41 Two rollers of diameters D_1 (in mm) and D_2 (in mm) are used to measure the internal taper angle in the V-groove of a machined component. The heights H_1 (in mm) and H_2 (in mm) are measured by using a height gauge after inserting the rollers into the same V-groove as shown in the figure.

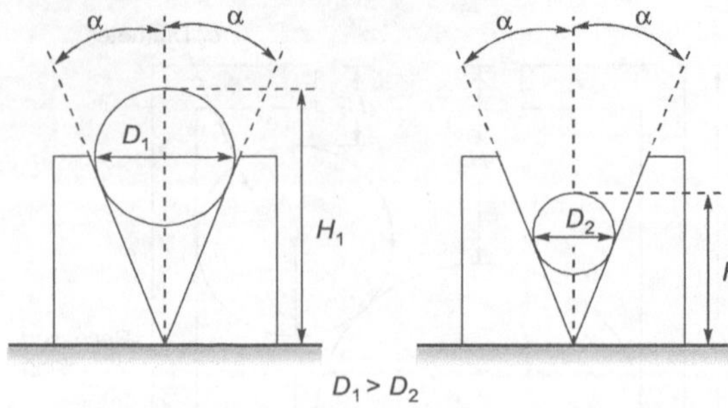

Which one of the following is the correct relationship to evaluate the angle α as shown in the figure?

(a) $\sin\alpha = \dfrac{(H_1 - H_2)}{(D_1 - D_2)}$

(b) $\sin\alpha = \dfrac{(D_1 - D_2)}{2(H_1 - H_2) - (D_1 - D_2)}$

(c) $\cos\alpha = \dfrac{(D_1 - D_2)}{2(H_1 - H_2) - 2(D_1 - D_2)}$

(d) $\operatorname{cosec}\alpha = \dfrac{(H_1 - H_2) - (D_1 - D_2)}{2(D_1 - D_2)}$

[2020 : 2 M, Set-2]

7.42 The following data applies to basic shaft system:
tolerance for hole = 0.002 mm,
tolerance for shaft = 0.001 mm,
allowance = 0.003 mm, basic size = 50 mm
The maximum hole size is _____ mm (round off to 3 decimal places). [2020 : 2 M, Set-1]

7.43 Consider the surface roughness profile as shown in the figure

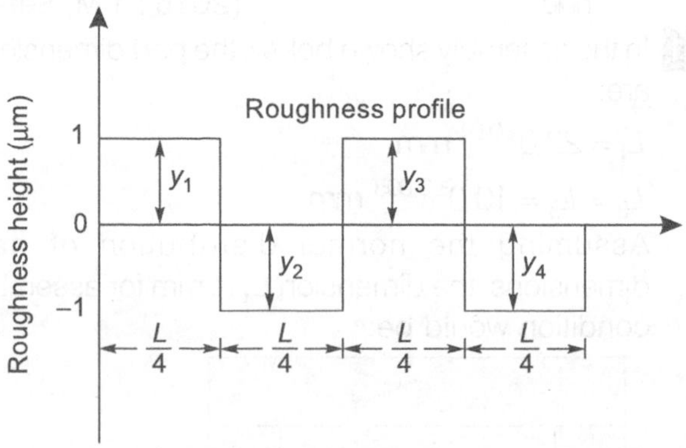

The center line average roughness (R_a in μm) of the measured length (L) is

(a) 1 (b) 4
(c) 2 (d) 0 [2021 : 1 M, Set-1]

7.44 A 76.2 mm gauge block is used under one end of a 254 mm sine bar with roll diameter of 25.4 mm. The height of gauge blocks required at the other end of the sine bar to measure an angle of 30° is_____mm (round off to two decimal places).

[2021 : 2 M, Set-2]

7.45 The allowance provided in between a hole and a shaft is calculated from the difference between
(a) upper limit of the shaft and the lower limit of the hole
(b) lower limit of the shaft and the lower limit of the hole
(c) upper limit of the shaft and the upper limit of the hole
(d) lower limit of the shaft and the upper limit of the hole [2021 : 1 M, Set-2]

7.46 The type of fit between a mating shaft of diameter $25.0^{+0.010}_{-0.010}$ mm and a hole of diameter $25.015^{+0.015}_{-0.015}$ mm is ________.
(a) Clearance (b) Transition
(c) Interference (d) Linear
[2022 : 1 M, Set-1]

7.47 A shaft of diameter $25^{-0.04}_{-0.07}$ mm is assembled in a hole of diameter $25^{0.02}_{-0.00}$ mm. Match the allowance and limit parameter in **Column I** with its corresponding quantitative value in **Column II** for this shaft-hole assembly.

Column I	Column II
P. Allowance	**1.** 0.09 mm
Q. Maximum clearance	**2.** 24.96 mm
R. Maximum material limit of hole	**3.** 0.04 mm
	4. 25 mm

(a) P-3, Q-1, R-4 (b) P-1, Q-3, R-2
(c) P-1, Q-3, R-4 (d) P-3, Q-3, R-2
[2022 : 1 M, Set-2]

7.48 The best size wire is fitted in a groove of a metric screw such that the wire touches the flanks of the thread on the pitch line as shown in the figure. The pitch (p) and included angle of the thread are 4 mm and 60°, respectively. The diameter of the best size wire is ____mm (round off to 2 decimal places).

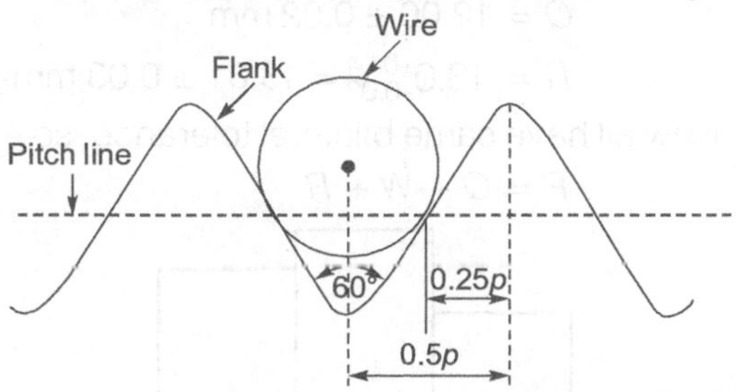

[2022 : 2 M, Set-2]

7.49 A cuboidal part has to be accurately positioned first, arresting six degrees of freedom and then clamped in a fixture, to be used for machining. Locating pins in the form of cylinders with hemi-spherical tips are to be placed on the fixture for positioning. Four different configurations of locating pins are proposed as shown. Which one of the options given is correct?

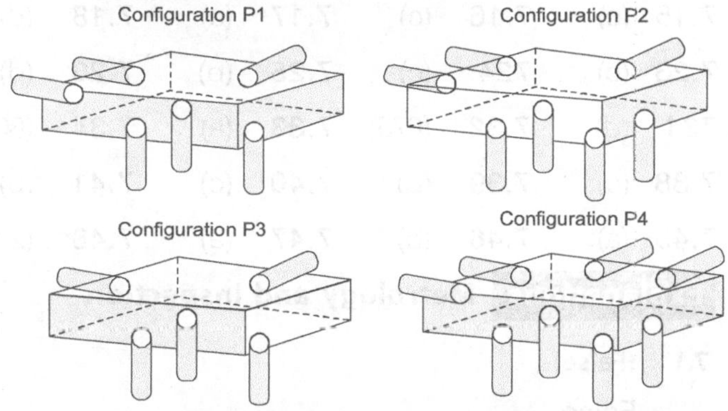

(a) Configuration P1 arrests 6 degrees of freedom, while Configurations P2 and P4 are over-constrained and Configuration P3 is under-constrained.
(b) Configuration P2 arrests 6 degrees of freedom, while Configurations P1 and P3 are over-constrained and Configuration P4 is under-constrained.
(c) Configuration P3 arrests 6 degrees of freedom, while Configurations P2 and P4 are over-constrained and Configuration P1 is under-constrained.
(d) Configuration P4 arrests 6 degrees of freedom, while Configurations P1 and P3 are over-constrained and Configuration P2 is under-constrained. [2023 : 1 M]

7.50 An optical flat is used to measure the height difference between a reference slip gauge *A* and a slip gauge *B*. Upon viewing via the optical flat using a monochromatic light of wavelength 0.5 μm, 12 fringes were observed over a length of 15 mm of gauge *B*. If the gauges are placed 45 mm apart, the height difference of the gauges is ________ μm. (Answer in integer)

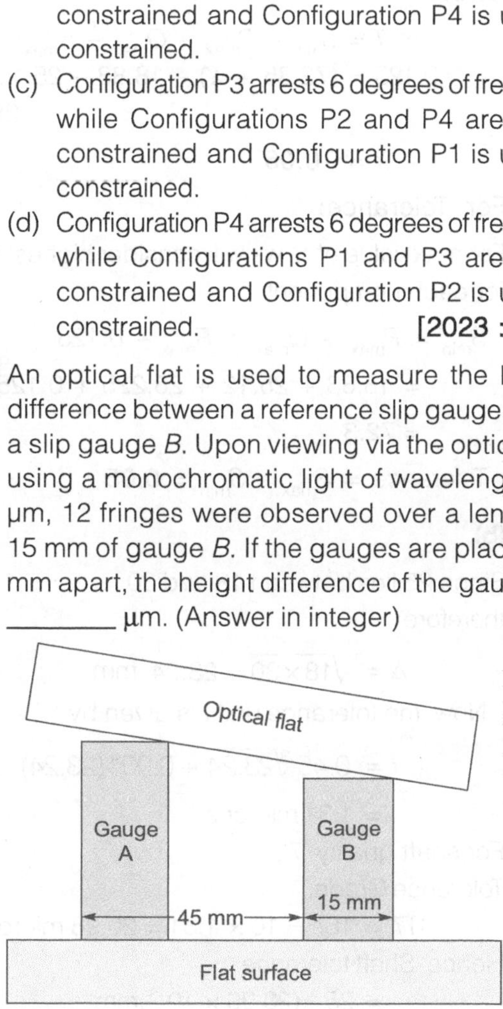

[2023 : 2 M]

Answers Metrology and Inspection

7.1	(False)	7.2	(Attributes)	7.3	(a)	7.4	(0.05)	7.5	(b)	7.6	(b)				
7.7	(b)	7.8	(a)	7.9	(b)	7.10	(d)	7.11	(None)	7.12	(c)	7.13	(a)	7.14	(c)
7.15	(c)	7.16	(c)	7.17	(a)	7.18	(d)	7.19	(a)	7.20	(c)	7.21	(c)	7.22	(d)
7.23	(c)	724	(c)	7.25	(b)	7.26	(d)	7.27	(d)	7.28	(93.30)	7.29	(b)	7.30	(c)
731	(d)	7.32	(675)	7.33	(a)	7.34	(800)	7.35	(a)	7.36	(29.030)	7.37	(57.782)		
7.38	(c)	7.39	(b)	7.40	(c)	7.41	(b)	7.42	(50.005)	7.43	(a)	7.44	(203.2)		
7.45	(a)	7.46	(b)	7.47	(a)	7.48	(2.309)	7.49	(a)	7.50	(9)				

Explanations Metrology and Inspection

7.1 (False)

False

It is false, on the contrary micrometers are calibrated using slip gauges.

7.2 (Attributes)

7.4 (0.05)

$$T = S_{min} - P_{max} - Q_{max} - R_{max}$$

$$0.125 = (72.35 + X) - 18.83 - 25.12 - 28.225$$

$$X = \mathbf{-0.05}$$

For Tolerance:

For max value of S, we will consider highest value for each component

$$S_{min} = P_{max} + Q_{max} + R_{max} + 0.125$$

$$= 18.83 + 25.12 + 28.225 + 0.125$$

$$= 72.3$$

$$\text{Tolerance} = S_{max} - S_{min} = \mathbf{0.05}$$

7.6 (b)

Since 25 lies between 18 and 30 therefore

$$\Delta = \sqrt{18 \times 30} = 23.24 \text{ mm}$$

Now, the tolerance unit i is given by

$$i = 0.45\sqrt[3]{23.24} + 0.001(23.24)$$

$$= 1.31 \text{ microns}$$

For shaft quality '7',

Tolerance Grade,

$$IT7 = 16i = 16 \times 1.31 = 20.96 \text{ microns}$$

Hence, Shaft tolerance

$$= 25 - (20.96 \times 10^{-3})\text{mm}$$

$$= \mathbf{24.979 \text{ mm}}$$

7.7 (b)

Given Metric thread M 16

In case of threads always rollers (wires) are used.

Pitch: P = 2 mm

Pitch diameter = 14.701 mm

In case of Metric thread $2\theta = 60°$

$\Rightarrow \theta = 30°$

For a best size wire

$$d_w = \frac{P}{2}\sec\theta = \frac{2}{2}\sec 30° = \mathbf{1.155 \text{ mm}}$$

7.8 (a)

$$\Delta h = \frac{N}{2} \times \frac{\lambda}{2}$$

Δh = difference in the height of slip gauges.

$\Delta h = 1.002 - 1.000 = 0.002$ mm

λ = wavelength

Total number of straight fringes,

$$N = \frac{4 \times \Delta h}{\lambda} = \frac{4 \times 0.002}{0.0058928} = 1.357 \approx \mathbf{2}$$

7.9 (b)

$$P = 35 \pm 0.08 \text{ mm}$$

$$Q = 12.00 \pm 0.02 \text{ mm}$$

$$R = 13.0^{+0.04}_{-0.02} = 13.01 \pm 0.03 \text{ mm}$$

Now all have same bilateral tolerance, so

$$P = Q + W + R$$

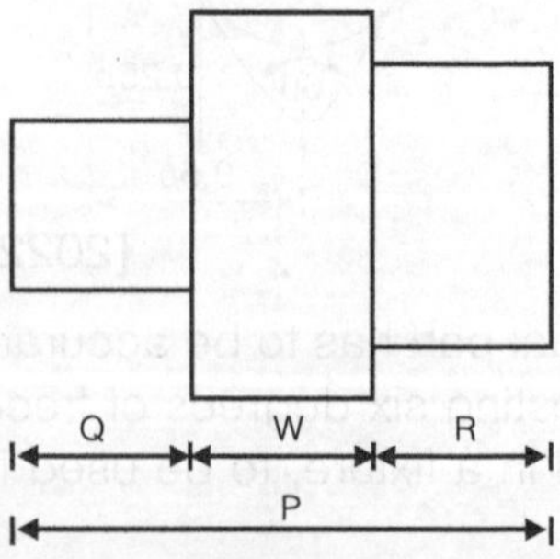

$\therefore \quad P = Q + W + R$

Considering dimension

$\Rightarrow \quad 35 = 12 + W + 13.01$

$W = 9.99$ mm

Tolerance are probabilities and not the absolute value on any part, at least one section must be there that treated as sink, and tolerance of sink will be cumulative sum of all tolerances.

$\therefore$ Tolerance= 0.08 + 0.02 + 0.03 = **0.13 mm**

7.10 (d)

Minimum shaft size = 25.000 – 0.04 = 24.96 mm

Maximum hole size = 25.000 + 0.02 = 25.02 mm

$\therefore$ maximum possible clearance

= maximum size of hole – Minimum size of shaft

= 25.02 – 24.96 = 0.060 mm = **60 micron**

7.11 (None)

For general purpose ISO system of gauge:

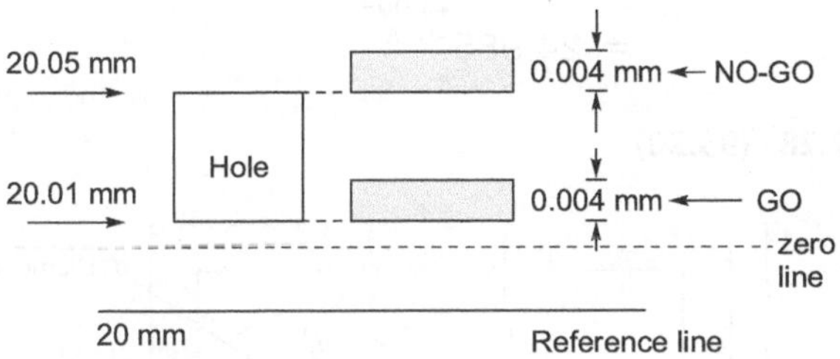

BS of GO gauge = 20.01 mm

LL of GO gauge = 20.01 mm

UL of GO gauge = 20.014 mm

GO gauge will be = $20.01^{+0.004}_{0}$ mm

BS of NO GO gauge = 20.05 mm

LL of NO GO gauge = 20.05 mm

UL of GO gauge = 20.054 mm

NO GO gauge will be = $20.05^{+0.004}_{+0}$ mm

In the question only UL of GO and NOGO is given, but it's impossible to produced. So answer will be none of this.

7.13 (a)

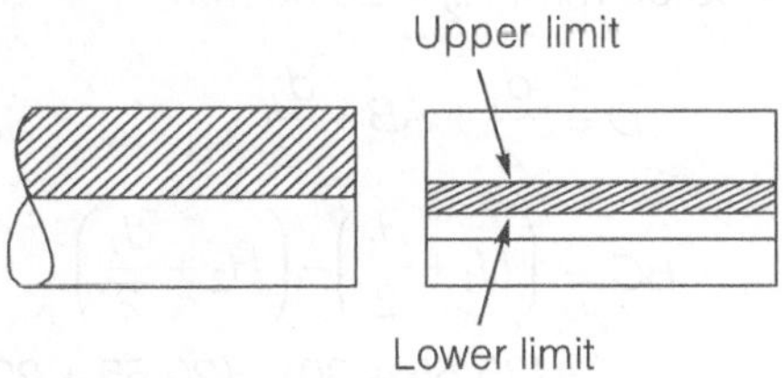

From the figure, it is essential that lower limit of shaft should be greater than upper limit of hole.

7.15 (c)

Low limit of hole = 40.000 mm

High limit of hole = 40.000 + 0.050

= 40.0.50 mm

High limit of shaft = 40.000 – 0.01

= 39.990 mm

Low limit of shaft = 39.990 – 0.04 = 39.950 mm

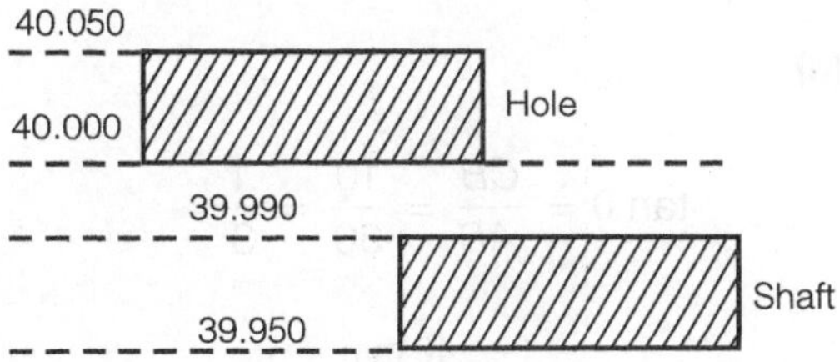

Maximum clearance

= 40.050 – 39.950 = **0.10 mm**

7.17 (a)

60 mm diameter lies in the diameter step of 50–80 mm. Therefore, geometric mean diameter,

$$D = \sqrt{D_{min} \times D_{max}} = \sqrt{50 \times 80} = 63.246 \text{ mm}$$

Fundamental tolerance,

$$i = (0.45\, D^{1/3} + 0.001\, D)\ \mu\text{m}$$

$$= [0.45\,(63.246)^{1/3} + 0.001\,(63.246)]$$

$$= 1.859\ \mu\text{m} = 0.00186 \text{ mm}$$

$$IT8 = 25\,i = 25 \times 0.00186 = 0.04646 \text{ mm}$$

Fundamental deviation for 'f' shaft,

$$\epsilon_d = -5.5\, D^{0.41}$$

$$= -5.5\,[63.246]^{0.41}$$

$$= -0.030115 \text{ mm}$$

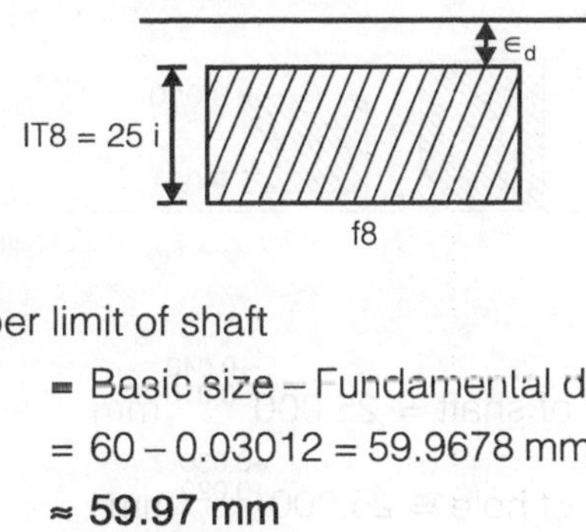

Upper limit of shaft

= Basic size – Fundamental deviation

= 60 – 0.03012 = 59.9678 mm

≈ **59.97 mm**

Lower limit of shaft

= Upper limit – Tolerance

= 59.97 – 0.0464 = 59.9236 mm

≈ **59.924 mm**

7.18 (d)

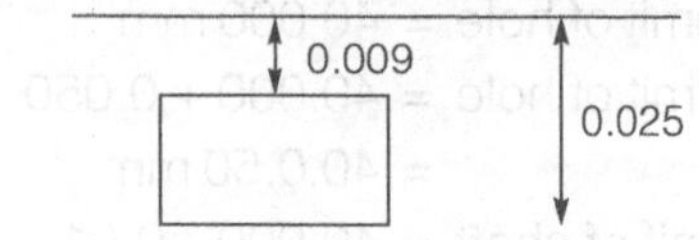

Tolerance = 0.025 − 0.009

= **0.016**

Fundamental deviation = **−0.009**

7.19 (a)

$$\tan\theta = \frac{CB}{AB} = \frac{10}{30} = \frac{1}{3}$$

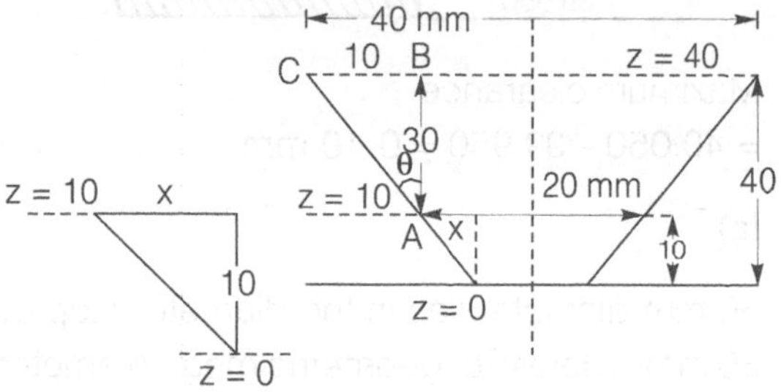

Now, $\tan\theta = \frac{x}{10}$

$$\frac{1}{3} = \frac{x}{10}$$

$$\Rightarrow \quad x = \frac{10}{3}\text{ mm} = 3.33\text{ mm}$$

So, dia at $z = 0$,

= 20 − 2x = 20 − 2 × 3.33 = **13.334 mm**

7.20 (c)

It is case of transition fit

Since,

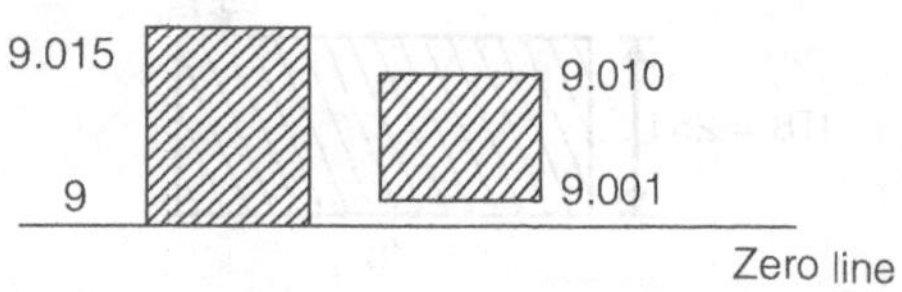

7.21 (c)

Size of shaft = $25.000^{+0.040}_{-0.010}$ mm

Size of hole = $25.000^{+0.030}_{+0.020}$ mm

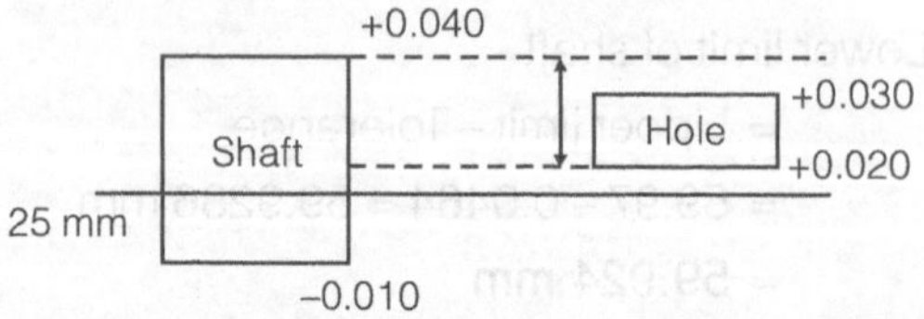

Maximum interference

= 0.040 − 0.020 = 0.020 mm = **20 μm**

7.22 (d)

Size of the Go gauge

$= (25 + 0.20) + [(30 + 2) \times 10^{-3}] \times 2$

= **25.084**

7.23 (c)

Pitch: $P = 2$ mm

Thread angle: $2\theta = 60°$ for metric thread

$\Rightarrow \quad \theta = 30°$

For best size wire

$$d_w = \frac{P}{2}\sec\theta = \frac{2}{2}\sec 30° = \mathbf{1.154\ mm}$$

7.27 (d)

Hole dimensions are 24.985mm and 25.015mm

Total tolerance = 0.015 + 0.015 = 0.03 mm

Work tolerance = 0.03 × 0.1 = 0.003 mm

Dimension of GO plug gauge

$$= 24.985^{+0.003}_{-0.000}$$

7.28 (93.30)

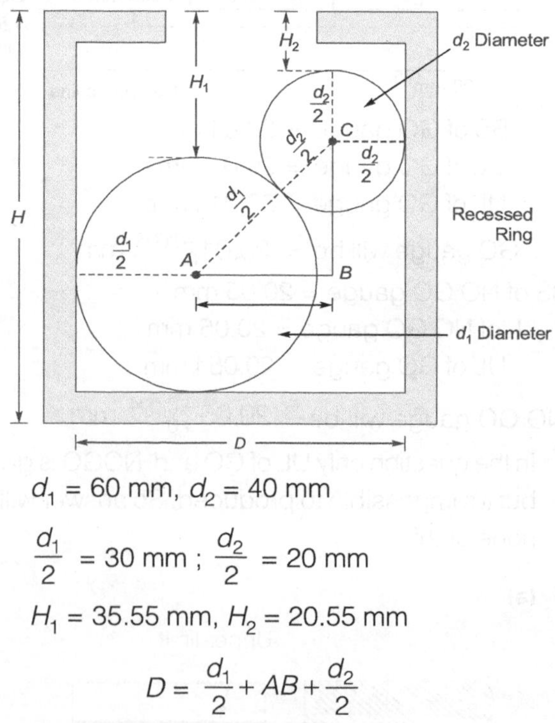

$d_1 = 60$ mm, $d_2 = 40$ mm

$\frac{d_1}{2} = 30$ mm ; $\frac{d_2}{2} = 20$ mm

$H_1 = 35.55$ mm, $H_2 = 20.55$ mm

$$D = \frac{d_1}{2} + AB + \frac{d_2}{2}$$

$$BC = \left(H_1 + \frac{d_1}{2}\right) - \left(H_2 + \frac{d_2}{2}\right)$$

$= (35.55 + 30) - (20.55 + 20)$

$= (65.55) - (40.55) = 25$ mm

In right angle ΔABC

$$AC = \frac{d_1}{2} + \frac{d_2}{2} = 30 + 20 = 50 \text{ mm}$$

$$AB = \sqrt{AC^2 - BC^2}$$

$$= \sqrt{50^2 - 25^2} = 25\sqrt{3} \text{ mm}$$

$$D = 30 + 25\sqrt{3} + 20 = \mathbf{93.30\ mm}$$

7.29 (b)

Minimum clearances = (Hole)$_{minimum}$ − (Shaft)$_{maximum}$

$= 25.02 - 25.005 = \mathbf{0.015\ mm}$

7.30 (c)

'GO' gauge controls the lower limit of a hole.

7.31 (d)

Given: $L_1 = 22.0^{\pm 0.01}$ mm;

$L_2 = L_3 = 10^{\pm 0.005}$ mm

$L_4 = ?$

Method I:

$$(L_4)_{max} = (L_1)_{max} - (L_2)_{min} - (L_3)_{min}$$
$$= 22.01 - (9.995) - (9.995)$$
$$= 2.02$$
$$(L_4)_{min} = (L_1)_{min} - (L_2)_{max} - (L_3)_{max}$$
$$= 21.99 - 10.005 - 10.005$$
$$= 1.98$$

Dimension L_4 is $2^{\pm 0.02}$ mm

Method II:

$$L_4 = L_1 - (L_2 + L_3)$$
$$= 22^{\pm 0.01} - (10^{\pm 0.005} + 10^{\pm 0.005})$$
$$= 22^{\pm 0.01} - (20^{\pm 0.01})$$
$$= 2^{\pm 0.02}$$

7.32 (675)

Given: 1 mm = 450 mm, Spacing = 1.5 mm

$$n = \frac{2}{\lambda} \quad \text{As given in question}$$

$$1.5 = n \times \frac{\lambda_2}{2}$$

$$\lambda_2 = \frac{1.5 \times 2}{n} = \frac{1.5 \times 2}{\frac{2}{450}} = 1.5 \times 450 = \mathbf{675\ nm}$$

7.33 (a)

Feeler gauge is used to measure clearance or gap between components. Fillet gauge is also known as radius gauge. Snap gauge is a limit gauge used to measure diameter within limits by comparison.

7.34 (800)

Pitch of thread, $P_t = 0.8$ mm

Speed of spindle, $N_t = 1500$ rpm

$$P_s = 1.5 \text{ mm}$$
$$N_s \times P_s \times Z_z = N_t \times P_t \times Z_t \quad [\because Z_s = Z_t = 1]$$
$$N_s \times 1.5 \times 1 = 1500 \times 0.8 \times 1$$
$$N_s = \mathbf{800\ rpm}$$

7.35 (a)

$$R_a = \frac{h}{4} = \frac{20}{4} = 5\ \mu\text{m}$$

7.36 (29.030)

GO ring gauge will inspect maximum metal conditions i.e. UL of shaft i.e. Largest size after platting.

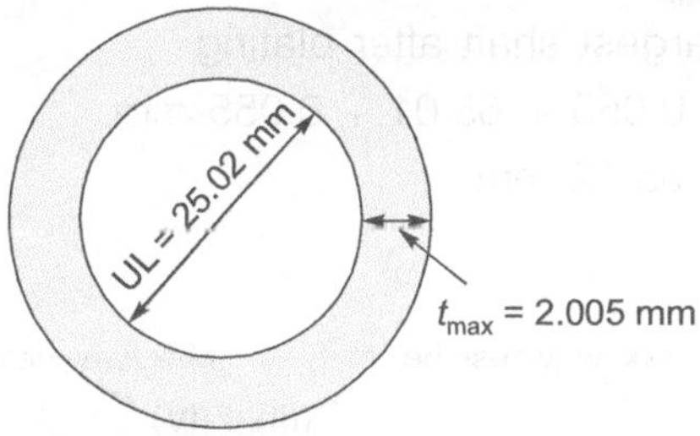

UL after platting

$$= 25.02 + 2.005 + 2.005 \text{ mm}$$
$$= \mathbf{29.030\ mm}$$

7.37 (57.782)

$$\theta = 27°32' = 27 + \left(\frac{32}{60}\right)° = 27.533°$$

In sine bar,

$$\sin\theta = \frac{H}{L} \quad \therefore \sin 27.533° = \frac{H}{125}$$

or $\quad H = \mathbf{57.782\ mm}$

7.38 (c)

Shaft $65^{+0.01}_{-0.05}$ mm

Coating thickness 50 ± 5 μm

Clearance fit

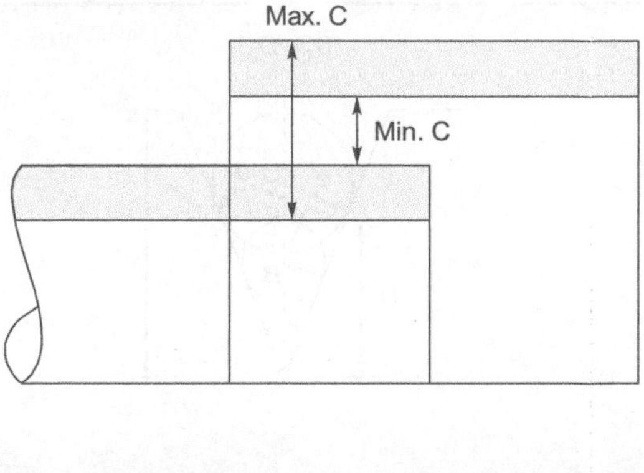

This is a question if clearance fit, for just clearance fit min. C will be zero. Therefore UL of shaft = LL of hole.

After coating largest shaft will be product if we start with largest shaft and add maximum coating.

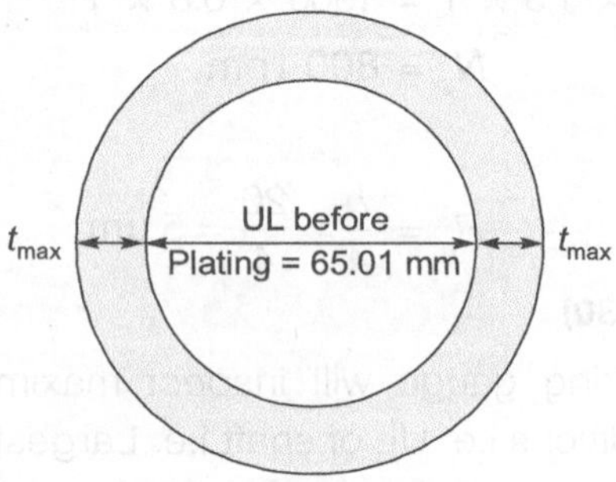

t_{max} = 55 μm = 0.055 mm

Largest shaft after plating

= 0.055 + 65.01 + 0.055 mm

= 65.12 mm

7.40 (c)

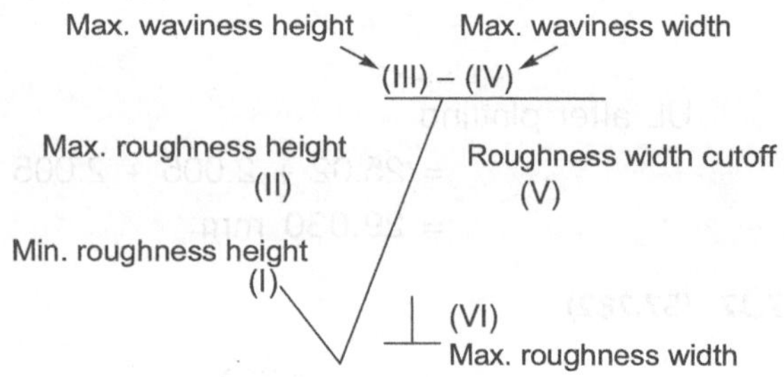

I-R, II-Q, III-P, IV-S, V-U, VI-T

7.41 (b)

As per given data:

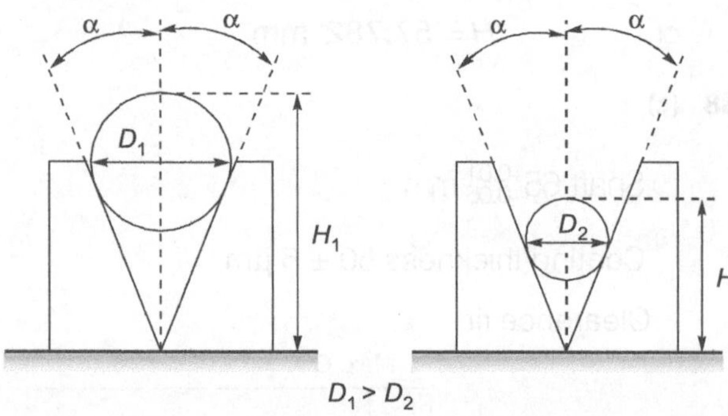

$D_1 > D_2$

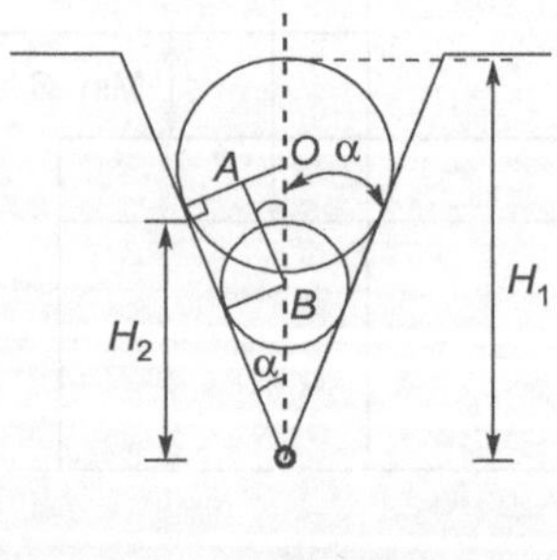

$$OB = \left(H_1 - \frac{D_1}{2}\right) - \left(H_2 - \frac{D_2}{2}\right)$$

$$= (H_1 - H_2) - \left(\frac{D_1 - D_2}{2}\right)$$

$$OA = \frac{D_1 - D_2}{2}$$

$$\sin\alpha = \frac{\frac{(D_1 - D_2)}{2}}{(H_1 - H_2) - \frac{(D_1 - D_2)}{2}}$$

$$\sin\alpha = \frac{(D_1 - D_2)}{2(H_1 - H_2) - (D_1 - D_2)}$$

7.42 (50.005)

Given: tolerance for hole = 0.002 mm, tolerance for shaft = 0.001 mm,

allowance = 0.003 mm, basic size = 50 mm

Using diagram method

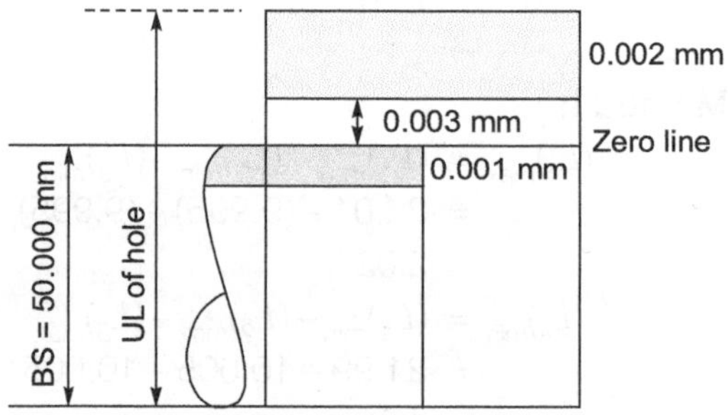

UL of hole = BS + 0.003 + 0.002 mm

= 50.005 mm

7.43 (a)

$$R_G = \frac{\sum_{i=1}^{n} y}{n} = \frac{4}{4} = 1$$

7.44 (203.2)

Case-I:

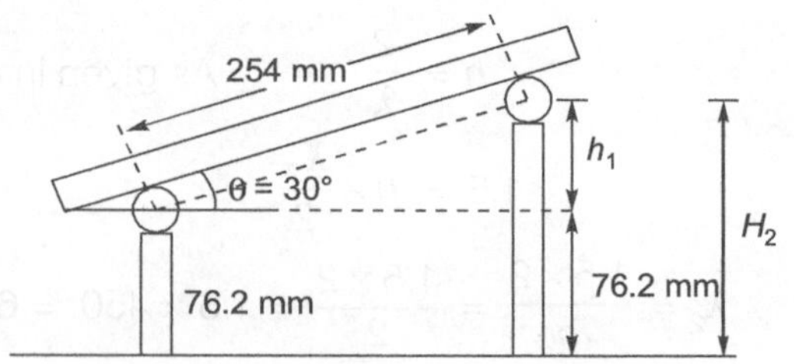

$$\sin 30° = \frac{h_1}{254}$$

or $h_1 = 254 \sin 30° = 127$ mm

$H_2 = 76.2 + 127 = 203.2$ mm

Case-II:

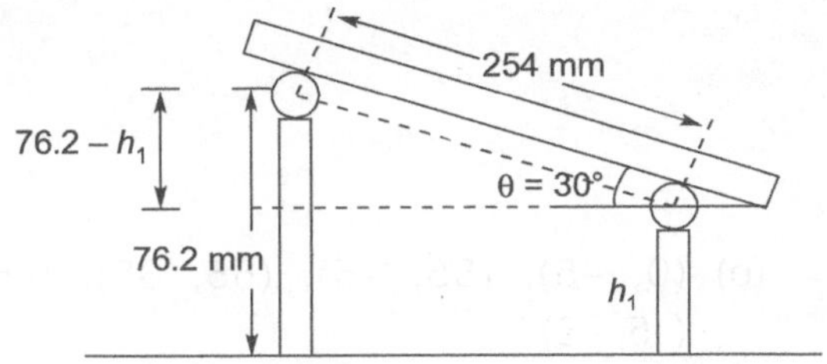

$$\sin 30° = \frac{76.2 - h_1}{254}$$

or $h_1 = 76.2 - 254 \sin 30°$

$= -50.8$ mm

Which is not feasible.

Therefore final answer 203.2 mm.

7.45 (a)

It is minimum clearance or maximum interference. It is the intentional difference between the basic dimensions of the mating parts. The allowance may be positive or negative.

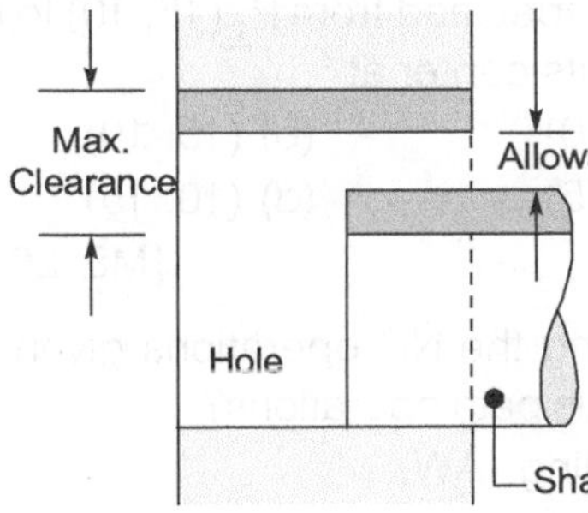

7.46 (b)

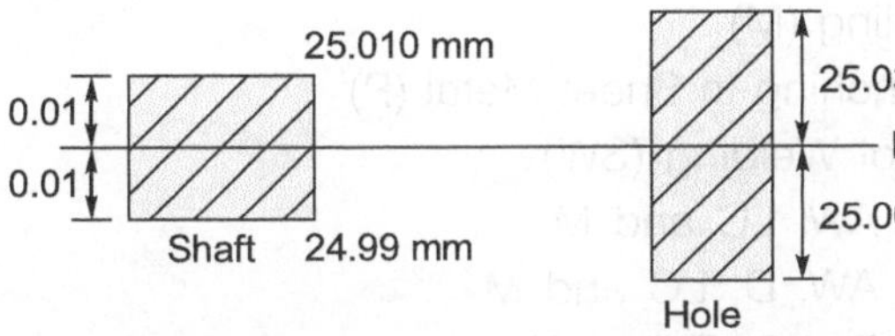

So, here we will have the transition fit which is the option (b).

7.47 (a)

Given: Shaft: $25^{-0.04}_{-0.07}$ and Hole = $25^{-0.02}_{-0.00}$

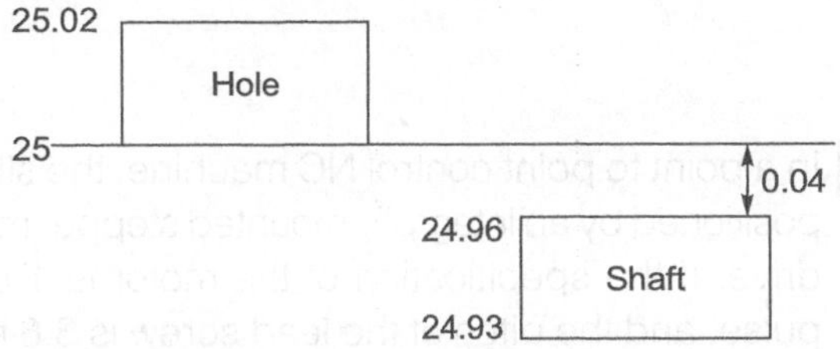

Allowance is the difference in the maximum material limit of hole and shaft.

- Allowance = Maximum material limit
  = 25 – 24.96 = 0.04 mm
- Max. clearance = $\text{Hole}_{max} - \text{Shaft}_{min}$
  = 25.02 – 24.93 = 0.09 mm
- Maximum material limit of hole = Minimum size of hole = 25 mm

7.48 (2.309)

Given: p = 4 mm, Thread angle, $\alpha = 60°$

$$\text{So, Best wire size} = \frac{p}{2}\sec\frac{\alpha}{2} = \frac{4}{2}\sec\frac{60}{2}$$

$$= 2 \sec 30$$

$$= 2 \times 1.154 \text{ mm} = 2.309 \text{ mm}$$

7.49 (a)

As per 3 - 2 - 1 principle by providing six pins, six degrees of freedom are arrested.

So, configuration P1 arrests six degrees of freedom. Configuration P2 and P4 are over constrained while configuration 3 is under constrained.

7.50 (9) (8.999 to 9.001)

Given

Number of fringes, n = 12 per 15 mm

Wavelength = 0.5 μm

Distance between gauges = 45 mm

Height difference between gauges,

$$h = \frac{n\lambda}{2} = \frac{12}{15} \times \frac{0.5}{2} \times 45 = 9 \ \mu\text{m}$$

■■■■

8 NC, CNC, DNC, APT, Automation

8.1 In a point to point control NC machine, the slides positioned by an integrally mounted stepper motor drive. If the specification of the motor is 1 deg/pulse, and the pitch of the lead screw is 3.6 mm. What is the expected positioning accuracy?

(a) 1 micron (b) 10 micron
(c) 50 micron (d) 100 micron

[ME 1997 : 2 M]

8.2 A cylinder of 155 mm diameter is to be reduced to 150 mm diameter in one turning cut with a feed of 0.15 mm/rev and a cutting speed of 150 m/min on a NC lathe. What are the programmed feed rate and the material removal rate.

[ME 1998 : 2 M]

8.3 In computer aided drafting practice, an arc is defined by

(a) two end points only
(b) centre and radius
(c) radius and one end point
(d) two end points and centre [ME 2000 : 1 M]

8.4 Cellular manufacturing is suitable for

(a) a single production in large volumes
(b) one-off production of several varieties
(c) production with similar features made in batches
(d) large variety of products in large volumes

[ME 2000 : 1 M]

8.5 Machining of complex shapes on CNC machines requires

(a) simultaneous control of x, y and z axes
(b) simultaneous control of x and y axes
(c) independent control of x and y axes
(d) independent control of x, y and z axes

[ME 2002 : 1 M]

8.6 In finish machining of an island on a casting with CNC milling machine, an end mill with 10 mm diameter is employed. The corner points of the island are represented by (0, 0), (0, 30), (50, 30) and (50, 0). By applying cutter radius compensation, the trajectory of the cutter will be

(a) (–5, 0), (–5, 35), (55, 35), (55, –5), (–5, 5)
(b) (0, –5), (55, –5), (55, 35), (–5, 35), (–5, –5)
(c) (5, 5), (5, 25), (45, 25), (45, 5), (5, 5)
(d) (5, 5), (45, 5), (45, 25), (5, 25), (5, 5)

[ME 2000 : 2 M]

8.7 During the execution of a CNC part program block N020 G02 X45.0 Y25.0 R5.0 the type of tool motion will be

(a) circular Interpolation – clockwise
(b) circular Interpolation – counterclockwise
(c) linear Interpolation
(d) rapid feed [ME 2004 : 1 M]

8.8 a 2-D CAD package, clockwise circular arc of radius 5, specified from P_1 (15, 10) to P_2 (10, 15) will have its center at

(a) (10, 10) (b) (15, 10)
(c) (15, 15) (d) (10, 15)

[ME 2004 : 2 M]

8.9 hich among the NC operations given below are continuous path operations?

Arc Welding (AW)
Drilling (D)
Laser Cutting of Sheet (LC)
Milling (M)
Punching in Sheet Metal (P)
Spot Welding (SW)

(a) AW, LC and M
(b) AW, D, LC and M
(c) D, LC, P and SW
(d) D, LC, and SW [ME 2005 : 1 M]

8.10 The tool of an NC machine has to move along a circular arc from (5,5) to (10,10) while performing an operation. The centre of the arc is at (10,5). Which one of the following NC tool path commands performs the above mentioned operation?

(a) N010 G02 X10 Y10 X5 Y5 R5
(b) N010 G03 X10 Y10 X5 Y5 R5
(c) N010 G01 X5 Y5 X10 Y10 R5
(d) N010 G02 X5 Y5 X10 Y10 R5

[ME 2005 : 2 M]

8.11 NC contouring is an example of

(a) continuous path positioning
(b) point-to-point positioning
(c) absolute positioning
(d) incremental positioning [ME 2006 : 1 M]

8.12 Which type of motor is NOT used in axis or spindle drives of CNC machine tools?
(a) induction motor (b) dc servo motor
(c) stepper motor (d) linear servo motor
[ME 2007 : 1 M]

8.13 For generating a Coon's surface we require
(a) a set of grid points on the surface
(b) a set of grid control points
(c) four bounding curves defining the surface
(d) two bounding curves and a set of grid control points [ME 2008 : 1 M]

Linked Answer Questions 8.14 and 8.15:

In the feed drive of a Point-to-Point open loop CNC drive, a stepper motor rotating at 200 steps/rev drives a table through a gear box and lead screw-nut mechanism (pitch = 4 mm, number of starts =1). The gear ratio

$$\left(= \frac{\text{Output rotational speed}}{\text{Input rotational speed}}\right) \text{is given by } U = \frac{1}{4}.$$

The stepper motor (driven by voltage pulses from a pulse generator) executes 1 step/pulse of the pulse generator .The frequency of the pulse train from the pulse generator is f = 10,000 pulses per minute.

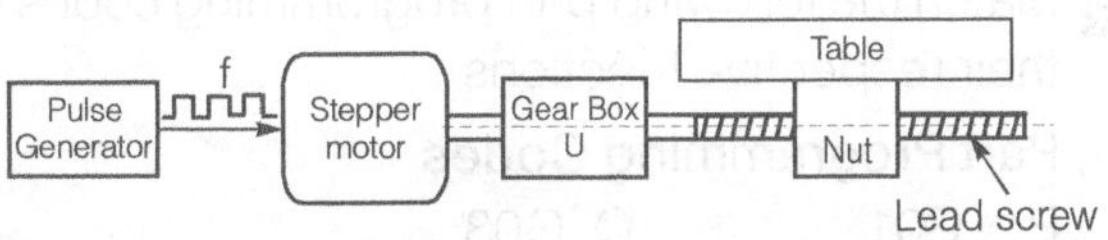

8.14 The Basic Length Unit (BLU), *i.e.*, the table movement corresponding to 1 pulse of the pulse generator, is
(a) 0.5 microns (b) 5 microns
(c) 50 microns (d) 500 microns
[ME 2008 : 2 M]

8.15 A customer insists on a modification to change the BLU of the CNC drive to 10 microns without changing the table speed. The modification can be accomplished by
(a) changing U to $\frac{1}{2}$ and reducing f to $\frac{f}{2}$
(b) changing U to $\frac{1}{8}$ and increasing f to $2f$
(c) changing U to $\frac{1}{2}$ and keeping f unchanged
(d) keeping U unchanged and increasing f to $2f$.
[ME 2008 : 2 M]

8.16 Match the following:

NC Code	Definition
P. M05	1. Absolute coordinate system
Q. G01	2. Dwell
R. G04	3. Spindle stop
S. G90	4. Linear interpolation

(a) P-2, Q-3, R-4, S-1
(b) P-3, Q-4, R-1, S-2
(c) P-3, Q-4, R-2, S-1
(d) P-4, Q-3, R-2, S-1 [ME 2009 : 2 M]

8.17 In a CNC program block, N002 G02 G91 X40 Z40..., G02 and G91 refer to
(a) circular interpolation in counterclockwise direction and incremental dimension
(b) circular interpolation in counterclockwise direction and absolute dimension
(c) circular interpolation in clockwise direction and incremental dimension
(d) circular interpolation in clockwise direction and absolute dimension [ME 2010 : 1 M]

8.18 A CNC vertical milling machine has to cut a straight slot of 10 mm width and 2 mm depth by a cutter of 10 mm diameter between points (0, 0) and (100, 100) on the XY plane (dimensions in mm). The feed rate used for milling is 50 mm/min. Milling time for the slot (in seconds) is
(a) 120 (b) 170
(c) 180 (d) 240 [ME 2012 : 1 M]

8.19 Which of the following is the correct data structure for solid models?
(a) solid part → face → edges → vertices
(b) solid part → edges → faces → vertices
(c) vertices → edges → faces → solid parts
(d) vertices → faces → edges → solid parts
[ME 2009 : 1 M]

8.20 For machining a rectangular island represented by coordinates P(0,0), Q(100,0), R(100,50) and S(0,50) on a casting using CNC milling machine, an end mill with a diameter of 16 mm is used. The trajectory of the cutter centre to machine the island $PQRS$ is
(a) (–8, – 8), (108, – 8), (108,58), (– 8,58), (– 8, – 8)
(b) (8,8), (94,8), (94,44), (8,44), (8,8)
(c) (– 8,8), (94,0), (94,44), (8,44), (– 8,8)

(d) (0,0), (100,0), (100,50), (50,0), (0,0)

[ME 2014 : 1 M, Set-1]

8.21 A robot arm PQ with end coordinates $P(0,0)$ and $Q(2,5)$ rotates counter clockwise about P in the XY plane by 90°. The new coordinate pair of the end point Q is

(a) (– 2, 5) (B) (– 5,2)
(c) (– 5, – 2) (D) (2, – 5)

[ME 2014 : 1 M, Set-2]

8.22 A minimal spanning tree in network flow models involves

(a) all the nodes with cycle/loop allowed
(b) all the nodes with cycle/loop not allowed
(c) shortest path between start and end nodes
(d) all the nodes with directed arcs

[ME 2014 : 1 M, Set-3]

8.23 For the CNC part programming, match Group A with Group B :

Group A

P. Circular interpolation, counter clock wise
Q. dwell
R. circular interpolation, clock wise
S. point to point countering

Group B

I. G02
II. G03
III. G04
IV. G00

(a) P - II, Q-III, R-I, S-IV
(b) P - I, Q-III, R-II, S-IV
(c) P - I, Q-IV, R-II, S-III
(d) P - II, Q-I, R-III, S-IV [ME 2014 : 2 M, Set-2]

8.24 The function of interpolator in a CNC machine controller is to

(a) control spindle speed
(b) coordinate feed rates of axes
(c) control tool rapid approach speed
(d) perform Miscellaneous (M) functions (tool change, coolant control etc.) [2015 : 1 M, Set-1]

8.25 A drill is positioned at point P and its has to proceed to point Q. The coordinates of point Q in the incremental system of defining position of a point in CNC part program will be

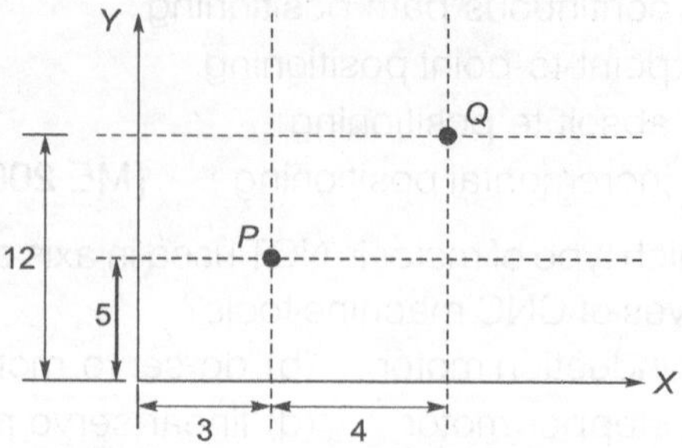

(a) (3, 12) (b) (5, 7)
(c) (7, 12) (d) (4, 7)

[2015 : 1 M, Set-3]

8.26 A triangular face in a CAD model has vertices: $P1(0, 0, 0)$; $P2(1, 1, 0)$ and $P3(1, 1, 1)$. The area of the facet is

(a) 0.500 (b) 0.707
(c) 1.414 (d) 1.732

[2015 : 2 M, Set-1]

8.27 In a CNC milling operation, the tool has to machine the circular arc from point (20, 20) to (10, 10) at sequence number 5 of the CNC part program. If the center of the arc is at (20, 10) and the machine has incremental mode of defining position coordinates, the correct tool path command is

(a) N 05 G90 G01 X–10 Y–10 R10
(b) N 05 G91 G03 X–10 Y–10 R10
(c) N 05 G90 G03 X20 Y20 R10
(d) N 05 G91 G02 X20 Y20 R10

[2015 : 2 M, Set-3]

8.28 Match the following part programming codes with their respective functions

Part Programming Codes

P. G01 Q. G03
R. M03 S. M05

Functions

I. Spindle stop
II. Spindle rotation, clockwise
III. Circular interpolation, anticlockwise
IV. Linear interpolation

(a) P–II, Q–I, R–IV, S–III
(b) P–IV, Q–II, R–III, S–I
(c) P–IV, Q–III, R–II, S–I
(d) P–III, Q–IV, R–II, S–I [2016 : 1 M, Set-3]

8.29 A point P (1, 3, –5) is translated by $2\hat{i} + 3\hat{j} - 4\hat{k}$ and then rotated counter clockwise by 90° about the z-axis. The new position of the point is:

(a) (– 6, 3, – 9) (b) (– 6, – 3, – 9)
(c) (6, 3, – 9) (d) (6, 3, 9)

[2016 : 2 M, Set-3]

8.30 For the situation shown in the figure below the expression for H in terms of r, R and D is

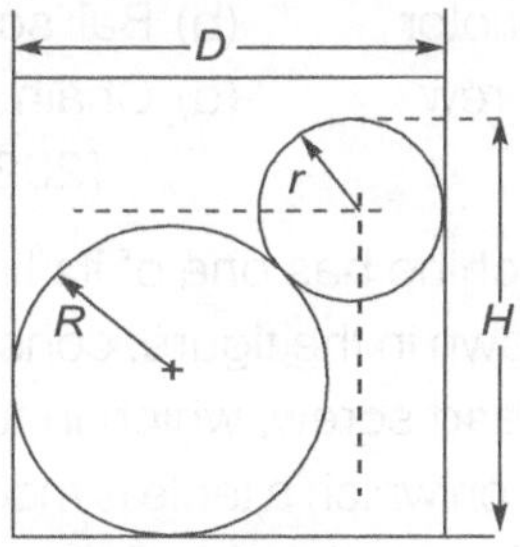

(a) $H = D + \sqrt{r^2 + R^2}$

(b) $H = (R + r) + (D + r)$

(c) $H = (R + r) + \sqrt{D^2 - R^2}$

(d) $H = (R + r) + \sqrt{2D(R + r) - D^2}$

[2016 : 2 M, Set-2]

8.31 The figure below represents a triangle PQR with initial coordinates of the vertices as P(1,3), Q(4, 5) and R(5.3.5). The triangle is rotated in XY plane about the vertex P by angle θ in clockwise direction. If $\sin\theta = 0.6$ and $\cos\theta = 0.8$, the new coordinates of the vertex Q are

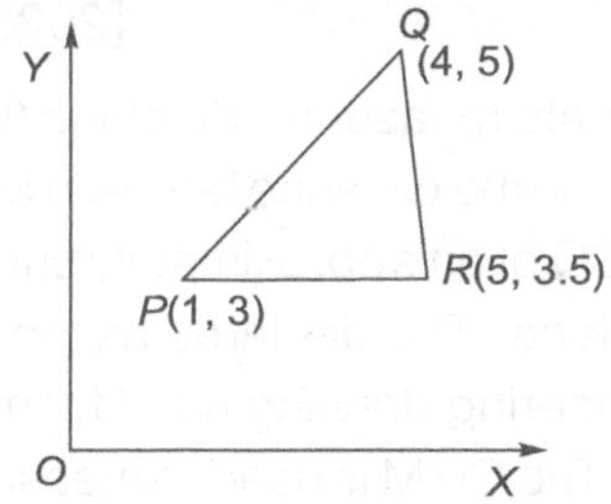

(a) (4.6, 2.8) (b) (3.2, 4.6)
(c) (7.9, 5.5) (d) (5.5, 7.9)

[2016 : 2 M, Set-1]

8.32 Circular arc on a part profile is being machined on a vertical CNC milling machine, CNC part program using metric units with absolute dimensions is listed below:

....

```
N60 G01 X 30 Y 55 Z 5 F50
N70 G02 X 50 Y 35 R 20
N80 G01 Z 5
```

....

The coordinates of the centre of the circular arc are:

(a) (30, 55) (b) (50, 55)
(c) (50, 35) (d) (30, 35)

[2017 : 2 M, Set-1]

8.33 Interpolator in a CNC machine

(a) controls spindle speed
(b) coordinates axes movements
(c) operates tool changer
(d) commands canned cycle

[2018 : 1 M, Set-1]

8.34 The position vector $\overrightarrow{OP}$ of P(20, 10) is rotated anti-clockwise in X-Y plane by an angle $\theta = 30°$ such that the point P occupies position Q, as shown in the figure. The coordinates (x, y) of Q are

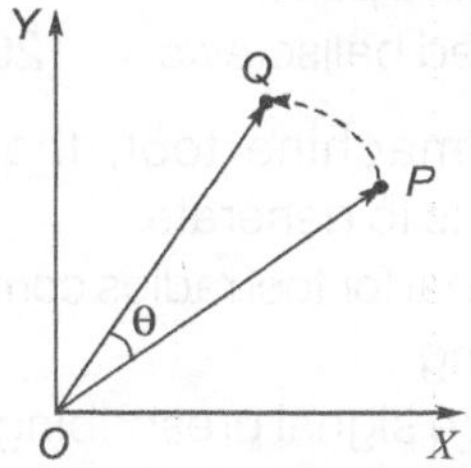

(a) (13.40, 22.32) (b) (18.66, 12.32)
(c) (22.32, 8.26) (d) (12.32, 18.66)

[2019 : 1 M, Set-1]

8.35 The transformation matrix for mirroring a point in x-y plane about the line $y = x$ is given by

(a) $\begin{bmatrix} -1 & 0 \\ 0 & 1 \end{bmatrix}$ (b) $\begin{bmatrix} 0 & 1 \\ 1 & 0 \end{bmatrix}$

(c) $\begin{bmatrix} 0 & -1 \\ -1 & 0 \end{bmatrix}$ (d) $\begin{bmatrix} 1 & 0 \\ 0 & -1 \end{bmatrix}$

[2019 : 1 M, Set-2]

8.36 A point P on a CNC controlled XY-stage is moved to another point 'Q' using the coordinate system shown in the figure below and rapid positioning command (G00).

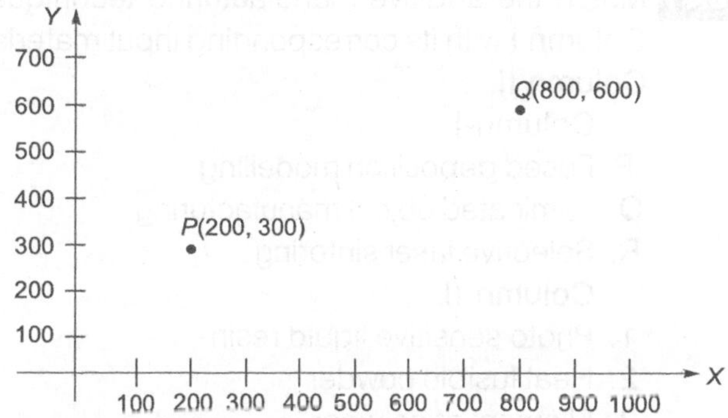

A pair of stepping motors with maximum speed of 800 rpm, controlling both the X and Y motion of the stage, are directly coupled to a pair of lead screw, each with a uniform pitch of 0.5 mm. The time needed to position the point 'P' to the point 'Q' is ____ minutes. (round off to 2 decimal places).

[2020 : 2 M, Set-2]

8.37 The XY table of a NC machine tool is to move from P(1, 1) to Q(51, 1); all coordinates are in mm. The pitch of the NC drive leadscrew is 1 mm. If the backlash between the leadscrew and the nut is 1.8°, then the total backlash of the table on moving from P to Q is _______ mm [Round off to two decimal places] [2021 : 2 M, Set-1]

8.38 In modern CNC machine tools, the backlash has been eliminated by
(a) slider crank mechanism
(b) rack and pinion
(c) ratchet and pinion
(d) preloaded ballscrews [2021 : 1 M, Set-1]

8.39 In a CNC machine tool, the function of an interpolator is to generate
(a) error signal for tool radius compensation during machining
(b) reference signal prescribing the shape of the part to be machined
(c) signal for the lubrication pump during machining
(d) NC code from the part drawing during post processing [2021 : 1 M, Set-2]

8.40 A CNC worktable is driven in a linear direction by a lead screw connected directly to a stepper motor. The pitch of the lead screw is 5 mm. The stepper motor completes one full revolution upon receiving 600 pulses. If the worktable speed is 5 m/minute and there is no missed pulse, then the pulse rate being received by the stepper motor is
(a) 20 kHz (b) 10 kHz
(c) 3 kHz (d) 15 kHz
[2022 : 1 M, Set-1]

8.41 Match the additive manufacturing technique in **Column I** with its corresponding input material in **Column II**.

Column-I
P. Fused deposition modelling
Q. Laminated object manufacturing
R. Selective laser sintering

Column-II
1. Photo sensitive liquid resin
2. Heat fusible powder
3. Filament of polymer
4. Sheet of thermoplastic or green compacted metal sheet

(a) P-3, Q-4, R-2
(b) P-1, Q-2, R-4
(c) P-2, Q-3, R-1
(d) P-4, Q-1, R-4
[2022 : 1 M, Set-2]

8.42 Which one of the following CANNOT impart linear motion in a CNC machine?
(a) Linear motor (b) Ball screw
(c) Lead screw (d) Chain and sprocket
[2022 : 1 M, Set-2]

8.43 A CNC machine has one of its linear positioning axes as shown in the figure, consisting of a motor rotating a lead screw, which in turn moves a nut horizontally on which a table is mounted. The motor moves in discrete rotational steps of 50 steps per revolution. The pitch of the screw is 5 mm and the total horizontal traverse length of the table is 100 mm. What is the total number of controllable locations at which the table can be positioned on this axis?

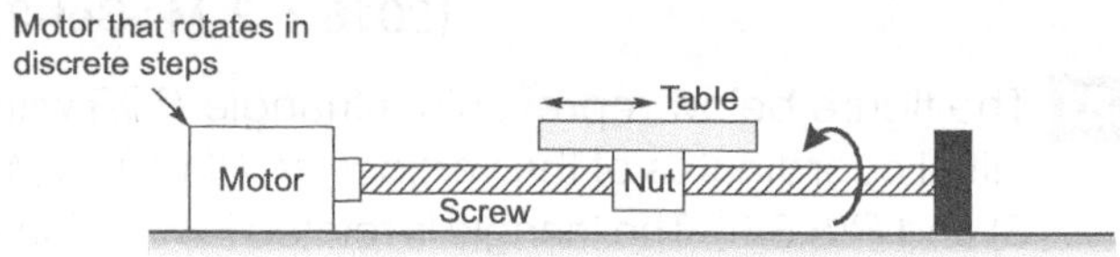

(a) 5000 (b) 2
(c) 1000 (d) 200
[2023 : 2 Marks]

8.44 A coordinate measuring machine (CMM) is used to determine the distance between Surface SP and Surface SQ of an approximately cuboidal shaped part. Surface SP is declared as the datum as per the engineering drawing used for manufacturing this part. The CMM is used to measure four points P1, P2, P3, P4 on Surface SP, and four points Q1, Q2, Q3, Q4 on Surface SQ as shown. A regression procedure is used to fit the necessary planes. The distance between the two fitted planes is_______ mm. (Answer in integer)

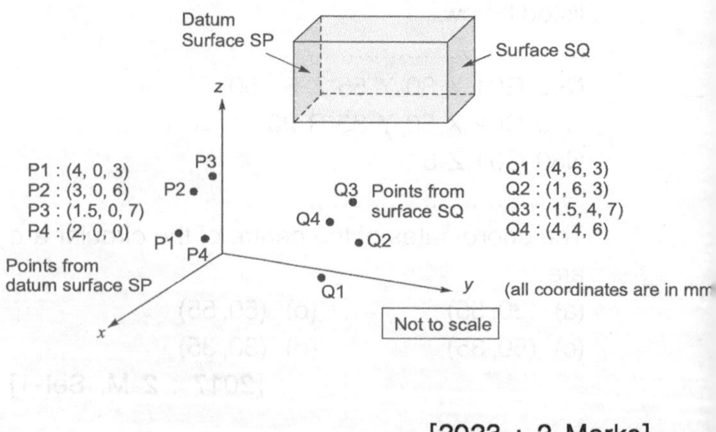

[2023 : 2 Marks]

8.45 A solid part (see figure) of polymer material is to be fabricated by additive manufacturing (AM) in square-shaped layers starting from the bottom of the part working upwards. The nozzle diameter of the AM machine is $a/10$ mm and the nozzle follows a linear serpentine path parallel to the sides of the square layers with a feed rate of $a/5$ mm/min.

Ignore any tool path motions other than those involved in adding material, and any other delays between layers or the serpentine scan lines.

The time taken to fabricate this part is ________ minutes. (Answer in integer)

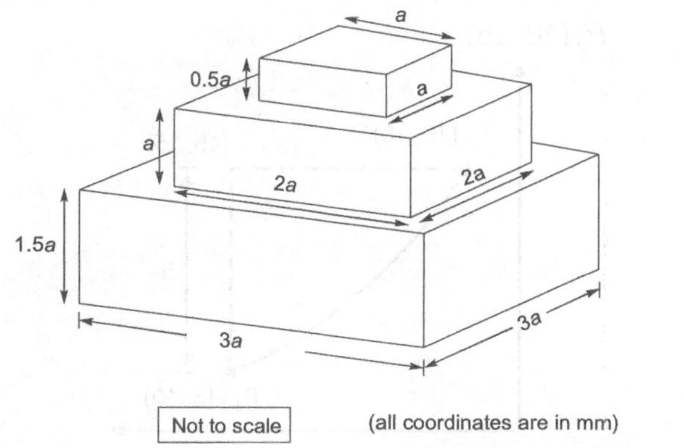

[2023 : 2 Marks]

8.46 The preparatory functions in Computer Numerical Controlled (CNC) machine programme are denoted by the alphabet

(a) P (b) O
(c) M (d) G **[2024 : 1 M]**

■■■■

Answers NC, CNC, DNC, APT, Automation

8.1	(b)	8.2	(56250)	8.3	(d)	8.4	(c)	8.5	(a)	8.6	(b)	8.7	(a)	8.8	(c)
8.9	(a)	8.10	(d)	8.11	(a)	8.12	(a)	8.13	(c)	8.14	(a)	8.15	(a)	8.16	(c)
8.17	(c)	8.18	(b)	8.19	(c)	8.20	(a)	8.21	(b)	8.22	(b)	8.23	(a)	8.24	(b)
8.25	(d)	8.26	(b)	8.27	(b)	8.28	(c)	8.29	(a)	8.30	(d)	8.31	(a)	8.32	(d)
8.33	(b)	8.34	(d)	8.35	(b)	8.36	(1.5)	8.37	(0.25)	8.38	(d)	8.39	(b)	8.40	(b)
8.41	(a)	8.42	(d)	8.43	(c)	8.44	(5)	8.45	(9000)	8.46	(d)				

Explanations NC, CNC, DNC, APT, Automation

8.1 (b)

For 1 rev. of lead screw slide will move by 3.6 mm

1 rev. = 360° = 360 pulses

360 pulses will cause motion of 3.6 mm

1 pulse will result in $\frac{3.6}{360}$ mm of motion

i.e., **0.01 mm or 10 μm.**

8.2 (56250)

$$V = \omega r = V = 2\pi Nr = \pi DN$$

$$N = \frac{150 \times 10^3}{\pi \times 155} = 308 \text{ rpm}$$

Programmed feed rate = $f \times N$ = 0.15 × 308 = **46.2 mm/min**

$$\text{Material removal rate} = f \times V \times \left(\frac{D_2 - D_1}{2}\right)$$

$$= 0.15 \times 150 \times 10^3 \times 2.5 = \mathbf{56250\ mm^3/min}$$

8.4 (c)

Cellular manufacturing is a part of group technology. Layout in the production facility is divided into various cell. Each cell is dedicated to a particular process.

8.6 (b)

Dia of end milling cutter = 10 mm

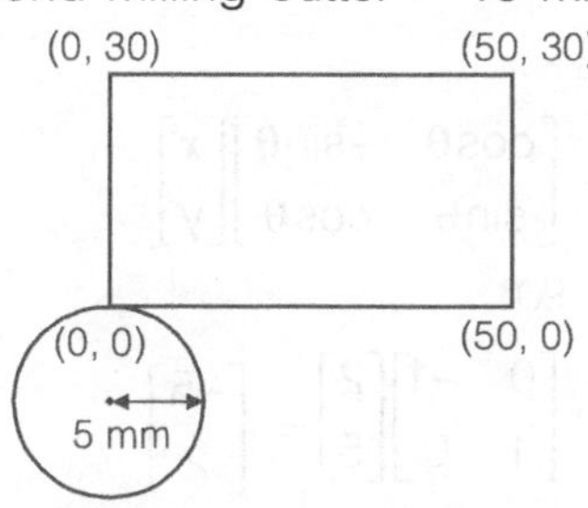

8.8 (c)

Given, P_1 (15, 10)

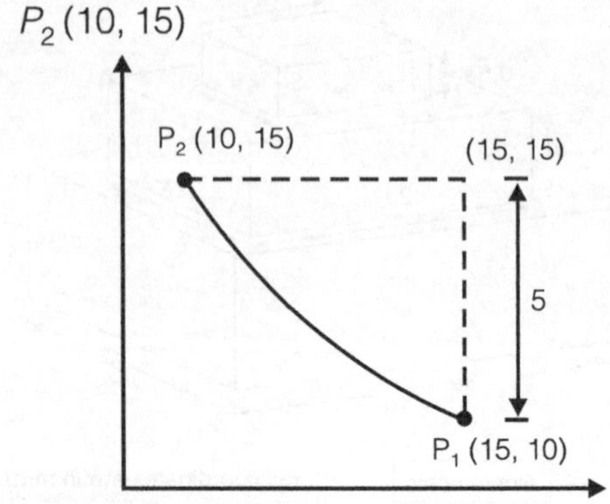

Clearly from figure

Centre of arc having radius 5 is
(15, (10+5)) = **(15, 15)**

8.15 (a)

Initially velocity was 10000 × 5 μm/min. Now velocity will be 5000 × 10 μm/min. So velocity of the table remains same.

8.19 (c)

Correct data structure for solid models is given by:

Vertices → edges → faces → solid parts

8.20 (a)

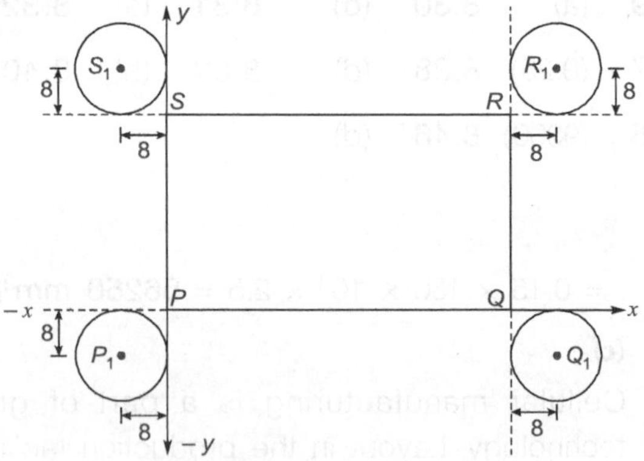

Normal Coordinates :

P(0, 0), Q(100, 0), R(100, 50), S(0, 50)

With cutter compensation :

P_1(–8, –8), Q(108, –8), R(108, 58), S(–8, 58)

8.21 (b)

$$\begin{bmatrix} x' \\ y' \end{bmatrix} = \begin{bmatrix} \cos\theta & -\sin\theta \\ \sin\theta & \cos\theta \end{bmatrix} \begin{bmatrix} x \\ y \end{bmatrix}$$

Here $\theta = 90°$

$$\begin{bmatrix} x' \\ y' \end{bmatrix} = \begin{bmatrix} 0 & -1 \\ 1 & 0 \end{bmatrix} \begin{bmatrix} 2 \\ 5 \end{bmatrix} = \begin{bmatrix} -5 \\ 2 \end{bmatrix}$$

8.25 (d)

∵ Incremental system of defining position of a point in *CNC* part program is used.

$\therefore$ Q(4, (12 – 5))

Q (4, 7)

8.26 (b)

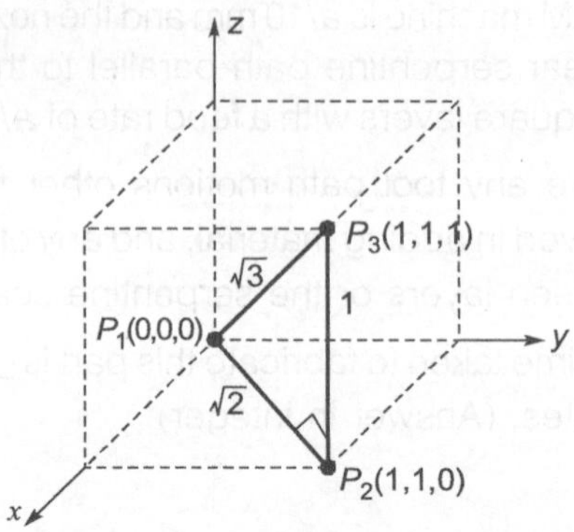

$$\text{Area} = \frac{1}{2} \times \sqrt{2} \times 1 = \frac{\sqrt{2}}{2} = \mathbf{0.707}$$

8.27 (b)

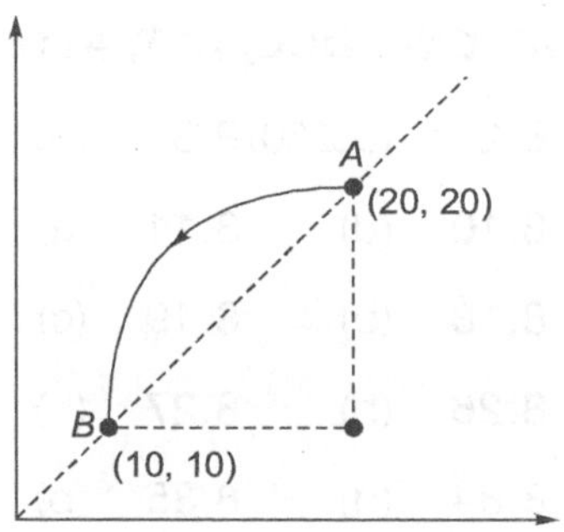

Radius of circle = 10

Incremental coordinate system is used. Anti-clockwise is the motion from *A* to *B*.

∴ (b) is the correct answer.

8.29 (a)

$$P = (1, 3, -5)$$

First translated by $2\hat{i} + 3\hat{j} - 4\hat{k}$

and then rotated about *z* axis by 90°,

Homogeneous Translation Matrix.

$$\left[\begin{array}{ccc|c} 1 & 0 & 0 & 2 \\ 0 & 1 & 0 & 3 \\ 0 & 0 & 1 & -4 \\ \hline 0 & 0 & 0 & 1 \end{array}\right] \begin{bmatrix} 1 \\ 3 \\ -5 \\ 1 \end{bmatrix} = \begin{bmatrix} 3 \\ 6 \\ -9 \end{bmatrix}$$

Then rotated about *z* axis,

$$\begin{bmatrix} \cos\theta & -\sin\theta & 0 \\ \sin\theta & \cos\theta & 0 \\ 0 & 0 & 1 \end{bmatrix} \begin{bmatrix} 3 \\ 6 \\ -9 \end{bmatrix}$$

$$\begin{bmatrix} 0 & -1 & 0 \\ 1 & 0 & 0 \\ 0 & 0 & 1 \end{bmatrix}\begin{bmatrix} 3 \\ 6 \\ -9 \end{bmatrix} = \begin{bmatrix} -6 \\ 3 \\ -9 \end{bmatrix} = [-6, 3, -9]$$

8.30 (d)

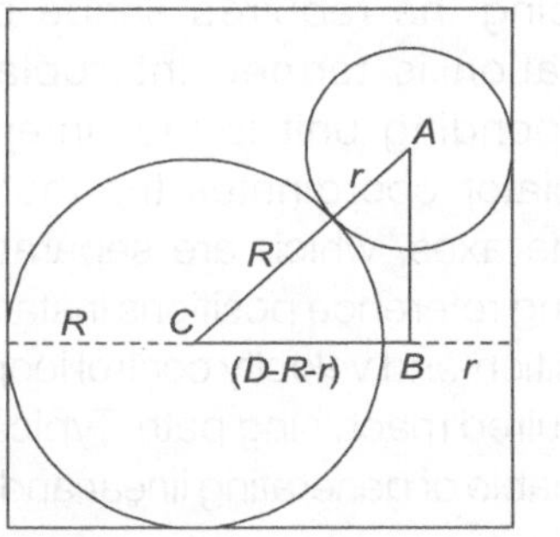

$$AB = \sqrt{(R+r)^2 - (D-R-r)^2}$$

$$H = R + AB + r$$

$$= R + r + \sqrt{(R+r+D-R-r)(R+r-D+R+r)}$$

$$= R + r + \sqrt{\begin{matrix}(R+r+D-R-r) \\ (R+r-D+R+r)\end{matrix}}$$

$$= R + r + \sqrt{D[2(R+r)-D]}$$

$$= R + r + \sqrt{2D(R+r) - D^2}$$

8.31 (a)

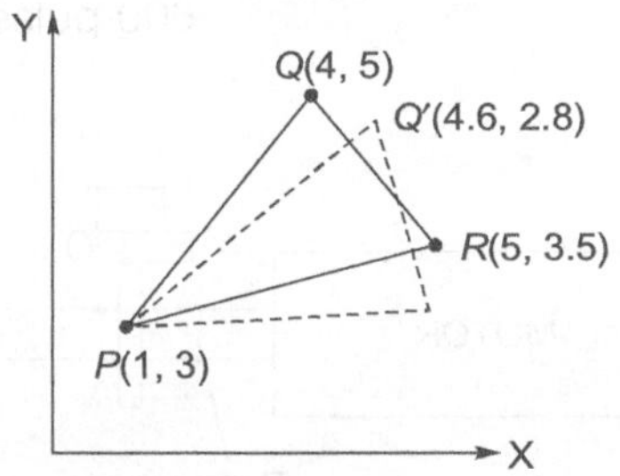

Step 1: Translate p to origin i.e., (0, 0)

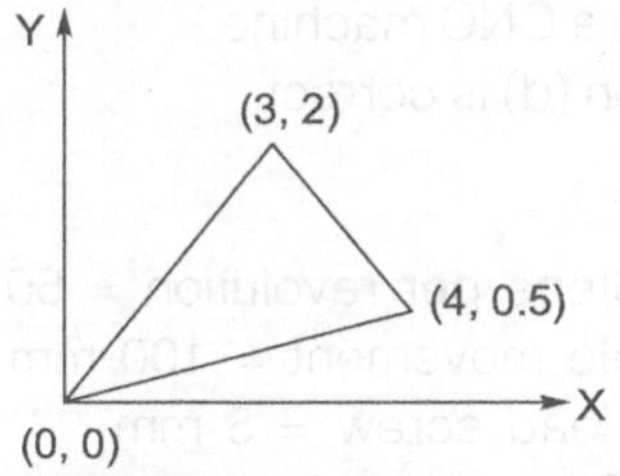

Step 2: Rotate about P

$$\begin{bmatrix} \cos\theta & \sin\theta \\ -\sin\theta & \cos\theta \end{bmatrix}\begin{bmatrix} 3 \\ 2 \end{bmatrix} = \begin{bmatrix} 0.8 & 0.6 \\ -0.6 & 0.8 \end{bmatrix}\begin{bmatrix} 3 \\ 2 \end{bmatrix} = \begin{bmatrix} 3.6 \\ -0.2 \end{bmatrix}$$

Step 3 : Translate it back,

$Q_X = 3.6 + 1 = 4.6$

$Q_Y = -0.2 + 3 = 2.8$

8.32 (d)

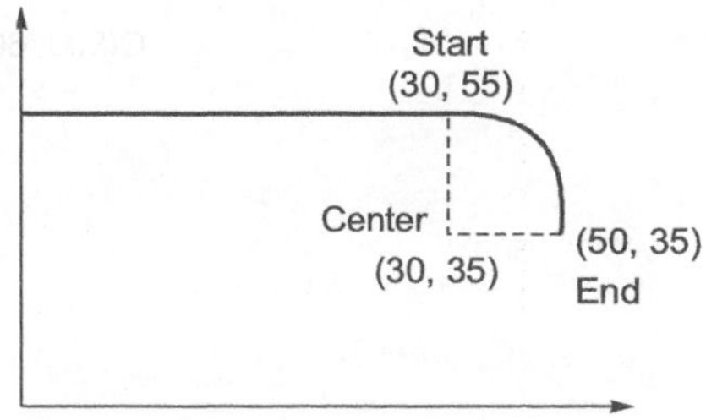

8.33 (b)

An interpolator provides two functions:

1. It computes individual axis velocities to drive the tool along the programmed path at given feed rate.
2. It generates intermediate coordinate positions along the programmed path.

8.34 (d)

From given diagram,

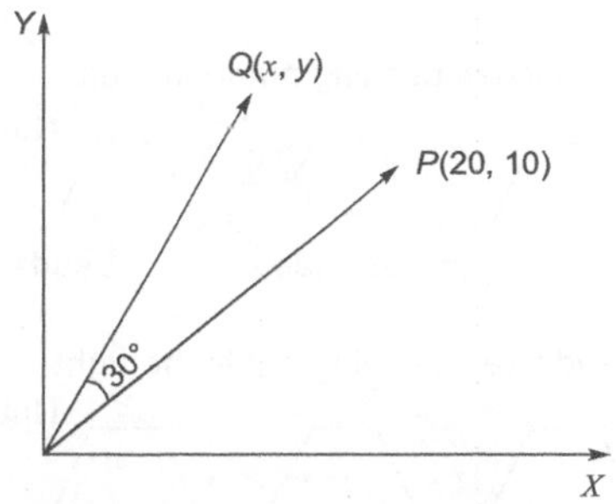

Rotation about x-y plane -counterclockwise

$$= \begin{bmatrix} \cos\theta & -\sin\theta \\ \sin\theta & \cos\theta \end{bmatrix}\begin{bmatrix} 20 \\ 10 \end{bmatrix} = \begin{bmatrix} \cos 30 & -\sin 30 \\ \sin 30 & \cos 30 \end{bmatrix}\begin{bmatrix} 20 \\ 10 \end{bmatrix}$$

$$= \begin{bmatrix} 0.866 & -0.5 \\ 0.5 & 0.866 \end{bmatrix}\begin{bmatrix} 20 \\ 10 \end{bmatrix} = \begin{bmatrix} 12.32 \\ 18.66 \end{bmatrix}$$

8.35 (b)

The transformation matrix for mirroring a point in x-y plane about the line $y = x$ is

$$\begin{bmatrix} 0 & 1 \\ 1 & 0 \end{bmatrix}$$

Option (b) is correct.

8.36 (1.5)

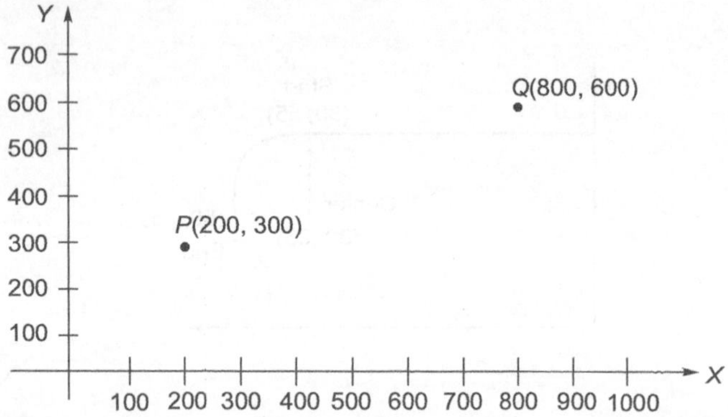

N = 800 rpm, P = 0.5 mm/rev

$$V = N \times P = \text{rev/min} \times \text{mm/rev}$$
$$= 400 \text{ mm/min}$$
$$\Delta t_x = \frac{600}{400} = 1.5 \text{ min}$$
$$\Delta t_y = \frac{300}{400} = 0.75 \text{ min}$$

There are two stepper motors so both will work till 0.75 min then *y* axis motor will stop then only *x* axis motor will run for 0.75 more, so total time will be 1.5 min.

8.37 (0.25)

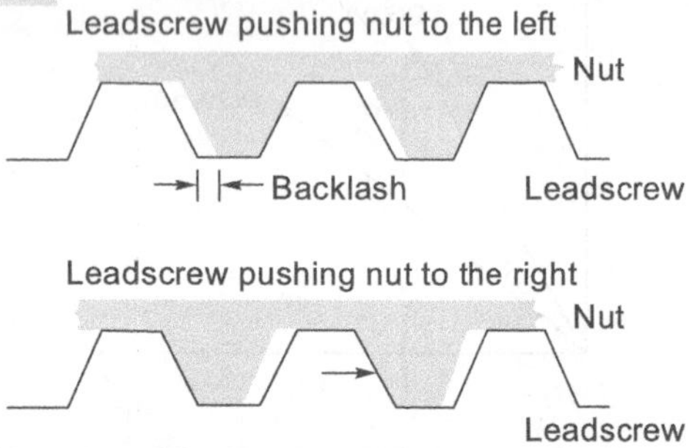

P(1, 1) to P(51, 1) is 50 mm distance. As pitch of the leadscrew is 1 mm. It has to move 50 rotation and in each rotation, backlash is 1.8 degree. Therefore total backlash = 1.8 × 50 degree.

In one rotation i.e. 360° rotation - 1 mm movement take place

$$\text{In } 1.8 \times 50° = \frac{1.8 \times 50}{360} \text{ mm} = 0.25 \text{ mm}$$

8.38 (d)

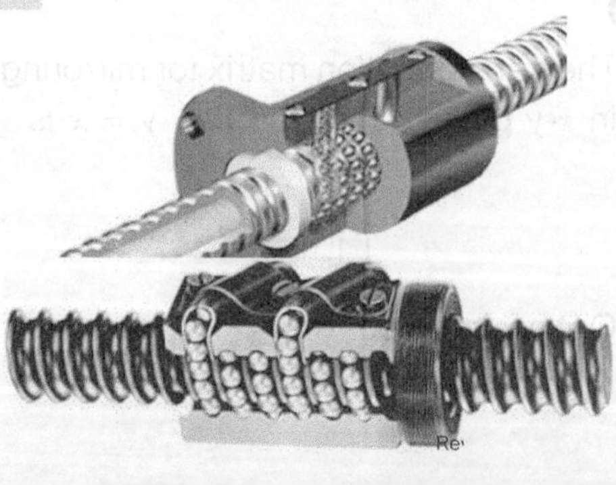

8.39 (b)

In contouring systems the machining path is usually constructed from a combination of linear and circular segments. It is only necessary to specify the coordinates of the initial and final points of each segment, and the feed rate. The operation of producing the required shape based on this information is termed interpolation and the corresponding unit is the "interpolator". The interpolator coordinates the motion along the machine axes, which are separately driven, by providing reference positions instant by instant for the position-and velocity control loops, to generate the required machining path. Typical interpolators are capable of generating linear and circular paths.

8.40 (b)

Table speed = BLU × Frequency (f)

$$\Rightarrow \quad f = \frac{\text{Table Speed}}{\text{B.L.U.}}$$

Given 600 pulses = 5 mm

$$\Rightarrow \quad 1 \text{ pulse} = \frac{5}{600} \text{ mm/pulses} = \text{B.L.U.}$$
$$\therefore \quad f = \frac{\text{T.S}}{\text{B.L.U}}$$
$$\Rightarrow \quad f = \frac{5000 \text{ mm}}{60 \sec \times \dfrac{5 \text{ mm}}{600 \text{ pulses}}} = 10 \text{ kHz}$$

8.42 (d)

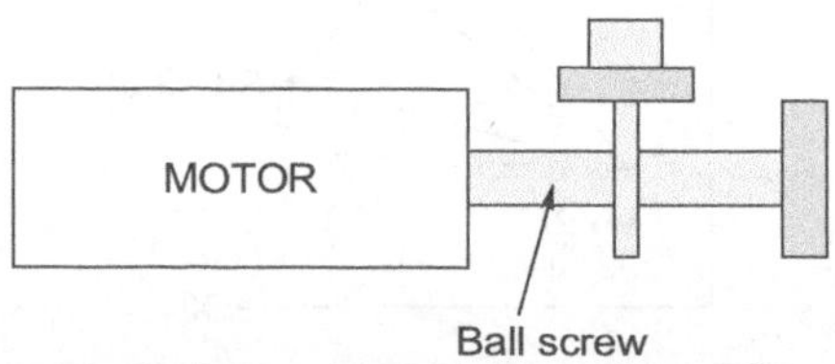

Except chain and sprocket, all can impact linear motion in a CNC machine.

So, option (d) is correct.

8.43 (c)

Given, Steps per revolution = 50

Total table movement = 100 mm

Pitch of lead screw = 5 mm

So, In 50 steps lead screw moves 5 mm

50 steps = 5 mm

hence 10 steps are required to move 1 mm.

So, number of steps required are = 100 × 10 = 1000

8.44 (5) (4.999 to 5.001)

Points P1, P2, P3, P4 lie on *xz* plane as their y-coordinate is zero. Thus distances of Q1,

Q2, Q3, Q4 from the SP surface are as described below.

Distance of Q1 from x-z plane = 6

Distance of Q2 from x-z plane = 6

Distance of Q3 from x-z plane = 4

Distance of Q4 from x-z plane = 4

Thus, Avarage distance between the two fitted planes = 5 mm

8.45 (9000) (8999.999 to 9000.001)

Given, feed rate, $f = a/5$ mm/min

$d = a/10$ mm

Total volume, $V = (3a)^2 \times 1.5a + (2a)^2 \times a + a^2 \times 0.5a$

$= 18a^3$

$$T = \frac{V}{fd^2} = \frac{18a^3}{\left(\frac{a}{10}\right)^2 \times \left(\frac{a}{5}\right)}$$

$$= \frac{18a^3}{a^3} \times 100 \times 5 = 9000 \text{ min}$$

8.46 (d)

Preparatory functions of CNC programming are denoted by G-codes. Preparatory functions are the G-codes that identify the type of activities the machine will execute. The preparatory functions describe the way in which the machine axes have to move, the method of interpolation, the dimension system, the time delay of program execution and the activation of specific operational modes in the control.

■■■■

9 Nonconventional Machining (EDM, High Energy Rate Forming)

9.1 Selection of electrolyte for ECM is as follows
(a) non-passivating electrolyte for stock removal and passivating electrolyte for finish control
(b) passivating electrolyte for stock removal and non-passivating electrolyte for finish control
(c) selection of electrolyte is dependent on current density
(d) electrolyte selection is based on tool work electrodes **[ME 1997 : 1 M]**

9.2 Inter electrode gap in ECG is controlled by
(a) controlling the pressure of electrolyte flow
(b) controlling the applied static load
(c) controlling the size of diamond particle in the wheel
(d) controlling the texture of the work
[ME 1997 : 1 M]

9.3 Match the following lists

List-I
A. ECM
B. EDM
C. USM
D. LBM

List-II
1. Plastic shear
2. Erosion/brittle fracture
3. Corrosive reaction
4. Melting vaporization
5. Ion displacement
6. Plastic shear and ion displacement
[ME 1998 : 2 M]

9.4 Match the following lists

List-I
A. Aluminium brake shoe
B. Plastic water bottles
C. Stainless steel cups
D. Soft drink can (aluminum)

List-II
1. Deep drawing
2. Blow moulding
3. Centrifugal casting
4. Impact extrusion
5. Sand casting
6. Upset forging
[ME 1998 : 2 M]

9.5 In EDM, the tool is made of
(a) Copper
(b) HSS
(c) Cast iron
(d) Plain carbon steel **[ME 1999 : 1 M]**

9.6 Choose the correct statement
(a) Fixture is to guide the tool as well as to locate and clamp the work piece
(b) Jig is to guide the tool as well as to locate and clamp the work piece
(c) Jigs are used on CNC machines to locate and clamp the workpiece and guide the tool
(d) No arrangement to guide the tool is provided in jig **[ME 1999 : 1 M]**

9.7 Deep hole drilling of small diameter, say 0.2 mm is done with EDM by selecting the tool material as
(a) Copper wire (b) Tungsten wire
(c) Brass wire (d) Tungsten carbide
[ME 2000 : 1 M]

9.8 In ECM the MRR is due to
(a) Corrosion (b) Erosion
(c) Fusion (d) Ion displacement
[ME 2001 : 1 M]

9.9 3-2-1 method of location in jig or fixture would collectively restrict the workpiece in '*n*' degree of freedom, where the value of '*n*' is
(a) 6 (b) 8
(c) 9 (d) 12 **[ME 2001 : 2 M]**

9.10 Estimate the MRR (in cc/sec) of an alloy containing 18% cobalt, 62% nickel and 20% chromium during ECM with current of 500 ampere. The density of the alloy is 8.28 g/cc. The following data is available.

Metal	Gram atomic weight	Velency
Cobalt	58.93	2
Nickel	58.71	2
Chromium	51.99	6

Assume Faraday's constant is 96500 coulombs/mole. **[ME 2002 : 2 M]**

9.11 As tool and work are not in contact in EDM process
(a) no relative motion occurs between them
(b) no wear of tool occurs
(c) no power is consumed during metal cutting
(d) no force between tool and work occurs
[ME 2003 : 1 M]

9.12 The mechanism of material removal in EDM process is
(a) melting and evaporation
(b) melting and corrosion
(c) erosion and cavitation
(d) cavitation and evaporation [ME 2004 : 1 M]

9.13 Typical machining operations are to be performed on hard-to-machine materials by using the processes listed below. Choose the best set of Operation- Process combinations

Operation
P. Deburring (internal surface)
Q. Die sinking
R. Fine hole drilling in thin sheets
S. Tool sharpening

Process
1. Plasma Arc Machining
2. Abrasive Flow Machining
3. Electric Discharge Machining
4. Ultrasonic Machining
5. Laser beam Machining
6. Electrochemical Grinding

(a) P–1, Q–5, R–3, S–4
(b) P–1, Q–4, R–1, S–2
(c) P–5, Q–1, R–2, S–6
(d) P–2, Q–3, R–5, S–6 [ME 2004 : 2 M]

9.14 A zigzag cavity in a block of high strength alloy is to be finish machined. This can be carried out by using

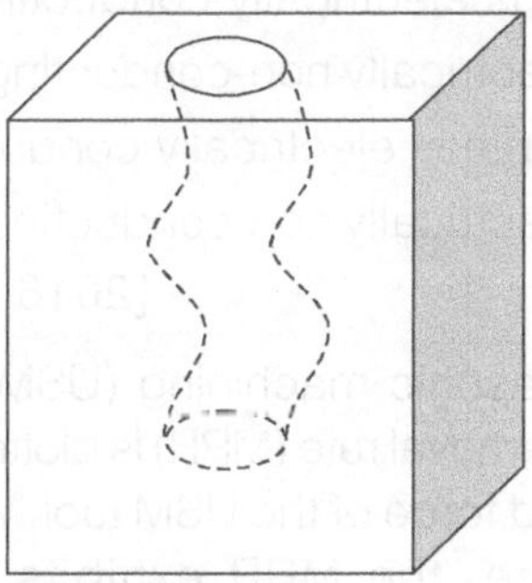

(a) electric discharge machining
(b) electro-chemical machining
(c) laser beam machining
(d) abrasive flow machining [ME 2005 : 1 M]

9.15 Arrange the processes in the increasing order of their maximum material removal rate.
Electrochemical Machining (ECM)
Ultrasonic Machining (USM)
Electron Beam Machining (EBM)
Laser Beam Machining (LBM)
Electric Discharge Machining (EDM)
(a) USM, LBM, EBM, EDM, ECM
(b) EBM, LBM, USM, ECM, EDM
(c) LBM, EBM, USM, ECM, EDM
(d) LBM, EBM, USM, EDM, ECM
[ME 2006 : 2 M]

9.16 In Electrodischarge Machining (EDM), if the thermal conductivity of tools is high and the specific heat of work piece is low, then the tool wear rate and material removal rate are expected to be respectively
(a) high and high (b) low and low
(c) high and low (d) low and high
[ME 2007 : 2 M]

9.17 Match the most suitable manufacturing processes for the following parts.

Parts
P. Computer chip
Q. Metal forming dies and moulds
R. Turbine blade
S. Glass

Manufacturing Processes
1. Electrochemical Machining
2. Ultrasonic Machining
3. Electrodischarge Machining
4. Photochemical Machining

(a) P–4, Q–3, R–1, S–2
(b) P–4, Q–3, R–2, S–1
(c) P–3, Q–1 R–4, S–2
(d) P–1, Q–2, R–4, S–3 [ME 2007 : 2 M]

9.18 A researcher conducts electrochemical machining (ECM) on a binary alloy (density 6000 kg/m^3) of iron (atomic weight 56, valency 2) and metal *P* (atomic weight 24, valency 4). Faraday's constant = 96500 coulomb/mole. Volumetric material removal rate of the alloy is 50 mm^3/s at a current of 2000 A. The percentage of the metal *P* in the alloy is closest to
(a) 40 (b) 25
(c) 15 (d) 79 [ME 2008 : 2 M]

9.19 Electrochemical machining is performed to remove material from an iron surface of 20 mm × 20 mm

under the following conditions:
Inter electrode gap = 0.2 mm
Supply voltage (DC) = 12 V
Specific resistance of electrolyte = 2 Ω cm
Atomic weight of iron = 55.85
Valency of Iron = 2
Faraday's constant = 96540 Coulombs
The material removal rate (in g/s) is
(a) 0.3471 (b) 3.471
(c) 34.71 (d) 347.1
[ME 2009 : 2 M]

9.20 In abrasive jet machining, as the distance between the nozzle tip and the work surface increases, the material removal rate
(a) increases continuously
(b) decreases continuously
(c) decreases, becomes stable and then increases
(d) increases, becomes stable and then decreases
[ME 2012 : 1 M]

9.21 Match the following non-traditional machining processes with the corresponding material removal mechanisms:

Machining process
P. Chemical machining
Q. Electro-chemical machining
R. Electro-discharge machining
S. Ultrasonic machining

Mechanism of material removal
1. Erosion
2. Corrosive reaction
3. Ion displacement
4. Fusion and vaporization

(a) P–2, Q–3, R–4, S–1
(b) P–2, Q–4, R–3, S–1
(c) P–3, Q–2, R–4, S–1
(d) P–2, Q–3, R–1, S–4 [ME 2011 : 2 M]

9.22 During the electrochemical machining (ECM) of iron (atomic weight = 56, valency = 2) at current of 1000 A with 90% current efficiency, the material removal rate was observed to be 0.26 cc/s. If Titanium (atomic weight = 48, valency = 3) is machined by the ECM process at the current of 2000 A with 90% current efficiency, the expected material removal rate in gm/s will be
(a) 0.11 (b) 0.23
(c) 0.30 (d) 0.52
[ME 2013 : 2 M]

9.23 The following four unconventional machining processes are available in a shop floor. The most appropriate one to drill a hole of square cross section of 6 mm × 6 mm and 25 mm deep is
(a) Abrasive Jet Machining
(b) Plasma Arc Machining
(c) Laser Beam Machining
(d) Electro Discharge Machining
[ME 2014 : 1 M, Set-2]

9.24 The process utilizing mainly thermal energy for removing material is
(a) Ultrasonic Machining
(b) Electrochemical Machining
(c) Abrasive Jet Machining
(d) Laser Beam Machining
[ME 2014 : 1 M, Set-3]

9.25 The principle of material removal in Electrochemical machining is
(a) Fick's law (b) Faraday's laws
(c) Kirchhoff''s laws (d) Ohm's law
[ME 2014 : 1 M, Set-4]

9.26 The primary mechanism of material removal in electrochemical machining (ECM) is
(a) chemical corrosion (b) etching
(c) ionic dissolution (d) spark erosion
[2015 : 1 M, Set-2]

9.27 A resistance-capacitance relaxation circuit is used in an electrical discharge machining process. The discharge voltage is 100 V. At a spark cycle time of 25 μs, the average power input required is 1 kW. The capacitance (in μF) in the circuit is
(a) 2.5 (b) 5.0
(c) 7.5 (d) 10.0
[2015 : 2 M, Set-2]

9.28 In a wire-cut EDM process the necessary conditions that have to be met for making a successful cut are that
(a) wire and sample are electrically non-conducting
(b) wire and sample are electrically conducting
(c) wire is electrically conducting and sample is electrically non-conducting
(d) sample is electrically conducting and wire is electrically non-conducting
[2016 : 1 M, Set-3]

9.29 In an ultrasonic machining (USM) process, the material removal rate (MRR) is plotted as a function of the feed force of the USM tool. With increasing feed force, the MRR exhibits the following behavior:
(a) increases linearly
(b) decreases linearly
(c) does not change

(d) first increases and then decreases [2016 : 1 M, Set-2]

9.30 The surface irregularities of electrodes used in an electrochemical machining (ECM) process are 3 μm and 6 μm as shown in the figure. If the work-piece is of pure iron and 12V DC is applied between the electrodes, the largest feed rate is ____mm/min.

Conductivity of the electrolyte	0.02 $ohm^{-1} mm^{-1}$
Over-potential voltage	1.5 V
Density of iron	7860 kg/m^3
Atomic weight of iron	55.85 gm

Assume the iron to be dissolved as Fe^{+2} and the Faraday constant to be 96500 Coulomb.

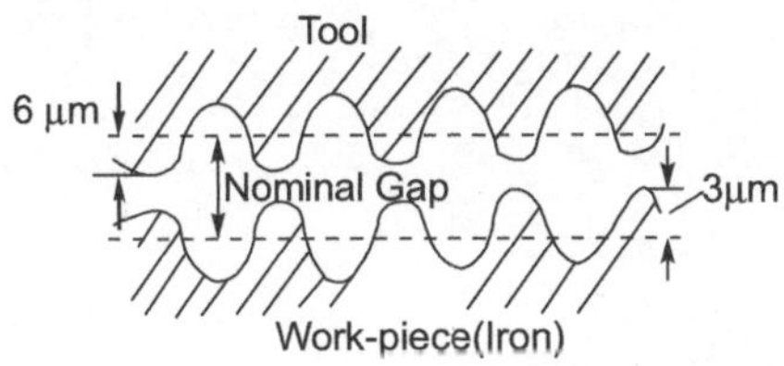

[2016 : 2 M, Set-2]

9.31 The non-traditional machining process that essentially requires vacuum is
(a) electron beam machining
(b) electro chemical machining
(c) electro chemical discharge machining
(d) electro discharge machining
[2016 : 1 M, Set-1]

9.32 Match the processes with their characteristics.

Process	Characteristics
P : Electrical Discharge Machining	1. No residual stress
Q : Ultrasonic machining	2. Machining of electrically conductive materials
R : Chemical machining	3. Machining of glass
S : Ion Beam Machining	4. Nano-machining

(a) P-2, Q-3, R-1, S-4
(b) P-3, Q-2, R-1, S-4
(c) P-3, Q-2, R-4, S-1
(d) P-2, Q-4, R-3, S-1 [2017 : 1 M, Set-1]

9.33 Which one of the following statements is TRUE for the ultrasonic machining (USM) process?
(a) In USM, the tool vibrates at subsonic frequency.
(b) USM does not employ magnetostrictive transducer.
(c) USM is an excellent process for machining ductile materials.
(d) USM often uses a slurry comprising abrasive-particles and water. [2017 : 1 M, Set-2]

9.34 An electrochemical machining (ECM) is to be used to cut a through hole into a 12 mm thick aluminium plate. The hole has a rectangular cross-section, 10 mm × 30 mm. The ECM operation will be accomplished in 2 minutes, with efficiency of 90%. Assuming specific removal rate for aluminium as 3.44×10^{-2} $mm^3/(A\ s)$, the current (in A) required is ________(correct to two decimal places).
[2018 : 2 M, Set-1]

9.35 Metal removal in electric discharge machining takes place through
(a) ion displacement (b) melting and vaporization
(c) corrosive reaction (d) plastic shear
[2018 : 1 M, Set-2]

9.36 A circular hole of 25 mm diameter and depth of 20 mm is machined by EDM process. The material removal rate (in mm^3/min) is expressed as $4 \times 10^4\ IT^{-1.23}$ where I = 300 A and the melting point of the material, T = 1600°C. The time (in minutes) for machining this hole is _______ (correct to two decimal places).
[2018 : 2 M, Set-2]

9.37 In an electrical discharge machining process, the breakdown voltage across inter electrode gap (IEG) is 200 V and the capacitance of the RC circuit is 50 μF. The energy (in J) released per spark across the IEG is _______ . [2019 : 1 M, Set-2]

9.38 The process, that uses a tapered horn to amplify and focus the mechanical energy for machining of glass, is
(a) electrochemical machining
(b) ultrasonic machining
(c) abrasive jet machining
(d) electrical discharge machining
[2020 : 1 M, Set-2]

9.39 A steel part with surface area of 125 cm^2 is to be chrome coaled through an electroplating process using chromium acid sulphate as an electrolyte. An increasing current is applied to the part according to the following current time relation:

$$I = 12 + 0.2t$$

where, I = current (A) and t = time (minutes). The part is submerged in the plating solution for a duration of 20 minutes for plating purpose. Assuming the cathode efficiency of chromium to be 15% and the plating constant of chromium acid sulphate to be 2.50×10^{-2} mm³/A·s, the resulting coating thickness on the part surface is ____ μm. (round off to one decimal place). [2020 : 2 M, Set-1]

9.40 The machining process that involves ablation is
(a) Laser Beam Machining
(b) Abrasive Jet Machining
(c) Chemical Machining
(d) Electrochemical Machining [2021 : 1 M, Set-2]

9.41 Electrochemical machining operations are performed with tungsten as the tool, and copper and aluminum as two different workpiece materials. Properties of copper and aluminum are given in the table below.

Material	Atomic mass (amu)	Valency	Density (g/cm³)
Copper	63	2	9
Aluminium	27	3	2.7

Ignore over potentials, and assume that current efficiency is 100% for both the workpiece materials. Under identical conditions, if the material removal rate (MRR) of copper is 100 mg/s, the MRR of aluminum will be ________ mg/s (round-off to two decimal places).
[2022 : 1 M, Set-1]

9.42 Which of these processes involve(s) melting in metallic workpieces?
(a) Electrochemical machining
(b) Electric discharge machining
(c) Laser beam machining
(d) Electron beam machining
[2022 : 1 M, Set-2]

■■■■

Answers Nonconventional Machining

9.1 (a) 9.2 (c) 9.3 (*) 9.4 (*) 9.5 (a) 9.6 (b, c) 9.7 (b) 9.8 (d)
9.9 (c) 9.10 (0.0124) 9.11 (d) 9.12 (a) 9.13 (d) 9.14 (a) 9.15 (d) 9.16 (d)
9.17 (a) 9.18 (b) 9.19 (a) 9.20 (d) 9.21 (a) 9.22 (c) 9.23 (d) 9.24 (d)
9.25 (b) 9.26 (c) 9.27 (b) 9.28 (b) 9.29 (d) 9.30 (51.542) 9.31 (a) 9.32 (a)
9.33 (d) 9.34 (968.992) 9.35 (b) 9.36 (7.1431) 9.37 (1) 9.38 (b)
9.39 (5.0) 9.40 (a) 9.41 (28.57) 9.42 (b, c, d)

Explanations Nonconventional Machining

9.1 (a)

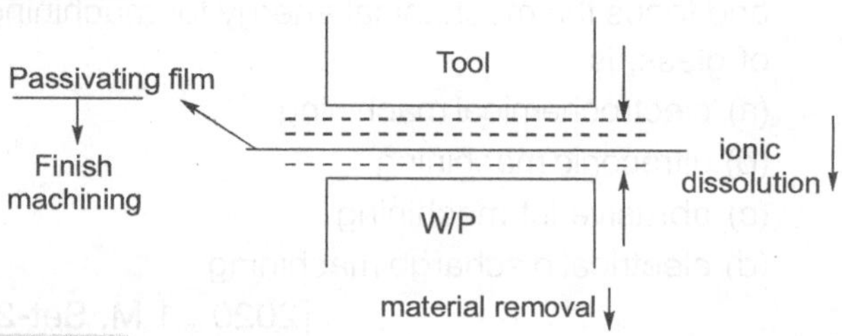

9.5 (a)

Tool for EDM are made of highly conductive material like copper, Brass, graphite etc.

9.6 (b, c)

In jigs provision for guiding tool is there along with locating and clamping arrangement for workpiece while fixtures are for locating and clamping workpiece only.

9.7 (b)

Tungsten has very high melting point therefore is more suitable.

9.9 (c)

Out of 12 degree of freedom, 3-2-1 method restricts 9 degree of freedoms leaving it with 3 degrees of freedom for mounting and dismounting workpiece.

9.10 (0.0124)

$$\text{MRR} = \frac{AI}{\rho ZF}$$

where, A = Atomic weight; I = Current
ρ = Density; Z = Valency
F = Faraday's constant

$$\left(\frac{Z}{A}\right)_{\text{equivalent}} = \frac{Z_1 x_1}{A_1} + \frac{Z_2 x_2}{A_2} + \frac{Z_3 x_3}{A_3}$$

$$= \frac{2\times0.18}{58.93}+\frac{2\times0.62}{58.71}+\frac{6\times0.2}{51.99}$$
$$= 5.03 \times 10^{-2}$$
$$\text{MRR} = \frac{500}{8.28\times5.03\times10^{-2}\times96500}$$
$$= \mathbf{1.24 \times 10^{-2}\ cc/s}$$

9.11 (d)

In EDM material is removed by erosion of repeated spark.

9.12 (a)

Material removal takes place due to erosion by repeated spark. After that melting and evaporation takes place.

9.15 (d)

Process	Maximum metal removal rate
1. ECM	10×10^3 mm^3/min
2. USM	180 mm^3/min
3. EBM	10 mm^3/min
4. LBM	5 mm^3/min
5. EDM	5×10^3 mm^3/min

9.18 (b)

For iron, atomic weight,

$$A_{tFe} = 56,\ \text{Valency: } r_{Fe} = 2$$

For metal P, at weight

$$A_{tP} = 24,\ \text{Valency: } r_P = 4$$

For Alloy, density,

$$\rho = 6000\ \text{kg/m}^3$$
$$= 6\ \text{gm/cc}$$
$$\text{MRR: } Q = 50\ \text{mm}^3/\text{s}$$
$$= 50 \times 10^{-3}\ \text{cc/s}$$
$$\text{MRR} = \frac{eI}{F\rho}$$
$$\Rightarrow 50\times10^{-3} = \frac{e\times2000}{96500\times6}$$
$$\Rightarrow \quad e = 14.475$$

Now, let % of P in alloy = x

$$\therefore \quad \frac{1}{e} = \left(\frac{100-x}{100}\right)\frac{V_{Fe}}{A_{tFe}}+\frac{x}{100}\times\frac{V_P}{A_{tP}}$$
$$\Rightarrow \quad \frac{1}{14.475} = \left(\frac{100-x}{100}\right)\frac{2}{56}+\frac{x}{100}\times\frac{4}{24}$$
$$\Rightarrow \quad x = \mathbf{24.48}$$

9.19 (a)

Electrochemical machining
Inter electrode gap, y = 0.2 mm
Supply voltage (DC), V = 12 V
Specific resistance of electrolyte,
$1/K = 2\ \Omega$ cm
Atomic weight of iron A = 55.85
Valency of iron: Z = 2
Faraday's constant: F = 96540 coulombs
Material is removed from iron surface of 20 mm × 20 mm
Surface area: a = 20 × 20 mm^2
= 2 × 2 cm^2

Rate of mass removal in the form

$$\dot{m} = \frac{AI}{ZF}$$

Current density with this gap is

$$J = \frac{K(V-\Delta V)}{y}$$

Assume total over voltage to be 0 volt ($\Delta V = 0$)

$$J = \frac{KV}{y} = \frac{V}{y\left(\frac{1}{K}\right)}$$
$$= \frac{12V}{\left(\frac{0.2}{10}\text{cm}\right)(2\ \Omega\ \text{cm})} = 300\ \text{amp/cm}^2$$

Current: I = (Current Density) × (Surface Area)
= 300 Amp/cm^2 × (2 × 2 cm^2)
= 1200 Amp

Rate of mass removal is

$$\dot{m} = \frac{AI}{ZF} = \frac{(55.85)(1200)}{2\times96540} = \mathbf{0.3471\ gm/s}$$

9.20 (d)

In abrasive jet machining the variation is as shown

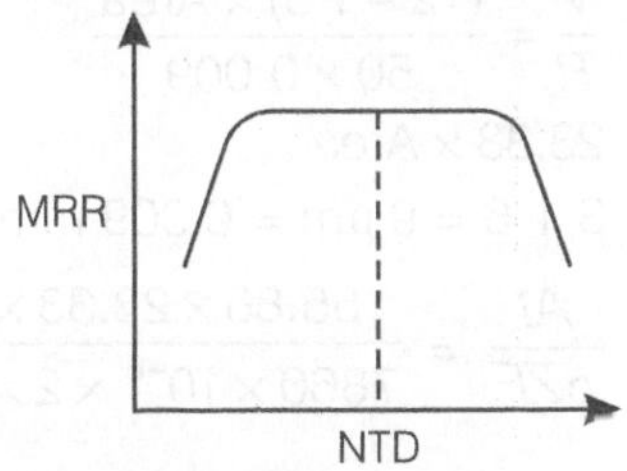

MRR = Metal removal rate
NTD = Nozzle tip distance

9.22 (c)

$$\text{MRR}_{(\text{Titanium})} = \frac{AI}{zF}\ \text{gm/s} = \frac{48\times2000\times0.9}{3\times96500}$$
$$= 0.298 \approx \mathbf{0.3\ gm/s}$$

9.23 (d)

EDM is most suitable process of square holes, intricate triangular holes etc.

9.24 (d)

Machining by laser beam is achieved through thermo-optic interaction between the beam and work material. Following phases are involved:
(*i*) interaction of laser beam with work material
(*ii*) heat conduction and temperature rise
(*iii*) melting, vaporization and ablation.

9.27 (b)

$$\text{Power} = \frac{\text{Work}}{\text{Time}}$$

$$P = \frac{1}{2}\frac{CV^2}{t}$$

$$\Rightarrow \quad C = \frac{P \times 2t}{V^2} = \frac{1000 \times 2 \times 25 \times 10^{-6}}{(100)^2}$$

$$= 5\ \mu\text{F}$$

9.29 (d)

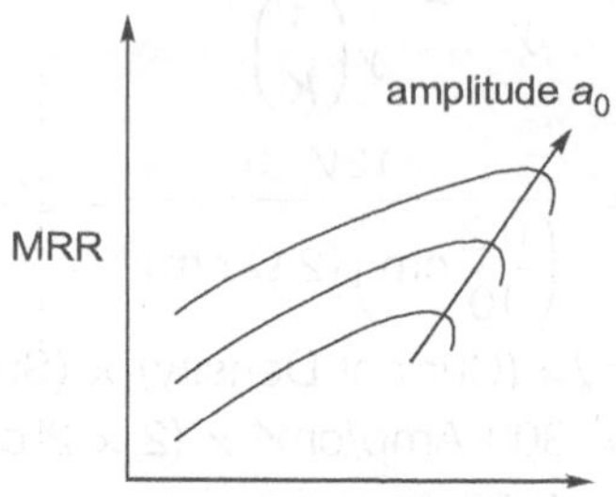

9.30 (51.542)

$$R = \frac{\rho L}{\text{Area}} = \frac{\frac{1}{0.02} \times 0.009}{\text{Area}} = \frac{50 \times 0.009}{\text{Area}}$$

$$I = \frac{V}{R} = \frac{(12 - 1.5) \times \text{Area}}{50 \times 0.009}$$

$$= 23.33 \times \text{Area}$$

$$L = 3 + 6 = 9\ \mu\text{m} = 0.009\ \text{mm}$$

$$\text{MRR} = \frac{AI}{\rho ZF} = \frac{55.85 \times 23.33 \times \text{Area}}{7860 \times 10^{-6} \times 2 \times 96500}$$

$$\frac{\text{MRR}}{\text{Area}} = 0.8590\ \text{mm/s} = \mathbf{51.542\ mm/min}$$

9.31 (a)

Electron beam machining requires vacuum chamber and therefore collision of accelerated electrons is avoided with the air molecules to obtain fine machining.

9.32 (a)

Process	Characteristics
P : Electrical Discharge Machining	2. Machining of electrically conductive materials
Q : Ultrasonic machining	3. Machining of glass
R : Chemical machining	1. No residual stress
S : Ion Beam Machining	4. Nano-machining

9.33 (d)

Ultrasonic Machining uses an abrasive slurry which transfers the impact load provided by the vibrating tool to the work piece.

9.34 (968.992)

Volume of metal to be removed

$$= (10 \times 30) \times 12\ \text{mm}^3 = 3600\ \text{mm}^3$$

Ideal energy required

$$= \frac{3600\ \text{mm}^3}{3.44 \times 10^{-2}\ \frac{\text{mm}^3}{(\text{A}s)}}$$

$$= 104651.163\ \text{As}$$

Actual energy required

$$= \frac{104651.163}{0.9}\ \text{As}$$

$$= 116279.07\ \text{As}$$

Time is 2 min = 120 s

$$\therefore \quad \text{Current} = \frac{116279.07}{120}\ \text{A} = 968.992\ \text{A}$$

9.35 (b)

In EDM, electric spark is the source due to which material undergoes fusion and vaporization.

9.36 (7.1431)

$$\text{MRR} = 4 \times 10^4\ IT^{-1.23}\ \text{mm}^3/\text{min}$$

Volume required to remove

$$= \frac{\pi}{4}D^2L = \frac{\pi}{4} \times 25^2 \times 20$$

$$= 9817.477\ \text{mm}^3$$

Now, $\quad \text{MRR} = 4 \times 10^4 \times I \times T^{-1.23}$

$$= 4 \times 10^4 \times 300 \times (1600)^{-1.23}$$

$$= 3 \times 4 \times 10^6 \times (1600)^{-1.23}$$

$$= 1374.40\ \text{mm}^3/\text{min}$$

$$\text{Time required} = \frac{9817.477}{1374.4} = \mathbf{7.1431\ min.}$$

9.37 (1)

$$E = \frac{1}{2}CV_d^2 = \frac{1}{2} \times 50 \times 10^{-6} \times 200^2$$

$= 1$ J

9.39 (5.0)

$I = 12 + 0.2t$

After time, 't', Next infinitly small time 'dt' let heat deposited 'dQ'.

$\therefore$ $dQ = 2.50 \times 10^{-2}$ (mm^3/A.s) $\times$ 12 + 0.2$t \times dt$

As we have to convert this 's' to 'min'

$\therefore$ $dQ = 2.50 \times 10^{-2}$ (mm^3/A $\times$ min) $\times$ 12 + 0.2$t \times dt$

Considering cathode efficiency of 15%

$dQ = 2.50 \times 10^{-2} \times 60 \times (12 + 0.2t)dt \times 0.15$ mm^3

$\therefore$ In 20 min,

$$Q = \int dQ = \int_0^{20} 2.50 \times 10^{-2} \times 60 \times (12 + 0.2t)dt \times 0.15$$

$$= 0.225\left[12 + 0.1t^2\right]_0^{20}$$

$$= 0.225\left[12 \times 20 + 0.1 \times 20^2\right] \text{mm}^3$$

$= 63$ mm^3

As area os 125 cm^2

Plating thickness,

$$t = \frac{63}{125 \times (100)} = 0.00504 \text{ mm} = 5.04 \ \mu\text{m}$$

(As 1 cm^2 = 100 mm^2)

9.41 (28.57) (27.00 to 30.00)

Material removal rate in ECM is given as MRR

$$\text{(gm/sec)} = \frac{eI}{F}$$

$\Rightarrow$ Using the equation for copper

$$\Rightarrow \quad \text{MRR (gm/sec)} = \frac{eI}{F}$$

$$\Rightarrow \quad 100 \times 10^{-3} = \frac{\frac{63}{2} \times I}{96500}$$

$\Rightarrow$ $I = 306.35$ Amperes

Using the equation for aluminium

$$\text{MRR (gm/sec)} = \frac{eI}{F} = \frac{\frac{27}{3} \times 306.35}{96500} \text{ gm/sec}$$

$= 0.028$ gm/sec $= 28.57$ mg/sec

9.42 (b, c, d)

For thermal energy methods of non-conventional machining, the mechanism of material removal is melting, fusion and evaporating. Since ECM is not a thermal energy method, it does not involve melting of workpiece.

For thermal methods mechanism will be same $\rightarrow$ Melting, fusion and vaporisation. So, except ECM all involves melting of metal workpieces.

■■■■

UNIT

XIII

Industrial Engineering

CONTENTS

Industrial Engineering

Syllabus

Production Planning and Control : Forecasting models, aggregate production planning, scheduling, materials requirement planning; lean manufacturing.

Inventory Control : Deterministic models; safety stock inventory control systems.

Operations Research : Linear programming, simplex method, transportation, assignment, network flow models, simple queuing models, PERT and CPM.

Analysis of Previous GATE Papers

Exam Year	1 Mark Ques.	2 Marks Ques.	3 Marks Ques.	5 Marks Ques.	Total Marks
1995	4	2	–	2	**18**
1996	1	1	–	2	**13**
1997	4	1	–	4	**26**
1998	2	–	–	1	**7**
1999	–	2	–	–	**4**
2000	–	2	–	4	**24**
2001	2	–	–	1	**7**
2002	1	2	–	2	**15**
2003	1	6	–	–	**13**
2004	3	6	–	–	**15**
2005	3	7	–	–	**17**
2006	2	7	–	–	**16**
2007	–	4	–	–	**8**
2008	2	6	–	–	**14**
2009	2	5	–	–	**12**
2010	4	4	–	–	**12**
2011	2	2	–	–	**6**
2012	1	2	–	–	**5**
2013	2	1	–	–	**4**
2014 Set-1	2	1	–	–	**4**
2014 Set-2	2	2	–	–	**6**

Exam Year	1 Mark Ques.	2 Mark Ques.	Total Marks
2014 Set-3	1	3	**7**
2014 Set-4	2	3	**8**
2015 Set-1	1	2	**5**
2015 Set-2	3	1	**5**
2015 Set-3	1	2	**5**
2016 Set-1	–	2	**4**
2016 Set-2	1	2	**5**
2016 Set-3	1	1	**3**
2017 Set-1	1	2	**5**
2017 Set-2	3	3	**9**
2018 Set-1	1	2	**5**
2018 Set-2	2	2	**6**
2019 Set-1	1	2	**5**
2019 Set-2	0	2	**4**
2020 Set-1	1	3	**7**
2020 Set-2	2	2	**6**
2021 Set-1	1	4	**9**
2021 Set-2	1	3	**7**
2022 Set-1	2	3	**8**
2022 Set-2	2	3	**8**
2023	2	1	**4**
2024	2	3	**8**

1 Forecasting

1.1 The most commonly used criteria for measuring forecast error is
(a) Mean absolute deviation
(b) Mean absolute percentage error
(c) Mean standard error
(d) Mean square error [1997 : 1 M]

1.2 In a forecasting model, at the end of period 13, the forecasted value for period 14 is 75. Actual value in the periods 14 to 16 are constant at 100. If the assumed simple exponential smoothing parameter is 0.5, then the MSE at the end of period 16 is
(a) 820.31 (b) 273.44
(c) 43.75 (d) 14.58 [1997 : 2 M]

1.3 When using a simple moving average to forecast demand, one would
(a) given equal weight to all demand data
(b) assign more weight to the recent demand data
(c) include new demand data in the average without discarding the earlier data
(d) include new demand data in the average after discarding some of the earlier demand data
[2001 : 1 M]

1.4 In a time series forecasting model, the demand for five time periods was 10, 13, 15, 18 and 22. A linear regression fit resulted in an equation $F = 6.9 + 2.9\,t$ where F is the forecast for period t. The sum of absolute deviations for the five data is
(a) 2.2 (b) 0.2
(c) –1.2 (d) 24.3 [2000 : 2 M]

1.5 A regression model is used to express a variable Y as a function of another variable X, this implies that
(a) There is a causal relationship between Y and X
(b) A value of X may be used to estimate a value of Y
(c) Values of X exactly determine values of Y
(d) There is no causal relationship between Y and X [2002 : 1 M]

1.6 The sale of cycles in a shop in four consecutive months are given as 70, 68, 82, 95. Exponentially smoothing average method with a smoothing factor of 0.4 is used in forecasting. The expected number of sales in the next month is
(a) 59 (b) 72
(c) 86 (d) 136 [2003 : 2 M]

1.7 For a product, the forecast and the actual sales for December 2002 were 25 and 20 respectively. If the exponential smoothing constant (α) is taken as 0.2, the forecast sales for January 2003 would be
(a) 21 (b) 23
(c) 24 (d) 27 [2004 : 1 M]

1.8 The sales of a product during the last four years were 860, 880, 870 and 890 units. The forecast for the fourth year was 876 units. If the forecast for the fifth year, using simple exponential smoothing, is equal to the forecast using a three period moving average the value of the exponential smoothing constant α is
(a) 1/7 (b) 1/5
(c) 2/7 (d) 2/5 [2005 : 1 M]

1.9 In an MRP system, component demand is
(a) forecasted
(b) established by the master production schedule
(c) calculated by the MRP system from the master production schedule
(d) ignored [2006 : 1 M]

1.10 A moving average system is used for forecasting weekly demand. $F_1(t)$ and $F_2(t)$ are sequences of forecasts with parameters m_1 and m_2, respectively, where m_1 and m_2 ($m_1 > m_2$) denote the numbers of weeks over which the moving averages are taken. The actual demand shows a step increase from d_1 to d_2 at a certain time. Subsequently,
(a) neither $F_1(t)$ nor $F_2(t)$ will catch up with the value d_2
(b) both sequences $F_1(t)$ and $F_2(t)$ will reach d_2 in the same period
(c) $F_1(t)$ will attain the value d_2 before $F_2(t)$
(d) $F_2(t)$ will attain the value d_2 before $F_1(t)$
[2008 : 2 M]

1.11 Which of the following forecasting methods takes a fraction of forecast error into account for the next period forecast?
(a) simple average method

(b) moving average method
(c) weighted moving average method
(d) exponential smoothening method

[2009 : 1 M]

1.12 The demand and forecast for February are 12000 and 10275, respectively. Using single exponential smoothening method (smoothening coefficient = 0.25), forecast for the month of March is
(a) 431 (b) 9587
(c) 10706 (d) 11000 [2010 : 1 M]

1.13 In simple exponential smoothing forecasting, to give higher weightage to recent demand information, the smoothing constant must be close to
(a) – 1 (b) zero
(c) 0.5 (d) 1.0 [2013 : 1 M]

1.14 In exponential smoothening method, which one of the following is true?
(a) $0 \le \alpha \le 1$ and high value of α is used for stable demand
(b) $0 \le \alpha \le 1$ and high value of α is used for unstable demand
(c) $\alpha \ge 1$ and high value of α is used for stable demand
(d) $\alpha \le 0$ and high value of α is used for unstable demand. [2014 : 1 M, Set-1]

1.15 The actual sales of a product in different months of a particular year are given below:

September	October	November	December	January	February
180	280	250	190	240	?

The forecast of the sales, using the 4 months moving average method, for the month of February is ______. [2014 : 1 M, Set-3]

1.16 Sales data of a product is given in the following table:

Month	January	February	March	April	May
Number of units sold	10	11	16	19	25

Regarding forecast for the month of June, which one of the following statements is TRUE?
(a) Moving average will forecast a higher value compared to regression
(b) Higher the value of order *N*, the greater will be the forecast value by moving average.
(c) Exponential smoothing will forecast a higher value compared to regression.
(d) Regression will forecast a higher value compared to moving average.

[2015 : 1 M, Set-2]

1.17 For a canteen, the actual demand for disposable cups was 500 units in January and 600 units in February. The forecast for the month of January was 400 units. The forecast for the month of March considering smoothing coefficient as 0.75 is ____.

[2015 : 2 M, Set-1]

1.18 The demand for a two-wheeler was 900 units and 1030 units in April 2015 and May 2015, respectively. The forecast for the month of April 2015 was 850 units. Considering a smoothing constant of 0.6, the forecast for the month of June 2015 is
(a) 850 units (b) 927 units
(c) 965 units (d) 970 units

[2016 : 2 M, Set-3]

1.19 The time series forecasting method that gives equal weightage to each of the *m* most recent observation is
(a) Moving average method
(b) Exponential smoothing with linear trend
(c) Triple Exponential smoothing
(d) Kalman Filter [2018 : 1 M, Set-1]

1.20 The table presents the demand of a product. By simple three months moving average method, the demand-forecast of the product for the month of September is

Month	Demand
January	450
February	440
March	460
April	510
May	520
June	495
July	475
August	560

(a) 536.67 (b) 490
(c) 530 (d) 510

[2019 : 1 M, Set-1]

1.21 The forecast for the monthly demand of a product is given in the table below.

Month	Forecast	Actual sales
1	32.00	30.00
2	31.80	32.00
3	31.82	30.00

The forecast is made by using the exponential smoothing method. The exponential smoothing coefficient used in forecasting the demand is
(a) 1.00 (b) 0.50
(c) 0.40 (d) 0.10

[2020 : 2 M, Set-2]

1.22 The demand and forecast of an item for five months are given in the table.

Month	Demand	Forecast
April	225	200
May	220	240
June	285	300
July	290	270
August	250	230

The Mean Absolute Percent Error (MAPE) in the forecast is____%. (round off to two decimal places)

[2021 : 2 M, Set-2]

1.23 An electric car manufacturer underestimated the January sales of car by 20 units, while the actual sales was 120 units. If the manufacturer uses exponential smoothening method with a smoothing constant of $\alpha = 0.2$, then the sales forecast for the month of February of the same year is _____ units (in integer).

[2022 : 1 M, Set-2]

Answers Forecasting

1.1 (d)	1.2 (b)	1.3 (d)	1.4 (a)	1.5 (a)	1.6 (c)	1.7 (c)	1.8 (c)
1.9 (c)	1.10 (d)	1.11 (d)	1.12 (c)	1.13 (d)	1.14 (b)	1.15 (240)	1.16 (d)
1.17 (569)	1.18 (d)	1.19 (a)	1.20 (d)	1.21 (d)	1.22 (8.07)	1.23 (104)	

Explanations Forecasting

1.1 (d)

Most commonly used criteria for measuring the forecast error is Mean square error

Sequence according to the uses is:

(i) Mean square error

(ii) Mean absolute deviation

(iii) Mean forecast error or Bias

(iv) Mean absolute percentage error

1.2 (b)

$$F_t = F_{t-1} + \alpha(D_{t-1} - F_{t-1})$$

$$F_{14} = 75\,; \quad D_{14} = 100\,; \quad \alpha = 0.5$$

$$F_{15} = 75 + 0.5(100 - 75) = 87.5$$

$$F_{16} = 87.5 + 0.5(100 - 87.5) = 93.75$$

Now,

$$\text{error}^2 = (D_t - F_t)^2$$

$$(\text{error})^2_{14} = (25)^2 = 625$$

$$(\text{error})^2_{15} = (12.5)^2 = 156.25$$

$$(\text{error})^2_{16} = (6.25)^2 = 39.06$$

$$\text{MSE} = \frac{\Sigma(\text{error})^2}{n} = \frac{625 + 156.25 + 39.06}{3}$$

$$= \mathbf{273.44}$$

1.3 (d)

Moving average method: In this method, past data are used fresh average is computed at the end of each period by adding the actual demand data of the most recent period and deleting the data of last period.

Note: In this method, as data changes from period to period so it is called as moving average method.

1.4 (a)

Period	D_i	$F_i = 6.9 + 2.9\,t$	$(D_i - F_i)$	$\lvert D_i - F_i \rvert$
1	10	9.8	0.2	0.2
2	13	12.7	0.3	0.3
3	15	15.6	–0.6	0.6
4	18	18.5	–0.5	0.5
5	22	21.4	0.6	0.6

$$\Sigma \lvert D_i - F_i \rvert = \mathbf{2.2}$$

1.5 (a)

A regression model is a part of a causal or economic method which is used for making forecasts for planning production schedules in the short range.

Note: The main aim of the causal method is to establish a relation so that changes in one variable will be useful for the prediction of other.

1.6 (c)

Taking forecast for the first period equal to demand.

$$F = F_{t-1} + \alpha e_{t-1}$$

D_i	F_i	e_i
70	70	0
68	70	–2
82	69.2	12.8
95	74.32	20.68
	82.59	

Hence the closest answer is **86**.

1.7 (c)

Given x_t = actual sale = 20

F_t = forecast sale = 25

Now, $F_{t+1} = F_t + \alpha(x_t - F_t)$

$= 25 + 0.2\ (20 - 25) = \mathbf{24}$

1.8 (c)

By three period moving average method,

d	f
860	
880	
870	
890	876

forecast of fifth year,

$$f_5 = \frac{880 + 870 + 890}{3} = 880 \text{ units}$$

By exponential smoothing method

$$f_5 = f_4 + \alpha(d_4 - f_4)$$

$$880 = 876 + \alpha(890 - 876)$$

$$14\alpha = 4$$

or $\alpha = \frac{4}{14} = \frac{2}{7}$

1.9 (c)

MRP is a simple system of calculating arithmetically the requirements of the input materials at different points of time based on actual production plan. MRP can also be defined as a planning and scheduling system to meet the time-phased materials requirements for production operations.

MRP always tries to meet the delivery schedule of end products as specified in the master production schedule.

MPS (Master proudction schedule): MPS is designed to meet the market demand (both the firm orders and forecasted demand) in future in the taken planning horizon. MPS mainly depicts the detailed delivery schedule of the end products. However, orders for replacement components can also be included in it to make it more comprehensive.

1.10 (d)

Since, $m_1 > m_2$ the weightage of the latest demand would be more in $F_2(t)$. Hence $F_2(t)$ will attain the value d_2 before $F_1(t)$.

1.11 (d)

Exponential smoothing methods of forecasting takes a fraction of forecast error into account for next period forecast.

$$F_t = F_{t-1} + \alpha(D_{t-1} - F_{t-1})$$

$$F_t = F_{t-1} + \alpha e_t$$

where, F_t is the forecast at time t

D_t is the actual demand at time t

F_{t-1} is the forecast at time $t - 1$

α is the smoothing coefficient

$e_t = D_{t-1} - F_{t-1}$, is forecast error

The value of α varies between 0 and 1, with commonly used value of 0.01 to 0.30.

New estimate= Old estimate of latest actual demand + α [latest actual demand – old estimate of latest actual demand]

Exponential smoothing provides a convenient, systematic and recursive method of revisiting the forecast for the next period whenever discrepancy exists between the previously forecast demand for current period and the actual demand for current period.

1.12 (c)

$F_t = F_{t-1} + \alpha(D_{t-1} - F_{t-1})$

F_t = current month forecast

F_{t-1} = last month forecast

D_{t-1} = last month demand

α = smoothening constant

$F_t = 10275 + 0.25(12000 - 10275)$

$F_t = 10706.25$, $F_t = \mathbf{10706}$

1.13 (d)

Simple exponential smoothing forecasting method

$F_{T+1} = F_T + \alpha(D_T - F_T)$ If $\alpha = 1$

$\therefore$ $F_{T+1} = D_T$

$\Rightarrow$ Higher weightage is given to recent demand if smoothing constant $\alpha = \mathbf{1}$.

Note: As we know, $\alpha = \frac{2}{n+1}$ i.e.

the value of α close to 1 has less smoothing effect and gives higher weightage to the recent data, while value of a close to 0 has greater smoothing effect and demand is less responsite to recent data.

1.14 (b)

In exponential smoothing forecast,

$$F_t = F_{t-1} + \alpha(D_{t-1} - F_{t-1})$$

where, $\alpha = \frac{2}{n+1}$ (n = number of periods)

Case I: For $n \to \infty$, $\alpha = 0$, i.e. stable demand.

Case II: For $n \to 1$, $\alpha = 1$ i.e. unstable demand.

So, $0 \le \alpha \le 1$ and high value of α is used for unstable demand.

POINTS TO REMEMBER

- Responsiveness indicates that forecast has a fluctuating and swinging pattern. It is preferred for new product and for that number of period is kept two i.e. $n \to 1$
- Stability represents that a forecast pattern is flat or has less fluctuation. It is preferred for old existing product and for that number of period kept large i.e. $n \to \infty$

1.15 Sol.

$$F = \frac{280+250+190+240}{4} = 240$$

1.16 (d)

As regression follow a pattern

$(\text{forecast})_{\text{June}}$ according to regression > 25

but according to moving average,

$$T_{\text{June}} < 25$$

1.17 (569)

$$F_{\text{Feb}} = F_{\text{Jan}} + \alpha[D_{\text{Jan}} - F_{\text{Jan}}]$$
$$= 400 + 0.75\,[500 - 400]$$
$$= 475 \text{ units}$$
$$F_{\text{March}} = F_{\text{Feb}} + \alpha[D_{\text{Feb}} - F_{\text{Feb}}]$$
$$= 475 + 0.75\,[600 - 475]$$
$$= 568.75 \approx \mathbf{569 \text{ units}}$$

1.18 (a)

Month	Demand	Forecast
April	900	850
May	1030	
June		??

Given: $\alpha = 0.6$,

Forecast for May, 2015

$$= 850 + 0.6\,(900 - 850) = 880$$

Forecast for June, 2015

$$= 880 + 0.6\,(1030 - 880)$$
$$= 880 + 90 = 970 \text{ units}$$

1.19 (a)

It gives equal weight to the previous data for a fixed period.

1.20 (d)

Three month moving average forecast for month of September.

$$F_{\text{sep}} = \frac{D_{\text{June}} + D_{\text{July}} + D_{\text{Aug}}}{2} = \frac{495+475+560}{3}$$

$$\text{Forecast for September} = \frac{1530}{3} = 510$$

1.21 (d)

From given data in table:

$$F_t = F_{t-1} + \alpha(D_{t-1} - F_{t-1})$$

For 2nd month $F_t = 31.8$,

for 1st month

$$F_{t-1} = 32 \text{ and } D_{t-1} = 30$$
$$31.8 = 32 + \alpha(30 - 32)$$
$$2\alpha = 32 - 31.8$$
$$\alpha = 0.1$$

1.22 (8.0712)

March	Di	Fi	ei	$\left\lvert\frac{ei}{Di}\right\rvert \times 100$
April	225	200	25	11.11%
May	220	240	-20	9.09%
June	285	300	-15	5.26%
July	290	270	20	6.896%
August	250	230	20	8.0%
				$\sum\left\lvert\frac{ei}{Di}\right\rvert \times 100 = 40.356$

$$\text{MAPE} = \frac{\sum\left\lvert\frac{ei}{Di}\right\rvert \times 100}{n} = 8.0712\%$$

1.23 (104) (104 to 104)

Given: $\alpha = 0.1$, Demand, $D_{\text{Jan}} = 120$,

Forecast, $F_{\text{Jan}} = 100$

So, error, $e_{\text{Jan}} = 20$

We know that, $F_t = F_{t-1} + \alpha e_{t-1}$

$$= 100 + 0.2 \times 20 = 104 \text{ units}$$

■■■■

2 Inventory Control & Break-Even Analysis

2.1 If the demand for an item is doubled and the ordering cost halved, the economic order quantity
(a) remains unchanged
(b) increases by factor of $\sqrt{2}$
(c) is doubled
(d) is halved **[1995 : 1 M]**

2.2 Setup costs do not include
(a) Labour cost of setting up machines
(b) Ordering cost or raw material
(c) Maintenance cost of the machines
(d) Cost of processing the work piece
[1997 : 1 M]

2.3 One of the following statements about PRS (Periodic Reordering System) is not true identify.
(a) PRS requires continuous monitoring of inventory levels
(b) PRS is useful in control of perishable items
(c) PRS provides basis for adjustments to account for variations in demand
(d) In PRS, inventory holding costs are higher than in Fixed Reorder Quantity System
[1998 : 1 M]

2.4 In inventory planning, extra inventory is unnecessarily carried to the end of the planning period when using one of the following lot size decision polices:
(a) Lot for lot production
(b) Economic Order Quantity (EOQ) lot size
(c) Period Order Quantity (POQ) lot size
(d) Part period total cost balancing **[1998 : 1 M]**

2.5 In computing Wilson's economic lot size for an item, by mistake the demand rate estimate used was 40% higher than the true demand rate. Due to this error in the lot size computation, the total cost of setup plus inventory holding per unit time. Would rise above the true optimum by approximately
(a) 1.4% (b) 3.3%
(c) 18.3% (d) 8.7% **[1999 : 2 M]**

2.6 Market demand for springs is 8,00,000 per annum. A company purchases these spring in lots and sells them. The cost of making a purchase order is ₹ 1,200. The cost of storage of springs is ₹ 120 per stored piece per annum. The economic order quantity is
(a) 400 (b) 2,828
(c) 4,000 (d) 8,000 **[2003:2 M]**

2.7 There are two products *P* and *Q* with the following characteristics

Product	Demand (units)	Order cost (Rs / order)	Holding Cost (Rs. unit / year)
P	100	50	4
Q	400	50	1

The Economic Order Quantity (EOQ) of products *P* and *Q* will be in the ratio
(a) 1 : 1 (b) 1: 2
(c) 1 : 4 (d) 1 : 8 **[2004 : 1 M]**

2.8 A company has an annual demand of 1000 units, ordering cost of Rs.100/ order and carrying cost of Rs. 100/unit-year. If the stock-out costs are estimated to be nearly Rs. 400 each time the company runs out-of-stock, the safety stock justified by the carrying cost will be
(a) 4 (b) 20
(c) 40 (d) 100 **[2004 : 2 M]**

2.9 Consider the following data for an item.
Annual demand : 2500 units per year
Ordering cost : Rs. 100 per order
Inventory holding rate: 25% of unit price

Price quoted by a supplier

Order quantity (units)	Unit price (Rs.)
< 500	10
> 500	9

The optimum order quantity (in units) is
(a) 447 (b) 471
(c) 500 (d) $\geq$ 600 **[2006 : 2 M]**

2.10 A stockist wishes to optimize the number of perishable items he needs to stock in any month in his store. The demand distribution for this perishable item is

Demand (in units)	2	3	4	5
Probability	0.10	0.35	0.35	0.20

The stockist pays Rs. 70 for each item and he sells each at Rs. 90. If the stock is left unsold in any month, he can sell the item at Rs. 50 each. There is no penalty for unfulfilled demand. To maximize the expected profit, the optimal stock level is

(a) 5 units (b) 4 units
(c) 3 units (d) 2 units [2006 : 2 M]

2.11 The maximum level of inventory of an item 100 and it is achieved with infinite replenishment rate. The inventory becomes zero over one and half month due to consumption at a uniform rate. This cycle continues throughout the year. Ordering cost is Rs. 100 per order and inventory carrying cost is Rs. 10 per item per month. Annual cost (in Rs.) of the plan, neglecting material cost, is

(a) 800 (b) 2800
(c) 4800 (d) 6800 [2007 : 2 M]

2.12 In machine shop, pins of 15 mm diameter are produced at a rate of 1000 per month and the same is consumed at a rate of 500 per month. The production and consumption continue simultaneously till the maximum inventory is reached. Then inventory is allowed to reduce to zero due to consumption. The lot size of production is 1000. If backlog is not allowed, the maximum inventory level is

(a) 400 (b) 500
(c) 600 (d) 700 [2007 : 2 M]

2.13 The net requirements of an item over 5 consecutive weeks are 50–0–15–20–20. The inventory carrying cost and ordering cost are ₹ 1 per item per week and ₹ 100 per order respectively. Starting inventory is zero. Use "Least Unit Cost Technique" for developing the plan. The cost of the plan (in ₹) is

(a) 200 (b) 250
(c) 225 (d) 260 [2007 : 2 M]

2.14 A company uses 2555 units of an item annually. Delivery lead time is 8 days. The reorder point (in number of units) to achieve optimum inventory is

(a) 7 (b) 8
(c) 56 (d) 60 [2009 : 2 M]

2.15 Annual demand for window frames is 10000. Each frame costs Rs. 200 and ordering cost is Rs. 300 per order. Inventory holding cost is Rs. 40 per frame per year. The supplier is willing to offer 2% discount if the order quantity is 1000 or more, and 4% if order quantity is 2000 or more. If the total cost is to be minimized, the retailer should

(a) order 200 frames every time
(b) accept 2% discount
(c) accept 4% discount
(d) order Economic Order Quantity [2010 : 2 M]

2.16 A component can be produced by any of the four processes I, II, III and IV. The fixed cost and the variable cost for each of the processes are listed below. The most economical process for producing a batch of 100 pieces is

Process	Fixed cost (in Rs.)	Variable cost per piece (in Rs.)
I	20	3
II	50	1
III	40	2
IV	10	4

(a) I (b) II
(c) III (d) IV [2014 : 1 M, Set-2]

2.17 Consider the following data with reference to elementary deterministic economic order quantity model

Annual demand of an item	100000
Unit price of the item (in Rs.)	10
Inventory carrying cost per unit per year (in Rs.)	1.5
Unit order cost (in Rs.)	30

The total number of economic orders per year to meet the annual demand is _____.

[2014 : 2 M, Set-2]

2.18 A manufacturer can produce 12000 bearings per day. The manufacturer received an order of 8000 bearings per day from a customer. The cost of holding a bearing in stock is Rs. 0.20 per month. Setup cost per production run is Rs. 500. Assuming 300 working days in a year, the frequency of production run should be

(a) 4.5 days (b) 4.5 months
(c) 6.8 days (d) 6.8 months

[2014 : 2 M, Set-3]

2.19 Annual demand of a product is 50000 units and the ordering cost is Rs. 7000 per order. Considering the basic economic order quantity model, the economic order quantity is 10000 units. When the annual inventory cost is minimized, the annual inventory holding cost (in Rs.) is _____.

[2015 : 1 M, Set-2]

2.20 A manufacturer has the following data regarding a product:
Fixed cost per month = Rs. 50000
Variable cost per unit = Rs. 200
Selling price per unit = Rs. 300
Production capacity = 1500 units per month
If the production is carried out at 80% of the rated capacity, then the monthly profit (in Rs.) is _____.
[2015 : 2 M, Set-2]

2.21 The annual requirement of rivets at a ship manufacturing company is 2000 unit. The rivets are supplied in units of 1 kg costing Rs. 25 each. If the costs Rs. 100 to place an order and the annual cost of carrying one unit is 9% of its purchase cost, the cycle length of the order (in days) will be_____. **[2015 : 2 M, Set-3]**

2.22 The annual demand for an item is 10,000 units. The unit cost is Rs. 100 and inventory carrying charges are 14.4% of the unit cost per annum. The cost of one procurement is Rs. 2000. the time between two consecutive orders to meet the above demand is ____month(s). **[2016 : 2 M, Set-1]**

2.23 A food processing company uses 25,000 kg of corn flour every year. The quantity-discount price of corn flour is provided in the table below.

Quantity (kg)	Unit Price(Rs/kg)
1-749	70
750-1499	65
1500 and above	60

The order processing charges are Rs. 500/ order. The handling plus carry-over charge on an annual basis is 20% of the purchase price of the corn flour per kg. The optimal order quantity (in kg) is _____. **[2016 : 2 M, Set-2]**

2.24 A local tyre distributor expects to sell approximately 9600 steel belted radial tyres next year. Annual carrying cost is Rs. 16 per tyre and ordering cost is ₹ 75. The economic order quantity of the tyres is
(a) 64 (b) 212
(c) 300 (d) 1200
[2018 : 1 M, Set-2]

2.25 Annual demand of valves per year in a company is 10,000 units; the current order quantity is 400 valves per order. The holding cost is ₹ 24 per valve per year & the ordering cost is ₹ 400 per order. If the current order quantity is changed to Economic Order Quantity, then saving in the total cost of inventory per year will be ₹_______. (Round off to two decimal places) **[2019 : 2 M, Set-2]**

2.26 Consider two cases as below.
Case 1: A company buys 1000 pieces per year of a certain part from vendor 'X'. The changeover time is 2 hours and the price is Rs. 10 per piece. The holding cost rate per part is 10% per year.
Case 2: For the same part, another vendor 'Y" offers a design where the changeover time is 6 minutes, with a price of Rs. 5 per piece, and a holding cost rate per part of 100% per year. The order size is 800 pieces per year from 'X' and 200 pieces per year from 'Y'.
Assume the cost of downtime as Rs. 200 per hour. The percentage reduction in the annual cost for Case 2, as compared to Case 1 is_________ (round off to 2 decimal places).
[2020 : 2 M, Set-1]

2.27 Robot Ltd. wishes to maintain enough safety stock during the lead time period between starting a new production run and its completion such that the probability of satisfying the customer demand during the lead time period is 95%. The lead time period is 5 days and daily customer demand can be assumed to follow the Gaussian (normal) distribution with mean 50 units and a standard deviation of 10 units.
Using $\phi^{-1}(0.95) = 1.64$, where ϕ represents the cumulative distribution function of the standard normal random variable, the amount of safety stock that must be maintained by Robot Ltd. to achieve this demand fulfilment probability for the lead time period is _______ units. [Round off to two decimal places] **[2021 : 1 M, Set-1]**

2.28 Daily production capacity of a bearing manufacturing company is 30000 bearings. The daily demand of the bearing is 15000, The holding cost per year of keeping a bearing in the inventory is ₹ 20. The setup cost for the production of a batch is ₹ 1800. Assuming 300 working days in a year, the economic batch quantity in number of bearings is _______ (in integer). **[2021 : 2 M, Set-2]**

2.29 Which one of the following is NOT a form of inventory?
(a) Raw materials
(b) Work-in-process materials
(c) Finished goods
(d) CNC Milling Machines **[2022 : 1 M, Set-1]**

2.30 The demand of a certain part is 1000 parts/ year and its cost is ₹1000/part. The orders are placed based on the economic order quantity

(EOQ). The cost of ordering is ₹100/order and the lead time for receiving the orders is 5 days. If the holding cost is ₹20/part/year, the inventory level for placing the orders is _____ parts. (round off to the nearest integer)

[2022 : 1 M, Set-2]

2.31 With reference to the Economic Order Quantity (EOQ) model, which one of the options given is correct?

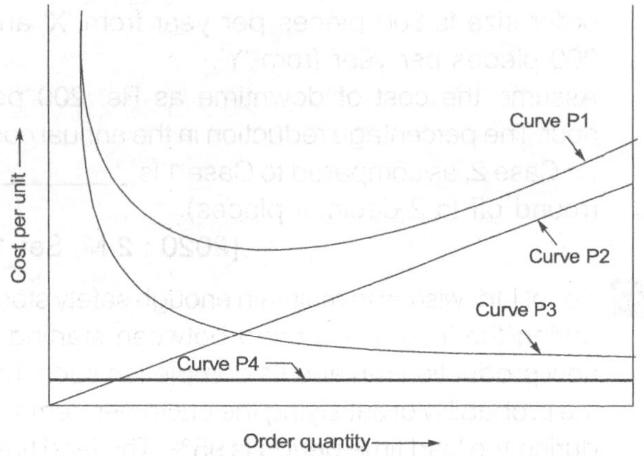

(a) Curve P1: Total cost, Curve P2: Holding cost, Curve P3: Set up cost, and Curve P4: Production cost.

(b) Curve P1: Holding cost, Curve P2: Set up cost, Curve P3: Production cost, and Curve P4: Total cost.

(c) Curve P1: Production cost, Curve P2: Holding cost, Curve P3: Total cost, and Curve P4: Set up cost.

(d) Curve P1: Total cost, Curve P2: Production cost, Curve P3: Holding cost, and Curve P4: Set up cost.

[2023 : 1 M]

2.32 A company orders gears in conditions identical to those considered in the economic order quantity (EOQ) model in inventory control. The annual demand is 8000 gears, the cost per order is 300 rupees, and the holding cost is 12 rupees per month per gear. The company uses an order size that is 25% more than the optimal order quantity determined by the EOQ model. The percentage change in the total cost of ordering and holding inventory from that associated with the optimal order quantity is

(a) 0 (b) 5
(c) 12.5 (d) 2.5

[2024 : 2 M]

■■■■

Answers Inventory Control & Break-Even Analysis

2.1	(a)	2.2	(d)	2.3	(a)	2.4	(c)	2.5	(c)	2.6	(c)	2.7	(c)	2.8	(c)
2.9	(c)	2.10	(b)	2.11	(d)	2.12	(b)	2.13	(b)	2.14	(c)	2.15	(c)	2.16	(b)
2.17	(50)	2.18	(c)	2.19	(35000)	2.20	(70000)	2.21	(76.948)	2.22	(2)	2.23	(1500)	2.24	(c)
2.25	(943.59)	2.26	(5.32)	2.27	(36.67)	2.28	(40250)	2.29	(d)	2.30	(14)	2.31	(a)	2.32	(d)

Explanations Inventory Control & Break-Even Analysis

2.1 (a)

$$\text{Economic order quantity, EOQ} = \sqrt{\frac{2DC_o}{C_h}}$$

Economic order quantity, when demand is doubled and ordering cost halved,

$$EOQ' = \sqrt{\frac{2(2D)\left(\frac{C_o}{2}\right)}{C_h}} = EOQ$$

So, Economic order quantity remains unchanged.

2.2 (d)

When the units are produced within the production system, then the cost associated with bringing shut down system, again into play is called setup cost. It includes maintenance cost, cost associated with arrangement of workers, raw material, tool extra, schedule chart preparation cost.

POINTS TO REMEMBER

Ordering cost and setup cost are same, both are denoted by the C_o.

2.3 (a)

In the periodic reordering system, Inventory level

is reviewed after a fixed period of time and fresh order is placed at that time. So, PRS dosn't require continuous monitoring of inventory levels.

POINTS TO REMEMBER

Two types of review system is used:
(i) Fixed order syste or reorder point system.
(ii) Periodic reordering system or fixed period system.

2.5 (c)

Method I:

$$Q^* \propto \sqrt{D}$$

$$Q_1^* \propto \sqrt{1.4D} \propto \sqrt{1.4}\sqrt{D} \propto \sqrt{1.4}\,Q^*$$

Now initial inventory cost = $\frac{D}{Q^*}C_o + \frac{Q^*}{2}C_n$

Now total inventory cost

$$= \frac{1.4D}{\sqrt{1.4}\,Q^*}C_o + \sqrt{1.4}\frac{Q^*}{2}C_n$$

$$= \sqrt{1.4}\left(\frac{D}{Q^*}C_o + \frac{Q^*}{2}C_n\right)$$

$= \sqrt{1.4}$ (initial inventory cost)

= 1.183 (initial inventory cost)

Total rise = **18.3%**

Method II:

Economic order quantity, EOQ = $\sqrt{\frac{2DC_o}{C_h}}$

Total inventory cost at EOQ,

$$(\text{TIC})_{\text{actual}} = \sqrt{2DC_oC_h}$$

$$(\text{TIC})_{\text{estimate}} = \sqrt{2(1.4D)C_oC_h} = 1.183\sqrt{2DC_oC_h}$$

Error in estimate,

$$= \frac{(\text{TIC})_{\text{estimate}} - (\text{TIC})_{\text{actual}} \times 100}{(\text{TIC})_{\text{actual}}}$$

$$= \left(\frac{1.183-1}{1}\right)\times 100 = 18.3\%$$

2.6 (c)

Given, Demand (D) = 8,00,000 per annum
Order cost C_o = Rs. 1200 per order
Holding cost C_h= Rs. 120 per stored piece per annum
We know that,

$$\text{EOQ} = \sqrt{\frac{2C_oD}{C_h}} = \sqrt{\frac{2\times 800000\times 1200}{120}}$$

$$= 4000$$

2.7 (c)

Given that, $D_P = 100$, $D_Q = 400$

$$(\text{EOQ})_p = \sqrt{\frac{2DC_o}{C_h}} = \sqrt{\frac{2\times 50\times 100}{4}} = 50$$

$$(\text{EOQ})_Q = \sqrt{\frac{2DC_o}{C_h}} = \sqrt{\frac{2\times 50\times 400}{1}} = 200$$

$$\therefore \frac{(\text{EOQ})_P}{(\text{EOQ})_Q} = \frac{50}{200} = \frac{1}{4}$$

2.8 (c)

Method I:

We need to find stock justified by carrying cost *i.e.*, M^*

$$M^* = Q^* - S^*$$

$$Q^* = \sqrt{\frac{2DC_o}{C_h}} \times \sqrt{\frac{C_h + C_b}{C_b}}$$

$$= \sqrt{\frac{2\times 100\times 1000}{100}} \times \sqrt{\frac{500}{400}} = 50$$

$$S^* = \frac{C_h}{C_b + C_h} \times Q^* = \frac{100}{100+400}\times 50$$

$$S^* = 10$$

and $M^* = 50 - 10 = \mathbf{40}$

Method II:

Given: Demand, D = 1000 units/year,
Ordering cost, C_o = ₹ 100/unit-year
Stock out cost, C_b = ₹ 400
Optimum level for stock out condition,

$$M^* = Q^*\left(\frac{C_b}{C_b + C_n}\right)$$

$$= \sqrt{\frac{2DC_o}{C_h}\times\left(\frac{C_o + C_h}{C_b}\right)}\left(\frac{C_b}{C_b + C_h}\right)$$

$$= 50\times\frac{400}{500} = 40 \text{ units}$$

2.9 (c)

Given: Annual demand: D = 2500 Units per year
Ordering cost: C_0 = Rs. 100 per order
Inventory holding rate: i = 0.25 of units prices
Case 1 : Let economic order quality is less than 500 for the inventory holding cost

$$C_C = \frac{25}{100}\times 10 = 2.5$$

for this case

$$EOQ = \sqrt{\frac{2DC_0}{C_C}} = \sqrt{\frac{2\times2500\times100}{2.5}} = 447.2$$

Inventory cost (TIC) = $\sqrt{2\times C_0\times C_C\times D}$

$$= \sqrt{2\times100\times0.25\times10\times2500}$$
$$= 1118.03$$

Case 2 : Let EQQ is greater than 500 (EOQ > 500) for this case

$$C_c = \frac{25}{100}\times9 = 2.25$$

$$EOQ = \sqrt{\frac{2\times2500\times100}{2.25}} = 471.4$$

This is against the assumption made so correct

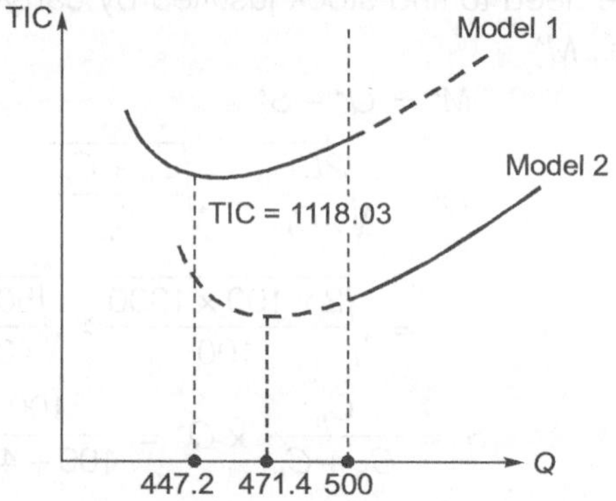

$$EOQ = 447.2 \approx 447 \text{ Units.}$$

When price break occurs

TIC at $Q = 500$

TIC = (Average inventory) × (unit inventory carrying cost) + (No. of order per year) × (cost of order)

$$= \frac{Q}{2}\times C_C + \frac{D}{Q}\times C_O$$

$$= \frac{500}{2}\times9\times.25 + \frac{2500}{500}\times100 = 1062.5$$

Hence, quantity to be ordered is **500**.

2.10 (b)

$p(d\geq Q)$ = Probability that the demand for Q units or more

M_p = Marginal profit per unit sold
=90 – 70 = Rs. 20

M_L = Marginal loss from each unit that is left unsold
= Rs. 70 – 50 = Rs. 20

$$p(d\geq Q) = \frac{M_P}{M_P+M_L} = \frac{20}{20+20} = 0.5$$

Demand	Probability that demand at this level	Cumulative Probability
2	0.1	0.1
3	0.35	0.45
4	0.35	0.8
5	0.20	1.00

Hence to maximize profit, optimal stock level is **4 units.**

2.11 (d)

Total Annual cost = Annual carrying cost + Annual ordering cost.

$$= \frac{Q}{2}\times C_h + \frac{D}{Q}C_o$$

Now, Annual carrying cost = $\frac{Q}{2}\times C_h$

where

Q = Units per order = 100

C_h = Inventory carrying cost is Rs. 10 per item per month

Hence, C_h = 12 × 10 per item per year
= Rs 120 per item per year

∴Annual carrying cost

$$= \frac{100}{2}\times120 = 6000$$

Now cost of an order

C_o = Rs. 100

and no of orders per year, $N = \frac{D}{Q} = \frac{1}{T}$

$$N = \frac{1}{T} = \frac{12\text{ months}}{1.5\text{ months}} = 8$$

Here, T = 1.5 months

Hence, annual ordering cost = 8 × 100 = 800

Hence, total Annual cost = 6000 + 800 = **₹ 6800**

2.12 (b)

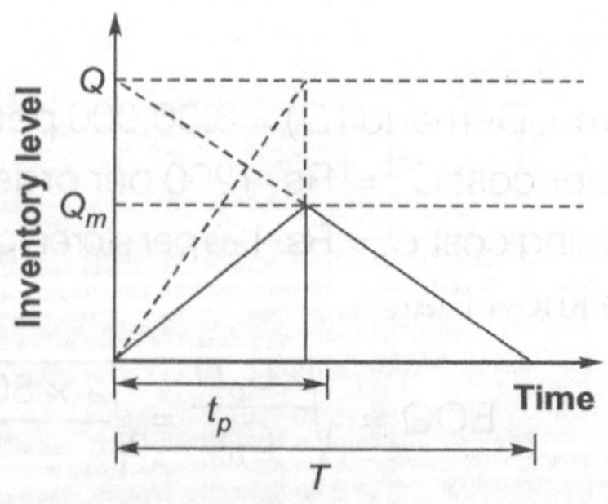

Maximum Inventory level,

$$Q_m = (p-d)\frac{Q}{P}$$

Here p = 1000 per month
d = 500 per month
Q = 1000 pins

∴ Maximum Inventory level

$$Q_m = (1000-500) \times \frac{1000}{1000} = \textbf{500 pins}$$

2.13 (b)

D	Single order	Cost
50	105 – 50	100
0	55	55
15	55 – 15 = 40	55
20	40 – 20	40
20	20 – 20	20
		Total Cost 270

D	Double order	Cost
50	65 – 50 = 15	100
0	15	15
15	15 – 15 = 0	15 + 100
20	40 – 20	20
20	20 – 20 = 0	Total cost 250

Hence, cost of plan Rs. **250**

2.14 (c)

Method I:

Using Wilson Harris model (Deterministic Model)

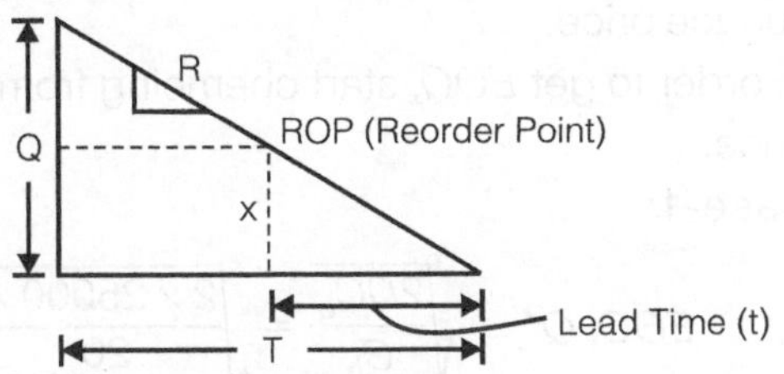

R = Consumption Rate = 2555 units annually

Q = Ordered Quantity

T = Cycle Time

t = Lead Time = 8 days = $\frac{8}{365}$ years

x = Reorder Quantity (Point)

Using Geometry of triangle

$$R = \frac{Q}{T} = \frac{x}{t}$$

$$2555 = \frac{x}{(8/365)}$$

or x = 56 unit

Therefore, the reorder point to achieve optimum inventory is **56 units**.

Method II:

Reorder point,

ROP = Lead time × Demand + Safety stock
= $Lt \times d + SS$

Since, $SS = 0$

So, ROP = $LT \times d$

Given: LT = 8 days

$$d = \frac{2555}{365} = 7 \text{ units/day}$$

∴ ROL = $LT \times d$ = 8 × 7 = 56 units

2.15 (c)

Total cost = Purchase cost + holding cost + ordering cost

It is a case of inventory with price breaks.

$$\text{EOQ, } Q^* = \sqrt{\frac{2C_oD}{C_h}} = \sqrt{\frac{2 \times 300 \times 10000}{40}}$$

= 387.298 units ≈ 388 units

For $EOQ\ Q^*$ = 388 units

$$T(C) = D \times C + \frac{Q^*}{2} \times h + \frac{D}{Q^*} \times A$$

$$= 10000 \times 200 + \frac{388}{2} \times 40 + \frac{10000}{388} \times 300$$

= 2015491.959 = ₹ 2015492

For Q = 1000

$T(C)$ = 10000 × (200 × 0.98) + 500 × 40 + 10 × 300
= ₹ 1983000

For Q = 2000

$T(C)$ = 10000 × (200 × 0.96) + 1000 × 40 + 5 × 300
= ₹ 1961500

Accept 4% discount as lowest total cost at Q = **2000**

2.16 (b)

Process	Fixed cost (in ₹)	Variable cost per piece(in ₹)	Total cost (₹)
I	20	3	20 + 3 × 100 = 320
II	50	1	50 + 1 × 100 = 150
III	40	2	40 + 2 × 100 = 240
IV	10	4	10 + 4 × 100 = 410

Total cost is minimum for process II, hence **process II** is most economical.

2.17 (50)

$$\text{EOQ, } Q^* = \sqrt{\frac{2C_oD}{C_h}} = \sqrt{\frac{2\times100000\times30}{1.5}}$$

$= 2000$ units

Number of orders,

$$N = \frac{D}{Q^*} = \frac{D}{\text{EOQ}} = \frac{100000}{2000} = \mathbf{50}$$

2.18 (c)

Given: $D = 8000$ per day; $P = 12000$ per day

$C_h = 0.2$ Rs. per month

$C_o = 500$ per production run

$$\text{EOQ, } Q^* = \sqrt{\frac{2C_oD}{C_h\left(1-\frac{D}{P}\right)}} = \sqrt{\frac{2\times500\times8000\times300}{0.2\times12\left(1-\frac{8000}{12000}\right)}}$$

$= 54772.25$

Frequency of production run

$$= \frac{54772.25}{8000} = \mathbf{6.8\ days}$$

POINTS TO REMEMBER

The frequency of production run is also known as total cycle time. And it is calculated as

$$T = \frac{Q^*}{D} = \frac{1}{N}$$

where, N = Number of orders.

2.19 Sol.

Method I:

For minimum annual inventory cost,

Ordering cost = Holding cost

7000 × number of order = holding cost

or Holding cost $= 7000\times\frac{50000}{10000} =$ **₹ 35,000**

Method II:

Given: EOQ, $Q^* = 10000$ units, $D = 50000$ units

$C_o =$ ₹ 7000/order

As we know, $Q^* = \sqrt{\frac{2DC_o}{C_h}}$

$$10000 = \sqrt{\frac{2\times50000\times7000}{C_h}}$$

$C_h =$ ₹ 7/unit/year

Annual carrying cost,

$$= \frac{Q^*}{2}\times C_h = \frac{10000}{2}\times 7 = ₹\ 35000$$

2.20 Sol.

Production $= 0.8\times1500 = 1200$

Profit $= S - V - F = 1200\times300 - 1200\times200 - 50000$

$=$ **₹ 70,000**

2.21 (76.948)

$$Q^* = \sqrt{\frac{2DC_o}{C_h}} = \sqrt{\frac{2\times2000\times100}{0.09\times25}} = 421.637$$

No. of orders, $N = \frac{D}{Q^*} = 4.7434$

$\therefore T = \frac{1}{N} = 0.2108$ year $=$ **76.948 days**

2.22 (2)

$$\text{EOQ, } Q^* = \sqrt{\frac{2DC_o}{C_h}} = \sqrt{\frac{2\times10000\times2000}{14.4}}$$

$= 1666.67$

(N) No. of orders per year $= \frac{D}{Q^*} = \frac{10000}{1666.67} = 6$

Time between two consecutive orders,

$= \frac{1}{N}\times12 = \frac{12}{6}$ months

$=$ **2 months**

2.23 (1500)

Annual demand, $D = 25000$ kg,

Ordering cost, $C_o =$ ₹ 500/order

Carrying cost or holding cost, $C_h = 20\%$ of unit surface price.

In order to get *EOQ*, start chemcing from lowest price.

Case-1:

$$EOQ, Q^* = \sqrt{\frac{2DC_o}{C_h}} = \sqrt{\frac{2\times25000\times500}{\frac{20}{100}\times60}}$$

$= 1443.376$ kg (not feasible)

Case-2:

$$\text{EOQ, } Q^* = \sqrt{\frac{2DC_o}{C_h}} = \sqrt{\frac{2\times25000\times500}{\frac{20}{100}\times65}}$$

$= 1386.75$ kg (feasible)

Total annual cost,

$= DC + \sqrt{2DC_oC_h}$

$$= 65\times25000 + \sqrt{2\times25000\times500\times0.2\times65}$$

$= 162500 + 18027.76 = ₹\,1643027.76$

For 1500 kg,

$$\text{Annual cost} = DC + \frac{D}{Q}C_0 + \frac{C_h}{2}Q$$

$$= 25000 \times 60 + \frac{1500 \times 20 \times 60}{2 \times 100} + \frac{25000}{1500} \times 500$$

$= ₹\,1517320.51$

So, at 1500 kg purchase, cost is minium then

EOQ = 1500 kg

2.24 (c)

Demand, $D = 9600$

Carrying cost, C_h = Rs. 16/year

Ordering cost, C_0 = Rs. 75/order

$$EOQ,\ Q^* = \sqrt{\frac{2DC_o}{C_h}} = \sqrt{\frac{2 \times 9600 \times 75}{16}}$$

$$= \sqrt{1200 \times 75} = \textbf{300 lyres}$$

2.25 (943.594)

As per given data

Annual demand, D = 10000 units

Q = 400/order

Holding cost (C_h) = ₹24 unit/year

Ordering cost (C_o) = ₹400/order

$$(\text{TIC})_I = \frac{D}{Q} \times C_o + \frac{Q}{2} \times C_h$$

$$= \frac{10000}{400} \times 400 + \frac{400}{2} \times 24$$

= ₹14800/year

If order quantity changed to EOQ,

$$\text{TIC}^* = \sqrt{2DC_oC_h}$$

$$= \sqrt{2 \times 10000 \times 400 \times 24}$$

= ₹13856.406/year

Hence total cost saving per year

= $(\text{TIC})_I - \text{TIC}^*$

= 14800–13856.406 = ₹943.594/year

2.26 (5.32)

Given Data : 1000 pieces/year from 'X'.

Changeover time = 2 hrs.

Cost of downtime = Rs. 200/hour

So, Total cost of downtime

2 × 200 = Rs. 400/downtime

C = Rs. 10/piece

Holding cost, C_h = 10% of Rs. 10

C_h = Rs. 1/unit/year

So, total cost for Case I :

= Material Cost + Downtime Cost + Inventory Holding Cost

$$= 1000 \times 10 + (1 \times 400) + \frac{1000}{2} \times 1$$

Total cost for case I = Rs. 10,900/-

Case II : Order quantity 800 units from X and 200 units from Y.

For Y :Change overtime = 6 min. = 0.1 hour

Downtime cost = 0.1 × 200 = Rs. 20/-

Unit cost, C = Rs. 5/piece

Holding cost, C_h = 100% of unit cost = Rs. 5

So, total cost for case II :

= Cost for 'X' + Cost for 'Y'

$$= \left(800 \times 10 + 400 + \frac{800}{2} \times 1\right) + \left(200 \times 5 + 20 + \frac{200}{2} \times 5\right)$$

= 8000 + 800 + 1000 + 520

Total cost for case II = Rs. 10,320/-

So, percentage reduction in total cost of case II

$$= \frac{10900 - 10320}{10900} \times 100$$

$$= \frac{580}{10900} \times 100 = 5.32\%$$

2.27 (36.67)

Given: $z = 1.64$

Safety stock = $z\sigma$

$$\sigma = \sqrt{10^2 + 10^2 + 10^2 + 10^2 + 10^2} = 10\sqrt{5}$$

Safety stock = $1.64 \times 10\sqrt{5} = 36.67$ units

2.28 (40250)

$$Q^* = \sqrt{\frac{2D \times C_0}{C_h} \times \frac{P}{P-d}}$$

$$= \sqrt{\frac{2 \times 15000 \times 300 \times 1800}{20} \times \left(\frac{30000}{30000 - 15000}\right)}$$

= 40249.2 ~ 40250 units

2.29 (d)

CNC milling machine does not make a part of the inventory. Machines come under asset category.

2.30 (14) (13 to 15)

Given: Demand, $D = 1000$ parts/year, Cost, $C =$ ₹1000 per part
Ordering cost, $C_o =$ ₹100 per order
Lead time, $LT = 5$ days
Holding cost, $C_h =$ ₹20 per part per year
Inventory level for placing orders

$$= LT \times \text{Demand per day} = \frac{LT \times D}{365}$$

Assuming 365 days in a year.

So, $$\text{ROL} = \frac{5 \times 1000}{365} = 13.698 \approx 14 \text{ parts}$$

2.31 (a)

- Since, holding cost is directly proportional to order quantity, P2 is holding cost curve.
- At EOQ, holding and set up costs are same. Also set up cost is inversely proportional to order quantity and thus, P3 is set up cost curve.

2.32 (d)

Given: $D = 8000$ units/yr, $C_o =$ Rs. 300/order, $C_h =$ Rs. $12 \times 12 =$ Rs. 144 per year

Now, $$Q^* = \sqrt{\frac{2C_oD}{C_h}} = \sqrt{\frac{2 \times 300 \times 8000}{144}}$$

$\therefore$ $$Q^* = 182.5 \text{ units}$$

Now, Actual quantity,

$$Q = 1.25Q^* = 1.25 \times 182.5 = 228.125 \text{ units}$$

Total cost at EOQ,

$$\text{TIC}_1 = \sqrt{2C_oC_hD} = \sqrt{2 \times 300 \times 144 \times 8000} = \text{Rs. } 26290.68$$

Now, Total cost at $Q = 228.125$ units,

$$\text{TIC}_2 = \frac{Q}{2} \times C_h + \frac{D}{Q} \times C_o$$

$$= \frac{228.125}{2} \times 144 + \frac{8000}{228.125} \times 300$$

$$= \text{Rs. } 26945.54$$

$$\therefore \text{ \% increase} = \frac{26945.54 - 26290.68}{26290.68} \times 100 = 2.5\%$$

Alternative solution

We know, if $Q = kQ^*$

then, $$\frac{TIC(Q)}{TIC(Q^*)} = \frac{1}{2}\left[k + \frac{1}{k}\right]$$

$\therefore$ Percentage increase,

$$\frac{TIC(Q) - TIC(Q^*)}{TIC(Q^*)} \times 100 = \left(\frac{TIC(Q)}{TIC(Q^*)} - 1\right) \times 100$$

$$= \left(\frac{1}{2}\left[k + \frac{1}{k}\right] - 1\right) \times 100$$

$$= \left(\frac{1}{2}\left[1.25 + \frac{1}{1.25}\right] - 1\right) \times 100$$

$$= 0.025 \times 100$$

$$= 2.5\%$$

■■■■

3 Linear Programming & Transportation

3.1 If at the optimum in a linear programming problem, a dual variable corresponding to a particular primal constraint is zero, then it means that

(a) Right hand side of the primal constraint can altered without affecting the optimum solution
(b) Changing the right hand side of the primal constraint will disturb the optimum solution
(c) The objectives function is unbounded
(d) The problem is degenerate **[1996 : 1 M]**

3.2 A manufacturer produces two types of products, 1 and 2, at production levels of x_1 and x_2 respectively. The profit is given is $2x_1 + 5x_2$. The production constraints are:

$x_1 + 3x_2 \leq 40$
$3x_1 + x_2 \leq 24$
$x_1 + x_2 \leq 10$
$x_1 > 0$
$x_2 > 0$

The maximum profit which can meet the constraints is

(a) 29 (b) 38
(c) 44 (d) 75 **[2003 : 2 M]**

3.3 A company produces two types of toys : *P* and *Q*. Production time of *Q* is twice that of *P* and the company has a maximum of 2000 time units per day. The supply of raw material is just sufficient to produce 1500 toys (of any type) per day. Toy type *Q* requires an electric switch which is available @ 600 pieces per day only. The company makes a profit of ₹ 3 and ₹ 5 on type *P* and *Q* respectively. For maximization of profits, the daily production quantities of *P* and *Q* toys should respectively be

(a) 1000, 500 (b) 500, 1000
(c) 800, 600 (d) 1000, 1000
[2004 : 2 M]

3.4 A component can be produced by any of the four processes, I, II, III and IV. Process I has fixed cost of Rs. 20 and variable cost of Rs. 3 per piece. Process II has a fixed cost of Rs. 50 and variable cost of Re. 1 per piece. Process III has a fixed cost of Rs. 40.00 and variable cost of Rs. 2 per piece. Process IV has fixed cost of Rs. 10 and Variable cost Rs. 4 per piece. If company wishes to produce 100 pieces of the component, from economic point of view it should choose

(a) Process I (b) Process II
(c) Process III (d) Process IV **[2005 : 2 M]**

3.5 A company has two factories S_1, S_2 and two warehouses D_1, D_2. The supplies from S_1 and S_2 are 50 and 40 units respectively. Warehouse D_1 requires a minimum of 20 units and a maximum of 40 units. Warehouse D_2 requires a minimum of 20 units and, over and above, it can take as much as can be supplied. A balanced transportation problem is to be formulated for the above situation. The number of supply points, the number of demand points, and the total supply (or total demand) in the balanced transportation problem respectively are

(a) 2, 4, 90 (b) 2, 4, 110
(c) 3, 4, 90 (d) 3, 4, 110 **[2005 : 2 M]**

Linked Answer Questions 3.6 and 3.7

Consider a linear programming problem with two variable and two constraints. The objective function is maximize $x_1 + x_2$. The corner points of the feasible region are (0, 0), (0, 2) (2, 0) and (4/3, 4/3)

3.6 If an additional constraint $X_1 + X_2 \leq 5$ is added, the optimal solution is

(a) (5/3, 5/3) (b) (4/3, 4/3)
(c) (5/2, 5/2) (d) (5, 0) **[2005 : 2 M]**

3.7 Let Y_1 and Y_2 be the decision variables of the dual and v_1 and v_2 be the slack variables of the dual of the given linear programming problem. The optimum dual variables are

(a) Y_1 and Y_2 (b) Y_1 and v_1
(c) Y_1 and v_2 (d) v_1 and v_2 **[2005 : 2 M]**

3.8 A firm is required to procure three items (*P*, *Q*, and *R*). The prices quoted for these items (in Rs.) by suppliers S_1, S_2 and S_3 are given in table. The management policy requires that each item has to be supplied by only one supplier and one supplier supply only one item. The minimum total cost (in Rs.) of procurement to the firm is

Item	Supplier		
	S_1	S_2	S_3
P	110	120	130
Q	115	140	140
R	125	145	165

(a) 350 (b) 360
(c) 385 (d) 395 [2006 : 2 M]

3.9 For the standard transportation linear program with m sources and n destinations and total supply equaling total demand, an optimal solution (lowest cost) with the smallest number of non-zero x_{ij} values (amounts from source i to destination j) is desired. The best upper bound for this number is
(a) mn (b) $2(m + n)$
(c) $m + n$ (d) $m + n - 1$ [2008 : 2 M]

Common Data for Questions 3.10 and 3.11 :
Consider the linear Programme (LP)
Max $4x + 6y$
Subject to
$3x + 2y \le 6$
$2x + 3y \le 6$
$x, y \ge 0$

3.10 After introducing slack variables s and t, the initial basic feasible solution is represented by the table below (basic variables are $s = 6$ and $t = 6$, and the objective function value is 0).

	–4	–6	0	0	0
s	3	2	1	0	6
t	2	3	0	1	6
	x	y	s	t	RHS

After some simplex iterations, the following table is obtained

	0	0	0	2	12
S	5/3	0	1	–1/3	2
y	2/3	1	0	1/3	2
	x	y	s	t	RHS

From this, one can conclude that
(a) the LP has a unique optimal solution
(b) the LP has an optimal solution that is not unique
(c) the LP is infeasible
(d) the LP is unbounded [2008 : 2 M]

3.11 The dual for the LP in **Q. 3.10** is
(a) Min $6u + 6v$
subject to
$3u + 2v \ge 4$
$2u + 3v \ge 6$
$u, v \ge 0$
(b) Max $6u + 6v$
subject to
$3u + 2v \le 4$
$2u + 3v \le 6$
$u, v \ge 0$
(c) Max $4u + 6v$
subject to
$3u + 2v \ge 6$
$2u + 3v \ge 6$
$u, v \ge 0$
(d) Min $4u + 6v$
subject to
$3u + 2v \le 6$
$2u + 3v \le 6$
$u, v \ge 0$
[2008 : 2 M]

3.12 Consider the following Linear Programming Problem (LPP):
Maximize $z = 3x_1 + 2x_2$
Subject to $x_1 \le 4$
$x_2 \le 6$
$3x_1 + 2x_2 \le 18$
$x_1 \ge 0, x_2 \ge 0$
(a) The LPP has a unique optimal solution
(b) The LPP is infeasible
(c) The LPP is unbounded
(d) The LPP has multiple optimal solution
[2009 : 2 M]

3.13 Simplex method of solving linear programming problem uses
(a) all the points in the feasible region
(b) only the corner points of the feasible region
(c) intermediate points within the infeasible region
(d) only the interior points in the feasible region
[2010 : 1 M]

Common Data Questions 3.14 and 3.15.
One unit of product P_1 requires 3 kg of resource R_1 and 1 kg resource R_2. One unit of product P_2 requires 2 kg of resource R_1 and 2 kg of resource R_2. The profits per unit by selling product P_1 and P_2 and ₹ 2000 and ₹ 3000 respectively. The manufacturer has 90 kg of resource R_1 and 100 kg of resource R_2.

3.14 The manufacturer can make a maximum profit of Rs.
(a) 60000 (b) 135000
(c) 150000 (d) 200000 [2011 : 2 M]

3.15 The unit worth of resource R_2, *i.e.* dual price of resource R_2 in ₹ per kg is
(a) 0 (b) 1350
(c) 1500 (d) 2000 [2011 : 2 M]

3.16 A linear programming problem is shown below:
Maximize $3x + 7y$
Subject to $3x + 7y \le 10$
$4x + 6y \le 8$
$x, y \ge 0$
It has
(a) an unbounded objective function
(b) exactly one optimal solution
(c) exactly two optimal solution
(d) infinitely many optimal solution [2013 : 2 M]

3.17 Consider an objective function $Z(x_1, x_2) = 3x_1 + 9x_2$ and the constraints
$x_1 + x_2 \le 8$
$x_1 + 2x_2 \le 4$
$x_1 \ge 0, x_2 \ge 0$

The maximum value of the objective function is ______.

[2014 : 2 M, Set-3]

3.18 If there are m sources and n destinations in a transportation matrix, the total number of basic variables in a basic feasible solution is

(a) $m + n$ (b) $m + n + 1$
(c) $m + n - 1$ (d) m

[2014 : 1 M, Set-2]

3.19 The total number of decision variables in the objective function of an assignment problem of size $n \times n$ (n jobs and n machines) is

(a) n^2 (b) $2n$
(c) $2n - 1$ (d) n [2014 : 1 M, Set-4]

3.20 For the linear programming problem:

$$\text{Maximize } z = 3x_1 + 2x_2$$
Subject to
$$-2x_1 + 3x_2 \le 9$$
$$x_1 - 5x_2 \ge -20$$
$$x_1, x_2 \ge 0$$

The above problem has

(a) unbounded solution
(b) infeasible solution
(c) alternative optimum solution
(d) degenerate solution [2015 : 2 M, Set-3]

3.21 Maximize $z = 15x_1 + 20x_2$

Subject to
$$12x_1 + 4x_2 \ge 36$$
$$12x_1 - 6x_2 \ge 24$$
$$x_1, x_2 \ge 0$$

The above linear programming problem has

(a) infeasible solution
(b) unbounded solution
(c) alternative optimum solutions
(d) degenerate solution [2016 : 2 M, Set-1]

3.22 Two models, P and Q of a product earn profits of ₹ 100 and ₹. 80 per piece, respectively. Production times for P and Q are 5 hours and 3 hours, respectively, while the total production time available is 150 hours. For a total batch size of 40, to maximize profit, the member of units of P to be produced is ____. [2017 : 2 M, Set-1]

3.23 Maximize $Z = 5x_1 + 3x_2$,

subject to
$$x_1 + 2x_2 \le 10$$
$$x_1 - x_2 \le 8$$
$$x_2, x_2 \ge 0$$

In the starting simplex tableau, x_1 and x_2 are non-basic variables and the value of Z is zero. The value of Z in the next simplex tableau is _______.

[2017 : 2 M, Set-2]

3.24 A product made in two factories, P and Q, is transported to two destinations, R and S. The per unit costs of transportation (in Rupees) from factories to destinations are as per the following matrix:

Factory \ Destination	R	S
P	10	7
Q	3	4

Factory P produces 7 units and factory Q produces 9 units of the product. Each destination requires 8 units. If the north-west corner method provides the total transportation cost X (in Rupees) and the optimized (the minimum) total transportation cost is Y (in Rupees), then $(X - Y)$, in Rupees, is

(a) 0 (b) 15
(c) 35 (d) 105

[2017 : 2 M, Set-2]

3.25 The minimum value of $3x + 5y$

such that:
$$3x + 5y \le 15$$
$$4x + 9y \le 8$$
$$13x + 2y \le 2$$
$$x \ge 0,\ y \ge 0$$

is__________. [2018 : 2 M, Set-1]

3.26 The problem of maximizing $Z = x_1 - x_2$ subject to constrains $x_1 + x_2 \le 10$, $x_1 \ge 0$, $x_2 \ge 0$ and $x_2 \le 5$ has

(a) no solution (b) one solution
(c) two solutions (d) more than two solutions

[2018 : 2 M, Set-2]

3.27 Five jobs (J1, J2, J3, J4 and J5) need to be processed in a factory. Each job can be assigned to any of the five different machines (M1, M2, M3, M4 and M5). The time durations taken (in minutes) by the machines for each of the jobs, are given in the table. However, each job is assigned to a specific machine in such a way that the total processing time is minimum. The total processing time is ________ minutes.

	M1	M2	M3	M4	M5
J1	40	30	50	50	58
J2	26	38	60	26	38
J3	40	34	28	24	30
J4	28	40	40	32	48
J5	28	32	38	22	44

[2019 : 2 M, Set-1]

3.28 Two business owners Shveta and Ashok run their businesses in two different states. Each of them, independent of the other, produces two products *A* and *B*, sell them at Rs. 2,000 per kg and Rs. 3,000 per kg, respectively, and uses Linear Programming to determine the optimal quantity of *A* and *B* to maximize their respective daily revenue. Their constraints are as follows: (i) for each business owner, the production process is such that the daily production of *A* has to be at least as much as *B*, and the upper limit for production of *B* is 10 kg per day, and (ii) the respective state regulations restrict Shveta's production of *A* to less than 20 kg per day and Ashok's production of *A* to less than 15 kg per day. The demand of both *A* and *B* in both the states is very high and everything produced is sold.

The absolute value of the difference in daily (optimal) revenue of Shveta and Ashok is _____ thousand Rupees (round off to 2 decimal places)

[2020 : 2 M, Set-1]

3.29 A factory produces m (i = 1,2, ... , m) products each of which requires processing on *n* (j = 1, 2, ... *n*) workstations. Let a_{ij} be the amount of processing time that one unit of the ith product requires on the jth workstation. Let the revenue from selling one unit of the ith product be r_i and h_i be the holding cost per unit per time period for the ith product. The planning horizon consists of T (t = 1, 2, ...T) time periods. The minimum demand that must be satisfied in time period *t* is d_{it}, and the capacity of the jth workstation below, with decision variables S_{it} (amount of product *i* sold in time period t), X_{it} (amount of product *i* manufactured in time period *t*) and I_{it} (amount of product *i* held in inventory at the end of time period *t*).

$$\max \sum_{t=1}^{T}\sum_{i=1}^{m}(r_i S_{it} - h_i I_{it})$$

$$S_{it} \geq d_{it} \quad \forall\ i,t$$

< capacity constraint >

< inventory balance constraint >

$$X_{it},\ S_{it},\ I_{it} \geq I_{io} = 0$$

The capacity constraints and inventory balance constraints for this formulation respectively are

(a) $\sum_{i}^{m} a_{ij} X_{it} \leq d_{it} \quad \forall\ i,t$ and

$I_{it} = I_{i,t-1} + X_{it} - S_{it} \quad \forall\ i,t$

(b) $\sum_{i}^{m} a_{ij} X_{it} \leq d_{it} \quad \forall\ i,t$ and

$I_{it} = I_{i,t-1} + S_{it} - X_{it} \quad \forall\ i,t$

(c) $\sum_{i}^{m} a_{ij} X_{it} \leq c_{jt} \quad \forall\ j,t$ and

$I_{it} = I_{i,t-1} + X_{it} - S_{it} \quad \forall\ i,t$

(d) $\sum_{i}^{m} a_{ij} X_{it} \leq c_{jt} \quad \forall\ i,t$ and

$I_{it} = I_{i,t-1} + X_{it} - d_{it} \quad \forall\ i,t$ **[2021 : 2 M, Set-2]**

3.30 In a linear programming problem, if a resource is not fully utilized, the shadow price of that resource is

(a) positive (b) negative
(c) zero (d) infinity

[2022 : 1 M, Set-1]

3.31 An assignment problem is solved to minimize the total processing time of four jobs (1, 2, 3 and 4) on four different machines such that each job is processed exactly by one machine and each machine processes exactly one job. The minimum total processing time is found to be 500 minutes. Due to a change in design, the processing time of Job 4 on each machine has increased by 20 minutes. The revised minimum total processing time will be ________ minutes (in integer). **[2022 : 2 M, Set-1]**

3.32 A manufacturing unit produces two products P_1 and P_2. For each piece of P_1 and P_2, the table below provides quantities of materials M_1, M_2, and M_3 required, and also the profit earned. The maximum quantity available per day for M_1, M_2 and M_3 is also provided. The maximum possible profit per day is ₹________.

	M_1	M_2	M_3	Profit per piece (₹)
P_1	2	2	0	150
P_2	3	1	2	100
Maximum quantity available per day	70	50	40	

(a) 5000 (b) 4000
(c) 3000 (d) 6000

[2022 : 2 M, Set-2]

3.33 Which one of the options given represents the feasible region of the linear programming model:
Maximize $45X_1 + 60X_2$

$$X_1 \leq 45$$
$$X_2 \leq 50$$
$$10X_1 + 10X_2 \geq 600$$
$$25X_1 + 5X_2 \leq 750$$

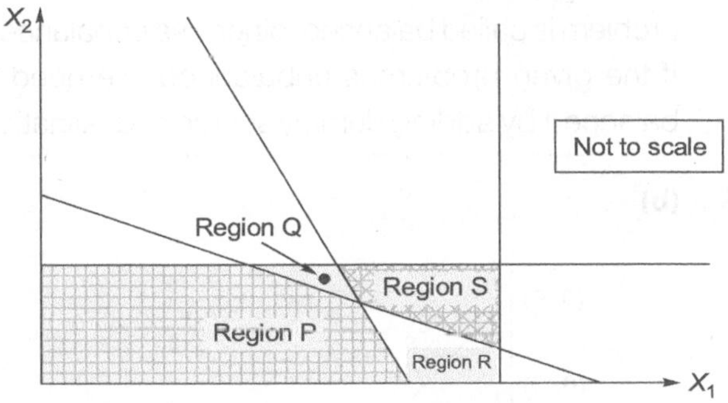

(a) Region P (b) Region Q
(c) Region R (d) Region S

[2023 : 1 Mark]

3.34 At the current basic feasible solution (bfs) $v_o (v_o \in \mathbb{R}^5)$, the simplex method yields the following form of a linear programming problem in standard form.

Minimize, $z = -x_1 - 2x_2$

st. $x_3 = 2 + 2x_1 - x_2$

$x_4 = 7 + x_1 - 2x_2$

$x_5 = 3 - x_1$

$x_1, x_2, x_3, x_4, x_5 \geq 0$

Here the objective function is written as a function of the non-basic variables. if the simplex method moves to the adjacent bfs $v_1 (v_1 \in \mathbb{R}^5)$ that best improves the objective function, which of the following represents the objective function at v_1, assuming that the objective function is written in the same manner as above?

(a) $z = -4 - 5x_1 + 2x_4$
(b) $z = -4 - 5x_1 + 2x_3$
(c) $z = -6 - 5x_1 + 2x_3$
(d) $z = -3 - x_5 - 2x_2$

[2024 : 2 M]

Answers Linear Programming & Transportation

3.1 (a) 3.2 (*) 3.3 (a) 3.4 (b) 3.5 (c) 3.6 (b) 3.7 (d)
3.8 (c) 3.9 (d) 3.10 (b) 3.11 (a) 3.12 (d) 3.13 (b) 3.14 (b) 3.15 (a)
3.16 (b) 3.17 (18) 3.18 (c) 3.19 (a) 3.20 (a) 3.21 (b) 3.22 (15)
3.23 (40) 3.24 (c) 3.25 (0) 3.26 (b) 3.27 (146) 3.28 (10) 3.29 (c)
3.30 (c) 3.31 (520) 3.32 (b) 3.33 (b) 3.34 (b)

Explanations Linear Programming & Transportation

3.2 (*)

Given system are

$$x_1 + 3x_2 \leq 40 \quad \text{...(I)}$$
$$3x_1 + x_2 \leq 24 \quad \text{... (II)}$$
$$x_1 + x_2 \leq 10 \quad \text{... (III)}$$
$$x_1 > 0, x_2 > 0$$

Objective function,

$z = 2x_1 + 5x_2$

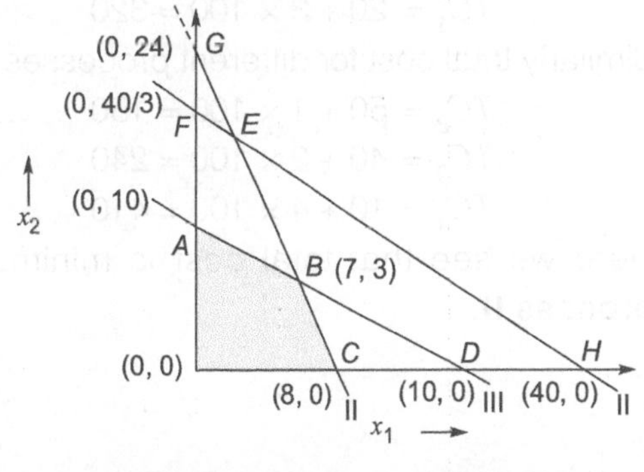

Solving the eqs. (ii) and (iii); we get

At A(0, 10), $z_{max} = 2 \times 0 + 5 \times 10 = 50$

At C(8, 0), $z_{max} = 8 \times 2 + 5 \times 0 = 16$

At B,(7, 3) $z_{max} = 2 \times 7 + 5 \times 3 = 29$

So maximum value of $z = 50$

No option is available.

3.3 (a)

Given $P + 2Q \leq 2000$ ...(i)

$P + Q \leq 1500$ ... (ii)

$Q \leq 600$

Objective function

$Z_{max} = 3P + 5Q$

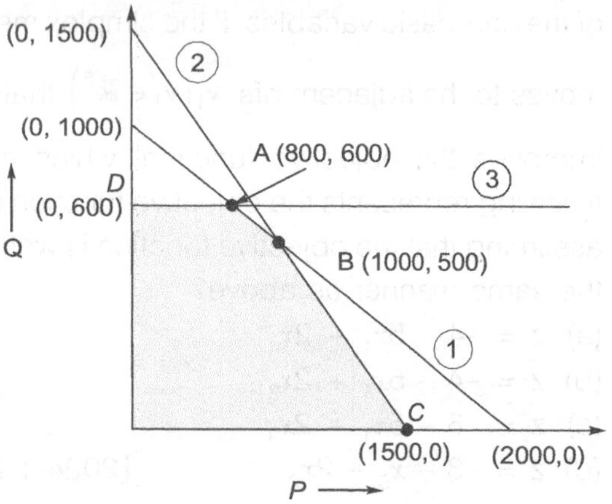

At point A, $z = 3 \times 800 + 600 \times 5 = 5400$

At point B, $z = 3 \times 1000 + 5 \times 500 = 5500$

At point C, $z = 3 \times 1500 + 5 \times 0 = 4500$

At point D, $z = 3 \times 0 + 5 \times 600 = 3000$

Hence z to be maximum at **B(1000, 500)**

3.4 (b)

No. of pieces produced

$N = 100$ piece

We know that,

Total cost: $TC = F + VN$

F = fixed cost

V = variable cost

For process I, total cost,

$TC_1 = 20 + 3 \times 100 = 320$

Similarly total cost for different processes,

$TC_2 = 50 + 1 \times 100 = 150$

$TC_3 = 40 + 2 \times 100 = 240$

$TC_4 = 10 + 4 \times 100 = 410$

Here we see that total cost is minimum for **process II**.

3.5 (c)

	S_1	S_2	
D_1			(20 – 40)
D_2			≥ 20
	50	40	

For balance transportation problem

no. of supply points $= m + n - 1 = 2 + 2 - 1 =$ **3**

No. of demand point = **4**

Total demand = 50 + 40 = **90**

POINTS TO REMEMBER

If the total supply from all the factories is equal to the total demand in all the destination, then the problem is called balanced, otherwise unbalanced. If the given problem is unbalanced, we need to balance it by adding dummy source or destination.

3.6 (b)

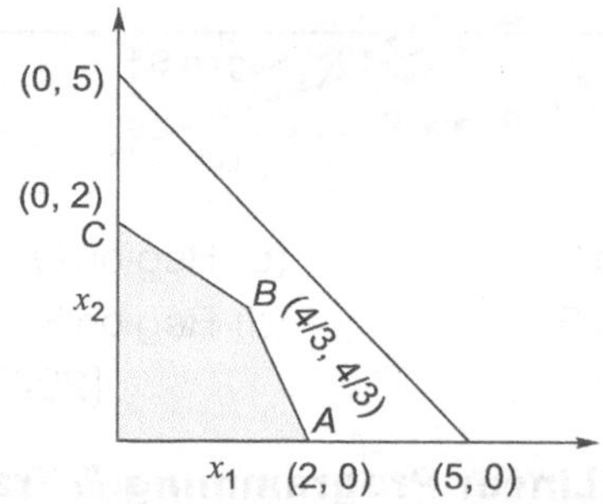

At point A (2, 0), $z = 1 \times 2 + 1 \times 0 = 2$

At point B, $\left(\frac{4}{3}, \frac{4}{3}\right)$, $z = 1 \times \frac{4}{3} + 1 \times \frac{4}{3}$

At point C, (0, 2) $= 1 \times 0 + 1 \times 2 = 2$

Constant equation $x_1 + x_2 \leq 5$ has no effect on optimal region.

So, maximum value of z occurs at point $B\left(\frac{4}{3}, \frac{4}{3}\right)$

Hence, optimal solution, $\left(\frac{4}{3}, \frac{4}{3}\right)$

3.8 (c)

	S_1	S_2	S_3
P	110	120	130
Q	115	140	140
R	125	145	165

Step 1 : Substract minimum entry in each column from all the entries on that column,

	S_1	S_2	S_3
P	0	0	0
Q	5	20	10
R	15	25	35

Step 2 : Substract minimum entry in each row of job-opportunity cost matrix from all the entries of that row.

0	0	0
0	15	5
0	10	20

Step 3 : Draw minimum number of horizontal and vertical lines to cover all zeros. This can be done in 2 lines, which is one less than the number of rows (which is 3). Thus the solution is yet not fixed. Hence go to step 4

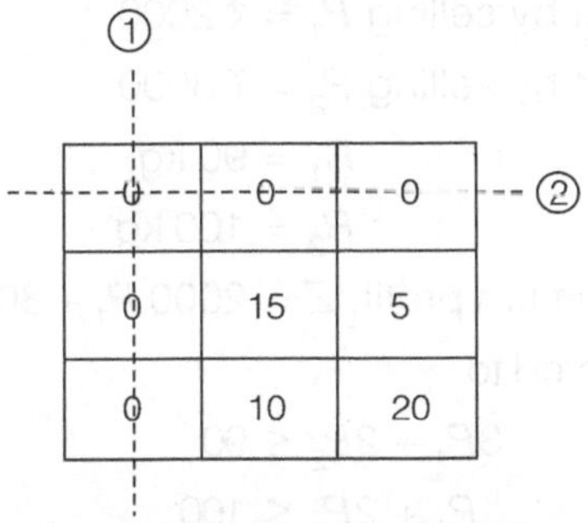

Step 4 : From the uncovered entries, the minimum is 5. Thus, substract 5 from all the entries, which are uncovered. Add 5 at junction of lines. We get the following matrix now, as the revised opportunity cost matrix.

	S_1	S_2	S_3
P	5	[0]	~~0~~
Q	~~0~~	10	[0]
R	[0]	5	15

Thus final assignment is

Item	Suppliers	Cost
P	S_2	120
Q	S_3	140
R	S_1	125

The minimum cost

$$=120 + 140 + 125 = ₹\,\mathbf{385}$$

POINTS TO REMEMBER

- It is a special case of transporation problem where in each row and column, 1-allocation is possible and the matrix should be a square matrix i.e. ($m = n$)
- The Hungarian method is used to solve assignment problem and it is developed only for minimization.

3.9 (d)

Degeneracy arises if number of allocation are less than $m + n - 1$

3.10 (b)

Method I:

Since zero has appeared in the optimal solution below a non-basic variable, it means there are alternate solutions. The LP has an optimal solution that is not unique.

Method II:

Given problem can be modified in terms of linear programming problem as,

$$\text{Max } z = 4x + 6y$$

$$3x + 2y \le 6 \quad ,,,(1)$$

$$2x + 3y \le 6 \quad ...(2)$$

$$x, y \le 0$$

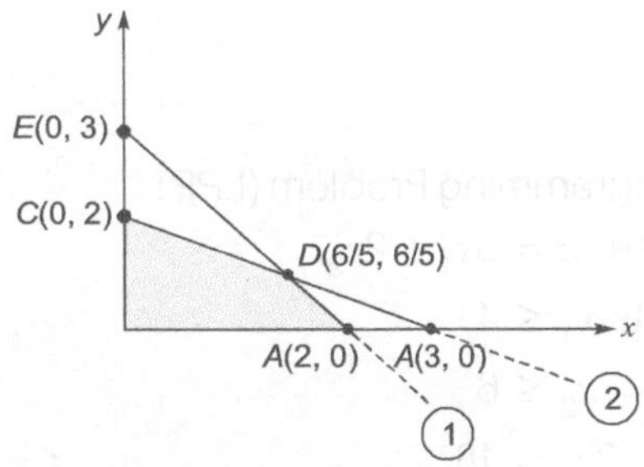

At point B (6/5, 6/3),

$$\text{Max } z = 4 \times \frac{6}{5} + 6 \times \frac{6}{5} = 12$$

At point C(0, 2)

$$\text{Max } z = 4 \times 0 + 6 \times 2 = 12$$

POINTS TO REMEMBER

It is also important to note that maximization z and constraint (2) $2x + 3y \le 6$ have the same slope. So we can also say that linear programing has multiple optimal solution.

3.11 (a)

Primal is

$\text{Max } Z = (4 \;\; 6)\begin{pmatrix} x \\ y \end{pmatrix}$ i.e Max $Z = CX$

Subject to $\begin{pmatrix} 3 & 2 \\ 2 & 3 \end{pmatrix}(x \;\; y) \le \begin{pmatrix} 6 \\ 6 \end{pmatrix}$ i.e $AX \le B$

Dual is

Min $Z = B^T Y$

$\text{Min } Z = (6 \;\; 6)\begin{pmatrix} u \\ v \end{pmatrix} \Rightarrow \text{Min } Z = 6u + 6v$

$A^T Y \ge C^T$

$$\begin{pmatrix} 3 & 2 \\ 2 & 3 \end{pmatrix}\begin{pmatrix} u \\ v \end{pmatrix} \ge \begin{pmatrix} 4 \\ 6 \end{pmatrix} \Rightarrow \begin{matrix} 3u + 2v \ge 4 \\ 2u + 3v \ge 6 \end{matrix}$$

and u, v ≥ 0

Note:

$$\left[\begin{array}{l} \text{If Primal is given by Min } Z = CX \\ AX \ge B \\ \text{the corresponding dual for LP is given by} \\ \text{Max } Z = B^T Y \\ A^T Y \le CT \end{array}\right]$$

$$\left[\begin{array}{l} \text{And if the Primal is given by Max } Z = CX \\ AX \le B \\ \text{Then corresponding dual is given by} \\ \text{Min } Z = B^T Y \\ A^T Y \ge C^T \end{array}\right]$$

3.12 (d)

Linear Programming Problem (LPP)

Maximize, $z = 3x_1 + 2x_2$

Constraints $x_1 \le 4$ ... (i)

$x_2 \le 6$ ... (ii)

$3x_1 + 2x_2 \le 18$ ... (iii)

$x_1 \le 0, x_2 \le 0$ ... (iv)

Using graphical method

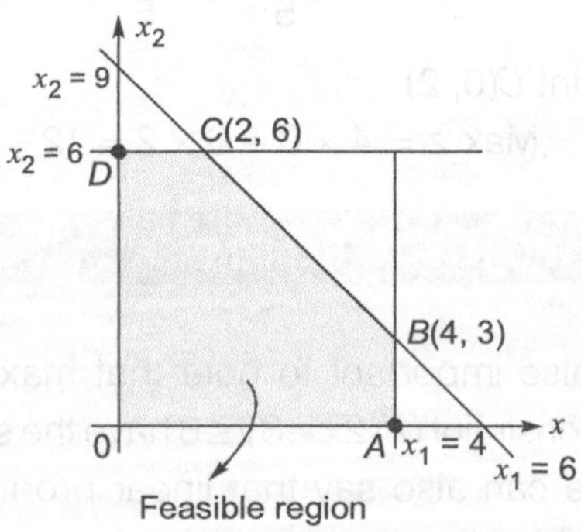

Because objective function have slope same as constraint (iii) *i.e.* objective function is parallel to constraint. Therefore the LPP has multiple optimal solution.

For example at Point B,

Maximum, $z = 3x_1 + 2x_2$

$= 3(4) + 2(3) = 18$

and at point C,

Maximum, $z = 3(2) + 2(6) = $ **18**

3.13 (b)

Linear programming solved by simplex method uses only the corner points of the feasible region because at one of these points give the value of objective function either maximum or minimum.

3.14 (b)

For product:

$P_1 = 3$ kg of R_1 and 1 kg of R_2 is required.

For product:

$P_2 = 2$ kg of R_1 and 2 kg of R_2 is required.

Profit by selling P_1 = ₹ 2000

Profit by selling P_2 = ₹ 3000

R_1 = 90 kg

R_2 = 100 kg

Maximum profit, $Z = 2000\, P_1 + 3000\, P_2$

Subjected to

$3P_1 + 2P_2 \le 90$ ... (i)

$P_1 + 2P_2 \le 100$ ... (ii)

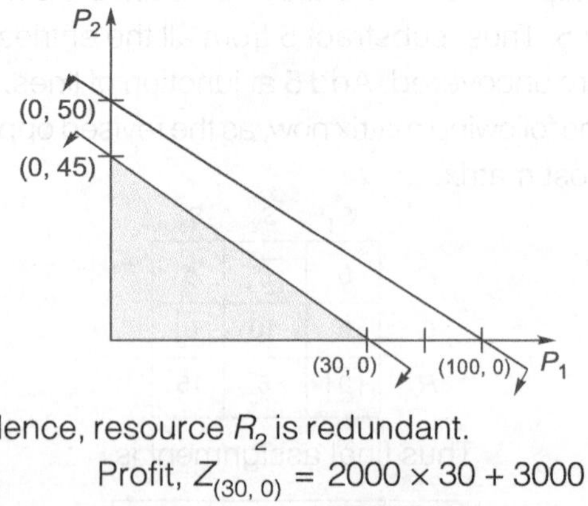

Hence, resource R_2 is redundant.

Profit, $Z_{(30, 0)} = 2000 \times 30 + 3000 \times 0$

$=$ ₹ 60000

$Z_{(0, 45)} = 2000 \times 0 + 3000 \times 45$

$=$ **₹135000**

3.15 (a)

The unit worth of resources R_2 i.e. dual price of resources R_2 in ₹ per kg is zero because the constraiht (2) is redundant. Therefore resources R_2 has no effect on the feasible solution.

3.16 (b)

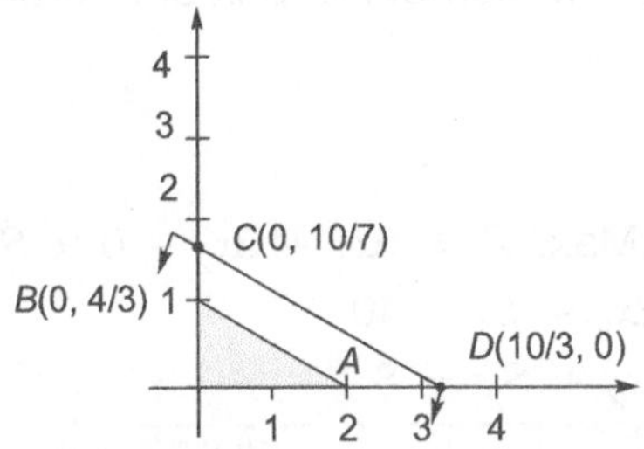

Max. $3x + zy$

Sub to $3x + 7y \le 10$

$4x + 6y \le 8$

$x, y \ge 0$

At point A, (2, 0),

$$z = 3 \times 2 + 7 \times 0 = 6$$

At point $B\left(0, \frac{4}{3}\right)$

$$z = 3 \times 0 + 7 \times \frac{4}{3} = \frac{28}{3} = 9.33$$

Hence, it has exactly one optimal solution.

3.17 Sol.

Max $Z = 3x_1 + 2x_2$

Subjected to:

$x_1 + x_2 \le 8$

$x_1 + 2x_2 \le 4$

$x_1, x_2 \ge 0$

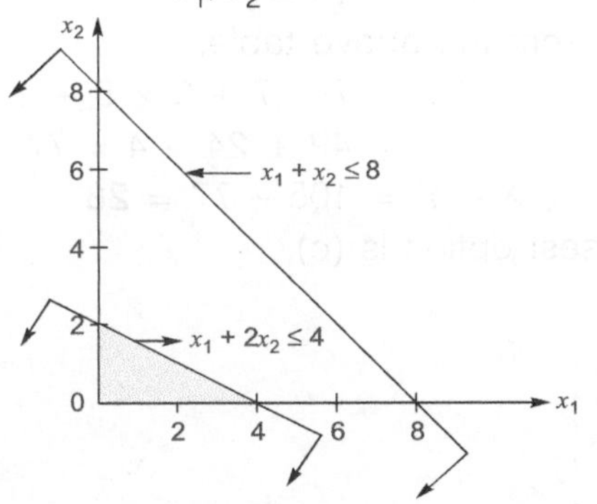

At point (4, 0), $(Z)_{(4,\,0)} = 3 \times 4 + 9 \times 0 = 12$

At pooint (0, 2), $(Z)_{(0,\,2)} = 3 \times 0 + 9 \times 2 = 18$

$Z_{max} = 18$

3.18 (c)

Basic feasible solution: For a maximum transportation problem, if the total number of allocatio is exactly equal to $m + n - 1$, then the solution is called basic feasible solution.

POINTS TO REMEMBER

Non-degenerate basic feasible solutions: For a maximum transportation problem, feasible solution is not degenerate, if the following two conditions are satisfied;

- Total number of allocation is exactly equal to $m + n - 1$
- These $m + n - 1$, allocations must be at independent position.

3.19 (a)

- Number of decision variables = $n \times n = n^2$
- Number of constraint = $n + n = 2n$
- Number of assignment = n
- Possible ways to make assignment = $n!$

3.20 (a)

Maximize $z = 3x_1 + 2x_2$

Subjected to

$$-2x_1 + 3x_2 \le 9 \quad ...(1)$$

$$x_1 - 5x_2 \ge -20 \quad ...(2)$$

Converting constraint (2) in ≤ form,

$$-2x_1 + 3x_2 \le 9 \quad ...(1)$$

$$-x_1 + 5x_2 \le 20 \quad ...(2)$$

Again,

$$\frac{x_1}{\left(\frac{-9}{2}\right)} + \frac{x_2}{3} \le 1 \quad ...(1)$$

$$\frac{x_1}{(-20)} + \frac{x_2}{4} \le 1 \quad ...(2)$$

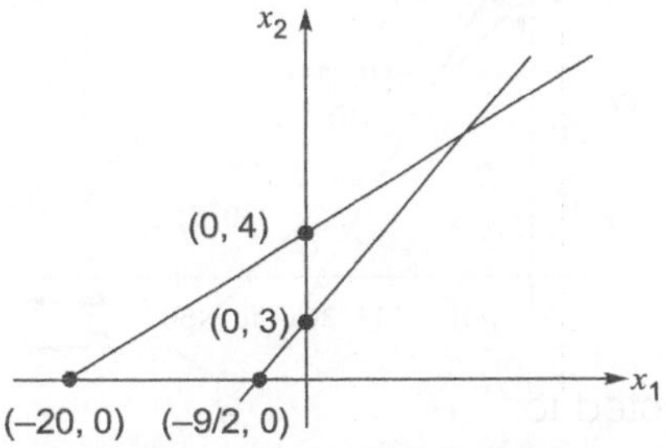

Since this is a problem for Max. z and according to the given constant, the maximum value of objective function lies at infinity. So it simply means that problem has unbounded solution.

POINTS TO REMEMBER

If according to the given condition, the greatest value of the objective function lies at infinity, then it simply means that the common feasible region is not bounded by limit on the constraint and the solution will be unbounded solution.

3.21 (b)

Maximize $z = 15x_1 + 20x_2$

Subjected to,

$12x_1 + 4x_2 \geq 36$ or $3x_1 + x_2 \geq 9$

$12x_1 - 6x_2 \leq 24$ or $2x_1 - x_2 \leq 4$

Let

$3x_1 + x_2 = 9$ or $2x_1 - x_2 = 4$

$x_1 = 0, x_2 = 9$ or $x_1 = 0, x_2 = -4$

$x_1 = 3, x_2 = 0$ or $x_1 = 2, x_2 = 0$

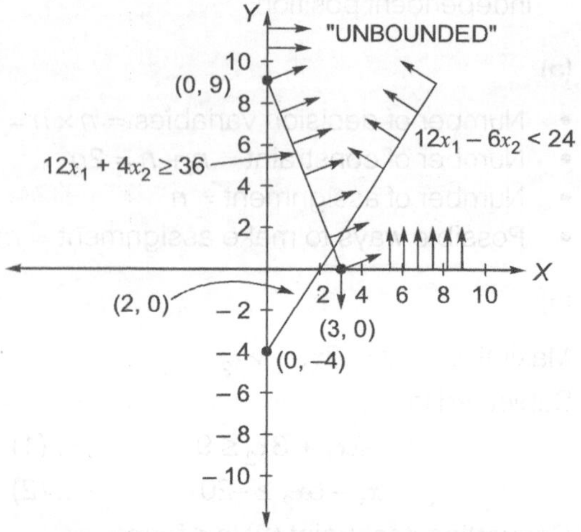

Since maximum value of objective function lies at infinity. So, problem has unbounded solution.

3.22 Sol.

Max. $Z = 100 \times P + 80 \times Q$

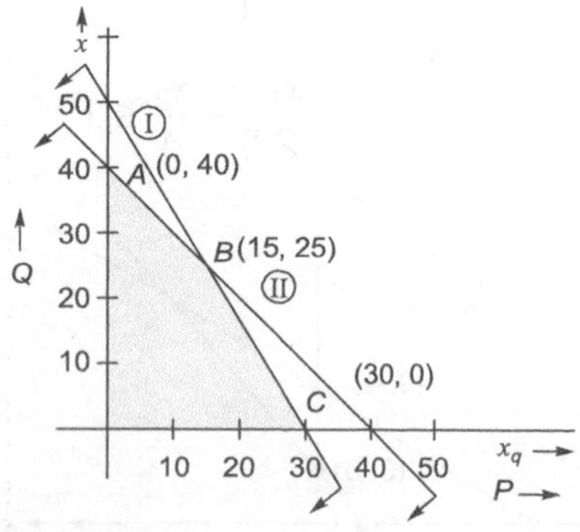

Subjected to:

$5P + 3Q \leq 150$

$P + Q \leq 40$

$$\frac{P}{30} + \frac{Q}{50} = 1$$

$$\frac{P}{40} + \frac{Q}{40} = 1$$

Putting value of corner point in objective function

$Z(A) = 3200$

$Z(B) = 3500$

$Z(C) = 3000$

Maximum at B, $P = 15$, $Q = 25$

Hence the member of unit of P to be produced is 15.

3.23 Sol.

Max. $Z = 5x_1 + 3x_2 + 0 \times S_1 + 0 \times S_2$

$x_1 + 2x_2 + S_1 = 10$

$x_1 - x_2 + S_2 = 8$

e_i	x_1	x_2	S_1	S_2	b_i	$\theta_i = b_i/a_{ij}$
0 S_1	1	2	1	0	10	10
0 S_2	(1)*	–1	0	1	8	8
C_j	5	3	0	0		
Z_j	0	0	0	0		
Δ_j	5	3	0	0		

Key Row (the S_2 row); Key column (the x_1 column)

Objective function will increase by ` 5 per unit of x_1 and bring 8 units.

So, total increase (8 × 5 = ` 40)

3.24 (c)

$X = 105$

Using penalty corner method and following modi method, we get

	R	S	Penalty		
P	[10]	[7] 7	7	3	0
Q	[3] 8	[4] 1	9	1	–3
	8	8			
Penalty	7	3			
v_j	6	7			

(4 marked in the P–R cell)

So, from the above table,

$Y = 7 \times 7 + 3 \times 8 + 4$

$= 49 + 24 + 4 = 77$

$X - Y = 105 - 77 = \mathbf{28}$

Closest option is (c).

3.25 Sol.

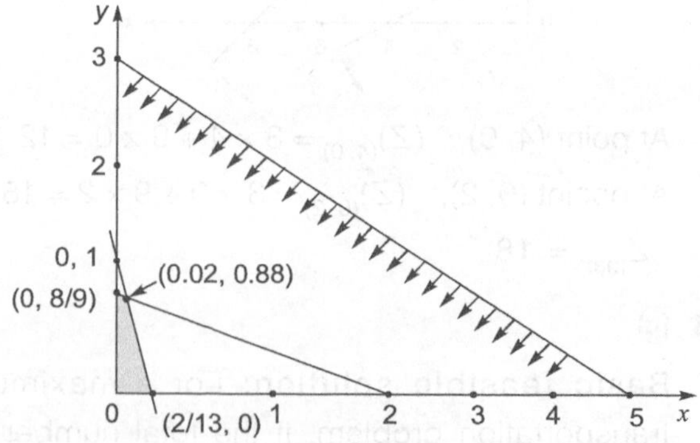

$Z = 3x + 5y$

Let us consider $3x + 5y \leq 15$ ...(i)

If $3x + 5y = 15$

$x = 0, y = 3$

$x = 5, y = 0$

Let us consider $4x + 9y \le 8$ ...(ii)

If $4x + 9y = 8$

$$x = 0,\ y = \frac{8}{9}$$

$$x = 2,\ y = 0$$

Let us consider $13x + 2y \le 2$ ...(iii)

If $13x + 2y = 2$

$$x = 0,\ y = 1$$

$$x = \frac{2}{13},\ y = 0$$

Comparing (ii) and (iii)

$4x + 9y = 8$ ...(ii)

$13x + 2y = 2$ ...(iii)

From (ii) $x = \frac{8-9y}{4}$

Putting in (iii)

$$\frac{13(8-9y)}{4} + 2y = 2$$

$$104 - 117y + 8y = 8$$

$$109y = 96$$

$\therefore$ $y = 0.88$

Hence, $x = 0.018$

Checking at corner points:

$$\therefore\quad Z\left(0, \frac{8}{9}\right) = 3(0) + 5\left(\frac{8}{9}\right) = 4.44$$

$$Z(0.018, 0.88) = 3(0.18) + 5(0.88) = 4.454$$

$$Z\left(\frac{2}{13}, 0\right) = 3\left(\frac{2}{13}\right) + 5(0) = 0.46$$

At (0, 0) $Z = 0$ so minimum value will be **0**.

3.26 (b)

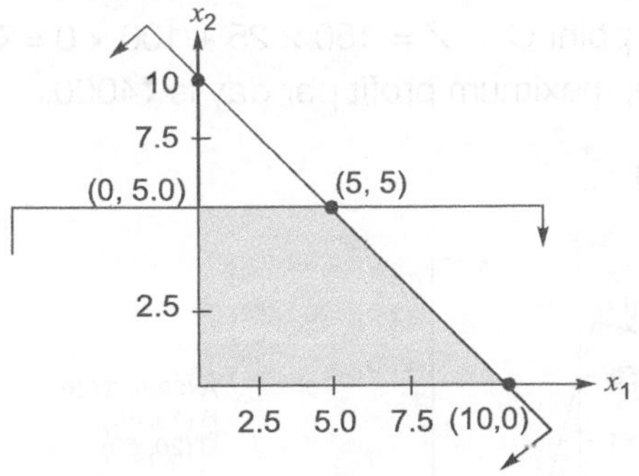

Max., $Z = x_1 - x_2$

Constrains : $x_1 + x_2 \le 10$;

$x_1 \ge 0$; $x_2 \ge 0$ and $x_2 \le 5$

$$Z(0, 5) = 0 - 5 = -5$$

$$Z(5, 5) = 5 - 5 = 0$$

$$Z(10, 0) = 10 - 0 = 10$$

$$Z_{max} = 10 \text{ at } (10, 0)$$

$\therefore$ **The problem has one solution**

3.27 Sol.

	M1	M2	M3	M4	M5
J1	40	30	50	50	58
J2	26	38	60	26	38
J3	40	34	28	24	30
J4	28	40	40	32	48
J5	28	32	38	22	44

Subtracting minimum value of each row from corresponding row elements.

	M1	M2	M3	M4	M5
J1	10	0	20	20	28
J2	0	12	34	0	12
J3	16	10	4	0	6
J4	0	12	12	4	20
J5	6	10	16	0	22

Subtracting minimum value of each column from corresponding column.

	M1	M2	M3	M4	M5
J1	10	[0]	16	20	22
J2	[0]	12	30	~~0~~	6
J3	16	10	~~0~~	~~0~~	[0]
J4	~~0~~	12	8	4	14
J5	6	10	12	[0]	16

Now number of assignments is less than the number of machines. Hence current solution is not optimal. Now adding minimum uncovered value at all junctions and subtracting from uncovered values.

	M1	M2	M3	M4	M5
J1	10	[0]	10	20	16
J2	~~0~~	12	24	~~0~~	[0]
J3	22	16	[0]	6	~~0~~
J4	[0]	12	2	4	8
J5	6	10	6	[0]	10

Hence, Job 1 $\rightarrow$ Machine 2

Job 2 $\rightarrow$ Machine 5

Job 3 $\rightarrow$ Machine 3

Job 4 $\rightarrow$ Machine 1

Job 5 $\rightarrow$ Machine 4

Total time = 30 + 38 + 28 + 28 + 22 = 146

3.28 Sol.

Maximum $z = 2000x_1 + 3000x_2$

$A \rightarrow x_1$ units $x_1 \ge x_2$

$B \rightarrow x_2$ units $x_2 \geq 10$

$x_1 < 20$

$x_1 < 15$

Shveta's profit = Rs. 70000 at (20, 10)

Ashok's profit = Rs. 60000 at (15, 10)

Difference Rs. 10000

3.29 (c)

$m \rightarrow i$ ... m ← product

$n \rightarrow i$ ... n ← workstation

$a_{ij} \rightarrow$ time

$r_i \rightarrow$ selling price

$h_i \rightarrow$ holding cost

$T \rightarrow t = 1, 2, \ldots T$

$d_{it} \rightarrow$ demand of product in time t

$c_{jt} \rightarrow$ capacity of workstation in time t

$S_{it} \rightarrow$ Number of product sold in time t

$x_{it} \rightarrow$ Number of product produced in time t

$I_{it} \rightarrow$ Number of product i hold in inventory at end of period t

Capacity constraint

$$a_{ij}x_{it} \leq c_{jt}$$

Inventory constraint

$$I_{it} = I_{i,\,t-1} + x_{it} - S_{it}$$

3.30 (c)

The shadow price indicates an additional unit of the resources in order to maximize profit under the resource constraints. If a resource is not completely used, i.e. there is slack then its marginal return is zero.

3.31 (520) (520 to 520)

If due to design change the time of one job increases by 20 minutes, then the total minimum time would also increase by 20 minutes.

New minimum total processing time = 500 + 20 = 520 minutes

3.32 (b)

	M_1	M_2	M_3	Profit per piece (₹)
P_1	2	2	0	150
P_2	3	1	2	100
Maximum quantity available per day	70	50	40	

$$2X_1 + 3X_2 \leq 70 \quad \ldots \text{(i)}$$

$$2X_1 + X_2 \leq 50 \quad \ldots \text{(ii)}$$

$$2X_2 \leq 40 \quad \ldots \text{(iii)}$$

$$\text{Max.}Z = 150X_1 + 100X_2$$

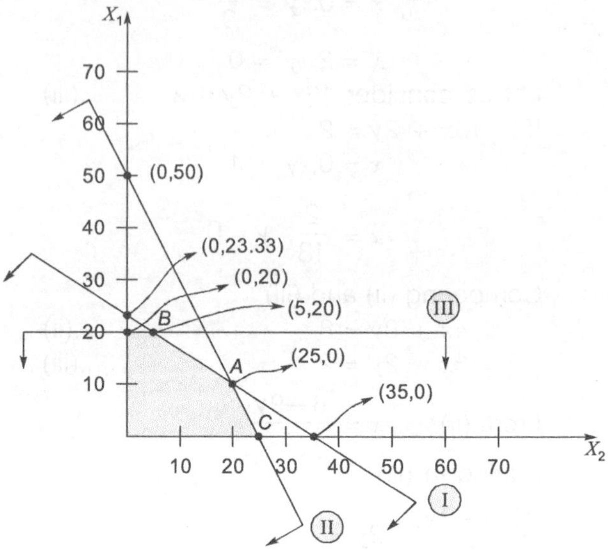

Solution of equation (i) and (iii) gives point B

$$2X_1 + 3 \times 20 = 70$$

$$X_1 = \frac{70-60}{2} = 5$$

So, coordinates of point B is (5, 20)

Solution of equation (i) and (ii) gives point A

$$(50 - X_2) + 2X_2 = 70$$

$$50 + 2X_2 = 70$$

$$X_2 = 10$$

$$\Rightarrow \quad X_1 = \frac{50-10}{2} = 20$$

So, coordinates of point A is (20, 10)

At point A, $Z = 150 \times 20 + 100 \times 10 = ₹4000$

At point B, $Z = 150 \times 5 + 100 \times 20 = ₹2750$

At point C, $Z = 150 \times 25 + 100 \times 0 = ₹3750$

So, maximum profit per day is ₹4000.

3.33 (b)

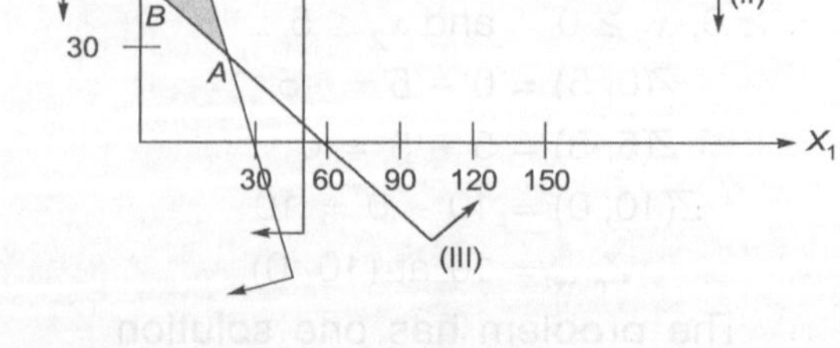

$$Z = 45X_1 + 60X_2$$

$$X_1 \le 45 \Rightarrow \frac{X_1}{45} = 1 \quad \text{... (i)}$$

$$X_2 \le 50 \Rightarrow \frac{X_2}{50} = 1 \quad \text{... (ii)}$$

$$10X_1 + 10X_2 \ge 600 \Rightarrow \frac{X_1}{60} + \frac{X_2}{60} = 1 \quad \text{... (iii)}$$

$$25X_1 + 5X_2 \le 750 \Rightarrow \frac{X_1}{30} + \frac{X_2}{150} = 1 \quad \text{... (iv)}$$

Point A: $X_1 = 60 - X_2$ (From (iii))

$25(60 - X_2) + 5X_2 = 750$ (From (iv))

$X_2 = 37.5$

$X_1 = 60 - 37.5 = 22.5$

Point B: $X_2 = 50$ (From (ii))

$10X_1 + 10(50) = 600$ (From (iii))

$X_1 = 10$

Point C: $X_2 = 50$ (From (ii))

$25X_1 + 5 \times 50 = 750$ (From (iv))

$X_1 = 20$

$$Z(A) = 45 \times 22.5 + 60 \times 37.5 = 3262.5$$

$$Z(B) = 45 \times 10 + 60 \times 50 = 3450$$

$$Z(C) = 45 \times 20 + 60 \times 50 = 3900$$

3.34 (b)

$$z = -x_1 - 2x_2$$

$$x_3 = 2 + 2x_1 - x_2 \quad \text{...(i)}$$

$$x_4 = 7 + x_1 - 2x_2 \quad \text{...(ii)}$$

$$x_5 = 3 - x_1 \quad \text{...(iii)}$$

$$x_1, x_2, x_3, x_4, x_5 \ge 0 \quad \text{...(iv)}$$

From equation (i);

$$x_2 = 2 + 2x_1 - x_3$$

On substituting this value in the objective function,

$$z = -x_1 - 2(2 + 2x_1 - x_3)$$

$$\Rightarrow \quad z = -4 - 5x_1 + 2x_3$$

Hence option (b) is correct.

Also from equation (iii);

$$x_1 = 3 - x_5$$

On substituting this value in the objective function,

$$z = -(3 - x_5) - 2x_2$$

$$\Rightarrow \quad z = -3 + x_5 - 2x_2$$

Hence option (d) is correct.

As per the given options, (b) and (d) both are correct. But since it is an MCQ problem, we will mark option (b) as correct option.

■■■■

4 PERT and CPM

4.1 In PERT, the distribution of activity time is assumed to be
(a) Normal (b) Gamma
(c) Beta (d) Exponential
[1995 : 1 M]

4.2 A dummy activity is used in PERT network to describe
(a) Precedence relationship
(b) Necessary time delay
(c) Resource restriction
(d) Resource idleness [1997 : 1 M]

4.3 The symbol used for Transport in work study is
(a) ⇒ (b) T
(c) ▭ (d) ∇ [2003 : 1 M]

4.4 A project consists of activities A to M shown in the net in the following figure with the duration of the activities Med in days

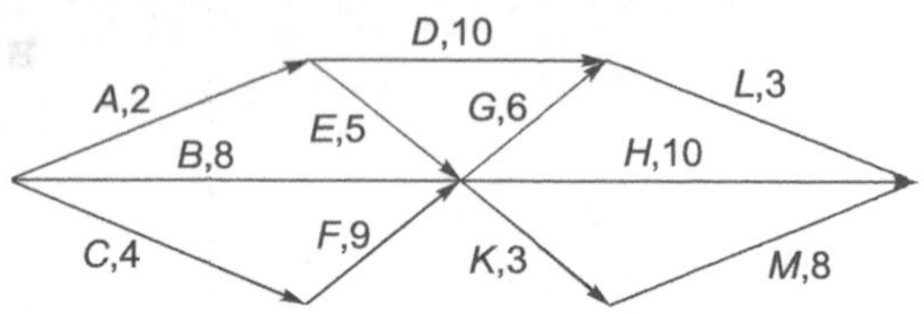

The project can be completed
(a) between 18, 19 days
(b) between 20, 22 days
(c) between 24, 26 days
(d) between 60, 70 days [2003 : 2 M]

4.5 In PERT analysis a critical activity has
(a) maximum float (b) zero float
(c) maximum cost (d) minimum cost
[2004 : 1 M]

4.6 An assembly activity is represented on an Operation Process Chart by the symbol
(a) □ (b) A
(c) D (d) O [2005 : 1 M]

4.7 A project has six activities (A to F) with respective activity durations 7, 5, 6, 6, 8, 4 days. The network has three path A-B, C-D and E-F. All the activities can be crashed with the same crash cost per day. The number of activities that need to be crashed to reduce the project duration by 1 day is
(a) 1 (b) 2
(c) 3 (d) 6 [2005 : 2 M]

Linked Answer Questions 4.8 and 4.9

Consider a PERT network for a project involving six tasks (a to f)

Task	Predecessor	Expected task time (in days)	Variance of the task time (in days^2)
a	–	30	25
b	a	40	64
c	a	60	81
d	b	25	9
e	b,c	45	36
f	d,e	20	9

4.8 The expected completion time of the project is
(a) 238 days (b) 224 days
(c) 171 days (d) 155 days [2006 : 2 M]

4.9 The standard deviation of the critical path of the project is
(a) $\sqrt{151}$ days (b) $\sqrt{155}$ days
(c) $\sqrt{200}$ days (d) $\sqrt{238}$ days
[2006 : 2 M]

4.10 A set of 5 jobs is to be processed on a single machine. The processing time (in days) is given in the table below. The holding cost for each job is ₹ K per day.

Job	Processing time
P	5
Q	2
R	3
S	2
T	1

A schedule that minimizes the total inventory cost is
(a) $T-S-Q-R-P$
(b) $P-R-S-Q-T$
(c) $T-R-S-Q-P$
(d) $P-Q-R-S-T$ [2008 : 1 M]

4.11 For the network below, the objective is to find the length of the shortest path from node P to node G. Let d_{ij} be the length of directed arc from node i to node j.

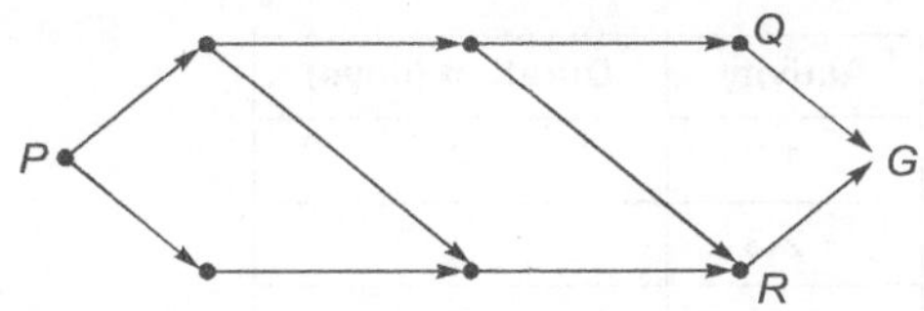

Let s_j be the length of the shortest path from P to node j. Which of the following equations can be used to find S_G?

(a) $S_G = \text{Min}\{S_Q, S_R\}$
(b) $S_G = \text{Min}\{S_Q - d_{QG}, S_R - d_{RG}\}$
(c) $S_G = \text{Min}\{S_Q + d_{QG}, S_R + d_{RG}\}$
(d) $S_G = \text{Min}\{d_{QG}, d_{RG}\}$ [2008 : 2 M]

4.12 The expected time (t_e) of a PERT activity in terms of optimistic time (t_o), pessimistic time (t_p) and most likely time (t_l) is given by

(a) $t_e = \dfrac{t_o + 4t_l + t_p}{6}$ (b) $t_e = \dfrac{t_o + 4t_p + t_l}{6}$

(c) $t_e = \dfrac{t_o + 4t_l + t_p}{3}$ (d) $t_e = \dfrac{t_o + 4t_p + t_l}{3}$

[2009 : 1 M]

Common data for Questions 4.13 and 4.14

Consider the following PERT network:

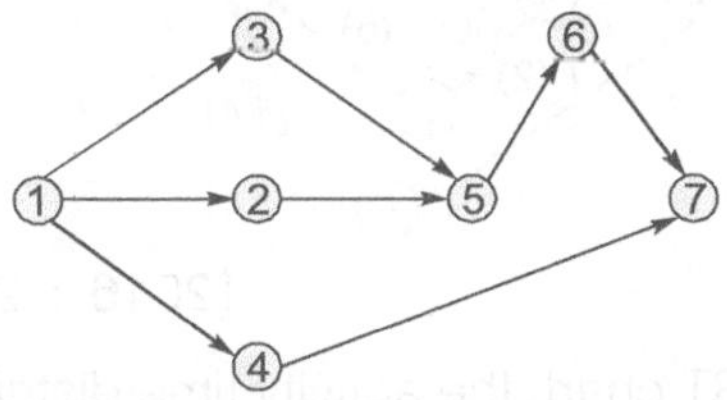

The optimistic time, most likely time and pessimistic time of all the activities are given in the table below:

Activity	Optimistic time (days)	Most likely time (days)	Pessimistic time (days)
1 – 2	1	2	3
1 – 3	5	6	7
1 – 4	3	5	7
2 – 5	5	7	9
3 – 5	2	4	6
5 – 6	4	5	6
4 – 7	4	6	8
6 – 7	2	3	4

4.13 The critical path duration of the network (in days) is

(a) 11 (b) 14
(c) 17 (d) 18 [2009 : 2 M]

4.14 The standard deviation of the critical path is

(a) 0.33 (b) 0.55
(c) 0.88 (d) 1.66 [2009 : 2 M]

4.15 The project activities, precedence relationships and durations are described in the table. The critical path of the project is

Activity	Precedence	Duration (in days)
P	-	3
Q	-	4
R	*P*	5
S	*Q*	5
T	*R*, *S*	7
U	*R*, *S*	5
V	*T*	2
W	*U*	10

(a) *P–R–T–V* (b) *Q–S–T–V*
(c) *P–R–U–W* (d) *Q–S–U–W*

[2010 : 2 M]

Linked Answer Questions 4.16 and 4.17.

For a particular project, eight activities are to be carried out. Their relationships with other activities and expected durations are mentioned in the table below.

Activity	Predecessors	Duration (days)
a	-	3
b	*a*	4
c	*a*	5
d	*a*	4
e	*b*	2
f	*d*	9
g	*c, e*	6
h	*f, g*	2

4.16 The critical path for the project is

(a) a–b–e–g–h (b) a–c–g–h
(c) a–d–f–h (d) a–b–c–f–h

[2012 : 2 M]

4.17 If duration of activity f is changed to 10 days, then the critical path for the project is

(a) Critical path remains the same and the total duration to complete the project changes to 19 days.
(b) Critical path and the total duration to complete the project remains the same.
(c) Critical path changes but the total duration to complete the project remains the same.
(d) Critical path changes and the total duration to complete the project changes to 17 days.

[2012 : 2 M]

4.18 A project has four activities *P*, *Q*, *R* and *S* as shown below.

Activity	Normal duration (days)	Predecessor	Cost slope (Rs./day)
P	3	–	500
Q	7	P	100
R	4	P	400
S	5	R	200

The normal cost of the project is ₹ 10,000/- and the overhead cost is ₹ 200/- per day. If the project duration has to be crashed down to 9 days, the total cost (in Rupees) of the project is ______.

[2014 : 2 M, Set-2]

4.19 Consider the given project network, where numbers along various activities represent normal time. The free float on activity 4-6 and the project duration, respectively, are

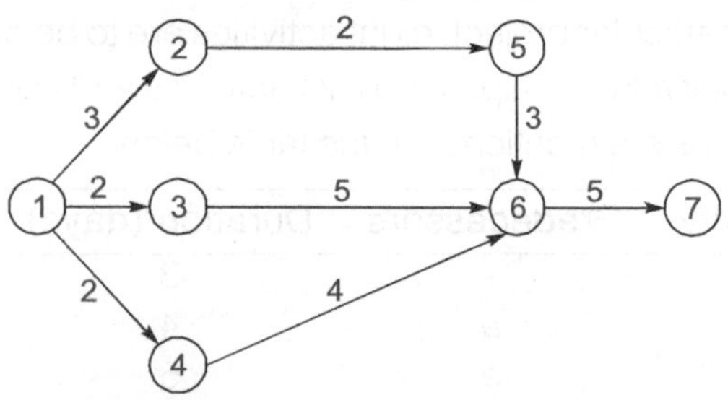

(a) 2 and 13 (b) 0 and 12
(c) –2 and 13 (d) 2 and 12

[2014 : 2 M, Set-3]

4.20 The precedence relations and duration (in days) of activities of a project network are given in the table. The total float (in days) of activities e and f, respectively, are

Activity	Predecessors	Duration (days)
a	–	2
b	–	4
c	a	2
d	b	3
e	c	2
f	c	4
g	d, e	5

(a) 0 and 4 (b) 1 and 4
(c) 2 and 3 (d) 3 and 1

[2014 : 2 M, Set-4]

4.21 Following data refers to the activities of a project, where, node 1 refers to the start and node 5 refers to the end of the project

Activity	Duration (days)
1-2	2
2-3	1
4-3	3
1-4	3
2-5	3
3-5	2
4-5	4

The critical path (CP) in the network is
(a) 1–2–3–5 (b) 1–4–3–5
(c) 1–2–3–4–5 (d) 1–4–5

[2015 : 2 M, Set-1]

4.22 A project consists of 14 activities, A to N. The duration of these activities (in days) are shown in brackets on the network diagram. The latest finish time (in days) for node 10 is______.

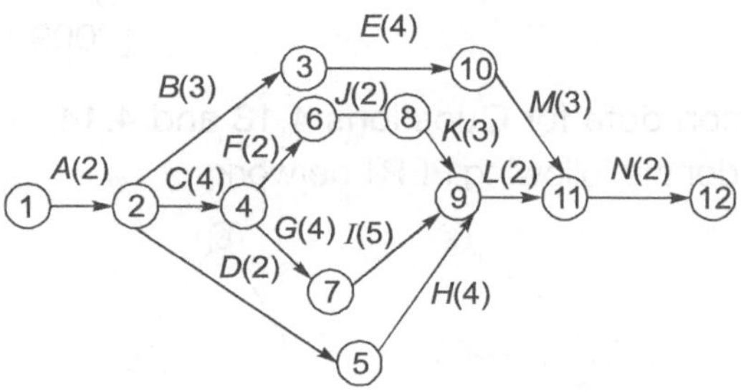

[2016 : 2 M, Set-2]

4.23 In PERT chart, the activity time distribution is
(a) Normal (b) Binomial
(c) Poisson (d) Beta

[2016 : 1 M, Set-3]

4.24 The standard deviation of linear dimensions P and Q are 3 μm and 4 μm, respectively. When assembled, the standard deviation (in μm) of the resulting linear dimension $(P + Q)$ is ______.

[2017 : 1 M, Set-2]

4.25 A project starts with activity A and ends with activity F. The precedence relation and durations of the activities are as per the following table:

Activity	Immediate predecessor	Duration (days)
A	–	4
B	A	3
C	A	7
D	B	14
E	C	4
F	D, E	9

The minimum project completion time (in days) is ___.

[2017 : 2 M, Set-2]

4.26 The arc lengths of a directed graph of a project are as shown in the figure. The shortest path length from node 1 to node 6 is ________.

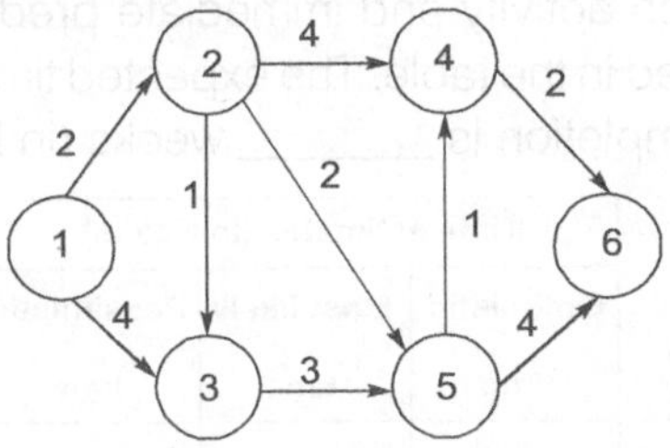

[2018 : 2 M, Set-2]

4.27 A project consists of six activities. The immediate predecessor of each activity and the estimated duration is also provided in the table below:

Activity	Immediate predecessor	Estimated duration (weeks)
P	-	5
Q	-	1
R	Q	2
S	P,R	4
T	P	6
U	S,T	3

If all the activities other than S take the estimated amount of time, the maximum duration (in weeks) of the activity *S* without delaying the completion of the project is ____.

[2019 : 2 M, Set-1]

4.28 The activities of a project, their duration and the precedence relationships are given in the table. For example, in a precedence relationship "X < Y, Z" means that X is predecessor of activities Y and Z. The time to complete the activities along the critical path is ________ weeks.

Activity	Duration (Weeks)	Precedence Relationship
A	5	$A < B,C,D$
B	7	$B < E,F,G$
C	10	$C < I$
D	6	$D < G$
E	3	$E < H$
F	9	$F < I$
G	7	$G < I$
H	4	$H < I$
I	2	....

(a) 25
(b) 21
(c) 17
(d) 23 [2019 : 2 M, Set-2]

4.29 Consider the following network of activities, with each activity named A–L, illustrated in the nodes of the network

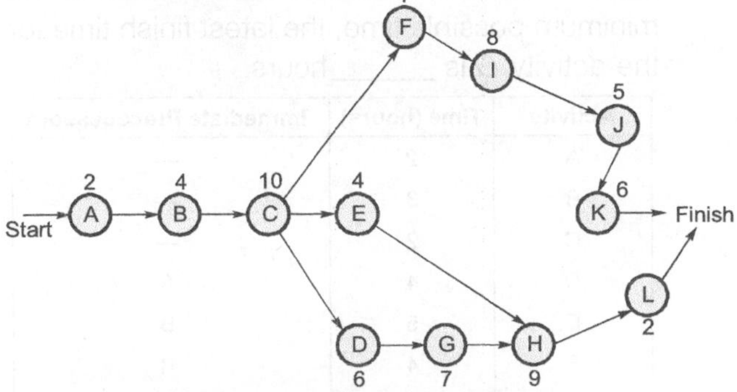

The number of hours required for each activity is shown alongside the nodes. The slack on the activity *L*, is ________ hours.

[2020 : 1 M, Set-2]

4.30 In the Critical Path Method (CPM), the cost-time slope of an activity is given by

(a) $\dfrac{\text{Crash Cost} - \text{Normal Cost}}{\text{Crash Time}}$

(b) $\dfrac{\text{Normal Cost}}{\text{Crash Time} - \text{Normal Time}}$

(c) $\dfrac{\text{Crash Cost} - \text{Normal Cost}}{\text{Crash Time} - \text{Normal Time}}$

(d) $\dfrac{\text{Crash Cost}}{\text{Crash Time} - \text{Normal Time}}$

[2020 : 1 M, Set-1]

4.31 Activities A, B, C and D form the critical path for a project with a PERT network. The means and variances of the activity duration for each activity are given below. All activity durations follow the Gaussian (normal) distribution, and are independent of each other.

Activity	A	B	C	D
Mean (days)	6	11	8	15
Variance (days^2)	4	9	4	9

The probability that the project will be completed within 40 days is ______. [Round off to two decimal places]

[Note : Probability is a number between 0 and 1]

[2021 : 2 M, Set-1]

4.32 A PERT network has 9 activities on its critical path. The standard deviation of each activity on the critical path is 3. The standard deviation of the critical path is

(a) 27
(b) 9
(c) 81
(d) 3 [2021 : 1 M, Set-2]

4.33 Activities *A* to *K* are required to complete a project. The time estimates and the immediate predecessors of these activities are given in the table. If the project is to be completed in the minimum possible time, the latest finish time for the activity *G* is _______ hours.

Activity	Time (hours)	Immediate Prececessors
A	2	—
B	3	—
C	2	—
D	4	A
E	5	B
F	4	B
G	3	C
H	10	D, E
I	5	F
J	8	G
K	3	H, I, J

(a) 5 (b) 10
(c) 8 (d) 9

[2022 : 2 M, Set-1]

4.34 A project consists of five activities (A, B, C, D and E). The duration of each activity follows beta distribution. The three time estimates (in weeks) of each activity and immediate predecessor(s) are listed in the table. The expected time of the project completion is _______ weeks (in integer).

Activity	Time estimates (in weeks) Optimistic time	Most likely time	Pessimistic time	Immediate predecessor(s)
A	4	5	6	None
B	1	3	5	A
C	1	2	3	A
D	2	4	6	C
E	3	4	5	B, D

[2022 : 2 M, Set-2]

■■■■

Answers PERT and CPM

4.1 (c)	4.2 (a)	4.3 (a)	4.4 (c)	4.5 (b)	4.6 (d)	4.7 (c)	4.8 (d)
4.9 (a)	4.10 (a)	4.11 (c)	4.12 (a)	4.13 (d)	4.14 (c)	4.15 (d)	4.16 (c)
4.17 (a)	4.18 (12500)	4.19 (a)	4.20 (b)	4.21 (b)	4.22 (14)	4.23 (d)	4.24 (5)
4.25 (30)	4.26 (7)	4.27 (6)	4.28 (d)	4.29 (2)	4.30 (c)	4.31 (50)	4.32 (b)
4.33 (b)	4.34 (15)						

Explanations PERT and CPM

4.1 (c)

In PERT, activity, time is assumed to be Beta distribution and project is assumed to be normal distribution.

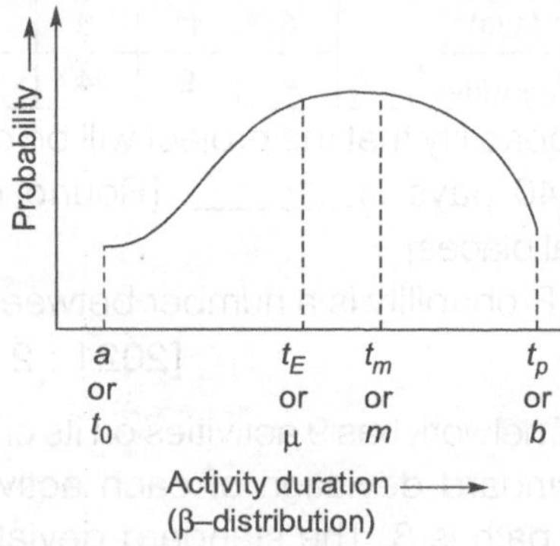

where, a or t_0 = optimistic time

t_E or μ = mean time or expected time

t_m or μ = most likely time

t_p or b = pesimistic time

expected or mean time,

$$t_E \text{ or } \mu = \frac{t_0 + 4t_m + t_p}{6} = \frac{a + 4m + b}{6}$$

4.3 (a)

The symbol used in work study is:

Operation	Storage	Inspection	Transportation	Delay or temporary storage
O	Δ	□	⇒	D

POINTS TO REMEMBER

- Above given symbols are used for process chart in the work study.
- The are also 18 Gilberth's therbligs in the micromotion study (work study).

4.4 (c)

Given network diagram as:

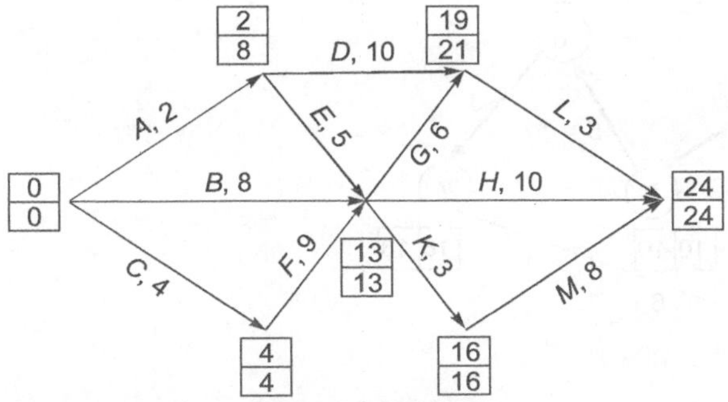

Maximum time along path CFKM.

So this is the critical path.

∴ The project can be completed along C-F-K-M in 4 + 9 + 3 + 8 = **24 days**

4.5 (b)

Critical activities decide the project time and there is no float for such activities.

4.6 (d)

Refer solution 4.4

4.7 (c)

The network diagram for the given project is,

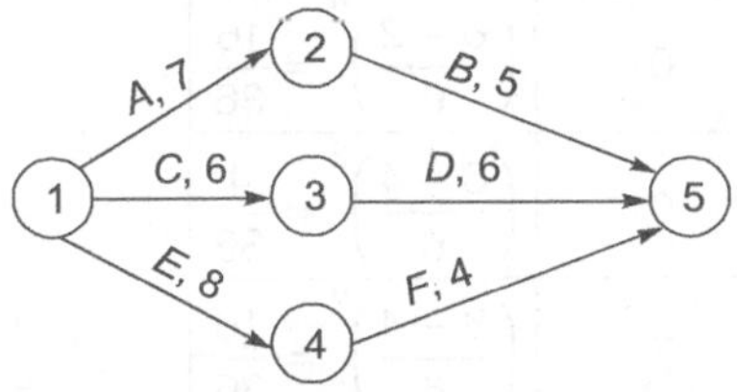

Since, all the paths have 12 days duration, so all three paths will be critical path.

To reduce the project duratio by 1 day, one activity from each path should be crashed. Hence total number of activities that should be crashed are 3.

4.8 (d)

PERT network are:

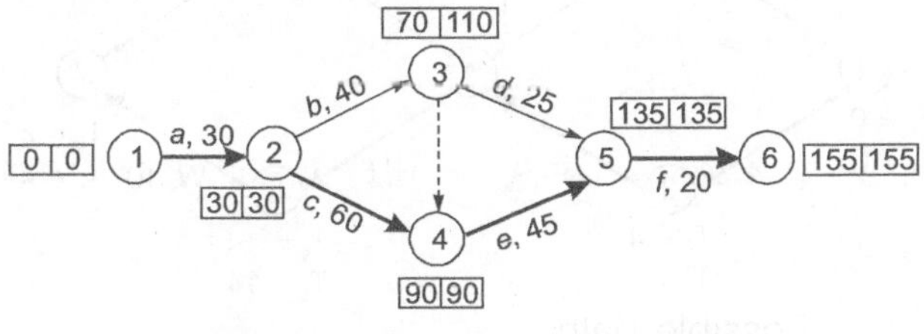

Here we see that expected completion time of the project is **155 days** and critical path is *a–c–e–f*

4.9 (a)

For critical path *a–c–e–f*, the variance for each task is shown below

Task	Variance (σ^2)
a	25
c	81
e	36
f	09

Total variance = 25 + 81 + 36 + 9 = 151

We know that

Standard deviation, $\sigma = \sqrt{\text{Variance}}$

∴ S.D. of critical path, $\sigma_{cp} = \sqrt{151}$ **days**

4.10 (a)

The total inventory cost is minimized when the jobs are processed according to the shortest processing time, *i.e.* job with least processing time should be done first.

POINTS TO REMEMBER

- Total tardiness is minimized by the earliest due date (EDD) rule.
- When we minimize the total completion time by shortest processing time (SPT) or earliest due date (EDD) rule then the minimum completion time will be obtained by the shortest processing time (SPT) rule.

4.11 (c)

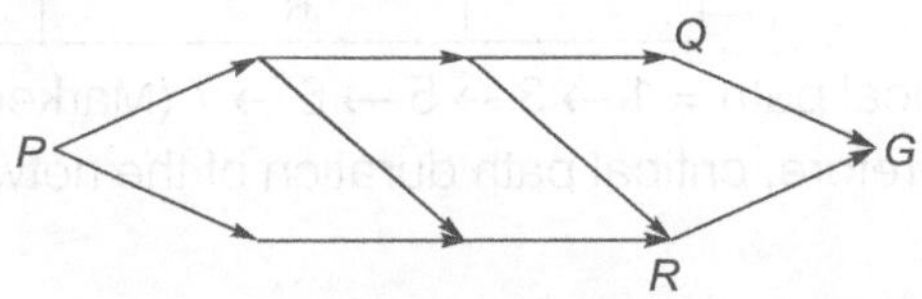

d_{ij} = length of directed are from node i to node j.

S_j = length of shortest path from P to node j.

There will be two possible paths to reach from P to G.

1. By reaching (P to Q) then (Q to G)
2. By reaching (P to R) then (R to G)

then, $S_G = \{S_Q + d_{QG}, S_R + d_{RG}\}$

4.12 (a)

The expected time (t_e) of PERT activity in terms of optimistic time (t_o), pessimistic time (t_p) and most likely time (t_l) is given by,

$$t_e = \frac{t_o + 4t_l + t_p}{6}$$

4.13 (d)

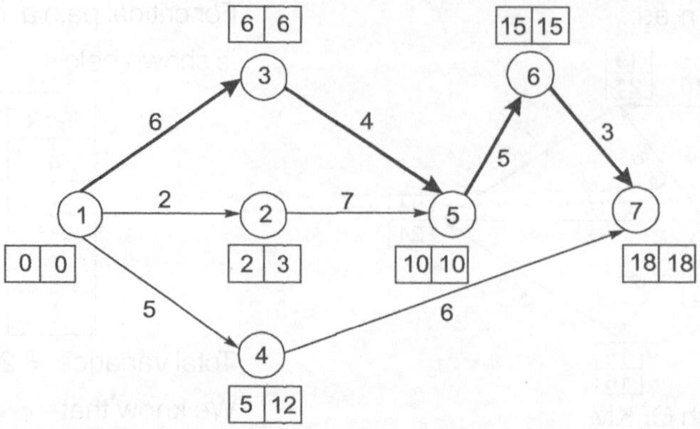

Activities	Duration $t_e = \dfrac{t_0 + 4t_m + t_p}{6}$	EST	LST	EFT	LFT	LST – EST = Float	Variance σ^2 $\left(\dfrac{t_p - t_0}{6}\right)^2$
1–2	$\dfrac{1+4\times2+3}{6} = 2$	0	0	2	3	0	$\left(\dfrac{3-1}{6}\right)^2 = \dfrac{4}{36}$
1–3	$\dfrac{5+6\times4+7}{6} = 6$	0	0	6	6	0	$\left(\dfrac{7-5}{6}\right)^2 = \dfrac{4}{36}$
1–4	$\dfrac{3+4\times5+7}{6} = 5$	0	0	5	12	7	$\left(\dfrac{7-3}{6}\right)^2 = \dfrac{16}{36}$
2–5	$\dfrac{5+4\times7+9}{6} = 7$	2	3	10	10	1	$\left(\dfrac{9-5}{6}\right)^2 = \dfrac{16}{36}$
3–5	$\dfrac{2+4\times4+6}{6} = 4$	6	6	10	10	0	$\left(\dfrac{6-2}{6}\right)^2 = \dfrac{16}{36}$
5–6	$\dfrac{4+4\times5+6}{6} = 5$	10	10	15	15	0	$\left(\dfrac{6-4}{6}\right)^2 = \dfrac{4}{36}$
4–7	$\dfrac{4+4\times6+8}{6} = 6$	5	12	18	18	7	$\left(\dfrac{8-4}{6}\right)^2 = \dfrac{16}{36}$
6–7	$\dfrac{2+4\times3+4}{6} = 3$	15	15	18	18	0	$\left(\dfrac{4-2}{6}\right)^2 = \dfrac{4}{36}$

Critical path = 1 → 3 → 5 → 6 → 7 (Marked by double lines)

Therefore, critical path duration of the network is = 6 + 4 + 5 + 3 = **18 days**

4.14 (c)

$\Sigma\sigma^2$ along the critical path

$$\Sigma\sigma^2 = \sigma^2_{1\text{–}3} + \sigma^2_{3\text{–}5} + \sigma^2_{5\text{–}6} + \sigma^2_{6\text{–}7}$$

$$= \frac{4}{36} + \frac{16}{36} + \frac{4}{36} + \frac{4}{36} = \frac{28}{36} = 0.882$$

Standard deviation of critical path is

$$\sigma = \sqrt{\frac{28}{36}} = \mathbf{0.88}$$

4.15 (d)

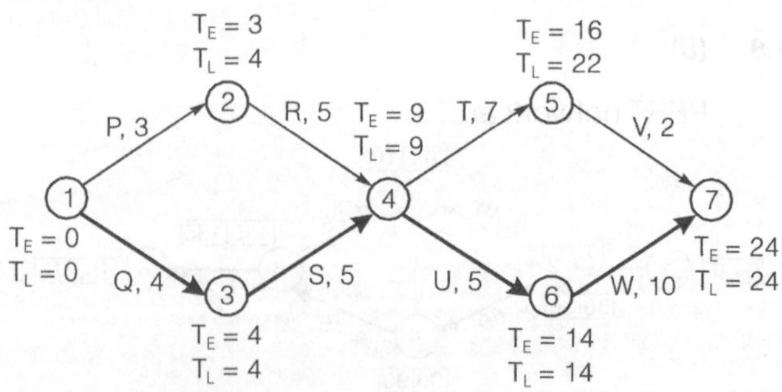

Possible path:

1. QSUW = 24 days ⇒ critical path
2. QSTV = 18 days
3. PRTV = 17 days
4. PR VW = 23 days

4.16 (c)

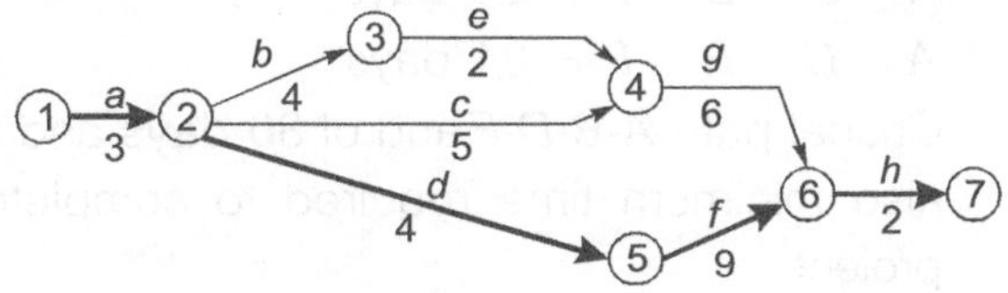

Various path

abegh — 17 days

acgh — 16 days

adfh — 18 days

∴ ***a–d–f–h*** is critical path.

4.17 (a)

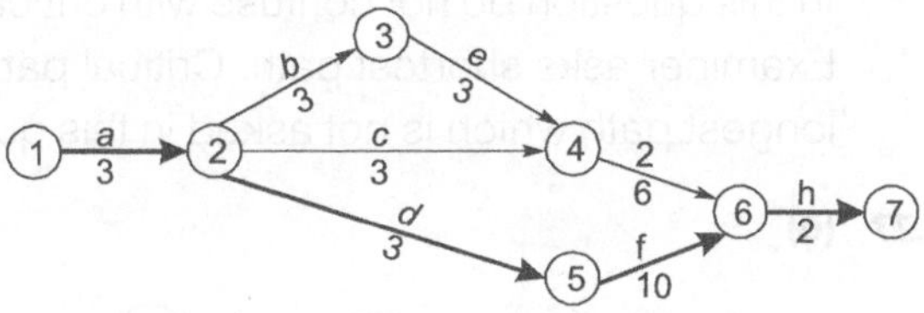

Critical path ⇒ *adfh* = 19 days

So, when the duration of activity, f is changed from 9 to 10 days, critical path remains same and the project duration will increase to 19 days.

4.18 (12500)

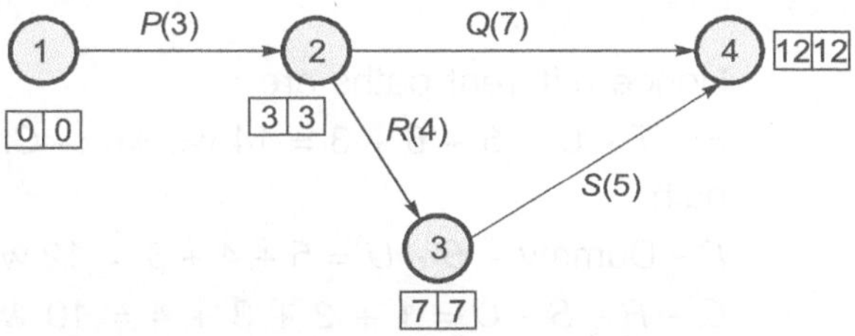

Before crashing:

TC = 10000 + 200 × 12 = ₹12400

Since cost slope of activity *S* is minimum.

So, activity *S* will be crashed by 2 days.

1st crashing:

$$TC = 12400 - \underbrace{200 \times 2}_{\text{overhead cost}} + \underbrace{200 \times 2}_{\text{crashing cost}}$$

= ₹12400

2nd crashing:

From various combinations/options, we get, least cost slope in *Q* + *S* activity

∴ crashing, *Q* and *S* by 1 day each

TC = 12400 – 1 × 200 + 1 × 300

= ₹12500

4.19 (a)

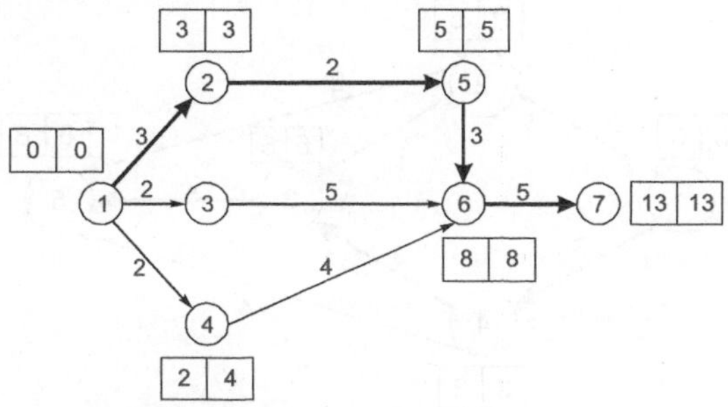

Critical path is 1 – 2 – 5 – 6 – 7

Project duration,

$$t_E = 3 + 2 + 3 + 5 = \mathbf{13}$$

$$\text{Free float } (4-6) = E_j - (E_i + t_E^{ij})$$

$$= 8 - (4 + 2) = \mathbf{2}$$

POINTS TO REMEMBER

Free float that part of total float which can be utilised without affecting the float of successive activities, i.e. if an activity is delayed by this value then also the successor activity starts at, it's earliest start time.

4.20 (b)

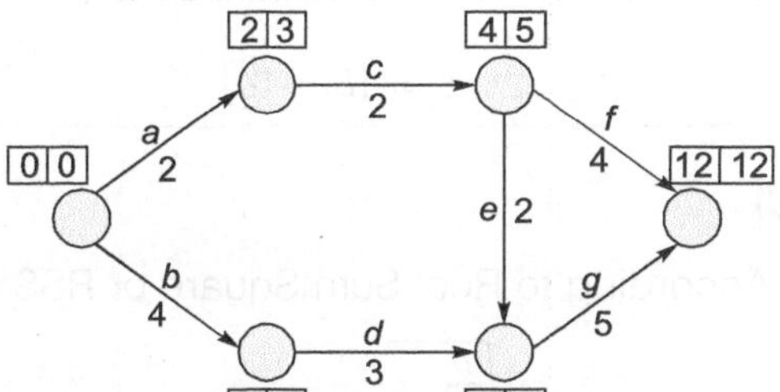

$$\text{Total float} = L_j - (E_i + t_E^{ij})$$

Total float for activity: $e = 7 - 2 - 4 = \mathbf{1}$

Total float for activity: $f = 12 - 4 - 4 = \mathbf{4}$

POINTS TO REMEMBER

Total float represents the amount of time by which a particular activity can be delayed without delaying the project completion time.

4.21 (b)

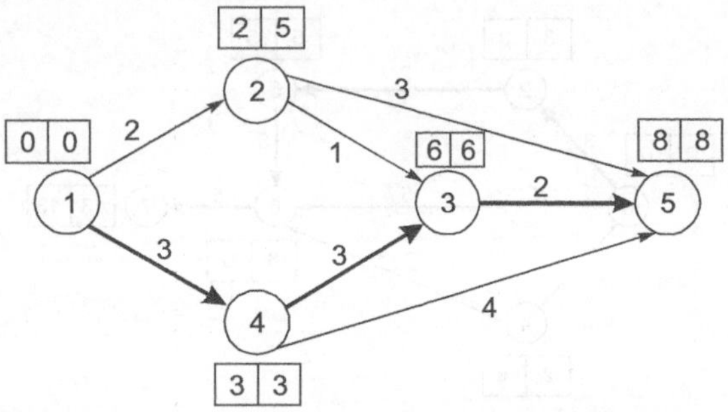

Critical path = **1–4–3–5**

4.22 Sol.

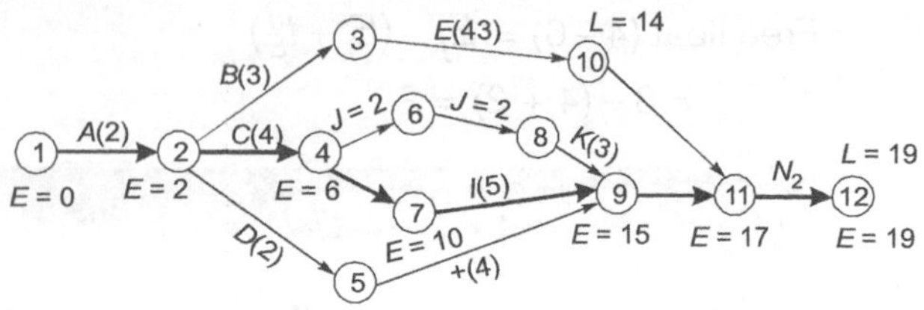

Critical path = 1 – 2 – 4 – 7 – 9 – 11 – 12 = 19 days

The latest finish time (in days) for node 10 is 14 days.

POINTS TO REMEMBER

The latest finish time is calculated as,

$$L_i = [L_j - t_E^{ij}]$$

4.24 (5)

According to Root Sum Square or RSS model

$$\sigma_a = \sqrt{\sum_{i=1}^{n} \sigma_i^2}$$

where, σ_a = assembly tolerance standard deviation

σ_i = i^{th} component tolerance standard deviation

Here $\sigma_P = 3\ \mu m$ and $\sigma_Q = 4\ \mu m$

$$\sigma_a = \sqrt{\sigma_P^2 + \sigma_Q^2} = \sqrt{3^2 + 4^2} = 5\ \mu m$$

4.25 (30)

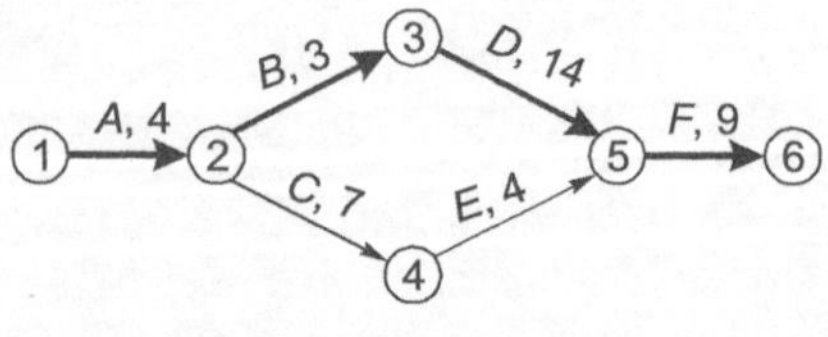

There are two path,

$A - B - D - F$ = 30 days

$A - C - E - F$ = 24 days

Critical path *A-B-D-F* and of **30 days** and it is also minimum time required to complete a project.

4.26 (7)

Shortest path is

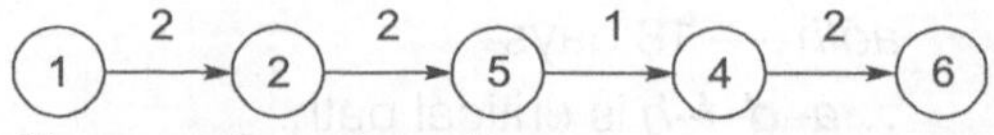

Shortest length = **7**

It is the problem of shortest path which will be 7. In this question do not confuse with critical path. Examiner asks shortest path. Critical path is the longest path which is not asked in this question.

4.27 (6)

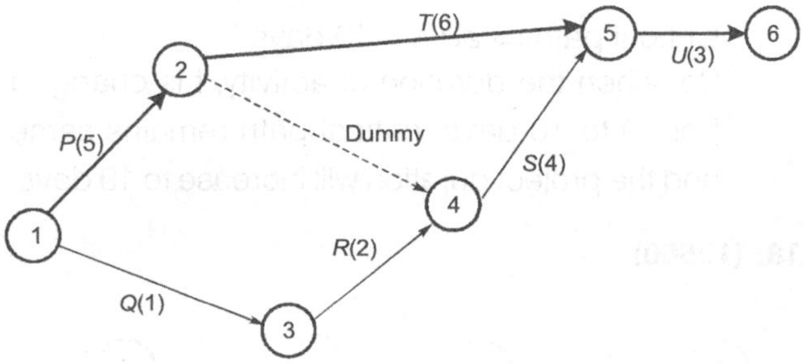

Hence different paths are:

P - T - U = 5 + 6 + 3 = 14 weeks ← Critical path

P - Dummy - *S - U* = 5 + 4 + 3 = 12 weeks

Q - R - S - U = 1 + 2 + 3 + 4 = 10 weeks

∵ Only '*S*' is delayed and no other activity is delayed.

So, maximum activity time for activity '*S*' can be 6 weeks without delaying the project.

4.28 (d)

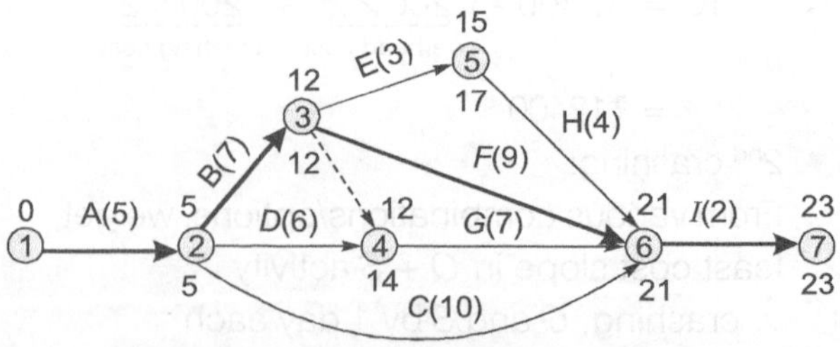

So critical path is *A - B - F - I*

Critical path duration = 23 weeks

4.29 (2)

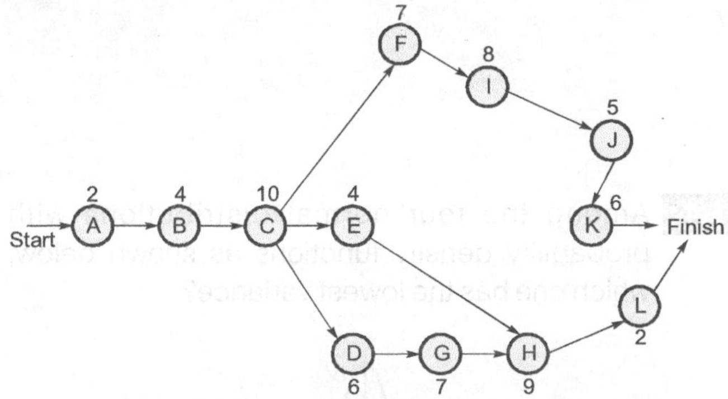

Time along path A-B-C-F-I-J-K = 42 hours
Time along path A-B-C-E-H-L = 31 hours
Time along path A-B-C-D-G-H-L = 40 hours
Slack for L = 42 – 40 = 2 hours

4.31 (50)

PERT-CPM

$$T_S = 40 \text{ days}, \quad T_E = 6 + 11 + 8 + 15$$
$$T_E = 40 \text{ days},$$
$$2 = \frac{T_S - T_E}{\sigma} = 0 \rightarrow 50\%$$

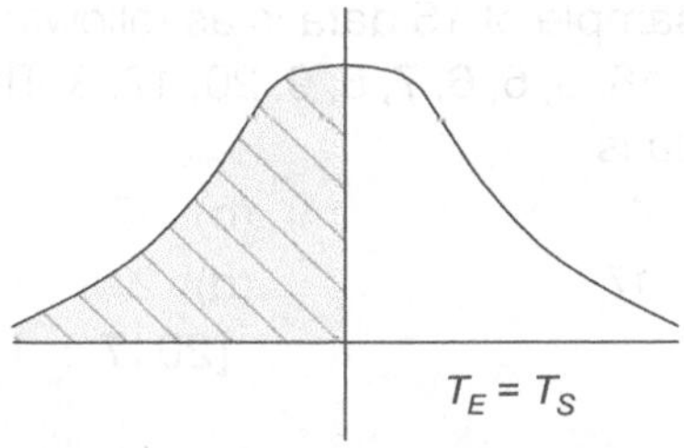

Probability of completing project in expected time is always 50%.

4.32 (b)

In CPM,

$$\sigma = \sqrt{\text{sum of variance along critical path}}$$

$$\sigma = \sqrt{\sigma^2 + \sigma^2 + ... + \sigma^2}$$

$$\sigma = \sqrt{9\sigma^2} = \sqrt{9 \times 9} = 9$$

4.33 (b)

Network diagram can be formed as:

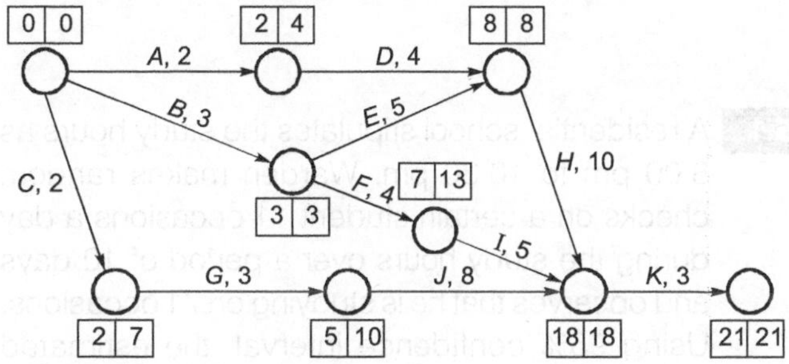

For activity G, latest finish time is 10.

4.34 (15) (15 to 15)

Activity	Time estimates (in weeks) Optimistic time (t_o)	Most likely time (t_m)	Pessimistic time (t_p)	Estimated time $t_e = \frac{t_o + 4t_m + t_p}{6}$
A	4	5	6	5
B	1	3	5	3
C	1	2	3	2
D	2	4	6	4
E	3	4	5	4

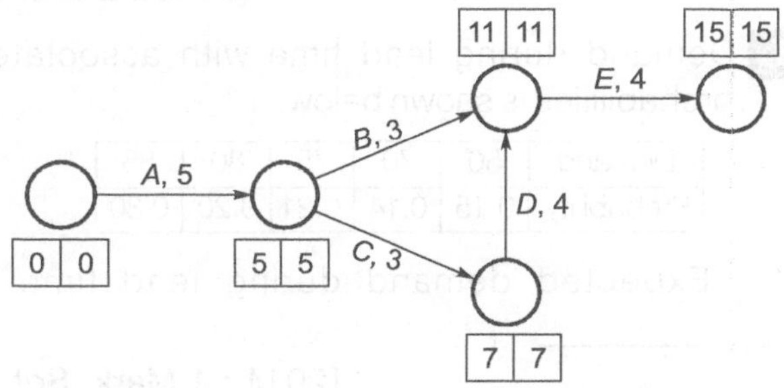

Expected time of the project completion is 15 weeks.

5 Probability and Statistics

5.1 A residential school stipulates the study hours as 8.00 pm to 10.30 pm. Warden makes random checks on a certain student 11 occasions a day during the study hours over a period of 10 days and observes that he is studying on 71 occasions. Using 95% confidence interval, the estimated minimum hours of his study during that 10 day period is

(a) 8.5 hours (b) 13.9 hours
(c) 16.1 hours (d) 18.4 hours

[2003 : 2 Marks]

5.2 The distribution of lead time demand for an item is as follows:

Lead time demand	Probability
80	0.20
100	0.25
120	0.30
140	0.25

The reorder level is 1.25 times the expected value of the lead time demand. The service level is

(a) 25% (b) 50%
(c) 75% (d) 100%

[2005 : 2 Marks]

5.3 Demand during lead time with associated probabilities is shown below:

Demand	50	70	75	80	85
Probability	0.15	0.14	0.21	0.20	0.30

Expected demand during lead time is ________

[2014 : 1 Mark, Set-4]

5.4 Among the four normal distributions with probability density functions as shown below, which one has the lowest variance?

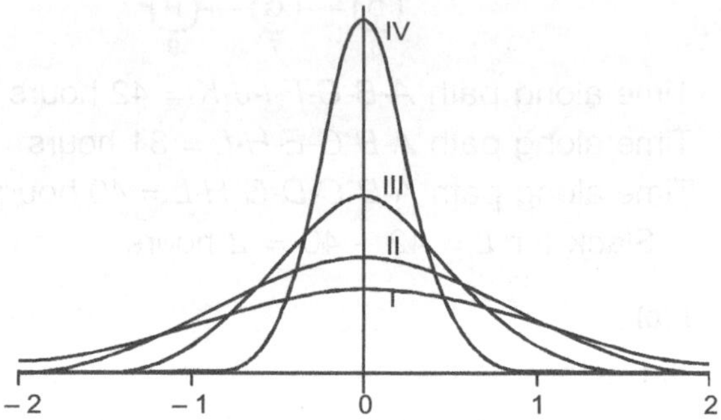

(a) I (b) II
(c) III (d) IV

[2015 : 1 Mark, Set-1]

5.5 A sample of 15 data is as follows: 17, 18, 17 17, 13, 18, 5, 5, 6, 7, 8, 9, 20, 17, 3. The mode of the data is

(a) 4 (b) 13
(c) 17 (d) 20

[2017 : 1 Mark, Set-2]

■■■■

Answers Probability and Statistics

5.1 (b) 5.2 (d) 5.3 (74 to 75) 5.4 (d) 5.5 (c)

Explanations **Probability and Statistics**

5.1 (b)

Total duration = 25 hr

$$P = \frac{71}{110} = 0.645$$

Average = 16.1 hr

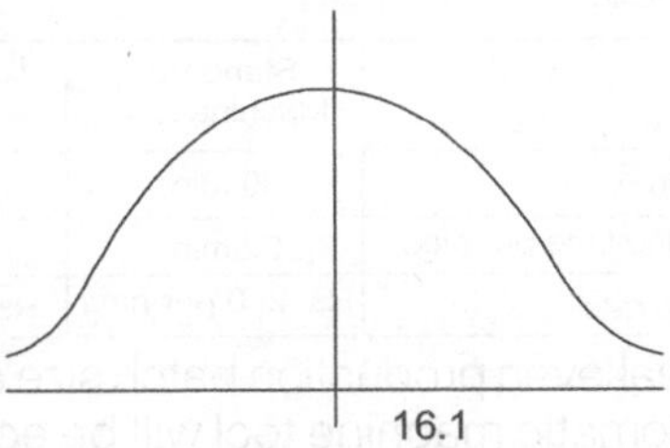

$$n = \frac{Z^2}{L^2} \cdot \frac{(1-P)}{P}$$

$$110 = \frac{1.96^2}{L^2} \cdot \frac{(1-0.645)}{0.645}$$

$$L = 0.1386$$

$$\text{Lower limit} = 16.1 \times (1-0.1386)$$

$$= \mathbf{13.89\ hrs.}$$

5.2 (d)

Expected value,

$$\bar{X} = 80 \times 0.2 + 100 \times 0.25 + 120 \times 0.3 + 140 \times 0.25$$

$$= 112$$

Lead time Demand	Probability	Cumulative probability
80	0.20	0.20
100	0.25	0.45
120	0.30	0.75
140	0.25	1.0

$$\text{Reorder level} = 1.25\bar{X}$$

$$= 1.25 \times 112 = 140$$

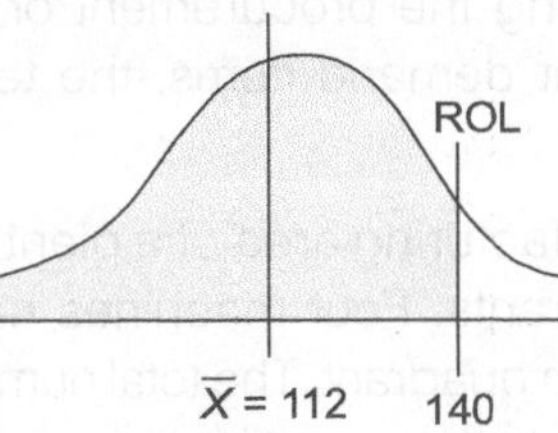

Demand in never more than 140, so service lvel is 100%.

5.3 Sol.

(74 to 75)

Expected Demand = 50 × 0.15 + 70 × 0.14+ 75 × 0.21 + 80 × 0.20 + 85 × 0.3 = **74.55**

5.4 (d)

Among the four normal distributions, as we see in the figure, distribution IV shows closeness towards the mean. So distribution IV will have less variation.

5.5 (c)

Mode means highest number of observations or occurrence of data most of the time as data 17, occurs four times, i.e., highest time. So mode is **17**.

■■■■

6 Products and Process, Planning and Control & Sequencing

6.1 For planning the procurement or production of dependent demand items, the technique most suited is [1995 : 1 M]

6.2 The manufacturing area of a plant is divided into four quadrants. Four machines have to located one in each quadrant. The total number of possible layouts is

(a) 4 (b) 8
(c) 16 (d) 24 [1995 : 1 M]

6.3 Match the following

List-I (Problem areas)	List-II (Techniques)
A. JIT	1. CRAFT
B. Computer assisted layout	2. PERT
C. Scheduling	3. Johnson's rule
D. Simulation	4. Kanbans
	5. EOQ rule
	6. Monte Carl

[1995 : 2 M]

6.4 In an assembly line for assembling toys, five workers are assigned tasks which take times of 10, 8, 6, 9 and 10 minutes respectively. The balance delay for line is

(a) 43.3% (b) 14.8%
(c) 14.0% (d) 16.3% [1996 : 2 M]

6.5 Production flow analysis (PFA) is a method of identifying part families that uses data from

(a) engineering drawings
(b) production schedule
(c) bill of materials
(d) route sheets [2001 : 1 M]

6.6 Two machines of the same production rate are available for use. On machine 1, the fixed cost is Rs. 100 and the variable cost is Rs. 2 per piece produced. The corresponding numbers for the machine 2 are Rs. 200 and Re. 1 respectively. For certain strategic reasons both the machines are to be used concurrently. The sale price of the first 800 units is Rs. 3.50 per unit and subsequently it is only Rs. 3.00. The breakeven production rate for each machine is

(a) 75 (b) 100
(c) 150 (d) 600 [2003 : 2 M]

6.7 A standard machine tool and an automatic machine tool are being compared for the production of a component. Following data refers to the two machines.

	Standard Machine tool	Automatic Machine tool
Setup time	30 min.	2 hours
Machining time per piece	22 min.	5 min
Machine rate	Rs. 200 per hour	Rs. 800 per hours

The breakeven production batch size above which the automatic machine tool will be economical to use, will be

(a) 4 (b) 5
(c) 24 (d) 225 [2004 : 2 M]

6.8 A soldering operation was work-sampled over two days (16 hours) during which an employee soldered 108 joints, actual working time was 90% of the total time and the performance rating was estimated to be 120 percent. If the contract provides allowance of 20 percent of the total time available, the standard time for the operation would be

(a) 8 min. (b) 8.9 min
(c) 10 min (d) 12 min [2004 : 2 M]

6.9 An electronic equipment manufacturer has decided to add a component subassembly operation that can produce 80 units during a regular 8-hour shift. This operation consists of three activities as below

Activity	Standard time (min)
M. Mechanical assembly	12
E. Electric wiring	16
T. Test	3

For line balancing the number of work stations required for the activities *M*, *E* and *T* would respectively be

(a) 2, 3, 1 (b) 3, 2, 1
(c) 2, 4, 2 (d) 2, 1, 3 [2004 : 2 M]

6.10 A welding operation is time-studied during which an operator was pace-rated as 120%. The operator took, on an average, 8 minutes for producing the weld-joint. If a total of 10% allowances are allowed for this operation. The expected standard production rate of the weld-joint (in units per 8 hour day) is

(a) 45 (b) 50
(c) 55 (d) 60 [2005 : 2 M]

6.11 A manufacturing shop processes sheet metal jobs, wherein each job must pass through two machines ($M1$ and $M2$, in that order). The processing time (in hours) for these jobs is

Machine	Jobs					
	P	Q	R	S	T	U
$M1$	15	32	8	27	11	16
$M2$	6	19	13	20	14	7

The optimal make-span (in hours) of the shop is
(a) 120 (b) 115
(c) 109 (d) 79 **[2006 : 2 M]**

6.12 The table gives details of an assembly line.

Work station	I	II	III	IV	V	VI
Total task time at the workstation (in minutes)	7	9	7	10	9	6

What is the line efficiency of the assembly line ?
(a) 70% (b) 75%
(c) 80 % (d) 85% **[2006 : 2 M]**

6.13 Capacities of production of an item over 3 consecutive months in regular time are 100, 100 and 80 and in overtime are 20, 20 and 40. The demands over those 3 months are 90, 130 and 110. The cost of production in regular time and overtime are respectively Rs. 20 per item and Rs. 24 per item. Inventory carrying cost is Rs. 2 per item month. The levels of starting and final inventory are nil. Backorder is not permitted. For minimum cost of plan, the level of planned production in overtime in the third month is
(a) 40 (b) 30
(c) 20 (d) 0 **[2007 : 2 M]**

6.14 The Product structure of an assembly P is shown in the figure.

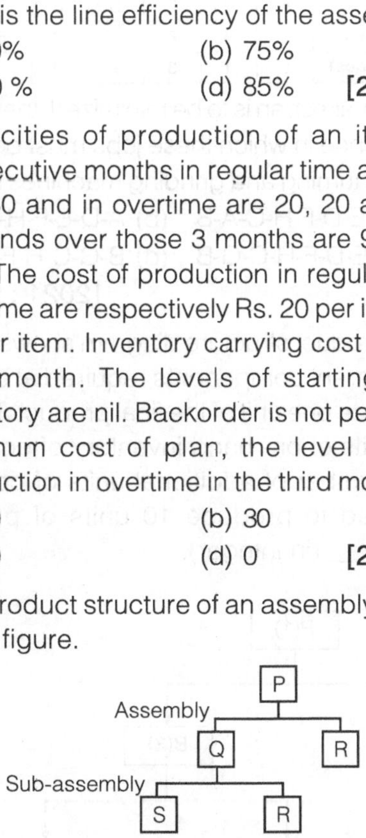

Estimated demand for end product P is as follows:

Week	1	2	3	4	5	6
Demand	1000	1000	1000	1000	1200	1200

Ignore lead times for assembly and subassembly. Production capacity (per week) for component R is the bottleneck operation. Starting with zero inventory, the smallest capacity that will ensure a feasible production plan up to week 6 is
(a) 1000 (b) 1200
(c) 2200 (d) 2400 **[2008 : 2 M]**

6.15 Six jobs arrived in a sequence as given below:

Jobs	Processing Time (days)
I	4
II	9
III	5
IV	10
V	6
VI	8

Average flow time (in days) for the above jobs using Shortest Processing Time rule is
(a) 20.83 (b) 23.16
(c) 125.00 (d) 139.00 **[2009 : 2 M]**

Common Data for Questions 6.16 and 6.17

Four jobs are to be processed on a machine as per data listed in the table.

Job	Processing time (in days)	Due date
1	4	6
2	7	9
3	2	19
4	8	17

6.16 If the Earliest Due Date (EDD) rule is used to sequence the jobs, the number of jobs delayed is
(a) 1 (b) 2
(c) 3 (d) 4 **[2010 : 2 M]**

6.17 Using the Shortest Processing Time (SPT) rule, total tardiness is
(a) 0 (b) 2
(c) 6 (d) 8 **[2010 : 2 M]**

6.18 Vehicle manufacturing assembly line is an example of
(a) product layout (b) process layout
(c) manual layout (d) fixed layout
[2010 : 1 M]

6.19 The word **kanban** is most appropriately associated with
(a) economic order quantity
(b) just-in-time production
(c) capacity planning
(d) product design **[2011 : 1 M]**

6.20 Which one of the following is NOT a decision taken during the aggregate production planning stage?
(a) Scheduling of machines

(b) Amount of labour to be committed
(c) Rate at which production should happen
(d) Inventory to be carried forward [2012 : 1 M]

6.21 During the development of a product, an entirely new process plan is made based on design logic, examination of geometry and tolerance information. This type of process planning is known as
(a) retrieval (b) generative
(c) variant (d) group technology based
[2015 : 1 M, Set-2]

6.22 Cylindrical pins of diameter $15^{\pm 0.020}$ mm are being produced on a machine. Statistical quality control tests show a mean of 14.995 mm and standard deviation of 0.004 mm. The process capability index *Cp* is
(a) 0.833 (b) 1.667
(c) 3.333 (d) 3.750
[2017 : 1 M, Set-1]

6.23 Following data refers to the jobs (P, Q, R, S) which have arrived at a machine for scheduling. The shortest possible average flow time is _____ days.

Job	Processing Time (days)
P	15
Q	9
R	22
S	12

[2017 : 2 M, Set-1]

6.24 Processing times (including step times) and due dates for six jobs waiting to be processed at a work centre are given in the table. The average tardiness (in days) using shortest processing time rule is ______(correct to two decimal places).

Job	Processing time (days)	Due date (days)
A	3	8
B	7	16
C	4	4
D	9	18
E	5	17
F	13	19

[2018 : 2 M, Set-1]

6.25 In Materials Requirement Planning, if the inventory holding cost is very high and the setup cost is zero, which one of the following lot sizing approaches should be used?
(a) Lot-for-Lot
(b) Fixed Period Quantity, for 2 periods
(c) Economic Order Quantity
(d) Base Stock Level [2020 : 1 M, Set-2]

6.26 For an assembly line, the production rate was 4 pieces per hour and the average processing time was 60 minutes. The WIP inventory was calculated. Now, the production rate is kept the same, and the average processing time is brought down by 30 percent. As a result of this change in the processing time, the WIP inventory
(a) increases by 25% (b) increases by 30%
(c) decreases by 25% (d) decreases by 30%
[2020 : 2 M, Set-1]

6.27 A set of jobs A, B, C, D, E, F, G, H arrive at time $t = 0$ for processing on turning and grinding machines. Each jobs needs to be processed in sequence - first on the turning machine and second on the grinding machine, and the grinding must occur immediately after turning. The processing times of the jobs are given below

Job	A	B	C	D	E	F	G	H
Turning (minutes)	2	4	8	9	7	6	5	10
Grinding (minutes)	6	1	3	7	9	5	2	4

If the makespan is to be minimized, then the optimal sequence in which these jobs must be processed on the turning and grinding machines is
(a) G-E-D-F-H-C-A-B (b) A-D-E-F-H-C-G-B
(c) A-E-D-F-H-C-G-B (d) B-G-C-H-F-D-E-A
[2021 : 2 M, Set-1]

6.28 The product structure diagram shows the number of different components required at each level to produce one unit of the final product *P*. If there are 50 units of on- hand inventory of component *A*, the number of additional units of component *A* needed to produce 10 units of product *P* is ______ (in integer).

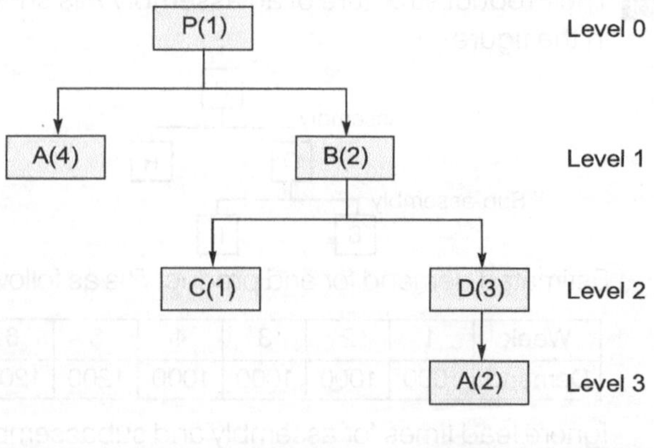

[2022 : 2 M, Set-1]

6.29 Parts P1-P7 are machined first on a milling machine and then polished at a separate machine. Using the information in the following table, the minimum total completion time required for carrying out both the operations for all 7 parts is ____ hours.

Part	Milling Machine	Polishing machine
P1	8	6
P2	3	2
P3	3	4
P4	4	6
P5	5	7
P6	6	4
P7	2	1

(a) 31 (b) 33
(c) 30 (d) 32

[2022 : 2 M, Set-2]

6.30 A part, produced in high volumes, is dimensioned as shown. The machining process making this part is known to be statistically in control based on sampling data. The sampling data shows that D1 follows a normal distribution with a mean of 20 mm and a standard deviation of 0.3 mm, while D2 follows a normal distribution with a mean of 35 mm and a standard deviation of 0.4 mm. An inspection of dimension C is carried out in a sufficiently large number of parts.

To be considered under six-sigma process control, the upper limit of dimension C should be ____ mm. (Rounded off to one decimal place)

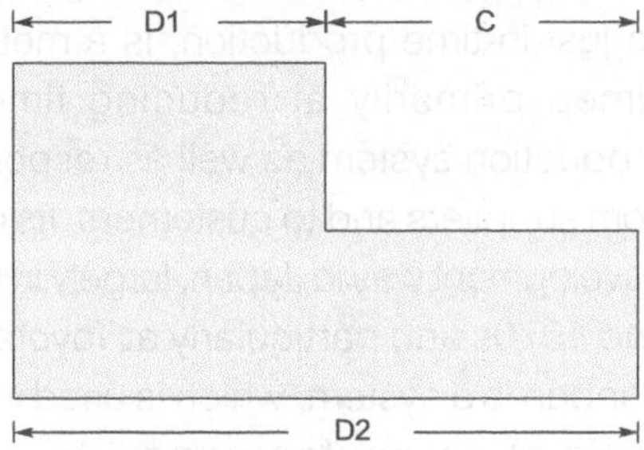

[2023 : 2 M]

6.31 A set of jobs U, V, W, X, Y, Z arrive at time t = 0 to a production line consisting of two workstations in series. Each job must be processed by both workstations in sequence (i.e., the first followed by the second). The process times (in minutes) for each job on each workstation in the production line are given below:

Job	U	V	W	X	Y	Z
Workstation 1	5	7	3	4	6	8
Workstation 2	4	6	6	8	5	7

The sequence in which the jobs must be processed by the production line if the total makespan of production is to be minimized is

(a) U-Y-V-Z-X-W (b) W-X-Z-V-Y-U
(c) W-X-V-Z-Y-U (d) W-U-Z-V-Y-X

[2024 : 1 M]

■■■■

Answers Products and Process, Planning and Control

6.1	(MRP)	6.2	(d)	6.4	(c)	6.5	(d)	6.6	(a)	6.7	(d)	6.8	(d)	6.9	(a)
6.10	(a)	6.11	(b)	6.12	(c)	6.13	(b)	6.14	(c)	6.15	(a)	6.16	(c)	6.17	(d)
6.18	(a)	6.19	(b)	6.20	(b)	6.21	(d)	6.22	(b)	6.23	(31)	6.24	(6.33)	6.25	(a)
6.26	(d)	6.27	(c)	6.28	(110)	6.29	(b)	6.30	(16.5)	6.31	(*)				

Explanations Products and Process, Planning and Control

6.1 Sol.

MRP is a computer based information system for scheduling, production and purchasing of dependent demand items MRP is a method of working out the production plan in a multi-stage production system that requires many product, sub-assembly and raw-material, so that all things needed are available at an appropriate time.

6.2 (d)

Arrangement of 4 machines into 4 quadrants = 4 ! = 24

6.3 Sol.

A–4, *B*–1, *C*–3, *D*–6

POINTS TO REMEMBER

- Just-in-time (JIT) manufacturing, also known as just-in-time production, is a methodology aimed primarily at reducing times within production system as well as response times from suppliers and to customers. Its origin and development was in Japan, largely in the 1960's and 1970s and particularly at Toyota.
- Kanban is a system, which is used in JIT (just-in-time) manufacturing to improve manufacturing efficiency.
- One of the main benefits of Kanban is to establish an upper limit to work in process inventory to avoid over capacity.

6.4 (c)

Assuming cycle time, $t_c = 10$ min

$\Sigma_i t_i = 10 + 8 + 6 + 9 + 10 = 43$

Balance delay,

$$= 1 - \frac{\Sigma t_i}{n \times t_c} = 1 - \frac{43}{5 \times 10} = 0.14 = \mathbf{14\%}$$

6.5 (d)

Bill of materials is a method is used in MRP (Material requirement planning). Production flow analysis (PFA) is a method of identifying part families that uses data from route sheets.

6.6 (a)

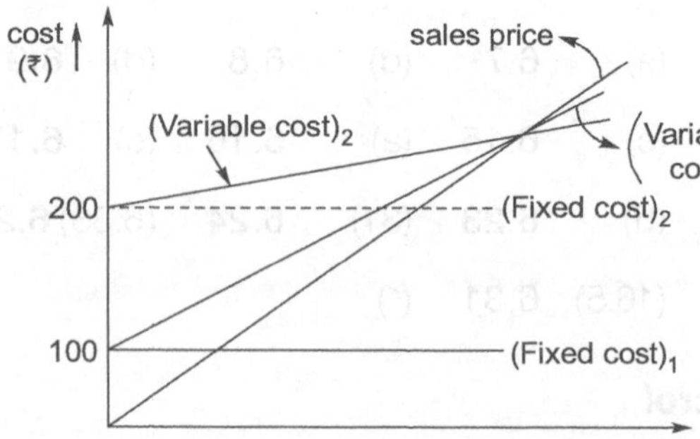

Let x be the quantity produced by each machine.

$\therefore$ The total cost in production using machine –1,

$$TC_1 = \text{Fixed cost + Variable cose}$$
$$= 100 + 2(x) = 100 + 2x$$

$\therefore$ Total cost in production using machine–2,

$$TC_2 = \text{Fixed cost + Variable cost}$$
$$= 200 + 1(x) = 200 + x$$

$\because$ At break-even point,

Total sales price = Total cost in production

$$3.5 \times 2x = TC_1 + TC_2$$
$$7x = 100 + 2x + 200 + x$$
$$= 300 + 3x$$

Now, quantity,

x = **75 units** (< 800 units) if value of x comes out > 800 units then for next unit sale price will be ₹3.

6.7 (d)

Total cost of x_1 component by using standard machine tool

$$(TC)_1 = \left(\frac{30}{60} + \frac{22x_1}{60}\right) \times 200$$
$$= 100 + \frac{2200}{30} x_1$$

Total cost x_2 component by using Automatic machine tool.

$$(TC)_2 = \left(2 + \frac{5}{60} \times x_2\right) \times 800$$
$$= 1600 + \frac{2000}{30} x_2$$

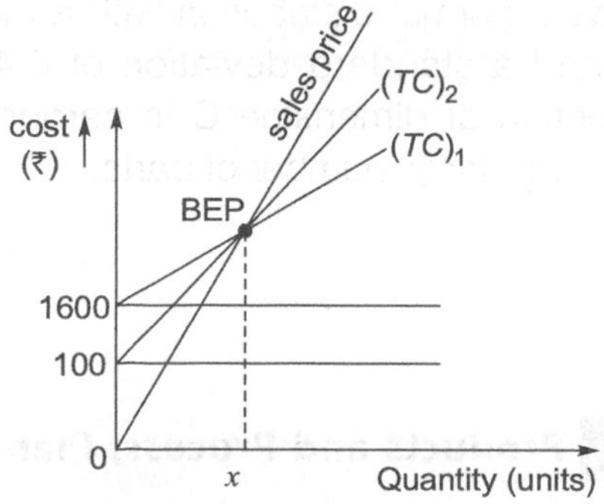

Let break even quantity be x.

At break even point

$$(TC)_1 = (TC)_2$$
$$\therefore 100 + \frac{2200}{30} x = 1600 + \frac{2000}{30} x$$

or $6.667x = 1500$

$\therefore$ x = **225**

6.8 (d)

Total time = 16 hours

$$\text{Actual time per joint} = \frac{90}{100} \times \frac{16}{108} \times 60 = 8 \text{ min}$$

Normal time = Actual time × Rating factor

$$\text{Normal time} = \text{Actual time} \times \frac{120}{100}$$
$$= 8 \times 1.2 = 9.6 \text{ min}$$

Standard time = Normal time + Allowances

$$S.T. = N.T. + 0.2S.T.$$

$$\Rightarrow \quad S.T. = \frac{9.6}{0.8} = 12 \text{ min}$$

6.9 (a)

Step - I

Activity	Standard time
M	12
E	16
T	3

Cycle time = 16 min

then no. of units produce

$$= \frac{8 \times 60}{16} = \textbf{30 units in 8 hrs.}$$

Step - II

Activity	Standard time
M	12
2*E*	8
T	3

Cycle time = 12 min

then no. of units produce

$$= \frac{8 \times 60}{12} = \textbf{40 units in 8 hrs.}$$

Step - III

Activity	Standard time
2*M*	6
2*E*	8
T	3

Cycle time = 8 min

then no. of units produce

$$= \frac{8 \times 60}{8} = \textbf{60 units in 8 hrs.}$$

Step - IV

Activity	Standard time
2*M*	6
3*E*	5.33
T	3

Cycle time = 6 min

then no. of units produce

$$= \frac{8 \times 60}{6} = \textbf{80 units in 8 hrs.}$$

So, number of work stations required **2, 3, 1 respectively.**

6.10 (a)

Rating factor = 1.2

Normal time = Element time × Rating factor

= 8 × 1.2 = 9.6 min

S.T. = Normal time + (% Allowance) × S.T.

$$S.T. = 9.6 + \frac{10}{100} \times S.T. = 10.667 \text{ min}$$

Standard production rate

$$= \frac{8 \times 60}{10.667} \text{ units} = 45$$

6.11 (b)

According to Johnson's algorithm, sequencing of jobs is

R	*T*	*S*	*Q*	*U*	*P*

Now,

Job	M_1		M_2	
	In	Out	In	Out
R	0	8	8	21
T	8	19	21	35
S	19	46	46	66
Q	46	78	78	97
U	78	94	97	104
P	94	109	109	115

We see that optimal makes span of the shop is **115 hours.**

6.12 (c)

Total time used = TWC = 7 + 9 + 7 + 10 + 9 + 6

= 48 min

T_c = Cycle time = 10 min

No. of work station = 6

$$\text{Line efficiency} = \frac{TWC}{n \times T_c} \times 100$$

$$= \frac{\text{Total time used}}{\text{No. of work station} \times \text{cycle time}} \times 100$$

$$= \frac{48}{6 \times 10} \times 100 = \textbf{80\%}$$

6.13 (b)

Since, it is given that starting and final inventory are nil. For min. cost of plan, the entire regular time production capacity will be used.

Month	Normal Production capacity	Overtime Production capacity	Demand	Unused capacity
1	100	20	90	10
2	100	20	130	–
3	80	40	110	–

Combined demand for first two months

$= 90 + 130 = 220$ units

Units producted by using normal production capacity $= 100 + 100 = 200$ units

Since by the use of overtime production capacity, we can easily fulfill the demand of remaining 20 units for the periods.

In 3rd month, there is a 30 units difference betwen regular production capacity and demand. So by making 30 units with the help of overtime production capacity, we can meet the deficit and there will be no inventory at the end of 3rd month.

6.14 (c)

Let 'Q' be the smallest capacity per week required to ensure a feasible production plan.

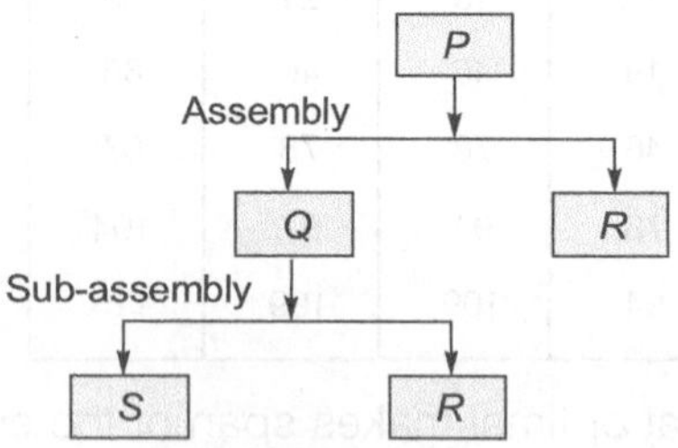

Week	Demand R(2)	Inventory level at the end of week
1	2000	Q–2000
2	2000	2Q–4000
3	2000	3Q– 6000
4	2000	4Q–8000
5	2400	5Q–10400
6	2400	6Q–12800

$\therefore$ For feasible production plan,

$$6Q - 12800 \geq 0$$

$$Q \geq \frac{12800}{6} \geq 2133.333$$

$\therefore$ The smallest capacity, $Q = 2200$ units

6.15 (a)

According to SPT rule, arranging the processing time in ncreasing.

Jobs	Processing time (days)
I	4
III	5
V	6
VI	8
II	9
IV	10

Jobs	Flow time (days)
I	0 + 4 = 4
III	4 + 5 = 9
V	4 + 5 + 6 = 15
VI	4 + 5 + 6 + 8 = 23
II	4 + 5 + 6 + 8 + 9 = 32
IV	4 + 5 + 6 + 8 + 9 + 10 = 42
	Total flow time = 125

$$\text{Avg. flow time} = \frac{\text{Total flow time}}{\text{Number of jobs}} = \frac{125}{6} = 20.83 \text{ days}$$

6.16 (c)

Sequence by Earliest Due Date (EDD)

Job	Time	Job flow time	Due date	Delay / Days or tardiness
1	4	0 + 4 = 4	6	+2
2	7	4 + 7 = 11	9	−2
4	8	11 + 8 = 19	17	−2
3	2	19 + 2 = 21	19	−2

No. of delay = **3**

6.17 (d)

Sequence by Shortest Processing Time (SPT)

Job	Process time	Job flow time	Due date	Days / Delays or Tardiness
3	2	0 + 2 = 2	19	0
1	4	2 + 4 = 6	6	0
2	7	6 + 7 = 13	9	4
4	8	13 + 8 = 21	17	4

Total tardiness= 0 + 0 + 4 + 4 = **8**

6.18 (a)

Product layout: A product layout arranges activities in a line according to the sequence of operations for a particular product or service. This would be an assembly line. In vehicle manufacturing assembly line, product layout is used.

POINTS TO REMEMBER

Process layout: A process layout is when you group similar activities together according to the process or function they perform. An example would be a machine shop where all the drills are located in one area of the shop.

6.19 (b)

Kanban is related to quality and this concept is used in just in time production *i.e.* zero inventory.

6.21 (d)

Group technolog is a manufacturing technique in which parts having similarities in geometry, manufacturing process an/or functions are manufactured in one location using a small number of machines or processes.
Group technology is based on a general principle that many problems are similar and by grouping similar problems, a single solution can be found to a set of problems, thus saving time and effort.

6.22 (b)

Process, capability index,

$$Cp = \frac{USL - LSL}{6\sigma}$$

$$= \frac{15.02 - 14.98}{6 \times 0.004} = 1.667$$

6.23 Sol.

SPT rule given shortest average

Jobs	Time	Job flow time
Q	9	0 + 9 = 9
S	12	9 + 12 = 21
P	15	21 + 15 = 36
R	22	36 + 22 = 58
		Total = 124

Job flow time = 124 days
So, applying SPT

$$\text{Average job flow time} = \frac{124}{4} = 31\,\text{days}$$

POINTS TO REMEMBER

When it is not given in problem that which (shortest possible time or earliest due date) rule is to be used. Then it is preferable to use SPT (shortest possible time) rule because SPT minimizes the total completion time more than the earliest due date (EDD) rule.

6.24 Sol.

By SPT Rule

Job	P.T.	D.D.	Job Flow Time	Tardiness
A	3	8	0 + 3 = 3	0
C	4	4	3 + 4 = 7	3
E	5	17	7 + 5 = 12	0
B	7	16	12 + 7 = 19	3
D	9	18	19 + 9 = 28	10
F	13	19	28 + 13 = 41	22
			Job flow time = 110	38

Total tardiness= 38
Average tardiness per job

$$= \frac{\text{Total tardiness}}{\text{No. of Jobs}} = \frac{38}{6} = 6.33 \text{ days}$$

6.25 (a)

For Lot-for-Lot approach inventory is minimum, hence inventory holding cost is reduced.

6.27 (c)

Sequencing,

A	(2)	6
B	4	(1)
C	8	(3)
D	9	(7)
E	(7)	9
F	6	(5)
G	5	(2)
H	10	(4)

A E D F H C G B

6.28 (110) (110 to 110)

Writing the different values required far making 10 units of *P*.

$\Rightarrow \quad n(A)_{(i)} = 4 \times 10 = 40$

$n(B) = 20$

$\Rightarrow \quad n(D) = 60$

$n(A)_{(ii)} = 120$

So, the total units of A required is 40 + 120

$= 160$ units

Now, we have 50 units of A in hand, therefore,

160 – 50 = 110 units more would be required.

6.29 (b)

By Johnson's Algorithm

	Milling machine	Polishing machine	
A	8	6	
B	3	2 ✓	X
X C	3 ✓	4	
D	4 ✓	6	
E	5 ✓	7	
F	6	4 ✓	
G	2	1 ✓	X

Sequence: $C — D — E — A — F — B — G$

Sequence	Milling machine		Polishing machine	
	In	Out	In	Out
C	0	3	3	7
D	3	7	7	13
E	7	12	13	20
A	12	20	20	26
F	20	26	26	30
B	26	29	30	32
G	29	31	32	33

So, MST = 33

6.30 (16.5) (16.4 to 16.6)

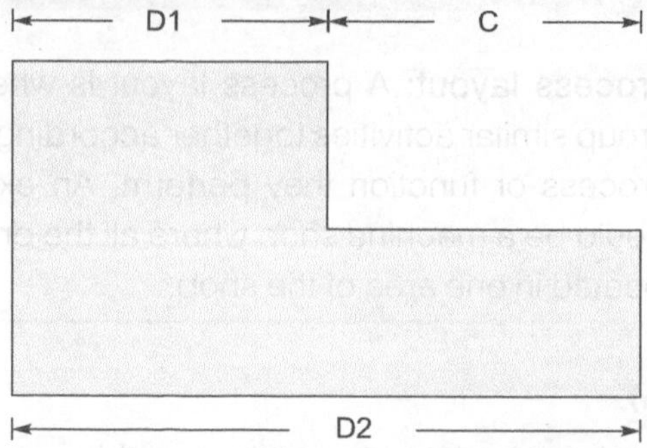

D1 and D2 follow normal distribution.

So, Upper limit = $\mu + 3\sigma$

Lower limit = $\mu - 3\sigma$

$\bar{D}1 = 20$ mm,

$\bar{D}2 = 35$ mm, $\sigma_{D2} = 0.4$ mm, $\sigma_{D1} = 0.3$ mm

Mean value, $\bar{C} = \bar{D}2 - \bar{D}1 = 35 - 20 = 15$ mm

Standard deviation of C

$$\sigma_C = \sqrt{\sigma_{D2}^2 + \sigma_{D1}^2} = \sqrt{0.4^2 + 0.3^2}$$

Upper limit of $C = \bar{C} + 3 \times \sigma_C = 15 + 3 \times 0.5 = 16.5$ mm

6.31 (*)

As per Johnson's algorithm, the optimum sequence is W-X-Z-V-Y-U.

Note: The question did not mention which method is to be used and hence it was marked correct for all students.

■■■■

7 Queuing Theory & Transportation

7.1 On an average 100 customers arrive at a place each hour, and on the average the server can process 120 customers per hour. What is the proportion of time the server is idle? **[1995 : 2 M]**

7.2 The cost of providing service in a queuing system increases with

(a) Increased mean time in the queue
(b) Increased arrival rate
(c) Decreased mean time in the queue
(d) Decreased arrival rate **[1997 : 1 M]**

7.3 At a production machine, parts arrive according to a Poisson process at the rate of 0.35 parts per minute. Processing time for parts have exponential distribution with mean of 2 minutes. What is the probability that a random part arrival finds that there are already 8 parts in the system (in machine + in queue)?

(a) 0.0247 (b) 0.0576
(c) 0.0173 (d) 0.082 **[1999 : 2 M]**

7.4 In a single server infinite population queuing model. Arrivals follow a Poisson distribution with mean λ = 4 per hour. The service times are exponential with mean service time equal to 12 minutes. The expected length of the queue will be

(a) 4 (b) 3.2
(c) 1.25 (d) 24.3 **[2000 : 2 M]**

7.5 The supply at three sources is 50, 40 and 60 units respectively while the demand at the four destinations is 20, 30, 10 and 50 units. In solving this transportation problem.

(a) a dummy source of capacity 40 units is needed
(b) a dummy destination of capacity 40 units is needed
(c) no solution exists as the problem is infeasible
(d) none solution exists as the problem is degenerate **[2002 : 2 M]**

7.6 Arrivals at a telephone booth are considered to be Poisson, with an average time of 10 minutes between successive arrivals. The length of a phone call is distributed exponentially with mean 3 minutes. The probability that an arrival does not have to wait before service is

(a) 0.3 (b) 0.5
(c) 0.7 (d) 0.9 **[2002 : 2 M]**

7.7 A maintenance service facility has Poisson arrival rates, negative exponential service time and operates on a 'first come first served' queue discipline. Break-downs occur on an average of 3 per day with a range of zero to eight. The maintenance crew can service an average of 6 machines per day with a range of zero to seven. The mean waiting time for an item to be serviced would be

(a) 1/6 day (b) 1/3 day
(c) 1 day (d) 3 day **[2004 : 2 M]**

7.8 Consider a single server queuing model with Poisson arrivals (λ = 4/hour) and exponential service (μ = 4/hour). The number in the system is restricted to a maximum of 10. The probability that a person who comes in leaves without joining the queue is

(a) 1/11 (b) 1/10
(c) 1/9 (d) 1/2 **[2005 : 1 M]**

7.9 The number of customers arriving at a railway reservation counter is Poisson distributed with an arrival rate of eight customers per hour. The reservation clerk at this counter takes six minutes per customer on an average with an exponentially distributed service time. The average number of the customers in the queue will be

(a) 3 (b) 3.2
(c) 4 (d) 4.2 **[2006 : 1 M]**

7.10 In an M/M/1 queuing system, the number of arrivals in an interval of length T is a Poisson random variable i.e. the probability of there being n arrivals in an interval of length T is $\dfrac{e^{-\lambda T}(\lambda T)^n}{n!}$.
The probability density function $f(t)$ of the inter-arrival time is given by

(a) $\lambda^2\left(e^{-\lambda^2 t}\right)$ (b) $\dfrac{e^{-\lambda^2 t}}{\lambda^2}$
(c) $\lambda e^{-\lambda t}$ (d) $\dfrac{e^{-\lambda t}}{\lambda}$
[2008 : 1 M]

7.11 Little's law is a relationship between

(a) stock level and lead time in an inventory system
(b) waiting time and length of the queue in a queuing system

(c) number of machines and job due dates in a scheduling problem
(d) uncertainty in the activity time and project completion time [2010 : 1 M]

7.12 Cars arrive at a service station according to Poisson's distribution with a mean rate of 5 per hour. The service time per car is exponential with a mean of 10 minutes. At steady state, the average waiting time in the queue is
(a) 10 minutes (b) 20 minutes
(c) 25 minutes (d) 50 minutes
[2011 : 1 M]

7.13 Customers arrive at a ticket counter at a rate of 50 per hour and tickets are issued in the order of their arrival. The average time taken for issuing a ticket is 1 min. Assuming that customer arrivals form a Poisson process and service times are exponentially distributed, the average waiting time in queue in min is
(a) 3 (b) 4
(c) 5 (d) 6 [2013 : 1 M]

7.14 At a work station, 5 jobs arrive every minute. The mean time spent on each job in the work station is 1/8 minute. The mean steady state number of jobs in the system is ______.[2014 : 2 M, Set-4]

7.15 The jobs arrive at a facility, for service, in a random manner. The probability distribution of number of arrivals of jobs in a fixed time interval is
(a) Normal (b) Poisson
(c) Erlang (d) Beta
[2014 : 1 M, Set-1]

7.16 Jobs arrive at a facility at an average rate of 5 in an 8 hour shift. The arrival of the jobs follows Poisson distribution. The average service time of a job on the facility is 40 minutes. The service time follows exponential distribution. Idle time (in hours) at the facility per shift will be
(a) $\frac{5}{7}$ (b) $\frac{14}{3}$
(c) $\frac{7}{3}$ (d) $\frac{10}{3}$
[2014 : 2 M, Set-1]

7.17 In the notation $(a/b/c) : (d/e/f)$ for summarizing the characteristics of queueing situation, the letters *b* and *d* stand respectively for
(a) service time distribution & queue discipline
(b) number of servers & size of calling source
(c) number of servers & queue discipline
(d) service time distribution and maximum number allowed in system [2015 : 1 M, Set-3]

7.18 In a single-channel queuing model, the customer arrival rate is 12 per hour and the serving rate is 24 per hour. The expected time that a customer is in queue is _____ minutes. [2016 : 1 M, Set-2]

7.19 For a single server with Poisson arrival and exponential service time, the arrival rate is 12 per hour. Which one of the following service rates will provide a steady state finite queue length?
(a) 6 per hour (b) 10 per hour
(c) 12 per hour (d) 24 per hour
[2017 : 1 M, Set-2]

7.20 The arrival of customers over fixed time intervals in a bank follow a Poisson distribution with an average of 30 customers/hour. The probability that the time between successive customer arrival is between 1 and 3 minutes is _______ (correct to two decimal places). [2018 : 1 M, Set-2]

7.21 Customers arrives at a shop according to the Poisson distribution with a mean of 10 customers/hour. The manager notes that no customer arrives for the first 3 minutes after the shop opens. The probability that a customer arrives within the next 3 minutes is
(a) 0.50 (b) 0.61
(c) 0.86 (d) 0.39
[2021 : 2 M, Set-1]

7.22 Consider a single machine workstation to which jobs arrive according to a Poisson distribution with a mean arrival rate of 12 jobs/hour. The process time of the workstation is exponentially distributed with a mean of 4 minutes. The expected number of jobs at the workstation at any given point of time is _____ [round off to the nearest integer]
[2021 : 2 M, Set-1]

7.23 A queueing system has one single server workstation that admits an infinitely long queue. The rate of arrival of jobs, to the queueing system follows the Poisson distribution with a mean of 5 jobs/hour. The service time of the server is exponentially distributed with a mean of 6 minutes. In steady state operation of the queueing system, the probability that the server is not busy at any point in time is
(a) 0.83 (b) 0.50
(c) 0.17 (d) 0.20
[2024 : 1 M]

7.24 In a supplier-retailer supply chain, the demand of each retailer, the capacity of each supplier, and the unit cost in rupees of material supply from each supplier to each retailer are tabulated below. The supply chain manager wishes to minimize the total cost of transportation across the supply chain.

	Retailer I	Retailer II	Retailer III	Retailer IV	Capacity
Supplier A	11	16	19	13	300
Supplier B	5	10	7	8	300
Supplier C	12	14	17	11	300
Supplier D	8	15	11	9	300
Demand	300	300	300	300	

The optimal cost of satisfying the total demand from all retailers is ____ rupees. (answer in integer)

[2024 : 2 M]

Answers Queuing Theory & Transportation

7.1 (0.167) 7.2 (c) 7.3 (c) 7.4 (b) 7.5 (b) 7.6 (c) 7.7 (a) 7.8 (a)
7.9 (b) 7.10 (c) 7.11 (b) 7.12 (d) 7.13 (c) 7.14 (1.67) 7.15 (b) 7.16 (b)
7.17 (a) 7.18 (2.5) 7.19 (d) 7.20 (0.383) 7.21 (d) 7.22 (4) 7.23 (b)
7.24 (12300)

Explanations Queuing Theory & Transportation

7.1 Sol.

λ = 100 customer per hour,
μ = 120 customer per hour,
Probability when the server is idle,

$$P_o = 1-\rho = 1-\frac{\lambda}{\mu}$$

$$P_o = 1-\frac{\lambda}{\mu} = 1-\frac{100}{120} = \frac{1}{6} = 0.167$$

7.3 (c)

Arrival rate (Poisson distribution)

$$\lambda = 0.35 \text{ parts/minute} = 0.35 \times 60 = 21 \text{ parts/hour}$$

Service rate (exponential distribution)

$$\mu = \frac{60}{2} = 30 \text{ parts/hour}$$

$$\rho = \frac{\lambda}{\mu} = \frac{21}{30} = \frac{7}{10} = 0.7$$

Probability of having exactly n-customers in the system,

$$P_n = \rho^n P_0 = \rho^n(1-\rho)$$

∴ Probabbility of having exactly 8-customers in the system,

$$P_8 = \rho^8(1-\rho) = (0.7)^8(1-0.7)$$
$$= (0.7)^8(0.3 = 0.01729 = 0.0173)$$

POINTS TO REMEMBER

- Parts arriving according to a Poisson process means Arrival is random.
- Service rate is exponential distribution means some parts take less time for service and others take more time for service.

7.4 (b)

Arrival rate (Poisson distribution), λ =4 per hr.,
Service rate (Exponential distribution)

μ = 5 per hr., ρ = 0.8

Expected length of queue or average number of customer in the queue,

$$L_q = \frac{\rho^2}{1-\rho} = \frac{\left(\frac{\lambda}{\mu}\right)}{1-\left(\frac{\lambda}{\mu}\right)} = \frac{\left(\frac{4}{5}\right)^2}{1-\left(\frac{4}{5}\right)} = 3.2$$

POINTS TO REMEMBER

- Average number of customer in the queue doesn't count the customer being served.
- Average number of customers in the system tells on an average total number of customers in the system and it improves both the customers waiting in the queue, along with being serve.

7.5 (b)

If the total supply from all the units per factories is equal to the total demand from all the destination, the problem will be balanced otherwise it will be unbalanced.

If the given problem is not balanced then we need to balance it by addition of dummy source of destination.

Total supply = 50 + 40 + 60 = 150

Total demand = 20 + 30 + 10 + 50 = 110

Total supply – Total demand = 150 – 110 = 40

Since supply is greater than demand, then a dummy source of capacity 40 units is needed.

7.6 (c)

$\lambda = 0.1$ per minute,

Service rate, $\mu = 0.33$ per minute

$$\rho = \frac{0.1}{0.33} = 0.3$$

Probability that an arrival does not have to wait before service or probability that a system is idle,

$$P_o = 1 - \rho = 1 - 0.3 = 0.7$$

7.7 (a)

Given:

Arrival rate: $\lambda = 3$ per day

Service rate: $\lambda = 6$ per day

Mean waiting time for an item to be serviced

$$= \frac{\lambda}{\mu(\mu-\lambda)} = \frac{3}{6(6-3)} = \frac{1}{2} \times \frac{1}{3} = \frac{1}{6} \text{day}$$

7.8 (a)

Given: $\lambda = 4$ per hour, $\mu = 4$ per hour

Traffic intensity,

$$\rho = \frac{\lambda}{\mu} = \frac{4}{4} = 1$$

We know that,

$$\sum_{n=0}^{10} P_n = 1$$

$$\therefore P_0 + P_1 + P_3 + ... + P_{10} = 1$$

$$P_0 + \rho P_0 + \rho^2 P_0 + \rho^3 P_0 + ... + \rho^{10} P_0 = 1$$

$$P_0(1 + \rho + \rho^2 + ... + \rho^{10}) = 1$$

$$P_0(1 + 1 + ... + 1) = 1$$

$$\therefore \quad P_0 = \frac{1}{11}$$

Probability that a person who comes in leaves without joining the queue i.e.

$$P_{11} = \rho^{11} P_0 = (1)^{11} \times \frac{1}{11} = \frac{1}{11}$$

7.9 (b)

Arrival rate, $\lambda = 8$ per hour

Service rate: $\mu = \frac{60}{6} = 10$ per hour

Average number of customer in the queue

$$L_q = \frac{\lambda^2}{\mu(\mu-\lambda)} = \frac{(8)^2}{10(10-8)} = \mathbf{3.2}$$

7.10 (c)

Probability of n arrivals in an interval of length,

$$T = \frac{e^{-\lambda T}(\lambda T)^n}{n!}$$

$\therefore$ probability density function f(t) of the inter-arrival time $\quad P(t) = \lambda e^{-\lambda t}$

POINTS TO REMEMBER

Probability that more then T time is needed to service a customer = $(\exp)^{-\lambda t}$

7.11 (b)

Little's law: Little's law is a relationship between waiting time and length of the queue in a queuing system. According to Little's law, for a stable system, the average number of customer, in the queue or the system, is given by average arrival rate multiplied by average waiting time in the queue or in the system.

Average number of customer in the system,

$$L_S = L \times W_S$$

(W_S = waiting time for system)

Average number of customer in the queue,

$$L_q = \lambda \times W_q$$

(W_S = waiting time for queue)

(λ = arrival rate)

7.12 (d)

Given data:

λ = arrival rate = 5/hour

μ = service rate = 6/hour

$\therefore$ Average waiting time in queue

$$W_q = \frac{\lambda}{\mu(\mu-\lambda)} = \frac{5}{6(6-5)} \text{hr}$$

$$= 0.833 \text{ hr} = \mathbf{50 \text{ minutes}}$$

7.13 (c)

Arrival rate, λ = 50 customer/hour

$$= \frac{50}{60} = \frac{5}{6} \text{ customer/minute}$$

Service rate,

μ = 1 customer/minute

Average waiting time in queue,

$$W_q = \frac{\lambda}{(\mu-\lambda)} = \frac{\frac{5}{6}}{1\left(1-\frac{5}{6}\right)} = 5 \text{ min}$$

POINTS TO REMEMBER

- Average waiting time in queue,

$$W_q = \frac{L_q}{\lambda} = W_S - \frac{1}{\mu} = \frac{\lambda}{m(\mu-\lambda)}$$

- Average waiting time in system,

$$W_S = \frac{L_S}{\lambda} = \frac{1}{(\mu-\lambda)}$$

7.14 Sol.

λ (arrival rate) = 5 jobs/minute

μ (service rate) = 8 jobs/minute

No. of jobs in the system,

$$L_S = \frac{\lambda}{\mu-\lambda} = \frac{5}{8-5} = \frac{5}{3} = \mathbf{1.67}$$

7.16 (b)

$$\lambda = \frac{5}{8}$$

$$\mu = \frac{60}{40} = \frac{3}{2}$$

Probability of idleness $= 1 - \frac{\lambda}{\mu}$

$$= 1 - \frac{5}{8} \times \frac{2}{3} = \frac{7}{12}$$

Idle time (in hrs) per shift $= \frac{7}{12} \times 8$

$$= \frac{14}{3}$$

7.17 (c)

Queuing model generally represents as "Kendall & Lee" notation whose general form is (a/b/c) : (d/e/f).

Where a = probability distribution of arrival pattern.

b = probability distribution of service pattern.

c = number of servers in the system

d = service rule or servoe order.

e = size of system

f = size or capacity of calling population

7.18 Sol.

Arrival rate, $\lambda = 12 \text{ hr}^{-1}$

Service rate, $u = 24 \text{ hr}^{-1}$

Expected time that a customer is in queue,

$$w_q = \frac{\lambda}{\mu(u-\lambda)} = \frac{12}{24 \times 12} \times 60 = \mathbf{2.5 \text{ minutes}}$$

7.19 (d)

λ = Arrival rate, μ = Service rate

$\therefore$ When $\lambda > \mu$,

queue length will kept on increasing and alternatively system will fail.

When $\lambda < \mu$, system works and provides a steady state finite queue length.

So $\mu > \lambda$ then only option (d) μ = 24 per hour is correct.

7.20 Sol.

Given, arrival rate,

$$\lambda = 30/\text{hour}$$

$$\lambda = \frac{1}{2} \text{ min.}$$

$$P = \text{prob.} = 1 - e^{-\lambda t}$$

$$P(1) = 1 - e^{-\frac{1}{2}\times 1} = 0.393$$

$$P(3) = 1 - e^{-\lambda t} = 1 - e^{-\frac{1}{2}\times 3}$$

$$= 1 - e^{-1.5} = 0.7768$$

$$P(1 \le T \le 3 \text{ min}) = 0.7768 - 0.393 = \mathbf{0.383}$$

7.21 (d)

(i) Probability of zero in 3 min ($\lambda T = 0.5$)

$$P_1 \rightarrow P(0, 3) = \frac{(\exp)^{-0.5} \cdot (0.5)^0}{0!} = 0.606$$

Probability of atleast one customer in next 3 minutes = 1 – 0.606 = 0.39

7.22 Sol.

$$\lambda = 12 \text{ per hour}, \ \frac{1}{\mu} = 4 \text{ min/Jobs}, \ \mu = 15 \text{ Jobs/hr}$$

$$\rho = \frac{12}{15} = \frac{4}{5}$$

$$L_s = \frac{\rho}{1-\rho} = \frac{\frac{4}{5}}{1-\frac{4}{5}} = \frac{\frac{4}{5}}{\frac{1}{5}}$$

$$\frac{4}{5} \times \frac{5}{1} = 4 \text{ Jobs}$$

7.23 (b)

Given: λ = 5/hr, μ = 10/hr

$$\therefore \quad \rho = \frac{\lambda}{\mu} = \frac{5}{10} = \frac{1}{2}$$

Now, probability for the system to be idle,

$$P_o = 1 - \rho = 1 - 0.5 = 0.5$$

7.24 (12300) (12300 to 12300)

As per Hungarian method, subtracting the smallest value from each column from other elements of that column, we get

6	6	12	5
0	0	0	0
7	4	10	3
3	5	4	1

Similarly, for each row, we get the following matrix.

1	1	7	0
0	0	0	0
4	1	7	0
2	4	3	0

Since, the minimum number of lines to cover each zero of every row and column is less than the order of matrix. So to get the opportunity matrix, we will subtract smallest value from every element without lines and adding the same at intersection we will get the following matrix.

0	0	6	0
0	0	0	0
3	0	6	0
1	3	2	0

Since the minimum number of lines is equal to the order of matrix. Assignment can be done in the matrix as shown below.

[0]	0	6	~~0~~
~~0~~	~~0~~	[0]	0
3	[0]	6	~~0~~
1	3	2	[0]

So, optimal solution is

$$= (11 + 7 + 14 + 9) \times 300$$
$$= ₹12300$$

■■■■

General Aptitude

UNIT XIV

General Aptitude

Syllabus

Verbal Aptitude : Basic English grammar: tenses, articles, adjectives, prepositions, conjunctions, verb-noun agreement, and other parts of speech Basic vocabulary: words, idioms, and phrases in context Reading and comprehension Narrative sequencing.

Quantitative Aptitude : Data interpretation: data graphs (bar graphs, pie charts, and other graphs representing data), 2- and 3-dimensional plots, maps, and tables Numerical computation and estimation: ratios, percentages, powers, exponents and logarithms, permutations and combinations, and series Mensuration and geometry Elementary statistics and probability

Analytical Aptitude : Logic: deduction and induction Analogy Numerical relations and reasoning.

Spatial Aptitude : Transformation of shapes: translation, rotation, scaling, mirroring, assembling, and grouping

Paper folding, cutting, and patterns in 2 and 3 dimensions .

Analysis of Previous GATE Papers

Exam Year	1 Mark Ques.	2 Marks Ques.	3 Marks Ques.	5 Marks Ques.	Total Marks
1995	–	–	–	–	–
1996	–	–	–	–	–
1997	–	–	–	–	–
1998	–	–	–	–	–
1999	–	–	–	–	–
2000	–	–	–	–	–
2001	–	–	–	–	–
2002	–	–	–	–	–
2003	–	–	–	–	–
2004	–	–	–	–	–
2005	–	–	–	–	–
2006	–	–	–	–	–
2007	–	–	–	–	–
2008	–	–	–	–	–
2009	–	–	–	–	–
2010	–	–	–	–	–
2011	–	–	–	–	–
2012	–	–	–	–	–
2013	–	–	–	–	–
2014 Set-1	5	5	–	–	**15**

Exam Year	1 Mark Ques.	2 Marks Ques.	Total Marks
2014 Set-2	5	5	**15**
2014 Set-3	5	5	**15**
2014 Set-4	5	5	**15**
2015 Set-1	5	5	**15**
2015 Set-2	5	5	**15**
2015 Set-3	5	5	**15**
2016 Set-1	5	5	**15**
2016 Set-2	5	5	**15**
2016 Set-3	5	5	**15**
2017 Set-1	5	5	**15**
2017 Set-2	5	5	**15**
2018 Set-1	5	5	**15**
2018 Set-2	5	5	**15**
2019 Set-1	5	5	**15**
2019 Set-2	5	5	**15**
2020 Set-1	5	5	**15**
2020 Set-2	5	5	**15**
2021 Set-1	5	5	**15**
2021 Set-2	5	5	**15**
2022 Set-1	5	5	**15**
2022 Set-2	5	5	**15**
2023	5	5	**15**
2024	5	5	**15**

1 General Aptitude

1.1 Which of the following options is the closest in meaning to the word below?

Circuitous

(a) Cyclic (b) Indirect
(c) Confusing (d) Crooked **[2010 : 1 M]**

1.2 The question below consists of a pair of related words followed by four pairs of words. Select the pair that best expresses the relation in the original pair.

Unemployed: Worker

(a) Fallow : Land
(b) Unaware : Sleeper
(c) Wit : Jester
(d) Renovated : House **[2010 : 1 M]**

1.3 Choose the most appropriate word from the options given below to complete the following sentence:

If we manage to .... our natural resources, we would leave a better planet for our children.

(a) uphold (b) restrain
(c) cherish (d) conserve **[2010 : 1 M]**

1.4 Choose the most appropriate word from the options given below to complete the following sentence:

His rather casual remarks on politics .... his lack of seriousness about the subject.

(a) masked (b) belied
(c) betrayed (d) suppressed **[2010 : 1 M]**

1.5 25 persons are in a room. 15 of them play hockey, 17 of them play football and 10 of them play both hockey and football. Then the number of persons playing neither hockey nor football is

(a) 2 (b) 17
(c) 13 (d) 3 **[2010 : 1 M]**

1.6 Modern warfare has changed from large scale clashes of armies to suppression of civilian populations. Chemical agents that do their work silently appear to be suited to such warfare; and regretfully, there exist people in military establishments who think that chemical agents are useful fools for their cause.

Which of the following statements best sums up the meaning of the above passage?

(a) Modern warfare has resulted in civil strife.
(b) Chemical agents are useful in modern warfare.
(c) Use of chemical agents in warfare would be undesirable.
(d) People in military establishments like to use chemical agents in war. **[2010 : 2 M]**

1.7 If 137 + 276 = 435 how much is 731 + 672?

(a) 534 (b) 1403
(c) 1623 (d) 1531 **[2010 : 2 M]**

1.8 5 skilled workers can build a wall in 20 days; 8 semiskilled workers can build a wall in 25 days; 10 unskilled workers can build a wall in 30 days. If a team has 2 skilled, 6 semiskilled and 5 unskilled workers, how long will it take to build the wall?

(a) 20 days (b) 18 days
(c) 16 days (d) 15 days **[2010 : 2 M]**

1.9 Given digits 2, 2, 3, 3, 3, 4, 4, 4, 4 how many distinct 4 digit numbers greater than 3000 can be formed?

(a) 50 (b) 51
(c) 52 (d) 54 **[2010 : 2 M]**

1.10 Hari (H), Gita (G), Irfan (I) and Saira (S) are siblings (i.e. brothers and sisters). All were born on 1st January. The age difference between any two successive siblings (that is born one after another) is less than 3 years. Given the following facts:

1. Hari's age + Gita's age > Irfan's age + Saira's age.
2. The age difference between Gita and Saira is 1 year. However, Gita is not the oldest and Saira is not the youngest.
3. There are no twins.

In what order were they born (oldest first)?

(a) IISIG (b) SGHI
(c) IGSH (d) IHSG **[2010 : 2 M]**

1.11 Choose the word from the options given below that is most nearly opposite in meaning to the given word:

Amalgamate

(a) merge (b) split
(c) collect (d) separate **[2011 : 1 M]**

1.12 Which of the following options is the closest in the meaning to the word below:

Inexplicable

(a) Incomprehensible (b) Indelible
(c) Inextricable (d) Infallible **[2011 : 1 M]**

1.13 If Log(P) = (1/2)Log(Q) = (1/3)Log(R), then which of the following options is TRUE?

(a) $P^2 = Q^3R^2$ (b) $Q^2 = PR$
(c) $Q^2 = R^3P$ (d) $R = P^2Q^2$ **[2011 : 1 M]**

1.14 Choose the most appropriate word(s) from the options given below to complete the following sentence.

In contemplated__________ Singapore for my vacation but decided against it.

(a) to visit (b) having to visit
(c) visiting (d) for a visit **[2011 : 1 M]**

1.15 Choose the most appropriate word from the options given below to complete the following sentence.

If you are trying to make a strong impression on your audience, you cannot do so by being understated, tentative or

(a) hyperbolic (b) restrained
(c) argumentative (d) indifferent

[2011 : 1 M]

1.16 A container originally contains 10 litres of pure spirit. From this container 1 litre of spirit is replaced with 1 litre of water. Subsequently, 1 litre of the mixture is again replaced with 1 litre of water and this processes is repeated one more time. How much spirit is now left in the container?

(a) 7.58 litres (b) 7.84 litres
(c) 7 litres (d) 7.29 litres **[2011 : 2 M]**

1.17 **Few school curricula include a unit on how to deal with bereavement and grief, and yet all students at some point in their lives suffer from losses through death and parting.**

Based on the above passage which topic would not be included in a unit on bereavement?

(a) how to write a letter of condolence
(b) what emotional stages are passed through in the healing process
(c) what the leading causes of death are
(d) how to give support to a grieving friend

[2011 : 2 M]

1.18 The variable cost (V) of manufacturing a product varies according to the equation $V = 4q$, where q is the quantity produced. The fixed cost (F) of production of same product reduces with q according to the equation $F = 100/q$. How many units should be produced to minimize the total cost ($V + F$)?

(a) 5 (b) 4
(c) 7 (d) 6 **[2011 : 2 M]**

1.19 P, Q, R and S are four types of dangerous microbes recently found in a human habitat. The area of each circle with its diameter printed in brackets represents the growth of a single microbe surviving human immunity system within 24 hours of entering the body. The danger to human beings varies proportionately with the toxicity, potency and growth attributed to a microbe shown in the figure below:

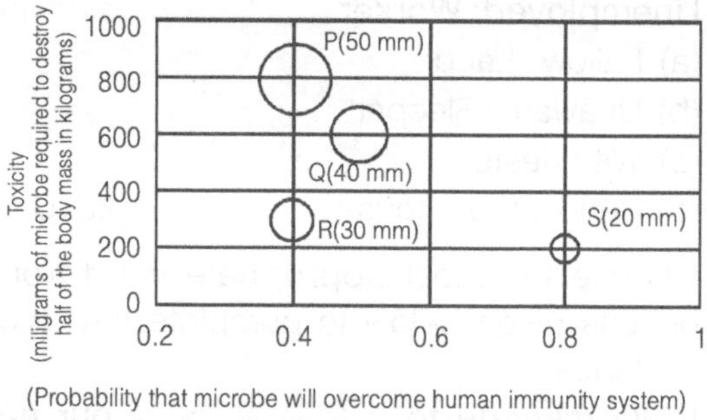

A pharmaceutical company is contemplating the development of a vaccine against the most dangerous microbe. Which microbe should the company target in its first attempt?

(a) P (b) Q
(c) R (d) S **[2011 : 2 M]**

1.20 A transporter receives the same number of orders each day. Currently, he has some pending orders (backlog) to be shipped. If he uses 7 trucks, then at the end of the 4th day he can clear all the orders. Alternatively, if he uses only 3 trucks, then all the orders are cleared at the end of the 10th day. What is the minimum number of trucks required so that there will be no pending order at the end of the 5th day?

(a) 4 (b) 5
(c) 6 (d) 7 **[2011 : 2 M]**

1.21 Choose the most appropriate alternative from the options given below to complete the following sentence.

Suresh's dog is the one _____ was hurt in the stampede.

(a) that (b) which
(c) who (d) whom **[2012 : 1 M]**

1.22 The cost function for a product in a firm is given by $5q^2$, where q is the amount of production. The firm can sell the product at a Met price of ₹ 50 per unit. The number of units to be produced by the firm such that the profit is maximized is

(a) 5 (b) 10
(c) 15 (d) 25 [2012 : 1 M]

1.23 Choose the most appropriate alternative from the options given below to complete the following sentence:

Despite several ____ the mission succeeded in its attempt to resolve the conflict.

(a) attempts (b) setbacks
(c) meetings (d) delegations
[2012 : 1 M]

1.24 Which one of the following options is the closest in meaning to the word given below?

Mitigate

(a) Diminish (b) Divulge
(c) Dedicate (d) Denote [2012 : 1 M]

1.25 Choose the grammatically INCORRECT sentence:

(a) They gave us the money back less the service charges to Three Hundred rupees.
(b) This country's expenditure is not less than that of Bangladesh.
(c) The committee initially asked for a funding of Fifty Lakh rupees, but later settled for a lesser sum.
(d) This country's expenditure on educational reforms is very less. [2012 : 1 M]

1.26 Given the sequence of terms, *AD CG FK JP*, the next term is

(a) *OV* (b) *OW*
(c) *PV* (d) *PW* [2012 : 2 M]

1.27 Wanted Temporary, Part-time persons for the post of Field Interviewer to conduct personal interviews to collect and collate economic data. Requirements: High School-pass, must be available for Day, Evening and Saturday work. Transportation paid, expenses reimbursed.

Which one of the following is the best inference from the above advertisement?

(a) Gender-discriminatory
(b) Xenophobic
(c) Not designed to make the post attractive
(d) Not gender-discriminatory [2012 : 2 M]

1.28 A political party orders an arch for the entrance to the ground in which the annual conventions is being held. The profile of the arch follows the equation $y = 2x - 0.1x^2$ where y is the height of the arch in meters. The maximum possible height of the arch is

(a) 8 meters (b) 10 meters
(c) 12 meters (d) 14 meters
[2012 : 2 M]

1.29 An automobile plant contracted to buy shock absorbers from two suppliers *X* and *Y*. *X* supplies 60% and *Y* supplies 40% of the shock absorbers. All shock absorbers are subjected to a quality test. The ones that pass the quality test are considered reliable. Of *X*'s shock absorbers, 96% are reliable. Of *Y*'s shock absorbers, 72% are reliable.

The probability that a randomly chosen shock absorber, which is found to be reliable, is made by *Y* is

(a) 0.288 (b) 0.334
(c) 0.667 (d) 0.720 [2012 : 2 M]

1.30 Which of the following assertions are CORRECT?

P: Adding 7 to each entry in a list adds 7 to the mean of the list
Q: Adding 7 to each entry in a list adds 7 to the standard deviation of the list
R: Doubling each entry n a list doubles the mean of the list
S: Doubling each entry in a list leaves the standard deviation of the list unchanged

(a) P, Q (b) Q, R
(c) P, R (d) R, S [2012 : 2 M]

1.31 Complete the sentence :

Universalism is to particularism as diffuseness is to _____.

(a) specificity (b) neutrality
(c) generality (d) adaptation
[2013 : 1 M]

1.32 Were you a bird, you ______ in the sky.

(a) would fly (b) shall fly
(c) should fly (d) shall have flown
[2013 : 1 M]

1.33 Which one of the following options is the closest in meaning to the word given below?

Nadir

(a) Highest (b) Lowest
(c) Medium (d) Integration
[2013 : 1 M]

1.34 Choose the grammatically INCORRECT sentence:

(a) He is of Asian origin.
(b) They belonged to Africa
(c) She is an European.
(d) They migrated from India to Australia.
[2013 : 1 M]

1.35 What will be the maximum sum of 44,. 42, 40, ...?
(a) 502 (b) 504
(c) 506 (d) 500 [2013 : 1 M]

1.36 Out of all the 2-digit integers between 1 and 100, a 2-digit number has to be selected at random. What is the probability that the selected number is not divisible by 7?
(a) 13/90 (b) 12/90
(c) 78/90 (d) 77/90 [2013 : 2 M]

1.37 A tourist covers half of this journey by train at 60 km/h, half of the remainder by bus at 30 km/h and the rest by cycle at 10 km/h. The average speed of the tourist in km/h during his entire journey is
(a) 36 (b) 30
(c) 24 (d) 18 [2013 : 2 M]

1.38 Find the sum of the expression

$$\frac{1}{\sqrt{1}+\sqrt{2}}+\frac{1}{\sqrt{2}+\sqrt{3}}+\frac{1}{\sqrt{3}+\sqrt{4}}+.....+\frac{1}{\sqrt{80}+\sqrt{81}}$$

(a) 7 (b) 8
(c) 9 (d) 10 [2013 : 2 M]

1.39 The current erection cost of a structure is ₹ 13,200. If the labour wages per day increase by 1/5 of the current wages and the working hours decrease by 1/24 of the current period, then the new cost of erection in ₹, is
(a) 16,500 (b) 15,180
(c) 11,000 (d) 10,120 [2013 : 2 M]

1.40 After several defeats in wars, Robert Bruce went in exile and wanted to commit suicide. Just before committing suicide, the came across a spider attempting tirelessly to have its net. Time and again, the spider failed but the did not deter it to refrain from making attempts. Such attempts by the spider made Bruce curious. Thus, Bruce started observing the near-impossible goal of the spider to have the net. Ultimately, the spider succeeded in having its net despite several failures. Such act of the spider encouraged Bruce not to commit suicide. And then, Bruce went back again and won many a battle, and the rest is history.
Which one of the following assertions is best supported by the above information?
(a) Failure is the pillar of success
(b) Honesty is the best policy
(c) Life begins and ends with adventures
(d) No adversity justifies giving up hope
[2013 : 2 M]

1.41 Choose the most appropriate phrase from the options given below to complete the following sentence. The aircraft ____ take off as soon as its flight plan was filed.
(a) is allowed to (b) will be allowed to
(c) was allowed to (d) has been allowed to
[2014 : 1 M, Set-1]

1.42 Read the statements:
All women are entrepreneurs.
Some women are doctors.
Which of the following conclusions can be logically inferred from the above statements?
(a) All women are doctors
(b) All doctors are entrepreneurs
(c) All entrepreneurs are women
(d) Some entrepreneurs are doctors
[2014 : 1 M, Set-1]

1.43 Choose the most appropriate word from the options given below to complete the following sentence.
Many ancient cultures attributed disease to supernatural causes. However, modern science has largely helped ______ such notions.
(a) impel (b) dispel
(c) propel (d) repel
[2014 : 1 M, Set-1]

1.44 The statistics of runs scored in a series by four batsmen are provided in the following table. Who is the most ***consistent*** batsman of these four?

Batsman	Average	Standard Deviation
K	31.2	5.21
L	46.0	6.35
M	54.4	6.22
N	17.9	5.90

(a) *K* (b) *L*
(c) *M* (d) *N*
[2014 : 1 M, Set-1]

1.45 What is the next number in the series?
12 35 81 173 357 _____.
[2014 : 1 M, Set-1]

1.46 Choose the most appropriate word from the options given below to complete the following sentence.
Communication and interpersonal skills are ______ important in their own ways.
(a) each (b) both
(c) all (d) either
[2014 : 1 M, Set-2]

1.47 Which of the options given below best completes the following sentence?

She will feel much better if she _________.

(a) will get some rest
(b) gets some rest
(c) will be getting some rest
(d) is getting some rest [2014 : 1 M, Set-2]

1.48 Choose the most appropriate pair of words from the options given below to complete the following sentence.

She could not _____ the thought of _______ the election to her bitter rival.

(a) bear, loosing (b) bare, loosing
(c) bear, losing (d) bare, losing

[2014 : 1 M, Set-2]

1.49 A regular die has six sides with numbers 1 to 6 Med on its sides. If a very large number of throws show the following frequencies of occurrence: $1 \rightarrow 0.167$; $2 \rightarrow 0.167$; $3 \rightarrow 0.152$; $4 \rightarrow 0.166$; $5 \rightarrow 0.168$; $6 \rightarrow 0.180$. We call this die

(a) irregular (b) biased
(c) Gaussian (d) insufficient

[2014 : 1 M, Set-2]

1.50 Fill in the missing number in the series.

2 3 6 15 _?_ 157.5 630

[2014 : 1 M, Set-2]

1.51 "India is a country of rich heritage and cultural diversity." Which one of the following facts best supports the claim made in the above sentence?

(a) India is a union of 28 states and 7 union territories.
(b) India has a population of over 1.1 billion.
(c) India is home to 22 official languages and thousands of dialects.
(d) The Indian cricket team draws players from over ten states. [2014 : 1 M, Set-3]

1.52 The value of one U.S. dollar is 65 Indian Rupees today, compared to 60 last year. The Indian Rupee has _________.

(a) depressed (b) depreciated
(c) appreciated (d) stabilized

[2014 : 1 M, Set-3]

1.53 'Advice' is _____________.

(a) a verb (b) a noun
(c) an adjective (d) both a verb and a noun

[2014 : 1 M, Set-3]

1.54 The next term in the series 81, 54, 36, 24, ... is________.

[2014 : 1 M, Set-3]

1.55 In which of the following options will the expression $P < M$ be definitely true?

(a) $M < R > P > S$ (b) $M > S < P < F$
(c) $Q < M < F = P$ (d) $P = A < R < M$

[2014 : 1 M, Set-3]

1.56 Which of the following options is the closest in meaning to the word underlined in the sentence below?

In a democracy, everybody has the freedom to <u>disagree</u> with the government.

(a) dissent (b) descent
(c) decent (d) decadent

[2014 : 1 M, Set-4]

1.57 After the discussion, Tom said to me, 'Please revert!'. He expects me to _______.

(a) retract (b) get back to him
(c) move in reverse (d) retreat

[2014 : 1 M, Set-4]

1.58 While receiving the award, the scientist said, "I feel vindicated". Which of the following is closest in meaning to the word 'vindicated'?

(a) punished (b) substantiated
(c) appreciated (d) chastened

[2014 : 1 M, Set-4]

1.59 Let $f(x, y) - x^n y^m = P$. If x is doubled and y is halved, the new value of f is

(a) $2^{n-m} P$ (b) $2^{m-n} P$
(c) $2(n-m)P$ (d) $2(m-n)P$

[2014 : 1 M, Set-4]

1.60 In a sequence of 12 consecutive odd numbers, the sum of the first 5 numbers is 425. What is the sum of the last 5 numbers in the sequence?

[2014 : 1 M, Set-4]

1.61 Find the odd one from the following group:

WEKO IQWA FNTX NVBD

(a) WEKO (b) IQWA
(c) FNTX (d) NVBD

[2014 : 2 M Set-1]

1.62 For submitting tax returns, all resident males with annual income below Rs 10 lakh should fill up Form *P* and all resident females with income below Rs 8 lakh should fill up Form *Q*. All people with incomes above Rs 10 lakh should fill up Form *R*, except non residents with income above ₹ 15 lakhs, who should fill up Form *S*. All others should fill Form *T*. An example of a person who should fill Form *T* is

(a) a resident male with annual income ₹ 9 lakh
(b) a resident female with annual income ₹ 9 lakh

(c) a non-resident male with annual income ₹ 16 lakh

(d) a non-resident female with annual income ₹ 16 lakh [2014 : 2 M Set-1]

1.63 A train that is 280 metres long, travelling at a uniform speed, crosses a platform in 60 seconds and passes a man standing on the platform in 20 seconds. What is the length of the platform in metres? [2014 : 2 M Set-1]

1.64 The exports and imports (in crores of ₹) of a country from 2000 to 2007 are given in the following bar chart. If the trade deficit is defined as excess of imports over exports, in which year is the trade deficit 1/5th of the exports?

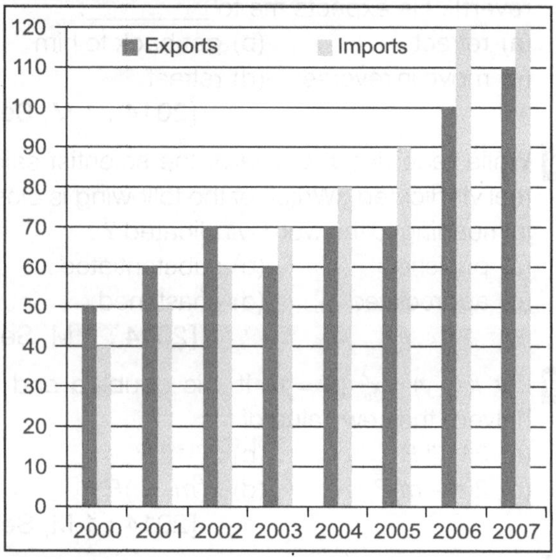

(a) 2005 (b) 2004
(c) 2007 (d) 2006

[2014 : 2 M Set-1]

1.65 You are given three coins: one has heads on both faces, the second has tails on both faces, and the third has a head on one face and a tail on the other. You choose a coin at random and toss it, and it comes up heads. The probability that the other face is tails is

(a) 1/4 (b) 1/3
(c) 1/2 (d) 2/3 [2014 : 2 M Set-1]

1.66 Find the odd one in the following group

QWZB, BHKM, WCGJ, MSVX,

(a) QWZB (b) BHKM
(c) WCGJ (d) MSVX

[2014 : 2 M Set-2]

1.67 Lights of four colors (red, blue, green, yellow) are hung on a ladder. On every step of the ladder there are two lights. If one of the lights is red, the other light on that step will always be blue. If one of the lights on a step is green, the other light on that step will always be yellow. Which of the following statements is not necessarily correct?

(a) The number of red lights is equal to the number of blue lights

(b) The number of green lights is equal to the number of yellow lights

(c) The sum of the red and green lights is equal to the sum of the yellow and blue lights

(d) The sum of the red and blue lights is equal to the sum of the green and yellow lights

[2014 : 2 M Set-2]

1.68 The sum of eight consecutive odd numbers is 656. The average of four consecutive even numbers is 87. What is the sum of the smallest odd number and second largest even number?

[2014 : 2 M Set-2]

1.69 The total exports and revenues from the exports of a country are given in the two charts shown below. The pie chart for exports shows the quantity of each item exported as a percentage of the total quantity of exports. The pie chart for the revenues shows the percentage of the total revenue generated through export of each item. The total quantity of exports of all the items is 500 thousand tonnes and the total revenues are 250 crore rupees. Which item among the following has generated the maximum revenue per kg?

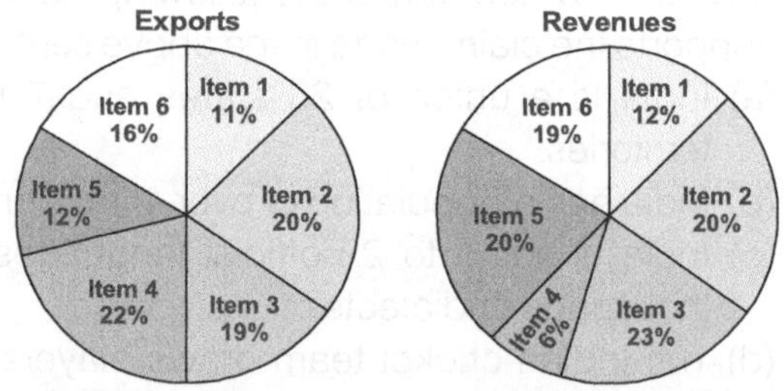

(a) Item 2 (b) Item 3
(c) Item 6 (d) Item 5

[2014 : 2 M Set-2]

1.70 It takes 30 minutes to empty a half-full tank by draining it at a constant rate. It is decided to simultaneously pump water into the half-full tank while draining it. What is the rate at which water has to be pumped in so that it gets fully filled in 10 minutes?

(a) 4 times the draining rate

(b) 3 times the draining rate

(c) 2.5 times the draining rate

(d) 2 times the draining rate [2014 : 2 M Set-2]

1.71 Find the next term in the sequence:

$7G$, $11K$, $13M$, _____.

(a) $15Q$ (b) $17Q$
(c) $15P$ (d) $17P$ **[2014 : 2 M Set-3]**

1.72 The multi-level hierarchical pie chart shows the population of animals in a reserve forest. The correct conclusions from this information are:

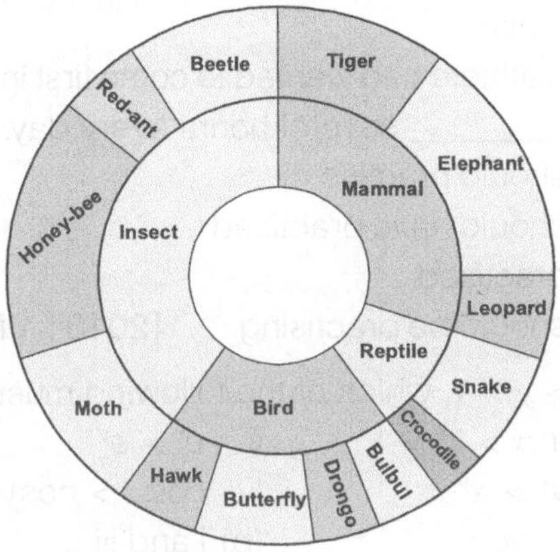

(i) Butterflies are birds
(ii) There are more tigers in this forest than red ants
(iii) All reptiles in this forest are either snakes or crocodiles
(iv) Elephants are the largest mammals in this forest

(a) (i) and (ii) only (b) (i),(ii), (iii) and (iv)
(c) (i), (iii) & (iv) only (d) (i),(ii) and (iii) only

[2014 : 2 M Set-3]

1.73 A man can row at 8 km per hour in still water. If it takes him thrice as long to row upstream, as to row downstream, then find the stream velocity in km per hour. **[2014 : 2 M Set-3]**

1.74 A firm producing air purifiers sold 200 units in 2012. The following pie chart presents the share of raw material, labour, energy, plant & machinery, and transportation costs in the total manufacturing cost of the firm in 2012. The expenditure on labour in 2012 is ₹ 4,50,000. In 2013, the raw material expenses increased by 30% and all other expenses increased by 20%. If the company registered a profit of ₹ 10 lakhs in 2012, at what price (in ₹) was each air purifier sold?

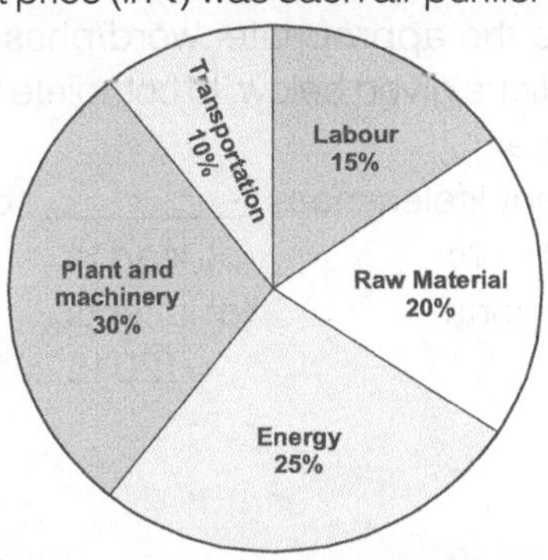

[2014 : 2 M Set-3]

1.75 A batch of one hundred bulbs is inspected by testing four randomly chosen bulbs. The batch is rejected if even one of the bulbs is defective. A batch typically has five defective bulbs. The probability that the current batch is accepted is_______. **[2014 : 2 M Set-3]**

1.76 Find the next term in the sequence:

$13M$, $17Q$, $19S$, ____.

(a) $21W$ (b) $21V$
(c) $23W$ (d) $23V$

[2014 : 2 M Set-4]

1.77 If 'KCLFTSB' stands for 'best of luck' and 'SHSWDG' stands for 'good wishes', which of the following indicates 'ace the exam'?

(a) MCHTX (b) MXHTC
(c) XMHCT (d) XMHTC

[2014 : 2 M Set-4]

1.78 Industrial consumption of power doubled from 2000-2001 to 2010-2011. Find the annual rate of increase in percent assuming it to be uniform over the years.

(a) 5.6 (b) 7.2
(c) 10.0 (d) 12.2

[2014 : 2 M Set-4]

1.79 A firm producing air purifiers sold 200 units in 2012. The following pie chart presents the share of raw material, labour, energy, plant & machinery, and transportation costs in the total manufacturing cost of the firm in 2012. The expenditure on labour in 2012 is ₹ 4,50,000. In 2013, the raw material expenses increased by 30% and all other expenses increased by 20%. What is the percentage increase in total cost for the company in 2013?

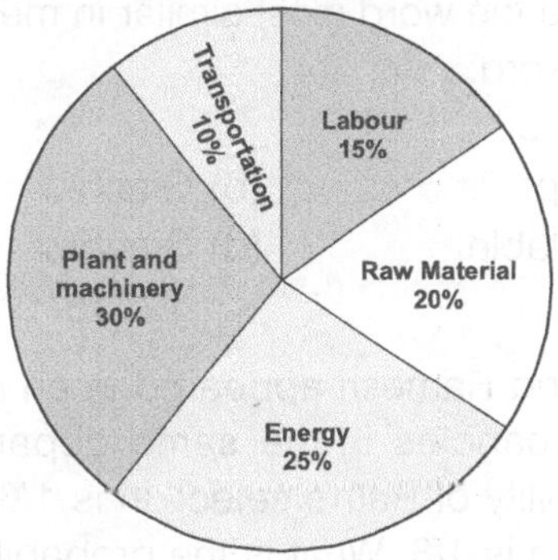

[2014 : 2 M Set-4]

1.80 A five digit number is formed using the digits 1,3,5,7 and 9 without repeating any of them. What is the sum of all such possible five digit numbers?

(a) 6666660 (b) 6666600
(c) 6666666 (d) 6666606

[2014 : 2 M Set-4]

1.81 An electric bus has onboard instruments that report the total electricity consumed since the start of the trip as well as the total distance covered. During a single day of operation, the bus travels on stretches *M*, *N*, *O* and *P*, in that order. The cumulative distances travelled and the corresponding electricity consumption are shown in the table below.

Stretch	Cumulative distance (km)	Electricity used (kWh)
M	20	12
N	45	25
O	75	45
P	100	57

The stretch where the electricity consumption per km is minimum is

(a) *M* (b) *N*
(c) *O* (d) *P*

[2015 : 1 M, Set-1]

1.82 What is the adverb for the given word below?
Misogynous

(a) Misogynousness (b) Misogynity
(c) Misogynously (d) Misogynous

[2015 : 1 M, Set-1]

1.83 Choose the appropriate word/phrase, out of the four options given below, to complete the following sentence:
Dhoni, as well as the other team members of Indian team, ____ present on the occasion.

(a) were (b) was
(c) has (d) have

[2015 : 1 M, Set-1]

1.84 Choose the word most similar in meaning to the given word:
Awkward

(a) Inept (b) Graceful
(c) Suitable (d) Dreadful

[2015 : 1 M, Set-1]

1.85 Ram and Ramesh appeared in an interview for two vacancies in the same department. The probability of Ram's selection is 1/6 and that of Ramesh is 1/8. What is the probability that only one of them will be selected?

(a) $\frac{47}{48}$ (b) $\frac{1}{4}$
(c) $\frac{13}{48}$ (d) $\frac{35}{48}$

[2015 : 1 M, Set-1]

1.86 Choose the correct verb to fill in the blank below:
Let us ___.

(a) introvert (b) alternate
(c) atheist (d) altruist

[2015 : 1 M, Set-2]

1.87 Choose the most appropriate word from the options given below to complete the following sentence.
If the athlete had wanted to come first in the race, he ________ several hours every day.

(a) should practise
(b) should have practised
(c) practised
(d) should be practising [2015 : 1 M, Set-2]

1.88 If $x > y > 1$, which of the following must be true?

i. $\ln x > \ln y$ ii. $e^x > e^y$
iii. $y^x > x^y$ iv. $\cos x > \cos y$

(a) i and ii (b) i and iii
(c) iii and iv (d) ii and iv

[2015 : 1 M, Set-2]

1.89 Choose the most suitable one word substitute for the following expression. Connotation of a road or way

(a) Pertinacious (b) Viaticum
(c) Clandestine (d) Ravenous

[2015 : 1 M, Set-2]

1.90 Find the missing sequence in the letter series below:
A, CD, GHI, ?, UVWXY

(a) LMN (b) MNO
(c) MNOP (d) NOPQ

[2015 : 1 M, Set-2]

1.91 Fill in the blank with the correct idiom/phrase.
That boy from the town was a ________ in the sleepy village.

(a) dog out of herd
(b) sheep from the heap
(c) fish out of water
(d) bird from the flock [2015 : 1 M, Set-3]

1.92 Choose the appropriate word/phase, out of the four options given below, to complete the following sentence:
Apparent lifelessness __________ dormant life.

(a) harbours (b) lead to
(c) supports (d) affects

[2015 : 1 M, Set-3]

1.93 Five teams have to compete in a league, with every team playing every other team exactly once, before going to the next round. How may matches will have to be held to complete the league round of matches?

(a) 20 (b) 10
(c) 8 (d) 5

[2015 : 1 M, Set-3]

1.94 Tanya is older than Eric.
Cliff is older than Tanya
Eric is older than Cliff.
If the first two statements are true, then the third statement is:

(a) True (b) False
(c) Uncertain (d) Data insufficient

[2015 : 1 M, Set-3]

1.95 Choose the statement where underlined word is used correctly.

(a) When the teacher eludes to different authors, he is being elusive
(b) When the thief keeps eluding the police, he is being elusive
(c) Matters that are difficult to understand, identify or remember are allusive
(d) Mirages can be allusive, but a better way to express them is illusory

[2015 : 1 M, Set-3]

1.96 Lamenting the gradual sidelining of the arts in school curricula, a group of prominent artists wrote to the Chief Minister last year, asking him to allocate more funds to support arts education in schools. However, no such increase has been announced in this year's Budget. The artists expressed their deep anguish at their request not being approved, but many of them remain optimistic about funding in the future.
Which of the statement(s) below is/are logically valid and can be inferred from the above statements?

i. The artists expected funding for the arts to increase this year.
ii. The Chief Minister was receptive to the idea of increasing funding for the arts.
iii. The Chief Minister is a prominent artist.
iv Schools are giving less importance to arts education nowadays.

(a) iii and iv (b) i and iv
(c) i, ii and iv (d) i and iii

[2015 : 2 M Set-1]

1.97 A tiger is 50 leap of its own behind a deer. The tiger takes 5 leaps per minute to the deer's 4. If the tiger and the deer cover 8 metre and 5 metre per leap respectively, what distance in meters will the tiger have a run before it catches the deer?

[2015 : 2 M Set-1]

1.98 Given below are two statements followed by two conclusions. Assuming these statements to be true, decide which one logically follows.

Statement:
i. All film stars are playback singers.
ii. All film directors are film stars.

Conclusions:
i. All film directors are playback singers.
ii. Some film stars are film directors.

(a) Only conclusion I follows
(b) Only conclusion II follows
(c) Neither conclusion I nor II follows
(d) Both conclusions I and II follow

[2015 : 2 M Set-1]

1.99 If $a^2 + b^2 + c^2 = 1$, then $ab + bc + ac$ lies in the interval

(a) $\left[1, \frac{2}{3}\right]$ (b) $\left[\frac{-1}{2}, 1\right]$
(c) $\left[-1, \frac{-1}{2}\right]$ (d) $[2, -4]$

[2015 : 2 M Set-1]

1.100 In the following sentence certain parts are underlined and Med P, Q and R. One of the parts may contain certain error or may not be acceptable in standard written communication. Select the part containing an error. Choose D as your answer if there is no error.

The student corrected $\underset{P}{\underline{\text{all the errors}}}$ that $\underset{Q}{\underline{\text{the instructor marked}}}$ on the $\underset{R}{\underline{\text{answer book}}}$

(a) P (b) Q
(c) R (d) No Error

[2015 : 2 M Set-1]

1.101 Ms. X will be in Bagdogra from 01/05/2014 to 20/05/2014 and from 22/05/2014 to 31/05/2014. On the morning of 21/05/2014, she will reach Kochi via Mumbai.
Which one of the statement below is logically valid and can be inferred from the above sentences?

(a) Ms. *X* will be in Kochi for one day, only in May.
(b) Ms. *X* will be in Kochi for only one day in May.
(c) Ms. *X* will be only in Kochi for one day in May.
(d) Only Ms. *X* will be in Kochi for one day in May.

[2015 : 2 M Set-2]

1.102 log tan 1° + log tan 2° + ... + log tan 89° is ___.

(a) 1 (b) $\frac{1}{\sqrt{2}}$
(c) 0 (d) − 1

[2015 : 2 M Set-2]

1.103 From a circular sheet of paper of radius 30 cm, a sector of 10% area is removed. If the remaining part is used to make a conical surface, then the ratio of the radius and height of the cone is ______.

[2015 : 2 M Set-2]

1.104 In the following questions, the first and the last sentence of the passage are in order and numbered 1 and 6. The rest of the passage is split into 4 parts and numbered as 2, 3, 4 and 5. These 4 parts are not arranged in proper order. Read the sentences and arrange them in a logical sequence to make a passage and choose the correct sequence from the given options.

1. On Diwali, the family rises early in the morning.
2. The whole family, including the young and the old enjoy doing this.
3. Children let off fireworks later in the night with their friends.
4. At sunset, the lamps are lit and the family performs various rituals.
5. Father, mother, and children visit relatives and exchanges gifts and sweets.
6. Houses look so pretty with lighted lamps all around.

(a) 2, 5, 3, 4 (b) 5, 2, 4, 3
(c) 3, 5, 4, 2 (d) 4, 5, 2, 3

[2015 : 2 M Set-2]

1.105 Ram and Shyam shared a secret and promised to each other that it would remain between them. Ram expressed himself in one of the following ways as given in the choices below. Identify the correct way as per standard English.

(a) It would remain between you and me
(b) It would remain between I and you
(c) It would remain between you and I
(d) It would remain with me

[2015 : 2 M Set-2]

1.106 Select the appropriate option in place of underlined part of the sentence.

<u>Increased productivity necessary</u> reflects greater efforts made by the employees.

(a) Increase in productivity necessary
(b) Increase productivity is necessary
(c) Increase in productivity necessarily
(d) No improvement required

[2015 : 2 M Set-3]

1.107 A coin is tossed thrice. Let *X* be the event that head occurs in each of the first two tosses. Let *Y* be the event that a tail occurs on the third toss. Let *Z* be the event that two tails occurs in three tosses. Based on the above information, which one of the following statements is TRUE?

(a) *X* and *Y* are not independent
(b) *Y* and *Z* are dependent
(c) *Y* and *Z* are independent
(d) *X* and *Z* independent [2015 : 2 M Set-3]

1.108 Right triangle *PQR* is to be constructed in the *xy*– plane so that the right angle is at *P* and line *PR* is parallel to the-axis. The *x* and *y* coordinates of *P*, *Q*, and *R* are to be integers that satisfy the inequalities:
$-4 \le x \le 5$ and $6 \le y \le 16$. How many different triangles could be constructed with these properties?

(a) 110 (b) 1,100
(c) 9,900 (d) 10,000

[2015 : 2 M Set-3]

1.109 In the given figure angle *Q* is a right angle, *PS*:*QS* = 3:1, *RT*:*QT* = 5:2 and *PU*:*UR* = 1:1. If area of triangle *QTS* is 20 cm, then the area of triangle *PQR* in cm^2 is ______.

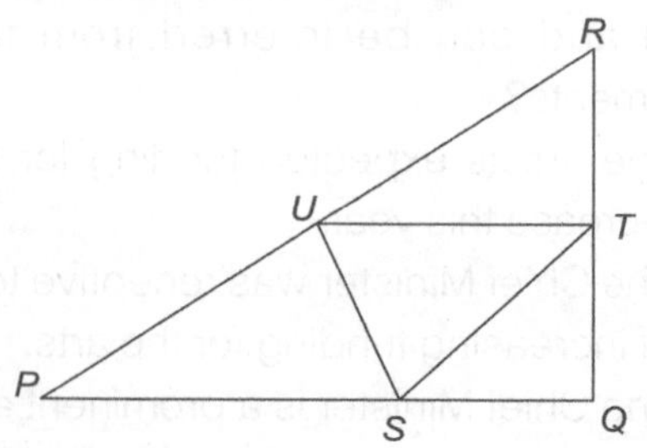

[2015 : 2 M Set-3]

1.110 Given below are two statements followed by two conclusions. Assuming these statements to be true, decide which one logically follows:

Statements:

I. No manager is a leader.

II. All leaders are executive.

Conclusions:

I. No manager is a executive.

II. All executive is a manager.

(a) Only conclusion I follows.

(b) Only conclusion II follows.

(c) Neither conclusion I nor II follows.

(d) Both conclusion I and II follow.

[2015 : 2 M Set-3]

1.111 Which of the following is CORRECT with respect to grammar and usage?

Mount Everest is____________.

(a) the highest peak in the world

(b) highest peak in the world

(c) one of highest peak in the world

(d) one of the highest peak in the world

[2016 : 1 M, Set-1]

1.112 The policeman asked the victim of a theft, "What did you______?"

(a) loose (b) lose

(c) loss (d) louse

[2016 : 1 M, Set-1]

1.113 Despite the new medicine's_____ in treating diabetes, it is not______ widely.

(a) effectiveness---prescribed

(b) availability---used

(c) prescription---available

(d) acceptance---proscribed

[2016 : 1 M, Set-1]

1.114 In a huge pile of apples and oranges, both ripe and unripe mixed together, 15% are unripe fruits. Of the unripe fruits, 45% are apples. Of the ripe ones, 66% are oranges. If the pile contains a total of 5692000 fruits, how many of them are apples?

(a) 2029198 (b) 2467482

(c) 2789080 (d) 3577422

[2016 : 1 M, Set-1]

1.115 Michael lives 10 km away from where I live. Ahmed lives 5 km away and Susan lives 7 km away from where I live. Arun is farther away than Ahmed but closer than Susan from where I live. From the information provided here, what is one possible distance (in km) at which I live from Arun's place?

(a) 3.00 (b) 4.99

(c) 6.02 (d) 7.01

[2016 : 1 M, Set-1]

1.116 A person moving through a tuberculosis prone zone has a 50% probability of becoming infected. However, only 30% of infected people develop the disease. What percentage of people moving through a tuberculosis prone zone remains infected but does not show symptoms of disease?

(a) 15 (b) 33

(c) 35 (d) 37

[2016 : 2 M Set-1]

1.117 In a world filled with uncertainty, he was glad to have many good friends. He had always assisted them in times of need and was confident that they would reciprocate. However, the events of the last week proved him wrong. Which of the following inference(s) is/are logically valid and can be inferred from the above passage?

(i) His friends were always asking him to help them.

(ii) He felt that when in need of help, his friends would let him down.

(iii) He was sure that his friends would help him when in need.

(iv) His friends did not help him last week.

(a) (i) and (ii) (b) (iii) and (iv)

(c) (iii) only (d) (iv) only

[2016 : 2 M Set-1]

1.118 Leela is older than her cousin Pavithra. Pavithra's brother Shiva is older than Leela. When Pavithra and Shiva are visiting Leela, all three like to play chess. Pavithra wins more often than Leela does.

Which one of the following statements must be TRUE based on the above?

(a) When Shiva plays chess with Leela and Pavithra, he often loses.

(b) Leela is the oldest of the three

(c) Shiva is a better chess player than Pavithra

(d) Pavithra is the youngest of the three

[2016 : 2 M Set-1]

1.119 If $q^{-a} = \frac{1}{r}$ and $r^{-b} = \frac{1}{s}$ and $s^{-c} = \frac{1}{q}$, the value of abc is______.

(a) $(rps)^{-1}$ (b) 0

(c) 1 (d) $r + q + s$

[2016 : 2 M Set-1]

1.120 P, Q, R and S are working on a project. Q can finish the task in 25 days, working alone for 12 hours a day. *R* can finish the task in 50 days, working alone for 12 hours per day. *Q* worked 12 hours a day but took sick leave in the beginning for two days, *R* worked 18 hours a day on all days. What is the ratio of work done by *Q* and *R* after 7 days from the start of the project?

(a) 10 : 11 (b) 11 : 10
(c) 20 : 21 (d) 21 : 20

[2016 : 2 M Set-1]

1.121 The volume of a sphere of diameter 1 unit is _____ than the volume of a cube of side 1 unit.

(a) least (b) less
(c) lesser (d) low

[2016 : 1 M, Set-2]

1.122 The unruly crowd demanded that the accused be __________ without trial.

(a) hanged (b) hanging
(c) hankering (d) hung

[2016 : 1 M, Set-2]

1.123 Choose the statement(s) where the underlined word is used correctly:

(i) A prone is a dried plum.
(ii) He was lying prone on the floor.
(iii) People who eat a lot of fat are prone to heart disease.

(a) (i) and (iii) only (b) (iii) only
(c) (i) and (ii) only (d) (ii) and (iii) only

[2016 : 1 M, Set-2]

1.124 **Fact:** If it rains, then the field is wet.
Read the following statements:

(i) It rains
(ii) The field is not wet
(iii) The field is wet
(iv) It did not rain

Which one of the options given below is **NOT** logically possible, based on the given fact?

(a) If (iii), then (iv). (b) If (i), then (iii).
(c) If (i), then (ii). (d) If (ii), then (iv).

[2016 : 1 M, Set-2]

1.125 A window is made up of a square portion and an equilateral triangle portion above it. The base of the triangular portion coincides with the upper side of the square. If the perimeter of the window is 6 m, the area of the window in m^2 is____.

(a) 1.43 (b) 2.06
(c) 2.68 (d) 2.88

[2016 : 1 M, Set-2]

1.126 Students taking an exam are divided into two groups, *P* and *Q* such that each group has the same number of students. The performance of each of the students in a test was evaluated out of 200 M. It was observed that the mean of group *P* was 105, while that of group *Q* was 85. The standard deviation of group *P* was 25, while that of group *Q* was 5. Assuming that the M were distributed on a normal distribution, which of the following statements will have the highest probability of being TRUE?

(a) No student in group *Q* scored less M than any student in group *P*.
(b) No student in group *P* scored less M than any student in group *Q*.
(c) Most students of group *Q* scored M in a narrower range than students in group *P*.
(d) The median of the M of group *P* is 100.

[2016 : 2 M Set-2]

1.127 A smart city integrates all modes of transport, uses clean energy and promotes sustainable use of resources. It also uses technology to ensure safety and security of the city, something which critics argue, will lead to a surveillance state. Which of the following can be logically inferred from the above paragraph?

(i) All smart cities encourage the formation of surveillance states.
(ii) Surveillance is an integral part of a smart city.
(iii) Sustainability and surveillance go hand in hand in a smart city.
(iv) There is a perception that smart cities promote surveillance.

(a) (i) and (iv) only (b) (ii) and (iii) only
(c) (iv) only (d) (i) only

[2016 : 2 M Set-2]

1.128 Find the missing sequence in the letter series.
B, FH, LNP, _ _ _ _.

(a) SUWY (b) TUVW
(c) TVXZ (d) TWXZ

[2016 : 2 M Set-2]

1.129 The binary operation is defined as $a \; b = ab + (a + b)$, where *a* and *b* are any two real numbers. The value of the identity element of this operation, defined as the number *x* such that $a \; x = a$, for any *a*, is ________.

(a) 0 (b) 1
(c) 2 (d) 10

[2016 : 2 M Set-2]

1.130 Which of the following curves represents the function $y = \ln\left(\left|e^{[\sin C(|x|)]}\right|\right)$

for $|x| < 2\pi$?

Here, x represents the abscissa and y represents the ordinate.

(a) 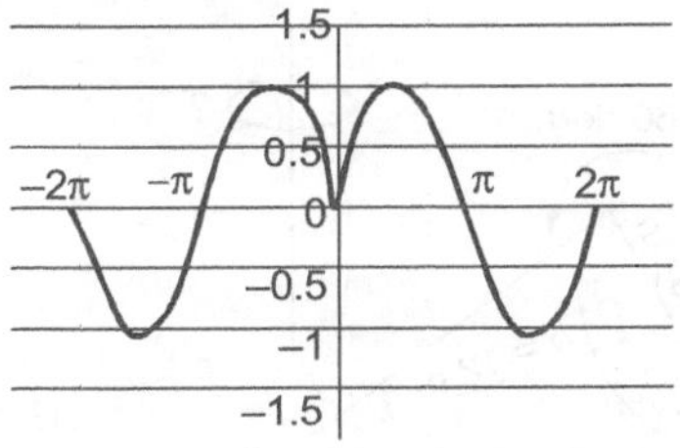

(b) 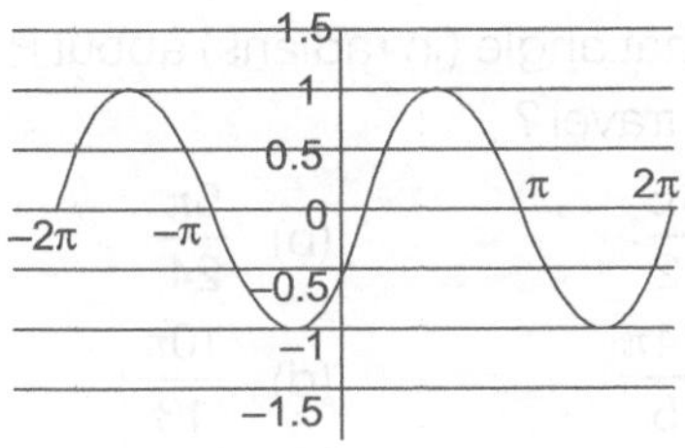

(c) 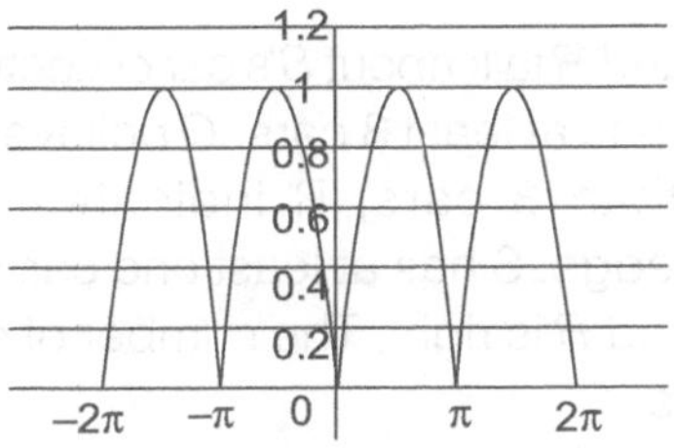

(d) 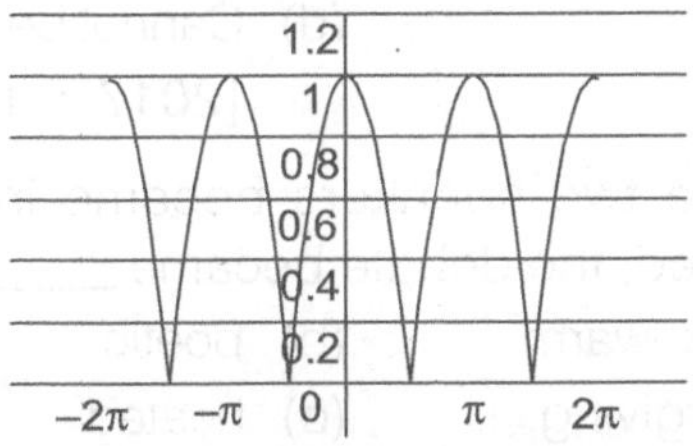

[2016 : 2 M Set-2]

1.131 Based on the given statements, select the appropriate option with respect to grammar and usage.

Statements

(i) The height of Mr. *X* is 6 feet.

(ii) The height of Mr. *Y* is 5 feet.

(a) Mr. *X* is longer than Mr. *Y*.
(b) Mr. *X* is more elongated than Mr. *Y*.
(c) Mr. *X* is tallor than Mr. *Y*.
(d) Mr. *X* is lengthier than Mr. *Y*.

[2016 : 1 M, Set-3]

1.132 The students __________ the teacher on teachers' day for twenty years of dedicated teaching.

(a) facilitated (b) felicitated
(c) fantasized (d) facilitated

[2016 : 1 M, Set-3]

1.133 After India's cricket world cup victory in 1985, Shrotria who was playing both tennis and cricket till then, decided to concentrate only on cricket. And the <u>rest is history.</u>

What does the underlined phrase mean in this context?

(a) history will rest in peace
(b) rest is recorded in history books
(c) rest is well known
(d) rest is archaic

[2016 : 1 M, Set-3]

1.134 Given (9 inches)$^{1/2}$ = (0.25 yards)$^{1/2}$, which one of the following statements is TRUE?

(a) 3 inches = 0.5 yards
(b) 9 inches = 1.5 yards
(c) 9 inches = 0.25 yards
(d) 81 inches = 0.0625 yards

[2016 : 1 M, Set-3]

1.135 *S*, *M*, *E* and *F* are working in shifts in a team to finish a project. *M* works with twice the efficiency of others but for half as many days as *E* worked. *S* and *M* have 6 hour shifts in a day, whereas *E* and *F* have 12 hours shifts. What is the ratio of contribution of *M* to contribution of *E* in the project?

(a) 1 : 1 (b) 1 : 2
(c) 1 : 4 (d) 2 : 1

[2016 : 1 M, Set-3]

1.136 The Venn diagram shows the preference of the student population for leisure activities.

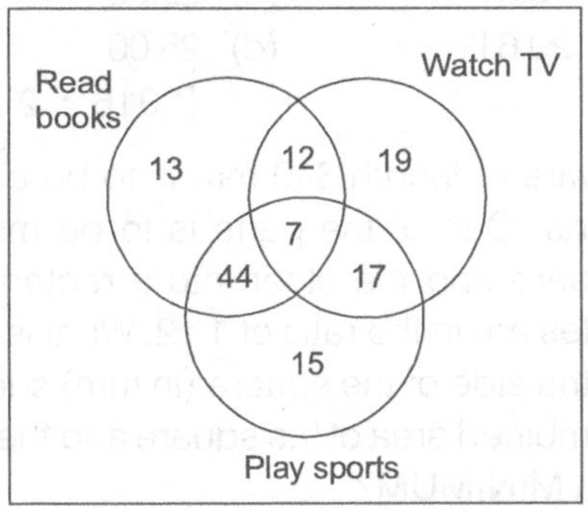

From the data given, the number of students who like to read books or play sports is____.

(a) 44 (b) 51
(c) 79 (d) 108

[2016 : 2 M, Set-3]

1.137 Social science disciplines were in an amorphous form until the colonial period when they were institutionalized. In varying degree, they were intended to further the colonial interest. In the time of globalization and the economic rise of postcolonial countries like India, conventional ways of knowledge production have become obsolete.

Which of the following can be logically inferred from the above statements?

(i) Social science disciplines have become obsolete

(ii) Social science disciplines had a pre-colonial origin.

(iii) Social science disciplines always promote colonialism.

(iv) Social science must maintain disciplinary boundaries.

(a) (ii) only (b) (i) and (iii) only
(c) (ii) and (iv) only (d) (iii) and (iv) only

[2016 : 2 M, Set-3]

1.138 Two and a quarter hours back, when seen in a mirror, the reflection of a wall clock without number Mings seemed to show 1 : 30. What is the actual current time shown by the clock?

(a) 8 : 15 (b) 11 : 15
(c) 12 : 15 (d) 12 : 45

[2016 : 2 M, Set-3]

1.139 *M* and *N* start from the same location. *M* travels 10 km East and then 10 km North-East. *N* travels 5 km South and then 4 km South-East. What is the shortest distance (in km) between *M* and *N* at the end of their level.

(a) 18.60 (b) 22.50
(c) 20.61 (d) 25.00

[2016 : 2 M, Set-3]

1.140 A wire of length 340 mm is to be cut into two parts. One of the parts is to be made into a square and the other into a rectangle where sides are in the ratio of 1 : 2. What is the length of the side of the square (in mm) such that the combined area of the square and the rectangle is a **MINIMUM**?

(a) 30 (b) 40
(c) 120 (d) 180

[2016 : 2 M, Set-3]

1.141 He was one of my best ________ and I felt his loss ________.

(a) friend, keenly (b) friends, keen
(c) friend, keener (d) friends, keenly

[2017 : 1 M, Set-1]

1.142 A right-angled cone (with base radius 5 cm and height 12 cm), as shown in the figure below, is rolled on the ground keeping the point P fixed until the point Q (at the base of the cone, as shown) touches the ground again.

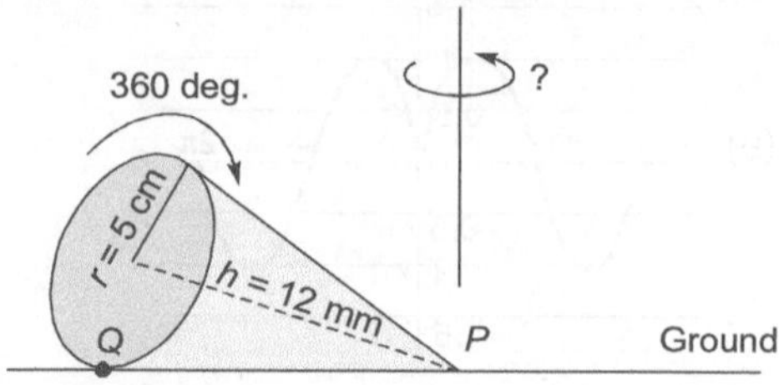

By what angle (in radians) about P does the cone travel?

(a) $\frac{5\pi}{12}$ (b) $\frac{5\pi}{24}$
(c) $\frac{24\pi}{5}$ (d) $\frac{10\pi}{13}$

[2017 : 1 M, Set-1]

1.143 *P*, *Q*, and *R* talk about *S*'s car collection, *P* states that *S* has at least 3 cars. *Q* believes that *S* has less than 3 cars, R indicates that to his knowledge. S has at least one car. Only one of *P*, *Q* and *R* is right. The number of cars owned by *S* is

(a) 0 (b) 1
(c) 3 (d) Cannot be determined

[2017 : 1 M, Set-1]

1.144 As the two speakers became increasingly agitated, the debate became ______.

(a) lukewarm (b) poetic
(c) forgiving (d) heated

[2017 : 1 M, Set-1]

1.145 In a company with 100 employees, 45 earn Rs. 20000 per month, 25 earn Rs. 30000, 20 earn Rs. 40000, 8 earn Rs. 60000, and 2 earn Rs. 150000. The median of the salaries is

(a) Rs. 20000 (b) Rs. 30000
(c) Rs. 32300 (d) Rs. 40000

[2017 : 1 M, Set-1]

1.146 What is the sum of the missing digits in the subtraction problem below?

```
  5..........
 −48_89
 ------
  1111
```

(a) 8 (b) 10
(c) 11 (d) Cannot be determined

[2017 : 2 M, Set-1]

1.147 Let S_1 be the plane figure consisting of the points (x, y) given by the inequalities $|x-1| \le 2$ and

$|y+2| \le 3$. Let S_2 be the plane figure given by the inequalities $x - y \ge -2$, $y \ge 1$, and $x \le 3$. Let S be the union of S_1 and S_2. The area of S is

(a) 26 (b) 28
(c) 32 (d) 34

[2017 : 2 M, Set-1]

1.148 Two very famous sportsmen M and Steve happened to be brothers, and played for country K. M teased James, and opponent from country E. There is no way you are good enough to play for your country. James replied, 'May be not, but at least I am the best player in my own family'.
Which one of the following can be inferred from this conversation.

(a) Mark was known to play better than James
(b) Steve was known to play better than M
(c) James and Steve were good friends
(d) James played better than Steve

[2017 : 2 M, Set-1]

1.149 Here, throughout the early 1820s, Stuart continued to fight his losing battle to allow his sepoys to year their caste-marks and their own choice of facial hair on parade, being again reprimanded by the commander-in-chief. His retort that A stronger instance than this of European prejudice with relation to this country has never come under my observations had no effect on his superiors.
According to this paragraph, which of the statements below is most accurate?

(a) Stuart's commander-in-chief was moved by this demonstration of his prejudice.
(b) The Europeans were accommodating of the sepoys desire to were their caste-marks
(c) Stuart's losing battle refers to his inability to succeed in enabling sepoys to wear caste-marks.
(d) The commander-in-chief was exempt from the European prejudice that dictated how the sepoys were to dress

[2017 : 2 M, Set-1]

1.150 The growth of bacteria (lactobacillus) in milk leads to curd formation. A minimum bacterial population density of 0.8 (in suitable) is needed to form curd. In the graph below, the population density of lactobacillus in 1 litre of milk is plotted as a function of time, at two different temperature, 25°C and 37°C.

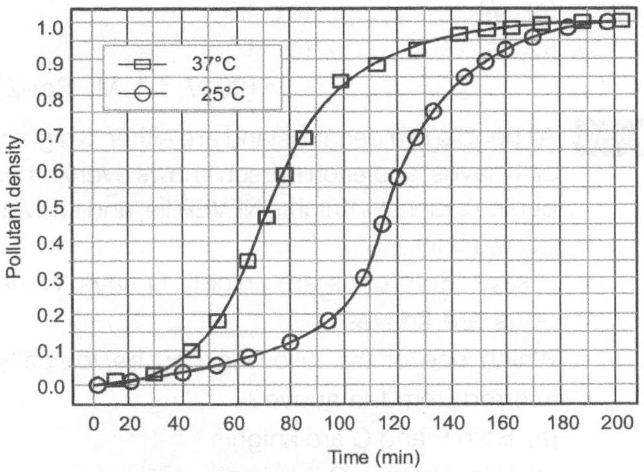

Consider the following statements based on the data shown above:

i. The growth in bacterial population stops earlier at 37°C as compared to 25°C
ii. The time taken for curd formation at 25°C is twice the time taken at 37°C

Which one of the following options is correct?

(a) only i (b) only ii
(c) both i and ii (d) neither i nor ii

[2017 : 2 M, Set-1]

1.151 If a and b are integers and $a - b$ is even, which of the following must always be even?

(a) ab (b) $a^2 + b^2 + 1$
(c) $a^2 + b + 1$ (d) $ab - b^2$

[2017 : 1 M, Set-2]

1.152 The ways in which this game can be played ________ potentially infinite.

(a) is (b) is being
(c) are (d) are being

[2017 : 1 M, Set-2]

1.153 If you choose plan P, you will have to _____ plan Q, as these two are mutually _____.

(a) forgo, exclusive
(b) forget, inclusive
(c) accept, exhaustive
(d) adopt, intrusive

[2017 : 1 M, Set-2]

1.154 P looks at Q while Q looks at R, P is married, R is not. The number of pairs of people in which a married person is looking at an unmarried person is

(a) 0 (b) 1
(c) 2 (d) cannot be determined

[2017 : 1 M, Set-2]

1.155 A couple has 2 children. The probability that both children are boys if the older one is a boy is

(a) 1/4 (b) 1/3
(c) 1/2 (d) 1

[2017 : 1 M, Set-2]

1.156 All people in a certain island are either 'Knights' or 'Knaves' and each person knows every other person's identity. Knights NEVER lie, and knaves ALWAYS lie.

P says 'Both of us are knight'. Q says 'None of us are knaves'

Which one of the following can be logically inferred from the above?

(a) Both P and Q are knights
(b) P is a knight; Q is a knave
(c) Both P and Q are knaves
(d) The identities of P, Q cannot be determined

[2017 : 2 M, Set-2]

1.157 There are 4 women P, Q, R, S and 5 men V, W, X, Y, Z in a group. We are required to form pairs each consisting of one woman and one man. P is not to be paired with Z, and Y must necessarily be paired with someone. In how many ways can 4 such pairs be formed ?

(a) 74 (b) 76
(c) 78 (d) 80

[2017 : 2 M, Set-2]

1.158 In the graph below, the concentration of a particular pollutant in a lake is plotted over (alternate) days of a month in winter (average temperature 10°C) and a month in summer (average temperature 30°C)

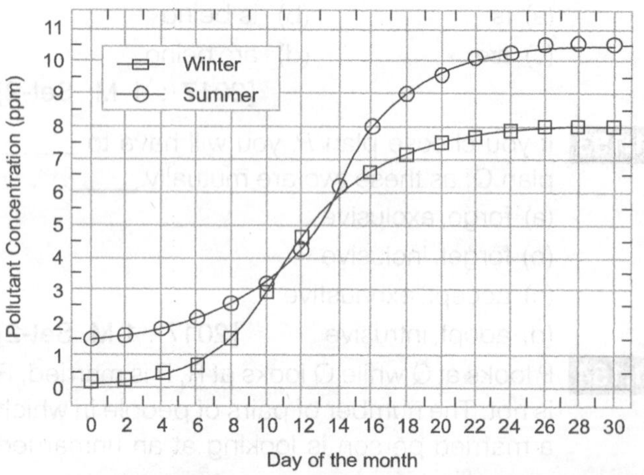

Consider the following statements based on the data shown above:

i. Over the given month, the difference between the maximum and the minimum pollutant concentrations is the same in both winter and summer.

ii. There are at least four days in the summer month such that the pollutant concentrations on those days are within 1 ppm of the pollutant concentrations on the corresponding days in the winter month.

Which one of the following options is correct?

(a) only i (b) only ii
(c) both i and ii (d) Neither i nor ii

[2017 : 2 M, Set-2]

1.159 X bullocks and Y tractors take 8 days to plough a field. If we halve the number of bullocks and double the number of tractors, it takes 5 days to plough the same field. How many days will it take X bullocks alone to plough the field?

(a) 30 (b) 35
(c) 40 (d) 45

[2017 : 2 M, Set-2]

1.160 If you are looking for a history of India, or for an account of the rise and fall of the British Raj, or for the reason of the cleaving of the subcontinent into two mutually antagonistic parts and the effects this mutilation will have in the respective sections, and ultimately on Asia, you will not find it in these pages; for though I have spent a lifetime in the country, I lived too near the seat of events, and was too intimately associated with the actors, to get the perspective needed for the impartial recording of these matters."

Which of the following is closest in meaning to 'cleaving'?

(a) deteriorating (b) arguing
(c) departing (d) splitting

[2017 : 2 M, Set-2]

1.161 A rectangle becomes a square when its length and breadth are reduced by 10 m and 5 m, respectively. during this process, the rectangle loses 650 m^2 of area. What is the area of the original rectangle in square meters?

(a) 1125 (b) 2250
(c) 2924 (d) 4500

[2018 : 1 M, Set-1]

1.162 A number consists of two digits. The sum of the digits is 9. If 45 is subtracted from the number, its digits are interchanged. What is the number?

(a) 63 (b) 72
(c) 81 (d) 90

[2018 : 1 M, Set-1]

1.163 Seven machines take 7 minutes to make 7 identical toys. At the same rate, how many minutes would it take for 100 machines to make 100 toys?

(a) 1 (b) 7

(c) 100 (d) 700

[2018 : 1 M, Set-1]

1.164 "Her ______ should not be confused with miserliness; she is ever willing to assist those in need."

The word that best fills the blank in the above sentence is:

(a) cleanliness (b) punctuality

(c) frugality (d) greatness

[2018 : 1 M, Set-1]

1.165 Going by the _____that many hands make light work, the school______involved all the students in the task."

(a) principle, principal (b) principal, principle

(c) principle, principle (d) principal, principal

[2018 : 1 M, Set-1]

1.166 Given that a and b are integers and $a + a^2b^3$ is odd, which one of the following statements is correct?

(a) a and b are both odd

(b) a and b are both even

(c) a is even and b is odd

(d) a is odd and b is even [2018 : 2 M, Set-1]

1.167 Consider the following three statements:

(i) Some roses are red

(ii) All red flowers fade quickly.

(iii) Some roses fade quickly

Which of the following statements can be logically inferred from the above statements?

(a) If (i) is true and (ii) is false, then (iii) is false.

(b) If (i) is true and (ii) is false, then (iii) is true.

(c) If (i) and (ii) are true, then (iii) is true.

(d) If (i) and (ii) are false, then (iii) is false.

[2018 : 2 M, Set-1]

1.168 Which of the following functions describe the graph shown in the below figure.

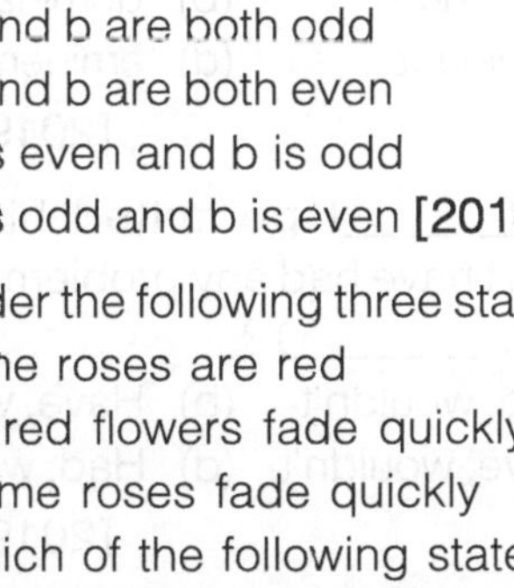

(a) $y = ||x|+1|-2$ (b) $y = ||x|-1|-1$

(c) $y = ||x|+1|-1$ (d) $y = ||x-1|-1|$

[2018 : 2 M, Set-1]

1.169 For integers a, b and c, what would be the minimum and maximum values respectively of $a + b + c$ if log úaú + log úbú + log úcú = 0

(a) −3 and 3 (b) −1 and 1

(c) −1 and 3 (d) 1 and 3

[2018 : 2 M, Set-1]

1.170 From the time the front of a train enters a platform, it takes 25 seconds for the back of the train to leave the platform, while travelling at a constant speed of 54 km/h. At the same speed, it takes 14 seconds to pass a man running at 9 km/h in the same direction as the train. What is the length of the train and that of the platform in meters, respectively?

(a) 210 and 140 (b) 162.5 and 187.5

(c) 245 and 130 (d) 175 and 200

[2018 : 2 M, Set-1]

1.171 "The dress ______ her so well that they all immediately ______ her on her appearance."

The words that best fill the blanks in the above sentence are

(a) complemented, complemented

(b) complimented, complemented

(c) complimented, complimented

(d) complemented, complimented

[2018 : 1 M, Set-2]

1.172 "The judge's standing in the legal community, though shaken by false allegations of wrongdoing, remained _______."

The word that best fill the blanks in the above sentence is

(a) undiminished (b) damaged

(c) illegal (d) uncertain

[2018 : 1 M, Set-2]

1.173 Find the missing group of letters in the following series:

BC, FGH, LMNO, _____

(a) UVWXY (b) TUVWX

(c) STUVW (d) RSTUV

[2018 : 1 M, Set-2]

1.174 The perimeters of a circle, a square and an equilateral triangle are equal. Which one of the following statements is true?

(a) The circle has the largest area

(b) The square has the largest area
(c) The equilateral triangle has the largest area
(d) All the three shapes have the same area

[2018 : 1 M, Set-2]

1.175 The value of the expression

$$\frac{1}{1+\log_u vw}+\frac{1}{1+\log_v wu}+\frac{1}{1+\log_w uv} \text{ is\_\_\_\_ .}$$

(a) –1 (b) 0
(c) 1 (d) 3

[2018 : 2 M, Set-2]

1.176 An unbiased coin is tossed six times in a row and four different such trials are conducted. One trial implies six tosses of the coin. If *H* stands for head and *T* stands for tail, the following are the observations from the four trials:

(1)HTHTHT (2) TTHHHT (3) HTTHHT (4) HHHT ____.

Which statement describing the last two coin tosses of the fourth trial has the highest probability of being correct?

(a) Two T will occur
(b) One H and one T will occur
(c) Two H will occur
(d) One H will be followed by one T

[2018 : 2 M, Set-2]

1.177 A contract is to be completed in 52 days and 125 identical robots were employed, each operational for 7 hours a day. After 39 days, five-seventh of the work was completed. How many additional robots would be required to complete the work on time, if each robot is now operational for 8 hours a day?

(a) 50 (b) 89
(c) 146 (d) 175

[2018 : 2 M, Set-2]

1.178 Forty students watched films, A, B and C over a week. Each student watched either only one film or all three. Thirteen students watched film A, sixteen students watched film B and nineteen students watched film C. How many students watched all three films?

(a) 0 (b) 2
(c) 4 (d) 8

[2018 : 2 M, Set-2]

1.179 A wire would enclose an area of 1936 m^2, if it is bent into a square. The wire is cut into two pieces. The longer piece is thrice as long as the shorter piece. The long and the short pieces are bent into a square and a circle, respectively. Which of the following choices is closest to the sum of the areas enclosed by the two pieces in square meters?

(a) 1096 (b) 1111
(c) 1243 (d) 2486

[2018 : 2 M, Set-2]

1.180 A house has a number which needs to be identified. The following three statements are given that can help in identifying the house number.

1. If the house number is a multiple of 3, then it is a number from 50 to 59.
2. If the house number is NOT a multiple of 4, then it is a number from 60 to 69.
3. If the house number is NOT a multiple of 6, then it is a number from 70 to 79.

Which is the house number?

(a) 54 (b) 65
(c) 66 (d) 76

[2018 : 2 M, Set-2]

1.181 John Thomas, an ________writer, passed away in 2018.

(a) prominent (b) dominant
(c) imminent (d) eminent

[2019 : 1 M, Set-1]

1.182 ____________I permitted him to leave, I wouldn't have had any problem with him being absent, ________I?

(a) Had, wouldn't (b) Have, would
(c) Have, wouldn't (d) Had, would

[2019 : 1 M, Set-1]

1.183 A worker noticed that the hour hand on the factory clock had moved 225 degrees during her stay at the factory. For how long does she stay in the factory?

(a) 8.5 hours (b) 7.5 hours
(c) 3.75 hours (d) 4 hours and 15 minutes

[2019 : 1 M, Set-1]

1.184 The sum and product of two integers are 26 and 165 respectively. The difference between these two integers is ______.

(a) 2 (b) 6
(c) 3 (d) 4

[2019 : 1 M, Set-1]

1.185 The minister avoided any mention of the issue of women's reservation in the private sector. He was accused of ________ the issue.

(a) tying (b) collaring
(c) skirting (d) belting

[2019 : 1 M, Set-1]

1.186 A firm hires employees at five different skill levels P, Q, R, S, T. The shares of employment at these skill levels of total employment in 2010 is given in the pie chart as shown. There were a total of 600 employees in 2010 and the total employment increased by 15% from 2010 to 2016. The total employment at skill levels P, Q, and R remained unchanged during this period. If the employment at skill level S increased by 40% from 2010 to 2016, how many employees were there at skill level T in 2016?

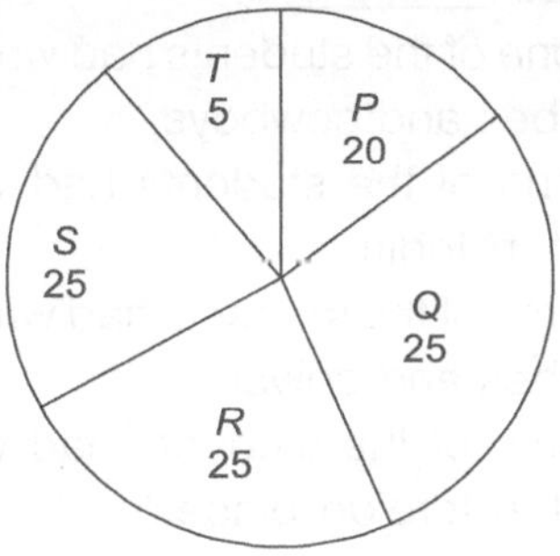

Percentage share of skills in 2010

(a) 35 (b) 30
(c) 72 (d) 60

[2019 : 2 M, Set-1]

1.187 M and N had four children P, Q, R and S. Of them, only P and R were married. They had children X and Y respectively. If Y is a legitimate child of W, which one of the following statements is necessarily FALSE?

(a) W is the wife of R
(b) R is the father of Y
(c) M is the grandmother of Y
(d) W is the wife of P [2019 : 2 M, Set-1]

1.188 Congo was named by Europeans. Congo's dictator Mobuto later changed the name of the country and the river to Zaire with the objective of Africanising names of persons and spaces. However, the name Zaire was a Portuguese alternation of *Nzadi o Nzere*, a local African term meaning 'River that swallows Rivers'. Zaire was the Portuguese name for the Congo river in the 16th and 17th centuries.

Which one of the following statements can be inferred from the paragraph above?

(a) The term *Nzadi o Nzere* was of Portuguese origin
(b) As a dictator Mobuto ordered the Portuguese to alter the name of the river to Zaire
(c) Mobuto's desire to Africanise names was prevented by the Portuguese
(d) Mobuto was not entirely successful in Africanising the name of his country

[2019 : 2 M, Set-1]

1.189 Under a certain legal system, prisoners are allowed to make one statement. If their statement turns out to be true then they are hanged. If the statement turns out to be false then they are shot. One prisoner made a statement and the judge had no option but to set him free. Which one of the following could be that statement?

(a) I committed the crime
(b) You committed the crime
(c) I will be shot
(d) I did not commit the crime

[2019 : 2 M, Set-1]

1.190 A person divided an amount of Rs. 100,000 into two parts and invested in two different schemes. In one he got 10% profit and in the other he got 12%. If the profit percentages are interchanged with these investments he would have got Rs. 120 less. Find the ratio between his investments in the two schemes.

(a) 9 : 16 (b) 37 : 63
(c) 47 : 53 (d) 11 : 14

[2019 : 2 M, Set-1]

1.191 Are there enough seats here? There are ________ people here than I expected.

(a) most (b) more
(c) least (d) many

[2019 : 1 M, Set-2]

1.192 If IMHO = JNIP; IDK = JEL; and SO = TP, then IDC = ________.

(a) JDE (b) JED
(c) JDC (d) JCD

[2019 : 1 M, Set-2]

1.193 Once the team of analysts identify the problem, we ________ in a better position to comment on the issue.

Which one of the following choices CANNOT fill the given blank?

(a) were to be (b) are going to be
(c) will be (d) might be

[2019 : 1 M, Set-2]

1.194 A final examination is the ______ of a series of evaluations that a student has to go through.

(a) insinuation (b) desperation
(c) culmination (d) consultation

[2019 : 1 M, Set-2]

1.195 The product of three integers X, Y and Z is 192. Z equal to 4 and P is equal to the average of X and Y. What is the minimum possible value of P?

(a) 8 (b) 7
(c) 9.5 (d) 6 [2019 : 1 M, Set-2]

1.196 X is an online media provider. By offering unlimited and exclusive online content at attractive prices for a loyalty membership, X is almost forcing its customers towards its loyalty membership. If its loyalty membership continues to grow at its current rate, within the next eight years more households will be watching X than cable television.
Which one of the following statements can be inferred from the above paragraph?

(a) Most households that subscribe to X's loyalty membership discontinue watching cable television
(b) Cable television operators don't subscribe to X's loyalty membership
(c) The X is cancelling accounts of non-members
(d) Non-members prefer to watch cable television [2019 : 2 M, Set-2]

1.197 Fiscal deficit was 4% of the GDP in 2015, and that increased to 5% in 2016. If the GDP increased by 10% from 2015 to 2016, then percentage increase in the actual fiscal deficit is_____

(a) 25.00 (b) 35.70
(c) 10.00 (d) 37.50

[2019 : 2 M, Set-2]

1.198 Mola is a digital platform for taxis in a city. It offers three type of rides-Pool, Mini and Prime. The table below presents the number of rides for the past four months. The platform earns one US dollar per ride. What is the percentage share of revenue contributed by prime to the total revenues of Mola, for the entire duration?

Type	Month			
	January	February	March	April
Pool	170	320	215	190
Mini	110	220	180	70
Prime	75	180	120	90

(a) 38.74 (b) 23.97
(c) 25.86 (d) 16.24

[2019 : 2 M, Set-2]

1.199 While teaching a creative writing class in India. I was surprised at receiving stories from the students that were all set in distant places: in the American West with cowboys and in Manhattan penthouses with clinking ice cubes. This was, till an eminent Caribbean writer gave the once-colonized countries the confidence to see the shabby lives around them as worthy of being "told".
The writer of this passage is surprised by the creative writing assignments of his students, because ________ .

(a) None of the students had written about ice cubes and cowboys.
(b) None of the students had written stories set in India.
(c) Some of the students had written about ice cubes and cowboys.
(d) Some of the students had written stories set in foreign places.

[2019 : 2 M, Set-2]

1.200 Two pipes P and Q can fill a tank in 6 hours and 9 hours respectively, while a third pipe R can empty the tank in 12 hours. Initially, P and R are open for 4 hours. Then P is closed and Q is opened. After 6 more hours R is closed. The total time taken to fill the tank (in hours) is _____

(a) 16.50 (b) 15.50
(c) 14.50 (d) 13.50

[2019 : 2 M, Set-2]

1.201 He is known for his unscrupulous ways. He always sheds ______ tears to deceive people.

(a) crocodile's (b) fox's
(c) crocodile (d) fox

[2020 : 1 M, Set-1]

1.202 Crowd funding deals with mobilisation of funds for a project from a large number of people, who would be willing to invest smaller amounts through web-based platforms in the project.
Based on the above paragraph, which of the following is correct about crowd funding?

(a) Funds raised through unwilling contributions on web-based platforms.
(b) Funds raised through coerced contributions on web-based platforms.

(c) Funds raised through large contributions on web-based platforms.
(d) Funds raised through voluntary contributions on web-based platforms.

[2020 : 1 M, Set-1]

1.203 Jofra Archer, the England fast bowler, is _____ than accurate.

(a) faster (b) more faster
(c) more fast (d) less fast

[2020 : 1 M, Set-1]

1.204 Select the word that fits the analogy:
Build : Building : : Grow : _______

(a) Grew (b) Growed
(c) Grown (d) Growth

[2020 : 1 M, Set-1]

1.205 I do not think you know the case well enough to have opinions. Having said that, I agree with your other point. What does the phrase "having said that" mean in the given text?

(a) in addition to what I have said
(b) despite what I have said
(c) contrary to what I have said
(d) as opposed to what I have said

[2020 : 1 M, Set-1]

1.206 The bar graph shows the data of the students who appeared and passed in an examination for four schools *P*, *Q*, *R* and *S*. The average of success rates (in percentage) of these four schools is _______.

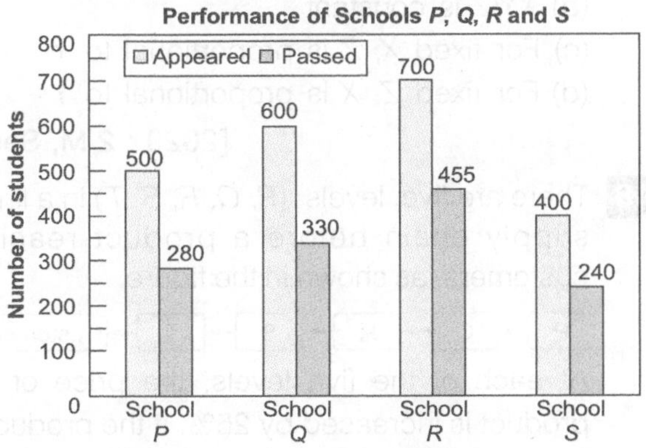

(a) 58.8 (b) 59.0
(c) 58.5 (d) 59.3

[2020 : 2 M, Set-1]

1.207 Select the graph that schematically represents BOTH $y = x^m$ and $y = x^{1/m}$ properly in the interval $0 \le x \le 1$. For integer values of m, where $m > 1$.

(a)

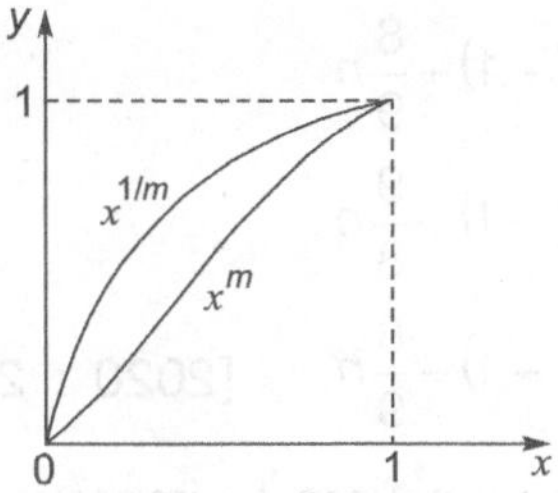

(b)

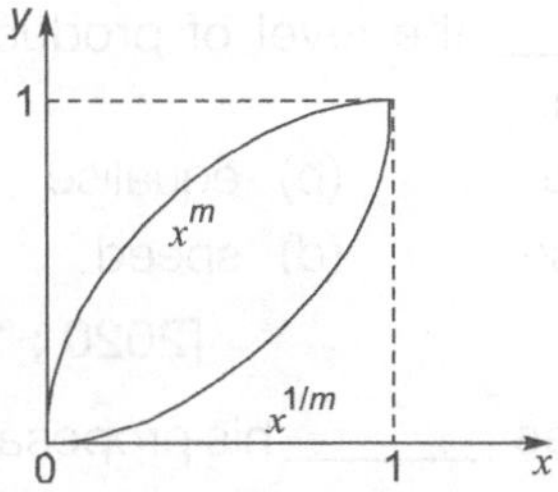

(c)

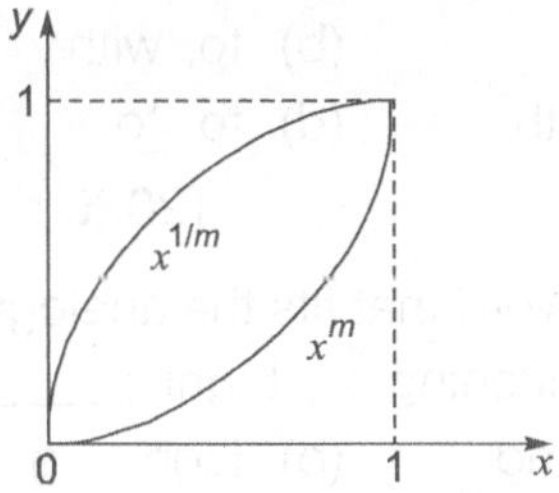

(d)

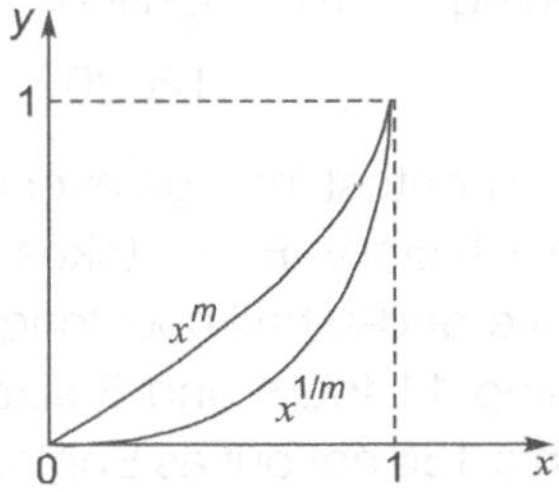

[2020 : 2 M, Set-1]

1.208 Define $[x]$ as the greatest integer less than or equal to x, for each $x \in (-\infty, \infty)$. If $y = [x]$, then area under y for $x \in [1, 4]$ is _______.

(a) 3 (b) 1
(c) 6 (d) 4

[2020 : 2 M, Set-1]

1.209 *P*, *Q*, *R* and *S* are to be uniquely coded using α and β. If P is coded as αα and *Q* as αβ, then R and S, respectively, can be coded as _______.

(a) βα and ββ (b) αβ and ββ
(c) βα and αβ (d) ββ and αα

[2020 : 2 M, Set-1]

1.210 The sum of the first *n* terms in the sequence 8, 88, 888, 8888, ... is ______.

(a) $\frac{81}{80}(10^n - 1) + \frac{9}{8}n$

(b) $\frac{80}{81}(10^n - 1) + \frac{8}{9}n$

(c) $\frac{81}{80}(10^n - 1) + \frac{9}{8}n$

(d) $\frac{80}{81}(10^n - 1) - \frac{8}{9}n$ [2020 : 2 M, Set-1]

1.211 The recent measures to improve the output would ______ the level of production to our satisfaction.

(a) increase (b) equalise
(c) decrease (d) speed

[2020 : 1 M, Set-2]

1.212 While I agree ________ his proposal this time, I do not often agree ______ him.

(a) with, to (b) to, with
(c) with, with (d) to, to

[2020 : 1 M, Set-2]

1.213 Select the word that fits the analogy:
White : Whitening : : Light : ________.

(a) Lightening (b) Lightning
(c) Enlightening (d) Lighting

[2020 : 1 M, Set-2]

1.214 In one of the greatest innings ever seen in 142 years of Test history. Ben Stokes upped the tempo in a five-and-a-half hour long stay of 219 balls including 11 fours and 8 sixes that saw him finish on a 135 not out as England squared the five-match series.

Based on their connotations in the given passage, which one of the following meanings DOES NOT match?

(a) squared = lost
(b) tempo = enthusiasm
(c) upped = increased
(d) saw = resulted in [2020 : 1 M, Set-2]

1.215 Climate change and resilience deal with two aspects - reduction of sources of nonrenewable energy resources and reducing vulnerability of climate change aspects. The terms 'mitigation' and 'adaptation' are used to refer to these aspects, respectively.

Which of the following assertions is best supported by the above information?

(a) Mitigation deals with actions taken to reduce the use of fossil fuels.
(c) Adaptation deals with causes of climate change.
(b) Mitigation deals with consequences of climate change.
(d) Adaptation deals with actions taken to combat green-house gas emissions

[2020 : 1 M, Set-2]

1.216 It was estimated that 52 men can complete a strip in a newly constructed highway connecting cities *P* and *Q* in 10 days. Due to an emergency, 12 men were sent to another project. How many number of days, more than the original estimate, will be required to complete the strip?

(a) 3 days (b) 5 days
(c) 10 days (d) 13 days

[2020 : 2 M, Set-2]

1.217 An engineer measures THREE quantities *X*, *Y* and *Z* in an experiment. She finds that they follow a relationship that is represented in the figure below: (the product of *X* and *Y* linearly varies with *Z*)

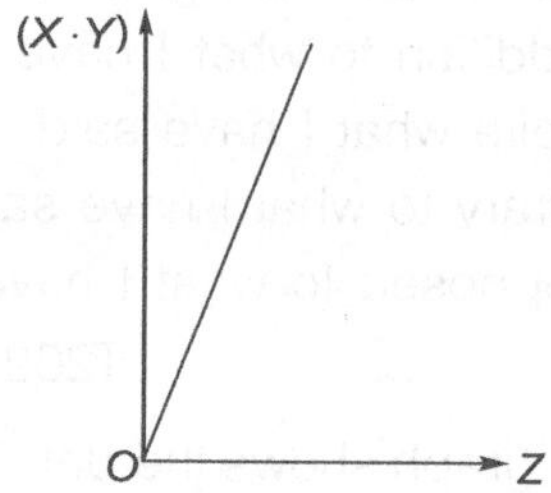

Then, which of the following statements is FALSE?

(a) For fixed *Y*; *X* is proportional to *Z*
(b) *XY*/*Z* is constant
(c) For fixed *X*; *Z* is proportional to *Y*
(d) For fixed *Z*; *X* is proportional to *Y*

[2020 : 2 M, Set-2]

1.218 There are five levels {*P*, *Q*, *R*, *S*, *T*} in a linear supply chain before a product reaches customers, as shown in the figure.

At each of the five levels, the price of the product is increased by 25%. If the product is produced at level *P* at the cost of Rs. 120 per unit, what is the paid (in rupees) by the customers?

(a) 292.96 (b) 187.50
(c) 234.38 (d) 366.21

[2020 : 2 M, Set-2]

1.219 The two pie-charts given below show the data of total students and only girls registered in different streams in a university. If the total number of students registered in the university is 5000, and the total number of the registered girls is 1500; then, the ratio of boys enrolled in Arts to the girls enrolled in Management is ____.

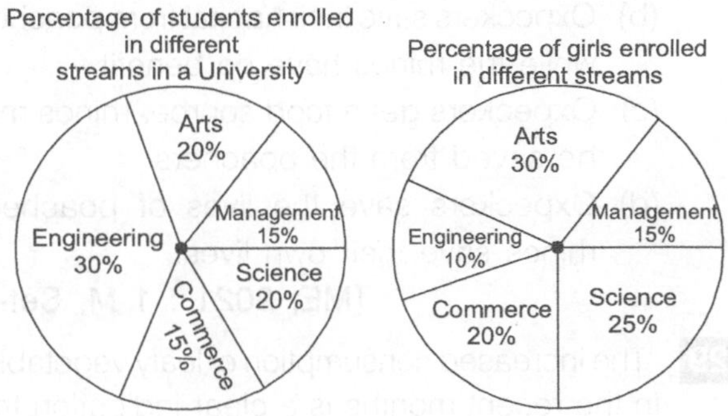

(a) 9 : 22 (b) 11 : 9
(c) 2 : 1 (d) 22 : 9

[2020 : 2 M, Set-2]

1.220 Find the missing element in the following figure:

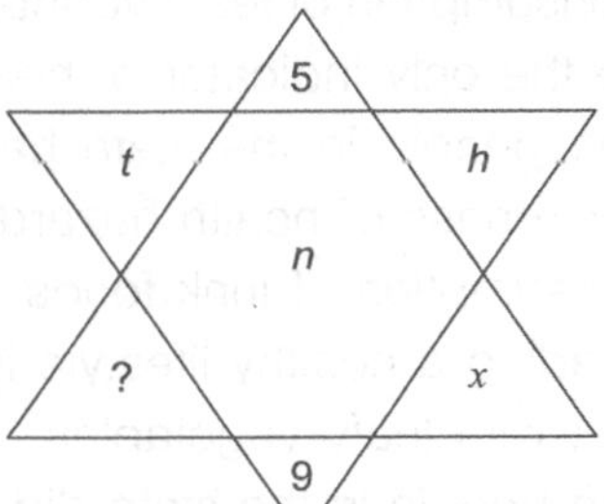

(a) *d* (b) *w*
(c) *e* (d) *y*

[2020 : 2 M, Set-2]

1.221

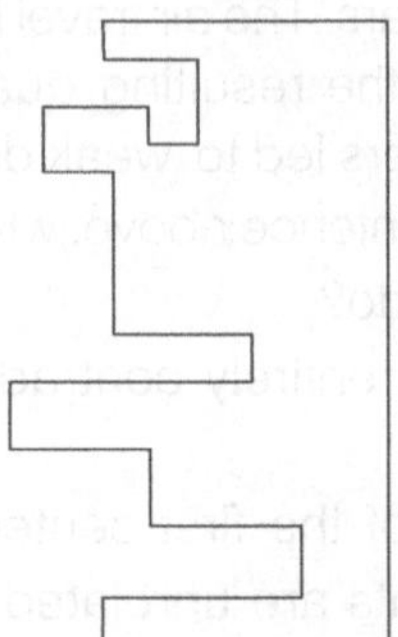

A jigsaw puzzle has 2 pieces. One of the pieces is shown above. Which one of the given options for the missing piece when assembled will form a rectangle? The piece can be moved, rotated or flipped to assemble with the above piece.

(a) 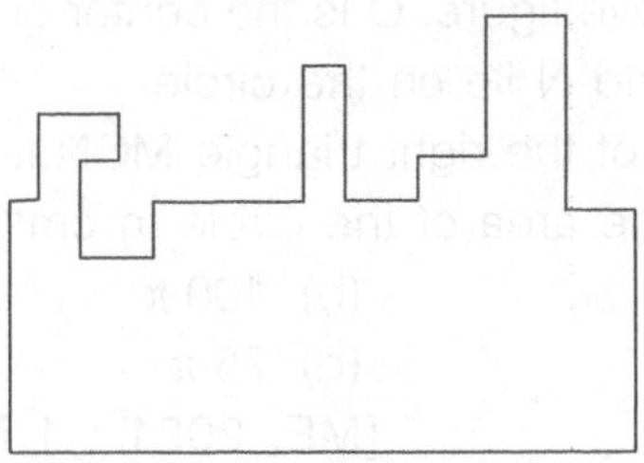

(b) 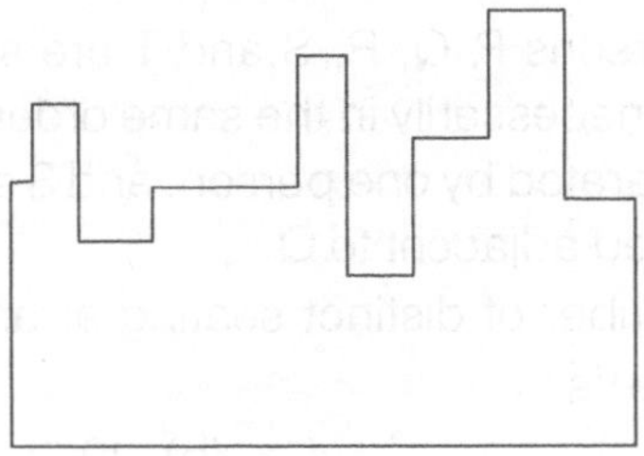

(c) 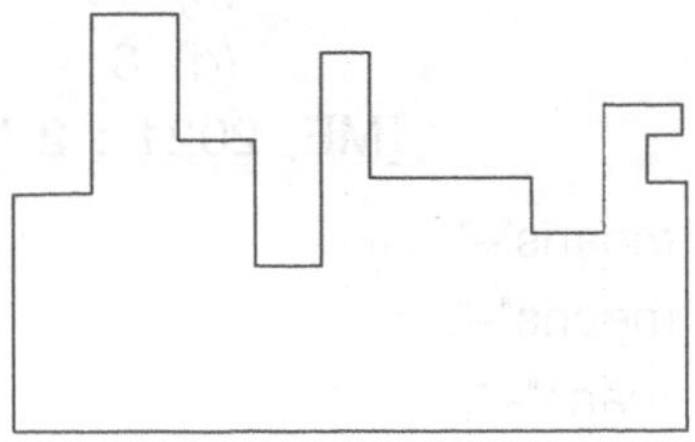

(d) 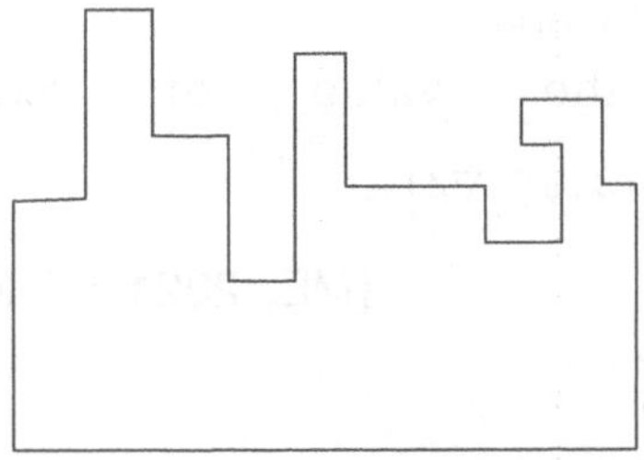

[ME, 2021 : 2 M, Set-1]

1.222 The number of hens, ducks and goats in farm P are 65, 91 and 169, respectively. The total number of hens, ducks and goats in a nearby farm Q is 416. The ratio of hens : ducks : goats in farm Q is 5 : 14 : 13. All the hens, ducks and goats are sent from farm Q to farm P.

The new ratio of hens : ducks : goats in farm P is ______.

(a) 5 : 7 : 13 (b) 21 : 10 : 26
(c) 5 : 14 : 13 (d) 10 : 21 : 26

[ME, 2021 : 2 M, Set-1]

1.223

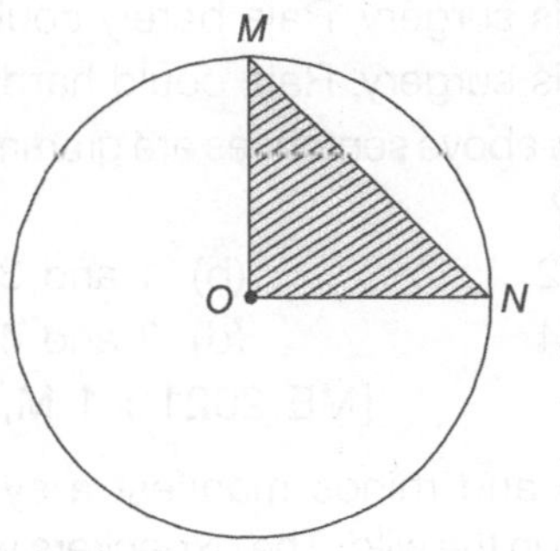

In the above figure, O is the center of the circle and, M and N lie on the circle.

The area of the right triangle MON is 50 cm^2.

What is the area of the circle in cm^2?

(a) 2π (b) 100π

(c) 50π (d) 75π

[ME, 2021 : 1 M, Set-1]

1.224 Five persons P, Q, R, S and T are sitting in a row not necessarily in the same order. Q and R are separated by one person, and S should not be seated adjacent to Q.

The number of distinct seating arrangements possible is

(a) 4 (b) 10

(c) 16 (d) 8

[ME, 2021 : 2 M, Set-1]

1.225 If $\begin{cases} "\oplus" \text{ means} "-", \\ "\otimes" \text{ means} "\div", \\ "\Delta" \text{ means} "+", \\ "\nabla" \text{ means} "\times", \end{cases}$

then, the value of expression

$\Delta 2 \oplus 3 \Delta ((4 \otimes 2) \nabla 4) =$

[ME, 2021 : 1 M, Set-1]

1.226

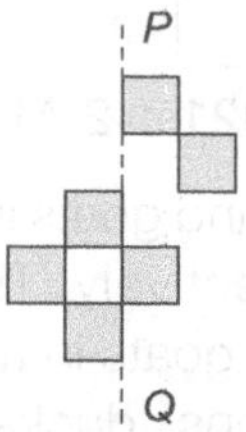

The least number of squares that must be added so that the line *P-Q* becomes the line of symmetry is ______.

[ME, 2021 : 1 M, Set-1]

1.227 Consider the following sentences:

1. After his surgery, Raja hardly could walk.
2. After his surgery, Raja could barely walk.
3. After his surgery, Raja barely could walk.
4. After his surgery, Raja could hardly walk.

Which of the above sentences are grammatically CORRECT?

(a) 1 and 2 (b) 1 and 3

(c) 2 and 4 (d) 3 and 4

[ME 2021 : 1 M, Set-1]

1.228 Oxpeckers and rhinos manifest a symbiotic relationship in the wild. The oxpeckers warn the rhinos about approaching poachers, thus possibly saving the lives of the rhinos. Oxpeckers also feed on the parasitic ticks found on rhinos.

In the symbiotic relationship described above, the primary benefits for oxpeckers and rhinos respectively are,

(a) Oxpeckers get a food source, rhinos have no benefit.

(b) Oxpeckers save their habitat from poachers while the rhinos have no benefit.

(c) Oxpeckers get a food source, rhinos may be saved from the poachers.

(d) Oxpeckers save the lives of poachers, rhinos save their own lives.

[ME, 2021 : 1 M, Set-1]

1.229 "The increased consumption of leafy vegetables in the recent months is a clear indication that the people in the state have begun to lead a healthy lifestyle".

Which of the following can be logically inferred from the information presented in the above statement?

(a) Consumption of leafy vegetables may not be the only indicator of healthy lifestyle.

(b) The people in the state have increased awareness of health hazards causing by consumption of junk foods.

(c) Leading a healthy lifestyle is related to a diet with leafy vegetables.

(d) The people in the state did not consume leafy vegetables earlier.

[ME 2021 : 2 M Set-1]

1.230 The world is going through the worst pandemic in the past hundred years. The air travel industry is facing a crisis, as the resulting quarantine requirement for travelers led to weak demand.

In relation to the first sentence above, what does the second sentence do?

(a) Second sentence entirely contradicts the first sentence.

(b) States an effect of the first sentence.

(c) The two statements are unrelated.

(d) Restates an idea from the first sentence.

[ME 2021 : 1 M Set-2]

1.231 Consider the following sentences:

(i) The number of candidates who appear for the GATE examination is staggering.

(ii) A number of candidates from any class are appearing for the GATE examination.

(iii) The number of candidates who appear for the GATE examination are stagging.
(iv) A number of candidates from my class is appearing for the GATE examination.

Which of the above sentences are grammatically CORRECT?

(a) (i) and (iii) (b) (i) and (ii)
(c) (ii) and (iii) (d) (ii) and (iv)

[ME 2021 : 1 M, Set-2]

1.232 If $\oplus \div \odot = 2$; $\oplus \div \Delta = 3$; $\odot + \Delta = 5$; $\Delta \times \otimes = 10$,

Then, the value of $(\otimes - \oplus)^2$, is:

(a) 4 (b) 16
(c) 1 (d) 0

[ME, 2021 : 1 M, Set-2]

1.233 Consider a square sheet of side 1 unit. The sheet is first folded along the main diagonal. This is followed by a fold along its line of symmetry. The resulting folded shape is again folded along its line of symmetry. The area of each face of the final folded shape, in square units, equal to _____.

(a) $\frac{1}{8}$ (b) $\frac{1}{32}$
(c) $\frac{1}{4}$ (d) $\frac{1}{16}$

[ME, 2021 : 2 M, Set-2]

1.234 Five persons P, Q, R, S and T are to be seated in a row, all facing the same direction, but not necessarily in the same order. P and T cannot be seated at either end of the row. P should not be seated adjacent to S. R is to be seated at the second position from the left end of the row. The number of distinct seating arrangements possible is:

(a) 2 (b) 3
(c) 4 (d) 5

[ME, 2021 : 1 M, Set-2]

1.235 The front door of Mr. X's house faces East. Mr. X leaves the house, walking 50 m straight from the back door that is situated directly opposite to the front door. He then turns to his right, walks for another 50 m and stops. The direction of the point Mr. X is now located at with respect to the starting point is _____.

(a) South-East (b) North-West
(c) North-East (d) West

[ME, 2021 : 2 M, Set-2]

1.236 A box contains 15 blue balls and 45 black balls. If 2 balls are selected randomly, without replacement, the probability of an outcome in which the first selected is a blue ball and the second selected is a black ball, is _____.

(a) $\frac{45}{236}$ (b) $\frac{1}{4}$
(c) $\frac{3}{16}$ (d) $\frac{3}{4}$

[ME, 2021 : 2 M, Set-2]

1.237 A digital watch X beeps every 30 seconds while watch Y beeps every 32 seconds. They beeped together at 10 AM.

The immediate next time that they will beep together is _____.

(a) 11.00 AM (b) 10.08 AM
(c) 10.42 AM (d) 10.00 PM

[ME, 2021 : 1 M, Set-2]

1.238 The ratio of the area of the inscribed circle to the area of circumscribed circle of an equilateral triangle is _____.

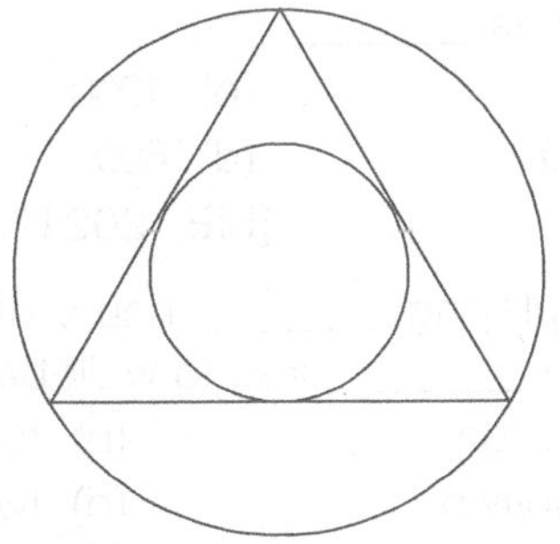

(a) $\frac{1}{4}$ (b) $\frac{1}{2}$
(c) $\frac{1}{6}$ (d) $\frac{1}{8}$

[ME, 2021 : 2 M, Set-2]

1.239 Given below are two statements 1 and 2, and two conclusions I and II.

Statement 1: All entrepreneurs are wealthy.
Statement 2: All wealthy are risk seekers.
Conclusion I: All risk seekers are wealthy.
Conclusion II: Only some entrepreneurs are risk seekers.

Based on the above statements and conclusions, which one of the following options is CORRECT?

(a) Only conclusion I is correct
(b) Only conclusion II is correct
(c) Neither conclusion I nor II is correct
(d) Both conclusions I and II are correct

[ME, 2021 : 2 M, Set-2]

1.240

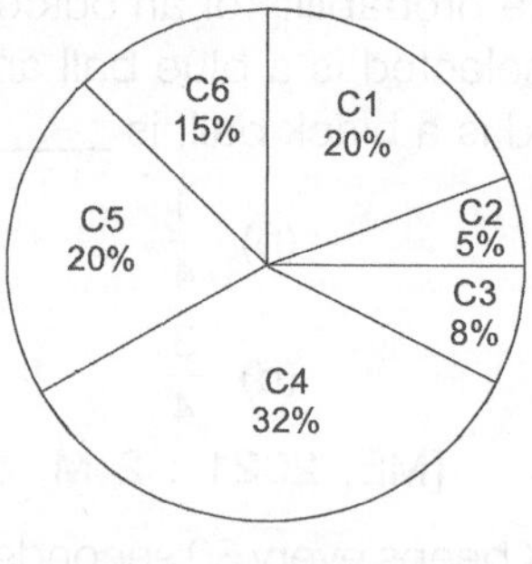

Company	Ratio
C1	3 : 2
C2	1 : 4
C3	5 : 3
C4	2 : 3
C5	9 : 1
C6	3 : 4

The distribution of employees at the rank of executives, across different companies C1, C2, ..., C6 is presented in the chart given above. The ratio of executives with a management degree to those without a management degree in each of these companies is provided in the table above. The total number of executives across all companies is 10000.

The total number of management degree holders among the executives in companies C2 and C5 together is ______.

(a) 225 (b) 1900
(c) 2500 (d) 600

[ME, 2021 : 2 M, Set-1]

1.241 After playing _______ hours of tennis, I am feeling _______ tired to walk back.

(a) too/too (b) too/two
(c) two/two (d) two/too

[2022 : 1 M, Set-1]

1.242 The average of the monthly salaries of M, N and S is ₹4000. The average of the monthly salaries of N, S and P is ₹5000. The monthly salary of P is ₹6000.

What is the monthly salary of M as a percentage of the monthly salary of P?

(a) 50% (b) 75%
(c) 100% (d) 125%

[2022 : 1 M, Set-1]

1.243 A person travelled 80 km in 6 hours. If the person travelled the first part with a uniform speed of 10 kmph and the remaining part with a uniform speed of 18 kmph. What percentage of the total distance is travelled at a uniform speed of 10 kmph?

(a) 28.25 (b) 37.25
(c) 43.75 (d) 50.00

[2022 : 1 M, Set-1]

1.244 Four girls P, Q, R and S are studying languages in a University. P is learning French and Dutch. Q is learning Chinese and Japanese. R is learning Spanish and French. S is learning Dutch and Japanese.

Given that: French is easier than Dutch; Chinese is harder than Japanese; Dutch is easier than Japanese, and Spanish is easier than French.

Based on the above information, which girl is learning the most difficult pair of languages?

(a) P (b) Q
(c) R (d) S

[2022 : 1 M, Set-1]

1.245

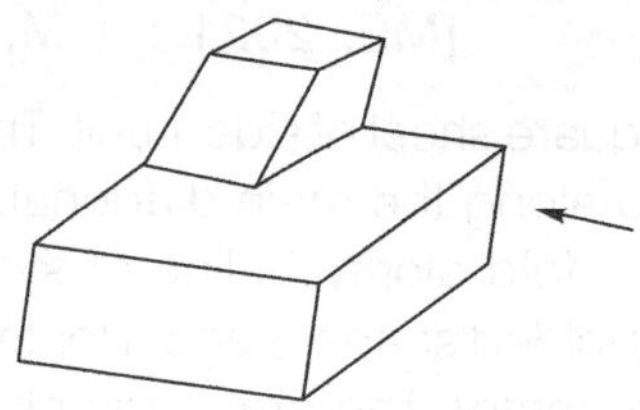

A block with a trapezoidal cross-section is placed over a block with rectangular cross section as shown above. Which one of the following is the correct drawing of the view of the 3D object as viewed in the direction indicated by an arrow in the above figure?

(a) 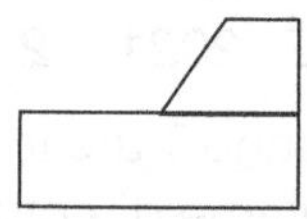 (b) 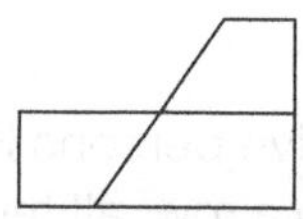

(c) 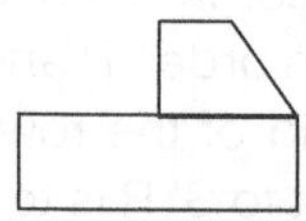 (d) 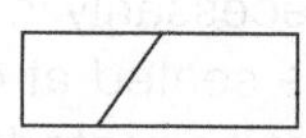

[2022 : 1 M, Set-1]

1.246 Humans are naturally compassionate and honest. In a study using strategically placed wallets that appear "lost", it was found that wallets with money are more likely to be returned than wallets without money. Similarly, wallets that had a key and money are more likely to be returned than wallets with the same amount of money alone. This suggests that the primary reason for this behavior is compassion and empathy.

Which one of the following is the CORRECT logical inference based on the information in the above passage?

(a) Wallets with a key are more likely to be returned because people do not care about money.

(b) Wallets with a key are more likely to be returned because people relate to suffering of others.
(c) Wallets used in experiments are more likely to be returned than wallets that are really lost.
(d) Money is always more important than keys.

[2022 : 2 M, Set-1]

1.247 A rhombus is formed by joining the midpoints of the sides of a unit square. What is the diameter of the largest circle that can be inscribed within the rhombus?

(a) $\frac{1}{\sqrt{2}}$ (b) $\frac{1}{2\sqrt{2}}$
(c) $\sqrt{2}$ (d) $2\sqrt{2}$

[2022 : 2 M, Set-1]

1.248 An equilateral triangle, a square and a circle have equal areas. What is the ratio of the perimeters of the equilateral triangle to square to circle?

(a) $3\sqrt{3}:2:\pi$ (b) $\sqrt{(3\sqrt{3})}:2:\sqrt{\pi}$
(c) $\sqrt{(3\sqrt{3})}:4:2\sqrt{\pi}$ (d) $\sqrt{(3\sqrt{3})}:2:2\sqrt{\pi}$

[2022 : 2 M, Set-1]

1.249 Given below are three conclusions drawn based on the following three statements.

Statement 1: All teachers are professors.

Statement 2: No professor is a male.

Statement 3: Some males are engineers.

Conclusion I: No engineer is a professor.

Conclusion II: Some engineers are professors.

Conclusion III: No male is a teacher.

Which one of the following options can be logically inferred?

(a) Only conclusion III is correct
(b) Only conclusion I and conclusion II are correct
(c) Only conclusion II and conclusion III are correct
(d) Only conclusion I and conclusion III are correct

[2022 : 2 M, Set-1]

1.250 In a 12-hour clock that runs correctly, how many times do the second, minute, and hour hands of the clock coincide, in a 12-hour duration from 3 PM in a day to 3 AM the next day?

(a) 11 (b) 12
(c) 144 (d) 2

[2022 : 2 M, Set-1]

1.251 Writing too many things on the _____ while teaching could make students get _____.

(a) bored/board (b) board/bored
(c) board/board (d) bored/bored

[2022 : 1 M, Set-2]

1.252 Which one of the following is a representation (not to scale and in bold) of all values of x satisfying the inequality $2-5x \le -\left(\frac{6x-5}{3}\right)$ on the real number line?

(a) 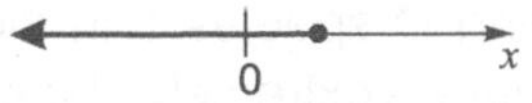

(b) 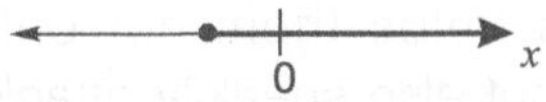

(c) 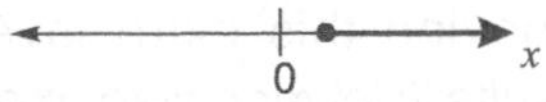

(d) 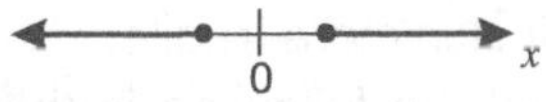

[2022 : 1 M, Set-2]

1.253 If $f(x) = 2\ln\left(\sqrt{e^x}\right)$, what is the area bounded by $f(x)$ for the interval [0, 2] on the x-axis?

(a) $\frac{1}{2}$ (b) 1
(c) 2 (d) 4

[2022 : 1 M, Set-2]

1.254 A person was born on the fifth Monday of February in a particular year. Which one of the following statements is correct based on the above information?

(a) The 2nd February of that year is a Tuesday.
(b) There will be five Sundays in the month of February in that year
(c) The 1st February of that year is Sunday.
(d) All Mondays of February in that year have even dates.

[2022 : 1 M, Set-2]

1.255 Which one of the groups given below can be assembled to get the shape that is shown above using each piece only once without overlapping with each other? (rotation and translation operations may be used).

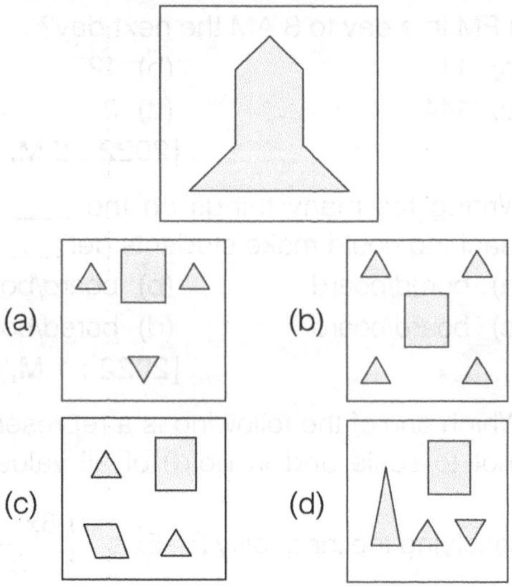

[2022 : 1 M, Set-2]

1.256 Fish belonging to species *S* in the deep sea have skins that are extremely black (ultra-black skin). This helps them not only to avoid predators but also sneakily attack their prey. However, having this extra layer of black pigment results in lower collagen on their skin, making their skin more fragile.

Which one of the following is the CORRECT logical inference based on the information in the above passage?

(a) Having ultra-black skin is only advantageous to species *S*.
(b) Species *S* with lower collagen in their skin are at an advantage because it helps them avoid predators.
(c) Having ultra-black skin has both advantages and disadvantages to species *S*.
(d) Having ultra-black skin is only disadvantageous to species *S* but advantageous only to their predators.

[2022 : 2 M, Set-2]

1.257 For the past *m* days, the average daily production at a company was 100 units per day. If today's production of 180 units changes the average to 110 units per day, what is the value of *m*?

(a) 18 (b) 10
(c) 7 (d) 5

[2022 : 2 M, Set-2]

1.258 Consider the following functions for non-zero positive integers, *p* and *q*.

$$f(p,q) = \underbrace{p\times p\times p\times \ldots\ldots\times p}_{q\text{ terms}} = p^q;\quad f(p,1) = p$$

$$g(p,q) = p^{p^{p^{p^{\cdot^{\cdot^{\cdot}}}}}}\ \text{(upto to } q \text{ terms)};\quad g(p,1) = p$$

Which one of the following options is correct based on the above?

(a) $f(2, 2) = g(2, 2)$
(b) $f(g(2, 2), 2) < f(2, g(2, 2))$
(c) $g(2, 1) \neq f(2, 1)$
(d) $f(3, 2) > g(3, 2)$ [2022 : 2 M, Set-2]

1.259 Four cities *P*, *Q*, *R* and *S* are connected through one-way routes as shown in the figure. The travel time between any two connected cities is one hour. The boxes beside each city name describe the starting time of first train of the day and their frequency of operation. For example, from city *P*, the first trains of the day start at 8 AM with a frequency of 90 minutes to each of *R* and *S*. A person does not spend additional time at any city other than the waiting time for the next connecting train.

If the person starts from *R* at 7 AM and is required to visit *S* and return to *R*, what is the minimum time required?

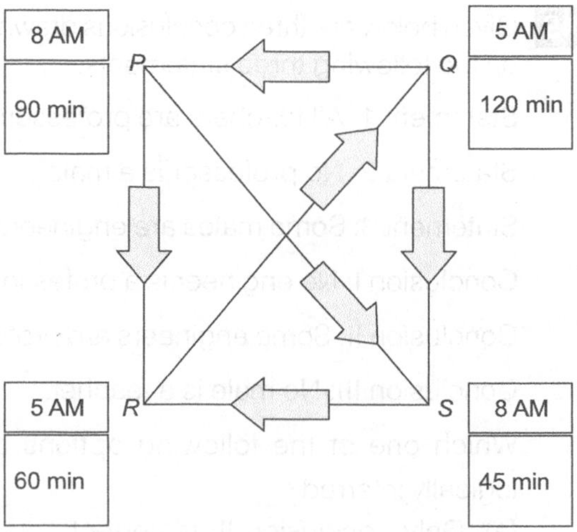

(a) 6 hours 30 minutes
(b) 3 hours 45 minutes
(c) 4 hours 30 minutes
(d) 5 hours 15 minutes [2022 : 2 M, Set-2]

1.260 Equal sized circular regions are shaded in a square sheet of paper of 1 cm side length. Two cases, case *M* and case *N*, are considered as shown in the figures below. In the case *M*, four circles are shaded in the square sheet and in

the case *N*, nine circles are shaded in the square sheet as shown.
What is the ratio of the areas of unshaded regions of case *M* to that of case *N*?

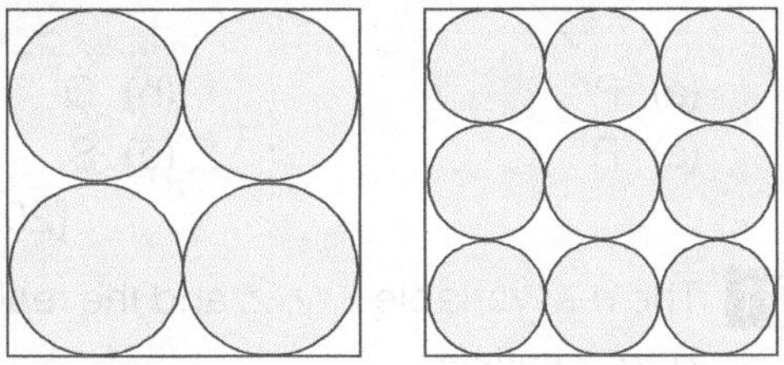

Case M　　Case N

(a) 2 : 3　　(b) 1 : 1
(c) 3 : 2　　(d) 2 : 1

[2022 : 2 M, Set-2]

1.261 He did not manage to fix the car himself, so he ____ in the garage.
(a) got it fixed　　(b) getting it fixed
(c) gets fixed　　(d) got fixed

[2023 : 1 M]

1.262 Planting : Seed : : Raising : ______. [By word meaning]
(a) Child　　(b) Temperature
(c) Height　　(d) Lift

[2023 : 1 M]

1.263 A certain country has 504 universities and 25951 colleges. These are categorised into Grades I, II, and III as shown in the given pie charts.
What is the percentage, correct to one decimal place, of higher education institutions (colleges and universities) that fall into Grade III?

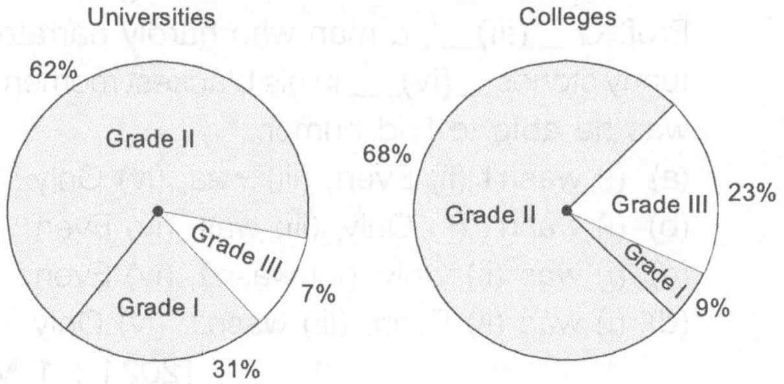

(a) 22.7　　(b) 23.7
(c) 15.0　　(d) 66.8

[2023 : 1 M]

1.264 The minute hand and second-hand of a clock cross each other ______ times between 09:15:00 AM and 09:45:00 AM on a day.
(a) 30　　(b) 15
(c) 29　　(d) 31 **[2023 : 1 M]**

1.265 The symbols ○, ∗, △, and □ are to be filled, one in each box, as shown below.
The rules for filling in the four symbols are as follows:
1. Every row and every column must contain each of the four symbols.
2. Every 2 × 2 square delineated by bold lines must contain each of the four symbols.

Which symbol will occupy the box Med with '?' in the partially filled figure?

	?		△
	○	∗	
		□	∗
∗		△	○

(a) ○　　(b) ∗
(c) △　　(d) □

[2023 : 1 M]

1.266 In a recently held parent-teacher meeting, the teachers had very few complaints about Ravi. After all, Ravi was a hardworking and kind student. Incidentally, almost all of Ravi's friends at school were hardworking and kind too. But the teachers drew attention to Ravi's complete lack of interest in sports. The teachers believed that, along with some of his friends who showed similar disinterest in sports, Ravi needed to engage in some sports for his overall development.
Based only on the information provided above, which one of the following statements can be logically inferred with certainty?
(a) All of Ravi's friends are hardworking and kind.
(b) No one who is not a friend of Ravi is hardworking and kind.
(c) None of Ravi's friends are interested in sports.
(d) Some of Ravi's friends are hardworking and kind.

[2023 : 2 M]

1.267 Consider the following inequalities:

$p^2 - 4q < 4$

$3p + 2q < 6$

where p and q are positive integers.

The value of $(p + q)$ is

(a) 2 (b) 1

(c) 3 (d) 4 **[2023 : 2 M]**

1.268 Which one of the sentence sequences in the given options creates a coherent narrative?

(i) I could not bring myself to knock.

(ii) There was a murmur of unfamiliar voices coming from the big drawing room and the door was firmly shut.

(iii) The passage was dark for a bit, but then it suddenly opened into a bright kitchen.

(iv) I decided I would rather wander down the passage.

(a) (iv), (i), (iii), (ii) (b) (iii), (i), (ii), (iv)

(c) (ii), (i), (iv), (iii) (d) (i), (iii), (ii), (iv)

[2023 : 2 M]

1.269 How many pairs of sets (S,T) are possible among the subsets of {1, 2, 3, 4, 5, 6} that satisfy the the condition that S is a subset of T?

(a) 729 (b) 728

(c) 665 (d) 664

[2023 : 2 M]

1.270 An opaque pyramid (shown below), with a square base and isosceles faces, is suspended in the path of a parallel beam of light, such that its shadow is cast on a screen oriented perpendicular to the direction of the light beam. The pyramid can be reoriented in any direction within the light beam. Under these conditions, which one of the shadows *P*, *Q*, *R* and *S* is NOT possible?

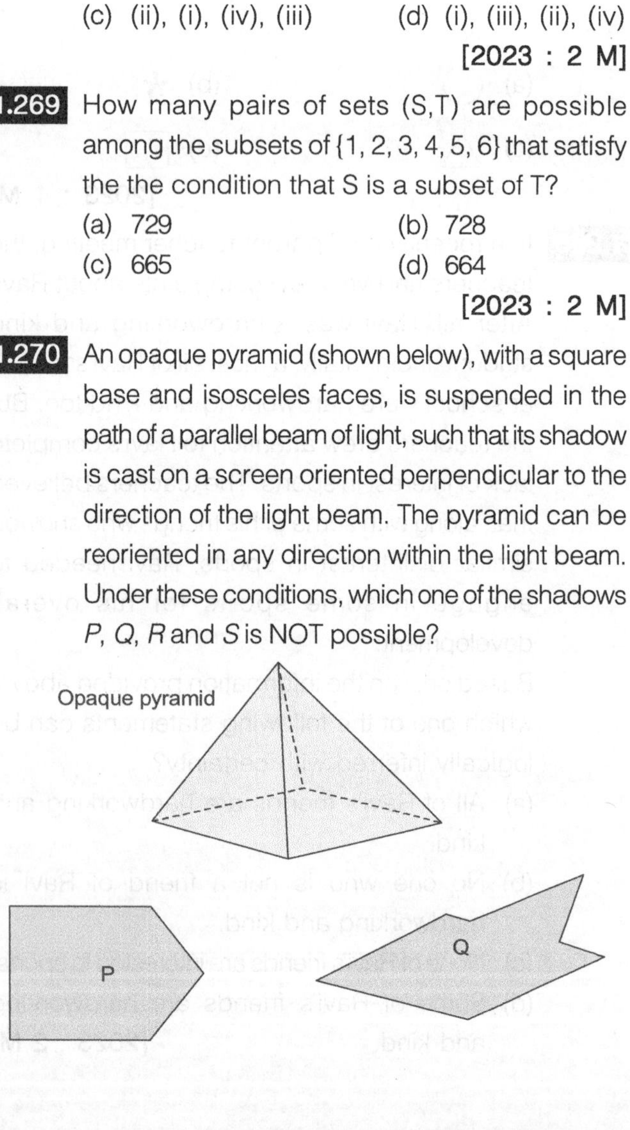

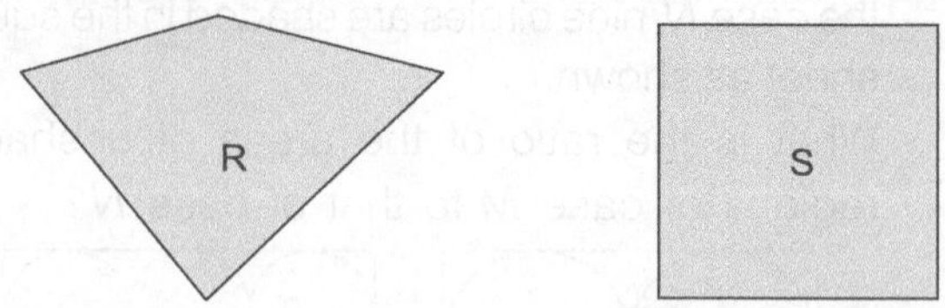

(a) P (b) Q

(c) R (d) S

[2023 : 2 M]

1.271 The real variable x, y, z, and the real constants p, q, r satisfy

$$\frac{x}{pq - r^2} = \frac{y}{qr - p^2} = \frac{z}{rp - q^2}$$

Given the denominator are non-zero, the value of $px + qy + rz$.

(a) pqr (b) 1

(c) $p^2 + q^2 + r^2$ (d) 0 **[2024 : 1 M]**

1.272 Four equilateral triangles are used to form a regular closed three-dimensional object by joining along the edges. The angle between any two faces is

(a) 45° (b) 30°

(c) 60° (d) 90°

[2024 : 1 M]

1.273 In the given text, the blanks are numbered (i)-(iv). Select the best match for all the blanks.

Prof. P **__(i)___** merely a man who narrated funny stories **__(ii)___** in his blackest moments he was capable of self-depreciating humor.

Prof. Q **__(iii)___** a man who hardly narrated funny stories **__(iv)___** in his blackest moments was he able to find humor.

(a) (i) wasn't (ii) Even, (iii) was, (iv) Only

(b) (i) wasn't (ii) Only, (iii) was, (iv) Even

(c) (i) was (ii) Only, (iii) wasn't, (iv) Even

(d) (i) was (ii) Even, (iii) wasn't, (iv) Only

[2024 : 1 M]

1.274 How many combinations of non-null sets *A*, *B*, *C* are possible from the subsets of {2, 3, 5} satisfying the condition: (i) A is a subset of B, and (ii) B is a subset of C?

(a) 18 (b) 27

(c) 28 (d) 19 **[2024 : 1 M]**

1.275 The bar chart gives the batting averages of VK and RS for 11 calender years from 2012 to 2022. Considering that 2015 and 2019 are world cup years, which one of the following options is true?

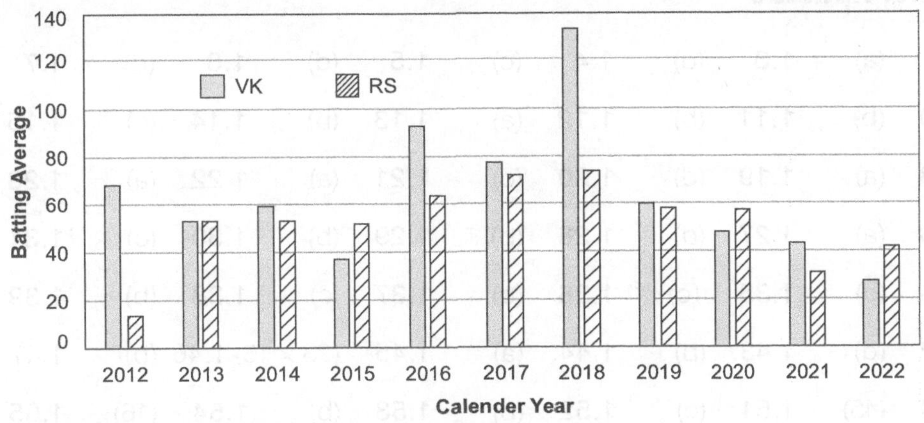

(a) VK has a higher yearly batting average than that of RS in every world cup year.
(b) VK's yearly batting average is consistently higher than that of RS between the two world cup years.
(c) RS's yearly batting average is consistently higher than that of VK in the last three years.
(d) RS has a higher yearly batting average than that of VK in every world cup year.

[2024 : 1 M]

1.276 Take two long dice (rectangular parallelepiped), each having four rectangular faces labelled as 2, 3, 5, and 7. If thrown, the long dice cannot land on the square faces and has 1/4 probability of landing on any of the four rectangular faces. The lebel on the top face of the dice is the score of the throw.

If thrown together, what is the probability of getting the sum of the two long dice scores greater than 11?

(a) $\frac{3}{16}$ (b) $\frac{1}{16}$

(c) $\frac{1}{8}$ (d) $\frac{3}{8}$

[2024 : 2 M]

1.277 Find the odd one out in the set: {19, 37, 21, 17, 23, 29, 31, 11}

(a) 23 (b) 21
(c) 29 (d) 37

[2024 : 2 M]

1.278 A planar rectangular paper has two V-shaped pieces attached as shown below:

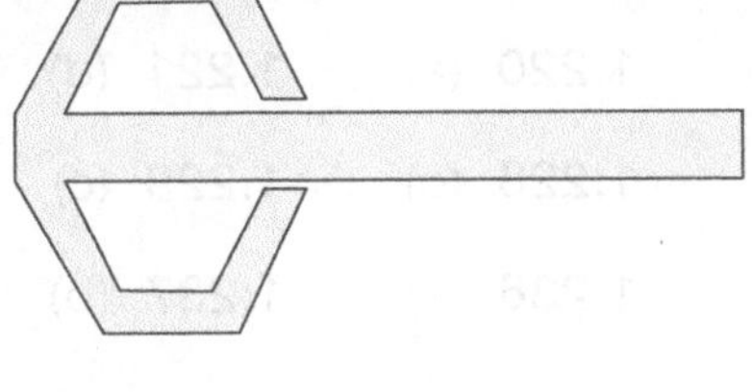

This piece of paper is folded to make the following closed three-dimensions object.

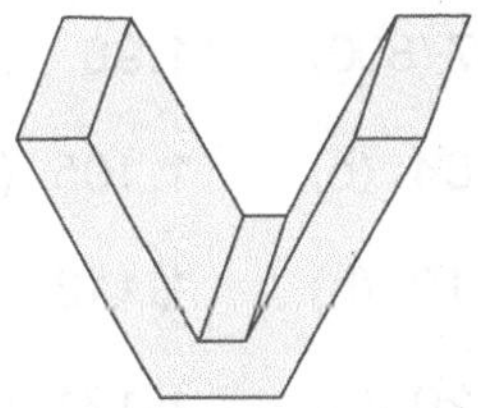

The number of folds required to form the above object is

(a) 8 (b) 7
(c) 9 (d) 11

[2024 : 2 M]

1.279 If '→' denotes increasing order of intensity, then the meaning of the words [smile → giggle → laugh] is analogous to [disapprove → ______ → chide].

Which one of the given options is appropriate to fill the blank?

(a) reprove (b) grieve
(c) reprise (d) praise

[2024 : 2 M]

1.280 In the following series, identify the number that needs to be changed to form the Fibonacci series.

1, 1, 2, 3, 6, 8, 13, 21, ...

(a) 13 (b) 8
(c) 21 (d) 6

[2024 : 2 M]

■■■■

Answers General Aptitude

Q	Ans	Q	Ans	Q	Ans	Q	Ans	Q	Ans	Q	Ans	Q	Ans	Q	Ans
1.1	(b)	1.2	(a)	1.3	(d)	1.4	(c)	1.5	(d)	1.6	(d)	1.7	(c)	1.8	(d)
1.9	(b)	1.10	(b)	1.11	(b)	1.12	(a)	1.13	(b)	1.14	(c)	1.15	(b)	1.16	(d)
1.17	(c)	1.18	(a)	1.19	(d)	1.20	(c)	1.21	(a)	1.22	(a)	1.23	(b)	1.24	(a)
1.25	(a)	1.26	(a)	1.27	(d)	1.28	(b)	1.29	(b)	1.30	(c)	1.31	(a)	1.32	(a)
1.33	(b)	1.34	(c)	1.35	(c)	1.36	(d)	1.37	(c)	1.38	(b)	1.39	(b)	1.40	(d)
1.41	(c)	1.42	(d)	1.43.	(b)	1.44.	(a)	1.45	(23 × 16)	1.46	(b)	1.47	(b)	1.48	(c)
1.49	(b)	1.50	(45)	1.51	(c)	1.52	(b)	1.53	(b)	1.54	(16)	1.55	(d)	1.56	(a)
1.57	(b)	1.58	(b)	1.59	(a)	1.60	(495)	1.61	(d)	1.62	(b)	1.63	(560)	1.64	(d)
1.65	(b)	1.66	(c)	1.67	(d)	1.68	(163)	1.69	(d)	1.70	(a)	1.71	(b)	1.72	(d)
1.73	(4)	1.74	(20,000)	1.75	(0.81)	1.76	(c)	1.77	(b)	1.78	(b)	1.79	(MXHTC)		
1.80	(b)	1.81	(d)	1.82	(c)	1.83	(b)	1.84	(a)	1.85	(b)	1.86	(b)	1.87	(b)
1.88	(a)	1.89	(b)	1.90	(c)	1.91	(c)	1.92	(a)	1.93	(b)	1.94	(b)	1.95	(b)
1.96	(b)	1.97	(800)	1.98	(d)	1.99	(b)	1.100	(b)	1.101	(b)	1.102	(c)		
1.103	(2.064)	1.104	(b)	1.105	(a)	1.106	(c)	1.107	(b)	1.108	(c)	1.109	(280)	1.110	(c)
1.111	(a)	1.112	(b)	1.113	(a)	1.114	(a)	1.115	(c)	1.116	(c)	1.117	(b)	1.118	(d)
1.119	(c)	1.120	(c)	1.121	(b)	1.122	(a)	1.123	(d)	1.124	(c)	1.125	(b)	1.126	(c)
1.127	(c)	1.128	(c)	1.129	(a)	1.130	(c)	1.131	(c)	1.132	(b)	1.133	(c)	1.134	(c)
1.135	(b)	1.136	(d)	1.137	(a)	1.138	(d)	1.139	(c)	1.140	(b)	1.141	(d)	1.142	(d)
1.143	(a)	1.144	(d)	1.145	(b)	1.146	(d)	1.147	(c)	1.148	(b)	1.149	(c)	1.150	(a)
1.151	(d)	1.152	(c)	1.153	(a)	1.154	(b)	1.155	(c)	1.156	(d)	1.157	(c)	1.158	(b)
1.159	(a)	1.160	(d)	1.161	(b)	1.162	(b)	1.163	(b)	1.164	(c)	1.165	(a)	1.166	(d)
1.167	(c)	1.168	(b)	1.169	(a)	1.170	(d)	1.171.	(d)	1.172	(a)	1.173	(b)	1.174	(a)
1.175	(c)	1.176	(b)	1.177	(∗)	1.178.	(c)	1.179	(c)	1.180	(d)	1.181	(d)	1.182	(d)
1.183	(b)	1.184	(d)	1.185	(c)	1.186	(d)	1.187	(d)	1.188	(d)	1.189	(c)	1.190	(c)
1.191	(b)	1.192	(b)	1.193	(a)	1.194	(c)	1.195	(b)	1.196	(a)	1.197	(d)	1.198	(b)
1.199	(b)	1.200	(c)	1.201	(c)	1.202	(d)	1.203	(c)	1.204	(d)	1.205	(b)	1.206	(b)
1.207	(c)	1.208	(c)	1.209	(a)	2.210	(d)	1.211	(a)	1.212	(b)	1.213	(a)	1.214	(a)
1.215	(a)	1.216	(a)	1.217	(d)	1.218	(d)	1.219	(d)	1.220	(a)	1.221	(d)	1.222	(d)
1.223	(b)	1.224	(c)	1.225	(7)	1.226	(d)	1.227	(c)	1.228	(c)	1.229	(c)	1.230	(b)
1.231	(b)	1.232	(c)	1.233	(a)	1.234	(b)	1.235	(b)	1.236	(a)	1.237	(b)	1.238	(a)

1.239 (c)	1.240 (b)	1.241 (d)	1.242 (a)	1.243 (c)	1.244 (b)	1.245 (a)	1.246 (b)
1.247 (a)	1.248 (b)	1.249 (a)	1.250 (a)	1.251 (b)	1.252 (c)	1.253 (c)	1.254 (a)
1.255 (b)	1.256 (c)	1.257 (c)	1.258 (a)	1.259 (a)	1.260 (b)	1.261 (a)	1.262 (a)
1.263 (a)	1.264 (a)	1.265 (b)	1.266 (d)	1.267 (a)	1.268 (c)	1.269 (a)	1.270 (b)
1.271 (d)	1.272 (*)	1.273 (a)	1.274 (*)	1.275 (b)	1.276 (a)	1.277 (b)	1.278 (c)
1.279 (a)	1.280 (d)						

Explanations General Aptitude

1.1 (b)

Circuitous: Deviating from a straight course ⇒ Indirect

(A) Cyclic: Recurring in cycle

(B) Indirect: Not leading by straight line

(C) Confusing: Lacking clarity

(D) Crooked: For shapes (irregular in shape)

1.2 (a)

Unemployed: Worker ⇒ Here one is opposite to other.

(A) Fallow: Land ⇒ Fallow means undeveloped land.

(B) Unaware: sleeper ⇒ Both are same unaware or asleep.

(C) Wit: Jester ⇒ Wit means ability to make jokes and jester is a joker.

(D) Renovated : House ⇒ Renovate means to make better and house can be renovated.

1.3 (d)

(A) Uphold: cause to remain ⇒ not appropriate

(B) Restrain: keep under control ⇒ not appropriate

(C) Cherish: be fond of ⇒ not related

(D) Conserve: Keep in safety and protect from harm, decay, loss, or destruction ⇒ most appropriate.

1.4 (c)

(A) Masked: Hide under a false appearance ⇒ opposite

(B) Belied: Be in contradiction with ⇒ not appropriate

(C) Betrayed: Reveal unintentionally⇒ most appropriate

(D) Suppressed: To put down by force or authority ⇒ irrelevant

1.5 (d)

Using the set theory formula

$n(A)$: Number of people who play hockey = 15

$n(B)$: Number of people who play football = 17

$n(A \cap B)$: Persons who play both hockey and football = 10

$n(A \cup B)$: Persons who play either hockey or football or both

Using the formula

$$n(A \cup B) = n(A) + n(B) - n(A \cap B)$$

$$n(A \cup B) = 15 + 17 - 10 = 22$$

Thus people who play neither hockey nor football = 25 − 22 = **3**

Alternative Method

Refer to Venn diagram given below:

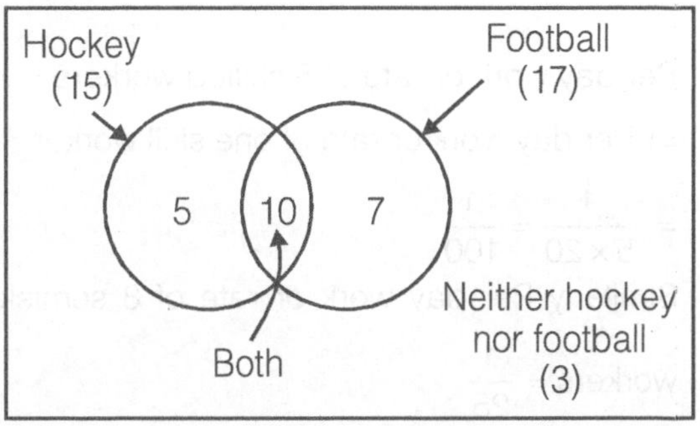

1.6 (d)

(A) Modern warfare has resulted in civil strife: There is no direct consequence of warfare given, so it is not appropriate.

(B) Chemical agents are useful in modern warfare: Passage does not say whether chemical agents are useful or not, so not appropriate.

(C) Use of chemical agents in warfare would be undesirable: Given that people in military think these are useful, undesirable is wrong

(D) People in military establishments like to use chemical agents in war; Correct choice as last statement tells that military people think that chemical agents are useful tools for their cause (work silently in warfare).

1.7 (c)

137 + 276 = 435

This is a octahedral addition thus

731 + 672(8) = **1623**

Alternative Method

7 and 6 added is becoming five means the given two numbers are added on base 8.

$$\begin{array}{r} (137)_8 \\ +(276)_8 \\ \hline (435)_8 \end{array}$$

Hence we have to add the another two given set of numbers also on base 8.

$$\begin{array}{r} (731)_8 \\ +(672)_8 \\ \hline (1623)_8 \end{array}$$

Hence the overall problem was based on identifying base which was 8 and adding number on base 8.

1.8 (d)

Per day work or rate of 5 skilled workers = $\frac{1}{20}$

⇒ Per day work or rate of one skill worker

$$= \frac{1}{5 \times 20} = \frac{1}{100}$$

Similarly Per day work or rate of 8 semiskilled workers = $\frac{1}{25}$

⇒ Per day work or rate of one semi-skill worker

$$= \frac{1}{8 \times 25} = \frac{1}{200}$$

And per day work or rate of 10 unskilled workers

$$= \frac{1}{30}$$

⇒ Per day work or rate of one semi-skill worker

$$= \frac{1}{10 \times 30} = \frac{1}{300}$$

Thus total per day work of 2 skilled, 6 semiskilled and 5 unskilled workers

$$= \frac{2}{100} + \frac{6}{200} + \frac{5}{300} = \frac{12+18+10}{600} = \frac{40}{600} = \frac{1}{15}$$

Thus time to complete the work is **15 days.**

Alternative Method

Let one day work of skilled semiskilled and unskilled worker be *a*, *b*, *c* units respectively.

$$5a \times 20 = 8b + 25 = 10c \times 30 = \text{Total unit of work}$$

$$100a = 200b = 300c$$

$$a = 2b = 3c$$

$$\Rightarrow \quad b = \frac{a}{2} \text{ and } c = \frac{a}{3}$$

Given that 2 skilled, 6 semiskilled and 5 unskilled workers are working. Let they finish the work in x days.

$$(2a + 6b + 5c)x = 5a + 20$$
$$= \text{Total units of work}$$

$$\left(2a + 3a + \frac{5}{3}a\right)x = 5a \times 20$$

$$\frac{20a}{3}x = 5a \times 20$$

$$x = \textbf{15 days}$$

1.9 (b)

We have to make 4 digit numbers, so the number should be start with 3 or 4, two cases possible;

Case (1) thousands digit is 3

Now other three digits may be any of 2, 2, 3, 3, 4, 4, 4, 4.

(a) Using 2, 2, 3
⇒223, 232, 322 -------

$\left(\frac{3!}{2!} = 3 \text{ numbers are possible}\right)$

(b) Using 2, 2, 4 ⇒ 224, 242, 422

$\left(\frac{3!}{2!} = 3 \text{ numbers are possible}\right)$

(c) Using 2, 3, 3 ⇒ 233, 323, 332

$\left(\frac{3!}{2!} = 3 \text{ numbers are possible}\right)$

(d) Using 2, 3, 4 ⇒ 234, 243, 324, 342, 423, 432

(3! = 6 numbers are possible)

(e) Using 2, 4, 4 ⇒ 244, 424, 442

$\left(\frac{3!}{2!} = 3 \text{ numbers are possible}\right)$

(f) Using 3, 3, 4 ⇒ 334, 343, 433

$\left(\frac{3!}{2!} = 3 \text{ numbers are possible}\right)$

(g) Using 3, 4, 4 ⇒ 344, 434, 443

$\left(\frac{3!}{2!} = 3 \text{ numbers are possible}\right)$

(h) Using 4, 4, 4 ⇒ 444

$\left(\frac{3!}{3!} = 1 \text{ numbers are possible}\right)$

Total 4 digit numbers in case 1 = 3 + 3 + 3 + 6 + 3 + 3 + 3 + 1 = 25

Case (2) thousands digit is 4 ; Now other three digits may be any of 2, 2, 3, 3, 3, 4, 4, 4.

(a) Using 2, 2, 3 ⇒ 223, 232, 322

$\left(\frac{3!}{2!} = 3 \text{ numbers are possible}\right)$

(b) Using 2, 2, 4 ⇒ 224, 242, 422

$\left(\frac{3!}{2!} = 3 \text{ numbers are possible}\right)$

(c) Using 2, 3, 3 ⇒ 233, 323, 332

$\left(\frac{3!}{2!} = 3 \text{ numbers are possible}\right)$

(d) Using 2, 3, 4 ⇒ 234, 243, 324, 342, 423, 432

$\left(\frac{3!}{2!} = 3 \text{ numbers are possible}\right)$

(e) Using 2, 4, 4 ⇒ 244, 424, 442

$\left(\frac{3!}{2!} = 3 \text{ numbers are possible}\right)$

(f) Using 3, 3, 3 ⇒ 333

$\left(\frac{3!}{3!} = 1 \text{ numbers are possible}\right)$

(g) Using 3, 3, 4 ⇒ 334, 343, 433

$\left(\frac{3!}{2!} = 3 \text{ numbers are possible}\right)$

(h) Using 3, 4, 4 ⇒ 344, 434, 443

$\left(\frac{3!}{2!} = 3 \text{ numbers are possible}\right)$

(i) Using 4, 4, 4 ⇒ 444

$\left(\frac{3!}{3!} = 1 \text{ numbers are possible}\right)$

Total 4 digit numbers in case 2 = 3 + 3 + 3 + 6 + 3 + 3
= 1 + 3 + 1 = 26

Thus total 4 digits numbers using case (1) and case (2)
= 25 + 26 = **51**

*** Alternative Method / Shortcut method**

As the number is greater than 3000.

So thousand's place can be there 3 or 4.

Let's consider the following two cases

Case (I) When thousand's place is 3.

3̲ a̲ b̲ c̲

If there is no restriction on number of two's, three's and four's. Then each of a, b, c can be filled with 2 or 3 or 4 each in 3 ways.

So 3 × 3 × 3 = 27 numbers are there. Out of which 3222, 3333 are invalid as 2 can be used twice & three thrice only so number of such valid numbers beginning with 3 are 27 – 2 = 25.

Case (II) When thousand's place is 4

4̲ a̲ b̲ c̲

Without restriction on number of 2's, 3's and 4's *a*, *b*, *c* (as explained in case I) can be filled in 27 ways.

Out of these 27 numbers, 4 2 2 2 is only invalid as two have to be used twice only.

So valid numbers are 27 – 1 = 26.

Total numbers from Case (I) & Case (II) 25 + 26 = **51**.

1.10 (b)

Suppose:Hari's age : H, Gita's age : G,
Saira's age : S, Irfan's age : I

- H + G > I + S
- Using Statement (2) both G – S = 1 or S – G = 1; G can't be oldest and S can't be youngest.
- There are no twins thus using statement (2) either GS or SG possible.

(A) HSIG: not possible as there is I between *S* and *G* which is not possible using statement (3)

(B) SGHI: SG order is possible, S > G > H > I and G + H > S + I (possible)
Because if { S = G + 1; and G = H + 1 and H = I + 2 then G + (I + 2) > (G + 1) + I}

(C) IGSH: according to this I > G and S > H thus adding these both inequalities we get I + S > G + H which is opposite of statement (2) thus not possible.

(D) IHSG: according to this I > H and S > G thus adding both inequalities I + S > H + G which is opposite of statement (2) thus not possible.

1.13 (b)

$$\log(P) = \frac{1}{2}\log(Q) = \frac{1}{3}\log(R)$$

$\Rightarrow \log(P) = \log(Q)^{1/2}$

$= \log(R)^{1/3} = K$

$\Rightarrow \quad P = (Q)^{1/2} = (R)^{1/3} = K$

$\Rightarrow \quad P = K,\ Q = K^2,\ R = K^3$ ...(i)

Now, only option (*B*) $Q^2 = PR$ satisfies

$Q^2 = (K^2)^2 = K^4$ From ...(i)

$P.R = K * K^3 = K^4$ From...(i)

Here, $Q^2 = PR$ holds true

1.16 (d)

Shortcut Method

Every time if we take 1 litre of mixture out and replace with water, content of pure spirit will keep on reducing by 10%.

So, final quantity of spirit after 3 such operations are

$10 \times 0.9 \times 0.9 \times 0.9 =$ **7.29 litres**

Alternate Solution

$$\frac{\text{Quantity of spirit left after } n^{th} \text{ operation}}{\text{Intial quantity of spirit}}$$

$$= \left(\frac{a-b}{a}\right)^n = \left(1-\frac{b}{a}\right)^n$$

where *a* is initial quantity of pure spirit and *b* is quantity taken out and replaced every time.

Hence, quantity of spirit left after 3rd operation

$$= \text{initial quantity} \times \left(1-\frac{1}{10}\right)^3$$

$= 10 \times 0.9 \times 0.9 \times 0.9$

$=$ **7.29 litres**

1.18 (a)

(T.C.) Total cost $= V + f$

$$\boxed{T.C = 4q + \frac{100}{q}}$$

As we have to minimize total cost.

Using options

$$(a)\ q = 5,\ T.C. = 4 \times 5 + \frac{100}{5} = 40$$

$$(b)\ q = 4,\ T.C. = 4 \times 4 + \frac{100}{4} = 41$$

$$(c)\ q = 7,\ T.C. = 4 \times 7 + \frac{100}{7} = 42.285$$

$$(d)\ q = 6,\ T.C. = 4 \times 6 + \frac{100}{6} = 40.66$$

Hence, T.C. is minimum at q = **5, (a) ans.**

∗ always put options to get answer fast.

Alternate Solution

$$\text{or} \quad T.C. = 4q + \frac{100}{q}$$

$$\text{for minimum} \quad \frac{d}{dq}\left(4q + \frac{100}{q}\right) = 0$$

$\Rightarrow \quad q^2 = 25$ or $q = 5$

$$\frac{d}{dq}\left(4q + \frac{100}{q}\right)_{q=5} > 0$$

Hence, TC (Total cost) is minimum at $q = 5$ **(a) ans.**

1.19 (d)

As we can understand that the danger of a microbe to human being will be directly proportional to potency and growth. At the same time, it will be inversely proportional to toxicity defined (more dangerous will a microbe be if lesser of its milligram is required). So level of dangerous $\alpha \frac{P\uparrow \times G\uparrow}{T\downarrow}$ where *P*, *G* and *T* are the potency, growth and toxicity as defined in question.

$$\text{So } D_i = \frac{KPG}{T} \quad \text{...(i)}$$

Where *K* is constant of proportionality.

So level of dangerous of *S* will be maximum given

$$\text{by } D_S = \frac{0.8 \times \pi(10\text{mm})^2}{200}.$$

Similar calculation for D_P, D_Q, D_R can be done based on (i) to find out that D_S is maximum and so most dangerous among them.

1.20 (c)

Let *y* we the backlog with transporter and *x* be the number of orders each day. So, as per conditions given in question

$4x + y = 28$ ...(i)

$10x + y = 30$ ...(ii)

Solving (i) and (ii),

$$x = \frac{1}{3} \text{ and } y = \frac{80}{3}$$

Since, we need to find out number of trucks so that no pending order will be there at the end of

5th day.

$5x + y = n \times 5$

So we need to find n, where n is the number of trucks required.

$$n = \frac{5x+y}{5} = \frac{5\times\frac{1}{3}+\frac{80}{3}}{5} = \frac{\frac{85}{3}}{5}$$

Hence, $5.\overline{66}$ truck will be required. As number of trucks have to be natural number. Hence, **6 trucks** will be required.

1.21 (a)

Suresh's dog is the one that was hurt in the stampede.

That is used with restrictive clauses.

1.22 (a)

Total cost $C - 5q^2$

Total Sales Revenue,

$S = 50q$

Profit $(P) = (S - C) = 50q - 5q^2$

As the profit have to be maximized

$P = 50q - 5q^2$

Using options

(a) $q = 5$

$P = 50 \times 5 - 5 \times 5^2 = 125$

(b) $q = 10$

$P = 50 \times 10 - 5 \times 10^2 = 0$

(c) $q = 15$

$P = 50 \times 15 - 5 \times 15^2 = $ –ive

(d) $q = 25$

$P = 50 \times 25 - 5 \times 25^2 = $ –ive

Hence the maximum profit will happen at $\mathbf{q = 5}$

Alternative solution:

$P = 50q - 5q^2$

For max/min $\frac{dp}{dq} = 0,$

$\Rightarrow 50 - 10q = 0$

$q = 10$

$\left.\frac{d^2p}{dq^2}\right]_{q=5} = -10$ which is negative

Hence maximum profit will happen at

$q = 5$

1.23 (b)

Setbacks

Despite several setbacks the mission succeeded in its attempt to resolve the conflict.

The word 'Despite' indicates that there has to be a contrast in the sentence, use of the word 'Setbacks' in the blank indicates that despite many problems the mission was successful.

1.24 (a)

Diminish

Mitigate means to reduce, to lessen etc. So only the word Diminish is close. Rest all choices have no link with the given word.

Divulge means to disclose or reveal which has no link with the given word.

1.26 (a)

AD, CG, FK, JP are given terms. The first letter in each term *A, C, F, J* are respectively 1, 3, 6, 10 letter of alphabetical sequence.

$A = 1$; 1st letter of alphabetical sequence

$C = 1 + 2 = 3$; 3rd letter of alphabetical sequence

$F = 1 + 2 + 3 = 6$; 6th letter of alphabetical sequence

$G = 1 + 2 + 3 + 4 = 10$; 10th letter of alphabetical sequence

The next term should begin with

$1 + 2 + 3 + 4 + 5 = 15$th letter of alphabetical sequence. So next term should begin with *O*.

Also in the series *AD, CG, FK, JP*, two letters in each term are separated by 2, 3, 4, 5 letters in between them respectively. Hence next term should be *OV* as there should be a gap of 6 alphabets between *O* and *V*.

Hence answer is ***OV***.

1.27 (d)

Not gender-discriminatory.

Choice (A) cannot be considered since there is no gender discrimination mentioned in the argument.

Choice (B) Xenophobic is one who has fear of foreigners, no link with the given argument.

Choice (C) It is wrong to say that the profile has not been designed to make the post attractive, since there are certain features which have been added to make the profile lucrative (which are given towards the end of the advertisement, like Transportation paid, expenses reimbursed).

1.28 (b)

$$y = 2x - 0.1x^2$$

For y (height) to be maximum $\frac{dy}{dx} = 0$

$$\frac{dy}{dx} = 2x - 2x = 0$$

$$\Rightarrow \quad x = 10$$

$$\left.\frac{d^2y}{dx^2}\right]_{x=10} = -0.2 \text{ (–ive)}, \quad \frac{d^2y}{dx^2} < 0$$

Hence maximum height will be at $x = 10$

$$y = 2 \times 10 - 0.1 \times (10)^2 = 10$$

Maximum height = **10 meters.**

1.29 (b)

Let total 100 shock absorbs are supplied

so X supplies = 60 (40% of 100)

Y supplies = 40 (60% of 100)

Reliable supply by X = 96% of 60 = 57.6

Reliable supply by Y = 72% of 40 = 28.8

So the required probability that random chosen reliable shock absorber is made by Y is

Required probability

$$= \frac{28.8}{57.6 + 28.8} = 0.333 \approx \mathbf{0.334}$$

1.30 (c)

Adding each and every observation by constant increases the arithmetic mean by same constant. Also multiplying each and every observation by constant multiplies the arithmetic mean by same constant. While the standard deviation will not show similar effect.

Alternative Solution:

$$\frac{x_1 + x_2 + x_3 \cdots x_N}{N} = \bar{x}$$

$$\frac{(x_1 + 7) + (x_2 + 7) + \cdots + (x_n + 7)}{N} = \bar{x} + 7$$

and $\frac{2(x_1 + x_2 \cdots x_N)}{N} = 2\bar{x}$

Hence only P, R are correct.

1.31 (a)

Specificity

Universalism and Particularism are opposite words. Similarly, we require a word that indicates an opposite to diffuseness.

Diffuseness refers to spreading or causing to spread in all directions. Specificity is an appropriate opposite.

1.32 (a)

Would fly

This is a sentence based on Conditional Sentences. In the first clause imagination (Were) has been used, so the main clause should comprise of 'would + I form of verb' usage.

1.33 (b)

Lowest

Nadir which is opposite to the word Zenith (the highest point) means the lowest point.

1.34 (c)

She is an European.

The usage of articles a and an is not decided by the first alphabet of the word but the sound / pronunciation of the word. European though starts with *E* which is a vowel but it does not yield a vowel sound. When read aloud, it leaves the sound of *Y* which is a consonant sound.

So the correct sentence should be, "She is a European".

1.35 (c)

For sum to be maximum, we will consider only positive terms

44 + 42 + ...2

or 2 + 4 + 6 .... 42 + 44

2[1 + 2 + 3 .... 22]

$$= \frac{2[22 \times 23]}{2} = \mathbf{506}$$

Alternatively:

Sum of even natural numbers

$2 + 4 + 6 \ldots 44 = n(n + 1)$

$22 \times 23 = \mathbf{506}$

1.36 (d)

There are Ninety (90) two digited numbers between 1 and 100 (typically 10 to 99). Among these numbers which are divisible by 7 are (14, 21, ...98) only 13 numbers.

$$\left[\because \left(\frac{100}{7}\right) - 1 = 13\right]$$

So, the required probability that the number is not divisible by

$$1-\frac{13}{90}=\frac{77}{90}$$

1.37 (c)

$$\text{Average speed} = \frac{\text{Total distance}}{\text{Total time}}$$

$$= \frac{D}{\frac{D/2}{60}+\frac{D/4}{30}+\frac{D/4}{10}} = 24 \text{ kmph}$$

1.38 (b)

The question can be written as..

$$\frac{1}{\sqrt{2}+\sqrt{1}}+\frac{1}{\sqrt{3}+\sqrt{2}}+\frac{1}{\sqrt{4}+\sqrt{3}}\cdots\frac{1}{\sqrt{8}+\sqrt{80}}$$

Rationalize each term by multiplying it with its conjugate in both numerator and denominator.

$$\left[\frac{1}{\sqrt{2}+\sqrt{1}}\times\frac{\sqrt{2}-\sqrt{1}}{\sqrt{2}-\sqrt{1}}\right]\times\left[\frac{1}{\sqrt{3}+\sqrt{2}}\times\frac{\sqrt{3}-\sqrt{2}}{\sqrt{3}-\sqrt{2}}\right]\times\ldots\ldots$$

$$\ldots\ldots\left[\frac{1}{\sqrt{8}\times\sqrt{80}}\times\frac{\sqrt{81}-\sqrt{80}}{\sqrt{81}-\sqrt{80}}\right]$$

Which can be resolved as

$$\left[\sqrt{2}-\sqrt{1}\right]+\left[\sqrt{3}-\sqrt{2}\right]\ldots\ldots\ldots\left[\sqrt{81}-\sqrt{80}\right]$$

(as denominator in each term will become 1)

Now cancelling like terms we will be left with

$$\sqrt{81}-\sqrt{1}=(9-1)=8$$

1.39 (b)

$$\text{Now cost} = 13200\left[1+\frac{1}{5}\right]\left[1-\frac{1}{24}\right]$$

$$= 13200\left[\frac{6}{5}\right]\times\left[\frac{23}{24}\right] = ₹\ 15{,}180$$

1.40 (d)

No adversity justifies giving up hope.

Options (b) and (c) can be clearly discarded since honesty and adventure making have no relationship with the given argument.

Option (a) may appear close. But remember, every failure may not result into success as it happens in the case of spider.

1.44. (a)

As batsman *K* have lowest standard deviation, he is the most consistent batsmen of these four.

1.45 Sol.

12, 35, 81, 173, 357, _____

Difference between consecutive terms of the series are : 13, 46, 92, 184, _____.

$35-12 = 23 = 23\times 1$

$81-35 = 46 = 23\times 2$

$173-81 = 92 = 23\times 4$

$357-173 = 184 = 23\times 8$

So the next difference have to be $23\times 8 = 368$, therefore next term have to $357+368$

$= \mathbf{725}$

$= 23\times 16$

$725-357 = 368 = 23\times 16$

1.49 (b)

As the numbers appearing on the faces are not equally likely because their probability of occurrence over a large number of throws are different (as given) so we can term it to be **biased** die.

1.50 Sol.

2, 3, 6, 15, _____, 157.5, 630

$2\times 1.5 = 3$

$3\times 2 = 6$

$6\times 2.5 = 15$

$15\times 3 = 45$

$45\times 3.5 = 157.5$

$157.5\times 4 = 630$

Hence the answer is **45**.

1.54 Sol.

81, 54, 36, 24, ____

Alternate terms are perfect squares 9^2, 6^2, 4^2 and in between terms are products 9×6, 6×4m _____ and so on missing terms is

4^2 = **16**

1.55 (d)

$$P = A < R < M$$

P is surely less than *M* only in option (d) else in any other option it cannot be conveyed surely.

1.59 (a)

$$f(x, y) = (2x)^n \times \left(\frac{y}{2}\right)^m = 2^{n-m}\times P$$

1.60 Sol.

Let the 12 consecutive odd numbers be

$a, a + 2, a + 4, a + 6 \ldots\ldots a + 22$

Sum of first five $(a) + (a + 2) + (a + 4) + (a + 6) + (a + 8) = 425$

$5a + 20 = 425 \Rightarrow a = 81$

Sum of last five $(a + 14) + (a + 16) + (a+18) + (a + 20) + (a + 22)$

$$5a + 90 = 5 \times 81 + 90 = \mathbf{495}$$

1.61 (d)

Difference between 1st and 4th, 2nd and 3rd letters in each series respectively is same except NVBD.

1.62 (b)

A resident female with annual income ₹ 9 Lakh should satisfy eligibility for filling form *T*. So answer is option B is answer. Option A, C , D corresponds to filling form P, S, D respectively.

1.63 Sol.

Let the length of platform is x and velocity of train is v.

Distance = Vel. × time

$(280 + x) = V \times 60$ ... (i)

$280 = V \times 20$ ... (ii)

From equation (i) and (ii), we get

x = **560 m**

1.64 (d)

Deficit should be equal to (1/5)th of exports which is possible in 2006

$$\text{Deficit} = \frac{\text{Imports} - \text{Exports}}{\text{Exports}}$$

$$= \frac{120-100}{100} = \frac{20}{100} = \frac{1}{5}$$

1.65 (b)

As one of the face is already shown as head of the possibility of second coin is not there (because second coin has both tail). Now sample space will contain now events (*H, H, T*). So the probability for another face as tail will be **1/3.**

1.66 (c)

Adding 6, 3 and 2 to the consecutive letters gives the only (c) does not follow.

1.67 (d)

Sum of the read and green lights as equal to sum of yellow and blue lights are red and blue will be numbered same and green and yellow will be numbered same.

So, $R + H = Y + B$

1.68 Sol.

Let eight consecutive odd numbers be

$a, a + 2 \ldots\ldots a + 14$

and four consecutive over numbers be

$b, b + 2, b + 14, b + 16$

Given $(a) + (a + 2), \ldots\ldots + (a + 14) = 656$

$8a + 56 = 656$

$a = 75$

Also given

$(b) + (b + 2) + (b + 4) + (b + 6) = 87 \times 4$

$= 348$

$4b + 12 = 348$

$b = 84$

Sum of smallest odd no (a) and second largest even no $(b + 4)$ will be

$= (a) + (b + 4)$

$= 75 + (84 + 4) = \mathbf{163}$

1.69 (d)

For maximum revenue per kg. Let us solve through options

A. Item 2 = $\dfrac{20\% \text{ of } 250}{20\% \text{ of } 500} = 0.5$

B. Item 3 = $\dfrac{23\% \text{ of } 250}{19\% \text{ of } 500} = 0.6$

C. Item 6 = $\dfrac{19\% \text{ of } 250}{16\% \text{ of } 500} = 0.59$

D. Item 5 = $\dfrac{20\% \text{ of } 250}{12\% \text{ of } 500} = 0.83$

Clearly item **5** is largest.

1.70 (a)

Drainage pipe drains half of the tank in 30 minutes or full tank is 60 minutes. Now the tank have to be filled in 10 minutes (when both drainage and filling pipes are in action)

Drainage pipe would drained 1/6th of capacity of tank in 10 minutes. So filling pipe have to fill 1/2 of capacity of 1/6 of capacity more i.e. 2/3rd of the tank in 10 minutes. or complete tank is 15 minutes. That's why the filling tank should have rate 4 times the draining rate.

Alternatively :

Drainage pipe can drain complete tank in 60 min. Let the filling pipe can fill the tank in 'x' min. Now both are operational and tank is half filled. Also remaining half needs to get filled in 10 minute.

$$10\left[\frac{1}{x}-\frac{1}{60}\right] = \frac{1}{2}$$

$$x = \textbf{15 minutes}$$

So filling pipe needs to have rate **4** times drainage pipe's rate.

1.71 (b)

7 *G*, 11 *K*, 13 *M*, _____

Next term will be *Q* which is 17 letter of English alphabet.

Among option only (b) (i.e. 17 *Q*) is correct.

1.72 (d)

On seeing the given figure in question i, ii and iii can be concluded.

1.73 Sol.

Let the velocity of stream is *V* time taken during downstream is *t* and distance is x.

$$t = \frac{x}{8+V} \quad \text{...(i)}$$

During upstream

$$3t = \frac{x}{8-V} \quad \text{...(ii)}$$

Dividing equation (ii) by (i), we get

$$V = \textbf{4 km/hr}$$

1.74 Sol.

Since,

$$\because \quad 15\% = 4{,}50{,}000$$

$$\therefore \quad 100\% = \frac{4{,}50{,}000\times 100}{15} = ₹\ 30{,}00{,}000$$

Selling price of purifier

$$= \frac{30{,}00{,}000+10{,}00{,}000}{200} = \textbf{₹ 20,000}$$

1.75 Sol.

Probability of bulb being defective $= \frac{5}{100}$

$= 0.05$

Probability of bulb being non defective

$= 1 - 0.05$

$= 0.95$

Probability that the batch is accepted

= None of the four bulb being defective

$= (0.95)^4 = 0.81450625$

$\approx$ **0.81**

1.76 (c)

13 *M*, 17 *Q*, 19 *S*

among options only ***W*** is correct as *W* is **23rd** letter in English language.

1.77 (b)

BEST OF LUCK

B–ST –F L–CK

GOOD WISHES

G—D W–SH–S

ACE THE EXAM

–C– TH – X–M

So answer is **MXHTC** (vowel is missing).

1.78 (b)

The value doubled over 10 years

$$A = P\left(1+\frac{R}{100}\right)^n$$

$$2P = P\left(1+\frac{R}{100}\right)^{10}$$

$$R \approx \textbf{7.2}$$

Alternate :

Simple growth rate for a quantity to double in 10 years will be 10%. For compounded growth answer have to be less than 10. Among options 7.2 (Ans. (b)) is correct as ((a) (5.6)) will be very much less than 10.

((c) (10)), ((d) (12) cannot be answer as compared growth rate have to be less than 10.

1.79 Sol.

In 2012, total cost

$$= \frac{4{,}50{,}000\times 100}{15} = 30{,}00{,}000$$

Cost of raw material in 2012

$$= 30{,}00{,}000\times\frac{20}{100} = 6{,}00{,}000$$

Cost of raw material in 2013

$= 600000 + 600000 \times \frac{30}{100} = 7,80,000$

other expenditure in 2012

$= 3000000 - 600000 = 2400000$

Increased cost of other expenditure in 2013

$= 2400000 + 2400000 \times \frac{20}{100}$

$= 2880000$

Total cost in 2013

$= 2880000 + 780000 = 3660000$

% increase in total cost in 2013

$= \frac{3660000 - 3000000}{3000000} \times 100 =$ **22%**

1.80 (b)

Overall there will be 5! = 120 numbers. Digit I will appear at unity, tens, hundred, thousand and ten thousand place 4! times each. So sum received due to all 1's will be 4! × (1 + 10 + 100 + 1000 + 10000) = 4! × 11111

Similarly sum received due to all 3's

4! × (3 + 30 + 300 + 3000 + 30000) = 4! × 33333

Similarly for digits 3, 4 and 5

So, overall sum received due to all digits

= 4! × 11111 [1 + 3 + 5 + 7 + 9]

= 24 × 11111 [25] = **6666600**

1.81 (d)

Following chart can be drawn from cumulative chart

Stretch	Cumulative distance (km)	Electricity used (kWh)	Electricity consumed per km
M	20	12	12/20 = 0.6
N	45	25	13/25 = 0.52
O	75	45	20/30 = 0.66
P	100	57	12/25 = 0.48

Hence minimum for *P*.

1.85 (b)

Required probability $= \frac{1}{6} \times \left(\frac{7}{8}\right) + \frac{1}{8} \times \left(\frac{5}{6}\right)$

$= \frac{12}{48} = \frac{1}{4}$

1.88 (a)

If $1 < y < x$ so $\log x > \log y$ as well $e^x > e^y$ other two will not always be true.

1.90 (c)

A, CD, GHI ?, UVWXY

Series is as follows: Write first letter A, leave aside nearby one B, write next two CD. Leave aside net two EF. Write next three GHI leave aside next JKL then we have to write **MNOP**.

1.91 (c)

Fish out of water : implies uncomfortable situation.

1.92 (a)

Harbours means to give shelter.

1.93 (b)

Total number of matches $= {}^5C_2 = \frac{5 \times 4}{2} =$ **10.**

1.94 (b)

$T > E, C > T \Rightarrow C > T > E$

So, Eric is actually younger than Cliff so third statement is false.

1.95 (b)

Elusive means difficult to catch.

1.97 Sol.

Tiger is 50 leaps of its own behind deer. Given tiger's one leap is 8 m. So initial separation of Tiger and deer = 50 × 8 = 400 m.

Tiger goes 5 leaps in a minute which is equal to 5 × 8 = 40 m/s

Deer goes 4 leaps in a minute which is equal to 4 × 5 = 20 m/s

Now relative distance of 400 m has to be covered with (40 – 20) = 20 m/min

Required time $= \frac{400\text{m}}{(40-20)\text{ m/min}}$

$\frac{\text{Distance}}{\text{Relative speed}} = 20 \text{ min}$

So, distance travelled by tiger @ 40 m/min × 20 min = **800 m.**

1.98 (d)

Clearly both the conclusion follows.

1.99 (b)

We know $(a + b + c)^2$

$= a^2 + b^2 + c^2 + 2[ab + bc + ca]$

$(a + b + c)^2 - 1 = 2[ab + bc + ca]$

[as $a^2 + b^2 + c^2 = 1$] for $ab + bc + ca$ to be min $(a + b + c) = 0$

min $(ab + bc + ca)$

$= -\frac{1}{2}$ [as putting $a + b + c = 1$]

Only option (b) has min value as $-\frac{1}{2}$.

1.102 (c)

$\log \tan 1° + \log \tan 2° + ... + \log \tan 89°$

$= \log(\tan 1° \times \tan 2° \times \cdots \tan 89°)$

As $[\log m + \log n = \log mn]$

$= \log(\tan 1° \times \tan 2° \cdots \tan(90° - 2°)) \tan(90° - 1°)$

$= \log(\tan 1° \times \tan 2° \times \cdots \tan 45° \times \cot 44 \cdots \cot 2° \cot 1°)$

$= \log(\tan 45°) = \log(1) = 0$

As $\tan(90° - \theta) = \cot\theta$ and

$\tan 1° \times \cot 1° = \tan 2° \times \cot 2° = \cdots 1$

1.103 Sol.

The shaded part of circular sheet is golded to form a cone (Figure below).

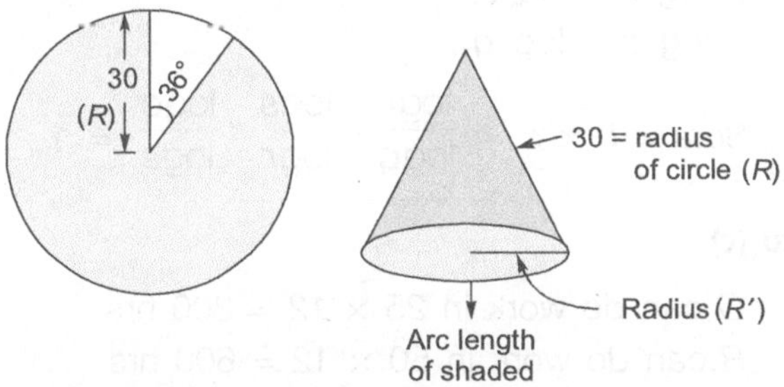

Slant height of conc, $l = 30$

$2\pi R' = \frac{9}{10} 2\pi \times 30 \quad R' = 27$

$h = \sqrt{30^2 - 27^2} = \sqrt{900 - 729}$

$= \sqrt{171}$

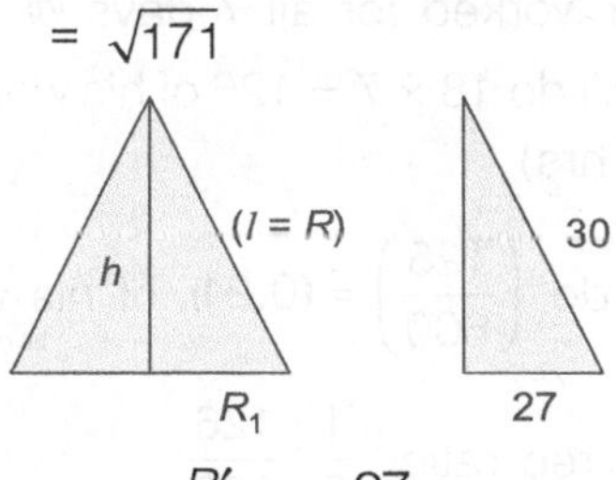

Required ratio $\frac{R'}{h} = \frac{27}{\sqrt{171}} = 2.064$

1.106 (c)

Increase in productivity necessarily reflects greater efforts made, by the employees.

1.107 (b)

Ambiguous question both A and B seem to be correct but question is ambiguous.

1.108 (c)

PR is parallel to x axis

$-4 \le x \le 5 \qquad 6 \le y \le 16$

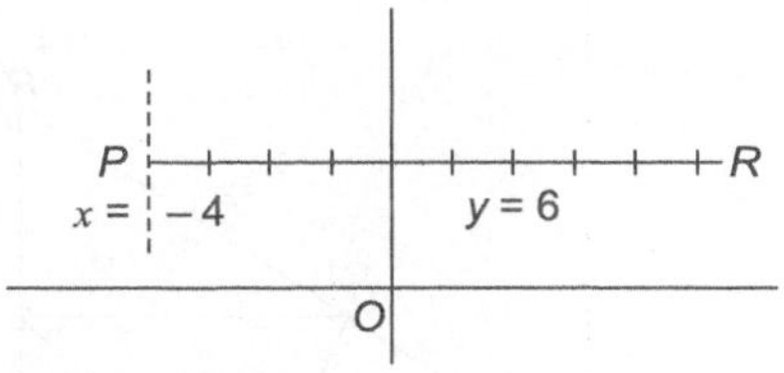

P at $x = -4$

Δ's = with base 1 unit y can take values (7, 8, ..., 16) : 10 values

Δ's = with base 2 unit y varies (7.. 16):10 values

Δ's = with base 9 unit y varies (7.... 16) : 10 values

Similarly P at $x = -3$ we will get 80 and 50 on

So [90 + 80 + ... 10] = 9[1 + 2 .... 10] = 10 $\sum_{n=1}^{9} n$

Similarly changing ordinate of P we will respectively get

[81 + 725 .... 9] Δ's = 9 [1 + ...9] = $9\sum_{n=1}^{9} n$

and further (64 + 56 .... 8) Δ's = $8\sum_{n=1}^{8} n$ and so on

If we sum all of the above

$$10\sum_{n=1}^{9} n + 9\sum_{n=1}^{9} n + 8\sum_{n=1}^{8} n . \Sigma n ....$$

$$9\sum_{n=1}^{9} n[1+2+3.....10] = \sum_{n=1}^{9} n \times \sum_{n=1}^{10} n$$

But there are overall four orientation for rt.

____ as shown 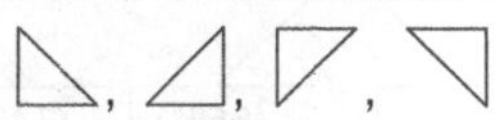, so actual answer will be $4 \times \sum_{n=1}^{9} n \times \sum_{n=1}^{10} n$

$$4 \times \left(\frac{9 \times 10}{2}\right) \times \left(\frac{10 \times 11}{2}\right) = 9900.$$

Alternate Method :

$-4 \le x \le 5,\ 6 \le y \le 16$

$PQ \parallel$ y axis and $PR \parallel x$ axis $\angle QPR$ is 90°

As all co-ordinates of P, Q, R, are integers so, we can choose (x, y) integer co-ordinates for vertex P in 10 × 11 = 110 ways.

Now with respect to fixed y co-ordinate for R we can choose its x-coordinate in only 9 ways as one point is already occupied (two points of a side of

Δ cannot coincide) overall 9 ways of doing it.

Now choosing y-coordinate for Q because x-coordinate of Q is already fixed (it has same x-coordinate as R) it can take 10 different values.

SO, overall $110 \times 9 \times 10 =$ **9900 ways** 9900 such Δ's will be possible.

1.109 Sol.

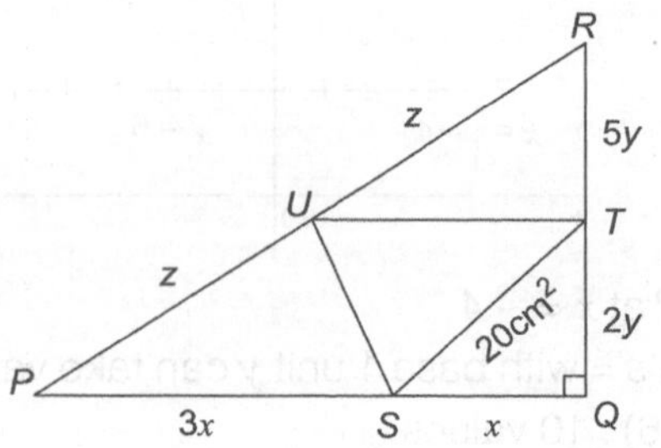

$$\text{Area } \Delta QTS = \frac{1}{2} SQ \times QT$$

$$20 = \frac{1}{2} \times x \times 2y \times 1$$

$$\Rightarrow \quad xy = 20$$

$$\Delta = \frac{1}{2} bh$$

$$\text{Now Area for } \Delta PQR = \frac{1}{2} PQ \times QR \times \sin 90°$$

$$= \frac{1}{2} \times 4x \times 7y = 14\,xy = \mathbf{280\ cm^2}$$

1.114 (a)

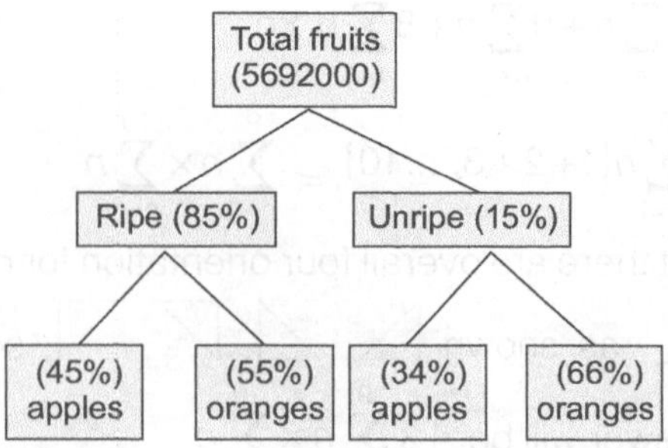

Total number of apples = Ripe apples + Unripe apples

$= (0.85 \times 0.45 + 0.15 \times 0.34) \times 569200$

$=$ **2029198**

1.115 (c)

Following line with respective distances can be drawn

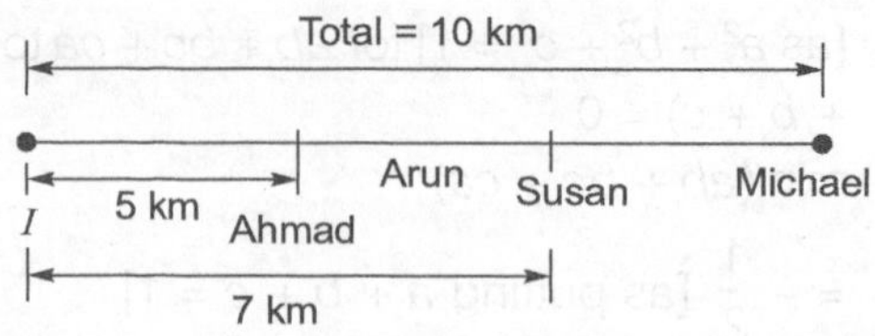

Arun can reside anywhere between Ahmed and Susan *i.e.* between 5 km and 7 km from I. Hence ans (c) **6.02**

$5 < 6.02 < 7$

1.116 (c)

The required probability

$$0.5 \times 0.7 = 0.35 \approx \frac{35}{100} = 35\%$$

1.118 (d)

$L > P$ (Leela is older than Pavithra)

$S > L$ (Shiv is older than Leela)

So Pavithra is youngest

1.119 (c)

$q^a = r,\ r^b = s,\ s^c = q$

$a \log q = \log r,$

$b \log r = \log s,$

$c \log s = \log q$

$$\text{So, } a \times b \times c = \frac{\log r}{\log q} \times \frac{\log s}{\log r} \times \frac{\log q}{\log s} = 1$$

1.120 (c)

Q can do work in $25 \times 12 = 300$ hrs

R can do work in $50 \times 12 = 600$ hrs

So we can say Q is twice efficient as R

Now Q worked only for 5 days @ 12 hrs/day. So he will do 60 units of his work (Total work for Q *i.e.* 300 hrs) he will do only $\frac{1}{5}$ of work $\left(\frac{60}{300} = \frac{1}{5}\right)$

While R worked for all 7 days @ 18 hrs/day

So he will do $18 \times 7 = 126$ of his work (Total work for 600 hrs)

He will do $\left(\frac{126}{600}\right) = (0.21)$ of his work

$$\text{So required ratio } \frac{1}{5} : \frac{126}{600}$$

$$120 : 126$$

$$20 : 21$$

1.125 (b)

Following figure can be drawn:

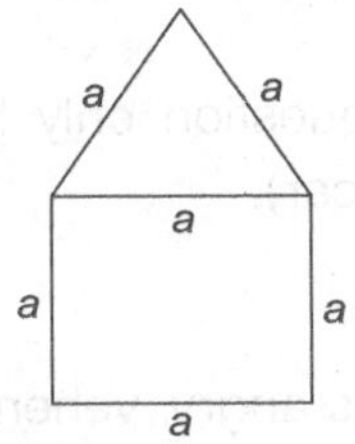

Perimeter of window,

$5a = 6$ m

$a = 1.2$ m

Complete area for window = Area of square + Area of equi Δ

$$= (1.2)^2 + \frac{\sqrt{3}}{4}(1.2)^2 \approx \mathbf{2.06\ m^2}$$

1.126 (c)

I esser standard deviation for group *Q* will mean that the observation sin group *Q* are closer and hence narrower is the range of marks scored in group *Q* as compared to group *P*. Other options (a), (b), (d) remains unconfined.

1.128 (c)

B	F	H	L	N	P	T	V	X	Z
2	2	8	12	14	16	20	22	24	26

4   6   8

So the required alphabets have to start with 20 is *T* and there after common difference in term will be 2 each hence answer is *TVXZ*

1.129 (a)

Use options only **(*a*) → 0**, satisfies

$a \quad b = ab - (a + b)$

$a \quad b = a \times 0 + (a + 0) = \mathbf{a}$

Given $a \quad x = a$ which satisfied for ($x = 0$)

1.135 (b)

M works with twice efficiency as *E* but worked for half as many days. So in this respect they will do equal work if their shifts would have been for same timings. But *M*'s shift is for hrs, while *E*'s shift for 12 hrs. Hence *E* will do twice the work as *M*.

Ratio of contribution of *M* : *E* in work, **1 : 2**

1.136 (d)

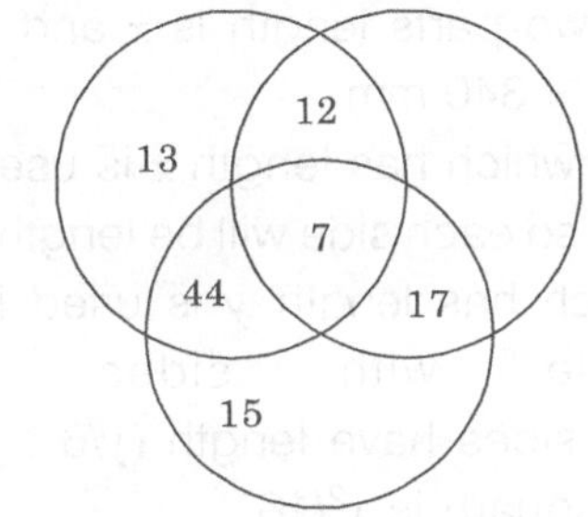

The number of student who like to read books or play sports have been shown

$= 13 + 12 + 44 + 7 + 15 + 1$

$= \mathbf{108}$

1.138 (d)

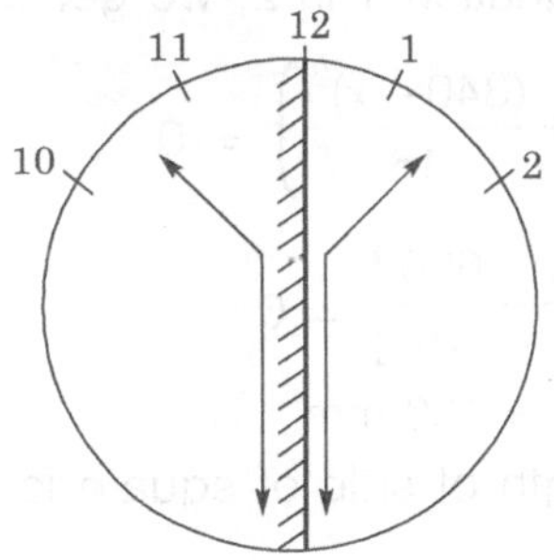

Mirror image of 1 : 30 is 10 : 30

10 : 30 was the time two and quarter hour back so time now will be **12 : 45**

1.139 (c)

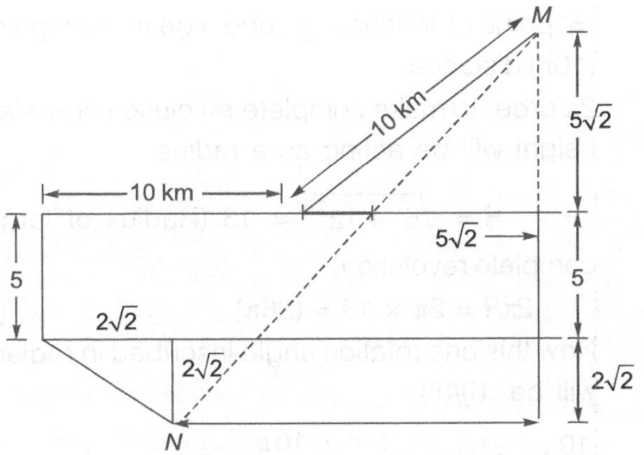

See the adjoining figure for solution

$$MM' = 5\sqrt{2} + 5 + 2\sqrt{2} = 5 + 7\sqrt{2}$$

$$NM' = 10 + 5\sqrt{2} - 2\sqrt{2} = 10 + 3\sqrt{2}$$

$$MN = \sqrt{(MM')^2 + (NM')^2}$$

$$= \sqrt{(5 + 7\sqrt{2}) + (10 + 3\sqrt{2})^2} \approx \mathbf{20.61}$$

1.140 (b)

Let the two parts length is x and y

so $x + y = 340$ mm ...(1)

The part which has length x is used in making a square so each side will be length $x/4$, and the part which has length y is used in making a rectangle with sides in ratio 1 : 1 so sides have length ($y/6$: $y/3$)

Area of square is $x^2/16$

Area of rectangle is $y^2/16$

so combined area is $\left(\frac{x^2}{16}+\frac{y^2}{18}\right)$ ...(2)

since the combined area should be minimum so using equation 1 in 2, we get

$$\frac{d}{dx}\left(\frac{x^2}{16}+\frac{(340-x)^2}{18}\right) = 0$$

$$\text{so}\left[\frac{x}{8}+\frac{x}{9}-\frac{680}{18}\right] = 0$$

or $x = 160$ mm

and length of side of square is

$$\frac{x}{4} = 40 \text{ mm}$$

1.142 (d)

Distance travelled by cone in one rotation

$= 2\pi r = 2\pi\,(5) = 10\pi$ ... (i)

$\Rightarrow$ point Q touches ground again after going (10π) distance.

2π order to make complete revolution now slant height will be acting as a radius

$R = \sqrt{5^2+12^2} = 13$ (Radius of larger complete revolution)

$2\pi R = 2\pi \times 13 = (26\pi)$ ... (ii)

Now this one rotation angle inscribed in radians will be (i)/(ii)

$$\frac{10\pi}{20\pi}\times 2\pi \text{ radians} = \frac{10\pi}{13} \text{ radians}$$

1.143 (a)

P says $\rightarrow$ S has atleast 3 Cars

Q says $\rightarrow$ S has less than 3 Cars

R says $\rightarrow$ S has atleast 1 Car

Solve by options

A (0)		B (1)		C (3)	
	P ×		P ×		P ✓
	Q ✓		Q ✓		Q ×
	R ×		R ✓		R ✓

as given in question only has to be right so answer is **0** (car).

1.144 (d)

Heated means angry, vehement

Which is reflected by speakers becoming agitated.

1.145 (b)

$45\text{p} \rightarrow 20\text{K}$

$25\text{p} \rightarrow 30\text{K}$

$20\text{p} \rightarrow 40\text{K}$

$8\text{p} \rightarrow 60\text{K}$

$2\text{p} \rightarrow 150\text{K}$

Median will be arithmetic mean 50th and 51st term, so both are in 30 K bracket.

So average or arithmetic mean of,

$$\frac{30000+30000}{2} = \textbf{Rs. 30000}$$

1.146 (d)

The following two possibilities exists.

$$\begin{array}{r} 5\ 0\ 1\ 0\ 0 \\ -\ 4\ 8\ 9\ 8\ 9 \\ \hline 0\ 1\ 1\ 1\ 1 \end{array} \quad \text{and} \quad \begin{array}{r} 5\ 0\ 0\ 0\ 0 \\ -\ 4\ 8\ 8\ 8\ 9 \\ \hline 0\ 1\ 1\ 1\ 1 \end{array}$$

Sum = 1 + 9 = 10 Sum = 0 + 8 = 8

1.147 (c)

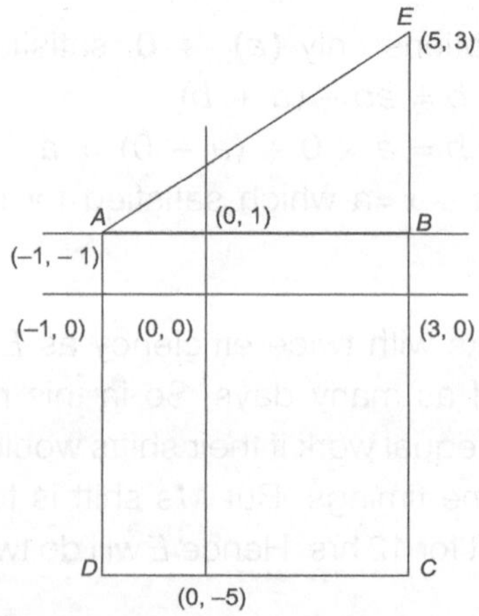

$|x-1| \le 2$ $\quad |y+2| \le 3$ $\quad x - y \ge -2$

$x \le 3$ $\quad y \le 1$

$x \ge -1$ $\quad -5 \le y$

Total area = Area of rectangle ABCD + Area of ΔABE

$$= 4\times6+\frac{1}{2}(4\times4)=32$$

1.148 (b)

Steve was known to be better player than Mark.

1.150 (a)

We can see graph only (i) follows.

1.152 (c)

Subject – The ways
Verb – are

1.153 (a)

Forgo –Means to go without something/refrain from

Exclusive– Excluding or not admitting other things

1.154 (b)

P $\rightarrow$ Married
R $\rightarrow$ Unmarried
Q $\rightarrow$ Status Not given
P $\rightarrow$ Looks @ Q
Q $\rightarrow$ Looks @ R

Now two possibility arise,

1. Let Q be married

$P^{married} \rightarrow Q^{married}$

$Q^{married} \rightarrow R^{unmarried}$

only one pair in which married person is looking @ unmarried.

2. Let Q be unmarried

$P^{married} \rightarrow Q^{unmarried}$

$Q^{unmarried} \rightarrow R^{unmarried}$

Then also only **one pair** exists.

So, only one pair exists in which married person is looking @ unmarried.

1.159 (a)

$x \rightarrow$ Bullocks
y $\rightarrow$ Tractors

$$(x + y)8 = \left(\frac{x}{2}+2y\right)5$$

$$8x + 8y = \frac{5x}{2}+10y$$

$$\Rightarrow\ 5.5x = 2y \text{ or } y = 2.75x$$

Now, x bullocks are alone working or 'd' days.

$$x \times d = (x + y)8$$
$$x \times d = (x + 2.75x)8$$
$$x \times d = 3.75x \times 8$$
$$d = \mathbf{30\ days}$$

1.160 (d)

Cleave – means to divide/split

1.161 (b)

Rectangle initial length is l and breath is b. If l is reduced by 10 m and breath reduced by 5 m then it becomes square.

So, 1st condition,

$$l - 10 = b - 5$$
$$l - b = 5$$

Given, initial, $l \times b = A$ ...(i)

2nd condition,

$$(l - 10)(b - 5) = A - 650$$
$$lb - 10b - 5l + 50 = A - 650$$
$$A - 10b - 5l = A - 700$$
$$10b + 5l = 700$$
$$10b + 5(b + 5) = 700$$
$$15b + 25 = 700$$
$$15b = 700 - 25$$
$$15b = 675 \Rightarrow b = 45 \text{ m}$$
$$l = 45 + 5 = 50 \text{ m}$$

Area of original rectangle = 45 × 50 = **2250 m²**

1.162 (b)

Let two digit no is xy

$$x + y = 9 \quad \text{...(i)}$$
$$10x + y - 45 = 10y + x$$
or
$$x - 5 = y$$
$$x - y = 5 \quad \text{...(ii)}$$

Adding (i) and (ii)

$$x = 7$$

Subtracting (i) and (ii)

$$y = 2$$

Therefore the number is **72**.

1.163 (b)

7 machines $\rightarrow$ 7 toys $\rightarrow$ 7 minutes

1 machine $\rightarrow$ 1 toy $\rightarrow$ 7 minutes

Because one machine takes 7 minute for making 1 toy.

So, 100 machines will take **7 minute** for making 100 toys.

1.164 (c)

Frugality is synonyms to miserliness
Frugality carries a positive connotation. It refers to the quality of being economical with money or food.
Miserliness is a negative word. It means excessive desire to save money or extreme meanness.

1.165 (a)

Principle - A moral rule / belief
Principal - The person in-charge of an educational institution.

1.166 (d)

Given, a and b are integer

$a + a^2b^3$ is odd

$a(1 + ab^3)$ is odd

Multiplication of odd and odd number is odd.

So, a is odd and $1 + ab^3$ is odd.

$1 + ab^3$ is odd, so ab^3 will be even.

Because a is odd so for ab^3 to be even b must be even.

So, a is odd and b is even.

1.167 (c)

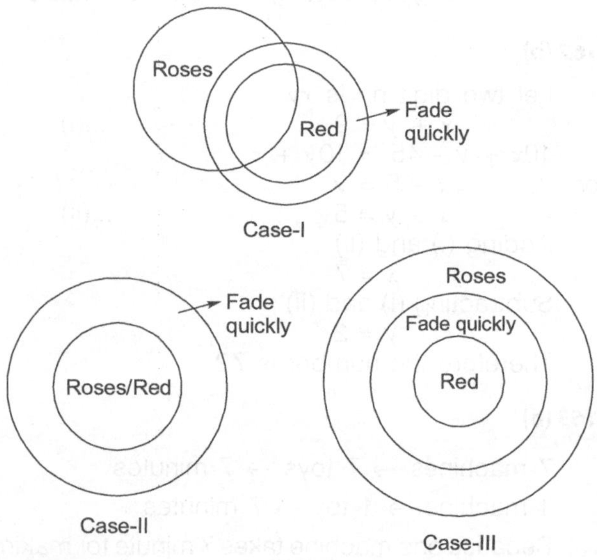

1.168 (b)

x	0	±1	±2
y	0	−1	0

Solving through options,

(a) $y = ||x|+1|-2$

Putting, $x = 1$

$y = 2 - 2 = 0$ Not satisfied

(b) $y = ||x|-1|-1$

Putting, $x = 1$

$y = |1 - 1| - 1$

$y = -1$ Satisfied

(c) $y = ||x|+1|-1$

Putting, $x = 1$

$y = |1 + 1| - 1$

$y = +1$ Not satisfied

(d) $y = ||x-1|-1|$

Putting, $x = 1$

$y = ||1 - 1| - 1|$

$y = |0 - 1|$

$y = 1$ Not satisfied

1.169 (a)

$\log|a| + \log|b| + \log|c| = 0$
It is possible only,
if $|a|, |b|$ and $|c|$ all are equal to 1.
So, a, b, c may be ±1, ±1, ±1 respectively.]
For minimum value all three will be − 1.
So, minimum value = **−3**
For maximum value all three will be + 1,
So, maximum value = **+3.**

1.170 (d)

Train speed = 54 km/h
Time = 25 sec for travelling length of train and length of platform
Man speed = 9 km/h
Relative speed of train with respect to man = 45 km/h
Time to cross the man = 14 sec
So, length of train = time × speed

$$= 14 \times 45 \times \frac{5}{18}$$

Length of train = 35 × 5 m = **175 m**
Length of platform + length of train = speed × time

$$= 54 \times \frac{5}{18} \times 25 = 15 \times 25 = 375\text{ m}$$

Length of platform = 375 − 175 = **200 m**

1.171 (d)

Complemented - Complete

Complimented - Remark of admiration

Complement - To contribute extra features to something so as to improve or emphasize their qualities.

Compliment - Politely congratulate or praise someone/something

1.172 (a)

Undiminished - Unambitious / unskilled.

Undiminished means not diminished, reduced or lessened. It is integrated, thorough and complete in nature.

1.173 (b)

BC, FGH, LMNO, TUVWX

1.174 (a)

$$P_C = P_S = P_T$$
$$\pi D = 4a = 3s$$

now, Area of circle $= \frac{\pi}{4}D^2$

Area of square $= a^2$ $[\pi D = 4a]$

$$= \frac{\pi^2 D^2}{16} = \frac{\pi^2}{16}D^2$$

Area of triangle $= \frac{\sqrt{3}}{4}s^2 = \frac{\sqrt{3}}{4}\left(\frac{\pi D}{3}\right)^2$ $\{\because \pi D = 3s\}$

$$= \frac{(\sqrt{3})\pi^2 D^2}{36}$$

$$\because \quad \frac{\pi}{4} > \frac{\pi^2}{16} > \frac{\sqrt{3}\pi^2}{36}$$

So, circle has the largest area.

1.175 (c)

$$\frac{1}{1+\log_u vw} + \frac{1}{1+\log_v wu} + \frac{1}{1+\log_w uv}$$

$$= \frac{1}{\log_u u + \log_u vw} + \frac{1}{\log_v v + \log_v wu} + \frac{1}{\log_w w + \log_w uv}$$

$$= \frac{1}{\log_u uvw} + \frac{1}{\log_v uvw} + \frac{1}{\log_w uvw}$$

$$= \log_{uvw} u + \log_{uvw} v + \log_{uvw} w = \log_{uvw} uvw = 1$$

1.176 (b)

For remaining last two tosses possible cases are,

H H

H T

T H

T T

Out of 4 possible cases one H and T will have the highest probability of occurrence.

1.177 (*)

(This question is to be challenged)

$$125 \times 7 \times 39 = \frac{5}{7}W$$

$W = 47775$ Robot hrs.

Left work $= 47775 - 125 \times 7 \times 39$

$= 13650$ Robot hrs.

$13 \times x \times 8 = 13650$

$x = 131.25$

Addition robot required

$= 131.25 - 125 = 6.25 \approx 7$

1.178 (c)

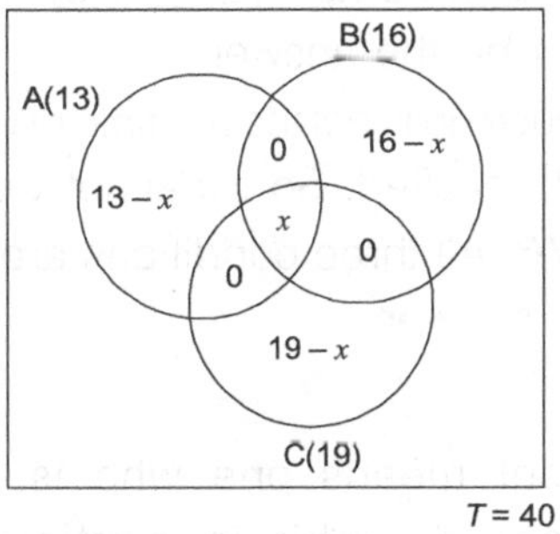

Total student $= 40$

$13 - x + 16 - x + 19 - x + x = 40$

Number of students who watched all three movies,

$\boldsymbol{x = 4}$

1.179 (c)

Initial square Area $= 1936$ m^2

$a^2 = 1936$ m^2

$a = 44$ m

Length of wire $= 4a = 4 \times 44 = 176$ m

Now,

part-1 length $= 3 \times 44 = 132$ m

part-2 length $= 1 \times 44 = 44$ m

Long wire is bent in square,

So, $4a' = 132$

$a' = 33$ m

Area of square $= 33^2 = 1089$ m^2

Now, small wire is bent in circle,

So, circle perimeter, $\pi D = 44$

$$\frac{22}{7} \times D = 44$$

$\boldsymbol{D = 14}$ **m**

Area of circle $= \frac{\pi}{4} \times D^2 = \frac{\pi}{4} \times 14^2 = 153.94 \text{ m}^2$

Total area enclosed = Area of square + Area of circle

$= 1089 + 153.94$

$= \mathbf{1242.97 \approx 1243 \text{ m}^2}$

1.180 (d)

By condition 1 ⇒

Possible numbers are → 51, 54, 57

By condition 2 ⇒ Possible numbers are → 61, 62, 63, 65, 66, 67, 69

By condition 3 ⇒ Possible numbers are → 70, 71, 73, 74, 75, 76, 77, 79

⇒ 66 is multiple of 3 and it does not belong to 50 to 59. So it will not be the answer.

⇒ 54 is multiple of 3 and 6. But it is not the multiple of 4. So according to second condition it cannot be the answer.

⇒ Because 65 is not the multiple of 6. So condition 3 is not satisfied. So it cannot be the answer.

⇒ For **76**. All three conditions are satisfied. So it is the answer.

1.181 (d)

Eminent means one who is famous and respected within a particular sphere; distinguished or renowned.

1.182 (d)

Had, I permitted him to leave, I wouldn't have had any problem

Past perfect conditional sentence pattern.

$\underbrace{\text{Had} + V_3 \text{ form,}}_{\text{If clause}} \ \underbrace{\text{would have} + V_3 \text{ form,}}_{\text{main clause}}$

Negative statement requires affirmative question tag so would I?

1.183 (b)

We know that watch hour hand complete one rotation (360°) in12 hours.

So, for 225°, corresponding time

$= \left(\frac{225}{360}\right) \times 12 = 7.5 \text{ hours}$

1.184 (d)

Let two numbers are x and y

Given, $(x + y) = 26$

$xy = 165$

We know that, $(x - y)^2 = (x + y)^2 - 4xy$

$= (26)^2 - 4 \times 165$

$(x - y)^2 = 16$

$x - y = 4$

1.185 (c)

He was accused of skirting.

Skirt as a verb means to avoid discussing a subject or problem, usually because there are difficulties that you do not want to deal with.

1.186 (d)

In 2010 total employees = 600

In 2016 there is 15% increment in number of employees.

In 2016 total employees = 1.15 × 600 = 690

Given, there is no change in P, Q and R type employees.

There is 40% increment in 'S' type employees.

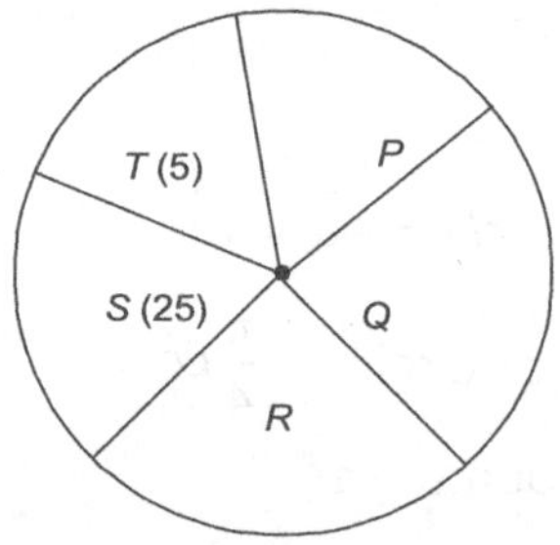

Now,Total employees in 2016

$= \left(600 \times \frac{25}{100}\right) \times 1.4 = 150 \times 1.4$

$= 210 \text{ employees}$

Increase in 'S' type employees

$= 210 - \left(600 \times \frac{25}{100}\right) = 60$

Initially, 'T' type employees $= \frac{5}{100} \times 600 = 30$

In 2016, increase in 'T' type employees = (690 – 600 – 60) = 30

In 2016, total 'T' type employees = (30 + 30) = 60 employees

1.187 (d)

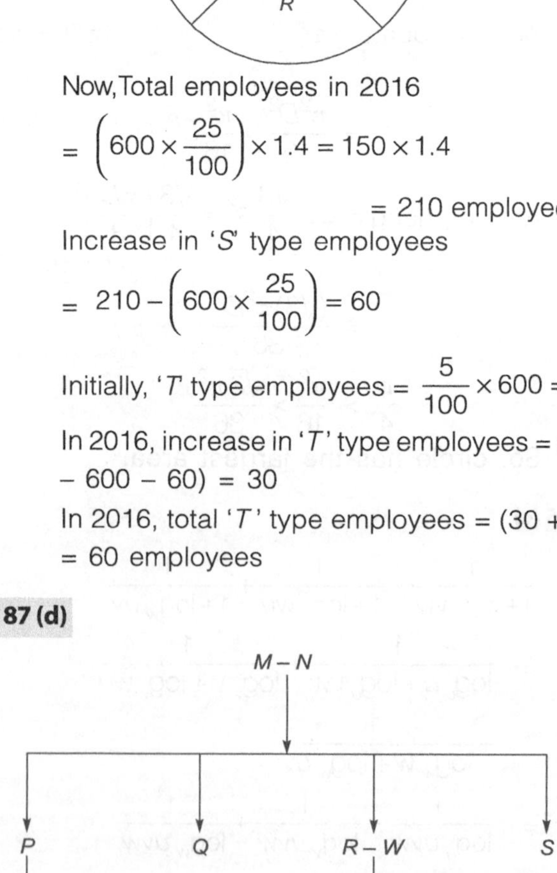

W can't be wife of P.

W can be husband or wife of R.

R can be father or mother of *Y*.
M can be grandfather of *Y*.

1.188 (d)

As the paragraph states, dictator Mobuto wanted to Africanise the name of Congo but could not succeed as the new name given to the country was Portuguese alteration of some other term.

1.189 (c)

Remaining three options would lead to the prisoner being hanged or shot. This answer choice creates two contradictory situations.

1.190 (c)

1st condition

Total amount = 100000

Assume, A has Invested Rs.x at 10% rate and Rs $(100000 - x)$ at 12% interest.

$$\text{Total interest} = 0.1x + (100000 - x) \times 0.12$$
$$(TI)_I = 12000 - 0.02x$$

2nd condition

Assume, A has invested Rs x at 12% rate and Rs $(100000 - x)$ at 10% interest.

$$\text{Total interest} = 0.12x + (100000 - x) \times 0.1$$
$$(TI)_{II} = 10000 + 0.02x$$

Given, $(TI)_I - (TI)_{II} = 120$

$$(12000 - 0.02x) - (10000 + 0.02x) = 120$$
$$2000 - 0.04x = 120$$
$$x = \frac{1880}{0.04} = \text{Rs. } 47000$$

Now, Another part $= 100000 - x$
$= 100000 - 47000$
$= \text{Rs. } 53000$

$$\frac{x}{(100000 - x)} = \frac{47}{53}$$

1.191 (b)

Comparative degree adjective is required since comparison is indicated (use of 'than').

1.192 (b)

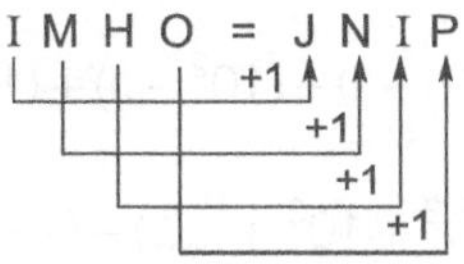

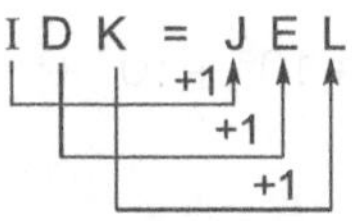

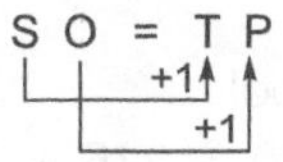

So,

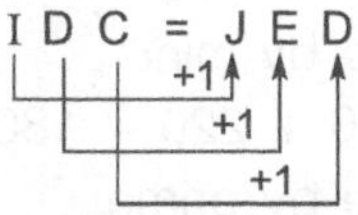

1.193 (a)

Since the 1st clause is in present tense, the second clause will not use past tense.
are going to be, will be, might be fill the given blank appropriately.

1.194 (c)

Culmination refers to the end point or the final result of a process.

1.195 (b)

Given, XYZ = 192, Z = 4

Now, XYZ = 192

$$XY = \frac{192}{4} = 48$$

XY = 48 → 1 × 48, 2 × 24, 3 × 16, 4 × 12, 6 × 8

Hence,
minimum possible value of

$$P = \frac{X+Y}{2} = \frac{6+8}{2} = 7$$

1.196 (a)

Most households that subscribe to X's loyalty membership discontinue watching cable television.
It is estimated that if X continues to offer loyalty membership, the number of subscribers of X will outgrow that of cable television. It is based on the assumption that most of the households subscribing to X don't watch cable television any longer.

1.197 (d)

Let GDP In 2015 = 100
GDP in 2016 = 100 + 10 = 110
Deficit in 2015 = 4
Deficit in 2016 = 5% of 110 = 5.5

Percentage increase in fiscal deficit

$$= \frac{5.5 - 4}{4} \times 100 = \frac{1.5}{4} \times 100 = 37.5\,\%$$

1.198 (b)

Revenue by pool

$=170 + 320 + 215 + 190 = 895$

Revenue by mini

$=110 + 220 + 180 + 70 = 580$

Revenue by prime

$=75 + 180 + 120 + 90 = 465$

Total revenue by Mola

$=895 + 580 + 465 = 1940$

Percentage of prime

$$=\frac{465}{1940}\times 100 = 23.97\%$$

1.199 (b)

None of the students had written stories set in India.

It is explicitly stated that the writer was surprised at receiving stories from the students that were all set in distant places.

1.200 (c)

In 1 hour P can fill $= \frac{1}{6}$

In 1 hour Q can fill $= \frac{1}{9}$

In 1 hour R can empty $= \frac{1}{12}$

In 4 hour $(P + R)$ can fill $= 4\times\left(\frac{1}{6}-\frac{1}{12}\right)=\frac{1}{3}$

In 6 hour $(Q + R)$ can fill $= 6\left(\frac{1}{9}-\frac{1}{12}\right)=\frac{1}{6}$

In 10 hour tank can fill $= \frac{1}{3}+\frac{1}{6}=\frac{1}{2}$

Remaining part of tank $= 1-\frac{1}{2}=\frac{1}{2}$

$\frac{1}{2}$ tank Q can fill $= \frac{1}{2}\times 9 = 4.5$ hour

Total time $=10 + 4.5 = 14.5$ hours

1.201 (c)

The term crocodile tears refers to hyporite crying. It is a false or insincere display of grief.

1.202 (d)

Crowd funding has been defined as a large number of people making small contribution for a project. Only option and is implied.

1.203 (c)

When two qualities of the same noun are compared, more + positive degree adjective is used. Use more fast and not 'faster'.

1.204 (d)

Build : Building
(verb) *(noun)*
Grow : Growth
(verb) *(noun)*

1.206 (b)

Average percentage of passing students

$$=\frac{\dfrac{\text{Total no. of passed students}}{\text{Total no. of students}}}{4}\times 100$$

$$=\frac{\dfrac{280}{500}+\dfrac{330}{600}+\dfrac{455}{700}+\dfrac{240}{400}}{4}\times 100$$

$$=\frac{0.56+0.55+0.65+0.60}{4}\times 100$$

$= 59\%$

1.208 (c)

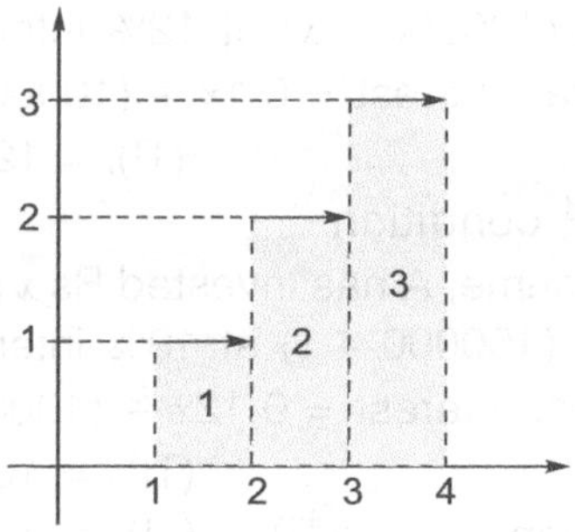

Area $= 1 \times 1 + 1 \times 2 + 1 \times 3$

$= 1 + 2 + 3 = 6$

1.209 (a)

$\beta\alpha$ and $\beta\beta$

Given, code of $P = \alpha\alpha$

code of $Q = \alpha\beta$

thesecode of $R = \beta\alpha$

andcode of $S = \beta\beta$

2.210 (d)

Using throw options put $n = 1$ and $n = 2$

Actual method:

$8 + 88 + 888 + ...$

$8[1 + 11 + 111 + ...]$

$\frac{8}{9}[9+99+999+......]$

$\frac{8}{9}[(10-1)+(10^2-1)+(10^3-1)+.....]$

$\frac{8}{9}[(10+10^2+10^3+.....)-(1+1+1+......+n)]$

$\frac{8}{9}[(10+10^2+10^3+.....)-n]$

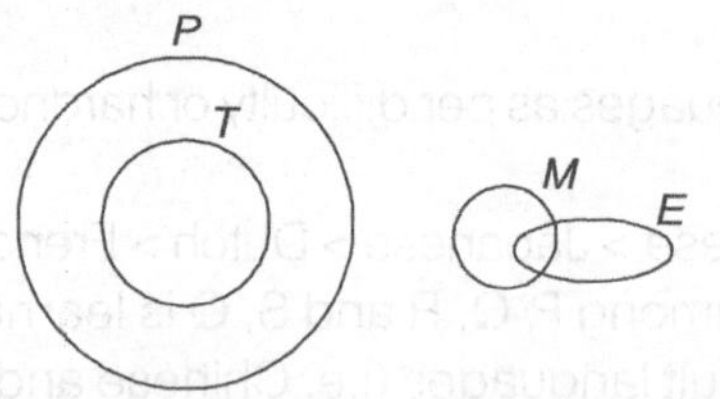

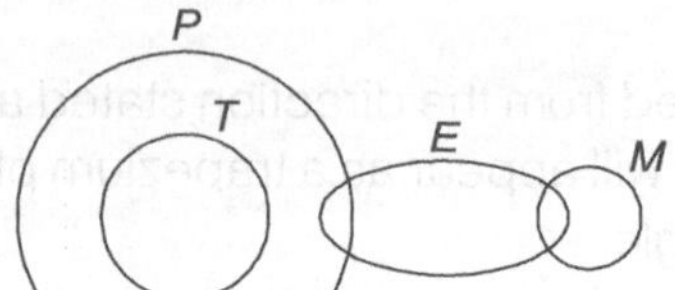

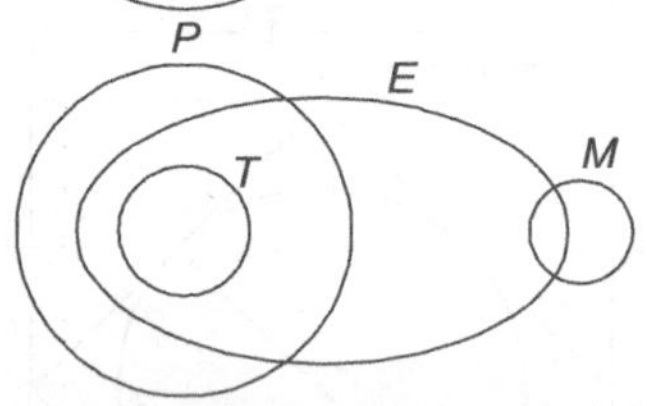

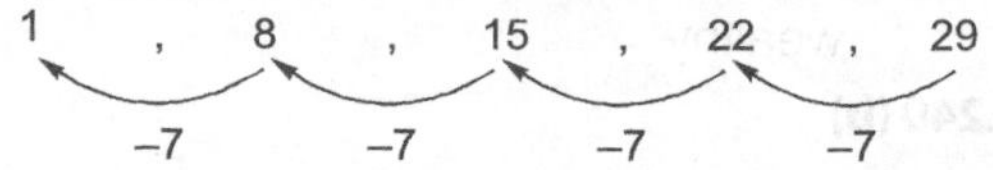

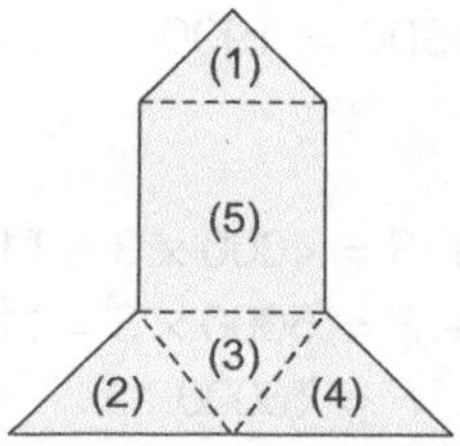

$$S_n = \frac{80}{81}(10^n - 1) - \frac{8n}{9}$$

1.212 (b)

Agree with - a person,
Agree to - an idea, proposal

1.216 (a)

52 men can do in 10 days.
Since 12 men were sent out
Remaining men left = 52 − 12 = 40
We know $M_1D_1 = M_2D_2$

$52 \times 12 = 40 \times x$

$$x = \frac{52 \times 10}{40} = 13 \text{ days}$$

Total number of days taken = 13 days
3 days more than the original estimate.

1.218 (d)

P → Q → R → S → T → Customer
↓
120 at each level increased 25% price paid by customer

$$= 120 \times \frac{125}{100} \times \frac{125}{100} \times \frac{125}{100} \times \frac{125}{100} \times \frac{125}{100} = 366.21$$

1.219 (d)

$$\frac{(\text{Boys})_{\text{Arts}}}{(\text{Girls})_{\text{Management}}} = \frac{20\% \text{ of } 5000 - 30\% \text{ of } 1500}{15\% \text{ of } 1500}$$

$$= \frac{1000 - 450}{225} = \frac{550}{225}$$

$$= \frac{110}{45} = \frac{22}{9} = 22 : 9$$

1.220 (a)

This is the log i.e. (n = 4)

5 + 4 = 9

t = 20, x = 24

20 + 4 = 24

Similarly, h = 8

Case-1: 8 + 4 = 12(l)

? + 4 = 8

? = 4 (d)

1.222 (d)

In farm P,
Hens = 65, Ducks = 91, Goats = 169
In farm Q,
Hens : Ducks : Goats
5 : 14 : 13
Number of Hens in farm Q

$$= \frac{5}{32} \times 416 = 65$$

Number of Ducks in firm Q

$$= \frac{14}{32} \times 416 = 182$$

Number of Goats in farm Q

$$= \frac{13}{32} \times 416 = 169$$

∵ From farm Q, hens, ducks and goats are sent to farm P,

∴ Total hens = 65 + 65 = 130
Total ducks = 91 + 182 = 273
Total goats = 169 + 169 = 338
New ratio = 130 : 273 : 338
= 10 : 21 : 26

1.223 (b)

∵ Area of ΔMON = 50 cm²

$$\frac{1}{2} \times r \times r = 50$$

$$r^2 = 100$$

Area of circle $= \pi r^2 = \pi \times 100$

$= 100\pi$

1.224 (c)

The possible seating arrangements as per the given condition are
QPRST, QTRSP
QPRTS, QTRPS
PQTRS, TQPRS
SPQTR, STQPR
RPQTS, RTQPS
SRPQT, SRTQP
SPRTQ, STRPQ
PSRTQ, TSRPQ
Hence, total seating arrangements are 16.

1.225 (7)

$$\Delta 2 \oplus 3 \Delta ((4 \otimes 2) \nabla 4) = +2 - 3 + ((4 \div 2) \times 4)$$
$$= +2 - 3 + (2 \times 4)$$
$$= +2 - 3 + 8$$
$$= 7$$

1.226 (d)

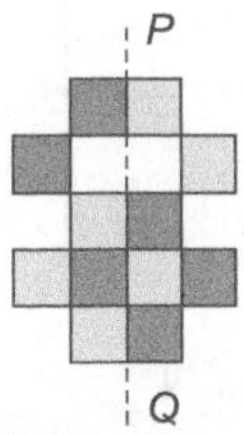

To make P-Q as symmetric line, minimum number of ▩ square added = 6.

1.227 (c)

Hardly/Scarcely/Barely have same sense that is negative and they are used after the verb (could barely and could hardly).

1.228 (c)

Option (a) and (b) are weekend by expression 'rhinos have no benefit'. Oxpeckers do not save life of poachers, so option (d) is incorrect. Hence, option (c) is correct.

1.229 (c)

The last sentence of the passage is reflecting in option (c) only.

1.230 (b)

First option is wrong because second sentence does not contradict the first sentence. Third option is wrong because two sentences are related. Fourth option is wrong because the second sentence does not repeat the first one. Hence second option is correct which shows the result of the cause.

1.231 (b)

"The number of" is singular and it takes singular verb. "A number of " is plural and it takes plural verb.

1.232 (c)

$$\frac{\oplus}{\odot} = 2, \quad \frac{\oplus}{\Delta} = 3$$

$$\therefore \quad \frac{\Delta}{\odot} = \frac{2}{3} \quad \text{...(i)}$$

$$\odot + \Delta = 5 \quad \text{...(ii)}$$

From eq. (i) and (ii)

$$\Delta = 2, \ \odot = 3$$

And

$$\oplus = 6, \ 2 \times \otimes = 10$$

$$\otimes = 5$$

$$\Rightarrow \quad (\otimes - \oplus)^2 = (5-6)^2 = 1$$

1.233 (a)

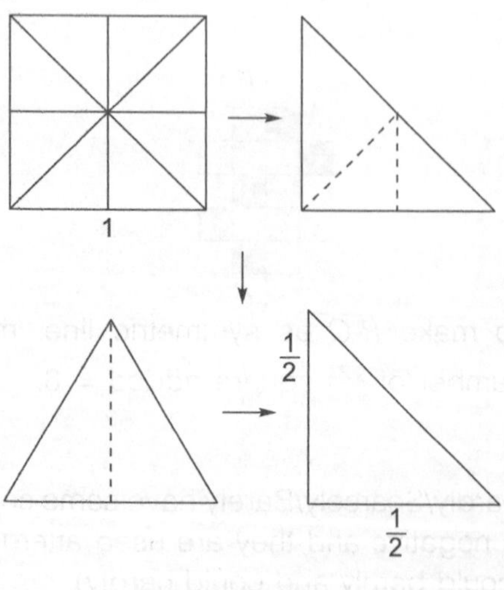

$$\text{Area} = \frac{1}{2} \times \frac{1}{2} \times \frac{1}{2} = \frac{1}{8}$$

1.234 (b)

The possible distinct arrangement are
S R P T Q, Q R P T S, S R T P Q
Hence, number of distinct seating arrangement.
= 3

1.235 (b)

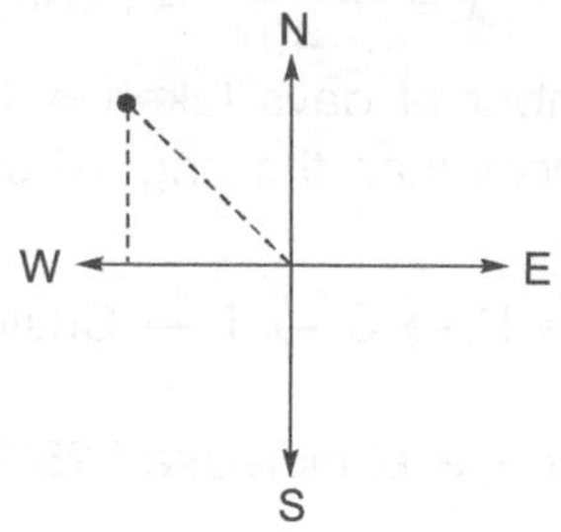

1.236 (a)

The probability of first ball is blue and second ball is black is given as,

$$P = \frac{15}{60} \times \frac{45}{59} = \frac{45}{236}$$

1.237 (b)

LCM of (30 and 32) is 480

480 seconds = 8 minutes

Hence, time will be 10.08 pm when they will beep together.

1.238 (a)

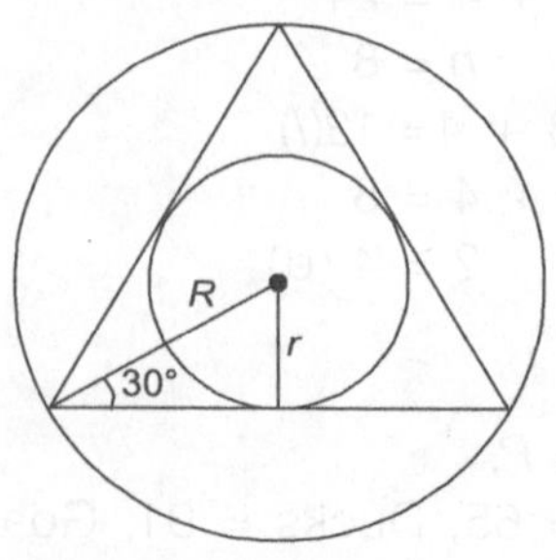

$$\sin 30° = \frac{r}{R}$$

$$\text{Area ratio} = \frac{\pi r^2}{\pi R^2} = \sin^2 30 = \frac{1}{4}$$

1.239 (c)

Possible cases are:

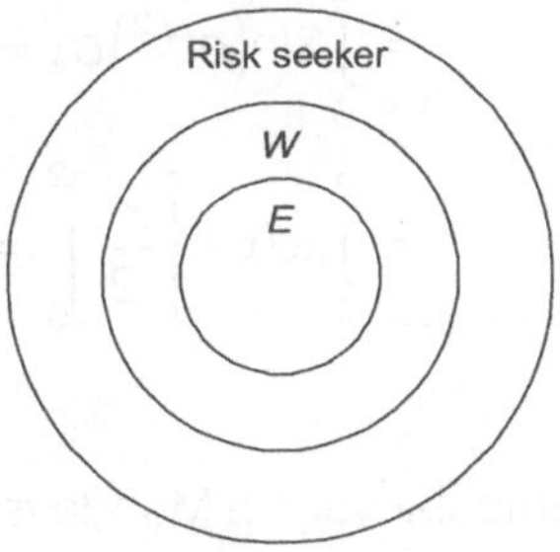

Conclusion-I is incorrect because some risk seeker are wealthy.
Conclusion-II is also incorrect because all the entrepreneurs are risk seeker as well as wealthy.

1.240 (b)

Number of employees in C2 company

$= \frac{5}{100} \times 10000 = 500$

Number of management degree holder employees in C2 $= \frac{1}{5} \times 500 = 100$

Number of employees in C5 company

$= \frac{20}{100} \times 10000 = 2000$

Number of management degree holder employees in C5 $= \frac{9}{10} \times 2000 = 1800$

Total management degree holder employees $= 100 + 1800 = 1900$

1.242 (a)

$$M + N + S = 4000 \times 3 = ₹12000 \quad \ldots (i)$$
$$N + S + P = 5000 \times 3 = ₹15000 \quad \ldots (ii)$$
$$\because \quad P = ₹6000$$

From (ii), $N + S = ₹9000$

From (i), $M + 9000 = 12000$

$$M = 3000$$
$$\frac{M}{P} = \frac{3}{6} = \frac{1}{2} \approx 50\%$$

1.243 (c)

Let x be the distance covered with 10 kmph so, $80 - x$ will be covered with 18 kmph.
$\because$ Total time to cover distance is 6 hr.

$$\Rightarrow \quad \left(\frac{x}{10}\right) + \left(\frac{80-x}{18}\right) = 6$$
$$\Rightarrow \quad \left(\frac{9x}{90}\right) + \frac{5(80-x)}{90} = 6$$
$$\Rightarrow \quad 9x + 400 - 5x = 540$$
$$\Rightarrow \quad x = 35$$

So, % Distance covered by 10 kmph $= \frac{35}{80} \times 100 = 43.75\%$

1.244 (b)

Languages as per difficulty or hardness w.r.t each other
Chinese > Japanese > Dutch > French > Spanish
So, among P, Q, R and S, Q is learning the most difficult languages (i.e. Chinese and Japanese).

1.245 (a)

If viewed from the direction stated above, the 2-D view will appear as a trapezium placed over a rectangle.

1.247 (a)

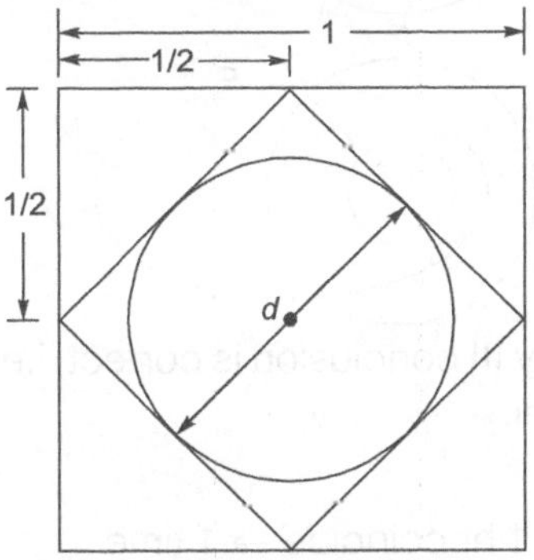

Diameter of circle = Side of rhombus

$$d = \sqrt{\left(\frac{1}{2}\right)^2 + \left(\frac{1}{2}\right)^2} = \frac{\sqrt{2}}{2} = \frac{1}{\sqrt{2}}$$

1.248 (b)

Let x = Side of equilateral D
a = Side of square
R = Radius of circle

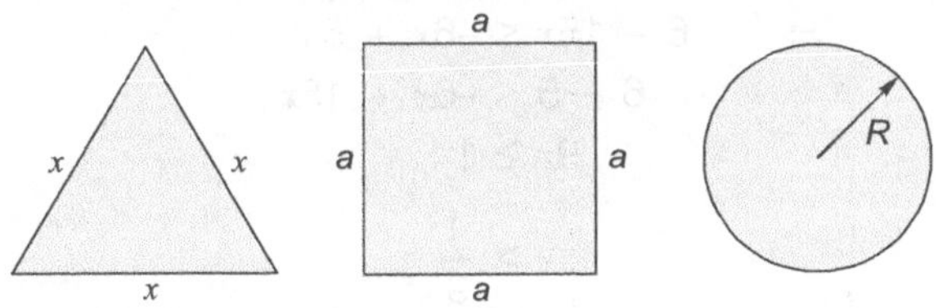

Area:

$$\frac{\sqrt{3}}{4}x^2 = a^2 = \pi R^2$$

or $$\frac{3^{1/4}}{2}x = a = \sqrt{\pi}R = k$$

Perimeter:

$3x : 4a : 2\pi R$

$$\frac{3.2k}{3^{1/4}}:4k:2\pi\frac{k}{\sqrt{\pi}}$$

$$\sqrt{(3\sqrt{3})}:2:\sqrt{\pi}$$

1.249 (a)

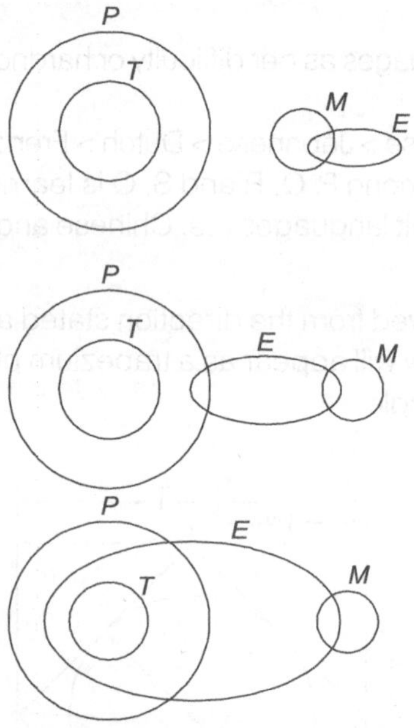

$\therefore$ Only III conclusion is correct, i.e. no male is a teacher.

1.250 (a)

Every 1 hr coincide → 1 time and in 12 h coincides → 11 times because between 11 to 1 they only meet one time.

1.251 (b)

Writing too many things on the board while teaching could make students get bored.

1.252 (c)

$$2 - 5x \le -\frac{(6x-5)}{3}$$

$$\Rightarrow \quad 6 - 15x \le -6x + 5$$

$$6 - 5 \le -6x + 15x$$

$$9x \ge 1$$

$$x \ge \frac{1}{9}$$

So, option (c) is correct.

1.253 (c)

Given: $f(x) = 2\ln\left(\sqrt{e^x}\right)$

Area bounded by $f(x)$ within $x \in [0, 2]$ on x-axis is given by,

$$A = \int_0^2 f(x)dx = \int_0^2 2\ln\left(\sqrt{e^x}\right)dx$$

$$= \int_0^2 2\ln\left(e^{x/2}\right)dx = \int_0^2 2\times\frac{x}{2}\ln e\, dx$$

$$= \int_0^2 x dx = \left[\frac{x^2}{2}\right]_0^2 = 2$$

1.254 (a)

In a particular year, 5 Mondays can only come in the month of February if that year is a leap year.

So, on 29th February there is a Monday.

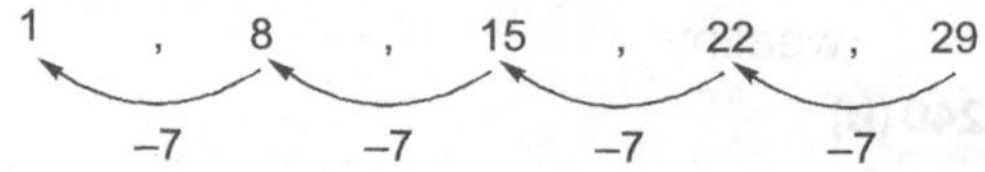

- Option (a) is correct, since 1st February is a Monday and hence 2nd February is a Tuesday.
- Option (b) is incorrect, since only Mondays can be five in February. So, only four Sundays are there.
- Option (c) is incorrect since 1st February is a Monday.
- Option (d) is incorrect as Mondays are on 1th, 8th, 15th, 22nd, 29th February which are all not even dates.

1.255 (b)

The given shape can be obtained by using 4 triangles (1, 2, 3 and 4) and 1 rectangle (5) as shown below.

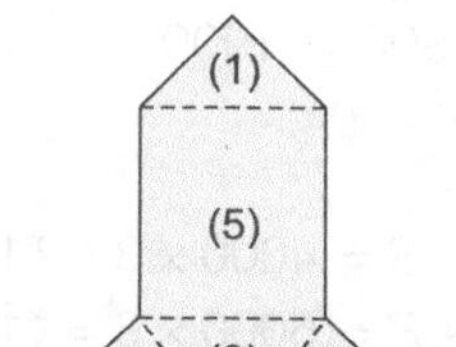

1.257 (c)

Total production in m days = $100 \times m$

Total production after today's day i.e. in $(m + 1)$ days = $100 \times (m + 1)$

The average production after todays' production = 110

$$\Rightarrow \quad \frac{100m+180}{(m+1)} = 110$$

$$100m + 180 = 110(m + 1)$$

$$100m + 180 = 110m + 110$$

$$70 = 10m$$

$$\Rightarrow \quad m = 7$$

1.258 (a)

Given: $f(p, q) = p^q$, $g(p, q) = p^{p^{p^{p^{\cdot^{\cdot^{\cdot}}}}}}$

Here, $f(2, 2) = 2^2 = 4$
$f(2, 1) = 2^1 = 2$
$f(3, 2) = 3^2 = 9$
$g(2, 2) = 2^2 = 4$
$g(2, 1) = 2 = 2$
$g(3, 2) = 3^3 = 9$

Also, $f(g(2, 2), 2) = f(4, 2) = 4^2 = 16$
$f(2, g(2, 2)) = f(2, 4) = 2^4 = 16$

On checking options, the correct option is (a).

1.259 (a)

The person starts from R at 7 AM and visits S and then returns to R if he follows the routes $R \to Q \to P \to S \to R$.

Since, travel time between any two connected cities is one hour, the total travel time for the route will be 4 hours as shown below.

$$R \xrightarrow[+1]{} Q \xrightarrow[+1]{} P \xrightarrow[+1]{} S \xrightarrow[+1]{} R$$

Now, the person reaches the city Q at 8 AM but the next train is at 9 AM. At 9 AM, he starts from Q and reaches P at 10 AM. The next train at P is at 11 AM, hence the waiting time at P is 1 hour.

He starts from P at 11 AM and reaches S at 12 AM but the next train at S is at 1:30 PM. Hence, waiting time at R is 1 hour and 30 Minutes.

Thus, the minimum time required is 6 hours 30 minutes.

1.260 (b)

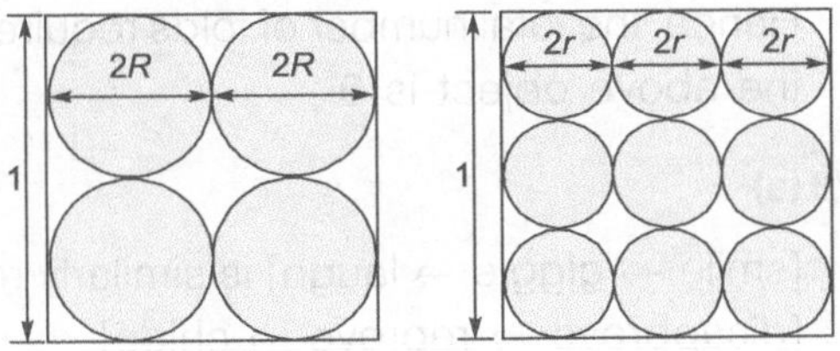

Case M Case N

Here, $6r = 1$ and $4R = 1$

$$\Rightarrow r = \frac{1}{6} \text{ and } R = \frac{1}{4}$$

Area of square in case M = Area of square in case N = 1 cm^2

Total area of all circles in case

$$M = 4(\pi R^2) = 4\pi \times \left(\frac{1}{4}\right)^2 = \frac{\pi}{4} \text{ cm}^2$$

Total area of all circles in case

$$N = 9(\pi r^2) = 9\pi \times \left(\frac{1}{6}\right)^2 = \frac{\pi}{4} \text{ cm}^2$$

Thus, area of unshaded region in case M,

$$A_1 = \left(1 - \frac{\pi}{4}\right) \text{cm}^2$$

Thus, area of unshaded region in case N,

$$A_2 = \left(1 - \frac{\pi}{4}\right) \text{cm}^2$$

Thus, the required ratio is

$$\frac{A_1}{A_2} = \frac{1 - \frac{\pi}{4}}{1 - \frac{\pi}{4}} = 1:1$$

1.263 (a)

Given,

Number of colleges, $C = 25951$

Number of universities, $U = 504$

Number of students with grade 3 in colleges

$$G_3(C) = 0.23 \text{ of } 25951 = 0.23 \times 25951$$
$$G_3(C) = 5968.73$$

Number of students with grade 3 in universities

$$G_3(U) = 0.07 \text{ of } 504 = 0.07 \times 504$$
$$G_3(U) = 35.28$$

Thus, required percentage

$$= \frac{\text{Total number of students with Grade 3}}{\text{Total number of students}}$$

$$= \frac{G_3(C) + G_3(U)}{c + u} = \frac{35.28 + 5968.73}{25951 + 504}$$

$= 0.2269$ or 22.7%

1.264 (a)

The minute hand and second hand will cross each other 1 time every minute after 9:15 i.e. 1 time each in 9:16 AM, 9:17 AM, 9.18 AM and so on upto 9:45 AM. Thus, they will cross each other 30 times.

1.265 (b)

□	*	○	△
△	○	*	□
○	△	□	*
*	□	△	○

1.267 (a)

Given inequalities,

$$p^2 - 4q < 4 \quad \text{...(i)}$$

$$3p + 2q < 6 \quad \text{...(ii)}$$

Since p and q are positive integers and satisfy the given inequalities, only possibility is

$$p = q = 1$$

This can be easily verified by putting values of p and q in inequality (ii),

For $p = q = 1$; $3(1) + 2(1) < 6$

For $p = 1$ and $q = 2$; $3(1) + 2(2) > 6$

And, similarly any other value of p and q will not satisfy the inequality.

Hence, $p + q = 1 + 1 = 2$

Alternate solution:

Multiplying inequality (ii) with 2,

$$6p + 4q < 12$$

or, $$4q < 12 - 6p \quad \text{...(iii)}$$

From (i) and (iii),

$$p^2 - 4 < 12 - 6p$$

$$p^2 + 6p - 16 < 0$$

$$(p + 8)(p - 2) < 0$$

Using wavy curve method,

$$\therefore \quad p \in (-8, 2)$$

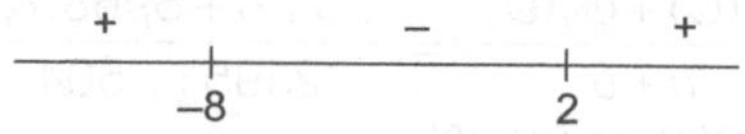

But it is given that p is positive integer,

$$\therefore \quad p = 1$$

From (iii), $$4q < 12 - 6(1)$$

$$4q < 6$$

$$\Rightarrow \quad q < \frac{3}{2}$$

But again q is positive integer,

$$\therefore \quad q = 1$$

Hence, $$p + q = 1 + 1 = 2$$

1.269 (a)

Since there are a total of 6 elements in the given set, the number of pairs of S & T possible are 3^6 i.e. 729. Note that these pairs also include the empty set 'ϕ'.

1.271 (d)

Let, $$\frac{x}{pq - r^2} = \frac{y}{qr - p^2} = \frac{z}{rp - q^2} = K$$

$$\frac{px}{p^2q - pr^2} = K$$

$$\frac{qy}{q^2r - qp^2} = K$$

$$\frac{rz}{r^2p - rq^2} = K$$

$$px + qy + rz = kp^2q - kpr^2 + kq^2r - kqp^2 + kr^2p - krq^2 = 0$$

1.276 (a)

Total outcomes : (2, 2), (2, 3), (2, 5), (2, 7)
(3, 2), (3, 3), (3, 5), (3, 7)
(5, 2), (5, 3), (5, 5), (5, 7)
(7, 2), (7, 3), (7, 5), (7, 7)

Favourable outcomes (sum > 11)
: (5, 7), (7, 5), (7, 7)

$$\text{Probability} = \frac{\text{Number of favourable outcomes}}{\text{Number of total outcomes}} = \frac{3}{16}$$

1.277 (b)

Except 21 all are prime number.

1.278 (c)

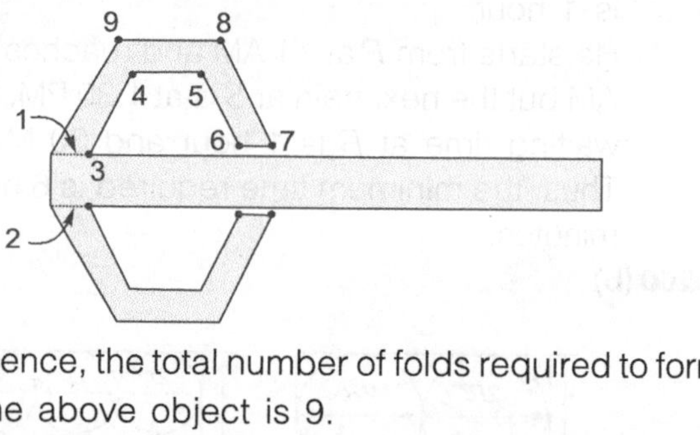

Hence, the total number of folds required to form the above object is 9.

1.279 (a)

[smile → giggle → laugh] is similarly related as [disapprove → reprove → chide]

1.280 (d)

In Fibonacci series, every number is equal to the sum of previous two numbers of the series. i.e. 0, 1, 1, 2, 3, 5, 8, 13, 21, ...

$$0 + 1 = 1$$
$$1 + 1 = 2$$
$$2 + 1 = 3$$
$$3 + 2 = 5$$
$$5 + 3 = 8$$
$$8 + 5 = 13$$
$$13 + 8 = 21$$

■■■■